Handbook of
Thin Film Materials

Handbook of Thin Film Materials

Volume 1

Deposition and Processing of Thin Films

Edited by

Hari Singh Nalwa, M.Sc., Ph.D.
Stanford Scientific Corporation
Los Angeles, California, USA

Formerly at
Hitachi Research Laboratory
Hitachi Ltd., Ibaraki, Japan

ACADEMIC PRESS

A Division of Harcourt, Inc.

San Diego   San Francisco   New York   Boston   London   Sydney   Tokyo

ACADEMIC PRESS
A division of Harcourt, Inc.
525 B Street, Suite 1900, San Diego, CA 92101-4495, USA
http://www.academicpress.com

Academic Press
Harcourt Place, 32 Jamestown Road, London, NW1 7BY, UK
http://www.academicpress.com

Library of Congress Catalog Card Number: 00-2001090614
International Standard Book Number, Set: 0-12-512908-4
International Standard Book Number, Volume 1: 0-12-512909-2

Printed in the United States of America
01 02 03 04 05 06 07 MB 9 8 7 6 5 4 3 2 1

To my children
Surya, Ravina, and Eric

Preface

Thin film materials are the key elements of continued technological advances made in the fields of electronic, photonic, and magnetic devices. The processing of materials into thinfilms allows easy integration into various types of devices. The thin film materials discussed in this handbook include semiconductors, superconductors, ferroelectrics, nanostructured materials, magnetic materials, etc. Thin film materials have already been used in semiconductor devices, wireless communication, telecommunications, integrated circuits, solar cells, lightemitting diodes, liquid crystal displays, magneto-optic memories, audio and video systems, compact discs, electro-optic coatings, memories, multilayer capacitors, flat-panel displays, smart windows, computer chips, magneto-optic disks, lithography, microelectromechanical systems (MEMS) and multifunctional protective coatings, as well as other emerging cutting edge technologies. The vast variety of thin film materials, their deposition, processing and fabrication techniques, spectroscopic characterization, optical characterization probes, physical properties, and structure-property relationships compiled in this handbook are the key features of such devices and basis of thin film technology.

Many of these thin film applications have been covered in the five volumes of the *Handbook of Thin Film Devices* edited by M. H. Francombe (Academic Press, 2000). The *Handbook of Thin Film Materials* is complementary to that handbook on devices. The publication of these two handbooks, selectively focused on thin film materials and devices, covers almost every conceivable topic on thin films in the fields of science and engineering.

This is the first handbook ever published on thin film materials. The 5-volume set summarizes the advances in thin film materials made over past decades. This handbook is a unique source of the in-depth knowledge of deposition, processing, spectroscopy, physical properties, and structure–property relationship of thin film materials. This handbook contains 65 state-ofthe-art review chapters written by more than 125 world-leading experts from 22 countries. The most renowned scientists write over 16,000 bibliographic citations and thousands of figures, tables, photographs, chemical structures, and equations. It has been divided into 5 parts based on thematic topics:

> Volume 1: Deposition and Processing of Thin Films
> Volume 2: Characterization and Spectroscopy of Thin Films
> Volume 3: Ferroelectric and Dielectric Thin Films
> Volume 4: Semiconductor and Superconductor Thin Films
> Volume 5: Nanomaterials and Magnetic Thin Films

Volume 1 has 14 chapters on different aspects of thin film deposition and processing techniques. Thin films and coatings are deposited with chemical vapor deposition (CVD), physical vapor deposition (PVD), plasma and ion beam techniques for developing materials for electronics, optics, microelectronic packaging, surface science, catalytic, and biomedical technological applications. The various chapters include: methods of deposition of hydrogenated amorphous silicon for device applications, atomic layer deposition, laser applications in transparent conducting oxide thin film processing, cold plasma processing in surface science and technology, electrochemical formation of thin films of binary III–V compounds, nucleation, growth and crystallization of thin films, ion implant doping and isolation of GaN and related materials, plasma etching of GaN and related materials, residual stresses in physically vapor deposited thin films, Langmuir–Blodgett films of biological molecules, structure formation during electrocrystallization of metal films, epitaxial thin films of intermetallic compounds, pulsed laser deposition of thin films: expectations and reality and b''-alumina single-crystal films. This volume is a good reference source of information for those individuals who are interested in the thin film deposition and processing techniques.

Volume 2 has 15 chapters focused on the spectroscopic characterization of thin films. The characterization of thin films using spectroscopic, optical, mechanical, X-ray, and electron microscopy techniques. The various topics in this volume include: classification of cluster morphologies, the band structure and orientations of molecular adsorbates on surfaces by angle-resolved electron spectroscopies, electronic states in GaAs-AlAs short-period superlattices: energy levels and symmetry, ion beam characterization in superlattices, *in situ* real time spectroscopic ellipsometry studies: carbon-based materials and metallic TiNx thin films growth, *in situ* Faraday-modulated fast-nulling single-wavelength ellipsometry of the growth of semiconductor, dielectric and metal thin films, photocurrent spectroscopy of thin passive films, low frequency noise spectroscopy for characterization of polycrystalline semiconducting thin films and polysilicon thin film transistors, electron energy loss spectroscopy for surface study, theory of low-energy electron diffraction and photoelectron spectroscopy from ultra-thin films, *in situ* synchrotron structural studies of the growth of oxides and metals, operator formalism in polarization nonlinear optics and spectroscopy of polarization inhomogeneous media, secondary ion mass spectrometry (SIMS) and its application to thin films characterization, and a solid state approach to Langmuir monolayers, their phases, phase transitions and design.

Volume 3 focuses on dielectric and ferroelectric thin films which have applications in microelectronics packaging, ferroelectric random access memories (FeRAMs), microelectromechanical systems (MEMS), metal–ferroelectric–semiconductor field-effect transistors (MFSFETs), broad band wireless communication, etc. For example, the ferroelectric materials such as barium strontium titanate discussed in this handbook have applications in a number of tunable circuits. On the other hand, high-permittivity thin film materials are used in capacitors and for integration with MEMS devices. Volume 5 of the *Handbook of Thin Film Devices* summarizes applications of ferroelectrics thin films in industrial devices. The 12 chapters on ferroelectrics thin films in this volume are complimentary to Volume 5 as they are the key components of such ferroelectrics devices. The various topics include electrical properties of high dielectric constant and ferroelectrics thin films for very large scale integration (VLSI) integrated circuits, high permittivity (Ba, Sr)TiO$_3$ thin films, ultrathin gate dielectric films for Si-based microelectronic devices, piezoelectric thin films: processing and properties, fabrication and characterization of ferroelectric oxide thin films, ferroelectric thin films of modified lead titanate, point defects in thin insulating films of lithium fluoride for optical microsystems, polarization switching of ferroelecric crystals, high temperature superconductor and ferroelectrics thin films for microwave applications, twinning in ferroelectrics thin films: theory and structural analysis, and ferroelectrics polymers Langmuir–Blodgett films.

Volume 4 has 13 chapters dealing with semiconductor and superconductor thin film materials. Volumes 1, 2, and 3 of the *Handbook of Thin Film Devices* summarize applications of semiconductor and superconductors thin films in various types of electronic, photonic and electro-optics devices such as infrared detectors, quantum well infrared photodetectors (QWIPs), semiconductor lasers, quantum cascade lasers, light emitting diodes, liquid crystal and plasma displays, solar cells, field effect transistors, integrated circuits, microwave devices, SQUID magnetometers, etc. The semiconductor and superconductor thin film materials discussed in this volume are the key components of such above mentioned devices fabricated by many industries around the world. Therefore this volume is in coordination to Volumes 1, 2, and 3 of the *Handbook of Thin Film Devices*. The various topics in this volume include; electrochemical passivation of Si and SiGe surfaces, optical properties of highly excited (Al, In)GaN epilayers and heterostructures, electical conduction properties of thin films of cadmium compounds, carbon containing heteroepitaxial silicon and silicon/germanium thin films on Si(001), germanium thin films on silicon for detection of near-infrared light, physical properties of amorphous gallium arsenide, amorphous carbon thin films, high-T_c superconducting thin films, electronic and optical properties of strained semiconductor films of group V and III-V materials, growth, structure and properties of plasma-deposited amorphous hydrogenated carbon–nitrogen films, conductive metal oxide thin films, and optical properties of dielectric and semiconductor thin films.

Volume 5 has 12 chapters on different aspects of nanostructured materials and magnetic thin films. Volume 5 of the *Handbook of Thin Film Devices* summarizes device applications of magnetic thin films in permanent magnets, magneto-optical recording, microwave, magnetic MEMS, etc. Volume 5 of this handbook on magnetic thin film materials is complimentary to Volume 5 as they are the key components of above-mentioned magnetic devices. The various topics covered in this volume are; nanoimprinting techniques, the energy gap of clusters, nanoparticles and quantum dots, spin waves in thin films, multi-layers and superlattices, quantum well interference in double quantum wells, electro-optical and transport properties of quasi-two-dimensional nanostrutured materials, magnetism of nanoscale composite films, thin magnetic films, magnetotransport effects in semiconductors, thin films for high density magnetic recording, nuclear resonance in magnetic thin films, and multilayers, and magnetic characterization of superconducting thin films.

I hope these volumes will be very useful for the libraries in universities and industrial institutions, governments and independent institutes, upper-level undergraduate and graduate students, individual research groups and scientists working in the field of thin films technology, materials science, solid-state physics, electrical and electronics engineering, spectroscopy, superconductivity, optical engineering, device engineering nanotechnology, and information technology, everyone who is involved in science and engineering of thin film materials.

I appreciate splendid cooperation of many distinguished experts who devoted their valuable time and effort to write excellent state-of-the-art review chapters for this handbook. Finally, I have great appreciation to my wife Dr. Beena Singh Nalwa for her wonderful cooperation and patience in enduring this work, great support of my parents Sri Kadam Singh and Srimati Sukh Devi and love of my children, Surya, Ravina and Eric in this exciting project.

Hari Singh Nalwa
Los Angeles, CA, USA

Contents

Chapter 1. METHODS OF DEPOSITION OF HYDROGENATED AMORPHOUS SILICON FOR DEVICE APPLICATIONS

Wilfried G. J. H. M. van Sark

Chapter 2.   ATOMIC LAYER DEPOSITION

Mikko Ritala, Markku Leskelä

Chapter 3.   LASER APPLICATIONS IN TRANSPARENT CONDUCTING OXIDE THIN FILMS PROCESSING

Frederick Ojo Adurodija

Chapter 4. COLD PLASMA PROCESSES IN SURFACE SCIENCE AND TECHNOLOGY

Pierangelo Gröning

Chapter 5. ELECTROCHEMICAL FORMATION OF THIN FILMS OF BINARY III–V COMPOUNDS

L. Peraldo Bicelli, V. M. Kozlov

Chapter 6.   FUNDAMENTALS FOR THE FORMATION AND STRUCTURE CONTROL OF THIN FILMS: NUCLEATION, GROWTH, SOLID-STATE TRANSFORMATIONS

Hideya Kumomi, Frank G. Shi

Chapter 7.   ION IMPLANT DOPING AND ISOLATION OF GaN AND RELATED MATERIALS

S. J. Pearton

Chapter 8. PLASMA ETCHING OF GaN AND RELATED MATERIALS

S. J. Pearton, R. J. Shul

Chapter 9. RESIDUAL STRESSES IN PHYSICALLY VAPOR-DEPOSITED THIN FILMS

Yves Pauleau

Chapter 10. LANGMUIR–BLODGETT FILMS OF BIOLOGICAL MOLECULES

Victor Erokhin

Chapter 11. STRUCTURE FORMATION DURING ELECTROCRYSTALLIZATION OF METAL FILMS

V. M. Kozlov, L. Peraldo Bicelli

Chapter 12. EPITAXIAL THIN FILMS OF INTERMETALLIC COMPOUNDS

Michael Huth

Chapter 13. PULSED LASER DEPOSITION OF THIN FILMS: EXPECTATIONS AND REALITY

Leonid R. Shaginyan

Chapter 14. SINGLE-CRYSTAL β″-ALUMINA FILMS

Chu Kun Kuo, Patrick S. Nicholson

About the Editor

Dr. Hari Singh Nalwa is the Managing Director of the Stanford Scientific Corporation in Los Angeles, California. Previously, he was Head of Department and R&D Manager at the Ciba Specialty Chemicals Corporation in Los Angeles (1999–2000) and a staff scientist at the Hitachi Research Laboratory, Hitachi Ltd., Japan (1990–1999). He has authored over 150 scientific articles in journals and books. He has 18 patents, either issued or applied for, on electronic and photonic materials and devices based on them.

He has published 43 books including *Ferroelectric Polymers* (Marcel Dekker, 1995), *Nonlinear Optics of Organic Molecules and Polymers* (CRC Press, 1997), *Organic Electroluminescent Materials and Devices* (Gordon & Breach, 1997), *Handbook of Organic Conductive Molecules and Polymers*, Vols. 1–4 (John Wiley & Sons, 1997), *Handbook of Low and High Dielectric Constant Materials and Their Applications*, Vols. 1–2 (Academic Press, 1999), *Handbook of Nanostructured Materials and Nanotechnology*, Vols. 1–5 (Academic Press, 2000), *Handbook of Advanced Electronic and Photonic Materials and Devices*, Vols. 1–10 (Academic Press, 2001), *Advanced Functional Molecules and Polymers*, Vols. 1–4 (Gordon & Breach, 2001), *Photodetectors and Fiber Optics* (Academic Press, 2001), *Silicon-Based Materials and Devices*, Vols. 1–2 (Academic Press, 2001), *Supramolecular Photosensitive and Electroactive Materials* (Academic Press, 2001), *Nanostructured Materials and Nanotechnology*–Condensed Edition (Academic Press, 2001), and *Handbook of Thin Film Materials*, Vols. 1–5 (Academic Press, 2002). The *Handbook of Nanostructured Materials and Nanotechnology* edited by him received the 1999 Award of Excellence in Engineering Handbooks from the Association of American Publishers.

Dr. Nalwa is the founder and Editor-in-Chief of the *Journal of Nanoscience and Nanotechnology* (2001–). He also was the founder and Editor-in-Chief of the *Journal of Porphyrins and Phthalocyanines* published by John Wiley & Sons (1997–2000) and serves or has served on the editorial boards of *Journal of Macromolecular Science-Physics* (1994–), *Applied Organometallic Chemistry* (1993–1999), *International Journal of Photoenergy* (1998–) and *Photonics Science News* (1995–). He has been a referee for many international journals including *Journal of American Chemical Society, Journal of Physical Chemistry, Applied Physics Letters, Journal of Applied Physics, Chemistry of Materials, Journal of Materials Science, Coordination Chemistry Reviews, Applied Organometallic Chemistry, Journal of Porphyrins and Phthalocyanines, Journal of Macromolecular Science-Physics, Applied Physics, Materials Research Bulletin*, and *Optical Communications*.

Dr. Nalwa helped organize the First International Symposium on the Crystal Growth of Organic Materials (Tokyo, 1989) and the Second International Symposium on Phthalocyanines (Edinburgh, 1998) under the auspices of the Royal Society of Chemistry. He also proposed a conference on porphyrins and phthalocyanies to the scientific community that, in part, was intended to promote public awareness of the *Journal of Porphyrins and Phthalocyanines*, which he founded in 1996. As a member of the organizing committee, he helped effectuate the First International Conference on Porphyrins and Phthalocyanines, which was held in Dijon, France in 2000. Currently he is on the organizing committee of the BioMEMS and Smart Nanostructures, (December 17–19, 2001, Adelaide, Australia) and the World Congress on Biomimetics and Artificial Muscles (December 9–11, 2002, Albuquerque, USA).

Dr. Nalwa has been cited in the *Dictionary of International Biography, Who's Who in Science and Engineering, Who's Who in America,* and *Who's Who in the World.* He is a member of the American Chemical Society (ACS), the American Physical Society (APS), the Materials Research Society (MRS), the Electrochemical Society and the American Association for the Advancement of Science (AAAS). He has been awarded a number of prestigious fellowships including a National Merit Scholarship, an Indian Space Research Organization (ISRO) Fellowship, a Council of Scientific and Industrial Research (CSIR) Senior fellowship, a NEC fellowship, and Japanese Government Science & Technology Agency (STA) Fellowship. He was an Honorary Visiting Professor at the Indian Institute of Technology in New Delhi.

Dr. Nalwa received a B.Sc. degree in biosciences from Meerut University in 1974, a M.Sc. degree in organic chemistry from University of Roorkee in 1977, and a Ph.D. degree in polymer science from Indian Institute of Technology in New Delhi in 1983. His thesis research focused on the electrical properties of macromolecules. Since then, his research activities and professional career have been devoted to studies of electronic and photonic organic and polymeric materials. His endeavors include molecular design, chemical synthesis, spectroscopic characterization, structure-property relationships, and evaluation of novel high performance materials for electronic and photonic applications. He was a guest scientist at Hahn-Meitner Institute in Berlin, Germany (1983) and research associate at University of Southern California in Los Angeles (1984–1987) and State University of New York at Buffalo (1987–1988). In 1988 he moved to the Tokyo University of Agriculture and Technology, Japan as a lecturer (1988–1990), where he taught and conducted research on electronic and photonic materials. His research activities include studies of ferroelectric polymers, nonlinear optical materials for integrated optics, low and high dielectric constant materials for microelectronics packaging, electrically conducting polymers, electroluminescent materials, nanocrystalline and nanostructured materials, photocuring polymers, polymer electrets, organic semiconductors, Langmuir-Blodgett films, high temperature-resistant polymer composites, water-soluble polymers, rapid modeling, and stereolithography.

List of Contributors

Numbers in parenthesis indicate the pages on which the author's contribution begins.

FREDERICK OJO ADURODIJA (161)
Inorganic Materials Department, Hyogo Prefectural Institute of Industrial Research,
3–1–12 Yukihira-cho, Suma-ku, Kobe, Japan

VICTOR EROKHIN (523)
Fondazione El.B.A., Corso Europa 30, Genoa, 16132 Italy

PIERANGELO GRÖNING (219)
Department of Physics, University of Fribourg, Fribourg, CH-1700 Switzerland

MICHAEL HUTH (587)
Institute for Physics, Johannes Gutenberg-University Mainz, 55099 Mainz, Germany

V. M. KOZLOV (261, 559)
Department of Physics, National Metallurgical Academy of Ukraine, Dniepropetrovsk, Ukraine

HIDEYA KUMOMI (319)
Canon Research Center, 5-1 Morinosato-Wakamiya, Atsugi-shi, Kanagawa 243-0193, Japan

CHU KUN KUO (675)
Ceramic Engineering Research Group, Department of Materials Science and Engineering,
McMaster University, Hamilton, Ontario L8S 4L7, Canada

MARKKU LESKELÄ (103)
Department of Chemistry, University of Helsinki, FIN-00014 Helsinki, Finland

PATRICK S. NICHOLSON (675)
Ceramic Engineering Research Group, Department of Materials Science and Engineering,
McMaster University, Hamilton, Ontario L8S 4L7, Canada

YVES PAULEAU (455)
National Polytechnic Institute of Grenoble, CNRS-UJF-LEMD,
B.P. 166, 38042 Grenoble Cedex 9, France

S. J. PEARTON (375, 409)
Department of Materials Science and Engineering, University of Florida,
Gainesville, Florida, USA

L. PERALDO BICELLI (261, 559)
Dipartimento di Chimica Fisica Applicata del Politecnico, Centro di Studio sui Processi
Elettrodici del CNR, 20131 Milan, Italy

MIKKO RITALA (103)
Department of Chemistry, University of Helsinki, FIN-00014 Helsinki, Finland

LEONID R. SHAGINYAN (627)
Institute for Problems of Materials Science, Kiev, 03142 Ukraine

FRANK G. SHI (319)
Department of Chemical and Biochemical Engineering and Materials Science,
University of California, Irvine, California, USA

R. J. SHUL (409)
Sandia National Laboratories, Albuquerque, New Mexico, USA

WILFRIED G. J. H. M. VAN SARK (1)
Debye Institute, Utrecht University, NL-3508 TA Utrecht, The Netherlands

Handbook of Thin Film Materials

Edited by H.S. Nalwa

Volume 1. DEPOSITION AND PROCESSING OF THIN FILMS

Volume 2. CHARACTERIZATION AND SPECTROSCOPY OF THIN FILMS

Volume 3. FERROELECTRIC AND DIELECTRIC THIN FILMS

Volume 4. SEMICONDUCTOR AND SUPERCONDUCTING THIN FILMS

Volume 5. NANOMATERIALS AND MAGNETIC THIN FILMS

Chapter 1

METHODS OF DEPOSITION OF HYDROGENATED AMORPHOUS SILICON FOR DEVICE APPLICATIONS

Wilfried G. J. H. M. van Sark

Debye Institute, Utrecht University, NL-3508 TA Utrecht, The Netherlands

Contents

Handbook of Thin Film Materials, edited by H.S. Nalwa
Volume 1: Deposition and Processing of Thin Films

ISBN 0-12-512909-2/$35.00

1. INTRODUCTION

This chapter describes the deposition of hydrogenated amorphous silicon (a-Si:H) and related materials by employing a low-temperature, low-density plasma. The method basically is a special form of chemical vapor deposition (CVD), which is known as plasma-enhanced chemical vapor deposition (PECVD) or plasma CVD. Essentially, silane gas (SiH_4) is excited by a radiofrequency (RF, 13.56 MHz) plasma, which causes silane molecules to dissociate. Subsequently, dissociation products are deposited on heated substrates and form a layer. Most research and industrial reactor systems consist of two parallel electrodes in a stainless steel chamber. Because of the relative ease of depositing a-Si:H uniformly over large areas, the original parallel plate geometry with RF excitation frequency is commonly used in industry, and has not changed much over the past two or three decades [1–13]. Material and device optimization is mostly done empirically, and so-called *device quality* a-Si:H layers having excellent uniformity are made by PECVD.

Nevertheless, several modifications have evolved since the first demonstration of deposition of a-Si:H, such as the use of higher excitation frequencies [from VHF (50–100 MHz) up to the gigahertz range], the use of remote excitation of the plasma, plasma beams, and modulation of the plasma in time or frequency. Even methods without the assistance of a plasma have evolved, such as the hot-wire CVD (HWCVD) method.

The material properties of layers deposited in a PECVD reactor strongly depend on the interaction between the growth flux and the film surface. Therefore, a central theme in this chapter is the relation between material properties and deposition parameters. Considering the plasma as a reservoir of species, we can distinguish neutrals, radicals, and ions, which can be either positive or negative. In a typical RF discharge, which is weakly ionized, the neutral species are the most abundant, having a concentration of about 10^{16} cm^{-3}, while the concentrations of radicals and ions are only about 10^{14} and 10^{10} cm^{-3}, respectively. The energies of these species may differ considerably. The neutrals, radicals and ions *within* the plasma are not energetic at all (below 0.1 eV). The ions that reach substrates and reactor walls are much more energetic (1–100 eV). This can have enormous consequences for the effect of species on a growing film. For example, the amount of energy that is present in the plasma amounts to 10^{13} eV/cm^3 for radicals and 10^{12} eV/cm^3 for ions. This shows that although ions are much less present in the plasma, their effect may be comparable to that of radicals.

In this chapter we will treat the common RF PECVD method for a-Si:H deposition, with emphasis on intrinsic material. First, a short introduction on the material properties of hydrogenated amorphous silicon is given. Subsequently, details are given on experimental and industrial deposition systems, with special emphasis on the UHV multichamber deposition system ASTER (Amorphous Semiconductor Thin Film Experimental Reactor) at Utrecht University [14, 15]. This is done not only because many experimental results presented in this chapter were obtained in that system, but because in our opinion it can also be seen as a generic multichamber deposition system. Then, a thorough description of the physics and the chemistry of the discharge is presented, followed by plasma modeling and plasma analysis results. Subsequently, relations will be formulated between discharge parameters and material properties, and models for the deposition of a-Si:H are presented.

In further sections extensions or adaptations of the PECVD method will be presented, such as VHF PECVD [16], the chemical annealing or layer-by-layer technique [17], and modulation of the RF excitation frequency [18]. The HWCVD method [19] (the plasmaless method) will be described and compared with the PECVD methods. The last deposition method that is treated is expanding thermal plasma CVD (ETP CVD) [20, 21]. Other methods of deposition, such as remote-plasma CVD, and in particular electron cyclotron resonance CVD (ECR CVD), are not treated here, as to date these methods are difficult to scale up for industrial purposes. Details of these methods can be found in, e.g., Luft and Tsuo [6].

As all these methods are used in research for improving material properties with specific applications in mind, a summary of important applications for which a-Si:H is indispensable is given in the last section.

It will be clear throughout this chapter that it is biased toward research performed at or in collaboration with the research group at the Debye Institute at Utrecht University. However, numerous references to other work are presented in order to put this research in a much broader perspective.

For further reading one may find excellent material in the (edited) books by (in chronological order) Pankove [1], Tanaka [2], Street [3], Kanicki [4, 5], Luft and Tsuo [6], Bunshah [7], Bruno et al. [8], Machlin [9, 10], Schropp and Zeman [11], Searle [12], and Street [13].

1.1. Historical Overview

The research on amorphous semiconductors in the 1950s and 1960s was focused on the chalcogenides, i.e. materials containing group VI elements (sulfur, selenium, and tellurium), such as As_2Se_3. These glasses are formed by cooling from the melt, their structure being similar to oxide glasses. Of particular interest was the relation between disorder of the structure and its electronic properties. This still is a question not fully answered; see for example Overhof in his Mott Memorial Lecture [22].

Amorphous silicon (and germanium) was prepared in those days by thermal evaporation or sputtering (see e.g. [23]). This unhydrogenated material was highly defective, which inhibited its use as a semiconductor. Research on incorporating hydrogen as a passivating element was pursued by introducing hydrogen in the sputtering system, which indeed improved the electronical properties [24–26].

In 1965 it was discovered that deposition of amorphous silicon employing the glow discharge technique yielded a material with much more useful electronic properties [27, 28]. The deposition occurs on a moderatedly heated substrate (200–300°C) through reactions of gas radicals with the substrate. At that time the infrared absorption bands of silicon–hydrogen bonds present in the deposited material were observed, but they were not recognized as such. Some years later, Fritsche and co-workers in Chicago confirmed that a-Si produced from a glow discharge of SiH_4 contains hydrogen [29, 30]. A recent personal account of the early years in amorphous silicon research shows the struggle to reach this conclusion [31]. Spear and co-workers in Dundee succeeded in improving the electrical properties, and finally, in 1975, a boost in research activities occurred when it was shown that a-Si:H could be doped n- or p-type by introducing phosphine (PH_3) or diborane (B_2H_6) in the plasma [32]: a range of resistivity of more than 10 orders of magnitude could be reached by adding small amounts of these dopants.

The achievement of Spear and LeComber [32] immediately paved the way for practical amorphous silicon devices. In fact, it is argued that the research field became "polluted by the applications" [33]. Carlson and Wronski at RCA Laboratories started in 1976 with the development of photovoltaic devices [34]. The first p–i–n junction solar cell was reported by the group of Hamakawa [35, 36]. In 1980, Sanyo was the first to market devices: solar cells for hand-held calculators [37]. Also, considerable research effort was directed towards amorphous silicon photoconductors for application in photocopying machines and laser printers [38, 39]. The a-Si:H photoconductor is, amongst other materials, used as the light-sensitive component in the electrophotographic process.

The first field effect transistors were also reported [40–42] at about this time. These thin film devices take advantage of the capability to deposit and process a-Si:H over large areas. It took only a few years before these thin-film transistors (TFTs) were utilized in active matrix liquid crystal displays (AMLCDs) by various companies. Active matrix addressing can also be used in printer heads. Combining the photoconductive properties and the switching capabilities of a-Si:H has yielded many applications in the field of linear sensor arrays, e.g., 2D image sensors and position-sensitive detectors of charged particles, X-rays, gamma rays, and neutrons [43–47].

1.2. Material Aspects of Hydrogenated Amorphous Silicon

1.2.1. Atomic Structure

Hydrogenated amorphous silicon is a disordered semiconductor whose optoelectronic properties are governed by the large number of defects present in its atomic structure. The covalent bonds between the silicon atoms in a-Si:H are similar to the bonds in crystalline silicon. The silicon atoms have the same number of neighbors and *on average* the same bond lengths and bond angles. One can represent the disorder by the atom pair distribution function, which is the probability of finding an atom at a distance r from another atom. A perfect crystal is completely ordered to large pair distances, while an amorphous material only shows short-range order. Because of the short-range order, material properties of amorphous semiconductors are similar to their crystalline counterparts.

Amorphous silicon is often viewed as a continuous random network (CRN) [48, 49]. In the ideal CRN model for amorphous silicon, each atom is fourfold coordinated, with bond lengths similar (within 1% [50]) to that in the crystal. In this respect, the short-range order (<2 nm) of the amorphous phase is similar to that of the crystalline phase. Amorphous silicon lacks long-range order because the bond angles deviate from the tetrahedral value (109.5°). The average bond-angle variation $\Delta\Theta$ reflects the degree of structural disorder in the random network. Raman spectroscopy is used to determine the vibrational density of states. The transverse optical (TO) peak region has been used to determine $\Delta\Theta$. Beeman et al. [51] have related the half width at half maximum, $\Gamma/2$, of the TO peak to $\Delta\Theta$. Values for $\Gamma/2$ range from 33 to 50 cm^{-1}, which translates into values of 8–13° for the average bond-angle variation.

The CRN may contain defects, but the crystalline concepts of interstitials or vacancies are not valid here. Instead, in the CRN one identifies a *coordination defect* when an atom has too few or too many bonds. In a-Si:H a silicon atom can have too few bonds to satisfy its outer sp^3 orbital. It is the common view that the dominant defect in amorphous silicon is a threefold-coordinated silicon atom. This structural defect has an unpaired electron in a nonbonding orbital, called a *dangling bond*. Pure amorphous silicon has a high defect density, 10^{20} cm^{-3} (one dangling bond for every ~500 Si atoms), which prevents photoconductance and doping. The special role of hydrogen with regard to amorphous silicon is its ability to passivate defects. Hydrogenation to a level of ~10 at.% reduces the defect density by four to five orders of magnitude.

The incorporation of phosphorus yields fourfold-coordinated P atoms, which are positively charged, as phosphorus normally is threefold coordinated. This substitutional doping mechanism was described by Street [52], thereby resolving the apparent discrepancy with the so-called $8 - N$ rule, with N the number of valence electrons, as originally proposed by Mott [53]. In addition, the incorporation mechanism, because charge neutrality must be preserved, leads to the formation of deep defects (dangling bonds). This increase in defect density as a result of doping explains the fact that a-Si:H photovoltaic devices are not simple p–n diodes (as with crystalline materials): an intrinsic layer, with low defect density, must be introduced between the p- and n-doped layers.

1.2.2. Microstructure and Hydrogen Content

The structure on a scale from the atomic level to about 10 nm is called *microstructure*. Many structural aspects are represented there, such as hydrogen bonding configurations (SiH, SiH$_2$, SiH$_3$, and polysilane [(SiH$_2$)$_n$] groups), multivacancies, internal surfaces associated with microvoids, density fluctuations, the unbonded hydrogen distribution (isolated and bulk molecular hydrogen), and the bonded hydrogen distribution (clustered and dispersed). Columnar structure, void size, and volume fraction are also included in this concept of microstructure, although these are of much larger scale.

The hydrogen content C_H greatly influences structure and consequently electronic and optoelectronic properties. An accurate measurement of C_H can be made with several ion-beam-based methods; see e.g. Arnold Bik et al. [54]. A much easier accessible method is Fourier-transform infrared transmittance (FTIR) spectroscopy. The absorption of IR radiation is different for different silicon–hydrogen bonding configurations. The observed absorption peaks have been indentified [55–57] (for an overview, see Luft and Tsuo [6]). The hydrogen content can be determined from the absorption peak at 630 cm^{-1}, which includes the rocking modes of all possible silicon–hydrogen bonding configurations. The hydrogen content is now defined as

$$C_H = A_{630} \int_{-\infty}^{+\infty} \frac{\alpha_{630}(\omega)}{\omega} \, d\omega \tag{1}$$

with $A_{630} = 2.1 \times 10^{19}$ cm^{-2} the proportionality constant determined by Langford et al. [58]. Alternatively, one can determine the hydrogen content by integration of the absorption peaks of the stretching modes (2000–2090 cm^{-1}), albeit by using the proportionality constant $A_{2000–2090}$ determined by Beyer and Abo Ghazala [59], $A_{2000–2090} = 1.1 \times 10^{20}$ cm^{-2}. Typical hydrogen contents are 9–11%.

In order to distinguish between isolated silicon–hydrogen bonds in a dense network and other bonding configurations, such as clustered monohydride and dihydride bonds, bonds on internal void surfaces, and isolated dihydride bonds, Mahan et al. [60] have defined the microstructure factor R^* as

$$R^* = \frac{I_{2070–2100}}{I_{2070–2100} + I_{2000}} \tag{2}$$

where I_{2000} is the integated absorption band due to isolated Si$-$H bonds. For all other bonds the IR absorption shifts to 2070–2100 cm^{-1}. Isolated bonds are preferred, so R^* is close to zero.

The refractive index of amorphous silicon is, within certain limits, a good measure for the density of the material. If we may consider the material to consist of a tightly bonded structure containing voids, the density of the material follows from the void fraction. This fraction f can be computed from the relative dielectric constant ϵ. Assuming that the voids have a spherical shape, f is given by Bruggeman [61]:

$$f = \frac{(\epsilon_d - \epsilon)(2\epsilon_d + 1)}{(\epsilon_d - 1)(2\epsilon_d + \epsilon)} \tag{3}$$

with the effective refractive index $n = \sqrt{\epsilon}$, and $n_d = \sqrt{\epsilon_d}$ the refractive index of the dense fraction.

The presence of a dense material with a varying void fraction results in compressive stress, with typical values of 500 MPa. Compressive stress can be determined conveniently by comparing the curvature of a crystalline silicon wafer before and after deposition of an a-Si:H film.

Hydrogenated amorphous silicon is not a homogeneous material. Its structure is thought to consist of voids embedded in an amorphous matrix [62, 63]. The size and number density of the voids depend on the deposition conditions. Poor-quality material can have a void fraction around 20%, while *device quality* a-Si:H has been shown to contain fewer voids, ~1%, with a diameter of ~10 Å [64–66]. The surfaces of the voids may be partly covered with hydrogen atoms [62, 67], and the voids are also thought to be filled with H$_2$ [68–72]. The influence of voids on the defect density in a-Si:H is the subject of much debate.

1.2.3. Electronic Structure and Transport

The preservation of the short-range order results in a similar electronic structure of the amorphous material to that of the crystalline one: bands of extended mobile states are formed (defined by the conduction and valence band edges, E_C and E_V), separated by the energy gap E_g, more appropriately termed the *mobility gap*. Figure 1 shows schematically the density of

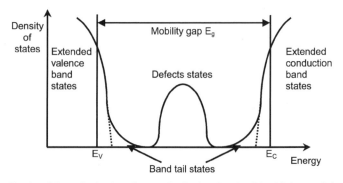

Fig. 1. Schematic density of states distribution. Bands of (mobile) extended states exist due to short-range order. Long-range disorder causes tails of localized states, whereas dangling bonds show up around midgap. The dashed curves represent the equivalent states in a crystal.

states distribution. The long-range atomic disorder broadens the densities of energy states, resulting in band tails of localized states that may extend deep into the bandgap. Coordination defects (dangling bonds) result in electronic states deep in the bandgap, around midgap. As electronic transport mostly occurs at the band edges, the band tails largely determine the electronic transport properties. The deep defect states determine electronic properties by controlling trapping and recombination.

Normally the defect density is low, and electronic transport is considered to occur predominantly at the mobility edges. For electron transport one can write the following expression for the conductivity σ:

$$\sigma(T) = \sigma_0 \exp\left(-\frac{E_C - E_F}{k_B T}\right) \quad (4)$$

where E_F denotes the Fermi energy, σ_0 the prefactor, k_B Boltzmann's constant, and T the temperature. Of particular interest for devices is the conductivity in the dark, σ_d, and the photoconductivity σ_{ph}. For typical undoped material E_F lies at midgap and σ_d equals about 10^{-10} Ω^{-1} cm^{-1} or less.

The activation energy E_A, defined as $E_C - E_F$ for the conduction band (and analogously for the valence band), can be used to assess the presence of impurities. Due to their presence, either intentional (B or P dopant atoms) or unintentional (O or N), the Fermi level shifts several tenths of an electron volt towards the conduction or the valence band. The activation energy is determined from plots of $\log\sigma(T)$ versus $1/T$, with $50 < T < 160°C$. For undoped material E_A is about 0.8 eV. The Fermi level is at midgap position, as typically E_g is around 1.6 eV.

The photoconductivity usually is determined by illuminating the material with light of known spectral content (AM1.5, 100 mW/cm^2). It has been shown [73] that the photoconductivity can be expressed as

$$\sigma_{ph} = e\mu\Delta n = \frac{e\eta_g\mu\tau(1 - R_f)F_0}{d}\left[1 - \exp(-\alpha(\lambda)d)\right] \quad (5)$$

with e the elementary charge, μ the mobility of photogenerated majority carriers (i.e. electrons), Δn their density, η_g the quantum efficiency for generation of carriers, τ their lifetime, R_f and d the reflectance and thickness of the material under study, F_0 the illumination intensity (photons/cm^2 s), and α the wavelength-dependent absorption coefficient. For good intrinsic a-Si:H one usually finds $\sigma_{ph} = 10^{-5}$ Ω^{-1} cm^{-1} or higher, which is 5–6 orders of magnitude higher than the dark conductivity. One uses the photoresponse (or photosensitivity) σ_{ph}/σ_d as a measure of optoelectronic quality.

Besides σ_{ph} and the photoresponse, the quantum-efficiency–mobility–lifetime product, $\eta_g\mu\tau$, is used as a figure of merit. Usually this product is measured at a wavelength of 600 nm, and typical values of $(\eta_g\mu\tau)_{600}$ are 10^{-7} cm^2/V or higher.

1.2.4. Optical Properties

The optical properties of a-Si:H are of considerable importance, especially for solar-cell applications. Because of the absence of

long-range order, the momentum $\vec{k}$ is not conserved in electronic transitions. Therefore, in contrast to crystalline silicon, a-Si:H behaves as though it had a direct bandgap. Its absorption coefficient for visible light is about an order of magnitude higher than that of c-Si [74]. Consequently, the typical thickness (sub-micrometer) of an a-Si:H solar cell is only a fraction of that of a c-Si cell.

In general one can divide the absorption behavior into three ranges:

1. For absorption coeffient α larger than 10^3 cm^{-1}, absorption takes place between extended states and is described by [75]

$$\left(\alpha(E)n(E)E\right)^{1/(1+p+q)} = B(\hbar\omega - E_g) \quad (6)$$

   with $\alpha(E)$ the energy-dependent absorption coefficient, $n(E)$ the refractive index, p and q constants related to the shape of the band edges, B a proportionality constant, $\hbar$ Planck's constant, and ω the frequency. Tauc has argued that the density of states near the band edges has a square-root dependence on energy, as is the case for crystalline semiconductors [75]. This results in $p = q = 1/2$. Thus, extrapolating $(\alpha(E)n(E)E)^{1/2}$ versus the photon energy to $\alpha(E) = 0$ for $\alpha(E) \geq 10^3$ cm^{-1} yields the Tauc gap. Klazes et al. [76] have proposed the density of states near the band edges to be linear, i.e. $p = q = 1$. Plotting $(\alpha(E)n(E)E)^{1/3}$ versus E yields a so-called cubic gap, which is about 0.1–0.2 eV smaller than the Tauc gap. The cubic plot is linear over a larger energy range than the Tauc gap, and fitting it yields more accurate values. Nevertheless, the Tauc gap is still widely used. For intrinsic a-Si:H the Tauc gap typically is 1.7 eV or smaller. The cubic gap is 1.6 eV or smaller. The refractive index at 600 nm is around 4.3; the absorption coefficient 4×10^4 cm^{-1}.

2. In the range of 1–$10 < \alpha < 10^3$ cm^{-1} absorption takes place between subbandgap states. An exponential dependence of α exists [23]:

$$\alpha = \alpha_0 \exp\left(\frac{E}{E_0}\right) \quad (7)$$

   with α_0 a prefactor, and E_0 the so-called Urbach energy. This reciprocal logarithmic slope of the edge, E_0, depends both on the temperature and on the disorder in the material. The slope of the absorption edge is mainly determined by the slope of the valence band tail. Typically, E_0 amounts to 50 meV. The slope of the conduction band tail is about half this value [77].

3. In the low-energy range α depends on the defect density, doping level, and details of the preparation process. Sensitive subbandgap spectroscopy is used to measure α and relate it to the defect density in the material [78, 79].

The optical properties of a-Si:H are influenced both by the hydrogen concentration and bonding in the film, and by the disorder in the a-Si:H network [80]. A linear relation between bandgap and hydrogen concentration is reported [81]: $E_g = 1.56 + 1.27C_H$. However, Meiling et al. [63, 82] have identified three regions in the relation between E_g and C_H. The bandgap first increases linearly as a function of hydrogen content, up to a value of about 0.1. A further increase up to $C_H \approx 0.22$ does not lead to a change in the bandgap. For higher values of C_H the bandgap further increases linearly, but with a smaller slope than in the first region. This is explained as follows: in the first region ($C_H < 0.1$) hydrogen is present only in monohydride bonds. The microstructure parameter is low; hence there are no voids in the material [62, 83]. In the middle region ($0.1 < C_H < 0.22$) the increasing amount of hydrogen is no longer incorporated as isolated SiH bonds. The microstructure parameter increases from 0 to 0.6, along with a reduction in silicon density. The material in this region consists of two phases, one which contains only SiH bonds, and one with SiH_2 bonds (chains) and voids. In the third region ($C_H > 0.22$) the material mainly consists of chains of SiH_2 bonds, and the material density is much lower.

Berntsen et al. [84, 85] have separated the effect of hydrogen content and bond-angle variation. The structural disorder causes broadening of the valence and conduction bands and a decrease of the bandgap by 0.46 eV. Hydrogenation to 11 at.% results in an independent increase of the bandgap with 0.22 eV.

For undoped a-Si:H the (Tauc) energy gap is around 1.6–1.7 eV, and the density of states at the Fermi level is typically 10^{15} eV^{-1} cm^{-3}, less than one dangling bond defect per 10^7 Si atoms. The Fermi level in n-type doped a-Si:H moves from midgap to approximately 0.15 eV from the conduction band edge, and in p-type material to approximately 0.3 eV from the valence band edge [32, 86].

1.2.5. Metastability

An important drawback of a-Si:H is its intrinsic metastability: the electronic properties degrade upon light exposure. This was discovered by Staebler and Wronski [87, 88], and is therefore known as the Staebler–Wronski effect (SWE). This effect manifests itself by an increase in the density of neutral dangling bonds, N_{db}, upon illumination, according to $N_{db}(t) \propto G^{0.6}t^{1/3}$, where G is the generation rate and t the illumination time [89]. The excess defects are metastable; they can be removed by annealing the material at temperatures above $\sim$150°C for some hours. The presence of these excess defects in concentrations up to 10^{17} cm^{-3} leads to a reduction of free carrier lifetime, and hence to a lower conversion effiency for solar cells.

The SWE is an intrinsic material property; it is also observed in very pure a-Si:H, with an oxygen concentration as low as 2×10^{15} cm^{-3} [90]. At impurity concentrations above 10^{18} cm^{-3} a correlation between SWE and impurity level has been established [91]. The hydrogen concentration, the bonding structure of hydrogen in the silicon network, and the disorder in the silicon network together affect the SWE. It has been long assumed that the origin of the SWE was local. Stutzmann et al. [89] have proposed a model in which photogenerated charge carriers recombine nonradiatively at weak silicon–silicon bonds

(e.g., strained bonds). The release of the recombination energy can be used to break the bond, and the two dangling bonds thus formed are prevented from recombining by the passivation of one of the bonds by a back-bonded hydrogen atom. Recent experimental results contradict the local nature of this model. Electron spin resonance (ESR) experiments have revealed that there is no close spatial correlation between the presence of hydrogen and light-induced dangling bonds [92]. Further, it has been found that the rate of dangling-bond creation is independent of temperature in the range of 4 to 300 K [93]. As hydrogen is immobile at 4 K, diffusion can be ruled out.

The SWE has been found to depend on the hydrogen microstructure. The amount of hydrogen bound to silicon in the dilute phase, i.e., monohydride bonds, determines the saturated density of metastable defects [94]. In addition, only in regions of low hydrogen density have light-induced defects been found in ESR experiments [92]; it may well be that in these regions only monohydride bonds up to the solubility limit of 2–4% of hydrogen in silicon [69, 95] are present. The presence of clustered hydrogen (high microstructure parameter R^*) leads to faster defect creation kinetics [96, 97]. These and other observations have led to the realization that the SWE extends over large regions, because configurational defects are transported over long distances. Branz proposed the hydrogen collision model [98]: the defect-creating recombination takes place at a Si$-$H bond, and the recombination energy is used to lift the hydrogen to a mobile energy level. The mobile hydrogen atom diffuses interstitially through the material. This mobile hydrogen atom is represented by a mobile complex of a Si$-$H bond and a dangling bond [99]. As the complex diffuses through the silicon network, it breaks one Si$-$Si bond after another, but each broken bond is re-formed after the complex has passed it. If two mobile hydrogen complexes collide, an immobile metastable two-hydrogen complex is formed. The net result is that two dangling bonds are created that are not spatially correlated to hydrogen atoms.

1.2.6. Alloys

The glow discharge technique is especially suitable for controlling various material properties by introducing other precursor gases in the plasma. As was first demonstrated by Anderson and Spear, incorporation of carbon or nitrogen in a-Si:H results in material with a large bandgap [100]. A linear relation between bandgap and carbon fraction x_C in a-SiC:H has been reported [101, 102]: $E_g = 1.77 + 2.45x_C$. On the other hand, by diluting the plasma with germane (GeH$_4$), material is obtained with small bandgaps [103]: from 1.7 eV (a-Si:H) to 1.0 eV (a-Ge:H). Here also a linear relation between bandgap and Ge fraction (x_{Ge}) has been reported [104], in which in addition the hydrogen fraction x_H is included: $E_g = 1.6 + x_H - 0.7x_{Ge}$.

The incorporation of elements as carbon, nitrogen, or germanium, however, leads to material with a low mobility and lifetime of charge carriers. This would limit the application of these alloys, and a large research effort has been undertaken to find ways around this problem.

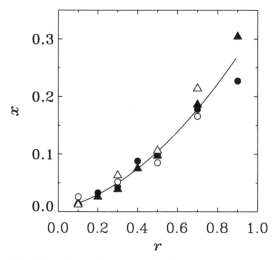

Fig. 2. The dependence of the carbon fraction $x = [C]/([Si] + [C])$ on the gas-flow ratio $r = [CH_4]/([SiH_4] + [CH_4])$ for films deposited in the ASTER system [AST1 (filled circles) and AST2 (filled triangles)] and for films deposited in a similar system (ATLAS) [ATL1 (open circles) and ATL2 (open triangles)]. (From R. A. C. M. M. van Swaaij, Ph.D. Thesis, Universiteit Utrecht, Utrecht, the Netherlands, 1994, with permission.)

The addition of an extra feedstock gas (GeH$_4$ or CH$_4$) in large quantities (compared to dopant gases) adds an extra degree of freedom to the already complex chemistry of the discharge. It is therefore even more complicated to relate material properties to deposition conditions [105].

As an example, GeH$_4$ is less stable than SiH$_4$ in the glow discharge, and Ge is preferentially deposited from SiH$_4$–GeH$_4$ mixtures: even a low GeH$_4$ fraction in the gas results in a high Ge fraction in the solid. Moreover, optimum conditions for a-Si:H deposition (low power density) cannot be translated simply to a-Ge:H deposition, in which case high power density is required to obtain good-quality material [106]. The deposition of a-SiGe:H requires elaborate fine tuning of deposition parameters. Other possibilities have been pursued to improve the properties of a-SiGe alloys, such as the use of strong dilution of SiH$_4$–GeH$_4$ mixtures with hydrogen [107], and the use of fluorinated reactants such as SiF$_4$ [108]. Modulation of the discharge (see also Section 8.3) has been demonstrated to increase the amount of Ge in the alloy by a factor of 2–10, depending on discharge conditions, compared to a continuous discharge [109].

As SiH$_4$ is less stable than CH$_4$, a low C fraction in the solid is obtained for a high CH$_4$ fraction in the gas. This is illustrated in Figure 2, which shows the dependence of the carbon fraction $x = [C]/([Si]+[C])$ on the gas flow ratio $r = [CH_4]/([SiH_4] + [CH_4])$ for films deposited in the ASTER system and a system with a similar reactor (ATLAS) [101, 102, 110]. These films were deposited in the so-called *low-power* deposition regime [111, 112], which is defined as the regime in which the applied power density is lower than the threshold power density required for the decomposition of methane [112]. In this regime the deposition of films is dominated by the decomposition of silane and is not dependent on the methane concentration

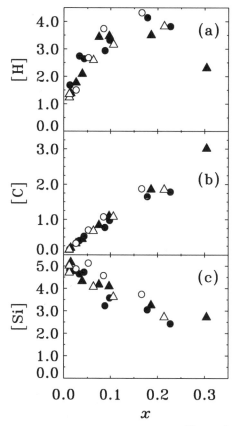

Fig. 3. Absolute atomic concentrations in units of 10^{22} at./cm^3, determined by ERD, RBS, and optical reflection and transmission spectroscopy, of (a) hydrogen, (b) carbon, and (c) silicon as a function of the carbon fraction x. Results are presented for the series AST1 (filled circles), AST2 (filled triangles), ATL1 (open circles), and ATL2 (open triangles). (From R. A. C. M. M. van Swaaij, Ph.D. Thesis, Universiteit Utrecht, Utrecht, the Netherlands, 1994, with permission.)

[111, 112]. Incorporation of carbon results from chemical reactions between methane molecules and silane species that are created by the plasma. The deposition rate in this regime is almost independent of the gas flow ratio r, provided the gas in the reactor is not depleted in the discharge [101, 102, 113].

The dependence of the absolute atomic concentration of hydrogen, silicon, and carbon on the carbon fraction is shown in Figure 3 [101, 102]. The atomic concentrations were determined by using elastic recoil detection (ERD) and Rutherford backscattering spectrometry (RBS) [114]. The most striking observation is the rapid increase of the hydrogen concentration upon carbon alloying (Fig. 3a), which can be ascribed to the incorporation of CH$_n$ groups in the material during deposition [112, 115]. It can be inferred from the data that up to a carbon fraction of about 0.1, three hydrogen atoms are incorporated per carbon atom. Above $x = 0.15$ the rate of increase of the hydrogen concentration becomes smaller. For higher x-values the hydrogen concentration even tends to decrease.

In Figure 3b and c the absolute atomic concentrations of carbon and silicon, respectively, are shown as a function of the carbon fraction. As expected, the carbon concentration

Table I. Selected Properties of *Device Quality* Hydrogenated Amorphous
Silicon Films

Property	Symbol	Value	Unit
Optical bandgap (Tauc)	E_g	1.8	eV
Optical bandgap (cubic)	E_g	1.6	eV
Refractive index	n	4.3	
Absorption coefficient (600 nm)	α_{600}	4×10^4	cm^{-1}
Urbach energy	E_0	50	meV
Dark conductivity	σ_d	10^{-10}	Ω^{-1} cm^{-1}
Photoconductivity (AM1.5)	σ_{ph}	10^{-5}	Ω^{-1} cm^{-1}
Activation energy	E_A	0.8	eV
Mobility–lifetime product	$\mu\tau_{600}$	10^{-7}	cm^2 V^{-1}
Hydrogen content	C_H	8–12	%
Microstructure factor	R^*	0–0.1	
HWHM of TO Raman peak	$\Gamma/2$	33	cm^{-1}
Intrinsic stress	σ_i	400–500	MPa
Defect density	N_s	10^{15}	cm^{-3} eV^{-1}

increases upon alloying. In contrast, the silicon content decreases rapidly, which implies that the material becomes less dense. As it was reported that the Si—Si bond length does not change upon carbon alloying [116], it thus can be inferred that the *a*-SiC:H material contains microscopic voids, which is in agreement with small-angle X-ray scattering (SAXS) results [62].

Much more can be said about Ge and C alloying, but that is not within the scope of this chapter. We refer to the books by Kanicki [5] and Luft and Tsuo [6].

1.2.7. *Device Quality Characteristics*

Over the past decades the term *device quality* has come to refer to intrinsic PECVD hydrogenated amorphous silicon that has optimum properties for application in a certain device. Of course, depending on the type of device, different optimum values are required; nevertheless the properties as listed in Table I are generally accepted, e.g. [6, 11]. Many of these properties are interrelated, which has to borne in mind when attempting to optimize only one of them.

Optimum properties for *p*- and *n*-type doped *a*-Si:H have been identified [6, 11]. For *p*-type doping boron is used as the dopant element. Due to alloying with silicon, the bandgap is reduced. This can be compensated by adding carbon. Typically silane (SiH_4), methane (CH_4), and diborane (B_2H_6) are used, with silane and carbon in about equal amounts and diborane a factor of thousand lower. This yields *p*-type material with $E_g = 2.0$ eV, $E_A = 0.5$ eV, $\sigma_d = 10^{-5}$ Ω^{-1} cm^{-1}, and $\alpha_{600} = 10^4$ cm^{-1}. Adding phosphine (PH_3) to silane in the ratio 0.025 : 1 yields *n*-type material of good quality: $E_g = 1.8$ eV, $E_A = 0.3$ eV, $\sigma_d = 10^{-3}$ Ω^{-1} cm^{-1}, and $\alpha_{600} = 4 \times 10^4$ cm^{-1}.

2. RESEARCH AND INDUSTRIAL EQUIPMENT

2.1. General Aspects

A plasma deposition system usually consists of several subsystems, each providing different functions [117]. The gas handling system includes process gas storage in high-pressure cylinders, mass flow controllers to measure and control the different gases to the reactor, and tubing. The vacuum system comprises pumps and pressure controllers. The plasma reactor is operated between 10^{-4} and 10 Torr.[1] A much lower background pressure, in the ultrahigh vacuum (UHV) range (10^{-9} mbar), is often required to ensure cleanliness of the process. High-vacuum rotary pumps are used in combination with turbomolecular pumps [118].

The deposition setup as shown in Figure 4a is the central part of the most commonly used planar diode deposition system. The power to the reactor system is delivered by means of a power supply connected to the reactor via appropriate dc or RF circuitry (matchboxes). Power supplies can consist of generator and amplifier combined in one apparatus, with a fixed RF frequency. More flexible is to have an RF generator coupled to a broadband amplifier [119, 120].

Using the planar diode geometry, *p*–*i*–*n* devices have been made in single-chamber reaction systems, either with [121] or without load lock [122]. In such single-chamber systems cross-contamination occurs. After deposition, small amounts of (dopant) gases will remain in the reactor due to adsorption on the walls. If no load lock is used, water vapor and oxygen are also adsorbed upon sample loading. These residual gases will desorb during subsequent depositions and may be built in as impurities in the amorphous layer [123, 124]. Additionally, the desorption of gases after a deposition may cause contaminated interfaces, e.g., a graded boron concentration profile may be present in the first monolayers of the *i*-layer of a p^+–*i* interface [125, 126]. Contaminated interfaces combined with (too) high impurity levels in the layers lead to worse electrical properties of the layer [127, 128], which results in low efficiency of solar cells.

By using a multichamber system [129], exchange of residual gases between successive depositions will be strongly decreased, and very sharp interfaces can be made. Furthermore, the use of a load-lock system ensures high quality of the background vacuum, and thus low levels of contaminants in the bulk layers. Multichamber reactor systems have been used for the fabrication of solar cells, and considerable improvements in energy conversion efficiency have been achieved [130, 131].

Most of the gases used are hazardous: they can be corrosive, flammable, explosive, and/or highly toxic. Silane is pyrophoric; the dopant gases diborane, phosphine, and arsine are extremely toxic, with TLV (threshold limited value) values in the low parts-per-million range. Therefore extreme care has to be taken

---

[1]Throughout this chapter the torr, millibar, and pascal are used as units of pressure, according to the original data rather then converting the first two to the SI unit. Note than 1 Pa = 0.01 mbar = 0.0076 Torr = 7.6 mTorr.

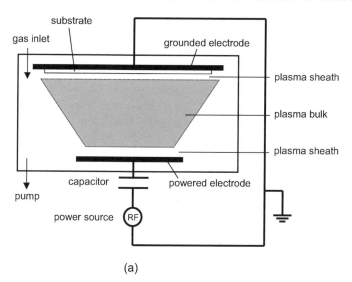

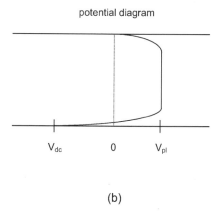

Fig. 4. Schematic representation (a) of a parallel-plate, capacitively coupled RF-discharge reactor, with unequal-size electrodes. The potential distribution (b) shows the positive plasma potential V_{pl} and the negative dc self-bias voltage V_{dc}.

in dealing with these gases. One generally installs, flow limiters (between the valve of the supply cylinder and pressure regulator, to prevent excessive gas flow in case of breakdown of the regulator), cross-purge assemblies [in order to purge the regulators and prevent release of the gas when exchanging (empty) gas cylinders], scrubbers or diluters to avoid above-TLV levels at the exhaust, detectors mounted at several critical locations, etc. More on safety issues can be found in, e.g., the proceedings of the 1988 Photovoltaic Safety Conference [132].

In the design of the reactor chamber one needs to address issues such as electrode geometry, gas flow patterns, heater design, and reactor volume (discharge and discharge-free regions). Hot walls will influence the gas temperature and the gas phase chemistry. Electrode size and electrode spacing as well as dark-space shields directly influence the discharge properties. Asymmetry introduces a bias, which may be controlled externally. Also the driving frequency is an important parameter. Over the years many reactor designs have been developed. Studies on the effects of process parameters on material qual-

ity do not always take geometric causes into account. This has prompted the development of a standard type of reactor, the GEC (Gaseous Electronics Conference) reference cell [133]. Nevertheless generalizations on the effect of process parameters on material quality have been made. For an overview see, e.g., Luft and Tsuo [6].

2.2. Reactor Configurations

Diode reactors can be powered by an RF or dc electric field. In case of RF excitation the deposition typically is on the grounded electrode. As the combined area of grounded electrode and reactor walls usually is larger than the area of the powered electrode, a dc self-bias is developed, as shown in Figure 4b. The potential drop at the powered electrode is much larger than at the grounded electrode. The powered electrode is more negative with respect to ground, and therefore is often called the cathode. The grounded electrode then is the anode. For a dc glow discharge, the potential distribution is similar to the one shown for the RF discharge in Figure 4b. In both cases ion bombardment at the cathode is larger than at the anode. Due to this, deposition of films on the cathode or anode leads to different microstructural properties. Deposition rates at the cathode are usually higher than at the anode. Films deposited at the cathode are dense, but also stressed. Anodic films are more porous. In dc discharges one sometimes uses a mesh positioned above the cathode, which has the same potential as the cathode [134]. Ions are thus slowed down by gas phase collisions in the region between mesh and cathode, and much better material is obtained [135]. Moreover, the mesh acts as a screen for reactive radicals. The SiH_2 radical has a large sticking probability, and it will stick to the mesh easily. As a consequence, the SiH_2 radical will be filtered out, and SiH_3 will dominate the deposition. The resulting material quality has been shown in RF triode discharges to have been improved [136–139].

Planar triode RF discharges used in research generally are asymmetric; the area of the powered electrode is much smaller than the area of all grounded parts taken together (the grounded electrode may be just a small part of the grounded area). The dc self-bias therefore is large. One can reduce the asymmetry by confining the discharge with a grounded mesh [140, 141] or wall [142]. Such confinement also allows for a higher power density in the discharge, which leads to enhanced deposition rates.

An external magnetic field has also been used to confine the plasma [143]. An arrangement where electromagnets are located under the cathode is known as the *controlled plasma magnetron method* [144]. The diffusion of electrons to the walls is prevented by the magnetic field between cathode and anode. This results in an increase in electron density, and consequently in a faster decomposition of silane and a higher deposition rate. At a deposition rate of 1 nm/s, device quality material is obtained [144]. In addition, a mesh is located near the anode, and the anode can by biased externally, both in order to confine the plasma and in order to control ion bombardment.

Hot reactor walls are sometimes used as a means to increase the density of the films that are deposited on the walls. This reduces the amount of adsorbed contaminants on the walls, and leads to lower outgassing rates. A hot wall is particularly of interest for single-chamber systems without a load-lock chamber. Material quality is similar to the quality obtained with a cold reactor wall [145].

Other configurations that are used include an concentric electrode setup in a tubular reactor, where the discharge still is capacitivily coupled. Also, inductive coupling has been used, with a coil surrounding the tubular reactor [146, 147].

2.3. Scale-Up to Systems of Industrial Size

A plasma process that has been demonstrated to yield good quality materials in the laboratory will one day need to be scaled up to a technology that can produce the materials in larger sizes and larger quantities. Such a transfer is not straightforward, and many technological difficulties will have to be overcome before a scaled-up process is commercially viable.

A plasma process is characterized by many parameters, and their interrelations are very complex. It is of paramount importance to understand, at least to a first approximation, how the plasma parameters have to be adjusted when the geometrical dimensions of the plasma system are enlarged. Especially of use in scaling up systems are scaling laws, as formulated by Goedheer et al. [148, 149] (see also Section 3.2.2).

In general, the substrate temperature will remain unchanged, while pressure, power, and gas flow rates have to be adjusted so that the plasma chemistry is not affected significantly. Grill [117] conceptualizes plasma processing as two consecutive processes: the formation of reactive species, and the mass transport of these species to surfaces to be processed. If the dissociation of precursor molecules can be described by a single electron collision process, the electron impact reaction rates depend only on the ratio of electric field to pressure, E/p, because the electron temperature is determined mainly by this ratio. The mass transport is pressure-dependent, because the product of diffusion constant and pressure, Dp, is constant. Hence, preservation of both plasma chemistry and mass transport is obtained by keeping both E and p constant during scale-up. As the area of the scaled-up reactor is larger (typically about 1 m^2) than its laboratory equivalent (1–100 cm^2), the total power or current supplied to the discharge must be enlarged in order to keep E constant. For a parallel plate reactor the power simply scales linearly with the area increase, while for other configurations this will not be a simple relation, because the plasma is not confined between the electrodes. Note that changing the area ratio of grounded to powered electrode will affect the dc self-bias voltage and the ion bombardment. The gas flow rate is the only external parameter that needs adjustment. It should be scaled up so that the average flow velocity is identical for laboratory- and industrial-scale reactors. Alternatively, one can keep the average residence time constant, which is defined as $\tau_r = pV/Q$, with V the volume of the reaction zone and Q the total mass flow rate. The residence time is a measure of the distance over which the reactive species diffuse in the reaction zone. With the pressure constant, the requirement of constant residence time scales the gas flow rate proportionally to the reaction volume.

The scale-up from a small to a large plasma reactor system requires only linear extrapolations of power and gas flow rates. However, in practice, the change in reactor geometry may result in effects on plasma chemistry or physics that were unexpected, due to a lack of precise knowledge of the process. Fine tuning, or even coarse readjustment, is needed, and is mostly done empirically.

A critical issue in scaling up a process is the uniformity in deposition rate and material quality. In general, once the deposition rate is constant within 5% over the whole substrate area, the material properties also do not vary much. After fine-tuning the power and gas flow rates, operators still may face in homogeneity issues. These can be caused by local changes in temperature, RF voltage, and gas composition, due to various causes. As an example, it has been reported that improper attachment of the substrate to the grounded electrode results in a local decrease of the deposition rate [150, 151].

Low contamination levels are readily achieved in laboratory scale UHV systems. Very high costs inhibit the use of UHV in industrial scale systems, however, so another, "local-UHV" approach has been proposed, viz. the plasma box reactor [152]. The substrate is mounted in a box, which is surrounded by a shell, which is pumped to a low pressure. The process pressure in the box is maintained by a throttle valve. As the pressure in the box is larger than the pressure in the surrounding shell, contaminants diffuse outwards and the incorporation of contaminants in the deposited layer is low.

Schropp and Zeman [11] have classified current production systems for amorphous silicon solar cells. They argue that cost-effective production of solar cells on a large scale requires that the product of the deposition time needed per square meter and the depreciation and maintenance costs of the system be small. Low deposition rates must be accompanied by low costs. The costs requirement is the main drive for ongoing investigations into the question how to increase the deposition rates while maintaining device quality material. Current production system configurations can be divided into three classes: (1) single-chamber systems, (2) multichamber systems, and (3) roll-to-roll systems [11].

In single-chamber systems there is no transport of substrates under vacuum conditions, which makes the system simple, but contamination levels may be relatively high. The addition of a load lock lowers contamination levels and requires only a linear transport mechanism. During the deposition process temperature and geometry cannot be changed easily, but changes are required for different layers in the solar cell. Substrates are loaded consecutively, and for every substrate a complete pump-down and heating are needed. The actual deposition time is much smaller than the total processing time. A higher throughput can be achieved by loading substrates in cassettes, as is general practice in the semiconductor industry. The investment is considered low, but so is the flexibility of the system.

Industrial scale multichamber systems offer many of the advantages that are characteristic of laboratory-scale systems. Typically, the deposition of each layer in a solar cell is performed in a separate chamber, and process parameter optimizations can be done for each layer individually. However, this high flexibility comes with high investment costs. Two types of multichamber systems can be distinguished, the cluster configuration and the in-line configuration. The cluster configuration is more or less an enlarged copy of a laboratory-scale system, and offers the greatest flexibility. Transport and isolation chambers can be shared by many deposition chambers, and, depending on the actual configuration, production can be continued even while some reactors are down. A cassette of substrates (batch) can be processed completely automatically, which increases throughput of the production system.

The in-line configuration consists of deposition chambers that are separated by isolation chambers [153]. The layer sequence of a solar cell structure prescribes the actual sequence of deposition chambers. The flexibility is much less than with a cluster configuration, and costs are generally much higher, but the throughput can also be much larger. In an in-line system the substrates can move while deposition takes place, which leads to very uniformly deposited layers, as uniformity of deposition is required only in one dimension (perpendicular to the moving direction).

In the roll-to-roll configuration the substrate consists of a continuous roll, e.g. stainless steel [154], SnO$_2$:F-coated aluminum [155], or plastic [156–158], which is pulled through a sequence of deposition chambers.

2.4. ASTER, a Research System

As a large part of the experimental and simulation results given in this chapter were obtained with or specifically calculated for the UHV multichamber deposition system ASTER (Amorphous Semiconductor Thin Film Experimental Reactor) at Utrecht University [14, 15], it is described in this section. It also is a more or less generic parallel plate multichamber research system. Similar commercially available multichamber equipment has been developed and described by Madan et al. [159]. It has also been installed at Utrecht University and is named Process Equipment for Amorphous Silicon Thin-Film Applications (PASTA).

The ASTER system is intended for use in research and consists of three identical plasma reactors, a spare chamber, and a load lock, all having the same outer dimensions. These chambers are connected radially via gate valves to a central transport chamber, in which a transport system or robot arm is located. All chambers are made of stainless steel, and all seals with valves, windows, gas-supply lines, and measuring devices are Conflat connections, ensuring that the vacuum will be of UHV quality. The ASTER system was completely designed and partly built in-house; the chambers and transport arm were manufactured by Leybold companies. A schematic drawing is shown in Figure 5, which illustrates the situation around 1990 [14, 15]. The spare chamber was designed first to

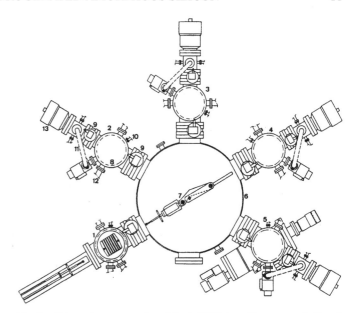

Fig. 5. Schematic representation of the ASTER deposition system. Indicated are: (1) load lock, (2) plasma reactor for intrinsic layers, (3) plasma reactor for p-type layers, (4) plasma reactor for n-type layers, (5) metal-evaporation chamber (see text), (6) central transport chamber, (7) robot arm, (8) reaction chamber, (9) gate valve, (10) gas supply, (11) bypass, (12) measuring devices, and (13) turbomolecular pump.

be used as an evaporation chamber, with which it was possible to evaporate contacts without exposing the sample to air. This was stopped for practical reasons. It was used then as a chamber where samples could be parked, to enhance productivity. As research shifted in the direction of plasma analysis, this fourth chamber was retrofitted with a quadrupole mass spectrometer [160]. After having built a separate system for plasma analysis in which the mass spectrometer was integrated [161], the fourth chamber now is in use for deposition of silicon nitride.

The central transport chamber is an 80-cm-diameter stainless steel vessel, and is pumped by a 1000-l/s turbomolecular pump, which is backed by a small (50 l/s) turbomolecular pump to increase the compression ratio for hydrogen, and by a 16-m^3/h rotating-vane pump. UHV is obtained after a bake-out at temperatures above 100°C (measured with thermocouples at the outside surface) of the whole system for about a week. A pressure in the low 10^{-11}-mbar range is then obtained. With a residual gas analyzer (quadrupole mass spectrometer, QMS) the partial pressures of various gases can be measured. During use of the system, the pressure in the central chamber is in the low 10^{-10}-mbar range due to loading of samples. Water vapor then is the most abundant species in the chamber.

The robot arm can transport a substrate holder for substrate sizes up to 10×10 cm^2. The arm can both rotate and translate, and is driven mechanically by two external dc motors, one for each movement. The translation mechanism was specially designed for fast transport of the samples. The mechanism is based on an eccentrically moving hinge joint, much like a stretching human arm, and it combines high speed with a shock-free start and stop of the movement. It takes only 4 s to translate the substrate holder from the transport arm to the outermost position in the reaction chamber. A gear transmission inside the transport chamber is used to decrease the torque to be applied to the feedthroughs. WSe_2 is used as lubricant for the gear wheels. The arm position is measured by potentiometers. The position of the arm in angular direction is reproducible to within $0.03°$ of the preset position.

A clear advantage of the central transport chamber is that a sample can be transported to the reaction chambers in any arbitrary sequence, without breaking the vacuum in the reaction chambers. Due to the very low pressure in the central chamber, the vacuum in the reaction chambers in fact is improved when a gate valve is opened. After a complete stretch of the robot arm, the substrate holder can be taken off the arm by a lift mechanism and clamped to the upper (grounded) electrode. Experienced operators are able to manually remove a substrate holder from one chamber and deliver it to any other within 30 s, including the time needed to open and close the gate valves. Automatic control by means of a programmable logic controller, coupled to a personal computer, only slightly shortens this transfer time. Because of the large heat capacity of the 12-mm-thick titanium substrate holder and the short transfer time, the cooling of the sample is limited to a few degrees. Fast transport and a low background pressure in the central chamber ensure that clean interfaces are maintained. Transport at 10^{-10} mbar for 30 s will give rise to a surface contamination of only about 0.01 monolayer of oxygen.

New substrates are mounted on the substrate holder, which then is loaded in the load-lock chamber, which subsequently is evacuated down to 10^{-6}–10^{-7} mbar with a 150-l/s turbomolecular pump in combination with a 25-m^3/h rotating vane pump. In this chamber the substrates are preheated to the desired temperature for the deposition, and at the same time they can be cleaned by dc argon sputtering. Usually the heating time (about 1 h) is much longer than the pumpdown time (less than 10 min). The substrate is transferred to the central chamber by a sequence of operations: opening the gate valve, translating the arm, lowering the substrate holder with the lift mechanism, retracting the arm (with the substrate holder), closing the gate valve. The pumping capacity of the central chamber is such that the 10^{-10}-mbar background pressure is reestablished within 10 s after closing the gate valve.

Each plasma reactor consist of a reaction chamber, an individual pumping unit, pressure gauges, and a dedicated gas-supply system with up to four different gases, according to the layer type to be deposited. Residual gas exchange is prevented by the separation of the pumping units, as well as by the presence of the central chamber. The background pressure in the plasma reactors is lower than 10^{-9} mbar, with partial pressures of water vapor and oxygen lower than 10^{-10} mbar, as a result of the capacity of the pumping unit, which is a 360-l/s turbomolecular pump backed with a 40-m^3/h rotating-vane pump. The process pressure typically is between 0.1 and 0.5 mbar. This can be accomplished by closing the gate valve between reaction chamber and pumping unit (9 in Figure 5). The chamber

then is pumped only via the bypass (11 in Figure 5), in which a butterfly valve regulates the process pressure.

A vertical cross section of the reaction chamber is shown in Figure 6. The inside diameter of the chamber amounts to 200 mm. The diameters of the grounded and powered electrodes are 180 and 148 mm, respectively, with a fixed interelectrode distance of 36.5 mm. The interelectrode distance has also been changed to 27 mm, and recently a modified powered electrode assembly has been retrofitted, with which it is possible to vary the interelectrode distance from 10 to 40 mm from the outside, i.e., without breaking the vacuum [162]. With process pressures in the range of 0.1–0.6 mbar the product of pressure and interelectrode distance, pL, may range from about 0.1 to 6 mbar cm. In practice, pL values are between 0.4 and 1.5 mbar cm, i.e., around the Paschen law minimum (see Section 3.2.4).

The lower electrode is coupled via a Π-type matching network to a 13.56-MHz generator. This network provides power matching between the RF power cable (50 Ω) and the plasma. Power levels are between 1 and 100 W, or between 6 and 600 mW/cm^2, using the area of the powered electrode.

The substrate holder is positioned face down (6 in Figure 6) on the upper electrode; thus deposition is upward, which prevents dust formed in the plasma from falling onto the substrate. The plasma is confined between the powered and the grounded electrode and the reactor walls, which in addition can be water-cooled. The substrate temperature can be varied up to about 500°C by means of fire rods in the grounded electrode. The temperature is regulated by a temperature controller, which measures the temperature with a thermocouple, also mounted in the upper electrode. The actual substrate temperature deviates from the temperature set by the controller, due to heat losses in the system. Therefore the actual substrate temperature has been measured by thermocouples on the substrate over a wide range of controller temperatures, and a calibration graph is used. Throughout this chapter, the actual substrate temperature is given.

The appropriate gas mixture can be supplied to the center of the reactor (4 in Fig. 6) via holes in the lower electrode (2 in Fig. 6), and is pumped out through the space between substrate electrode and the reactor wall to the exhaust (5 in Fig. 6). Alternatively, the gas mixture can be supplied horizontally, parallel to the electrodes, through a flange in the reactor wall, positioned between the electrodes (perpendicular to the plane of the cross section in Fig. 6, not shown). In this case, the gas is pumped out at the opposite side of the supply.

The volume of the reactor is about 10 l. At a typical process pressure of 0.2 mbar and a total gas flow rate of 60 sccm, the average residence time of molecules in the reactor amounts to about 1.3 s.

Different gases experience different conductances towards the pump. The conductance is a combined result of the manually adjustable butterfly valve in the bypass line and the pumping system. This results in gas-dependent pressures for identical gas flows, as the flow Q is related to pressure p and conductance C_{gas} as $Q = pC_{gas}$. In Figure 7 the pressures of

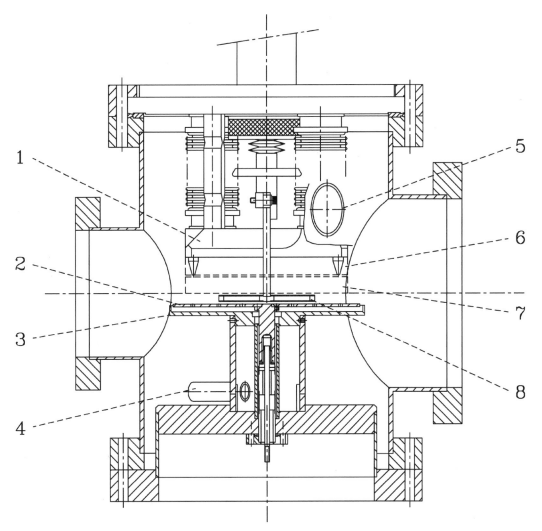

Fig. 6. Vertical cross section of the reaction chamber. Indicated are: (1) the grounded electrode, (2) the RF electrode, (3) the dark space shield, (4) the gas supply, (5) the gas exhaust, (6) the position of the sample holder during deposition, (7) the position of the sample holder when loaded, and (8) the lift mechanism.

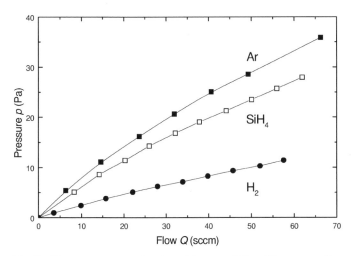

Fig. 7. The pressure p as a function of flow rate Q for different gases (Ar, SiH_4, H_2) at one and the same setting of the pressure regulating butterfly valve in the bypass (11 in Fig. 5). (Redrawn from E. A. G. Hamers, Ph.D. Thesis, Universiteit Utrecht, Utrecht, the Netherlands, 1998, with permission.)

pure argon, silane, and hydrogen are shown as a function of flow rate [163]. The setting of the butterfly valve was the same for all cases. As can be seen, the conductance for hydrogen is about a factor of 2–3 larger than for argon and silane. From the fact that the rate $dp/dQ = C^{-1}$ decreases with increasing flow, it is inferred that the flow of the gases leaving the reactor is a mixed laminar–molecular (Knudsen) flow [118].

As a consequence of gas-dependent conductances, the composition of a gas mixture may differ from the one expected on the basis of flow ratios. As an example, the partial pressures (measured with a QMS) of a range of silane–hydrogen mixtures at a total flow of 30 sccm is shown in Figure 8a as a function of the silane flow fraction $r_Q = Q_{SiH_4}/(Q_{SiH_4} + Q_{H_2})$ [163]. A flow of hydrogen yields a lower pressure (about 3 times lower) than the same flow of silane; cf. Figure 7. The actual partial pressure ratio $r_p = p_{SiH_4}/(p_{SiH_4} + p_{H_2})$ for these mixtures is shown in Figure 8b. It is clear that equal flow rates of different gases (in this case hydrogen and silane) do not lead to

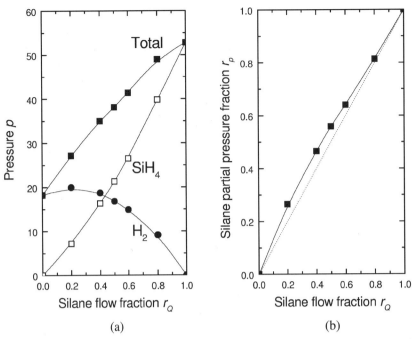

Fig. 8. (a) The total and partial pressures p and (b) the partial pressure ratio r_p of silane and hydrogen in a silane–hydrogen mixture, at different flow ratios r_Q. The total flow rate is 30 sccm. (Adapted from E. A. G. Hamers, Ph.D. Thesis, Universiteit Utrecht, Utrecht, the Netherlands, 1998, with permission.)

equal partial pressure ratios. This is important when comparing experimental and simulation data.

Each of the reaction chambers has its own gas supply system, consisting of two channels 3–5 m in length. It also includes a nitrogen purge system. One channel is connected to a gas manifold for three types of gases; the other channel is connected to a single gas bottle. With this configuration it is possible to supply the reaction chamber with either a continuous flow of different gases or a mixture of gases, which may change rapidly during deposition (cf. chemical annealing [164, 165]). The gases are connected via pressure regulators to the supply channels. They can be switched to go via a mass flow controller to the reaction chamber or to go to a drain, which is continuously purged with argon and pumped with an 8-m³/h rotating-vane pump. When the pressure in the drain or the exhaust lines is too high, a safety exhaust is opened with a much larger diameter than the regular exhaust line from the reaction chamber. Nitrogen is available for pressurizing the system, and also is used to dilute the exhaust lines from the reaction chambers, to ensure below-TLV values on the roof of the building. If one needs to clean a reaction chamber, a separate channel with NF_3 is available. Etching with a NF_3 discharge has been found to reduce considerably the amount of Si_xH_y compounds adsorbed on the wall [166].

In the plasma reactor dedicated for intrinsic material deposition (2 in Fig. 5), only hydrogen and silane are used, along with argon. A mixture of trimethylboron (5% TMB in H_2), SiH_4, and methane (CH_4) is used in the p-plasma reactor (3 in Fig. 5). Diborane can also be used. A mixture of phosphine [PH_3 (1% in H_2)] and SiH_4 is used in an n-plasma reactor (4 in Fig. 5). All

gases are of 6.0 quality (99.9999% pure) if available from manufacturers, and otherwise as pure as possible.

The low background pressure (10^{-9} mbar) together with the purity of the gases used ensures a low concentration of contaminants. Amorphous silicon films made in the intrinsic reactor have been analyzed by using ERD, which is available in our laboratory [114]. The determined oxygen content in these films typically is lower than 3×10^{18} cm^{-3}, which is somewhat lower than the values required for obtaining *device quality* films reported by Morimoto et al. [167].

It goes without saying that safety is very important in working with such a setup. A risk analysis has been performed. Safe operation procedures have been formulated. Personnel have been trained to work safely with the setup, which includes exchanging gas bottles using pressurized masks. Interlocks are installed, as well as emergency switches. Emergency power is available in case of power failures. Gases are stored in gas cabinets that were designed and installed with assistence of gas manufacturers. All valves in the gas-supply lines are normally closed valves: if compressed air (6 bar) is not available, all valves are by definition closed. Valves for nitrogen supply are normally open valves. Gate valves are double-action valves: compressed air is needed to either open or close the valve.

3. PHYSICS AND CHEMISTRY OF PECVD

3.1. General Introduction

A plasma can be defined as a partially ionized, quasineutral gas, consisting of about equal numbers of positive and negative charges, and a different number of un-ionized neutral molecules. An external source of energy is needed to sustain the plasma for a sufficiently long time. The simplest and most widespread method that is used is the electrical discharge, dc or RF. High electric fields applied at millibar pressures yield nonequilibrium plasmas: free electrons are accelerated to 1–10 eV, while ions and neutrals have low energy (0.1 eV). These *hot* electrons initiate chemical reactions through collisions with the *cold* neutrals. Therefore, processing temperatures can be much lower than in thermal CVD, which has tremendous consequences for the applicability of PECVD. Without PECVD modern-day electronic chips would never have been possible. For general references see, e.g., Bruno et al. [8], Grill [117], Chapman [134], and Ricard [168].

This section treats the plasma physics and plasma chemistry of the typical silane–hydrogen RF discharge, with occasional examples that employ a somewhat higher excitation frequency. Electrical characterization of the discharge is followed by an analysis of the silane chemistry. An appropriate set of gas phase species is presented, which are then used in the modeling of the plasma. A comparison is made between modeling results and experimental work in ASTER. Extension to 2D modeling is presented as well.

Plasma analysis is essential in order to compare plasma parameters with simulated or calculated parameters. From the optical emission of the plasma one may infer pathways of chemical reactions in the plasma. Electrical measurements with electrostatic probes are able to verify the electrical properties of the plasma. Further, mass spectrometry on neutrals, radicals, and ions, either present in or coming out of the plasma, will elucidate even more of the chemistry involved, and will shed at least some light on the relation between plasma and material properties. Together with ellipsometry experiments, all these plasma analysis techniques provide a basis for the model of deposition.

3.2. Plasma Physics

3.2.1. Plasma Sheath

A schematic layout of a typical parallel-plate RF-discharge system is depicted in Figure 4a. The RF power is capacitively coupled to the discharge between the two electrodes. Silane is introduced between the electrodes, and reaction products and unreacted gas are pumped away from the reactor. The substrate onto which a-Si:H is deposited is mounted on the grounded (top) electrode (anode). The RF voltage $V(t) = V_{RF} \sin \omega t$ is applied to the cathode, with $\omega/2\pi = 13.56$ MHz. This particular frequency had been chosen because radiating energy at this frequency would not interfere with communications [134]. However, higher harmonics do, and one should be careful to design proper shielding.

Accelerated electrons in the applied electric field ionize gas molecules, and in these ionization processes extra electrons are created. In the steady state the loss of charged particles is balanced by their production. Due to their much lower mass, electrons move much faster than ions. As a result, charge separation creates an electric field, compensating the differences in electron and ion velocities. The potential in the central region, or bulk plasma, is slightly positive with respect to the electrodes, due to the small surplus of positive ions.

The time-averaged potential profile is shown in Figure 4b. As ions cannot follow the oscillations in the applied electric field, it is this profile that ions experience. The bulk plasma is characterized by a constant potential, V_{pl}. In both *sheaths* (regions between plasma bulk and the electrodes), the ions experience a potential difference and are accelerated towards the electrodes. This leads to energetic ion bombardment of the electrodes. Electrons are expelled from the sheaths, so all ionization and dissociation processes must occur in the plasma bulk. Plasma light, resulting from emission from excited molecules, is emitted only from the plasma bulk; the sheaths are dark.

The electrons follow the oscillations in the electric field, and experience the time-dependent plasma potential. Due to the capacitor through which the RF power is coupled to the electrodes, no dc current flows through the plasma. The ion and electron currents towards each of the electrodes balance each other over one RF period.

In most systems the substrate electrodes are larger than the powered electrodes. This asymmetric configuration results in a negative dc self-bias voltage V_{dc} on the powered electrode. Without that, the difference in electrode areas would result in a net electron current per RF period [134, 169]. It has been shown that the ratio of the time-averaged potential drops for the sheaths at the grounded (V_{sg}) and the powered electrode (V_{sp}) are inversely proportional to a power of the ratio of the areas of the two electrodes (A_g, A_p) [134, 170–172]:

$$\frac{V_{sg}}{V_{sp}} = \left(\frac{A_p}{A_g}\right)^n \quad \text{with} \quad 1 < n < 4 \qquad (8)$$

For PECVD conditions one usually finds $n \leq 2$ [173]. The plasma potential V_{pl} equals the voltage drop at the sheath of the grounded electrode. It has been reported [170] that $V_{pl} = V_{sg} = (V_{RF} + V_{dc})/2$. In the case of an RF frequency of 13.56 MHz, typical values are 10–30 V for the plasma potential, and with $n = 2$ and with $1 < A_{sg}/A_{sp} < 2$ typical dc self-bias voltages are -50 V for low power (a few watts) to -250 V for high power (up to 100 W).

The plasma potential is the maximum value with which ions can be accelerated from the edge of the sheath towards the substrate, located at the grounded electrode. This may cause ion bombardment, which may induce ion–surface interactions such as enhancement of adatom diffusion, displacement of surface atoms, trapping or sticking of incident ions, sputtering, and implantation; see Section 6.2.1.

Meijer and Goedheer [174] have developed a 1D model with which, among other things, the dc self-bias and the sheath voltages can be calculated. It is assumed that the potential variation over the *complete* discharge is harmonic:

$$V_{sp}(t) - V_{sg}(t) = V_{RF} \sin \omega t + V_{dc} \qquad (9)$$

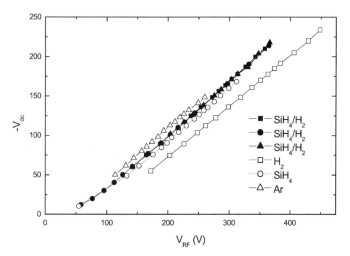

Fig. 9. Relation between dc self-bias voltage V_{dc} and applied RF voltage V_{RF} for different gases (Ar, H_2, SiH_4) and gas mixtures $[0.1 < [SiH_4]/([H_2] + [SiH_4]) < 0.9]$. Note that the slope is independent of the gas used.

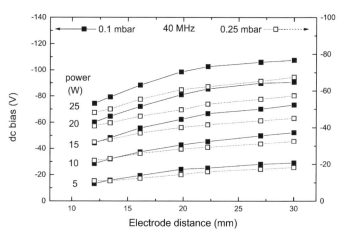

Fig. 10. Dc self-bias voltage V_{dc} as a function of electrode distance for two pressures and six power levels at 40 MHz.

A further assumption is that the time-averaged current is zero, because of the capacitive coupling of the power supply to the electrodes. With this model unequal-size electrode systems can be modeled, by using an area ratio $\alpha = A_g/A_p$. The ratio between V_{dc} and V_{RF} has been calculated for $1 < \alpha < 10$, and is in excellent agreement [174, 175] with experimental data [176, 177].

Meiling et al. [151] have presented measured data on the relation between V_{dc} and V_{RF} in the ASTER system with an electrode area ratio $\alpha = A_g/A_p$ of 2.1 for various silane–hydrogen mixtures and different applied power levels. It was found that $V_{dc} = aV_{RF} + C$, with $a = 0.64$ and $C = 31$, independent of power and independent of silane–hydrogen mixture. The values of α and a are in excellent agreement with the calculated results from Meijer and Goedheer [174]. Calculations by Goedheer et al. [149] show that this relation also is independent of frequency. The slope a only depends on the geometry (viz. the area ratio) of the reactor system. The intercept C has a different value for pure argon or pure hydrogen plasmas, as can be inferred from Figure 9. The intercept is not predicted in the 1D model by Meijer and Goedheer [174], on account of the low dimensionality of the model. In a 2D model, the intercept is correctly calculated [178].

3.2.2. Scaling Laws

Goedheer et al. [148, 149] have attempted to formulate scaling laws. They varied the power P, pressure p, and frequency ω and measured the dc self-bias V_{dc}. From their experimental results (frequency range 60–100 MHz, pressure range 20–60 Pa) a scaling law was formulated [148]:

$$P^{exp} \propto \omega^2 p^{2/3} V_{dc} \qquad (10)$$

Additional data (frequency range 13.56–100 MHz, pressure range 10–60 Pa) resulted in a slightly changed scaling law [149], $P^{exp} \propto \omega^2 p^{0.5} V_{dc}$, which was refined using modeling data

to [149, 179]

$$P^{model} \propto \omega^{1.5} p^{0.7} V_{dc} \qquad (11)$$

In addition a scaling law was derived in which also the electrode distance and radius were introduced:

$$P^{model} \propto \omega^{1.5} p^{0.7} \left(\frac{L}{R}\right)^{-0.4} R^{2.2} V_{dc} \qquad (12)$$

In the ASTER system (see Section 2.4), experiments were performed in order to test this scaling law. To this end, a newly designed RF electrode assembly was retrofitted to a deposition chamber. With this electrode setup, it was possible to change the electrode distance from the outside, without breaking the vacuum. A large data set was taken, consisting of 420 data points [162]: at three values of the pressure ($0.1 \leq p \leq 0.45$ mbar), five of the RF power ($5 \leq P \leq 25$ W), seven of the electrode distance ($12 \leq L \leq 30$ mm), and four of the RF frequency ($13.56 \leq \omega/2\pi \leq 50$ MHz).

A selection of the data set is shown in Figure 10. The dc self-bias voltage is shown as a function of electrode distance, for different values of the RF power, at two pressures and a frequency of 40 MHz. It is clear from the experimental data that the dc self-bias voltage is not strongly dependent on electrode distance. Rauf and Kushner [180] have reported the effect of electrode distance on dc self-bias voltage for the Gaseous Electronics Conference reference cell (GECRC) [133]. They observe a gradual increase of dc self-bias versus electrode distance much like the one presented here, but only above a certain value of L. They attribute this to a change in current flows. Bringing the electrodes, which are of equal size, closer together caused a larger part of the current to flow out through the grounded electrode. Hence, the discharge became more symmetric, and as a consequence the dc self-bias decreased.

Comparing published data on the relation between the dc self-bias and RF voltage between Rauf and Kushner [180] and our group [151] shows that the proportionality constant (slope) in the GECRC is nearly twice the one observed in the ASTER reactor. In other words, the ASTER reactor is less asymmetric

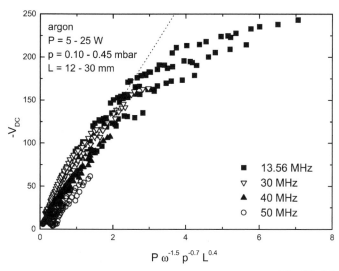

Fig. 11.   Scaling law between dc self-bias voltage V_{dc} and $P\omega^{-1.5}p^{-0.7}L^{0.4}$ for an argon discharge.

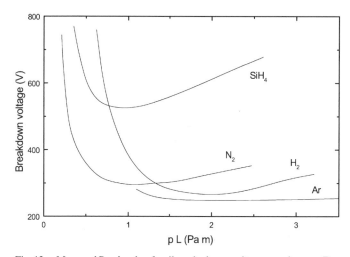

Fig. 12.   Measured Paschen law for silane, hydrogen, nitrogen, and argon. (Redrawn from K. M. H. Maessen, Ph.D. Thesis, Universiteit Utrecht, Utrecht, the Netherlands, 1988, with permission.)

than the GECRC. This may be an additional explanation why the V_{dc} data presented here do not show a sudden change as a function of electrode distance.

In Figure 11 the complete data set is plotted using the scaling law

$$P^{exp} \propto \omega^{1.5}p^{0.7}L^{-0.4}V_{dc} \qquad (13)$$

It is clear that the scaling law, which was deduced from modeling results, also holds for the experimental results. However, above a value for the dc self-bias of about -150 V, deviations occur. A closer look at the data reveals that the 13.56 MHz data are responsible for this. Following the arguments discussed by Rauf and Kushner [180], this may well be explained by the fact that two different matching networks were employed, viz. a Π-network for the 13.56-MHz data, and an L-network for the other frequencies. These circuits are not included in the model. Another possible explanation is the presence of higher harmonics at high power levels, and large heat dissipation in the network.

3.2.3. Effective Discharge Power

It is difficult to determine the true power that is dissipated in the discharge. Usually, the reading of a directional power meter in the 50-Ω line just before the matching network is reported as the RF power. The effective or true power is not simply the difference between input and reflected power, which is usually also measured with the same power meter. Godyak and Piejak have presented three methods to determine the true power: a subtractive method, a phase shift method, and an integration method [181]. All three methods give the same results. In the phase shift method the discharge power is determined from the discharge voltage and current and the phase shift ϕ between them. In the integration method the product of voltage and current waveform is integrated over one RF cycle.

The subtractive method was adapted from Horwitz [182], and is easiest in use. The principle is to measure the power delivered to the system, including the tuned matching network, in the case that the discharge is on (P_{tot}) and in the case that it is off, i.e. when the system is evacuated (P_{vac}), with the constraint that in both cases P_{tot} and P_{vac} are measured for the same electrode voltage V_{pp}. The matcher efficiency [181] or power transfer efficiency η_P [183] then is defined as

$$\eta_P = \left(\frac{P_{tot} - P_{vac}}{P_{tot}}\right)_{V_{pp}=const} = \frac{P_{eff}}{P_{tot}} \qquad (14)$$

where P_{eff} is the effective power dissipated in the discharge. The power transfer efficiency η_P has been reported to vary between 40% and 90% as a function of pressure [181] and excitation frequency [183].

Both the subtractive and the phase shift method are valid as long as the RF waveform remains sinusoidal [184]. For asymmetrical discharges the integral method is to be used.

3.2.4. Paschen Law

In order to sustain a discharge between two parallel plate electrodes, the product of pressure p and interelectrode distance L has to satisfy the Paschen law for the gas mixture used. This law relates the product pL and the breakdown voltage V_b (or minimum sustaining voltage) of the discharge:

$$V_b = \frac{C_1 pL}{C_2 + \ln pL} \qquad (15)$$

with C_1 and C_2 constants that are dependent on the gas mixture. For large pL the breakdown voltage is proportional to pL. For most gases the minimum breakdown voltage is between 100 and 500 V for pL in the range of 0.1–10 Torr cm [117]. Measurements on argon, silane, hydrogen, and nitrogen [185, 186] are shown in Figure 12. Note that 1 Torr cm equals 1.33 Pa m.

Table II.   Electron Impact Collisions[a]

Acronym	Electron impact reaction	Threshold energy (eV)	Reference	Comment
SE1	$SiH_4 + e^- \rightarrow SiH_2^+ + 2H + 2e^-$	11.9	[206]	Lumped
S_2E1	$Si_2H_6 + e^- \rightarrow Si_2H_4^+ + 2H + 2e^-$	10.2	[206]	Lumped
SV1	$SiH_4 + e^- \rightarrow SiH_4^{(1-3)} + e^- \rightarrow SiH_4 + e^-$	0.271	[212]	Vibr. exc. 1–3
SV2	$SiH_4 + e^- \rightarrow SiH_4^{(2-4)} + e^- \rightarrow SiH_4 + e^-$	0.113	[212]	Vibr. exc. 2–4
SE2	$SiH_4 + e^- \rightarrow SiH_2 + 2H + e^-$	8.3	[199, 201, 202]	Fragment. 83%
SE3	$SiH_4 + e^- \rightarrow SiH_3 + H + e^-$	8.3	[199, 201, 202]	Fragment. 17%
S_2E2	$Si_2H_6 + e^- \rightarrow SiH_3 + SiH_2 + H + e^-$	7.0	[201]	Fragment. 91% [204]
SE4	$SiH_4 + e^- \rightarrow SiH_3^- + H$	5.7	[208]	Lumped
HE1	$H_2 + e^- \rightarrow H_2^+ + 2e^-$	15.7	[207]	Lumped
HV	$H_2^{v=0} + e^- \rightarrow H_2^{v\neq0} + e^- \rightarrow H_2^{v=0} + e^-$	0.54	[213]	$v = 1$
		1.08		$v = 2$
		1.62		$v = 3$
HE2	$H_2 + e^- \rightarrow H + H + e^-$	8.9	[205]	$b^3\Sigma_u^+$ exc.

[a] Adapted from G. J. Nienhuis, Ph.D. Thesis, Universiteit Utrecht, Utrecht, the Netherlands, 1998, with permission.

At low pL the energy of the electrons is limited by collisions with the electrodes, and a large voltage is required to compensate the energy lost to the walls. At high pL electrons will collide with gas phase species, and the voltage again is large to compensate for the energy lost to collisions within the plasma, which however promotes polymerization or powder formation. Usually one employs a voltage just above the Paschen minimum to be able to deposit device quality material. Interelectrode distances typically are between 10 and 50 mm. Ishihara et al. [187] have reported only monohydride bonds to be present in the material for $pL \approx 1$ Torr cm. Dihydride bonds begin to appear above $pL = 1.5$ Torr cm.

3.3. Plasma Chemistry

For low-pressure plasmas containing mainly inert gases the electrons can be characterized by a Maxwellian electron energy distribution function (EEDF). However, deviations have been reported [188]. The shape of the EEDF depends on the energy lost via electron collisions and the energy gained by the time-dependent electric field in the plasma. Adding silane and/or hydrogen to the plasma results in a non-Maxwellian EEDF [189]. Electrons determine the plasma characteristics, as they are responsible for elastic collisions and inelastic collisions, which lead to rovibrational and electronic excitation, ionization, and dissociation of the neutral gas molecules. In elastic collisions the internal energies of the neutrals are unchanged, whereas the energies, momenta, and directions of the electrons undergo large changes. In inelastic collisions the energy that is transferred from the electron to the neutral species causes many processes to occur, such as dissociation, (dissociative) ionization, (dissociative) attachment, recombination, and vibrational, electronic, and dissociative excitation. As a result of vibrational excitation of the molecules in the plasma, the collisions between these excited molecules allow for redistribution

of vibrational states, charge exchange between molecules, excitation of molecules, dissociation, and dissipation of vibrational energy to walls of the system. Each of these processes is characterized by its own (energy-dependent) cross section $\sigma(E)$. The rate constant k of the collisional processes is proportional to the product of EEDF [$f(E)$] and cross section, according to $k = \int E^{1/2} f(E)\sigma(E) \, dE$. The onset of these processes is at 10 eV or higher, while the EEDF peaks around 1–2 eV. Hence, the tail of the EEDF mainly determines which processes occur.

3.3.1. Species Considered

In a silane–hydrogen discharge the feedstock gases SiH_4 and H_2 take part in all the processes that occur. A large number of reactions have been proposed (see e.g. Kushner [190]). Nienhuis et al. [191] have performed a sensitivity analysis in their self-consistent fluid model, from which a minimum set of reactions have been extracted for a typical low-pressure RF discharge. Tables II and III list these reactions. They will be used in the plasma models described in subsequent sections. The review articles on silane chemistry by Perrin et al. [192] and on hydrogen by Phelps [193] and Tawara et al. [194] have been used. The electron collision data are compiled in Figure 13 [189].

3.3.2. Electron Impact Collisions

Dissociation of the gases SiH_4 and H_2 by electron impact will create reactive species (radicals) and/or neutrals (Si_2H_6 and even higher-order silanes [195–198]). Atomic hydrogen is an important particle because it is formed in nearly all electron impact collisions, and the H-abstraction reaction [199, 200] of (di)silane is an important process, as is seen from sensitivity study. Dissociation of SiH_4 can create different SiH_x (with $x = 0, 1, 2, 3$) radicals. Only silylene (SiH_2) and silyl (SiH_3)

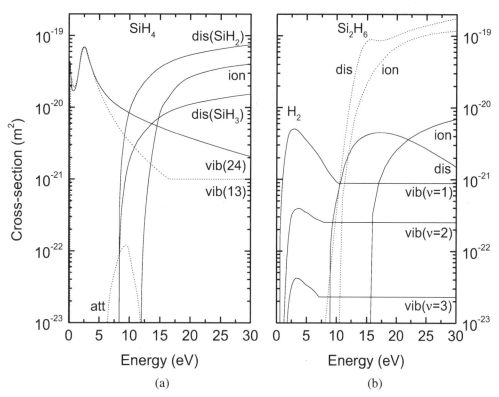

Fig. 13. Cross sections for electron collisions for SiH_4–H_2: (a) SiH_4, (b) Si_2H_6 (dotted lines) and H_2 (solid lines). Abbreviations are: ion, ionization; dis, dissociation; vib, vibrational excitation; att, attachment. See Table II for details and references. (Adapted from G. J. Nienhuis, Ph.D. Thesis, Universiteit Utrecht, Utrecht, the Netherlands, 1998, with permission.)

Table III.   Chemical Reactions[a]

Acronym	Chemical reaction	Reaction rate ($cm^3 s^{-1}$)	Reference	Comment
HS	$H + SiH_4 \rightarrow SiH_3 + H_2$	$2.8 \times 10^{-11} \exp(-1250/T_{gas})$	[192, 199]	
HS_21	$H + Si_2H_6 \rightarrow Si_2H_5 + H_2$	$1.6 \times 10^{-10} \exp(-1250/T_{gas})$	[192, 199, 200]	
HS_22	$H + Si_2H_6 \rightarrow SiH_3 + SiH_4$	$0.8 \times 10^{-10} \exp(-1250/T_{gas})$	[192, 199, 200]	
HS_n	$H + Si_nH_{2n+2} \rightarrow Si_nH_{2n+5} + H_2$	$2.4 \times 10^{-10} \exp(-1250/T_{gas})$	[192]	Est., $n > 2$
H_2S1	$H_2 + SiH_2 \rightarrow SiH_4$	$3.0 \times 10^{-12}[1 - (1 + 0.00023 p_0)^{-1}]$	[192]	p_0 (Pa)
SS1	$SiH_4 + SiH_2 \rightarrow Si_2H_6$	$2.0 \times 10^{-10}[1 - (1 + 0.0032 p_0)^{-1}]$	[192]	p_0 (Pa)
S_2S1	$Si_2H_6 + SiH_2 \rightarrow Si_3H_8$	$4.2 \times 10^{-10}[1 - (1 + 0.0033 p_0)^{-1}]$	[192]	p_0 (Pa)
S_nS	$Si_nH_{2n+2} + SiH_2 \rightarrow Si_{n+1}H_{2n+4}$	$4.2 \times 10^{-10}[1 - (1 + 0.0033 p_0)^{-1}]$	[192]	Est., p_0 (Pa), $n > 2$
SS2	$SiH_3 + SiH_3 \rightarrow SiH_4 + SiH_2$	1.5×10^{-10}	[192, 215]	
SS3	$SiH_3^- + SiH_2^+ \rightarrow SiH_3 + SiH_2$	1.2×10^{-7}	[214]	
S_2S2	$SiH_3^- + Si_2H_4^+ \rightarrow SiH_3 + 2 SiH_2$	1.0×10^{-7}	[214]	
H_2S2	$SiH_3^- + H_2^+ \rightarrow SiH_3 + H_2$	4.8×10^{-7}	[214]	
S_2S_2	$Si_2H_5 + Si_2H_5 \rightarrow Si_4H_{10}$	1.5×10^{-10}	[192]	Est.

[a] Adapted from G. J. Nienhuis, Ph.D. Thesis, Universiteit Utrecht, Utrecht, the Netherlands, 1998, with permission.

radicals are included in the plasma chemistry, because these two radicals are considered to be the most important ones [190]. The SiH_3 radicals are assumed to be responsible for the film growth (they are *growth precursors*), and the SiH_2 radicals are assumed to be precursors of higher silanes, i.e., Si_nH_{2n+2}. Because the H-abstraction reaction is important, it is necessary to include the Si_nH_{2n+1} radical if Si_nH_{2n+2} is included. Polymers larger than Si_8H_{18} can be ignored [191]. Dissociation of Si_2H_6 pro-

duces mainly SiH_3 and Si_2H_5. The latter is also assumed to be a growth precursor.

The total dissociation cross sections of silane and disilane have been taken from Perrin et al. [201]. An uncertainty in the present knowledge of the silane chemistry is the branching ratio of the silane dissociation channels [192]. Here, the branching ratio is taken from Doyle et al. [197], who suggest using the branching ratio determined by Perkins et al. [202] for photolysis, viz., a branching of 83% into SiH_2 and 17% into SiH_3 (reactions SE2 and SE3 in Table II). However, Perrin et al. [192] suggest that the formation of SiH is also of importance at low electron energies (9.9 eV): it should be at least 20%. In contrast, Hertl et al. [203] have measured SiH densities in typical SiH_4–H_2 discharges, and they show that the SiH density is at least one order of magnitude lower than the SiH_2 density. Therefore, the SiH radical is not included. The fragmentation pattern of disilane dissociation (reaction S_2E2 in Table II) is taken from Doyle et al. [204]; it is 91% $SiH_3 + SiH_2 + H$ and 9% $H_3SiSiH + 2H$. As a simplification the second branching path is neglected. Dissociation of H_2 by electron impact is included [205].

Ionization of SiH_4, Si_2H_6, and H_2 can produce a large number of different positive ions [206, 207]. For simplicity all the ionization cross sections of each background neutral leading to different ions have been lumped and attached to the ion that is most likely to be formed: for SiH_4, Si_2H_6, and H_2 these ions are SiH_2^+, $Si_2H_4^+$, and H_2^+, respectively (reactions SE1, S_2E1, and HE1 in Table II).

Similarly to the lumping of the positive ionization cross sections, the dissociative attachment cross sections of each background neutral leading to different anions have been lumped to one cross section (reaction SE4 in Table II). The most abundant anions of SiH_4, and H_2 are SiH_3^- and H^-, respectively [208, 209]. Since calculations have shown that dissociative attachment of H_2 has a negligible influence, it is not included. The sum of the dissociative attachment cross sections of Si_2H_6 is comparable in magnitude to that of SiH_4 [210]. However, as the partial pressure of Si_2H_6 is at most a tenth that of SiH_4, the dissociative attachment of Si_2H_6 is neglected.

Vibrational excitation by electron impact of the background neutrals is an important process, because it is a major cause of energy loss for the electrons [reactions SV1 (SiH_4 stretching mode), SV2 (SiH_4 bending mode), and HV in Table II]. Moreover, the density of the vibrationally excited molecules has been reported to be important [211]. However, information about reaction coefficients of vibrationally excited molecules is scarce [192]. Here, only the vibrational excitation of SiH_4 and H_2 is included [212, 213].

3.3.3. Chemical Reactions

Table III lists the set of chemical reactions, with reaction rates and references. The rate constants for mutual neutralization of a positive ion and SiH_3^- (reactions SS3, S_2S2, and H_2S2 in Table III) have been taken from a scaling formula due to Hickman [214], where the electron affinities are taken from Perrin et al. [192] and the ion temperature is assumed equal to the gas temperature. The end products of the mutual neutralization reaction of SiH_3^- and $Si_2H_4^+$ are taken to be SiH_3 and $2SiH_2$ (not SiH_3 and S_2H_4); hence no extra species (Si_2H_4) has to be introduced. The radical–radical reaction $SiH_3 + SiH_3$ can lead to Si_2H_6 or $SiH_2 + SiH_4$ [215]. However, Si_2H_6 will be an excited association product, and collisional deexcitation will only occur

at high pressures [192]. This excited Si_2H_6 is likely to dissociate into SiH_2, and SiH_4, and therefore the only end product of $SiH_3 + SiH_3$ is assumed to be $SiH_2 + SiH_4$.

Chemical reactions of positive ions with background neutrals [193, 216], e.g., $SiH_2^+ + SiH_4 \rightarrow SiH_3^+ + SiH_3$ or $H_2^+ + H_2 \rightarrow H_3^+ + H$, have not been included in the model. The reaction coefficients of these reactions are of equal magnitude to those of some radical–neutral reactions, but the densities of the ions are much lower than the radical densities, making ion–neutral reaction of negligible influence on the partial pressure of the background neutrals and the growth rate. For a detailed analysis of the densities and fluxes of ions, the lumped sum of ion densities should be unraveled again.

3.3.4. Plasma–Wall Interaction

The plasma–wall interaction of the neutral particles is described by a so-called *sticking model* [136, 137]. In this model only the radicals react with the surface, while nonradical neutrals (H_2, SiH_4, and Si_nH_{2n+2}) are reflected into the discharge. The surface reaction and sticking probability of each radical must be specified. The nature (material, roughness) and the temperature of the surface will influence the surface reaction probabilities. Perrin et al. [136] and Matsuda et al. [137] have shown that the surface reaction coefficient β of SiH_3 is temperature-independent at a value of $\beta = 0.26 \pm 0.05$ at a growing a-Si:H surface in a temperature range from room temperature to 750 K. The surface sticking coefficient s of SiH_3 is $s = 0.09 \pm 0.02$ in a temperature range from room temperature to 575 K. Above 575 K the ratio s/β increases to 1 (saturation) at a temperature of 750 K. Note that typical deposition temperatures are below 575 K.

Information about the surface reaction coefficients of radicals Si_nH_{2n+1} where $n > 1$ is scarce. Because the structure of these radicals is similar to that of SiH_3, the same surface reaction coefficients are used. It is assumed that if Si_nH_{2n+1} radicals recombine at the surface with a hydrogen atom, a Si_nH_{2n+2} neutral is formed and is reflected into the discharge. Another possibility is the surface recombination of Si_nH_{2n+1} radicals with physisorbed Si_mH_{2m+1} radicals at the surface. Matsuda et al. [137] have shown that the probability of surface recombination of SiH_3 with physisorbed SiH_3 decreases with increasing substrate temperature. Doyle et al. [204] concluded that at a typical substrate temperature of 550 K, SiH_3 radicals mainly recombine with physisorbed H atoms.

For the SiH_2 radicals the surface reaction coefficients have been taken as $s = \beta = 0.8$ [192]. This sticking coefficient is large because there is no barrier for insertion of this species into the a-Si:H surface. Kae-Nune et al. [217] specify a surface recombination probability of about 1 for atomic hydrogen on an a-Si:H surface during deposition that results mainly in recombination of H with an H-atom bounded to the surface.

If a Si_nH_{2n+1} radical sticks to the surface, not all hydrogen will be incorporated in the layer, and therefore molecular hydrogen will be reflected into the discharge. The amount of molecular hydrogen thus reflected is taken such that the percentage of hydrogen in the deposited layer is 10%. The growth rate is calculated by summing the flux of Si-containing neutrals and ions toward the electrodes and assuming a constant density of 2.2 g/cm^3 of the deposited layer [218].

3.3.5. Transport Coefficients

The diffusion constant D_j of neutral particle j is calculated in two steps. First, the binary diffusion coefficient D_{ij} in each of the background gas species (SiH$_4$, Si$_2$H$_6$, H$_2$) is calculated, following Perrin et al. [192]. Then D_j is approximated using Blanc's law [219]:

$$\frac{p_{\text{tot}}}{D_j} = \sum_{i=\text{background}} \frac{p_i}{D_{ij}} \qquad (16)$$

where p_{tot} is the total pressure and p_i is the partial pressure of background species i.

The ion mobility coefficients μ_j are calculated similarly. First, the ion mobility of ion j in background neutral i is calculated using the low-E-field Langevin mobility expression [219]. Then Blanc's law is used to calculate the ion mobility in the background mixture. The ion diffusion coefficient is calculated with Einstein's relation

$$D_i = \frac{k_B T_{\text{ion}}}{e} \mu_i \qquad (17)$$

where T_{ion} is the ion temperature, which is assumed to be equal to the gas temperature T_{gas}.

4. PLASMA MODELING

All properties of a-Si:H are a result of operating conditions of the discharge (pressure, RF frequency, RF power, gas mixture, geometry). Optimization of properties mostly is done empirically, due to the complex chemistry of the silane–hydrogen discharge, including plasma–wall (= substrate) interaction. Nevertheless, serious attempts have been made to model the discharge, thus providing knowledge of the chemistry and the deposition process, with the ultimate goal to find the optimum parameter space of the specific reactor system used.

Various types of models have been used to describe and study glow discharges. These models differ in the approximations that are made [e.g., the dimension, self-consistency, approach (fluid or kinetic), and number of chemical processes], which will influence the physical relevance of the modeling results. However, a higher accuracy (fewer approximations) will require a larger computational effort. Much effort has been put into the development of self-consistent models for glow discharges. Examples are fluid models [220, 221], particle-in-cell/Monte Carlo (PIC/MC) models [222–224], and hybrid models [190, 225, 226]. More extensive overviews have been presented elsewhere [195, 196, 227, 228]. Most of these self-consistent models have been used for discharges with a relatively simple chemistry and no plasma–wall interaction of the

neutrals. Other models include the silane–hydrogen chemistry and the surface interaction to a relatively high degree of completeness, but these models are not self-consistent with respect to the rates of the electron impact collisions (i.e., ionization, dissociation, excitation, and attachment); see for instance Perrin [211]. Models developed for plug-flow silane–hydrogen discharges [190, 195, 196] are self-consistent with respect to the silane–hydrogen chemistry, the electron impact collisions, the plasma–wall interaction, and the transport phenomena. However, in this type of models the composition of the neutral background gas is taken equal to the gas feedstock composition, that is, the depletion of silane will be low.

4.1. 1D Fluid Discharge Model

4.1.1. Model Description

Nienhuis et al. [189, 191] have developed a self-consistent fluid model that describes the electron kinetics, the silane–hydrogen chemistry, and the deposition process of a perfectly stirred reactor. Due to the discharge processes, the composition of the background neutrals will differ from the composition of the feedstock gases, because hydrogen and higher-order silanes are formed, while silane is consumed. Fluid models describe the discharge by a combination of balances for the particle, momentum, and energy densities of the ions, electrons, and neutrals obtained from moments of the Boltzmann transport equation, see e.g. [174, 229, 230]. A limitation of a fluid model is the requirement that the mean free path of *all* particles must be less than the characteristic dimensions of the discharge. The balance equations are coupled to the Poisson equation to calculate the electric field, where only electrostatic forces are taken into account. Solving these equations yields the behavior in space and time of the particle densities and the electric potential during one cycle of the periodic steady state RF discharge.

Rates for electron impact collisions as well as the electron transport coefficients depend on the EEDF. The collision rates, the electron transport coefficients, and the average electron energy are obtained by solving the Boltzmann equation for the EEDF. This EEDF has been expanded in two terms with respect to the velocity [231]. The EEDF is calculated as a function of the electric field for a given composition of the neutral background density. A lookup table has been constructed to obtain the collision rates and electron transport coefficients as functions of the average electron energy, which are used in the fluid model. The deposition process is modeled by the use of sticking coefficients, which actually specify the boundary conditions of the particle density balance equations.

The 1D fluid model describes a discharge created and sustained between two parallel plates, where the time-dependent discharge characteristics only vary along the z-axis, i.e., the direction normal to the plates. Plasma parameters such as the dc self-bias voltage, uniformity of the deposition, and radial transport cannot be studied. Self-consistent models in more than one dimension can model these transport phenomena more appropriately [220, 221, 232]. In reality radial dependences do exist in the discharge, due to radial transport processes. Silane is introduced and is transported from the inlet into the discharge, where it may be transformed into other species via chemical reactions. This new species may be transported outside the discharge toward the pump outlet. The model described here corrects for this by means of additional source terms. The reactor is divided into two volumes, the discharge volume (defined as the volume between the two electrodes), and the discharge-free volume; see also Figure 4a. The charged particles and the radicals are confined to the discharge volume. Radical reaction times or diffusion times are in the millisecond range, shorter than the diffusion time to the outside wall, and shorter than the residence time (~ 1 s) in the reactor. Hence, radial dependences are neglected, and the reactor is considered perfectly stirred. The radical densities are spatially resolved in the z-direction, while the nonradical neutral densities are assumed to be homogeneous throughout the whole reactor volume. In the remaining discharge-free volume only nonradical neutrals are considered.

The main reason for using a 1D model is the reduction of the computational effort compared to the higher-dimensional models. This reduction to one dimension is acceptable because it is possible to study the sustaining mechanisms and the chemistry in the discharge with a 1D self-consistent model.

To verify the model and to establish in which process parameter space it is valid, a comparison is made with experimental data. These experimental data are the partial pressures of silane, disilane, and hydrogen and the growth rate, obtained during deposition of amorphous silicon. Data are compared for various combinations of the total pressure in the reactor, the electrical power, and the frequency of the power source.

A sensitivity study of the influence of the elementary data (i.e., reaction coefficients, cross sections, and transport coefficients) has been performed to determine the importance of specific elementary data, so to guide further research in this area [189].

4.1.2. Particle Balances

For each nonradical neutral j in the discharge volume V_D, the balance equation of their total number N_j^D can be written as

$$\frac{\partial N_j^D}{\partial t} = \int_D S_{\text{reac},j}\, dV + Q_j^{F \to D} \qquad (18)$$

where the source term $S_{\text{reac},j}$ represents the creation or destruction of corresponding species j by electron impact collisions and/or by chemical reactions, e.g., $A + B \to C + D$, where A, B, C, and D indicate the reaction species. The source term for the particles in such a reaction is

$$S_{\text{reac,C}} = S_{\text{reac,D}} = -S_{\text{reac,A}} = -S_{\text{reac,B}} = n_A n_B k_r \qquad (19)$$

where k_r is the reaction rate. For electron collisions the reaction rate depends on the local (averaged) electron energy; see Figure 13 and Table II. Other chemical reaction rate coefficients are constant; see Table III. The term $Q_j^{F \to D}$ represents the total number of neutrals j per second that are transported from the discharge-free (F) volume into the discharge (D) volume.

For each nonradical neutral j in the discharge-free volume V_F, the balance equation of their total number N_j^F can be written as

$$\frac{\partial N_j^F}{\partial t} = Q_j^{\text{in}} + Q_j^{\text{pump}} - Q_j^{F \to D} \qquad (20)$$

where the term Q_j^{in} represents the number of particles that enter the reactor volume per second as feedstock gas. The term Q_j^{pump} represents the number of pumped particles per second, and is given by

$$Q_j^{\text{pump}} = -\frac{N_j^F}{\tau} \qquad (21)$$

with τ the average residence time, which is computed so that the total pressure in the discharge p_{tot}, given by the ideal gas law, equals the preset pressure.

Only the steady-state situation is considered; therefore the partial time derivatives are set equal to zero. Then, combining the above equations and solving for N_j^F yields (with $n_j = N_j^F / V_F$)

$$n_j = \frac{\tau}{V_F}\left(Q_j^{\text{in}} + \int_D S_{\text{reac},j}\, dV \right) \qquad (22)$$

For each particle j in the discharge volume (electrons, ions, and radicals) the density balance can be written as

$$\frac{\partial n_j}{\partial t} + \frac{\partial \Gamma_j}{\partial z} = S_{\text{reac},j} \qquad (23)$$

where n_j is the density and Γ_j is the flux of particle j.

4.1.3. Particle Fluxes

In the fluid model the momentum balance is replaced by the drift–diffusion approximation, where the particle flux Γ consists of a diffusion term (caused by density gradients) and a drift term (caused by the electric field E):

$$\Gamma_j = \mu_j n_j E - D_j \frac{\partial n_j}{\partial z} \qquad (24)$$

where μ_j and D_j are the mobility and diffusion transport coefficients, respectively, which are charge-sign-dependent (see Section 3.3.5). The mobilities of electrons and negative ions are negative; that of positive ions is positive. For the neutral particles the mobility is set equal to zero. The use of the drift–diffusion approximation is allowed when two conditions are satisfied. The first condition is that the characteristic time between momentum transfer collisions (of the charged particles) is much smaller than the RF period. In other words, the drift–diffusion approximation assumes that a charged particle reacts instantaneously to a change of the electric field, because the moment equations of the fluid model are simplified by neglecting the particle inertia. The typical momentum transfer frequency for electrons is about 100 MHz. For ions the characteristic momentum transfer frequency is only a few megahertz. To use the drift–diffusion approximation for RF frequencies higher than a few megahertz the electric field is replaced by an effective electric field [233], which takes account of the fact that the ions do not respond instantaneously to the actual electric field, unless the momentum transfer frequency is much larger than the applied frequency.

The second condition for the use of the drift–diffusion approximation specifies that the mean free path of these collisions is much smaller than the characteristic gradient lengths in the discharge. In other words the motion of the particles must be collision-dominated. The mean free path of the electrons is the largest among all particles, and their description will become less accurate in the low pressure regime, i.e., at pressures lower than about 10 Pa. The cross section for momentum transfer for the electrons in silane is about 10^{-19} m^2 [212]. At 10 Pa, this gives a mean free path of about 5×10^{-3} m, which is of the order of the length scale of the electron density gradients in the discharge. For ions also this pressure limit is valid.

4.1.4. Poisson Equation

The electric field E and potential V are calculated using the Poisson equation,

$$\frac{\partial^2 V}{\partial z^2} = -\frac{e}{\epsilon_0}\left(\sum_{i=\text{ions}} q_i n_i - n_e\right) \qquad E = -\frac{\partial V}{\partial z} \qquad (25)$$

with e the elementary charge, ϵ_0 the vacuum permitivity, n_e the electron density, n_i the ion density, and q_i the sign of the ion charge i, i.e., $+1$ or -1.

4.1.5. Energy Balances

The electron energy density $w_e = n_e\epsilon$ (the product of the electron density n_e and average electron energy ϵ) is calculated self-consistently from the second moment of the Boltzmann equation:

$$\frac{\partial w_e}{\partial t} + \frac{\partial \Gamma_w}{\partial z} = -e\Gamma_e E + S_w \qquad (26)$$

where Γ_w is the electron energy density flux:

$$\Gamma_w = \frac{5}{3}\left(\mu_e w_e E - D_e \frac{\partial w_e}{\partial z}\right) \qquad (27)$$

and μ_e and D_e are the electron mobility and electron diffusion coefficients. The term S_w in the electron energy balance equation is the loss of electron energy due to electron impact collisions. The various source terms for electron impact collisions depend on the local average electron energy ϵ, which is calculated by dividing the energy density w_e by the electron density n_e. The energy loss due to elastic collisions of electrons with neutrals [234] amounts to only about 1% of the total energy dissipation at typical discharge conditions, and this term is therefore neglected.

No energy balances for the ions and neutrals are included. The energy balance for neutrals is complex, due to the large

range of possible energy exchange reactions in a typical discharge. Moreover, most of the reaction coefficients are unknown [211]. It is assumed that the ions have nearly the same energy as the neutral atoms, because of the efficient energy transfer between ions and neutrals due to their nearly equal mass. Further, the energy gain of the ions from the electric field is low because of the high inertia of the ions; the ions cannot respond to the fast oscillations of the electric field. This assumption is correct for the bulk of the discharge, where the time-averaged electric field is nearly zero. However, the time-averaged electric field in the sheaths is *not* negligible and will increase the energy of the ions far above the thermal energy of the neutrals, as is experimentally shown for silane–hydrogen plasmas [235, 236].

4.1.6. Boundary Conditions

In order to solve the differential equations, the boundary conditions of the potential and density profiles must be specified. The potential at the grounded electrode ($z = L$, radius R) is set equal to zero. The potential at the driven electrode ($z = 0$) is set equal to

$$V(z = 0, t) = V_{\text{RF}} \sin(2\pi \nu_{\text{RF}} t) \qquad (28)$$

where V_{RF} is the amplitude of the potential and ν_{RF} is the frequency of the oscillating potential. The value of V_{RF} is adjusted until the dissipated power in the discharge equals the preset electrical power. The dissipated power (expressed in watts) is given by the ohmic heating of the charged particles:

$$P_{\text{dis}} = \frac{e\pi R^2}{\tau_{\text{RF}}} \sum_{\substack{j=\text{charged}\\\text{particles}}} \int_0^{\tau_{\text{RF}}} \int_0^L E(z,t)\Gamma_j(z,t)\,dz\,dt \qquad (29)$$

where τ_{RF} is the RF period.

As the electric field always points in the direction of the electrode, the densities of the electrons and negative ions are set equal to zero at the electrode. It is assumed that the ion flux at the electrodes has only a drift component, i.e., the density gradient is set equal to zero. The conditions in the sheath, which depend on pressure, voltage drop, and sheath thickness, are generally such that secondary electrons (created at the electrodes as a result of ion impact) will ionize at most a few molecules, so no ionization avalanches will occur. Therefore, secondary electrons can be neglected.

The boundary conditions for the density balance equation of neutrals are given by the plasma–wall interaction (for instance by deposition processes), which is described by a sticking model; see Figure 14. This surface interaction is modeled by using surface reaction coefficients β_j, which specify the probability that neutral j will react at the surface. The coefficient $R_j = 1 - \beta_j$ represents the fraction of the incident neutrals that are reflected into the gas phase keeping their initial identity. It is assumed that the nonradical neutrals do not react at the wall, whereas the radical neutrals may do so, and form a nonradical neutral or contribute to a-Si:H layer formation.

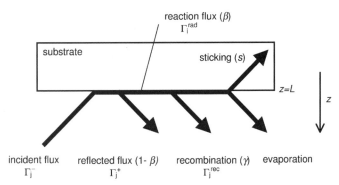

Fig. 14. Schematic representation of plasma–wall interaction for neutrals. (Adapted from G. J. Nienhuis, Ph.D. Thesis, Universiteit Utrecht, Utrecht, the Netherlands, 1998.)

For the radical neutrals, boundary conditions are derived from diffusion theory [237, 238]. One-dimensional particle diffusion is considered in gas close to the surface at which radicals react (Figure 14). The particle fluxes in the two z-directions can be written as

$$\Gamma_j^- = \frac{1}{4} n_j v_{\text{th},j} + \frac{D_j}{2} \frac{\partial n_j}{\partial z} \quad (30)$$

$$\Gamma_j^+ = \frac{1}{4} n_j v_{\text{th},j} - \frac{D_j}{2} \frac{\partial n_j}{\partial z} \quad (31)$$

where $v_{\text{th},j}$ is the thermal velocity of neutral j, given by $v_{\text{th},j} = \sqrt{8k_B T_{\text{gas}}/\pi m_j}$ with m_j the mass of particle j. The first terms in Eqs. (30) and (31) are the random contributions, and the second terms are the contributions due to the gradient in the concentration of diffusing particles. The resulting net flux Γ_j is given by $\Gamma_j = \Gamma_j^+ - \Gamma_j^-$. The fraction R_j of the flux incident on the wall (Γ_j^-) is equal to the flux coming from the wall, $\Gamma_j^+ = R_j \Gamma_j^-$, when no radicals are produced at the wall. The boundary condition of radical j, which reacts at the surface of the electrode, then is given by

$$\Gamma_j^{\text{rad}} = -D_j \frac{\partial n_j^{\text{rad}}}{\partial z}\bigg|_{z=0} = -\frac{\beta_j}{1-\beta_j/2} \frac{v_{\text{th},j}}{4} n_j^{\text{rad}}\bigg|_{z=0} \quad (32)$$

where Γ_j^{rad} and $n_j^{\text{rad}}|_{z=0}$ are the flux of radicals j reacting at the surface, and their density at the surface, respectively.

The recombination probability γ_{ij} is defined as the probability that a neutral radical j reacts on the surface, forming a volatile stable product, i.e., a nonradical neutral i, which is reflected into the discharge (e.g. $SiH_{3(gas)} + H_{(surface)} \rightarrow SiH_{4(gas)}$). This causes a flux into the discharge:

$$\Gamma_j^{\text{rec}} = \sum_{j \neq i} \frac{\gamma_{ij}}{\beta_j} D_j \frac{\partial n_j^{\text{rad}}}{\partial z}\bigg|_{z=0} \quad (33)$$

The sticking coefficient $s_i = \beta_i - \gamma_i$ (where $\gamma_i = \sum_{j \neq i} \gamma_{ij}$) is the probability that a neutral i sticks to the surface. In case of silane radicals Si_nH_m, this means film growth by Si incorporation. At the same time atomic hydrogen is released or evaporated by the layer. The evaporation flux is chosen such that the amount of hydrogen in the layer is 10%, i.e., the typical

device quality hydrogen content (the possibility of a hydrogen-rich overlayer is neglected).

For each neutral particle the boundary conditions are formulated; one then obtains a coupled set of equations, which is solved to obtain the boundary conditions for all neutral density balance equations.

4.1.7. Electron Energy Distribution Function

In order to calculate the rates for electron impact collisions and the electron transport coefficients (mobility μ_e and diffusion coefficient D_e), the EEDF has to be known. This EEDF, $f(\vec{r}, \vec{v}, t)$, specifies the number of electrons at position $\vec{r}$ with velocity $\vec{v}$ at time t. The evolution in space and time of the EEDF in the presence of an electric field is given by the Boltzmann equation [231]:

$$\frac{\partial f}{\partial t} + \vec{\nabla}_{\vec{r}} \cdot (\vec{v}f) - \vec{\nabla}_{\vec{v}} \cdot \left(\frac{e}{m_e} \vec{E}(\vec{r}, t)f\right) = \left(\frac{\partial f}{\partial t}\right)_{\text{coll}} \quad (34)$$

where $(\partial f/\partial t)_{\text{coll}}$ is the collision term. Some approximations have to be made, as this equation is complex (details can be found in [189]).

First, it is assumed that the EEDF is spatially uniform and temporally constant, which is allowed if the energy relaxation time of the EEDF is much shorter than the RF-cycle duration, and if the relaxation length is much smaller than the typical gradient scale length. This assumption implies a spatially and temporally constant electric field. It reduces the Boltzmann equation to a problem exclusively in the velocity space.

Second, in the collision term only electron–neutral collisions are considered, because RF plasmas are weakly ionized. The inelastic collisions considered are ionization, dissociation, excitation, and attachment (see also Table II). The crude first approximation is relaxed in the fluid model by the introduction of the spatial and temporal dependence of the average electron energy. The collision rates, transport coefficients, and average electron energy are calculated in a range of electric fields E for a given composition of the background neutrals. The results are then tabulated to obtain a lookup table of the rates and transfer coefficients as a function of the averaged electron energy. In the fluid model the (time- and space-dependent) average electron energy is calculated with the energy balance equation. The validity of this assumption has been demonstrated [175, 221, 230, 234, 239].

Third, a further simplification of the Boltzmann equation is the use of the two-term spherical harmonic expansion [231] for the EEDF (also known as the Lorentz approximation), both in the calculations and in the analysis in the literature of experimental data. This two-term approximation has also been used by Kurachi and Nakamura [212] to determine the cross section for vibrational excitation of SiH_4 (see Table II). Due to the magnitude of the vibrational cross section at certain electron energies relative to the elastic cross sections and the steep dependence of the vibrational cross section, the use of this two-term approximation is of variable accuracy [240]. A Monte Carlo calculation is in principle more accurate, because in such a model the spatial and temporal behavior of the EEDF can be included. However, a Monte Carlo calculation has its own problems, such as the large computational effort needed to reduce statistical fluctuations.

The method to calculate the rates of electron impact collisions as used in this model has been studied by Meijer et al. [230] and has also been used by others [221, 234].

4.1.8. Calculation Scheme

The system of equations of the fluid model as well as the Boltzmann equation of the EEDF must be untangled numerically. The balance equations are spatially discretized (129 grid points in this study) using the Sharfetter–Gummel exponential scheme [241]. An implicit second-order method is used to numerically treat the time evolution of the balance equations [220]. The electron transport equation is solved simultaneously with Poisson's equation to avoid numerical instabilities. A Newton method is used to solve the resulting set of nonlinear equations. This approach limits the required number of time discretization steps (here 80 time steps per RF cycle). Doubling the number of grid points and time steps resulted in 10^{-2} and 10^{-4} for the maximum relative changes of the partial pressures (SiH_4, H_2, and Si_2H_6) and of the deposition rate, respectively. A multigrid technique is used to enhance the convergence of the fluid model [242]. Further details about the numerical techniques of the fluid model and the Boltzmann solver can be found elsewhere [175, 243].

The fluid model calculates the density profiles based on the electron impact collision rates calculated by the Boltzmann solver. This Boltzmann solver needs the densities of the background neutrals as input. Therefore, the fluid model and the Boltzmann solver are run alternately until the changes in the densities of the background neutrals are less than 10^{-4}. Convergence of the fluid model is obtained when the relative change ϵ_n of the discharge parameters (density profiles and potential profile) at the beginning of two subsequent RF periods (n and $n+1$) is less than 10^{-6}. To check whether this criterion is sufficient, some calculations were continued until ϵ_n reached a value 10^{-7}. In those cases, the partial pressures of SiH_4, H_2, and Si_2H_6 and the deposition rate differed relatively by at most 10^{-5} from the results obtained with ϵ_n equal to 10^{-6}.

4.1.9. Simulation Results

The 1D fluid discharge model has been applied to the ASTER deposition system (see Section 2.4). The deposition reactor has an inner volume of 10 l and an inner diameter of 20 cm. The upper electrode is grounded (see Fig. 4a), and the powered electrode is located 2.7 cm lower. Other typical silane–hydrogen discharge parameters are summarized in Table IV.

Using these discharge settings in the model yields the following partial pressures for the three background neutral particles considered: 7.72 Pa for SiH_4, 11.51 Pa for H_2, and 0.55 Pa for Si_2H_6. With these partial pressures, and the data on electron collisions as given in Table II, first the EEDF can be calculated,

Table IV. Typical Settings for a Silane–Hydrogen Discharge in the ASTER System

Parameter	Value	Unit
Electrode spacing	0.027	m
Electrode radius	0.08	m
Reactor volume	0.010	m^3
Silane flow	30.0	sccm
Hydrogen flow	30.0	sccm
Pressure	20.0	Pa
Gas temperature	400	K
Substrate temperature	550	K
RF frequency	50	MHz
RF power	5	W

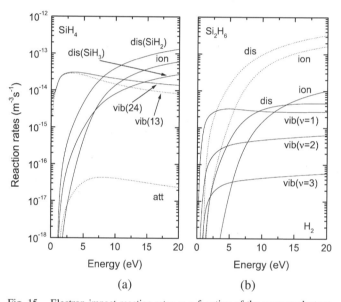

Fig. 15. Electron impact reaction rates as a function of the average electron energy. A 1 : 1 mixture of SiH_4 and H_2 was used, at a total pressure of 83 Pa. (a) Reaction rates for SiH_4, (b) reaction rates for Si_2H_6 (dotted lines) and H_2 (solid lines). Abbreviations are: ion, ionization; dis, dissociation; vib, vibrational excitation; att, attachment. See Table II for details and references. (Adapted from G. J. Nienhuis, Ph.D. Thesis, Universiteit Utrecht, Utrecht, the Netherlands, 1998.)

followed by the calculation of the electron transport coefficients and electron impact rates. From a comparison of a Maxwellian EEDF and the two-term Boltzmann EEDF as calculated here, it follows that the Maxwellian EEDF underestimates the EEDF at lower energies and overestimates it at higher energies [189].

In Figure 15 the electron impact rates for the various reactions (Table II) are shown as functions of the average electron energy. In a typical silane–hydrogen discharge the average electron energy amounts to about 5 eV. As can be seen in Figure 15, at an energy of 5 eV the dissociation of silane and disilane and the vibrational excitation of silane have the highest rates.

Ohmic heating of the electrons amounts to about 80% of the plasma power. This dissipated power is mainly used for

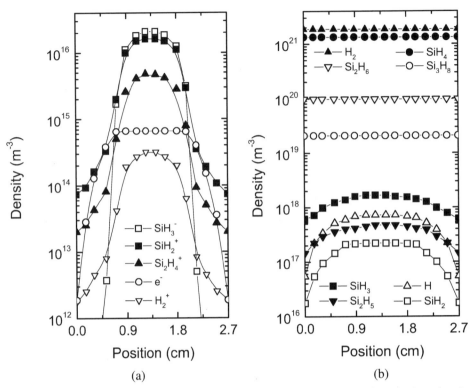

Fig. 16.   Time-averaged densities of (a) charged and (b) neutral species for the discharge settings given in Table IV. Note the difference in density scales. (Compiled from G. J. Nienhuis, Ph.D. Thesis, Universiteit Utrecht, Utrecht, the Netherlands, 1998.)

vibrational excitation of the molecules, in contrast to a inert gas discharge (e.g. Ar), where a much larger fraction of the power is used for the acceleration of the positive ions in the sheaths. Another remarkable outcome is the high density of negative ions (2.38×10^{16} m^{-3}) compared to the electron density (7.19×10^{14} m^{-3}). This is typical for an electronegative discharge, where the rate coefficient for recombination of a negative ion with positive ions is low.

Figure 16 shows the calculated time-averaged densities of charged (Fig. 16a) and neutral species (Fig. 16b). The SiH$_3^-$ density is the highest in this case, followed by the SiH$_2^+$ density: quasineutrality is preserved. It can be clearly seen that the SiH$_3^-$ ions are confined in the discharge. The electrons are mobile and therefore much less confined. The temporal variation of the ion density profile is small, except in the sheaths. The mass of the ions is too high to respond instantaneously to the changing electric field. Ion transport is only influenced by the time-averaged electric field. The total (positive) ion flux to the electrode is 4.06×10^{18} m^{-3} s^{-1}.

From Figure 16b it is clear that the density of the nonradical neutrals (H$_2$, SiH$_4$, Si$_2$H$_6$, and Si$_3$H$_8$) is much higher than that of radicals (SiH$_3$, H, Si$_2$H$_5$, and SiH$_2$). Moreover, from the decreasing density near the electrodes, the species that react can be identified. Note that the SiH$_3$ density is about one order of magnitude larger than the SiH$_2$ density.

4.1.10. Comparison with Experiments

In this sub-subsection the results of the 1D model are shown together with data taken from experiments in the ASTER system. Both the partial pressures of the background neutrals (H$_2$, SiH$_4$, and Si$_2$H$_6$) and the deposition rates are accessible in experiment and model. The comparison is made as a function of one of the discharge parameters: the RF frequency, the RF power, or the total pressure in the reactor. Other process parameters have been kept constant; see Table IV.

Of course, care should be taken when comparing the experimental data and the modeling results; the model is an approximation, and the experiment has its uncertainties. The most important approximation is that the model is one-dimensional. A main uncertainty of the experiment is the relation between *source* power and *plasma* power (see Section 3.2.3). The source power is defined as the power delivered by the power generator. In experiments, the source power is a discharge setting (Table IV). The plasma power is the power dissipated by the plasma. As a rule, this plasma power is smaller than the source power, due to losses in the matching network, which matches the plasma impedance to the impedance of the source power generator, viz., 50 Ω [180]. From comparison of experimental data with modeling results (e.g., [220]) in a wide range of discharge settings it is inferred that the plasma power is approximately 50% of the source power. This is a crude approximation, as in fact this efficiency depends on discharge settings

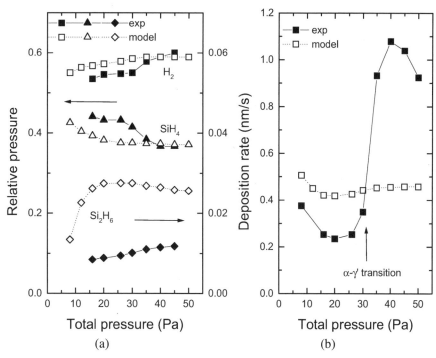

Fig. 17.   (a) The relative pressures (i.e., the ratio of the partial pressure to the total pressure) of H_2, SiH_4, and Si_2H_6, and (b) the deposition rate, as a function of total pressure at an RF frequency of 50 MHz and a plasma power of 5 W. Other discharge settings are given in Table IV. Modeling results are in dotted lines and open symbols, experimental data in solid lines and filled symbols. Note the sudden increase at 30 Pa, i.e., the transition from the α- to the γ'-regime. (Compiled from G. J. Nienhuis, Ph.D. Thesis, Universiteit Utrecht, Utrecht, the Netherlands, 1998.)

and the resulting discharge impedance. Agreement, therefore, is not expected between experimental and modeling results. Tendencies, however, will be clear from the results.

Another important problem to address is the regime in which the discharge is operating, whether the α (dust-free) regime or the γ'-regime, where dust plays an important role in the discharge [244] (see Section 6.1.6). Dust particles consist of large clusters of Si and H atoms, and are negatively charged by attachment of free electrons. They can be considered as an additional wall surface. Consequently, the free electron density decreases. In order to maintain the discharge, i.e., the amount of ionization, the average electron energy has to increase. This also leads to an enhanced dissociation of SiH_4 and thus to a larger radical production.

Generally, at low pressures and at low power levels the discharges operate in the α-regime. At a certain critical pressure, which depends more or less upon the other discharge parameters, the discharge makes a transition from the α- to the γ'-regime. The modeling does not take dust effects into account. The discharge settings used are interesting, because at increasing total pressure the discharge regime is changed from α to γ', at a critical pressure of about 30 Pa.

4.1.10.1. Pressure Variation

In Figure 17 are shown the effects of total pressure on the relative pressures (i.e., the ratio of the partial pressure to the total pressure) of silane, hydrogen, and disilane (Fig. 17a) and on the

deposition rate (Fig. 17b). The RF frequency is 50 MHz, and the plasma power is 5 W. The relative pressure of hydrogen slowly increases, and the relative pressure of silane slowly decreases, both in model as well as in experiment. This is caused by an increase in silane depletion at higher total pressures, which results from a higher power dissipation by the electrons. At 16 Pa the fraction of dissipated power that is used to heat electrons is 79%, at 50 Pa this increases to 97%. This increase leads to a higher dissociation rate of silane, and explains the increased depletion.

The relative pressure of disilane increases as a function of total pressure, due to the increased production of radicals, which is a result of increased dissociation of silane, as well as to the shorter gas volume reaction times and longer diffusion times to the walls, which result from increasing the pressure.

Below 30 Pa the trends agree between model and experiment. The model slightly underestimates the silane partial pressure, by about 10%, and overestimates the hydrogen pressure by about 6%. Possible causes for this discrepancy are the simple 1D geometry of the model, the assumed relation between plasma power and source power, and the approximation for the pumping by using an average residence time. The discrepancy for the disilane data might be explained e.g. by questioning the dissociation branching ratio of silane. Layeillon et al. [195, 196] have discussed this, and conclude that dissociation of silane should result in a larger fraction of SiH_3.

At about 30 Pa, the experimental deposition rate (Fig. 17b) clearly shows a sudden increase, which is not revealed by the

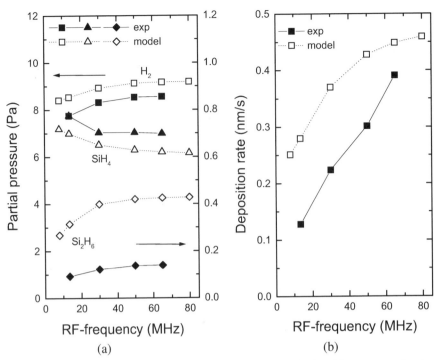

Fig. 18. (a) The partial pressures of H_2, SiH_4, and Si_2H_6 and (b) the deposition rate, as a function of RF frequency at a total pressure of 16 Pa and a plasma power of 5 W. Other discharge settings are given in Table IV. At these settings the discharge is in the α-regime. Modeling results are in dotted lines and open symbols, experimental data in solid lines and filled symbols. (Compiled from G. J. Nienhuis, Ph.D. Thesis, Universiteit Utrecht, Utrecht, the Netherlands, 1998.)

model. In addition, the partial pressure of silane decreases (and of hydrogen increases). These features have been observed in other experiments as well [245–248], and are caused by the transition from the α- to the γ'-regime.

4.1.10.2. RF Frequency Variation

In Figure 18 are shown the effects of RF frequency on the partial pressures of silane, hydrogen, and disilane (Fig. 18a) and on the deposition rate (Fig. 18b). The total pressure is 16 Pa, and the plasma power is 5 W. The discharge is in the α-regime.

Both model and experiment show the same tendency: the dissociation of silane increases as a function of RF frequency. This can be explained following the arguments from Heintze and Zedlitz [236]. They found that the power dissipated by the ions in the sheaths decreases with increasing RF frequency. This is confirmed by the modeling results: at 13.56 MHz about 50% of the power is used for the acceleration of the ions, whereas at 80 MHz this is reduced to about 15%. Hence, at higher RF frequencies more power is used for the heating of electrons, which in turn leads to a higher dissociation of silane. As a result, more hydrogen and silane radicals are produced, and more molecular hydrogen is formed. A saturation is observed when all power is consumed by the electrons.

The experimentally found linear increase of the deposition rate as a function of frequency is not seen in the modeling results that show saturation; see Figure 18b. The linear increase has also been measured by others [119, 120, 249], up to an RF frequency of 100 MHz. Howling et al. [250] have measured this

linear relationship, while taking special care that the effective power is independent of frequency.

A possible explanation for the difference in tendencies of the deposition rate between experiment and model is that in the model the surface reaction and sticking coefficients of the radicals are taken to be independent of the discharge characteristics. In fact, these surface reaction coefficients may be influenced by the ions impinging on the surface [251]. An impinging ion may create an active site (or dangling bond) at the surface, which enhances the sticking coefficient. Recent experiments by Hamers et al. [163] corroborate this: the ion flux increases with the RF frequency. However, Sansonnens et al. [252] show that the increase of deposition rate cannot be explained by the influence of ions only.

The discrepancy may also be caused by the approximations in the calculation of the EEDF. This EEDF is obtained by solving the two-term Boltzmann equation, assuming full relaxation during one RF period. When the RF frequency becomes comparable to the energy loss frequencies of the electrons, it is not correct to use the time-independent Boltzmann equation to calculate the EEDF [253]. The saturation of the growth rate in the model is not caused by the fact that the RF frequency approaches the momentum transfer frequency ν_{me} [254]. That would lead to less effective power dissipation by the electrons at higher RF frequencies and thus to a smaller deposition rate at high frequencies than at lower frequencies.

Another possible explanation is the approximation that the vibrationally excited silane molecules are treated in the model

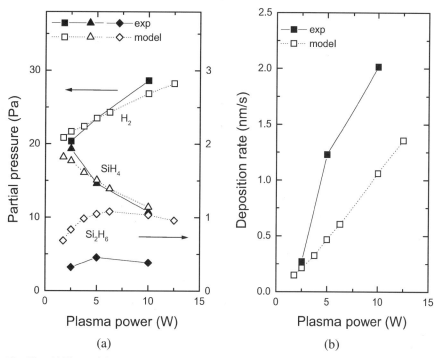

Fig. 19. (a) The partial pressures of H_2, SiH_4, and Si_2H_6 and (b) the deposition rate, as a function of plasma power at a total pressure of 40 Pa and an RF frequency of 50 MHz. Other discharge settings are given in Table IV. At these settings the discharge is in the γ'-regime. Modeling results are in dotted lines and open symbols, experimental data in solid lines and filled symbols. (Compiled from G. J. Nienhuis, Ph.D. Thesis, Universiteit Utrecht, Utrecht, the Netherlands, 1998.)

as ground state molecules. This is based on the assumption that collisional deexcitation in the gas phase and at the walls is fast enough. The density of vibrationally excited molecules is thus neglected. The fraction of the total power used for vibrationally exciting silane molecules increases with the RF frequency, which may result in a nonnegligible excited state density. Inclusion of vibrationally excited silane molecules leads to a higher radical production and deposition rate. Vibrationally excited silane molecules have a lower threshold for electron dissociation than a ground state molecule. Also, multivibrational excitation may lead to homogeneous pyrolysis ($e^- + SiH_4 \rightarrow SiH_4^{vib} \rightarrow SiH_2 + H_2$). Perrin et al. [192, 211] have ruled out this pyrolysis channel because of the high rates of relaxation for vibrationally excited silane molecules, but their observation was made for an RF frequency of 13.56 MHz. In addition, vibrationally excited silane molecules have a drastically increased cross section for dissociative attachment [192, 255, 256], which increases the production of negative ions and the loss of electrons. In order to sustain the discharge, a higher average electron energy is required, which also leads to a higher dissociation.

4.1.10.3. Plasma Power Variation

In Figure 19 are shown the effects of RF power on the partial pressures of silane, hydrogen, and disilane (Fig. 19a) and on the deposition rate (Fig. 19b). The total pressure is 40 Pa, and the RF frequency 50 MHz. The discharge is in the γ'-regime.

Results are shown as a function of the supposed effective RF power, i.e., 50% of the power set at the power source.

The decrease of the silane partial pressure and the concomitant increase of the hydrogen partial pressure as a function of plasma power can be understood in terms of the increased electron density and electron energy. Both lead to a higher dissociation of silane and hydrogen. The silane radicals and atomic hydrogen thus created diffuse to the electrodes, where deposition takes place and molecular hydrogen is formed.

The partial pressure of disilane as a function of plasma power first increases due to the higher production of silane radicals. Above a certain power the increase in disilane dissociation is higher than the increase in disilane production; hence the disilane partial pressure decreases again. Similar behavior has been observed by Kae-Nune et al. [217].

In Figure 19b the deposition rate is seen to increase with the plasma power. Andújar et al. [246] also measured an increase of the deposition rate with increasing total power. They found that the relation between deposition rate and electrical power depends on the pressure. At pressures lower than about 20 Pa, their measured deposition rate increases linearly with power, which is the same relation as shown by the 1D discharge model. However, at pressures higher than 20 Pa there is a saturation of the deposition rate as a function of the power, which is similar to the experiments shown here (Fig. 19b). Andújar et al. [246] conclude that at 20 Pa the discharge regime is changed from the α- to the γ'-regime. This value of 20 Pa depends on discharge parameters.

4.1.10.4. Sensitivity Analysis

A possible cause for the discrepancy between experiment and model is an error in the elementary parameters (reaction coefficients, cross sections, and transport coefficients) which are obtained from the literature. With a sensitivity study it is possible to identify the most important processes [189].

To calculate the sensitivity with respect to one of the elementary parameters, its value is changed slightly and the resulting change in the plasma characteristic (i.e., the partial pressure of SiH_4, Si_2H_6, or H_2, or the deposition rate) is calculated. A measure $\Lambda = \Delta Y / \Delta X$ of the sensitivity is the relative change ΔY of the plasma characteristic Y divided by the relative change ΔX of the elementary parameter X. A sensitivity study has been made for a discharge at a pressure of 16 Pa effective power of 5 W, and RF frequency of 50 MHz. The partial pressures of SiH_4 and H_2 are not sensitive ($-0.01 < \Lambda < +0.01$) to any change in the chemical parameters. The partial pressure of Si_2H_6 is most sensitive to the reactions $SiH_4 + SiH_2$ ($\Lambda = +0.18$) and $Si_2H_6 + SiH_2$ ($\Lambda = -0.14$) and to the dissociation branching ratio SiH_3/SiH_2 of SiH_4 ($\Lambda = -0.20$). The correctness of the silane dissociation branching ratio has also been discussed by Layeillon et al. [195, 196]. They concluded that the branching ratio should result in a larger fraction of SiH_3 in order to improve modeling results. This is also observed here. The growth rate is most sensitive to the surface reaction ($\Lambda = -0.20$) and sticking probability ($\Lambda = +0.29$) of SiH_3 and to the diffusion coefficients of SiH_2 ($\Lambda = +0.12$) and H ($\Lambda = -0.12$).

Nienhuis [189] has used a fitting procedure for the seven most sensitive elementary parameters (reactions $SiH_4 + SiH_2$ and $Si_2H_6 + SiH_2$, dissociation branching ratio of SiH_4, surface reaction coefficient and sticking probability of SiH_3, and diffusion coefficients of SiH_2 and H). In order to reduce the discrepancy between model and experiment significantly, some values had to be changed by more than two orders of magnitude, which clearly is unrealistic. Uncertainties in the literature never are larger than 50%.

4.2. 2D Fluid Discharge Model

4.2.1. Model Description

In the previous section a 1D discharge model was formulated and results were compared with experimental data. Although this discharge model is sophisticated, there are some disadvantages related to its low dimensionality. Therefore, this model is extended to a 2D discharge model, the extension being only in geometry.

The self-consistent fluid model was developed by Nienhuis [189]. This model is two-dimensional (cylindrically symmetric). It is an extension, purely in geometry, of a previously published 1D model [191]. Only the important aspects of the model are summarized here. A fluid model consists of balance equations for the densities of the various species in the discharge (i.e. electrons, positive and negative ions, and neutrals) and an energy balance equation for the electrons. The

heavy species (ions and neutrals) are assumed to have a uniform temperature close to the wall temperature, because of the good energy transfer both between the heavy species and between the heavy species and the wall. Heating or cooling of the electrodes is not considered. The electric field in the discharge is calculated using Poisson's equation. For the charged particle fluxes the drift-diffusion approximation is used, with the actual electric field for the electrons and an effective field for the ions, in order to allow for their inertia [233]. In the 1D model [191] the nonradical neutrals were treated with a model based on perfect mixing between the discharge volume and the discharge-free volume of the reactor vessel. Here, diffusive transport of all neutrals is taken into account, while modeling only the discharge volume.

In a fluid model the correct calculation of the source terms of electron impact collisions (e.g. ionization) is important. These source terms depend on the EEDF. In the 2D model described here, the source terms as well as the electron transport coefficients are related to the average electron energy and the composition of the gas by first calculating the EEDF for a number of values of the electric field (by solving the Boltzmann equation in the two-term approximation) and constructing a lookup table.

A sticking model is used for the plasma–wall interaction [137]. In this model each neutral particle has a certain surface reaction coefficient, which specifies the probability that the neutral reacts at the surface when hitting it. In case of a surface reaction two events may occur. The first event is sticking, which in the case of a silicon-containing neutral leads to deposition. The second event is recombination, in which the radical recombines with a hydrogen atom at the wall and is reflected back into the discharge.

The chemistry (i.e. the species and reactions included) is the same as described for the 1D model. However, here the higher-order silanes Si_nH_{2n+2} and silane radicals Si_nH_{2n+1} are limited to $n < 4$ to reduce the computational effort. Thus, Si_3H_7 and Si_3H_8 are representative for all silanes with $n > 2$. The formation of powder (large silane clusters) is not taken into account in this model. The discharge settings for the calculations shown here are a total pressure of 20 Pa, a power input of 250 W m^{-2}, an RF frequency of 50 MHz, and an inlet flow of 30 sccm of SiH_4 and 30 sccm of H_2. This parameter set is chosen because it results in a situation where most of the silane is consumed in a large reactor. This situation is required for economic reasons in industrial applications.

4.2.2. Comparison with Experiments: Small Reactor

The 2D model has been validated and compared with the 1D model and with the experimental data shown in the previous section. Details can be found in [189]. It was found that the partial pressures (averaged in the case of the 2D model) differed by at most 9%. On the symmetry axis in the 2D model the partial pressures of SiH_4, H_2, and Si_2H_6 are 7.77, 11.23, and 0.64 Pa, respectively. For the 1D model the values are 7.72, 11.51, and 0.55 Pa, respectively. Averaged over the whole discharge,

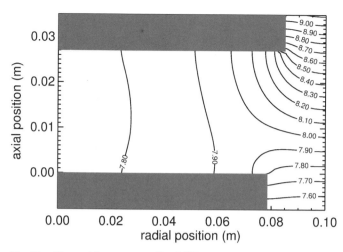

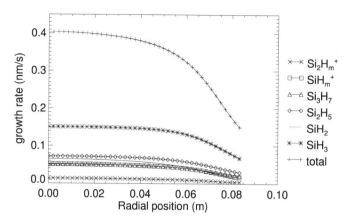

Fig. 20. The partial-pressure profile of SiH₄ calculated with the 2D model. Discharge settings are a total pressure of 20 Pa, a power of 5 W, and an RF frequency of 50 MHz. (From G. J. Nienhuis, Ph.D. Thesis, Universiteit Utrecht, Utrecht, the Netherlands, 1998, with permission.)

Fig. 21. Deposition rate at the grounded electrode, with the contribution of each silicon-containing species. Discharge settings are a total pressure of 20 Pa, a power of 5 W, and an RF frequency of 50 MHz. (From G. J. Nienhuis, Ph.D. Thesis, Universiteit Utrecht, Utrecht, the Netherlands, 1998, with permission.)

the 2D model calculates the partial pressures of SiH_4, H_2, and Si_2H_6 to be 8.05, 11.16, and 0.56 Pa, respectively, which is closer to the experimental data (see Fig. 17a). A partial-pressure profile of SiH_4 as calculated with the 2D model for both the volume between the electrodes (discharge volume) and the volume next to the electrodes (discharge-free volume) is shown in Figure 20. The dimensions are chosen to mimic the experimental reactor, with an electrode distance L of 2.7 cm (axial coordinate z) and electrode radius R of 7.6 cm (radial coordinate r). The inlet of SiH_4 is a (cylindrically symmetric) ring inlet, shown at the top right of the figure, where the partial pressure is at its highest value. The pump outlet is at the bottom right. It can be seen that the silane partial pressure varies only 10% as a function of the axial position, and it decreases slightly towards the axis of the discharge. This reflects the consumption of silane. Most of the silane flows from the inlet to the outlet; only a small amount diffuses to the discharge region. Similar profiles for the partial pressures of hydrogen and disilane are found.

Other discharge characteristics calculated with the 2D model differ only slightly from the ones calculated with the 1D model. The calculated growth rate is 0.419 nm/s, versus 0.426 nm/s for the 1D model. Electron and ion densities are lower for the 2D model: the electron density is 5.78×10^{14} m^{-3} (2D) and 7.19×10^{14} m^{-3} (1D); the negative-ion density is 2.00×10^{16} m^{-3} (2D) and 2.38×10^{16} m^{-3} (1D). The ion fluxes also are lower: the ion flux to the grounded electrode is 3.38×10^{18} m^{-2} s^{-1} (2D) and 4.06×10^{18} m^{-2} s^{-1} (1D); the ion flux to the powered electrode is 3.82×10^{18} m^{-2} s^{-1} (2D) and 4.06×10^{18} m^{-2} s^{-1} (1D). This is due to the radial losses in the 2D model. The difference in ion fluxes to the grounded and powered electrode results in a dc self-bias voltage of -13.96 V, whereas the experimental V_{dc} is -22 V.

From the profile of the average electron energy it is seen that a relatively high average electron energy exists near the edge of the powered electrode ($z = 2$–5 mm, $r = 7$ cm), which is

independent of the phase in the RF cycle [189]. This is caused by the high electric field between the outside of the powered electrode and the grounded shield around it (known as the dark space shield).

The partial pressure profile of SiH_3 shows a maximum between the electrodes, much like the 1D profile (see Fig. 16). The SiH_2 partial-pressure profile shows a maximum towards the edge of the discharge volume ($z \approx 8$ mm, $r \approx 6.5$ cm). Both the SiH_3 and SiH_2 source term also peak in this region. The high average electron energy in this region of the discharge is responsible for the formation of these radicals by electron dissociation of silane.

The total deposition rate at the grounded electrode and the contribution of each individual species to the total deposition rate is shown in Figure 21 as a function of radial position. The deposition rate is related to the flux into the surface of silicon-containing species, SiH_3, SiH_2, Si_2H_5, Si_3H_7, SiH_m^+, and $Si_2H_m^+$. For the discharge conditions used here it follows that SiH_3 is the radical with the highest contribution to the deposition rate, 37%. Other radicals contribute 16% (SiH_2), 17% (Si_2H_5), and 13% (Si_3H_7). The total contribution of ions to the deposition rate is 17%: 13% for SiH_m^+, and 4% for $Si_2H_m^+$.

The deposition rate is not uniform over the grounded electrode. A common requirement for large-scale reactors is a 5% difference in deposition rate over a certain electrode area, preferably the complete electrode. Using this, the 2D model would predict a useful electrode area of about 30%, i.e., an electrode radius of about 4 cm.

In the ASTER reactor deposition experiments were performed in order to compare with the 2D model results. Normalized deposition rates are plotted in Figure 22 as a function of radial position for data taken at 25 and 18 Pa. The deposition takes place on a square glass plate. For each pressure two profile measurements were performed, each profile perpendicular to the other (a and b in Fig. 22). A clear discrepancy is present. The use of the simplified deposition model is an explanation for this. Another recent 2D fluid model also shows discrepancies between the measured and calculated deposition rate [257],

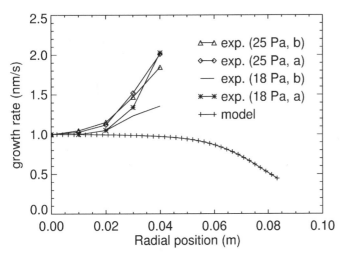

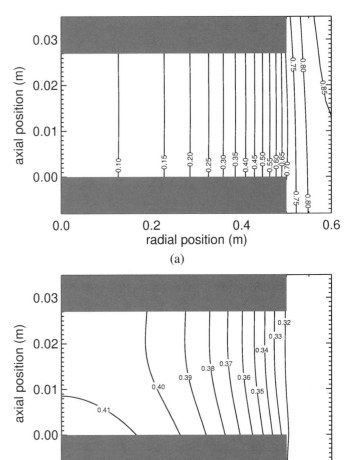

(a)

(b)

Fig. 22.   Normalized deposition rate at the grounded electrode of the 2D model compared with experiments. Discharge settings for the model are a total pressure of 20 Pa, a power of 5 W, and an RF frequency of 50 MHz. The experiments were performed at 18 and 25 Pa. The measurements were done for two perpendicular directions, a and b. (From G. J. Nienhuis, Ph.D. Thesis, Universiteit Utrecht, Utrecht, the Netherlands, 1998, with permission.)

which are attributed to the relative simplicity of the deposition model.

Another explanation for the observed discrepancy might be the influence of the electrode material. In all model calculations it is assumed that the potential at the grounded electrode is zero. In reality, as one generally uses insulating substrates such as glass, other boundary conditions are needed. Nienhuis [189] has reformulated boundary conditions taking into account the actual square 10×10-cm^2 glass substrate mounted on the cylindrically symmetric grounded electrode. It was found that the deposition rate at the center of the grounded electrode (covered with the glass) was reduced by about 15%. Going towards the edges of the glass, the deposition rate increases up to the level that is calculated for the full metal electrode. For larger radial positions the deposition rate equals the one for the full metal electrode. In addition, calculation of the contribution of individual species to the deposition rate show that the contributions of SiH$_3$ and SiH$_2$ increase around the glass–metal transition. It is concluded that the effect of a mixed boundary condition may at least partially explain the observed increase of deposition rate towards the edges of the glass.

4.2.3. Comparison with Experiments: Large Reactor

In a small reactor the depletion of silane is usually low, and consequently the background gas composition can be considered homogeneous. In a large reactor the composition varies throughout the discharge volume, as a result of the much higher consumption of silane. For commercial reasons the flow of silane is kept as low as possible, so as to obtain full consumption; this, however, may lead to loss of homogeneity. Large reactor systems have recently been modeled using the 2D model by Nienhuis and Goedheer [232]. A reactor is modeled with the same electrode spacing as above (2.7 cm), but

Fig. 23.   The partial pressure of silane (Pa) for a large-scale reactor with (a) ring inlet and (b) showerhead inlet. Note the difference in axial and radial scales. Discharge settings are a total pressure of 20 Pa, a power of 5 W, and an RF frequency of 50 MHz. (From G. J. Nienhuis, Ph.D. Thesis, Universiteit Utrecht, Utrecht, the Netherlands, 1998, with permission.)

with electrode radius of 50 cm and reactor radius 60 cm, which are typical dimensions used in reactors for industrial applications [252]. Two types of gas inlet are modeled: a showerhead, i.e., a powered electrode with small pores through which the gas is introduced [258], and a ring inlet, i.e., an inlet through the ring-shaped area between the grounded electrode and the reactor wall. In both cases, pumping occurs through the area between the powered electrode and the wall.

The discharge settings for the calculations shown here are a total pressure of 20 Pa, a power input of 250 W m^{-2}, an RF frequency of 50 MHz, and an inlet flow of 30 sccm of SiH$_4$ and 30 sccm of H$_2$. The resulting silane pressure distribution is shown in Figure 23 as a function of radial position for both the ring inlet (Fig. 23a) and the showerhead inlet (Fig. 23b). In case of the ring inlet the silane pressure distribution is very inhomogeneous; for the showerhead inlet it is relatively homogeneous. A strong depletion of silane is seen for the ring inlet on go-

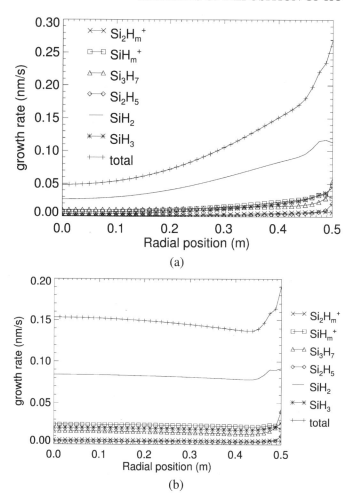

Fig. 24. The deposition rate at the grounded electrode with the contribution of each silicon-containing species, for a large scale reactor with (a) ring inlet and (b) showerhead inlet. Note the difference in uniformity. Discharge settings are a total pressure of 20 Pa, a power of 5 W, and an RF frequency of 50 MHz. (From G. J. Nienhuis, Ph.D. Thesis, Universiteit Utrecht, Utrecht, the Netherlands, 1998, with permission.)

ing from the outside to the center of the discharge. In fact, the discharge consists of an almost pure hydrogen discharge at the center and a mixture of silane and hydrogen towards the edges. For the showerhead inlet the situation is reversed, and the effect is much smaller.

The depletion of silane has large consequences for the uniformity of the deposition, as is shown in Figure 24. In the case of the showerhead the deposition is very homogeneous, and decreases slowly from the center to the edge of the discharge, which can be related to the small decrease in silane partial pressure as seen in Figure 23b. The ring inlet shows a large increase in the deposition rate from the electrode center to the outside. This reflects the high silane consumption in the opposite direction. The sharp increase of the deposition rate at the very edge of the electrode has also been observed experimentally [259].

The large depletion of silane has consequences for the chemistry. In the electron impact dissociation of silane the two radicals SiH_2 and SiH_3 are formed with branching ratio 83%:17% [197]; the other product is H (two atoms or one, re-

spectively). When there is enough silane available, the SiH_2 radical is lost by the insertion reaction $SiH_2 + SiH_4 \rightarrow Si_2H_6$, and via the H-abstraction reaction $H + SiH_4 \rightarrow SiH_3 + H_2$ the radical SiH_3 is produced. Deposition of SiH_2 and SiH_3 is the loss process at the walls.

The loss and production depend on silane partial pressure. For a high partial pressure of silane most SiH_2 radicals react with silane before reaching the wall. As the production of SiH_3 radical is also high, the SiH_3 radical is the main precursor for deposition. This is the case in the small reactor (see Fig. 21). For a low partial pressure, the reaction time of SiH_2 with SiH_4 is longer than the diffusion time to the wall. Together with the lower production of SiH_3, this causes SiH_2 to be the main precursor for deposition (see Fig. 23). The main precursor for deposition thus depends on the silane partial pressure. This will have serious consequences for the quality of the deposited material. The SiH_2 radical has a much larger sticking probability than the SiH_3 radical. The structure of a layer deposited with SiH_2 will be different from that of layer deposited with SiH_3. SiH_2 leads to columnar deposition, which is unwanted. *Device quality* material is obtained when deposition occurs via the SiH_3 radical. Scaling up of a small to a large reactor therefore is not straightforward.

4.3. Particle-in-Cell Discharge Models

4.3.1. Particle-in-Cell Model Principles

A complete model for the description of plasma deposition of *a*-Si:H should include the kinetic properties of ion, electron, and neutral fluxes towards the substrate and walls. The particle-in-cell/Monte Carlo (PIC/MC) model is known to provide a suitable way to study the electron and ion kinetics. Essentially, the method consists in the simulation of a (limited) number of computer particles, each of which represents a large number of physical particles (ions and electrons). The movement of the particles is simply calculated from Newton's laws of motion. Within the PIC method the movement of the particles and the evolution of the electric field are followed in finite time steps. In each calculation cycle, first the forces on each particle due to the electric field are determined. Then the equations of motion for electrons and ions are integrated. Subsequently collisional processes are treated by the Monte Carlo method. The chance of a collision for each particle during one time step is calculated using the cross sections for the various processes, and random number generation controls the actual occurrence of the collision. Finally the particle distribution is interpreted into electron and ion densities, and the Poisson equation for the electric field is solved. Thus the ions and electrons move in an electric field that is itself directly determined by the particle densities (via the Poisson equation). The result is referred to as the *consistent* solution for the electric field.

Despite the simplifications and the incomplete information with respect to the cross sections, the PIC method gives remarkably good results in RF modeling [260, 261]. The advantage of a fully kinetic description is accompanied by the disadvantage of very time-consuming computations. The PIC method allows only for a fully explicit treatment of the particle transport. Numerical instabilities due to large particle velocities and the coupling between particle densities and the electric field can only be avoided by reducing the time step in the simulation and so increasing the computational effort. Large numerical fluctuations in the calculated quantities as a result of the small number of computer particles (small with respect to the real physical numbers of ions and electrons) can also not be avoided in PIC simulations.

Some authors have made their code for PIC calculations in RF discharges available for a larger public. In particular, Birdsall et al. [261–263] developed the XPDP1 code, which is an implementation of the PIC method for RF discharges, written in C for use in UNIX/X-window environments. This XPDP1 code is developed for simulations in 1D RF discharges with plasmas consisting of one type of ion (argon, hydrogen), which have, as collisional processes, elastic collisions of electrons and ions with background-gas neutrals, excitation, and ionization. The code is built up in a modular way, which makes it easy to modify or add parts of the code, e.g., to make it suitable for use with other gases with more collisional processes. The code is intended to run interactively on workstations or PCs using X-windows, and communicates with the outside world via windows and dialog boxes.

Treatment of the SiH_4–H_2 discharge requires the addition of extra processes (ionization, dissociation, and attachment for the electrons, recombination for the ions) in the collision component. This resulted in major changes in this part of the code [264, 265]. Especially the incorporation of collision processes with target particles, which do not form a constant background density, but have a space- and time-varying density, needed some significant modifications. They were necessary to handle ion–ion recombination, but may also be used in future, e.g. for ion–electron collisions and for electron–radical collisions. In the original code partial collision probabilities were determined via cross sections; in the modified one, inverse free paths had to be applied. The cross sections for the electrons were obtained from the literature [266]. The cross sections for ion–neutral scattering had to be estimated. For the energy dependence of the scattering cross section an ion-induced dipole interaction has been assumed [267]. The recombination cross section is directly estimated from rate coefficients [190], and results in a total recombination sink (integrated over velocity space) identical to the fluid one.

4.3.2. Hybrid PIC/MC–Fluid Model

As a first attempt to modify the code to be able to run simulations on SiH_4–H_2 discharges, a hybrid PIC/MC–fluid code was developed [264, 265]. It turned out in the simulations of the silane–hydrogen discharge that the PIC/MC method is computationally too expensive to allow for extensive parameter scans. The hybrid code combines the PIC/MC method and the fluid method. The electrons in the discharge were handled by the fluid method, and the ions by the PIC/MC method. In this way

a large gain in computational effort is achieved, whereas kinetic information of the ions is still obtained.

The fluid model is a description of the RF discharge in terms of averaged quantities [268, 269]. Balance equations for particle, momentum, and/or energy density are solved consistently with the Poisson equation for the electric field. Fluxes described by drift and diffusion terms may replace the momentum balance. In most cases, for the electrons both the particle density and the energy are incorporated, whereas for the ions only the densities are calculated. If the balance equation for the averaged electron energy is incorporated, the electron transport coefficients and the ionization, attachment, and excitation rates can be handled as functions of the electron temperature instead of the local electric field.

Especially for the electrons, the fluid model has the advantage of a lower computational effort than the PIC/MC method. Their low mass (high values of the transport coefficients) and consequent high velocities give rise to small time steps in the numerical simulation ($v\Delta t < \Delta x$) if a so-called explicit method is used. This restriction is easily eliminated within the fluid model by use of an implicit method. Also, the electron density is strongly coupled with the electric field, which results in numerical instabilities. This requires a simultaneous implicit solution of the Poisson equation for the electric field and the transport equation for the electron density. This solution can be deployed within the fluid model and gives a considerable reduction of computational effort as compared to a nonsimultaneous solution procedure [179]. Within the PIC method, only fully explicit methods can be applied.

The disadvantage of the fluid model is that no kinetic information is obtained. Also, transport (diffusion, mobility) and rate coefficients (ionization, attachment) are needed, which can only be obtained from experiments or from kinetic calculations in simpler settings (e.g. Townsend discharges). Experimental data on silane–hydrogen Townsend discharges are hardly available; therefore PIC/MC simulations have been performed.

The quasi-Townsend discharge is an infinitely extended plasma, in which the electrons move in a constant, spatially homogeneous electric field E. The electrons collide with neutrals in a background gas of infinite extent. The constant electric field replaces the solution of the Poisson equation and makes the PIC method applicable for the electrons within reasonable computational limits. During the particle simulation in the quasi-Townsend discharge, averaged values, over all electron particles i, of the energy ϵ_i, diffusion constant D_i, mobility μ_i, ionization rate $k_{\mathrm{ion},i}$, and attachment rate $k_{\mathrm{att},i}$ are calculated. This results in values for the transport and rate coefficients: the electron transport coefficients D_e and μ_e, the ionization and attachment rates k_{ion} and k_{att} and the averaged electron energy $\langle \epsilon \rangle = 3k_B T_e/2$. All parameters are obtained for several values of the electric field E.

Subsequently, because the electron temperature is known as a function of the electric field E, the temperature T_e can be used instead of E as a parameter for coefficients and rates, by elimination of E. Thus, the coefficients are available both as a function of E and of T_e. Both the local electric field [225,

269] and the electron temperature [239, 268] have been used as parameters in fluid modeling.

A Townsend PIC simulation was performed for several gas mixtures ($SiH_4 : H_2 = 100 : 0, 50 : 50, 20 : 80$, and $0 : 100$) [264, 265]. It was found that the dependence of the electron temperature on the electric field strongly depends on the mixture ratio. As a result, the rate coefficient parametrized by the electric field, $k = k(E)$, also strongly depends on the mixture ratio. In contrast, it was found that the rate as a function of the electron energy, $k = k(T_e)$, shows only minor changes ($\leq 2\%$) for different mixing ratios. In Figure 25 the ionization (Fig. 25a) and attachment (Fig. 25b) rates are given. Here, two often used coefficients in RF fluid modeling are also shown for reference: an argon ionization rate derived from Townsend experiments (Fig. 25a), and a CF_4 attachment coefficient [179] (Fig. 25b).

For the following reasons the choice of the temperature T_e as the experimental parameter is more appropriate [268, 270]: (a) the high-energy tail of the energy distribution function of the electrons determines ionization versus attachment; (b) both in pure silane and hydrogen and in mixtures, T_e is a good pa-rameter for the distribution function; (c) T_e as a function of the electric field varies strongly with the mixture ratio SiH_4/H_2.

Methods that compensate for nonequilibrium effects in the situation of E-parametrized coefficients are very complicated, and are sometimes not firmly grounded. Because the electron temperature also gives reasonable results without correction methods, the rate and transport coefficients were implemented as a function of the electron energy, as obtained from the PIC calculations presented in Figure 25.

4.3.3. Simulation of RF Discharges in Silane–Hydrogen Mixtures with the Hybrid Model

In the deposition of amorphous silicon the RF frequency plays an important role. Therefore, in the simulations the frequency dependence was emphasized. Calculations were performed for frequencies $\nu_{RF} = 10, 40$, and 80 MHz.

The $SiH_4 : H_2$ ratio was $20\% : 80\%$ in all simulations. As most experiments in the ASTER deposition system are performed at a constant total power, the calculations were also performed at constant power instead of a given RF voltage. This is especially important in that the power has a large influence on the ion energy spectra at the electrodes. Calculations were performed at a power of $P = 12$ W, with an electrode area of 0.010 m^2 and an electrode distance of 0.04 m. The pressure was $p = 250$ mTorr. Also a calculation was performed at $p = 150$ mTorr for the standard RF frequency of 10 MHz.

In Figure 26 density profiles for two frequencies, 10 MHz (Fig. 26a) and 80 MHz (Fig. 26b), are shown. The densities of the species (electrons, positive ions, negative ions) and power dissipated in these species are presented in Table V, as well as the maximum electric field over the sheath and the outflux of positive ions.

It can be clearly observed that the electron density increases with frequency. The increase of frequency at a constant power results in lower RF voltages and, consequently, lower electric fields in the sheath. The maximum electric field E_{max} in the sheath diminishes with increasing frequency. The field decrease is too small to prevent the displacement current, $\epsilon_0 \partial E/\partial t = \epsilon_0 \nu_{RF} \partial E/\partial \tau$, with $\tau = t/\tau_{RF}$, from increasing. As a result, the electron conduction current in presheath and bulk increases, and the heating of electrons becomes stronger, whereas the lower potential drop across the sheath causes a lower power dissipation into the ions. The increase of the electron heating results in a higher ionization of silane (SiH_m^+) and hydrogen (H_2^+). This production is partly compensated for by higher transport losses (outfluxes) The net result is an increasing electron and positive ion density with increasing frequency. On the other hand the negative ion density decreases due to the enhanced recombination between positive ions and SiH_m^-. As a result of the different curves of $k_{ion}(\epsilon)$ for H_2 and SiH_4, the ionization of H_2 is enhanced relatively more by the increasing electron heating than that of SiH_4. The increased ion fluxes lead to a shrinking of the sheath.

A further result of the increase of power dissipation in the electrons has consequences for the plasma chemistry. Besides

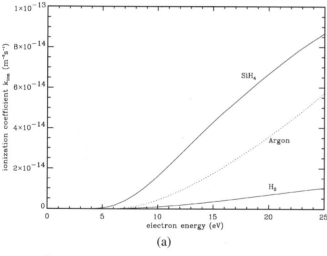

(a)

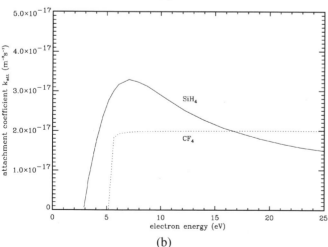

(b)

Fig. 25. Simulation results for quasi-Townsend discharges: (a) ionization co-efficients for SiH_4, H_2, and Ar; (b) attachment coefficients for SiH_4 and CF_4.

Table V. Simulation Results for the Hybrid PIC/MC–Fluid Model for a SiH$_4$–H$_2$ Discharge Driven with a Power of 12 W

p (mTorr)	ν_{RF} (MHz)	Density (10^{16} m^{-3})				Power (W)				E_{max} (10^4 V/m)	Outflux (10^{18} m^{-2} s^{-1})	
		e^-	SiH$_m^+$	SiH$_m^-$	H$_2^+$	e^-	SiH$_m^+$	SiH$_m^-$	H$_2^+$		SiH$_m^+$	H$_2^+$
250	10	1.8	6.3	4.7	0.18	1.3	7.2	0.0	3.5	11	15	1.1
250	40	4.8	9.4	4.9	3.2	2.7	5.9	0.0	3.4	8.5	28	2.8
250	80	6.9	10.1	4.5	3.9	3.7	5.0	0.0	3.3	7.1	36	3.9
150	10	1.2	6.9	5.7	0.048	1.1	7.5	0.0	3.4	9.4	14	0.9

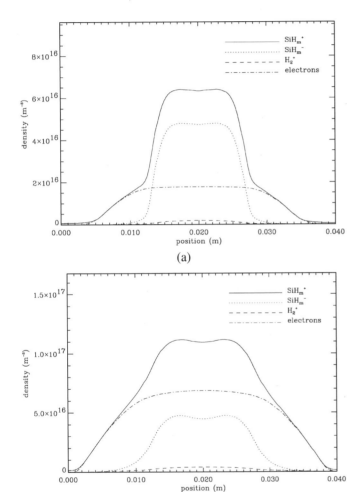

Fig. 26. Density profiles in a SiH$_4$–H$_2$ discharge at 250 mTorr for an RF frequency of (a) 10 MHz and (b) 80 MHz.

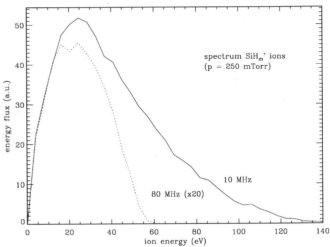

Fig. 27. Ion energy distribution of SiH$_m^+$ ions in a SiH$_4$–H$_2$ discharge at 250 mTorr for an RF frequency of 10 MHz (solid line) and 80 MHz (dotted line).

the increased ion densities, also the production of radicals will be increased, which may lead to higher deposition rates.

It is observed that a decrease of the pressure (from $p = 250$ to 150 mTorr) mainly results in a decrease of the densities due to higher transport losses, and in an extension of the sheath due to a higher ion mobility. The electric field and the electron heating diminish slightly for lower pressure. The electron and H$_2^+$

density, and consequently the H$_2^+$ outflux, are much more influenced by the pressure decrease than the SiH$_m^+$ and SiH$_m^-$ ion densities and the SiH$_m^+$ outflux.

The ion energy distribution of SiH$_m^+$ is presented in Figure 27 for 10 and 80 MHz. Because the total power deposition was kept constant at 12 W, they give insight into the positive ions' behavior in deposition experiments. It is seen that for silane ions (Fig. 27)—and also for hydrogen ions (not shown)—the ion energy distribution shifts to lower energies for increasing frequency, due to the decreasing electric field and sheath thickness. The saddle structure due to the RF frequency, which can be seen at very low pressures (<10 mTorr) [271], has totally disappeared.

For one specific set of discharge parameters, in a comparison between the hybrid approach and a full PIC/MC method, the spectra and the ion densities of the hybrid model showed some deviations from those of the full particle simulation. Nevertheless, due to its computational advantages, the hybrid model is appropriate for gaining insight into plasma behavior under varying parameters, especially with respect to the ion energy distribution at the deposition layer.

4.3.4. Simulation of RF Discharges in Silane–Hydrogen Mixtures with a Full PIC/MC Model

The full PIC/MC method has not been used frequently to study SiH_4–H_2 discharges [223], because of the large amount of computer time required. Yan and Goedheer [224] have developed a 2D PIC/MC code with some procedures to speed up the calculation. In the code, electrons and positive and negative ions are described in a kinetic way. The capacitively coupled RF discharge in a mixture of SiH_4 and H_2 has been simulated for pressures below 300 mTorr and frequencies from 13.56 to 65 MHz. Yan and Goedheer [224] studied the effects of frequency on (1) the EEDF in the discharge area, (2) the ion energy distribution (IED) at the substrate, (3) the ion flux to the substrate, and (4) the power dissipation at constant power density.

The PIC/MC method is based on a kinetic description of the particle motion in velocity–position space. Positive and negative *superparticles* move in the self-consistent electric field they generate. The electric field is obtained by solving the Poisson equation. A Monte Carlo formalism is used to describe the effect of collisions. The displacement between collisions is ruled by Newton's law. There are four kinds of charged particles in the 2D model that is used by Yan and Goedheer [224]: electrons, positive ions (SiH_3^+, H_2^+), and negative ions (SiH_3^-). The negative hydrogen ion (H^-) is not considered, because its density is much lower than the other ion densities. For a computationally stable simulation, the distance between adjacent points of the grid on which the Poisson equation is solved is taken to be less than several Debye lengths, and the product of the time step and the electron plasma frequency is taken to be much smaller than one. The number of superparticles for each species is around 10^4. The superparticle size for species with a low density (electrons and H_2^+) is between 10^6 and 10^7; for species with a high density (SiH_3^+ and SiH_3^-) it is between 10^8 and 10^9.

The simulation of an electronegative gas discharge converges much more slowly than that of an electropositive discharge. This is mainly caused by the slow evolution of the negative-ion density, which depends only on attachment (to create negative ions) and ion–ion recombination (to annihilate negative ions), both processes with a very small cross section. In addition to the common procedures adopted in the literature [222, 223, 272, 273], such as the null collision method, and different superparticle sizes and time steps for different types of particle, two other procedures were used to speed up the calculation [224].

The first procedure is to use the rate equilibrium equations for electrons and ions at the quasi-steady state for a new guess of the particle densities. In the final result, quasi-steady balances of the positive ions and negative ions integrated over the whole discharge region will be reached, i.e., the balance for the creation of positive ions (SiH_3^+) by ionization and their loss to the reactor boundary and by recombination, and the balance for the creation of negative ions by attachment and their loss by recombination. These balances are used in the following way. After a number of cycles, the average ionization, recombina-

tion, and attachment rates are known, as well as the average loss of positive ions to the wall. Also the deviation from the balances is known. The new densities of electrons and positive and negative ions can be guessed by combining charge neutrality with the balance equations in a global model. These densities will be different from the values obtained in the previous simulation, and the number of simulation particles thus has to be adapted. The easiest way to implement the changes is through a change of the size of the superparticles. To do this, only the interior part of the discharge is used, where the negative ions are present. The changes are spatially distributed according to the profile of the negative ions. After the correction the standard PIC/MC calculation is restarted. This procedure is repeated several times during a full simulation and reduces the computational effort considerably.

The second procedure is to use a small number of large particles in the beginning of the simulation. The results of that are used as the initial conditions for a simulation with a large number of small particles. Thus, in the final results, the statistical fluctuations are reduced. Also, results from a previous simulation can be used as initial conditions for a new simulation with changed parameters, which saves a large amount of time during parameter scans.

Most of the electron and ion impact reactions that are included in this code are listed in Table II. Added to this are the elastic collisions $SiH_4 + e^-$ [274] and $H_2 + e^-$ [275], the recombination $SiH_3^- + SiH_3^+ \rightarrow 2SiH_3$ [214], and the charge exchange $H_2^{+*} + H_2 \rightarrow H_2^+ + H_2^*$ [193]. Note that disilane is not included. The cross sections are taken from the references in Table II. Ion–neutral elastic collision cross sections are obtained according to the Langevin formula [192]. The total number of ion–ion recombinations in the discharge is obtained from the integral of the product of the recombination rate coefficient k_{rec} and the ion density profiles. The recombination event is attributed to the negative ion; it recombines with the nearest positive ion. In this model the focus is on the energy dissipation, EEDF, IED, and ion flux to the wall, but not on the details of chemical reactions. Chemical reactions are not included in our code. The experiments described above showed that the background pressure and the gas composition can be regarded as being constant for the dimensions used. In the simulation, the pressure and the gas composition are assumed to be constant.

The frequency effects are studied in the cylindrical geometry ($R = 0.08$ m, $L = 0.027$ m) at a constant power of 25 W, which corresponds to a volume power density 46 mW cm^{-3}. The pressure is 120 mTorr with 45% SiH_4 and 55% H_2. It is found that the RF voltage at this power scales with the frequency as $V_{RF} \nu_{RF}^{1.15} = C$, with C a constant. Because the induced displacement current increases with the frequency, a lower RF voltage is needed at high frequencies to keep the same power density. This relation can explain the increase of the deposition rate with frequency at a constant RF voltage [276], because in that case the dissipated power increases strongly.

The simulations show further that the average electron energy decreases slightly as a function of frequency, from about 5.8 to 5 eV on going from 13.56 to 65 MHz. The low voltage applied at high frequencies leads to a weaker sheath electric field. Consequently positive ions are pulled out less efficiently, so their density increases. Due to the charge neutrality in the bulk, the electron density also increases, from 1.8×10^{15} m^{-3} at 13.56 MHz to 3.2×10^{15} m^{-3} at 65 MHz. On the other hand, the current density $j \approx e\mu_e E$ in the bulk is constant because of the constant power. Therefore a higher electron density leads to a weaker electric field in the bulk, which is responsible for heating electrons. This causes the decrease of the average energy. The explanation of the increase of the deposition rate with frequency as observed by Curtins et al. [277] is that there is a larger relative population of high-energy electrons. These high-energy electrons would yield higher ionization and dissociation rates. The simulation described here shows that the density of low-energy electrons increases strongly while the density of high-energy electrons even goes down as a result of the combination of high-frequency effects on the EEDF and on the electron density. Stochastic heating [278] does not seem to lead to an increase of the electron energy, probably because its effect is canceled by the decreasing voltage and by the efficient energy loss to vibrational excitation.

The flux of SiH$_3^+$ to the substrate and the average ion energy in this flux at different frequencies are shown in Figure 28a. The ion flux and ion energy are sampled over a disk one-quarter of the electrode radius around the center of the grounded electrode, which avoids the influence of the edge. The average ion energy decreases quickly with increasing frequency due to the decrease of the RF voltage. The ion flux to the wall remains approximately the same (1.7×10^{19} m^{-2} s^{-1}) when the frequency is varied. This result is quite different from the results presented by Heintze and Zedlitz [279], where the ion flux showed a significant increase with frequency. Their results, however, are not in agreement with the Bohm criterion, because the electron density did not change much while the ion flux changed one order of magnitude [224]. This may be caused by dust formation. Another inconsistency is that the dissociation rate measured by the same authors [280] only increased by a factor of 1.5 when the frequency was increased from 50 to 250 MHz. The dissociation cross section is larger than the ionization cross section and with lower threshold; see also Figure 13. This implies that dissociation has higher probability than ionization. The obvious increase in ion flux should therefore be accompanied by at least the same amount of dissociation.

The IED of SiH$_3^+$ on the substrate at different frequencies is shown in Figure 28b. With increasing frequency, the energy distribution profile becomes more peaked. That is attributed to a decrease in the number of collisions experienced by ions, because at higher frequencies the lower applied RF voltage leads to a narrower sheath. The same reason obviously causes the maximum ion energy to decrease with increasing frequency. This result is consistent with the evolution of the estimated maximum ion energy by Dutta et al. [281]. But in other experiments by Hamers [163] it was found that the maximum ion energy did not change much.

The power dissipated at two different frequencies has been calculated for all reactions and compared with the energy loss

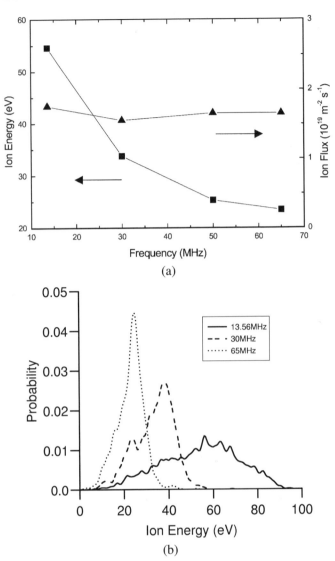

Fig. 28. The effect of RF frequency on (a) ion energy and ion flux to the substrate, and (b) the ion energy distribution for SiH$_3^+$. The power is 25 W, and the pressure is 120 mTorr. [From M. Yan and W. J. Goedheer, *Plasma Sources Sci. Technol.* 8, 349 (1999), © 1999, Institute of Physics, with permission.]

to the walls. It is shown that at 65 MHz the fraction of power lost to the boundary decreases by a large amount compared to the situation at 13.56 MHz [224]. In contrast, the power dissipated by electron impact collision increases from nearly 47% to more than 71%, of which vibrational excitation increases by a factor of 2, dissociation increases by 45%, and ionization stays approximately the same, in agreement with the product of the ionization probability per electron, the electron density, and the ion flux, as shown before. The vibrational excitation energy thresholds (0.11 and 0.27 eV) are much smaller than the dissociation (8.3 eV) and ionization (13 eV) ones, and the vibrational excitation cross sections are large too. The reaction rate of processes with a low energy threshold therefore increases more than those with a high threshold.

The simulated rate of vibrational excitation at 30 MHz is already twice that at 13.56 MHz. At 65 MHz, this rate is three times the rate at 13.56 MHz. The dissociation and ionization rates do not change much. The increase of the vibrational excitation leads to an increase of the density of excited SiH_4 molecules. This agrees with an N_2 discharge experiment where the vibrational temperature of N_2 was observed to rise with the applied frequency [282].

The excited SiH_4 can participate in various reactions. For example, one can speculate that the excited SiH_4 molecules could enhance pyrolysis ($SiH_4 \rightarrow SiH_2 + H_2$) [283] and consequently contribute to the deposition rate. SiH_2 can further dissociate to SiH, and it was observed that the emission of SiH shows a similar increase to the deposition rate [250]. This explanation of an increased deposition via excited SiH_4 molecules, however, is not supported by the results obtained when the pressure is varied. Simulations were performed at 13.56 MHz and a power of 16 W, at pressures ranging from 150 down to 90 mTorr. Although the behavior of the vibrational excitation, dissociation, and ionization are quite similar to the behavior as a result of frequency variation, the deposition rate in the experiments remains the same at low pressures [163, 245]. At high pressures the discharge will reach the dusty (γ') regime. This implies that the vibrational excitation does not influence the deposition directly.

The IED reacts quite differently to pressure changes than to frequency changes. Although the IEDs are somewhat different due to the larger number of collisions in the sheath at higher pressures, they cover approximately the same energy range. It has been reported by Dutta et al. [284] that the quality of the deposited material is better at high frequency than at lower frequency, because the intrinsic stress goes down when the frequency is increased. Since the displacement energy of atoms in the solid is about 20 eV [285], there most likely is an energy range of ions that effectively assists the radical deposition on the surface by influencing the reactivity of the films. At high frequencies more ions are in the effective (lower) energy range than that at a lower frequency (Fig. 28b). This could well cause the higher deposition rate. At different pressures, the energy spectrum of the ions is almost the same, and hence the deposition rate is the same. Chemical reactions are not included here, so the influence of changes in the radical fluxes on the deposition process is not studied. The simulation and experimental results [191, 247], however, indicate that the composition of the SiH_4–H_2 mixture does not change dramatically.

5. PLASMA ANALYSIS

In order to relate material properties with plasma properties, several plasma diagnostic techniques are used. The main techniques for the characterization of silane–hydrogen deposition plasmas are optical spectroscopy, electrostatic probes, mass spectrometry, and ellipsometry [117, 286]. Optical emission spectroscopy (OES) is a noninvasive technique and has been developed for identification of Si, SiH, Si^+, and H^+ species in the plasma. Active spectroscopy, such as laser induced fluorescence (LIF), also allows for the detection of radicals in the plasma. Mass spectrometry enables the study of ion and radical chemistry in the discharge, either *ex situ* or *in situ*. The Langmuir probe technique is simple and very suitable for measuring plasma characteristics in nonreactive plasmas. In case of silane plasma it can be used, but it is difficult. Ellipsometry is used to follow the deposition process *in situ*.

5.1. Optical Emission

A plasma emits light, as a result of excited species that undergo a radiative transition from a high to a lower level state. By measuring the wavelenghts and intensities of the emitted spectral lines, it is possible to identify and quantify neutrals and ions in the plasma. This spectral fingerprint allows for the study of physical and chemical processes that occur in the plasma. OES is a nonintrusive technique, which can provide spatial and temporal information. By its nature, OES is limited to the detection of light-emitting species.

Optical emision spectra nowadays are simply measured using a fiber optic cable that directs the plasma light to a monochromator, which is coupled to a photodetector. By rotating the prism in the monochromator a wavelength scan of the emitted light can be obtained. Alternatively, an optical multichannel analyzer can be used to record (parts of) an emission spectrum simultaneously, allowing for much faster acquisition. A spectrometer resolution of about 0.1 nm is needed to identify species.

Quantitative analysis of emission spectra is difficult. As a first assumption the emission intensity from a species is proportional to its concentration, but the proportionality constant depends among other things on the electron density and on the quantum efficiency of the excitation process. As the electron density depends on the RF power and the gas mixture, and the quantum efficiency on the total pressure and the gas mixture, the proportionality "constant" never really is constant [117].

Optical emission is a result of electron impact excitation or dissociation, or ion impact. As an example, the SiH radical is formed by electron impact on silane, which yields an excited or superexcited silane molecule ($e^- + SiH_4 \rightarrow SiH_4^* + e^-$). The excess energy in SiH_4^* is released into the fragments: $SiH_4^* \rightarrow SiH^* + H_2 + H$. The excited SiH^* fragments spontaneously release their excess energy by emitting a photon at a wavelength around 414 nm, the bluish color of the silane discharge. In addition, the emission lines from Si^*, H^*, and H_2^* have also been observed at 288, 656, and 602 nm, respectively.

Matsuda and Hata [287] have argued that the species that are detectable using OES only form a very small part (<0.1%) of the total amount of species present in typical silane deposition conditions. From the emission intensities of Si^* and SiH^* the number density of these excited states was estimated to be between 10^6 and 10^7 cm^{-3}, on the basis of their optical transition probabilities. These values are much lower than radical densities, 10^{12} cm^{-3}. Hence, these species are not considered to partake in the deposition. However, a clear correlation between the emission intensity of Si^* and SiH^* and the deposition rate has been observed [288]. From this it can be concluded that the emission intensity of Si* and SiH* is proportional to the concentration of deposition precursors. As the Si* and SiH* excited species are generated via a one-electron impact process, the deposition precursors are also generated via that process [123]. Hence, for the characterization of deposition, discharge information from OES experiments can be used when these common generation mechanisms exist [286].

LIF has been used to determine spatial profiles of SiH and Si radicals at various deposition conditions. These radicals are excited by a tuned pulsed laser, and the fluorescent intensities are a measure of the density of the particles studied. Spatial profiles can be obtained by translating incident laser beam and detection optics. LIF is highly selective and highly sensitive, but it is a complex technique. Mataras et al. [289, 290] have compared OES and LIF profiles of SiH* and SiH. From the differences observed it is concluded that the generation processes of these radicals are different. The SiH* concentration is related to the magnitude of both cathode and anode sheath potentials, whereas the SiH concentration is related to the density of energetic electrons in the plasma bulk. As the generation of SiH_2 and of SiH_3 radicals follow similar pathways, it is inferred that the LIF profile of SiH is a good measure of the generation profile of SiH_2 and SiH_3 too [286]. This seems to be confirmed by data obtained by infrared laser absorption spectroscopy (IR-LAS) [215, 291, 292], where the SiH_3 profile is measured to be similar to the LIF SiH profile.

Profiles of SiH measured by OES as a function of RF voltage have been reported by Böhm and Perrin [184]. In their nearly symmetrical discharge a clear relation exists between SiH emission intensity and RF voltage, at low pressure (55 mTorr). Further, the SiH profiles are M-shaped, i.e., they peak around the time-averaged sheath boundary and are lower at the center of the plasma, reflecting the time-averaged electron density profile. Going to high pressures (180 mTorr) the so-called α–γ' transition occurs, and the SiH emission intensity is highest at the center of the discharge. LIF experiments at these pressures are impossible due to the presence of powder, which causes strong light scattering.

5.2. Electrostatic Probes

An electrostatic, or Langmuir, probe is a metallic electrode, insulated except at the tip, which is introduced into the plasma. The metal mostly used is tungsten, because of its high melting point. The insulator of the probe should be chemically inert. The probe is inserted into a discharge chamber via an electrically insulating vacuum seal, within which the probe is either fixed or movable, usually in one direction, via a bellows. This *in situ* technique locally perturbs the plasma by changing the electric field. The magnitude of the disturbance depends on the dimensions of the probe relative to the Debye length λ_D, which in turn depends on the plasma properties. Usually λ_D is in the micrometer range, and probe diameters in the millimeter range, so disturbances are very small.

Employing a Langmuir probe, it is possible to determine the plasma densities (n_i and n_e), electron temperature T_e, plasma potential V_{pl}, and floating potential V_{fl} [117, 134, 293, 294]. It is often used as a complementary technique to optical and/or mass spectrometry. A Langmuir probe is operated by applying an external variable potential (usually between -100 and $+100$ V) and measuring the current (in the milliampere range) through the probe. The sweep rate usually is 1–10 V/μs, and multiple I–V characteristics are acquired and averaged. From such an averaged I–V characteristic all plasma parameters mentioned above can in principle be determined.

When a probe is inserted into a plasma, it will experience electrons and ions colliding with its tip. Due to the high mean speed of electrons, the flow of electrons is higher than the flow of ions. Consequently, the tip will charge up negatively until the electrons are repelled, and the net current then is zero. The probe potential then is the floating potential, V_{fl}. The electron current density J_e then balances the ion current density J_i. At potentials lower than V_{fl} the ion current cannot increase further—in fact, only ions are collected from the plasma—and the ion saturation current I_{is} is measured. The plasma potential V_{pl} is defined as the potential at which all electrons arriving near the probe are collected and the probe current equals the electron current. Note that the plasma assumes the plasma potential in the absence of a probe; hence probe perturbation at V_{pl} is minimal. At potentials higher than V_{pl} the electron saturation current I_{es} is measured.

The I–V characteristic thus is divided into three regions: (I) $V < V_{fl}$, (II) $V_{fl} < V < V_{pl}$, and (III) $V > V_{pl}$. In region I the ion saturation current dominates; in region III, the electron saturation current. They are derived to be [117, 294]:

$$I_{is} = 0.4 n_i e A_p \sqrt{\frac{2k_B T_e}{m_i}} \tag{35}$$

$$I_{es} = n_e e A_p \sqrt{\frac{8k_B T_e}{m_e}} \sqrt{1 + \frac{eV}{k_B T_e}} \tag{36}$$

with A_p the probe area. The expression for I_{es} is derived for a cylindrical probe tip.

In the intermediate region II, assuming a Maxwellian EEDF, the electron retarding current is given by

$$I_{er} = n_e e A_p \sqrt{\frac{k_B T_e}{2\pi m_e}} \exp\left(\frac{eV}{k_B T_e}\right) \tag{37}$$

The electron temperature is derived from the I–V characteristic by first subtracting the measured ion current from the measured current in region II. From the slope of the logarithm of I_{er} versus V, T_e is determined [Eq. (37)]. Plotting the square of the probe current versus voltage in region III yields a straight line [Eq. (36)], and the electron density is derived from its slope (and the known T_e). The ion density then is easily calculated with Eq. (35). The plasma potential is determined from the intersection between the extrapolated currents from regions II and III. The floating potential is the potential at which the current is zero. Other methods for the extraction of plasma properties from I–V data, not assuming a Maxwellian EEDF, are described by Awakowicz [294].

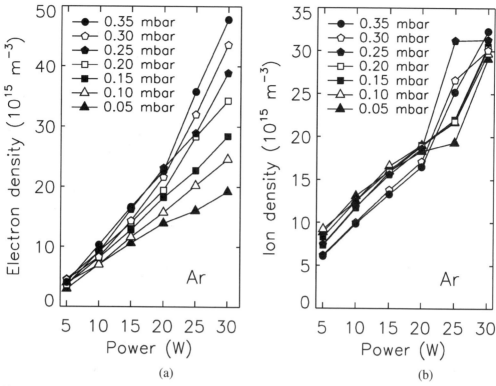

Fig. 29. Electron (a) and ion (b) density as a function of power for different pressures in the case of an argon discharge.

In RF discharges one has to compensate for the RF voltage component across the probe sheath. This is done with a pickup element, situated close to the probe tip, which should be in the plasma bulk as well. Compensating electronics within the probe is usually designed to work at a prescribed RF frequency, and cannot be used at other frequencies.

In the ASTER system a data series is measured for an argon and a hydrogen plasma running at 13.56 MHz, in which the power (5–30 W) and pressure (5–50 Pa) are varied [265, 295]. The probe tip is positioned exactly between powered and grounded electrode, at the center of the discharge.

The plasma potential determined from I–V data for argon increases linearly from about 24 to 27 V with increasing power at a pressure of 0.05 mbar. At the highest pressure of 0.35 mbar these values have shifted downwards by about 2 V only. For the hydrogen discharge similar behavior is observed, with an increase from 25 to 32 V with power, at 0.10 mbar. These values shift by about 3 V on going to a pressure of 0.5 mbar. The variation of the plasma potential in the case of hydrogen is larger than in the case of argon. This difference in behavior is also seen in the electron temperature and the electron and ion density. In the case of argon, T_e varies between 1.8 and 2.8 eV. The effect of pressure variation is larger than the effect of power variation. At low pressure (0.05 mbar) T_e is around 2.5 eV, whereas at high pressure (0.35 mbar) T_e is around 2 eV. In the case of hydrogen, T_e varies between 1.5 and 4.2 eV. At low pressure (0.10 mbar) T_e varies slightly around 4 eV, whereas at high pressure (0.5 mbar) T_e changes from 4 eV at low power (5 W) to 1.5 eV at high power (30 W).

The electron and ion densities for argon and hydrogen are shown in Figure 29 and Figure 30, respectively. The absolute values of n_i for argon (Fig. 29b) are about 5–8 times larger than the ones for hydrogen (Fig. 30b). Further, the increase of n_i as a function of applied power is only slightly pressure-dependent in the case of argon. The dependence is much clearer in the case of hydrogen. Similar observations can be made for the electron density.

Employing a extendable bellows, the probe tip can be moved towards the edge of the discharge over a distance of about 5 cm. Measurements performed in an argon discharge at 0.15 mbar at three different power levels show that V_{pl}, T_e, n_e, and n_i do not vary more than 5–10%. This shows that the discharge is homogeneous.

The Langmuir probe method is easily employed in discharges; however, especially in deposition plasmas, care has to be taken in interpreting I–V data. The presence of negative ions in silane plasmas complicates the analysis. Their contribution should be taken into account. Further, (thin) insulating layers may form on the probe, which can result in short-circuiting of the probe. These contamination effects distort the probe characteristics, especially around V_{pl}. The probe surface may be cleaned *in situ* by ion bombardment, i.e., by biasing the probe to a large negative potential in a *non*depositing plasma such as argon. Melting the deposited contaminations off the probe should be possible by passing a high electron current through the probe.

Probe measurements in silane discharges have been reported [296, 297]. Apparently, no difficulties were experienced,

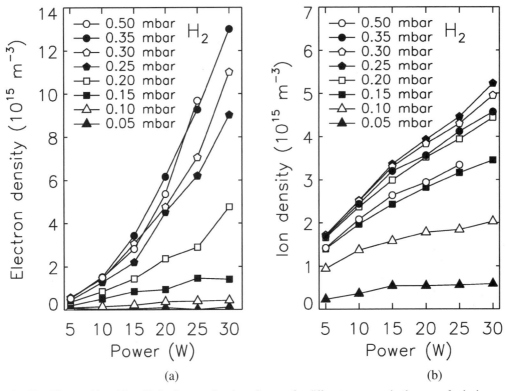

Fig. 30. Electron (a) and ion (b) density as a function of power for different pressures in the case of a hydrogen discharge.

as the deposited amorphous silicon layer on the tip was sufficiently photoconductive. For typical silane discharge conditions values for T_e are found to be between 2 and 2.5 eV. Electron densities are around 1×10^9 cm^{-3} [296]. Probe measurement in the ASTER system failed due to strong distortions of the probe current, even after following cleaning procedures.

5.3. Mass Spectrometry

Mass spectrometry is used to obtain information on the neutral and ionic composition of the deposition discharge. Three modes of sampling are commonly used: analysis of discharge products downstream, line-of-sight sampling of neutrals, and direct sampling of ions from the discharge. Mass spectrometry is a simple method to measure and quantify neutral species in a silane discharge, although calibrations are required. For ionic species, mass spectrometry is the only method available. A quadrupole mass filter is a proven instrument that is essential in mass spectrometry of low-pressure discharges.

5.3.1. Analysis of Neutrals

The analysis of the neutral gas composition in a discharge yields useful information on the mechanisms and kinetics of silane dissociation. However, it should be borne in mind that with mass-spectrometric analysis one only detects the final products of a possibly long chain of reactions.

The partial pressures of the stable neutral molecules in the discharge (silane, hydrogen, disilane, trisilane) can be measured

by a quadrupole mass spectrometer (QMS). The QMS usually is mounted in a differentially pumped chamber, which is connected to the reactor via a small extraction port [286]. In the ASTER system a QMS is mounted on the reactor that is used for intrinsic material deposition. The QMS background pressure (after proper bake-out) is between 10^{-12} and 10^{-13} mbar. The controllable diameter in the extraction port is adjusted so that during discharge operation the background pressure never exceeds 10^{-11} mbar.

The gas that enters the QMS is ionized by electron impact at a factory-preset electron energy of 70 eV (or 90 eV), and subsequently mass-analyzed. The ion currents at different mass-to-charge ratios (m/e) need to be converted to partial pressures by careful calibrations, as reported by Hamers [163]. Gas X is admitted to the reactor and the corresponding pressure p is measured, as well as the ion current I at a specific ratio $m/e = \mu_X$ of the gas, e.g., for argon $m/e = 40$ amu/e. The background signal at the same $m/e = \mu_X$ that results from residual gases in the QMS chamber is subtracted from the measured ion current. The calibration factor γ_X is defined as the ratio between the corrected ion current and the reactor pressure. Calibrations are performed typically at three pressures before and after measurements on the discharge, as the sensitivity of the channeltron in the QMS may rapidly change. The calibrations are performed with hydrogen, argon, silane, and disilane, with the main contributions at $m/e = 2, 40, 30$, and 60 amu/e, respectively.

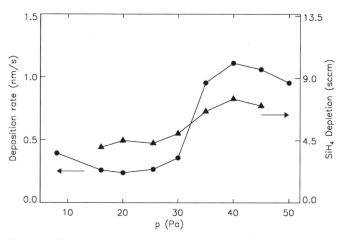

Fig. 31. The deposition rate and the corresponding silane depletion as a function of the total process pressure. Other conditions are 50 MHz, 30 : 30 SiH$_4$: H$_2$. (From E. A. G. Hamers, Ph.D. Thesis, Universiteit Utrecht, Utrecht, the Netherlands, 1998.)

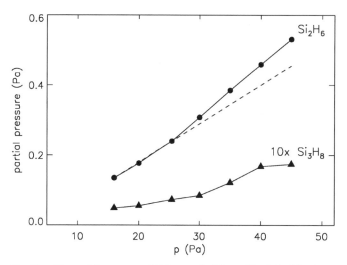

Fig. 32. The partial pressures of disilane and trisilane. The dashed line is an extrapolation of the disilane partial pressure in the α-regime. (From E. A. G. Hamers, Ph.D. Thesis, Universiteit Utrecht, Utrecht, the Netherlands, 1998.)

In the deposition of a-Si:H the dissociation of the process gas silane leads to the formation of hydrogen, disilane, trisilane, higher-order silanes, and a solid film. Silane is depleted. From the consumption of silane, one may estimate the deposition rate of the solid film. The difference in the silane partial pressure between the discharge-on and the discharge-off state is often used as an indication of the amount of silane that is depleted [298–300]. However, the relative change in pressure in general is not equal to the relative change in flow, due to gas-dependent conductances [163, 301] (see also Section 2.4). For example, in the ASTER deposition system equal partial pressures of silane and hydrogen are observed for a flow ratio of silane to hydrogen of about 1.25. Hence, another way of determining the silane depletion is used [163, 301]. First, the partial pressures of silane and hydrogen are measured in the discharge-on state. Subsequently, the discharge is switched off, and the silane and hydrogen flow rates are adjusted so that the same partial pressures as in the discharge-on state are reached. These flow rates then are the flow rates of silane and hydrogen that leave the reactor in the discharge-on state. The silane depletion ΔQ_{SiH_4} is the difference between the admitted silane flow rate and the flow rate of silane that leaves the reactor when in the discharge-on state. Additionally, an estimate can be made of the deposition rate, under the assumptions that all depleted silane is used for the deposition (no large amount of higher-order silanes or dust is formed). Also the deposition is assumed to occur homogeneously over the surface of the reactor. The estimated deposition rate is $r_d^{\text{est}} = \Delta Q_{\text{SiH}_4} / k_B T_0 \rho A_r$, with ρ the atomic density of the amorphous silicon network (5×10^{22} cm^{-3}), T_0 the standard temperature (300 K), and A_r the reactor surface (0.08 m^2). This yields an estimated deposition rate of 0.11 nm/s per sccm SiH$_4$ depletion [163]. In Figure 31 the deposition rate is compared with the corresponding silane depletion as a function of process pressure for a silane–hydrogen discharge at 50 MHz and 10 W. For these conditions the α–γ' transition occurs at 30 Pa: the deposition rate is increased by a factor of 4,

whereas the depletion is increased only by a factor of 2. The right-hand axis is scaled using the value 0.11 nm/s·sccm. In the α-regime the deposition rate is underestimated, whereas in the γ'-regime it is overestimated. This is in part due to the assumed uniformity of deposition. In the α-regime the deposition rate at the center of the substrate is lower than at the edges; in the γ'-regime this is reversed. Moreover, from the low dc bias voltages observed in the γ'-regime it is inferred that the discharge is more confined.

The partial pressures of disilane and trisilane are shown in Figure 32, for the same process conditions as in Figure 31. Both partial pressures increase as a function of pressure. Around the α–γ' transition at 30 Pa the disilane partial pressure increases faster with increasing pressure, as can be seen from the deviation from the extrapolated dashed line. The disilane partial pressure amounts to about 1% of the total pressure, and the trisilane partial pressure is more than an order of magnitude lower. Apparently, in the γ'-regime the production of di- and trisilane is enhanced.

Not only the silane depletion, but also the hydrogen production can be used to obtain information on the reaction products of the decomposition process. In Figure 33 the silane depletion and the corresponding hydrogen production are shown for a number of experiments, with process parameters such as to cover both the α- and the γ'-regime [163, 301]. A clear correlation exists. In addition, the solid line in Figure 33 relates the hydrogen production and silane depletion in the case that only a-Si:H is formed with 10 at.% H, i.e., a-Si:H$_{0.1}$ and H$_2$. The dashed line represents the case where 30% of the silicon from the consumed amount of silane would leave the reactor as disilane, instead of being deposited as a-Si:H$_{0.1}$. The data can well be explained by these two cases. In fact, they are an indication that part of the consumed silane leaves the reactor with a larger amount of H per Si atom than a-Si:H, which may be caused by the formation of higher silanes and/or powders.

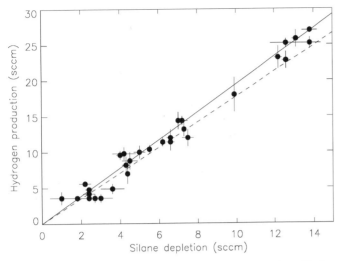

Fig. 33. The depletion of silane and the corresponding production of hydrogen for several process conditions, covering both the α- and the γ'-regime. The solid line represents the case where all the consumed silane is converted into a-Si:H$_{0.1}$ and 1.95H$_2$. The dashed line represents the case where 30% of the consumed silane is converted into disilane instead of being deposited. (From E. A. G. Hamers, Ph.D. Thesis, Universiteit Utrecht, Utrecht, the Netherlands, 1998, with permission.)

Mass-spectrometric research on silane decomposition kinetics has been performed for flowing [298, 302–306] and static discharges [197, 307]. In a dc discharge of silane it is found that the reaction rate for the depletion of silane is a linear function of the dc current in the discharge, which allows one to determine a first-order reaction mechanism in electron density and temperature [302, 304]. For an RF discharge, similar results are found [303, 305]. Also, the depletion and production rates were found to be temperature-dependent [306]. Further, the depletion of silane and the production of disilane and trisilane are found to depend on the dwell time in the reactor [298]. The increase of di- and trisilane concentration at short dwell times (<0.5 s) corresponds to the decrease of silane concentration. At long dwell times, the decomposition of di- and trisilane produces a much greater decrease in silane concentration. Moreover, a strong correlation is observed between deposition rate and di- and trisilane concentration. This has led to a proposal [283, 298, 308] of the main electron impact dissociation channel of silane, viz., e$^-$ + SiH$_4$ $\rightarrow$ SiH$_2$ + H$_2$ + e$^-$. The higher silanes are partially decomposed into radicals, which are responsible for deposition [308].

In a static discharge, first silane is introduced and subsequently all valves are closed. Then the discharge is ignited, and the mass-spectrometer signals for silane ($m/e = 30$ amu/e), disilane ($m/e = 62$ amu/e), trisilane ($m/e = 92$ amu/e), and tetrasilane ($m/e = 120$–128 amu/e) are followed as a function of time [197]. Rate equations can be formulated that explain the decrease in silane concentration and the increase in di-, tri-, and tetrasilane. From these data another proposal [197] of the main electron impact dissociation channel of silane is formulated: e$^-$ + SiH$_4$ $\rightarrow$ SiH$_2$ + 2H + e$^-$, in agreement

with others [201, 309, 310]. The abstraction reaction H + SiH$_4$ $\rightarrow$ SiH$_3$ + SiH$_3$ explains the large production of SiH$_3$ radicals in this case. The discrepancy as reflected in the hydrogen production (either H or H$_2$) cannot be resolved solely by mass-spectrometric data.

5.3.2. Analysis of Radicals

It is possible but difficult to detect directly, by mass spectrometry, the neutral radicals produced in a silane discharge. To discrimate between the radicals and the much more abundant neutrals, ionization of the species by low-energy electrons can be used [307], and the term *threshold ionization mass spectrometry* (TIMS) is used to denote this technique. This method is based on the principle of ionization of neutrals by electrons that have a well-defined energy. These ionized neutrals then are detected by means of a quadrupole mass spectrometer. In general, the radical A can be detected when the electron impact ionization process leads to the formation of an ion A$^+$. In order for this to occur, the energy of the electrons should be larger than the ionization potential E_i. Also, dissociative ionization of a molecule containing A may lead to A$^+$, following AB + e$^-$ $\rightarrow$ A$^+$ + B + 2e$^-$. Here, the energy of the electrons should be larger than the appearance potential E_a. It holds that $E_a > E_i$, and $E_a \approx E_i + E_{A-B}$, where E_{A-B} is the energy of the A$-$B bond [311]. When the electron energy is larger than E_a, the dissociative ionization process dominates, because the density of the molecule AB is much larger than that of the radical A. When the electron energy has a value between E_i and E_a, then only the radicals are ionized, and one is able to determine the radical density.

In silane discharges, one observes the following: when the discharge is off, the mass spectrometric signal at $m/e = 31$ amu/e (SiH$_3^+$) as a function of electron energy is due to dissociative ionization of SiH$_4$ in the ionizer of the QMS, with an ionization potential of 12.2 eV [312]. The signal with the discharge on is due to ionization of the radical SiH$_3$ plus the contribution from dissociative ionization of silane (the ions from the discharge are repelled by applying a positive voltage to the extraction optics of the QMS). The appearance potential of the SiH$_3$ radical is 8.4 eV [312], and therefore a clear difference between discharge on and off is observed. The corresponding threshold energies for SiH$_2$ are 11.9 and 9.7 eV (these four numbers are also given by Kae-Nune et al. [311], as 12.0, 8.0, 11.5, and 9.0 eV, respectively). The net radical contribution to the ion signal is given by [311]

$$\Delta I(\text{A}^+) = (I_{\text{on}} - I_{\text{bg}} - I_{\text{ions}}) - (I_{\text{off}} - I_{\text{bg}}) \frac{n_{\text{SiH}_4,\text{on}}}{n_{\text{SiH}_4,\text{off}}} \quad (38)$$

where I_{on} (I_{off}) is the signal when the discharge is on (off), I_{bg} the background signal, and I_{ions} the signal due to ions (which should be zero). The factor $n_{\text{SiH}_4,\text{on}}/n_{\text{SiH}_4,\text{off}}$ takes account of the depletion of the density of SiH$_4$, which can be obtained from the ratio $I_{\text{on}}/I_{\text{off}}$ at high electron energies. The concentration of radical SiH$_x$ can be expressed as [311]

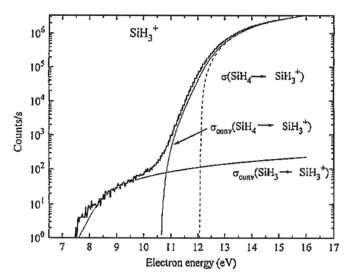

Fig. 34. Measured and fitted curves for the ionization of the SiH$_3$ radical. [From P. Kae-Nune, J. Perrin, J. Guillon, and J. Jolly, *Plasma Sources Sci. Technol.* 4, 250 (1995), © 1995, Institute of Physics, with permission.]

$$\frac{n(\text{SiH}_x)}{n(\text{SiH}_4)} = C \frac{Q(\text{SiH}_4 \rightarrow \text{SiH}_x^+, E_2 > E_a)}{I(\text{SiH}_x^+, E_2 > E_a)}$$

$$\times \frac{1}{E_1'' - E_1'} \int_{E_i - E_1'}^{E_1'' < E_a} \frac{\Delta I(\text{SiH}_x^+, E)}{Q(\text{SiH}_x \rightarrow \text{SiH}_x^+, E)} \, dE \tag{39}$$

where $I(\text{SiH}_x^+, E_2 > E_a)$ is the signal due to dissociative ionization of SiH_4, $\Delta I(\text{SiH}_x^+, E)$ the net radical signal [Eq. (38)], $Q(\text{SiH}_4 \rightarrow \text{SiH}_x^+, E_2 > E_a)$ the electron impact cross section for the dissociative ionization of SiH_4, $Q(\text{SiH}_x \rightarrow \text{SiH}_x^+, E)$ the electron impact cross section of the ionization of the radical SiH_x, and C a geometrical correction factor. Clearly, the knowledge of the electron impact cross sections is essential for the exact determination of the radical densities. Kae-Nune et al. [311] have used data for deuterated methane radicals (CD_3 and CD_2 [313]) to determine $Q(\text{SiH}_3 \rightarrow \text{SiH}_3^+, E)$ and $Q(\text{SiH}_2 \rightarrow \text{SiH}_2^+, E)$ by scaling. The value of $Q(\text{SiH} \rightarrow \text{SiH}^+, E)$ is determined by taking the average of $Q(\text{SiH}_2 \rightarrow \text{SiH}_2^+, E)$ and the measured $Q(\text{Si} \rightarrow \text{Si}^+, E)$ [314]. In Figure 34 the measured SiH_3^+ signal is compared with fitted curves for radical and dissociative ionization [311].

Nowadays, a commercially available QMS system (Hiden EQP300) is used by many groups, with which detection of radicals, and also of ions (see Section 5.3.3), is possible [315]. Kae-Nune et al. [198, 217, 311] have mounted this QMS system in the grounded electrode of their parallel plate PECVD reactor. Neutral (and ionic) species are sampled from the discharge via a 0.3-mm-diameter hole in the grounded electrode. Two-stage differential pumping ensures a low background pressure ($<10^{-6}$ Torr) in the ionization chamber of the QMS.

With this setup a series of measurements was taken in which the power was varied from 5 to 30 W. The pressure was low, 0.06 Torr, while the temperature was 250°C. The SiH_x ($0 < x < 3$) radical densities near the surface were measured as a function of effective power delivered to the discharge. It was

found that the SiH$_3$ radical is the most abundant one, with a concentration varying between 1.8×10^{11} and 3.2×10^{11} cm^{-3}. The SiH$_2$ radical concentration was about a factor of 20 lower. Both radical concentrations increase with increasing power.

From the radical concentrations it is possible to derive the radical flux to the surface:

$$\Phi_{\text{SiH}_x} = n_{\text{SiH}_x} \frac{v_{\text{SiH}_x}}{4} \frac{\beta_{\text{SiH}_x}}{1 - \beta_{\text{SiH}_x}/2} \tag{40}$$

where v_{SiH_x} is the thermal velocity ($= \sqrt{8k_B T/\pi m}$) of the radical SiH_x, β_{SiH_x} the surface reaction probability (see also Fig. 14 in Section 4.1.6). The quantity β is the adsorbed fraction of radicals, which either stick (sticking probability s_{SiH_x}) or recombine as SiH_4 or Si_2H_6 and go to the gas phase (recombination probability γ_{SiH_x}). By definition, $\beta = s + \gamma$. An amount $r = 1 - \beta$ is reflected from the surface directly.

The contribution of radical SiH_x to the total deposition rate follows from

$$R_{d, \text{SiH}_x} = \Phi_{\text{SiH}_x} \frac{s_{\text{SiH}_x}}{\beta_{\text{SiH}_x}} \frac{M_{\text{Si}}}{N_A \rho_{a\text{-Si:H}}} \tag{41}$$

where N_A is Avogadro's number, M_{Si} the molar mass of silicon, and $\rho_{a\text{-Si:H}}$ the mass density of a-Si:H (2.2 g/cm^3). For SiH$_3$ Matsuda et al. have determined β and s [137]: $\beta_{\text{SiH}_3} = 0.26$ and $s_{\text{SiH}_3} = 0.09$. For SiH$_2$, SiH, and Si large β and s are assumed, viz., $\beta_{\text{SiH}_2} = s_{\text{SiH}_2} = 0.8$ [311], $\beta_{\text{SiH}} = s_{\text{SiH}} = 0.95$ [316], and $\beta_{\text{Si}} = s_{\text{Si}} = 1$ [311]. Using these values for β and s, Kae-Nune et al. [311] have determined the contribution of radicals to the deposition rate. At low power levels, the deposition rate can fully be accounted for by the SiH$_3$ and SiH$_2$ radicals in a 60 : 40 ratio. At high power levels the other radicals also become important. At their highest power level of about 50 mW/cm^2, the sum of all radical contributions to the deposition rate is only 65%, indicating that dimer and trimer radicals, as well as ions, also contribute to the deposition.

Using the same threshold ionization mass spectrometry setup, Perrin et al. [317] have measured the temporal decay of radical densities in a discharge afterglow. From these experiments the coefficient β for the radical SiH$_3$ has been determined to be 0.28, which is in agreement with already known results from other (indirect) experimental approaches [136, 137, 318]. For the Si$_2$H$_5$ radical β is determined to be between 0.1 and 0.3. The coefficient β for atomic hydrogen on a-Si:H lies between 0.4 and 1, and is thought to represent mainly surface recombination to H$_2$.

5.3.3. Analysis of Ions

Detection of ions from a discharge is done by direct sampling through an orifice. In order to extract the ions collisionlessly the dimensions of the sampling orifice should be smaller than the sheath thickness; they are typically of the order of 100 μm. Moreover, the detected ions and their energies are representative of the plasma bulk situation only when the sheath is collisionless, i.e., at low pressures [286]. One generally is interested in the interaction of ions with a growing surface; hence normal operating pressures are used.

In RF discharges of silane, SiH_3^+ usually is the most abundant ion, but others (SiH_2^+, and $Si_nH_m^+$ with $1 < n < 9$) also are present [319]. The relative abundance of SiH_3^+ ions increases with increasing pressure, while that of SiH_2^+ decreases [319]. The ionization cross section of SiH_2^+ is higher than that of SiH_3^+ [320], but SiH_2^+ is lost via the reaction $SiH_2^+ + SiH_4 \rightarrow SiH_3^+ + SiH_3$ [305]. At pressures lower than 0.1 Torr SiH_2^+ becomes the dominant ion.

The ion clusters ($Si_2H_m^+$, $Si_3H_n^+$, ...) are also present, and larger clusters (positive and neutral) can be formed through reactions with silane molecules. Negative ions have been detected [321], which are thought responsible for the powders in the discharge [322].

Ion energy distributions (IEDs) are measured by several groups [323–326]. The reliability of IEDs depends strongly on the knowledge of the transmission function of the instruments, which most likely is energy-dependent. Improper adjustment of the various potential levels throughout the instruments can result in "ghost" structures in the IED, and they reflect the physics of the instrument rather than of the discharge. Hamers et al. [161, 163] have presented a method to determine the proper transmission function of an electrostatic lens system of a commercially available ion energy and mass spectrometer, the Hiden EQP [315]. This system is used by many groups to study plasmas [160, 198, 311, 321, 327, 328]. It is very versatile with respect to the control of voltages on the many electrostatic lenses.

The EQP is mounted in a plasma reactor identical to the ones in ASTER; see Figure 35. The plasma is generated between the two parallel electrodes. In the grounded electrode, where normally the substrate is placed for the deposition of a-Si:H, a small orifice is located to sample particles that arrive from the discharge. The orifice is made in a 20-μm-thick stainless steel foil, and has a diameter of 30 μm. The foil is integrated in a stainless steel flange. The ratio of the thickness of the foil to the diameter of the orifice results in a physical acceptance angle of 55° with respect to the normal to the electrode. The small size of the orifice compared to the sheath thickness and the mean free path of the particles in the plasma ensures that the orifice does not influence the discharge [323]. Using this orifice in deposition plasmas leads to deposition of a-Si:H on the surface of the orifice, but also on its inner sides, thereby narrowing the sampling diameter. This can be observed by monitoring the pressure rise in the mass spectrometer during experiments. Typically, a 50% reduction in orifice area takes 3 h. The a-Si:H film on the orifice then needs to be etched away, which is done *ex situ* by a KOH etch.

The design of the instrument, together with the pumping capacity, ensures a low background pressure ($<10^{-9}$ mbar). Under process conditions the pressure directly behind the orifice is about a factor of 10^5 lower than the process pressure; in the mass filter, even a factor of 10^6. The mean free path of particles that have entered the EQP therefore is several meters.

The ion optics (IO in Figure 35) consists of a number of electrostatic lenses, which direct the sampled ions through the drift

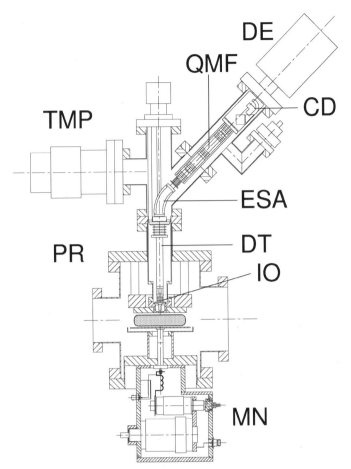

Fig. 35. Vertical cross section of the reaction chamber equipped with the mass spectrometer system. Indicated are: QMF, the quadrupole mass filter; ESA, the electrostatic analyzer; CD, the channeltron detector; DE, the detector electronics; DT, the drift tube; IO, the ion optics; TMP, the turbomolecular pump; PR, the plasma reactor; and MN, the matching network.

tube (DT) to the electrostatic analyzer (ESA). The ESA transmits only those ions with a specific energy-to-charge ratio ϵ_{pass}. An ESA has a constant relative energy resolution $\Delta\epsilon/\epsilon_{pass}$. Because this ESA operates at a constant $\Delta\epsilon$, the pass energy ϵ_{pass} is constant. Here, $\epsilon_{pass} = 40$ eV, and $\Delta\epsilon = 1.5$ eV (FWHM). The measurement of IEDs requires that sampled ions be decelerated or accelerated to this pass energy. This is also done in the ion optics part.

The other very important function of the ion optics is to shape the ion beam. Voltages on the various lenses should be set to avoid chromatic aberration, which causes energy-dependent transmission of ions in the instrument, and as a result erroneous IEDs [161, 163]. The correct lens settings have been found by simulations of ion trajectories in the EQP using the simulation program SIMION [329]. In addition, an experimental method to find the correct settings has been presented [161, 163].

The acceptance angle depends on the lens settings and on the ion energy. Low-energy ions are deflected more towards the optical axis than high-energy ions, which results in a larger acceptance angle. The acceptance angle varies between about 6° and 1° for ion energies between 1 and 100 eV [161, 163].

For lower energies the acceptance angle can reach values above 20°, which causes drastic increases of the low-energy part of the measured IED.

The measured ion energy distributions are affected by the processes that ions experience during their passage through the plasma sheath. In a collisionless situation, the IED is solely determined by the RF modulation of the plasma potential. The extent to which an ion will follow these modulations depends on its mass. Electrons are able to follow the electric field variations instantaneously, while ions experience a time-averaged electric field. The energy of an ion that arrives at the electrode depends on the value of the plasma potential at the time that the ion entered the sheath. Hence, the ion energy is modulated in much the same way that the plasma potential is modulated. Many ions arrive at the electrode with energies close to the extreme values of the harmonically varying voltage, and this results in a saddle structure in the IED, in which the two peaks reflect the minimum and maximum plasma potential [134]. The heavier the ion, the more it experiences a time-averaged plasma potential, and the narrower the saddle structure is.

Elastic collisions and chemical reactions in the sheath lead to a broad angular distribution of the ion velocity at the electrode. Examples for SiH_3^+ and SiH_2^+ are the H^- abstraction reactions $SiH_3^+ + SiH_4 \rightarrow SiH_4 + SiH_3^+$ [330] and $SiH_2^+ + SiH_4 \rightarrow SiH_3^+ + SiH_3$ [331]. Due to the small acceptance angle of the EQP, these processes are only in part reflected in the IED [163, 332].

5.3.3.1. Charge Exchange

Much more important for the shape of the IED are charge exchange processes. In a charge exchange process an electron is exchanged between a neutral and an ion. In the plasma sheath a charge exchange between a neutral of thermal energy and a fast ion leads to the formation of a thermal ion and a fast neutral. These newly created thermal ions are accelerated towards the electrode, and the kinetic energy that they gain depends on the phase in the RF period and their position at the time of creation. This leads to distinct peaks in the IED, and the origin of these peaks can be explained as follows [325]. Electrons respond to the RF modulation and move to and from the electrodes. The movement of the electron front is a function of the phase ϕ in the RF period. When an ion is created (by a charge exchange process) behind the electron front, i.e., in the plasma bulk, it will not experience an electric field. Once the electron front moves inward to the plasma bulk, the ion will be accelerated to the electrode. Depending on the place of creation x, the ion will be able to reach the electrode before the electron front is at the position of the ion again. Once in the plasma bulk, the ion will not be accelerated until the electron front moves inward again. This may be repeated several times. Wild and Koidl [325] have presented a description of the origin of the maxima in IEDs. They argue that these maxima only occur if the ion energy ϵ at the electrode is independent of the phase ϕ and the position x of creation of the newly formed ion. Two conditions are to be met: $d\epsilon/d\phi = 0$ and $d\epsilon/dx = 0$. As shown by Wild and

Koidl [325], the energy ϵ varies between $\epsilon_{\min}$ and $\epsilon_{\max}$, and for every x, the values of $\epsilon_{\min}$, $\epsilon_{\max}$, and $\epsilon_{\phi=0}$ fulfill the condition $d\epsilon/d\phi = 0$. Further, $\epsilon_{\phi=0}$ exhibits a series of extrema as a function of x, and at these extrema is alternately equal to $\epsilon_{\min}$ and $\epsilon_{\max}$. At these extrema both conditions are met, and a saddle structure is observed in the IED.

In addition, ions that arrive with the mean energy of the saddle structure have reached the electrode after an integer number of periods. Thus, ions in successive charge exchange saddle structures have needed an increased number of periods to reach the electrode, on average. Hamers has very nicely illustrated this by performing Monte Carlo simulations of charge exchange processes [163], using a time- and position-dependent electric field, which is calculated from the sheath model as described by Snijkers [333]. About 300,000 SiH_2^+ ions were followed. The ion density at the electrode was 3×10^8 cm^{-3}, the plasma potential 25 V, and the excitation frequency 13.56 MHz. The main saddle structure is found to be around 25 V (23–27 V). In the IED four charge exchange peaks are clearly distinguished, at about 2, 7.5, 14, and 19 eV. From a comparison with the transit time, i.e. the time an ion needs to cross the sheath, it is clear that the peaks in the IED correspond to ions that needed an integer number of periods to cross the sheath. The first peak at 2 eV corresponds to ions that needed one period, and so on. In general, ions in charge exchange peak p needed p periods to cross the sheath.

The energy position ϵ_p of the peak and the corresponding number p of periods of time T are used by Hamers et al. [163, 332] to reconstruct the time-averaged potential profile $V(x)$ in the sheath. An ion that arrives with energy ϵ_p at the electrode has a velocity that follows from $\frac{1}{2}mv^2 = \epsilon_p - eV(x)$. With $v = dx/dt$ one derives

$$\int_0^{x_p} \sqrt{\frac{m/2}{\epsilon_p - eV(x)}}\, dx = \int_0^{pT} dt = pT \qquad (42)$$

where x_p is the mean position of creation of the ions that arrive at the electrode with energy ϵ_p. The unknown $V(x)$ now is found by further assuming that $V(x)$ can be represented by a parabolic function [328, 333]. A parabolic potential profile corresponds to a constant (net) charge carrier density n as a function of distance x in the sheath, according to $-\vec{\nabla}^2 V = \vec{\nabla} \cdot \vec{E} = en/\epsilon_0$. The electric field $E(x)$ in the sheath then is a linear function of x:

$$E(x) = -E_0 + \frac{enx}{\epsilon_0} \qquad (43)$$

with E_0 the electric field at the electrode ($x = 0$). Integrating $E(x)$ from the electrode ($V = 0$) over a distance x gives the potential

$$V(x) = E_0 x + \frac{en}{2\epsilon_0}x^2 \qquad (44)$$

The sheath thickness d_s is found from the position where the electric field vanishes: $d_s = \epsilon_0 E_0/(en)$. Here the potential is equal to the time-averaged plasma potential $V_{\rm pl}$. We then find, with Eq. (44), for the electric field at the electrode

$$E_0 = \sqrt{\frac{2enV_{pl}}{\epsilon_0}} \tag{45}$$

and for the sheath thickness

$$d_s = \sqrt{\frac{2\epsilon_0 V_{pl}}{en}} \tag{46}$$

Rewriting Eq. (44) yields

$$V(x) = V_{pl}\left[\frac{2x}{d_s} - \left(\frac{x}{d_s}\right)^2\right] \tag{47}$$

Substitution of $V(x)$ in Eq. (42) and solving for ϵ_p then yields

$$\epsilon_p = eV_{pl}\left\{1 - \left[\cosh\left(pT\sqrt{\frac{e^2 n}{m\epsilon_0}}\right)\right]^{-2}\right\} \tag{48}$$

The energy position ϵ_p of peak p in the IED of an ion with mass m is seen to be dependent on the plasma potential V_{pl}, the RF period T, and the ion plasma frequency $\omega_i = \sqrt{e^2 n/(m\epsilon_0)}$. Equation (48) can be used to determine the (net) charge carrier density in the sheath and the time-averaged potential V_{pl} from measured IEDs. The mean position x_p follows from combining Eq. (47) and Eq. (48):

$$x_p = d_s\left\{1 - \left[\cosh\left(pT\sqrt{\frac{e^2 n}{m\epsilon_0}}\right)\right]^{-1}\right\} \tag{49}$$

In silane discharges several ions are observed to be involved in a charge exchange process, and therefore maxima in their ion energy distribution at distinct energies are observed. The charge carrier density and the plasma potential that result from the fit of the IED allow for the quantification of the related parameters sheath thickness and ion flux. This method has been be used to relate the material quality of a-Si:H to the ion bombardment [301, 332]; see also Section 6.2.3.

In the following, IEDs measured in silane–argon and silane–hydrogen discharges are shown, and V_{pl} and n are determined from fitting the data using Eq. (48). In fact, V_{pl} is determined from an IED that is not affected by collisions in the sheath; this is the case for $Si_2H_4^+$, where only one peak is observed in the IED. It should be noted that in general the width of the saddle structure is smaller than the energy resolution (1.5 eV) of the instrument, and therefore cannot be distinguished.

5.3.3.2. Silane–Argon Discharges

Silane–argon discharges are very illustrative, for no less than four different ions have charge exchange maxima in their ion energy distribution. In Figure 36 IEDs are shown that were measured in a plasma that was created in a mixture of 13 sccm Ar and 13 sccm SiH_4, at a pressure of 0.1 mbar. The applied RF frequency was 13.56 MHz, at a power of 10 W. The dc self-bias voltage that developed was -135 V. The substrate temperature was 250°C. At least the first six peaks in the IED of Ar^+ and the first five peaks in the IED of SiH_2^+

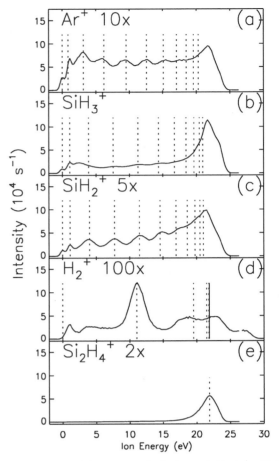

Fig. 36. The measured ion energy distributions of (a) Ar^+, (b) SiH_3^+, (c) SiH_2^+, (d) H_2^+, and (e) $Si_2H_4^+$ from a SiH_4/Ar plasma. The conditions are 13 sccm SiH_4, 13 sccm Ar, 13.56 MHz, 10 W, 0.2 mbar, 250°C, $V_{dc} = -135$ V. The dotted lines in (a), (b), (c), and (d) indicate the position of the first 10 charge exchange peaks based on a plasma potential of 21.8 V and a charge carrier density n of 1.66×10^8 cm^{-3}. The dotted line in the IED of $Si_2H_4^+$ (e) indicates the value of the plasma potential. Note the different scaling factors of the various IEDs. (From E. A. G. Hamers, Ph.D. Thesis, Universiteit Utrecht, Utrecht, the Netherlands, 1998, with permission.)

can clearly be discerned. From the IED of $Si_2H_4^+$ (Fig. 36e) a time-average plasma potential V_{pl} of 21.8 ± 0.3 V is derived. Using now only *one* fitting parameter, i.e., the charge carrier density n, all distinct peaks in the IEDs of Ar^+, SiH_3^+, SiH_2^+, and H_2^+ can be fitted with Eq. (48). Performing the fit yields $n = (1.66 \pm 0.05) \times 10^8$ cm^{-3}. The energy positions are fitted well, as can be seen by comparing the data in Figure 36 with the dotted vertical lines. The sheath thickness as calculated from Eq. (46) is 3.81 ± 0.05 mm. The first charge exchange peak in the Ar^+ IED originates at a distance of 100 μm from the electrode, as calculated from Eq. (49). The sixth peak, which is still visible, originates at a distance of about half the sheath thickness from the electrode.

It is observed that the SiH_3^+ ion is the most abundant ion, see Figure 36b. The peak around 23 V is the main saddle structure and is asymmetric due to collision processes. SiH_2^+ (see Figure 36c) exhibits a 5 times lower peak intensity. The first five charge exchange maxima in the IED of SiH_2^+ (Fig. 36c) are clearly visible. The IED intensity is about 5 times lower than for SiH_3^+. Although the SiH_2^+ ion is the main product of dissociative ionization of SiH_4 [320], the measured IED intensity of SiH_3^+ is always larger. The H^- transfer reaction causes SiH_3^+ to be more abundant than SiH_2^+. Also, in the IED of Ar^+ (Fig. 36a) about six of the charge exchange maxima are clearly visible. The broad overall IED shows that many ions underwent a charge exchange reaction in the sheath. Even H_2^+ (Fig. 36d) IEDs can be measured. The low intensity is due to the low partial pressure of hydrogen (it is created by silane dissociation only), and to the fact that the ionization potential of hydrogen is, like that of argon, higher than that of silane. Therefore the number of created hydrogen ions will be small.

The maximum in the IED at 11 eV is the first charge exchange maximum of H_2^+. The energy of this maximum is always somewhat lower than the energy of the fourth charge exchange maximum in the SiH_2^+ IED. This is due to the almost four times higher plasma frequency of the H_2^+ ion ($m/e = 2$ amu/e) than that of the SiH_2^+ ion ($m/e = 30$ amu/e).

The H_2^+ saddle structure is broader than the saddle structures of the other ions, due to the larger ion plasma frequency of H_2^+. The ions with an energy larger than 25 eV are part of the high-energy side of the main saddle structure.

$Si_2H_4^+$ is an ion that is created in the plasma by polymerization reactions. Several pathways may lead to this ion. The first pathway is the dissociative ionization of Si_2H_6 that is formed in a radical–neutral reaction. The second pathway is the direct formation in the ion–molecule reaction [192] $SiH_2^+ + SiH_4 \rightarrow Si_2H_4^+ + H_2$.

5.3.3.3. Silane–Hydrogen Discharges

The IEDs of several ions in a typical silane–hydrogen plasma are shown in Figure 37. The plasma was created in a mixture of 26 sccm SiH_4 and 24.5 sccm H_2, at a pressure of 0.2 mbar. The applied RF frequency is 13.56 MHz, at a power of 10 W. The self-bias voltage that developed was -118 V. The substrate temperature was 250°C. The discharge is in the α-regime. The IEDs of H_2^+ and SiH_2^+ show distinct peaks. The positions of the peaks are again described with the plasma potential, as deduced from the $Si_2H_4^+$ IED, and the charge carrier density n. The value for V_{pl} is 23.8 ± 0.2 V, and the value for n is $(1.11 \pm 0.04) \times 10^8$ cm^{-3}. The sheath thickness is 4.88 ± 0.10 mm. Under these and under most other conditions, the SiH_3^+ IED shows no distinguishable charge exchange maxima.

The measured intensity in the SiH_3^+ IED again is higher than that in the SiH_2^+ IED. There are two distinct peaks at low energies in the SiH_3^+ IED, and such peaks are observed in most SiH_3^+ IEDs. Their energy positions cannot be explained by charge exchange processes. The intensity in the H_2^+ IED is higher than in the silane–argon plasma, simply because the partial pressure of hydrogen in this silane–hydrogen mixture is higher.

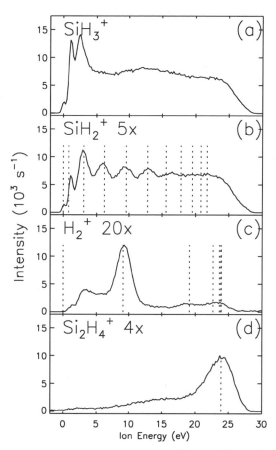

Fig. 37. The ion energy distributions of (a) SiH_3^+, (b) SiH_2^+, (c) H_2^+, and (d) $Si_2H_4^+$ from a SiH_4–H_2 plasma. The conditions are 26 sccm SiH_4, 24.5 sccm H_2, 13.56 MHz, 10 W, 0.2 mbar, 250°C, $V_{dc} = -118$ V. The dotted lines in (b) and (c) indicate the position of the first 10 charge exchange peaks based on a plasma potential of 23.8 V and a charge carrier density n of 1.11×10^8 cm^{-3}. The dotted line in the IED of $Si_2H_4^+$ (d) indicates the value of the plasma potential. Note the different scaling factors of the various IEDs. (From E. A. G. Hamers, Ph.D. Thesis, Universiteit Utrecht, Utrecht, the Netherlands, 1998, with permission.)

Charge exchange maxima are in general pronounced in the SiH_2^+ IED in discharges at 13.56 MHz. The successive charge exchange peaks tend to overlap each other at higher frequencies. The IED of H_2^+ exhibits charge exchange peaks that are still well separated at higher frequencies due to the almost four times higher ion plasma frequency of H_2^+ than that of SiH_2^+.

Very different IEDs can be observed in discharges that are operated in the γ'-regime and that contain powder. The discharge, in this case, is created in a mixture of 30 sccm SiH_4 and 30 sccm H_2 at a pressure of 0.60 mbar. The RF frequency is 13.56 MHz, and the applied power is 33.7 W. The dc self-bias voltage is -32.5 V. The substrate temperature is 250°C. The measured IEDs are shown in Figure 38. The plasma potential is determined from the IED of $Si_2H_4^+$, and is 49.2 ± 0.5 V, which is about twice the value in discharges operated in the α-regime. The charge carrier density is $(9.9 \pm 0.3) \times 10^8$ cm^{-3}. The sheath thickness is 2.34 ± 0.05 mm.

The peaks in both the SiH_2^+ (Fig. 38a) and the SiH_3^+ (Fig. 38b) IED are well separated, and the intensity in the region between the charge exchange peaks is relatively low. The

Table VI. Sheath Properties of an Argon–Silane and Two Silane–Hydrogen Discharges[a]

Property	Ar/SiH$_4$	H$_2$/SiH$_4$ (α)	H$_2$/SiH$_4$ (γ')	Unit
V_{pl}	21.8 ± 0.3	23.8 ± 0.2	49.2 ± 0.5	V
n	1.66 ± 0.05	1.11 ± 0.04	9.9 ± 0.3	10^8 cm^{-3}
d_s	3.81 ± 0.05	4.88 ± 0.10	2.34 ± 0.05	mm
E_0	11.4 ± 0.2	9.8 ± 0.2	42.0 ± 0.8	kV/m
Γ_{max}	1.93 ± 0.05	1.35 ± 0.04	17.3 ± 0.7	10^{18} m^{-2} s^{-1}
$(\epsilon\Gamma)_{max}$	6.7 ± 0.2	5.2 ± 0.2	136 ± 4	W m^{-2}

[a] α and γ' regimes; plasma potential V_{pl}, charge carrier density n, sheath thickness d_s, electric field at the electrode E_0, ion flux Γ_{max}, and ion energy flux $(\epsilon\Gamma)_{max}$. Compiled from E. A. G. Hamers, Ph.D. Thesis, Universiteit Utrecht, Utrecht, the Netherlands, 1998.

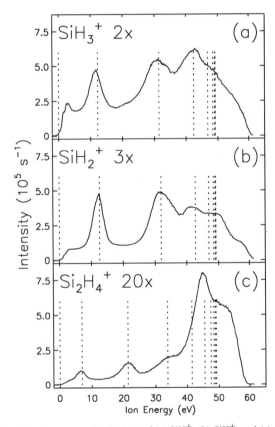

Fig. 38. The ion energy distributions of (a) SiH$_3^+$, (b) SiH$_2^+$, and (c) Si$_2$H$_4^+$ from a SiH$_4$–H$_2$ plasma in the γ'-regime. The dotted lines in (a), (b), and (c) indicate the position of the first 10 charge exchange peaks based on a plasma potential of 49.2 V and a charge carrier density n of 9.9×10^8 cm^{-3}. Note the different scaling factors of the various IEDs. (From E. A. G. Hamers, Ph.D. Thesis, Universiteit Utrecht, Utrecht, the Netherlands, 1998, with permission.)

IED of Si$_2$H$_4^+$ also shows charge exchange maxima on the low-energy side of the IED, as well as an asymmetrical main saddle structure around 50 eV.

The sheath characteristics of the argon–silane and the two silane–hydrogen discharges of which the IEDs were shown above are summarized in Table VI. The sheath characteristics of the 10-W silane–argon discharge and the 10-W silane–hydrogen discharge in the α-regime are rather similar. The high-power, high-pressure discharge (γ'-regime) has a twice larger plasma potential, a twice smaller sheath distance, a four times larger electric field E_0, and a much larger ion flux and ion energy flux (see Section 6.2.2) than the discharge in the α-regime.

5.3.3.4. Charge Exchange Reactions

The charge exchange reactions between Ar and Ar$^+$ and between H$_2$ and H$_2^+$ are well known [325, 328]:

$$\text{Ar}^{+(f)} + \text{Ar}^{(t)} \rightarrow \text{Ar}^{(f)} + \text{Ar}^{+(t)} \qquad (50)$$

$$\text{H}_2^{+(f)} + \text{H}_2^{(t)} \rightarrow \text{H}_2^{(f)} + \text{H}_2^{+(t)} \qquad (51)$$

where (f) and (t) denote a fast ion or neutral and an ion or neutral with thermal energy. These reactions are called symmetrical resonant charge exchange reactions, as they occur between an ion and the corresponding neutral.

The ionization energies of hydrogen (15.4 eV) and argon (15.8 eV) are higher those of silane (11.6 eV) and disilane (9.9 eV). Therefore, ion–molecule reactions of H$_2^+$ or Ar$^+$ with silane or disilane will result in electron transfer from silane to the ion [190], which leads to dissociative ionization of silane. These known asymmetric charge exchange reactions result in SiH$_x^+$ ($0 < x < 3$) ions with thermal energy in the sheath; see Table VII [334, 335]. The charge exchange of H$_2^+$ with SiH$_4$ is a major loss process for the hydrogen ions in typical silane plasmas.

The cross sections of these asymmetric charge exchange reactions are in general lower than the cross sections of resonant charge exchange processes [134]. For comparison, the cross section of the Ar$^+$–Ar charge exchange reaction is 4×10^{-15} cm^2, about one order of magnitude larger than the reactions listed in Table VII. The total cross section of charge exchange reactions between Ar$^+$ and SiH$_4$ is much smaller than the total cross section of charge exchange reactions between H$_2^+$ and SiH$_4$.

The discharge in which the charge exchange peaks of Si$_2$H$_4^+$ are observed is in the γ'-regime. A considerable amount of silane will be depleted and gas phase polymerizations occur, which means that Si$_2$H$_6$ is likely to be present in large quantities. Therefore, Hamers et al. [235] have suggested that Si$_2$H$_6$

Table VII. Charge Exchange Reactions of Ar^+ and H_2^+ with $SiH_4{}^a$

Reaction	σ $(10^{-16}\ cm^2)$
$Ar^+ + SiH_4 \rightarrow Ar + Si^+ + 2H_2$	0.2
$Ar^+ + SiH_4 \rightarrow Ar + SiH^+ + H_2 + H$	0.3
$Ar^+ + SiH_4 \rightarrow Ar + SiH_2^+ + H_2$	0.4
$Ar^+ + SiH_4 \rightarrow Ar + SiH_3^+ + H$	2
$H_2^+ + SiH_4 \rightarrow H_2 + Si^+ + 2H_2$	2
$H_2^+ + SiH_4 \rightarrow H_2 + SiH^+ + H_2 + H$	2
$H_2^+ + SiH_4 \rightarrow H_2 + SiH_2^+ + H_2$	11
$H_2^+ + SiH_4 \rightarrow H_2 + SiH_3^+ + H$	34

^{a}The cross sections σ for reactions with Ar^+ are determined at thermal energies of the reactants [334]; the cross sections for reactions with H_2^+ are determined at a kinetic energy of the reactants of 1 eV [335]. (Compiled from E. A. G. Hamers, Ph.D. Thesis, Universiteit Utrecht, Utrecht, the Netherlands, 1998.)

is involved in the charge exchange reaction where $Si_2H_4^+$ is created:

$$H_2^+ + Si_2H_6 \rightarrow H_2 + Si_2H_4^+ + H_2 \qquad (52)$$

5.4. Ellipsometry

Another *in situ* technique that is used to study deposition in real time is ellipsometry. This optical technique is non-invasive and does not perturb the discharge. In ellipsometry one measures the amplitude and phase of the reflected light from a surface. Spectroscopic ellipsometry (SE) allows for the wavelength-dependent measurement of the dielectric function of the material, at submonolayer sensitivity [336]. Spectroscopic phase-modulated ellipsometry (SPME) gives the possibility of fast data recording, and therefore is exploited for kinetic studies of the deposition of a-Si:H [337]. Spectroscopic ellipsometry is a very sensitive tool to study surface and interface morphology in the UV–visible range. The use of IR light allows for the identification of Si–H bonding configurations [338].

Polarization modulation can be done by rotating the polarizer, as is done in rotating-element ellipsometers [339], or by using photoelastic devices [337]. Data acquisition nowadays is fast, which makes real-time measurements of film and interface formation possible: a full spectrum ranging from 1.5 to 5 eV can be measured in less than a second [340–342].

In a typical ellipsometry experiment a sample is irradiated with polarized light, which subsequently is reflected from the sample surface and detected after passing an analyzer. The ratio ρ of complex reflectances for perpendicularly (s) and parallelly (p) polarized light usually is represented as follows:

$$\rho = \frac{r_p}{r_s} = \tan \Psi \exp(i\Delta) \qquad (53)$$

with r_p and r_s the complex reflectances for p- and s-polarized light, and Ψ and Δ two convenient angles. This ratio can be related directly to the complex dielectric function $\epsilon = \epsilon_1 + i\epsilon_2$ in the special case of light incident at an angle ϕ on the surface

of a semiinfinite medium [336]:

$$\epsilon = \sin^2 \phi \left(1 + \tan^2 \phi \left[\frac{1 - \rho}{1 + \rho} \right]^2 \right) \qquad (54)$$

In case of multilayer structures much more complex expressions result.

In the IR one usually presents spectroellipsometric measurements using the optical density $D = \ln(\overline{\rho}/\rho)$, where $\overline{\rho}$ refers to the substrate before deposition of a film [343]. The presence of a vibrational mode (rocking, bending, stretching) of a silicon–hydrogen bond, for instance, is revealed by a peak at the corresponding energy of that mode in the real part of D versus energy (in cm^{-1}). The contribution of a specific vibration mode to the dielectric function of the film is estimated from the Lorentz harmonic-oscillator expression [344]:

$$\epsilon = \epsilon_\infty + \frac{F}{\omega_0^2 - \omega^2 - i\Gamma\omega} \qquad (55)$$

with ω_0 the frequency of the mode, F its oscillator strength, and Γ a damping constant. The oscillator strength can be expressed as $F = \omega_0^2(\epsilon_0 - \epsilon_\infty)$, where ϵ_0 and ϵ_∞ are the high- and low-frequency dielectric constants, respectively. The vibration mode of a chemical bond will show up as a maximum or minimum in the real part of D versus energy, depending on the dielectric function of the substrate.

The interpretation of ellipsometric data is based on the description of the optical properties of the material under study. Effective-medium theories (EMTs) are widely used for this purpose, as they can be applied when materials are heterogeneous and when the size of inhomogeneities is small compared to the wavelength of the light, so that scattering of light can be neglected. In a-Si:H the presence of voids will have an effect on the dielectric function. Also, during deposition the surface will roughen more or less, depending on deposition conditions, which will have a perturbing effect on the polarization of the incident light. An optical model of a substrate with an a-Si:H film could consist of a substrate, an a-Si:H film with voids, and a surface layer of a-Si:H with a large number of voids, which reflects the surface roughness.

The general form that is used in EMTs to describe the dielectric function ϵ of an a-Si:H layer with a certain fraction of voids is given by [345]

$$\epsilon = \frac{\epsilon_a \epsilon_v + \kappa \epsilon_h (f_a \epsilon_a + f_v \epsilon_v)}{\kappa \epsilon_h + (f_a \epsilon_v + f_v \epsilon_a)} \qquad (56)$$

where κ is the screening parameter [$\kappa = (1/q) - 1$ and $0 \leq q \leq 1$], ϵ_h is the host dielectric function, ϵ_a and f_a are the dielectric function and volume fraction of the a-Si:H component, and ϵ_v and f_v are the dielectric function and volume fraction of the void component. Several approximations can be made. At the end of the seventies Aspnes et al. [346] found that the Bruggeman approximation provided the best description for the data available. Recently Fujiwara et al. [345], with new spectroellipsometric data, confirmed this fact. The Bruggeman approximation [61] defines $\epsilon_h = \epsilon$. For spherical inclusions $q = 1/3$, i.e., $\kappa = 2$, and the dielectric function is isotropic. For the effect of voids present in the a-Si:H layer we further simplify Eq. (56) by taking $\epsilon_v = 1$, and substituting $f_a = 1 - f_v$.

In a real-time spectroellipsometric measurement in which the kinetics of a-Si:H deposition is studied, trajectories are recorded in the $\Delta-\Psi$ plane at various photon energies between 2 and 4 eV. These trajectories can be simulated and fitted to models that represent the growing a-Si:H layer. Canillas et al. [347] have made a detailed study of the deposition of the first few layers of a-Si:H on a NiCr/glass substrate. Similar results are obtained for a c-Si substrate. They have proposed several models to explain the data. One possible model is the *hemispherical* nucleation model, which describes a hexagonal network of spherical a-Si:H nuclei located at an average distance d between them. The growth is represented by an increase in the radii of the nuclei until the nuclei make contact. This results in an a-Si:H layer with a large void fraction, of $f_v = 0.39$ [286]. From the start of the deposition ($f_v = 1$) to the point where the nuclei make contact ($f_v = 0.39$), the dielectric function is calculated using the Bruggeman EMT, and a $\Delta-\Psi$ trajectory can be simulated. In another model, the *columnar* nucleation model, columns of a-Si:H material start to grow from the initial hexagonal network, where the growth is represented by an increase in column radius and height. Again, f_v is calculated, and the dielectric function can be deduced from that. A further refinement is that the columns coalesce, and form a dense layer, with a layer of a certain thickness on top, which consists of free standing columns. This extra layer, typically 5 nm in thickness, represents the roughness, and increases with film thickness for typical deposition conditions. A recent study supports the concept of coalescing 3D islands of a-Si:H [340]: directly from the start of the deposition, the thickness of the surface roughness layer increases to 2 nm. At that thickness the islands start to combine into a continuous layer, and the thickness of the surface roughness layer decreases to remain constant at 1.5 nm.

Andújar et al. [246] have used SE to study the effects of pressure and power on the properties of a-Si:H. In their system the $\alpha-\gamma'$ transition occurs just above 0.2 mbar at a power of 10 W. They found that the density of the material is decreased on going from the α- to the γ'-regime, as is deduced from the decrease in the imaginary part of the dielectric function ϵ_2.

Collins et al. [342] have used SE to study the effects of hydrogen dilution. On native oxide c-Si substrates they have found that the thickness of the roughness layer is nearly constant at 1.5–2 nm in the case of deposition with pure silane, for a total deposited layer thickness up to 1000 nm. At a dilution factor $R = [H_2]/[SiH_4]$ of 10, the thickness of the roughness layer decreases from 1.5 to just below 1 nm. In case of $R = 20$, a similar decrease is observed (with a minimum of 0.3 nm at a bulk layer thickness of 20 nm), but followed by a steep increase to 4.5 nm at a bulk layer thickness of 100 nm, which is consistent with the formation of a microcrystalline film. At values above $R = 25$, the deposited layer is microcrystalline. These deposition experiments have also been carried out starting on a 300-nm-thick a-Si:H film, prepared with $R = 0$. Here, the amorphous-to-microcrystalline transition, as evidenced by

the steep increase of the thickness of the roughness layer, occurs at 200 nm for $R = 15$ and at 60 nm for $R = 30$. Based on these observations, a phase diagram was constructed, showing the amorphous-to-microcrystalline transition as a function of the bulk layer thickness and the dilution factor R. The importance of this phase diagram lies in the fact that a deposition process that is close to the amorphous-to-microcrystalline transition yields material that has excellent optoelectronic properties [348–352].

Spectroscopic ellipsometry has also been applied in the characterization of compositionally graded a-SiC:H alloys [353], where the flow ratio $z = [CH_4]/([CH_4] + [SiH_4])$ was varied during deposition. The analysis of the SE data on the graded layer prompted the development of a new four-medium model, which consists of the ambient, a surface roughness layer, an outer layer, and the pseudo-substrate. A virtual interface approximation is applied at the interface between outer layer and pseudo-substrate. This model gives the near-surface carbon content and the instantaneous deposition rate, as well as the thickness of the roughness layer. Besides a-Si:H–a-SiC:H interfaces, also p–i interfaces have been characterized, which is possible because the dielectric function changes at the interface [354, 355].

The use of IR in SE allows for the investigation of hydrogen incorporation in a-Si:H. The identification of vibrational modes in nanometer thin films is difficult, due to the weak signals. Nevertheless, Blayo and Drévillon have shown that monohydride and dihydride bonding configurations can be discerned in 0.5- to 1-nm-thick a-Si:H layers [338], using IR phase-modulated ellipsometry (IRPME). In *in situ* IRPME studies the real part of D is presented in the stretching mode region of the IR, i.e., 1900–2200 cm^{-1}. Three stretching modes can be revealed in this range, typically around 2000, 2100, and 2160 cm^{-1}, corresponding to SiH, SiH$_2$, and SiH$_3$ bonding configurations. From the evolution of the SiH– and the SiH$_2$ stretching mode with increasing deposited thickness, it was found [356] that the deposited layer is built up from two layers. The bottom layer contains SiH bonds, and linearly increases in thickness with deposition time. The top layer has a nearly constant thickness of 1–2 nm, and contains SiH$_2$ bonds. At the start of the deposition a layer consisting of SiH$_2$ is formed, and its thickness increases with time. At a thickness of about 1 nm, another layer between substrate and the SiH$_2$-containing layer is formed, which contains SiH. The deposition proceeds as the thickness of this latter layer increases with time. Attenuated total reflection IR spectroscopy has revealed that SiH$_2$ and SiH$_3$ surface modes are present even before the SiH$_2$-containing layer is formed [340].

The position of the SiH and SiH$_2$ stretching modes varies with the thickness of the deposited layer. In the first few monolayers the SiH peak shifts upwards by about 20 cm^{-1}, as a result of the very high hydrogen content [340]. At the end of the coalescence phase, the SiH peak is at 1995 cm^{-1}, and its position does not change any more. The substrate used was c-Si with a native oxide layer. In contrast, another study shows an increase in the SiH peak position from 1960 to 1995 cm^{-1}, together with an increase in the SiH_2 peak position from 2060 to 2095 cm^{-1}, for thicknesses of the deposited layer between 0 and 500 nm [344]. Moreover, these shifts depend on the deposition temperature and on the substrate. These findings were related to the change in disorder as a function of deposited thickness. The disorder, characterized by the width of the Raman TO peak, decreased with increasing thickness.

6. RELATION BETWEEN PLASMA PARAMETERS AND MATERIAL PROPERTIES

6.1. External Parameters

Luft and Tsuo have presented a qualitative summary of the effects of various plasma parameters on the properties of the deposited a-Si:H [6]. These generalized trends are very useful in designing deposition systems. It should be borne in mind, however, that for each individual deposition system the optimum conditions for obtaining device quality material have to be determined by empirical fine tuning. The most important external controls that are available for tuning the deposition processs are the power (or power density), the total pressure, the gas flow(s), and the substrate temperature. In the following the effects of each parameter on material properties will be discussed.

6.1.1. Plasma Power

When the plasma power (density) is increased, the deposition rate is increased monotonically up to the point where the gas flow rate becomes the limiting factor [81, 357] (see Figure 39). This deposition rate increase comes with a number of disadvantages, including poor film quality and powder formation. At low power levels a large part of the SiH_4 remains undissociated [303], and the corresponding films contain only silicon monohydrides, inferred from FTIR measurements. At high power levels the SiH_4 is strongly dissociated, and the films contain a large amount of SiH_2 [358]. The microstructure parameter R^* is high. The high power levels also lead to columnar microstructure, which is accompanied by high spin densities [359, 360]. Experimental results obtained in ASTER show similar trends, as is shown in Section 6.2.3.

The electrical properties (dark conductivity and photoconductivity) are reported to first decrease and then increase upon increasing power [361]. The optical bandgap increases with increasing power, due to the increase of the hydrogen content [63, 82, 362, 363]. However, at very high power levels, microcrystalline silicon is formed [364], which causes the hydrogen content (and, consequently, the bandgap) to decrease.

As a result of the deterioration of film properties with increasing power levels, to obtain *device quality* material the use of low power is required, albeit with a concomitant low deposition rate. Increasing the deposition rate without altering the *device quality* material properties is a large research challenge.

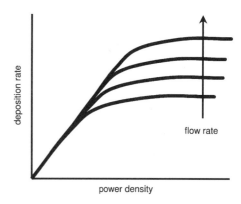

Fig. 39. Schematic representation of the influence of power density and flow rate on the deposition rate. [After A. Matsuda, *J. Vac. Sci. Technol. A* 16, 365 (1998).]

6.1.2. Total Pressure

The deposition rate as a function of total pressure shows two different dependences. In the low-pressure, or supply-limited, regime the deposition rate is proportional to the pressure, while in the power-limited regime it is constant [365]. At low pressure the film quality is not good. The supply of SiH_3 radicals, needed for high quality films, is depleted easily, and SiH_2 and SiH radicals can reach the substrate. Moreover, ion bombardment may be too severe. Therefore, higher pressure levels are required. Usually the operating pressure (in combination with electrode distance and power lever) is set so that one works just above the Paschen curve (see Section 3.2.4). The pressure should not be too high, otherwise gas-phase polymerization occurs [55]. These yellow powders (which turn white upon oxygen exposure) are to be avoided, as they increase the downtime of deposition systems due to clogging of the pumps. At normal working pressures it has been found that the hydrogen content and dihydride content increase with the pressure [81].

6.1.3. Gas Flow

The gas flow rate is usually presented as a deposition parameter; however, it is much more instructive to report the gas residence time [6], which is determined from the flow rate and the geometry of the system. The residence time is a measure of the probability of a molecule to be incorporated into the film. The gas depletion, which is determined by the residence time, is a critical parameter for deposition. At high flow rates, and thus low residence times and low depletion [303], the deposition rate is increased [357, 365] (see Figure 39) and better film quality is obtained, as is deduced from low microstructure parameter values [366].

6.1.4. Hydrogen Dilution

The addition of a hydrogen flow to the silane flow often is used to improve the material quality of the deposited film. Moderate hydrogen dilution of silane ($0.15 \leq [SiH_4]/([SiH_4] + [H_2]) \leq 1$) has been found to yield a lower optical bandgap, a lower activation energy, a lower total hydrogen content, a lower microstructure factor, and a higher photoresponse [15, 82, 367–369]. Unfortunately, also the deposition rate is reduced. Device quality material is obtained for values of $[SiH_4]/([SiH_4]+[H_2])$ between about 0.3 and 0.7. Moreover, an improvement of the uniformity of deposition over a 10×10-cm^2 substrate area has been found as well [370]: for $[SiH_4]/([SiH_4]+[H_2]) = 1/3$ the variation in thickness amounted to only 1.5%.

In the intermediate regime of hydrogen dilution ($0.05 \leq [SiH_4]/([SiH_4]+[H_2]) \leq 0.15$) the hydrogen content increases again, as well as the bandgap, while the microstructure factor remains low [369]. In this regime wide-bandgap a-Si:H can be obtained with better optoelectronic properties than a-SiC:H.

Highly diluted silane ($[SiH_4]/([SiH_4]+[H_2]) \leq 0.05$) causes the deposited films to be crystalline rather than amorphous, due to selective etching of strained and weak bonds by atomic hydrogen [371, 372]. Interestingly, using hydrogen dilution to deposit a-Si:H films "on the edge of crystallinity" [154, 349] resulted in improved material stability. These materials are also referred to as polymorphous silicon (pm-Si) [373, 374].

The admixture of a small amount of silane to a pure hydrogen plasma drastically changes the plasma potential, the dc self-bias voltage, and the charge carrier density. Hamers [163] has found a doubling of the carrier density by changing $[SiH_4]/([SiH_4]+[H_2])$ from 0 to 0.1. The dc self-bias becomes more negative by 25%, and the plasma potential is lowered by about 10–15%. For fractions from 0.1 to 0.5 these parameters remain about constant. The parameters change again when $[SiH_4]/([SiH_4]+[H_2])$ is larger than 0.5.

6.1.5. Substrate Temperature

The substrate temperature is a very important deposition parameter, as it directly affects the kinetics of ad- and desorption of growth precursors, surface diffusion, and incorporation. Actual substrate temperatures may differ from substrate heater setpoints. Calibration of temperature readings is needed, so as to report the correct substrate temperature.

The deposition rate is nearly independent of temperature, while the total hydrogen content, the microstructure parameter, and the disorder decrease with increasing temperature [375]; see also Figure 40 [84, 85]. As a consequence, the optical bandgap decreases as well. An optimum deposition temperature around 250°C exists, such that the material contains only SiH bonds. At higher temperatures the hydrogen evolution from the film causes lowering of the hydrogen content, and leads to higher defect densities. The photoconductivity and photoresponse also have an optimum value around 250°C.

6.1.6. α–γ′ Transition

As was shown above, the properties of the discharge will change with process parameters such as the process pressure, the RF power, and the excitation frequency. Two different plasma regimes can be distinguished: the α- and the γ′-regime [184, 244–246, 248]. In the transition from the α-regime to

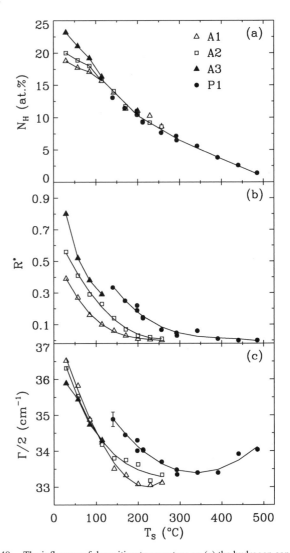

Fig. 40. The influence of deposition temperature on (a) the hydrogen concentration, (b) the microstructure parameter, and (c) the Raman half width $\Gamma/2$. The labels A and P refer to the ASTER and the PASTA deposition system. Series A1 was prepared from a SiH_4–H_2 mixture at 0.12 mbar. Series A2 and A3 were deposited from undiluted SiH_4 at 0.08 and 0.12 mbar, respectively. Series P1 was deposited from undiluted SiH_4. (From A. J. M. Berntsen, Ph.D. Thesis, Universiteit Utrecht, Utrecht, the Netherlands, 1998, with permission.)

the γ′-regime a change occurs in plasma properties. This includes the plasma impedance, the optical emission from the plasma [184], and the dc self-bias voltage at the powered electrode [245] in a reactor with electrodes of unequal size.

An increase in the deposition rate of the film and a change of the film properties is observed near the transition from the α- to the γ′-regime [245]. As an example, Figure 41 shows the deposition rate as a function of pressure, for various hydrogen dilution ratios [376]. It can be seen that higher hydrogen dilution pushes the α–γ′ transition away, as it were. Other examples are shown in Figure 46 and Figure 64. The transition is assumed to be initiated by the formation of larger particles in the gas phase [246, 322]. Negative ions [377] are trapped in the plasma bulk, and recombine with silane radicals to become large negative ions (or small clusters). The charge of these particles can

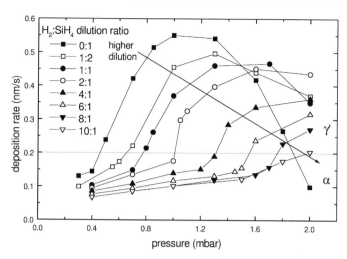

Fig. 41. The deposition rate as a function of deposition pressure for different ratios of hydrogen dilution ($0 : 1 < H_2 : SiH_4 < 10 : 1$). The deposition temperature was 240°C, the SiH_4 flow was 600 sccm, and the excitation frequency was 13.56 MHz. The arrow indicates increasing hydrogen dilution. The α-regime is below the dotted line, the γ' above it. [After R. B. Wehrspohn, S. C. Deane, I. D. French, I. Gale, J. Hewett, M. J. Powell, and J. Robertson, *J. Appl. Phys.* 87, 144 (2000).]

fluctuate, and can become positive. The clusters coalesce, and when the size of the coagulates is sufficiently large the charge builds up and remains negative. They grow further by deposition of *a*-Si:H on their surface. These particles are often referred to as powder or dust [378]. Due to the large amount of negative charge (up to hundreds of electrons) on them, the free electron density drops. Therefore, the electron temperature has to increase in order to sustain the discharge. The discharge becomes more resistive, causing more efficient power coupling into the plasma [322]. Most of these phenomena are observed in discharges with the conventional 13.56-MHz excitation frequency, but they can also be observed at higher excitation frequencies [301].

Interestingly, it has been argued that nanoparticulate formation might be considered as a possibility for obtaining new silicon films [379]. The nanoparticles can be crystalline, and this fact prompted a new line of research [380–383]. If the particles that are suspended in the plasma are irradiated with, e.g., an Ar laser (488 nm), photoluminescence is observed when they are crystalline [384]. The broad spectrum shifts to the red, due to quantum confinement. Quantum confinement enhances the bandgap of material when the size of the material becomes smaller than the radius of the Bohr exciton [385, 386]. The broad PL spectrum shows that a size distribution of nanocrystals exists, with sizes lower than 10 nm.

Moreover, it was found that incorporation of nanoparticles about 8 nm in diameter in *a*-Si:H led to improved properties, the most important one being enhanced stability against light soaking and thermal annealing [387]. A later study revealed a typical crystallite size of 2–3 nm, with a hexagonal close-packed structure [388]. Diamond structures can also be observed [389]. Hence the name polymorphous silicon is justified.

In view of the above, instead of avoiding powder formation regimes in discharges, careful powder management involving optimization of the ratio of radicals and silicon nanoparticles arriving on the substrate has been proposed [379].

6.1.7. Other Deposition Parameters

Depending on the flow pattern in the reactor, the depletion of gases can cause nonuniform deposition across the surface of the substrate. As is depicted in Figure 4, the gas usually is introduced at the top of the cylindrically symmetric reactor, and it flows into (and out of) the discharge region from the sides. Many other configurations exist, e.g., a radial flow reactor, where the gas flow is introduced underneath the powered electrode, and pumped away through the center of this electrode. A reverse radial flow reactor, where the gas flow is introduced from the center of the powered electrode and pumped away at the sides, has also been proposed [117, 370]. The best solution to overcome depletion is to inject the gas directly via a showerhead at the grounded electrode [173].

Meiling et al. [82] have investigated the effect of electrode shape. They found that in the case of a ring electrode, less dense material with a larger bandgap was obtained than in the case of a flat-plate electrode. In addition, the ring electrode induces an extra nonuniformity in thickness, as a result of the nonuniform electric field between the powered and the grounded electrode. Further, the area ratio of grounded to powered electrode in the case of the ring electrode is much higher, which yields higher dc bias voltages and higher ion bombardment energies. Hence, a ring electrode should not be used.

Many different substrates are used for *a*-Si:H deposition. Usually Corning 7059 glass [390] and crystalline silicon are used for materials research, as both have similar thermal expansion coefficients to *a*-Si:H. Devices are mostly made on glass coated with transparent conductive oxide (TCO). As TCO coatings one uses indium tin oxide (ITO), fluorine-doped tin oxide (SnO_2:F), and zinc oxide (ZnO). Ion bombardment may lead to the reduction of the ITO coating [391], while ZnO-coated glass is much more resistant to this. Polymer films have been used as substrates for flexible solar cell structures. They require lower deposition temperatures than glass or stainless steel.

It has been found that various material properties are thickness-dependent. Raman experiments show a dependence on the type of substrate (glass, *c*-Si, stainless steel, ITO on glass) and on the thickness (up to 1 μm) of the films [392, 393]. Recent transmission electron microscopy (TEM) results also show this [394]. This is in contrast to other results, where these effects are negligible for thicknesses larger than 10 nm [395, 396], as is also confirmed by ellipsometry [397] and IR absorption [398] studies.

6.2. Internal Parameters

6.2.1. The Role of Ions

There are many effects possible due to the interaction of ions with a surface during deposition, e.g., the enhancement of

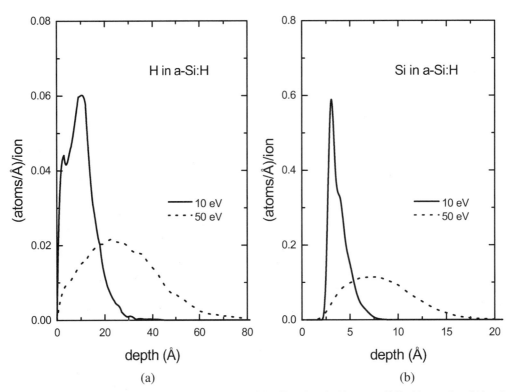

Fig. 42.   Simulated depth distributions of (a) hydrogen and (b) silicon ions incident on *a*-Si:H with energies of 10 and 50 eV. Note the difference in depth scale. Simulations were performed with TRIM92.

adatom diffusion, ion-induced desorption, displacement of surface or subsurface atoms, sticking or subsurface trapping of the ions, sputtering, and implantation in subsurface layers (subplantation). Light and slow ions are able to excite SiH surface bonds vibrationally, which may enhance the surface migration or desorption of weakly physisorbed species, such as SiH_3. For the nonmobile chemisorbed species, such as SiH_2, to be desorbed, a higher ion mass and/or energy is needed. At the same time, however, this may cause subplantation and resulting collision cascades or thermal spikes, which will lead to local defects and poor electronic properties [173].

Several views exist on the exact processes that occur during the deposition of *a*-Si:H. Ganguly and Matsuda [399] explain the growth of *a*-Si:H using a surface diffusion model, in which ions are ignored. Perrin [211] has formulated a model of the RF discharge in which, also, ions are not thought to be of importance to the growth. On the other hand, Veprek et al. [400] introduce the term *ion-induced dehydrogenation*, and Heintze and Zedlitz [249, 280] show the importance of ion fluxes in plasmas, which are excited with very high frequency (VHF) electrode voltages. In addition, molecular dynamics studies of low-energy (10 eV) particle bombardment on crystalline silicon surfaces show surface dimer bond breaking [401] and the formation of interstitials [402], among other effects. Few measurements are available on the ion energies and fluxes in the sheath of a discharge under typical deposition conditions [280, 403–405]. The influence of ion bombardment on defect density [405], mobility [404], and electronic properties [403, 406] has been reported. The dependence of stress on the ion bombardment has been suggested by many authors [284, 376, 407, 408]. In the field of PECVD of amorphous or diamondlike carbon, it is assumed that ions are of prime importance for the formation of dense material [409]. Others also stress the importance of ion bombardment on film properties [410–412]. In addition, in electron beam deposition of amorphous silicon it was shown that the formation of microvoids is inhibited when Ar^+ assistance is used during deposition [413]. It was found that the energy per atom is the decisive parameter instead of the ion/atom ratio. Supplying 12 eV per Ar^+ ion is sufficient for annihilation of all microvoids. As a further example, calculation of projected ranges of low-energy particles in a solid can be performed with the Monte Carlo simulation program TRIM (transport of ions in matter) developed and supplied by Ziegler et al. [414]. The results of simulations with TRIM92 (TRIM Version 1992) of hydrogen and silicon ions incident on *a*-Si:H ($C_H = 0.1$) with energies of 10 and 50 eV, respectively, are presented in Figure 42. The projected ranges of hydrogen (1 nm at 10 eV, 2.5 nm at 50 eV) are much larger than those of silicon (0.3 nm at 10 eV, 0.75 nm at 50 eV). Clearly, these low-energy ions influence the surface and subsurface layers of the deposited film.

It is difficult to isolate the effect of ion bombardment on material properties. An external bias can be applied to the discharge in an attempt to study the effect of varying acceleration voltages over the sheath. However, this modifies the whole potential profile, and consequently the discharge chemistry. Conclusions drawn from such experiments obviously are misleading. It has been demonstrated in a multipole discharge

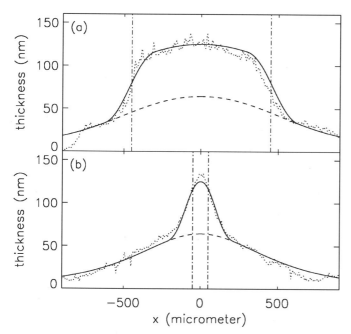

Fig. 43. Measured and simulated thickness profiles on the substrate behind the slit for a film deposited under discharge conditions that typically yield good material properties: (a) along the length of the slit, (b) across the slit. The vertical dashed–dotted lines indicate the boundaries of the apertures. The dotted lines represent the measured profiles, the solid lines the simulated profiles. The dashed lines are the simulated deposition profiles of the radicals. (From E. A. G. Hamers, Ph.D. Thesis, Universiteit Utrecht, Utrecht, the Netherlands, 1998, with permission.)

at low pressure (<5 mTorr) that the ion flux can be controlled without disturbing the plasma properties and radical flux [358, 415–418]. At such low pressure the ion flux is monoenergetic and the relative reactive ion flux (Γ_+/Γ_0) can be as large as 0.8. Perrin [173] has summarized the effects of ion bombardment for $\Gamma_+/\Gamma_0 \approx 0.2$ and ion energy up to 150 eV: the deposited films show a densification of the microstructure, a disappearance of columnar growth, and a reduction of surface roughness. Further, the monohydride bonds in the network are favored over dihydride bonds: R^* is lowered to about zero, and the compressive stress present in the material is increased. Also the optoelectronic properties are improved.

In an RF discharge the relative reactive ion flux is much lower (0.03 [358]) and the IED is considerably broadened. Kae-Nune et al. [311] have estimated, from measurements on radicals, that ions could account for about 10% of the deposition rate. A straightforward method to directly determine the contribution of ions to the deposition process has been presented by Hamers et al. [163, 419]. The ions and radicals are separated by exploiting the difference in angular distribution of their velocities. As ions are accelerated towards the electrode, their angular distribution of ion velocities will be narrowed. Radicals diffuse through the sheath, and their angular velocity distribution will be isotropic. If one places an aperture in front of a substrate, the ions will mainly reach the substrate directly behind the aperture, while radicals will spread over a much larger area on the substrate. A similar configuration has been used to determine the reaction probability of radicals [420]. The

effect is clearly demonstrated in Figure 43: the thickness (deposition rate) is enhanced by about a factor of two just behind the aperture. The deposition conditions were chosen to yield good quality material. From geometrical modeling it was found that under these discharge conditions the contribution of ions to the deposition was 10% [419], which confirms the estimate from Kae-Nune et al. [311].

6.2.2. The Ion Flux

In order to study the influence of ions on the deposition process, a reliable quantification of the ion flux and energy is imperative. This flux cannot be determined directly from the detected number of ions in an IED as measured by means of QMS, for three reasons [332]. First, the orifice size decreases during subsequent measurements due to deposition of a-Si:H on the edges of the orifice. Second, due to the limited acceptance angle of the mass spectrometer system, only a fraction of the ions that arrive at the substrate is actually detected. This fraction depends on the type and number of interactions that an ion experiences while traversing the sheath, and also on the ion species itself. In addition, this implies that the mean kinetic energy of the impinging ions cannot be determined. Third, the ion flux from a typical silane plasma consists of many different ions. All ions must be taken into account in the quantification of the ion flux.

Another approach is needed [163, 301, 332], and one follows from the definition of the ion flux Γ, i.e. the product of the density of the ions n_i and their mean velocity $\bar{v}_i$: $\Gamma = n_i \bar{v}_i$. The contribution of electrons to the time-averaged charge carrier concentration at the electrode n can be neglected, as the electron current towards the electrode is peaked in a very short time during the RF period. We thus have $n_i = n$. A further assumption is that there are no collisions in the sheath. As a result of this, the velocity $\bar{v}_i$ is equal to the maximum velocity, v_{max}, that the ions gain. The value of v_{max} follows from the time-averaged plasma potential eV_{pl}: $v_{max} = \sqrt{2eV_{pl}/\overline{m}}$, with $\overline{m}$ the average mass of the ions. If the ions reach the electrode without collisions in the sheath, the IED will show a distinct peak at eV_{pl}; see e.g. the IED of $Si_2H_4^+$ in an α-silane discharge (Figure 37).

The maximum ion flux Γ_{max} at the electrode is thus estimated to be

$$\Gamma_{max} = nv_{max} = n\sqrt{\frac{2eV_{pl}}{\overline{m}}} \qquad (57)$$

The maximum *kinetic energy* flux of ions, $(\epsilon\Gamma)_{max}$ (W m^{-2}), can be defined as

$$(\epsilon\Gamma)_{max} = eV_{pl}\Gamma_{max} \qquad (58)$$

where ϵ denotes the ion energy.

The collisionless-sheath assumption leads to an overestimation of the velocity. The use of an average mass $\overline{m}$, which is taken to be 31 amu/e (i.e., the mass of SiH_3^+), in addition may lead to an overestimation of the velocity, as ions with more than one silicon atom are present, albeit in lower concentrations. Hamers et al. [163, 301, 332] have argued that the velocity of

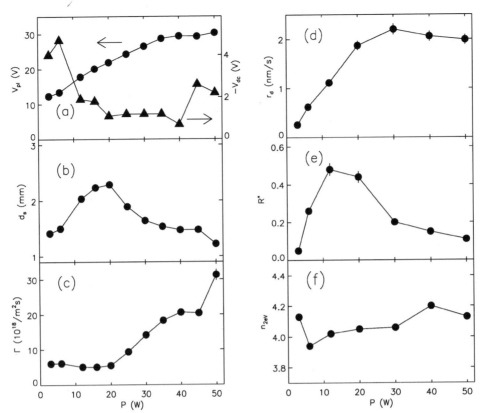

Fig. 44. Plasma parameters as deduced from the IEDs and material properties as a function of power delivered to the SiH$_4$–Ar discharge at an excitation frequency of 50 MHz and a pressure of 0.4 mbar: (a) the plasma potential V_{pl} (circles) and dc self bias V_{dc} (triangles), (b) the sheath thickness d_s, (c) the maximum ion flux Γ_{max}, (d) the growth rate r_d, (e) the microstructure parameter R^*, and (f) the refractive index n_2 eV. (Compiled from E. A. G. Hamers, Ph.D. Thesis, Universiteit Utrecht, Utrecht, the Netherlands, 1998.)

the ions (and the flux) is overestimated by at most a factor of two. Also, the ion kinetic energy flux overestimates the real energy flux if collisions take place in the sheath. However, in many collisions the kinetic energy is distributed over the different particles. For instance, in a charge exchange process, the original ion is neutralized, but retains its kinetic energy. The newly formed ion will gain the energy that the neutralized ion would have gained in the rest of the trajectory. The two particles will transport the total amount of kinetic energy corresponding to the plasma potential to the surface [163].

In order to express the importance of the ions to the growth process quantitatively, two related quantities can be defined: the fraction of arriving *ions* per deposited *atom*, R_i, and the kinetic energy transferred by *ions* per deposited *atom*, E_{max}. These quantities are used in ion-beam-assisted deposition in order to relate material properties to ion flux and energy [421]. Their definition is

$$R_i = \frac{\Gamma_{max}}{\rho r_d} \qquad (59)$$

$$E_{max} = \frac{(\epsilon \Gamma)_{max}}{\rho r_d} \qquad (60)$$

where the number of deposited atoms per unit time and per unit area is ρr_d, with ρ the atomic density of the amorphous network (5×10^{22} cm^{-3}) and r_d the growth rate.

6.2.3. Relation between Ion Flux and Material Quality

A systematic study of the role of the ions in the deposition process and their influence on the quality of the layers has been performed by Hamers et al. [163, 301, 332] in the ASTER deposition system. More specifically, a study has been made on the relation between the plasma parameters and the material properties in both the α- and the γ'-regime at typical deposition conditions. Here, the results for power and pressure variation are summarized. Details can be found elsewhere [163, 301, 332].

First, results on power variation are described, for two different discharges: (1) a silane–hydrogen discharge in the powder-free α-regime, and (2) a silane–argon discharge in both the α- and the powder-producing γ'-regime.

The process conditions of the series in silane–hydrogen are an excitation frequency of 13.56 MHz, a pressure of 0.20 mbar, gas flows of 30 sccm SiH$_4$ and 30 sccm H$_2$, and a substrate temperature of 250°C. The power was varied between 5 and 20 W. It was found that the absolute magnitude of the dc self-bias voltage increases linearly with increasing power, and the trend of the ion flux is similar (0.8×10^{18} m^{-2} s^{-1} at 5 W, and 2.1×10^{18} m^{-2} s^{-1} at 20 W). This linear relationship between V_{dc} and growth rate had been observed earlier [15, 151, 422], and in practice can even be used as a calibration curve. The

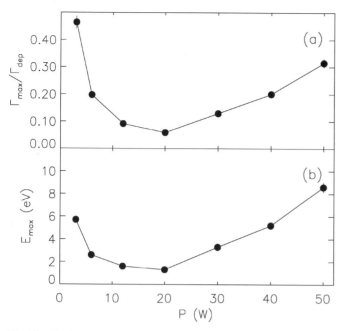

Fig. 45. (a) The ratio of the ion flux to the deposition flux, R_i, and (b) the ratio of the energy flux to the deposition flux, E_{max}, as a function of power delivered to the SiH_4–Ar discharge at an excitation frequency of 50 MHz and a pressure of 0.4 mbar. (Compiled from E. A. G. Hamers, Ph.D. Thesis, Universiteit Utrecht, Utrecht, the Netherlands, 1998.)

plasma potential was 22.7 V at 5 W and was only 3 V higher at 20 W. In this series only two films were deposited, one at 10 W and one at 15 W. The refractive index in both cases was 4.33, representative of good material. The fractions of arriving ions per deposited atom, R_i, as defined in Section 6.2.2, are 0.26 and 0.20, respectively, while the kinetic energy transferred by ions per deposited atom, E_{max} is about 5 eV in both cases.

The process conditions of the series in silane–argon are an excitation frequency of 50 MHz, a pressure of 0.40 mbar, gas flows of 30 sccm SiH_4 and 60 sccm Ar, and a substrate temperature of 250°C. The power is varied between 3 and 50 W. Plasma parameters deduced from measurements of the IED and material properties are shown in Figure 44 and Figure 45. The discharge changes from the α-regime to the γ'-regime around 10 W, which can best be observed by the decrease of the dc self-bias and an increase in sheath thickness (see Figure 44a and b). The bias is very low in the γ'-regime, typically between -2 and -1 V, indicative of a more symmetrical discharge. The plasma potential increases with increasing power and tends to saturate at the highest powers. Visual inspection of the discharge shows that the optical emission from the plasma becomes brighter and more homogeneously distributed on going from the α- to the γ'-regime. In the α-regime the emission intensity is highest at the boundary between plasma bulk and sheaths.

The growth rate, plotted in Figure 44d, increases with increasing power, which is ascribed to the higher degree of dissociation in the discharge. The growth rate reaches a maximum around 30 W, and decreases somewhat at higher powers: the growth rate then is limited by the flow rate. It has been estimated that about 60% of the silane is used for deposition [163];

the rest may well (partially) partake in the processes that lead to the creation of powder. This yellow powder is always observed in the reactor after experiments performed in the γ'-regime. It is always found outside the plasma confinement, and never on the substrate.

At the transition from the α- to the γ'-regime the enhancement of the growth rate (Fig. 44d) is larger than the change in ion flux (Fig. 44c), as shown in Figure 45a, where R_i is plotted. An R_i-value of 0.25 or larger is commonly found in the α-regime, whereas R_i is typically 0.10 or lower in the γ'-regime. In addition, the amount of kinetic ion energy per deposited atom, E_{max}, shows a minimum of about 2 eV between 10 and 20 W; see Figure 45b.

The microstructure parameter is low in the material deposited at the lowest power (Fig. 44e); it increases rapidly with increasing power up to 20 W, and then decreases again with further increasing power. The opposite holds for the refractive index (Fig. 44f), although that is less clear. A high value of the microstructure indicates a large fraction of Si–H_2 bonds in the material, corresponding to an open material structure and a low refractive index.

From a comparison between the behavior of the microstructure parameter R^* (Fig. 44e) and the ion kinetic energy per deposited atom, E_{max} (Fig. 45b), it can be concluded that a one-to-one relation appears to exist between the relative strength of the ion bombardment, expressed in terms of E_{max}, and the microstructure parameter. This has also been suggested by others [246, 422].

Both the dark conductivity and the photoconductivity of the sample deposited at 40 W have been measured. The photoconductivity has been measured under AM1.5 conditions. The dark conductivity and photoconductivity of the material deposited at 40 W are 1×10^{-10} and 1.2×10^{-4} $\Omega^{-1} cm^{-1}$, respectively. These values are indicative of good electronic properties, such as a low defect density and a high carrier mobility. The improvement of electrical properties of the deposited films at high RF power densities has also been reported by Nishikawa et al. [361].

The influence of power variation on the material properties is in agreement with the trends observed by Andújar et al. [246], who studied the α–γ' transition in pure silane discharges at 13.56 MHz. Further, this has also been observed for pure silane discharges at 50 MHz by Meiling et al. [423].

The α–γ' transition can also be induced by changing the process pressure. At low pressures the discharge is in the α-regime. At a certain pressure it changes to the γ'-regime on increasing the pressure by only 0.02–0.04 mbar. The experiments presented here span a pressure range from 0.08 to 0.5 mbar, and, as above, plasma properties that are deduced from IED measurements are compared with material properties. The other process conditions are an excitation frequency of 50 MHz, a power level of 10 W, gas flows of 30 sccm SiH_4 and 30 sccm H_2, and a substrate temperature of 250°C. Data are summarized in Figure 46 and Figure 47. The self-bias voltage decreases quite rapidly in magnitude with increasing pressure, as shown in Figure 46a. The plasma potential slowly increases with pressure

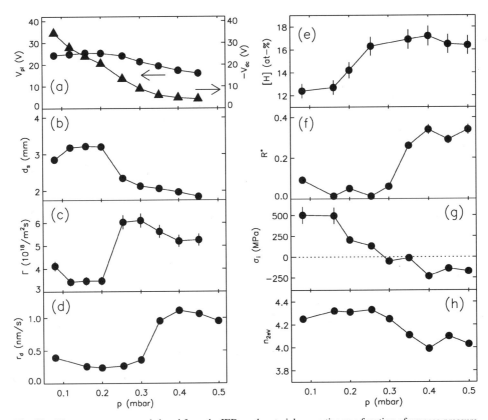

Fig. 46. Plasma parameters as deduced from the IEDs and material properties as a function of process pressure of a SiH_4–H_2 discharge at an excitation frequency of 50 MHz and a power of 10 W: (a) the plasma potential V_{pl} (circles) and dc self-bias V_{dc} (triangles), (b) the sheath thickness d_s, (c) the maximum ion flux Γ_{max}, (d) the growth rate r_d, (e) the hydrogen content, (f) the microstructure parameter R^*, (g) the internal stress σ, and (h) the refractive index $n_{2\,eV}$. (Compiled from E. A. G. Hamers, Ph.D. Thesis, Universiteit Utrecht, Utrecht, the Netherlands, 1998.)

up to a pressure of 0.20 mbar and decreases towards the highest pressures (Fig. 46a). As both dc self-bias and plasma potential decrease, V_{rf} will also decrease [134]. Such a decrease has been measured by others [184, 245]. The sheath thickness falls from around 3 to 2 mm at around 0.2 mbar (Fig. 46b). The ion flux (Fig. 46c) is rather constant in the region up to 0.2 mbar, whereas it is nearly doubled on the high-pressure side. A slight decrease of the ion flux in the α-regime has been observed in 13.56-MHz discharges by Roca i Cabarrocas [403]. The growth rate increases by about a factor of 4 upon a pressure change from 0.2 to 0.3 mbar (Fig. 46d). The fraction R_i is shown in Figure 47a to be around 0.25 in the α-regime, whereas it drops to about 0.10 at pressures higher than 0.3 mbar, because the deposition rate increases a factor of 2.5 more than the ion flux. Since the plasma potential decreases with increasing pressure, the ions arriving at the film surface are expected to have a lower energy per ion in the γ'-regime than in the α-regime. This is clearly seen in Figure 47b, where E_{max} changes from about 5 eV at pressures below 0.3 mbar to about 2 eV at the higher pressures.

As can most clearly be seen in Figure 46d, the α–γ' transition occurs at a pressure of about 0.3 mbar for these experimental conditions. The impedance of the plasma, as well as the optical emission from the plasma, changes on going through

the transition. The depletion of SiH_4 during deposition was already shown and compared with the deposition rate in Figure 31. The effect of the α–γ' transition on the partial pressures of disilane and trisilane was already presented in Figure 32 (see Section 5.3.1). The amount of silicon in these higher-mass neutral species is only about 5% of the total amount of silicon in SiH_4, as can be concluded from the measurements of the partial pressures of these gases. The transition however effects the production of these higher silanes. The increase of the partial pressures is larger than the increase in depletion. This means that the amount of these higher silanes produced per consumed quantity of silane increases at the higher pressures.

The material properties are also affected on going through the α–γ' transition. The hydrogen content is about 12% up to 0.2 mbar, and increases to over 16% at higher pressures (Fig. 46e). The increase in the hydrogen content occurs at a lower pressure than the increase in deposition rate, much like the behavior of the ion flux. The microstructure parameter is lower than 0.1 up to a pressure of 0.3 mbar, and is around 0.3 above this pressure (Fig. 46f). The internal stress in the layers is about 500 MPa compressive at low pressures, changes to a tensile stress around 0.3 mbar, and remains tensile at 200 MPa at higher pressures (Fig. 46g). The refractive index is around 4.30 up to a pressure of 0.3 mbar, but decreases towards the high-

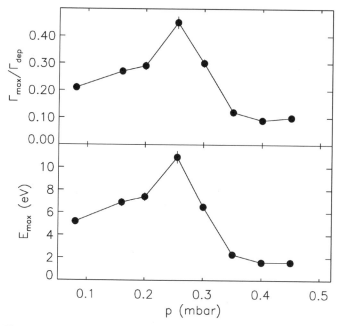

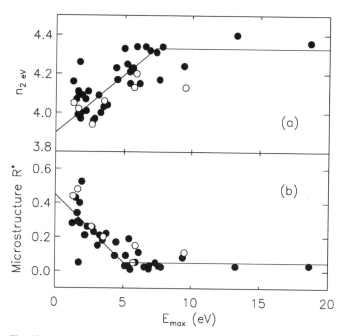

Fig. 47. (a) The ratio of the ion flux to the deposition flux, R_i, and (b) the ratio of the energy flux to the deposition flux, E_{max}, as a function of process pressure of a SiH_4–H_2 discharge at an excitation frequency of 50 MHz and a power of 10 W. (Compiled from E. A. G. Hamers, Ph.D. Thesis, Universiteit Utrecht, Utrecht, the Netherlands, 1998.)

Fig. 48. Material properties as functions of E_{max} at 250°C: (a) refractive index $n_{2\,eV}$, and (b) microstructure parameter R^*. The closed circles represent pure silane and silane–hydrogen plasmas. The open circles refer to silane–argon plasmas. Lines are to guide the eye. (Adapted from E. A. G. Hamers, Ph.D. Thesis, Universiteit Utrecht, Utrecht, the Netherlands, 1998.)

est pressures (Fig. 46h). In summary, the material becomes of poorer quality at higher pressures. The γ'-regime is associated with the formation of particulates in the plasma [322]. However, these particles are assumed not to be incorporated in the film and thus not to be the direct cause of the different material properties.

In the presented range of pressure variation, Hamers [163] also has studied the influence of the substrate temperature on the plasma and the material. It was found that in the temperature range of 200 to 300°C the trends of the bias voltage, the plasma potential, and the growth rate as functions of pressure all are the same, while the absolute magnitude depends on the temperature. The trends in material properties are similar to the ones reported above: at a temperature of 200°C the material quality is worse than at higher temperatures. The α–γ' transition occurs at a lower pressure than at a temperature of 250°C. This has been observed before [248].

In all the experiments reported here it is observed that the value of R_i is rather constant at a value of 0.25 in the α-regime, whereas in the γ'-regime it typically amounts to 0.10. In other words, if all the Si atoms that originate from ions contribute to the deposition, the contribution of ions to the deposition in the α-regime is 25%, and in the γ'-regime 10%. The observation that these values are rather constant in each regime strongly indicates that the deposition rate is limited by the ion flux.

In an attempt to relate ion bombardment to material structure it is very illustrative to correlate the refractive index $n_{2\,eV}$ and the microstructure parameter R^* with the kinetic ion energy per deposited atom, E_{max}. The data presented above for various discharges (pure silane, silane–argon, and silane–hydrogen) at

a temperature of 250°C are thus summarized in Figure 48. It is very clear that the structural properties of the layers are poor (small $n_{2\,eV}$ and large R^*) for small E_{max}. The structural properties improve rapidly with increasing E_{max} up to a value of about 5 eV. All samples with E_{max} below 5 eV are deposited in the γ'-regime. Above 5 eV a dense network with only a small fraction of Si−H_2 bonds is produced. The structural properties do not change further with increasing E_{max}.

The role of the substrate temperature can be inferred from a plot of $n_{2\,eV}$ and R^* versus E_{max} at the three temperatures mentioned: 200, 250, and 300°C (see Figure 49). At a substrate temperature of 200°C the refractive index is lower at every E_{max} than at a substrate temperature of 250°C. Further, the threshold at which dense material is obtained is observed to be a few electron volts higher than at 250°C. The refractive index at 300°C is high and independent of E_{max}. The microstructure parameter R^* as a function of E_{max} behaves similarly for material deposited at 200 and at 250°C. At 300°C the value of R^* is less than 0.1 and independent of E_{max}. It is noteworthy to show the relation between the internal stress and E_{max} as a function of temperature (Fig. 50). The stress is linearly dependent on E_{max} (between 1 and 7 eV) with a slope of about 175 MPa/eV for material deposited at 200 and at 250°C. For the 300°C data series this relation is shifted upwards by about 600 MPa, and at values of the stress larger than about 1000 MPa (E_{max} was about 4.5 eV) the deposited layers started to peel off from the silicon substrate directly after exposure to air. In summary, the ion bombardment clearly is needed to create or promote a dense amorphous network at temperatures up to 250°C. At higher

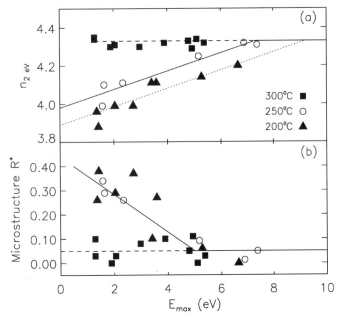

Fig. 49. Material properties as function of E_{max} at three temperatures: (a) refractive index $n_{2 eV}$, and (b) microstructure parameter R^*. Lines are to guide the eyes. (Adapted from E. A. G. Hamers, Ph.D. Thesis, Universiteit Utrecht, Utrecht, the Netherlands, 1998.)

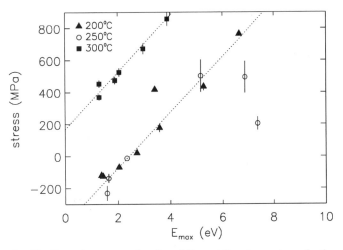

Fig. 50. Internal stress as a function of E_{max} at three temperatures, for the pressure series in Figure 46 in the SiH_4–H_2 discharge running at an excitation frequency of 50 MHz and a power of 10 W. (From E. A. G. Hamers, Ph.D. Thesis, Universiteit Utrecht, Utrecht, the Netherlands, 1998, with permission.)

temperatures energetic ion bombardment results in too high intrinsic stress.

6.2.4. Ion–Surface Interactions

The different processes in which the ions might well be involved at the surface can be indentified on the basis of their chemical reactivity at the surface and the amount of energy they transport towards the surface.

It is assumed that SiH_3 radicals are responsible for the production of deposition sites at the hydrogen-covered surface when no ions are present [138]. If ions contribute to the depo-

sition process, a higher deposition rate has been reported; see e.g. [419]. The fraction of arriving ions per deposited silicon atom varies between 0.1 and 0.3, depending on the discharge regime (see Section 6.2.3). Assuming equal sticking probabilities of ions and radicals (0.3 for the SiH_3 radical [192]), up to about 10% of the deposition is attributed directly to the ions. This value has also been found by Kae-Nune et al. [311]. On the other hand, ions can penetrate into the subsurface layers, and hence their sticking probability will be closer to one. In addition, direct incorporation of ions in the amorphous network can only partly explain the increase in deposition rate. The creation of deposition sites by the ions possibly is of greater importance. It has been suggested that a direct relation between the ion flux and the deposition rate exists, if the availability of deposition sites is the limiting factor in the deposition process [236, 400, 424]. Also, partial dissociation of a polyatomic ion upon impact may occur [412, 425–427]. The dissociation of a hydrogen-containing ion may lead to the production of more than one hydrogen atom per ion. These atoms create molecular hydrogen and dangling bonds upon reaction with hydrogen bonded to the surface. The probability γ of the recombination process of atomic hydrogen has been estimated to be between 0.4 and 1 [317].

In discharges operating in the γ'-regime large particles consisting of several ions are present, in contrast to discharges in the α-regime. In both regimes the same relative amount of ions contributes to the deposition, as is deduced from the measured amount of radicals that contribute to the deposition, which is regime-independent [317]. These larger ions in the γ'-regime are thought to create more deposition sites per arriving ion than the smaller ions in the α-regime [163]. This results in a smaller number of ions per deposited atom in the γ'-regime.

The kinetic ion energy flux, $(\epsilon\Gamma)_{max}$, which is typically 20 W m^{-2} [163, 301], will raise the substrate temperature by only a few degrees. Therefore, the influence of ions will be limited to the vicinity of impact. Furthermore, typical ion energies are below the sputtering threshold of silicon [134].

Enhancement of surface diffusion of the growth precursors is considered as one of the beneficial effects of ion bombardment [246, 428]. The potential energy of ions, which is released when the ion is neutralized, is typically 10 eV. This energy can be a substantial fraction of the total energy transferred. The release of this ionization energy is sufficient to excite atoms into excited electronic states, thereby weakening their bonds and enhancing their mobilities [429].

The influence of ion bombardment on the structure of deposited materials has been studied by Müller [428, 430] by performing molecular dynamics calculations on ion-assisted vapor phase growth of nickel. Nickel, of course, is very different from hydrogenated amorphous silicon, but the resemblance of the two deposition processes and some material properties is remarkable. In both cases, the growth precursors stick to the surface or diffuse along the surface, and the material can have a columnar-like structure or a structure with voids. Ion bombardment during the deposition of Ni promotes the formation of a dense film. The ion energies and fluxes in the molecular dynamics studies were similar to those in typical a-Si:H deposition conditions. Two important processes modify the structure of the nickel film, viz., forward sputtering and diffusion enhancement of surface species. Similar processes are likely to occur in a-Si:H deposition. In the deposition of amorphous germanium by ion-assisted electron-beam evaporation, it was also found that a kinetic energy of about 5 eV per deposited atom is needed to produce a dense amorphous germanium network, irrespective of the amount of kinetic energy of the ions: the effect of five ions with 20 eV was the same as that of one ion of 100 eV [431]. These two additional examples clearly indicate the beneficial role that ion bombardment apparently has on the density of the deposited film.

In the formation of a hydrogenated amorphous silicon network, the presence of the bonded hydrogen must be taken into account, as it was shown in the preceding that both the hydrogen content and the microstructure parameter are influenced by the ion bombardment. The high value of the microstructure parameter and the concomitant low density of the material for low values of E_{max} at 250°C (Fig. 48) indicate that the cross-linking is not sufficiently fast to create a dense network. This cross-linking process is activated by the locally released energy of chemisorption [432, 433], and involves the temperature-activated evolution of H_2 from the surface. Hydrogen desorption due to cross-linking, which is known to occur from thermal desorption measurements [434], starts to become important at temperatures higher than 300°C. At this temperature a drastic change in the importance of ion bombardment for achieving a dense network was observed (see Fig. 49). At lower deposition temperatures the cross-linking during the deposition is temperature-limited and the kinetic energy of the ions is needed to achieve cross-linking. If the amount of ion kinetic energy is insufficient, cross-linking will be incomplete and will result in an open network structure with a large amount of Si–H_2 bonds. At larger values of the ion kinetic energy the cross-linking is stimulated and a dense network is formed. Ion bombardment rather than the temperature induces cross-linking, as the microstructure parameter, the stress, and (to a lesser extent) the refractive index have the same dependence on E_{max} at both 200 and 250°C. At 300°C the thermally activated cross-linking process is becoming important.

Tensile stress can be related to the presence of voids in the material. The large microstructure parameter at low E_{max} is indicative of voids. A kinetic energy of the ions of 20 eV is typically enough to implant some of these ions a few monolayers below the surface of the growing film [400]. The extra Si atoms deposited in the layer at this depth will cause a compressive stress. Light hydrogen ions, which are less abundant than e.g. SiH_3^+ ions, penetrate deeper into the material (see also the TRIM simulation results in Fig. 42). The stress in the films deposited at 200 and 250°C depends similarly on E_{max}. At these temperatures the ion energy is needed to densify and to cross-link the network. At a temperature of 300°C the stress is much higher, which is explained by the fact that the ions reach an already dense network and some will be implanted, resulting in excess-stress in the material.

7. DEPOSITION MODELS

A conceptual view of the processes that occur in PECVD of a-Si:H has been given by Perrin [173] and is shown in a slightly adapted form in Figure 51. Primary gas phase reactions are electron-impact excitation, dissociation, and ionization of the source gas molecules (SiH_4), thereby producing radicals and positive and negative ions. Secondary reactions in the gas phase between (charged) molecules and radicals produce other species. Diffusion of reactive neutral species leads to material deposition. Positive ions are accelerated to the substrate and bombard the growing film. Negative ions are confined in the bulk of the discharge, which can lead to particulate formation. On the surface, species diffuse to growth sites and contribute to the film after sticking to the underlying material. Also, surface species recombine with other species and desorb from the surface (see also Fig. 14). Subsurface reactions are the release of hydrogen from this hydrogen-rich layer and the relaxation of the silicon network.

Nowadays, it is generally accepted that *device quality a-Si:H* is obtained under PECVD conditions where the SiH_3 radical is the predominant growth precursor [123, 137, 192]. At typical deposition temperatures and pressures the a-Si:H surface is almost completely terminated by hydrogen atoms. The SiH_3 radical is thought to physisorb on this hydrogen-terminated surface before it is incorporated into the film [136, 137, 317, 435–438]. The excess hydrogen is eliminated from the hydrogen-rich (40–50%) subsurface, so as to form "bulk" a-Si:H with a hydrogen content of about 10%.

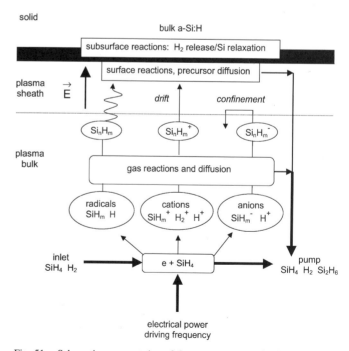

Fig. 51. Schematic representation of the processes occurring in a SiH_4–H_2 discharge and the various particles present in the energy and material balances. [After J. Perrin, in "Plasma Deposition of Amorphous Silicon-Based Materials," (G. Bruno, P. Capezzuto, and A. Madan, Eds.), Chap. 4, p. 177. Academic Press, Boston (1995).]

Robertson has summarized the three recent classes of models of a-Si:H deposition [439]. In the first one, proposed by Ganguly and Matsuda [399, 440], the adsorbed SiH$_3$ radical reacts with the hydrogen-terminated silicon surface by abstraction or addition, which creates and removes dangling bonds. They further argue that these reactions determine the bulk dangling bond density, as the surface dangling bonds are buried by deposition of subsequent layers to become bulk defects.

The second class of models was formulated by Winer [441] and Street [442, 443]. Here the notion that hydrogen atoms are more mobile than silicon atoms forms the basis of the model. The silicon network is fixed up to a temperature near the crystallization temperature (650°C). Hydrogen is mobile above a temperature of about 200°C, and its motion allows for the interconversion of defects, weak bonds, and strong bonds. These processes occur in the subsurface layer, and surface processes are not explicitly included. In this model, the optimum deposition temperature is the one at which hydrogen atoms can diffuse one atomic spacing as to remove the weak bonds. After deposition, the weak bond distribution can only be changed irreversibly while the defect distribution can be changed reversibly, as described by the defect pool models [444–446].

In the third class of models, computer simulations try to fully incorporate all processes in the discharge, the interaction of species created in the discharge with the wall (i.e., the substrate), and the network formation [190, 191, 232, 447–449]. These models to date do not treat the formation of disorder or defects, but aim at the understanding of the deposition rate, hydrogen content, and other macroscopic properties in relation to the discharge conditions (see also Section 4).

Robertson has combined and extended the first and second classes of models [439] by focusing on the origin of weak bonds. A surface adsorption model is used to describe surface coverage of the silyl radical SiH$_3$. Then, the processes that cause hydrogen to be expelled from the subsurface layer lead to the formation of weak bonds, which are frozen in. The dangling bond defects arise from these weak bonds by the defect pool process. In the following, a description is given of the Robertson deposition model (details can be found in [439, 450]).

7.1. Surface Adsorption

On the hydrogen-terminated surface, the surface species are mainly ≡SiH, but at lower temperature (<200°C) =SiH$_2$ and —SiH$_3$ become increasingly important. The SiH$_2$ and SiH$_3$ radicals react differently with this surface. The SiH$_2$ radical can insert directly into Si—H or Si—Si bonds. The sticking coefficient is close to 1, which leads to columnar growth (physical vapor deposition regime [451]). Also, the SiH$_2$ can react with SiH$_4$ to form Si$_2$H$_6$. The SiH$_3$ radical cannot react directly with the surface. It has a small sticking coefficient, and it diffuses over the surface to find a reactive site. This reactive site has been suggested to be a dangling bond at kink and step edges [452]. As a result, this leads to conformal coverage (chemical vapor deposition regime [451]). The SiH$_3$ radical is physisorbed onto the surface [436]. A contrasting view has been proposed by

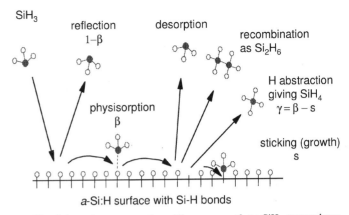

Fig. 52. Schematic respresentation of the processes that a SiH$_3$ may undergo at a hydrogen-terminated a-Si:H surface. [After J. Robertson, *J. Appl. Phys.* 87, 2608 (2000).]

Table VIII.   Reactions of Surface Sites ≡SiH, ≡Si·, and ≡SiHSiH$_3$ with the Radical ·SiH$_3$, with Corresponding Rate Coefficients

Symbol	Reaction		Rate coefficient
R1	≡SiH + ·SiH$_3$	→ ≡SiHSiH$_3$	s_1
R2	≡SiHSiH$_3$	→ ≡SiH + ·SiH$_3$	v_d
R3	≡SiHSiH$_3$ + ≡SiH	→ ≡SiH + ≡SiHSiH$_3$	v_h
R4	≡SiH + ·SiH$_3$	→ ≡Si· + SiH$_4$	v_a
R5	≡Si· + ·SiH$_3$	→ ≡Si–SiH$_3$	s_0
R6	2 ≡SiHSiH$_3$	→ 2 ≡SiH + Si$_2$H$_6$	v_r
R7	2 ≡SiH	→ 2 ≡Si· + H$_2$	v_b

Von Keudell and Abelson [453], who conclude from their experiments that SiH$_3$ can insert into strained Si—Si bonds on the surface to form a fivefold-coordinated bond. This was also observed by Ramalingam et al. [454] in molecular dynamics simulations of SiH$_3$ impinging on a 2×1 Si(100) reconstructed surface. Nevertheless, as these observations do not as yet explain film deposition, they are considered at least questionable.

The possible reactions with a ≡SiH surface site are schematically shown in Figure 52, and summarized together with their reaction rates in Table VIII. As a simplification, the surface site is ≡SiH, the dangling bond is ≡Si·, and the adsorbed SiH$_3$ radical is ≡SiHSiH$_3$. The possible reactions include adsorption (R1 in Table VIII), desorption (R2), hopping to an adjacent ≡SiH site (R3), abstraction of an hydrogen atom from ≡SiH to create ≡Si· (R4), addition to ≡Si· (R5), surface recombination with another SiH$_3$ radical to form a desorbing Si$_2$H$_6$ molecule (R6), and elimination of hydrogen molecules from two adjacent ≡SiH sites to yield surface dangling bonds (R7).

Several simplifications are further made [439]: (1) the SiH$_3$ adsorption (R1) is temperature-independent, and s_1 is constant; (2) the SiH$_3$ addition to a dangling bond is temperature-independent, and $s_0 = 1$, but the SiH$_3$ must hop to this dangling bond site from a neighboring site, so the net rate is $v_h s_0$; (3) the abstraction reaction R4 is faster than the desorption reaction R2, so the latter is ignored; and (4) the recombination reaction R6 is ignored.

Perrin et al. have denoted the fractional coverages of the surface sites as θ_1, θ_0, and θ_4 for $\equiv$SiH, $\equiv$Si, and $\equiv$SiHSiH$_3$, respectively [317], with $\theta_1 + \theta_0 + \theta_4 = 1$. These three fractional coverages can be found by expressing them as rates in terms of creation and destruction. This gives, with ϕ the incident SiH$_3$ flux density per surface site [439],

$$\frac{d\theta_0}{dt} = \nu_0\theta_4(1 - \theta_0) - \nu_4\theta_4\theta_0 + \nu_b(1 - \theta_0) \quad (61)$$

$$\frac{d\theta_4}{dt} = \phi - \nu_a\theta_4(1 - \theta_0) - \nu_h\theta_4\theta_0 \quad (62)$$

In the steady state, and ignoring multiple hopping, the coverages are approximated as

$$\theta_0 \approx \frac{\nu_a}{\nu_h} + \frac{1}{1 + \phi/\nu_b} \quad (63)$$

$$\theta_4 \approx \frac{\phi}{2\nu_a(1 + \nu_b/\phi)} \quad (64)$$

At temperatures lower than 350–400°C the expressions for θ_0 and θ_4 reduce to $\theta_0 \approx \nu_a/\nu_h$ and $\theta_4 \approx \phi/(2\nu_a)$. The residence time of SiH$_3$ radicals at the surface, τ_4, is given by the coverage θ_4 divided by the arrival flux ϕ: $\tau_4 \approx 1/(2\nu_a)$. The SiH$_3$ surface diffusion length is $L = \sqrt{D\tau_4}$, where $D = \nu_h a^2$ is the SiH$_3$ diffusion coefficient, with a the Si$-$Si bond length. We thus have $L = a\sqrt{\nu_h/2\nu_a} = a/\sqrt{\theta_0}$.

The reaction rates ν_i ($i = h, a, b$) for each reaction are thermally activated as $\nu_i = \nu_{i0}\exp(-E_i/k_BT)$. Robertson has critically assessed the values ν_{i0} and E_i for each reaction [439], and used for the model $\nu_{h0} = 10^{13}$ s^{-1}, $E_h = 0.2$ eV, $\nu_{a0} = 3 \times 10^{11}$ s^{-1}, $E_a = 0.4$ eV, $\nu_{b0} = 10^{13}$ s^{-1}, and $E_b = 2.2$ eV. These values give a surface dangling bond density that increases monotonically with temperature from about 10^{-5} at room temperature to about 10^{-3} at 400°C. Thermal desorption of hydrogen then sets in, and the surface dangling bond density is suddenly increased and approaches unity around 600°C. This increase in θ_0 is due to the fact that $E_a > E_h$. The critical parameter is the difference $E_a - E_h$, not their absolute values. At typical deposition temperatures of 200–250°C the surface dangling bond density is about 3×10^{-4}, which is two orders of magnitude larger than the bulk dangling bond density. The SiH$_3$ surface diffusion length decreases from about 30 nm at room temperature to about 8 nm at 250°C.

These findings from Robertson [439, 450] are in contrast to the ones from the model proposed by Ganguly and Matsuda [399, 440]. They found a minimum in the surface dangling bond density, and a maximum in the surface diffusion coefficient, simultaneously occurring at about 250°C ($\theta_0 \approx 10^{-5}$, $L \approx 400$ nm). This would explain the optimum in material quality around 250°C. Surface defect densities have been measured during deposition by *in situ* electron spin resonance by Yamasaki et al. [455]. They found that the surface defect density was much higher than the bulk defect density. Therefore, the hypothesis of surface dangling bonds that are buried by deposition of subsequent layers is questionable. Robertson argues

that as most of the hydrogen is eliminated from the hydrogen-rich a-Si:H surface layer, about half of the bonds in this layer are changed, from Si$-$H to Si$-$Si. It is very unlikely that this enormous change in bonding configurations would leave a surface dangling bond density of 10^{-6} unchanged [450].

7.2. Solubility of Hydrogen in Silicon

The elimination of excess hydrogen from the surface layer seems counterintuitive, the hydrogen leaves the a-Si:H network towards the surface, where the hydrogen content is larger. This can be understood by considering the thermodynamics of hydrogen in silicon [439]. In general, the Gibbs free energy of a mixture of two components A and B, with mole fractions x_A and x_B (with $x_A + x_B = 1$), is given as [456]

$$G = x_A G_A + x_B G_B + x_A x_B \Omega_{AB} + RT(x_A \ln x_A + x_B \ln x_B) \quad (65)$$

with G_A and G_B the Gibbs free energies of the pure species, Ω_{AB} the heat of mixing, and $RT(x_A \ln x_A + x_B \ln x_B)$ the entropy of mixing. The chemical potential μ_A is given by

$$\mu_A = \left(\frac{\partial G}{\partial x_A}\right)_{T,B} = G_A + \Omega_{AB}(1 - x_A)^2 + RT \ln x_A \quad (66)$$

and similarly for μ_B. If Ω_{AB} is negative, mixing is favored. If Ω_{AB} is positive, two stable mixtures (A-rich and B-rich) exist with a miscibility gap in between. Any mixture with a composition between the two stable phases will decrease its free energy by decomposing into an A-rich and a B-rich phase. Such a process is known as spinoidal decomposition, and in it species diffuse up a concentration gradient [456].

In the case of a-Si:H, it turns out that Ω_{AB} is positive [439]. This can be inferred from the bond energies of Si$-$Si (2.35 eV), Si$-$H (3.3 eV), and H$-$H (4.5 eV). A Si$-$H network is thus unstable compared to pure Si and H$_2$. A mixture of Si$-$Si and Si$-$H bonds will be driven towards their two stable mixture phases, which leads to the solubility limit of H in Si. Acco et al. [69, 70] have determined this limit to be about 4%. Excess hydrogen in a-Si:H forms SiH$_2$ or SiH around microvoids.

7.3. Elimination of Hydrogen from a-Si:H

Successive burial of hydrogen-rich surface layers leads to the formation of the a-Si:H material. The large amount of hydrogen at the surface is to saturate the surface dangling bonds. The much lower hydrogen content in the bulk is due to the solubility limit of hydrogen in silicon.

The hydrogen content in the bulk a-Si:H depends on deposition temperature and plasma conditions (e.g. [375, 457]; see also Section 6.1). In the γ'-regime the hydrogen content increases considerably with decreasing deposition temperature, while in the α-regime the hydrogen content also increases with decreasing temperature, but this increase sets in at lower temperatures. Robertson has defined the hydrogen elimination rate R_H, expressed as the number of hydrogen atoms per unit cell, as the sum of a thermal (R_T) and an athermal, plasma-driven (R_P)

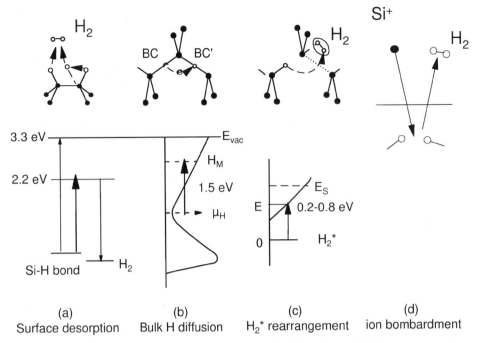

Fig. 53. Possible hydrogen elimination processes: (a) surface loss from adjacent Si—H groups by prepairing on a single Si site, (b) bulk diffusion of atomic H by activation to a H mobility level μ_M, as described by the hydrogen density of states (HDOS), (c) rearrangement of H_2^*, and the lowering of the barrier energy below the normal value E_s by excitation to a weak bond, expressed in the HDOS, (d) subsurface ion-induced displacement and desorption. [After J. Robertson, *J. Appl. Phys.* 87, 2608 (2000).]

elimination process [450]: $R_H = R_T + R_P$. Also, R_H equals the growth rate r_d multiplied by the fractional loss of hydrogen from the surface layer to the bulk: $R_H = r_d(C_{H,s} - C_{H,b})$, where $C_{H,s}$ and $C_{H,b}$ are the hydrogen concentration at the surface and in the bulk, respectively.

7.3.1. Thermal Reactions

The thermal processes can be derived starting with the behavior of hydrogen content as a function of reciprocal temperature [439, 450]. There are two regimes: a high-temperature regime ($T_s > 400°C$), characterized by an apparent activation energy of about 1.6 eV, and a low-temperature regime ($T_s < 250°C$) with an apparent activation energy of about 0.15 eV. Various hydrogen elimination processes are summarized in Figure 53 [439].

In the high-temperature regime it is most likely that hydrogen is eliminated to form H_2 at the surface, with an activation energy of 2.1 eV (Fig. 53a). This is similar to H evolution from crystalline silicon [439, 458]. At intermediate temperatures ($250°C < T_s < 400°C$) the elimination of hydrogen occurs via diffusion of atomic hydrogen towards the surface with an activation energy of 1.5 eV to recombine as H_2, i.e., the standard effusion process [459–461]. This is described with the hydrogen density of states (HDOS) diagram [442, 443, 446, 459, 462]; see Figure 53b. The HDOS diagram displays the energy of a hydrogen atom in *a*-Si:H compared to that of a free hydrogen atom in vacuum. A hydrogen bound as Si—H has an energy of −3.3 eV with respect to the vacuum level, while a

hydrogen at a Si—Si bond center (BC in Fig. 53b) has an energy of about −1 eV. In *a*-Si:H hydrogen is more stable at the bond center of a weak bond, which is represented as a tail in the HDOS below −1 eV. In the HDOS there exists a mobility edge (H_M), above which the hydrogen states form a percolation path. Hydrogen diffusion occurs by excitation from μ_H to H_M. The difference $\mu_H - H_M$ amounts to 1.5 eV which is less than 2.2 eV, but much larger than the apparent activation energy of 0.15 eV.

Robertson proposed that the local bond rearrangement close to the surface can account for this low activation energy [439]. It is assumed that hydrogen exists largely in pairs, analogous to the H_2^* in crystalline silicon [463]. This H_2^* consists of two Si—H bonds in the same direction [464, 465]. It can undergo a local rearrangement with one hydrogen atom passing through two bond center positions, at a cost of 1 eV [245]. Another local rearrangement is possible in which the H passes through the bond center to combine with the other hydrogen atom to form the interstitial molecule H_2 (Figure 53c). The Si—Si bond center site is the transition state of the reaction. As hydrogen is more strongly bound at a weak Si—Si bond than at a normal Si—Si bond [466, 467], the energy barrier is lowered. As a consequence, the elimination of hydrogen to form H_2 is eased in the presence of weak bonds.

An alternative hydrogen elimination process has been proposed by Severens et al. [432, 468]. They argue that hydrogen can be eliminated by a cross-linking step [469] immediately after a physisorbed SiH_3 radical has chemisorbed on a surface dangling bond. This cross-linking probability is thermally activated, but due to the energy released at the chemisorption of SiH$_3$ (about 2 eV), the activation is quite low [432]. If this cross-linking process does not occur immediately after chemisorption, it is assumed that hydrogen is incorporated into the film. An activation energy of 0.15 eV is deduced from fitting hydrogen incorporation data [468].

A detailed description of the local bond rearrangement has been derived [439], using the concept of the HDOS with a low-energy tail that corresponds to the H present at weak Si—Si bonds. The width of this tail is $2E_{v0}$, i.e., twice the width of the valence band tail in the electronic density of states, which in turn is about equal to the Urbach energy E_0 [442, 443]. The HDOS then is [439]

$$N(E) = \frac{N_0}{2E_{v0}} \exp\left(\frac{E - E_s}{2E_{v0}}\right) \tag{67}$$

with E_s at the top of the HDOS, i.e., the energy barrier for rearrangement of the ideal H$_2^*$ (1 eV). The rearrangement occurs by activation of H to a state E in the tail. The reaction rate in atoms per unit surface area is the product of this HDOS, the Maxwell–Boltzmann factor, and an attempt frequency ν_1 ($\sim 10^3$ s^{-1}), and is expressed in the low-temperature regime as [439]

$$R_T = \frac{\nu_1 k_B T}{2E_{v0}} \exp\left(-\frac{E_s}{2E_{v0}}\right) \tag{68}$$

The values obtained from this equation compare well with the experimental values of E_{v0} as a function of temperature.

7.3.2. Athermal Reactions

Next, athermal, plasma-driven hydrogen elimination reactions are considered. Ions play an important role in the deposition. They can account for 10–25% of the total deposition rate [301, 311], depending on the regime of the discharge. Moreover, the energy that they carry to the (sub)surface [163] can be used to overcome energy barriers (see also Section 6.2.1). Ions penetrate the film and eliminate hydrogen by displacing hydrogen atoms from subsurface Si—H bonds. They subsequently recombine as hydrogen molecules and effuse back to the surface (see Fig. 53d). The displacement yield Y_D is given by the modified Kirchin–Pease equation $Y_D = E_i/2E_D$, where E_i is the ion energy and E_D the displacement energy [412]. Here, E_D is equal to the Si—H bond energy, i.e., 3.3 eV. In the α-regime the contribution of ions to the deposition rate is 25% [301], and a typical average ion energy is 10 eV (see Section 6.2.1). This gives a displacement yield Y_D of about 40% of the deposition rate. This illustrates the beneficial role of ions in improving the material quality [284, 301, 358, 417, 470]. For the γ'-regime about 10% of the ions contribute to the deposition rate, and typical average ion energies are 20–30 eV [301]. This gives a displacement energy Y_D of about 30–45% of the deposition rate. In this regime average ion energies are larger than the bulk displacement energy of silicon of 22 eV [285, 376], which degrades the material quality.

It is this athermal dehydrogenation process that reduces the excess weak-bond density associated with a high polymeric content, rather than ions' densifying the material by removing polymeric SiH$_x$ groups. Low energy ions are especially effective at dehydrogenation because the displacement energy E_d is low. This favors the use of VHF PECVD reactors, where the ion energies are much lower and the ion densities much higher than at 13.56 MHz.

Further, atomic hydrogen in hydrogen-diluted discharges can contribute to abstraction of surface and subsurface hydrogen [471]. The atomic hydrogen reduces the fraction of higher SiH$_x$ groups [466]. Only a small amount is incorporated directly [472]. Atomic hydrogen does not reduce the surface hydrogen content below 50%, because this would result in surface dangling bonds that would be passivated by incident hydrogen [472, 473]. Atomic hydrogen can also diffuse into the a-Si:H with an activation energy of 0.4–0.7 eV [461], which allows subsurface hydrogen abstraction. However, this would be less important at low temperatures, due to the considerable activation energy.

7.4. Dangling-Bond and Weak-Bond Density

The hydrogen elimination processes described above create the distribution of weak bonds. The dangling-bond distribution is formed from this by interconversion from weak bonds to dangling bonds. For substrate temperatures above the hydrogen equilibration temperature of 200°C, hydrogen is mobile, and the defect pool operates. This allows the weak-bond and dangling-bond distributions to equilibrate [439]. At temperatures below 200°C hydrogen motion is inhibited, and equilibration should not be possible. Nevertheless, Stutzmann [474] reported that the dangling-bond density as a function of temperature shows the the same behavior as the weak-bond density, for all temperatures. This suggests that a defect pool of sorts operates below 200°C in the surface layer [439].

It is assumed that a large athermal, plasma-driven hydrogen flux is present in the hydrogen elimination zone at temperatures below 200°C, which mediates the equilibration of the dangling-bond and weak-bond distributions. The dangling-bond density and weak-bond density thus remain linked in the elimination zone by a largely temperature-independent relationship, given by the defect pool [445, 446]. The dangling-bond density in the elimination layer becomes the bulk defect density as the layer is buried deeper and becomes frozen in. This proposed process allows the weak-bond density to determine the dangling-bond density at all temperatures [439].

8. MODIFICATIONS OF PECVD

Of the great number of possible and reported modifications, in this section only the most important ones are described, viz., the use of higher excitation frequency (VHF), the use of gas flow modulation (known as chemical annealing or layer-by-layer deposition), and the use of RF modulation.

8.1. VHF

8.1.1. General

Deposition of hydrogenated amorphous silicon employing the VHF PECVD technique (typical frequency range 20–110 MHz) has been reported to yield an increase in deposition rate by one order of magnitude over the conventionally used frequency of 13.56 MHz [16, 146, 250, 280], without adversely affecting material quality [183, 280, 475]. This is of great importance for lowering the production cost of a-Si:H solar cells.

The explanation of the VHF influence on the deposition rate is still a topic of discussion. It has been proposed theoretically that the high-energy tail in the EEDF is increased with an increasing ratio ω/ν of the excitation frequency to the energy collision frequency [276, 476]. This leads to an increased ionization rate [146, 250, 280]. However, mass spectrometry results on the decomposition of silane as a function of frequency show that the increase of the deposition rate cannot solely be attributed to the enhancement of radical production [119, 120]. It has also been found or deduced that the flux of ions towards the surface is increased with increasing frequency [146, 183, 279, 284] while at the same time the ion energy is decreased. This low-energy, high-flux ion bombardment enhances surface mobilities of adsorbed species. It was shown in Section 6.2.3 that a kinetic energy of 5 eV per deposited atom is needed to produce good quality a-Si:H [301].

Because the wavelength of the RF signal is of the order of the substrate dimensions (3 m at 100 MHz), it can be expected that uniform deposition is more difficult at these high frequencies [477]. In fact, a practical optimum frequency around 60–70 MHz is used [478, 479], which provides a good compromise between high deposition rate and attainability of uniform deposition. Further, the use of a distributed RF electrode network where all nodes have identical amplitude and phase improves the homogeneity of deposition [480].

8.1.2. Optimization of Deposition Conditions

A change of excitation frequency prompts for optimization of other deposition parameters, such as pressure, power, and geometry. Chatham and Bhat have shown that the maximum of SiH* emission at 414 nm is shifted towards lower pressure upon increasing the frequency [146]. This means that the maximum in radical production also shifts to lower pressure. Experimental results obtained in the ASTER deposition system corroborate this [247].

A 20-fold increase in deposition rate was reported for an increase in frequency from 13.56 to 100 MHz [119, 120]. Here the electrode distance was about 4 cm, while the pressure was 0.12 mbar. A reasonable thickness uniformity was observed. For a smaller electrode distance (2.7 cm) higher pressures are needed in order to maintain good homogeneity in thickness. This is demonstrated in Figure 54, where the conditions to obtain a homogeneity deviating less than 5% over a 10×10-cm^2 area are presented. The power density in this case was 57 mW/cm^2. The shaded area represents the process

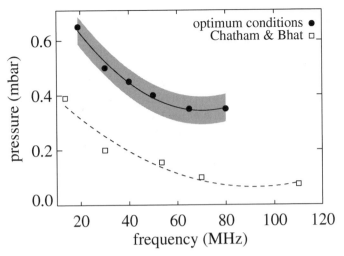

Fig. 54. Optimum values for pressure and frequency for uniform (thickness variation <5%) film deposition. The shaded area signifies the process window. Also shown are similar data by Chatham and Bhat [146].

window. It is clear that for this electrode distance the homogeneity requirement can be met, if one decreases the pressure according to the increase in frequency. This observation also explains the improvement in homogeneity as reported by Howling et al. [250]: at constant pressure the deposited layers become more homogeneous upon increasing the frequency. A similar effect was observed by Chatham and Bhat [146]. For comparison, their data are also shown in Figure 54. The electrode distance in their work amounted to about 1 cm, while the reported power density was 31 mW/cm^3.

The fact that pressure and frequency are related by the need to obtain homogeneous layers leads to the suggestion that the sheath thickness d_s between the grounded electrode and plasma bulk plays an important role. It is known that an increase of frequency yields a decrease in sheath thickness, while a decrease in pressure enlarges it. The combination of these effects suggests that an optimum sheath thickness is required.

The dependence of sheath thickness on frequency or pressure has been reported by several authors [179, 328, 481–485]. Their results are based on both theoretical studies of argon and experimental studies of argon, helium, and silane, using parallel plate configurations. It was found in these studies that $d_s \propto p^a$, where a ranges from -0.3 to -0.55, and $d_s \propto \omega^b$, where b ranges from -0.5 to -1.0. The data represented in Figure 46 and Figure 63 (Section 8.1.3) give $a = -0.28 \pm 0.08$ and $b = -0.26 \pm 0.02$.

Fitting the data in Figure 54 using the relationship $p^a f^b = c$, with c a constant that could represent d_s, shows that several combinations of the powers a and b can be used, as long as the exponents a and b are related as $b \approx 0.45a$. These fit results are inconsistent with the mentioned values for a and b; hence the sheath thickness probably is not constant. This was also deduced from direct measurement of the IEDs [161, 235, 486, 487]. In contrast, a recently developed method to determine the sheath thickness directly from deposition experiments [150,

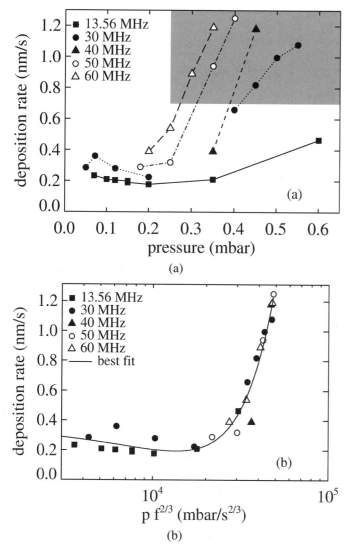

Fig. 55. (a) Deposition rate as a function of pressure at various frequencies. The shaded area shows the 5% uniformity conditions. (b) Best fit of deposition rate versus $pf^{2/3}$, clearly showing the α–γ' transition.

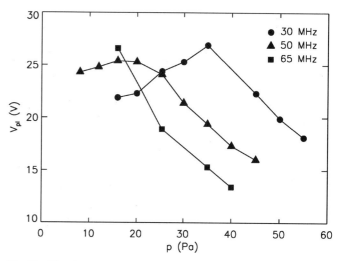

Fig. 56. The plasma potential as a function of pressure for a frequency of 30, 50, and 65 MHz. (From E. A. G. Hamers, Ph.D. Thesis, Universiteit Utrecht, Utrecht, the Netherlands, 1998, with permission.)

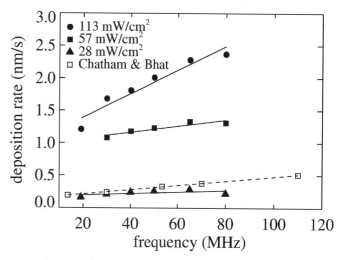

Fig. 57. Deposition rate as function of frequency at optimum pressure at each frequency (see Fig. 54) at three power densities. Also shown are data by Chatham and Bhat [146].

151, 488] revealed that the value of d_s is constant for the conditions where $p\omega^{1/2}$ is constant [487].

The deposition rate as a function of pressure at various frequencies and at a power density of 57 mW/cm^2 is shown in Figure 55a. The shaded area represents the conditions for which uniform deposition is observed. At all frequencies the deposition rate at low pressure is lower than at high pressure by a factor of 5 to 6. This effect is well known, and it is explained by the transition from the α- to the γ'-regime of the plasma [245]. It follows from the figure that the transition region shifts to lower pressure with higher frequencies. Fitting the deposition rate as a function of the scaling parameter $p^a f^b$ shows that a good fit is obtained for $a = 1$ and $b = 2/3$; see Figure 55b. This holds for all frequencies. The same behavior has been found for the plasma potential [163]. As is illustrated in Figure 56, the plasma potential slightly increases in the low-pressure α-regime, but clearly decreases with increas-

ing pressure at all three frequencies. The pressure at which the plasma potential starts to decrease occurs at a lower pressure at the higher frequencies. The rapid increase of deposition rate as a function of frequency, as reported in the literature, e.g. [16, 119, 120, 280, 475], is observed in the α-regime. In or near the γ'-regime the enhancement in deposition rate as a function of frequency is not pronounced. This is illustrated in Figure 57 by results at three power densities. Chatham and Bhat's data [146] also show only a slight increase with frequency.

In Figure 58 optical and structural material properties are shown. The refractive index at 2.07 eV (600 nm), $n_{2.07\ eV}$, decreases with increasing applied RF power P, and is more or less independent of the frequency (Fig. 58a). The absorption coefficient at 2.07 eV behaves similarly. Values are around 2 to 3×10^4 cm^{-1}. The cubic bandgap E_g increases with power, and is also not dependent on frequency (Fig. 58b). The hydro-

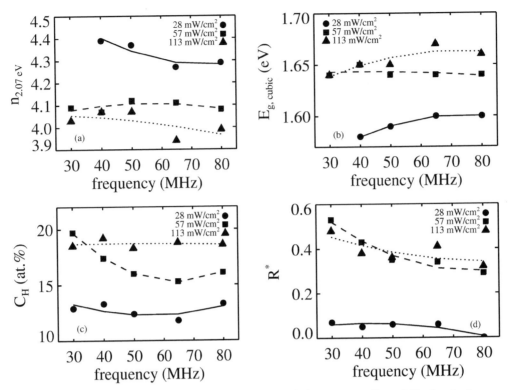

Fig. 58. (a) Refractive index $n_{2.07\,\text{eV}}$, (b) cubic bandgap E_g, (c) hydrogen content C_H, and (d) microstructure parameter R^* as function of frequency at optimum pressures (see Fig. 54) at three power densities.

gen content C_H increases from 13 at.% at low power to 19 at.% at high power, again being nearly independent of frequency (Fig. 58c). The microstructure parameter R^* increases with power, with only a slight dependence on frequency (Fig. 58d). The character of the intrinsic stress is changed from compressive (about 600 MPa) at low power to tensile (100 MPa) at high power. Only a slight influence of frequency is observed, and the dependence on power is much clearer. Small–angle X-ray spectroscopy (SAXS) data show that voids are absent in the samples deposited at low power densities [66]. For higher powers voids are observed, and there appears to be a sudden change between 27 and 58 mW/cm^2.

Electrical data are shown in Figure 59 as a function of deposition rate for all frequencies, using the relation between deposition rate and power density as depicted in Figure 54. Both dark conductivity and photoconductivity decrease exponentially with increasing deposition rate. The data in this range of deposition rates can be fitted with $\sigma_d = 9 \times 10^{-12} \exp(-1.5 r_d)$ and $\sigma_{ph} = 8 \times 10^{-6} \exp(-0.5 r_d)$, with σ_d and σ_{ph} expressed in Ω^{-1} cm^{-1}, and r_d in nanometers per second. Consequently the photoresponse σ_{ph}/σ_d increases with deposition rate as about $10^6 \exp(r_d)$. Activation energies amounted typically to 0.7–1.0 eV. From thermally stimulated conductivity (TSC) measurements [489–492] a midgap density of states (DOS) of 1.5×10^{16} cm^{-3} eV^{-1} is determined. The product $\mu\tau$ at 300 K is 9×10^{-5} cm^2 V^{-1}. Both DOS and $\mu\tau$ are independent of frequency.

Summarizing the results, it is clear that most of the material properties are not strongly dependent on frequency in the range

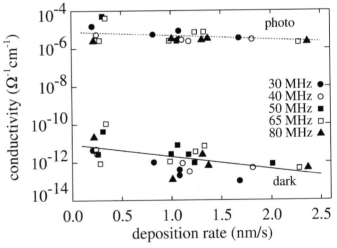

Fig. 59. Dark conductivity (σ_d) and photoconductivity (σ_{ph}) as functions of deposition rate for all frequencies.

of 30 to 80 MHz. The effect of power density is much more important. The material deposited at the lowest power is of *device quality*. Using larger power densities results in less dense material. This can be inferred from the decrease of the refractive index and the increase of the microstructure parameter, in combination with the increase in hydrogen content.

Perrin et al. [245] have shown that the increase in deposition rate is more pronounced above a certain threshold power density. This power density is close to the lowest power density

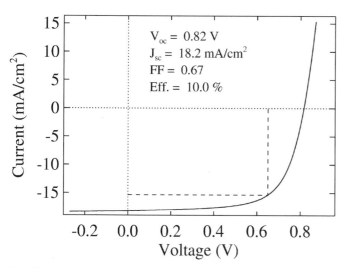

Fig. 60. Current–voltage characteristics of a solar cell made at 65 MHz and 42 mW/cm². The dashed line indicates the maximum-power point.

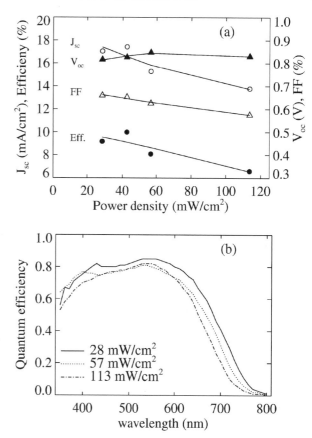

Fig. 61. Solar cell performance parameters as functions of power density for cells made at 65 MHz: (a) J_{sc}, efficiency, V_{oc}, and fill factor (FF); (b) spectral response.

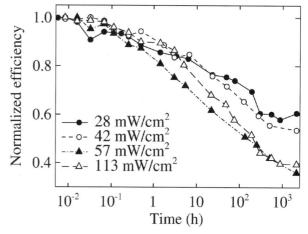

Fig. 62. Normalized solar cell efficiency as a function of illumination time for different power densities as obtained by continuous illumination of 1000-W/m² AM1.5 light. The initial efficiencies of the four cells were 9%, 10%, 8%, and 6% for 28-, 42-, 57-, and 113-mW/cm² power density, respectively.

in the experiments shown here. The deposition rate is probably the main parameter causing the observed difference between the low-power and high-power material and the deterioration of layers at higher power levels. This confirms the known observation that in general, the quality of material deposited in the γ'-regime is worse than in the α-regime [245].

The presented material quality results are similar to results reported by others [119, 120, 280, 475], although those frequency series were obtained in the α-regime at constant low pressures. Here, good quality homogeneous layers were deposited *at the α–γ' transition region*. Especially the low power results compare well with results by others.

The intrinsic material fabricated at the frequencies reported above was incorporated in p^+–i–n^+ solar cells [493]. The p- and n-layer were prepared by the conventional 13.56-MHz discharge. The *device quality* films indeed yield good solar cells, of 10% efficiency, as is shown in Figure 60. This cell is manufactured with a 500-nm-thick i-layer made at 65 MHz with a power density of 42 mW/cm², resulting in excellent properties. The deposition rate still is 2–3 times higher than at the conventional 13.56 MHz; 0.2–0.5 nm/s versus 0.1–0.15 nm/s [494]. Similar results were reported by Jones et al. [495].

The effects of power density on the material quality are also reflected in the cell quality, e.g. the efficiency. Figure 61a shows a decrease in efficiency with increasing power density or deposition rate. This is also seen in the spectral response measurements (Fig. 61b). A shift in the red from long (700 nm) to shorter (650 nm) wavelength is observed if one compares cells deposited at low and high power densities. This can be explained by the fact that the material is less dense, with a higher hydrogen content, yielding a larger bandgap. The observed decrease in the blue part of the curve as a function of power density may well be due to changes in the p–i interface, which are induced by a different ion bombardment condition.

The differences in material properties are also reflected in the degradation behavior of the cells. Figure 62 shows the nor-

malized efficiencies as a function of illumination time. The efficiencies of the cells with the i-layer deposited at low power densities stabilize at around 60% of their initial value, while the cells with the i-layer deposited at high power densities stabilize at 40%. This correlates with the hydrogen content of the i-layers, which is around 12% and 19% for the low and the

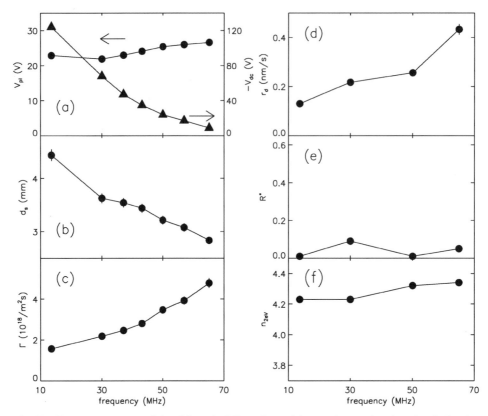

Fig. 63. Plasma parameters as deduced from the IEDs and material properties as a function of excitation frequency of the SiH_4–H_2 discharge at a power of 10 W and a pressure of 0.16 mbar: (a) the plasma potential V_{pl} (circles) and dc self-bias V_{dc} (triangles), (b) the sheath thickness d_s, (c) the maximum ion flux Γ_{max}, (d) the growth rate r_d, (e) the microstructure parameter R^*, and (f) the refractive index $n_{2\,eV}$. (Compiled from E. A. G. Hamers, Ph.D. Thesis, Universiteit Utrecht, Utrecht, the Netherlands, 1998.)

high power densities (see Fig. 58) and which is indicative for the low-density material at obtained at these high power densities.

Also a frequency dependence was observed, especially in V_{oc}, which was not expected from the material study. As only the deposition conditions of the i-layers are varied, a change at the p–i interface must be responsible for the change in V_{oc}. The lower value of V_{oc} at low frequency was attributed to the difference in ion bombardment at the p–i interface [493].

8.1.3. Discharge Analysis

In the ASTER deposition system, experiments have been carried out in which the excitation frequency was varied between 13.56 and 65 MHz [169]. The other process conditions were kept constant at a power of 10 W, a pressure of 0.16 mbar, gas flows of 30 sccm SiH_4 and 30 sccm H_2, and a substrate temperature of 250°C. As in Section 6.2.3, plasma properties that are deduced from IED measurements are compared with material properties in Figure 63. The IEDs of SiH_2^+ at four frequencies are shown in Figure 64.

The plasma potential is about 25 V (Figure 63a). This value of the plasma potential is typical for the silane plasmas in the *asymmetric* capacitively coupled RF reactors as used in the

ASTER deposition system, and is also commonly found in argon or hydrogen plasmas [170, 280, 327]. From the considerable decrease of the dc self-bias with increasing frequency (Figure 63a) it is inferred that the potential drop over the sheath of the grounded electrode (V_{pl}) and the one over the sheath of the powered electrode ($V_{pl} - V_{dc}$), become comparable in magnitude. Hence, the discharge is becoming more symmetric with increasing frequency.

In contrast, Heintze and Zedlitz [236] also presented data on the plasma potential as function of frequency in silane plasmas: the plasma potential varies from about 27 V at 35 MHz to about 20 V at 180 MHz. Moreover, Dutta et al. [284] used a *symmetric* capacitively coupled RF reactor and estimated the plasma potential in their system from the applied voltage at the powered electrode. A decrease of the plasma potential from 45 V at 13.56 MHz to only 15 V at 70 MHz is observed. This difference in behavior is thought to be solely due to the different reactor geometries.

The charge carrier density in the sheath increases by a factor of 3 on increasing the excitation frequency from 13.56 to 65 MHz. As a consequence, the sheath thickness d_s decreases from 4.4 mm at 13.56 MHz to 2.8 mm at 65 MHz (Fig. 63b), which results in a reduced number of collisions in the sheath.

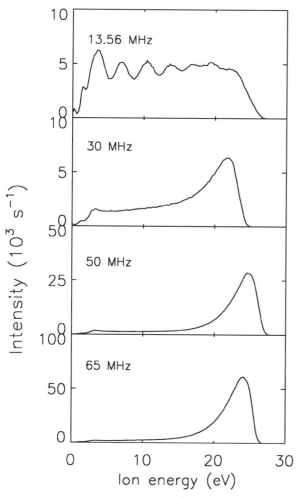

Fig. 64. The ion energy distributions of SiH_2^+ at several frequencies. (From E. A. G. Hamers, Ph.D. Thesis, Universiteit Utrecht, Utrecht, the Netherlands, 1998, with permission.)

This is nicely illustrated in Figure 64, where the IEDs of SiH_2^+ are shown for several frequencies. Clearly, the contribution of low-energy ions decreases upon increasing the frequency.

The increase in the deposition rate r_d (Fig. 63d) corresponds to the increase in the ion flux (Fig. 63c): the fraction of arriving ions per deposited atom, R_i, is constant at about 0.25. Such observations have also been reported by Heintze and Zedlitz [236], who furthermore suggested that the deposition rate may well be controlled by the ion flux. The kinetic ion energy per deposited atom, E_{max}, is also constant and amounts to about 5 eV. As was shown in Section 6.2.3, the material quality as reflected in the refractive index $n_{2\,eV}$ (Fig. 63e) and the microstructure parameter R^* (Fig. 63f) is good: $n_{2\,eV}$ is around 4.25, and R^* is low (<0.1). The depletion of the silane stays constant at a value of 4.0 ± 0.4 sccm in this frequency range. The partial pressures of silane, hydrogen, disilane (1.3×10^{-3} mbar), and trisilane (2×10^{-5} mbar) in the plasma are also independent of frequency. Similar results have been obtained by Heintze and Zedlitz [280]. As a result of varying the frequency from 50 to 250 MHz, the dissociation rate

only increased by a factor of 1.5, while the deposition rate on the electrodes increased by nearly an order of magnitude.

From the constant depletion, assuming a homogeneous deposition throughout the reactor, a frequency-independent deposition rate of 0.44 nm/s is deduced [163]. At 13.56 MHz the observed deposition rate is a factor of 3 lower, whereas at 65 MHz the measured deposition rate equals the estimated value of 0.44 nm/s. This suggests that the increased deposition rate at higher frequencies, as measured at the center of the grounded electrode, is partly due to a more homogeneous deposition profile throughout the reactor at higher frequencies [163, 280]. This suggestion is further supported by the fact that the discharge is more symmetric at higher frequencies, as deduced from the low dc self-bias at high frequencies.

8.2. Chemical Annealing

It was recognized by, e.g., Tsai et al. [496] that high hydrogen dilution resulted in microcrystalline films as a result of atomic hydrogen, which preferentially etches weak bonds during deposition. This prompted the use of fluorine-containing precursors, with which enhanced etching is achieved during deposition. Etching by means of periodically interrupting the deposition process has been reported by Tsuo et al. [497], who used a constant XeF_2 flow and a periodic SiH_4 flow. The use of an RF hydrogen discharge for the etching cycles and a diluted silane discharge for the deposition cycles was tried by Xu, as reported by Luft and Tsuo [6]. With both methods good material is obtained.

The *chemical annealing* technique was proposed by Shirai et al. [17, 471, 498] as a means to tackle the metastability problem in a-Si:H solar cells. It aims at the lowering of the hydrogen content in a-Si:H by alternating deposition and exposure to atomic hydrogen. It was demonstrated that a lower hydrogen content indeed leads to more stable material. A two-discharge-zone apparatus was used by Asano et al. [499, 500]. In one zone SiH_4 and H_2 were mixed to deposit a-Si:H, while in the other zone pure H_2 was used for the etching process. The substrate was rotated between the two zones. Chemical annealing alternatively is termed alternately repeating deposition and hydrogen plasma treatment (ADHT) and layer by layer (LBL).

The application of this technique in conventional RF deposition equipment revealed the possibility of producing microcrystalline silicon at relatively low power densities and temperatures [501]. Depending on deposition conditions, (i.e., deposition cycle time, hydrogen exposure time, and power density), different microcrystalline fractions are found; see [165, 251, 502].

As an example, the Raman spectra of three films deposited in the ASTER deposition system are shown in Figure 65, for which only the hydrogen exposure time was varied, 10–30 s. In the deposition cycle a 1 : 1 mixture of silane and hydrogen was used. Conditions were chosen so that *device quality* material was obtained [370]. The duration of the deposition cycle was kept constant at 10 s. At the start of the H_2-plasma cycle the SiH_4 flow was stopped and extra H_2 was introduced to keep

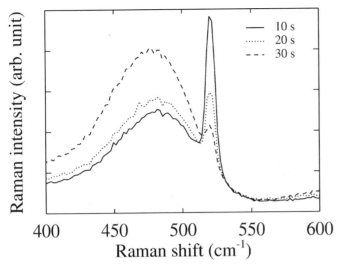

Fig. 65. Raman spectra for films deposited employing the chemical annealing technique for different hydrogen exposure times, 10, 20, and 30.5 seconds.

the pressure constant. A practical problem to be addressed is the matching of the discharge when changing from H_2 to SiH_4 and vice versa. A sudden change in dc self-bias is observed, as well as a change in reflected power. Fast automated matching networks are to be used to minimize these effects.

The broad TO peak, characteristic for a-Si:H, is observed at 480 cm^{-1}, while the crystalline peak is at 520 cm^{-1}. The ratio between integrated peak intensities clearly depends on the hydrogen exposure time. The hydrogen content increases linearly with exposure time; the extra hydrogen forms SiH_2 bonds. These effects were thought to originate from the hydrogen bombardment during the hydrogen exposure cycle [165].

Layadi et al. have shown, using *in situ* spectroscopic ellipsometry, that both surface and subsurface processes are involved in the formation of μc-Si [502, 503]. In addition, it was shown that the crystallites nucleate in the highly porous layer below the film surface [502, 504], as a result of energy released by chemical reactions [505, 506] (chemical annealing). In this process four phases can be distinguished: incubation, nucleation, growth, and steady state [507]. In the incubation phase, the void fraction increases gradually while the amorphous fraction decreases. Crystallites start to appear when the void fraction reaches a maximum (at a thickness of about 10 nm). Subsequently, the crystalline fraction increases exponentially in the nucleation phase. Then a linear increase in the crystalline fraction sets in, corresponding to the growth of crystallites in the whole film. At a certain thickness, the crystalline fraction remains constant, and a steady state is reached. *In situ* conductivity measurements revealed that the dark conductivity increases by more than an order of magnitude at the onset of the growth phase [507]. These films have been employed in thin-film transistors, and excellent stability was demonstrated [508].

It was argued by Saito et al. [509] that a hydrogen plasma treatment as used in the LBL technique causes *chemical transport*: the a-Si:H film deposited on the cathode prior to the hydrogen treatment is etched and transferred to the anode.

Hence, in the hydrogen plasma silane-related molecules and radicals are present. In fact, material is deposited on the anode under the same conditions as in the case of high dilution of silane with hydrogen.

8.3. RF Modulation

Modulation of the RF excitation has been used in an attempt to increase the deposition rate. Increasing the gas pressure or raising the power generally leads to dust formation and deterioration of material properties. To overcome this problem, one can pulse the plasma by modulating the RF signal in amplitude with a square wave (SQWM). Depending on the regime (α or γ'), different effects are observed.

A considerable decrease both in the deposition rate and in powder formation was found when the 13.56-MHz excitation was modulated with a square wave of 2 Hz [510, 511]. In the γ'-regime, Biebericher et al. [512] have observed a decrease in deposition rate from 1.0 nm/s in a continuous-wave (cw) 50-MHz SiH_4–H_2 plasma, to 0.2 nm/s in a similar (i.e. with the same *average* power of 10 W) plasma, modulated by a frequency of 100 kHz.

When modulating the power in the α-regime, an increase in deposition rate has been observed with SQWM, with an optimum at a frequency of about 100 kHz [513–515]. It was reported that the time-averaged electron density for a modulated discharge was higher than in the case of a continuous discharge [18]. Moreover, the amount of dust generated in the discharge was much smaller, due to the low density of negative ions.

At the start of each modulation pulse, a sharp peak in optical emission is seen. Similar SiH* emission peaks in pulsed plasmas have been found by Scarsbrook et al. [516] and Howling et al. [321]. The sharp peak was claimed to be caused by a pulse of high-energy electrons. Overzet and Verdeyen [517] measured electron densities at a 2.9-MHz excitation frequency and modulation frequencies up to 20 kHz. The optical emission of a SQWM argon plasma was measured by Booth et al. [518], who also performed particle-in-cell modeling.

Here, results are shown from experiments performed in ASTER, reported by Biebericher et al. [512, 519]. A SiH_4–H_2 (50 : 50 flow ratio, total flow 60 sccm) plasma was generated at an RF excitation frequency of 50 MHz. The substrate temperature was 250°C. The RF signal was ampitude modulated (AM) by a square wave. The modulation frequency has been varied in a range of 1–400 kHz. The modulation depth was always 90%. The duty cycle was fixed at 50%. The pressure amounted to 0.2 mbar, and the *average* power was kept at 10 W. With a duty cycle of 50%, this leads to a power of 20 W during the plasma-on period.

The measured dependence of the deposition rate on the modulation frequency is shown in Figure 66a. It can be clearly seen that the deposition rate at the optimum of 100 kHz is about three times higher than the deposition rate of a cw discharge of 10 W (dotted line). In addition, there is a clear correlation between

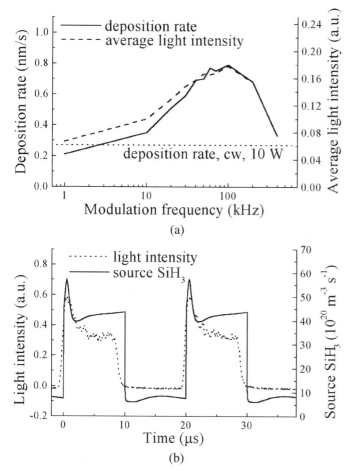

Fig. 66. Effects of modulating the RF excitation frequency: (a) deposition rate and average light intensity as a function of the modulation frequency, with the deposition rate at cw conditions indicated by the dotted line, (b) measured spectrally integrated emission and calculated production of SiH₃ radicals as a function of time, at a modulation frequency of 50 kHz and a 50% duty cycle. [From A. C. W. Biebericher, J. Bezemer, W. F. van der Weg, and W. J. Goedheer, *Appl. Phys. Lett.* 76, 2002 (2000), © 2000, American Institute of Physics, with permission.]

the deposition rate and the time-averaged optical emission intensity of the pulsed discharge. This average intensity is related to the electron temperature: a higher light intensity can be attributed to an increase in the electron temperature. This results in an increased production of radicals, which is reflected in an increased deposition rate.

The spectrally integrated (185–820 nm) emission of a 50-MHz discharge, modulated at 50 kHz, is shown in Figure 66b. Similar results are obtained for the SiH^* emission. The peak appearing at the onset is generated by the plasma. The rise time of about 1 μs and the FWHM of the peak of 6 μs are much longer than the 40-ns detection limit of the electronics. When the power is turned off, the light intensity decreases very fast and the plasma fully extinguishes.

Upon changing the modulation frequency it was observed that only the length of the plateau varied, whereas the height and FWHM of the peak at plasma onset remained almost constant. Therefore, the relative importance of the peak in the time-averaged light intensity is higher at higher modulation

frequencies, and consequently, the average light intensity increases with increasing modulation frequency. This holds up to the observed optimum of 100 kHz. At this frequency the plasma-on period is 5 μs, i.e. about equal to the FWHM of the peak. A further increase in modulation frequency leads to a reduction of the peak and the average light intensity.

In order to be able to explain the observed results plasma modeling was applied. A one-dimensional fluid model was used, which solves the particle balances for both the charged and neutral species, using the drift–diffusion approximation for the particle fluxes, the Poisson equation for the electric field, and the energy balance for the electrons [191] (see also Section 4.1).

The calculated production of SiH_3 radicals during two modulation cycles at a modulation frequency of 50 kHz is also shown in Figure 66b. Good correlation with the behavior of average light intensity versus time is observed, including the sharp peak at plasma onset. The shape of the curve is determined by the instantaneous sharp rise of dissociation and by chemical reactions. The nonzero production of SiH_3 radicals during the plasma-off period is due to chemical reactions. SiH_3 is produced not only by dissociation of SiH_4 through electron impact, but also by an abstraction reaction of SiH_4 and H. These hydrogen atoms can originate from the dissociation of SiH_4 into SiH_2 and 2H [512].

The behavior of the electrons as a function of time can explain the observed results. Simulation results for a discharge operated at a pressure of 0.267 mbar, a frequency of 50 MHz, an electrode distance of 27 mm, a voltage modulated between 134 V (power 20 W) and 40 V (power <1 W), and a modulation frequency of 5 kHz are shown in Figure 67. In this figure the electron energy and density are shown as a function of the distance between the electrodes. The numbers indicate the numbers of passed RF cycles. A modulation frequency of 5 kHz corresponds to 10,000 RF cycles. The plasma is turned off at cycle 0 and is turned on at cycle 5000. It can be seen in Figure 67a that the average electron energy decreases very fast. This is consistent with the observed decay of the emission of the discharge (see Fig. 66b). The electron density also decreases, albeit with a longer decay time. The electron density is about halved after 100 cycles. Thus, very soon after shutoff, within a few cycles, no new radicals are produced, apart from those generated in neutral–neutral reactions. The electron (and ion) density decreases by the loss due to ambipolar diffusion to a value one order of magnitude lower than the value at cycle 0. When the power is switched on at cycle 5000, the electron energy increases to a value nearly two times larger than the value at cycle 0 (and 10,000). This fast increase is attributed to the low electron density at cycle 5000, which absorbs the power. The increase in electron energy results in an even steeper increase in reaction rates per electron, due to the energy dependence of the cross-sections. Thus, a peak in light intensity as well as a peak in SiH₃ radical production is observed. This leads to an increase of the average deposition rate to a value above the rate corresponding to the average deposited power (10 W); cf. Figure 66a. At a low modulation frequency (1 kHz) the decay of the profiles

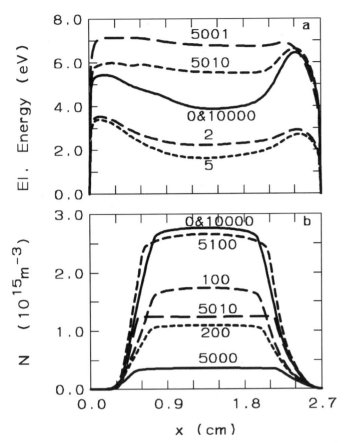

Fig. 67. The development in time of (a) the average electron energy and of (b) the density of electrons between the electrodes in the case of a 50-MHz plasma and a modulation frequency of 5 kHz. The RF electrode is located at $x = 0$, and the grounded electrode at $x = 2.7$ cm. The numbers near the curves in both graphs represent the number of RF cycles. One RF cycle $\equiv 20$ ns. The plasma is turned off at cycle 0, turned on at cycle 5000, and turned off again at cycle 10,000. [From A. C. W. Biebericher, J. Bezemer, W. F. van der Weg, and W. J. Goedheer, *Appl. Phys. Lett.* 76, 2002 (2000), © 2000, American Institute of Physics, with permission.]

and the overshoot of the rates cover only a small fraction of the modulation period. In other words, the discharge is in a quasi steady state with respect to the applied power, and the deposition rate is only slightly above the corresponding deposition rate at cw conditions. As the modulation frequency increases, the overshoot becomes more important, which leads to an increase in deposition rate. When the period of the modulation becomes shorter than the decay time of the electron density, the power is distributed over too many electrons to yield considerable heating. The heating becomes less, and less of the radicals is produced, as is shown in Figure 66a. It is therefore concluded that an optimum in the modulation frequency exists, which corresponds to the decay time of the electron density. From Figure 67b it can be inferred that this optimum should be around 100 kHz, as at 200 cycles the electron density is reduced to $1/e$ of its value at cycle 0, i.e., the lifetime is 200 RF cycles, or 4 μs.

Another beneficial effect of modulation of the discharge is the observed improvement of uniformity of deposition. At cw discharge conditions that normally result in thickness variations

of 70%, modulation lowers the variation to about 10% [519]. This was attributed to the fact that, as a result of modulation, reactions are not confined to the sheaths, and radicals are produced in the entire volume between the electrodes.

Madan et al. [515] have presented the effect of modulation on the properties of the material (dark conductivity and photoconductivity) and of solar cells. They also observe an increase in deposition rate as a function of modulation frequency (up to 100 kHz) at an excitation frequency of 13.56 MHz, in their PECVD system [159]. The optimum modulation frequency was 68 kHz, which they attribute to constraints in the matching networks. Increasing the deposition rate in cw operation of the plasma by increasing the RF power leads to worse material. Modulation with a frequency larger than 60 kHz results in improved material quality, for material deposited with equal deposition rates. This is also seen in the solar cell properties. The intrinsic a-Si:H produced by RF modulation was included in standard p–i–n solar cells, without buffer or graded interface layers. For comparison, solar cells employing layers that were deposited under cw conditions were also made. At a low deposition rate of about 0.2 nm/s, the cw solar cell parameters are: $V_{oc} = 0.83$ V, $J_{sc} = 15.8$ mA/cm^2, FF = 0.67, and $\eta = 8.8\%$ (at AM1.5 illumination conditions). Solar cells produced with the intrinsic layer deposited at 68-kHz modulation frequency are of the same quality: $V_{oc} = 0.81$ V, $J_{sc} = 17.1$ mA/cm^2, FF = 0.63, and $\eta = 8.7\%$.

Biebericher et al. [519] report that beyond a modulation frequency of about 40 kHz in their 50-MHz discharge the material is slightly worse than at standard cw conditions. However, they argue that due to the enhancement in deposition rate, lower partial pressures of silane can be used, which results in more efficient silane consumption. Indeed, they have shown that lowering the gas flows by a factor of 3 leads to a reduction in deposition rate of only a factor of 1.5 at 100-kHz modulation frequency [520]. Clearly, the efficiency of silane consumption is of great importance in production.

9. HOT WIRE CHEMICAL VAPOR DEPOSITION

9.1. General Description

The hot wire chemical vapor deposition (HWCVD) method was introduced as early as 1979 by Wiesmann et al. [521]. The principle of HWCVD is based on the thermal decomposition of silane at a heated tungsten (or tantalum) filament, foil, or grid. At temperatures of 1400–1600°C the silane is decomposed into a gaseous mixture of silicon and hydrogen atoms. Matsumura [522–524] further developed HWCVD, but termed it catalytical CVD (CTL-CVD); for he showed that the decomposition of silane at a heated tungsten filament is a catalytic process. The use of a much higher pressure (>0.1 mbar) than the one Wiesmann et al. used ($< 5 \times 10^{-4}$ mbar) led to a high deposition rate (0.5 nm/s) [521]. Doyle et al. showed that the silicon atoms deposited on the filament are then thermally evaporated onto the substrate, which is located within a few centimeters of the filament; hence the term *evaporative surface decomposition* [525].

At the same time, thermal and catalytic dissociation of silane and hydrogen lead to the generation of hydrogen atoms at the filament surface [526].

Device quality a-Si:H made by HWCVD (as they termed it) was first reported by Mahan et al. [19, 527]. They obtained a-Si:H with hydrogen concentrations as low as 1%. Deposition rates as high as 5 nm/s [528] and 7 nm/s [529] have been achieved for a-Si:H of high quality. In order to obtain *device quality* material it was shown by Doyle et al. [525] that the radicals that are generated at the filament (atomic Si and atomic H) must react in the gas phase to yield a precursor with high surface mobility. Hence, the mean free path of silane molecules should be smaller than the distance between filament and substrate, d_{fs}. Too many reactions between radicals and silane molecules, however, result in worse material. In fact, optimal film properties are found for values of pd_{fs} of about 0.06 mbar·cm [530, 531].

The combination of high deposition rate and the ability to produce *device quality* material is of particular interest for solar cell production and TFT fabrication [532–535]. Further, the low hydrogen content was expected to yield improved stability against light-induced degradation [527], as the Staebler–Wronski effect is related to the hydrogen content in the material (see also Section 1.2.5). This was demonstrated by Crandall et al. [536], who incorporated an HWCVD-deposited layer in solar cells, and observed reduced degradation upon light soaking as compared to devices with a conventional PECVD layer. In these *hybrid* solar cells only the intrinsic layer was made by HWCVD; all other (doped) layers were deposited by PECVD. Wang et al. [537] have reported solar cells that were completely made using HWCVD.

HWCVD-deposited a-Si:H layers have also successfully been used as the semiconductor layer in inverted–staggered TFTs [538, 539]. Moreover, it was demonstrated that these TFTs have excellent electrical properties, they do not suffer from a shift of the threshold voltage upon prolonged application of a gate voltage and these HWCVD TFTs are stable [538, 540]. In a TFT, the metastable character of a-Si:H manifests itself as a reversible threshold-voltage shift, upon prolonged application of gate voltage (see Section 11.2).

Dilution of silane with hydrogen leads to the formation of polycrystalline silicon films (poly-Si:H) that still contain a small amount (<1%) of hydrogen [529, 541–544]. This is believed to occur as a result of increased etching by atomic hydrogen. Molecular hydrogen is decomposed at the filament to reactive atomic hydrogen, which is able to etch silicon atoms from disordered or strained bonding sites in the amorphous (sub)surface network. The amorphous network thus is transformed to a polycrystalline one, without amorphous tissue. Large grains are obtained (70-nm average grain size), and the crystalline fraction in these highly compact materials amounts to 95% [544]. The grain boundaries are passivated by the small amount of hydrogen.

Doping is achieved by adding dopant gases such as phosphine, diborane, or trimethylboron to the silane [545–547]. The

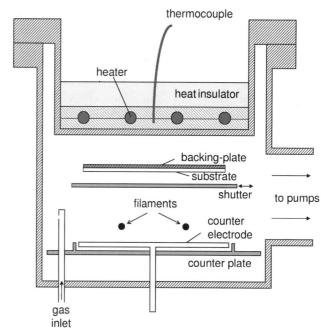

Fig. 68. Cross-sectional view of a HWCVD deposition reactor. (From K. F. Feenstra, Ph.D. Thesis, Universiteit Utrecht, Utrecht, the Netherlands, 1998, with permission.)

doped material is of electronic quality comparable to PECVD-doped material.

9.2. Experimental Setup

A cross-sectional overview of a HWCVD setup is shown in Figure 68 [548]. This particular reactor is part of a multichamber, ultrahigh-vacuum deposition system, PASTA, described by Madan et al. [159]. The background pressure of the reactor is 10^{-8} mbar. In this HWCVD reactor two coiled tungsten filaments are mounted parallel to each other in order to improve thickness uniformity. The filaments have a diameter of 0.5 mm and a length of 15 cm. The distance between the filaments is 4 cm, and the distance from the filaments to the substrate (d_{fs}) is 3 cm. The filaments are heated by sending a dc current through the wires. Calibration of the filament temperature as a function of dc current is needed, and can be done with a pyrometer. Temperatures exceeding about 1800°C should be used as to avoid silicide formation on the filament. On the other hand, the temperature should not be so high that tungsten is evaporated from the filament and is incorporated in the film [549].

A shutter can be placed in front of the substrate to prevent deposition during filament preheating. The reaction gases are injected on the lower-left side of the reactor, and directed towards the reactor wall in order to achieve more homogeneous gas flows. The (unreacted) gas is pumped out at the right side, so the gas flow is perpendicular to the filaments.

The substrate is radiatively heated by heaters that are placed outside the vacuum. A backing plate ensures a laterally homogenous temperature profile. In the same chamber also PECVD can be carried out. The backing plate then is the grounded electrode, and the RF voltage is applied to the counter electrode.

The high filament temperature used causes additional radiative heating of the substrate [530, 531]. Feenstra et al. [531, 548] have developed a heat transport model of their setup (Figure 68). All heat exchange is assumed to occur via radiation. They found, e.g., that at a heater temperature of 100°C and a filament temperature of 1900°C the substrate temperature reached a value of 350°C after a typical stabilization time of 25 min. With the shutter closed the stable substrate temperature was 50°C lower. This can be exploited to preheat the substrate, when the shutter is closed, to the desired temperature. Just before the deposition is started, the filament temperature is set to its correct value, and the shutter can be opened. This procedure avoids any temperature drift during deposition.

9.3. Material Properties and Deposition Conditions

With the setup described, a series of depositions was carried out [531, 548], in which the substrate temperature was varied between 125 and 650°C, the pressure between 0.007 and 0.052 mbar, the gas flow rate between 15 and 120 sccm, and the dilution of the silane gas with hydrogen ($[SiH_4]/([SiH_4] + [H_2])$) between 0.1 and 1. Under these conditions the deposition rate varied between 1 and 2.5 nm/s [531]. Molenbroek et al. [530] reported a variation of the deposition rate between 1 and 9 nm/s for similar process conditions, at a filament temperature of 2000°C.

The variation of deposition temperature has similar effects on the material properties to those on PECVD-deposited material. With increasing temperature (125–650°C), the material becomes more dense (the refractive index extrapolated to 0 eV increases from 3.05 to 3.65), and the hydrogen content is decreased (15 to 0.3 at.%), as well as the microstructure factor (0.4 to 0). The activation energy is 0.83 eV up to a deposition temperature of 500°C. The dark conductivity and AM1.5 photoconductivity are about 5×10^{-11} and 5×10^{-6} Ω^{-1} cm^{-1}, respectively, up to a deposition temperature of 500°C, i.e., the photoresponse is 10^5.

The deposition rate increases upon increasing the pressure. This is explained by noting that the impingement rate per unit area, r_i, of molecules on the filament is linearly dependent on the pressure as $r_i = p/\sqrt{2\pi k_B T_g}$, with T_g the gas temperature. However, as the pressure becomes higher, the collisional mean free path of the silane becomes smaller, and the silane supply to the filaments becomes restricted. Moreover, the transport of deposition precursors to the substrate is restricted as well. The mean free path of silane was estimated to be 2.5 cm at a pressure of 0.02 mbar [531], i.e., the mean free path about equals the distance between filament and substrate. Indeed, a maximum in deposition rate is observed at this pressure. This corresponds to a value of pd_{fs} of 0.06 (cf. [530]). The microstructure parameter plotted as a function of pd_{fs} has a minimum around $pd_{fs} = 0.06 \pm 0.02$ [530].

The silane flow rate determines to a large extent the possible deposition rate. If the supply of silane is too low, the envi-

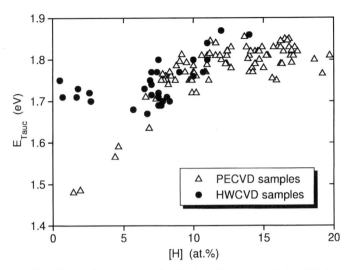

Fig. 69. Relation between the optical bandgap (Tauc convention) and the hydrogen content in a-Si:H deposited by HWCVD (closed circles) and PECVD (open triangles). (From K. F. Feenstra, Ph.D. Thesis, Universiteit Utrecht, Utrecht, the Netherlands, 1998, with permission.)

ronment in the immediate vicinity of the filament will become depleted of silane. The generation of deposition precursors (and also the deposition rate) is limited by the silane supply. Above a certain flow, the dissociation reaction rate is the limiting factor, and as a consequence the deposition rate will remain constant upon a further increase of the silane flow.

Dilution of silane with hydrogen to moderate amounts ($0.75 < [SiH_4]/([SiH_4] + [H_2]) < 1$) is not expected to have a large influence on material properties, as even in the pure silane case a large amount of atomic hydrogen is present. Every SiH_4 molecule dissociates to give four H atoms [525]. Also, the mean free path of the atomic H is larger than that of Si, being about 8 cm at 0.02 mbar [531], and consequently all hydrogen will reach the substrate. It is found that the deposition rate depends linearly on the dilution fraction, as it depends on the partial pressure of silane. Significant changes in the refractive index, the hydrogen content, the microstructure parameter, and electrical properties are observed for values of $[SiH_4]/([SiH_4]+[H_2])$ lower than about 0.3–0.4. A closer look at these data reveals that moderate hydrogen dilution slightly improves the dark conductivity and photoresponse, whereas dilution fractions smaller than 0.3–0.4 lead to deterioration of the material. Similar trends were also observed by Bauer et al. [550] and Molenbroek et al. [528, 530].

The common linear dependence of the bandgap on the hydrogen content as observed for PECVD-deposited material is not observed for HWCVD-deposited material. A large dataset is shown in Figure 69, in which the relation between bandgap and hydrogen content is compared for HWCVD- and PECVD-deposited material [531]. It is seen that the bandgap of HWCVD-deposited material follows a similar behavior, as a function of hydrogen content, to that of PECVD-deposited material, for values of the hydrogen content of 7% and higher: the bandgap varies between 1.7 and 1.85 eV [531]. The bandgap of PECVD-deposited materials decreases to 1.5 eV

Table IX. Exothermic Gas Phase Reactions in the HWCVD Process, Their Enthalpies, and Their Rates [192], Estimated for Typical Operating Conditions

Chemical reaction	Enthalpy (eV)	Reaction rate ($cm^3 s^{-1}$)
$H + SiH_4 \rightarrow SiH_3 + H_2$	-0.54	2.68×10^{-12}
$H + SiH_3 \rightarrow SiH_2 + H_2$	-1.46	2×10^{-11}
$SiH_3 + SiH_3 \rightarrow SiH_4 + SiH_2$	-0.91	1.5×10^{-10}
$Si + SiH_4 \rightarrow Si_2H_2 + H_2$	-0.34	3.5×10^{-10}
$Si_2H_2 + SiH_4 \rightarrow Si_3H_4 + H_2$	$-0.4/-0.05$	2×10^{-10}

at a hydrogen concentration of 2%, whereas the bandgap of HWCVD-deposited material remains unchanged at 1.7 eV, even at hydrogen concentrations of 1%. This difference in behavior is due to the difference in the number of voids present in the material. A large number of voids reduces the effective density of the material, and increases the average $Si-Si$ bond distance. As a result, the bandgap remains high. Crandall et al. [71] found that the number of voids increased with decreasing hydrogen content in HWCVD-deposited material. This observation is also supported by small-angle X-ray scattering (SAXS) [551] and nuclear magnetic resonance (NMR) [552] data.

9.4. Deposition Model

The deposition mechanism in HWCVD of a-Si:H can be divided into three spatially separated processes. First, silane is decomposed at the tungsten filament. Second, during the diffusion of the generated radicals (Si, H) from the filament to the substrate, these radicals react with other gas molecules and radicals, and new species will be formed. Third, these species arrive at the substrate and contribute to the deposition of a-Si:H.

The filament material acts as a catalyzer in the silane decomposition process. This is clear by comparing the $Si-H$ bond energy (3–4 eV) in silane with the thermal energy at the filament temperature of 2000°C (0.25 eV): they differ by an order of magnitude. The dissociation of silane proceeds in two steps [525]. First, the silane reacts with the tungsten to form a tungsten silicide W_xSi_y at the surface of the filament. Hydrogen atoms leave the surface before they can recombine to form molecular hydrogen. Subsequently, the silicon atoms evaporate from the surface, which is an activated process with an activation energy of 3.6 eV [525]. At a filament temperature of about 1700°C the silane decomposition rate and the evaporation rate balance each other. For lower temperatures a tungsten–silicon alloy is formed on the filament.

A fraction of the generated radicals react with gas molecules before they are able to reach the substrate directly. This fraction depends on the process pressure. A number of reactions may take place, and only the exothermic ones are listed in Table IX, as endothermic reactions are not very likely to occur. As can be seen in Table IX, the Si and H radicals are highly reactive with silane. Most reactions will occur with the omnipresent silane. Quantum chemistry computations, confirming the findings of Molenbroek et al. [530], have shown that the reaction of

Si with SiH_4 proceeds via insertion of Si into the SiH_4 to form Si_2H_4, followed by a rearrangement of the thus formed species and the elimination of H_2, which yields Si_2H_2 [553]. An increasing amount of SiH_3 will be created via the reaction of H with SiH_4 upon increasing the pressure. As a product of the reaction $SiH_3 + SiH_3$, SiH_2 is created which at higher pressures may polymerize further to Si_3H_4. Holt et al. [554] have performed Monte Carlo simulations to study gas phase and kinetic processes in HWCVD. They showed that under the conditions for obtaining *device quality* material SiH_3 is the most abundant species at the substrate. At a pressure of about 0.08 mbar the fluxes of SiH_3, Si_2H_2, and Si to the substrate were calculated to be 5×10^{17}, 1×10^{16}, and 5×10^{13} $cm^{-2} s^{-1}$, respectively.

The HWCVD deposition process is more or less the same as for PECVD, and was described in Section 7. Important differences between the two is the absence of ions, and the limited number of different species present in the gas phase, in the former. At low pressure atomic Si is the main precursor. This yields void-rich material with a high microstructure factor. Increasing the pressure allows gas phase reactions with Si and H to create more mobile deposition precursors (SiH_3), which improves the material quality. A further increase leads to the formation of higher silanes, and consequently to a less dense film.

Using threshold ionization mass spectrometry and *in situ* ellipsometry, Schröder and Bauer [555] have shown that the Si_2H_4 radical may well be the species responsible for deposition, rather than SiH_3 as in PECVD. This larger and less mobile precursor is thought to be the cause of the observed differences in the deposition conditions required in HWCVD and PECVD to obtain *device quality* material.

10. EXPANDING THERMAL PLASMA CHEMICAL VAPOR DEPOSITION

10.1. General Description

The expanding thermal plasma chemical vapor deposition (ETP CVD) technique has been developed in the group of Schram [20, 556] and has been used for the deposition of several materials, such as hydrogenated amorphous silicon [21] and carbon [557], and even diamond [558]. The technique is a remote-plasma deposition method. The generation of the plasma, the transport of the plasma, and the deposition are geometrically separated. In remote-plasma CVD the substrate holder is not a necessary electrode for the plasma. The absence of direct plasma contact with the substrate allows for better control of ion bombardment, which is advantageous. Also, the properties of the plasma can be varied independently, which makes optimization of the whole process simpler.

An argon–hydrogen plasma is created in a dc thermal arc (cascaded arc) operated at high pressure (≈ 0.5 bar) [556, 559, 560] (the cascaded arc is also employed in IR ellipsometry, providing a well-defined source of intense IR radiation; see Section 5.4 [343]). As the deposition chamber is at much lower pressure (0.1–0.3 mbar), a plasma jet is created, expanding into

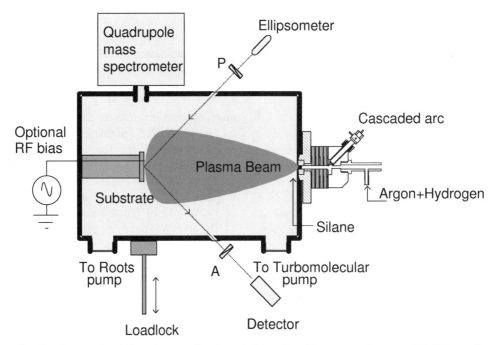

Fig. 70.   Cross-sectional view of a expanding thermal plasma deposition reactor. (Courtesy of M.C.M. van de Sanden, Eindhoven University of Technology, Eindhoven, The Netherlands.)

the deposition chamber. Near the plasma source silane is injected, and the active plasma species dissociate the silane into radicals and ions. These species can deposit on the substrate, which is positioned further downstream.

The main advantage of this technique is the very high deposition rate that can be obtained for the different materials. However, this high deposition rate may not always be compatible with good material quality. A large effort has been made and has resulted in the deposition of good-quality a-Si:H at deposition rates as high as 10 nm/s [432, 561, 562].

10.2. Experimental Setup

The ETP CVD setup is schematically shown in Figure 70. It consists of a thermal plasma source, a cascaded arc, and a low-pressure deposition chamber [563, 564]. The cascaded arc consists of three cathodes, a stack of ten copper plates with a central bore of 4 mm, and an anode plate with a conical nozzle [556]. The length of the cascade is about 10 cm. All parts are water-cooled. Pure argon or an argon–hydrogen mixture is introduced into the arc. The argon flow is varied between 1800 and 6000 sccm, and the hydrogen flow between 0 and 1200 sccm. The arc pressure is between 0.4 and 0.6 bar. The dc discharge is current-controlled. The arc current and voltage are 30–75 A and 45–120 V, respectively. Typical arc plasma parameters are a electron density and temperature of 10^{16} cm^{-3} and 1 eV [565], respectively.

The deposition chamber is a cylindrical vessel with a diameter of 50 cm and a length of 80 cm. At about 5 cm from the arc outlet, silane can be introduced via an injection ring (7.5-cm diameter) that contains eight holes of 1-mm diameter each. The distance between arc outlet and substrate is 38 cm.

The substrates are heated via the substrate holder, of which the temperature can be controlled between 100 and 500°C. Samples can be loaded via a load lock equipped with a magnetic transfer arm. The substrate can be optionally RF-biased. A residual gas analyzer (QMS) and an ellipsometer complete the setup. The typical pressure is in the range of 0.15–0.5 mbar. The deposition chamber has a volume of 180 l. During processing it is pumped by a stack of two Roots blowers and one forepump (total pumping capacity is about 1500 m^3/h); otherwise it is pumped by a turbo pump (450 l/s), with which a base pressure of 10^{-6} mbar is reached.

As a result of the large pressure gradient between arc and deposition chamber the plasma expands supersonically from the nozzle into the deposition chamber. After a stationary shock front at about 5 cm from the nozzle, the plasma expands subsonically. As a consequence, the electron density and temperature are drastically reduced [566]. In this downstream region the plasma is recombining, much as in an afterglow plasma. The low electron temperature in the region where the silane is introduced implies that electron-induced reactions are ineffective for the ionization and dissocation of silane. It is found that the admixture of H$_2$ to Ar in the arc is the cause of the strong reduction in the ion and electron density [567]. For pure Ar, the ion density is 10^{13} cm^{-3}, and consists mainly of Ar$^+$. At high H$_2$ flows (1800 sccm), the ion density is about 4×10^{10} cm^{-3}, and is mainly H$^+$. At low hydrogen flows, ArH$^+$ is the most abundant ion. The reason for this drastic change in density is that fast molecular processes become an important recombination channel for the ions [560, 568]. The atomic H concentration increases from about 2×10^{11} cm^{-3} at low (240 sccm) H$_2$ flow to about 2×10^{12} cm^{-3} at high (1800 sccm) H$_2$ flow. The arc changes basically from an argon ion source to an atomic hydrogen source on going from pure argon to admixture of a large amount of hydrogen.

10.3. Material Properties and Deposition Conditions

The deposition rate is found to be independent of the deposition temperature, for low (0.3 nm/s) to high (30 nm/s) rates [472, 473]. The deposition rate and the silane consumption increase with increasing partial pressure of silane [569]. A higher arc current yields a higher silane depletion and deposition rate [563]. The bond-angle distortion, as determined from Raman spectroscopy measurements, varied between 8.3° at low deposition rate and 9.1° at high deposition rate [570], indicating a slightly increased disorder.

The incorporated amount of hydrogen depends on the deposition temperature. The hydrogen content decreases with increasing temperature. For films deposited at a high rate, the hydrogen content always is higher than for the films that were deposited at a low rate. The hydrogen content varies from 60 at.% at a deposition temperature of 100°C to 10 at.% at a deposition temperature of 400°C for high-rate deposited films. The corresponding values for low-rate deposited films are 22 at.% (100°C) and 5 at.% (400°C). The microstructure factor also decreases with increasing temperature. The refractive index increases with increasing temperature [468]. Both dark conductivity and photoconductivity show an increase with increasing temperature, and have a maximum at 400–450°C [571] of 4×10^{-6} and 10^{-9} Ω^{-1} cm^{-1}, respectively.

The deposition rate and the silane consumption decrease sharply by a factor of about four upon adding 60–120 sccm hydrogen to pure argon in the arc, as in shown in Figure 71a. Further addition of hydrogen does not influence these parameters much [569, 572]. Similar effects are observed for the refractive index, microstructure factor, activation energy, and dark conductivity and photoconductivity: adding a small amount of hydrogen leads to a large change, which is saturated upon further hydrogen admixture.

Using ion and threshold ionization mass spectrometry (TIMS), Langmuir probe measurements, and cavity ring down absorption spectroscopy (CRDS), the influence of hydrogen admixture on the film growth precursors has been investigated [567, 569, 573]. For the argon–hydrogen plasma, it was found that electron temperature and ion fluence decrease with increasing hydrogen flow. At low hydrogen flow the electron temperature is 0.3 eV, whereas at high hydrogen flow it is 0.1 eV. The ion fluence is decreased from about 120 sccm at low hydrogen flow to about 4 sccm at high flow.

Adding silane to the argon–hydrogen plasma leads to the formation of large, hydrogen-poor positive silicon ions [567]. With increasing flow the cluster size increases; in fact, the higher silane clusters $Si_nH_m^+$ ($n > 5$) are much more abundant than the lower silane clusters. The total ion flux towards the substrate scales linearly with the silane flow, while it decreases with increasing hydrogen flow. The maximum contribution of the cationic clusters to film deposition ranges from 4% to 7%

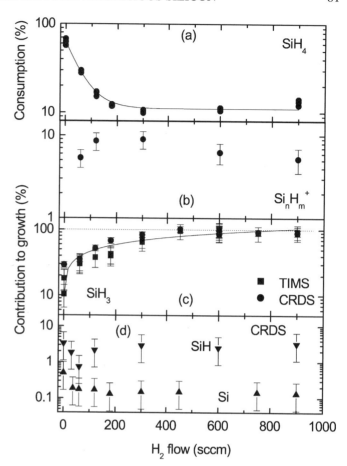

Fig. 71. The influence of hydrogen admixture on (a) the silane consumption, and the contribution of (b) $Si_nH_m^+$, (c) SiH_3, (d) SiH and Si to the deposition, as measured by TIMS and CRDS. (From W. M. M. Kessels, Ph.D. Thesis, Technische Universiteit Eindhoven, Eindhoven, the Netherlands, 2000, with permission.)

for varying silane flows, and from 5% to 9% for varying hydrogen flows (Fig. 71b), and is nearly independent of both the silane and the hydrogen flow [567].

As a consequence, the contribution of neutral species to the deposition is more than 90%, and consists mainly of SiH_3. This contribution increases from about 20% at low hydrogen flow to about 90% at intermediate hydrogen flow, and remains constant with further increase (Fig. 71c). A contribution of SiH_2 larger than 0.1% could only be measured at zero hydrogen flow, and amounted to about 5%. Other radicals, viz., SiH and Si, contribute about 2% and 0.2%, respectively, to the deposition (Fig. 71d), irrespective of hydrogen flow.

The optimized conditions for the deposition of good-quality a-Si:H have been found to be: an argon flow of 3300 sccm, a hydrogen flow of 600 sccm, a silane flow of 600 sccm, an arc current of 45 A, an arc voltage of 140 V, an arc pressure of 0.4 bar, a process pressure of 0.2 mbar in the downstream region, and a substrate temperature of 400°C [571, 572]. At this high deposition temperature the hydrogen content is 6–7 at.%. Consequently the (cubic) bandgap is low, 1.51 eV. Also, the microstructure factor still is nonzero, viz. 0.2, and what is more, the dark conductivity is a factor of 10 higher than needed for *device quality* material. The electron drift mobility is somewhat smaller than usual for device quality films, but the hole drift mobility is one order of magnitude larger. The activation energy, Urbach energy, and defect density are 0.75 eV, 50–55 meV, and 10^{16} cm^{-3}.

10.4. Deposition Model

Kessels has summarized the information on the reactive species emanating from the Ar–H$_2$-operated cascaded arc [571], and has formulated a global reaction scheme. For a hydrogen flow between 0 and 120 sccm, the dissociation of silane is governed by dissociative charge transfer of Ar$^+$ with SiH$_4$, which generates SiH$_n^+$ ($n \leq 3$) ions: Ar$^+$ + SiH$_4 \rightarrow$ Ar + SiH$_n^+$ + $\cdots$. Then, as the electron density is high, these reactions are quickly followed by dissociative recombination reactions with electrons, which generates SiH$_m^+$ ($m \leq 2$) radicals: SiH$_n^+$ + e$^- \rightarrow$ SiH$_m$ + $\cdots$. These highly reactive radicals react with silane molecules, forming polysilane radicals: SiH$_m$ + SiH$_4 \rightarrow$ Si$_2$H$_p$ + $\cdots$. When the electron density is decreased as a result of the dissociative recombination reactions, ion–molecule reactions between silane ions and silane will become significant as well. These reactions (SiH$_n^+$ + SiH$_4 \rightarrow$ Si$_2$H$_m^+$ + $\cdots$) lead to sequential ion–SiH$_4$ reactions (Si$_p$H$_q^+$ + SiH$_4 \rightarrow$ Si$_{p+1}$H$_r^+$ + $\cdots$), which generate cationic clusters (up to Si$_{10}$H$_r^+$ have been observed).

At a hydrogen flow larger than 300 sccm the ion fluence from the arc is greatly reduced. Reactions with atomic H emanating from the arc now are more effective. The dissociation of silane is governed by hydrogen abstraction, which generates the SiH$_3$ radical (H + SiH$_4 \rightarrow$ H$_2$ + SiH$_3$). The small flow of H$^+$ from the arc is responsible for a charge exchange reaction with silane, which creates SiH$_n^+$ ($n \leq 3$), which initiates sequential ion–molecule reactions as above.

The deposition of good-quality a-Si:H using ETP CVD is a result of the dominant presence of the SiH$_3$ radical under conditions of high hydrogen flow in the cascaded arc. The deposition process itself is more or less the same as for PECVD, and was described in Section 7. An important difference between the ETP CVD and the PECVD process is the absence of ions with considerable energy (>1 eV) and the larger number of cationic clusters in the former. Nevertheless, good-quality a-Si:H is deposited at high rates.

11. APPLICATIONS

The first practical device demonstrating the use of a-Si:H as a photovoltaic material was the 2.4%-efficient solar cell reported by Carlson and Wronski [34]. Since then, interest in a-Si:H has been growing rapidly, more or less prompted by the many possible applications. It is used in solar cells and optical sensors based on the photovoltaic effect. Especially the possibility of uniform large-area deposition has been exploited in a-Si:H thin film transistors (TFTs) that are used in controlling active-matrix liquid crystal displays (LCDs). The high photoconductivity and

fast photoresponse of a-Si:H is of great importance for use in large-size linear image sensors.

A selection of applications is presented in the following subsections: solar cells, TFTs, light sensors (visible, IR, X-ray), and chemical sensors. Also, light-emitting devices, in particular utilizing erbium incorporation in a-Si:H, are presented. Finally, electrostatic loudspeakers in which an a-Si:H film is incorporated are described. Details of various applications described here, as well as many other applications, can be found in the excellent edited books [4, 5, 11, 13, 574].

11.1. Solar Cells

11.1.1. Operation Principle

The operation principle of a solar cell is based on charge separation. Photogenerated electrons and holes must be spatially separated in order to contribute to the net current of the device. Charge separation is done by an internal electric field. In crystalline semiconductors this can be achieved by stacking a p-type doped material on an n-type doped material. The internal field at the p–n junction prevents the recombination of photogenerated electrons and holes. Once separated, however, recombination may occur, and is an important loss process in a solar cell [575]. In amorphous semiconductors a p–n junction hardly shows photovoltaic action. Photogenerated electrons and holes cannot diffuse over long distances, as the defect density in p- and n-type doped material is high. Therefore an undoped (intrinsic) layer has to be introduced between the p- and the n-layer, with a thickness smaller than the mean free path of the slower carriers, viz. the holes.

In the presence of an electric field the drift length is the mobility–lifetime product times the electric field: $\lambda_{\mathrm{mfp}} = \mu \tau E$ [576]. With typical values of $\mu \tau$ and E the mean free path usually exceeds by far the thickness of the solar cell, and virtually all photogenerated carriers can be collected. However, under certain operating conditions, field-free regions in the i-layer may exist, and the collection efficiency is decreased because the diffusion lengths of the carriers are much smaller than the thickness of the solar cell [11, 577].

The p and n layers provide the built-in potential but do not contribute to the collection of carriers. Therefore these layers need to be only as thick as the depletion layer (5–20 nm), as any additional thickness will unnecessarily decrease the collection efficiency by absorbing light. The intrinsic layer should be as thick as possible to absorb the maximum fraction of photons, but the depletion width at operating conditions of about 1 μm sets an upper limit. Typically, one uses a thickness of 500 nm.

A schematic cross-section of a p–i–n a-Si:H solar cell [11] is shown in Figure 72a. In this so-called superstrate configuration (the light is incident from above), the material onto which the solar cell structure is deposited, usually glass, also serves as a window to the cell. In a substrate configuration the carrier onto which the solar cell structure is deposited forms the back side of the solar cell. The carrier usually is stainless steel, but flexible materials such as metal-coated polymer foil (e.g. polyimid) or a very thin metal make the whole structure flexible [11].

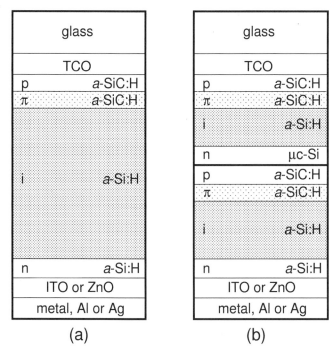

Fig. 72. Schematic cross-section of (a) a single junction p–i–n a-Si:H superstrate solar cell and (b) a tandem solar cell structure. (From R. E. I. Schropp and M. Zeman, "Amorphous and Microcrystalline Silicon Solar Cells—Modeling, Materials and Device Technology," Kluwer Academic Publishers, Boston, 1998, with permission.)

The current–voltage (I–V) characteristic of a solar cell is given by

$$I(V) = I_{ph}(V) - I_0 \left[\exp\left(\frac{e(V + I(V)R_s)}{Ak_BT} \right) - 1 \right]$$
$$- \frac{V + I(V)R_s}{R_{sh}} \qquad (69)$$

where $I(V)$ is the current I at voltage V, $I_{ph}(V)$ the photogenerated current, I_0 the reverse saturation current, A the diode ideality factor, R_s the series resistance, and R_{sh} the shunt resistance parallel to the junction. Usually the current density $J(V)$ rather than the current $I(V)$ is shown, e.g., see Figure 60. An I–V measurement is performed by connecting a variable load or by applying an external voltage [578], in the dark or under illumination with an AM1.5 spectrum. Four parameters determine the performance of a solar cell: the efficiency η, the fill factor FF, the open circuit voltage V_{oc}, and the short circuit current I_{sc}. The efficiency is defined as the ratio of the maximum delivered power to the power of the incident light. The maximum-power point is the point where $I_{max}V_{max}$ has the maximum value. The open circuit voltage V_{oc} and the short circuit current I_{sc} are defined as the values of the voltage and current at $I(V) = 0$ and $V = 0$, respectively. The fill factor FF is defined as the ratio of the maximum power to the product of I_{sc} and V_{oc}. Typical values of these parameters were shown in Figure 60. A characteristic resistance R_{ch} is sometimes defined as V_{oc}/I_{sc} in order to assess the effect of R_s and R_{sh} on the fill factor.

When $R_s \ll R_{ch}$ or $R_{sh} \gg R_{ch}$ little effect is seen on the fill factor [575].

The short circuit current is the product of the photon flux $\Phi(\lambda)$ of the incident solar spectrum and the wavelength-dependent spectral response or collection efficiency $Q(\lambda)$ integrated over all wavelengths: $J_{sc} = \int \Phi(\lambda)Q(\lambda)\,d\lambda$ (see Fig. 61b). The collection efficiency is about 80% between 450 and 600 nm, demonstrating that there is little loss due to recombination (the i-layer is of *device quality*). The decreasing collection efficiency at the red side is due to the decreasing absorption coefficient of a-Si:H. In the blue, the decreasing collection efficiency is due to absorption in the p-layer and/or buffer layer.

11.1.2. Single-Junction Solar Cells

A schematic cross section of a single-junction p–i–n a-Si:H solar cell in the superstrate configuration was shown in Figure 72a. The front electrode consists of a transparent conductive oxide (TCO). Examples are tin oxide (SnO$_2$:F), indium tin oxide (ITO, In$_2$O$_3$:Sn), and doped zinc oxide (ZnO:Al) [579]. A high transparency in the spectral range of the solar cell is required, as well as a low sheet resistance, to minimize series resistance losses. Further, the contact resistance with the p-layer should be low. The surface of the TCO should be textured to reduce optical reflection and to increase the optical path length by scattering, thereby increasing the absorption in the active i-layer [577, 580, 581]. Shunting paths and pinholes should be avoided by a proper morphology. The TCO should also be PECVD-compatible, i.e., it should be highly resistant to the reducing H ambient during PECVD deposition.

ITO, SnO$_2$:F, and ZnO:Al share high transparency (>90%), large bandgap (>3.5 eV), and low sheet resistance (<10 $\Omega/\square$). Tin oxide is made by atmospheric CVD, which inherently produces a native texture. Zinc oxide and ITO, both made by sputtering or evaporation, usually are flat. Zinc oxide can be etched by HCl to produce a rough surface [11]. Both tin oxide and zinc oxide are chemically resistant. ITO can be made chemically resistant by adding a thin titanium oxide layer as a protective coating [582]. Zinc oxide has a high contact resistance with doped a-Si:H, which can be circumvented by using a microcrystalline doped layer [583].

Absorption in the p-layer can be reduced by using an a-SiC:H alloy with a bandgap of about 2 eV [584]. Carbon profiling within the p-layer further improves the window properties [585]. An intentionally graded p–i interface (*buffer layer*) 10 nm in thickness enhances the spectral response in the blue [125, 494, 586], which can be attributed to a reduced interface recombination.

The properties of the i–n interface and the n-layer are not critical, as long as the conductivity is high. Introducing a TCO layer between n-layer and metal back contact enhances the light trapping even further. Absorption mainly is enhanced in the long-wavelength region [587]. The optimum thickness is found to be 70 nm [588].

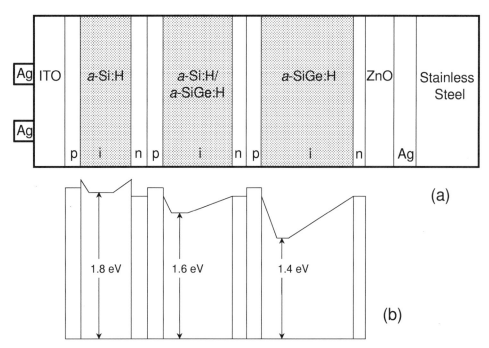

Fig. 73.   Schematic cross section of a triple-junction *a*-Si:H substrate solar cell on stainless steel (a), and the corresponding schematic band diagram (b). (From R. E. I. Schropp and M. Zeman, "Amorphous and Microcrystalline Silicon Solar Cells—Modeling, Materials and Device Technology," Kluwer Academic Publishers, Boston, 1998, with permission.)

In the substrate configuration the stainless steel carrier is coated with a Ag–ZnO bilayer in order to enhance the back reflection of the back contact; see Figure 73 [11]. An increase in J_{sc} of about 50% was achieved by Banerjee and Guha [589] by using a textured Ag–ZnO bilayer, which further enhances the optical path length and consequently the absorption. As at this stage no *a*-Si:H has been deposited, there are virtually no restrictions on the process temperature.

After depositing the *n*-layer, a similar *i*–*n* buffer layer can be deposited as in *p*–*i*–*n* cells. Bandgap profiling, i.e., increasing the bandgap towards the top *p*-layer, is much easier for *n*–*i*–*p* structures, as here simply decreasing the deposition temperature during deposition automatically yields a larger bandgap [11]. In superstrate *p*–*i*–*n* cells, deposition of the *i*-layer is from the *p*- to the *n*-layer. In this case, a decrease in the bandgap is achieved by increasing the temperature during deposition, which may not always be desirable for the already deposited *p*-layer and *p*–*i* interface. Another advantage is that the most critical top layers are deposited last.

The transparent top contact is deposited last of all, which imposes restrictions on the process temperature. Thermally evaporated ITO and ZnO deposited by metal–organic CVD (MOCVD) are most suitable. At a typical thickness of 70 nm the ITO serves as a good antireflection coating as well. Due to the somewhat high sheet resistance, a metal (Ag) grid is necessary to reduce the series resistance [11].

11.1.3. Multiple-Junction Solar Cells

The performance of *a*-Si:H can be improved by stacking two (or more) cells with different optical bandgaps on top of each other; see Figure 72b [11]. In such a *tandem* solar cell, the larger-bandgap material, of which the top cell is made, collects the high-energy photons, while the lower-energy photons are collected in the lower cell. While theoretical efficiencies of single-junction *a*-Si:H solar cells are around 15%, values above 20% are calculated for multijunction cells. A two-terminal (top and bottom contact) double-junction cell consisting of a top cell with a bandgap of 1.75 eV and a bottom cell with a bandgap of 1.15 eV is predicted to have a conversion efficiency of 21%, while a triple stack of 2.0-, to 1.7-, 1.45-eV cells is predicted to have an efficiency of 24% [590]. In the latter cell carbon alloying of *a*-Si:H would make the top cell, and germanium alloying the bottom cell. However, due to high defect densities present in *a*-SiC:H and *a*-SiGe:H, other ways of obtaining good-quality high- and low-bandgap material have been developed. Dilution of SiH$_4$ with hydrogen in the regime $0.05 \leq$ [SiH$_4$]/([SiH$_4$] + [H$_2$]) ≤ 0.15 yields material of better optoelectronic properties than *a*-SiC:H [369], and is therefore used for the top cell (Fig. 72b).

Hydrogen dilution also improves the quality of *a*-SiGe:H alloys [369]. An *a*-SiGe:H bottom cell must be designed to operate at long wavelengths, with relatively weakly absorbed light. The concept of bandgap grading at the *p*–*i* interface, as in *a*-Si:H cells, has been demonstrated to be inappropriate for *a*-SiGe:H cells [591]. Several bandgap profiles were investigated. The best solar cell performance was obtained for a bandgap that first decreased but then increased; see also Zimmer et al. [592].

A triple junction structure based on 1.79-, 1.55-, and 1.39-eV materials [593] was reported to have an initial efficiency of 14.6% and a stabilized (see Section 11.1.4) efficiency of

12.1% [154]. In these series-connected structures the open circuit voltage is the sum of the V_{oc}'s of the individual cells (here $V_{oc} = 2.3$ V), while the currents are matched to be equal in all cells: $J_{sc} = 7.56$ mA/cm^2. The fill factor is 0.7 [154]. A schematic cross section of such a triple-junction p–i–n a-Si(Ge):H solar cell structure in the substrate configuration is shown in Figure 73 [11], together with the corresponding band diagram. This band diagram shows the intricate bandgap profiling scheme that is needed to obtain these high-efficiency solar cells.

A critical part in the structure of multijunction solar cells is the n–p junction that separates the individual cells. Here, electrons that are generated in the top cell (Fig. 72b) flow to the n–p junction, where they must recombine with holes from the bottom cell. As a result of the very high doping levels used, recombination takes place via tunneling. When the recombination rate is not balanced with the supply of carriers, space-charge accumulation will occur, which negatively influences the electric field in the adjacent cell that has the highest generation rate [11]. Often one of the layers of the tunnel junction is microcrystalline [594].

Another important issue in multijunction cells is current matching. The individual currents must exactly balance; otherwise a loss in efficiency occurs. A current mismatch can be easily revealed by measuring the spectral response [595]. If the currents are matched, then the quantum efficiency is flat over a wide range of wavelengths. If one of the cells is limiting the current, then the observed quantum efficiency is not flat, and in fact is the quantum efficiency of the current-limiting cell.

Empirical optimization of the thicknesses of the individual cells within the structure is to be combined with computer modeling. A comprehensive model is required in which an optical and an electrical model of a-Si:H are integrated [11]. Since such a model contains about 100 input parameters for a single-junction solar cell, a careful calibration procedure is needed to extract input parameters from measured layer properties [596, 597]. Current matching in tandem cells has been investigated using a model that not only well describes recombination via tunneling but also takes into account the optical enhancement due to the textured TCO [598]. It was found that for an a-Si:H–a-Si:H stack the optimum thickness of the bottom cell amounted to 300 nm, while for an a-Si:H–a-SiGe:H stack it amounted to 150 nm. The optimum thickness of the top cell was 50 and 60 nm for the a-Si:H–a-Si:H and the a-Si:H–a-SiGe:H stack, respectively [11, 598].

11.1.4. Stability

Illumination of solar cells causes a reduction of efficiency and fill factor, as a result of light-induced creation of defects (Staebler–Wronski effect, Section 1.2.5). This reduction is halted after several hundred hours of illumination. The reduction is correlated with solar cell thickness. A large intrinsic layer thickness leads to a large reduction of efficiency and fill factor compared to a small intrinsic layer thickness. The solar cell properties can be completely recovered by annealing at about 150°C. The open circuit voltage and short circuit current decrease only slightly.

Material properties cannot always be correlated with degradation behavior [493, 495, 599, 600]. Lee et al. [600] have shown the degradation kinetics of cells in which the intrinsic layer was deposited by highly diluting the silane. A clear correlation was observed between the decrease in efficiency and the increase of $\mu\tau$. Another example that demonstrates a correlation between degradation behavior and material quality was shown in Figure 62. Here the normalized cell efficiency as a function of illumination time was depicted for solar cells, where the intrinsic layer is deposited at different discharge power levels [493]. At the lower power levels the efficiency stabilizes at about 60% of its initial value, whereas at the higher power levels it stabilizes at 40%. This has been attributed to the fact that at the higher power densities the density of the material is lower and the hydrogen content is higher.

As a result of the creation of defects, trapping of electrons or holes is enhanced. Thus, the created defects reduce the product $\mu\tau$, which depresses charge collection and solar cell efficiency. The charge collection length, i.e., the average distance that a carrier travels before it is trapped, is defined as $d_c = \mu\tau V_{bi}/d$, with V_{bi} the built-in voltage (typically around 1.2 V [3]) and d the thickness of the cell. An empirical relationship between fill factor and charge collection length d_c was reported by Faughnan and Crandall [601]:

$$\text{FF} = 0.35 + 0.15 \ln\left(\frac{d_c}{d}\right) \tag{70}$$

Smith et al. [602] have derived the dependence of FF on illumination time. They combined the empirical relationship between FF and d_c [Eq. (70)] with the time dependence of the defect density ($N_{db}(t) \propto G^{0.6}t^{1/3}$ [89]) and the relation between $\mu\tau$ and defect density ($N_{db} \propto 1/\mu\tau$ [3]). They arrived at

$$\text{FF}(t) = \text{FF}_i - \frac{k}{3} \log\left(\frac{t}{t_i}\right) \tag{71}$$

where FF_i is the initial FF, k a kinetic constant for degradation, and t_i the initial time. Their data were fitted with $\text{FF}_i = 0.68$ and $k = 0.25$.

Catalano et al. [603] have introduced the device thickness into an empirical time dependence of the efficiency:

$$\eta = \eta_i \left[1.1 - K \log\left(\frac{t}{t_i}\right)\right] \tag{72}$$

where η_i is the initial efficiency, and K a rate constant. The removal of defects was found to be a temperature-activated process [604] represented by $K = K' \exp(E_A/k_B T)$, where K' is a constant increasing with the thickness ($\propto d^{0.54}$), and E_A an activation energy (0.2 eV [603]). At 35°C Eq. (72) can be rewritten as [603]

$$\eta = \eta_i \left[1.1 - 0.165 d^{0.54} \log\left(\frac{t}{t_i}\right)\right] \tag{73}$$

From the increase of the rate constant with thickness it is clear that thick cells will degrade deeper than thin cells. However, because the initial efficiency increases with thickness, an optimum thickness of 200–300 nm is found [577].

The degradation of solar cell properties can be circumvented by proper device design [11]. The parameter of interest here is the electric field profile after degradation, which should be optimized for carrier collection. A thinner intrinsic layer combined with enhanced light confinement leads to a higher electric field after degradation. Also, bandgap profiling may assist the carrier transport in the low-field region that is present in the intrinsic layer after degradation [605]. Further hydrogen dilution during deposition of the intrinsic layer has been reported to improve the stability [606].

Individual cells in multijunction cells are more stable, due to their reduced thickness compared to single-junction cells. Moreover, as the amount of light that is absorbed in the bottom cell is reduced and as the degradation rate is reduced at lower light levels, the stability is improved [606]. Therefore the multijunction cell is expected to be more stable than a single-junction cell of the same thickness, which indeed has been demonstrated [577, 606–608].

Outdoor operating temperatures of solar cells are around 60°C. At this temperature significant annealing of defects occurs [606], and the stability is better than at 25°C. Interestingly, seasonal measurements of the performance of solar cells show that the efficiency is higher in summer than in winter [609], which was attributed to annealing of defects [610]. However, it was subsequently reported that these seasonal effects are due to the spectral differences in summer and winter, rather than the increased operating temperature in summer [611, 612].

The research effort in many laboratories around the world on the optimization of laboratory-scale (1-cm^2 area) cells has led to the development of large-area (1 m^2) solar cell modules with an efficiency exceeding 10% [613, 614]. In a module individual cells can be connected in series or parallel, depending on the desired output voltage and current. Details on this topic can be found elsewhere, e.g., [11, 157, 577, 607, 615, 616].

11.2. Thin Film Transistors

11.2.1. Operation Principle

The most common a-Si:H TFT structure is the so-called inverted staggered transistor structure [40], in which silicon nitride is used as the gate insulator. A schematic cross section is shown in Figure 74. The structure comprises an a-Si:H channel, a gate dielectric (SiN$_x$), and source, drain, and gate contacts.

The operation principle of these TFTs is identical to that of the metal–oxide–semiconductor field-effect transistor (MOSFET) [617, 618]. When a positive voltage V_G is applied to the gate, electrons are accumulated in the a-Si:H. At small voltages these electrons will be localized in the deep states of the a-Si:H. The conduction and valence bands at the SiN$_x$–a-Si:H interface bend down, and the Fermi level shifts upward. Above a certain threshold voltage V_{th} a constant proportion of the electrons will be mobile, and the conductivity is increased linearly

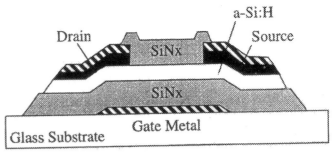

Fig. 74. Schematic cross section of an inverted staggered TFT structure.

with $V_G - V_{th}$. As a result the transistor switches on, and a current flows from source to drain. The source–drain current I_{SD} can be expressed as [619]

$$I_{SD} = \mu_{FE} C_G \frac{W}{L} \left[(V_G - V_{th}) - \frac{V_D}{2} \right] V_D \qquad (74)$$

where μ_{FE} is the effective carrier or field-effect mobility, C_G the gate capacitance, V_D the drain voltage, and W and L the channel width and length, respectively. Saturation of the source–drain current occurs when $dJ_{SD}/dV_D = 0$, and it is then expressed as

$$I_{SD,sat} = \mu_{FE} C_G \frac{W}{2L} (V_G - V_{th})^2 \qquad (75)$$

The maximum current in the on state is about 10^{-4} A, while the current at $V_G = 0$ is 10^{-11} A, which shows the large dynamic range. The threshold voltage is around 1 V, typically, and μ_{FE} can amount to 1 cm^2 V^{-1} s^{-1} [618].

At the interface of the nitride ($E_g = 5.3$ eV) and the a-Si:H the conduction and valence band line up. This results in band offsets. These offsets have been determined experimentally: the conduction band offset is 2.2 eV, and the valence band offset 1.2 eV [620]. At the interface a small electron accumulation layer is present under zero gate voltage, due to the presence of interface states. As a result, band bending occurs. The voltage at which the bands are flat (the flat-band voltage V_{FB}) is slightly negative.

11.2.2. Stability

TFTs are not perfectly stable. After prolonged application of a gate voltage, the threshold voltage starts to shift. Two mechanisms can account for this [621]. One is the charge trapping of electrons in the nitride, where the silicon dangling bond is the dominant electron trap. When electrons are transferred from the a-Si:H to the nitride, they can be trapped close to the interface, and a layer of fixed charge results. This fixed charge Q_F reduces the charge induced in the a-Si:H, and the transfer characteristic shifts by an amount Q_F/C_G. In the second mechanism deep defect states are created by the breaking of strained or weak Si—Si bonds. Electrons are trapped in these states and act as fixed negative charges, which results in band bending. This gives a decrease in source–drain current or a positive shift of V_{th} of several volts. It has been found that the time dependence of the threshold current follows a stretched exponential behavior [622]:

$$\Delta V_{th} = (V_{th}(\infty) - V_{th}(0)) \left[\left(\exp\left(-\frac{t}{t_0} \right)^{\beta} \right) - 1 \right] \quad (76)$$

where $V_{th}(\infty)$ and $V_{th}(0)$ are the threshold voltages at infinity and at $t = 0$, respectively. The factor β has been related to the time-dependent $(\propto t^{\beta-1})$ hydrogen diffusion coefficient in a-Si:H [623, 624].

HWCVD-deposited a-Si:H layers incorporated in inverted–staggered TFTs have been reported to result in stable behavior of the TFT [538, 539]. The threshold voltage does not shift upon prolonged application of a gate voltage. It has been suggested that not only the H content but also the H bonding structure is responsible for the difference in metastable character for PECVD and HWCVD layers [94, 96, 97, 625].

Another disadvantageous phenomenon in TFTs is the photoconductivity of a-Si:H [626]. Electrons and holes are photogenerated and recombine at the back surface (gate insulator). The photocurrent reduces the on/off ratio of the TFT. Illumination, however, cannot always be avoided, e.g., in active matrix displays. A way of circumventing this is to make the a-Si:H as thin as possible.

11.2.3. Fabrication

The TFTs are made on transparent glass substrates, onto which gate electrodes are patterned. Typically, the gate electrode is made of chromium. This substrate is introduced in a PECVD reactor, in which silane and ammonia are used for plasma deposition of SiN_x as the gate material. After subsequent deposition of the a-Si:H active layer and the heavily doped n-type a-Si:H for the contacts, the devices are taken out of the reactor. Cr contacts are evaporated on top of the structure. The transistor channel is then defined by etching away the top metal and n-type a-Si:H. Special care must be taken in that the etchant used for the n-type a-Si:H also etches the intrinsic a-Si:H. Finally the top passivation SiN_x is deposited in a separate run. This passivation layer is needed to protect the TFT during additional processing steps.

The thickness of the active layer is about 100–300 nm, while the source–drain distance (channel length) amounts to a few micrometers. The channel length is determined by the current requirements and usually exceeds 10 μm. Other manufacturing schemes as well as alternative structures are described elsewhere [619, 621]. Technology developments for the next generation TFTs that are to be used for high-resolution displays have been summarized by Katayama [627].

11.2.4. Application in Active Matrix Displays

One of the most important applications of TFTs is in active matrix addressing of liquid crystal displays (AMLCD) [619, 621]. The light transmission of a liquid crystal between two glass plates is controlled by the voltage on the pixel electrode. Nematic liquid crystals consist of rod-shaped molecules that align themselves parallel to an applied electrical field. When no field is present, they align themselves according to the surfaces in between which they are contained. By orienting the crystals, a change in polarization of the light passing through the crystals can be achieved. In this way, the crystals act as light switches. The voltage on the pixel electrode is controlled by the transistors. These act as switches that pass the voltage on their source to the drain, which is, the pixel electrode. When the gate is switched off, the pixel electrode is electrically isolated from the source, and the information remains on the pixel electrode. An $n \times m$ matrix of source and gate lines is able to address $n \times m$ pixels. This allows for high-definition large-area television displays. As an example, a 640×480 pixel display with three subpixels for each color consists of over one million transistors.

Active matrix addressing can also be used in printer heads [628] and linear sensor arrays [629]. a-Si:H TFTs are used to address a linear array of output or input transducers, respectively. In the linear sensor array, a-Si:H photodiodes are used.

11.3. Light Sensors

11.3.1. Linear Arrays

Linear arrays of a-Si:H photodiodes are widely used in optical page scanning applications such as fax machines and document scanners [630]. The large linear dimension of the array (as wide as the page to be scanned) allows a much simpler design. No optics are needed for image size reduction, which is required when a CCD camera is used as the detecting element. The matrix addressing is similar to that in AMLCD technology, but confined to one dimension. For each pixel a p–i–n photodiode is connected to a transistor. During illumination of the pixel, charge is transferred to the bottom electrode, and accumulates when the TFT is off. Upon switching on of the TFT, the charge will flow out and can be read by external electronic circuitry. In this way, the sensor integrates the signal during the time that it is not addressed. Readout times are very short in comparison with accumulation times.

11.3.2. Photoreceptors

a-Si:H photoconductors can be used as the light-sensitive components in electrophotographic or xerographic processes [39, 631, 632]. In contrast with conventional materials such as As_2Se_3, CdS, or ZnO, a-Si:H is nontoxic and provides a hard surface. Typically, a three-layer structure is used on top of a glass substrate: n-a-Si:H/i-a-Si:H/a-SiC:H.

The electrophotographic process consists of a number of steps [633]. First, the photoconductor is charged in the dark via a corona discharge. Then the photoconductor is illuminated by projecting the image to be copied onto the surface. In the areas that are exposed to light the material becomes conductive and charge flows to the substrate. The nonilluminated parts still contain charge. The photoconductor then is moved under a bias electrode, and an electric field is present in the gap between electrode and charged surface. Now, liquid toner (ink) is introduced in this gap, and the pigment particles from the toner drift to the surface of the photoconductor due to the electrostatic force. The last step is the transfer of the toner to paper, where it is fused by heat or pressure to make the image permanent.

A charged photoconductor discharges slowly in the dark, as is evident from the small decrease in time of the surface voltage. The discharge during illumination is much faster. When the discharge is not complete, a small residual voltage remains on the surface. A low dark current is required, which limits the dark photoconductivity of a-Si:H to $10^{-12}\ \Omega^{-1}\ cm^{-1}$. The residual voltage can be minimized by optimizing the $\mu\tau$ of the material.

The purpose of the n-type doped a-Si:H layer is to prevent injection of charge from the substrate into the photoconductor. Thus it serves as a blocking layer. Injection of surface charge into the photoconductor is prevented by the surface blocking layer. The a-SiC:H is of low quality, with a high midgap density of states. Surface charge will be trapped in these midgap states.

The corona discharges produces oxygen ions and ozone, which may react with the photoconductor [634]. As a means to circumvent possible degradation of the surface layer, an extra, protective thin layer was proposed, with high carbon content [101, 635, 636]. This would reduce silicon–oxygen reactions at the surface. Excellent electrophotographic characteristics have been obtained with a thin device comprising a 0.1-μm-thick n-type a-Si:H layer, a 1.0-μm intrinsic a-Si:H layer, a 0.1-μm undoped a-SiC$_{0.1}$:H layer, and a 0.014-μm undoped a-SiC$_{0.3}$:H layer [101].

11.3.3. Position Sensor

An a-Si:H-based position sensor consists of an intrinsic film sandwiched between two transparent conductive electrodes [637]. Two line contacts on the top are perpendicular to two on the bottom. When a light spot is incident on the device, carriers are generated, and a photocurrent flows to the contacts. The contacts form resistive dividers, so that from the ratio of the photocurrents the lateral position relative to the top or bottom contacts can be determined. The top contacts give the x-position, and the bottom contacts the y-position.

A p-i-n diode has been used on glass and on polyimid as a position-sensitive detector [638, 639]. The position of an incident light spot is measured by means of the lateral photovoltage.

11.3.4. Color Sensors

Similar to multijunction solar cells are color detectors, which have been designed to detect two [640, 641] or three [642–644] colors. A so-called adjustable threshold color detector (ATCD) consists of an a-Si:H–a-SiC:H p-i-n-i-p-i-n stack [642]. This ATCD discriminates between the three fundamental colors, red, green, and blue. As a result of proper thickness and bandgap design, the blue is absorbed in the top p-i-n junction, the green in the middle n-i-n junction, and the red in the bottom n-i-p junction. A change in polarity and value of the voltage selects the color that can be detected. The operation of the device is based on adjusting the photocurrent detection in the middle n-i-n junction.

A TCO/p-i-n-i-p/TCO/p-i-n/metal stack has been designed as a three-color sensor [643, 644]. An extra contact is made to the middle TCO. With appropriate bandgaps the peak detection is at 450, 530, and 635 nm for the blue, green, and red, respectively.

An active matrix of two-color a-Si:H photodetectors has been reported [645], where a n-i-p-i-n switching device is stacked on a two-color p-i-n-i-p structure.

11.3.5. IR Sensor

The absorption values of a-Si:H in the IR are much higher than of c-Si, due to transitions between extended states and deep levels. The IR sensitivity increases with increasing defect density. However, as the collection efficiency decreases drastically with increasing defect density, photodiodes cannot be used. Another approach is based on the change in photocapacitance due to IR radiation. A structure similar to a p-i-n a-Si:H solar cell has been used as an infrared detector [646]. The intrinsic layer in fact is a so-called *microcompensated* layer with very low concentrations of boron and phosphorus, but with a high defect density. In this layer the IR sensitivity is enhanced, due to transitions between extended states and hole traps. The increase in trapped charge modifies the electric field in the device, which is observed as a change in capacitance. A high responsivity is observed between 800 and 1400 nm, as well as between 3500 and 4500 nm.

11.3.6. X-Ray Sensor

High-energy radation can be imaged with a-Si:H, either directly or via a converter [3]. A thick film is required for direct detection, due to the weak interaction of the radiation with the material. A converter usually is a phosphor, which emits in the visible, and thin a-Si:H films are needed. X-rays with an energy up to 100 keV eject the electrons from the inner atomic core levels to high levels in the conduction band. The emitted electrons create electron–hole pairs due to ionization. These pairs can be detected in the same way as in p-i-n photodiodes.

X-ray imagers consist of a phosphor in direct contact with the surface of an array of a-Si:H photodiodes. The device is a matrix-addressed array, in which each imaging pixel consists of a photodiode and a TFT [647–649]. These X-ray imagers are very suitable for static and dynamic imaging in medical diagnosis.

11.4. Chemical Sensors

11.4.1. pH ISFET

An ion-sensitive field effect transistor (ISFET) with a reference electrode is similar to a MOSFET, with the gate exposed to an electrolyte. Selected ion concentrations can be measured. The threshold voltage of an ISFET depends on the potential of the electrolyte–insulator surface, which in turn is dependent on temperature and on the pH of the electrolyte. The threshold voltage has been found to shift down linearly by about 0.5 V on going from pH 7 to pH 1 [650]. Also, a drift of the gate voltage in time is found; it is small (1 mV/h) for low pH and large (7 mV/h) for high pH [651]. This is related to a small temporal change in capacitance of the insulator [652].

11.4.2. Hydrogen Sensor

A metal–insulator–semiconductor (MIS) device where palladium is used as the metal is sensitive to hydrogen. This sensitivity is a result of dissociation of hydrogen on the Pd surface, which induces an interface dipole layer that changes the barrier height and dark diode current (reverse bias current) [653]. Using a MIS structure with a-Si:H, i.e., glass/Cr/n-a-Si:H/ i-a-Si:H/SiO$_x$/Pd, Fortunato et al. [654] reported that the presence of 400 ppm of hydrogen leads to an increase of the reverse current of this device by three orders of magnitude. When these devices are illuminated during operation, 400 ppm of hydrogen induces a shift of the open circuit voltage to lower values. Hence, these photochemical sensors can operate as a gas sensor either in the dark or under illumination.

11.5. Other Applications

11.5.1. Loudspeakers

The capability of depositing a-Si:H uniformly over large areas has been of special interest for application in electrostatic loudspeakers, where the a-Si:H layer is suitable for the retention layer on the vibrating foils in the loudspeaker. One aims at obtaining a large-area thin film that can be either positively or negatively charged by applying a high voltage, but at the same time can hold large surface charge densities that do not displace laterally. Here, the high photosensitivity is not important, and in fact should be quenched for proper operation.

Electroacoustic transducers can be divided into electrodynamic and electrostatic transducers. The most commonly used loudspeakers convert electrical energy to acoustical energy by electrodynamical means: driving an electromagnetic coil that activates a diaphragm. At low frequencies, however, the diaphragm acts as a high-pass filter, with a cutoff frequency that depends on the diameter of the diaphragm. The mechanical construction of large-area loudspeakers adversely influences the amplitude and phase behavior, especially at high frequencies. These drawbacks usually are overcome by using different-size loudspeakers for different frequency ranges.

The electrostatic loudspeaker is used for high-end sound reproduction. The operation principle of an electrostatic loudspeaker is illustrated in Figure 75. Essentially, the electrostatic loudspeaker is a capacitor with air as the dielectric. One of the electrodes is a thin, electrically conducting foil (a in Fig. 75), and the other is a perforated flat fixed plate (b in Fig. 75) that

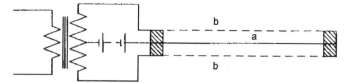

Fig. 75. Schematic drawing of an electrostatic loudspeaker: a, thin flexible foil coated with a-Si:H (or graphite); b, perforated flat fixed plates.

also is conductive. The electrodes are oppositely charged by a high voltage, 5 kV. Sound is produced by superimposing audio-frequency voltages that drive the movement of the foil [655].

The following issues need to be addressed. The amplitude of the foil motion should be very small to avoid nonlinear distortion. A large-area foil therefore is used. Further, the electrical force between the electrodes can be as large as 20 kV/cm, which may lead to electrical breakdown. Another issue is the presence of an electric field component over the surface of the foil, as a result of clamping the foil to the edges of the support. This field will vary with the amplitude of the foils, which leads to charge displacements and associated power dissipation. This effect especially may be present at low frequencies, where amplitudes are large.

The thin film should meet the following requirements. Areas in excess of 1 m^2 need to be deposited continuously, reproducibly, and homogeneously. The use of plastic foils sets a limit to the deposition temperature. The thin film should be capable of holding large surface charge concentrations, and have a large lateral resistivity. It should be mechanically stable, capable of accommodating mechanical deformations, and resistant to humidity. Usually, the foils are coated with a graphite layer as the semiconducting thin film. However, these layers are frequently unstable, and suffer from charge discplacement effects, which eventually lead to electrical breakdown [656].

Prototype electrostatic loudspeakers where the graphite is replaced by a-Si:H have been made, where a Mylar foil (area 10×10 cm^2, thickness 6 μm) is used [657]. Deposition of the a-Si:H layer was carried out in the ASTER deposition system. Uniform deposition (standard deviation of thickness, 1.5%) was achieved by diluting the SiH$_4$ with H$_2$ with SiH$_4$: H$_2$ = 1 : 2 [370]. The deposition was at room temperature. The hydrogen content amounted to 18 at.%, and the bandgap was 1.81 eV. The dark conductivity and AM1.5 photoconductivity were 7.5×10^{-9} and 1.8×10^{-8} Ω^{-1} cm^{-1}, respectively. In practice the film would not be exposed to light.

The frequency reponse characteristics of the foil with a-Si:H were measured in the range from 2.9 Hz to 5.46 kHz, by making use of generated pink noise with a continuous energy distribution. In Figure 76 the frequency response characteristics of the a-Si:H-coated foil are compared with those of a conventional graphite-coated foil of the same area. It can be concluded from this figure that the output of the a-Si:H-coated loudspeaker is lower than that of the graphite-coated one from 60 Hz upwards. In spite of the lower efficiency in this range, the a-Si:H-coated loadspeaker outperforms the graphite-coated one in the low range (10–60 Hz). The response is more extended towards low

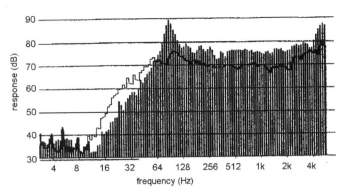

Fig. 76. Frequency response characteristics of a conventional, graphite-coated diaphragm (bar graph) and an *a*-Si:H-coated diaphragm (line graph).

frequencies, and the cutoff behavior (dB/octave) also seems improved. The frequency response is stable, even after prolonged application of low-frequency signals. The improvement was attributed to the reduced power dissipation at low frequencies due to the absence of charge displacement currents at the surface of the *a*-Si:H coated foil [657].

11.5.2. Erbium in a-Si:H

Erbium, one of the rare-earth materials, has been very important in the development of optical communication technology in the past decades. The trivalent erbium ion (Er^{3+}) has an incomplete $4f$ shell, which is shielded from the outside by $5s$ and $5p$ shells. Due to this, in erbium-doped materials a sharp optical transition from the first excited state to the ground state of Er^{3+} is observed, which occurs at an energy of 0.8 eV. The corresponding wavelength is 1.54 μm, which is of great importance in that standard silica-based optical fibers have the highest transparency at this wavelength. Indeed, Er-doped silica optical fibers have been demonstrated to operate as optical amplifiers at 1.54 μm [658, 659]. Planar amplifiers in which Er-doped channel waveguides are manufactured on a planar substrate have also been demonstrated [660]. These are of particular importance in that they can be integrated with other waveguide devices on a single chip. If efficient light emission from Er-doped silicon were possible, integration of optical and electrical functions on a single silicon chip would be within reach.

Due to the small emission and absorption cross sections of Er^{3+}, a high Er density is needed to reach reasonable values of optical gain. Typically Er densities are between 0.1% and 1.0% (10^{19}–10^{20} Er/cm^3). These values are far beyond the equilibrium solubility limits of Er in silicon. Therefore, nonequilibrium methods have to be used, such as ion implantation. Er implantation in crystalline silicon leads to amorphization, and additional annealing (600°C) is required to recrystallize the silicon. Optical activation of the Er may even require subsequent annealing at higher temperatures. Impurities such as oxygen or carbon have been found to enhance the luminescence. More can be found in the reviews on Er in silicon that have been published by Polman [661] and Priolo et al. [662, 663].

In this sub-subsection, the Er doping of amorphous silicon is discussed. The problem of limited solubility of Er in crystalline silicon has been circumvented. However, the electrical properties of pure *a*-Si are poor compared to *c*-Si. Therefore, hydrogenated amorphous silicon is much more interesting. Besides, the possibility of depositing *a*-Si:H directly on substrates, i.e., optical materials, would make integration possible. Both low-pressure chemical vapor deposition (LPCVD) [664] and PECVD [665, 666] have been used to make the *a*-Si:H into which Er is implanted. In both methods oxygen is intentionally added to the material, to enhance the luminescence.

The *a*-Si:H is deposited on Si(100) by LPCVD from SiH$_4$ and N$_2$O at 620°C, with hydrogen and oxygen contents of 10% and 31%, respectively. The *a*-Si:H layer was 340 nm thick. Erbium was implanted at 500 keV to a dose of 1×10^{15} Er/cm^2. The peak concentration of Er at a depth of 150 nm was 0.2%. Upon annealing at 400°C, room-temperature photoluminescence is observed [664], with the characteristic Er^{3+} peak at 1.54 μm. Temperature-dependent luminescence measurements in the range from 77 to 300 K show quenching of the peak luminescence by only a factor of 3. In the case of to Er-implanted crystalline silicon, this quenching is 10–100 times larger. Moreover, the luminescence intensity for the amorphous material is higher over the whole range of measured temperatures.

Luminescence lifetime measurements reveal a double-exponential decay, with lifetimes around 160 and 800 μs, independent of temperature in the range of 77–300 K. This behavior differs from the lifetime quenching as observed for oxygen- or nitrogen-codoped *c*-Si, where the lifetime is quenched by one to nearly three orders of magnitude [667, 668]. This has been attributed to nonradiative deexcitation of excited Er^{3+} that occurs at higher temperatures and depends on the bandgap of the material [669]: the larger the bandgap, the less quenching should be observed. Indeed, the large bandgap of the LPCVD *a*-Si:H (about 2 eV) seems to inhibit quenching.

With PECVD *a*-Si:H is deposited on a Corning glass substrate from SiH$_4$ at 230°C, employing the PASTA deposition system described by Madan et al. [159]. The hydrogen and oxygen contents were of 10% and 0.3%, respectively. The presence of this background concentration of oxygen probably was due to postponed maintenance; however, it was advantageous for this particular experiment. The *a*-Si:H layer was 250 nm thick. Erbium was implanted at 125 keV to a dose of 4×10^{14} Er/cm^2. The peak concentration of Er at a depth of 35 nm was 0.2%. In some samples additional oxygen was implanted at 25 keV to a dose of 7×10^{15} Er/cm^2, which resulted in an oxygen peak concentration of 1.7%, which overlapped the Er depth profile. Upon annealing at 400°C, room-temperature photoluminescence is observed [665, 666], with the characteristic Er^{3+} peak at 1.54 μm, as shown in Figure 77. The inset shows the behavior of the peak intensity as a function of annealing temperature. Temperatures between 300 and 400°C clearly are optimum annealing temperatures. Both the as-implanted samples and the samples annealed at 500°C show no erbium-related luminescence. In order to remove irradiation-induced

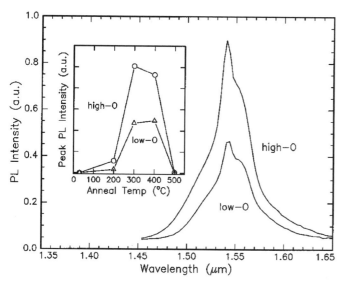

Fig. 77. Room-temperature photoluminescence spectra of Er-implanted PECVD a-Si:H, annealed at 400°C. The implantation energy and dose were 125 keV and 4×10^{14} Er/cm^2, respectively, which resulted in peak concentration of 0.2 at.%. "Low-O" and "high-O" denote a peak oxygen concentration in a-Si:H of 0.3 and 1.3 at.%, respectively. The inset shows the 1.54-μm peak intensity as a function of annealing temperature, for both oxygen concentrations. [From J. H. Shin, R. Serna, G. N. van den Hoven, A. Polman, W. G. J. H. M. van Sark, and A. M. Vredenberg, *Appl. Phys. Lett.* 68, 697 (1996), © 1996, American Institute of Physics, with permission.]

effects, high annealing temperatures are needed; however, that increases the outdiffusion of hydrogen.

Temperature-dependent luminescence measurements in the range from 77 to 300 K show quenching of the peak luminescence by a factor of about 15. Similar behavior is observed in the lifetime quenching [665, 666]. As the band gap of the PECVD a-Si:H is about 1.6 eV, nonradiative deexcitation of Er may occur at elevated temperatures. The amount of quenching lies in between that of c-Si and LPCVD a-Si:H, just like the bandgap.

Electroluminescent devices were made to demonstrate the possible application of these a-Si:H materials. Er-doped p–n diodes in c-Si show electroluminescence, both in forward and reverse bias [670–672]. Under forward bias the electroluminescence signal is attributed to electron–hole recombination, which results in excitation of the Er^{3+}. Under reverse bias, in particular beyond the Zener breakdown, the erbium is excited by impact excitation by hot electrons that are accelerated across the junction.

Incorporated in a device, the LPCVD a-Si:H material shows electroluminescence only in reverse bias [673]. The mechanism is similar to the one described for c-Si. The PECVD a-Si:H material was incorporated in a p–i–n solar cell structure, with a thickness of the intrinsic layer of 500 nm (see Section 11.1). Oxygen was coimplanted at 80 keV (3.2×10^{15} O/cm^2) and at 120 keV (5.5×10^{15} O/cm^2), which resulted in a roughly constant oxygen concentration of 1.0% in the Er projected range in the middle of the intrinsic a-Si:H layer. Electroluminescence is observed under forward bias [674].

Incorporation of Er *during* deposition of a-Si:H has also been achieved using molecular beam epitaxy [675], magnetron-assisted decomposition of silane [676], PECVD with a metal–organic source of solid erbium (tris(2,2,6,6-tetramethyl-2,5-heptadionato)Er(III) [677, 678] and tris(hexafluoroacetylacetate)Er(III)·1,2-dimethoxyethane [679]), electron cyclotron resonance PECVD in combination with sputtering of erbium [680], and catalytic CVD combined with pulsed laser ablation [681]. Explanations of the quenching behavior seem to be dependent on the sample preparation technique; recently, however, a comprehensive energy transfer model based on a Förster mechanism (resonant dipole coupling [682]) has been proposed [683].

11.5.3. Miscellaneous

Photodiode arrays have been used as retinal implants [684]. These arrays of p–i–n diodes are fabricated on a thin titanium layer bonded to a glass plate. The total thickness of this flexible structure is 1.5 μm. The microphotodiode array (MPDA) is used to replace photoreceptors (rods and cones) that have become defective due to disease.

Near-field optical microscopy (NSOM) [685, 686] has been used as a tool for nanolithography on the submicron length scale. Here, the fiber optic probe is used as a light source to expose a resist by scanning the surface. Dithering (oscillating the probe tip parallel to the surface) is needed to keep a constant probe–sample distance. As NSOM is not diffraction-limited, it is in principle possible to generate features with smaller dimensions (<100 nm [687, 688]) than those that are achievable with far-field optical lithography. Patterning of a-Si:H involves oxidizing the surface by local removal of hydrogen passivation. The thus formed oxide layer is the mask for subsequent etching. Herndon et al. [689] have reported on NSOM patterning of PECVD a-Si:H in open air, by using an Ar ion laser as the light source. Subsequent etching in a hydrogen plasma revealed an etch selectivity of 500 : 1. The feature widths are found to be dependent on the dither amplitude

A monolithic, amorphous-silicon-based, photovoltaic-powered electrochromic window has been reported by Gao et al. [690]. A thin, wide-bandgap a-SiC:H n–i–p solar cell on top of a glass substrate is used as a semitransparent power supply. An electrochromic device consisting of Li$_y$WO$_3$/LiAlF$_4$/V$_2$O$_5$ is deposited directly on top of the solar cell. The operation principle is based on the change of optical absorption by inserting and extracting Li$^+$ in and out of the WO$_3$ film. A prototype 16-cm^2 device is able to modulate the transmittance of the stack by more than 60% over a large part of the visible spectrum.

12. CONCLUSION

In this chapter the deposition of a-Si:H by PECVD has been described. The chapter covers material as well as discharge issues. It tries to relate material and discharge properties in various ways. Plasma modeling provides a means to study in detail the physical and chemical processes that occur in the plasma. The presented models show a high degree of sophistication, but from the comparison with experimental data it is clear that especially the deposition model needs improvement. Also, a full 2D model most probably is not needed, as differences between 1D and 2D modeling results are not very large.

Plasma analysis reveals information on the products of chemical processes, and can be used to good effect as a feedback to plasma modeling. The role of ions has been thoroughly illustrated, and the important result that ion bombardment with moderate energy is beneficial for material quality has been quantified.

The many modifications to the conventional RF PECVD method show that one still is trying to find methods that will in the end lead to improved material properties. This is especially the case for the intrinsic metastability of a-Si:H. In this respect, the stable material that is obtained at discharge conditions "at the edge of crystallinity" is very promising. Also, the quest for higher deposition rates while at least maintaining *device quality* material properties shows the industrial drive behind the research. Faster deposition allows for more solar cells to be produced in the same time.

In the three deposition methods PECVD, HWCVD, and ETP CVD one can see certain similarities. In one region in the reactor system, the generation of growth precursors takes place. For PECVD this occurs in the discharge bulk, for HWCVD at and around the wire, and for ETP CVD in the vicinity of the expanding beam. A transport region exists where precursors are transported to the substrate. For HWCVD and ETP CVD this region is between wire/source and substrate. For PECVD this region is the sheath. The third region is the deposition region, in all cases the substrate. Looking at these techniques from this perspective, one can conclude that the deposition mechanism of a-Si:H is universal. The best material quality is obtained when the SiH_3 radical is abundantly present.

In the last part of this chapter a summary was given of some applications. These are the driving force for the large research effort that continues. From these applications and the many more that were not treated here, it will be clear that a-Si:H is a material that is omnipresent and contributes greatly to our well-being.

Acknowledgments

It goes without saying that the work presented here could not have been done without the help and support of many people. First, many thanks go to John Bezemer, my supervisor, who taught me all there is to know about plasma deposition of a-Si:H, and now enjoys his retirement. Second, I gratefully acknowledge the colleagues with whom I performed and discussed numerous deposition and plasma analysis experiments: Hans Meiling, Edward Hamers, Stefania Acco, Martin van de Boogaard, Arjan Berntsen, Ernst Ullersma, René van Swaaij, Meine Kars, Gerard van der Mark, and students Renée Heller, Remco van der Heiden, Pieter van de Vliet, Micha Kuiper, Tjitske Kooistra, Jan-Wijtse Smit, Edward Prendergast, Tom Thomas, and Lenneke Slooff. Further, I would gratefully like to thank the following persons for various types of continuous support during my active time of research on a-Si:H deposition: Wim Arnold Bik, Frans Habraken, Arjen Vredenberg, Dirk Knoesen, Wim de Kruif, Albert Polman, Gerhard Landweer, Jeroen Daey Ouwens, Kees Feenstra, Maarten van Cleef, Jatindra Rath, Wim Sinke, Bert Slomp, Jeike Wallinga, Karine van der Werf, Ruud Schropp, and Werner van de Weg, who headed the group for over a decade.

I very much enjoyed the fruitful collaboration with Wim Goedheer and his group—Gert Jan Nienhuis, Peter Meijer, and Diederick Passchier—at the FOM Institute for Plasmaphysics "Rijnhuizen," and thank him for supplying me with published and additional data used in the section on plasma modeling.

Also, the biannual Dutch meetings on solar cell research always allowed for many discussions with colleagues from Eindhoven Technical University (Daan Schram, Gijs Meeusen, Richard van de Sanden) and Delft Technical University (Miro Zeman, Wim Metselaar).

Further I thank Hans Gerritsen and Gijs van Ginkel for allowing and stimulating me to write this chapter.

And last but not least, I thank my wife Joni, son Guido, and daughter Sophie for the enormous joy they give me every day.

Parts of the work presented here were financially supported by the Netherlands Agency for Energy and the Environment (NOVEM), the Foundation for Fundamental Research on Matter (FOM), the Royal Netherlands Academy of Art and Sciences (KNAW), the Netherlands Technology Foundation (STW), and the Netherlands Organization for Scientific Research (NWO).

REFERENCES

1. J. I. Pankove, Ed., "Hydrogenated Amorphous Silicon, Semiconductors and Semimetals, Vol. 21, Parts A-D." Academic Press, Orlando, FL, 1984.
2. K. Tanaka, Ed., "Glow-Discharge Hydrogenated Amorphous Silicon." KTK Scientific Publishers, Tokyo, 1989.
3. R. A. Street, "Hydrogenated Amorphous Silicon" Cambridge Solid State Science Series. Cambridge University Press, Cambridge, U.K., 1991.
4. J. Kanicki, Ed., "Amorphous and Microcrystalline Semiconductor Devices—Optoelectronic Devices." Artech House, Norwood, MA, 1991.
5. J. Kanicki, Ed., "Amorphous and Microcrystalline Semiconductor Devices—Materials and Device Physics." Artech House, Norwood, MA, 1992.
6. W. Luft and Y. S. Tsuo, "Hydrogenated Amorphous Silicon Alloy Deposition Processes." Marcel Dekker, New York, 1993.
7. R. F. Bunshah, Ed., "Handbook of Deposition Technologies for Films and Coatings—Science, Technology and Applications." Second Edition. Noyes Publications, Park Ridge, NJ, 1994.
8. G. Bruno, P. Capezzuto, and A. Madan, Eds., "Plasma Deposition of Amorphous Silicon-Based Materials." Academic Press, Boston, 1995.
9. E. S. Machlin, "Materials Science in Microelectronics—the Relationships between Thin Film Processing and Structure." Giro Press, Croton-on-Hudson, NY, 1995.
10. E. S. Machlin, "Materials Science in Microelectronics—Volume 2, The Effects of Structure on Properties in Thin Films." Giro Press, Croton-on-Hudson, NY, 1998.

11. R. E. I. Schropp and M. Zeman, "Amorphous and Microcrystalline Silicon Solar Cells—Modeling, Materials and Device Technology." Kluwer Academic Publishers, Boston, 1998.

12. T. M. Searle, Ed., "Properties of Amorphous Silicon and its Alloys, EMIS Datareview Series, No. 19." IEE, London, 1998.

13. R. A. Street, Ed., "Technology and Applications of Amorphous Silicon." Springer-Verlag, New York, 2000.

14. C. A. M. Stap, H. Meiling, G. Landweer, J. Bezemer, and W. F. van der Weg, "Proceedings of the Ninth E.C. Photovoltaic Solar Energy Conference, Freiburg, F.R.G., 1989" (W. Palz, G. T. Wrixon, and P. Helm, Eds.), p. 74. Kluwer Academic, Dordrecht, 1989.

15. H. Meiling, Ph.D. Thesis, Universiteit Utrecht, Utrecht, 1991.

16. H. Curtins, N. Wyrsch, and A. V. Shah, *Electron. Lett.* 23, 228 (1987).

17. H. Shirai, J. Hanna, and I. Shimizu, *Japan. J. Appl. Phys.* 30, L881 (1991).

18. J. T. Verdeyen, L. Beberman, and L. Overzet, *J. Vac. Sci. Technol. A* 8, 1851 (1990).

19. A. H. Mahan, J. Carapella, B. P. Nelson, R. S. Crandall, and I. Balberg, *J. Appl. Phys.* 69, 6728 (1991).

20. G. M. W. Kroesen, D. C. Schram, and M. J. van de Sande, *Plasma Chem. Plasma Proc.* 10, 49 (1990).

21. G. J. Meeusen, E. A. Ershov-Pavlov, R. F. G. Meulenbroeks, M. C. M. van de Sanden, and D. C. Schram, *J. Appl. Phys.* 71, 4156 (1992).

22. H. Overhof, *J. Non-cryst. Solids* 227–230, 15 (1998).

23. J. Tauc, Ed., "Amorphous and Liquid Semiconductors." Plenum, New York, 1974.

24. A. J. Lewis, G. A. N. Connell, W. Paul, J. R. Pawlik, and R. J. Temkin, "Tetrahedrally Bonded Amorphous Semiconductors," AIP Conference Proceedings, Vol. 20 (M. H. Brodsky, S. Kirkpatrick, and D. Weaire, Eds.), p. 27. American Institute of Physics, New York, 1974.

25. E. C. Freeman and W. Paul, *Phys. Rev. B* 18, 4288 (1978).

26. W. Paul and H. Ehrenreich, *Phys. Today* 38 (8), 13 (1985).

27. H. F. Sterling and R. C. G. Swann, *Solid State Electron.* 8, 653 (1965).

28. R. C. Chittick, J. H. Alexander, and H. F. Sterling, *J. Electrochem. Soc.* 116, 77 (1969).

29. A. Triska, D. Dennison, and H. Fritzsche, *Bull. Am. Phys. Soc.* 20, 392 (1975).

30. H. Fritzsche, "Proceedings of the 7th International Conference on Amorphous and Liquid Semiconductors" (W. E. Spear, Ed.), p. 3. CICL, Edinburgh, 1977.

31. H. Fritsche, *Mater. Res. Soc. Proc. Symp.* 609, A17.1.1 (2000).

32. W. Spear and P. LeComber, *Solid State Commun.* 17, 1193 (1975).

33. I. Solomon, *Mater. Res. Soc. Proc. Symp.* 609, A17.3.1 (2000).

34. D. E. Carlson and C. R. Wronski, *Appl. Phys. Lett.* 28, 671 (1976).

35. H. Okamoto, Y. Nitta, T. Yamaguchi, and Y. Hamakawa, *Sol. Energy Mater.* 3, 313 (1980).

36. Y. Hamakawa, *Mater. Res. Soc. Proc. Symp.* 609, A17.2.1 (2000).

37. Y. Kuwano, T. Imai, M. Ohnishi, and S. Nakano, "Proceedings of the 14th IEEE Photovoltaic Specialists Conference, San Diego, 1980," p. 1408. IEEE, New York, 1980.

38. I. Shimizu, T. Komatsu, T. Saito, and E. Inoue, *J. Non-cryst. Solids* 35 & 36, 773 (1980).

39. S. Oda, Y. Saito, I. Shimizu, and E. Inoue, *Philos. Mag. B* 43, 1079 (1981).

40. P. G. LeComber, W. E. Spear, and A. Ghaith, *Electron. Lett.* 15, 179 (1979).

41. A. J. Snell, K. D. MacKenzie, W. E. Spear, P. G. LeComber, and A. J. Hughes, *J. Appl. Phys.* 24, 357 (1981).

42. M. J. Powell, B. C. Easton, and O. F. Hill, *Appl. Phys. Lett.* 38, 794 (1981).

43. R. L. Weisfield, *J. Non-cryst. Solids* 164–166, 771 (1993).

44. S. N. Kaplan, J. R. Morel, T. A. Mulera, V. Perez-Mendez, G. Schlurmacher, and R. A. Street, *IEEE Trans. Nucl. Sci.* NS-33, 351 (1986).

45. T. C. Chuang, L. E. Fennel, W. B. Jackson, J. Levine, M. J. Thompson, H. C. Tuan, and R. Weisfield, *J. Non-cryst. Solids* 97 & 98, 301 (1987).

46. L. E. Antonuk, C. F. Wild, J. Boundry, J. Jimenez, M. J. Longo, and R. A. Street, *IEEE Trans. Nucl. Sci.* NS-37, 165 (1990).

47. C. van Berkel, N. C. Bird, C. J. Curling, and I. D. French, *Mater. Res. Soc. Symp. Proc.* 297, 939 (1993).

48. W. H. Zachariasen, *J. Am. Chem. Soc.* 54, 3841 (1932).

49. D. E. Polk, *J. Non-cryst. Solids* 5, 365 (1971).

50. M. A. Paesler, D. E. Sayers, R. Tsu, and J. G. Hernandez, *Phys. Rev. B* 28, 4550 (1983).

51. D. Beeman, R. Tsu, and M. F. Thorpe, *Phys. Rev. B* 32, 874 (1985).

52. R. A. Street, *Phys. Rev. Lett.* 49, 1187 (1982).

53. N. F. Mott, *Philos. Mag.* 19, 835 (1969).

54. W. M. Arnold Bik, C. T. A. M. de Laat, and F. H. P. M. Habraken, *Nucl. Instrum. Methods Phys. Res. B* 64, 832 (1992).

55. M. H. Brodsky, *Thin Solid Films* 40, L23 (1977).

56. G. Lucovsky, R. J. Nemanich, and J. C. Knights, *Phys. Rev. B* 19, 2064 (1979).

57. H. Shanks, C. J. Fang, L. Ley, M. Cardona, F. J. Demond, and S. Kalbitzer, *Phys. Status Solidi B* 100, 43 (1980).

58. A. A. Langford, M. L. Fleet, B. P. Nelson, W. A. Lanford, and N. Maley, *Phys. Rev. B* 45, 13367 (1992).

59. W. Beyer and M. S. Abo Ghazala, *Mater. Res. Soc. Symp. Proc.* 507, 601 (1998).

60. A. H. Mahan, P. Raboisson, and R. Tsu, *Appl. Phys. Lett.* 50, 335 (1987).

61. D. A. G. Bruggeman, *Ann. Phys. (Leipzig)* 24, 636 (1935).

62. A. H. Mahan, D. L. Williamson, B. P. Nelson, and R. S. Crandall, *Sol. Cells* 27, 465 (1989).

63. H. Meiling, M. J. van den Boogaard, R. E. I. Schropp, J. Bezemer, and W. F. van der Weg, *Mater. Res. Soc. Symp. Proc.* 192, 645 (1990).

64. M. J. van den Boogaard, S. J. Jones, Y. Chen, D. L. Williamson, R. A. Hakvoort, A. van Veen, A. C. van der Steege, W. M. Arnold Bik, W. G. J. H. M. van Sark, and W. F. van der Weg, *Mater. Res. Soc. Symp. Proc.* 258, 407 (1992).

65. M. J. van den Boogaard, Ph.D. Thesis, Universiteit Utrecht, Utrecht, 1992.

66. S. Acco, D. L. Williamson, W. G. J. H. M. van Sark, W. C. Sinke, W. F. van der Weg, A. Polman, and S. Roorda, *Phys. Rev. B* 58, 12853 (1998).

67. A. H. Mahan, D. L. Williamson, B. P. Nelson, and R. S. Crandall, *Phys. Rev. B* 40, 12024 (1989).

68. P. C. Taylor, in "Semiconductors and Semimetals" (J. I. Pankove, Ed.), Vol. 21C, p. 99. Academic Press, Orlando, FL, 1984.

69. S. Acco, D. L. Williamson, P. A. Stolk, F. W. Saris, M. J. van den Boogaard, W. C. Sinke, W. F. van der Weg, and S. Roorda, *Phys. Rev. B* 53, 4415 (1996).

70. S. Acco, W. Beyer, E. E. Van Faassen, and W. F. van der Weg, *J. Appl. Phys.* 82, 2862 (1997).

71. R. S. Crandall, X. Liu, and E. Iwaniczko, *J. Non-cryst. Solids* 227–230, 23 (1998).

72. S. Acco, D. L. Williamson, S. Roorda, W. G. J. H. M. van Sark, A. Polman, and W. F. van der Weg, *J. Non-cryst. Solids* 227–230, 128 (1998).

73. P. J. Zanzucchi, C. R. Wronski, and D. E. Carlson, *J. Appl. Phys.* 48, 5227 (1977).

74. G. D. Cody, B. Abeles, C. Wronski, C. R. Stephens, and B. Brooks, *Sol. Cells* 2, 227 (1980).

75. J. Tauc, in "Optical Properties of Solids" (F. Abelès, Ed.), Chap. 5, p. 277. North-Holland, Amsterdam, the Netherlands, 1972.

76. R. H. Klazes, M. H. L. M. van den Broek, J. Bezemer, and S. Radelaar, *Philos. Mag. B* 45, 377 (1982).

77. T. Tiedje, J. M. Cebulka, D. L. Morel, and B. Abeles, *Phys. Rev. Lett.* 46, 1425 (1981).

78. K. Pierz, B. Hilgenberg, H. Mell, and G. Weiser, *J. Non-cryst. Solids* 97/98, 63 (1987).

79. K. Pierz, W. Fuhs, and H. Mell, *Philos. Mag. B* 63, 123 (1991).

80. N. Maley and J. S. Lannin, *Phys. Rev. B* 36, 1146 (1987).

81. R. C. Ross and J. Jaklik, *J. Appl. Phys.* 55, 3785 (1984).

82. H. Meiling, W. Lenting, J. Bezemer, and W. F. van der Weg, *Philos. Mag. B* 62, 19 (1990).

83. R. S. Crandall, *Sol. Cells* 24, 237 (1988).

84. A. J. M. Berntsen, W. F. van der Weg, P. A. Stolk, and F. W. Saris, *Phys. Rev. B* 48, 14656 (1993).

85. A. J. M. Berntsen, Ph.D. Thesis, Universiteit Utrecht, Utrecht, 1993.

86. W. E. Spear and P. G. LeComber, *Adv. Phys.* 26, 811 (1977).

87. D. L. Staebler and C. R. Wronski, *Appl. Phys. Lett.* 31, 292 (1977).

88. D. L. Staebler and C. R. Wronski, *J. Appl. Phys.* 51, 3262 (1980).

89. M. Stutzmann, W. B. Jackson, and C. C. Tsai, *Phys. Rev. B* 32, 23 (1985).

90. T. Kamei, N. Hata, A. Matsuda, T. Uchiyama, S. Amano, K. Tsukamoto, Y. Yoshioka, and T. Hirao, *Appl. Phys. Lett.* 68, 2380 (1996).

91. C. C. Tsai, J. C. Knights, R. A. Lujan, W. B. B. L. Stafford, and M. J. Thompson, *J. Non-cryst. Solids* 59 & 60, 731 (1983).

92. S. Yamasaki and J. Isoya, *J. Non-cryst. Solids* 164–166, 169 (1993).

93. P. Stradins and H. Fritzsche, *Philos. Mag. B* 69, 121 (1994).

94. C. Godet, P. Morin, and P. Roca i Cabarrocas, *J. Non-cryst. Solids* 198–200, 449 (1996).

95. J. Daey Ouwens and R. E. I. Schropp, *Phys. Rev. B* 54, 17759 (1996).

96. S. Zafar and E. A. Schiff, *Phys. Rev. B* 40, 5235 (1989).

97. C. Manfredotti, F. Fizzotti, M. Boero, P. Pastorino, P. Polesello, and E. Vittone, *Phys. Rev. B* 50, 18046 (1994).

98. H. M. Branz, *Solid State Commun.* 105, 387 (1998).

99. R. Biswas, Q. Li, B. C. Pan, and Y. Yoon, *Mater. Res. Soc. Symp. Proc.* 467, 135 (1997).

100. D. A. Anderson and W. E. Spear, *Philos. Mag.* 35, 1 (1977).

101. R. A. C. M. M. van Swaaij, Ph.D. Thesis, Universiteit Utrecht, Utrecht, 1994.

102. R. A. C. M. M. van Swaaij, A. J. M. Berntsen, W. G. J. H. M. van Sark, H. Herremans, J. Bezemer, and W. F. van der Weg, *J. Appl. Phys.* 76, 251 (1994).

103. J. Chevallier, H. Wieder, A. Onton, and C. R. Guarnieri, *Solid State Commun.* 24, 867 (1977).

104. S. Wagner, V. Chu, J. P. Conde, and J. Z. Liu, *J. Non-cryst. Solids* 114, 453 (1989).

105. C. M. Fortmann, in "Plasma Deposition of Amorphous Silicon-Based Materials" (G. Bruno, P. Capezzuto, and A. Madan, Eds.), Chap. 3, p. 131. Academic Press, Boston, 1995.

106. W. Paul, S. J. Jones, F. C. Marques, D. Pang, W. A. Turner, A. E. Wetsel, P. Wickboldt, and J. H. Chen, *Mater. Res. Soc. Symp. Proc.* 219, 211 (1991).

107. A. Matsuda and K. Tanaka, *J. Non-cryst. Solids* 97, 1367 (1987).

108. G. Bruno, P. Capezzuto, G. Cicala, and F. Cramarossa, *J. Mater. Res.* 4, 366 (1989).

109. G. Bruno, P. Capezzuto, M. Losurdo, P. Manodoro, and G. Cicala, *J. Non-cryst. Solids* 137&138, 799 (1991).

110. H. Herremans, W. Grevendonk, R. A. C. M. M. van Swaaij, W. G. J. H. M. van Sark, A. J. M. Berntsen, W. M. Arnold Bik, and J. Bezemer, *Philos. Mag. B* 66, 787 (1992).

111. M. P. Schmidt, I. Solomon, H. Tran-Quoc, and J. Bullot, *J. Non-cryst. Solids* 77–78, 849 (1985).

112. I. Solomon, M. P. Schmidt, and H. Tran-Quoc, *Phys. Rev. B* 38, 9895 (1988).

113. M. N. P. Carreño, I. Pereyra, M. C. A. Fantini, H. Takahashi, and R. Landers, *J. Appl. Phys.* 75, 538 (1994).

114. W. M. Arnoldbik and F. H. P. M. Habraken, *Rep. Prog. Phys.* 56, 859 (1993).

115. M. L. Oliveira and S. S. Camargo, Jr., *J. Appl. Phys.* 71, 1531 (1992).

116. F. Evangelisti, *J. Non-cryst. Solids* 164–166, 1009 (1993).

117. A. Grill, "Cold Plasma in Materials Fabrication—from Fundamentals to Applications." IEEE Press, Piscataway, NJ, 1994.

118. J. F. O'Hanlon, "A User's Guide to Vacuum Technology." Second Edition. Wiley, New York, 1989.

119. J. Bezemer, W. G. J. H. M. van Sark, M. B. von der Linden, and W. F. van der Weg, "Proceedings of the Twelfth E.C. Photovoltaic Solar Energy Conference, Amsterdam, the Netherlands, 1994" (R. Hill, W. Palz, and P. Helm, Eds.), p. 327. H.S. Stephens & Associates, Bedford, U.K., 1994.

120. J. Bezemer and W. G. J. H. M. van Sark, "Electronic, Optoelectronic and Magnetic Thin Films, Proceedings of the 8th International School on Condensed Matter Physics (ISCMP), Varna, Bulgaria, 1994," p. 219. Wiley, New York, 1995.

121. M. Böhm, A. E. Delahoy, F. B. Ellis, Jr., E. Eser, S. C. Gau, F. J. Kampas, and Z. Kiss, "Proceedings of the 18th IEEE Photovoltaic Specialists Conference, Las Vegas, 1985," p. 888. IEEE, New York, 1985.

122. C. R. Dickson, J. Pickens, and A. Wilczynski, *Sol. Cells* 19, 179 (1986).

123. K. Tanaka and A. Matsuda, *Mater. Sci. Rep.* 2, 139 (1987).

124. F. Jansen and D. Kuhman, *J. Vac. Sci. Technol.* 6, 13 (1988).

125. W. Y. Kim, H. Tasaki, M. Konagai, and K. Takahashi, *J. Appl. Phys.* 61, 3071 (1987).

126. A. Catalano and G. Wood, *J. Appl. Phys.* 63, 1220 (1988).

127. Y. Kuwano, M. Ohnishi, S. Tsuda, Y. Nakashima, and N. Nakamura, *Japan. J. Appl. Phys.* 21, 413 (1982).

128. R. A. Street, J. Zesch, and M. J. Thompson, *Appl. Phys. Lett.* 43, 672 (1983).

129. M. Ohnishi, H. Nishiwaki, E. Enomoto, Y. Nakashima, S. Tsuda, T. Takahama, H. Tarui, M. Tanaka, H. Dojo, and Y. Kuwano, *J. Non-cryst. Solids* 59/60, 1107 (1983).

130. S. Nakano, S. Tsuda, H. Tarui, T. Takahama, H. Haku, K. Watanabe, M. Nishikuni, Y. Hishikawa, and Y. Kuwano, *Mater. Res. Soc. Symp. Proc.* 70, 511 (1986).

131. S. Tsuda, T. Takahama, M. Isomura, H. Tarui, Y. Nakashima, Y. Hishikawa, N. Nakamura, T. Matsuoka, H. Nishiwaki, S. Nakano, M. Ohnishi, and Y. Kuwano, *Japan. J. Appl. Phys.* 26, 33 (1987).

132. W. Luft, Ed., "Photovoltaic Safety," AIP Conference Proceedings, Vol. 166. American Institute of Physics, New York, 1988.

133. P. J. Hargis, Jr., K. E. Greenberg, P. A. Miller, J. B. Gerardo, J. R. Torczynski, M. E. Riley, G. A. Hebner, J. R. Roberts, J. K. Olthoff, J. R. Whetstone, R. J. Van Brunt, M. A. Sobolewski, H. M. Anderson, M. P. Splichal, J. L. Mock, P. Bletzinger, A. Garscadden, R. A. Gottscho, G. Selwyn, M. Dalvie, J. E. Heidenreich, J. W. Butterbaugh, M. L. Brake, M. L. Passow, J. Pender, A. Lujan, M. E. Elta, D. B. Graves, H. H. Sawin, M. J. Kushner, J. T. Verdeyen, R. Horwath, and T. R. Turner, *Rev. Sci. Instrum.* 65, 140 (1994).

134. B. Chapman, "Glow Discharge Processes—Sputtering and Plasma Etching." Wiley, New York, 1980.

135. A. Gallagher, *Mater. Res. Soc. Symp. Proc.* 70, 3 (1986).

136. J. Perrin, Y. Takeda, N. Hirano, Y. Takeuchi, and A. Matsuda, *Surf. Sci.* 210, 114 (1989).

137. A. Matsuda, K. Nomoto, Y. Takeuchi, A. Suzuki, A. Yuuki, and J. Perrin, *Surf. Sci.* 227, 50 (1990).

138. A. Matsuda and K. Tanaka, *J. Appl. Phys.* 60, 2351 (1986).

139. T. Ichimura, T. Ihara, T. Hama, M. Ohsawa, H. Sakai, and Y. Uchida, *Japan. J. Appl. Phys.* 25, L276 (1986).

140. T. Hamasaki, M. Ueda, A. Chayahara, M. Hirose, and Y. Osaka, *Appl. Phys. Lett.* 44, 600 (1984).

141. T. Hamasaki, M. Ueda, A. Chayahara, M. Hirose, and Y. Osaka, *Appl. Phys. Lett.* 44, 1049 (1984).

142. P. Roca i Cabarrocas, J. B. Chevrier, J. Huc, A. Lloret, J. Y. Parey, and J. P. M. Schmitt, *J. Vac. Sci. Technol. A* 9, 2331 (1991).

143. M. Taniguchi, M. Hiorse, T. Hamasaki, and Y. Osaka, *Appl. Phys. Lett.* 37, 787 (1980).

144. M. Ohnishi, H. Nishiwaki, K. Uchihashi, K. Yoshida, M. Tanaka, K. Ninomiya, M. Nishikuni, N. Nakanura, S. Tsuda, S. Nakano, T. Yazaki, and Y. Kuwano, *Japan. J. Appl. Phys.* 27, 40 (1988).

145. F. N. Boulitrop, N. Proust, J. Magarino, E. Criton, J. F. Peray, and M. Dupre, *J. Appl. Phys.* 58, 3494 (1985).

146. H. Chatham and P. K. Bhat, *Mater. Res. Soc. Symp. Proc.* 149, 447 (1989).

147. H. Chatham, P. Bhat, A. Benson, and C. Matovich, *J. Non-cryst. Solids* 115, 201 (1989).

148. P. M. Meijer, J. D. P. Passchier, W. J. Goedheer, J. Bezemer, and W. G. J. H. M. van Sark, *Appl. Phys. Lett.* 64, 1780 (1994).

149. W. J. Goedheer, P. M. Meijer, J. Bezemer, J. D. P. Passchier, and W. G. J. H. M. van Sark, *IEEE Trans. Plasma Sci.* PS-23, 644 (1995).

150. W. G. J. H. M. van Sark, H. Meiling, J. Bezemer, M. B. von der Linden, R. E. I. Schropp, and W. F. van der Weg, *Sol. Energy Mater. Sol. Cells* 45, 57 (1997).

151. H. Meiling, W. G. J. H. M. van Sark, J. Bezemer, R. E. I. Schropp, and W. van der Weg, "Proceedings of the 25th IEEE Photovoltaic Specialists Conference, Washington, D.C., 1996," p. 1153. IEEE, New York, 1996.

152. J. P. M. Schmitt, J. Meot, P. Roubeau, and P. Parrens, "Proceedings of the Eighth E.C. Photovoltaic Solar Energy Conference, Florence, Italy, 1988" (I. Solomon, B. Equer, and P. Helm, Eds.), p. 964. Kluwer Academic, Dordrecht, 1989.

153. S. Tsuda, S. Sakai, and S. Nakano, *Appl. Surf. Sci.* 113/114, 734 (1997).

154. J. Yang, A. Banerjee, S. Sugiyama, and S. Guha, *Appl. Phys. Lett.* 70, 2975 (1997).

155. R. E. I. Schropp, C. H. M. van der Werf, M. Zeman, M. C. M. van de Sanden, C. I. M. A. Spee, E. Middelman, L. V. de Jonge-Meschaninova, A. A. M. van der Zijden, M. M. Besselink, R. J. Severens, J. Winkeler, and G. J. Jongerden, *Mater. Res. Soc. Symp. Proc.* 557, 713 (1999).

156. S. Fujikake, K. Tabuchi, T. Yoshida, Y. Ichikawa, and H. Sakai, *Mater. Res. Soc. Symp. Proc.* 377, 609 (1995).

157. T. Yoshida, S. Fujikake, S. Kato, M. Tanda, K. Tabuchi, A. Takano, Y. Ichikawa, and H. Sakai, *Sol. Energy Mater. Sol. Cells* 48, 383 (1997).

158. E. Middelman, E. van Andel, P. M. G. M. Peters, L. V. de Jonge-Meschaninova, R. J. Severens, G. J. Jongerden, R. E. I. Schropp, H. Meiling, M. Zeman, M. C. M. van de Sanden, A. Kuipers, and C. I. M. A. Spee, "Proceedings of the 2nd World Conference on Photovoltaic Energy Conversion, Vienna, Austria" (J. Schmid, H. A. Ossenbrink, P. Helm, H. Ehmann, and E. D. Dunlop, Eds.), p. 816. European Commision Joint Research Centre, Ispra, Italy, 1998.

159. A. Madan, P. Rava, R. E. I. Schropp, and B. von Roedern, *Appl. Surf. Sci.* 70/71, 716 (1993).

160. W. G. J. H. M. van Sark, J. Bezemer, M. Kars, M. Kuiper, E. A. G. Hamers, and W. F. van der Weg, "Proceedings of the Twelfth E.C. Photovoltaic Solar Energy Conference, Amsterdam, the Netherlands, 1994" (R. Hill, W. Palz, and P. Helm, Eds.), p. 350. H. S. Stephens & Associates, Bedford, U.K., 1994.

161. E. A. G. Hamers, W. G. J. H. M. van Sark, J. Bezemer, W. J. Goedheer, and W. F. van der Weg, *Int. J. Mass Spectrom. Ion Processes* 173, 91 (1998).

162. W. G. J. H. M. van Sark, J. Bezemer, and W. J. Goedheer, unpublished results.

163. E. A. G. Hamers, Ph.D. Thesis, Universiteit Utrecht, Utrecht, 1998.

164. H. Shirai, D. Das, J. Hanna, and I. Shimizu, *Appl. Phys. Lett.* 59, 1096 (1991).

165. W. G. J. H. M. van Sark, J. Bezemer, P. G. van de Vliet, M. Kars, and W. F. van der Weg, "Proceedings of the Twelfth E.C. Photovoltaic Solar Energy Conference, Amsterdam, the Netherlands, 1994" (R. Hill, W. Palz, and P. Helm, Eds.), p. 335. H.S. Stephens & Associates, Bedford, U.K., 1994.

166. H. Kausche, M. Möller, and R. Plättner, "Proceedings of the Fifth E.C. Photovoltaic Solar Energy Conference, Athens, Greece, 1983." Reidel, Dordrecht, 1983.

167. A. Morimoto, M. Matsumoto, M. Kumeda, and T. Shimizu, *Japan. J. Appl. Phys.* 29, L1747 (1990).

168. A. Ricard, "Reactive Plasmas." Société Française du Vide, Paris, 1996.

169. H. S. Butler and G. S. Kino, *Phys. Fluids* 6, 1346 (1963).

170. K. Köhler, J. W. Coburn, D. E. Horne, E. Kay, and J. H. Keller, *J. Appl. Phys.* 57, 59 (1985).

171. M. A. Liebermann and S. E. Savas, *J. Vac. Sci. Technol. A* 16, 1632 (1990).

172. M. V. Alves, M. A. Liebermann, V. Vahedi, and C. K. Birdsall, *J. Appl. Phys.* 69, 3823 (1991).

173. J. Perrin, in "Plasma Deposition of Amorphous Silicon-Based Materials" (G. Bruno, P. Capezzuto, and A. Madan, Eds.), Chap. 4, p. 177. Academic Press, Boston, 1995.

174. P. M. Meijer and W. J. Goedheer, *IEEE Trans. Plasma Sci.* PS-19, 170 (1991).

175. P. M. Meijer, Ph.D. Thesis, Universiteit Utrecht, Utrecht, 1991.

176. K. Köhler, D. E. Horne, and J. W. Coburn, *J. Appl. Phys.* 58, 3350 (1985).

177. A. D. Kuypers and H. J. Hopman, *J. Appl. Phys.* 63, 1894 (1988).

178. W. J. Goedheer, private communication.

179. J. D. P. Passchier, Ph.D. Thesis, Universiteit Utrecht, Utrecht, 1994.

180. S. Rauf and M. J. Kushner, *J. Appl. Phys.* 83, 5087 (1998).

181. V. A. Godyak and R. B. Piejak, *J. Vac. Sci. Technol. A* 8, 3833 (1990).

182. C. M. Horwitz, *J. Vac. Sci. Technol. A* 1, 1795 (1983).

183. F. Finger, U. Kroll, V. Viret, A. Shah, W. Beyer, X.-M. Tang, J. Weber, A. A. Howling, and C. Hollenstein, *J. Appl. Phys.* 71, 5665 (1992).

184. C. Böhm and J. Perrin, *J. Phys. D* 24, 865 (1991).

185. K. M. H. Maessen, Ph.D. Thesis, Universiteit Utrecht, Utrecht 1988.

186. M. J. M. Pruppers, Ph.D. Thesis, Universiteit Utrecht, Utrecht 1988.

187. S. Ishihara, K. Masatoshi, H. Takashi, W. Kiyotaka, T. Arita, and K. Mori, *J. Appl. Phys.* 62, 485 (1987).

188. V. A. Godyak and R. B. Piejak, *Phys. Rev. Lett.* 65, 996 (1990).

189. G. J. Nienhuis, Ph.D. Thesis, Universiteit Utrecht, Utrecht, 1998.

190. M. J. Kushner, *J. Appl. Phys.* 63, 2532 (1988).

191. G. J. Nienhuis, W. J. Goedheer, E. A. G. Hamers, W. G. J. H. M. van Sark, and J. Bezemer, *J. Appl. Phys.* 82, 2060 (1997).

192. J. Perrin, O. Leroy, and M. C. Bordage, *Contrib. Plasma Phys.* 36, 3 (1996).

193. A. V. Phelps, *J. Phys. Chem. Ref. Data* 19, 653 (1990).

194. H. Tawara, Y. Itikawa, H. Nishimura, and M. Yoshino, *J. Phys. Chem. Ref. Data* 19, 617 (1990).

195. L. Layeillon, P. Duverneuil, J. P. Couderc, and B. Despax, *Plasma Sources Sci. Technol.* 3, 61 (1994).

196. L. Layeillon, P. Duverneuil, J. P. Couderc, and B. Despax, *Plasma Sources Sci. Technol.* 3, 72 (1994).

197. J. R. Doyle, D. A. Doughty, and A. Gallagher, *J. Appl. Phys.* 68, 4375 (1990).

198. P. Kae-Nune, J. Perrin, J. Guillon, and J. Jolly, *Japan. J. Appl. Phys.* 33, 4303 (1994).

199. E. R. Austin and F. W. Lampe, *J. Phys. Chem.* 81, 1134 (1977).

200. T. L. Pollock, H. S. Sandhu, A. Jodhan, and O. P. Strausz, *J. Am. Chem. Soc.* 95, 1017 (1973).

201. J. Perrin, J. P. M. Schmitt, G. De Rosny, B. Drévillon, J. Huc, and A. Lloret, *Chem. Phys.* 73, 383 (1982).

202. G. C. A. Perkins, E. R. Austin, and F. W. Lampe, *J. Am. Chem. Soc.* 101, 1109 (1979).

203. M. Hertl, N. Dorval, O. Leroy, J. Jolly, and M. Péalet, *Plasma Sources Sci. Technol.* 7, 130 (1998).

204. J. R. Doyle, D. A. Doughty, and A. Gallagher, *J. Appl. Phys.* 71, 4771 (1992).

205. A. G. Engelhardt and A. V. Phelps, *Phys. Rev.* 131, 2115 (1963).

206. E. Krishnakumar and S. K. Srivastava, *Contrib. Plasma Phys.* 35, 395 (1995).

207. H. Tawara and T. Kato, *At. Data Nucl. Data Tables* 36, 167 (1987).

208. P. Haaland, *J. Chem. Phys.* 93, 4066 (1990).

209. J. M. Wadehra and J. N. Bardsley, *Phys. Rev. Lett.* 41, 1795 (1978).

210. E. Krishnakumar, S. K. Srivastava, and I. Iga, *Int. J. Mass Spectrom. Ion Processes* 103, 107 (1991).

211. J. Perrin, *J. Phys. D* 26, 1662 (1993).

212. W. Kurachi and Y. Nakamura, *J. Phys. D* 22, 107 (1989).

213. H. Ehrhardt, L. Langhans, F. Linder, and H. Taylor, *Phys. Rev.* 173, 222 (1968).

214. A. P. Hickman, *J. Chem. Phys.* 70, 4872 (1979).

215. N. Itabashi, K. Kato, N. Nishiwaki, T. Goto, C. Yamada, and E. Hirota, *Japan. J. Appl. Phys.* 28, L325 (1989).

216. J. M. S. Henis, G. W. Steward, M. K. Tripodi, and P. P. Gaspar, *J. Chem. Phys.* 57, 389 (1972).

217. P. Kae-Nune, J. Perrin, J. Jolly, and J. Guillon, *Surf. Sci.* 360, L495 (1996).

218. T. Fuyuki, B. Allain, and J. Perrin, *J. Appl. Phys.* 68, 3322 (1990).

219. E. W. McDaniel and E. A. Mason, "The Mobility and Diffusion of Ions in Gases." Wiley, New York, 1973.

220. J. D. P. Passchier and W. J. Goedheer, *J. Appl. Phys.* 74, 3744 (1993).

221. J. P. Boeuf and L. C. Pitchford, *Phys. Rev. E* 51, 1376 (1995).

222. C. K. Birdsall, *IEEE Trans. Plasma Sci.* 19, 65 (1991).

223. J. P. Boeuf and P. Belenguer, *J. Appl. Phys.* 71, 4751 (1992).

224. M. Yan and W. J. Goedheer, *Plasma Sources Sci. Technol.* 8, 349 (1999).

225. N. Sato and H. Tagashira, *IEEE Trans. Plasma. Sci.* PS-19, 10102 (1991).

226. N. V. Mantzaris, A. Boudouvis, and E. Gogolides, *J. Appl. Phys.* 77, 6169 (1995).

227. M. J. Kushner, *Solid State Technol.* 6, 135 (1996).

228. G. G. Lister, *Vacuum* 45, 525 (1994).

229. E. Gogolides and H. H. Sawin, *J. Appl. Phys.* 72, 3971 (1992).

230. P. M. Meijer, W. J. Goedheer, and J. D. P. Passchier, *Phys. Rev. A* 45, 1098 (1992).

231. I. P. Shkarofsky, T. W. Johnston, and M. P. Bachynski, "The Particle Kinetics of Plasmas." Addison-Wesley, Reading, MA 1966.

232. G. J. Nienhuis and W. J. Goedheer, *Plasma Sources Sci. Technol.* 8, 295 (1999).

233. A. D. Richards, B. E. Thompson, and H. H. Sawin, *Appl. Phys. Lett.* 50, 492 (1987).

234. D. P. Lymberopoulos and D. J. Economou, *J. Phys. D* 28, 727 (1995).

235. E. A. G. Hamers, W. G. J. H. M. van Sark, J. Bezemer, W. F. van der Weg, and W. J. Goedheer, *Mater. Res. Soc. Symp. Proc.* 420, 461 (1996).

236. M. Heintze and R. Zedlitz, *J. Non-cryst. Solids* 198–200, 1038 (1996).

237. E. W. McDaniel, "Collision Phenomena in Ionized Gases." Wiley, New York, 1964.

238. P. J. Chantry, *J. Appl. Phys.* 62, 1141 (1987).

239. M. S. Barnes, T. J. Cotler, and M. E. Elta, *J. Comput. Phys.* 77, 53 (1988).

240. W. L. Morgan, *Plasma Chem. Plasma Process.* 12, 477 (1992).

241. D. L. Sharfetter and H. K. Gummel, *IEEE Trans. Electron Devices* ED-16, 64 (1969).

242. W. Hackbusch, "Multigrid Methods and Applications." Springer-Verlag, Berlin, 1985.

243. J. D. P. Passchier, Ph.D. Thesis, Universiteit Utrecht, Utrecht, 1994.

244. A. A. Fridman, L. Boufendi, T. Hbid, B. V. Potapkin, and A. Bouchoule, *J. Appl. Phys.* 79, 1303 (1996).

245. J. Perrin, P. Roca i Cabarrocas, B. Allain, and J.-M. Friedt, *Japan. J. Appl. Phys.* 27, 2041 (1988).

246. J. L. Andújar, E. Bertran, A. Canillas, C. Roch, and J. L. Morenza, *J. Vac. Sci. Technol. A* 9, 2216 (1991).

247. W. G. J. H. M. van Sark, J. Bezemer, E. M. B. Heller, M. Kars, and W. F. van der Weg, *Mater. Res. Soc. Symp. Proc.* 377, 3 (1995).

248. G. Oversluizen and W. H. M. Lodders, *J. Appl. Phys.* 83, 8002 (1998).

249. M. Heintze and R. Zedlitz, *Prog. Photovolt. Res. Appl.* 1, 213 (1993).

250. A. A. Howling, J.-L. Dorier, C. Hollenstein, U. Kroll, and F. Finger, *J. Vac. Sci. Technol. A* 10, 1080 (1992).

251. W. G. J. H. M. van Sark, J. Bezemer, and W. F. van der Weg, *Surf. Coat. Technol.* 74–75, 63 (1995).

252. L. Sansonnens, A. A. Howling, and C. Hollenstein, *Plasma Sources Sci. Technol.* 7, 114 (1998).

253. M. Capitelli, C. Gorse, R. Winkler, and J. Wilhelm, *Plasma Chem. Plasma Process.* 8, 399 (1988).

254. H. Curtins, N. Wyrsch, N. Favre, and A. V. Shah, *Plasma Chem. Plasma Process.* 7, 267 (1987).

255. L. A. Pinnaduwga, W. Ding, and D. L. McCorkle, *Appl. Phys. Lett.* 71, 3634 (1997).

256. L. A. Pinnaduwga and P. G. Datskos, *J. Appl. Phys.* 81, 7715 (1997).

257. O. Leroy, G. Gousset, L. L. Alves, J. Perrin, and J. Jolly, *Plasma Sources Sci. Technol.* 7, 348 (1998).

258. J. P. M. Schmitt, *Thin Solid Films* 174, 193 (1989).

259. L. Sansonnens, D. Franz, C. Hollenstein, A. A. Howling, J. Schmitt, E. Turlot, T. Emeraud, U. Kroll, J. Meier, and A. Shah, "Proceedings of the 13th European Photovoltaic Solar Energy Conference, Nice, France, 1995" (W. Freiesleben, W. Palz, H. A. Ossenbrink, and P. Helm, Eds.), p. 276. H.S. Stephens & Associates, Bedford, U.K., 1995.

260. D. Vender and R. W. Boswell, *IEEE Trans. Plasma Sci.* 18, 725 (1990).

261. C. K. Birdsall, *IEEE Trans. Plasma. Sci.* 19, 65 (1991).

262. C. K. Birdsall and A. B. Langdon, "Plasma Physics via Computer Simulation." Adam Hilger, Bristol, 1991.

263. V. Vahedi, "XPDP1 Manual." University of California, Berkeley, CA, 1994.

264. J. D. P. Passchier, Technical Report No. GF95.08, Utrecht University, Utrecht, 1995.

265. W. G. J. H. M. van Sark and J. Bezemer, Technical Report, Utrecht University, Utrecht, 1995.

266. Y. Ohmori, M. Shimozuma, and H. Tagashira, *J. Phys. D* 19, 1029 (1986).

267. E. A. Mason and E. W. McDaniel, "Transport Properties of Ions in Gases." Wiley, New York, 1988.

268. D. B. Graves, *J. Appl. Phys.* 62, 88 (1987).

269. J.-P. Boeuf, *Phys. Rev. A* 36, 2782 (1987).

270. E. Gogolides and H. H. Sawin, *J. Appl. Phys.* 72, 3971 (1992).

271. A. Manenschijn and W. J. Goedheer, *J. Appl. Phys.* 69, 2923 (1991).

272. V. Vahedi, C. K. Birdsall, M. A. Lieberman, G. Dipeso, and T. D. Rognlien, *Phys. Fluids B* 5, 2719 (1993).

273. K. Nanbu and Y. Kitatani, *Vacuum* 47, 1023 (1996).

274. A. K. Jain and D. G. Thompson, *J. Phys. B* 20, L389 (1987).

275. H. Tawara, Y. Itikawa, H. Nishimura, and M. Yoshino, *J. Phys. Chem. Ref. Data* 19, 621 (1990).

276. M. R. Wertheimer and M. Moisan, *J. Vac. Sci. Technol. A* 3, 2643 (1985).

277. H. Curtins, N. Wyrsch, M. Favre, and A. V. Shah, *Plasma Chem. Plasma Process.* 7, 267 (1987).

278. V. A. Godyak, *Sov. Phys. Tech. Phys.* 16, 1073 (1973).

279. M. Heintze and R. Zedlitz, *J. Phys. D* 26, 1781 (1993).

280. M. Heintze and R. Zedlitz, *J. Non-cryst. Solids* 164–166, 55 (1993).

281. J. Dutta, W. Bacsa, and C. Hollenstein, *J. Appl. Phys.* 72, 3220 (1992).

282. P. Mérel, M. Chaker, M. Moisan, A. Ricard, and M. Tabbal, *J. Appl. Phys.* 72, 3220 (1992).

283. S. Veprek and M. G. J. Veprek-Heijman, *Appl. Phys. Lett.* 56, 1766 (1990).

284. J. Dutta, U. Kroll, P. Chabloz, A. Shah, A. A. Howling, J.-L. Dorier, and C. Hollenstein, *J. Appl. Phys.* 72, 3220 (1992).

285. T. Takagi, *J. Vac. Sci. Technol. A* 2, 282 (1983).

286. G. Turban, B. Drévillon, D. S. Mataras, and D. E. Rapakoulias, in "Plasma Deposition of Amorphous Silicon-Based Materials" (G. Bruno, P. Capezzuto, and A. Madan, Eds.), Chap. 2, p. 63. Academic Press, Boston, 1995.

287. A. Matsuda and N. Hata, in "Glow-Discharge Hydrogenated Amorphous Silicon" (K. Tanaka, Ed.), Chap. 2, p. 9. KTK Scientific Publishers, Tokyo, 1989.

288. A. Matsuda, T. Kaga, H. Tanaka, L. Malhotra, and K. Tanaka, *Japan. J. Appl. Phys.* 22, L115 (1983).

289. D. Mataras, S. Cavadias, and D. Rapakoulias, *J. Appl. Phys.* 66, 119 (1989).

290. D. Mataras, S. Cavadias, and D. Rapakoulias, *J. Vac. Sci. Technol. A* 11, 664 (1993).

291. N. Itabashi, K. Kato, N. Nishiwaki, T. Goto, C. Yamada, and E. Hirota, *Japan. J. Appl. Phys.* 27, L1565 (1988).

292. N. Itabashi, N. Nishiwaki, M. Magane, S. Naito, T. Goto, A. Matsuda, C. Yamada, and E. Hirota, *Jpn. J. Appl. Phys.* 29, L505 (1990).

293. N. Hershkowitz, M. H. Cho, C. H. Nam, and T. Intrator, *Plasma Chem. Plasma Process.* 8, 35 (1988).

294. P. Awakowicz, *Mater. Sci. Forum* 287–288, 3 (1998).

295. R. van der Heijden, Master's Thesis, Utrecht University, Utrecht, 1996.

296. E. R. Mosburg, R. C. Kerns, and J. R. Abelson, *J. Appl. Phys.* 54, 4916 (1983).

297. G. Bruno, P. Capezzuto, G. Cicala, P. Manodoro, and V. Tassielli, *IEEE Trans. Plasma Sci.* PS-18, 934 (1990).

298. S. Veprek and M. Heintze, *Plasma Chem. Plasma Process.* 10, 3 (1990).

299. J. Cárabe, J. J. Gandía, and M. T. Gutiérrez, *J. Appl. Phys.* 73, 4618 (1993).

300. N. Spiliopoulos, D. Mataras, and D. E. Rapakoulias, *J. Electrochem. Soc.* 144, 634 (1997).

301. E. A. G. Hamers, W. G. J. H. M. van Sark, J. Bezemer, H. Meiling, and W. F. van der Weg, *J. Non-cryst. Solids* 226, 205 (1998).

302. G. Nolet, *J. Electrochem. Soc.* 122, 1030 (1975).

303. G. Turban, Y. Catherine, and B. Grolleau, *Thin Solid Films* 60, 147 (1979).

304. J. J. Wagner and S. Veprek, *Plasma Chem. Plasma Process.* 2, 95 (1982).

305. G. Turban, Y. Catherine, and B. Grolleau, *Plasma Chem. Plasma Process.* 2, 61 (1982).

306. P. A. Longeway, H. A. Weakliem, and R. D. Estes, *J. Appl. Phys.* 57, 5499 (1985).

307. R. Robertson, D. Hils, H. Chatham, and A. Gallagher, *Appl. Phys. Lett.* 43, 544 (1983).

308. M. Heintze and S. Vepřek, *Appl. Phys. Lett.* 54, 1320 (1989).

309. G. Turban, Y. Catherine, and B. Grolleau, *Thin Solid Films* 77, 287 (1981).

310. G. Turban, *Pure Appl. Chem.* 56, 215 (1984).

311. P. Kae-Nune, J. Perrin, J. Guillon, and J. Jolly, *Plasma Sources Sci. Technol.* 4, 250 (1995).

312. R. Robertson and A. Gallagher, *J. Appl. Phys.* 59, 3402 (1986).

313. F. A. Baiocchi, R. C. Wetzel, and R. S. Freund, *Phys. Rev. Lett.* 53, 771 (1984).

314. R. S. Freund, R. C. Wetzel, R. J. Shul, and T. R. Hayes, *Phys. Rev. A* 41, 3575 (1990).

315. Hiden Analytical Ltd., 240 Europa Boulevard, Gemini Business Park, Warrington WA5 5UN, England.

316. P. Ho, W. G. Breiland, and R. J. Buss, *J. Chem. Phys.* 91, 2627 (1989).

317. J. Perrin, M. Shiratani, P. Kae-Nune, H. Videlot, J. Jolly, and J. Guillon, *J. Vac. Sci. Technol. A* 16, 278 (1998).

318. J. Perrin, C. Böhm, R. Eternadi, and A. Lloret, *Plasma Sources Sci. Technol.* 3, 252 (1994).

319. G. Turban, Y. Catherine, and B. Grolleau, *Thin Solid Films* 67, 309 (1980).

320. H. Chatham, D. Hils, R. Robertson, and A. Gallagher, *J. Chem. Phys.* 81, 1770 (1984).

321. A. A. Howling, L. Sansonnens, J.-L. Dorier, and C. Hollenstein, *J. Appl. Phys.* 75, 1340 (1994).

322. L. Boufendi and T. Bouchoule, A. Hbid, *J. Vac. Sci. Technol. A* 14, 572 (1996).

323. B. E. Thompson, K. D. Allen, A. D. Richards, and H. H. Sawin, *J. Appl. Phys.* 59, 1890 (1986).

324. J. Liu, G. L. Huppert, and H. H. Sawin, *J. Appl. Phys.* 68, 3916 (1990).

325. C. Wild and P. Koidl, *J. Appl. Phys.* 69, 2909 (1991).

326. R. J. M. M. Snijkers, M. J. M. Sambeek, G. M. W. Kroesen, and F. J. de Hoog, *Appl. Phys. Lett.* 63, 308 (1993).

327. J. K. Olthoff, R. J. van Brunt, S. B. Radovanov, J. A. Rees, and R. Surowiec, *J. Appl. Phys.* 75, 115 (1994).

328. M. Fivaz, S. Brunner, W. Swarzenbach, A. A. Howling, and C. Hollenstein, *Plasma Sources Sci. Technol.* 4, 373 (1995).

329. D. A. Dahl, J. E. Delmore, and A. D. Appelhans, *Rev. Sci. Instrum.* 61, 607 (1990).

330. M. L. Mandich, W. D. Reents, and K. D. Kolenbrander, *J. Chem. Phys.* 92, 437 (1990).

331. W. D. Reents and M. L. Mandich, *J. Chem. Phys.* 96, 4429 (1992).

332. E. A. G. Hamers, J. Bezemer, H. Meiling, W. G. J. H. M. van Sark, and W. F. van der Weg, *Mater. Res. Soc. Symp. Proc.* 467, 603 (1997).

333. R. J. M. M. Snijkers, Ph.D. Thesis, Technische Universiteit Eindhoven, Eindhoven, The Netherlands, 1993.

334. E. R. Fisher and P. B. Armentrout, *J. Chem. Phys.* 93, 4858 (1990).

335. W. N. Allen, T. M. H. Cheng, and F. W. Lampe, *J. Chem. Phys.* 66, 3371 (1977).

336. R. M. A. Azzam and N. M. Bashara, "Ellipsometry and Polarized Light." North Holland, Amsterdam, 1977.

337. B. Drévillon, J. Perrin, R. Marbot, A. Violet, and J. L. Dalby, *Rev. Sci. Instrum.* 53, 969 (1982).

338. N. Blayo and B. Drévillon, *Appl. Phys. Lett.* 59, 950 (1991).

339. R. W. Collins, *Rev. Sci. Instrum.* 61, 2029 (1990).

340. H. Fujiwara, Y. Toyoshima, M. Kondo, and A. Matsuda, *Phys. Rev. B* 60, 13598 (1999).

341. R. Brenot, B. Drévillon, P. Bulkin, P. Roca i Cabarrocas, and R. Vanderhaghen, *Appl. Surf. Sci.* 154–155, 283 (2000).

342. R. W. Collins, J. Koh, A. S. Ferlauto, P. I. Rovira, Y. Lee, R. J. Koval, and C. R. Wronski, *Thin Solid Films* 364, 129 (2000).

343. B. Drévillon, *Prog. Cryst. Growth Charact. Mater.* 27, 1 (1993).

344. R. Ossikovski and B. Drévillon, *Phys. Rev. B* 54, 10530 (1996).

345. H. Fujiwara, J. Koh, P. I. Rovira, and R. W. Collins, *Phys. Rev. B* 61, 10832 (2000).

346. D. E. Aspnes, J. B. Theeten, and F. Hottier, *Phys. Rev. B* 20, 3292 (1979).

347. A. Canillas, E. Bertran, J. L. Andújar, and B. Drévillon, *J. Appl. Phys.* 68, 2752 (1990).

348. Y. Lu, S. Kim, M. Gunes, Y. Lee, C. R. Wronski, and R. W. Collins, *Mater. Res. Soc. Symp. Proc.* 336, 595 (1994).

349. D. V. Tsu, B. S. Chao, S. R. Ovshinsky, S. Guha, and J. Yang, *Appl. Phys. Lett.* 71, 1317 (1997).

350. J. Koh, Y. Lee, H. Fujiwara, C. R. Wronski, and R. W. Collins, *Appl. Phys. Lett.* 73, 1526 (1998).

351. S. Guha, J. Yang, D. L. Williamson, Y. Lubianiker, J. D. Cohen, and A. H. Mahan, *Appl. Phys. Lett.* 74, 1860 (1999).

352. A. H. Mahan, J. Yang, S. Guha, and D. L. Williamson, *Phys. Rev. B* 61, 1677 (2000).

353. H. Fujiwara, J. Koh, C. R. Wronski, and R. W. Collins, *Appl. Phys. Lett.* 70, 2150 (1997).

354. H. Fujiwara, J. Koh, C. R. Wronski, and R. W. Collins, *Appl. Phys. Lett.* 72, 2993 (1998).

355. H. Fujiwara, J. Koh, C. R. Wronski, and R. W. Collins, *Appl. Phys. Lett.* 74, 3687 (1999).

356. N. Blayo and B. Drévillon, *J. Non-cryst. Solids* 137 & 138, 771 (1991).

357. A. Matsuda, *J. Vac. Sci. Technol. A* 16, 365 (1998).

358. B. Drévillon, J. Perrin, J. M. Siefert, J. Huc, A. Lloret, G. de Rosny, and J. P. M. Schmitt, *Appl. Phys. Lett.* 42, 801 (1983).

359. R. A. Street, J. C. Knights, and D. K. Biegelsen, *Phys. Rev. B* 19, 3027 (1978).

360. J. C. Knights and R. A. Lujan, *Appl. Phys. Lett.* 35, 244 (1979).

361. S. Nishikawa, H. Kakinuma, T. Watanabe, and K. Nihei, *Japan. J. Appl. Phys.* 24, 639 (1985).

362. G. D. Cody, T. Tiedje, B. Abeles, B. Brooks, and Y. Goldstein, *Phys. Rev. Lett.* 47, 1480 (1981).

363. K. M. H. Maessen, M. J. M. Pruppers, J. Bezemer, F. H. P. M. Habraken, and W. F. van der Weg, *Mater. Res. Soc. Symp. Proc.* 95, 201 (1987).

364. K. Tanaka, K. Nakagawa, A. Matsuda, M. Matsumura, H. Yamamoto, S. Yamasaki, H. Okushi, and S. Izima, *Japan. J. Appl. Phys.* 20, 267 (1981).

365. P. Sichanugrist, M. Konagai, and K. Takahashi, *Japan. J. Appl. Phys.* 25, 440 (1986).

366. M. Hirose, *J. Phys. (Paris), Colloq.* 42, C4 (1981).

367. P. Chaudhuri, S. Ray, and A. K. Barua, *Thin Solid Films* 113, 261 (1984).

368. J. Shirafuji, S. Nagata, and M. Kuwagaki, *J. Appl. Phys.* 58, 3661 (1985).

369. S. Okamoto, Y. Hishikawa, and S. Tsuda, *Japan. J. Appl. Phys.* 35, 26 (1996).

370. H. Meiling, R. E. I. Schropp, W. G. J. H. M. van Sark, J. Bezemer, and W. F. van der Weg, "Proceedings of the Tenth E.C. Photovoltaic Solar Energy Conference, Lisbon, Portugal, 1991" (A. Luque, G. Sala, W. Palz, G. Dos Santos, and P. Helm, Eds.), p. 339. Kluwer Academic, Dordrecht, 1991.

371. J. P. M. Schmitt, *J. Non-cryst. Solids* 59 & 60, 649 (1983).

372. C. C. Tsai, G. B. Anderson, and R. Thompson, *J. Non-cryst. Solids* 137–138, 637 (1991).

373. P. Roca i Cabarrocas, S. Hamma, S. N. Sharma, G. Viera, E. Bertran, and J. Costa, *J. Non-cryst. Solids* 227–230, 871 (1998).

374. C. Longeaud, J. P. Kleider, P. Roca i Cabarrocas, S. Hamma, R. Meaudre, and M. Meaudre, *J. Non-cryst. Solids* 227–230, 96 (1998).

375. J. C. Knights, *Japan. J. Appl. Phys.* 18, 101 (1979).

376. R. B. Wehrspohn, S. C. Deane, I. D. French, I. Gale, J. Hewett, M. J. Powell, and J. Robertson, *J. Appl. Phys.* 87, 144 (2000).

377. A. A. Howling, J.-L. Dorier, and C. Hollenstein, *Appl. Phys. Lett.* 62, 1341 (1993).

378. A. Bouchoule, Ed., "Dusty Plasmas: Physics, Chemistry, and Technological Impact in Plasma Processing." Wiley, New York, 1999.

379. P. Roca i Cabarrocas, *J. Non-cryst. Solids* 266–269, 31 (2000).

380. M. Otobe, T. Kanai, T. Ifuku, H. Yajima, and S. Oda, *J. Non-cryst. Solids* 198–200, 875 (1996).

381. D. M. Tanenbaum, A. L. Laracuente, and A. Gallagher, *Appl. Phys. Lett.* 68, 705 (1996).
382. P. Roca i Cabarrocas, *Mater. Res. Soc. Symp. Proc.* 507, 855 (1998).
383. E. Stoffels, W. W. Stoffels, G. M. W. Kroesen, and F. J. de Hoog, *Electron Technol.* 31, 255 (1998).
384. C. Courteille, J.-L. Dorier, J. Dutta, C. Hollenstein, A. A. Howling, and T. Stoto, *J. Appl. Phys.* 78, 61 (1995).
385. S. Schmitt-Rink, D. A. B. Miller, and D. S. Chemla, *Phys. Rev. B* 35, 8113 (1987).
386. A. J. Read, R. J. Needs, K. J. Nash, L. T. Canham, P. D. J. Calcott, and A. Qteish, *Phys. Rev. Lett.* 69, 1232 (1992).
387. P. Roca i Cabarrocas, P. Gay, and A. Hadjadj, *J. Vac. Sci. Technol. A* 14, 655 (1996).
388. A. Fontcuberta i Morral, R. Brenot, E. A. G. Hamers, R. Vanderhaghen, and P. Roca i Cabarrocas, *J. Non-cryst. Solids* 266–269, 48 (2000).
389. R. Butté, S. Vignoli, M. Meaudre, R. Meaudre, O. Marty, L. Saviot, and P. Roca i Cabarrocas, *J. Non-cryst. Solids* 266–269, 263 (2000).
390. Corning 7059 Material Information, Corning Glass Works, NY.
391. J. Wallinga, D. Knoesen, E. A. G. Hamers, W. G. J. H. M. van Sark, W. F. van der Weg, and R. E. I. Schropp, *Mater. Res. Soc. Symp. Proc.* 420, 57 (1996).
392. M. B. Schubert and G. H. Bauer, "Proceedings of the Twenty-First IEEE Photovoltaic Specialists Conference, Kissimee, U.S.A., 1990," p. 1595. IEEE, New York, 1990.
393. Y. Hishikawa, M. Ohnishi, and Y. Kuwano, *Mater. Res. Soc. Symp. Proc.* 192, 3 (1990).
394. A. Shin and S. Lee, *J. Non-cryst. Solids* 260, 245 (1999).
395. N. Maley and J. S. Lannin, *Phys. Rev. B* 31, 5577 (1985).
396. A. J. M. Berntsen, W. G. J. H. M. van Sark, and W. F. van der Weg, *J. Appl. Phys.* 78, 1964 (1995).
397. R. W. Collins and J. M. Cavese, *J. Appl. Phys.* 61, 1869 (1987).
398. Y. Toyoshima, K. Arai, A. Matsuda, and K. Tanaka, *Appl. Phys. Lett.* 56, 1540 (1990).
399. G. Ganguly and A. Matsuda, *J. Non-cryst. Solids* 164–166, 31 (1993).
400. S. Veprek, F.-A. Sarrott, S. Rambert, and E. Taglauer, *J. Vac. Sci. Technol. A* 7, 2614 (1989).
401. B. J. Garrison, M. T. Miller, and D. W. Bremmen, *Chem. Phys. Lett.* 146, 513 (1988).
402. M. Kitabatake, P. Fons, and J. E. Greene, *J. Vac. Sci. Technol. A* 8, 3726 (1990).
403. P. Roca i Cabarrocas, *Mater. Res. Soc. Symp. Proc.* 149, 33 (1989).
404. G. Ganguly, T. Ikeda, I. Sakata, and A. Matsuda, *Mat. Res. Soc. Proc. Symp.* 420, 347 (1996).
405. A. I. Kosarev, A. S. Smirnov, A. S. Abramov, A. J. Vinogradov, A. Y. Ustavschikov, and M. V. Shutov, *J. Vac. Sci. Technol. A* 15, 298 (1997).
406. A. S. Abramov, A. Y. Vinogradov, A. I. Kosarev, M. V. Shutov, A. S. Smirnov, and K. E. Orlov, *Tech. Phys.* 43, 180 (1998).
407. J. P. Harbison, A. J. Williams, and D. V. Lang, *J. Appl. Phys.* 55, 946 (1984).
408. K. H. Müller, *J. Appl. Phys.* 62, 1796 (1987).
409. J. Robertson, *J. Non-cryst. Solids* 164–166, 1115 (1993).
410. J. R. Abelson, *Appl. Phys. A* 56, 493 (1993).
411. R. Biswas, *J. Appl. Phys.* 73, 3295 (1993).
412. W. Möller, *Appl. Phys. A* 56, 527 (1993).
413. M. Hoedemaker, J. van der Kuur, E. J. Melker, and B. J. Thijsse, *Nucl. Instrum. Methods Phys. Res. B* 127/128, 888 (1997).
414. J. F. Ziegler, J. P. Biersack, and U. Littmarck, "The Stopping and Range of Ions in Solids." Pergamon, New York, 1985.
415. B. Drévillon, J. Huc, and N. Boussarssar, *J. Non-cryst. Solids* 59/60, 735 (1983).
416. B. Drévillon and M. Toutlemonde, *Appl. Phys. Lett.* 59, 950 (1985).
417. J. Perrin, Y. Takeda, N. Hirano, H. Matsuura, and A. Matsuda, *Japan. J. Appl. Phys.* 28, 5 (1989).
418. P. Roca i Cabarrocas, P. Morin, V. Chu, J. P. Conde, J. Z. Liu, H. R. Park, and S. Wagner, *J. Appl. Phys.* 69, 2942 (1991).
419. E. A. G. Hamers, J. Bezemer, and W. F. van der Weg, *Appl. Phys. Lett.* 75, 609 (1999).
420. D. A. Doughty, J. R. Doyle, G. H. Lin, and A. Gallagher, *J. Appl. Phys.* 67, 6220 (1990).
421. J. K. Hirvonen, in "Materials and Processes for Surface and Interface Engineering" (Y. Pauleau, Ed.), Chap. 9, p. 307. Kluwer Academic, Dordrecht, 1995.
422. J. E. Potts, E. M. Peterson, and J. A. McMillan, *J. Appl. Phys.* 52, 6665 (1981).
423. H. Meiling, J. Bezemer, R. E. I. Schropp, and W. F. van der Weg, *Mater. Res. Soc. Proc. Symp.* 467, 459 (1997).
424. S. Habermehl and G. Lucovsky, *J. Vac. Sci. Technol. A* 14, 3024 (1996).
425. S. R. Kasi, H. Kang, C. S. Sass, and J. W. Rabalais, *Surf. Sci. Rep.* 10, 1 (1989).
426. L. Hanley, H. Lim, D. G. Schultz, S. B. Wainhaus, P. de Sainte Claire, and W. L. Hase, *Nucl. Instrum. Methods Phys. Res. B* 125, 218 (1997).
427. H. Sugai, Y. Mitsuoka, and H. Toyoda, *J. Vac. Sci. Technol. A* 16, 290 (1998).
428. K. H. Müller, *Phys. Rev. B* 35, 7906 (1987).
429. J. W. Rabalais, A. H. Al-Bayati, K. J. Boyd, D. Marton, J. Kulik, Z. Zhang, and W. K. Chu, *Phys. Rev. B* 53, 10781 (1996).
430. K. H. Müller, *J. Appl. Phys.* 59, 2803 (1986).
431. J. E. Yehoda, B. Yang, K. Vedam, and R. Messier, *J. Vac. Sci. Technol. A* 6, 1631 (1988).
432. R. J. Severens, M. C. M. van de Sanden, H. J. M. Verhoeven, J. Bastiaansen, and D. C. Schram, *Mater. Res. Soc. Symp. Proc.* 420, 341 (1996).
433. M. C. M. van de Sanden, R. J. Severens, W. M. M. Kessels, F. van de Pas, L. van Ijzendoorn, and D. C. Schram, *Mater. Res. Soc. Proc. Symp.* 467, 621 (1997).
434. E. H. C. Ullersma, Ph.D. Thesis, Universiteit Utrecht, Utrecht, 1998.
435. R. Robertson and A. Gallagher, *J. Chem. Phys.* 85, 3623 (1986).
436. A. Gallagher, *J. Appl. Phys.* 63, 2406 (1988).
437. J. L. Guizot, K. Nomoto, and A. Matsuda, *Surf. Sci.* 244, 22 (1991).
438. A. J. Flewitt, J. Robertson, and W. I. Milne, *J. Appl. Phys.* 85, 8032 (1999).
439. J. Robertson, *J. Appl. Phys.* 87, 2608 (2000).
440. G. Ganguly and A. Matsuda, *Phys. Rev. B* 47, 3661 (1993).
441. K. Winer, *Phys. Rev. B* 41, 7952 (1990).
442. R. A. Street, *Phys. Rev. B* 43, 2454 (1991).
443. R. A. Street, *Phys. Rev. B* 44, 10610 (1991).
444. R. A. Street and K. Winer, *Phys. Rev. B* 40, 6236 (1989).
445. M. J. Powell and S. C. Deane, *Phys. Rev. B* 48, 10815 (1993).
446. M. J. Powell and S. C. Deane, *Phys. Rev. B* 53, 10121 (1996).
447. K. Gleason, K. S. Wang, M. K. Chen, and J. A. Reimer, *J. Appl. Phys.* 61, 2866 (1987).
448. M. J. McCaughy and M. J. Kushner, *J. Appl. Phys.* 65, 186 (1989).
449. S. Shirafuji, S. Nakajima, Y. F. Wang, T. Genji, and K. Tachibana, *Japan. J. Appl. Phys.* 32, 1546 (1993).
450. J. Robertson, *Mater. Res. Soc. Proc. Symp.* 609, A1.4.1 (2000).
451. C. C. Tsai, J. C. Knights, G. Chang, and B. Wacker, *J. Appl. Phys.* 59, 2998 (1986).
452. A. J. Flewitt, J. Robertson, and W. I. Milne, *J. Non-cryst. Solids* 266–269, 74 (2000).
453. A. von Keudell and J. R. Abelson, *Japan. J. Appl. Phys.* 38, 4002 (1999).
454. S. Ramalingam, D. Maroudas, and E. S. Aydill, *J. Appl. Phys.* 86, 2872 (1999).
455. S. Yamasaki, T. Umeda, J. Isoya, and K. Tanaka, *Appl. Phys. Lett.* 70, 1137 (1997).
456. D. A. Porter and K. E. Easterling, "Phase Transformations in Metals and Alloys." Chapman & Hall, London, 1992.
457. P. Roca i Cabarrocas, Y. Bouizem, and M. L. Theye, *Philos. Mag. B* 65, 1025 (1992).
458. C. M. Greenlief, S. M. Gates, and P. A. Holbert, *J. Vac. Sci. Technol. A* 7, 1845 (1989).
459. W. B. Jackson, *J. Non-cryst. Solids* 164–166, 263 (1993).
460. W. Beyer, *J. Non-Cryst. Solids* 198–200, 40 (1996).
461. W. Beyer and U. Zastrow, *Mater. Res. Soc. Symp. Proc.* 420, 497 (1996).
462. C. G. Van de Walle, *Phys. Rev. B* 49, 4579 (1994).

463. J. Robertson, C. W. Chen, M. J. Powell, and S. C. Deane, *J. Non-cryst. Solids* 227–230, 138 (1998).

464. K. J. Chang and J. D. Chadi, *Phys. Rev. Lett.* 62, 937 (1989).

465. K. J. Chang and J. D. Chadi, *Phys. Rev. B* 40, 11644 (1990).

466. Y. Miyoshi, Y. Yoshida, S. Miyazaki, and M. Hirose, *J. Non-cryst. Solids* 198–200, 1029 (1996).

467. J. Robertson and M. J. Powell, *Thin Solid Films* 337, 32 (1999).

468. W. M. M. Kessels, R. J. Severens, M. C. M. van de Sanden, and D. C. Schram, *J. Non-cryst. Solids* 227–230, 133 (1998).

469. G. H. Lin, J. R. Doyle, M. Z. He, and A. Gallagher, *J. Appl. Phys.* 64, 341 (1988).

470. P. Roca i Cabarrocas, *Appl. Phys. Lett.* 65, 1674 (1995).

471. H. Shirai, J. Hanna, and I. Shimizu, *Japan. J. Appl. Phys.* 30, L679 (1991).

472. M. C. M. van de Sanden, W. M. M. Kessels, R. J. Severens, and D. C. Schram, *Plasma Phys. Controlled Fusion* 41, A365 (1999).

473. M. C. M. van de Sanden, W. M. M. Kessels, A. H. Smets, B. A. Korevaar, R. J. Severens, and D. C. Schram, *Mater. Res. Soc. Symp. Proc.* 557, 13 (1999).

474. M. Stutzmann, *Philos. Mag. B* 60, 531 (1989).

475. A. Shah, J. Dutta, N. Wyrsch, K. Prasad, H. Curtins, F. Finger, A. Howling, and C. Hollenstein, *Mater. Res. Soc. Symp. Proc.* 258, 15 (1992).

476. C. M. Ferreira and J. Loureiro, *J. Phys. D* 17, 1175 (1984).

477. J. Kuske, U. Stephan, O. Steinke, and S. Röhlecke, *Mater. Res. Soc. Symp. Proc.* 377, 27 (1995).

478. U. Kroll, J. Meier, M. Goetz, A. Howling, J.-L. Dorier, J. Dutta, A. Shah, and C. Hollenstein, *J. Non-cryst. Solids* 164–166, 59 (1993).

479. H. Meiling, J. F. M. Westendorp, J. Hautala, Z. Saleh, and C. T. Malone, *Mater. Res. Soc. Symp. Proc.* 345, 65 (1995).

480. S. Röhlecke, R. Tews, A. Kottwitz, and K. Schade, *Surf. Coat. Technol.* 74–75, 259 (1995).

481. C. Beneking, *J. Appl. Phys.* 68, 4461 (1990).

482. C. Beneking, *J. Appl. Phys.* 68, 5435 (1990).

483. C. Beneking, F. Finger, and H. Wagner, "Proceedings of the Eleventh E.C. Photovoltaic Solar Energy Conference, Montreux, Switzerland 1992" (L. Guimarães, W. Palz, C. de Reyff, H. Kiess, and P. Helm, Eds.), p. 586. Harwood, Chur, Switzerland, 1993.

484. M. J. Colgan, M. Meyyappan, and D. E. Murnick, *Plasma Sources Sci. Technol.* 3, 181 (1994).

485. U. Kroll, Y. Ziegler, J. Meier, H. Keppner, and A. Shah, *Mater. Res. Soc. Symp. Proc.* 336, 115 (1994).

486. E. A. G. Hamers, W. G. J. H. M. van Sark, J. Bezemer, W. F. van der Weg, G. J. Nienhuis, and W. J. Goedheer, "Proceedings of the Thirteenth European Sectional Conference on the Atomic and Molecular Physics of Ionized Gases," ESCAMPIG 96, Poprad, Slovakia, *Europhys. Conf. Abs.* 20E, 27 (1996).

487. W. G. J. H. M. van Sark, H. Meiling, E. A. G. Hamers, J. Bezemer, and W. F. van der Weg, *J. Vac. Sci. Technol. A* 15, 654 (1997).

488. H. Meiling, W. G. J. H. M. van Sark, J. Bezemer, and W. van der Weg, *J. Appl. Phys.* 80, 3546 (1996).

489. G. E. N. Landweer and J. Bezemer, in "Amorphous Silicon and Related Materials, Advances in Disordered Semiconductors 1" (H. Fritzsche, Ed.). World Scientific, Singapore, 1989.

490. M. Zhu, M. B. von der Linden, J. Bezemer, R. E. I. Schropp, and W. F. van der Weg, *J. Non-cryst. Solids* 137–138, 355 (1991).

491. M. Zhu, M. B. von der Linden, and W. F. van der Weg, *Mater. Res. Soc. Symp. Proc.* 336, 449 (1994).

492. M. B. von der Linden, Ph.D. Thesis, Universiteit Utrecht, Utrecht, 1994.

493. W. G. J. H. M. van Sark, J. Bezemer, and W. F. van der Weg, *J. Mater. Res.* 13, 45 (1998).

494. R. E. I. Schropp, H. Meiling, W. G. J. H. M. van Sark, J. Stammeijer, J. Bezemer, and W. F. van der Weg, "Proceedings of the Tenth E.C. Photovoltaic Solar Energy Conference, Lisbon, Portugal, 1991" (A. Luque, G. Sala, W. Palz, G. Dos Santos, and P. Helm, Eds.), p. 1087. Kluwer Academic, Dordrecht, 1991.

495. S. J. Jones, D. L. Williamson, T. Liu, X. Deng, and M. Izu, *Mater. Res. Soc. Proc. Symp.* 609, A7.4.1 (2000).

496. C. C. Tsai, R. Thompson, C. Doland, F. A. Ponce, G. B. Anderson, and B. Wacker, *Mater. Res. Soc. Symp. Proc.* 118, 49 (1988).

497. Y. S. Tsuo, R. Weil, S. Asher, A. Nelson, Y. Xu, and R. Tsu, "Proceedings of the 19th IEEE Photovoltaic Specialists Conference, New Orleans, 1987," p. 705. IEEE, New York, 1987.

498. H. Shirai, D. Das, J. Hanna, and I. Shimizu, *Appl. Phys. Lett.* 59, 1096 (1991).

499. A. Asano, T. Ichimura, and H. Sakai, *J. Appl. Phys.* 65, 2439 (1989).

500. A. Asano, *Appl. Phys. Lett.* 56, 533 (1990).

501. G. Parsons, *IEEE Electron Device Lett.* 13, 80 (1992).

502. N. Layadi, P. Roca i Cabarrocas, B. Drévillon, and I. Solomon, *Phys. Rev. B* 52, 5136 (1995).

503. P. Roca i Cabarrocas, N. Layadi, B. Drévillon, and I. Solomon, *J. Non-cryst. Solids* 198–200, 871 (1996).

504. P. Roca i Cabarrocas, S. Hamma, A. Hadjadi, J. Bertomeu, and J. Andreu, *Appl. Phys. Lett.* 69, 529 (1996).

505. N. Shibata, K. Fukada, H. Ohtoshi, J. Hanna, S. Oda, and I. Shimizu, *Mater. Res. Soc. Symp. Proc.* 95, 225 (1987).

506. Y. H. Yang, M. Katiyar, G. F. Feng, N. Maley, and J. R. Abelson, *Appl. Phys. Lett.* 65, 1769 (1994).

507. S. Hamma and P. Roca i Cabarrocas, *J. Appl. Phys.* 81, 7282 (1997).

508. P. Roca i Cabarrocas, R. Brenot, P. Bulkin, R. Vanderhaghen, B. Drévillon, and I. French, *J. Appl. Phys.* 86, 7079 (1999).

509. K. Saito, M. Kondo, M. Fukawa, T. Nishimiya, A. Matsuda, W. Futako, and I. Shimizu, *Appl. Phys. Lett.* 71, 3403 (1997).

510. C. Anandan, C. Mukherjee, T. Seth, P. N. Dixit, and R. Bhattacharyya, *Appl. Phys. Lett.* 66, 85 (1995).

511. C. Mukherjee, C. Anandan, T. Seth, P. N. Dixit, and R. Bhattacharyya, *Appl. Phys. Lett.* 68, 194 (1996).

512. A. C. W. Biebericher, J. Bezemer, W. F. van der Weg, and W. J. Goedheer, *Appl. Phys. Lett.* 76, 2002 (2000).

513. H. Kirimura, H. Maeda, H. Murakami, T. Nakahigashi, S. Ohtani, T. Tabata, T. Hayashi, M. Kobayashi, Y. Mitsuda, N. Nakamura, H. Kuwahara, and A. Doi, *Japan. J. Appl. Phys.* 33, 4389 (1994).

514. S. Morrison, J. Xi, and A. Madan, *Mater. Res. Soc. Symp. Proc.* 507, 559 (1998).

515. A. Madan, S. Morrison, and H. Kuwahara, *Sol. Energy Mater. Sol. Cells* 59, 51 (1999).

516. G. Scarsbrook, I. P. Llewellyn, S. M. Ojha, and R. A. Heinecke, *Vacuum* 38, 627 (1988).

517. L. J. Overzet and J. T. Verdeyen, *Appl. Phys. Lett.* 48, 695 (1986).

518. J. P. Booth, G. Cunge, N. Sadeghi, and R. W. Boswell, *J. Appl. Phys.* 82, 552 (1997).

519. A. C. W. Biebericher, J. Bezemer, W. F. van der Weg, and W. J. Goedheer, *Mater. Res. Soc. Proc. Symp.* 609, A4.1.1 (2000).

520. A. C. W. Biebericher, J. Bezemer, W. F. van der Weg, and W. J. Goedheer, unpublished results.

521. H. Wiesmann, A. K. Ghosh, T. McMahon, and M. Strongin, *J. Appl. Phys.* 50, 3752 (1979).

522. H. Matsumura, *Japan. J. Appl. Phys.* 25, L949 (1986).

523. H. Matsumura, *Appl. Phys. Lett.* 51, 804 (1987).

524. H. Matsumura, *J. Appl. Phys.* 65, 4396 (1989).

525. J. R. Doyle, R. Robertson, G. H. Lin, M. Z. He, and A. Gallagher, *J. Appl. Phys.* 64, 3215 (1988).

526. F. Jansen, I. Chen, and M. A. Machonkin, *J. Appl. Phys.* 66, 5749 (1989).

527. A. H. Mahan and M. Vaněček, "Amorphous Silicon Materials and Solar Cells," AIP Conference Proceedings, Vol. 234 (B. Stafford, Ed.), p. 195. American Institute of Physics, New York, 1991.

528. E. C. Molenbroek, A. H. Mahan, E. Johnson, and A. C. Gallagher, *J. Appl. Phys.* 79, 7278 (1996).

529. M. Heintze, R. Zedlitz, H. N. Wanka, and M. B. Schubert, *J. Appl. Phys.* 79, 2699 (1996).

530. E. C. Molenbroek, A. H. Mahan, and A. C. Gallagher, *J. Appl. Phys.* 82, 1909 (1997).

531. K. F. Feenstra, R. E. I. Schropp, and W. F. van der Weg, *J. Appl. Phys.* 85, 6843 (1999).

532. P. Papadopoulos, A. Scholz, S. Bauer, B. Schröder, and H. Öchsner, *J. Non-cryst. Solids* 164–166, 87 (1993).

533. B. P. Nelson, E. Iwaniczko, R. E. I. Schropp, H. Mahan, E. C. Molenbroek, S. Salamon, and R. S. Crandall, "Proceedings of the Twelfth E.C. Photovoltaic Solar Energy Conference, Amsterdam, the Netherlands, 1994" (R. Hill, W. Palz, and P. Helm, Eds.), p. 679. H. S. Stephens & Associates, Bedford, U.K., 1994.

534. A. H. Mahan, E. Iwaniczko, B. P. Nelson, R. C. Reedy, Jr., R. S. Crandall, S. Guha, and J. Yang, "Proceedings of the 25th IEEE Photovoltaic Specialists Conference, Washington DC, 1996," p. 1065. IEEE, New York, 1996.

535. R. E. I. Schropp, K. F. Feenstra, E. C. Molenbroek, H. Meiling, and J. K. Rath, *Philos. Mag. B* 76, 309 (1997).

536. R. S. Crandall, A. H. Mahan, B. Nelson, M. Vaněček, and I. Balberg, "AIP Conference Proceedings" (R. Noufi, Ed.), Vol. 268, p. 81. American Institute of Physics, Denver, 1992.

537. Q. Wang, E. Iwaniczko, Y. Xu, W. Gao, B. P. Nelson, A. H. Mahan, R. S. Crandall, and H. M. Branz, *Mater. Res. Soc. Proc. Symp.* 609, A4.3.1 (2000).

538. H. Meiling and R. E. I. Schropp, *Appl. Phys. Lett.* 70, 2681 (1997).

539. V. Chu, J. Jarego, H. Silva, T. Silva, M. Reissner, P. Brogueira, and J. P. Conde, *Appl. Phys. Lett.* 70, 2714 (1997).

540. H. Meiling, A. M. Brockhoff, J. K. Rath, and R. E. I. Schropp, *J. Non-cryst. Solids* 227–230, 1202 (1998).

541. J. Cifre, J. Bertomeu, J. Puigdollers, M. C. Polo, J. Andreu, and A. Lloret, *Appl. Phys. A* 59, 645 (1994).

542. A. R. Middya, S. Hazra, S. Ray, C. Longeaud, and J. P. Kleider, "Proceedings of the Thirteenth European Photovoltaic Solar Energy Conference, Nice, France, 1995" (W. Freiesleben, W. Palz, H. A. Ossenbrink, and P. Helm, Eds.), p. 679. H. S. Stephens & Associates, Bedford, U.K., 1995.

543. P. Brogueira, J. P. Conde, S. Arekat, and V. Chu, *J. Appl. Phys.* 79, 8748 (1996).

544. J. K. Rath, H. Meiling, and R. E. I. Schropp, *Sol. Energy Mater. Sol. Cells* 48, 269 (1997).

545. H. N. Wanka, R. Zedlitz, M. Heintze, and M. B. Schubert, "Proceedings of the Thirteenth European Photovoltaic Solar Energy Conference, Nice, France, 1995" (W. Freiesleben, W. Palz, H. A. Ossenbrink, and P. Helm, Eds.), p. 1753. H.S. Stephens & Associates, Bedford, U.K., 1995.

546. V. Brogueira, P. Chu, A. C. Ferro, and J. P. Conde, *J. Vac. Sci. Technol. A* 15, 2968 (1997).

547. P. Alpuim, V. Chu, and J. P. Conde, *Mater. Res. Soc. Proc. Symp.* 609, A22.6.1 (2000).

548. K. F. Feenstra, Ph.D. Thesis, Universiteit Utrecht, Utrecht, 1998.

549. C. Horbach, W. Beyer, and H. Wagner, *J. Non-cryst. Solids* 137 & 138, 661 (1991).

550. S. Bauer, B. Schröder, and H. Oechsner, *J. Non-cryst. Solids* 227–230, 34 (1998).

551. A. H. Mahan, Y. Chen, D. L. Williamson, and G. D. Mooney, *J. Non-cryst. Solids* 137–138, 65 (1991).

552. Y. Wu, J. T. Stephen, D. X. Han, J. M. Rutland, R. S. Crandall, and A. H. Mahan, *Phys. Rev. Lett.* 77, 2049 (1996).

553. R. P. Muller, J. K. Holt, D. G. Goodwin, and W. A. Goddard, III, *Mater. Res. Soc. Proc. Symp.* 609, A6.1.1 (2000).

554. J. K. Holt, M. Swiatek, D. G. Goodwin, and H. A. Atwater, *Mater. Res. Soc. Proc. Symp.* 609, A6.2.1 (2000).

555. B. Schröder and S. Bauer, *J. Non-cryst. Solids* 266–269, 1115 (2000).

556. A. T. M. Wilbers, G. M. W. Kroesen, C. J. Timmermans, and D. C. Schram, *Meas. Sci. Technol.* 1, 1326 (1990).

557. J. W. A. M. Gielen, P. R. M. Kleuskens, M. C. M. van de Sanden, L. J. van IJzendoorn, D. C. Schram, E. H. A. Dekempeneer, and J. Meneve, *J. Appl. Phys.* 80, 5986 (1996).

558. J. J. Beulens, A. J. M. Buuron, and D. C. Schram, *Surf. Coat. Technol.* 47, 401 (1991).

559. M. J. de Graaf, R. J. Severens, R. P. Dahiya, M. C. M. van de Sanden, and D. C. Schram, *Phys. Rev. E* 48, 2098 (1993).

560. R. F. G. Meulenbroeks, M. F. M. Steenbakkers, Z. Qing, M. C. M. van de Sanden, and D. C. Schram, *Phys. Rev. E* 49, 2272 (1994).

561. W. M. M. Kessels, A. H. M. Smets, B. A. Korevaar, G. J. Adriaenssens, M. C. M. van de Sanden, and D. C. Schram, *Mater. Res. Soc. Symp. Proc.* 557, 25 (1999).

562. B. A. Korevaar, G. J. Adriaenssens, A. H. M. Smets, W. M. M. Kessels, H.-Z. Song, M. C. M. van de Sanden, and D. C. Schram, *J. Non-cryst. Solids* 266–269, 380 (2000).

563. M. C. M. van de Sanden, R. J. Severens, W. M. M. Kessels, R. F. G. Meulenbroeks, and D. C. Schram, *J. Appl. Phys.* 84, 2426 (1998).

564. M. C. M. van de Sanden, R. J. Severens, W. M. M. Kessels, R. F. G. Meulenbroeks, and D. C. Schram, *J. Appl. Phys.* 85, 1243 (1999).

565. J. J. Beulens, M. J. de Graaf, and D. C. Schram, *Plasma Sources Sci. Technol.* 2, 180 (1993).

566. M. C. M. van de Sanden, J. M. de Regt, and D. C. Schram, *Plasma Sources Sci. Technol.* 3, 501 (1994).

567. W. M. M. Kessels, C. M. Leewis, M. C. M. van de Sanden, and D. C. Schram, *J. Appl. Phys.* 86, 4029 (1999).

568. R. F. G. Meulenbroeks, R. A. H. Engeln, M. N. A. Beurskens, R. M. J. Paffen, M. C. M. van de Sanden, J. A. M. van der Mullen, and D. C. Schram, *Plasma Sources Sci. Technol.* 4, 74 (1995).

569. W. M. M. Kessels, M. C. M. van de Sanden, and D. C. Schram, *J. Vac. Sci. Technol. A* 18, 2153 (2000).

570. W. M. M. Kessels, M. C. M. van de Sanden, R. J. Severens, L. J. van Ijzendoorn, and D. C. Schram, *Mater. Res. Soc. Symp. Proc.* 507, 529 (1998).

571. W. M. M. Kessels, Ph.D. Thesis, Technische Universiteit Eindhoven, Eindhoven, the Netherlands, 2000.

572. W. M. M. Kessels, R. J. Severens, A. H. M. Smets, B. A. Korevaar, G. J. Adriaenssens, D. C. Schram, and M. C. M. van de Sanden, *J. Appl. Phys.* 89, 2404 (2001).

573. W. M. M. Kessels, J. P. M. Hoefnagels, M. G. H. Boogderts, D. C. Schram, and M. C. M. van de Sanden, *J. Appl. Phys.* 89, 2065 (2001).

574. H. S. Nalwa, Ed., "Handbook of Advanced Electronic and Photonic Materials and Devices." Academic Press, Boston, 2000.

575. M. A. Green, "Solar Cells, Operating Principles, Technology and System Applications." University of New South Wales, Kensington, NSW, Australia, 1992.

576. M. Hack and M. Shur, *J. Appl. Phys.* 58, 997 (1985).

577. A. Catalano, in "Amorphous and Microcrystalline Semiconductor Devices—Optoelectronic Devices" (J. Kanicki, Ed.), Chap. 2, p. 9. Artech House, Norwood, MA, 1991.

578. A. Polman, W. G. J. H. M. van Sark, W. C. Sinke, and F. W. Saris, *Sol. Cells* 17, 241 (1986).

579. H. Hartnagel, A. Dawar, A. Jain, and C. Jagadish, "Semiconducting Transparent Thin Films." Institute of Physics Publishing, Bristol, 1995.

580. E. Yablonovitch and G. D. Cody, *IEEE Trans. Electron. Dev.* ED-29, 300 (1982).

581. H. W. Deckman, C. R. Wronski, H. Witzke, and E. Yablonovitch, *Appl. Phys. Lett.* 42, 968 (1983).

582. J. Daey Ouwens, R. E. I. Schropp, J. Wallinga, W. F. van der Weg, M. Ritala, M. Leskelä, and M. Hyrvärinen, "Proceedings of the Twelfth E.C. Photovoltaic Solar Energy Conference, Amsterdam, the Netherlands, 1994" (R. Hill, W. Palz, and P. Helm, Eds.), p. 1296. H.S. Stephens & Associates, Bedford, U.K., 1994.

583. M. Kubon, E. Böhmer, F. Siebke, B. Rech, C. Beneking, and H. Wagner, *Sol. Energy Mater. Sol. Cells* 41/42, 485 (1996).

584. Y. Tawada, H. Okamoto, and Y. Hamakawa, *Appl. Phys. Lett.* 39, 237 (1981).

585. K. Miyachi, N. Ishiguro, T. Miyashita, N. Yanagawa, H. Tanaka, M. Koyama, Y. Ashida, and N. Fukuda, "Proceedings of the Eleventh E.C. Photovoltaic Solar Energy Conference, Montreux, Switzerland, 1992" (L. Guimarães, W. Palz, C. de Reyff, H. Kiess, and P. Helm, Eds.), p. 88. Harwood, Chur, Switzerland, 1993.

586. R. R. Arya, A. Catalano, and R. S. Oswald, *Appl. Phys. Lett.* 49, 1089 (1986).

587. C. Beneking, B. Rech, T. Eickhoff, Y. G. Michael, N. Schultz, and H. Wagner, "Proceedings of the Twelfth E.C. Photovoltaic Solar Energy

Conference, Amsterdam, the Netherlands, 1994," p. 683. Kluwer Academic, Dordrecht, 1994.

588. G. E. N. Landweer, B. S. Girwar, C. H. M. van der Werf, J. W. Metselaar, and R. E. I. Schropp, "Proceedings of the Twelfth E.C. Photovoltaic Solar Energy Conference, Amsterdam, the Netherlands, 1994," p. 1300. Kluwer Academic, Dordrecht, 1994.

589. A. Banerjee and S. Guha, *J. Appl. Phys.* 69, 1030 (1991).

590. Y. Kuwano, M. Ohnishi, H. Nishiwaki, S. Tsuda, T. Fukatsu, K. Enomoto, Y. Nakashima, and H. Tarui, "Proceedings of the 16th IEEE Photovoltaic Specialists Conference, San Diego, 1982," p. 1338. IEEE, New York, 1982.

591. S. Guha, J. Yang, A. Pawlikiewicz, T. Glatfelter, R. Ross, and S. R. Ovshinsky, *Appl. Phys. Lett.* 54, 2330 (1989).

592. J. Zimmer, H. Stiebig, and H. Wagner, *J. Appl. Phys.* 84, 611 (1998).

593. R. E. Rochelau, M. Tun, and S. S. Hegedus, "Proceedings of the 26th IEEE Photovoltaic Specialists Conference, Anaheim, 1996," p. 703. IEEE, New York, 1997.

594. J. K. Rath, R. E. I. Schropp, and W. Beyer, *J. Non-cryst. Solids* 227–230, 1282 (1998).

595. J. Bruns, M. Choudhury, and H. G. Wagemann, "Proceedings of the Thirteenth European Photovoltaic Solar Energy Conference, Nice, France, 1995" (W. Freiesleben, W. Palz, H. A. Ossenbrink, and P. Helm, Eds.), p. 230. H.S. Stephens & Associates, Bedford, U.K., 1995.

596. M. Zeman, J. A. Willemen, S. Solntsev, and G. W. Metselaar, *Sol. Energy Mater. Sol. Cells* 34, 557 (1994).

597. M. Zeman, R. A. C. M. M. van Swaaij, E. Schroten, L. L. A. Vosteen, and J. W. Metselaar, *Mater. Res. Soc. Symp. Proc.* 507, 409 (1998).

598. M. Zeman, J. A. Willemen, L. L. A. Vosteen, G. Tao, and G. W. Metselaar, *Sol. Energy Mater. Sol. Cells* 46, 81 (1997).

599. X. Xu, J. Yang, and S. Guha, *Appl. Phys. Lett.* 62, 1399 (1993).

600. Y. Lee, L. Jiao, Z. Lu, R. W. Collins, and C. R. Wronski, *Sol. Energy Mater. Sol. Cells* 49, 149 (1997).

601. B. W. Faughnan and R. Crandall, *Appl. Phys. Lett.* 44, 537 (1984).

602. Z. E. Smith, S. Wagner, and B. W. Faughnan, *Appl. Phys. Lett.* 46, 1078 (1985).

603. A. Catalano, R. Bennett, R. Arya, K. Rajan, and J. Newton, "Proceedings of the 18th IEEE Photovoltaic Specialists Conference, Las Vegas, 1985," p. 1378. IEEE, New York, 1985.

604. M. Stutzmann, W. B. Jackson, and C. C. Tsai, *Phys. Rev. B* 34, 63 (1986).

605. R. E. I. Schropp, A. Sluiter, M. B. von der Linden, and J. Daey Ouwens, *J. Non-cryst. Solids* 164–166, 709 (1993).

606. D. E. Carlson, K. Rajan, R. R. Arya, F. Willing, and L. Yang, *J. Mater. Res.* 13, 2754 (1998).

607. Y. Hamakawa, W. Ma, and H. Okamoto, in "Plasma Deposition of Amorphous Silicon-Based Materials" (G. Bruno, P. Capezzuto, and A. Madan, Eds.), Chap. 6, p. 283. Academic Press, Boston, 1995.

608. L. L. Kazmerski, *Renew. Sustain. Energy Rev.* 1, 71 (1997).

609. A. V. Shah, R. Platz, and H. Keppner, *Sol. Energy Mater. Sol. Cells* 38, 501 (1995).

610. T. Strand, L. Mrig, R. Hansen, and K. Emery, *Sol. Energy Mater. Sol. Cells* 41/42, 617 (1996).

611. R. Ruther and J. Livingstone, *Sol. Energy Mater. Sol. Cells* 36, 29 (1996).

612. J. Merten and J. Andreu, *Sol. Energy Mater. Sol. Cells* 52, 11 (1998).

613. J. Yang, A. Banerjee, T. Glatfelter, K. Hoffman, X. Xu, and S. Guha, Proceedings of the 24th IEEE Photovoltaic Specialists Conference, Waikoloa HI, 1994," p. 380. IEEE, New York, 1995.

614. S. Guha, J. Yang, A. Banerjee, T. Glatfelter, and S. Sugiyama, *Sol. Energy Mater. Sol. Cells* 48, 365 (1997).

615. S. Kiyama, S. Nakano, Y. Domoto, H. Hirano, H. Tarui, K. Wakisaka, M. Tanaka, S. Tsuda, and S. Nakano, *Sol. Energy Mater. Sol. Cells* 48, 373 (1997).

616. Y. Hamakawa, *Appl. Surf. Sci.* 142, 215 (1999).

617. S. M. Sze, "Semiconductor Devices, Physics and Technology." Wiley, New York, 1985.

618. M. J. Powell, *IEEE Trans. Elec. Dev.* ED-36, 2753 (1989).

619. K. Suzuki, in "Amorphous and Microcrystalline Semiconductor Devices—Optoelectronic Devices" (J. Kanicki, Ed.), Chap. 3, p. 77. Artech House, Norwood, MA, 1991.

620. A. Iqbal, W. B. Jackson, C. C. Tsai, J. W. Allen, and C. W. Bates, *J. Appl. Phys.* 61, 2947 (1987).

621. C. van Berkel, in "Amorphous and Microcrystalline Semiconductor Devices—Materials and Device Physics" (J. Kanicki, Ed.), Chap. 8, p. 397. Artech House, Norwood, MA, 1992.

622. W. B. Jackson and M. Moyer, *Mater. Res. Soc. Symp. Proc.* 118, 231 (1988).

623. J. Kakalios, R. A. Street, and W. B. Jackson, *Phys. Rev. B* 59, 1037 (1987).

624. W. B. Jackson, J. M. Marshall, and M. D. Moyer, *Phys. Rev. B* 39, 1164 (1989).

625. A. M. Brockhoff, E. H. C. Ullersma, H. Meiling, F. H. P. M. Habraken, and W. F. van der Weg, *Appl. Phys. Lett.* 73, 3244 (1998).

626. C. van Berkel and M. J. Powell, *J. Appl. Phys.* 60, 1521 (1986).

627. M. Katayama, *Thin Solid Films* 341, 140 (1999).

628. H. C. Tuan, *J. Non-cryst. Solids* 115, 132 (1989).

629. H. Matsumura, H. Hayama, Y. Nara, and K. Ishibashi, *Japan. J. Appl. Phys.* 20, 311 (1980).

630. R. L. Weisfield, *IEEE Trans. Electron. Dev.* ED-36, 2935 (1989).

631. I. Shimizu, S. Shirai, and E. Inoue, *J. Appl. Phys.* 52, 2776 (1981).

632. I. Shimizu, in "Semiconductors and Semimetals" (J. I. Pankove, Ed.), Vol. 21D, p. 55. Academic Press, Orlando, FL, 1984.

633. R. Schaffert, "Electrophotography." Focal, London, 1975.

634. J. Kodama, S. Araki, M. Kimura, and T. Inagaki, *Japan. J. Appl. Phys.* 29, L867 (1990).

635. R. A. C. M. M. van Swaaij, W. P. M. Willems, J. Bezemer, H. J. P. Lokker, and W. F. van der Weg, *Mater. Res. Soc. Symp. Proc.* 297, 559 (1993).

636. R. A. C. M. M. van Swaaij, S. J. Elmer, W. P. M. Willems, J. Bezemer, J. M. Marshall, and A. R. Hepburn, *J. Non-cryst. Solids* 164–166, 533 (1993).

637. T. Takeda and S. Sano, *Mater. Res. Soc. Symp. Proc.* 118, 399 (1988).

638. E. Fortunato, M. Vieira, I. Ferreira, C. N. Carvalho, G. Lavareda, and R. Martins, *Mater. Res. Soc. Symp. Proc.* 297, 981 (1993).

639. E. M. C. Fortunato, D. Brida, I. M. M. Ferreira, H. M. B. Águas, P. Nunes, A. Cabrita, F. Guiliani, Y. Nunes, M. J. P. Maneira, and R. Martins, *Mater. Res. Soc. Proc. Symp.* 609, A12.7.1 (2000).

640. H. Stiebig and N. Bohm, *J. Non-cryst. Solids* 164–166, 789 (1993).

641. R. Brüggemann, T. Neidlinger, and M. B. Schubert, *J. Appl. Phys.* 81, 7666 (1997).

642. G. de Cesare, F. Irrera, F. Lemmi, and F. Palma, *Appl. Phys. Lett.* 66, 1178 (1995).

643. M. Topič, F. Smole, J. Furlan, and W. Kusian, *J. Non-cryst. Solids* 227–230, 1326 (1998).

644. J. Krč, M. Topič, F. Smole, D. Knipp, and H. Stiebig, *Mater. Res. Soc. Proc. Symp.* 609, A12.5.1 (2000).

645. D. Caputo and G. de Cesare, *Sensors and Actuators A* 78, 108 (1999).

646. D. Caputo, G. de Cesare, A. Nascetti, F. Palma, and M. Petri, *Appl. Phys. Lett.* 72, 1229 (1998).

647. R. A. Street, X. D. Wu, R. Weisfield, S. Ready, R. Apte, M. Nguyen, and P. Nylen, *Nucl. Instrum. Methods Phys. Res. A* 380, 450 (1996).

648. R. A. Street, R. B. Apte, T. Granberg, M. P., S. E. Ready, K. S. Shah, and R. L. Weisfield, *J. Non-cryst. Solids* 227–230, 1306 (1998).

649. J.-P. Moy, *Nucl. Instrum. Methods Phys. Res. A* 442, 26 (2000).

650. J. C. Chou, Y.-F. Wang, and J. S. Lin, *Sensors and Actuators B* 62, 92 (2000).

651. J. C. Chou, H. M. Tsai, C. N. Shiao, and J. S. Lin, *Sensors and Actuators B* 62, 97 (2000).

652. S. Jamash, S. D. Collins, and R. L. Smith, *IEEE Trans. Electron. Dev.* ED-45, 1239 (1998).

653. P. F. Ruths, S. Askok, S. J. Fonash, and J. M. Ruths, *IEEE Trans. Electron. Dev.* ED-28, 1003 (1981).

654. E. Fortunato, A. Malik, A. Sêco, I. Ferreira, and R. Martins, *J. Non-cryst. Solids* 227–230, 1349 (1998).

655. F. V. Hunt, "Electroacoustics, the Analysis of Transduction and Its Historical Background." American Institute of Physics, New York, 1982.

656. M. Smits, private communication.

657. R. E. I. Schropp, M. Smits, H. Meiling, W. G. J. H. M. van Sark, M. M. Boone, and W. F. van der Weg, *Mater. Res. Soc. Symp. Proc.* 219, 519 (1991).

658. P. J. Mears, L. Reekie, I. M. Jauncey, and D. N. Payne, *Electron. Lett.* 23, 1026 (1987).

659. E. Desurvire, R. J. Simpson, and P. C. Becker, *Opt. Lett.* 12, 888 (1987).

660. G. N. van den Hoven, A. Polman, C. van Dam, J. W. M. van Uffelen, and M. K. Smit, *Appl. Phys. Lett.* 68, 1886 (1996).

661. A. Polman, *J. Appl. Phys.* 82, 1 (1997).

662. F. Priolo, G. Franzò, S. Coffa, and A. Carnera, *Mat. Chem. Phys.* 54, 273 (1998).

663. F. Priolo, G. Franzò, S. Coffa, and A. Carnera, *Phys. Rev. B* 57, 4443 (1998).

664. G. N. van den Hoven, J. H. Shin, A. Polman, S. Lombardo, and S. U. Campisano, *J. Appl. Phys.* 78, 2642 (1995).

665. J. H. Shin, R. Serna, G. N. van den Hoven, A. Polman, W. G. J. H. M. van Sark, and A. M. Vredenberg, *Appl. Phys. Lett.* 68, 697 (1996).

666. A. Polman, J. H. Shin, R. Serna, G. N. van den Hoven, W. G. J. H. M. van Sark, A. M. Vredenberg, S. Lombardo, and S. U. Campisano, *Mater. Res. Soc. Symp. Proc.* 422, 239 (1996).

667. S. Coffa, G. Franzò, A. Polman, and R. Serna, *Phys. Rev. B* 49, 16313 (1994).

668. P. G. Kik, M. J. A. de Dood, K. Kikoin, and A. Polman, *Appl. Phys. Lett.* 70, 1721 (1997).

669. F. Priolo, G. Franzò, S. Coffa, A. Polman, S. Libertino, R. Barklie, and D. Carey, *J. Appl. Phys.* 78, 3874 (1995).

670. G. Franzò, F. Priolo, S. Coffa, A. Polman, and A. Carnera, *Appl. Phys. Lett.* 64, 2235 (1994).

671. B. Zheng, J. Michel, F. Y. G. Ren, L. C. Kimerling, D. C. Jacobson, and J. M. Poate, *Appl. Phys. Lett.* 64, 2842 (1994).

672. J. Palm, F. Gan, B. Zheng, J. Michel, and L. C. Kimerling, *Phys. Rev. B* 54, 17603 (1996).

673. S. Lombardo, S. U. Campisano, G. N. van den Hoven, and A. Polman, *J. Appl. Phys.* 77, 6504 (1995).

674. L. H. Slooff, Master's thesis, FOM—Institute for Atomic and Molecular Physics, Amsterdam, 1996.

675. J. Stimmer, A. Reittinger, J. F. Nützel, G. Abstreiter, H. Holzbrecher, and C. Buchal, *Appl. Phys. Lett.* 68, 3290 (1996).

676. W. Fuhs, I. Ulber, G. Weiser, M. S. Bresler, O. B. Gusev, A. N. Kuznetsov, V. K. Kudoyarova, E. I. Terukov, and I. N. Yassievich, *Phys. Rev. B* 56, 9545 (1997).

677. E. I. Terukov, O. I. Kon'kov, V. K. Kudoyarova, O. B. Gusev, and G. Weiser, *Semiconductors* 24, 884 (1998).

678. E. I. Terukov, M. M. Kazanin, O. I. Kon'kov, V. K. Kudoyarova, K. V. Kougiya, Y. A. Nikulin, and Kazanskii, *Semiconductors* 34, 829 (2000).

679. V. B. Voronkov, V. G. Golubev, A. V. Medvedev, A. B. Pevtsov, N. A. Feoktistov, N. I. Gorshkov, and D. N. Suglobov, *Phys. Solid State* 40, 1301 (1998).

680. J. H. Shin and M. Kim, *J. Vac. Sci. Technol. A* 17, 3230 (1999).

681. A. Masuda, J. Sakai, H. Akiyama, O. Eryu, K. Nakashima, and H. Matsumura, *J. Non-cryst. Solids* 226–229, 136 (2000).

682. T. Förster, *Discuss. Faraday Soc.* 27, 7 (1959).

683. H. Kühne, G. Weiser, E. I. Terukov, A. N. Kuznetsov, and V. K. Kudoyarova, *J. Appl. Phys.* 86, 1896 (1999).

684. M. B. Schubert, A. Hierzenberger, H. J. Lehner, and J. H. Werner, *Sensors Actuators A* 74, 193 (1999).

685. E. Betzig and J. K. Trautmann, *Science* 257, 189 (1992).

686. E. Betzig, P. L. Finn, and J. S. Weiner, *Appl. Phys. Lett.* 60, 2484 (1992).

687. S. Madsen, M. Mullenborn, K. Birkelund, and F. Grey, *Appl. Phys. Lett.* 69, 544 (1996).

688. S. Madsen, S. I. Bozhevolnyi, K. Birkelund, M. Mullenborn, J. M. Hvam, and F. Grey, *J. Appl. Phys.* 82, 49 (1997).

689. M. K. Herndon, R. T. Collins, R. E. Hollingsworth, P. R. Larson, and M. B. Johnson, *Appl. Phys. Lett.* 74, 141 (1999).

690. W. Gao, S. H. Lee, J. Bullock, Y. Xu, D. K. Benson, S. Morrison, and H. M. Branz, *Sol. Energy Mater. Sol. Cells* 59, 243 (1999).

Chapter 2

ATOMIC LAYER DEPOSITION

Mikko Ritala, Markku Leskelä

Department of Chemistry, University of Helsinki, FIN-00014 Helsinki, Finland

Contents

1. INTRODUCTION

Atomic layer deposition (ALD) is a chemical gas phase thin film deposition method based on alternate saturative surface reactions. As distinct from the other chemical vapor deposition techniques, in ALD the source vapors are pulsed into the reactor alternately, one at a time, separated by purging or evacuation periods. Each precursor exposure step saturates the surface with a monomolecular layer of that precursor. This results in a unique self-limiting film growth mechanism with a number of advantageous features, such as excellent conformality and uniformity, and simple and accurate film thickness control.

ALD was developed and world widely introduced with a name of atomic layer epitaxy (ALE) (for other names, see Section 2) in the late 1970s by Suntola and co-workers in Finland [1–4]. The motivation behind developing ALD was the desire to make thin film electroluminescent (TFEL) flat panel displays [2, 5–9]. This is a demanding application since electric fields in the range of megavolts per centimeter are applied across

Handbook of Thin Film Materials, edited by H.S. Nalwa
Volume 1: Deposition and Processing of Thin Films

ISBN 0-12-512909-2/$35.00

polycrystalline or amorphous luminescent and insulator films (Sections 6.1. and 6.2). Nevertheless, ALD was successful in meeting the requirements of high dielectric strength, low pinhole density, and uniformity over the large-area substrates, and it has been employed in TFEL production since the early 1980s [7–11].

Soon after the successful introduction of ALD, its applicability to epitaxial compound semiconductors was demonstrated by several groups [12–15], and since that semiconductors, especially the III–V compounds, have been the most extensively examined materials [16–25]. However, though many outstanding results have been achieved, the overall success in that field has remained limited with no reported commercial applications. Meanwhile, new nonepitaxial applications were taken into research more slowly but steadily. Solar cells, microelectronics, optics, protective applications, and gas sensors have been among the areas examined.

Since the mid 1990s, rapidly increasing interest toward ALD has arisen from the silicon-based microelectronics. This increase is a direct consequence of the ever-decreasing device dimensions and the increasing aspect ratios in the integrated circuits (IC). The thin film deposition techniques used for a long time in the IC industry are foreseen to meet major conformality problems during the next few years, and ALD is currently considered as one of the most potential substitutes for them. At the same time, the films are shrinking so thin that the major concern of ALD, the low deposition rate, is becoming less important.

In addition to thin films on planar substrates, ALD has also been examined in surface processing of porous materials. A great deal of this work has concentrated on chemical modification of high surface area silica and alumina powders for heterogeneous catalysts [26–33] but also nanoporous silicon layers [34, 35] and alumina membranes [36, 37] have been processed.

As the current interest in ALD is largely centered on nonepitaxial films, this review chapter mainly focuses on these materials and the related chemistry. The growth of epitaxial semiconductors has been thoroughly reviewed elsewhere [16–25] and is touched here only from the chemistry point of view. In addition, the ALD processing of high surface area powders has been summarized in a number of articles [30–33] and is considered here only as a valuable source of information on the growth mechanisms (Section 7.2.2). In addition to these and other, more general review articles [5–9, 38–42], the proceedings of the biannually held topical international symposia [43–46] give a good overview of the development of the ALD method.

We start with a few notes on the alternative names of ALD, followed by an introduction of the basic features, benefits, and limitations of the method. After examining ALD reactors, precursors will be discussed. This is followed by a summary of ALD made film materials. Finally, the various ways of optimizing and characterizing ALD processes are surveyed.

2. ALTERNATIVE NAMES

At the time of its introduction, the method was given a name atomic layer epitaxy (ALE) [1, 2] where the term "epitaxy," coming from the Greek and meaning "on-arrangement," was used to emphasize the sequentially controlled surface reactions upon the previously deposited layer [7, 38]. However, as "epitaxy" is more commonly used in describing a growth of a single crystalline film on a single crystalline substrate with a well-defined structural relationship between the two, unfortunate confusions have arisen when applying the name ALE in the case of amorphous or polycrystalline films. For this and other reasons, many synonyms, all basically referring to the same method, have been suggested for ALE during the years (Table I). Among these, ALD appears to have become the most widely used one and is to be understood as a general name of this method covering all kinds of films, whereas ALE should now be reserved for epitaxial films only. Though the words "atomic" and "layer" have also been questioned (Table I), their continuous use is strongly motivated by the desire to keep the terminology consistent, thereby assisting the follow-up of the literature.

3. BASIC FEATURES OF ALD

3.1. ALD Cycle

In ALD, the film growth takes place in a cyclic manner. In the simplest case, one cycle consists of four steps: (i) exposure of the first precursor, (ii) purge or evacuation of the reaction chamber, (iii) exposure of the second precursor, and (iv) purge or evacuation. This cycle is repeated as many times as necessary

Table I. Alternative Names to the ALD Method

Name	Acronym	Comments
Atomic layer deposition	ALD	General, covers all kinds of films
		In a close connection with the original name
Atomic layer epitaxy	ALE	The original name, but should be reserved for epitaxial films only
Atomic layer growth	ALG	Like ALD but less used
Atomic layer chemical vapor deposition	ALCVD	Emphasizes the relation to CVD
Molecular layer epitaxy	MLE	Emphasizes molecular compounds as precursors
Digital layer epitaxy	DLE	Emphasizes the digital thickness control
Molecular layering	ML	Dates back to old Russian literature
Successive layerwise chemisorption		
Sequential surface chemical reaction growth		
Pulsed beam chemical vapor deposition		

a)   b)

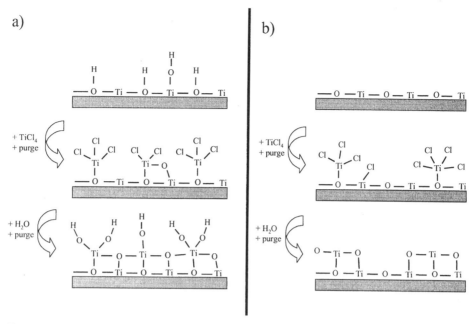

Fig. 1.   An ALD film deposition cycle shown schematically with the TiCl$_4$-H$_2$O process as an example. To illustrate the importance of surface species possibly remaining after the water pulse, or more generally, after the nonmetal precursor pulse, two hypothetical reaction routes have been sketched: (a) hydroxyl group terminated and (b) dehydroxylated surface after the water pulse.

to obtain the desired film thickness. Depending on the process, one cycle deposits 0.1- to 3-Å film thickness.

The actual reactions taking place under each exposure step depend largely on the presence or the absence of reactive functional groups on the surface of the growing film. In the case of oxides, for example, surface hydroxyls often terminate the surface after exposure to water. Here, the ALD process is schematically illustrated by an example of the growth of TiO$_2$ from TiCl$_4$ and H$_2$O (Fig. 1). To emphasize the importance of the hydroxyl groups, two alternative reaction routes have been sketched out, one with a hydroxyl group terminated surface (Fig. 1a) and the other with a completely dehydroxylated surface (Fig. 1b). In reality, the hydroxyl coverage is temperature dependent, and thus the two routes may be mixed.

When dosed onto the surface terminated with functional (here hydroxyl) groups, the precursor molecules react with these groups (Fig. 1a) releasing some of their ligands. If there are no functional groups on the surface, the incoming molecule can only chemisorb, either intact or dissociatively (Fig. 1b). As a consequence of a finite number of the reaction or chemisorption sites on the surface, in maximum only a monomolecular layer of the precursor becomes firmly bound to the surface. The following purge (or evacuation) period removes all the excess precursor molecules and volatile byproducts leaving only the monolayer on the surface. For simplicity, from here on, this monolayer will be called a chemisorption layer regardless of whether it is formed via exchange reactions (Fig. 1a) or true chemisorption (Fig. 1b). Subsequently, the other precursor is dosed in and reacts with the chemisorption layer liberating the ligands and producing the desired solid. At the same time, the surface is converted back to its original state, i.e., in our example either hydroxyl terminated (Fig. 1a) or bare oxide sur-

face (Fig. 1b). The second purge period completes the deposition cycle which is then repeated.

Ideally, each exposure and purge step is complete. The precursor molecules chemisorb or react with the surface groups saturatively so that after the formation of the chemisorption monolayer no further adsorption takes place, and the purge periods remove all the excess precursor and volatile byproduct molecules. Under these conditions, the film growth is self-limiting: the amount of the film material deposited during each cycle is the same and is determined only by the density of the chemisorption or reaction sites at the surface. The self-limiting growth mechanism gives ALD a number of advantageous features (Table II) which are discussed in more detail in the next section.

Most of the ALD processes are based on the above described exchange reactions between molecular precursors. Another possible reaction type is additive with elemental precursors but because only a few metals are volatile enough, the applicability of these reactions is limited. The third reaction type, quite rare as well, involves a self-limiting adsorption of a precursor followed by its decomposition by an appropriate energy pulse (Section 5.3.4).

In a majority of the ALD processes reported, the reactions are activated only thermally under isothermal conditions. However, additional activation methods have also been examined. These include light irradiation with lasers [15, 17, 19, 47–51] or lamps [52–57] for promoting photolytic or photothermal reactions, and upstream generation of reactive radicals by thermal cracking [58–69] or with plasma discharges [70–75].

The alternate pulsing is definitely the most characteristic feature of ALD but almost as distinctive is the self-limiting growth mechanism. However, some deviations from the absolutely self-

Table II.   Characteristic Features of ALD with the Consequent Advantages

Characteristic feature of ALD	Inherent implication on film deposition	Practical advantage
Self-limiting growth process	Film thickness is dependent only on the number of deposition cycles	Accurate and simple thickness control
	No need of reactant flux homogeneity	Large-area capability
		Large-batch capability
		Excellent conformality
		No problems with inconstant vaporization rates of solid precursors
		Good reproducibility
		Straightforward scale-up
	Atomic level control of material composition	Capability to produce sharp interfaces and superlattices
		Possibility to interface modification
Separate dosing of reactants	No gas phase reactions	Favors precursors highly reactive toward each other, thus enabling effective material utilization
	Sufficient time is provided to complete each reaction step	High quality materials are obtained at low processing temperatures
Processing temperature windows are often wide	Processing conditions of different materials are readily matched	Capability to prepare multilayer structures in a continuous process

limited growth conditions may be accepted with certain precautions. The reactions may be somewhat incomplete leaving the chemisorption layer short from saturation, or there may be some precursor decomposition in addition to the chemisorption. As long as the reactions causing these nonidealities proceed in a surface controlled rather than a mass transport controlled manner, the coverage of the chemisorption layer and thereby the deposition rate remain everywhere the same, thus maintaining the advantageous characteristics of ALD. Alternatives to accepting these nonidealities would be to increase the exposure time to complete the reactions, or to lower the deposition temperature to avoid the decomposition, but these would be accompanied by negative effects of increased process time and reduced film quality.

A common misconception is that ALD growth always proceeds in a layer-by-layer manner but this is often not the case as only a fraction of a monolayer may be deposited in each cycle. Reasons for the less than a monolayer per cycle growth are the limited number of reactive surface sites [33, 42] and the steric hindrances between bulky ligands in the chemisorption layer [8, 76]. As a consequence, even if saturatively formed, the chemisorption layer contains too few metal atoms for forming a full monolayer of the film material. On the other hand, also surface reconstructions may cause a decrease in the growth rate [77, 78], though in the apparently exceptional case of the 2 ML cycle^{-1} growth of AlAs they appear to have the opposite effect [19, 79–81].

Yet another misconception is that ALD would produce atomically smooth films. This indeed may often be the case with epitaxial or amorphous films, but the nucleation and grain growth involved in the formation of polycrystalline films usually lead to a measurable surface roughness which increases along with film thickness (Section 7.2.1) [39, 82–87].

3.2. Benefits of ALD

Table II summarizes the relationships between the characteristic features of the ALD process and the resulting benefits. The benefits are now discussed with selected examples. It is worth emphasizing already here that the achievement of these benefits requires appropriate chemistry fulfilling the special requirements set for ALD processes (Section 5).

As the film growth proceeds in a self-limiting manner, each cycle deposits exactly the same amount of material, and thus the film thickness may be accurately controlled simply by the number of deposition cycles. However, it must be recognized that during the very first cycles when the surface is converted from the substrate to the film material, the surface density of the chemisorption or reactive sites and, accordingly, the growth rate may change.

The self-limiting growth mechanism also ensures that the precursor fluxes do not need to be uniform over the substrate. What is only needed is a flux which is large enough so that the chemisorption layer becomes saturated, all the excess will be subsequently purged away. As a result, ALD provides perfect conformality and trench-fill capability (Fig. 2) [88] as well as good large-area uniformity [88, 89] and large-batch capability. Batches as large as 82 substrates of a size 15.5×26.5 cm^2 [41] or 42 times 40×50 cm^2 substrates [90] may be processed in the existing ALD reactors, and this is certainly not the upper limit. The self-limiting growth mechanism also ensures good reproducibility and relatively straightforward scale-up.

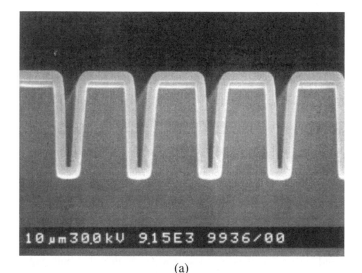

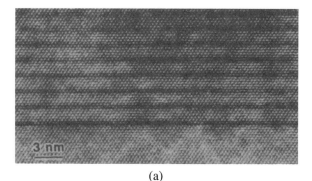

(a)

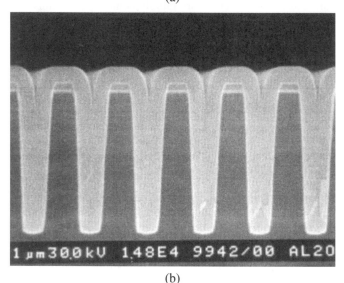

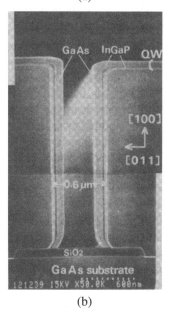

(a)

(b)

(b)

Fig. 2. Cross-sectional SEM images of a 300-nm Al_2O_3 film deposited onto a patterned silicon substrate showing (a) perfect conformality and (b) trench-fill capability [88]. Note that on the top surface of the silicon wafer there is a thermal silicon oxide layer below the Al_2O_3 film. Reprinted with permission from M. Ritala et al., Perfectly conformal TiN and Al_2O_3 films deposited by atomic layer deposition, *Chem. Vapor Deposition* 5, 7 (1999), © 1999, Wiley-VCH Verlag GmbH.

Fig. 3. (a) (110) cross-sectional transmission electron microscopy (TEM) micrograph of a 1-ML InAs – 5-ML GaAs superlattice. (b) Cross-sectional scanning electron microscopy (SEM) micrograph of an $In_{1-x}Ga_xP/GaAs$ single quantum well structure on a (011) sidewall of a GaAs substrate [18]. Reprinted with permission from A. Usui, Atomic layer epitaxy of III–V compounds: Chemistry and applications, *Proc. IEEE* 80, 1641 (1992), © 1992, IEEE.

Precursor flux homogeneity is not required in timescale either. Therefore, solid sources which often have inconstant sublimation rates are more easily adopted in ALD than in chemical vapor deposition (CVD). The separate dosing of the precursors, in turn, naturally eliminates all possible detrimental gas phase reactions.

As the film is deposited in a layerwise manner, either a full monolayer or a fraction of a monolayer at a time, the material composition may be controlled down to a nanometer level, and in the most ideal cases even to an atomic level. This offers a simple way to create superlattices (Fig. 3a) [7, 18, 91, 92] and other multilayer structures with accurately controlled layer thicknesses, such as optical multilayers tailored for soft X-ray or visible range (Section 6.8), or nanolaminate insula-

tors (Section 6.2). An outstanding demonstration of the accurate film thickness control and perfect conformality is shown in Figure 3b where a single quantum well structure has been grown into a trench. Note also the selective growth between the GaAs and SiO_2 surfaces.

Practice has shown that ALD made films often, though of course not always, possess superior quality as compared with films made by other methods at the corresponding temperatures. This can be related to the fact that in ALD each monomolecular layer reaction step is given enough time to reach completion while in other methods the continuous growth may prevent this by covering the unreacted species with new deposits.

Many ALD processes may be performed over a relatively wide temperature range. Therefore, a common growth temperature is often found for different materials, thereby making it possible to deposit multilayer structures in a continuous manner. This is utilized, for example, in manufacturing of the TFEL displays where the insulator-luminescent material-insulator three-layer structure (Section 6.1) is grown in one continuous process.

3.3. Limitations of ALD

The major limitation of ALD is evidently its slowness since at best a monolayer of the film is deposited during one cycle. Typically, deposition rates of 100–300 nm h^{-1} are obtained and 1 μm h^{-1} appears to be the upper limit for most processes, though a record rate of 2 μm h^{-1} was achieved for epitaxial GaAs in a specially designed reactor (Section 4.3) [93]. However, the low growth rate does not necessarily mean low productivity when the latter is considered in terms of film volume (= thickness × area) produced per time unit. By making use of the good large-batch and large-area processing capabilities of ALD, the low growth rate may be effectively compensated for. On the other hand, there are certain application areas, in particular integrated circuits, where single substrate processing is increasingly preferred because of the cost risks related to losing a batch of valuable substrates. Luckily, at the same time, the film thicknesses have shrunk down to a level where the low growth rate of ALD is losing its significance when weighted against the potential benefits.

One limitation to a widespread use of ALD has been the lack of good and cost-effective processes for some important materials. Among these are, for example, metals, Si, SiO_2, Si_3N_4, and several ternary and multicomponent materials. Many of these materials have been made by ALD (Table V in Section 6) but the processes are still far from ideal in a sense that long cycle times are required.

4. ALD REACTORS

4.1. Overview

Just like CVD, ALD processes may also be performed in many kinds of reactors over a wide pressure range from atmospheric to ultrahigh vacuum (UHV). ALD reactors can be divided into two groups: inert gas flow reactors operating under viscous or transition flow conditions at pressures higher than about 1 Torr, and high- or ultrahigh-vacuum reactors with molecular flow conditions. The former resemble CVD reactors while the latter are like molecular beam epitaxy (MBE) reactors. In any case, when the special features and the needs of ALD are taken into consideration in the reactor design, some significant differences emerge as discussed in this section.

The main parts of an ALD reactor are:

1. transport gas supply (if any),
2. sources of one or several of the following types: gas, liquid, and solid sources,
3. flow and sequencing control of the sources,
4. reaction chamber,
5. temperature control of the heated sources and the reaction chamber (and the substrate, if heated separately),
6. vacuum pump and related exhaust equipment.

Provisional items include:

7. *in situ* surface and gas phase analysis equipments, such as an ellipsometer and mass spectrometer, for process characterization and control (Section 7.2),
8. load-locks for changing substrates or sources,
9. separate preheating chamber for the substrates.

The last one is especially effective in maximizing throughput in large-batch processing. Besides equipping with a load-lock, often it may be desirable to make the ALD reactor a part of a larger cluster tool for an inert transfer of the substrates from one process step to another.

As already noted, in ALD no strict precursor flux homogeneity is required. This gives freedom in the reactor design and assists in constructing large-batch reactors. On the other hand, cost-effectiveness of the process requires that both exposure and purge sequences are rapidly completed. For this reason, flow-type reactors are preferred especially in production scale use: purging of a properly designed reactor under viscous flow conditions is much more rapid than evacuation of a high-vacuum chamber. Another crucial aspect related to purging or evacuation is that the chamber and the lines which are to be purged or evacuated should have small volume and uniformly heated walls. Otherwise, long purge periods may be required to desorb precursor molecules adsorbed on "cold spots" on the walls. This is especially important in the growth of oxides where water is often used as an oxygen source.

The desire for high precursor utilization efficiency is also an important aspect for the reactor design. In high-vacuum reactors, the precursor molecules make at best only a few collisions with the substrate surface before being pumped out and thus have a limited probability to react. On the other hand, in the flow-type reactors, and in the traveling-wave reactors (Section 4.2.4) in particular, a precursor molecule makes multiple hits while being transported through the reactor. As a consequence, reaction probability and thereby precursor utilization efficiency are increased, and the saturation of the chemisorption layer is accelerated which makes the process faster.

MBE types of reactors are basically easy to operate in the ALD mode since they are usually equipped with shutters by means of which precursor fluxes from the Knudsen effusion cells can be chopped. When volatile reactants are used, their pulsing is realized with valves. What is still needed to make an ALD process is a proper choice of substrate temperature and evacuation periods between the alternate precursor pulses to ensure the self-limiting growth conditions. The major benefit of the MBE type of reactors is the rich variety of *in situ* analytical techniques which may be implemented for detailed reaction mechanism studies. Large chambers also make it easy to employ various sources of extra energy for activating the reactions (Fig. 4). On the other hand, the need of long exposure and evacuation times limits the throughput.

CVD reactors with liquid and gas sources only are also relatively easy to operate in the ALD mode. For pulsing the precursors, simple solenoid or pneumatic valves may be used. However, if the valves are far from the reaction chamber and there

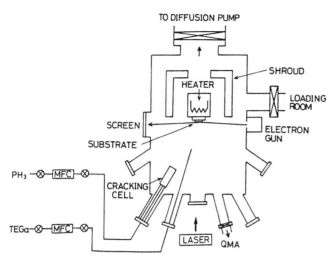

Fig. 4. An example of a high-vacuum ALD reactor incorporating RHEED (electron gun and screen) and a quadrupole mass analyzer (QMA) for *in situ* characterization, and a laser and a thermal cracking cell for activating the reactions [49]. Reprinted with permission from M. Yoshimoto, A. Kajimoto, and H. Matsunami, Laser-assisted atomic layer epitaxy of GaP in chemical beam epitaxy, *Thin Solid Films* 225, 70 (1993), © 1993, Elsevier Science.

Fig. 5. A photograph of an ASM Microchemistry F-950 ALD reactor for production of flat panel displays and solar cells. (Courtesy ASM Microchemistry, Espoo, Finland.)

are long tubings in between, their effective purging calls for special attention. Even more problematic is the pulsing of solid sources which need to be heated far above room temperature where no mechanical valves can be used. In the flow-type ALD reactors described in the next two sections, these issues have been specially addressed.

4.2. Flow-Type ALD Reactors with Inert Gas Valving

The ALD reactors commercially available [90] and those in industrial use in TFEL manufacturing are all of a flow type with inert gas valving [3, 4, 38]. These reactors are carefully designed to minimize the pulse and purge times and to maximize the precursor utilization efficiency. They are available in both small size for research and large size for production [90] (Fig. 5). The main points of these reactor designs are now examined. The discussion is limited mainly to those features which are special to ALD whereas more common issues, like mass flow controllers, pumps, heaters, and their controllers etc., are not dealt with. For a detailed model of mass transport in the flow-type ALD reactors, see Ref. [94].

4.2.1. Transport Gas

Inert gas, most often nitrogen, is used for transporting the precursor vapors and purging the reaction chamber. A continuous flow of the inert gas is regulated with mass flow controllers. The flow rate is optimized together with the total pressure and the reactor geometry to give the best combination of speed, transportation and utilization efficiency of the reactants, inert gas valving (Section 4.2.3), and the consumption of the inert gas itself. In small research reactors, the flow rate is about 0.5 slm whereas the largest production reactors require 20–50 slm. The operating pressures are typically in the range of 1–10 Torr.

A crucial aspect of the transport gas is purity because it is the main source of impurities in an ALD process. In the minimum, 99.999% purity should be used. In Section 4.2.4, a calculation of the effects of the impurities is presented.

4.2.2. Sources

The source chemicals may be divided into two groups depending on if their vapor pressures at room temperature, or close to it, are higher or lower than the total pressure inside the reactor (1–10 Torr). The high vapor pressure sources; i.e., gases and highly volatile liquids, are led into the reactor from their external cylinders or vessels by pulsing with fast valves. No bubblers or transport gases are necessary since the source vapors enter the reactor as a consequence of the pressure difference. Once entered into the reactor, the vapor is transported further by the carrier gas flow inside the reactor. However, as large scale manufacturing requires large doses, carrier gas transportation already from the source vessel may become necessary.

For low vapor pressure source chemicals, two options exist. The first one is to have them in external vessels which, together with the source lines and the required valving systems, are properly heated to ensure effective transportation. The second option is to place the source chemicals inside the reactor where they are heated to temperatures where appropriate vapor pressures are obtained. In research reactors, for instance, a vapor pressure around 0.1 Torr has been found to be a good starting point for a new precursor. To realize pulsing of solid sources inside the reactor, a clever inert gas valving system (next section) was developed [3, 4].

4.2.3. Inert Gas Valving

An exact mathematical description of the inert gas valving system has been given elsewhere [3, 4, 38], so here only the basic principle is introduced. The inert gas valving is based on a

a

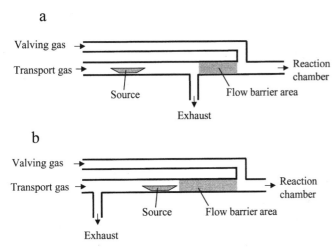

b

Fig. 6. A schematic of the inert gas valving system where the direct connection to the exhaust is (a) between the source and the reaction chamber, or (b) behind (upstream of) the source. The shading shows the area where the flow barrier is formed when set on.

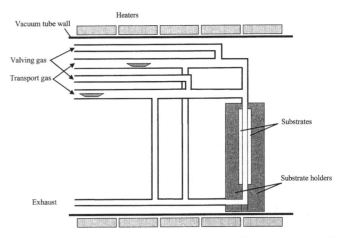

Fig. 7. Schematics of a small research ALD reactor with two inert gas valved source lines and a traveling-wave reaction chamber inserted to a main vacuum tube (not to scale). The intermediate space contains inert gas at a pressure slightly higher than that inside the tubings and the reaction chamber.

design where each source tube is connected both to a reaction chamber and directly to an exhaust so that the exhaust connection is either between the source and the chamber (Fig. 6a) or behind the source (Fig. 6b). Each tube is also equipped with two inert gas flows: one is the transport gas and the other is the valving gas. The valving gas enters the source tube between the source and the reaction chamber. A practical way to realize the two flows is to employ coaxial tubes [3, 4].

When the source is in its off state, the valving gas flow is on and the transport gas flow is off. The valving gas flow divides into two parts, one purging the reaction chamber and the other one setting up a laminar flow barrier (shaded areas in Fig. 6). The flow rate of the valving gas in the flow barrier is adjusted to counterbalance the diffusion rate of the precursor toward the chamber so that the precursor vapor pressure in the chamber does not exceed a predetermined level, for example, one part per million (ppm) of its pressure in the source. In other words, the velocity of the valving gas in the barrier is set equal to or greater than the velocity by means of which the specified precursor isobar diffuses. When the source is turned on, the valving gas is turned off and the transport gas is turned on, thereby breaking the flow barrier. All this can be done rapidly with valves located at room temperature, no matter how hot the source itself is.

The source vapors are led into the reaction chamber through separate lines which all, i.e., also those through which vapors from the external gas or liquid sources are led in, are equipped with the inert gas valving system. Usually, the valving points are close to the chamber and the separate lines merge just in front of it so that the volume to be purged is minimized and well heated and there is hardly any film growth before the chamber.

The operation of the inert gas valving system is easily tested by pulsing only one precursor into the chamber while keeping the other one behind the flow barrier. No film deposition is observed when the barrier holds.

4.2.4. Reaction Chamber

The reactors may be equipped with two kinds of reaction chambers: a compact, cassette-like, traveling-wave reactor chamber, and a more open tubular chamber. They both may be combined easily with the inert gas valving system. The open chambers have the benefit that they can incorporate any arbitrarily shaped substrate. However, they are not as fast and as cost-effective as the traveling-wave reactors which therefore are preferred with planar substrates in both research and large scale production.

In the traveling-wave reactor geometry, the substrates are located face to face close to each other. The distance between two substrates is typically only a few millimeters leaving a narrow flow channel in between. As an example, Figure 7 shows schematics of a small reactor where there is only one substrate pair. A connection to two inert gas valved source lines is shown as well. In practice, there are usually more source lines to facilitate deposition of multicomponent films or multilayer structures. Favorably, the substrates are located close to the point where the source lines merge since this minimizes the chamber wall area subjected to the film growth. The whole assembly of the source lines and reaction chamber is inserted into a vacuum tube outside which heaters are located. The temperature increases stepwise from the gas inlets toward the reaction chamber so that each solid source may be set independently to an appropriate temperature.

In large-batch reactors, there may be tens of substrates arranged in pairs so that the gases flow over the front but not on the backsides of the substrates (Fig. 8). Instead of the one flow channel in the small reactor, in the large reactor there are several of them parallel with each other. Connections to the sources are basically similar to the small reactor. Together with the self-limiting film growth mechanism, these similarities between the research and large scale reactors assist in scale-up of ALD processes.

When pulsed into the reaction chamber, the precursor molecules are transported by the carrier gas along the flow

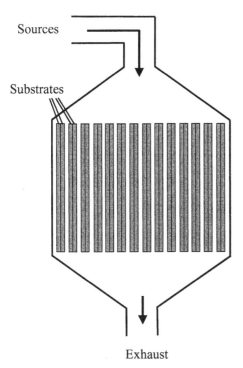

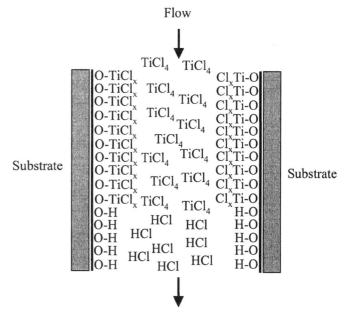

Fig. 8. Reaction chamber of a large scale ALD reactor incorporating several substrates in pairs so that narrow flow channels are left between each pair.

Fig. 9. Propagation of a TiCl$_4$ pulse along a flow channel over substrate surfaces which the preceding water pulse has left hydroxyl group terminated. The multiple hit conditions ensure that the hydroxyl groups become effectively consumed in exchange reactions at the TiCl$_4$ pulse front. HCl molecules formed as reaction byproducts travel in front of the TiCl$_4$ pulse and thus may readsorb and block reaction sites from TiCl$_4$.

channel between the substrates. The flow velocity and pressure are adjusted to maintain essentially pluglike flow conditions which keep the adjacent precursor pulses separated—like waves—and ensure effective and rapid purging of the small volume left between the substrates. In the small research reactors with two 5×5 cm^2 substrates (Fig. 7), purge times as short as 0.1 s are enough for separating the precursor pulses. If the pulses were not well separated, films with nonuniform thickness would be obtained, and in the worst case particles might form in gas phase reactions.

While traveling along the narrow flow channel, the precursor molecules undergo multiple collisions with the substrates. As a result of these multiple hit conditions, the precursor molecules have a high probability to find an open chemisorption or reaction site and thus become effectively utilized. At the same time, the surface becomes rapidly saturated: often 0.2 s is enough for the saturation in the previously described research reactor. A low operation pressure favors high hitting frequency but reduces the precursor transport capability of the carrier gas. The optimal pressure is thus a compromise of the two factors.

The special conditions in the traveling-wave reactors have a number of interesting consequences. In the beginning of the precursor pulse, its vapor pressure in the flow channel is very much location and time dependent [95]. At first, only the leading edge of the substrate; i.e., the edge closest to the inlet, receives precursor molecules. Because of their high hitting frequency, and preferably also high reactivity, the precursor molecules become rapidly exhausted from the pulse front to the open surface sites. Thus, the latter parts of the substrate do not receive any precursor molecules before the leading parts

have become covered. In other words, the precursor pulse front rolls like a wave over the substrate saturating the surface on its way (Fig. 9). Experimentally, this can easily be seen and verified by underdosing one precursor and by observing two distinct regions on the substrate: one covered with a uniform film and another one covered with no film. The higher the precursor reactivity, the sharper the border between the two areas. In research reactors, borders as sharp as a few millimeters are observable. In principle, the steepness of the border could be used in studying the reactivities of the precursors. Quantitative evaluation would, however, require constant doses throughout the experiment which, especially with solid sources, may often be difficult to realize.

The sharp border between the growth and nongrowth areas points out the possibility of maximizing the precursor utilization. The closer the border is set to the substrate, the less precursor is wasted to the growth on the downstream side chamber walls or to the exhaust. However, at the same time the risks related to variations in precursor doses are increased.

In the traveling-wave reactor geometry, the surface hitting densities estimated from the partial pressures with the basic equations of the kinetic gas theory lose their meaning quite a lot. This is illustrated by considering the effects of impurities in the carrier gas. Taking again the research reactor (Fig. 7) as an example, we have, approximately, a total pressure of 5 Torr and 200 stdcm^3 min^{-1} carrier gas flow rate to the flow channel limited by the two 5×5 cm^2 substrates. A typical moisture content of 1 ppm in the carrier gas means a partial pressure of 5×10^{-6} Torr which at room temperature implies a hitting density of 2.4×10^{15} water molecules cm^{-2} s^{-1}. For water, a high reactivity toward a surface covered with metal precursors may be assumed and therefore the sticking coefficient should approach unity. Then, the amount of oxygen corresponding to a typical ALD oxide film growth rate (3.0×10^{14} oxygen atoms cm^{-2} cycle^{-1}) would become adsorbed in about 0.1 s already from the residues in the carrier gas. However, from the flow rate of the carrier gas, one can calculate that in the same time (0.1 s), only 9×10^{12} H$_2$O molecules enter the flow channel. When averaged over the 50-cm^2 substrate area, this gives only 1.8×10^{11} oxygen atoms cm^{-2} which is more than 3 orders of magnitude less than that estimated from the hitting density. Clearly, the ordinary hitting density calculations cannot be applied.

What actually happens is that the moisture impurities become adsorbed at the upstream edge of the substrate, i.e., right after the point where the precursor flow lines merge and film growth starts (cf. Fig. 7). Assuming a coverage of 1×10^{14} molecules cm^{-2} and a cycle time of 1.0 s, this means that 0.9-cm^2 area adsorbs the incoming moisture impurities. In practice, a somewhat larger area, typically 5 cm^2, may be disturbed close to the leading edge but besides the impurities other factors also contribute to this area. In any case, the upstream edge of the substrate serves as an effective internal getter for the impurities. By moving the substrates a bit further away from the merging point of the separate precursor lines, the getter region would be eliminated from the substrate but then films would grow more extensively on the chamber walls. Therefore, in research it is more convenient to accept the leading edge area and to consider only the area outside that as representative for the true ALD growth.

A less favorable consequence of the multiple hit conditions is that also the reaction byproducts, like HCl, have a high prob-

ability to adsorb. As they are formed in the reactions at the front edge of the precursor pulse, the byproducts travel in front of this pulse (Fig. 9) and thereby they have a possibility to decrease the growth rate by blocking reaction sites from the precursor molecules [96, 97]. This effect may explain the differences which are sometimes observed in the growth rates obtained with the traveling-wave and other kinds of reactors.

In addition to the limitations on substrate shape, the traveling-wave reactors are also restricted in respect to the utilization of extra energy sources like plasma, hot filaments, or light. Because of the compact construction of the reaction chamber, there hardly is any room for the implementation of the extra energy sources, especially in the large-batch reactors. The multiple wall collision conditions, in turn, make it difficult to transport reactive radicals from their remote sources over the entire substrates. Therefore, these reactors are quite limited to thermally activated reactions only.

4.3. Flow-Type ALD Reactors with Moving Substrates

The alternate precursor dosing as required in ALD may be realized not only by chopping precursor fluxes to the stationary substrates as above, but also by moving substrates between separate precursor fluxes which are either constantly on or accordingly pulsed. While a high-vacuum version of such a reactor was used in the very first ALD experiments in 1974 [1, 5, 7], also with this approach flow-type systems have subsequently become dominating. In these reactors, inert gases are used for transporting the precursors and for setting up shrouds between the zones with the different precursors.

The most popular type of moving substrate reactors is based on the rotating susceptor design [3, 4, 13, 20, 63, 93, 98–100], an example of which is shown in Figure 10. A fixed part with

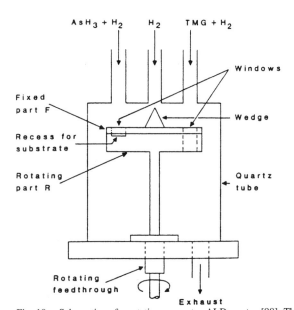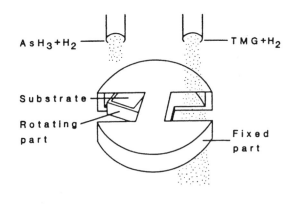

Fig. 10. Schematics of a rotating susceptor ALD reactor [98]. The susceptor consists of fixed and rotating parts. The precursor streams are directed into windows in the fixed part below which the substrate rotates and cuts through the streams. A wedge and a large H$_2$ flow from the center tube prevent mixing of the two precursors. Reprinted with permission from M. A. Tischler and S. M. Bedair, Self-limiting mechanism in the atomic layer epitaxy of GaAs, *Appl. Phys. Lett.* 48, 1681 (1986), © 1986, American Institute of Physics.

two openings covers the rotating susceptor. The source gases are continuously directed to the openings and flow unimpeded through the fixed part except when the substrate cuts through the streams. The exposure times are controlled by the rotational speed and, if needed, they may be elongated by pausing the substrate. A carrier gas flown with a high rate at the center and guided by the wedge on the fixed part prevents mixing of the source gases. The fixed part also shears off a part of the boundary layer above the substrate. With a small clearance between the fixed and moving parts, the rotation speed could have been increased up to 120 rpm which corresponds to 85 ms exposure times and gives a high, self-limited (1 ML cycle^{-1}) growth rate of 2 μm h^{-1} for GaAs [20, 93]. On the other hand, the rotating susceptor reactors have limitations in that they are difficult to adopt into processes which require complex sequencing of several sources, like a growth of multilayer structures or compositionally varied multicomponent films.

5. ALD PRECURSORS

Chemistry, especially the choice of proper precursors, is certainly the key issue in a successful design of an ALD process. In this section, we first examine the properties required for the ALD precursors. Next, some practical aspects of choosing the precursors are discussed. Finally, a comprehensive survey of the precursors used in ALD is carried out.

5.1. Requirements for ALD Precursors

Requirements for ALD precursors are summarized in Table III. The precursor chemistry of ALD is convenient to compare with CVD. There are many similarities but because of the certain characteristics of ALD, differences exist as well. However, the unique features of ALD are not reflected that much in the individual compounds which are usually the same as those used in CVD (see Table IV). Rather, it is how they are combined together which makes the difference. For ALD, the most aggressively reacting precursor combinations are looked for while in CVD too aggressive reactions must be avoided.

5.1.1. Volatility

The volatility requirement is of course common to ALD and CVD but as the self-limiting growth mechanism makes ALD less susceptible to small variations in precursor fluxes, solid precursors are more easily adopted to ALD than to CVD. Solids do possess some disadvantages, however. First, because solids are often loaded inside the ALD reactor in a limited source volume, the sources must be reloaded frequently, possibly after each run, which is laborious especially in industry. Second, if the particle size is too small, the particles may be transported by the carrier gas, even if operated under reduced pressure, to the substrates where they cause detrimental defects. For these reasons, liquid or gaseous precursors should be preferred, if available.

Table III. Requirements for ALD Precursors

Requirement	Comments
Volatility	For efficient transportation, a rough limit of 0.1 Torr at the applicable maximum source temperature
	Preferably liquids or gases
No self-decomposition	Would destroy the self-limiting film growth mechanism
Aggressive and complete reactions	Ensure fast completion of the surface reactions and thereby short cycle times
	Lead to high film purity
	No problems of gas phase reactions
No etching of the film or substrate material	No competing reaction pathways
	Would prevent the film growth
No dissolution into the film or substrate	Would destroy the self-limiting film growth mechanism
Unreactive volatile byproducts	To avoid corrosion
	Byproduct readsorption may decrease the growth rate
Sufficient purity	To meet the requirements specific to each process
Inexpensive	
Easy to synthesize and handle	
Nontoxic and environmentally friendly	

The required precursor vapor pressure is reactor specific and depends on many factors, like source geometry, substrate area, and flow rates. In research scale flow-type reactors, for example, a vapor pressure of 0.1 Torr is usually appropriate and ensures effective transportation (cf. Section 4.2.2). In large scale production, higher doses and accordingly higher precursor vapor pressures are often desired. The minimum requirements to the precursor volatility are thus set by these vapor pressures and the highest applicable source temperatures (e.g., 500°C).

Precursor vapor pressure data are not always available, and especially with low volatility solids such measurements may be quite troublesome. In these cases, thermogravimetric (TG) measurements can be used for examining the volatility [101] and for estimating the required source temperatures [102]. One option is to use the conventional constant heating rate measurements. From the weight loss curves of compounds which have already been used in ALD, one can identify certain temperatures, such as those corresponding to 10 or 50% weight losses, and one can plot these as a function of the source temperatures used in a given type of an ALD reactor (Fig. 11). On basis of this data set and a similar TG measurement, a source temperature may be estimated for any new compound to be used in a similar reactor. Another possibility is to perform isothermal weight loss measurements over a range of temperatures. After calibration with a compound with a known vapor pressure, vapor pressures of the compounds of an interest may be estimated [102].

Table IV.   A Summary of the Precursor Combinations Used and Examined
in ALD Thin Film Processes[a]

Metal precursor	Nonmetal and other precursors	Film material	Reference
Elements			
Zn	S	ZnS	[1, 2, 116]
	Se	ZnSe	[116–120]
	Se(C$_2$H$_5$)$_2$, prepyrolyzed	ZnSe	[121]
	Te	ZnTe	[118, 122]
Cd	S	CdS	[123]
	Se	CdSe	[116]
	Te	CdTe	[122, 124–127]
Halides			
AlCl$_3$	H$_2$O	Al$_2$O$_3$	[2–4, 107, 128–136]
	O$_2$	Al$_2$O$_3$	[137, 138]
	ROH	Al$_2$O$_3$	[129, 139]
	H$_2$O/ROH + P$_2$O$_5$/(CH$_3$O)$_3$PO	Al$_2$O$_3$: P	[140, 141]
	NH$_3$	AlN	[142]
	AsH$_3$	AlAs	[143]
GaCl	NH$_3$	GaN	[144]
	PH$_3$	GaP	[145]
	P$_4$ + As$_4$	GaP$_{1-x}$As$_x$	[146, 147]
	AsH$_3$	GaAs	[14, 145, 148–153]
GaCl$_3$	NH$_3$	GaN	[154]
	AsH$_3$	GaAs	[143, 155, 156]
	CuCl + H$_2$S	CuGaS$_2$	[157]
GaBr	AsH$_3$	GaAs	[158]
GaI	AsH$_3$	GaAs	[158]
InCl	PH$_3$	InP	[145]
	((CH$_3$)$_3$C)PH$_2$	InP	[129]
	AsH$_3$	InAs	[148, 153]
InCl$_3$	H$_2$O/H$_2$O$_2$	In$_2$O$_3$	[86, 115, 159]
	H$_2$O/H$_2$O$_2$ + SnCl$_4$	In$_2$O$_3$: Sn	[86, 159, 160]
	H$_2$S	In$_2$S$_3$	[161]
SiCl$_4$	H$_2$O	SiO$_2$	[162–167]
	NH$_3$	Si$_3$N$_4$	[168]
	H$_2$S	Cl dopant in SrS and CaS	[169]
Si$_2$Cl$_6$	Si$_2$H$_6$	Si	[60, 170]
	Atomic H[b]	Si	[61]
	N$_2$H$_4$	Si$_3$N$_4$	[171]
GeCl$_4$	Atomic H[b]	Ge	[62]
SnCl$_4$	H$_2$O	SnO$_2$	[34, 35, 172–174]
	H$_2$O + SbCl$_5$	SnO$_2$: Sb	[175]
SbCl$_5$	H$_2$O	Sb$_2$O$_5$	[175]
TiCl$_4$	H$_2$-plasma	Ti	[74]
	H$_2$O	TiO$_2$	[82, 96, 107, 176–182]
	H$_2$O$_2$	TiO$_2$	[183, 184]
	NH$_3$ (+ Zn)	TiN	[88, 111, 112, 185, 186]
	(CH$_3$)$_2$NNH$_2$	TiN	[114]
TiI$_4$	H$_2$O$_2$	TiO$_2$	[187–190]
	NH$_3$	TiN	[186, 191]
ZrCl$_4$	H$_2$O	ZrO$_2$	[83, 133, 192–195]
ZrI$_4$	H$_2$O$_2$	ZrO$_2$	[196]

Table IV.   (Continued)

Metal precursor	Nonmetal and other precursors	Film material	Reference
$HfCl_4$	H_2O	HfO_2	[84, 131, 195, 197, 198]
$NbCl_5$	NH_3 (+ Zn)	NbN	[103, 112, 185]
	$(CH_3)_2NNH_2$	NbN	[114]
$TaCl_5$	H_2-plasma	Ta	[74]
	H_2O	Ta_2O_5	[3, 4, 104–106, 130, 131, 135, 195, 199–202]
	$NH_3 + H_2O$	TaO_xN_y	[113]
	NH_3	Ta_3N_5	[113, 185]
	$NH_3 + Zn$	TaN	[113]
	$(CH_3)_2NNH_2$	Ta_3N_5	[114]
TaI_5	H_2O_2	Ta_2O_5	[203]
$MoCl_5$	Zn	Mo	[204]
	NH_3 (+ Zn)	MoN, Mo_2N	[114, 185]
	$(CH_3)_2NNH_2$	MoN, Mo_2N	[114]
WO_xF_y	H_2O	WO_3	[205]
WF_6	Si_2H_6	W	[206]
	NH_3	W_2N	[207]
$MnCl_2$	H_2S	Dopant in ZnS	[3, 4, 8]
MnI_2	H_2S	Dopant in ZnS	[208]
CuCl	H_2	Cu	[209, 210]
	Zn	Cu	[110]
	$GaCl_3 + H_2S$	$CuGaS_2$	[157]
$ZnCl_2$	H_2S	ZnS	[2–4, 85, 211–213]
	H_2Se	ZnSe	[214]
	$H_2S + Se$	$ZnS_{1-x}Se_x$	[108]
	HF	ZnF_2	[215]
$CdCl_2$	H_2S	CdS	[123, 213]
Alkyl compounds			
$Al(CH_3)_3$	H_2O	Al_2O_3	[36, 37, 88, 89, 133, 134, 162, 194, 216–224]
	H_2O_2	Al_2O_3	[184, 225–229]
	N_2O	Al_2O_3	[228]
	NO_2	Al_2O_3	[217]
	O_2-plasma	Al_2O_3	[75]
	NH_3	AlN	[230–233]
	PH_3	AlP	[234]
	AsH_3	AlAs	[13, 17, 47, 58, 79, 80, 235, 236]
$Al(CH_3)_2Cl$	H_2O	Al_2O_3	[237]
	NH_3	AlN	[238]
$Al(CH_3)_2H$	PH_3	AlP	[239]
	AsH_3	AlAs	[79, 240]
$Al(C_2H_5)_3$	H_2O	Al_2O_3	[241]
	NH_3	AlN	[242, 243]
	AsH_3	AlAs	[17, 47]
$Al(CH_2CH_2(CH_3)_2)_3$	AsH_3	AlAs	[244]
$Ga(CH_3)_3$	NH_3	GaN	[245–249]
	N_2-H_2-plasma	GaN	[71]
	PH_3	GaP	[234, 239, 250, 251]
	AsH_3	GaAs	[12, 13, 15, 17, 48, 70, 78, 79, 93, 98, 99, 235, 236, 252–256]
	$((CH_3)_3C)AsH_2$	GaAs	[99, 257, 258]
$Ga(CH_3)_2(C_2H_5)$	AsH_3	GaAs	[252]

Table IV.   (Continued)

Metal precursor	Nonmetal and other precursors	Film material	Reference
$Ga(C_2H_5)_3$	NH_3	GaN	[243, 259, 260]
	NH_3, thermally cracked	GaN	[63]
	PH_3	GaP	[49]
	AsH_3	GaAs	[17, 19, 47, 50, 252]
	$((CH_3)_3C)AsH_2$	GaAs	[258, 261]
	$((CH_3)_2N)_3As$	GaAs	[262]
$Ga(C_2H_5)_2Cl$	AsH_3	GaAs	[263]
$Ga(CH_2CH_2(CH_3)_2)_3$	AsH_3	GaAs	[19]
$Ga(CH_2C(CH_3)_3)_3$	$((CH_3)_3C)AsH_2$	GaAs	[258]
$In(CH_3)_3$	H_2O	In_2O_3	[264]
	PH_3	InP	[78, 265, 266]
	$((CH_3)_3C)PH_2$	InP	[267, 268]
	AsH_3	InAs	[48, 269]
	$((CH_3)_3C)AsH_2$	InAs	[270]
$((CH_3)_2(C_2H_5)N)In(CH_3)_3$	AsH_3	InAs	[257]
$In(CH_3)_2Cl$	AsH_3	InAs	[92]
$In(CH_3)_2(C_2H_5)$	NH_3	InN	[246–248]
$In(C_2H_5)_3$	PH_3	InP	[250, 251]
	AsH_3	InAs	[91]
$Sn(CH_3)_4$	NO_2	SnO_2	[271]
$Sn(C_2H_5)_4$	NO_2	SnO_2	[271]
$Zn(CH_3)_2$	H_2O	ZnO	[272]
	$H_2O + Al(CH_3)_3$	ZnO : Al	[272]
	$H_2S + H_2O$	$ZnO_{1-x}S_x$	[273]
	H_2S	ZnS	[274–276]
	$H_2S + H_2Se$	$ZnS_{1-x}Se_x$	[276–278]
	H_2Se	ZnSe	[276, 279–282]
	$(C_2H_5)_2Te$	ZnTe	[283]
	$(CH_3)(CH_2CHCH_2)Te$	ZnTe	[283]
$Zn(C_2H_5)_2$	H_2O	ZnO	[52, 89, 272, 284–293]
	$H_2O + B_2H_6$	ZnO : B	[289–292, 294]
	$H_2O + Al(CH_3)_3$	ZnO : Al	[272]
	$H_2O + Ga(CH_3)_3$	ZnO : Ga	[293]
	$H_2S + H_2O$	$ZnO_{1-x}S_x$	[288]
	H_2S	ZnS	[288, 295, 296]
	$(C_2H_5)_2S_2$	ZnS	[297]
	$(C_2H_5)_2Se_2$	ZnSe	[297]
$Cd(CH_3)_2$	H_2S	CdS	[298–302]
	H_2Se	CdSe	[279]
	$(C_2H_5)_2Te$	CdTe	[283]
	$(CH_3)(CH_2CHCH_2)Te$	CdTe	[283, 303]
	$((CH_3)_2CH)_2Te$ + temperature modulation	CdTe	[53]
$Hg(CH_3)_2$	$CH_3(allyl)Te$	HgTe	[303]
Alkoxides			
$Al(OC_2H_5)_3$	H_2O/ROH	Al_2O_3	[129]
$Al(OCH_2CH_2CH_3)_3$	H_2O/ROH	Al_2O_3	[129]
	$H_2O/ROH + P_2O_5$	Al_2O_3 : P	[141]
$Ti(OC_2H_5)_4$	H_2O	TiO_2	[304, 305]
$Ti(OCH(CH_3)_2)_4$	H_2O	TiO_2	[306–308]
	$Bi(C_6H_5)_3 + H_2O$	$Bi_xTi_yO_z$	[309]
$Zr(OC(CH_3)_3)_4$	H_2O	ZrO_2	[310]

Table IV.   (Continued)

Metal precursor	Nonmetal and other precursors	Film material	Reference
$Nb(OC_2H_5)_5$	H_2O	Nb_2O_5	[136, 192, 202, 311]
$Ta(OC_2H_5)_5$	H_2O	Ta_2O_5	[133, 136, 192, 197, 202, 312, 313]
	NH_3	TaO_xN_y	[314]
$Pb(OC(CH_3)_3)_2$	H_2S	PbS	[315]
$Pb_4O(OC(CH_3)_3)_6$	H_2S	PbS	[315]
β-diketonato complexes			
$Mg(thd)_2{}^c$	H_2O_2	MgO	[316]
	O_3	MgO	[317]
$Ca(thd)_2$	H_2S	CaS	[87, 169, 318–321]
	HF	CaF_2	[215]
$Sr(thd)_2$	O_3	SrO ($SrCO_3$)	[322]
	$O_3 + Ti(OCH(CH_3)_2)_4 + H_2O$	$SrTiO_3$	[322]
	H_2S	SrS	[169, 296, 318, 323–325]
	$H_2S + Se$	$SrS_{1-x}Se_x$	[109]
	HF	SrF_2	[215]
$Ba(thd)_2$	H_2S	BaS	[318, 326]
$Ga(acac)_2{}^d$	O_3/H_2O	Ga_2O_3	[35, 327]
$In(acac)_3$	H_2S	In_2S_3	[288]
$Pb(thd)_2$	H_2S	PbS	[315]
$Y(thd)_3$	O_3, O_2	Y_2O_3	[328, 329]
	H_2S	Y_2O_2S	[330]
$La(thd)_3$	H_2S	La_2S_3, La_2O_2S	[331]
	O_3	La_2O_3	[332, 333]
	$Co(thd)_2 + O_3$	$LaCoO_3$	[332]
	$Ni(thd)_2 + O_3$	$LaNiO_3$	[333]
	$Mn(thd)_3 + O_3$	$LaMnO_3$	[334]
$Ce(thd)_4$	O_3	CeO_2	[335]
	H_2S	Dopant in CaS, SrS and BaS	[169, 296, 326, 336–338]
$Ce(thd)_3(phen)^e$	H_2S	Dopant in SrS	[339]
$Ce(tpm)_3{}^f$	H_2S	Dopant in SrS	[339]
$Ce(tpm)_4$	H_2S	Dopant in SrS	[340]
$Ce(tfa)_4{}^g$	H_2S	Dopant in SrS	[340]
$Mn(thd)_3$	O_3	MnO_x	[334]
	H_2S	Dopant in ZnS	[295, 296]
$Co(thd)_2$	O_3	Co_3O_4	[332]
$Ni(acac)_2$	O_3	NiO	[341, 342]
$Ni(thd)_2$	O_3	NiO	[333]
$Cu(thd)_2$	H_2	Cu	[186, 210, 343, 344]
$Ln(thd)_3{}^h$	H_2S	Dopants in CaS, SrS and ZnS	[296, 319, 336, 345–352]
Cyclopentadienyl compounds			
$Mg(C_5H_5)_2$	H_2O	MgO	[241, 353–355]
$Sr(C_5{}^iPr_3H_2)_2{}^i$	$Ti(OCH(CH_3)_2)_4 + H_2O$	$SrTiO_3$	[356–358]
	H_2S	SrS	[359]
$Sr(C_5Me_5)_2{}^j$	H_2S	SrS	[359]
$Ba(C_5Me_5)_2$	$Ti(OCH(CH_3)_2)_4 + H_2O$	$BaTiO_3$	[356, 357]
	H_2S	BaS	[359]
$Ba(C_5{}^tBu_3H_2)_2{}^k$	$Ti(OCH(CH_3)_2)_4 + H_2O$	$BaTiO_3$	[357]
$Mn(C_5H_5)_2$	H_2S	Dopant in ZnS	[295]
$Mn(C_5MeH_4)(CO)_3$	H_2S	Dopant in ZnS	[295]
$Zr(C_5H_5)_2Cl_2$	O_3	ZrO_2	[322]
$Ce(C_5Me_4H)_3$	H_2S	Dopant in SrS	[360]

Table IV. (Continued)

Metal precursor	Nonmetal and other precursors	Film material	Reference
Carboxylates			
$Zn(CH_3COO)_2{}^l$	H_2S	ZnS	[212, 361, 362]
	H_2O	ZnO	[361, 363]
Hydrides			
$((CH_3)_3N)AlH_3$	NH_3 + temperature modulation	AlN	[365]
	AsH_3	AlAs	[366]
	$((CH_3)_2N)_3As$	AlAs	[262]
$((CH_3)_2(C_2H_5)N)AlH_3$	NH_3	AlN	[367–369]
	AsH_3	AlAs	[81, 370–372]
SiH_4	Temperature modulation	Si	[54, 55]
$SiCl_2H_2$	H_2	Si	[373, 374]
	Atomic H^b	Si	[60, 64–66, 170]
	UV light	Si	[375]
	C_2H_2	SiC	[376, 377]
	NH_3-plasma	Si_3N_4	[72, 73]
	NH_3, thermally cracked	Si_3N_4	[59]
Si_2H_6	Temperature modulation	Si	[51, 57, 378]
	Thermal precracking and temperature modulation	Si	[67]
	Si_2Cl_6	Si	[60]
	C_2H_2	SiC	[379, 380]
	C_2H_4	SiC	[100]
Si_3H_8	Temperature modulation	Si	[65, 381]
$Si(C_2H_5)_2H_2$	Temperature modulation	Si	[382]
GeH_4	Temperature modulation	Ge	[55, 56]
$Ge(CH_3)_2H_2$	Atomic H^b	Ge	[68, 69]
$Ge(C_2H_5)_2H_2$	Temperature modulation	Ge	[383]
Alkyl- and silylamides			
$Ti(N(CH_3)_2)_4$	NH_3	TiN	[384, 385]
	NH_3 + SiH_4	$Ti_xSi_yN_z$	[384, 385]
$Ti(N(C_2H_5)(CH_3))_4$	NH_3	TiN	[386]
$Ce(N(Si(CH_3)_3)_2)_3$	H_2S	Dopant in SrS	[387]
Others			
$Si(NCO)_4$	H_2O	SiO_2	[388]
	$N(C_2H_5)_3$	SiO_2	[389]
$CH_3OSi(NCO)_3$	H_2O_2	SiO_2	[390, 391]
$Pb((C_2H_5)_2NCS_2)_2$	H_2S	PbS	[315, 392]
$Ni(apo)_2{}^m$	O_3	NiO	[341]
$Ni(dmg)_2{}^n$	O_3	NiO	[341]
Two metal compounds			
$AlCl_3$	$Al(OC_2H_5)_3$	Al_2O_3	[393]
	$Al(OCH(CH_3)_2)_3$	Al_2O_3	[393]
	$Ti(OC_2H_5)_4$	$Al_xTi_yO_z$	[393]
	$Ti(OCH(CH_3)_2)_4$	$Al_xTi_yO_z$	[393]
$Al(CH_3)_3$	$Al(OCH(CH_3)_2)_3$	Al_2O_3	[393]
	$Ti(OCH(CH_3)_2)_4$	$Al_xTi_yO_z$	[393]
	CrO_2Cl_2	$Al_xCr_yO_z$	[394]
$ZrCl_4$	$Al(OC_2H_5)_3$	$Al_xZr_yO_z$	[393]
	$Ti(OCH(CH_3)_2)_4$	$Ti_xZr_yO_z$	[393]

Table IV.   (Continued)

Metal precursor	Nonmetal and other precursors	Film material	Reference
	$Si(OC_2H_5)_4$	$Zr_xSi_yO_z$	[393]
	$Si(OC_4H_9)_4$	$Zr_xSi_yO_z$	[393]
$HfCl_4$	$Al(OC_2H_5)_3$	$Al_xHf_yO_z$	[393]
	$Ti(OCH(CH_3)_2)_4$	$Ti_xHf_yO_z$	[393]
$TaCl_5$	$Ta(OC_2H_5)_5$	Ta_2O_5	[393, 395]

[a] Classification goes according to the type of the metal precursor. Processes with two metal precursors with no separate nonmetal source are listed separately in the end of the table.

[b] Obtained by thermal dissociation of H_2.

[c] thd is 2,2,6,6,-tetramethyl-3,5-heptanedione. Alkaline earth and yttrium thd-complexes used may also contain a neutral adduct molecule or they may have been slightly oligomerized.

[d] acac is acetyl acetonate.

[e] phen is 1,10-phenanthroline.

[f] tpm is 1,1,1-trifluoro-5,5-dimethyl-2,4-hexanedione.

[g] tfa is 1,1,1-trifluoro-2,4-pentanedione.

[h] Ln is Ce, Pr, Nd, Sm, Eu, Gd, Tb, Tm.

[i] ^{i}Pr is $-CH(CH_3)_2$.

[j] Me is $-CH_3$.

[k] ^{t}Bu is $-C(CH_3)_3$.

[l] Converts to $Zn_4O(CH_3COO)_6$ when annealed [364].

[m] apo is 2-amino-pent-2-en-4-onato.

[n] dmg is dimethylglyoximato.

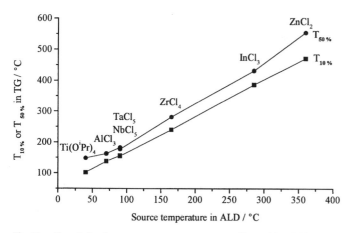

Fig. 11.   Correlation between temperatures corresponding to 10 and 50% mass losses in TG measurements and source temperatures used in research scale ALD reactors [102]. The TG measurements were carried out in 1-atm nitrogen atmosphere with a heating rate of $10°C\,min^{-1}$.

TG measurements also reveal if the volatilization takes place in one step, or if there are decomposition reactions involved [101].

5.1.2. Stability Against Self-Decomposition

The self-limiting film growth via the surface exchange reactions is obviously achievable only under conditions where the precursors do not decompose on their own. A simple way to test if a precursor is thermally decomposing is to pulse only that compound in the reactor and to see if any film grows. As the decom-

position reactions may often be catalyzed by the film material, it is advised to carry out the decomposition experiments not only on bare substrates but also on previously deposited films and to see if their thicknesses increase.

Because the decomposition is a thermally activated reaction, its rate increases exponentially with temperature. On the other hand, the longer the exposure time, the more pronounced the decomposition in comparison to the desired exchange reaction (Section 7.1.1). This explains why in processing of porous materials and in many surface chemistry studies, both with long exposure times of minutes, decomposition is observed at significantly lower temperatures than in film growth experiments with fast pulsing. Clearly, it is a material-, process-, and application-dependent choice where to set the uppermost limit for the acceptable decomposition rate.

In fortuitous cases, the product of the decomposition reaction is the same as that of the exchange reactions, like, for example, in the growth of oxides from metal alkoxides and water where also the thermal decomposition of the alkoxide deposits the same oxide. In such a case, quite a substantial contribution of the decomposition may be accepted, provided of course that the decomposition proceeds in a surface reaction rate limited manner, thereby maintaining the good uniformity and conformality (Section 7.1.2). On the other hand, if the decomposition reaction causes incorporation of contaminants into the film, it must be minimized to a level where the resulting contamination can be tolerated.

5.1.3. Aggressive and Complete Reactions

The alternate pulsing means that there are no risks of gas phase reactions in ALD. Therefore, precursors which react aggressively with each other can be used. In fact, such combinations should be preferred to ensure rapid completion of the reactions and thereby short cycle times as well as effective precursor utilization. This is in marked contrast with CVD where too reactive precursor combinations must be avoided. Thermodynamically, this means that the reactions used in ALD should have as negative a Gibbs free energy change ΔG as possible [39], while in CVD ΔG should still be negative but close to zero.

The reactions also need to be completed so that no impurities are left in the films. The aggressiveness of the reaction does not necessarily guarantee the completion. Especially at low temperatures, kinetic reasons may prevent the completion and thus some unreacted ligands from the metal precursors may be incorporated into the films. Likewise, nonmetal precursors such as H_2O or NH_3, for example, may leave $-OH$ or $-NH_x$ residues if the reactions are incomplete. The amount of these residues in the films decreases as the temperature is increased but if that is increased too much, the precursor decomposition may start to produce new kinds of contaminants.

5.1.4. No Etching Reactions

A negative consequence of the alternate supply of the precursors is that there are no competing reaction pathways for possible etching reactions where the film itself, substrate, or an underlying film is etched by one of the precursors. Such etching reactions have, for example, prevented the growth of Nb_2O_5 from $NbCl_5$ and H_2O and have been interpreted to involve a formation of volatile oxochlorides [103]:

$$Nb_2O_5(s) + 3NbCl_5(g) \rightarrow 5NbOCl_3(g)$$

Similar etching is also observed in the analogous $TaCl_5$-H_2O process but only above a certain threshold temperature of about 300°C, and even then the etching is slow enough not to totally prevent the film growth but just to cause a decrease of the deposition rate with longer $TaCl_5$ pulse times [104–106]. A simple test to verify an existence of etching reactions is to expose a film to one precursor only and to observe a disappearance of the film or part of it.

A somewhat less detrimental type of gas–solid reaction is the one which involves an exchange of the film constituents, like a replacement of a metal cation in an oxide with another metal having a higher affinity toward oxygen [107]:

$$4AlCl_3(g) + 3TiO_2(s) \rightarrow 2Al_2O_3(s) + 3TiCl_4(g)$$

Such reactions are often self-terminated once the surface becomes covered by the exchange reaction product (Al_2O_3 in the preceding example), and therefore they may be tolerated unless very sharp interfaces are required. On the other hand, the same kind of replacement reactions may also be utilized in making, for example, metal sulfoselenides where sulfur atoms deposited through reactions between metal precursors and H_2S

are in the next step partially replaced by elemental selenium, thereby avoiding the need to use the highly toxic H_2Se [108, 109] (cf. Section 5.3.2).

5.1.5. No Dissolution to the Film or the Substrate

The precursors should chemisorb on the surface upon which they are dosed but they are not allowed to dissolve into that material. Examples of precursor dissolution are rare but one is the attempt to deposit metallic copper from CuCl and elemental zinc, the latter being used as a reducing agent [110]. Metallic copper with only about 3 at.% zinc was indeed obtained but the film growth was evidently not self-limiting and growth rates were high, above 5 Å cycle^{-1}. The results were explained with the following mechanism in which the copper film is formed as desired by the reductive exchange reaction:

$$CuCl(ads) + 0.5Zn(g) \rightarrow Cu(s) + 0.5ZnCl_2(g)$$

However, once copper has been formed on the surface, the dissolution of zinc into copper starts to complicate the process. When dosed onto a copper surface with a CuCl chemisorption layer, Zn first reduces CuCl as before but then, rather than just adsorbing on the surface, zinc also dissolves into the copper,

$$Cu(s) + Zn(g) \rightarrow Cu(Zn)(s)$$

The resulting Cu(Zn) alloy is not stable and during the following purge period Zn outdiffuses and evaporates

$$Cu(Zn)(s) \rightarrow Cu(s) + Zn(g)$$

The outdiffusion is slow, however, and also continues during the next CuCl pulse. This leads to a CVD-type of growth and thereby to a loss of the self-limiting growth mechanism:

$$CuCl(g) + 0.5Zn(g) \rightarrow Cu(s) + 0.5ZnCl_2(g)$$

The precursor may dissolve not only into the film material but also into the substrate. Processes where this can possibly take place are the growths of transition metal nitride films onto silicon substrates from the metal halides and ammonia with elemental zinc as an additional reducing agent (Section 5.3.6) [103, 111–114]. If dissolved into silicon, zinc forms detrimental electrically active defects. In fact, this has been regarded as such a major concern that the zinc-based processes have not been taken into a more detailed study even if they produce the highest quality nitrides (Section 6.5) and the occurrence of zinc dissolution into silicon still remains to be verified.

5.1.6. Unreactive Byproducts

Preferably, the precursors should produce unreactive byproducts which would be easily purged out of the reactor. Reactive byproducts may cause corrosion problems in the reactor or in the exhaust. In addition, they may readsorb on the film surface and block adsorption sites from the precursor molecules, thereby decreasing the growth rate (Section 4.2.4). In the worst case, the byproducts may etch the film. Nonetheless, in many successfully implemented ALD processes (e.g., $ZnCl_2$-H_2S,

AlCl$_3$-H$_2$O, TiCl$_4$-H$_2$O) hydrogen chloride is formed as a byproduct and routinely handled in properly designed reactors. Thus, the unreactivity of the byproducts may be considered only as a secondary, though still an important requirement while planning an ALD process.

5.1.7. Other Requirements

Precursor purity requirements are specific to each process and application, and also depend on the nature of the impurities. Naturally, semiconductors are more sensitive to impurities than, for example, protective coatings. If nonvolatile and unreactive with the precursor, the impurities are of less concern because they remain in the source.

The other requirements, such as low cost, easy synthesis and handling, nontoxicity and environmental friendliness, are all common to the CVD precursor requirements. However, all these and the above requirements may not always be fulfilled simultaneously, and often only these last ones can be sacrificed while developing a new ALD process. If toxic or otherwise dangerous compounds cannot be avoided, special care must of course be given to their handling before and after the process, and appropriate safety systems must be installed.

While considering the cost issue, it must be kept in mind that a price of any chemical is largely determined by its demand. Therefore, no compound should be rejected from research just because of its current price. Rather, one should have a look at its synthesis procedure and try to estimate how much the price might be lowered if the precursor would be taken into a large scale use. Furthermore, in many cases the contribution of the precursor chemicals to the final price of a thin film device is rather small.

5.2. Choice of Precursors

After examining the desired properties of the ALD precursors, we now present some basic ideas about how to proceed in selecting precursors for a new process. The necessary background information can be looked for from basically two sources: literature and thermodynamic calculations.

5.2.1. Literature Survey

For planning a new ALD process and choosing precursors for it, looking at the literature makes a good start. While searching the literature, all the alternative names of the method (Table I) should be taken into account. If no ALD process for the particular film material has been reported, processes of any material containing the constituent elements should be looked for to see which kind of precursors have been used. To assist the literature survey, ALD processes reported so far have been summarized in Table IV in Section 5.3.

In addition to ALD literature, CVD publications also contain valuable information, such as volatility data and reaction and decomposition temperatures. For example, notations of precursor combinations too reactive for CVD are often especially promising. Other potential sources of volatility data are publications on synthesis or thermal analysis of the compounds

and catalogues and data sheets of commercial suppliers. Any comments on moisture or air sensitivity give hope for fast reactions, especially in the growth of oxides but also in more general terms, though at the same time they imply that special care must be devoted to the precursor handling.

5.2.2. Thermodynamic Consideration

Thermodynamics refers to equilibrium conditions while ALD is certainly a nonequilibrium process: the precursors are dosed into the reactor alternately and the byproducts are constantly pumped out. In addition, thermodynamics tells nothing about reaction rates. In any case, thermodynamic calculations provide valuable information about the compounds under consideration for ALD, provided of course that relevant thermodynamic data are available. Today, these calculations are easily made with relatively inexpensive commercial PC programs with integrated databases, but while evaluating the results the previously mentioned limitations of thermodynamics must carefully be kept in mind.

The simplest calculation is the vapor pressure of the compound. Additionally, the stable form, e.g., monomer vs dimer, in the gas phase at each temperature of an interest is easily checked.

To estimate a reactivity of a pair of precursors, ΔG for a suggested net reaction is calculated and is compared with other alternatives, all balanced to deposit an equal amount of the film. If otherwise possible and reasonable, the reaction with the most negative ΔG should be chosen. However, in some cases, like in the growth of In$_2$O$_3$ from InCl$_3$ and H$_2$O [115], reactions with even slightly positive ΔG have been used and have been explained to be possible because of the continuous feed of one precursor and the pump out of the byproducts. On the other hand, it also must be emphasized that in ALD the two precursors can meet each other only if there is a certain mechanism by means of which at least one of them can adsorb on the surface. Thermodynamic estimations of such mechanisms are of course impossible because of a lack of thermodynamic data for the intermediate surface species.

The occurrence of possible side reactions, like etching (Section 5.1.4), may also be estimated by thermodynamic calculations. One may either calculate ΔG for the suggested side reaction or perform equilibrium composition calculations for a system where the film or the substrate material is exposed to one precursor vapor.

5.3. Overview of Precursors and Their Combinations Used in ALD

Table IV summarizes the precursors and their combinations used in ALD as comprehensively as the authors know (compiled in June 2000). For the materials of the greatest interest in nonepitaxial applications (oxides, nitrides, sulfides, fluorides, metals) rather extensive referencing has been attempted, whereas for the widely examined epitaxial semiconductors, especially for the III–V compounds, only a few examples have been given for each precursor combination and review articles

[7, 16–25, 38] are recommended for broader coverage including also the semiconductor dopant precursors which are not dealt with here. Basically, all the processes which have resulted in film growth are included in Table IV, even if some of them are slow requiring unpractically long exposure times.

The evolution of ALD precursor chemistry has been reviewed [396] so only a rather concise summary is presented. Other recommended review articles can be found in Refs. [7, 8, 18, 19, 39], for example.

5.3.1. Metal Precursors

Elemental zinc and sulfur were used in the first ALD experiments for growing ZnS [1, 2]. Subsequently, zinc and cadmium have been used in reactions with sulfur, selenium, and tellurium (Table IV), but except for these and mercury the use of metal elements is limited because of their low vapor pressures.

Soon after the first demonstrations of ALD, metal chlorides were taken under study. Among the first processes examined were also two of the most successful ALD processes, namely, $ZnCl_2$-H_2S for ZnS and $AlCl_3$-H_2O for Al_2O_3 [2–4]. Though both these processes are based on solid chlorides and are therefore somewhat laborious in production, they have been used in industry from the very beginning of the ALD manufacturing of the TFEL displays. For doping of the ZnS : Mn phosphor, $MnCl_2$ is used, even if it has quite a low vapor pressure. Among the first chloride precursors examined were also the liquids $TiCl_4$ and $SnCl_4$, the former of which has also been adopted to the TFEL production for making $Al_xTi_yO_z$ insulators (Section 6.2.1). Subsequently, basically all possible volatile chlorides have been examined (Table IV). By contrast, of the other halides only TiI_4, ZrI_4, TaI_5, WF_6, and MnI_2 have been examined so far. Especially the iodides should be examined in more detail because the low metal-iodine bond energies suggest that iodide residues could be removed from the films more completely than chlorides.

As the interest toward ALE (here the name ALE is used to emphasize the epitaxial deposits) of the III–V materials arose, a natural starting point was to choose metal alkyls, like $Ga(CH_3)_3$, as precursors because at that time they were already well known and were broadly used in metal organic vapor phase epitaxy (MOVPE). However, this choice is not completely in line with the previously discussed requirements to ALD precursors because the reactions with the group V hydrides usually require such high temperatures that pyrolysis of the organometallic compounds occurs and complicates the control of the self-limiting growth conditions [18, 19, 22, 48, 255, 397]. Indeed, a mechanism which would explain the different results obtained under different GaAs growth conditions has been speculated about for a long time. The current opinion is that the two key requirements for the self-limiting growth are an avoidance of complete gas phase decomposition of $Ga(CH_3)_3$ in the boundary layer above the substrate and a high enough methyl radical flux to compensate for the methyl radicals which desorb from the surface [22]. On the other hand, with the aid of laser assistance the growth temperature may be lowered so

that the thermal decomposition is significantly decreased, and at the same time the width of the temperature range for the self-limiting film growth is increased [15, 17, 48, 49]. Another consequence of the metal alkyl pyrolysis is carbon residual incorporation causing relatively high p-type carrier concentrations [18, 19]. However, with careful control of the growth conditions, in particular by reducing the flow rate and the exposure time of the metal alkyl and by increasing the corresponding parameters for the group V hydride, low carrier concentrations and even n-type conductivity could have been achieved [19, 93, 253, 254]. In any event, because of these reasons, chloride-based processes emerged as another main route to ALE of III–V compounds [18]. Here the precursors have been either binary chlorides, often formed in situ, or alkyl chlorides which pyrolyze in the reactor into the binary chlorides. With the chloride-based processes, the heavy carbon contamination is avoided and the n-type materials are easily obtained.

The early work on the III–V compounds in the late 1980s was focused on compounds of Ga, Al, In, As, and P. Somewhat later, in the early 1990s, ALE of gallium and aluminum nitrides from the trialkyls was examined [242, 243, 259, 260], and later InN was also grown [246–248]. In addition to the III–V compounds, metal alkyl precursors have successfully been used also in the ALD of epitaxial or nonepitaxial II–VI compounds and polycrystalline or amorphous oxides. Here ZnS, CdS, Al_2O_3, and ZnO have been the most thoroughly examined materials (Table IV). Reactions with H_2S and H_2O are facile and thus proceed at low temperatures, occasionally even at room temperature [274, 298–302], where the pyrolysis of the metal alkyls is much weaker than at the temperatures typically used in the III–V processes.

Metal alkoxides offer chlorine-free alternatives to the growth of oxide thin films. The first ALD experiments were made on aluminum alkoxides using water or alcohols as oxygen sources [129], and subsequently titanium, tantalum, and niobium alkoxides (mainly ethoxides) have successfully been used in reactions with water (Table IV). For tantalum and especially for niobium, alkoxides are important precursors because of the etching problems in the chloride processes (Section 5.1.4). Alkoxides decompose thermally at elevated temperatures and therefore the ALD processes are limited to temperatures below 400°C. On the other hand, the decomposition product often is a rather pure oxide and therefore the decomposition may be accepted to a certain extent. All the previously mentioned alkoxide processes produce good quality films, but zirconium tetra-tert-butoxide [310] resulted in ZrO_2 films which were clearly of a lower quality than those obtained from $ZrCl_4$. This is unfortunate because the liquid alkoxide was hoped to solve a problem encountered in the $ZrCl_4$-H_2O process where very fine particles from the solid $ZrCl_4$ source are often transported by the carrier gas and are incorporated into the films.

After the success of the ZnS : Mn TFEL phosphor, the strive toward full color displays turned the ALD research on other TFEL phosphors, most of which have alkaline earth metal sulfides as the host materials and rare-earth metals as the dopants (Section 6.1). These electropositive metals have few volatile compounds, β-diketonates being the well known and the most thoroughly examined exception. Though the β-diketonates, especially those of the alkaline earth metals, suffer somewhat from instability, they have served as reasonably good precursors for the TFEL phosphors (Table IV). For example, the $Sr(thd)_2$-$Ce(thd)_4$-H_2S process has been used in a pilot scale growth of $SrS:Ce$ [338]. On the other hand, while reactive with H_2S, most of the β-diketonates do not react with water at temperatures where they would not decompose already on their own. Here, $Ga(acac)_3$ has been an exception and reacted with water but the Ga_2O_3 films obtained were heavily carbon contaminated [327]. $Mg(thd)_2$, in turn, has been used together with H_2O_2 to grow MgO films but the deposition rate was low, only 0.1–0.2 Å cycle^{-1} [316]. Thus, oxides are best grown from the β-diketonates with ozone as the oxygen source (Table IV). In reactions with ozone, the large β-diketonato ligands are probably burned into smaller molecules and thus these reactions essentially differ from most of the other ALD processes where the ligands are removed more or less intact, either after protonation by the hydrogen containing nonmetal precursor or as a consequence of radical desorption.

Cyclopentadienyl compounds (metallocenes) have been examined as alternatives to alkaline earth and rare-earth metal β-diketonates with a number of promising results (Table IV). The use of cyclopentadienyls in ALD was first demonstrated in the growth of MgO from $Mg(C_5H_5)_2$ and H_2O [241, 353, 354], and subsequently $Mn(C_5H_5)_2$ and $Mn(C_5MeH_4)(CO)_3$ were examined as dopant sources for $ZnS:Mn$ [295] and $Ce(C_5Me_4H)_3$ for $SrS:Ce$ [360]. In any event, so far the greatest progress with the cyclopentadienyls has been made in the growth of $SrTiO_3$ and $BaTiO_3$ [356–358] (Section 6.2.2), because these can be deposited from the β-diketonates only with ozone and then the resulting film is amorphous which requires high temperature annealing to get crystallized [322]. Using $Sr(C_5{}^iPr_3H_2)_2$ together with $Ti(O^iPr)_4$ and H_2O, crystalline $SrTiO_3$ was obtained already at 250°C [356–358]. Despite some decomposition of $Sr(C_5{}^iPr_3H_2)_2$, a good control of film stoichiometry was achieved by varying the pulsing ratio of the two metal compounds. $BaTiO_3$ films were obtained with an analogous process with $Ba(C_5Me_5)_2$ [356, 357] or $Ba(C_5{}^tBu_3H_2)_2$ [357] as the barium precursor. These Sr and Ba compounds also react with hydrogen sulfide forming the corresponding sulfides at remarkably low temperatures: well-crystallized SrS was obtained already at 120°C [359].

Overall, cyclopentadienyl compounds form a large family of potential precursors since these are known for many metals and the ligands can be varied by substitutions in the carbon 5-ring, largening the ring system (indene, fluorene) and by linking two rings together by a bridge. Cyclopentadienyls have often been considered too air sensitive but the foregoing experiences demonstrate that with a reasonably careful handling they can be used without difficulty. The first experiments on these compounds have been promising but the full potential still remains to be explored and may provide interesting findings in the near future.

5.3.2. Nonmetal Precursors

For the nonmetals, the simple hydrides have mostly been used: H_2O, H_2O_2, H_2S, H_2Se, H_2Te, NH_3, N_2H_4, PH_3, AsH_3, SbH_3, and HF. These are all sufficiently reactive but some of them are very poisonous and therefore their alkyl derivatives have been examined as substitutes. Hydrogen peroxide is oxidizing by its nature while most of the others, hydrazine in particular, are reducing. The oxidizing power of H_2O_2 is seldom made use of, however. By contrast, the reducing action is highly needed in the growth of transition metal nitrides where the metals have higher oxidation states in their precursors than in the nitrides. Ammonia is powerful enough to reduce most of the transition metal chlorides to the desired metallic nitrides, though films with better conductivity are usually obtained when elemental zinc is used as an additional reducing agent (Section 5.3.6). However, $TaCl_5$ is not reduced by ammonia and insulating Ta_3N_5 is obtained [113]. A hydrazine derivative $(CH_3)_2NNH_2$ does not reduce tantalum either [114], and TaN is obtained only with zinc as a reducing agent [113].

Molecular oxygen, O_2, is usually too inert to react with the typical metal compounds under reasonable growth temperatures. Instead, ozone O_3 has been used to grow oxides from many β-diketonate compounds (Table IV) which do not react with the most common oxygen sources water and hydrogen peroxide. Other less used oxygen sources include N_2O and alcohols. A new approach applicable to many oxides is to use metal alkoxides as both metal and oxygen sources (Section 5.3.5).

Similarly to oxygen, also molecular nitrogen, N_2, and hydrogen, H_2, are quite inert toward the common metal sources at low temperatures. They are indeed often used as carrier gases, but while nitrogen is in most cases truly inert, hydrogen sometimes participates the surface reactions and removes surface terminating ligands (Section 7.2.1) [18]. Hydrogen has also been used as a reducing agent in ALD but this requires either catalytically active surfaces or high temperatures (Section 5.3.6). On the other hand, when dissociated by thermal cracking or plasma discharge, reactive atomic species are produced from hydrogen and nitrogen [60–62, 64–66, 68–71, 74]. Also oxygen may be dissociated in a plasma discharge but until now atomic-oxygen-based ALD processes have only briefly been demonstrated [75].

For the chalcogens (S, Se, Te), elements can be used as precursors, though usually only when the metal source also is an element (Table IV). It has also been demonstrated that elemental Se replaces sulfur atoms on the surface of the growing ZnS and SrS films. With this replacement reaction, the corresponding sulfoselenides with as much as 90 anion-% Se were deposited from $ZnCl_2$ or $Sr(thd)_2$, H_2S, and Se [108, 109], thereby avoiding the use of the highly toxic H_2Se. Finally, elemental As_4 and P_4 have been employed in the growth of GaAs and GaP from GaCl with hydrogen as a carrier gas [146, 147].

5.3.3. In situ Synthesized Precursors

A majority of the precursors used in ALD are synthesized beforehand but in some cases precursors generated *in situ* have been used as well. *In situ* synthesis makes it possible to use compounds which are otherwise difficult to handle because of their high reactivity or which easily age when stored for long times.

The most common way for the *in situ* synthesis is to employ gas–solid reactions. As in CVD, also in ALD this has most often been used with chlorides, especially with gallium and indium chlorides, but also strontium and barium β-diketonates and tungsten oxyfluorides have been synthesized *in situ*. Gallium and indium chlorides were synthesized from metallic gallium and indium and hydrogen chloride at around 750°C [18]. For *in situ* synthesis of $[Sr(thd)_2]_n$, reactions between solid Sr, SrO, $Sr(OH)_2$, or $SrCO_3$ and Hthd vapor were examined and the best results were obtained with Sr and SrO [325]. $[Ba(thd)_2]_n$ was synthesized from $Ba(OH)_2$ and Hthd [326], and WO_xF_y from WO_3 and WF_6 [205]. A risk related to the *in situ* precursor synthesis by the gas–solid reactions is the possibility that part of the gaseous reactant passes the source to the reaction chamber and etches the film. Therefore, the souce must be properly designed to ensure complete reactions.

The *in situ* generation of HF differs from the preceding solid–gas reactions in that HF was obtained by decomposing NH_4F thermally [215]. The ammonia liberated in the decomposition apparently had no effect in that study since the fluoride films deposited (ZnF_2, CaF_2, SrF_2) are inert toward ammonia.

In addition, gas–gas reactions have been employed for *in situ* generation of desired precursor compounds too. Chlorides of gallium and indium were formed by mixing $Ga(C_2H_5)_3$ and $In(CH_3)_3$ vapors with HCl in a reaction zone heated around 100°C, thereby avoiding problems related to the low vapor pressures of the binary chlorides and alkyl chlorides [18].

5.3.4. Single Source Precursors

In general, ALD processes employ two or more precursors which each contain one of the film constituent elements (see Section 5.3.5, however). Single source precursors which contain all the film constituents are actively examined in CVD, but in ALD they are not directly applicable because the self-limiting growth mechanism by its nature requires exchange reactions between different precursors and stability against self-decomposition. On the other hand, single source ALD processes can be realized by ramping the temperature repeatedly so that the adsorption takes place at a low temperature and the decomposition takes place at a higher temperature. Simple thermal heating and cooling of the substrate is inevitably too slow to be practical except for very thin films, but heating with laser or lamp or photodissociation with high energy photons makes the process faster. However, so far such processes have not been examined for compound films for which the single source CVD precursors are usually aimed, but only for deposition of elemental silicon [51, 54, 55, 57, 65, 67, 378, 381] and germanium [55, 56, 383]. For example, to deposit germanium, $Ge(C_2H_5)_2H_2$ was adsorbed at 220°C with a release of hydrogen and a formation of a monolayer of $Ge(C_2H_5)_2$, and then the ethyl groups

were desorbed by heating above 400°C [383]. Similarly, silicon was deposited in a self-limiting manner at 180–400°C by first adsorbing Si_2H_6 and then by heating the substrate with a UV laser to desorb the surface terminating hydrogen atoms [51, 378].

The combination of using only one precursor and short heating (or dissociating) light pulses also offers an interesting possibility to have the precursor continuously present in the reactor and omit the purging. Within the duration of short light pulses, adsorption of new molecules in place of the decomposed chemisorption monolayer is negligible, and thus self-limiting film growth is achieved. This approach has been used for ALD of silicon and germanium from SiH_4 and GeH_4 by heating with a Xe flash lamp [54–56]. In fact, with these particular precursors the continuous presence is vital because the surface density of the chemisorbed species is determined by the balance between adsorption and desorption; i.e., the chemisorbed species have a high desorption probability and might thus be lost during a purge period.

5.3.5. Combinations of Two Metal Compounds

Quite recently a novel approach to the ALD of oxides was introduced [393]. In this process, two metal compounds, at least one of which is an alkoxide and thus contains a metal–oxygen bond, were used (Table IV). The metal alkoxide serves as both the metal and the oxygen source while the other metal compound, typically a metal chloride, acts as the other metal source:

$$bM(OR)_a + aM'X_b \rightarrow M_bM'_aO_{ab} + abRX$$

Depending on whether M and M' are similar or different, binary or mixed oxides are obtained. The major benefit of not using separate oxygen sources like water or hydrogen peroxide is the less susceptible oxidation of the substrate surface. This is especially important when thin high permittivity dielectric layers are to be deposited directly on silicon without creating an interfacial silicon oxide layer (Section 6.2.2).

5.3.6. Reducing Agents and Other Additional Reagents

Under this heading, reagents which take part in the film formation reactions but do not leave any constituents to the film are considered. The most obvious of these are the reducing agents in the growth of elemental films. For this purpose, both molecular and atomic hydrogen as well as elemental zinc and disilane have been examined. Atomic hydrogen which may be produced by either plasma discharge or thermal cracking is very reactive and facilitates even the deposition of metallic titanium and tantalum from their chlorides [74]. Also $SiCl_2H_2$ [60, 64–66, 170], Si_2Cl_6 [61], $GeCl_4$ [62], and $Ge(CH_3)_2H_2$ [68, 69] have been reduced with atomic hydrogen. Molecular hydrogen, by contrast, is quite inert and reduces CuCl [209, 210] and $Cu(thd)_2$ [210, 343, 344] only on appropriate metal surfaces (Section 6.6). For instance, the $Cu(thd)_2$-H_2 process resulted in a film growth only when the surface was seeded with a predeposited platinum–palladium layer [210, 343, 344], and even then the copper deposition could not be repeated in a different reactor where high partial pressures of hydrogen could not be used and where the residence time of the precursors was much shorter [186]. The ALD studies on Ni(acac)$_2$, Cu(acac)$_2$, and Pt(acac)$_2$ reduction by H$_2$ also indicated the low reactivity of molecular hydrogen; metallic deposits seemed to be obtained only through interactions with substrates or as a consequence of thermal decomposition of the metal precursors [342]. Finally, high temperatures of 815–825°C were required to obtain monomolecular layer growth when SiCl$_2$H$_2$ was reduced by H$_2$ [373, 374].

Elemental zinc is a powerful reducing agent and has been employed in the ALD of copper [110] and molybdenum [204] but it suffers from its tendency to dissolve into the metallic films. Though the films contain only a few atomic percentages of zinc, the dissolution and subsequent outdiffusion during the process destroy the self-limiting growth mechanism (Section 5.1.5). In the growth of transition metal nitrides, the dissolution of zinc into the films is not a problem [112] and intermediate zinc pulses have been employed in improving the properties of the films deposited from metal chlorides and ammonia (Section 6.5). In addition to acting as a reducing agent, zinc may also assist in removing chlorides from the surface by forming ZnCl$_2$. In the TaCl$_5$-NH$_3$ process, the use of zinc is vital because otherwise semiconducting Ta$_3$N$_5$ is obtained instead of the desired TaN [113], but in the other cases the positive effect of zinc is not necessarily just simply due to a reducing action but also other chemical and structural factors appear to be involved [112]. However, as zinc forms electrically active states in silicon, the concern that zinc could dissolve into a silicon substrate has limited further interest toward these zinc-based ALD nitride processes. The most recently examined reducing agent is disilane Si$_2$H$_6$ which was successfully used in the ALD of metallic tungsten from WF$_6$ [206] (Sections 6.6 and 7.2.2). No silicon or fluorine could be detected in the films by X-ray photoelectron spectroscopy.

Moving to other additional reagents, pyridine [165, 166] and ammonia [167] have been used in catalyzing reactions between the alternately pulsed SiCl$_4$ and H$_2$O. In this way, the deposition temperature of SiO$_2$ was decreased remarkably from above 300°C to room temperature and at the same time the reactant exposure required to complete the surface reactions was decreased from 10^9 to 10^4–10^5 L (1 L = 10^{-6} Torr s). Unfortunately, also these catalyzed reactions required tens of seconds to get completed and thus the main problem of SiO$_2$-ALD, i.e., the long cycle times causing low deposition rates per time unit, still remains. The mechanism proposed for explaining the catalytic effect of pyridine and ammonia involves hydrogen bonding of the nitrogen atom in the catalyst molecule to surface hydroxyl groups (during the SiCl$_4$ pulse) or water molecules (during the H$_2$O pulse) with a concomitant weakening of the O—H bond and an increase of the nucleophilicity of the oxygen atom. The catalytic effect decreased as the temperature was increased above room temperature which may be related to decreased surface coverages of SiCl$_4$ and, perhaps more importantly, of catalyst molecules. As good film properties were achieved with

these catalyzed processes, they are clearly an interesting approach to very low temperature ALD and they deserve further studies to clarify their applicability to other materials.

Additional reagents may also be used for assisting the completion of the reactions. Occasionally, the complete ligand removal from the metal precursors is hard and may thus lead to impurity incorporation. This has been a severe problem in the epitaxial growth of the III–V semiconductors where low impurity levels are required; especially aluminum containing compounds are prone to carbon incorporation. In the growth of AlAs from Al(CH$_3$)$_2$H and AsH$_3$, dimethylamine ((CH$_3$)$_2$NH) supplied intermediately after Al(CH$_3$)$_2$H was used to remove methyl groups from the surface [240]. A mechanism which involves reactions between the methyl groups and amine and hydrogen radicals was suggested to explain the observed decrease of the carbon content from 6 × 10^{20} to 8 × 10^{19} cm^{-3} as determined by secondary ion mass spectrometry (SIMS). Nitrogen contents were not reported, however. Anyhow, the idea of using separate reagents for assisting surface ligand removal is evidently worth further consideration.

6. FILM MATERIALS AND APPLICATIONS

Table V summarizes the film materials deposited by ALD thus far. As in Table IV, all the reported materials have been included regardless of how effective the processes actually are. For references, see Table IV. In this section, the most important nonepi-

Table V.   Thin Film Materials Deposited by ALD^a

II–VI compounds	ZnS, ZnSe, ZnTe, ZnS$_{1-x}$Se$_x$,
	CaS, SrS, BaS, SrS$_{1-x}$Se$_x$,
	CdS, CdTe, MnTe, HgTe, Hg$_{1-x}$Cd$_x$Te,
	Cd$_{1-x}$Mn$_x$Te
II–VI-based TFEL phosphors	ZnS : M (M = Mn, Tb, Tm),
	CaS : M (M = Eu, Ce, Tb, Pb),
	SrS : M (M = Ce, Cu, Tb, Pb)
III–V compounds	GaAs, AlAs, AlP, InP, GaP, InAs,
	Al$_x$Ga$_{1-x}$As, Ga$_x$In$_{1-x}$As, Ga$_x$In$_{1-x}$P
Nitrides	
semiconductors/dielectric	AlN, GaN, InN, SiN$_x$, Ta$_3$N$_5$
metallic	TiN, Ti-Si-N, TaN, NbN, MoN, W$_2$N
Oxides	
dielectric	Al$_2$O$_3$, TiO$_2$, ZrO$_2$, HfO$_2$, Ta$_2$O$_5$,
	Nb$_2$O$_5$, Y$_2$O$_3$, MgO, CeO$_2$, SiO$_2$, La$_2$O$_3$,
	SrTiO$_3$, BaTiO$_3$, Bi$_x$Ti$_y$O$_z$
Transparent conductors/	In$_2$O$_3$, In$_2$O$_3$: Sn, SnO$_2$, SnO$_2$: Sb, ZnO,
semiconductors	ZnO : Al, ZnO : B, ZnO : Ga, Ga$_2$O$_3$, WO$_3$,
	NiO, Co$_3$O$_4$, MnO$_x$
Ternary oxides	LaCoO$_3$, LaNiO$_3$, LaMnO$_3$
Fluorides	CaF$_2$, SrF$_2$, ZnF$_2$
Elements	Si, Ge, Cu, Mo, Ta, W
Others	La$_2$S$_3$, PbS, In$_2$S$_3$, CuGaS$_2$, SiC

^{a}For references, see Table IV.

taxial thin film materials, grouped according to their applications, are discussed.

As the epitaxial semiconductors were limited outside the main scope of this presentation, they are not dealt with here but rather review articles [19–23] are suggested for the III–V compounds, and [24, 25] for the II–VI compounds. In contrast to the compound semiconductors, little has been published thus far on the material properties of the ALD made Si and Ge films, the focus in these studies has been on the film growth and the related chemistry. Also deposits other than thin films are skipped here; the use of ALD in processing of porous materials [30–33] and in nanotechnology [398] have been reviewed in the references cited.

6.1. Electroluminescent Display Phosphors

As already noted, ALD was originally developed by Suntola and co-workers for making thin film electroluminescent (TFEL) displays [2, 5–9]. In this application, ALD has been very successful and has been used for nearly 20 years in manufacturing [7–11]. In addition, research on using ALD in making thin films for the TFEL displays has been active all the time and has formed a solid basis for the more recent ALD research, particularly that on insulators (Section 6.2). Therefore, it is instructive to first have a closer look at the TFEL display itself.

A schematic of a conventional TFEL display is shown in Figure 12. A more recently developed, so-called inverted structure is otherwise similar but the places of the transparent and metal electrodes have been exchanged. In the conventional structure, glass is used as a substrate because viewing is through the substrate. Also in the inverted structure glass is often used but also opaque ceramics may be applied as substrates because the viewing is on the opposite side. The benefit of glass is its low price whereas ceramics make it possible to use high annealing temperatures for improving the phosphor crystallinity. If soda lime glass is used, it is passivated by an ALD made Al_2O_3 to prevent sodium outdiffusion (Section 6.4). The ac-

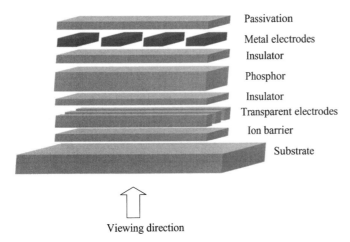

Fig. 12. Schematics of a TFEL display with a conventional structure. In the inverted structure the metal and transparent electrodes have changed places and the viewing is from the film stack side of the substrate.

tual TFEL device structure consists of an electrode-insulator-semiconducting phosphor-insulator-electrode film stack where the electrodes are patterned into stripes perpendicular to each other while the other films are continuous. The electrode on the viewing side must of course be transparent and thus indium–tin oxide (ITO) is usually used. The other electrode is metal; in the conventional structure it is aluminum but in the inverted structure molybdenum or tungsten with better thermal properties must be used. A passivating layer is finally deposited on the top of the structure. At present, ALD is used in industrial scale for all the other films except the electrodes which are sputtered. $ZnS:Mn$, Al_2O_3 and Al_2O_3/TiO_2(ATO) are the dominant ALD made materials used by the industry. The thickness of the phosphor layer is in the range of 500–1000 nm while the insulators are about 200-nm thick, and the three layers are deposited in one continuous ALD process.

In the TFEL displays, each crossing point of the bottom and top electrodes defines a picture element (pixel). A pixel is lit by applying an ac (typically 60 Hz) voltage to the two electrodes. At low voltages, the insulator-phosphor-insulator structure acts like three capacitors in series, but when the electric field in the phosphor layer exceeds a certain threshold value, electrons begin to flow through the phosphor from one insulator-phosphor interface to the other. When arriving at the interface, the electrons are trapped in the interface states from which they are emitted when the polarity of the electric field reverses. For an effective operation of the typically used phosphor materials, the threshold field must be in the range of 1 to 2 MV cm^{-1}, so that once released, the electrons are rapidly accelerated to energies high enough to impact excite the luminescent centers which then emit light while returning to the ground state. The phosphor layer must withstand these high fields without destructive breakdown. Likewise, the insulators must possess high breakdown strength and low leakage currents at the operation voltages. Pinhole-freeness over the large-area substrate is a key aspect in meeting these requirements. In the late 1970s, few techniques existed for making high quality insulator-phosphor-insulator structures, and that was the main motivation for developing ALD.

Unlike the other flat panel displays, the TFEL displays have complete solid-state structures which gives them many advantages like wide operating temperature, ruggedness, exceptionally broad viewing angle, and fast response. On the other hand, operation voltages around 200 V are typically needed which means that the driving electronics and thus the whole display is rather expensive as compared to its main competitors, especially liquid crystal displays. Therefore, TFEL displays are not found in laptops but rather they are found in applications like medical, instrumentation, and transportation where their special characteristics are valued.

Quite recently a new kind of TFEL display, active matrix EL (AMEL) display has been developed [399]. Here, the insulator-phosphor-insulator-transparent top electrode film stack is deposited directly on single crystal silicon-on-insulator (SOI) wafers to which all the required driving circuitry has been integrated. This makes it possible to make small high resolution displays for head mounted applications, for example. On the other hand, the deposition on top of the driving circuitry sets strict demands on the film conformality, thus favoring the use of ALD (Fig. 13).

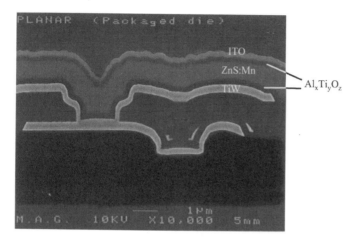

Fig. 13. Cross-sectional image of an AMEL display. Only the films of the TFEL structure are shown, the driving circuitry below is not shown. The ZnS : Mn and Al$_x$Ti$_y$O$_z$ films are made by ALD. (Courtesy B. Aitchison, Planar Systems Inc.)

Comprehensive review articles on TFEL phosphor materials can be found in Refs. [11, 400, 401], so here they are discussed chiefly from the ALD point of view. A summary of the ALD made TFEL phosphors is presented in Table VI. The most important of the TFEL phosphors is the yellow–orange emitting ZnS : Mn which is made by ALD primarily from ZnCl$_2$, MnCl$_2$, and H$_2$S [3, 4, 8, 402], though alternatives have been used for both zinc (Zn(C$_2$H$_5$)$_2$) and manganese (Mn(thd)$_3$, Mn(C$_5$H$_5$)$_2$, Mn(C$_5$MeH$_4$)(CO)$_3$) [295, 296, 402]. The concentration of Mn is typically 0.5–2 mol%. In ALD, the doping is realized most simply by replacing a certain number of the zinc precursor pulses with manganese precursor pulses. Though this seems to lead to delta doping depth profiles, concentration quenching has not been found to be a major problem. Apparently, surface roughness and diffusion smooth the doping profile. On the other hand, due to the forbidden transitions in the Mn luminescence the Mn–Mn energy transfer is not very probable. The other option for doping is to supply the dopant simultaneously with the matrix cation but in that approach differences in reactivities may cause nonuniformity over the substrate.

ZnS : Mn is the most efficient TFEL phosphor ever found (3–8 lm W^{-1}). No major differences in the performance can be found between the ALD films [402, 403] and those made

Table VI. Luminescent Thin Films Made by ALD for TFEL Devices

Phosphor material	Emission color	Luminance cd m^{-2} at 60 Hz[a]	CIE coordinates		Reference
			x	y	
ZnS : Mn^{2+} (chloride-process)	Yellow	440	0.52	0.48	[402]
ZnS : Mn^{2+} (organometallic precursors)	Yellow	430	0.54	0.46	[402, 403]
ZnS : Tb^{3+}	Green	35			[350]
ZnS : Tb^{3+}(O, Cl)	Green	75	0.28	0.64	[351]
ZnS : Tm^{3+}	Blue	<1 (300 Hz)			[345]
CaS : Eu^{2+}	Red	2–6	0.68	0.31	[404, 405]
CaS : Tb^{3+}	Green	20			[319]
CaS : Pb^{2+}	Blue	2.5 (300 Hz)	0.17	0.13	[406]
		80	0.14–0.15	0.07–0.15	[407]
SrS : Ce^{3+}	Bluish-green	130	0.30	0.54	[408]
SrS : Ce^{3+} (filtered)	Blue	8	0.08	0.20	[408]
SrS : Ce^{3+}, Y^{3+}	Bluish-green	100	0.3	0.5	[360]
SrS : Ce^{3+}, Y^{3+} (filtered)	Blue	20	0.21	0.39	[360]
SrS : Pr^{3+}	White	30 (300 Hz)			[345]
SrS : Tb^{3+}	Green	5.5			[347]
SrS : Mn^{2+}	Green	3.5 (300 Hz)	0.37–0.44	0.55–0.61	[409]
SrS : Mn^{2+}, Pb^{2+}	White	7 (300 Hz)	0.28	0.39	[409]
SrS : Cu$^+$	Blue (green)	25	0.17	0.30	[410]
SrS : Pb^{2+}	Bluish-green	17 (300 Hz)	0.26	0.33	[406]
SrS : Pb^{2+} (filtered)	Blue	1.8 (300 Hz)	0.14	0.09	[406]
CaF$_2$: Eu^{2+}	Blue (450 nm)	2 (1 kHz)	0.2	0.1	[215]
ZnS : Mn^{2+}/SrS : Ce^{3+}	White	480[b]	0.49	0.48	[338, 402, 408]

[a]Note that the voltages in the luminance measurements vary in different references from 25 to 50 V above the threshold.

[b]Value for one pixel; in patterned full devices values of 21 and 70 cd m^{-2} with 60 and 350 Hz, respectively, have been achieved for areal luminance [411, 412].

by other methods, like thermal and electron beam evaporation or sputtering. A clear advantage of ALD is, however, a larger grain size at the beginning of the film growth which ensures that in the ALD made films the dead layers, i.e., layers with no emission due to poor crystallinity, are thinner than in the films made by the other techniques [413, 414]. The emission band of ZnS : Mn is so broad that with proper filters it may also be used as a red or a green phosphor for multi- and full color displays. To avoid parallax, the filters need to be placed close to the film stack and because this is not possible in the conventional structure (Fig. 12) where the substrate would be left in between, the inverted structure was developed by replacing the positions of the metal and the transparent electrodes [10].

ZnS : Tb is the other zinc sulfide-based material which has shown EL properties good enough for applications. In the deposition, Tb(thd)$_3$ has been used as a precursor and good results have been obtained when several subsequent Tb(thd)$_3$/H$_2$S cycles have been used instead of one. Thus, the actual structure has been of a TbS$_x$/ZnS sandwich type [350, 351]. The rare-earth ions (Ln^{3+}) are large as compared to the zinc ion and therefore in practical rare-earth concentrations (few percent) the rare-earth ions when homogeneously distributed cannot locate at the zinc site in ZnS but rather a Ln$_2$O$_2$S center is formed [415].

Due to the excellent performance of ZnS : Mn in the long wavelength part of the visible spectrum, quite a lot of the more recent phosphor research has been devoted to the blue phosphors which are needed to complement ZnS : Mn in making a full color display. Here, the most intensively studied material has been bluish-green emitting SrS : Ce for which the basic process uses Sr(thd)$_2$, Ce(thd)$_4$, and H$_2$S as the precursors [337, 338], but also different fluorinated β-diketonate Ce complexes have been studied as source materials for the Ce dopant [339, 340]. The use of cyclopentadienyl compounds as precursors began with Ce compounds [360] but also the host material SrS has been made from metalorganic precursors, from Sr(C$_5{}^i$Pr$_3$H$_2$)$_2$ for instance [359, 416]. Despite all the efforts, the performance of SrS : Ce after blue filtering (10–20 cd m^{-2} depending on the filter) is still somewhat lacking the level required (20–30 cd m^{-2}) for commercializing the full color displays. The total EL brightness of SrS : Ce is good (>100 cd m^{-2}) but unfortunately the majority of the emission falls in the green region (Table VI). Quite recently, SrS : Cu and CaS : Pb have been taken under study as new potential blue phosphors. Especially for CaS : Pb, deposited from Ca(thd)$_2$, Pb(C$_2$H$_5$)$_4$, and H$_2$S, very promising results have been reported: deep blue emission, CIE (Comission Internationale de l'Eclairage) color coordinates ranging from (0.14, 0.07) to (0.15, 0.15), and luminance of 80 cd m^{-2} with low driving voltage but the efficiency is still an issue [407].

In addition to the previously mentioned, many other potential phosphors have been deposited by ALD in the search of materials for full color TFEL displays (Table VI). The materials studied are rare-earth doped zinc and alkaline earth sulfides. The rare-earth ions, besides cerium and terbium, include europium (red), samarium (red), praseodymium (white), and thulium (blue). Most of the materials have been examined already in the 1980s and the EL performance levels reached by ALD are also here quite similar to those obtained by the other film deposition methods. Unfortunately, the EL properties of these materials are far below the level needed in TFEL devices.

6.2. Insulators

6.2.1. Insulators for TFEL Displays

In the TFEL displays (Section 6.1, Fig. 12), the insulator films limit current transport across the TFEL device. Since high electric fields of about 2 MV cm^{-1} are typically used, high dielectric field strength (high breakdown field, E_{BD}) and pinhole-freeness over the whole display area is required. Therefore, the insulators seem to be perhaps the most critical part in the preparation of reliable TFEL displays. A detailed discussion on the role of the insulators and their requirements has been presented in Refs. [11, 400, 401], so here we just summarize the most important properties of the insulator films:

   (i) high electric field strength,
  (ii) high relative permittivity, ε_r,
 (iii) pinhole-free structure,
  (iv) self-healing breakdown mode,
   (v) convenient and stable interface-state distribution from which electrons are emitted into the phosphor at proper electric fields,
  (vi) good thickness uniformity and conformality,
 (vii) good adhesion and stability with the adjacent electrode and phosphor layers,
  (ix) stress-free.

The insulator films should preferably be amorphous because polycrystalline films lead to rough interfaces and they contain grain boundaries through which electrons can flow and ions can migrate. The requirements of high field strength and high permittivity are contradictory since, in general, insulators with high permittivity suffer from low breakdown fields. In addition, their breakdown mechanism is usually propagating rather than self-healing. Therefore, insulators with moderate permittivity but high E_{BD} have usually been preferred. A convenient figure of merit for the insulators is the charge storage factor, $\varepsilon_0\varepsilon_r E_{BD}$, which shows the maximum charge density that can be stored in a capacitor made of a given insulator. Here, it must be noted, however, that while ε_r is usually well defined, E_{BD} is a statistical quantity and depends on the measurement method and the definition of breakdown; i.e., whether the breakdown field is that causing a destructive breakdown or a certain leakage current density, like 1 μA cm^{-2}. Table VII summarizes dielectric properties of ALD made insulators potential for TFEL displays. In addition to those listed in Table VII, several other potential insulator films have also been grown by ALD (Y$_2$O$_3$, MgO, CeO$_2$, AlN, see Tables IV and V) but their dielectric properties have not been reported in enough detail. On the other hand, some of the high permittivity insulators (TiO$_2$, Nb$_2$O$_5$, SrTiO$_3$, BaTiO$_3$) are considered too leaky to be applied in the TFEL

Table VII.   Dielectric Properties of ALD Made Insulators Potential for TFEL Displays[a]

Material	ε_r	E_{BD} (MV cm^{-1})	$\varepsilon_0 \varepsilon_r E_{BD}$ (nC mm^{-2})	Reference
Al_2O_3	7–9	3–8	16–50	[129, 133, 135, 136, 216, 220, 225, 227, 237]
$Nb_x Al_y O_z$	8	5	42	[136]
Al_2O_3/TiO_2	9–18	5–7	40–80	[107, 417]
Al_2O_3/Ta_2O_5	10–20	2–6.6	20–55	[130, 131, 133, 135]
HfO_2	13–16	1–5	8	[131, 197]
Ta_2O_5/HfO_2	19–23	2.5–5	40–66	[131, 197]
ZrO_2	20	1	20	[133]
Ta_2O_5	23–25	0.5–1.5	5–10	[131, 133, 197]
Ta_2O_5/ZrO_2	25–28	2–2.5	40–60	[133]
$Ta_x Ti_y O_z$	27–28	1	33	[418]
$Nb_x Ta_y O_z$	25–35	0.5–1	25	[202]
$Nb_x Ta_y O_z/ZrO_2$	31–33	3	83	[192]
Ta_2O_5/Nb_2O_5	38	0.5	17	[202]

[a]Typically the best results or a range of the best values are given for each material. The notation of AB/CD refers to a stacked insulator (nanolaminate) with a variable number of layers. Note that the E_{BD} values depend on its definition, and consequently also $\varepsilon_0\varepsilon_r E_{BD}$ depend on the definition of E_{BD}. Most of the results are based on E_{BD} corresponding to a leakage current density of 1 μA cm^{-2}.

displays as such, without a combination with higher resistivity materials, and are therefore not included in Table VII.

In the commercial TFEL displays, the ALD made insulators are Al_2O_3/TiO_2 (ATO) or just Al_2O_3 which are made from the corresponding chlorides and water. For aluminum, $Al(CH_3)_3$ may also be used. The Al_2O_3/TiO_2 insulator [107, 399, 402, 417, 419] is a good representative of composite structures where advantageous characteristics of two or more materials are combined in realizing insulators which are at the same time reliable (high E_{BD}) and efficient (high ε_r). In other words, the $\varepsilon_0\varepsilon_r E_{BD}$ value of the composite exceeds those of its binary constituents (Table VII).

Later, Ta_2O_5 based composite insulators have been examined in great detail (Table VII). Ta_2O_5 has a relatively high permittivity of 25 but in the as-deposited state it is usually quite leaky due to oxygen deficiency. To improve this shortcoming, a concept of nanolaminate was introduced. Nanolaminates consist of alternating layers of two or more insulator materials so that each separate layer has a thickness in a range of 1–20 nm (Fig. 14a) [197]. Due to the sequential film deposition in ALD, the preparation of nanolaminates with accurately varying composition depth profiles is straightforward. The ALD made nanolaminates studied so far have consisted of stacked layers of Ta_2O_5, ZrO_2, HfO_2, Al_2O_3, Nb_2O_5, and their solid solutions deposited from $Ta(OC_2H_5)_5$, $TaCl_5$, $ZrCl_4$, $Zr(OC(CH_3)_3)_4$, $HfCl_4$, $AlCl_3$, $Nb(OC_2H_5)_5$, and H_2O at 325°C and below. Figure 14 compares the dielectric properties of HfO_2-Ta_2O_5 nanolaminates to their binary constituents. The labels describe the nanolaminate configuration as $N \times (d_{HfO_2} + d_{Ta_2O_5})$ where N is the number of the HfO_2-Ta_2O_5 bilayers, and d_{HfO_2} and $d_{Ta_2O_5}$ are the thicknesses of the corresponding single layers in nanometers. A number of conclusions can be drawn:

(i) the leakage current in the nanolaminates is lowered in relation to both constituents and is strongly dependent on the actual nanolaminate configuration (Fig. 14b),

(ii) the permittivity is nearly, though not entirely a linear function of the relative thickness of the binary constituents (Fig. 14c),

(iii) the charge storage factor has a maximum, about 10-fold of that of the binaries, at a certain relative thickness of the constituents (Fig. 14d).

An interesting observation is that in the nanolaminates the leakage current density is lowered not only in relation to the more leaky component, Ta_2O_5, but also in relation to the higher resistivity HfO_2 (Fig. 14b). In other words, the addition of layers of the high leakage current material Ta_2O_5 to HfO_2 has a positive effect on the leakage current properties. This improvement, which at first sight seems somewhat surprising, is attributed to the elimination of grain boundaries extending through the whole insulator from one electrode to the other. Polycrystalline insulator films, like HfO_2 here, usually exhibit extra conductivity along the grain boundaries. In the nanolaminate structure, the continuous grain growth of HfO_2, and thus the continuous grain boundaries, are interrupted by the amorphous Ta_2O_5 layers. Another potential explanation for the improved leakage current properties of the nanolaminates is the trapping of electrons at an interface next to a weak point in one sublayer, thereby decreasing the injecting electric field in the vicinity of this point. Nevertheless, if these two mechanisms were the only ones responsible for the reduced leakage current, the layer thicknesses should not have such a significant effect as observed (Fig. 14b). Therefore, it appears that also some other, layer thickness-dependent factors are involved. Especially the crystal structures and crystallite sizes of the poly-

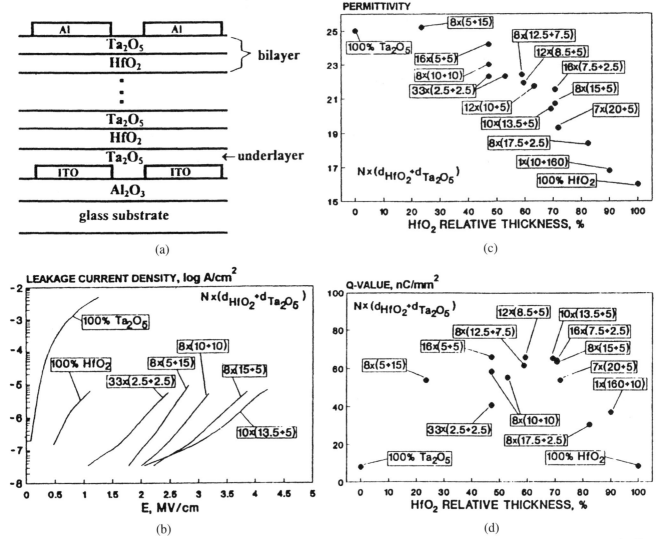

Fig. 14. (a) Schematic of a HfO$_2$-Ta$_2$O$_5$ nanolaminate insulator between ITO and Al electrodes. (b) Leakage current densities of HfO$_2$ and Ta$_2$O$_5$ films and various HfO$_2$-Ta$_2$O$_5$ nanolaminates, all having a total thickness of about 170 nm. The nanolaminate configurations are described with a notation $N \times (d_{HfO_2} + d_{Ta_2O_5})$ where N is the number of bilayers, and d_{HfO_2} and $d_{Ta_2O_5}$ are thicknesses of the single HfO$_2$ and Ta$_2$O$_5$ layers, expressed in nanometers. (c) Relative permittivities and (d) charge storage factors $Q = \varepsilon_0 \varepsilon_r E_{BD}$ of the HfO$_2$-Ta$_2$O$_5$ nanolaminates vs the relative thickness of the HfO$_2$ layers. E_{BD} is taken as a field which induces a leakage current density of 1 μA cm^{-2} [197]. Reprinted with permission from K. Kukli et al., Tailoring the dielectric properties of HfO$_2$-Ta$_2$O$_5$ nanolaminates, *Appl. Phys. Lett.* 68, 3737 (1996), © 1996, American Institute of Physics.

crystalline layers seem to affect both leakage current and permittivity [133, 197].

The other nanolaminate structures examined have shown rather similar basic behavior as the HfO$_2$-Ta$_2$O$_5$ nanolaminates, though some interesting differences have been identified as well. In the Al$_2$O$_3$-Ta$_2$O$_5$ nanolaminates where the both constituents are amorphous, no improvement of the leakage current was observed in comparison with Al$_2$O$_3$ which is the component having the better resistivity [133]. This supports the idea that the improvements observed in the HfO$_2$-Ta$_2$O$_5$ nanolaminates are indeed due to the grain boundary interruption. In the ZrO$_2$-Ta$_2$O$_5$ nanolaminates, on the other hand, the permittivity did not decrease smoothly as a function of the increasing ZrO$_2$ relative thickness but had values (27 to 28) which were higher than those of either of the constituents (25 for Ta$_2$O$_5$, 20

for ZrO$_2$) when as thick (150–200 nm) as the total thickness of the nanolaminate [133]. This was attributed to the dominance of the higher permittivity tetragonal phase of ZrO$_2$ in the thin layers in the nanolaminates as opposite with the monoclinic phase in the thicker binary films. Such an effect could also possibly explain the observed nonlinearities in the permittivity of the HfO$_2$-Ta$_2$O$_5$ nanolaminates (Fig. 14c) since these films were found to consist of a mixture of monoclinic and tetragonal HfO$_2$. By contrast, in the completely amorphous Al$_2$O$_3$-Ta$_2$O$_5$ nanolaminates the permittivity decreased linearly from 25 for Ta$_2$O$_5$ to 8 for Al$_2$O$_3$ [133].

The highest charge storage factor obtained so far is 84 nC mm^{-2} measured for the Nb$_x$Ta$_y$O$_z$-ZrO$_2$ nanolaminates where solid solutions were formed from Ta$_2$O$_5$ and Nb$_2$O$_5$, the latter of which has a high permittivity (about 50) but is very leaky in the binary form [192]. When mixed with Ta_2O_5, the leakage is decreased to a reasonable level, and even if also the permittivity decreases to 25–35, it remains higher than that of Ta_2O_5. As a consequence, the $Nb_xTa_yO_z$ solid solution shows improved dielectric properties in comparison to both of its constituents, and hence the nanolaminates are also improved by replacing Ta_2O_5 with $Nb_xTa_yO_z$.

6.2.2. Insulators for Microelectronics

The success in the demanding application of the TFEL displays clearly demonstrated the capabilities of ALD in making high quality insulators for large-area applications. This, together with the other beneficial features of ALD, has subsequently encouraged ALD insulator research also in other areas, particularly in microelectronics. While the TFEL insulator research serves as a good starting point for microelectronic applications, there also are some important differences which somewhat limit the direct applicability of the results from the TFEL insulators. First, the insulators in the TFEL displays are usually about 200 nm thick while the present and future integrated circuits require much thinner films down to a few nanometers. Second, the electrodes are quite different: in TFEL displays Al, Mo, and ITO are typically used whereas in the integrated circuits the insulators are combined with Si, Al, Pt, and perhaps with also some other conductors yet to be defined.

In the silicon-based integrated circuits, SiO_2 and SiO_xN_y have been the insulators of choice for decades. Unfortunately, no effective ALD processes for these materials have been found so far. In all the processes reported by now (Table IV), lengthy exposure times are required. On the other hand, an interesting room temperature ALD process for depositing SiO_2 from $SiCl_4$ and H_2O using pyridine [165, 166] or ammonia [167] as a catalyst has been reported, though also there long exposure times were needed (Section 5.3.6).

As is well known, the continuous decrease of the device sizes in the integrated circuits will soon lead to a situation where SiO_2 and SiO_xN_y can no longer be used as the insulators. In both metal-oxide-semiconductor-field-effect transistors (MOSFET) and dynamic random access memory (DRAM) capacitors silicon-based insulators are projected to be scaled so thin in the coming years that direct tunneling currents through them become detrimentally large. Therefore, insulators with higher permittivity and methods for their controlled deposition are urgently looked for. As the new materials are required for replacing SiO_2 in devices where the latter would need to have a thickness of 1.5 nm or below, the substitutes must accordingly have an equivalent thickness ($d_{eq} = (\varepsilon_{SiO_2}/\varepsilon)d = (3.9/\varepsilon)d$) of 1.5 nm or less. Such equivalent thicknesses must be achieved repeatedly over the 200- or 300-mm wafers. Clearly, this is an application to which ALD is ideally suited: very thin films with accurately controlled thickness over large areas. In addition, in DRAM capacitors excellent conformality is also highly appreciated as the memory capacitors will have three-dimensional structures.

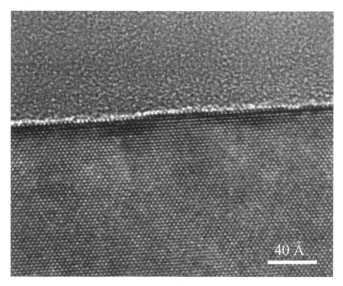

Fig. 15. High resolution TEM image of an Al_2O_3-Si interface where the Al_2O_3 film was deposited from $AlCl_3$ and $Al(OCH(CH_3)_2)_3$.

In MOSFETs, the gate insulator is deposited directly on silicon. An achievement of an equivalent thickness of 1.5 nm and below requires that the high permittivity insulator is deposited on silicon so that only very thin, preferably only a one-to-two monolayers thick silicon oxide layer forms at the interface. If the interface oxide becomes of a considerable thickness, it will start to dominate the overall capacitance of the resulting insulator stack which may be understood to consist of two capacitors in a series giving a total equivalent thickness $d_{eq, tot} = d_{SiO_2} + d_{eq}$. A very thin interface layer is preferred, however, as it is beneficial for good interface characteristics, i.e., low interface trap density. The interfacial SiO_2 layer may form by two mechanisms: by a reaction between silicon and the high permittivity oxide, and by oxidation of the silicon surface during the oxide deposition. To avoid the first mechanism, oxides which appear to be thermodynamically stable in contact with silicon should be chosen, such as Al_2O_3, ZrO_2, and HfO_2 and their silicates. ALD processes have been reported for all these oxides (Table IV).

Amorphous Al_2O_3 films have already been grown directly on HF etched silicon without creating an interfacial layer by both $Al(CH_3)_3$-H_2O [194, 222] and $AlCl_3$-$Al(OCH(CH_3)_2)_3$ (Fig. 15) [393] ALD processes. The latter belongs to the new group of ALD oxide processes where no separate oxygen sources are used (Section 5.3.5). These were developed to eliminate the possibility that the conventional oxygen sources used in the ALD processes, like water and hydrogen peroxide, could oxidize the silicon surface and thereby create the interfacial oxide layer. Detailed electrical characterization of these films are still in progress, however. The absence of the interface layer below the films deposited by the $Al(CH_3)_3$-H_2O process is somewhat surprising as water might be expected to oxidize the silicon surface (see the following). Perhaps this is counterbalanced by the reducing nature of $Al(CH_3)_3$. These Al_2O_3 films have shown low leakage currents, and C–V measurements have indicated relatively low midgap interface-state densities (one estimate being 10^{11} states eV^{-1} cm^{-2} [216]) but have at the same time given indications of some mobile and fixed charges, especially when large bias voltages have been applied [216, 220, 222].

Ultrathin ZrO_2 films have been deposited by ALD from $ZrCl_4$ and H_2O at 300°C on differently treated silicon surfaces [193, 194, 420, 421]. Contrary to the previously mentioned Al_2O_3 films, on native oxide-free (HF etched) silicon the ZrO_2 films exhibited rather poor morphology with islands of crystalline ZrO_2 dispersed in an amorphous matrix and roughened interfaces [193, 194]. In addition, an interfacial SiO_2 layer was also formed. On thermally oxidized silicon, the crystallites were better developed and interfaces were abrupt, and more ZrO_2 was deposited than on the HF stripped silicon. These findings suggest that the initial nucleation of ZrO_2 on oxide-free, hydrogen terminated silicon is suppressed, but the possibility of detrimental reactions between $ZrCl_4$ and silicon must be kept in mind as well. The ZrO_2 films on thermal SiO_2 were stable against intermixing and interfacial reactions up to approximately 900°C but higher temperature annealing in vacuum resulted in a formation of $ZrSi_2$ islands.

On chemically oxidized silicon, $ZrCl_4$ interacted with the oxide layer resulting in a Zr-Si-O interface layer with a permittivity twice as high as that of SiO_2 [420]. This difference in comparison with the thermal oxide was attributed to the presence of silicon suboxides in the chemical oxide. The resulting Zr-Si-O interface layer was not very stable, however, and decomposed into ZrO_2 and SiO_2 at 500°C. Electrical characterization of SiO_x/ZrO_2 dielectric stacks obtained by depositing 5 to 10-nm ZrO_2 on a chemical oxide indicated both a presence of fixed negative charge (effective density of 5.2×10^{12} cm^{-2}) and an electron trapping (1.5×10^{12} cm^{-2}) [421]. Leakage through such a stack was concluded to proceed by a direct tunneling of electrons through SiO_x and trap-assisted tunneling through ZrO_2. A similar mechanism was also observed for an analogous SiO_x/Ta_2O_5 dielectric stack [421].

A good quality ZrO_2 was obtained on an ammonia treated (nitrided) silicon. A 4.0-nm thick ZrO_2 film on a 1.3-nm interface layer resulted in an equivalent thickness $d_{eq} = 1.4$ nm and a low leakage current of 10^{-8} A cm^{-2} [420].

On a bare silicon substrate, the $TaCl_5$-H_2O process caused a formation of a 1.5-nm SiO_2 interface layer at 300°C [195]. When the silicon was passivated with 2.8-nm silicon nitride, no oxide layer was observed at the interface.

A common observation on many thin ALD oxide films on silicon is that their permittivity decreases considerably as the thicknesses are scaled down. Such a decrease cannot be explained solely in terms of a lower permittivity interface layer but is also observed when the contribution of the interface is taken into account. For example, in the study of the SiO_x/ZrO_2 dielectric stacks a permittivity of 15 was evaluated for the 5- to 10-nm ZrO_2 layers after subtracting the contribution of SiO_x [421] while values around 20 have usually been obtained for thicker films (Table VII). A similarly decreased permittivity of 15 was also obtained for 10-nm ZrO_2 in another study where,

in addition, parallel decreases were observed with Ta_2O_5 (from 27 to 15 for a 16-nm film) and HfO_2 (from 16 to 11 for an 11-nm film) [195]. On the other hand, rapid thermal annealing at 700°C increased the permittivity of Ta_2O_5 to 21, obviously due to a densification of the films. In fact, studies on the effects of postdeposition treatments on the dielectric properties of ALD made oxide films have so far remained in their infancy and should be explored in detail in the future.

In the DRAM capacitors, the insulator film is deposited on three-dimensional electrodes. For the near future DRAMs, $Ba_{1-x}Sr_xTiO_3$ is the most extensively examined insulator. Recently ALD processes were developed for both $SrTiO_3$ and $BaTiO_3$ using novel Sr and Ba cyclopentadienyl compounds as precursors together with $Ti(OCH(CH_3)_2)_4$ and H_2O [356–358]. The use of these new precursors was crucial because the β-diketonates $Sr(thd)_2$ and $Ba(thd)_2$ that have been widely used in CVD, and that have also been used in the ALD of sulfides (cf. Table IV), do not react with water at temperatures which would be low enough so that they or $Ti(OCH(CH_3)_2)_4$ would not decompose thermally. The β-diketonates do react with ozone, e.g., $Sr(thd)_2$ reacts with ozone at around 300°C but the resulting film contains $SrCO_3$, or when combined with titanium, an amorphous mixture which needs to be annealed at 700°C to get $SrTiO_3$ crystallized [322]. By contrast, using strontium bis(tri-isopropylcyclopentadienyl) $(Sr(C_5{}^iPr_3H_2)_2)$, crystalline $SrTiO_3$ films with excellent conformality were obtained at 250–325°C (Fig. 16a and b) [356–358]. The Sr/Ti ratio in the films was well correlated to the pulsing ratio of the metal precursors (Fig. 16c), and after annealing at 500°C in air, the films with the optimized composition had permittivities up to 180 (Fig. 16c) as measured in a capacitor structure $ITO/SrTiO_3/Al$. However, because the leakage of $SrTiO_3$ is largely governed by the Schottky emission, large leakage current densities were measured with this electrode configuration. Recent studies have indeed verified that the application of Pt electrodes significantly reduces the leakage current level [357]. On the other hand, as common to these materials, a decrease of permittivity with film thickness was observed: for a 50-nm $SrTiO_3$ a permittivity of only 100 was measured [356]. Clearly, high temperature annealings will be needed to increase the permittivity of the thinnest films. Likewise, solid solutions of $Ba_{1-x}Sr_xTiO_3$ are to be examined.

Though clearly beneficial in the case of thicker insulators (Section 6.2.1), the nanolaminate concept is quite problematic to apply in MOSFETs and DRAMs where the insulators are much thinner. Note, for example, that in Figure 14 the lowest leakage current was achieved with a configuration of $10 \times (13.5 + 5$ nm), for which the permittivity was 20. Thus, for an equivalent thickness of, e.g., 2 nm, a single layer pair would already be too thick. On the other hand, taking a thickness of 2.5 nm for each layer, the permittivity is 22 (Fig. 14), and thus a configuration of $2 \times (2.5 + 2.5$ nm) should give an equivalent thickness of 1.8 nm. However, in an attempt to make thin Ta_2O_5-ZrO_2 nanolaminates consisting of either 2.5- or 1.0-nm sublayers, it was observed that although permittivities of 22 (on Si) and 24 (on ITO) were achieved with 100-nm films, the per-

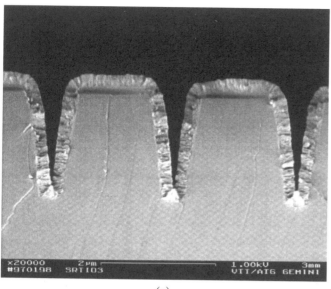

(a)

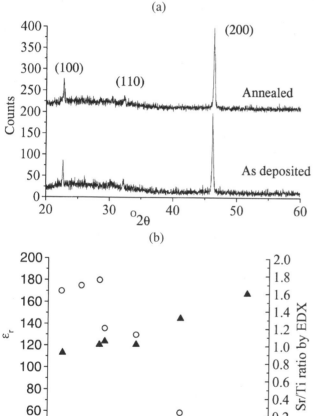

(b)

(c)

Fig. 16. (a) SEM image of a cross section of an ALD made SrTiO₃ film on a substrate with trenches. (b) X-ray diffraction patterns of a SrTiO₃ film as deposited at 325°C and after annealing in air at 500°C. (c) Sr/Ti ratio (▲) and permittivity (○), the latter being measured after the 500°C postdeposition annealing in air, as a function of the Sr(C₅ⁱPr₃H₂)₂-H₂O to Ti(OCH(CH₃)₂)₄-H₂O cycle ratio for SrTiO₃ thin films deposited at 325°C [356]. Reprinted with permission from M. Vehkamäki et al., Growth of SrTiO₃ and BaTiO₃ thin films by atomic layer deposition, *Electrochem. Solid-State Lett.* 2, 504 (1999). © 1999, The Electrochemical Society Inc.

mittivities decreased substantially when the films were made thinner, 7- to 10-nm thick [357]. The decrease was more severe on silicon ($\varepsilon = 12$ for 2.5 nm and $\varepsilon = 8$ for 1.0-nm sublayers) than on ITO ($\varepsilon = 15$ for 2.5 nm and $\varepsilon = 10$ for 1.0-nm sublayers), but as in both cases a decrease was observed, also other factors than the interfacial SiO$_x$ layer must be involved. These films were in their as-deposited state, however, and hence a further decrease of the obtained equivalent thicknesses of about 3.5 nm may be expected after postannealing treatments. Similar permittivities of 12–14 have also been obtained in another study where thin (10–12 nm, including the underlying 2.8-nm Si$_3$N$_4$) Ta$_2$O$_5$-HfO$_2$, Ta$_2$O$_5$-ZrO$_2$, and ZrO$_2$-HfO$_2$ nanolaminates were examined, and the resulting equivalent thicknesses were accordingly from 3.0 to 3.3 nm [195].

Another concern in applying nanolaminates in microelectronic applications is charge trapping at the internal interfaces, especially when one of the components shows significant leakage like Ta$_2$O$_5$. Such a trapping may manifest itself as hysteresis in the C–V characteristics, for example. In any event, the possibility offered by ALD to straightforward preparation of accurately controlled multilayer insulator structures is clearly an opportunity worth further studies.

Ta$_2$O$_5$-HfO$_2$ nanolaminates have also been examined in integrated passive capacitors [422, 423]. A 48-nm thick film made up of 3-nm layers exhibited a permittivity of 23, so the capacitance density was 4 to 5 nF mm^{-2}. The breakdown strength was high, 6.5 MV cm^{-1}, but the defect density was relatively high because of particles from the HfCl$_4$ source.

In the integrated circuit metallization schemes the situation is the reverse of that in MOSFETs and DRAMs. The intermetal and interlevel SiO$_2$ insulators have shrunken so narrow and thin that capacitances between the adjacent conductors have become large enough to be the limiting factor for the overall speed of the circuit. In addition, crosstalk between the closely spaced conductor lines has become a concern. To minimize these problems, insulators with permittivity lower than that of SiO$_2$, such as fluorinated SiO$_2$ and polymers, are actively examined. Until now, no studies on the ALD of these materials have been reported, however.

6.3. Transparent Conductors

Transparent conducting oxides (TCO) are used in many large-area applications like flat panel displays, solar cells, windows, and antistatic coatings. A good ALD TCO process would be especially important in the TFEL display manufacturing where it could be combined with the preceding ALD process, namely, ion barrier deposition in the conventional structure (cf. Fig. 12) or insulator-phosphor-insulator process in the inverted structure. Nevertheless, though ALD processes have been developed and examined for many TCO films (Table VIII), none of them has so far replaced the well-established sputtered indium–tin oxide (In$_2$O$_3$: Sn, ITO) in commercial TFEL displays.

In general, the properties of the ALD made TCO films (Table VIII) are rather similar to those obtained by other methods. The lowest resistivity (2.4×10^{-4} Ω cm) has been achieved

Table VIII.   Electrical Properties of Transparent Conducting Oxide Thin Films Prepared by ALD[a]

Material	ρ (10^{-4} Ω cm)	μ (cm^2 V^{-1} s^{-1})	n (10^{19} cm^{-3})	Reference
In$_2$O$_3$	35	72	2.5	[115, 159]
In$_2$O$_3$: Sn	2.4	34	78	[159, 160]
SnO$_2$	1400	20	0.2	[173, 175]
	120			[271]
SnO$_2$: Sb	9	12	58	[175]
SnO$_2$: F	3.3			[271]
ZnO	45	33	4.3	[272, 284, 289, 292]
ZnO with UV irradiation	6.9	20	39	[52]
ZnO : Al	8.0	30	26	[272]
ZnO : Ga	8.0	40	20	[293]
ZnO : B	5.0	26	49	[289, 292]

[a] ρ is resistivity, μ is electron mobility, and n is electron concentration.

with ITO deposited from InCl$_3$, SnCl$_4$, and H$_2$O or H$_2$O$_2$. The ITO process suffers, however, from a relatively low deposition rate of 0.2–0.4 Å cycle^{-1} so that the effective growth rate is at best only 2 nm min^{-1} [159]. Attempts to find indium precursors more reactive than InCl$_3$ have all failed so far [424]. Also the SnCl$_4$-SbCl$_5$-H$_2$O process used for depositing SnO$_2$: Sb is rather slow (0.35 Å cycle^{-1}) [173]. By contrast, ZnO doped with boron, aluminum, or gallium can be deposited from the corresponding metal alkyls, diborane, and water with high rates from 0.5 to 2.8 Å cycle^{-1}, and effective growth rates up to 13 nm min^{-1} are achievable [424]. However, the resistivities are remarkably higher than in ITO (Table VIII), and thus a sheet resistance of 5 Ω $\square^{-1}$ which is typically required in the TFEL displays could be achieved only with about 1600 nm thick ZnO : Al films as compared to the presently used 300- to 500-nm ITO films. An interesting observation on nondoped ZnO films was that their electron concentration could be increased 10-fold by UV irradiation during the growth, thereby resulting in a resistivity of 6.9×10^{-4} Ω cm [52].

As compared to the other ALD TCO processes, the zinc alkyl-based processes have a major benefit in that they may be run well below 200°C. This has been utilized in making front contacts to the temperature sensitive CdS/CuInSe$_2$ solar cell absorber structures: an efficiency of 14.3% was achieved using ZnO : Al made by ALD at 150°C [425]. ZnO : B, in turn, has been examined as a transparent conductor in amorphous-silicon (a-Si) solar cells. Because the ALD made ZnO : B films are only weakly textured, their light scattering properties are poor for the a-Si solar cells. Therefore, a bilayer structure combining a texture of a metal organic chemical vapor deposited (MOCVD) ZnO : B and the low resistivity and good stability of the ALD ZnO : B was developed resulting in an efficiency of 8.2% in a 1.0-cm^2 single p–i–n junction a-Si solar cell [292].

Though not an actual conductor, TiO$_2$ is often so oxygen deficient that its conductivity may reach 1 (Ω cm)$^{-1}$. The utiliza-

tion of this moderate conductivity and the good chemical stability of TiO$_2$ has been examined in a-Si solar cells in protecting ITO from a hydrogen plasma used for depositing the a-Si layer [426]. A 30-nm TiO$_2$ layer made by ALD from TiCl$_4$ and H$_2$O increased the cell efficiency from 4.9 to 6.6%. In addition to its protecting action, TiO$_2$ also served as an antireflection layer for the blue light.

6.4. Passivating and Protecting Layers

The pinhole-free and dense structure of the ALD made films make them interesting also for many passivating and protecting applications. The first one of these is the use of ALD Al$_2$O$_3$ as an ion barrier layer in the TFEL displays (Fig. 12). Usually, the sodium out-diffusion from soda lime glass prevents this inexpensive glass from being used as a substrate for the TFEL and other devices sensitive to sodium. It has been proved, however, that the ALD made Al$_2$O$_3$ films effectively prevent the sodium migration [427], thus facilitating the use of soda lime glass instead of the more expensive near-zero alkali glasses. ALD made Al$_2$O$_3$ is also used as a final passivation layer on top of the TFEL thin film stack (Fig. 12). The excellent conformality and the amorphous structure with no grain boundaries are clearly the key benefits of the ALD Al$_2$O$_3$ in blocking the migration of ions and moisture.

Yet another use where the protecting properties of the ALD Al$_2$O$_3$ have been examined in the TFEL displays is an etch stop for patterning the phosphor layers [428]. This was needed when a stacked dual substrate structure was examined for multicolor displays. In this structure, one substrate had the blue phosphor as a continuous layer while the other substrate incorporated the red and green phosphors, ZnS : Mn and ZnS : Tb, respectively, patterned to separate pixels. After depositing and patterning ZnS : Tb, that was covered by an Al$_2$O$_3$ etch stop layer and ZnS : Mn so that during the wet etching of the latter, the etch stop protected the ZnS : Tb layer. Remarkably, an as thin as 60-nm Al$_2$O$_3$ etch stop layer was effective for this purpose. However, because of the complexity of the stacked dual substrate TFEL structure, it was subsequently rejected as a commercial display.

A key aspect in corrosion protection is a prevention of material transportation across the metal-electrolyte interface. Therefore, the good barrier properties discussed earlier clearly point the potential of the ALD made films in corrosion protection. Preliminary studies have indicated that the metal surface has a crucial effect on the film density: dense films were obtained on stainless steel but ordinary steel was much more difficult to protect [429]. The best results were achieved with Ta$_2$O$_5$ which at the same time is amorphous and chemically stable. Al$_2$O$_3$ is also amorphous and thus free of grain boundaries, but being amphoteric it dissolves to strong acids and bases. TiO$_2$, in turn, is chemically stable but because it is polycrystalline, it contains grain boundaries along which ions can migrate, though slowly. Similar to the nanolaminate insulators (Section 6.2.1), better results were achieved also in this case by using Al$_2$O$_3$-TiO$_2$ multilayers where the grain boundaries extending through the whole film were interrupted. Nevertheless, as erosion is usually strongly involved in corrosion, it is doubtful if it is economic to make ALD films thick enough. Rather, it may be envisioned that ALD films could be used as underlayers below thicker but less dense layers made by faster methods. In such a structure, the thicker overlayer would take the erosion load whereas the ALD made layer would give the chemical protection at the interface.

6.5. Transition Metal Nitride Diffusion Barriers

Transition metal nitride layers are used in the integrated circuits as barriers in preventing interdiffusion and reactions between metals and silicon or insulators. As the device dimensions have been aggressively scaled down, the cross sections of metal lines have decreased accordingly. Consequently, the importance to be able to fill the available space with metal as fully as possible has continuously increased. At the same time, the aspect ratios of trenches and vias have increased. These trends have led to such requirements for the future diffusion barriers that will clearly favor the use of ALD: the barriers should be thin, a few nanometers only, uniformly all around the deep trenches and vias, and still have dense and void-free structure to effectively prevent the detrimental interactions. Low resistivity is also required but the thinner the films can be made, the less stringent this requirement apparently becomes. Finally, as the future low permittivity intermetal insulators may be polymers with limited thermal stability, 400°C is often cited as the maximum deposition temperature.

A majority of the ALD transition metal nitride processes examined so far are based on the metal chloride–ammonia precursor combinations (Table IX). In general, these nitrides have acceptably low resistivities which are comparable to those reported for CVD nitrides. The highest quality films have been obtained using zinc as an additional reducing agent, but because of concerns related to zinc these processes hardly become accepted to the semiconductor applications. Nonetheless, also without zinc reasonable film properties have been obtained, tantalum nitride being an exception because without zinc the insulating Ta_3N_5 was obtained even when dimethyl hydrazine was used as the nitrogen source [114]. Excellent film conformality extending down to the sharp corners at the trench bottoms has been verified (Fig. 17), and trench filling as an ultimate conformality test has also been proved [88]. Preliminary barrier studies on TiN between Cu and Si were encouraging as well, the failure temperatures were around 650°C as judged on the basis of surface silicon content and etch pit tests [186]. Grain boundary diffusion was concluded to be the dominant mechanism for the barrier failure.

One concern of the metal chloride–ammonia processes, also shared by the only metal iodide (TiI_4)-ammonia process examined so far and the dimethyl hydrazine-based processes, is the low deposition rate which has typically been only about 0.2 Å cycle^{-1}, thereby implying low saturation coverage of the metal precursor. It still remains open whether this is due to a small number of reactive $-NH_x$ groups left on the surface after the ammonia pulse or the adsorption site blockage by a readsorbed reaction byproduct HCl. For instance, the stability

Table IX.   Properties of ALD Made Transition Metal Nitrides

Material	Precursors	Growth temperature (°C)	Resistivity ($\mu\Omega$ cm)	Reference
TiN	$TiCl_4 + NH_3$	500	250	[111, 112, 185]
	$TiCl_4 + Zn + NH_3$	500	50	[111, 112]
		400	200	[111]
	$TiI_4 + NH_3$	500	70	[191]
		450	150	[191]
		400	380	[191]
	$TiCl_4 + (CH_3)_2NNH_2$	350	500	[114]
	$Ti(N(CH_3)_2)_4 + NH_3$	180	5000	[384]
$Ti_xSi_yN_z$	$Ti(N(CH_3)_2)_4 + SiH_4 + NH_3$	180	3×10^4	[384]
NbN	$NbCl_5 + NH_3$	500	200, 550	[103, 112, 185]
	$NbCl_5 + Zn + NH_3$	500	200	[103, 112]
	$NbCl_5 + (CH_3)_2NNH_2$	400	2900	[114]
TaN	$TaCl_5 + Zn + NH_3$	400–500	900	[113]
Ta_3N_5	$TaCl_5 + NH_3$	400–500	4.1×10^4, 5×10^5	[113, 185]
	$TaCl_5 + (CH_3)_2NNH_2$	400	NM^a	[114]
MoN_x	$MoCl_5 + NH_3$	500	100, 260	[114, 185]
	$MoCl_5 + Zn + NH_3$	500	500	[114]
		400	3600	[114]
	$MoCl_5 + (CH_3)_2NNH_2$	400	930	[114]
W_2N	$WF_6 + NH_3$	330–530	4500	[207]

^{a}Not measurable.

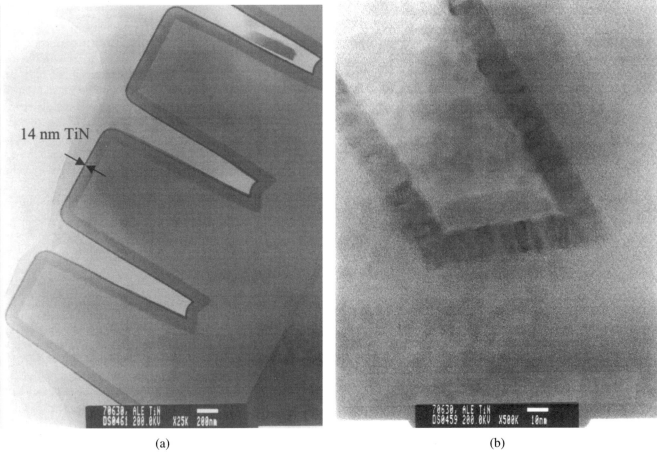

(a)   (b)

Fig. 17.   (a) Cross-sectional TEM image of a 14-nm TiN film deposited on deep trenches. (b) Detail from the bottom of a trench.

of species formed on a TiN surface in an exposure to NH$_3$ is known to be low [430] at temperatures which are required to get the ALD nitride deposition reactions to proceed. The readsorption of HCl is also well possible because most of these processes have been examined in the traveling-wave reactors (Fig. 7) where the byproduct readsorption probability is especially high because these travel in front of the precursor pulse (cf. Fig. 9). Indeed, a related WF$_6$-NH$_3$ process examined at similar temperatures in a more open reaction chamber resulted in a remarkably higher deposition rate of 2.55 Å cycle^{-1} [207]. In any event, still a direct comparison with one process examined in two different reactors is lacking. Other concerns include halide contamination and possible incompatibility problems with metal surfaces due to the reactions between metals and metal chlorides or HCl. The halide contents are strongly dependent on the deposition temperature increasing with decreasing temperature. At 500°C, they are typically well below 1 at.% but at 400°C a few at.% chlorine may be found which is just at the acceptance limit for the microelectronic applications. On the other hand, the iodine content in the TiN films deposited from TiI$_4$ was 2 at.% already at 350°C. Here, it must be noted that until now no postdeposition treatments have been applied to these films. On basis of the CVD literature, it may be expected that both contamination levels and resistivities could be decreased by, for example, N$_2$-H$_2$ plasma treatments.

Metal alkyl amides have been extensively examined as CVD precursors in an effort to decrease nitride deposition temperatures. In ALD, however, these compounds have been only scarcely examined. According to our experiences [431], Ti(N(C$_2$H$_5$)$_2$)$_4$ starts to decompose below temperatures which would be needed for effective reactions with ammonia, and thus films with unacceptably high resistivities were obtained. Nevertheless, in the Ti(N(C$_2$H$_5$)(CH$_3$))$_4$-NH$_3$ process a self-limiting growth was observed at 170–210°C by using high NH$_3$ doses with a flow rate of 250 sccm and 10 s or longer exposure times [386]. The deposition rate was high, 5 to 6 Å cycle^{-1}, more than twice the distance (2.45 Å) between the subsequent Ti layers in the [111] direction of TiN, thereby implying a complicated reaction mechanism. The films were amorphous and contained 4 at.% carbon and 6 at.% hydrogen but resistivities were not reported. A high deposition rate of 4.4 Å cycle^{-1} was also reported for the Ti(N(CH$_3$)$_2$)$_4$-NH$_3$ process at 180°C [384, 385]. With the addition of silane, either as a separate pulse between Ti(N(CH$_3$)$_2$)$_4$ and NH$_3$ or simultaneously with ammonia, Ti$_x$Si$_y$N$_z$ films with up to 23 at.% Si were deposited with a rate which decreased with increasing silicon content from 2.4 Å cycle^{-1} (18 at.% Si) to 1 Å cycle^{-1} (23 at.% Si) [384, 385]. A 10-nm Ti$_{0.32}$Si$_{0.18}$N$_{0.50}$ film prevented a diffusion of copper in a barrier test where a metal-oxide-semiconductor (MOS) structure Cu/Ti$_{0.32}$Si$_{0.18}$N$_{0.50}$/SiO$_2$/Si was annealed at 800°C for 60 min in $H_2(10\%)$-Ar(90%) [385]. These films showed excellent conformality but they suffered from high resistivities of 5000 $\mu\Omega$ cm (TiN) and 30,000 $\mu\Omega$ cm ($Ti_{0.32}Si_{0.18}N_{0.50}$) [384]. Also here the effects of postdeposition treatments remain to be examined.

6.6. Metals

Effective ALD metal processes could find use in many applications. A copper ALD process would offer an interesting alternative for the integrated circuit metallizations. Even if the entire copper layer was not made by ALD, a uniform seed layer for electrochemical copper deposition would already be of interest, especially if directly combined to a diffusion barrier ALD nitride process (see the previous section). Similar to the barriers, also here excellent conformality in high aspect ratio structures is required. On the other hand, some metals like Ta and Ti are considered as potential barriers for copper. For the TFEL displays, a metal ALD process might be interesting because it could be combined with the preceding ALD process step, i.e., deposition of the insulator-phosphor-insulator multilayer in the conventional device (Fig. 12) or the ion barrier in the inverted structure.

ALD of metals is, however, a chemically challenging task with only limited success so far. An effective reducing agent must be found and the metal precursor must be chosen so that it does not etch the metal film by forming volatile lower oxidation state compounds, and while being easily reduced, the metal precursor must be stable against decomposition and disproportionation. Yet another question is the mechanism by means of which the metal precursor can adsorb on the film surface, especially if the surface is free of functional groups.

The WF_6-Si_2H_6 process [206] addresses the previous issues elegantly as the growth involves exchange reactions during both the WF_6 and Si_2H_6 exposures (Section 7.2.2). Tungsten films were deposited at 150–330°C with a high rate per cycle, about 2.5 Å cycle^{-1}, but the cycle time was not specified. No silicon or fluorine could be detected in the films with X-ray photoelectron spectroscopy (XPS). The films were nearly amorphous and thus had smooth surfaces and a relatively high resistivity of 120 $\mu\Omega$ cm.

Cu films have been deposited by ALD using hydrogen as a reducing agent both for CuCl [209, 210] and for Cu(thd)$_2$ [210, 343, 344]. However, both processes seem to be quite specific about the substrate material which suggests that the substrate actively participates or catalyzes the film growth reactions, especially at the beginning of the growth. The CuCl-H_2 process was studied only on tantalum substrates which reduced CuCl causing an initial film thickness of about 30 nm. The Cu(thd)$_2$-H_2 process was examined on many surfaces (Pt/Pd seeded glass, bare glass, Ta, Fe, TiN, Ni, In_2O_3 : Sn) but film deposition occurred only on the glass surface seeded with a few angstroms thick layer of sputtered Pt/Pd alloy. A self-limiting growth with a rate of 0.3 Å cycle^{-1} was observed within a temperature range of 190–260°C. The catalytic effect of the Pt/Pd seed was explained in terms of a more complete dissociative adsorption of Cu(thd)$_2$ on an oxide-free metal surface as compared with hydroxyl terminated oxide and oxidized metal surfaces where one of the thd ligands remains bound with the adsorbing Cu atom and blocks further reactions. However, even more probably the Pt/Pd seed assists in dissociating the otherwise quite inert molecular hydrogen into reactive hydrogen atoms. In any event, even with the seed layer the process did not result in a film growth in another kind of ALD reactor with lower hydrogen partial pressure and shorter residence times [186].

The Cu films deposited from CuCl and H_2 around 400°C consisted of large grains, typically 1–4 μm in a lateral diameter, some even 10 μm, and showed good adhesion to the tantalum substrate [209]. They contained only 0.5–1.0 at.% chlorine as determined by XPS. No resistivity was reported, however. With Cu(thd)$_2$, a smaller grain size of 0.1 to 0.3 μm was obtained, possibly because of lower deposition temperatures below 350°C [343, 344]. No carbon could be detected with XPS in the highest purity films which had been cooled under hydrogen atmosphere but, interestingly, those cooled under nitrogen contained about 12 at.% carbon. Oxygen concentration was low in all the films, and below 3 at.% hydrogen was detected with nuclear reaction analysis. A resistivity of 8 $\mu\Omega$ cm was measured for a 60-nm film but this high value was attributed to the small thickness. Indeed, lower resistivities down to 3.5 $\mu\Omega$ cm were obtained later with 80- to 120-nm thick films [210].

As already noted in Sections 5.1.5 and 5.3.6, the CuCl-Zn [110] and $MoCl_5$-Zn [204] processes suffered from a reversible dissolution-outdiffusion of zinc into the Cu and Mo films. Further complications to the latter process arose from an etching of the Mo film by $MoCl_5$ through a formation of lower molybdenum chlorides $MoCl_{5-x}$. As a consequence, self-limiting growth conditions were hard to achieve and reproducibility was quite poor. Especially Cu was deposited with a high rate of above 5 Å cycle^{-1} (500°C). The Cu films consisted of large grains which stayed isolated so long that after a formation of a continuous film the sheet resistance was too low for a reliable measurement. About 3 at.% zinc was left in the films while no chlorine could be detected with energy dispersive X-ray spectroscopy (EDX). When Mo films were deposited at 420°C, the growth rate and the zinc residual content decreased from 0.8 to 0.4 Å cycle^{-1} and 2.5 to 0.5 at.%, respectively, with increasing purge time after the zinc pulse. Above 460°C the film growth practically stopped, obviously because of the etching reactions. The lowest resistivites of about 15 $\mu\Omega$ cm were obtained with the films with the lowest zinc contents.

Metallic Ta and Ti films with a conformality approaching 100% have been deposited at 25–400°C using a plasma enhanced ALD process where $TaCl_5$ and $TiCl_4$ were reduced by atomic hydrogen generated upstream with an inductively coupled RF plasma discharge [74]. Ta was deposited with a rate varying from 0.16 Å cycle^{-1} at 25°C to 1.67 Å cycle^{-1} at 250–400°C, and Ti was deposited with a rate of 1.5 to 1.7 Å cycle^{-1}. Because of oxidation during air exposure after the film growth, no meaningful analysis of the Ti films could be done and the Ta films also appeared to be strongly oxidized since the measured oxygen levels were high, from a few percent up to 26 at.%. On the other hand, chlorine residues were below 3 at.%. At deposition temperatures of 250°C and below, no crystallinity could be observed in the Ta films up to thicknesses of about 40 nm, but at higher temperatures crystalline Ta was obtained. These preliminary results on obtaining metallic films of the highly electropositive elements of titanium and tantalum are strongly encouraging and motivate further studies on atomic hydrogen-based ALD metal processes.

Another approach toward utilization of the advantages of ALD in depositing metal films involves a two step process where an oxide film is first made by ALD and then reduced by hydrogen. As a first example of this approach, a 135-nm NiO film deposited from $Ni(acac)_2$ and ozone at 250°C was converted to a metallic, well-adherent nickel at 260°C in a 5% H_2–95% Ar atmosphere (1 atm) [342]. Though in this case the reduction caused pinholes to the originally dense films, this approach is evidently worth further examination, perhaps with thinner films or by dividing the process into several deposition-reduction steps.

6.7. Solar Cell Absorbers

Studies on ALD deposition of thin film solar cell absorbers have been quite scarce but some promising results have been obtained. An efficiency of 14.0% was achieved using a CdS/CdTe heterojunction made by ALD from elemental sources in a flow-type reactor [432–434] but because of the general concerns related to CdTe solar cells; i.e., the stability of the contacts and the environmental issues, the interest to this ALD application has subsequently turned down.

For $CuInSe_2$-based solar cells, ALD has been examined in depositing not only the transparent front contacts (Section 6.3) but also the buffer layers between $CuInSe_2$ and TCO. As a standard, chemical bath deposited CdS is used as this buffer but for environmental reasons cadmium-free substitutes are actively looked for. With ALD made nondoped ZnO, efficiencies of 11.4–13.2% have been obtained using $Cu(In,Ga)Se_2$ as an absorber [285, 286, 288] while In_2S_3 gave an even higher efficiency of 13.5% [288]. Importantly, both processes were carried out at 160–165°C. A 10-nm ZnSe buffer layer deposited on $Cu(In,Ga)Se_2$ from elemental sources at 250°C by an MBE system operated in an ALD mode resulted in an efficiency of 11.6% [119].

Also epitaxial solar cell absorbers have been made by ALD. With an $Al_{1-x}Ga_xAs/GaAs$ tandem solar cell structure, an open circuit voltage of 1.945 V and a fill factor of 0.812 were achieved under 10 suns-AM 1.5 conditions [236].

6.8. Optical Coatings

The accurate film thickness control and good uniformity over large-area substrates offered by ALD are valuable characteristics also in making multilayer structures for optics. Due to the wide variety of dielectric materials made by ALD, refractive indexes from 1.43 to 2.6 are available [435]. In addition, a suitable mixing of pulsing sequences of binaries with different refractive indexes should make it rather straightforward to create films with graded refractive index profiles. On the other hand, the slowness of the method is a limiting factor especially when thick structures with tens of layers are needed for sophisticated visible range components.

Optical multilayers have been made by ALD for both visible [435] and soft X-ray [184, 239] wavelength ranges. The visible range components consisted of polycrystalline ZnS ($ZnCl_2$-H_2S) and amorphous Al_2O_3 ($AlCl_3$-H_2O) layers with thicknesses of 62 and 86 nm, respectively. Various structures ranging from simple two- or three-layer antireflection coatings and neutral beam splitters to 19-layer Fabry–Perot filters were made. They all accurately reproduced the spectral responses calculated for the designed multilayer structures, thus evidencing that both the thicknesses and optical properties of the films were well controlled.

In the soft X-ray range, the layers are only a few nanometers thick, thus setting rigorous requirements to the film thickness accuracy and interface smoothness, but at the same time making the deposition rate of less concern as compared to the visible range components. In good agreement with the model calculations, a high reflectance of over 30% at a wavelength of 2.734 nm and an incidence angle of 71.8°C from the surface normal was achieved with an ALD made multilayer consisting of 20 pairs of amorphous Al_2O_3 ($Al(CH_3)_3$-H_2O_2) and TiO_2 ($TiCl_4$-H_2O_2) layers which had a layer-pair thickness of 4.43 nm [184]. Soft X-ray multilayer mirrors have also been made by depositing AlP ($Al(CH_3)_2H$-PH_3) and GaP ($Ga(CH_3)_3$-PH_3) layers on a GaP substrate with a rate of an exact one monolayer per cycle. A structure with 50 bilayers of $(AlP)_{22ML}(GaP)_{13ML}$ gave a maximum reflectance in excess of 10% at a wavelength of about 17 nm and an incidence angle of 35° from the surface normal [239].

7. CHARACTERIZATION OF ALD PROCESSES

ALD processes may be examined by many different ways as summarized in Table X. Among these, the film growth experiments are the most common as they focus on the final goal, i.e., the films themselves, but they give only a limited chemical information. Therefore, other methods are required to get a better understanding of the chemistry involved.

As discussed in Section 4, the most important ALD reactors are the flow-type ones operating at pressures above 1 Torr. In addition, the reaction chambers are often of a traveling-wave design with tightly packed substrates (Figs. 7 and 8). Only a few *in situ* characterization methods may be added to these kinds of reactors, and therefore many alternative approaches have been taken for examining the chemistry relevant to ALD (Table X). A general trend is, however, that the more detailed chemical information the method gives, the farther are the experimental conditions from the flow-type reactors, thus arising an issue of the representativeness of the results for the actual

Table X.   Summary of Methods Used to Characterize ALD Processes

Method	Advantages	Limitations
Film growth experiments	Focus on the overall goal, the optimization of the film growth process	Limited chemical information, interpretation requires additional data
In situ measurements under real growth conditions:	Relevancy to the actual film growth processes	Difficulties in sampling
• Microgravimetry	Direct measurement of (relative) surface mass changes	Interpretation requires certain assumptions of the surface species
• Mass spectrometry	Identification of volatile byproducts at various stages of an ALD cycle	Cracking and reactions during ionization complicate the interpretation
• Optical methods	Sensitive observation of changes in surface termination	Chemical interpretation is difficult
Measurements after an inert transfer from the growth conditions to the analysis chamber	A large number of surface analytical techniques applicable ensures thorough characterization	Possible changes during the sample transfer
Reactions under high-vacuum conditions	A large number of surface analytical techniques applicable ensures thorough characterization	Representativity is questionable because of differences in pressures (pressure gap) and reaction times
Reactions on high surface area substrates	The large amount of products makes it possible to use routine chemical techniques, like IR, NMR, and elemental analysis	Representativity is questionable because of very long reaction times Possible changes during the sample transfer

growth processes. The most important differences lie usually in reaction times and operation pressures. Therefore, to attain the most comprehensive understanding of the chemistry in the flow-type ALD reactors, one should combine detailed surface chemistry studies under specifically optimized conditions and *in situ* studies under the real growth conditions—or conditions as close to them as possible. The latter either verify that the reaction mechanisms observed in the surface chemistry studies also prevail under the actual growth conditions or, if not, point out the differences.

In this section, the most common approaches for characterizing the ALD processes are introduced and discussed. The main emphasis is put on the methods which lead to a thorough understanding of the chemistry in the flow-type reactors.

7.1. Film Growth Experiments

The aim of the film growth experiments is of course to find the best conditions for growing high quality films. These studies involve the examination of the effects of the experimental parameters (precursor fluxes (vapor pressures), growth temperatures, pulse and purge times) on the growth rate and film properties, like composition, structure, and uniformity. In addition to revealing the best growth conditions, the observed interdependencies provide a great deal of information about the growth reactions as discussed in this section.

7.1.1. Examination of Growth Rate

ALD growth rate is usually expressed in terms of thickness increment per cycle, e.g., nanometers per cycle or angstroms per

cycle. Normally, the growth rate is determined after the film growth by dividing the measured film thickness by the number of deposition cycles applied, but sometimes optical *in situ* measurements are used for real time growth rate measurements (see Section 7.2.1).

7.1.1.1. Pulse Times

The effects of pulse times are usually examined by keeping the other parameters constant and by varying the exposure time of one precursor at the time. As the pulse time is increased, the density of the chemisorbed species on the surface increases toward the saturation level. The fulfilling of the chemisorption sites is basically governed by three factors:

- the number of precursor molecules arriving on the substrate, being determined by the precursor flux and the exposure time,
- the rate of the chemisorption reactions, whether true chemisorption or exchange reactions,
- the rate of further reactions on the surface which reopen chemisorption sites that may have become temporarily blocked by the nearby chemisorbed species, e.g., bulky ligands of a chemisorbed precursor molecule may impose steric hindrances until reordered or desorbed.

If the pulse time is so short that not enough precursor molecules are transported to the reaction chamber, severe thickness nonuniformities result. Especially in the traveling-wave reactors (Section 4.2.4) the precursor is consumed at the substrate area closest to the inlet while the areas further away receive only few molecules, if any. Therefore, a certain delay period

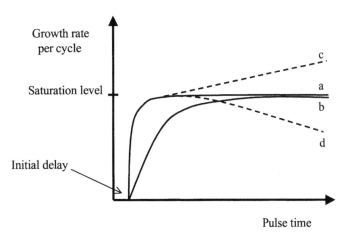

Fig. 18. Different growth rate vs pulse time curves observable in ALD processes: (a) fast and (b) slow chemisorption reactions with no decomposition or etching, (c) chemisorption reactions followed by precursor decomposition, and (d) chemisorption reactions followed by etching reactions. The initial delay period is due to the limited rate of precursor transportation and its length is dependent on the measurement position in the reactor.

is observed in the growth rate vs exposure time curve (Fig. 18). The length of this delay depends on the distance of the measurement point from the reactant inlet and on the precursor flux (see what follows). With increasing pulse time, the whole substrate becomes covered and different kinds of growth rate vs pulse time curves are observed depending on the chemistry (Fig. 18). As the self-limiting film growth mechanism is the characteristic feature of ALD, the growth rate is expected to saturate with long enough exposure times. Ideally, the reactions are fast and thus the saturation level is rapidly achieved (curve a in Fig. 18) but sometimes; e.g., at lower temperatures or with less reactive precursors, the reactions proceed slower (curve b in Fig. 18).

If the precursor is decomposing thermally, the growth rate does not saturate but increases quite linearly with increasing pulse time (Fig. 18, curve c). If the decomposition rate is only modest, the growth rate increases first more rapidly corresponding to the desired surface reactions, and only after that the decomposition dominated region with a smaller slope begins. Naturally, the higher the temperature, the faster the decomposition, and thus the steeper the slope in the decomposition region.

Occasionally, it is observed that the growth rate first increases but then starts to decrease as the pulse time is increased further (Fig. 18, curve d). This may be taken as an indication of an etching reaction which begins after the growth reactions have been completed; i.e., at first the incoming metal precursor reacts with the surface species (e.g., —OH groups) left from the previous pulse, but when they are consumed, the precursor starts to react with the film material itself.

Curve b in Figure 18 also points out why it may often be more effective not to strive for the fully saturated growth rate but instead to satisfy with 90% or so of it. Reaching the last 10% may require significant elongation of the pulse time and thereby decrease the productivity, i.e., deposition rate per time unit. In fact, in the most cases the highest productivities are achieved

with the shortest exposure times, the minimum being set by the time required to get the whole substrate covered. Quite often also film properties, such as uniformity, conformality, and purity are acceptable even if the growth rate is not fully saturated, and thus the slightly undersaturated growth conditions may be chosen to maximize the productivity.

7.1.1.2. Precursor Fluxes

The precursor fluxes are usually controlled by either the source temperatures (solids and liquids) or the mass flow controllers (gases). As the first approximation, the precursor flux may be assumed to have a similar effect as the pulse time because the precursor dose is a product of the two. Thus, with insufficiently small fluxes thickness nonuniformities result, and with increasing flux the growth rate behaves quite similarly as with increasing pulse time (Fig. 18). Therefore, a need for long exposure times, causing long cycle times and thereby slow deposition rates, may be circumvented, at least to some extent (see below), by increasing the precursor flux.

However, the relation between exposure time and precursor flux is not that trivial that only their product would determine the growth rate. There are many examples that the growth rate per cycle may be increased only by increasing the exposure time, but not by increasing the precursor flux [19, 269, 270, 307]. This has been attributed to the slowness of the surface reactions which lead to the reopening of the temporarily blocked chemisorption sites. On the other hand, studies on ALD of oxide films showed that the growth rate could be increased by increasing the water flux more than by increasing the water exposure time, though in both cases a saturation was observed [224]. This was explained by an increased surface hydroxylation under high water flux: because of the increased surface hydroxyl group density, more reactive sites were available for the subsequently pulsed metal precursor. Though one might expect that with increased densities the hydroxyl groups could become unstable against dehydroxylation and thereby make the growth dependent on the purge time following the water pulse, this was not observed. Apparently, if dehydroxylation occurred, it was fast enough to reach saturation in less than 0.2 s.

7.1.1.3. Purge Length

Insufficient purging causes overlapping of the precursor pulses and thereby CVD-like growth which manifests itself as increased growth rate and thickness nonuniformity. With increasing purge time, the overlapping is reduced and the growth rate decreases and finally saturates to the constant level. However, if desorption of the chemisorbed precursor molecules occurs, the growth rate continues to decrease even if the pulses are well separated.

In practice, surface species are usually stable and no desorption related behavior can be observed at temperatures used. This is especially true when the chemisorption involves exchange reactions between the incoming precursor and the surface species (Fig. 1a) because by nature these reactions are irreversible as part of the precursor molecule ligands are immediately released and purged away. On the other hand, desorption may occur if the precursor is molecularly or dissociatively chemisorbed (Fig. 1b), or if the surface species formed in the exchange reactions can combine with each other. In ALD of oxides, for example, hydroxyl groups are produced on the surface in the exchange reactions between the chemisorbed metal compounds and water, and also dissociative adsorption of water may add them. Subsequently, the hydroxyl groups may combine with each other and thereby dehydroxylate the surface, and if this occurs reasonably slowly, it may cause a decrease of the growth rate with increasing purge time after the water pulse [83]. Yet another possibility for desorption at high temperatures is that the film constituents themselves have reasonable equilibrium pressures.

In general, the purge times required to separate the precursor pulses are very much reactor dependent but do not depend that much on the precursors, water with a highly polar molecule perhaps being an exception. Thus, once appropriate purge times are determined for a given reactor with one process, they are usually applicable to also other processes in the same reactor. As discussed in Section 4, the minimization of purge times is one of the key aspects in a design of a productive ALD reactor.

7.1.1.4. Temperature

Different kinds of growth rate vs temperature dependencies which may be observed in ALD processes are schematically depicted in Figure 19. The key temperature range is the one in the middle where the growth proceeds in the self-limiting manner. Depending on if the surface density of the chemisorbed species is temperature dependent or not, the growth rate is also temperature dependent or constant. In the former case, the growth rate usually decreases with increasing temperature because of a decreased density of reactive surface species, a typical example being the dehydroxylation of oxide surfaces (e.g., [436]). In any event, the temperature dependency is usually so weak that small variations in the growth temperature have only minor effects. Temperature-independent growth rates are often observed in the growth of II–VI compounds, for example, [38].

In the low temperature side of the self-limiting region, the growth rate may either decrease or increase with decreasing temperature (Fig. 19). The decrease appears to be much more often observed and may be related to kinetic reasons; i.e., the growth reactions become so slow that they are not completed within the given pulse time. In principle, they might be completed if the pulse times were increased but that would increase the cycle time substantially. The increase of the growth rate with decreasing temperature may occur due to a multilayer adsorption and the condensation of low vapor pressure precursors.

Also at the high temperature side of the self-limiting region the growth rate may either decrease or increase (Fig. 19). The decrease of the growth rate with increasing temperature is related to a desorption of the precursors while the increase is due to a decomposition.

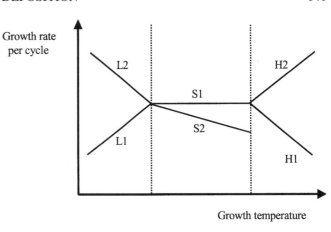

Fig. 19. Various growth rate vs temperature curves observable in ALD processes:

S1: self-limiting growth with temperature-independent rate,
S2: self-limiting growth with temperature-dependent rate,
L1: self-limitation not reached because of slow reactions,
L2: self-limiting growth rate exceeded because of multilayer adsorption or condensation,
H1: self-limitation not maintained because of precursor desorption,
H2: self-limiting growth rate exceeded because of precursor decomposition.

7.1.2. Examination of Film Properties

7.1.2.1. Film Composition

Film compositions, especially impurities, give indications of incomplete reactions or precursor decomposition. At low temperatures, the growth reactions are often somewhat incomplete and thus constituent elements of the precursor molecules may be found as impurities in the films. With increasing temperature, the reactions become more complete and thus the impurity contents decrease (Fig. 20a). This is also reflected in other film properties, like refractive index and dielectric permittivity of insulators (Fig. 20b), or conductivity in the case of conducting materials. On the other hand, if the temperature is increased too much, precursor decomposition may begin and cause an increase of the impurity contents.

7.1.2.2. Uniformity and Conformality

In the ideal ALD process, both uniformity and conformality are excellent. However, as noted earlier, the real processes may differ from the ideal one in two ways: the growth reactions are not entirely saturated or some precursor decomposition occurs in addition to the exchange reactions. The extent to which these nonidealities may be accepted is largely determined by how they affect the uniformity and conformality; i.e., if these two properties are not deteriorated, the reactions responsible for the nonideal behavior may be concluded to proceed in a surface reaction limited manner. For the incomplete saturation, this means that the increase of the surface density of the chemisorbed precursor molecules toward the saturation value is restricted by the rate of surface reactions, i.e., either the chemisorption reaction itself or the further reactions which reopen the momentarily blocked adsorption sites. For the decom-

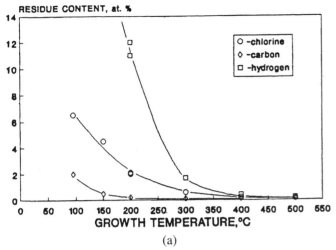

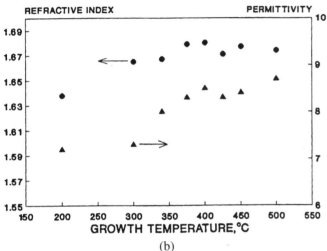

Fig. 20. The effect of growth temperature on (a) impurity contents and (b) refractive index and permittivity of Al_2O_3 films deposited from $Al(CH_3)_2Cl$ and water [237]. Reprinted with permission from K. Kukli et al., Atomic layer epitaxy growth of aluminium oxide thin films from a novel $Al(CH_3)_2Cl$ precursor and H_2O, *J. Vac. Sci. Technol. A* 15, 2214 (1997), © 1997, American Vacuum Society.

position, surface control means that the decomposition reaction rate is slow in comparison to the rate at which new precursors are adsorbed on the sites which are opened in the place of the decomposed molecules.

Thickness nonuniformity and nonconformality may also result because of insufficient purging. In deep and narrow trenches and holes, material transportation and thus also purging is slower than on planar surfaces and therefore purge periods that are too short should first be observed as reduced conformality. In practice, however, 0.5-s purging periods have already resulted in excellent conformality and complete filling of trenches with dimensions comparable to microelectronic devices [88].

Yet another potential reason for thickness nonuniformity is readsorption of reaction byproducts (cf. Section 4.2.4). If this happens, the largest disturbance in the thickness uniformity occurs at the areas closest to the merging point of the precursor flow channels because there the amount of the reaction byprod-

ucts varies the most. The surface at the point where the film growth begins receives only the precursor molecules, so the chemisorption density is the highest there. At the same time, this surface and any surface downstream produce the reaction byproducts, the concentration of which thereby increases in the carrier gas until a steady state is reached. As a consequence, the film thickness decreases over the range where this concentration develops [96, 97].

7.1.2.3. Crystallinity and Morphology

Film crystallinity is not related that much to the growth reactions but rather reflects the mobility of the film material and is thus strongly affected by the growth temperature—not to forget the possible crystallization promoting effect of the substrate, however. In general, the temperature limit between amorphous and crystalline deposits and the overall crystallinity are in the first place material dependent, and not largely affected by the precursors. There are some exceptions, however. For example, the crystallinity of TiO_2 films appears to be better when deposited from titanium alkoxides than when deposited from $TiCl_4$. This has been explained by an ordering effect of bridging alkoxide groups on the surface of the growing film [304, 307]. In contrast, Ta_2O_5 films obtained from $Ta(OC_2H_5)_5$ at around 300°C are completely amorphous [312] whereas the films obtained from $TaCl_5$ show some crystallinity [106, 200] which is ascribable to the rather exceptional formation of tantalum oxychlorides which, on the one hand, appear to favor crystallization by increasing mobility but, on the other hand, being volatile lead to etching of Ta_2O_5 above 300°C [104–106].

The formation of a polycrystalline film is accompanied by surface roughening which increases with increasing film thickness [82–87]. In this respect, the polycrystalline deposits made by ALD differ from the amorphous and epitaxial films which usually preserve rather well the morphology of the substrate surface. In any case, even if within the dimensions of the crystallites the film does not grow uniformly, in larger scale, like that of microelectronic device structures (Figs. 2 and 16) or on macroscopic level, the uniformity is still excellent.

Ex situ AFM studies on polycrystalline films have revealed that during the early stages of the growth agglomerates are formed on the surface (Fig. 21). As more ALD cycles are applied, the agglomerates grow both laterally and vertically, and once they grow into contact with each other, a continuous film is formed but with a roughened surface as compared with the substrate surface (Fig. 21). The roughness of a film with a given thickness is largely determined by the areal density of the agglomerates formed in the beginning of the growth.

Basically, agglomeration might be explained either as a process by means of which the system tries to minimize the substrate-film interface energy, or as a consequence of a limited number of reactive sites on the starting surface. However, though it has indeed been suggested that roughening would be an inherent consequence of a less than a monolayer per cycle growth, the growth rate is hardly the determining factor because rough films are obtained also with a full monolayer per

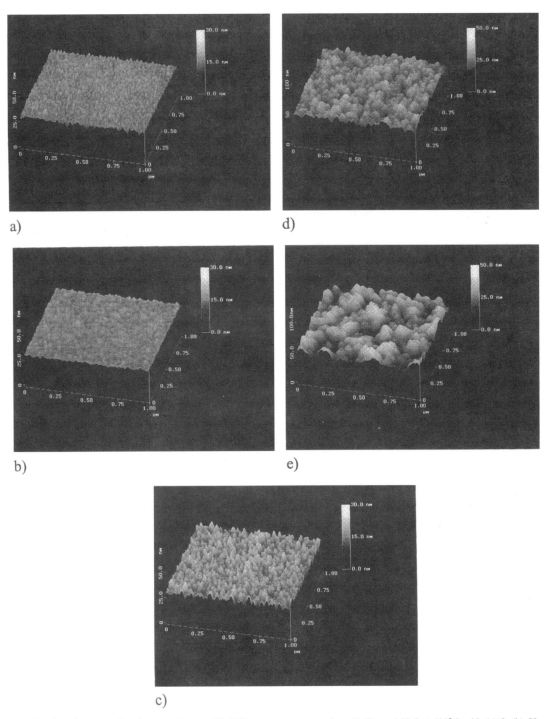

Fig. 21. Development of surface roughness of ZnS films grown on glass from $ZnCl_2$ and H_2S at 500°C with (a) 8, (b) 50, (c) 200, (d) 800, and (e) 3000 deposition cycles [85]. Reprinted with permission from J. Ihanus et al., AFM studies on ZnS thin films grown by atomic layer epitaxy, *Appl. Surf. Sci.* 120, 43 (1997), © 1997, Elsevier Science.

cycle growth rates [52, 284] and smooth films with lower rates [132]. Therefore, the interface energy minimization appears to be a more valid explanation. Evidently, in the case of epitaxial films, the driving force for the agglomeration is much lower because of the well-matched crystal structures of the film and the substrate.

Since polycrystalline structure and roughening are nearly always observed together, a connection between them appears evident. Crystallization and the interface energy driven agglomeration are both thermodynamically favored processes which, however, require sufficient mobility to occur. At low temperatures, the mobility is low, and amorphous and smooth films are obtained, but as the temperature is increased, the mobility increases and agglomeration and crystallization occur. Most probably, the agglomerates are preferably formed at certain special surface sites which on one substrate may be more numerous than on the other. For example, on mica the initial ZnS agglomerate density was much lower than on glass [85].

The final and apparently the hardest question is the atomic mechanism which at the same time can explain the agglomeration and maintain the self-limitation of the film growth. By now, the identity of the migrating species can mainly be just speculated about. The simplest possibility is that the agglomerates are formed as a consequence of a migration and the nucleation of the film material itself. Once an ALD cycle has been completed and a layer or a sublayer of the film material has been deposited, the material may be mobile enough to nucleate, i.e., to form agglomerates with a crystalline structure. To estimate if this is possible, one can try to look for a similar film-substrate combination made by reactive evaporation at the same substrate temperature: if a deposit formed in such a process has sufficient mobility for crystallization and agglomeration, the ALD made material should have that too. In other words, there should not be any reason why the nucleation would be restricted in ALD.

Another mechanistic explanation for the agglomeration is that one or several of the species formed as intermediates in the ALD reactions are able to migrate on the surface or in the gas phase. Such species are hard to identify, but in the $TiCl_4$-H_2O process, for example, $Ti(OH)_xCl_y$ and $TiOCl_2$ have been suggested [27, 28, 82]. Importantly, the possible formation of such species does not hurt the self-limitation of the film growth, since that would occur through reactions which passivate the reaction or chemisorption sites against further interaction with $TiCl_4$, for example:

$$-OH + TiCl_4(g) \rightarrow -Cl + Ti(OH)Cl_3(g)$$
$$\Rightarrow O + TiCl_4(g) \rightarrow 2-Cl + TiOCl_2(g)$$

7.2. Reaction Mechanism Studies

ALD reaction mechanisms may be investigated by analyzing either the surface or the gas phase, preferably both. The ultimate goal is to specify the species existing on the surface at each phase of an ALD cycle and to identify the volatile byproducts and the time of their release. From this information, the prevailing reaction mechanism can be determined. Obviously, the most definite conclusions would be obtained via a detailed chemical characterization of the surface but such studies are often hard to realize and, instead, less direct methods need to be applied. For example, the two alternative reaction mechanisms suggested in Figure 1 for the deposition of TiO_2 may be distinguished by observing whether gaseous HCl is released during both $TiCl_4$ and H_2O pulses (Fig. 1a) or only during the latter (Fig. 1b). Alternatively, or complementary, these mechanisms also may be distinguished by measuring surface mass changes during the various steps of an ALD cycle, the ratio of which is characteristic for each suggested mechanism (Section 7.2.1).

7.2.1. Flow-type Reactor Conditions

The reaction mechanism studies discussed here are those that have been carried out under conditions which are the most representative for the practical flow-type ALD thin film reactors in what comes to the pressure (a few Torrs), the exposure time (a few seconds in maximum), and the surface area of the substrates. For these studies, a differentially pumped quadrupole mass spectrometer (QMS) has been the choice for analyzing the gas phase whereas microgravimetry and optical techniques have been used for the surface characterization. Even these techniques are quite hard to integrate to the compact traveling-wave reactor geometries, and therefore certain modifications have been needed for their implementation. Nevertheless, the required modifications are after all quite straightforward, and because all these measurements are done in real time, they might also be applicable for process control in industry.

7.2.1.1. Analysis of the Gas Phase

For the gas phase analysis, it is important to note that the amount of byproducts released is directly related to the reactive surface area: an order of magnitude estimation is 10^{14} molecules released per 1 cm^2. In addition, the reactions are often fast and completed in fractions of a second. Therefore, the analysis method, also including the sampling arrangement, and the reactive surface area must be carefully designed and adjusted in accordance with each other to ensure meaningful and time-resolved detection of the byproducts. Until now, only QMS with differentially pumped sampling has been used [211, 223, 437–439].

Because mass spectrometers need to be operated below 10^{-4} Torr, a pressure reduction of at least 4 orders of magnitude is needed from the flow-type ALD reactor conditions. This has been realized by sampling with either a capillary [211, 223, 437] or an orifice [438, 439]. To minimize the contributions of reactions in the sampling setup, there must be as small a surface area as possible and therefore an orifice is the best choice, though capillaries can also be made small enough, e.g., 0.2 mm in diameter and 2 mm in length [223, 437]. As many of the precursors used in ALD are solids, the sampling setup must be heated; most preferably it is held at the reaction temperature. To keep the detector and the associated electronics close to room temperature, the QMS is inserted into a temperature gradient so that only the sampling setup and the ionizer are at the reaction chamber temperature. An ALD-QMS setup, also incorporating a quartz crystal microbalance (QCM, see the following) [439] is shown in Figure 22.

Figure 23 shows an example of data acquired with the instrument shown in Figure 22 on the $TiCl_4$-D_2O process. The dominant reaction byproduct is DCl which is followed as $D^{37}Cl^+$ ($m/z = 39$). The ratio of DCl liberated during the $TiCl_4$ pulse to that liberated during the D_2O pulse is about 1/3 which implies that each chemisorbing $TiCl_4$ molecule reacts with one surface $-OD$ group releasing one of its Cl ligands. The other three Cl ligands are released during the D_2O pulse. The simultaneously measured QCM data supports this conclusion (see the following).

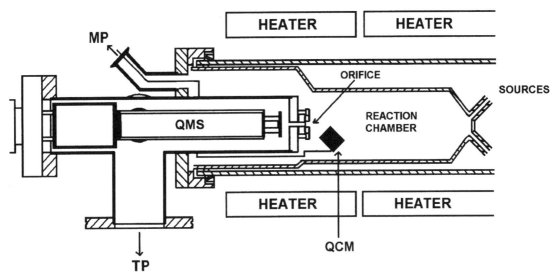

Fig. 22. Schematics of an ALD-QMS-QCM setup (after [438]). The precursors are transported by a carrier gas from their sources at right (not shown) through the reaction chamber to the mechanical pump (MP). The reaction chamber is quite tightly packed with glass substrates to have enough surface area (about 3500 cm^2) for producing detectable amounts of reaction byproducts. A small fraction of the total flow is pumped through the sampling orifice and the QMS chamber by the turbomolecular pump (TP). The pressure drop across the orifice is from about 2 Torr in the reaction chamber to below 10^{-6} Torr in the QMS chamber. The QCM is located in the vicinity of the orifice so that they both are at the reaction chamber temperature.

7.2.1.2. Analysis of the Surface

As the detailed identification of surface species is difficult under the flow-type reactor conditions, the surface studies have focused quite a lot on observing changes in surface termination during each ALD process step. Rates of exchange reactions, chemisorption, desorption, and decomposition, and their effects on self-limitation have been of interest. Because all these processes affect the mass of the film-substrate system, they may be followed by real time mass measurements. In addition, when only well-defined exchange reactions occur, the relative mass changes provide information of the average species terminating the surface at each stage of an ALD cycle.

Real time mass measurements may be done with either a quartz crystal microbalance (QCM) [96, 104, 105, 128, 179, 188, 198, 305, 308, 313, 438, 439] (Fig. 22) or with a sensitive electrobalance [144, 151, 152]. An example of the latter is shown in Figure 24. The electrobalance itself is outside the reaction chamber and is protected with a carrier gas flow. The substrate is connected to the electrobalance with a fused quartz fiber. The sensitivity of the microbalance is better than 0.025 μg, so with a 2.6-cm^2 substrate the smallest detectable mass change corresponds to less than 0.05 ML of GaAs or GaN, for example. The main problem of this system appears to be its sensitivity to flow rate changes associated with pulsing of the precursors [151]. In addition, a reactor specially designed for a connection to the electrobalance is evidently needed.

Quartz crystal microbalances are routinely used as real time film thickness monitors in evaporation, for example. The operation of QCM is based on the relation between the resonance oscillation frequency of the piezoelectric quartz crystal and its mass, also including the film mass. QCM is rather easily integrated to any ALD reactor into a space of a few cubic centimeters, but its sensitivity to temperature variations may cause problems. Therefore, during the measurements the temperature of the crystal must either stay constant or vary with a constant rate so that its contribution can be subtracted from the measured frequency changes. Any temperature variations due to the precursor pulsing must be eliminated. In any case, above 300–350°C noise in the QCM signal becomes largely disturbing. The primary data of QCM, i.e., the changes in crystal oscillation frequency, directly give the relative surface mass changes which are usually enough for examining the ALD reaction mechanisms. For this reason and because of uncertainties related to the absolute mass measurements, the latter are rarely done.

Figure 23 shows QCM data measured on the TiCl$_4$-D$_2$O process [439] with the ALD-QMS-QCM setup shown in Figure 22. Basically, similar data are also obtained with electromicrobalances. During each precursor pulse, a fast mass change followed by a saturation is observed as expected for an effective and self-limiting ALD process. Any decomposition or etching reactions causing deviations from the self-limiting growth conditions could easily be distinguished as continuous mass increase or decrease, respectively, [104, 105, 313]. During the TiCl$_4$ pulse, the mass increases by m_1 as adsorbates $-$TiCl$_x$ are formed, and during the D$_2$O pulse the mass decreases as Cl ligands are replaced by $-$OD groups or oxide ions. The mass increase m_0 brought about by the complete cycle corresponds to TiO$_2$. The ratio m_1/m_0 can be calculated for any suggested reaction mechanism, and thereby the correct mechanism may be distinguished from the measured ratio. In Figure 23, the m_1/m_0 ratio is about 1.9 which, in good agreement with the QMS results (see the foregoing text), points to the following reaction

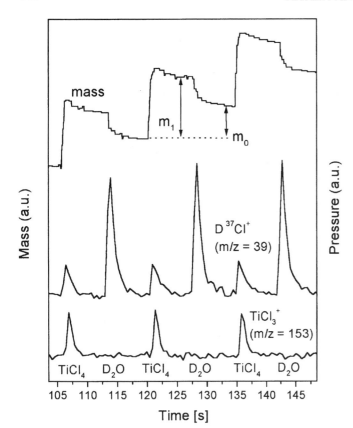

Fig. 23. QMS and QCM measurements on a TiCl₄-D₂O process done with the ALD-QMS-QCM setup shown in Figure 22. The pulse times are 0.5 s for both precursors and purge periods are 6.0 s. The reaction temperature is 200°C [439]. The QMS data shows the time variation of the ions $TiCl_3^+$ ($m/z = 153$) and $D^{37}Cl^+$ ($m/z = 39$). From the $D^{37}Cl^+$ signal the background observed during subsequent reference D₂O pulses applied after the ALD cycles has been subtracted. The mass signal was obtained from the negative resonance oscillation frequency change, $-\Delta f$, which is in a direct relationship to the surface mass load. m_1 and m_0 denote the mass changes during the TiCl₄ pulse and during one completed ALD cycle, respectively, and they are related to the intermediate surface species $-TiCl_x$ and TiO₂, respectively.

mechanism:

$$-OD + TiCl_4(g) \rightarrow -O-TiCl_3 + DCl(g)$$
$$-O-TiCl_3 + 2D_2O \rightarrow -O-Ti(O)-OD + 3DCl(g)$$

$$
\begin{aligned}
m_1 &\propto M(-TiCl_3) - M(D) \\
&= (154 - 2)\ \text{g mol}^{-1} = 152\ \text{g mol}^{-1} \\
m_0 &\propto M(TiO_2) = 80\ \text{g mol}^{-1} \\
\frac{m_1}{m_0} &= 1.9
\end{aligned}
$$

Note that the surface species formed in the latter reaction is $-O-Ti(O)-OD$ rather than $-O-Ti(-OD)_3$ because otherwise the $-OD$ group density would be threefold in the beginning of the next cycle whereas the cycle-to-cycle repeatability (Fig. 23) indicates that such variation does not exist. On the other hand, the reaction mechanism was found to be clearly temperature dependent: the higher the temperature, the lower the amount of DCl released during the TiCl₄ pulse. This behavior can be correlated to the density of hydroxyl groups on the TiO₂ surface which decreases with increasing temperature [436]. Obviously, as the reaction temperature is increased, few hydroxyl groups are left on the surface after the water pulse and the dominating mechanism become that shown in Figure 1b; i.e., TiCl₄ chemisorbs without exchange reactions and all the Cl ligands are released during the water pulse. It must be emphasized, however, that both QMS and QCM give information of only the net reactions during each pulse, and the actual mechanism may consist of several elementary reactions. For example, even if $-OD$ groups existed on the surface in significant amounts and reacted with TiCl₄, a release of DCl would not necessarily be observed during the TiCl₄ pulse as DCl might readsorb on the surface and stay there until the next water pulse.

Optical techniques form another group of methods which are applicable for analyzing surface processes in the flow-type ALD reactors. In fact, these techniques have been applied over the entire pressure range from UHV to atmospheric. Only one or two windows and an open light path to the sample surface are needed. If film growth on the window constitutes a problem, it may be prevented with proper inert gas flow curtains. Optical methods are well suited for following changes in surface termination and for examining dynamics of these changes. However, the chemical interpretation of the data is difficult without supplementary methods.

Surface photoabsorption (SPA) [18, 149–152, 156, 266, 397, 440–444] and reflectance difference spectroscopy (RDS) [244, 257, 258, 270, 445] have been the most extensively used optical techniques in the characterization of ALD processes. In SPA, the reflectivity of p-polarized light incident on the surface at the Brewster angle is measured [446]. With this geometry, the bulk contribution is minimized, enabling sensitive detection of the reflectivity changes caused by surface reactions. Under the self-limiting ALD growth conditions, the reflectivity varies periodically between two constant levels corresponding to the two differently terminated surfaces. Deviations from these reflectivity levels give indications of other kinds of surface terminations caused by, for example, incomplete reactions, decomposition, or desorption. For instance, SPA measurements have indicated that the GaCl-AsH₃ process proceeds differently in hydrogen and helium carrier gases, and it was suggested that in H₂ chlorides are removed from the surface as HCl already during the purge following the GaCl pulse whereas in the inert carrier the chlorides are removed during the AsH₃ pulse [150]. The suggested mechanism in H₂ was later verified by gravimetric studies [151] whereas further SPA investigations revealed that part of the Ga atoms left on the surface after the chloride removal in H₂ are covered with hydrogen atoms [443]. Additionally, arsenic desorption during elongated purge times after the AsH₃ pulse was revealed and kinetically measured by both SPA and microgravimetry [444].

In RDS, nearly normal incident light polarized along two certain crystallographic axes of the substrate is used, and the difference in their reflectivity is measured [447]. The alternate pulsing in ALD causes changes in the RDS signal similar to the SPA. Because the RDS signal is derived from the anisotropy between two crystallographic directions in the surface plane,

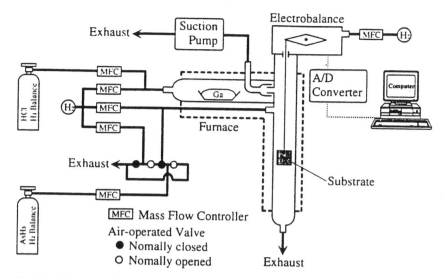

Fig. 24. Schematic of an *in situ* gravimetric monitoring system used to examine ALD processes in which GaCl was formed in situ from a metallic gallium and HCl vapor [151]. Reprinted with permission from A. Koukitu, N. Takahashi, and H. Seki, *In situ* monitoring of the growth process in GaAs atomic layer epitaxy by gravimetric and optical methods, *J. Cryst. Growth* 146, 467 (1995), © 1995, Elsevier Science.

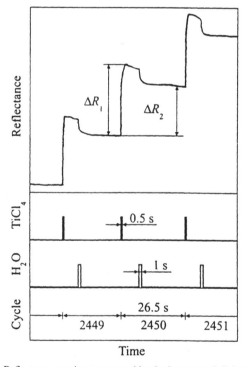

Fig. 25. Reflectance transients measured by the incremental dielectric reflection technique during three subsequent TiO_2 growth cycles with $TiCl_4$ and H_2O as precursors [182]. Reprinted with permission from A. Rosental et al., Surface of TiO_2 during atomic layer deposition as determined by incremental dielectric reflection, *Appl. Surf. Sci.* 142, 204 (1999), © 1999, Elsevier Science.

RDS is applicable only to epitaxial growth. SPA has also mainly been used in epitaxy processes, though in principle it is also applicable to nonepitaxial cases. In fact, a similar technique, called incremental dielectric reflection (IDR) [182, 448, 449] was applied for following the ALD growth of TiO_2 on glass (Fig. 25). As in SPA, also in IDR the reflection of *p*-polarized

light incident at the Brewster angle is measured but here the reflectivity variations originate from interference effects rather than from absorption. If not only the film but also the substrate is transparent to the probing light, the substrate may be used as the window for the light [182, 449]. This makes it possible to apply IDR also to the traveling-wave reactor geometries. Another technique similarly based on the effects of the surface layer on light interference, named as surface photointerference (SPI), has been applied for the *in situ* characterization of ALD growth of epitaxial ZnSe and CdSe on GaAs [279]. As compared with the methods based on light absorption, IDR and SPI have a benefit in that they employ lower energy photons which apparently have less effect on the growth reactions.

Ellipsometer is a common tool for *in situ* characterization of thicknesses and optical properties of thin films. Its suitability to the analysis of the separate ALD reaction steps has been quite scarcely examined, however. A spectroscopic ellipsometer has been used for following the $Al(CH_3)_3$-H_2O_2 ALD process [229]. Using predetermined optical properties and a single layer model for the film, thickness increments of 0.17 nm cycle^{-1} were resolved during the $Al(CH_3)_3$ pulse though the data was quite noisy, peak-to-peak variations in the thickness were at best 0.1 nm.

7.2.2. High Surface Area Substrates

High surface area substrates have been a subject of an ALD research both because of their practical importance, like in heterogeneous catalysts, and because they offer plenty of material for detailed chemical characterization. Nanoporous silica and alumina powders have high specific surface areas of several hundred square meters per gram, silica even up to 1000 $m^2\,g^{-1}$, which ensures that already a monolayer of reaction products or intermediates is enough for routine chemical analysis techniques, such as IR, NMR, and elemental analysis. For example, if 1 cm^2 planar substrate holds about 10^{14} surface species, 100-mg nanoporous powder with a surface area of 300 m^2 g^{-1} contains 3 × 10^{19} of them.

The major difference of reactions on porous powders in comparison with thin film growth comes from the slow transportation into and out of the porous particles and from the need to deliver much larger quantities of precursors in each pulse. Therefore, the exposure and purge times must be much longer than on planar substrates, typical values being a few hours. As a consequence, the reactions may proceed quite differently than during the second and subsecond range exposures in thin film growth. Slower reactions may become dominating when more time is given, especially decomposition is observed at much lower temperatures than on planar surfaces. In addition, because only a few cycles are applied, the reactions often represent more an interface formation than a steady-state film growth, though sometimes this may be taken as an advantage as well. From a practical point of view, it is important to note that prior to and during the analysis the sample must be kept under an inert atmosphere, hence requiring appropriate means for transferring the sample powder from the reactor to the analysis.

Studies on porous powders have clearly pointed out the importance of the surface terminating groups with respect to the metal precursor chemisorption density [30–33]. For example, when the hydroxyl group density on silica is decreased by preheating it at increasing temperatures, the number of metal atoms bound to silica decreases accordingly. At the same time, a consumption of the hydroxyl groups in reactions with the metal precursor may be verified with IR and NMR measurements. These observations imply that the chemisorption mainly occurs via a reaction with the hydroxyl groups.

The chemisorption density is also affected by steric factors: the larger the precursor molecule of a given metal, the lower the number of metal atoms chemisorbed. For example, Ni(acac)$_2$ results in a saturation density of 2.5 Ni nm^{-2} whereas the more bulky Ni(thd)$_2$ gives only 0.92 Ni nm^{-2} [30, 33]. If the precursors are significantly different in their character, their binding modes to the surface may differ. For instance, Al(CH$_3$)$_3$ has a high affinity toward oxygen and thus reacts on silica not only with the —OH groups but also with siloxane bridges (Si—O—Si), whereas Al(acac)$_3$ reacts only with the —OH groups and, unlike Al(CH$_3$)$_3$, is not able to consume them all because of its larger size [33]. On the other hand, the chemical nature of the powder may also have significant effects. Alumina, for example, catalyzes the decomposition of Ti(OCH(CH$_3$)$_2$)$_4$ already above 100°C whereas on silica self-limiting adsorption occurs at least up to 200°C (Fig. 26a) [32]. The elemental C/Ti ratios measured after the Ti(OCH(CH$_3$)$_2$)$_4$ exposure (Fig. 26b) imply that on silica two ligands are consumed in reactions with the surface —OH groups and the surface is terminated with (—Si—O—)$_2$Ti(—OCH(CH$_3$)$_2$)$_2$ species.

An intermediate case between loose beds of porous powders and planar substrates is formed by thinner porous objects, such as porous silicon surface layers, high surface area mem-

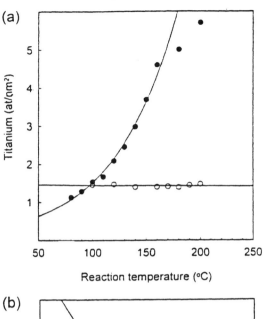

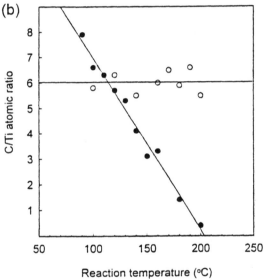

Fig. 26.   (a) Titanium surface density and (b) C/Ti atomic ratio on high surface area alumina (•) and silica (○) powders after Ti(OCH(CH$_3$)$_2$)$_4$ exposure at different temperatures [32]. Reprinted with permission from M. Lindblad et al., Processing of catalysts by atomic layer epitaxy: Modification of supports, *Appl. Surf. Sci.* 121–122, 286 (1997), © 1997, Elsevier Science.

branes, and powders pressed into thin grids or in the form of thin pellets. These resemble powders in that IR characterization is possible but the required exposure times are shorter, tens of seconds, and thus closer to thin film growth, though still the difference is notable, 1 to 2 orders of magnitude. In any event, the reactions are fast enough so that several deposition cycles may be repeated for observing the cyclic changes in surface termination. Also these studies have supported the view that surface terminating groups play a decisive role in each chemisorption step in ALD of oxides [36, 42, 162, 163, 219, 264] and nitrides [168, 207, 231, 232]. The same also applies to the novel tungsten process where WF$_6$ and Si$_2$H$_6$ are used as precursors [206]. The increase and the decrease, respectively, in the Si—H and W—F absorption bands in the IR spectra after the Si$_2$H$_6$ exposure im-

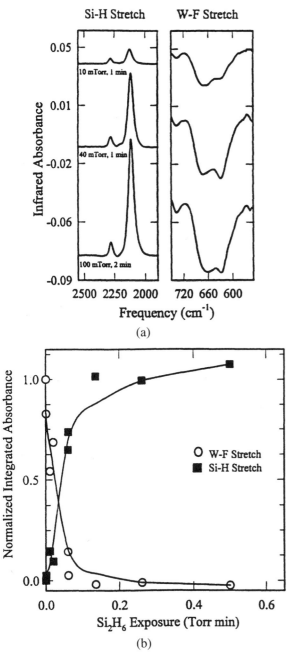

(b)

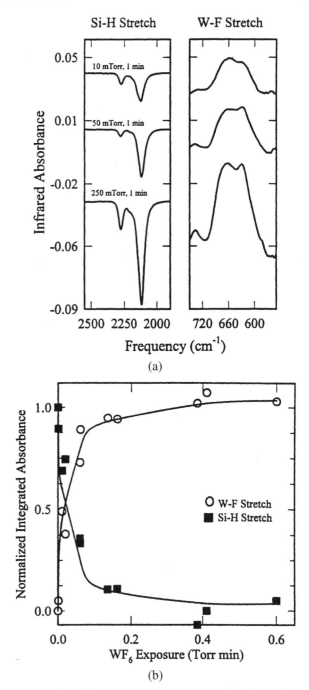

(a)

Fig. 27. (a) Fourier transform infrared (FTIR) difference spectra recorded in the tungsten ALD process after various Si_2H_6 exposures on a surface previously saturated with WF_6. Each spectrum is referenced to the initial surface left after the WF_6 pulse. The reaction temperature is 150°C. (b) The Si_2H_6 exposure dependence of the normalized integrated absorbances of the W–F stretching vibration at ~680 cm^{-1} and the Si–H stretching vibrations at 2115 and 2275 cm^{-1} [206]. Reprinted with permission from J. W. Klaus, S. J. Ferro, and S. M. George, Atomic layer deposition of tungsten using sequential surface chemistry with a sacrificial stripping reaction, *Thin Solid Films* 360, 145 (2000), © 2000, Elsevier Science.

Fig. 28. (a) FTIR difference spectra recorded in the WF_6-Si_2H_6 ALD process vs WF_6 exposure at 150°C. Each spectrum is referenced to the initial, Si_2H_6 saturated surface. (b) Normalized integrated absorbances of the W–F (680 cm^{-1}) and Si–H (2115 and 2275 cm^{-1}) stretching vibrations vs the WF_6 exposure at 150°C [206]. Reprinted with permission from J. W. Klaus, S. J. Ferro, and S. M. George, Atomic layer deposition of tungsten using sequential surface chemistry with a sacrificial stripping reaction, *Thin Solid Films* 360, 145 (2000), © 2000, Elsevier Science.

ply that disilane reduces –WF_x surface species into metallic tungsten and forms –SiH_xF_y species instead (Fig. 27). Subsequently, WF_6 reacts with the –SiH_xF_y surface groups and regenerates the –WF_x terminated surface (Fig. 28). Since the films deposited with this process contained no silicon or flu-

orine detectable with XPS, it could be concluded that the intermediate SiH_xF_y groups were removed as volatile SiH_aF_b species. Both reactions appear self-limiting, as the absorbances saturate above certain exposure levels (Figs. 27b and 28 b).

7.2.3. High Vacuum Conditions

As ALD processes proceed via saturating surface reactions, it is evident that techniques having the highest surface sensitivity should be applied for their characterization. Among these techniques the most valuable are perhaps XPS which provides information of not only the elements present on the surface but also of their chemical environment, and low energy ion scattering (LEIS) which probes only the topmost layer of atoms. Other applicable surface analysis techniques include the well-known Auger electron spectroscopy (AES), SIMS, reflection high energy electron diffraction (RHEED), and low energy electron diffraction (LEED), for example. However, the ultrahigh-vacuum conditions required by these techniques are in a clear contradiction with the practical ALD reactors. Therefore, either sample transfer or reactions performed in high vacuum are needed.

Sample transfer from a higher pressure reactor to the UHV analysis chamber may be realized in several ways. If the two are separate, a transfer container with inert atmosphere or vacuum must be used. More conveniently, a reaction chamber is connected directly to the analysis chamber through a gate valve. The reaction chamber may be just another high-vacuum chamber which during the reactions is filled alternately to certain pressures of the precursors and then evacuated to a similar level as the analysis chamber. Such systems are well suited for reasonably volatile precursors but low vapor pressure sources may constitute problems because of condensation. To avoid these problems, a special mini-ALD reactor built in connection to a UHV surface analysis chamber is apparently advantageous [450]. It is also possible to build inside the analysis chamber a small high pressure reaction chamber which after evacuation is opened for the surface studies [163]. Occasionally, the reactions have been carried out in the analysis chamber itself.

These kinds of high-vacuum studies are not unique to ALD but they have also been performed in an attempt to deepen the understanding of CVD chemistry. There also alternate dosing is often used to better identify the potential surface intermediates. No matter what the original aim of the studies, their eventual outcome is obvious: the elemental composition of the surface intermediates terminating the surface after each reaction step (Fig. 29). Light elements, especially hydrogen, are unfortunately difficult to detect with most of the methods. On the other hand, as XPS shows information of the chemical environment of the elements, it can indirectly reveal also the presence of hydrogen. RHEED and LEED, in turn, give structural information and are usually used in connection with epitaxial growth processes on single crystal substrates.

It must be recognized, however, that both sample transfer and different reaction conditions may effect the chemistry in comparison with the practical ALD processes. Sample transfer obviously takes time and involves cooling of the sample, thus raising a question if the surface intermediates stay unchanged. Likewise, reactions often behave quite differently under high-vacuum conditions as compared with higher pressures. In fact, sometimes reactions known from higher pressure conditions appear not to proceed at all in high vacuum. This effect is well

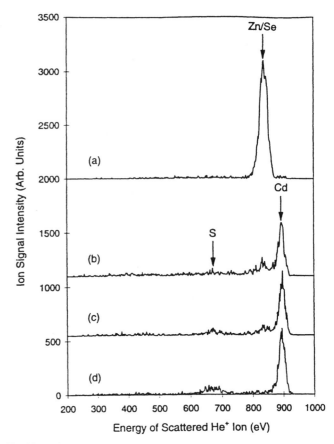

Fig. 29. Low energy ion scattering (LEIS) spectra from (a) a clean ZnSe surface, and after (b) one, (c) two, and (d) four CdS deposition cycles [299]. The energy of the incident He$^+$ ions was 1000 eV and the peaks at 679, 840, and 896 eV correspond to He$^+$ ions scattered off S, Zn or Se, and Cd, respectively. Reprinted with permission from Y. Luo et al., *Langmuir* 14, 1493 (1998), © 1998, American Chemical Society.

known in heterogeneous catalyst studies, for instance, and is often called a pressure gap. To explain the pressure gap, both a too small number of molecules having enough thermal energy to overcome the reaction barrier and a lack of collision-induced chemistry have been suggested [451]. In any case, application of conclusions drawn from the high-vacuum experiments to practical, higher pressure ALD reactors clearly requires critical evaluation and, preferably, comparison with film growth and/or reaction mechanism studies under the reactor conditions.

Temperature programmed desorption (TPD, also called thermal desorption spectroscopy, TDS) [18, 299–301, 375, 382, 383, 442] may also be considered as a high-vacuum method though high vacuum is not necessarily required but often used to obtain supplementary information with the surface analysis techniques. TPD is an effective method for examining the identity, relative quantity, and thermal stability of surface species. In TPD, the sample is first prepared at a chosen temperature by exposing the surface to the compound being studied and then, after evacuation, the temperature is steadily increased and the desorbed species are monitored with a mass spectrometer. When applied to ALD studies, a number of alternate precursor exposures may be given before starting the heating.

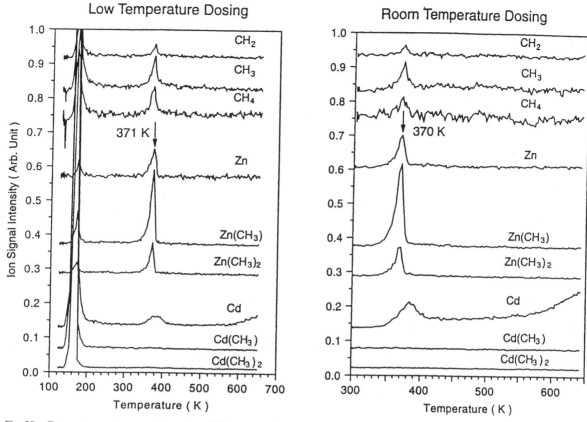

Fig. 30.   Temperature programmed desorption (TPD) spectra measured after dosing of $Cd(CH_3)_2$ on to ZnSe at 100 K (low temperature dosing) and room temperature [299]. Reprinted with permission from Y. Luo et al., *Langmuir* 14, 1493 (1998), © 1998, American Chemical Society.

An illustrative example of TPD studies on ALD processes is the examination of CdS deposition from $Cd(CH_3)_2$ and H_2S on a ZnSe substrate [299–301]. Figure 30 shows two TPD spectra measured after dosing $Cd(CH_3)_2$ onto ZnSe at 100 K (low temperature dosing) and room temperature [299]. In the case of the low temperature dosing, desorption of physisorbed $Cd(CH_3)_2$ multilayer is observed at 160 K, otherwise the two spectra are essentially similar. At 370 K, desorption of $Zn(CH_3)_2$ is observed (the lower mass species come from cracking of $Zn(CH_3)_2$) which indicates that $Cd(CH_3)_2$ was dissociatively adsorbed on ZnSe, and upon heating the surface terminating methyl groups combined preferably with zinc, thereby causing zinc to cadmium replacement in the surface layer. Remarkably, the dissociative adsorption occurred also during the low temperature dosing before the desorption of the physisorbed species.

When H_2S was dosed at 100 K upon the methyl terminated surface, left from the preceding exposure of $Cd(CH_3)_2$ onto ZnSe, no reaction occurred and the physisorbed H_2S desorbed at 150 K and the methyl groups at 370 K (Fig. 31). By contrast, when dosed at room temperature, H_2S reacted completely with the $-CH_3$ groups which were not detected in TPD anymore (Fig. 31). After this reaction, the surface was terminated with $-SH$ groups which were stable up to 485 K where they combined and desorbed as H_2S. LEIS measurements showed that already after the first deposition cycle the ZnSe substrate was

nearly completely covered by CdS (Fig. 29). As a consequence, the desorption of $Zn(CH_3)_2$ ceased rapidly with an increasing number of deposition cycles [301]. However, at this phase no substantial $Cd(CH_3)_2$ desorption was observed either, but rather the $-CH_3$ groups desorbed in the form of radicals [301].

The $Cd(CH_3)_2$-H_2S precursor combination is apparently an exceptionally efficient one since the reactions go into completion already at room temperature. In any case, procedures similar to that outlined above are applicable for analyzing other ALD processes as well but the dosing temperatures must be carefully chosen according to the kinetic limitations of the chemistry under study. A serious concern is that the surface species forming during the low temperature exposure are different from those which form at higher temperatures in an ALD process. Likewise, if the initial surface in TPD differs from that in an ALD process, different kinds of surface species are likely to form. For example, if an oxide surface is carefully cleaned under high vacuum prior to a TPD experiment, it most probably has no $-OH$ groups and therefore no exchange reactions are possible in an adsorption of a metal compound. By contrast, as discussed before, in ALD oxide processes surface $-OH$ groups are often present and thus some of the metal precursor ligands may become lost in reactions with them. The surface species formed in these two alternative cases apparently exhibit different decomposition and desorption behavior in TPD. For

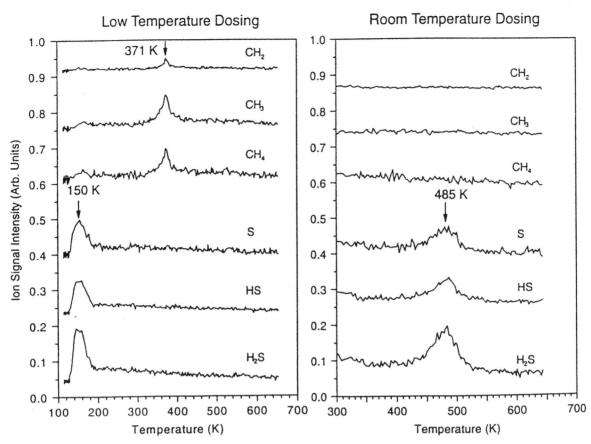

Fig. 31.   TPD spectra after dosing of H_2S at 100 K (low temperature dosing) and room temperature on a ZnSe surface which was previously exposed to $Cd(CH_3)_2$ and was thus methyl terminated [299]. Reprinted with permission from Y. Luo et al., *Langmuir* 14, 1493 (1998), © 1998, American Chemical Society.

these reasons, the sample preparation in TPD must be carefully planned to be the most representative for the ALD process conditions. It might also be valuable to compare TPD spectra of surfaces prepared at various temperatures.

8. SUMMARY

ALD is a unique process based on alternate surface reactions which are accomplished by dosing the gaseous precursors on the substrate alternately. Under the ideal conditions, these reactions are saturative ensuring many advantageous features like excellent conformality, large-area uniformity, accurate, and simple film thickness control, repeatability, and large-batch processing capability. As these advantages can often be achieved also under growth conditions which are not fully saturated, in many cases it is beneficial to employ short exposure times to maximize the effective growth rate even if the process is not then strictly self-limited.

While ALD has been employed in TFEL display manufacturing for many years, its broader utilization has been limited by its slowness as evaluated in terms of thickness increment per time unit. At the same time, the high productivity achievable through large-batch processing has perhaps not gained enough attention. In the near term future, the conformality demands and

film thicknesses in the integrated circuits are approaching levels which favor ALD as a highly potential manufacturing method, provided of course that proper processes can be developed for the materials needed. As a consequence, the main focus of the ALD research is foreseen to change from the thus far dominated epitaxial semiconductors to oxides, nitrides, and metals.

Among the various reactor types used for ALD, the flow-type traveling-wave geometry has been the most common one, at least in industry. It has a number of benefits, like fast pulsing and purging, effective precursor utilization and scalability to large batches. The major limitation is a difficulty to implement extra energy sources. Methods for *in situ* process control should also be developed.

Precursor chemistry is definitely the key issue in developing effective ALD processes. Aggressively reacting precursor combinations should be looked for to ensure fast completion of the reactions and effective precursor utilization. Solid sources are otherwise well applicable but they are quite laborious, and if in the form of a fine powder, may cause particle defects into the films. Therefore, high vapor pressure compounds, preferably liquids, should be looked for in the first place but often such choices are not available. By now, development of precursors specifically tailored for ALD has been quite limited but also in this area increasing efforts may be expected.

ALD processes activated by light or radicals have been demonstrated, but so far this has remained quite an unexplored area and will certainly open up new possibilities for ALD. Lowered deposition temperatures and higher purity films may be expected, and metal precursors which have so far been quite inert toward the common nonmetal sources may turn reactive. In addition, completely new film materials like metals are likely to appear in the ALD material selection. On the other hand, while suitable for single substrate processing, the activated processes may be hard to scale-up to large batches which are often required to make ALD productive enough.

Many approaches have been taken toward a better understanding of the ALD chemistry. The reaction mechanism studies all point out the importance of surface groups as intermediates in ALD processes.

REFERENCES

1. T. Suntola and J. Antson, U.S. Patent 4,058,430, 1977.
2. T. Suntola, J. Antson, A. Pakkala, and S. Lindfors, *SID 80 Dig.* 11, 108 (1980).
3. T. S. Suntola, A. J. Pakkala, and S. G. Lindfors, U.S. Patent 4,389,973, 1983.
4. T. S. Suntola, A. J. Pakkala, and S. G. Lindfors, U.S. Patent 4,413, 022, 1983.
5. T. Suntola and J. Hyvärinen, *Annu. Rev. Mater. Sci.* 15, 177 (1985).
6. C. H. L. Goodman and M. V. Pessa, *J. Appl. Phys.* 60, R65 (1986).
7. T. Suntola, *Mater. Sci. Rep.* 4, 261 (1989).
8. M. Leskelä and L. Niinistö, in "Atomic Layer Epitaxy" (T. Suntola and M. Simpson, Eds.), p. 1. Blackie, Glasgow, 1990.
9. T. Suntola, *Thin Solid Films* 216, 84 (1992).
10. R. Törnqvist, *Displays* 13, 81 (1992).
11. Y. A. Ono, "Electroluminescent Displays." World Scientific, Singapore, 1995.
12. J. Nishizawa, H. Abe, and T. Kurabayashi, *J. Electrochem. Soc.* 132, 1197 (1985).
13. S. M. Bedair, M. A. Tischler, T. Katsuyama, and N. A. El-Masry, *Appl. Phys. Lett.* 47, 51 (1985).
14. A. Usui and H. Sunakawa, *Japan. J. Appl. Phys.* 25, L212 (1986).
15. A. Doi, Y. Aoyagi, and S. Namba, *Appl. Phys. Lett.* 49, 785 (1986).
16. M. A. Tischler and S. M. Bedair, in "Atomic Layer Epitaxy" (T. Suntola and M. Simpson, Eds.), p. 110. Blackie, Glasgow, 1990.
17. Y. Aoyagi, T. Meguro, S. Iwai, and A. Doi, *Mater. Sci. Eng. B* 10, 121 (1991).
18. A. Usui, *Proc. IEEE* 80, 1641 (1992).
19. M. Ozeki, *Mater. Sci. Rep.* 8, 97 (1992).
20. S. M. Bedair and N. A. El-Masry, *Appl. Surf. Sci.* 82–83, 7 (1994).
21. Y. Sakuma, M. Ozeki, N. Ohtsuka, Y. Matsumiya, H. Shigematsu, O. Ueda, S. Muto, K. Nakajima, and N. Yokoyama, *Appl. Surf. Sci.* 82–83, 46 (1994).
22. S. M. Bedair, *J. Vac. Sci. Technol. B* 12, 179 (1994).
23. J. Nishizawa and T. Kurabayashi, *Thin Solid Films* 367, 13 (2000).
24. T. Yao, in "Atomic Layer Epitaxy" (T. Suntola and M. Simpson, Eds.), p. 155. Blackie, Glasgow, 1990.
25. M. Konagai, Y. Takemura, H. Nakanishi, and K. Takahashi, *Mater. Res. Soc. Symp. Proc.* 222, 233 (1991).
26. E.-L. Lakomaa, S. Haukka, and T. Suntola, *Appl. Surf. Sci.* 60–61, 742 (1992).
27. S. Haukka, E.-L. Lakomaa, and A. Root, *J. Phys. Chem.* 97, 5085 (1993).
28. S. Haukka, E.-L. Lakomaa, O. Jylhä, J. Vilhunen, and S. Hornytzkyj, *Langmuir* 9, 3497 (1993).
29. A. Kytökivi, E.-L. Lakomaa, A. Root, H. Österholm, J.-P. Jacobs, and H. H. Brongersma, *Langmuir* 13, 2717.
30. E.-L. Lakomaa, *Appl. Surf. Sci.* 75, 185 (1994).
31. S. Haukka and T. Suntola, *Interface Sci.* 5, 119 (1997).
32. M. Lindblad, S. Haukka, A. Kytökivi, E.-L. Lakomaa, A. Rautiainen, and T. Suntola, *Appl. Surf. Sci.* 121–122, 286 (1997).
33. S. Haukka, E.-L. Lakomaa, and T. Suntola, *Stud. Surf. Sci. Catal.* 120, 715 (1998).
34. C. Dücsö, N. Q. Khanh, Z. Horváth, I. Bársony, M. Utriainen, S. Lehto, M. Nieminen, and L. Niinistö, *J. Electrochem. Soc.* 143, 683 (1996).
35. M. Utriainen, S. Lehto, L. Niinistö, C. Dücsö, N. Q. Khanh, Z. E. Horváth, I. Bársony, and B. Pécz, *Thin Solid Films* 297, 39 (1997).
36. A. W. Ott, J. W. Klaus, J. M. Johnson, S. M. George, K. C. McCarley, and J. D. Way, *Chem. Mater.* 9, 707 (1997).
37. B. S. Berland, I. P. Gartland, A. W. Ott, and S. M. George, *Chem. Mater.* 10, 3941 (1998).
38. T. Suntola, in "Handbook of Crystal Growth" (D. T. J. Hurle, Ed.), Vol. 3b, p. 601. Elsevier, Amsterdam, 1994.
39. M. Leskelä and M. Ritala, *J. Phys. IV* 5, C5-937 (1995).
40. L. Niinistö, M. Ritala, and M. Leskelä, *Mater. Sci. Eng. B* 41, 23 (1996).
41. M. Ritala, *Appl. Surf. Sci.* 112, 223 (1997).
42. S. M. George, A. W. Ott, and J. W. Klaus, *J. Phys. Chem.* 100, 13,121 (1996).
43. L. Niinistö, Ed., *Acta Polytechnol. Scand., Ser. Chem. Technol. Metall.* 195 (1990).
44. S. M. Bedair, Ed., *Thin Solid Films* 225 (1993).
45. M. Ozeki, A. Usui, Y. Aoyagi, and J. Nishizawa, Eds., *Appl. Surf. Sci.* 82–83 (1994).
46. H. Sitter and H. Heinrich, Eds., *Appl. Surf. Sci.* 112 (1997).
47. T. Meguro, S. Iwai, Y. Aoyagi, K. Ozaki, Y. Yamamoto, T. Suzuki, Y. Okano, and A. Hirata, *J. Cryst. Growth* 99, 540 (1990).
48. P. D. Dapkus, S. P. DenBaars, Q. Chen, W. G. Jeong, and B. Y. Maa, *Prog. Cryst. Growth Charact.* 19, 137 (1989).
49. M. Yoshimoto, A. Kajimoto, and H. Matsunami, *Thin Solid Films* 225, 70 (1993).
50. Q. Chen and P. D. Dapkus, *Thin Solid Films* 225, 115 (1993).
51. D. Lubben, R. Tsu, T. R. Bramblett, and J. E. Greene, *J. Vac. Sci. Technol. A* 9, 3003 (1991).
52. K. Saito, Y. Watanabe, K. Takahashi, T. Matsuzawa, B. Sang, and M. Konagai, *Solar Energy Mater. Solar Cells* 49, 187 (1997).
53. R. M. Emerson, J. L. Hoyt, and J. F. Gibbons, *Appl. Phys. Lett.* 65, 1103 (1994).
54. J. Murota, M. Sakubara, and S. Ono, *Appl. Phys. Lett.* 62, 2353 (1993).
55. M. Sakuraba, J. Murota, T. Watanabe, Y. Sawada, and S. Ono, *Appl. Surf. Sci.* 82–83, 354 (1994).
56. M. Sakuraba, J. Murota, N. Mikoshiba, and S. Ono, *J. Cryst. Growth* 115, 79 (1991).
57. J. Nishizawa, A. Murai, T. Ohizumi, T. Kurabayashi, K. Ohtsuka and T. Yoshida, *J. Cryst. Growth* 209, 327 (2000).
58. H. Yokoyama, M. Shinohara, and N. Inoue, *Appl. Phys. Lett.* 60, 377 (1992).
59. S. Yokoyama, N. Ikeda, K. Kajikawa, and Y. Nakashima, *Appl. Surf. Sci.* 130–132, 352 (1998).
60. D. D. Koleske and S. M. Gates, *J. Appl. Phys.* 76, 1615 (1994).
61. D. D. Koleske and S. M. Gates, *Appl. Phys. Lett.* 64, 884 (1994).
62. S. Sugahara, Y. Uchida, T. Kitamura, T. Nagai, M. Matsuyama, T. Hattori, and M. Matsumura, *Jpn. J. Appl. Phys.* 36, 1609 (1997).
63. J. Sumakeris, Z. Sitar, K. S. Ailey-Trent, K. L. More, and R. F. Davis, *Thin Solid Films* 225, 244 (1993).
64. S. Imai, T. Iizuka, O. Sugiura, and M. Matsumura, *Thin Solid Films* 225, 168 (1993).
65. S. Imai and M. Matsumura, *Appl. Surf. Sci.* 82–83, 322 (1994).
66. E. Hasunuma, S. Sugahara, S. Hoshino, S. Imai, K. Ikeda, and M. Matsumura, *J. Vac. Sci. Technol. A* 16, 679 (1998).
67. Y. Suda, Y. Misato, and D. Shiratori, *Jpn. J. Appl. Phys.* 38, 2390 (1999).
68. S. Sugahara, T. Kitamura, S. Imai, and M. Matsumura, *Appl. Surf. Sci.* 82–83, 380 (1994).
69. S. Sugahara, M. Kadoshima, T. Kitamura, S. Imai, and M. Matsumura, *Appl. Surf. Sci.* 90, 349 (1995).

70. M. de Keijser and C. van Opdorp, *Appl. Phys. Lett.* 58, 1187 (1991).

71. C.-Y. Hwang, P. Lu, W. E. Mayo, Y. Lu, and H. Liu, *Mater. Res. Soc. Symp. Proc.* 326, 347 (1994).

72. H. Goto, K. Shibahara, and S. Yokoyama, *Appl. Phys. Lett.* 68, 3257 (1996).

73. S. Yokoyama, H. Goto, T. Miyamoto, N. Ikeda, and K. Shibahara, *Appl. Surf. Sci.* 112, 75 (1997).

74. S. M. Rossnagel, A. Sherman, and F. Turner, *J. Vac. Sci. Technol. B* 18, 2016 (2000).

75. A. Sherman, private communication.

76. M. Ylilammi, *Thin Solid Films* 279, 124 (1996).

77. T. Suntola, *Thin Solid Films* 225, 96 (1993).

78. K. Kodama, M. Ozeki, Y. Sakuma, K. Mochizuki, and N. Ohtsuka, *J. Cryst. Growth* 99, 535 (1990).

79. M. Ishizaki, N. Kano, J. Yoshino, and H. Kukimoto, *Thin Solid Films* 225, 74 (1993).

80. M. Ozeki and N. Ohtsuka, *Appl. Surf. Sci.* 82–83, 233 (1994).

81. S. Hirose, M. Yamaura, A. Yoshida, H. Ibuka, K. Hara, and H. Munekata, *J. Cryst. Growth* 194, 16 (1998).

82. M. Ritala, M. Leskelä, L.-S. Johansson, and L. Niinistö, *Thin Solid Films* 228, 32 (1993).

83. M. Ritala and M. Leskelä, *Appl. Surf. Sci.* 75, 333 (1994).

84. M. Ritala, M. Leskelä, L. Niinistö, T. Prohaska, G. Friedbacher, and M. Grasserbauer, *Thin Solid Films* 250, 72 (1994).

85. J. Ihanus, M. Ritala, M. Leskelä, T. Prohaska, R. Resch, G. Friedbacher, and M. Grasserbauer, *Appl. Surf. Sci.* 120, 43 (1997).

86. T. Asikainen, M. Ritala, M. Leskelä, T. Prohaska, G. Friedbacher, and M. Grasserbauer, *Appl. Surf. Sci.* 99, 91 (1996).

87. S. Dey and S. J. Yun, *Appl. Surf. Sci.* 143, 191 (1999).

88. M. Ritala, M. Leskelä, J.-P. Dekker, C. Mutsaers, P. J. Soininen, and J. Skarp, *Chem. Vapor Deposition* 5, 7 (1999).

89. J. I. Skarp, P. J. Soininen, and P. T. Soininen, *Appl. Surf. Sci.* 112, 251 (1997).

90. *Eur. Semicond.* April 1999, p. 69.

91. B. T. McDermott, N. A. El-Masry, M. A. Tischler, and S. M. Bedair, *Appl. Phys. Lett.* 51, 1830 (1987).

92. K. Mori, S. Sugou, Y. Katol, and A. Usui, *Appl. Phys. Lett.* 60, 1717 (1992).

93. A. Dip, G. M. Eldallal, P. C. Colter, N. Hayafuji, and S. M. Bedair, *Appl. Phys. Lett.* 62, 2378 (1993).

94. M. Ylilammi, *J. Electrochem. Soc.* 142, 2474 (1995).

95. J. Aarik and H. Siimon, *Appl. Surf. Sci.* 81, 281 (1994).

96. H. Siimon, J. Aarik, and T. Uustare, *Electrochem. Soc. Proc.* 97-25, 131 (1997).

97. H. Siimon and J. Aarik, *J. Phys. D: Appl. Phys.* 30, 1725 (1997).

98. M. A. Tischler and S. M. Bedair, *Appl. Phys. Lett.* 48, 1681 (1986).

99. H. Liu, P. A. Zawadski, and P. E. Norris, *Thin Solid Films* 225, 105 (1993).

100. J. J. Sumakeris, L. B. Rowland, R. S. Kern, S. Tanaka, and R. F. Davis, *Thin Solid Films* 225, 219 (1993).

101. T. Ozawa, *Thermochim. Acta* 174, 185 (1991).

102. A. Niskanen, T. Hatanpää, M. Ritala, and M. Leskelä, *J. Thermal. Anal. Calorim.* 64, 955 (2001).

103. K.-E. Elers, M. Ritala, M. Leskelä, and E. Rauhala, *Appl. Surf. Sci.* 82–83, 468 (1994).

104. J. Aarik, A. Aidla, K. Kukli, and T. Uustare, *J. Cryst. Growth* 144, 116 (1994).

105. J. Aarik, K. Kukli, A. Aidla, and L. Pung, *Appl. Surf. Sci.* 103, 331 (1996).

106. K. Kukli, M. Ritala, R. Matero, and M. Leskelä, *J. Cryst. Growth* 212, 459 (2000).

107. J. Skarp, U.S. Patent 4,486,487, 1984.

108. J. Ihanus, M. Ritala, M. Leskelä, and E. Rauhala, *Appl. Surf. Sci.* 112, 154 (1997).

109. J. Ihanus, M. Ritala, and M. Leskelä, *Electrochem. Soc. Proc.* 97-25, 1423 (1997).

110. M. Juppo, M. Ritala, and M. Leskelä, *J. Vac. Sci. Technol. A* 15, 2330 (1997).

111. M. Ritala, M. Leskelä, E. Rauhala, and P. Haussalo, *J. Electrochem. Soc.* 142, 2731 (1995).

112. M. Ritala, T. Asikainen, M. Leskelä, J. Jokinen, R. Lappalainen, M. Utriainen, L. Niinistö, and E. Ristolainen, *Appl. Surf. Sci.* 120, 199 (1997).

113. M. Ritala, P. Kalsi, D. Riihelä, K. Kukli, M. Leskelä, and J. Jokinen, *Chem. Mater.* 11, 1712 (1999).

114. M. Juppo, M. Ritala, and M. Leskelä, *J. Electrochem. Soc.* 147, 3377 (2001).

115. T. Asikainen, M. Ritala, and M. Leskelä, *J. Electrochem. Soc.* 141, 3210 (1994).

116. W. Faschinger, P. Juza, S. Ferreira, H. Zajicek, A. Pesek, H. Sitter, and K. Lischka, *Thin Solid Films* 225, 270 (1993).

117. Y. Takemura, M. Konagai, K. Yamasaki, C. H. Lee, and K. Takahashi, *J. Electron. Mater.* 22, 437 (1993).

118. M. Konagai, Y. Takemura, K. Yamasaki, and K. Takahashi, *Thin Solid Films* 225, 256 (1993).

119. M. Konagai, Y. Ohtake, and T. Okamoto, *Mater. Res. Soc. Symp. Proc.* 426, 153 (1996).

120. A. Szczerbakow, E. Dynowska, K. Swiatek, and M. Godlewski, *J. Cryst. Growth* 207, 148 (1999).

121. R. Kimura, M. Konagai, and K. Takahashi, *J. Cryst. Growth* 116, 283 (1992).

122. F. Hauzenberger, W. Faschinger, P. Juza, A. Pesek, K. Lischka, and H. Sitter, *Thin Solid Films* 225, 265 (1993).

123. A. Rautiainen, Y. Koskinen, J. Skarp, and S. Lindfors, *Mater. Res. Soc. Symp. Proc.* 222, 263 (1991).

124. M. Ahonen, M. Pessa, and T. Suntola, *Thin Solid Films* 65, 301 (1980).

125. M. Pessa, P. Huttunen, and A. Herman, *J. Appl. Phys.* 54, 6047 (1983).

126. A. Kytökivi, Y. Koskinen, A. Rautiainen, and J. Skarp, *Mater. Res. Soc. Symp. Proc.* 222, 269 (1991).

127. H. Sitter and W. Faschinger, *Thin Solid Films* 225, 250 (1993).

128. J. Aarik, A. Aidla, A. Jaek, A.-A. Kiisler, and A.-A. Tammik, *Acta Polytechnol. Scand., Ser. Chem. Technol. Metall.* 195, 201 (1990).

129. L. Hiltunen, H. Kattelus, M. Leskelä, M. Mäkelä, L. Niinistö, E. Nykänen, P. Soininen, and M. Tiitta, *Mater. Chem. Phys.* 28, 379 (1991).

130. H. Kattelus, M. Ylilammi, J. Saarilahti, J. Antson, and S. Lindfors, *Thin Solid Films* 225, 296 (1993).

131. H. Kattelus, M. Ylilammi, J. Salmi, T. Ranta-aho, E. Nykänen, and I. Suni, *Mater. Res. Soc. Symp. Proc.* 284, 511 (1993).

132. M. Ritala, H. Saloniemi, M. Leskelä, T. Prohaska, G. Friedbacher, and M. Grasserbauer, *Thin Solid Films* 286, 54 (1996).

133. K. Kukli, M. Ritala, and M. Leskelä, *J. Electrochem. Soc.* 144, 300 (1997)

134. S. J. Yun, K.-H. Lee, J. Skarp, H. R. Kim, and K.-S. Nam, *J. Vac. Sci. Technol. A* 15, 2993 (1997).

135. Y. S. Kim, K.-L. Cho, and S. J. Yun, "5th International Conference on the Science and Technology of Display Phosphors," San Diego, 1999, p. 97, Extended Abstracts.

136. K. Kukli, M. Ritala, and M. Leskelä, *J. Electrochem. Soc.* 148, F35 (2001).

137. G. Oya, M. Yoshida, and Y. Sawada, *Appl. Phys. Lett.* 51, 1143 (1987).

138. G. Oya and Y. Sawada, *J. Cryst. Growth* 99, 572 (1990).

139. M. Leskelä, L. Niinistö, E. Nykänen, P. Soininen, and M. Tiitta, *Acta Polytechnol. Scand., Ser. Chem. Technol. Metall.* 195, 193 (1990).

140. M. Nieminen, L. Niinistö, and R. Lappalainen, *Mikrochim. Acta* 119, 13 (1995).

141. M. Tiitta, E. Nykänen, P. Soininen, L. Niinistö, M. Leskelä, and R. Lappalainen, *Mater. Res. Bull.* 33, 1315 (1998).

142. K.-E. Elers, M. Ritala, M. Leskelä, and L.-S. Johansson, *J. Phys. IV* 5, C5-1021 (1995).

143. M. Akamutsu, S. Narahara, T. Kobayashi, and F. Hagesawa, *Appl. Surf. Sci.* 82–83, 228 (1994).

144. A. Koukitu, Y. Kumagai, T. Taki, and H. Seki, *Japan. J. Appl. Phys.* 38, 4980 (1999).

145. A. Usui, H. Sunakawa, F. J. Stützler, and K. Ishida, *Appl. Phys. Lett.* 56, 289 (1990).

146. H. Ikeda, Y. Miura, N. Takahashi, A. Koukitu, and H. Seki, *Appl. Surf. Sci.* 82–83, 257 (1994).

147. T. Taki, T. Nakajima, A. Koukitu, and H. Seki, *J. Cryst. Growth* 183, 75 (1998).

148. J. Ahopelto, H. P. Kattelus, J. Saarilahti, and I. Suni, *J. Cryst. Growth* 99, 550 (1990).

149. K. Nishi, A. Usui, and H. Sakaki, *Thin Solid Films* 225, 47 (1993).

150. N. Takahashi, M. Yagi, A. Koukitu, and H. Seki, *Japan. J. Appl. Phys.* 32, L1277 (1993).

151. A. Koukitu, N. Takahashi, and H. Seki, *J. Cryst. Growth* 146, 467 (1995).

152. A. Koukitu, T. Taki, K. Norita, and H. Seki, *J. Cryst. Growth* 198–199, 1111 (1999).

153. A. Koukitu, H. Nakai, A. Saegusa, T. Suzuki, O. Nomura, and H. Seki, *Japan. J. Appl. Phys.* 27, L744 (1988).

154. H. Tsuchiya, M. Akamutsu, M. Ishida, and F. Hagesawa, *Japan. J. Appl. Phys.* 35, L748 (1996).

155. R. Kobayashi, K. Ishikawa, S. Narahara, and F. Hagesawa, *Japan. J. Appl. Phys.* 31, L1730 (1992).

156. R. Kobayashi, S. Narahara, K. Ishikawa, and F. Hagesawa, *Japan. J. Appl. Phys.* 32, L164 (1992).

157. N. Tsuboi, T. Isu, N. Kakuda, T. Terasako, and S. Iida, *Japan. J. Appl. Phys.* 33, L244 (1994).

158. T. Taki and A. Koukitu, *Appl. Surf. Sci.* 112, 127 (1997).

159. M. Ritala, T. Asikainen, and M. Leskelä, *Electrochem. Solid-State Lett.* 1, 156 (1998).

160. T. Asikainen, M. Ritala, and M. Leskelä, *J. Electrochem. Soc.* 142, 3538 (1995).

161. T. Asikainen, M. Ritala, and M. Leskelä, *Appl. Surf. Sci.* 82–83, 122 (1994).

162. S. M. George, O. Sneh, A. C. Dillon, M. L. Wise, A. W. Ott, L. A. Okada, and J. D. Way, *Appl. Surf. Sci.* 82–83, 460 (1994).

163. O. Sneh, M. L. Wise, A. W. Ott, L. A. Okada, and S. M. George, *Surf. Sci.* 334, 135 (1995).

164. J. W. Klaus, A. W. Ott, J. M. Johnson, and S. M. George, *Appl. Phys. Lett.* 70, 1092 (1997).

165. J. W. Klaus, O. Sneh, and S. M. George, *Science* 278, 1934 (1997).

166. J. W. Klaus, O. Sneh, A. W. Ott, and S. M. George, *Surf. Rev. Lett.* 6, 435 (1999).

167. J. W. Klaus and S. M. George, *Surf. Sci.* 447, 81 (2000).

168. J. W. Klaus, A. W. Ott, A. C. Dillon, and S. M. George, *Surf. Sci.* 418, L14 (1998).

169. P. Soininen, L. Niinistö, E. Nykänen, and M. Leskelä, *Appl. Surf. Sci.* 75, 99 (1994).

170. D. D. Koleske and S. M. Gates, *Appl. Surf. Sci.* 82–83, 344 (1994).

171. S. Morishita, S. Sugahara, and M. Matsumura, *Appl. Surf. Sci.* 112, 198 (1997).

172. M. Ylilammi, M.Sc. Thesis, Helsinki University of Technology, 1979.

173. H. Viirola and L. Niinistö, *Thin Solid Films* 249, 144 (1994).

174. M. Utriainen, K. Kovács, J. M. Campbell, L. Niinistö, and F. Réti, *J. Electrochem. Soc.* 146, 189 (1999).

175. H. Viirola and L. Niinistö, *Thin Solid Films* 251, 127 (1994).

176. S. B. Desu, *Mater. Sci. Eng. B* 13, 299 (1992).

177. M. Ritala, M. Leskelä, E. Nykänen, P. Soininen, and L. Niinistö, *Thin Solid Films* 225, 288 (1993).

178. J. Aarik, A. Aidla, T. Uustare, and V. Sammelselg, *J. Cryst. Growth* 148, 268 (1995).

179. J. Aarik, A. Aidla, V. Sammelselg, H. Siimon, and T. Uustare, *J. Cryst. Growth* 169, 496 (1996).

180. J. Aarik, A. Aidla, A.-A. Kiisler, T. Uustare, and V. Sammelselg, *Thin Solid Films* 305, 270 (1997).

181. V. Sammelselg, A. Rosental, A. Tarre, L. Niinistö, K. Heiskanen, K. Ilmonen, L.-S. Johansson, and T. Uustare, *Appl. Surf. Sci.* 134, 78 (1998).

182. A. Rosental, A. Tarre, P. Adamson, A. Gerst, A. Kasikov, and A. Niilisk, *Appl. Surf. Sci.* 142, 204 (1999).

183. H. Kumagai, M. Matsumoto, K. Toyoda, M. Obara, and M. Suzuki, *Thin Solid Films* 263, 47 (1995).

184. H. Kumagai, K. Toyoda, K. Kobayashi, M. Obara, and Y. Iimura, *Appl. Phys. Lett.* 70, 2338 (1997).

185. L. Hiltunen, M. Leskelä, M. Mäkelä, L. Niinistö, E. Nykänen, and P. Soininen, *Thin Solid Films* 166, 149 (1988).

186. P. Mårtensson, M. Juppo, M. Ritala, M. Leskelä, and J.-O. Carlsson, *J. Vac. Sci. Technol. B* 17, 2122 (1999).

187. K. Kukli, M. Ritala, M. Schuisky, M. Leskelä, T. Sajavaara, J. Keinonen, T. Uustare, and A. Hårsta, *Chem. Vapor Deposition* 6, 297 (2000).

188. K. Kukli, A. Aidla, J. Aarik, M. Schuisky, A. Hårsta, M. Ritala, and M. Leskelä, *Langmuir* 16, 8122 (2000).

189. M. Schuisky, A. Hårsta, A. Aidla, K. Kukli, A.-A. Kiisler, and J. Aarik, *J. Electrochem. Soc.* 147, 3319 (2000).

190. M. Schuisky, K. Kukli, A. Aidla, J. Aarik, M. Ludvigsson, and A. Hårsta, *Electrochem. Soc. Proc.* 2000-13, 637 (2000).

191. M. Ritala, M. Leskelä, E. Rauhala, and J. Jokinen, *J. Electrochem. Soc.* 145, 2914 (1998).

192. K. Kukli, M. Ritala, and M. Leskelä, *Nanostruct. Mater.* 8, 785 (1997).

193. M. Copel, M. Gibelyuk, and E. Gusev, *Appl. Phys. Lett.* 76, 436 (2000).

194. E. P. Gusev, M. Copel, E. Cartier, D. Buchanan, H. Okorn-Schmidt, M. Gribelyuk, D. Falcon, R. Murphy, S. Molis, I. J. R. Baumvol, C. Krug, M. Jussila, M. Tuominen, and S. Haukka, *Electrochem. Soc. Proc.* 2000-2, 477 (2000).

195. H. Zhang, R. Solanki, B. Roberds, G. Bai, and I. Banerjee, *J. Appl. Phys.* 87, 1921 (2000).

196. K. Kukli, K. Forsgren, J. Aarik, T. Uustare, A. Aidla, A. Niskanen, M. Ritala, M. Leskelä, and A. Hårsta, to be published.

197. K. Kukli, J. Ihanus, M. Ritala, and M. Leskelä, *Appl. Phys. Lett.* 68, 3737 (1996).

198. J. Aarik, A. Aidla, A.-A. Kiisler, T. Uustare, and V. Sammelselg, *Thin Solid Films* 340, 110 (1999).

199. M. Pessa, R. Mäkelä, and T. Suntola, *Appl. Phys. Lett.* 38, 131 (1981).

200. K. Kukli, J. Aarik, A. Aidla, O. Kohan, T. Uustare, and V. Sammelselg, *Thin Solid Films* 260, 135 (1995).

201. H. Siimon and J. Aarik, *J. Phys. IV* 5, C5-277 (1995).

202. K. Kukli, M. Ritala, and M. Leskelä, *J. Appl. Phys.* 86, 5656 (1999).

203. K. Forsggren, J. Sundqvist, A. Hårsta, K. Kukli, J. Aarik, and A. Aidla, *Electrochem. Soc. Proc.* 2000-13, 645 (2000).

204. M. Juppo, M. Vehkamäki, M. Ritala, and M. Leskelä, *J. Vac. Sci. Technol. A* 16, 2845 (1998).

205. P. Tägtsröm, P. Mårtensson, U. Jansson, and J.-O. Carlsson, *J. Electrochem. Soc.* 146, 3139 (1999).

206. J. W. Klaus, S. J. Ferro, and S. M. George, *Thin Solid Films* 360, 145 (2000).

207. J. W. Klaus, S. J. Ferro, and S. M. George, *J. Electrochem. Soc.* 147, 1175 (2000).

208. M. Tammenmaa, M.Sc. Thesis, Helsinki University of Technology, 1983.

209. P. Mårtensson and J.-O. Carlsson, *Chem. Vapor Deposition* 3, 45 (1997).

210. P. Mårtensson, "Acta Universitatis Upsaliensis, Comprehensive Summaries of Uppsala Dissertations from the Faculty of Science and Technology," Vol. 421, 1999.

211. J. Hyvärinen, M. Sonninen, and R. Törnqvist, *J. Cryst. Growth* 86, 695 (1988).

212. M. Oikkonen, *Mater. Res. Bull.* 23, 133 (1988).

213. A. Szczerbakow, M. Godlewski, E. Dynowska, V. Yu. Ivanov, K. Swiatek, E. M. Goldys, and M. R. Phillips, *Acta Phys. Pol. A* 94, 579 (1998).

214. C. D. Lee, B. K. Kim, J. W. Kim, H. L. Park, C. H. Chung, S. K. Chang, J. I. Lee, and S. K. Noh, *J. Cryst. Growth* 138, 136 (1994).

215. M. Ylilammi and T. Ranta-aho, *J. Electrochem. Soc.* 141, 1278 (1994).

216. G. S. Higashi and C. G. Flemming, *Appl. Phys. Lett.* 55, 1963 (1989).

217. V. E. Drozd, A. P. Baraban, and I. O. Nikiforova, *Appl. Surf. Sci.* 82–83, 583 (1994).

218. A. W. Ott, K. C. McCarley, J. W. Klaus, J. D. Way, and S. M. George, *Appl. Surf. Sci.* 107, 128 (1996).

219. A. W. Ott, J. W. Klaus, J. M. Johnson, and S. M. George, *Thin Solid Films* 292, 135 (1997).

220. P. Ericsson, S. Begtsson, and J. Skarp, *Microelectron. Eng.* 36, 91 (1997).

221. Y. Kim, S. M. Lee, C. S. Park, S. I. Lee, and M. Y. Lee, *Appl. Phys. Lett.* 71, 3604 (1997).

222. E. P. Gusev, M. Copel, E. Cartier, I. J. R. Baumvol, C. Krug, and M. A. Gribelyuk, *Appl. Phys. Lett.* 76, 176 (2000).

223. M. Juppo, A. Rahtu, M. Ritala, and M. Leskelä, *Langmuir* 16, 4034 (2000).

224. R. Matero, A. Rahtu, M. Ritala, M. Leskelä, and T. Sajavaara, *Thin Solid Films* 368, 1 (2000).

225. J.-F. Fan, K. Sugioka, and K. Toyoda, *Japan. J. Appl. Phys.* 30, L1139 (1991).

226. J.-F. Fan and K. Toyoda, *Appl. Surf. Sci.* 60–61, 765 (1992).

227. J.-F. Fan and K. Toyoda, *Japan. J. Appl. Phys.* 32, L1349 (1993).

228. H. Kumagai, K. Toyoda, M. Matsumoto, and M. Obara, *Japan. J. Appl. Phys.* 32, 6137 (1993).

229. H. Kumagai and K. Toyoda, *Appl. Surf. Sci.* 82–83, 481 (1994).

230. T. M. Mayer, J. W. Rogers, Jr., and T. A. Michalske, *Chem. Mater.* 3, 641 (1993).

231. M. E. Bartram, T. A. Michalske, J. W. Rogers, Jr., and R. T. Paine, *Chem. Mater.* 5, 1424 (1993).

232. H. Liu, D. C. Bertolet, and J. W. Rogers, Jr., *Surf. Sci.* 340, 88 (1995).

233. D. Riihelä, M. Ritala, R. Matero, M. Leskelä, J. Jokinen, and P. Haussalo, *Chem. Vapor Deposition* 2, 277 (1996).

234. J. R. Gong, S. Nakamura, M. Leonard, S. M. Bedair, and N. A. El-Masry, *J. Electron. Mater.* 21, 965 (1992).

235. J. R. Gong, P. C. Colter, D. Jung, S. A. Hussien, C. A. Parker, A. Dip, F. Hyuga, W. M. Duncan, and S. M. Bedair, *J. Cryst. Growth* 107, 83 (1991).

236. N. Hayafuji, G. M. Eldallal, A. Dip, P. C. Colter, N. A. El-Masry, and S. M. Bedair, *Appl. Surf. Sci.* 82–83, 18 (1994).

237. K. Kukli, M. Ritala, M. Leskelä, and J. Jokinen, *J. Vac. Sci. Technol. A* 15, 2214 (1997).

238. A. Niskanen, M. Ritala, and M. Leskelä, unpublished results.

239. M. Ishii, S. Iwai, H. Kawata, T. Ueki, and Y. Aoyagi, *J. Cryst. Growth* 180, 15 (1997).

240. S. Hirose, N. Kano, M. Deura, K. Hara, H. Munekata, and H. Kukimoto, *Japan. J. Appl. Phys.* 34, L1436 (1995).

241. R. Huang and A. H. Kitai, *J. Electron. Mater.* 22, 215 (1993).

242. M. Asif Khan, J. N. Kuznia, R. A. Skogman, D. T. Olson, M. MacMillan, and W. J. Choyke, *Appl. Phys. Lett.* 61, 2539 (1992).

243. M. Asif Khan, J. N. Kuznia, D. T. Olson, T. George, and W. T. Pike, *Appl. Phys. Lett.* 63, 3470 (1993).

244. D. E. Aspnes, I. Kamiya, H. Tanaka, R. Bhat, L. T. Florez, J. P. Harbison, W. E. Quinn, M. Tamargo, S. Gregory, M. A. A. Pudensi, S. A. Schwarz, M. J. S. P. Brasil, and R. E. Nahory, *Thin Solid Films* 225, 26 (1993).

245. N. H. Karam, T. Parodos, P. Colter, D. McNulty, W. Rowland, J. Schetzina, N. El-Masry, and S. M. Bedair, *Appl. Phys. Lett.* 67, 94 (1995).

246. K. S. Boutros, F. G. McIntosh, J. C. Roberts, S. M. Bedair, E. L. Piner, and N. A. El-Masry, *Appl. Phys. Lett.* 67, 1856 (1995).

247. E. L. Piner, M. K. Behbehani, N. A. El-Masry, F. G. McIntosh, J. C. Roberts, K. S. Boutros, and S. M. Bedair, *Appl. Phys. Lett.* 70, 461 (1997).

248. F. G. McIntosh, E. L. Piner, J. C. Roberts, M. K. Behbehani, M. E. Aumer, N. A. El-Masry, and S. M. Bedair, *Appl. Surf. Sci.* 112, 98 (1997).

249. S.-C. Huang, H.-Y. Wang, C.-J. Hsu, J.-R. Gong, C.-I. Chiang, S. L. Tu, and H. Chang, *J. Mater. Sci. Lett.* 17, 1281 (1998).

250. B. T. McDermott, K. G. Reid, N. A. El-Masry, S. M. Bedair, W. M. Duncan, X. Yin, and F. H. Pollak, *Appl. Phys. Lett.* 56, 1172 (1990).

251. B. T. McDermott, N. A. El-Masry, B. L. Jiang, F. Hyuga, S. M. Bedair, and W. M. Duncan, *J. Cryst. Growth* 107, 96 (1991).

252. Y. Sakuma, M. Ozeki, N. Ohtsuka, and K. Kodama, *J. Appl. Phys.* 68, 5660 (1990).

253. M. Ozeki, K. Mochizuki, N. Ohtsuka, and K. Kodama, *Appl. Phys. Lett.* 53, 1509 (1988).

254. K. Mochizuki, M. Ozeki, K. Kodama, and N. Ohtsuka, *J. Cryst. Growth* 93, 557 (1988).

255. J. Nishizawa and T. Kurabayashi, *Appl. Surf. Sci.* 106, 11 (1996).

256. K. Mukai, N. Ohtsuka, H. Shoji, and M. Sugawara, *Appl. Surf. Sci.* 112, 102 (1997).

257. B. Y. Maa and P. D. Dapkus, *Appl. Phys. Lett.* 58, 2261 (1991).

258. R. Arès, S. P. Watkins, P. Yeo, G. A. Horley, P. O'Brien, and A. C. Jones, *J. Appl. Phys.* 83, 3390 (1998).

259. M. Asif Khan, R. A. Skogman, J. M. Van Hove, D. T. Olson, and J. N. Kuznia, *Appl. Phys. Lett.* 60, 1366 (1992).

260. M. Asif Khan, J. N. Kuznia, D. T. Olson, J. M. Van Hove, M. Blasingame, and L. F. Reitz, *Appl. Phys. Lett.* 60, 2917 (1992).

261. M. Ait-Lhouss, J. L. Castaño, B. J. Garcia, and J. Piqueras, *J. Appl. Phys.* 78, 5834 (1995).

262. I. Suemune, *Appl. Surf. Sci.* 82–83, 149 (1994).

263. K. Mori, M. Yoshida, A. Usui, and H. Terao, *Appl. Phys. Lett.* 52, 27 (1988).

264. A. W. Ott, J. M. Johnson, J. W. Klaus, and S. M. George, *Appl. Surf. Sci.* 112, 205 (1997).

265. W. K. Chen, J. C. Chen, L. Anthony, and P. L. Liu, *Appl. Phys. Lett.* 55, 987 (1989).

266. Y. Kobayashi and N. Kobayashi, *Japan. J. Appl. Phys.* 31, L71 (1992).

267. N. Pan, J. Carter, S. Hein, D. Howe, L. Goldman, L. Kupferberg, S. Brierley, and K. C. Hsieh, *Thin Solid Films* 225, 64 (1993).

268. N. Otsuka, J. Nishizawa, H. Kikuchi, and Y. Oyama, *J. Cryst. Growth* 205, 253 (1999).

269. W. G. Jeong, E. P. Menu, and P. D. Dapkus, *Appl. Phys. Lett.* 55, 244 (1989).

270. C. A. Tran, R. Ares, S. P. Watkins, G. Soerensen, and Y. Lacroix, *J. Electron. Mater.* 24, 1597 (1995).

271. V. E. Drozd and V. B. Aleskovski, *Appl. Surf. Sci.* 82–83, 591 (1994).

272. V. Lujala, J. Skarp, M. Tammenmaa, and T. Suntola, *Appl. Surf. Sci.* 82–83, 34 (1994).

273. B. W. Sanders and A. Kitai, *Chem. Mater.* 4, 1005 (1992).

274. A. Hunter and A. H. Kitai, *J. Cryst. Growth* 91, 111 (1988).

275. C. H. Liu, M. Yokoyama, and Y. K. Su, *Japan. J. Appl. Phys.* 35, 5416 (1996).

276. C. T. Hsu, *Thin Solid Films* 335, 284 (1998).

277. C. T. Hsu, *J. Cryst. Growth* 193, 33 (1998).

278. C. T. Hsu, *Mater. Chem. Phys.* 58, 6 (1999).

279. A. Yoshikawa, M. Kobayashi, and S. Tokita, *Appl. Surf. Sci.* 82–83, 316 (1994).

280. I. Bhat and S. Akram, *J. Cryst. Growth* 138, 127 (1994).

281. C.-T. Hsu, *Japan. J. Appl. Phys.* 35, 4476 (1996).

282. M. Yokoyama, N. T. Chen, and H. Y. Ueng, *J. Cryst. Growth* 212, 97 (2000).

283. W.-S. Wang, H. Ehsani, and I. Bhat, *J. Electron. Mater.* 22, 873 (1993).

284. B. Sang and M. Konagai, *Japan. J. Appl. Phys.* 35, L602 (1996).

285. S. Chaisitsak, T. Sugiyama, A. Yamada, and M. Konagai, *Japan. J. Appl. Phys.* 38, 4989 (1999).

286. A. Shimizu, S. Chaisitsak, T. Sugiyama, A. Yamada, and M. Konagai, *Thin Solid Films* 361–362, 193 (2000).

287. E. B. Yousfi, J. Fouache, and D. Lincot, *Appl. Surf. Sci.* 153, 223 (2000).

288. E. B. Yousfi, T. Asikainen, V. Pietu, P. Cowache, M. Powalla, and D. Lincot, *Thin Solid Films* 361–362, 183 (2000).

289. A. Yamada, B. Sang, and M. Konagai, *Appl. Surf. Sci.* 112, 216 (1997).

290. B. Sang, A. Yamada, and M. Konagai, *Solar Energy Mater. Solar Cells* 49, 19 (1997).

291. B. Sang, A. Yamada, and M. Konagai, *Japan. J. Appl. Phys.* 37, L1125 (1998).

292. B. Sang, K. Dairiki, A. Yamada, and M. Konagai, *Japan. J. Appl. Phys.* 38, 4983 (1999).

293. A. W. Ott and R. P. H. Chang, *Mater. Chem. Phys.* 58, 132 (1999).

294. B. Sang, A. Yamada, and M. Konagai, *Japan. J. Appl. Phys.* 37, L206 (1998).

295. E. Soininen, G. Härkönen, and K. Vasama, "3rd International Conference on the Science and Technology of Display Phosphors," Huntington Beach, CA, 1997, p. 105, Extended Abstracts.

296. S. Moehnke, M. Bowen, S.-S. Sun, and R. Tuenge, "4th International Conference on the Science and Technology of Display Phosphors," Bend, OR, 1998, p. 203, Extended Abstracts.

297. H. Fujiwara, H. Kiryu, and I. Shimizu, *J. Appl. Phys.* 77, 3927 (1995).

298. Y. Luo, D. Slater, M. Han, J. Moryl, and R. M. Osgood, Jr., *Appl. Phys. Lett.* 71, 3799 (1997).

299. Y. Luo, D. Slater, M. Han, J. Moryl, R. M. Osgood, Jr., and J. G. Chen, *Langmuir* 14, 1493 (1998).

300. M. Han, Y. Luo, J. E. Moryl, R. M. Osgood, Jr., and J. G. Chen, *Surf. Sci.* 415, 251 (1998).

301. M. Han, Y. Luo, J. E. Moryl, and R. M. Osgood, Jr., *Surf. Sci.* 425, 259 (1999).

302. Y. Luo, M. Han, D. A. Slater, and R. M. Osgood, Jr., *J. Vac. Sci. Technol. A* 18, 438 (2000).

303. N. H. Karam, R. G. Wolfson, I. B. Bhat, H. Ehsani, and S. K. Ghandhi, *Thin Solid Films* 225, 261 (1993).

304. M. Ritala, M. Leskelä, and E. Rauhala, *Chem. Mater.* 6, 556 (1994).

305. J. Aarik, A. Aidla, V. Sammelselg, M. Ritala, and M. Leskelä, *Thin Solid Films* 370, 163 (2000).

306. H. Döring, K. Hashimoto, and A. Fujishima, *Ber. Bunsenges. Phys. Chem.* 96, 620 (1992).

307. M. Ritala, M. Leskelä, L. Niinistö, and P. Haussalo, *Chem. Mater.* 5, 1174 (1993).

308. J. Aarik, A. Aidla, T. Uustare, M. Ritala, and M. Leskelä, *Appl. Surf. Sci.* 161, 385 (2000).

309. M. Schuisky, K. Kukli, M. Ritala, A. Hårsta, and M. Leskelä, *Chem. Vapor Deposition* 6, 139 (2000).

310. K. Kukli, M. Ritala, and M. Leskelä, *Chem. Vapor Deposition* 6, 297 (2000).

311. K. Kukli, M. Ritala, M. Leskelä, and R. Lappalainen, *Chem. Vapor Deposition* 4, 29 (1998).

312. K. Kukli, M. Ritala, and M. Leskelä, *J. Electrochem. Soc.* 142, 1670 (1995).

313. K. Kukli, J. Aarik, A. Aidla, H. Siimon, M. Ritala, and M. Leskelä, *Appl. Surf. Sci.* 112, 236 (1997).

314. H.-J. Song, W. Koh, and S.-W. Kang, *Mater. Res. Soc. Symp. Proc.* 567, 469 (1999).

315. E. Nykänen, J. Laine-Ylijoki, P. Soininen, L. Niinistö, M. Leskelä, and L. G. Hubert-Pfalzgraf, *J. Mater. Chem.* 4, 1409 (1994).

316. T. Hatanpää, J. Ihanus, J. Kansikas, I. Mutikainen, M. Ritala, and M. Leskelä, *Chem. Mater.* 11, 1846 (1999).

317. M. Putkonen, L.-S. Johansson, E. Rauhala, and L. Niinistö, *J. Mater. Chem.* 9, 2449 (1999).

318. M. Tammenmaa, H. Antson, M. Asplund, L. Hiltunen, M. Leskelä, and L. Niinistö, *J. Cryst. Growth* 84, 151 (1987).

319. M. Leskelä, M. Mäkelä, L. Niinistö, E. Nykänen, and M. Tammenmaa, *Chemtronics* 3, 113 (1988).

320. J. Rautanen, M. Leskelä, L. Niinistö, E. Nykänen, P. Soininen, and M. Utriainen, *Appl. Surf. Sci.* 82–83, 553 (1994).

321. T. Hänninen, I. Mutikainen, V. Saanila, M. Ritala, M. Leskelä, and J. C. Hanson, *Chem. Mater.* 9, 1234 (1997).

322. M. Putkonen, A. Kosola, and L. Niinistö, private communication.

323. M. Leskelä, L. Niinistö, E. Nykänen, P. Soininen, and M. Tiitta, *Mater. Res. Soc. Symp. Proc.* 222, 315 (1991).

324. J. Aarik, A. Aidla, A. Jaek, M. Leskelä, and L. Niinistö, *J. Mater. Chem.* 4, 1239 (1994).

325. P. Soininen, E. Nykänen, L. Niinistö, and M. Leskelä, *Chem. Vapor Deposition* 2, 69 (1996).

326. V. Saanila, J. Ihanus, M. Ritala, and M. Leskelä, *Chem. Vapor Deposition* 4, 227 (1998).

327. M. Nieminen, L. Niinistö, and E. Rauhala, *J. Mater. Chem.* 6, 27 (1996).

328. H. Mölsä, L. Niinistö, and M. Utriainen, *Adv. Mater. Opt. Electron.* 4, 389 (1994).

329. M. Putkonen, T. Sajavaara, L.-S. Johansson, and L. Niinistö, *Chem. Vapor Deposition* 7, 44 (2001).

330. K. Kukli, M. Peussa, L.-S. Johansson, E. Nykänen, and L. Niinistö, *Mater. Sci. Forum* 315–317, 216 (1999).

331. K. Kukli, H. Heikkinen, E. Nykänen, and L. Niinistö, *J. Alloys Compd.* 275–277, 10 (1998).

332. H. Seim, M. Nieminen, L. Niinistö, H. Fjellvåg, and L.-S. Johansson, *Appl. Surf. Sci.* 112, 243 (1997).

333. H. Seim, H. Mölsä, M. Nieminen, H. Fjellvåg, and L. Niinistö, *J. Mater. Chem.* 7, 449 (1997).

334. O. Nilsen, M. Peussa, H. Fjellvåg, L. Niinistö, and A. Kjekshus, *J. Mater. Chem.* 9, 1781 (1999).

335. H. Mölsä and L. Niinistö, *Mater. Res. Soc. Symp. Proc.* 335, 341 (1994).

336. M. Tammenmaa, M. Leskelä, T. Koskinen, and L. Niinistö, *J. Less-Common Met.* 126, 209 (1986).

337. M. Leppänen, M. Leskelä, L. Niinistö, E. Nykänen, P. Soininen, and M. Tiitta, *SID 91 Dig.* 22, 282 (1991).

338. M. Leppänen, G. Härkönen, A. Pakkala, E. Soininen, and R. Törnqvist, "Conference Proceedings Eurodisplay 93," Starsbourg, 1993, p. 229.

339. T. Leskelä, K. Vasama, G. Härkönen, P. Sarkio, and M. Lounasmaa, *Adv. Mater. Opt. Electron.* 6, 169 (1996).

340. E. Nykänen, P. Soininen, L. Niinistö, M. Leskelä, and E. Rauhala, "Proceedings of the 7th International Workshop on Electroluminescence," Beijing, 1995, p. 437.

341. M. Utriainen, M. Kröger-Laukkanen, and L. Niinistö, *Mater. Sci. Eng. B* 54, 98 (1998).

342. M. Utriainen, M. Kröger-Laukkanen, L.-S. Johansson, and L. Niinistö, *Appl. Surf. Sci.* 157, 151 (2000).

343. P. Mårtensson and J.-O. Carlsson, *J. Electrochem. Soc.* 145, 2929 (1998).

344. P. Mårtensson and J.-O. Carlsson, *Electrochem. Soc. Proc.* 97-25, 1529 (1997).

345. P. Soininen, E. Nykänen, M. Leskelä, and L. Niinistö, unpublished results.

346. K. Swiatek, M. Godlewski, M. Leskelä, and L. Niinistö, *Acta Phys. Pol. A* 79, 255 (1991).

347. M. Leskelä, L. Niinistö, E. Nykänen, P. Soininen, and M. Tiitta, *J. Less-Common Met.* 153, 219 (1989).

348. K. Swiatek, M. Godlewski, M. Leskelä, and L. Niinistö, *J. Appl. Phys.* 74, 3442 (1993).

349. K. Swiatek, A. Suchocki, A. Stapor, L. Niinistö, and M. Leskelä, *J. Appl. Phys.* 66, 6048 (1989).

350. G. Härkönen, K. Härkönen, and R. Törnqvist, *SID 90 Dig.* 21, 232 (1990).

351. S. J. Yun, S.-D. Nam, J. S. Kang, K.-S. Nam, and H.-M. Park, "3rd International Conference on the Science and Technology of Display Phosphors," Huntington Beach, CA, 1997, p. 167, Extended Abstracts.

352. M. Asplund, J. Hölsä, M. Leskelä, L. Niinistö and E. Nykänen, *Inorg. Chim. Acta* 139, 261 (1987).

353. R. Huang and A. H. Kitai, *Appl. Phys. Lett.* 61, 1450 (1992).

354. R. Huang and A. H. Kitai, *J. Mater. Sci. Lett.* 12, 1444 (1993).

355. M. Putkonen, T. Sajavaara, and L. Niinistö, *J. Mater. Chem.* 10, 1857 (2000).

356. M. Vehkamäki, T. Hatanpää, T. Hänninen, M. Ritala, and M. Leskelä, *Electrochem. Solid-State Lett.* 2, 504 (1999).

357. M. Ritala, K. Kukli, M. Vehkamäki, T. Hänninen, T. Hatanpää, P. I. Räisänen, and M. Leskelä, *Proc. Electrochem. Soc.* 2000-13, 597 (2000).

358. M. Vehkamäki, T. Hänninen, M. Ritala, M. Leskelä, T. Sajavaara, E. Rauhala, and J. Keinonen, *Chem. Vapor Deposition* 7, 75 (2001).

359. J. Ihanus, T. Hänninen, T. Hatanpää, T. Aaltonen, I. Mutikainen, T. Sajavaara, J. Keinonen, M. Ritala, and M. Leskelä, submitted for publication.

360. G. Härkönen, M. Lahonen, E. Soininen, R. Törnqvist, K. Vasama, K. Kukli, L. Niinistö, and E. Nykänen, "4th International Conference on the Science and Technology of Display Phosphors," Bend, OR, 1998, p. 223, Extended Abstracts.

361. M. Tammenmaa, T. Koskinen, L. Hiltunen, L. Niinistö, and M. Leskelä, *Thin Solid Films* 124, 125 (1985).

362. M. Oikkonen, M. Blomberg, T. Tuomi, and M. Tammenmaa, *Thin Solid Films* 124, 317 (1985).

363. K. Kobayashi and S. Okudaira, *Chem. Lett.* 511 (1997).

364. L. Hiltunen, M. Leskelä, M. Mäkelä, and L. Niinistö, *Acta Chem. Scand. A* 41, 548 (1987).

365. H. Liu and J. W. Rogers, Jr., *J. Vac. Sci. Technol. A* 17, 325 (1999).

366. K. Kitahara, N. Ohtsuka, T. Ashino, M. Ozeki, and K. Nakajima, *Japan. J. Appl. Phys.* 32, L236 (1993).

367. J. N. Kidder, Jr., J. S. Kuo, A. Ludviksson, T. P. Pearsall, J. W. Rogers, Jr., J. M. Grant, L. R. Allen, and S. T. Hsu, *J. Vac. Sci. Technol. A* 13, 711 (1995).

368. A. Ludviksson, D. W. Robinson, and J. W. Rogers, Jr., *Thin Solid Films* 289, 6 (1996).

369. J. N. Kidder, Jr., H. K. Yun, J. W. Rogers, Jr., and T. P. Pearsall, *Chem. Mater.* 10, 777 (1998).

370. M. Nagano, S. Iwai, K. Nemoto, and Y. Aoyagi, *Japan. J. Appl. Phys.* 33, L1289 (1994).

371. N. Kano, S. Hirose, K. Hara, J. Yoshino, H. Munekata, and H. Kukimoto, *Appl. Surf. Sci.* 82–83, 132 (1994).

372. S. Hirose, N. Kano, K. Hara, H. Munekata, and H. Kukimoto, *J. Cryst. Growth* 172, 13 (1997).

373. J. Nishizawa, K. Aoki, S. Suzuki, and K. Kikuchi, *J. Electrochem. Soc.* 137, 1898 (1990).

374. J. Nishizawa, K. Aoki, and S. Suzuki, *J. Cryst. Growth* 99, 502 (1990).

375. Y. Takahashi and T. Urisu, *Japan. J. Appl. Phys.* 30, L209 (1991).

376. H. Nagasawa and Y. Yamaguchi, *Thin Solid Films* 225, 230 (1993).

377. H. Nagasawa and Y. Yamaguchi, *Appl. Surf. Sci.* 82–83, 405 (1994).

378. Y. Suda, D. Lubben, T. Motooka, and J. E. Greene, *J. Vac. Sci. Technol. B* 7, 1171 (1989).

379. T. Fuyuki, T. Yoshinobu, and H. Matsunami, *Thin Solid Films* 225, 225 (1993).

380. S. Hara, T. Meguro, Y. Aoyagi, M. Kawai, S. Misawa, E. Sakuma and S. Yoshida, *Thin Solid Films* 225, 240 (1993).

381. S. Imai, S. Takagi, O. Sugiura, and M. Matsumura, *Japan. J. Appl. Phys.* 30, 3646 (1991).

382. P. A. Coon, M. L. Wise, A. C. Dillon, M. B. Robinson, and S. M. George, *J. Vac. Sci. Technol. B* 10, 221 (1992).

383. Y. Takahashi, H. Ishii, and K. Fujinaga, *J. Electrochem. Soc.* 136, 1826 (1989).

384. J.-S. Min, H. S. Park, W. Koh, and S.-W. Kang, *Mater. Res. Soc. Symp. Proc.* 564, 207 (1999).

385. J.-S. Min, H. S. Park, and S.-W. Kang, *Appl. Phys. Lett.* 75, 1521 (1999).

386. J.-S. Min, Y.-W. Son, W.-G. Kang, S.-S. Chun, and S.-W. Kang, *Japan. J. Appl. Phys.* 37, 4999 (1998).

387. W. S. Rees, Jr., O. Just, and D. S. Van Derveer, *J. Mater. Chem.* 9, 249 (1999).

388. W. Gasser, Y. Uchida, and M. Matsumura, *Thin Solid Films* 250, 213 (1994).

389. K. Yamaguchi, S. Imai, N. Ishitobi, M. Takemoto, H. Miki, and M. Matsumura, *Appl. Surf. Sci.* 130–132, 202 (1998).

390. S. Morishita, W. Gasser, K. Usami, and M. Matsumura, *J. Non-Cryst. Solids* 187, 66 (1995).

391. S. Morishita, Y. Uchida, and M. Matsumura, *Japan. J. Appl. Phys.* 34, 5738 (1995).

392. M. Leskelä, L. Niinistö, P. Niemelä, E. Nykänen, P. Soininen, M. Tiitta, and J. Vähäkangas, *Vacuum* 41, 1457 (1990).

393. M. Ritala, K. Kukli, A. Rahtu, P. I. Räisänen, M. Leskelä, T. Sajavaara, and J. Keinonen, *Science* 288, 319 (2000).

394. V. E. Drozd, A. A. Tulub, V. B. Aleskovski, and D. V. Korol'kov, *Appl. Surf. Sci.* 82–83, 587 (1994).

395. K. Kukli, M. Ritala, and M. Leskelä, *Chem. Mater.* 12, 1914 (2000).

396. M. Leskelä and M. Ritala, *J. Phys. IV* 9, Pr8-837 (1999).

397. N. Kobayashi, Y. Kobayashi, Y. Yamauchi, and Y. Horikoshi, *Appl. Surf. Sci.* 60–61, 544 (1992).

398. M. Ritala and M. Leskelä, *Nanotechnology* 10, 19 (1999).

399. M. Aguilera and B. Aitchison, *Solid State Technol.*, Nov., 109 (1996).

400. M. Leskelä, W.-M. Li, and M. Ritala, "Electroluminescence," *Semicond., Semimet.* 64, 413 (1999).

401. P. D. Rack and P. H. Holloway, *Mater. Sci. Eng.* R21, 171 (1998).

402. A. Pakkala, EL displays based on ALE grown phosphors and insulator films, www.planar.com.

403. E. Soininen, G. Härkönen, and K. Vasama, "3rd International Conference on the Science and Technology of Display Phosphors," Huntington Beach, CA, 1997, p. 105, Extended Abstracts.

404. L. Hiltunen, M. Leskelä, L. Niinistö, E. Nykänen, P. Soininen, and M. Tiitta, *Acta Polytechnol. Scand. Ser. Appl. Phys.* 170, 233 (1990).

405. P. Soininen, M.Sc. Thesis, Helsinki University of Technology, Espoo, 1990.

406. E. Nykänen, S. Lehto, M. Leskelä, L. Niinistö, and P. Soininen, "Electroluminescence, Proceedings of the 6th International Workshop on Electroluminescence," El-Paso, 1992, p. 199.

407. S. J. Yun, Y. S. Kim, J. S. Kang, S.-H. K. Park, K. I. Cho, and D. S. Ma, *SID 99 Dig.* 30, 1142 (1999).

408. R. Törnqvist, *SID 97 Dig.* 28, 855 (1997); TFEL color by white, www.planar.com.

409. P. Soininen, M. Leskelä, L. Niinistö, E. Nykänen, and E. Rauhala, "Electroluminescence, Proceedings of the 6th International Workshop on Electroluminescence," El-Paso, 1992, p. 217.

410. E. Soininen, private communication.

411. T. Harju, G. Härkönen, M. Lahonen, J. Määttänen, E. Soininen, R. Törnqvist, K. Vasama, J. Viljanen, and T. Flegal, *SID 97 Dig.* 28, 859 (1997); Bright 320 (x3). 240 RGB TFEL display based on "color by white", www.planar.com.

412. J. Haaranen, T. Harju, P. Heikkinen, G. Härkönen, M. Leppänen, T. Lindholm, J. Maula, J. Määttänen, A. Pakkala, E. Soininen, M. Sonninen, R. Törnqvist, and J. Viljanen, *SID 95 Dig.* 26, 883 (1995).

413. R. Törnqvist, J. Antson, J. Skarp, and V. Tanninen, *IEEE Trans. Electron Devices* ED-30, 468 (1983).

414. D. Theis, H. Oppolzer, G. Ebbinghaus, and S. Schild, *J. Cryst. Growth* 63, 47 (1983).

415. Y. Charreire, A. Marbeuf, G. Tourillon, M. Leskelä, L. Niinistö, E. Nykänen, P. Soininen, and O. Tolonen, *J. Electrochem. Soc.* 139, 619 (1992).

416. J. Ihanus, T. Hänninen, M. Ritala, and M. Leskelä, "4th International Conference on the Science and Technology of Display Phosphors," Bend, OR, 1998, p. 263, Extended Abstracts.

417. S.-S. Sun and R. Khormaei, *Proc. SPIE* 1664, 48 (1992).

418. K. Kukli, M. Ritala, and M. Leskelä, *Electrochem. Soc. Proc.* 97-25, 1137 (1997).

419. J. Antson, *SID 82 Dig.* 13, 124 (1982).

420. S. Haukka, M. Tuominen, and E. Granneman, Semicon Europa/Semieducation, Munich, 5th of April, 2000.

421. M. Houssa, M. Tuominen, M. Naili, V. Afanas'ev, A. Stesmans, S. Haukka, and M. M. Heyns, *J. Appl. Phys.* 87, 8615 (2000).

422. H. Kattelus, H. Ronkainen, and T. Riihisaari, *Int. J. Microcircuits Electron. Packag.* 22, 254 (1999).

423. H. Kattelus, H. Ronkainen, T. Kanniainen, and J. Skarp, "Proceedings of the 28th European Solid-State Device Research Conference ESSDERC'98," Bordeaux, France, 1998, p. 444.

424. M. Ritala, T. Asikainen, M. Leskelä, and J. Skarp, *Mater. Res. Soc. Symp. Proc.* 426, 513 (1996).

425. V. Lujala, J. Skarp, M. Tammenmaa, T. Suntola, and J. Wallinga, "Proceedings of the 12th European Photovoltaic Solar Energy Conference," Amsterdam, The Netherlands, 1994, p. 1511.

426. J. D. Ouwens, R. E. I. Schropp, J. Wallinga, W. F. van der Weg, M. Ritala, M. Leskelä, and J. Hyvärinen, "Proceedings of the 12th European Photovoltaic Solar Energy Conference," Amsterdam, The Netherlands, 1994, p. 1296.

427. H. Antson, M. Grasserbauer, M. Hamilo, L. Hiltunen, T. Koskinen, M. Leskelä, L. Niinistö, G. Stingeder, and M. Tammenmaa, *Fresenius Z. Anal. Chem.* 322, 175 (1985).

428. W. Barrow, R. Coovert, E. Dickey, T. Flegal, M. Fullman, C. King, and C. Laakso, "Conference Record of the 1994 International Display Research Conference," 1994, p. 498.

429. R. Matero, M. Ritala, M. Leskelä, T. Salo, J. Aromaa, and O. Forsen, *J. Phys. IV* 9, Pr8-493 (1999).

430. L. A. Okada and S. M. George, *Appl. Surf. Sci.* 137, 113 (1999).

431. M. Juppo. M. Ritala, and M. Leskelä, unpublished results.

432. J. Skarp, Y. Koskinen, S. Lindfors, A. Rautiainen, and T. Suntola, "Proceedings of the 10th European Photovoltaic Solar Energy Conference," Lisbon, Portugal, 1991, p. 567.

433. J. Skarp, E. Anttila, A. Rautiainen, and T. Suntola, *Int. J. Solar Energy* 12, 137 (1992).

434. T. Suntola, *MRS Bull.* Oct. 45 (1993).

435. D. Riihelä, M. Ritala, R. Matero, and M. Leskelä, *Thin Solid Films* 289, 250 (1996).

436. G. D. Parfitt, *Prog. Surf. Membr. Sci.* 11, 181 (1976).

437. M. Ritala, M. Juppo, K. Kukli, A. Rahtu, and M. Leskelä, *J. Phys. IV* 9, Pr8-1021 (1999).

438. A. Rahtu and M. Ritala, *Electrochem. Soc. Proc.* 2000-13, 105 (2000).

439. R. Matero, A. Rahtu, and M. Ritala, submitted for publication.

440. N. Kobayashi and Y. Kobayashi, *Thin Solid Films* 225, 32 (1993).

441. J. P. Simko, T. Meguro, S. Iwai, K. Ozasa, Y. Aoyagi, and T. Sugano, *Thin Solid Films* 225, 40 (1993).

442. A. Usui, *Thin Solid Films* 225, 53 (1993).

443. A. Koukitu and T. Taki, *Appl. Surf. Sci.* 112, 63 (1997).

444. A. Koukitu, T. Taki, K. Narita, and H. Seki, *J. Cryst. Growth* 198–199, 1111 (1999).

445. B. Y. Maa and P. D. Dapkus, *Thin Solid Films* 225, 12 (1993).

446. N. Kobayashi and Y. Horikoshi, *Japan. J. Appl. Phys.* 28, L1880 (1989).

447. D. E. Aspnes, J. P. Harbison, A. A. Studna, and L. T. Florez, *Phys. Rev. Lett.* 59, 1687 (1987).

448. A. Rosental, P. Adamson, A. Gerst, and A. Niilisk, *Appl. Surf. Sci.* 107, 178 (1996).

449. A. Rosental, P. Adamson, A. Gerst, H. Koppel, and A. Tarre, *Appl. Surf. Sci.* 112, 82 (1997).

450. R. G. van Welzenis, R. A. M. Bink, and H. H. Brongersma, *Appl. Surf. Sci.* 107, 255 (1996).

451. S. T. Ceyer, *Science* 249, 133 (1990).

Chapter 3

LASER APPLICATIONS IN TRANSPARENT CONDUCTING OXIDE THIN FILMS PROCESSING

Frederick Ojo Adurodija

Inorganic Materials Department, Hyogo Prefectural Institute of Industrial Research, 3–1–12 Yukihira-cho, Suma-ku, Kobe, Japan

Contents

Handbook of Thin Film Materials, edited by H.S. Nalwa
Volume 1: Deposition and Processing of Thin Films
Copyright © 2002 by Academic Press
All rights of reproduction in any form reserved.

ISBN 0-12-512909-2/$35.00

1. INTRODUCTION

1.1. History of Transparent Conducting Films and Applications

The history of transparent conducting oxide (TCO) semiconductor films goes as far back as 1907, when Bädeker first prepared cadmium oxide (CdO) films by the thermal oxidation of sputtered cadmium [1]. Following this, there has been tremendous interest in TCO films, with the advancement in optoelectronic technology. Over the past decades, several new materials, including tin oxide (SnO_2), indium oxide or tin-doped indium oxide (In_2O_3:Sn or ITO), zinc oxide (ZnO), and cadmium stannate have emerged. Numerous new techniques for the preparation of TCO films have also evolved. The importance of TCO films is based on a combination of important properties, including high luminous transmittance, high infrared reflectance, and good electrical resistivity.

Presently, TCO films are found in a host of optoelectronic, mechanical, and architectural systems, including flat panel displays, organic light emitting devices [2–4], solar cells, heat mirror, energy-efficient windows, and gas sensors [5, 6]. In flat panel displays, a TCO layer is needed to transmit the back light through the device to the viewer, while in solar cells a TCO layer is used both as a transparent electrode or an n-type layer in the case of a p–n heterojunction system. In addition, a TCO layer coated on solar cells also serves as an antireflection coating. With increasing sophistication in the areas of optoelectronic applications such as flat panel displays, TCO thin films of much lower resistivity ($<10^{-4}$ Ω cm) are required. These display devices now utilize substrates such as polymers and organic color filter films that cannot survive temperatures over 200°C. Therefore, such a low resistivity should be achieved possibly at room temperature.

1.2. Indium Tin Oxide Thin Films

Among the class of TCO films, indium tin oxide (ITO) thin film is highly favored because of its combination of technologically important properties such as low electrical resistivity, high luminous transmittance, high near infrared reflectance, excellent substrate adherence, hardness, and chemical inertness [5–11]. These unique properties make ITO the most studied among TCO materials since it was first discovered in 1954, by Rupprecht [12]. It is not surprising that relentless efforts have been directed toward the development of techniques to produce ITO thin films and such developments are contained in several review papers and textbooks [5, 6].

1.3. Deposition Techniques

Some of the early developed techniques that have been to grow TCO films include thermal or electron-beam evaporation in a reactive atmosphere [6, 13–29], spray pyrolysis [30–33], and sputtering in reactive [34–59] and nonreactive environments. Others are reactive ion plating [60–63], chemical vapor deposition [64–69], dip coating, or sol–gel processes [70–76]. Among these techniques for ITO growth, dc and rf sputtering, spray pyrolysis, and chemical vapor deposition are currently in use even at a commercial scale. Other techniques such as ion-cluster-beam [77, 78] and pulsed laser deposition (PLD) [79–92] have emerged as promising techniques for the growth of materials of technological importance. Although earlier applications of PLD were for the growth of metallic and multicomponent superconducting oxide films, the last few years have seen an upsurge of the use of PLD for the growth of TCO films.

The achievement of the lowest resistivity and highly transparent ITO films is the bull's eye of all technologies. It is well known that the properties of ITO films are strongly dependent on the deposition conditions. Efforts at understanding the physical phenomena governing the growth conditions are often retarded due to the too many methods available. In spite of this, tremendous effort has been directed at improving all the processes available for ITO film deposition. ITO film of extremely low resistivity of 4.4×10^{-5} Ω cm has been obtained by thermal evaporation at temperature of 300°C and to date no one else has been able to reproduce such a low value [27, 28]. The commercially produced ITO films have resistivity of the order of 2×10^{-4} Ω cm.

1.4. General Remarks

In the past decade the deposition techniques for ITO films have passed through series of changes and this is due to the increasing sophistication of applications. The major changes that occurred in the deposition techniques until 1983 have been reviewed by Vossen, Jarzenbski, Manifacier, Dawar and Joshi, and Chopra [5, 8–10, 93]. These reviews covered the preparation methods, the properties of TCO films, and applications in thin film devices. The fundamental optical properties of ITO films and their application in energy-efficient windows have been reviewed by Hamberg and Granqvist [6]. Tahar et al. [73] reviewed the electrical properties of ITO films. In the review Tahar et al. also identified important areas of further research in order to enhance the electrical properties. Due to its newness, PLD has often been excluded from major review papers as a potential deposition method for TCO films. However, substantial work done on TCO films existed in the literature as early as 1989. In spite of the short history of PLD, remarkable progress has been made on the enhancement of the optoelectronic properties of ITO films on small substrate areas by taking advantage of the versatility of the technique. This chapter provides an up to date and comprehensive review on PLD for the growth of TCO films, particularly ITO films.

This chapter is organized as follows: Sections 1 and 2 cover the history, deposition techniques, and fundamental electrical properties of ITO films. Section 3 provides a brief review on the principles, applications, and benefits of excimer lasers. The fundamentals of the PLD technique are presented in Section 4, together with approaches that have been used to enhance the quality of TCO films. The use and optimization of the background oxygen gas is also considered. In this section, the initial growth of ITO film is also discussed. The methods of film characterization form the final part of this section. Sections 5 and 6 cover the deposition and properties In_2O_3 and ITO films. Included in the discussion are the electrical, structural, optical, and chemical-states properties of the films. Section 7 provides a review on a novel excimer laser irradiation of the surfaces of ITO films during growth. The topics addressed in this section include the structural, electrical, and optical properties of laser-irradiated ITO films. Discussion on the electronic transport properties of ITO films is also included. A brief review on the properties of other TCO films such as ZnO films deposited by PLD are covered in Section 8. Section 9 covers some applications of ITO films deposited by PLD in device fabrication. Finally, the review is concluded in Section 10.

2. GENERAL ELECTRICAL PROPERTIES OF TCO FILMS

Section 2 concerns mainly the fundamental electrical properties of In_2O_3 and ITO films. The three fundamental requirements for the electronic structure of a TCO material are outlined as follows [94]:

(a) Optical transparency in the visible region requires the bandgap of the TCO to be greater than 3.1 eV (also the TCO must be an intrinsic insulator).

(b) Electric current is transported by electron introduction to the conduction band, and the conduction band must constitute an extended state so that doped electrons can migrate in the lattice under a weak applied electric field.

(c) Electron doping in the oxide crystals is obtained using methods such as substitution of cations or anions or by the addition of oxygen vacancies or excess cations.

2.1. Conduction Mechanism

Tahar et al. [73] has recently reviewed the electrical properties of ITO films, including the conduction mechanism and defect models. The discussion in this section is intended to follow a similar path. The electrical conductivity is expressed as the product of the carrier concentration and mobility and is represented as

$$\sigma = 1/\rho = N\mu e \qquad (1)$$

where e is the electronic charge.

Therefore, higher conductivity is expected for larger carrier concentration and mobility. The electrical properties of the oxide films rely solidly on the oxidation state of the metal component and the amount of impurity added to the films, deliberately or unintentionally. Perfectly stoichiometric In_2O_3 film is either an insulator or an ionic conductor and the latter is of little importance as a transparent conductor due to the high activation energy necessary for ionic conduction [8].

In_2O_3 has a cubic bixbyite Mn_2O_3 (I) structure originating from an array of unoccupied tetrahedral oxygen anion sites [6]. The electrical properties of In_2O_3 depend mainly on the oxidation state of the indium metal. Hence, the creations of oxygen vacancies (V_o) contribute significant conduction electrons responsible for the high conductivity (low resistivity) in undoped In_2O_3, oxygen deficient material. Stoichiometric In_2O_3 single crystal can exhibit resistivities as low as 10^{-4} Ω cm [73]. Basically, In_2O_3 should have a filled O^{2-} $2p$ that has the oxygen $2p$ in characteristic [95]. Below the valence band edge E_v lies the In $3d$ core. The conduction band is the metal-$5s$ state with an edge $E_c \approx 3.5$ eV above the valence band. The following higher band is the metal-$5p$ state [96, 97]. As-deposited In_2O_3 films lack stoichiometry, because of the oxygen-array vacancies V_o. A high V_o density creates an impurity band that overlaps the lower part of the conduction band (E_c), resulting in a degenerate semiconductor [8]. V_o provides a maximum of two electrons for the conduction process and is represented by

$$O_o \rightarrow V_o + \frac{1}{2}O_2(g) \qquad (2)$$

$$V_o \rightarrow V_o^{\cdot\cdot} + 2e^- \qquad (3)$$

The chemical composition for In_2O_3 material is denoted by In_2O_{3-x}. Besides the supply of conduction electrons, the oxygen vacancies (V_o) also enable the mobility of O^{2-} ions. Hence, In_2O_{3-x} is considered as a mixed conductor possessing both electronic and O^{2-} ion conduction [95], but the latter is insignificant when compared to electronic conduction.

Alternatively, free electrons are created in In_2O_3 by doping with elements of valence electron equal to or greater than four. This doping technique could be important, especially if we consider the relationship between the carrier concentration and the mobility. Elements such as Sn, Ge, Cu, Te, S, Zn Pb, Er, F, and Mg (see [5, 6, 98–100] and references therein) have been investigated as dopant in In_2O_3. But Sn is the most effective among these elements and lots of work has been done on Sn-doped In_2O_3 or ITO films [56, 73, 101]. The Sn atom with four valence electrons substitutes for an In atom having three valence electrons in the crystal lattice, thus giving an extra electron to the conduction band as represented by

$$Sn_{In}' \rightarrow Sn_{In} + e^- \qquad (4)$$

Thus, there is an overall preservation of the charge neutrality. In actual fact, in ITO both substitutional Sn and oxygen vacancies contribute to the high conductivity [95, 102]. The chemical composition for Sn-doped In_2O_3 material is represented as $In_{2-x}Sn_xO_{3-x}$ [95]. The effective doping effects are obtained when the ionic radius of the dopant is the same as or less than the host ion, and furthermore the dopant oxide does not form compounds or solid solutions with the host oxide [5]. The theoretical maximum carrier density contributed by only Sn is $3 \times 10^{20} \times C_{Sn}$ (at.%) where C_{Sn} is the Sn concentration. In practice, the creation of free carriers via Sn doping is self-limiting and does not increase as anticipated.

2.2. Defect Models

The majority of the research efforts has been focused on improving the conductivity of ITO films by increasing the effective number of free electrons through Sn doping. It is well known that as the carrier concentration increases, the mobility of free electrons is impaired. Hence, improvement in the resistivity is a compromise between the carrier concentration and the mobility of free electrons. The above phenomenon shows that the mobility (μ) and carrier concentration (n) are controlled by the relation, $\mu \propto n^{-2/3}$, which is a deviation from the relation $\mu = (4e/h)(\pi/3)^{1/3}n^{-2/3} = 9.916 \times 10^{14}n^{-2/3}$ ($cm^2\,V^{-1}\,s^{-1}$) earlier deduced by Johnson and Lark-Horovitz [103] for a complete degeneracy. The later relation should result in a resistivity of $2 \times 10^{-4}\,\Omega\,cm$. However, the resistivity of ITO films is much lower than that predicted supposing every soluble Sn atom contributes one free electron [73]. This shows that some of the Sn atoms in the crystal lattice are inactive. Several investigations on the doping mechanism and chemical state of Sn atoms in ITO films using major deposition techniques like spray pyrolysis and sputtering have been reported [102, 104, 105]. In spray pyrolysis films, Frank and Köstlin [105] observed that the carrier concentration increases almost in a linear form with increasing Sn doping content up to 5 wt% and then stabilizes with further increase in Sn doping level. The electrically inactive Sn produces a deleterious effect on the electrical conductivity by way of contributing neutral scattering centers. Several reports on the chemical state of Sn in In_2O_3 films are available [95]. In addition, structural models for the inactive Sn dopant in In_2O_3 films, where the deactivation of the Sn dopant is associated with the chemical reaction between substitutional (Sn/In) and interstitial oxygen, have been documented [105, 106]. Two models have been proclaimed in the literature to explain the compensation of Sn donors by the inclusion of oxygen anions that trap the free electrons as follows [107, 108]:

(i) Anion interstitial model: Excess interstitial oxygen anions $\delta/2$ (O_i) occupy some of the quasianion sublattice, thus neutralizing the δ ionized Sn donors, represented by $In_{2-\delta}Sn_\delta(O_i)_{\delta/2}O_3$.

(ii) Cation vacancy model: Production of cation $\delta/3$ vacancies (V_{In}) to preserved electrical neutrality, represented by $In_{2-4\delta/3}(V_{In})_{\delta/3}Sn_\delta O_3$.

Frank and Köstlin discarded model (ii) because they observed a minor change of ~0.4% in the ITO film thickness after the oxidation process. Based on model (i), they suggested the formation of the following neutral defects:

(a) (Sn_2O_i''): Two Sn^{4+} ions that are not on nearest-neighbor locations weakly bound to an interstitial oxygen anion. The interstitial oxygen ion dissociates on annealing under reducing conditions: $Sn_2O_i'' \leftrightharpoons 1/2O_2^{(g)}$.

(b) (Sn_2O_4)^x: Two nearest-neighbor Sn^{4+} ions bound on regular anion locations and extra interstitial oxygen ion on nearest quasianion position. This neutral defect was already reported [104, 105].

(c) (Sn_2O_i'') and (Sn_2O_4)^x: This is a combination of the first two defects mentioned above.

Other ways to regulate the Sn doping efficiency have been discussed in [73]. In another study, Nadaud et al. [109], using Rietvelt analysis, has observed that quasianion sites are partly occupied by oxygen atoms which is also in agreement with model (i). Yamada et al. [110] have also reported the doping mechanisms and the donor compensation effects in ITO powder with various Sn concentrations, using ^{119}Sn transmission Mossbauer spectroscopy and precise X-ray diffraction analyses. From the TMS analysis, they could distinguish between the electrically active and inactive Sn dopants, which revealed that deactivation of dopants is the result of the inclusion of excess oxygen anions (O_i'') into the In_2O_3 host lattice. Yamada et al. has been able to relate the chemical state of Sn to the electrical properties of ITO. They clearly observed that the steady increase in deactivated Sn^{4+} that is coordinated by seven or eight oxygen anions (one or two excess interstitial oxygen anions) with increasing Sn doping concentration is the result of the low doping efficiency of the heavily doped ITO of more than 5 wt%.

3. EXCIMER LASERS

The interaction of intense laser radiation with condensed matter is the primary event in all applications of lasers for material processing. These interactions involve aspects of optics, heat and mass transfer, and plasma physics. This part provides only a brief review on the types of excimer lasers that are used in PLD work. Numerous specialized textbooks and review papers on the principles and applications of lasers have been published. Therefore, further reading on this subject may be obtained from these sources (see [111] and references therein).

The commonest laser systems that have been used in deposition work involve excimer and Nd^{3+}:YAG lasers. The Nd^{3+}:YAG laser is a solid state laser system where neodymium (Nd) ions act as the active medium and are impurities in the yttrium aluminum garnet (YAG) composite (host) [112]. YAG lasers have emissions at a long wavelength (infrared region) of 1064 nm, which is far outside the range desired for thin film deposition. In YAG lasers the emitted infrared radiation yields thermal energy that is used to vaporize the material. A good description on the subject of YAG lasers is found in a book written by Yariv [113].

Excimer lasers are gas lasers that emit their radiation in the ultraviolet unlike YAG lasers. Hence, excimer lasers can

Table I. Excimer Lasers and Their Corresponding Operating Wavelengths

Excimer	Wavelength (nm)
F_2	157
ArF	193
KrCl	222
KrF	248
XeCl	308
XeF	351

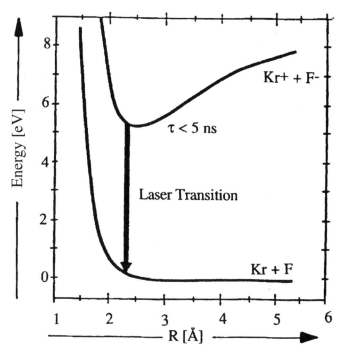

Fig. 1. Schematic diagram of the electronic potential of the KrF excimer. (Reprinted with permission from S. M. Green, A. Pique, and S. Harshavardhan, in "Equipment: Pulsed Laser Deposition of Thin Films" (D. B. Chrisey and G. K. Hubler, Eds.), p. 25. Wiley, New York, 1994.)

transfer very thin layers from the ablation target composition onto the growing film (solid–vapor ablation). By this process, the high-energy radiation breaks down the chemical bonds of the material target into its chemical components in the solid state. This scenario is different from the YAG lasers that physically melt the source target before evaporation takes place. The solid–vapor ablation effect gives excimer lasers an edge over the other laser systems as well as conventional deposition methods like sputtering. This benefit has made it possible for the processing of organic compound films using excimer PLD.

The excimer lasers utilize noble gas compounds for the lasing action. Noble gases such as Ar, Kr, Xe, etc., which under normal conditions do no react with other elements when excited in a laser cavity with the aid of an electrical discharge can be ionized. The ionized noble atoms can attract neutral atoms like fluorine (F_2) or chlorine (Cl_2) to form active excimer molecules. Examples of commercially developed excimer complex molecules include fluoride (F_2), argon fluoride (ArF), krypton chloride (KrCl), krypton fluoride (KrF), xenon fluoride (XeF), and xenon chloride (XeCl). Table I shows a summary of these laser systems and their analogous excimer wavelengths. Among the excimer lasers listed above, KrF, ArF, and XeCl have been widely used for PLD work. But the KrF laser is found to yield the highest gain among the electrical discharged pumped excimer laser [112]. KrF and ArF excimer lasers are commonly used among researchers engaged in PLD work, including deposition of TCO films [79–92].

Presently, commercially available excimer laser systems can achieve output in excess of 1 J/pulse. Systems that can accomplish pulse frequencies of hundreds of hertz with energy output of about 500 mJ/pulse are available [112]. Recently, new excimer lasers with large aperture and capable of delivery up to 20 J/pulse and average power in kilowatt range have been developed [114].

3.1. Principles of Excimer Lasers

The basics of excimer lasers are discussed below according to Green et al. [112]. The light output from an excimer laser is derived from a molecular gain medium in which the lasing action takes place between a bound upper electronic state and a repulsive or loosely bound ground electronic state. Since the ground state is repulsive, the excimer molecule can detach quickly (on

the order of a vibrational period $\sim$10–13 s) as it emits a photon during passage from the upper state to the ground state. The high ratio of the upper state lifetime to lower state lifetime makes the excimer, somehow, the ideal laser medium because population inversion and hence high gain are easily attained. A typical molecular potential illustration of an excimer system is shown in Figure 1 [112]. The excimer molecules are formed in a gaseous mixture of their constituent gases, such as Xe, HCl, and Ne in the case of the XeCl laser. Energy is pumped into the gas mixture via fast avalanche electric discharge. The pumping originates ionic and electronically excited species that undergo chemical reaction and produce the excimer molecules. Other discharge methods such as electron-beam and microwave methods have also been used to pump the gas mixture.

The kinetic and chemical reactions that lead to the formation of the excimer molecules are quite complicated and can involve very many steps. In the case of KrF, some of the more important steps are outlined as follows [112]:

$$Kr + e^- \rightarrow Kr^+, Kr^*, Kr_2^+$$
$$F_2 + e^- \rightarrow F + F^-$$
$$Kr^+ + F^-X \rightarrow KrF^* + X$$
$$Kr_2^+ + F^- \rightarrow KrF^* + Kr$$
$$Kr^* + F_2 \rightarrow KrF^* + F$$

where the * represents an electronically excited species and X represents a third body (He, Ne).

Once the excimer is created, it decays through spontaneous emission and collisional deactivation, giving the molecule a lifetime of ~2.5 ns. Modest output energies of several hundred milijoules per pulse dictate an excimer population density requirement on the order of 10^{15} cm^{-3}. Hence, in order for lasing action to happen, the generation rate of the ionic and excited precursors must be fast enough to produce excimers at a rate of several 10^{23} cm^{-3} s^{-1} [112]. Since an excimer laser is stabilized by a third body, the fast kinetics involved in producing the excimers requires total gas pressure over a range of 2 to 4 atmospheres within the discharge volume. Other discharge parameters are electron densities, current densities, and electron temperature of the order of 10^{15} cm^{-3}, 10^3 A cm^{-2}, and 1200 K, respectively. These conditions are satisfied with electric discharge field strengths of 10–15 kV cm^{-1}. Consequently, discharge electrode spacing is limited to 2–3 cm; hence discharge voltages can be in the range of 20–45 kV [112].

3.2. Principles of Excimer PLD

In excimer PLD, laser light in the ultraviolet and the near ultraviolet (wavelength of 193 nm to 400 nm) is of interest in thin-film formation. This short wavelength region produces high photon energy that is required for the ablation process. The generality of the materials used for PLD research has a very high absorption coefficient in this wavelength region. For example, ITO has an absorption coefficient of ~3×10^5 cm^{-1} in the 248 nm wavelength [115]. The shorter the wavelength of the excimer laser, the higher the absorption. The importance of this feature is that it minimizes the interaction time between the laser radiation and the target material, hence making it possible to deposit nanolayers of the target material during PLD.

The basic components of an excimer PLD system include a vacuum system equipped with a target and substrate holder as well as deposition monitoring systems. The evaporating source is a high-power laser located outside of the vacuum chamber. An optical system (lenses and mirrors) is used to focus the high-power laser beam on the substrate. The mechanism of PLD leading to material transfer to the substrate involves many stages, although it depends mainly on the type of laser, optics, and properties of the target used. First, a target is heated by short, concentrated burst of laser radiation and when the laser radiation is absorbed by the solid surface, electromagnetic energy is converted first into electronic excitation and then into thermal and the material is ablated. Subsequently, the evaporants form a plume which consists of a mixture of high-energy species including atoms, molecules, electrons, ions, clusters, and even micron-sized particulates. The plume that is formed propagates through a background of working gas toward the substrate. The plume is characterized by numerous collisions that may affect the velocities and the mean-free paths of the ablated atoms and ions as well as initiating reactions. The reduction of the mean-free path caused by collision leads to the fast expansion of the plume from the target surface to form a narrow forward angular distribution of the evaporants [116]. During this process, some material is redeposited onto the target or chamber walls, but most reaches the substrate where nucleation and growth occur. The eventual formation of the film is not only governed by the ablation and plume propagation but also by the substrate properties such as structure and topography, working gas, and the laser power. To study these effects and how they influence the efficiency of laser material processing, a variety of spectroscopic and image analysis techniques are employed.

3.3. Major Applications of Excimer Lasers

Apart from thin-film deposition, excimer lasers are used in industrial applications such as micromachining, planarization, and UV lithographic patterning [114]. Other uses include surgical operations. Over the past decade the application of excimer lasers for thin-film deposition and other applications has increasingly multiplied in both research institutions and industries. In actual fact, the market for excimer lasers is gradually shifting toward large-scale industrial applications.

3.4. Advantages and Disadvantages of PLD

The PLD process for thin-film formation has its merits and demerits. The major merits include the ability to reproduce the stoichiometry of the material from a multicomponent target in thin-film form during ablation (congruent evaporation). Other advantages are flexibility, fast response, and energetic evaporants. However, such favorable results do not occur in all materials. For some metallic alloys, semiconductors, complex superconducting oxides, and ferroelectrics it is difficult to maintain the stoichiometric composition during laser ablation [117–122]. Other demerits are related to the formation of clusters or particulate in the growing films of some materials and the narrow forward angular distribution of the evaporants that may pose problems for large area deployment. The methods to overcome these problems are continuously being addressed [111, 122, 123].

For TCO thin-film deposition, the advantages of PLD are following:

(i) stoichiometric removal of elemental species from the bulk target during deposition,
(ii) nonparticulate formation in the growing film,
(iii) low growth temperatures of high quality films,
(iv) stable and possible standardization of deposition parameters,
(v) good adhesion of the film to the substrate,
(vi) good reproducibility of film properties.

A major disadvantage of PLD for TCO thin-film formation is the coating of the quartz-glass window admitting the laser beam into the chamber during deposition. It is known that TCO films have high absorption to UV light in the ultraviolet range; hence the laser energy is drastically reduced as deposition progresses. This problem is mostly resolved by constant cleaning or replacement of the windows after a couple of deposition runs. In terms of the film properties, the merits of PLD outweigh those of conventional methods such as sputtering, evaporation, chemical vapor deposition, or spray pyrolysis.

4. PLD DEPOSITION TECHNIQUE

So far we have discussed the types and principles of excimer lasers. In this section we shall look at the laser system setup and conditions of thin-film growth. PLD, even though it was first demonstrated in 1965 as a potential thin-film deposition technology, did not receive widespread attention until the discovery of high-T_c superconductors in 1986 [111]. Therefore, PLD is still a relatively new technology compared to other deposition technologies such as sputtering and physical vapor evaporation. It is noteworthy to mention that considering the short history of PLD, many research groups have achieved acceptable thin-film quality of complex superconductors and semiconductors, including transparent conducting oxides, with minimum initial research effort. The strides in research and development in this area have been rapid within the last five years, but there are still many questions to address in terms of commercial deployment of this rather complex technology. However, like any other new technology time will provide answers to some of these questions.

The type of application of TCO films often dictates the choice of the deposition technique. Novel devices require coating of films on flexible and less weighty substrates such as polymer films. These novel applications have led to an increase in the demand for low temperature deposition technologies for TCO films. For example, for novel flat panel displays which require the formation of ITO films on heat sensitive polymer sheets or polymer based color filters, the conservation of low substrate temperature (<200°C), perhaps through low kinetic energy deposition techniques, is desirable. In heterojunction solar cells, where an n-type ITO film is deposited onto a p-type absorber layer, high substrate temperature (>200°C) contributes to degradation of the junction properties due to interdiffusion of the elemental species. Therefore, the development of low temperature techniques for the growth of ITO films is essential to realize optimal performance of such devices.

It is well established that a low resistivity of $\sim 2 \times 10^{-4}$ Ω cm is readily obtained when ITO films are deposited on glass substrate at temperatures greater than 300°C [73]. Using conventional techniques like sputtering and evaporation, amorphous films with inferior optoelectronic properties are mostly observed at low temperature (<150°C) [124]. The conditions that yield amorphous or partly amorphous ITO films depend on the deposition parameters such as substrate temperature, oxygen partial pressure, or film thickness [125–129]. Using reactive evaporation, Muranaka [128, 129] observed that ITO films deposited at temperatures less than 150°C are amorphous from the outset of the deposition but change to the crystalline phase with increasing film thickness. The change from amorphous to crystalline phase has been witnessed with a dramatic improvement in the optoelectronic properties of the films. Therefore, the amorphous–crystalline phase transition is bound to play a major role in determining the microstructural, electrical, and optical properties of crystalline ITO films deposited on heated substrates. Paine et al. have observed that upon annealing of the as-deposited amorphous ITO film at low temperature

(125–165°C), it undergoes a structural relaxation that includes local ordering and crystallization. The central focus of their study was to establish conditions for forming crystalline ITO films with low resistivity at low temperature [124]. In recent years PLD has been found to be a suitable low-temperature deposition tool for low-resistivity ITO films [79–92]. In processes like electron-beam evaporation and sputtering, postdeposition heat treatment at 300–550°C in air is required to oxidize the residual metal component in order to enhance the electrical resistivity of ITO films. However, for PLD the high-energy plume produced during deposition is often substituted for high substrate temperature to form low-resistivity films even at room temperature. This unique feature may be beneficial for low-temperature applications.

4.1. ITO Target Ablation and Modification

During PLD, highly energetic atoms and ions are ejected from the target as a result of laser–target interaction. These atoms, ions, and clusters approach the substrate to form a high-density laser plume. The laser-produced plume is visible in size and shape and is dependent on the background oxygen pressure and the laser fluence [92]. During the transport process between the target and the substrate, many collisions occur among the energetic particles, ions, and oxygen atmosphere. At the time of the target–substrate transport, the energetic particles collide with the oxygen molecules at a frequency that is dependent on the mean-free paths of the ablated particles (i.e., oxygen pressure). At high oxygen pressure, collisions between the laser-ejected particles and the oxygen atmosphere increase. The major consequences of the increased collisions are as follows: (i) reduction of the mean-free paths and consequently the number of particles reaching the substrate and (ii) change in the angular and velocity distribution of the plume that fits the $\cos^n \theta$, where θ is determined from the surface perpendicular to the target and n is a variable of increasing oxygen pressure [92]. Figure 2a shows a schematic diagram of the ablation plume during deposition of ITO films by PLD. The various particles and ion species present during ablation are shown. The diagnosis of the plume by plasma emission monitoring using an optical spectrum analyzer is shown in Figure 2b. The major identified radicals (In*, Sn*, O_2^*, and In$^+$) present in the plume around the surface of the target are indicated. The spectrum reflects the particles present in the ITO target that are transferred via the plume to the substrate to form the thin film.

Target surface modification is a common feature observed during deposition using techniques like sputtering and PLD. The shape and pattern of the surface often depend on the type of material. For example, formation of cones has been observed in Y–Ba–Cu–O targets after exposure to laser energy pulses [130]. In an overview provided by Foltyn [130], vaporization-resistant impurities are said to be responsible for laser-cone formation. Particulate formation during deposition of the film has been found to be break off tips of cones which ultimately end up in the deposited film. The size and density of the particulate are dependent on the deposition conditions, such as laser

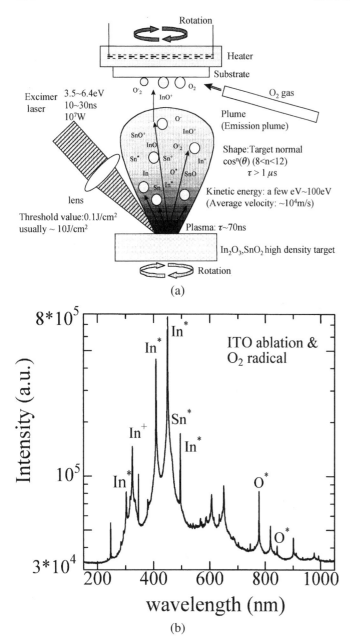

Fig. 2. (a) Deposition scheme for pulsed laser deposition. (b) Emission spectrum of the laser plume obtained from optical spectrum analyzer.

shows the surface scanning electron microscopy (SEM) micrographs of ITO targets with packing densities of 70 and 98% (low and high density) before and after irradiation with several laser pulses at a fluence of ~ 4 J cm^{-2}, respectively. Before laser irradiation, the low-density target consists of microvoids, whereas the high-density target appeared to be void free, as shown in Figure 3a and b. After exposure to several laser pulses, SEM revealed ITO target surfaces that have undergone a typical melting and solidification process. The surface of the high-density target indicated smooth and uniform erosion of the materials during the ablation process, while the surface of the low-density target showed lateral grooves that suggested the formation of *low-degree* cones. Although the surface roughness of the low- and high-density targets after exposure to laser irradiation differs, no particulate formation has been observed in the deposited films.

4.2. Background Gas

The basic PLD process usually requires the use of a background gas during deposition. A background gas is required for optimal growth of complex superconducting and semiconductor oxide films [123]. The addition of a background gas assists not only in the slow down and thereby the moderation of the energy of the ablated species, but it also causes a significant reduction in the concentration of ejected atomic and ionic components from the target that are transmitted to the substrate [91, 92, 132, 133]. Furthermore, during the formation of compound oxides and nitrides thin films, the interaction of the plume with a reactive background gas plays an important role in producing the atomic and molecular precursors required for the growth of the compound phase [133].

During the growth of TCO films by PLD and other deposition methods, it is commonly required to maintain an oxidizing atmosphere. The common arrangement for the use of a reactive oxygen gas is to blend the gas into the vacuum chamber continually at a steady flow rate, while the chamber is concurrently pumped to preserve a constant ambient pressure. The oxidizing environment helps to form and stabilize the desired compound phase during deposition. ITO films deposited in vacuum by PLD, sputtering, and evaporation methods have been found to lack stoichiometry, showing a semimetallic mirrorlike appearance [6]. The ITO films also display poor optical and electrical properties. Most of the works reported so far on ITO films deposited by PLD have utilized molecular oxygen (O_2) as the oxidizing gas [79–92].

Depending on the oxygen pressure during growth, both gas-phase (plume-oxygen gas) and surface reactions are important for the oxidation process, with the latter contributing nearly only at lower pressures. The deposition scheme for ITO film and the emission spectrum of the plume during deposition by PLD are shown in Figure 2a and b. The possible gas-phase and surface oxidation reactions that occur during growth are also shown. The gas phase includes neutral cation and anion components. A similar particle–gas interaction during growth

wavelength, laser power, laser spot size, and background gas pressure [120, 121, 123, 131]. The subject of cone formation continues to receive considerable attention, since it tends to affect the deposition parameters, particularly the deposition rate and the quality of the deposited film [120]. In novel ultrahigh performance optoelectronic applications like flat panel displays where strict constraints exist for surface smoothness, the tolerance of particulate formation is very low.

In excimer PLD of TCO films, no critical problems of particulate formation in the deposited films has been experienced. However, the dependence of the film quality on the deposition conditions, including background gas, substrate temperature, and deposition rate, has been reported [79–92]. Figure 3a–d

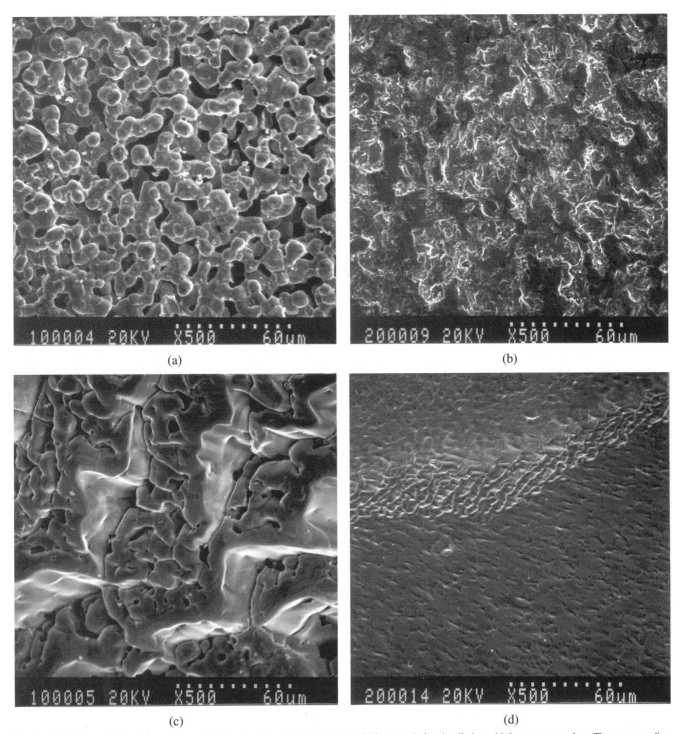

Fig. 3.   SEM micrographs of the surfaces of (a) low-density and (b) high-density ITO targets before irradiation with laser energy pulses. The corresponding surfaces after laser irradiation are shown in (c) and (d).

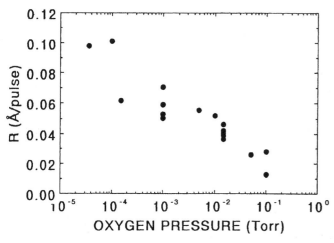

Fig. 4. Dependence of deposition rate on the oxygen pressure. The target–substrate distance was 7 cm. (Reprinted with permission from J. P. Zheng and H. S. Kwok, *Thin Solid Films* 232, 99 (1993), © 1993, Elsevier Science.)

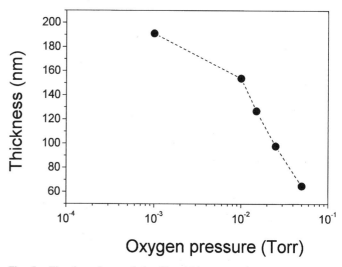

Fig. 5. The dependence of the film thickness on the oxygen pressure. (Reprinted with permission from F. O. Adurodija, H. Izumi, T. Ishihara, H. Yoshioka, H. Matsui, and M. Motoyama, *Japan. J. Appl. Phys. 1* 38, 2710 (1999), © 1999, Japan Society of Applied Physics.)

leads to the formation of superconducting and ferroelectric oxide films. According to Gupta [133], during deposition substantial interactions between the oxygen gas and the ablated particle constituents occur. The oxygen gas can experience photon and electron-impact decomposition during the ablation process, primarily in the locality of the target to produce atomic oxygen. This is in addition to the atomic oxygen directly released from the target along with molecular oxygen and the cation species. Some of the generated atomic oxygen is lost during transfer to the substrate due to chemical reactions with the oxygen gas [133]. Similarly, the cation species are attenuated due to reactions and elastic scattering during the target–substrate transition. The unattenuated atomic and molecular species that arrive and are adsorbed by the substrate experience a series of oxidation, recombination, and desorption steps that finally leads to the formation of the film.

The properties of ITO films have been observed to be highly sensitive to the oxygen pressure during deposition and the effects of oxidizing gas on the properties of ITO films are well documented in the literature [79–92]. Specifically, the effects of oxygen pressure over a wide range of 1×10^{-5} to 1×10^{-1} Torr on the properties of ITO films have been reported [81–83, 88, 91, 92]. For apparent practical reasons, most of the high-quality ITO films deposited by PLD are obtained within a narrow oxygen pressure of $(1–1.5) \times 10^{-2}$ Torr [79–92]. At low pressures ($<5 \times 10^{-3}$ Torr), poor quality ITO films have been observed and the cause has been associated with insufficient oxidizing gas. In addition, the low oxygen pressure causes the kinetic energy of the ablated species to increase, leading to an expanded *c*-lattice parameter and sometimes splitting of diffraction peaks for films deposited on heated glass substrate [82]. At high oxygen pressures, there exists excessive scattering of the gas phase during transport to the substrate with a consequential degradation of the film properties. The effects of oxygen pressure on the properties of ITO films are discussed in Sections 5 and 6.

4.3. Growth Rate and Film Thickness

Zheng and Kwok [92] have found from the optical time-of-flight study that during the growth of ITO films by PLD, the laser-produced atoms and ions approach the substrate within a time interval of 10 μs and at a high velocity of 10^6 cm s^{-1}. Therefore, at high oxygen pressure, the velocities of the different species in the laser plume are drastically reduced because of the high collision rates with the oxygen gas molecules. The dependence of the growth rate on the oxygen pressure is shown in Figure 4 [92]. In a separate study, Adurodija et al. [82] also showed the dependence of the film thickness on the oxygen pressure for ITO films deposited at a deposition time length of approximately 15 minutes as shown in Figure 5 [82]. A significant decrease in the film thickness with increasing oxygen pressure can be seen, implying that oxygen pressure has a profound effect on the deposition rate and film thickness due to increased collisions and scattering rates.

4.4. Target to Substrate Distance

The source target to substrate distance is another major factor that affects the properties of thin films during deposition. For the PLD technique, Zheng and Kwok [92] have studied the effect of the target–substrate distance on growth rate of ITO films deposited at an oxygen pressure of 1.5×10^{-2} Torr. Figure 6 shows the variations of the deposition rate with $1/d^2$, where d is the target–substrate distance. They observed that the growth rate reduces quicker than $1/d^2$ which is a characteristic of a point-source evaporation in vacuum. This feature was attributed to possible collisions between the atoms, ions, and clusters with the oxygen atmosphere. Zheng and Kwok have also shown that the target–substrate distance influences the electrical properties of ITO films during deposition by PLD. In a previous investigation, the group obtained the minimum resistivity at a target–substrate distance of 7 cm for a small-scale laboratory setup.

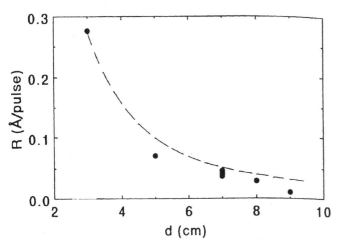

Fig. 6. Variations of deposition rate on the target–substrate distance. The dashed line shows the $1/d^2$ law. (Reprinted with permission from J. P. Zheng and H. S. Kwok, *Thin Solid Films* 232, 99 (1993), © 1993, Elsevier Science.)

They also reported in a separate study that a similar target–substrate distance is required for producing better quality superconducting films following the pressure–distance scaling law [134]. However, they noted that the target–substrate distance of 7 cm only applied well to an oxygen pressure of around 1.5×10^{-2} Torr, and there may exist an optimal target–substrate distance for various oxygen pressures. A majority of the laser experimental setups for the deposition of ITO films, including the setup described in this section, is based on a target–substrate spacing of 7 cm. In other studies, few groups have maintained a smaller target–substrate distance of 3–4 cm in their PLD system configuration and low-resistivity films have also been attained [88–91].

4.5. Optimization of Deposition Conditions (Background Gas)

To achieve high-quality ITO films, it is important to optimize the deposition parameters of the PLD system. Among these parameters the effect of oxygen pressure on the properties of the films is of fundamental importance. Since oxygen pressure moderates the energetic atomic and ionic species emerging from the target, it is essential to understand its effect on the properties of the film during growth. For this reason, Wu et al. [91] have developed a gas-dynamic model to explain the effect of the oxygen pressure on the electrical resistivity of ITO films based on a simple gas-phase collision representation. In their study, Wu et al. have been able to attain the following. (a) First, measure accurately the stoichiometry as a function of the chamber oxygen pressure. (b) Second, correlate the stoichiometry with the resistivity and the transmissivity of the films. (c) Finally, show that the resistivity can be modeled by a simple gas-phase collision representation which describes the resistivity as a function of oxygen gas pressure. The summary of their experimental and theoretical evaluations is presented as follows.

From Rutherford backscattering (RBS) measurements, Wu et al. have shown that the ratio of metal (In + Sn) to oxygen

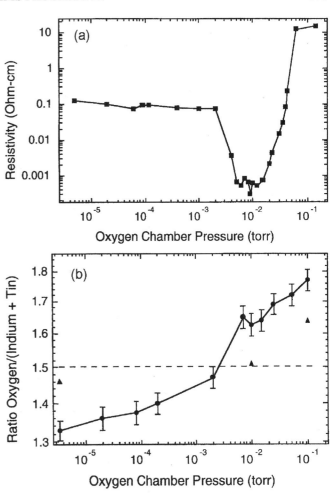

Fig. 7. (a) and (b) show the resistivity and oxygen-to-metal (indium + tin) ratio as a function of oxygen chamber pressure for ITO films deposited by PLD. ■ Resistivity, ● O/(In + Sn) ratio for room temperature deposit, ▲ O/(In + Sn) ratio after 400°C anneal, "dashed": ideal ratio for In$_{2-x}$Sn$_x$O$_3$. (Reprinted with permission from Y. Wu, C. H. M. Marée, R. F. Haglund, Jr., J. D. Hamilton, M. A. Morales Paliza, M. B. Huang, L. C. Feldman, and R. A. Weller, *J. Appl. Phys.* 86, 991 (1999), © 1999, American Institute of Physics.)

content during deposition can affect the resistivity of the films. RBS uses a monoenergetic beam of alpha particles to probe the near-surface layer of the material to be analyzed. The number and energy of backscattered particles contain information about the elemental composition of the sample versus thickness. Figure 7a and b shows the resistivity and oxygen-to-metal ratio as a function of oxygen pressure for the ITO films, respectively. It is obvious that the films deposited at oxygen pressure higher than 3×10^{-3} Torr contained excess oxygen. Below this oxygen pressure, the films were oxygen deficient according to the stoichiometric composition. In agreement with previous studies, Wu et al. observed that the resistivity is sensitive to the change in oxygen pressure [81–83, 88, 90, 92]. Films deposited at oxygen pressures below 3×10^{-3} Torr (i.e., oxygen deficient) yielded high resistivity, but at oxygen pressures of around 2×10^{-3} Torr, an onset abrupt decrease in resistivity occurred, just at the point where the oxygen-to-metal ratio showed a steep increase. ITO films with minimal resistivities were observed when the chamber pressure was over the range 5×10^{-3} to 1.5×10^{-2} Torr. From RBS measurements, the group observed that an optimum oxygen-to-metal ratio of ~ 1.6 produces the lowest resistivity films and this ratio coincided with oxygen pressure of around 1×10^{-2} Torr, shown in Figure 7b. This oxygen pressure range has been previously found by other groups to produce the best resistivity ITO films [81–83, 88, 90, 92]. Higher oxygen pressures result in increased oxygen-to-metal ratio with a corresponding sharp increase in the resistivity of the films.

4.5.1. Background Gas-Dynamic Model

The simple gas-phase collision model used by Wu et al. [91] to explain the behavior of the resistivity with change in the oxygen pressure for room temperature deposited ITO films represented in Figure 7a as shown in Figure 8 is discussed next. The model has been based on the following assumptions:

(i) Target atoms (ions and clusters) are emitted with a relatively high kinetic energy. When such atoms impinge on the substrate surface at room temperature, they cannot undergo surface diffusion in order to form "low-resistivity" stoichiometric films.

(ii) With increasing oxygen pressure, the energetic atoms equilibrate through atom–oxygen collisions.

(iii) The equilibrated fraction gently falls on the surface and is able to undergo chemical reactions to form stoichiometric, low-resistivity films. Stoichiometric clusters may also be formed in the "plume phase."

(iv) The fraction of equilibrated particles, f_{eq}, is represented by a simple multiple-collision scattering mechanism. The f_{eq} is calculated by assuming an energy independent, elastic scattering cross-section, σ, while arbitrarily defining f_{eq} as the fraction that has undergone m or more collisions, $f_{\geq m}$, given by [91, 135]

$$f_{\geq m} = e^{-(\sigma N)} \sum_{n=m}^{\infty} \frac{(\sigma N)^n}{n!} \quad (5)$$

where N represents the real density of the background oxygen molecules in the plume and is directly proportional to the background oxygen pressure.

(v) The resistivity is calculated as the parallel combination of the film containing equilibrated ($f_{\geq m}$) and the unequilibrated ($1 - f_{\geq m}$) film. The resistivity of the unequilibrated component is taken as $\rho_0 = 1 \times 10^{-1} \, \Omega \, \text{cm}$, the experimental value achieved at low oxygen pressure. For the equilibrated component, the resistivity ρ_i is taken as $1 \times 10^{-4} \, \Omega \, \text{cm}$, the "ideal value" for ITO film. The measured resistivity, ρ_m, is given by [91]

$$\rho_m = \frac{\rho_i \rho_0}{\rho_0 f_{\geq m} + \rho_i (1 - f_{\geq m})} \quad (6)$$

where $f_{\geq m}$ is a function of the oxygen pressure, as shown in Eq. (5).

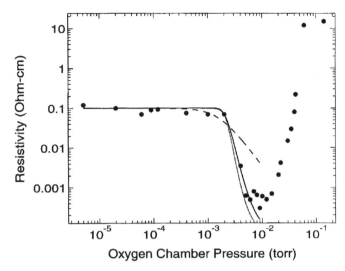

Fig. 8. Fitting of the gas dynamic model to the resistivity dependence on low chamber oxygen pressure for m or more collisions with cross section σ. ● Experimental; "dashed": $m = 2$, $\sigma = 2.3 \times 10^{-16} \, \text{cm}^2$; "solid": $m = 9$, $\sigma = 9.8 \times 10^{-15} \, \text{cm}^2$. (Reprinted with permission from Y. Wu, C. H. M. Marée, R. F. Haglund, Jr., J. D. Hamilton, M. A. Morales Paliza, M. B. Huang, L. C. Feldman, and R. A. Weller, *J. Appl. Phys.* 86, 991 (1999), © 1999, American Institute of Physics.)

From the model, Wu et al. were able to precisely determine the relationship between the resistivity and oxygen pressure during the growth of ITO films by PLD, as shown in Figure 8. Their theoretical model strongly agreed with the experimental data, particularly in the low oxygen pressure regime (i.e., $<1 \times 10^{-2}$ Torr). The best fit was obtained for the equilibrated particles, $f_{\geq m}$, for $m = 9$ and cross-section $\sigma = 9.8 \times 10^{-15} \, \text{cm}^2$. A similar model of equilibration has been reported by Wood et al. [135] using a scattering and hydrodynamic model to determine the energy distribution of PLD particles.

At higher oxygen pressure ($>1 \times 10^{-2}$ Torr), the rapid increase in the resistivity is controlled by the nonstoichiometric (oxygen-rich) property of the film. In this oxygen pressure region, the model could not explain the behavior of the resistivity of ITO films due to the lack of an adequate quantitative/physical model. Generally, the study has provided a clear understanding on the role of oxygen pressure on the composition as well as the resistivity of ITO films during deposition by PLD.

4.6. Initial Growth of ITO Films

During growth of thin films, atoms amassed on the substrate may diffuse laterally until they are lost by reevaporation, used up in formation of critical-size nuclei, seized by existing clusters, or caught at specific sites. The growth of thin film is generally divided into three growth modes. First, a three-dimensional (Volmer–Weber) growth mode, where film growth proceeds via formation and growth of separated islands; second, the layer-by-layer or two-dimensional growth (Frank–van der Merwe) mode which consists of the deposition of a monolayer at a time; third, the two-dimensional growth mode that translates into a three-dimensional growth (Stranski–Krastinov) mode where

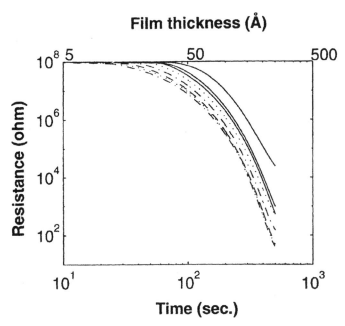

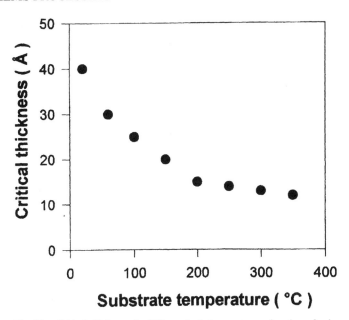

Fig. 9. Resistance of ITO films as a function of film thickness for various substrate temperatures. To avoid cluttering, the temperature values are not indicated in the figure. They are, starting from the top curve, 20, 50, 100, 150, 200, 250, 300, and 350°C, respectively. (Reprinted with permission from X. W. Sun, H. C. Huang, and H. S. Kwok, *Appl. Phys. Lett.* 68, 2663 (1996), © 1996, American Institute of Physics.)

Fig. 10. Critical thickness for ITO conductivity onset as a function of substrate temperature. (Reprinted with permission from X. W. Sun, H. C. Huang, and H. S. Kwok, *Appl. Phys. Lett.* 68, 2663 (1996), © 1996, American Institute of Physics.)

the monolayer growth is proceeded by island formation. Detailed discussions on the concepts of film growth can be found in [136, 137].

These initial growth modes determine the end properties of the films. The properties of the film are also susceptible to the nature of the supporting substrate during growth. In the case of ITO films, the initial stages of growth using PLD have been studied by Sun et al. [86] and their findings will constitute the majority of the discussion in this part. When ITO film is deposited on glass, no lattice matching is expected. Hence, under low-temperature conditions the initial growth proceeds through island formation from nucleation sites [138]. How the islands grow and coalesce was the focus in the study conducted by Sun et al. [86] ITO films were deposited on the glass substrates at substrate temperatures over the range 20 to 350°C. The initial growth stages were monitored using *in situ* resistance measurement, a method that has been used to study the properties of thin metallic films and dynamics oxidation of high-temperature superconducting oxide films [139, 140]. The same method has been used to prove that the growth of ITO films passes through stages of tunneling, percolation, and linear ohmic growth [19]. Sun et al. have shown from *in situ* resistance measurements that at temperatures below 150°C, the islands grow in three dimensions similar to the Volmer–Weber mechanism. On the other hand, above 150°C, the growth is two dimensional, similar to the Frank–van der Merwe mechanism. They found that the change in the growth attributes is related to the crystallinity of the deposited films.

Figure 9 shows the resistance of ITO films measured at various temperatures ranging from 20 to 350°C. It was observed that in all the cases there was an initial period where the film did not conduct and followed by a sudden onset of conductivity. In the experiment Sun et al. defined the onset of conductivity as the time when the measured resistance was 95 MΩ, which was affirmed from the maximum resistance of 100 MΩ the instrument could measure. It is apparent from Figure 9 that the region of no conductivity was due to formation of islands that were well separated. Figure 10 shows plots of the critical thickness of the film versus substrate temperature. The critical thickness has been defined as the nominal thickness of the film at the commencement of conductivity. It is obvious that the critical thickness decreased with increasing temperature before stabilizing at 1.2 nm, which was close to the lattice constant of ITO. By extrapolating the low-temperature-decreasing portion of the curve to intercept the 1.2 nm horizontal line at 150°C in Figure 10, Sun et al. predicted that the growth characteristics of ITO were different below and above the substrate temperature of 150°C. This temperature coincided with the transition temperature from amorphous to crystalline film during growth. The group interpreted their results as follows. At the low-temperature region, the initial growth proceeds via island formation on the glass substrate. These islands grow and coalesce via either a two-dimensional or a three-dimensional route. Below 150–200°C, the islands grow in a three-dimensional mode similar to Volmer–Weber mechanism, while above 150–200°C the islands become typically one unit cell thick and then grow sideways in a two-dimensional mode conforming to the Frank–van der Merwe mechanism.

As mentioned above, in the Frank–van der Merwe thin-film growth mechanism, the building blocks grow laterally in a two-dimensional form. Therefore, according to Sun et al., the volume of the film scales as the total area of the film. In this case the critical thickness will be constant regardless of the separation of the nucleation sites. Assuming that the temperature only affected the density of nucleation site, the critical thickness would be independent of thickness. This was the situation for the films deposited at temperature higher than 150°C. Hence, the growth mechanism at high temperature followed the Frank–van der Merve mechanism, and the critical thickness was similar to that of the blocks which was equal to one lattice constant. At lower temperature (<150°C), the films were amorphous, and the critical thickness was temperature dependent. This representation conformed to a three-dimensional growth of islands. In three-dimensional growth, the volume of the film ascends as L^3, while the area ascends as L^2, where L is the linear dimension of the islands. Consequently, the critical thickness of the film when the islands start to merge would depend on the separation of the nucleation sites. Assuming that the nucleation sites are arranged in a hexagonal form with a separation of d, and the islands are half spheres, when the islands contact, the average thickness of the film is easily shown to be $\pi d/6\sqrt{3}$. Using this model, Sun et al. showed that the nucleation sites were separated by 13.2 nm at 20°C and 6.6 nm at 150°C. Thus, these results implied that during growth of ITO films by PLD, the growth begins with the formation of islands on activated nucleation sites. At low temperatures, the islands are amorphous and they grow laterally as well as in height in a three-dimensional mode. At high temperatures, the islands are polycrystalline and they grow to a thickness of about one lattice parameter before growing laterally in a two-dimensional mode. The transition temperature between the two modes exists at 150–200°C, which also coincides with the transition temperature from amorphous to polycrystalline ITO film.

4.7. Film Deposition and Characterization

The discussion in this part is focused on the fundamental laser system setup for thin-film work. The excimer laser system layout described here is based on the system we have used in our PLD work [79–83]. The basic operation of the laser setup described below is relatively similar to those that have been used by other groups [84–92]. Therefore, we begin by describing this process in some detail.

The films were deposited using a KrF (248 nm) excimer laser (Lambda Physik, COMPex 102) system, as shown in Figure 11. The layout consisted of a stainless steel cylindrical chamber that was equipped with a turbomolecular pump. The chamber was capable of pumping down to 1×10^{-6} Torr. The target holder was a multiple type that can accommodate a maximum of six targets. The laser beam was focused onto the target located within the vacuum chamber through a quartz glass window using a spherical lens with a diameter of 50 cm. The substrate holder was a stainless steel plate that could be electrically heated up to 850°C. The distance between the target and the substrate was fixed at 7 cm. Figure 12 shows a schematic diagram of the PLD system. Depending on the deposition con-

Fig. 11.   Photograph of a typical PLD system.

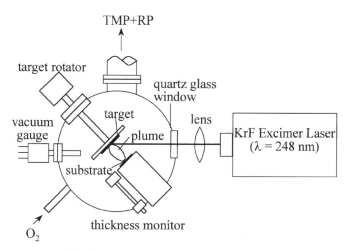

Fig. 12.   A schematic diagram of the PLD system.

ditions such as the oxygen pressure, the laser operating at a pulse rate of 20 Hz can produce energy densities between 1 and 10 J cm^{-2}.

Targets consisting of high-purity (99.99%) undoped In$_2$O$_3$ and Sn-doped (2–15 wt%) In$_2$O$_3$ hot-pressed sintered ceramics have been investigated for film deposition by PLD [79–92]. Glass substrate is widely used to support ITO film during PLD, although the substrate used depends on the type of application. Glass enables the optoelectronic properties of the films to be examined. However, other substrates that have been used are silicon and yttrium-stabilized zirconia; the later is used to investigate the epitaxial growth of ITO films by PLD. It is well established that the properties of the films such as hardness, optical, electrical, and microstructural properties depend on the surface conditions of the supporting substrate. Therefore, prior to mounting in the deposition system the substrates were first cleaned to free the surfaces from contaminants such as dusts, clothing or skin particles, solvent impurities, and skin oils, which could alter the optoelectronic properties of the deposited films.

ITO films have been studied over a wide range of chamber oxygen pressures ranging from 1×10^{-5} to 5×10^{-1} Torr, and substrate temperatures ranging from room temperature to 500°C [79–92]. Oxygen could be released from the source material or target during deposition process [6]. The oxygen released causes a nonstoichiometry in the films and the magnitude is dependent on the deposition method. Therefore, in order to grow films of superior quality with good reproducibility it is essential to carefully monitor the oxygen pressure. In the PLD system, this was achieved by controlling the oxygen pressure using a mass flow controller, so that stoichiometric In_2O_3 and ITO films were produced via creation of the desired oxygen vacancies. This is particularly important for In_2O_3 films where doping is entirely dependent on the creation of oxygen vacancies. In this way, the film stoichiometry were maintained.

The substrate temperature is another important parameter that requires crucial control during deposition. This is due to the fact that the reactivity of the oxygen with the ablated particles at the substrate is essential. The substrate temperature has been found to influence the crystallinity of ITO films [5, 6, 124]. The deposition rate is a most important parameter for the film quality which in turn can affect the thickness of the film. The deposition rate was monitored by the use of a vibrating quartz microbalance mounted near the substrate. The frequency response was converted into deposition rate and the resultant film thickness was obtained on a microprocessor. The growth rates mostly depended on the oxygen pressure and the energy density used during film growth and varied from 2 to 15 nm min^{-1}. A typical deposition rate of 12 nm min^{-1} was achieved at oxygen pressure of 1×10^{-2} Torr using an energy density of ~5 J cm^{-2} [82]. It has also been reported that the properties of ITO films depend on the thickness [8, 141, 142] and could result from a display of the deviation in the deposition conditions [6]. The thickness of the deposited films can vary depending on the type of application. However, for flat panel displays, the film thickness is usually limited to 100 ± 20 nm. Hence, during deposition, the substrate temperature, deposition rate, oxygen pressure, and laser output energy were monitored continuously in order to maintain the growth parameters.

Proper characterization of the films is essential for the understanding of their electrical, optical, structural, and chemical properties. The thickness of the ITO films was measured by a stylus profilometer. For such measurements, a well-defined edge across the substrate is required. Hence a stainless steel strip was clamped over the substrate to create a step for accurate determination of the film thickness. The grain sizes, crystallinity, and surface roughness were determined from the top surface of the films using high resolution SEM and atomic force microscopy (AFM). The X-ray diffraction (XRD) patterns of the ITO films were generated using a computer controlled diffractometer with Bragg–Bretano focusing geometry and Cu $K\alpha$ ($\lambda = 1.5405$ Å) radiation. Wide-angle 2θ scans from 15 to 70° taken at 0.05 steps were used to ascertain the identity of the phases present in the films.

It has been reported that during the growth of TCO films, strain could be induced in the films due to the difference in the thermal conductivity of the substrate and the film [89, 143]. In addition, high-energy processes have often generated strain

in thin film during growth [143]. It has been observed from XRD analysis that ITO films deposited, particularly, under low oxygen pressure displayed peak splitting as a result of induced strain during growth [82]. The induced strain in the ITO films deposited by PLD has been determined using grazing incidence X-ray diffraction. The electrical properties of the films deposited on glass were examined by Hall measurements using van der Pauw geometry. The electrical resistivities of the films deposited on silicon substrates were determined using the four-point-probe method. Information on the optical properties was obtained from the optical spectrophotometer, operating at wavelengths between ultraviolet (150 nm) and near infrared (2500 nm) regions. The results from the electrical, optical, and structural properties of the ITO films are presented in Sections 5 to 7.

5. PROPERTIES OF PLD INDIUM OXIDE FILMS

Undoped In_2O_3 film has been the foundation for the many studies conducted on ITO films. Although much attention is now focused on ITO, In_2O_3 is still studied for fundamental understanding of the material properties of ITO films. Knowing that the conduction mechanism of ITO is governed by intrinsic doping with oxygen vacancies and addition of Sn dopants, it is essential first to understand the intrinsic properties of undoped films deposited by any technique. This will help to relate properly the effect of extrinsic dopants on the film properties with respect to changes in deposition conditions. In this regard, Adurodija et al. [83] have studied the intrinsic properties of undoped In_2O_3 films deposited by PLD. Depositions were performed using the laser system described in Section 4.7. During deposition, the energy density was maintained at ~8 J cm^{-2}, while the substrate temperature and the oxygen pressure were varied from room temperature (20°C) to 350°C, and 1×10^{-3} to 5×10^{-2} Torr, respectively. The thickness of the deposited films was of the order of 100 nm. The experimental results of the structural and electrooptic properties of the films are discussed in this section.

5.1. Electrical Properties

It has been shown in Section 2 that the electrical properties of undoped In_2O_3 rely completely on the oxidation state of the indium metal constituent. The properties of the films were investigated in relation to the changes in the oxygen pressure and substrate temperature. Figure 13a–c shows the respective resistivity, carrier concentration, and Hall mobility of In_2O_3 films deposited at room temperature, 100°C, 150°C, and 200°C as a function of oxygen pressure [83]. High resistivities were observed at low oxygen pressure of 1×10^{-3} Torr for all the substrate temperatures investigated. With increasing oxygen pressure to between 1×10^{-2} and 1.5×10^{-2} Torr, the resistivities decreased to minimal values of 3.6×10^{-4}, 3.2×10^{-4}, 2.8×10^{-4}, and 3.7×10^{-4} Ω cm at room temperature, 100°C, 150°C, and 200°C, respectively. A subsequent abrupt rise in the

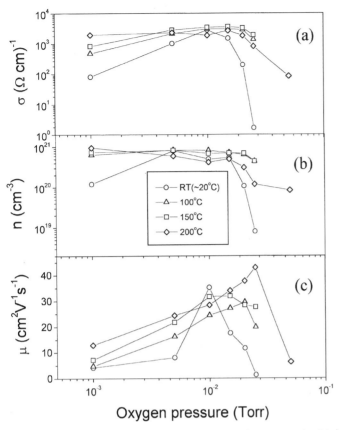

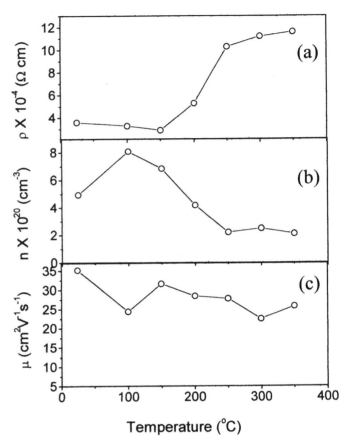

Fig. 13. (a)–(c) show the respective resistivity (ρ), carrier concentration (n), and Hall mobility (μ) of In$_2$O$_3$ films deposited at different substrate temperature as a function of oxygen pressure. (Reprinted with permission from F. O. Adurodija, H. Izumi, T. Ishihara, H. Yoshioka, H. Matsui, and M. Motoyama, *Appl. Phys. Lett.* 74, 3059 (1999), © 1999, American Institute of Physics.)

Fig. 14. Plots of (a) resistivity (ρ), (b) carrier concentration (n), and (c) Hall mobility (μ) against substrate temperature for In$_2$O$_3$ films. (Reprinted with permission from F. O. Adurodija, H. Izumi, T. Ishihara, H. Yoshioka, H. Matsui, and M. Motoyama, *Appl. Phys. Lett.* 74, 3059 (1999), © 1999, American Institute of Physics.)

resistivity was observed with further increase in oxygen pressure, irrespective of the substrate temperature used. For example, at 200°C, the resistivity was already more than an order of magnitude higher than the least value as the oxygen pressure increases from 1.5×10^{-2} to 5×10^{-2} Torr. The room temperature resistivity value marked the lowest reported for undoped In$_2$O$_3$ films deposited by any technique [73, 144].

From Figure 13b, at room temperature carrier concentration of $\sim 10^{20}$ cm^{-3} was obtained at an oxygen pressure of 1×10^{-3} Torr. It then increased to around $(5-7) \times 10^{20}$ cm^{-3} and remained fairly stable at oxygen pressures between 5×10^{-3} and 1.5×10^{-2} Torr, before decreasing rapidly to about 3×10^{18} cm^{-3} as the oxygen pressure increased to 5×10^{-2} Torr. However, for heated substrates (100–200°C), the carrier concentration remained roughly consistent at $(4-8) \times 10^{20}$ cm^{-3} at oxygen pressures between 1×10^{-3} and 1.5×10^{-2} Torr. It then dropped freely to $\sim 10^{18}$ cm^{-3} at an oxygen pressure of 5×10^{-2} Torr with a consequential increase in the resistivity. The films with the low resistivity yielded higher carrier concentrations (up to 8×10^{20} cm^{-3}) than previously reported for pure In$_2$O$_3$ films [16]. In Figure 13c, modest Hall mobility, less than 45 cm^2 V^{-1} s^{-1}, was observed in all the films and these values are much lower than earlier reported for In$_2$O$_3$ films

[16, 144]. The low resistivity observed in the films was essentially a result of the high carrier concentration. It is not understood why the films exhibited high carrier concentration and only a shallow maximum of Hall mobility. However, this maximum yielded the lowest resistivity within the oxygen pressure range of 1×10^{-2} to 1.5×10^{-2} Torr. These observations corresponded closely to those earlier made by Fan et al. [48] in their sputtered ITO films. These results showed that the number of oxygen vacancies required for obtaining reasonably low resistivity In$_2$O$_3$ films during PLD process falls within oxygen pressures of 1×10^{-2} and 1.5×10^{-2} Torr which is in agreement with what has been reported for ITO [79–92].

The effects of the substrate temperature on the resistivity, the carrier concentration, and the Hall mobility of the In$_2$O$_3$ films are shown in Figure 14a–c. It is seen in Figure 14a that an increase in the substrate temperature led to a significant increase in the resistivity of the In$_2$O$_3$ films. This implied an increase in the compensation for the oxygen vacancies during growth. This may be due to loss of free carriers contributed by the creation of oxygen vacancies during the formation and crystallization of the films at high substrate temperature, as shown in Figure 14b. The Hall mobility remained moderately low at high substrate

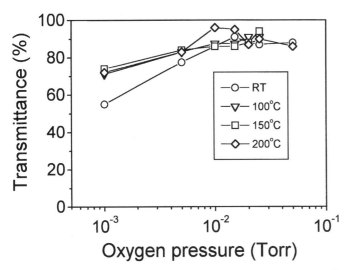

Fig. 15. Optical transmittance of In$_2$O$_3$ films averaged over a wavelength range 450–800 nm, as a function of oxygen pressure. (Reprinted with permission from F. O. Adurodija, H. Izumi, T. Ishihara, H. Yoshioka, H. Matsui, and M. Motoyama, *Appl. Phys. Lett.* 74, 3059 (1999), © 1999, American Institute of Physics.)

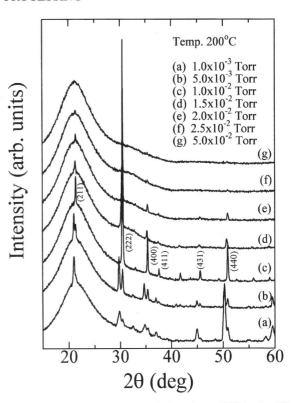

Fig. 16. XRD spectra of ITO films deposited on glass at 200°C under different oxygen pressures. (Reprinted with permission from F. O. Adurodija, H. Izumi, T. Ishihara, H. Yoshioka, H. Matsui, and M. Motoyama, *Japan. J. Appl. Phys. 1* 38, 2710 (1999), © 1999, Japan Society of Applied Physics.)

temperature and decreased even further with increasing substrate temperature, as shown in Figure 14c.

5.2. Optical Properties

The optical transmittance of the In$_2$O$_3$ films averaged over a wavelength range of 450 to 800 nm (visible range) plotted against oxygen pressure for temperature over the range 20–200°C is shown in Figure 15 [83]. At oxygen pressures below 5×10^{-3} Torr, the films appeared brownish in color and translucent (transmittance <80%), irrespective of the deposition temperature. With increasing oxygen pressure, the transmittance also increased before stabilizing between 85 and 95%. For the films deposited at room temperature, in addition to increasing transmittance with increasing oxygen pressure, a shift of the absorption edge to shorter wavelengths also occurred. This shift correlated to the decrease in carrier concentration, as shown Figure 13a, and is related to the Burstein–Moss effect [145]. Further results on the optical properties of In$_2$O$_3$ or ITO films are presented in Section 6.

5.3. Structural Properties

The structural properties of the In$_2$O$_3$ films were examined by XRD. XRD analyses showed that the films deposited at substrate temperature below 100°C were amorphous, whereas from a substrate temperature of 150°C, crystalline phases corresponding to the cubic bixbyite structure appeared. A dependence of the structure on the oxygen pressure was also observed. It has been found that films grown at high oxygen pressure, over 2×10^{-2} Torr, indicated weak reflections, while as the oxygen pressure decreases, peak reflections also became sharper. Detailed results on the structural properties of In$_2$O$_3$ and ITO films are discussed in Section 6.

6. PROPERTIES OF PLD ITO FILMS

In this section the general properties of ITO films deposited by PLD are discussed. The areas covered include the structural, electrical, and optical properties of the films.

6.1. Structural and Other Properties

6.1.1. Structural Properties

Next we discuss properties of ITO films containing different Sn doping content. The properties of ITO films are compared with those of undoped In$_2$O$_3$ films. The crystalline structures of the films are studied by X-ray diffraction as a function of the oxygen pressure and substrate temperature. XRD analysis of the films deposited at room temperature showed no diffraction peaks, irrespective of the changes in Sn doping content. The absence of diffraction peaks showed that the films were amorphous. The XRD spectra shown in Figure 16 indicated that the films deposited at 200°C were polycrystalline [83]. XRD profiles showed only standard ITO peaks with strong (222) peaks, thus suggesting a ⟨111⟩ preferred orientation. However, the ⟨111⟩ preferred orientation of the films reduced with changes in oxygen pressure. It is seen that the diffraction peaks decreased with an increase or decrease in the oxygen pressure during growth. The deterioration in the crystallinity of the films, especially at high oxygen pressure, has been attributed to an increase

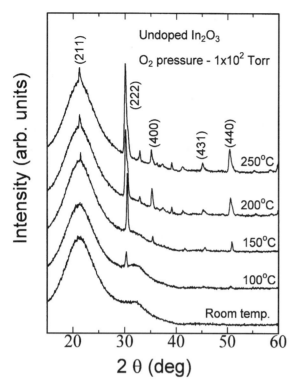

Fig. 17. Dependence of XRD spectra on the substrate temperature for undoped In$_2$O$_3$ films deposited on glass. (Reprinted with permission from F. O. Adurodija, H. Izumi, T. Ishihara, H. Yoshioka, M. Motoyama, and K. Murai, *J. Vac. Sci. Technol.* A 18, 814 (2000), © 2000, American Institute of Physics.)

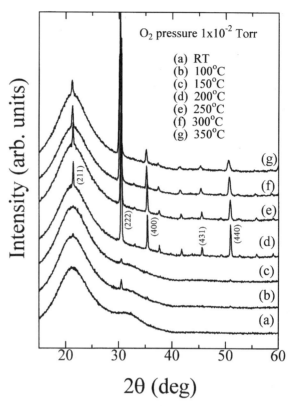

Fig. 18. Dependence of XRD patterns on the substrate temperature for 5 wt% Sn-doped In$_2$O$_3$ films deposited on glass. (Reprinted with permission from F. O. Adurodija, H. Izumi, T. Ishihara, H. Yoshioka, M. Motoyama, and K. Murai, *J. Vac. Sci. Technol.* A 18, 814 (2000), © 2000, American Institute of Physics.)

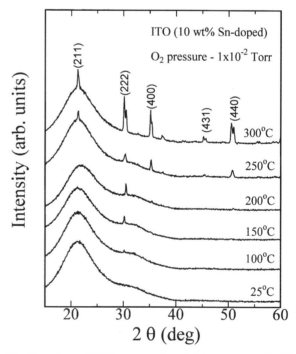

Fig. 19. Dependence of XRD spectra on the substrate temperature for 10 wt% Sn-doped In$_2$O$_3$ films deposited on glass. (Reprinted with permission from F. O. Adurodija, H. Izumi, T. Ishihara, H. Yoshioka, M. Motoyama, and K. Murai, *J. Vac. Sci. Technol.* A 18, 814 (2000), © 2000, American Institute of Physics.)

in the adsorption of oxygen into the films which could prevent the surface migration of In or Sn atoms, leading to a structural disorder [44, 146]. Kim et al. [88] also observed a ⟨111⟩ preferred orientation in the ITO films deposited at 300°C under oxygen pressure of 1×10^{-2} Torr. However, they noted that as the oxygen pressure increases from 1×10^{-3} to 1×10^{-1} Torr the (222) peak intensity also increases.

From Figure 16 it is also found that upon decreasing the oxygen pressure to less than 1×10^{-2} Torr, the splitting of diffraction peaks appeared, thus indicating the formation of layered structures. Splitting of the peaks has been reported for sputtered and electron-beam-evaporated ITO films deposited at 185°C and the cause has been associated with the formation of two different strained layers resulting from thermally assisted (solid-phase) crystallization and vapor-phase crystallization [126]. In PLD ITO films, peak splitting due to strain has been observed mostly in the films deposited on heated substrates at low oxygen pressures and having a thickness greater than 150 nm. Detailed analysis on induced strain in ITO films during growth by PLD is discussed in the later part of this section.

Undoped In$_2$O$_3$ was mostly observed to be crystalline at substrate temperature above 150°C and exhibit structures similar to that of Sn-doped In$_2$O$_3$ films deposited by PLD. Figures 17 to 19 show XRD patterns of undoped In$_2$O$_3$ and 5 wt% and 10 wt% Sn-doped In$_2$O$_3$ films as a function of substrate temperatures, respectively [83]. The films were deposited at the optimal oxygen pressure of $(1-1.5) \times 10^{-2}$ Torr. From substrate temperatures as low as 100°C, crystalline films started to form in the undoped In_2O_3 and ITO films [80]. This temperature is less than the crystallization temperature (T_c = 150–160°C) previously reported for ITO films [44, 147, 148]. A T_c of 150°C has also been reported for 10 wt% Sn-doped In_2O_3 films prepared by Sun et al. [86] using PLD with a laser fluence of 1 J cm^{-2}. Kim et al. [88] also observed crystalline films from a temperature of 100°C for ITO films deposited by PLD using a laser fluence of 2 J cm^{-2}. The group observed a strong (400) diffraction peak, indicating a ⟨100⟩ preferred orientation, but as the substrate temperature increased the orientation changed to ⟨111⟩ plane.

It has been documented in the literature that ablated particles possess high kinetic energy that increases with increase in laser fluence [92, 149]. Therefore, a combined use of higher laser fluence of 2–5 J cm^{-2} (more energetic photon-ejected particles) and low thermal energy (100°C) could initiate crystallization of the films, with a consequential enhancement in the quality of the films. In undoped In_2O_3 films, strong XRD peak intensities were already formed at a substrate temperature of 150°C unlike Sn-doped In_2O_3 films, as shown in Figures 18 and 19. It should be mentioned that the growth of polycrystalline films at room temperature or substrate temperature below 50°C at high sputtering power and reduced target–substrate distance has been reported [150, 151]. However, these films exhibited high resistivity and low optical transmittance. The attainment of crystalline ITO films by sputtering at low substrate temperature was related to the high kinetic energy of the sputtered particles [44, 124, 151]. Therefore, the crystallization of ITO films observed at 100°C by PLD could be associated with the energetic photon-ejected particles impinging on the film surface coupled with a secondary thermal effect.

At substrate temperatures above 200°C, XRD showed crystalline films with strong (222) peak intensity and ⟨111⟩ preferred orientation for undoped In_2O_3 and 5 wt% Sn-doped In_2O_3 films. In the case of the 10 wt% Sn-doped In_2O_3 films, weak diffraction peaks with a slight ⟨100⟩ preferred orientation were observed at a substrate temperature of 250°C. At 300°C, splitting of the diffraction peaks occurred in the 10 wt% Sn-doped In_2O_3 films. This feature is similar to that observed for 5 wt%-doped In_2O_3 films grown at oxygen pressure below 5×10^{-3} Torr, as shown in Figure 18.

6.1.2. Morphological Properties

Figure 20 shows the surface SEM micrographs of ITO (5 wt% Sn-doped In_2O_3) films deposited at room temperature and oxygen pressures of (a) 1×10^{-2} Torr and (b) 2×10^{-2} Torr. The films show smooth surface morphology with no defined granular structures, thus implying an amorphous state. This feature confirmed the XRD data shown in Figure 18. No change in the microstructure with changes in Sn doping content was observed. It is also noticed that the films grown at low and high oxygen pressures ($< 5 \times 10^{-3}$ and $> 2 \times 10^{-2}$ Torr) were nonuniform as shown by islands present on the surfaces.

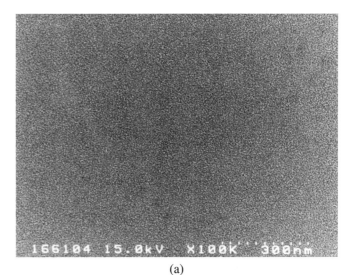
(a)

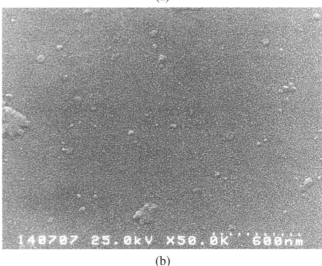
(b)

Fig. 20. Surface SEM micrographs of ITO films (5 wt% Sn-doped In_2O_3) deposited at room temperature and oxygen pressures of (a) 1×10^{-2} Torr and (b) 2×10^{-2} Torr.

Figure 21 shows the surface SEM micrographs of (a) an undoped In_2O_3 film deposited at 1×10^{-2} Torr and (b) to (d) 5 wt% Sn-doped In_2O_3 films deposited at 5×10^{-2} Torr, 1×10^{-2} Torr, and 1×10^{-3} Torr, respectively. All the films were deposited at temperature of 200°C. The SEM micrographs clearly showed that the microstructure of the films differed with changes in oxygen pressure. At oxygen pressures of 1×10^{-3} and 5×10^{-2} Torr, the surfaces of the films showed poor microstructures with smooth and spotting materials across the surfaces. In contrast, the SEM micrographs of the undoped In_2O_3 and 5 wt% Sn-doped In_2O_3 deposited at the optimum oxygen pressure of 1×10^{-2} Torr showed distinctive microstructure with very dense crystallites and average sizes larger than 300 nm. The SEM analysis of the films closely agreed with the results of the XRD measurements discussed earlier.

The surface roughness (R_a) of the films was studied as a function of the substrate temperature using AFM. The films were deposited under oxygen pressure of 1×10^{-2} Torr and substrate temperature ranging from room temperature to 200°C.

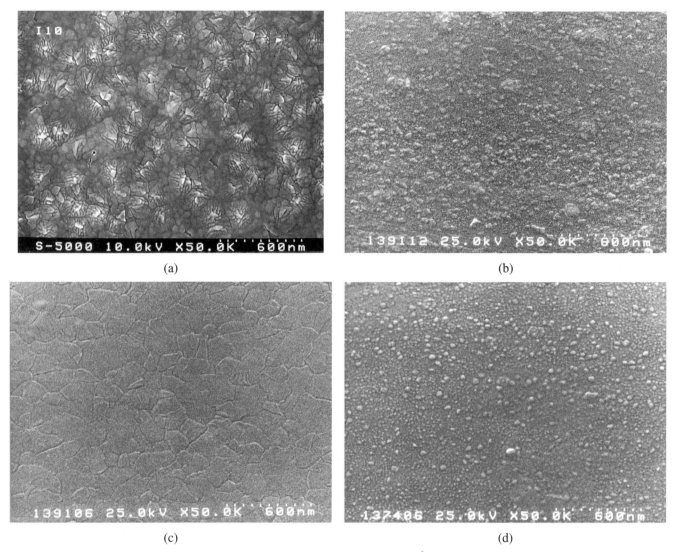

(a)

(b)

(c)

(d)

Fig. 21. Surface SEM micrographs of (a) an undoped In_2O_3 film deposited at 200°C under 1×10^{-2} Torr, and (b) to (d) 5 wt% Sn-doped In_2O_3 films deposited under 5×10^{-2}, 1×10^{-2}, and 1×10^{-3} Torr, respectively.

Figure 22 shows the average R_a of the films plotted against substrate temperature [83]. The R_a for an unheated bare SiO_2 fused quartz glass substrate was determined as $\sim$1.5 nm. The films deposited at room temperature and 200°C were smoother, as indicated by the low R_a values ($<$1.3 nm). Between temperatures of 100 and 150°C, the films appeared textured (as observed from the AFM pictures, not shown) with surface roughness ranging from 2.5 to 4.9 nm. These results show that oxygen pressure and substrate temperature could significantly affect the microstructure of ITO films.

6.1.3. Effects of Strain on the Structure of ITO Films

This section concerns further investigations on the splitting of diffraction peaks earlier mentioned in Section 6.1.1. As discussed previously in Section 2, In_2O_3 has a $\langle 111 \rangle$ texture, exhibiting a cubic structure and a lattice constant of 10.118 Å [5, 152]. However, the lattice constant can vary from 10.122

to 10.238 Å depending on the amount of Sn dopant added into the films [88, 153]. The increase in the lattice constant due to Sn doping is explained as the substitutional inclusion of Sn^{2+} ions into In^{3+} sites and/or the inclusion of Sn ions in the interstitial locations [88]. Because the radius (0.93 Å) of Sn^{2+} ions is greater than that (0.79 Å) of In^{3+}, the substitution of Sn^{2+} for In^{3+} could result in lattice distortion [88, 154]. The increase in the lattice constant also has been related to oxygen deficiency and strain effect due to mismatch in the thermal expansion coefficient between the film (7.2×10^{-6} °C) [88, 155] and the glass substrate (4.6×10^{-6} °C) [88, 156, 157].

It has been reported that the existence of strain in thin film is dependent on the growth conditions such as working gas pressure, substrate temperature, and deposition rate [143, 158] and these changes have been found to affect the behavior of systems like SiGe and SiO_2/Si [159, 160]. This study was carried out to shed light on the probable cause of the peak splitting in the ITO films. In recent years, grazing incidence X-ray

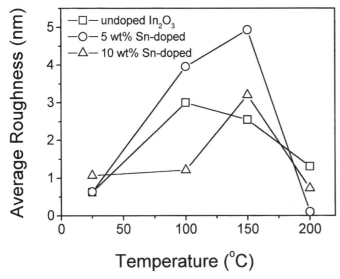

Fig. 22. Average surface roughness (R_a) of In$_2$O$_3$ and ITO films plotted against substrate temperature. (Reprinted with permission from F. O. Adurodija, H. Izumi, T. Ishihara, H. Yoshioka, M. Motoyama, and K. Murai, *J. Vac. Sci. Technol. A* 18, 814 (2000), © 2000, American Institute of Physics.)

diffraction (GIXRD) has been used to analyze the surface and the bulk structures of both crystalline and polycrystalline thin films [161]. Using GIXRD, the surface sensitivity is enhanced, thereby enabling the surface and the bulk properties of the film to be determined. GIXRD combined with conventional θ–2θ XRD method was employed to examine the structure and the evolution of strain in ITO layers deposited by PLD [162]. GIXRD with step scan of $0.04°$ using a Cu $K\alpha$ ($\lambda = 1.5405$ Å) radiation was used. During measurements, the incidence angle (γ) was varied from 0.3 to $5°$. Small γ provided information on the surface properties of the film, while larger γ gave information on the bulk and near the substrate.

Undoped In$_2$O$_3$ and Sn-doped (3 wt%, 5 wt%, and 10 wt%) In$_2$O$_3$ films with thickness of about 100 nm were deposited on fused quartz SiO$_2$ glass substrates at $200°$C and oxygen pressure of 1×10^{-2} Torr using a KrF (248 nm) excimer laser with a fluence of 3–5 J cm^{-2}. In Figure 18 it was shown that apart from the changes in the diffraction peak intensities with decreasing oxygen pressure, splitting of the peaks also occurred at oxygen pressures less than 5×10^{-3} Torr. This feature has been associated with the increased energetic ablated particles incident on the substrate due to a decrease in the scattering rates by the low oxygen ambient.

In the PLD method, laser-ejected particles were reported to carry high velocities of $\sim10^6$ cm/s (high kinetic energy) [92]. Hence, at low oxygen pressure ($< 5 \times 10^{-3}$ Torr) repeated bombardment of the growing film by the energetic particles could generate internal strain in the films with a consequential splitting of the diffraction peaks. At high oxygen pressure ($> 5 \times 10^{-3}$ Torr), because the kinetic energy of the ablated particles was significantly reduced due to the high scattering rates, no splitting occurred in the films. The kinetic energy of ablated particles was considered to be several electron volts (eV) high,

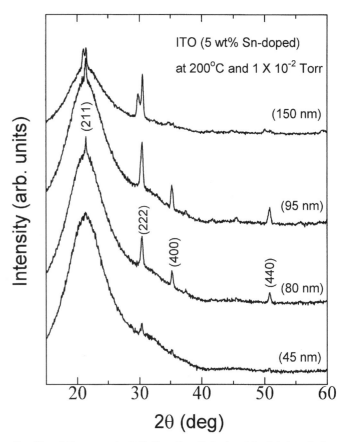

Fig. 23. XRD spectra for ITO films (5 wt% Sn-doped In$_2$O$_3$) deposited at $200°$C and oxygen pressure of 1×10^{-2} Torr as a function of thickness. (Reprinted with permission from F. O. Adurodija, H. Izumi, T. Ishihara, H. Yoshioka, and M. Motoyama, *J. Mater. Sci.: Mater. Electronics* 12, 57 (2001), © 2000, Kluwer Academic Publishers.)

and in the case of the deposition of YBa$_2$Cu$_3$O$_{7-\delta}$ superconducting films by PLD, the kinetic energy of the particles has been found to be greater than 13 eV [149]. Besides, the energy of the flying particles has been reported to affect the properties of thin films [44, 150]. For example, using sputtering techniques, the effect of high-energy particles has been found to severely affect the structure and the electrical properties of ITO films during growth [44, 150, 163].

Furthermore, the film thickness was varied from 45 to 175 nm in order to study the development of peak splitting. Figure 23 shows XRD spectra for ITO films (5 wt% Sn-doped In$_2$O$_3$) deposited at $200°$C and oxygen pressure of 1×10^{-2} Torr as a function of thickness [162]. Films with thickness less than 100 nm showed only a single diffraction peak, but on increasing the film thickness, above 150 nm, splitting of the diffraction peaks became apparent. This suggested that the existence of strain in the ITO layers was predominant in thicker films. Residual strain can be classified into intrinsic and thermal components. The former which is often referred to as growth strain is the cumulative result of chemical and microstructural defects included during deposition, while the latter is due to thermal expansion of the film and the substrate during cooling [157].

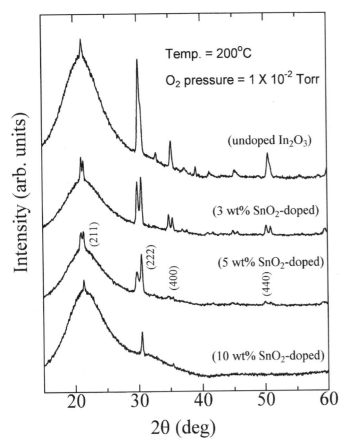

Fig. 24. XRD pattern of ITO films with Sn doping content between 0 and 10 wt% showing splitting of the peaks for the 3 and 5 wt% Sn doped films. (Reprinted with permission from F. O. Adurodija, H. Izumi, T. Ishihara, H. Yoshioka, and M. Motoyama, *J. Mater. Sci.: Mater. Electronics* 12, 57 (2001), © 2001, Kluwer Academic Publishers.)

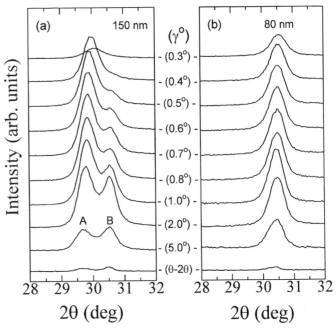

Fig. 25. XRD spectra of a (222) peak for 150 and 80 nm ITO (5 wt% Sn-doped In₂O₃) films measured at different GIXRD angles (γ). (Reprinted with permission from F. O. Adurodija, H. Izumi, T. Ishihara, H. Yoshioka, and M. Motoyama, *J. Mater. Sci.: Mater. Electronics* 12, 57 (2001), © 2001, Kluwer Academic Publishers.)

Intrinsic strain resulting from chemical and microstructural defects is assumed to be dominant in the ITO films in view of the low deposition temperature of 200°C used. The ablated particles carrying high energies are suspected to induce microstructural defects in the films during growth.

The XRD pattern of ITO films with Sn doping content between 0 and 10 wt% is shown in Figure 24 [162]. It is seen that peak splitting only occurred in the ITO films with Sn-doping content between 3 and 5 wt% and of thickness ≥150 nm. However, peak splitting has also be observed in undoped In₂O₃ and ITO films (10 wt% Sn-doped) deposited at lower oxygen pressures and temperatures of 250°C and above, as shown in Figure 19. A similar strain-induced distortion in the structure has also been reported for ITO films (4 wt% Sn-doped In₂O₃) prepared by sputtering and electron-beam evaporation at 185°C [126]. Figure 25a and b shows the XRD spectra of a (222) peak for 150 and 80 nm thick ITO (5 wt% Sn-doped In₂O₃) films measured at different GIXRD angles (γ) [162]. A steady shift in the (222) diffraction peak position toward higher 2θ angle with increasing γ occurred, particularly for the 150-nm-thick film. The shift in the peak position was associated mainly with intrinsic strain (microstructural defect) that resulted from repeated

bombardment of the growing film by the high-energy particles. Strain has been reported to be a common cause of peak shift in many thin-film systems, including ZnO films [143].

From Figure 25a, a single (222) peak that corresponded to higher 2θ angle (peak B) was formed near the surface of the film, but as γ (X-ray penetration depth) increases, two (222) peaks (A and B) emerged. It should be noted that the split affected all the peaks. It is speculated that peak A resulted primarily from the intrinsic strain induced in the film during growth. This observation is similar to that reported by Yi et al. on sputtered and electron-beam-deposited ITO films [126]. They attributed this phenomenon to the presence of two strained layers that was caused by thermal-assisted (solid-phase) crystallization and vapor-phase crystallization [126]. The formation of two layers could not be established in the ITO films deposited by PLD. However, the dual effects of energetic-particle bombardment of the growing film and substrate heating could exert intrinsic strain with a resultant change in the lattice constants (peak shifts), splitting of XRD peaks, and the presence of two layers in the films. In the case of the 80-nm-thick ITO film, no apparent splitting of the (222) diffraction peak was observed and the shift in the peak positions with increasing γ was minimal.

Assuming that the induced-strain is uniform, the percentage of the strain in the ITO layers (150 and 80 nm) was determined as a function of the grazing incidence angle using the relation [143]

$$\text{Strain [\%]} = (c_{\text{film}} - c_{\text{bulk}})/c_{\text{bulk}} \qquad (7)$$

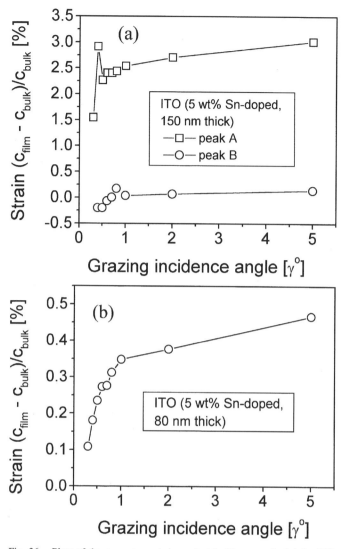

Fig. 26. Plots of the percentage strain against incidence angle (γ) for ITO films with thickness of (a) 150 nm and (b) 80 nm. (Reprinted with permission from F. O. Adurodija, H. Izumi, T. Ishihara, H. Yoshioka, and M. Motoyama, *J. Mater. Sci.: Mater. Electronics* 12, 57 (2001), © 2001, Kluwer Academic Publishers.)

where c_{film} and c_{bulk} are the lattice parameters of ITO film and In_2O_3 powder, respectively. The lattice constant of In_2O_3 was taken as 10.118 Å as per JCPDS files [153].

The dependence of the percentage intrinsic strain on γ for 80- and 150-nm-thick ITO films as determined from peak (222) is given in Figure 26a and b [162]. A deconvolution analysis using a psuedo-Voigt function with Gauss ratio between 0 and 1 was used to verify the positions of the doublet peaks on the 2θ axis [126]. Higher lattice constants varying from 10.413 to 10.441 Å were observed for peak A (resulting from intrinsic strain). The lattice constants obtained for peaks B at different GIXRD angle varied from 10.098 to 10.135 Å. From Figure 26a, intrinsic strain of nearly 3% was observed for peak A, compared to peak B with 0.25%. Such a high percentage of strain is not unusual considering the high laser energy fluence of 5 J cm^{-2} used during the PLD process. In the case of the

80-nm-thick ITO film, a maximum of 0.5% was observed, as shown in Figure 26b. It was found that strain in ITO films could be curtailed by use of low energy fluence ≤ 2 J cm^{-2} during growth. However, such a low fluence drastically reduces the deposition rates. For example, a deposition of ~ 12 nm/min for 5 J cm^{-2} could be reduced to ~ 2 nm/min when the fluence is lowered to 1 J cm^{-2}. The above results showed that the structure of ITO films could be distorted due to the effect of stain when high fluence is used. However, despite the splitting of the peaks caused by strain, no detrimental effects on the electrical properties were detected in the films. The ITO films yielded low resistivities of $(1.8-2.8) \times 10^{-4}$ Ω cm.

6.1.4. Epitaxial Growth of ITO Films by PLD

One way to use metallic electrodes is to deposit the ferroelectric layer on TCO films such as In_2O_3 [84, 164, 165]. The chemical resemblance between the TCO and ferroelectric materials can safeguard interdiffusion and provide a more suitable condition for oriented growth [84]. Oriented ferroelectric films have been demonstrated on textured YBCO contacting layers formed on LaAlO$_3$, SrTiO$_3$, and epitaxial yittria-stabilized zirconia (YSZ)-buffered silicon substrate [166, 167]. Earlier work involving the deposition of ferroelectrics on In_2O_3 has concentrated on the use of sputtered polycrystalline films deposited on glass substrate by sputtering [164, 165]. It has been found that the polycrystalline characteristics of such films are not suitable for epitaxial growth. Attempts have been made to grow oriented In_2O_3 films on MgO and YSZ substrates using several deposition techniques, including PLD [84, 87, 168, 169]. Tarsa et al. [84] has studied the growth of oriented In_2O_3 films on InAs, GaAs, MgO, and YSZ substrates. The In_2O_3 films were deposited at substrate temperature between 400 and 450°C using a KrF (248 nm) excimer laser. Figure 27 shows typical $\theta-2\theta$ XRD analysis and rocking curve measurements for epitaxial growth of In_2O_3 films on MgO and YSZ substrate [84]. The tendency of orientation as deduced from these spectra symbolized cubic-on-cubic texturing, i.e., $(001)_{In_2O_3} \| (001)_{MgO}$, $[100]_{In_2O_3} \| [100]_{MgO}$, $(001)_{In_2O_3} \| (001)_{YSZ}$, and $[100]_{In_2O_3} \| [100]_{YSZ}$. The group observed a high degree of orientation (rocking curve full width at half maximum (FWHM) of 0.29°) for YSZ compared to MgO with 1.5°. The high degree of crystalline orientation of the film on YSZ has been associated with a smaller mismatch of the 2:1 epitaxial relation for the YSZ [84]. The lattice constant of YSZ ranges from 5.11 to 5.18 Å (depending on the content of Y_2O_3), hence giving a mismatch of $\sim 2\%$ for the cubic-on-cubic orientation. Recently, Kwok et al. [87] have demonstrated an even higher degree of orientation for epitaxial formation of ITO films on YSZ substrate at temperature of 400°C and oxygen pressure of 2×10^{-2} Torr by PLD. They obtained XRD rocking curve FWHM as narrow as 0.08°. The reason for the excellent crystallinity of the ITO was also ascribed to the small lattice mismatch between the YSZ and In_2O_3.

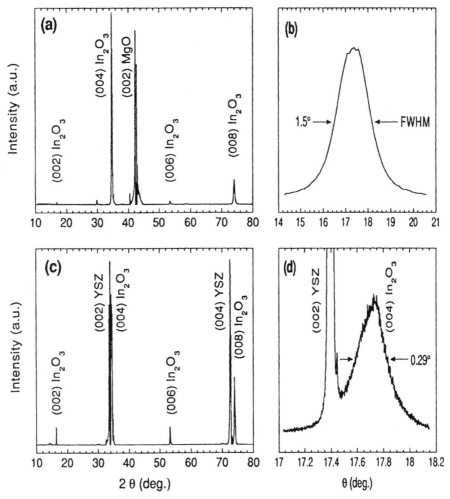

Fig. 27.   X-ray θ–2θ scan of In_2O_3 films grown on (a) MgO and (c) YSZ substrates. The corresponding (004) In_2O_3 rocking curves are shown in (b) and (d). The obvious splitting of the higher intensity substrate peaks in (a) and (b) is due to saturation of the X-ray detector. (Reprinted with permission from E. J. Tarsa, J. H. English, and J. S. Speck, *Appl. Phys. Lett.* 62, 2332 (1993), © 1993, American Institute of Physics.)

6.2. Electrical Properties of ITO Films

6.2.1. Effects of Oxygen Pressure on the Electrical Properties of the Films Deposited at Room Temperature

The electrical properties of ITO films deposited at room temperature are considered next. As explained in Section 6, all the films deposited at room temperature were amorphous. Figure 28 shows plots of resistivities versus oxygen pressure for 0 wt%, 5 wt%, and 10 wt% Sn-doped In_2O_3 films [81–83]. The resistivities of the films were strongly dependent on the oxygen pressure and followed a similar tendency. High resistivity was achieved at low oxygen pressure of 1×10^{-3} Torr for all the ITO films. With increasing oxygen pressure to between 1×10^{-2} and 1.5×10^{-2} Torr, the resistivities reduced to their minimal values of 1.8×10^{-4}, 4.8×10^{-4}, and 5.4×10^{-4} Ω cm for 0 wt% (undoped In_2O_3), 5 wt%, and 10 wt% Sn-doped ITO films, respectively. From oxygen pressure greater than 1.5×10^{-2} Torr, an abrupt increase in the resistivities of the films was noticed. The optimal oxygen pressure that yielded the lowest resistivity films

lies within a narrow range of $(1–1.5) \times 10^{-2}$ Torr [79–92]. The poor resistivity obtained at low oxygen pressure (1×10^{-3} Torr) and high oxygen pressure ($> 1.5 \times 10^{-2}$ Torr) was ascribed to nonstoichiometry resulting from a deficit of oxygen vacancies. Composition wise, an increase in the resistivity of the films with increasing Sn doping content was witnessed at the optimum oxygen pressure of 1×10^{-2} Torr. The increase in the resistivity of the films at this oxygen pressure with increasing Sn doping content was attributed to scattering caused by Sn atoms [73]. The results suggested that at room temperature, lower resistivity films could be obtained from an undoped In_2O_3 target compared to a Sn-doped In_2O_3 target by the PLD technique.

Figure 29 shows (a) the carrier concentration and (b) the Hall mobility plotted against oxygen pressure, respectively [81–83]. At an oxygen pressure of 1×10^{-3} Torr, the carrier concentration varied from 4.5×10^{18} to 3.5×10^{20} cm^{-3} and was minimum for 10 wt% Sn-doped In_2O_3 film. It increased to about 8×10^{20} cm^{-3} at an oxygen pressure of 5×10^{-3} and then decreased slightly to around $(5–6) \times 10^{20}$ cm^{-3} at an oxygen pressure of 1×10^{-2} Torr for all the films, irrespective of the change

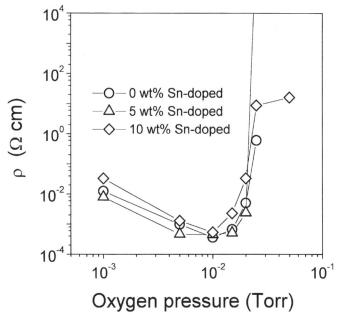

Fig. 28.   Plots of resistivities (ρ) versus oxygen pressure for 0 wt%, 5 wt%, and 10 wt% Sn-doped In_2O_3 films deposited at room temperature.

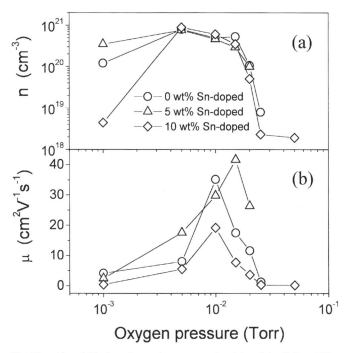

Fig. 29.   (a) and (b) show the carrier concentration (n) and the Hall mobility (μ) plotted against oxygen pressure for the ITO films deposited at room temperature.

in Sn doping content. In the region of optimal oxygen pressure of $(1–1.5) \times 10^{-2}$ Torr, little change of the carrier concentration with changes in Sn doping content occurred. From oxygen pressure above 1.5×10^{-2} Torr, the carrier concentrations of all the films decreased very sharply and this decrease corresponded to the poor resistivity observed (Fig. 28). The carrier concentration of the undoped and Sn-doped In_2O_3 films responded in a

similar way to the changes in oxygen pressure.

As described in Section 2, for undoped In_2O_3 films free carriers are generated by the creation of oxygen vacancies [46, 170], while in ITO films generation of free carriers occurs through the creation of oxygen vacancies and substitutional four-valence Sn atoms. The creation of an oxygen vacancy supplies a maximum of two electrons, while the substitution of an In atom with a Sn atom provides an electron to the conduction band as represented in Eqs. (3) and (4) in Section 2. Equation (3) applies mainly to In_2O_3, while Eqs. (3) and (4) are true for crystalline ITO films. However, in amorphous state (i.e., films deposited at room temperature), it has been reported that the increasing presence of Sn atoms in ITO films does not contribute free carriers to the conduction band, rather they act as scattering centers [73, 144]. In Figure 30 is illustrated the structures of amorphous and crystalline ITO material, portraying a distorted and an orderly structure, respectively. Therefore, the high carrier concentrations observed in the undoped and Sn-doped In_2O_3 amorphous films at the optimal oxygen pressure were a consequence of the creations of oxygen vacancies. On the other hand, the low carrier concentrations achieved at low ($< 1 \times 10^{-3}$ Torr) and high ($> 1.5 \times 10^{-2}$ Torr) oxygen pressures could be due to a defect in oxygen vacancies leading to nonstoichiometric films [46].

In Figure 29b, low Hall mobility (<5 cm^2 V^{-1} s^{-1}) was obtained for the films deposited at low oxygen pressure of 1×10^{-3} Torr, in spite of the changes in Sn doping content. With increasing oxygen pressure, the mobility also increased and peaked at the region of optimal oxygen pressure of $(1–1.5) \times 10^{-2}$ Torr. A further increase in oxygen pressure led to a sharp decrease in carrier mobility. At the optimum oxygen pressure of 1×10^{-2} Torr, mobilities of 41.9, 29.8, and 19.2 cm^2 V^{-1} s^{-1} were measured for the 0 wt%, 5 wt%, 10 wt% Sn-doped In_2O_3 films, respectively. However, in the case of 5 wt% Sn-doped In_2O_3 the highest mobility of 41.7 cm^2 V^{-1} s^{-1} was obtained at 1.5×10^{-2} Torr. At oxygen pressure of 1×10^{-2} Torr, it is seen that the mobility decreases with increasing Sn doping content and it corresponds to the increase of resistivity observed in Figure 28. The mobility of free carriers in ITO films is affected by the disorder due to the structure of In_2O_3 and modification of the network resulting from Sn doping [73, 88, 144]. Hence, at a Sn doping content of 10 wt%, the observed low mobility could be caused by scattering of conduction electrons resulting from the excess Sn atoms.

6.2.2. Effects of Oxygen Pressure on the Electrical Properties of the Films Deposited at a Temperature of 200°C

Figure 31a–c shows the resistivity, carrier concentration, and Hall mobility plotted against oxygen pressure for films deposited from 0 wt%, 5 wt%, and 10 wt% Sn-doped In_2O_3 targets at 200°C [81, 82]. The films indicated a weak dependence of the resistivity on the oxygen pressure, particularly below 1.5×10^{-2} Torr. The lowest resistivity of 1.8×10^{-4} Ω cm was observed at oxygen pressure of 1×10^{-2} Torr for 5 and 10 wt%

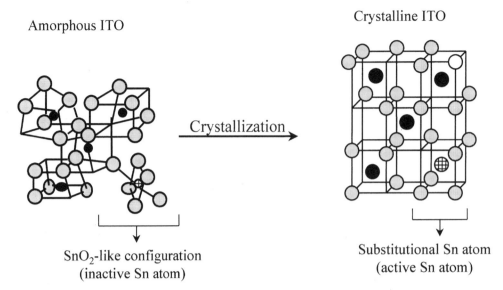

Fig. 30.   Illustrations of the structures of amorphous and crystalline ITO material.

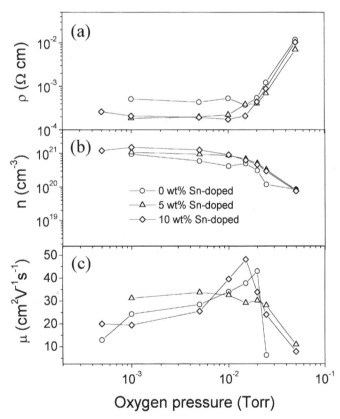

Fig. 31.   Resistivity (ρ), carrier concentration (n), and Hall mobility (μ) plotted against oxygen pressure for 0, 5, and 10 wt% Sn doped In_2O_3 films deposited at 200°C. (Adapted from F. O. Adurodija, H. Izumi, T. Ishihara, H. Yoshioka, H. Matsui, and M. Motoyama, *Appl. Phys. Lett.* 74, 3059 (1999); F. O. Adurodija, H. Izumi, T. Ishihara, H. Yoshioka, H. Matsui, and M. Motoyama, *Japan. J. Appl. Phys. 1* 38, 2710 (1999); F. O. Adurodija, H. Izumi, T. Ishihara, H. Yoshioka, K. Yamada, H. Matsui, and M. Motoyama, *Thin Solid Films* 350, 79 (1999) from American Institute of Physics, Japan Society of Applied Physics, and Elsevier Science.)

Sn-doped In_2O_3 films. This observation was not unusual since the combined effect of energetic vapor and thermally induced crystallization was expected to aid surface migration of In and Sn cations with a consequential enhancement in the film resistivity [47, 171, 172]. The ITO conduction mechanism is partly controlled by the intrinsic doping by oxygen vacancies; hence at the optimum oxygen pressure, resistivity was lowered. Undoped In_2O_3 films showed a somewhat higher resistivity above 3.6×10^{-4} Ω cm at this region of oxygen pressure. The reason for the high resistivity in the undoped In_2O_3 films has been discussed in Section 5. From oxygen pressure above 1×10^{-2} Torr, a rapid increase in the resistivities occurred in all the films, irrespective of the changes in the Sn doping content. The resistivity of the films deposited at 200°C was better than or comparable to the values reported for ITO films deposited by other techniques [6, 73, 85, 90, 171–173]. Several groups have observed that oxygen pressure in the region of $(1–1.5) \times 10^{-2}$ Torr is optimal for achieving highly conducting ITO films deposited on heated substrates by the PLD process [90, 92, 171]. It is especially interesting to note that low resistivity films were achieved within a narrow oxygen pressure range of $(1–1.5) \times 10^{-2}$ Torr. This is important for the optimization and standardization of the PLD growth conditions.

Figure 31b shows a plot of the carrier concentration as a function of oxygen pressure for films grown at 200°C. A high carrier concentration of around 10^{21} cm^{-3} was obtained for Sn-doped In_2O_3 films deposited at oxygen pressures over the range 1×10^{-3} to 1×10^{-2} Torr. As expected, the carrier concentration was maximal for the samples that yielded the lowest resistivities. Lower carrier concentrations of $(2–4) \times 10^{20}$ cm^{-3} were observed for the undoped In_2O_3 films as described in Section 5. At higher oxygen pressures ($> 1 \times 10^{-2}$ Torr), the carrier concentrations of the films declined rapidly to around 8×10^{19} cm^{-3}. This behavior is similar to that observed in the ITO films deposited at room temperature. Carl et al. [47] also observed a similar behavior for sputtered ITO films. The decrease in the carrier concentrations with increasing oxygen pressure has been associated with an increase in the absorption of free electrons [73].

Plots of the Hall mobilities versus oxygen pressure for films deposited at temperature of 200°C shown in Figure 31c. A similarity in the behavior of the carrier mobility for the undoped and 10 wt% Sn-doped In_2O_3 films with changes in oxygen pressure is apparent. Moderately low mobilities were observed at low oxygen pressure over the range 1×10^{-3} to 5×10^{-3} Torr. This was followed by an increase in the mobility with increasing oxygen pressure to about 1×10^{-2} Torr before decreasing with a further increase in oxygen pressure. In contrary, the Hall mobility of the 5 wt% Sn-doped In_2O_3 films tended to stabilize at about 33 cm^2 V^{-1} s^{-1} cm for oxygen pressures between 1×10^{-3} and 1.5×10^{-2} Torr before decreasing with further increase in oxygen pressure. The best Hall mobilities of 43–47 cm^2 V^{-1} s^{-1} were observed for the films deposited under oxygen pressure of $(1–1.5) \times 10^{-2}$ Torr, particularly for the undoped and 10 wt% Sn-doped In_2O_3 films. At low and high oxygen pressures, the low mobilities exhibited by the films could be related to the poor crystallinity, as discussed in Section 6. On the other hand, the high carrier concentration ($\sim 10^{21}$ cm^{-3}) observed in the films deposited at low oxygen pressures suggested that scattering by impurity centers due to excess cations (Sn and In) was responsible for the low Hall mobility [48, 55, 73, 174]. In general, the sharp drop in the resistivity at higher oxygen pressures was caused by the drastic reduction of the carrier concentration and the Hall mobility.

6.2.3. Effects of Substrate Temperature on the Electrical Properties of the Films

The influence of the substrate temperature on the electrical properties of the films is considered next. Figure 32a–c shows the dependence of resistivity, carrier concentration, and Hall mobility on the substrate temperature for undoped In_2O_3 and 5 wt% and 10 wt% Sn-doped In_2O_3 films deposited at an oxygen pressure of 1×10^{-2} Torr, respectively [80]. The error (experimental and instrumental) in the data presented here was less than 5%. At low substrate temperature ($\leq 100°C$), the least resistivity of $\sim 2 \times 10^{-4}$ Ω cm was obtained for undoped In_2O_3 film. The creation of oxygen vacancies contributed to the high carrier concentrations shown in Figure 32b. Moderate Hall mobility (<45 cm^2 V^{-1} s^{-1}) was observed for In_2O_3 films which was much lower than earlier reported [16, 171]. In contrast, amorphous Sn-doped In_2O_3 films deposited at substrate temperatures below 150°C yield lower carrier concentrations, consistent with the claim that Sn doping does not donate free carriers to amorphous films [44, 86, 124, 129, 144, 147, 148, 170]. Therefore, it could be deduced that at low substrate temperature, the creation of oxygen vacancies dominated the free carrier generation mechanism resulting in the low resistivity of the films [16, 144]. These results were in agreement with those

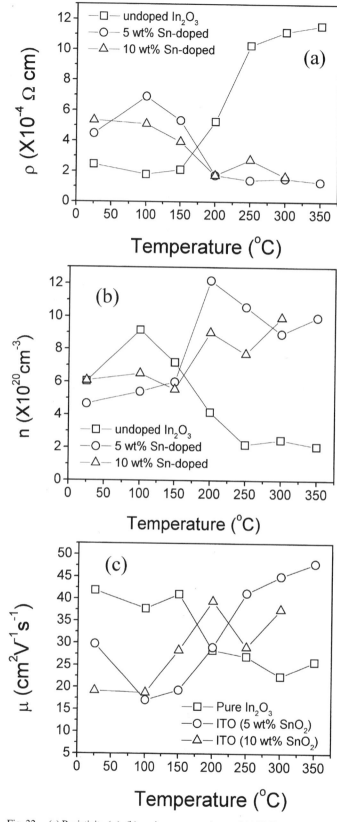

Fig. 32. (a) Resistivity (ρ), (b) carrier concentration, and (c) Hall mobility as a function of substrate temperature for undoped In_2O_3 and ITO films containing different Sn doping contents. (Reprinted with permission from F. O. Adurodija, H. Izumi, T. Ishihara, H. Yoshioka, M. Motoyama, and K. Murai, *J. Vac. Sci. Technol. A* 18, 814 (2000), © 2000, American Institute of Physics.)

previously reported for amorphous Sn-doped In$_2$O$_3$ films deposited by other techniques such as sputtering [124, 144, 148, 170]. In general, this study has shown that low resistivity In$_2$O$_3$ films comparable to ITO films grown at high substrate temperature (200–400°C) by other deposition techniques [73] can be obtained at low substrate temperature using PLD.

On increasing the substrate temperature above 150°C, a sudden increase in resistivity was noticed for undoped In$_2$O$_3$ films as shown in Figure 32a. The degradation in the resistivity of undoped In$_2$O$_3$ resulted from the lowering of the carrier concentration and the Hall mobility. The low carrier concentration may be due to the absence of Sn cations to compensate the depleting oxygen vacancies during growth of In$_2$O$_3$ films at high substrate temperature. The decrease in Hall mobility of undoped In$_2$O$_3$ films deposited at high substrate temperature may be due to grain boundary scattering. On the other hand, a marked improvement in the resistivity of Sn-doped In$_2$O$_3$ films occurred at substrate temperatures above 150°C. The lowest resistivity of 1.3×10^{-4} Ω cm was obtained for 5 wt% Sn-doped In$_2$O$_3$ at 350°C (maximum substrate temperature investigated) [80]. The reduction in the resistivity of Sn-doped In$_2$O$_3$ films resulted from an increase in the carrier concentration and Hall mobility, as shown in Figure 32b and c. This behavior is not unexpected, since at high substrate temperature (crystalline films) thermal activation of Sn into Sn^{4+} occurs which provides free electrons to the conduction band [73]. The results further ascertained that Sn doping is effective at high substrate temperature.

6.3. Optical Properties

The optical properties of the ITO films produced under different growth conditions are discussed next. This section concerns the experimental data and theoretical evaluations for amorphous and crystalline ITO films. The optical transparency of ITO film in the visible and near-infrared regions of the solar spectrum depends on its wide optical bandgap. The optical bandgap of ITO films varies from 3.75 to 4.2 eV and is dependent on the film composition (Sn doping content) and deposition conditions such as substrate temperature, oxygen pressure, and film thickness. The absorption edge falls in the ultraviolet (UV) region and shifts to shorter wavelengths (i.e., narrowing of the bandedge) with increasing carrier concentration [5]. Figure 33 shows the band structures of (a) an undoped and (b) a highly Sn-doped In$_2$O$_3$ film. A partial filling of the conduction band in addition to shifts in energy of the bands is apparent. This shift is associated with the filling of the states near the bottom of the conduction band and is known as the Burstein–Moss shift [145]. Figure 34 shows the transmission, reflection, and absorption spectra for an ITO film deposited from a 5 wt% Sn-doped In$_2$O$_3$ target on glass at temperature of 200°C and oxygen pressure of 1×10^{-2} Torr [88]. The transparent regions of the film are the visible and the near infrared. The transparency in these regions is limited by: (a) reflection losses which include specular and scattered components (it usually results from surface roughness and could increase with film thickness), (b) free car-

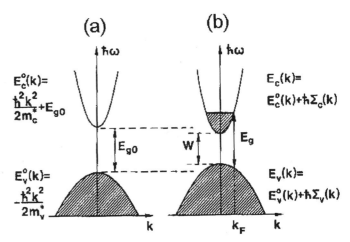

Fig. 33. The assumed band structure of undoped In$_2$O$_3$ in the vicinity of the top of the valence band and the bottom of the conduction band. (b) The effect of Sn doping. A shift of the band is apparent. Shaded areas denote occupied states. Bandgaps, Fermi wave number, and dispersion relations are indicated. (Reprinted with permission from I. Hamberg and C. G. Granqvist, *J. Appl. Phys.* 60, R123 (1986), © 1986, American Institute of Physics.)

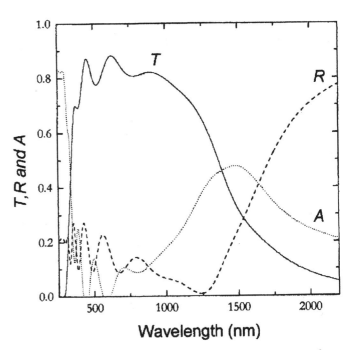

Fig. 34. Transmittance (T), reflectance (R), and absorption (A) spectra for an ITO film deposited from a 5 wt% Sn doped In$_2$O$_3$ target on glass at a temperature of 200°C and oxygen pressure of 1×10^{-2} Torr. The film thickness was ~300 nm. (Reprinted with permission from H. Kim, C. M. Gilmore, A. Piqué, J. S. Horwitz, H. Mattoussi, H. Murata, Z. H. Kafafi, and D. B. Chrisey, *J. Appl. Phys.* 86, 6451 (1999), © 1999, American Institute of Physics.)

rier absorption in the film, and (c) change in the transmittance due to change in the film thickness [5].

The transmittance curves for ITO film deposited at room temperature and 200°C under oxygen pressures over the range 1×10^{-3} to 5×10^{-2} Torr are shown in Figure 35a and b. In Figure 35c is shown the transmittance of the films deposited at different temperature under oxygen pressure of 1×10^{-2} Torr [83].

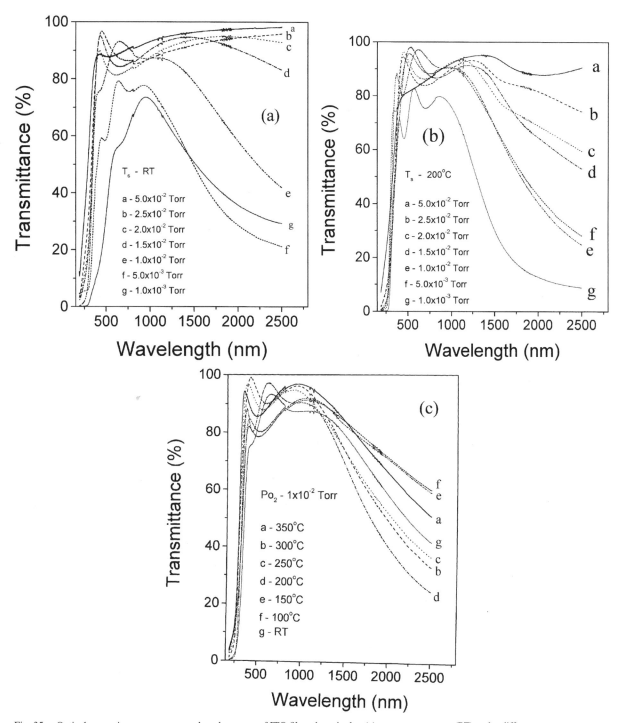

Fig. 35. Optical transmittance versus wavelength spectra of ITO films deposited at (a) room temperature (RT) under different oxygen pressure. (b) Optical transmittance versus wavelength spectra of ITO films deposited at 200°C under different oxygen pressure. (c) Optical transmittance versus wavelength spectra of ITO films deposited at different temperature under oxygen pressure of 1×10^{-2} Torr. (Reprinted with permission from F. O. Adurodija, H. Izumi, T. Ishihara, H. Yoshioka, H. Matsui, and M. Motoyama, *Japan. J. Appl. Phys. 1* 38, 2710 (1999), © 1999, Japan Society of Applied Physics.)

The changes in the transmittance curves with respect to changes in oxygen pressure for the films deposited at room temperature and 200°C followed a similar tendency. At oxygen pressure of 1×10^{-3} Torr, the films appeared brownish in color, hence exhibiting a low optical transparency of less than 60%, for the room temperature deposited films.

A salient feature observed in the films was the increasingly high near-infrared reflectance (NIR) with decreasing oxygen pressure. This feature is particularly important for a window layer coating. The high NIR displayed by the films deposited at room temperature is interesting in view of the fact that the films were not post-heat-treated. This may have contributed to the moderately high electron concentrations evidenced in the films deposited at room temperature and 200°C, as shown in Figure 29b in Section 6.2. High NIR has been observed by other groups [82, 88, 90]. From Figure 35a and b, a shift in the absorption edge (bandedge) toward the ultraviolet (short-wavelength range) with increasing oxygen pressure is apparent. This phenomenon called the Burstein–Moss (B-M) shift is discussed later in this section. This feature has been observed in ITO films deposited by other methods when the composition or the deposition conditions were changed [5, 6, 88]. ITO films with a high transmittance and a high NIR were obtained at oxygen pressures in the region of 5×10^{-3} and 1.5×10^{-2} Torr. This region of oxygen pressure has been reported as the optimum for creating suitable oxygen vacancies necessary for high-quality ITO films.

ITO films grown by PLD at the optimal deposition conditions have shown high transmissivity in the visible and high reflectivity in the infrared regions of the solar spectrum. The films also indicated an abrupt absorption edge in the UV region. A plot of the absorption coefficient (α) versus photon energy for such an abrupt absorption edge usually shows a linear proportionality which implies that direct allowed transitions occurs in the films. Following the law of conservation of energy, α is determined from the relations [174]

$$T(\lambda) + R(\lambda) + A_f(\lambda) = 1 \qquad (8)$$

where $A_f(\lambda)$ is the absorption of the film. For ITO films with reflectance (visible region) less than 20% the transmittance can be written as

$$T = (1 - R)^2 e^{-\alpha t} \qquad (9)$$

where t is the film thickness. The absorption coefficient is given by

$$\alpha = \ln\left[\frac{(1-R)^2}{T}\right]\frac{1}{t} \qquad (10)$$

$R(\lambda)$ is determined from the part of the transmittance where $\alpha \approx 0$.

For such direct transitions the theory of fundamental absorption between the simple parabolic bands illustrated in Figure 29 leads to a photon energy dependence, near the absorption edge [175–177], of

$$\alpha(h\nu) = \frac{A}{h\nu}(h\nu - E_g)^{1/2} \qquad (11)$$

where E_g is the optical bandgap, $h\nu$ is photon energy, and A is a constant.

Therefore, for a direct bandgap semiconductor, the plot of $\alpha(h\nu)^2$ versus $h\nu$ is linear and the intercept on the $h\nu$ axis gives the direct optical bandgap, E_g. Several groups have used this relation to calculate the optical bandgap of ITO films [5, 6]. Figure 32a shows plots of $\alpha(h\nu)^2$ versus $h\nu$ for ITO films deposited at substrate temperature over the range 25 to 300°C, thus confirming direct bandgap transitions, as determined by Kim et al. [88]. The group observed that the direct bandgap increases from 3.89 to 4.21 eV with an increase in the substrate

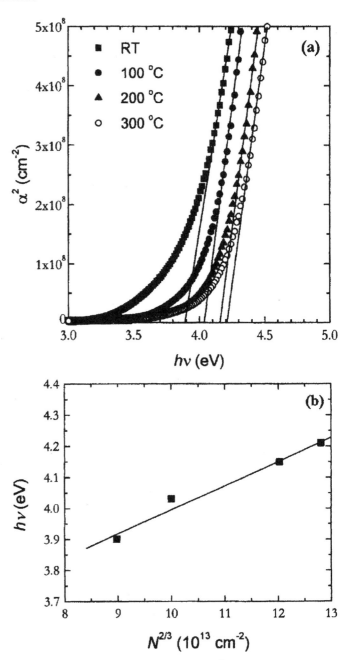

Fig. 36. (a) Dependence of photon energy on α^2 for ITO films deposited at different substrate temperatures. (b) Variation of bandgap as a function of $N^{2/3}$ for the same film shown in (a). (Reprinted with permission from H. Kim, C. M. Gilmore, A. Piqué, J. S. Horwitz, H. Mattoussi, H. Murata, Z. H. Kafafi, and D. B. Chrisey, *J. Appl. Phys.* 86, 6451 (1999), © 1999, American Institute of Physics.)

temperature from 25 to 300°C. This increase in the bandgap (B-M shifts) has been associated with an increase in the carrier concentration of the films.

The relationship between the optical bandgap (eV) and $N^{2/3}$ regarding the ITO films represented in Figure 36a is shown in Figure 36b [88]. The linear relationship exhibited between $N^{2/3}$ and the direct optical bandgap was attributed to the B-M shift. In the case of parabolic band edges, where the B-M shift is predominant, the energy bandgap for a free-electron Fermi gas is given by the relation [5, 6, 145]

$$E_g = E_{g0} + \Delta E_g^{\text{B-M}} \qquad (12)$$

where E_{g0} represents the intrinsic bandgap (3.75 eV) [6] and $\Delta E_g^{\text{B-M}}$ is the bandgap shift due to the filling of the states near the bottom of the conduction band given by the B-M theory [5, 6, 104, 145, 178],

$$\Delta E_g^{\text{B-M}} = \frac{\hbar^2}{2m_{vc}^*}\left(3\pi^2 N\right)^{1/2} \qquad (13)$$

where m_{vc}^* is the reduced effective mass which also includes the contribution of carriers from the valence band curvature given as

$$\frac{1}{m_{vc}^*} = \frac{1}{m_v^*} + \frac{1}{m_c^*} \qquad (14)$$

where m_v^* and m_c^* represent the effective mass of the carriers in the conduction and valence bands, respectively.

In a more complete theoretical approach, the band shifts due to electron interaction and impurity scattering, particularly at very high carrier concentrations, are also considered. Hence the bandgap given in Eq. (11) is written as [6]

$$E_g = E_{g0} + \Delta E_g^{\text{B-M}} + \hbar\Sigma_c - \hbar\Sigma_v \qquad (15)$$

where $\hbar\Sigma_c$ and $\hbar\Sigma_v$ are self-energies resulting from electron–electron and electron–impurity scattering. The linear proportionality shown in Figure 36b indicated that the B-M shift was more important than the $\hbar\Sigma$ in ITO films.

Furthermore, Kim et al. [88] have observed that the refractive index of ITO films (300-nm-thick) grown by PLD at a substrate temperature of 250°C was dependent on the wavelength, as shown in Figure 37 [88]. They observed an average value of 1.95 ± 0.21 for the visible range (550 nm) of the solar spectrum. However, the refractive index is of little concern in the areas of low resistivity applications. The extinction coefficient k, however, varies with conductance.

In the near-infrared region, absorption due to free carriers becomes important for TCO films. The plasma resonance of the charge carrier can be simplified on the basis of the classical Drude's theory and has been successfully used in the case of ITO and SnO_2 films [62, 88, 163]. According to Drude's classical theory, the real and the imaginary parts of the dielectric constants are given by

$$\varepsilon = (n - ik)^2 = \varepsilon' + \varepsilon'' \qquad (16)$$

$$\varepsilon' = n^2 - k^2 = \varepsilon_\infty\left(\frac{1 - \omega_p^2}{\omega^2 + \gamma^2}\right) \qquad (17)$$

$$\varepsilon'' = 2nk = \frac{\omega_p^2 \gamma \varepsilon_\infty}{\omega(\omega^2 + \gamma^2)} \qquad (18)$$

The plasma frequency ω_p is given as

$$\omega_p^2 = \frac{4\pi N e^2}{\varepsilon_0 \varepsilon_\infty m_e^*} \qquad (19)$$

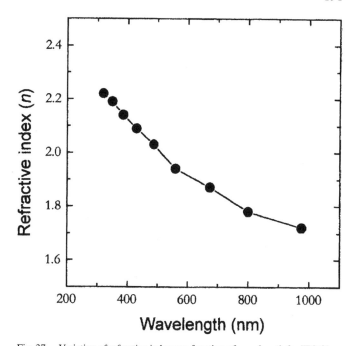

Fig. 37. Variation of refractive index as a function of wavelength for ITO films deposited at 250°C and 1×10^{-2} Torr. (Reprinted with permission from H. Kim, C. M. Gilmore, A. Piqué, J. S. Horwitz, H. Mattoussi, H. Murata, Z. H. Kafafi, and D. B. Chrisey, *J. Appl. Phys.* 86, 6451 (1999), © 1999, American Institute of Physics.)

where ε_∞ and ε_0 represent the high-frequency dielectric constant of the film and free space, respectively, and γ is equivalent to $1/\tau$, where τ is the relaxation time. τ is considered to be independent of ω_p but is related to the mobility as follows:

$$\gamma = \frac{1}{\tau} = -\frac{e}{m_e^* \mu} \qquad (20)$$

Equation (19) shows that the resonance wavelength is dependent on the mobility. Therefore, a large $m_e^* \mu$ implies a large NIR and also a sharper transition from low to high reflection. This feature has been found to be very useful in the case of applications in heat mirrors [5].

6.4. Chemical States Analysis of ITO Films

At the beginning of this chapter it was mentioned that the conductivity of TCO films is directly related to the concentration and mobility of free electrons. ITO films with high carrier concentration ($\sim 10^{21}$ cm^{-3}) are achieved by the substitution of Sn^{4+} for In^{3+} which provides in an ideal situation, one carrier electron to the conduction band [73]. It has been reported that the carrier concentration of ITO films increases as the Sn doping content increases to 5 wt% and beyond this doping level, no further increase in the carrier concentration occurs [105]. The dependence of the resistivity, carrier concentration, and Hall mobility of ITO films deposited by PLD is shown in Figure 38 [88]. The critical limit of Sn doping of 5 wt% in the ITO film is apparent. Sn can exist as either SnO (valence of 2) or SnO_2 (valence of 4). Since indium has a valence of 3, the presence of

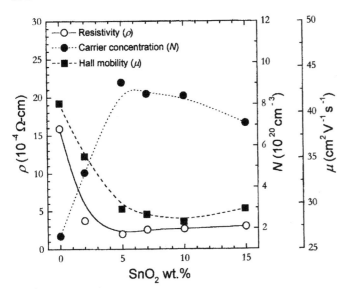

Fig. 38. Dependence of resistivity (ρ), carrier concentration (N), and Hall mobility (μ) on Sn doping content for ITO films deposited at temperature of 250°C and oxygen pressure of 1×10^{-2} Torr. (Reprinted with permission from H. Kim, C. M. Gilmore, A. Piqué, J. S. Horwitz, H. Mattoussi, H. Murata, Z. H. Kafafi, and D. B. Chrisey, *J. Appl. Phys.* 86, 6451 (1999), © 1999, American Institute of Physics.)

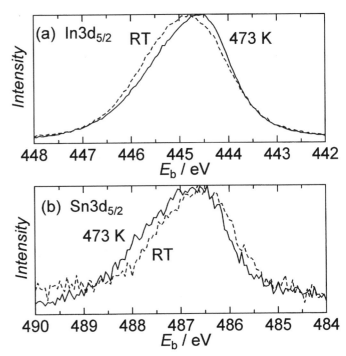

Fig. 39. X-ray photoelectron spectra of (a) In $3d_{5/2}$ and (b) Sn $3d_{5/2}$ peaks of 5 wt% Sn-doped In$_2$O$_3$ film grown at room temperature (dashed lines) and 200°C (solid lines). (Reprinted with permission from H. Yoshioka, F. O. Adurodija, H. Izumi, T. Ishihara, and M. Motoyama, *Electrochem. Solid State Lett.* 3, 540 (2000), © 2000, Electrochemical Society Inc.)

SnO$_2$ would contribute two electrons to the conduction band. It has been reported that at low deposition temperature Sn is present in ITO films as SnO resulting in low carrier concentration, but on annealing in air the SnO is converted to SnO$_2$, leading to an increase in carrier concentration [35, 105, 179]. Therefore, in order to improve the electrical properties of ITO films, it is important to understand the chemical state of Sn dopants as a function of the film composition and deposition conditions. For this reason, X-ray photoelectron spectroscopy (XPS) measurements were performed on the ITO films deposited by PLD. XPS allows measurement of the binding energy with a high resolution, which makes it possible to detect the different chemical states of an element. Studies on the chemical state of Sn in sputtered ITO films have been reported [35, 96, 180, 181]. However, for sputtered ITO films, the change in Sn chemical states with changes in the electrical properties and deposition conditions was hardly detected due to the high response from the Sn segregated on the surface [161, 182].

XPS analysis was performed with an ULVAC-phi 5500MT spectrometer [183]. The spectra were taken with Mg $K\alpha$ (1254 eV) radiation, operating at 15 kV–200 W with an analyzer pass energy of 23.5 eV. The total instrumental resolution was about 1.0 eV. Since a very slight charging was noticed during measurements of the ITO films, the binding energy was calibrated using a C $1s$ peak of adventitious carbon as 285.0 eV. The take-off angle of the photoelectrons was varied from 15 to 90° to determine the depth uniformity of the films. The $3d_{5/2}$ peak area ratio of In and Sn in the films showed very little change with the take-off angle in contrast to sputtered ITO films with Sn segregation on the surface layer [161, 182]. Also, no significant difference in the peak position and shape was observed in the angle resolved measurements, implying that the

chemical states across the depth of the ITO films were uniformly distributed. Therefore, only the spectra measured at the take-off angle of 45° were analyzed.

Figure 39 shows (a) In $3d_{5/2}$ and (b) Sn $3d_{5/2}$ spectra for 5 wt% Sn-doped In$_2$O$_3$ films deposited at room temperature and 200°C, respectively [183]. The film deposited at room temperature indicated a Sn $3d_{5/2}$ binding energy of 486.6 eV which agreed very well with the binding energy of SnO$_2$, and this is a typical value for tetravalent Sn ions incorporated in the oxide [184]. On the other hand, the peak position of the film grown at 200°C shifted by about 0.3 eV to the higher binding energy side. An increase in the peak width and a tailing in the curve at higher binding energy side were also evidenced. As for the In $3d_{5/2}$ spectra of the films deposited at 200°C, asymmetry in the peak shape was larger, although no significant change in the peak width and position was observed.

In Figures 29a, 31b, and 38 were shown the carrier concentration as a function of Sn doping content for the films deposited at room temperature, 200°C, and 250°C. It was found that at Sn doping content higher than 5 wt%, no increase in the effective carrier concentration of the films occurred. At room temperature, we saw that amorphous ITO films with low carrier concentration that is independent Sn doping level were formed, indicating that Sn^{4+} ions are inactive and do not contribute to free electron generation. On the other hand, high carrier concentration was observed in the ITO films that were deposited at 200°C. The reason has been ascribed to thermal activation of Sn into Sn^{4+} ions in the host ITO crystal lattice, thus releasing free electrons to the conduction band. The change observed in the Sn spectra of the films deposited at room temperature and 200°C (Fig. 39b) reflected the difference between the chemical states of the inactive and the active Sn.

The chemical state analysis of Sn in rf-sputtered ITO films using XPS was reported earlier by Fan and Goodenough [95] where they observed asymmetry in the Sn $3d$ peaks and attributed it to a mixture of the chemical states of SnO and SnO$_2$. In the present study, the asymmetry and chemical shift which increased with increasing carrier concentration could not be attributed to the presence of SnO, since Sn^{2+} substitution of In^{3+} would reduce the carrier concentration. The asymmetric shape observed in both In and Sn spectra probably resulted from the loss of photoelectron energy due to excitation of electron–hole pairs near the Fermi surface and has been represented by the Doniach–Sunjic equation [185]. This tends to explain better the results shown in Figure 39.

The effects of Sn doping content on the structural properties and chemical states were also compared. From XRD data shown in Figures 17 to 19 in Section 6, it is apparent that the structural properties of the films were affected by Sn doping. Better crystallinity was achieved for the 5 wt% Sn-doped In$_2$O$_3$ films compared to 10 wt% Sn-doped In$_2$O$_3$ films deposited at 200°C. Relating the crystallinity to the carrier concentration, it was found that the 10 wt% Sn-doped (Fig. 18) film yielded a lower carrier concentration of 6×10^{20} cm^{-3} versus 1.2×10^{21} cm^{-3} for the 5 wt% doped film (Fig. 19). At 300°C, the carrier concentration of the 10 wt% Sn-doped In$_2$O$_3$ film was found to increase to 1.0×10^{21} cm^{-3} due to improvement in the crystallinity. The lowering of the doping efficiency at higher Sn doping content above 5 wt% (Fig. 38) has been explained for ITO films deposited by spray pyrolysis on the basis of the inactivity of Sn due to the formation of complexes such as Sn$_2$O and Sn$_2$O$_4$ [105].

Figure 40 shows Sn $3d_{5/2}$ XPS spectra for 5 and 10 wt% Sn-doped In$_2$O$_3$ films deposited at 200 and 300°C [183]. At 200°C, the 5 wt% Sn-doped In$_2$O$_3$ film displayed higher binding energy and larger peak width than the 10 wt% Sn-doped In$_2$O$_3$ films, as expected from the large difference in the carrier concentration of 1.2×10^{21} and 6×10^{20} cm^{-3}, respectively. At 300°C, the peak position and peak shape of the films are comparable as anticipated. However, the binding energy of the 5 wt% Sn-doped In$_2$O$_3$ film was higher than that of the 10 wt% Sn-doped In$_2$O$_3$ film. This suggested that the amount of inactive Sn in the 10 wt% Sn-doped In$_2$O$_3$ film is larger than that of the 5 wt% Sn-doped In$_2$O$_3$ film, hence the lower average binding energy, even though the carrier concentrations are similar. It is difficult to specify the origin of the chemical shift between active and inactive Sn from the present XPS theory. However, one of the explanations given for the chemical shift is the formation of Sn complexes accompanied by oxygen incorporation into vacant sites, leading to an increase in the Sn coordinated oxygen vacancies.

In general, the chemical shift between active and inactive Sn from XPS could be observed for the ITO films deposited by PLD. The amount of the chemical shift corresponded well with

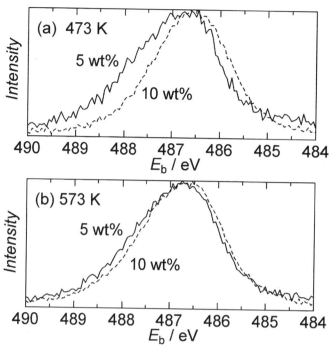

Fig. 40. X-ray photoelectron spectra of Sn $3d_{5/2}$ peak of the ITO films prepared at (a) 200°C and (b) 300°C with Sn concentrations of 5 and 10 wt% Sn-doped In$_2$O$_3$ films (solid lines) and (dashed lines), respectively. (Reprinted with permission from H. Yoshioka, F. O. Adurodija, H. Izumi, T. Ishihara, and M. Motoyama, *Electrochem. Solid State Lett.* 3, 540 (2000), © 2000, Electrochemical Society Inc.)

the carrier concentration and doping efficiency that changes with the substrate temperature and Sn doping content.

7. LASER IRRADIATION

7.1. Excimer Laser Irradiation of Thin Films

In ion beam thin-film processing, low-energy (<100 eV) and high-density bombardment by gas (clusters) is focused on the substrate surface. The low-energy bombardment of the substrate during film growth can produce unusual effects on the film properties which are associated with ion–solid interactions [186]. The clusters are made up of a few hundred to thousands of atoms generated from various kinds of gas materials. Multiple collisions during the impact of accelerated cluster ions onto the surface of the substrate essentially create low-energy bombarding effects in the range of few electron volts (eV) to hundreds of eV per atom at very high density. These bombarding characteristics have been used among others for surface modification, including sputtering, smoothing, cleaning, and low-temperature thin-film formation. Sometimes the bombarding effect enhances the surface reactivity with a consequential improvement of the crystal growth and orientation. Riesse and Weissmentel [187] have shown that better adhesion can be obtained in boron nitride films by neutral ion bombardment of the growing films during PLD.

Some of the useful effects of ion beam assisted deposition like increasing the nucleation sites, improvements in the film morphology, crystallinity, and preferred crystal orientation, and low growth temperature can be realized by laser irradiation of the growing films. UV photon energy excitation can cause significant improvements in the film properties during growth. These unique characteristics have been applied to PLD utilizing its beneficial features, including nonequilibrium heating which makes it possible to heat only the surface of the film. In addition, the flexibility in the positioning of the energy source makes it easy to channel part of the laser energy beam at the substrate surface. In thin-film formation, high quality films can be achieved by focusing a beam of excimer pulsed laser energy emanating from a single source at the substrate surface during deposition. The laser beam can produce (a) high quality films through high-density, low-energy (<1 eV) bombardment, (b) a very smooth surface through control of activated nucleation and growth process, and (c) compound formation through high reactivity. Laser beams have been used in semiconductor wafer annealing to remove lattice defects and to restore electrical properties in ion-implanted devices [188]. Because of the very short pulse width, a depth of a few microns is heated; the overall device can be maintained at low temperature, hence preventing redistribution of implanted dopants [188].

The basic characteristics of laser irradiation or annealing of thin films are as follows [188]:

(a) The absorption of energy is predictable with less dependence on uncontrollable parameters, including doping density and surface reflectivity, than the absorption of energy light beams. Some of the energy is lost due to scattering of the photons with the ablated particles and the plume.

The limited penetration depth of the laser beam and reflectivity of the melted surface do not permit reaction of films with thickness of the order of 100 nm. However, in the case of film formation, the bombardment of the growing film by pulsed laser energy has been found to be efficient in enhancing the overall quality of the films. Laser annealing of the growing films during PLD process has been demonstrated for a variety of superconducting, ferroelectric, and TCO films, as discussed in the later part of this section.

(b) It is possible to adjust the penetration depth of the process from a simple range–energy relation [189, 190] or spatial distribution of energy loss [191].

(c) It is possible to treat large areas with appreciable uniformity by raster of the laser beam over the entire substrate area.

(d) There is the possibility of varying the pulse width of the laser beam and the energy fluence on the substrate surface. These features are also applicable to pulsed electron irradiation and annealing [188].

The heating effect from the UV laser irradiation could contribute to epitaxial growth of thin films on suitable substrates at lower temperatures. A study conducted by Merli [191] on the spatial distribution of heat in a silicon wafer has shown that the heating is "quasi-adiabatic"; i.e., the melting temperature is reached in the underlayer prior to the surface layer. The energy densities required for laser irradiation are somewhat dependent on the material. For ITO films, the energy density of interest is less than 100 mJ cm^{-2}. Higher energy density may lead to damage or reablation of the deposited film. The recent status on excimer laser annealing of amorphous silicon for increasing the quality of poly-thin-film transistors from an industrial point of view has been presented by Stehle [114]. The article compared the physical and economic factors of the scanning single-shot excimer laser crystallization for achieving large area surface treatment. He noted that single-shot excimer laser annealing is more reliable and could be used at room temperature and atmospheric pressure.

Laser irradiation of the film during growth can be effective in improving the surface mobility of the adatoms due to photochemical and photothermal effects. This is in contrast to ion bombardment, where the kinetic energy is basically transferred to the adatoms via momentum-transfer collision effects. Increased adatom surface mobility ordinarily results in enhancement of the film-surface morphology due to improved nucleation and annealing of defects during deposition. During depositions within a background gas, laser irradiation could lead to selective excitation and photodissociation of the gas in the proximity of the substrate, or it could be adsorbed on the surface to produce precursor atoms for the film growth [133]. Enhancements in the properties of the film due to photoexcitation of the growing film have been observed using UV lamps or laser irradiation during molecular beam epitaxy or chemical vapor deposition [192, 193]. Alternatively, laser induced chemical vapor deposition (LCVD) of GaAs has been reported to enhance the crystallinity during growth using Ar^{+} ($\lambda = 514$ nm) lasers [194]. The enhancement process in the LCVD was a result of photocatalytic reaction on the locally heated substrate surface. Hence, the LCVD was based on the photodecomposition of molecules near the substrate surface with subsequent deposition of the desired species.

Initial works on surface irradiation during PLD utilize a single laser for both target ablation and substrate irradiation. The laser beam was split into two parts using a beam splitter and part of the beam is concentrated onto the target, while the other half with energy density ranging from 10–100 mJ cm^{-2} was directed at the surface of the substrate. In this experimental arrangement, the arrival time of the ablated beam and the substrate irradiation beam was almost the same and no delay time essentially existed. The interaction of the laser beam with the material deposited from the previous pulse occurred, since the laser beam arrived at the substrate surface before the arrival of the ablated species [133]. Using this experimental configuration, Otsubo et al. [195] observed an enhancement of the crystallinity and surface morphology of superconducting Y–Ba–Cu–O (YBCO) films deposited by an ArF (193 nm) excimer laser at a lower substrate temperature of 600°C in oxygen atmosphere. The observed improvement in the films was connected with a photochemically induced temperature rise in the growing films, resulting in local heating of the films. Minamikawa et al. and Kanai et al. [196, 197] also observed similar effects of laser irradiation on the properties of YBCO and $Bi_2Sr_2CuO_6$ films in N_2O environment.

Successive experimental setups included the use of a separate laser system for excimer laser irradiation of the film during growth. The objective was to optimize the delay time between the ablation plume and the laser irradiation pulses via re-excitation of the ablated species, thus leading to further improvement in the crystallinity of the film. Using this arrangement, Estler et al. [198] deposited YBCO superconducting films on YSZ substrates using a 308 nm laser system with delay times varying from 0 to 100 μs and energy densities of 10–20 $mJ\,cm^{-2}$. They observed no significant improvements in the films with changes in delay time. However, a later study by Morimoto et al. [199] on YBCO films deposited on (100) MgO substrates using a ArF (193 nm) excimer laser with an energy density of 50 $mJ\,cm^{-2}$ and delay times of 10–20 μs has indeed shown noticeable effects on the crystal orientation of the films deposited at 600°C. Tabata et al. [200, 201] have also shown that epitaxial growth temperature of $PbTiO_3$ films on $SrTiO_3$ substrate can be reduced from 500°C to 300°C by concurrent laser irradiation of the growing film during PLD process. The experiment was performed by introducing a delay time of 1 μs during laser irradiation, which conformed to the mean arrival time of the ablated constituents on the substrate. Tabata et al. [202] also observed that $PbTiO_3$ films grown on Pt or $(La,Sr)_2CuO_4$ superconducting substrates by ArF (193 nm) excimer laser irradiation of the growing films with energy densities of 30–100 $mJ\,cm^{-2}$ caused a significant improvement on the crystalline structure of the films. For fluences of 50–100 $mJ\,cm^{-2}$, films with similar crystalline quality were observed, but at a higher energy density of 120 $mJ\,cm^{-2}$ an increase in the surface roughness of the films occurred due to damage caused by the irradiation effect. Laser irradiation of the growing film has been found to reduce the growth temperature of $BaFe_{12}O_{19}$ and $BaTiO_3$ epitaxial films to 600°C and 650°C, respectively [133, 203]. The films were found to exhibit significant improvements in the structure and surface morphology. These results have suggested that laser irradiation of the growing films influences surface migration of atoms, thus enabling the removal of defects leading to high quality films.

Although most of the works reported on laser irradiation involve the use of lasers in the UV range which cause surface excitation, a purely radiative source of energy for heating of the substrate during the PLD process has also been studied. The use of a continuous wave CO_2 laser has been studied for substrate heating during deposition [133]. The advantages of this heating method compared to resistive heaters were identified as avoidance of the problems of outgassing and oxidation of heating elements that were common with resistive heaters. Rapid heating and cooling could be achieved by using light materials as substrate holders. Studies in this area have been performed by Dyer et al. and Wu et al. [204, 205]. Good quality YBCO films have been realized by use of this heating method.

7.2. Laser Irradiation of ITO Films

Systematic study on KrF excimer laser ablation of sputter-deposited ITO films of various thicknesses ranging from 70 to 500 nm on glass substrates has been carried out by Szörényi et al. [206]. Excimer lasers that emit UV pulses of several 10 ns duration are considered to offer a direct approach to high-resolution surface structuring with a minimum thermal or mechanical stress compared with lithography and wet or reactive ion etching [206]. Szörényi et al. have shown through thermal model calculations a framework for explaining the notably different single-shot ablation characteristics of these films when the thickness is within the thermal diffusion length. The low-fluence ablation characteristics of ITO films of different thicknesses are shown in Figure 41 [206]. They observed that for film thickness of 70 and 160 nm complete ablation of the film from the glass substrate occurred when laser fluences of 100–160 and 150–390 $mJ\,cm^{-2}$ were used, respectively. The ablation characteristic was observed to differ with increasing ITO film thickness. From mathematical modeling using the finite-difference method (i.e., the local thermodynamic equilibrium approximation) they have shown that the ablation depth decreased with increasing film thickness. The calculated ablation depth as a function of fluence is shown in Figure 42 [206]. This study showed that for fluence less than 100 $mJ\,cm^{-2}$, the film thickness essentially remains unaffected.

Little has been done on laser irradiation of TCO film during growth by PLD. However, scattered works exist on annealing of ZnO and ITO films by UV lamp and laser energy pulses [111, 207–210]. Laser annealing of amorphous ITO films has been reported by Hosono et al. [111, 210]. They observed that

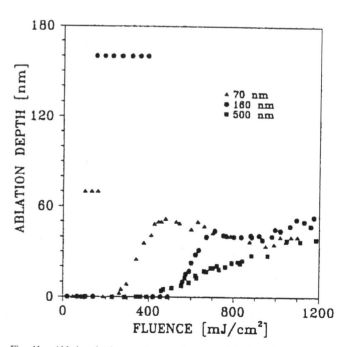

Fig. 41. Ablation depth per pulse as a function of the laser fluence for 70, 160, and 500 nm thick ITO films. (Reprinted with permission from T. Szörényi, Z. Kántor, and L. D. Laude, *Appl. Surf. Sci.* 86, 219 (1995), © 1995, Elsevier Science.)

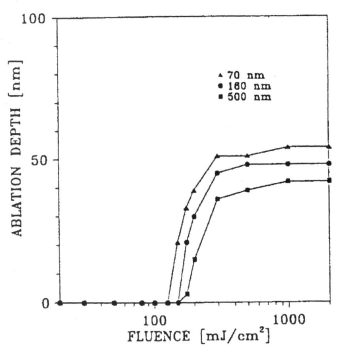

Fig. 42. Calculated ablation depth as a function of laser fluence for 70, 160, and 500 nm thick ITO films. (Reprinted with permission from T. Szörényi, Z. Kántor, and L. D. Laude, *Appl. Surf. Sci.* 86, 219 (1995), © 1995, Elsevier Science.)

sputter-deposited amorphous ITO films could be converted to crystalline films at room temperature by *in situ* pulsed laser annealing with a fluence of around 40 mJ cm^{-2}. They also enumerated the benefits of pulsed laser annealing of ITO films as follows: (i) The UV photon energy is effectively absorbed by ITO film due to its high absorption coefficient of 3×10^5 cm^{-1} at $\lambda = 248$ nm which corresponds to a penetration depth of 33 nm. In the case of a 193 nm laser wavelength, an absorption coefficient of 4.5×10^5 cm^{-1} produced a penetration depth of 22 nm. (ii) A UV laser beam with a pulse interval of 20 ns does not cause a permanent substrate temperature rise, thus making it possible to define the optimal conditions of annealing. In their study using KrF (248 nm) and ArF (193 nm) lasers, Hosono et al. observed that the decisive factor leading to the crystallization of amorphous ITO films was the energy density (>40 mJ cm^{-2} per pulse) for both excimer lasers and not the pulsed frequency. Evidence of improvements in the crystalline structure and optoelectronic properties was observed in the laser-annealed films.

We have carried out studies on excimer laser irradiation of ITO films during growth by PLD [79, 211, 212]. Remarkable improvements on the overall properties of the films were observed even at room temperature. The experimental procedure as well as the results of the ITO films will be considered in subsequent sections.

7.3. Film Preparation by Laser Irradiation

Depositions were carried out using the same KrF (248 nm) laser system described earlier in Section 4.5. The only addition to the arrangement was the laser irradiation assembly consisting of a beam splitter, reflecting mirrors, and focusing lens to direct part of the laser beam at the substrate surface. A schematic diagram of the experimental setup is shown in Figure 43 [79]. The laser system was operated at a total energy of 290–310 mJ and a pulsed frequency of 20 Hz. The laser beam from the system was split into two equal parts using a beam splitter. Half of the energy was focused on the target using a spherical lens with a focal length of 25 cm, thereby yielding an energy density of approximately 3 J cm^{-2}. The other half of the energy was directed at the substrate surface via an independent quartz glass window using a single convex lens with a focal length of 70 cm. The energy density of the laser beam incident on the substrate was roughly 70 mJ cm^{-2}.

During depositions, undoped In$_2$O$_3$ (0 wt% Sn-doped) and Sn-doped In$_2$O$_3$ (3 wt%, 5 wt%, 8 wt%, and 10 wt%) sintered ceramic targets were used. The objective was to determine the effect of target composition on the properties of the laser-irradiated films. Films were deposited at different oxygen pressures, from 1×10^{-3} to 4.5×10^{-2} Torr, and substrate temperatures ranging from room temperature (20°C) to 400°C. The substrates were clamped onto the holder using stainless steel strips, thus producing a well-defined step in the nonirradiated part of the film close to the focused laser beam on the substrate in order to enable the film thickness to be properly determined. Films with reasonable uniformity were deposited on SiO$_2$ fused-quartz glass with substrate areas of 6.25 cm^2. The middle part of the substrate that was irradiated with the laser energy pulses was oval in shape with a total area of about 1 cm^2. The above conditions produced deposition rates of approximately 15 nm/min. The thickness of the films was around 80 ± 20 nm.

The thickness of the films was measured from the well-defined edge by the stainless strip in the nonirradiated part (close to the laser-irradiated portion) using a stylus profilometer. Because the energy density of the irradiation beam was lower than the threshold of 120 mJ cm^{-2} necessary to cause reablation of the deposited film, no major change in the film thickness was expected [205]. Therefore, the measured thickness on the nonirradiated parts of the films was considered to be similar to the laser-irradiated parts, in view of the reasonable uniformity attained over the substrate during deposition. The films were characterized for electrical, optical, and structural properties using the techniques already described in Section 4.7. Both the laser-irradiated and the nonirradiated parts of the films were analyzed in order to compare their properties directly.

7.4. Effect of Sn-Doping on the Electrical Properties of Laser-Irradiated ITO Films

This part consists of the electrical properties of the laser-irradiated ITO films deposited by PLD. The films were deposited from ITO targets of different Sn doping content. The properties of the laser-irradiated and nonirradiated parts of the films are compared. From XRD analysis it was found that the

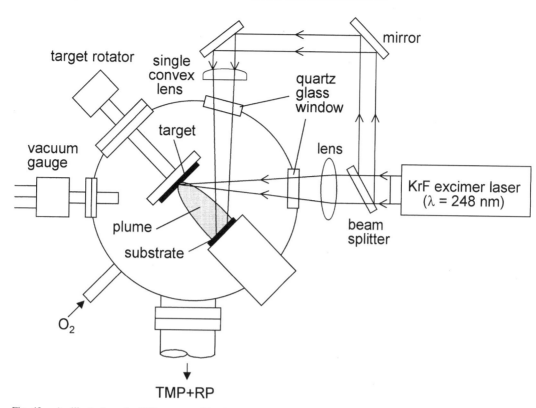

Fig. 43. An illustration of a PLD system with a laser irradiation wing. (Reprinted with permission from F. O. Adurodija, H. Izumi, T. Ishihara, H. Yoshioka, M. Motoyama, and K. Murai, *Japan. J. Appl. Phys. Lett. 2* 39, L377 (2000), © 2000, Japan Society of Applied Physics.)

laser-irradiated films deposited at room temperature and those deposited at 200°C were crystalline with strong ⟨111⟩ preferred orientation, while the nonirradiated portions of the films deposited at room temperature were amorphous [79]. The experimental results of the microstructural properties of the laser-irradiated films are discussed in Sections 7.5 and 7.6.

Figure 44 shows the resistivity of ITO films with Sn doping over the range 0 to 10 wt% deposited at room temperature and 200°C for the laser-irradiated and nonirradiated parts. At room temperature, the nonirradiated parts of the films showed reasonably low resistivity which increased slightly from 2.2×10^{-4} to 2.6×10^{-4} Ω cm with increasing Sn doping content from 0 to 10 wt%. This small increase in the resistivity at high Sn doping content was attributed to the electrical inactivity of Sn at high doping levels [104, 110]. This observation is consistent with observations made for amorphous films, as discussed in Section 6.2. In contrast, the laser-irradiated (crystalline) parts of the undoped In_2O_3 (0 wt% Sn-doped) films deposited at room temperature and 200°C exhibited higher resistivity values of $(4–6.5) \times 10^{-4}$ Ω cm. The high resistivity observed in the In_2O_3 films was due to a deficit of oxygen vacancies leading to nonstoichiometric compositions, as discussed in Section 5 [88, 144]. Within Sn doping content over the range 3 to 10 wt%, the resistivity reduced to between 8.9×10^{-5} and 1.3×10^{-4} Ω cm for the laser-irradiated parts of the films deposited at room temperature and 200°C. The minimum resistivity of 8.9×10^{-5} Ω cm

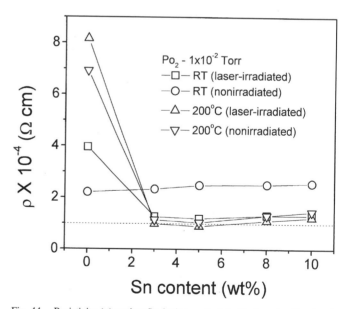

Fig. 44. Resistivity (ρ) against Sn doping content for the laser irradiated and the nonirradiated parts of ITO films deposited at room temperature and 200°C, respectively. (Reprinted with permission from F. O. Adurodija, H. Izumi, T. Ishihara, H. Yoshioka, and M. Motoyama, *J. Appl. Phys.* 88, 4175 (2000), © 2000, American Institute of Physics.)

was observed for the laser-irradiated part of an ITO film deposited from a 5 wt% Sn-doped In_2O_3 target at 200°C. With a

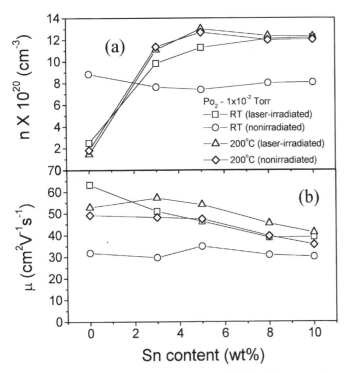

Fig. 45. Plots of (a) carrier concentration (n) and (b) Hall mobility (μ) against Sn doping content for the laser irradiated and the nonirradiated parts of ITO films deposited at RT and 200°C, respectively. (Reprinted with permission from F. O. Adurodija, H. Izumi, T. Ishihara, H. Yoshioka, and M. Motoyama, *J. Appl. Phys.* 88, 4175 (2000), © 2000, American Institute of Physics.)

further increase in the Sn doping content from 5 to 10 wt%, the resistivity increased slightly and the cause was associated with the increasing presence of impurity scattering centers [73, 144, 110].

Figure 45 shows (a) the carrier concentration and (b) the Hall mobility measured from the laser-irradiated and the nonirradiated parts of ITO films deposited at room temperature and 200°C, respectively. The nonirradiated parts of the films deposited at room temperature yielded steady carrier concentrations of $(7.4–8.8) \times 10^{20}$ cm^{-3} for Sn doping content over the range 0 to 10 wt%. This feature suggested that Sn doping had little effect on the creation of free carriers in the amorphous ITO films [143].

For the laser-irradiated parts of the films deposited at room temperature and 200°C (crystalline films), the carrier concentrations indicated a strong dependence on Sn doping content, as shown in Figure 45a. In the case of the undoped In$_2$O$_3$ films, low carrier concentrations of $(1.5–2.5) \times 10^{20}$ cm^{-3} were observed at both substrate temperatures. The carrier concentrations increased sharply to about $(1.1–1.3) \times 10^{21}$ cm^{-3} with increasing Sn doping content to 5 wt%. A further increase in the Sn doping content from 5 to 10 wt% resulted in a shallow increase from 1.1×10^{21} to 1.2×10^{21} cm^{-3} for the films deposited at room temperature. However, for the films deposited at 200°C, the carrier concentrations saturated at 1.3×10^{21} cm^{-3} for 5 wt% doping content, before decreasing slightly to about 1.2×10^{21} cm^{-3} with further increase in Sn doping content to

10 wt%. The carrier concentrations of the laser-irradiated parts of the ITO films deposited at room temperature and the non-irradiated parts of the films deposited at 200°C showed close resemblance.

From the experimental data reported above, the marked increase in the carrier concentration was attributed to the diffusion of Sn atoms from interstitial sites or grain boundaries into the In cation sites due to improvement in the crystalline structure caused by photochemical and/or thermal crystallization effects [213]. This is expected since a Sn atom has a valence of four electrons compared to In with a valence of three electrons. However, the slight reduction in the carrier concentration of the ITO films with increasing Sn doping content within 5 and 10 wt% implied an increase in the electrical inactivity of Sn atoms at higher doping levels. Yamada et al. [110] and Köstlin et al. [104] have shown in their respective studies on the doping mechanism of Sn in In$_2$O$_3$ powder and ITO films deposited by spray pyrolysis that the doping efficiency decreases with increasing Sn content, above 5 wt%. They reported that for Sn doping content above 5 wt%, complexes that do not donate free electrons but only existed as impurity scattering centers in the ITO material are formed. They have shown that the impurity scattering centers reduce the doping efficiency and the mobility of free carriers [104, 110]. The results on the relationship between Sn doping and carrier concentration for the laser-irradiated films closely agreed with those reported by Köstlin et al. and Yamada et al. A similar observation has also been made by Kim et al. on laser deposited ITO films at 250°C [88].

Apparently, the Hall mobility of the nonirradiated parts of the films deposited at room temperature remained unaffected ($\sim$32 cm^2 V^{-1} s^{-1}) by the change in Sn doping content from 0 wt% to 10 wt%, as shown in Figure 45b. This behavior is consistent with the stable resistivity and carrier concentration shown in Figures 44 and 45a. For the laser-irradiated parts of the films deposited at room temperature, high Hall mobilities of 39–63.3 cm^2 V^{-1} s^{-1} were observed for Sn doping over the range 0 to 10 wt%. At room temperature, the laser-irradiated part of an undoped In$_2$O$_3$ produced a maximum mobility of 63.3 cm^2 V^{-1} s^{-1}, which gradually decreased to 39 cm^2 V^{-1} s^{-1} with increasing Sn doping content to 10 wt%. The Hall mobilities of the laser-irradiated parts of the films deposited at room temperature are comparable to those of the nonirradiated parts deposited at 200°C.

In contrast, the laser-irradiated parts of the films deposited at 200°C showed a remarkable increase in the Hall mobility, irrespective of the changes in Sn doping content. A similar relationship between the Hall mobility and Sn doping has been reported for ITO films deposited by electron-beam evaporation and PLD [88, 144]. The observed steady decrease in the mobility has been attributed to the presence of complexes that act as scattering centers. In general, laser irradiation of the growing ITO films on unheated and heated glass substrates by PLD was found to enhance the electrical properties. The increase in the Hall mobility of the laser-irradiated films even at 200°C is associated with the improvement in the crystallinity of the films. The moderately low mobility measured at high Sn doping con-

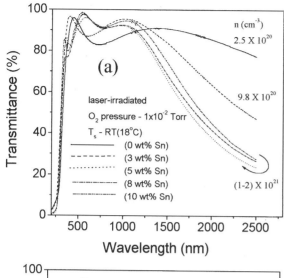

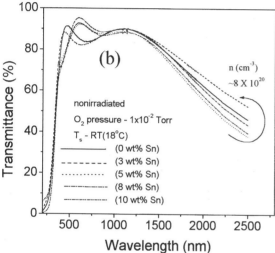

Fig. 46. Transmittance curves as a function of wavelength for (a) laser-irradiated and (b) nonirradiated parts of ITO films containing different Sn doping contents. (Reprinted with permission from F. O. Adurodija, H. Izumi, T. Ishihara, H. Yoshioka, and M. Motoyama, *J. Appl. Phys.* 88, 4175 (2000), © 2000, American Institute of Physics.)

tent is explained on the basis of the interaction of scattering centers and formation of neutral scattering defect [73, 144, 214].

Figure 46a and b shows the respective optical transmittance measured at wavelengths over the range 190 to 2500 nm for the laser-irradiated and nonirradiated parts of the 0–10 wt% Sn-doped films deposited at room temperature. The films showed high optical transmittance greater than 85% in the visible range of the solar spectrum. From Figure 46a and b, the Sn doping could be correlated to the optical properties of the laser-irradiated and nonirradiated parts of the films. For the laser-irradiated parts of the films, the changes in the NIR and absorption edge could be related directly to the carrier concentration shown in Figure 45b. It is seen that the Sn doping content (carrier concentration) increased with increasing NIR, while the optical absorption edge shifted (Burstein–Moss shift) toward the shorter wavelength, thus implying bandgap widening. The shift in the absorption edge could also be related to the increase in

the filling up of the lower energy levels in the conduction band by electrons released from the Sn atoms. A similar feature has been reported for ITO films deposited by electron-beam evaporation and PLD [5, 88] as well as fluorine doped tin oxide (SnO_2:F) films [5].

7.4.1. Electronic Transport Properties

Carrier mobilities of ITO films are reported to be greatly affected by disorder due to the structure of In_2O_3 as well as modification of the network resulting from Sn doping [73, 144]. It is well established that with increasing Sn doping content, complexes that do not contribute free electrons, but only exist as scattering centers that reduce the mobilities of free electrons, are formed in the ITO films [73, 104, 110, 144, 214]. Electron scattering sources such as grain boundaries, acoustical phonons, and neutral and ionized impurity centers have been found to affect the electrical and optical properties of ITO films [73]. However, in most of the cases that have been studied, grain boundaries and acoustical phonons played secondary roles, since the mean-free path of the electron is smaller than the average crystallite sizes and also shows little dependence on growth temperatures over the range 100 to 500°C [73, 215].

The experimental data on the laser-irradiated and nonirradiated parts of the films were studied further in order to clarify the dominant scattering process. Therefore, the mean-free path (l) was calculated using a sufficiently degenerate gas model, given by [73, 144, 216]

$$V_f = \left(3\pi^2\right)^{1/3}\left(\hbar/m^*\right)n^{1/3} \tag{21}$$

$$l = V_f\tau = \left(3\pi^2\right)^{1/3}\left(\hbar/e^2\right)\rho^{-1}n^{-2/3} \tag{22}$$

where V_f, m^*, and n represent the electron velocity at the Fermi surface, electron effective mass, and carrier concentration, respectively. The relaxation time and the resistivity are denoted by τ and ρ. Within the limits of experimental error (<5%), the calculated mean-free paths plotted against Sn doping content are shown in Figure 47, for the laser-irradiated and nonirradiated parts of the ITO films deposited at room temperature and 200°C, respectively. Apparently, no significant change in the mean-free paths was observed for both the amorphous films ($\approx$6 nm) and the crystalline ITO films (<12 nm) deposited at room temperature with laser irradiation and at 200°C. The values of the mean-free paths were generally less than 12 nm for all the films. For the crystalline films, the calculated l was much smaller than the average crystallite sizes of more than 200 nm, as shown later in Sections 7.5 and 7.6. Therefore, it was concluded that electron scattering at grain boundaries was not important.

Scattering of the conduction electrons in ITO films by neutral and ionized impurity centers appeared to be predominant in most cases that have been studied [73, 144]. Erginsoy has calculated the contribution to conductivity due to neutral impurities, particularly for semiconductors with small degrees of ionization [217]. The theoretical evaluation on the contribution to conductivity due to ionized scattering has been performed

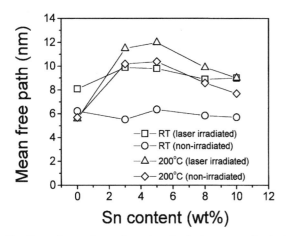

Fig. 47. Plots of mean free paths against Sn doping content for the laser-irradiated and the nonirradiated parts of ITO films deposited at room temperature and 200°C, respectively. (Reprinted with permission from F. O. Adurodija, H. Izumi, T. Ishihara, H. Yoshioka, and M. Motoyama, *J. Appl. Phys.* 88, 4175 (2000), © 2000, American Institute of Physics.)

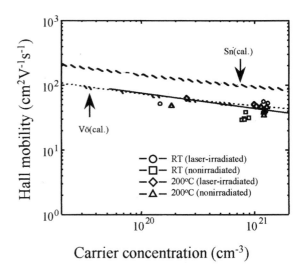

Fig. 48. Hall mobility versus carrier for the laser-irradiated and nonirradiated parts of ITO films deposited at RT and 200°C, respectively: experimental data (circles, squares, triangles, and diamonds) and calculated mobilities based on scattering by ionized impurities and the assumption that all the carriers are due to singly charged tin (Sn·) donors (dashed line) or doubly charged oxygen vacancies ($V_O^{..}$) (dotted line). (Reprinted with permission from F. O. Adurodija, H. Izumi, T. Ishihara, H. Yoshioka, and M. Motoyama, *J. Appl. Phys.* 88, 4175 (2000), © 2000, American Institute of Physics.)

by Conwell and Weisskopf [218] and Dingle [219]. The above-named theories as given below were used to further explain the electronic transport properties of the laser-irradiated ITO films. These theories have been used by many others to describe the effect of neutral and charge scattering centers on the carrier mobility in degenerate semiconductors as [73, 104, 144, 170, 220, 221]

$$\mu_N = \frac{(m^*e^3)}{20\varepsilon_0\varepsilon_r\hbar^3 n_N} \tag{23}$$

$$\mu_I = \frac{24\pi^3(\varepsilon_0\varepsilon_r)^2\hbar^3 n}{e^3 m^{*2} g(x) Z^2 n_I} \tag{24}$$

where the screening function $g(x)$ is given by

$$g(x) = \ln\left(1 + \frac{4}{x}\right) - \left(1 + \frac{4}{x}\right)^{-1} \tag{25}$$

where

$$x = \frac{4e^2 m^*}{4\pi\varepsilon_0\varepsilon_r\hbar^2(3\pi^5)^{1/3} n^{1/3}} \tag{26}$$

μ_N and μ_I denote the mobilities due to neutral and charged centers, respectively. ε_0 is the permittivity of free space. m^* and ε_r are the effective mass of the electron and low frequency of permittivity. For ITO films the values m^* and ε_r are taken as $0.3m_e$ [73, 144, 220] and 9 [6, 73, 144], respectively. n and Z represented the carrier concentration and the charge of the ionized centers, respectively. n_N and n_I denote neutral and ionized scattering centers, respectively. The measured total mobility μ_T is stated as the sum of μ_N and μ_I as follows:

$$\frac{1}{\mu_T} = \frac{1}{\mu_N} + \frac{1}{\mu_I} \tag{27}$$

In crystalline ITO films, the behavior of the electronic properties is known to be regulated by electron donors originating from substitutional four-valence Sn atoms (supply of one electron or Sn·) and oxygen vacancies (supply of two electrons or

$V_O^{..}$). In essence, the scattering caused by doubly charged oxygen vacancies ($V_O^{..}$) is greater than that of single charge from Sn (Sn·), since the values of $n/n_I Z^2$ (from Eq. (25)), ($V_O^{..}$) and (Sn·) are 1 and 0.5, respectively [73, 144]. Figure 48 shows plot of the measured Hall mobility against carrier concentration for the nonirradiated and laser-irradiated parts of the ITO films deposited at room temperature and 200°C. Using Eq. (25), Figure 48 also shows the calculated μ_I plotted as a function of n assuming that the measured mobility was totally due to either charged vacancies ($V_O^{..}$ or dotted line) or singly charged Sn (Sn· or dashed line). It was observed that the gradients of the measured and the calculated mobilities versus the carrier concentrations were in good agreement. The results concluded that scattering at ionized centers appeared to be responsible for the decrease in the mobility of the ITO films. However, from Figure 45a, since the increase in carrier concentration was principally associated with a rise in Sn doping contents, some Sn atoms present may also contribute to neutral scattering centers.

7.5. Effects of Oxygen Pressure on the Properties of Laser-Irradiated ITO Films

ITO films discussed here were deposited at room temperature (18°C) and various oxygen pressures. The films were characterized for structural, electrical, and optical properties and the results of the various analyses are presented.

7.5.1. Structural and Morphological Properties

A typical XRD pattern of the nonirradiated part of an ITO film deposited under an oxygen pressure of 1×10^{-2} Torr is

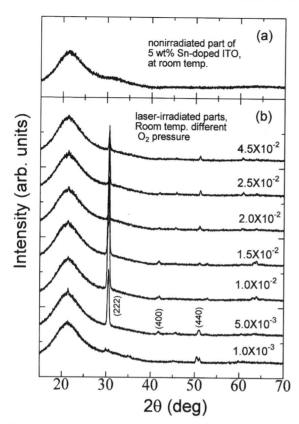

Fig. 49. (a) XRD spectrum of a nonirradiated part of an ITO film deposited at RT and P_{O_2} of 1×10^{-2} Torr, (b) XRD spectra of the laser-irradiated parts of ITO films deposited at room temperature and oxygen pressure over the range 1×10^{-3} to 4.5×10^{-2} Torr.

shown in Figure 49a. In Figure 49b the XRD patterns of the laser-irradiated parts of ITO films deposited at oxygen pressures over the range 1×10^{-3} to 4.5×10^{-2} Torr are shown. The nonirradiated parts of the films prepared at room temperature were amorphous as indicated by the absence of diffraction peaks as expected. In the case of the laser-irradiated parts, crystalline phases indicated by the presence of diffraction peaks were observed. XRD patterns show only standard ITO peaks with a strong ⟨111⟩ preferred orientation, except for the films deposited at low oxygen pressures of 1×10^{-3} Torr. At the low oxygen pressure region, the XRD spectrum of the film appeared to contain double peaks due to induced strain during film growth. At higher oxygen pressure ($> 1.5 \times 10^{-2}$ Torr), the crystalline quality deteriorated in a similar way earlier described in Section 6 for films deposited by the conventional PLD method.

Figure 50a shows typical surface AFM micrographs of a nonirradiated part of a film deposited at 1×10^{-2} Torr, and Figure 50b–d represents the laser-irradiated parts of ITO films deposited at oxygen pressure of 1×10^{-2}, 1×10^{-3}, and 2.5×10^{-2} Torr, respectively. The nonirradiated part of the films exhibited smooth surface features, indicating an amorphous phase. In the case of the laser-irradiated parts, contrasting morphologies were displayed with changes in oxygen pressure. At low oxygen pressures of 1×10^{-3} Torr, the films appeared

very rough, while at high oxygen pressures ($> 2 \times 10^{-2}$ Torr), uniform films containing small crystallites were obtained, thus confirming the poor crystalline structures observed from the XRD analysis (Fig. 49b). However, at oxygen pressures of 1×10^{-2} Torr, ITO films with densely packed crystallites and sizes of around 200 nm were evidenced. It is noteworthy that the optimal oxygen pressure of 1×10^{-2} Torr also yielded the best crystallinity for the laser-irradiated part of the films [79–92, 171]. The distinct improvement in the structural and morphological properties of the films has been associated with the photocrystallization effect since no thermal heating was involved during deposition.

7.5.2. Electrical Properties

The dependences of the resistivity on the oxygen pressure for the laser-irradiated and the nonirradiated parts of ITO films deposited at room temperature are shown in Figure 51. In both cases, a strong influence of oxygen pressure on the resistivity, particularly at oxygen pressures above 1.5×10^{-2} Torr, is apparent. At low oxygen pressure (1×10^{-3} Torr), the laser-irradiated part of the ITO film yielded a resistivity of 3.1×10^{-4} Ω cm compared to 4.7×10^{-3} Ω cm for the nonirradiated part. With increasing oxygen pressure to the optimal region of $(1–1.5) \times 10^{-2}$ Torr, the resistivity of the laser-irradiated and the nonirradiated parts of the films decreased to minimal values of 1.2×10^{-4} and 2.3×10^{-4} Ω cm, respectively. However, a further increase in oxygen pressure ($> 1.5 \times 10^{-2}$ Torr) resulted in a sharp rise in the resistivities of both parts of the films. The lower resistivity observed in the laser-irradiated parts of the films was associated with the improvement in the crystallinity of the films due to the photocrystallization effect. Table II shows the electrical properties of ITO films deposited by a multitude of techniques. It is seen that at low-temperature PLD ITO films displayed lower resistivities, higher carrier concentrations, and higher mobilities compared to the other techniques.

The carrier concentration and the Hall mobility plotted against the oxygen pressure for the laser-irradiated and the nonirradiated parts of ITO films deposited at room temperature are shown in Figure 52. At low oxygen pressures of 1×10^{-3} Torr, carrier concentrations of 7.7×10^{20} and 4.7×10^{20} cm^{-3} were measured for the laser-irradiated and the nonirradiated parts of the films, respectively. With increasing oxygen pressure to 1.5×10^{-2} Torr, the carrier concentrations increased to optimal values of 1.3×10^{21} and 8.8×10^{20} cm^{-3} for the laser-irradiated and the nonirradiated parts of the films, respectively. This oxygen pressure region also produced the lowest resistivity, as shown in Figure 51. Further increases in oxygen pressure ($> 1.5 \times 10^{-2}$ Torr) led to a rapid decrease in the carrier concentration. This behavior is similar to what was reported for ITO films deposited by conventional PLD [82]. The carrier concentrations of the laser-irradiated and the nonirradiated parts of the films were lower, particularly at oxygen pressures below 1.5×10^{-2} Torr. At all oxygen pressures, the carrier concentrations of the laser-irradiated parts were much higher than those of the nonirradiated parts of the films. The improvement in the

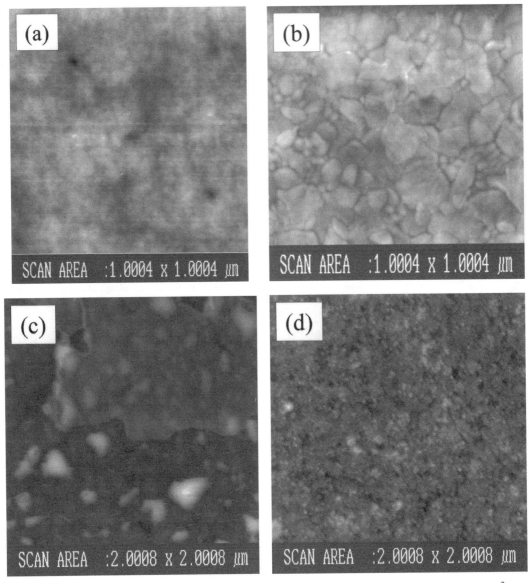

Fig. 50. Typical surface AFM pictures of (a) a nonirradiated part of the films deposited at oxygen pressure of 1×10^{-2} Torr. (b) to (d) show the laser-irradiated part of ITO films deposited at oxygen pressures of 1×10^{-2}, 1×10^{-3}, and 2.5×10^{-2} Torr, respectively. All films were deposited at room temperature.

carrier concentration was a consequence of the activation of Sn atoms into Sn^{4+} within the host In_2O_3 crystal lattice during the photocrystallization process. The migration of Sn atoms was enhanced by the laser irradiation; hence more Sn atoms could substitute for In atoms in the lattice, leaving one electron each more than the requirement for bonding.

The dependence of Hall mobility on oxygen pressure for the laser-irradiated and the nonirradiated parts of the ITO films are shown in Figure 52b. The effects of oxygen pressure on the Hall mobility in both parts of the films followed similar tendencies. At low oxygen pressures of 1×10^{-3} Torr, the laser-irradiated part of a film exhibited a moderately low Hall mobility of $26.2 \text{ cm}^2 \text{ V}^{-1} \text{ s}^{-1}$ that rose to 40.4–$45.2 \text{ cm}^2 \text{ V}^{-1} \text{ s}^{-1}$ with increasing oxygen pressure to between 5×10^{-3} and 2×10^{-2} Torr.

Further increases in oxygen pressure to 4.5×10^{-2} Torr resulted in a sharp decrease in the carrier mobility to $2.2 \text{ cm}^2 \text{ V}^{-1} \text{ s}^{-1}$. In the case of the nonirradiated parts of the films, a very low Hall mobility of $4.5 \text{ cm}^2 \text{ V}^{-1} \text{ s}^{-1}$ was observed at a low oxygen pressure of 1×10^{-3} Torr followed by a rapid increase to $30.4 \text{ cm}^2 \text{ V}^{-1} \text{ s}^{-1}$ with increasing oxygen pressure to 1×10^{-2} Torr. Between oxygen pressures of 1×10^{-2} and 2×10^{-2} Torr, no appreciable change in Hall mobility of the nonirradiated part of the films occurred. However, at an oxygen pressure of 4.5×10^{-2} Torr, the mobility and the carrier concentration of the nonirradiated part of the films could not be determined due to the very high resistivity. The best Hall mobility was obtained for the laser-irradiated and the nonirradiated parts of the ITO films deposited at the optimal oxy-

Table II. Compilation of the Electrical Properties of In_2O_3 and ITO Films Deposited by a Multitude of Techniques

Film material	Deposition technique	Deposition temperature (°C)	Resistivity ($\times 10^{-4}$ Ω cm)	Carrier concentration ($\times 10^{20}$ cm^{-3})	Hall mobility (cm^2 V^{-1} s^{-1})	Comment	References
In_2O_3	PLD	RT	3.6	5	35	–	[83]
In_2O_3	PLD	RT	2.5	6	30	–	[80]
In_2O_3	PLD	100	1.8	9	47	–	[80]
In_2O_3	PLD	200	5	4	28	–	[80]
In_2O_3	PLD	350	11.5	2	25	–	[83]
In_2O_3	TE	400	4	4	72	–	[222]
In_2O_3	TE	320–350	2	4.7	70	–	[23]
In_2O_3	TE	150	8–16	0.5–1.1	60–95	–	[16]
In_2O_3	RE	200–400	20–30	0.35	25–60	–	[126]
ITO	PLD	RT	2.8	–	–	–	[92]
ITO	PLD	RT	4.5	5	30	–	[82]
ITO	PLD	RT	4	–	–	–	[89]
ITO	PLD	RT	4	6	12	–	[91]
ITO	PLD	200	1.7	9	33	–	[82]
ITO	PLD	310	2	–	–	–	[85]
ITO	PLD	350	1.3	~10	~48	–	[80]
ITO	PLD	RT	1.2	10	40	PLI	[79]
ITO	PLD	200	0.89	~12	55	PLI	[212]
ITO	PLD	300	0.84	~12	57	PLI	[212]
ITO	DCMS	RT	5.5	7.1	16	–	[38]
ITO	RFMS	370	0.6–0.8	36	27	–	[41]
ITO	RFMS	Unheated	~4	–	–	–	[155]
ITO	RFMS	130	~4	10	~10	–	[49]
ITO	RFMS	200	1.59	10	~40	–	[224]
ITO	RFMS	300	2.47–1.4	11.5–12.4	22–36	–	[225]
ITO	RFMS	Unheated	2.55	15.9	15.4	Annealed 400°C	[55]
ITO	EBE	300	0.44	13.8	103	–	[27, 28]
ITO	ARE	200	1.64	>10	30	–	[226]
ITO	ARE	370	0.7	~10	~30	–	[15]
ITO	HDPE	280	1.23	10.7	47.5	–	[227]
ITO	HDPE	180	1.7	–	–	–	[147]
ITO	CVD	260	28.3	–		–	[64]
ITO	CVD	450	~4.5	10	5–22	–	[228]
ITO	Spray	480	1.5	11	43	–	[229]
ITO	Spray	350–450	1.6–1.8	8.8	43	–	[230]
ITO	Sol–gel	–	30–50	–	–	Annealed 550°C	[231]

TE—thermal evaporation, RE—reactive evaporation, PLI—pulsed laser irradiation, DCMS—dc magnetron sputtering, RFMS—rf magnetron sputtering, EBE—electron beam evaporation, ARE—activated reactive evaporation, HDPE—high density plasma-enhanced evaporation, CVD—chemical vapor deposition.

gen pressure. The higher Hall mobility observed for the laser-irradiated part was associated with the improvement in the crystallinity of the films. The relationship between the Hall mobility and oxygen pressure is consistent with those earlier reported for ITO films deposited at room temperature and (200°C) by PLD, as discussed earlier in Section 6.2 [81]. In general, the poor electrical properties obtained in the films deposited at low oxygen pressure (1×10^{-3} Torr) and high oxygen pressures ($>1.5 \times 10^{-2}$ Torr) was due to the poor crystalline quality, as shown in Figures 49b and 50.

7.5.3. Optical Properties of the Films

The optical properties of the films ($\sim$100 nm) were determined by measuring the transmittance at wavelengths over the range 190 to 2500 nm. Figure 53a and b shows the transmittance spectra versus the wavelength for the laser-irradiated and the nonir-

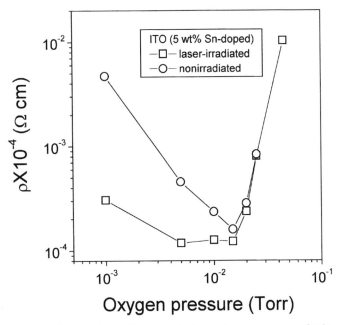

Fig. 51. The dependence of the resistivity (ρ) on the oxygen pressure for the laser-irradiated and the nonirradiated parts of the ITO films deposited at room temperature.

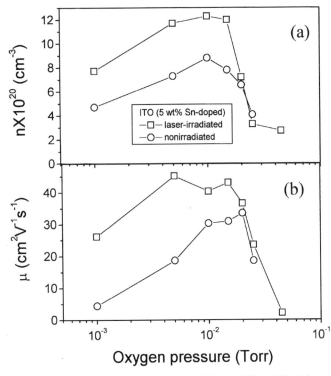

Fig. 52. (a) The carrier concentration (n) and (b) the Hall mobility (μ) as a function of oxygen pressure for the laser-irradiated and the nonirradiated parts of the ITO films deposited at room temperature.

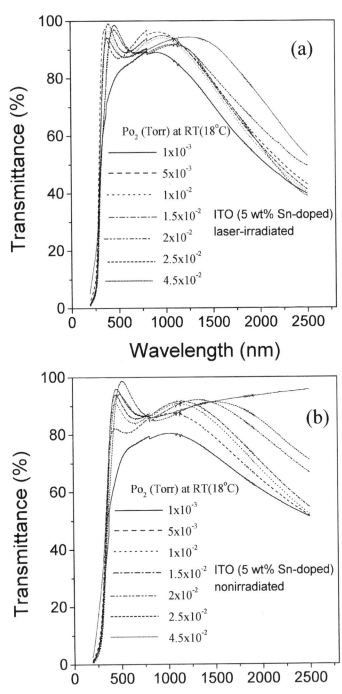

Fig. 53. Optical transmittance spectra versus the wavelength for the laser-irradiated (a) and the nonirradiated (b) parts of the ITO films deposited under oxygen pressure over the range 1×10^{-3} to 4.5×10^{-2} Torr at room temperature.

radiated parts of ITO films deposited at oxygen pressures over the range 1×10^{-3} to 4.5×10^{-2} Torr. The films deposited at low oxygen pressure ($= 1 \times 10^{-3}$ Torr) were visually brownish in color and less transparent ($<85\%$) to the visible light (400–

800 nm). The transmittance (visible) of both the laser-irradiated and the nonirradiated parts of ITO films increased with increasing oxygen pressure to 4.5×10^{-2} Torr. At the optimal oxygen pressure region, $(1–1.5) \times 10^{-2}$ Torr, transmittance above 90% was measured for the laser-irradiated films, while that of the nonirradiated films was slightly less than 90%.

The NIR was also affected by the laser irradiation and the change in oxygen pressure during growth of the ITO films. With increasing oxygen pressure, the NIR of the nonirradiated part of films decreased accordingly, while that of the laser-irradiated parts remained reasonably high. In addition, the absorption edge was found to change with changes in oxygen pressure.

7.6. Effect of Substrate Temperature on the Properties of Laser-Irradiated ITO Films

So far it has been shown that at room temperature crystalline ITO films with improved optoelectrical properties can be achieved by laser irradiation of the substrate during PLD process. In this section the effect of the substrate temperature on the material properties of the laser-irradiated films is discussed. The discussion covers the properties of the laser-irradiated and nonirradiated parts of the films.

7.6.1. Structural and Morphological Properties

The XRD spectra of the laser-irradiated and the nonirradiated portions of the ITO films deposited at substrate temperatures ranging from 18°C to 400°C under oxygen pressure of 1×10^{-2} Torr are shown in Figure 54 [223]. Crystalline films with a strong $\langle 111 \rangle$ preferred orientation were observed for the laser-irradiated films deposited at all temperatures. In contrast, the nonirradiated parts of the ITO films deposited at substrate temperatures lower than 150°C were amorphous, while above 150°C the films were crystalline exhibiting strong $\langle 111 \rangle$ preferred orientation as expected. Details on the structural properties of ITO films deposited without laser irradiation have been described in Section 6. With increasing substrate temperature, a slight shift in the diffraction peak positions to lower 2θ angles were observed and the cause was associated with induced strain in the films during growth. In addition, the (222) diffraction peak intensities of the laser-irradiated films were much stronger than those of the nonirradiated parts of the films.

The surface morphology of the films deposited at various temperatures was also analyzed using the AFM. Figure 55 shows the surface AFM micrographs of the laser-irradiated and nonirradiated parts of ITO films deposited at 100°C [223]. Figure 56 shows the surface AFM micrographs of the laser-irradiated and nonirradiated parts of ITO films deposited at 300°C [223]. These samples represented the low and high deposition temperature regimes where amorphous and crystalline films are achieved without laser irradiation. At 100°C, the nonirradiated films were amorphous as confirmed from XRD analysis Figure 54a. At 300°C, the nonirradiated and the laser-irradiated part of the films were crystalline. The crystallinity of the ITO films was highly improved with increasing substrate temperature as shown from both the XRD and AFM analyses. These data indicated that laser irradiation of the substrate during growth even at higher temperature is effective in enhancing the structural properties of ITO films. Below a substrate temperature of 100°C, crystallization of the films was supposed

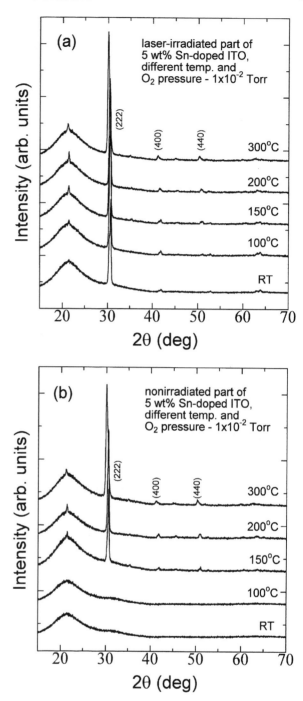

Fig. 54. XRD spectra of the (a) laser-irradiated and (b) nonirradiated parts of ITO films deposited at oxygen pressure of 1×10^{-2} Torr, as a function of substrate temperature. (Reprinted with permission from F. O. Adurodija, H. Izumi, T. Ishihara, H. Yoshioka, M. Motoyama, and K. Murai, *Appl. Surf. Sci.* 177, 114 (2001), © 2001, Elsevier Science.)

to proceed through a photochemical mechanism (photocrystallization). On the other hand, above a substrate temperature of 150°C, film crystallization occurred through the combined effects of photo and thermal mechanisms (photothermal crystallization) [133, 213].

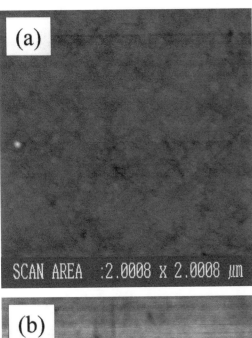

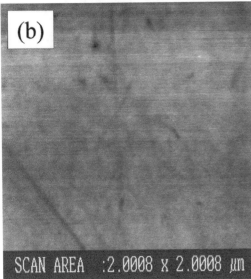

Fig. 55. Surface AFM micrographs of the (a) laser-irradiated and (b) nonirradiated parts of an ITO film deposited at a temperature of 100°C. (Reprinted with permission from F. O. Adurodija, H. Izumi, T. Ishihara, H. Yoshioka, M. Motoyama, and K. Murai, *Appl. Surf. Sci.* 177, 114 (2001), © 2001, Elsevier Science.)

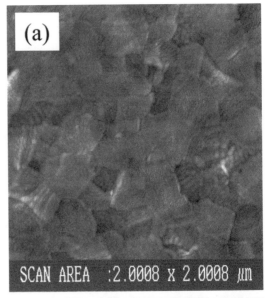

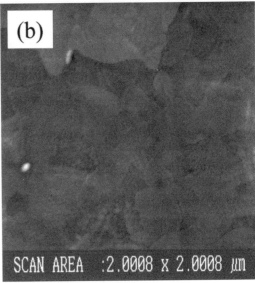

Fig. 56. Surface AFM micrographs of the (a) laser-irradiated and (b) nonirradiated parts of an ITO film deposited at temperature of 300°C. (Reprinted with permission from F. O. Adurodija, H. Izumi, T. Ishihara, H. Yoshioka, M. Motoyama, and K. Murai, *Appl. Surf. Sci.* 177, 114 (2001), © 2001, Elsevier Science.)

7.6.2. Electrical Properties

The dependences of the resistivity on the substrate temperature for the laser-irradiated and the nonirradiated parts of the ITO films are shown in Figure 57 [223]. All films were deposited at the optimum oxygen pressure of 1×10^{-2} Torr. At room temperature, the lowest resistivity of 1.2×10^{-4} Ω cm was achieved for the laser-irradiated part of the film compared to 2.3×10^{-4} Ω cm for the nonirradiated part as already explained in Sections 7.4 and 7.5. With increasing substrate temperature the resistivities of both parts of the films decreased linearly, while the differences between the resistivity values narrowed and coincided at 300°C. The measured resistivity at 300°C was 8.4×10^{-5} Ω cm. When the substrate temperature was raised to 400°C, the resistivities of the laser-irradiated and the nonirradiated parts of the

films increased to 1.1×10^{-4} and 1.2×10^{-4} Ω cm, respectively. The increase in the resistivity was associated with film contamination originating from the components of the substrate holder.

The carrier concentration and the Hall mobility as a function of the substrate temperature for the laser-irradiated and the nonirradiated parts of the ITO films are shown in Figure 58a and b [223]. With increasing substrate temperature from room temperature to 200°C, the carrier concentration of the nonirradiated part of the films increased from 8.8×10^{20} to 1.2×10^{21} cm^{-3} and saturated with further increase in temperature to 400°C. The initial increase in the carrier concentration at temperature between room temperature and 200°C (amorphous–crystalline) was a consequence of the thermal excitation of the Sn atoms into Sn^{4+} cations [73]. In the case of the laser-irradiated parts

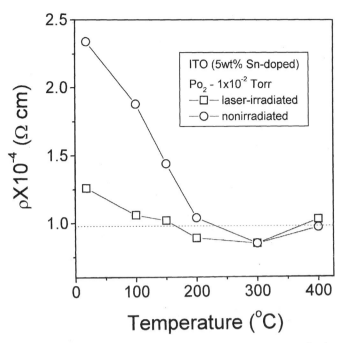

Fig. 57. Dependence of the resistivity (ρ) on substrate temperature for the laser-irradiated and the nonirradiated parts of ITO films. (Reprinted with permission from F. O. Adurodija, H. Izumi, T. Ishihara, H. Yoshioka, M. Motoyama, and K. Murai, *Appl. Surf. Sci.* 177, 114 (2001), © 2001, Elsevier Science.)

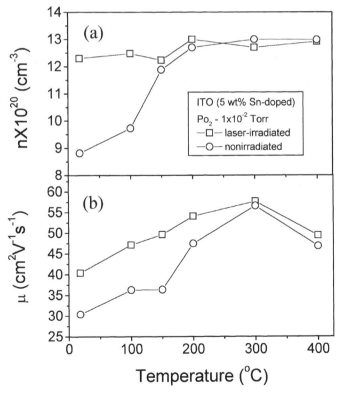

Fig. 58. (a) The carrier concentration (n) and (b) the Hall mobility (μ) as a function of temperature for the laser-irradiated and the nonirradiated parts of ITO films, respectively. (Reprinted with permission from F. O. Adurodija, H. Izumi, T. Ishihara, H. Yoshioka, M. Motoyama, and K. Murai, *Appl. Surf. Sci.* 177, 114 (2001), © 2001, Elsevier Science.)

of the ITO films, no appreciable change in the carrier concentration was observed at substrate temperatures over the range 18 to 400°C. The high carrier concentration obtained in the laser-irradiated parts of the films even at low substrate temperature was attributed to photothermal crystallization effect [213]. However, a slight increase in the carrier concentration with increasing substrate temperature from room temperature to 400°C was observed and could be due to further improvement in the crystallinity of the films. From 200°C, the carrier concentrations of the laser-irradiated and nonirradiated parts of the ITO films were $\sim 1.2 \times 10^{21}$ cm^{-3}. These results further ascertain the effectiveness of Sn atoms as carrier contributors in crystalline ITO films [73]. The carrier concentration of the ITO films was strongly enhanced at low substrate temperature by the laser irradiation effect.

The relationships between the Hall mobility and the substrate temperature for the laser-irradiated and the nonirradiated parts the ITO films are shown in Figure 58b. The Hall mobilities of the laser-irradiated and nonirradiated parts of the films increased appreciably from 40 and 30 cm^2 V^{-1} s^{-1} at room temperature to 57 and 56 cm^2 V^{-1} s^{-1} at 300°C, respectively. A further increase in the substrate temperature to 400°C resulted in a decrease in the Hall mobility to 47 cm^2 V^{-1} s^{-1} for the laser-irradiated and 45 cm^2 V^{-1} s^{-1} for the nonirradiated parts of the films. At temperature of 400°C, the observed increase in the resistivity was primarily a consequence of the reduced Hall mobility. The high Hall mobilities of 57 and 56 cm^2 V^{-1} s^{-1} measured for the laser-irradiated and the nonirradiated parts of the ITO film deposited at 300°C also contributed to the lowering of the resistivities of the films. The relationship between the Hall mobility and the substrate temperature (from 18 to 300°C) is consistent with those previously reported for sputtered ITO and In$_2$O$_3$ films [48, 55]. The thermal effects on the Hall mobility were noticed from the narrowing of the narrowing of the difference between the values of mobilities of the laser-irradiated and the nonirradiated parts of the films with increasing temperature. The higher Hall mobility values observed in the laser-irradiated parts of the films compared to the nonirradiated parts were chiefly due to the enhancement of the crystallinity.

7.6.3. Optical Properties

The effects of the substrate temperature on the optical properties of the laser-irradiated and the nonirradiated parts of the films were also manifested. Figure 59a and b shows the optical transmittance curves as a function of wavelength for the laser-irradiated and nonirradiated parts of ITO films deposited at substrate temperatures over the range 18 to 300°C [223]. In comparison, no significant change in transmittance ($\sim$90% in the visible region) occurred in the laser-irradiated parts of the films, while the nonirradiated parts showed an increase from 85 to about 90% with increasing substrate temperature. In addition, higher NIR that changed slightly with changes in the substrate temperature was noticed for the laser-irradiated parts

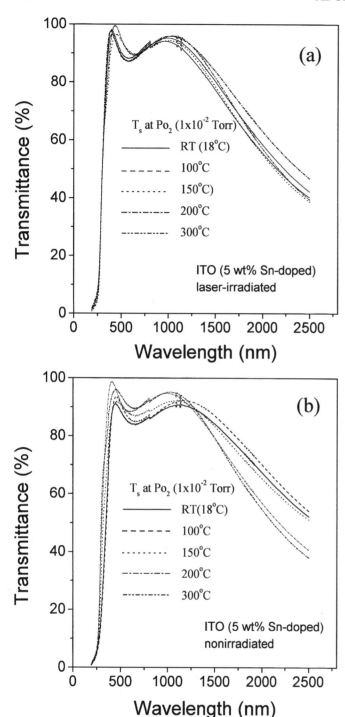

Fig. 59. Optical transmittance curves as a function of the wavelength for (a) the laser-irradiated and (b) the nonirradiated parts of ITO films deposited at temperature over the range room temperature to 300°C. (Reprinted with permission from F. O. Adurodija, H. Izumi, T. Ishihara, H. Yoshioka, M. Motoyama, and K. Murai, *Appl. Surf. Sci.* 177, 114 (2001), © 2001, Elsevier Science.)

of the films, while the nonirradiated parts indicated lower NIR that increases with increasing substrate temperature.

Furthermore, the absorption edge of the nonirradiated parts of the ITO films shifted (B-M shift) to shorter wavelength, im-

plying a widening of the optical bandgap with increasing substrate temperature, as shown in Figure 59b. This feature is similar to what was observed in ITO films deposited by conventional PLD (i.e., without laser irradiation), sputtering, and evaporation methods [80–92, 171]. In the case of the laser-irradiated parts of the films, no apparent change in the absorption edge occurred for substrate temperature over the range 18 to 300°C, as shown in Figure 59a. These results suggested that the optical properties of ITO films could be enhanced by laser irradiation of the substrates even on heated substrate by laser irradiation during growth by PLD.

8. OTHER TCO MATERIALS—ZINC OXIDE (ZnO) THIN FILMS

This section focuses on other TCO thin films that have been prepared by PLD with particular emphasis on ZnO films. ZnO is well known for its high conductivity, piezoelectricity, and high optical transparency; hence it is ideally suited for applications such as transparent electrodes, piezoelectric transducers, tribology, and surface acoustic devices [232–243]. Like ITO, ZnO has been the subject of numerous studies and several techniques, including PLD, have been used to grow ZnO thin films on a variety of substrates [6, 207–209, 234, 244–258]. ZnO has a hexagonal crystal structure of the wurtzite group [244, 259–264]. Basically, the microstructure, crystal orientation, and defect density control the properties of ZnO films. The presence of dopants can also affect the properties of ZnO films [264]. Generally ZnO exhibits *n*-type conduction resulting from oxygen deficiencies and interstitial Zn ions which act as donors in the ZnO crystal lattice.

High quality ZnO thin films have been deposited by the PLD technique since 1983 [207–209, 248–258]. More recently, PLD has been used to grow epitaxial ZnO films on sapphire substrate in order to study the nonlinear optical and piezoelectric properties and their use as a buffer layer for epitaxial GaN growth [249, 250, 265–267]. Like ITO films, the relatively low deposition temperature of high quality ZnO films has been the argument in favor of PLD technique. For example Narasimhan et al. have demonstrated that strong *c*-axis-oriented ZnO films can be deposited at room temperature [251].

Different kinds of lasers such as Nd:YAG ($\lambda = 1064$ nm) [249], Cu-vapor ($\lambda = 510$ and 578 nm) [250], and excimers (KrF and ArF) [208, 209, 248, 249] have been used to deposit ZnO films. The pulsed excimer lasers appeared to be a popular choice because of their high excitation energy of the ionized or ejected species in the plume. Apart from the conventional PLD, UV-assisted irradiation during deposition has been investigated [207]. Therefore, in this section we intended to provide an overview on the electrical, optical, and structural properties of ZnO films deposited by PLD as drawn from the works of Suzuki et al. [253, 254] and Hayamizu et al. [252]. ZnO is usually doped with impurity atoms in order to enhance the electrical properties. Aluminum is the most used element to doped ZnO to form Al-doped or ZnO:Al. Gallium (Ga) has

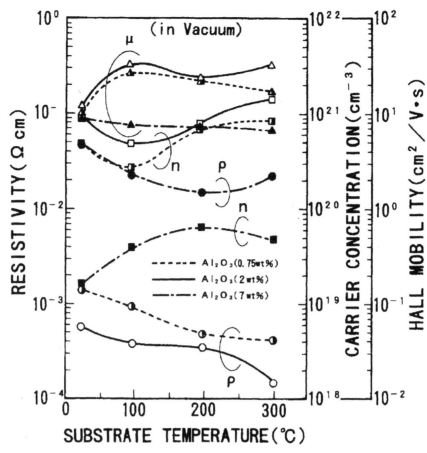

Fig. 60. Dependence of the electrical properties of Al-doped ZnO films deposited in vacuum on glass substrates. (Reprinted with permission from A. Suzuki, T. Matsushita, N. Wada, Y. Sakamoto, and M. Okuda, *Japan. J. Appl. Phys. 2* 35, L56 (1996), © 1996, Japan Society of Applied Physics.)

also been found as an efficient dopant in ZnO films, thus forming Ga-doped ZnO or ZnO:Ga. The effects of the Al and Ga dopant elements in ZnO films will be discussed in detail in subsequent paragraphs. Prior to this, the effect of oxygen pressure on the properties of undoped ZnO films deposited at substrate temperature over the range 25–300°C has been reported by Narasimhan et al. [251]. They observed that optimal oxygen pressure of around 1×10^{-2} Torr yielded the lowest resistivity of $(2–3) \times 10^{-3}$ Ω cm for the ZnO films. This oxygen pressure agreed with that used to produce high quality ITO films as discussed in previous sections of this chapter.

The dependence of the resistivity, carrier concentration, and Hall mobility on the substrate temperature for ZnO films containing different Al doping contents is shown in Figure 60 [254]. These films were deposited on glass substrates using an ArF (193 nm) excimer laser with a fluence of 1 J cm^{-2} and a pulse rate of 10 Hz in a vacuum of 1×10^{-8} Torr. The optimum doping efficiency was achieved at 2 wt%, similar to other deposition techniques like sputtering [254]. The lowest resistivities obtained at room temperature and 300°C were 5.6×10^{-4} and 1.4×10^{-4} Ω cm, respectively. Apparently, the low resistivity resulted from the higher carrier concentration and mobility compared to 0.75 and 7 wt% Al doping content. Figure 61

shows the optical transmittance of ZnO:Al films deposited at 300°C, indicating high transmittance to the visible light [254]. The 2 wt% Al-doped ZnO film displayed strikingly high near infrared reflectance, thus confirming the improvement in the electrical properties compared with the 0.75 and 7 wt% Al doping content. Analyzing the microstructure of the films using XRD, Suzuki et al. noticed that crystalline film could be produced at substrate temperatures over the range 25 to 300°C. Another salient feature was that the surface of the films deposited at 300°C was very flat.

The same group (Suzuki et al.) also reported interesting results on the doping effect of Ga on the properties of ZnO films. The dependence of the resistivity, carrier concentration, and Hall mobility on the substrate temperature for ZnO films containing different Ga doping content is shown in Figure 62 [253]. All the films were deposited on glass substrates. The deposition conditions are similar to those used for growing ZnO:Al films as described above. The best resistivities of 2.1×10^{-4} and 2.9×10^{-4} Ω cm were obtained for the 7 wt% Ga doped ZnO films at 200 and 300°C, respectively. The low resistivity was also due to the high carrier concentration and Hall mobility. Table III shows a compilation of the electrical properties of ZnO films deposited by various techniques at different de-

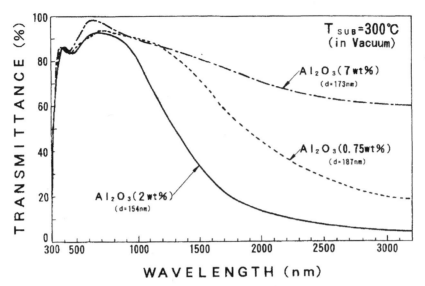

Fig. 61.   Optical transmittance spectra of ZnO films with 0.75, 2, 7 wt% Al$_2$O$_3$ doping content. (Reprinted with permission from A. Suzuki, T. Matsushita, N. Wada, Y. Sakamoto, and M. Okuda, *Japan. J. Appl. Phys. 2* 35, L56 (1996), © 1996, Japan Society of Applied Physics.)

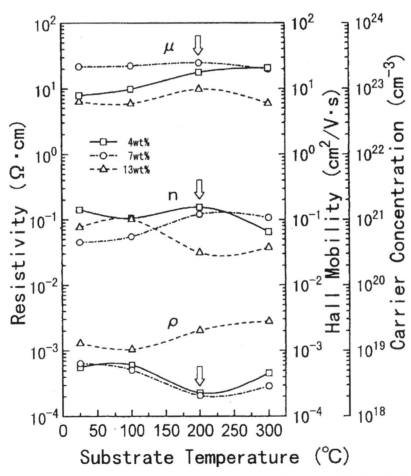

Fig. 62.   Dependence of the substrate temperature on the electrical properties of Ga-doped ZnO films deposited on glass substrates. (Reprinted with permission from A. Suzuki, T. Matsushita, Y. Sakamoto, N. Wada, T. Fukuda, H. Fujiwara, and M. Okuda, *Japan. J. Appl. Phys.* 35, 5457 (1996), © 1996, Japan Society of Applied Physics.)

Table III.   Compilation of the Electrical Properties of ZnO Films Deposited by Various Techniques

Film material	Deposition technique	Deposition temperature (°C)	Resistivity ($\times 10^{-4}$ Ω cm)	Carrier concentration ($\times 10^{20}$ cm^{-3})	Hall mobility (cm^2 V^{-1} s^{-1})	References
ZnO	PLD	RT	20–30	–	–	[251]
ZnO:Al	PLD	300	9	5.8	12	[248]
ZnO:Al	PLD	300	1.4	15	30	[252]
ZnO:Ga	PLD	200	2.1	1	10	[253]
ZnO:Ga	PLD	300	3.6	8.7	18	[268]
ZnO	RFMS	90	5	1	120	[269]
ZnO	RFMS	–	4.6	5	27	[270]
ZnO	DCMS	–	40	–	–	[271]
ZnO:Al	DCMS	RT	7.7	4.2	19.5	[272]
ZnO:Al	DCRMS	300	4.2	2.6	57	[273]
ZnO:Al	ACRMS	300	4	4.9	32	[273]
ZnO	Spray	450	20	–	–	[274]
ZnO:Al	Spray	425	35	0.2	60	[275]
ZnO:In	Spray	375	8	4	16	[276]
ZnO:F	Spray	700	6.8	–	–	[277]
ZnO:Al	DCRMS	250	1.9	11	30	[278]
ZnO:Al	RFMS	RT	6.5	6.5	15	[279]
ZnO:Al	RFMS	90	1.9	15	22	[280]
ZnO:B	RFMS	90	6.4	2.5	39	[281]
ZnO:In	RFMS	90	8.1	4	20	[281]
ZnO:Ga	RFMS	90	5.1	2.5	39	[281]
ZnO:Al	MOCVD	420	3	8.8	23	[282]
ZnO:Al	CVD	350	71	–	–	[283]
ZnO:B	CVD-ALD	150	5.3	–	–	[284]
ZnO:Al	Sol–gel	450	7	6	10	[285]
ZnO:Al	Electroless	225	2.1	1.8	17	[286]

RFMS—rf magnetron sputtering, DCMS—dc magnetron sputtering, DCRMS—dc reactive magnetron sputtering, ACRMS—ac reactive magnetron sputtering, MOCVD—metal organic chemical vapor deposition.

position conditions. Films deposited by PLD produced the best resistivity at low temperatures.

The transmittance (T) and reflectance (R) of the ZnO:Ga films deposited at temperature of 200°C are shown in Figure 63 [253]. The films displayed high transmittance in excess of 90% in the visible and high reflectance in the infrared regions of the spectrum. From the studies on the microstructure of the films using XRD and SEM, Suzuki et al. also observed that crystalline films with very smooth surfaces could be deposited at substrate temperatures over the range 25 to 300°C, especially for the 7 wt% ZnO:Ga films.

Highly c-axis-oriented ZnO films have been observed on glass substrates by PLD even at room temperature [251, 252]. Figure 64a shows a typical XRD spectrum of a ~200-nm-thick ZnO film deposited on the glass substrate at a temperature of 500°C by an ArF (193 nm) excimer laser [252]. Apparently, only two diffraction peaks of (002) and (004) were present, implying a very strong c-axis of ZnO growth on the glass substrate. In Figure 64b is shown the rocking curve of the (002)

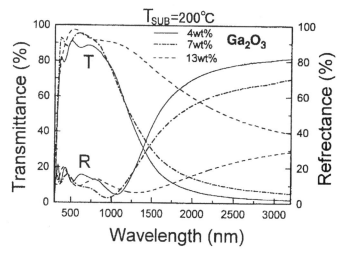

Fig. 63.   Optical transmittance (T) and reflectance (R) of Ga-doped ZnO films deposited on quartz glass substrates. (Reprinted with permission from A. Suzuki, T. Matsushita, Y. Sakamoto, N. Wada, T. Fukuda, H. Fujiwara, and M. Okuda, *Japan. J. Appl. Phys.* 35, 5457 (1996), © 1996, Japan Society of Applied Physics.)

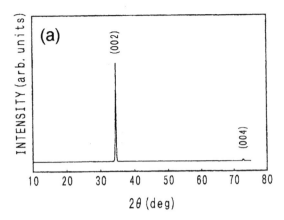

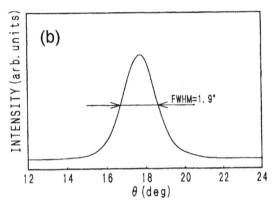

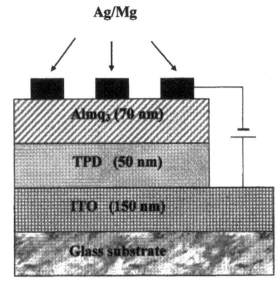

Fig. 65. Schematic structure of an OLED with an ITO anode deposited by PLD. ITO film, deposited at room temperature in oxygen pressure of 1×10^{-2} Torr, is used in the device. (Reprinted with permission from H. Kim, C. M. Gilmore, A. Piqué, J. S. Horwitz, H. Mattoussi, H. Murata, Z. H. Kafafi, and D. B. Chrisey, *J. Appl. Phys.* 86, 6451 (1999), © 1996, American Institute of Physics.)

Fig. 64. (a) XRD spectra of ZnO films grown on glass substrates. (b) X-ray rocking curve of the (002) peak. (Reprinted with permission from S. Hayamizu, H. Tabata, H. Tanaka, and T. Kawai, *J. Appl. Phys.* 80 (1996), © 1996, American Institute of Physics.)

plane. The rocking curve at FWHM was found to be 1.9° and the fluctuation of the c-axis of the ZnO film was very little.

The effects of UV irradiation on the electrical, structural, and optical properties of ZnO films during deposition by sputtering and PLD have been studied [207–209]. It has been found that UV irradiation during sputter deposition of ZnO films in the presence of extra Zn atoms produced some improvements in the optoelectronic properties via improvement in the crystallinity of the films. On the other hand, Craciun et al. [207] have obtained highly textured ZnO films on Si (100) and sapphire substrates by *in situ* UV-assisted PLD. Using this method, they were able to grow highly textured ZnO films on Si substrates at low temperature of 100°C. On the sapphire substrates epitaxial films with $[001]_{ZnO} \| [001]_{sap}$ and $[100]_{ZnO} \| [110]_{sap}$ were obtained at 400°C. The FWHM of the rocking curve of the (002) diffraction peak was found to be 0.168°. These experimental data have shown that PLD is an effective deposition tool for producing high quality ZnO films.

9. APPLICATIONS OF PLD ITO FILMS

As enumerated in Section 1.1, a number of applications require TCO (ITO) film in the form of a transparent conducting electrodes. Some of the these devices, including solar cells, liquid

crystal flat panel displays, and organic light emitting diodes (OLED), are very sensitive to high-temperature (>200°C) treatment. Hence, low-temperature deposition techniques of high quality films are desirable. In solar cells the commonly used technique for depositing ITO films include sputtering and spray pyrolysis. ITO films deposited by PLD have been used to fabricate ITO/InP solar cells by Jia et al. [85]. The ITO films were deposited at room temperature and 310°C. The results of their studies have shown that ITO/InP cells fabricated from ITO films deposited at room temperature and 310°C yielded open circuit voltages of 660 and 610 mV and short circuit currents of ∼23.5 and ∼23 mA, respectively. The improvement in the device affected mainly the open circuit voltage, implying a preservation of the junction properties via low-temperature deposition. Higher spectra response and acceptable dark current characteristics were also confirmed from the device fabricated from ITO deposited at room temperature [85].

ITO films deposited by PLD have also been used to fabricate OLED devices [88, 89]. ITO film is usually chosen as an anode to these devices, because of the high efficiency for hole injection into the organic materials. An illustration of an OLED structure is shown in Figure 65 [88, 89]. The device consisted of a hole transport layer (∼50 nm thick) of N,N'-diphenyl-N,N-bis(3-methyl)-1,1'-diphenyl-4,4'-diamine and an electron transport layer (ETL/EML, 70 nm thick) of tris(4-methyl-8-hydroxyquinolinolato)aliminum (III) (Almq3). An alloy of Mg:Ag (ratio = 12 : 1 and thickness of 100 nm) was used as the cathode contact on the ETL layer. The device had an active area of ∼4 mm². The current–voltage–luminance characteristic of the device measured in nitrogen atmosphere is shown in Figure 66 [88, 89]. The current–voltage and luminance–

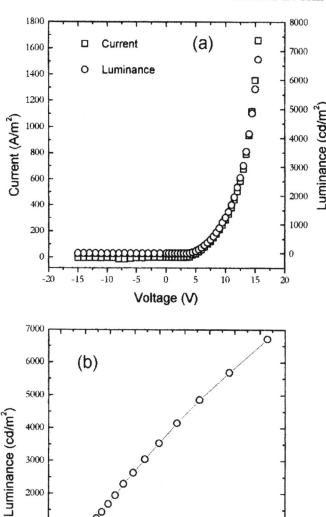

Fig. 66. (a) Current–voltage–luminance (*I–V–L*) and (b) luminance–current (*L–I*) characteristics of a heterostructure device with PLD ITO film as the hole injecting layer to TPD layer. ITO film, grown at room temperature in oxygen pressure of 1×10^{-2} Torr, was used in the device. (Reprinted with permission from H. Kim, C. M. Gilmore, A. Piqué, J. S. Horwitz, H. Mattoussi, H. Murata, Z. H. Kafafi, and D. B. Chrisey, *J. Appl. Phys.* 86, 6451 (1999), © 1999, American Institute of Physics.)

10. CONCLUSION

In conclusion, experimental investigations of PLD for TCO films depositions have shown that high quality films can be produced even at room temperature. A direct relation between the quality of the In_2O_3, ITO, and ZnO films and experimental conditions (laser energy density, substrate distance, background oxygen pressure, substrate temperature, and growth rates) was established. The experimental studies also showed a good qualitative agreement with the theoretical evaluations. Compared to other deposition techniques it is observed that low resistivity and highly transparent ITO and ZnO films could be prepared at low substrate temperature by PLD. The qualities of the films were highly susceptible to the changes in the oxygen pressure and the substrate temperature. In particular, oxygen pressure is found to exert significant influence on the overall properties of the films. ITO films deposited at substrate temperatures below 100°C even though they were amorphous indicated low resistivity. Crystalline films were only produced at substrate temperatures above 100°C and the films exhibited higher conductivity.

Furthermore, irradiating the surfaces of the growing ITO films by laser energy pulses during the PLD process the microstructural and electro-optical properties of the films were improved. Electrical resistivity below 10^{-4} Ω cm was attained at temperatures as low as 200°C. The enhancement in the electrical properties was primarily due to photochemical and photothermal induced crystallization during the laser irradiation of the growing films. Applications of the laser deposited ITO films at room temperature to solar cells and OLED device fabrications have shown some encouraging performances.

Acknowledgments

The author acknowledges the immense contributions of Drs. M. Motoyama, T. Ishihara, H. Izumi, H. Yoshioka, and K. Murai to this work. The results of some of the works reported in this chapter were partly supported by the New Energy and Industrial Technology Development Organization (NEDO), Japan.

REFERENCES

1. K. Bädeker, *Ann. Phys. Leizpzig* 22, 749 (1907).
2. C. C. Wu, C. I. Wu, J. C. Sturm, and A. Kahn, *Appl. Phys. Lett.* 70, 1348 (1997).
3. F. Li, H. Tang, and J. Shinar, *Appl. Phys. Lett.* 70, 2741 (1997).
4. S. A. Van Slyke, C. H. Chen, and C. W Tang, *Appl. Phys. Lett.* 69, 2160 (1996).
5. K. L. Chopra, S. Major, and D. K. Pandya, *Thin Solid Films* 102, 1 (1983).
6. I. Hamberg and C. G. Granqvist, *J. Appl. Phys.* 60, R123 (1986).
7. E. Ritter, "Progress in Electro-Optics" (E. Camatini, Ed.). Plenum, New York, 1975.
8. J. L. Vossen, *Phys. Thin Films* 9, 1 (1977).
9. J. C. Manifacier, *Thin Solid Films* 90, 297 (1982).
10. Z. M. Jarzebski, *Phys. Status Solidi A* 71, 13 (1982).
11. G. Haacke, *Annu. Rev. Mater. Sci.* 7, 73 (1977).
12. G. Rupprecht, *Z. Phys.* 139, 504 (1954).
13. N. Rucher and K. J. Becker, *Thin Solid Films* 53, 163 (1978).
14. P. Nath and R. F. Bunshah, *Thin Solid Films* 69, 63 (1980).

voltage curves showed a typical diode characteristic with the power output observed only in the forward bias. In addition, the current and luminance data superimpose reasonably well, in agreement with what has been achieved with commercial ITO films. The external quantum efficiency for such a heterostructure device was reported as $\eta_{ext} = 1.5\%$ at 100 A m^{-2} compared to 1.5–2.5% for commercial ITO films [287–289]. The results indicated that ITO films deposited at room temperature by PLD are of good quality and also present the potential for improvement of the performances of these devices.

15. P. Nath, R. F. Bunshah, B. M. Basol, and O. M. Staffsud, *Thin Solid Films* 72, 463 (1980).

16. Z. Ovadyahu, B. Ovryn, and H. W. Kraner, *J. Electrochem. Soc.* 130, 917 (1983).

17. P. Manivannan and A Subrahmanyam, *J. Phys. D* 26, 1510 (1993).

18. J. L. Yao, S. Hao, and J. Wilkinson, *Thin Solid Films* 189, 227 (1990).

19. K. Korobov, M. Leibovitch, and Y. Shapira, *Appl. Phys. Lett.* 65, 2290 (1994).

20. M. Abbas, I. A. Qazi, A. Samina, M. Masaood, A. Majeed, and A. ul Haq, *J. Vac. Sci. Technol. A* 13, 152 (1995).

21. P. Thilakan and J. Kumar, *Thin Solid Films* 292, 50 (1997).

22. S.-W. Jan and S.-C. Lee, *J. Electrochem. Soc.* 134, 2061 (1987).

23. C. A. Pan and T. P. Ma, *Appl. Phys. Lett.* 37, 163 (1980).

24. Z. Ovadyahu and H. Wiesmann, *J. Appl. Phys.* 52 (1981).

25. I. Hamberg, A. Hjortsberg, and C. G. Granvist, *Appl. Phys. Lett.* 40, 362 (1982).

26. M. Penza, S. Cozzi, M. A. Tagliente, L. Mirenghi, C. Matucci, and A. Quirini, *Thin Solid Films* 349, 71 (1999).

27. I. A. Rauf, *J. Mater. Sci. Lett.* 12, 1902 (1993).

28. I. A. Rauf, *J. Appl. Phys.* 79, 4057 (1995).

29. I. A. Rauf, *Mater. Lett.* 23, 73 (1995).

30. H. Haitjema and J. J. Ph. Elich, *Thin Solid Films* 205, 93 (1991).

31. C. Goebbert, R. Nonniger, M. A. Aegerter, and H. Schmidt, *Thin Solid Films* 351, 79 (1999).

32. H. Bisht, H.-T. Eun, A. Mehrtens, and M. A. Aegerter, *Thin Solid Films* 351, 109 (1999).

33. Ph. Parent, H. Dexpert, G. Tourillon, and J.-M. Grimal, *J. Electrochem. Soc.* 139, 282 (1992).

34. I. Baía, M. Quintela, L. Mendes, P. Nunes, and R. Martins, *Thin Solid Films* 337, 171 (1999).

35. L. J. Meng, A. Maçrarico, and R. Martins, *Vacuum* 46, 673 (1995).

36. T. Karasawa and Y. Miyata, *Thin Solid Films* 223, 135 (1993).

37. L. Davis, *Thin Solid Films* 236, 1 (1993).

38. S. Ishibashi, Y. Higuchi, Y. Ota, and K. Nakamura, *J. Vac. Sci. Technol. A* 8, 1399 (1990).

39. M. Higuchi, S. Uekusa, R. Nakano, and K. Yokogawa, *J. Appl. Phys.* 74, 6710 (1993).

40. B. S. Chiou and S. T. Hsieh, *Thin Solid Films* 229, 146 (1993).

41. S. Ray, R. Benerjee, N. Basu, A. K. Batabyal, and A. K. Barua, *J. Appl. Phys.* 54, 3497 (1983).

42. S. B. Lee, J. C. Pincenti, A. Cocco, and D. L. Naylor, *J. Vac. Sci. Technol. A* 11, 2742 (1993).

43. S. Kulkarni and M. Bayard, *J. Vac. Sci. Technol. A* 9, 1193 (1991).

44. P. K. Song, Y. Shigesato, I. Yasui, C. W. Ow-Yang, and D. C. Paine, *Japan. J. Appl. Phys.* 37, 1870 (1998).

45. Y. Shigesato and D. C. Paine, *Thin Solid Films* 238, 44 (1994).

46. P. K. Song, H. Akao, M. Kamei, Y. Shigesato, and I. Yasui, *Japan. J. Appl. Phys.* 38, 5224 (1999).

47. K. Carl, H. Schmitt, and I. Friedrich, *Thin Solid Films* 295, 151 (1997).

48. J. C. C. Fan, F. J. Bachner, and G. H. Foley, *Appl. Phys. Lett.* 31, 773 (1977).

49. M. Buchanan, J. B. Webb, and D. F. Williams, *Thin Solid Films* 80, 373 (1981).

50. K. Ishibashi, K. Watanabe, T. Sakarai, O. Okada, and N. Hosokawa, *J. Non-Cryst. Solids* 218, 354 (1997).

51. R. W. Moss, D. H. Lee, K. D. Voung, R. A. Condrate, Sr., X. W. Wang, M. DeMarco, and J. Stuckey, *J. Non-Cryst. Solids* 218, 105 (1997).

52. J. C. C. Fan and F. J. Bachner, *J. Electrochem. Soc.* 122, 1719 (1975).

53. L.-J. Meng and M. P. dos Santos, *Thin Solid Films* 289, 65 (1996).

54. H. M. Zahirul Alam, P. K. Saha, T. Hata, and K. Sasaki, *Thin Solid Films* 352, 133 (1999).

55. H. Nanto, T. Minami, S. Orito, and S. Takata, *J. Appl. Phys.* 63, 2711 (1988).

56. R. Aitchison, *Austral. J. Appl. Sci.* 5, 10 (1954).

57. J. A. Thornton and V. L. Hedgcoth, *J. Vac. Sci. Technol.* 13, 117 (1976).

58. S. Yoshida, *Appl. Opt.* 17, 145 (1978).

59. R. R. Mehta and S. F. Vogel, *J. Electrochem. Soc.* 119, 752 (1972).

60. Y. Suzuki, F. Niino, and K. Katoh, *J. Non-Cryst. Solids* 218, 30 (1997).

61. Y. Murayama, *J. Vac. Sci. Technol.* 12, 818 (1975).

62. R. P. Howson, J. N. Avaritsiotis, M. I. Ridge, and C. A. Bishop, *Thin Solid Films* 58, 379 (1978).

63. T. Ishida, H. Kobayashi, and Y. Nakato, *J. Appl. Phys.* 73, 4344 (1993).

64. T. Maruyama and Kitamura, *Japan. J. Appl. Phys. 2* 28, L1096 (1989).

65. M. Privman, J. Berger, and D. S. Tannhauser, *Thin Solid Films* 102, 117 (1983).

66. J.-W. Bae, S.-W. Lee, K.-H. Song, J.-I. Park, K.-J. Park, Y.-W. Ko, and G.-Y. Yeom, *Japan. J. Appl. Phys. 1* 38, 2917 (1999).

67. T. Maruyama and K. Fukui, *J. Appl. Phys.* 70, 3848 (1991).

68. T. Maruyama and K. Fukui, *Thin Solid Films* 203, 297 (1991).

69. R. G. Gordon, *Mater. Res. Symp. Proc.* 426, 419 (1996).

70. Y. Djaoued, V. H. Phong, S. Badilescu, P. V. Ashrit, F. E. Girouard, and V.-V. Truong, *Thin Solid Films* 293, 108 (1997).

71. A. Gurlo, M. Ivanovskaya, A. Pfau, U. Weimar, and W. Göpel, *Thin Solid Films* 307, 288 (1997).

72. Y. Takahashi, S. Okada, R. B. H. Tahar, K. Nakano, T. Ban, and Y. Ohya, *J. Non-Cryst. Solids* 218, 129 (1997).

73. R. B. H. Tahar, T. Ban, Y. Ohya, and Y. Takahashi, *J. Appl. Phys.* 83, 2631 (1998).

74. G. Yi and M. Sayer, *Ceram. Bull.* 70, 1173 (1991).

75. H. Dislich, *J. Non-Cryst. Solids* 57, 371 (1988).

76. D. M. Mattox, *Thin Solid Films* 204, 25 (1991).

77. W. Quin, R. P. Howson, M. A. Akizuki, J. Matsuo, G. H. Takaoka, and I. Yamada, "Extended Abstracts of the 44th Spring Meeting of the Japan Society of Appl. Phys., Vol. 2," 1997, p. 466.

78. R. P. Howson, J. Matsuo, and I. Yamada, "Extended Abstracts of the 44th Spring Meeting of the Japan Society of Appl. Phys.," 1997, p. 463.

79. F. O. Adurodija, H. Izumi, T. Ishihara, H. Yoshioka, and M. Motoyama, *Japan. J. Appl. Phys. Lett. 2* 39, L377 (2000).

80. F. O. Adurodija, H. Izumi, T. Ishihara, H. Yoshioka, M. Motoyama, and K. Murai, *J. Vac. Sci. Technol. A* 18, 814 (2000).

81. F. O. Adurodija, H. Izumi, T. Ishihara, H. Yoshioka, K. Yamada, H. Matsui, and M. Motoyama, *Thin Solid Films* 350, 79 (1999).

82. F. O. Adurodija, H. Izumi, T. Ishihara, H. Yoshioka, H. Matsui, and M. Motoyama, *Japan. J. Appl. Phys. 1* 38, 2710 (1999).

83. F. O. Adurodija, H. Izumi, T. Ishihara, H. Yoshioka, H. Matsui, and M. Motoyama, *Appl. Phys. Lett.* 74, 3059 (1999).

84. E. J. Tarsa, J. H. English, and J. S. Speck, *Appl. Phys. Lett.* 62, 2332 (1993).

85. Q. X. Jia, J. P. Zheng, H. S. Kwok, and W. A. Anderson, *Thin Solid Films* 258, 260 (1995).

86. X. W. Sun, H. C. Huang, and H. S. Kwok, *Appl. Phys. Lett.* 68, 2663 (1996).

87. H. S. Kwok, X. W. Sun, and D. H. Kim, *Thin Sold Films* 335, 229 (1998).

88. H. Kim, C. M. Gilmore, A. Piqué, J. S. Horwitz, H. Mattoussi, H. Murata, Z. H. Kafafi, and D. B. Chrisey, *J. Appl. Phys.* 86, 6451 (1999).

89. H. Kim, A. Piqué, J. S. Horwitz, H. Mattoussi, H. Murata, Z. H. Kafafi, and D. B. Chrisey, *Appl. Phys. Lett.* 74, 3444 (1999).

90. F. Hanus, A. Jadin, and L. D. Laude, *Appl. Surf. Sci.* 96–98, 807 (1996).

91. Y. Wu, C. H. M. Maree, R. F. Haglund, Jr., J. D. Hamilton, M. A. Morales Paliza, M. B. Huang, L. C. Feldman, and R. A. Weller, *J. Appl. Phys.* 86, 991 (1999).

92. J. P. Zheng and H. S. Kwok, *Thin Solid Films* 232, 99 (1993).

93. A. L. Dawar and J. C. Joshi, *J. Mater. Sci.* 19, 1 (1984).

94. H. Kawazoe and K. Ueda, *J. Amer. Ceram. Soc.* 82, 3330 (1999).

95. J. C. C. Fan and J. B. Goodenough, *J. Appl. Phys.* 48, 3524 (1977).

96. V. M. Vainshtein and V. I. Fistul, *Sov. Phys. Semicond.* 1, 104 (1967).

97. H. K. Müller, *Phys. Status Solidi* 27, 723 (1968).

98. S. J. Wen, G. Campet, J. Portier, G. Couturier, and B. Goodenough, *Mater. Sci. Eng. B* 14, 115 (1992).

99. T. Maruyama and T. Tago, *Appl. Phys. Lett.* 64, 1395 (1994).

100. S. J. Wen, G. Couturier, G. Campet, J. Portier, and J. Claverie, *Phys. Status Solidi A* 130, 407 (1992).

101. S. Takaki, K. Matsumoto, and K. Suzuki, *Appl. Surf. Sci.* 33/34, 919 (1988).

102. J. C. Manifacier, L. Szepessy, J. F. Bresse, M. Perotin, and R. Stuck, *Mater. Res. Bull.* 14, 163 (1979).

103. V. A. Johnson and K. Lark-Horovitz, *Phys. Rev.* 71, 374 (1947).

104. H. Köstlin, R. Jost, and W. Rems, *Phys. Status Solidi A* 29, 87 (1975).

105. G. Frank and H. Köstlin, *Appl. Phys. A* 27, 197 (1982).

106. F. T. J. Smith and S. L. Lyu, *J. Electrochem. Soc.* 128, 2388 (1981).

107. J. H. de Wit, *J. Solid State Chem.* 20, 143 (1977).

108. E. C. Subbarao, P. H. Sutter, and J. Hirizo, *J. Amer. Ceram. Soc.* 48, 443 (1965).

109. N. Nadaud, N. Lequeux, M. Nanot, J. Jove, and T. Roisnel, *J. Solid State Chem.* 135, 140 (1998).

110. N. Yamada, I. Yasui, Y. Shigesato, H. Li, Y. Ujihira, and K. Nomura, *Japan. J. Appl. Phys. 1* 38, 2856 (1999).

111. D. B. Chrisey and G. K. Hubler, Eds., "Pulsed Laser Deposition of Thin Films." Wiley, New York, 1994.

112. S. M. Green, A. Piqué, K. S. Harshavardhan, and J. S. Bernstein, in "Pulsed Laser Deposition of Thin Films" (D. B. Chrisey and G. K. Hubler, Eds.), p. 24. Wiley, New York, 1994.

113. A. Yariv, "Introduction to Optical Electronics," 2nd ed. Holt, Riehart & Winston, New York, 1976.

114. M. Stehle, *J. Non-Cryst. Solids* 218, 218 (1997).

115. H. Hosono, M. Kurita, and H. Kawazoe, *Thin Solid Films* 351, 137 (1999).

116. J. T. Cheung, in "Pulsed Laser Deposition of Thin Films" (D. B. Chrisey and G. K. Hubler, Eds.), p. 4. Wiley, New York, 1994.

117. R. P. van Ingen, R. H. J. Fastenau, and E. J. Mittemeijer, *J. Appl. Phys.* 76, 1871 (1994).

118. H. U. Kreds, S. Fahler, and O. Bremert, *Appl. Surf. Sci.* 86, 86 (1995).

119. F. Antoni, E. Fogarassy, C. Fuchs, J. J. Grob, B. Prevot, and J. P. Stoquert, *Appl. Phys. Lett.* 67, 2072 (1995).

120. E. Sobol, "Phase Transformations and Ablation in Laser Treated Solids." Wiley, New York, 1995.

121. E. van der Riet, J. C. S. Kools, and J. Dieleman, *J. Appl. Phys.* 73, 8290 (1993).

122. D. B. Chrisey, J. S. Horowitz, and K. S. Grabowski, *Mater. Res. Soc. Symp. Proc.* 191, 25 (1990).

123. T. Yano, T. Oosie, M. Yoneda, and M. Katsumura, *J. Mater. Sci. Lett.* 15, 1994 (1993).

124. D. C. Paine, T. Whitson, D. Janiac, R. Beresford, C. Ow-Yang, and B. Lewis, *J. Appl. Phys.* 85, 8445 (1999).

125. T. J. Vink, W. Walrave, J. L. C. Daams, P. C. Baarslag, and J. E. A. van der Meerakker, *Thin Solid Films* 266, 145 (1995).

126. C. H. Yi, Y. Shigesato, I. Yasui, and S Takaki, *Japan. J. Appl. Phys. 2* 34, L244 (1995).

127. H. P. Klug and L. E. Alexander, "X-Ray Diffraction Procedures for Polycrystalline and Amorphous Materials," 2nd ed. Wiley, New York, 1974.

128. S. Muranaka, *Japan. J. Appl. Phys. 2* 30, L2062 (1991).

129. S. Muranaka, Y. Bando, and T. Takada, *Thin Solid Films* 25, 355 (1987).

130. S. R. Foltyn, in "Pulsed Laser Deposition of Thin Films" (D. B. Chrisey and G. K. Hubler, Eds.), p. 89. Wiley, New York, 1994.

131. L.-C. Chen, in "Pulsed Laser Deposition of Thin Films" (D. B. Chrisey and G. K. Hubler, Eds.), p. 167. Wiley, New York, 1994.

132. D. B. Geohegan, *Mater. Res. Soc. Symp. Proc.* 201, 557 (1991).

133. A. Gupta, in "Pulsed Laser Deposition of Thin Films," p. 265. Wiley, New York, 1994.

134. H. S. Kim and H. S. Kwok, *Appl. Phys. Lett.* 61, 2234 (1992).

135. R. F. Wood, K. R. Chen, J. N. Leboeuf, A. A. Puretsky, and D. B. Geohegan, *Phys. Rev. Lett.* 79, 1571 (1997).

136. J. A. Venebles, G. D. T. Spiller, and M. Hanbucken, *Rep. Prog. Phys.* 47, 399 (1984).

137. B. Lewis and J. C. Anderson, "Nucleation and Growth of Thin Films." Academic Press, New York, 1978.

138. L. Mao, R. E. Benoit, and J. Proscia, *Mater. Res. Soc. Symp. Proc.* 317, 191 (1994).

139. Q. Y. Ying, H. S. Kim, D. T. Shaw, and H. S. Kwok, *Paal. Phys. Lett.* 55, 1041 (1989).

140. H. S. Kwok and Q. Y. Ying, *Physica C* 177, 122 (1991).

141. K. Saxena, S. P. Singh, R. Thangaraj, and O. P. Agnihotri, *Thin Solid Films* 117, 95 (1984).

142. H. Hoffman, J. Pickl, M. Schmidt, and D. Krause, *Appl. Phys.* 16, 239 (1978).

143. K. Ellmer, K. Diesner, R. Wendt, and S. Fiechter, in "Polycrystalline Semiconductors IV—Physics, Chemistry and Technology" (S. Pissini, H. P. Strunk, and J. H. Werner, Eds.), p. 541. Trans Tech, Zug, Switzerland.

144. Y. Shigesato and D. C. Paine, *Appl. Phys. Lett.* 62, 1268 (1993).

145. E. Burstein, *Phys. Rev.* 93, 632 (1952).

146. J. A. Thornton and W. D. Hoffman, *Thin Solid Films* 171 (1989).

147. T. Oyama. N. Hashimoto, J. Shimizu, Y. Akao, H. Kojima, K. Aikawa, and K. Suzuki, *J. Vac. Sci. Technol. A* 10, 1682 (1992).

148. M. Kamei, Y. Shigesato, I. Yasui, N. Taga, and S. Takaki, *J. Non-Cryst. Solids* 218, 267 (1997).

149. H. Izumi, K. Ohata, T. Sawada, T. Morishita, and S. Tanaka, *Japan. J. Appl. Phys.* 30, 1956 (1991).

150. E. Terzini, G. Nobile, S. Loreti, C. Minarini, T. Ploichetti, and P. Thilakan, *Japan. J. Appl. Phys.* 38, 3448 (1999).

151. P. K. Song, Y. Shigesato, M. Kamei, and I. Yasui, *Japan. J. Appl. Phys.* 38, 2921 (1999).

152. R. W. G. Wyckoff, in "Crystal Structures," 2nd ed., Vol. 2, p. 4. Krieger, Malabar, FL, 1986.

153. JCPDS Card No. 06-0416.

154. W.-F. Wu, B.-S. Chiou, and S.-T. Hsieh, *Semicond. Sci. Technol.* 9, 1242 (1994).

155. W.-F. Wu and B.-S. Chiou, *Thin Solid Films* 293, 244 (1997).

156. J. A. Thornton, in "Annual Review of Materials Science" (R. A. Huggins, R. H. Bube, and R. W. Roberts, Eds.), p. 239. Annual Rev. Inc., Palo Alto, CA, 1977.

157. J. Michler, M. Mermoux, Y. von Kaenel, A. Haouni, G. Lucazeau, and E. Blank, *Thin Solid Films* 357, 189 (1999).

158. T. C. Huang, *Adv. X-ray Anal.* 33, 91 (1990).

159. V. Craciun, I. W. Boyd, P. Andreazza, and C. Boulmer-Leborgne, *J. Appl. Phys.* 83, 1770 (1997).

160. E. Hasegawa, A. Ishtani, K. Akimoto, M. Tsukiji, and N. Ohta, *J. Electrochem. Soc.* 142, 273 (1995).

161. R. Feidenhans'l, *Surf. Sci. Rep.* 10, 105 (1989).

162. F. O. Adurodija, H. Izumi, T. Ishihara, H. Yoshioka, and M. Motoyama, *J. Mater. Sci.: Mater. Electronics* 12, 57 (2001).

163. C. V. R. Vasant Kumar and A. Mansingh, *J. Appl. Phys.* 65, 1270 (1989).

164. G. Yi, Z. Wu, and M. Sayer, *J. Appl. Phys.* 64, 2717 (1989).

165. K. Sreenivas, M. Sayer, T. Laursen, J. L. Whitton, R. Pascual, D. J. Johnson, D. T. Amm, G. I. Sproule, D. F. Mitchell, M. J. Graham, S. C. Gujrathi, and K. Oxorn, *Mater. Res. Symp. Proc.* 200, 255 (1990).

166. R. Ramesh, A. Inam, W. K. Chan, F. Tillerot, B. Wilkens, C. C. Chang, T. Sands, J. M. Tarascon, and V. G. Keramidas, *Appl. Phys. Lett.* 59, 3542 (1991).

167. Y. A. Boikov, S. K. Esayan, Z. G. Ivanov, G. Brorsson, T. Cleaeson, J. Lee, and S. Safari, *Appl. Phys. Lett.* 61, 528 (1992).

168. M. Kamei, T. Yagami, S. Takaki, and Y. Shigesato, *Appl. Phys. Lett.* 64, 2712 (1994).

169. M. Kamei, Y. Shigesato, S. Takaki, Y. Hayashi, M. Sasaki, and T. E. Haynes, *Appl. Phys. Lett.* 65, 546 (1994).

170. J. R. Bellingham, W. A. Phillips, and C. J. Adkins, *J. Phys. Condens. Matter* 2, 6207 (1990).

171. J. P. Zheng and H. S. Kwok, *Appl. Phys. Lett.* 63, 1 (1993).

172. R. Latz, K. Michael, and M. Scherer, *Japan. J. Appl. Phys.* 30, L149 (1991).

173. K. L Narasimhan, S. P Pai, V. R. Palkar, and R. Pinto, *Thin Solid Films* 295, 104 (1997).

174. S. Noguchi and H. Sakata, *J. Phys.* 13, 1129 (1980).

175. H. Daves, *Proc. IEE* 101, 209 (1954).

176. S. A. Agnihotry, *J. Phys. D* 18, 2087 (1985).

177. R. A. Smith, "Wave Mechanics of Crystalline Solids." Chapman and Hall, London, 1961.

178. T. S. Moss, *Proc. Phys. Soc. London Sect. B* 67, 775 (1954).

179. C. H. L. Weijtens, *J. Electrochem. Soc.* 138, 3432 (1991).

180. M. D. Stoev, J. Touskova, and J. Tousek, *Thin Solid Films* 299, 67 (1997).

181. T. L. Barr and Y. L. Liu, *J. Phys. Chem. Solids* 50, 657 (1989).

182. Y. Shigesato, S. Takaki, and T. Haranoh, *J. Appl. Phys.* 71, 1 (1992).

183. H. Yoshioka, F. O. Adurodija, H. Izumi, T. Ishihara, and M. Motoyama, *Electrochem. Solid State Lett.* 3, 540 (2000).

184. C. D. Wagner, W. M. Riggs, L. E. Davis, J. F. Moulder, and G. E. Mullenberg, "Handbook of X-ray Photoelectron Spectroscopy." Perkin-Elmer, Eden Prairie, MN, 1979.

185. S. Doniach and M. Sunjic, *J. Phys. C* 3, 285 (1970).

186. I. Yamada, *Mater. Sci. Forum* 264, 239 (1997).

187. G. Riesse and S. Weissmentel, *Thin Solid Films* 355–356, 105 (1999).

188. A. Luches, in "Physical Processes in Laser–Materials Interactions" (M. Bertolotti, Ed.), p. 443. Plenum, New York, 1981.

189. K. Kanaya and S. Okayama, *J. Phys. D* 5, 43 (1972).

190. L. Katz and A. S. Penfold, *Rev. Mod. Phys.* 24, 28 (1952).

191. P. G. Merli, *Optik* 56, 205 (1980).

192. R. N. Brickell, N. C. Giles, and J. F. Schetzina, *Appl. Phys. Lett.* 49, 1095 (1986).

193. R. L. Abber, in "Handbook of Thin-Film Deposition Processes and Technique" (K. K. Schuegraf, Ed.), p. 270. Noyes, Park Ridge, NJ, 1988.

194. S. A. Hussein, N. H. Karam, S. M. Bedair, A. A. Fahmy, and N. A. El-Masry, *Mater. Res. Soc. Symp. Proc.* 129, 165 (1988).

195. S. Otsubo, T. Minamikawa, Y. Yonezawa, T. Maeda, A. Morimoto, and T. Shimizu, *Japan. J. Appl. Phys.* 28, 2211 (1989).

196. T. Minamikawa, Y. Yonezawa, S. Otsubo, Y. Yonewawa, A. Morimoto, and T. Shimizu, in "Advances in Superconductivity II" (T. Ishiguro and K. Kajimura, Eds.), p. 857. Springer-Verlag, Tokyo, 1990.

197. M. Kanai, K. Horiuchi, T. Kawai, and S. Kawai, *Appl. Phys. Lett.* 57, 2716 (1990).

198. R. C. Estler, N. S. Nogar, R. E. Muenchausen, R. C. Dye, C. Flamme, J. A. Martin, A. Garcia, and S. Fotlyn, *Mater. Lett.* 9, 342 (1990).

199. A. Morimoto, S. Otsubo, T. Shimizu, T. Minamikawa, and Y. Yonezawa, *Mater. Res. Soc. Symp. Proc.* 275, 371 (1990).

200. H. Tabata, T. Kawai, and S. Kawai, *Appl. Phys. Lett.* 58, 1443 (1991).

201. H. Tabata, T. Kawai, S. Kawai, O. Murata, J. Fujioka, and S. Minakata, *Appl. Phys. Lett.* 59, 2354 (1991).

202. H. Tabata, O. Murata, T. Kawai, S. Kawai, and M. Okuyama, *Japan. J. Appl. Phys.* 31, 2968 (1992).

203. A. Ito, A. Machida, and M. Obara, *Japan. J. Appl. Phys. 2* 36, L805 (1997).

204. P. E. Dyer, A. Issa, P. H. Key, and P. Monk, *Supercond. Sci. Technol.* 3, 472 (1990).

205. K. H. Wu, C. L. Lee, J. Y. Juang, T. M. Uen, and Y. S. Gou, *Appl. Phys. Lett.* 58, 1089 (1991).

206. T. Szörényi, Z. Kántor, and L. D. Laude, *Appl. Surf. Sci.* 86, 219 (1995).

207. V. Craciun, R. K. Singh, J. Perriere, J. Spear, and D. Craciun, *J. Electrochem. Soc.* 147, 1077 (2000).

208. V. Craciun, J. Elders, J. G. E. Gardeniers, and I. W. Boyd, *Appl. Phys. Lett.* 65, 2963 (1994).

209. V. Craciun, J. Elders, J. G. E. Gardeniers, J. Geretovsky, and I. W. Boyd, *Thin Solid Films* 259 (1995).

210. H. Hosono, M. Kurita, and H. Kawazoe, *Japan. J. Appl. Phys. 2* 37, L1119 (1998).

211. F. O. Adurodija, H. Izumi, T. Ishihara, H. Yoshioka, and M. Motoyama, *Mater. Sci. Lett.* 19, 1719 (2000).

212. F. O. Adurodija, H. Izumi, T. Ishihara, H. Yoshioka, and M. Motoyama, *J. Appl. Phys.* 88, 4175 (2000).

213. H. Tabata, O. Murata, T. Kawai, S. Kawai, and M. Okuyama, *Japan. J. Appl. Phys.* 31, L2968 (1998).

214. E. Gerlach and Rautenberg, *Phys. Status Solidi B* 86, 479 (1978).

215. R. B. H. Tahar, T. Ban, Y. Ohya, and Y. Takahashi, *J. Appl. Phys.* 82, 865 (1997).

216. C. Kittel, "Introduction to Solid State Physics," 5th ed., Chap. 6. Maruzen, Tokyo, 1985.

217. C. Erginsoy, *Phys. Rev.* 79, 1013 (1950).

218. E. Conwell and V. F. Weiskopf, *Phys. Rev.* 77, 388 (1950).

219. R. B. Dingle, *Philos. Mag.* 46, 831 (1955).

220. R. Clanget, *Appl. Phys.* 2, 247 (1973).

221. Fujisawa, T. Nishino, and Y. Hamakawa, *Japan. J. Appl. Phys.* 27, 552 (1988).

222. M. Mizuhashi, *Thin Solid Films* 70, 91 (1980).

223. F. O. Adurodija, H. Izumi, T. Ishihara, H. Voshioka, M. Motoyama, and K. Murai, *Appl. Surf. Sci.* 177, 114 (2001).

224. K. Utsumi, O. Matsunaga, and T. Takahata, *Thin Solid Films* 334, 30 (1998).

225. X. W. Sun, L. D. Wang, and H. S. Kwok, *Thin Solid Films* 360, 75 (2000).

226. K. Aikawa, Y. Asai, and T. Hosaka, *Proc. ISSP* 195–198 (1993).

227. Y. Shigesato, S. Takai, and T. Haranoh, *J. Appl. Phys.* 71, 3356 (1992).

228. E. Kawamata and K. Ohshima, *Japan. J. Appl. Phys.* 18, 205 (1979).

229. G. Blandenet, M. Court, and Y. Lagarde, *Thin Solid Films* 77, 81 (1982).

230. L. A. Ryabova, V. S. Salun, and I. A. Serbinor, *Thin Solid Films* 92, 327 (1982).

231. T. Maruyama and A. Kojima, *Japan. J. Appl. Phys. 2* 27, L1829 (1993).

232. S. V. Prasad, S. D. Walck, and J. S. Zabinski, *Thin Solid Films* 360, 107 (2000).

233. J. Exarhos and S. K. Sharma, *Thin Solid Films* 270, 27 (1995).

234. I. Petrov, V. Orlonov, and A. Misiuk, *Thin Solid Films* 120, 55 (1984).

235. L. I. Maissel and R. Glang, "Handbook of Thin Film Technology," p. 124. McGraw–Hill, New York, 1970.

236. D. L. Raimondi and E. Kay, *J. Vac. Sci. Technol.* 7, 96 (1970).

237. F. S. Hickernell, *J. Appl. Phys.* 44, 1061 (1973).

238. K. Machida, M. Shibutani, Y. Maruyama, and T. R. Matsumoto, *IECE* 62, 358 (1979).

239. T. Shiosaki, S. Ohnishi, and A. Kawabata, *J. Appl. Phys.* 50, 3113 (1979).

240. J. S. Zabinski, J. Corneille, S. V. Prasad, N. T. McDevitt, and J. B. Bultman, *J. Mater. Sci.* 32, 5313 (1997).

241. S. V. Prasad and J. S. Zabinski, *Wear* 203–204, 498 (1997).

242. H. Nanto, T. Minami, S. Shooji, and S. Tanaka, *J. Appl. Phys.* 55, 1029 (1984).

243. R. M. White and F. W. Voltmer, *Appl. Phys. Lett.* 7, 314 (1965).

244. T. Yamamoto, T. Shiosaki, and A. Kawabata, *J. Appl. Phys.* 51, 3113 (1980).

245. K. Matsubara, I. Yamada, N. Nagao, K. Tominaga, and T. Takagi, *Surf. Sci.* 86, 290 (1979).

246. M. Kadota, T. Kasanami, and M. Minakata, *Japan. J. Appl. Phys.* 31, 3013 (1992).

247. W. T. Seeber, M. O. Abou-Helal, S. Barth, D. Beil, T. Höche, H. H. Afify, and S. E. Demain, *Mater. Sci. Semicond. Proc.* 2, 45 (1999).

248. Z. Y. Ning, S. H. Cheng, S. B. Ge, Y. Chao, Z. G. Gang, Y. X. Zhang, and X. G. Liu, *Thin Solid Films* 307, 50 (1997).

249. X. W. Sun and H. S. Kwok, *J. Appl. Phys.* 86, 408 (1999).

250. L. N. Dinh, M. A. Schildbach, M. Balooch, and W. McLean II, *J. Appl. Phys.* 86, 1149 (1999).

251. K. L. Narasimhan, S. P. Pai, V. R. Palkar, and R. Pinto, *Thin Solid Films* 295, 104 (1997).

252. S. Hayamizu, H. Tabata, H. Tanaka, and T. Kawai, *J. Appl. Phys.* 80, (1996).

253. A. Suzuki, T. Matsushita, Y. Sakamoto, N. Wada, T. Fukuda, H. Fujiwara, and M. Okuda, *Japan. J. Appl. Phys.* 35, 5457 (1996).

254. A. Suzuki, T. Matsushita, N. Wada, Y. Sakamoto, and M. Okuda, *Japan. J. Appl. Phys. 2* 35, L56 (1996).

255. H. Sankur and J. T. Cheng, *J. Vac. Sci. Technol. A* 1, 1806 (1983).

256. T. Nakayama, *Surf. Sci.* 133, 101 (1983).

257. N. J. Ianno, L. McConville, N. Shaikh, S. Pittal, and P. G. Snyder, *Thin Solid Films* 220, 92 (1992).

258. M. Joseph, H. Tabata, and T. Kawai, *Appl. Phys. Lett.* 74, 2534 (1999).

259. Y. J. Kim and H. J. Kim, *Mater. Lett.* 21, 351 (1994).

260. S. Takada, *J. Appl. Phys.* 73, 4739 (1993).

261. T. Hata, E. Noda, O. Morimoto, and T. Hada, *Appl. Phys. Lett.* 37, 633 (1980).

262. N. Fujimura, T. Nishihara, S. Goto, J. Xu, and T. Ito, *J. Cryst. Growth* 130, 269 (1993).

263. W. Gopel and U. Lampe, *Phys. Rev. B* 22, 6447 (1980).

264. Y. Morinaga, K. Sakuragi, N. Fujimura, and T. Ito, *J. Cryst. Growth* 174, 691 (1997).

265. J.-M. Liu and C. K. Ong, *Appl. Phys. A* 67, 493 (1998).

266. H. Cao, J. Y. Wu, H. C. Ong, J. Y. Day, and R. P. H. Chang, *Appl. Phys. Lett.* 73, 572 (1998).

267. J. Narayan, K. Dovidenko, A. K. Sharma, and S. Oktyabrsky, *J. Appl. Phys.* 84, 2597 (1998).

268. G. A. Hirata, J. McKittick, J. Siqueiros, O. A. Lopez, T. Cheeks, O. Contreras, and J. Y. Yi, *J. Vac. Sci. Technol. A* 14, 791 (1996).

269. T. Minami, H. Nanto, and S. Takata, *Appl. Phys. Lett.* 41, 958 (1982).

270. R. E. I. Schropp and A. Madan, *J. Appl. Phys.* 66, 2027 (1984).

271. M. Brett, R. Mcmahon, J. Affinito, and R. Parsons, *J. Vac. Sci. Technol. A* 1, 162 (1984).

272. R. Menner, R. Schäffler, B. Sprecher, and B. Dimmler, "Proc. 2nd World Conf. Exhib. Photovoltaic Solar Energy Conv.," Vienna, 1998, p. 660.

273. S. Jäger, B. Szyszka, J. Szczyrbowski, and G. Bräuer, *Surf. Coat. Technol.* 98, 1304 (1998).

274. J. Aranovich, A. Ortiz, and R. H. Bube, *J. Vac. Sci. Technol.* 16, 994 (1979).

275. J. Song, I.-J. Park, and K.-H. Yoon, *J. Korean Phys. Soc.* 29, 219 (1996).

276. S. Major, A. Banerjee, and K. L. Chopra, *Thin Solid Films* 108, 333 (1983).

277. S. Fujihara, J. Kusakado, and T. Kimura, *J. Mater. Sci. Lett.* 17, 781 (1998).

278. T. Tominaga, N. Umezu, I. Mori, T. Ushiro, T. Moriga, and I. Nakabayashi, *Thin Solid Films* 334, 35 (1998).

279. R. Cebulla, R. Wendt, and K. Ellmer, *J. Appl. Phys.* 83, 1087 (1998).

280. T. Minami, H. Nanto, and S. Takata, *Japan. J. Appl. Phys.* 23, L280 (1984).

281. T. Minami, H. Sato, H. Nanto, and S. Takata, *Japan. J. Appl. Phys.* 24, L781 (1985).

282. J. Hu and R. G. Gordon, *J. Appl. Phys.* 71, 880 (1992).

283. T. Minami, H. Sato, H. Sonohara, S. Takata, T. Miyata, and I. Fukuda, *Thin Solid Films* 253, 14 (1994).

284. B. Sang, A. Yamada, and M. Konagai, *Japan. J. Appl. Phys.* 37, L1125 (1998).

285. W. Tang and D. C. Cameron, *Thin Solid Films* 238, 83 (1994).

286. D. Raviendra and J. K. Sharma, *J. Appl. Phys.* 58, 838 (1985).

287. J. Kido and Y. Iizumi, *Chem. Lett.* 10, 963 (1997).

288. J. Kido and Y. Iizumi, *Appl. Phys. Lett.* 73, 2721 (1998).

289. H. Murata, C. D. Merritt, H. Mattoussi, and Z. H. Kafifi, *Proc. SPIE* 3476, 88 (1998).

Chapter 4

COLD PLASMA PROCESSES IN SURFACE SCIENCE AND TECHNOLOGY

Pierangelo Gröning

Department of Physics, University of Fribourg, Fribourg, CH-1700 Switzerland

Contents

1. INTRODUCTION

1.1. Plasma the Fourth State of Matter

Plasmas are ubiquitous, comprising more than 99% of the known matter in the universe. Taking into consideration the energy of the particles constituting the plasma, plasma is energetically the fourth and highest state of the matter, apart from the solid, liquid, and gaseous states. Irving Langmuir and his collaborators at General Electric were the first to study phenomena in plasma in the early 1920s while working on the development of vacuum tubes for high currents. It was Langmuir [1] who in 1929 used the term "plasma" for the first time to describe ionized gases.

In our everyday life encounters with plasmas are limited to a few examples: the aurora borealis (Fig. 1) at the polar regions, the flash of a lightning bolt, the conducting gas inside a fluorescent neon tube, and the small amount of ionization in the flame of a welding torch. The reason plasmas are so rare in our everyday life can be seen from the Saha equation, which describes the amount of ionization to be expected in a gas in thermal equilibrium:

$$\frac{n_i}{n_n} \approx 2.4 \times 10^{15} \frac{T^{3/2}}{n_i} e^{-U_i/kT} \qquad \text{Saha equation}$$

Here n_i and n_n are, respectively, the density of ions and neutral atoms, T is the temperature, k is the Boltzmann constant, and U_i is the ionization energy of the gas. For dry air at room temperature the fractional ionization $n_i/n_n \approx 10^{-122}$ predicted by the Saha equation is negligibly low. As the temperature increases and U_i is only a few times kT the degree of ionization rises abruptly, and the gas goes into the plasma state. Not any ionized gas can be called a plasma. Of course, there is always some small degree of ionization in any gas [1, 2]. A useful and common definition for the plasma state is as follows:

The plasma is a quasineutral gas of charged and neutral particles which exhibits collective behavior.

"Quasineutral" means that the plasma is neutral enough so that the electromagnetic forces do not vanish. Then one can take the

Handbook of Thin Film Materials, edited by H.S. Nalwa
Volume 1: Deposition and Processing of Thin Films

ISBN 0-12-512909-2/$35.00

Fig. 1.   Aurora borealis (from the website http://gedds.pfrr.alaska.edu).

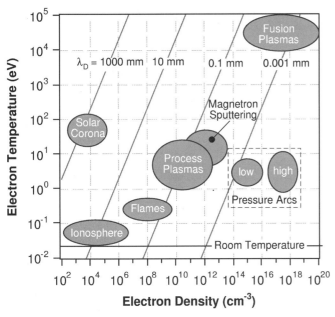

Fig. 2.   Classification of plasmas as a function of electron density and temperature (1 eV $\sim$ 11,600 K). The Debye length λ_D characterizes the interaction distance between charges.

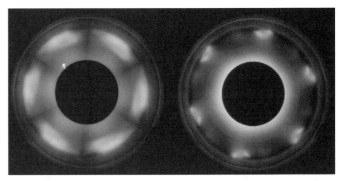

Fig. 3.   Low pressure ($p = 10^{-2}$ mbar) microwave plasma excited at the ECR condition. Left O_2 and right Ne plasma.

ion and electron densities equal $n_i \cong n_e =: n$, where n is a common density called the plasma density. In a plasma the motion of the particles can cause local concentrations of positive and negative electric charges which create long-range Coulombic fields that affect the motion of the charged particles far away from the charge concentrations. Thus elements of the plasma affect each other, even at large separations, giving the plasma its characteristic "collective behavior." The local charge concentrations in a plasma are confined to small volumes of size λ_D, called the Debye length (Fig. 2). The Debye length describes quantitatively the ability of the plasma to shield out electric potentials that are applied to it. The formula for the Debye length λ_D can be deduced from the Poisson equation and it is

$$\lambda_D = \left(\frac{\varepsilon_0 k T_e}{n_e e^2} \right)^{1/2} \quad \text{Debye length}$$

where T_e is the electron temperature, the quantity to characterize the kinetic energy of the electrons. The electron temperature T_e is used in the formula of λ_D because the electrons, being more mobile than the ions, generally do the shielding by moving so as to create a surplus or deficit of negative charge. Finally, a criterion for an ionized gas to be a plasma is that it is dense enough that λ_D is much smaller than the dimensions of the system.

The plasma contains a multitude of different neutral and charged particles, as well as ultraviolet (UV) and vacuum ultraviolet (VUV) radiation. It is broadly characterized by the following basic parameters:

- the gas pressure (p) or the density of the neutral particles, n_n (Fig. 3),
- the ion and electron density, n_i and n_e, which are equal in the quasineutral state of the plasma ($n_i \cong n_e =: n$, n is called the plasma density),
- the energy distributions of the neutral particles, $f_n(E)$, the ions, $f_i(E)$, and the electrons, $f_e(E)$.

One of the physical parameters defining the state of a neutral gas in thermodynamic equilibrium is its temperature T, representing the mean kinetic energy of the particles:

$$\frac{1}{2} m \langle \vec{v} \rangle^2 = \frac{3}{2} k T$$

The plasma is a mixture of particles with different electric charges and masses. The electrons and the heavy particles (neutrals and ions) in the plasma can be approximately considered, thermally, as two or better three subsystems, each in its own thermal equilibrium. Therefore, in analogy to the temperature of a neutral gas, the plasma can be characterized by the three temperatures, the electron, T_e, the ion, T_i, and the gas temperature (neutral), T_g. Thermodynamic equilibrium in the plasma will only exist if all temperatures of the different subsystems are equal.

The essential mechanisms in the plasma are excitation and relaxation, dissociation, ionization, and recombination. To maintain a steady state in the plasma density, the recombination's must be balanced by an ionization process; i.e., an external energy source is required. Usually an electric field is used as energy source, acting directly on the charged particles only. The energy transfer W from the electric field $\vec{E}$ to a simple charged particle with mass m is given by

$$W = e\vec{E} \cdot \vec{r} = \frac{(eEt)^2}{2m}$$

where $\vec{r}$ is the distance travelled in time t.

The expression for the energy transfer W from the electric field $\vec{E}$ to a charged particle shows clearly that the field primarily gives energy to the electrons since the mass of the ions m_i is much larger than those of the electrons m_e ($m_i \gg m_e$). From the electrons the energy is transferred to atoms and ions by collisions. Therefore, at low pressures, the electron temperature T_e is much higher than those of the ions T_i and neutrals (gas) T_g, respectively. The three temperatures converge to similar values at a pressure around 100 mbar and the plasma becomes arclike. Atmospheric pressure plasmas have temperatures of a few thousand Kelvin. Such plasmas, where local thermodynamic equilibrium ($T_e = T_i = T_g$) is achieved, in volumes of the order of the mean free path length for collision, are called local thermodynamic equilibrium (LTE) plasma or thermal plasma. In low pressure plasmas, where the collision rate and therefore the energy transfer from electrons to neutrals is reduced, the electron temperature is much higher than the temperature of the ions and the gas ($T_e \gg T_i \cong T_g$). In such plasmas the LTE conditions are not achieved and therefore are called non-LTE plasmas or cold plasmas, because the ion and gas temperatures are at about

room temperature ($T_i \cong T_g \cong$ RT), whereas the electron can reach temperatures of 10^4–10^5 K (1–10 eV).

1.2. Cold Plasma

Cold plasma processes are nowadays used in various technological applications. Cold plasmas are used for coating, polymerisation, activation and cleaning of workpieces, for etching of microelectronic and micromechanical devices, and for finishing of textiles. In the microelectronic industry plasma surface cleaning is established as the cleaning technique for the nanometer scale. In comparison to wet chemical processes plasma processes are much environmental friendly, cheaper and a higher cleaning degree can be achieved. This makes plasma processes very attractive for the industry and is the driving force that the field of applications increases permanently.

There are many excellent books reviewing the physics and chemistry [4, 5], the diagnostic [6], and the applications [7, 8] of cold plasmas. Here we will only introduce briefly the principles and a few important parameters of cold plasmas, without pretension on completeness (Fig. 4).

1.2.1. Electron Temperature T_e

In cold plasmas the electrons and the heavy particles are not in thermodynamic equilibrium, even at local scale in the order of the Debye length λ_D. Such plasmas can be excited and

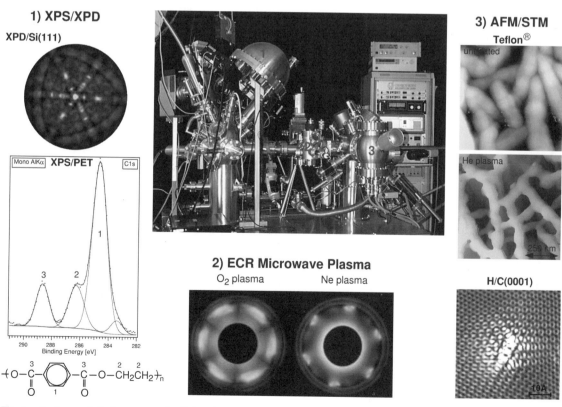

Fig. 4. Surface analytical system at the University of Fribourg (Switzerland) to investigate cold plasma processes. The system is equipped with an ESCA system (1) for photoelectron spectroscopy (XPS, UPS) and photoelectron diffraction (XPD), with an ECR microwave plasma (2) and a room temperature AFM/STM (3). The whole system is operating at a base pressure of 5×10^{-11} mbar.

sustained by direct current (DC), radio frequency (RF), or microwave (MW) electric fields applied to the gas. From the electric field the energy is transferred to the available free electrons by accelerating them, as mentioned before. Concomitantly the electrons lose energy in collisions with the gas. As long as the energy of the electrons is too low to excite or ionise the gas, the collisions will necessarily be elastic and therefore heat the gas. Since the energy transfer from the electron to the gas is extremely small in the elastic collision the electron gains energy from the electric field until it is sufficiently high to cause ionization or excitation through inelastic collisions. The electrons produced in the ionization process are in turn accelerated by the electric field and produce further ionization up to the plasma state.

In thermodynamic equilibrium the velocity distribution and consequently the energy distribution of the particles in the gas is Maxwellian. In cold plasma the electrons are in a nonequilibrium state ($T_e \gg T_g \cong$ RT). The slower electrons make elastic collisions only, whereas electrons with higher energies are liable to lose a much larger fraction of their energy by inelastic collisions. Therefore the energy distribution of the electrons in cold plasmas is shifted to higher energy compared to the Maxwell–Boltzmann distribution (Fig. 5). Instead of using the Maxwell–Boltzmann distribution $f_{MB}(E)$ it is better to approximate the electron energy distribution in cold plasma by the Druvestyn distribution $f_D(E)$:

$$f_D(E) = 1.04 \cdot \langle E \rangle^{-3/2} E^{1/2} \exp\left(\frac{-0.55E^2}{\langle E \rangle^2}\right)$$

Druvestyn distribution

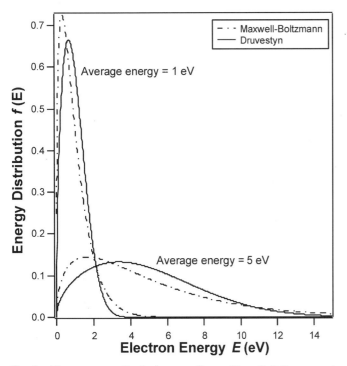

Fig. 5. Electron energy distribution according to Maxwell–Boltzmann and Druvestyn.

The Druvestyn distribution considers the motion of electrons in a weak electric field, such as exist in cold plasmas. Characteristic for both energy distribution functions is the high-energy tail. For reactions in the plasma requiring relevant electrons energies between 3 and 10 eV the Druvestyn distribution predicts a higher reaction rate than the Maxwell–Boltzmann distribution.

1.2.2. Plasma Potential V_p

A very important parameter in cold plasma processing is the plasma potential V_p which defines the kinetic energy of the ions impinging on the surface of a solid in contact with the plasma. The plasma potential is a consequence of the enormous difference in the temperature, i.e., speed $\langle v \rangle$, between the electrons and the ions in the cold plasma. A sample inserted on an electrically insulated sample holder into the plasma will be struck by electrons and ions with currents densities predicted from Avogadro's law to be

$$j_e = \frac{-e \cdot n_e \cdot \langle v_e \rangle}{4}$$
$$j_i = \frac{e \cdot n_i \cdot \langle v_i \rangle}{4}$$

Since $\langle v_e \rangle \gg \langle v_i \rangle$ the electron current density is much larger than the ion one ($|j_e| \gg |j_i|$). Hence the sample will be immediately negatively charged, until the electron current is reduced by electrostatic repulsion just enough to balance the ion current. The negative potential, which balances the electron and ion current, is called the floating potential V_{fl}. The same charging effect occurs also with the plasma container walls, so that the plasma is positively charged with respect to the container. This potential is called the plasma potential V_p. The plasma potential is in fact the electric potential of the plasma due to the ion excess caused by the higher recombination rate of the electrons at the plasma container walls compared to the ions. The Debye shielding effect causes a positive space charge region, called the plasma sheath, in front of all surfaces (V_{fl}) in contact with the plasma (V_p). Therefore the plasma potential V_p is always the highest positive potential in a plasma system. At low pressures where the collision mean free path is much larger than the plasma sheath d_s, the latter is approximately given by

$$d_s \approx \lambda_D \cdot \left[\frac{e(V_p - V_{fl})}{kT_e}\right]^{3/4}$$

plasma sheath

The plasma sheath is for the electrons a potential barrier in front of the substrate. Only electrons with enough kinetic energy can overcome this barrier. However, the ions are accelerated by the plasma sheath toward the substrate. According to the voltage drop ($V_p - V_{fl}$) in this space charge region, the kinetic energy E_{kin} of the ions impinging on the substrate is given by

$$E_{kin} = e(V_p - V_{fl})$$

As suggested above the plasma potential is a consequence of the finite volume of the plasma device, i.e., determined by the recombination rates in the plasma and at the plasma container walls. With increasing gas pressure, the electron–ion recombination probability in the gas increases relative to that of the electrons at the container walls. This means that the plasma potential is pressure dependent and decreases with increasing gas pressure. Typical values for the plasma potential V_p are between 5 and 30 V. Accordingly the kinetic energies of the ions impinging the substrate are several electronvolts, which is about thousand times higher than the ion temperature T_i and high enough to induce significant physical and chemical effects on the substrate surface.

1.2.3. Self-Bias

The physical etching effect of the ions during plasma treatment can be increased by negatively biasing the sample to increase the kinetic energy of the ions hitting the surface. Whereas for metallic samples a simple DC supply can be used to charge it negatively, an RF source has to be applied for insulating samples to use the self-biasing effect [9] of the plasma (Fig. 6).

The enormous difference in speed between the electrons and the ions in the cold plasma causes, by applying an RF potential, initially ($t = 0$) an excess electron current on the substrate (Fig. 7a). Therefore the substrate begins to charge negatively until (few cycles) the DC-offset voltage (U_{bias}) on the RF potential balances the electron and ion current on the substrate (Fig. 7b).

By self-biasing, negative DC potentials (U_{bias}) up to a few hundreds of volts can be obtained, depending on the RF power (P_{RF}) and the gas pressure (p):

$$U_{bias} \propto \left(\frac{P_{RF}}{p} \right)^{1/2}$$

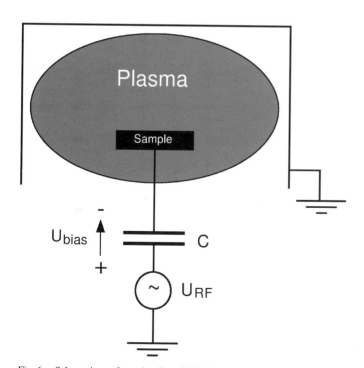

Fig. 6. Schematic configuration for self-biasing a sample (C: blocking capacitor).

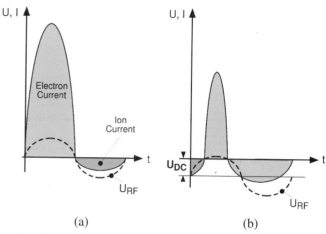

Fig. 7.   Self-biasing of a substrate (a) initial potential ($t = 0$), (b) steady state (in practice the asymmetry between electron and ion current is much larger than indicated).

In the steady state the substrate potential is positive only during a very short fraction of each cycle. Therefore the substrate will be almost continuously bombarded with ions of an average kinetic energy ($E_{kin} \cong e \cdot U_{bias}$) defined by the bias potential U_{bias}. Self-biasing can be used for ion etching and sputter deposition of insulators.

1.2.4. Plasma Excitation

To excite and sustain cold plasmas an electric field has to be applied to the gas. This can be done by DC, RF, or MW power, whereby the electric energy is transferred via the electrons to the plasma. In the situation without collisions the electron energy W in an electric field ($E(t)$) is given by the equation of motion and

$$W = \frac{m_e \dot{x}^2}{2} = \frac{(\widehat{E}et)^2}{2m_e\omega^2} \sin^2(\omega t) \qquad \left(E(t) = \widehat{E}\cos(\omega t) \right)$$

The formula suggests that besides a high electric field strength ($\widehat{E}$), a low frequency (ω) is required to reach a maximal electron energy (W). This is in opposition to the practice where the MW discharge is more efficient than the RF discharge, and the RF discharge is more efficient than the DC discharge in promoting ionization and sustaining the discharge. MacDonald and Tetenbaum [10] explained this behavior by the model in which the electron making an elastic collision at an atom, reversing its motion at the time the electric field changes the direction, will continue to gain velocity and energy. In this way, electrons could reach ionization energies for quite weak electric fields. Taking into account the elastic electron atom collisions at a frequency ν the mean power $\langle P \rangle$ absorbed by an electron from the electric field is according to Venugopalan [11] given by

$$\langle P \rangle = \frac{(\widehat{E}e)^2}{2m_e} \frac{\nu}{\nu^2 + \omega^2}$$

The formula shows that the power absorption reaches a maximum when the collision frequency ν equals the electric field

Table I.   Characteristics of the Various Cold Plasmas

	DC	RF	MW	ECR
Coupling of the electric power	Internal and conductive electrodes	Internal as well as external electrodes, conductive or nonconductive	No electrodes	No electrodes
Pressure (mbar)	0.5–10	10^{-2}–1	0.5–10	10^{-5}–10^{-2}
Ionization degree	≪0.1%	<0.1%	<1%	1–10%
Ion density (cm^{-3})	$<10^{10}$	$\sim10^{10}$	$\sim10^{11}$	$\sim10^{11}$
Kinetic energy of ions hitting the substrate (eV)	Very high 100–600	Can be controlled by a negative bias	Can be controlled by a negative bias	Can be controlled by a negative bias >10 eV

frequency ω. This means that the power adsorption efficiency depends on the gas pressure. For gas pressures in the mbar range the optimal excitation frequency is in the MW range. Since the energy acquired by an electron during one cycle decreases with increasing excitation frequency it is favorable to use RF instead of MW for pressures less than 1 mbar. The coupling of the MW power into the plasma for pressures $p < 1$ mbar can be improved by operating the excitation at the electron cyclotron resonance (ECR) condition. Such plasmas called ECR plasmas can be sustained over a wide pressure range from few mbar down to extremely low pressures of 10^{-6} mbar. ECR plasmas make good use of the gyromotion of charged particles in a magnetic field $\vec{B}$. The radius r_c, called the cyclotron or Larmor radius, and the frequency ω_c, called the cyclotron or Larmor frequency, are given by

$$\omega_c = \frac{eB}{m} \qquad r_c = \frac{mv_\perp}{eB} \qquad \begin{array}{l}\text{cyclotron or Larmor frequency}\\ \text{and radius, respectively}\end{array}$$

$v_\perp$ is the velocity component of the charged particle normal to the magnetic field $\vec{B}$. The cyclotron frequency ω_c is independent of the velocity of the charged particles and therefore characteristic for the individual charged particles. If now an electric field $\vec{E}$ is rotating in the same direction (right hand polarized) as the gyromotion, the charged particles have a reduced relative frequency of $\omega - \omega_c$. Therefore, the formula for the mean power $\langle P \rangle$ transferred from the electric field to the electron gyrating around the magnetic field lines is given by

$$\langle P \rangle = \frac{(\widehat{E}e)^2}{2m_e} \frac{\nu}{\nu^2 + (\omega - \omega_{ce})^2}$$

ω_{ce} is the cyclotron frequency of the electron.

If the frequency ω of the electric field equals the cyclotron frequency ω_{ce} a resonant situation ($\omega - \omega_{ce} = 0$) is achieved and the electrons are continuously accelerated in the magnetic field $\vec{B}$ and the mean power $\langle P \rangle$ absorbed by the electrons is maximal. For the commonly used microwave frequency of 2.45 GHz the ECR condition is satisfied for a magnetic field of 87.5 mT. An excellent review of the physics of ECR microwave discharges has been given by Asmussen [12].

Usually ECR plasma sources are operating in the pressure range of 10^{-5} to 10^{-2} mbar. In this pressure range the ECR microwave coupling is the most efficient one, resulting in a high ionization degree. The characteristics of the various cold plasmas are summarized in Table I.

1.3. Plasma Chemistry

As mentioned above, in all cold plasmas the electrons are the principal sources for transferring electrical energy to the gas. Elastic electron–molecule collisions will increase the temperatures of the neutrals (T_g) and ions (T_i), while inelastic collisions (Table II) generally result in ionization, excitation, or fragmentation of the molecules. In addition, deexcitation processes produce the glow, i.e., the electromagnetic radiation, which also contributes to these reactions.

In the plasma chemical reactions can take place either in the gaseous phase or at the substrate surface; the former are called homogeneous and the latter are heterogeneous reactions.

The rate of homogeneous reaction R between two types of particles (A, B) depends on the cross section σ_{AB} of the reaction, the mean free path between interactions λ_{AB}, the densities n_A, n_B of the two particles, and the relative velocity v_{AB} between the particles:

$$R = \frac{1}{\lambda_{AB}} \cdot n_A \cdot v_{AB} = \sigma_{AB} \cdot n_B \cdot n_A \cdot v_{AB} \quad (cm^{-3}\,s^{-1})$$

Since the rate of homogeneous reaction R depends on the product of the particle densities these reactions are generally performed in the pressure range of 0.1 to 10 mbar. At much lower pressures the mean free path is too long for gas collisions to be significant. On the other hand chemical reactions with relative high activation energies require lower gas pressures to secure reactions even at a slow rate which otherwise would not occur at limiting electron temperatures T_e. Therefore a number of reactions are most efficient in MW plasmas which usually have relatively high electron temperatures.

The important aspect of plasma reactions is the variety of different reactions that may be produced simultaneously, including electron impact reactions (ionization, dissociation, ex-

Table II.   Types of Inelastic Collisions

Deexcitation	$A + e^- \Longleftrightarrow A^* + e^-$	Excitation
Photoemission	$A + h\nu \Longleftrightarrow A^*$	Photoexcitation
Recombination	$A + e^- \Longleftrightarrow A^+ + 2e^-$	Ionization
Recombination	$A + h\nu \Longleftrightarrow A^+ + 2e^-$	Photoionization
Photodetachment	$A + e^- \Longleftrightarrow A^- + h\nu$	Radiative attachment
Recombination	$A^* + B \Longleftrightarrow A + B^+ + e^-$	Penning ionization
Charge exchange	$A + B^+ \Longleftrightarrow A^+ + B$	Charge exchange

$+$ Ions.

$*$ Excited species (not in the ground state).

Table III.   Typical Heterogeneous Reactions

$A(g) + S \rightarrow A(s)$	Adsorption
$A^*(g) + S \rightarrow A^*(s)$	Adsorption
$S + A(s) + A(g) \rightarrow S + A_2(g)$	Recombination
$S + B(s) + A^+(g) \rightarrow S^+ + B(g) + A(g)$	Sputtering
$A^*(g) + S \rightarrow S + A(g)$	Deexcitation
$R(g) + R(s) \rightarrow P(s)$	Polymerization
$A(g) + R(s) \rightarrow P(s)$	Polymerization
$R(s) + R(s) \rightarrow P(s)$	Polymerization

Labelling: A, B for molecule, S for surface, R radical, P for polymer, $*$ for excited, g for gaseous, and s for solid.

citation), radical–radical, ion–molecule, ion–radical, molecule–excited species, etc. Some of these reactions create new chemical bonds and new molecules; others form secondary ions and secondary radicals or new excited species, etc. Since excitation as well as dissociation energies ($< 10 \, eV$) are considerably lower than ionization energies ($> 10 \, eV$) the densities of the ions n_i of the radicals n_r and neutrals n_n are related by

$$n_i \ll n_r \ll n_n$$

With respect to these partial particle densities radical–molecule reactions will be generally more important than ion–molecule reactions in homogeneous reactions. The reaction rates as well as the reaction channels can be changed by the presence of a so-called carrier gas, which furnishes additional electrons. Typical carrier gases are N_2 and the noble gases.

The second class of reactions in the plasma is the heterogeneous reactions, which occur at the solid surface exposed to the plasma species. Typical heterogeneous reactions are adsorption, recombination, deexcitation, polymerization, and sputtering. Typical heterogeneous reactions are summarized in Table III.

Homogeneous reactions in the plasma are dominated by chemical reactions. In contrast to that, physical reactions must be considered in heterogeneous reactions at surfaces in contact with the plasma. The presence of energetic particles and electromagnetic radiation in the plasma causes physical reactions at any surface in contact with the plasma. If a molecular ion with enough kinetic energy ($> 10 \, eV$) collides with a solid surface, a part of the kinetic energy is transferred into internal energy (vibrational and/or electronic excitation), which generally is much larger than the chemical bond energy. This causes the dissociation of the molecule into its constituents, which afterward can react with the solid. Positive ions accelerated toward the solid surface with sufficient high kinetic energy can remove an atom from the surface by inducing a collision cascade in the first nanometer of the surface. This process is called sputtering. The threshold energy E_{th} for sputtering a target atom with mass M_1 by a projectile ion with mass M_2 has been calculated by Behrisch et al. [13] and interpreted by Bohdansky et al. [14]. The threshold energy E_{th}, calculated in computer simulations for light ions, can be expressed as follows:

$$E_{th} = \frac{U_0}{\gamma(1 - \gamma)} \qquad \text{for } M_2/M_1 \leq 0.3$$
$$E_{th} = 8U_0\left(\frac{M_2}{M_1}\right)^{2/5} \qquad \text{for } 0.3 < M_2/M_1 < 1$$

with

$$\gamma = \frac{4M_1 M_2}{(M_1 + M_2)^2} \qquad \text{energy transfer coefficient}$$

U_0 \qquad binding energy of the surface atom which is approximative to the heat of sublimation

The threshold energy E_{th} for sputtering is typically between 25 and 40 eV. Ions with lower kinetic energies cause phonon excitation in the solid, i.e., heat up the substrate. In the topic of thin film deposition the most important effects which can occur by ion–solid interactions can be identified as follows:

- desorption or sputtering of adsorbed impurities from the substrate surface, often used for cleaning the substrate before coating or accompanied by forward recoil implantation of impurities into the substrate,
- the implantation and entrapment of coating and support gas atoms initially into the substrate and subsequently into the coating,
- displacement of substrate and coating atoms and generation of lattice defects. Atomic displacements can lead to intermixing of the substrate and coating atoms whilst enhanced defect densities can promote rapid interdiffusion. The net effects of such processes are a denser and more homogenous structure of the coating and the broadening of the substrate–coating interfaces generally results in a better adhesion of the coating.

The scanning electron microscope images in Fig. 8 illustrate on two examples the effect of energetic particle bombardment on the morphology of physical vapor deposited thin films. The by classical physical vapor deposition (PVD) produced Cr and Al_2O_3 films shows a columnar grain structure with crystal of about 0.1 µm in diameter.

The condensation from the vapor involves incident atoms becoming bonded adatoms which then diffuse over the film surface until they desorb or are trapped at low energy lattice sites. The incorporated atoms reach their equilibrium positions in the lattice by bulk diffusive motion. This growth mode involves

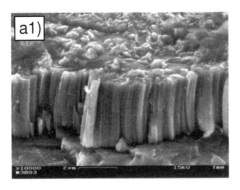

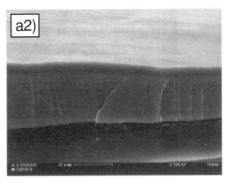

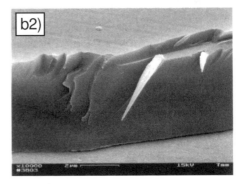

Fig. 8. Comparison of the morphology of thin films deposited by conventional PVD and plasma activated PVD (PAPVD). (a1) Cr film deposited by PVD. (a2) Cr film deposited by PAPVD ($U_{\text{bias}} = -100$ V). (b1) Al_2O_3 film deposited by reactive PVD. (b2) Al_2O_3 film deposited by PAPVD ($U_{\text{bias}} = -100$ V). Reprinted with permission from [15], © 2000, Elsevier Science.

four basic processes: shadowing, surface diffusion, bulk diffusion, and desorption. The last three are quantified by the characteristic diffusion and sublimation activation energies whose magnitudes scale directly with the melting point T_m of the condensate. Shadowing is a phenomenon arising from the geometric constraint imposed by the roughness of the growing film and the line-of-sight impingement of the arriving atoms.

In 1969 Movchan and Demchishin [16] proposed a zone structure model which describes the morphology of PVD films as a function of the relative substrate temperature T/T_m with respect to the melting point T_m of the condensate. They identified three temperature zones forming characteristic film morphologies. In zone 1 ($T/T_m < 0.3$), the structure which appears is the result of the shadowing effects that overcome limited adatom surface diffusion. The film has needlelike crystallites and is porous. In zone 2 ($0.3 < T/T_m < 0.45$), the structure is the result of surface diffusion controlled growth. The film structure is columnar with increasing column diameter and decreasing porosity for higher substrate temperatures. In the last, the high temperature zone 3 ($T/T_m > 0.45$), the film growth is bulk diffusion controlled, giving rise to an equiaxed recrystallized grain structure. A similar zone schema was introduced by Thornton [17, 18] but with four zones, an additional transition zone T between zone 1 and zone 2. In 1978, Lardon et al. [19] extended the structure zone schema by the bias potential (U_{bias}) parameter by taking into account the influence of energetic particle bombardment on the morphology of the condensate. As mentioned above energetic particle bombardment leads to dis-

placement of coating atoms and generation of lattice defects, which increases the mobility of the adatoms, resulting a denser and more homogenous structure of the coating as shown by the images (a2) and (b2) in Fig. 8. The bombardment of the surface with energetic ions can be interpreted as a local heating, which increases the mobility of the atom in the impact zone. Therefore energetic particle bombardment influences the morphology of a coating in the sense that the boundaries in the zone structure model are shifted to lower T/T_m values.

An excellent review, on the interaction of low energy ions with surfaces, has been given by Carter and Armour [20].

2. APPLICATIONS

2.1. Carbon Thin Films

The versatility of cold plasma in thin film technology can impressively be demonstrated by the deposition of carbon thin films.

The electronic structure of the carbon atom is $1s^2 2s^2 p^2$. Due to its three possible hybridization states (sp, sp^2, sp^3), carbon can exist in various stable allotropic forms (Fig. 9). The thermodynamically stable form of carbon at normal pressure and temperature is graphite. The free-energy difference between the different allotropic phases is relatively small, for instance, 0.03 eV/atom between diamond and graphite, which is slightly higher than kT at room temperature (0.025 eV). However, there is always a large energy barrier between the different

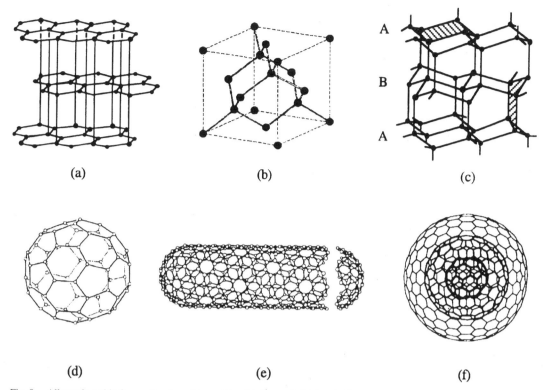

Fig. 9. Allotropic stable forms of carbon: (a) graphite, (b) diamond, (c) lonsdaelite, (d) fullerene C_{60}, (e) single-walled nanotube (SWNT), (f) onion C_{540}.

allotropic phases, which makes them all very stable at normal conditions. For instance, 15 GPa static pressure and a temperature of 1800 K is needed to transform graphite into diamond. With plasmas all stable allotropic as well as amorphous carbon phases can be synthesized or deposited as thin films, depending on the plasma and substrate conditions.

2.1.1. Amorphous Carbon Films

Amorphous carbon films can be produced by sputter deposition, a PVD process, and by chemical vapor deposition (CVD) with hydrocarbon-containing gases. Amorphous carbon thin films may be broadly classified as:

(i) amorphous carbon films, usually deposited by PVD processes,
(ii) hydrogenated carbon films, a-C : H, usually deposited by plasma-assisted CVD (PACVD).

Both types of films contain different amounts of sp^2 and sp^3 bonded carbon. The amount of sp^2 bonded carbon can be measured directly using nuclear magnetic resonance spectroscopy. The classification of amorphous carbon according to carbon bond type and hydrogen content can be represented in a triangular phase diagram [21] (Fig. 10). The corners at the base of the triangle correspond to diamond (100% sp^3 carbon) and graphite (100% sp^2 carbon). The upper limit for formation of solid films is defined by the tie line between the compositions of polyethylene $-(CH_2)_n-$ and polyethyne $-(CH)_n-$.

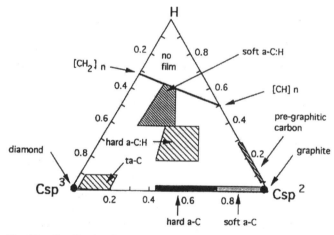

Fig. 10. Classification diagram for amorphous carbon films. Reprinted with permission from [21], © 1996, John Wiley & Sons.

Hard forms of amorphous hydrogenated carbon (a-C : H), also known as "diamondlike carbon" (DLC), were first produced by Aisenberg and Chabot [22] in 1971. The expression diamondlike refers to the properties of the coating like high hardness (up to 6000 HV), high thermal conductivity, high chemical inertness, low wear, and low friction [23, 24]. Due to their outstanding properties DLC coatings are industrially used for a variety of applications, i.e., protective coatings on magnetic hard discs, sliding bearings, medical implants, forming tools, as well as many special applications. Figure 11 shows an

Fig. 11. Aluminium yarn storage disk coated with 3 μm DLC for applications in spinning machines. (Photo: Berna/Bernex AG, CH-4600 Olten, Switzerland.)

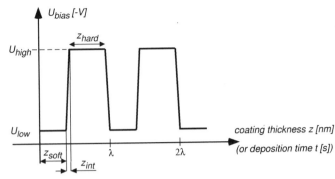

Fig. 13. Bias voltage curve for DLC multilayer structure. λ is the repeat unit of the multilayer structure. Reprinted with permission from [133], © 1987, Elsevier Science.

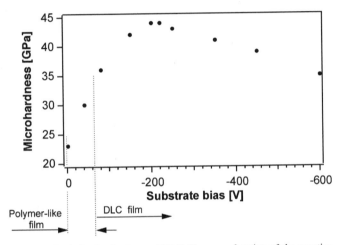

Fig. 12. Typical microhardness of DLC films as a function of the negative substrate bias (deposition: MW CH_4 plasma, $p = 20$ mbar).

aluminium yarn storage disk from a spinning machine coated with 3 μm DLC.

DLC consists of an amorphous network of sp^3 and sp^2 hybridized carbon and can be synthesized at room temperature by PACVD [25–27] or other ion-assisted processes. Any hydrocarbon with sufficient vapor pressure can in principle be used as precursor for PACVD of DLC films. Commonly used process gases are acetylene (C_2H_2) and methane (CH_4). The DLC deposition is a nonequilibrium process characterized by the interaction of energetic ions with the surface of the growing film. The deposition of DLC by PACVD has to be done on negatively biased substrates to actuate reactions like thermal and pressure spikes at the growth surface by energetic ions. Tsai and Bogy [28] calculated for 100 eV ions thermal spikes of 3300 K and pressures of 1.3×10^{10} Pa for a period of 10^{-10} s. The lifetime of these spikes is much longer than the vibrational period ($\sim 10^{-14}$ s) for diamond and thus well beyond what is required to allow bonding. The DLC process forms a metastable amorphous material whose structure is conserved due to extremely high quenching rates of the thermal spikes. The hardness of the DLC films can be varied by adjusting the substrate bias, as shown in Figure 12.

The characteristic process of DLC deposition is the interaction of the impinging energetic ions with the hydrocarbons physisorbed on the surface of the growing film. This causes an increased deposition rate with increasing ion flux and decreasing substrate temperature. Further the properties of the DLC films are independent of the chemical precursor used if the negative bias is high enough ($U_B < -100$ V), as found by Koidl et al. [29]. For lower bias voltages the impact energy is to small for efficient hydrocarbon dissociation and the film becomes soft and polymerlike, as indicated in Figure 12.

In recent years many research groups have reported new types of coating, consisting of multiphase materials with structures in the nanometer range, that exhibit outstanding mechanical properties [30–32].

The fact that the hardness of DLC films depends only on the applied negative substrate bias allows one to deposit DLC as nanoscaled multilayer coatings consisting of alternative hard and soft a-C:H layers [33]. The thickness of the multilayer repeat unit and the interface width between these layers can be controlled in the nanometer range by the substrate bias. A typical bias voltage function for a DLC multilayer structure is shown in Figure 13.

Hauert et al. [33] have deposited DLC multilayer coatings with different repeat unit thicknesses λ and interface widths z_{int}. They found that the wear resistance of the different multilayer coatings depends strongly on the repeat unit thickness λ and especially on the interface width z_{int} as shown in Figure 14. The friction coefficient of all samples was between 0.07 and 0.18.

Though the main cause of the DLC multilayer structure on the tribological behavior is not fully understood, the experiments demonstrate the enormous potential that DLC multilayer structures can have in tribology.

PACVD in combination with magnetron sputtering allows the deposition of DLC with different dopands. The electrical conductivity of metal-containing hydrogenated amorphous carbon (M-a-C:H) films, for example, can be varied over many orders of magnitude [34]. Also, depending on the type of metal and the amount used, the M-a-C:H film may exhibit different hardness, friction, and wear [35]. The excellent friction behavior combined with the chemical inertness makes DLC coatings interesting for biomedical implants. Moreover, the release of

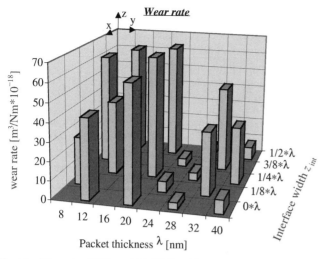

Fig. 14. Wear rate of 300 nm thick DLC multilayer coatings as a function of the repeat unit thickness λ and interface with z_{int}. Reprinted with permission from [33].

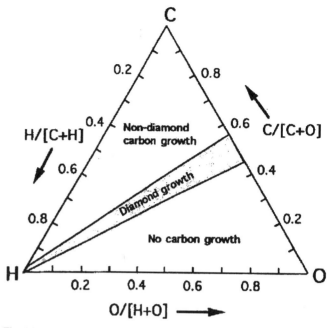

Fig. 15. Bachmann diagram for CVD diamond film growth. Reprinted with permission from [41], © 1991, Elsevier Science.

metal ions or wear particles from metallic orthopedic implants into the surrounding tissue, which is believed to lead to bone resorption and consequently to failure and loss of the implant, can be minimized or prevented by using DLC as a protective coating [36]. Francz et al. [37] have shown that by additional incorporation of different metals into the a-C:H matrix, thin film coatings with new biological properties can be generated. They found that by adding Ti to the carbon matrix, cellular reactions such as tendential differentiation of osteoblasts together with a reduced osteoclast activity could be obtained.

2.1.2. Diamond Films

In 1971 Deryagin and Fedoseev reported successful continuos growth of diamond at low pressures by methods which they later described [38]. The outstanding properties of diamond, including its hardness, chemical inertness, good optical transparency, and highest thermal conductivity, triggered enormous research activity in diamond thin film deposition. The book "Low-Pressure Synthetic Diamond" edited by Dischler and Wild [39] gives a nice overview on the manufacturing and applications of diamond thin films.

Today, diamond thin films can be deposited by various low pressure techniques [40] such as hot-filament, MW plasma, DC discharge, or plasma jet techniques. Plasma is among the most widely used techniques (PECVD) for gas activation, permitting the deposition of diamond thin films.

The low pressure diamond deposition process is fundamentally different from those of a-C:H. Characteristic for the a-C:H deposition process is the physical effect of the impinging energetic ions with the hydrocarbons physisorbed on the surface of the growing film. In opposition to that the diamond deposition process is the result of various chemical reactions on the surface of the growing film. The difference in the two deposition processes clearly finds expression by the precursor gas needed. While for diamond deposition the hydrocar-

bon precursor must be strongly diluted with hydrogen (typically $CH_4/H_2 = 1/99$), the a-C:H deposition process does not need the hydrogen. Further differences of the two deposition processes are the pressure, which is a few ten of millibars for diamond and below 1 millibar for a-C:H, and the needed substrate temperature, 600 to 1100°C for diamond and room temperature for a-C:H. The hydrogen chemistry is essential for the diamond growth at low pressures. The triangular CHO diagram of Bachmann et al. [41] shows with which gas compositions CVD diamond can be grown (see Fig. 15).

The growth surface of diamond can be represented schematically by a layer of carbon atoms terminated with atomic hydrogen. Individual hydrogen atoms are removed and provide a chemically active site for addition of carbon. Hydrogen abstraction has been estimated to occur every 70 μs on average, with each such site being refilled after 15 μs [42, 43]. Growth occurs when a hydrocarbon radical is attached at an active site of the lattice and loses its hydrogen atoms through hydrogen abstraction. Among the different radicals (methane, methyl, acetylene, acetyl), methyl radicals are usually accepted as the main growth species, especially for the (100) surface [44].

First, a surface hydrogen atom recombines with atomic hydrogen of the plasma to H_2 molecules, producing a radical site on the surface (Fig. 16b). This site can react with a hydrogen atom (Fig. 16a) or a methyl radical (Fig. 16c). The next step is the abstraction by atomic hydrogen of the plasma of a hydrogen atom of the methyl group or of another surface atom (Fig. 16d). Finally, another hydrogen atom is abstracted to leave two adjacent radical carbon atoms. They react to form a carbon–carbon bond (Fig. 16e) forming an adamantane molecule. In parallel, a methyl group can add to the CH_2 radical site to form an ethyl group (C_2H_5), diversifying the growth mechanisms [45]. Frenklach et al. [46, 47] proposed a growth mechanism based on the addition of acetyl to a radical site.

Characteristic of CVD diamond deposition is the need for nucleation centers. On commercial silicon wafer without any treatment the CVD diamond process leads to the deposition of individual nanocrystallites with a density lower than 10^5 nuclei per cm^2. The main reason for this extremely low nucleation density is the high surface energy of diamond relative to that of the substrate material. Surface nucleation enhancement is necessary in order to get a continuous polycrystalline film. Three main mechanisms are known for diamond nucleation:

(a) nucleation on dislocation ledges, kinks, or intentional scratches,
(b) diffusion barrier enhanced nucleation,
(c) nucleation on a molecular precursor.

The corresponding techniques are:

(a) scratching the substrate with diamond, nitrides or other ceramic powders, to create defects which act as nucleation centers [48]; the nucleation density is

between 10^5 to 10^{10} cm^{-2}. No oriented diamond film has been observed yet on such prepared substrates,

(b) enhanced nucleation on a precursor such as thin metal films, graphite fibers, C_{60}, or others [48, 49]; the nucleation density is between 10^6 and 10^{10} cm^{-2} depending on the precursor.

A further technique is the bias enhanced nucleation by applying a DC or RF voltage during the first minutes of the deposition [48, 50, 51]. The nucleation density obtained with this technique is between 10^8 and 10^{11} cm^{-2}. At present it is the only nucleation technique involving the oriented growth of the diamond film relative to the silicon substrate as shown in Figure 17.

The mechanisms of oriented bias enhanced nucleation are still not well understood and one crucial question is the structure of the interlayer between the silicon substrate and the diamond film formed during the growth and especially during the bias enhanced nucleation process. The nature of this interlayer has been investigated by a number of analytical techniques including X-ray photoelectron spectroscopy (XPS) [53], transmission electron microscopy (TEM) [54], and IR spectroscopy [55]. No clear picture can be drawn whether oriented diamond nucleates on the β-SiC interlayer that is formed during the nucleation process or directly on silicon. TEM investigations show nucleation on all materials [56, 57]. Gerber et al. [58, 59] have proposed a subplantation model for the oriented nucleation. They claimed that ion bombardment of the a-C:H layer formed on top of the SiC layer leads to subplantation effects in the a-C:H layer which cause a high compressive stress. Locally this stress transforms the sp^2 to an sp^3 structure, giving rise to a diamond nucleation center.

We investigated the silicon surface during the first minutes of bias enhanced nucleation by using X-ray photoelectron diffraction (XPD) [60, 61]. XPD is a natural extension to regular XPS [62]. A spectral feature is selected out of a XPS spectrum from a single-crystalline sample. The intensity modulation of this particular signal is measured as a function of the electron emission angle. Based on the forward focusing effect (Fig. 18)

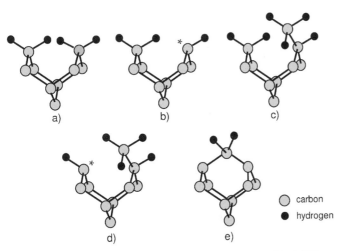

carbon
hydrogen

Fig. 16.   Methyl based diamond growth model according to Harris [45].

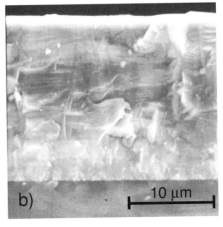

Fig. 17.   SEM pictures of (100)-oriented polycrystalline diamond film grown by PECVD on silicon (100): (a) top view, (b) side view. Reprinted with permission from [52].

of the electrons by atoms the measured intensity modulation gives information on the surface atomic structure.

Figure 19 shows the deconvoluted C1s XPS spectrum together with XPD pattern of the C—Si and C—C components. The diffractograms show for both components the typical structure of a (100) surface. The C—Si diffractogram proves the oriented growth of a β-SiC layer with respect to the silicon (100) substrate (Si—Si diffractogram not shown). The formation of oriented β-SiC is not surprising because the structure can be formed by substitution of Si atoms with carbon atoms from the plasma. The formation of oriented β-SiC is independent of the

bias and appears also without bias [61]. This proves that the presence of oriented β-SiC is not a sufficient condition for oriented diamond growth!

The diffractogram of the C—C component is identical to that of the C—Si component, indicating the presence of small carbon clusters in the β-SiC matrix. The clusters are oriented with respect to the silicon (100) substrate and act as nucleation centers for the oriented diamond growth. The small diamond crystallites which are formed during the bias-enhanced nucleation (Fig. 20) do not significantly contribute to the analysis because their coverage is below 1% and therefore below the XPS detection limit. In opposition to the β-SiC layer, oriented carbon clusters are only formed at sufficiently high negative bias voltage ($U_{bias} < -100$ V).

Unfortunately, the initially euphoric expectations in diamond thin films as the universal solution for all kind of problems in wear, tribology, electronics, etc., have not yet come true. Nevertheless diamond thin films are nowadays used in many applications such as surface acoustic wave filters (Sumitomo Electric), thermal heat sinks in electronic devices, inert electrodes in electrochemical disinfection of fresh water and purification of waste water [63], conductive atomic force microscope tips (Fig. 21), etc.

Also very promising is the field electron emission behavior of diamond films. Since the first observations and investigations of the low field electron emission properties of natural and chemical vapor deposited diamond in the early 1990s [65, 66], the interest in this field has steadily increased until the present day. Over 150 patents issued on diamond field emitters prove the big economic interest for that application [67]. The estimated annual turnover of 20 billion dollar for flat panel displays in the year 2002 may be the strong driving force to de-

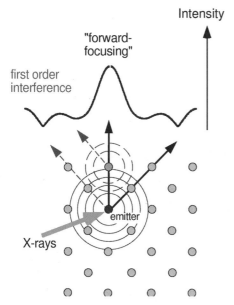

Fig. 18.   Schematic illustration of the forward focusing effect associated with photoelectron diffraction.

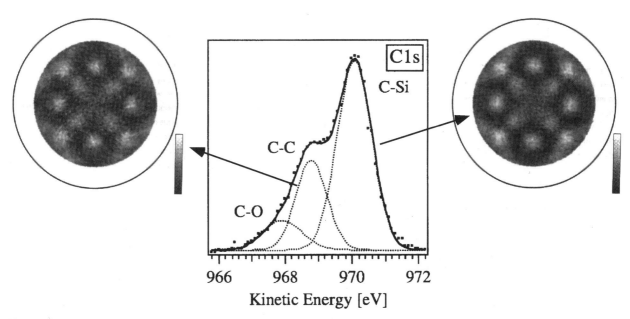

Fig. 19.   Fitted C1s XPS signal together with XPD patterns of the C—Si and C—C components of the silicon (100) surface after 8 min of bias-enhanced nucleation and 10 min deposition. Reprinted with permission from [60], © 1997, American Physical Society.

velop micrometer sized electron emission structures for which carbon thin films are good candidates.

Field emission is usually due to the tunnelling of electrons through the narrow surface potential barrier created by an intense electric field. In the case of a typical metal having a work function of 5 eV, fields in the range of 2500 V μm^{-1} are required to get detectable field emission currents (>1 nA). For technological applications such high fields are very difficult to generate on flat surfaces. In order to create sufficiently high fields for electron field emission one usually has to rely on the field enhancing effect of tiplike structures. The local electric field $\vec{F}$ at the apex of a tip exposed to an electric field $\vec{F}_0$ can be expressed by $\vec{F} = \vec{F}_0 \cdot \beta$, where β is the field enhancement factor. It depends on the tip geometry and on the tip orientation in the field. In the case of a needle shaped tip, with radius r and height h, parallel to the field $\vec{F}_0$, the field enhancement factor β is equal, in first approximation, to the aspect ratio h/r. Many research groups have shown that the electron field emission behavior of CVD diamond films depends dominantly on the surface morphology. Films exhibiting bad crystalline quality show the best electron emission properties, especially the lowest threshold field. Usually these are films with nanocrystalline structures which are characterized by a Raman spectrum showing a very weak or even absent 1332 cm^{-1} diamond line and by a strong feature of sp^2 bonded carbon (Fig. 22).

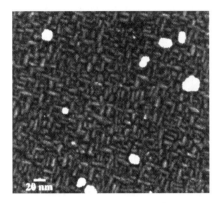

Fig. 20. HRSEM picture of the silicon surface after the bias-enhanced nucleation. The white spots are diamond crystallites. Reprinted with permission from [60], © 1997, American Physical Society.

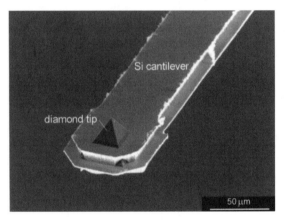
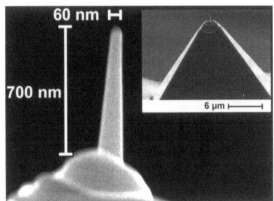

Fig. 21. CVD pyramidal diamond tip on silicon cantilever. The diamond tip has been modified by focused ion beam etching [64] (from CSEM "Scientific and Technical Report 1999" [57]).

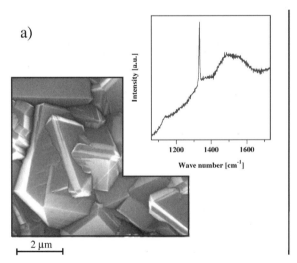
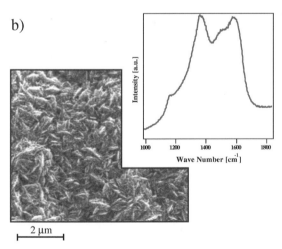

Fig. 22. SEM images and Raman spectra of: (a) weakly emitting CVD diamond grown at 850°C and 1% CH$_4$ in H$_2$, (b) good emitting CVD diamond grown at 950°C and 5% CH$_4$ in H$_2$. Reprinted with permission from [68]. Reprinted with permission from [69], © 1999, American Vacuum Society.

These films can be grown at relatively high substrate temperature $T_s > 950°C$ and high carbon precursor concentration (e.g., >3% CH_4).

The electron field emission phenomenon was first reported by Wood in 1897. In 1928 Fowler and Nordheim delivered the first generally accepted explanation of electron field emission in terms of the newly developed theory of quantum mechanics [70]. They treated the conduction electrons in the metal as a gas of free particles obeying the Fermi–Dirac statistic. The electrons are confined in a metal by the surface potential barrier, the shape of which is considered to be determined by the potential inside the metal (work function ϕ), the image charge potential, and the applied external potential as schematically illustrated in Figure 23.

The result of this quantum mechanic calculation is the well-known Fowler–Nordheim relation for the emission current density j_e,

$$j_e = \frac{e^3}{4(2\pi)^2 \hbar \phi} F^2 \exp\left(-\frac{4\sqrt{2m_e}\phi^{1.5}}{3\hbar e F}\right)$$

Fowler–Nordheim relation

where j_e is in $A\,m^{-2}$ and F is the field in $V\,m^{-1}$.

The Fowler–Nordheim relation shows that the emission current density depends only on the work function ϕ and the local electric field F present at the emission site. Because usually neither ϕ nor F is known it is impossible to determine these emission characterizing values from a simple I–V measurement. To get the second necessary equation to determine ϕ and F we measured the field emitted electron energy distribution (FEED). Figure 24 shows the electron energy distribution expected for the classical Fowler–Nordheim field emission. The energy distribution shows a peak centered at the Fermi energy E_F of the emitter. At 0 K temperature the peak would show a sharp cutoff toward the high energy side, showing the position of the emitter's Fermi level E_F. With increasing temperature this cutoff broadens due to the thermal excitation of electrons at the Fermi level into states of higher energy.

The low energy side shows the exponentially decreasing tunneling probability with decreasing energy due to the increasing width of the tunneling barrier. The transmission probability $D(E)$ for an electron with energy E to tunnel through a potential barrier is given in the WKB (Wentzel–Kramers–Brillouin) approximation by [71]

$$D_{\mathrm{WKB}}(E) = \exp\left\{-\frac{A_v}{F} + \left[C_v(E - E_F)\right]\right\}$$

transmission probability

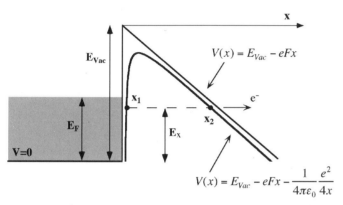

Fig. 23. Schematic illustration of the surface potential barrier $V(x)$ under the action of an external electric field.

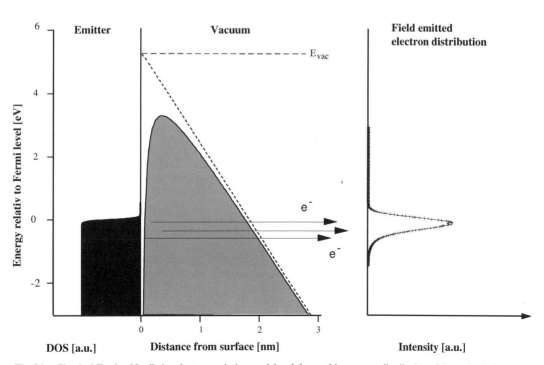

Fig. 24. Classical Fowler–Nordheim electron emission model and the resulting energy distribution of the emitted electrons.

a)

b)

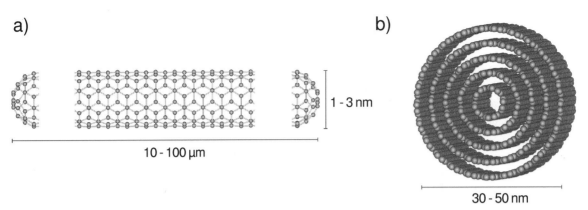

1 - 3 nm

10 - 100 μm

30 - 50 nm

Fig. 25.   Schematic illustration of (a) single-wall carbon nanotube, (b) multi-wall carbon nanotube.

with

$$A_v = \frac{4}{3}\sqrt{\frac{2m_e}{\hbar^2}}\frac{\phi^{3/2}}{F} \qquad C_v = 2\sqrt{\frac{2m_e}{\hbar^2}}\frac{\phi^{1/2}}{F}$$

From the Fowler–Nordheim relation j_e (I–V curve) and the tunneling transmission probability $D(E)$ (fit of the low energy side of the FEED) it is now possible to determine the work function ϕ, the local electric field F, and the field enhancement factor β at the emission site. With this approach we proved that the electron field emission of diamond and DLC films is classical Fowler–Nordheim tunnelling [68, 69, 72]. We demonstrated that the excellent field electron emission behavior of CVD diamond films is due to very high field enhancement at the film asperities. This explains why diamond films with nanocrystalline structures (Fig. 22b) show the best field electron emission behavior. For nanocrystalline diamond films we found typically a work function of 5 eV, characteristic for carbon allotropes, and field enhancement factors up to 800. The corresponding local electric field at the emission site is typically 2500 V μm^{-1}, which is equal to the required field for electron emission of flat metal surfaces. Since the excellent electron field emission behavior of CVD diamond is only based on a geometrical effect, it is evident that carbon nanotubes (see next section) with their needle shaped structure are optimal electron field emitters [73–75].

2.1.3. Carbon Nanotubes

Carbon nanotubes (CNT) are ultrathin carbon fibers with nanometer-size diameter and micrometer-size length and were discovered by Iijima in 1991 [76] in the carbon cathode used for arc-discharge processing of fullerenes (C$_{60}$). The structure of CNT may be viewed as enrolled cylindrical graphene sheets and closed by fullerenoid end caps. There are single-wall carbon CNT (SWCNT) and multiwall CNT (MWCNT), consisting of several nested coaxial single wall tubules. Typical dimensions of MWCNT are outer diameter: 2–20 nm, inner diameter: 1–3 nm, and length: 1–100 μm. The intertubular distance is 0.340 nm, which is slightly larger than the interplanar distance in graphite. Excellent reviews on the synthesis and the physical properties can be found in [77–79] (see Fig. 25).

CNTs can be prepared by arc evaporation [80], laser ablation [81], pyrolysis [82], electrochemical methods [83, 84], and also PECVD [75]. The CNT formation by PECVD and pyrolysis as well is a catalytic process using metallic catalysts. In the arc method metallic catalysts improve the production yield and control the diameter of the CNTs. The catalytic methods for making CNTs have their origin in the corresponding work on carbon fibers [85].

Baker and Harris [86] proposed a model for the catalytic filament growth where hydrocarbon is decomposed on the surface of the metal particle, producing hydrogen and carbon, which then dissolves in the metal. The dissolved carbon then diffuses through the particle, to be deposited on the trailing face, forming the filament. This model seems also to be applicable for general CNT growth (Fig. 26). However the growth mechanisms responsible for the structure type of the growing CNT (SWCNT or MWCNT) are still unknown.

For the PECVD technique we found that CNTs grow exactly at the same process conditions as used for diamond films. In the PECVD the substrate pretreatment is the only step which defines the structure of the carbon deposit (diamond or CNT). In the case of CNT a metallic catalyst has to be deposited. We performed sputter-coating (Ni, Fe) or spin-coating (Fe(NO$_3$)$_3$) for the catalyst deposition.

Figure 27 shows scanning electron microscope (SEM) pictures of CNT brush grown perpendicular to the substrate. The backscattered electron image (Fig. 27b) reveals small metal clusters (white spots) on top of the CNT bundle. This image supports emphatically that the growth model for carbon fibers proposed by Baker and Harris is also valid for CNTs. Figure 28b shows a TEM picture of a nanotube end with the encapsulated metal catalyst.

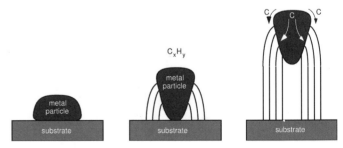

Fig. 26.   Proposed model for catalytic growth of CNTs according to the model of Baker and Harris [86] proposed for carbon filaments.

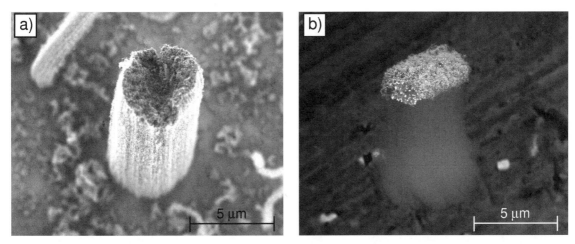

Fig. 27. SEM pictures of a CNT brush: (a) secondary electron image and (b) backscattered electron image. Catalyst: 0.0025 mol Fe(NO$_3$)$_3$ in ethanol, $T = 700°C$, 2% CH$_4$ in H$_2$, $p = 50$ mbar (from [87]).

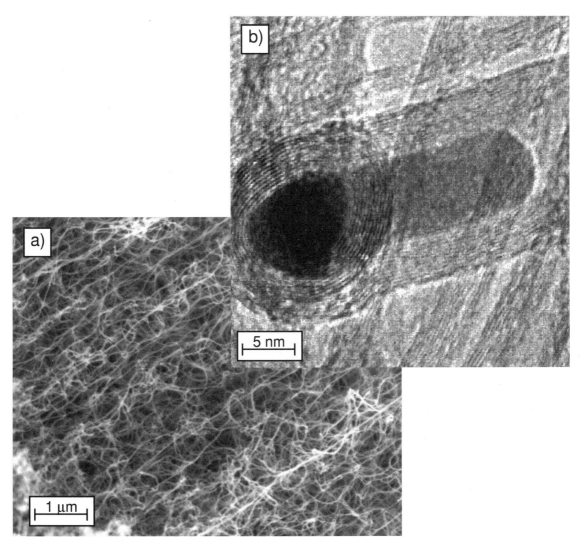

Fig. 28. (a) SEM image of nanotubes inside the CNT brush in Figure 27 (from [87]). (b) TEM image from the end of a CNT with the encapsulated metal catalyst (from Anke Weidenkaff, Universität Augsburg, Germany).

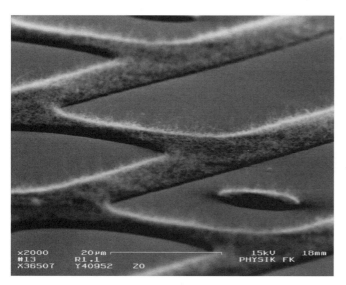

Fig. 29.   SEM image of patterned CNT film grown on a silicon wafer. The Fe(NO₃)₃ catalyst was deposited by microcontact printing. From a 0.01 mol Fe(NO₃)₃ ethanol solution (from [87]).

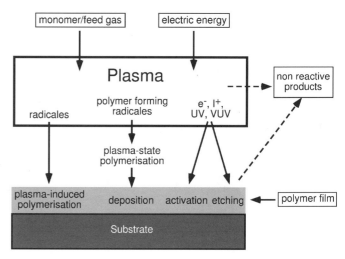

Fig. 30.   Reaction diagram of the plasma polymerization process according to Poll et al. [96].

The oriented growth of the CNT inside the brush is forced by the tube density. Above a critical density the direction of the growing nanotube is confined by the surrounding growing tubes. A zoom with the SEM into the CNT brush (Fig. 28a) shows straight as well as twisted nanotubes forming a kind of tissue. The straight tubes may have the lowest growth rate and may limit the growth rate of the whole brush by anchoring the "head" of the brush to substrate. This "head" (Fig. 27) is composed by clusters of the metal catalyst and probably by amorphous carbon. Fast growing nanotubes cannot get through this "head" and are forced to twist. This explains the constant height of the nanotubes in the brush.

Since the CVD process needs a catalyst for the CNT growth it is relatively easy to prepare patterned CNT films. It has been shown recently that patterned CNT can be prepared by standard lithographic techniques [88, 89]. We used microcontact printing to pattern the catalyst onto the substrate [90]. For this a polydimethylsiloxane stamp was first hydrophilized to considerably increase the affinity between the stamp and the metal catalyst–ethanol solution. The stamp was subsequently inked and after drying the catalyst was printed on a silicon wafer. Figure 29 shows a SEM image of patterned CNT film produced by the described microcontact printing technique.

2.2. Plasma Polymerization

2.2.1. Generals

The development of plasma polymerization processes began in the 1950s [91, 92] and since the 1960s plasma polymerization processes have been studied intensively [93–95]. Plasma polymerization is essentially a PECVD process. It refers the deposition of polymer films through reactions of the plasma with an organic monomer gas. Plasma polymerization is a specific type of plasma chemistry and involves homogeneous (Table I)

and heterogeneous (Table III) reactions; the former are reactions between plasma species and the latter are reactions between plasma and surface species and the surface species itself. The two types of reactions are often called "plasma-state polymerization" and "plasma-induced polymerization" [96]. Several kinetic models of plasma polymerization have been proposed. The most popular are the models of Yasuda [8], Poll et al. [96], and Lam et al. [97]. These models involve ablation and polymerization mechanisms in a competitive process (see Fig. 30).

The expression "plasma polymerization" is strictly spoken not correct because the process results in the preparation of a new type of material and is not a kind of polymerization in the classical sense. In contrast to conventional polymerization, based on molecular processing, the plasma polymerization is an atomic process with rearrangements of the atoms within the monomer. Consequently the materials formed by plasma polymerization are very different from conventional polymers. Its structures are more like highly crosslinked oligomers. This special structure makes plasma polymers generally chemically and physically different from conventional polymers. Plasma polymers are generally:

(1) chemically inert,
(2) very adherent to a variety of substrates including polymer, glass, and metal surfaces,
(3) pinhole-free,
(4) easily formed with thicknesses from microns down to nanometers, and as multilayer films (Section 2.1.1) or films with grading of chemical or/and physical properties.

Plasma polymers find applications as adhesion promoters for all kind of substrates, membranes for gas separation or vapor barriers, UV protection coatings, antireflection coatings, photoresist, antiadhesive film, etc.

Plasma polymerization is a field somewhere between physics and chemistry. The choice, excitation, and characteriza-

a) 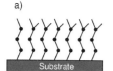 b) 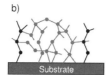 c)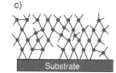

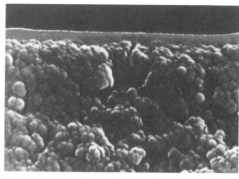

Fig. 31. Schematic illustration of the different film structures of plasma polymers: (a) grafted monomers, (b) low crosslinked (polymerlike), (c) highly crosslinked, compact (hard coating).

Fig. 32. Polyethersulphone membrane with a 0.4 μm thick plasma polymer coating for permeability selective gas separation. Reprinted with permission from [99].

tion of suitable plasmas presume a well-founded knowledge of plasma physics, but their applications to relevant technical problems require the experience and intuition of chemistry. Systematic investigations on a large scale of plasma polymerization began in the 1970s on three classes of monomer, hydrocarbons, fluorocarbons, and siloxanes.

As mentioned in Section 2.1.1 the plasma polymerization process allows formation of films with a variety of physical and chemical properties, starting from the same monomer, only by changing the plasma parameters. In general the energy flux into the growing plasma polymer film is decisive for the structure and the properties of the film. With increasing energy flux the film usually becomes harder, more disordered, and crosslinked with reduced hydrogen content. In opposition a low energy flux causes a plasma polymer that retains more the molecular structure of the monomer. The approach to create soft plasma polymers instead of hard coatings in the PECVD process is the minimization of the plasma–precursor and plasma–film interactions. Summarized, the following plasma conditions may aspire to achieve enhanced control of the plasma polymerization chemistry [98]:

(1) minimization of the W/FM parameter, called the *Yasuda parameter* [8] (W: power coupled into the plasma; F: flow rate; M: monomer molecular weight),
(2) use of monomers with polymerisable double bonds,
(3) use of a Faraday cage around the substrate,
(4) sample position downstream from the plasma zone,
(5) use of cold substrate,
(6) pulsed plasma excitation to reduce the plasma on time (periodicity $T \sim 10^{-3}$ s).

It is obvious that precautions (3)–(6) reduce the demanded plasma–precursor and plasma–film interactions. The W/FM parameter is a value associated with the molecular weight of the polymer forming radical. It contains the dissociation probability, represented by the plasma power P, the dwell-time of the monomer in the plasma zone, represented by the flow rate F, and the molecular weight of the precursor M. The smaller the W/FM parameter, the more the plasma polymer retains the chemical structure of the precursor. Figure 31 shows schematically the different film structures which can be created by plasma polymerization.

Plasma polymerization conditions with very weak plasma interactions lead to the formation of an ultrathin film of grafted monomers (Fig. 31a). Such films are of great interest in biosensor surface engineering. Adsorption is an ubiquitous phe-

nomenon that occurs when molecules and materials meet at surfaces. In biosensors applications, the nonspecific adsorption of biomolecules may lead to reduced sensitivity or to a loss of sensitivity with time. Grafted monomer films can be used as spacer layers for specific immobilization of biomolecules. Crosslinked plasma polymers show excellent gas barrier properties, which make them interesting as membranes for permeability selective gas separation (Fig. 32).

This application of plasma polymers was discussed in great detail by Nomura et al. [100] who have described the gas separation behavior of a variety of plasma polymers. They found that the separation of H_2/N_2 and H_2/CO_2 is molecular sieve type, i.e., based on difference in molecular size, and for O_2/N_2, CO_2/CH_4 solution-diffusion type separation, whereby sieve type separations were more effective. The permeability ratio for the above mentioned gas mixtures is between 3 and 50 [100–102]. In highly crosslinked plasma polymers (Fig. 31c) the chemical structure of the precursor is no longer retained. The chemical and physical properties of these films differ completely from those of conventional polymers. The films are generally chemically inert, compact, pinhole-free, and hard. Well known examples of such films are diamondlike carbon synthesized from hydrocarbons discharges, SiO_2 and Si_3N_4 synthesized from silane, and siloxane plasmas. Such films are used to impart abrasion resistance to softer, especially optical, substrates. An excellent review on plasma polymerization of organosilicones has been given by Wróbel and Wertheimer [8].

A promising application of plasma polymers is as solid lubricant or antiadhesive film in surface micromachining. Surface micromachining, defined as the fabrication of micromechanical structures from deposited thin films, is one of the core technologies underlying microelectromechanical systems (Fig. 33), which promises to extend the benefits of microelectronic fabrication technology to sensing and actuating functions. Such microstructures typically range from 0.1 to several μm in thickness, with lateral dimensions of 3–300 μm, and lateral and vertical gaps to other structures or to the substrate of around 1 μm. The large surface area and the small offset from adjacent surfaces makes microstructures especially vulnerable to adhesion upon contact.

The adhesion of microstructures to adjacent surfaces can occur either during the final steps of the micromachining process (release-related adhesion) or after packaging of the device, due to overranging of input signals or electromechanical instability (in-use adhesion). The causes of strong adhesion can be traced to the interfacial forces existing at the dimensions of microstructures. These include capillary, electrostatic, van der Waals, and chemical forces. An effective treatment of microstructures to reduce stiction must provide a hydrophobic, conductive surface with a low surface energy in order to avoid electrostatic forces and the formation of water layers on the surface, thereby eliminating van der Waals and capillary forces altogether. Plasma polymers in the form of grafted monomers or ultrathin polymerlike films [103] are effective to reduce stiction. The Centre Suisse d'Electronique et de Microtechnique (CSEM) used such a film to prevent stiction in an electrostatic microshutter array for high-speed light switching (Fig. 34).

Each microshutter consists of a shutter blade, a suspension beam, and several electrodes and stoppers. The microshutter is

a typical bistable microelectromechanical device where the active position of the movable shutter is confined by mechanical stoppers. These stoppers are exposed to very large mechanical loads by adsorbing the momentum and the kinetic energy of the shutter. It was observed that after millions of cycles shutters could become immobilized against the stoppers. A solution for this in-use stiction problem is a hydrophobic antistiction coating of the shutter and stopper walls.

2.2.2. Plasma Polymerized Fluorinated Monomer Coatings

As already mentioned, the plasma polymerization involves deposition and ablation mechanisms in a competitive process. This behavior can be observed particularly well in fluorocarbon sustained discharges. Pioneering work on fluorocarbon sustained discharges was done by Coburn and Winters [104] in the late 1970s.

They found that fluorocarbon sustained discharges can be operated in overall polymerization or etching conditions depending on the F/C ratio of the precursor (Fig. 35). Typically, a high F/C ratio of the precursor, e.g., CF_4, leads to efficient etching, whereas a low F/C ratio, e.g., C_2F_4, favors easy polymerization.

The ability to tune the fluorocarbon sustained discharge for polymerization or etching demands the presence of three competitive active species in the discharge, namely CF_x radicals as reactive fragments for the polymer deposits, the F atoms to trigger the etching by forming volatile fluorides and charged particles. The excess of one of these species defines the polymerizing or etching behavior of the discharge. If we consider the CF_2 radicals as the main polymerizing species then the characteristic of the fluorocarbon discharge can be simply determined by the corresponding chemical equation:

$$CF_4 \xrightarrow{plasma} CF_2 + 2F \qquad \text{etching condition}$$
$$C_2F_6 \xrightarrow{plasma} 2CF_2 + 2F \qquad \text{boundary condition}$$
$$C_4F_{10} \xrightarrow{plasma} 4CF_2 + 2F \qquad \text{polymerizing condition}$$

Fig. 33. SEM image from a microelectromechanical system (from Sandia National Lab Server (http://www.sandia.gov)).

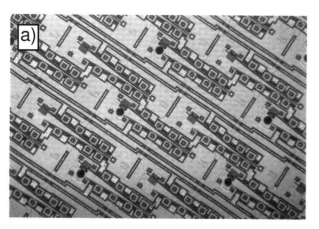
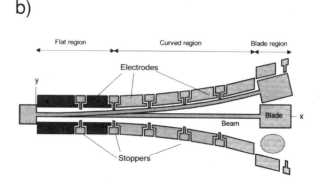

Fig. 34. Electrostatic microshutter array in polysilicon. (a) Close-up picture of the shutter array. (b) Layout of the shutter design (length 100 μm) (from CSEM "Scientific and Technical Report 1999" [64]).

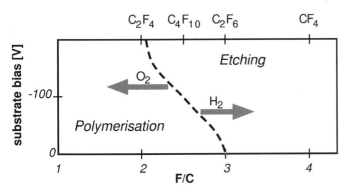

Fig. 35. Boundary between polymerization and etching conditions as a function of the fluorine to carbon (F/C) ratio and the negative substrate bias. Additional oxygen displaces the boundary to lower F/C ratio, hydrogen to higher F/C ratio (according to Coburn and Winters [104]).

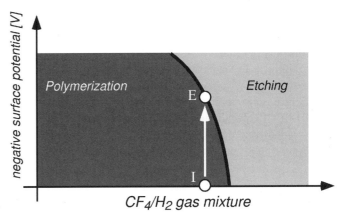

Fig. 36. Schematic illustration of the a "self-thickness-limited" plasma polymerization process. Reprinted with permission from [103], © 1996, American Vacuum Society.

Therefore, a better criterion for characterizing the plasma behavior is the ratio of the active species $[F]/[CF_x]$ rather than the F/C ratio of the precursor. The active species in excess defines the etching or polymerizing behavior of the plasma. The relative ratio of the active species can be varied by changing precursor gas or by varying the type and the amounts of additive gases. Oxygen as additive gas increases the etching behavior of the fluorocarbon discharge, while the addition of hydrogen increases the polymerization behavior. In fact, hydrogen reacts with F atoms, leading to unreactive HF resulting in a reduced etching activity. Obviously similar behavior can be obtained with a precursor containing hydrogen, such as CHF_3. On the other hand, active carbon sites can be saturated by O atoms and converted to inactive CO, CO_2, and COF_2. F atoms are no longer recombined and their content therefore increases, resulting in a higher etching activity.

Positively charged particles are the third active species influencing the characteristic of the fluorocarbon discharge. These particles are accelerated to the substrate and affect etching and heterogeneous polymerization processes. With increasing particle energy (substrate bias) the etching process is enhanced, as indicated by the bias dependence of the polymerization/etching boundary in Figure 35.

We take advantage of the behavior illustrated in Figure 35 to develop a "self-thickness-limited" plasma polymerization process to create an ultrathin antiadhesive film [103]. With the process we were able to produce films with an extremely low surface energy of 4 mJ m^{-2} and 5 nm thickness, autocontrolled by the plasma. The developed deposition process takes place in two steps, a growth and a treatment step. With the CF_4/H_2 gas mixture used, we tuned the plasma conditions such that the polymerization starts close to the polymerization/etching boundary (Fig. 36, point I). Due to the mobile electrons in the plasma the growing insulating polymer film becomes negatively charged until it reaches the polymerization/etching boundary (Fig. 36, point E). At this point, the etching process balances the polymerization process and the deposition changes from the growth to the treatment step. During the growth step ($t \leq 2$ min) the film is characterized by small CF_2

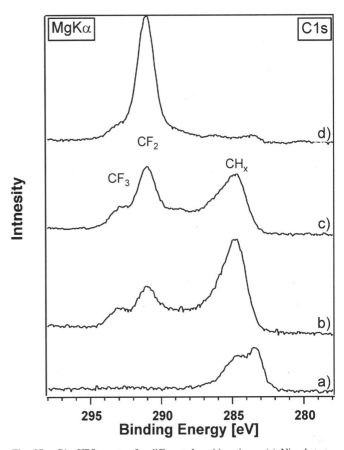

Fig. 37. C1s XPS spectra for different deposition times. (a) Ni substrate, (b) 2 min, (c) 6 min, (d) 10 min. Reprinted with permission from [103], © 1996, American Vacuum Society.

and high CH_x content (Fig. 37b). In this phase of the deposition process the antiadhesive property of the film is not extraordinary but the adhesion to the Ni substrate is excellent. During the treatment step the CH_x groups at the film surface are transformed to CF_2 groups (Fig. 37c and d). At the end of the deposition process the plasma polymer film contains 85% CF_2 groups (Fig. 38) and shows the extremely low surface energy of 4 mJ m^{-2}, which is 4 times lower than that of Teflon®. The excellent hydrophobic behavior of this film is demonstrated in Figure 39 by a water droplet on a Swiss coin covered with such a plasma polymer film.

The film was tested as an antiadhesive coating for the replication of micro-optical structures, using a Ni shim with a very fine surface relief grating (400 nm periodicity) to hot emboss ($T = 180°C$) polycarbonate. The tests demonstrate the excellent antiadhesive property of the film against the polycarbonate and reveal a good adhesion of film with the Ni shim [105, 106].

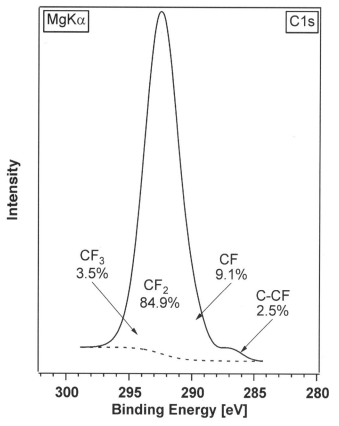

Fig. 38.   Deconvoluted C1s XPS spectrum of the PPFM film. Reprinted with permission from [103], © 1996, American Vacuum Society.

2.3. Surface Treatments

The most widespread industrial application of cold plasmas is in cleaning and treatment of surfaces, particularly of polymer surfaces. Plasma treatment of polymer surfaces to increase their adhesive properties regarding painting, printing, or metallization is nowadays a well established and successful processes in the automobile, food packaging, capacitor (Fig. 40), and electronic industries.

The images in Figure 41 show two applications of metallized polymer foils where cold plasma treatment of the polymer is used to improve the adhesion to the metal film. One example is Al/Zn metallized ($d = 15$ nm) polypropylene (PP) for high energy density capacitors [108]. Such capacitors are used in high power electronics. On the new Swiss high power locomotive of the "type 465," 3000 kg or 4000 km of this segmented metallized polymer foils ($d = 15$ μm, $b = 15$ cm) are installed as capacitors. Compared to the conventional high power capacitor technology where metal and paper sheets are wound to a capacitor, the new technology with the segmented

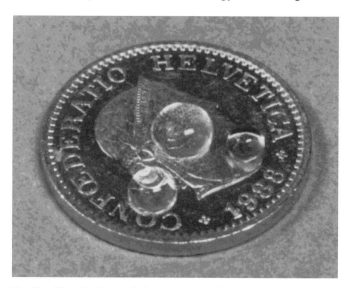

Fig. 39.   Water droplet on a Swiss coin covered with a 5 nm thick antiadhesive plasma polymer film. Reprinted with permission from [103], © 1996, American Vacuum Society.

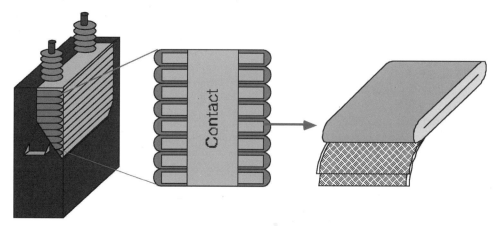

Fig. 40.   Schematic drawing of the high energy density capacitor manufactured by Montena SA, CH-1728 Rossens, Switzerland. Reprinted with permission from [107].

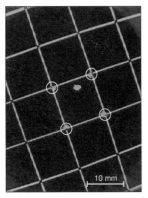

10 mm

100 μm

Fig. 41. On the left: mosaic structure of Al/Zn metallized PP used as high energy density capacitors. The picture illustrates the self-healing effect of the mosaic structure after an electrical breakdown in a segment. Reprinted with permission from [107]. On the right: laser microstructured test device of Cu metallized PI for the application as a flexible electronic circuit board.

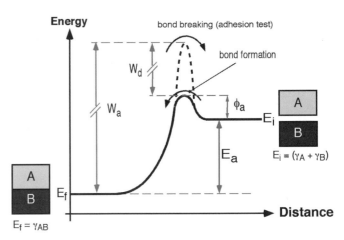

Fig. 42. Illustration of the two different notions of the "adhesion."

electrode metallized on PP allows one to increase the energy density by a factor of more than two. The segmentation of the electrode is important for the self-healing effect in the case of a breakdown through the dielectric (PP), which makes the capacitor defect-tolerable and safe. In case of such a breakdown in one of the segments the small current gates interconnecting the segments serve as fuses; they isolate the segment and therefore the breakdown from the rest of the electrode [109].

The other example is Cu metallized polyimide (PI) which will be used as microstructured flexible electronic circuit boards. The patterning is made by laser delamination irradiating the metallized PI with UV light through a mask.

In the domain of surface cleaning the cold plasma has been established as the cleaning technique for the nanometer scale. In comparison to wet chemical processes cold plasmas are environmentally friendly and cheaper, and a higher cleaning degree of the surface can be achieved.

2.3.1. Plasma Treatment to Improve Adhesive Properties

The notion of adhesion can be misleading, because it refers to several meanings depending on the context [110, 111]. In physics or chemistry, the adhesion is directly connected to the intermolecular forces acting across an interface. From a technological point of view, the adhesion corresponds to the mechanical force or energy that is required to separate two bodies in intimate contact. The two different meanings of "adhesion" are illustrated in Figure 42.

When two bodies are brought into intimate contact (path: "bond formation" in Fig. 42) a new interface is formed at the expense of two free surfaces. The nature of the interaction at the interface determines the adhesion energy (E_a). Thermodynamically, this energy is defined as the difference between the energy of the initial state (E_i), when the two bodies are separated, and the final state (E_f), when they are in intimate contact. The adhesion energy (E_a) is then, according to Dupré,

$$E_a := E_i - E_f = (\gamma_A + \gamma_B) - \gamma_{AB} \qquad \text{Dupré equation}$$

Here, γ_A, γ_B are the surface energies of the bodies A and B, and γ_{AB} is the interface energy of the two bodies in contact. From the Dupré equation it is clear that the adhesion energy E_a is in the order of the surface energies of the bodies, also mJ m^{-2}. In this thermodynamic sense the adhesion is a pure interface phenomenon determined by a few monolayers of atoms at the interface. Apart from the formation of a mechanical bond we may examine the process to break this bond (path: "adhesion test"). While the adhesion energy from the Dupré equation is on the order of mJ m^{-2} the energy W_a measured in an adhesion test amounts to several hundreds of J m^{-2}. The reason for this huge difference in the adhesion energy is the deformation energy W_d, which has to be brought up in all adhesion tests. This indicates that the measurement of the practical adhesion is actually a superposition of adhesive and elasticity properties at the interface.

The following factors or theories contribute to mechanism of adhesion:

- mechanical interlocking,
- diffusion theory,
- electronic theory,
- adsorption theory.

2.3.1.1. Mechanical Interlocking

This theory associates adhesion to mechanical interlocking around the irregularities at the substrate surface. However, the effects of topography on adhesion are much more complex. The potential bonding area of a rough surface is clearly greater than for a smooth surface. However, the surface topography has to be appropriate. For example, if the viscosity of an adhesive is high, the irregularities on the surface have not to be deep and narrow, because then the wetting may be far from complete. The Washburn equation, which describes the time t for a liquid to move a distance z in a capillary of radius r, gives an indication of the wetting rate for irregularities by the adhesive:

$$t = \frac{2\eta}{\gamma \cos\theta} \frac{z^2}{r} \qquad \text{Washburn equation}$$

Here η and γ are the viscosity and surface tension of the adhesive, respectively. θ is the wetting angle of the adhesive on the substrate. The wetting rate may therefore be much lower for small capillaries and for low energy surfaces such as for nonpolar polymers.

2.3.1.2. Diffusion Theory

The intrinsic adhesion between polymers can be affected by the mutual diffusion of polymer molecules across the interface. But this requires that the macromolecules or chain segments of the polymers possess sufficient mobility and are mutually soluble. The latter requirement may be restated by the condition that they possess similar values of solubility parameters. The condition of similar solubility parameters is only valid if the polymers are amorphous. If the polymer possesses a significant degree of crystallinity then the free energy of crystallization makes it more resistant to form a solution and the diffusion process can be neglected for the adhesion.

2.3.1.3. Electronic Theory

If the two bodies (A and B) in contact have different electronic band structures it is likely to have some electron transfer on contact to balance the Fermi levels: This will result in the formation of a double layer of electrical charges at the interface. Deryaguin's theory essentially treats the interface between two bodies as a capacitor where the electrostatic pressure p_C of this double layer capacitor corresponds to the adhesion force:

$$p_C = \frac{\varepsilon_0}{d^2}\left(\frac{\phi_A - \phi_B}{e}\right)^2 \quad \text{electrostatic pressure}$$

$$\frac{\phi_A - \phi_B}{e} =: V_C \quad \text{contact voltage}$$

$$\phi_A, \phi_B \quad \text{work functions}$$

$$e \quad \text{elementary charge}$$

$$d \quad \text{charge separation}$$

For a contact potential of 1 V and a charge separation of 0.5 nm the electrostatic pressure p_C is 20 MPa, which is only a fifth of the pressure created by van der Waals forces under the same conditions. The effect of charge transfer at the interface on the adhesion is secondary at best.

2.3.1.4. Adsorption Theory

The adsorption theory of adhesion is the most widely applicable theory. It proposes that, if sufficiently intimate molecular contact is achieved at the interface, the materials will adhere because of the interatomic forces, which are established between the atoms at the interface of the bodies in contact. The strength of these bonds and their range can be estimated using classic approaches of electrostatics and electrodynamics if the spatial structure and distribution of molecules and atoms are known. The adsorption theory allows the definition of the thermodynamic work of adhesion E_a (Dupré equation) required to create two surfaces by breaking a solid.

Nowadays it has become generally accepted that while the adsorption theory has the widest applicability, each of the other theories may be appropriate under certain circumstances and often make a contribution to the intrinsic adhesion forces which are acting across the interface. Therefore, the following well known basic requirements have to be considered for good adhesion:

- absence of weak boundary layers,
- good contact between the bodies,
- avoidance of stress concentration at the interface which could lead to disbonding.

2.3.2. Surface Cleaning

Maybe the most important requirement for good adhesion is that the weak boundary layer can be removed. In some cases one has to prevent the formation of a weak boundary layer. In practice, it cannot be considered that the bulk structure of the solid is preserved at the surface. Technical solids show in general a surface structure with a reaction layer and a contamination layer on top, as sketched in Figure 43. These layers can act as a weak boundary layer and prevent adequate adhesion. For this reason it is very important to proceed to a surface cleaning process to remove the contamination from these layers, and if necessary also the oxide layer, to obtain optimal adhesion.

The crucial effect that the contamination layer may have on adhesion and tribology will be shown here using the example of the thermosonic wire bonding process.

2.3.2.1. Thermosonic Wire Bonding

Thermosonic wire bonding is the most widely used assembly technique in semiconductors to connect the internal semiconductor die to the external leads (Fig. 44). To prevent high cost and unreliable microelectronic devices wire bonding failures must be eliminated.

The microelectronic industry accepts a failure rate of 1 ppm for the wire bonding process. Approximately 27% of failures in microelectronic devices are caused by wire bonding failure [112]. Therefore the wire bonding process is the most expensive step in microelectronic packaging. The continuous increase in microelectronic packing density demands a permanent

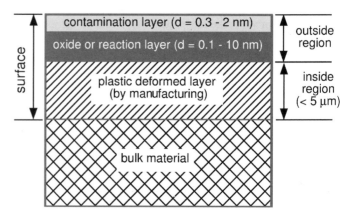

Fig. 43.   Schematic illustration of the surface structure of a technical surface.

(a)

(b)

Fig. 44. (a) Capillary with the bond wire and the microelectronic device. (b) Wire bond on 80 μm large contact (from ESEC SA, CH-6330 Cham, Switzerland).

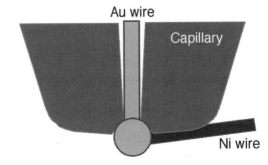

Fig. 45. Thermocouple in the bondability analyzer (diameter of Au and Ni wire 25 μm) (from [107], reprinted with permission).

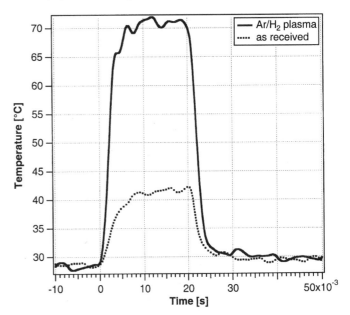

Fig. 46. Thermosignals of the bondability analyzer performed on an Ag plated Cu lead frame before and after Ar/H_2 plasma cleaning (friction time 20 ms). Reprinted with permission from [107].

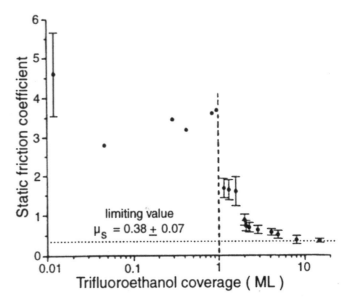

Fig. 47. The static friction coefficient μ_s as a function of triflouroethanol coverage on Cu(111). Reprinted with permission from [118], © 1995, American Chemical Society.

progress in wire bonding, which is only possible by a detailed understanding of the physics of the bonding process.

The thermosonic wire bonding process begins with a gold ball at the end of the wire, centered within the inside chamfer of the bonding capillary. During the bonding process the gold ball is vibrated at the ultrasonic frequency over the bond pad. The heat generated by this friction process causes microwelding at the interface [113], yielding to a stable mechanical and electrical bond. Since the bonding process is a kind of welding process it is crucial to dissipate maximal friction heat Q at contact surface. This demands a maximal friction coefficient μ. Several investigations have shown that the wire bond quality of Au and Ag contacts is determined decisively by the organic contamination layer on the bond pad [114–116]. One origin of the organic contamination is the die bonding process, where the die is hot glued on the substrate. During this process organic material can be evaporated and deposited on the contacts. From XPS and pull force measurements, we found that good bond contact quality is only achieved if the organic contamination layer on the bond pad is thinner than 4 nm.

To study the influence of the contamination on the tribosystem of the thermosonic wire bonding process we developed an apparatus which allows one to characterize the bonding behavior of the bond pad and to study bonding process in detail and in real time [117]. The principle of the developed bondability analyzer is the measurement of the temperature at the bond ball during the thermosonic wire bonding process by using a thermocouple instead of the bond wire (Fig. 45). The temperature at the bond ball is directly correlated to the dissipated heat during the bonding process. With the bondability analyzer we found that the thin contamination layer acts as a very efficient lubricant. Removing the contamination layer by plasma cleaning increases the friction coefficient μ and consequently the dissipated heat by a factor of four, as indicated by the thermosignals in Figure 46. This result absolutely conforms to measurements of the static friction coefficient μ_s of Cu(111) sliding on Cu(111) (Fig. 47), both surfaces covered with trifluoroethanol (CF$_3$CH$_2$OH) [118]. The measurements show that below monolayer coverage the molecular adsorbates have no influence on the static friction coefficient μ_s, while a sharp drop in μ_s occurs at monolayer coverage over a thickness of 10 monolayers. Systematic investigations have shown that plasma cleaning (Ar/H$_2$ plasma) of bond pads improves the pull force and reproducibility of thermosonic wire bonds significantly. The cleaning process of Ag contacts by Ar/H$_2$ plasma is described in Section 2.3.4. Plasma cleaning of bond pads of different materials (Cu, Ag, Au) to improve the quality and the reproducibility of wire bond contacts is nowadays a successful and well established process in IC packaging.

2.3.3. Plasma Treatment of Polymers

Owing to the low intrinsic adhesive properties of polymers, many applications of these materials require a surface pretreatment. Therefore, surface pretreatments of polymers in order to enhance their adhesive properties are of great technological importance. Plasma treatment of polymers is one of the more recent methods to achieve improved adhesion in systems such as polymer–adhesive [119, 120] of polymer–metal interfaces [121]. The merits of plasma surface treatment to improve the macroscopic adhesion strengths are generally accepted. Different effects of the plasma treatment on polymers are known. They include cleaning by ablation of low-molecular-weight material, activation of the surface, dehydrogenation, change in surface polarity and wetting characteristics, crosslinking and chain scission, and structural modification. These effects depend very much on the considered polymer and plasma used.

2.3.3.1. Chemical and Structural Modifications of Surfaces

Plasma treatment is a very efficient process to increase the surface energy γ, i.e., wettability of polymers, dramatically. After plasma treatment polymer surfaces can easily show water contact angles ϑ smaller than 15°, as illustrated in Figure 48. The increase of the surface energy γ of polymers by plasma treatment is related to the surface cleaning and the formation of polar bonds at the surface. For example, O$_2$ plasma is very effective to increase the surface energy of polyolefines as PP or polyethylene (PE) by forming oxygen functionalities (Fig. 49). The oxidation of PP ($-$CHCH$_3$–CH$_2-$) in the O$_2$ plasma occurs via dissociation of the methyl group ($-$CH$_3-$) [122], as indicated by the relative intensity decrease of the associated peak in the valence band (Fig. 50).

As discussed at the beginning of this section the practical adhesion W_a is related to the thermodynamic work of adhesion E_a defined by the Dupré equation. Under the assumption of an ideal surface treatment (bulk properties not affected), the practical adhesion is completely determined by thermodynamic work of adhesion E_a, i.e., the surface energies γ_A, γ_B of the two bodies and the interface energy γ_{AB} of the bodies in intimate contact. Therefore it makes sense to maximize the surface energy γ of the bodies to get optimal practical adhesion. In reality the plasma treatment of polymers is not limited to the topmost surface atoms and is therefore not an ideal surface treatment. For this reason plasma treatment on polymers has to be carried out very carefully. Plasma "overtreatment" is probably the most frequent cause for adhesive failures of metal films on polymers.

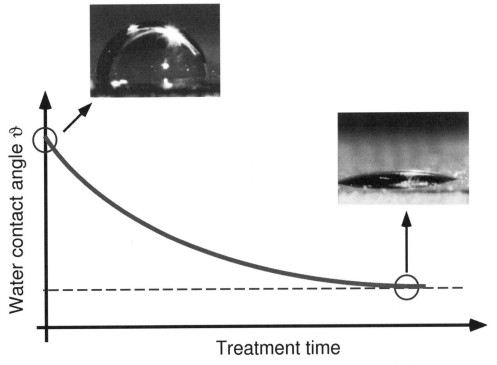

Fig. 48.   Schematic illustration of the surface energy γ of a plasma treated polymer as a function of the plasma exposure time.

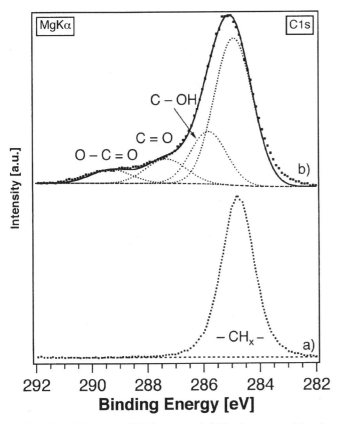

Fig. 49. C1s XPS spectra of PP (a) as received, (b) O_2 plasma treated ($t = 5$ s, $p = 1 \times 10^{-2}$ mbar). Reprinted with permission from [123].

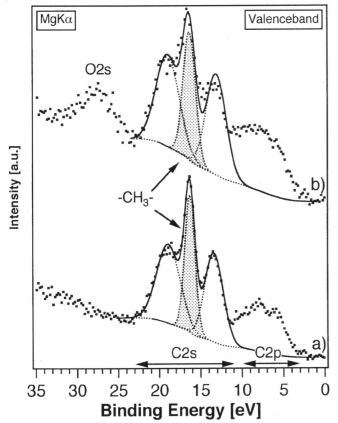

Fig. 50. XPS valence band spectra of PP (a) as received, (b) O_2 plasma treated ($t = 5$ s, $p = 1 \times 10^{-2}$ mbar). Reprinted with permission from [123].

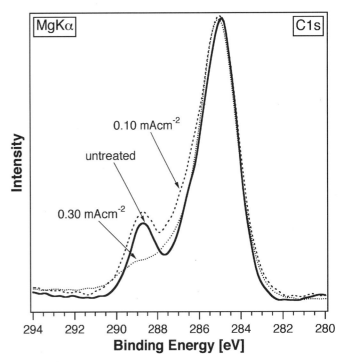

Fig. 51. Illustration of the side-chain scission in PMMA due to low energy ions. Reprinted with permission from [124], © 1995, Elsevier Science.

Fig. 52. C1s XPS spectra of PMMA, untreated and after O_2 plasma treated with to different ion fluxes. Reprinted with permission from [124], © 1995, Elsevier Science.

The susceptibility of a polymer on structural changes induced by plasma treatment depends strongly on its chemical structure. For example, polymethylmethacrylate (PMMA) is very susceptible for structural changes during plasma treatment especially in plasma with high ionization degree [124]. Positive ions from the plasma react and neutralize at the partially negatively charged carbonyl oxygen (Fig. 51). After that, the created electron hole in the carbonyl group becomes filled by an electron transfer from the neighboring nonpolar C–C bond, cleaving side chains from the polymer backbone. In noble gas plasmas, where no incorporation of chemical active species is possible, all side chains within a thickness of 10 nm are removed after few seconds of plasma treatment. The ion induced degradation process is in competition to the aspired formation of polar bonds and must be suppressed.

The C1s XPS spectra in Figure 52 illustrate the effect of the ion flux during O_2 ECR plasma treatment on the chemical composition of the PMMA surface. The ion flux was varied by changing the position of the sample relative to the plasma zone.

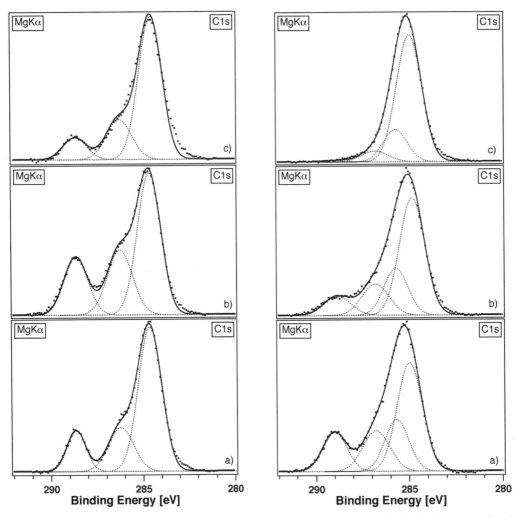

Fig. 53.   Deconvoluted C1s XPS spectra of PMMA (left) and PET (right): (a) as received, (b) O₂ plasma treated, and (c) Ar plasma treated. Reprinted with permission from [126], © 1999, American Institute of Physics.

In this way the influence of the UV/VUV radiation is the same for both treatments and differences in surface modifications are correlated to the ion flux.

Polymers with delocalized electrons (phenyl ring) in their chemical structure are generally less susceptible to ion induced damage [125–127]. This is because electron transfer from the delocalized electrons system to electron hole, created by the ion neutralization, can occur without breaking a bond. This is the reason plasma treatment on polyethyleneterephtalate (PET) is in general very successful, while it is very difficult on PMMA. The C1s XPS spectra in Figure 53 illustrate the type of chemical modifications on the PMMA and PET surface undergoing O₂ and Ar ECR plasma treatment. The spectra reveal as already mentioned the etching of polar oxygen groups indicated by the lower intensity of the corresponding peaks (peaks located above $E_B = 285$ eV). The etching of oxygen functionalities is much stronger on the PMMA surface than on the PET surface, indicating that the plasma reaction on the polymer surface depends on the whole chemical structure of the monomer.

The stability of a polymer in the plasma is primarily determined by the response of the polymer on the plasma induced damage and less on the primary damage itself. Generally, surface modifications of polymers by plasma are always accompanied with degradation reactions of the polymer in the near surface layer. The thickness of the modified surface layer is typically 10 nm. This plasma modified surface layer is initially in a highly excited state and likes to relax. In the presence of a reactive metal film (adhesive layer, e.g., Cr) the functional groups introduced by the plasma tend to segregate to the metal, forming a covalent bond [123]. This process becomes equal to a phase separation process between the plasma modified surface layer and the bulk polymer, resulting in the formation of a weak boundary a few nanometers underneath the surface. Plasma pretreated metallized polymer foils show therefore usually cohesive fractures in peel strength measurements. Gong et al. [128] studied the effect of sticker groups (−COOH) on the fracture energy of Al–cPBD–Al interfaces (cPBD: carboxylated polybutadiene (PBD)) and found that a small amount of sticker groups improves the fracture energy considerably. They found a crit-

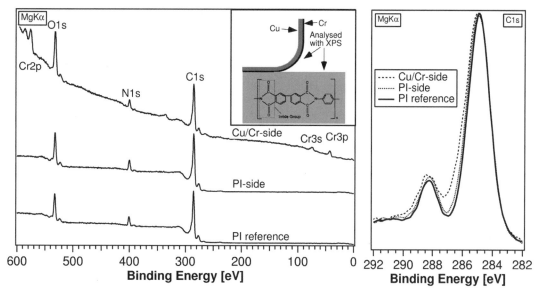

Fig. 54. XPS spectra from the fracture surfaces from peel test fragments of Cu metallized polyimide (UPILEX) with a Cr adhesive layer peel strength of $12\,\mathrm{N\,cm^{-1}}$. Reprinted with permission from [123].

ical concentration around 3 mol% to give a maximum bond strength. They showed that the sticker groups tend to segregate to the metal surface, resulting in a large concentration gradient of the sticker groups in the lattice layer close to the metal substrate. This investigation suggests that plasma modified polymers are very susceptible to reorganization processes, which may limit the cohesion in a few nanometers thick surfaces of the polymer, as seen above. These processes may be important for the aging behavior of the plasma modified polymer/metal interface.

Because the plasma induced phase separation process occurs a few nanometers underneath the polymer/metal interface, surface sensitive techniques must be used to detect this effect, as shown in Figure 54 for Cu metallized PI. The small Cr signals (adhesive layer) in the XPS spectrum of the "Cu/Cr-side" peel test fragment indicate the cohesive failure in the polymer. The identical C1s spectra of the "PI-side" fragment and the untreated PI and its difference to the spectra of the "Cu/Cr-side" fragment reveals that the cohesive fracture occurs along the interface of the plasma modified surface layer and the bulk polymer.

To avoid phase separation processes between the plasma modified polymers and the bulk polymers degradation reactions must be minimized in the plasma treatment.

As discussed, plasma surface modifications on polymers are mainly due to the formation of functional groups and degradation reactions. Radicals in the plasma remove hydrogen atoms on the polymer surface to form carbon radicals. Successively, these radicals combine with other radicals in the plasma to form new functional groups. On the other hand electrons and ions bombard the polymer surface making chain scission to form carbon radicals at the end of the polymer fragments. This may result in degradation reactions of the polymer chain to yield degradation products with low molecular weight. These low

Table IV. Reaction Constants of the Dominant Recombination Processes in an O_2 plasma [129]

Reaction	Rate constant k
$e + O_2^+ \rightarrow O^* + O$	$<10^{-7}$ (cm^3 s^{-1})
$e + O^+ \rightarrow O^*$	$<10^{-7}$ (cm^3 s^{-1})
$2O + O_2 \rightarrow 2O_2$	2.3×10^{-33} (cm^6 s^{-1})

molecular weight polymer chains are then subjected to reorientation at the polymer/metal interface, which may result in the formation of a weak boundary layer. To escape from these degradation effects the plasma parameters must be tuned in the direction that the charged particle concentration in the treatment zone is minimal so that radicals alone are predominantly used as active species for the surface modification. This is possible because the lifetime of radicals in the plasma is generally much longer than those of the electrons and ions (Table IV). In addition electron–ion recombination takes place generally via radical formation. As a result, radicals will live much longer than ions and electrons outside the plasma excitation zone or when the plasma is switched off. This opens now the possibility to reduce the electron and ion reaction during the plasma treatment by operating the plasma either in the remote mode (separation in space) or as pulsed plasma (separation in time).

Inagaki and co-workers [130–132] compared remote, pulsed, and continuous H_2 plasma treatments on polytetrafluoroethylene (PTFE). They found similar chemical modifications on the PTFE surface for all plasma operation modes, observed by XPS and contact angle measurements. Further they found 2 to 4 times better adhesion between PTFE and copper by using remote or pulsed plasma treatments instead of a continuous plasma treatment. The authors associate this behavior to

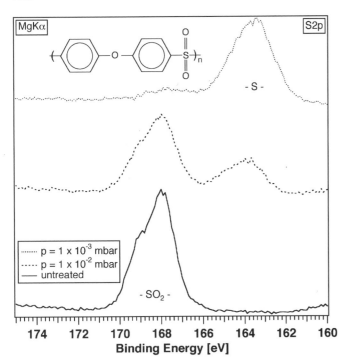

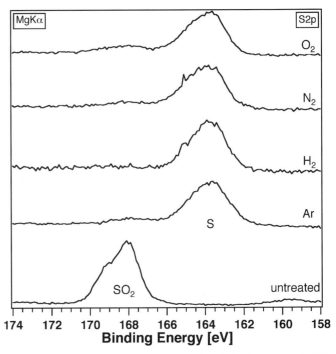

Fig. 55. S2p XPS spectra of the untreated and Ar ECR plasma treated PES surface. Reprinted with permission from [127], © 1996, Elsevier Science.

Fig. 56. S2p XPS spectra of (Ar, H₂, N₂, O₂) ECR plasma treated PES. $p = 3 \times 10^{-4}$ mbar, $t = 1$ s. Reprinted with permission from [127], © 1996, Elsevier Science.

the amount of degradation products present on the surface. To check the presence of low molecular weight degradation products on the PTFE surface they performed contact angle measurements of the treated surfaces before and after acetone rinsing. They suppose, if some degradation products are generated on the polymer surface by the plasma treatment, there will be some differences in the contact angle on the treated polymer surface with and without degradation products. The measurements reveal a much larger difference in contact angle for the PTFE surface treated in continuous plasma than for pulsed plasma treated surfaces (different modulations).

Another approach to minimize degradation reactions on the polymer surface by the ions of the plasma is to limit their kinetic energy. As mentioned in the Introduction the mean kinetic energy or temperature T_i of the ions in cold plasmas is at about room temperature, hence in the range of 25 meV. But due to the positive plasma potential V_p (Section 1.2.2) the kinetic energy of the negative ions impinging on the substrate can be several eV, which is high enough to penetrate the surface and to induce significant physical degradation effects on the polymer surface. The plasma potential V_p as a consequence of the ion excess in the plasma is strongly gas pressure dependent. With increasing gas pressure, the probability for electron–ion recombination in the plasma increases at the expense of electron trapping at the container walls. This decreases the ion excess and consequently the plasma potential V_p and thus the kinetic energy of the ions impinging on the substrate. How dramatic the plasma polymer reactions can change as a function of the gas pressure is shown here by the example of polyethersulfone (PES) [127]. In Figure 55 the S2p XPS spectra illustrate the chemical modification on the PES surface by

Ar plasma treatment. The spectra reveal that sulphone groups ($-SO_2-$; $E_B = 167.6$ eV) are reduced to sulphide ($-S-$; $E_B = 163.7$ eV) by the Ar plasma treatment. The XPS spectra show further that the reduction of the sulphone to sulphide is more efficient at lower gas pressure.

The S2p XPS spectra in Figure 56 show that the reduction of sulphone to sulphide is independent on the feed gas used. Apparently, gas-specific plasma properties (radicals, UV, VUV) are not responsible for the reduction process. At the very low gas pressure of 3×10^{-4} mbar, all sulphone groups at least within a surface layer of 10 nm (XPS probing depth) are reduced to sulphide within 1 s treatment time. In contrast to that, at pressures $p > 10^{-1}$ mbar the reduction process is completely suppressed. These observations indicate a threshold behavior in the reaction of the plasma with the sulphone group of the polymer. This threshold behavior and the fact that at low pressures the reduction process is independent of gas-specific plasma properties indicate that the ions in the plasma are responsible for the process. The threshold behavior in the reaction can be explained by the kinetic energy of the ions.

In the used ECR plasma the mean kinetic energy of the ions hitting the substrate decreases progressively from 22 eV at $p = 3 \times 10^{-4}$ mbar to 3 eV at $p = 10^{-1}$ mbar. The threshold energy E_{th} to penetrate polymer surfaces is typically higher than 10 eV (Section 1.3). With decreasing kinetic energy the number of ions overcoming the threshold energy necessary to interact with the polymer decreases. This explains the higher efficiency of reaction for plasmas operating at low pressures. For pressures $p > 10^{-1}$ mbar the mean kinetic energy of 3 eV is far from the threshold energy E_{th}. Consequently, the specific ion–

polymer reaction is suppressed. This example on the pressure sensitive chemical modifications on PES illustrates the variety of chemical activity of plasmas depending on the operating parameters.

2.3.3.2. Topographic and Physical Modifications of Surfaces

Besides chemical modifications also physical and topographic modifications can be made on polymer surfaces by plasma treatments. The ability to change the topography of polymers by plasma is for example very interesting for the fabrication of soft contact lenses. The eye tolerability and the wearing comfort of such contact lenses are determined considerably by its surface topography (Fig. 57) and wettability. Both properties can be optimized by plasma treatment.

The surface topography of polymers also has influence on the adhesion of metallized films. Several publications report on investigations on the topography of treated polymer surfaces with atomic force microscopy (AFM) or SEM analysis. The applied treatments were ion bombardment [133], excimer laser treatment [134], or various plasma treatments such as corona

discharge [135, 136] and RF or DC plasmas [137–140]. AFM is well suited to investigate the topography of polymer surfaces. For example, the all-trans conformation order of cold-extruded PE [141] or the presence of nonofibrile patterns oriented perpendicular to the fibers of oriented ultrahigh molecular weight PE [142] can be observed by AFM techniques.

Our AFM investigations on PP [143] and PMMA [144] have shown that completely different surface topographies can be created by plasma treatment depending on the gas used. Plasma treatments with reactive gases (O_2, N_2, NH_3) induce weak morphology changes. Moreover, the modifications of the surface roughness are very sensitive to the plasma conditions, so that polymer surfaces treated in reactive gas plasmas can be either smoother or rougher than the untreated polymer. For the untreated PP we measured for 6 μm $\times$ 6 μm area a root mean square (RMS) roughness of 45 nm. After N_2 plasma treatment the RMS roughness is between 5 and 100 nm, depending on the treatment conditions. The situation is totally different by using noble gas plasmas. Noble gas plasmas (He, Ar, Ne, Xe) are able to create complete new morphologies at polymer surfaces. The AFM images in Figure 58 show the morphology changes on PP induced by Ar and He plasma. The He treated PP surface is covered by a dense network of macroscopic chains, while the Ar plasma treated surface presents a smaller granular structure. In fact the granular structure of the Ar plasma treated surface is only the first stage in the process of chain formation as observed on the He treated surface. Actually, all noble gas plasmas form first the granular and afterward the chain structure at the PP surface, depending on the treatment time. The formation of chainlike structures finds application in the manufacturing of soft contact lenses (Fig. 57).

Overney et al. [135] found that corona treated PP (energy dose 112.5 J cm^{-2}) presents dropletlike protrusions of about 450 nm in diameter. The protrusion size increases (up to 700 nm) with increasing energy dose. They attribute the protrusion formation to local surface melting. Similar morphologies such as on the noble gas plasma treated PP and PMMA surfaces have been found on polyethersulphone (PES) and polyarylsulphone after XeCl or KeF excimer laser ablation in air and under vacuum [145]. A minimal energy deposition of 0.5 to

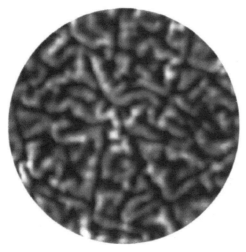

Fig. 57.   AFM image of plasma coated soft contact lens (www.solvias.com).

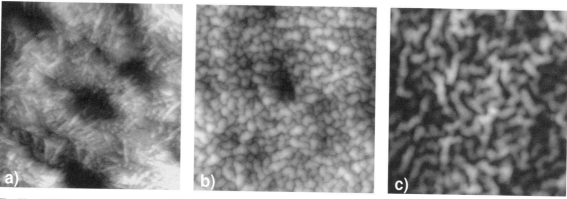

Fig. 58.   AFM images of (1 μm $\times$ 1 μm) of PP (a) untreated, (b) Ar ECR plasma, and (c) He ECR plasma, treated for 120 s and -120 V of RF bias. Reprinted with permission from [143], © 1996, Elsevier Science.

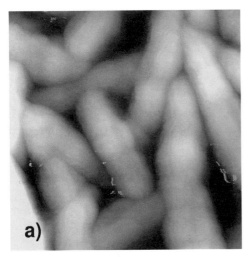

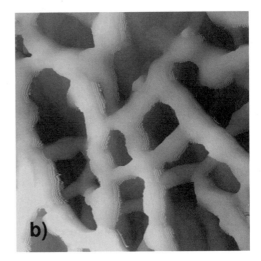

Fig. 59.   AFM images of (1 μm × 1 μm) illustrating the etching effect of He plasma on polytetrafluoroethylene (PTFE): (a) untreated, (b) He ECR plasma treated. Reprinted with permission from [144].

1.0 J cm^{-2} is needed to obtain the characteristic chain structure with diameter ranging from 0.7 to 1.3 μm. They attribute this phenomenon to photoinduced thermal processes of high mobility and solidification at the etched polymer surface during and after ablation. These results indicate that the difference between noble gas and reactive gas plasma treatments on the surface morphology is associated to a much higher mobility of noble gas plasma modified polymer chains. The main difference between reactive gas plasmas and noble gas plasmas is the lack of chemically active species in the latter. In reactive gas plasmas the radicals formed on the polymer can be saturated by chemical active species from the plasma forming new chemical groups or crosslinks. Due to the lack of chemical active species this is not possible in noble gas plasmas. In noble gas plasma the formation of radicals on the polymer is always associated to molecular mass loss (etching effect, Fig. 59) of the polymer. This effect conjugated with chain scission and local temperature increase leads to a high mobility of the polymer chains and consequently to recrystallization or reconstruction of the plasma modified surface region.

To characterize the topography of a surface accurately it is necessary to measure the surface roughness σ as a function of the horizontal length scale L. The RMS roughness $\sigma(L)$ as a function of the horizontal length scale L is given by

$$\sigma(L) = \sqrt{\left\langle \frac{\sum_{i,j=1}^{N}(h_{i,j} - \overline{h})^2}{N^2} \right\rangle}$$

where $N = L/l_0$ is the number of raster points in a length L and l_0 is the raster, h_{ij} is the height of the coordinate ij, $\overline{h} = (1/N^2) \sum_{i,j=1}^{N} h_{ij}$ is the average height, and the average $\langle \rangle$ is calculated over all squares if the size $L \times L$ fits into the image. If a surface is a self-affine fractal, one expects that $\sigma(L)$ is given by

$$\sigma(L) \approx L^\chi$$

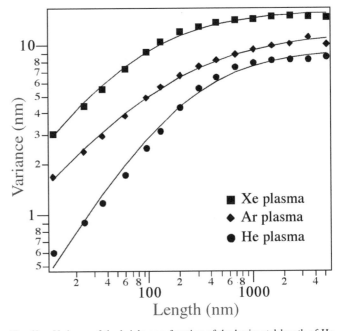

Fig. 60.   Variance of the height as a function of the horizontal length of He, Ar, Xe ECR plasma treated PP ($t = 60$ s, RF bias $= -120$ V). The continuous lines are the fits with the above given crossover function (from [143], reprinted with permission).

where $\chi (< 1)$ is the roughening exponent related to the fractal dimension d_f by $d_f = 3 - \chi$. In practice, surfaces are not fractal on all length scales L and $\sigma(L)$ is described by the crossover function

$$\sigma(L) = \frac{\sigma_0}{(1 + (L_0/L))^\chi}$$

where L_0 is a crossover length. L_0 can be interpreted as a characteristic size of the surface structures with roughness σ_0.

Figure 60 displays the average variance of the height as a function of the horizontal length scale L of an AFM investigated PP surface after He, Ar, and Xe plasma treatment. The fit

Table V. Results of the Variance Curves in Figure 60 Fitted with the Above Given Crossover Function (from [143])

	σ_0 (nm)	L_0 (nm)	χ
He plasma	9.4	236	0.96
Ar plasma	11.9	153	0.72
Xe plasma	16	70	0.86

Table VI. Resistance of PP after ECR Plasma Treatments with Several Gases [146]

Treatment	Resistance (Ω)	RF bias (V)	Treatment time (s)
As-received	$>10^{16}$	—	—
N_2	$>10^{16}$	−200	600
O_2	$>10^{16}$	−200	600
H_2	$>10^{16}$	−200	600
He	2×10^{7}	−230	30
Ne	1×10^{10}	−220	120
Ar	1×10^{10}	−150	120

results of variance curves with the crossover function are given in Table V. The roughness analysis shows that the macroscopic RMS roughness of the plasma treated PP, which corresponds to the saturation value of the variance $\sigma(L)$, is between 10 and 20 nm, which is half of the value for the untreated PP surface. The size of the characteristic surface structure appears to be smaller but higher for the heavy atom plasma (Xe) than for the light atom plasma (He).

As mentioned above, due to the lack of chemically active species, noble gas plasma treatments never lead to an incorporation of new chemical species. Radicals formed at the polymer surface are therefore forced to saturate themselves on the polymer. In the particular case of polyolefins, containing only carbon and hydrogen, noble gas plasma treatments induce hydrogen desorption, resulting in the formation of carbon double bonds, clearly evidenced by XPS measurements [146] (see Fig. 61).

The formation of carbon double bonds in polyolefines by noble gas plasma treatments leads to a dramatic decrease in the electrical resistance [146]. Actually, a resistance drop of up to ten orders of magnitude was measured on He plasma treated PP. Table VI shows the resistance of PP after plasma treatments with several gases. The resistance values are steady state values obtained by the individual plasma treatment. The results in Table VI show that the decrease in resistance by noble gas plasma treatments becomes smaller with increasing mass of the gas atoms, and reactive gas plasma treatments do not affect the resistance. The change in resistance by noble gas plasma treatments can be traced to ions in the plasma. The saturation of the resistance decrease occurs at ion doses between 10^{17} and 10^{18} ions cm^{-2} and the higher the ion energy (RF bias: 0–230 V), the lower the resistance. Taking into account the penetration

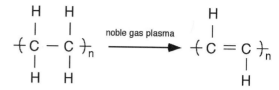

Fig. 61. Illustration of the carbon double bonds on the example of PE.

depth of low energetic ions in polymers, which is between 1 and 10 nm, the resistivity of the He plasma modified surface layer can be estimated to be around 0.1 Ω cm.

2.3.4. Plasma Chemical Diffusion Treatments

It has been 70 years since the Swiss engineer Bernhard Berghaus filed his original patents on the plasma processing on metals [147, 148]. With these patents Berghaus provided the "foundation stone" for two modern classes of plasma surface engineering:

(a) plasma-assisted diffusion methods (nitriding, carburizing, etc.),
(b) plasma-assisted deposition (sputttering, plasma activated evaporation).

Surface grafting by thermal controlled diffusion is a very old process where metal surfaces are alloyed with interstitial elements (B, C, N, O). Classical thermal controlled diffusion processes in metallurgy are oxidation, nitriding, and carburizing for surface hardening. Compared to conventional thermal controlled diffusion methods, plasma diffusion treatments have several significant advantages:

- environmental cleanliness,
- economy,
- reduced cycle times,
- reliability,
- flexible, easy masking,
- minimal distortion,
- lower process temperatures.

Although thermochemical diffusion treatments can be carried out using solid, liquid, or gaseous media, the technical, economic, and environmental constraints imposed by their use cause many heat treaters to change to plasma technology. Enterprises are facing increased pressure from the government legislation to "clean up" the environment and minimize industrial pollution. In this respect, the near mature technologies of plasma diffusion treatments are very attractive.

From the technical point of view the main advantages of plasma diffusion treatments compared to conventional thermal controlled diffusion methods are reduced treatment times and therefore the possibility of lower process temperature. The additional formation of chemically active species (ions, atoms, electrons) in the plasma causes a far greater number of chemical active species to arrive at the workpiece surface per unit time in the plasma than for a gaseous system of equivalent pressure.

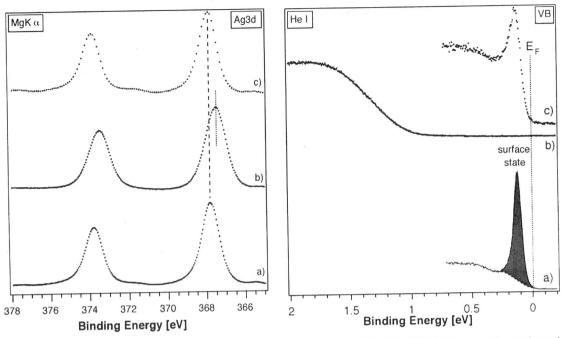

Fig. 62.   Ag3*d* XPS spectra and valence band UPS spectra (recorded at normal emission) of Ag(111): (a) after several sputtering and annealing cycles, (b) after 5 min O$_2$ plasma treatment, (c) sample (b) after 60 min H$_2$ plasma treatment (from [153]).

2.3.4.1. Oxidation of Ag by O$_2$ Plasma Treatment

Noble-metal monoxides are compounds of widespread technological interest for experimental chemistry and material science, as they possess unique catalytic properties and have been used as sensors, as photocathodes in water electrolysis [149], for thermal decomposition of inorganic salts [150], and in other applications. The exceptional catalytic activity of silver for a number of partial oxidation reactions has been known for nearly a century. The ethylene epoxidation and formaldehyde synthesis reactions are the two industrially relevant reactions [151].

With regards to the formation of a compound exhibiting a minimum Gibbs energy of formation, it is known that silver oxides are inherently unstable. The stoichiometric oxides Ag$_2$O and AgO decompose at 503 and 373 K, respectively [153–155]. Therefore it is very difficult to form thick AgO films by thermal diffusion. We have shown [156] that relatively thick ($d \sim 50$ nm) AgO films can be prepared by plasma assisted diffusion within 5 min and at room temperature. The formation of AgO is clearly indicated by the chemical shift of the Ag3$d_{5/2}$ XPS signal from 367.8 to 367.45 eV binding energy (Fig. 62). The valence band spectra (Fig. 62b) reveal for the plasma oxidized silver an electronic gap of 1.2 eV which indicates that the silver oxide formed is an insulator. In fact the plasma oxidized silver is transparent as proved by plasma oxidizing a 100 nm thick Ag film evaporated on glass.

2.3.4.2. Reduction and Recrystallization of AgO to Ag(111) by H$_2$ Plasma Treatment

For studying phase transitions during plasma assisted diffusion we chose an Ag(111) single crystal as substrate. The Ag(111) single crystal is suitable because it shows a Shockley-type surface state around the center $\overline{\Gamma}$ of the surface Brillouin zone [157]. As the surface state is a collective electronic phenomenon at the surface of a crystal, it is inherently sensitive to the perfection of the crystal order. Thus it offers a perfect probe for investigating surface modifications. After plasma oxidation no discernible low energy electron diffraction (LEED) patterns are observable, indicating no long range in the oxide layer formed. XPD of the Ag3$d_{5/2}$ signal shows some residual structure of the original Ag(111) orientation, indicating that some regions with short range order (next nearest neighbors) still remain intact by the oxidation.

H$_2$ plasma treatment (at room temperature) of the oxidized Ag(111) single crystal leads to a complete reduction of the oxide layer ($d = 100$ nm), as observable by the chemical shift of the Ag3$d_{5/2}$ signal back to the original binding energy (367.8 eV) of Ag^0 (Fig. 62).

XPD patterns as well as sharp (1x1) LEED patterns can be observed (Fig. 63), showing that the short and long range order of the Ag crystal have been restored by the H$_2$ plasma treatment. The reappearance of the surface state in the valence band (Fig. 62) indicates a remarkable crystalline quality of the re-established Ag(111) surface. The result that silver oxide can be reduced over a thickness of 100 nm by H$_2$ plasma treatment is very interesting in connection with the application to use Ar/H$_2$ plasma to clean bond pads (Ag, Cu) to improve the quality and the reproducibility of the thermosonic wire bonding process (Section 2.3.2). It may explain the long term effect (typically 10 days even in air) of the Ar/H$_2$ plasma cleaning on the thermosonic wire bonding process [158].

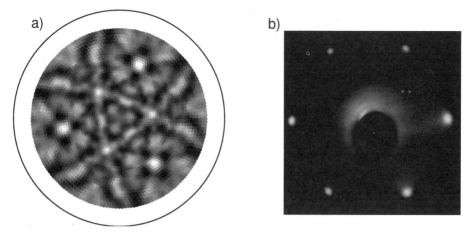

Fig. 63. (a) XPD pattern and (b) (1x1) LEED pattern of Ag(111) after O_2 plasma oxidation and H_2 plasma reduction ($T = 300$ K) (from [153]).

2.4. Surface Termination by H_2 Plasma Treatment

2.4.1. H_2 Plasma Treatment of Diamond

The interaction of hydrogen with semiconductor surfaces is a subject of considerable importance in surface and material science and technology. Hydrogen adsorption can unreconstruct surfaces and passivate surface dangling bonds, and thus can change the surface properties dramatically. Formation of various hydride CVD processes plays a dominant role in thin film structures and surface morphologies (e.g., diamond deposition, Section 2.1.2). As an example of a hydrogen induced surface reconstruction, the one of the hydrogen terminated and the hydrogen free diamond (100) surface is shown schemati-

cally in Figure 64. The influence of the hydrogen termination on the electronic property of the diamond (100) surface is the appearance of the negative electron affinity (NEA). NEA is defined if the vacuum level (E_{vac}) lies below the conduction band minimum at the surface and hence electrons excited into the conduction band can easily escape into vacuum. Himpsel et al. [159] first demonstrated NEA for hydrogen terminated diamond (111) surface (C(111) − (1x1) : H) in 1979. In 1994, Van der Weide et al. [160] found that NEA on diamond is not limited to the (111) surface but also is present for the hydrogen terminated (100) surface. The hydrogen free diamond surfaces C(100) − (2x1) and C(111) − (1x1) show always positive electron affinity (PEA). The NEA property is used in p–n junction electron emitter devices [161] (see Fig. 65).

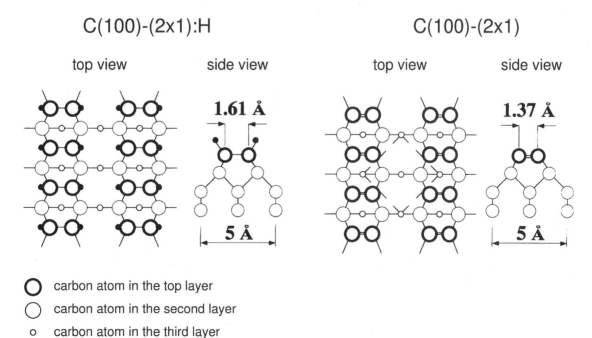

C(100)-(2x1):H   C(100)-(2x1)

top view   side view   top view   side view

1.61 Å   1.37 Å

5 Å   5 Å

○   carbon atom in the top layer

○   carbon atom in the second layer

○   carbon atom in the third layer

●   hydrogen atom

Fig. 64. Schematic illustration of the surface reconstruction of the hydrogen terminated and hydrogen free diamond (100) surface. Reprinted with permission from [163], © 1996, Elsevier Science.

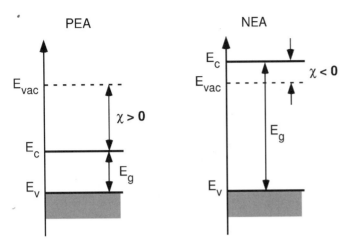

Fig. 65. Energy diagrams of a semiconductor showing positive electron affinity (PEA: $\chi > 0$) and a large bandgap semiconductor showing "true" negative electron affinity (NEA: $\chi < 0$). Reprinted with permission from [163], © 1996, Elsevier Science.

Küttel et al. [162] have shown that the diamond (100) as well as the diamond (111) surface can be polished and hydrogen terminated in H_2 plasma at 870°C and 40 mbar. The initial surface roughness of 7 nm (RMS) can be reduced to 1 nm by this plasma treatment (Fig. 66). This smooth and hydrogen terminated $C(100) - (2x1):H$ surface shows a sharp $(2x1)$ LEED pattern and a strong NEA peak in the UPS spectrum upon annealing to 300°C [163]. At higher temperatures the hydrogen desorbs and the diamond (100) surface shows PES behavior.

Few research groups have shown that O_2 plasma produces moderately high erratic rates of reactive ion etching of diamond although accompanied by an increase of the surface roughness [165–168]. The addition of Ar results in a greater uniformity and better reproducibility than pure O_2 plasma [166, 168]. In contrast, the etching of diamond in SF_6 [167] and CF_4 [169] plasmas produces round asperities on the diamond surface but without a decrease in overall roughness. The plasma etching of CVD diamond is attributed to ion-enhanced chemical etching. Steinbruchel et al. [170] have shown that reactive ion etching of

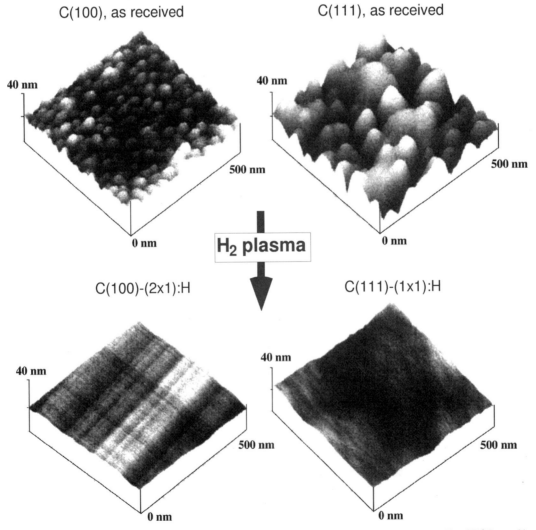

Fig. 66. AFM images of the diamond C(100) and C(111) surfaces before and after H_2 plasma treatment. $T = 870°C$, $t = 6$ h, $p = 40$ mbar. Reprinted with permission from [163], © 1996, Elsevier Science.

a wide range of materials occurs by a process of either physical sputtering or ion-enhanced chemical etching. In both processes, the etch yield $Y(E)$ can be described by the following expression:

$$Y(E) = A\left(\sqrt{E_i} - \sqrt{E_{th}}\right)$$

E_i is the kinetic energy of the ion, E_{th} is the threshold energy for ion etching, and A is a constant, which is for ion-enhanced chemical etching significantly higher than for physical sputtering [171].

2.4.2. H₂ Plasma Treatment of Graphite

The interaction of hydrogen atoms with graphitic surfaces is of outstanding interest for material science and combustion, as well as fundamental research. It is, e.g., a hot topic in astrophysics where the understanding of adsorption and diffusion processes of hydrogen on interstellar dust grain is believed to be a key issue for the explanation of H_2 formation. Further, defects control many physical properties of solids and dominate the electronic behavior of nanoscale metallic and semiconducting systems, particularly at low dimensions. Graphite was one of the first materials for which long range order electronic effects caused by surface defects have been predicted and experimentally observed using scanning tunneling microscopy [172, 173]. Graphite is a semimetal composed of stacked hexagonal planes of sp^2 hybrid bonded carbon atoms. The ABAB stacking of these planes in a three-dimensional crystal creates two inequivalent sites with different properties with regard to the electronic structure: α site atoms are exactly located above atoms of the underlying plane, whereas β site atoms are located above the center of the hexagonal rings of the underlying plane. The weak Van der Waals interaction between adjacent planes leads to a suppression of the charge density at the Fermi level E_F at α sites [174].

The study of changes in the electronic structure of graphite due to adsorbates is relevant for other carbon materials (such as nanotubes, fullerenes, etc.) having the six membered ring as a structural unit.

Recent theoretical work showed the existence of a chemisorbed state and a partial tetrahedrization (sp^3 hybridization) of the carbon atom in the graphite basal plane [175]. Different models used for the simulation of hydrogen adsorption on graphite yield qualitatively the same results but quantitative discrepancies regarding activation energies and chemisorption energies are current debated. Chen et al. [176] found the chemisorption of hydrogen on the basal plane of graphite to be endothermic. A metastable state exits in all examined configurations, the most stable one being the on-top site. The on-top configuration also shows the lowest activation barrier of ~ 1.8 eV. This high energy barrier makes it impossible to populate this state with atomic hydrogen formed in a thermal source [177], where the energy of the atoms is only kT. We showed that in cold H_2 plasma the energy of the ions hitting the substrate (typically ~ 10 eV) is large enough to populate the hydrogen chemisorption state on graphite [178]. By performing a combined mode of scanning tunneling and atomic force microscopy we found long range electronic effects caused by hydrogen–carbon interaction at the graphite surface. Two types of surface modifications could be distinguished by this method: chemisorption of hydrogen on the basal plane and atomic vacancy formation.

2.4.2.1. Chemisorbed Hydrogen

Scanning the cold plasma treated graphite surface in the AFM contact mode with simultaneous acquisition of the current signal reveals clear superlattice-type modifications of the electronic properties over a distance of 25 lattice constants (Fig. 67). However, no modifications were detected in the topography im-

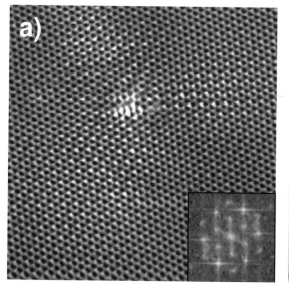 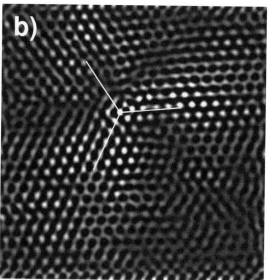

Fig. 67. (a) Current image of a H_2 plasma treated graphite surface recorded in AFM contact mode (94 Å × 94 Å). The inset shows the center of the fast Fourier transform of the image. (b) The same data as (a) after applying a two-dimensional band pass filter to isolate the superlattice component. Reprinted with permission from [175], © 1999, Elsevier Science.

age. The inset of Figure 67a shows the center of the Fourier transform of the current image. The six inner spots correspond to the $(\sqrt{3} \times \sqrt{3})R30°$ superstructure.

The six outer spots with the same orientation are due to the second harmonic of the superlattice, whereas the six outer spots rotated by 30° originate from the regular graphite lattice. The threefold symmetry of the scattering pattern is clearly seen in Figure 67b, which shows the same data as (a) but after applying a two-dimensional band pass filter to isolate the superlattice contribution. Kelly et al. [179] discussed the dependence of the orientation of the threefold scattering pattern on the type of atom (α or β site atom), which is affected by an atomic vacancy. Scattering patterns originating from α site defects are rotated by 60° relative to the ones from β site atoms, reflecting the symmetry of the dangling bonds of the nearest neighbors. These bonds are also affected if the hybridization of a carbon atom is changed by forming an additional bond to a chemisorbed hydrogen atom. Only one type of carbon atom seems to be affected. If α and β sites were involved, one would expect a sixfold symmetric scattering pattern. A more detailed analysis of the bandpass filtered current image shows that the

measured scattering pattern results from two interfering scattering patterns with the same orientation, but with an origin shift of 2.46 Å, i.e., the distance between two carbon atoms of the same type.

2.4.2.2. Atomic Vacancies

Typical images of the simultaneously recorded current and topography images of a local defect are shown in Figure 68. The topographic structure of this defect consists of three local depressions with a corrugation of 0.3 Å. This strong enhancement of the corrugation, as compared to the 10 times weaker corrugation of the defect free zones, is due to atomic vacancies in the first carbon monolayer, which were formed upon the H_2 plasma treatment. The vacancy formation probability is found to be five times smaller than hydrogen chemisorption. The depressions in Figure 68b are separated by 4.1 Å. The direction and the distance between the topographic minima show that carbon atoms of the same type (α or β site atoms) are affected (distance between second nearest neighbors of the same type is $\sqrt{3} \cdot 2.46$ Å $= 4.26$ Å). The current image shows the lo-

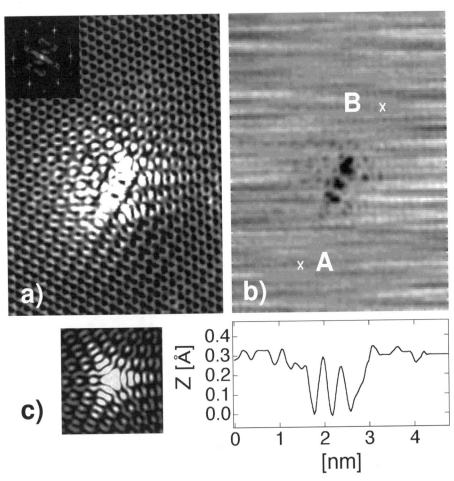

Fig. 68. Simultaneously recorded current and topography image of a H_2 plasma treated graphite surface in AFM contact mode (58 Å $\times$ 78 Å). (a) Current image with fast Fourier transform (inset). Current range: 2 nA (black)–40 nA (white). (b) Topography image with line profile from A to B. (c) Calculated LDOS in the vicinity of a single atomic vacancy using a tight binding model. Reprinted with permission from [175], © 1999, Elsevier Science.

cal density of states (LDOS) in the vicinity of the defect and again the $(\sqrt{3} \times \sqrt{3})R30°$ superlattice, which is visible over 25 lattice constants. The calculated LDOS for a single atomic vacancy [172] (Fig. 68c) shows in all details the same features as the experiment.

The threshold energies for sputtering off carbon atoms have been calculated by Bohdansky and Roth [14] (see Section 1.3). For the interaction of hydrogen ions (proton) with graphite an energy threshold of $E_{th} \sim 36$ eV is estimated. The maximum ion energy in ECR plasma treatment is given by the plasma potential. The maxium ion energies in the plasma chamber used for this experiment range from ~ 22 eV ($p = 10^{-4}$ mbar) to about 9 eV ($p = 10^{-1}$ mbar) [180], which is below the estimated energy threshold for vacancy formation by sputtering. A possible channel for vacancy formation is the neutralization process of the ions. The charge transfer from the surface to the approaching ion leads to a weakening and a finite destruction cross section of the sp^2 bonds [181].

3. OUTLOOK

In the past 15 years many cold plasma processes have been established in various technological applications. Many innovative new products have only become possible by using cold plasma processes. Many industries, e.g., optical, microelectronic, food packaging, and mechanical engineering industries, use cold plasma processes in various areas of applications (Fig. 69). On an industrial scale plasma processes are relatively cheap so that even for low cost products (e.g., food packaging) plasma processes are widely used. Compared to wet chemical processes the consumption of chemicals is much lower for plasma processes, which make them much more ecologically safe. The ecological pressure makes plasma processes very attractive for the industry and is a strong driving force that the field of applications increases permanently.

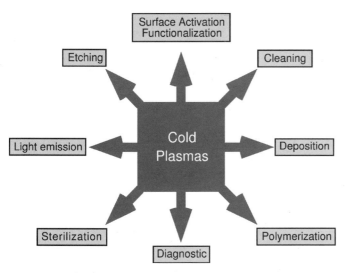

Fig. 69. Areas of application for cold plasma processes.

With the present contribution we touched briefly on a few cold plasma processes important in surface science and technology. The selection of examples is by no means complete. Important and well established industrial applications, for example, reactive ion etching of semiconductors especially silicon or plasma polymerization of organosilicones and organometallics, are not discussed in this chapter. Nevertheless we hope that the present contribution gives an illustrative introduction of cold plasma processes in surface science and technology.

For the future, efforts in plasma diagnostics (not discussed in this chapter) have to be done for a better understanding of the particular plasma processes. This will allow design of new and better plasma sources with characteristics optimized for intended processes, which guarantee that the areas of applications increase permanently.

Acknowledgments

I thank M. Collaud Coen, A. Schneuwly, P. Ruffieux, P. Schwaller, O. Nilsson, M. Bielmann, and O. Gröning for a lot of excellent work done on the investigations presented here. Generous financial support was received from the Swiss National Science Foundation (Programs NFP 36 and NFP 47 and PPM) and the Commission of Technology and Innovation (CTI) of Switzerland.

REFERENCES

1. I. Langmuir, *Phys. Rev.* 33, 954 (1929).
2. F. F. Chen, "Introduction to Plasma Physics." Plenum Press, New York, 1974.
3. S. Ichimaru, "Basic Principles of Plasma Physics." Benjamin–Cummings, Redwood City, CA, 1973.
4. B. Chapman, "Glow Discharge Processes." Wiley, New York, 1980.
5. A. Grill, "Cold Plasma in Materials Fabrication." IEEE Press, New York, 1994.
6. A. Auciello and D. L. Flamm, "Plasma Diagnostic," Vol. 1. Academic Press, San Diego, 1990.
7. R. d'Agostino, "Plasma Deposition, Treatment, and Etching of Polymers." Academic Press, San Diego, 1990.
8. H. Yasuda, "Plasma Polymerization." Academic Press, Orlando, 1985.
9. H. S. Butler and G. S. Kino, *Phys. Fluid* 6, 1346 (1963).
10. A. D. MacDonald and S. J. Tetenbaum, "Gaseous Electronics," Vol. 1. Academic Press, New York, 1978.
11. M. Venugopalan, "Reactions under Plasma Conditions," Vol. 1. Wiley-Interscience, New York, 1971.
12. J. Asmussen, *J. Vac. Sci. Technol. A* 7, 883 (1989).
13. R. Behrisch, G. Maderlechner, B. M. U. Scherzer, and M. T. Robinson, *Appl. Phys.* 18, 391 (1979).
14. J. Bohdansky, J. Roth, and H. L. Bay, *J. Appl. Phys.* 51, 2861 (1980).
15. S. Schiller, Chr. Metzner, and O. Zywitzki, *Surf. Coat. Technol.* 125, 240 (2000).
16. B. A. Movchan and A. V. Demchshin, *Phys. Met. Metallogr.* 28, 83 (1969).
17. J. A. Thornton, *J. Vac. Sci. Technol.* 11, 666 (1974).
18. J. A. Thornton, *Annu. Rev. Mater. Sci.* 7, 239 (1977).
19. M. Lardon, R. Buhl, H. Singer, H. K. Pukler, and E. Moll, *Thin Solid Films* 54, 317 (1978).

20. G. Carter and D. G. Armour, *Thin Solid Films* 80, 13 (1981).

21. P. K. Bachmann, in "Ullman's Encyclopaedia of Industrial Chemistry," Vol. A 26, p. 720, 1996.

22. S. Aisenberg and R. Chabot, *J. Appl. Phys.* 42, 2953 (1971).

23. K. Enke, H. Dimigen, and H. Hübsch, *Appl. Phys. Lett.* 36, 291 (1980).

24. C. Donnet, T. LeMogne, L. Ponsonnet, M. Belin, A. Grill, V. Patel, and C. Jahnes, *Tribol. Lett.* 4, 259 (1998).

25. J. A. Thornton, in "Deposition Technologies for Films and Coatings" (R. F. Bunshah et al., Eds.). Noyes, Park Ridge, NJ, 1982.

26. J. Robertson, *Progr. Solid State Chem.* 21, 199 (1991).

27. S. R. P. Silva, J. Robertson, and G. A. J. Amaratunga, "Amorphous Carbon: State of the Art." World Scientific, Singapore, 1997.

28. H. Tsai and D. B. Bogy, *J. Vac. Sci. Technol. A* 5, 3287 (1987).

29. P. Koidl, C. Wild, R. Locher, and R. E. Sah, in "Diamond and Diamond-like Films and Coatings" (R. E. Clausing, L. L. Horton, J. C. Angus, and P. Koidl, Eds.), NATO-ASI Series B: Physics, Vol. 266, p. 243. Plenum, New York, 1991.

30. W. D. Sproul, *Science* 273, 889 (1996).

31. U. Helmersson, S. Todorova, L. Market, S. A. Barnett, J.-E. Sundren, and J. E. Greene, *J. Appl. Phys.* 62, 481 (1987).

32. W. D. Sproul, P. J. Rudnik, M. E. Grahan, and S. L. Rhode, *Surf. Coat. Technol.* 43/44, 270 (1990).

33. R. Hauert, J. Patscheider, L. Knoblauch, and M. Diserens, *Adv. Mater.* 11, 175 (1999).

34. B. Meyerson and F. W. Smith, *Solid State Comm.* 34, 531 (1980).

35. H. Dimigen, H. Hübsch, and R. Memming, *Appl. Phys. Lett.* 50, 1056 (1987).

36. T. L. Jacobs, J. H. Spence, S. S.Wagal, and H. J. Orien, "Applications of Diamond Films and Related Materials: Third International Conference," 1995, p. 753.

37. G. Francz, A. Schroeder, and R. Hauert, *Surf. Interface Anal.* 28, 3 (1999).

38. B. V. Deryagin and D. V. Fedoseev, *Science* 233, 102 (1975).

39. B. Dischler and C. Wild, "Low-Pressure Synthetic Diamond." Springer-Verlag, Berlin/Heidelberg, 1998.

40. P. K. Bachmann, in "Properties and Growth of Diamond" (G. Davies, Ed.), EMIS Datareviews Series, p. 349. Short Run, Exeter, 1994.

41. P. K. Bachmann, D. Leers, and H. Lydt, *Diamond Relat. Mater.* 1, 1 (1991).

42. J. E. Butler and R. L. Woodlin, *Philos. Trans. Roy. Soc. London Ser. A* 342, 209 (1993).

43. L. S. Plano, in "Diamond: Electronic Properties and Applications" (L. S. Pan and D. R. Kania, Eds.), p. 65, 1995.

44. R. Gat and J. C. Angus, in "Properties and Growth of Diamond" (G. Davies, Ed.), EMIS Datareviews Series, p. 325. Short Run, Exeter, 1994.

45. S. J. Harris, *Appl. Phys. Lett.* 56, 2298 (1990).

46. S. Skotov, B. Weimer, and M. Frenklach, *J. Phys. Chem.* 99, 5616 (1995).

47. M. Frenklach and K. E. Spear, *J. Mater. Res.* 3, 133 (1998).

48. H. Liu and D. S. Dandy, *Diamond Relat. Mater.* 4, 535 (1995).

49. M. Ece, B. Oral, J. Patscheider, and K.-H. Ernst, *Diamond Relat. Mater.* 4, 720 (1995).

50. B. R. Stoner, G. H. M. Ma, S. D. Wolter, and J. T. Glass, *Phys. Rev. B* 45, 11067 (1992).

51. C. Wild, P. Koidl, W. Müller-Sebert, H. Walcher, R. Kohl, N. Herres, R. Locher, and R. Brenn, *Diamond Relat. Mater.* 2, 158 (1993).

52. E. Maillard-Schaller, Study of the Interface between Highly Oriented Diamond and Silicon, Ph.D. Thesis No. 1145, Physics Department, University of Fribourg, Switzerland, 1996.

53. S. D. Wolter, B. R. Stoner, J. T. Glass, P. J. Jenkins, D. S. Buhaenko, C. E. Jenkins, and P. Southworth, *Appl. Phys. Lett.* 62, 1215 (1993).

54. N. Jiang, B. W. Sun, Z. Zhang, and Z. Lin, *J. Mater. Res.* 9, 2695 (1994).

55. W. Kulisch, B. Sobisch, M. Kuhr, and R. Beckmann, *Diamond Relat. Mater.* 4, 401 (1994).

56. S. D. Wolter, T. H. Boert, A. Vescan, and E. Kohn, *Appl. Phys. Lett.* 68, 3558 (1996).

57. C. L. Jia, K. Urban, and X. Jiang, *Phys. Rev. B* 52, 5164 (1995).

58. J. Gerber, M. Weiler, O. Sohr, K. Jung, and H. Ehrhardt, *Diamond Relat. Mater.* 3, 506 (1994).

59. S. Sattel, N. Weiler, J. Greber, H. Roth, M. Scheib, K. Jung, H. Ehrhard, and J. Robertson, *Diamond Relat. Mater.* 4, 333 (1995).

60. E. Maillard-Schaller, O. M. Küttel, P. Gröning, O. Gröning, R. G. Agostino, P. Aebi, L. Schlapbach, P. Wurzinger, and P. Pongartz, *Phys. Rev. B* 55, 15895 (1997).

61. E. Maillard-Schaller, O. M. Küttel, P. Gröning, and L. Schlapbach, *Diamond Relat. Mater.* 6, 282 (1997).

62. C. S. Fadley, in "Synchrotron Radiation Research" (R. Z. Bachrach, Ed.). Plenum, New York, 1989.

63. A. Olbrich, B. Ebersberger, C. Boit, Ph. Niederdermann, W. Hänni, J. Vancea, and H. Hoffmann, *J. Vac. Sci. Technol. B* 17, 1570 (1999).

64. Centre Suisse d'Electronique et de Microtechnique SA (CSEM), CH-2007 Neuchâtel, Switzerland.

65. M. W. Geis, N. M. Elfremov, J. D. Woodhouse, M. D. McAleese, M. Marywka, D. G. Socker, and J. F. Hochedez, *IEEE Trans. Electron. Device Lett.* 12, 456 (1991).

66. C. Wang, A. Garcia, D. Ingram, M. Lake, and M. E. Kordesch, *Electron. Lett.* 27, 1459 (1991).

67. IBM Patent Server, www.patents.ibm.com.

68. O. Gröning, Field Emission Properties of Carbon Thin Films and Carbon Nanotubes, Ph.D. Thesis No. 1258, Physics Department, University of Fribourg, Switzerland, 1999.

69. O. Gröning, O. M. Küttel, P. Gröning, and L. Schlapbach, *J. Vac. Sci. Technol. B* 17, 1970 (1999).

70. R. H. Fowler and L. W. Nordheim, *Proc. Roy. Soc. London Ser. A* 119, 173 (1928).

71. The WKB method was published in 1926 independently by the following three authors: G. Wentzel, *Z. Phys.* 38, 518 (1926); H. A. Kramers, *Z. Phys.* 39, 828 (1926); L. Brillouin, *C. R. Acad. Sci. Paris* 183, 24 (1926).

72. O. Gröning, O. M. Küttel, P. Gröning, and L. Schlapbach, *J. Vac. Sci. Technol. B* 17, 1064 (1999).

73. W. A. de Heer, A. Châtelain, and D. Ugarte, *Science* 270, 1179 (1995).

74. O. M. Küttel, O. Gröning, Ch. Emmenegger, and L. Schlapbach, *Appl. Phys. Lett.* 73, 2113 (1998).

75. O. Gröning, O. M. Küttel, Ch. Emmenegger, P. Gröning, and L. Schlapbach, *J. Vac. Sci. Technol. B* 18, 665 (2000).

76. S. Iijima, *Nature* 354, 56 (1991).

77. P. J. F. Harris, "Carbon Nanotubes and Related Structures, New Materials for the Twenty-First Century." Cambridge Univ. Press, Cambridge, UK, 1999.

78. K. Tanaka, T. Yamabe, and K. Fukui, "The Science and Technology of Carbon Nanotubes." Elsevier, New York, 1999.

79. D. Tomanek and R. J. Enbody, "Science and Application of Nanotubes," Proccedings of Nanotube 99. Kluwer Academic/Plenum, New York, 2000.

80. T. W. Ebbesen, *Annu. Rev. Mater. Sci.* 24, 235 (1994).

81. T. Guo, P. Nikolaev, A. Thess, D. T. Colbert, and R. E. Smalley, *J. Phys. Chem.* 55, 10694 (1995).

82. M. Endo, K. Takeuchi, S. Igarashi, K. Kobori, M. Shiraishi, and H. W. Kroto, *J. Phys. Chem. Solids* 54, 1841 (1993).

83. W. K. Hsu, J. P. Hare, M. Terrones, H. W. Kroto, D. R. M. Walton, and P. J. F. Harris, *Nature* 377, 687 (1995).

84. W. K. Hsu, M. Terrones, J. P. Hare, H. Terrones, H. W. Kroto, and D. R. M. Walton, *Chem. Phys. Lett.* 262, 161 (1996).

85. M. S. Dresselhaus, G. Dresselhaus, I. L. Spain, and H. A. Goldberg, Eds., in "Graphite Fibers and Filaments." Springer Verlag, New York, 1988.

86. R. T. K. Baker and P. S. Harris, *Chem. Phys. Carbon* 14, 83 (1978).

87. P. Gröning et al., Physics Department, University of Fribourg, Pérolles, CH-1700 Fribourg, Switzerland.

88. J. Kong, H. T. Soh, A. M. Cassell, C. F. Quate, and H. Dai, *Nature* 395, 878 (1998).

89. S. Fan, M. G. Chapline, N. R. Franklin, T. W. Tombler, A. M. Cassell, and H. Dai, *Science* 283, 512 (1999).

90. H. Kind, J.-M. Bonard, Ch. Emmenegger, L.-O. Nilsson, K. Hernadi, E. Maillard-Schaller, L. Schlapbach, L. Forro, and K. Kern, *Adv. Mater.* 11, 1285 (1999).

91. H. König and G. Helwig, *Z. Phys.* 129, 491 (1951).

92. H. Schmellenmeier, *Exp. Techn. Phys.* 1, 49 (1953).

93. J. Goodmann, *J. Polym. Sci.* 44, 551 (1960).

94. A. P. Bradley and J. P. Hammes, *J. Electrochem. Soc.* 110, 15 (1963).

95. T. Williams and M. W. Hayes, *Nature* 209, 769 (1966).

96. H. U. Poll, M. Artz, and K. H. Wickleder, *European Polym. J.* 12, 505 (1976).

97. D. K. Lam, R. F. Baddour, and A. F. Stancell, in "Plasma Chemistry of Polymers" (M. Shen, Ed.). New York, 1976.

98. B. D. Ratner, *Polymer* 34, 643 (1993).

99. Ch. Oehr and H. Brunner, *Vakuum Forsch. Praxis* 1, 35 (2000).

100. H. Nomura, P. W. Kramer, and H. Yasuda, *Thin Solid Films* 118, 187 (1984).

101. J. Sakata, M. Yammamoto, and M. Hirai, *J. Appl. Polym. Sci.* 31, 1999 (1986).

102. A. F. Stancell and A. T. Spencer, *J. Appl. Polym. Sci.* 16, 1505 (1972).

103. P. Gröning, A. Schneuwly, L. Schlapbach, and M. T. Gale, *J. Vac. Sci. Technol. A* 14, 3043 (1996).

104. J. W. Coburn and H. F. Winters, *J. Vac. Sci. Technol.* 16, 391 (1979).

105. R. Jaszewski, H. Schift, P. Gröning, and G. Margaritondo, *Micro. Nano-Eng.* 35, 381 (1997).

106. R. W. Jaszewski, H. Schift, B. Schnyder, A. Schneuwly, and P. Gröning, *Appl. Surf. Sci.* 143, 301 (1999).

107. A. Schneuwly, Engineering Relevant Surfaces: Treatment, Characterisation and Study of Their Properties, Ph.D. Thesis No. 1225, Physics Department, University of Fribourg, Switzerland, 1998.

108. A. Schneuwly, P. Gröning, L. Schlapbach, C. Irrgang, and J. Vogt, *IEEE Trans. Dielectrics Electrical Insulation* 5, 862 (1998).

109. A. Schneuwly, P. Gröning, and L. Schlapbach, *Mater. Sci. Eng. B* 55, 210 (1998).

110. A. J. Kinloch, "Adhesion and Adhesives." Chapman and Hall/Cambridge Univ. Press, London, 1990.

111. L.-H. Lee, "Fundamentals of Adhesion." Plenum, New York, 1991.

112. S. Trigwell, *Solid State Technol.* 45 (1993).

113. A. Carrass and V. P. Jäcklin, *Dtsch. Verb. Schweisstech.* 173, 135 (1995).

114. A. Schneuwly, P. Gröning, L. Schlapbach, and V. Jäcklin, *IEEE J. Electron. Mater.* 27, 990 (1998).

115. N. Onda, A. Dommann, H. Zimmermann, C. Lüchinger, V. Jäcklin, D. Zanetti, E. Beck, and J. Ramm, "Semicon Singapore Proc.," 1996.

116. E. Wandtke, *Transfer* 40, 34 (1996).

117. A. Schneuwly, P. Gröning, L. Schlapbach, and G. Müller, *IEEE J. Electron. Mater.* 27, 1254 (1998).

118. C. F. McFadden and A. F. Gellmann, *Langmuir* 11, 273 (1995).

119. E. M. Liston, *J. Adhes.* 30, 199 (1989).

120. E. Occhiello, M. Morra, G. Morini, F. Garbassi, and D. Johnson, *J. Appl. Polym. Sci.* 42, 2045 (1991).

121. L. J. Gerenser, *J. Vac. Sci. Technol. A* 8, 3682 (1990).

122. U. W. Gedde, K. Pellfolk, M. Braun, and C. Rodehed, *J. Appl. Polym. Sci.* 39, 477 (1990).

123. P. Gröning, S. Berger, M. Collaud Coen, O. Gröning, and B. Hasse, "5th European Adhesion Conference EURADH'2000," Lyon, 2000, p. 299.

124. P. Gröning, O. M. Küttel, M. Collaud Coen, G. Dietler, and L. Schlapbach, *Appl. Surf. Sci.* 89, 83 (1995).

125. G. Marletta, S. M. Catalano, and S. Pignataro, *Surf. Interface Anal.* 16, 407 (1990).

126. P. Gröning, M. Collaud, G. Dietler, and L. Schlapbach, *J. Appl. Phys.* 76, 887 (1994).

127. P. Gröning, M. Collaud Coen, O. M. Küttel, and L. Schlapbach, *Appl. Surf. Sci.* 103, 79 (1996).

128. L. Gong, A. D. Friend, and R. P. Wool, *Macromolecules* 31, 3706 (1998).

129. A. T. Bell, Techniques and Applications of Plasma Chemistry (J. R. Hollahan and A. T. Bell, Eds.), p. 1. Wiley, New York, 1974.

130. Y. Yamada, T. Yamada, S. Tasaka, and N. Inagaki, *Macromolecules* 29, 4331 (1996).

131. N. Inagaki, S. Tasaka, and T. Umehara, *J. Appl. Polym. Sci.* 71, 2191 (1999).

132. N. Inagaki, S. Tasaka, and K. Mochizuki, *Macromolecules* 32, 8566 (1996).

133. R. Michael and D. Stulik, *Nucl. Instrum. Methods B* 28, 259 (1987).

134. H. Niino, M. Nakano, S. Nakano, A. Yabe, H. Moriya, and T. Miki, *Japan. J. Appl. Phys.* 28, L2225 (1989).

135. R. M. Overney, R. Lüthi, H. Haefke, J. Frommer, E. Meyer, H.-J. Güntherodt, S. Hild, and J. Fuhrmann, *Appl. Surf. Sci.* 64, 197 (1993).

136. R. M. Overney, H.-J. Güntherodt, and S. Hild, *J. Appl. Phys.* 75, 1401 (1994).

137. K. Harth and S. Hibst, *Surf. Coat. Technol.* 59, 363 (1994).

138. Y. Khairallah, F. Arefi, J. Amouroux, D. Leonard, and P. Bertrand, *J. Adhes. Sci. Technol.* 8, 363 (1994).

139. N. Inagaki, S. Tasaka, and K. Hibi, *J. Adhes. Sci. Technol.* 8, 395 (1994).

140. H. Yasuda, *J. Macromol. Sci. Chem. A* 10, 383 (1976).

141. S. N. Magonov, K. Qvarnström, V. Elings, and H.-J. Cantow, *Polym. Bull.* 25, 689 (1991).

142. A. Wawkuschewski, H.-J. Cantow, and S. N. Magonov, *Adv. Mater.* 6, 476 (1994).

143. M. Collaud Coen, G. Dietler, S. Kasas, and P. Gröning, *Appl. Surf. Sci.* 103, 27 (1996).

144. M. Collaud Coen, P. Gröning, and R. Lehmann, submitted for publication.

145. H. Niino and A. Yabe, *J. Photochem. Photobiol. A* 65, 303 (1992).

146. M. Collaud Coen, P. Gröning, G. Dietler, and L. Schlapbach, *J. Appl. Phys.* 77, 5595 (1995).

147. B. Berghaus, German Patent DPR 668.639, 1932.

148. B. Berghaus, German Patent DPR 851.560, 1939.

149. K. R. Thampi and M. Gratzel, *J. Mol. Catal.* 60, 31 (1990).

150. A. A. Said, *J. Mater. Sci. Lett.* 11, 1903 (1992).

151. "Ullman's Encyclopedia of Industrial Chemistry," 5th ed., Vol. A 11, p. 619, 1988.

152. A. Nagy and G. Mestl, *Appl. Catal.* 188, 337 (1999).

153. X. Bao and J. Deng, *J. Catal.* 99, 391 (1986).

154. X. Bao, M. Muhler, Th. Schedel-Niedrig, and R. Schlögl, *Phys. Rev. B* 54, 2249 (1996).

155. X. Bao, B. Pettinger, G. Ertl, and R. Schlögel, *Ber. Bunsenges. Phys. Chem.* 97, 97 (1993).

156. M. Bielmannm, P. Ruffieux, P. Schwaller, L. Schlapbach, and P. Gröning, submitted for publication.

157. R. Paniago, R. Matzdorf, G. Meister, and A. Goldmann, *Surf. Sci.* 336, 113 (1995).

158. Plasma cleaner LFC 150, Balzers Instruments, PO 1000, FL-9496 Balzers, Lichtenstein.

159. F. J. Himpsel, J. A. Knapp, J. A. VanVechten, and D. E. Eastmann, *Phys. Rev. B* 20, 624 (1979).

160. J. van der Weide, Z. Zhang, P. K. Baumann, M. G. Wensell, J. Bernholc, and R. J. Nemanich, *Phys. Rev. B* 50, 5803 (1994).

161. R. L. Bell, "Negative Electron Affinity Devices." Clarendon, Oxford, 1973.

162. O. M. Küttel, L. Diederich, E. Schaller, O. Carnal, and L. Schlapbach, *Surf. Sci. Lett.* 337, L812 (1995).

163. L. Diederich, O. M. Küttel, E. Schaller, and L. Schlapbach, *Surf. Sci.* 349, 176 (1996).

164. L. Diederich, Negative Electron Affinity of Hydrogen-Terminated Diamond Surfaces Studies by Photoelectron Spectroscopy, Ph.D. Thesis No. 1210, Physics Department, University of Fribourg, Switzerland, 1998.

165. G. B. Sandhu and W. K. Chu, *Appl. Phys. Lett.* 55, 437 (1989).

166. A. Vescan, W. Ebert, T. H. Borst, and E. Kohn, *Diamond Relat. Mater.* 5, 774 (1996).

167. C. Vivensang, L. Ferlazzo-Manin, M. V. Ravet, G. Turban, F. Rosseaux, and A. Gicquel, *Diamond Relat. Mater.* 5, 840 (1996).

168. S. A. Grot, R. A. Ditzio, G. Sh. Gildenblat, A. R. Badzian, and S. J. Fonash, *Appl. Phys. Lett.* 61, 2326 (1992).

169. K. Kobayashi, N. Mutsukura, and Y. Machi, *Thin Solid Films* 200, 139 (1991).

170. Ch. Steinbruchel, H. W. Lehmann, and K. Frick, *J. Electrochem. Soc.* 132, 180 (1985).

171. Ch. Steinbruchel, *Appl. Phys. Lett.* 55, 1960 (1989).

172. H. A. Mizes and J. S. Foster, *Science* 244, 559 (1989).

173. J. Xhie, K. Sattler, U. Muller, N. Venkateswaran, and G. Raina, *Phys. Rev. B* 43, 8917 (1991).

174. D. Tomanek, S. G. Louie, H. J. Mamin, D. W. Abraham, R. E. Thomson, E. Ganz, and J. Clarke, *Phys. Rev. B* 35, 7790 (1987).

175. L. Jeloaica and V. Sidis, *Chem. Phys. Lett.* 300, 157 (1999).

176. J. P. Chen and R. T. Yang, *Surf. Sci.* 216, 418 (1989).

177. U. Bischler and E. Bertel, *J. Vac. Sci. Technol. A* 11, 458 (1993).

178. P. Ruffieux, O. Gröning, P. Schwaller, L. Schlapbach, and P. Gröning, *Phys. Rev. Lett.* 84, 4910 (2000).

179. K. F. Kelly and N. J. Halas, *Surf. Sci.* 416, L1085 (1998).

180. S. Nowak, P. Gröning, O. M. Küttel, M. Collaud, and G. Dietler, *J. Vac. Sci. Technol. A* 10, 3419 (1992).

181. V. V. Khvostov, M. B. Guseva, V. G. Babaev, and E. A. Osherovich, *Surf. Sci.* 320, L123 (1994).

Chapter 5

ELECTROCHEMICAL FORMATION OF THIN FILMS OF BINARY III–V COMPOUNDS

L. Peraldo Bicelli

Dipartimento di Chimica Fisica Applicata del Politecnico, Centro di Studio sui Processi Elettrodici del CNR, 20131 Milan, Italy

V. M. Kozlov

Department of Physics, National Metallurgical Academy of Ukraine, Dniepropetrovsk, Ukraine

Contents

Handbook of Thin Film Materials, edited by H.S. Nalwa
Volume 1: Deposition and Processing of Thin Films

ISBN 0-12-512909-2/$35.00

1. INTRODUCTION

Currently, electronic and optoelectronic industries provide some of the largest markets and challenges for thin-film semiconductors. Current techniques for the growth of these materials include physical methods (e.g., molecular beam epitaxy (MBE), liquid-phase epitaxy (LPE), liquid encapsulated Czochralski (LEC), sputtering), and chemical methods (e.g., chemical vapor deposition (CVD), metalorganic chemical vapor deposition (MOCVD), metalorganic vapor phase epitaxy (MOVPE), electrodeposition). Physical methods are expensive but give relatively more reliable and reproducible results. Most of the chemical methods are more economical, but their full potential for obtaining device-quality films has not been completely explored in many cases. Moreover, several authors (e.g., [1]) projected in recent time a future in which solid-state synthesis will be conducted in so-called round-bottom flasks, the strategies being more similar to those of organic chemistry than to those of the high-temperature methods currently employed. Emerging syntheses of technologically important semiconductors, including those of group III–V family, have been outlined [2]. They are based on solution-chemical processes performed in mild reaction conditions, making it possible to turn down the heat on semiconductor growth and to control crystallite size and morphology. Among these processes, electrodeposition has also been considered (this mainly refers to II–VI chalcogenides).

Electrochemical synthesis is one of the oldest methods for the formation of thin films and may offer a simple and viable alternative to the very complicated and highly expensive high-temperature and/or high-vacuum processes. It offers especially attractive features, such as simplicity, low cost, high output, and scalability compared with the competing methods of deposition, such as MBE and CVD. Moreover, it is generally performed near room temperature, and it is thus appealing as a low-temperature and isothermal technique that avoids the deleterious effects of interdiffusion, contamination, and dopant redistribution, which are typical of high-temperature processes.

The main advantages of electrodeposition include the following:

1. It affords precise process control by means of current or voltage modulation and thus the control of film thickness, morphology, stoichiometry, doping, etc. This is due to its electrical nature, as electrical parameters are inherently easier to control than thermal parameters.
2. It makes possible the growth of uniform films on different substrates and over large and/or irregular areas, from cm^2 to m^2.
3. It occurs closer to equilibrium conditions than many high-temperature vacuum deposition methods, but, on the other hand, it allows the electrodeposition of compositionally nonequilibrium materials or modulated structures.

4. It is particularly suited to the fabrication of heterojunctions simply through a change in the deposition electrolyte.
5. It allows an efficient use of chemical precursors, because deposition occurs on the working electrode only and not on other reactor surfaces.
6. It is less polluting, as it does not involve the use of toxic or gaseous precursors, such as, e.g., AsH_3, $AsCl_3$, or metalorganic chemicals.
7. It provides the possibility of a technological transfer from the established plating industry and, thus, a wide range of industrial experience.

The electrochemical deposition technique is particularly suited to the preparation of polycrystalline and amorphous thin-film semiconductors for applications in which the use of high-performance materials is not necessary and cost effectiveness is the most important aspect. For example, advanced optoelectronic devices for mid-infrared detection imply the use of both sources and detectors of low band-gap semiconductors, which, in principle, do not need to be monocrystalline but may be polycrystalline, 1–2-μm-thick films. Polycrystalline materials are also used for solar energy conversion, and in several cases good quality materials have been achieved that were appropriate for efficient devices. For example, high efficiencies (10.6% under 100 mW cm^{-2} illumination) were obtained with an all-electroplated CdS/p-MCT (mercury cadmium telluride) heterojunction solar cell, in which p-MCT was a ~ 1.2-μm thin film of polycrystalline Cd-rich mercury cadmium telluride [3]. A remarkable achievement of the electrochemical method is the preparation of polycrystalline thin films of materials with considerable molecular complexity, such as ternary chalcogenides. Further exciting developments have been anticipated for superlattices and other nanostructured semiconductors, as well as in microdiode arrays and quantum dots.

In addition, a variety of electrodeposition techniques, either single or in combination with others (i.e., direct current, pulsed current, direct plus pulsed current, electroless deposition, etc.), have been reported to produce better quality materials, but they have mainly been applied to the electrodeposition of metals and alloys.

Notwithstanding the numerous attractive features of the electrodeposition technique in the preparation of thin-film semiconductors, there are also several drawbacks, as described elsewhere in this chapter. Here, we only recall that the target materials often suffer from contamination with impurity phases, have a poor morphology, and are nonhomogeneous in their thickness, and their stoichiometry is far from the expected one.

Although the processes involved in compound semiconductor electrodeposition are quite complex because of the need to maintain stoichiometry, electrodeposition has successfully been applied to several II–VI compounds, notably binary and ternary Cd and Zn chalcogenides. In contrast, the electrodeposition of III–V compound semiconductors has not received as much attention, despite the increasing interest in them for a series of applications. Indeed, III–V compounds are found in several important areas of application of modern microelectronic techniques because of their use, e.g., in the production of advanced optoelectronic devices, infrared detectors and optical communication systems, photovoltaic and photoelectrochemical solar cells, semiconductor laser and light-emittting diodes (LEDs), very high-speed integrated circuits, high-speed computers, and magnetoresistive sensors.

Purpose of this review chapter is to chart the progress of the science of the formation by electrochemical deposition of III–V compound thin films. The challenging problems presented by the physicochemical characterization of these electrosynthesized compounds is also addressed, along with the valuable information gained on the thermodynamic and kinetic aspects of their formation and growth. Strategies for improving the compositional purity and morphological quality of these materials are considered, too. It is also pertinent to underline again the relatively advanced level of the existing knowledge base for II–VI compounds, whereas the synthesis of III–V compounds has been much less successful. The use of less conventional electrolytes (e.g., low-temperature organic melts) and more sophisticated techniques (e.g., electrochemical atomic layer epitaxy), however, may offer an attractive way out of these difficulties.

2. GROUP III–V COMPOUNDS

The typical III–V family consists of the nine 1 : 1 binary semiconductor compounds formed by the elements of group IIIA, Al, Ga, and In, and those of group VA, P, As, and Sb, that is, AlP, AlAs, and AlSb; GaP, GaAs, and GaSb; InP, InAs, and InSb. They crystallize with the zinc-blende (cubic) structure, and their atoms are tetrahedrally coordinated, forming approximately covalent sp^3 bonds, as shown in Figure 1a [4].

The zinc-blende structure belonging to the B-3 structural type may be related to the diamond lattice belonging to the A-4 type. In this latter structural type, two fcc lattices are shifted with respect to each other by 1/4 along the space diagonal. In the zinc-blende structure (e.g., ZnS), the sites of the first of the two fcc lattices are occupied by the atoms of one element (e.g., Zn), and those of the second of the two fcc lattices, by the atoms of the other element (e.g., S). The space group is $F\bar{4}3m$, and there are four molecules per unit cell.

As for the other two IIIA and VA elements, Tl and Bi, respectively, the situation is completely different. Indeed, no compounds are formed between Tl and P or As, as well as between Bi and Al or Ga. The phase diagram [5] shows the existence of the two compounds Tl_7Sb_2 [6] and $TlBi_2$ (with a homogeneity range between 34 and 46 at.% Tl [6]) both with a very low melting point, 226°C (but with a question mark [5]) and 213°C [5], respectively, as well as the existence of three In–Bi compounds, InBi, In_5Bi_3, and $InBi_2$, with even lower melting points (110°C for the former and around 90°C for the latter two). Notwithstanding its 1 : 1 composition, InBi does not belong to the typical III–V family, owing to its completely different structure, i.e.,

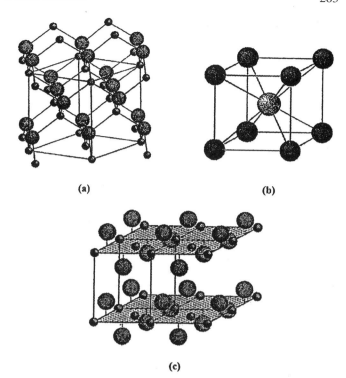

Fig. 1. Atomic arrangement in (a) zinc-blende (B-3 structural type), (b) cesium chloride (B-2), and (c) lithium hydroxide (B-10). Four unit cells are depicted in the last case to show the layered structure [4]. Reprinted from [89] with the kind permission of Elsevier Science.

a tetragonal unit cell with the atoms arranged in layers (Fig. 1c). Moreover, it has metallic bonds with fourfold coordination.

We will now describe the properties of typical III–V compounds in more detail. First of all, it is interesting to examine their phase diagrams, which are very simple. The main results may be summarized as follows.

All nine binary phase diagrams show the presence of a 1 : 1 binary compound that melts congruently. It does not form solid solutions with the two pure elements and has, with one exception (InSb), a higher melting point than both of them. Note, however, that the phase diagrams of Al–P and Al–As systems are speculative, owing to the few data available. They were assumed to be topologically similar to those of other III–V systems, such as, for example, GaAs. Moreover, the phase diagram of Al–Sb has been assessed from the data of several authors. Finally, the Ga–P diagram is incomplete, only ranging to a 50 at.% P content.

In some cases, there are one or, exceptionally, two eutectics between the compound and the pure elements:

Al–Sb, two eutectics at 657 ± 2.5°C and 0.4 at.% Sb and 627 ± 2.5°C and 98.5 at.% Sb
Ga–Sb, one eutectic at 588.5°C and 88.4 at.% Sb
In–Sb, one eutectic at 494°C and 68.2 at.% Sb
In–As, one eutectic at 731°C and 87.5 at.% As

These eutectics appear in all antimonides and, in the case of the In compounds, in arsenide, mainly at the side of the group V element.

Table I.   Percentage of the Ionic Character of III–V Bonds

	P	As	Sb
Al	9	6.5	4
Ga	6.5	4	2.5
In	4	2.5	1

Data are from [7].

Table II.   Standard Gibbs Free Energy (kJ mol^{-1}) of Formation of III–V Compounds

	P	As	Sb
Al	>−166.5	>−116.3	—
Ga	>−88	−67.8	−38.9
In	−77.0	−53.6	−25.5

Data are from [8].

Table III.   Energy Gap (eV) of III–V Compounds

	P	As	Sb
Al	—	2.15 (ID)	1.6 (ID)
Ga	2.3 (ID)	1.43 (D)	0.68 (D)
In	1.34 (D)	0.36 (D)	0.17 (D)

D and ID, direct and indirect gap transition, respectively. Data are from [10].

Table IV.   Melting Point (°C) of III–V Compounds and of Their Elements

	P (white, 44.14)	As (≈815)	Sb (630.755)
Al (660.452)	≈2500	1760	1058 ± 10
Ga (29.774)	1467 ± 3	1238	709.6
In (156.634)	1071	943	525

Melting points of elements are in parentheses. Data are from [5].

The chemical bond in the III–V compounds is not completely covalent, because group V elements are slightly more electronegative than those of group III elements, and this imparts a weak ionic character to the prevailing covalent bond. The percentage of ionic character of the bond in each compound, determined from the difference in the electronegativities of V and III elements, is collected in Table I [7]. In this and in the following tables of this type, each compound is virtually presented at the intersection point of the horizontal line, corresponding to its group III element, with the vertical line corresponding to its group V element. At this same point, the value of the considered quantity is reported, in the present case of the percentage of ionic character.

The ionic character of the bonds decreases from left to right and from top to bottom, so that AlP is the most heteropolar and InSb the least heteropolar compound. In this same direction moves the stability of the compounds toward their bond rupture, as expressed, e.g., by their Gibbs free energy of formation from the elements in standard conditions [8], which is reported in Table II. Unfortunately, not all experimental data are available, as for AlP, AlAs, and GaP the enthalpy of formation only is known, as often occurs. However, the entropy contribution is expected to decrease during the formation in solid-state conditions of the compound from the elements, thus implying that the Gibbs free energy of the process is higher (less negative) than the related enthalpy, as indicated in Table II. The correlation between the values connected to the difference in electronegativity and the thermodynamic values is easy to understand.

InSb has the weakest bond among the typical III–V compounds but InBi has an even weaker bond; its Gibbs free energy is only −3.72 kJ mol^{-1} [9]. So, InBi is considered to be an intermetallic compound, and all typical compounds of the III–V family are semiconductors, but the values of their band gap energy decrease from left to right and from top to bottom, just like the ionic character of their bonds. This is shown in Table III [10]. GaP has the highest value (2.3 eV), and InSb, the smallest one (0.17 eV). Owing to its very low band gap, InSb, too, is often considered as an intermetallic compound [11]. The gap transition is mainly a direct one (D in Table III); those of the Al compounds and GaP, however, are indirect ones (ID in Table III).

Although the structure is the same, the melting points of the nine compounds are different and reflect the strength of the bonds, as evident is from a comparison of the values reported in Table IV to those of Table I. The melting points of the single elements are also shown in the table. They are listed in parentheses close to the chemical symbol of the pure element, to clearly show the differences in the values of the compounds and their elements. Indeed, the melting points of both elements and III–V compounds are important, because several of them are prepared in molten salt electrolytes at high temperatures or are deposited at room temperature and then submitted to thermal annealing to complete the process of formation through diffusion and reaction (mainly in the solid state) and/or to improve the crystallinity and morphology of the samples. Moreover, In and, more especially, Ga have low melting points, and this may create several problems during the electrochemical process of compound formation and the subsequent heat treatment. Furthermore, about half of the value of the activation energy of the diffusion process is due to the energy of formation of the vacancies [12], which is strictly related to the melting temperature of the phases where diffusion occurs, namely that of the elements and of the compound. The comparison of the melting temperature of the III group element with that of the V group element forming a compound through diffusion and reaction may indicate which of them is prevailingly diffuse. So, for example, Ga and In are expected and observed to be the diffusing species, respectively, in the case of GaSb and InSb formation.

Finally, it is also interesting to compare the lattice constants of the nine III–V compounds that have the same cubic struc-

Table V.   Lattice Constants (Å) of the Cubic Zinc-Blende Structure of III–V Compounds

	P (1.10)	As (1.18)	Sb (1.36)
Al (1.26)	5.451	5.662	6.1347
Ga (1.26)	5.4506	5.6538	6.095
In (1.44)	5.869	6.058	6.4782

Atomic radii for tetrahedral coordination of the elements are in parentheses. Data are from [6, 13].

Table VI.   Possible Modulation Ranges of the Energy Gap and of the Lattice Constant of III–V Compounds

Ternary compounds	Binary compounds	Energy gap (eV)	Lattice constant (Å)
Ga–As–Sb	GaAs–GaSb	1.43–0.68	5.6538–6.095
Ga–In–Sb	GaSb–InSb	0.68–0.17	6.095–6.4782
In–As–P	InAs–InP	0.36–1.34	6.058–5.869
In–As–Sb	InAs–InSb	0.36–0.17	6.058–6.4782

Data are from [10] and [6].

ture [6]. In this case, there are at least two factors influencing such parameters, the steric and chemical interactions. The former is expressed by the dimension of the elements as given by their atomic radii for tetrahedral coordination [13], the latter by the bond strength. They influence the lattice constant in opposite ways, but the former is expected to be the prevailing factor, as experimentally observed. Indeed, as shown in Table V, the lattice constants increase from left to right and from top to bottom. To more clearly demonstrate this correlation, the values of the atomic radii have also been listed in the table, close to the symbol of each element. Because Al and Ga have the same atomic radius, the lattice constants of the compounds in the first two lines have practically the same value, which increases from left to right. The mass densities ($g \, cm^{-3}$) follow the same trend as that of the lattice constants.

Interesting correlations among binary III–V semiconductor compounds regarding some of their physical properties and parameters (e.g., lattice dynamics, piezoelectric and optical properties, electron affinity, effective mass, and variation with pressure of the band gap energy) are reported in a review by Adachi [14].

An attractive opportunity provided by the electrochemical approach is the ease with which substitutional alloys may be generated, not only in the case of metal alloys but alloy compounds. Some examples of group III–V combinations that have been electrosynthesized are those resulting from the GaAs–GaSb, GaSb–InSb, InAs–InP, and InAs–InSb systems, which will be examined in the following. They are interesting because both the band gap and the lattice constant can in principle be modulated over a wide range, as shown in Table VI. However, according to Rajeshwar [15], the greater number of active

species in the case of the electrosynthesis of semiconductor alloys with respect to that of metal alloys renders the composition modulation a little more difficult. Adachi [14] suggested an interpolation scheme as a useful tool for estimating some physical parameters of ternary alloy compounds from those of the corresponding binary compounds (e.g., $Al_xGa_{1-x}As$ from AlAs and GaAs), whereas Tomashik [16] proposed a composite tetrahedral heptahedron enabling graphic representation of all possible systems formed by 16 III–V compounds consisting of four cationic and four anionic species. As a matter of fact, multicomponent systems of compound semiconductors are of great interest in the search for new semiconducting materials with superior properties. However, the investigation of such systems is a very complex and labor-consuming task, and such a graphic representation may be particularly useful when the number of constituent components increases.

In addition to alloys, superstructures (e.g., superlattices) may be considered, too. They differ from the former because their composition is modulated along one direction, whereas the plane perpendicular to it is compositionally homogeneous. The current interest in nanomodulated superstructures lies in their unique optoelectronic properties and interesting quantization effects [17]. A typical example is the CdSe-ZnSe superlattice, with alternating layers of CdSe and ZnSe along the z reference axis. This ternary system may be prepared by potential or current pulsing between the limits characteristic of CdSe and ZnSe formation [15, 18, 19].

The next challenge lies in securing truly nanomodulated thin films. The recent electrochemical technique called ECALE (electrochemical atomic layer epitaxy) makes it possible to achieve nanometer-thick epitaxial films of compound semiconductors from aqueous solutions through layer-by-layer deposition under ultrahigh purity conditions [20, 21]. It is based on a thin-layer methodology and on underpotential deposition to alternately lay down atop a suitable substrate a monolayer of each species of a binary compound, e.g., Ga–As or In–As, as discussed below. Such a technique is also potentially applicable to the electrosynthesis of superstructures and has recently been applied in the case of the CdS-ZnS system [22].

Another possible new technique is the electrochemical fabrication of microdiode arrays, which consists of the electrochemical synthesis within the pores of special membranes of an array containing, e.g., more than 10^9 CdSe or graded CdSe/CdTe cylinders with a diameter of 200 nm. After removal of the membrane (e.g., by dissolution in aqueous NaOH), an array of wires freely standing on a metal substrate is obtained. Current–voltage data show that the Ni-CdSe array is rectifying, with a rectification ratio of 1000 at ±2 V [23].

The unique size-dependent, optical, photocatalytic, and nonlinear optical properties of colloidal nanocrystalline semiconductor particles, called quantum dots, continue to attract considerable interest. In contrast to the preparation of high-quality quantum dots of II–VI semiconductors, that of III–V semiconductors has proved to be problematic. However, the synthesis and properties of well-crystallized GaP and InP quantum dots have recently been reported; the diameter of the former ranges from 20 to 65 Å, and that of the latter is about 25 Å [24]. They were obtained by heating a solution containing the precursor species (GaP and InP) with a mixture of trioctylphosphine oxide and trioctylphosphine as a colloidal stabilizer. Instead, the Argonne National Laboratory [25] used an electrochemical technique to produce quasi-periodic quantum dot arrays, with excellent control over dot size and interdot spacing to be used in microelectronic applications. The manufacturing costs are expected to be orders of magnitude lower than those of conventional nanofabrication.

3. ELECTRODEPOSITION

Fundamental thermodynamic and kinetic considerations related to the electrodeposition of alloy and compound semiconductors are presented, with emphasis on the processes of interest in the case of binary III–V semiconducting compounds.

Although the main aspects of metal electrodeposition may be applied to semiconductor electrodeposition, semiconductors present some typical properties that cannot be ignored:

1. Inside a semiconductor, at its interface with the electrolyte, a typical space-charge layer is formed that is practically nonexistent at the metal–electrolyte junction.
2. Semiconductors have a much higher resistivity than metals, and the interfacial potential and charge distribution may change dramatically during electrodeposition, once some semiconducting layers have been deposited.
3. Semiconductors are sensitive to several defects that influence their resistivity, so that, in extreme cases, they may even become a degenerate solid, losing their characteristic properties.
4. Owing to other properties, electrochemical behavior, particularly the current density–potential characteristics of semiconductors, is more complex than that of metals, as shown by Gerischer (e.g., [26]).

Characteristic aspects of compound semiconductor electrodeposition are (i) the very negative Gibbs free energy of compound formation from the elements, which may be beneficial for simultaneous electrodeposition, and (ii) the great difference in the standard reduction potentials of the different components, owing to which several difficulties are encountered, particularly in the case of III–V compounds formed from a typical metallic (Al, Ga, In) and nonmetallic (P, As, Sb) element. Indeed, for simultaneous deposition the reduction potentials have to be equal.

The most common methods used to prepare thin films of III–V group compounds are simultaneous cathodic electrodeposition (codeposition), a pure electrochemical method, and sequential (consecutive) electrodeposition of precursor multilayers with subsequent annealing treatment, a mixed electrochemical and thermal method. Another mixed method, typically applied to phosphides (such as InP), is the preliminary electrodeposition of the metal on an inert substrate followed by a thermal treatment in the presence of the other component (P or PH_3) in the gaseous phase. Several other methods are applied, mainly to prepare II–VI group semiconductors, which have been studied more extensively, and reference is made to [27].

More details on semiconductor electrodeposition, including deposit characterization techniques, are reported in a recent handbook [10].

4. CODEPOSITION: BASIC CONSIDERATIONS

Cathodic codeposition makes it possible to directly obtain an alloy or a compound in a single electrochemical step and, when the control of the material resistivity is possible, to prepare even high-thickness films. This method has a wide application for the preparation of several members of the III–V semiconductor family, from both aqueous and molten salt solutions, as well as of many binary and ternary chalcogenides.

The codeposition of alloy and compound semiconductors is quite difficult, as the conditions favorable for deposition of one component normally differ from those necessary for the other one. Moreover, electrodeposition is complicated if a desired stoichiometry is requested, which is the rule for semiconductors.

The factors governing codeposition at a given current density and temperature are (e.g., [28–31]) (i) the deposition potential of the individual ions in the electrolyte, (ii) the cathodic polarization caused by the difference in deposition potentials, (iii) the relative ion concentration in the bath, (iv) the dissolving tendency of the deposited compound, and (v) the hydrogen overpotential on the deposit surface at the cathode. The role of these parameters has been discussed by Brenner [28], mainly for alloy electrodeposition.

4.1. Thermodynamic Aspects

Kröger [29] analyzed the influence of both the characteristic thermodynamic parameters of the codeposition method and the experimental variables on the deposit composition. He based his analysis on a model that was applied to the case of the cathodic deposition of alloys or of binary compounds with a well-defined stoichiometry and metallic or semiconducting properties. This theory was then shown to also be applicable to the case of the cathodic deposition of the ternary semiconducting compound $CuInSe_2$ [32]. In the following discussion, the formation of binary compounds only will be examined.

Following Kröger's theory [29], let us consider the simultaneous electrodeposition of the two species M and N to prepare the $M_r N_s$ compound, where N is the more noble of the two elements, i.e., that forming first. This means that its potential in standard conditions, E_N°, is greater (more positive) than that of the other one.

According to the well-known Nernst equation, the equilibrium potentials of M and N in a solution of their ions, E_M and E_N, respectively, depend on the activities of the same ions in the electrolyte, $a(M^{m+})$ and $a(N^{n+})$, and on the activities of M and N in the deposit, $a(M)$ and $a(N)$:

$$E_M = E_M^\circ + (RT/mF)\ln\left[a(M^{m+})/a(M)\right]$$

$$E_N = E_N^\circ + (RT/nF)\ln\left[a(N^{n+})/a(N)\right]$$

The latter activities are equal to 1 if the deposit is a pure, perfect element and are smaller than 1 if the deposit is a compound. R and F are the gas and Faraday constants, respectively, and T is the temperature.

As true reversible potentials are rarely met in practice, Kröger [29] introduced the concept of quasi-rest potentials to help explain zero current conditions that otherwise only approximate true thermodynamic reversibility. The electrode potentials at which codeposition is possible are equal to the quasi-rest potentials of the compound plus a polarization increasing with the current density. Because polarization is established with a relaxation time not higher than 10^{-8} s, the quasi-rest potential can be determined by opening the electrolysis circuit, so that overpotentials and ohmic drops are absent, and measuring the electrode potential within about 10^{-3} s after interruption of the deposition current density.

Quasi-rest potentials are important for characterization of the bulk of the deposit. They originate in four factors: (i) the equilibrium potentials of the components, which may differ widely; (ii) the changes in the activities of the components in the deposit when the compound is formed; (iii) the interfacial activities of the ionic species during electrodeposition, which may significantly differ from their bulk activities; and (iv) the relative magnitudes of the exchange current densities of the components in the deposit.

Kröger assumed that the quasi-rest potential is an equilibrium potential of the deposit relative to the electrolyte with activities of the potential-determining species, as they are at the solid–electrolyte interface during electrodeposition. With this assumption, the first three factors can be taken into account by expressing previous potentials, E_M and E_N, in terms of the activities of the ionic species at the deposit–electrolyte interface during electrodeposition and of the activities of the components in the deposit. The former activities can often be approximated by the concentrations as they are at the interface, whereas this is usually not possible for the latter activities. E_M and E_N are now the quasi-rest potentials.

Electrodeposition of M and N usually occurs at potentials more negative than the quasi-rest potentials, the difference being the overpotential, η_M and η_N. The latter is related to the kinetic factors influencing the different steps of the deposition process, such as charge transfer, crystallization, mass transport, and so on, and increases at increasing current density. The conditions necessary to obtain the simultaneous deposition of the two kinds of ions at the cathode can be written as

$$E_M + \eta_M = E_N + \eta_N$$

This equation may be fulfilled in different ways, but two common cases may be considered. If the deposition potentials are not far apart, they may be brought together by decreasing the ionic activity of the more noble species, N, with respect to that of the less noble one, M. However, there is a limit to the application of this method related to the logarithmic dependence of E_N on the activity of N^{n+} ions, also because extremely diluted electrolytes in an ionic species would soon be depleted. Instead, if E_N is markedly different from E_M. Three kinds of approach are possible. (i) A chemical approach, e.g., by introducing a complexant into the bath, decreasing the activity of the discharging N^{n+} ions through the formation of a complex. (ii) An electrochemical approach, e.g., by increasing the cathodic overpotential of N by adding some particular agents. (iii) An instrumental approach, e.g., by performing a potentiostatic electrolysis in conditions of the limiting current density of N, so that E_N may be decreased as desired. A practical rule is that codeposition requires that the deposition potentials differ by less than 200 mV. Another way to solve this problem is to take advantage of the favorable Gibbs free energy of formation of the compound from the elements, ΔG.

The activities $a(M)$ and $a(N)$ of M and N appearing in the Nernst equations change their values during electrodeposition as a consequence of a change not only in the concentration of M and N in the deposit, but also in their physicochemical state, e.g., whether they are part of a compound or not. So, the activities in the deposit also reflect the environment of its constituents. They are related through the solid-state reaction

$$rM + sN = M_rN_s + \Delta G$$

where ΔG (G in Kröger's paper [29]) is the connected Gibbs free energy and $r = s = 1$ in the case of III–V compounds. Assuming that such a reaction occurs in equilibrium conditions, it results in

$$a(M_rN_s)/\left[a(M)^r \times a(N)^s\right] = \exp(-\Delta G/RT)$$

where $a(M_rN_s) \approx 1$ for all practical purposes. Hence,

$$\left[a(M)^r \times a(N)^s\right] = \exp(\Delta G/RT)$$

so a high value of $a(M)$ gives rise to a small value of $a(N)$, and vice versa, whereas the activities of M and N in the compound are determined by their concentrations and by the thermodynamic stability of the deposit [11].

The limiting values of $a(M)$ and $a(N)$ are determined by the activities of the coexisting phases in the phase diagram. Assuming that M_rN_s is the only compound in the system, as in the case of the considered III–V compounds, almost pure M and N are in coexistence with the compound at the M_rN_s–M and N–M_rN_s phase boundaries, respectively, and the limiting values are

$$a(M) = 1, \quad a(N) = \exp(\Delta G/sRT)$$

and

$$a(N) = 1, \quad a(M) = \exp(\Delta G/rRT)$$

respectively.

The variations of $a(M)$ and $a(N)$ over the existence range of the compound produces a related variation in the quasi-rest potential of M and N; the total variations are

$$\Delta E_M = -\Delta G/mrF \quad \text{and} \quad \Delta E_N = -\Delta G/nsF$$

respectively.

Owing to the negative value of the Gibbs free energy of formation of the compound, ΔG, the quasi-rest potentials of both components shift in the anodic direction when their activity in the deposit decreases, that is, the formation of the compound produces a positive shift of the individual potentials.

To proceed further, it is important to consider whether the deposition rate constants appearing in the exchange current densities of the two individual components, M and N, are of the same order of magnitude or differ considerably.

In the former case, the two species have equal weights in determining the potential. The species with the larger value of the product of the discharge rate constant and the activity in the electrolyte at the deposit–electrolyte interface is believed to be dominant in determining the potential.

In the latter case, the component with the highest rate constant may be determing the potential under all conditions and is favored over the other one. The treatment of this last case requires the exact knowledge of the kinetics of the charge transfer process and cannot be generalized. Therefore, the first case only, the most interesting for compound electrodeposition, will be discussed, as is usually done (e.g., [10, 27]).

Because a deposit of one composition can have only one quasi-rest potential, the necessary condition for codeposition to form the compound is

$$E(M_rN_s) = E_M = E_N$$

Taking this equality of the quasi-rest potentials into account, a relation is obtained among the activities, both of the ionic species in solution at the electrolyte–deposit interface and of the components in the solid, and the standard potentials,

$$(RT/F)\ln\left[a\left(M^{m+}\right)^{1/m}/a\left(N^{n+}\right)^{1/n}\right]$$
$$= \left(E_N^\circ - E_M^\circ\right) + (RT/F)\ln\left[a(M)^{1/m}/a(N)^{1/n}\right]$$

or, at the M_rN_s–M phase boundary, where $a(M) = 1$,

$$(RT/F)\ln\left[a\left(M^{m+}\right)^{1/m}/a\left(N^{n+}\right)^{1/n}\right]$$
$$= \left(E_N^\circ - E_M^\circ\right) - \Delta G/nsF \quad (1)$$

and at the N–M_rN_s phase boundary, where $a(N) = 1$,

$$(RT/F)\ln\left[a\left(M^{m+}\right)^{1/m}/a\left(N^{n+}\right)^{1/n}\right]$$
$$= \left(E_N^\circ - E_M^\circ\right) + \Delta G/mrF \quad (2)$$

The last term in both equations has been obtained by introducing the activities of the components in the deposits expressed as a function of ΔG, which was previously considered.

Equations (1) and (2) determine the ratio in which the various species contribute to the exchange current density, that is, which of them is the main potential-determining species.

Because it has been assumed that N is more noble than M, $(E_N^\circ - E_M^\circ)$ is positive, and the thermodynamic quantity is negative. Hence, the second member of Eq. (1) is always positive.

This means that at the M_rN_s–M phase boundary, $a(M^{m+}) \gg a(N^{n+})$, and M is the potential-determining species for all possible compositions of the compound. It is worth noting that the activities of the ionic species in the electrolyte are those close to those at the deposit surface; the bulk are not. At the N–M_rN_s phase boundary [Eq. (2)], two cases arise, depending on whether $(E_N^\circ - E_M^\circ)$ is greater or smaller than $(-\Delta G/mrF)$, namely, whether the difference in the reversible deposition potentials of the individual components N and M is larger (class I) or smaller (class II) than the shift in deposition potential of the less noble component M, as a result of compound formation.

4.1.1. Class I: $(E_N^\circ - E_M^\circ) > (-\Delta G/mrF)$; $a(M^{m+}) \gg a(N^{n+})$

Although the ratio between the two activities is not as large as at the M boundary, M remains the potential-determining species over the entire deposition range. The condition for codeposition can be realized by maintaining a low N^{n+} ion concentration in the electrolysis solution. This means that the rate of the electrodeposition process is controlled by the diffusion of the more noble species of the two. An increase in the mass transport in the solution (e.g., by stirring the bath) increases the surface activity of the N^{n+} ions, thus increasing the molar fraction of N in the deposit. In these systems, the codeposition potential shifts monotonically with composition and varies between a minimum value equal to $E_M^\circ + (RT/mF)\ln[a(M^{m+})]$, where the deposit, rich in M, is formed by the compound and an excess of M, and a maximum value differing from the minimum one by $(-\Delta G/mrF)$, where the deposit is rich in N and contains an excess of N over the compound composition. This is shown in Figure 2.

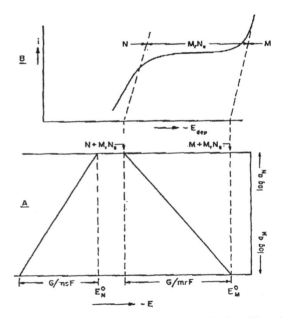

Fig. 2. (A) Equilibrium potentials E of M and N as a function of the activities and (B) corresponding current density vs. cathode potential curve for deposition of class I compounds. Reproduced from [29] with the kind permission of the Electrochemical Society, Inc.

Figure 2A depicts the variation in the M and N potentials, with the activities of M and N in the deposit for constant values of $a(M^{m+})$ and $a(N^{n+})$ equal to one. This figure does not represent the situation in the electrolyte at the deposit–electrolyte interface during electrodeposition, but it indicates how differences in the nobility of the two components are affected by their interaction in the deposits. It shows which species must be expected to be potential determining and helps in selecting an electrolyte composition that allows codeposition. For example, codeposition of the noble N and the less noble M is only possible if $a(M^{m+}) \gg a(N^{n+})$, so that the curve due to N in Figure 2A is shifted toward more cathodic potentials to intersect that due to M. For electrolytes with ion activities differing from those of the figure, the curves have to be shifted accordingly.

Figure 2B depicts the corresponding current density–cathode potential curve for M_rN_s deposition from an electrolyte with $a(M^{m+}) \approx 1$ and $a(N^{n+}) \ll 1$. Moving in the cathodic direction, pure N is deposited first, whereas the value of $a(N^{n+})$ at the interface gradually decreases. When it is close to zero, a plateau is reached where the M_rN_s compound is formed. The current density at the plateau is proportional to $a(N^{n+})$. A further sharp increase in the current density occurs at the potential where deposition of pure M becomes possible. Owing to the relatively high M^{m+} concentration in the electrolyte, the condition $a(M^{m+}) \approx 0$ at the interface is reached at extremely large values of current density, that is, at large cathodic potentials not shown in the figure. Moreover, because polarization (overpotentials plus ohmic drop) increases when the current density increases, the deposition potentials are shifted in Figure 2B to the negative side relative to the quasi-rest potentials by polarization.

The formation of the compound involves the deposition of both M and N, and their interaction makes the deposition of M possible at potentials more positive than that required for pure M. According to Brenner [28], N, the more noble species, which deposits first, has induced the deposition of the less noble one. Examples of this so-called induced codeposition or underpotential deposition have been observed in several cases, such as for GaAs formation in molten salts [33] and formation of Cd and Zn chalcogenides [29].

4.1.2. Class II: $(E_N^\circ - E_M^\circ) < (-\Delta G/mrF)$; $a(M^{m+}) \ll a(N^{n+})$

At the M phase boundary, M is still the potential-determining species, whereas at the N boundary, it is now N, the more noble species. In this case, the potential-determining species changes from the M to the N phase boundary. The cross-over takes place at intermediate activities of the components in the deposit near the point where the contributions of M and N to the exchange current density are of the same order. At or near this point, both species are involved in determining the potential. For class II, codeposition is also possible for comparable values of the activities of the two ionic species in solution. In this case, the deposit contains the two components at comparable quantities,

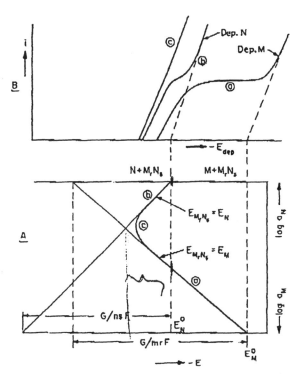

Fig. 3. (A) Equilibrium potentials E of M and N as a function of the activities and (B) corresponding current density vs. cathode potential curves for deposition of class II compounds from electrolytes of different compositions. Reproduced from [29] with the kind permission of the Electrochemical Society, Inc.

and the codeposition potential is more positive than that of either M or N when they are deposited individually. So, owing to the very negative value of the Gibbs free energy of formation of the compound, codeposition is thermodynamically favored over the deposition of either species, M or N. Figure 3 illustrates the new situation.

Figure 3A shows that the two curves intersect for class II, so that codeposition is also possible when $a(M^{m+}) = a(N^{n+}) = 1$. As both activities are high, no appreciable exhaustion occurs, as indicated by the corresponding current density–potential curve c in Figure 3B. Curve a in the same figure refers to the case already considered for class I, where $a(M^{m+}) \approx 1$ and $a(N^{n+}) \ll 1$ and, therefore, is similar to the only curve in Figure 2B. Hence, the deposition of M-rich products requires an electrolyte with $a(M^{m+}) \gg a(N^{n+})$, as for class I, and the rate of the deposition process is controlled by diffusion of the more noble species, N. A change in current density or rate of stirring makes it possible to obtain deposits of various compositions but all at the M-rich side of the system. Instead, curve b refers to the new case $a(M^{m+}) \ll 1$ and $a(N^{n+}) \approx 1$, where N-rich products are formed. So, the deposition of N-rich products (which occurs over a limited deposition range) requires an electrolyte with $a(M^{m+}) \ll a(N^{n+})$. N is now the potential-determining species, and the rate of the deposition process is controlled by the diffusion of the less noble species, M. Because for class I, codeposition is only possible when the ratio $a(M^{m+})/a(N^{n+}) \gg 1$, for class II, it is also possible when this ratio is much lower than (case b of Fig. 3) or even equal to (case c) 1. In the latter case, $a(M^{m+}) \approx a(N^{n+}) \approx 1$. For class I, quasi-rest potentials uniquely characterize the deposits, whereas for class II, different deposits may have the same quasi-rest potential.

The class II behavior has been observed for several metal alloys, e.g., Ni–Sn [34], Co–Sn [35], and Co–Sb [36].

Finally, it is worth recalling that in 1988 Engelken [37] named the class I and class II electrodepositions of Kröger's pioneering work conventional and pure underpotential depositions, respectively (i.e., CUD and PUD). This was done to distinguish the first case, in which only the less noble species can be deposited in underpotential conditions, owing to compound formation, from the second case, in which both species can be deposited.

4.2. Kinetic Aspects

In addition to the thermodynamic aspects of the electrodeposition process, kinetic aspects are also important in controlling the quality of the deposits.

In general, the current–potential characteristic (voltammetric curve) of the codeposition process is not the resultant of the characteristics of the two partial processes. So, knowledge of these processes does not automatically lead to a forecast of the nature (eutectic, solid solution, chemical compound) and composition of the deposit. Moreover, although similar in some respect to metallic alloy electroplating, as described by Brenner [28], the very negative Gibbs free energy of formation of III–V (and II–VI) compounds, the prevailing semiconducting nature of the deposits, the large difference in the standard reduction potentials of the nonmetallic and of most metallic species, the changing nature of the deposit during deposition, and the possible presence of metastable states introduce additional complexities into any theoretical description of the process [38]. Therefore, few theoretical contributions have been published on compound semiconductor eletrodeposition [28, 29, 39, 40]. However, Engelken et al., in a series of papers [37, 38, 41, 42], developed a simple kinetic model for the electrodeposition of CdTe, a typical class I system, from acidic solutions containing reducible ions of both constituents. This model is also applicable to other II–VI or to III–V binary compounds. The electrochemical reactions involved in the case of CdTe deposition are

$$Cd^{2+} + 2e = Cd$$
$$3H^+ + HTeO_2^+ + 4e = Te + 2H_2O$$

and finally,

$$Cd + Te = CdTe$$

where e indicates an electron, the ions are in the aqueous phase, and the elements and the compound are in the solid phase. This is a typical example of induced codeposition, according to the classification by Brenner [28], who divided the reaction mechanisms into induced and regular codeposition.

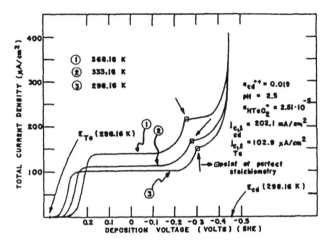

Fig. 4. Plots of total current density vs. deposition potential with temperature as a parameter. Reproduced from [41] with the kind permission of the Electrochemical Society, Inc.

The model of Engelken et al. is based upon a generalized Butler–Volmer equation that considers ion transport limitations near the cathode. Although deposition itself is a nonequilibrium process, the reaction between the deposited Cd and Te is assumed to be sufficiently rapid that any infinitesimally small volume in the deposit remains in quasi-chemical equilibrium conditions. Notwithstanding other simplifying assumptions [38], the model yielded a convenient computational algorithm that was used to numerically simulate the voltammetric curves related to the codeposition of CdTe from Cd^{2+} and $HTeO_2^+$ solutions, for different values of the experimental variables, particularly the deposition temperature, and to calculate the mole fractions and activities of Cd, Te, and CdTe at the deposit surface for any value of the deposition potential.

The simulated curves were similar to those already described by Kröger [29] for class I systems, presenting the typical current density plateau due to the diffusion of the $HTeO_2^+$ ions, as shown in Figure 4 [41]. The onset of the cathodic current density took place at the Nernst potential of Te. So, Te only was initially deposited, and because of ion transport limitations, the current density soon saturated, reaching the limiting value for Te deposition (main plateau region). Then, the current density increased again, owing to Cd underpotential deposition and formation of the first CdTe crystals. This deposition proceeded, and the deposit became richer and richer in Cd, finally achieving the theoretical composition of the compound at the potential of perfect stoichiometry. At this potential, equal numbers of moles of Cd and Te were deposited. Cd deposition continued in underpotential conditions reaching saturation.

Figure 5 [41] depicts the dependence of the deposit composition (mole fraction of Cd, Te, and CdTe) on the deposition potential for curve 3 reported in Figure 4. A shift of the potential in the cathodic or anodic direction from the potential of perfect stoichiometry involved a significant abrupt increase in the Cd or Te atomic fraction, respectively, whereas the mole fraction of the compound changed only slightly around the potential of perfect stoichiometry. So, pure CdTe was practically formed at

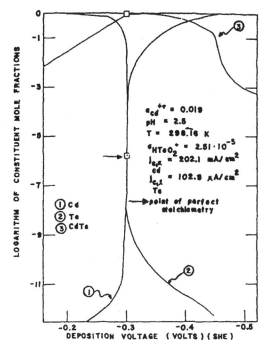

$a_{Cd}^{++} = 0.019$
pH = 2.5
T = 298.16 K
$a_{HTeO_2^+} = 2.51 \cdot 10^{-5}$
$i_{c,L}^{Cd} = 202.1$ mA/cm^2
$i_{c,L}^{Te} = 102.3$ μA/cm^2

← point of perfect stoichiometry

① Cd
② Te
③ CdTe

Fig. 5. Logarithmic plots of constituent mole fractions vs. deposition potential for curve 3 in Figure 4. Reproduced from [41] with the kind permission of the Electrochemical Society, Inc.

the potential of perfect stoichiometry only, whereas CdTe containing an excess of either Cd or Te was formed at the more or less negative potentials, respectively, and the material became *n*- or *p*-type, consequently.

According to the authors, the model allows computer simulation of the electrodeposition process for metal selenides, tellurides, arsenides, and antimonides and leads to several interesting predictions [41]. Following this same kinetic model, computer simulation was also performed for CdTe electrodeposition at 160°C from an ethylene glycol-based bath, and the results were discussed within the framework of those obtained in aqueous solutions [43].

These simulations consider the charge transfer process with ionic species diffusion into the electrolyte and formation of the compound in the solid state, but they ignore effects such as surface adsorption, limitations in nucleation, and crystal growth, as well as second-order effects, such as hydrogen evolution, generation or reduction of impurities, the possible semiconducting nature of the deposit over a narrow stoichiometric range, and incomplete or slow metal–nonmetal reactions.

5. CODEPOSITION FROM AQUEOUS SOLUTIONS

Typical examples of regular codeposition according to Brenner [28] are the synthesis of InSb and InAs [44, 45] in the presence of buffered solutions (citric acid and citrate) containing InCl$_3$ and SbCl$_3$ or AsCl$_3$, respectively. The citrate ions act as the complexing agent of group V ions, so that their reduction potential becomes closer to that of the In ions and codeposition

is possible. The following is a schematic representation of the electrochemical reactions (citrate representing the citrate ions):

$$In^{3+} + Sb^{3+}\text{-citrate} + 6e = InSb + citrate$$
$$In^{3+} + As^{3+}\text{-citrate} + 6e = InAs + citrate$$

The resulting process is the simultaneous deposition of the two types of ions under diffusion control of the more noble species, Sb or As. The stoichiometry of the compound is thus obtained by controlling the composition of the bath and the mass transport in the solution.

Critical technical parameters for semiconductor codeposition are the relative concentrations of the two ionic species in the electrolyte, the pH of the solution, temperature, current density, and deposition time. Research typically investigates the influence of these same parameters on the deposit properties.

Cathodic codeposition from aqueous media has scarcely been performed in the case of III–V compounds, because it was often ineffective. So, the relevant literature, particularly the earlier papers, is not as extensive as that for II–VI compounds. A significant analysis of this subject has recently been done [46]. It concludes that phosphides cannot be electrodeposited because of the P resistance toward cathodic deposition, except in induced deposition, which is not observed with group III elements. However, direct InP electrodeposition from aqueous electrolytes containing hexafluorophosphates has recently been reported [47, 48], although with unsatisfactory results. The deposition of Al compounds is precluded by the impossibility of depositing this metal from aqueous media. GaAs and GaSb present several difficulties, owing to the great difference in the deposition potentials of Ga and As or Sb, respectively, as the standard potential of the first element is very negative, and those of the latter two are too positive. So, the only III–V compounds that appear to be directly accessible by cathodic codeposition from aqueous baths are InSb [44, 45] and, to a lower extent, InAs [45].

5.1. Pourbaix's Equilibrium Diagrams

The thermodynamic stability of III–V semiconductors in aqueous solutions is usually evaluated with the help of the potential–pH equilibrium diagrams by Pourbaix [49] for the constituent–water systems, at 25°C. These curves show the dependence of the electrode potential [vs. standard hydrogen electrode (SHE)] on the pH value of the electrolyte (as the ion activity is pH dependent) for a constant concentration of the dissolved substances as well as for a constant partial pressure of the gaseous substances. They are valid only in the absence of substances with which the elements can form soluble complexes or insoluble salts. They have been used, for example, as a guide to fix the preliminary electrodeposition conditions to obtain corrosion-free, pure electrodeposits. As an example, Figure 6 shows a simplified representation of the Pourbaix diagrams for the Ga–H$_2$O, In–H$_2$O, As–H$_2$O, and Sb–H$_2$O systems, separately reflecting the corrosion, passivation, and immunity domains, but not indicating the different thermodynamically possible chemical and electrochemical reactions. During corrosion,

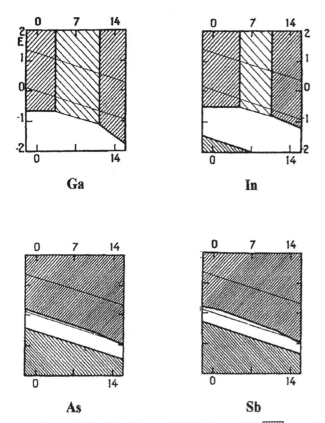

Fig. 6. Simplified Pourbaix diagrams of Ga, In, As, and Sb. ▨▨, corrosion by dissolution; ▨▨, corrosion by gasification; ▨▨, passivation by oxide or hydroxide layer; ☐, immunity. Reproduced from [49] with the kind permission of CEBELCOR.

soluble or gaseous products are formed (corrosion by dissolution or gasification, respectively), whereas passivation occurs by an oxide or hydroxide or even hydride film covering the material surface. In the immunity domain, the solid stable form is the element itself; the corrosion reaction is thermodynamically impossible. In the passivity domain, the solid stable form is not the metal but an oxide, hydroxide, hydride, or a salt. For the definition of the corrosion threshold in these representations, Pourbaix arbitrarily adopted a value of 10^{-6} M for the concentration of the soluble species (of Ga, In, As, or Sb, in our figure) and of 10^{-6} atm for the partial equilibrium pressure of the gaseous species (hydrides of In, AsH_3, and SbH_3). As shown in Figure 6, the system $Ga-H_2O$ presents two corrosion domains, one at low pH (with formation of Ga^{3+} ions) and the other at high pH (with formation of GaO_3^{3-}), between which there is a domain of passivation by an oxide or hydroxide layer. A similar diagram is shown by the $In-H_2O$ system, which has, in addition, a domain due to the existence of In hydrides at the more cathodic potential values and at low pH. Instead, the diagrams of $As-H_2O$ and $Sb-H_2O$ are different from the previous ones because they still present two corrosion domains, extending all over the considered pH range (from −2 to 16). The anodic one is due to corrosion and dissolution of the scarcely protective oxides, particularly those of Sb, and the cathodic one is due to

gasification to AsH_3 and SbH_3. Between these two domains lies a small immunity domain.

To attain the codeposition of the two species and thus the formation of the pure binary compound (e.g., GaAs), both potential and pH values must be selected, providing a common immunity domain for Ga and As. Indeed, for codeposition to occur, the concentration of the soluble species and the pH of the electrolyte should be adjusted so that the electrode potentials of all of the individual deposits lie in this region. If the experiment is performed in galvanostatic conditions, the suitable current density will be that corresponding to a potential lying in the common immunity domain. Such a domain may be found by simply superimposing the two Pourbaix diagrams and finding the common immunity region between Ga and As. This rapid and intuitive method (often applied, e.g., by Singh et al. [50], Chandra et al. [51], and Rajeshwar [15]) makes it possible to determine by inspection whether codeposition is possible. For example, in the case of GaAs it shows that the conditions for codeposition are generally rather critical, as experimentally found. Therefore, with some reservations, which will be pointed out in the following, this method is applied to the III–V compounds, in a joint analysis of the Pourbaix diagrams of their two components with the help of Figure 6. The resulting conclusions are as follows.

1. Aluminum has a diagram similar to that of Ga, but it is considerably more reactive. Its immunity domain lies below about −1.8 V vs. SHE, so that it is practically impossible to bring Al in a state of immunity because of its very low equilibrium potential. For the same reason, it is practically impossible to electrodeposit Al from aqueous solutions, although, under very special conditions, it has been deposited electrochemically [49]. No common stability domain can be found for the Al–As and Al–Sb systems.

2. Phosphorus is always unstable in the presence of aqueous solutions. However, InP has been electrodeposited from $InCl_3$ and NH_4PF_6, at low pH, with limited success [47, 48].

3. As shown in Figure 6, Ga has an immunity domain lying below −0.7 V vs. SHE, which overlaps that of As in a limited pH range, as already discussed. Instead, practically no overlapping occurs with the immunity domain of Sb, which has an equilibrium diagram similar to that of As, but it is slightly shifted toward less negative potentials. However, GaSb thin films in mixture with Sb have recently been produced, the stoichiometric control being limited by parasitic hydrogen and SbH_3 evolution [52].

4. Indium, with a diagram similar to that of Ga but with an immunity region extending toward less negative values, presents a common immunity domain with both As and Sb, as well as with Bi, in a wide pH range (from about 2 to 14).

This method of using the Pourbaix diagrams to determine whether codeposition of more elements is possible suffers from several limitations when applied to practical cases without consideration of their specificity, and may be misleading [21]. Indeed, there are several points to be taken into account. First of all, the diagrams have to be recalculated for each particular system, to take into account the effective concentration of the soluble species and the effective partial pressure of the gaseous ones, rather than assuming values of 10^{-6} M and 10^{-6} atm, respectively. Second, the interaction between the components in the deposit and formation of the compound shifts the deposition potential of the less noble species (e.g., Ga) to more positive values, as already discussed (underpotential deposition). This is expected to enlarge the common immunity domain in the case of GaAs and to allow superposition in the case of GaSb. Finally, similarly to all thermodynamic treatments, kinetic effects are not considered, and it may be that, because of a high irreversibility of the process, the real lines are shifted with respect to the thermodynamic lines, while concurrent processes (e.g., hydrogen evolution and hydride formation) may become important. Moreover, microscopic effects such as substrate surface structure and coverage are not considered, although they are important for an accurate prediction of the experimental underpotential deposition potential, for example, in the case of the ECALE technique.

A more accurate application of Pourbaix's diagrams than that previously discussed for GaAs in contact with aqueous solutions containing Ga and As soluble species was carried out by Perrault [53]. He performed ab initio calculations also considering the reactions involving GaAs, but determined the region of stability of GaAs in solutions where all of the soluble species including the products of GaAs decomposition had a unit activity. When the more stringent conditions are used as a criterion for corrosion, that is, 10^{-6}, as done by Pourbaix, the stable region becomes narrower. Moreover, in most practical cases, the decomposition products of the compound are not present in the electrolyte where the electrode is being immersed or their concentrations are very low. This made Perrault's results scarcely comparable with experimental data. Instead, Villegas and Stickney [21] evaluated the stability of GaAs exactly under the conditions to be used for ECALE deposition. They calculated several potential vs. pH diagrams, assuming that the concentration of the soluble species in contact with GaAs were those expected when Ga or As was deposited as well as when the electrode was rinsed with a practically pure electrolyte (Ga and As species present with a concentration of 10^{-6} M). The so calculated potential vs. pH plots showed that, because of GaAs formation, the potential of Ga deposition on As was, for a given pH, more positive by 200 mV than on bulk Ga, and a similar effect was observed for As deposition on Ga.

Earlier studies [54, 55] have also discussed the application of Pourbaix's diagrams to the prediction and understanding of the thermodynamic stability of semiconductors when employed in electrochemical cells with aqueous electrolytes for the conversion of solar energy into chemical (photoelectrolysis cells) and electrical (photoelectrochemical cells) energy. Park and Barber [55] refer to the photostability, in the presence of the electrochemical reaction of interest, of n- and p-type semiconductors (Cd chalcogenides, GaAs and GaP) with regard to anodic and cathodic decomposition, respectively. Their investigation is mainly based on the theoretical work by Gerischer

[56] and Bard and Wrighton [57], which, however, was limited to a given pH.

The standard Pourbaix diagrams were recalculated by Park and Barber, taking the compound formation into account, assuming a concentration of 10^{-6} M of the soluble species and reporting in the same diagrams the level of the semiconductor flat-band potential referred to the electrochemical scale. From these diagrams, several important conclusions were drawn regarding (i) the electrode stability against anodic dissolution and cathodic decomposition, (ii) the interaction of the redox reaction of interest (e.g., reduction and oxidation of water) with the semiconductor, and (iii) the appropriate pH region for a given redox couple to be chosen.

The analysis of GaAs and GaP indicated the redox reactions responsible for the very small region of stability of the former compound and for the absence of a stability region of the latter one. Both materials are expected to undergo anodic dissolution, and, therefore, the n-type doped electrodes may not be stable under operating conditions. Experimental results agree with this prediction [58–60]. As for the photoassisted electrolysis of water to hydrogen, the reduction of H^+ will always compete with the reduction of the p-type GaAs electrode, whereas it may be accomplished at the p-type GaP electrode, because its cathodic dissolution potential is much more negative than that of H^+ reduction. The photoassisted water electrolysis was shown to occur on p-type GaP without a significant amount of detected dissolution products [60]. More systematic papers have recently been published on the photostability of semiconductors in several electrolytes, but they mainly refer to chalcogenides (e.g., [61–63]).

5.2. Parasitic Reactions

During compound semiconductor electrodeposition from aqueous baths, the possibility of simultaneous evolution of hydrogen should also be considered. In general, the complete Pourbaix diagrams also define the stability domain of water, that is, the region where water is thermodynamically stable at 1 atm. This region is delimited by two parallel straight lines, shown in Figure 6, which represent the equilibrium conditions of water oxidation (or of its OH^- ions) to gaseous oxygen and of water reduction (or of its H^+ ions) to gaseous hydrogen, when the partial pressure of oxygen or hydrogen is 1 atm, at 25°C. The equations of these two lines are, respectively,

$$E = 1.228 - 0.0591\,\text{pH} \quad \text{in V vs. SHE}$$
$$E = 0.000 - 0.0591\,\text{pH} \quad \text{in V vs. SHE}$$

If the stability domains of a compound or of its components lie below the stability domain of water, hydrogen evolution is expected to be superimposed on the electrodeposition process, unless the hydrogen overpotential of the deposited material is sufficiently high to hinder it.

From the corresponding Pourbaix diagrams [49] it is concluded that the whole of the stability domain of Al, Ga, In, and P lies entirely below that of water, whereas a small portion of the stability domain of As and Sb overlaps that of water. Hence,

1. Aluminum decomposes water, forming hydrogen, in the presence of sufficiently acid or alkaline solutions, dissolving mainly as Al^{3+} or as AlO_2^-, respectively.

2. In the presence of water, Ga becomes covered with an oxide film, the protective effect of which seems not to be perfect. Ga is appreciably attacked, with possible hydrogen evolution, by very acid or very alkaline solutions, mainly forming Ga^{3+} or GaO_3^{3-}, respectively. It can be electrodeposited from sufficiently acid or sufficiently alkaline solutions, with simultaneous hydrogen evolution. According to Perrault [53], electrodeposition seems to be easier from alkaline solutions.

3. Indium is not corroded by water at room temperature, owing to a protective oxide film whereas it tends to dissolve with hydrogen evolution, mainly as In^{3+} or as InO_2^- in the presence of definitely acid or alkaline solutions, respectively. The baths for its electrodeposition are usually chloride-, sulfate-, and sulfammate-based. With increased current density, hydrogen evolution occurs mainly in indium sulfate solutions [64, 65].

4. Phosphorus is always unstable when in contact with aqueous electrolytes. It can theoretically reduce water with the formation of hydrogen, hypophosphites, phosphites, phosphates, and gaseous phosphine.

5. Arsenic is a fairly noble element that is stable in the presence of water and aqueous solutions of all pH's, free from oxidizing agents. It can be electrodeposited either by the reduction of arsenites (AsO^+, $HAsO_2$, AsO_2^-) or arsenates (AsO_4^{3-}) or by the limited oxidation of arsine (AsH_3). Depending on the nature of the electrolyte and on the operating conditions, it is possible to obtain either As alone or As simultaneously with its hydrides (mainly AsH_3) and with hydrogen [66]. The limits of the domain of the relative stability of As and AsH_3 are given by

$$E = -0.490 - 0.0591 \, pH \quad \text{in V vs. SHE}$$
$$[\text{for } p(AsH_3) = 10^{-6} \text{ atm}]$$

6. Antimony is a slightly noble element that is unaffected in the presence of nonoxidizing aqueous solutions free from complexing substances. In the trivalent and pentavalent states, Sb forms a large number of complexes. Similarly to As, Sb can be electrodeposited from solutions of increasing pH values containing the SbO^+, $HSbO_2$, and SbO_2^- species. Gaseous SbH_3 can be obtained together with hydrogen by the electrolysis of acid and alkaline solutions. The limits of the domain of the relative stability of Sb and SbH_3 are given by

$$E = -0.392 - 0.0591 \, pH \quad \text{in V vs. SHE}$$
$$[\text{for } p(SbH_3) = 10^{-6} \text{ atm}]$$

Taking these observations into account and recalling that during the formation of a compound, the potential of the less noble species (Ga, In) shifts in the positive direction, it may be possible to avoid the parasitic hydrogen evolution reaction during the codeposition of Ga–As, Ga–Sb, In–As, and In–Sb. However, successful codeposition will be achieved only if a potential control is kept at all times, thus also controlling the overpotentials of both electrodeposition processes due to kinetic limita-

tions. At very negative codeposition potentials, hydrogen evolution may be paralleled by other parasitic processes, such as AsH_3 or SbH_3 formation from As- or Sb-containing baths, as already stated.

Hydrogen evolution during deposition has several negative effects. First of all, the parasitic process decreases the growth rate of the deposit, lowering the overall coulombic efficiency of the main electrodeposition process. Moreover, the physical, chemical, and mechanical properties of the deposit may be affected. Indeed, even hydrogen gas bubbles may induce defects in the deposited layers. Much more important is hydrogen incorporation into the deposit, which gives rise to stress [67] and cracking [68] in thin films. So, cracks or large cavities can be generated at the defective sites when the pressure of the accumulated gas is sufficiently high. Furthermore, hydrogen may preferentially be adsorbed on certain crystallographic planes of the deposit, thus changing the surface coverage and producing or changing the preferred surface orientation.

As for the electrolyte properties, hydrogen discharge increases the pH of the electrolyte close to the electrode surface, and if the electrolyte becomes sufficiently alkaline, precipitation of solid products may occur with the formation of a film on the deposit surface.

5.3. Classification of Cathodic Codeposition Processes

As previously shown, the cathodic deposition of M^{m+} and N^{n+} ions to form the $M_r N_s$ compound was distinguished in two classes, depending on whether the difference in equilibrium potentials of the two pure components ($E_N^\circ - E_M^\circ$) is greater (class I) or smaller (class II) than the shift in the deposition potential of the less noble component as a result of compound formation ($-\Delta G / mr F$). In the case of the considered group III–V compounds, $r = s = 1$ and $m = n = 3$.

Table VII reports the data relevant to GaAs, GaSb, InAs, and InSb, which are prevailingly deposited from acid solutions. Depending on the pH of the electrolyte, different reduction reactions have to be considered, so the involved ionic species and the related pH ranges are also indicated in the table. Whereas Ga and In are present in the electrolyte as Ga^{3+} and In^{3+} ions, the As or Sb ionic species are AsO^+ and $HAsO_2$ or SbO^+ and $HSbO_2$, respectively, for increasing pH values but still in the acid region. The values of the equilibrium potentials are taken from Pourbaix's equilibrium data [49] and those of the Gibbs free energy of the compound formation from the elements from the NBS Tables [8].

The data in Table VII show that the above compounds in acid solutions have to be classified under class I, with Ga and In being the potential-determining species within the entire deposition range considered. Their behavior is similar to that observed for CdTe [29, 69].

GaAs has also been codeposited from highly alkaline solutions (pH > 9.21), so that in this case the ionic species are GaO_3^{3-} and AsO_2^- [70]. As shown in the table, GaAs has now to be classified under class II, and the potential-determining species changes from Ga at the Ga side of the system to As at

Table VII. Classification of Some III–V Compounds According to Kröger [29]

MN	M species	pH	E_M° (V vs. SHE)	N species	pH	E_N° (V vs. SHE)	$E_N^\circ - E_M^\circ$ (V)	$-\Delta G/3F$ (V)	Class
GaAs	Ga^{3+}	≤ 2.56	-0.529	AsO^+	≤ -0.34	0.254	0.783	0.234	I
GaAs	Ga^{3+}	≤ 2.56	-0.529	$HAsO_2$	≥ -0.34 and ≤ 9.21	0.248	0.777	0.234	I
GaAs	GaO_3^{3-}	≥ 11.74	0.319	AsO_2^-	≥ 9.21	0.429	0.110	0.234	II
GaSb	Ga^{3+}	≤ 2.56	-0.529	SbO^+	≤ 0.87	0.212	0.741	0.134	I
GaSb	Ga^{3+}	≤ 2.56	-0.529	$HSbO_2$	≥ 0.87 and ≤ 11	0.230	0.759	0.134	I
InAs	In^{3+}	≤ 3.88	-0.342	AsO^+	≤ -0.34	0.254	0.596	0.185	I
InAs	In^{3+}	≤ 3.88	-0.342	$HAsO_2$	≥ -0.34 and ≤ 9.21	0.248	0.590	0.185	I
InSb	In^{3+}	≤ 3.88	-0.342	SbO^+	≤ 0.87	0.212	0.554	0.088	I
InSb	In^{3+}	≤ 3.88	-0.342	$HSbO_2$	≥ 0.87 and ≤ 11	0.230	0.572	0.088	I

Electrochemical and thermodynamic data are from [49] and [8], respectively.

the As side, whereas near the cross-over point, both Ga and As are involved in determining the potential, as already discussed.

6. CODEPOSITION FROM MOLTEN SALTS

Cathodic codeposition was also carried out from high-temperature molten salts and more recently from low-temperature melts, mainly organic chlorides. Several III–V compounds, such as GaP, InP, GaAs, and InSb, were prepared from these electrolytes. One of the major problems to be faced in molten salt electrodeposition is often the proper choice of a suitable solute–solvent system [71]. The choice of the solute is usually less difficult, because it has just to be stable in the selected solvent and to be highly soluble therein. Several stringent properties are needed in the solvent. It should have (i) low volatility, (ii) low viscosity to favor the diffusion of the dissolved species, (iii) a low melting point, (iv) a low reactivity with the electrodes and the electrochemical cell materials, (v) high stability at the operating potentials, and (vi) high solubility in some aqueous or organic solvents that can be easily removed from the deposit. Other important aspects are its toxicity and compatibility with the environment, availability at high chemical purity, and low cost.

To obtain high-quality materials (e.g., single crystals) it is necessary that the interface between the growing phase and the electrolyte be stable against arbitrary perturbation. Otherwise, protuberances develop, cellular growth results, and impurity entrapment usually occurs. A large departure from stable growth conditions leads to powder formation ahead of the crystal surface.

There have been several theoretical approaches to the stability of a crystal interface during crystallization from molten salts (e.g., [72–78]). In addition, Huggins and Elwell [79] considered the particular case of crystal growth during electrocrystal-

lization where the driving force is faradaic rather than thermal. These authors defined the conditions for stable growth generalizing a previous model [80] based on the variation in chemical potential ahead of the interface of a growing crystal. Such a criterion was applied to solute crystallization with either a thermal or an electrical driving force, and an equation was deduced that gave the conditions for the maximum stable growth rate during electrocrystallization.

Stable growth requires a high concentration of solute ions; a high ionic mobility, which is directly proportional to the solute diffusion coefficient, is also desirable. It also requires a high effective electric potential gradient at the crystallization surface to prevent the occurrence of a supersaturated region ahead of the interface. Such a gradient may be enhanced by stirring the solution, which is expected to reduce the thickness of the boundary layer and may be controlled independently of the growth rate. If growth occurs at a constant cell voltage (or overpotential), stirring is likely to increase the growth rate, perhaps without affecting the stability of the interface. The temperature dependence of the maximum stable growth rate is difficult to predict, as it is governed by the ratio between the diffusion coefficient of the solute and the absolute temperature, but generally a temperature increase slightly enhances the stability.

In an optimized system, growth should proceed at a rate close to the maximum stable value. Beyond this value, protrusions are formed where the supersaturation is highest, typically at edges and corners of faceted crystals. According to Huggins and Elwell [79], operating in non-steady-state conditions, for example, in alternating potentials, which provides further flexibility, may be advantageous in the preparation of perfect, inclusion-free deposits [81, 82].

In conclusion, the factors affecting the stability of deposition are a high solute concentration, a high solute diffusion coefficient, and a low growth rate. It has been confirmed experimentally that stable growth occurs only below a critical current density, or, in other words, the maximum stable growth rate is the critical factor, as in conventional solution growth [11, 71].

7. SEQUENTIAL ELECTRODEPOSITION

This method is more easily applicable and makes it possible to prepare thin films of binary and ternary III–V (and II–VI) compounds with well-defined composition and thickness. The starting point is the sequential deposition of each component in the desired quantity or thickness, followed by an annealing treatment. Note that, very often, thermal treatments do not present many difficulties and that, in several cases, they are also performed after the electrochemical process to improve the deposit properties and performance, whichever method is followed, and, therefore, even after codeposition. This occurs particularly in the case of semiconducting thin films. Post-electrodeposition annealing usually results in an increase in the grain size, and it may create vacancies, helping solid-state diffusion. Moreover, laser annealing was shown to enhance the carrier mobility.

The sequential electrodeposition process may schematically be described as

$$M^{m+} + me = M$$
$$B^{n+} + ne = N$$

and the thermal process as

$$rM + sN = M_rN_s$$

Owing to the sequential character of the process, optimal experimental conditions for each element may be selected, such as bath composition, type of process, galvanostatic or potentiostatic process, temperature, and deposition time. If the current efficiency of each electrochemical process is known, it is possible to exactly evaluate the deposited quantity of each element and, thus, the final compound thickness, if the densities of all of the relevant materials are known. Indeed, in contrast to codeposition, the exact coupling of the fluxes of each element is not necessary to obtain the correct deposit stoichiometry. In several cases, it is also feasible to dope a deposit during electrodeposition, in an attempt to control the concentration of mobile charge carriers. However, a problem often of no easy solution is the choice of the deposition sequence, which depends on the chemical and electrochemical stability of the initially deposited element in the particular operating conditions for the deposition of the following element. Indeed, the underlayer may become unstable in the new electrochemical situation and dissolve at least partly or form gaseous compounds, or it may react with the subsequently deposited element, forming undesired products. For example, difficulties in the choice of the operative sequence have more particularly been encountered when Ga or As was involved. So, the sequence As first and Ga second was not practicable, owing to AsH_3 formation at the highly cathodic potentials at which Ga is deposited. On the other hand, even the reverse sequence encountered difficulties, owing to the Ga tendency to alloy with many metal substrates.

The heat treatment following the preparation of the deposited multilayer is a fundamental step in the sequential method. It has to be carried out in accurately selected and controlled conditions that allow the solid-state diffusion process with chemical reaction to occur, so that only the compound of interest is formed, possibly with the desired stoichiometry. Evaporation of some components and their reaction with the substrate or with some species of the heating atmosphere (when heating is not performed in vacuum or in an inert gas) must carefully be avoided. Furthermore, when the phase diagram of the system under investigation is not a simple one, for example, when more than one compound may be formed (as in the case of the In–Bi system) or miscibility gaps are present, the process may become critical.

Several III–V compounds were produced according to this method, in particular GaSb [83] and InSb [84]. They were prepared by electrodeposition from aqueous solutions of their inorganic salts. However, the sequence Ga first and Sb second was not possible, because Ga was stripped from the substrate; that is, it was reoxidized at the potentials at which Sb was deposited. Instead, the reverse sequence Sb first and Ga second was successful, as well as the sequence Sb first and In second, in obtaining InSb.

The mechanism of the diffusion and reaction process occurring during the annealing treatment of the deposit was also investigated, but in a few cases only, as research in this field is still scarce. The systems examined were In electrodeposits on Bi [85–87] or on Sb [88–90] to form the binary InBi and InSb compounds, respectively, as well as Ga [91] electrodeposits on Sb to form GaSb. The aim was to determine the diffusion coefficient, in the specific case of In and Ga, initially into the other element, and then, when the chemical reaction has already taken place, into the compound. From measurements made at different temperatures, it was also also possible to evaluate the activation energy and the frequency factor of the process and thus to more deeply analyze the kinetic aspects. This part will be presented in Sections 9 and 10.

A typical sequential method is the ECALE technique, which was developed in 1990–1991 by Stickney and co-workers [20, 92, 93] as a low-cost procedure for forming ultrathin layers of compound semiconductors with a well-ordered structure, mainly II–VI and III–V compounds. It was proposed as the electrochemical analog of atomic layer epitaxy (ALE), for example, molecular beam and chemical vapor atomic layer epitaxy (MBALE and CVALE, respectively). The ALE methodology was developed in the middle of the 1970s for the production of compounds, one layer at the time, with the use of surface-limited reactions [94]. These reactions, which favor layer-by-layer growth, are well known in electrochemistry as underpotential deposition (UPD) [95].

UPD is a process in which an atomic layer of one element is deposited on a surface of a different element at a potential prior to (i.e., under) that necessary to deposit an element on itself. UPD was mainly observed in systems such as metal electrodeposition on both polycrystalline (e.g., [96]) and single-crystal (e.g., [97, 98]) materials. It has already been discussed in connection with Kröger's theory of codeposition [29]. In the ECALE method, the UPD electrooxidation of a nonmetallic element is alternated with the UPD electroreduction of a metallic element. This makes it possible to keep the two elements on the surface simultaneously. So, one of the elements has to be stable and water soluble in a negative state, generally as an anion. Different solutions and potentials are used for each element, and they are alternated in a cycle, with each deposition followed by rinsing the cell with an inert electrolyte to prevent possible precipitation of the compound. Every deposition cycle forms a monolayer of the compound, so problems with three-dimensional nucleation and growth processes are avoided. Moreover, when UPD is performed on a single-crystal face, the resulting deposit is generally epitaxial; that is, it is strongly affected by the crystallographic orientation of the substrate. The thickness of the deposit depends on the number of cycles. These consist of a series of individual and controllable electrochemical steps, which are investigated and optimized independently from each other, resulting in increased control over the deposit at the atomic level. The degrees of freedom available in compound electrodeposition are thus expanded, allowing the growth of nanometric materials particularly suitable for use in technological applications. A growth of this type is certainly not possible with conventional methodologies, such as simultaneous or consecutive electrodeposition followed by annealing. Moreover, contamination was minimized by the use of thin-layer electrodeposition cells and ultrahigh purity chemicals. Indeed, such a special configuration reduces the amount of contaminants having access to the electrode surface because of its high ratio between the surface area and the volume of the solution, such as 1 cm^2 to 4 μl [20].

The ECALE method requires specific electrochemical conditions, namely deposition potentials and times, reactants and their concentrations, supporting electrolytes, solution pH, and the possible use of complexing agents to shift the ion potential in the desired direction. These conditions are strictly connected to the compounds one wants to prepare and to the substrate to be used.

Stickley et al. extensively applied the ECALE method to prepare II–VI (Cd and Zn chalcogenides, e.g., [92, 93, 99]) and III–V (GaAs [21, 100] and InAs [101]) compound semiconductors, mainly on polycrystalline and single-crystal Au electrodes. Recently, such a technique was extended to Ag substrates to deposit II–VI chalcogenides [102, 103] and InAs [104].

The ECALE method seems to work well for II–VI compounds but not for III–V compounds, for which the deposition conditions are more critical [105].

8. ELECTRODEPOSITION OF GROUP III–V COMPOUNDS

To preliminarily consider a brief history of the subject, it is worth starting with the observations by Hobson and Leidheiser [88]. These authors performed kinetic studies to prepare InSb by electrodepositing In on an Sb substrate, to obtain quantitative information on the rate of formation of this representative III–V compound under a gradient potential. In this same paper dating back to 1965, they mentioned two electrochemical studies [106, 107] in which they observed that intermetallic compounds may be formed on the surface of polarized electrodes. They investigated the possibility of using this technique for the electrochemical preparation of intermetallic compounds and subsequently applied it to the synthesis of III–V semiconducting compounds. They claimed that GaSb, InAs, InSb, and InBi were successfully formed by electrodeposition of the group III element from a boiling acid solution onto an electrode made of the group V element.

The first paper, a brief communication published in 1962, describes a simple electrochemical method for the preparation of AuSn and its applicability to other intermetallic compounds [106]. In a typical experiment, in which a gold cylinder was polarized in boiling 2 M HCl containing 0.001 M Sn^{2+} to a potential about 15 mV more noble than the Sn potential in this medium, they observed the formation of a deposit identified as AuSn by electron and X-ray diffraction. Based on these results and on those on the Pd–Sn system (PdSn$_4$ was recognized as the major constituent) as well as on preliminary studies, it appeared that intermetallic compounds of Sn, Pb, and Cd with noble metals could be prepared by electrochemical deposition of the former on the latter metals in the solid state.

The second paper was published in 1964 [107]; it dealt with the poisoning of the surface of Pt metals by intermetallic compound formation. The lowered effectiveness of particular platinum metals in catalyzing the corrosion of less noble metals in boiling acids could be explained by the formation of an intermetallic compound on the platinum surface when it was polarized to a potential close to the corrosion potential of the less noble metal. This observation was supported by the results of their quantitative experiments on Pt, Rh, or Pd and Sn in boiling 2 M HCl. Thus neither of these papers refers to the possible formation of III–V compounds.

To our knowledge, the first researchers to prepare a binary III–V compound were Styrkas et al. [108], who in 1964 recognized by X-ray diffraction the formation of InSb, together with In and Sb crystals, during the codeposition of In and Sb from solutions of their sulfates in ethylene glycol. The two elemental species completely disappeared during the subsequent heat treatment in hydrogen, and pure InSb was obtained. However, in the current literature, e.g., [31, 71, 109], the first electrodeposition of III–V semiconductors is credited to Cuomo and Gambino [110], who in 1968 achieved the epitaxial growth of GaP by electrolysis at 800–1000°C of a melt containing Ga in the form of an oxide and P as a metaphosphate. As a matter of fact, they were the first to carry out systematic research on different melts, cathodic materials (some of them single crystals with different crystallographic orientations of their surface), as well as bath temperature and electrodeposition conditions. Moreover, the reaction products were submitted to X-ray diffraction, metallographic investigation, and, in some cases, to spectrochemical analysis.

The research was further deepened by De Mattei et al. [111], who 10 years later investigated the conditions necessary for the stable growth of GaP epitaxial layers by molten salt electrodeposition on Si and GaP substrates. This investigation represents the first attempt to ascertain the identity of the variables affecting deposit morphology and homogeneity. Other III–V semiconductors, InP (in 1981 [112]) and GaAs (in 1978 [113] and 1982 [33]) were electrodeposited from molten salts, but none of them could be obtained as epitaxial films.

In 1986, Wicelinski and Gale [114] described the formation of GaAs thin films from low temperature (40°C) molten salts, that is, from organic chloroaluminate melts, a mixture of $AlCl_3$ and alkylpyridinium halides. Other molten salt systems were subsequently investigated: organic chlorogallate [109, 115] and chloroindate [116] melts in 1990 and 1994, respectively. They allowed the electrodeposition of III–V semiconductors (GaAs and InSb, respectively) at temperatures close to room temperature.

It was not until 1985 that a III–V compound was prepared by electrochemical deposition from aqueous solutions when Sadana and Singh [44] prepared InSb from acid citrate solutions. However, the electrodeposition of Sb–In alloys of practical interest, mainly to obtain hard corrosion-resistant coatings, was investigated much earlier, and the first mention of these alloys, not of the InSb compound, was made in 1955–1956 in patents issued to Smart [117, 118].

Several other authors electrodeposited the III–V semiconductor compounds in these last 15 years, from both aqueous and nonaqueous electrolytes, and their research, together with the earlier work, is examined in detail in the next part of this chapter, which specifically deals with each compound.

8.1. Aluminum Compounds

The group III–V compounds of Al (i.e., AlP, AlAs, and AlSb) are used for several applications, mainly in connection with other III–V materials, and were prepared following different physical and chemical methods. Instead, electrochemical methods are practically nonexistent (with one exception), because, as already stated, Al deposition from aqueous electrolytes is practically impossible. This was confirmed for AlAs by Perrault [53], who performed the thermodynamic calculations for AlAs electrodeposition from aqueous solutions and could not find a stability domain.

AlP is of interest, for example, in AlP-GaP superlattices prepared by vapor phase epitaxial growth. Their optical properties suggest that superlattices composed of two indirect band gap semiconductors offer great potential for application in optical devices [119]. Recently, a gas-source migration-enhanced epitaxial growth of GaP, AlP, and AlAs has specifically been studied, and it was observed that the growth rates of the layers depend on the flux of the group III source [120].

AlAs and its multinary compounds are potentially important in heterostructure systems, for example, with GaAs as a useful material for application in many high-speed electronic and optoelectronic devices. This is because the lattice parameter difference between, for example, GaAs and $Al_xGa_{1-x}As$ ($0 \leq x \leq 1.0$) is very small, promising an insignificant concentration of undesirable interface states [14]. An overview of a numerically stable and efficient method for computing transmission coefficients in semiconductor heterostructures with the use of multiband band structure models has recently been published. The AlAs/GaAs double barrier heterostructure and an AlSb/GaSb/InAs interband tunnel device are used to illustrate the applications of this method to several phenomena, such as X-point tunneling, hole tunneling, and interband magneto-tunneling [121].

AlSb is studied, for example, for the development of quantum confined structures based on the AlSb-GaSb-InAs family of semiconductors. A primary emphasis is the flexibility of this material system as the basis for a wide variety of electrooptical modulators, frequency doublers, infrared diode lasers, resonant tunneling circuits, and other devices [122]. Moreover, theoretical work using the ab initio pseudopotential theory has recently been carried out to investigate the properties of a variety of imperfect AlSb/InAs heterostructures [123].

AlSb seems to be the only compound for which an electrochemical synthesis is known. Indeed, Chandra et al. [124] published a paper on photoelectrochemical solar cells using electrodeposited GaAs and AlSb semiconductor films. These films were prepared under various deposition conditions, and a reasonable stoichiometry could be achieved, as revealed by energy dispersive X-ray analysis (EDX).

AlSb was prepared by dissolving aluminum sulfate and antimony oxide in concentrated HCl. With a well-polished Ti cathode in this electrolyte, the electrochemical codeposition reaction of Al and Sb took place according to

$$Al^{3+} + 3e = Al$$
$$HSbO_2 + 3H^+ + 3e = Sb + 2H_2O$$

the standard potentials of which are -1.663 and 0.230 V vs. SHE, respectively. Moreover, the potential of the second reaction decreases with increasing pH. Owing to the great difference in the values of the standard potentials (Sb being the more noble of the two species), the authors used electrolytes with a much higher concentration of the Al salt than of the Sb oxide, typically in the range of 1000 : 1, to obtain codeposition. They investigated the relative composition of Al and Sb in the electrodeposited AlSb films at different pH and deposition times for different ratios of the two salts. All of the electrodeposition experiments were performed at a current density of 2 mA cm^{-2} at 32°C. The pH of the solution played an important role in this system, and a value ranging from 3.0 to 4.0 allowed good quality deposits. The best film had the stoichiometry $Al_{1.12}Sb_{0.88}$, and its behavior was investigated in photoelectrochemical solar cells in which the film deposited on the Ti substrate was the working electrode and a Pt plate was the counterelectrode. Both electrodes were dipped in either aqueous or nonaqueous media containing a suitable redox electrolyte. However, AlSb films quickly hydrolyzed in aqueous electrolytes. The mechanism of the process proposed by Rudorff and Kohlmeyer [125] assumes a fast reaction with hydrogen evolution,

$$AlSb + 2H_2O = AlSbO \cdot H_2O + H_2$$

followed by a slow reaction, again with hydrogen evolution,

$$AlSbO \cdot H_2O + H_2O = Al(OH)_3 + Sb + 1/2H_2$$

Therefore, a nonaqueous electrolyte (i.e., acetonitrile) was used with either the iodine/iodide or the nitrobenzene redox couple. Under 100 mW cm^{-2} illumination, the open-circuit photovoltage was 210 and 240 mV, respectively, for the two photoelectrochemical cells, and the short-circuit photocurrent density was 0.35 and 0.1 mA cm^{-2}, respectively. Both systems were stable, but the photoresponse was low. The band gap was around 1.6 eV. The authors concluded that the AlSb system is not suitable for terrestrial use, as it is likely to hydrolyze if perfect sealing is not attained.

8.2. Gallium Phosphide

Literature on III–V phosphides is rather scarce, because electrodeposition of P from aqueous media is not feasible [49]. Indeed, its induced codeposition is possible with Fe group but not with group III elements [28, 126]. So, other methods were mainly followed (e.g., electrodeposition from melts). As a matter of fact, the synthesis of III–V compounds which have a more marked covalent character is more difficult and needs higher temperatures and, consequently, the use of a molten salt as the electrolyte. So, GaP, but also InP, was first prepared from these baths.

Cuomo and Gambino [110] were among the first researchers who electrodeposited III–V semiconductors. They obtained epitaxial GaP deposits on (111), (110), and (100) oriented single-crystal Si cathodes from a fused salt melt that contained 2.0 NaPO$_3$, 0.5 NaF, and 0.125–0.25 Ga$_2$O$_3$ in molar ratios. The sodium metaphosphate simultaneously acted as a solvent and as a P source. The melt operated between 800°C and 1000°C. The lower limit was determined by the liquidus temperature of the mixture, which became a clear liquid at about 750°C. The melt began to vaporize quite rapidly at about 1000°C. So the latter temperature was selected as the upper limit. Within this temperature range, the quality of the GaP deposits appeared to be fairly independent of temperature. Two alkali halide baths were also examined that contained, respectively (in molar ratios), 1.2 LiCl, 0.8 KCl (eutectic composition), 0.05 Ga$_2$O$_3$, and 0.1 NaPO$_3$; and 1.0 NaCl, 1.0 KCl, 0.1 Ga$_2$O$_3$, and 0.2 NaPO$_3$. They had both lower melting temperatures than the previous one (the former working from 520°C to 600°C and the latter at 800°C), but they were unsatisfactory because they yielded multiphase deposits comprising a Ga-GaP mixture. Therefore, they will not be considered further.

As for the metaphosphate melt, owing to its corrosive nature, a graphite crucible (which also served as the anode) had to be used in an inert atmosphere. It was housed inside a resistively heated quartz chamber. The cathodic polarization characteristics of the melt were preliminarily studied to determine the optimum deposition potential. Below a cell voltage of 0.6 V,

no solid deposits were observed, but a gas evolved that was believed to be P vapor. At higher voltages (around 0.6 V) GaP was electrodeposited at the cathode, as was P vapor. The cell current increased rapidly with increasing cell voltage. After electrodeposition, the samples were slowly cooled before being removed from the electrolysis chamber to prevent film cracking caused by temperature-induced stresses.

The mechanism of the GaP electrodeposition process was not explicitly studied; however, some tentative conclusions were drawn from experimental results. GaP was suggested to be formed by codeposition and reaction of Ga and P at the cathode surface. Ga was probably formed by direct reduction of Ga^{3+} ions, whereas the reduction of the metaphosphate ions to elemental P was that already proposed by Yocom [127] to be

$$4PO_3^- + 5e = 3PO_4^{3-} + P$$

This reaction was based on the observation that the electrolysis products at the cathode were P and sodium ortophosphate.

Epitaxial GaP deposits as thick as 100 μm were obtained at a current density of 50 mA cm^{-2}. Typically, the deposits consisted of an epitaxial layer about 25 μm thick that was covered with a rather dendritic, polycrystalline overgrowth. With the addition of adequate doping elements to the deposition bath, the material could be grown either as p- or n-type. Doping was achieved by adding small concentrations of Zn as ZnO (for p-type) and of Se as Na$_2$SeO$_4$ (for n-type) in the solution. Electroluminescent p–n junctions were prepared by a two-step synthesis, that is, by growing a layer of one carrier type, then overgrowing a layer of the other carrier type. In a few cases, studies of electrical properties and spectrochemical analyses were made. The resulting room temperature resistivity, carrier density, and mobility of the p-type film were 0.0703 Ω cm, 3.3 × 10^{18} cm^{-3}, and 27 cm^2 V^{-1} s^{-1}, respectively. Orange-red light-emitting diodes were successfully fabricated from the electrodeposited material.

Cuomo and Gambino also briefly mentioned the synthesis of InP and AlP from the metaphosphate melt when Ga$_2$O$_3$ was replaced with In$_2$O$_3$ and Al$_2$O$_3$, respectively, but no details were reported. However, Eldwell [71] claimed not to be able to confirm these results and that the solubility of these alternative oxides in the bath was very low. Cuomo and Gambino [110] also deposited ZnSe and ZnTe epitaxially on Si from the halide melts. The success of these initial experiments suggested that most of the III–V (and II–VI) compounds could be deposited by molten salt electrolysis.

According to De Mattei et al. [111], Cuomo and Gambino described the first synthesis of a compound semiconductor, GaP, by molten salt electrodeposition. They prepared epitaxial layers and measured the characteristics of a p–n junction diode, but did not systematically investigate the experimental conditions to deposit uniform epitaxial layers without polycrystalline overgrowths. So, as already stated, De Mattei et al. were the first to study the conditions for stable growth and to attempt to identify the variables that critically determine the morphology and uniformity of the deposited layers during GaP growth, notably substrate material, temperature, and electrochemical parameters.

GaP was synthesized between 750°C and 900°C from the metaphosphate melt of Cuomo and Gambino, which contained (in wt.%) 17.2 Ga_2O_3, 75.1 $NaPO_3$, and 7.7 NaF. GaP was produced at the cathode according to the direct reaction of Ga^{3+} and PO_3^-,

$$Ga^{3+} + 4PO_3^- + 8e = GaP + 3PO_4^{3-}$$

or

$$Ga^{3+} + PO_3^- + 8e = GaP + 3O^{2-}$$

that is, following a mechanism different from that suggested by Cuomo and Gambino.

The anode was a graphite crucible that helped to scavenge the electrodeposited oxygen, forming carbon dioxide. Three different substrate materials were studied: graphite; P-doped, n-type (100) Si; and S-doped, n-type (111) GaP. However, GaP layers were produced on the last two of these only. Indeed, the deposits on graphite were needle-like. Silicon, which was chosen as the substrate because of its availability and close lattice match to GaP, presented two major disadvantages for the growth of epitaxial layers: the large difference in the thermal expansion coefficient and its slow reaction with the melt producing GaP and SiO_2. As for the GaP substrate, a series of depositions lasting 20 min was made at constant current densities from 6 to 48 mA cm^{-2}, at 800°C. A systematic change in the deposit morphology was observed. The results clearly indicated that a current density below 10–20 mA cm^{-2} was necessary for the growth of uniform coherent layers, whereas a current density of 40 mA cm^{-2} led to a rapid breakdown in uniformity of the initial coherent layer and to the formation of craters and dendrites. These changes were associated with the generation of excess P at the cathode surface, which limited the stable growth rate to one below 23 μm h^{-1}, corresponding to 20 mA cm^{-2}. This value was typical of multicomponent growth processes and was comparable to rates normally used in liquid and vapor phase epitaxy. No investigations were made of the impurities in the deposits or of the techniques for decreasing them. Studies of this type would clearly be necessary to prepare materials for electronic applications. However, the authors maintained that the establishment of conditions for the reproducible deposition of uniform epitaxial layers was a major step toward this goal.

8.3. Indium Phosphide

Because of its high electron mobility, InP is a very interesting material and is used in important applications, for example, in lasers, photodiodes, and optical switches. It also has a very favorable band gap for solar cell applications, but it was remarked that it is not well suited to this purpose because of its sensitivity to water vapor [128].

Early deposition of InP from melts was obtained by a method similar to that used to obtain GaP [111]. However, InP was subject to more constraints, especially the higher volatility of In_2O_3

and the higher vapor pressure of P above the solid. Thus, Eldwell et al. [112] explored many possible solute–solvent systems as well as electrodeposition conditions on several substrates. The selection of melt composition was critical. The major difficulties in finding a melt to deposit InP were that the most appropriate sources of In, In_2O_3, and InF_3 were relatively volatile above 650°C and that InP significantly decomposed in this region. These problems limited the maximum temperature to the range of 600°C to 650°C. The difficulties encountered with the $NaPO_3$-NaF melts used for GaP preparation were the very low solubility of In_2O_3 and InF_3 and the very high viscosity of the bath at 600°C. Therefore, a systematic search was undertaken for an alternative system involving an alkali metal salt, owing to the electrochemical stability of the cationic species. The best results of all systems investigated were achieved with a melt of composition $Na_{0.814}K_{0.186}(PO_3)_{0.75}F_{0.25}$ as the solvent (a quaternary $NaPO_3$-KPO_3-NaF-KF eutectic previously proposed by Bukhalova and Mardirosova [129]) and In_2O_3 as the solute. However, at the operating temperature (600°C), the In_2O_3 solubility in this bath was still poor (i.e., only about 1 mol%), and the viscosity was high. These conditions were unfavorable to stable growth, and no epitaxial growth could be obtained.

The reactions proposed for the formation of InP were as follows: at the cathode, the simultaneous reduction of In^{3+}, and of PO_3^- to PO_4^{3-}, and at the anode, oxygen evolution according to the equations

$$In^{3+} + 4PO_3^- + 8e = InP + 3PO_4^{3-}$$
$$4PO_4^{3-} = 4PO_3^- + 2O_2 + 8e$$

The process proceeded only if a high potential difference between anode and cathode was applied, as shown in Figure 7. The minimum value obtained by extrapolating the linear part of the curve vs. potential was 0.85 ± 0.05 V vs. graphite. This limiting potential value was found to be insensitive to the In_2O_3 concentration.

Several materials were tried as cathodes (e.g., graphite, Ge, Au, Mo, Nb, Pt, and Ta), but they suffered from a relatively

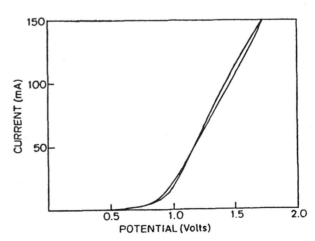

Fig. 7. Variation of current with potential vs. graphite for InP deposition on CdS from an $Na_{0.814}K_{0.186}(PO_3)_{0.75}F_{0.25}$ bath at 600°C. Reprinted from [112] with the kind permission of Elsevier Science.

poor adhesion of InP. Instead, Ni turned out to be the most suitable material. Epitaxial growth was also attempted on single crystals of CdS and InP; the former was chosen in part because of its application in CdS/InP heterojunction solar cells [130]. The CdS samples were oriented according to the hexagonal plane and were slightly soluble in the melt, which produced a light etching of the sulfide surface before InP deposition. The morphology of the deposits prepared at potentials from 0.85 to 1.30 V vs. graphite was examined by scanning electron microscopy (SEM). The deposits at the lower potentials were irregular and nonuniform, whereas a stable growth of coherent InP layers was observed in the range from 0.90 to 1.10 V vs. graphite while the corresponding current density was between 1 and 3 mA cm^{-2}. These last values were much lower than those found in the case of GaP deposition on n-type GaP, where current densities as high as 20 mA cm^{-2} were found [111]. The InP deposits were polycrystalline, probably except for an initial layer of submicron thickness, and their thickness was typically 2 to 5 μm. Similar results were achieved with (111) and (100) oriented InP wafers. The electrodeposited In films from a new melt contained 0.5% Si, 0.05% Al, 0.03% Cu, and 0.01% Fe as major impurities. The impurity concentration decreased with prolonged electrolysis, so that the only impurities detected in the 11th film deposited from the same bath were 30 ppm Fe and 2 ppm Ca.

InP layers of constant thickness were obtained on both CdS and InP, substrates but difficulties were encountered in the reproducible deposition of epitaxial layers, even at current densities below 1 mA cm^{-2}. This behavior was attributed to the electrolytic solution. Indeed, in agreement with theoretical results [79], the conditions for stable growth are a high concentration of the solute and a low viscosity of the melt. Neither these conditions were verified in the case of InP deposition, because of the limitations in the highest operating temperature (not more than 600–650°C), in contrast to what occurred in the case of GaP deposition, where the temperature was 800–900°C [111].

According to the authors, the most promising practical application of electrodeposited InP was its use for polycrystalline layers on inexpensive substrates. So, InP layers of good morphology were deposited on Ni and could be utilized in heterojunction solar cells, perhaps by a subsequent application of a CdS layer by spray pyrolysis.

Sahu investigated the formation of InP films from aqueous [47, 48] and nonaqueous [131] solutions. The aqueous bath contained InCl$_3$ and NH$_4$PF$_6$ in various proportions, and its pH was adjusted to low values by the addition of HCl drops to prevent hydrolysis of the salt. The film composition was sensitive to the solution concentration, pH, temperature, and electrolysis current density.

InP films that were 0.45 μm thick were obtained on a Ti cathode by electrodeposition, under stirred conditions, at a constant current density of 20 mA cm^{-2} for 7 min, at 22°C. The atomic composition of the as-deposited films obtained from electrolytes of different compositions was determined by EDX analysis. The P content in the deposits increased with the NH$_4$PF$_6$ concentration, at constant InCl$_3$ concentration, whereas a slight increase in InCl$_3$ could drastically increase the In content. A solution of 1.25 mM InCl$_3$ and 58.3 mM NH$_4$PF$_6$ yielded a nearly stoichiometric In-to-P atomic ratio of 49.0 to 51.0.

Hydrogen evolution prevailed when the current density was higher than 20 mA cm^{-2} or the pH was lower than 2.0, and In only was deposited. At pH values higher than 2.8 or at temperatures higher than 50°C, the amount of P in the deposit decreased. This decrease was possibly due to an increase of (i) the cathode current efficiency (not determined here); (ii) the basicity close to the electrode–electrolyte interface, owing to hydrogen evolution; and (iii) the deposition potential. A similar behavior of decreasing P content with increasing basicity, temperature, and current efficiency was also observed during the deposition of other P-containing alloys [132, 133].

SEM analysis showed a poor morphology with a grain size smaller than 10 nm. The X-ray diffraction pattern presented the peaks of InP, but also those of In, and P. When the samples were annealed for 2 h in vacuum at 300°C, the intensities of the peaks increased and their half-width decreased, probably as a result of the improved crystallinity, but the films still remained multiphase.

The band gap of both as-grown and annealed deposits, estimated from optical absorption spectra, was 1.22 eV. Moreover, measurements of the Hall effect showed the conductivity to be n-type and the carrier concentration to be 5.8×10^{18} cm^{-3}. In a more recent paper on the preparation of InP films by a PH$_3$ treatment of electrodeposited In, Cattarin et al. [134] observed that these claims of a direct InP electrodeposition did not seem to be well substantiated.

Owing to the unsatisfactory results achieved in aqueous solutions, an attempt was made by Sahu [131, 135] to electrodeposit InP from nonaqueous electrolytes. The solutions were prepared by dissolving InCl$_3$ and NH$_4$PF$_6$ in different ratios in dimethylformamide. Cathodic electrodeposition was carried out at 23°C with constant stirring, at a current density ranging from 1.5 to 3.0 mA cm^{-2}. The characterization was performed on a nearly stoichiometric film, 0.4 μm thick, obtained from an In-poor electrolyte, whose InCl$_3$-to-NH$_4$PF$_6$ molar ratio was 2.0 : 38.2. X-ray diffraction studies showed the presence of two peaks due to InP in the as-deposited films, along with numerous peaks of In and P. However, after annealing for 2 h at 300°C or 400°C, the InP peaks disappeared, and those of In and P were observed exclusively. The morphology of the deposit drastically changed after the heat treatment, from a characteristic needle- and thread-like structure to a globular-spherical grain structure that was like molten In or P dissolved in molten In. In several regions, even the bare Ti substrate was seen.

Notwithstanding the poor quality of the deposit, the as-grown thin film was analyzed for its optical and electrical properties. The former investigation showed it to be a semiconductor with a band gap of 1.35 eV (close to that of InP) and an absorption threshold of less than 1.3 eV; the latter corresponded to a polycrystalline semiconducting film with a high electron concentration, 6.5×10^{18} cm^{-3} and a low electronic mobility,

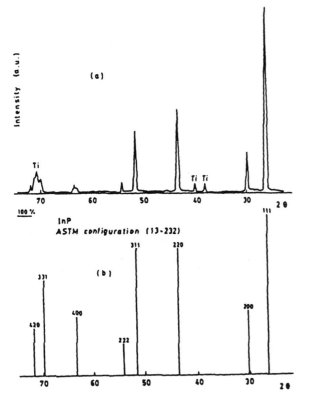

Fig. 8. (a) X-ray diffraction pattern of an electroplated In thin film after heating in a P atmosphere and (b) ASTM lines for InP. Reproduced from [45] with the kind permission of the Electrochemical Society, Inc.

$228.0 \text{ cm}^{-2} \text{ V}^{-1} \text{ s}^{-1}$. The Hall measurements indicated n-type conductivity. On the basis of the overall results, the InP deposit could be regarded as an n-type degenerate semiconductor.

The author maintained that the research clearly demonstrated the feasibility of InP deposition from nonaqueous solutions and that films of good quality could be prepared from a nonaqueous bath with a higher temperature [136], but, to our knowledge, no other paper was published by Sahu on the subject.

In 1989, the year of Sahu's first paper [47], Ortega and Herrero [45] prepared InP thin films by a phosphorization treatment of metallic In films electrodeposited from a citric bath [137]. This method consisted of submitting the metallic precursor of the semiconductor electrodeposit on an inert substrate (in this case Ti) to a heat treatment in the presence of the other component in the gaseous phase. The In-plated Ti electrode was inserted into a Pyrex tube that contained 50–100 mg of red P. After the tube was evacuated to a moderate vacuum of 0.01 mm Hg (around 1.33 Pa), it was sealed, placed in a muffle furnace, and heated. A heating treatment at 130°C for at least 1 h was necessary to avoid In globule formation. After this pretreatment, the samples were heated for about 3 h at 400°C and subsequently cooled to room temperature.

The characterization of the In films after phosphorization was accomplished by X-ray diffraction. As shown in Figure 8, the pattern matched the American Society for Testing and Materials (ASTM) data for the InP sphalerite phase and showed very sharp peaks in the correct intensity ratio, indicating that the material was crystalline with no preferred orientation. The authors also tested the photoelectrochemical behavior of the InP films in a HCl solution and observed that the photoconductivity was n-type. The physical parameters, band gap (from the spectral response) and flat-band potential (from the Mott–Schottky plot), were also obtained in this electrolyte. They were 1.29 eV and −0.42 V vs. saturated calomel electrode (SCE), respectively.

Recently, Cattarin et al. [134] followed a method similar to that of Ortega and Herrero [45] to prepare InP. They electrodeposited In at −1.0 V vs. SCE on a Ti substrate from a pH 2, 0.1 M indium sulfate solution according to literature data [138] and annealed the deposit in the presence of a PH_3 flow instead of red P. The In films were exposed to PH_3 vapors in an Aixron AIX 200 reactor, with Pd-diffused hydrogen as a carrier gas. The PH_3 pressure was 1.18 mbar (118 Pa) in the reactive chamber, the pressure of which was kept at 100 mbar (10^4 Pa). The samples were heated at 400–600°C for 20–80 min, inducing the reaction

$$2\text{In} + 2PH_3 = 2\text{InP} + 3H_2$$

The authors described in particular the advantages of their method with respect to the dominant preparation route of InP thin films by MOCVD. This technique is based on the reaction occurring at 600–700°C between indium alkyls and PH_3,

$$\text{InR}_3 + PH_3 = \text{InP} + 3\text{RH}$$

Indeed, the method devised by Cattarin et al. required lower temperatures. Furthermore, no trialkylindium derivatives had to be handled, which are pyrophoric and sensitive to oxygen and moisture, considerably increasing the cost of preparation facilities [139].

Different techniques were employed to characterize the obtained material, scanning electron microscopic and EDX analyses, and X-ray diffraction. Several conditions related to the annealing temperature in PH_3 and the In substrate thickness were tested to prepare InP layers of optimal morphology and composition.

Thin In substrates of 2.5-μm average thickness (deposited charge 5 C cm^{-2}) were heat-treated for 40 min in PH_3 at temperatures from 400°C to 600°C, every 50°C. At the lower temperatures, part of In did not react, whereas at the higher temperatures, crystallinity improved but the surface coverage became quite irregular, owing to the formation of large In globules. The best material was obtained at 450°C. Quite regular InP was formed at this temperature, although In-rich regions remained; the morphology was fairly good because the formation of In globules was limited. As for this fastidious globule formation, it is worth noting that it could perhaps have been avoided by submitting the In deposit to the short (1 h) and mild (130°C) heat pretreatment in a vacuum performed by Ortega and Herrero [45].

To improve the surface coverage, electrodes with a deposited charge between 2.5 and 15 C cm^{-2} were annealed at 450°C for 40 min in a PH_3 flow. Low deposition charges produced only a partial In coverage of the substrate, whereas high charges increased the tendency of molten In to form globules. In this case (10 or 15 C cm^{-2}), a fairly large amount of unreacted In was recognized by X-ray diffraction. The best coverage was obtained when the deposition charge was 5 C cm^{-2}. A sequence of two identical steps was also tried, each step involving the deposition of 5 C cm^{-2} of In, followed by annealing in a PH$_3$ flow for 40 min, at 450°C. The coverage was quite good, but In globules were still present. However, a final annealing treatment at 600°C lasting 20 min improved the crystallinity of the InP layer and further decreased the In quantity, and the X-ray diffraction pattern matched the ASTM spectrum for InP [6].

An important observation was that even after a prolonged annealing treatment in a PH$_3$ flow, the samples showed an In-to-P atomic ratio between 1 and 0.97, close to the stoichiometric value. This means that the stoichiometry was easily obtained and that no strict control of the parameters of the PH$_3$ treatment was necessary. Such behavior was tentatively attributed to a loss in the catalytic activity of the sample surface toward PH$_3$ decomposition once the whole In amount had reacted.

The thin InP films were submitted to a basic photoelectrochemical (PEC) investigation in an acidic polyiodide electrolyte. The photocurrent–voltage curves recorded under illumination with chopped white light showed that the samples obtained from nominally pure I$_3^-$ by a single deposition step were n-type. They presented a significant anodic photoactivity together with a fairly large dark current that changed in sign around the redox potential of the I$_3^-$/I$^-$ redox couple. Because n-type semiconductors cannot sustain large anodic currents in the dark, such behavior was attributed to oxidation processes occurring on bare metallic areas, presumably on the Ti substrate exposed to the electrolyte, owing to the uneven substrate coverage. The films obtained by the two deposition steps showed a much smaller dark current, indicating an improved surface coverage, but a worse PEC performance, which was related to increased recombination losses and to bulk resistance in the thicker films.

Interesting enough was the preparation of p-type InP by depositing, at 40 mA cm^{-2}, 5 or 50 mC cm^{-2} of Zn from a literature bath on the top of In layers of 5 mC cm^{-2} before annealing. The bath contained ZnCl$_2$, NH$_4$Cl, and citric acid, and its pH was adjusted to 7 with NH$_4$OH. This is a well-established method for achieving p-type conductivity, and Zn is the most frequently used acceptor-type dopant among bivalent metals such as Zn, Cd, and Mg. The PEC response of the p-type InP was similar, *mutatis mutandis*, to that observed for the n-type InP. The current–voltage curves presented a large cathodic dark current that was attributed to reduction processes of species in solution, again taking place on the metallic regions of the electrode.

The surface of the n-InP electrode polarized at 0.0 V vs. SCE in the polyiodide medium was scanned with a laser spot of 10-μm radius to obtain laser beam-induced current (LBIC) imaging. Poorly photoactive regions were indicated that were more numerous on electrodes prepared by the one-step method. In no areas of the surface were changes in the sign of the pho-

tocurrent observed; this was an indication that the as-grown material, although inhomogeneous, was n-type without exception.

With the intention of comparing the results, the authors also deposited an n-InP film on a Ti substrate by the conventional MOCVD technique. These deposits appeared quite homogeneous, despite the presence of significant cracks, and had an In-to-P atomic ratio of 0.98, and their X-ray diffraction pattern matched the ASTM patterns [6]. Their PEC performance was somewhat better but quite comparable to that of the films obtained by the phosphorization of electrodeposited In layers; the photocurrent was typically 1.2 to 1.4 times higher and the dark current 0.4 to 0.6 times lower. The analysis of the photoaction spectrum indicated a direct allowed transition with a band gap of 1.33 eV, close to the value of 1.34 eV for single-crystal electrodes [140]. A smaller value (1.31–1.32 eV) was found for n-InP films obtained from electrodeposited In layers, probably because of the sub-band gap response of the polycrystalline material. The estimated quantum efficiency of these films was around 0.2 electrons in the external circuit per incident quantum, a value that was considered to be reasonable for polycrystalline thin film deposits.

The method for the preparation of InP films proposed by Cattarin et al. [134] easily produced n-InP layers of correct stoichiometry and good crystallinity, whose photoactivity was higher than that presented by the films prepared by Ortega and Herrero [45]. Their overall light-to-energy conversion efficiency in PEC cells was rather poor (0.25%), but it might be regarded as significant when we consider that the preparation procedure had not been optimized and that comparable values were shown by undoped films prepared by the MOCVD technique. Indeed, the authors expected a marked improvement of the material performance by a doping treatment with S or Si. However, at present, the main limit of these n-InP films lies in their irregular surface coverage and morphology, which are a consequence of In being in the molten state during annealing in a PH$_3$ flow so that In globules are formed. The authors felt confident that more satisfactory results could be obtained by applying their method to metals whose melting temperature is higher than that of the heat treatment to which they have to submit their deposits to induce the chemical reaction with the gaseous partner.

8.4. Gallium Arsenide

GaAs is possibly the most interesting compound of the III–V semiconductor class and is becoming increasingly important in electronics and opto-electronics, owing to its unique optical and electronic properties. Indeed, GaAs has a higher carrier mobility than Si, resulting in faster electronics and a direct band gap matching the solar spectrum. It is of interest for many applications, including very high-speed integrated circuits, microwave devices, optical fibers, laser and LEDs, and for efficient solar cells, owing to its favorable band gap and high absorption coefficient [141]. For example, it is well known for its use in photovoltaic, photoelectrochemical, and photoelectrolysis cells. Indeed, considerable progress has been made in the state of the art of these cells. Recently, Khaselev and Turner [142] developed a photoelectrolysis cell that can split water into hydrogen and oxygen. A device with 12.4% efficiency has been demonstrated, an efficiency almost twice as high as previously reported. It is a unique combination of a GaAs photovoltaic cell and a gallium indium phosphide photoelectrochemical cell. The photovoltaic component provides the additional voltage needed to electrolyze water efficiently.

Despite the properties of GaAs, its electrodeposition was not pursued intensively, owing to several difficulties encountered in the synthesis of III–V compounds with more marked covalent character. Higher temperatures were needed, and, consequently, high temperature molten salts were preliminarily investigated as electrolytes.

GaAs was first synthesized by De Mattei et al. [113] at 720–760°C in molten NaF and B_2O_3 containing Ga_2O_3 and $NaAsO_2$. The melt composition that has given the best results was 20.3, 67.4, 4.2 and 8.1 wt%, respectively, and the molar concentration of Ga_2O_3 and $NaAsO_2$ was 1.4 and 4.1%, respectively; the excess of As was required to deposit GaAs alone. A vitreous carbon crucible and an Au anode were employed because of the high reactivity of the bath. Electrodeposition was carried out on Ni and GaAs single-crystal substrates. The minimum cell voltage was 2.4 V on Ni and 1.7 V on GaAs. The films obtained on the latter substrate were epitaxial and 10 μm thick. Attempts to deposit GaAs from solutions of Ga_2O_3 and $NaAsO_2$ in a Li_2O-B_2O_3-LiF solvent used to grow LaB_6 [143] were unsuccessful.

According to Dioum et al. [33], the high temperatures used were partly due to the necessity of dissolving a nonvolatile Ga donor (e.g., Ga_2O_3). So, an attempt was made to electrodeposit GaAs from molten $GaCl_3$–KCl (potassium tetrachlorogallate, $KGaCl_4$), previously synthesized by the authors [144], and AsI_3 at a relatively low temperature (300°C). The addition of $AsCl_3$ did not change the current–voltage curves, owing to the volatility of $AsCl_3$ (whose boiling point is 130.2°C), which evaporated before being solvated by the melt. The higher boiling point of AsI_3 (403°C) allowed the introduction of greater amounts of As. Electrodeposits were obtained either potentiostatically or galvanostatically on Au-electroplated Ni electrodes. The electrodeposited material was separated from the substrate, reduced to powder, washed with water and acetone, and analyzed by X-ray diffractometry with Ag powder as an internal standard. The X-ray diffraction analysis of the electrodeposits showed that, under potentiostatic conditions, only As was formed, whereas under galvanostatic conditions, some GaAs was also formed; this occurred, however, at very negative values of the electrode potential. So apparently, stoichiometric deposits of useful quality, particularly those free of As, could not be obtained.

In 1986, Wicelinski and Gale [114] reported the codeposition of GaAs at even lower temperatures (40°C) from acidic chloroaluminate melts composed of $AlCl_3$ and 1-butylpyridinium chloride (PyCl). A mixture of $GaCl_3$ and $AsCl_3$ was added to the melts, and their electrochemical behavior was investigated by voltammetry and coulometry. Reduction under potentiostatic control led to the formation of films containing both Ga and As, which were analyzed by X-ray diffraction and X-ray fluorescence spectroscopy. Inductive coupled plasma and electron dispersive scattering analyses verified their composition.

The melt used by Wicelinski and Gale as an electrolyte was one of a number of $AlCl_3$ organic chloride salts with a low melting point, allowing molten salt electrochemistry to take place at or around room temperature [145–148]. These solvents presented a large potential window without decomposition, so that the parasitic hydrogen evolution process did not interfere with the cathodic codeposition process of interest, contrary to what was observed during GaAs electrodeposition from aqueous solutions [51]. Furthermore, by changing the ratio between the moles of $AlCl_3$ and the moles of the organic chloride, a wide range of acidity was offered. So, if this ratio was less than unity, the melts were basic, and the reaction of their formation was

$$RCl + AlCl_3 = R^+ + AlCl_4^-$$

R^+ is the pyridinium or imidazolium ion. If this ratio was greater than 1 (i.e., $AlCl_3$ was in excess), the melts were acidic and an aluminum dimer ion was formed:

$$AlCl_4^- + AlCl_3 = Al_2Cl_7^-$$

The potential window of the melts was limited on the anodic side by the oxydation of Cl^- and on the cathodic side by the reduction of either the organic cation or the Al(III) species in the basic or acidic melt, respectively.

In another paper, Wicelinski and Gale [149] studied the electrochemistry of the Ga species in $AlCl_3$ and 1-butylpyridinium chloride melts and observed that Ga deposition was possible from acidic but not basic melts. Furthermore, the prospects for depositing pure GaAs films were briefly discussed.

A particularly attractive salt among the chloroaluminates for its wider potential window [150] and a wide acidic–basic range [151] was a mixture of $AlCl_3$ and 1-methyl-3-ethylimidazolium chloride (ImCl). Preliminary results obtained by Carpenter and Verbrugge [109] suggested that Ga and As codeposition could be achieved in acidic melts at room temperature after the addition of $GaCl_3$ and $AsCl_3$ as solutes.

In their continuing work to find media suitable for thin film fabrication of III–V semiconductors, Wicelinski et al. [152] reported for the first time the existence of low-temperature chlorogallate molten salt systems formed by the addition of $GaCl_3$ to either ImCl or PyCl, to control the exothermic reaction to avoid melt decomposition. When properly prepared, these melts were clear, colorless liquids. The $GaCl_3$–ImCl melts were liquid at or below 20°C when the molar fraction of $GaCl_3$ was in the composition range from 0.3 to 0.7. Although most of the $GaCl_3$–PyCl melts in the acidic regime and some in the basic regime were liquid at 25°C, certain melts near the neutral region were not liquid until approximately 45°C.

The electrochemical windows of electrolyte stability obtained by cyclic voltammetry were similar for the two melts. In the acidic regime, it was reasonably expected that the anodic and cathodic limits corresponded to chlorine evolution and to Ga deposition, respectively. In the basic regime, whereas the anodic limit was still the same, the cathodic limit appeared to correspond to the reduction of the organic cation, Im^+ or Py^+. Differential scanning calorimetry [153], nuclear magnetic resonance, fast atom bombardment mass spectrometry [154], and Raman spectroscopy [155], along with density measurements were pursued to develop an understanding of the physical and chemical properties of these melts. These studies were intended to assist in the determination of the usefulness of the new chlorogallate melts as solvents for electrochemical and spectroscopic research.

GaAs deposited from a chloroaluminate melt would be expected to contain some residual Al contamination. This was undesirable, because even traces of Al contamination could significantly modify the electronic properties of the semiconductor. Therefore, to minimize contamination, Carpenter and Verbrugge [109] chose the chlorogallate melt $GaCl_3$–ImCl, reported by Wicelinski et al. [152], as an electrolyte for GaAs electrodeposition. This melt was presumed to have a behavior similar to that of the $AlCl_3$–ImCl melt with regard to the reaction of its own formation,

$$ImCl + GaCl_3 = Im^+ + GaCl_4^-$$

and the reactions determining the anodic and cathodic sides of the potential window when such a melt was acidic or basic had already been discussed by Wicelinski and Gale [114], as mentioned before. In particular, Ga(III) species reduction was expected to occur in acidic melts only. In contrast to this, Carpenter and Verbrugge [109] found that Ga(III) species could also be reduced in basic chlorogallate melts, yielding Ga deposits without apparent electrolyte degradation. So, they added $AsCl_3$ as the solute to the basic $GaCl_3$–ImCl melt and codeposited Ga and As at room temperature, forming GaAs.

They conducted the experiments in a simple electrochemical cell housed in a glove box containing dry and extremely pure nitrogen. The reference electrode was an Al wire placed in a $GaCl_3$–ImCl melt with a 40–60 molar ratio, which was found to be stable in the basic electrolytes. The working electrode was a glassy carbon disk, and the counterelectrode was a Ga anode. The electrochemical behavior of the melt before and after $AsCl_3$ addition was investigated by cyclic voltammetry, and the deposits were characterized by EDX analysis and X-ray photoelectron spectroscopy (XPS) to gain information concerning their elemental composition and chemical state, respectively.

The authors based the interpretation of their results on the already discussed formation of Im^+ and $GaCl_4^-$. The reduction of the latter ionic species resulted in Ga metal deposition, according to the reaction

$$GaCl_4^- + 3e = Ga + 4Cl^-$$

Moreover, similar to what was observed in $GaCl_3$–KCl melts [33], the addition of $AsCl_3$ to the basic $GaCl_3$–ImCl melt presumably allowed the formation of $AsCl_4^-$,

$$ImCl + AsCl_3 = Im^+ + AsCl_4^-$$

which was then reduced to As.

Table VIII. Current Densities and Deposit Compositions

Potential (V)	Cathodic current density (mA/cm^2)		Deposit composition Potential step (mole fraction As)
	30 mV/s	Potential step	
−0.4	0.9	0.8	0.80
−0.8	4.3	4.0	0.68
−1.0	8.0	6.6	0.13
−1.4	6.9	4.0	0.38
−2.0	2.1	1.0	0.16

Reproduced from [109] with the kind permission of the Electrochemical Society, Inc.

Table VIII synthesizes some typical electrochemical results. It reports the cathodic currrent density at decreasing potential values both during cyclic voltammetry at a slow scan rate and during potential-step experiments, in the latter case also showing the deposit composition determined by EDX analysis. These results indicated that As was the more noble species of the two, as in aqueous solutions, and that, therefore, small concentrations of $AsCl_3$ with respect to those of $GaCl_3$ had to be added to the bath if equimolar deposits were desired. They also showed that the deposition of Ga was prevailing over that of As at potentials below −1 V vs. the Al reference electrode. The best potential range for yielding equimolar Ga–As deposits was between about −0.4 and −1 V vs. the Al reference electrode. The XPS analysis confirmed the formation of the GaAs compound. However, the stoichiometry and morphology of the deposits were unsatisfactory, and free Ga and As were found in addition to GaAs.

The proposed electrochemical route for GaAs codeposition from room temperature molten salts was encouraging, also because the high rates of film growth by electrodeposition were promising for the preparation of large-area thin films. Nevertheless, further work is needed to avoid the nonuniform current distribution resulting from the inadequate geometry of the disk electrode employed in the system, to achieve a 1 : 1 Ga-to-As composition all over the cathode surface and to obtain uniform and optimized GaAs deposits.

In a subsequent paper, Verbrugge and Carpenter [115] looked for more details on the process of Ga electrodeposition from chlorogallate melts. They based the behavior of these melts on the known chemistry [156] and electrochemistry [157] of the $AlCl_3$–ImCl system.

Regarding Al electrodeposition, it was believed that the $AlCl_4^-$ ions present in the $AlCl_3$–ImCl melt were too stable to be reduced, and that only the reduction of the $Al_2Cl_7^-$ species could take place according to

$$Al_2Cl_7^- + 3e = Al + AlCl_4^- + 3Cl^-$$

as supported by the work of several authors [157, and the references quoted therein].

As previously indicated, the $Al_2Cl_7^-$ species was only formed in the acidic bath owing to the presence of an excess of

AlCl$_3$, and this was why Al deposition did from the basic melts not occur. However, in the case of the GaCl$_3$–ImCl system, Ga was also deposited from basic melts [109], and, therefore, Verbrugge and Carpenter [115] assumed that Ga$_2$Cl$_7^-$ was present in the basic melts, too. The following equation was then used to describe the charge transfer process:

$$Ga_2Cl_7^- + 3e = Ga + GaCl_4^- + 3Cl^-$$

instead of the one already considered based on the direct reduction of the GaCl$_4^-$ species. However, because very little GaCl$_3$ was thought to exist in the basic melt, Ga$_2$Cl$_7^-$ was not expected to be formed, according to the reaction

$$GaCl_4^- + GaCl_3 = Ga_2Cl_7^-$$

as in acid melts of the similar AlCl$_3$–ImCl system. On the contrary, Ga$_2$Cl$_7^-$ was expected to be formed according to the GaCl$_4^-$ dimerization reaction,

$$2GaCl_4^- = Ga_2Cl_7^- + Cl^-$$

The equations describing the Ga$_2$Cl$_7^-$ formation (through dimerization) and its reduction to Ga constituted the description of the salient chemistry and electrochemistry of the Ga deposition process from the GaCl$_3$–ImCl melt. Its comprehension was considered to be necessary for the analysis of the GaAs deposition process. Therefore, it was simulated numerically.

Microelectrodes were shown to be useful in the acquisition of the thermodynamic, transport, and kinetic data necessary for the calculations. They presented several advantages over the traditional (planar) macroelectrodes for the investigation of charge transfer processes, among which was the reduction of ohmic and capacitive effects. They made it possible to obtain high flux rates without significant mass transport resistance. So, the greatest part of the measured potential drop could be associated to charge transfer resistance, permitting more accurate measurements of the related rate constant.

The mathematical model first considered the relevant phenomenological equations and subsequently the conservation equations (of mass, energy, charge, and electroneutrality) satisfying the natural laws of classical physical chemistry. The nonlinear transport equations were solved for all mobile species in the chlorogallate melt, and a general model was developed for microcylinder (a Pt wire of 12.5-μm radius) and microhemisphere (12.5-μm radius) electrodes whose results were compared with those of planar microelectrodes (a Pt disk of 5-μm radius).

From a simulation of the experimental results for the Ga deposition process, the physicochemical constants were obtained, and a plausible interpretation of the Ga deposition mechanism was given.

Migration was found to be an important mode for the transport of the Ga$_2$Cl$_7^-$ ions, and their concentration profile was simulated during current-step experiments as a function of time and of the radial position, in the case of the microcylinder electrode. Far from the electrode surface, the Ga$_2$Cl$_7^-$ concentration never changed its bulk value. Instead, at the electrode surface,

the Ga$_2$Cl$_7^-$ concentration, initially equal to the bulk value, decreased with time nearly to zero (transition time). Soon after this time, the current was switched to zero, and the potential was driven to a more cathodic value, because another electrode reaction was forced to proceed, and the surface concentration of Ga$_2$Cl$_7^-$ was increased by the transport from the bulk to the surface.

The slight current oscillations observed in the current–potential curves at the onset and at the end of the Ga deposition reaction were investigated with a microdisk electrode, the area of which was far smaller than that of the microcylinder electrode. Therefore, it was particularly useful for the observation of single events, like those associated with Ga crystallization and not obscured by the noise of other similar events. Such oscillation phenomena appeared to be related to metal crystallization processes.

The analysis of the kinetic data showed that Ga deposition from the basic chlorogallate melt was not as easy as that of Al deposition from the similar acidic chloroaluminate melt.

Before the above two papers, Verbrugge and Carpenter [158] presented a patent on the electrochemical codeposition of uncontaminated so-called metal–Ga films (e.g., GaAs) from a melt essentially consisting of GaCl$_3$-dialkylimidazolium chloride and a salt of the other metal. In these melts (i) the alkyl groups comprised not more than four C atoms, (ii) the molar ratio of dialkylimidazolium chloride to GaCl$_3$ was not less than 1 but lower than 20 (i.e., the bath was basic), and (iii) the molar ratio of the metal salt to GaCl$_3$ was less than 0.5.

Compared with other conventional methods of preparing GaAs thin films, the electrochemical deposition from nonmolten salts, mainly aqueous solutions, has the advantages of low operating and equipment costs, relatively easy control of film properties, and nontoxic, nonvolatile raw material. However, earlier literature data suggest that this process is subject to marked drawbacks [53, 159], sometimes with problems [53]. Indeed, the codeposition of Ga–As thin films is severely hindered by the very negative deposition potential of Ga. According to Musiani et al. [160], As is an element that is not easy to deposit from aqueous solutions, and not much is known about plating onto As, whereas its tendency to inhibit some electrochemical reactions is well documented [161]. Several researchers agree that As electrodeposits cannot be grown as desired (their maximum attainable thickness is on the order of 10 μm), because both current and current efficiency markedly decline with prolonged potentiostatic electrolysis [66, 162–164]. This behavior has been explained as an effect of either the high resistivity of the amorphous As deposits [162] or the high overvoltage for further As deposition on the initially formed layer [163]. The intrinsic difficulty in depositing As is probably one of the reasons for the relative scarcity of the literature on the plating of As alloys.

In a 1984 patent, Astier et al. [165] claimed the electrodeposition of GaAs from aqueous solutions of Ga$_2$(SO$_4$)$_3$, As$_2$O$_3$, and H$_2$SO$_4$. Murali and co-workers [166] first deposited Ga from a sodium gallate solution on a Cu foil substrate and subsequently As from an arsenite solution; then they annealed the samples in a high vacuum at temperatures ranging from 247°C to 527°C, but no film characterization was reported.

Chandra and Khare [51] were among the first to investigate the preparation from aqueous chloride solutions of GaAs for solar cells, following the codeposition method. The authors preliminarily compared the potential–pH equilibrium diagrams of Pourbaix for Ga and As indicating the common immunity domain (as already shown in Section 5.1). A pH between 0.3 and 4.1 was found by overlapping the Pourbaix diagrams of Ga and As without considering the positive shift due to compound formation. The hydrogen evolution process created some problems in these acid solutions, and it was eliminated, at least partially, by stirring of the electrolyte. The latter was prepared by dissolving metallic Ga and As_2O_3 in concentrated HCl and then adding the two solutions in the appropriate ratio. The working electrode was a Ti plate. The electrochemical reactions were

$$Ga^{3+} + 3e = Ga$$
$$AsO^+ + 2H^+ + 3e = As + H_2O$$

and the related standard potentials were -0.529 and 0.254 V vs. SHE, respectively.

The electrolysis parameters were controlled in a series of experiments in which one parameter was changed while the others remained fixed. So, the influence of (i) the relative concentration of Ga and As ions in the electrolyte, (ii) the solution pH, (iii) the electrolysis current density, (iv) the temperature of the electrolyte, and (v) the deposition time was investigated to determine the optimal conditions for obtaining a deposit with nearly the stoichiometric GaAs composition. The results of this systematic research suggested that the best GaAs film could be obtained from a pH 0.7 solution containing 10 g liter^{-1} of both Ga and As_2O_3 and performing the electrodeposition at 2.5 mA cm^{-2} for 7 min, at a temperature of 22°C. The average film composition was $Ga_{1.04}As_{0.96}$, whereas EDX analysis indicated that the deposit stoichiometry was not uniform in the entire deposit thickness. So, the films were multiphase. The electron diffraction patterns of the film were recorded with an electron microscope, and they showed a good correspondence with the standard pattern for polycrystalline GaAs.

The band gap and the type of transition of the electrodeposited film were determined by optical absorption spectrophotometry in the transmission mode. For this purpose, the film was prepared on a Ti-coated semitransparent glass plate. A band gap around 1.5 eV was found, and GaAs was confirmed to be a direct band gap material. The authors also determined the resistivity (about 10^3 Ω cm) and the trap density (10^{14} cm^{-3}) of the electrodeposited material from the current–voltage characteristics of a metal–electrodeposited GaAs film–metal junction.

In another paper [167], the photoelectrochemical behavior at different light intensities of the same GaAs film was investigated in the presence of the sulfide–polysulfide redox system. The current–voltage characteristics of the interface in the dark and under illumination indicated that the charge transport at the electrolyte interface occurred via holes involving surface states and deep traps. Photoelectrochemical solar cells fabricated with the electrodeposited films had a conversion efficiency of about

0.3%, only, possibly owing to the poor quality of the film. The authors claimed that their studies were the first exploratory studies to obtain GaAs electrolytically and suffered from the limitation of impurity control, use of a proper complexing agent during deposition, annealing, etc.

In a subsequent paper [124], Chandra et al. again presented and discussed the experimental results reported in their previous papers [51, 167] but added new results mainly on the photoelectrochemical behavior of their $Ga_{1.04}As_{0.96}$ thin film. Moreover, they also reported the preparation and properties of AlSb (see Section 8.1).

As for GaAs, the authors gave a more systematic arrangement of the procedure for fixing the starting conditions of their trial to electrodeposit the compound: the Pourbaix diagram and the polarization curve approach. The former made it possible to obtain the common immunity region for Ga and As (pH range 0.3–4.2 at about -0.8 V vs. SHE). As for the latter approach, a sudden increase in the current at one electrode potential indicates the onset of the deposition of one species. The same occurs for the other species at another potential. In between, a plateau is observed, which indicates the region of the likely simultaneous deposition of both. Two steps, located at about 0.7 and 0.3 V vs. SCE, corresponding to As and Ga deposition, respectively, were noted in the polarization curve of a pH 7 solution containing 10 g liter^{-1} each of Ga and As_2O_3. A plateau was observed between these two values, and in this same range the electrode potential for electrodeposition was fixed.

The optical absorption spectrum of the best GaAs film ($Ga_{1.04}As_{0.96}$) was investigated again to determine the band gap. The experimental data were now treated according to the theoretical equation relating the absorption coefficient to the light frequency and not to an approximated one, as previously done [51]. However, the same value of 1.5 eV was obtained. The photoelectrochemical cells with GaAs photoanodes were studied with the use of three electrolytes containing different redox systems: (i) the sulfide–disulfide redox couple, (ii) the ferrocene–ferrocenium redox couple in acetonitrile, and (iii) the $MV^{2+/+}$ (methyl viologen) redox couple still in acetonitrile. GaAs was unstable in the aqueous electrolyte and deteriorated with time, whereas it was stable in both nonaqueous electrolytes, but the response was low.

The detailed photocurrent–photovoltage characteristics of the photoelectrochemical cells showed that the GaAs films had an efficiency of about 1% and a fill factor of about 0.5 (the redox electrolyte was not indicated). The authors hoped that by improving the stoichiometry and uniformity of the deposits, a competitive and economically viable system could be obtained.

Perrault [53] performed the thermodynamic calculations of GaAs electrodeposition, considering two possible mechanisms for GaAs electrosynthesis: direct electrodeposition from aqueous solutions (mechanism I) and electroprecipitation as the result of the preliminary reduction of the As compound to AsH$_3$ and its chemical reaction with the Ga ions,

$$AsH_3 + Ga^{3+} = GaAs + H^+$$

(mechanism II). He first determined the Ga–H$_2$O and the As–H$_2$O potential–pH equilibrium diagrams, considering only the monomeric three-valent Ga compounds for which data were available and not considering the pentavalent compounds of As, as he was looking for a way to prepare of arsenides. Then he calculated the complete diagram for the GaAs–Ga–As–H$_2$O system, taking account of the equilibrium reactions of GaAs with all of the species involved in the Ga–H$_2$O and the As–H$_2$O diagrams. According to his thermodynamic results, the complete GaAs diagram showed a potential–pH domain existing between pH 1.3 and 13.3, where GaAs could be stable in contact with aqueous solutions containing both As(III) and Ga(III) ionic species. This provided that the hydrogen evolution overvoltage was high enough for the particular set of experimental conditions, and GaAs could be electrodeposited by the simultaneous reduction of these two species. However, as already discussed in Section 5.1, the choice to assume a unit activity for the soluble species and to consider the GaAs decomposition products, too, made the results of Perrault scarcely comparable to the experimental results.

Electrochemical experiments in various pH ranges were also carried out with solutions obtained by dissolving Ga$_2$O$_3$ and As$_2$O$_3$ in alkaline media and adjusting the pH value by adding either hydrochloric or phosphoric acid.

In highly alkaline solutions (pH near 14) Ga only was deposited. Drops of the molten metal were formed on the cathode surface when the measurements were carried out at temperatures above the Ga melting point (29.774°C), whereas no Ga deposition occurred below this temperature, probably owing to a slower kinetics of the process. The deposits contained only traces of As. GaAs was not observed, as it was not thermodynamically stable at these pH values.

Deposits from very acid solutions (pH from 0 to 1) contained both Ga and As in various proportions (the latter was predominant), but GaAs was not found by X-ray diffraction analysis. Mechanism I (direct GaAs electrodeposition) was not possible, owing to the instability of GaAs in these solutions, and, as for mechanism II (electroprecipitation), the author could assume, on the basis of his experimental results, that it was not possible, either.

For solutions above pH 10, the results were comparable with those achieved in highly alkaline solutions, whereas at pH 4 or 5, they were similar to those obtained at pH 1. In neutral solutions (pH from 6 to 8) a very smooth, mirror-like As deposit formation was noted, but again GaAs electrodeposition could not be achieved. The deposition of As was tentatively explained, at least partially, by some of the reactions considered in the As thermodynamic diagram. Indeed, it might be inferred that, owing to the high overpotential for As deposition, AsH$_3$ was first formed and reacted with arsenious compounds in the electrolyte according to the antidisproportionation reaction

$$AsH_3 + HAsO_2 = 2As + 2H_2O \qquad (3)$$

thus chemically forming As.

The standard Gibbs free energy of this reaction has a very low value (-140.72 kJ mol^{-1}), which is much lower than that of the possible chemical reactions between AsH$_3$ and the various gallium ions in the electrolyte leading to the formation of GaAs. Indeed, these values are not lower than -23.24 kJ mol^{-1}. This might explain why mechanism II for GaAs electrosynthesis did not occur.

In conclusion, all results clearly showed that the electrochemical formation of GaAs, either by direct electrodeposition or by electroprecipitation, could not be achieved in aqueous solutions of arsenious ionic species. According to Perrault's opinion, the main problem is to prevent the arsenious compounds from reacting with AsH$_3$ and consuming it. AsH$_3$ was very easily formed through the electrochemical reduction of the same As(III) species and appeared to be the probable intermediate in the GaAs electrosynthesis. However, additional experimental work would have to be performed to confirm this last point. The author also suggested some possible solutions of the problem, such as the use of additives able to complex the As(III) compounds or to replace them with other arsenious compounds not containing oxygen, to prevent their reaction with AsH$_3$. Of course, these substantial modifications of the electrolyte composition might considerably affect the electrochemical process of AsH$_3$ formation and, consequently, that of GaAs.

Calculations for AlAs were also performed, but it appeared that a thermodynamic stability domain could not be found for this compound in contact with aqueous solutions. Perrault also intended to consider in the future other III–V compounds for which thermodynamic data were available (i.e., GaSb, InSb, InAs, or InP), but, to our knowledge, no other results have been published.

Yang et al. [70] could not duplicate the electrochemical results of Chandra and Khare's work [51, 124, 167]. The former authors electrodeposited GaAs from both alkaline (pH not lower than 12) and acid (pH not higher than 3) aqueous electrolytes on several cathodes. For their research, they selected highly alkaline and highly acid solutions because Ga oxides and hydroxides have a minimum solubility at about pH 8.

The authors started from previous considerations that according to the Ga–H$_2$O and the As–H$_2$O potential–pH equilibrium diagrams, electrodeposition of both Ga and As was possible from acid as well as from alkaline aqueous electrolytes. Despite the large difference in their standard potentials, the kinetic behavior might change the required potential for codeposition. Once Ga and As had been codeposited, they were expected to react to form GaAs, owing to the much higher stability of the compound with respect to the two elements, as indicated by its large negative standard Gibbs free energy of formation (-83.67 kJ mol^{-1}, but -67.8 kJ mol^{-1}, according to more reliable data [8]). So, a mixture was expected to be formed that contained GaAs with an excess of either Ga or As, depending on the atomic ratio between the two deposited elements. Based on these thermodynamic–kinetic considerations, the authors performed a systematic investigation of GaAs deposition from aqueous baths, the influence of the operating parameters on the deposit properties, and, possibly, the electroplating mechanism.

The working electrode was a Ti rotating disk, and the counterelectrode was made of Pt. During the preparation of the acid electrolyte, 1 M GaCl$_3$ was mixed in an appropriate ratio with a 1 M As$_2$O$_3$ solution (in concentrated HCl) in deionized water. The pH value was then adjusted with the addition of a reagent-grade NH$_4$OH or KOH solution. The alkaline solutions were prepared by adding As$_2$O$_3$ to a Ga$_2$O$_3$ solution and adjusting the pH value with concentrated KOH. Three different reference electrodes were used: Ag–AgCl in saturated KCl (saturated silver chloride (SSC), 0.197 V vs. SHE), saturated calomel (SCE, 0.241 V vs. SHE), and Hg–HgO in 0.1 M NaOH (0.157 V vs. SHE). The authors characterized the effect on the deposit properties of the applied potential, current density, electrolyte composition, substrate material, deposition temperature, pH, complexing agents, and stirring.

GaAs films were obtained from both alkaline and acid baths. The X-ray diffraction pattern indicated that the as-deposited films on Ti cathodes were amorphous or microcrystalline GaAs, which transformed to a completely crystalline phase with some loss of As content because of its volatility, after annealing at 250–300°C for as little as 6 h in a nitrogen atmosphere. Mixtures of the electrodeposited Ga and As yielded crystalline GaAs after annealing. The secondary ion mass spectrometry (SIMS) analysis also confirmed the existence of GaAs in the as-deposited films from alkaline baths.

As a general behavior, the deposits from alkaline electrolytes were thick, porous, powdery, and relatively uniform in depth, and they showed a scarce adhesion to the substrate. The deposits from acid electrolytes were thinner, more compact, and adherent with a relatively nonuniform composition. Both deposits, especially those obtained from acid baths, contained some oxygen, which was recognized by EDX analysis and Auger electron spectroscopy (AES).

The results are now examined in more detail. The deposits from alkaline and acid solutions are considered separately.

8.4.1. Alkaline Solutions

A strong dependence of the deposit composition on the deposition potential was observed in alkaline solutions, as shown in Figure 9. Around 62 at.% Ga, the curve presents a typical plateau between −1.65 and −1.73 V vs. Hg–HgO, which is not affected by agitation (300 rpm). At increasing cathodic potentials, AsH$_3$ was produced by As reduction during the deposition process, so that deposits of nearly pure Ga containing only traces of As were obtained, as found by Perrault [53]. When the Ga or As concentration in the solution was increased, their content in the deposited film also increased. Special research was performed to investigate the possible routes followed by As after electrodeposition on the Ti substrate. Arsenic could (i) stay on the electrode surface with Ga, (ii) leave the cathode and form a black powder in the solution, (iii) be reduced to AsH$_3$ and either evolve into air or (iv) react with gallate ions (GaO$_3^{3-}$) to form fine As or GaAs. A combination of the first two routes prevailed at potentials where a constant Ga content was observed as a function of the deposition potential, whereas

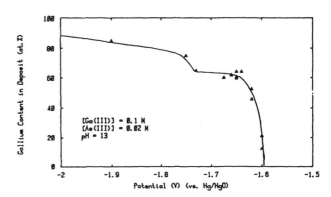

Fig. 9. Relationship between potential and Ga content in deposited Ga–As films. Reproduced from [70] with the kind permission of the Electrochemical Society, Inc.

Table IX. Ga Contents in Ga–As films on Various Substrates, Electrolyte Compositions and Deposition Potentials

Substrate	[Ga(III)] (M)	[As(III)] (M)	pH	Potential (V)	Ga (%)
Silicon	0.251	0.051	14	−2.99 (Ag–AgCl)	100
	0.251	0.051	14	−2.90 (Ag–AgCl)	77.8
	0.251	0.051	14	−2.80 (Ag–AgCl)	53.9
Lead	0.251	0.051	14	−1.73 (Ag–AgCl)	100
	0.251	0.051	14	−1.71 (Ag–AgCl)	55.6
	0.251	0.051	14	−1.70 (Ag–AgCl)	37.3
	0.251	0.051	14	−1.65 (Ag–AgCl)	29.0
	0.251	0.051	14	−1.60 (Ag–AgCl)	19.9
Tin	0.100	0.020	13	−1.845 (SCE)	49.8
	0.020	0.020	13	−1.73 (Hg–HgO)	32.0
Graphite	0.177	0.044	13	−1.70 (SCE)	68.7
	0.202	0.077	13	−1.70 (SCE)	48.0

Reproduced from [70] with the kind permission of the Electrochemical Society, Inc.

a combination of the latter two mainly occurred at the more cathodic potentials, where Ga-rich deposits were formed.

Because Ga has a low melting point (29.774°C), only deposits formed at low temperatures were uniform in their thickness. When the deposition temperature was increased, the Ga content in the sample decreased, and at least part of the Ga(III) ions were believed to penetrate through the porous deposit and were directly deposited on the Ti substrate.

As for the effect of the nature of the substrate, materials different from Ti were also investigated. They needed to have a high hydrogen overvoltage, because deposition was typically performed at very cathodic potentials. Moreover, no alloys or intermetallic compounds had to be formed with either Ga or As. So, Cu cathodes could not be used because of the formation of the intermetallic CuGa$_2$ compound. Table IX lists the Ga content in the deposits obtained on Si, Pb, Sn, and graphite cathodes under different experimental conditions of electrolyte composition and pH, as well as deposition potential. Note the very negative values of the deposition potential on Si, depending on its semiconducting properties. According to the table, by the selection of an appropriate electrolyte composition and deposition potential (i.e., by carrying out a systematic investigation as done with Ti cathodes), a 50 at.% Ga content in the deposit could also be obtained on these substrates.

The Auger depth profiles of Ga, As, O, and C were obtained by Ar^+ sputtering a deposit formed at -1.62 V vs. Hg–HgO from a pH 13 solution containing 0.1 M Ga(III) and 0.02 M As(III). They showed changes in the deposit composition occurring during the first minute, that is within 0.06 μm from the surface, because the sputtering rate was estimated to be about 0.06 μm min^{-1}. Then, the deposit composition was reasonably uniform. The origin of the slightly higher concentration of oxygen at the sample surface was related to the possible reaction of air oxygen with the as-deposited film or to its adsorption on it. Instead, the more deeply located oxygen must have mainly come from the electrolyte permeating the cathode during the electrodeposition of Ga and As. The lower concentration of Ga at the deposit surface was attributed to its dissolution in the electrolyte a few seconds before removal of the sample from the electrochemical cell. Indeed, the deposit was observed to dissolve into the bath, and Ga disappeared more quickly than As.

As stated, the deposits were submitted to an annealing treatment to improve their crystalline properties. Pure crystalline GaAs was obtained from alkaline solutions after the annealing, at 240°C for 21 h, of a sample containing 68.1 at.% of Ga deposited at -1.67 V vs. Ag–AgCl from a solution containing 0.25 M Ga(III), 0.05 M As(III), and 2 M KOH.

Yang et al. [70] also tried two different consecutive deposition methods to prepare GaAs: deposition of As on Ga previously deposited on the Ti substrate, and deposition of Ga on predeposited As on Ti. The two baths for either Ga or As deposition had the composition 0.75 M Ga(III) in 6 M KOH and 0.06 M As(III) in 0.3 M KOH, respectively. However, the former sequence only was successful. Before annealing, the As substrate was so thick in comparison with the electrodeposited Ga film that only small Ga quantities were detected by EDX investigation. Moreover, no peaks due to Ga, As, or GaAs were recognized in the X-ray diffraction pattern. After a heat treatment of the deposit at 300°C for 10 h, crystalline GaAs was obtained. The Ga content determined by EDX analysis was 49 at.%. As for the latter sequence (Ga deposition on As), the As film detached from the cathode surface during Ga deposition and fell onto the cell bottom, possibly because of the formation of gas bubbles. The authors suggested two possible explanations for this phenomenon: (i) As was reduced to AsH$_3$ and lost adhesion, and (ii) Ga penetrated through the As film, lifting it. These facts made it difficult to electrodeposit a Ga and As multilayer. It is worth noting that both sequences have been investigated, perhaps slightly earlier than by Yang et al. [70], even by Mengoli et al. [168], who also experienced negative results, as shown below.

8.4.2. Acid Solutions

Acid solutions had a pH value between 0.7 and 3.0 to avoid gallium oxide and hydroxide precipitation. The deposition of

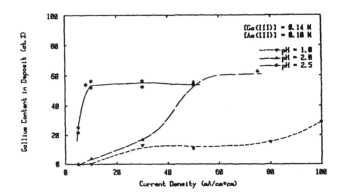

Fig. 10. Relationship between current density and Ga content in Ga–As films deposited from solutions of different pH. Reproduced from [70] with the kind permission of the Electrochemical Society, Inc.

Ga started at -1.2 V vs. Ag–AgCl, and its content in the film continuously increased at more cathodic potentials, reaching a value of 65 at.% at -1.5 to -1.6 V vs. Ag–AgCl. At these negative potentials, possibly owing to a pH increase due to hydrogen evolution, white gallium hydroxides precipitated on the surface of the black deposit.

The effect of the solution pH on the Ga content in the deposits is depicted in Figure 10 as a function of the current density. First of all, to avoid precipitation, the pH could not be higher than 2.5. Below this limit, higher current densities (or very negative potentials) were needed to reach a constant value of the Ga content in the deposit, and this Ga content was higher. Relatively thick deposits with about 50 at.% Ga content, the best ones, were obtained from a pH 2.5 electrolyte for current densities between 10 and 50 mA cm^{-2}.

Because the deposition potentials of Ga were more cathodic than those of As by at least 0.65 V, the possibility of finding a complexing agent for As that did not complex Ga was explored but not found. In aqueous solutions, both As and Ga form strong bonds with oxygen, so that few chemical species other than hydroxide can form complexes with them. Moreover, the bond with oxygen is expected to be stronger for As than for Ga, because the former does not exist as a free ion in any solution, whereas free Ga ions are present at low pH. Therefore, a complexing agent may more easily react with Ga than with As, as confirmed by the values of the stability constants of complexes. As a matter of fact, the experimental results showed the difficulties in finding an optimal complexing agent. Indeed, some of the most frequently used agents for III–V compounds (citric acid and sodium citrate solutions) inhibited both Ga and As deposition; others (acetic acid, tartrate, and succinate) did not improve the quality of the deposits, which were rougher and more porous than those prepared from baths not containing complexing agents.

The former complexing agents were more closely investigated under several conditions. Solutions of a variety of different compositions were examined: 0.057, 0.263, and 0.14 M Ga(III) plus 0.02, 0.017, and 0.10 M As(III), respectively, at pH values between 0.7 and 3. The total concentrations of citric acid and citrate were at least equal to the sum of Ga(III) and As(III),

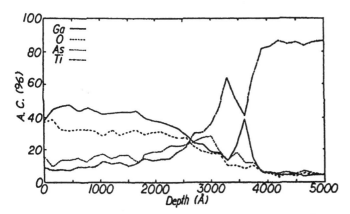

Fig. 11. Auger depth profile for a Ga–As film deposited at 30 mA cm^{-2} from a pH 2.5 electrolyte containing 0.14 M Ga(III) and 0.10 M As(III). Reproduced from [70] with the kind permission of the Electrochemical Society, Inc.

and the current densities ranged from 2.5 to 100 mA cm^{-2}. However, no deposit formation was observed on the cathode.

The films deposited from acid solutions showed no X-ray diffraction peaks of crystalline GaAs. They were so thin (the estimated value was 0.3 μm) that some samples presented the peaks of the Ti substrate only. However, GaAs peaks were observed after an annealing treatment at 260°C for 8 h. The grain size determined from the Scherrer equation was at least 300 Å, as also found for deposits from alkaline solutions.

The Auger depth profile (Fig. 11) for a Ga–As deposit obtained at 30 mA cm^{-2} from an acid solution (pH 2.5) containing 0.14 M Ga(III) and 0.10 M As(III) was more complex than that for a deposit obtained from alkaline solutions. It showed that the thin film was about 3000 Å thick, as indicated by the steep increase in the Ti curve at this value, and that the oxygen content was very high, about 30 at.%. The oxygen profile showed a trend similar to that of the Ga content and could more likely be associated with Ga than with As, in agreement with the results of EDX measurements. Indeed, whereas the As content increased slightly with depth, strongly decreasing at the deposit–Ti interface, the Ga and oxygen contents decreased monotonically. To explain such oxygen incorporation, three hypotheses were formulated: (i) a partial reduction of Ga(III) to Ga$_2$O, (ii) the constant coprecipitation on the cathode surface of Ga species containing oxygen owing to the high pH near the electrode, and (iii) a promotion of the Ga oxidation due to As, which prevented Ga from forming a continuous phase, thus increasing its reacting surface. However, in our opinion, this last hypothesis seems to be less probable, because an EDX peak of oxygen, although a weak one, was also observed in Ga deposits obtained at 30 mA cm^{-2} from an electrolyte (0.14 M Ga(III), pH 2.5) not containing As. In contrast, it was not observed in As deposits obtained at the same current density from an electrolyte (0.10 M As(III), pH 2.5) not containing Ga. It is worth noting that the deposition conditions (current density, electrolyte composition of each single species, pH) chosen for these last two experiments matched at best those used during the simultaneous deposition of Ga and As.

In conclusion, the results of the systematic research by Yang et al. [70] showed once more the difficulties in obtaining GaAs films from aqueous electrolytes of interest for practical applications.

According to Murali et al. [159], problems such as limited thickness, nonuniform thickness, and possibly large-scale cracking have usually been observed during the electrodeposition of semiconductors in direct current conditions. These problems could be overcome by the pulse plating technique, which has distinct advantages such as improved deposit distribution, improved adhesion, lower impurity content, and improved conductivity. It is possible to control deposition by varying the current density, the pulse frequency, or the duty cycle [169, 170]. This technique was found to yield crack-free hard deposits [171] and fine-grained films of lower porosity than that of films obtained by electrodeposition under direct current conditions [172–174]. So the authors [159] investigated for the first time the preparation of GaAs films by the pulse plating technique. The substrates were SnO$_2$, Ti, and stainless steel. A two-electrode system was employed; the counterelectrode was a Pt foil. The deposition was carried out at room temperature, with stirring, with the use of an aqueous solution with the composition 0.15 M GaCl$_3$ and 0.1 M AsCl$_3$. Different potentials were applied in the range of 2–6 V, with an on time of 10 s and an off time of 20 s, and the current density was in the range of 50–100 μA cm^{-2}. The deposition process lasted 30 min.

X-ray diffraction studies showed that the films were polycrystalline. They exhibited peaks corresponding to the GaAs phase alone, for pulse amplitudes below 2.5 V, and additional peaks due to Ga and As, for pulse amplitudes above 2.5 V.

The film thickness was around 1.0 μm, and the optical absorption spectrum indicated an energy gap of 1.43 eV. Owing to the high resistivity of the films, which was around 10^8 Ω cm and increased to around 10^{10} Ω cm on heat treating at 550°C for 5 min, the films did not yield measurable photoelectrochemical outputs. In situ doping of the GaAs films and variation of the parameters of pulse plating might yield low-resistivity films. Therefore, efforts in this direction to prepare low-resistivity GaAs useful for photoelectrochemical and photovoltaic devices have been under way since the authors maintained that this new technique opens up great potentials. However, as far as we know, no new data have been published.

Gao et al. [175] prepared polycrystalline GaAs thin films after having optimized the critical technological parameters of electrodeposition to obtain the correct stoichiometry. Aqueous HCl solutions containing Ga and As$_2$O$_3$ in an appropriate ratio were used, and the cathode was either a SnO$_2$-coated glass or a TiO$_2$ substrate. Different operating conditions were selected for the two substrates. The optimal ones were the following: Ga(III) and As(III) concentration in the electrolyte: 7×10^{-4} and 5.6×10^{-3} M, respectively, or 1.4×10^{-1} and 1.7×10^{-3}, respectively; pH: 0.93 or 0.64; current density: 5.56 or 2.50 mA cm^{-2}; deposition time: 3.5 or 7 min, respectively, whereas the electrolyte temperature was 22°C in both cases. The stoichiometry of the deposits on the SnO$_2$-coated glass substrate was $Ga_{0.91}As_{1.09}$, and that on the TiO_2 substrate was $Ga_{0.93}As_{1.07}$.

The current density had a critical value beyond which the deposit adhered poorly to the SnO_2-coated glass. Cowache et al. [176] previously showed that a pretreatment of the surface of this substrate improved the quality of electrodeposited films (in the specific case of CdTe). Moreover, Gao et al. observed that this same substrate was partly corroded under a prolonged deposition time at constant current density, bringing the process to an end. The X-ray diffraction spectra of the deposits presented a good correspondence with standard data for GaAs, but additional peaks due to SnO_2, Sn, or Ti were observed.

The $Ga_{0.91}As_{1.09}$ samples were submitted to optical characterization and to Mott–Schottky measurements [175, 177]. They showed a direct-gap nature, where the band gap was 1.43 eV, as for GaAs. Compared with the absorption curve of intrinsic GaAs, that of the deposits presented an adsorption edge shift toward higher wavelengths, which generated an exponential tail. This tail was suggested to be caused by the presence of unoccupied shallow impurity states, lattice microstrains, and the nonuniform distribution of charged centers. From the dependence on the applied bias of the depletion layer capacitance of the film dipped in a sulfide–polysulfide electrolyte, the flatband potential and the donor concentration of the film were determined. They were observed to be equal to -1.4 V vs. SCE and 6.9×10^{15} cm^{-3}, respectively, and from all previous values the positions of the energy band edges were also obtained. The film turned out to be n-type. Moreover, the photocurrent density, as a function of the applied bias of the $Ga_{0.91}As_{1.09}$–electrolyte junction, was investigated, and the efficiency was found to be 0.51%. The authors maintained that lower-cost and higher-efficiency GaAs photoelectrochemical cells could be prepared by pretreating the substrate surface and improving the electrodeposition technique further.

Tests for the synthesis of GaAs from aqueous solutions following the sequential method were performed by Mengoli et al. [168]. Two synthetic sequences were tried, both revealing some weakness in either the electrochemical or the thermal annealing step. In the first sequency, Ga was electrodeposited on Ni-plated Cu sheets at -2.0 V vs. SCE from an alkaline gallate bath (0.6 M $GaCl_3$ in 5 M KOH) according to paper [178], and As was then electrodeposited on the Ga layer at -1.0 V vs. SCE from a 5 wt.% solution of As_2O_3 in 7 M HCl, at 50°C, following reference [66]. Under these conditions, 2–3 μm-thick As deposits were formed with a small current decrease during electrolysis, a result that the authors maintained is possibly due to the formation of an amalgam with the liquid phase of Ga (the melting point of which is 29.774°C).

The as-deposited Ga–As two-layer samples were amorphous; their thermal annealing at 400°C for 5 h induced the formation of crystalline phases identified by X-ray diffraction as Ga–Cu and Ga–Ni alloys, together with small amounts of GaAs. Indeed, the first and the third most intense diffraction peaks only were observed, the second being hidden by the peaks of the substrate metals and their alloys with Ga. The formation of the two Ga alloys probably occurred because Ga was de-

posited as the inner layer. A further difficulty with this first sequence was that, owing to the high volatility of As, the desired 1 : 1 Ga-to-As ratio could be maintained only by performing the annealing treatment under an As overpressure.

The second synthetic sequence (i.e., As first and Ga second) was severely hindered by the electrochemical instability of As, which was extensively converted to AsH_3 [161] at the very negative potentials required for Ga deposition. A possible way out of these difficulties, which was suggested by the authors but not tested, might be Ga electrodeposition from an adequate aprotic medium. Instead, the authors [168] yielded nearly pure polycrystalline GaAs films by following a different procedure, that is, by annealing at 350°C for 5 h the two-layer samples obtained by vacuum evaporation of a Ga film onto electrodeposited As. These films contained only traces of the products of the reaction of As with the underlying metals.

Subsequently, the same research team, with the intention of preparing ternary Ga–As–Sb thin films, preliminarily synthesized GaAs films by deposition of As on electrodeposited Ga [179]. They followed the same electrochemical conditions as in their previous research [168]. Again, the first problem was the tendency of Ga to alloy with many metal substrates during the annealing step, but Ti seemed to be a convenient one for Ga deposition, because it showed both a high hydrogen overvoltage and a limited tendency, if any, to alloy with Ga, as already observed by Yang et al. [70]. However, at the temperature of 50°C recommended for Ga electrodeposition (owing to the higher coulombic efficiency [178] and the fairly homogeneous coverage of many substrates, particularly Sb [83]), Ga was in the molten state, and its adhesion to the Ti substrate was critical. So, Ga deposition was performed at 20°C, the yield being 30–60% only, and the Ga deposit was microcrystalline, irregularly covering the substrate. The subsequent As deposition, with an efficiency close to 100%, produced a powdery deposit, and an effective control of the local Ga-to-As ratio was not possible. Nevertheless, after a 4-h annealing process at 450°C, in the presence of a small amount of As powder to limit losses from the samples, the X-ray diffraction pattern of the two-layer deposit presented the GaAs peaks in their correct ratio and minor Ti substrate peaks. These data confirmed previously reported results of Yang et al. [70], but, according to the authors, they were comparably better, indicating a more homogeneous substrate coverage. However, local deviations in the Ga-to-As ratio from 1 : 1 with both Ga-rich and As-rich areas were indicated by EDX analysis, but apparently these areas did not cause significant changes in the X-ray diffraction pattern.

As for the ternary compounds, it is interesting to note that the preparation of $GaAs_xSb_{1-x}$ by sequential deposition of Ga on an As-Sb alloy and subsequent annealing at 500°C for 8 h gave rise to the formation of two microcrystalline phases, an As-poor ($GaAs_{0.19}Sb_{0.81}$) and an As-rich ($GaAs_{0.82}Sb_{0.18}$) phase [179]. These results are in agreament with literature data showing a miscibility gap between GaAs and GaSb at all temperatures below 860°C [180] and could be compared with similar results obtained during the synthesis of In–As–Sb thin films by In plating an electrodeposited As–Sb alloy and submitting the sample to a thermal treatment [168], as will be discussed under InAs.

GaAs was also prepared by Stickney and co-workers [21, 100] in 1992, following the ECALE technique. The aim was to prepare high-quality semiconducting thin films by taking advantage of surface-limited processes such as UPD to circumvent problems associated with nucleation and growth leading to the convoluted morphologies observed during direct codeposition.

Deposition was performed on polycrystalline and single-crystal Au electrodes oriented according to the (100), (110), and (111) planes [21]. Gold was selected as the substrate, owing to its stability in aqueous solutions and to its extended double layer window, which allowed UPD studies with minimum interference from substrate oxidation and hydrogen evolution. It was fortuitous that a low lattice mismatch of only 3.5% exists between GaAs and twice the Au lattice constant.

Preliminary investigations [100] were carried out with the polycrystalline Au electrode in a thin-layer electrochemical cell to limit contamination. They determined the potentials for oxidative As UPD deposition and reductive Ga UPD deposition with different solution concentrations and pH values. The reference electrode was Ag–AgCl, 1.0 M NaCl. Arsenic was shown to be reductively deposited from a $HAsO_2$ solution with pH 4 between -0.3 and -1.5 V vs. Ag–AgCl, 1.0 M NaCl:

$$HAsO_2 + 3H^+ + 3e = As + 2H_2O$$

At very negative potentials, the As surface coverage decreased because of the formation of arsine:

$$As + 3H^+ + 3e = AsH_3$$

These results were interpreted in two different ways: (i) deposited As was reduced at very negative potentials, except for a monoatomic layer stabilized by bonding with Au surface atoms; alternatively, (ii) $HAsO_2$ was quantitatively converted to arsine according to the reaction

$$HAsO_2 + 6H^+ + 6e = AsH_3 + 2H_2O$$

and a monoatomic As layer was deposited by oxidative UPD from AsH_3.

The subsequent electrochemical studies described in the same paper [100] were performed in a ultrahigh-vacuum antechamber interfaced to the main analysis chamber, where AES and low-energy electron diffraction (LEED) characterization were used to obtain information about the surface composition and structural analysis, respectively. The surface composition analysis was also carried out by coulometry. This configuration made it possible to perform the electrochemical treatments in an argon atmosphere, at ambient pressure, and the subsequent analysis in an ultrahigh vacuum, protecting the samples from contact with air. Scanning tunneling microscopy investigation was performed under nitrogen at atmospheric pressure, inside a glove box.

ECALE deposition on the surface of the Au single-crystal electrodes was achieved by alternative deposition of atomic layers of As and Ga. Arsenic was first deposited from a 1.0 mM

$HAsO_2$ and 1.0 mM H_2SO_4 solution with a pH of 3.2 and subsequently reduced at -1.25 V vs. Ag–AgCl, 1.0 M NaCl, in a 10.0 mM Cs_2SO_4 solution with a pH of 4.7, containing 1 mM acetate buffer. So, low-coverage, ordered layers of As were obtained by reductive UPD from $HAsO_2$ and by oxidative UPD from AsH_3 on the Au (100) and Au (110) substrates with one-forth and one-half surface coverage, respectively. The next step was Ga deposition on the As-covered surfaces from a 0.5 mM $Ga_2(SO_4)_3$ solution with a pH of 2.7, at potentials between -0.71 and -0.56 V vs. Ag–AgCl, 1.0 M NaCl. At these values, stoichiometric Ga quantities, with respect to the already deposited As, were deposited, and the Ga and As coverages resulted in (2×2) and $c(2 \times 2)$ LEED patterns on the Au (100) and (110) surfaces, respectively.

The stability of GaAs in aqueous solutions was also evaluated; the results indicated that GaAs can be electrodeposited from aqueous solutions by following the ECALE methodology, avoiding emersion of the substrate and loss of potential control.

According to Chandra and co-workers [10], more theoretical studies will have to be made before the potential of the ECALE technique can be fully exploited.

8.5. Indium Arsenide

In 1989, Ortega and Herrero [45] described electrochemical synthesis by the codeposition of InAs, in addition to that of InSb, which is further examined here. InAs was obtained from citric acid solutions of $InCl_3$ and $AsCl_3$ to complex the group V element, shifting its electrodeposition potential toward a more negative value, that is, closer to that of the less noble In.

To clarify the mechanism of As electrodeposition, Figure 12 compares the curve giving the current density as a function of the applied potential of a solution containing 0.04 M $AsCl_3$ and 0.3 M citric acid with that of a solution containing citric acid only. Two electrochemical reduction reactions took place, form-

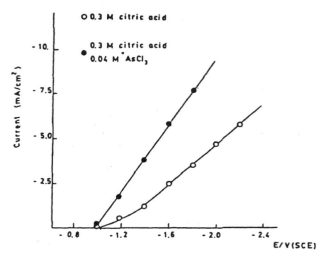

Fig. 12. Current density vs. potential plot for As deposition from a citric bath. Reproduced from [45] with the kind permission of the Electrochemical Society, Inc.

ing AsH_3 as the final product, according to the equations

$$AsO_2^- + 4H^+ + 3e = As + 2H_2O$$
$$As + 3H^+ + 3e = AsH_3$$

The standard potentials are 0.194 and -0.843 V vs. SCE, respectively. Note the similarity of these reactions to those considered during GaAs preparation following the ECALE method. Because of the acid pH of the solution, a more appropriate reagent would have been $HAsO_2$ instead of AsO_2^-, whose domain of relative predominance is above pH 9.21 [49]. The standard potential of the $HAsO_2$–As redox system is 0.013 V vs. SCE. As a matter of fact, Ortega and Herrero also considered the presence of $HAsO_2$ in the bath. Indeed, they maintained that the arsine produced on the cathode could be assumed to interact chemically with the $HAsO_2$ contained in the electrolytic solution, so that As was finally produced:

$$AsH_3 + HAsO_2 = 2As + 2H_2O$$

which in turn reacted with In^{3+} to form the InAs compound when $InCl_3$ was also added to the bath:

$$As + In^{3+} + 3e = InAs$$

According to the authors, this mechanism of codeposition is similar to that proposed by Kröger [29, 69] and simulated by Engelken et al. [38, 41] for the formation of CdTe from aqueous solutions of Cd^{2+} and H_2TeO_3, which has already been considered.

Adherent InAs thin films could be obtained on a Ti cathode at -1.40 V vs. SCE from a citric bath. The as-deposited InAs was amorphous. After being annealed at 400°C for 30 min, its X-ray diffractogram showed the peaks of fcc InAs, but with a low crystallinity. With illumination of the electrode during InAs electrodeposition, a cathodic photocurrent was observed, which indicated that the growing semiconducting layers were p-type, as shown in Figure 13 by the current–potential curves for darkness and illumination.

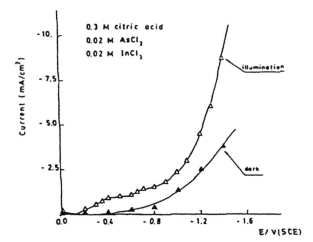

Fig. 13. Current density vs potential plot for InAs deposition in darkness and under illumination. Reproduced from [45] with the kind permission of the Electrochemical Society, Inc.

InAs thin films were also prepared by Mengoli et al. [168], who In-plated electrodeposited As and submitted the two-layer samples thus obtained to thermal annealing. Indeed, of the two possible deposition sequences, that involving As deposition as the first step was chosen because previous experience during the preparation of Ga–As deposits showed that As losses by evaporation during the annealing treatment was much more severe when the group V element formed the outer layer.

The As layers were electrodeposited from stirred 5 wt.% solutions of As_2O_3 in concentrated (7 M) HCl at -1.0 V vs. SCE and 50°C. The procedure was that proposed by Menzies and Owen [66], who observed that with increased cathodic current density and deposition time, the current efficiency with respect to elementary As deposition strongly decreased. However, the deposition of As was essentially quantitative for the short electrolysis durations (3–4 min) and low current densities (3–15 mA cm^{-2}) employed by Mengoli et al. [168] to grow As layers up to 2 μm thick at most. This result is in keeping with the well-known inhibition of the hydrogen evolution process by arsenious compounds [181].

The In layers were deposited from unstirred 0.1 M $InCl_3$, 0.13 M ethanolamine, and 0.035 M ammonia solutions whose pH was adjusted to 2 with HCl. The deposition conditions, taken from the paper of Herrero and Ortega regarding the preparation of In_2S_3 and In_2Se_3[182], were -1.0 V vs. SCE at 25°C. The In current efficiency was 95 $\pm$ 4%. To prepare InAs thin films, the As and In deposition charges were controlled to obtain a 1 : 1 stoichiometry. Mengoli et al. [168] also compared the cyclic voltammograms recorded on Ni- or As-plated Ni electrodes in the In-containing solution (here only 0.05 M $InCl_3$). By sweeping toward negative potentials, In deposition on As was clearly delayed by about 400 mV. This shift was independent of the As layer thickness, and it was concluded to arise from a larger overpotential of In on As than on Ni, rather than from the resistance of the As layer.

The two-layer samples were then annealed at 325°C for 4 h under either a nitrogen or an As atmosphere. In both cases, no weight losses were found, and the EDX analysis showed no change in the overall composition. X-ray diffraction indicated the formation of polycrystalline InAs without a marked preferred orientation, as found by comparing the experimental spacings and relative intensities of the InAs diffraction peaks with those of literature data. The authors also determined the average crystallite size according to the Scherrer relation. The crystallite dimension turned out to be in the range of 40–50 nm, which is comparable to the values obtained for antimonides, InSb (35 nm) and GaSb (60 nm), prepared following the same method: Sb first, either In or Ga second [84]. In conclusion, the difficulties expected when As is used, on the basis of its low electrical conductivity and high volatility, did not actually hinder the electrochemical or the annealing step in the case of this InAs synthesis.

Then the same authors studied the synthesis of In–As–Sb thin films involving electrochemical steps (In deposition on an electrodeposited As–Sb layer) followed by thermal annealing at 300°C for 4 h. The ternary compounds consisted of a mixture of two phases, an As-rich and a Sb-rich phase [168], the lattice parameters of which were determined by X-ray diffraction analysis. In a previous paper [84], Mengoli et al. observed that also the ternary $Ga_x In_{1-x}Sb$ compound (obtained by coating an In–Sb layer with Ga, which will be examined under InSb) was formed as mixtures of two phases. This result might be related to the annealing temperatures, which in both cases are not sufficiently high to induce the formation of a homogeneous melt.

The attempt to prepare ternary electrodeposits by simultaneous In, As, and Sb deposition from a solution containing $InCl_3$, As_2O_3, and $SbCl_3$ in citric acid was not successful, because the deposits were heterogeneous [168]. SEM analysis showed the presence of In crystals in a matrix prevailingly consisting of an As–Sb alloy, and X-ray diffraction analysis showed the presence of In peaks and of one peak due, perhaps, to some ternary compound, $InAs_xSb_{1-x}$. No As or Sb diffraction peaks were observed; they were probably in the amorphous state.

In a subsequent paper [46], the same group of researchers simultaneously deposited an In–As–Sb alloy from pH 2 tartaric acid solutions, obtaining a homogeneous deposit of a composition suitable for achieving $InAs_xSb_{1-x}$. Indeed, poorly crystalline Sb-rich ternary phases were formed at room temperature, the crystallinity of which improved after annealing at 250°C. The results of X-ray and EDX analyses suggested that the entire deposit had been converted into the $InAs_xSb_{1-x}$ crystalline phase.

Wishing to also prepare the ternary $InAs_xP_{1-x}$ compound by PH_3 treatment of In–As precursors, Dalchiele et al. of the previous research group [183] preliminarily prepared (by codeposition) In–As alloys and (by consecutive deposition) As–In bilayer films. This research was directly connected to preceding work by the same authors already discussed, on the formation of n- and p-type InP films by a high-temperature PH_3 treatment of electrodeposited In layers [134]. During such treatment, In melted and formed microdrops because of its low melting point (156.634°C), so that a somewhat heterogeneous film was obtained that presented an unsatisfactory photoelectrochemical behavior. The In–As precursor was expected to have a lower tendency to form such drops than pure In, and, therefore, a final ternary In–As–P film with a better morphology and surface coverage was expected.

Both In–As alloys and As–In bilayer films were potentiostatically electrodeposited at 50°C from an unstirred and deaerated aqueous solution on a Ti working electrode. They were a few micrometers thick, as estimated from their weight. The codeposition of the In–As alloy films was performed at a potential ranging from −0.800 to −1.250 V vs. SCE from a sulfate bath with a typical composition of 20 mM $NaAsO_2$, 10–50 mM $In_2(SO_4)_3$, 25 mM potassium citrate, and 100 mM K_2SO_4. The solution pH was adjusted to 2.0 with H_2SO_4. The As–In bilayers were obtained by consecutively plating (according to methods in the often quoted literature) As at −1.0 V vs. SCE from a 5 wt.% As_2O_3 solution in 7 M HCl at 50°C [66] and In at −1.05 V vs. SCE from a 0.1 M $In_2(SO_4)_3$ solution of pH 2 [138].

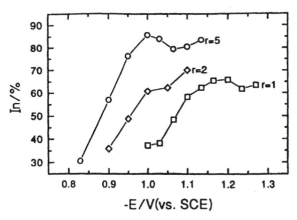

Fig. 14. Potential dependence of the composition of In–As alloys deposited from pH 2 sulfate–citrate solutions containing 20 mM $NaAsO_2$ and 10 (□), 20 (◊) or 50 (○) mM $In_2(SO_4)_3$. Reprinted from [183] with the kind permission of Elsevier Science.

As previously shown, Ortega and Herrero [45] already investigated In and As codeposition to prepare InAs. They used a chloride bath containing citric acid, whereas Dalchiele et al. [183] considered an acid sulfate bath containing citrate and investigated the effect of the electrodeposition parameters on the composition and morphology of the films.

The influence of the deposition potential on the film composition is shown in Figure 14 for three different electrolytes with an increasing In(III) concentration, so that the ratio between the In and the As atomic concentrations, r, was 1, 2, and 5, respectively. For each curve, for more negative potentials, the In at.% in the deposit increased, as expected on a thermodynamic basis, because In is the less noble metal of the two. The standard potential of the AsO^+–As and In^{3+}–In redox systems is 0.254 and −0.342 V vs. SHE, respectively, with a difference of 0.596 V. The citrate ions were added to the electrolytic solution to limit such difference, because they were expected to form complexes with the As(III) ions but not with the In(III) ones. The curves attained a quasi-constant value when the process became diffusion-limited, because the electrolyte was not stirred.

Note that here again the standard potential to be referred to is that of the $HAsO_2$–As redox system, because the AsO^+ domain is below the pH value −0.34. In this case, however, the value of the standard potential (0.248 V vs. SHE) is close to that of the AsO^+–As redox system [49].

The limiting value of the In-to-As ratio in the deposits increased with r, so deposits with a value close to that in the solution were obtained when r was equal to 5, whereas they contained an excess of In when $r = 1$. Indeed, by decreasing r, the curves shifted toward more cathodic values, where As may be reduced to AsH_3 [184]. This In percentage increase was confirmed by SEM investigation, which showed the deposits from the $r = 1$ bath to become less compact and somewhat dendritic at increasing cathodic potentials. Comparing the three curves of Figure 14, it appears that for decreasing r values, deposits with a lower In-to-As ratio were obtained at the same potential, whereas deposits with the same composition were obtained

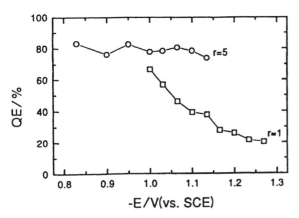

Fig. 15. Potential dependence of the current efficiency for deposition of In–As alloys from pH 2 sulfate–citrate solutions containing 20 mM NaAsO$_2$ and 10 (□) or 50 (○) mM In$_2$(SO$_4$)$_3$. Reprinted from [183] with the kind permission of Elsevier Science.

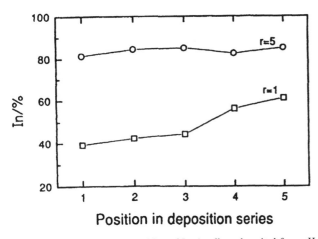

Fig. 16. Evolution of the composition of In–As alloys deposited from pH 2 sulfate–citrate solutions containing 20 mM NaAsO$_2$ and 10 (□) or 50 (○) mM In$_2$(SO$_4$)$_3$ in a series of depositions from the same bath. Reprinted from [183] with the kind permission of Elsevier Science.

at increasingly more cathodic potentials. So, stoichiometric deposits could be prepared at −0.880, −0.950, and −1.065 V vs. SCE from electrolytes with r equal to 5, 2, and 1, respectively.

The coulombic efficiency, QE, during each codeposition of In and As was estimated by EDX analysis. Its trend as a function of the applied potential for two r values is depicted in Figure 15. The parasitic hydrogen and AsH$_3$ evolution processes were competitive with the main process at the more cathodic potentials [184] when the In^{3+} ion concentration in the electrolyte was low ($r = 1$). They became less important at higher concentrations. Indeed, for $r = 5$, the coulombic efficiency was practically constant around 80 at.%. The changes in the composition of films consecutively deposited from the same bath at the same potential (−1.1 V vs. SCE) were also examined (Fig. 16). Once again, these changes were practically negligible for $r = 5$; the In content remained nearly constant, around 80 at.%. In contrast, a significant and continuous increase of the In content in the In–As alloy was observed for the smallest r value. The authors maintained that this behavior was prob-

ably due to a continuous increase in the In-to-As ratio in the solution because of AsH$_3$ formation and its reaction with the As(III) species to form As, according to the antidisproportionation reaction (3) described by Perrault [53], as confirmed by the increasing presence of a brownish suspension in the solution.

The X-ray diffraction pattern of the 1 : 1 deposits obtained from a solution with $r = 1$ presented weak and broad peaks due to InAs and sharp peaks due to In. The corresponding peaks of As were not observed, probably because As was in the amorphous state. After a heat treatment at 400°C for 4 h, the deposits presented sharp peaks due to the InAs compound only. Their relative intensities and the observed distances of the lattice planes were in good agreement with literature data [6].

As–In bilayers prepared following the consecutive electrodeposition method were also considered as a precursor for the InAs$_x$P$_{1-x}$ films formation because of their easy conversion to InAs by a thermal treatment, as already discussed [168].

Both of these bilayers and the electrodeposited In–As alloys were annealed under PH$_3$ flow in an Aixtron AIX 200 reactor at 500°C for 15 min and subsequently at 600°C for 30 min and then were submitted to EDX analysis. After the PH$_3$ treatment of the In–As alloy, two polycrystalline phases were detected by X-ray diffraction. One of these phases was almost pure InP (with a very small As content, not more than a few at.%); the other one presented peaks intermediate between those of InP and InAs and was, therefore, identified as InAs$_x$P$_{1-x}$. The ratio of In to (As + P) was very close to 1 for all ternary compounds, thus indicating spontaneous control of the stoichiometry occurring during the PH$_3$ treatment, as previously observed in the case of the binary InP compound formation [134]. The x value in InAs$_x$P$_{1-x}$ ranged from 0.25 to 1. The ternary phase was observed for all of the As-to-In ratios, even starting from an As-rich In–As alloy with a ratio equal to 1.70. In this case, the InP amount that accompanied the InAs$_x$P$_{1-x}$ phase was very small. A special X-ray diffraction analysis at grazing angle incidence and SIMS depth profile analysis indicated that the InP percentage increased closer to the surface.

Despite the absence of a miscibility gap for the InAs–InP system, the formation of two phases was similar to what was found for the ternary InAs–InSb system prepared by In plating of an As–Sb alloy [168]. According to the authors [183], this might be a consequence of synthesis conditions where the phase formation was controlled by kinetic factors.

As for to the morphology of the deposits, bilayers, and In–As alloys, the electrodes were regularly coated with small grains whose size increased after PH$_3$ annealing, as found by SEM investigation.

The photoelectrochemical research performed in an aqueous acidic iodide solution showed that the InP phase only was photoactive. It was found to be n-type with a band gap with an upper value of 1.32–1.33 eV, as found for InP samples prepared by the reaction of In with PH$_3$ [134], and a lower value of 1.27 eV. The transition was a direct one.

Recently, InAs films were directly electrodeposited at room temperature by the ECALE method [101]. Up to 160-nm-thick films were formed by the alternating electrodeposition of As and In atomic layers in underpotential (UPD) conditions. The instrument for the measurements consisted of a series of solution reservoirs, pumps, a selection valve, a potentiostat, and a 0.10-ml thin-layer flow cell containing the substrate. The latter was a (100) Si wafer onto which a 5-nm-thick Ti film was first vapor deposited, followed by a 200-nm Au film. The electrolytes were a 2.5 mM solution of As_2O_3 with a pH of 4.8, buffered with 50.0 mM sodium acetate, and a 0.3 mM solution of In_2O_3 with a pH of 3. The pH of the solution was adjusted with H_2SO_4, and the supporting electrolyte was 0.5 M $NaClO_4$.

The cyclic voltammograms of the Au deposited on Si showed a peak due to UPD electrodeposition at −0.15 V vs. Ag–AgCl in the As-containing solution and one at −0.20 V vs. Ag–AgCl in the In-containing solution. From these results, a program was suggested for the deposition cycle, with −0.25 V for As UPD and −0.3 V vs. Ag–AgCl for In UPD. After the first few cycles, the program was adjusted to more negative values. The initial deposits appeared gray and metallic; however, nonmetallic deposits were obtained with the use of significantly more negative potentials for the deposition of both elements.

The films were characterized by several techniques: SEM, AFM (atomic force microscopy), X-ray diffraction (showing strong peaks of InAs but also of In and Au), ellipsometry, and infrared absorption. Moreover, the band gap was estimated from the experimentally determined absorption coefficient dependence on the energy of the radiation, here in the infrared region. The band gap of the electrodeposited film was 0.45 eV, whereas that of an InAs single-crystal wafer was significantly lower (0.36 eV). The authors attributed this blue shift to the $1/d^2$ dependence of the band gap on the diameter, d, of the crystallites in the film [185–187]. Comparing the 0.1-eV shift to those observed for InAs nanocrystals, the authors estimated a d value on the order of 20 nm for their InAs films, in agreement with the values found by X-ray diffraction (15 nm) and the AFM technique (40 nm).

The growth of InAs on Ag single-crystal electrodes was recently investigated with the ECALE method by Innocenti et al. [104]. This research group mainly worked on Ag substrates oriented according to low-index planes. The use of Ag instead of Au might be advantageous for its lower cost and for the often better defined UPD voltammetric peaks. On the other hand, Ag is less noble and, therefore, more reactive than Au. As a consequence, the double-layer region in the positive direction is narrower than on Au, and different experimental conditions had to be selected. Furthermore, the Ag surface required more severe treatments before use, for example, chemical polishing followed by cathodic polarization to produce mild hydrogen evolution [102].

On the basis of previous experiences and chemical considerations, it was established that the first monoatomic layer to be deposited was most conveniently that of As. Cyclic voltammograms of the Ag substrate in an alkaline solution containing $NaAsO_2$ showed that the bulk As deposition peak was preceded by a peak indicative of surface-limited processes. However, because of the high irreversibility of the system, this last peak occurred at a potential more cathodic than the reversible potential

for bulk As deposition. Therefore, it was not possible to deposit an As monolayer without the concomitant bulk As deposition. To circumvent this situation, an excess of As was deposited at −1.1 V vs. Ag–AgCl, then a potential of −1.6 V vs. Ag–AgCl was applied that was sufficiently cathodic to reduce the massive As to AsH_3 without reducing the first As overlayer in direct contact with the Ag substrate. This As could be considered to be obtained in UPD conditions, As_{upd}. Chronocoulombometric measurements confirmed the two-step AsO_2^- reduction mechanism,

$$AsO_2^- + 2H_2O + 3e = As + 4OH^-$$
$$As + 3H_2O + 3e = AsH_3 + 3OH^-$$

This mechanism is similar to the first of the two mechanisms already considered by Stickney et al. [100] for GaAs ECALE deposition from acid solutions containing $HAsO_2$.

As for In ECALE deposition, the following considerations had to be taken into account: (i) solubilizing the In compounds in neutral and alkaline electrolytes and (ii) finding a potential where the deposition rate was sufficiently high to be compatible with the typical working times but not high enough to yield bulk In deposition. Because of these limitations, In was deposited from a 0.5 mM In^{3+} sulfuric buffer (pH 2.2) at −0.58 V vs. Ag–AgCl. At this value, In could be considered to be obtained in UPD conditions (In_{upd}). So, the first ECALE deposition cycle of InAs consisted of depositing As_{upd} and replacing the electrolyte with a solution of In_2O_3 in sulfuric buffer followed by In UPD deposition under a strict potential control. A special electrochemical cell made it possible to automatically make substitutions for the electrolytes and control the potential. Of course, the use of solutions of such different pH values introduced additional technical difficulties. The first cycle was followed by other cycles, to obtain deposits of consistent thickness. The presence of one or more deposition cycles of In and As was verified by anodic stripping of the two deposited species. A direct proportionality between the stripping charge and the number of cycles confirmed the validity of the procedure and gave an indication of the stoichiometry of the InAs compound.

8.6. Gallium Antimonide

The electrochemical synthesis of this III–V compound semiconductor has not received much attention until recently, despite its use in applications such as solar cells, photodetectors, and substrates for epitaxial films. A Ga–Sb alloy was first deposited in a single electrochemical step from a nonaqueous (glycerol) bath containing the two elements [188]. The effect of the cathode material, the current density, the relative concentrations of the Ga and Sb species in the bath (their total concentration was 1 g-equiv. liter^{-1}), temperature, and stirring on the deposition of both species and their alloys was examined. The nature of the cathode material influenced the deposition potential of the components and of the alloys as well as the deposit quality. The best results were achieved with Cu cathodes. At the current densities of 2–2.5 mA cm^{-2}, the deposition potential of the components, Ga and Sb, leveled off. With a further increase in the current density, the total polarization curve shifted to a more positive region than the polarization curves of the components, leading to a depolarization factor that favored the formation of either a solid solution or an intermetallic compound. A rise in temperature and agitation reduced the region of the limiting current density and shifted the deposition potential toward positive values.

More recently, Mengoli and co-workers [83] explored the two alternative electrochemical strategies of simultaneous and sequential deposition of Sb and Ga to synthesize GaSb thin films. They tried the former approach first, but they failed, because the large difference in the standard potentials of the Sb(III)–Sb (0.212 V vs. SHE) and of the Ga(III)–Ga (−0.529 V vs. SHE) redox systems could not be compensated for with the use of solutions containing excess Ga ions and complexing agents (citrate) for Sb ions. Instead, they succeeded in preparing GaSb polycrystalline films by the alternative method, involving sequential deposition in potentiostatic conditions of Sb and Ga films from an acid $SbCl_3$ solution and an alkaline $GaCl_3$ solution, respectively, and subsequent annealing. The substrate was an approximately 5-μm-thick Ni film galvanostatically deposited on a Cu sheet from a Watts-type bath. The choice of the deposition sequence depends on the stability of the initially deposited metal under the deposition conditions of the subsequent metal layer. So, the sequence Ga first–Sb second was not possible because the deposited Ga anodically dissolved at a potential greater than or equal to −0.6 V vs. SCE and was displaced from the substrate during Sb deposition at these values. To avoid Ga reoxidation, Sb electrodeposition had to be performed at potentials smaller than −0.6 V vs. SCE, where the process was diffusion-controlled and where inhomogeneous, scarcely adherent Sb films, formed by dendrites, were obtained.

These difficulties did not exist when Sb was deposited first. However, potentials had to be chosen for the subsequent Ga deposition that were not too negative, otherwise Sb was converted to SbH_3 [189]. The authors performed a test to determine the Sb losses due to this reaction by polarizing for 100 s a 0.5-μm-thick Sb deposit in a 5 M KOH solution at increasingly negative values. Whereas this loss was lower than 3% at −1.85 V vs. SCE, it was higher than 75% at −2.1 V vs. SCE. This result imposed a potential limit for Ga deposition.

Sb was deposited at −0.25 V vs. SCE with 100% coulombic efficiency from a solution of 0.3 M $SbCl_3$, 1.7 M HCl, and 1.6 M H_2SO_4, as suggested in the literature [190], whereas Ga was potentiostatically deposited from strongly alkaline solutions (5 M KOH), where it was present as GaO_3^-. Several different conditions were examined to improve its coulombic efficiency because of the parasitic hydrogen evolution process. The coulombic efficiency of Ga deposition, QE, was measured as the ratio between the charge spent for stripping Ga, Q_s, and the total charge transferred during deposition, Q_d, and its value was the average over the deposition time. A window of the potential was found at −1.85 V vs. SCE (Fig. 17), where this efficiency was fairly high and Sb degradation to SbH_3 was negligible. Therefore, Ga was deposited at this same potential from a bath of 0.3 M $GaCl_3$ and 5 M KOH for various times to reach

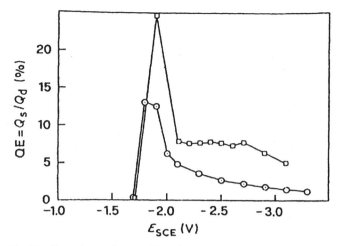

Fig. 17. Dependence of coulombic efficiency of Ga deposition on the potential. Electrolysis duration, 30 s; electrolyte composition, 0.11 M (o) or 0.22 M (□) $GaCl_3$, 5 M KOH. Reproduced from [83] with the kind permission of Kluwer Academic Publishers.

the wanted film thickness. Moreover, the authors forced a flow of electrolyte tangential to the electrode surface with a peristaltic pump, obtaining two results: an increase in the efficiency and the removal of hydrogen gas bubbles, which otherwise induced defects in the Ga layer.

To produce the interdiffusion and reaction of Ga with Sb, electrodeposition was followed by thermal annealing in an inert atmosphere, generally at 100°C for 4 h. A comparison of the stripping voltammograms of the individual metal layers (Ga and Sb) with those of a two-layer sample (Ga–Sb) as deposited and after annealing allowed the authors to conclude that no reaction between Ga and Sb occurred at room temperature both during and after Ga deposition, but that GaSb was formed during a mild heat treatment. So, the two-layer samples were heated in the stated conditions, and the formation of GaSb was observed by X-ray diffraction. The diffractogram relevant to a 40 μm Sb + 5 μm Ga sample submitted to such treatment showed almost four peaks due to GaSb and another two due to GaSb only partially, because the latter overlapped with some Sb peaks. The remaining peaks were due to the Sb underlayer. Reflections due to GaSb and Sb only were observed, suggesting that Ga which was in defect with respect to Sb had completely reacted. The characterization by SEM, EDX, and SIMS of the annealed deposits showed that the very large majority of their surfaces was homogeneously covered by small crystallites, typically of 0.1–2-μm diameter, although a few large features up to 40 μm in diameter were detected, too, possibly because of residual unreacted Ga not detected by X-ray diffraction. Moreover, the EDX analysis showed that most of the sample surface was chemically homogeneous at least at the scale of the area sampled by the electron beam (around 25 μm^2). A SIMS depth profile analysis of the surface of a 1 μm Sb and 1 μm Ga annealed sample sputtered with an Ar^+ beam indicated that the deposit was homogeneous in the direction perpendicular to its surface. In contrast, in the case of an as-deposited two-layer sample sputtered for the same time, a sharp interface between Ga and Sb was detected.

The proposed electrochemical route made it possible to prepare GaSb thin films from aqueous solutions, although difficulties in the control of the Ga layer thickness were encountered because of hydrogen evolution simultaneously occurring with Ga electrodeposition. According to the authors, such difficulties might be overcome if an adequate aprotic electrolyte is found.

Finally, the authors maintained that this was the first report on the electrochemical synthesis of a III–V semiconductor from aqueous solutions carried out by the sequential deposition method followed by thermal annealing, which appeared to be suitable for the preparation of other related materials. This synthesis was then followed by the same group of authors to prepare thin films of InSb and the ternary Ga–In–Sb [84] compound; InAs [160, 168], In–As–Sb [46, 160, 168], and In–As–P [183]; and GaAs [168, 179] and Ga–As–Sb [179]. However, it is worth recalling that 3 years earlier, in 1987, Murali et al. [166] had already applied the sequential deposition method to prepare a III–V compound (GaAs) from an aqueous bath.

In 1996, McChesney et al. [52], intending to prepare III–V compounds for applications in infrared photodetectors and optical communication systems, developed suitable electrochemical methods for the production of GaSb and InSb deposits. The InSb deposits are discussed under InSb.

The GaSb samples were grown at a constant potential and at 80°C from an aqueous solution containing $SbCl_3$ and either $Ga_2(SO_4)_3$ or $GaCl_3$. The pH value was lower than 1. The explored molar ratio of the Ga-to-Sb species ranged from 1 : 1 to 1000 : 1. Several different substrates were tested, such as Cu and Ti sheets and glass coated with indium tin oxide (ITO). The deposits were investigated by X-ray diffraction, SEM, EDX, and LA-ICP (laser ablation inductively coupled plasma mass spectroscopy). Spongy Sb deposits, poorly adherent to the substrate, were formed under most experimental conditions. However, at potentials more negative than -0.8 V vs. SSC (Ag–AgCl, 0.197 V vs. SHE) with a solution 0.04 M $GaCl_3$, 0.001 M $SbCl_3$, and ITO substrates, mixtures of cubic GaSb and rhombohedral Sb were formed that adhered well to the substrate. The authors asserted that to their knowledge this was the first report of the direct codeposition of GaSb from aqueous electrolytes. As a matter of fact, a previous attempt by Mengoli and co-workers [83] was unsuccessful, as already discussed.

McChesney et al. also tried to prepare stoichiometric GaSb films but encountered the same difficulties as Mengoli et al.: the large difference in the standard potentials of Ga^{3+} and SbO^+ ions could not be compensated for by increasing the concentration of the former in the bath or by decreasing that of the latter. The use of an electrolyte also containing a 0.3 M citric acid complexant did not have the desired effect, because the deposits were poorly adherent and consisted solely of Sb. An analysis of the Pourbaix diagrams [49] showed that Ga^{3+}, SbO^+, and GaSb can coexist only at very negative potentials, more negative than those at which concurrent hydrogen and SbH_3 evolution took place. At these potentials, the current efficiency for Ga deposition was minimal. So, since then, the authors have concentrated their work on the electrochemical deposition of suitable precursors from nonaqueous electrolyte systems, but to our knowledge no new paper on the subject has been published yet.

8.7. Indium Antimonide

As already discussed in the historical part of this chapter, in 1964 Styrkas et al. [108] prepared InSb by codeposition of In and Sb from a nonaqueous bath. In 1965, kinetic studies were undertaken by Hobson and Leidheiser [88] to obtain quantitative information on the growth rate of InSb and on the In diffusion coefficient during In electrodeposition on an Sb cathode. This paper will be examined in Section 9.2 in connection with the results of recent research on the subject [89]. Subsequently in 1970, Belitskaya et al. [191] studied In and Sb codeposition from four glycerol solutions containing a total metal ion concentration of 0.4 M with In as $InCl_3$ (0.35, 0.3, 0.2, and 0.1 M) and Sb as potassium antimony tartrate as the remainder and 70 g liter^{-1} of KOH. Solutions containing a high concentration of Sb gave deposits of essentially pure Sb. With lower Sb ion concentration and increased cathodic potential, the individual discharge currents of Sb and In were approximately equal; thus the deposit could be assumed to contain In–Sb alloys. However, changing the deposition rate did not play an important role in the formation of the intermetallic InSb compound.

Because of the commercial applications of In alloys with many metals, several reviews have been published regarding their electrochemical preparation by codeposition from both aqueous and nonaqueous solutions and the process conditions. In particular, In–Sb alloys have been produced electrochemically for anticorrosion coatings.

In 1981, Walsh and Gabe [192] discussed in some detail the major developments for the preparation of In alloy electrodeposits, mainly for overlay alloy bearings and special solders with low melting points and briefly considered In–Sb alloys. They mentioned In electrodeposition from tartrate baths (usually containing In and Sb as chlorides) and suggested that when a low-pH (0.5–3.0) version of this solution was used together with the addition of polyethylene-polyamine (PEPA), a fine-grained smooth deposit could be produced. PEPA was reported to have a very high throwing power, presumably promoted by its viscosity. Walsh and Gabe also suggested that the deliberate production of InSb might be advantageous in the production of infrared detectors and magnetoresistors.

In a paper published in three parts in 1982, Walker and Duncan [193] described the industrial use of In and its alloys, including those with Sb, and examined the operating characteristics as well as the advantages and disadvantages of the different plating conditions. Although In alloy electrodeposition from fluoborate solutions appeared not to have been extensively studied, they recommended this electrolyte, as it produced dense, silvery, fine crystalline In–Sb deposits [193, Part II]. The pH of the bath in particular had to be controlled because a value below 1 caused a decrease in the Sb content of the deposits [194]. A nonaqueous, glycerol-based electrolyte was also suggested for In–Sb alloy deposition [193, Part III]. It contained In as chloride and Sb as potassium antimony tartrate and worked at 85°C. Either In-rich (50–70 wt.%) or In-poor (20–40 wt.%) deposits were obtained, depending on whether electrodeposition was performed without or with agitation [195]. However, the fluoborate bath was preferred over the glycerol bath because it worked at room temperature and did not need such high temperatures.

In 1985, Sadana et al. [196] surveyed the literature up to the end of 1983 regarding the electrodeposition of Sb and its alloys, including those with In. Numerous authors, prevailingly Russian researchers, investigated Sb and In codeposition mainly to obtain hard corrosion-resistant coatings. A variety of plating solutions were employed, including aqueous solutions (sulfate, chloride, fluoride–sulfate, chloride–tartrate–glycerol, and tartrate–chloride baths) and nonaqueous solutions (ethylene glycol). For more details, reference is made to the review by Sadana et al., who in several cases also reported the composition of the electrolytic baths and the plating conditions (temperature and current density). They also mentioned the paper by Styrkas et al. [108] and that by Belitskaya et al. [191], which have already been discussed, as they refer to the formation of the InSb compound.

In 1985, Sadana and Singh [44] systematically investigated In–Sb alloy codeposition from aqueous solutions containing citrate complexes of the two metals and were able to bring the deposition potentials closer. Indeed, the standard electrode potentials of the SbO^+–Sb and In^{3+}–In redox systems are 0.212 and -0.342 V vs. SHE, respectively [49], and their potential difference (0.554 V) is too large to permit a satisfactory codeposition from acidic solutions of the simple salts. Therefore, solutions containing complexing agents were investigated. The authors studied six different plating solutions that varied in In, Sb, potassium citrate, and citric acid content, so that the weight % of Sb in the solution ranged from 20.00 to 93.33. The plating solutions were prepared by mixing in adequate proportions solutions of In and Sb obtained by dissolving In metal in sulfuric acid and Sb_2O_3 in a solution of potassium citrate and citric acid, respectively. To adjust the pH value (normally equal to 1.7 or 2) either potassium citrate or sulfuric acid solutions were added.

Figure 18 shows a voltammogram representing the circulating current as a function of the applied potential of In^{3+} and Sb^{3+} ions in the same citrate electrolyte of pH 2, under identical conditions. As the difference in their half-wave potentials was 0.1 V only, the authors maintained that alloy deposition from such a solution was feasible, as experimentally verified. From this starting point, they investigated the influence of several bath compositions, pH values, current densities, temperatures on the deposit composition, and cathodic current efficiency, as well as the effect of moderate or fast stirring.

Electrodeposition of In–Sb alloys from an unstirred bath was shown to be a diffusion-controlled process in which the In-to-Sb concentration ratio in the electrolyte had a great influence on the composition of the electrodeposited thin film [44], as also subsequently observed by Okubo and Landau [197]. Agitation increased both the limiting current density and the Sb content of the deposit especially at high current densities (greater than

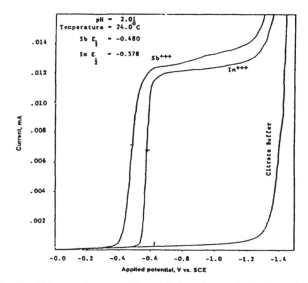

Fig. 18. Voltammetric curve of In and Sb in a citrate solution. Reproduced from [44] with the kind permission of *Plating and Surface Finishing*.

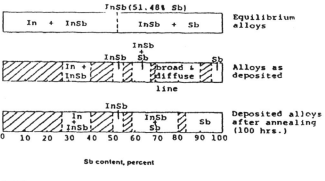

Fig. 19. X-ray structure of In–Sb deposits. Reproduced from [44] with the kind permission of *Plating and Surface Finishing*.

4 A dm^{-2}), where the diffusion effect was expected to be large. This was due to the fact that Sb depletion in the cathode layer was greater than that of In, because of its preferential deposition.

The X-ray pattern of the deposits examined before and after annealing at 350°C, for 100 h, showed the formation of InSb around the stoichiometric composition and an excess of either In or Sb on its sides. An example of the results of the structural investigation is depicted in Figure 19, which compares the phase structure of the alloys, both as deposited and after 100 h annealing, with that of the equilibrium alloys. The InSb phase existed in the electrodeposited film in the composition range of 49.9–54.1 wt.% Sb.

The authors also examined the appearance of the deposits, which could be bright with a bluish tinge (those containing 80 wt.% or more of Sb), dull, or gray to yellowish black, again exploring the influence of the bath composition, current density, pH, and temperature. The Sb-rich alloys were bright and tarnish-resistant and were expected to be promising as decorative and protective finishes.

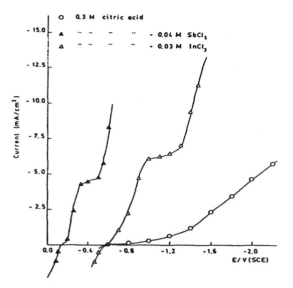

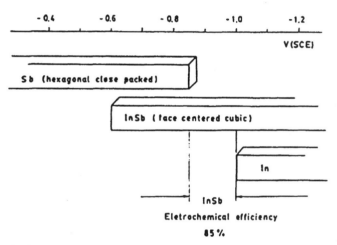

Fig. 20. Current–potential curves for In and Sb deposition from citric acid baths. Reproduced from [45] with the kind permission of the Electrochemical Society, Inc.

Fig. 21. Potential dependence of the X-ray structure of In–Sb films obtained by electrodeposition. Reproduced from [45] with the kind permission of the Electrochemical Society, Inc.

Ortega and Herrero [45], too, observed that diffusion limited the electrodeposition of both In and Sb from citric acid solutions, as shown in Figure 20, which compares the current density vs. potential curves for In and Sb deposition from a citric acid bath. At the end of the plateau, a current density rise is observed because of hydrogen evolution, the hydrogen overvoltage depends on the cathodic process over the electrode surface. At the more cathodic potentials, SbH_3 formation could occur in parallel with hydrogen evolution in the case of Sb electrodeposition. These authors examined in particular the potential dependence of the X-ray structure of the films (Fig. 21). Polycrystalline InSb thin films, free of In and Sb and adherent to the Ti substrate, were electrodeposited between -0.85 and -1.0 V vs. SCE from a bath of 0.3 M citric acid, 0.033 M $InCl_3$, and 0.047 M $SbCl_3$. The current efficiency was higher than 85%, at -0.90 V vs. SCE, but decreased at more negative values, because of the hydrogen evolution process, as stated.

A completely different approach was followed by Mengoli et al. [84], who prepared InSb thin films by the sequential deposition method involving thermal annealing, which they first applied to the synthesis of GaSb [83]. However, the interdiffusion and reaction process presented a higher activation energy in the case of the In–Sb system, and, therefore, the latter required higher temperatures than the Ga–Sb system [27]. Because of the good stability of Sb in the electrolytic solution at the In deposition potential, the sequence Sb first–In second was followed. Sb was deposited with a 94 wt.% efficiency at -0.25 V vs. SCE from the bath containing 0.6 M $SbCl_3$, 1.6 M H_2SO_4, and 1.7 M HCl proposed by Iyer and Deshpande [190]. Then, an In layer was deposited on it at -0.85 V vs. In from an electrolyte containing 0.3 M $InCl_3$ and 1 M KCl, the pH of which was adjusted to 1.5–2.0 with HCl. The two-layer samples, consisting of an inner Sb and an outer In layer (about 1 μm each) deposited on a Pt substrate, were then annealed in nitrogen at 150, 175, and 185°C for times increasing from 1 to 15 h. No weight losses were observed during this treatment.

The conversion percentage of the two metals to InSb was determined by stripping voltammetry performed in a strongly acidic solution (1.7 M HCl, 1.6 M H_2SO_4). The stripping voltammogram of the as-deposited two-layer samples presented two anodic peaks at -0.640 and -0.024 V vs. SCE, which the authors attributed to the oxidation of In to In(III) [198, 199] and of Sb to Sb(III) [200, 201], respectively (Fig. 22). Because InSb could be converted to SbH_3 at the more cathodic values while it was stable at -0.3 V vs. SCE, the stripping voltammograms of the annealed samples were performed starting from this value. Going toward the anodic direction, a peak was observed at -0.3 and another at -0.025 V vs. SCE. The charge flowing at the former peak, Q_1, was ascribed to the dissolution of the unreacted In, and that at the latter one, Q_2, to both InSb and unreacted Sb oxidation. The peak due to InSb oxidation, which involved six electrons per molecule [202], was confirmed by the anodic dissolution curve of an authentic InSb sample. The recovered total charge was equal to the charge initially spent in separately depositing the two species, In and Sb. The converted fraction of these two metals into InSb was estimated as $(Q_2 - Q_1)/(Q_2 + Q_1)$.

The authors determined the percentage of conversion of the two-layer In–Sb samples to the InSb compound as a function of the annealing conditions, investigating the already quoted three temperatures and four heating times (1, 2, 5, and 15 h). The results are collected in Table X. They were qualitatively confirmed by X-ray diffraction investigation. The In and Sb peaks only were detected after annealing at 150°C, a temperature not far from the In melting point (156.634°C), and both the typical peaks of InSb and of the unreacted components were present at 175°C, their relative amount increasing with the annealing time. The single component peaks completely disappeared by annealing at temperatures equal to or higher than 185°C for 5–10 h.

The data in Table X were subsequently utilized [89] to estimate the value of the In diffusion coefficient in InSb, as discussed in Section 9.2.

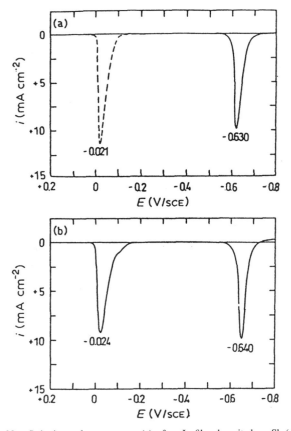

Fig. 22. Stripping voltammograms (a) of an In film deposited on Sb (solid line) and an Sb film deposited on Pt (dashed line) and (b) of the as-deposited In–Sb two-layer sample deposited on Pt Sweep rate, 1 mV s^{-1}; electrolyte composition, 1.7 M HCl and 1.6 M H$_2$SO$_4$. Reproduced from [84] with the kind permission of Kluwer Academic Publishers.

Table X.   Conversion of Two-Layer In–Sb Samples to InSb as a Function of Annealing Conditions

T (°C)	Percentage conversion after			
	1 h	2 h	5 h	15 h
150	—	—	—	0
175	67	69	78	87
185	87	—	99	—

Reprinted from [84] with kind permission of Kluwer Academic Publishers.

Mengoli et al. [84] synthesized InSb by the sequential method, not only as an alternative route to the codeposition method by Sadana and Singh [44] and by Ortega and Herrero [45], but also as a preliminary step for the preparation of the ternary Ga-In–Sb compound by deposition of a Ga layer on two-layer In–Sb samples and annealing at 400°C for 5 h. They obtained Ga$_x$In$_{1-x}$Sb films consisting of a single phase for $x < 0.09$ and of a mixture of two phases, one rich in In and the other rich in Ga, for $x > 0.16$. The average crystallite sizes determined by X-ray diffraction, of the electrochemically prepared GaSb and InSb were 60 and 35 nm, respectively; those of

the Ga-rich and of the In-rich phase ranged from 28 to 88 and from 20 to 49 nm, respectively.

In 1996, McChesney et al. [52] carried out InSb electrodeposition from the bath (0.3 M citric acid, 0.033 M InCl$_3$, and 0.047 M SbCl$_3$) used by Ortega and Herrero [45]. They worked at room temperature and without electrolyte stirring, to limit the preferential Sb deposition, because Sb discharge at the cathode was diffusion-limited, as discussed above. The substrate was a Ti sheet, and the applied potential ranged from -0.4 to -1.4 V vs. SSC (Ag–AgCl, 0.197 vs. SHE). As already observed by Ortega and Herrero, the deposit stoichiometry depended on the potential. At increasing negative values, the deposits consisted of Sb, Sb + InSb, Sb + In + InSb, and In + InSb mixtures, as was shown by X-ray diffraction [52]. The films obtained between -0.750 and -0.950 V vs. SSC were amorphous; all of the other ones were polycrystalline. The amorphous films were crystallized by heating at 250°C for 15 min in nitrogen, producing an In + InSb deposit. The results duplicated those by Ortega and Herrero [45], with one substantial exception: stoichiometric InSb could not be obtained.

Substrates of Cu and ITO were also investigated, and it was noted that the type of substrate greatly affected the deposit composition. For example, on both substrates, mixtures of In and InSb were deposited under most conditions, as no correlation was observed between applied potential and deposit stoichiometry, contrary to what occurred in the case of the Ti substrate. This result was attributed to the different cathodic overpotentials associated with the various materials. Because thin films for electronic devices often require a particular substrate (e.g., ITO in the case of photodetectors), this point could be a limitation for the electrochemical method. In any case, it could not be neglected.

In an attempt to deposit pure InSb, McChesney et al. [52] also tried solutions with different In-to-Sb ratios as well as different deposition conditions (higher temperatures and bath stirring), but they mostly obtained Sb or In plus InSb thin films. According to the authors, these results indicate the existence of kinetic barriers to the formation of the compound, barriers that might be surmounted by operating at high temperatures, that is, in nonaqueous baths.

Carpenter and Verbrugge [116] recently prepared InSb by electrochemical codeposition of In and Sb from a novel molten salt electrolyte at 45°C. This electrolyte consisted of an organic chloroindate obtained from a mixture of InCl$_3$ and 1-methyl-3-ethylimidazolium (ImCl) in the appropriate ratio, which was heated with stirring at 50–65°C. The resulting melt was left for several hours at 45°C to equilibrate before use. For the codeposition measurements, SbCl$_3$ was added with stirring immediately before deposition. The organic molten salt was formed by the reaction of the two solids as follows:

$$ImCl + InCl_3 = Im^+ + InCl_4^-$$

in analogy to the behavior of the previously reported low-melting chloroaluminates [148, 203] and chlorogallates [109, 152] involving AlCl$_3$ or GaCl$_3$, respectively, instead of InCl$_3$.

This was the first report of a low-temperature organic chloroindate melt, besides the high-temperature molten salt system, $InCl_3$–KCl, where the presence of $InCl_4^-$ was recognized by Raman spectroscopy in 1969 [204].

Either acidic or basic melts of $InCl_3$–ImCl were expected to be formed, depending on whether the $InCl_3$-to-ImCl molar ratio was greater or smaller than 1, respectively, because of the presence of excess metal chloride (a Lewis acid) or the presence of chloride ions from the dissociation of excess ImCl. However, only the basic melts formed a clear liquid without solid particle suspension at 65°C, so the basic 45 to 55 chloroindate melt was utilized in all experiments, in analogy to a similar basic clorogallate melt from which Ga was successfully electrodeposited [109].

The availability of a series of group III metal (Al, Ga, In) chloride–ImCl melts provided an interesting opportunity to perform comparative chemical and electrochemical research. Moreover, Carpenter and Verbrugge observed that $SbCl_3$–ImCl melts, too, could be prepared at 65°C (possibly because of the formation of $SbCl_4^-$). This was the first time that the existence of a melt similar to previous ones but involving a group V metal was reported.

However, in the considered paper, the $InCl_3$–ImCl melt only was investigated in several electrochemical experiments (such as voltammetric, cyclic voltammetric, and deposition–corrosion) to determine the electrochemical reactions accessible in the melt, particularly In electrodeposition. Although the exact mechanism of the process was not clear, it appeared that both In(I) and In(III) species were involved in the electrochemistry of the melt. Moreover, current oscillation phenomena were observed at the more negative potentials (−1.3 to −1.5 V vs. In) during experiments performed at the lower scan rates (e.g., at 50 instead of 100 mV s^{-1}). Similar oscillation phenomena have been observed for a number of electrochemical systems, and, in general, they are representative of complex chemical processes. Despite of their interest, the authors did not discuss them further.

The codeposition experiments were carried out on a Pt microcylinder electrode in the 45 to 55 bath containing 0.1 M $SbCl_3$, where the Sb(III) concentration was much lower than that of In(III). The counterelectrode and reference electrode consisted of In and were made by dip-coating a Pt wire in molten In. Cyclic voltammetry showed that for scanning of the potential to large negative values, from 1.00 to −1.00 V vs. In, the voltammogram for the codeposition process was similar to that for In deposition. In contrast, when the minimum value was −0.50 V vs. In only, Sb(III) deposited before In(III), indicating that codeposition of In and Sb was possible.

A preliminary characterization was carried out on deposits obtained by codeposition on Pt and glassy carbon disks at potentials ranging from −0.3 to −1.2 V vs. In. The deposits were often not homogeneous and were poorly adherent. One of the most homogeneous deposits was yielded with the use of a square-pulse potential source to control the potential of the working electrode with respect to the reference electrode between 0 and −1.2 V vs. In with a half-cycle period of 100 ms.

The atomic ratios of In to Sb in the deposits were determined by EDX analysis. They ranged from 14 to 0.7, and those most rich in In were obtained from Sb-poor electrolytes. The deposition potential influenced the composition ratio in a complex way, and, according to the authors' opinion, further studies were required to fully explain this behavior.

Interesting results were obtained by XPS investigation, which clearly showed the presence of InSb and made it possible to determine the oxidation states of Sb and In. Those of Sb, corresponding to Sb $3d$ electrons, were obtained by a peak-fitting analysis of the spectra from samples deposited with the use of a pulsed potential. The largest peak was assigned to InSb, but smaller peaks due to Sb metal [Sb(0)] and Sb oxides (Sb_2O_3 and Sb_2O_5) were also observed. A semiquantitative analysis of the data showed that about 67% of Sb in the examined layer (about 20 Å thick) was due to InSb, and the remaining 33% was due to Sb metal and its oxides. The authors assumed it likely that process optimization and deposit annealing at relatively low temperatures (350°C) would increase this percentage and improve the deposit quality. They concluded that the series of group III metal chloride–organic cloride melts could, in principle, be used as electrolytes for the deposition of almost any III–V compound, both binary and ternary (AlGaAs, GaInSb, etc.); ternary compounds are of great interest for electronic and/or optoelectronic applications.

In a patent [205] preceding the paper discussed above [116], the same authors described the method of preparation thin films by codeposition from an organic chloroindate melt containing a salt of at least one metal selected from the group consisting of P, As, and Sb, and an $InCl_3$–dialkylimidazolium chloride wherein each alkyl group contained no more than four C atoms, and the molar ratio of the $InCl_3$ to the organic chloride ranged from about 45–55 to 2–3. Substitution of small amounts of $InCl_3$ with a trichloride salt of another group III metal could be employed to obtain deposits containing other group III metals. For molar ratios of the metal salt to $InCl_3$ other than 45–55, the melt was heated to 45°C or greater.

8.8. Indium–Bismuth Compounds

Research on codeposition of In and Bi to obtain InBi thin films is scarce [206]. Sadana and Wang [207] showed that In–Bi alloys can be obtained from aqueous baths containing diethylenetriamine pentaacetate complexes of the two metals. Alloys containing up to 90 wt.% In were deposited by varying the current density and bath composition. The authors observed that increasing the current density, and In content of the bath decreased the percentage of Bi in the deposit. Moreover, the current efficiency decreased with increasing current density, but it was not significantly affected by a change in the In content of the bath. Deposits containing 63–89 wt.% In were of good visual quality and were promising decorative finishes. Later, Sadana and Wang examined the effect of pH, temperature, and stirring on the composition and properties of the deposits [208]. X-ray analysis of the original and annealed (at 50°C for 200 h) deposits showed the existence of two intermediate phases. The first was InBi, and the second had a much lower Bi content than that required for In$_2$Bi (the phase expected, according to the phase diagram, in addition to InBi and In$_5$Bi$_3$ [5]) and was assumed to be In$_3$Bi. However, the structure and lattice parameters were the same as those reported for In$_2$Bi [6]. Note that the authors did not consider In$_5$Bi$_3$, which, on the other hand, is difficult to recognize in the presence of the other two In–Bi intermetallic compounds [85], as discussed below.

In 1993, a bath based on Trilon B (dinatrii-ethylendiamine-tetraacetate) containing indium nitrate, bismuth nitrate (total concentration 0.25 M), and ammonium nitrate was developed for the deposition of In–Bi alloys having good weldability and high corrosion resistance [209]. The electrodeposition conditions were investigated, as were the structure and some properties of the deposits, such as their microhardness, internal stress, and corrosion resistance. Polarization measurements showed that the deposition of In in the alloy began at a potential more positive than the equilibrium potential of pure In in the same electrolyte. The alloys containing 10–15 at.% In showed a sharp decrease in the corrosion rate, and this behavior was explained by the presence in their structure of an excess of the intermetallic phase and the smoothed profile of the deposit surface. Some decrease in the corrosion resistance of alloys obtained at low cathodic polarization was apparently due to the deterioration of the quality of the deposits and their inclusion of indium hydroxide.

9. DIFFUSION PROCESS AND FORMATION OF GROUP III–V COMPOUNDS

Because of their different aims, another group of research papers on III–V compounds will be considered separately. They examine the electrochemical preparation of some of these compounds by electrodeposition of a group III element either on a bulk substrate of a group V element (In on Bi, In on Sb) or on a previously electrodeposited thin film of a group V element, followed by an annealing treatment (In on Sb, Ga on Sb). In these papers, the process of diffusion (mainly in the solid state) and chemical reaction at different temperatures to form the III–V compound was more closely examined, as was the influence of the substrate structure and morphology on the same process (In or Ga on crystalline and amorphous Sb thin films).

9.1. The Indium–Bismuth System

Already in 1969 [210], it was shown that the plating of In on Bi cathodes at room temperature leads to the formation of intermetallic compounds (In$_2$Bi and InBi), because In diffusion inside the Bi substrate is fast not only at high [211] but also at low [210] temperature. The availability of more accurate techniques and instrumentation made it possible to identify a third intermetallic compound (In$_5$Bi$_3$) in addition to In$_2$Bi and InBi [85] and to outline the In diffusion process [86], as shown in the following. Such a compound never emerged on the electrode surface and, therefore, could not be observed during the electrochemical measurements.

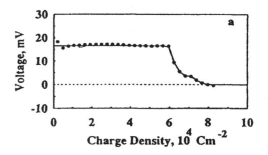

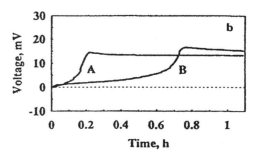

Fig. 23. (a) Open-circuit potential vs. In of the Bi cathode as a function of the charge circulated during In deposition at 4.8 A m^{-2}. (b) Time dependence of the cathode potential vs. In after In deposition at 7.2 A m^{-2} for 2 h (curve A) and 2.7 h (curve B). Reprinted from [86] with the kind permission of Elsevier Science.

Indium exhibits an electrochemical behavior that is practically reversible on both the cathodic and anodic sides. At higher current densities, anomalies may be observed on the cathodic side, particularly in indium sulfate baths. They are always associated with a strong increase in the overvoltage and a decrease in the current efficiency due to hydrogen evolution [64, 65]. Therefore, experimental conditions (electrolyte composition, pH, current density, etc.) were selected in which the parasitic process did not occur.

Indium was electrodeposited at 25°C in galvanostatic conditions from an aqueous solution of InCl$_3$ (0.67 N, pH 1.3) on a bulk Bi electrode. The anode was made of In, and an In wire inserted in a Luggin glass capillary was used as the reference electrode. The potential of the Bi electrode vs. In was measured during and after electrodeposition at current densities from 2.4 to 21.4 A m^{-2}. The deposition time was varied from 1.2 to 8.3 h to obtain 2.8–12.6-μm-thick deposits.

Figure 23a depicts the open-circuit potential vs. In of the Bi cathode as a function of the circulated charge. It was obtained by periodically opening the circuit for about 30 s during galvanostatic electrodeposition. Thermodynamic and X-ray diffraction results showed that the three plateaus observed around 15, 5, and 0 mV vs. In on the open-circuit curve were due to the formation of InBi, In$_2$Bi, and In, respectively. The cathode potential was monitored even after In electrodeposition to obtain information on the changes occurring in the composition of the electrode surface. As shown in Figure 23b, a double transition took place in most cases, because In and In$_2$Bi as well were not stable on the electrode surface. Thus, curve B presents an extended plateau whose value corresponds to In$_2$Bi followed

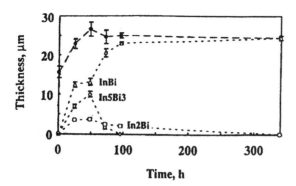

Fig. 24. Cumulative thickness of the deposit (o) and thickness of the In$_2$Bi (□), In$_5$Bi$_3$ (◇), and InBi (△) layers as a function of time after In deposition at 21.4 A m^{-2} for 3 h. The standard deviation from the average value is also indicated. Reprinted from [85] with the kind permission of Elsevier Science.

by a potential increase up to the typical value of InBi, whereas curve A (due to a thinner deposit obtained at the same current density as curve B) had a faster evolution showing the plateau of InBi only. This time evolution was due to In diffusion into and reaction with Bi, as shown by SEM and EDX analyses of the sample surface and cross section. Indeed, in all of the cases where the final electrode potential vs. In was equal to zero, isolated In microcrystals were observed on the sample surface immediately after electrodeposition. These microcrystals emptied out and collapsed on the sample surface, just as though In diffused into the bulk of the electrode. The smaller the deposited In quantity, the faster was the time evolution.

Simultaneously with the SEM observations, the In surface concentration of the 12.6-μm-thick deposits was analyzed by EDX as a function of time. Because the emission depth of the backscattered electrons was smaller than 1 μm, only the deposit surface was investigated. Because of the time needed to perform the analysis, the In surface concentration was initially not higher than 90 wt.% and then regularly decreased to 35.6 wt.%, very close to that of InBi (35.46 wt.%).

Interesting details on changes in the morphology and composition of the deposits were obtained by analyzing their inner structure as a function of time. To obtain the cross section, the samples were cut after immersion in liquid nitrogen for a few seconds. The results showed that the deposit had a multilayered structure, that inside each layer one of the three intermetallic compounds was present, and that their sequence from the surface to the bulk was In$_2$Bi/In$_5$Bi$_3$/InBi.

By SEM-EDX analysis, the thickness of each layer was also determined, as was the cumulative thickness of the deposit as a function of time. Figure 24 shows the trend of these thicknesses and, therefore, that of the diffusion and reaction process, for a film deposited at 21.4 A m^{-2} for 3 h. It clearly indicates that In$_5$Bi$_3$ never emerged on the electrode surface. After a certain time (around 100 h in the example) the previous sequence of the intermetallic compounds (In$_2$Bi/In$_5$Bi$_3$/InBi) changed to In$_2$Bi/InBi, the latter being the unique compound at the end of the analyzed time scale (1500 h). So, In$_5$Bi$_3$ exclusively appeared as a separation phase.

To emphasize the difficulty of observing In$_5$Bi$_3$ when the other In–Bi intermetallic compounds are present, it is worth noting that In$_5$Bi$_3$ was recognized in 1967, in addition to In$_2$Bi and InBi [212, 213], in a revised In–Bi equilibrium phase diagram [214]. A more recent In–Bi phase diagram is reported in Massalski's handbook [5], and the presence of In$_5$Bi$_3$ has also been confirmed by thermodynamic calculations [9].

To give further support to the SEM-EDX results, the deposits were submitted to X-ray diffraction analysis. The results mimicked those already discussed: In entirely disappeared from the surface, the two transient compounds In$_2$Bi and In$_5$Bi$_3$ evolved, the latter disappearing before the former, and the InBi layer continuously increased.

To evaluate the In diffusion coefficient in the three In–Bi compounds, the mechanism of the process was investigated, with the intention of creating an interpretative model [86].

First, assuming that InBi only was formed, two subsequent steps were envisaged during electrodeposition: (i) the In^{3+} charge transfer process and (ii) In diffusion into the InBi layer toward the InBi–Bi interface to further form InBi. The kinetics of the first step was determined by the current density, i, which was constant in the electrochemical experiments. The moles of In deposited on the unit surface area, m_{dep}, increased linearly with time, t, according to Faraday's law,

$$m_{dep} = it/zF$$

where z is the number of charges transferred during the electrochemical process ($z = 3$) and F is Faraday's constant, as usual.

Schmalzried [215, pp. 53, 95] had already analyzed the case where two metals M (In) and N (Bi) react to form the intermetallic compound M$_r$N$_s$ (InBi), separated from the reactants by two phase boundaries, so that the reaction proceeds by diffusion of the participating components through the reaction product. The overall driving force for the reaction is the difference in Gibbs free energy between the reactants and the reaction product. For a very low solubility of the reactants in the reaction product, the particle fluxes are locally constant. As long as the local thermodynamic equilibrium is maintained within the reaction layer and at the phase boundaries, a parabolic growth rate law results. Consequently, the number of moles of In that could be removed from the unit surface area to form the intermetallic compound, m_{rem}, also shows a parabolic time dependence:

$$m_{rem} = \left(D^* t\right)^{1/2}$$

where D^* is proportional to the In diffusion coefficient, D, and to the difference between the In chemical potential at the two phase boundaries of the reaction product, that is, to the standard Gibbs free energy of formation of InBi from the elements. Generally, D will depend to a greater or lower extent upon the component activities. However, as a first approximation, its average value over the reaction layer can be used, especially when the activities of M and N within the region of homogeneity of phase M$_r$N$_s$ do not vary by more than a power of 10. For intermetallic compounds with very small values of the standard Gibbs free energy of formation from the elements (e.g., for

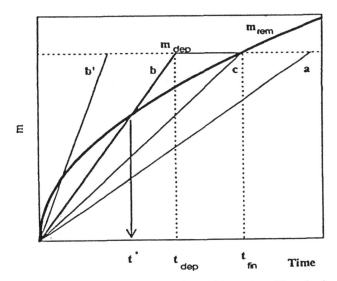

Fig. 25. Comparison of the time dependence of the number of In moles deposited on the unit surface at several current densities, m_{dep} (straight lines), and the number of In moles removed from the unit surface, m_{rem} (parabolic curve). In case b, the arrow indicates the time at which $m_{rem} = m_{dep}$. Reprinted from [86] with the kind permission of Elsevier Science.

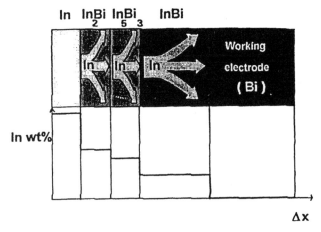

Fig. 26. Scheme of the cathode composition and of In diffusion into the In–Bi compounds at the end of the electrodeposition process (upper part), and In wt.% as a function of the deposit thickness (lower part). Reprinted from [86] with the kind permission of Elsevier Science.

NiAl, it is lower than $-100\ \text{kJ mol}^{-1}$), this simplification is not valid. However, this is not the case for the In–Bi intermetallic compounds, whose values are higher than $-15\ \text{kJ mol}^{-1}$ [9], as shown below. Indium only was assumed to diffuse, because of a mobility higher than that of Bi, as shown by its lower melting point ($156.634°C$) in comparison with that of Bi ($271.442°C$).

Figure 25 indicates the typical time dependence of both m_{dep} and m_{rem} at room temperature. m_{dep} depends on the electrodeposition rate, and, therefore, different straight lines are obtained at the different (constant) current densities. In contrast, just one curve depicts the removed In quantity, which depends, at constant temperature and pressure, on the material properties only (particularly on the In diffusion coefficient). Three different situations may occur:

1. $m_{rem} > m_{dep}$. All of the electrodeposited In diffuses into the cathode, forming an InBi layer, the growth rate of which is determined by the current density.
2. $m_{rem} < m_{dep}$. In accumulates on the electrode surface because the diffusion and reaction process is not able to remove all of the electrodeposited material.
3. $m_{rem} = m_{dep}$. The maximum possible growth rate of the layer is observed without In surface accumulation. The intersection point of the $m_{rem}(t)$ curve with each of the $m_{dep}(t)$ straight lines gives the limiting time, t^*, above which In surface accumulation takes place at each current density.

The experimental curves (e.g., Fig. 23a) made it possible to determine this time, because just at this same time, the open-circuit potential vs. In decreased from the typical value for InBi toward zero, indicating a progressive surface enrichment with In. So, by equating m_{rem} to m_{dep}, and introducing the experimental t^* values for several runs, it was possible to determine

the average D^* value and from this the average In diffusion coefficient in InBi, at $25°C$, $D \approx 1 \times 10^{-15}\ \text{m}^2\ \text{s}^{-1}$.

For this calculation, the value of Chevalier [9] for the InBi standard Gibbs free energy of formation at $25°C$ was utilized. Chevalier determined the free energies for InBi, In_5Bi_3, and In_2Bi by averaging the experimental data of different authors [216–218]. They were found to be small: -3.72, -14.2, and $-5.54\ \text{kJ mol}^{-1}$, respectively.

It is worth examining the very high value of $D(\text{InBi})$ at room temperature. Much lower values, usually between 10^{-20} and $10^{-50}\ \text{m}^2\ \text{s}^{-1}$, were obtained for common metals and for Ge and Si. Values comparable with that for In in InBi were only observed during the diffusion of very small atoms (e.g., around $10^{-15}\ \text{m}^2\ \text{s}^{-1}$ for Li in Ge and Si, and $10^{-17}\ \text{m}^2\ \text{s}^{-1}$ for C in body-centered cubic (bcc) Fe at $70°C$) or at high temperatures (e.g., around $10^{-15}\ \text{m}^2\ \text{s}^{-1}$ for Zn in Cu at $734°C$ and $10^{-17}\ \text{m}^2\ \text{s}^{-1}$ for Fe in Fe at $727°C$) [12].

In the most general case, when the diffusion process was not able to remove all of the deposited In to form InBi, the deposit was composed of In, In_2Bi, In_5Bi_3, and InBi at the end of the electrodeposition process, as schematically shown in Figure 26, and the deposit composition evolved with time. The behavior of such multiphase product layers with respect to In diffusion and reaction was evaluated by applying previous treatments to each of the individual phases as their range of homogeneity was sufficiently narrow, and it was assumed that local equilibrium was maintained [215, p. 95]. This resulted in a parabolic growth rate for each reaction product and, therefore, for the total thickness of the multiphase layer. However, the difference between the In chemical potential at the In_rBi_s interfaces, $\Delta\mu(In_rBi_s)$, was no longer proportional to the standard free energy of formation of In_rBi_s from the elements, as was the case when only one intermetallic compound (InBi) was formed. This difference was calculated by the method suggested by Schmalzried [215, p. 124], taking into account that, as shown in Figure 26, each intermetallic compound was produced at its respective right-hand interface by reaction of the diffusing species (In) with the intermetallic compound that had a lower In content, while it was consumed at its left-hand interface by reaction with In to form the compound that had a higher In content. By applying the equilibrium conditions at the two interfaces of each reaction product layer, the difference $\Delta\mu_{In}$ between the In chemical potentials at the left- and right-hand boundaries was evaluated from thermodynamic data. This difference at either side of the In_2Bi and $InBi$ layers had a positive sign, whereas for In_5Bi_3 it was negative. Therefore, In diffusion through In_2Bi and $InBi$ was favorable, whereas it was not through In_5Bi_3; this explains why the latter compound was never observed on the electrode surface and no additional plateau could be found in the In electrodeposition curve.

As long as metallic In was present on the sample surface, the ratios of the thicknesses, Δx, of the different intermetallic compounds were independent of time and are given by

$$\left[\Delta x(In_rBi_s)/\Delta x(InBi)\right]$$
$$= \left\{\left[D(In_rBi_s) \times \Delta\mu(In_rBi_s)\right]/\left[D(InBi) \times \Delta\mu(InBi)\right]\right\}^{1/2}$$

where In_rBi_s refers to In_2Bi or In_5Bi_3.

The thicknesses of the layers of the different In–Bi intermetallic compounds were experimentally determined as a function of time, as were their ratios to that of $InBi$. Time was measured starting from the end of the electrodeposition process. With the introduction of the average experimental values in the previous equation, together with the estimated values of $D(InBi)$ and $\Delta\mu(In_rBi_s)$, the average In diffusion coefficients in the In_2Bi and In_5Bi_3 layers (at 25°C) were finally obtained:

$$D(In_2Bi) \approx 1 \times 10^{-16} \text{ m}^2\,\text{s}^{-1}$$
$$D(In_5Bi_3) \approx 3 \times 10^{-16} \text{ m}^2\,\text{s}^{-1}$$

The average In diffusion coefficient also turned out to be very high in In_2Bi and In_5Bi_3 compared with the values of other metals at room temperature.

The presence of a multiphase product layer between In and Bi made it possible to explain the observed potentiometric behavior. After electrodeposition, In diffusion with reaction occurred, utilizing metallic In on the deposit surface. As soon as In was no longer available at the sample surface, In_2Bi and In_5Bi_5 decomposed. As previously observed, the chemical potential of In at the In_2Bi–In_5Bi_3 interface increased within In_5Bi_3, whereas that at the In_5Bi_3–$InBi$ interface decreased within $InBi$. Therefore, In_5Bi_3 would be expected to be more readily decomposed than In_2Bi, as was observed experimentally (Fig. 24). Following a similar argument, it may be expected that even during In electrodeposition, In_5Bi_3 did not appear on the electrode surface, because, otherwise, the In chemical potential would strongly increase within the outermost deposit layer (of In_5Bi_3), and this is an unfavorable condition for In diffusion. Hence, the surface would immediately be covered by In or In_2Bi. So, the assignment of the quasi-plateau at 5 mV vs. In to In_2Bi and not to In_5Bi_3 seems to be reasonable.

To summarize the results, during In electrodeposition on Bi cathodes, In diffused and reacted with Bi, forming overlayers of In–Bi intermetallic compounds ordered according to their relative composition. The solid-state process continued after electrodeposition until a layer of $InBi$ only formed. The exceptionally high mobility at room temperature of In in In–Bi intermetallic compounds could explain this singular behavior.

The influence of temperature on the process of In–Bi intermetallic compound formation, in particular of $InBi$, was also investigated [87]. Indium was electrodeposited on Bi electrodes for 1.5–4.5 h at current densities from 5.0 to 14.3 A m^{-2} and at temperatures from 30°C to 70°C, to observe the time evolution of the deposit composition. The coulometric thickness of the deposited In films ranged from 2.8 to 6.1 μm. For the SEM-EDX analysis, thicker deposits (12.6 μm) were prepared, at 25°C and 70°C, by electrodeposition at 21.4 A m^{-2} for 3 h.

At the higher electrodeposition temperatures, the plateau in the electrochemical curves due to $InBi$ formation was more extended and was observed at slightly higher potentials. Moreover, no intermediate quasi-plateaus were noted, the In diffusion rate being so high that the two In–Bi compounds less stable than $InBi$ could not be formed.

This different behavior was clearly indicated by SEM and EDX analyses of the surface and cross section of the thicker deposits during time evolution after electrodeposition at 25°C and at 70°C. Moreover, X-ray diffraction confirmed the results.

By considering the charge transfer and mass transfer phenomena, the authors also estimated the coefficient of In diffusion into $InBi$. The $InBi$ Gibbs free energy of formation from the elements at different temperatures, $\Delta G(T)$, was evaluated at 273 K from literature data [216–219], according to the Gibbs–Helmholtz equation. However, because of the spread in the values of the different authors, the equation

$$\Delta G(T) = -1464.4 - 7.58128 \times T \qquad \text{(in J mol}^{-1}\text{)}$$

given by Chevalier [9] was preferred. Chevalier optimized the experimental data of other authors during his thermodynamic evaluation of the In–Bi phase diagram. The results showed a relatively small difference in the Gibbs free energies at the different temperatures, which implied a very small difference in the plateau voltages of $InBi$, within the range of the experimental error.

The values obtained for the diffusion coefficient at 30–70°C ranged from 0.79×10^{-15} to 3.77×10^{-15} m^2 s^{-1}. Again, these values were unusually high.

Applying an Arrhenius-type relation, $\ln D(InBi)$ as a function of $1/T$, the values of 32.8 kJ mol^{-1} and 3.7×10^{-10} m^2 s^{-1} were estimated for the activation energy and the frequency factor of the process, respectively. The very small value of the activation energy (usually about 100–400 kJ mol^{-1} for many materials of interest [12]) was certainly a factor contributing to the high diffusion coefficient observed for In in $InBi$, notwithstanding the exceptionally small frequency factor. The theoretically predicted and experimentally observed values are often between 10^{-5} and 10^{-7} m^2 s^{-1} [12].

9.2. The Indium–Antimony System

The investigation of the In–Bi system was extended to the In–Sb system [89]. The Sb cathode was prepared by melting high-purity Sb. The other experimental conditions, counterelectrode, reference electrode, and electrolyte, were the same as those of the In–Bi system.

The electrodes (In anode and Sb cathode) were polarized for 10 min at a constant current density of 2 A m^{-2} to deposit In. After a very small potential transient, which was related to the nucleation process on a foreign substrate, the cathode potential attained a nearly constant value. The current was then switched off and the electrode potential vs. In immediately decreased to zero, indicating the presence of In on the electrode surface. A similar experiment was carried out by periodically opening the circuit for about 60 s during electrodeposition to monitor the changes occurring in the surface composition. The open-circuit potentials did not reach constant values, as found in the similar case of In deposition on Bi. As a matter of fact, at the beginning of the experiment, their values were positive and close to that of bare Sb, whereas the following showed a progressively decreasing value until they reached 0 V vs. In. Hence, the cathode behaved like a mixed electrode, and no plateau due to the formation of InSb was observed.

After the current was switched off, the electrode potential was still monitored. After about 15 min, the electrode potential showed a typical inflection, an extremely short quasi-plateau, which, on the basis of thermodynamic data, was attributed to InSb emerging on the electrode surface. Then, the electrode potential increased again, because InSb immediately dissolved into the electrolyte, and the potential returned toward the initial open-circuit value because of the bare Sb substrate.

To verify the above InSb and In dissolution, a special experiment was performed. Two In electrodeposits on Sb were prepared under the same conditions (10 A m^{-2}, 300 s). At the end of the process, one of the samples was immediately taken from the bath, while the other was left in rest conditions. After 66 h, the former sample showed a very small X-ray diffraction peak due to InSb in addition to the peaks of the unreacted In, whereas the latter sample showed neither InSb nor In peaks, in agreement with electrochemical results [220].

As In disappeared not only because of the solid-state diffusion and formation of InSb, but prevailingly because of its dissolution into the electrolyte, it was impossible to obtain reliable values of the In diffusion coefficient from electrochemical results. Therefore, another method was followed based on X-ray data, which also made it possible to estimate the D values at high temperatures [89].

For this purpose, thin (0.8 μm) and thick (18 μm) In deposits were prepared for observation of the formation of the intermetallic compound. At the end of the deposition process, the electrodes were immediately removed from the electrolyte to avoid In dissolution into the bath, rinsed in water, and dried. To study the influence of temperature, they were annealed at 70°C, 110°C and 135°C for increasing times (up to several hundred hours).

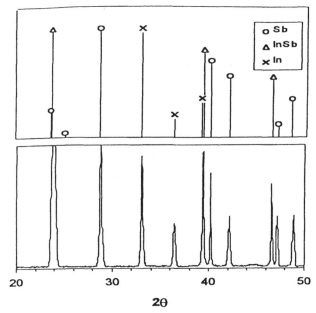

Fig. 27. Comparison of the schematic ASTM lines of In, Sb, and InSb [6] with the X-ray diffraction pattern of a 0.8-μm-thick deposit heated at 135°C for 51.5 h. Reprinted from [89] with the kind permission of Elsevier Science.

Thin In deposits obtained at room temperature consisted of isolated microcrystals with an average size of 5 μm and showing quite regular crystal planes. EDX investigation indicated that these crystals were due to In, whereas on the remaining regions of the surface Sb only was found. When the deposits were heated, the surface of the microcrystals became rounded, especially at the highest temperature (135°C), not far from the In melting point.

Thick deposits were rough and imperfect with typical holes, but the In dot map analysis of their cross section showed their practical continuity. The SEM-EDX analysis was performed at several points of this section from the surface toward the bulk of the specimen 2 years after electrodeposition. It showed the continuous decrease in the In atomic percentage, which reached a value close to 50% at the deposit–substrate interface, indicating the formation of InSb. These results indicated the much slower In diffusion and reaction process for the Sb substrate than for the Bi substrate. At higher temperatures, the deposits were continuous and InSb was more easily observed, for example, after the samples were heated for 334 h at 110°C. From scans of their cross section, they showed the three-layer structure In–InSb–Sb from the surface to the bulk.

The X-ray diffraction pattern of a thin deposit heated at 135°C for 51.5 h is compared in Figure 27 with the schematic ASTM peaks of In, Sb, and InSb [6]. Although some lines are superimposed on the ASTM spectra, it was possible to select for each phase at least one typical reflection peak not influenced by the nearby peaks. This peak was at $2\theta = 28.6°$ for Sb (because of the {012} planes), at 33.0° for In ({101}), and at 46.6° for InSb ({311}). The comparison of the sample peaks with the ASTM identification peaks showed that the deposit contained In and InSb.

Both EDX and X-ray diffraction analyses indicated the increasing formation of the InSb phase with heating time and temperature. However, because of the very low rate of the process, InSb could be detected either after a very long aging of the deposits at room temperature or after a severe annealing.

The In diffusion coefficient was estimated at different temperatures from the decrease with time of the most intense X-ray diffraction peak of In ({101}). Because the same sample area was always examined, it was assumed that the intensity of this line was proportional to the In quantity in the deposit. So, from the In moles deposited on and removed by diffusion from the unit surface area,

$$m_{rem} = m_{dep}(1 - I_1/I_0)$$

where I_0 and I_1 are the intensities of the In line immediately after deposition and after deposit heating, respectively. As previously shown, m_{dep} and m_{rem} could be determined from the deposition and the diffusion time (in this case the heating time), respectively. The Schmalzried theory, discussed in detail for the similar case of InBi formation [86], could also be applied in this case, because of the relatively high standard Gibbs free energy of formation of InSb from the elements, equal to -25.5 kJ mol^{-1} [8]. Again, In only was assumed to diffuse, because of its higher mobility, as shown from the much lower melting point: 156.634°C and 630.755°C for In and Sb, respectively.

The average D values and their standard deviation, estimated from the experimental I_1/I_0 ratio, ranged from $(0.10 \pm 0.04) \times 10^{-19}$ m^2 s^{-1} at 25°C to $(2.59 \pm 0.33) \times 10^{-19}$ m^2 s^{-1} at 135°C. From the linear dependence of $\ln D$ on $1/T$, the activation energy $(30 \pm 2$ kJ mol$^{-1})$ and the frequency factor $[(1.8 \pm 0.5) \times 10^{-15}$ m^2 s$^{-1}]$ of the diffusion process were determined.

The obtained D values were discussed in connection with the results of other authors. As stated in Section 8.7, Mengoli et al. [84] determined the time dependence of the percent conversion of the In–Sb samples to InSb at high temperatures. As the initial quantities of In and Sb were known, it was possible to evaluate the InSb moles formed on the unit surface area as a function of time. From the observed square root dependence, the diffusion coefficient was estimated. The values turned out to be around 3×10^{-18} and 7×10^{-18} m^2 s^{-1}, at 175°C and 185°C, respectively, in reasonable agreement with the previous values, considering that at these temperatures In was in the molten state. In contrast, no agreement was found with the D values obtained by Hobson and Leidheiser [88]. Performing a chemical analysis of the deposits, these authors determined the InSb quantity formed both after In electrodeposition at 45°C on bulk Sb and during a subsequent heat treatment in boiling water. The D values were around 1.5×10^{-16} and lower than 1.7×10^{-18} m^2 s^{-1}, respectively. The authors maintained that the data available when the paper was written were insufficient to explain such a great difference in the diffusion coefficients. So far, their highest value remains unexplainable and perhaps not convincing, also because when the Schmalzried relation-

ship [215] was applied to their data, an even higher diffusion coefficient was found (10^{-12} instead of 10^{-16} m^2 s^{-1}) [89].

Eisen and Birchenall [221] investigated the In and Sb self-diffusion in InSb at 478–521°C with a tracer technique, determining the activation energy and frequency factor. The data yielded slightly different diffusion coefficients for group III and V atoms, 8.3×10^{-19} and 5.9×10^{-19} m^2 s^{-1}, respectively, at 478°C. The values are on the order of those of the substitutional elements in Ge, thus tending to support a vacancy mechanism in InSb, too, which has a similar strong tetrahedral coordination. A reasonable mechanism for self-diffusion accounting for the above data seemed to be the vacancy diffusion in each of the two fcc sublattices formed by the two constituents of InSb. On the basis of bond strength considerations, the authors could also assume that the diffusing entities in InSb were neutral atoms and could explain the higher diffusion coefficient of In with respect to Sb.

Contrary to what was observed in the case of In electrodeposition on Bi cathodes, the In diffusion coefficient in InSb is very low, about five orders of magnitude lower than that in InBi [87]. This different behavior was related to the different crystalline structures of the two compounds. As already discussed, the majority of the 1 : 1 compounds formed by elements of group IIIA and VA of the periodic table, to which In, Sb, and Bi belong, crystallize with the zinc-blende (cubic) structure with tetrahedrically coordinated atoms (Fig. 1a). Exceptions in this series are the compounds involving the more metallic elements, particularly InBi, which has a tetragonal unit cell and metallic bonds with fourfold coordination. Its B-10 structural type (Fig. 1c) consists of layers of like atoms normal to the c axis, with adjacent In layers separated by two Bi layers (In–Bi$_1$–Bi$_2$–In stacking). Each In layer is bonded to the Bi layer to either side of it, and the closest In–Bi interatomic distance (3.13 Å) is appreciably greater than the value expected for a covalent bond (2.92 Å) or for an ionic structure (2.86 Å) and much greater than the size of the closest approach of In and Sb in InSb (2.80 Å). Moreover, the size of the closest approach of Bi atoms in neighboring Bi layers (3.68 Å) is somewhat larger than the interatomic separation of Bi atoms in adjacent layers of metallic Bi (3.47 Å). This arrangement gives rise to a marked cleavage plane normal to the c axis [222]. A structure of this type can also be described as a distorted cesium chloride (cubic) arrangement (Fig. 1b) when the axial ratio c/a approaches the value of $\sqrt{2}/2$ and the atoms in the layers Bi$_1$ and Bi$_2$ lie in the same plane (In–Bi–In stacking). However, in this case, the distortion is so great that a fourfold coordination results and only a formal resemblance exists between InBi and a distorted cesium chloride arrangement.

The typically layered structure of InBi is expected to be highly favorable to diffusion, particularly in the case of metals, such as In, that have a low melting point and, therefore, a high mobility. Instead, a compound such as InSb with strong bonds and a more compact atomic arrangement is found to be more "impervious."

It is now worth recalling that the In diffusion process is accompanied by a chemical reaction to form the intermetallic compound. The In quantity removed from the unit surface area of the deposit in the time unit depends not only on the diffusion coefficient but also on the Gibbs free energy of formation from the elements of the intermetallic compound. The absolute value of this thermodynamic quantity is much higher in InSb than in InBi but not high enough to compensate for the difference in the diffusion coefficients. So, the structural factor prevailed over the physicochemical one, and it was not possible to observe the deposited In microcrystals collapsing on and disappearing from the sample surface, as occurred in the case of In deposits on Bi cathodes [85].

10. INFLUENCE OF THE SUBSTRATE STRUCTURE AND MORPHOLOGY ON THE DIFFUSION PROCESS

10.1. The Indium–Antimony System

In general, diffusion and reaction processes in the solid state are scarcely reproducible phenomena, because of the consistent influence of the crystal geometry (surface coverage, material morphology) and structure (mono- or polycrystalline, with preferred orientation, amorphous, with grain and dislocation boundaries and other structural defects), so that only the order of magnitude of the average diffusion coefficient is significant. Because the geometry and structure of the In deposit and the Sb substrate depend on the way they were prepared, the values obtained for the In diffusion coefficient typically refer to In electrodeposits on a bulk Sb substrate [89]. To deepen the understanding of these aspects, which to the authors' knowledge have never been examined before, the influence of the substrate geometry and structural properties on the diffusion coefficient was investigated [90]. For this purpose, crystalline and amorphous Sb deposits were prepared by electrochemical deposition under different experimental conditions. Indium was then electrodeposited on the two different substrates, and the formation of InSb was examined at several temperatures.

Sb was electrodeposited on an Fe disk from an aqueous solution of 0.3 M SbCl$_3$, 1.6 M H$_2$SO$_4$, and 1.7 M HCl. Two optimized conditions were found to produce either crystalline or amorphous Sb deposits. Crystalline deposits were obtained by electrodeposition at 440 A m^{-2} and 50°C, and the amorphous ones were obtained at 40 A m^{-2} and 22°C. The deposition time was 20 min in both cases. The average thickness of the deposits was 33 and 3 μm, respectively. Moreover, two different types of crystalline deposits were prepared: those obtained from the electrolyte either not containing any special addition agent or containing the organic compound Trilon B at a concentration of 40 g liter^{-1}.

Before being utilized, the structures of the Sb deposits were checked by X-ray diffraction. Whereas the crystalline deposits presented the typical diffraction peaks of rhombohedral Sb, the amorphous ones showed a characteristic broad band in the range $2\theta = 24$–$36°$ and minor bands at higher values, but no peaks due to the crystalline phase. As for the deposit morphology, Figure 28 compares the SEM micrograph and the rough-

ness profile of the crystalline Sb deposit obtained from the electrolyte not containing Trilon B (c-Sb) with those of amorphous Sb (a-Sb). The crystalline deposits presented regularly shaped rosette-like crystals with a mean size of 15–20 μm (Fig. 28a), whereas the morphology of the amorphous deposits was completely different (Fig. 28c). Typical spheroidal particles were observed to have an average size of about 2 μm and were arranged along the etching lines of the Fe disk. This is due to the fact that the thinner amorphous deposits were still influenced by the surface inhomogeneities of the same Fe disk. The corresponding roughness profiles along 200 μm of a central line are shown in Figure 28b and d, together with the average value of the maximum individual roughness depth, R_{max}. Both curves as well as the numerical values dramatically indicate the great difference in the surface profiles, as also expected on the basis of the morphological aspects of the deposits. Crystalline deposits obtained in the presence of Trilon B (ct-Sb) were similar to c-Sb deposits. However, EDX results showed a high quantity of carbon on their surface because of the adsorption of the organic salt.

Indium was deposited from the usual chloride bath at 35°C, at a current density of 177 A m^{-2} for 1.5 min on the Sb substrates with the different structures. The deposits completely covered the substrate surface and their thickness (about 0.9 μm) did not hinder the X-ray observation of the growing InSb intermetallic compound. Those on c-Sb presented a much smoother profile than the original substrate, whereas those on a-Sb did not show any practical difference. After In deposition on Sb, the samples were heated at 40°C, 70°C, 110°C, and 140°C in vacuum for increasing times. Of course, the deposits on the amorphous substrate were heated only up to the time at which the Sb transition from the amorphous to the crystalline phase did not occur. Although according to Hashimoto and Nohara [223], very thin (nm-sized) amorphous Sb layers inherently in an unstable supercooled state may easily be crystallized at room temperature by some trivial perturbation; micrometer-sized layers are much more stable. So, a specific investigation was made of the crystallization process of a-Sb, whose main lines will be discussed later.

The In diffusion coefficient was estimated from X-ray results according to the Schmalzried theory [215]. The values obtained for the different substrates and temperatures are collected in Table XI, and the dependence of $\ln D$ on $1/T$ is depicted in Figure 29. The observed linear behavior made it possible to determine the activation energy and the frequency factor of the In diffusion process in crystalline InSb on the c-Sb substrate: 63 ± 4 kJ mol^{-1} and $(8.20 \pm 2.50) \times 10^{-10}$ m^2 s^{-1}, respectively; on the ct-Sb substrate: 65 ± 4 kJ mol^{-1} and $(2.64 \pm 1.50) \times 10^{-10}$ m^2 s^{-1}, respectively; and on the a-Sb substrate: 53 ± 4 kJ mol^{-1} and $(2.06 \pm 0.80) \times 10^{-10}$ m^2 s^{-1}, respectively.

As shown in Table XI and Figure 29, the In diffusion coefficient strongly depended on the substrate and increased according to the sequence D(ct-Sb) $< D$(c-Sb) $< D$(a-Sb). Moreover, from previous results [89] for bulk Sb substrates, b-Sb, D(b-Sb) was practically equal to or, at the two highest temperatures, even lower than D(ct-Sb).

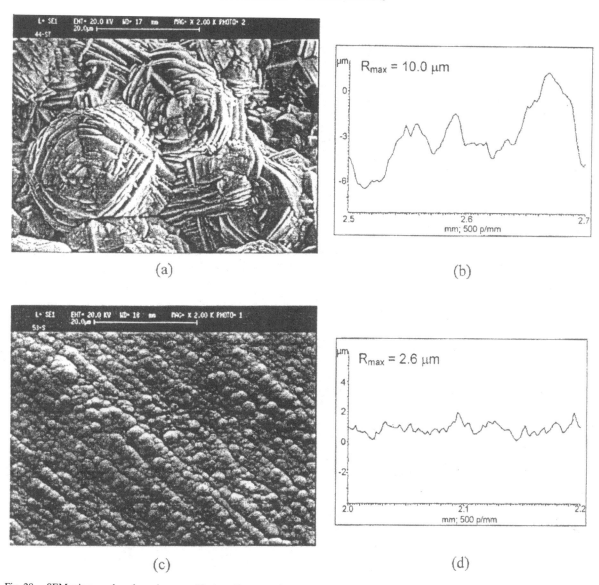

Fig. 28. SEM micrograph and roughness profile (a and b, respectively) of the c-Sb substrate and (c and d, respectively) of the a-Sb substrate. Reprinted from [90] with the kind permission of Elsevier Science.

Table XI. Indium Diffusion Coefficient as a Function of Temperature for the Different Sb Substrates

Substrate	40°C	70°C	110°C	140°C
Crystalline Sb without Trylon B $D \times 10^{19}$ $(m^2 s^{-1})$	0.204 ± 0.150	2.14 ± 1.05	23.1 ± 8.5	69.2 ± 21.0
Crystalline Sb with Trylon B $D \times 10^{19}$ $(m^2 s^{-1})$	—	0.352 ± 0.245	2.91 ± 1.15	17.6 ± 6.5
Amorphous Sb without Trylon B $D \times 10^{19}$ $(m^2 s^{-1})$	3.44 ± 1.25	17.1 ± 6.5	112 ± 30	492 ± 90

Reprinted from [90] with the kind permission of Elsevier Science.

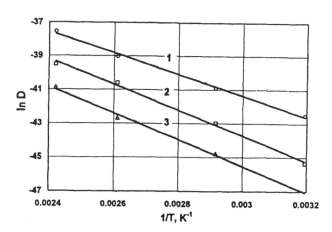

Fig. 29. Dependence of the logarithm of the In diffusion coefficient on the reverse of the absolute temperature (Arrhenius plot) for In electrodeposits on a-Sb (curve 1), c-Sb (curve 2), and ct-Sb (curve 3). Reprinted from [90] with the kind permission of Elsevier Science.

The diffusion coefficient is well known to be influenced by at least two main factors, geometrical and structural, as mentioned above. The former depends on the surface morphology of the Sb substrate, because in comparison with a smooth surface profile, a rough one presents a higher effective area for diffusion. However, the D(ct-Sb) and D(c-Sb) values were smaller than the value of D(a-Sb), notwithstanding the higher surface roughness of the crystalline substrates with respect to the amorphous one. So, the structural factor was prevailing. To give direct evidence of this hypothesis, an a-Sb sample was heated for 3 h at 110°C to induce complete crystallization. The so obtained crystalline phase (ca-Sb) showed a very smooth morphology similar to that of the original amorphous material. On this substrate, In was deposited, as usual, and the sample was heated at 110°C for 1 h, 30 min to determine the In diffusion coefficient. The value of D(ca-Sb) was found to be practically equal to that for the c-Sb substrate; that is, it was smaller than for the a-Sb substrate. This confirmed that the differences observed between D(a-Sb) and D(c-Sb) or D(ca-Sb) were due entirely to the structural and not to the geometrical factor.

The two different crystalline substrates had practically the same morphology, but that obtained in the presence of the addition agent was shown to have foreign particles selectively adsorbed on its surface, possibly mainly localized at the grain boundaries. This certainly hindered the In intergranular diffusion, thus explaining why D(ct-Sb) was smaller than D(c-Sb), as also confirmed by the difference in the pre-exponential factor and not in the activation energy for diffusion into the two substrates. Finally, the small D values of the bulk Sb substrate were tentatively ascribed to its high defectiveness.

To explain the results, the path of In was considered during its diffusion from the In–InSb to the InSb–Sb interface inside the growing InSb layer. Because the latter was always polycrystalline, this part of the overall process could not be a cause of the observed differences in the D(c-Sb) and D(a-Sb) values. Instead, a chemical reaction between In and Sb occurred at the InSb–Sb interface, where In further diffused into the either crystalline or amorphous Sb substrate. So, after nucleation and growth of the new InSb phase, the InSb–Sb interface was shifted more and more into the bulk of the Sb deposit. Hence, this last step, strictly depending on the Sb structure, was at the heart of the observed behavior.

As for the substrate structure, in crystalline (rhombohedral) Sb the atoms are incorporated into two-dimensional networks of puckered sixfold rings forming typical layers. The threefold coordinated atoms in the same layer are at a distance of 2.87 Å, and the nearest atoms in adjacent layers are at 3.37 Å [4]. The structure of amorphous Sb was investigated by several authors, and the experimental interatomic distances were found to be generally larger (e.g., [224]) and the density lower (e.g., [225]) than those in rhombohedral Sb. Moreover, the interlayer correlation existing in the crystalline form was minimized in the amorphous structure, a feature probably accounting for the semiconducting properties and low density of a-Sb [226].

The electrodeposited a-Sb samples were submitted to structural characterization by X-ray diffraction, and the radial distribution function was determined [227]. The results of the analysis could be interpreted on the basis of the model reported in the literature for rhombohedral Sb. The prevailing structural disorder in the amorphous sample was due to the average distance between neighboring double layers, which was higher by 27% than in crystalline Sb deposits. So, even electrodeposited Sb followed the rule that some interatomic distance strongly increased from the crystalline to the amorphous phase. These results, indicating the more open structure of a-Sb than of c-Sb, might explain the more favorable conditions for In diffusion in the former material. These more favorable conditions were due entirely to the lower activation energy for the a-Sb substrate, as the pre-exponential factor was higher for the c-Sb phase. Because the latter represents the frequency of successful attempts to overcome the energy barrier, the higher value found for the crystalline material was related to its more regular structure.

The authors concluded that the amorphous substrate was more suitable than the crystalline one for the preparation of InSb by solid-state diffusion and reaction, because the process was moderately fast, even at low temperatures. Indeed, in contrast to the In diffusion coefficient for c-Sb, that for a-Sb lies slightly beyond the upper limit of the often mentioned range normally considered for common metals and semiconductors [12].

10.2. The Gallium–Antimony System

The influence of the amorphous and the crystalline structure of Sb on Ga diffusion into the same Sb substrate has also been studied, and interesting results are already available [91].

The Ga diffusion coefficient was determined at 50°C, 75°C, and 100°C, and the formation of the GaSb semiconductor compound was observed. The D values were on the order of 10^{-18} to 10^{-16} m^2 s^{-1}. At these temperatures, Ga was at least partially in the molten state, and the geometrical factor prevailed over the structural one, because, owing to its larger surface area, the c-Sb substrate was more suitable than the a-Sb substrate for promoting the Ga diffusion and reaction process: D(c-Sb) $>$ D(a-Sb). However, in the case of a crystalline substrate with the same morphology as amorphous Sb obtained by a-Sb crystallization, the reverse occurred: D(ca-Sb) $<$ D(a-Sb). The primary role played by the surface morphology when Ga was in the molten state was explained by high Ga fluidity. Indeed, Ga in the liquid phase is expected to be able to easily reach surface regions where the diffusion process is less difficult, as also shown by the very high value of the frequency factor for both c-Sb and a-Sb substrates (about 10^{-5} and 10^{-6} m^2 s^{-1}, respectively).

The diffusion coefficients were also determined at 20°C, a temperature at which Ga is in the solid state. Values around three orders of magnitude lower were achieved, and the structural factor prevailed over the geometrical one, that is, D(c-Sb) $<$ D(a-Sb), despite the larger surface area for diffusion of the c-Sb substrate. This behavior was similar to what was previously observed for In diffusion into a-Sb and c-Sb [90] and was attributed, in this case too, to the high structural disorder in a-Sb, which also presented higher interatomic distances than c-Sb.

10.3. Amorphous Antimony Crystallization

As already stated, to select the annealing treatment that best avoided crystallization during the diffusion measurements, specific research was carried out on a-Sb crystallization [227].

Several authors studied amorphous Sb films of several thicknesses prepared in different ways and at various temperatures, indicating the strong dependence of their stability against crystallization on the preparation conditions [223, 228–234]. The results indicated the important role of the thickness in the crystallization of amorphous Sb films and, therefore, the difficulties in obtaining amorphous samples some micrometers in thickness. In particular, they showed that it was not possible to avoid crystallization during vacuum deposition at room temperature. On the contrary, this was possible by electrodeposition of Sb on Fe substrates and adequately selection of the electrolyte composition and the current density, as shown above [90].

To investigate the influence of temperature on a-Sb crystallization and the mechanism of the process, the deposits were heated at 70°C and 110°C for increasing times, and the changes in the intensity of the main X-ray diffraction peak of c-Sb were analyzed. As pointed out by several authors [235], the complex process of amorphous material crystallization usually occurs in several more or less distinguishable steps. They can consist of nucleation and crystal growth controlled either by short-range diffusion or by interfacial chemical reactions, and the slowest step is rate-determining. So, the experimental results were discussed on the basis of the classical JMAYK (Johnson–Mehl–Avrami–Yerofeev–Kolmogorov) nucleation–growth equation, and the kinetic parameters were determined. They were interpreted according to Šesták's treatment [235], and it was found that the a-Sb crystallization process occurred through instantaneous bulk nucleation and subsequent three-dimensional growth controlled by diffusion. The three-dimensional growth of the nuclei was directly verified by SEM observation and indirectly by X-ray determination of the crystallite size. The activation energy for Sb diffusion in a-Sb during crystallization was about 25.1 kJ mol^{-1}. It is worth noting that this value is smaller, but not far from that determined in the same temperature range for In diffusion in a-Sb with formation of crystalline InSb: 53 kJ mol^{-1} [90].

The mechanism of the process is in agreement with current ideas that in so-called amorphous materials small clusters of the crystalline phase are often dispersed that are not detectable by usual X-ray analysis and may act as nucleation centers.

11. CONCLUSIONS

In this review chapter, we examined the general advantages and disadvantages of the electrochemical deposition process used to obtain the different III–V semiconductors, focusing on principles, methods, and results. We examined in detail the particular electrodeposition conditions, discussing the goal already reached, what remains open, and the future expectations of the researchers in the field. Another point that was stressed concerns the diffusion and reaction process mainly occurring in the solid state during the formation of III–V compounds. The mechanism of the process was investigated, as was the influence of the structure and morphology of the substrate on this same process.

The short introduction (Section 1) lists the main aspects of the electrochemical methodology used to prepare III–V compounds, comparing them with those relevant to the physical and chemical processes more frequently used.

Section 2 analyzes the typical properties of III–V semiconductors that make them so unique and of interest for practical applications. They include, for example, the ionic character of their bonds, the Gibbs free energy of formation from the elements, the band gap, the melting point, and the lattice constant. The formation of multinary compounds, truly nanomodulated thin films, microdiode arrays, and quantum dots (of increasing interest) is also briefly stressed.

Generally speaking, the electrodeposition of III–V semiconducting thin films involves three consecutive stages, which have to be carefully controlled. The first stage involves the formation of stable precursor ions in solution(s) or melt to be submitted to the desired redox reactions. The second concerns the simultaneous or consecutive discharge of the compound components at the cathode. The last stage is the generation of a homogeneous thin film with the desired stoichiometry, structure, morphology and with a low defectivity, which may eventually be obtained by a final thermal treatment.

Although the deposition parameters may "easily" be optimized in the case of a single element deposition, the situation becomes more complicated when two or more chemical species have to be codeposited. Kröger [29] developed the theoretical basis for simultaneous deposition, the standard deposition methodology, and his theory is presented in Section 4. A deposition potential is usually chosen that optimizes the stoichiometry of the deposit. In general, it is such that one of the elements (the more noble one) is deposited at a rate controlled by mass transfer toward the cathode surface, while the other element (which is present in excess) reacts with it to form the compound.

Another essential aspect of codeposition is the thermodynamic stability of the compound to be electrodeposited in the electrolyte. This stability was discussed in the case of aqueous electrolytes by several authors with the aid of the classical potential–pH plots due to Pourbaix [49]. These equilibrium diagrams also make it possible to determine the parasitic reactions that may join the main processs, typically hydrogen evolution and/or arsine or stibine formation (Section 5).

An argument that deserved a separate analysis is codeposition from molten salts, particularly the high-temperature ones, because the operative conditions needed to achieve deposits of some quality are severe. So, an important aspect is the stability against arbitrary perturbations of the growing deposit–electrolyte interface, which was investigated by Huggins and Elwell [79], as reported in Section 6.

Research on the sequential electrodeposition method of III–V compound semiconductors started in the earlier 1990s. Although less straightforward than codeposition, it presents several advantages, because it does not require a compromise between diverging requirements, because each element may be deposited under appropriate conditions. So, for example, it is insensitive to large differences in the deposition potentials of the two elements; on the other hand, the choice of the deposition sequence is often problematic. Moreover, this method requires a final thermal treatment to favor interdiffusion and chemical reaction (Section 7).

Section 8 surveys the electrochemical formation of thin films of the different III–V compounds, giving problems and strategies and critically analyzing the state of the art of the subject.

Research on GaP dates back to the work by Cuomo and Gambino [110] and by De Mattei et al. [111], in 1968 and 1978, respectively. Epitaxial layers were electrodeposited at about 800–1000°C from a highly corrosive metaphosphate melt containing Ga_2O_3 on single-crystal cathodes. InP, too, was codeposited from a bath containing P as a metaphosphate, but more difficulties were encountered, and no epitaxial growth could be observed [112]. Codeposition from an aqueous electrolyte containing $InCl_3$ and NH_4PF_6 was tried by Sahu [47, 48], but with unsatisfactory results. A method that seems to work better is the heat treatment of electrodeposited In films with gaseous P [45] or PH_3 vapors [134]. The pioneering research by Ortega and Herrero [45] has recently been deepened by Cattarin et al. [134], who prepared both *p*- and *n*-type polycrystalline InP films. However, these films presented cathodic and anodic dark currents, respectively, attributed to processes that occurred on metallic areas, presumably of the bare Ti substrate, owing to the irregular surface and poor morphology of the InP film. This morphology was a consequence of the formation of In globules during the phosphorization treatment performed at a temperature (between 400°C and 600°C) much higher than the In melting temperature.

Because of the considerable interest in GaAs films with controlled properties, several authors studied their electrodeposition from molten salts as well as from aqueous solutions.

Ten-micrometer-thick epitaxial films were obtained by De Mattei et al. in 1978 [113] on GaAs single-crystal electrodes by electrodeposition at 720–760°C from molten NaF and B_2O_3, a highly reactive bath also containing Ga_2O_3 and $NaAsO_2$. Of growing importance in more recent times is a series of low-temperature melts (40°C or even 20°C) formed by an organic chloride salt and either $AlCl_3$ or $GaCl_3$, from which Wicelinsky and Gale in 1986 [114] and Carpenter and Verbrugge in 1990 [109], respectively, codeposited GaAs. Although the results were encouraging, the stoichiometry and morphology of the deposits were still unsatisfactory. However, the potential of these melts must not be underestimated.

A systematic investigation of GaAs codeposition from aqueous baths was first reported by Chandra and Khare [51] in 1987, but the films obtained were multiphase. Moreover, these results could not be duplicated by Yang et al. [70]. On the basis of his thermodynamic calculations, Perrault [53] found a large potential–pH compatibility domain existing between GaAs and Ga(III) and As(III) ionic species in aqueous solutions. Nevertheless, GaAs was not obtained from acid or alkaline and neutral solutions.

Yang et al. [70] again reported the problems encountered in codepositing GaAs films of practical interest from aqueous baths. Indeed, the deposits obtained from alkaline solutions were thick, powdery, and scarcely adherent to the substrate, whereas those from acid solutions were thinner, more compact, but not uniform in composition. Both types of deposits were microcrystalline and yielded crystalline GaAs, with some loss of As content, after annealing.

Thin GaAs films approximating the stoichiometric composition were codeposited by Gao et al. [175] from acid solutions. The photoelectrochemical response of the electrodeposited films in contact with a redox electrolyte was poor because of their nonuniform composition and porosity as well as the presence of impurities and defects.

Not much better results were obtained by preparing GaAs following the consecutive deposition method, and additional difficulties had to be faced in determining the deposition sequence, which presented minor drawbacks.

In 1995, Andreoli et al. [179], following this method, deposited As on Ga predeposited on a Ti substrate, which practically did not alloy with Ga. After the two-layer samples were annealed, GaAs films were obtained that more homogeneously covered the substrate than did those prepared by Yang et al. [70], although they presented Ga-rich and As-rich areas. The reverse sequence (i.e., As first and Ga second) was not practicable, because As was reduced to AsH_3 during Ga electrodeposition [168].

In 1990, Sticking and co-workers [20] developed a typical sequential method, called electrochemical atomic layer epitaxy, and, in 1992 [21, 100], prepared nanometer-thick GaAs films of high purity as well as high structural and morphological perfection. They were obtained by alternately depositing layers of As and Ga on single-crystal Au electrodes, while strictly controlling the electrolyte composition and electrodeposition conditions. The same technique was followed by previous authors [101] and by Foresti et al. [104] to prepare InAs. In the latter case, single-crystal Ag substrates were used for their lower cost, although they are more reactive than the Au ones.

In 1989, Ortega and Herrero [45] first achieved InAs thin films by codeposition from citric acid solutions containing the chlorides of both In and As. The as-electrodeposited samples were amorphous and presented low crystallinity, even after annealing. They showed a *p*-type conductivity. InAs was later (1992) also obtained by Mengoli et al. [168] by In plating of electrodeposited As and submitting the samples to a subsequent thermal treatment. No overall composition changes occurred; the X-ray diffraction spectra practically matched the standard spectrum of polycrystalline InAs with respect to peak positions and related intensities. The average grain size was in the range of 40–50 nm. So, none of the expected difficulties were encountered with As, because of its low electrical conductivity and volatility.

Mengoli and co-workers [83] also succeeded in preparing GaSb by electrodepositing Ga on Sb at not too negative potentials to avoid Sb degradation to SbH_3. However, it was not easy to control the thickness of the Ga layer because of the simultaneous hydrogen evolution. After a mild heat treatment, the deposits were chemically homogeneous on their surface and perpendicular to it. More recently (1996), McChesney et al. [52] attempted to grow GaSb following the codeposition method, but the control of stoichiometry was limited by the concurrent evolution of H_2 and SbH_3.

Much research was carried out on InSb formation. In 1985, Sadana and Singh [44] systematically studied the achievement of In–Sb alloys from acid citrate baths of several compositions and recognized the InSb phase in the electrodeposits both before and after annealing. Subsequently, Ortega and Herrero [45] investigated In and Sb codeposition from citric acid solutions, mainly examining InSb formation as a function of the deposition potential and determining the conditions needed to obtain deposits free from the In and Sb phases. This research was practically duplicated by McChesney et al. [52], but the formation of stoichiometric InSb could not be repeated.

In 1991, InSb was prepared according to the sequential method of depositing Sb as the first element. The efficiency of conversion of the two-layer film to InSb was investigated as a function of the annealing conditions [84]. The binary compound was obtained when annealing was carried out at temperatures slightly higher than the In melting point, whereas pure InSb was achieved when the heat treatment was performed at 185°C for 5 h.

In 1994, a novel low-temperature molten salt electrolyte was reported by Carpenter and Verbrugge [116] that consisted of $InCl_3$ and an organic chloride from which codeposition of In and Sb was possible at 45°C, after the addition of $SbCl_3$. Electrochemical experiments showed that the chemical process leading to InSb deposit formation from this chloroindate melt is quite complex. In the authors' opinion, it is likely that improved-quality deposits containing a large fraction of InSb are attainable through an optimized electrochemical control and after low-temperature (350°C) annealing. Moreover, in a patent [205], the same authors also referred to the existence of organochloroindate melts comprising a salt of at least one group V element (P, As, Sb), which, in principle, could be used as electrolytes for the electrodeposition of almost any III–V compound after the replacement of a small amount of $InCl_3$ with a trichloride salt of another group III metal.

The above sections show the importance of the diffusion and reaction process in obtaining III–V compounds, and in the last two sections of the chapter, these arguments are discussed in more detail. So, the diffusion and reaction process of In into Bi [85–87] was compared with that of In into Sb [89] (Section 9). A diffusion coefficient five orders of magnitude higher was observed in the former case, and it was related to the different structural types of the two binary compounds: InBi has a typically layered structure favorable to diffusion, and InSb has a zinc-blende structure with tetrahedrically coordinated atoms and practically covalent bonds.

In Section 10, a more subtle difference is considered in comparing the In diffusion and reaction process into either crystalline or amorphous Sb [90]. In this case, too, the main role was played by the structure. Indeed, diffusion was easier into amorphous Sb with a more open structure and larger interatomic distances than crystalline Sb, although this latter presented a higher surface area (i.e., an area more favorable to diffusion). So, the structural factor prevailed over the geometrical one.

In conclusion, the systematic part of this chapter details the practical difficulties encountered in preparing high-quality thin films of III–V semiconductors following the electrochemical methodology. As a matter of fact, electrodeposition of compound semiconductors is complicated because of the need to maintain stoichiometry. Investigation is still at the level of fundamental research, and even the most recent work often lacks of a comprehensive film characterization that could be achieved with the use of the most advanced techniques, which could be used to systematically control the material quality.

Clearly, electrodeposition is a fascinating and promising method for preparing III–V semiconductors as thin films, but considerable development is required before it can be established whether such a technique has a place in future technology, as in the case of II–VI compounds. Perhaps the immense potential of electrosynthesis science and technology for the production of advanced semiconducting materials and structures has scarcely been explored, and the numerous application possibilities uniquely suitable for electrosynthesized thin films have not yet been fully examined.

As for the near future, the results of this survey leave us under the impression that the electrodeposition of III–V thin films can hardly compete with its vacuum counterpart, but that it can play a role in yielding advanced products with highly specific properties, such as superlattices, well-ordered atomic layer epitaxial films, and microdiode arrays.

Acknowledgments

V. M. Kozlov gratefully acknowledges the Cariplo Foundation, "A. Volta" Center for Scientific Culture, for financial support. L. Peraldo Bicelli thanks Prof. M. L. Foresti (University of Florence, Italy) for fruitful discussions regarding electrochemical atomic layer epitaxy.

REFERENCES

1. A. Stein, S. W. Keller, and T. E. Mallouk, *Science* 259, 1558 (1993).
2. W. E. Buhro, K. M. Hickman, and T. J. Trentler, *Adv. Mater.* 8, 685 (1996).
3. B. M. Basol and E. S. Tseng, *Appl. Phys. Lett.* 48, 946 (1986).
4. T. Ernst, in "Landolt-Boernstein. Zahlenwerte und Funktionen" (K. H. Hellwege, Ed.), Vol. 6, Band I, Teil 4, pp. 19 (A-7 type), 24 (B-2 type and B-3 type), 27 (B-10 type). Springer-Verlag, Berlin, 1955.
5. T. B. Massalski, "Binary Alloy Phase Diagrams," Vol. 1, pp. 88 (Al–As), 95 (Al–Bi), 145 (Al–P), 161 (Al–Sb), 200 (Ga–As), 206 (In–As), 231 (Tl–As), 501 (Ga–Bi), 509 (In–Bi), 547 (Tl–Bi); Vol. 2, pp. 1149 (Ga–P), 1157 (Ga–Sb), 1386 (In–P), 1391 (In–Sb), 1831 (Tl–P), 2023 (Tl–Sb). American Society for Metals, Metals Park, OH, 1986–1987.

6. JCPDS, Powder diffraction file, No. 12-470, 1972 (AlP); 17-915, 1974 (AlAs); 6-0233, 1967 (AlSb); 32-397, 1990 (GaP); 32-389, 1990 (GaAs); 7-215, 1967 (GaSb); 32-452, 1990 (InP); 15-869, 1972 (InAs); 6-0208, 1967 (InSb); 32-113, 1990 (InBi); 23-850, 1983 (In_5Bi_3); 11-566, 1972 (In_2Bi); 31-106, 1990 (Tl_7Sb_2); 14-5, 1972 ($TlBi_2$); 5-0642, 1974 (In); 35-732, 1990 (Sb); 5-0519, 1974 (Bi).

7. C. A. Coulson, "Valence," 2nd ed., p. 139. Oxford Univ. Press, Oxford, 1961.

8. D. D. Wagman, V. H. Evans, V. B. Parker, R. H. Schumm, I. Halow, S. M. Bailey, K. L. Churney, and R. L. Nuttall, *J. Phys. Chem. Ref. Data* **11**, 2-130 (AlP, AlAs), 2-132 (GaP, GaAs, GaSb), 2-134 (InP, InAs, InSb) (1982).

9. P. Y. Chevalier, *Calphad* 12, 383 (1988).

10. R. K. Pandey, S. N. Sahu, and S. Chandra, "Handbook of Semiconductor Electrodeposition," Vol. 5, p. 261. Dekker, New York, 1996.

11. C. D. Lockhande and S. H. Pawar, *Phys. Status Solidi A* 111, 17 (1989).

12. C. A. Wert and R. M. Thomson, "Physics of Solid," p. 54. McGraw-Hill, New York, 1970.

13. S. Koritnig, in "Landolt-Boernstein. Zahlenwerte und Funktionen" (K. H. Hellwege, Ed.), Vol. 6, Band I, Teil 4, p. 527. Springer-Verlag, Berlin, 1955.

14. S. Adachi, *J. Appl. Phys.* 58, R1 (1985).

15. K. Rajeshwar, *Adv. Mater.* 4, 23 (1992).

16. V. N. Tomashik, *Neorg. Mater.* 32, 1178 (1996) [*Inorg. Mater. (USSR)* 32, 1030 (1996)].

17. S. D. Lester and H. G. Strectman, *Superlattices Microstruct.* 2, 33 (1986).

18. J. Yahalom and O. Zadok, *J. Mater. Sci.* 22, 499 (1987).

19. V. Krishnan, D. Ham, K. K. Mishra, and K. Rajeshwar, *J. Electrochem. Soc.* 139, 23 (1992).

20. B. W. Gregory, M. L. Norton, and J. L. Stickney, *J. Electroanal. Chem.* 293, 85 (1990).

21. I. Villegas and J. L. Stickney, *J. Electrochem. Soc.* 139, 686 (1992).

22. F. Loglio, M. Tugnoli, F. Forni, and M. L. Foresti, in "Giornate dell'Electrochimica Italiana," p. 10. Divisione di Elettrochimica, Società Chimica Italiana, Como, Italy, 1999.

23. J. D. Klein, R. D. Herrick II, D. Palmer, M. J. Sailor, C. J. Brumlik, and C. R. Martin, *Chem. Mater.* 5, 902 (1993).

24. O. I. Micic, J. R. Sprague, C. J. Curtis, K. M. Jones, J. L. Machol, A. J. Nozik, H. Giessen, B. Fluegel, G. Mobs, and N. Peyghambarian, *J. Phys. Chem.* 99, 7754 (1995).

25. S. Bandyopadhyay, A. E. Miller, D. F. Yue, G. Banerjee, and R. E. Ricker, *J. Electrochem. Soc.* 142, 3609 (1995).

26. H. Gerischer, in "Physical Chemistry" (H. Eyring, D. Henderson, and W. Jost, Eds.), Vol. 9A, Chap. 5. Academic Press, New York, 1970.

27. G. Mengoli, M. M. Musiani, and F. Paolucci, *Chim. Ind.* 73, 477 (1991).

28. A. Brenner, "Electrodeposition of Alloys: Principles and Practice." Academic Press, New York, 1963.

29. F. A. Kröger, *J. Electrochem. Soc.* 125, 2028 (1978).

30. R. Weil, *Plating Surf. Finish.* 12, 46 (1982).

31. G. F. Fulop and R. M. Taylor, *Annu. Rev. Mater. Sci.* 15, 197 (1985).

32. D. Pottier and G. Maurin, *J. Appl. Electrochem.* 19, 361 (1989).

33. I. G. Dioum, J. Vedel, and B. Tremillon, *J. Electroanal. Chem.* 139, 329 (1982).

34. J. W. Cuthbertson, N. Parkinson, and H. P. Rooksby, *J. Electrochem. Soc.* 100, 107 (1953).

35. G. Tammann and A. Koch, *Z. Anorg. Chem.* 133, 179 (1924).

36. F. Ducelliz, *Bull. Soc. Chim.* 7, 606 (1910).

37. R. D. Engelken, *J. Electrochem. Soc.* 135, 834 (1988).

38. R. D. Engelken and T. P. Van Doren, *J. Electrochem. Soc.* 132, 2904 (1985).

39. M. Takahashi, K. Uosaki, and H. Kita, *J. Electrochem. Soc.* 131, 2305 (1984).

40. M. W. Verbrugge and C. W. Tobias, *J. Electrochem. Soc.* 132, 1298 (1985).

41. R. D. Engelken and T. P. Van Doren, *J. Electrochem. Soc.* 132, 2910 (1985).

42. R. D. Engelken, *J. Electrochem. Soc.* 134, 832 (1987).

43. M. Fracastoro-Decker and L. Peraldo Bicelli, *Mater. Chem. Phys.* 37, 149 (1993).

44. Y. N. Sadana and J. P. Singh, *Plating Surf. Finish.* 12, 64 (1985).

45. J. Ortega and J. Herrero, *J. Electrochem. Soc.* 136, 3388 (1989).

46. S. Cattarin, M. M. Musiani, U. Casellato, P. Guerriero, and R. Bertoncello, *J. Electroanal. Chem.* 380, 209 (1995).

47. S. N. Sahu, *J. Mater. Sci. Lett.* 8, 533 (1989).

48. S. N. Sahu, *Sol. Energy Mater.* 20, 349 (1990).

49. M. Pourbaix, "Atlas of Electrochemical Equilibria in Aqueous Solutions," 2nd ed., pp. 504 (P), 516 (As), 524 (Sb), 168 (Al), 428 (Ga), 436 (In), 533 (Bi). NACE, Houston, TX, and CEBELCOR, Brussels, 1974.

50. R. P. Singh, S. L. Singh, and S. Chandra, *J. Phys. D: Appl. Phys.* 19, 1299 (1986).

51. S. Chandra and N. Khare, *Semicond. Sci. Technol.* 2, 214 (1987).

52. J.-J. McChesney, J. Haigh, I. M. Dharmadasa, and D. J. Mowthorpe, *Opt. Mater.* 6, 63 (1996).

53. G. G. Perrault, *J. Electrochem. Soc.* 136, 2845 (1989).

54. J. O'M. Bockris and K. Uosaki, *J. Electrochem. Soc.* 124, 1348 (1977).

55. S.-M. Park and M. E. Barber, *J. Electroanal. Chem.* 99, 67 (1979).

56. H. Gerischer, *J. Electroanal. Chem.* 82, 133 (1977).

57. A. J. Bard and M. S. Wrighton, *J. Electrochem. Soc.* 124, 1706 (1977).

58. R. L. Meek and N. E. Schumber, *J. Electrochem. Soc.* 119, 1148 (1972).

59. A. Yamamoto and S. Yano, *J. Electrochem. Soc.* 122, 260 (1975).

60. M. Tomkiewicz and J. M. Woodall, *Science* 196, 990 (1977).

61. A. B. Ellis, S. W. Kaiser, J. M. Bolts, and M. S. Wrighton, *J. Am. Chem. Soc.* 99, 2839 (1977).

62. D. Cahen and Y. Mirovsky, *J. Phys. Chem.* 89, 2818 (1985).

63. L. Peraldo Bicelli, *J. Phys. Chem.* 96, 9995 (1992).

64. R. Piontelli and G. Serravalle, *Electrochim. Metal.* 1, 149 (1966).

65. A. N. Campbell, *Can. J. Chem.* 55, 1710 (1977).

66. I. A. Menzies and L. W. Owen, *Electrochim. Acta* 11, 251 (1966).

67. J. Scarminio, S. N. Sahu, and F. Decker, *J. Phys. E: Sci. Instrum.* 22, 755 (1989).

68. S. Chandra and S. N. Sahu, *J. Phys. D: Appl. Phys.* 17, 2115 (1984).

69. M. P. R. Panicker, M. Knaster, and F. A. Kröger, *J. Electrochem. Soc.* 125, 566 (1978).

70. M.-C. Yang, U. Landau, and J. C. Angus, *J. Electrochem. Soc.* 139, 3480 (1992).

71. D. Elwell, *J. Cryst. Growth* 52, 741 (1981).

72. G. P. Ivantsov, *Dokl. Akad. Nauk SSSR* 81, 179 (1951).

73. J. W. Rutter and B. Chalmers, *Can. J. Phys.* 31, 15 (1953).

74. C. Wagner, *J. Electrochem. Soc.* 101, 225 (1954).

75. W. W. Mullins and R. F. Sekerka, *J. Appl. Phys.* 35, 444 (1964).

76. W. A. Tiller, *J. Cryst. Growth* 2, 69 (1968).

77. H. J. Scheel and D. Elwell, *J. Electrochem. Soc.* 120, 818 (1973).

78. A. A. Chernov, *J. Cryst. Growth* 24–25, 11 (1974).

79. R. A. Huggins and D. Elwell, *J. Cryst. Growth* 37, 159 (1977).

80. R. A. Huggins, *J. Electrochem. Soc.* 122, 90C (1975).

81. A. R. Despic and K. I. Popov, *J. Appl. Electrochem.* 1, 275 (1971).

82. U. Cohen and R. A. Huggins, *J. Electrochem. Soc.* 122, 245C (1975).

83. F. Paolucci, G. Mengoli, and M. M. Musiani, *J. Appl. Electrochem.* 20, 868 (1990).

84. G. Mengoli, M. M. Musiani, F. Paolucci, and M. Gazzano, *J. Appl. Electrochem.* 21, 863 (1991).

85. S. Canegallo, V. Demeneopoulos, L. Peraldo Bicelli, and G. Serravalle, *J. Alloys Comp.* 216, 149 (1994).

86. S. Canegallo, V. Demeneopoulos, L. Peraldo Bicelli, and G. Serravalle, *J. Alloys Comp.* 228, 23 (1995).

87. S. Canegallo, V. Agrigento, C. Moraitou, A. Toussimi, L. Peraldo Bicelli, and G. Serravalle, *J. Alloys Comp.* 234, 211 (1996).

88. M. C. Hobson, Jr., and H. Leidheiser, Jr., *Trans. Metal. Soc. AIME* 233, 482 (1964).

89. V. M. Kozlov, V. Agrigento, D. Bontempi, S. Canegallo, C. Moraitou, A. Toussimi, L. Peraldo Bicelli, and G. Serravalle, *J. Alloys Comp.* 259, 234 (1997).

90. V. M. Kozlov, V. Agrigento, G. Mussati, and L. Peraldo Bicelli, *J. Alloys Comp.* 288, 255 (1999).

91. V. M. Kozlov and L. Peraldo Bicelli, *J. Alloys Comp.* 313, 161 (2000).
92. B. W. Gregory, D. W. Suggs, and J. L. Stickney, *J. Electrochem. Soc.* 138, 1279 (1991).
93. B. W. Gregory and J. L. Stickney, *J. Electroanal. Chem.* 300, 543 (1991).
94. S. Bedair, in "Atomic Layer Epitaxy" (S. Bedair, Ed.), p 304. Elsevier, Amsterdam, 1993.
95. A. A. Gewirth and B. K. Niece, *Chem. Rev.* 97, 1129 (1997).
96. D. M. Kolb, in "Advances in Electrochemistry and Electrochemical Engineering" (H. Gerischer and C. W. Tobias, Eds.), Vol. 11. Wiley, New York, 1978.
97. K. Juttner and W. J. Lorenz, *Z. Phys. Chem. N. F.* 122, 163 (1980).
98. J. L. Stickney, S. D. Rosasco, B. C. Schardt, and A. T. Hubbard, *J. Phys. Chem.* 88, 251 (1984).
99. C. K. Rhee, B. M. Huang, E. M. Wilmer, S. Thomas, and J. L. Stickney, *Mater. Manuf. Processes* 10, 283 (1995).
100. I. Villegas and J. L. Stickney, *J. Vac. Sci. Technol. A* 10, 3032 (1992).
101. T. L. Wade, L. C. Ward, C. B. Maddox, U. Happek, and J. L. Stickney, *Electrochem. Solid-State Lett.* 2, 616 (1999).
102. M. L. Foresti, G. Pezzatini, M. Cavallini, G. Aloisi, M. Innocenti, and R. Guidelli, *J. Phys. Chem. B* 102, 7413 (1998).
103. G. Pezzatini, S. Caporali, M. Innocenti, and M. L. Foresti, *J. Electroanal. Chem.* 475, 164 (1999).
104. M. Innocenti, F. Forni, E. Bruno, G. Pezzatini, and M. L. Foresti, in "Giornate dell'Elettrochimica Italiana," p. 8. Divisione di Elettrochimica, Società Chimica Italiana, Como, Italy, 1999.
105. M. L. Foresti, private communication.
106. T. G. Carver and H. Leidheiser, Jr., *J. Electrochem. Soc.* 109, 68 (1962).
107. W. R. Buck III and H. Leidheiser, Jr., *Nature (London)* 204, 177 (1964).
108. A. D. Styrkas, V. V. Ostroumov, and G. V. Anan'eva, *Zh. Prikl. Khim.* 37, 2431 (1964).
109. M. K. Carpenter and M. W. Verbrugge, *J. Electrochem. Soc.* 137, 123 (1990).
110. J. J. Cuomo and R. J. Gambino, *J. Electrochem. Soc.* 115, 755 (1968).
111. R. C. De Mattei, D. Elwell, and R. S. Feigelson, *J. Cryst. Growth* 44, 545 (1978).
112. D. Elwell, R. S. Feigelson, and M. M. Simkins, *J. Cryst. Growth* 51, 171 (1981).
113. R. C. De Mattei, D. Elwell, and R. S. Feigelson, *J. Cryst. Growth* 43, 643 (1978).
114. S. P. Wicelinski and R. J. Gale, *Proc. Electrochem. Soc.* 86-1, 144 (1986).
115. M. W. Verbrugge and M. K. Carpenter, *AIChE J.* 36, 1097 (1990).
116. M. K. Carpenter and M. W. Verbrugge, *J. Mater. Res.* 9, 2584 (1994).
117. C. F. Smart, German Patent DE 929,103, 1955.
118. C. F. Smart, U.S. Patent US 2,755,537, 1956.
119. M. Kumagai, T. Takagahara, and E. Hanamura, *Phys. Rev. B: Condens. Matter* 37, 898 (1988).
120. M. Nagano, *Japan. J. Appl. Phys.* 38, 3705 (1999).
121. D. Z.-Y. Ting, *Microelectron. J.* 30, 985 (1999).
122. C. A. Hoffman, J. R. Meyer, F. J. Bartoli, J. R. Waterman, B. V. Shanabrook, B. R. Bennet, and R. J. Wagner, *Proc. SPIE* 2554, 124 (1995).
123. M. J. Shaw, *Phys. Rev. B: Condens. Matter* 58, 7834 (1998).
124. S. Chandra, N. Khare, and H. M. Upadhyaya, *Bull. Mater. Sci.* 10, 323 (1988).
125. W. Rudorff and E. J. Kohlmeyer, *Z. Metallkl.* 45, 608 (1954).
126. A. P. Tomilov and N. E. Chomutov, "Encyclopedia of the Electrochemistry of the Elements," Vol. 3. Dekker, New York, 1975.
127. P. N. Yocom, The preparation of transition-metal phosphides by fused salt electrodes, Ph.D. thesis, Univ. of Michigan, 1958.
128. K. W. Böer, *Phys. Status Solidi A* 40, 355 (1977).
129. G. A. Bukhalova and I. V. Mardirosova, *Russ. J. Inorg. Chem.* 11, 497 (1966).
130. K. J. Bachmann, E. Buehler, J. L. Shay, and S. Wagner, *Appl. Phys. Lett.* 29, 121 (1976).
131. S. N. Sahu, *J. Mater. Sci. Electron.* 3, 102 (1992).
132. J. Bielinski and A. Bielinska, *Surf. Technol.* 24, 219 (1985).
133. X. Rajnarayan and M. N. Mungole, *Surf. Technol.* 24, 233 (1985).
134. S. Cattarin, M. Musiani, U. Casellato, G. Rossetto, G. Razzini, F. Decker, and B. Scrosati, *J. Electrochem. Soc.* 142, 1267 (1995).
135. S. N. Sahu and A. Bourdillon, *Phys. Status Solidi A* 111, K 179 (1989).
136. R. B. Gore, R. K. Pandey, and S. Kulkarni, *J. Appl. Phys.* 65, 2693 (1989).
137. J. Herrero and J. Ortega, *Sol. Energy Mater.* 17, 357 (1988).
138. R. Piercy and N. A. Hampson, *J. Appl. Electrochem.* 5, 1 (1975).
139. P. Zanella, G. Rossetto, N. Brianese, F. Ossola, M. Porchia, and J. O. Williams, *Chem. Mater.* 3, 225 (1991).
140. G. Harbeke, O. Madelung, and U. Rössler, in "Landolt-Boernstein. Numerical Data and Functional Relationships in Science and Technology" (O. Madelung, Ed.), Vol. 17a, p. 281. Springer-Verlag, Berlin, 1982.
141. S. Chandra, "Photoelectrochemical Solar Cells," Vol. 5, p. 175. Gordon and Breach, New York, 1985.
142. O. Khaselev and J. A. Turner, *Science* 280, 425 (1998).
143. I. V. Zubeck, R. S. Feigelson, R. A. Huggins, and P. A. Petit, *J. Cryst. Growth* 34, 85 (1976).
144. I. G. Dioum, J. Vedel, and B. Tremillon, *J. Electroanal. Chem.* 137, 219 (1982).
145. H. L. Chum, V. R. Koch, L. L. Miller, and R. A. Osteryoung, *J. Am. Chem. Soc.* 97, 3264 (1975).
146. R. A. Carpio, L. A. King, R. E. Lindstrom, J. C. Nardi, and C. L. Hussey, *J. Electrochem. Soc.* 126, 1644 (1979).
147. C. L. Hussey, L. A. King, and J. S. Wilkes, *J. Electroanal. Chem.* 102, 321 (1979).
148. J. S. Wilkes, J. A. Levisky, C. L. Hussey, and M. L. Druelinger, *Proc. Electrochem. Soc.* 81-9, 245 (1981).
149. S. P. Wicelinski and R. J. Gale, *Proc. Electrochem. Soc.* 87-7, 591 (1987).
150. M. Lipsztajn and R. A. Osteryoung, *J. Electrochem. Soc.* 130, 1968 (1983).
151. J. S. Wilkes, J. A. Levisky, R. A. Wilson, and C. L. Hussey, *Inorg. Chem.* 21, 1263 (1982).
152. S. P. Wicelinski, R. J. Gale, and J. S. Wilkes, *J. Electrochem. Soc.* 134, 262 (1987).
153. S. P. Wicelinski, R. J. Gale, and J. S. Wilkes, *Thermochim. Acta* 126, 255 (1988).
154. S. P. Wicelinski, R. J. Gale, K. M. Pamidimukkala, and R. A. Laine, *Anal. Chem.* 60, 2228 (1988).
155. S. P. Wicelinski, R. J. Gale, S. D. Williams, and G. Mamantov, *Spectrochim. Acta, Part A* 45, 759 (1989).
156. C. L. Hussey, *Adv. Molten Salt Chem.* 5, 185 (1983).
157. P. K. Lay and M. Skyllas-Kazacos, *J. Electroanal. Chem.* 248, 431 (1988).
158. M. W. Verbrugge and M. K. Carpenter, U.S. Patent US 4,883,567, 1989.
159. K. R. Murali, V. Subramanian, N. Rangarajan, A. S. Lakshmanan, and S. K. Rangarajan, *J. Mater. Sci. Mater. Electron.* 2, 149 (1991).
160. M. M. Musiani, F. Paolucci, and P. Guerriero, *J. Electroanal. Chem.* 332, 113 (1992).
161. A. P. Tomilov and N. E. Chomutov, "Encyclopedia of the Electrochemistry of the Elements," Vol. 2, p. 21. Dekker, New York, 1975.
162. G. Wranglen, *J. Electrochem. Soc.* 108, 1069 (1961).
163. R. Piontelli and G. Poli, *J. Electrochem. Soc.* 109, 551 (1962).
164. T. E. Dinan, W. F. Jon, and H. Y. Cheh, *J. Electrochem. Soc.* 136, 3284 (1989).
165. L. Astier, R. Michelis-Quiriconi, and M.-L. Astier, French Patent F 2,529,713, 1984.
166. K. R. Murali, M. Jayachandran, and N. Rangarajan, *Bull. Electrochem.* 3, 261 (1987).
167. S. Chandra and N. Khare, *Semicond. Sci. Technol.* 2, 220 (1987).
168. G. Mengoli, M. M. Musiani, and F. Paolucci, *J. Electroanal. Chem.* 332, 199 (1992).
169. N. Ibl, *Surf. Technol.* 10, 81 (1980).
170. T. L. Lam, A. Kaike, I. Ohno, and S. Naruyama, *J. Metal. Finish Soc. Jpn.* 24, 425 (1983).
171. J. Houlston, *Product. Finish.* 42, 20 (1989).
172. G. Holmbom and B. E. Jacobson, *J. Electrochem. Soc.* 135, 2720 (1989).
173. G. Devaraj, K. I. Vasu, and S. K. Seshadri, *Bull. Electrochem.* 5, 333 (1989).

174. G. Devaraj, G. N. K. Ramesh Bapu, J. Ayyapparaju, and S. Guruviah, *Bull. Electrochem.* 5, 448 (1989).
175. Y. Gao, A. Han, Y. Lin, Y. Zhao, and J. Zhang, *J. Appl. Phys.* 75, 549 (1994).
176. P. Cowache, D. Lincot, and J. Vedel, *J. Electrochem. Soc.* 136, 1646 (1989).
177. Y. Gao, A. Han, Y. Lin, Y. Zhao, and J. Zhang, *Thin Solid Films* 232, 278 (1993).
178. R. Dorin and E. J. Frazer, *J. Appl. Electrochem.* 18, 134 (1988).
179. P. Andreoli, S. Cattarin, M. Musiani, and F. Paolucci, *J. Electroanal. Chem.* 385, 265 (1995).
180. K. Onabe, *Japan. J. Appl. Phys.* 21, 964 (1982).
181. J. O'M. Bockris and B. E. Conway, *Trans. Faraday Soc.* 45, 989 (1949).
182. J. Herrero and J. Ortega, *Sol. Energy Mater.* 16, 477 (1987).
183. E. Dalchiele, S. Cattarin, M. Musiani, U. Casellato, P. Guerriero, and G. Rossetto, *J. Electroanal. Chem.* 418, 83 (1996).
184. J. L. Valdes, G. Cadet, and J. W. Mitchell, *J. Electrochem. Soc.* 138, 1654 (1991).
185. O. Alvarez-Fregoso, J. G. Mendoza-Alvarez, and O. Zelaya-Angle, *J. Appl. Phys.* 82, 708 (1997).
186. L. Levy, D. Ingert, N. Fekin, and M. P. Pileni, *J. Cryst. Growth* 184–185, 377 (1998).
187. T. van Buuren, L. N. Dinh, L. L. Chase, W. J. Siekhaus, and L. J. Terminello, *Phys. Rev. Lett.* 80, 3803 (1998).
188. Sh. Z. Khamundkhanova, G. F. Rubinchik, and A. M. Murtazaev, *Dokl. Akad. Nauk Uzb. SSR* 30, 47 (1973).
189. A. L. Pitman, M. Pourbaix, and N. Zoubov, *J. Electrochem. Soc.* 104, 594 (1957).
190. R. K. Iyer and S. G. Deshpande, *J. Appl. Electrochem.* 17, 936 (1987).
191. T. B. Belitskaya, V. M. Kochegarov, and Yu. I. Chernov, *Elektrokhimiya* 6, 215 (1970).
192. F. C. Walsh and D. R. Gabe, *Surf. Technol.* 13, 305 (1981).
193. R. Walker and S. J. Duncan, *Metal Finish.* 9, 21 (1982), Part 1; 10, 77 (1982), Part 2; 11, 59 (1982), Part 3.
194. V. M. Mogilev and A. I. Falicheva, *Zashchita Metallov* 10, 192 (1974).
195. L. Domnikov, *Metal Finish.* 71, 50 (1973).
196. Y. N. Sadana, J. P. Singh, and R. Kumar, *Surf. Technol.* 24, 319 (1985).
197. T. Okubo and M. Landau, *Proc. Electrochem. Soc.* 88-2, 547 (1988).
198. V. V. Losev and A. P. Pchelnikov, *Electrochim. Acta* 18, 589 (1973).
199. G. Gunawardena, D. Pletcher, and A. Razaq, *J. Electroanal. Chem.* 164, 363 (1984).
200. I. A. Ammar and A. Saad, *J. Electroanal. Chem.* 34, 159 (1972).
201. L. L. Wikstrom, N. T. Thomas, and K. Nobe, *J. Electrochem. Soc.* 122, 1201 (1975).
202. M. E. Straumanis and L. Hu, *J. Electrochem. Soc.* 119, 818 (1972).
203. T. A. Zawodzinski, Jr., and R. A. Osteryoung, *Inorg. Chem.* 28, 1710 (1989).
204. J. H. R. Clarke and R. E. Hester, *J. Chem. Phys.* 50, 3106 (1969).
205. M. K. Carpenter and M. W. Verbrugge, U.S. Patent US 5,264,111, 1993.
206. V. M. Kochegarov and V. D. Samuilenkova, *Elektrokhimiya* 1, 1470 (1965).
207. Y. N. Sadana and Z. Z. Wang, *Surf. Technol.* 25, 17 (1985).
208. Y. N. Sadana and Z. Z. Wang, *Metal Finish.* 83, 23 (1985).
209. V. V. Povetkin and T. G. Shibleva, *Zashchita Metallov* 29, 518 (1993).
210. L. Peraldo Bicelli, C. Romagnani, and G. Serravalle, *Electrochim. Metal.* 4, 233 (1969).
211. C. Sunseri and G. Serravalle, *Metall. Ital.* 7–8, 373 (1976).
212. O. H. Henry and E. L. Badwick, *Trans. Metall. Soc. AIME* 171, 389 (1947).
213. E. A. Peretti and S. C. Carapella, *Trans. Am. Soc. Metall.* 41, 947 (1949).
214. B. C. Giessen, A. Morris, and N. J. Grant, *Trans. Metall. Soc. AIME* 239, 883 (1967).
215. H. Schmalzried, "Solid State Reactions," pp. 53, 95, 124. Academic Press, New York, 1974.
216. R. Boom, P. C. M. Vendel, and F. R. De Boer, *Acta Metall.* 21, 807 (1973).
217. H. P. Singh, M. H. Rao, and S. Misra, *Scr. Metall.* 6, 621 (1972).
218. H. P. Singh, *Scr. Metall.* 6, 519 (1972).
219. P. M. Robinson and M. B. Bever, *Trans. Metall. Soc. AIME* 233, 1908 (1965).
220. V. M. Kozlov, V. Agrigento, D. Bontempi, S. Canegallo, C. Moraitou, A. Toussimi, L. Peraldo Bicelli, and G. Serravalle, unpublished observations.
221. F. H. Eisen and C. Birchenall, *Acta Metall.* 5, 265 (1957).
222. P. Binnie, *Acta Crystallogr.* 9, 686 (1956).
223. M. Hashimoto and T. Nohara, *Thin Solid Films* 199, 71 (1991).
224. L. A. Zhukova and O. Yu. Sidorov, *Fiz.-Khim. Issled. Metall. Protsessov* 13, 53 (1985).
225. H. Krebs, F. Schultze-Gebhardt, and R. Thees, *Z. Anorg. Chem.* 282, 177 (1955).
226. N. F. Mott and E. A. Davis, "Electronic Processes in Non-Crystalline Solids," 2nd ed., p. 439. Clarendon, Oxford, 1979.
227. V. M. Kozlov, B. Bozzini, V. Licitra, M. A. Lovera, and L. Peraldo Bicelli, *Int. J. Nonequilibrium Processing*, in press.
228. J. J. Hauser, *Phys. Rev. B: Condens. Matter* 9, 2623 (1974).
229. N. Kaiser, H. Müller, and Ch. Gloede, *Thin Solid Films* 85, 293 (1981).
230. M. Hashimoto, H. Sugibuchi, and K. Kambe, *Thin Solid Films* 98, 197 (1982).
231. N. Kaiser, *Thin Solid Films* 115, 309 (1984).
232. M. Hashimoto and M. Matui, *Appl. Surf. Sci.* 33–34, 826 (1988).
233. M. Hashimoto, K. Umezawa, and R. Murayama, *Thin Solid Films* 188, 95 (1990).
234. P. Jensen, P. Melinon, M. Treilleux, A. Hoareau, J. X. Hu, and B. Cabaud, *Appl. Phys. Lett.* 59, 1421 (1991).
235. J. Šesták, in "Kinetic Phase Diagrams Nonequilibrium Phase Transitions" (Z. Chvoj, J. Šesták, and A. Tříska, Eds.), Vol. 10, p. 219. Elsevier, Amsterdam, 1991.

Chapter 6

FUNDAMENTALS FOR THE FORMATION AND STRUCTURE CONTROL OF THIN FILMS: NUCLEATION, GROWTH, SOLID-STATE TRANSFORMATIONS

Hideya Kumomi

Canon Research Center, 5-1 Morinosato-Wakamiya, Atsugi-shi, Kanagawa 243-0193, Japan

Frank G. Shi

Department of Chemical and Biochemical Engineering and Materials Science, University of California, Irvine, California, USA

Contents

1. INTRODUCTION

1.1. Structures and Properties

Thin films can be treated as being three-dimensional condensed matter with the dimension in one direction much smaller than the dimensions in the other perpendicular directions. The basic structure consists of the two surfaces and the intermediate bulk between them. Unless the thin film is freestanding, it touches the other adjacent layer(s) like a substrate, and at least one surface becomes an interface.

In addition to these basic components, thin films may include the finer structures. Atoms composing the film and the chemical bonds among them provide the (atomic) nanostructure. The defects in the atomic structure are also the issues. They are vacancies, interstitial atoms, impurities, broken bonds,

Handbook of Thin Film Materials, edited by H.S. Nalwa
Volume 1: Deposition and Processing of Thin Films

ISBN 0-12-512909-2/$35.00

deformed bonds, atomic surface roughness, etc. The chain or group of these defects makes the microstructures. They are voids, stacking faults, spatial fluctuation of the atomic composition, segregation of the impurities, dislocations and twin boundaries in crystalline films, surface undulation, spatial variation of the film thickness, and so on. In the case of polycrystalline thin films, the granular structures should also be considered. They are the variation of crystallographic orientation among the grains, the variation of the grain size, and the grain boundaries where the adjacent grains meet.

These structures could be connected closely with the mechanical, chemical, electrical, and optical properties or functions of the thin films, and their spatial uniformity. In other words, the properties or their uniformity of thin films would be limited by their structures. For instance, the point defects and the dislocations in the epitaxial thin films always limit their performance. The surface roughness of polycrystalline Si (poly-Si) thin films used as the gate electrodes of metal oxide silicon transistors inhibits one from decreasing the film thickness down to be comparable to the surface roughness, and hence from reducing the device dimensions. When the poly-Si thin films serve as the active layers of thin film transistors, the electric potential barrier localized at the grain boundaries hinders the carrier transport and reduces the performance of the devices. For piezoelectric polycrystalline thin films, the fluctuation of the normal orientation among the grains attenuates the overall piezoelectric efficiency of the films. Thus it is often necessary to control the structures of thin films for improving their properties, functions, or uniformity. The control of the thin films structures has long been one of the central problems in modern materials science and technology.

1.2. Structures and Formation Process

Thin films could be prepared through one or more steps of the processes. Except certain kinds of organic membranes or soap bubbles, thin films need to be formed over the substrates. The preparation and the treatment of the substrates must be the initial step. The subsequent step involves the deposition of materials on the substrates from vapor, liquid, or solution. There are many possibilities that these formation processes provide the origins of the thin films structures. For example, if there is imperfect smoothness at the substrate surface, it directly causes the interfacial roughness or becomes the origin of the defects which could propagate in the deposited film. If the deposition over substrates without any epitaxial relation to the deposited films evolves by island growth of crystallites like Volmer–Weber or Stranski–Krastanov modes, the grain boundaries must be formed where the adjacent islands meet.

The as-deposited thin films may be further treated with the postdeposition processes which not only improve the as-deposited structures but also introduce some new structures. For instance, the surface roughness of the as-deposited thin films can be reduced by etching or polishing techniques, while these processes bring some defects into the thin films. Amorphous thin films deposited onto a single crystalline substrate can be crystallized, layer by layer in solid phase, with energies applied to the films. If there are some impurities like oxides at the interface to the substrate, the dislocations or the stacking faults propagate from the impurities through the crystallized layer. If the substrate is also amorphous, the crystallization evolves with random nucleation of crystallites and their growth in the deposited films. The grain boundaries are formed at the meeting points of the adjacent crystallites. Polycrystalline thin films deposited on an insulating amorphous substrate can be melted and recrystallized by sweeping a stripe heater over the thin films, which is known as zone melting technique. The length of the achieved single crystalline regions along the sweeping direction is limited by thermal stress and undesirable random nucleation of the crystallites.

Hence most of the thin films microstructures originate in the formation processes. For the purpose of controlling the structures, it is only one approach, or at least one of the best approaches, to control the formation processes. Motivated by this fact, a number of scientific investigations have been devoted to elucidate the mechanisms in the formation processes of the thin films, and many engineering efforts have been spent to introduce novel techniques to the formation processes.

1.3. Nucleation, Growth, and Solid-State Transformations

The elementary process responsible for the thin film formation is the phase transformation of matters. The deposition process starts with the phase transformation of deposited species from vapor or liquid phase in the free space over the substrate to adherent states at the substrate surface. The postdeposition process also involves the phase transformation such as amorphous–crystalline or liquid–crystalline phase transition.

The phase transformation is generally initiated by nucleation and growth of clusters of new phases. For example, the dislocations in crystals are formed by the nucleation and growth of the embryo dislocations. Without continuous supply of atomic steps containing kinks, the deposition of epitaxial thin films proceeds layer by layer with the nucleation of two-dimensional clusters and their lateral growth. The deposition of polycrystalline thin films and the crystallization of amorphous thin films on amorphous substrates are just the processes of crystallite nucleation and growth. It is important to understand the nucleation, the growth, and the resultant solid-state transformations for investigating the formation of the thin films.

1.4. Scope of This Chapter

In view of the above discussion, this chapter aims at providing the fundamentals for the formation and the structure control of thin films focusing on nucleation and growth processes. The formation of polycrystalline thin films and their granular structures is mainly dealt with as a typical example.

Section 2 describes the state-of-the-art theories for nucleation and growth, basic parameters of nucleation and growth, and the theoretical basis for the observables in the nucleation and growth. Section 3 describes the experimental methods for measuring the nucleation and growth and their theoretical bases with some examples shown. Section 4 describes the strategies and the methods for controlling the grain size, the size distribution of grains, the location of grains in the formation of polycrystalline thin films by the control of the nucleation and growth by showing some examples.

At the end of this introductory section, some of the important or recent reviews on thin films formation [1–16] are cited, which might be helpful to those who are interested in the general issues beyond the scope of this chapter.

2. THEORY OF NUCLEATION AND GROWTH

Theories of the nucleation and growth consist of both thermodynamic and kinetic parts. The thermodynamic part relates to the potential for the formation of a new phase: it describes the formation free energy of the clusters in the new phase and the size distribution of the cluster population in the equilibrium state. The kinetic part presents the formation rate of a new phase: it describes the growth and shrinkage of clusters, the dynamic evolution of size distribution of the clusters, and the rates of nucleation, growth, and the volume fraction transformed into the new phase.

In the followings, the theories of nucleation and growth are reviewed and the theoretical bases for the observables in the nucleation and growth are summarized. It is noticed in this chapter that the terms "particle," "domain," "grain," "cluster," and "crystallite" are used synonymously but in different ways according to the situations.

2.1. Thermodynamics of Nucleation and Growth

The thermodynamics study of nucleation and growth dates back to the 19th century [17, 18]. The pioneering efforts in the early 20th century [19–24] laid down the firm foundation for our current thermodynamic understanding of nucleation and growth.

2.1.1. Free Energy of Nucleus

Nucleation starts with formation of small clusters of the new phase in a matrix of the mother phase. The thermodynamical stability of the cluster is evaluated by the free energy of the cluster, W, which is defined as the difference in the free energy between the cluster and the equivalent in the matrix. Generally, W is a function of the cluster size, g, and has a peak in the size space. As illustrated in Figure 1, $W(g)$ increases from 0 with g, shows a maximum at $g = g_*$, and then decreases monotonically to become negative after $g = g_0$.

The size of g_* is called the *critical size*, which can be derived as a solution of

$$\left[\frac{\partial W(g)}{\partial g} \right]_{g=g_*} = 0 \qquad (2.1)$$

for g_*. The g_*-sized cluster is called a *critical* cluster or simply a *nucleus*. The clusters in the regions of $g < g_*$ and $g > g_*$

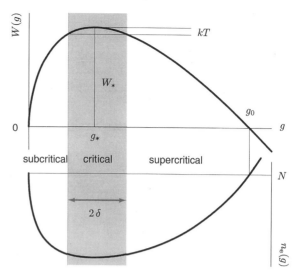

Fig. 1. Dependence of the free energy of a cluster, $W(g)$, on the cluster size, g, and the equilibrium distribution of the cluster size, $n_e(g)$.

are often called *subcritical* or *embryo* and *supercritical* clusters, respectively. The value of the maximum, $W(g_*) \equiv W_*$, is called the free-energy barrier to nucleation.

Up to the size of g_0, $W(g)$ is positive and the formation of the cluster is unfavorable. Particularly in the region of $g < g_*$, the derivative, $\partial W(g)/\partial g$, is positive, and the subcritical clusters prefer to shrink and dissolve back into the matrix of the mother phase. In the region of $g > g_*$, $\partial W(g)/\partial g$ is negative, and the supercritical clusters tend to grow thereafter. There is a special region around g_*, called the *critical region*, the *nucleation barrier layer*, or the *nucleation boundary layer*, in which the cluster is hardly influenced by the drift field made by the gradient of $W(g)$ and readily diffuses in the size space by a small fluctuation less than one quantum unit. The left and right boundaries of the critical region are derived as the two solutions of

$$W_* - W(g) \leq kT \qquad (2.2)$$

for g, where k denotes Boltzmann constant and T is the absolute temperature. Instead of the exact form of $W(g)$ in Eq. (2.2) for calculating the boundaries, its approximation by the quadratic expansion around $g = g_*$,

$$W(g) \simeq W_* + \frac{1}{2} \left[\frac{\partial^2 W(g)}{\partial g^2} \right]_{g=g_*} (g - g_*)^2 \qquad (2.3)$$

is frequently used for solving Eq. (2.2). Substituting Eq. (2.3) for $W(g)$ in Eq. (2.2), the boundary sizes are approximated as

$$g = g_* \pm \delta \quad \text{with } \delta = \left[-\frac{1}{2kT} \frac{\partial^2 W(g)}{\partial g^2} \right]_{g=g_*}^{-1/2} \equiv \frac{1}{\sqrt{\pi} Z} \qquad (2.4)$$

where 2δ corresponds to the width of the critical region, and Z is called the Zeldovich factor. With δ, Eq. (2.3) can be rewritten as

$$W(g) \simeq W_* - kT\left(\frac{g - g_*}{\delta}\right)^2$$

Nucleation is the process in which the cluster grows beyond the critical size, g_*, crossing the free-energy barrier to nucleation, W_*. If g_* is as small as 1 or $W_* \leq kT$, the nucleation occurs without any other conditions. It is frequent, however, that $g_* \gg 1$ and $W_* \gg kT$. In such conditions, if there is initially no cluster in the vicinity of the critical region, it seems that the nucleation scarcely occurs within the thermodynamic aspect of nucleation. Practically, there is another driving force to nucleation other than the drift field of $W(g)$, which makes the cluster nucleate in any condition. This issue will be explained in detail later in Section 2.2.1.

2.1.2. Classical Model for the Nucleation Free Energy

The early developed model for the nucleation free energy is based upon the capillary approximation to clusters, which is today called the classical model.

Provided a cluster is formed in a matrix, the cluster consists of the surface (or interface) to the matrix and the body wrapped by the surface. The Gibbs free-energy change by the existence of this cluster, W, can be a sum of contributions from the volume energy and the surface energy as

$$W(g) = -g\Delta\mu + \sigma\kappa g^{1-1/d} \qquad (2.5)$$

where g is the number of monomers such as atoms or molecules in the cluster and represents the size of the cluster, $\Delta\mu \, (> 0)$ is the chemical potential standing for the difference of the free energy per monomer between the phases of the cluster and the matrix, σ denotes the interfacial energy density (i.e., per unit areas) at the interface between the cluster and the matrix, κ is the geometrical factor with which $\kappa g^{1-1/d}$ becomes the area of the interface between the cluster and the matrix, and d stands for the effective geometrical dimension of the cluster. If the cluster has a compact spherical shape, $d = 3$ and $\kappa = (36\pi V_1^2)^{1/3}$ where V_1 is the volume that a monomer occupies in the cluster.

The plot of $W(g)$ in Figure 1 is drawn based on Eq. (2.5). In the case of the classical model, the g-dependence of $W(g)$ is due to the competition between the volume energy term, $-g\Delta\mu$, and the surface energy term, $\sigma\kappa g^{1-1/d}$. For $W(g)$ of Eq. (2.5), the critical size is calculated from Eq. (2.1) as

$$g_* = \left(\frac{d-1}{d}\frac{\sigma\kappa}{\Delta\mu}\right)^d = g\left[\frac{\partial}{\partial g}\frac{W(g)}{\Delta\mu} + 1\right]^d$$

the gradient of the free energy can be expressed as

$$\frac{\partial}{\partial g}W(g) = \Delta\mu\left[\left(\frac{g_*}{g}\right)^{1/d} - 1\right]$$

the free-energy barrier to nucleation is

$$W_* = \frac{\Delta\mu}{d-1}g_* = \left(\frac{\sigma\kappa}{d}\right)^d\left(\frac{d-1}{\Delta\mu}\right)^{d-1} \qquad (2.6)$$

the half width of the critical region is

$$\delta = \sqrt{2dkT\frac{g_*}{\Delta\mu}} = \sqrt{2kTd^{1-d}(d-1)^d(\sigma\kappa)^d\Delta\mu^{-(d+1)}}$$

and the size at which $W(g)$ changes its sign is

$$g_0 = \left(\frac{\sigma\kappa}{\Delta\mu}\right)^d$$

The model expressed by Eq. (2.5) is valid when the nucleating cluster has a single interface to the matrix, such as homogeneous nucleation of isotropic clusters. If the nucleation is heterogeneous, e.g., formation of clusters on a substrate, the clusters have the other interface to the substrate in addition to that to the matrix. For such cases where the cluster has multiple interfaces, Eq. (2.5) should be slightly modified into

$$W(g) = -g\Delta\mu + \sum_{i=1}^{N}\sigma_i\kappa_i g^{1-1/d} \qquad (2.7)$$

where N is the number of the kinds of the interfaces, and σ_i and κ_i are the interfacial energy density and the geometrical factor of the ith interface, respectively. It is noted that the basic features in the size dependence of Eq. (2.7) do not differ from those of Eq. (2.5).

2.1.3. Nonclassical Models for the Nucleation Free Energy

The classical model for the free energy to nucleation has long been used for explaining and investigating phenomena related to nucleation. In fact, the theoretical predictions of the classical model can successfully describe some kinds of nucleation in undercooled liquids and glasses. However, a few discrepancies have been also found between the classical model and experimental results. The classical model have been criticized as well from theoretical points of view, mainly on the applicability of the model to the small clusters. Since the number of monomers composing the surface of the cluster gets close to the total number of monomers in the cluster as the cluster size decreases, the linear decomposition of the free energy into the volume energy term and the surface energy term becomes questionable. In the theoretical prediction using the classical model, the values measured for the bulk materials are often used for the thermodynamic parameters (i.e., $\Delta\mu$ and σ) in Eqs. (2.5) and (2.7), since it is difficult to measure the values for the small clusters. These values for the small clusters would change from those for the bulk materials. For example, because of the large curvature, the surface bond structures of small spherical clusters must be deformed from those of the flat surface, and thus the surface or interfacial energy density could be different between the bulk and the small clusters.

To improve or transcend the classical model, a number of attempts have been devoted to establishing nonclassical or modern models. One of the typical attempts is to compensate the interfacial energy for the small clusters considering its dependence on the curvature. The other typical attempt is the field-theoretical approach to the free-energy barrier to nucleation.

Other than these, many new models have been still proposed. Since it is out of the scope of this chapter to review all of the modern models, they are not mentioned in any more detail here. The readers are referred to the brilliant reviews [25–35] for more details. It should be noted, however, that many models have not been sufficiently tested by experimental studies, as will be discussed in detail later in Section 3.5.1.

2.1.4. Equilibrium Size Distribution of Clusters

If the system is in the equilibrium state, the size distribution of clusters is determined only by the size dependence of the free energy, according to the statistical consideration [36, 37]. Suppose that the monomers in the system are distributed into the clusters with various size. The total number of the monomers, N, is the sum of the monomers composing the clusters,

$$N = \sum_{g=1}^{N} g n(g) \qquad (2.8)$$

where $n(g)$ is the number of g-sized clusters. Considering all the possible combinations for distributing the monomers into the clusters, the partition function can be written as

$$Z = A \sum_{n(g)} \prod_{g} \frac{1}{n(g)!} \left[N \exp\left(-\frac{\epsilon_g}{kT}\right) \right]^{n(g)} \qquad (2.9)$$

where A is a constant and ϵ_g represents the potential energy of g-sized clusters. Since the most probable distribution is the largest term in the sum of Eq. (2.9), this can be derived by seeking the maximum at $\partial Z / \partial n(g) = 0$, which is

$$n(g) = C^g N \exp\left(-\frac{\epsilon_g}{kT}\right) \qquad (2.10)$$

where C is a constant. Considering the number of monomers ($g = 1$) in Eq. (2.10), one obtains

$$C = \frac{n(1)}{N} \exp\left(\frac{\epsilon_1}{kT}\right)$$

When the system is in the initial stage of nucleation and growth, the number of clusters, $n(g)$, is much smaller than N, and $N \approx n(1)$. Thus the equilibrium distribution, $n_e(g)$, is approximated by

$$n_e(g) = n(1) \exp\left(\frac{g\epsilon_1 - \epsilon_g}{kT}\right) \qquad (2.11)$$

Noting that $W(g)$ is defined as the free energy of g-sized clusters with that of untransformed monomers being set to zero, and if the potential energy of monomers, ϵ_1, can be approximated by that of the untransformed monomers, Eq. (2.11) can be written as

$$n_e(g) = n(1) \exp\left[-\frac{W(g)}{kT}\right] \qquad (2.12)$$

The above result is derived for homogeneous nucleation in which all the monomers can play the role of a nucleation site. The nucleation sites, however, are not necessarily equivalent to

the available monomers. For the case of heterogeneous nucleation where something other than monomers provides the nucleation sites, the number of the sites, n_s, could be smaller or larger than that of monomers, $n(1)$. Even if $n_s > n(1)$, the maximum number of clusters, n_{max}, which is achieved when all the clusters are monomers, does not exceed $n(1)$, and hence, Eq. (2.12) is valid. If $n_s < n(1)$, n_{max} is equal to n_s, and thus, $n(1)$ in Eq. (2.12) should be replaced by n_s.

The g-dependence of Eq. (2.12) is plotted together in Figure 1 using the classical model for $W(g)$ of Eq. (2.5). As shown in the figure, $n_e(g)$ exhibits a minimum at $g = g_*$ and diverges infinitely after that, which is never observed experimentally. The equilibrium size distribution of Eq. (2.12) is physically meaningless in the size range far larger than the critical region. However, this is useful in deriving the actual size distribution of clusters around the critical region, as will be described in Section 2.2.

2.2. Kinetics of Nucleation and Growth

The kinetics study of nucleation and growth was initiated also in the early 20th century. Since then, a number of studies have been devoted to establishing and improving the kinetic theory.

In the first half of the 20th century, the master difference equation describing the kinetic process was reduced to a continuous partial differential equation by approximating the free energy of clusters in the critical region with some analytical functions [20, 24, 38–41], based upon the Szilard model [42]. The reduced equation is today called the "Farkas–Becker–Döring–Zeldovich–Frenkel equation" or simply the "Zeldovich–Frenkel equation." This is a kind of Fokker–Planck equation describing general diffusion processes and was expected to be solved. It had not been easy, however, to obtain a complete form of the analytical solution, and the solution techniques have been investigated for over half a century introducing the further approximations.

The first asymptotic form of the analytical solution was given by Zeldovich [24] with a quadratic approximation of the free energy. Similar approaches are adopted by Wakeshima [43], Lifshitz and Slyozov [44], and Feder et al. [45] and lead to simple approximated solutions. Probstein [46], Kantrowitz [47], and Chakraverty [48] got a start on a technique of expanding the solution into a series and reducing the equation into an eigenvalue problem. Collins [49] and Kashchiev [50–53] developed this methodology and obtained a series solution by approximating the critical region with a square potential well. Until recently, Kashchiev's solution had been the most accurate and the most utilized for analyzing experimental data [54–57]. Kelton et al. [58] compared all of the analytical solutions which had been obtained by 1983 and concluded that Kashchiev's result best coincided with their numerical solution of the master equation. On the other hand, Binder and Stauffer [25] indicated that mathematical and physical errors in Kashchiev's approach happened to be canceled out in the final result and thus lead to the conclusion of Kelton et al. Shizgal and Barrett [59] enumerated every problem in Kashchiev's treatment. Shneidman [60] pointed out that the time lag for nucleation Kashchiev obtained counted only the relaxation time for clusters to diffuse across the critical region. Wu [61–63] reported that the coincidence Kelton et al. found between Kashchiev's solution and the numerical results could be observed only within the investigated range of temperature. For the purpose of obtaining more accurate and problem-free solutions, Shneidman [60, 64–67] and Shi et al. [68–73] proposed new methods using a singular perturbation approach based on boundary-layer theory. Demo and Kožíšek [74] further introduced a method which is based upon the boundary-layer theory combined with the Green function technique. These methods provide the analytical solutions with the fewest approximations and hypotheses at present.

About that time when Farkas–Becker–Döring–Zeldovich–Frenkel equation was established, the "Kramers–Moyal expansion" [75, 76] was introduced to another approximation of the master equation. A set of equations obtained by this expansion without relying on the Szilard model is called the "Kramers–Moyal equation" which is equivalent to the master equation except for continuity. Matkowsky et al. [77] derived an asymptotic solution of this equation considering dynamical aspects of Markov process. Shizgal and Barrett [59] reduced the Kramers–Moyal equation to a kind of Fokker–Planck equation and further to an equivalent Schrödinger equation to which one could apply various methods developed in quantum physics for obtaining an approximated solution. Gitterman et al. [78–82] approximated the free energy in the critical region with the quadratic potential well of a harmonic oscillator and solved the corresponding Schrödinger eigen value problem. Demeio and Shizgal [83] applied the Wentzel–Kramers–Brillouin (WKB) approximation to the Schrödinger equation. Trinkaus and Yoo [84] did not adopt the Schrödinger equation but directly solved the Fokker–Planck equation using the Green function method. However, these solutions provide very complicated expressions and have not been tested sufficiently by experiments.

Recently, Shi proposed a new approximation of the master equation [85] by replacing the Kramers–Moyal expansion with a mesoscopic description of inhomogeneous and nonequilibrium Markov process [86, 87]. The obtained equation is valid for all cluster sizes much greater than 1, while Farkas–Becker–Döring–Zeldovich–Frenkel equation is valid only within the critical region. The equation can be analytically solved by the singular perturbation method based on the boundary-layer theory, yielding an analytical expression for the size distribution of clusters for both in the critical regions and beyond [88, 89]. In the following, the fundamental concept of the kinetic approach, the master equation, the approximated equation, and the solution will be described in detail.

2.2.1. Basic Concept: Diffusion Field in the Size Space

The essence of the kinetic approach to nucleation and growth is to consider the rate of transition between the different cluster sizes and the size distribution of clusters. The transition rate is

Fig. 2. Illustration for showing the transition among the clusters at the sizes of g_1, g_2, and g_3.

defined as $\Gamma(g_i \mid g_j)$ which is the probability of changing the cluster size from g_j to g_i per unit time. The size distribution of clusters is described by the number concentration of clusters as a function of their size, $f(g)$, namely, the number of g-sized clusters per unit volume.

Suppose that there are clusters at three different sizes, $g_1 < g_2 < g_3$, and they grow or shrink to one of these sizes as illustrated in Figure 2. The change of the concentration of g_2-sized clusters, $f(g_2)$, during a unit time should be a sum of four kinds of contributions,

$$\frac{\partial f(g_2)}{\partial t} = \Gamma(g_2 \mid g_1)f(g_1) - \Gamma(g_1 \mid g_2)f(g_2) \\ - \Gamma(g_3 \mid g_2)f(g_2) + \Gamma(g_2 \mid g_3)f(g_3) \quad (2.13)$$

As will be explicitly described later, the size dependence of the transition rate is determined by the driving force from the drift field. In the subcritical region of nucleation, $\Gamma(g_j \mid g_i) < \Gamma(g_i \mid g_j)$ for $g_i < g_j$, while the sign of the first inequality reverses in the supercritical region.

Let us consider here that the clusters at the three sizes are all in the subcritical region; i.e., $g_1 < g_2 < g_3 < g_* - \delta$. If the size distribution of the clusters, $f(g)$, is uniform or monotonically increasing so that $f(g_1) \leq f(g_2) \leq f(g_3)$, the sign of Eq. (2.13) must be negative. Furthermore, if this is realized for any possible combinations of the three cluster sizes in a successive domain of the subcritical region, the clusters cannot grow on average over the domain. On the other hand, if the size distribution is monotonically decreasing, the sign of Eq. (2.13) could be positive even if the net rate of the transition is backward. Moreover, if the size distribution keeps decreasing throughout a size space with satisfying the condition that Eq. (2.13) is positive for any combination of the three cluster sizes, the size distribution evolves toward the right and above, as shown in Figure 3b. Therefore, the net growth of the clusters is possible also in the subcritical region.

Thus the nucleation and growth of clusters can be considered as a process of diffusion accompanied by drift of clusters in the size space. The diffusion is driven by the diffusion field made by the gradient of the size distribution, while the drift is due to the free-energy barrier to nucleation. As illustrated in Figure 3a, the clusters in the subcritical region diffuse to grow resisting the backward force of the drift, then pass diffusing across the critical region without the drift, and after that, stably grow by both the diffusion and the drift in the supercritical region. The diffusion and drift process is theoretically derived in the following, by the kinetic description of the nucleation and growth.

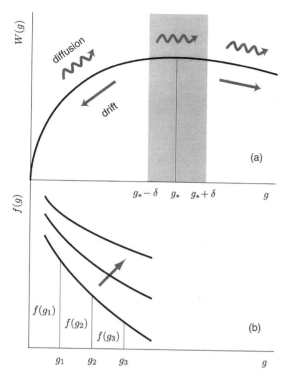

Fig. 3. Size dependence of (a) the free energy of a cluster, $W(g)$, and (b) the size distribution of the clusters, $f(g)$.

2.2.2. Master Equation

Considering the balance between inflow and outflow of monomers and clusters into and from g-sized clusters like Eq. (2.13), the dynamic evolution of the cluster concentration should be described by

$$\frac{\partial f(1,t)}{\partial t} = \sum_{g'=1}^{N(t)} \left[\Gamma\left(1 \mid g',t\right) f\left(g',t\right) - \Gamma\left(g' \mid 1,t\right) f(1,t) \right]$$
$$+ K(1,t) - L(1,t)$$

$$\frac{\partial f(2,t)}{\partial t} = \sum_{g'=1}^{N(t)} \left[\Gamma\left(2 \mid g',t\right) f\left(g',t\right) - \Gamma\left(g' \mid 2,t\right) f(2,t) \right]$$
$$+ K(2,t) - L(2,t)$$

$$\vdots$$

$$\frac{\partial f(g,t)}{\partial t} = \sum_{g'=1}^{N(t)} \left[\Gamma\left(g \mid g',t\right) f\left(g',t\right) - \Gamma\left(g' \mid g,t\right) f(g,t) \right] \quad (2.14)$$
$$+ K(g,t) - L(g,t)$$

$$\vdots$$

$$\frac{\partial f(N,t)}{\partial t} = \sum_{g'=1}^{N(t)} \left[\Gamma\left(N \mid g',t\right) f\left(g',t\right) - \Gamma\left(g' \mid N,t\right) f(N,t) \right]$$
$$+ K(N,t) - L(N,t)$$

where $f(g,t)$ is the number concentration of g-sized clusters at a time of t, $N = N(t)$ is the total number of the monomers distributed to 1- through N-sized clusters at t, $\Gamma(g \mid g',t)$ is the rate of transition from g'-sized clusters to g-sized clusters at t,

$K(g,t)$ is the inflow rate of g-sized clusters introduced into the system from the outside at t, and $L(g,t)$ is the outflow rate of g-sized clusters discharged from the system at t. Equation (2.14) is the master equation that describes all of the processes related to nucleation, growth, and coarsening of clusters with any possible transitions between the cluster sizes. Mathematically, the master equation is equivalent to Pauli's equation in quantum statistics [90] or to the description of Markov process for random walk.

If the clusters are hardly introduced from or discharged to the outside of the system, $K(g,t) \simeq L(g,t) \simeq 0$ for $g > 1$. Furthermore, if the system is completely closed, $K(g,t) = L(g,t) = 0$ for all g. In such cases, the gth equation of Eq. (2.14) can be rewritten as a recurrence formula,

$$\frac{\partial f(g,t)}{\partial t} = J(g-1,t) - J(g,t) \quad (2.15)$$

where

$$J(g,t) \equiv \sum_{i=g+1}^{N(t)} \sum_{j=1}^{g} \left[\Gamma(i \mid j,t) f(j,t) - \Gamma(j \mid i,t) f(i,t) \right]$$
$$(2.16)$$

is the rate for clusters to pass the size of g at t and is called the "flux" of clusters in the size space.

As mentioned in Section 2.2.1, clusters are made to drift by the gradient of the free energy in the size space. Since the forward transition (addition) rate should be different from the backward (dissociation) one, variables in the transition rate are not commutative; i.e., $\Gamma(i \mid j) \neq \Gamma(j \mid i)$. It has been a custom in reaction rate theory to distinguish them as

$$\Gamma(i \mid j) = \begin{cases} \alpha(i \mid j) & \text{for } i < j \\ \beta(i \mid j) & \text{for } i > j \end{cases} \quad (2.17)$$

Since Eq. (2.14) is a set of difference equations, it cannot be analytically solved. Instead, the numerical solutions have been often attempted to obtain using numerical calculation of computers [9, 25, 58, 59, 82, 91–107]. These numerical results are used for evaluating analytical solutions of continuous equations approximated from the master equation and for directly estimating experimental results. It should be noted, however, the numerical results are also limited by the finite space and time for the computation and are never exactly valid. Moreover, it is impossible with the numerical solutions to quantitatively estimate any parameter of the nucleation and growth from experimental data. The proper analytical solutions are indispensable to the comprehensive study of nucleation and growth, and the proper approximation of the master equation is necessary for obtaining an analytically solvable equation.

2.2.3. Kinetic Description of Inhomogeneous Nonequilibrium Process

For proper approximation of the master equation, it is useful to start with general kinetic description of inhomogeneous and nonequilibrium processes. Provided that the variable g can be treated as continuous in the size space, the system is closed, and the transition rate does not explicitly depend on time, Eq. (2.14) is equivalent to the master equation of the Markov process [108–110],

$$
\begin{aligned}
\frac{\partial p(x,t)}{\partial t} &= \int \left[\Gamma(x \mid x') p(x',t) - \Gamma(x' \mid x) p(x,t) \right] dx' \\
&= \int_{-\infty}^{\infty} \left[\Gamma(x \mid x+r) p(x+r,t) \right. \\
&\qquad \left. - \Gamma(x+r \mid x) p(x,t) \right] dr
\end{aligned}
\tag{2.18}
$$

where x denotes a continuous state variable, $p(x,t)$ is the state probability to be found at x and a time of t, $\Gamma(x \mid x') \, dx'$ is the rate of transition from one state x' to the other state x, and $r \equiv x' - x$ is the transition length. When the transition rates of a process depend on its state variable, x, whether linearly or nonlinearly, the Markov process must be *inhomogeneous*, or else it is *homogeneous*. The process of nucleation and growth is essentially inhomogeneous because the transition rate depends on the cluster size except in its critical region.

For reducing Eq. (2.18) into solvable forms, it is necessary to approximate the right hand side, and hence to obtain $\Gamma(x \mid x')$ and $\Gamma(x' \mid x)$ for all x'. In the actual process, however, one can only know a small transition,

$$
\Gamma(x + \Delta x \mid x) \equiv \Gamma(x, \Delta x)
$$

rather than $\Gamma(x \mid x')$, where Δx is a small but finite transition length. A mesoscopic treatment for the transition [86, 87] allows one to obtain $\Gamma(x \mid x')$ from the known $\Gamma(x, \Delta x)$. Let us first divide a transition length of $r = x' - x$ into m small regions with a same length of $\Delta x = r/m \ll x$. For the ith region with $i = 0, 1, 2, \ldots, m-1$, the rate of transition from $x + i\Delta x$ to $x + (i+1)\Delta x$ is

$$
\Gamma(x + (i+1)\Delta x \mid x + i\Delta x) = \Gamma(x + i\Delta x, \Delta x)
$$

Provided here that the transition from x to x' can be treated as a chain of small transitions passing each successive small region and that failure to pass any of the small regions immediately causes a failure of the whole transition from x to x', the rate of the whole transition is expressed by the product of rates of each small transition,

$$
\begin{aligned}
\Gamma(x' \mid x) = \big[\Gamma(x, \Delta x) &\times \Gamma(x + \Delta x, \Delta x) \times \Gamma(x + 2\Delta x, \Delta x) \\
&\cdots \times \Gamma(x + (m-1)\Delta x, \Delta x) \big]^{1/m}
\end{aligned}
\tag{2.19}
$$

Taking a limit of $\Delta x / x \to 0$ in Eq. (2.19) leads to

$$
\Gamma(x' \mid x) = \exp\left[\frac{1}{x' - x} \int_{x}^{x'} \ln \Gamma(x, dx) \, dx \right]
\tag{2.20a}
$$

$$
\Gamma(x \mid x') = \exp\left[\frac{1}{x - x'} \int_{x'}^{x} \ln \Gamma(x, -dx) \, dx \right]
\tag{2.20b}
$$

Substituting Eqs. (2.20) into Eq. (2.18), one obtains an equation

$$
\begin{aligned}
\frac{\partial p(x,t)}{\partial t} \\
= \int_{-\infty}^{\infty} \Bigg\{ &\exp\left[\frac{1}{r} \int_{x}^{x+r} \ln \Gamma(x, dx) \, dx \right] p(x+r,t) \\
&- \exp\left[-\frac{1}{r} \int_{x+r}^{x} \ln \Gamma(x, -dx) \, dx \right] p(x,t) \Bigg\} dr
\end{aligned}
\tag{2.21}
$$

for the observable transition rate, $\Gamma(x, dx)$, where all possible transitions of the elementary processes with any transition length, r, are allowable. The time evolution of $p(x,t)$ at x results from all the possible contributions of $-\infty < r < \infty$. It is possible with Eq. (2.20) to describe the kinetics of an inhomogeneous nonequilibrium process whose transition lengths are comparable to or even larger than the inhomogeneity of the system.

Considering further the limit of $x' \to x + r$ with small transition lengths as $|r/x| \ll 1$, Eq. (2.20) is reduced to

$$
\begin{aligned}
\Gamma(x' \mid x) &= \exp\left[\frac{1}{r} \ln \Gamma\left(\frac{x + x + r}{2}, dx \right) \Big|_{dx = x + r - x} \times r \right] \\
&= \Gamma\left(x + \frac{r}{2}, r \right)
\end{aligned}
\tag{2.22a}
$$

$$
\begin{aligned}
\Gamma(x \mid x') \\
= \exp&\left[-\frac{1}{r} \ln \Gamma\left(\frac{x + r + x}{2}, -dx \right) \Big|_{dx = x - (x+r)} \times (-r) \right] \\
= \Gamma&\left(x + \frac{r}{2}, -r \right)
\end{aligned}
\tag{2.22b}
$$

Then using Eq. (2.22), Eq. (2.21) can be expanded as

$$
\frac{\partial p(x,t)}{\partial t} = \frac{\partial}{\partial x} a_0(x) p(x,t) + \sum_{i=1}^{\infty} \frac{1}{2^i i!} \frac{\partial^i}{\partial x^i} a_i(x) \frac{\partial p(x,t)}{\partial x}
\tag{2.23}
$$

with

$$
\begin{aligned}
a_i(x) &= \int_{-\epsilon}^{\epsilon} (-r)^{i+1} \Gamma(x, r) \, dr \\
\Gamma(x, r) &\equiv \Gamma(x, dx)|_{dx = r}
\end{aligned}
$$

where ϵ is a small quantity satisfying $-\epsilon < r < \epsilon$. If it is a sufficient approximation to take the first two terms of the expansion in Eq. (2.23), one obtains

$$
\frac{\partial p(x,t)}{\partial t} = -\frac{\partial}{\partial x} A(x) p(x,t) + \frac{\partial}{\partial x} D(x) \frac{\partial p(x,t)}{\partial x}
\tag{2.24}
$$

with

$$
\begin{aligned}
A(x) &= \int_{-\epsilon}^{\epsilon} r \Gamma(x, r) \, dr \\
D(x) &= \frac{1}{2} \int_{-\epsilon}^{\epsilon} r^2 \Gamma(x, r) \, dr
\end{aligned}
$$

2.2.4. Kinetic Description of Nucleation and Growth

Nucleation and growth are typical examples of the inhomogeneous nonequilibrium processes described by Eqs. (2.23) and (2.24). Neglecting degree of freedom associated with the shape of clusters, the state variable the clusters can take is just the number of monomers composing the clusters. In a closed system, the concentration of the clusters is the probability of finding the cluster at its size in unit volume. Thus we obtain a kinetic description of nucleation and growth in the inhomogeneous nonequilibrium system by replacing the state variable, x, and the state probability, $p(x,t)$, with the cluster size, g, and the cluster concentration, $f(g, t)$, respectively in Eqs. (2.23) and (2.24). Additionally, in the case of cluster growth, it is supposed that the multistep transition less occurs than the one step transition, $r = 1$; that is, Szilard model [42] is effective, which is expressed using the notation of Eq. (2.17) by

$$\Gamma(x' \mid x) = \beta(x' \mid x)\delta_{x,x'-1} + \alpha(x' \mid x)\delta_{x,x'+1} \quad (2.25a)$$

$$\Gamma(x \mid x') = \beta(x \mid x')\delta_{x'+1,x} + \alpha(x \mid x')\delta_{x'-1,x} \quad (2.25b)$$

where $\delta_{i,j}$ stands for Kronecker delta. For sufficiently large $g \gg 1$, i.e., $r/g \ll 1$, Eqs. (2.22) and (2.25) give

$$\beta(g + 1 \mid g) = \beta(g + 1/2, 1) \quad (2.26a)$$

$$\alpha(g \mid g + 1) = \alpha(g + 1/2, -1) \quad (2.26b)$$

Thus by replacing the variables and the function, and using Eqs. (2.25) and (2.26), the full expansion form of Eq. (2.23) is transformed into

$$\frac{\partial f(g, t)}{\partial t} = \frac{\partial}{\partial g} a_1(g) f(g, t)$$
$$+ \sum_{i=1}^{\infty} \frac{(-1)^{i+1}}{2^i \, i!} \frac{\partial^i}{\partial g^i} a_{i+1}(g) \frac{\partial f(g, t)}{\partial g} \quad (2.27)$$

with

$$a_i(g) = \Gamma(g, 1) + (-1)^i \Gamma(g, -1)$$
$$= \beta(g, 1) + (-1)^i \alpha(g, -1)$$

and the second-order expansion form of Eq. (2.24) becomes

$$\frac{\partial f(g, t)}{\partial t} = -\frac{\partial}{\partial g} A(g) f(g, t) + \frac{\partial}{\partial g} D(g) \frac{\partial f(g, t)}{\partial g} \quad (2.28a)$$

with

$$A(g) = \beta(g, 1) - \alpha(g, -1) \quad (2.28b)$$

$$D(g) = \frac{1}{2}[\beta(g, 1) + \alpha(g, -1)] \quad (2.28c)$$

where $\beta(g, 1) \equiv \beta(g, dg)|_{dg=1}$ and $\alpha(g, -1) \equiv \alpha(g, dg)|_{dg=-1}$. Here the flux of clusters defined by Eq. (2.16) becomes

$$J(g, t) = A(g) f(g, t) - D(g) \frac{\partial f(g, t)}{\partial g} \quad (2.29)$$

The kinetic description of nucleation and growth by Eqs. (2.27) or (2.28) is valid for all $g \gg 1$. When the critical size, g_*, is sufficiently larger than 1, these equations are valid in the critical region and beyond. It should be emphasized that Eqs. (2.27) and (2.28) differ from the Kramers–Moyal expansion [75, 76] and the Fokker–Planck equation [112–114], respectively, which cannot correctly describe inhomogeneous processes.

It is also noted that Eqs. (2.27) and (2.28) can describe the relatively early stage of the phase transformation by nucleation and growth of clusters. In the early stage, the clusters are small and their concentration is low so that they scarcely collide with one another, while in the later stage, the grown clusters at high concentration could impinge upon the adjacent clusters and coalesce. Since the coalescence of grown clusters corresponds to multistep transitions of the cluster size (i.e., $r > 1$), it contradicts the assumption of a one-step restriction of the Szilard

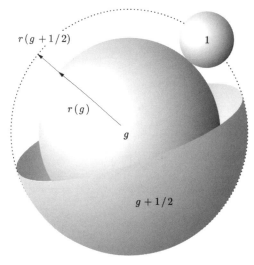

Fig. 4.   Illustration of a molecule condensing into a g-sized spherical cluster.

model. Thus Eqs. (2.27) and (2.28) would not be applicable for describing the nucleation and growth in the later stages.

The cluster size of $g + 1/2$ appearing in Eq. (2.26) might seem a little strange. It would not be easy to understand the physical meanings of the half size of a monomer, although g is considered to be continuous. Let us consider that a spherical g-sized cluster is formed in the atmosphere, and a vapor molecule condenses into the cluster. According to the kinetic theory of gas, the collision rate of the molecules is proportional to the surface area of the cluster, and Dufour and Defay [111] pointed out "when molecules condense on a curved surface, the area available for condensation is not the area of the surface itself but the area of an exterior of one molecule semidiameter," as illustrated in Figure 4. This idea is exactly reflected in the physical meanings of Eq. (2.26). In spite of the crude picture with the spherical cluster, such an approximation captures the essential physics of an inhomogeneous process for $g \gg 1$, as manifested by the fact that the kinetic description by Eq. (2.28) is expressed in terms of the known $\beta(g, dg)$ and $\alpha(g, dg)$.

2.2.5. Nucleation Equation as a Limit of the New Kinetic Description

The kinetic description of nucleation and growth given by Eq. (2.28) could be further approximated within the critical region. The restriction of the one-step transition by the Szilard model [Eq. (2.25)] reduces the flux of clusters [Eq. (2.16)] to

$$J(g, t) = \beta(g+1 \mid g) f(g, t) - \alpha(g \mid g+1) f(g+1, t) \quad (2.30)$$

If the system is in the equilibrium state, $J(g, t) = 0$ for any g, and one can expect that the *detailed balance* expressed by

$$\beta(g + 1 \mid g) f_e(g, t) = \alpha(g \mid g + 1) f_e(g + 1, t) \quad (2.31)$$

is kept, where $f_e(g, t)$ is the equilibrium number concentration of g-sized clusters at t, i.e., the equilibrium size distribution of

Eq. (2.11) per unit volume, which is given by

$$f_{e}(g, t) = f(1, t) \exp\left[-\frac{W(g)}{kT}\right] \quad (2.32)$$

Considering $g \gg 1$, Eqs. (2.12), (2.26), and (2.31) give a ratio

$$
\begin{aligned}
\frac{\alpha(g + 1/2, -1)}{\beta(g + 1/2, 1)} &= \frac{\alpha(g \mid g + 1)}{\beta(g + 1 \mid g)} \\
&= \frac{f_{e}(g, t)}{f_{e}(g + 1, t)} = \frac{n_{e}(g)}{n_{e}(g + 1)} \\
&= \exp\left[\frac{W(g + 1) - W(g)}{kT}\right] \\
&\approx \exp\left[\frac{\partial}{\partial g}\frac{W(g)}{kT}\right]_{g = g + 1/2} \quad (2.33)
\end{aligned}
$$

which is equivalent to

$$\alpha(g, -1) = \beta(g, 1) \exp\left[\frac{\partial}{\partial g}\frac{W(g)}{kT}\right] \quad (2.34)$$

Then, from Eqs. (2.28b), (2.28c), and (2.34), one obtains

$$
\begin{aligned}
\frac{A(g)}{D(g)} &= 2\frac{\beta(g, 1) - \alpha(g, -1)}{\beta(g, 1) + \alpha(g, -1)} \\
&= 2\frac{1 - \exp\left[\frac{\partial}{\partial g}\frac{W(g)}{kT}\right]}{1 + \exp\left[\frac{\partial}{\partial g}\frac{W(g)}{kT}\right]} \\
&\approx -\frac{\partial}{\partial g}\frac{W(g)}{kT} \quad (2.35)
\end{aligned}
$$

since $|\partial W(g)/\partial g|/kT \ll 1$ in the critical region. Substituting Eq. (2.35) into Eq. (2.28) and considering $D(g) = [\beta(g, 1) + \alpha(g, -1)]/2 \approx \beta(g, 1)$ in the critical region lead to the nucleation limit of Eq. (2.28),

$$\frac{\partial f(g, t)}{\partial t} = \frac{\partial}{\partial g}\left[\beta(g, 1)\left\{\frac{\partial f(g, t)}{\partial g} + f(g, t)\frac{\partial}{\partial g}\frac{W(g)}{kT}\right\}\right] \quad (2.36)$$

where the flux of clusters [Eq. (2.29)] becomes

$$J(g, t) = -\beta(g, 1)\left\{\frac{\partial f(g, t)}{\partial g} + f(g, t)\frac{\partial}{\partial g}\frac{W(g)}{kT}\right\} \quad (2.37)$$

Equation (2.36) is quite similar to Farkas–Becker–Döring–Zeldovich–Frenkel equation,

$$\frac{\partial f(g, t)}{\partial t} = \frac{\partial}{\partial g}\left[\beta(g + 1 \mid 1)\left\{\frac{\partial f(g, t)}{\partial g} + f(g, t)\frac{\partial}{\partial g}\frac{W(g)}{kT}\right\}\right] \quad (2.38)$$

except for a subtle difference between the terms of $\beta(g, 1)$ and $\beta(g + 1 \mid g)$. Equation (2.38) was originally obtained by a simple quadratic approximation of Eq. (2.15) with Eqs. (2.30) and (2.31) and had been most frequently used for investigating nucleation and growth. Fortunately, in most previous studies, the incorrectness of involving $\beta(g + 1 \mid g)$ used to be compensated by mistaking $\beta(g, 1)$ for $\beta(g + 1 \mid g)$ without awareness of their subtle but significant difference. This fact also con-

firms that the Farkas–Becker–Döring–Zeldovich–Frenkel equation could be valid only within the critical region.

2.2.6. Size Distribution of Clusters in the Critical Region

The proper kinetic description of nucleation and growth expressed by Eq. (2.28) can be analytically solved for $f(g, t)$ by employing the Laplace transformation and a singular perturbation approach with the asymptotic matching method [115]. For simplicity in the present description of the solving processes, we adopt the following situation as an example:

(1) The capillary model given by Eq. (2.5) is a good approximation to the free energy of clusters.
(2) The addition rate of monomers to g-sized clusters is proportional to the surface density of monomers and the surface area of the clusters [42, 116] and is given by

$$\beta(g, t) \equiv \beta(g, 1) = \omega\kappa_{vd}\{f_{m}(t)\}^{1-1/d}g^{1-v/d} \quad (2.39)$$

where ω stands for the jumping frequency of a monomer into clusters, $\kappa_{vd} \equiv \pi^{v/d}(6V_{1})^{1-v/d}$ is the geometrical factor with V_{1} being the volume per monomer in the clusters, $f_{m}(t)$ is the number concentration of monomers available to the growth of the clusters, and v denotes the index determining the mode of the monomer addition. When $f(1, t)$ is the monomer concentration and is spatially uniform, $f_{m}(t) = f(1, t)$. If the monomer addition is controlled by the chemical reaction or the formation of the chemical bonds at the interface, $v = 1$, and if controlled by the diffusion of monomers, $v = 2$. Here we take $f_{m}(t) = f(1, t)$, $v = 1$, and $d = 3$ in Eqs. (2.5) and (2.39) for a while.

(3) The initial and the boundary conditions are

$$
\begin{aligned}
f(g, 0) &= 0 && \text{for } g > 1 && (2.40a) \\
f(1, t) &= f_{e}(1, t) && \text{for all } t && (2.40b) \\
\frac{f(g, t)}{f_{e}(g, t)} &\to 0 && \text{for } g_{*}/g \to 0 && (2.40c)
\end{aligned}
$$

In the following, the solutions are obtained in the critical region first. The solutions beyond the critical region will be shown later in Sections 2.2.7 and 2.2.8.

Let us first introduce the scaled concentration of clusters,

$$y(g, t) \equiv \frac{f(g, t)}{f_{er}(g, t)} \quad (2.41)$$

where

$$
\begin{aligned}
f_{er}(g, t) &\equiv f_{e}\left[g, t - t_{g}(1, g)\right] \\
&= f_{1}(g, t) \exp\left[-\frac{W(g)}{kT}\right] \quad (2.42)
\end{aligned}
$$

is the equilibrium number concentration of g-sized clusters at a retroactive time when the cluster whose size is g at t was a monomer. Here,

$$f_1(g,t) \equiv \begin{cases} f(1,0) & \text{for } t \leq t_g(1,g) \\ f[1, t - t_g(1,g)] & \text{for } t > t_g(1,g) \end{cases} \quad (2.43)$$

is the retroactive concentration of monomers or nucleation sites (hence, $f_{er}(x,0) = f_e(x,0)$), where

$$t_g(1,g) \equiv \int_1^g \frac{dg'}{\dot{g}'} \quad (2.44)$$

is the time for clusters to grow from monomer to g. The deterministic growth rate of the cluster, $\dot{g} \equiv dg/dt$, will be defined later by Eq. (2.92). In addition to the transformation by Eq. (2.41), the transformations by scaled variables and parameters,

$$\tau \equiv \frac{\delta^2}{2\beta(g_*,t)} \quad x \equiv \frac{g}{g_*} \quad \text{and} \quad \epsilon \equiv \frac{\delta}{g_*} \quad (2.45)$$

lead the nucleation limit of the kinetic description given by Eq. (2.36) into

$$\frac{\tau}{x^{2/3}}\left[\frac{\partial y(x,t)}{\partial t} + \frac{y(x,t)}{f_{er}(x,t)}\frac{\partial f_{er}(x,t)}{\partial t}\right]$$
$$= \frac{\epsilon^2}{2}\frac{\partial^2 y(x,t)}{\partial x^2} + \left[\frac{\epsilon^2}{3x} + 3\left(1 - x^{-1/3}\right)\right]\frac{\partial y(x,t)}{\partial x} \quad (2.46)$$

under a condition of $\partial \tau/\partial t \ll \tau/t$.

Under another condition of

$$\left|\frac{\partial y(x,t)}{\partial t}\right| \gg \left|\frac{y(x,t)}{f_{er}(x,t)}\frac{\partial f_{er}(x,t)}{\partial t}\right| \quad (2.47)$$

Eq. (2.46) is reduced to

$$\frac{\tau}{x^{2/3}}\frac{\partial y(x,t)}{\partial t} = \frac{\epsilon^2}{2}\frac{\partial^2 y(x,t)}{\partial x^2} + \left[\frac{\epsilon^2}{3x} + 3\left(1 - x^{-1/3}\right)\right]\frac{\partial y(x,t)}{\partial x} \quad (2.48)$$

Then, the Laplace transform of the scaled concentration,

$$y(x,s) = \int_0^\infty y(x,t)\exp(-st)\,dt \quad (2.49)$$

leads Eq. (2.48) into an ordinary differential equation about x,

$$\epsilon^2\frac{\partial^2 y(x,s)}{\partial x^2} + \left[\frac{2}{3x}\epsilon^2 + 6\left(1 - x^{-1/3}\right)\right]\frac{\partial y(x,s)}{\partial x}$$
$$- \frac{\tau}{x^{2/3}}\left[y(x,s)s - y(x,0)\right] = 0 \quad (2.50)$$

and the boundary conditions described by Eqs. (2.40b) and (2.40c) into

$$y(1/g_*,s) = 1/s \quad y(x,s) = 0 \quad \text{for } x \gg 1 \quad (2.51)$$

Since the principal term of Eq. (2.48), $\partial y(x,s)/\partial x$, changes its sign at $x = 1$ in the interval of $[1/g_*, \infty]$, there exists a "boundary layer" around $x = 1$. Following the boundary layer theory, the size space can be divided into three successive intervals: the left outer region, the inner region, and the right outer region, which correspond to the subcritical, critical, and supercritical regions, respectively. Then Eq. (2.48) can be entirely solved by obtaining the solutions in these regions and by asymptotically matching them.

Suppose that the left and right outer solutions, $y_{out}(x,s)$, can be asymptotically expanded to a perturbation series with respect to the power of ϵ,

$$y_{out}(x,s) \rightarrow \sum_{n=0}^\infty \epsilon^n y_n(x,s) \quad \text{as} \quad \epsilon \rightarrow 0+ \quad (2.52)$$

The boundary conditions for each coefficient are

$$y_0^\ell(1/g_*,s) = 1 \quad y_n^\ell(1/g_*,s) = 0 \quad \text{for } n \geq 1 \quad (2.53a)$$
$$y_n^r(\infty,s) = 0 \quad \text{for } n \geq 0 \quad (2.53b)$$

By substituting Eq. (2.52) into Eq. (2.48) and equating the coefficients for the same power of ϵ, one obtains a set of equations:

$$\frac{\partial y_0(x,s)}{\partial x} - \frac{s\tau}{3}\frac{x^{-2/3}}{1-x^{-1/3}}y_0(x,s) = 0$$
$$\frac{\partial y_1(x,s)}{\partial x} - \frac{s\tau}{3}\frac{x^{-2/3}}{1-x^{-1/3}}y_1(x,s) = 0$$
$$\qquad\qquad\qquad\qquad\qquad\qquad\qquad (2.54)$$
$$\frac{\partial^2 y_{n-2}(x,s)}{\partial x^2} + \frac{2}{3x}\frac{\partial y_{n-2}(x,s)}{\partial x} + 6\left(1 - x^{-1/3}\right)\frac{\partial y_n(x,s)}{\partial x}$$
$$- 2s\tau x^{-2/3}y_n(x,s) = 0 \quad \text{for } n \geq 2$$

It is found in the detailed analysis by Hoyt and Sunder [117] that the perturbation terms of $n \geq 2$ are negligible for the solutions of Eq. (2.54) under the boundary conditions of Eq. (2.53). It is apparent that the first order solutions vanish ($y_1(x,s) = 0$) for both the left and the right outer regions. Therefore, only the leading order solution ($n = 0$) remains in Eq. (2.52), and we obtain the left and right outer solutions:

$$y_{out}^\ell(x,s) = \frac{1}{s}\left(\frac{1 - x^{1/3}}{1 - g_*^{-1/3}}\right)^{s\tau}$$
$$\times \exp\left[s\tau\left(x^{1/3} - g_*^{-1/3}\right)\right] \quad (2.55a)$$
$$y_{out}^r(x,s) = 0 \quad (2.55b)$$

For convenience to obtain the inner solution, we introduce a variable transformation of $y_{in}(x,s) \rightarrow Y_{in}(X,s)$ with an inner variable, $X \equiv (x-1)e^\xi$ ($\xi > 0$), and transform Eq. (2.48) into

$$\frac{\epsilon^{2(1-\zeta)}(1+\epsilon^\zeta X)}{2}\frac{\partial^2 Y_{in}(X,s)}{\partial X^2}$$
$$+ 3\epsilon^{-\zeta}(1+\epsilon^\zeta X)\left[1 - \left(1+\epsilon^\zeta X\right)^{-1/3}\right]\frac{\partial Y_{in}(X,s)}{\partial X}$$
$$+ \frac{\epsilon^{2-\zeta}}{3}\frac{\partial Y_{in}(X,s)}{\partial X}$$
$$- \tau\left(1+\epsilon^\zeta X\right)^{-1/3}sY_{in}(X,s) = 0 \quad (2.56)$$

and the boundary condition into

$$Y_{in}(\infty,s) = 0 \quad (2.57)$$

Here if $\epsilon \rightarrow 0$ with a variable X being fixed, $\lambda \rightarrow 1$. As well as the outer solutions, the asymptotic expansion of $Y_{in}(X,s)$ gives a perturbation series:

$$Y_{in}(X,s) \rightarrow \sum_{n=0}^\infty \epsilon^n Y_n(X,s) \quad \text{as } \epsilon \rightarrow 0+ \quad (2.58)$$

By substituting Eq. (2.58) into Eq. (2.56) and equating the coefficients for the same power of ϵ, one obtains a set of equations:

$$\frac{\partial^2 Y_0(X, s)}{\partial X^2} - 2s\tau(1 + \epsilon X)^{-1/3} Y_0(X, s) = 0$$

$$\frac{\partial^2 Y_n(X, s)}{\partial X^2} + X\frac{\partial^2 Y_{n-1}(X, s)}{\partial X^2}$$
$$- 2s\tau(1 + \epsilon X)^{-1/3} Y_n(X, s) = 0 \quad \text{for } n \geq 1 \quad (2.59)$$

Here again, only the leading order solution can remain. Thus the inner solution satisfying the boundary condition [Eq. (2.57)] is obtained as

$$Y_{\text{in}}(X, s) = \frac{1}{2}\big[A(s)\, i^{s\tau} \text{erfc}(X) + B(s)\, i^{s\tau} \text{erfc}(-X)\big] \quad (2.60)$$

where

$$i^{s\tau} \text{erfc}(X) = \frac{2}{\sqrt{\pi}} \int_X^\infty \frac{(t - X)^{s\tau}}{(s\tau)!} e^{-t^2} dt$$

is the repeated error function. The coefficients $A(s)$ and $B(s)$ in Eq. (2.60) are determined by asymptotically matching Eq. (2.60) to the outer solutions of Eq. (2.55) as

$$\lim_{x \to 1-} y_{\text{out}}^\ell(x, s) = \lim_{X \to -\infty} Y_{\text{in}}(X, s)$$
$$\lim_{X \to +\infty} Y_{\text{in}}(X, s) = \lim_{x \to 1+} y_{\text{out}}^r(x, s) \quad (2.61)$$

which give

$$A(s) = \frac{1}{s}\left[\frac{\epsilon/3}{1 - g_*^{-1/3}} \exp\big(1 - g_*^{-1/3}\big)\right]^{s\tau} \Gamma(1 + s\tau)$$

$$B(s) = 0 \quad (2.62)$$

where $\Gamma(z)$ stands for the gamma function. By substituting Eq. (2.62) into Eq. (2.60) and restoring the variables, one obtains

$$y_{\text{in}}(x, s) = \frac{1}{2s}\left[\frac{\epsilon}{3(1 - g_*^{-1/3})} \exp\big(1 - g_*^{-1/3}\big)\right]^{s\tau}$$
$$\times \Gamma(1 + s\tau)\, i^{s\tau} \text{erfc}\left(\frac{x - 1}{\epsilon}\right) \quad (2.63)$$

Finally, considering the initial condition of Eq. (2.40a), the inverse Laplace transform of Eqs. (2.55) and (2.63) gives the outer and inner solutions in real g space,

$$y_{\text{out}}^\ell(g, t) = \Theta\big[t - \mu(g)\tau\big] - \Theta\big[-\mu(g)\tau\big] \quad (2.64a)$$

$$y_{\text{in}}(g, t) = \frac{1}{2}\big[\eta(g, t, \tau) - \eta(g, 0, \tau)\big] \quad (2.64b)$$

$$y_{\text{out}}^r(g, t) = 0 \quad (2.64c)$$

with

$$\mu(g) = -\frac{t_g(1, g)}{\tau} = g_*^{-1/3} - x^{1/3} - \ln\left[\frac{1 - x^{1/3}}{1 - g_*^{-1/3}}\right] \quad (2.64d)$$

$$\eta(g, t, \tau) = \text{erfc}\left[\frac{g - g_*}{\delta} + \exp\left(\frac{\lambda\tau - t}{\tau}\right)\right] \quad (2.64e)$$

$$\lambda = \lim_{g \to g_* - \delta} \mu(g) = g_*^{-1/3} - 1 + \ln\left[\frac{3}{\epsilon}\big(1 - g_*^{-1/3}\big)\right] \quad (2.64f)$$

where $\Theta(z)$ is the unit step function and

$$\text{erfc}(z) = 1 - \frac{2}{\sqrt{\pi}} \int_0^z e^{-t^2} dt$$

is the complementary error function.

Three kinds of time constants are considered here. The first is $\mu(g)\tau$ which corresponds to the time lag for the concentration of subcritical clusters to approach its steady state. The second is $\lambda\tau$ which is the relaxation time for the inner solution, and thus the relaxation time for the cluster concentration in the critical region to establish its steady state. The last is τ which stands for the duration for clusters to pass diffusing across the critical region. It is noted in Eqs. (2.64) that the size distribution of clusters $f(g, t)$ is dynamically scaled with $f_{\text{er}}(g, t)$ since $y(g, t) = f(g, t)/f_{\text{er}}(g, t)$.

Equations (2.64) are valid also for arbitrary values of d and ν, for which Eqs. (2.64d) and (2.64f) are generalized as

$$\mu(g) \equiv \frac{1}{1 + \nu}\Big[g^{(1+\nu)/d}\,_2F_1\big(1, 1 + \nu; 2 + \nu; x^{1/d}\big)$$
$$- g_*^{-(1+\nu)/d}\,_2F_1\big(1, 1 + \nu; 2 + \nu; g_*^{-1/d}\big)\Big]$$
$$= x^{(1+\nu)/d}\Phi\big(x^{1/d}, 1, 1 + \nu\big)$$
$$- g_*^{-(1+\nu)/d}\Phi\big(g_*^{-1/d}, 1, 1 + \nu\big) \quad (2.64g)$$

$$\lambda = \ln\big(1 - g_*^{-1/d}\big) + \frac{1}{2}\left[\frac{d(d - 1)}{2}\frac{W_*}{kT}\right] - \int_{g_*^{-1/d}}^1 \frac{z^\nu - 1}{z - 1} dz \quad (2.64h)$$

where $_2F_1(a, b; c; z)$ denotes the hypergeometric function, and $\Phi(z, s, a)$ stands for the Lerch transcendent. Furthermore, even if the free energy of clusters is altered from its capillary approximation, the results obtained are the same as Eqs. (2.64), and only $\mu(g)$ and λ change. Thus the dynamic scaling is essential to the dynamic evolution of the cluster-size distribution, independent of the actual form of the free energy. With Eq. (2.64h) we obtain the universal form of the size distribution in the critical region,

$$f(g, t) = \frac{1}{2} f_{\text{er}}(g, t)\big[\eta(g, t, \tau) - \eta(g, 0, \tau)\big] \quad (2.65a)$$

with

$$\eta(g, t, \tau) = \text{erfc}\left[\frac{g - g_*}{\delta} + \exp\left(\frac{\lambda\tau - t}{\tau}\right)\right] \quad (2.65b)$$

For future convenience, the size distribution without regarding the depletion of monomers and nucleation sites is defined here as

$$f_0(g, t) = \frac{1}{2} f_{\text{er}}(g, 0)\big[\eta(g, t, \tau_0) - \eta(g, 0, \tau_0)\big] \quad (2.66)$$

where $f_{\text{er}}(g, 0) = f_e(g, 0)$ and

$$\tau_0 = \frac{\delta^2}{2\beta(g_*, 0)} \quad (2.67)$$

2.2.7. Size Distribution of Clusters in the Supercritical Region

Equation (2.64c), $y^r_{out}(g, t) = 0$, gives the right outer solution of Eq. (2.28), which does not mean that $f(g, t)$ is exactly zero but negligibly smaller than $f_{er}(g, t)$ in the supercritical region. This result does not provide, however, the exact form of $f(g, t)$ and was obtained starting from the nucleation limit of the kinetic description by Eq. (2.46). For obtaining a solution of Eq. (2.28) in the supercritical region, it does not help to introduce $y(g, t) = f(g, t)/f_{er}(g, t)$ unlike the treatment in the critical region, since $f(g, t)/f_{er}(g, t) \ll 1$. Thus only with the scaling of $x \equiv g/g_*$ and $\epsilon \equiv \delta/g_*$ can Eq. (2.28) be transformed into

$$\tau \frac{\partial f(x, t)}{\partial t} = \frac{\epsilon^2 x^{2/3}}{2} \frac{\partial^2 f(x, t)}{\partial x^2}$$
$$+ \left[\frac{x^{-1/3}}{3} \epsilon^2 - 3x^{1/3}(x^{1/3} - 1) \right] \frac{\partial f(x, t)}{\partial x}$$
$$- x^{-1/3}(2 - x^{-1/3}) f(x, t) \quad (2.68)$$

Since Eq. (2.28) is valid in the early stage before the coalescence of clusters becomes significant, where the growth of clusters continues only by monomer addition, $f(g, t)$ in the supercritical region rapidly decreases with g and the following relation can be suggested:

$$\left| \frac{\partial^2 f(x, t)}{\partial x^2} \right| \ll \left| \frac{\partial f(x, t)}{\partial x} \right| \ll f(x, t)$$

Therefore, the term including the second derivative of $f(x, t)$ can be discarded in Eq. (2.68). It is possible, moreover, to neglect $\epsilon^2 x^{-1/3}/3$ in comparison to $3x^{1/3}(x^{1/3} - 1)$ in the coefficient of $\partial f(x, t)/\partial x$, since $x \gg 1$ in the supercritical region and $\epsilon^2 = 3kT \ll 1$. Consequently, Eq. (2.68) is reduced to a simple equation:

$$\tau \frac{\partial f(x, t)}{\partial t} = -3x^{1/3}(x^{1/3} - 1) \frac{\partial f(x, t)}{\partial x}$$
$$- x^{-1/3}(2 - x^{-1/3}) f(x, t) \quad (2.69)$$

The solution of Eq. (2.69) must asymptotically match the solution in the critical region [Eq. (2.65)] at the left boundary of the supercritical region, i.e., the right boundary of the critical region, $g = g_* + \delta$. Thus the left boundary condition for Eq. (2.69) should be

$$f(g_* + \delta, t) = \frac{f_{er}(g_* + \delta, t)}{2} \left\{ \text{erfc}\left[1 + \exp\left(\frac{\lambda \tau - t}{\tau} \right) \right] \right.$$
$$\left. - \text{erfc}[1 + e^\lambda] \right\} \quad (2.70a)$$

and the right boundary condition and the initial condition are

$$f(\infty, t) = 0 \quad (2.70b)$$
$$f(g, 0) = 0 \quad (2.70c)$$

As mentioned earlier, the equilibrium size distribution of clusters loses its physical meanings in the supercritical region,

and one cannot scale $f(g, t)$ with $f_{er}(g, t)$ for the same g. Instead, the cluster concentration should be scaled with the equilibrium concentration at the right boundary of the critical region as

$$y(x, t) = \frac{f(x, t)}{f_{e\delta}(x, t)}$$
$$\text{with } f_{e\delta}(x, t) = f_1(x, t) \exp\left[-\frac{W(g_* + \delta)}{kT} \right] \quad (2.71)$$

where $f_{e\delta}(1 + \delta/g_*, t) = f_{er}(1 + \delta/g_*, t)$. With this scaling, Eq. (2.69) can be solved by an ordinary method using the Laplace transformation, under the initial and boundary conditions of Eq. (2.70). After the inverse transformation of the variables, the solution for $f(g, t)$ is thus obtained as

$$f(g, t) = f_{e\delta}(g, t)\phi(g)\left[\psi(g, t, \tau) - \psi(g, 0, \tau) \right] \quad (2.72a)$$

with

$$\phi(g) = \frac{\rho(g)}{2} \left(\frac{g_* + \delta}{g} \right)^{1/3} \quad (2.72b)$$

$$\psi(g, t, \tau)$$
$$= \text{erfc}\left[1 + \exp\left[\frac{\lambda \tau - t + t_g(g_* + \delta, g, \tau)}{\tau} \right] \right] \quad (2.72c)$$

$$t_g(g_* + \delta, g, \tau)$$
$$= \frac{\tau}{g_*} \left[g^{1/3} - (g_* + \delta)^{1/3} - g_* \ln \rho(g) \right] \quad (2.72d)$$

$$\rho(g) = \frac{(g_* + \delta)^{1/3} - g_*^{1/3}}{g^{1/3} - g_*^{1/3}} \quad (2.72e)$$

where λ is given by Eq. (2.64h). It is obvious from Eqs. (2.65) and (2.72c) that the concentration of supercritical clusters approaches the steady state with a similar time dependence to that in the critical region except for the shift of time, $t_g(g_* + \delta, g, \tau)$. The assumptions of $d = 3$ and $\nu = 1$ reflect only on $\phi(g)$ and $t_g(g_* + \delta, g, \tau)$ through $\rho(g)$. They can be readily generalized for any values of d and ν as

$$\phi(g) = \frac{\rho(g)}{2} \left(\frac{g_* + \delta}{g} \right)^{(d-\nu-1)/d} \quad (2.72f)$$

$$t_g(g_* + \delta, g, \tau)$$
$$= \tau \left[-\ln \rho(g) + \int_{(1+\delta/g_*)^{1/d}}^{(g/g_*)^{1/d}} \frac{z^\nu - 1}{z - 1} dz \right] \quad (2.72g)$$

$$\rho(g) = \frac{(g_* + \delta)^{1/d} - g_*^{1/d}}{g^{1/d} - g_*^{1/d}} \quad (2.72h)$$

For the same purpose of defining Eq. (2.66), the size distribution without regarding the depletion of monomers and nucleation sites is defined as

$$f_0(g, t) = f_e(g, 0)\phi(g)\left[\psi_0(g, t, \tau_0) - \psi_0(g, 0, \tau_0) \right] \quad (2.73)$$

Figure 5 shows the dynamic evolution of the cluster-size distribution around the critical region, at $t/\tau = 0.5, 1, 2, 3, 4, 5, 6$, and ∞. The plots are calculated from Eqs. (2.65) and (2.72) employing Eqs. (2.5) and (2.39) with $d = 3$, $\nu = 1$, $\Delta\mu = 0.1$ eV,

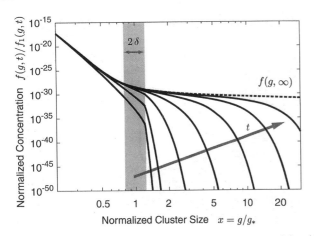

Fig. 5.   Dynamic evolution of the size distribution of clusters around the critical region, which is based upon Eqs. (2.64), (2.65), and (2.72). Note that the size distribution asymptotically approaches the steady-state distribution, $f(g, \infty)$, with the time and the cluster size.

$\sigma = 1.93 \times 10^{18}$ eV m^{-2}, and $a = 3.58 \times 10^{-19}$ m^2. The concentrations of clusters are normalized by the monomer concentration, $f_1(g, t)$, and are plotted as a function of the normalized cluster size, $x = g/g_*$. It is seen that the size distribution approaches the steady-state distribution, $f(g, \infty)$, with time. For each distribution, the steady state is being established from a smaller size, and the cluster concentration exhibits a rapid drop beyond the region where the steady state has been already established.

2.2.8. Size Distribution of Clusters Far Beyond the Critical Region

While Eqs. (2.72) are valid in the entire supercritical region (i.e., $g \geq g_* + \delta$), they can be approximated to more simple and useful forms in the region far beyond the critical region. It is suggested from Eqs. (2.72f) and (2.72h) that the steady-state term of Eq. (2.72) should obey a power-law dependence on the size as

$$\phi(g, t) \propto g^{\nu/d - 1} \qquad (2.74)$$

for the sufficiently large cluster size of $g \gg g_* + \delta$. Thus one may expect

$$\left| \frac{\partial \phi(g, t)}{\partial g} \middle/ \phi(g, t) \right| \to 0 \quad \text{as } g \to \infty \qquad (2.75)$$

As will be shown by Eqs. (2.92) and (2.93) in Section 2.3.2, the rate of cluster growth, $\dot{g}$, can be nearly equal to $\beta(g, t)$ for $g \gg g_*$, and hence, the time for clusters to grow, Eq. (2.72d), can be approximated as

$$t_g(g_* + \delta, g, \tau) = \int_{g_* + \delta}^{g} \frac{dg'}{\dot{g}'} \approx \int_{g_* + \delta}^{g} \frac{dg'}{\beta[g', t(\tau)]} \qquad (2.76)$$

where $t(\tau)$ is the time when $\beta(g_*, t) = \delta^2/\tau$, and hence, $t(\tau_0) = 0$. Since $t_g(g_* + \delta, g, \tau)$ of Eq. (2.76) monotonously

increases with g, one may also expect from Eq. (2.72c),

$$\left| \frac{\partial \psi(g, t)}{\partial g} \middle/ \psi(g, t) \right| \to 0 \quad \text{as } g \to \infty \qquad (2.77)$$

Considering the estimates of Eqs. (2.75) and (2.77), one can conclude

$$|f(g, t)| \gg \left| \frac{\partial f(g, t)}{\partial g} \right|$$

generally for sufficiently large clusters in the supercritical region. Therefore, noting that $A(g) \approx 2D(g) \approx \beta(g)$ for $g \gg g_*$ and from Eq. (2.29), one obtains a relation

$$J(g, t) = \beta(g, t) f(g, t) \qquad (2.78)$$

Provided here that the steady-state distribution has been established up to a certain size, $g_s \gg g_* + \delta$, the following relation can be derived from Eqs. (2.28) and (2.29),

$$\frac{\partial f(g, t)}{\partial t} = -\frac{\partial J(g, t)}{\partial g} = 0$$

which indicates that the flux of clusters is constant throughout the size region, $g < g_s$. Then the constant flux can be represented by the steady-state flux at the critical size:

$$J(g, t) = J_{ss}(g_*, t) \quad \text{for } g < g_s \qquad (2.79)$$

Equations (2.78) and (2.79) lead to the steady-state distribution of cluster size for $g \gg g_* + \delta$ as

$$\varphi(g, t) \equiv \lim_{t \to \infty} f(g, t) = \frac{J_{ss}(g_*, t)}{\beta(g, t)} \qquad (2.80)$$

Comparing Eq. (2.80) to the infinite time limit of Eq. (2.72), and using Eq. (2.76), the overall cluster concentration for $g \gg g_* + \delta$ is finally obtained as

$$f(g, t) = \varphi(g, t) \frac{\psi(g, t, \tau) - \psi(g, 0, \tau)}{\psi(g, \infty, \tau) - \psi(g, 0, \tau)} \qquad (2.81a)$$

with

$$\varphi(g, t) = \frac{J_{ss}(g_*, t)}{\beta(g, t)} \qquad (2.81b)$$

$$\psi(g, t, \tau)$$
$$= \text{erfc}\left[1 + \exp\left[\frac{\lambda \tau - t + t_g(g_* + \delta, g, \tau)}{\tau} \right] \right] \qquad (2.81c)$$

$$t_g(g_* + \delta, g, \tau) = \int_{g_* + \delta}^{g} \frac{dg'}{\beta[g', t(\tau)]} \qquad (2.81d)$$

It is noted that Eq. (2.81) is applicable to any forms of $W(g)$ and $\beta(g, t)$. Of course, if employing Eq. (2.39) for $\beta(g, t)$, one finds that the steady-state distribution, $\varphi(g, t)$, obeys a power-law size dependence,

$$\varphi(g, t) \propto g^{\nu/d - 1}$$

as was predicted in Eq. (2.74).

Similarly to Eqs. (2.66) and (2.73), the size distribution without regarding the depletion of monomers and nucleation sites is defined here as

$$f_0(g, t) = \varphi(g, 0) \frac{\psi(g, t, \tau_0) - \psi(g, 0, \tau_0)}{\psi(g, \infty, \tau_0) - \psi(g, 0, \tau_0)} \quad (2.82)$$

2.2.9. Beyond the Early Stages of Nucleation and Growth

The theoretical results presented in Sections 2.2.4–2.2.8 are valid under the condition of the one-step transition (Szilard model) expressed by Eq. (2.25). The multistep transition, in other words, coalescence or coagulation of clusters, is not negligible when the concentration of clusters is high and (1) if the clusters are spatially mobile so that the probability of their collision is large, or (2) when the clusters have grown and the average distance between the surfaces of adjacent clusters (i.e., growth front) becomes small so that they readily impinge upon the neighboring clusters by their own further growth. Even though the clusters are not mobile, the coalescence is inevitable in the late stages of phase transformations where the accumulated number concentration of clusters becomes high and the clusters having early nucleated have grown large. Thus the theoretical results are applicable only to the early precoalescence stages of phase transformations driven by nucleation and growth.

There is another reason that the applicability is restricted to the early stage. The nucleation sites and/or the monomers should be consumed by nucleating and growing clusters and should be depleted with time unless they are continuously supplied faster than or at least as fast as their consumption. It is found in Eqs. (2.12), (2.64) or (2.65), (2.72) with (2.71), and (2.81) with (2.78) and (2.79) that the size distribution of clusters is proportional to the number concentration of nucleation sites or that of the monomers, $f_1(g, t)$. This result originates in deriving Eq. (2.46) under the condition of $\partial \tau/\partial t \ll \tau/t$ and reducing it to Eq. (2.48) under the condition of Eq. (2.47). These conditions can be interpreted as

$$\frac{f_s(1, t)}{f_s(1, 0)} > 0 \quad (2.83)$$

$$\left| \frac{f_1(g, t)}{f(g, t)} \frac{\partial f(g, t)}{\partial t} - \frac{\partial f_1(g, t)}{\partial t} \right| \gg \left| \frac{\partial f_1(g, t)}{\partial t} \right| \quad (2.84)$$

respectively. The first condition [Eq. (2.83)] is obviously satisfied before complete depletion of monomers, i.e., $f_s(1, t) = 0$. The second condition [Eq. (2.84)] requires that the depletion of nucleation sites or monomers is sufficiently slow and the nucleation is still in the transient period where the steady-state distribution has not been established and hence $\partial f(g, t)/\partial t > 0$ in the critical region. After the transient period, the effects of the nucleation site and/or monomer depletion have to be considered. Letting the time scale for the depletion be

$$t_{\mathrm{d}} \equiv -\left[\frac{1}{f_1(g, t)} \frac{\partial f_1(g, t)}{\partial t} \right]^{-1} \quad (2.85)$$

and if t_{d} varies slowly with time, Eq. (2.48) can be solved under the initial and boundary conditions of Eq. (2.40) to obtain the

size distribution in the critical region as

$$f(g, t) = \frac{f_{\mathrm{er}}(g, t)}{\sqrt{\pi}} \exp\left(\frac{\lambda \tau}{t_{\mathrm{d}}} \right) \int_{X + \exp(\lambda - t/\tau)}^{X + \exp(\lambda)} \frac{\exp(-z^2)}{(z - X)^{\tau/t_{\mathrm{d}}}} \, dz \quad (2.86)$$

which is readily reduced to Eq. (2.65) for $t_{\mathrm{d}} \gg \tau$. For rare cases with $t_{\mathrm{d}} \lesssim \tau$, Eq. (2.65) should be replaced by Eq. (2.86) which influences the size distribution in the supercritical region through the boundary condition at $g = g_* + \delta$ [corresponding to Eq. (2.70a)].

Consequently, the solutions of the kinetic equations shown in this chapter are applicable to the early stage of nucleation and growth before the complete depletion of nucleation sites or monomers and the occurrence of coalescence become considerable.

Except the earliest stage where $f(1, t)/f(1, 0) \sim 1$, it is necessary to take the depletion of nucleation sites or monomers into account. The dependence of $f_1(g, t)$ on g and t is given by

$$f_1(g, t) = \left\{ 1 - \chi\left[t - t_{\mathrm{g}}(g) \right] \right\} f(1, 0) \quad (2.87)$$

where $\chi(g, t)$ is the transformed or clustered volume fraction when g-sized clusters at t were monomers and is generally defined as

$$\chi(t) = \frac{1}{V_1} \int_2^\infty g f(g, t) \, dg \quad (2.88)$$

If Eqs. (2.87) and (2.88) are explicitly included into the kinetic equations, Eqs. (2.46) and (2.68) become differential-integral equations of $f(g, t)$ which, unfortunately, have not been completely solved. It is possible, however, to draw a qualitative picture of the cluster-size distributions in the late stages, if the coalescence is still negligible. Since $\chi(t)$ increases with time approaching 1 and $t_{\mathrm{g}}(g)$ increase with g, $f_1(g, t)$ of Eq. (2.87) must monotonously increase with g. On the other hand, the scaled size distribution [Eqs. (2.64) and (2.72)] always decreases with g. Therefore, the resultant size distribution could have a plateau or a peak so as to be a lognormal-like distribution in the supercritical region. For the purpose of analyzing the experimental data with the theoretical results for the size distribution, $f_1(g, t)$ could be treated as a predetermined quantity, because $\chi(t)$ can be measured independently.

2.3. Observables in Nucleation and Growth

Although the theoretical expressions for the size distribution of clusters are obtained, they cannot always be applied to experimental observations directly. Experimentally observable quantities are shown in the following with their theoretical basis.

2.3.1. Size Distribution of Clusters

The kinetic theory of nucleation and growth provides the size distribution of clusters both in the critical and supercritical regions. Since the size of observable large clusters can be precisely measured, the result far beyond the critical region [Eq. (2.81)] is applicable directly to the data measured. It is usually impossible, however, to compare the result for the critical region with the experimental results. Even with the up-to-date techniques for experimental investigations, the clusters around the critical size are generally too small to observe quantitatively. Furthermore, since clusters which have not exceeded the critical region are not stable, they could be surely detected only by *in situ* observations, which makes the experimental approach more difficult. Instead of directly using Eq. (2.65), the nucleation rate which is one of the typical observables can be derived from it as described in Section 2.3.3.

There is a point which one should pay attention to in the expression of the size distribution. The kinetic theory originally provides the size distribution expressed in the number of monomers composing the clusters, g. When the observed cluster sizes are described by g, one can directly apply $f(g, t)$ to them. On the other hand, the cluster size is usually measured in a unit of length and is described by a characteristic length, r, such as the radius of spherical clusters and the side length of cubic clusters. The thermodynamic theory of nucleation and growth had been originally described with r. When the size distribution measured in the length unit needs to be treated as is, it is necessary to transform $f(g, t)$ into $f(r, t)$. However, even if $r = r(g)$ is a function of g and the inverse function $g = g(r)$ exists, $f(g(r), t)$ is not simply equivalent to $f(r, t)$. Instead, the transformation is derived from the conservation of clusters. First, the accumulated number concentration of clusters should be the same whichever expression is adopted,

$$N(g_s, t) = \int_{g_s}^{\infty} f(g, t) \, dg = \int_{r_s}^{\infty} f(r, t) \, dr \qquad (2.89)$$

where $r_s = r(g_s)$. Equation (2.89) is satisfied if

$$\int_{g-\Delta g/2}^{g+\Delta g/2} f(g, t) \, dg = \int_{r-\Delta r/2}^{r+\Delta r/2} f(r, t) \, dr \qquad (2.90)$$

is valid for arbitrary g, Δg, $r = r(g)$, and $\Delta r = r(g + \Delta g/2) - r(g - \Delta g/2)$. If $f(g, t)$ and $f(r, t)$ are continuous throughout the size range, the limit of Δg, $\Delta r \to 0$ leads Eq. (2.90) to $f(g, t)\Delta g = f(r, t)\Delta r$. Thus the transformation is

$$f(r, t) = f(g(r), t) \frac{dg(r)}{dr} \qquad (2.91)$$

It should be noted in Eq. (2.91) that dimensionality of $f(r, t)$ differs from that of $f(g, t)$. Since g is a dimensionless number and r has a dimension of length $[L]$, the dimension of $f(r, t)$ is smaller than $f(g, t)$ by 1 with respect to $[L]$. If the size distribution is observed in a space with a Euclidean dimension of D_e, the dimension is $[L^{-D_e}]$ for $f(g, t)$, and $[L^{-D_e-1}]$ for $f(r, t)$, respectively.

2.3.2. Growth Rate

The growth rate or the growth velocity defined as the change of the size of a cluster per unit time is one of the important parameters that control the nucleation and growth. The net growth rate should be the difference between the forward transition (addition) rate and the backward transition (dissolution) rate, which

is expressed using Eq. (2.34) as

$$v_g(g, t) \equiv \dot{g} = \frac{dg}{dt} = \beta(g, t) - \alpha(g, t)$$
$$= \beta(g, t) \left\{ 1 - \exp\left[\frac{\partial}{\partial g} \frac{W(g)}{kT} \right] \right\} \qquad (2.92)$$

Equation (2.92) is called the *deterministic* growth rate which is equivalent to the result of the reaction rate theory [118]. The growth rate, $v_g(g, t)$, changes its sign from negative to positive at the critical size. Since $\alpha(g, t)/\beta(g, t) \to 0$ as $g \to \infty$,

$$v_g(g, t) \approx \beta(g, t) \qquad (2.93)$$

for $g \gg g_* + \delta$.

The addition rate of monomers into clusters is generally given by Eq. (2.39) in which the jumping frequency of monomers into clusters, ω, usually exhibits a temperature dependence of thermal activation processes; i.e.,

$$\omega = \omega_0 \exp\left(-\frac{E_g}{kT} \right) \qquad (2.94)$$

where ω_0 is the jumping frequency at $E_g = 0$ or $T \to \infty$, and E_g is the activation energy for the growth of clusters. Thus the addition rate of monomers and hence the growth rate are also thermally activated as

$$v_g(g, t) = \beta(g, t) \propto \exp\left(-\frac{E_g}{kT} \right) \qquad (2.95)$$

2.3.3. Nucleation Rate

The nucleation rate is also one of the essential parameters for nucleation and growth, which is the number of clusters nucleating per unit volume and unit time. If the nucleation is defined as the passage of clusters across the nucleus size, g_n, the nucleation rate is just the flux of clusters at $g = g_n$. Thus the definition of the nucleation rate depends on that of the nucleus.

2.3.3.1. Nucleation Rate of Nuclei in the Critical Region

Clusters having passed through the critical region are often regarded as the stable clusters. The flux of clusters in the critical region, given by Eq. (2.96), can be rewritten as

$$J(g, t) = -\beta(g, t) \left[f_{er}(g, t) \frac{\partial}{\partial g} \frac{f(g, t)}{f_{er}(g, t)} \right.$$
$$\left. + \frac{f(g, t)}{f_1(g, t)} \frac{\partial f_1(g, t)}{\partial g} \right] \qquad (2.96)$$

Since in the critical region,

$$\left| \frac{\partial}{\partial g} \ln \frac{f(g, t)}{f_{er}(g, t)} \right| \gg \left| \frac{\partial \ln f_1(g, t)}{\partial g} \right|$$

for the early stage before the complete depletion of monomers or nucleation sites, the second term of Eq. (2.96) is negligible, and hence

$$J(g, t) = -\beta(g, t) f_{er}(g, t) \frac{\partial}{\partial g} \frac{f(g, t)}{f_{er}(g, t)} \qquad (2.97)$$

Substituting Eq. (2.65) for $J(g, t)$ in Eq. (2.97), the flux of clusters in the critical region is given by

$$J(g, t) = \frac{1}{\sqrt{\pi}\delta}\beta(g, t)f_{er}(g, t)[\zeta(g, t, \tau) - \zeta(g, 0, \tau)] \tag{2.98a}$$

with

$$\zeta(g, t, \tau) = \exp\left[-\left\{\frac{g - g_*}{\delta} + \exp\left(\frac{\lambda\tau - t}{\tau}\right)\right\}^2\right] \tag{2.98b}$$

which is often called the *transient* nucleation rate. Equation (2.98) increases from 0 with time and asymptotically approaches

$$J_{ss}(g, t) = \frac{1}{\sqrt{\pi}\delta}\beta(g, t)f_{er}(g, t)$$
$$\times \left\{\exp\left[-\left(\frac{g - g_*}{\delta}\right)^2\right] - \zeta(g, 0, \tau)\right\} \tag{2.99}$$

which is called the *steady-state* nucleation rate. The flux without regarding the depletion of monomers and nucleation sites, as well as Eqs. (2.66), (2.73), and (2.82), can be also defined as

$$J_0(g, t) = \frac{1}{\sqrt{\pi}\delta}\beta(g, 0)f_{er}(g, 0)$$
$$\times [\zeta(g, t, \tau_0) - \zeta(g, 0, \tau_0)] \tag{2.100}$$
$$J_{s0}(g) = J_{ss}(g, 0) \tag{2.101}$$

While, strictly in terms of thermodynamics, the smallest stable cluster should be the cluster at the right boundary of the critical region, the nucleation rate has been often calculated at the critical size conventionally. With Eqs. (2.98) and (2.99), the nucleation rate at the critical size and its steady-state value are given by

$$J(g_*, t) = \frac{1}{\sqrt{\pi}\delta}\beta(g_*, t)f_{er}(g_*, t)$$
$$\times \left\{\exp\left[-\exp\left(2\frac{\lambda\tau - t}{\tau}\right)\right] - \exp[-e^{2\lambda}]\right\} \tag{2.102}$$
$$J_{ss}(g_*, t) = \frac{1}{\sqrt{\pi}\delta}\beta(g_*, t)f_{er}(g_*, t)\{1 - \exp[-e^{2\lambda}]\} \tag{2.103}$$

When g_* is sufficiently larger than δ, $\exp[-\exp(2\lambda)] \ll 1$ and hence

$$J_{ss}(g_*, t) \approx \frac{1}{\sqrt{\pi}\delta}\beta(g_*, t)f_{er}(g_*, t) \tag{2.104}$$

The conventionally used steady-state nucleation rate corresponds to

$$J_{ss}(g_*) \equiv J_{ss}(g_*, 0) = \frac{1}{\sqrt{\pi}\delta}\beta(g_*, 0)f_e(g_*, 0) \tag{2.105}$$

2.3.3.2. Nucleation Rate of Nuclei in the Supercritical Region

The nucleus may be defined as the stable cluster in the supercritical region, or as the smallest detectable cluster which is usually far beyond the critical region. As shown by Eqs. (2.92) and (2.93), $\beta(g, t)/\alpha(g, t) \ll 1$ for $g \geq g_* + \delta$, and hence

$$A(g, t) = \beta(g, t) - \alpha(g, t) \approx \beta(g, t)$$
$$D(g, t) = \frac{1}{2}[\beta(g, t) + \alpha(g, t)] \approx \frac{1}{2}\beta(g, t)$$

Thus from Eq. (2.29), the flux of clusters in the supercritical region is

$$J(g, t) = \beta(g, t)\left[f(g, t) - \frac{1}{2}\frac{\partial f(g, t)}{\partial g}\right] \tag{2.106}$$

where $f(g, t)$ is given by Eq. (2.72) or (2.81), which, as well as the flux of clusters in the critical region, approaches the steady-state value with time. If $\partial f_1(g, t)/\partial g \approx 0$, the steady-state flux for $g \gtrsim g_* + \delta$ is derived from Eq. (2.72) as

$$J_{ss}(g, t) = \beta(g, t)f_{e\delta}(g, t)$$
$$\times \left[\{erfc(1) - erfc(1 + e^\xi)\}\left\{\phi(g) - \frac{1}{2}\frac{\partial\phi(g)}{\partial g}\right\}\right.$$
$$\left. - \frac{\phi(g)}{\sqrt{\pi}\tau\dot{g}}\exp[\xi - (1 + e^\xi)^2]\right]$$

where $\xi = \lambda + t_g(g_* + \delta, g, \tau)/\tau$. For $g \gg g_* + \delta$,

$$J_{ss}(g, t) = \beta(g, t)\left[\varphi(g, t) - \frac{1}{2}\frac{\partial\varphi(g, t)}{\partial g}\right]$$
$$= J_{ss}(g_*, t)\left[1 + \frac{1}{2\beta(g, t)}\frac{\partial\beta(g, t)}{\partial g}\right]$$
$$\approx J_{ss}(g_*, t) \tag{2.107}$$

is obtained from Eq. (2.81), since $\partial\beta(g, t)/\partial g/\beta(g, t) = (1 - \nu/d)/g \ll 1$ for $g \gg 1$. Also in the supercritical region, the flux without regarding the depletion of monomers and nucleation sites can be defined as

$$J_0(g, t) = \beta(g, 0)\left[f_0(g, t) - \frac{1}{2}\frac{\partial f_0(g, t)}{\partial g}\right] \tag{2.108}$$

2.3.3.3. Activation Energy of the Nucleation Rate

Equations (2.98)–(2.107), and Eqs. (2.4), (2.42) and (2.95), suggest that the flux of clusters or the nucleation rate is thermally activated as

$$J(g, t) \propto J_{ss}(g_*, t) \propto T^{-1/2}\exp\left(-\frac{E_g + W_*}{kT}\right) \tag{2.109}$$

at any time at any size in the critical region or far beyond the critical region. This indicates that both the nucleation and growth contribute to the nucleation rate.

2.3.4. Transformed Fraction

The transformed or clustered fraction is often measured in the experiments for studying the formation of thin films. This is also called the crystallized fraction for the thin films formed by crystallization of noncrystalline materials, or it corresponds to the ratio of surface coverage for the thin films formed by deposition over substrates. The transformed fraction is defined as the ratio of the volume transformed at a time to the entire volume which finally can be transformed. If clusters whose sizes are larger than g are regarded as transformed clusters, their fraction is

$$\chi(g, t) = \frac{1}{V_1} \int_g^\infty g' f(g', t) \, dg' \qquad (2.110)$$

where $\chi(2, t)$ corresponds to Eq. (2.88). The fraction $\chi(g, t)$ should increase with time from 0 and converge at 1 for $t \to \infty$. Unfortunately, Eq. (2.110) with $f(g, t)$ of Eqs. (2.64) or (2.65) and (2.72) or (2.81) cannot be integrated into a simple formula expressed by elementary functions. Moreover, when $\chi(g, t)$ is not negligible, the concentration of clusters is high and the early nucleated clusters grow large, and hence the coalescence of clusters could considerably occur, where the theoretical results for $f(g, t)$ are not applicable.

A simple model was proposed by Johnson and Mehl [119] and Avrami [120–122], for describing the time evolution of the transformed fraction taking the effects of the coalescence and the depletion of monomers or nucleation sites into account. Their result, called the Johnson–Mehl–Avrami formalism, can be generalized as

$$\chi_{\text{JMA}}(g_s, t) = 1 - \exp\left[-V_1 \int_{g_s}^\infty g' f_0(g', t) \, dg'\right]$$
$$= 1 - \exp\left[-V_1 \int_0^t g_m(g_s, t - t') J_0(g_s, t') \, dt'\right] \qquad (2.111)$$

where $f_0(g, t)$ is given by Eq. (2.66), (2.73), or (2.82), where g_s is the size of the smallest detectable clusters, $g_m(g, t)$ is the size for g-sized clusters to reach during a period of t, and $J_0(g, t)$ is given by Eq. (2.100) or (2.108). If $\partial\beta(g, t)/\partial t \approx 0$ and $g \gg g_* + \delta$, Eq. (2.111) can be reduced to

$$\chi_{\text{JMA}}(g_s, t) = 1 - \exp\left[-V_1 \int_0^t J_0(g_s, t')\right.$$
$$\left. \times \left\{\frac{\nu}{d}\beta_0(t - t') + g_s^{\nu/d}\right\}^{d/\nu} dt'\right] \qquad (2.112)$$

where $J_0(g_s, t)$ is given by Eq. (2.108) and

$$\beta_0 \equiv \beta(1, 0) = \omega \kappa_{\nu d}\{f_m(0)\}^{1-1/d} \qquad (2.113)$$

Equation (2.112) is often approximated by approximating $J_0(g, t)$ as

$$J_0(g_s, t) = \begin{cases} 0 & \text{for } t < \tau_{\text{tr}} \\ J_{\text{ss}}(g_*, 0) & \text{for } t \geq \tau_{\text{tr}} \end{cases} \qquad (2.114)$$

where τ_{tr} is the transient period and usually is taken to be proportional to τ. With this approximation, Eq. (2.112) for $t \geq \tau_{\text{tr}}$ can be reduced to

$$\chi_{\text{JMA}}(g_s, t) = 1 - \exp\left[-\frac{V_1 J_{\text{ss}}(g_*, 0)}{\beta_0(1 + \nu/d)}\right.$$
$$\left. \times \left[\left\{g_s^{\nu/d} + \frac{\beta_0 \nu}{d}(t - \tau_{\text{tr}})\right\}^{1+d/\nu} - g_s^{1+\nu/d}\right]\right] \qquad (2.115)$$

Except the initial stage of the nucleation and growth, it is a rather good approximation to take g to be 0 in Eq. (2.115). For $\nu = 1$ and $g_s \to 0$, one obtains from Eq. (2.115) a popular form of the Johnson–Mehl–Avrami formalism:

$$\chi_{\text{JMA}}(t) = 1 - \exp\left[-\frac{V_1}{108} J_{\text{ss}}(g_*, 0)\beta_0^d(t - \tau_{\text{tr}})^{1+d}\right] \qquad (2.116)$$

2.3.5. Accumulated Number Concentration of Clusters

In the experimental studies of nucleation and growth, the nucleation rate has been measured often by the time evolution of the number concentration of all the stable clusters having various sizes. The total number concentration of clusters at a time, t, is equal to the number concentration of clusters having been accumulated until t, which is calculated by

$$N(g_s, t) = \int_{g_s}^\infty f(g, t) \, dg = \int_0^t J(g_s, t') \, dt' \qquad (2.117)$$

It is generally difficult to obtain a simple expression of Eq. (2.117) for arbitrary g_s and t. A few examples for presenting the basic behaviors are shown in the following.

2.3.5.1. Precoalescence Stage

In the precoalescence stage, the theoretical results for $f(g, t)$ or $J(g_*, t)$ can be used for calculating the accumulated number concentration of clusters larger than the critical size ($g_s = g_*$), which is given by

$$N(g_*, t) = \frac{J_{\text{ss}}(g_*, 0)}{\{f_m(0)\}^{1-1/d} f_1(g_*, 0)} \int_0^t \{f_m(t')\}^{1-1/d} f_1(g_*, t')$$
$$\times \left\{\exp\left[-\exp\left(2\frac{\lambda\tau - t'}{\tau}\right)\right] - \exp(-e^{2\lambda})\right\} dt' \qquad (2.118)$$

Equation (2.118) gradually increases with time from 0, exhibits a steep climb at $t \sim \lambda\tau - \tau \ln[-\ln\{1/2 + \exp(-e^{2\lambda})\}]$, and afterward asymptotically approaches a constant,

$$N(g_*, t) \to \frac{J_{\text{ss}}(g_*, 0)}{\{f_m(0)\}^{1-1/d} f_1(g_*, 0)}$$
$$\times \int_0^t \{f_m(t')\}^{1-1/d} f_1(g_*, t') \, dt' \qquad (2.119)$$

for $t \gg \lambda\tau$ when the monomers and the nucleation sites have been almost depleted.

If the depletion of monomers and/or nucleation sites is negligible (i.e., $J(g_*, t) \approx J_0(g_*, t)$) in the early stage, Eq. (2.118) can be reduced to

$$N(g_*, t) = \frac{J_{\text{ss}}(g_*, 0)\tau}{2} \left\{E_1\left[\exp\left(2\frac{\lambda\tau - t}{\tau}\right)\right] - E_1(e^{2\lambda})\right.$$
$$\left. - \exp(-e^{2\lambda})\frac{2t}{\tau}\right\} \qquad (2.120)$$

where $E_1(z) \equiv \int_1^\infty e^{zt}/t \, dt$ is the exponential integral. Furthermore, if the depletion of monomers and/or nucleation sites has been negligible until $t > \lambda\tau$, $N(g_*, t)$ of Eq. (2.120) asymptotically approaches a linear dependence on time with a slope of

$$
\begin{aligned}
\frac{dN(g_*, t)}{dt} &= J_{ss}(g_*, 0)\big[1 - \exp\big(-e^{2\lambda}\big)\big] \\
&\approx J_{ss}(g_*, 0)
\end{aligned} \tag{2.121}
$$

Otherwise, $N(g_*, t)$ approaches the constant [Eq. (2.119)] without showing the asymptotic behavior to the linear dependence on time.

Generally, for $g_s \neq g_*$, $N(g_s, t)$ does not possess any simple expression like Eq. (2.120). It is possible, however, to draw a picture of the basic feature for sufficiently large g_s. If the steady-state flux has been established at the size of g_s ($\gg g_* + \delta$) since a time of $\tau_{tr}(g_s)$, and if the depletion of monomers and/or nucleation sites is still negligible at t ($> \tau_{tr}(g_s)$), the accumulated number concentration of clusters larger than g_s can be approximated to

$$
\begin{aligned}
N(g_s, t) &\approx \int_{\tau_{tr}(g_s)}^{t} J_{ss}(g_s, 0)\, dt' + \int_0^{\tau_{tr}(g_s)} J_0\big(g_s, t'\big)\, dt' \\
&= J_{ss}(g_*, 0)\big[t - \tau_{tr}(g_s)\big] + C
\end{aligned} \tag{2.122}
$$

where $\partial C/\partial t = 0$. Thus $N(g_s \gg g_* + \delta, t)$ also approaches the linear dependence on time with the same slope as $N(g_*, t)$.

2.3.5.2. *Up to the Late Stage*

If the Johnson–Mehl–Avrami formalism is employed to consider the depletion of monomers and nucleation sites and/or the effect of coalescence, one can obtain an approximated expression for the accumulated number concentration of the cluster,

$$
N_{JMA}(g, t) = \int_0^t \big[1 - \chi_{JMA}\big(g, t'\big)\big] J_0\big(g, t'\big)\, dt' \tag{2.123}
$$

which is approximately valid up to the late stage where the effect of the coalescence becomes significant. When Eqs. (2.114) and (2.116) are good approximations to $J_0(g, t')$ and $\chi_{JMA}(g, t')$, the right hand side of Eq. (2.123) can be analytically integrated into

$$
\begin{aligned}
N_{JMA}(t) &= \frac{1}{1+d}\left[\frac{108}{V_1}\left(\frac{J_{ss}}{\beta_0}\right)^d\right]^{1/(1+d)} \\
&\quad \times \Bigg[\Gamma\left(\frac{1}{1+d}\right) \\
&\qquad - \Gamma\left(\frac{1}{1+d}, \frac{V_1}{108} J_{ss}(g_*, 0)\beta_0^d(t - \tau_{tr})^{1+d}\right)\Bigg]
\end{aligned} \tag{2.124}
$$

for $t \geq \tau_{tr}$, where $\Gamma(a, z) = \int_z^\infty t^{a-1}e^{-t}\, dt$ is the incomplete gamma function. $N_{JMA}(t)$ of Eq. (2.124) increases from $t = \tau_{tr}$ and converges to a constant.

2.3.6. *Final Grain Size*

The grain size of granular structures like polycrystalline thin films has been one of the parameters that attract the most attention in their industrial applications, because the grain size often dominates their performance as the materials. The grain size in question is generally the final grain size that is achieved after complete transformation and never changes any more. The transformation is completed when the nucleation and growth stops. In the case of thin films deposited as polycrystalline from vapor or liquid sources, the nucleation of new clusters is terminated only by stopping the supply of species for the clusters, while the clusters cease growing when they impinge upon the adjacent clusters. In the case of thin films formed by the transformation of the starting thin films, the nucleation and growth come to the end when the nucleation sites and the monomers surrounding the clusters are completely depleted, respectively. If the clusters are formed by precipitation of the monomers dispersed in the matrix and are separately located in the completely transformed thin films, the nucleation stops when either of the monomers surrounding the clusters or the nucleation sites is completely depleted, while the growth can last as long as the monomers surrounding the clusters exist. If the thin film is formed by filling up its untransformed region with the clusters, the nucleation ceases at the complete depletion of the nucleation sites, and the growth of each cluster is terminated by their impingement upon the adjacent clusters, but the growth of the last cluster continues until the untransformed region disappears.

For any case, the mean of the final grain size, $\overline{g_f}$, is just the number of monomers available to compose the clusters per the number of the clusters, which is given by dividing the initial concentration of monomers by the finally accumulated number concentration of clusters. Although Eq. (2.118) predicts that the accumulated number concentration becomes constant as $t \to \infty$, this equation cannot be applied to the present situation where the coalescence of clusters would play a crucial role in controlling the final grain size. Instead, the Johnson–Mehl–Avrami formalism shown in Section 2.3.4 can be utilized for estimating the final concentration of clusters [123]. If one adopts the same situation and approximation as those for Eq. (2.124), the mean of the final grain size can be estimated from Eq. (2.124) as

$$
\begin{aligned}
\overline{g_f} &= \frac{f_m(0)}{\lim_{t \to \infty} N_{JMA}(t)} \\
&= f_m(0)\left(\frac{V_1}{108}\right)^{1/(1+d)}\left[\Gamma\left(\frac{2+d}{1+d}\right)\right]^{-1} \\
&\quad \times \left[\frac{\beta_0}{J_{ss}(g_*, 0)}\right]^{d/(1+d)}
\end{aligned} \tag{2.125a}
$$

Here, $\beta_0 = \omega\kappa_{vd}\{f_m(0)\}^{1-1/d}$ vanishes in

$$
\begin{aligned}
\frac{\beta_0}{J_{ss}(g_*, 0)} &= \frac{\beta_0}{\beta_0 g_*^{1-1/d} f_e(g_*, 0)/\sqrt{\pi}\delta} \\
&= \frac{\sqrt{2\pi d k T}}{f(1, 0)\Delta\mu^{1/d}}\big[(d-1)W_*\big]^{1/d - 1/2} \exp\left(\frac{W_*}{kT}\right)
\end{aligned} \tag{2.125b}
$$

Hence, $\overline{g_f}$ is independent of the growth rate and its components such as the activation energy of growth, E_g. It is remarkable that

$\overline{g_f}$ is determined only by the initial concentration of monomers available to compose the clusters, $f_m(0)$, the initial concentration of nucleation sites, $f(1, 0)$, the effective dimension, d, the chemical potential, $\Delta\mu$, the temperature, T, and the free-energy barrier to nucleation, W_*.

3. MEASUREMENT OF NUCLEATION AND GROWTH

For understanding the formation of thin films or for controlling the microstructures of thin films, it is indispensable to precisely measure the characteristics of the thin films and the physical quantities or the parameters that control the nucleation and growth in the formation of thin films. Among the various parameters, the energy barriers to nucleation and growth are most essential and should be the final destination in the measurement. However, it is necessary and useful to measure preliminary quantities such as sizes and concentrations, not only for attaining the energy barriers, but also for evaluating the thin films. This section provides the fundamentals for measuring these quantities and parameters in the experiments.

3.1. Dimensions

3.1.1. Size and Shape of Clusters

The kinetic theory describes the dynamic evolution of nucleation and growth in the size space. The size of clusters must be measured for analyzing experimental data with the theoretical results for the size distribution of clusters. Even when one attempts to measure the nucleation rate, it is of importance to strictly regulate the size of the smallest sampled clusters, since the flux of clusters depends on the cluster size. We cannot avoid paying attention to the cluster size in any cases.

As shown in Section 2.2, the size of clusters is originally defined as the number of monomers composing the clusters, g, in the kinetic theory of nucleation and growth. Although high resolution microscopy enables one to directly observe the atomic images of the clusters, it is impossible to precisely count all the monomers included in a cluster unless the clusters are two-dimensional islands with uniformly layered thickness of monomers. Furthermore, it is in practice difficult to count more than hundreds or thousands of monomers in a cluster. Instead, the size of clusters is usually measured in a unit of length and then transformed into the number of monomers. Otherwise, the theoretical results are transformed into those expressed by the cluster size in a unit of length, if possible. For both approaches, the transformation should be given by

$$gV_1 = V \tag{3.1}$$

where V is the volume of a g-sized cluster measured in the unit of length. Thus we have to measure the volume of clusters for measuring their sizes.

For clusters having a single compact shape such as a sphere, a cube, and a crystallographic habit, it is satisfactory to measure a characteristic length of clusters, r, like the diameter or the side length. The volume of the cluster can be calculated by $V = \kappa r^3$ for three-dimensional clusters or $V = \kappa r^2 \ell$ for two-dimensional ones, where κ denotes the geometrical factor and ℓ stands for the film thickness. Otherwise, we have to record all the information about the shape of each cluster besides the characteristic length. This is possible for two-dimensional clusters, while it is not always possible for three-dimensional ones. It is often necessary to regard their variational three-dimensional shapes as an approximate compact one.

The sizes and shapes of clusters can be measured by various experimental methods. The Raman spectroscopy can be applied to estimate the averaged size of crystalline clusters in the thin films [124, 125], which also involves the other structural information such as strain or stress. This method by itself could not always provide the absolute dimensions of the clusters without relying on the empirical conversion from the Raman spectra to the average size of clusters. X-ray diffraction has been also used for estimating the average size of crystallites at each crystallographic orientation [126–130], neglecting the influence from the random stress. This also relies upon the empirical Sherrer's formula [131] which relates the half-width of the crystalline spectra to the average size. Furthermore, since these spectroscopy or diffraction methods consider the crystalline region having perfectly single crystallinity to be a single cluster, the clusters involving imperfectness of crystallinity such as stacking faults and twin boundaries could not be correctly measured. On the other hand, one can measure the absolute dimensions of clusters from the real images of thin films in which the shapes of the clusters are clearly recognizable.

The images of thin films can be captured by using various microscopy techniques. For three-dimensional clusters dispersed over the substrates, optical microscopy is available for capturing their plan-view images if they are larger than several micron meters. However, with the optical microscope, we can measure only the outlines of clusters projected onto the image planes, because of its shallow focus depth of field. It is thus necessary to make some approximation to the three-dimensional shapes of the clusters by regarding them as hemispheres and so on. For the smaller clusters, scanning electron microscopy (SEM) is suitable for capturing their plan-view and slanted-view images [132–135]. With SEM, the shapes of the clusters can be observed in detail due to the sufficient focus depth of field. It is not easy, however, to quantitatively measure the complete three-dimensional shapes unless their two-dimensional images are captured from various directions. For two-dimensional clusters which epitaxially grow over the substrates under high vacuum conditions and have the thickness of atomic steps, the reflection electron microscopy is effective in determining their size and shapes [136]. For two-dimensional clusters which are formed by partial transformation of the starting thin films and thus are embedded in the untransformed matrices at the same thickness, optical microscopy is also available for observing the plan views [57, 137], if the optical indices are sufficiently different between the clusters and the matrices, and if the clusters are larger than several micron meters. Surface tracing microscopy such as scanning tunneling microscopy [138] and atomic force microscopy [139, 140] can

image analysis
at 4.66 nm/pixel

area: 5.966 μm²
perimeter: 18.10 μm
max length: 3.063 μm
min length: 2.480 μm

1 μm

Fig. 6. Plan-view bright-field TEM image of a two-dimensional crystalline Si cluster embedded in amorphous Si thin films with the thickness of 0.1 μm, and the results of the image processing and analysis. The volume of this crystallite is determined to be $V = 0.5966$ μm³, and the number of monomers (i.e., Si atoms) is estimated to be $g = 2.953 \times 10^{10}$.

capture the images of the embedded clusters, when the structural or electrical properties of the film surfaces are different between the cluster and the matrices. If the matrices can be removed by etching techniques leaving only the clusters on the substrates, the optical microscope [137, 141], SEM [142, 143], and the above-mentioned surface tracing microscopy are able to capture the embossed shapes of the clusters. For the smaller two-dimensional clusters embedded in the films thinner than micron meters, transmission electron microscopy (TEM) is suitable for measuring their size from the plan-view images [123, 126, 127, 144–163]. With the plan-view images of these two-dimensional clusters, their volume can be precisely determined, regardless of their shapes, by measuring the areas of the clusters and multiplying them by the film thickness. Figure 6 shows the example of measuring the size of a crystalline two-dimensional Si cluster embedded in amorphous Si thin films from its plan-view bright-field TEM image. The plan-view TEM is also applicable to three-dimensional clusters embedded within thin films, if their diameters are not much smaller than the film thickness. In this case, one can only measure the outlines of the clusters projected onto the film plane, and one needs some assumption to their three-dimensional shapes. For the smaller three-dimensional clusters embedded in thin films, only the cross-sectional-view TEM works. However, in this case, the observable cross-section of the clusters is not always equal to their maximum cross-section, and their sizes would be less estimated than they are.

It is generally more difficult to measure the size of clusters which are densely packed and impinge upon the adjacent clusters. When the thin film has a columnar structure composed by the two-dimensional clusters, and the surface of the film is not flat so that the grain boundaries are seen where the two surfaces of the adjacent clusters meet at a certain contained angle, SEM is able to capture the images of the grain boundaries and hence the shapes of cluster. When the surface of the columnar thin films is flat, but ditches are engraved along the grain boundaries, SEM or surface tracing microscopy are available for capturing the images of the clusters. Otherwise, TEM works if the scattering of the incident electron beam inside the cluster is uniform and different from that at the grain boundaries. If the film is polycrystalline and composed by the randomly oriented crystallites, the dark-field TEM images are able to distinguish the

area of the crystallites with a specific crystallographic orientation from the others and to measure their sizes [123, 145–148, 158], although it is not easy to measure all the crystallites with various orientations. Unless the clusters are periodically placed and uniform in size, the location of the grain boundaries must be random, and their shapes widely varies. Thus it is indispensable to record the shape of each cluster for measuring their sizes.

3.1.2. Effective Dimension of Clusters

The kinetic theory of nucleation and growth provides the expressions of the number concentration or the flux of clusters and the other observables derived from them. All of these theoretical expressions include the effective geometrical dimension of clusters, d, through the addition rate of monomers, $\beta(g, t)$, if $\beta(g, t)$ is given by the general form of Eq. (2.39). Moreover, if the free energy of clusters, $W(g)$, includes a term or a factor relating to the geometry of the clusters like the capillary model described by Eq. (2.5), the theoretical expressions for the observables further include d through $W(g)$.

The effective geometrical dimension, or simply the effective dimension d, can be integer and equivalent to the Euclidean dimension of clusters, if the clusters keep a compact shape at any size. Otherwise, d is a noninteger. Furthermore, if the geometry of the cluster is fractal, d should be irrational [164, 165]. Generally in the case of noninteger effective dimensions, the value of d is not *a priori* given and has to be experimentally determined.

If the fractal objects have a self-similar shape, their effective dimension can be and has been often determined using the box-counting method [166, 167]. In this method, the dimension is measured from the slope on the log-log plot of number of boxes needed to cover the entire surface of the object as a function of the size of the boxes. For the precision of the determination, the range of the box size should be as wide as possible. In other words, the image of the object should have much great detail at the boundary. Such an image is obtained only by constructing that of the whole object with a number of the partial images captured at high magnification, since the spatial resolution in imaging systems increases with the magnification in general. This method is thus effective in investigating a small number of similar objects, but it is not always appropriate when the size of the objects is widely distributed and it is unknown whether the same d governs them all. One would often be faced with such a situation when treating the clusters in thin films formed by nucleation and growth. An alternative method [168] is proposed for experimentally determining arbitrary d, which is suitable for many clusters with a variety of sizes as shown in the following.

3.1.2.1. Fractal Dimension of the Self-Similar Surface of Objects

Let us first consider the surface area of a g-sized cluster, which is expressed by

$$S(g) = \kappa g^{1-1/d} \tag{3.2}$$

where κ is the geometrical factor. Equation (3.2) suggests that one can determine d from the g-dependence of $S(g)$. Provided here that the surface of the clusters has a self-similar fractal structure, the observed surface area of the clusters, $s(r)$, should be scaling with the cluster size expressed in a unit of length, r, as

$$s(r) = qr^D \qquad (3.3)$$

where q denotes the undetermined coefficient and D stands for the fractal dimension with respect to the surface area of the cluster. It is noted that $s(r)$ differs from $S(g)$ because $s(r)$ is the surface area observed with a certain finite resolution while $S(g)$ is the actual one. On the other hand, the observed surface area of the self-similar fractal cluster can be scaled also with the resolution of the measurement. Thus the undetermined coefficient q, and hence $s(r)$, also should be a function of the resolution expressed by the characteristic length scale of the measurement, R, as

$$s(r, R) = q(R)r^D \qquad (3.4)$$

Thus the surface area of a cluster at a size which is equal to the resolution of the measurement (i.e., $r = R$) becomes

$$\lim_{r \to R} s(r, R) = s(R, R) = q(R)R^D \qquad (3.5)$$

Taking the minimum curvature of the observable curved surface to be the resolution, R, the observed surface area of the R-sized cluster should be the actual surface area of the perfect sphere with a radius of R,

$$s(R, R) = \pi(2R)^{D_e - 1} \qquad (3.6)$$

where D_e is the Euclidean dimension of the space the cluster belongs to. Equations (3.5) and (3.6) lead to the coefficient, q, as

$$q(R) = \pi 2^{D_e - 1} R^{D_e - D - 1} \qquad (3.7)$$

What the above formulation means can be understood by the example of the Koch snowflake shown in Figure 7. Since $D_e = 2$ for the Koch snowflake, the surface corresponds to the perimeter. Let us consider first that an r-sized (Fig. 7c) and an $r \times p$-sized (Fig. 7a) Koch snowflakes are observed at the same resolution of R. The observed surface areas of both are related by Eqs. (3.4) and (3.7) as

$$s(pr, R) = 2\pi R^{1-D}(pr)^D = p^D 2\pi R^{1-D} r^D = p^D s(r, R) \qquad (3.8)$$

If one observes the r-sized Koch snowflake at the higher resolution of R/p (Fig. 7d), the observed surface area is calculated using Eq. (3.8) as

$$\begin{aligned} s(r, R/p) &= 2\pi(R/p)^{1-D}r^D \\ &= \frac{1}{p}p^D(2\pi R^{1-D}r^D) \\ &= \frac{1}{p}s(pr, R) \end{aligned} \qquad (3.9)$$

Therefore, the homothetic magnification by p makes the Koch snowflake of Figure 7d fit in that of Figure 7a.

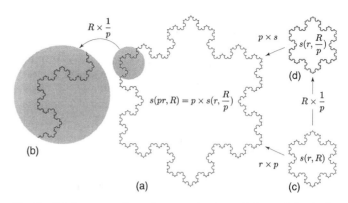

Fig. 7. Relation among the surface area, the size, and the resolution in the measurement of the surface area (i.e., the perimeter) of the self-similar Koch snowflakes. The identical structure (b) is reproduced after the magnification of the original (a) by enhancing the resolution to $R \times 1/p$. The increase in the size by p from (c) to (a) results in the same surface area as that resulting from the identical magnification of (d) which is as large as (c) but observed with the finer resolution of $R \times 1/p$.

Equations (3.4) and (3.7) indicate that, for the self-similar fractals, the increase in size and the enhancement of the resolution are in the one-to-one correspondence, and the observed surface area is scaled with the same fractal dimension, D, to both the size and the resolution. This observation suggests that one can determine the fractal dimension not only by changing the resolution (e.g., the size of boxes in the box-counting method) but also by investigating the size dependence of the surface area.

3.1.2.2. Effective Dimension of the Self-Similar Surface of Clusters

The effective dimension, d, in Eq. (3.2) is then derived by relating the observed surface area, $s(r, R)$, to the actual one, $S(g)$. Considering a case of $D_e = 3$ for simplicity, the observed surface area is given by Eqs. (3.4) and (3.7) as

$$s(r, R) \equiv s_3(r, R) = 4\pi r^D R^{2-D} \qquad (3.10)$$

If the resolution could be made infinitely small, Eq. (3.10) would diverge to infinity, which indicates that $s_3(r, R)$ is never the actual surface area. Reminded that clusters are the condensed matters of monomers, one may take the radius of the monomer to be the resolution limit for a proper estimate of the actual surface area from the observed one. Then let us define the effective radius of g-sized cluster as the radius of a perfect sphere whose volume is equal to that of the cluster by

$$r(g) = \left(\frac{3gV_1}{4\pi}\right)^{1/3} \qquad (3.11)$$

If the limit of Eq. (3.10) by $R \to r(1)$ is regarded as the actual surface area, this is given by

$$\begin{aligned} S(g) \equiv S_3(r(g)) &= \lim_{R \to r(1)} s_3(r, R) \\ &= 4\pi r^D \left(\frac{3V_1}{4\pi}\right)^{(2-D)/3} \\ &= \left(36\pi V_1^2\right)^{1/3} g^{D/3} \end{aligned} \qquad (3.12)$$

Comparing Eq. (3.12) to Eq. (3.2), we obtain the relation between the effective dimension and the fractal dimension, and the geometrical factor:

$$d = \frac{3}{3 - D} \qquad \kappa \equiv \kappa_3 = \left(36\pi V_1^2\right)^{1/3} \qquad (3.13)$$

In conclusion, for determining the effective dimension, we only need to measure the surface area of clusters at a fixed resolution as a function of their size and then evaluate d from the index of the power-law dependence using Eqs. (3.10) and (3.13).

3.1.2.3. Effective Dimension of Pseudo Two-Dimensional Clusters

However, it is not easy to measure the surface area of complicated three-dimensional objects. In reality, the three-dimensional clusters are observed by their two-dimensional images. For instance, the three-dimensional clusters deposited over substrates are usually captured by the plan-view or slanted-view images, and the clusters embedded in matrices are also observed by their cross-section. In the case of crystallization of thin films, the crystalline clusters are essentially restricted within the pseudo two-dimensional space of the starting thin films. Thus, practically, the method for measuring the effective dimension of clusters should treat the pseudo two-dimensional clusters.

Let us suppose that the pseudo two-dimensional cluster has a disklike shape with a thickness of ℓ and the planar dimensions of r ($\gg \ell$), as shown in Figure 8a. One may regard this cluster as a thin foil sliced from the three-dimensional cluster to which the cluster could originally grow without any restriction. If it is possible to measure the surface area of the flank of the cluster, the observed surface area of the flank can be approximately related to that of the imaginary three-dimensional cluster as

$$s_2(r, R) \approx s_3(r, R) \times \frac{\ell}{2r} \qquad (3.14)$$

Here, r is the effective radius of this disklike cluster when observed from the normal direction to the disk, which is expressed by

$$r(g) = \left(\frac{g V_1}{\pi \ell}\right)^{1/2} \qquad (3.15)$$

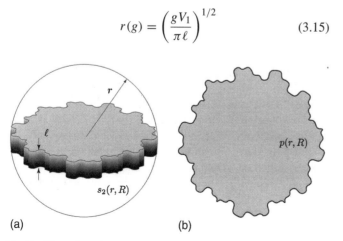

(a)   (b)

Fig. 8.   Illustration of a pseudo two-dimensional cluster having a disklike shape. (a) Slanted view. (b) Plan view from the normal direction.

Similarly to Eq. (3.12), the actual surface area of the flank is given by

$$S(g) \equiv S_2(r(g)) = \lim_{R \to r(1)} s_2(r, R)$$
$$= 2\pi \ell r^{D-1} \left(\frac{V_1}{\pi \ell}\right)^{-D/2}$$
$$= 2(\pi \ell V_1)^{1/2} g^{(D-1)/2} \qquad (3.16)$$

Thus the geometrical factor becomes

$$\kappa \equiv \kappa_2 = 2(\pi \ell V_1)^{1/2} \qquad (3.17)$$

Even with Eq. (3.16), it is still difficult to measure the surface area of the flank. Instead, it is possible to measure the perimeter of the plan view as shown by Figure 8b. Since the fractal dimension of the perimeter should be $D - 1$ according to the dimension analysis, the observed perimeter is given by

$$p(r, R) = 2\pi r^{D-1} R^{2-D}$$
$$= p_0 g^{(D-1)/2}$$
$$= p_0 g^{1-3/(2d)} \qquad (3.18a)$$

where

$$p_0 \equiv 2\pi R^{2-D} \left(\frac{V_1}{\pi \ell}\right)^{(D-1)/2} \qquad (3.18b)$$

The effective dimension can be thus determined from the power-law correlation observed between the perimeter of the pseudo two-dimensional clusters and their size.

3.1.2.4. Example of Determining the Effective Dimension

Figures 9 and 10 show an example of determining the effective dimension by measuring the perimeter of pseudo two-dimensional clusters. The clusters are Si crystallites formed by the solid-phase crystallization of amorphous Si (a-Si) thin films over SiO_2 substrates [169]. The a-Si thin films are deposited on the substrates by low-pressure chemical-vapor deposition using SiH_4 gas at a temperature of 823 K, under a pressure of 40 Pa, and at a rate of 2.8×10^{-5} μm s^{-1}. The substrates are crystalline Si wafers coated by amorphous SiO_2 films which are formed by thermal oxidation of the wafers. The thickness of the a-Si films is $\ell = 0.1$ μm. Then, Si$^+$ ions accelerated to 70 keV are implanted into the a-Si films at room temperature and at a dose of 1×10^{13} mm^{-2}. The a-Si thin films are crystallized in solid phase by the isothermal annealing at a temperature of 873 K in nitrogen atmosphere for 1.08×10^5 s [171]. The crystallization is initiated by the continuous nucleation of crystalline clusters at random locations of the thin films, and it is advanced by their epitaxial growth at the interfaces between the crystallites and the amorphous matrix. When the crystallites are smaller than the film thickness ($\ell = 0.1$ μm, in this example) they may grow three-dimensionally. After their growth front reaches the top surface of the thin film or the bottom interface to the SiO_2 substrate, their growth is restricted in the direction of the film plane. Consequently, the crystallites larger than ~0.1 μm become the pseudo two-dimensional clusters, as shown by the schematic slanted view of the partially crystallized film in Figure 9a. In this type of crystallization, the crystallites grow involving the

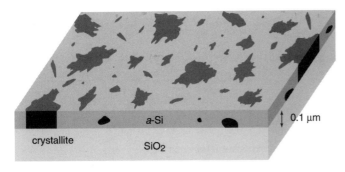

(a)

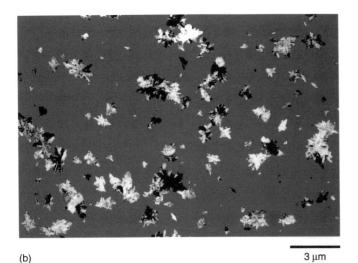

(b)   3 μm

Fig. 9. Planar Si crystallites formed by the solid-phase crystallization of a
0.1 μm thick a-Si thin film over the SiO$_2$ substrate. The schematic slanted-
view (a) and a plan-view bright-field TEM image of the partially crystallized
a-Si thin film, in which dendritic crystallites are embedded at random locations
of the amorphous matrix.

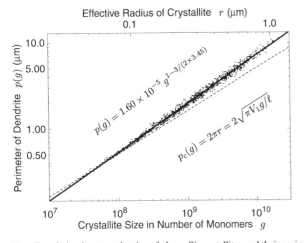

Fig. 10. Correlation between the size of planar Si crystallites and their perimeter, which is measured from the plan-view TEM images as shown by Figure 9b.
The crystallites are formed at 873 K by the solid-phase crystallization of a
0.1 μm thick a-Si thin film over the SiO$_2$ substrate. The correlation consisting
of 1095 crystallites exhibits a power-law dependency which apparently differs
from that of the compact circular objects represented by the dotted line with
$p_c(g) = 2\pi r = 2\sqrt{\pi V_1 g/\ell}$.

twin boundaries which provide the preferential growth points at
their surface [172, 173]. The multiple twinning thus makes the
shape of the planar crystallites look like dendrites. As shown
in the plan-view TEM images of Figure 9b, the dendritic crystallites are formed at random locations with a wide spread of
size from submicron to a few micron meters. Figure 10 plots
the correlation between the size and the perimeter, which consists of 1095 dendritic crystallites sampled from several tens of
the images like that shown in Figure 9b. Each small dot corresponds to a sampled crystallite. The perimeter is measured by
the image analysis as shown in Figure 6, but at the resolution of
0.02×0.02 μm^2 per pixel. In Figure 10, the plots for the dendritic crystallites exhibit an apparent power-law correlation and
definitely differ from the ideal correlation for compact circular objects which is expressed by $p_c(g) = 2\pi r = 2\sqrt{\pi V_1 g/\ell}$
and plotted by a dotted line together. This fact confirms that the
dendritic Si crystallites have a fractal surface. By the fitting of a
power-law function to all of the 1095 plots, the perimeter is estimated as $p(g) = 1.60 \times 10^{-5} g^{1-3/(2\times 3.45)}$. Thus the effective
dimension is determined to be $d = 3.45$. With this coefficient
and Eq. (3.18b), the actual resolution is estimated to be $R \sim
0.05$ μm, which is comparable to but larger than the pixel size
of 0.02 μm, and hence consistent. It is additionally noted that
the fitted solid line for the dendrites, $p(g)$, and the dotted line
for the compact circular objects, $p_c(g)$, cross at $g \sim 4.35 \times 10^7$,
because $p(g)$ is measured at the limited resolution of $R \sim
0.05$ μm, while $p_c(g)$ is the actual perimeter with $R = r(1)$.

3.2. Ratios

3.2.1. Transformed Fraction

The transformed fraction has been measured for investigating
the formation processes of thin films for a long time. In thin
films engineering, the transformed fraction is used for detecting
the completion of the transformation of the starting thin films,
or the complete coverage of the substrate surface by the deposited thin films. Also, in thin films science, the transformed
fraction could be utilized for investigating the thermodynamics
and kinetics of the nucleation and growth.

There are various experimental methods available for measuring or estimating the transformed fraction. If taking examples in polycrystalline thin films formed by solid-phase crystallization of amorphous thin films, a calorimetric technique such
as differential scanning calorimetry [174, 175] may be applied
to detect the complete transformation. The transformed fraction
can be otherwise monitored by the X-ray diffraction [176, 177],
optical reflectivity or transmission [178–181], electrical resistivity or conductivity [182–187, 189, 190], and, for some cases,
luminescence, electron spin resonance, photocurrent measurements, and so on. Although the measured quantities by these
methods reflect the transformed fraction, they do not directly
provide the actual values of the transformed fraction. In addition to the average size of clusters, Raman spectroscopy is able
to estimate the transformed fraction [125, 188, 191, 192]. It is
also necessary, however, to employ the empirical conversion
from the Raman spectra to the transformed fraction. Thus the

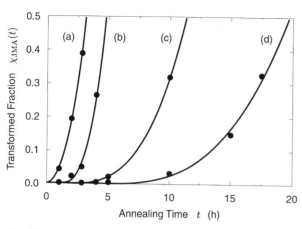

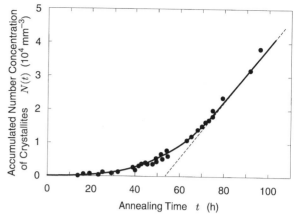

Fig. 11. Time evolution of the transformed fraction observed in the solid-phase crystallization of a-Si thin films with various conditions of Si$^+$ ion implantation prior to the isothermal annealing at 873 K. The ion dose and the acceleration energy are (a) 2×10^{12} mm^{-2} and 70 keV, (b) 1×10^{13} mm^{-2} and 40 keV, (c) 1×10^{13} mm^{-2} and 55 keV, and (d) 1×10^{13} mm^{-2} and 65 keV [171]. The solid lines stand for $\chi_{\rm JMA}(t)$ of Eq. (2.116) fitted to the experimental plots.

Fig. 12. Time evolution of the accumulated number concentration of crystallites formed by the solid-phase crystallization of lithium disilicate (Li$_2$O·SiO$_2$) glass at 703 K. The plot data are collected from Figure 1a of [197]. The solid line stands for the theoretical $N(t)$ of Eq. (2.120) fitted to the experimental plots.

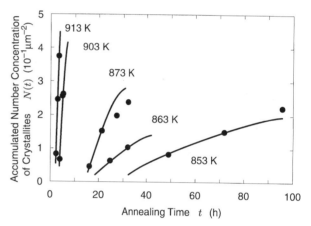

Fig. 13. Time evolution of the accumulated number concentration of Si crystallites formed by the solid-phase crystallization of a-Si thin films at various temperatures. The plot data are collected from from Figure 4 of [123]. The solid lines stand for the theoretical $N(t)$ of Eq. (2.124).

most direct method would be microscopy and its image analysis [123, 132–134, 143, 144, 151, 156, 157, 162, 168, 193]. For instance, by applying the image analysis as shown in Figure 6 to the samples shown in Figure 9b, the transformed fraction is determined to be $\chi(t) = 0.1099$.

If the time evolution of the transformed fraction is measured, one may fit the theoretical expression like Eq. (2.115) or (2.116) to the experimental data set regarding the parameters for nucleation and growth as the unknown fitting parameters. Figure 11 shows an example of the time evolution of the transformed fraction measured using the TEM and the image analysis in the solid-phase crystallization of a-Si thin films formed under various conditions [171] and the results of the fitting with $\chi_{\rm JMA}(t)$. In principle, these parameters could be simultaneously determined by this method using only the transformed fraction [159], while the transformed fraction can be combined with the other observables like the accumulated number concentration of clusters for determining these parameters [123, 151, 193].

3.2.2. Accumulated Number Concentration of Clusters

The accumulated number concentration of clusters can be measured only by counting the clusters on the real images of thin films. The images have been captured for this purpose by optical micrographs [57, 137, 141], SEM [143], and TEM [55, 56, 123, 144, 147, 151–153, 163, 193–196]. As described in Section 2.3.5, the accumulated number concentration of clusters, $N(g_{\rm s}, t)$, depends on the size of the smallest sampled clusters, $g_{\rm s}$, which was confirmed also by the numerical computer simulation based on the master differential equation of Eq. (2.14) [58]. It is thus important to fix the size of the smallest sampled clusters in the measurement. Moreover, the sensitivity of detecting clusters is not always constant throughout the size range from the detection limit but generally is reduced near the detection limit. It is also necessary to set $g_{\rm s}$ at the smallest size where the sensitivity of the detection is constant.

The time evolution of $N(g_{\rm s}, t)$ has been one of the important clues to the kinetics and thermodynamics of nucleation and growth. According to Eqs. (2.121) and (2.122), $N(g_{\rm s}, t)$ approaches a linear dependence on time with the slope of $J_{\rm ss}(g_*, 0)$ after a certain transient period, if the steady-state flux has been established before the depletion of monomers or nucleation sites becomes significant. Figure 12 shows an example of such an observation reported by Fokin et al. [197]. However, $N(g_{\rm s}, t)$ does not always exhibit such an apparent approach to the linear dependency. In order to confirm the linearity, one needs to observe $N(g_{\rm s}, t)$ for a sufficiently long period during which the depletion of monomers or nucleation sites has to be negligible. Before the linearity with the slope of $J_{\rm ss}(g_*, 0)$ is established, the influence of the depletion of monomers or nucleation sites may be superposed on the observed $N(g_{\rm s}, t)$. Therefore, even if $N(g_{\rm s}, t)$ seems to approach the linearity, there is no guarantee that its slope is $J_{\rm ss}(g_*, 0)$. No linear region can be even observed, especially when the depletion of monomers or nucleation sites becomes significant before the steady-state flux is established. Figure 13 shows a typical example reported by

Iverson and Reif [123]. For such a situation, instead of finding the linearity, one can apply $N_{JMA}(g, t)$ of Eq. (2.123) or (2.124) for evaluating the observed $N(g_s, t)$. In another attempt, Im and Atwater [152] normalize the observed $N(g_s, t)$ with the fraction of the untransformed region, $1 - \chi(t)$, measured at the same time, t, and observed a linear dependence of $N(g_s, t)/[1 - \chi(t)]$ on t. This method may be utilized for the preliminary evaluation of the time evolution of $N(g_s, t)$ but could not always be accurate. As formulated by Eqs. (2.42)–(2.44), the number concentration of clusters should be normalized with the untransformed fraction when the clusters were formed or nucleated, $1 - \chi[t - t_g(g)]$, which depends on the cluster size, g. Therefore, it is essentially invalid to normalize $N(g_s, t)$ including clusters at various sizes with a single $1 - \chi(t)$.

3.2.3. Size Distribution of Clusters

The size distribution of clusters has been measured mostly for the purpose of observing itself and was scarcely used for investigating the kinetics of nucleation and growth [56, 135, 141, 151, 162, 195, 196, 198], until the proper kinetic description of the nucleation and growth up to the observable cluster size which is usually far beyond the critical region was established [85, 88, 89]. As described in Section 2.2 and as will be demonstrated in Section 3.5, the size distribution of clusters should be considered also as an important and direct clue to the kinetics of nucleation and growth.

Unless the distribution is obtained directly, the measurement of the size distribution starts with measuring the size of each cluster by the methods described in Section 3.1.1. Generally, the cluster size is measured initially in a unit of length, r, and then can be transformed into the number of monomers, g, if necessary. The frequency distribution of clusters is produced by sampling a sufficient number of clusters and by giving an appropriate set of subregions in the whole range of the cluster sizes so as to obtain a smooth distribution curve. If the cluster size is expressed by the number of monomers, the number of the sampled clusters whose sizes are within a subregion of $[g - \Delta g/2, g + \Delta g/2]$ is related to the actual number concentration by

$$n(g, \Delta g, t) = V_s \int_{g-\Delta g/2}^{g+\Delta g/2} f(g', t)\, dg' \approx V_s f(g, t)\Delta g$$

(3.19)

if $|\partial f/\partial g|\Delta g \ll 1$, where V_s is the volume of the whole sampled space. Thus the actual number concentration is obtained by normalizing the number of clusters in the subregion with V_s and the Δg. It is not always necessary that the intervals of the subregions, Δg, are uniform throughout the size range. The size dependent Δg is often useful in keeping the accuracy almost constant in the observed range of size, when the number concentration widely varies.

When the size distribution needs to be treated in the size space represented in a unit of length, the number of the sampled

clusters within a subregion of $[r - \Delta r/2, r + \Delta r/2]$ is

$$n(r, \Delta r, t) = V_s \int_{r-\Delta r/2}^{r+\Delta r/2} f(r', t)\, dr' \approx V_s f(r, t)\Delta r \quad (3.20)$$

where $f(r, t)$ is given by Eq. (2.91). Thus normalizing $n(r, \Delta r, t)$ with V_s and Δr results in $f(r, t)$ whose dimensionality is $[L^{-D_e-1}]$, where D_e is the Euclidean dimension of the sampled space.

3.2.3.1. Example of Measuring and Analyzing the Size Distribution of Clusters

Figure 14 shows an example of the measured size distribution of dendritic Si crystallites embedded in a-Si thin films. The sample preparation and evaluation are the same as those for Figures 9 and 10 except that the a-Si film was annealed at 863 K for 1.80×10^5 s [199]. The 1060 crystallites are sampled from the plan-view TEM images as shown in Figure 9b, and the crystallite sizes are measured by the method shown in Figure 6, which provides the projected area, A_p, of the crystallite. Then the projected area is transformed into the effective radius, $r \equiv \sqrt{A_p/\pi}$, that is, the radius of a circle whose area is A_p. As shown in Figure 9a, the crystallites grow three-dimensionally before their growth front reaches the surfaces of the a-Si thin film. It is thus necessary to divide the growth processes into the three-dimensional mode and the two-dimensional one for deriving the actual cluster size from the measured effective radii, which is accomplished by the following consideration. The solid-phase crystallization of a-Si progresses by rearranging the disordered atomic bonds into the crystalline one, which occurs just at the interface between the crystallites and the surrounding amorphous matrices [200, 201]. The single Si atom plays a role of the monomer for the nucleation and growth. There is no

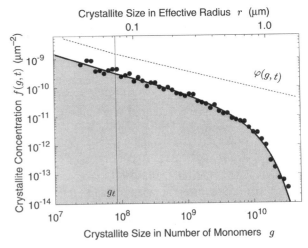

Fig. 14. Size distribution of dendritic Si crystallites embedded in an a-Si thin film. The film is partially crystallized in solid phase as shown in Figure 9. The closed circles are the experimental plots made from the 1060 sampled crystallites. The solid line stands for the theoretical distribution fitted to the experimental plots. The dotted line is the steady-state distribution, $\varphi(g, t)$, extracted by the fitting. The vertical straight line indicates $g = g_\ell$ where the crystallites change the growth mode from the three-dimensional to the two-dimensional.

long-range transport of monomers which affects the rearrangement. The displacement of the monomers is no longer than the bond length. Thus the addition mechanism of monomers into the crystallites is controlled by the chemical reaction at the interface, i.e., $\nu = 1$, and the monomer concentration around the crystallites is constant, i.e., $f_m(t) = f_m(0)$. Hence, the addition rate of monomers given generally by Eq. (2.39) is described as

$$\beta(g) = \omega' S(g) \qquad (3.21)$$

where $\omega' \equiv \omega\{f_m(0)\}^{1-1/d}$ is the frequency of monomer addition per unit area with ω being given by Eq. (2.94). The surface area of the g-sized crystallite, $S(g)$, should be the area of the interface between the crystallites and a-Si matrices. Equations (3.12), (3.13), and (3.16) indicate that $S(g)$ should be described as

$$S(g) = \begin{cases} \kappa_3 g^{1-1/d} & g < g_\ell \\ \kappa_2 g^{1-3/2d} & g \geq g_\ell \end{cases} \qquad (3.22)$$

where g_ℓ is the size at which the three-dimensionally growing crystallites reach the film surfaces; κ_3 and κ_2 are given by Eqs. (3.13) and (3.17), respectively. With Eqs. (2.81b), (2.105), (3.21), and (3.22), the steady-state size distribution is obtained as

$$\varphi(g,t) = \begin{cases} Cg^{1-1/d} & g < g_\ell \\ \dfrac{\kappa_3}{\kappa_2} Cg^{1-3/2d} & g \geq g_\ell \end{cases} \qquad (3.23)$$

if $g_* < g_\ell$, where $C = f_1(g,t) \exp(-W_*/kt)/\sqrt{\pi}$ and the estimate of $\ln(g_*^{1-1/d}/\delta) \ll \ln C$ [168] is adopted. The requirement for the continuity of $\varphi(g,t)$ at $g = g_\ell$ gives

$$g_\ell = \left(\frac{2}{9}\right)^{2d/3} \left(\frac{\pi \ell^3}{V_1}\right)^{d/3}$$

With the effective dimension, $d = 3.45$, predetermined by the method described in Section 3.1.2 and $\ell = 0.1$ μm, $g_\ell = 8.283 \times 10^7$ is estimated. Assuming that the crystallites instantly change their growth mode from the three dimensional to the two dimensional, the cluster size expressed in the number of monomers is calculated from the measured effective radius by

$$g = \begin{cases} \dfrac{4\pi}{3V_1} r^3 & r < r_\ell \\ \dfrac{\pi \ell}{V_1} r^2 & r \geq r_\ell \end{cases} \qquad (3.24)$$

where $r_\ell \equiv \sqrt{g_\ell V_1/(\pi \ell)}$. After transforming the measured effective radius into the cluster size in the number of monomers by Eq. (3.24), the frequency distribution is produced with the variable intervals of $\Delta g = 4.500 \times 10^7 \times 1.162^{m-3}$ for $m = 0, 1, 2, \ldots, 48$ and then is normalized by Eq. (3.19). The closed circles in Figure 14 are thus obtained and represent the actual concentrations of clusters. The lower horizontal axis indicates the crystallite size expressed in the number of monomers, g, and the upper one indicates the corresponding effective radius, r. The thin straight line vertically penetrating the plots indicates the boundary at $g = g_\ell$. The measured plots exhibit a monotonous decrease with g and a rapid drop for large g, which

is just predicted by the kinetic theory of nucleation and growth as shown in Figure 5. The solid line outlining the closed circles is the theoretical size distribution obtained as follows. By substituting Eqs. (3.21) and (3.22) into Eq. (2.81d), the time to grow can be calculated as

$$t_g(g_* + \delta, g) \approx \begin{cases} \dfrac{d}{\omega' \kappa_3} g^{1/d} & g < g_\ell \\ \dfrac{d}{\omega'} \left\{ \dfrac{g_\ell^{1/d}}{\kappa_3} + \dfrac{2}{3\kappa_2} \left(g^{3/2d} - g_\ell^{3/2d}\right) \right\} & g \geq g_\ell \end{cases} \qquad (3.25)$$

Then the theoretical size distribution of Eq. (2.81) with Eqs. (2.105), (3.23), (3.25) and the predetermined d is numerically fitted to the experimentally obtained plots, regarding a set of quantities $\{W_*, \tau, \lambda, \omega'\}$ as the undetermined parameters. As seen in Figure 14, the fitted theoretical size distribution well reproduces the experimentally obtained plots. The dotted straight line bending at $g = g_\ell$ indicates the steady-state distribution, $\varphi(g,t)$, of Eq. (3.23) which is extracted from the overall size distribution, $f(g,t)$, by the fitting. The overall distribution has not reached the extracted steady-state distribution, which suggests that the steady state has not been established in the observed range of the cluster size.

3.3. Rates

3.3.1. Nucleation Rate

As can be found in Eq. (2.103) or (2.105) with Eqs. (2.39), (2.42), and (2.95), the steady-state nucleation rate, J_{ss}, brings the information about the energy barriers to nucleation and growth. It is found in Eq. (2.102) with Eqs. (2.39) and (2.45) that the transient nucleation rate, $J(t)$, includes the information of the growth also in its relaxation time (i.e., τ). Thus the nucleation rate has been measured as a clue to the thermodynamics and the kinetics of nucleation and growth.

The nucleation rate has been measured or estimated by several methods. Wu et al. [202, 203] attempted a preliminary estimate of J_{ss} using the onset time of the intensity of the X-ray diffraction observed in a-Si thin films crystallized in solid phase, while this method has no theoretical basis.

The time evolution of the transformed fraction can be used for estimating the nucleation rate and the growth rate simultaneously [159, 184–187]. Although it is generally impossible to obtain an analytical nonintegral expression of the transformed fraction, i.e., $\chi_{JMA}(g,t)$ of Eq. (2.111) or (2.112) with $J_0(g,t)$ of Eq. (2.100) or (2.108), the power-series expansion of $J_0(g,t)$ about a certain time, t', enables us to derive the analytical series expression of $\chi_{JMA}(g,t)$, which is effective if the observed range of time is restricted around t'. Otherwise, one could employ the simple expression of Eq. (2.115) relying on the two-stage approximation of $J_0(g,t)$ expressed by Eq. (2.114). In both cases, the highest degree of t in such expressions of $\chi_{JMA}(g,t)$ must be higher than d/ν. There is thus no way but to numerically fit the analytical $\chi_{JMA}(g,t)$ to the set of the experimental data regarding a set of $\{J_{ss}, \beta_0, \tau, \lambda\}$ or

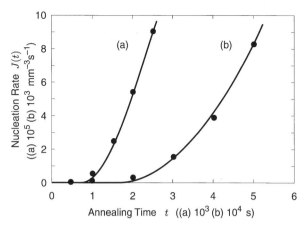

Fig. 15.   Time evolution of the nucleation rate observed in the solid-phase crystallization of a-Si deposited by the electron beam evaporation. The annealing temperature is 920 K for (a) and 861 K for (b). The plot data are collected from Figures 3a and c of [55]. The solid lines stand for the theoretical $J(t)$ of Eq. (2.102) fitted to the experimental plots.

$\{J_{ss}, \beta_0, \tau_{tr}\}$ as the undetermined fitting parameters. If the numerical fitting is indispensable, one may numerically calculate the integral term of Eq. (2.111) or (2.112) at the same time in the fitting process, instead of expanding $J_0(g, t)$ to the power series of t or relying on the two-stage approximation. Note that if one employs the further simple expression for $\chi_{JMA}(t)$ of Eq. (2.116) in which the smallest cluster size is taken to be 0, it is impossible to separate J_{ss} and β_0 only from the transformed fraction.

The most direct method is to observe the derivative in the time evolution of the accumulated number concentration of clusters, $N(g_s, t)$, which provides the transient nucleation rate, $J(g_s, t) = dN(g_s, t)/dt$. Figure 15 shows an example, which is a part of the results reported by Köster [55] for the solid-phase crystallization of a-Si thin films. In this example, $J(g_s, t)$ increases all the time during the observation and does not seem to converge at a constant which is to be the steady-state nucleation rate, J_{ss}. Even if one goes on observing, $J(g_s, t)$ might exhibit no apparent plateau and decrease after showing a peak because of the depletion of monomers or nucleation sites. Instead of waiting for the establishment of the steady state, Köster fitted a theoretical expression of $J(g_*, t)$ to the observed transient rate of nucleation and estimated the values of J_{ss}. If an apparent linear dependence of $N(g_s, t)$ on t is observed, its slope could be regarded as the steady-state nucleation rate J_{ss} [57, 143, 152, 194], although in the observed time evolution of $N(g_s, t)$, there is no available evidence that the steady state is actually established.

It is naturally possible to directly fit the theoretical expression of the accumulated number concentration to the observed time evolution for estimating J_{ss}. Iverson and Reif [123, 193] assumed the two-stage process for nucleation in the solid-phase crystallization of a-Si thin films, which is described by Eq. (2.114). Then they fitted a theoretical expression for $N_{JMA}(t)$, which is basically the same as Eq. (2.124), to the observed time evolution of $N(g_s, t)$ as shown in Figure 13. As

seen in Eq. (2.124), it is impossible in principle to separate J_{ss} and β_0 only from the accumulated number concentration of clusters derived considering the Johnson–Mehl–Avrami formalism. Therefore, they also fitted the theoretical expression for $\chi_{JMA}(t)$, which is basically the same as Eq. (2.116), to the time evolution of $\chi(g_s, t)$ observed together with $N(g_s, t)$. Finally, they found the simultaneous best fit of $N_{JMA}(t)$ and $\chi_{JMA}(t)$ to both the observed time evolutions and estimated J_{ss} and β_0.

As another approach to the steady-state nucleation rate, a method using the size distribution of cluster has been proposed and demonstrated [162]. However, as will be explained in Section 3.5, the size distribution was found to be a direct clue to the thermodynamics and the kinetics of nucleation and growth, and a method for measuring the essential parameters such as the free-energy barrier to nucleation directly from the size distribution was developed. Thus this method has no more utility value today.

3.3.2. Growth Rate

The growth rate or the growth velocity is not only a clue to the kinetics of nucleation and growth, but also an important parameter in the industrial usage of thin films, since the growth rate often controls the total processing time needed to fabricate the devices using the thin films. When the growth on average progresses in the direction perpendicular to the surface of substrates or the films, which may be called the "one-dimensional" growth, the growth rate can be directly measured from the time evolution of the film thickness. In the layer-by-layer epitaxial growth of thin films from vapor or liquid phase, the growth rate is equivalent to the deposition rate, which can be measured with various *in situ* or *ex situ* methods. This is valid also when the deposition goes on by the cluster growth and the nucleation of the clusters on top of deposited clusters, except in the initial stage of the deposition. Even when the growth front is hidden under the surface of the starting thin films, such as the solid-phase epitaxial growth of amorphous thin films over crystalline substrates, the movement of the growth front can be *in situ* monitored by the ion channeling and the backscattering techniques [204–212], or by the oscillation of the optical reflectivity [213–216].

On the other hand, it is generally difficult to measure the growth rate of clusters growing two- or three-dimensionally. Wu et al. [202, 203] made a preliminary estimate regarding the increasing rate of the X-ray diffraction intensity as the growth rate, in the solid-phase crystallization of a-Si thin films, but this treatment is not theoretically valid. The direct measurement is accomplished only by tracking the sizes of the same clusters during the growth, employing *in situ* observations. A special TEM apparatus equipped with hot-stage sample holders has been used [55, 144, 152, 153, 194, 217–219] for this purpose. Figure 16 shows such an example for the *in situ* TEM observation of the same crystallite grown by the solid-phase crystallization of a-Si thin film. The plot data are calculated from the brilliant experimental result reported by Im et al. [153]. However,

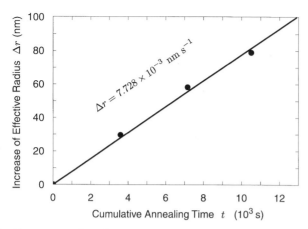

Fig. 16. Increase of the effective radius of a single crystallite with the cumulative annealing time for the solid-phase crystallization of an *a*-Si thin film. The plot data are calculated from the result in Figure 2b of [153]. The size of the same crystallite is measured by the *in situ* TEM observation. The solid line indicates the linear function of *t* fitted to the experimental plots. The growth rate is determined to be 7.728×10^{-3} nm s^{-1} from the fitting.

for employing this type of *in situ* observation, the TEM specimen has to be either thin cut foils exposing the cross-section of the thin film, the self- (or free-) standing thin films, or at least the thin films on the substrates which are as thin as the thin films. The growth conditions are also restricted such as in ultrahigh vacuum or for relatively short periods. Furthermore, it is not easy to sample a number of the clusters, which is large enough for ensuring statistical accuracy.

Without employing the *in situ* observations, one has to estimate the growth rate from the time evolution of the transformed fraction by the numerical fitting of its theoretical expressions, $\chi_{\mathrm{JMA}}(g_s, t)$, like Eq. (2.111), (2.112), or (2.115) [159, 176, 184, 185]. Otherwise, the theoretical expressions for $\chi_{\mathrm{JMA}}(t)$ of Eq. (2.116) and $N_{\mathrm{JMA}}(t)$ of Eq. (2.124) are numerically fitted to their experimental data simultaneously, and the growth rate is estimated together with the nucleation rate [123, 193]. Using the analysis of the size distribution as shown in Figure 14, one can determine not the growth rate itself, but the frequency of monomer addition, ω [168].

3.4. Characteristic Time

The characteristic times for the nucleation and growth have also been measured as a clue to their kinetics. One of the often observed characteristic times is the transformation time, τ_c, in the formation of thin films [123, 143, 178, 179, 220], which is defined as the time when the transformed fraction reaches a certain value. For instance, if one takes the value at $\chi_{\mathrm{JMA}}(\tau_c) = 1 - 1/e$ and employs Eq. (2.116) for the transformed fraction, $\chi_{\mathrm{JMA}}(t)$, the transformation time is given by

$$\tau_c = \left[\frac{108}{V_1 J_{\mathrm{ss}}(g_*, 0) \beta_0^d} \right]^{1/1+d} + \tau_{\mathrm{tr}}$$

Thus the transformation time includes information about the steady-state nucleation rate and the growth rate. However, this

is basically the same as the time evolution of $\chi_{\mathrm{JMA}}(t)$ and $N_{\mathrm{JMA}}(t)$ provides. Since one has to observe the time evolution of $\chi_{\mathrm{JMA}}(t)$ for determining τ_c anyhow, it would be better to directly use $\chi_{\mathrm{JMA}}(t)$ than τ_c.

The time constant for establishing the steady-state nucleation is the characteristic time that has most attracted attention in both the theoretical investigations [43, 46, 49, 62, 78, 81, 95, 221–225] and the experimental investigations [56, 57, 123, 142, 143, 151, 162, 177–179, 183, 189, 220]. This time constant theoretically corresponds to τ and $\lambda\tau$ given by Eqs. (2.45) and (2.64h). Particularly, the former time constant, τ, has long been investigated and is called the *time lag*, although its definition has slightly varied with the development of the kinetic theory of nucleation. In the experiments, however, τ is not directly accessible. Instead, the related characteristic times have been measured, which are expected to be proportional to τ and are called with various names such as incubation, induction, relaxation, or transient time (or period). The transient time, τ_{tr}, appearing in Eqs. (2.114)–(2.116) and (2.124) corresponds to one such characteristic time. These characteristic times related to τ have been often measured either from the onset of nucleation where the formation of clusters becomes detectable, or from the intersection of the extrapolated linear line along the estimated steady-state region and the horizontal axis, in the time evolution of the accumulated number concentration of clusters, as indicated by the dotted line in Figure 12. Otherwise, the absolute value of τ can be determined by fitting the theoretical expressions for the transformed fraction, the accumulated number concentration of clusters, or the size distribution of clusters to their experimental data. In any cases, as seen in Eqs. (2.45), the time lag, τ, includes the information about the nucleation and growth of clusters, which can be obtained also from the other observables. Another time constant, $\lambda\tau$, was first introduced in the latest kinetic theory of nucleation and growth and has not been used for investigating the kinetics of nucleation and growth.

3.5. Energy Barriers

3.5.1. Free-Energy Barrier to Nucleation

The measurement of the free-energy barrier to nucleation, W_*, is one of the destinations in the experimental investigations of nucleation and growth. It may be said that all the observables discussed in the previous sections are just the means to derive W_* from them. As briefly introduced in Section 2.1.3, a number of models for the free energy of nucleating clusters have been proposed. They should be tested by comparing their predictions for the magnitude or the dependency of W_* to the experimentally determined ones. The magnitude of W_* is also an important parameter for controlling the structure of thin films. As discussed in Section 2.3.6 and shown by Eq. (2.125), the final grain size of granular thin films is controlled mainly by W_*. Thus the measurement of W_* is desired from both the scientific and engineering interests in the formation of thin films. Three methods for the experimental determination of W_* are discussed in the following.

3.5.1.1. Arrhenius Method

The Arrhenius method has long been only one means available for experimentally estimating W_*. As shown by Eqs. (2.95) and (2.109), both the growth rate and the nucleation rate are thermally activated as

$$v_g(g) = \beta(g) \propto \exp\left(-\frac{E_g}{kT}\right)$$

$$J(g) \propto J_{ss}(g_*) \propto T^{-1/2} \exp\left(-\frac{E_g + W_*}{kT}\right)$$

Thus, if the exponents of $v_g(g)$ and $J_{ss}(g_*)$ are obtained from their Arrhenius plots, one can expect that their difference is W_*. Iverson and Reif [123] employed this method and estimated $W_* = 2.4 \pm 0.15$ eV in the solid-phase crystallization of a-Si thin films for the temperature range of 853–913 K. Masaki et al. [159], following the same approach, estimated $W_* = 2.0 \pm 1.6$ eV in similar thin films for the temperature range of 823–923 K and also calculated $W_* = 2.1$ eV from the results reported by Zellama et al. [187]. There is, however, a restriction on the applicability of this Arrhenius method.

Practically, the exponent of $J_{ss}(g_*)$ is derived from the slope on its Arrhenius plot. On the other hand, since W_* is a Gibbs free energy, it should be decomposed into

$$W_* = H_* - TS_* \tag{3.26}$$

where H_* and S_* are the differences in the enthalpy and entropy between a critical cluster and a monomer, and hence H_* and $-TS_*$ correspond the enthalpic and entropic barriers to nucleation, respectively. Thus the slope in the Arrhenius plot of $J_{ss}(g_*)$ should be

$$-\frac{d \ln J_{ss}(g_*)}{d(1/kT)} = H_* + E_g - \frac{kT}{2} \tag{3.27}$$

Consequently, when E_g is subtracted from the slope in the Arrhenius plot of $J_{ss}(g_*)$, what is obtained is not W_* but H_*. The Arrhenius method for measuring W_* is valid only if the entropic contribution to W_* is negligibly small; i.e., $|H_*| \gg |TS_*|$. In other words, the Arrhenius method is not valid if W_* changes with the temperature, T. When the Arrhenius method is once employed, a model for W_* in which W_* does not depend on T automatically presupposed. Hence the Arrhenius method is not always valid for testing any theoretical model for W_*.

3.5.1.2. Model-Independent Method Using the Ratio of the Growth Rate to the Steady-State Nucleation Rate

A simple and direct method which does not rely on the Arrhenius plot is proposed and demonstrated [226–231]. The method utilizes the absolute magnitudes of the growth rate of large clusters (i.e., $g \gg g_* + \delta$), $v_g(g) = \beta(g)$, and the steady-state nucleation rate, $J_{ss}(g_*)$, which are observed at the same temperature. The basic idea is that the energy barrier to growth, E_g, is

canceled in their ratio as

$$\frac{\beta(g)}{J_{ss}(g_*)} \propto \frac{\exp[-E_g/kT]}{T^{-1/2} \exp[-(E_g + W_*)/kT]} \propto \exp\left(\frac{W_*}{kT}\right)$$

if E_g is common to the critical clusters and the observable g-sized clusters. Actually, from Eqs. (2.94) and (2.105), the ratio is exactly given by

$$\frac{\beta(g)}{J_{ss}(g_*)} = \frac{\sqrt{\pi}\delta}{f(1,0)} \exp\left(\frac{W_*}{kT}\right) \frac{\beta(g)}{\beta^*(g_*)} \tag{3.28}$$

where $\beta^*(g)$ and $\beta(g)$ are the addition rate of monomers at $g = g_*$ and $g \gg g_* + \delta$, respectively, and they do not always have the same functional form with respect to g. Thus the free-energy barrier to nucleation should be determined by

$$\frac{W_*}{kT} = \ln\left[f(1,0)\frac{\beta_0^*}{J_{ss}(g_*)}\right] + \ln\left[\frac{g_*^{1-\nu/d}}{\sqrt{\pi}\delta}\right]$$

$$\approx \ln\left[f(1,0)\frac{\beta_0^*}{J_{ss}(g_*)}\right] \tag{3.29}$$

where β_0^* is β_0 given by Eq. (2.113) for $\beta^*(g)$, and the approximation is always applicable considering the second term, $\ln[g_*^{1-\nu/d}/\sqrt{\pi}\delta]$, is negligibly small compared with the first one. When $\beta^*(g) = \beta(g)$ including the common E_g, the magnitude of β_0^* is readily derived by

$$\beta_0^* = \left[\frac{v_g(r)}{g^{1-\nu/d}}\frac{dg}{dr}\right]_{g=g(r)} \tag{3.30}$$

where r is the effective radius of clusters and $v_g(r) \equiv dr/dt$ is the growth rate with respect to r, i.e., in a unit of length. For instance, in the case of $g(r) = (4\pi/3V_1)r^3$ and $\nu = 1$ for spherical clusters, Eq. (3.30) becomes

$$\beta_0^* = \left(\frac{36\pi}{V_1}\right)^{1/3} v_g(r) \tag{3.31}$$

Even when $\beta^*(g) \neq \beta(g)$, β_0^* can be determined from $v_g(r)$, as long as the frequency of monomer addition, ω, including E_g, is common to the critical cluster and the observable g-sized cluster, and if the effective dimension, d, and the index for the growth mode, ν, are known for the both clusters. As an example of this case, if $\beta(g)$ is given by Eqs. (3.21) and (3.22) which are considered for the analysis in Figure 14, $\beta_0^*(g)$ is given by

$$\beta_0^* = 2\left(\frac{\pi\ell}{V_1}\right)^{3/2d}\frac{\kappa_3}{\kappa_2}r^{3/d-1}v_g(r) \tag{3.32}$$

With this method, one can determine W_* directly from only one set of $\{v_g(r), J_{ss}(g_*)\}$ measured at the same temperature, independent of any model for W_* and independent of E_g. Consequently, this method enables us to measure the temperature dependence of W_* and provides an opportunity to test any models for W_*.

Figure 17 shows an example of applying the present method to the nucleation and growth of FeBSi nanocrystallites in the metallic glass annealed at various temperatures. The results are

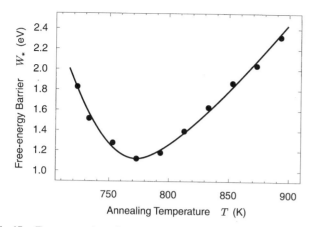

Fig. 17. Temperature dependence of the free-energy barrier to nucleation, W_*, directly determined by the method using the ratio of the growth rate, $v_g(r)$, and the steady-state nucleation rate, $J_{ss}(g_*)$, for the solid-phase crystallization of a-FeBSi alloy. The data of $v_g(r)$ and $J_{ss}(g_*)$ used for calculating W_* are collected from Figure 4 of [232]. The solid line is drawn just for a guide to the dependency. Note that W_* strongly depends on the annealing temperature.

calculated using the values of $\{v_g(r), J_{ss}(g_*)\}$ reported by Tong et al. [232]. It is seen that W_* strongly depends on temperature and exhibits an apparent valley in the observed range of the temperature. Thus the Arrhenius method is not definitely applicable to this system. Table I shows another example of applying the present method to the nucleation and growth in the solid-phase crystallization of a-Si thin films under various experimental conditions. The results are collected from those reported in [227] and are calculated using the numerical reports on $v_g(r)$ and $J_{ss}(g_*)$, which are cited also in Table I. It is found for the isothermal annealing that the directly determined W_* is around 2 eV regardless of the preparation conditions of the starting a-Si thin films. Also it seems that W_* does not exhibit an apparent dependency on the temperature, which suggests that the entropic barrier to nucleation could be small. For observing the temperature dependence of W_*, one needs more precise measurement under fixed conditions except the temperature.

Table I. Directly Determined Free-Energy Barrier to Nucleation in Solid-Phase Crystallization of a-Si Thin Films under Various Conditions

Ref.[a]	Starting material	Substrate	ℓ (nm)[b]	Crystallization	T (K)[c]	W_* (eV)[d]
[55]	EB a-Si[e]	NaCl, free	150–350	isothermal	824	2.04
					861	2.11
					920	2.08
[233]	↑	Si_3N_4	40	↑	983	2.08
[143]	↑	SiO_2	200	↑	895	2.05
			400			2.00
[159]	PECVD a-Si:H[f]	quartz	100	↑	827	1.98
					848	2.19
					840	2.09
					895	2.17
					920	1.99
[123]	LPCVD poly-Si[g]	↑	100	↑	903	1.98
[193]	↑	↑	150	↑	903	2.10
[234]	↑	↑	100	↑	873	2.10
					888	2.12
					903	2.05
				ion irradiated[h]	873	1.15
					888	1.20
					903	1.19

[a]The reference number for the reported source of $v_g(r)$ and $J_{ss}(g_*)$.

[b]The thickness of the a-Si thin film.

[c]The annealing temperature.

[d]The free-energy barrier to nucleation.

[e]a-Si deposited using an electron-beam gun.

[f]Hydrogen-rich a-Si deposited by plasma-enhanced chemical-vapor deposition.

[g]Low-pressure chemical-vapor deposition poly-Si amorphized by self-ion implantation.

[h]Xe^{2+}-ion-beam induced crystallization during the isothermal annealing.

3.5.1.3. Model-Independent Method Using the Size Distribution of Clusters

While the method using the ratio of $v_g(r)$ to $J_{ss}(g_*)$ is suitable for directly measuring W_*, these quantities are not always accessible, or at least, it is more delicate to precisely measure them than has been regarded in general. As an alternative to this method, a method using the size distribution of clusters is proposed and demonstrated [168, 235]. The alternative method is based upon the same kinetic basis as that of the method using the ratio of $v_g(r)$ to $J_{ss}(g_*)$, but it uses the size distribution of clusters instead.

Recall the size distribution of clusters far beyond the critical region, which is expressed by Eq. (2.81). It is found in this equation that the steady-state size distribution of Eq. (2.81b), $\varphi(g) = J_{ss}(g_*)/\beta(g)$, is just equivalent to the inverse of Eq. (3.28). Therefore, if it is possible to observe the steady-state size distribution and precisely measure its dependence on the cluster size, W_* can be determined in the similar manner of the method using the absolute values of $v_g(r)$ and $J_{ss}(g_*)$. According to Eqs. (2.39) and (2.81b), the steady-state size distribution should be a power-law function of the cluster size as

$$\varphi(g) = C g^{1-v/d} \tag{3.33}$$

with

$$C = \frac{\kappa_{v'd'}}{\kappa_{vd}} \frac{[f(1,0)]^{1-1/d'}}{[f(1,0)]^{1-1/d}} \frac{g_*^{1-v/d}}{\sqrt{\pi}\delta} f(1,0) \exp\left(-\frac{W_*}{kT}\right) \tag{3.34}$$

where d' and v' are for the critical cluster. Consequently, from the coefficient of the power-law size distribution, W_* is determined by

$$\begin{aligned}
\frac{W_*}{kT} &= -\ln C + \ln\left(\frac{\kappa_{v'd'}}{\kappa_{vd}}\right) \\
&\quad + \left(1 + \frac{1}{d} - \frac{1}{d'}\right) \ln f(1,0) + \ln\left(\frac{g_*^{1-v/d}}{\sqrt{\pi}\delta}\right) \\
&\approx -\ln C + \ln\left(\frac{\kappa_{v'd'}}{\kappa_{vd}}\right) + \left(1 + \frac{1}{d} - \frac{1}{d'}\right) \ln f(1,0)
\end{aligned} \tag{3.35}$$

where the approximation is as valid as in Eq. (3.29). Note that if the steady-state size distribution has been actually established and observable, it is not always necessary to determine the effective dimension, d, in advance, for the purpose of determining W_* by this method.

Even when the steady state has not been established in the observed range of the cluster size, one can extract $\varphi(g)$ from the overall transient size distribution, $f(g, t)$, by the proper numerical fitting to the measured data and then follow the above procedure. Otherwise, without the intermediate process of determining the coefficient, C, one may directly fit the overall $f(g, t)$ with W_* being regarded as one of the fitting parameters. Figure 14 shows an example of such a fitting based on $f(g, t)$ with Eqs. (3.21)–(3.25), by which the free-energy barrier to nucleation is determined to be $W_* = 2.149$ eV. In this figure, the extracted $\varphi(g)$ is also plotted by the dotted line. It is

apparently seen that the overall $f(g, t)$ is still more than several times distant from the extracted $\varphi(g)$ in the observed range of the cluster size. On the other hand, if one only glances at the shapes of the observed $f(g, t)$, it may appear as if there were a power-law range in a smaller size than $g \sim 10^9$. If we mistakenly regarded this pseudo-power-law range as the steady-state distribution, which is not true in this case, and directly analyzed it using Eqs. (3.33)–(3.35) without fixing d in advance, we must have obtained a quite different (i.e., wrong) value for W_*. These observations indicate that, for the precise measurement of W_*, the analysis of the size distribution with the overall $f(g, t)$ and the predetermined d is preferable to that with only $\varphi(g)$.

Since this method is based on the same kinetic basis as that of the method using the absolute values of $v_g(r)$ and $J_{ss}(g_*)$, the former possesses all of the advantages the latter has. One can determine W_* directly from only one size distribution measured at one temperature and at one time and hence is able to measure the temperature dependence of W_*. Also, the method is independent of any model for W_* and independent of the energy barrier to growth and hence suitable to test any models for W_*. Unlike the method using $v_g(r)/J_{ss}(g_*)$, the present method is free from the experimental difficulties in precisely measuring their absolute values, such as fixing the detection limit of small clusters, the estimates of $v_g(r)$, and the often relying on the inevitable use of Johnson–Mehl–Avrami formalism. These difficulties can be easily avoided in the present method by using only the reliable range of the cluster size, by not measuring the unnecessary growth rate, by observing the thin films before the depletion of monomers or nucleation sites becomes significant, etc.

3.5.2. Enthalpic and Entropic Barriers to Nucleation

The free-energy barrier to nucleation, W_*, essentially changes with the temperature, T, as described by Eq. (3.26), unless the entropic barrier is negligibly smaller than the enthalpic barrier. When the temperature dependence of W_* is observed in their determination by the direct methods, the enthalpic and entropic barriers can be estimated by

$$\frac{H_*}{kT^2} = \frac{d}{dT}\left(\frac{W_*}{kT}\right) \tag{3.36a}$$

$$S_* = -\frac{dW_*}{dT} \tag{3.36b}$$

respectively. These enthalpic and entropic barriers to nucleation can provide various information about the nucleation and growth of clusters and the phase transformation as shown in the following examples.

If W_* exhibits a linear dependence on T, H_* and S_* could be constant within the observed range of the temperature. Figure 18 shows such an example of the temperature dependence of W_* which is measured in the solid-phase crystallization of amorphous $CoSi_2$ thin films by the direct method using the ratio of $v_g(r)$ to $J_{ss}(g_*)$ [228]. With this dependence on T and Eq. (3.36), the enthalpic and entropic barriers are determined to

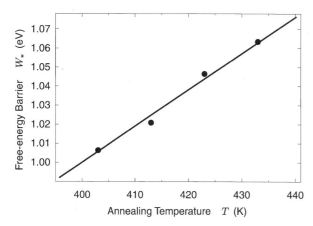

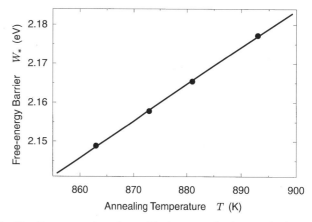

Fig. 18.   Temperature dependence of the free-energy barrier to nucleation, W_*, observed in the solid-phase crystallization of 100 nm thick amorphous $CoSi_2$ thin films. The plots of W_* are recalculated by the direct method using the absolute values for v_g and $J_{ss}(g_*)$ reported in Table I of [228]. The solid line indicates the linear dependence based on the enthalpic and entropic barriers determined by Eq. (3.36).

Fig. 19.   Temperature dependence of the free-energy barrier to nucleation, W_*, observed in the solid-phase crystallization of a-Si thin films. The results are obtained by the direct method using the size distribution of clusters as shown in the analysis of Figure 14. The plot data of W_* originate in the same source used in Figure 6 of [168]. The solid line indicates the linear dependency based on the determined enthalpic and entropic barriers by Eq. (3.36)

be $H_* = 0.240$ eV and $-TS_* = 1.899 \times 10^{-3}T$ eV, respectively. The positive value obtained for H_* is a manifestation of the small size of the critical clusters. Since the enthalpic barrier is the difference in the enthalpy between the critical cluster and the monomer, it should be described as $H_* \equiv h_{cn}(g_*) - g_* h_a$, where $h_{cn}(g_*)$ represents the enthalpy of g_*-sized crystalline critical cluster, and h_a is the enthalpy of a monomer in the bulk amorphous phase. It is noted that, under the same conditions of the crystallization, the enthalpy of a bulk crystalline phase, h_c, should be less than that of a bulk amorphous phase, h_a. Thus the positive H_* indicates that there is another positive contribution to $h_{cn}(g_*)$ which is most likely due to the interface creation associated with the nucleation. If the crystallization temperature is raised higher, the critical size, g_*, increases in general. For such a sufficiently large critical crystallite, the positive contribution may become negligibly small compared to h_c, resulting in a negative H_*. It is more remarkable here that the entropic barrier, $-TS_* \sim 0.76$–0.82 eV, is significantly larger than the enthalpic barrier, $H_* = 0.240$ eV. In other words, the free-energy barrier to nucleation comes mainly from the entropic barrier, which corresponds to the difference in the ordering between the crystalline and amorphous $CoSi_2$. Conversely, the relatively small enthalpic barrier indicates that the difference in the bond strength between the two phases is not large. These observations suggest that the nucleation event is essentially a process of topological rearrangement of the disordered amorphous structure into the ordered crystalline one.

Figure 19 shows another example of the temperature dependence of W_* which is measured in the solid-phase crystallization of a-Si thin films by the direct method using the size distribution of clusters [199] as shown in the analysis of Figure 14. The enthalpic and entropic barriers are determined to be $H_* = 1.332$ eV and $-TS_* = 9.458 \times 10^{-4}T$ eV, respectively. As for the positive value obtained for H_*, the same discussion is valid here as that for the solid-phase crystallization of a-$CoSi_2$. However, unlike the result for $CoSi_2$, both

the enthalpic barrier ($H_* = 1.332$ eV) and the entropic barrier ($-TS_* \sim 0.81$–0.85 eV) are significant parts of the free-energy barrier to nucleation in the a-Si thin films. This observation suggests that the nucleation of crystallites in the a-Si thin films used for this experiment progresses not only by the topological rearrangement of a simply disordered amorphous structure like a continuous random network model, but also with some enthalpic changes accompanied by constructing new atomic bonds from dangling bonds or broken bonds.

3.5.3. Energy Barrier to Growth

The energy barrier to growth, E_g, has been central to the study of the growth mechanisms of thin films and the control of the growth rate. Since E_g is just the enthalpic barrier associated with the atomic jump, displacement, or diffusion, its value can be determined simply as the activation energy of the growth rate, v_g, or the frequency of monomer addition per unit area, ω', by their Arrhenius plots.

Figure 20 shows the Arrhenius plot of the growth rate, v_g, of $CoSi_2$ crystallites grown by the solid-phase crystallization of a-$CoSi_2$ thin films. The data for v_g are collected from Table I of [228] and are the same as those used for calculating W_* shown in Figure 18. With this Arrhenius slope, the energy barrier to growth is determined to be $E_g = 1.134$ eV. Figure 21 shows the Arrhenius plot of the frequency of the monomer addition per unit area, ω', observed in the solid-phase crystallization of a-Si thin films. The data for ω' are obtained by the analysis of the size distribution of clusters as shown in Figure 14, in which the values of ω are determined as one of the optimized fitting parameters. With this Arrhenius slope, the energy barrier to growth is determined to be $E_g = 3.42$ eV, which is close to the estimates by the Arrhenius plots of v_g, for the similar a-Si thin films, e.g., $E_g = 3.3$ eV [123] and 3.1 ± 0.6 eV [159].

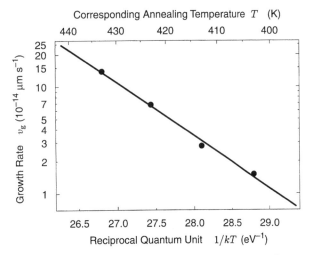

Fig. 20. Arrhenius plot of the growth rate, v_g, of CoSi$_2$ crystallites grown by the solid-phase crystallization of a-CoSi$_2$ thin films. The data for v_g are collected from Table I of [228] and are the same as those used for calculating W_* shown in Figure 18. The upper horizontal axis represents the corresponding annealing temperature.

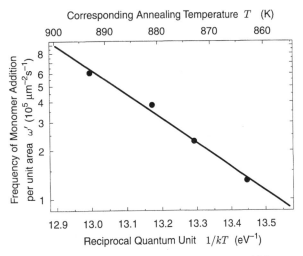

Fig. 21. Arrhenius plot of the frequency of the monomer addition per unit area, ω', into Si crystallites grown by the solid-phase crystallization of a-Si thin films. The data for ω' originate in the same source used in Figure 7 of [168], which are obtained by the analysis of the size distribution of clusters as shown in Figure 14 regarding ω' as one of the optimized fitting parameters.

4. CONTROL OF NUCLEATION AND GROWTH

The control of the nucleation and growth of clusters is the initial approach to the control of the microstructures in thin films formed by this process. In this section, the controls of three microstructures in granular thin films are discussed, which are the grain size, the size distribution of grain, and the grain locations. The term "grain" or "crystallite" is mainly used here instead of "cluster."

4.1. Grain Size

The grain size is one of the greatest concerns in the microstructures of granular materials such as polycrystalline thin films.

If the desired properties of the thin films come from the large surface-to-volume ratio of the grains, such as quantum confinement, the grain size has to be smaller than the threshold size for the appearance of the small-size effect. If the bulk properties inside the grains are desired, the grain size should be as large as possible because of the following reason. In the case of densely packed granular thin films, their structure consists of the grains and the grain boundaries. The grain boundaries which are the interface between the adjacent grains generally possess different properties from those inside the grains and often impede the appearance of the expected bulk properties of the grains on average. The total area of the grain boundaries per unit volume is reciprocally proportional to the mean grain size, when they are represented in a unit of length. Thus if the mean grain size increases, the density of the grain boundaries decreases, and consequently the bulk properties of the grains become manifest. For instance, the transport of electric charges in polycrystalline Si thin films strongly depends on their mean crystallite size [236–238]. In any case, it is crucial to control the grain size.

The grain size in question is generally the final grain size, since thin films are used after the complete transformation of the thin films. As given by Eq. (2.125) in Section 2.3.6, the mean of the final grain size, $\overline{g_f}$, depends only on the initial concentration of monomers available to compose the clusters, $f_m(0)$, the initial concentration of nucleation sites, $f(1, 0)$, the effective dimension, d, the chemical potential, $\Delta\mu$, the temperature, T, and the free-energy barrier to nucleation, W_*. In the following, several cases are introduced for showing the instances of controlling the grain size by some of these parameters.

4.1.1. Control by Formation Temperature

Among the control parameters, the mean of the final grain size depends most strongly on those at the index of the exponential in Eq. (2.125b). Thus one can consider the dependence of the mean of the final grain size as

$$\overline{r_f} \propto \exp\left(\frac{1}{d+1}\frac{W_*}{kT}\right) \qquad (4.1)$$

where $\overline{r_f}$ ($\propto \overline{g_f}^{1/d}$) stands for the mean of the final grain size represented by the effective radius in a unit of length. If d is fixed for the growth of the grains, the dimensionless quantity of W_*/kT is the most dominant parameter over $\overline{r_f}$.

Equation (4.1) suggests that $\overline{r_f}$ could be controlled by the temperature for the formation of the thin films, unless $d(W_*/kT)/dT = 0$. The temperature dependence of $\overline{r_f}$ is controlled by that of W_*. If $dW_*/dT > W_*/T$ (> 0), $\overline{r_f}$ is expected to increase with T. Inversely if $dW_*/dT < W_*/T$ which is satisfied always when $dW_*/dT \leq 0$, $\overline{r_f}$ may decrease with T. For example, Iverson and Reif [123] estimate the temperature dependence of $\overline{r_f}$ for the solid-phase crystallization of a-Si thin films using the data set of the growth rate and the steady-state nucleation rate they measured and a counterpart of Eq. (2.125), and regarding W_* as constant with respect to T.

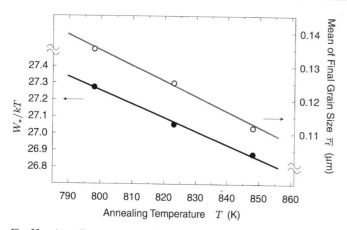

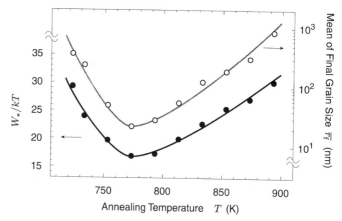

Fig. 22. Annealing temperature dependence of W_*/kT (●) and the mean of the final grain size, $\overline{r}_f$ (○), for the solid-phase crystallization of a-SiGe thin films. The plots for W_*/kT are calculated from the data of the growth rate and the steady-state nucleation rate reported in Table 2 of [231]. The data for the plots of $\overline{r}_f$ are collected from Figure 6 of the same reference. The solid linear lines are drawn as guides.

Fig. 23. Annealing temperature dependence of W_*/kT (●) and the mean of the final grain size, $\overline{r}_f$ (○), for the solid-phase crystallization of a-FeBSi alloy. The plots for W_*/kT are calculated from the data for W_* shown in Figure 17 where the method for determining W_* is described. The data for the plots of $\overline{r}_f$ are collected from the results reported in Figure 3 of [232]. The solid lines are drawn as guides.

As expected from the above discussion, they observe the decrease of $\overline{r}_f$ with T and an Arrhenius dependence with the activation energy of 0.6 eV which is consistent with their estimation of $W_* = 2.4$ eV according to Eq. (4.1) with $d = 3$. Unlike such an estimation, two examples below will show the temperature dependence of actually measured $\overline{r}_f$ and directly determined W_*/kT.

4.1.1.1. Solid-Phase Crystallization of SiGe Thin Films

Figure 22 shows the temperature dependence of W_*/kT and $\overline{r}_f$ for the formation of polycrystalline SiGe thin films by the solid-phase crystallization of a-SiGe thin films. The 130 nm thick a-SiGe thin films are deposited over SiO_2 substrates by low-pressure chemical-vapor deposition using a mixture gas of SiH_4 and GeH_4, and then they are isothermally annealed for the crystallization in a temperature range of 798–848 K. The plots for W_*/kT are calculated from the data for the growth rate and the steady-state nucleation rate reported in Table 2 of [231], with the direct method for determining W_* using their ratio as described in Section 3.5.1. The data for the plots of $\overline{r}_f$ are collected from the results reported in Figure 6 of the same reference. It is found that the directly determined W_* increases with T from 1.88 to 1.96 eV for the observed temperature range. However, as shown in Figure 22, the resultant W_*/kT decreases with T because $d(W_*/kT)/dT$ (= 1.77×10^{-3} eV K^{-1} on average over the temperature range) is smaller than W_*/T (= 2.32–2.35 × 10^{-3} eV K^{-1}). Thus $\overline{r}_f$ also decreases with T within the observed range of temperature. It is predicted from the measured decrease of W_*/kT and Eq. (4.1) that $\overline{r}_f$ should decrease by 14% which is in agreement with the actually measured decrease of $\overline{r}_f$ by 17%.

4.1.1.2. Solid-Phase Crystallization of FeBSi Alloy

Figure 23 shows the annealing temperature dependence of W_*/kT and $\overline{r}_f$ for the formation of FeBSi nanocrystallites in

the metallic glass. The plots for W_*/kT are calculated from the data for W_* shown in Figure 17 where the method for determining W_* is described. Since in this case W_* changes much with T, exhibiting an apparent valley at around 775 K as shown in Figure 17, the resultant W_*/kT reflects its dependence as shown in Figure 23. The data for the plots of $\overline{r}_f$ are collected from the results reported in Figure 3 of [232] and are represented in log scale. It is apparently seen that $\overline{r}_f$ exhibits a quite similar dependence on T to that of W_*/kT. For the quantitative evaluation, if one compares the plots at $T = 723$ K to those at $T = 773$ K, the ratio of $\overline{r}_f$ expected from the ratio of W_*/kT and Eq. (4.1) is about 20 which agrees approximately with about 17 determined from the actually measured $\overline{r}_f$.

4.1.2. Control by Composition of Thin Films

Equation (4.1) also suggests that $\overline{r}_f$ could be controlled only by the free-energy barrier to nucleation, W_*, without changing the formation temperature. The magnitude of W_* can vary with the materials of thin films in general. If it is acceptable to alter the atomic composition of the thin films made from compound materials, W_* could be controlled by shifting the composition ratio. Below is shown the example of controlling $\overline{r}_f$ of polycrystalline $Si_{1-x}Ge_x$ thin films by the x.

In the solid-phase crystallization of $Si_{1-x}Ge_x$ amorphous compounds, the crystallites nucleating in the amorphous matrix suffer an amount of strain because of the difference in the lattice constant between crystalline Si and Ge. If the capillary model described by Eq. (2.5) is regarded as a good approximation for this system, the free energy of the $Si_{1-x}Ge_x$ crystallites can be expressed as [239]

$$W(g) = -g\Delta\mu + \sigma\kappa g^{1-1/d} + E_s g$$

where E_s is the strain energy per monomer associated with the difference in the lattice constant. Thus the free-energy barrier

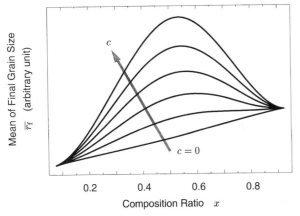

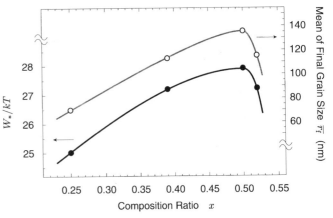

Fig. 24. Simulated dependence of the mean of the final grain size, $\overline{r}_f$, on the composition ratio, x, of polycrystalline $Si_{1-x}Ge_x$ thin films formed by the solid-phase crystallization at 823 K, with the undetermined scaling parameter, c, being varied from 0. The simulation is based on the theoretical description by Eqs. (4.1)–(4.6) with the known constants for crystalline Si, Ge, and SiGe.

Fig. 25. Dependence of W_*/kT (•) and the mean of the final grain size, $\overline{r}_f$ (○) on the composition ratio, x, of polycrystalline $Si_{1-x}Ge_x$ thin films formed by the solid-phase crystallization at 823 K. The plot data are collected and calculated from the results reported in Figure 3 of [243]. The solid lines are drawn as guides.

to nucleation is derived as

$$W_* = \left(\frac{\sigma\kappa}{d}\right)^d \left(\frac{\Delta\mu - E_s}{d-1}\right)^{1-d} \tag{4.2}$$

Here, E_s should be a function of the composition ratio, x, and should be given by [240, 241]

$$E_s(x) = c\left[(1-x)^2\left(\frac{a_x - a_{Si}}{a_{Si}}\right)^2 + x(1-x)\left(\frac{a_x - a_{SiGe}}{a_{SiGe}}\right)^2 + x^2\left(\frac{a_x - a_{Ge}}{a_{Ge}}\right)^2\right] \tag{4.3}$$

where c is the unknown scaling parameter corresponding to the elastic energy density, a_{Si}, a_{Ge}, a_{SiGe}, and a_x are the lattice constants of crystalline Si, Ge, SiGe (i.e., $x = 0.5$ for $Si_{1-x}Ge_x$), and $Si_{1-x}Ge_x$ at arbitrary x, respectively. The lattice constant of crystalline $Si_{1-x}Ge_x$ is generally undetermined but can be estimated by [241]

$$a_x = (1-x)a_{Si} + xa_{Ge} \tag{4.4}$$

Similarly, the free-energy difference per monomer between the amorphous and the crystalline $Si_{1-x}Ge_x$, $\Delta\mu$, can be obtained from

$$\Delta\mu(x) = (1-x)\Delta\mu_{Si} + x\Delta\mu_{Ge} \tag{4.5}$$

where $\Delta\mu_{Si}$ and $\Delta\mu_{Ge}$ are the free-energy difference for pure Si and Ge, respectively. The free-energy difference can be generally given by [242]

$$\Delta\mu_X = \left[(\ln 2 - 1)T + \frac{T_{mX}}{2}\right]\frac{\Delta H_X}{T_{mX}} \tag{4.6}$$

where X denotes Si or Ge, T is the crystallization temperature, T_{mX} is the melting temperature of the X element, and ΔH_X stands for the enthalpy difference between the amorphous and the crystalline X.

The theoretical description by Eqs. (4.2)–(4.6) suggests that W_* can be dependent on the composition ratio, x, and hence the mean of the final grain size, $\overline{r}_f$, can be controlled by x. Figure 24

shows the dependence of $\overline{r}_f$ on x, which is simulated based upon Eqs. (4.1)–(4.6) at $T = 823$ K with $d = 3$ and the known constants for crystalline Si, Ge, and SiGe. The unknown scaling parameter, c, which is the magnitude of the elastic contribution to $W(g)$, is varied from 0. As seen in the figure, the theory predicts that at small c, $\overline{r}_f$ monotonously increases with x, but when c exceeds a certain value, $\overline{r}_f$ exhibits a peak around $x = 0.5$–0.6. Such a dependence of $\overline{r}_f$ on x is actually observed in the experimental study. Figure 25 shows the experimentally observed dependence of $\overline{r}_f$ and that of W_*/kT on x. The a-$Si_{1-x}Ge_x$ thin films are prepared by the same procedures of those for Figure 22 and are then isothermally annealed at 823 K. The plot data are collected and calculated from the results reported in Figure 3 of [243]. The magnitude of W_* for calculating W_*/kT is determined by the direct method using the ratio of the growth rate to the steady-state nucleation rate. The observed $\overline{r}_f$ exhibits a peak at $x \sim 0.5$, which agrees with the theoretical prediction shown in Figure 24. The observed W_*/kT also exhibits a quite similar dependence on x. These observations with the theoretical consideration present a typical example of controlling $\overline{r}_f$ by the composition ratio in compound thin films through the change of the free-energy barrier to nucleation.

4.1.3. Control by States of Starting Materials

The method described in Section 4.1.2 for controlling the mean of the final grain size, $\overline{r}_f$, is applicable only to the compound thin films and if the variation of the composition is acceptable for the use of the transformed thin films. If the thin films are made from a single element or if the composition ratio of compound thin films may not be changed, the states of the starting thin films can control the free-energy barrier to nucleation, W_*, and $\overline{r}_f$. Actually, the control of the grain size in the thin films made from a single element has been often attempted by modifying the starting thin films. For example, the grain size of polycrystalline Si thin films formed by the isothermal annealing of a-Si thin films has been enlarged mainly by introducing various formation methods of the starting a-Si and by optimizing the conditions of the formation, which could alter the states of the a-Si [244]. The introduced formation methods are vacuum evaporation deposition [245, 246], electron-beam vacuum deposition [178], radio-frequency [144] or dc [220] magnetron sputtering deposition, atmospheric-pressure [175] or the low-pressure (LP) [126, 128] chemical-vapor deposition (CVD) using SiH$_4$ source gas, amorphization of polycrystalline Si thin films by ion bombardment [145, 247, 248], plasma-enhanced CVD [249], LPCVD using Si$_2$H$_6$ source gas [155, 250], ion implantation into a-Si thin films prepared by LPCVD using SiH$_4$ source gas [163, 202, 203], rapid-thermal CVD using Si$_2$H$_6$ source gas [251], and so on. The optimized formation conditions are the deposition rate, the deposition temperature, the deposition pressure, the dopant concentration, the conditions for the ion implantation, the source gas species for the deposition, etc. These formation methods and conditions could control the states of the a-Si such as bonding structure, impurities, voids, residual strain, surface roughness, and so on. Although there are few comprehensive and quantitative studies for the comparison among the various preparation methods of the starting a-Si thin films, a few examples of controlling W_* and $\overline{r_f}$ by the formation conditions are shown in the following.

4.1.3.1. Low-Pressure Chemical-Vapor Deposition using Si$_2$H$_6$ Source Gas

In the solid-phase crystallization of a-Si thin films prepared with CVD systems using silanes as the source gas, it has been observed that the final grain size of the crystallized poly-Si thin films apparently depends on the deposition temperature, T_d, for the formation of the starting a-Si thin films [148, 154, 155, 158, 161]. Particularly, the final grain size exhibits a peak at a certain T_d in the LPCVD systems using SiH$_4$ [148, 158] and Si$_2$H$_6$ [155, 158, 161]. Nakazawa and Tanaka observed the strong correlation between the peak intensity ratio of the Raman spectra and the nucleation rate in the SiH$_4$-PECVD (plasma enhanced chemical upor deposition) system [154] and in the Si$_2$H$_6$-LPCVD system [155]. They attributed the dependence of the final maximum grain size on T_d to the difference in the structural disorder of the a-Si. Voutsas et al. observed in their detailed Raman study [125] that the degree of the structural disorder estimated from Raman spectra monotonically decreased with T_d and could not account by itself for the non-monotonic dependence of $\overline{r_f}$ on T_d. Furthermore, Voutsas and Hatalis [161] found at a fixed T_d that $\overline{r_f}$ increased only with the deposition rate. This finding contradicts the result of the Raman study [125] in which the degree of the structural disorder also increases with the deposition rate, because the large structural disorder should enhance the free-energy difference between the amorphous and the crystalline phases, $\Delta\mu$, then reduce W_*, and consequently reduce $\overline{r_f}$ according to Eqs. (2.6) and (4.1). Regardless of this contradiction, however, the dependence of $\overline{r_f}$ can be understood in the theoretical framework for Eq. (4.1) if one observes the dependence of W_*/kT as follows.

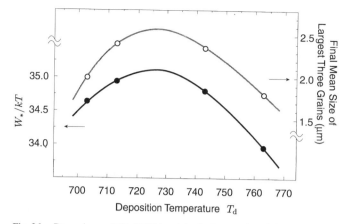

Fig. 26. Dependence of W_*/kT (•) and the final mean size of the largest three grains (○) on the deposition temperature of the starting a-Si thin film, T_d, in the solid-phase crystallization of a-Si thin films at 823 K, deposited by LPCVD using Si$_2$H$_6$ source gas at T_d. The plots for W_*/kT are calculated from the data of the growth rate and the steady-state nucleation rate reported in Figure 6 of [155]. The data for the plots of the final grain size are collected from Figure 4 of the same reference. The solid lines are drawn as guides.

Figure 26 shows the dependence of W_*/kT and the final grain size on T_d. The plots for W_*/kT are calculated from the data of the growth rate, v_g, and the steady-state nucleation rate, J_{ss}, reported in Figure 6 of [155], with the direct method for determining W_* using their ratio as described in Section 3.5.1. The data for the plots of the final grain size are collected from the results reported in Figure 4 of the same reference. The a-Si thin films are deposited by LPCVD using Si$_2$H$_6$ source gas and then crystallized in solid phase at 873 K. The deposition rate is presumably varied with T_d so that the dependence on T_d results from both T_d and the deposition rate. The final grain size is not equal to $\overline{r_f}$, but the mean radius of the largest three grains sampled in the completely crystallized films, which is expected to reflect $\overline{r_f}$. It is seen that both W_*/kT and the final grain size exhibit quite similar dependence on T_d showing a peak at around $T_d = 730$ K.

Figure 27 shows the results of the same analyses as those in Figure 26, using the data set reported in Figures 5, 7, and 9 of [161]. The 100 nm thick a-Si thin films are deposited at a constant deposition rate of 0.06 nm s^{-1} for all T_d by adjusting the other deposition conditions and are isothermally annealed at 873 K. Thus the effect of the deposition rate variation is eliminated, while the influence of changing the other deposition conditions is unknown. The grain size here is measured as being $\overline{r_f}$ and is different from that in Figure 26. There is a considerable difference between the magnitude of W_*/kT in Figure 26 and that in Figure 27, which comes from the significant difference in the magnitudes of v_g and J_{ss} that the original investigators estimated. Nevertheless, it is observable also in Figure 27 that both W_*/kT and $\overline{r_f}$ exhibit a quite similar dependence on T_d, showing a peak at around $T_d = 735$ K as well as those in Figure 26.

Figures 26 and 27 indicate that $\overline{r_f}$ changes, obeying W_*/kT for the two different experiments in spite of the significant difference in their magnitudes of v_g and J_{ss}. This observation con-

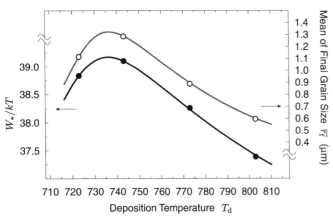

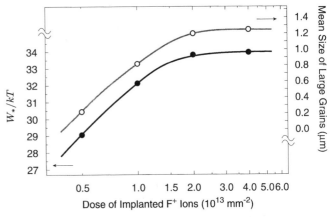

Fig. 27. Dependence of W_*/kT (●) and the mean of the final grain size, $\overline{r_f}$ (○), on the deposition temperature of the starting a-Si thin film, T_d, in the solid-phase crystallization of a-Si thin films at 823 K, deposited by LPCVD using Si_2H_6 source gas at T_d. The plots for W_*/kT are calculated from the data of the growth rate and the steady-state nucleation rate reported in Figures 7 and 9 of [161]. The data for the plots of the grain size are collected from Figure 5 of the same reference. The solid lines are drawn as guides.

Fig. 28. Dependence of W_*/kT (●) and the mean size of large grains (○) on the implanted dose of the F^+ ions, which is observed in the solid-phase crystallization of 150 nm thick a-Si thin films. The a-Si thin films are deposited at 818 K in a SiH_4-LPCVD system, then implanted with the F^+ ions accelerated to 60 keV, and finally annealed at 873 K. The plots for W_*/kT are calculated from the data of the growth rate and the steady-state nucleation rate reported in Figure 3 of [163]. The data for the plots of the grain size are collected from Figure 4 of the same reference. The solid lines are drawn as guides.

firms that $\overline{r_f}$ is controlled mostly by W_*/kT whatever the origins of the change of W_*/kT are. They also demonstrate that $\overline{r_f}$ can be controlled by the state of the starting thin films.

4.1.3.2. Ion Implantation into a-Si Thin Films Deposited in SiH₄-LPCVD Systems

The solid-phase crystallization of a-Si thin films formed in SiH_4-LPCVD systems yields the poly-Si thin films whose typical grain size is smaller than half a micron meter. Wu et al. [202, 203] invented a method for enlarging the grain size obtained from the SiH_4-LPCVD by self-ion implantation into the as-deposited a-Si thin films prior to the isothermal annealing for their crystallization. Park et al. [163] observed almost the same effect replacing the silicon ions by fluorine ions. Under the optimum conditions of the ion implantation, the resultant grain size is enlarged over micron meters. When the silicon ions are implanted into the a-Si, both the as-deposited and the implanted thin films are a-Si. Even with the fluorine ions, their content is no more than that of the other impurities in the as-deposited a-Si thin films since the dose is no larger than $\sim 10^{14}$ mm^{-2}. Hence, this technique only modifies the state of the a-Si thin films. Some of the analyses in terms of the change of W_*/kT are given below.

The dependence of W_*/kT and the final grain size on the dose of fluorine ions is shown in Figure 28, which is observed in the following experiment and the analysis. The 150 nm thick a-Si thin films are deposited at $T_d = 818$ K over SiO_2 substrates by LPCVD using 100% SiH_4; then F^+ ions accelerated to 60 keV are implanted into the as-deposited a-Si thin films at various dose, and the implanted films are annealed at 873 K. The grain size is the mean size of typical large grains observed in the completely crystallized films, which is expected to reflect $\overline{r_f}$. The data for the plots of the mean size of large grains are collected from the results reported in Figure 4 of [163]. The

plots for W_*/kT are calculated from the data of the growth rate, v_g, and the steady-state nucleation rate, J_{ss}, reported in Figure 3 of the same reference, with the direct method for determining W_* using their ratio as described in Section 3.5.1. It is seen that both W_*/kT and the the mean size of large grains increase with the dose of F^+ ions, exhibiting a quite similar dependence, and seem to saturate above $\sim 2 \times 10^{13}$ mm^{-2}. When comparing the results at the doses of 5×10^{12} and 5×10^{13} mm^{-2}, the increase of W_*/kT is about 4.9 which agrees with about a 5.6 times increase of the grain size. Thus the enlargement of the grain size can be attributed to the enhancement of W_*/kT.

For the case of the self-ion implantation, the measured data of the final grain size is not available, but the dependence of W_* on the ion conditions is precisely measured. Figures 29 and 30 show the dependence of W_*/kT on the dose and the acceleration energy of the Si^+ ions, respectively, which are observed in the following experiment and analysis [252]. The a-Si thin films are deposited up to a thickness of 100 nm by LPCVD at a deposition temperature of 823 K, under a pressure of 40 Pa, and at a deposition rate of 2.8×10^{-2} nm s^{-1}. Then the Si^+ ions are uniformly implanted into the films at a fixed acceleration energy of 70 keV with the ion dose being varied (Fig. 29), or at a fixed ion dose of 1×10^{13} mm^{-2} with the acceleration energy being varied (Fig. 30). Finally, the self-implanted a-Si thin films are crystallized by the isothermal annealing in nitrogen atmosphere at 873 K. The magnitudes of W_* for obtaining W_*/kT are determined by the direct method using the size distribution of the crystallites as shown in Figure 14 and Section 3.5.1.

As seen in Figures 29 and 30, W_*/kT remarkably increases as the ion dose or the acceleration energy increases. Such large enhancement of W_*/kT suggests that the grain size is expected to greatly enlarge by the self-ion implantation. The TEM images in Figure 31 show an example of comparison between the crystalline grains obtained with two different conditions of the

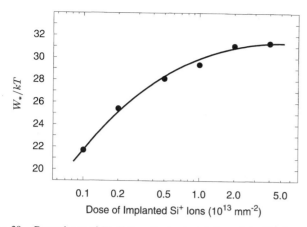

Fig. 29. Dependence of W_*/kT on the implanted dose of the Si$^+$ ions accelerated to 70 keV, which is observed in the solid-phase crystallization of a-Si thin films. The 100 nm thick a-Si thin films are deposited at 823 K in a SiH$_4$-LPCVD system, then implanted with the Si$^+$ ions, and finally annealed at 873 K. The data of W_* for the plots of W_*/kT are determined directly from the size distribution of the crystallites as in [235]. The solid line is drawn as a guide.

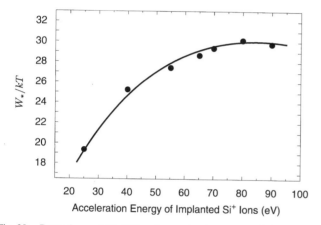

Fig. 30. Dependence of W_*/kT on the acceleration energy of the implanted Si$^+$ ions at a fixed dose of 1×10^{14} mm^{-2}, which is observed in the solid-phase crystallization of a-Si thin films. The 100 nm thick a-Si thin films are deposited at 823 K in a SiH$_4$-LPCVD system, then implanted with the Si$^+$ ions, and finally annealed at 873 K. The data of W_* for the plots of W_*/kT are determined directly from the size distribution of the crystallites as in [235]. The solid line is drawn as a guide.

self-ion implantation. The a-Si thin films are partially crystallized for appropriate annealing durations so as to see the individual grains in the bright-field plan-view TEM images. The conditions of the self-ion implantation are picked out of those on Figure 30 so that the difference in the grain size is visible in the TEM images presented at the same magnification: the accelerating energies are 40 keV (Fig. 31a) and 70 keV (Fig. 31b), and the dose of 1×10^{14} mm^{-2} is common to both. It is seen in these TEM images that most grains obtained at 40 keV are as small as submicron meters while the typical grains obtained at 70 keV exceed 1 μm in diameter. According to Figure 30, the enhancement of W_*/kT between these accelerating energies is about 4 which is in agreement of the observed enlargement of the grain size. Although the result of this comparison is just

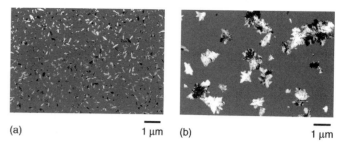

(a)   1 μm   (b)   1 μm

Fig. 31. Bright-field plan-view TEM images of partially crystallized a-Si thin films presented at the same magnification. The a-Si thin films are prepared by LPCVD using SiH$_4$ source gas, and Si$^+$ ions are implanted prior to the isothermal annealing at 873 K. The conditions of ion implantation and the annealing duration are (a) 40 keV, 1×10^{14} mm^{-2}, 1.44×10^4 s, and (b) 70 keV, 1×10^{14} mm^{-2}, 6.30×10^4 s.

for the partially crystallized films, the similar enlargement of $\overline{r}_f$ may be observed also in the completely crystallized films.

The observations and analyses in Figures 28–31 indicate that the ion implantation into the a-Si thin films prepared in SiH$_4$-LPCVD systems is effective in enhancing W_* and suppressing the nucleation in the subsequent crystallization. The dependence of W_*/kT on the acceleration energy of the implanted ions could provide a clue to the reason for the suppression of the nucleation. As seen in Figure 30, the increase of W_*/kT decelerates for the larger acceleration energy and W_*/kT seems to show a broad peak at around 80 keV. In the present case of the self-ion implantation into 100 nm thick a-Si thin films, the projected range of the implanted Si$^+$ ions reaches the interface to the SiO$_2$ underlayer when the Si$^+$ ions are accelerated to about 70 keV. The range of the energy deposit by the collision of Si$^+$ ions becomes maximum at the interface when the acceleration energy is about 80 keV. These observations suggest that the suppression of the nucleation results mainly from the modification near the interface between the a-Si and the SiO$_2$. There have been some speculations as to what is modified at the interface, such as the oxygen recoil from the SiO$_2$ to the a-Si by the impact of the ions [202, 203], the increase in the structural disorder of the amorphous network [163], and the amorphization of the preexisting crystalline clusters localized near the interface [9]. In any case, the self-ion implantation must modify the state of the a-Si especially in the vicinity of the interface to the underlayer, which enhances W_* and hence W_*/kT for the nucleation of crystallites, and consequently enlarge the grain size.

4.1.4. Control by Coarsening of Grains

In the previous sections, we have discussed the control of grain size by controlling nucleation through W_*/kT. There are some other approaches to the control of grain size. One of the promising approaches is to utilize a phenomenon known as "coarsening" of grains, which explosively enlarges the size of grains having already nucleated, i.e., independent of the nucleation process. In this phenomenon, relatively large grains grow while the others do not grow or shrink, and consequently the mean grain size increases. Such a coarsening phenomenon has been observed in various systems for the formation of condensed matter. The coarsening can occur from the beginning of the formation of thin films, halfway through the formation after a certain size distribution of grains is established, or in the post-formation processes after the primary transformation for the formation of thin films is completed. For instance, in the vapor phase growth of grains, the high mobility of small grains in free space often enables them to coalesce resulting in their coarsening. Phenomena called "normal" and "secondary grain growth" [253] correspond to the coarsening in the postformation processes. When the columnar polycrystalline thin films are annealed at considerably higher temperature than their primary formation temperature, the grain boundaries of relatively large grains quickly move to expand their regions, incorporating the surrounding smaller grains due to the excess free energy resulting from the grain boundaries or the crystallographic anisotropy in the surface energy of grains. Consequently, the number concentration of smaller grains decreases while that of larger grains increases, and the mean of the grain size increases. In the case of thin films which are not densely packed with grains, a well-known phenomenon called "Ostwald ripening" can affect similarly on the grain size. Below is shown an example of the coarsening observed halfway through the formation.

When crystalline Si thin films are deposited by CVD under the conditions that single crystalline Si epitaxially grows, three-dimensional islands of crystalline Si grains are formed over insulating substrates such as silicon oxides and nitrides. In the CVD using the gases of SiH_4 or SiH_2Cl_2, HCl, and H_2 at around 1200 K under low pressure conditions, it had been known that the all Si grains are formed instantly after a short incubation period and only grow afterward [254]. It is found, however, under conditions with high HCl partial pressures, that an anomalous coarsening of Si grains occurs in halfway through film formation [132]. Figure 32 shows the dynamic evolution of the size distribution of the grains and their slanted-view SEM micrographs observed during the coarsening phenomenon. Here, Si is deposited over SiO_x ($x < 2$) substrates at 1223 K under 2×10^4 Pa, using gases of SiH_2Cl_2, HCl, and H_2 at a flow rate of 8.83×10^{-3}, 0.03, and $1.67\,\ell\,s^{-1}$, respectively. The grain size is represented by the effective radii of hemispheres whose areas projected onto the substrate are equivalent to those of the Si grains on their plan-view SEM images. It was observed that the Si grains were formed as soon as the deposition started. As shown in Figure 32 and its inset micrographs, the size distribution of grains has been restricted to the submicron range and looked lognormal with a single peak up to about 240 s. During this time region, the grains seem to gradually increase in number with time but do not appear to grow in size. At around 360 s, some micron-sized large grains emerge among the preexisting small ones, resulting in a kind of bimodal size distribution. The large grains rapidly grow and increase in number, retaining the bimodal size distribution until 420 s. At 480 s, most of the substrate surface is occupied by the large grains while the small ones disappear with their concentration being reduced by three orders of magnitude. As a result, the size distribution becomes broad over several micron meters.

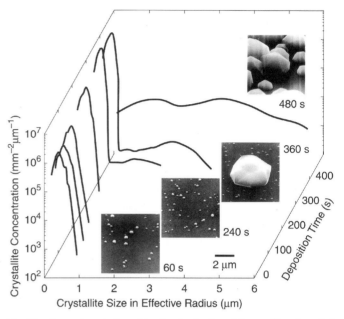

Fig. 32. Dynamic evolution of the size distribution of Si crystallites deposited over SiO_x ($x < 2$) by chemical vapor deposition at 1223 K under 2×10^4 Pa using SiH_2Cl_2, HCl, and H_2 gases at a flow rate of 8.83×10^{-3}, 0.03, and $1.67\,\ell\,s^{-1}$, respectively. The SiO_x is formed by Si^+ ion implantation into SiO_2 at an accelerating energy of 20 keV and at a dose of 4×10^{14} mm^{-2}. The insets are the SEM micrographs of the deposited surfaces at the deposition times of 60, 240, 360, and 480 s.

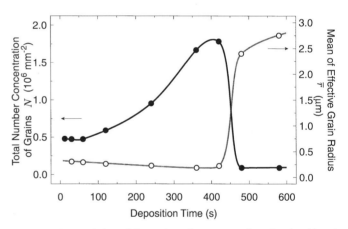

Fig. 33. Time evolution of the total number concentration of grains (●) and the mean grain size expressed in the effective radius (○), with the deposition time. Both are calculated from the size distributions shown in Figure 32. The experimental conditions are the same as those for Figure 32. The solid lines are drawn as guides.

The effects of this coarsening phenomenon on the total number concentration of grains, N, and the mean grain size, $\bar{r}$, are evident in their time evolution shown by Figure 33, both of which are calculated from the size distributions as shown in Figure 32. Until the coarsening starts around 360 s, N gradually increases with time, while $\bar{r}$ slightly decreases. The coarsening reduces the increasing rate of N around 400 s and then causes a steep drop of N at about 450 s down to a value even lower than the initial concentration. At the same time, $\bar{r}$ suddenly increases by nearly 10 times. Afterward, N changes little and only $\bar{r}$ increases. Consequently, the final concentration of grains and hence the mean of the final grain size are controlled by the coarsening process and not by the initial nucleation process.

It is elucidated by the analysis using the surface coverage ratio [132] and by the experiment using selective deposition [133] that most of the initially formed small grains disappear during the coarsening. This fact indicates that the growth of large grains and the etching of small grains simultaneously occur. The etching itself is certainly enhanced by chlorides in the formation gas, which is known as the etching agent of Si. Hwang et al. [255, 256] proposed the likeliest mechanism based on the "charged cluster model" for the most part of this puzzling phenomenon. According to the model, Si clusters first nucleate in the vapor above the substrate and then are deposited on the substrate. The nucleation of the clusters in the vapor reduces the supersaturation and makes the driving force to etching by the atomic unit at the substrate. If the clusters in the vapor and the deposited grains are charged with a same polarity, there is Coulomb repulsion between them when they are comparable in size. It is known, however, that net attraction could exist between a small particle and a nearby large particle due to image charge forces [257]. The attraction increases with size difference between two particles and with the amount of charges in the larger particle. Therefore, once some grains on the substrate survive the etching and continue to build up charges, they will progressively attract the small charged clusters in the vapor and will exclusively grow, while the others will be etched away by the atomic unit. The coarsening of CVD-Si grains is thus explained by the charged cluster model at least qualitatively.

4.1.5. Control by Other Means

There are a number of potential approaches to the control of grain size other than those described in Sections 4.1.1–4.1.4. Below are briefly shown some of them.

4.1.5.1. Formation Conditions

Section 4.1.1 treated the control by formation temperature which is just one of the formation conditions. There are many other conditions which can control grain size, depending upon the formation methods. For example, in the vapor phase deposition of thin films, one can adjust various formation parameters such as the choice and the partial pressures of gas species, and they often influence the final grain size. In the liquid phase formations or the deposition from solutions, the supercooling ratio or rate is one of the crucial parameters. In any case, the formation conditions control the final grain size through the change of W_*/kT as far as the final number concentration of grains is controlled by the nucleation process.

4.1.5.2. Nonisothermal Processes

The previous sections mainly treated the isothermal processes for the formation of thin films. Regarding the thermal processes for crystallization of thin films except the isothermal ones, various methods have been developed, such as zone melting, pulse-, continuous-, or excimer-laser annealing, electron beam annealing, rapid thermal annealing, and so on. As to nonthermal processes for crystallization of thin films, there have been ion-beam induced crystallization, pressure enhanced crystallization, X-ray enhanced crystallization, etc. For the deposition of thin films over substrates, electrochemical deposition, photochemical deposition, sputtering deposition, electrically biased deposition, etc., can be enumerated. The techniques using energy or particle beams have potential of local enhancement of the formation processes. For instance, the local irradiation of laser beams onto a-Si thin films causes the local crystallization [139, 258–260]. The nonthermal techniques enable one to control W_* independent of the formation temperature and can be used together with the thermal treatment. For instance, as shown in Table I, the irradiation of high energy ion beam largely reduces W_* for the solid-phase nucleation of Si crystallites in a-Si thin films and significantly reduces the formation temperature for the crystallization [152, 261, 262].

4.1.5.3. Introduction of Preferential Nucleation Sites

Equation (2.125) suggests that the mean of the final grain size also depends on the initial monomer concentration, $f_m(0)$, and the initial concentration of nucleation sites, $f(1, 0)$, as $\bar{r}_f \propto f_m(0)^{1/d}/f(1, 0)^{1/(1+d)}$. Since any monomer can act as nucleation site in the homogeneous nucleation, i.e., $f_m(0) = f(1, 0)$, the change in $f(1, 0)$ hardly influences $\bar{r}_f$. On the other hand, if it is possible to introduce preferential nucleation sites, $f(1, 0)$ can be changed independently of $f_m(0)$, and $\bar{r}_f$ could be controlled by $f(1, 0)$. Although the dependence of $\bar{r}_f$ on $f(1, 0)$ is not as strong as that on W_*/kT, the control of grain size with $f(1, 0)$ would be effective if $f(1, 0)$ is exponentially variable. Particularly when $f(1, 0) \ll f_m(0)$, and when the nucleation sites are depleted mostly by the nucleation of grains and not by their growth, the finally accumulated number concentration of grains becomes comparable to $f(1, 0)$, and hence the final grain size is controlled mainly by $f(1, 0)$ as $\bar{r}_f \approx [f_m(0)/f(1, 0)]^{1/(1+d)}$.

The preferential nucleation sites are often introduced by manipulating heterogeneous nucleation sites in the starting thin films or on the substrates. The free-energy barrier to the heterogeneous nucleation at these introduced sites has to be sufficiently lower than that to the homogeneous nucleation so that the former dominates over the latter. There have been various attempts for introducing such heterogeneous nucleation sites. A method called "graphoepitaxy" [263–265] may be a typical example. In this method, thin films are formed over SiO_2 substrates with gratinglike shallow steps at their surfaces. The thin films are KCl polycrystalline films deposited onto the substrate from their liquid solution [263], poly-Si thin films formed by melting recrystallization of a-Si over the substrate [264], etc. The top, bottom, or edge of the steps serves as the heterogeneous nucleation site, yielding large grain size of the polycrystalline films. In the chemical vapor deposition of polycrystalline diamond thin films, it is known that the scratches over substrates become the heterogeneous sites. The grain density is often controlled by scratching the substrate surface. Besides these structurally introduced sites, impurities can sometimes work as the heterogeneous nucleation sites. For example, Sb ions implanted into a-Si thin films provide the heterogeneous sites in their solid-phase crystallization and induce preferential nucleation over the spontaneous nucleation [266]. Zhao et al. [267] crystallize Er-doped a-Si thin films in solid phase, in which the doped Er serves as a heterogeneous nucleation site, and control the size of Si nanocrystallites for controlling the quantum size effect on their photoluminescence. Metal-contact-induced crystallization of amorphous thin films [268–270] can be regarded as one of the variants.

4.1.5.4. Introduction of Seeds

Since the final number concentration of grains controls the final grain size, one may prepare seeds for the growth of grains in thin films before their formation, instead of introducing the heterogeneous nucleation sites. The seeds can be either provided externally from the surface of the thin films or internally embedded in the thin films. If the seeds are larger than the stable grains, exceeding the critical size for nucleation, they can grow without the nucleation process like seed crystals for the growth of bulk crystals. Techniques called "selective epitaxial growth" [271], "epitaxial lateral overgrowth" [272], "lateral solid-phase epitaxy" [273–275], etc., are models for introducing the external seeds to the thin films. In these techniques, crystalline thin films are formed over the mask layers on single crystalline substrates. The surfaces of Si substrates are locally exposed through the openings of the mask layer. The crystalline grains can homoepitaxially grow from the exposed seed regions over the mask layer and impinge on the adjacent grains having grown from the adjacent seeds. A similar idea is adopted in the solid-phase epitaxial growth of a-Si thin films over quartz substrates [276]. In this method, single crystalline Si tips are pressed on the surface of the a-Si thin films during the crystallization. The a-Si thin films are crystallized preferentially from the regions where the Si tips are pressed. A method called "seed selection through ion channeling" [146, 247, 277] is a typical example of the internal seeding. In this method, poly-Si thin films on SiO_2 substrates are partially amorphized by self-ion implantation, so as to leave small crystallites of a specific crystallographic orientation in the amorphous matrix through the ion channeling effect, and then are recrystallized in the solid phase. The small crystallites having survived the ion bombardment serve as the seed crystals and grow to crystallize the amorphous matrix before the spontaneous nucleation of crystallites takes place. A similar configuration of starting materials can be obtained also in as-deposited Si thin films formed in LPCVD systems [278, 279]. The phase of Si thin films deposited by LPCVD varies with the deposition conditions from the crystalline to the amorphous. Under an intermediate condition, mixed-phase thin films are formed, in which the small crystallites coexist in the amorphous matrix.

The preexisting crystallites in the as-deposited thin films serve as the seed crystals in the crystallization. Epitaxial growth of Si crystallites formed by agglomeration of poly-Si thin films over SiO_2 substrates [280], the local crystallization of amorphous thin thin films by laser-beam irradiation [139, 258–260], and the local crystallization of a-Si thin films by excimer-laser annealing with local cooling techniques [281, 282] are also regarded as the seeding methods.

4.1.5.5. Substrate Conditions

Substrate is not only a supporting component of thin films but also a condition which can control the formation of thin films and the final grain size in various ways. The properties of substrate surfaces such as material, structure, roughness, mechanical and electrical states, adsorbates, etc., control the surface energy density of the substrate. In classical heterogeneous models of droplet nucleation on the substrates, the surface energy density of the substrate surface controls the interfacial energy density at the contact to the droplet, which controls the wetting angle of the droplet to the substrate. The wetting angle controls W_* and hence controls the final grain size. In the case of depositions in vapor phase, the surface properties also affect surface migration of adatoms, which controls the supersaturation at the surface and hence controls W_*. Some of the surface properties might introduce preferential nucleation sites to the substrates. The differences in the thermomechanical properties between the starting thin film and the substrate often introduce a substantial stress into the thin films during their transformation. The stress may affect the nucleation of grains during the formation.

For instance, the surface treatment of substrates often changes the nucleation behaviors and the final grain size. For the solid-phase crystallization of a-Si thin films, Voutsas and Hatalis [283] observed that the surface treatment of substrates before the deposition of the starting a-Si significantly influenced the final grain size. Song et al. [284] compared the nucleation behaviors in the solid-phase crystallization of a-Si thin films on quartz, LPCVD-SiO_2, and thermally grown SiO_2 substrates and found that the surface roughness of the substrate strongly affected the nucleation mechanism. Also in the same formation system, it is observed that the volume shrinkage of a-Si thin films by the crystallization causes an amount of stress in the thin films since the substrate does not shrink [285], which could influence the nucleation during the crystallization. In CVD systems, it is known that the final number concentration of grains depends on the substrate materials. For example, in Si–H–Cl CVD systems like that for Figures 32 and 33, the final number concentration of crystalline Si grains over Si_3N_4 can be more than 100 times larger than that over SiO_2 [254, 286, 287], resulting in the difference in the final grain size. This phenomenon is also explained by the difference in the amount of charges which can be built up on the substrates and by the interaction between the charged substrates and incoming charged clusters, in the charged cluster model [255, 256, 288–298]. Thus the final grain size can be controlled by the choice of the materials for substrates and by changing the states of the substrate surface.

Finally, it is noted that the change of grain size by the methods except those simply changing W_*/kT could be accompanied by the change in the size distribution of grains. In other words, the modification of the size distribution could be attributed to the change of the grain size. The next section treats the control of the size distribution in detail.

4.2. Size Distribution of Grains

When the properties inside grains are desired for granular thin films, the mean grain size should be as large as possible since the larger grain size reduces the density of grain boundaries. This strategy is always valid as far as the properties averaged over a volume containing a sufficiently large number of grains is concerned. However, if the volume contains just a small number of grains, another problem arises in uniformity among multiple thin films formed by the same procedure or among multiple parts of a single thin films. Such a situation confronts us when the grain size is enlarged comparably up to the volume or when the volume is reduced comparably down to the volume of the grain. Thin film transistors (TFT) fabricated on poly-Si thin films provide a typical example of this situation [170]. Since the poly-Si TFT is mostly used as a part of integrated circuits (IC) for devices such as active-matrix liquid-crystal displays, the uniformity among a large number of the TFTs in the IC is a critical demand. The performance of the poly-Si TFTs has been improved by enlarging the grain size of the poly-Si thin films and by reducing the size of the TFTs. As the grain size gets close to the size of TFTs, the fluctuation of performance among the TFTs progressively increases although the highest and the average performances observed among the TFTs also increase [299, 300]. Due to the broad distribution of the performance, the TFTs made of a small number of Si grains have not been unpracticed.

The uniformity problem arises from spatial distribution of grain size in the thin films. Even if one region of a thin film happens to contain a small number of grains, the others having the same volume could contain a large number of grains. One of the approaches to the improvement of the spatial uniformity is to narrow the range of the overall size distribution of grains itself.

As shown in Sections 2.2 and 3.2.3, the size distribution of grains formed through random nucleation processes should exhibit a continuous decrease from the monomer size to the largest size without coalescence in the initial stage of formation. In the later stage, the size distribution is deformed by both the depletion of nucleation sites and the impingement on the adjacent grains. The depletion of nucleation sites suppresses the nucleation of new grains and decreases the number concentration of smaller grains with time. The impingement between the adjacent grains reduces their growth rate or stops their growth and hinders further expansion of the size distribution toward the larger size. Consequently, the final size distribution in completely transformed thin films could be lognormal-like, showing a peak at a certain size, as mentioned in Section 2.2.9. Actually, Kouvatsos et al. [301] observed the lognormal-like distributions in completely crystallized a-Si thin films. To further narrow such a final size distribution, there is nothing but to make the size distribution lognormal-like from the early stage when the transformed fraction is still small. Below are shown some of the methodologies for controlling the early size distributions.

4.2.1. Control by Coarsening of Grains

There have been few reports on the control of size distribution of grains in thin films by controlling the coarsening of grains. Here we note just the possibilities to be discussed.

Lognormal-like size distributions are often observed in vapor phase or vacuum deposition of nanometer-sized metallic clusters from the early stages when the coverage of substrate is still negligible and the clusters are discrete. This type of lognormal-like size distribution has been attributed to the coarsening of clusters through coalescence of small clusters. The coalescence is possible for the clusters because they are mobile over substrates [302]. Conversely, the size distribution could be controlled through the coalescence processes by the interactions among the clusters and the mobility of clusters, which are generally controlled by the formation conditions and the states of the substrate surface. It is observed, however, that the coalescence process often produces a bimodal size distribution resulting in a broader distribution than the lognormal-like one. The lognormal-like size distributions are observed also in CVD of larger Si crystallites, as shown in the early stage distributions of Figure 32. In this case, the coarsening further evolves the size distribution to the bimodal and finally to the broad one. On the other hand, it is reported by Ross et al. [303] that in CVD of Ge quantum dots on single crystalline Si substrates, the coarsening of strained Ge islands first leads to a bifurcation in the size distribution but ultimately to a narrow size range. They attributed this coarsening behavior to a shape transition of the islands with size, which gives rise to an abrupt change in chemical potential. Thus this type of coarsening could be controlled also by thermodynamic formation parameters.

Lognormal size distributions have been frequently studied in vapor phase growth of particles such as aerosols and powers which can be captured on substrates and constitute thin films. The lognormal distribution in this system has been explained in the framework of the coarsening by coalescence [304] or by Brownian coagulation [305–307] based upon the general theories [44, 308, 309]. As far as the lognormality originates in these mechanisms, it is basically controlled by the aerial mobility of particles through gas viscosity, etc. However, it is found that the lognormality can be explained by the distribution of flight time of particles in the growth space [310]. In this model, the random walk of particles with a constant gas flow in a finite space gives rise to a lognormal distribution of their flight time, and the distribution is controlled by the dimension of the growth space and the gas flow rate. Thus it may be possible to control the resultant size distribution of particles by these formation conditions.

4.2.2. Control by Formation Time of Stable Grains

Without coarsening processes, the size distribution of grains in the early stage of formation is controlled exclusively by the distribution of time when the stable grains are formed. The early size distribution of grains is narrowed by narrowing the time distribution of grain formation. Ultimately, if all the stable grains are formed at the same instant with a same size, the early size distribution shrinks into a single size. A typical example is the growth of thin films seeded with the stable grains having the same size as mentioned in Section 4.1.5. The CVD of Si crystallites using Si–H–Cl gas systems under moderate partial pressure of HCl, which was mentioned in Section 4.1.4, is virtually the case where all the grains nucleate instantaneously after a short incubation time, and no additional nucleation occurs afterward [254].

When the stable grains are formed by continuous nucleation, it is necessary to control the time distribution of nucleation for obtaining a lognormal-like size distribution, which is accomplished by controlling both the start and the end of nucleation. For the start of nucleation, faster nucleation can shrink the tail of the size distribution toward the large size. The faster nucleation is obtained by a higher steady-state nucleation rate, J_{ss}, and shorter characteristic times for transient nucleation such as the time lag. Both the conditions are satisfied with larger growth rate and narrower critical region. The magnitude of J_{ss} can be independently enhanced by increasing the concentration of nucleation sites, $f(1, 0)$, or more effectively by reducing the free-energy barrier to nucleation, W_*. However, as described by Eq. (2.125), both the increase of $f(1, 0)$ and the reduction of W_* lead to the reduction of the mean of the final grain size. For obtaining the left shoulder of lognormal-like size distribution, it is necessary to reduce the number concentration of smaller grains. This is achieved only by stopping nucleation before complete transformation of thin films. To stop the nucleation keeping the already formed grains growing, there is nothing but to reduce $f(1, 0)$ less than the concentration of monomers, as mentioned in Section 4.1.5. However, since the reduction of $f(1, 0)$ is accompanied by the reduction of J_{ss}, it is necessary at the same time to reduce W_* for maintaining high J_{ss}. Overall, it is effective in obtaining lognormal-like size distribution in continuous nucleation systems to (1) enhance the growth rate, (2) shrink the width of critical region, or (3) reduce W_* and $f(1, 0)$.

The third of the above approaches is often actualized by introducing preferential nucleation sites such as heterogeneous ones. As mentioned in Section 4.1.5, W_* for the heterogeneous nucleation is generally smaller than that for the homogeneous nucleation. Once such preferential nucleation sites are introduced at a concentration much lower than monomers, fast nucleation could occur, which is followed by the stop of nucleation due to the complete depletion of the preferential nucleation sites, and only initially formed grains could continue to grow to complete the transformation before the homogeneous nucleation starts. For example, in the study reported by Zhao et al. [267] on the solid-phase crystallization of Er-doped a-Si

thin films, a sharp blue photoluminescence was observed in the Er-doped nanocrystalline Si thin films, while the nondoped ones exhibited a quite different spectra, despite having the same mean size of crystallites. They attributed this difference to the broad size distribution of Si nanocrystallites formed without the Er doping compared with that of Er-doped ones. One can interpret through this observation that the doped Er atoms serve as the heterogeneous (i.e., preferential) nucleation sites and control the size distribution of Si nanocrystallites in addition to their mean size so that the quantum size effect becomes conspicuous. A lognormal-like size distribution is observed also in partially crystallized a-Si thin films in solid phase even at a very low transformed fraction of 0.028 [311]. This phenomenon might be also due to depletion of something like preferential nucleation sites. The introduction of preferential nucleation sites will be treated in more detail with another method combined in the next section.

4.3. Grain Locations

The size distribution of grains in the early stage of formation could be narrowed by the controls described in Section 4.2.2. However, in the case of continuous thin films, the final size distribution is controlled not only by the initial nucleation processes and the depletion of nucleation sites but also by the impingement among the adjacent grains. The grain size is thus finally fixed by grain boundaries where the growth fronts of adjacent grains meet. In other words, the grain size is determined by the location of the grain boundaries. If the growth rates are common to each grain, which is generally valid, the location of the grain boundaries is then controlled by the grain location, that is, the location where the grains are formed to stably grow. Thus, even though all the grains are formed ideally at the same instant and isotropically grow at a same growth rate, the resultant granular structure could be at most a Voronoï diagram as shown in Figure 34A, because the location of grain nucleation is randomly distributed in the space (Fig. 34a). The size distribution of these Voronoï polygons is exactly lognormal, where the standard deviation is determined by the spatial distribution of grain locations. Therefore, it is necessary to control the spatial distribution of grain locations by arranging them in order to further shrink the final size distribution of grains. In addition to the shrinkage of the size distribution, the location control of grains and grain boundaries will also enable us to fabricate small devices inside a single grain which are not affected by the grain boundaries [170]. Ultimately, if the location of grain nucleation is periodically arranged as shown in Figure 34b, where the grains are placed at square lattice points, the resultant grain boundaries in the completely transformed thin films constitute a checkered pattern as shown in Figure 34B. In such controlled textures, the size distribution of grains shrinks into a single size, and one can know where the grain boundaries are.

There are two approaches to the control of grain locations, which are the spatial arrangement of seeding thin films and that of preferential nucleation sites as mentioned in Section 4.1.5. In the case of seeding with stable (i.e., supercritical) grains, there

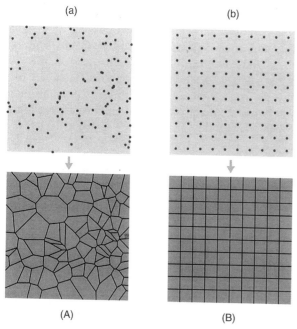

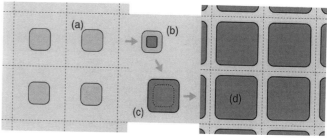

Fig. 35. Schematic representation of the nucleation and growth of grains at artificial nucleation sites. (a) Artificial nucleation sites having a finite size are placed at square lattice points. (b) A single grain nucleates exclusively in each site. (c) The single grain grows to occupy and get over the site before additional nucleation occurs in the site. (d) The grains having grown from the sites are to impinge upon one another and will form the grain boundaries around the median lines between the sites.

Fig. 34. Schematic representations of the final grain structures and their origins. Instant formation of stable grains at random locations (a) leads to a Voronoï diagram (A) where the grain size is lognormally distributed. If the location of the stable grain formation can be periodically arranged (b), the resultant grain boundaries constitute a checkered pattern (B), and the size distribution shrinks into a single size.

is no dispersion in their formation time, and all of them start growing as soon as the formation starts. If the arranged seed grains are of the same size, the resultant grain structure becomes just like Figure 34B. Such a control can be actually achieved by external seeds as in the epitaxial lateral overgrowth [272] and the lateral solid-phase epitaxy [273–275], if the seed regions exposed in the mask have the same size and are placed periodically. On the other hand, it is still difficult to place an internal single seed at an arbitrary location of a starting thin film. In both the techniques of seed selection through ion channeling [146, 247, 277] and mixed-phase thin films [278, 279], one cannot know where the crystallites are embedded, since they are originally dispersed at random in the thin films. In the local crystallization techniques by laser-beam irradiation [258–260], the crystallization is not yet controllable enough to form a single grain at a laser-irradiated portion of thin films.

It is not easy either to place the preferential nucleation sites at desired locations, since the actual nucleation site is considered to be as small as a monomer. With any modern techniques for microfabrication, the manipulatable dimension over a wide area is still far larger than such an atomic size. Furthermore, if one could manipulate an actual nucleation site, the free-energy barrier to the nucleation at the site must be greatly lower than the others in order to form a grain there without fail, since the nucleation is a stochastic process. Such control over a wide range of the free-energy barrier is not always available.

Instead of manipulating each actual nucleation site, it was proposed [312, 313] to periodically place "artificial nucleation sites." The artificial nucleation site is not identical to the ac-

tual nucleation site but is a finite region in which a number of actual nucleation sites can be included but only a single grain is formed. This idea is schematically represented in Figure 35. First, the artificial nucleation site is placed at, for instance, each square lattice point (Fig. 35a). Other dispositions of the sites such as a hexagonal arrangement are also allowable. In the early stage of the formation of a thin film, a single grain nucleates exclusively in each artificial site (Fig. 35b). Then the single grain grows to occupy the whole region of the site and to get over the site (Fig. 35c). Finally, the grown grain will impinge upon the adjacent grains which have grown from the next sites and will form the grain boundaries around the median lines between the adjacent sites (Fig. 35d). There are several conditions for controlling the grain locations by such a nucleation and growth process: (1) Grains nucleate at all the sites within a short period compared with the total duration for the complete transformation. (2) The first grain in a site grows to occupy the whole region of the site before the next nucleation occurs. (3) No nucleation occurs outside the sites' regions until the complete transformation. (4) The dimension of the artificial site is sufficiently smaller than the intersite distance. These conditions restrict the rates of nucleation and growth inside and outside the site, the dimensions of the site and the intersite distance, and the disposition of the sites, which are traded off for each other. The last condition controls the spatial dispersion of grain locations since the single grain can be formed anywhere within the site. The top three conditions can be summarized in the phrase "selective nucleation." The method of actualizing the selective nucleation depends on the materials and the methods of formation of thin films. Some of the examples are given in the following.

4.3.1. Control in Deposition Systems

For the formation of thin films in deposition systems, substrate conditions provide a method of selective nucleation. As briefly mentioned in Section 4.1.5, the final number concentration of grains can be widely changed by the choice of substrate materials or by the states of substrate surfaces. When two kinds of surfaces coexist on a substrate, the grains can be deposited exclu-

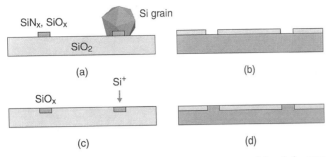

Fig. 36. Possible configurations of substrates represented by their cross-sectional schemes for the selective nucleation of Si crystallites in Si–H–Cl CVD. (a) Small portions made of SiN_x ($x \le 4/3$) or SiO_x ($x < 2$) for artificial nucleation sites are placed over SiO_2 surfaces. (b) The reversed configuration of (a). (c) Small SiO_x ($x < 2$) regions are formed by local Si doping such as Si^+ ion implantation into SiO_2 surfaces. (d) The reversed configuration of (c), formed by local oxidation of SiN_x ($x \le 4/3$) or SiO_x ($x < 2$).

Fig. 37. SEM micrographs showing the time evolution of selective nucleation on a SiN_x ($x < 4/3$) artificial nucleation site in a CVD using gases of SiH_2Cl_2, HCl, and H_2. At 480 s, a number of submicron-sized Si grains are formed exclusively over a site with an area of 4×4 μm^2. At 720 s, a micron-sized grain emerges among the preexisting fine ones by the coarsening phenomenon as shown in Figure 32. At 960 s, the emerged grain rapidly grow to almost occupy the site, while all the preexisting fine grains vanish.

sively on one surface with no grain on the other surface simultaneously. This is known as "selective deposition" which has been found in various systems for vapor-phase deposition. For example, in Si–H–Cl CVD systems, Si crystallites can be deposited exclusively on Si_3N_4 [254, 286, 287], SiN_x ($x < 4/3$) [314], and SiO_x ($x < 2$) [313] over SiO_2 substrates. Thus if small portions of these materials for preferential nucleation are manipulated and placed periodically over SiO_2 surfaces, as shown in Figure 36, they can serve as artificial nucleation sites.

Actually, the formation of a single Si grain is observed at a Si_3N_4 small portion manipulated as the substrate configuration of Figure 36b with the circular opening in diameter of 2 μm, in a CVD system using gases of $SiCl_4$ and H_2 [312]. However, the yield of single grain formation was unsatisfactory in this system due to the insufficient selectivity of nucleation. The yield was improved by replacing Si_3N_4 with SiN_x ($x = 0.56$) or SiO_x ($x < 2$), and $SiCl_4$ gas with SiH_2Cl_2 and HCl. The improved CVD system corresponds to that in Figure 32, where the coarsening of Si grains occurs on artificial nucleation sites [134] as shown in the slanted-view SEM micrographs of Figure 37. Here, the substrate configuration is that represented by Figure 36a. The artificial nucleation site is made by a square portion of a 100 nm thick SiN_x thin film which is deposited as

Si_3N_4 by LPCVD, then implanted with Si^+ ions at a dose of 1.1×10^{14} mm^{-2} and at an acceleration energy of 10 keV, and finally patterned into a region of 4×4 μm^2 in area. The deposition conditions are identical to those for Figure 32. It is seen at 480 s that a number of submicron-sized small grains are formed exclusively over the site. At 720 s, these small grains increase in number and a micron-sized large grain emerges among the small ones and rapidly grows. Finally, at 960 s, the site is monopolized by a single large grain while all the small grains vanish. Thus the coarsening phenomenon is responsible for the single grain formation at the artificial nucleation sites in this system.

The selective nucleation of a single grain is possible also at multiple sites as shown in Figure 38. Here, the substrate configuration is that represented by Figure 36c. The artificial nucleation sites of SiO_x ($x < 2$) are formed by the irradiation of a Si^{2+} focused-ion beam into 1.2×1.2 μm^2 areas of SiO_2 substrates at a dose of 4×10^{14} mm^{-2} and at an acceleration energy of 40 keV, and they are placed at square lattice points at 50 μm periods. For the purpose of recovering from the damage caused by the ion irradiation, the substrate is annealed at 1173 K for 10 min in H_2 atmosphere. Then, to expose the most Si-rich layer of SiO_x areas, about a 0.1 μm thick layer from the surface is removed by dipping the substrate in HF solution. Finally, Si is deposited for 1800 s at the same CVD conditions as those for Figure 32. As seen in the micrograph, Si grains whose diameters are tens of microns are arranged at the square lattice points at 50 μm periods with no grain on the SiO_2 surface. All the grains are furnished with clear facets which can be assigned to the indices of Si crystals. These crystallographic habits indicate that any grain can be classified into one of three types: the first is a part of the single crystal bounded by 8 {111} facets and 24 {311} facets, the second is the simple mirror-twin crystal of the first, and the last is a part of a multiply twinned icosahedron which is bounded by the 20 equivalent {111} facets. This observation, along with the other analyses such as micro-X-ray diffraction and TEM observation, confirms that the grains are grown from single nuclei formed by selective nucleation at artificial nucleation sites. The size distribution of these grains is narrow and lognormal-like as shown in Figure 39. The mean grain size in effective radius and the standard deviation are $\bar{r} = 21.87$ μm and $\sigma = 1.987$ μm, respectively, while $\bar{r} = 2.397$ μm and $\sigma = 1.388$ μm are derived for the size distribution without the location control of grains as shown by the data at 480 s in Figure 32. The small difference of these standard deviations, in spite of the large increase of the mean grain size, suggests that the initial size dispersion just after the coarsening is preserved in the large location-controlled grains shown in Figure 38.

The results shown in Figures 37–39 are the cases where a single grain is successfully formed at almost every site. When the deposition conditions or the designs of artificial nucleation sites are partly changed from these settings, one could encounter the failure of single grain formation. For instance, if the area of the sites is expanded or x of SiN_x or SiO_x decreases, multiple grains would be finally formed over a site. If these conditions are changed contrarywise, there would be some sites

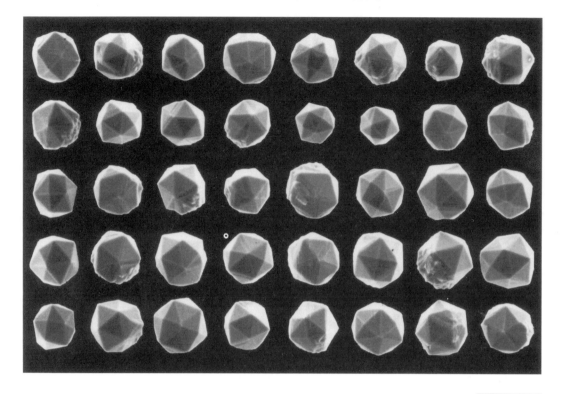

50 μm

Fig. 38. Plan-view SEM micrographs of Si grains whose locations are controlled by selective nucleation using artificial nucleation sites. The artificial nucleation sites are made of $1.2 \times 1.2 \ \mu\text{m}^2$ square regions of SiO_x ($x < 2$) and are placed at square lattice points at 50 μm periods. Si is deposited for 1800 s under deposition conditions identical to those of Figure 32.

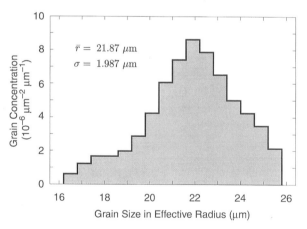

Fig. 39. Histogram for the size distribution of location-controlled grains which are partly shown in Figure 38. The distribution looks like lognormal with a standard deviation of $\sigma = 1.987 \ \mu$m at a mean size of $\bar{r} = 21.87 \ \mu$m. The magnitude of standard deviation indicates that the initial size dispersion just after the coarsening is almost preserved after the grains grow to be large.

without any grains finally. If x of SiN_x or SiO_x increases and the partial pressure of HCl gas is reduced, undesired grains nucleate on the SiO_2 surface before the controlled grains at the sites impinge upon the adjacent ones, although the selective nucleation of single grains succeeds.

Once a single grain is successfully formed at each site and the nucleation of grains is suppressed on SiO_2, the transfor-

mation can be completed by the impingement of the location-controlled grains as shown in a plan-view SEM micrograph of Figure 40. Here, $1.2 \times 1.2 \ \mu\text{m}^2$ square portions of SiN_x form the artificial nucleation sites and are placed in period of 80 μm. The deposition conditions are the same as those for Figures 37–39 except the substrate temperature of 1263 K. After the deposition for 5400 s, the surface of the grains is polished to be flat. It is seen in the micrograph that the grains constitute nearly a checkered pattern with unfilled voids at the corners.

4.3.1.1. Other Controls in Deposition Systems

The control of grain locations by artificial nucleation sites was first demonstrated in the poly-Si thin films formed by CVD systems as shown above. Ever since, the method has been extended to a wide variety of formation of thin films in deposition systems. Some examples are enumerated below.

For the formation of diamond thin films, Hirabayashi et al. [315] first reported the location control of diamond grains over crystalline Si substrates. They abraded the substrate surface with diamond dust and then smoothed it by Ar+ ion etching with a patterned mask overlayer, leaving a two-dimensional array of small regions unetched. After removing the mask, diamond was deposited on the substrate by CVD using a tungsten filament. Since diamond grains are preferentially deposited on rough or scratched surfaces over smooth surfaces, the masked

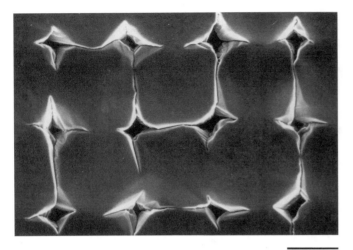

50 μm

Fig. 40. Plan-view SEM micrographs of impinging Si grains. Each grain has grown from an artificial nucleation site which was formed by $1.2 \times 1.2 \ \mu m^2$ square portions of SiN_x and placed in period of 80 μm. After the deposition for 5400 s, the surface of the grains was polished to be flat. The grains constitute nearly a checkered pattern with unfilled voids at the corners.

regions serve as the artificial nucleation sites. The local abrading approach to the formation of artificial nucleation sites has been developed in a variety of techniques [316–321]. Ma et al. reported another approach using an array of SiO_2 dots placed over Si substrates [322–324] or Si dots patterned on Si substrates [325] in combination with an oblique ion-beam irradiation. In these techniques, the unetched portions remain at one edge or a shadow of the dots where diamond grains preferentially nucleate. Roberts et al. [326] proposed a method named "selective area dc bias-induced nucleation" which is based on the selective deposition of diamond on Si over SiO_2 in a CVD system with applying a negative dc bias to substrates [327].

Tokunaga et al. [328] reported an attempt at selective nucleation in the formation of polycrystalline GaAs thin films deposited by metal–organic CVD. They choose the small islands of a poly-Si thin films for artificial nucleation sites over SiO_2 substrates. For sufficient selectivity, it was necessary to add HCl to the ordinary deposition gases composed of AsH_3, trimethylgallium, and H_2.

Aizenberg et al. [329] reported the selective nucleation of single calcite ($CaCO_3$) crystallites on metal substrates such as Ag, Au, or Pd in liquid-phase deposition. They first prepare a polydimethylsiloxane "stamp" by casting, on which an array of small terraces is periodically formed. Then the stamp is pressed on metal substrates with "inks" of $HS(CH_2)_nX$ (X = CO_2H, SO_3H, OH) ethanol solution. At the stamped portion of the substrate, X-terminated self-assembled monolayers are transcribed. Next, the substrate is washed with $HS(CH_2)_{15}CH_3$ to terminate the unstamped area with CH_3. Finally, the calcite crystallites are deposited exclusively on the stamped regions in a $CaCl_2$ solution dissolving CO_2 and NH_3. Thus the stamped regions terminated with the monolayers of X molecules serve as the artificial nucleation sites. It is reported that the orientation of calcite crystallites is also controlled in addition to their locations.

With the control of grain location, it is further possible to control the external shape of grains without their impingement. If an artificial nucleation site or a seed is placed at the bottom of the dimple on a substrate, the selectively formed grain at the site can grow to fill the dimple, exceeding the dimple region. By levelling the grain on the substrate at the top surface of the substrate, the grain embedded in the dimple remains on the substrate. By these processes, the external shape of grains can be molded into the predetermined one. Such a technique is often useful when a device is fabricated within one grain, since thin films are not always needed forming over the whole area of a substrate.

4.3.2. Control in Transformation of Starting Thin Films

The principle of the method for controlling grain locations by artificial nucleation sites can be applied also to the formation of thin films by the transformation of starting thin films such as crystallization or recrystallization of amorphous or polycrystalline materials. The selective nucleation of single grains in this system was first demonstrated in the solid-phase crystallization of a-Si thin films [156, 157]. In the case of solid-phase crystallization of amorphous thin films, the nucleation and growth of grains stops only by the depletion of nucleation sites or the impingement on the adjacent grains. There is neither coarsening of grains nor long-range transport of monomers which affect the addition of monomers into grains. Thus a quantitative description of the nucleation and growth with artificial nucleation sites [162] is readily derived based on the theoretical framework shown in Sections 2.1.4 and 3.2.3.

Let us consider that artificial nucleation sites having an area of a_s are placed two-dimensionally in a starting thin film at a concentration of $f_s(0)$. The primary requirement for selective nucleation is the preferential nucleation inside the artificial sites with no spontaneous nucleation outside, which can be estimated by

$$\int_0^{t_{ct}} \left[1/f_s(0) - a_s \right] J_{out}(t') \, dt' < 1 \qquad (4.7)$$

where t_{ct} denotes the time needed to completely transform the the starting thin film, and $J_{out}(t)$ is the nucleation rate outside the artificial sites. The second condition is to suppress the excess nucleation inside an artificial site until the first grain grows to occupy the whole area of the artificial site, which is approximately described as

$$\int_0^{t_{co}} f_s(t) \int_t^{t+\sqrt{a_s/\pi v_r^2}} \left[a_s - \pi \left\{ v_r(t'-t) \right\}^2 \right] J_{in}(t') \, dt' dt < 1 \qquad (4.8)$$

where t_{co} denotes the time when all the artificial sites are occupied by grains, $f_s(t)$ is the concentration of artificial site with no grain at t, v_r is the growth rate of grains with respect to their effective radius, and $J_{in}(t)$ is the nucleation rate inside the artificial sites. Equations (4.7) and (4.8) suggest that one has to

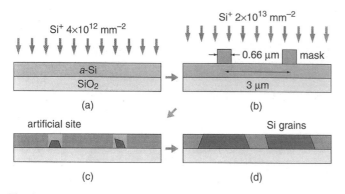

(a)   (b)

(c)   (d)

Fig. 41. Cross-sectional schemata representing a process of preparing artificial nucleation sites in solid-phase crystallization of a-Si thin films, which was employed in [156, 157]. (a) Si$^+$ ions are implanted uniformly into an as-deposited LPCVD a-Si thin film. (b) Local ion implantation with periodically placed masks. (c) In the crystallization, the selective nucleation of a single grain is expected at each masked portion which acts as an artificial nucleation site. (d) The grains laterally grow beyond the site region.

control the nucleation rates inside and outside the artificial sites for actualizing the selective nucleation.

In the first demonstration for the solid-phase crystallization of a-Si thin films, the nucleation rate was controlled through the free-energy barrier to nucleation, W_*, by Si$^+$ ion implantation into as-deposited amorphous films. As shown in Figures 29 and 30, W_* widely changes with the conditions of ion implantation. The artificial nucleation sites can be prepared by spatially altering these conditions in plane of the as-deposited thin films. Figure 41 shows the cross-sectional schemata representing one of the processes for giving the spatial inhomogeneity to the conditions of ion implantation, which was employed in [156, 157]. First, a 100 nm thick a-Si thin film is deposited over a SiO$_2$ substrate by LPCVD under the same conditions as those for Figure 31, and Si$^+$ ions are implanted uniformly into the as-deposited a-Si thin film at a dose of 4×10^{12} mm^{-2} and at an acceleration energy of 70 keV (Fig. 41a). Second, Si$^+$ ions are implanted again at a dose of 2×10^{13} mm^{-2} and at the same acceleration energy, locally into the a-Si thin film by placing masks with a diameter of 0.66 μm at square lattice points periods of 3 μm (Fig. 41b). By this two-step ion implantation, the masked portions of the a-Si thin film suffer a lower ion bombardment than the unmasked area and hence will act as artificial nucleation sites. In the isothermal annealing at 873 K in nitrogen atmosphere, a single grain is expected to nucleate at each artificial site (Fig. 41c), and the grains will laterally grow over the sites (Fig. 41d).

Figure 42 shows the result of the process in Figure 41, by plan-view TEM micrographs representing the time evolution of the selective nucleation. Each micrograph covers an area containing four artificial nucleation sites. It is seen at 2.0×10^4 s that submicron-sized single grains are formed at the artificial sites which are visible as dark circular areas. At 3.6×10^4 s, most of the grains have grown beyond the site's area to be micron-sized, and they are about to impinge upon the neighbors at 5.4×10^4 s. Figure 43 shows the plan-view TEM image of the thin films annealed for 3.6×10^4 s, covering a wider area than those in Figure 42. It is clearly seen that the grains are exclu-

2.0×10^4 s   3.6×10^4 s   5.4×10^4 s   2 μm

Fig. 42. Plan-view TEM micrographs showing the time evolution of the selective nucleation obtained by the process shown in Figure 41. At 2.0×10^4 s, submicron-sized single grains are formed at the artificial sites which are visible as dark circular areas. At 3.6×10^4 s, most of the grains have grown beyond the site's area to be micron-sized. At 5.4×10^4 s, the controlled grains are about to impinge upon the neighbors.

sively formed at any artificial sites and the grain locations are controlled as a square lattice. The selected-area transmission-electron diffraction of these controlled grains reveals that they have continuous crystalline structures [156, 162]. This observation also confirms, in addition to the time evolution shown in Figure 42, that the controlled grains have grown single nuclei selectively formed at the artificial nucleation sites.

In Figure 43, it is also seen that most of the grains have already grown beyond the site's areas, while a few small grains still stay within the sites. The existence of these small grains predicts that the size of the controlled grains is lognormally distributed over a finite range. If the conditions for selective nucleation expressed by Eqs. (4.7) and (4.8) are fulfilled here, the dynamic evolution of the size distribution of controlled grains can be theoretically formulated up to the complete depletion of nucleation sites [330]. Since only a single grain is allowed to nucleate at each artificial site, the artificial sites available to the subsequent nucleation are restricted to those with no grain and are being depleted one by one as a new grain nucleates. Therefore, the concentration of the available artificial sites, $f_s(t)$, at a time of t decreases with t, obeying

$$\frac{df_s(t)}{dt} + a_s f_s(t) \frac{dN(t)}{dt} = 0 \qquad (4.9)$$

where $N(t)$ is the accumulated number concentration of the stable grains and is given by Eq. (2.120) with Eq. (2.105) and Eqs. (3.21)–(3.25). Under the initial conditions of $f_s'(0) = 0$, $f_s(t) \to 0$, and $f_s'(t) \to 0$ as $t \to \infty$, Eq. (4.9) is solved, yielding

$$f_s(t) = f_s(0) \exp[-a_s N(t)] \qquad (4.10)$$

With Eq. (4.10), the concentration of actual nucleation sites over the entire thin film is derived as

$$f_1(g, t) = a_s f(1, 0) f_s[t - t_g(g)] \qquad (4.11)$$

where $f_1(g, t)$ corresponds to Eq. (2.43) and $t_g(g)$ is given by Eq. (3.25). The size distribution of large grains formed by the selective nucleation at artificial nucleation sites is thus given by Eq. (2.81) with Eqs. (2.42), (2.104), (3.21)–(3.25), and (4.11). It is noted that such a description is valid also after the complete depletion of actual nucleation sites until the impingement

3 µm

Fig. 43.   Plan-view TEM micrograph of a-Si thin films annealed for 3.6×10^4 s with the grain location being controlled by artificial nucleation sites. The image covers a wider area than those in Figure 42. It is seen that the dendritic Si grains are exclusively formed at any artificial nucleation sites, and the grain locations are controlled as a square lattice.

between adjacent location-controlled grains starts in the considerably later stage than the case of random nucleation. Since the nucleation sites are supposed to be depleted while the transformed fraction is still negligible, the size distribution of grains should become lognormal-like from the early stage of the transformation as mentioned in Section 2.2.9.

Actually, the size distributions of controlled grains are lognormal-like in both the experimentally measured results and the theoretical predictions, as shown by the dynamic evolution in Figure 44. Here, the plots indicated by three kinds of the discrete markers correspond to the size distributions at the three annealing times and are observed in the samples shown by the micrographs of Figure 42 at the corresponding annealing times, respectively. The solid lines represent the theoretical predictions calculated by the above formulations with an interpolated value of $W_* = 2.05$ eV at an ion dose of 4×10^{12} mm^{-2} in Figure 29, and the measured value for the effective dimension of $d = 3.41$. The theoretical predictions are in good agreement with the experimental results, particularly in the following features: the size distribution appears lognormal and shifts toward the larger size with time, preserving its lognormal-like form, but slightly broadens, reducing its peak height. The theory suggests that such a slight change of the distribution form with time is due to the dendritic shape of grains. If the shapes of grains are compact (i.e., $d = 3$), the size distribution should simply shift with time, keeping its form unchanged. Thus the good agree-

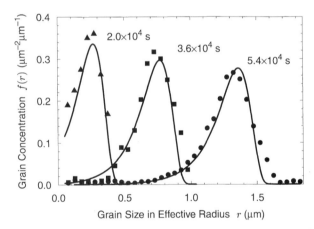

Fig. 44.   Dynamic evolution of the size distribution of grains in the controlled solid-phase crystallization of a-Si thin films with artificial nucleation sites. Three distributions marked with ▲, ■, and ● are observed in the samples shown by the micrographs in Figure 42, respectively. The solid lines are the theoretical predictions with the depletion of the nucleation sites and the noninteger effective dimension of dendritic grains taken into account.

ment between the theory and the experimental results also confirms the selective nucleation of a single grain at each artificial nucleation site, on which the theory is based.

The results shown in Figures 42–44 demonstrate not only a method for the control of grain locations but also an alternative origin of lognormal size distributions [330]. It has been long considered that the lognormal size distribution originates in the coarsening of grains through coalescence or coagulation. In the system with grain location being controlled, the lognormal-like size distributions can be observed without involving any coarsening process and long before the coalescence starts. Moreover, the experimentally observed results can be successfully explained by the simple theory taking no coalescence but only the depletion of nucleation sites into account. Thus these results elucidate that the formation of lognormal size distribution in phase transformations can be a consequence of the interplay between the dynamics of nucleation and growth and that of the nucleation site depletion.

4.3.2.1. Other Controls in Transformation of Starting Thin Films

The control of grain locations has been attempted also in the formation of various thin films by transformation of starting thin films. Some issues are shown below.

Asano et al. [331, 332] reported the location control of solid-phase nucleation at the surface steps. They deposited a-Si thin films by electron-beam evaporation over a SiO_2 substrate on which about 100 nm height steps are formed in a stripe pattern. In the solid-phase crystallization, preferential nucleation of grains is observed along the steps. Although such an one-dimensional control of grain locations does not yield the selective nucleation of a single grain at a predetermined position, a small terrace which has a short closed step could theoretically provide the artificial nucleation site for a single grain.

As another potential approach to selective nucleation in the transformation of thin films, one may consider metal-contact-induced crystallization of amorphous thin films. The metallic thin films on top of amorphous semiconductor surfaces provide heterogeneous nucleation sites in their solid-phase crystallization. Liu and Fonash [269] deposited ultrathin Pd islands on a selective area of a-Si thin films and crystallized them by rapid thermal annealing. They observed the exclusive formation of polycrystalline regions beneath the areas capped with Pd. Atwater et al. [270] reported a similar method which is more oriented toward the selection of single grains. They deposited a periodic array of 5 μm diameter In islands in period of 20–100 μm by mechanically masked evaporation on top of amorphous Ge film, and they accomplished the selective nucleation of approximately 5–8 grains at each site under the In island. Although the selective nucleation of single grains at these artificial nucleation sites was not achieved, it is theoretically possible if the size of the metal islands could be made smaller. Asano Makihara and Asano [333] introduces a metal imprint method in which an array of Ni-coated tips is pressed onto a-Si thin films prior to the crystallization. A small amount of Ni transferred to the a-Si surface could reduce the number of grains grown at the imprinted site, potentially down to one.

There are a few attempts also in the crystallization of Si thin films melted by laser annealing, where Si thin film is first melted by heating with laser exposure and then crystallized in liquid phase by cooling with the end of the exposure. All of the attempts are based on the spatial control of heating or cooling. Noguchi and Ikeda [258] prepared a reflecting overlayer with small openings on top of a-Si thin films and irradiated the substrate with an excimer laser. During laser irradiation, the a-Si regions under the small openings are exclusively heated so as to restrict the melting and liquid-phase crystallization there. Consequently, the crystallized regions are formed at predetermined positions in the a-Si thin film. Subsequent isothermal annealing at low temperatures induces solid-phase epitaxial growth from the seed grains formed by excimer-laser annealing. Toet et al. [139, 259, 260] reported a similar method without the reflecting overlayer. They irradiated a cw Ar^+-laser beam, which was focused into a spot of $\sim$1 μm diameter on a-Si thin films, and induced the local melting and liquid-phase crystallization there. Unfortunately, in the above two cases, the selective nucleation of a single grain has not been achieved, probably due to the difficulty in heat control within a small region. Kim and Im [334, 335] reported the effect of a stripe-patterned SiO_2 layer over a-Si thin films in the crystallization process only by excimer-laser annealing. The antireflective coating effect induces complete melting of a-Si exclusively under the SiO_2 caps, while the uncapped regions are partially melted. As a result, large grains grown from the uncapped regions into the capped one are aligned perpendicular to the SiO_2 stripes. This case is also just a one-dimensional control of grain locations, but the observed phenomenon might be used for manipulating artificial nucleation sites. Song and Im [336, 337] proposed a method using patterned square islands of a-Si thin films which are connected to small masked areas via narrow bridges of the a-Si. In the excimer-laser annealing, the a-Si islands are completely annealed except the masked areas. Thus the masked region provides seed grains whose growth propagates through the bridge to the entire island. If the propagating growth was restricted through the bridge to that from a single seed, the entire island becomes a single grain. Wilt and Ishihara [281, 282] developed another method for local cooling during excimer-laser annealing of a-Si thin films. They prepared a small bump by patterning crystalline Si substrate, then deposited a SiO_2 layer thicker than the height of the bump. After flattening the SiO_2 surface, an a-Si thin film was deposited over it. By these processes, the SiO_2 layer is thinner over the bump than elsewhere so that the heat conductance from the top Si layer to the substrate is better near the bump. Thus in the excimer-laser annealing, the melted Si near the underlaid bump is cooled faster than the other area, and preferential nucleation occurs there. It is reported that a single grain was formed at a bump by reducing its size and adjusting the thickness of the SiO_2 layer over it.

Acknowledgments

The authors of this chapter are grateful to Professor V. M. Fokin for the use of the data published in Figure 1a of [197] in Figure 12, Professor M. K. Hatalis for the use of the data published on Figures 5, 7, and 9 of [161] in calculating the plots of Figure 27, Professor U. Köster for the use of the data published in

Figure 3a and c of [55] in calculating the plots of Figure 15, Professor K. Lee for the use of the data published in Figures 3 and 4 of [163] in calculating the plots of Figure 28, Dr. K. Nakazawa for the use of the data published in Figures 4 and 6 of [155] in calculating the plots of Figure 26, Professor R. Reif for the use of the data published in Figure 4 of [123] in Figure 13, and Dr. H. Y. Tong for the use of the data published in Figures 3 and 4 of [232] in calculating the plots of Figures 17 and 23.

The authors also appreciate the permission of the following publishers to adopt or reproduce the published materials or to reuse the same sources used in the published materials, on which they hold copyrights: American Institute of Physics for the use of Figure 4 of [123] in Figure 13, Figure 2b of [153] in Figure 16, Figures 3 and 4 of [232] in Figures 17 and 23, Tables I and II of [227] in Table I, Figures 4 and 6 of [155] in Figure 26, and Figures 1, 2, and 3 of [134] in Figures 37, 33, and 32, respectively; Elsevier Science B. V. for the use of Figure 1a of [197] in Figure 12, Table 2 and Figure 6 of [231] in Figure 22, Figures 3 and 4 of [163] in Figure 28, and Figures 1, 2, and 3 of [132] in Figures 32 and 33; Materials Research Society for the use of Figure 1 of [235] in Figures 6 and 10, Figure 3 of [243] in Figure 25, Figures 2a, 3, and 4 of [133] in Figures 32 and 33; The American Physical Society for the use of Table I and Figure 1 of [228] in Figure 18, Figures 6 and 7 of [168] in Figures 19 and 21, respectively, and Figures 1 and 2 of [330] in Figures 42 and 44, respectively; the Electrochemical Society, Inc., for the use of Figures 5, 7, and 9 of [161] in Figure 27; the Institute of Pure and Applied Physics for the use of Figures 3, 4, 5, 6a, and 8a of [313] in Figures 38, 37, 32, 9, and 43, respectively; and Wiley-VCH Verlag GmbH for the use of Figure 3a and c of [55] in Figure 15.

REFERENCES

1. B. Lewis and J. C. Anderson, "Nucleation and Growth of Thin Films." Academic Press, New York, 1978.
2. H. Neddermeyer, *Crit. Rev. Solid State Mater. Sci.* 16, 309 (1990).
3. H. Liu and D. S. Dandy, "Diamond Chemical Vapor Deposition: Nucleation and Early Growth Stages." Noyes, Park Ridge, NJ, 1995.
4. F. Y. Genin, *Acta Metal. Mater.* 43, 4289 (1995).
5. I. V. Markov, "Crystal Growth for Beginners: Fundamentals of Nucleation, Crystal Growth, and Epitaxy." World Scientific, Singapore, 1995.
6. S. A. Kukushkin and A. V. Osipov, *Progr. Surf. Sci.* 51, 1 (1996).
7. I. Markov, *Phys. Rev. B* 56, 12544 (1997).
8. I. Markov, *Mater. Chem. Phys.* 49, 93 (1997).
9. C. Spinella, S. A. Lombardo, and F. Priolo, *J. Appl. Phys.* 84, 5383 (1998).
10. I. Kusaka, Z.-G. Wang, and J. H. Seinfeld, *J. Chem. Phys.* 108, 6829 (1998).
11. H. Brune, *Surf. Sci. Rep.* 31, 125 (1998).
12. D. M. Gruen, *Mater. Res. Soc. Bull.* 23, 32 (1998).
13. R. Lacmann, A. Herden, and C. Mayer, *Chem. Eng. Tech.* 22, 279 (1999).
14. S. T. Lee, Z. D. Lin, and X. Jiang, *Mater. Sci. Eng. R-Rep.* 25, 123 (1999).
15. M. Zinke-Allmang, *Thin Solid Films* 346, 1 (1999).
16. H. Rapaport, I. Kuzmenko, M. Berfeld, K. Kjaer, J. Als-Nielsen, R. Popovitz-Biro, I. Weissbuch, M. Lahav, and L. Leiserowitz, *J. Phys. Chem. B* 104, 1399 (2000).
17. J. W. Gibbs, *Trans. Connect. Acad.* 3, 108 (1875); 3, 343 (1878).
18. W. Ostwald, *Z. Phys. Chem. (Leipzig)* 22, 289 (1897).
19. M. Volmer and A. Weber, *Z. Phys. Chem.* 119, 227 (1926).
20. L. Farkas, *Z. Phys. Chem. A* 125, 236 (1927).
21. R. Kaischew and I. N. Stranski, *Z. Phys. Chem. B* 26, 317 (1934).
22. J. I. Frenkel, *J. Phys.* 1, 315 (1939).
23. W. Bank, *J. Chem. Phys.* 7, 324 (1939).
24. J. B. Zeldovich, *Acta Physiochim. URSS* 18, 1 (1943).
25. K. Binder and D. Stauffer, *Adv. Phys.* 25, 343 (1976).
26. K. Binder, *Rep. Progr. Phys.* 50, 783 (1987).
27. K. F. Kelton, *Solid State Phys.* 45, 75 (1991).
28. D. W. Oxtoby, *J. Phys. Condens. Matter* 4, 7627 (1992).
29. R. Pretorius, T. K. Marais, and C. C. Theron, *Mater. Sci. Eng. R-Rep.* 10, 1 (1993).
30. I. Gutzow and J. Schmelzer, "The Vitreous State: Thermodynamics, Structure, Rheology, and Crystallization." Springer-Verlag, Berlin, 1995.
31. D. W. Oxtoby, *Accounts Chem. Res.* 31, 91 (1998).
32. L. Gránásy and P. F. James, *J. Chem. Phys.* 111, 737 (1999).
33. L. Gránásy, *J. Mol. Struc.* 486, 523 (1999).
34. I. H. Leubner, *Current Opinion Colloid Interface Sci.* 5, 151 (2000).
35. D. Kashchiev, "Nucleation: Basic Theory with Applications Physics." Butterworth–Heinemann, Oxford, 2000.
36. D. Walton, *J. Chem. Phys.* 37, 2182 (1962).
37. L. D. Landau and E. M. Lifshitz, "Statistical Physics," Chap. 12. Pergamon, Oxford, 1969.
38. M. Volmer, *Z. Phys. Chem.* 25, 555 (1929).
39. R. Becker and W. Döring, *Ann. Phys.* 24, 719 (1935).
40. J. I. Frenkel, "Kinetic Theory of Liquids." Clarendon, Oxford, 1946.
41. D. Turnbull and J. C. Fisher, *J. Chem. Phys.* 17, 71 (1949).
42. W. J. Dunning, in "Nucleation" (A. C. Zettlemoyer, Ed.), p. 1. Dekker, New York, 1969.
43. H. Wakeshima, *J. Chem. Phys.* 22, 1614 (1954).
44. I. M. Lifshitz and V. V. Slyozov, *J. Phys. Chem. Solids* 19, 35 (1961).
45. J. Feder, K. C. Russell, J. Lothe, and G. M. Pound, *Adv. Phys.* 15, 111 (1966).
46. R. F. Probstein, *J. Chem. Phys.* 19, 619 (1951).
47. A. Kantrowitz, *J. Chem. Phys.* 19, 1097 (1951).
48. B. K. Chakraverty, *Surface Sci.* 4, 205 (1966).
49. F. C. Collins, *Z. Electrochem.* 59, 404 (1955).
50. D. Kashchiev, *Surface Sci.* 14, 209 (1969).
51. D. Kashchiev, *Surface Sci.* 18, 293 (1969).
52. D. Kashchiev, *Surface Sci.* 18, 389 (1969).
53. D. Kashchiev, *Surface Sci.* 22, 319 (1970).
54. S. Toschev, A. Milchev, and S. Stoyanov, *J. Cryst. Growth* 13/14, 123 (1972).
55. U. Köster, *Phys. Status Solidi* 48, 313 (1978).
56. U. Köster and M. Blank-Bewersdorff, *J. Less-Common Metals* 140, 7 (1988).
57. P. F. James, *Phys. Chem. Glasses* 15, 95 (1974).
58. K. F. Kelton, A. L. Greer, and C. V. Thompson, *J. Chem. Phys.* 79, 6261 (1983).
59. B. Shizgal and J. C. Barrett, *J. Chem. Phys.* 91, 6505 (1989).
60. V. A. Shneidman, *Sov. Phys. Tech. Phys.* 32, 76 (1987).
61. D. T. Wu, *J. Chem. Phys.* 97, 1922 (1992).
62. D. T. Wu, *J. Chem. Phys.* 97, 2644 (1992).
63. D. T. Wu, *J. Chem. Phys.* 99, 1990 (1993).
64. V. A. Shneidman, *Sov. Phys. Tech. Phys.* 33, 1338 (1989).
65. V. A. Shneidman and M. C. Weinberg, *J. Chem. Phys.* 95, 9148 (1991).
66. V. A. Shneidman, *Phys. Rev. A* 44, 2609 (1991).
67. V. A. Shneidman, *Phys. Rev. A* 44, 8441 (1991).
68. G. Shi and J. H. Seinfeld, *J. Colloid Interface Sci.* 135, 252 (1990).
69. G. Shi, J. H. Seinfeld, and K. Okuyama, *Phys. Rev. A* 41, 2101 (1990).
70. G. Shi, J. H. Seinfeld, and K. Okuyama, *Phys. Rev. A* 44, 8443 (1991).
71. G. Shi, J. H. Seinfeld, and K. Okuyama, *J. Appl. Phys.* 68, 4550 (1990).
72. G. Shi and J. H. Seinfeld, *J. Chem. Phys.* 92, 687 (1990).
73. G. Shi and J. H. Seinfeld, *J. Chem. Phys.* 93, 9033 (1990).
74. P. Demo and Z. Kožíšek, *Phys. Rev. B* 48, 3620 (1993).
75. H. A. Kramers, *Physica* 7, 284 (1940).
76. J. E. Moyal, *J. Roy. Statist. Soc. Ser. B* 11, 150 (1949).

77. B. J. Matkowsky, Z. Schuss, and C. Knessl, *Phys. Rev. A* 29, 3359 (1984).
78. M. Gitterman and Y. Rabin, *J. Chem. Phys.* 80, 2234 (1984).
79. Y. Rabin and M. Gitterman, *Phys. Rev. A* 29, 1496 (1984).
80. I. Edrei and M. Gitterman, *J. Chem. Phys.* 85, 190 (1986).
81. I. Edrei and M. Gitterman, *J. Phys. A* 19, 3279 (1986).
82. I. Edrei and M. Gitterman, *Phys. Rev. A* 33, 2821 (1986).
83. L. Demeio and B. Shizgal, *J. Chem. Phys.* 98, 5713 (1993).
84. H. Trinkaus and M. H. Yoo, *Philos. Mag. A* 55, 269 (1987).
85. F. G. Shi, *Scripta Metal. Materialia* 30, 1195 (1994).
86. F. G. Shi, *Phys. Lett. A* 183, 311 (1993).
87. F. G. Shi, *Chem. Phys. Lett.* 212, 421 (1993).
88. G. Shi and J. H. Seinfeld, *AIChE J.* 40, 11 (1994).
89. G. Shi and J. H. Seinfeld, *Mater. Chem. Phys.* 37, 1 (1994).
90. H. Huang, "Statistical Mechanics." New York, 1963.
91. W. G. Courtney, *J. Chem. Phys.* 36, 2009 (1962).
92. F. F. Abraham, *J. Chem. Phys.* 51, 1632 (1969).
93. W. J. Shugard and H. Reiss, *J. Chem. Phys.* 65, 2827 (1976).
94. B. T. Draine and E. E. Salpeter, *J. Chem. Phys.* 67, 2230 (1977).
95. I. K. Dannetschek and D. Stauffer, *J. Aerosol Sci.* 12, 105 (1981).
96. A. R. Despic and M. N. Djorovic, *Electrochim. Acta* 29, 131 (1984).
97. V. Volterra and A. R. Cooper, *J. Non-Cryst. Solids* 74, 85 (1985).
98. K. F. Kelton and A. L. Greer, *J. Non-Cryst. Solids* 79, 295 (1986).
99. H.-J. Müller and K.-H. Heinig, *Nuclear Instrum. Methods B* 22, 524 (1987).
100. Z. Kožíšek, *Cryst. Res. Technol.* 23, 1315 (1988).
101. O. T. Valls and G. F. Mazenko, *Phys. Rev. B* 42, 6614 (1990).
102. A. L. Greer, P. V. Evans, R. G. Hamerton, D. K. Shangguan, and K. F. Kelton, *J. Cryst. Growth* 99, 38 (1990).
103. N. Miloshev and G. Miloshev, *Atom. Res.* 25, 417 (1990).
104. B. Nowakowski and E. Ruckenstein, *J. Colloid Interface Sci.* 145, 182 (1991).
105. P. V. Evans and S. R. Stiffler, *Acta Metal. Mater.* 39, 2727 (1991).
106. Z. Kožíšek and P. Demo, *J. Chem. Phys.* 108, 9835 (1998).
107. C. Spinella, S. A. Lombardo, and F. Priolo, *Thin Solid Films* 337, 90 (1999).
108. G. Nicolis and I. Prigogine, "Self-Organization in Nonequilibrium Systems," Chaps. 9 and 10. Wiley, New York, 1977.
109. N. G. van Kampen, "Stochastic Processes in Physics and Chemistry," Chap. 5. North-Holland, New York, 1981.
110. C. W. Gardiner, "Handbook of Stochastic Methods." Springer-Verlag, Berlin, 1983.
111. L. Dufour and R. Defay, "Thermodynamics of Clouds," Chap. 13. Academic Press, New York, 1963.
112. M. Planck, *Sitzungsber. Preuss. Akad. Wiss.* (1917).
113. F. C. Goodrich, *Proc. Roy. Soc. London Ser. A* 227, 167 (1964).
114. I. Prigogine, "Introduction to Thermodynamics of Irreversible Processes," 3rd. ed. Interscience, New York, 1968.
115. C. M. Bender and S. A. Orszag, "Advanced Mathematical Methods for Scientists and Engineers," Part III. McGraw–Hill, New York, 1978.
116. K. C. Russel, *Adv. Colloid Interface Sci.* 13, 205 (1980).
117. J. J. Hoyt and G. Sunder, *Scripta Metal. Materialia* 29, 1535 (1993).
118. J. W. Christian, "The Theory of Transformations in Metals and Alloys." Pergamon, Oxford, 1975.
119. W. A. Johnson and R. F. Mehl, *Amer. Inst. Min. Metall. Pet. Eng.* 135, 416 (1939).
120. M. Avrami, *J. Chem. Phys.* 7, 1103 (1939).
121. M. Avrami, *J. Chem. Phys.* 8, 212 (1940).
122. M. Avrami, *J. Chem. Phys.* 9, 177 (1941).
123. R. B. Iverson and R. Reif, *J. Appl. Phys.* 62, 1675 (1987).
124. H. Richter, Z. P. Wang, and L. Ley, *Solid State Comm.* 39, 625 (1981).
125. A. T. Voutsas, M. K. Hatalis, J. Boyce, and A. Chiang, *J. Appl. Phys.* 78, 6999 (1995).
126. G. Harbeke, L. Krausbauer, E. F. Steigmeier, A. E. Widmer, H. F. Kappert, and G. Neugebauer, *Appl. Phys. Lett.* 42, 249 (1983).
127. G. Harbeke, L. Krausbauer, E. F. Steigmeier, A. E. Widmer, H. F. Kappert, and G. Neugebauer, *J. Electrochem. Soc.* 131, 675 (1984).
128. F. S. Becker, H. Oppolzer, I. Weitzel, H. Eichermüller, and H. Schaber, *J. Appl. Phys.* 56, 1233 (1984).
129. T. Aoyama, G. Kawachi, N. Konishi, T. Suzuki, Y. Okajima, and K. Miyata, *J. Electrochem. Soc.* 136, 1169 (1989).
130. S. Hasegawa, S. Sakamoto, T. Inokuma, and Y. Kurata, *Appl. Phys. Lett.* 62, 1218 (1993).
131. H. P. Klung and L. E. Alexander, "X-ray Diffraction Procedure." Wiley, New York, 1959.
132. H. Kumomi, T. Yonehara, Y. Nishigaki, and N. Sato, *Appl. Surf. Sci.* 41/42, 638 (1989).
133. H. Kumomi and T. Yonehara, *Mater. Res. Soc. Symp. Proc.* 202, 83 (1991).
134. H. Kumomi and T. Yonehara, *Appl. Phys. Lett.* 54, 2648 (1989).
135. E. Molinari, R. Polini, and M. Tomellini, *Appl. Phys. Lett.* 61, 1287 (1992).
136. A. A. Shklyaev, M. Shibata, and M. Ichikawa, *Phys. Rev. B* 58, 15647 (1998).
137. P. Joubert, *J. Appl. Phys.* 60, 2823 (1986).
138. K. Uesugi, T. Yao, T. Sato, T. Sueyoshi, and M. Iwatsuki, *Appl. Phys. Lett.* 62, 1600 (1993).
139. D. Toet, P. V. Santos, S. Eitel, and M. Heintze, *J. Non-Cryst. Solids* 198/200, 887 (1996).
140. K. M. Chen, D. E. Jesson, S. J. Pennycook, T. Thundat, and R. J. Warmack, *Phys. Rev. B* 56, R1700 (1997).
141. J. J. Hammel, *J. Chem. Phys.* 46, 2234 (1967).
142. D. G. Burnett and R. W. Douglas, *Phys. Chem. Glasses* 12, 117 (1971).
143. S. Roorda, D. Kammann, W. C. Sinke, G. F. A. van de Walle, and A. A. van Gorkum, *Mater. Lett.* 9, 259 (1990).
144. J. E. Greene and L. Mei, *Thin Solid Films* 37, 429 (1976).
145. Y. Komen and I. W. Hall, *J. Appl. Phys.* 52, 6655 (1981).
146. K. T.-Y. Kung and R. Reif, *J. Appl. Phys.* 62, 1503 (1987).
147. T. Katoh, *IEEE Trans. Electron Devices* 35, 923 (1988).
148. M. K. Hatalis and D. W. Greve, *J. Appl. Phys.* 63, 2260 (1988).
149. U. Köster and M. Blank-Bewersdorff, *J. Less-Common Metals* 140, 7 (1988).
150. I. Mizushima, W. Tabuchi, and H. Kuwano, *Japan. J. Appl. Phys.* 27, 2310 (1988).
151. S. Kambayashi, S. Onga, I. Mizushima, K. Higuchi, and H. Kuwano, "Extended Abstracts of the 21st Conference on Solid State Devices and Materials," Business Center for Academic Societies Japan, Tokyo, 1989, p. 169.
152. J. S. Im and H. A. Atwater, *Appl. Phys. Lett.* 57, 1766 (1990).
153. J. S. Im, J. H. Shin, and H. A. Atwater, *Appl. Phys. Lett.* 59, 2314 (1991).
154. K. Nakazawa and K. Tanaka, *J. Appl. Phys.* 68, 1029 (1990).
155. K. Nakazawa, *J. Appl. Phys.* 69, 1703 (1991).
156. H. Kumomi and T. Yonehara, *Mater. Res. Soc. Symp. Proc.* 202, 645 (1991).
157. H. Kumomi, T. Yonehara, and T. Noma, *Appl. Phys. Lett.* 59, 3565 (1991).
158. C. H. Hong, C. Y. Park, and H.-J. Kim, *J. Appl. Phys.* 71, 5427 (1992).
159. Y. Masaki, P. G. LeComber, and A. G. Fitzgerald, *J. Appl. Phys.* 74, 129 (1993).
160. S. Lombardo, S. U. Campisano, and F. Baroetto, *Phys. Rev. B* 47, 13561 (1993).
161. A. T. Voutsas and M. K. Hatalis, *J. Electrochem. Soc.* 140, 871 (1993).
162. H. Kumomi and T. Yonehara, *J. Appl. Phys.* 75, 2884 (1994).
163. J. W. Park, D. G. Moon, B. T. Ahn, H. B. Im, and K. Lee, *Thin Solid Films* 245, 228 (1994).
164. F. Hausdorff, *Math. Ann.* 79, 157 (1918).
165. B. B. Mandelbrot, "Fractals: Form, Chance and Dimension." Freeman, San Francisco, 1977.
166. G. A. Edgar, "Measure, Topology and Fractal Geometry." Springer-Verlag, New York, 1990.
167. K. Falconer, "Fractal Geometry, Mathematical Foundation and Applications." Wiley, Chichester, 1990.
168. H. Kumomi and F. G. Shi, *Phys. Rev. B* 52, 16753 (1995).
169. The extensive reviews on the solid-phase crystallization of *a*-Si thin films are found in [9, 170].

170. N. Yamauchi and R. Reif, *J. Appl. Phys.* 75, 3235 (1994),

171. The details of the sample preparation are described in [162].

172. A. Nakamura, F. Emoto, E. Fujii, Y. Uemoto, A. Yamamoto, K. Senda, and G. Kano, *Japan. J. Appl. Phys.* 27, L2408 (1988).

173. T. Noma, T. Yonehara, and H. Kumomi, *Appl. Phys. Lett.* 59, 653 (1991).

174. H. S. Chen and D. Turnbull, *J. Appl. Phys.* 40, 4214 (1969).

175. N. Nagasima and N. Kubota, *J. Vac. Sci. Technol.* 14, 54 (1977).

176. R. Bisaro, J. Magariño, N. Proust, and K. Zellama, *J. Appl. Phys.* 59, 1167 (1986).

177. R. Bisaro, J. Magariño, Y. Pastol, P. Germain, and K. Zellama, *Phys. Rev. B* 40, 7655 (1989).

178. N. A. Blum and C. Feldman, *J. Non-Cryst. Solids* 11, 242 (1972).

179. N. A. Blum and C. Feldman, *J. Non-Cryst. Solids* 22, 29 (1976).

180. M. Libera and M. Chen, *J. Appl. Phys.* 73, 2272 (1993).

181. R. B. Bergmann, *J. Cryst. Growth* 165, 341 (1996).

182. M. Hirose, M. Hayama, and Y. Osaka, *Japan. J. Appl. Phys.* 13, 1399 (1974).

183. K. P. Chik and P.-K. Lim, *Thin Solid Films* 32, 45 (1976).

184. P. Germain, S. Squelard, J. Bourgoin, and A. Gheorghiu, *J. Appl. Phys.* 48, 1909 (1977).

185. P. Germain, S. Squelard, J. C. Bourgoin, and A. Gheorghiu, *J. Non-Cryst. Solids* 23, 93 (1977).

186. P. Germain, K. Zellama, S. Squelard, J. C. Bourgoin, and A. Gheorghiu, *J. Appl. Phys.* 50, 6986 (1979).

187. K. Zellama, P. Germain, S. Squelard, J. C. Bourgoin, and P. A. Thomas, *J. Appl. Phys.* 50, 6995 (1979).

188. R. Tsu, J. Gonzalez-Hernandez, S. S. Chao, S. C. Lee, and K. Tanaka, *Appl. Phys. Lett.* 40, 534 (1982).

189. R. Bisaro, J. Magariño, K. Zellama, S. Squelard, P. Germain, and J. F. Morhange, *Phys. Rev. B* 31, 3568 (1985).

190. T. Mohammed-Brahim, K. Kis-Sion, D. Briand, M. Sarret, O. Bonnaund, J. P. Kleider, C. Longeaud, and B. Lambert, *J. Non-Cryst. Solids* 227–230, 962 (1998).

191. J. Gonzalez-Hernandez and R. Tsu, *Appl. Phys. Lett.* 42, 90 (1983).

192. H. Kakinuma, M. Mohri, M. Sakamoto, and T. Tsuruoka, *J. Appl. Phys.* 70, 7374 (1991).

193. R. B. Iverson and R. Reif, *Appl. Phys. Lett.* 52, 645 (1988).

194. Á. Barna, P. B. Barna, and J. F. Pócza, *J. Non-Cryst. Solids* 8–10, 36 (1972).

195. C. Spinella, S. Lombardo, and S. U. Campisano, *Appl. Phys. Lett.* 57, 554 (1990).

196. C. Spinella, S. Lombardo, and S. U. Campisano, *Phys. Rev. Lett.* 66, 1102 (1991).

197. V. M. Fokin, A. M. Kalinina, and V. N. Filipovich, *J. Cryst. Growth* 52, 115 (1981).

198. S. K. Heffels and E. J. de Jong, *Chem. Eng. Technol.* 13, 63 (1990).

199. The experimental details are found in [168].

200. T. Saito and I. Ohdomari, *Philos. Mag. B* 43, 673 (1981).

201. T. Saito and I. Ohdomari, *Philos. Mag. B* 49, 471 (1984).

202. I.-W. Wu, A. Chiang, M. Fuse, L. Öveçoglu, and T. Y. Huang, *J. Appl. Phys.* 65, 4036 (1989).

203. I.-W. Wu, A. Chiang, M. Fuse, L. Öveçoglu, and T. Y. Huang, *Mater. Res. Soc. Symp. Proc.* 182, 107 (1990).

204. L. Csepregi, J. W. Mayer, and T. W. Sigmon, *Appl. Phys. Lett.* 29, 92 (1976).

205. L. Csepregi, R. P. Küllen, J. W. Mayer, and T. W. Sigmon, *Solid State Comm.* 21, 1019 (1977).

206. L. Csepregi, E. F. Kennedy, T. J. Gallagher, J. W. Mayer, and T. W. Sigmon, *J. Appl. Phys.* 48, 4234 (1977).

207. E. F. Kennedy, L. Csepregi, J. W. Mayer, and T. W. Sigmon, *J. Appl. Phys.* 48, 4241 (1977).

208. L. Csepregi, E. F. Kennedy, J. W. Mayer, and T. W. Sigmon, *J. Appl. Phys.* 49, 3606 (1978).

209. I. Golecki, G. E. Chapman, S. S. Lau, B. Y. Tsaur, and J. W. Mayer, *Phys. Lett.* 71A, 267 (1979).

210. A. La Ferla, S. Cannavò, G. Ferla, S. U. Campisano, E. Rimini, and M. Servidori, *Nuclear Instrum. Methods B* 19/20, 470 (1987).

211. A. La Ferla, E. Rimini, and G. Ferla, *Appl. Phys. Lett.* 52, 712 (1988).

212. J. P. de Souza, L. Amaral, and P. F. P. Fichtner, *J. Appl. Phys.* 71, 5423 (1992).

213. F. Priolo, A. La Ferla, C. Spinella, E. Rimini, G. Ferla, F. Baroetto, and A. Licciardello, *Appl. Phys. Lett.* 53, 2605 (1988).

214. F. Priolo, C. Spinella, and E. Rimini, *Phys. Rev. B* 41, 5235 (1990).

215. G. Q. Lu, E. Nygren, M. J. Aziz, D. Turnbull, and C. W. White, *Appl. Phys. Lett.* 56, 137 (1990).

216. J. M. C. England, P. J. Timans, C. Hill, P. D. Augustus, and H. Ahmed, *Appl. Phys. Lett.* 61, 1670 (1992).

217. J. Morgiel, I.-W. Wu, A. Chiang, and R. Sinclair, *Mater. Res. Soc. Symp. Proc.* 182, 191 (1990).

218. A. S. Kirtikar, R. Sinclair, J. Morgiel, I.-W. Wu, and A. Chiang, *Mater. Res. Soc. Symp. Proc.* 202, 627 (1991).

219. R. Sinclair, J. Morgiel, A. S. Kirtikar, I. W. Wu, and A. Chiang, *Ultramicroscopy* 51, 41 (1993).

220. M. Aoucher, G. Farhi, and T. Mohammed-Brahim, *J. Non-Cryst. Solids* 227–230, 958 (1998).

221. R. P. Andres and M. Boudart, *J. Chem. Phys.* 42, 2057 (1965).

222. S. Toschev and I. Gutzow, *Phys. Status Solidi* 21, 683 (1967).

223. H. L. Frisch and C. C. Carlier, *J. Chem. Phys.* 54, 4326 (1971).

224. K. C. Russell, *Fundam. Appl. Ternary Diffusion* 59 (1990).

225. P. Demo and Z. Kožíšek, *Philos. Mag. B* 70, 49 (1994).

226. F. G. Shi, *Scripta Metal. Materialia* 31, 1227 (1994).

227. F. G. Shi, *J. Appl. Phys.* 76, 5149 (1994).

228. F. G. Shi and K. N. Tu, *Phys. Rev. Lett.* 74, 4476 (1995).

229. F. G. Shi, H. Y. Tong, and J. D. Ayers, *Appl. Phys. Lett.* 67, 350 (1995).

230. H. Y. Tong, F. G. Shi, E. J. Lavernia, and J. D. Ayers, *Scripta Metal. Materialia* 32, 511 (1995).

231. H. Y. Tong, T. J. King, and F. G. Shi, *Thin Solid Films* 290–291, 464 (1996).

232. H. Y. Tong, B. Z. Ding, H. G. Jiang, K. Lu, J. T. Wang, and Z. Q. Hu, *J. Appl. Phys.* 75, 654 (1994).

233. J. L. Batstone, *Philos. Mag. A* 67, 51 (1993).

234. J. S. Im and H. A. Atwater, *Nuclear Instrum. Methods Phys. B* 59, 422 (1991); J. H. Shin and H. A. Atwater, *ibid.* 80/81, 973 (1993).

235. H. Kumomi and F. G. Shi, *Mater. Res. Soc. Symp. Proc.* 472, 403 (1997).

236. T. I. Kamins, *J. Appl. Phys.* 42, 4357 (1971).

237. J. Y. W. Seto, *J. Appl. Phys.* 46, 5247 (1975).

238. S. Solmi, M. Severi, R. Angelucci, L. Baldi, and R. Bilenchi, *J. Electrochem. Soc.* 129, 1811 (1982).

239. G. Shi and J. H. Seinfeld, *J. Mater. Res.* 6, 2091 (1991).

240. M. R. Weidmann and K. E. Newman, *Phys. Rev. B* 45, 8388 (1992).

241. P. Kringhøj, R. G. Elliman, M. Fyhn, S. Y. Shiryaev, and A. N. Larsen, *Nuclear Instrum. Methods B* 106, 346 (1995).

242. C. V. Thompson and F. Spaepen, *Acta Metall.* 27, 1855 (1979).

243. H. Y. Tong, Q. Jiang, D. Hsu, T. J. King, and F. G. Shi, *Mater. Res. Symp. Proc.* 472, 397 (1997).

244. T. Mohammed-Brahim, M. Sarret, D. Briand, K. Kis-sion, L. Haji, O. Bonnaud, D. Louër, and A. Hadjaj, *Philos. Mag. B* 76, 193 (1997).

245. R. Grigorovici, *Mater. Res. Bull.* 3, 13 (1968).

246. M. H. Brodsky, R. S. Title, K. Weiser, and G. D. Petit, *Phys. Rev. B* 1, 2632 (1970).

247. R. Reif and J. E. Knott, *Electron. Lett.* 17, 587 (1981).

248. M.-K. Kang, T. Matsui, and H. Kuwano, *Japan. J. Appl. Phys.* 34, L803 (1995).

249. R. Kakkad, J. Smith, W. S. Lau, and S. J. Fonash, *J. Appl. Phys.* 65, 2069 (1989).

250. A. Nakamura, F. Emoto, E. Fujii, A. Yamamoto, Y. Uemoto, H. Hayashi, Y. Kato, and K. Senda, *IEDM Tech. Dig.* 847 (1990).

251. K. E. Violette, M. C. Öztürk, K. N. Christensen, and D. M. Maher, *J. Electrochem. Soc.* 143, 649 (1996).

252. The experimental details are reported in [162, 235].

253. C. V. Thompson, *J. Appl. Phys.* 58, 763 (1985), and references therein.

254. W. A. P. Claassen and J. Bloem, *J. Electrochem. Soc.* 127, 194 (1980).

255. N. M. Hwang, *J. Cryst. Growth* 205, 59 (1999).

256. N. M. Hwang, W. S. Cheong, and D. Y. Yoon, *J. Cryst. Growth* 206, 177 (1999).

257. D. B. Dove, *J. Appl. Phys.* 35, 2785 (1964).

258. T. Noguchi and Y. Ikeda, "Proc. Sony Research Forum," Sony, Tokyo, 1993, p. 200.

259. D. Toet, B. Koopmans, P. V. Santos, R. B. Bergmann, and B. Richards, *Appl. Phys. Lett.* 69, 3719 (1996).

260. D. Toet, B. Koopmans, R. B. Bergmann, B. Richards, P. V. Santos, M. Arbrecht, and J. Krinke, *Thin Solid Films* 296, 49 (1997).

261. T. Yamaoka, K. Oyoshi, T. Tagami, Y. Arima, K. Yamashita, and S. Tanaka, *Appl. Phys. Lett.* 57, 1970 (1990).

262. C. Spinella, S. Lombardo, F. Priolo, and S. U. Campisano, *Phys. Rev. B* 53, 7742 (1996), and references therein.

263. H. I. Smith and D. C. Flanders, *Appl. Phys. Lett.* 32, 349 (1978).

264. M. W. Geis, D. C. Flanders, and H. I. Smith, *Appl. Phys. Lett.* 35, 71 (1979).

265. M. W. Geis, D. A. Antoniadis, D. J. Silversmith, R. W. Mountain, and H. I. Smith, *Appl. Phys. Lett.* 37, 454 (1980).

266. H. Kumomi, unpublished.

267. X. Zhao, H. Isshiki, Y. Aoyagi, T. Sugano, S. Komuro, and S. Fujita, *Mater. Sci. Eng. B* 51, 154 (1998).

268. S. R. Herd, P. Chaudhari, and M. H. Brodsky, *J. Non-Cryst. Solids* 7, 309 (1972).

269. G. Liu and S. J. Fonash, *Appl. Phys. Lett.* 55, 660 (1989).

270. H. A. Atwater, J. C. M. Yang, and C. M. Chen, *Am. Inst. Phys. Conf. Proc.* 404, 345 (1997).

271. B. D. Joyce and J. A. Baldrey, *Nature* 195, 485 (1962).

272. L. Jastrzebski, *J. Cryst. Growth* 63, 493 (1983).

273. Y. Kunii, M. Tabe, and K. Kajiyama, *J. Appl. Phys.* 54, 2847 (1983).

274. H. Ishiwara, H. Yamamoto, S. Furukawa, M. Tamura, and T. Tokuyama, *Appl. Phys. Lett.* 43, 1028 (1983).

275. Y. Ohmura, Y. Matsushita, and M. Kashiwagi, *Japan. J. Appl. Phys.* 21, L152 (1983).

276. A. Doi, M. Kumikawa, J. Konishi, and Y. Nakamizo, *Appl. Phys. Lett.* 59, 2518 (1991).

277. K. Egami, A. Ogura, and M. Kimura, *J. Appl. Phys.* 59, 289 (1986).

278. A. T. Voutsas and M. K. Hatalis, *Appl. Phys. Lett.* 63, 1546 (1993).

279. A. T. Voutsas and M. K. Hatalis, *J. Appl. Phys.* 76, 777 (1994).

280. K. Yamagata and T. Yonehara, *Appl. Phys. Lett.* 61 2557 (1992).

281. P. Ch. van der Wilt and R. Ishihara, *Phys. Status Solidi A* 166, 619 (1998).

282. R. Ishihara and P. Ch. van der Wilt, *Japan. J. Appl. Phys.* 37, L15 (1998).

283. A. T. Voutsas and M. K. Hatalis, *J. Electrochem. Soc.* 140, 282 (1993).

284. Y.-H. Song, S.-Y. Kang, K. I. Cho, and H. J. Yoo, *ETRI J.* 19, 25 (1997).

285. H. Miura, H. Ohta, N. Okamoto, and T. Kaga, *Appl. Phys. Lett.* 60, 2746 (1992).

286. W. A. P. Claassen and J. Bloem, *J. Electrochem. Soc.* 127, 1836 (1980).

287. W. A. P. Claassen and J. Bloem, *J. Electrochem. Soc.* 128, 1353 (1981).

288. N. M. Hwang and D. Y. Yoon, *J. Cryst. Growth* 160, 98 (1996).

289. N. M. Hwang, J. H. Hahn, and D. Y. Yoon, *J. Cryst. Growth* 162, 55 (1996).

290. K. Choi, S.-J. L. Kang, H. M. Jang, and N. M. Hwang, *J. Cryst. Growth* 172, 416 (1997).

291. H. M. Jang and N. M. Hwang, *J. Mater. Res.* 13, 3527 (1998).

292. H. M. Jang and N. M. Hwang, *J. Mater. Res.* 13, 3536 (1998).

293. N. M. Hwang, *J. Cryst. Growth* 198/199, 945 (1999).

294. Jn.-D. Jeon, C. J. Park, D.-Y. Kim, and N. M. Hwang, *J. Cryst. Growth* 213, 79 (2000).

295. M. C. Barnes, D.-Y. Kim, H. S. Ahn, C. O. Lee, and N. M. Hwang, *J. Cryst. Growth* 213, 83 (2000).

296. W. S. Cheong, D. Y. Yoon, D.-Y. Kim, and N. M. Hwang, *J. Cryst. Growth* 218, 27 (2000).

297. N. M. Hwang, W. S. Cheong, D. Y. Yoon, and D.-Y. Kim, *J. Cryst. Growth* 218, 33 (2000).

298. N. M. Hwang and D.-Y. Kim, *J. Cryst. Growth* 218, 40 (2000).

299. T. Noguchi, T. Ohshima, and H. Hayashi, *Japan. J. Appl. Phys.* 28, L309 (1989).

300. N. Yamauchi, J.-J. Hajjar, and R. Reif, *IEEE Trans. Electron Devices* ED-38, 55 (1991).

301. D. N. Kouvatsos, A. T. Voutsas, and M. K. Hatalis, *IEEE Trans. Electron Devices* 43, 1399 (1996).

302. For reviews, see D. Kashchiev, *Surf. Sci.* 86, 14 (1979); K. Kinoshita, *Thin Solid Films* 85, 223 (1981), and references therein.

303. F. M. Ross, J. Tersoff, and R. M. Tromp, *Phys. Rev. Lett.* 80, 984 (1998).

304. C. G. Granqvist and R. A. Buhrman, *J. Appl. Phys.* 47, 2200 (1976).

305. S. K. Friedlander and C. S. Wang, *J. Colloid Interface Sci.* 22, 126 (1966).

306. K. W. Lee, *J. Colloid Interface Sci.* 92, 315 (1983).

307. K. W. Lee, Y. J. Lee, and D. S. Han, *J. Colloid Interface Sci.* 188, 486 (1997).

308. M. V. Smoluchowsky, *Z. Phys. Chem.* 92, 129 (1917).

309. C. Wagner, *Z. Elektrochem.* 65, 581 (1961).

310. J. Söderlund, L. B. Kiss, G. A. Niklasson, and C. G. Granqvist, *Phys. Rev. Lett.* 80, 2386 (1998).

311. R. B. Bergmann, F. G. Shi, and J. Krinke, *Phys. Rev. Lett.* 80, 1011 (1998).

312. T. Yonehara, Y. Nishigaki, H. Mizutani, S. Kondoh. K. Yamagata, T. Noma, and T. Ichikawa, *Appl. Phys. Lett.* 52, 1231 (1988).

313. H. Kumomi and T. Yonehara, *Japan. J. Appl. Phys.* 36, 1383 (1997).

314. N. Sato and T. Yonehara, *Appl. Phys. Lett.* 55, 636 (1989).

315. K. Hirabayashi, Y. Taniguchi, O. Takamatsu, T. Ikeda, K. Ikoma, and N. Iwasaki-Kurihara, *Appl. Phys. Lett.* 53, 1815 (1988).

316. T. Inoue, H. Tachibana, K. Kumagai, K. Miyata, K. Nishimura, K. Kobashi, and A. Nakaue, *J. Appl. Phys.* 67, 7329 (1990).

317. K. Kobashi, T. Inoue, H. Tachibana, K. Kumagai, K. Miyata, K. Nishimura, and A. Nakaue, *Vacuum* 41, 1383 (1990).

318. J. F. DeNatale, J. F. Flintoff, and A. B. Harker, *J. Appl. Phys.* 68, 4014 (1990).

319. R. Ramesham, T. Roppel, C. Ellis, D. A. Jawarske, and W. Baugh, *J. Mater. Res.* 6, 1278 (1991).

320. S. J. Lin, S. L. Lee, H. Hwang, C. S. Chang, and H. Y. Wen, *Appl. Phys. Lett.* 60, 1559 (1991).

321. S. Miyauchi, K. Kumagai, K. Miyata, K. Nishimura, K. Kobashi, A. Nakaue, J. T. Glass, and I. M. Buckley-Golder, *Surf. Coating Technol.* 47, 465 (1991).

322. J.-S. Ma, H. Kawarada, T. Yonehara, J.-I. Suzuki, J. Wei, Y. Yokota, H. Mori, H. Fujita, and A. Hiraki, *Appl. Surf. Sci.* 41/42, 572 (1989).

323. J.-S. Ma, H. Kawarada, T. Yonehara, J. Suzuki, J. Wei, Y. Yokota, and A. Hiraki, *Appl. Phys. Lett.* 55, 1070 (1989).

324. J.-S. Ma, H. Kawarada, T. Yonehara, J. Suzuki, J. Wei, Y. Yokota, and A. Hiraki, *J. Cryst. Growth* 99, 1206 (1990).

325. J.-S. Ma, H. Yagyu, A. Hiraki, H. Kawarada, and T. Yonehara, *Thin Solid Films* 206, 192 (1991).

326. P. G. Roberts, D. K. Milne, P. John, M. G. Jubber, and J. I. B. Wilson, *J. Mater. Res.* 11, 3128 (1996).

327. S. Katsumata and S. Yugo, *Diamond Relat. Mater.* 2, 1490 (1993).

328. H. Tokunaga, H. Kawasaki, and Y. Yamazaki, *Japan. J. Appl. Phys.* 31, L1710 (1992).

329. J. Aizenberg, A. J. Black, and G. M. Whitesides, *Nature* 398, 495 (1999).

330. H. Kumomi and F. G. Shi, *Phys. Rev. Lett.* 82, 2717 (1999).

331. T. Asano and K. Makihara, *Japan. J. Appl. Phys.* 32, 482 (1993).

332. T. Asano, K. Makihara, and H. Tsutae, *Japan. J. Appl. Phys.* 33, 659 (1994).

333. K. Makihara and T. Asano, *Appl. Phys. Lett.* 76, 3774 (2000).

334. H. J. Kim and J. S. Im, *Appl. Phys. Lett.* 68, 1513 (1996).

335. H. J. Kim and J. S. Im, *Mater. Res. Soc. Symp. Proc.* 397, 401 (1996).

336. H. J. Song and J. S. Im, *Appl. Phys. Lett.* 68, 3165 (1996).

337. H. J. Song and J. S. Im, *Mater. Res. Soc. Symp. Proc.* 397, 459 (1996).

Chapter 7

ION IMPLANT DOPING AND ISOLATION OF GaN AND RELATED MATERIALS

S. J. Pearton

Department of Materials Science and Engineering, University of Florida, Gainesville, Florida, USA

Contents

1. INTRODUCTION

Ion implantation is the preferred method for selected-area doping of semiconductors due to its high spatial and dose control and its high throughput capability [1–10]. In GaN there have been a number of systematic reports detailing the achievement of *n*- and *p*-type doping [11–15] (using Si and O for *n*-type, and Mg and Ca for *p*-type) and the application of ion implantation to improve ohmic contact resistance on heterostructure field-effect transistors [16] and to producing *p–n* junctions for fabrication of junction field-effect transistors [12] and for light-emitting diodes [4]. At high implant doses ($\geq 5 \times 10^{14}$ cm^{-2}) it is clear that conventional rapid thermal annealing at 1100–1200°C can activate the dopants but not remove the ion-induced structural damage [14, 15]. At higher annealing temperatures (≥ 1400°C), it is difficult to provide a sufficiently high N$_2$ pressure to prevent dissociation of the GaN surface. Three different approaches have been reported: the first is use of an NH$_3$ ambi-

ent in a metal-organic chemical vapor deposition (MOCVD) reactor [17], the second is use of a high N$_2$ overpressure (15 kbar) in a GaN bulk crystal growth apparatus [14], and the third is use of an AlN encapsulant to prevent nitrogen loss from the GaN [18]. The third approach is clearly the most convenient.

Implantation is used for two different reasons in wide bandgap nitrides. The first is to create doping (either *n*- or *p*-type) by introducing additional dopants that are activated by high temperature annealing to move the impurities onto substitutional lattice positions. The second is to create high resistivity material by introduction of deep trap states that remove carriers from either band and decrease the sample conductivity.

2. RANGE STATISTICS

Table I lists experimental and calculated range data for elements implanted into GaN. The first two columns give the element and implantation energy. The next three columns give the depth

Handbook of Thin Film Materials, edited by H.S. Nalwa
Volume 1: Deposition and Processing of Thin Films

ISBN 0-12-512909-2/$35.00

Table I.   Implantation Images for Dopants and Isolation Species in GaN

Element	Energy (keV)	Experiment			TRIM	
		R_m (μm)	R_p (μm)	ΔR_p (μm)	R_p (μm)	ΔR_p (μm)
H	40	0.428	0.428	0.172	0.383	0.089
H	50	0.571	0.517	0.157	0.570	0.137
Li	100	0.408	0.362	0.146	0.491	0.146
Be	100	0.360	0.332	0.130	0.357	0.112
C	260	0.538	0.460	0.185	0.543	0.130
F	100	0.137	0.189	0.110	0.170	0.067
Na	100	0.106	0.089	0.096	0.154	0.065
Mg	100	0.112	0.089	0.066	0.148	0.063
Si	150	0.173	0.159	0.173	0.190	0.075
S	200	0.170	0.103	0.290	0.199	0.074
Zn	300	0.104	0.13	0.08	0.159	0.063
Ge	500	0.107	0.126	0.077	0.260	0.096
Se	500	0.112	0.125	0.067	0.235	0.085

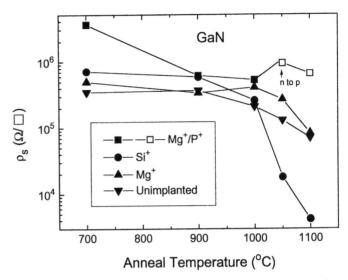

Fig. 1.   Sheet resistivity in Si, Mg, and Mg/P implanted GaN, as a function of annealing temperature. The implant dose was 10^{14} cm^{-2} at an energy of 100 keV in each case.

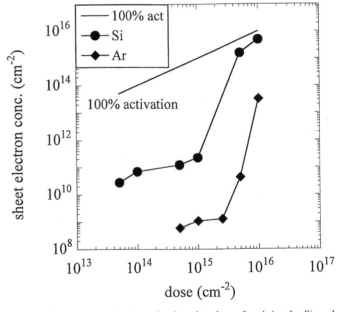

Fig. 2.   Sheet electron density vs implantation dose of each ion for Si- and Ar-implanted GaN annealed at 1100°C for 15 s. The top line represents 100% activation of the implanted dose assuming full ionization.

of the peak (R_m), the projected range (R_p) or first moment, and the range straggle (ΔR_p) or second moment of the depth distribution. The depth distributions were measured using secondary ion mass spectroscopy (SIMS) as described elsewhere [19, 20]. The values of R_m and ΔR_p were determined by applying a Pearson IV computer fitting routine to the experimentally measured depth profile. The final two columns are the values of range and range straggle determined from TRIM92 calculations [21], using the value 6.0 gm/cm^3 for the density of GaN.

The agreement between experiment and transport-of-ions-in-matter (TRIM) calculations appears to vary substantially with the mass of the implanted element. For the higher elements, Be and F, the agreement is good. For the lightest element, H, the experimental values are less than the calculated values, the discrepancy increasing with the mass and becoming as large as a factor of two for the range data. We have no explanation for this observation other than some unaccounted affect in the TRIM simulation for projectiles that have a much greater mass than the light N atoms in the GaN. The values of ΔR_p show more random variations of a small magnitude. It has been observed that it is more difficult to sputter craters and measure the resulting crater depths in GaN than in the other III-V semiconductors, so a larger experimental error is suggested as well as the need for more SIMS measurements to confirm or improve these preliminary results.

3. DONOR IMPLANTS (Si, O, S, Se, AND Te)

Most of the *n*-type doping in GaN by implantation has been performed with Si$^+$ implants. Figure 1 shows the sheet resistance of initially resistive GaN after Si$^+$ or Mg$^+$ implantation, as a function of annealing temperature. At temperatures above 1050°C, the Si becomes electrically active, producing

a sharp decrease in resistivity [22]. Variable temperature Hall measurements showed the ionization level to be ~30–60 meV. Subsequently ion channeling, nuclear reaction analysis, and particle-induced X-ray emission measurements showed that almost 100% of the Si goes onto the Ga site at 1100°C [23], creating the shallow donor state.

Figure 2 shows the sheet electron concentration in Si-implanted GaN for 1100°C annealing, as a function of implant dose. For the two highest dose Si-implanted samples (5 and 10×10^{15} cm^{-2}) 35% and 50%, respectively, of the implanted Si ions created ionized donors at room temperature. The possibility that implant damage alone was generating the free electrons can be ruled out by comparing the Ar-implanted samples at the same dose with the Si-implanted samples at the same dose which had over a factor of 100 times more free electrons. If the implantation damage was responsible for the carrier generation or for enhanced conduction by a hopping mechanism then the Ar-implanted samples as a result of Ar's heavier mass would have demonstrated at least as high a concentration of free electrons as the Si-implanted sample. Since this is not the case, implant damage cannot be the cause of the enhanced conduction and the implanted Si must be activated as donors. The significant activation of the implanted Si in the high-dose samples and not the lower dose samples is explained by the need for the Si concentration to exceed the background carbon concentration ($\sim 5 \times 10^{18}$ cm^{-3}) that was thought to be compensating the lower dose Si samples.

For samples implanted with Si at liquid nitrogen temperature to doses below about 3×10^{15} cm^{-2}, the GaN was not amorphous after implantation and exhibited a small decrease in damage with increasing annealing temperature. This behavior is illustrated in Figure 3a, where Rutherford backscattering channeling (RBS-C) spectra are shown for 2×10^{15} cm^{-2} Si-implanted GaN at liquid nitrogen temperature and annealed at 400, 800, and 1100°C. In this case, even at 1100°C, RBS-C shows that the damage is not completely removed. For lower Si doses down to 5×10^{13} cm^{-2}, damage removal is more extensive during annealing, but the damage removal is still not completely removed even at 1100°C. On the other hand, for Si doses greater than about 5×10^{15} cm^{-2}, annealing up to 1100°C does not result in any appreciable reduction in damage as observed in RBS-C spectra. Figure 3b, which shows the peak damage from such RBS-C spectra as a function of Si dose for as-implanted and 1100°C annealing, illustrates this behavior. Most of the data in Figure 3b refer to liquid nitrogen implants, but the limited data at room temperature, where the initial damage is usually lower, show similar trends.

For doses greater than 10^{16} cm^{-2} shown in Figure 3b, the RBS-C spectra show that the disorder reaches the 100% level, which indicates the formation of an amorphous layer after implantation. It is interesting to examine the annealing behavior of such layers. Cross sectional transmission electron microscopy (XTEM) micrographs show the as-implanted disorder for a Si dose of 4×10^{16} cm^{-2} (liquid nitrogen implant) and revealed a thick amorphous layer with a dense, deeper band of extended defects (D) in crystalline GaN. Misfit dislocations (M) thread toward the surface from the underlying GaN. When this high dose sample is annealed at 1100°C, the amorphous layer recrystallizes as polycrystalline GaN, with no detectable epitaxial growth. The underlying disorder is observed to coarsen but is otherwise unchanged. Thus the XTEM observations are consistent with the RBS-C data in Figure 3b, where the disorder is only marginally reduced at 1100°C. Further comparison of RBS-C data and cross sectional transmission electron microscopy (XTEM) (not shown) indicates the crystallization occurs between 800 and 1100°C.

Oxygen implantation also creates n-type doping in GaN [24]. Figure 4 shows an Arrhenius plot of the sheet resistivity of

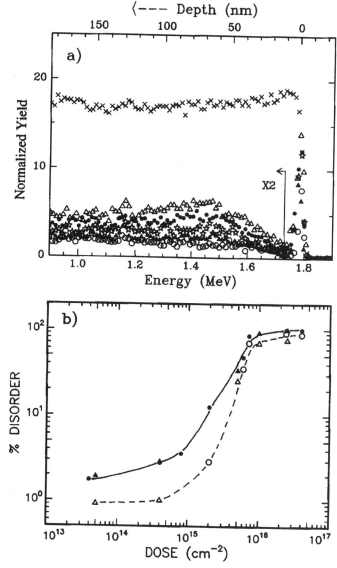

Fig. 3. (a) 2 MeV He$^+$ RBS-C spectra illustrating the annealing of 90 keV Si ion damage (2×10^{15} cm^{-2}) in GaN at temperatures of 400°C, 10 min (filled circles), 8000°C, 10 min (stars), and 1100°C, 30 s (filled triangles). The unimplanted (open circles), as-implanted (open triangles), and random (crosses) spectra are also shown. (b) Peak disorder from RBS-C spectra plotted as a function of dose for 90–100 keV Si ions; as-implanted, liquid nitrogen temperature (filled circles) and room temperature (filled triangles) implants; annealed 1100°C (30 s) for liquid nitrogen (open circles) and room temperature (open triangles) implants.

GaN before and after O$^+$ implantation and 1100°C annealing. The oxygen creates a shallow donor state with activation energy 29 meV, but there is relatively low efficiency, i.e., $\leq 10\%$ of the implanted oxygen becomes electrically active.

Figure 5 shows an Arrhenius plot of S$^+$ activation in GaN. The sheet carrier concentration measured at 25°C shows an activation energy of 3.16 eV for the annealing temperature range between 1000 and 1200°C and basically saturates thereafter. The maximum sheet electron density, $\sim 7 \times 10^{13}$ cm^{-2}, corresponds to a peak volume density of $\sim 5 \times 10^{18}$ cm^{-3}. This is well below that achieved with Si$^+$ implantation and anneal-

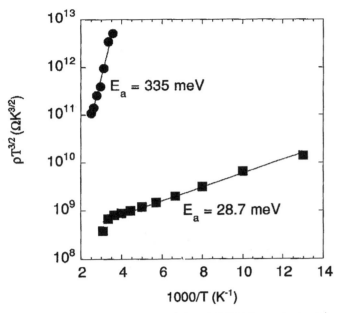

Fig. 4. Arrhenius plot of sheet resistivity of GaN before and after O⁺-implantation.

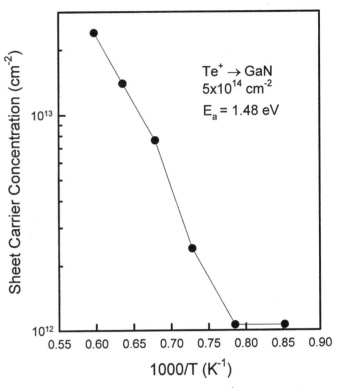

Fig. 6. Arrhenius plot of sheet electron density in Te⁺-implanted GaN versus annealing temperature.

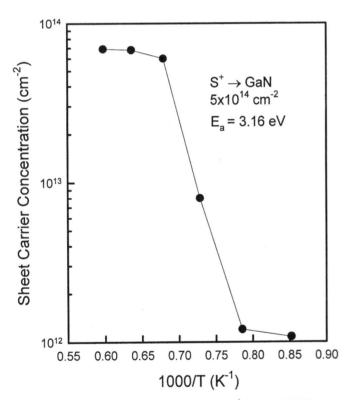

Fig. 5. Arrhenius plot of sheet electron density in S⁺-implanted GaN versus annealing temperature.

be annealed out. Even though implanted Si⁺ at the same dose showed evidence of site switching and self-compensation, it still produces a higher peak doping level than the nonamphoteric donor S, which is only slightly heavier (^{32}S vs ^{28}Si). From temperature-dependent Hall measurements, we find a S⁺ donor ionization level of 48 ± 10 meV, so that the donors are fully ionized at room temperature.

Similar data are shown in Figure 6 for Te⁺ implantation. The activation starts around the same temperature as for S, but much lower sheet electron densities are obtained, the activation energy is significantly lower (1.5 eV), and the carrier concentration does not saturate, even at 1400°C. It is likely that because of the much greater atomic weight of ^{128}Te, even higher annealing temperatures would be required to remove all its associated lattice damage, and that the activation characteristics are still being dominated by this defect removal process. Residual lattice damage from the implantation is electrically active in all III-V semiconductors, producing either high resistance behavior (GaAs) or residual n-type conductivity (InP, GaN). The only data available on group VI doping in epitaxial material are from Se-doped MOCVD material, where maximum electron concentrations of 2×10^{18}–6×10^{19} cm^{-3} were achieved [29, 30]. These are also below the values reported for Si doping and suggest that the group VI donors do not have any advantage over Si for creation of n-type layers in GaN. From limited temperature-dependent Hall data, we estimate the Te ionization level to be 50 ± 20 meV.

Data for activation of Si⁺ implants in GaN are shown in Figure 7. The activation energy of 5.2 eV is higher than that for the

ing ($>10^{20}$ cm^{-3}) [25–28]. In the latter case, the carrier density showed an activation energy of 5.2 eV. The physical origin of this activation energy contains several components—basically it is the energy required to move an implanted ion onto a substitutional lattice site and for it to show electrical activity. This latter requirement means that compensating defects must also

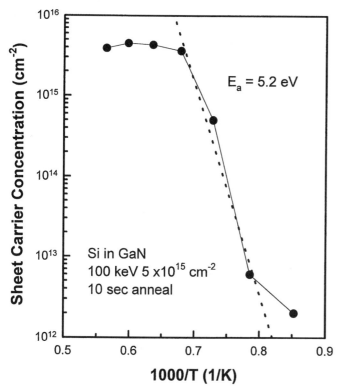

Fig. 7. Arrhenius plot of sheet electron density in Si$^+$-implanted GaN versus annealing temperature.

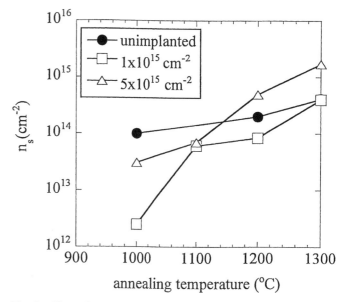

Fig. 8. Sheet electron concentration for unimplanted and Si-implanted (100 keV) AlGaN (15% AlN) at the doses shown versus annealing temperature.

group VI dopants discussed above, but overall Si is the best choice for creating n-type doping in GaN.

AlGaN layers will be employed in heterostructure transistors to realize a two-dimensional electron gas and to increase the transistor breakdown voltage. As discussed in the introduction, implantation can be used to reduce the transistor access resistance of the AlGaN barrier. One would anticipate that the addition of Al to the GaN matrix will increase the damage threshold as is the case for AlGaAs as compared to GaAs [31], but little work has been reported in this area. Recently the first implantation doping studies have been reported for AlGaN [32, 33].

For the work by Zolper et al. [34], the AlGaN layer used for the Si implantation was nominally 1.0 μm thick grown on a c-plane sapphire substrate. The Al composition was estimated to be 15% based on X-ray and photoluminescence measurements. The as-grown minimum backscattering yield measured by channeling Rutherford backscattering was 2.0% and is comparable to a high quality GaN layer.

The AlGaN samples were implanted with Si at room temperature at an energy of 100 keV at one of two doses, 1 or 5×10^{15} cm^{-2}. The higher Si dose has previously been shown not to amorphize GaN and produce an as-implanted channeling yield of 34% in GaN.

Samples were characterized by channeling Rutherford backscattering (C-RBS) with a 2 MeV beam with a spot size of 1 mm^2 at an incident angle of 155°. Aligned spectra are taken with the beam parallel to the c axis of the GaN film. Random spectra are the average of five off-axis, off-planar orientations.

Electrical characterization was performed using the Hall technique at room temperature.

Figure 8 shows the sheet electron concentration versus the annealing temperature for the Si-implanted AlGaN sample. Data for an unimplanted sample is included as a control. First of all it is clear that the unimplanted samples have significant donors produced by the annealing process alone. This may be due to the activation of unintentional impurities, such as Si or O, in the film. O may be a particular suspect due to the tendency of O to incorporate in Al-containing material. At the highest temperature, the high dose Si-implanted sample has four times higher free electron concentration (1.7×10^{15} cm^{-2}) than the unimplanted sample. This corresponds to 34% activation of the implanted Si.

Figure 9 shows aligned C-RBS spectra for 15% Al in AlGaN either as-grown (unimplanted), after Si implantation at a dose of 1 or 5×10^{15} cm^{-2}, and for the higher dose samples after annealing. As was the case for GaN, the 1×10^{15} cm^{-2} sample shows limited dechanneling while the higher dose sample shows a marked damage peak. The minimum channeling yield for the high dose sample was 26.67% which is lower than that seen for GaN which showed χ_{min} between 34% and 38% implanted under the same conditions. This means the addition of 15% aluminum to the GaN matrix increases its damage threshold as is the case for Al additions to GaAs to form AlGaAs [31]. The spectrum for the annealed sample shows limited damage removal, again consistent with that seen for GaN at this temperature [35]. There is evidence, however, of improvement in the near surface as seen by the reduction in the first surface peak. This peak has been suggested to be due to preferential sputtering of nitrogen from the film surface [36]. The reduction of this peak via annealing suggests the surface stoichiometry

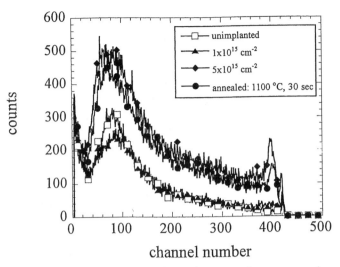

Fig. 9. Channeling Rutherford backscattering spectra for as-grown, unimplanted, Si-implanted (100 keV, 1 or 5×10^{15} cm^{-2}) and implanted (100 keV, 5×10^{15} cm^{-2}) annealed AlGaN.

is restored during the anneal. Further study is needed to better understand this effect.

The redistribution of implanted Si, Mg, and C in Al$_{0.123}$-Ga$_{0.88}$N was studied by Polyakov and co-workers [33]. While Si and C demonstrated no measurable diffusion by SIMS after at 1140°C, 1 h anneal, Mg did show appreciable profile broadening and indiffusion under these conditions. Activation of a modest Si dose (5×10^{14} cm^{-2}) was also achieved by annealing at 1140°C with a resulting peak electron concentration of 1.2×10^{18} cm^{-3}. More work is needed on the effect of Al composition on the implantation properties of AlGaN.

4. ACCEPTOR IMPLANTS

In Figure 1 we had also shown activation data for Mg and Mg/P implanted GaN. The purpose of the co-implant scheme is to enhance occupation of Ga substitutional sites by the group II acceptor Mg. An *n*-to-*p* conversion was observed after 1050°C annealing, corresponding to movement of the Mg atom onto substitutional sites.

Similar results were observed for Ca and Ca/P implanted GaN [13], as shown in Figure 10. The *n*-to-*p* conversion occurred at 1100°C and the *p*-type material showed an activation energy of ~16^9 meV for the Ca acceptor (Fig. 11). Ion channeling results found that 80% of the Ca was near Ga sites as-implanted, with a displacement of ~0.2 Å from the exact Ga site. Upon annealing at 1100°C the Ca moved onto the substitutional Ga lattice positions [37].

The effects of postimplant annealing temperature on the sheet carrier concentrations in Mg$^+$ and C$^+$ implanted GaN are shown in Figure 12. There are two important features of the data: first, we did not achieve *p*-type conductivity with carbon and, second, only ~1% of the Mg produces a hole at 25°C. Carbon has been predicted previously to have a strong self-compensation effect [38], and it has been found to produce

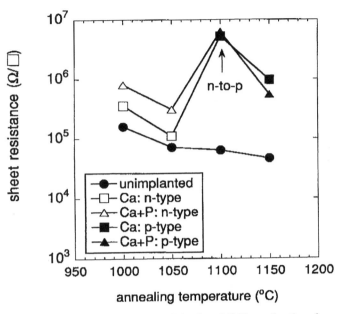

Fig. 10. Sheet resistivity in Ca or Ca/P implanted GaN, as a function of annealing temperature. The implant dose was 10^{14} cm^{-2} at an energy of 100 keV in each case.

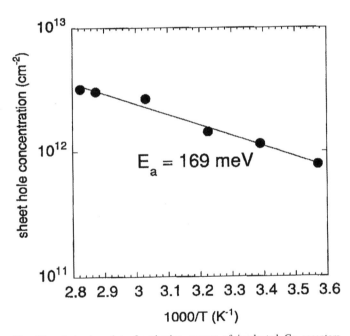

Fig. 11. Arrhenius plot of activation energy of implanted Ca acceptors in GaN.

p-type conductivity only in metal–organic molecular beam epitaxy where its incorporation on a N-site is favorable [39]. Based on an ionization level of ~170 meV, the hole density in Mg-doped GaN would be calculated to be ~10% of the Mg acceptor concentration when measured at 25°C. In our case, we see an order of magnitude less holes than predicted. This should be related to the existing *n*-type carrier background in the material and perhaps to residual lattice damage which is also *n*-type in GaN. At the highest annealing temperature (1400°C), the hole density falls, which could be due to Mg coming out of

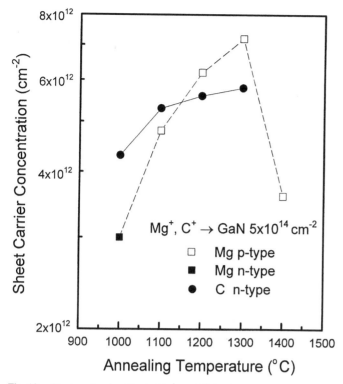

Fig. 12. Sheet carrier densities in Mg$^+$- or C$^+$-implanted GaN as a function of annealing temperature.

Theoretical evidence suggests that Be may be an effective acceptor in GaN when co-incorporated with oxygen [44]. Be solubility in GaN is predicted to be low, due to the formation of Be_3N_2, but the co-incorporation might enhance the solubility. There have been some experimental indications of such a result [45], but these have not been verified.

Table II summarizes the current level of understanding of maximum achievable doping levels, ionization energies, and diffusivities for implanted species in GaN.

5. DAMAGE REMOVAL

Previous work on damage accumulation during implantation had been limited to Si-implanted GaN at 77 K or room temperature [46, 47]. That work demonstrated that the amorphization dose for ~100 keV Si implantation was ~2×10^{16} cm^{-2}. Recent work has addressed the damage accumulating during Ca, Ar, and In implants in GaN at 77 K (Ca and Ar) and room temperature (In) [48, 49]. Additional work has also been reported from the damage annealing characteristics for high dose Si implantation [50]. Figure 13 shows the change in the Rutherford backscattering (RBS) minimum channeling yield (χ_{mm}) versus dose for Ca- and Ar-implanted GaN at 77 K [48]. The amorphization dose ($\chi_{mm} = 100\%$) is 6×10^{15} cm^{-2} for both elements. As expected, this dose is lower than that reported for Si-implanted GaN of 2×10^{16} cm^{-2} due to the higher mass of Ca and Ar as compared to Si. In the same study, it was determined that for a Ca dose as low as 3×10^{14} cm^{-2} an amorphous component to the X-ray diffraction spectra is created. This suggests that local pockets of amorphous material are formed prior to the complete amorphization of the implanted regions. Further study of the removal of such an amorphous

Table II. Characteristics of Different Implanted Dopants in GaN

	Max achievable doping level (cm^{-3})	Diffusivity (cm^2 s^{-1})	Ionization level (meV)
Donors			
Si	5×10^{20}	$< 2 \times 10^{-13}$ (1500°C)	28
S	5×10^{18}	$< 2 \times 10^{-13}$ (1400°C)	48
Se	2×10^{18}	$< 2 \times 10^{-13}$ (1450°C)	–
Te	1×10^{18}	$< 2 \times 10^{-13}$ (1450°C)	50
O	3×10^{18}	$< 2 \times 10^{-13}$ (1200°C)	30
Acceptors			
Mg	~5×10^{18a}	$< 2 \times 10^{-13}$ (1450°C)	170
Ca	~5×10^{18a}	$< 2 \times 10^{-13}$ (1450°C)	165
Be	$< 5 \times 10^{17}$	defect-assisted	–
C	n-type	$< 2 \times 10^{-13}$ (1400°C)	–

^{a}Acceptor concentration.

Note: It appears that Si is the best choice for n-doping, while either Mg or Ca is the best choice for p-type doping.

the solution or to the creation of further compensating defects in the GaN.

We also point out that unless most residual defects are removed by annealing, the implanted acceptors will remain compensated and not create p-type material [40, 41]. Improved optical activation of Zn acceptors in implanted GaN has been achieved using high pressure (10 kbar) N_2 annealing [42, 43].

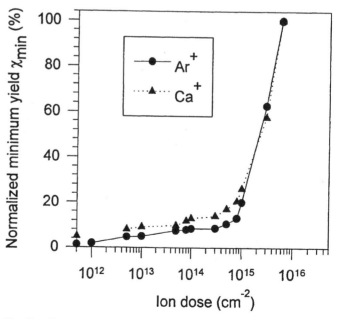

Fig. 13. Change in the RBS minimum channeling yield (χ_{min}) versus implant dose for 180 keV Ca and Ar implanted GaN at 77 K.

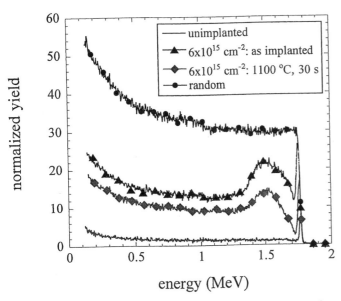

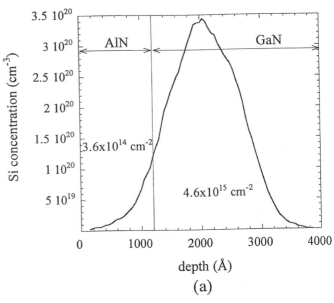

(a)

Fig. 14. Channeling Rutherford backscattering spectra for as-grown (random and aligned, unimplanted) and Si-implanted (90 keV, 6×10^{15} cm^{-2}) GaN (as-implanted and after a 1100°C, 30 s anneal). The implants were performed at room temperature.

material during annealing will be important to optimize the annealing process.

The In-implantation study examined the local lattice environment after implantation and subsequent annealing using the emission channeling technique and perturbed-$\gamma\gamma$-angular correlation. Ronning and co-workers [49] found that the majority of the In atoms were substitutional as-implanted but within a heavily defective lattice. A gradual recovery of the damage was seen between 600 and 900°C with about 50% of the In atoms occupying substitutional lattice sites with defect-free surroundings after a 900°C anneal. Results for higher temperature annealing were not reported. It remains to be determined at what point complete lattice recovery is achieved as determined by this technique.

Additional work on Si implantation had demonstrated significant implantation-induced damage remains in GaN even after an 1100°C activation anneal that produces electrically active donors [50, 51]. This is shown in the RBS spectra of Figure 14 and by XTEM images. The RBS spectra shows that even after an 1100°C anneal the high channeling yield is evidence of significant damage in the crystal. This is confirmed by the XTEM images before and after annealing where, despite some coarsening of the damage, no significant reduction in the damage concentration has occurred. In the following section, work on complete removal of the implantation damage by annealing up to 1500°C is presented.

As discussed in the previous section, an annealing temperature well above 1100°C is needed to remove the implantation-induced damage in GaN. However, since GaN will readily dissociate at these temperatures, special precautions must be taken. The approaches used to maintain the GaN stoichiometry at up to 1500°C were either to use an encapsulating overlayer such as the AlN described above or to perform the anneal under a high

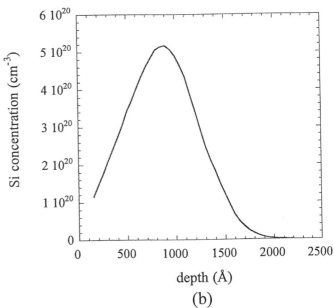

(b)

Fig. 15. Calculated Si-implantation profiles using TRIM92 code: (a) 210 keV, 5×10^{15} cm^{-2} Si in AlN/GaN and (b) 100 keV, 5×10^{15} cm^{-2} Si in GaN.

N$_2$ overpressure. Results from both approaches are presented below. Both electrical and structural data are presented to correlate the effect of the removal of implantation-induced damage to electrical activation of implanted Si donors.

To better understand the removal of implantation-induced damage, annealing experiments were formed in a MOCVD growth reactor capable of reaching 1400°C. While the anneal was performed in flowing ammonia to help stabilize the surface against decomposition, AlN encapsulation was also studied to further suppress N loss [51, 52]. The first set of samples was encapsulated with 120 nm of sputter deposited AlN. Si implantation was performed through the AlN at an energy of 210 keV and a dose of 5×10^{15} cm^{-2}. The results of Monte Carlo TRIM calculations, shown in Figure 15a, predict that ~7% of the Si ions come to rest in the AlN film with a dose of 4.6×10^{15} cm^{-2}

being placed in the GaN [53]. The Si peak range from the GaN surface is estimated to be approximately 80 nm. The second set of samples was unencapsulated and implanted with Si at an energy of 100 keV and dose of 5×10^{15} cm^{-2}. This also gives a range from the GaN surface of 80 nm as shown in Figure 15b. A sample from each set was annealed under one of four conditions. One pair of samples was annealed in a rapid thermal annealer (RTA) inside a SiC coated graphite susceptor and processed in flowing N$_2$ since this procedure was used previously to activate implanted Si in GaN. The remaining samples were annealed in a custom built MOCVD system that employed rf heating with the samples placed on a molybdenum holder on a SiC coated graphite susceptor. The stated temperatures were measured with an Accufiber Model-10 or a Minolta Cyclota-52 pyrometer, which were calibrated by the melting point of Ge at 934°C. The pressure in the MOCVD reactor was 630 mTorr with gas flows of 4 slm of N$_2$ and 3 slm of NH$_3$. The encapsulated and unencapsulated samples for a given temperature were annealed together.

Figure 16a and b shows the sheet electron concentration and electron Hall mobility versus the annealing conditions for the unencapsulated and AlN encapsulated samples, respectively. An unimplanted sample annealed at 1100°C for 15 s in the RTA remained highly resistive with $n \ll 1 \times 10^{15}$ cm^{-3} after annealing. The mobility of this undoped film could not be reliably measured with the Hall effect due to the low carrier concentration.

First looking at the data for the unencapsulated samples (Fig. 16a), the sample annealed at 1100°C for 15 s in the RTA has a sheet electron concentration of 6.8×10^{14} cm^{-2} or 13.6% of the implanted dose. This activation percentage is in the range reported for earlier Si-implanted GaN samples annealed in this way. After the 1100°C, 30 s MOCVD anneal the number of free electrons goes up to 40% of the implanted dose before decreasing to 24% for the 1300°C anneal. The decrease for the 1300°C sample was accompanied by a degradation of the surface of the sample as determined by observation under an electron microscope as shown by scanning electron microscope (SEM) micrographs. This point will be revisited when discussing the C-RBS spectra later, but it is believed that the GaN layer has started to decompose during this anneal. Therefore, the reduction in the electron concentration may be due to loss of material. The Hall mobility increased with increasing thermal treatment and is suggestive of improved crystalline quality. No data are given in Figure 16a for the unencapsulated sample annealed at 1400°C since the GaN film completely sublimed or evaporated during this anneal. This was confirmed by C-RBS data for this sample that showed only the substrate Al and O peaks with a slight Ga surface peak.

Now turning to the data for the AlN encapsulated samples (Fig. 16b). There is increasing sheet electron density with increasing thermal treatments for the encapsulated samples including the highest temperature anneal. The RTA sample has a lower activity than the comparable unencapsulated sample (3.5% vs 13.6%); however, all the other AlN encapsulated samples have higher electron concentrations than the compa-

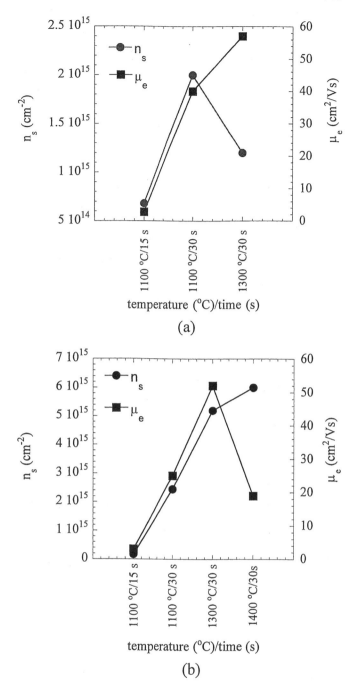

(a)

(b)

Fig. 16. Sheet electron concentration and electron Hall mobility versus annealing treatment for (a) unencapsulated and (b) AlN encapsulated Si-implanted GaN.

rable unencapsulated sample. The 1300°C sample has a sheet electron concentration of 5.2×10^{15} cm^{-2} that is 113% of the Si dose that should have been retained in the GaN layer (4.6×10^{15} cm^{-2}). The excess electron concentration may be due to small uncertainties in the Hall measurement due to the nonideal contact geometry (i.e., ideal point contacts were not used). The error in the Hall measurement is estimated to be $\sim$10%. The apparent excess electron concentration may also be due to indiffusion of the Si from the AlN encapsulant into the GaN substrate or to the activation of other native donor de-

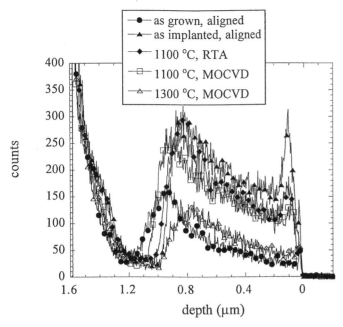

Fig. 17. Channeling Rutherford backscattering spectra for Si-implanted (100 keV, 5×10^{15} cm^{-2}) GaN either as-implanted or annealed as shown in the legend.

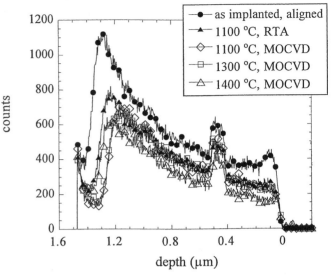

Fig. 18. Channeling Rutherford backscattering spectra for Si-implanted (210 keV, 5×10^{15} cm^{-2}), GaN encapsulated with 120 nm of AlN either as-implanted or annealed as shown in the legend.

fects in the GaN layer such as N vacancies. The sample annealed at 1400°C has a still higher free electron concentration (6×10^{15} cm^{-2}) that may also be partly due to measurement errors or to activation of native defects. This sample had visible failures in the AlN layer (cracks and voids) that allowed some degree of decomposition of the GaN layer. This was confirmed by SEM micrographs that showed regions of GaN loss (confirmed by Auger electron spectroscopy). Therefore, the formation of N vacancies in this sample is very likely and may contribute to the electron concentration. This would also contribute to the observed reduction in the electron mobility. The failure of the AlN film may be due to nonoptimum deposition conditions resulting in nonstoichiometric AlN or to evolution of hydrogen from the GaN epitaxial layer that ruptures the AlN during escape. If the failure is due to nonoptimum AlN properties, this can be rectified by examining the AlN deposition parameters. If hydrogen evolution is the cause of the failure an anneal (~900°C for 5 min) to drive out the hydrogen prior to AlN deposition should rectify this problem.

Figure 17 shows a compilation of the C-RBS spectra for the unencapsulated samples along with the aligned spectrum for an unimplanted sample. The as-implanted sample shows the damage peak near 100 nm with an additional peak at the surface. The surface peak may be due to preferential sputtering of the surface during Si implantation studies. The sample annealed in the RTA has improved channeling and an apparent reduction in the implant damage. It should be noted, however, that the surface peak is also diminished in the RTA sample and it has previously been reported that the change in the surface peak can account for the apparent reduction in the implantation damage peak. Upon annealing at higher temperatures or longer times in the MOCVD reactor, the channeling continues to improve and

approaches, but does not reach, the unimplanted aligned spectra. The 1300°C sample, however, appeared to show evidence of material loss as demonstrated by the change in position and abruptness of the substrate signal as well as the observation of surface roughing. Therefore, the reduction in the implantation damage peak in this sample may be due to sublimation or evaporation of the implanted region and not recovery of the original crystal structure. The loss of at least part of the Si-implanted region is consistent with the reduction in the free electron concentration in this sample shown in Figure 16a.

Figure 18 shows a compilation of the C-RBS spectra for the AlN encapsulated sample. The as-implanted sample has a damage peak at ~80 nm with no additional surface peak as seen in the unencapsulated sample. The lack of surface peak in this samples supports the hypothesis that this peak on the unencapsulated sample is due to preferential sputtering since the GaN surface of the encapsulated sample is protected from sputter loss during implantation. A significant reduction in the implantation-induced damage peak occurs after the RTA anneal with a further reduction with increasing thermal processing.

Table III summarizes the values for the minimum channeling yield (χ_{min}) of the GaN layer for the sample studied. The dechanneling for an unimplanted sample was estimated to be 2.5% and is close to the theoretical limit. For the AlN encapsulated samples the value of χ_{min} has had the effect of the AlN overlayer subtracted out based on spectra of unimplanted, unannealed AlN on GaN. The unencapsulated samples show a continuous reduction in χ_{min} with increasing annealing temperature. This reduction is at least partly due to removal of a surface damage peak as previously discussed. The χ_{min} value at 1300°C of 6.7% may be anomalously low due to sublimation of the damaged region in this sample. The 1400°C annealed encapsulated sample shows a significant reduction in χ_{min} from the as-implanted value of 38.6% to 12.6%. Since the RBS analysis was performed on regions of the sample with the lowest

Table III. Summary of C-RBS Results for Si-implanted, AlN-encapsulated, and Unencapsulated GaN Annealed Under the Conditions Shown

Sample	Anneal (temperature/time) (°C, s)	χ_{min} for GaN (%)	GaN thickness (μm)
GaN as-grown	none	2.5	1.12
AlN/GaN as-implanted	none	38.6	1.36
1	1400, 30	12.6	1.29
2	1300, 30	17.2	1.23
3	1100, 30	20.2	1.23
4	1100, 15	23.6	1.29
GaN as-implanted	none	34.1	0.90
5	1400, 30	a	–
6	1300, 30	6.7	0.90
7	1100, 30	17.2	0.98
8	1100, 15	20.8	0.94

a The GaN layer completely evaporated during the anneal.

Note: All 30 s anneals were done in MOCVD reactor under flowing N_2/NH_3 and the 15 s anneals were done in a RTA under flowing N_2.

density of craters, this χ_{min} value should be characteristic of the GaN that has not decomposed during the anneal. While the reduction in χ_{min} demonstrated partial recovery of the crystal lattice, the original channeling properties were not realized. This suggests still higher temperature anneals may be required. This would be consistent with a 2/3 rule for relating the melting point (T_{mp}) of a semiconductor to the implantation activation temperature (T_{act}) since for GaN $T_{mp} \sim 2500°C$ and therefore T_{act} should be $\sim 1650°C$. To reach this temperature in a controllable manner will require development of new annealing furnaces. Preferably, such a furnace should operate in a rapid thermal processing mode (i.e., with rapid heating and cooling) to minimize the thermal budget of the anneal process. This will require new equipment designs such as have been reported in the previous section.

6. HIGH TEMPERATURE ANNEALING

6.1. Surface Protection

The usual environment for high temperature annealing of III-nitrides is NH_3 [54], but this is inconvenient for processes such as rapid thermal annealing for implant activation, contact annealing for implant isolation. In those situations we would like to provide some form of N_2 overpressure to minimize loss of nitrogen from the semiconductor surface at high temperature [55]. With conventional III-V materials such as GaAs and InP this is achieved in several ways [56–69], namely by two methods: (i) placing the sample of interest face down on a substrate of the same type [60, 67], so that the onset of preferential As or P loss quickly suppresses further loss. The disadvantages of this method include the fact that some group V atoms are lost from the near surface. There is always a possibility of mechanical abrasion of the face of the sample of interest, and con-

tamination can easily be transferred from the dummy wafer to the one of interest. The second method involves (ii) placing the wafer in a SiC-coated graphite susceptor [68, 69], which either has had its internal surfaces coated with As or P by heating a sacrificial wafer within it, or in which granulated or powdered GaAs or InP is placed in reservoirs connected to the region in which the wafer is contained. In both cases subsequent heating of the susceptor produces an As or P vapor pressure above the surface of the process wafer, suppressing loss of the group V element.

The former approach is widely used in III-V research and is known as the proximity geometry. The latter approach is widely used in industry for anneal processes for GaAs and to a lesser extent InP.

A variety of authors previously published experimental and theoretical vapor pressure curves for AlN [70–73], GaN [74–76], and InN [77–80].

In all of this previous work the equilibrium N_2 pressures above the solid (or solid plus liquid) have been the focus. In many process steps RTA using the proximity geometry is employed, and this is a nonequilibrium situation. In this case the sample of interest is placed face down on a substrate that can provide a group V partial pressure, and heating by tungsten–halogen lamps produces temperature ramp rates of $100–200°C\,s^{-1}$. Zolper et al. [81] observed that the luminescence and surface morphologies of GaN grown on Al_2O_3 annealed in flowing N_2 actually improved for RTA temperatures up to 1100°C. Similar results were obtained at lower temperature by Lin et al. [82]. This is a common situation for lattice-mismatched systems (e.g., GaAs/Si), where postgrowth or even in situ annealing is generally found to improve structural and optical properties, provided that group V loss from the surface can be suppressed and that external impurities do not diffuse in during the anneal.

In this section we report on an investigation of the thermal stability of AlN, GaN, InN, InAlN, and InGaN during rapid thermal annealing. The electrical conductivity, surface morphology, and surface stoichiometry have all been measured as a function of annealing temperature. The In-containing materials are found to be substantially less thermally stable than either GaN or AlN, and loss of nitrogen creates thin n-type surface layer in all three binary nitrides.

The sheet resistance normalized to the as-grown value for all of the nitride samples is shown in Figure 19 as a function of annealing temperature. The values for the GaN, AlN, and InN are found to drop by approximately three orders of magnitude with annealing, up to 900°C. The material becomes strongly n-type in all cases, even in the GaN and AlN, which were resistive as grown. The sheet resistance for AlN continues to drop steadily with anneal temperature until 1100°C. As will be shown below, AlN shows only a small loss of N from the surface as determined by Auger electron spectroscopy (AES). However, the electrical measurements are more sensitive to small changes in the composition than the Auger. Here we believe the N vacancies created by the loss of N from the surface are creating shallow donors. This agrees with the theoretical prediction of

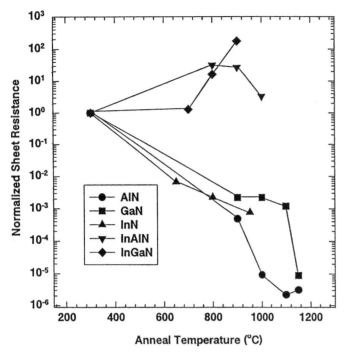

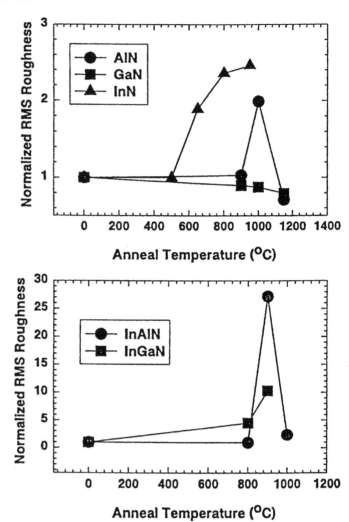

Fig. 19. Sheet resistance normalized to the as-grown value for AlN, GaN, InN, InAlN, and InGaN as a function of annealing temperature. The initial sheet resistances were $\sim 10^8$ Ω per square (AlN), 10^6 Ω per square (GaN), 1.6×10^5 Ω per square (InAlN), and 16 Ω per square (InN and InGaN).

Maruska and Tietjen [83]. The actual values of sheet resistance for AlN are much higher than the GaN up to 900°C, and significantly higher than InN at all temperatures. The data in Figure 19 are in agreement with the melting point and vapor pressure curves for these materials. AlN is predicted to be stable under N_2 gas up to ~ 2500°C and to melt at ~ 3700°C at atmospheric pressure. GaN is predicted to melt at ~ 3000°C and InN at only ~ 2400°C and to degrade at 600°C. AES has confirmed loss of N from the annealed GaN sample, which would suggest that N vacancies are contributing to the conductivity. Above 900°C Ga vacancies in the GaN may be creating compensating shallow acceptors, thus causing the sheet resistance to level off. We also may be getting Ga or N antisite defects that form deeper traps. At 1150°C the sheet resistance for the GaN drops sharply, indicating that N is being lost at a much greater rate than the Ga. Groh et al. [84] showed loss of nitrogen beginning at 710°C in vacuum annealed GaN, with significant loss at ≥ 980°C. The sheet resistance for the InN drops steadily over the temperature range, which correlates to the problems of nonstoichiometry in InN. The large size differences between the N and In make the material less stable.

The sheet resistance for InGaN and InAlN, on the other hand, increases with annealing. The InAlN sheet resistance increases by a factor of 10^2 from the value for the as-grown material when annealed at 800°C. Its resistance then remains constant to 900°C and then decreases slightly at 1000°C. For InGaN the sheet resistance remains constant up to 700°C and then increases rapidly with increasing temperature. This suggests that simple N vacancies are not the cause of the residual n-type conductivity in these samples since at the highest

Fig. 20. The rms data normalized to the as-grown roughness as a function of anneal temperature for AlN, GaN, and InN. The initial rms values were 4 nm (AlN), 8.9 nm (GaN), and 16.1 nm (InN).

temperatures we are losing N from the surface, as described below. However, these samples become less conducting, suggesting creation of compensating acceptors or annealing of the native donors is occurring. It is likely, in contrast to the binary nitrides, that the V_N have several different energy levels in the ternaries because of the differences in strength between In−N and Ga−N bonds. Some of these may be creating a deep acceptor which compensates the shallow donors, or a deep level electron trap, making the material more resistive. There could also be the presence of compensating V_{Ga}- or V_{In}-related defects, which might be easier to form in ternaries because of different sizes and bond strengths of the In and Ga. Finally, it is possible that at least some of the conductivity in the ternaries is due to carbon-related donors [34], and subsequent annealing might produce carbon self-compensation, as in other III-V's. At this point we cannot be more precise regarding the differences between the binaries and the ternaries and this is under close examination.

The rms data normalized to the as-grown roughness as a function of anneal temperature are shown in Figure 20 (top) for AlN, GaN, and InN. The AlN is still smooth at 900°C but becomes quite rough at 1000°C. Further surface reconstruction continues at higher annealing temperatures. At 1150°C the sample becomes smooth again—in fact slightly smoother than the as-grown sample. GaN shows no roughening, becoming smoother with annealing due to defect annealing and surface reconstruction. InN, on the other hand, is a factor of 2 rougher than for as-grown samples at 650°C, indicating the weaker bond strength of this material.

In Figure 20 (bottom) the rms roughnesses for InAlN and InGaN are shown as a function of rapid thermal anneal temperature. We see that the InAlN remained smooth until 800°C and at 900°C has increased an order of magnitude in roughness. At 1000°C the rms roughness returns to a value close to that of the value for the as-grown material. We found this to be a result of In droplets forming on the surface above 800°C and then evaporating above 900°C. The InGaN surface was unchanged at 700°C, with the roughness increasing above that temperature. In all cases we found that the nitride surface was deficient in nitrogen after high temperature annealing.

For uncapped annealing, the III-V nitrides were thermally stable to relatively high temperatures. AlN and GaN remain smooth and stoichiometric at 1000°C, InN and InGaN up to 800°C, and InN up to 600°C. Above these temperatures capping is necessary to prevent the loss of N and, sometimes, In. Consistent with the predicted melting temperatures and thermal stabilities of the nitrides, we found AlN to be somewhat more stable than GaN, and much more stable than InN. InAlN was found to be more stable than InGaN, as expected from a consideration of the binary component N_2 vapor pressures. AlN may prove to be a good capping material for the other nitrides, because of its high stability and the fact that it can be selectively removed by wet etching in KOH based solutions.

6.2. Susceptors

It would be convenient for GaN device processing if development of a similar process for rapid thermal processing of III nitrides occurred, in which an overpressure of N_2 is supplied to a susceptor. In this section we compare use of powdered AlN or InN as materials for use in the susceptor reservoirs and compare the results with those obtained by simple proximity annealing.

The GaN, AlN InN, InGaN, and InAlN samples were grown using metal–organic molecular beam epitaxy on semi-insulating, (100) GaAs substrates of Al_2O_3 c-plane substrates in an Intevac Gen II system as described previously. The group-III sources were triethylgallium, trimethylamine alane, and trimethylindium, respectively, and the atomic nitrogen was derived from an electron cyclotron resonance Wavemat source operating at 200 W forward power. The layers were single crystal with a high density (10^{11}–10^{12} cm^{-2}) of stacking faults and microtwins. The GaN and AlN were resistive as-grown, and the InN was highly autodoped n-type ($>10^{20}$ cm^{-3}) due to presence of native defects. InAlN and InGaN were found to contain both hexagonal and cubic forms. The $In_{0.75}Al_{0.25}N$ and

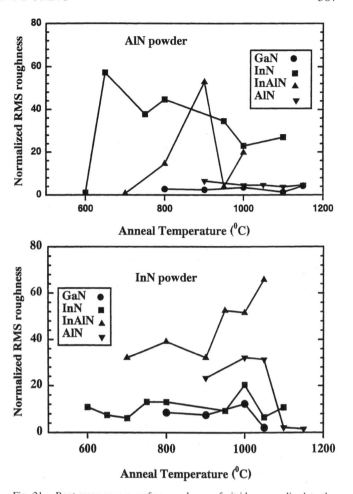

Fig. 21. Root-mean-square surface roughness of nitrides normalized to the as-grown value, as a function of anneal temperature using either AlN powder (top) or InN powder (bottom) in the susceptor reservoirs.

$In_{0.5}Ga_{0.5}N$ were conducting n-type grown ($\sim10^{20}$ cm^{-3}) due to residual autodoping by native defects.

The samples were annealed either (i) face down on samples of the same type, i.e., GaN when annealing GaN, InN for InN, etc., or (ii) within a SiC-coated graphite susceptor in which the reservoirs were filled with either powdered AlN or InN (average particle size ~10 μm). Annealing was performed for 10 s at peak temperatures between 650 and 1100°C in flowing N_2 gas. The sheet resistance was measured at room temperature on a van der Pauw Hall system with 1:1 InHg alloyed contacts (400°C, 2 min) on the corners. An atomic force microscope (AFM), operated in tapping mode with Si tips, was used to measure the root-mean-square (rms) roughness of the samples. The surface morphology was examined with a SEM. Energy dispersive X-ray spectroscopy (EDAX) was used to analyze the surface composition of some samples. AES was used to investigate near-surface stoichiometry before and after anneal.

A comparison of the annealing temperature dependence of nitride rms surface roughness for sample processed in the graphite susceptor with either AlN or InN powder in the reservoirs is shown in Figure 21. One would expect the InN powder to provide higher vapor pressure of N_2 at equivalent tem-

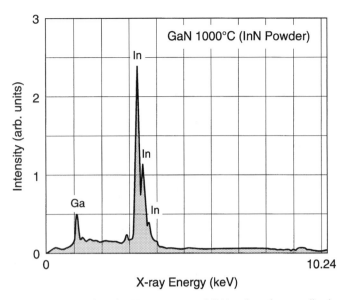

Fig. 22. Energy dispersive X-ray spectrum of GaN surface after annealing in susceptor containing InN powder in the reservoir.

peratures than AlN [70, 85, 86] and this appears to be evident in the rms data for InN, where the surface roughens dramatically above 600°C with AlN powder while the roughening is less obvious with InN powder. The data in Figure 21 need to be considered in the light of the results from the other materials. For example, the GaN and AlN rms values are consistently higher for the InN powder annealing. These results clearly indicate large-scale (~ 1 μm) roughness evident on the samples annealed with the InN powder.

SEM examination of all the samples revealed the cause of this roughening. After 1100°C annealing with AlN powder there is no change in morphology from the control samples. By sharp contrast, metallic droplets are visible on samples annealed with InN powder even at 800°C. Similar droplets were observed in all materials after annealing with InN powder at ≥ 750°C. EDAX measurements identified these droplets as In in each case (Fig. 22). Therefore, it is clear that the InN powder initially provides good surface protection for annealing temperatures ≥ 750°C through incongruent evaporation of In from the powder leads to condensation of droplets on the samples contained within the reservoir. At temperatures approaching 1100°C, these droplets evaporate from the surface of GaN or AlN, leading to an apparent surface smoothing when measured by AFM.

Some other features of the annealing are salient with respect to implant activation processes. First, if we employed 90% N_2 : 10% H_2 as the purge gas in the RTA system instead of pure N_2, we noticed that the temperature at which surface dissociation was evident by either AFM or SEM was lowered by 100–200°C for each of the nitrides. A similar effect was observed if O_2 was present in the annealing ambient, and thus, it is critical to avoid residual O_2 or H_2 in RTA systems during annealing of the nitrides. Second, under optimized ambient conditions (pure N_2 purge gas, and use of either the proximity

geometry or powdered AlN in the susceptor reservoirs), AES was able to detect N_2 loss from the surface of GaN even after 1000°C anneals, and from InAlN and InGaN after 800°C anneals [85]. However, N_2 loss for AlN was detectable only after anneals at 1150°C, emphasizing the extremely good thermal stability of this material. Indeed Zolper et al. [87] have recently reported use of sputtered AlN as an encapsulant for annealing GaN at temperatures up to 1100°C for Si^+ or Mg^+ ion implant activation. The AlN could be selectively removed with KOH solutions after the annealing process [88].

The loss of N_2 from binary nitride surfaces during annealing produced thin (<0.5 μm as determined by subsequent dry etching and remeasurement) highly conducting n^+ regions. These were evident for annealing temperatures above ~ 600°C in InN and at ~ 1125°C in GaN. For example, the sheet resistance increased by 2–4 orders of magnitude in both materials. This agrees with the theoretical prediction of Maruska and Tietjen [83] that N vacancies create a shallow donor state in binary nitrides, and the temperatures to which the materials are stable are in general agreement with the published vapor pressure and melting point data for the binary nitrides. In the ternary material, however, the sheet resistance of the epitaxial layers increased with annealing temperature, suggesting that simple nitrogen vacancies are not the only cause of the residual n-type conductivity in these samples, since at the highest temperatures there is clear loss of nitrogen. The annealing, in this case, appears to create compensating acceptors, or else annealing of the native donors is occurring [89]. The nature of the defects present in as-grown and annealed nitrides is currently under investigation with IR absorption and variable temperature Hall measurements.

In summary, several approaches to rapid thermal processing of binary and ternary nitrides have been investigated. In the proximity geometry AlN and GaN retain smooth stoichiometric surfaces to ≥ 1000°C, InAlN and InGaN to 800°C, and InN up to 600°C. Similar thermal stabilities were obtained for face-up annealing in graphite susceptors in which AlN powder provides a N_2 overpressure. An attempt to increase this overpressure through use of InN powder was unsuccessful because of In droplet condensation on all samples at temperatures ≥ 750°C. This could only be rectified if one could design a two-zone rapid thermal processing chamber in which a reservoir of InN powder was maintained at ≥ 750°C, while the samples to be annealed were separately heated to their required temperatures.

6.3. AlN Encapsulant

The existing commercial rapid thermal processing (RTP) equipment typically relies on a series of tungsten–halogen lamps as heat sources to rapidly heat up the semiconductor wafers [90]. However, this type of lamp-based RTP system suffers from many problems such as their point heat source nature, fluctuating lamp temperature during processing, and only modest temperature capacity (<1100°C). Recent interest in developing wide bandgap compound semiconductors has pushed the processing temperature requirements to much higher values (up to

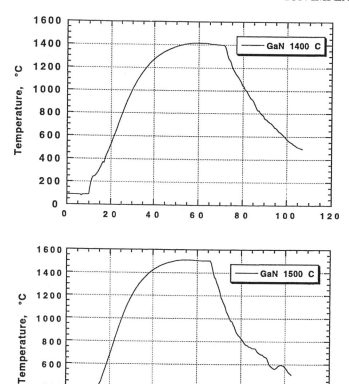

Fig. 23. Time–temperature profiles for RTP annealing of GaN at 1500°C.

1500°C). Presently, there are no specific RTP systems that can operate at such high temperatures. In the study of annealing of GaN up to 1400°C [91], a custom system (based on MOCVD system) that employed rf heating was built and utilized. There is an urgent need in GaN and SiC technology to develop new RTP systems which can provide uniform heating to very high temperatures (>1500°C).

Most existing RTP equipment utilizes either an array of 10 or more tungsten–halogen lamps or a single arc lamp as heat sources [17]. This lamp-based RTP equipment can achieve only modest processing temperatures (<1100°C), primarily because of the pointlike nature of the sources and large thermal mass of the systems. To realize higher temperature capacity, new types of heat sources have to be employed. In the past few years, there has been development of novel molybdenum intermetallic composite heaters that may be used in air at temperatures up to 1900°C [91]. These heaters have high emissivity (up to 0.9) and allow heat up time of the order of seconds and heat fluxes up to 100 W/cm^2. This novel RTP unit is capable of achieving much higher temperatures than the lamp-based RTP equipment. Figure 23 shows some typical time–temperature fluctuation to rapidly heat up and cool down the wafer, the Zapper™ unit relies on wafer movement (in/out of furnace horizontally) to achieve rapid ramp-up and ramp-down rates similar to those of conventional RTP systems.

A variety of undoped GaN layers ∼3 μm thick were grown at ∼1050°C by metal–organic chemical vapor deposition using trimethylgallium and ammonia. Growth was preceded by deposition of thin (∼200 Å) GaN or AlN buffers (growth temperature 530°C) on the Al$_2$O$_3$ substrates. Capacitance–voltage measurements on the GaN showed typical n-type background carrier concentrations of ≤3 × 10^{16} cm^{-3}. Si$^+$ was implanted to a dose of 5 × 10^{15} cm^{-2}, 100 keV, producing a maximum Si concentration of ∼6 × 10^{20} cm^{-3} at a depth of ∼800 Å. Some of the samples were encapsulated with 1000–1500 Å of AlN deposited in one of two ways. In the first, AlN was deposited by reactive sputtering of pure AlN targets in 300 mTorr of 20% N$_2$ Ar. The deposition temperature was 400°C. In the second method, AlN was grown by metal–organic molecular beam epitaxy (MOMBE) at 750°C using dimethylamine alane and plasma dissociated nitrogen.

The samples were sealed in quartz ampoules under N$_2$ gas. To ensure good purity of this ambient the quartz tube (with sample inside and one end predoped) was subjected to an evacuation/N$_2$ purge cycle for three repetitions before the other end of the tube was closed, producing a final N$_2$ pressure of ∼15 psi. This negative pressure was necessary to prevent blowout of the ampoule at elevated annealing temperatures. The samples were then annealed at 1100–1500°C, for a dwell time of ∼10 seconds (Fig. 23). The time difference for reaching the annealing temperature was between 4 and 6 seconds. To compensate for this heating time lag inside the ampoule, the dwell time was purposely extended to ∼15 seconds. Ramp rates were 80°C s^{-1} from 25–1000°C and 30°C s^{-1} from 1000–1500°C, producing an average ramp-up rate over the entire cycle of ∼50°C s^{-1}. The typical ramp-down rate was ∼25°C s^{-1}. Thermocouple measurements of temperature uniformity over a typical wafer size (4 inch diameter) were ±8°C at both 1400 and 1500°C. After removal of the samples from the ampoules they were examined by SEM, AFM, and van der Pauw geometry Hall measurements obtained with alloyed HgIn eutectic contacts.

SEM micrographs of unencapsulated GaN surfaces annealed at 1200, 1300, 1400, or 1500°C showed that the 1200°C annealing does not degrade the surface, and the samples retain the same appearance as the as-grown material. After 1300°C annealing, there is a high density (∼10^8 cm^{-2}) of small hexagonal pits which we believe is due to incongruent evaporation from the surface. The 1400°C annealing produces complete dissociation of the GaN, and only the underlying AlN buffer survives. Annealing at 1500°C also causes loss of this buffer layer, and a smooth exposed Al$_2$O$_3$ surface is evident.

By sharp contrast to the results for GaN, both the sputtered and MOMBE AlN were found to survive annealing above 1300°C. For the sputtered material we often observed localized failure of the film, perhaps due to residual gas agglomeration. For the MOMBE films this phenomenon was absent. Also in the sputtered material the rms surface roughness tended to go through a maximum, due to some initial localized bubbling, followed by the film densification.

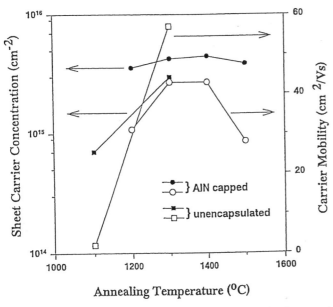

Fig. 24. Sheet carrier density and electron mobility in capped and uncapped Si-implanted GaN, as a function of annealing temperature.

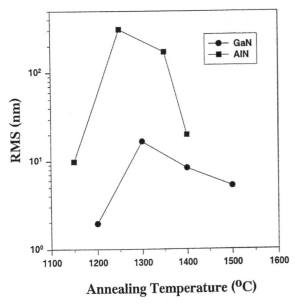

Fig. 25. Root-mean-square surface roughness of GaN and AlN after unencapsulated annealing at different temperatures.

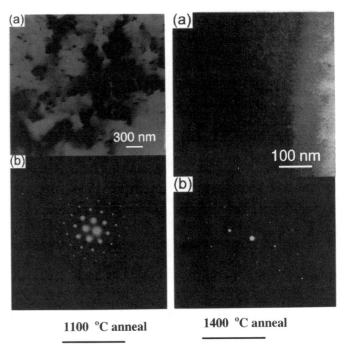

1100 °C anneal   1400 °C anneal

Fig. 26. TEM cross-sections of Si-implanted (5×10^{15} cm^{-2}, 150 keV) GaN after high temperature annealing.

The clear result from all this data is that the implanted GaN needs to be encapsulated with AlN in order to preserve the surface quality. We have previously shown that AlN is selectively removable from GaN using KOH-based solution. Figure 24 shows the sheet carrier concentration and electron mobility in the Si$^+$-implanted GaN, for both unencapsulated and AlN-encapsulated material, as a function of annealing temperature. For unencapsulated annealing we see an initial increase in electron concentration, but above 1300°C the GaN layer disintegrates (Fig. 25). By contrast, for AlN-encapsulated samples the Si$^+$ implant activation percentage is higher (~90%)

and peaks around 1400°C. This corresponds to a peak carrier concentration of $\geq 5 \times 10^{20}$ cm^{-3}. For 1500°C annealing both carrier concentration and mobility decrease, and this is consistent with Si-site switching as observed in Si$^+$-implanted GaAs at much lower temperatures [92]. The results in Figure 24 are compelling evidence of the need for high annealing temperatures and the concurrent requirement for effective surface protection of the GaN.

There is clear evidence from both ion channeling and TEM measurements that temperatures above 1300°C are required to completely remove implantation damage in GaN. Since the residual damage tends to produce n-type conductivity, it is even more imperative in acceptor-implanted material to completely remove its influence. However, a premium is placed on prevention of surface dissociation, because loss of nitrogen also leads to residual n-type conductivity in GaN. The combination of RTP annealing in the Zapper™ unit at 1400–1500°C and high quality AlN encapsulants produces metallic doping levels ($\sim 5 \times 10^{20}$ cm^{-3}) in Si$^+$-implanted GaN. Figure 26 shows that annealing Si-implanted GaN at 1400°C removes all of the extended defects, whereas annealing at 1100°C leaves a large density of these defects.

Since the activation anneal for GaN was initially done in the range of 1100°C, the formation of a Pt/Au Schottky contact on n-type GaN was studied after such a high temperature anneal with and without a AlN encapsulation layer. Since AlN has a higher dissociation temperature it should act to suppress the dissociation of the GaN. One set of samples was n-type as-grown with a background donor concentration of ~ 5–10×10^{16} cm^{-3} (samples A1, A2). The second set of

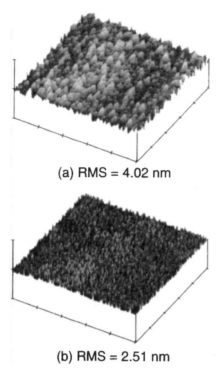

(a) RMS = 4.02 nm

(b) RMS = 2.51 nm

Fig. 27. Atomic force microscope images of GaN after an 1100°C, 15 s anneal either (a) uncapped or (b) capped with reactively sputtered AlN. The AlN film was removed in a selective KOH-based etch (AZ400K developer) at 60–70°C. On both images, the vertical scale is 50 nm per division and the horizontal scale is 2 μm per division.

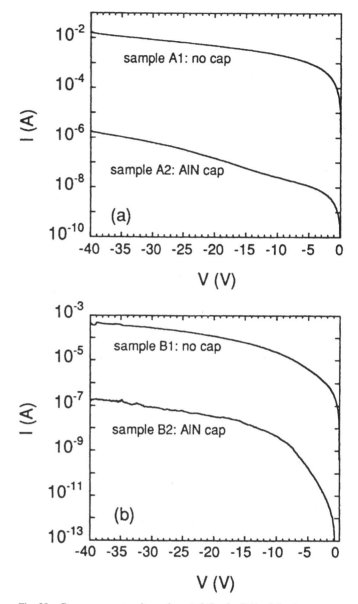

Fig. 28. Reverse current–voltage characteristics for Pt/Au Schottky contacts on GaN annealed at 1100°C, 15 s either uncapped (samples A1, B1) or capped with AlN (samples A2, B2). (a) is for unintentionally doped GaN with an as-grown donor concentration of 5–10 $\times$ 10^{16} cm^{-3} and (b) is for initially semi-insulating GaN implanted with ^{28}Si (100 keV, 5 $\times$ 10^{13} cm^{-2}) to simulate a MESFET channel implant.

samples (B1, B2) was semi-insulating as-grown and was implanted with ^{28}Si (100 keV, 5×10^{13} cm^{-2}) to simulate a metal-semiconductor field effect transistor (MESFET) channel implant ($n \sim 3 \times 10^{17}$ cm^{-3}). One sample from each set had 120 nm of AlN deposited in an Ar-plasma at 300 W using an Al-target and a 10 sccm flow of N$_2$. The film had an index of refraction of 2.1 ± 0.05 corresponding to stoichiometric AlN. All samples were annealed together in a SiC coated graphite crucible at 1100°C for 15 s in flowing N$_2$. Following annealing, the AlN was removed in a selective KOH-based etch (AZ400K developer) at 60–70°C. This etch has been shown to etch AlN at rates of 60 to 10,000 Å min^{-1}, depending on the film quality, while under the same conditions no measurable etching of GaN was observed. Ti/Al ohmic contacts were deposited and defined by conventional lift-off techniques on all samples and were annealed at 500°C for 15 s. Pt/Au Schottky contacts were deposited and defined by lift-off within a circular opening in the ohmic metal. Electrical characterization was performed on a HP4145 at room temperature on 48 μm diameter diodes. Samples prepared in the same way, except without any metallization, were analyzed with AES surface and depth profiles. The surface morphology was also characterized by AFM before and after annealing.

Figure 27 shows three-dimensional AFM images of the surface of samples A1 (uncapped) and A2 (AlN cap) after annealing at 1100°C. While both of the annealed samples displayed some increase in their rms surface roughness (A1: 4.02 nm, A2: 2.51 nm) over the as-grown sample (1.4 nm) the sample annealed without the AlN cap (A1) was markedly rougher with a dramatically different surface morphology. These differences were attributed to more N-loss in the uncapped sample as will be discussed later.

Figure 28 shows the reverse current/voltage characteristics of the Pt/Au Schottky diodes for the four samples studied. Samples A1 and B1, annealed without the AlN cap, demonstrated very leaky reverse characteristics with roughly a 3 to 4 order-of-magnitude increase in leakage current over the samples annealed with the AlN cap. In fact, sample A1 approached ohmic behavior. Samples A2 and B2, on the other hand, demonstrated very good rectification with reverse breakdown voltages in ex-

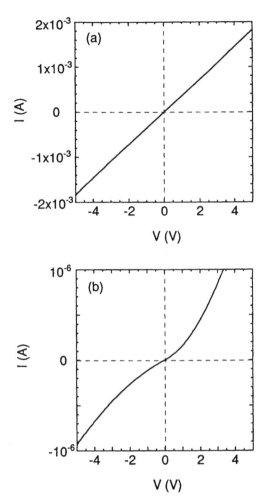

Fig. 29. Current–voltage characteristics for Ti/Al ohmic contacts on GaN after an 1100°C, 15 s anneal either (a) uncapped or (b) capped with reactively sputtered AlN. The as-grown GaN was *n*-type with a background concentration of $5–10 \times 10^{16}$ cm^{-3}. Note the change in scale for the current axis between the two plots.

N-loss and the formation of N-vacancies were proposed as the key mechanisms involved in changing the electrical properties of the Schottky diodes and ohmic contacts. Since N-vacancies are thought to contribute to the background *n*-type conductivity in GaN, an excess of N-vacancies at the surface should result in an n^+-region (possibly a degenerate region) at the surface. This region would then contribute to tunnelling under reverse bias for the Schottky diodes and explain the increase in the reverse leakage in the uncapped samples. Similarly, an n^+-region at the surface would improve the ohmic contact behavior as seen for the uncapped samples [93]. The effectiveness of the AlN cap during the anneal to suppress N-loss is readily understood by the inert nature of AlN and its extreme thermal stability thereby acting as an effective diffusion barrier for N from the GaN substrate.

In summary, the viability of using reactively sputtered AlN films as an encapsulating layer for GaN during 1100°C annealing was demonstrated. The AlN cap was selectively removed in a KOH-based etch. By employing an AlN cap, Pt/Au Schottky barriers with reverse breakdown voltages in excess of 40 V were realized on samples annealed at 1100°C. Samples annealed uncapped under the same conditions have 3 to 4 orders-of-magnitude higher reverse leakage than the capped samples while also displaying improved ohmic contact performance. These results are explained by the formation of an n^+-layer at the surface created by N-loss and N-vacancy formation in the uncapped samples. The AlN encapsulation, on the other hand, effectively suppressed N-loss from the GaN substrate. The use of AlN capping should allow the realization of all ion-implanted GaN MESFETs.

6.4. NH₃ Annealing

High pressure, high temperature annealing was used to study the fundamental limits of implantation induced damage removal in GaN [94]. By employing high N-overpressures (up to 15 kbar) sample decomposition is suppressed and the damage removal can be uncompromisingly examined [95, 96]. Figure 30 shows aligned C-RBS spectra for GaN implanted with 100 keV Si at dose of 5×10^{15} cm^{-2} and annealed under the conditions shown in the legend. Included in the legend in parentheses is the minimum channeling yield (χ_{min}) for each sample. An as-implanted sample (spectra not shown) had a χ_{min} of ∼34%; therefore significant damage removal has occurred for the 1250°C ($\chi_{min} = 14.28\%$) sample with continuing improvement with increased temperature. The 1500°C sample has a channeling yield equivalent to an unimplanted sample and demonstrated no macroscopic surface decomposition. This result suggests that implantation damage can in fact be removed in GaN given a high enough annealing temperature. The next step will be to find alternative ways, besides the extremely high N-overpressure, to maintain the sample stoichiometry.

The samples of Figure 30 were also characterized by the Hall technique, by photoluminescence, and by secondary ion mass spectroscopy (SIMS). Hall data suggest 46% electric activity of the implanted Si at 1250°C with increasing activity to

cess of 40 V where the breakdown voltage is taken at 1 μA/μm of diode perimeter current. Figure 29 shows the current/voltage characteristics for adjacent Ti/Al ohmic contacts for the two unintentionally doped samples A1 and A2. In this case, the samples annealed without the AlN cap demonstrated improved ohmic behavior as compared to the samples annealed with the AlN cap. This is consistent with the Schottky characteristics and suggests a change in the near surface stoichiometry of the GaN resulting from the uncapped anneal as will be discussed below.

In an attempt to quantify the change in the GaN surface resulting from annealing with and without the AlN cap, AES surface and depth profiles were performed. When comparing the Ga/N ratio for each case an increase was seen for the sample annealed without the AlN cap (Ga/N ratio = 2.34) as compared to the as-grown sample (Ga/N ratio = 1.73). This was explained by N-loss from the GaN during the annealing process. The AlN cap prevented this loss. AES depth profiles of the uncapped and annealed sample suggest that the N-loss occurred in the very near surface region (∼50 Å).

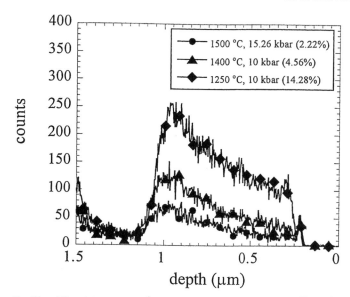

Fig. 30. Aligned C-RBS spectra for Si-implanted (100 keV, 5×10^{15} cm^{-2}) GaN annealed for 15 min under the conditions shown. In parentheses is the minimum channeling yield (χ_{min}) for each sample.

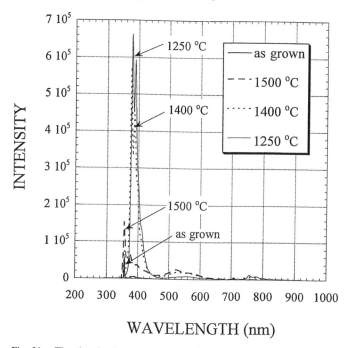

Fig. 31. The photoluminescence spectra of samples as-grown and after implantation (Si: 100 keV, 5×10^{15} cm^{-2}) and annealing as listed in the legend.

88% at 1500°C. However, the SIMS data show high levels of oxygen in the samples; therefore, the free donor concentration may also have a component due to O which is known to act as donor in GaN [97]. The source of the O is unclear; however, it may have diffused out from the sapphire substrate or in from the annealing ambient. The Hall mobility of the 1250°C was ~100 cm^2 V^{-1} s^{-1} and is very respectable for such a high donor level (on the order of 10^{20} cm^{-3}). The mobility was roughly constant for the higher annealing temperatures.

The photoluminescence spectra of samples as-grown and after implantation and annealing is shown in Figure 31. The as-

implanted samples (not shown) had no appreciable luminescence while the annealed samples had both near bandedge and donor/acceptorlike emission peaks. The 1500°C annealed sample has a stronger bandedge luminescence intensity than the as-grown material by a factor of three, but was worse than the 1250 and 1400°C samples. This may result from indiffusion of contaminants. The exact nature of this enhancement is under study, but an enhanced donor/acceptor recombination associated with the Si-doping along with removal of nonradiative centers during the annealing process was postulated.

7. DIFFUSIVITY OF IMPLANTED SPECIES

Figure 32 shows a SIMS profile of implanted Mg in GaN, both before and after annealing at 1450°C. Within the resolution of SIMS (~200 Å under these conditions) there is no motion of the Mg. Using a simple $2\sqrt{Dt}$ estimation of the diffusivity at this temperature gives a value of $\leq 2 \times 10^{-13}$ cm^2 s^{-1}. This is in sharp contrast to its behavior in GaAs [98–100], where the rapid diffusion of the Ga-site acceptors during annealing can only be suppressed by co-implanting group V elements to create a sufficient number of vacant sites for the initially interstitial acceptor ions to occupy upon annealing. This reduces the effective diffusivity of the acceptor and increases its electrical activation. The additional advantage gained from using a group V co-implant is that it maximizes occupation of the group III site by the acceptor [101]. In addition, implanted Mg in GaAs often displays outdiffusion toward the surface (in most case up, rather than down, the concentration gradient), leading to loss of

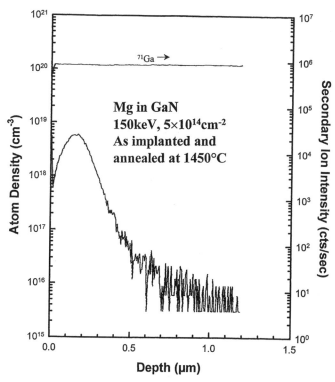

Fig. 32. SIMS profiles before and after annealing at 1450°C of implanted Mg (150 keV, 5×10^{14} cm^{-2}) in GaN. The profiles are essentially coincident.

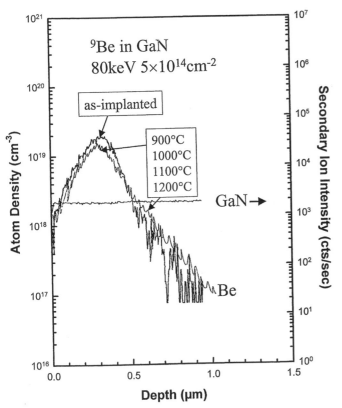

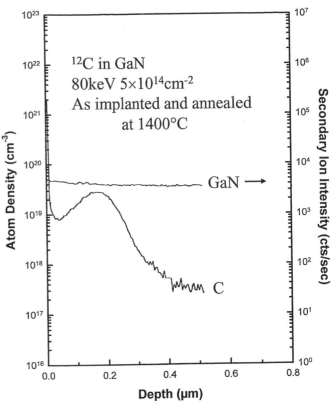

Fig. 33.   SIMS profiles before and after annealing at different temperatures of implanted Be (80 keV, 5 × 10^{14} cm^{-2}) in GaN.

Fig. 34.   SIMS profiles before and after annealing at 1400°C of implanted C (80 keV, 5 × 10^{14} cm^{-2}) in GaN. The profiles are essentially coincident.

dopant into the annealing cap [102]. This has been suggested to be due to nonequilibrium levels of Ga interstitials created by the implantation process. This mechanism is clearly absent for implanted Mg in GaN. Temperature-dependent Hall measurements showed an ionization level of ~170 meV, consistent with data in doped epitaxial material.

Figure 33 shows a series of profiles for ^{9}Be before and after annealing up to 1200°C. Note that there is an initial broadening of the profile at 900°C, corresponding to an effective diffusivity of ~5 × 10^{-13} cm^2 s^{-1} at this temperature. However, there is no subsequent redistribution at temperatures up to 1200°C. Implanted Be shows several types of anomalous diffusion in GaAs, including uphill diffusion and movement in the tail of the profile, in addition to normal concentration-dependent diffusion [100], which also result from the nonequilibrium concentrations of point defects created by the nuclear stopping process of the implanted ions. It appears that in GaN, the interstitial Be undergoes a type of transient-enhanced diffusion until these excess point defects are removed by annealing, at which stage the Be is basically immobile. We did not observe p-type conductivity in Be-implanted GaN.

Carbon is typically a very slow diffuser in all III-V compounds, since it strongly prefers substitutional lattice sites [98–102]. In GaN, it has been shown that it is possible to get p-type conductivity in carbon-doped material, albeit with low hole concentrations. In general, however, GaN containing high concentrations of carbon is self-compensated, suggesting that carbon is occupying both Ga and N sites. Figure 34 shows that

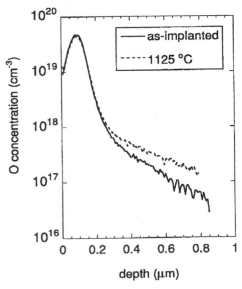

Fig. 35.   SIMS profiles before and after annealing at 1125°C of implanted Ca in GaN.

it is an extremely slow diffuser when implanted into GaN, with $D_{eff} \leq 2 \times 10^{-13}$ cm s^{-1} at 1400°C. Hall measurements showed that the C-implanted GaN remained n-type for all annealing conditions and did not convert to p-type conductivity.

Similar data are shown in Figure 35 for Ca implants in GaN. Once again there was no detectable motion of the acceptor dopant.

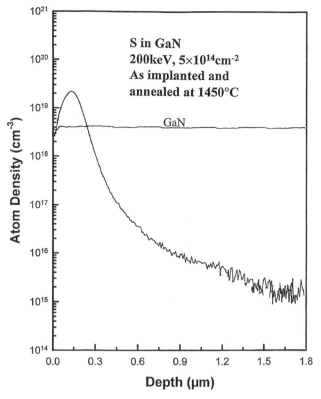

Fig. 36. SIMS profiles before and after annealing at 1450°C of implanted S (200 keV, 5×10^{14} cm^{-2}) in GaN. The profiles are essentially coincident.

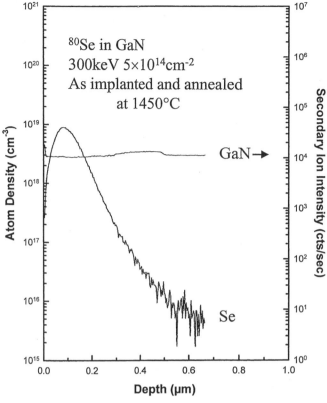

Fig. 37. SIMS profiles before and after annealing at 1450°C of implanted Se (300 keV, 5×10^{14} cm^{-2}) in GaN. The profiles are essentially coincident.

Our previous results showed that Si is an effective donor impurity in implanted GaN. Little work has been performed on the group VI donors, although Feng et al. [103] showed that Se displayed relatively high diffusion during growth by MOCVD at 1000°C. Wilson reported some redistribution of implanted S after annealing at 700–1000°C in relatively thin layers of GaN, which might have been influenced by the high crystalline defect density in the material. The ionization level of Si was measured as ~30 meV in implanted GaN samples. We have reported on the lattice quality of Si-implanted material elsewhere, but in brief, annealing at ≥ 1400°C is sufficient to completely remove extended defects, whereas 1100°C annealing leaves a high degree of disorder.

Figure 36 shows SIMS profiles before and after 1450°C annealing of implanted S in GaN. There is clearly no motion of the sulfur under these conditions, which suggests that, as expected, the structural quality of the GaN may have a strong influence on the apparent diffusivity of dopants. The samples in the present experiment are much thicker than those employed in the previous work, and the extended defect density will be correspondingly lower in the implanted region ($\sim 5 \times 10^8$ cm^{-2} compared to $\sim 10^{10}$ cm^{-2} in the thin samples). The Si ionization level was measured as 48 ± 10 meV from Hall data.

The other group VI donors, Se and Te, have low diffusion coefficients in all compounds semiconductors (for example, $D_{Se} = 5 \times 10^{-15}$ cm^2 s^{-1} at 850°C in GaAs) [100, 101], and we find a similar result for these species implanted into GaN. Figure 37 shows the SIMS profiles before and after 1450°C annealing for Se, while Figure 38 shows similar data for Te.

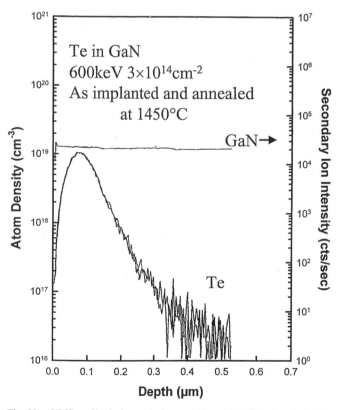

Fig. 38. SIMS profiles before and after annealing at 1450°C of implanted Te (600 keV, 3×10^{14} cm^{-2}) in GaN.

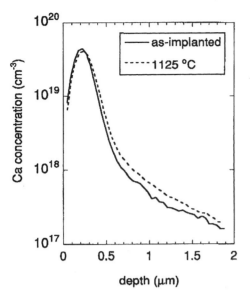

Fig. 39. SIMS profiles before and after annealing at 1125°C of implanted O in GaN.

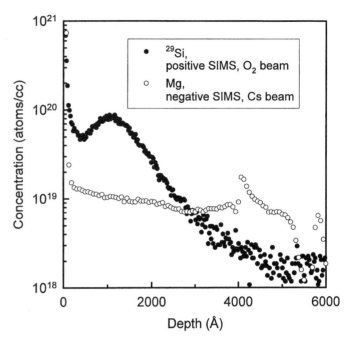

Fig. 40. SIMS profiles of Si and Mg in an implanted GaN junction.

In both cases, the effective diffusivity at this temperature is $\leq 2 \times 10^{-13}$ cm^2 s^{-1}. From limited temperature-dependent Hall data, we obtained an estimate of 50 ± 20 meV for the Te ionization level.

Figure 39 shows SIMS profiles of implanted O before and after 1125°C annealing. There was also no detectable redistribution of the oxygen.

Most of the common acceptor and donor species have been implanted into GaN at room temperature, and subsequently annealed up to 1450°C. With the exception of Be, which shows an apparent damage-assisted redistribution at 900°C, none of the species show detectable motion under these conditions. This bodes well for the fabrication of GaN-based power devices such as thyristors and insulated gate bipolar transistors which will require creation of doped well or source/drain regions by implantation. The low diffusivities of implanted dopants in GaN means that junction placement should be quite precise and there will be fewer problems with lateral diffusion of the source/drain regions toward the gate. Finally, the results show the effectiveness of the AlN cap in protecting the GaN surface from dissociation, since if any of the surface was degraded during annealing, the implant profiles would not longer overlap.

8. p–n JUNCTION FORMATION

In order to fabricate devices such as the enhancement lateral GaN MOSFET, there are several key process modules that must be developed, such as gate oxide deposition, p–n junction formation by implantation, ohmic contact improvement, and device passivation. There has already been significant progress in some of these areas, including the recent demonstration of the first GaN MOSFET using GdGaO$_x$ as the gate dielectric [104]. Moreover, the group of Murakami et al. [105] reported excellent p-ohmic specific contact resistivities of $\sim 10^{-5}$ Ω cm^2 using Ta-based metallization. The formation of p–n junctions by ion implantation is one of the key process steps in the fabrication

of the lateral MOSFET. Previous work has shown that low dose ($\leq 5 \times 10^{13}$ cm^{-2}) Si$^+$ implants can be efficiently activated in GaN by annealing at ≥ 1100°C. Careful attention must be paid to avoiding loss of N$_2$ from the GaN surface during annealing and various approaches such as AlN capping, N$_2$ overpressures, or use of powdered GaN in the reservoirs of susceptors containing the samples to be annealed have been reported [106–111]. A key requirement for good activation appears to be that the starting GaN have high resistivity. Higher dose implants of Si$^+$ require higher annealing temperatures for optimum activation, in the range ≥ 1400°C. Less success has been obtained with p-type implantation because of the generally high n-type residual background carrier concentration in GaN and the large ionization level (~ 170 meV) of Mg.

In this section we report the fabrication of $n^+ p$ junctions in GaN by implantation of ^{29}Si$^+$, followed by AlN-capped rapid annealing at 1100°C. Rectifying characteristics were obtained, but there was evidence of incomplete implant damage removal.

A ~ 5000 Å thick p-GaN layer was grown on c-plane Al$_2$O$_3$ by MOCVD at 1040°C. Mg doping was derived from a Cp$_2$Mg precursor. The total Mg concentration measured by SIMS was $\sim 10^{19}$ cm^{-3}, corresponding to a hole concentration at 25°C of $\sim 10^{17}$ cm^{-3}. ^{29}Si$^+$ ions were implanted in a nonchanneling direction to a dose of 2×10^{14} cm^{-2} at an energy of 120 keV. Part of each sample was masked with photoresist during the implant step. After removal of this resist, the sample was capped with sputtered AlN and annealing performed at 1100°C under a 1 atm N$_2$ ambient. The n-type regions were metallized with Ti/Al, while the p-regions were metallized with Ni/Au.

From past experiments, it is known that the diffusivity of implanted Si is $\leq 10^{-12}$ cm^{-2} s^{-1} at 1100°C. We also found that there was no detectable motion of the Mg during annealing as shown in the SIMS profiles of Figure 40. This suggests that

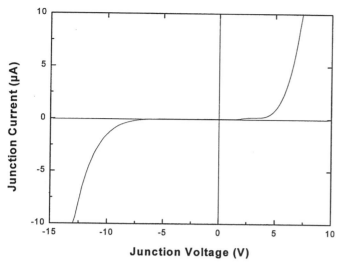

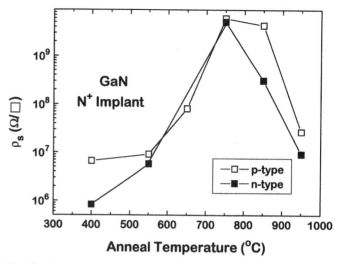

Fig. 41. *I–V* characteristics from an implanted GaN junction (Si-implant into Mg-doped *p*-GaN).

Fig. 42. Sheet resistance versus annealing temperature for N implanted initially *n*- and *p*-type GaN. The N was implanted at multiple energies to give an approximately uniform ion concentration of 4×10^{18} cm^{-3} across ~500 nm.

the majority of the Mg is incorporated substitutionally, or we should have expected significant redistribution during annealing. Note that the junction depth will be difficult to determine from the SIMS data, because the background sensitivity for Si is relatively high. From a TRIM simulation we estimate the junction depth to be ~4000 Å based on the distance at which the Si concentration falls to ~10^{17} cm^{-3} (the hole concentration due to the Mg acceptors). We assume that at 25°C essentially all of the Si donors are ionized ($E_d \sim 28$ meV) and we have activated ≥90% of the implanted donors.

The current–voltage (*I–V*) characteristic from a typical implanted junction is shown in Figure 41. The reverse breakdown voltage is 13 V at a current density of 5.1×10^{-4} A cm^{-2}. From the slope of the forward section of the characteristics, the junction ideality factor is ~2. This is indicative of generation–recombination being the dominant current conduction mechanism in the junction. It is shown that heteroepitaxial GaN contains a high density of threading dislocations, and these are a source of possible generation–recombination centers.

Plan view TEM was performed to examine the effectiveness of the damage removal by the annealing step. The micrographs showed that the GaN clearly contains a significant density of dislocations. These are not observed in unimplanted control samples, and thus we ascribe their origin to condensation of point defects created during the implant step. Separate measurements have shown that annealing temperatures ≥1400°C are necessary to remove these dislocations. We expect that creation of n^+p junctions will be simpler than p^+n junctions, because of the lower activation efficiency of *p*-implants and the more stringent requirement for avoidance of N$_2$ loss from the near-surface region. The *I–V* characteristics we obtained here can certainly be improved, both in terms of breakdown voltage and ideality factor.

9. ISOLATION

Implant isolation has been widely used in compound semiconductor devices for interdevice isolation such as in transis-

tor circuits or to produce current channeling such as in lasers [111–113]. The implantation process can compensate the semiconductor layer either by a damage or chemical mechanism. For damage compensation, the resistance typically goes through a maximum with increased postimplantation annealing temperature as the damage is annealed out and hopping conduction is reduced. At higher temperatures the defect density is further reduced below that required to compensate the material and the resistivity decreases. For chemical compensation, the postimplantation resistance again increases with annealing temperature with a reduction in hopping conduction but it then stabilizes at higher temperatures as a thermally stable compensating deep level is formed. Typically there is a minimum dose (dependent on the doping level of the sample) required for the chemically active isolation species to achieve thermally stable compensation [114]. Thermally stable implant isolation has been reported for *n*- and *p*-type AlGaAs where an Al−O complex is thought to form [114, 115] and for C-doped GaAs and AlGaAs where a C−N complex is postulated [116]. With this background, the implant isolation properties of the III-N materials are reviewed.

As shown in Figure 42, N-implantation (at doses of 10^{12}–10^{13} cm^{-3}) effectively compensates both *p*- and *n*-type GaN. For both doping types the resistance first increases with annealing temperature then reaches a maximum before demonstrating a significant reduction in resistance after a 850°C anneal for *n*-type and a 950°C anneal for *p*-type GaN. This behavior is typical of implant-damage compensation. The defect levels estimated from Arrhenius plots of the resistance/temperature product are 0.83 eV for initially *n*-type and 0.90 eV for initially *p*-type GaN (Fig. 43). These levels are still not at midgap, but are sufficiently deep to realize a sheet resistance >10^9 Ω/square. The implantation has also been reported to effectively isolate *n*-type GaN, with the material remaining compensated to over 850°C. Interestingly, H-implanted compensation of *n*-type GaN is reported to anneal out at ~400°C with

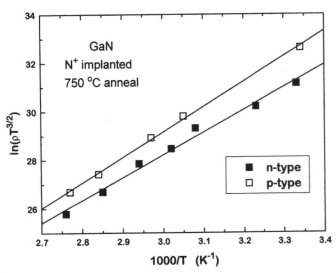

Fig. 43. Arrhenius plots of sheet resistance in N$^+$-implanted n- and p-GaN annealed at 750°C.

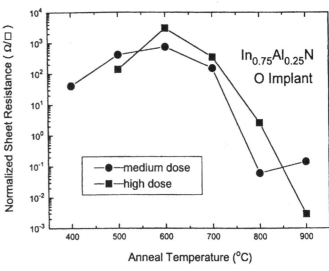

Fig. 44. Normalized sheet resistance (relative to the unimplanted values) versus anneal temperature for In$_{0.75}$AlN implanted with various doses of O$^+$ at multiple energies.

an anomalous dependence on implant energy. The reason for this is presently not known. In light of this result, however, H-implantation in GaN will require further study, as H is often the ion of choice for photonic device isolation applications that require deep isolation schemes. Moreover, both the He and N isolation appear to rely solely on implantation damage without any chemical compensation effects analogous to those in the O/Al/GaAs case [111, 114, 115]. However, the implantation-induced defects in GaN are more thermally stable than other III-V semiconductor materials, such as GaAs or InP, where the damage levels begin to anneal out below 700°C [111]. This may be a result of the higher bandgap of GaN or the more polar nature of the lattice causing more stable defects. Further work is still required to understand the nature of the implantation damage in GaN.

Figure 44 shows the annealing temperature dependence of sheet resistivity in O$^+$-implanted Al$_{0.2}$Ga$_{0.8}$N. Very high sheet resistances ($\sim 10^{12}$ Ω/square) can be obtained because of the large bandgap of this material, but once again the isolation is due to damage and not to chemical deep levels. Similar results were obtained with n^+ implantation. Deep level transient spectroscopy measurements showed states at $E_c - 0.67$, $E_c - 0.60$, and $E_c - 0.27$ eV in N$^+$-implanted GaN [117]. It is not known what effect the annealing ambient has, since the presence of oxygen or hydrogen is known to enhance the decomposition of the GaN surface and lead to preferential N$_2$ loss [118–122].

Figure 44 shows the evolution of normalized sheet resistance in O$^+$-implanted In$_{0.75}$Al$_{0.25}$N with subsequent annealing. The sheet resistance increases up to ~ 600°C and then is reduced to the original unimplanted values at ~ 800°C. This behavior is typically of that seen in other implanted III-V semiconductors and is caused by the introduction of deep acceptor states related to the implant damage that compensate the shallow native donors. This produces an increase in sheet resistance of the material, the magnitude of which is dose dependent. The usual situation is that all of the shallow levels are compensated, but

some residual conductivity remains due to the presence of inter-defect hopping of trapped carriers from one closely spaced trap site to another. In other words, there is an excess of deep states over that required for optimum compensation. Subsequent annealing removes these excess states, leading to an increase in sheet resistance of the material. Continued annealing above ~ 600°C for In$_{0.75}$Al$_{0.25}$N reduces the deep acceptor concentration below that required to trap all of the original free electrons, and the conductivity increases back to the preimplanted value. In the present case, annealing above 800°C actually produces a sheet resistance lower than in the as-grown samples due either to loss of nitrogen from the uncapped material or the existence of additional shallow donor states created by the implantation.

Similar results were obtained for N$^+$-implanted In$_{0.75}$-Al$_{0.25}$N. The data show the same basic trends as for oxygen implantation and lead to the conclusion that neither oxygen nor nitrogen produces a significant concentration of chemical deep states in InAlN or the high resistance values would be maintained even at the maximum annealing temperatures. Oxygen and nitrogen are known to create chemical deep levels in AlGaAs. Note that even in this relatively conducting InAlN, implantation and optimized annealing is capable of producing increases in sheet resistance of 3–8 × 10^3 times (absolute values >10^8 Ω/square). This is well above the values required for electronic device isolation ($\geq 10^6$ Ω/square).

Temperature-dependent measurements of the sheet resistance of an In$_{0.75}$Al$_{0.25}$N sample implanted with either a medium or high dose of O$^+$ ions and annealed at 600°C are shown in Figure 45. The activation energy obtained for the medium dose sample is ~ 0.29 eV, consistent with its lower sheet resistance relative to that of the higher dose material where the activation energy was 0.54 eV. Note that for optimum implant isolation it is desirable to create midgap states. In InAlN the states created are still relatively high in its band gap, similar to the behavior of InP and InGaAs.

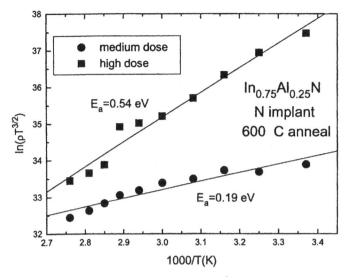

Fig. 45. Arrhenius plots of sheet resistance of N^+-implanted $In_{0.75}Al_{0.25}N$ after annealing at 600°C.

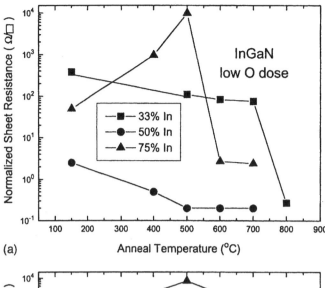

(a)

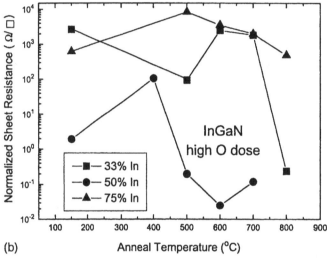

(b)

Fig. 46. Normalized sheet resistance versus anneal temperature for $In_xGa_{1-x}N$ implanted with (a) low doses and (b) doses of O^+ at multiple energies.

Turning to $In_xGa_{1-x}N$, Figure 46 shows the increase in sheet resistance of O^+-implanted material of three different compositions as a function of annealing temperature. The maximum increases in sheet resistance are between 10^2 and 10^4 for both low and high dose conditions, with consistently higher values for the higher doses. Once again there is no evidence for a chemically active deep acceptor state for oxygen.

Similar data were obtained for N^+ ion implantation in InGaN. The maximum increases in sheet resistance are less than a factor of 10^3, slightly lower than for the O^+ implant results. This is most likely a result of the slightly different growth conditions for this set of samples leading to higher initial conductivities. The same trends in sheet resistance with annealing temperature are evident as for O^+ implantation. In this highly conducting $In_xGa_{1-x}N$ the absolute sheet resistances achievable are borderline for electronic device isolation but are acceptable for applications such as current path delineation in photonic devices.

The results for F^+ implantation in InGaN also showed there is basically no difference from the results for N^+ or O^+ implants, indicating that none of these elements produce chemical deep states in $In_xGa_{1-x}N$ and that all of the changes in electrical properties are due to introduction of damage-related deep levels and their subsequent annealing. Examples of the measurement temperature dependence of sheet resistance, corrected for the temperature dependence of mobility, are shown in Figure 47 for $In_{0.33}Ga_{0.67}N$ implanted with a high dose of N^+, and annealed at either 500 or 800°C. In the former case, an activation energy of 0.40 eV is obtained. This is relatively high in the bandgap of this material ($\sim$2.8 eV). As discussed earlier, implant isolation is most effective when the damage-related states are at midgap, as in the case in the $Al_xGa_{1-x}As$ and $In_xGa_{1-x}P$ materials systems. The behavior of InAlN and InGaN is similar to that of InP and InGaAs, where the damage-related levels are relatively high in the gap. The result is that initially n-type material achieves only moderate resistivities upon implantation with nonchemically active ions. The other side of this situation

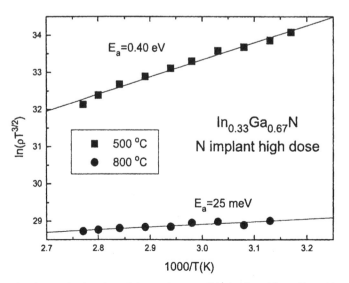

Fig. 47. Arrhenius plots of sheet resistance of N^+ implanted $In_{0.33}Ga_{0.67}N$ implanted with a high dose of N^+ and subsequently annealed at either 500 or 800°C.

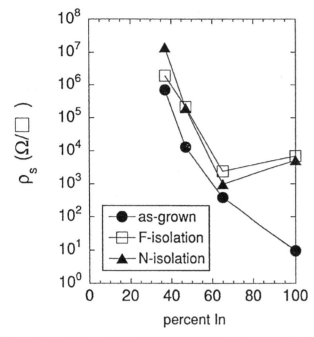

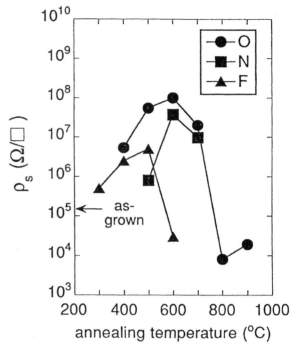

Fig. 48. Maximum sheet resistance versus percent In for InGaN either as-grown or implanted with F or N and annealed at the temperature for maximum compensation for each composition (ion concentration $\sim 5 \times 10^{19}$ cm^{-3}).

Fig. 49. Sheet resistance versus annealing temperature for O-, N-, or F-implanted In$_{0.75}$Al$_{0.25}$N (ion concentration $\sim 5 \times 10^{18}$ cm^{-3}).

is that if one starts with initially p-type InP one can achieve very high resistivities because the Fermi level moves from near the valance band and through midgap on its way to the states in the upper part of the band and therefore an optimum choice of dose can place the Fermi level at midgap. To date no one has produced p-type InGaN or InAlN, but it will be interesting to see if this behavior is observed in these materials. Upon annealing at 800°C, where it again becomes very conducting, the N$^+$-implanted In$_{0.33}$Ga$_{0.67}$N displays an activation energy of only $\sim$25 meV, consistent with the values obtained in as-grown samples.

To summarize the results for the alloys, implant isolation of the In-containing nitrides (InN, InGaN, and InAlN) was first reported using F-implantation [123]. That work showed that InN did not demonstrate significant compensation while the ternaries increased in sheet resistance by roughly an order of magnitude after a 500°C anneal. Data from a more extensive study of In$_x$G$_{1-x}$N implant isolation for varying In composition using N- and F-implantation is summarized in Figure 48 [124]. The InGaN ternaries only realize a maximum of a 100-fold increase in sheet resistance independent of ion species after a 550°C anneal. Pure InN shows a higher increase of 3 orders of magnitude but still only achieves a maximum sheet resistance of 10^4 Ω/square. This may be high enough for some photonic device current-guiding applications but is not sufficient for interdevice isolation in electronic circuits. The damage levels created by N-implantation are estimated from an Arrhenius plot of the resistance/temperature product to be a maximum of 390 meV below the conduction band [124]. The defect level is high in the energy gap, not near midgap, as is ideal for implant compensation. The position of the damage level is anal-

ogous to the defect position reported for implant compensated n-type InP and InGaAs [125] but different from the damage-associated, midgap states created in GaAs and AlGaAs.

As shown in Figure 49, In$_{0.75}$Al$_{0.25}$N, in contrast to InGaN, can be highly compensated with N- or O-implantation with over a 3 order-of-magnitude increase in sheet resistance after a 600–700°C anneal while F-implantation produces only a 1 order-of-magnitude increase in sheet resistance [123, 126]. The compensating level in InAlN is also high in the bandgap with the deepest level estimated from Arrhenius plots as being 580 meV below the conduction bandedge in high dose N-isolated material; however, it is sufficiently deep to achieve highly compensated material. The enhanced compensation for N- and O-implantation, as compared to F-implantation, in InAlN suggests some chemical component to the compensation process. For N-implantation a reduction in N-vacancies, which are thought to play a role in the as-grown n-type conduction, may explain the enhance compensation. For O-implantation, the enhanced compensation may be the result of the formation of an O—Al complex as is thought to occur in O-implanted AlGaAs.

Figure 50 schematically summarizes the present knowledge of the position in the bandgap of the compensating implanted defect levels in III-N materials and compares these to those in GaAs and InP. Although the levels are not at midgap, as is ideal for optimum compensation as occurs in GaAs and p-type InP, with the exception of InGaN, the levels are sufficiently deep to produce high resistivity material.

Very effective isolation of AlGaN/GaN heterostructure filed effect transistor structures has been achieved using a combined P$^+$/He$^+$ implant process [127]. The groups of Asbeck and Lau at UCSD demonstrated that a dual energy (75/180 keV) P$^+$ im-

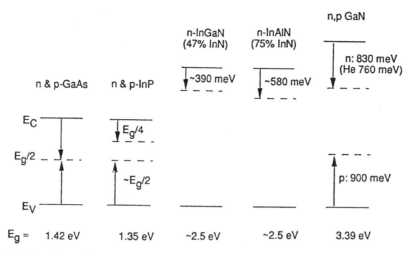

Fig. 50. Schematic representation of the position in the energy gap of compensating defect levels from implant isolation in GaAs, InP, $In_{0.47}Ga_{0.53}N$, $In_{0.75}Al_{0.25}N$, and GaN.

plant (doses of 5×10^{11} and 2×10^{12} cm^{-2}, respectively), followed by a 75 keV He$^+$ implant (6×10^{13} cm^{-2}), was able to produce sheet resistance of $\sim 10^{12}$ Ω/⊙ in AlGaN/GaN structures with 1 μm thick undoped GaN buffers. The temperature dependence of the resistivity showed an activation energy of 0.71 eV, consistent with past measurements of deep states induced in GaN by implant damage.

In this section we report on the defect levels created on n- and p-type GaN by Ti, O, Fe, or Cr implantation. The annealing temperature dependence of the sample resistance was measured up to 900°C. The thermally stable, electrically active concentration of deep states produced by these species was found to be $< 7 \times 10^{17}$ cm^{-3} in both conductivity types of GaN, with the sample resistivity approaching its original, unimplanted value by ~ 900°C in all cases. The defect levels created in the implanted material are within 0.5 eV of either bandedge.

0.3 μm thick n (Si-doped) or p (Mg-doped) type GaN layers were grown on 1 μm thick undoped GaN on (0001) sapphire substrates by rf plasma activated molecular beam epitaxy. The carrier concentration in the doped layers was 7×10^{17} cm^{-3} in each case. Ohmic contacts were formed in a transmission line pattern (gap spacings of 2, 4, 8, 16, and 32 μm) by e-beam evaporation and lift-off of Ti/Au (n-type) or Ni/Au (p-type) annealed at 700°C under N$_2$. The total metal thickness was 4000 Å, so that these regions could act as implanted masks. A schematic of the resultant structure is shown in Figure 51 (top).

The samples were then implanted at 25°C using multiple-energy Ti$^+$, O, Fe, or Cr ions. An example of the calculated ion distribution for the Ti$^+$ implant scheme is shown at the bottom of Figure 51. The doses and energies were chosen to create an average ion concentration of $\sim 10^{19}$ cm^{-3} throughout the 0.3 μm thick doped GaN layers.

In the case where the implanted species is chemically active in the GaN it is the ion profiles that are the important feature, since it is the electrically active fraction of these implanted species that determines the isolation behavior. In the case where the isolation simply results from damage-related deep levels,

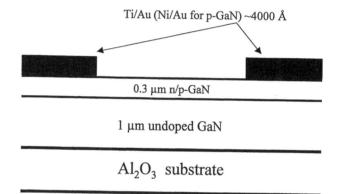

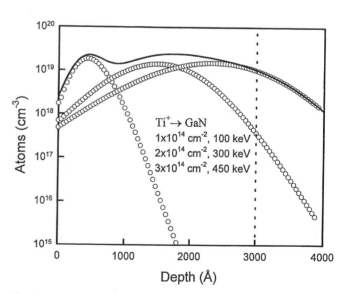

Fig. 51. Schematic of GaN structure for measurement of sheet resistance after implantation (top) and ion profiles for Ti$^+$ implant sequence (bottom).

then it is the profile of ion damage that is important. Figure 52 shows both the calculated ion profiles (top) and damage profiles (bottom) for the multiple energy O$^+$ implant scheme, obtained

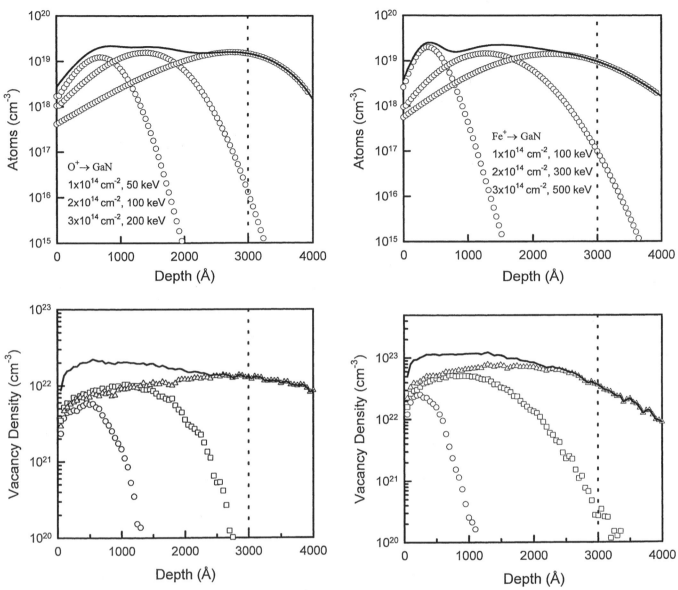

Fig. 52. Ion (top) and damage (bottom) profiles for multiple energy O^+ implant sequence into GaN.

Fig. 53. Ion (top) and damage (bottom) profiles for multiple energy Fe^+ implant sequence into GaN.

from TRIM simulations. Note that the defect density is generally overstated in these calculations due to recombination of vacancies and interstitials. In any case, the doses are below the amphorphization threshold for GaN.

Both Fe and Cr were implanted at 100 keV (10^{14} cm^{-2}), 300 keV (2×10^{14} cm^{-2}), and 500 keV (3×10^{14} cm^{-2}). The ion and vacancy profiles for the Fe^+ implants are shown at the top and bottom, respectively, of Figure 53.

The sheet resistance of the implanted regions was measured for annealing temperatures up to 900°C and for measurement temperatures up to ~250°C.

Figure 54 shows the annealing temperature dependence of sheet resistance for Cr^+ (top) and Fe^+ (bottom) implanted n- and p-type GaN. The trends in the sheet resistance are typical of those observed with damage-related isolation. The as-

implanted resistance is 6–7 orders of magnitude higher than that of the unimplanted material due to creation of deep traps that remove carriers from the conduction and valence bands. Subsequent annealing tends to further increase the sheet resistance, by reducing the probability for hopping conduction as the average distance between trap sites is increased. Beyond particular annealing temperatures (500–600°C in this case) the trap density begins to fall below the carrier concentration and carriers are returned to the conduction or valence bands. This produces a decrease in sheet resistance toward the original, unimplanted values. If Cr or Fe produced energy levels in the bandgap with concentrations greater than the carrier density in the material, then the sheet resistance would remain high for annealing temperatures above 600°C. For these two impurities it is clear that the electrically active concentration of deep states

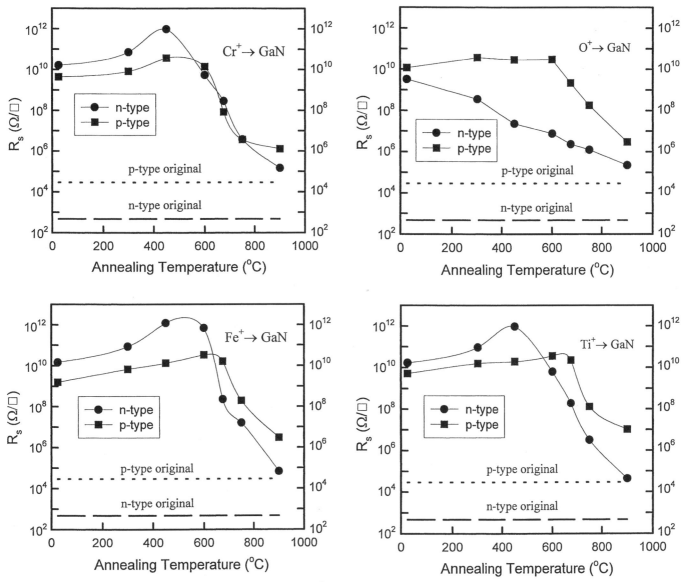

Fig. 54. Evolution of sheet resistance of GaN with annealing temperature after either Cr^+ (top) or Fe^+ (bottom) implantation.

Fig. 55. Evolution of sheet resistance of GaN with annealing temperature after either O^+ (top) or Ti^+ (bottom) implantation.

is $< 7 \times 10^{17}$ cm^{-3}; otherwise all the carriers would remain trapped beyond an annealing temperature of 600°C.

Similar data are shown in Figure 55 for O^+ (top) and Ti^+ (bottom) implants in both n- and p-type GaN. Basically the same trends are observed as for Cr^+ and Fe^+ implants. The sheet resistances tend to be lower for the O^+ implants because of the lower vacancy and interstitial concentrations created. We do not believe this is a result of O-related shallow donor states because these are not activated until annealing temperatures above 1100°C. Once again, the concentration of electrically active deep states related to the chemical nature of both Ti and O must be $< 7 \times 10^{17}$ cm^{-3} when they are introduced into GaN by ion implantation.

Figure 56 shows Arrhenius plots of the sheet resistance of Cr^+ (top) or Fe^+ (bottom) implanted n- and p-type GaN annealed at either 450°C (n-type) or 600°C (p-type). These an-

nealing temperatures were chosen to be close to the point where the maxima in the sheet resistances occur for the two different conductivity types. The activation energies derived from these plots represent the Fermi level position for the material at the particular annealing temperatures employed. Note that the values are far from midgap (1.7 eV for hexagonal GaN) but are still large enough to create very high sheet resistances in the ion damaged material. Within the experimental error (± 0.04 eV), the activation energies are the same for Cr^+ and Fe^+ implants for both conductivity types. This again suggests the defect states created are damage related and not chemical in nature.

Similar data are shown in Figure 57 for O^+ (top) and Ti^+ (bottom) implants in both n- and p-type GaN annealed at either 300°C (O^+, n-type), 450°C (Ti, n-type), or 600°C (O and

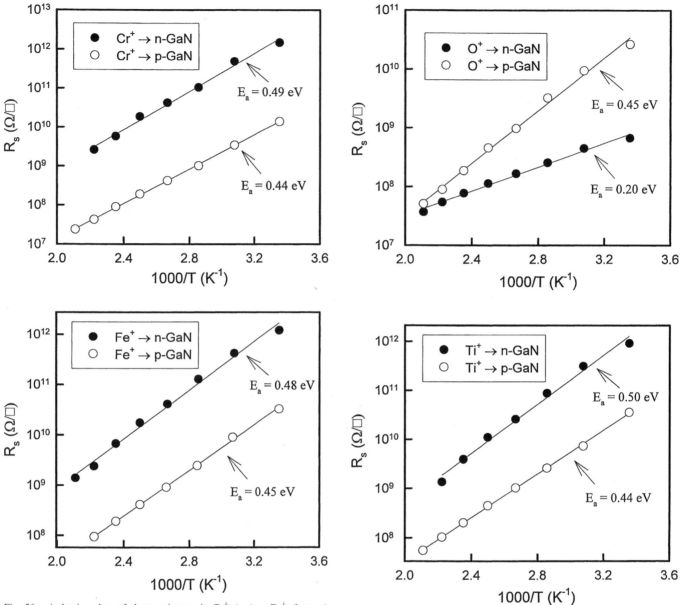

Fig. 56. Arrhenius plots of sheet resistance in Cr^+ (top) or Fe^+ (bottom) implanted n- and p-GaN, after annealing at either 450°C (n-type) or 600°C (p-type).

Fig. 57. Arrhenius plots of sheet resistance in O^+ (top) or Ti^+ (bottom) implanted n- and p-type GaN, after annealing at either 300°C (O^+, n-type). 450°C (Ti, n-type), or 600°C (O^+ and Ti^+, p-type).

Ti, p-type). The activation energies are again similar to those obtained with Cr^+ or Fe^+ implants, except for the case of O^+ into n-GaN. This difference may be related to the lower damage density with O^+ implantation described earlier. The defect states in the gap are most likely due to point defect complexes of vacancies and/or interstitials and the exact microstructure of these complexes and their resultant energy levels are expected to be very dependent on damage density and creation rate. This might also explain the differences reported in the literature for the activation energies obtained with different implant species.

Figure 58 shows a schematic of the energy level positions found in this work for Ti, Cr, Fe, and O implanted p- and n-type GaN annealed to produce the maximum sheet resistance. Al-

though the levels are not a midgap as is ideal for optimum compensation, they are sufficiently deep to produce high resistivity material. In GaN contaminated with transition metal impurities, no-phonon photoluminescence lines attributed to Fe^{3+} at 1.3 eV and Ti^{2+} at 1.9 eV have been reported, but to date there are no electrical measurements.

The main points of this investigation may be summarized as follows:

(i) Ti, O, Fe, and Cr do not produce electrically active deep energy levels in the GaN bandgap with concentrations approaching 7×10^{17} cm^{-3} when introduced by implantation.

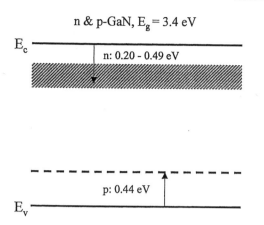

Fig. 58. Schematic representation of the position in the energy gap of defect levels from Fe, Cr, Ti, or O implant isolation in GaN.

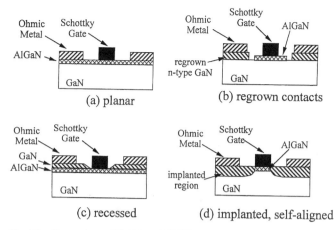

Fig. 59. Comparison of GaN-based HFET structures: (a) planar, (b) recessed gate, (c) regrown n^+ ohmic regions, and (d) self-aligned, implanted. Implantation is the most practical means to achieve the selective area doping required to reduce the transistor access resistance.

(ii) GaN implanted with these species displays typical damage-related isolation behavior, with no evidence of chemically induced thermally stable isolation.

(iii) Sheet resistances of $\sim 10^{12}$ Ω/square in n-GaN and $\sim 10^{10}$ Ω/square in p-GaN can be achieved by implantation of Cr, Fe, or Ti.

(iv) The activation energies for the sheet resistance of the implanted GaN are in the range 0.2–0.49 eV in n-type and ~ 0.44 eV for p-type.

10. DEVICES

The best example of how ion implantation can directly impact the performance of group III-nitride transistors is illustrated in Figure 59. The figure shows four device structures that could be employed to fabricate AlGaN/GaN high electron mobility transistors (HEMTs). To date, the majority of the AlGaN/GaN HEMTs transistors have been fabricated in a planar structure as shown in Figure 16a, where the ohmic source and drain contacts are placed directly on the wide bandgap AlGaN layer without any increased local doping to reduce the contact or access resistance. This leads to a high access resistance, reduced current capability, and a high transistor knee voltage. This in turn reduces the transistor power gain, power-added efficiency, and linearity. Figure 59b and c shows two approaches that have been taken to reduce the access resistance. One is to selectively etch away the wide bandgap material in the contact regions and then regrow highly doped GaN to improved access resistance; however, the manufacturability of this approach, as with any regrowth process, is questionable. The recess gate approach of Figure 16c has been widely used in other mature compound semiconductors such as GaAs and InP. Although this type of structure has been demonstrated in GaN-based devices, the unavailability of controlled wet etching of GaN requires the use of a plasma recess etch. Use of a plasma etch introduces surface damage in the semiconductor in the region under Schottky gate that degrades the rectifying properties of the gate. Finally, the self-aligned ion implanted structure of Figure 59d is used to create selective areas of highly doped regions for the source and drain contacts in a highly manufacturable fashion without any plasma etching of the gate region. To date, ion implantation has been used to realize a GaN junction field effect transistor but has not been applied to AlGaN/GaN HEMTs. As will be discussed later, one of the key challenges to applying ion implantation to AlGaN/GaN HEMTs is the avoidance of surface degradation that will negatively impact the Schottky gate formation during the high temperature implant activation anneal.

Progress in GaN-based electronics has been remarkably rapid due to several factors. One of these is that the experience gained in GaAs/AlGaAs HEMTs has been quickly applied to the GaN/AlGaN system. At one time it was thought that MBE and related techniques would be the best choice for growth of electronic device structures, and this may still be the case if GaN substrates become available. However, for heteroepitaxial growth there is still a need to grow thick buffer or epitaxial lateral growth (ELO) structures, which are best done with MOCVD. The rapid advances in material purity, ohmic contact quality, and gate contact stability have fueled the progress in HFETs in the GaN/AlGaN system. Much work remains to be done on vertical device structures such as thyristors and HBTs, where minority carrier lifetime, interface quality, and doping control are important factors. As with GaAs electronics, much of the impetus for nitride electronics is coming from defense applications and there is as yet no commercialization of these devices.

An example of how selective area implantation can be used to improve ohmic contact resistance on device structures is shown in Figure 60. In this case, W contact deposited on the implanted n-GaN showed minimum specific contact resistances of $\sim 10^{-4}$ Ω cm^2. After implantation of Si$^+$ followed by activation at 1100°C to produce n^+ GaN regions ($n \sim 5 \times 10^{20}$ cm^{-3}), subsequently deposited W contacts show a contact resistance almost 2 orders of magnitude lower than on unimplanted material.

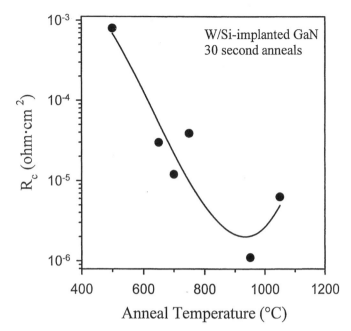

Fig. 60. Contact resistance versus annealing temperature for W on Si-implanted GaN that was initially activated at 1100°C to produce n^+ GaN.

Acknowledgment

This work is partially supported by grants from DARPA/EPRI and NSF.

REFERENCES

1. J. C. Zolper, *J. Cryst. Growth* 178, 175 (1997).
2. J. C. Zolper, in "GaN and Related Materials, Optoelectronic Properties of Semiconductors and Superlattices," Vol. 2. Gordon and Breach, New York, 1997.
3. R. G. Wilson, *Proc. Electrochem. Soc.* 95-21, 152 (1995).
4. H. P. Maruska, M. Lioubtchenko, T. G. Tetreault, M. Osinski, S. J. Pearton, M. Schurmann, R. Vaudo, S. Sakai, Q. Chen, and R. J. Shul, *Mater. Rev. Soc. Symp. Proc.* 483, 345 (1998).
5. J. C. Zolper, H. H. Tan, J. S. Williams, J. Zou, D. J. H. Cockayne, S. J. Pearton, M. H. Crawford, and R. F. Karlicek, Jr., *Appl. Phys. Lett.* 70, 2729 (1997).
6. S. C. Binari, H. B. Dietrich, G. Kelner, L. B. Rowland, K. Doverspike, and D. K. Wickenden, *J. Appl. Phys.* 78, 3008 (1995).
7. C. Liu, B. Mensching, M. Zeitler, K. Volz, and B. Rauschenbach, *Phys. Rev. B* 57, 2530 (1998).
8. C. Ronning, N. Dalmer, M. Deicher, M. Restle, M. D. Bremser, R. F. Davis, and H. Hofsass, *Mater. Res. Soc. Symp. Proc.* 468, 407 (1997).
9. H. Kobayashi and W. M. Gibson, *Appl. Phys. Lett.* 73, 1406 (1998).
10. T. Suski, J. Jun, M. Leszczynski, H. Teisseyre, I. Grzegopy, S. Prorwski, J. M. Baranowski, A. Rockett, S. Strite, A. Stanert, A. Turos, H. H. Tan, J. S. Williams, and C. Jagadish, *Mater. Res. Soc. Symp. Proc.* 482, 703 (1998).
11. S. J. Pearton, C. R. Abernathy, C. B. Vartuli, J. C. Zolper, C. Yuan, and R. A. Stall, *Appl. Phys. Lett.* 68, 2273 (1996).
12. J. C. Zolper, R. J. Shul, A. G. Baca, R. G. Wilson, S. J. Pearton, and R. A. Stall, *Appl. Phys. Lett.* 68, 2273 (1996).
13. J. C. Zolper, R. G. Wilson, S. J. Pearton, and R. A. Stall, *Appl. Phys. Lett.* 68, 1945 (1996).
14. J. C. Zolper, J. Han, S. B. Van Deusen, R. M. Biefeld, M. H. Crawford, J. Han, T. Suski, J. M. Baranowski, and S. J. Pearton, *Mater. Res. Soc. Symp. Proc.* 483, 609 (1998).
15. H. H. Tan, J. S. Williams, J. Zou, D. J. H. Cockayne, S. J. Pearton, and R. A. Stall, *Appl. Phys. Lett.* 69, 2364 (1996).
16. J. Burm, K. Chu, W. A. Davis, W. J. Schaff, L. F. Eastman, and T. J. Eustis, *Appl. Phys. Lett.* 70, 464 (1997).
17. J. C. Zolper, J. Han, R. M. Biefeld, S. B. Van Deusen, W. R. Wampler, D. J. Reiger, S. J. Pearton, J. S. Williams, H. H. Tan, and R. Stall, *J. Electron. Mater.* 27, 179 (1998).
18. J. C. Zolper, D. J. Reiger, A. G. Baca, S. J. Pearton, J. W. Lee, and R. A. Stall, *Appl. Phys. Lett.* 69, 538 (1996).
19. R. G. Wilson, F. A. Stevie, and C. W. Magee, "Secondary Ion Mass Spectrometry." Wiley, New York, 1989.
20. R. G. Wilson, *J. Electrochem. Soc.* 138, 718 (1991).
21. J. P. Biersack, *Nucl. Instrum. Methods B* 35, 205 (1988).
22. S. J. Pearton, C. R. Abernathy, C. B. Vartuli, J. C. Zolper, C. Yuan, and R. A. Stall, *Appl. Phys. Lett.* 67, 1435 (1995).
23. H. Kobayashi and W. M. Gibson, *Appl. Phys. Lett.* 73, 1406 (1998).
24. J. C. Zolper, R. G. Wilson, S. J. Pearton, and R. A. Stall, *Appl. Phys. Lett.* 68, 1945 (1996).
25. J. C. Zolper, J. Han, R. M. Biefeld, S. B. Van Deusen, W. R. Wampler, D. J. Reiger, S. J. Pearton, J. S. Williams, H. H. Tan, and R. Stall, *J. Electron. Mater.* 27, 179 (1998).
26. J. C. Zolper, D. J. Reiger, A. G. Baca, S. J. Pearton, J. W. Lee, and R. A. Stall, *Appl. Phys. Lett.* 69, 538 (1996).
27. M. Fu, V. Sarvepalli, R. K. Singh, C. R. Abernathy, X. A. Cao, S. J. Pearton, and J. A. Sekhar, *Mater. Res. Soc. Symp. Proc.* 483, 345 (1998).
28. X. A. Cao, C. R. Abernathy, R. K. Singh, S. J. Pearton, M. Fu, V. Sarvepalli, J. A. Sekhar, J. C. Zolper, D. J. Rieger, J. Han, T. J. Drummond, R. J. Shul, and R. G. Wilson, *Appl. Phys. Lett.* 73, 229 (1998).
29. M. S. Feng, J. D. Guo, and G. C. Chi, *Proc. Electrochem. Soc.* 95-21, 43 (1995).
30. C. C. Yi and B. W. Wessels, *Appl. Phys. Lett.* 69, 3206 (1996).
31. H. H. Tan, C. Jagadish, J. S. Williams, H. Zoa, D. J. H. Cockayne, and A. Sikorski, *J. Appl. Phys.* 77, 87 (1995).
32. J. C. Zolper, J. Han, S. B. Van Deusen, R. Biefeld, M. H. Crawford, J. Jun, T. Suski, J. M. Baranowski, and S. J. Pearton, *Mater. Res. Soc. Symp. Proc.* 482, 618 (1998).
33. A. Y. Polyakov, M. Shin, M. Skowronski, R. G. Wilson, D. W. Greve, and S. J. Pearton, *Solid-State Electron.* 41, 703 (1997).
34. J. C. Zolper, D. J. Rieger, A. G. Baca, S. J. Pearton, J. W. Lee, and R. A. Stall, *Appl. Phys. Lett.* 69, 538 (1996).
35. J. C. Zolper, M. H. Crawford, H. H. Tan, J. S. Williams, J. Zhou, D. J. H. Cockayne, S. J. Pearton, and R. F. Karlicek, Jr., *Appl. Phys. Lett.* 70, 2729 (1997).
36. H. H. Tan, J. S. Williams, J. Zou, D. J. H. Cockayne, S. J. Pearton, and C. Yuan, *J. Electrochem. Soc.* 96-11, 142 (1996).
37. H. Kobayashi and W. M. Gibson, *Appl. Phys. Lett.* 74, 2355 (1999).
38. P. Bogulawski, E. L. Briggs, and J. Bernholc, *Phys. Rev. B* 51, 17255 (1995).
39. C. R. Abernathy, J. D. MacKenzie, S. J. Pearton, and W. S. Hobson, *Appl. Phys. Lett.* 66, 1969 (1995).
40. B. Mensching, C. Liu, B. Rauschenbach, K. Kornitzer, and W. Ritter, *Mater. Sci. Eng. B* 50, 105 (1997).
41. J. S. Chan, N. W. Cheung, L. Schloss, E. Jones, W. S. Wong, N. Newman, X. Liu, E. R. Weber, A. Gassman, and M. D. Rubin, *Appl. Phys. Lett.* 68, 2702 (1996).
42. A. Pelzmann, S. Strite, A. Dommann, C. Kirchner, M. Kamp, K. J. Ebeling, and A. Nazzal, *MRS Internet J. Nitride Semicond. Res.* 2, 4 (1997).
43. S. Strite, A. Pelzmann, T. Suski, M. Leszczynski, J. Jun, A. Rockett, M. Kamp, and K. J. Ebeling, *MRS Internet J. Nitride Semicond. Res.* 2, 15 (1997).
44. F. Bernardini, V. Fiorentini, and A. Bosin, *Appl. Phys. Lett.* 70, 2990 (1997).

45. O. Brandt, H. Yang, H. Kostial, and K. Ploog, *Appl. Phys. Lett.* 69, 2707 (1996).

46. H. H. Tan, J. S. Williams, C. Yuan, and S. J. Pearton, *Mater. Res. Soc. Symp. Proc.* 395, 708 (1996).

47. H. H. Tan, J. S. Williams, J. Zou, D. J. H. Cockayne, S. J. Pearton, and R. A. Stall, *Appl. Phys. Lett.* 69, 2364 (1996).

48. C. Liu, B. Mecnsching, M. Zeitler, K. Volz, and B. Rauschenbach, *Phys. Rev. B* 57, 308 (1997).

49. C. Ronning, N. Dalmer, M. Deicher, M. Restle, M. D. Bremser, R. F. Davis, and H. Hofsass, *Mater. Res. Soc. Symp. Proc.* 468, 407 (1997).

50. J. C. Zolper, M. H. Crawford, H. H. Tan, J. S. Williams, J. Zhou, D. J. H. Cockayne, S. J. Pearton, and R. F. Karlicek, Jr., *Appl. Phys. Lett.* 70, 2729 (1997).

51. J. C. Zolper, in "GaN and Related Materials II" (S. J. Pearton, Ed.). Gordon and Breach, New York, 1998.

52. J. I. Pankove, E. A. Miller, and J. E. Berkeyheiser, "Luminescence of Crystals, Molecules and Solutions" (F. Williams, Ed.), p. 426. Plenum, New York, 1973.

53. J. F. Ziegler, J. P. Biersack, and U. Littmark, "The Stopping and Range of Ions in Solids," Vol. 1. Pergamon, New York, 1985.

54. S. Nakamura, M. Senoh, and T. Mukai, *Appl. Phys. Lett.* 62, 2390 (1993).

55. V. Swaminathan and A. T. Macrander, "Materials Aspects of GaAs and InP Based Structures." Prentice Hall, Englewood Cliffs, NJ, 1991.

56. D. E. Davies, *Nucl. Instrum. Methods* 7/8, 387 (1985).

57. K. G. Stevens, *Nucl. Instrum. Methods* 209/210, 589 (1983).

58. N. Nishi, T. Inada, and T. Saito, *Japan. J. Appl. Phys.* 122, 401 (1983).

59. J. D. Oberstar and B. G. Streetman, *Thin Solid Films* 103, 17 (1983).

60. B. Molnar, *Appl. Phys. Lett.* 36, 927 (1980).

61. C. A. Armiento and F. C. Prince, *Appl. Phys. Lett.* 48, 1623 (1986).

62. S. J. Pearton and K. D. Cummings, *J. Appl. Phys.* 58, 1500 (1985).

63. R. T. Blunt, M. S. M. Lamb, and R. Szweda, *Appl. Phys. Lett.* 47, 304 (1985).

64. A. Tamura, R. Uenoyama, K. Nishi, K. Inoue, and T. Onuma, *J. Appl. Phys.* 62, 1102 (1987).

65. T. E. Haynes, W. K. Chu, and S. T. Picraux, *Appl. Phys. Lett.* 50, 1071 (1987).

66. J. M. Woodall, H. Rupprecht, R. J. Chicotka, and G. Wicks, *Appl. Phys. Lett.* 38, 639 (1981).

67. S. Reynolds, D. W. Vook, W. C. Opyd, and J. E. Gibbons, *Appl. Phys. Lett.* 51, 916 (1987).

68. W. H. Haydl, *IEEE Electron. Device Lett.* EDL-5, 78 (1984).

69. S. J. Pearton and R. Caruso, *J. Appl. Phys.* 66, 663 (1989).

70. S. Porowski and I. Grzegory, in "Properties of Group III Nitrides" (J. Edgar, Ed.). INSPEC, London, 1994.

71. G. A. Slack and T. F. McNelly, *J. Cryst. Growth* 34, 263 (1976).

72. J. Pastrnak and L. Roskovcova, *Phys. Status Solidi* 7, 331 (1964).

73. J. A. Van Vechten, *Phys. Rev. B* 7, 1479 (1973).

74. J. Karpinski, J. Jun, and S. Porowski, *J. Cryst. Growth* 66, 1 (1984).

75. J. R. Madar, G. Jacob, J. Hallais, and R. Fruchart, *J. Cryst. Growth* 31, 197 (1975).

76. C. D. Thurmond and R. A. Logan, *J. Electrochem. Soc.* 119, 622 (1972).

77. J. B. MacChesney, P. M. Bridenbaugh, and P. B. O'Connor, *Mater. Res. Bull.* 5, 783 (1970).

78. R. D. Jones and K. Rose, *J. Phys. Chem. Solids* 48, 587 (1987).

79. A. M. Vorobev, G. V. Evseeva, and L. V. Zenkevich, *Russian J. Phys. Chem.* 47, 1616 (1973).

80. T. Matsuoka, H. Tanaka, T. Sasaki, and A. Katsui, *Inst. Phys. Conf. Ser.* 106, 141 (1990).

81. J. C. Zolper, M. Hagerott-Crawford, A. J. Howard, J. Rainer, and S. D. Hersee, *Appl. Phys. Lett.* 68, 200 (1996).

82. M. E. Lin, B. N. Sverdlov, and H. Morkoc, *Appl. Phys. Lett.* 63, 3625 (1993).

83. H. P. Maruska and J. J. Tietjen, *Appl. Phys. Lett.* 15, 327 (1969).

84. R. Groh, G. Gerey, L. Bartha, and J. I. Pankove, *Phys. Status Solidi A* 26, 363 (1974).

85. C. B. Vartuli, S. J. Pearton, C. R. Abernathy, J. D. MacKenzie, E. S. Lambers, and J. C. Zolper, *J. Vac. Sci. Technol. B* 14, 2761 (1996).

86. J. Karpinski, J. Jun, and S. Porowski, *J. Cryst. Growth* 66, 1 (1984).

87. J. C. Zolper, D. J. Rieger, A. G. Baca, S. J. Pearton, J. W. Lee, and R. A. Stall, *Appl. Phys. Lett.* 69, 538 (1996).

88. J. R. Mileham, S. J. Pearton, C. R. Abernathy, J. D. MacKenzie, R. J. Shul, and S. P. Kilcoyne, *Appl. Phys. Lett.* 67, 1119 (1995).

89. T. Matsuoka, H. Tanaka, T. Susaki, and A. Kasui, *Inst. Phys. Conf. Ser.* 106, 141 (1990).

90. F. Roozeboom, in "Rapid Thermal Processing: Science and Technology" (R. B. Fair, Ed.), pp. 349–423. Academic Press, New York, 1993.

91. J. A. Sekhar, S. Penumella, and M. Fu, in "Transient Thermal Processing Techniques in Electronic Materials" (N. M. Ravindra and R. K. Singh, Eds.), pp. 171–175. TMS, Warrendale, PA, 1996.

92. S. J. Pearton, *Internat. J. Mod. Phys. B* 7, 4687 (1993).

93. J. Brown, J. Ramer, K. Zheng, L. F. Lester, S. D. Hersee, and J. C. Zolper, *Mater. Res. Soc. Symp. Proc.* 395, 702 (1996).

94. J. C. Zolper, J. Han, S. B. Van Deusen, R. Biefeld, M. H. Crawford, J. Jun, T. Suski, J. M. Baranowski, and S. J. Pearton, *Mater. Res. Soc. Symp. Proc.* 482, 618 (1998).

95. J. A. Van Vechten, *Phys. Rev. B* 7, 1479 (1973).

96. J. Karpinski, J. Jun, and S. Porowski, *J. Cryst. Growth* 66, 1 (1984).

97. B.-C. Chung and M. Gershenzon, *J. Appl. Phys.* 72, 651 (1992).

98. M. D. Deal and H. G. Robinson, *Appl. Phys. Lett.* 55, 1990 (1989).

99. H. G. Robinson, M. D. Deal, and D. A. Stevenson, *Appl. Phys. Lett.* 58, 2000 (1991).

100. M. D. Deal, C. J. Hu, C. C. Lee, and H. G. Robinson, *Mater. Res. Soc. Symp. Proc.* 300, 365 (1993).

101. H. G. Robinson, M. D. Deal, P. B. Griffin, G. Amaratunga, P. B. Griffin, D. A. Stevenson, and J. D. Plummer, *J. Appl. Phys.* 71, 2615 (1992).

102. S. J. Pearton and C. R. Abernathy, *Appl. Phys. Lett.* 55, 678 (1989).

103. M. S. Feng, J. D. Guo, and G. C. Chi, *Proc. Electrochem. Soc.* 95-21, 43 (1995).

104. F. Ren, M. Hong, S. N. G. Chu, M. A. Marcus, M. J. Schurmann, A. Baca, S. J. Pearton, and C. R. Abernathy, *Appl. Phys. Lett.* 73, 3893 (1998).

105. M. Suzuki, T. Kawakami, T. Arai, S. Kobayashi, Y. Koide, T. Uemura, N. Shibata, and M. Murakami, *Appl. Phys. Lett.* 74, 275 (1999).

106. T. Suski, J. Jun, M. Leszczynski, H. Teisseyre, I. Grzegory, S. Porowski, J. M. Baranowski, A. Rockett, S. Strite, A. Stanert, A. Turos, H. H. Tan, J. S. Williams, and C. Jagadish, *Mater. Res. Soc. Symp. Proc.* 482, 703 (1998).

107. S. Strite, P. W. Epperlein, A. Dommann, A. Rockett, and R. F. Broom, *Mater. Res. Soc. Symp. Proc.* 395, 795 (1996).

108. J. Hong, J. W. Lee, C. B. Vartuli, J. D. MacKenzie, S. M. Donovan, C. R. Abernathy, R. Crockett, S. J. Pearton, J. C. Zolper, and F. Ren, *Solid-State Electron.* 41, 681 (1997).

109. J. Hong, J. W. Lee, C. B. Vartuli, C. R. Abernathy, J. D. MacKenzie, S. M. Donovan, S. J. Pearton, and J. C. Zolper, *J. Vac. Sci. Technol. A* 15, 797 (1997).

110. J. Hong, J. W. Lee, J. D. MacKenzie, S. M. Donovan, C. R. Abernathy, S. J. Pearton, and J. C. Zolper, *Semicond. Sci. Technol.* 12, 1310 (1997).

111. S. J. Pearton, *Mater. Sci. Rep.* 4, 313 (1990).

112. M. Orenstein, N. G. Stoffel, A. C. Von Lehmen, J. P. Harbison, and L. T. Florez, *Appl. Phys. Lett.* 59, 31 (1991).

113. K. L. Lear, R. P. Schneider, K. D. Choquette, S. P. Kilcoyne, J. J. Figiel, and J. C. Zolper, *IEEE Photon. Tech. Lett.* 6, 1053 (1994).

114. J. C. Zolper, A. G. Baca, and S. A. Chalmers, *Appl. Phys. Lett.* 62, 2536 (1993).

115. S. J. Pearton, M. P. Iannuzzi, C. L. Reynolds, Jr., and L. Peticolas, *Appl. Phys. Lett.* 52, 395 (1988).

116. J. C. Zolper, M. E. Sherwin, A. G. Baca, and R. P. Schneider, Jr., *J. Electron. Mater.* 24, 21 (1995).

117. D. Haase, M. Schmid, W. Kurner, A. Dornen, V. Harle, F. Scholz, M. Burkard, and H. Schweizer, *Appl. Phys. Lett.* 69, 2525 (1996).

118. A. Rebey, T. Boufaden, and B. El Jani, *J. Cryst. Growth* 203, 12 (1999).

119. C. J. Sun, P. Kung, A. Saxler, H. Ohsato, E. Bigan, M. Razeghi, and D. K. Gaskill, *J. Appl. Phys.* 76, 236 (1994).

120. H. Tanaka and A. Nakadaira, *J. Cryst. Growth* 189/190, 730 (1998).

121. J. Han, T. B. Ng, R. M. Biefeld, M. H. Crawford, and D. M. Follstaedt, *Appl. Phys. Lett.* 71, 3114 (1997).

122. Y. Kobayashi and N. Kobayashi, *J. Cryst. Growth* 189/190, 301 (1998).

123. S. J. Pearton, C. R. Abernathy, P. W. Wisk, W. S. Hobson, and F. Ren, *Appl. Phys. Lett.* 63, 1143 (1993).

124. J. C. Zolper, S. J. Pearton, C. R. Abernathy, and C. B. Vartuli, *Appl. Phys. Lett.* 66, 3042 (1995).

125. S. J. Pearton, C. R. Abernathy, M. B. Panish, R. A. Hamm, and L. M. Lunardi, *J. Appl. Phys.* 66, 656 (1989).

126. J. C. Zolper, S. J. Pearton, C. R. Abernathy, and C. B. Vartuli, *Mater. Res. Soc. Symp. Proc.* 378, 408 (1995).

127. G. Harrington, Y. Hsin, Q. Z. Liu, P. M. Asbeck, S. S. Lau, M. A. Khan, J. W. Yang, and Q. Chen, *Electron. Lett.* 34, 193 (1998).

Chapter 8

PLASMA ETCHING OF GaN AND RELATED MATERIALS

S. J. Pearton

Department of Materials Science and Engineering, University of Florida, Gainesville, Florida, USA

R. J. Shul

Sandia National Laboratories, Albuquerque, New Mexico, USA

Contents

1. INTRODUCTION

GaN and related alloys are being used in applications such as fabrication of blue/green/UV emitters (light-emitting diodes and lasers) and high-temperature, high-power electronic devices [1–15]. The emitter technology is relatively mature: light-emitting diodes have been commercially available since 1994, and blue laser diodes are also available. Electronic devices such as heterostructure field effect transistors (FETs), heterojunction bipolar transistors (HBTs), metal oxide semiconductor field effect transistors (MOSFETs), and diode rectifiers have all been realized in the AlGaInN system, with very promising high-temperature ($>300°C$) and high-voltage performance. The applications for the emitter devices lies in full color displays, optical data storage, white-light sources, and covert communications, and electronic devices are suited to high-power switches and microwave power generation.

Handbook of Thin Film Materials, edited by H.S. Nalwa
Volume 1: Deposition and Processing of Thin Films
Copyright © 2002 by Academic Press

ISBN 0-12-512909-2/$35.00

Because of limited wet chemical etch results for the group III nitrides, a significant amount of effort has been devoted to the development of dry etch processing [16–20]. Dry etch development was initially focused on mesa structures in which high etch rates, anisotropic profiles, smooth sidewalls, and equirate etching of dissimilar materials were required. For example, commercially available LEDs and laser facets for GaN-based laser diodes were patterned with the use of reactive ion etch (RIE). However, as interest in high-power, high-temperature electronics [21–24] increased, etch requirements expanded to include smooth surface morphology, low plasma-induced damage, and selective etching of one layer rather than. Dry etch development is further complicated by the inert chemical nature and strong bond energies of the group III nitrides as compared with other compound semiconductors. GaN has a bond energy of 8.92 eV/atom, InN a bond energy of 7.72 eV/atom, and AlN a bond energy of 11.52 eV/atom [25].

2. PLASMA REACTORS

Dry plasma etching has become the dominant patterning technique for the group III nitrides, because of the shortcomings of wet chemical etching. Plasma etching proceeds by physical sputtering, chemical reaction, or a combination of the two, often referred to as ion-assisted plasma etching. Physical sputtering is dominated by the acceleration of energetic ions formed in the plasma to the substrate surface at relatively high energies, typically >200 eV. Because of the transfer of energy and momentum to the substrate, material is ejected from the surface. This sputter mechanism tends to yield anisotropic profiles; however, it can result in significant damage, rough surface morphology, trenching, poor selectivity, and nonstoichiometric surfaces, thus minimizing device performance. Pearton and co-workers measured sputter rates for GaN, InN, AlN, and InGaN as a function of Ar^+ ion energy [26]. The sputter rates increased with ion energy but were quite slow, <600 Å/min, because of the high bond energies of the group III—N bond.

Chemically dominated etch mechanisms rely on the formation of reactive species in the plasma that absorb to the surface, form volatile etch products, and then desorb from the surface. Because ion energies are relatively low, etch rates in the vertical and lateral directions are often similar, thus resulting in isotropic etch profiles and loss of critical dimensions. However, because of the low ion energies used, plasma-induced damage is minimized. Alternatively, ion-assisted plasma etching relies on both chemical reactions and physical sputtering to yield anisotropic profiles at reasonably high etch rates. Provided the chemical and physical components of the etch mechanism are balanced, high-resolution features can be realized with minimal damage, and optimum device performance can be obtained.

2.1. Reactive Ion Etching

RIE utilizes both the chemical and physical components of an etch mechanism to achieve anisotropic profiles, fast etch rates, and dimensional control. RIE plasma is typically generated by applying a radio frequency (rf) power of 13.56 MHz between two parallel electrodes in a reactive gas (see Fig. 1a). The substrate is placed on the powered electrode, where a potential is induced and ion energies, defined as they cross the plasma sheath, are typically a few hundred electron volts. RIE is operated at low pressures, ranging from a few milliTorr up to 200 mTorr, which promotes anisotropic etching due to increased mean free paths and reduced collisional scattering of ions during acceleration in the sheath.

Adesida et al. were the first to report RIE of GaN in $SiCl_4$-based plasmas [19]. Etch rates increased with increasing dc bias and were >500 Å/min at −400 V. Lin et al. reported similar results for GaN in BCl_3 and $SiCl_4$ plasmas with etch rates of 1050 Å/min in BCl_3 at 150-W cathode (area 250 in^2) rf power [27]. Additional RIE results have been reported for HBr- [28], CHF_3-, and CCl_2F_2-based [29] plasmas with typical etch rates of <600 Å/min. The best RIE results for the group III nitrides have been obtained in chorine-based plasmas under high ion energy conditions, where the III—N bond breaking and the sputter desorption of etch products from the surface are most efficient. Under these conditions, plasma damage can occur and degrade both electrical and optical device performance. Lowering the ion energy or increasing the chemical activity in the plasma to minimize the damage often results in slower etch rates or less anisotropic profiles, which significantly limits the critical dimension. Therefore, it is necessary to pursue alternative etch platforms that combine high-quality etch characteristics with low damage.

2.2. High-Density Plasmas

The use of high-density plasma etch systems, including electron cyclotron resonance (ECR), inductively coupled plasma (ICP), and magnetron RIE (MRIE), has resulted in improved etch characteristics for the group III nitrides as compared with RIE. This observation is attributed to plasma densities that are 2 to 4 orders of magnitude higher than RIE, thus improving the III—N bond breaking efficiency and the sputter desorption of etch products formed on the surface. Furthermore, because ion energy and ion density can be more effectively decoupled as compared with RIE, plasma-induced damage is more readily controlled. Figure 2b shows a schematic diagram of a typical low-profile ECR etch system. High-density ECR plasmas are formed at low pressures with low plasma potentials and ion energies due to magnetic confinement of electrons in the source region. The sample is located downstream from the source to minimize exposure to the plasma and to reduce the physical component of the etch mechanism. Anisotropic etching can be achieved by superimposing an rf bias (13.56 MHz) on the sample and operating at low pressure (<5 mTorr) to minimize ion scattering and lateral etching. However, as the rf biasing is increased, the potential for damage to the surface increases. Figure 2 shows a schematic of the plasma parameters and sample position in a typical high-density plasma reactor.

a) RIE

b) ECR

c) ICP

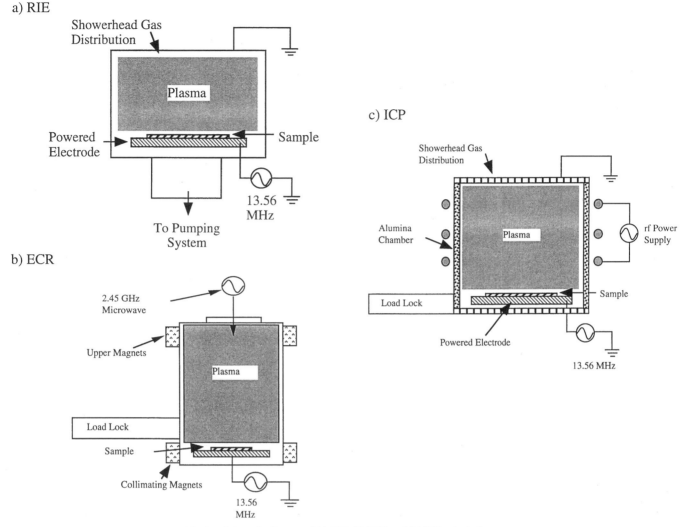

Fig. 1.   Schematic diagram of (a) RIE, (b) ECR, and (c) ICP etch platforms.

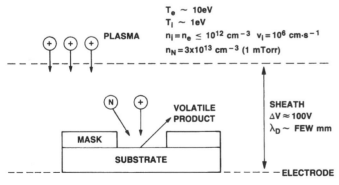

Fig. 2.   Schematic diagram of the high-density plasma etching process.

Pearton and co-workers were the first to report ECR etching of group III nitride films [30, 31]. Etch rates for GaN, InN, and AlN increased as either the ion energy (dc bias) or ion flux (ECR source power) increased. Etch rates of 1100 Å/min for AlN and 700 Å/min for GaN at a dc bias of −150 V in a Cl$_2$/H$_2$ plasma and 350 Å/min for InN in a CH$_4$/H$_2$/Ar plasma at a dc bias of −250 V were reported. The etched features were anisotropic, and the surface remained stoichiometric over a wide range of plasma conditions. GaN ECR etch data have been reported by several authors, including etch rates as high as 1.3 μm/min [32–43].

ICP offers another high-density plasma etch platform to pattern group III nitrides. ICP plasmas are formed in a dielectric vessel encircled by an inductive coil to which rf power is applied (see Fig. 1c). The alternating electric field between the coils induces a strong alternating magnetic field, trapping electrons in the center of the chamber and generating a high-density plasma. Because ion energy and plasma density can be effectively decoupled, uniform density and energy distributions are transferred to the sample while ion and electron energies are kept low. Thus, ICP etching can produce low damage while maintaining fast etch rates. Anisotropy is achieved by superimposition of rf bias on the sample. ICP etching is generally believed to have several advantages over ECR, including easier scale-up for production, improved plasma uniformity over a wider area, and lower cost of operation. The first ICP etch results for GaN were reported in a $Cl_2/H_2/Ar$ ICP-generated plasma with etch rates as high as ~6875 Å/min [44, 45]. Etch rates increased with increasing dc bias, and etch profiles were highly anisotropic, with smooth etch morphologies over a wide range of plasma conditions. GaN etching has also been reported for a variety of halogen- and methane-based ICP plasmas [46–55].

MRIE is another high-density etch platform that is comparable to RIE. In MRIE, a magnetic field is used to confine electrons close to the sample and minimize electron loss to the wall [56–58]. Under these conditions, ionization efficiencies are increased and high plasma densities and fast etch rates are achieved at much lower dc biases (less damage), as compared with RIE. GaN etch rates of ~3500 Å/min were reported in BCl_3-based plasmas at dc biases less than −100 V [60]. The etch was fairly smooth and anisotropic.

2.3. Chemically Assisted Ion Beam Etching

Chemically assisted ion beam etching (CAIBE) and reactive ion beam etching (RIBE) have also been used to etch group III nitride films [20, 59–63]. In these processes, ions are generated in a high-density plasma source and accelerated by one or more grids to the substrate. In CAIBE, reactive gases are added to the plasma downstream of the acceleration grids, thus enhancing the chemical component of the etch mechanism, whereas in RIBE, reactive gases are introduced in the ion source. Both etch platforms rely on relatively energetic ions (200–2000 eV) and low chamber pressures (<5 mTorr) to achieve anisotropic etch profiles. However, with such high ion energies, the potential for plasma-induced damage exists. Adesida and co-workers reported CAIBE etch rates for GaN as high as 2100 Å/min with 500-eV Ar^+ ions and Cl_2 or HCl ambients [19, 59–61]. Rates increased with beam current, reactive gas flow rate, and substrate temperature. Anisotropic profiles with smooth etch morphologies were observed.

2.4. Reactive Ion Beam Etching

The RIBE removal rates for GaN, AlN, and InN are shown in Figure 3 as a function of Cl_2 percentage in Cl_2/Ar beams at 400 eV and a current of 100 mA. The trend in removal rates basically follows the bond energies of these materials. At a fixed Cl_2/Ar ratio, the rates increased with beam energy, as shown in Figure 4. At very high voltages, one would expect the rates to saturate or even decrease, because of ion-assisted desorption of the reactive chlorine from the surface of the nitride sample before it can react to form the chloride etch products.

There was relatively little effect of either beam current (Fig. 5) or sample temperature (Fig. 6) on the RIBE removal rates of the nitride. The etch profiles are anisotropic, with light trenching at the base of the features. This is generally ascribed to ion deflection from the sidewalls, causing an increased ion flux at the base of the etched features.

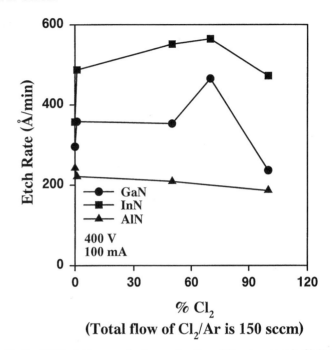

Fig. 3. RIBE nitride removal rates as a function of Cl_2 percentage in Cl_2/Ar beams.

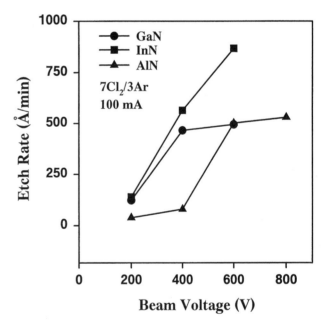

Fig. 4. RIBE nitride removal rates as a function of beam voltage.

2.5. Low-Energy Electron-Enhanced Etching

Low-energy electron-enhanced etching (LE4) of GaN has been reported by Gilllis and co-workers [17, 64–66]. LE4 is an etch technique that depends on the interaction of low-energy electrons (<15 eV) and reactive species at the substrate surface. The etch process results in minimal surface damage, because there is negligible momentum transferred from the electrons to the substrate. GaN etch rates of ~500 Å/min in a H_2-based LE4 plasma and ~2500 Å/min in a pure Cl_2 LE4 plasma have been reported [17, 66]. GaN has also been etched by photoassisted

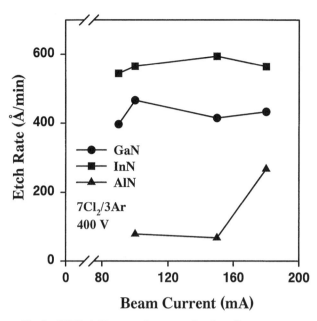

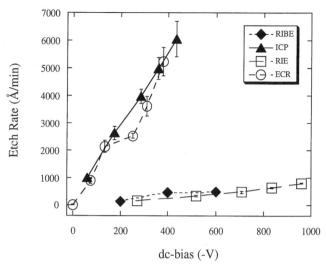

Fig. 7. GaN etch rates in RIE, ECR, ICP, and RIBE Cl$_2$-based plasmas as a function of dc bias.

Fig. 5. RIBE nitride removal rates as a function of beam current.

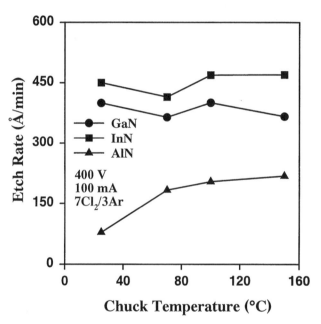

Fig. 6. RIBE nitride removal rates as a function of chuck temperature.

dry etch processes where the substrate is exposed to a reactive gas and ultraviolet laser radiation simultaneously. Vibrational and electronic excitations lead to improved bond breaking and desorption of reactant products. Leonard and Bedair obtained GaN etch rates of <80 Å/min in HCl with the use of a 193-nm ArF excimer laser [67].

GaN etch rates are compared in Figure 7 for RIE, ECR, and ICP Cl$_2$/H$_2$/CH$_4$/Ar plasmas as well as a RIBE Cl$_2$/Ar plasma. CH$_4$ and H$_2$ were removed from the plasma chemistry to eliminate polymer deposition in the RIBE chamber. Etch rates increased as a function of dc bias regardless of etch technique. GaN etch rates obtained in the ICP and ECR plasmas were

much faster than those obtained in RIE and RIBE. This was attributed to higher plasma densities (1–4 orders of magnitude higher), which resulted in more efficient breaking of the III—N bond and sputter desorption of the etch products. Slower rates observed in the RIBE may also be due to lower operational pressures (0.3 mTorr compared with 2 mTorr for the ICP and ECR) and/or lower ion and reactive neutral flux at the GaN surface due to high source-to-sample separation.

3. PLASMA CHEMISTRIES

3.1. Cl$_2$-Based

Etch characteristics are often dependent upon plasma parameters, including pressure, ion energy, and plasma density. As a function of pressure, plasma conditions, including the mean free path and the collisional frequency, can change, resulting in changes in both ion energy and plasma density. GaN etch rates are shown as a function of pressure for an ICP-generated BCl$_3$/Cl$_2$ plasma in Figure 8. Etch rates increased as the pressure was increased from 1 to 2 mTorr and then decreased at higher pressures. The initial increase in etch rate suggested a reactant limited regime at low pressure; however, at higher pressures the etch rates decreased because of lower plasma densities (ions or radical neutrals), redeposition, or polymer formation on the substrate surface. At pressures less than 10 mTorr, GaN etches were anisotropic and smooth, whereas at pressures greater than 10 mTorr the etch profile was undercut and poorly defined, because of a lower mean free path, collisional scattering of the ions, and increased lateral etching of the GaN.

GaN etch rates are plotted as a function of dc bias (which correlates to ion energy) for an ICP-generated BCl$_3$/Cl$_2$ plasma in Figure 9. The GaN etch rates increased monotonically as the dc bias or ion energy increased. Etch rates increased because of improved sputter desorption of etch products from the surface as well as more efficient breaking of the Ga—N bonds. Etch rates have also been observed to decrease under high ion bombardment energies due to sputter desorption of reactive species from the surface before the reactions occur. This is often referred to as an adsorption-limited etch regime. In Figure 10, scanning electron microscopy (SEM) micrographs are shown for -50 (Fig. 10a), -150 (Fig. 10b), and -300 V (Fig. 10c) dc bias. The etch profile became more anisotropic as the dc bias increased from -50 to -150 V dc bias because of the perpendicular path of the ions relative to the substrate surface, which maintained straight wall profiles. However, as the dc bias was increased to -300 V, a tiered etch profile with vertical striations in the sidewall was observed, because of erosion of the resist mask edge. The GaN may become rougher under these

conditions because of mask redeposition and preferential loss of N_2.

In Figure 11, GaN etch rates are shown as a function of ICP source power while the dc bias is held constant at -250 V. GaN etch rates increased as the ICP source power increased, because of higher concentrations of reactive species, which increases the chemical component of the etch mechanism and/or higher ion flux, which increases the bond breaking and sputter desorption efficiency of the etch. Etch rates have also been observed to stabilize or decrease under high plasma flux conditions due either to saturation of reactive species at the surface or sputter desorption of reactive species from the surface before the reactions occur. The etch profile was anisotropic and smooth up to

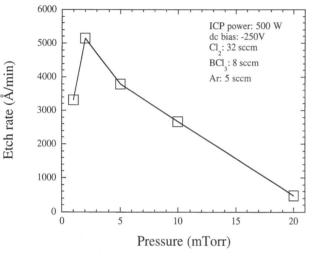

Fig. 8.   GaN etch rates as a function of pressure in an ICP-generated $BCl_3/Cl_2/Ar$ plasma at 32 sccm Cl_2, 8 sccm BCl_5, 5 sccm Ar, 500 W IPC source power, -150 V dc bias, and $10°C$ electrode temperature.

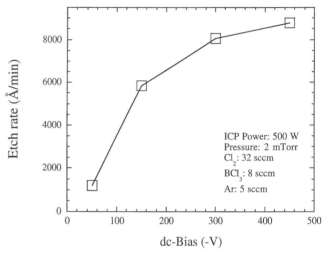

Fig. 9.   GaN etch rates as a function of dc bias in an ICP-generated $BCl_3/Cl_2/Ar$ plasma at 32 sccm Cl_2, 8 sccm BCl_3, 5 sccm Ar, 500 W ICP source power, 2 mTorr pressure, and $10°C$ electrode temperature.

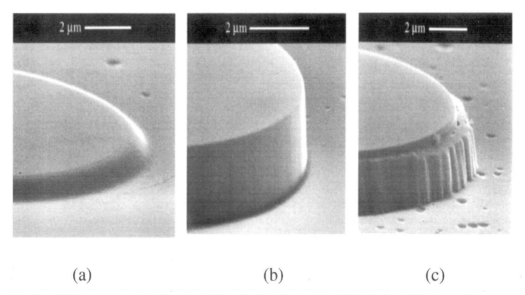

(a)   (b)   (c)

Fig. 10.   SEM micrographs for GaN etched at (a) -50, (b) -150, and (c) -300 V dc bias. ICP etch conditions were 32 sccm Cl_2, 8 sccm BCl_3, 5 sccm Ar, 500 W ICP source power, 2 mTorr pressure, and $10°C$ electrode temperature.

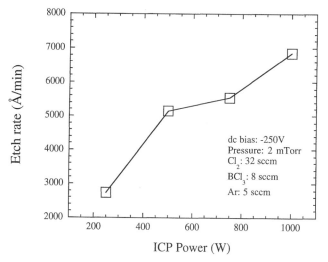

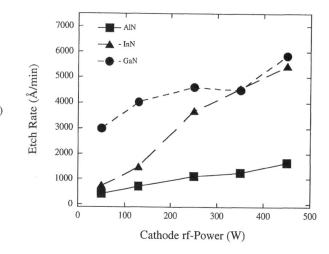

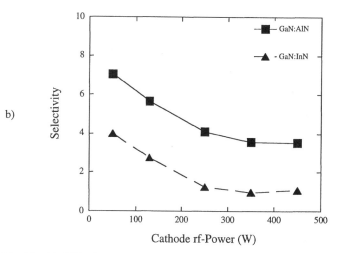

Fig. 11. GaN etch rates as a function of ICP source power in an ICP-generated $BCl_3/Cl_2/Ar$ plasma at 32 sccm Cl_2, 8 sccm BCl_3, 5 sccm Ar, −250 V dc bias, 2 mTorr pressure, and 10°C electrode temperature.

Fig. 12. (a) GaN, InN, and AlN etch rates and (b) GaN : AlN and GaN : InN etch selectivities as a function of dc bias in a Cl_2/Ar ICP plasma. Plasma conditions were 25 sccm Cl_2, 5 sccm Ar, 2 mTorr chamber pressure, 500 W ICP source power, and 25°C cathode temperature.

an ICP power of 1000 W, where the feature dimensions were lost and sidewall morphology was rough because of erosion of the mask edge under high plasma flux conditions.

In addition to etch rates, etch selectivity or the ability to etch one film at higher rates than another can be very important in device fabrication. For example, optimization of etch selectivity is critical to the control of threshold voltage uniformity for high electron mobility transistors (HEMTs), to accurately stop on either the emitter or collector regions for metal contacts for HBTs, and for low-resistivity n-ohmic contacts on InN layers. Several studies have recently reported etch selectivity for the group III nitrides [51, 54, 68–72]. For example, Figure 12 shows GaN, InN, and AlN etch rates and etch selectivities as a function of cathode rf power in an ICP-generated Cl_2/Ar plasma. Etch rates for all three films increased with increasing cathode rf power or dc bias because of improved breaking of the III−N bonds and more efficient sputter desorption of the etch products. Increasing InN etch rates were especially significant because $InCl_3$, the primary In etch product in a Cl-based plasma, has a relatively low volatility. However, under high dc bias conditions, desorption of the $InCl_3$ etch products occurred before coverage of the etch surface [69]. The GaN : InN and GaN : AlN etch selectivities were less than 8 : 1 and decreased as the cathode rf power or ion energy increased. Smith and co-workers reported similar results for a Cl_2/Ar ICP plasma where GaN : AlN and GaN : AlGaN selectivities decreased as the dc bias increased [54]. At a dc bias of −20 V, etch selectivities of ∼ 39 : 1 and ∼ 10 : 1 were reported for GaN : AlN and GaN : AlGaN, respectively.

Temperature-dependent etching of the group III nitrides has been reported for ECR and ICP etch systems [33, 43, 55]. Etch rates are often influenced by the substrate temperature, which can affect the desorption rate of the etch process, the gas–surface reaction kinetics, and the surface mobility of reactants.

The substrate temperature can be controlled and maintained during the etch process by a variety of clamping and backside heating or cooling procedures. GaN and InN etch rates are shown in Figure 13 as a function of temperature in $Cl_2/H_2/Ar$ ICP plasma. GaN etch rates were much faster than InN rates because of the higher volatility of the $GaCl_3$ etch products as compared with $InCl_3$ and showed little dependence on temperature. However, the InN etch rates showed a considerable temperature dependence, increasing at 150°C because of the higher volatilities of the $InCl_3$ etch products at higher substrate temperatures.

Several different plasma chemistries have been used to etch the group III nitrides. As established above, etch rates and profiles can be strongly affected by the volatility of the etch products formed. Table I shows the boiling points of possible etch products for the group III nitrides exposed to halogen- and hydrocarbon-based plasmas. For halogen-based plasmas, etch rates are often limited by the volatility of the group III halogen etch product. For Ga- and Al-containing films, chlorine-based plasmas typically yield fast rates with anisotropic, smooth etch

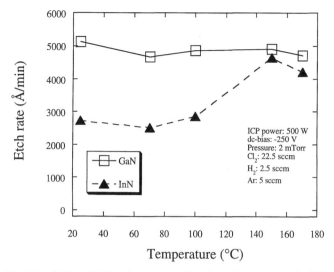

Fig. 13. GaN and InN etch rates as a function of temperature for ICP-generated $Cl_2/H_2/Ar$ plasmas. ICP etch conditions were 22.5 sccm Cl_2, 2.5 sccm H_2, 5 sccm Ar, 500 W ICP source power, -250 V dc bias, and 2 mTorr pressure.

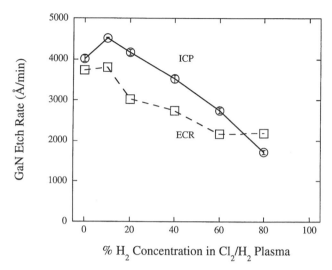

Fig. 14. GaN etch rates in an ICP and ECR Cl_2H_2/Ar plasma as a function of H_2 percentage.

Table I. Boiling Points for Possible Etch Products of Group III Nitride Films Etched in Halogen- or CH_4/H_2-Based Plasmas

Etch products	Boiling points (°C)
$AlCl_3$	183
AlF_3	NA
AlI_3	360
$AlBr_3$	263
$(CH_3)_3Al$	126
$GaCl_3$	201
GaF_3	1000
GaI_3	Sublimes 345
$GaBr_3$	279
$(CH_3)_3Ga$	55.7
$InCl_3$	600
InF_3	>1200
InI_3	NA
$InBr_3$	Sublimes
$(CH_3)_3In$	134
NCl_3	<71
NF_3	-129
NBr_3	NA
NI_3	Explodes
NH_3	-33
N_2	-196
$(CH_3)_3N$	-33

NA = not available.

profiles. CH_4/H_2-based plasma chemistries have also yielded smooth, anisotropic profiles for Ga-containing films, however, at much slower rates. Based only on a comparison of etch product volatility, slower etch rates in CH_4-based plasmas are unexpected because the $(CH_3)_3Ga$ etch product has a much

lower boiling point than $GaCl_3$. This observation demonstrates the complexity of the etch process, where redeposition, polymer formation, and gas-phase kinetics can influence the results. As shown above, etch rates for In-containing films obtained in room temperature chlorine-based plasmas tend to be slow, with rough surface morphology and overcut profiles due to the low volatility of the $InCl_3$ and preferential loss of the group V etch products. However, at elevated temperatures (>130°C), the $InCl_3$ volatility increases and the etch rates and surface morphology improve [33, 43, 55, 73–75]. Significantly better room temperature etch results are obtained in CH_4/H_2-based plasmas because of the formation of more volatile $(CH_3)_3In$ etch products [31, 76].

The source of reactive plasma species and the addition of secondary gases to the plasma can vary etch rates, anisotropy, selectivity, and morphology. The fragmentation pattern and gas-phase kinetics associated with the source gas can have a significant effect on the concentration of reactive neutrals and ions generated in the plasma, thus affecting the etch characteristics. Secondary gas additions and variations in gas ratios can change the chemical : physical ratio of the etch mechanism. The effect of Ar, SF_6, N_2, and H_2 additions to Cl_2- and BCl_3-based ICP and ECR plasmas for GaN etching has been reported [77]. In general, GaN etch rates were faster in Cl_2-based plasmas as compared with BCl_3, because of the higher concentration of reactive Cl. The addition of H_2, N_2, or SF_6 to either Cl_2- or BCl_3-based plasmas changed the relative concentration of reactive Cl in the plasma, which directly correlated with the GaN etch rate. For example, in Figure 14, GaN etch rates are shown as a function of the percentage H_2 concentration for ECR- and ICP-generated $Cl_2/H_2/Ar$ plasmas. GaN etch rates in the ECR and ICP increased slightly as H_2 was initially added to the Cl_2/Ar plasma, indicating a reactant-limited regime. Monitoring of the ECR plasma with a quadruple mass spectrometry (QMS) showed that the Cl concentration (indicated by $m/e = 35$) remained relatively constant at 10% H_2. As the H_2 concentration was increased above 10%, the Cl concentration decreased and a

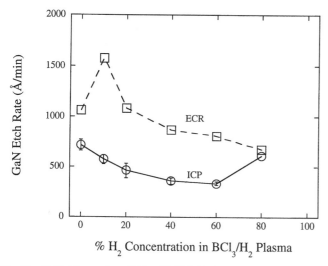

Fig. 15.   GaN etch rates in an ICP and ECR BCl$_3$/H$_2$/Ar plasma as a function of H$_2$ percentage.

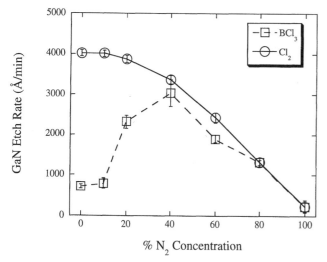

Fig. 16.   GaN etch rates as a function of N$_2$ percentage for ICP-generated Cl$_2$- and BCl$_3$-based plasmas.

peak corresponding to HCl increased. GaN etch rates decreased at H$_2$ concentrations greater than 10% in both the ECR and ICP, presumably because of the consumption of reactive Cl by hydrogen. In Figure 15, H$_2$ was added to BCl$_3$-based ECR and ICP plasmas. In the ECR plasma, the GaN etch rate increased at 10% H$_2$, corresponding to an increase in the reactive Cl concentration as observed by quadrupole mass spectrometry (QMS). As the H$_2$ concentration was increased further, GaN etch rates decreased, the Cl concentration decreased, and the HCl concentration increased, presumably because of the consumption of reactive Cl by hydrogen. In the ICP reactor, GaN etch rates were slow and decreased as hydrogen was added to the plasma up to 80% H$_2$, where a slight increase was observed.

Another example of plasma chemistry-dependent etching of GaN is shown in Figure 16 for Cl$_2$/N$_2$/Ar and BCl$_3$/N$_2$/Ar ICP-generated plasmas. In the Cl$_2$-based plasma, GaN etch rates decreased as the percentage N$_2$ increased, presumably because of a reduction in reactive Cl. In the BCl-based plasma GaN etch rates increased up to 40% N$_2$ and then decreased at higher N$_2$

concentration. This observation has also been reported for ECR and ICP etching of GaAs, GaP, and In-containing films [78–80]. Ren and co-workers first observed maximum etch rates for In-containing films (InGaN and InGaP) in an ECR discharge at a gas ratio of 75 : 25 for BCl$_3$/N$_2$ [78]. Using optical emission spectroscopy (OES), Ren reported maximum emission intensity for atomic and molecular Cl at 75% BCl$_3$ as well as a decrease in the BCl$_3$ intensity and the appearance of a BN emission line. The authors speculated that N$_2$ enhanced the dissociation of BCl$_3$, resulting in higher concentrations of reactive Cl and Cl ions and thus higher etch rates. Furthermore, the observation of BN emission suggested that less B was available to recombine with reactive Cl. This explanation may also be applied to the peak GaN etch rates observed at 40% N$_2$ in the ICP BCl$_3$/N$_2$/Ar plasmas. However, OES of the BCl$_3$/N$_2$/Ar ICP discharge did not reveal higher concentrations of reactive Cl or a BN peak emission. In Figure 17, OES spectra are shown for 100% BCl$_3$ (Fig. 17a), 75% BCl$_3$/25% N$_2$ (Fig. 17b), 25% BCl$_3$/75% N$_2$ (Fig. 17c), and 100% N$_2$ ICP plasmas (Fig. 17d). As N$_2$ was added to the BCl$_3$ plasma, the BCl$_3$ emission (2710 Å) and Cl emission (5443 and 5560 Å) decreased, whereas the BN emission (3856 Å) was not obvious.

BCl$_3$/Cl$_2$ plasmas have shown encouraging results in the etching of GaN films [46, 53]. The addition of BCl$_3$ to a Cl$_2$ plasma can improve sputter desorption due to higher mass ions and reduce surface oxidation by gettering H$_2$O from the chamber. In Figure 18, GaN etch rates are shown as a function of percentage Cl$_2$ in a BCl$_3$/Cl$_2$/Ar ICP plasma. As the percentage of Cl$_2$ increased, GaN etch rates increased up to 80% because of higher concentrations of reactive Cl. OES showed that the Cl emission intensity increased and the BCl emission intensity decreased as the percentage of Cl$_2$ increased. Slower GaN etch rates in a pure Cl$_2$ plasma were attributed to less efficient sputter desorption of etch products in the absence of BCl$_3$. Similar results were reported by Lee et al. [46]. The fastest GaN etch rates were observed at 10% BCl$_3$, where the ion current density and Cl radical density were the greatest as measured by OES and a Langmuir probe.

In general, GaN : AlN and GaN : InN etch selectivities are < 10 : 1 as a function of plasma chemistry for Cl$_2$- or BCl$_3$-based plasmas. GaN : AlN and GaN : InN etch selectivities were higher for Cl$_2$-based ICP plasmas as compared with BCl$_3$-based ICP plasma because of the higher concentration of reactive Cl produced in the Cl$_2$-based plasmas, thus resulting in faster GaN etch rates [72]. Alternatively, InN and AlN etch rates showed much less dependence on plasma chemistry and were fairly comparable in Cl$_2$- and BCl$_3$-based plasmas. An example of etch selectivity dependence on plasma chemistry is shown in Figure 19. GaN, AlN, and InN etch rates and etch selectivities are plotted as a function of percentage SF$_6$ for an ICP Cl$_2$/SF$_6$/Ar plasma. GaN and InN etch rates decreased as SF$_6$ was added to the plasma, because of the consumption of Cl by S and therefore lower concentrations of reactive Cl. The AlN etch rates increased with the addition of SF$_6$ and reached a maximum at 20% SF$_6$. As SF$_6$ was added to the Cl$_2$ plasma, slower AlN etch rates were expected because of the formation of low-

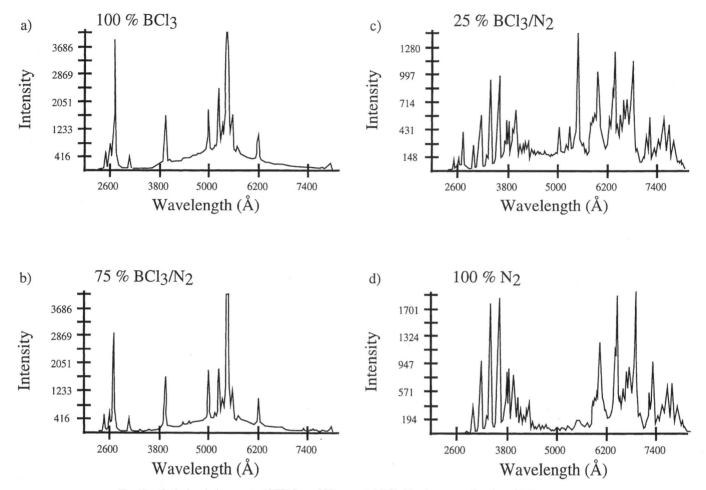

Fig. 17. Optical emission spectra (OES) for an ICP-generated BCl_3/N_2 plasma as a function of BCl_3 percentage.

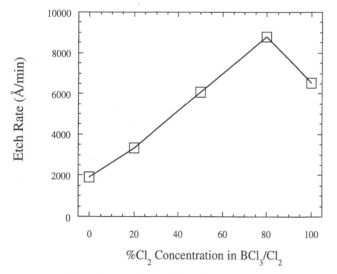

Fig. 18. GaN etch rates in ICP BCl_3/Cl_2 plasmas as a function of Cl_2.

volatility AlF_3 etch products. However, because of the high ion flux in the ICP, the sputter desorption of the AlF_3 may occur before passivation of the surface [69]. Therefore, the GaN : AlN selectivity decreased rapidly from $\sim 6:1$ to $<1:1$ with the ad-

dition of SF_6. The GaN : InN selectivity reached a maximum of $4:1$ at 20% SF_6.

The simple Cl_2/Ar chemistry works very well for most device fabrication processes, providing controllable etch rates (Fig. 20). Even at biases less than 90 V, the GaN etch rate is still typically ~ 1000 Å/min.

3.2. I_2- and Br_2-Based

Other halogen-containing plasmas including ICl/Ar, IBr/Ar, BBr_3/Ar, and BI_3/Ar have been used to etch GaN, with promising results [39, 68, 81–83]. Vartuli and co-workers reported GaN, InN, AlN, InN, InAlN, and InGaN etch rates and selectivities for CRS ICl/Ar and IBr/Ar plasmas [84]. In general, etch rates increased for all films as a function of dc bias because of improved III−N bond breaking and sputter desorption of etch products from the surface. GaN etch rates greater than 1.3 μm/min were obtained in the ICl/Ar plasma at a rf power of 250 W (bias of −200 V), whereas GaN etch rates were typically <4000 Å/min in IBr/Ar. Cho et al. reported typical GaN etch rates of <2000 Å/min in ICP-generated BI_3/Ar and BBR_3/Ar plasmas [83]. ICl/Ar and IBr/Ar ECR plasmas yielded GaN : InN, GaN : AlN, GaN : InGaN, and GaN : InAlN

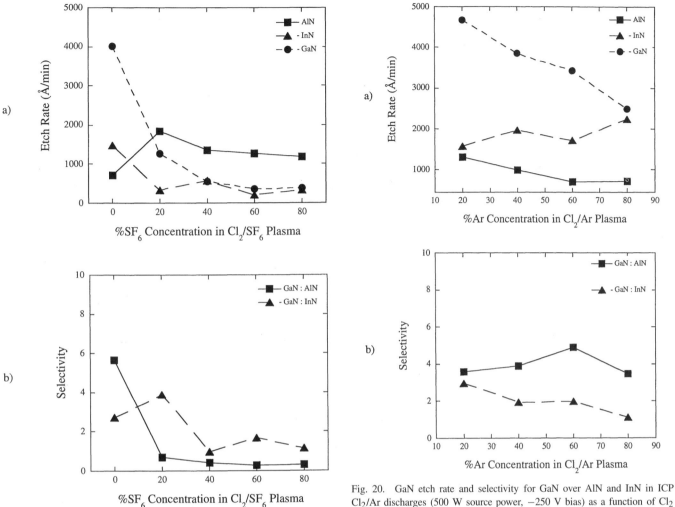

Fig. 19.   (a) GaN, InN, and AlN etch rates and (b) GaN : AlN and GaN : InN etch selectivities as a function of SF_6 percentage in a $Cl_2/SF_6/Ar$ ICP plasma.

Fig. 20.   GaN etch rate and selectivity for GaN over AlN and InN in ICP Cl_2/Ar discharges (500 W source power, −250 V bias) as a function of Cl_2 percentage.

selectivities less than 6 : 1; however, etch selectivities greater than 100 : 1 were obtained for InN : GaN and InN : AlN in BI_3/Ar plasmas [68, 69, 81–83]. Fast etch rates obtained for InN were attributed to the high volatility of the InI_3 etch products as compared with the GaI_3 and AlI_3 etch products, which can form passivation layers on the surface. Maximum selectivities of $\sim 100 : 1$ for InN : AlN and $\sim 7.5 : 1$ for InN : GaN were reported in the BBr_3/Ar plasma [83]. InI_x products have higher volatility than corresponding $InGl_x$ species, making iodine an attractive etchant for InGaN alloys. The interhalogen compounds are weakly bonded and therefore should easily break apart under plasma excitation to form reactive iodine, bromine, and chlorine.

Figure 21 shows etch rates for the binary nitrides and selectivities for InN over both GaN and AlN as a function of the boron halide percentage by flow in the gas load. The dc chuck self-bias decreases as the BI_3 content increases, suggesting that the ion density in the plasma is increasing. The InN etch rate is proportional to the BI_3 content, indicating the presence of a strong chemical component in its etching. In comparison, AlN

and GaN show very low rates up to $\sim 50\%$ BI_3 (~ 500 Å/min for AlN and ~ 1700 Å/min for GaN). An increase in the BI_3 content in the discharges actually produces a falloff in the etch rate for both AlN and GaN. We expect there are several possible mechanisms that may explain these data. First, the decrease in chuck self-bias and hence the ion energy under these conditions may more than compensate for the higher active iodine neutral flux. Second, the formation of the less volatile GaI_x and AlI_x etch products may create a selvege layer that suppresses the etch rate. This mechanism occurs in the Cl_2-reactive ion etching of InP. In this system, etching does not occur unless elevated sample temperatures or higher dc biases are used to facilitate removal of the $InCl_3$ etch product. In InN etch selectivity for both materials initially increases but also goes through a minimum. Note, however, that selectivities of greater than 100 can be achieved for both InN/AlN and InN/GaN.

Data for BBr_3/Ar discharges are also shown in Figure 21 for fixed source power (750 W) and rf chuck power (350 W). Higher rf powers were required to initiate etching with BBr_3 compared with BI_3, and the dc self-bias increased with the BBr_3 content. The etch rate of InN is again a strong function of

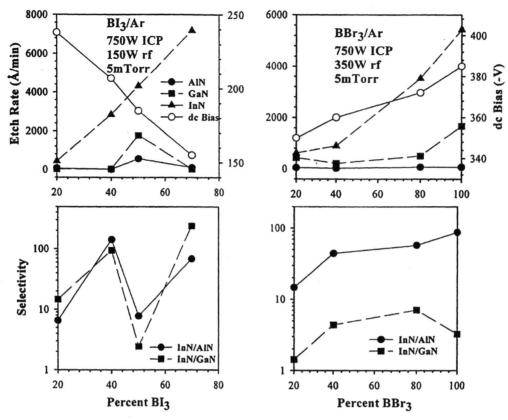

Fig. 21. Nitride etch rates (top) and etch selectivities for InN/AlN and InN/GaN (bottom) in BI$_3$/Ar discharges (750 W source power, 5 mTorr) as a function of the boron halide content.

the boron halide content, whereas GaN shows significant rates (∼1800 Å/min) only for pure BBr$_3$ discharges, and AlN shows very low etch rates over the whole range of conditions investigated. Maximum selectivities of ∼ 100 : 1 for InN/AlN and ∼ 7.5 : 1 for InN/GaN are obtained.

Based on the results in Figure 21, we chose fixed plasma compositions and varied the ion energy and flux through control of the source and chuck powers. Figure 22 shows that the source power had a significant effect only on the InN etch rate for both 4BI$_3$/6Ar and 4BBr$_3$/6Ar discharges at fixed rf power (150 W). The etch rate of InN continues to increase with source power, which controls the ion flux and dissociation of the discharge, whereas the GaN and AlN rates are low for both plasma chemistries. The InN etch rates are approximately a factor of 2 faster in BI$_3$/Ar than in BBr$_3$/Ar, even for lower rf chuck powers. This is expected when we take into consideration the relative stabilities of the respective In etch products (the InI$_3$ melting point is 210°C; InBr$_3$ sublimes at <600°C). The resultant selectivities are shown at the bottom of Figure 22; once again, a value of ∼ 100 : 1 for InN over GaN is achieved with BI$_3$, whereas BBr$_3$ produced somewhat lower values.

The dependence of the etch rate and InN/AlN and InN/GaN selectivities on rf chuck power for both plasma chemistries at a fixed source power (750 W) is shown in Figure 23. While the GaN and AlN etch rates (top left) increase only at the high-

est chuck powers investigated for 4BI$_3$/6Ar discharges, the InN etch rate increases rapidly to 250 W. This is consistent with a strong ion-assisted component for the latter under these conditions. The subsequent decrease in the etch rate at higher power produces corresponding maxima (≥100) in etch selectivity for chuck powers in the range of 150–250 W. This type of behavior is quite common for high-density plasma etching of III–V materials, where the etching is predominantly ion-assisted desorption of somewhat volatile products, with insignificant rates under ion-free conditions. In this scenario, at very high ion energies, the active etching species (iodine neutral in this case) can be removed by sputtering before they have a chance to complete the reaction with substrate atoms. Similar data for BBr$_3$/Ar mixtures are also shown in Figure 23. For this chemistry the InN etch rate saturates, and we did not observe any reduction in etch rate, although this might be expected to occur if higher powers could be applied (our power supply is limited to 450 W). GaN does show an etch rate maximum at ∼350 W, producing a minimum in the resultant InN/GaN selectivity. The etch selectivity of InN over the other two nitrides for BI$_3$/Ar is again much higher than for BBr$_3$/Ar.

The effect of plasma composition on etch rates and selectivities if GaN, AlN, and InN in ICl/Ar and IBr/Ar discharges at a source power of 750 W, a rf chuck power of 250 W, and 5 mTorr is shown in Figure 24. The etch rates of InN and AlN

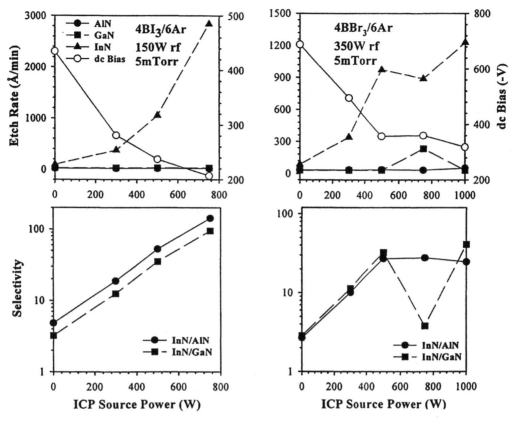

Fig. 22. Nitride etch rates (top) and etch selectivities for InN/AlN and InN/GaN (bottom) in BI_3/Ar discharges as a function of source power.

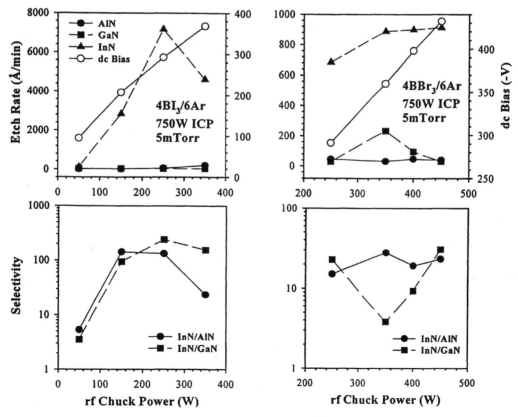

Fig. 23. Nitride etch rates (top) and etch selectivities for InN/AlN and InN/GaN (bottom) in BI_3/Ar or BBr_3/Ar discharges as a function of rf chuck power.

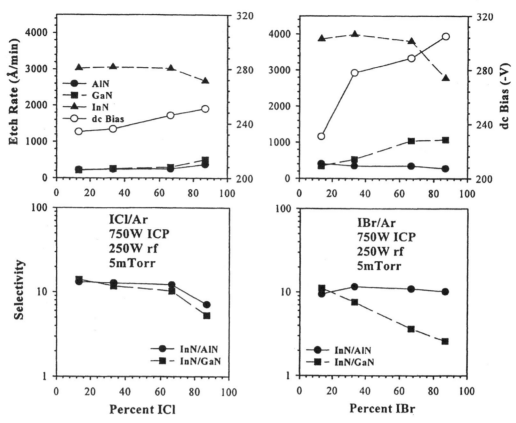

Fig. 24. Nitride etch rates (top) and etch selectivities for InN/AlN and InN/GaN (bottom) in ICl/Ar or IBr/Ar discharges (750 W source power, 250 W rf chuck power, 5 mTorr) as a function of interhalogen content.

are relatively independent of the plasma compositions of the two chemistries over a broad composition range, indicating that the etch mechanism is dominated by physical sputtering. The dc bias voltage increased with increasing interhalogen concentrations. The decrease in ion flux also implies an increase in the concentrations of neutral species such as Cl, Br, and I. The etch rate of GaN steadily increased with increasing ICl concentration. In contrast, the etch rate of GaN saturated beyond 66.7% IBr. These results indicate that etching of GaN in both chemistries can be attributed more to chemical etching by increased concentrations of reactive neutrals than to ion-assisted sputtering. The effect of plasma composition showed an overall trend of decrease in selectivities for InN over both AlN and GaN as the concentration of ICl and IBr increased.

The ICP source power dependence of the etch rates and selectivities is shown in Figure 25 for ICl/Ar and IBr/Ar discharges at a fixed plasma composition chuck power (250 W) and process pressure (5 mTorr). InN again showed higher etch rates than AlN and GaN. The increase in the etch rate with increased source power is due to the higher concentration of reactive species in the plasma, suggesting a reactant-limited regime, and to higher ion flux to the substrate surface. The relatively constant etch rate with a further increase in the ICP power is attributed to the competition between the ion-assisted etch reaction and the ion-assisted desorption of the reactive species

at the substrate surface before etch reaction. The selectivity of InN over AlN showed maximum values, depending on the ICP source power, whereas that of InN over GaN increased overall as the source power increased.

The effect of rf chuck power on the etch rates and selectivities of InN over AlN and over GaN is shown in Figure 26. The etch rates for all materials increased as the rf power or the ion-bombarding energy increased. The increase in etch rate with the chuck power can be attributed to enhanced sputter desorption of etch products as well as to physical sputtering of the InN surface. It is also interesting to see that the magnitude of the etch rate is on the order of the bond energies, InN (7.72 eV) < GaN (8.92 eV) < AlN (11.52 eV). Maximum values of selectivity of $\sim$30 for InN/AlN and of $\sim$14 for InN/GaN were obtained. The root-mean-square (rms) roughness of unetched GaN was 2.4 nm.

3.3. CH$_4$/H$_2$/Ar

Pearton and co-workers were the first to etch group III nitride films in an ECR-generated CH$_4$/H$_2$/Ar plasma [31]. Etch rates for GaN, InN, and AlN were <400 Å/min at a dc bias of $\sim$−250 V. Vartuli et al. reported ICP GaN, InN, and AlN etch rates approaching 2500 Å/min in CH$_4$/H$_2$Ar and CH$_4$/H$_2$/N$_2$ plasmas [84]. Etch rates increased with increasing dc bias or

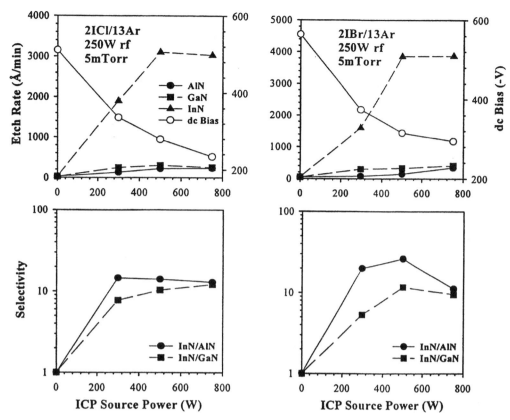

Fig. 25. Nitride etch rates (top) and etch selectivities for InN/AlN and InN/GaN (bottom) in ICl/Ar or IBr/Ar discharges (250 W rf chuck power, 5 mTorr) as a function of source power.

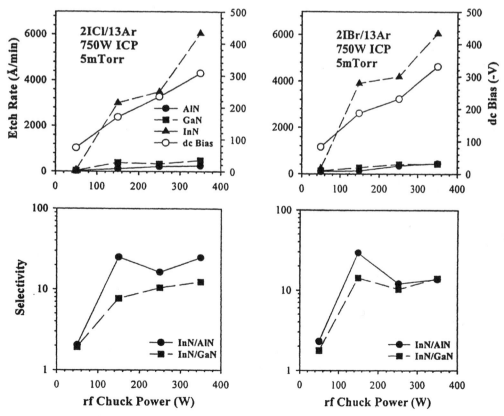

Fig. 26. Nitride etch rates (top) and etch selectivities for InN/AlN and InN/GaN (bottom) in ICl/Ar or IBr/Ar discharges (750 W source power, 5 mTorr) as a function of rf chuck power.

ion flux and were higher in CH$_4$/H$_2$/Ar plasmas. Anisotropy and surface morphology were good over a wide range of conditions. As compared with Cl-based plasmas, etch rates were consistently slower, which may make the CH$_4$/H$_2$-based processes applicable for devices where etch depths are relatively shallow and etch control is extremely important.

Vartuli and co-workers compared etch selectivities in CH$_4$/H$_2$/Ar and Cl$_2$/Ar plasmas in both RIE- and ECR-generated plasmas [69]. For CH$_4$/H$_2$/Ar plasmas, InN : GaN and InGaN : GaN etch selectivities ranged from 1 : 1 to 6 : 1, whereas etch selectivities of 1 : 1 or favoring GaN over the In-containing films were reported for Cl$_2$/Ar plasmas.

4. ETCH PROFILE AND ETCHED SURFACE MORPHOLOGY

Etch profile and etched surface morphology can be critical to postetch processing steps, including the formation of metal contacts, deposition of interlevel dielectric or passivation films, or epitaxial regrowth. Figure 27 shows SEM micrographs of GaN, AlN, and InN etched in Cl$_2$-based plasma. The GaN (Fig. 27a) was etched at a chamber pressure of 5 mTorr, an ICP power of 500 W, 22.5 standard cm^3 Cl$_2$, 2.5 standard cm^3 H$_2$, 5 standard cm^3 Ar, a temperature of 25°C, and a dc bias of -280 ± 10 V. Under these conditions, the GaN etch rate was $\sim$6880 Å/min with highly anisotropic, smooth sidewalls. The sapphire substrate was exposed during a 15% overetch. Pitting of the sapphire surface was attributed to defects in the substrate or growth process. The AlN (Fig. 27b) and InN (Fig. 27c) features were etched at a chamber pressure of 2 mTorr, an ICP power of 500 W, 25 sccm Cl$_2$, 5 standard cm^3 Ar, a temperature of 25°C, and a cathode rf power of 250 W. Under these conditions, the AlN etch rate was $\sim$980 Å/min and the InN etch rate

was $\sim$1300 Å/min. Anisotropic profiles were obtained over a wide range of plasma chemistries and conditions, with sidewall striations present.

Sidewall morphology is especially critical in the formation of laser mesas for ridge waveguide emitters or for buried planar devices. The vertical striations observed in the GaN sidewall in Figure 27a were due to striations in the photoresist mask, which were transferred into the GaN feature during the etch. The sidewall morphology and, in particular, the vertical striations were improved in an ICP Cl$_2$/BCl$_3$ plasma at a dc bias of -150 V. In Figure 28, a SEM micrograph of GaN etched in an ICP Cl$_2$/BCl$_3$ plasma shows highly anisotropic profiles and smooth sidewall morphology. The etch conditions were 2 mTorr chamber pressure, 500 W ICP power, 32 standard cm^3 Cl$_2$, 8 standard cm^3 BCl$_3$, 5 standard cm^3 Ar, 25°C, and a dc bias of -150 ± 10 V. Ren et al. have demonstrated improved GaN sidewall morphology etched in an ECR, with the use of a SiO$_2$ mask [85]. Vertical striations in the SiO$_2$ mask were reduced by optimizing the lithography process used to pattern the SiO$_2$. The SiO$_2$ was then patterned in a SF$_6$/Ar plasma in which a low-temperature dielectric overcoat was used to protect the resist sidewall during the etch.

In several studies atomic force microscopy (AFM) has been used to quantify the etched surface morphology as rms roughness. Rough etch morphology often indicates a nonstoichiometric surface due to preferential removal of either the group III or group V species. For example, in Figure 29, GaN and InN rms roughnesses are shown for as-grown samples and for samples exposed to an ECR Cl$_2$/H$_2$/CH$_4$/Ar plasma as a function of cathode rf power. The rms roughness for as-grown GaN and InN was 3.21 ± 0.56 and 8.35 ± 0.50 nm, respectively. The GaN rms roughness increased as the cathode rf power was increased, reaching a maximum of $\sim$85 nm at 275 W. The rms roughness for InN was greatest at a cathode rf power of 65 W, imply-

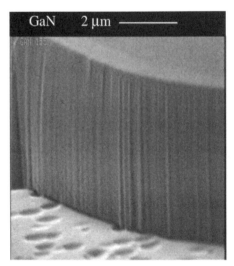

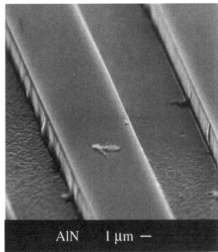

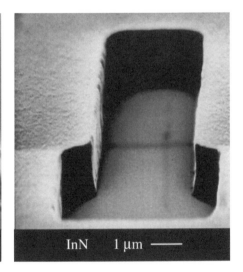

a)   b)   c)

Fig. 27.   SEM micrographs of (a) GaN, (b) AlN, and (c) InN etched in Cl$_2$-based ICP plasmas.

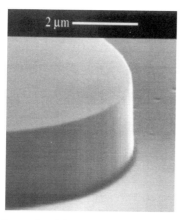

Fig. 28. SEM micrographs of GaN etched in a BCl₃/Cl₂-based ICP plasma.

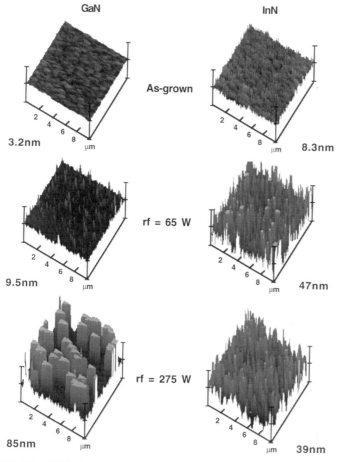

GaN   InN

As-grown

3.2nm   8.3nm

rf = 65 W

9.5nm   47nm

rf = 275 W

85nm   39nm

Fig. 29. AFM micrographs for (a) GaN and InN as grown, (b) GaN and InN etched at a rf cathode power of 65 W, and (c) GaN and InN etched at a rf cathode power of 275 W in an ECR-generated Cl₂/H₂/CH₄/Ar plasma. The Z scale is 100 nm/division.

5. PLASMA-INDUCED DAMAGE

Plasma-induced damage often degrades the electrical and optical properties of compound semiconductor devices. Because GaN is more chemically inert than GaAs and has higher bonding energies, more aggressive etch conditions (higher ion energies and plasma flux) may be used with potentially less damage to the material. Limited data have been reported for plasma-induced damage of the group III nitrides [86–89]. Pearton and co-workers reported increased plasma-induced damage as a function of ion flux and ion energy for InN, InGaN, and InAlN in an ECR [86]. The authors also reported (a) more damage in InN films as compared with InGaN, (b) more damage in lower doped materials, and (c) more damage under high ion energy conditions due to formation of deep acceptor states, which reduce the carrier mobility and increased resistivity. Postetch annealing processes removed the damage in the InGaN, whereas the InN damage was not entirely removed.

Ren and co-workers measured electrical characteristics for InAlN and GaN FET structures to study plasma-induced damage for ECR BCl₃, BCl₃/N₂, and CH₄/H₂ plasmas [87]. They reported (a) doping passivation is found in the channel layer in the presence of hydrogen; (b) high ion bombardment energies can create deep acceptor states that compensate for the material; and (c) preferential loss of N can produce rectifying gate characteristics. Ping and co-workers studied Schottky diodes for Ar and SiCl₄ RIE plasmas [88]. More damage was observed in pure Ar plasmas and under high dc bias conditions. Plasma-induced damage of GaN was also evaluated in ICP and ECR Ar plasma, with the use of photoluminescence (PL) measurements as a function of cathode rf power and source power [89]. The peak PL intensity decreased with increasing ion energy independent of etch technique. As a function of source power or plasma density, the results were less consistent. The PL intensity showed virtually no change at low ICP source power and then decreased as the plasma density increased. In the ECR plasma, the PL intensity increased by ~115% at low ECR source power and improved at higher ECR source powers, but at a lower rate. The effect of postetch annealing in Ar varied, depending on initial film conditions; however, annealing at temperatures above 440°C resulted in a reduction in the PL intensity.

Surface stoichiometry can also be used to evaluate plasma-induced damage. Nonstoichiometric surfaces can be created by preferential loss of one of the lattice constituents. This may be attributed to the higher volatility of the respective etch products, leading to enrichment of the less volatile species or preferential sputtering of the lighter element. Auger electron spectroscopy (AES) can be used to measure surface stoichiometry. Figure 30 shows characteristic Auger spectra for (a) as-grown GaN samples and samples exposed to an ECR plasma applied microwave power of 850 W and cathode rf powers of (b) 65 W and (c) 275 W. For the as-grown sample, the Auger spectrum showed a Ga:N ratio of 1.5 with normal amounts of adventitious carbon and native oxide on the GaN surface. After plasma

ing that the ion bombardment energy is critical to a balance of the chemical and sputtering effects of this plasma chemistry in maintaining smooth surface morphologies.

A summary of etch rate results for the nitrides with different chemistries and different techniques is shown in Table II.

Table II.   Summary of Etch Rate Results for GaN, AlN, and InN, with
Different Plasma Chemistries in Different Techniques

Gas chemistry	Etching technique	Etch rate (nm/min) at given bias						
		GaN		Ref.	AlN		Ref.	InN		Ref.
$SiCl_4$ [w/Ar, SiF_4]	RIE	55	−400 V	[19]	—	—		—	—	
BCl_3	RIE	105	−230 V	[27]	—	—		—	—	
HBr [w/Ar, H_2]	RIE	60	−400 V	[28]	—	—		—	—	
CHF_3, C_2ClF_5	RIE	45	500 W	[29]	—	—		—	—	
SF_6	RIE	17	−400 V	[28]	—	—		—	—	
CHF_3, C_2ClF_5	RIE	60	−500 V	[29]	—	—		—	—	
BCl_3/Ar	ECR	30	−250 V	[30]	17	−250 V	[30]	17	−300 V	[30]
CCl_2F_2/Ar	ECR	20	−250 V	[30]	18	−300 V	[30]	18	−300 V	[30]
CH_4/H_2/Ar	ECR	40	−250 V	[21]	2.5	−300 V	[30]	10	−300 V	[30]
Cl_2/H_2/Ar	ECR	200	−180 V	[33]	110	−150 V	[33]	150	−180 V	[33]
$SiCl_4$/Ar	ECR	95	−280 V	[36]	—	—		—	—	
HI/H_2	ECR	110	−150 V	[16]	120	−150 V	[16]	100	−150 V	[16]
HBr/H_2	ECR	70	−150 V	[16]	65	−150 V	[16]	17	−150 V	[16]
ICl/Ar	ECR	1300	−275 V	[39]	200	−272 V	[39]	1150	−275 V	[39]
IBr/Ar	ECR	300	−170 V	[68]	160	−170 V	[68]	325	−170 V	[68]
BCl_3	M-RIE	350	<−100 V	[58]	125	<−100 V	[16]	100	<−100 V	[16]
Cl_2/H_2/Ar	ICP	688	−280 V	[45]	—	—		—	—	
Cl_2/Ar	ICP	980	−450 V	[54]	670	−450 V	[54]	150	−100 V	[49]
Cl_2/N_2	ICP	65	−100 V	[49]	39	−100 V	[49]	30	−100 V	[49]
BBr_3	ICP	150	−380 V	[50]	50	−200 V	[50]	500	−380 V	[50]
BI_3	ICP	200	−175 V	[50]	100 V	−175	[50]	700	−240 V	[50]
ICl	ICP	30	−300 V	[50]	30	−300 V	[50]	600	−300 V	[50]
IBr	ICP	20	−300 V	[50]	30	−300 V	[50]	600	−300 V	[50]
Ar ion	Ion milling	110	500 eV	[20]	29	500 eV	[26]	61	500 eV	[26]
Cl_2 [Ar ion]	CAIBE	210	500 eV	[60]	62	500 eV	[27]	—	—	
HCl [Ar ion]	CAIBE	190	500 eV	[59]	—	—		—	—	
Cl_2	RIBE	150	500 eV	[51]	—	—		—	—	
HCl	RIBE	130	500 eV	[51]	—	—		—	—	
Cl_2/Ar	RIBE	50	−400 V	[51]	50	−400 V	[51]	80	−400 V	[51]
HCl	Photoassisted	0.04 Å/pulse		[67]	—	—		—	—	
H_2, Cl_2	LE4	50–70	1–15 eV	[65, 66]	—	—		—	—	

exposure, the Ga : N ratio increased as the cathode rf power increased with some residual atomic Cl from the plasma. Under high ion energy conditions, preferential removal of the lighter N atoms as observed resulted in Ga-rich surfaces.

5.1. n-GaN

The etching requirements for electronic devices are more demanding than those for photonic devices, at least from the viewpoint of electrical quality. One of the most sensitive tests of near-surface electrical properties is the quality of rectifying contacts deposited on the etched surface. There has been relatively little work in this area to date. Ren et al. [85, 87] found that rectifying contacts on electron cyclotron resonance plasma-etched GaN and InAlN surfaces were very leaky, though some improvement could be obtained by postetch annealing at

400°C. Ping et al. [88] found that reactively ion-etched n-GaN surfaces had poor Schottky contact properties, but that plasma chemistries with a chemical component (e.g., Cl_2-based mixtures) produced less degradation than purely physical etching.

The layer structure and contact metals are shown schematically in Figure 31. The GaN was grown by rf plasma-assisted molecular beam epitaxy on c-plane Al_2O_3 substrates. The Ti/Au ohmic contracts were patterned by lift-off and annealed at 750°C, producing contact resistances in the 10^{-5} $\Omega\,cm^{-2}$ range. Samples were exposed to either pure N_2 or H_2 discharges in a Plasma Therm 790 ICP system at a fixed pressure of 5 mTorr. The gases were injected into the ICP source at a flow rate of 15 standard cm^3/min. The experimentally varied parameters were source power (300–1000 W) and rf chuck power (40–250 W), which control ion flux and ion energy, respectively. In some cases the samples were either an-

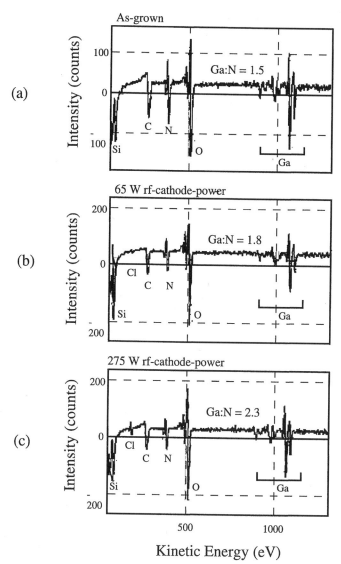

As-grown

Ga:N = 1.5

(a)

65 W rf-cathode-power

Ga:N = 1.8

(b)

275 W rf-cathode-power

Ga:N = 2.3

(c)

Kinetic Energy (eV)

Fig. 30. AES surface scans of GaN (a) before exposure to the plasma and at (b) 65 W (−120 V bias) and (c) 275 W of rf cathode power (−325 V bias), 1 mTorr, 170°C, and 850 W of microwave power in an ECR-generated Cl_2/H_2 discharge.

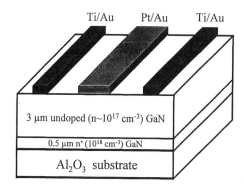

Fig. 31. Schematic of GaN Schottky diode structure.

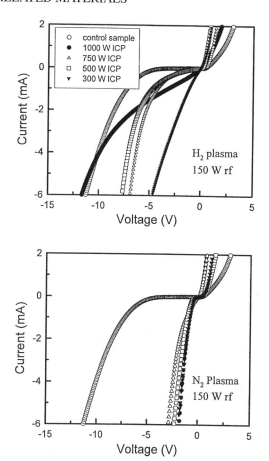

Fig. 32. $I-V$ characteristics from GaN diodes before and after H_2 (top) or N_2 (bottom) plasma exposure (150 W rf chuck power, 5 mTorr) at different ICP source powers.

nealed in N_2 for 30 s at 300–850°C or photoelectrochemically etched in 0.2 M KOH solutions at 25°C after plasma exposure. The Pt/Au Schottky metallization was then deposited through a stencil mass by e-beam evaporation. Current–voltage characteristics were recorded on a HP4145A parameter analyzer, and we defined the reverse breakdown voltage (V_B) as the voltage at which the leakage current was 10^{-3} A. We found in all cases that plasma exposure caused significant increases in forward and reverse current, with ideality factors increasing from typical values of 1.4–1.7 on control samples to >2. For this reason we were unable to extract meaningful values of either ideality factor or barrier height.

Figure 32 shows a series of $I-V$ characteristics from the GaN diodes fabricated on samples exposed to either H_2 or N_2 discharges at different source powers. It is clear that N_2 plasma exposure creates more degradation of the diode characteristics than does H_2 exposure. This implicates the ion mass ($^{28}N_2^+$, $^2H_2^+$ for the main positive ion species) as being more important in influencing the electrical properties of the GaN surface than a chemical effect, because H_2 would be likely to preferentially remove nitrogen from the GaN as NH_3.

The variations in V_B of the diodes with the source power during plasma exposure are shown in Figure 33. For any expo-

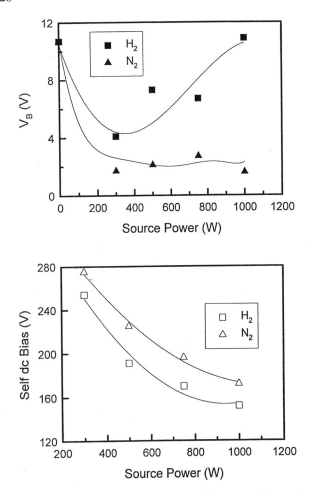

Fig. 33. Variation of V_B in GaN diodes (top) and dc chuck self-bias (bottom) as a function of ICP source power in H$_2$ or N$_2$ plasmas (150 W rf chuck power, 5 mTorr).

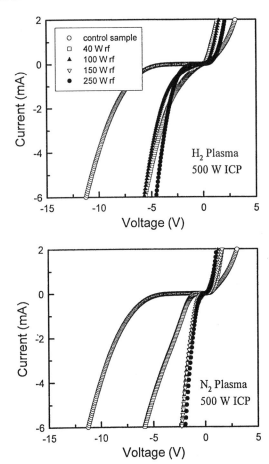

Fig. 34. I–V characteristics from GaN diodes before and after H$_2$ (top) or N$_2$ (bottom) plasma exposure (500 W source power, 5 mTorr) at different rf chuck powers.

sure to the N$_2$ discharges V_B is severely reduced. In contrast, there is less degradation with the H$_2$ plasma to the lower average ion energy under those conditions, as shown at the bottom of Figure 33. The average ion energy is approximately equal to the sum of dc self-bias and plasma potential, where the later is in the range of -22 to -28 V, as determined by Langmuir probe measurements. Ion-induced damage in GaN displays n-type conductivity, and, in addition, the heavy N$_2^+$ ions are also more effective in preferential sputtering of the N relative to Ga, compared with H$_2^+$ ions of similar energy.

Similar conclusions can be drawn from the data on the effect of increasing rf chuck power. Figure 34 shows the diode I–V characteristics from H$_2$ or N$_2$ plasma-exposed samples at a fixed source power (500 W) but varying rf chuck power. There are once again very severe decreases in breakdown voltage and increases in leakage current. The dependence of V_B on rf chuck power during the plasma exposures is shown in Figure 35, along with the dc self-bias. The V_B values fall by more than a factor of 2, even for very low self-biases, and emphasize how sensitive the GaN surface is to degradation by energetic ion bombardment. The degradation saturates beyond a chuck power of $\sim$100 W, corresponding to ion energies of $\sim$175 eV. We

assume that once the immediate surface becomes sufficiently damaged, the contact properties basically cannot be made any worse, and the issue is then whether the damage depth increases with the different plasma parameters. Because ion energy appears to be a critical factor in creating the near-surface damage, we would expect damage depth to increase with ion energy in a nonetching process. In the case of simultaneous etching and damage creation (e.g., in Cl$_2$/Ar etch processing), higher etch rates would lead to lower amounts of residual damage because the disordered region would be partially removed.

The damage depth was established by photoelectrochemically wet etching different amounts of the plasma-exposed GaN surfaces and then depositing the Pt/Au metal. Figure 36 (top) shows the effect on the I–V characteristics of this removal of different depths of GaN. There is a gradual restoration of the reverse breakdown voltage, as shown at the bottom of the figure. Note that the forward part of the characteristics worsens for the removal of 260 Å of GaN and shows signs of high series resistance. This would be consistent with the presence of a highly resistive region beneath the conducting near-surface layer, created by point defect diffusion from the surface. A similar model applies to ion-damaged InP, i.e., a nonstoichiometric near-surface region (deficient in P in that case) [90], followed

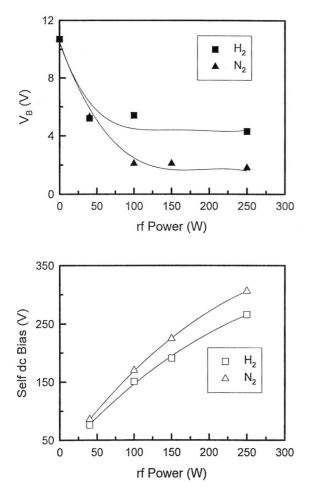

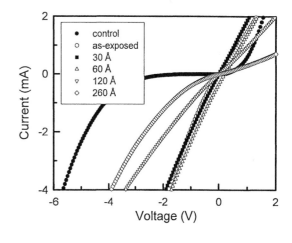

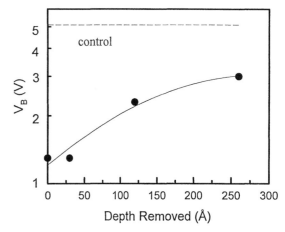

Fig. 35. Variation of V_B in GaN diodes (top) and dc chuck self-bias (bottom) as a function of rf chuck power in H_2 or N_2 plasmas (500 W source power, 5 mTorr).

Fig. 36. I–V characteristics from N_2 plasma-exposed GaN diodes before and after wet etch removal of different amounts of GaN before deposition of the Schottky contact (top) and variation of V_B as a function of the amount of material removed (bottom).

by a transition to a stoichiometric but point defect-compensated region, and finally to unperturbed InP.

The fact that plasma exposure severely degraded the surface is clear from the AFM data of Figure 37. Exposure to a source power of 500 W, a rf chuck power of 150 W (dc self-bias −221 V), and 5-mTorr N_2 discharge increased the rms surface roughness from 0.8 to 4.2 nm. Subsequent photoelectrochemical etching restored the initial morphology. However, we observed the onset of increasingly rough surfaces for deeper etch depths [91], reducing a relatively inaccurate measure of how much of the surface had to be removed to restore the diode breakdown voltage to its original value. We were able to estimate this depth as ∼600 ± 150 Å for the N_2 plasma conditions mentioned above.

Another method for trying to restore the electrical properties of the plasma-exposed surface is annealing. Figure 37 also shows AFM scan from samples after annealing at 550°C or 750°C, with no significant change in rms values. I–V data from annealed samples are shown in Figure 38. At the top are characteristics from samples that were plasma exposed (N_2, 500 W source power, 150 W rf chuck power, 5 mTorr) and then annealed and in which the contact was deposited. These samples

show that increasing the annealing temperature to 750°C brings a substantial improvement in V_B (Fig. 38, bottom). However, for annealing at 850°C the diode began to degrade, and this is consistent with the temperature at which N_2 begins to be lost from the surface. In the case where the samples were exposed to the N_2 plasma, and then the Pt/Au contact was deposited before annealing, the I–V characteristics show continued worsening upon annealing (Fig. 38, center). In this case, the Pt/Au contact is stable to 700°C on unetched samples. The poorer stability in etched samples could be related to the surface damage enhancing interfacial reaction between the Pt and GaN.

The main findings of this study can be summarized as follows:

1. There is a severe degradation in the electrical quality of GaN surfaces after ICP H_2 or N_2 discharge exposure. Under all conditions there is a strong reduction of V_B in diode structures to the point at which the Schottky contacts show almost ohmic-like behavior. These observations are consistent with the creation of a conducting n-type surface layer resulting from energetic ion bombardment. Heavier ions (N_2^+) create

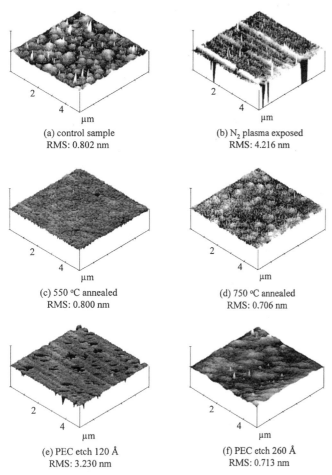

Fig. 37. AFM scans before and after N$_2$ plasma exposure (500 W source power, 150 W rf chuck power, 5 mTorr) and subsequent annealing or photochemical etching.

more damage than lighter ions (H$_2^+$) in this situation, where damage accumulates without any concurrent etching of the surface.

2. The depth of the damage is approximately 600 Å, as judged by the return of the diode characteristics to their control values.

3. Annealing at 750°C is also effective in helping to remove the effects of plasma exposure. Higher temperatures lead to degradation in GaN diode properties for uncapped anneals.

5.2. p-GaN

The layer structure consisted of 1 μm of undoped GaN ($n \approx 5 \times 10^{16}$ cm^{-3}) grown on a c-plane Al$_2$O$_3$ substrate, followed by 0.3 μm of Mg doped ($p \approx 10^{17}$ cm^{-3}) GaN. The samples were grown by rf plasma-assisted molecular beam epitaxy. Ohmic contacts were formed with Ni/Au deposited by e-beam evaporation, followed by lift-off and annealing at 750°C. The GaN surface was then exposed for 1 min to ICP H$_2$ or Ar plasmas in a Plasma-Therm 790 System. The 2-MHz ICP source power was varied from 300 to 1400 W, and the 13.56 MHz

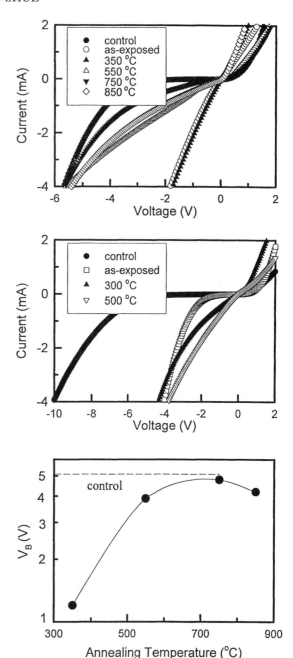

Fig. 38. I–V characteristics from GaN diodes before and after N$_2$ plasma exposure (500 W source power, 150 W rf chuck power, 5 mTorr) and subsequent annealing either before (top) or after (center) the deposition of the Schottky metallization. The variation of V_B in the samples annealed before metal deposition is shown at the bottom of the figure.

rf chuck power was varied from 20 to 250 W. The former parameter controls ion flux incident on the sample, and the latter controls the average ion energy. Before to deposition of 250-μm-diameter Ti/Pt/Au contacts through a stencil mask, the plasma-exposed surfaces were either annealed under N$_2$ in a rapid thermal annealing system or immersed in boiling NaOH solutions to remove part of the surface. As reported previously, it is possible to etch damaged GaN in a self-limiting fashion in hot alkali or acid solutions. The current–voltage (I–V) charac-

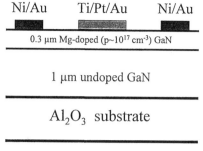

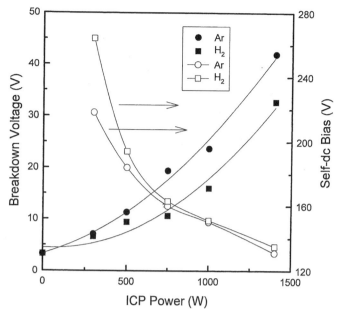

Fig. 41. Variation of diode breakdown voltage in samples exposed to H_2 or Ar ICP discharges (150 W rf chuck power) at different ICP source powers before deposition of the Ti/Pt/Au contact. The dc chuck self-bias during plasma exposure is also shown.

Fig. 39. Schematic of *p*-GaN Schottky diode structures.

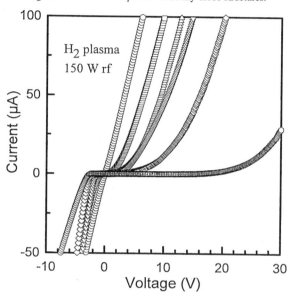

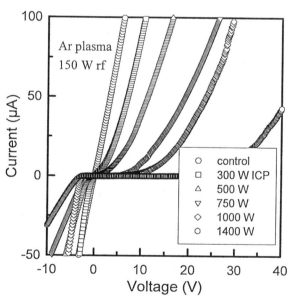

Fig. 40. *I–V* characteristics from samples exposed to either H_2 (top) or Ar (bottom) ICP discharges (150 W rf chuck power) as a function of ICP source power before deposition of the Ti/Pt/Au contact.

teristics of the diodes were recorded on an HP 4145A parameter analyzer. A schematic of the final test structures is shown in Figure 39. The unetched control diodes have reverse breakdown

voltages of ∼2.5–4 V, depending on the wafer—these values were uniform (±12%) across a particular water.

Figure 40 shows the *I–V* characteristics from samples exposed to either H_2 (top) or Ar (bottom) ICP discharges (150 W rf chuck, 2 mTorr) as a function of source power. In both cases there is an increase in both the reverse breakdown voltage and the forward turn-on voltage, with these parameters increasing monotonically with the source power during plasma exposure.

Figure 41 shows this increase in breakdown voltage as a function of source power and the variation of the chuck dc self-bias. As the source power increases, the ion density also increases and the higher plasma conductivity suppresses the developed dc bias. Note that the breakdown voltage of the diodes continues to increase even as this bias (and hence ion energy, which is the sum of this bias and the plasma potential) decreases. These results show that ion flux lays an important role in the change of diode electrical properties. The other key result is that Ar leads to a consistently greater increase in breakdown voltage, indicating that ion mass, rather than any chemical effect related to removal of N_2 or NH_3 in the H_2 discharges, is important.

The increase in breakdown voltage on the *p*-GaN is due to a decrease in hole concentration in the near-surface region through the creation of shallow donor states. The key question is whether there is actually conversion to an *n*-type surface under any of the plasma conditions. Figure 42 shows the forward turn-on characteristics of the *p*-GaN diodes exposed to different source power Ar discharge at low source power (300 W); the turn-on remains close to that of the unexposed control sample. However, there is a clear increase in the turn-on voltage at higher source powers, and in fact at ≥750 W the characteristics

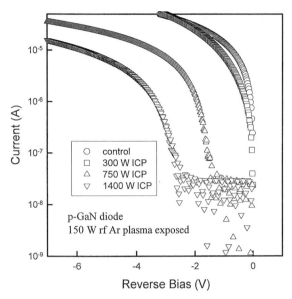

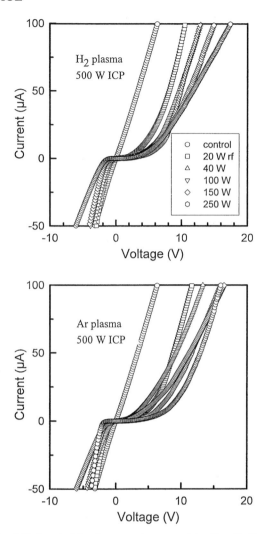

Fig. 42. Forward turn-on characteristics of diodes exposed to ICP Ar discharges (150 W rf chuck power) at different ICP source powers before deposition of the Ti/Pt/Au contact.

Fig. 43. I–V characteristics from samples exposed to either H₂ (top) or Ar (bottom) ICP discharges (500 W source power) as a function of rf chuck power before deposition of the Ti/Pt/Au contact.

are those of an n–p junction. Under these conditions the concentration of plasma-induced shallow donors exceeds the hole concentration, and there is surface conversion. In other words, the metal–p-GaN diode has become a metal–n-GaN–p-GaN junction. We always find that plasma-exposed GaN surfaces are N_2-deficient relative to their unexposed state, and therefore the obvious conclusion is that nitrogen vacancies create shallow donor levels. This is consistent with thermal annealing experiments in which N_2 loss from the surface produced increased n-type conduction.

The influence of rf chuck power on the diode I–V characteristics is shown in Figure 43 for both H_2 and Ar discharges at a fixed source power (500 W). A trend similar to that for the source power experiments is observed, namely the reverse breakdown voltage increases, consistent with a reduction in p-doping level near the GaN surface.

Figure 44 plots breakdown voltage and dc chuck self-bias as a function of the applied rf chuck power. The breakdown voltage initially increases rapidly with ion energy (the self-bias plus ∼25-V plasma potential) and saturates above ∼100 W, probably because the sputtering yield increases and some of the damaged region is removed. Note that these are very large changes in breakdown voltage even for low ion energies, emphasizing the need to carefully control both flux and energy. We should also point out that our experiments represent worst-case scenarios, because with real etching plasma chemistries such as Cl_2/Ar, the damaged region would be much shallower, because of the much higher etch rate. As an example, the sputter rate of GaN in a 300-W source power, 40-W rf chuck power, Ar ICP discharge is ∼40 Å/min, whereas the etch rate in a Cl_2/Ar discharge under the same conditions is ∼1100 Å/min.

An important question is the depth of the plasma-induced damage. We found that we were able to etch p-GaN very slowly

in boiling NaOH solutions, at rates that depended on the solution molarity (Fig. 45), even without any plasma exposure of the material. This enabled us to directly measure the damage depth in plasma-exposed samples in two different ways.

The first method involved measuring the etch rate as a function of depth from the surface. Defective GaN resulting from plasma, thermal, or implant damage can be wet chemically etched at rates much faster than those for undamaged material, because the acid or base solutions are able to attack the broken or strained bonds present. Figure 46 shows the GaN etch rate as a function of depth in samples exposed to a 750-W source power, 150-W rf chuck power Ar discharge. The etch rate is a strong function of the depth from the surface and saturates between ∼425 and 550 Å. Within this depth range the etch rate is returned to the "bulk" value characteristic of undamaged p-GaN.

The second method of establishing the damage depth, of course, is simply to measure the I–V characteristics after different amounts of material are removed by wet etching before de-

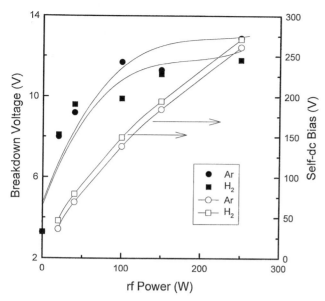

Fig. 44. Variation of diode breakdown voltage in samples exposed to H_2 or Ar ICP discharges (500 W source power) at different rf chuck powers before deposition of the Ti/Pt/Au contact. The dc chuck self-bias during plasma exposure is also shown.

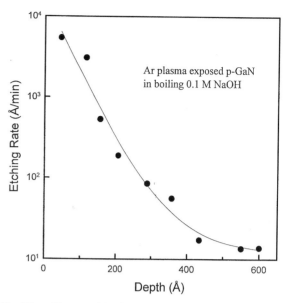

Fig. 46. Wet etching rate of Ar plasma-exposed (750 W source power, 150 W rf chuck power) GaN as a function of depth in the sample.

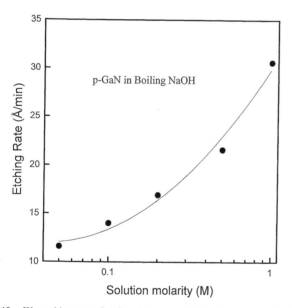

Fig. 45. Wet etching rate of *p*-GaN in boiling NaOH solutions as a function of solution molarity.

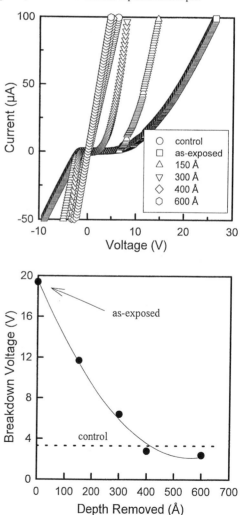

Fig. 47. *I–V* characteristics from samples exposed to ICP Ar discharges (750 W source power, 150 W rf chuck power) and subsequently wet etched to different depths before deposition of the Ti/Pt/Au contact (top) and breakdown voltage as a function of depth removed (bottom).

position of the rectifying contact. Figure 47 (top) shows the *I–V* characteristics from samples exposed to 750-W source power, 150-W rf chuck power (-160 V dc chuck bias) Ar discharges and subsequently wet etched to different depths with the use of 0.1 M NaOH solutions before deposition of the Ti/Pt/Au contact. Figure 47 (bottom) shows the effect of the amount of material removed at the diode breakdown voltage. Within the experimental error of $\pm 12\%$, the initial breakdown voltage is reestablished in the range of 400–450 Å. This is consistent with the depth obtained from the etch rate experiments described above. These values are also consistent with the damage depths

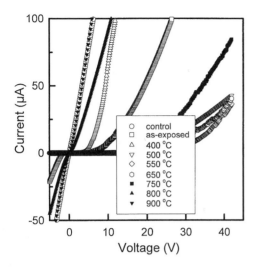

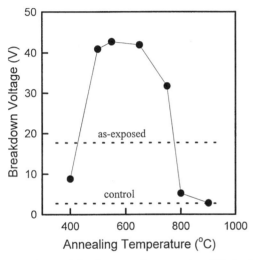

Fig. 48. I–V characteristics from samples exposed to ICP Ar discharges (750 W source power, 150 W rf chuck power) and subsequently annealed at different temperatures before deposition of the Ti/Pt/Au contact (top) and breakdown voltage as a function of annealing temperature (bottom).

we established in n-GaN diodes exposed to similar plasma conditions.

The other method of removing plasma-induced damage is annealing. In these experiments we exposed the samples to the same type of plasma (Ar, 750-W source power, 150-W rf chuck power) and then annealed under N_2 at different temperatures. Figure 48 (top) shows the I–V characteristics of these different samples and (bottom) the resulting breakdown voltages as a function of annealing temperature. On this wafer, plasma exposure caused an increase in breakdown voltage from ~2.5 to ~18 V. Subsequent annealing at 400°C initially decreased the breakdown voltage, but a higher temperature produced a large increase. At temperatures above 700°C, the diode characteristics returned to their initial values and were back to the control values by 900°C. This behavior is similar to that observed in implant-isolated compound semiconductors in which ion damage compensates for the initial doping in the material, producing higher sheet resistance. In many instances the

damage site density is larger than that needed to trap all of the free carriers, and trapped electrons or holes may move by hopping conduction. Annealing at higher temperatures removes some of the damage sites, but there are still enough to trap all of the conduction electrons/holes. Under these conditions the hopping conduction is reduced and the sample sheet resistance actually increases. At still higher annealing temperatures, the trap density falls below the conduction electron or hole concentration, and the latter are returned to their respective bands. Under these conditions the sample sheet resistance returns to its preimplanted value. The difference in the plasma-exposed samples is that the incident ion energy is a few hundred electron volts, compared with a few hundred kiloelectron volts in implant-isolated material. In the former case the main electrically active defects produced are nitrogen vacancies near the surface, whereas in the latter case there are vacancy and interstitial complexes produced in far greater numbers to far greater depths. In our previous work on plasma damage in n-GaN we found that annealing at ~750°C almost returned the electrical properties to their initial values. If the same defects are present in both n- and p-type material after plasma exposure, this difference in annealing temperature may be a result of a Fermi-level dependence of the annealing mechanism.

The main conclusions of this study may be summarized as follows:

1. The effect of either H_2 or Ar plasma exposure on p-GaN surfaces is to decrease the net acceptor concentration through the creation of shallow donor levels, most likely N_V. At high ion fluxes or ion energies there can be type conversion of the initially p-type surface. The change in electrical properties is more pronounced with Ar than with H_2 plasmas under the same conditions.

2. Two different techniques for measuring the damage depth find it to be in the range of 400–500 Å under our conditions. After the removed of this amount of GaN, both the breakdown voltage and wet chemical etch rates are returned to their initial values.

3. Postetch annealing in N_2 at 900°C restores the initial breakdown voltage on plasma-exposed p-GaN. Annealing at higher temperatures degraded the electrical properties, again most likely because of N_2 loss from the surface.

5.3. Schottky Diodes

Contrary to initial expectations, the surface of GaN is relatively sensitive to energetic ion bombardment or thermal degradation encountered during device processing. In particular, it can preferentially lose N_2, leaving strong n-type conducting regions. Whereas dry etching has been used extensively for patterning of photonic devices (light-emitting and laser diodes) and opto-electronic devices (UV detectors), there has been little work performed on understanding the electrical effects of ion-induced point defects or nonstoichiometric surfaces result-

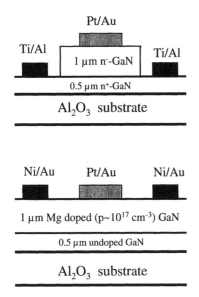

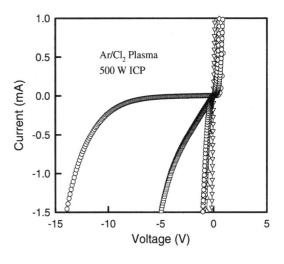

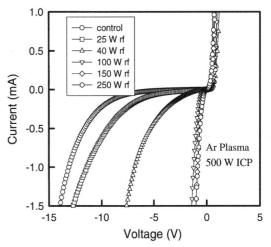

Fig. 49. Schematic of *n*- and *p*-GaN diode structures.

Fig. 50. *I–V* characteristics from *n*-GaN samples exposed to ICP Cl_2/Ar (top) or Ar (bottom) discharges (500 W source power) as a function of rf chuck power before deposition of the rectifying contact.

ing from the plasma exposure. Several groups have reported increases in the sheet resistance of GaN exposed to high-density plasmas, along with decreases in reverse breakdown voltage (V_B) and reductions in Schottky barrier height (ϕ_B) in diodes formed on *n*-type GaN. In this latter case, low-bias forward currents were increased by up to 2 orders of magnitude after exposure of the diode to pure Ar discharges. Conversely, whereas the rectifying contact properties were degraded by plasma exposure, the specific resistance of *n*-type ohmic contacts was improved. Similarly, in *p*-type GaN, the effect of Ar or H_2 high-density plasma exposure was to decrease the net acceptor concentration to depths of ~500 Å. At high ion fluxes or energies, there was type conversion of the initially *p*-GaN surface.

Dry etching is needed for a range of GaN electronic devices, including mesa diodes, rectifiers, thyristors, and HBTs for high-temperature, high-power operation. These applications include control of power flow in utility grids, radar, and electronic motor drives. It is critical to understand the depth and thermal stability of dry etch damage in both *n*- and *p*-type GaN and its effect on the current–voltage (*I–V*) characteristics of simple diode structures.

In this section we report on a comparison of the effects of Cl_2/Ar and Ar ICP exposure on the electrical properties of *n*- and *p*-GaN Schottky diodes. In some cases it was found that Cl_2/Ar discharges could produce even more damage than pure Ar, because of the slightly higher ion energies involved. The damage saturates after a short exposure to either Cl_2/Art or Ar discharges and is significant even for low ion energies. Annealing between 700°C and 800°C restored ≥70% of the reverse breakdown voltage on *n*-GaN, and the damage depth was again established to be ~500 Å in *p*-GaN.

The diode structures are shown schematically in Figure 49. The GaN layers were grown by rf plasma-assisted molecular beam epitaxy on *c*-plane Al_2O_3 substrates. The Ti/Al (for *n*-type) and Ni/Au (for *p*-type) ohmic contacts were patterned by liftoff and annealed at 750°C. The samples were exposed

to either $10Cl_2$/5Ar or 15Ar (where the numbers denote the gas flow rate in standard cubic centimeters per minute) ICP discharges in a Plasma-Therm ICP reactor at a fixed pressure of 3 mTorr. We investigated a range of rf chuck powers (25–250 W) and etch times (4–100 s), with a fixed source power of 500 W. In some cases, the samples were either annealed in N_2 for 30 s at 500–800°C or wet etched in 0.1 M NaOH solutions at ~100°C after plasma exposure. The Schottky metallization (Pt/Au in both cases) was then deposited through a stencil mask ($\phi = 70$ or 90 μm) by e-beam evaporation. Current–voltage characteristics were recorded on an HP 4145A parameter analyzer, and we defined the reverse breakdown voltage as that at which the leakage current was 10^{-3} A. The forward on-voltage (V_F) was defined as the voltage at which the forward current was 100 A cm^{-2}. In all cases the ideality factors increased from 1.3 to 1.6 on control samples to >2 after plasma exposure, and thus we were unable to extract meaningful values of either barrier height or ideality factor.

Figure 50 shows a series of *I–V* characteristics from *n*-type GaN diodes fabricated on samples exposed to either Cl_2/Ar

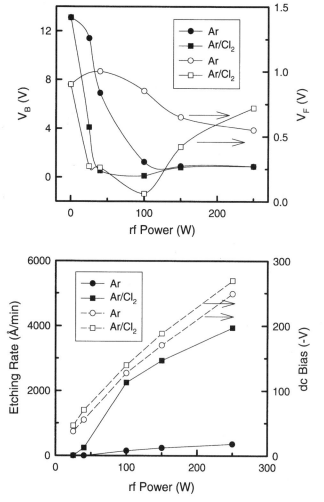

Fig. 51. Variations of V_B and V_F (top) and of n-GaN etching rate (bottom) as a function of rf chuck power for n-GaN diodes exposed to ICP Cl$_2$/Ar discharges (500 W source power).

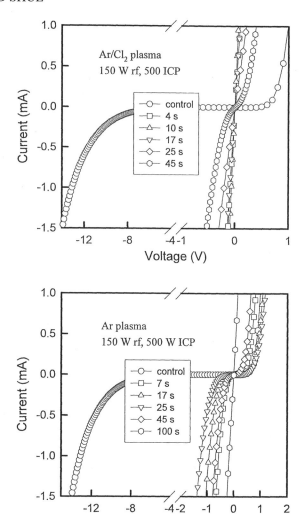

Fig. 52. I–V characteristics from n-GaN samples exposed to ICP Cl$_2$/Ar (top) or Ar (bottom) discharges (150 W rf chuck power, 500 W source power) as a function of plasma exposure time before deposition of the rectifying contact.

(top) or Ar (bottom) discharges at different rf chuck powers. There is a significant reduction in V_B under all conditions, with Ar producing less damage at low chuck powers. This is probably related to two factors: the slightly higher chuck bias with Cl$_2$/Ar due to the lower positive ion density in the plasma (Cl is more electronegative than Ar) and the heavier mass of the Cl$_2^+$ ions compared with Ar$^+$. This is consistent with our past data on the relative effects of N$_2$ and H$_2$ plasma exposure, in which ion mass was found to be more important in influencing the electrical properties of the GaN surface than any chemical effects.

The variations of V_B and V_F with the rf chuck power during plasma exposure are shown in Figure 51 (top). At powers of ≤ 100 W, the Cl$_2$/Ar creates more degradation of V_B, as discussed above, whereas at higher powers the damage saturates. The average ion energy is the sum of dc self-bias (shown at the bottom of the figure) and plasma potential (which is about 22–25 eV under these conditions). Thus for ion energies less than ~ 150 eV, Ar produces less damage than Cl$_2$/Ar, even though

the etch rate with the latter is much higher. This is also reflected in the variation of V_F with rf chuck power.

Figure 52 shows a series of I–V characteristics from n-type GaN diodes fabricated on samples exposed to the two different plasmas for different times at fixed rf chuck power (150 W) and source power (500 W). It is clear that the damage accumulates rapidly, with the I–V characteristics becoming linear at longer times. It should be remembered that this is damage accumulating ahead of the etch front.

Figure 53 shows the variations in V_B and V_F in n-type diodes with plasma exposure time to 500-W source power, 150-W rf chuck power Cl$_2$/Ar or Ar discharges (top), together with the etch depth versus etch time (bottom). As is readily apparent, V_B decreases dramatically after even short plasma exposures and then tends to recover slightly up to ~ 25 s. The V_B values are < 1 V for basically all plasma exposure times for both Cl$_2$/Ar and Ar. For V_F, there was more degradation with Cl$_2$/Ar for short exposure times.

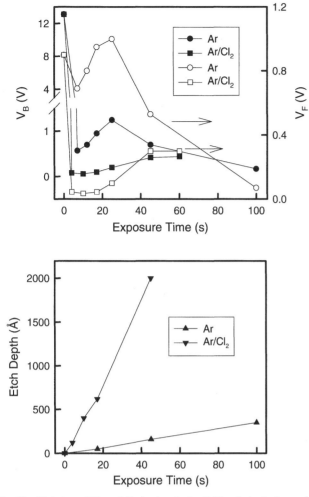

Fig. 53. Variation of V_B and V_F (top) and of n-GaN etch depth (bottom) as a function of plasma exposure time for n-GaN diodes exposed to ICP Cl$_2$/Ar discharges (500 W source power, 150 W rf chuck power).

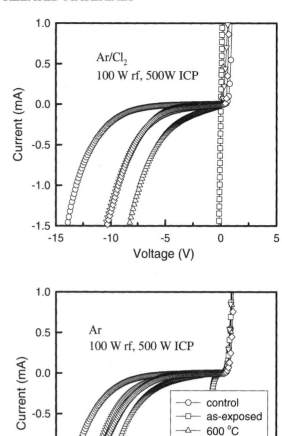

Fig. 54. I–V characteristics from n-GaN samples exposed to ICP Cl$_2$/Ar (top) or Ar (bottom) discharges (500 W source power, 100 W rf chuck power) as a function of annealing temperature before deposition of the rectifying contact.

To examine the thermal stability of the etch damage, n-type samples were exposed to Ar or Cl$_2$/Ar discharges at a fixed source power (500 W) and rf chuck power (150 W rf) and then annealed at different temperatures before deposition of the rectifying contact. Figure 54 shows I–V characteristics from control, plasma-exposed, and annealed diodes. The annealing produces a significant recovery of the electrical properties for samples exposed to either type of plasma. The V_B values are shown in Figure 55, as a function of post-plasma exposure annealing temperature. Annealing temperatures between 700°C and 800°C restore more than 70% of the original V_B value, but clearly annealing alone cannot remove all of the dry etch-induced damage. Annealing temperatures above 800°C were found to lead to preferential loss of N$_2$ from the surface, with a concurrent degradation in V_B.

Turning to p-GaN diodes, Figure 56 shows I–V characteristics from samples that were wet etched to various depths in NaOH solutions after exposure to either Cl$_2$/Ar or Ar discharges (500-W source power, 150-W rf chuck power, 1 min). For these plasma conditions we did not observe type conversion

of the surface. However, we find that the damaged GaN can be effectively removed by immersion in hot NaOH, without the need for photo- or electrochemical assistance of the etching. The V_B values increase on p-GaN after plasma exposure because of introduction of shallow donor states that reduce the wet acceptor concentration.

Figure 57 shows two methods for determining the depth of the damaged region in p-GaN diodes. At the top is a plot of the variation of V_F and V_B with the depth of material removed by NaOH etching. The values of both parameters are returned to their control values by depths of 500–600 Å. What is clear from these data is that the immediate surface is not where the p-doping concentration is most affected, because the maximum values peak at depths of 300–400 Å. This suggests that N_V or other compensating defects created at the surface diffuse rapidly into this region, even near room temperature. This is consistent with results in other semiconductors, where damage depths are typically found to be many times deeper than the projected range of incident ions. The bottom part of Fig-

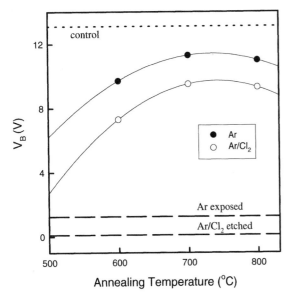

Fig. 55. Variation of V_B in n-GaN diodes exposed to ICP Cl_2/Ar or Ar discharges (500 W source power, 100 W rf chuck power) with annealing temperature before deposition of the rectifying contact.

ure 57 shows the wet etch depth in plasma-damaged p-GaN as a function of etching time. The etch depth saturates at depths of 500–600 Å, consistent with the electrical data. It has previously been shown that the wet etch depth on thermally or ion-damaged GaN was self-limiting. This is most likely a result of the fact that defective or broken bonds in the material are readily attached by the acid or base, whereas in undamaged GaN the etch rate is negligible.

The main findings of our study may be summarized as follows:

1. Large changes in V_B and V_F of n- and p-GaN Schottky diodes were observed after exposure to both Cl_2/Ar and Ar ICP discharges. In some cases the electrical properties are more degraded with Cl_2/Ar, even though this plasma chemistry has a much higher etch rate.
2. The damage accumulates near the surface, even for very short exposure times (4 s). The damage depth was established to be 500–600 Å from both the changes in electrical properties and the depth dependence of the wet etch rate.
3. Annealing in the range 700–800°C partially restores V_B in n-GaN diodes, but full recovery can only be achieved with an additional wet etch step for removal of the damaged material. The combination of annealing and a wet etch clean-up step looks very promising for GaN device fabrication.

5.4. p–n Junctions

Layer structures were grown by metal organic chemical vapor deposition (MOCVD) on c-plane Al_2O_3 substrates at 1040°C. The structure consisted of a low-temperature (530°C) GaN

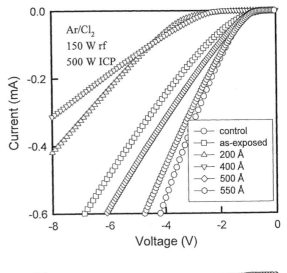

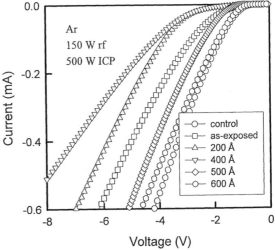

Fig. 56. I–V characteristics from p-GaN samples exposed to ICP Cl_2/Ar (top) or Ar (bottom) discharges (500 W source power, 150 W rf chuck power) and wet etched in boiling NaOH to different depths before deposition of the rectifying contact.

buffer, 1.2 μm of n (2×10^{17} cm^{-3}, Si-doped) GaN, 0.5 μm of nominally undoped ($n \approx 10^{16}$ cm^{-3}) GaN, and 1.0 μm of p ($N_A \approx 5 \times 10^{19}$ cm^{-3}, Mg-doped) GaN. The p-ohmic metal (Ni/Au) was deposited by e-beam evaporation and liftoff and then alloyed at 750°C. A mesa was then formed by BCl_3/Cl_2/Ar (8/32/5 standard cm^3) ICP etching to a depth of 1.6 μm under different plasma conditions to examine the effect of ion energy and ion flux, respectively. The ICP reactor was a load-locked Plasma-Therm SLR 770, which used a 2-MHz, three-turn coil ICP source. All samples were mounted with a thermally conductive paste on an anodized Al carrier that was clamped to the cathode and cooled with He gas. The ion energy or dc bias was defined by superimposing a rf bias (13.56 MHz) on the sample. The n-type ohmic metallization (Ti/Al) was then deposited, to produce the structure shown in Figure 58. Reverse I–V measurements were made on 300-μm-diameter diodes with a HP 4145B semiconductor parameter analyzer. In this study the reverse leakage current was measured at a bias of -30 V. Etch

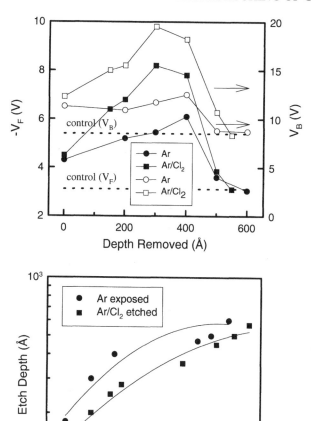

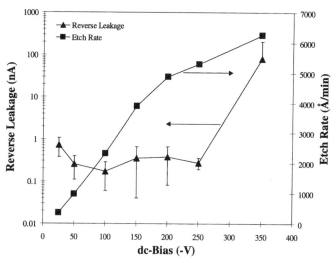

Fig. 59.   Reverse leakage current measured at −30 V for GaN p–i–n junctions etched in ICP 32Cl$_2$/8BCl$_3$/5Ar discharges (500 W source power, 2 mTorr), as a function of dc chuck self-bias.

Fig. 57.   Variation of V_B and V_F (top) with depth of p-GaN removed by wet etching before deposition of the rectifying contact, and wet etch depth versus etch time in boiling NaOH solutions for plasma-damaged p-GaN (bottom).

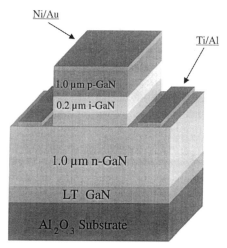

Fig. 58.   Schematic of GaN p–i–n junction formed by dry etching.

rates were calculated from bulk GaN samples patterned with AZ-4330 photoresist. The depth of etched features was measured with an Alpha-step stylus profilometer after the photore-

sist was removed. Etch profile and surface morphology were analyzed by SEM and AFM, respectively.

Figure 59 shows the effect of dc chuck bias on the reverse junction leakage current, along with the corresponding GaN etch rates. There is little effect on the current below chuck biases of −250 V. This corresponds to an ion energy of approximately −275 eV, because this energy is the sum of chuck bias and plasma potential (about −25 eV in this tool under these conditions). The reverse current decreases slightly as the dc self-bias is increased from −25 to −50 V. This may result from the sharp increase in etch rate, which leads to faster removal of near-surface damage. The reverse current increases rapidly above an ion energy of −275 V, which is a clear indication of severe damage accumulating on the sidewall. The damage probably takes the form of point defects such as nitrogen vacancies, which increase the n-type conductivity of the surface. The total reverse current density, J_R, is the sum of three components, namely diffusion, generation, and surface leakage, according to

$$J_R = \left(\frac{eD_h}{l_h N_D} + \frac{eD_e}{l_e N_A} \right) n_i^2 + \frac{eWn_i}{\tau_g} + J_{SL}$$

where e is the electronic charge, $D_{e,h}$ are the diffusion coefficients of electrons or holes, $l_{e,h}$ are the lengths of the n and p regions outside the depletion region in a p–n junction, $N_{D,A}$ are the donor/acceptor concentrations on either side of the junction, n_i is the intrinsic carrier concentration, W is the depletion with τ_g the thermal generation lifetime of carriers, and J_{SL} is the surface current component, which is bias-dependent. The latter component is most affected by the dry etch process and dominates the reverse leakage in diodes etched at high ion energies.

GaN sidewall profiles and etch morphologies have been evaluated from previous results as a function of dc bias. The etch becomes more anisotropic as the dc bias increases from −50 to −150 V dc bias because of the perpendicular nature of the ion bombardment energies. However, at −300 V dc bias a tiered etch profile with vertical striations in the sidewall was

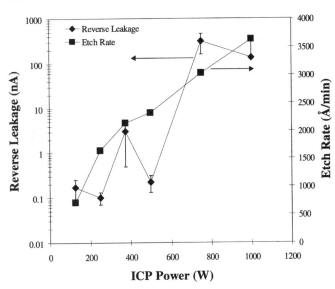

Fig. 60. Reverse leakage current measured at −30 V for GaN p–i–n junctions etched in ICP 32Cl₂/8BCl₃/5Ar discharges (−100 V dc chuck self-bias, 2 mTorr), as a function of source power.

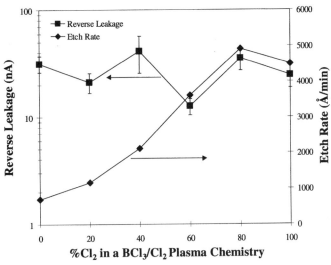

Fig. 61. Reverse leakage current measured at −30 V for GaN p–i–n junctions etched as a function of Cl₂ percentage in an ICP Cl₂/BCl₃/Ar plasma. Plasma conditions were −100 V dc chuck self-bias, 2 mTorr, 500 W ICP power, and 40 standard cm³ total gas flow.

observed because of erosion of the mask edge under high ion bombardment energies. The physical degradation (both profile and morphology) of the etched sidewall at −300 V could help explain higher reverse leakage currents above −250 V dc bias. Under high bias conditions, more energetic ions scattering from the surface could strike the sidewalls with significant momentum, thus increasing the likelihood of increased damage and higher reverse leakage currents. Under low bias conditions, the sidewall profile is less anisotropic, implying increased lateral etching of the GaN (undercutting of the mask). Under these conditions the etch process becomes dominated by the chemical component of the etch mechanism, which may account for the slightly higher reverse leakage observed at −25 V dc bias.

Figure 60 shows the effect of ICP source power on the junction reverse leakage current. The plasma flux is proportional to source power. In this experiment the ion energy was held constant at −100 V dc bias. There is a minimal effect on leakage current for source powers of ≤500 W, with severe degradation of the junction characteristics at higher powers, even though the GaN etch rate continues to increase. This is an important result because it shows that the conditions that produce the highest etch rate are not necessarily those that lead to the least damage. Increased sidewall damage under high plasma flux conditions may be due to increased ion scattering as well as more interactions of reactive neutrals with the sidewall of the mesa. SEM micrographs from bulk GaN samples also show a degradation of sidewall profile under high ICP source power conditions. At an ICP source power of 1000 W, the sidewall has a tiered profile with vertical striations possibly due to erosion of the mask edge. However, sidewall profiles at 250 and 500 W looked reasonably anisotropic and smooth.

Reverse leakage currents were relatively insensitive to chemistry effects in a Cl₂/BCl₃/Ar ICP discharge. As shown in Figure 61, the reverse leakage current ranged between ∼10 and

40 nA as the Cl₂ percentage changed from 0 to 100. This is not too surprising, given that BCl₃ ions will be the heaviest ions in the discharge under all of these conditions, and we expect ion damage to be dominated at this flux. The reverse leakage currents were measured from a different GaN wafer as compared with other samples used in this study. The surface morphology for the as-grown wafer was significantly higher for this sample and may account for higher reverse leakage currents measured under the standard conditions. Notice that the GaN etch rate increased as Cl₂ was added to the BCl₃/Ar plasma up to 80%. In Cl₂/Ar plasma the GaN etch rate decreased because of lower concentrations of reactive Cl neutrals. Etch profiles were relatively anisotropic and smooth, except for the Cl₂/Ar plasma, where the etch was slightly rough.

In Figure 62, reverse leakage currents and GaN etch rates are plotted as a function of chamber pressure. Under low-pressure conditions (1 mTorr) the reverse leakage was high, possibly because of higher mean free paths and more energetic collisions of the plasma ions with the sidewall. As the pressure was increased to 2 mTorr and higher, the reverse leakage currents decreased and remained relatively constant. The GaN etch rate increased at 2 mTorr and then decreased at 10 mTorr. This may be attributed to lower plasma densities, redeposition, or polymer formation on the substrate surface. Under low-pressure etch conditions the etch was anisotropic and smooth; however, at 10 mTorr the etch profile was undercut and poorly defined because of a lower mean free path, higher collisional scattering of the ions, and increased lateral etching of the GaN. For the most part, the rms surface roughness was <2 nm regardless of pressure, which is similar to the rms roughness for the as-grown GaN. This is expected because the ion energy is below the value where surface roughening occurs [14, 15].

Two samples were annealed in this study to determine if the defects caused by plasma-induced damage to the p–n junc-

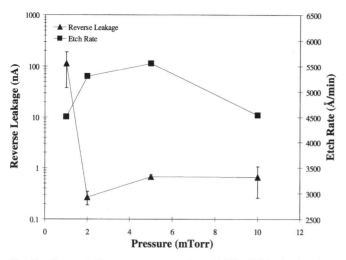

Fig. 62. Reverse leakage current measured at −30 V for GaN p–i–n junctions etched in ICP 32Cl$_2$/8BCl$_3$/5Ar discharges (−100 V dc chuck self-bias, 500 W ICP source power), as a function of chamber pressure.

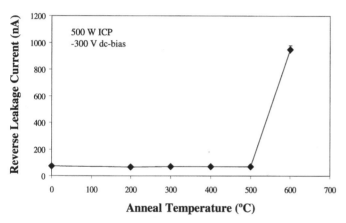

Fig. 63. Reverse leakage current measured at −30 V for GaN p–i–n junctions etched in ICP 32Cl$_2$/8BCl$_3$/5Ar discharges (−300 V dc chuck self-bias, 500 W ICP source power, 2 mTorr), as a function of anneal temperature.

tion could be removed and low reverse leakage currents recovered. The first sample was initially exposed to the following ICP conditions: 32 standard cm^3 Cl$_2$, 8 standard cm^3 BCl$_3$, 5 standard cm^3 Ar, 500 W ICP power, −300 V dc bias, and 2 mTorr pressure. The reverse leakage remained essentially constant up to 600°C, where the reverse leakage increased by more than an order of magnitude (see Fig. 63). (Note that all reverse leakage data were taken at −30 V, except for the 600°C data, which were taken at lower voltages because of breakdown at −30 V.) A similar trend was observed for the second sample (although there was much more scatter in the data, which were collected under the same ICP conditions, with the exceptions of 750 W ICP source power and −100 V dc bias). The inability to remove damage from these samples may be due to annealing temperatures that were not high enough. Cao et al. [33, 34] have reported improved breakdown voltages for dry-etched n- and p-GaN Schottky diodes annealed in the range

of 400–700°C; however, annealing temperatures greater than 800°C were needed to produce a near-complete recovery in breakdown voltage.

In summary, there are high-density plasma etching conditions for GaN where there is minimal degradation in the reverse leakage current of p–i–n mesa diodes. Both ion energy and ion flux are important in determining the magnitude of this current, and a high etch rate is not necessarily the best choice for minimizing dry etch damage.

6. DEVICE PROCESSING

6.1. Microdisk Lasers

A novel laser structure is the microdisk geometry, which does not require facet formation. These lasers should in principle have low thresholds because of their small active volume. Although microcylinder geometries are possible, superior performance is expected when the active disk region sits only on a thin support post or pedestal. To fabricate this latter geometry, it is necessary to have a selective wet etch for the material under the active layer. A schematic of the process is shown in Figure 64. A microcylinder is initially formed by anisotropic dry etching. We have employed ECR or ICP Cl$_2$/Ar discharges to produce the initial vertical etch. The undercut is then produced by the use of KOH solutions at ∼80°C to selectively etch the AlN buffer layer on which the InGaN/GaN quantum well is grown. SEM micrographs of two different lasers are shown in Figure 65. In both cases we used an upper cladding layer of Al-GaN, which was etched somewhat more slowly than the pure AlN bottom cladding layer.

6.2. Ridge Waveguide Lasers

The achievement of continuous-wave GaN-InGaN laser diodes has tremendous technological significance. For commercially acceptable laser lifetimes (typically ≥10,000 h), there is immediate application in the compact disk data storage market. The recording and reading of data on these disks are currently performed with near-infrared (∼780 nm) laser diodes. The switch to the much shorter wavelength (∼400 nm) GaN-based laser diodes will allow higher recording densities [by ∼$(780/400)^2$ or almost a factor of 4]. There is also a large potential market in projection displays, where laser diodes with the three primary colors (red, green, and blue) would replace the existing liquid crystal modulation system. The laser-based system would have advantages in terms of greater design simplicity, lower cost, and broader color coverage. The key development is the need to develop reliable green InGaN laser diodes. The high output power of GaN-based lasers and fast off/on times should also have advantages for improved printer technology, with higher resolution than existing systems based on infrared lasers. In underwater military systems, GaN lasers may have application for covert communications because of a transmission passband in water between 450 and 550 nm.

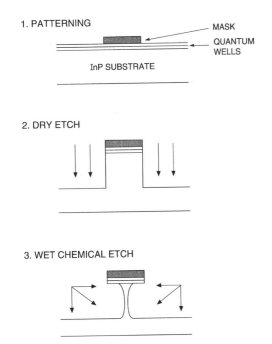

1. PATTERNING
MASK
QUANTUM
WELLS
InP SUBSTRATE

2. DRY ETCH

3. WET CHEMICAL ETCH

Fig. 64.   Schematic of microdisk laser fabrication process.

Although a number of groups have now reported room-temperature lasers in the InGaN/GaN/AlGaN heterostructure system under pulsed [92–105] and cw operation [106–116], the field has been completely dominated by Nakamura et al. [92–97, 99, 100, 105–115]. The growth is performed by MOCVD, generally at atmospheric pressure. Initial structures were grown on c-plane (0001) sapphire, with a low-temperature (550°C) GaN buffer, a thick n^+ GaN lower contact region, an n^+ InGaN strain-relief layer, an n^+ AlGaN cladding layer, a light-guiding region of GaN, and a multiquantum well region consisting of Si-doped $In_{0.15}Ga_{0.85}N$ wells separated by Si-doped $In_{0.02}Ga_{0.98}N$ barriers. The p side of the device consisted of sequential layers of p-AlGaN, p^+ GaN light-guiding, p-$Al_{0.09}Ga_{0.92}N$ cladding, and p^+ GaN contact. A ridge geometry was fabricated by dry etching in most cases (material removed down to the p-$Al_{0.08}Ga_{0.92}N$ layer), followed by dry etching, cleaving, or polishing to form a mirror facet. These facets are coated (with TiO_2/SiO_2 in the Nichia case) to reduce laser threshold, and Ni/Au (p type) and Ti/Al (n type) were employed for ohmic metallization. The typical Nichia structure is shown in Figure 66.

For this type of structure, threshold current densities are typically ≥ 4 kA cm^{-2}, with an operating voltage of ≥ 5 V at the threshold current. The emission mechanism is still the subject of intense study but may be related to the localization of excitons at compositional fluctuations (leading to potential minima in the band structure) in the InGaN wells [117–119]. These devices display relatively short lifetimes under cw operation, typically tens to hundreds of hours. The failure mechanism is most commonly short-circuiting of the p–n junction, a result of p-contact metallization punch-through. It is not that surprising that in this high-defect-density material that the metal can migrate down threading dislocations or voids under high drive-

Fig. 65.   SEM micrographs of GaN/InGaN/AlN microdisk laser structures.

p-electrode
p-GaN
p-Al $_{0.14}$Ga $_{0.86}$N/GaN MD-SLS
p-GaN
p-Al $_{0.2}$Ga $_{0.8}$N
$In_{0.02}Ga_{0.98}N/In_{0.15}Ga_{0.85}N$ MQW
n-GaN
n-Al $_{0.14}$Ga $_{0.86}$N/GaN MD-SLS
n-In $_{0.1}$Ga $_{0.9}$N
SiO$_2$
GaN buffer layer
SiO$_2$
n-electrode
n-GaN
(0001) sapphire substrate

Fig. 66.   Schematic of GaN/InGaN/AlGaN laser diodes grown on Al$_2$O$_3$.

current conditions. The threshold carrier densities of the laser diodes on sapphire are typically $\sim 10^{20}$ cm^{-3}, well above the theoretical values ($\sim 10^{19}$ cm^{-3}) [119–121].

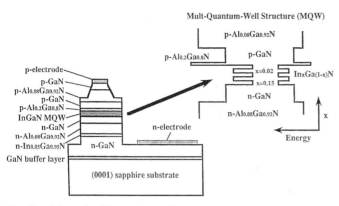

Fig. 67. Schematic of GaN/InGaN/AlGaN laser diodes grown on ELOG substrates.

A major breakthrough in laser diode lifetime occurred with two changes in the growth. The first was replacement of the AlGaN cladding layers with AlGaN/GaN strained-layer superlattices, combined with modulation doping. These changes had the effects of reducing the formation of cracks that often occurred in the AlGaN and reducing the diode operating voltage [111]. The second was the use of epitaxial lateral overgrowth (ELOG) [113, 122, 123]. In this technique GaN is selectively grown on a SiO_2-masked GaN/Al_2O_3 structure. After ~ 10 μm of GaN is deposited over the SiO_2 stripes, it coalesces to produce a flat surface [123]. For a sufficiently wide stripe width, the dislocation density becomes negligible, compared with $\geq 10^9$ cm^{-2} in the window regions. A typical laser diode structure grown by the ELOG method is shown in Figure 67. The laser itself is fabricated slightly off-center from the mask regions, because of gaps that occur there due to imperfect coalescence of the GaN. These devices have a lower threshold current density (≤ 4 kA cm^{-2}) and operating voltage (4–6 V) and much longer (10,000 h) room-temperature lifetimes. The reduction in threading dislocation density dramatically changes the lifetime, because the p metal no longer has a direct path for shorting out the junction during operation. The carrier density at threshold is also reduced to $\sim 3 \times 10^{19}$ cm^{-3}, not far above the expected values. An output power greater than 400 mW and a lifetime greater than 160 h at 30 mW constant output power have been reported.

Subsequent work from Nichia has focused on the growth of laser diodes on quasi-GaN substrates. Thick (100–200 μm) GaN is grown on ELOG structures by either MOCVD or hydride vapor phase epitaxy (VPE). The sapphire substrate is then removed by polishing, to leave a freestanding GaN substrate. The mirror facet can then be formed by cleaving. The GaN substrate has better thermal conductivity than sapphire.

One of the most important features of the etching of the ridge waveguide is the smoothness of the sidewall. Figure 68 shows SEM micrographs of features etched into pure GaN, with the use of a SiN_x mask and an ICP Cl_2/Ar discharge at moderate powers (500 W source power, 150 W rf chuck power). Although the sidewalls are reasonably vertical, one can see striations, which result from roughness on the photoresist mask used

Fig. 68. SEM micrograph of dry-etched GaN feature.

to pattern the SiN_x. Another problem than can occur is illustrated in the SEM micrograph at the top of Figure 69. In this a very high ion energy was employed during the etching, leading to roughening of the feature sidewall. This problem is absent when ion energies below approximately 200 eV are employed, as shown in the micrograph at the bottom of Figure 69.

When careful attention is paid to the lithography, the etching of the SiN_x mask, and the etching of the nitride laser structure, results like those shown in the SEM micrographs of Figure 70 are obtained. The active region of the laser is visible as the horizontal lines along the middle of the sidewall.

6.3. Heterojunction Bipolar Transistors

Wide bandgap semiconductor HBTs are attractive candidates for applications in high-frequency switching, communications, and radar. Although field effect transistors can be used for these same applications [124–126], HBTs have better linearity, higher current densities, and excellent threshold voltage uniformity. The GaN/AlGaN system is particularly attractive

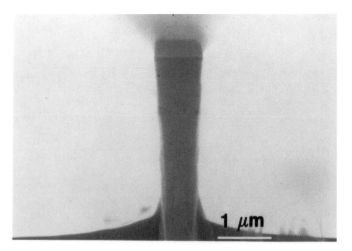

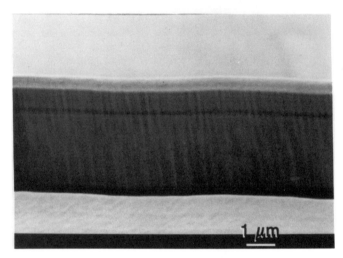

Fig. 69.   SEM micrographs of features etched into GaN at high (top) or moderate (bottom) ion energy.

Fig. 70.   SEM micrographs of dry-etched GaN/InGaN/GaN ridge waveguide laser structure.

because of its outstanding transport properties and the experience base that has developed as a result of the success of light-emitting diodes [127], laser diodes [128–130], and UV detectors [131] fabricated from AlGaInN materials. GaN/SiC HBTs with excellent high-temperature (535°C) performance have been reported [132]. Recently two reports have appeared on the operation of GaN/AlGaN HBTs [133–135]. In one case the extrinsic base resistance was reduced through the selective regrowth of GaN(Mg), and devices with 3×20 μm^2 emitters showed a dc current gain of ~ 3 at 25°C [133]. In work from our group, GaN/AlGaN HBTs have been fabricated with a non-self-aligned, low-damage dry etch process based on that developed for the GaAs/AlGaAs, GaAs/InGaP, and InGaAs/AlInAs systems [136]. The performance of GaN/AlGaN devices fabricated by that method also showed low gains at room temperature, typically ≤ 3. When operated at higher temperatures the gain improved, reaching ~ 10 at 300°C as more acceptors in the base region became ionized and the base resistance decreased.

In this section we review the fabrication process for GaN/AlGaN HBTs, examine the temperature dependence of the p-ohmic contacts, and report measurements of typical background impurity concentrations, determined by secondary ion mass spectrometry (SIMS).

Structures grown by two different methods were examined. In the first, rf plasma-assisted molecular beam epitaxy at a rate of ~ 0.5 $\mu m\,h^{-1}$ was used to grow the HBT structure on top of a 2-μm-thick undoped GaN buffer that was grown on c-plane (0001) sapphire [137]. An 8000-Å-thick GaN sub-collector (Si $\sim 10^{18}$ cm^{-3}) was followed by a 5000-Å-thick GaN collector (Si $\sim 10^{17}$ cm^{-3}), a 1500-Å-thick GaN base (Mg acceptor concentration $\sim 10^{18}$ cm^{-3}), a 1000-Å-thick $Al_{0.15}Ga_{0.85}N$ emitter (Si $\sim 5 \times 10^{17}$ cm^{-3}), and a 500-Å grade to a 2000-Å-thick GaN contact layer (Si $\sim 8 \times 10^{18}$ cm^{-3}).

The second structure was grown by MOCVD on c-plane sapphire [138, 139], with trimethylgallium, trimethylaluminum, and ammonia as the precursors and high-purity H_2 as the carrier gas. The growth process has been described in detail previously [139]. The basic layer structure is shown in Figure 71.

The process flow for device fabrication is shown schematically in Figure 72. First the emitter metal (Ta/Al/Pt/Au) is patterned by liftoff and used as an etch mask for the fabrication of the emitter mesa. The dry etching was performed in a Plasma Therm 770 ICP system with Cl_2/Ar discharges. The process

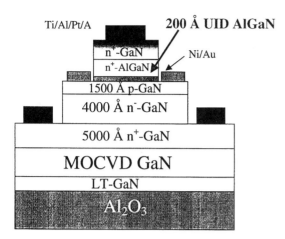

Fig. 71. Schematic of MOCVD-grown GaN/AlGaN HBT.

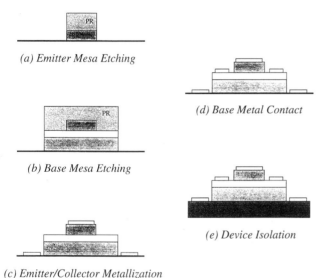

(a) Emitter Mesa Etching

(b) Base Mesa Etching

(c) Emitter/Collector Metallization

(d) Base Metal Contact

(e) Device Isolation

Fig. 72. Schematic process sequence for GaN/AlGaN HBT.

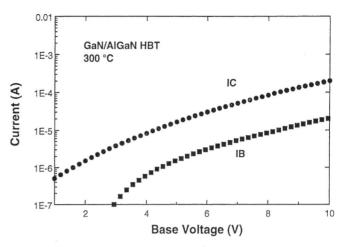

Fig. 73. Gummel plot measured at 300°C for GaN/AlGaN HBT.

pressure was 5 mTorr, and the source was excited with 300 W of 2-MHz power. This power controlled the ion flux and neutral density, and the incident ion energy was controlled by the application of 40 W of 13.56-MHz power to the sample chuck. Base metallization of Ni/Pt/Au was patterned by liftoff, and then the mesa was formed by dry etching. The etch rate of GaN under our conditions was ∼1100 Å/min and was terminated at the subcollector, where Ti/Al/Pt/Au metallization was deposited. The contacts were alloyed at 700–800°C.

It has been firmly established that high specific contact resistivities are a limiting factor in GaN-based device performance, and, in particular, for the p-ohmic contact. We examined the alloying temperature dependence of the current–voltage ($I–V$) characteristics for several different p-metal schemes. The as-deposited contacts are rectifying. Annealing at progressively higher temperatures produced a significant improvement. But even for 800°C anneals the contacts were not purely ohmic when measured at room temperature. This is consistent with past data, showing that p-metallization on GaN is often better described as a leaky Schottky contact.

As the measurement temperature is increased, the hole concentration in the p-GaN increases through higher ionization efficiency of the Mg acceptors. For example, the hole concentration would increase from ∼10% of the acceptor density at 25°C to ∼60% at 300°C, based on Fermi–Dirac statistics. The p-contact becomes truly ohmic at ≥300°C. From transmission line measurements, we found that $\rho_c \approx 2 \times 10^{-2} \ \Omega \, cm^{-2}$ at this temperature. This indicates that the GaN/AlGaN HBT will perform better at elevated temperatures, where the base contact resistivity is lower. The contact barrier is on the order of 0.5 eV, whereas the Mg acceptor has an ionization level of 0.18 eV.

The device performances of the molecular beam epitaxy- and MOCVD-grown devices were similar, namely a common-emitter current gain of ≤3 at 25°C, increasing to ∼10 at 300°C. A Gummel plot from the molecular beam epitaxy device is shown in Figure 73. In both devices the performance was still limited by the base resistance, and methods to increase the base doping and lower the extrinsic resistance in this region will be critical for future efforts in this area. The common base current gain, α, was in the range of 0.75 (25°C) to 0.9 (300°C), indicating that the base transport factor is close to unity and that I_B is dominated by reinjection to the emitter.

Another important aspect of the realization of GaN/AlGaN HBTs is confinement of the Mg doping to the base. If the p-type spills over into the relatively lightly doped emitter, then the junction is displaced and the advantage of the heterostructure is lost.

In summary, GaN/AlGaN HBTs have been fabricated on both molecular beam epitaxy- and MOCVD-grown material, and they display similar performances, i.e., a common-emitter current gain of ∼10 when operated at ∼300°C. However, junction leakage is also higher at elevated temperatures, which is a major drawback in most applications. The fabrication process developed previously for other compound semiconductor systems works well for the GaN/AlGaN system, with the main difference being higher annealing temperatures required for the ohmic contacts. The device performance is still limited by the base doping for both molecular beam epitaxy and MOCVD structures.

6.4. Field Effect Transistors

Much attention has been focused recently on the development of AlGaN/GaN heterostructure field effect transistors (HFETs) for high-frequency and high-power applications [140–145]. Both enhancement and depletion mode devices have been demonstrated, with gate lengths down to 0.2 μm. Excellent dc performance has been reported up to 360°C [145], and the best devices have a maximum frequency of oscillation f_{max} of 77 GHz at room temperature [142]. Even better speed performance could be expected from InAlN channel structures, because of the superior transport properties and the ability to use highly doped $In_xAl_{1-x}N$ ($x = 0 \rightarrow 1$) graded contact layers, which should produce low specific contact layers, which should produce specific contact resistivities. We have previously demonstrated that nonalloyed Ti/Pt/Au metal on degenerately doped InN ($n = 5 \times 10^{20}$ cm^{-3}) has a ρ_c of $\sim 1.8 \times 10^{17}$ Ω cm^2 [146]. Although MOCVD has generally been employed for the growth of nitride-based photonic devices and for most of the prototype electronic devices [147, 149], the ability of the molecular beam techniques to control layer thickness and incorporate higher In concentration in the ternary alloys is well suited to the growth of HFET structures [150, 151].

The exceptional chemical stability of the nitrides has meant that dry etching must be employed for patterning. To date most of the work in this area has concentrated on the achievement of higher etch rates with minimal mask erosion, particularly because a key application is the formation of dry-etched layer facets. In that case etch rate, etch anisotropy, and sidewall smoothness are the most important parameters, and little attention has been paid to the effect of dry etching on the stoichiometry and electrical properties of the nitride surface.

In these experiments, we used an InAlN and GaN FET structure as a test vehicle for measuring the effect of ECR BCl$_3$-based dry etching on the surface properties of InAlN and GaN. Preferential loss of N leads to roughened morphologies and the creation of a thin n^+ surface layer that degrades the rectifying properties of subsequently deposited metal contacts.

The InAlN samples were grown by metal organic molecular beam epitaxy (MOMBE) on 2″-diameter GaAs substrates with the use of WAVEMAT ECR N$_2$ plasma and metalorganic group III precursors (trimethylamine alane, triethylindium). A low-temperature ($\sim$400°C) AlN nucleation layer was followed by a 500-Å-thick AlN buffer layer grown at 700°C. The In$_{0.3}$Al$_{0.7}$N channel layer ($\sim$5 $\times$ 10^{17} cm^{-2}) was 500 Å thick, and then an ohmic contact layer was produced by grading to pure InN over a distance of $\sim$500 Å.

The GaN layer structure was grown on double side polished c-Al$_2$O$_3$ substrates prepared initially by HCl/HNO$_3$/H$_2$O cleaning and an in situ H$_2$ bake at 1070°C. A GaN buffer more than 300 Å thick was grown at 500°C and crystallized by ramping the temperature to 1040°C, where trimethylgallium and ammonia were again used to grow $\sim$1.5 μm of undoped GaN ($n < 3 \times 10^{16}$ cm^{-3}), a 2000-Å channel ($n = 2 \times 10^{17}$ cm^{-3}), and a 1000-Å contact layer ($n = 1 \times 10^{18}$ cm^{-3}).

FET surfaces were fabricated by the deposition of TiPtAu source/drain ohmic contacts, which were protected by photore-

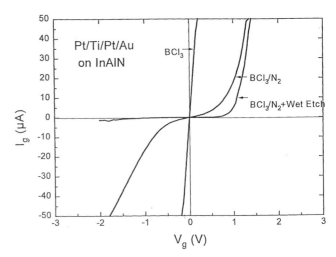

Fig. 74. *I–V* characteristics of Pt/TiPt/Au contacts on InAlN exposed to different ECR plasmas.

sists. The gate mesa was formed by dry etching down to the InAlN or *n*-GaN channel, with the use of ECR BCl$_3$ or BCl$_3$/N$_2$ plasma chemistry. During this process, we noticed that the total conductivity between the ohmic contacts did not decrease under some conditions. CH$_4$/H$_2$ etch chemistry was also studied. To simulate the effects of this process, we exposed the FET substrates to D$_2$ plasma and saw strong reductions in sample conductivity. The incorporation of D$_2$ into InAlN was measured by SIMS. Changes to the surface stoichiometry were measured by AES. All plasma processes were carried out in a Plasma Therm SLR 770 System with an Astex 7700 low-profile ECR source operating at 500 W. The samples were clamped to an rf-powered, He backside cooled chuck, which was left at floating potential (about -30 V) relative to the body of the plasma.

Upon dry-etch removal of the InAlN capping layer, a Pt/Ti/Pt/Au gate contact was deposited on the exposed InAlN to complete the FET processing. When pure BCl$_3$ was employed as the plasma chemistry, we observed ohmic and not rectifying behavior for the gate contact. When BCl$_3$/N$_2$ was used, there was some improvement in the gate characteristics. A subsequent attempt at a wet-etch clean-up with either H$_2$O$_2$/HCl or H$_2$O$_2$//HCl produced a reverse breakdown in excess of 2 V (Fig. 74). These results suggest that the InAlN surface becomes nonstoichiometric during the dry etch step, and that addition of N$_2$ retards some of this effect.

Figure 75 shows the I_{DS} values obtained as a function of dry etch time in ECR discharges of either BCl$_3$ or BCl$_3$/N$_2$. In the former case the current does not decrease as material is etched away, suggesting that a conducting surface layer is continually being created. In contrast, BCl$_3$/N$_2$ plasma chemistry does reduce the drain-source current as expected, even though the breakdown characteristics of gate metal deposited on this surface are much poorer than would be expected.

From the AFM studies of the InAlN gate contact layer surface after removal of the In$_x$Al$_{1-x}$N contact layer surface in BCl$_3$, BCl$_3$/N$_2$, or BCl$_3$/N$_2$ plus wet etch, we observed that the mean surface roughness is worse for the former two

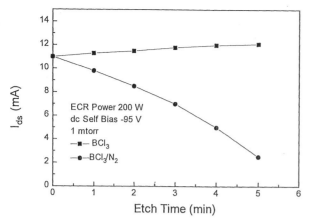

Fig. 75. I_{DS} values at 5-V bias for InAlN FETs etched for various times in BCl_3 or BCl_3/N_2 ECR plasmas.

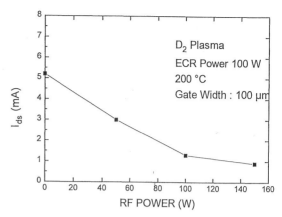

Fig. 76. I_{DS} values at 5-V bias, as a function of the rf power used during D_2 ECR plasma exposures at 200°C for 30 min.

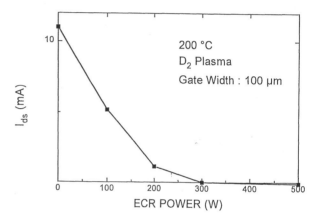

Fig. 77. I_{DS} values at 5-V bias as a function of the ECR source power used during D_2 plasma exposure at 200°C for 30 min.

chemistries, indicating that preferential loss of N is probably occurring during the dry etch step. To try to remove the group III-enriched region, the samples were rinsed in a 1 : 1 HCl/H_2O_2 solution at 50°C for 1 min. This step did increase the reverse breakdown voltage (~2.5 V) compared with dry-etched only samples, (<1 V), but did not produce a completely damage-free surface. This is because the HCl/H_2O_2 does not remove the InAlN immediately below the surface, which is closer to stoichiometry but is still defective.

Although BCl_3/N_2 produces less enrichment than pure BCl_3, there is clearly the presence of a defective layer that prevents the achievement of acceptable rectifying contacts. From the I–V measurements, we believe this defective layer is probably strong n-type, in analogy with the situation with InP described earlier. At this stage, there is no available wet etch solution for InAlN that could be employed to completely remove the non-stoichiometric layer in the type of clean-up step commonly used in other III–V materials. Other possible solutions to this problem include the use of a higher Al concentration in the stop layer, which should be more resistant to nitrogen loss, or employment of a layer structure that avoids the need for gate recess.

A study of simulating CH_4/H_2 etch chemistry was conducted on a full FET structure, which requires etching of the InAlN contact layer to expose the gate contact layer for deposition of the gate metallization. Figure 76 shows the I_{DS} as a function of applied bias before and after a D_2 plasma treatment at 200°C for 30 min. The loss of conductivity could be due to two different mechanisms. The first is hydrogen forming neutral complexes (D—H)—o—, with the donor extra electrons taken up by the formation of forming a bond with the hydrogen. The second mechanism is the creation of deep acceptor states that trap the electrons and remove them from the conduction process. These states might be formed by the energetic D^+ or D_2^+ ion bombardment from the plasma.

The decrease in I_{DS} in the InAlN FET was a strong function of the ion energy in the plasma and of the active neutral (D^0) and ion density (D^+, D_2^+). Figure 77 shows the dependence of I_{DS} on ECR microwave power. As this power is in-

creased, both dissociation of D_2 molecules into atoms and ionization of atomic and molecular species will increase, and thus it is difficult to separate out bombardment and passivation effects. At fixed ECR power, the I_{DS} values also decrease with rf power, which controls the ion energy. At 150 W, the ion energy increases to ~200 eV, with a corresponding decrease in I_{DS}. This is good evidence that creation of deep traps is playing at least some role, and the fact the hydrogen is found to diffuse all the way through the sample also implicates passivation as contributing to the loss of conductivity.

Figure 78 shows the gate current–voltage characteristics when the gate metal is deposited on the as-etched GaN surface. The Schottky contact is extremely leaky, with poor breakdown voltage. We believe this is caused by the presence of a highly conducting N-deficient surface, similar to the situation encountered on dry etching InP, where preferential loss of P produces a metal-rich surface, which precludes the achievement of rectifying contacts. AES analysis of the etched GaN surface showed an increasing Ga-to-N ratio (from 1.7 to 2.0 in terms of raw counts) upon etching. However, a 5-min anneal at 400°C under N_2 was sufficient to produce excellent rectifying contacts, with a gate breakdown of ~25 V (Fig. 79). We believe the presence of the conducting surface layer after etching is a strong con-

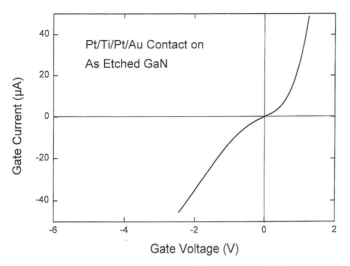

Fig. 78. *I–V* characteristic on ECR BCl₃-etched GaN.

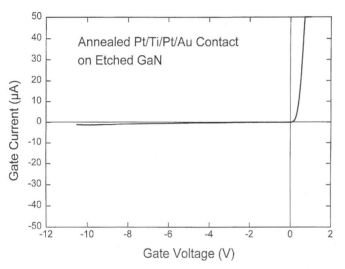

Fig. 79. *I–V* characteristic on ECR BCl₃-etched GaN annealed at 400°C before deposition of the gate metal.

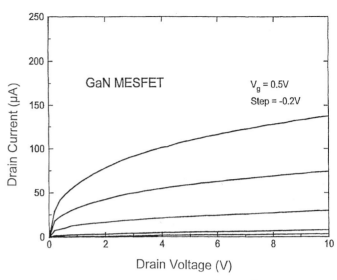

Fig. 80. Drain *I–V* characteristics of a $1 \times 50 \ \mu\text{m}^2$ MESFET.

tributing factor to the excellent ρ_c values reported by Lin et al. [150] for contacts on a reactively ion-etched *n*-GaN.

The drain *I–V* characteristics of the $1 \times 50 \ \mu\text{m}^2$ MESFET are shown in Figure 80. The drain-source breakdown was −20 V, with a threshold voltage of −0.3 V. The device displays good pinch-off and no slope to the *I–V* curves due to gate leakage, indicating that the anneal treatment is sufficient to restore the surface breakdown characteristics. We believe these devices are well suited for high power applications, because GaN is a robust material and the contract metallizations employed are also very stable.

III-nitride FET structures are sensitive to several effects during dry etching of the gate mesa. First, if hydrogen is present in the plasma there can be passivation of the doping in the channel layer. Second, the ion bombardment from the plasma can create deep acceptor states that compensate for the material. Third, even when these problems are avoided through the use of H-free plasma chemistries and low ion energies and fluxes, preferen-

tial loss of N can produce poor rectifying gate characteristics for metal deposited on the etched surface. Ping et al. observed that pure Ar etching produced more damage in Schottky diodes than did SiCl₄ RIE. The diode characteristics were strongly dependent on plasma self-bias, and annealing at 680°C removed much of the damage.

6.5. UV Detectors

Gallium nitride (GaN) and its alloys of aluminum gallium nitride (AlGaN) are the most promising semiconductors for the development of ultraviolet (UV) photodetectors for applications such as combustion monitoring, space-based UV spectroscopy, and missile plume detection. With a direct bandgap energy of approximately 3.39 eV (366 nm), GaN is an ideal material for the fabrication of photodetectors capable of rejecting near-infrared and visible regions of the solar spectrum while retaining near-unity quantum efficiency in the UV. The use of AlGaN materials in photodetector fabrication makes possible bandgap engineering of the peak responsivity to shorter wavelengths in the deep UV [152]. GaN is also an extremely robust semiconductor suitable for high-temperature (>200°C) applications.

Nitride-based UV photodetectors that have been reported include *p–n* photodiode devices with 0.05 mm² junction area and 0.07 A/W peak responsivity [153], 0.04 mm² junction area [154], 0.25 mm² junction area and 0.1 A/W peak responsivity [155], and 0.59 mm² junction area and 0.195 A/W peak responsivity [156]. Other reported photodetectors include semitransparent Schottky junction devices [157] and metal–semiconductor–metal devices [158, 159].

The GaN and AlGaN UV photodiodes were grown on (0001) basal-plane sapphire substrates by molecular beam epitaxy with the use of an rf atomic nitrogen plasma source [160]. Details of the growth process for nitride detectors have been reported [161]. The *p–i–n* detector epitaxial layers consisted of a $5 \times 10^{18} \ \text{cm}^{-3}$ *n*-GaN layer followed by a 5000-Å intrin-

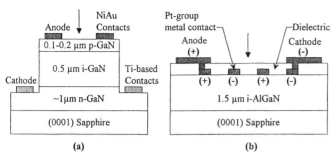

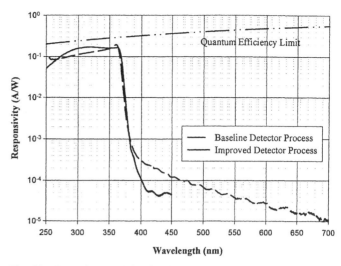

Fig. 81. Schematic of epitaxial and device structures for (a) GaN p–i–n UV photodetectors and (b) AlGaN MSM photodetectors.

sic region with unintentional n-type doping in the 10^{15} cm^{-3} decade. The topmost epitaxial layer consisted of 1000–2000 Å 1×10^{18} cm^{-3} p-GaN. Mesas reaching the n-GaN cathode contact layer were formed by ICP plasma etching with chlorine-based chemistry [162]. Ohmic contacts with the n-type and p-type GaN were made by Ti-based and Ni-based metallizations, respectively. All of the GaN p–i–n UV detectors were fabricated with an optical detection area of 0.5 mm^2 and a p–i–n junction area of 0.59 mm^2, which is considerably larger (>12.5 times) than other GaN p–n detectors reported with noise measurements.

In addition to p–i–n type detectors, shorter UV wavelength (MSM) photodetectors, operating in a quasi-photoconductive mode, were fabricated from 1.5-μm-thick silicon-doped ($\sim 1 \times 10^{17}$ cm^{-3}) n-AlGaN with a bandgap energy of approximately 320 nm. The MSMs were fabricated by first depositing a Pt-group metallization 1 μm wide with a 5-μm pitch to form the Schottky contacts. Next a dielectric was deposited to act as an insulator between the AlGaN semiconductor and the bond pads. The dielectric process was not optimized to function as an antireflection coating. Reported here are results for AlGaN MSMs with active areas of 0.25 mm^2. Schematics of the two structures are shown in Figure 81.

Shunt resistance and spectral responsivity data were collected with the use of on-wafer probing. The shunt resistance was determined by the linear trace of the current–voltage (I–V) characteristic from -10 mV to $+10$ mV. The spectral responsivities of the UV photodiodes were measured in photovoltaic mode (zero bias) for p–i–n devices and a photoconductive mode (biased) for MSM devices, with the use of a 75-W xenon arc lamp chopped at 700 Hz and filtered by a 1/8-m monochromator set to a 5-nm bandpass. The power of the monochromatic light was measured with a calibrated, National Institutes of Standards and Techniques (NIST)-traceable, silicon photodiode and then focused onto GaN wafers resting on a micropositioner stage.

The GaN p–i–n UV photodetector responsivity measurements reported were obtained with the devices operating in the unbiased, photovoltaic mode. Shown in Figure 82 is a 25°C spectral responsivity curve for a baseline UV photodetector with 0.194 A/W peak responsivity and a visible rejection of 4 orders of magnitude, which has been reported [157]. The maximum theoretical peak responsivity at the 360-nm bandgap

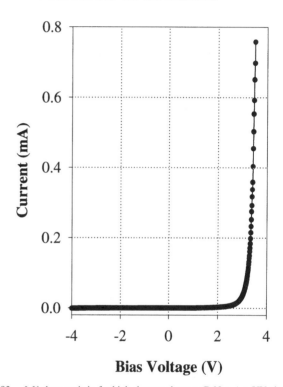

Fig. 82. Spectral responsivity for GaN p–i–n UV photodetectors plotted against the maximum theoretical value with no reflection.

Fig. 83. I–V characteristic for high-shunt-resistance GaN p–i–n UV photodiodes. The excellent forward bias characteristic was based on a 3.1-V junction potential.

is 0.28 A/W with no reflection and 0.23 A/W including reflection at the GaN surface. Also included on the plot is a trace for a GaN p–i–n UV photodetector with an improved p-type epitaxial process, which yields a greater visible rejection and more constant deep UV responsivity. The improved GaN p–i–n device was fabricated with a 1000-Å p-type cathode layer. The shunt resistances for these improved 0.59 mm^2 devices ranged from 200 MΩ to 50 GΩ, depending on the process. Shown in Figure 83 is the current–voltage (I–V) characteristic of the high shunt resistance photodetector, the responsivity of which

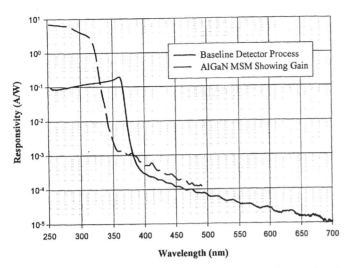

Fig. 84.   Spectral responsivity for an AlGaN MSM UV photodetector plotted against the baseline GaN *p–i–n* UV photodetector spectral responsivity.

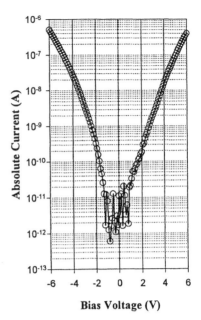

Fig. 85.   *I–V* characteristic for high-shunt-resistance AlGaN MSM photodetectors. The near-zero bias voltage measurements were limited by ambient room electrical noise.

is traced in Figure 82. The device exhibits a low dark current and excellent forward-biased diode *I–V* characteristics with a built-in potential of approximately 3.1 V.

The AlGaN MSM UV photodetector measurement shown in Figure 84 was obtained at a bias of 6 V between the anode and cathode contacts. The responsivity measurement is the absolute value, which includes a 20% reflection of light from the metal contacts. The AlGaN MSM exhibited a substantial photoconductive gain (>700×), which yielded a responsivity of 7 A/W at 250 nm. The AlGaN device band-edge response does not decay as rapidly as that seen in devices fabricated from GaN material. The MSM devices, fabricated from AlGaN, which exhibited a luminescent peak at 320 nm, exhibited a rejection of slightly more than 3 orders of magnitude (1278×) of 360-nm light over 320-nm light. The rejection ratio for 250-nm light over 360-nm light was more than 5.5 orders of magnitude (5263×). The increased gain at shorter wavelengths is believed to be the result of greater electron-hole generation near the high electric field regions at the surface of the device. As expected, the MSM devices exhibited excellent shunt resistance (>100 GΩ) and dark current characteristics at low bias, as shown in Figure 85.

GaN *p–i–n* UV photodetectors with an optically active surface area of 0.5 mm^2 and a junction area of 0.59 mm^2 have been fabricated on 3-inch-diameter GaN *p–i–n* epitaxial wafers and characterized. Wafer maps of photodetector peak responsivity (maximum of 0.194 A/W at 359 nm) indicated that more than 60% (±1σ) of all of the GaN UV photodetectors performed within a ±12% deviation from the average peak responsivity. Furthermore, the vast majority of GaN UV photodetectors were characterized with shunt resistances that were within one decade of each other.

High-temperature testing of the GaN *p–i–n* photodetectors up to 300°C indicated no significant increase in visible spectral responsivity or short-term degradation. The room-temperature spectral responsivity of the GaN photodetectors was fully recovered after 300°C testing. The 300°C GaN photodetector 1/*f*

noise power densities were measured to be 6.6×10^{-19} and 2.1×10^{-21} A^2/Hz at 100 Hz and 1 kHz, respectively. The room-temperature, 100-Hz, and 1-kHz noise power density of the GaN *p–i–n* photodetectors was extrapolated to be of the order of 10^{-30} A^2/Hz ($\sim 10^{-15}$ A/Hz$^{1/2}$ noise current density).

The AlGaN MSM photodetectors, which were fabricated from AlGaN with a near-bandgap luminescent peak of 320 nm, exhibited substantial photoconductive gain, resulting in 7 A/W responsivity at 250 nm and 1.7 A/W responsivity at 320 nm. The AlGaN MSMs were characterized by responsivity rejection ratios of 5263 and 1278 for 250-nm and 320-nm light, respectively, versus 360-nm light.

Acknowledgments

The work at the University of Florida is partially supported by a Defense Advanced Research Planning Agency/Electric Power Research Institute grant (MDA972-98-1-0006) (D. Radach and B. Damsky) monitored by the Office of Naval Research (J. C. Zolper) and a National Science Foundation grant (DMR97-32865) (L. D. Hess). Sandia is a multiprogram laboratory operated by Sandia Corporation, a Lockheed-Martin Company, for the U.S. Department of Energy under contract DEAC04-94-AC-85000.

REFERENCES

1. S. J. Pearton, Ed., "GaN and Related Materials." Gordon and Breach, New York, 1997.

2. J. I. Pankove and T. D. Moustakas, Eds., "GaN." Academic Press, San Diego, 1988.

3. M. G. Craford, in "IUVSTA Workshop on GaN," Hawaii, August 1997.

4. S. Nakamura, M. Senoh, S. Nagahama, N. Iwasa, T. Yamada, T. Matsushita, Y. Sugimoto, and H. Kikoyu, *Jpn. J. Appl. Phys., Part 2* 36, L1059 (1997).

5. S. Nakamura, *IEEE J. Sel. Areas Commun.* 3, 435 (1997).

6. S. Nakamura, in "3rd International GaN Conference," Tokushima, Japan, October 1997.

7. S. Nakamura, *IEEE J. Sel. Areas Commun.* 4, 483 (1998).

8. N. Morkoc, "Wide Band Gap Nitrides and Devices." Springer-Verlag, Berlin, 1998.

9. S. Nakamura and G. Fosol, "The Blue Laser Diode." Springer-Verlag, Berlin, 1998.

10. S. N. Mohammad and H. Morkoc, *Prog. Quantum Electron.* 20, 361 (1996).

11. S. N. Mohammad, A. Salvador, and H. Morkoc, *Proc. IEEE* 83, 1306 (1995).

12. H. Morkoc, S. Strite, G. B. Gao, M. E. Lin, B. Sverdlov, and M. Burns, *J. Appl. Phys.* 76, 1368 (1994).

13. F. A. Ponce, in "Encyclopedia of Applied Physics" (G. L. Trigg, Ed.). VCH, Weinheim, 1998.

14. F. A. Ponce and D. P. Bour, *Nature (London)* 386, 351 (1997).

15. I. Akasaki and H. Amano, *J. Electrochem. Soc.* 141, 2266 (1994).

16. S. J. Pearton and R. J. Shul, in "Gallium Nitride I" (J. I. Pankove and T. D. Moustakas, Eds.). Academic Press, San Diego, 1998.

17. P. Gillis, D. A. Choutov, P. A. Steiner, J. D. Piper, J. H. Crouch, P. M. Dove, and K. P. Martin, *Appl. Phys. Lett.* 66, 2475 (1995).

18. R. J. Shul, in "Processing of Wide Bandgap Semiconductors" (S. J. Pearton, Ed.). Noyes, Park Ridge, NJ, 1999.

19. I. Adesida, A. Mahajan, E. Andideh, M. Asif Khan, D. T. Olsen, and J. N. Kuznia, *Appl. Phys. Lett.* 63, 2777 (1993).

20. I. Adesida, A. T. Ping, C. Youtsey, T. Sow, M. Asif Khan, D. T. Olsen, and J. N. Kuznia, *Appl. Phys. Lett.* 65, 889 (1994).

21. O. Aktas, Z. Fan, S. N. Mohammad, A. Botcharev, and H. Morkoc, *Appl. Phys. Lett.* 69, 25 (1996).

22. M. A. Khan, Q. Chen, M. S. Shur, B. T. McDermott, J. A. Higgins, J. Burm, W. Schaff, and L. F. Eastman, *Electron. Lett.* 32, 357 (1996).

23. Y. F. Wu, S. Keller, P. Kozodoy, B. P. Keller, P. Parikh, D. Kapolnek, S. P. DenBaars, and V. K. Mishra, *IEEE Electron. Device Lett.* 18, 290 (1997).

24. M. A. Khan, J. N. Kuznia, M. S. Shur, C. Eppens, J. Burm, and W. Schaff, *Appl. Phys. Lett.* 66, 1083 (1995).

25. W. A. Harrison, "Electronic Structure and Properties of Solids." Freeman, San Francisco, 1980.

26. S. J. Pearton, C. R. Abernathy, F. Ren, and J. R. Lothian, *J. Appl. Phys.* 76, 1210 (1994).

27. M. E. Lin, Z. F. Zan, Z. Ma, L. H. Allen, and H. Morkoc, *Appl. Phys. Lett.* 64, 887 (1994).

28. A. T. Ping, I. Adesida, M. Asif Khan, and J. N. Kuznia, *Electron. Lett.* 30, 1895 (1994).

29. H. Lee, D. B. Oberman, and J. S. Harris, Jr., *Appl. Phys. Lett.* 67, 1754 (1995).

30. S. J. Pearton, C. R. Abernathy, F. Ren, J. R. Lothian, P. W. Wisk, A. Katz, and C. Constantine, *Semicond. Sci. Technol.* 8, 310 (1993).

31. S. J. Pearton, C. R. Abernathy, and F. Ren, *Appl. Phys. Lett.* 64, 2294 (1994).

32. S. J. Pearton, C. R. Abernathy, and F. Ren, *Appl. Phys. Lett.* 64, 3643 (1994).

33. R. J. Shul, S. P. Kilcoyne, M. Hagerott Crawford, J. E. Parmeter, C. B. Vartuli, C. R. Abernathy, and S. J. Pearton, *Appl. Phys. Lett.* 66, 1761 (1995).

34. C. B. Vartuli, J. D. MacKenzie, J. W. Lee, C. R. Abernathy, S. J. Pearton, and R. J. Shul, *J. Appl. Phys.* 80, 3705 (1996).

35. L. Zhang, J. Ramer, K. Zheng, L. F. Lester, and S. D. Hersee, *Mater. Res. Soc. Symp. Proc.* 395, 763 (1996).

36. L. Zhang, J. Ramer, J. Brown, K. Zheng, L. F. Lester, and S. D. Hersee, *Appl. Phys. Lett.* 68, 367 (1996).

37. B. Humphreys and M. Govett, *MRS Internet J. Nitride Semicond. Res.* 1, 6 (1996).

38. J. W. Lee, J. Hong, J. D. MacKenzie, C. R. Abernathy, S. J. Pearton, F. Ren, and P. F. Sciortino, *J. Electron. Mater.* 26, 290 (1997).

39. C. B. Vartuli, S. J. Pearton, J. W. Lee, J. Hong, J. D. MacKenzie, C. R. Abernathy, and R. J. Shul, *Appl. Phys. Lett.* 69, 1426 (1996).

40. R. J. Shul, A. J. Howard, S. J. Pearton, C. R. Abernathy, C. B. Vartuli, P. A. Barnes, and M. J. Bozack, *J. Vac. Sci. Technol., B* 13, 2016 (1995).

41. C. B. Vartuli, S. J. Pearton, C. R. Abernathy, R. J. Shul, A. J. Howard, S. P. Kilcoyne, J. E. Parmeter, and M. Hagerott Crawford, *J. Vac. Sci. Technol., A* 14, 1011 (1996).

42. R. J. Shul, C. I. H. Ashby, D. J. Rieger, A. J. Howard, S. J. Pearton, C. R. Abernathy, C. B. Vartuli, and P. A. Barnes, *Mater. Res. Soc. Symp. Proc.* 395, 751 (1996).

43. R. J. Shul, in "GaN and Related Materials" (S. J. Pearton, Ed.). Gordon and Breach, Amsterdam, the Netherlands, 1997.

44. R. J. Shul, G. B. McClellan, S. J. Pearton, C. R. Abernathy, C. Constantine, and C. Barratt, *Electron. Lett.* 32, 1408 (1996).

45. R. J. Shul, G. B. McClellan, S. A. Casalnuovo, D. J. Rieger, S. J. Pearton, C. Constantine, C. Barratt, R. F. Karlicek, Jr., C. Tran, and M. Schurmann, *Appl. Phys. Lett.* 69, 1119 (1996).

46. Y. H. Lee, H. S. Kim, G. Y. Yeom, J. W. Lee, M. C. Yoo, and T. I. Kim, *J. Vac. Sci. Technol., A* 16, 1478 (1998).

47. H. S. Kim, Y. H. Lee, G. Y. Yeom, J. W. Lee, and T. I. Kim, *Mater. Sci. Eng., B* 50, 82 (1997).

48. Y. B. Hahn, D. C. Hays, S. M. Donovan, C. R. Abernathy, J. Han, R. J. Shul, H. Cho, K. B. Jung, and S. J. Pearton, *J. Vac. Sci. Technol., B* 17, 1237 (1999).

49. H. Cho, C. B. Vartuli, S. M. Donovan, J. D. MacKenzie, C. R. Abernathy, S. J. Pearton, R. J. Shul, and C. Constantine, *J. Electron. Mater.* 27, 166 (1998).

50. H. Cho, C. B. Vartuli, S. M. Donovan, C. R. Abernathy, S. J. Pearton, R. J. Shul, and C. Constantine, *J. Vac. Sci. Technol., A* 6, 1631 (1998).

51. J. W. Lee, H. Cho, D. C. Hays, C. R. Abernathy, S. J. Pearton, R. J. Shul, G. A. Vawter, and J. Han, *IEEE J. Sel. Top. Quantum Electron.* 4, 557 (1998).

52. R. J. Shul, C. G. Willison, M. M. Bridges, J. Han, J. W. Lee, S. J. Pearton, C. R. Abernathy, J. D. MacKenzie, and S. M. Donovan, *Solid-State Electron.* 42, 677 (1999).

53. R. J. Shul, C. I. H. Ashby, C. G. Willison, L. Zhang, J. Han, M. M. Bridges, S. J. Pearton, J. W. Lee, and L. F. Lester, *Mater. Res. Soc. Symp. Proc.* 512, 487 (1998).

54. S. A. Smith, C. A. Wolden, M. D. Bremser, A. D. Hanser, and R. F. Davis, *Appl. Phys. Lett.* 71, 3631 (1997).

55. R. J. Shul, in "GaN and Related Materials II" (S. J. Pearton, Ed.). Gordon and Breach, New York, 1998.

56. G. F. McLane, M. Meyyappan, M. W. Cole, and C. Wrenn, *J. Appl. Phys.* 69, 695 (1991).

57. M. Meyyappan, G. F. McLane, H. S. Lee, E. Eckart, M. Namaroff, and J. Sasserath, *J. Vac. Sci. Technol., B* 10, 1215 (1992).

58. G. F. McLane, L. Casas, S. J. Pearton, and C. R. Abernathy, *Appl. Phys. Lett.* 66, 3328 (1995).

59. A. T. Ping, I. Adesida, and M. Asif Khan, *Appl. Phys. Lett.* 67, 1250 (1995).

60. A. T. Ping, A. C. Schmitz, and M. Asif Khan, *J. Electron. Mater.* 25, 825 (1996).

61. A. T. Ping, M. Asif Khan, and I. Adesida, *Semicond. Sci. Technol.* 12, 133 (1997).

62. R. J. Shul, G. A. Vawter, C. G. Willison, M. M. Bridges, J. W. Lee, S. J. Pearton, and C. R. Abernathy, *Solid-State Electron.* 42, 2259 (1998).

63. J. W. Lee, C. Vartuli, J. MacKenzie, J. R. Mileham, S. J. Pearton, R. J. Shul, J. C. Zolper, M. Crawford, J. Zavada, R. Wilson, and R. Schwartz, *J. Vac. Sci. Technol., B* 14, 3637 (1996).

64. H. P. Gillis, D. A. Choutov, K. P. Martin, and L. Song, *Appl. Phys. Lett.* 68, 2255 (1996).

65. H. P. Gillis, D. A. Choutov, and K. P. Martin, *J. Mater.* 50, 41 (1996).

66. H. P. Gillis, D. A. Choutov, K. P. Martin, S. J. Pearton, and C. R. Abernathy, *J. Electrochem. Soc.* 143, 251 (1996).

67. R. T. Leonard and S. M. Bedair, *Appl. Phys. Lett.* 68, 794 (1996).

68. C. B. Vartuli, S. J. Pearton, J. W. Lee, J. D. MacKenzie, C. R. Abernathy, and R. J. Shul, *J. Vac. Sci. Technol., A* 15, 638 (1997).

69. C. B. Vartuli, S. J. Pearton, C. R. Abernathy, R. J. Shul, and F. Ren, *Proc. Electrochem. Soc.* 97-34, 39 (1998).

70. J. W. Lee, J. Hong, and S. J. Pearton, *Appl. Phys. Lett.* 68, 847 (1996).

71. R. J. Shul, C. G. Willison, M. M. Bridges, J. Han, J. W. Lee, S. J. Pearton, C. R. Abernathy, J. D. MacKenzie, and S. M. Donovan, *Mater. Res. Soc. Symp. Proc.* 482, 802 (1998).

72. R. J. Shul, C. G. Willison, M. M. Bridges, J. Han, J. W. Lee, S. J. Pearton, C. R. Abernathy, J. D. MacKenzie, S. M. Donovan, L. Zhang, and L. F. Lester, *J. Vac. Sci. Technol., A* 16, 1621 (1998).

73. C. Constantine, C. Barratt, S. J. Pearton, F. Ren, and J. R. Lothian, *Electron. Lett.* 28, 1749 (1992).

74. C. Constantine, C. Barratt, S. J. Pearton, F. Ren, and J. R. Lothian, *Appl. Phys. Lett.* 61, 2899 (1992).

75. D. G. Lishan and E. L. Hu, *Appl. Phys. Lett.* 56, 1667 (1990).

76. T. R. Hayes, in "Indium Phosphide and Related Materials: Processing, Technology and Devices" (A. Katz, Ed.), Chap. 8, pp. 277–306. Artech House, Boston, 1992.

77. R. J. Shul, R. D. Briggs, J. Han, S. J. Pearton, J. W. Lee, C. B. Vartuli, K. P. Killeen, and M. J. Ludowise, *Mater. Res. Soc. Symp. Proc.* 468, 355 (1997).

78. F. Ren, J. R. Lothian, J. M. Kuo, W. S. Hobson, J. Lopata, J. A. Caballero, S. J. Pearton, and M. W. Cole, *J. Vac. Sci. Technol., B* 14, 1 (1995).

79. F. Ren, W. S. Hobson, J. R. Lothian, J. Lopata, J. A. Caballero, S. J. Pearton, and M. W. Cole, *Appl. Phys. Lett.* 67, 2497 (1995).

80. R. J. Shul, G. B. McClellan, R. D. Briggs, D. J. Rieger, S. J. Pearton, C. R. Abernathy, J. W. Lee, C. Constantine, and C. Barratt, *J. Vac. Sci. Technol., A* 15, 633 (1997).

81. C. B. Vartuli, S. J. Pearton, J. D. MacKenzie, C. R. Abernathy, and R. J. Shul, *J. Electrochem. Soc.* 143, L246 (1996).

82. C. B. Vartuli, S. J. Pearton, J. W. Lee, J. D. MacKenzie, C. R. Abernathy, and R. J. Shul, *J. Vac. Sci. Technol., B* 15, 98 (1997).

83. H. Cho, J. Hong, T. Maeda, S. M. Donovan, J. D. MacKenzie, C. R. Abernathy, S. J. Pearton, R. J. Shul, and J. Han, *MRS Internet J. Nitride Semicond. Res.* 3, 5 (1998).

84. C. B. Vartuli, S. J. Pearton, J. W. Lee, J. D. MacKenzie, C. R. Abernathy, R. J. Shul, C. Constantine, and C. Barratt, *J. Electrochem. Soc.* 144, 2844 (1997).

85. F. Ren, S. J. Pearton, R. J. Shul, and J. Han, *J. Electron. Mater.* 27, 175 (1998).

86. S. J. Pearton, J. W. Lee, J. D. MacKenzie, C. R. Abernathy, and R. J. Shul, *Appl. Phys. Lett.* 67, 2329 (1995).

87. F. Ren, J. R. Lothian, S. J. Pearton, C. R. Abernathy, C. B. Vartuli, J. D. MacKenzie, R. G. Wilson, and R. F. Karlicek, *J. Electron. Mater.* 26, 1287 (1997).

88. T. Ping, A. C. Schmitz, I. Adesida, M. A. Khan, O. Chen, and Y. W. Yang, *J. Electron. Mater.* 26, 266 (1997).

89. R. J. Shul, J. C. Zolper, M. Hagerott Crawford, S. J. Pearton, J. W. Lee, R. F. Karlicek, Jr., C. Tran, M. Schurmann, C. Constantine, and C. Barratt, *Proc. Electrochem. Soc.* 96-15, 232 (1996).

90. S. J. Pearton, U. K. Chakrabarti, and F. A. Baiocchi, *Appl. Phys. Lett.* 55, 1633 (1989).

91. H. Cho, K. A. Auh, R. J. Shul, S. M. Donovan, C. R. Abernathy, E. S. Lambers, F. Ren, and S. J. Pearton, *J. Electron. Mater.* 28, 288 (1999).

92. S. Nakamura, M. Senoh, S. Nagahama, N. Iwasa, T. Yamada, T. Matsushita, H. Kiyoku, and Y. Sugimoto, *Jpn. J. Appl. Phys., Part 2* 35, L74 (1996).

93. S. Nakamura, M. Senoh, S. Nagahama, N. Iwasa, T. Yamada, T. Matsushita, H. Kiyoku, and Y. Sugimoto, *Jpn. J. Appl. Phys., Part 2* 35, L217 (1996).

94. S. Nakamura, M. Senoh, S. Nagahama, N. Iwasa, T. Yamada, T. Matsushita, H. Kiyoku, and Y. Sugimoto, *Appl. Phys. Lett.* 68, 2105 (1996).

95. S. Nakamura, M. Senoh, S. Nagahama, N. Iwasa, T. Yamada, T. Matsushita, H. Kiyoku, and Y. Sugimoto, *Appl. Phys. Lett.* 68, 3269 (1996).

96. S. Nakamura, M. Senoh, S. Nagahama, N. Iwasa, T. Yamada, T. Matsushita, H. Kiyoku, Y. Sugimoto, and H. Kiyoku, *Appl. Phys. Lett.* 69, 1477 (1996).

97. S. Nakamura, M. Senoh, S. Nagahama, N. Iwasa, T. Yamada, T. Matsushita, H. Kiyoku, Y. Sugimoto, and H. Kiyoku, *Appl. Phys. Lett.* 69, 1568 (1996).

98. K. Itaya, M. Onomura, J. Nishino, L. Sugiura, S. Saito, M. Suzuki, J. Rennie, S. Nunoue, M. Yamamoto, H. Fujimoto, Y. Kokubun, Y. Ohba, G. Hatakoshi, and M. Ishikawa, *Jpn. J. Appl. Phys., Part 2* 35, L1315 (1996).

99. S. Nakamura, M. Senoh, S. Nagahama, N. Iwasa, T. Yamada, T. Matsushita, Y. Sugimoto, and H. Kiyoku, *Appl. Phys. Lett.* 69, 3034 (1996).

100. S. Nakamura, M. Senoh, S. Nagahama, N. Iwasa, T. Yamada, T. Matsushita, Y. Sugimoto, and H. Kiyoku, *Appl. Phys. Lett.* 70, 616 (1997).

101. G. E. Bulman, K. Doverspike, S. T. Sheppard, T. W. Weeks, H. S. Kong, H. M. Dieringer, J. A. Edmond, J. D. Brown, J. T. Swindell, and J. F. Schetzina, *Electron. Lett.* 33, 1556 (1997).

102. M. P. Mack, A. Abare, M. Aizcorbe, P. Kozodoy, S. Keller, U. K. Mishra, L. Coldren, and S. DenBaars, *MRS Internet J. Nitride Semicond. Res.* 2, 41 (1997) (available from http://nsr.mij.mrs.org/2/5/).

103. Kuramata, K. Domen, R. Soejima, K. Horino, S. Kubota, and T. Tanahashi, *Jpn. J. Appl. Phys., Part 2* 36, L1130 (1997).

104. T. Kobayashi, F. Nakamura, K. Naganuma, T. Toyjo, H. Nakajima, T. Asatsuma, H. Kawai, and M. Ikeda, *Electron. Lett.* 34, 1494 (1998).

105. S. Nakamura, M. Senoh, S. Nagahama, N. Iwasa, T. Yamada, T. Matsushita, Y. Sugimoto, and H. Kiyoku, *Appl. Phys. Lett.* 69, 4056 (1996).

106. S. Nakamura, M. Senoh, S. Nagahama, N. Iwasa, T. Yamada, T. Matsushita, Y. Sugimoto, and H. Kiyoku, *Appl. Phys. Lett.* 70, 1417 (1997).

107. S. Nakamura, M. Senoh, S. Nagahama, N. Iwasa, T. Yamada, T. Matsushita, Y. Sugimoto, and H. Kiyoku, *Appl. Phys. Lett.* 70, 2753 (1997).

108. S. Nakamura, *IEEE J. Sel. Top. Quantum Electron.* 3, 435 (1997).

109. S. Nakamura, M. Senoh, S. Nagahama, N. Iwasa, T. Yamada, T. Matsushita, Y. Sugimoto, and H. Kiyoku, *Jpn. J. Appl. Phys., Part 2* 36, L1059 (1997).

110. S. Nakamura, in "24th International Symposium on Compound Semiconductors," San Diego, CA, Plen-1, September 8–11, 1997.

111. S. Nakamura, M. Senoh, S. Nagahama, N. Iwasa, T. Kozaki, H. Umemoto, M. Sano, and K. Chocho, *Jpn. J. Appl. Phys., Part 2* 36, L1568 (1997).

112. S. Nakamura, M. Senoh, S. Nagahama, N. Iwasa, T. Yamada, T. Matsushita, H. Kiyoku, Y. Sugimoto, T. Kozacki, H. Umemoto, M. Sano, and K. Chocho, *Appl. Phys. Lett.* 72, 2014 (1998).

113. S. Nakamura, M. Senoh, S. Nagahama, N. Iwasa, T. Yamada, T. Matsushita, H. Kiyoku, Y. Sugimoto, T. Kozacki, H. Umemoto, M. Sano, and K. Chocho, *Jpn. J. Appl. Phys., Part 2* 37, L309 (1998).

114. S. Nakamura, M. Senoh, S. Nagahama, N. Iwasa, T. Yamada, T. Matsushita, H. Kiyoku, Y. Sugimoto, T. Kozacki, H. Umemoto, M. Sano, and K. Chocho, *Appl. Phys. Lett.* 73, 822 (1998).

115. S. Nakamura, M. Senoh, S. Nagahama, N. Iwasa, T. Yamada, T. Matsushita, H. Kiyoku, Y. Sugimoto, T. Kozacki, H. Umemoto, M. Sano, and K. Chocho, *Jpn. J. Appl. Phys., Part 2* 37, L627 (1998).

116. S. Chichibu, T. Azuhata, T. Sota, and S. Nakamura, *Appl. Phys. Lett.* 69, 4188 (1996).

117. Y. Narukawa, Y. Kawakami, S. Fujita, and S. Nakamura, *Phys. Rev. B: Condens. Matter* 55, 1938R (1997).

118. Y. Narukawa, Y. Kawakami, M. Funato, S. Fujita, and S. Nakamura, *Appl. Phys. Lett.* 70, 981 (1997).

119. M. Suzuki and T. Uenoyama, *Jpn. J. Appl. Phys., Part 1* 35, 1420 (1996).

120. M. Suzuki and T. Uenoyama, *Appl. Phys. Lett.* 69, 3378 (1996).

121. W. W. Chow, A. F. Wright, and J. S. Nelson, *Appl. Phys. Lett.* 68, 296 (1996).

122. O. H. Nam, M. D. Bremser, T. Zheleva, and R. F. Davis, *Appl. Phys. Lett.* 71, 2638 (1997).

123. Y. Kato, S. Kitamura, K. Hiramatsu, and N. Sawaki, *J. Cryst. Growth* 144, 133 (1994).

124. J. C. Zolper, *Solid-State Electron.* 42, 2153 (1998).

125. M. S. Shur and M. A. Khan, in "High Temperature Electronics" (M. Willander and H. L. Hartnagel, Eds.), pp. 297–321. Chapman and Hall, London, 1996.

126. E. R. Brown, *Solid-State Electron.* 42, 2119 (1998).

127. S. Nakamura, *IEEE J. Sel. Top. Quantum Electron.* 3, 345 (1997).

128. S. Nakamura, M. Senoh, S. Nagahama, N. Iwasa, T. Yamada, T. Matsushita, Y. Sugimoto, and H. Kiyoku, *Appl. Phys. Lett.* 70, 616 (1997).

129. G. E. Bulman, K. Doverspike, S. T. Sheppard, T. W. Weeks, H. S. Kong, H. Dieringer, J. A. Edmond, J. D. Brown, J. T. Swindell, and J. F. Schetzina, *Electron. Lett.* 33, 1556 (1997).

130. M. P. Mack, A. Abare, M. Aizcorbe, P. Kozodoy, S. Keller, U. K. Mishra, L. Coldren, and S. P. DenBaars, *MRS Internet J. Nitride Semicond. Res.* 2, 41 (1997).

131. J. M. Van Hove, R. Hickman, J. J. Klaassen, P. P. Chow, and P. P. Ruden, *Appl. Phys. Lett.* 70, 282 (1997).

132. J. I. Pankove, M. Leksono, S. S. Chang, C. Walker, and B. Van Zeghbroeck, *MRS Internet J. Nitride Semicond. Res.* 1, 39 (1996).

133. L. S. McCarthy, P. Kozodoy, S. P. DenBaars, M. Rodwell, and U. K. Mishra, in "25th International Symposium Compound Semiconductors," Nara, Japan, October 1998. *IEEE Electron. Dev. Lett. EDL* 20, 277 (1999).

134. F. Ren, C. R. Abernathy, J. M. Van Hove, P. P. Chow, R. Hickman, J. J. Klaassen, R. F. Kopf, H. Cho, K. B. Jung, J. R. LaRoche, R. G. Wilson, J. Han, R. J. Shul, A. G. Baca, and S. J. Pearton, *MRS Internet J. Nitride Semicond. Res.* 3, 41 (1998).

135. J. Han, A. G. Baca, R. J. Shul, C. G. Willison, L. Zhang, F. Ren, A. P. Zhang, G. T. Dang, S. M. Donovan, X. A. Cao, H. Cho, K. B. Jung, C. R. Abernathy, S. J. Pearton, and R. G. Wilson, *Appl. Phys. Lett.* 74, 2702 (1999).

136. F. Ren, J. R. Lothian, S. J. Pearton, C. R. Abernathy, P. W. Wisk, T. R. Fullowan, B. Tseng, S. N. G. Chu, Y. K. Chen, L. W. Yang, S. T. Fu, R. S. Brozovich, H. H. Lin, C. L. Henning, and T. Henry, *J. Vac. Sci. Technol., B* 12, 2916 (1994).

137. R. Hickman, J. M. Van Hove, P. P. Chow, J. J. Klaassen, A. M. Wowchack, and C. J. Polley, *Solid-State Electron.* 42, 2138 (1998).

138. J. Han, M. H. Crawford, R. J. Shul, J. J. Figiel, M. Banas, L. Zhang, Y. K. Song, H. Zhou, and A. V. Nurmikko, *Appl. Phys. Lett.* 73, 1688 (1998).

139. J. Han, T.-B. Ng, R. M. Biefeld, M. H. Crawford, and D. M. Follstaedt, *Appl. Phys. Lett.* 71, 3114 (1997).

140. M. A. Khan, J. N. Kuznia, D. T. Olson, W. Schaff, J. Burm, and M. S. Shur, *Appl. Phys. Lett.* 65, 1121 (1994).

141. S. C. Binari, L. B. Rowland, W. Kruppa, G. Kelner, K. Doverspike, and D. K. Gaskill, *Electron. Lett.* 30, 1248 (1994).

142. M. A. Khan, M. S. Shur, J. N. Kuznia, J. Burm, and W. Schuff, *Appl. Phys. Lett.* 66, 1083 (1995).

143. M. A. Khan, Q. Chen, C. J. Sun, J. W. Wang, M. Blasingame, M. S. Shur, and H. Park, *Appl. Phys. Lett.* 68, 514 (1996).

144. W. Kruppa, S. C. Binari, and K. Doverspike, *Electron. Lett.* 31, 1951 (1995).

145. S. C. Binari, L. B. Rowland, G. Kelner, W. Kruppa, H. B. Dietrich, K. Doverspike, and D. K. Gaskill, *Inst. Phys. Conf. Ser.* 141, 459 (1995).

146. F. Ren, C. R. Abernathy, S. N. G. Chu, J. R. Lothian, and S. J. Pearton, *Appl. Phys. Lett.* 66, 1503 (1995).

147. S. Strike, M. E. Lin, and H. Morkoc, *Thin Solid Films* 231, 197 (1993).

148. H. Morkoc, S. Strite, G. B. Bao, M. E. Lin, B. Sverdlov, and M. Burns, *J. Appl. Phys.* 76, 1363 (1994).

149. C. R. Abernathy, J. D. MacKenzie, S. R. Bharatan, K. S. Jones, and S. J. Pearton, *Appl. Phys. Lett.* 66, 1632 (1995).

150. M. E. Lin, Z. E. Fan, Z. Ma, L. H. Allen, and H. Morkoc, *Appl. Phys. Lett.* 64, 887 (1994).

151. A. T. Ping, A. C. Schmitz, I. Adesida, M. A. Khan, O. Chen, and Y. W. Yang, *J. Electron. Mater.* 26, 266 (1997).

152. J. M. Van Hove, P. P. Chow, R. Hickman, A. M. Wowchak, J. J. Klaassen, and C. J. Polley, *Mater. Res. Soc. Proc.* 449, 1227 (1996).

153. G. Xu, A. Salvador, A. E. Botchkarev, W. Kim, C. Lu, H. Tang, H. Morkoc, G. Smith, M. Estes, T. Dang, and P. Wolf, *Mater. Sci. Forum, Part 2* 264–268, 1441 (1998).

154. D. V. Kuksenkov, H. Temkin, A. Osinsky, R. Gaska, and M. A. Khan, "1997 International Electron Devices Meeting Technical Digest," p. 759. IEEE, Piscataway, NJ, 1997.

155. A. Osinsky, S. Gangopadhyay, R. Gaska, B. Williams, and M. A. Khan, *Appl. Phys. Lett.* 71, 2334 (1997).

156. R. Hickman, J. J. Klaassen, J. M. Van Hove, A. M. Wowchak, C. Polley, M. R. Rosamond, and P. P. Chow, *MRS Internet J. Nitride Semicond. Res.* 4S1, G7.6 (1999).

157. M. A. Khan, J. N. Kuznia, D. T. Olson, M. Blasingame, and A. R. Bhattaria, *Appl. Phys. Lett.* 63, 1781 (1993).

158. J. C. Carrano, T. Li, D. L. Brown, P. A. Grudowski, C. J. Eiting, R. D. Dupuis, and J. C. Campbell, *Appl. Phys. Lett.* 73, 2405 (1998).

159. Ferguson, S. Liang, C. A. Tran, R. F. Karlicek, Z. C. Feng, Y. Lu, and C. Joseph, *Mater. Sci. Forum, Part 2* 264–268, 1437 (1998).

160. J. M. Van Hove, G. J. Cosimini, E. Nelson, A. M. Wowchak, and P. P. Chow, *J. Cryst. Growth* 150, 908 (1995).

161. J. M. Van Hove, R. Hickman, J. J. Klaassen, and P. P. Chow, *Appl. Phys. Lett.* 70, 2282 (1997).

162. H. Cho, C. B. Vartuli, C. R. Abernathy, S. M. Donovan, S. J. Pearton, R. J. Shul, and J. Han, *Solid-State Electron.* 42, 2277 (1998).

Chapter 9

RESIDUAL STRESSES IN PHYSICALLY VAPOR-DEPOSITED THIN FILMS

Yves Pauleau

National Polytechnic Institute of Grenoble, CNRS-UJF-LEMD, B.P. 166, 38042 Grenoble Cedex 9, France

Contents

1. INTRODUCTION

Thin films of a wide variety of conductive, semiconductor, and insulating materials prepared by physical vapor deposition (PVD) techniques are used as active or passive components in a large number of devices for advanced technologies (microelectronics, optics, mechanics, etc.). Usually, three major categories of techniques can be distinguished to produce PVD thin films: evaporation, sputtering, and ion plating [1]. The growth mechanism of PVD thin films onto substrate surfaces at

Handbook of Thin Film Materials, edited by H.S. Nalwa
Volume 1: Deposition and Processing of Thin Films
Copyright © 2002 by Academic Press
All rights of reproduction in any form reserved.

ISBN 0-12-512909-2/$35.00

ambient or low temperatures (with respect to the melting point of the deposited material) involves the condensation of atoms or molecules physically generated in the vapor phase from a condensed (solid or liquid) source material; this step appears as the common characteristic of PVD processes. Additional common features of all PVD processes include: a high vacuum (base pressure $\leq 10^{-9}$ Pa) and gas delivery system, high purity source materials, a substrate mounting assembly, and often a number of diagnostic tools used to control and to characterize the deposition process (thickness monitors, mass, and optical emission spectrometers, residual gas analyzers, etc.).

For practical purposes, PVD thin films must be tightly adherent to substrate surfaces. Indeed, defects and delamination at the film-substrate interface may lead to poor performance and destruction of devices. Since the film is generally firmly bonded to a massive substrate (much thicker than the film), any change in length along the film plane which is not matched exactly by an equal change in length of the substrate results in a stress in the film. Excessive residual stresses in thin films may cause mechanical damage (fracture) and adhesion failures [2, 3] as well as defect formation in the substrate [4, 5]. On a less catastrophic scale, large stresses can affect the optical, electrical, and magnetic properties of films [6, 7]. The elimination and the release of residual stresses are also the cause of the undesirable formation of hillocks, whiskers, and holes in PVD thin films [8, 9]. Thus, there is a considerable interest in measuring the magnitude of residual stresses in PVD thin films prepared under different experimental conditions. Usually, the control of residual stresses is a critical point in a given PVD process to produce reliable thin films. The need to understand the origin of residual stresses is vital to the implementation of efficient PVD processes for the manufacturing production of high quality films for a wide variety of applications such as protective films, active and passive electronic devices, layered composite structures, etc.

Three component parts are currently distinguished in the overall residual stresses developed in thin films at a given temperature. The magnitude of residual stresses, σ, can be expressed as

$$\sigma = \sigma_{\text{th}} + \sigma_i + \sigma_e \qquad (1)$$

The thermal stress, σ_{th}, arises from the different thermal expansion coefficients of films and substrates combined with the difference between the deposition or substrate temperature, T_s, and the temperature of samples, T_r, during the determination of stresses which is commonly equal to room temperature. The intrinsic stress, σ_i, in thin films is introduced during the PVD process and is not directly related to thermal mismatch between the deposited material and the substrate; this stress may originate from incomplete structural-ordering processes occurring during the growth of films. The magnitude of the intrinsic stress can be related to the microstructure or morphology of films which depends on various process parameters. These physical features of films (microstructure, grain size, and morphology, texture) are directly dependent on the deposition parameters [10–12], i.e., in particular, on the substrate temperature and the

pressure in the deposition chamber which affects the kinetic energy of species impinging on the substrate or film surface during deposition of films; these incident species including molecules, radicals, atoms, ions can be either condensed and incorporated in the film lattice or simply reflected at the film surface. As a result, it is convenient to separate considerations on intrinsic stresses in thin films produced from nonenergetic particles (with an average kinetic energy of incident particles corresponding to the thermal energy at evaporation which is less than 0.5 eV) from those regarding films deposited by condensation of energetic species and/or under bombardment of particles having kinetic energies ranging from a few electron volts to a few hundreds of electron volts; these PVD processes such as sputtering, dual ion beam sputtering, ion beam-assisted deposition in which the simultaneous deposition and the energetic particle bombardment of films are involved are generally referred to as ion-assisted deposition (IAD) or ion beam-assisted deposition (IBAD) processes [11]. The third component in residual stresses, named extrinsic stress, σ_e, is induced by external factors and results from interactions between the deposited material and the environment which are subsequent to deposition, e.g., this type of stress may arise from adsorption of water vapor in porous films exposed to room air immediately after removing the samples from the deposition chamber. Then, interactions between the adsorbed species and the films may be responsible for changes in composition and volume of the deposited material; as a result, a slow and progressive variation of the extrinsic stress (and residual stresses) may be observed as a function of the aging time of adsorbed species containing films. The effects of residual stresses in PVD thin films are so crucial for the performance of devices that residual stresses have been the object of many articles and continue to attract attention. The literature related to the production, the release, and the determination of residual stresses in thin films or other types of stress data is so extensive that every article cannot be reviewed in detail.

The present chapter deals with the effects of PVD process parameters including physical factors which are specific to the PVD process on residual stresses in continuous films, i.e., with a thickness higher than that corresponding to a film formed by the simple coalescence of islands on flat substrate surfaces [10]. The average value of residual stresses in continuous films is observed to be independent of the film thickness. In addition, PVD thin films can grow on various types of substrates (single crystal, polycrystalline, or amorphous materials). The microstructure and the morphology of PVD films then, in turn, the residual stresses are affected by the substrate materials [13]. However, a large majority of PVD thin films are deposited onto nonepitaxial substrates; i.e., the substrate is amorphous or polycrystalline and may be covered with an adsorbed film of impurities providing weak bonds to the adatoms. This type of substrate is likely to induce an island mode of growth with no crystallographic relationship between the orientation of the islands (if crystalline) and the substrate; i.e., the islands are not epitaxial with the substrate. Hence, the chapter is essentially focused on continuous films deposited onto nonepitaxial substrates at low temperatures, i.e., at temperatures at which thermally activated annealing processes such as grain growth, defect annihilation, and mass transport are excluded.

The goal of this chapter is to review the major models or mechanisms proposed in the literature to explain the origin of residual stresses in PVD thin films and, then to present various stress data illustrating the validity and the limits of these models. Section 2 of this chapter gives a summary of the major features of the microstructure and the morphology of PVD thin films together with a presentation of the major structure-zone models proposed for films produced under various deposition conditions. Section 3 concentrates on the determination of residual stresses by various experimental techniques, calculation of the magnitude of residual stresses in thin films deposited on thick substrates and mechanical stability of films or film-substrate structures. Section 4 reviews the models invoked to explain the origin of residual stresses in thin films produced by PVD techniques from nonenergetic particles and with energetic particle bombardment. Section 5 discusses the effects of major process parameters (pressure, substrate bias voltage, substrate temperature) on intrinsic stress built up during deposition of films. Section 6 is related to stress data reported in the literature for silicon dioxide and amorphous carbon films prepared by various PVD techniques. These data were selected and were presented to illustrate the applicability, the validity, or the limits of major models for the origin of residual stresses in PVD films discussed in previous sections of the chapter.

2. MICROSTRUCTURE AND MORPHOLOGY OF PVD THIN FILMS

Various experimental parameters related to the deposition equipment and the process affect the nucleation on the substrate surface and the early stages of the growth of PVD films. Furthermore, the nucleation steps and the growth mechanisms of PVD films determine the microstructure and the morphology of the deposited materials which, ultimately, determine the physical properties of PVD films such as microhardness, residual stresses, surface roughness, mass density, reflectivity, etc. Therefore, it is important to understand the growth mechanisms involved in the development of grain structures and morphology of films.

The major process parameters which can affect the early stages of growth include: source and substrate materials purity, composition, and crystallographic orientation of substrate surfaces, deposition rate of films, impurity concentration, deposition, or substrate temperature, T_s, pressure, kinetic energy of species arriving at the film surface which condense or simply reflect at the surface of films during deposition. The nucleation and the growth of films generally occur via condensation of individual atoms or polyatomic species striking the substrate surface. The condensation of species onto the surface involves an energetic equilibrium step with the surface atoms and a diffusion step of adatoms across the surface up to a low energy adsorption site. Usually, the final resting sites of adatoms are

defect sites such as vacancies, growth ledges, or preexisting nucleus. A very short time of the order of a lattice vibration time of 10^{-12} s is required for equilibrating adatoms with the surface atoms; the corresponding quenching rate is about 10^{15} K s^{-1}. The adatoms accommodated on the surface may either desorb and return to the gas phase or chemisorb and become incorporated in the growing film. The kinetics of thin film nucleation are almost always controlled by the defect density on the surface; in addition, the role of defects increases as the adatom flux decreases [14].

2.1. Nucleation and Growth Modes of PVD Thin Films

In general, three primary growth modes of PVD thin films can be distinguished on the basis of the bonding strength between depositing atoms and substrate atoms (Fig. 1).

2.1.1. Three-Dimensional Island Growth Mode

This three-dimensional island growth referred to as the Volmer–Weber growth mode involves the nucleation of the condensed phase into distinct, small clusters directly on the substrate surface. Then, with elapsing time and arrival of additional adatoms, the clusters grow into islands which begin to touch and finally coalesce to form a continuous film. This type of growth mode is encountered with adatoms more strongly bound to each other than to the substrate atoms as is often the case for metal films deposited onto insulating or contaminated substrates. Clusters below the critical size dissolve and break apart whereas those larger than the critical size stabilize and grow in accordance with the heterogeneous three-dimensional nucleation theory using the capillarity model [15–17].

2.1.2. Two-Dimensional Layer-by-Layer Growth Mode

This type of two-dimensional growth corresponding to the Frank–van der Merwe growth mode occurs when the depositing atoms exhibit a greater bonding strength with the substrate atoms than with other depositing atoms. The free-energy barrier for nucleation or the free energy of clusters with the critical size is reduced to zero; thus, the adatom mobility is very high and layer-by-layer growth occurs. The two-dimensional

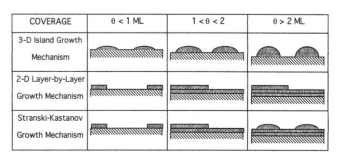

COVERAGE	$\theta < 1$ ML	$1 < \theta < 2$	$\theta > 2$ ML
3-D Island Growth Mechanism			
2-D Layer-by-Layer Growth Mechanism			
Stranski-Kastanov Growth Mechanism			

Fig. 1. Schematic of three growth modes of films. θ is the substrate surface coverage in monolayer (ML).

growth mode is involved in the homoepixatial growth on a clean substrate surface as is the case for metal–metal and semiconductor–semiconductor systems.

2.1.3. Mixed Growth Mode

The third growth mode or the Stranski–Kastanov mode is a combination of the first two modes described previously. This growth mode is characterized by a shift from the two-dimensional layer-by-layer growth mode during the initial stages of the growth to the three-dimensional island growth mode in the latter stages. In other words, after forming one to several two-dimensional monolayers, the two-dimensional layer-by-layer growth mode becomes unfavorable and the depositing adatoms begin to nucleate and to form islands on the epitaxial material. The origin of this transition from two-dimensional to three-dimensional growth is not completely understood. The driving force for this transition can be the release of elastic energy stored in the film caused by film–substrate lattice mismatch. The release of stresses between the substrate and the deposited material leads to the formation of defects in the film that act as heterogeneous nucleation sites. The mixed growth mode can be found in metal–metal and metal–semiconductor systems such as indium films grown on ⟨100⟩-oriented single crystal silicon substrates.

2.2. Effect of Energetic Particle Condensation and/or Bombardment on Nucleation and Early Stages of the Growth of Films

The bombardment of the substrate surface and the growing films with energetic (ionized and/or neutral) particles, i.e., with a kinetic energy ranging from a few electron volts to a few hundreds of electron volts, has been observed to produce beneficial modifications in a number of characteristics and properties which are critical to the performance of PVD thin films. The deposition of films with energetic particle bombardment performed under appropriate conditions can lead to improved adhesion of films to substrates, mass density enhancement, or densification of films grown at low substrate temperatures, and better control of texture (orientation), morphology, and grain sizes; thereby, modification of residual stresses can be observed. In addition to a sputter-etching effect on the materials surface, the energetic particle-solid surface interactions can induce fundamental changes in nucleation mechanisms and growth kinetics of films.

Various effects resulting from these interactions have been investigated and discussed in the literature; the major effects include the production of defects in the substrate surface which can act as preferential adsorption sites, trapping or implantation of incident species in the near-surface region, dissociation of small clusters during the early stages of the growth of films, enhanced adatom diffusion. No single mechanism can explain all observed results. For a given experiment, the dominating effects essentially depend on the film–substrate materials combination,

kinetic energy, flux and mass of the incident particles, and substrate temperature.

2.2.1. Three-Dimensional Island Growth Mode

The density of preferential nucleation sites created on the substrate surface under energetic particle bombardment is determined by the difference between their rate of production (which depends on the kinetic energy, flux, and mass of incident particles as well as on the mass of substrate atoms) and their rate of loss via thermal annealing phenomena (which increases with increasing substrate temperature). For instance, the nucleation rate of germanium films produced by IAD can be increased or decreased depending upon the choice of the substrate temperature and the substrate material [18]. A decrease in island density has been observed on substrate surfaces under energetic particle bombardment; as a result, islands can be produced with larger average sizes for a given nominal film thickness and ultimately the films exhibit larger grain sizes [19].

The major mechanism proposed to explain increased average island sizes in systems associating substrates with weakly bound adatoms is the depletion of small clusters caused by sputtering and particle-induced dissociation [20]. Small clusters with subcritical sizes are energetically unfavorable and dissociate spontaneously into adatoms which desorb and return to the gas phase or diffuse on the substrate surface to larger clusters or islands. Meanwhile, the energetic particle bombardment of the island-substrate system results in a minor sputter etching of large clusters. This growth mechanism leading to the formation of islands with larger sizes and films with larger grain sizes is expected to occur under experimental conditions corresponding to a high incident energetic particle to vapor flux ratio, ϕ_p/ϕ_c.

Furthermore, the diffusivity of adatoms can be enhanced during IAD through the initiation of shallow collision cascades and the excitation of surface phonons. However, the excess energy gained by near-surface atoms involved in collision cascades are transferred to the lattice and the atoms are thermalized within several vibrational periods. This mechanism which may contribute to a decrease in the epitaxial temperature would not be expected to result in enhanced diffusion of adatoms over lengths of more than several lattice distances. Consequently, various phenomena occurring in IAD processes are greatly involved in the growth kinetics and the mechanisms of three-dimensional islands; these phenomena affect the defect concentration, the density of adsorption sites, the island, and the grain sizes. However, much work remains to be done to understand in detail these effects occurring for complex deposition processes of thin films.

2.2.2. Two-Dimensional Layer-by-Layer Growth Mode

A decrease in the epitaxial temperature defined as the lowest substrate temperature at which a (1×1) reflection high-energy electron diffraction (RHEED) pattern was obtained from a 5-nm-thick film, was reported for silicon films deposited on single crystal Si and Al_2O_3 substrates by molecular beam epitaxy (MBE) from a flux of partially ionized and accelerated species [21]; the epitaxial temperature was reduced by more than 100°C on Si and Al_2O_3 substrates after ionizing less than 1% of the incident flux of silicon atoms and accelerating the ions to 200 and 100 eV, respectively.

Mechanisms based on sputter cleaning or etching effects on the surface can be invoked to explain the bombardment-induced enhancement of the film epitaxy [22, 23]. However, other mechanisms may also be involved in the MBE process from energetic species as demonstrated by molecular dynamic (MD) simulations [24]. The use of a flux of accelerated species for deposition of films increases the impact mobility of adatoms [25]; as a result, an increase in the average distance between the point of the first interaction with the substrate surface and the final adsorption site of the adatom was found to be of the order of a few lattice distances in MD simulations. In addition, local arrangements of lattice atoms occur in the first few hundred fentoseconds following the collision between an energetic particle and the surface as well as during the subsequent relaxation period. Finally, the energetic particle bombardment during deposition of films provides local atomic arrangements allowing atoms to relax into lower energy sites and leading to reduced epitaxial temperatures.

However, under certain experimental conditions, the energetic particle bombardment may damage the local atomic arrangement and may create defects having a detrimental effect on the structure of films. High quality films can be grown at reduced epitaxial temperatures provided that a careful balance is maintained between beneficial effects of the IAD process and harmful effects of near-surface residual defects which can be annealed during deposition under appropriate conditions. The most favorable deposition conditions would be low energy particles, relatively high accelerated particle-to-vapor flux ratios, and low deposition rates with no contamination from background impurities.

2.2.3. Mixed Growth Mode

The growth of InAs and GaAs films on single crystal silicon substrates occurs via the Stranski–Kastanov mode. However, an IAD process can be used to suppress the three-dimensional island nucleation, postpone the two-dimensional–three-dimensional transition to higher film thickness, and to decrease the defect concentration in the deposited material [26–29]. The dissociation of clusters by energetic particles was invoked to explain these results and an optimum energy range for incident particles was determined using a simple collision model [27, 29]. Particles with kinetic energies below this range are not efficient for cluster dissociation. In fact, the lower energy threshold is related to the energy of atom displacement on the surface [30]. Incident particles with too high kinetic energies contribute in the cluster dissociation but also transfer momentum to lateral atoms so that adatoms can reach other large clusters and the expected effect is totally neutralized.

2.3. Structure-Zone Models

The microstructure of a continuous film deposited by evaporation, i.e., from nonenergetic particles is mainly dependent on the substrate temperature and the substrate surface roughness. The substrate temperature plays a dominant role in determining the adatom surface mobility and the bulk diffusion rates since the kinetics of these diffusion processes are exponentially and thermally activated. Smooth substrate surfaces lead to more uniform fine grained microstructures. Rough surfaces with various defects (peaks, valleys) exhibit preferred adsorption, nucleation, and growth sites, lead to shadowing effects of the adatoms and more columnar open voided microstructures can be expected. However, at relatively high substrate temperatures, the kinetics of surface diffusion processes are sufficiently fast to overcome topological effects and thin films with relatively smooth surfaces can be produced.

2.3.1. Movchan and Demchishin Structure-Zone Diagram

The microstructure of metal and oxide films produced by electron beam evaporation was first categorized by Movchan and Demchishin [31] using a structure-zone diagram describing the major features of the microstructure or the morphology of films as a function of the normalized growth temperature T_s/T_m which is the ratio between the absolute substrate temperature and the absolute melting point of the deposited material (Fig. 2).

Films produced at low deposition temperatures, T_s, exhibit microstructures schematically represented in zone 1 ($T_s/T_m < 0.3$). These films contain tapered crystallites with domed tops separated by open voided boundaries. Each grain may extend through the thickness of films which consist of textured and "fibrous" grains. Analyses by high resolution transmission electron spectroscopy have shown that the film structure corresponding to zone 1 consists of very fine (20 nm in diameter) equiaxed grains, i.e., grains of equal extent in all three dimensions although these grains are often collected into bundles of approximately the same orientation [32]. The width of crystallites increases with increasing normalized growth temperature; however, the apparent activation energy in the range 0.1–0.2 eV

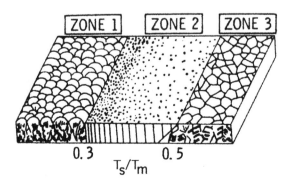

Fig. 2. Structure-zone diagram proposed by Movchan and Demchishin for metal and insulating films produced by thermal evaporation [31]. Reprinted from B. A. Movchan and A. V. Demchishin, *Phys. Met. Metallogr.* 28, 83 (1969), with permission.

is too low to explain the grain growth via bulk or surface diffusion mechanisms. The open voided columnar structure results from low adatom surface diffusion so that the incident atoms stick where they land [31]. This low adatom mobility can explain the porosity of relatively thick films deposited at high rates and also the shadowing effects that may control the relative heights of adjacent grains [33]. In spite of extensive studies, the initial stages of epitaxial growth and the condensation mechanisms controlling the size and the orientation of the first nuclei at very low substrate temperatures are not well understood [34].

In zone 2 with $0.3 < T_s/T_m < 0.5$, the microstructure of films consists of columnar grains separated by dense intercrystalline boundaries. In addition, the surface of films is relatively smoother than in zone 1. The average grain width is less than the thickness of films and increases with increasing normalized growth temperature. The apparent activation energies are of the order of that for surface diffusion. Surface and bulk diffusion processes may occur during deposition of films.

In zone 3 ($0.5 < T_s/T_m < 1$) when the substrate temperature is higher than half of the melting point of the deposited material, the films consist of squat uniform columnar grains with diameters which exceed the thickness of the films. The surface of the deposited material is smooth and shiny for metal films. Sometimes, the grains are misleadingly referred to as equiaxed. In fact, the columnar microstructure results from granular epitaxy operating at high substrate temperatures. The apparent activation energies of grain growth processes correspond to that of bulk diffusion [35]. The variation in grain size can be explained either by bulk diffusion processes or by recrystallization. The transition between zones 2 and 3 was observed to be gradual and the boundary was drawn with a positive slope (Fig. 2).

2.3.2. Thornton Structure-Zone Diagram

Later, Thornton [36, 37] added an additional axis for the extension of the Movchan and Demchishin structure-zone diagram and the description of the microstructure of metal films deposited by magnetron sputtering (Fig. 3); this axis is to account for the presence of the sputtering gas. The normalized growth temperature at which the zone boundaries occur increases with increasing gas pressure. In fact, the sputtering gas pressure affects the microstructure of sputter-deposited films through several indirect mechanisms. As the sputtering gas pressure is increased, the mean free path for elastic collisions between sputtered (or evaporated) species and the sputtering gas (or residual gas) atoms decreases down to the source or target–substrate distance. As a result, the oblique component of the flux of sputtered particles which condense on the substrate surface increases and films with a more open voided microstructure corresponding to zone 1 are obtained. Furthermore, a decrease in sputtering gas pressure during deposition results in larger mean free paths of particles and increased kinetic energy of species impinging on the substrate surface. Sputter-deposited films produced under low sputtering gas pressures exhibit relatively dense microstructures.

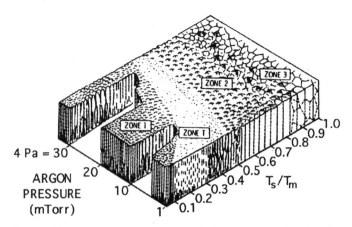

Fig. 3. Structure-zone diagram proposed by Thornton for metal films deposited by magnetron sputtering [30]. Reprinted from J. A. Thornton and D. W. Hoffman, *Thin Solid Films* 171, 5 (1989), with permission.

An additional region labeled zone T appears in the Thornton structure-zone diagram between zones 1 and 2 (Fig. 3). The microstructure consists of a dense array of poorly defined fibrous grains and the surface of films is relatively smoother than that of films corresponding to zone 1. The transition zone (zone T) as the substrate temperature is increased is a consequence of the onset of surface diffusion of adatoms [39]. Adatoms can migrate on the surface and arriving atoms are sufficiently mobile on the surface to join vacant lattice sites on the grain surface before being covered with additional layers of condensed atoms. Each initial grain continues to grow by local epitaxy or granular epitaxy as the film thickens. The first grains play a major role in controlling the final grain structure of films. Higher substrate temperatures would be required for granular epitaxy to occur if the deposition rate of films is very high.

2.3.3. Structure-Zone Diagram Proposed by Grovenor, Hentzell, and Smith

The structure and the morphology of metal films grown by thermal evaporation have been examined by Grovenor, Hentzell, and Smith [40] who further modified the earlier structure-zone diagram of Movchan and Demchishin. Analyses and transmission electron microscopy (TEM) examinations of films lead to classify the microstructure into four characteristic zones (Fig. 4).

At substrate temperatures below $0.15T_m$, films exhibit a homogeneous equiaxed grain structure with grain diameters in the range 5–20 nm. At these temperatures, the mobility of the deposited atoms is low and the atoms stick where they land. A calculation of the critical nucleus radius for homogeneous nucleation of solid Ni from the vapor at $0.15T_m$ indicates that the critical nucleus size corresponds to the size of a single atom and that the maximum diffusion length of a deposited atom before being covered with further deposited atoms is about one atomic distance. These results suggest that a noncrystalline or amorphous microstructure is favored and can be expected. However, crystalline grains and not amorphous films

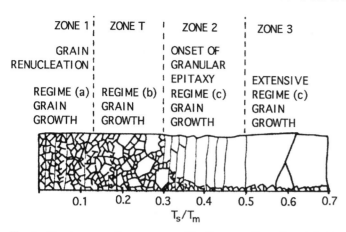

ZONE 1	ZONE T	ZONE 2	ZONE 3
GRAIN RENUCLEATION		ONSET OF GRANULAR EPITAXY	EXTENSIVE
REGIME (a) GRAIN GROWTH	REGIME (b) GRAIN GROWTH	REGIME (c) GRAIN GROWTH	REGIME (c) GRAIN GROWTH

0.1   0.2   0.3   0.4   0.5   0.6   0.7

T_s/T_m

Fig. 4. Structure-zone diagram proposed by Grovenor, Hentzell, and Smith for metal films produced by thermal evaporation [40]. Reprinted from C. R. M. Grovenor, H. T. G. Hentzell, and D. A. Smith, *Acta Metall.* 32, 773 (1984), with permission.

can be formed under these conditions; this experimental result suggests that collective transformations are important. It is tentatively suggested that the small equiaxed crystalline grains found in films deposited by evaporation at low substrate temperatures ($T_s < 0.15T_m$) are produced by some kind of athermal process occurring when the precursor phase reaches a critical thickness [40]. Small groups of atoms crystallize athermally, release the heat of crystallization and this energy stimulates further crystallization until neighboring nuclei impinge. The grain size is governed by the density of nuclei on the surface. This explanation is consistent with the existence of tensile residual stresses in films grown at low substrate temperatures [41]. Indeed, the precursor to crystalline transformation involves a decrease in volume for most metals; this volume change is constrained by the substrate so that the film is in a tensile state and the substrate is in a compressive state.

In zone T ($0.15T_m < T_s < 0.3T_m$), the microstructure can be described as a transition state between the equiaxed microstructure found in zone 1 and the columnar microstructure corresponding to zone 2. This transition zone is attributed to the onset of surface diffusion of adatoms which can migrate to energetically favorable sites before being covered with additional deposited atoms. In addition, under these conditions, certain grain boundaries become mobile and a grain growth process or recrystallization can be anticipated. In fact, the small-grained substructure consists of a bimodal distribution of sizes resulting from the process of secondary recrystallization. This process requires that the boundaries between the randomly oriented grains be immobile while those between the growing grains and their neighbors be mobile. Intercolumnar voids are responsible for preventing grain boundaries from moving. For most of the grain boundaries, the driving force due to their curvature tending to make them move is equal to the pinning force due to the voids. The secondary recrystallization process driven by a difference in the energies of surface planes is responsible for the development of the fiber texture corresponding to a film surface having the lowest free energy. For example, 50-nm-thick

gold films deposited onto amorphous SiO_2 substrates at room temperature ($T_s/T_m = 0.22$) exhibit a microstructure with a bimodal grain size distribution which consists of $\langle 111 \rangle$-oriented grains of several micrometers in diameter surrounded by grains having the as-deposited grain size of about 20–50 nm [42]. The boundaries surrounding the $\langle 111 \rangle$-oriented grains move because the extra driving force due to the difference in surface energy between the $\langle 111 \rangle$-oriented grains and their neighbors allows these boundaries to overcome the resistance to their motion due to the voids [43].

The microstructures of films in zones 2 and 3 are similar to the microstructures corresponding to zones 2 and 3 of the Movchan and Demchishin structure-zone diagram. The growth mechanisms which can be invoked are also similar. Briefly, in zone 2, the deposited atoms possess sufficient surface mobilities to diffuse and to increase the grain size. In this temperature range, films exhibit a uniform columnar grain structure with grain diameters lower than the film thickness; in addition, the grain diameters increase with increasing both substrate temperature and film thickness. In zone 3 at substrate temperatures higher than $0.5T_m$, the microstructure of films consists of squat uniform grains with diameters larger than the film thickness.

2.3.4. Effects of Incident Energetic Particles on the Microstructure of PVD Films

Incident hyperthermal particles impinging on film surfaces during deposition of films by IAD can produce displacement spikes, forward and inward sputtering. The kinetic energy of hyperthermal particles exceeds thermal energy and can vary from about 1 to 10^3 eV. Displacement spikes can yield a liquid-like structure resulting from a collision cascade in the deposited material [44]. The region is unstable. At low normalized growth temperatures corresponding to zone 1 of a structure-zone diagram (Fig. 4), the shear modulus of film materials tends to be zero at some points during simultaneous energetic particle bombardment and deposition. Transformation sequences such as solid to liquid-like to metastable solid phase can be promoted by incident hyperthermal particles and can result in microstructures with randomly oriented grains. At higher normalized growth temperatures, relaxation processes can occur during deposition and defects can be removed. These relaxation processes include diffusion of interstitial atoms to vacancies and other stable sites, diffusion of vacancies to stable vacancy sites, collapse of disordered atom arrays to form new crystalline grains of different orientations. In addition, a fraction of displacement generated vacancies can collide during diffusion and can form vacancy clusters which can then collapse to form dislocation loops. Segments of these dislocations can align to form tilt and twist boundaries which, in turn, yield misoriented grains. Voids between grains and columns can be eliminated by forward sputtering caused by incident energetic particles. The grain boundary migration is no longer pinned down by voids; the process of secondary recrystallization cannot occur and bimodal grain size distribution is not found in films deposited from incident hyperthermal particles.

Furthermore, the substrate surface can be altered by energetic particle bombardment during deposition of films. The substrate surface can be ion etched and cleaned by incident particle beams with appropriate energy and flux. As a result, a nonepitaxial substrate can be converted into an epitaxial substrate and an epitaxial monocrystalline film can be formed on a monocrystalline substrate after an appropriate surface ion-etching process. Moreover, energetic particle bombardment on clean substrate surfaces can promote the formation of nucleation sites and can increase the density of clusters. As a result, reduced grain sizes can be observed in films produced by IAD without epitaxial relationship between film and substrate materials. However, appropriate particle bombardment can also lead to the dissociation or lead to the elimination of small clusters by sputtering and, thereby, can yield an enhancement in grain size. In addition, energetic particle bombardment can promote adatom diffusion and grain boundary migration processes which can possibly affect or modify the normalized growth temperatures defining the boundaries in structure-zone diagrams. The prediction of the effect of a specific particle bombardment on the microstructure of films under particular deposition conditions is very difficult or even impossible for various systems.

In conclusion, these structure-zone diagrams provide a useful method for a qualitative description of observed film microstructures. However, no quantitative insight into the growth mechanism of films can be drawn from these diagrams. In addition, it is wise to be cautious in using such diagrams to predict microstructures and properties of films since the growth kinetics of films are known to be dependent on other factors than the substrate temperature. For instance, the surface roughness of substrates can promote microstructures of films corresponding to zone 1 at elevated substrate temperatures by enhancing shadowing effects caused by oblique deposition angles [39]. The surface contamination can also play a major role in determining film microstructures. The adatom mobility is reduced by surface impurities and the extent of zone 1 increases with increasing surface impurity content [12].

2.4. Major Physical Parameters Affecting the Microstructure of PVD Films

According to the structure-zone diagrams described previously, the substrate temperature, T_s, through the normalized growth temperature, T_s/T_m, and the gas pressure, P, in the deposition chamber appear as the major experimental parameters which affect the microstructure of PVD films. The effects of these deposition parameters, T_s and P, on the microstructure of films can be investigated or analyzed via a more quantitative approach than that corresponding to the simple morphological description of PVD films given by the structure-zone diagrams. Practically, the microstructure evolution is related to the surface mobility of adatoms as well as to the surface and the bulk diffusion steps of atoms allowing atomic rearrangements in the lattice of the deposited material. These diffusion characteristics are greatly dependent on two major physical parameters, i.e., substrate temperature and kinetic energy of particles impinging

on the surface; these incident particles (atoms and/or ions) can be either adsorbed on the surface and contribute to the growth of films or reflected at the surface and return to the gas phase. The kinetic energy of incident particles depends on several factors; one of them is the gas pressure in the deposition chamber.

2.4.1. Effect of the Substrate Temperature

At substrate temperatures corresponding to zone T in structure-zone diagrams, the surface mobility of adatoms is believed to be sufficiently large for allowing surface diffusion of adatoms up to energetically favorable sites before arrival of additional condensing atoms. This enhanced surface mobility and surface diffusion process of adatoms are recognized to be responsible for the microstructure of films grown by granular epitaxy. The substrate temperature corresponding to the onset of a significant surface diffusion of adatoms before arrival of additional condensing atoms can be evaluated from the diffusion length, X, given by [45]

$$X = \sqrt{4D\Delta t} \qquad (2)$$

where D is the diffusivity of adatoms and Δt is the time interval between deposition of successive monolayers. The diffusivity of adatoms depends on the substrate temperature, T_s,

$$D = D_0 \exp\left(-\frac{Q}{kT_s}\right) \qquad (3)$$

where k is the Boltzmann constant. For metals, the preexponential factor $D_0 \approx 1 \times 10^{-7}$ m^2 s^{-1}, and the activation energy is related to the melting point, i.e., $Q = 6.5kT_m$ [45, 46]. In addition, the time interval, Δt, depends on the deposition rate, V_R, of films as

$$\Delta t = N_0 \frac{\Omega}{V_R} \qquad (4)$$

where N_0 is the number of atoms per unit area in a monolayer of metal atoms on the surface and Ω is the atomic volume. Therefore, the normalized growth temperature, T_s/T_m, is given by

$$\frac{T_s}{T_m} = \frac{6.5}{\ln(4D_0 N_0 \Omega/(X^2 V_R))} \qquad (5)$$

For metals, N_0 and Ω are typically equal to about 1×10^{19} m^{-2} and 0.02×10^{-27} m^3, respectively. The deposition rate of metal films produced by thermal evaporation can be varied in the range $(1-30) \times 10^{-9}$ m s^{-1}. For a diffusion length $X = 1$ nm, corresponding to about 10 interatomic distances, the normalized growth temperature calculated from Eq. (5) is in the range 0.26–0.30. This value is in good agreement with the lower temperature limit of zone 2 determined experimentally by Movchan and Demchishin for various metal films [31]. Moreover, Eq. (5) suggests that the deposition rate of films also plays a significant role in determining the microstructure of films deposited at a given substrate temperature. This calculation of the normalized growth temperature can be faulted when impurity atoms on metal films impede the surface diffusion of adatoms. Thus, the transition temperatures in structure-zone diagrams are clearly also dependent on two additional physical parameters, namely, the deposition rate of films and the impurity concentration on the surface of growing films.

2.4.2. Effect of the Kinetic Energy of Atoms Condensed on the Substrate and the Film Surfaces

The surface mobility of adatoms is related to the kinetic energy of atoms condensed on the growing film and the incident species reflected at the film surface via elastic collisions. With sufficiently energetic atomic interactions between the film surface and the species coming from the gas phase, adatoms can escape from the potential energy well corresponding to an adsorption site and can diffuse on the surface from one site to another until a more favorable energetic site is reached. The resulting atomic rearrangements lead to the growth of films with larger mass densities.

The maximum kinetic energy, E_0, of atoms evolved from a thermal evaporation source at a temperature, T_0, is usually less than 0.5 eV ($E_0 = (3/2)kT_0$). In addition, the kinetic energy distribution of evaporated particles is very narrow as suggested by the velocity distribution of Cu evaporated atoms (Fig. 5). By contrast, the kinetic energy distribution of sputtered atoms is relatively broader and exhibits a large distribution tail in the high energy range; a significant number of sputtered atoms with kinetic energies ranging from 10 to 30 eV can be ejected from the target for sputtering gas–target material systems used in various technological applications. The average kinetic energy of sputtered atoms is currently 10 times higher than that of evaporated atoms. These data indicate that the sputtering mechanism is connected with momentum transfer and not connected with local heating.

The evaporated atoms lose kinetic energy by collisions with residual gas species occurring between the evaporation source and the substrate surface. For a given source–substrate distance, d, the kinetic energy of condensed atoms, E_c, depends on the

number of collisions, n_c, with residual gas species. Assuming a negligible effect of the angular scattering phenomena on the free path of the species, the average collision number of an evaporated atom traveling through a residual gas is given by

$$n_c = \frac{d}{\lambda} \tag{6}$$

where λ is the mean free path of the residual gas species which depends on the pressure, P, and temperature, T, in the deposition chamber,

$$\lambda = \frac{RT}{4\pi \sqrt{2} \rho_m^2 N_{Av} P} \tag{7}$$

where R is the ideal gas constant, N_{Av} is the Avogadro number, and ρ_m is the radius of the residual gas species (atoms or molecules). For oxygen as a major residual gas in the deposition chamber at room temperature, the mean free path, λ (in centimeters), is given by

$$\lambda = \frac{7.1 \times 10^{-3}}{P} \tag{8}$$

where P is the oxygen pressure expressed in millibars. Furthermore, the kinetic energy, E_1, of an evaporated atom of mass M_e, before collision with a residual gas molecule of mass M_g, is related to the kinetic energy, E_2, after collision by [48]

$$w = \ln\left(\frac{E_2}{E_1}\right) = \left[\frac{(1-M)^2}{2M} \ln\left(\frac{1+M}{1-M}\right)\right] - 1 \tag{9a}$$

with the atom mass ratio $M = M_g/M_e < 1$, or

$$w = \ln\left(\frac{E_2}{E_1}\right) = \left[\frac{(M-1)^2}{2M} \ln\left(\frac{M+1}{M-1}\right)\right] - 1 \tag{9b}$$

with $M > 1$.

The kinetic energy, E_c, of atoms condensed on the substrate surface which depends on the initial kinetic energy, E_0, of evaporated atoms can be calculated from Eqs. (6) to (9):

$$E_c = E_0 \exp(n_c w) = E_0 \exp\left(\frac{dw}{\lambda}\right) \tag{10}$$

For an evaporation process in oxygen as a residual gas in the evaporation chamber, the kinetic energy of condensed atoms is given by

$$E_c = E_0 \exp\left(\frac{dwP}{7.1 \times 10^{-3}}\right) \tag{11}$$

where the term w is given by Eq. (9a) or (9b) depending on the atom mass ratio M; the source–substrate distance, d, and the oxygen pressure, P, in the deposition chamber are expressed in centimeters and millibars, respectively.

For deposition of SiO_2 films by direct thermal evaporation of silica, silicon monoxide molecules, SiO, were found to be the major species in the gas phase evolved from the evaporation source [49]; this result was obtained from thermodynamic calculations based on the minimization of the free energy function of the system at equilibrium. For an SiO molecule colliding with an oxygen molecule, O_2, the factor w calculated from

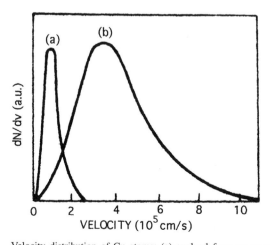

Fig. 5. Velocity distribution of Cu atoms: (a) evolved from an evaporation source, and (b) ejected from a target surface by sputtering [47]. Reprinted from L. Eckertova, "Physics of Thin Films," Chap. 2, p. 23. Plenum, New York, 1977, with permission.

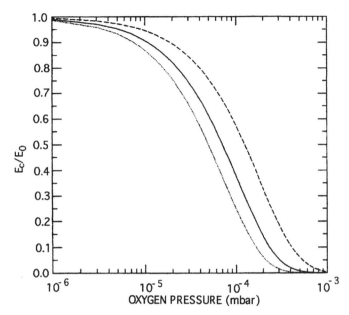

Fig. 6. Condensed energy-to-initial energy ratio versus oxygen pressure for SiO molecules traveling through oxygen for various source–substrate distances: 40 cm (dashed line), 70 cm (solid line), and 100 cm (dotted line).

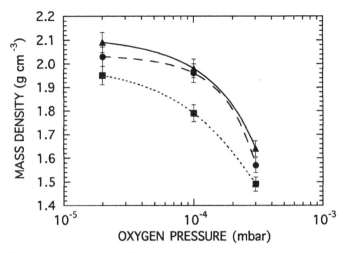

Fig. 7. Oxygen pressure effect on the mass density of stoichiometric SiO_2 films deposited by thermal evaporation on substrates at: ambient temperature (dotted line), 200°C (dashed line), and 250°C (solid line) [49]. Reprinted from H. Leplan, Ph.D. Thesis, National Polytechnic Institute of Grenoble, France, 1995.

Eq. (9a) is equal to -0.905. The energy ratio, E_c/E_0, for SiO molecules traveling through oxygen was dependent on the residual oxygen pressure in the deposition chamber (Fig. 6). The kinetic energy of SiO molecules condensed on the substrate, 70 cm far from the evaporation source, can vary from $0.91E_0$ to $0.07E_0$ under a residual gas pressure in the range $(1–30) \times 10^{-5}$ mbar. With an evaporation source at 2000 K, the average kinetic energy of evaporated species is about 0.26 eV and the kinetic energy of SiO molecules condensed on the substrate surface can vary from 0.236 to 0.018 eV depending on the oxygen pressure in the deposition chamber. The microstructure of amorphous SiO_2 films produced under these conditions was not fully dense. The mass density of stoichiometric SiO_2 films was greatly dependent on the deposition temperature, T_s, and the oxygen pressure in the deposition chamber although these parameters were varied in relatively narrow ranges (Fig. 7).

These results illustrate the effect of two major physical parameters (substrate temperature and kinetic energy of incident atoms condensed on the surface) affecting the microstructure and the properties of thin films produced by thermal evaporation, i.e., from nonenergetic species. Moreover, these results are quite compatible with those obtained from a two-dimensional simulation model for the deposition of nickel films at 300 and 400 K with kinetic energies of incident atoms of 0.1, 1, and 2 eV [50]. Both the substrate temperature and the incident energy of condensing atoms can affect the packing density of Ni films defined as the ratio between the mass density of films and the bulk metal density (Fig. 8); the substrate temperature has a more noticeable effect on the film structure than the incident energy does. Therefore, the effects of T_s and E_c on the microstructure of insulating and metal films deposited by thermal evaporation are quite similar.

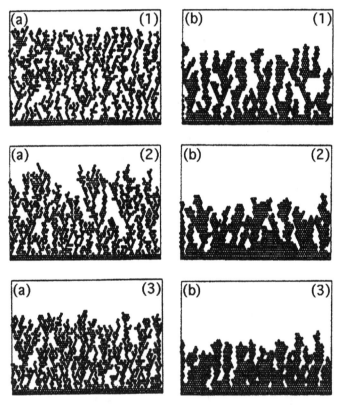

Fig. 8. Simulation results of thin film growth at a substrate temperature of: (a) 300 K and (b) 400 K with a kinetic energy of condensing Ni atoms of: (1) 0.1 eV, (2) 1 eV, and (3) 2 eV [50]. Reprinted from M. Xia, X. Liu, and P. Gu, *Appl. Opt.* 32, 5443 (1993), with permission.

A two-dimensional molecular dynamics (two-dimensional MD) approach was also used to investigate the microstructure, the average packing density, and the homoepitaxy of a simple Lennard-Jones system as a function of the incident kinetic energy of the arriving species in the low energy range between

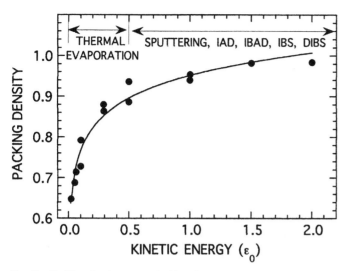

Fig. 9. Packing density versus incident kinetic energy of adatoms [25]. Reprinted from K.-H. Müller, *Surf. Sci.* 184, L375 (1987), with permission.

$0.02\varepsilon_0$ and $2\varepsilon_0$ where ε_0 is the depth of the Lennard-Jones potential well [25]. The deposition process was simulated at a substrate temperature of 0 K so that structural effects resulting from thermal adatom mobility can be excluded. As a result, the microstructure and the packing density of films were mainly determined by the adatom impact mobility which depends on the incident kinetic energy. The packing density of films increased with increasing incident kinetic energy E in units of ε_0 (Fig. 9). A relatively large scattering of data can be observed since a small number of atoms were considered for the simulation to reduce computational cost. However, the trends are clearly discernible. The incident kinetic energy in the range 0 to $0.5\varepsilon_0$ corresponds to the regime of thermal evaporation processes involving nonenergetic species condensed on the substrate surface. Incident particles with energies higher than $0.5\varepsilon_0$ can be found for deposition of thin films by sputtering. Thus, the voided microstructure and the porosity of PVD films are clearly related to low adatom impact mobilities.

2.4.3. Effect of the Momentum Transfer of Incident Energetic Particles

Hyperthermal particles with kinetic energies of a few hundreds of electron volts which impinge on the surface of films produced by IAD penetrate the material to an average depth of a few nanometers. Particle–film interactions induce various phenomena such as collision cascades in the deposited material, sputtering of atoms from the film surface, implantation of incident particles in the growing film, creation of vacancy-interstitial pairs and thermal spikes [51]. Energetic particle bombardment of films during deposition results in surface depletion due to sputtering and inwardly recoil atoms. In addition, film density enhancement deeper in the bulk can be observed because of particle and recoil implantation.

The improved mass density of films produced by IAD may be attributed to the distortion of the growing surface layer by energetic particles; these phenomena can be analyzed on the basis of the ion-peening model [52]. Energetic particles striking and penetrating the film surface randomly displace atoms from their equilibrium positions through a series of primary and recoil collisions which lead to a volumetric distortion. At low normalized growth temperatures ($T_s/T_m < 0.25$), mass transport phenomena and defect mobility are low enough to freeze the volumetric distortion in place. A model based on the knock-on linear cascade theory of transmission sputtering proposed by Sigmund [53] predicts the dependence of the volumetric film distortion on the flux and the energy of incident hyperthermal particles [54]. The volumetric distortion, d, is assumed to be proportional to the fractional number of atoms, n/N, displaced from equilibrium positions,

$$d = K \frac{n}{N} \qquad (12)$$

where K is the proportionality factor. The validity of this hypothesis was discussed in various articles and was evidenced by experimental data [55–57]. According to Sigmund's concept of forward sputtering, the number of displaced atoms per unit volume, n, in a growing film is given by the product of the forward sputtering yield Y (atoms/projectile) of the deposited material (target) and the flux, ϕ_p, of energetic particles (projectiles), i.e., $n = Y\phi_p$. For polycrystalline and amorphous targets, the sputtering yield Y is equal to the product of the deposited energy per unit depth, F, and a term G containing the target material properties, i.e., $Y = GF$. For interatomic collisions at low incident energies ($<$ kiloelectron volts), Born–Mayer interatomic potentials are assumed to be valid and the term G (expressed in $eV^{-1} \text{Å}^{-2}$) is equal to $0.042(NU_0)$ where U_0 (in eV at^{-1}) is the cohesive energy or sublimation energy of the target. As a result, the volumetric distortion can be written as

$$d = 0.042 \left(\frac{K\phi_p F}{N^2 U_0} \right) \qquad (13)$$

The deposited energy per unit depth, F, for normal incidence and moderate mass projectiles, M_p, is given by

$$F = \alpha S_n N \qquad (14)$$

where α is an energy-independent function representing the efficiency of the momentum transfer; from the data of Andersen and Bay [58], for $0.675 < M_t/M_p < 4.8$, the term α can be approximated to

$$\alpha \approx 0.07 \left(1 + \frac{M_t}{M_p} \right) \qquad (15)$$

where M_t and M_p are the target and projectile atomic masses, respectively. The second term, S_n (eV Å^2) in Eq. (14), is given by [53]

$$S_n = 84.75 \left[\frac{M_p Z_p Z_t}{(M_p + M_t)(Z_p^{2/3} + Z_t^{2/3})^{1/2}} \right] s_n(\varepsilon) \qquad (16)$$

where Z_t and Z_p are the target and projectile atomic numbers, respectively; $s_n(\varepsilon)$ is the reduced stopping power which can be approximated to

$$s_n(\varepsilon) \approx 3.33\varepsilon^{1/2} \qquad (17)$$

for low reduced energies, i.e., $\varepsilon = E_p/E_c < 0.02$ [59]. The term, ε, is the projectile energy normalized by the Coulomb energy, E_c (keV), of the projectile atom–target atom interaction which is given by

$$E_c = 0.03\left(1 + \frac{M_p}{M_t}\right)Z_t Z_p \left(Z_t^{2/3} + Z_p^{2/3}\right)^{1/2} \qquad (18)$$

From Eqs. (15) to (18), the deposited energy, F, can be expressed as

$$F = 114.05\left[\frac{M_t^{1/2}}{(M_p + M_t)^{1/2}}\right]\left[\frac{(Z_p Z_t)^{1/2}}{\left(Z_p^{2/3} + Z_t^{2/3}\right)^{3/4}}\right]N\sqrt{E_p} \qquad (19)$$

and the volumetric distortion is given by

$$d = 4.78\left(\frac{K\phi_p}{NU_0}\right)\left[\frac{M^{1/2}}{(M_p + M_t)^{1/2}}\right]\left[\frac{(Z_p Z_t)^{1/2}}{\left(Z_p^{2/3} + Z_t^{2/3}\right)^{3/4}}\right]\sqrt{E_p} \qquad (20)$$

or

$$d = 4.79\left(\frac{K\phi_p}{N}\right)\delta\sqrt{E_p} \qquad (21)$$

with

$$\delta = \frac{M_t^{1/2}(Z_p Z_t)^{1/2}}{U_0(M_p + M_t)^{1/2}\left(Z_p^{2/3} + Z_t^{2/3}\right)^{3/4}} \qquad (22)$$

From Eqs. (13) and (21), the number of atoms per unit volume, n, displaced by energetic particle bombardment from equilibrium positions in the deposited material during deposition of films can be written as

$$n = 4.79\delta\phi_p\sqrt{E_p} \qquad (23)$$

Assuming that the microstructure of films, i.e., morphology, texture, and grain size subjected to energetic particle bombardment, during deposition by thermal evaporation, is determined by the number of atoms per unit volume displaced from equilibrium positions, the bombardment effect is rather related to the momentum of incident particles (dependence of n on $(E_p)^{1/2}$).

Furthermore, various theoretical attempts were conducted to elucidate the densification mechanism of films produced by IAD. Monte Carlo calculations were developed to investigate the effect of thermal spikes on the film density [60]. For these ion-induced localized annealing events, the relevant parameter is the kinetic energy of ions. The major conclusion of this work was that the thermal spikes do not significantly improve the mass density of films under typical IAD conditions. In fact, the energetic particle bombardment of films during deposition results in surface depletion due to sputtering and inwardly recoil atoms, and also leads to an improved mass density deeper in the bulk because of particle and recoil implantation. Finally, a collision cascade model showed that the densification of films can be caused by the recoil implantation of near-surface atoms which is a momentum-dependent process [61].

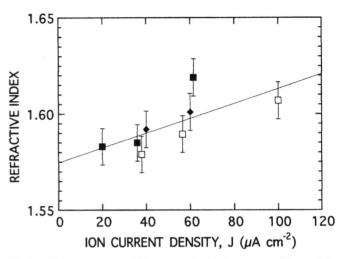

Fig. 10. Refractive index at 350 nm of LaF$_3$ films bombarded with neon ($\square$), argon ($\blacklozenge$), and krypton ($\blacksquare$) ions as a function of the ion current density [62]. Reprinted from J. D. Targove and H. A. Macleod, *Appl. Opt.* 27, 3779 (1988), with permission.

An experimental verification of momentum transfer as the dominant mechanism of the mass density enhancement of films produced by IAD was proposed by Targove and Macleod [62]. To separate the kinetic energy effect from the momentum effect of incident hyperthermal particles, lanthanum fluoride films deposited by thermal evaporation were bombarded with three different inert gas ions (Ne, Ar, and Kr) of 500 eV. The refractive indices of the films were used as a measure of their packing density [63]. According to the simple thermal spike model, the number of atoms rearranged during the lifetime of a spike, n_T, is proportional to $q^{5/3}$ where q is the energy deposited into the spike by the incident ions [64]. The rate of atomic rearrangements in the film is given by the product $(n_T J)$ where the ion current density at the substrate, J, is proportional to the ion or the projectile flux, ϕ_p. The spike lifetime being independent of the ion mass and the ion current density, the number of atoms rearranged, n_T, should be constant for all these experiments, and the rate of atomic rearrangements should be proportional to the ion current density J. In fact, the data given in Figure 10 show that the refractive indices and, therefore, packing densities of films are not proportional to the ion current density and depend on the mass of ions striking the film surface.

Furthermore, the theoretical model proposed by Müller [61] which takes advantage of previously developed fast methods for three-dimensional Monte Carlo cascade computation shows that the ion incorporation and the recoil implantation of surface atoms lead to an improvement of the film density slightly below the surface of a growing film. The densification of films depends on the ability of condensing atoms to refill surface vacancies which are created by sputtering and driven-in atoms. Therefore, the mechanism responsible for the film density improvement may involve a momentum transfer from incident particles to atoms in the deposited material. The implantation depth is about proportional to the momentum transferred to the target atom (film atom). With a negligible ion energy loss before an implantation event, the maximum momentum transfer per projectile, P_{max}, is given by [62]

$$P_{max} = \sqrt{2M_p \gamma E_p} \qquad (24)$$

where M_p and E_p are the mass and the energy of ions (projectiles), respectively; γ is the energy transfer coefficient which governs the maximum energy transferred from the incident ion to the film atom of mass M_t:

$$\gamma = \frac{4M_p M_t}{(M_p + M_t)^2} \qquad (25)$$

This expression of γ is derived from equations corresponding to the kinetic energy and the momentum conservation laws for elastic collisions assuming that an elastic head-on collision occurs between the incident ion and the film atom; i.e., the direction of the ion velocity after collision is reverse to the incident direction and the knocked atom moves after collision in the incident direction. The term, γ, reflects the efficiency of the energy transfer from incident ions to condensed atoms for elastic head-on collisions. A mean value of γ was adopted since two elements, La and F, are found in LaF_3 films,

$$\gamma_x = 0.25(\gamma_{x-La} + 3\gamma_{x-F}) \qquad (26)$$

where x represents Ne, Ar, or Kr; the terms γ_{x-La} and γ_{x-F} are calculated using Eq. (25). The total momentum transfer rate is approximated to

$$P_{tot} = P_{max}J = J\sqrt{2M_p \gamma E_p} \qquad (27a)$$

or for ions of charge q as projectiles,

$$P_{tot} = q\phi_p \sqrt{2M_p \gamma E_p} \qquad (27b)$$

where P_{tot} is expressed in $(\mu A\,cm^{-2})\,(uma)^{1/2}\,(eV)^{1/2}$. An excellent linear correlation was found between the refractive indices of films at 350 nm and the total momentum transfer (Fig. 11). These data suggest that the increase in refractive index and, therefore, the packing density of films produced by IAD are related to the momentum transfer rather than to the energy transfer.

However, with a given flux, ϕ_p, and the kinetic energy, E_p, of incident ions, the efficiency of the IAD process for the film density enhancement is expected to be dependent on the deposition rate of films. For a given flux of incident hyperthermal particles, the bombardment effect on the densification of films is expected to decrease with increasing deposition rate. As a result, the normalized momentum, P_n, appears to be a more relevant parameter governing the densification of films produced by IAD; this normalized momentum can be approximated to

$$P_n = \frac{\phi_p}{\phi_c} \sqrt{2M_p \gamma E_p} \qquad (28)$$

where ϕ_p and ϕ_c are the fluxes of incident energetic particles and condensing atoms on the film surface, respectively.

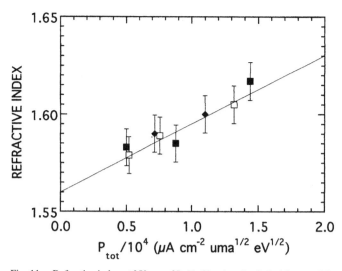

Fig. 11. Refractive index at 350 nm of LaF_3 films bombarded with neon ($\square$), argon ($\blacklozenge$), and krypton ($\blacksquare$) ions as a function of the total momentum transfer rate P_{tot} [62]. Reprinted from J. D. Targove and H. A. Macleod, *Appl. Opt.* 27, 3779 (1988), with permission.

2.4.4. Kinetic Energy of Species Impinging on the Film Surface and the Normalized Momentum Involved in the Deposition Process

2.4.4.1. Films Prepared by Thermal Evaporation and Ion-Assisted Deposition

As already mentioned, the maximum kinetic energy, E_0, of atoms evolved from a thermal evaporation source is usually less than 0.5 eV and the kinetic energy distribution is very narrow (Fig. 5). Depending on the residual gas and the pressure in the deposition chamber, the kinetic energy, E_c, of these atoms condensed on the substrate surface is a fraction of the initial kinetic energy; the kinetic energy ratio, E_c/E_0, can be derived from Eq. (10).

For direct evaporation processes carried out under pressures lower than 10^{-5} mbar, the maximum kinetic energy of atoms condensed on the substrate can be as high as 90% of the initial kinetic energy, E_0 (Fig. 6). However, for reactive evaporation processes performed under relatively high pressures of reactive gas such as oxygen or nitrogen, the kinetic energy of atoms condensed on the substrate surface can be lower than $0.3E_0$. Therefore, moderate substrate temperatures are needed to activate chemical surface reactions and to produce high quality oxide or nitride thin films with stoichiometric compositions.

On the basis of Eq. (28), the normalized momentum, P_n, involved in deposition processes by direct thermal evaporation is given by

$$P_n = \sqrt{2M_c E_c} \qquad (29)$$

where M_c and E_c are the atomic mass and the kinetic energy of condensing atoms, respectively. The flux ratio, ϕ_p/ϕ_c, and the energy transfer coefficient, γ, are equal to unity.

When the growth of films deposited by direct thermal evaporation is assisted by ion bombardment as for IAD processes, the normalized momentum, P_n, can be expressed as

$$P_n = \sqrt{2M_c E_c} + \frac{\phi_p}{\phi_c}\sqrt{2M_p \gamma E_p} \qquad (30)$$

However, as the kinetic energy of incident ions is increased, the first term in Eq. (30) becomes readily negligible with respect to the second one and the normalized momentum is approximately given by Eq. (28).

2.4.4.2. Films Deposited by Sputtering

The surface of films sputter deposited on grounded substrates is exposed to the effect of three types of energetic particles: (1) atoms ejected from the target which condense on the substrate or the film surface, (2) atoms of the sputtering gas implanted in the target as sputtering ions and subsequently ejected from the target surface together with target atoms, and (3) neutral atoms originating from sputtering ions which are neutralized and elastically reflected at the target surface. The initial kinetic energy distribution of atoms ejected from the target is typically similar to the velocity distribution of Cu atoms represented in Figure 5. The maximum energy lies in the range U_0 to $U_0/2$ where U_0 is the cohesive energy of atoms in the target. The average initial kinetic energy, E_0, of sputtered atoms which can be as high as 10 eV for a number of metal atoms increases with increasing atomic number, Z_t, of target atoms (Fig. 12). The energy distribution may broaden as the sputtered atoms travel through the plasma from the target to the substrate because of elastic collisions with sputtering gas atoms. The average kinetic energy, E_c, of sputtered atoms which condense on the substrate surface after n_c elastic collisions in the gas phase can be calculated using Eq. (10). In general, the average kinetic energy of condensing atoms is adopted as the relevant parameter to describe or to investigate the kinetic energy effect of atoms on the microstructure, the morphology, and the properties of films produced by sputtering; however, the small fraction

of atoms with kinetic energy values in the distribution tail may also affect the characteristics of sputter-deposited films.

A large fraction of sputtering ions striking the target surface are implanted in the target to an average depth of a few nanometers. Later, in the permanent sputtering regime, these atoms are ejected from the target surface together with target atoms with an average kinetic energy similar to those given in Figure 12. These atoms traveling through the plasma experience a kinetic energy loss caused by elastic collisions with gas phase species and these sputtering gas atoms impinge on the growing film surface with an average kinetic energy, E_{sga}.

A fraction of sputtering ions are neutralized and elastically backscattered from the target surface. The kinetic energy of these reflected neutral atoms impinging on the growing film surface may be as high as the kinetic energy of sputtering ions, i.e., more than 1 order of magnitude above the average kinetic energy of condensing atoms. The reflection coefficients in terms of ion flux, R_0, and ion energy, γ_0, of ions used for sputter deposition of films are defined by

$$R_0 = \frac{\phi_{si}}{\phi_{si}^0} \qquad (31)$$

$$\gamma_0 = \frac{E_{si}}{E_{si}^0} \qquad (32)$$

where ϕ_{si} and E_{si} are the flux and the kinetic energy of neutral sputtering gas atoms backscattered from the target surface, respectively; ϕ_{si}^0 and E_{si}^0 are the flux and the kinetic energy of sputtering ions striking the target surface, respectively. Usually, the initial kinetic energy of sputtering ions may vary in the range 0.5–2 keV.

These reflection coefficients, R_0 and γ_0, are dependent on the atom mass ratio, M_t/M_i, where M_t and M_i are the atomic masses of target atoms and sputtering ions, respectively. A strong dependence of the reflection coefficient, R_0, on the atom mass ratio is observed in Figure 13; i.e., the fraction of ions reflected at the target surface may vary from 0.5 to 20%. The reflection coefficient in terms of energy, γ_0, also depends on the atom mass ratio (Fig. 14). The neutral atom backscattered from the target surface can possess a relatively large kinetic energy in the range $0.08E_{si}^0$ to $0.30E_{si}^0$ with a typical kinetic energy of sputtering ions, E_{si}^0, of the order of 1 keV. In other words, these energetic neutral particles can impinge on the surface of growing films with an average kinetic energy of about 100 eV. The energy loss caused by elastic collisions in the gas phase can be minimized and even negligible under a sufficiently low sputtering gas pressure and with a short target–substrate distance.

Under these conditions, the normalized momentum, P_n, for sputter deposition of films on grounded substrates can be expressed approximately as

$$P_n = \sqrt{2M_c E_c} + \frac{R_0 \phi_{si}^0}{\phi_c}\sqrt{2M_{si}\gamma_0 E_{si}^0}$$
$$+ \frac{(1-R_0)\phi_{si}^0}{\phi_c}\sqrt{2M_{si}E_{sga}} \qquad (33)$$

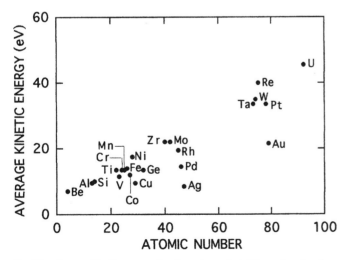

Fig. 12. Average kinetic energy of various atoms ejected from a target surface under 1200 eV krypton ion bombardment [65]. Reprinted from G. K. Wehner and G. S. Anderson, in "Handbook of Thin Film Technology" (L. I. Maissel and R. Glang, Eds.), p. 3–23. McGraw-Hill, New York, 1970, with permission.

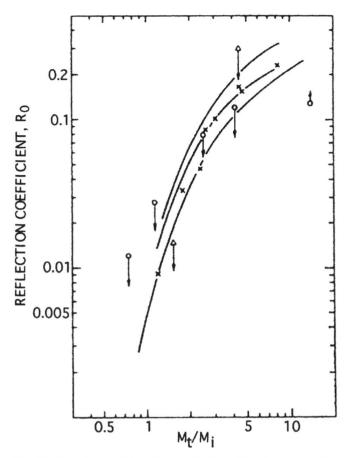

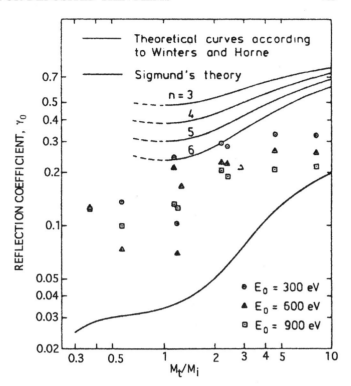

Fig. 13. Reflection coefficient, R_0, as a function of the atomic mass ratio, M_t/M_i, at a fixed reduced energy, ε, of 0.2. The uppermost curve is calculated using a Thomas–Fermi cross section and the correction for surface effects on backscattering coefficients, the second curve is the same without the surface correction, and the third curve is the uncorrected curve. Inelasticity is neglected. Experimental data points: ($\times$) from Bottiger et al. [66], ($\bigcirc$) from Brunneé [67], ($\triangle$) from Petrov [68]. Arrows indicate estimated extrapolations to $\varepsilon = 0.2$ by Bottiger et al. [66]. Reprinted from J. Bottiger, J. A. Davies, P. Sigmund, and K. B. Winterbon, *Radiat. Effects* 11, 69 (1971), with permission.

Fig. 14. Reflection coefficient, γ_0, as a function of the atomic mass ratio, M_t/M_i. The experimental data are compared with theoretical curves according to Sigmund [69], and Winters and Horne [70]. The term, n, is the average collision number of the backscattering process and E_0 is the kinetic energy of incident ions [71]. Reprinted from W. R. Gesang, H. Oechsner, and H. Schoof, *Nucl. Instrum. Methods* 132, 687 (1976), with permission.

reflection coefficients, R_0 and γ_0, can be estimated from data given in Figures 13 and 14.

2.4.4.3. Films Produced by Ion Beam Sputtering and Dual Ion Beam Sputtering

The residual gas pressure effect on the kinetic energy of sputtered atoms is negligible using ion beam sputtering (IBS) processes. Ion-assisted growth of films can be accomplished by using an additional ion source. The surface of films produced by dual ion beam sputtering (DIBS) processes is exposed to energetic particles already mentioned for films deposited by sputtering on grounded substrates. An additional type of energetic particles (projectiles) coming from the ion source for ion-assisted deposition is involved in the DIBS process. The normalized momentum for DIBS processes can be written as

$$P_n = \sqrt{2M_c E_c} + \frac{R_0 \phi_{si}^0}{\phi_c}\sqrt{2M_{si}\gamma_0 E_{si}^0}$$
$$+ \frac{(1 - R_0)\phi_{si}^0}{\phi_c}\sqrt{2M_{si}E_{sga}} + \frac{\phi_p}{\phi_c}\sqrt{2M_p\gamma E_p} \quad (35)$$

where the first three terms in the right-hand side member are similar to those given in Eqs. (33) and (34). The fourth term is similar to the second term of Eq. (30) where ϕ_p and ϕ_c are the fluxes of incident ions (projectiles coming from the additional

In Eq. (33), the fluxes of particles impinging on the film surface are assumed to be similar to fluxes of particles reflected at the target surface; strictly speaking, these fluxes are proportional.

The surface of films sputter deposited on negatively biased substrates is also bombarded with ions of charge q coming from the plasma. Therefore, the normalized momentum is given by

$$P_n = \sqrt{2M_c E_c} + \frac{R_0 \phi_{si}^0}{\phi_c}\sqrt{2M_{si}\gamma_0 E_{si}^0}$$
$$+ \frac{(1 - R_0)\phi_{si}^0}{\phi_c}\sqrt{2M_{si}E_{sga}} + \frac{\phi_i}{\phi_c}\sqrt{2M_{si}\gamma q V_B} \quad (34)$$

where ϕ_i is the flux of sputtering gas ions striking the growing film surface, γ is the energy transfer coefficient reflecting the efficiency of the energy transfer from incident ions to condensing atoms (γ can be calculated using Eq. (25)), and V_B is the substrate bias voltage. Various parameters involved in Eqs. (33) and (34) can be determined by computer calculation codes such as transport of ions in matter or TRIM code [72]. The values of ion source for IAD) and condensing atoms on the film surface, respectively; M_p and E_p are the atomic mass and the kinetic energy of incident ions, respectively.

3. MAGNITUDE OF RESIDUAL STRESSES IN PVD THIN FILMS

Various methods for measuring the magnitude of residual stresses in thin films can be categorized on the basis of the physical phenomena or parameters involved in the experiments such as X-ray diffraction methods, optical, electrical, and electromechanical methods. Detailed information about the instruments needed for these measurements was already extensively reported in the literature [73–76]. The choice of the appropriate method depends on various factors or needs such as *in situ* or *ex situ* measurements, type of substrate materials, type of film–substrate structures.

With a film tightly adherent to a thin substrate, a biaxial stress in the film of a sufficiently high intensity causes the substrate to bend elastically. A tensile stress (with a positive conventional sign) bends the film–substrate structure so that the film surface is concave whereas a compressive stress (with a negative conventional sign) bends the sample so that the film surface is convex. The most common methods for determining residual stresses in thin films are based on measurements of the deformation of film–substrate structures. The substrates can be in the form of either thin cantilevered beams or disks. Cantilevered beam type substrates are generally employed for *in situ* measurements of stresses in the films during deposition. The residual stresses are determined by measurements of the radius of curvature of the beam, the deflection of the free end of the beam, or by observing the displacement of the center of a circular disk type substrate. Then, the average value of residual stresses in the films can be calculated from the characteristic length value (radius of curvature, deflection, or displacement) using appropriate equations developed in the next sections.

Alternatively, the residual stresses in polycrystalline and single crystal films can be determined by X-ray diffraction techniques. If the film is deposited on a single crystal substrate, X-ray measurements can be used to determine the bending of the lattice planes, and the average residual stresses are calculated from the radius of curvature of the substrate. Furthermore, the position of a diffraction peak in the X-ray pattern of the deposited material gives the interplanar spacing of the set of corresponding lattice planes, and the strain in the film can be deduced directly from the lattice parameter of the deposited material. Then, the average value of residual stresses in the film is calculated from the strain in the film using appropriate elastic constants and Hooke's law [77]. It is worthwhile to note that X-ray diffraction techniques give the strain, and hence the average residual stresses in a crystallite lattice. These stress values are not necessarily similar to those determined from substrate bending measurements since the stress at the grain boundaries may be different from the stress in the crystallite.

The strain distribution through the thickness of polycrystalline films can be deduced from analyses by X-ray diffraction techniques [78–81]. This aspect corresponds to the major advantage of X-ray diffraction measurements over the substrate curvature techniques for determination of residual stresses in polycrystalline films. However, PVD films exhibit in general amorphous or nanocrystalline structures and the determination of residual stresses by X-ray diffraction techniques is often difficult or even impossible. Therefore, the major advantage of the substrate bending techniques over X-ray diffraction measurements is that residual stresses can be determined even in amorphous films.

3.1. Determination of Residual Stresses from the Radius of Curvature of Substrates

The most common methods for the determination of residual stresses in thin films deposited on beam or disk type substrates are based on the measurements of the deformation of film–substrate structures. This approach can be used for a wide range of materials including elementary metal, semiconductor, and insulating films or multilayer structures deposited on various types of substrates such as metal, semiconductor, glass.

3.1.1. Relation between Residual Stresses and Radius of Curvature of Substrates

The equation for stress can be derived on the basis of elementary considerations of the beam elasticity theory. First of all, it is necessary to clearly define various physical parameters involved in the calculations. The neutral axis is defined as that longitudinal axis of a beam which undergoes no additional strain, i.e., no change in length when the beam is bent (Fig. 15). Therefore, the neutral axis lies at the center of a simple beam. When a thin strip of metal (or other material) or a thin substrate restrained from bending is covered with a thin film, the strip is compressed by the tension in the film which is thereby also shortened and loses some of its stress. If the constraints are now released and the strip is allowed to bend, the stress in the film

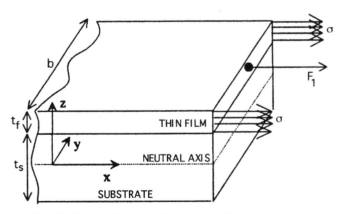

Fig. 15.   Distribution of stresses in the structure held rigid before contraction and bending of the substrate (the stresses in the nondeformed substrate are nil).

is still further relieved. These losses in stress depend on the dimensions of the substrate and, hence the final equilibrium stress in the film is not a constant quantity but depends on the experimental conditions. As a result, it is necessary to define the stress in the coating, S, which is independent of the mode of measurement. This stress referred to as the true stress is existing in the film when it is deposited on a rigid, incompressible surface or for practical purposes on a substrate thick enough to undergo no appreciable deformation. This term, S, is given by

$$S = \int_0^{t_f} \sigma(z)\, dz \tag{36}$$

where t_f is the thickness of the film and $\sigma(z)$ is the stress for a fiber at a distance z from the chosen axis. The average stress in the film can be calculated from the true stress and the thickness of the film:

$$\sigma = \frac{S}{t_f} \tag{37}$$

Usually, the stress data given in the literature are the values of the average stress, σ. Then, for a correct comparison of stress values obtained from samples with films of various thicknesses, it is necessary to verify that the stress is independent of the film thickness.

Two basic conditions must be satisfied by the internal longitudinal fiber stresses of a beam in equilibrium,

$$F = \int \sigma(z)\, dA = 0 \tag{38}$$

$$M = \int \sigma(z)z\, dA = 0 \tag{39}$$

These equations can be taken over any cross section of the beam. Equation (38) states that the sum, F, of the longitudinal forces within the beam is zero, i.e., that the internal compressive forces are equal to the internal tensile forces. Equation (39) states that at equilibrium the internal bending moment, M, of the beam is zero about any axis. The variable, z, is the distance of the fibers, of stress $\sigma(z)$, from the chosen axis, and dA is the element of area of the cross section.

Before considering the application of these general equations to the curvature of a substrate (typically, a strip of metal) covered with a thin film, various assumptions can be considered to simplify the calculations. It will be assumed that: (i) the properties of materials involved in the sample (substrate and thin film) fulfill the requirements for application of the mechanics of elastic, homogeneous, and continuous materials, (ii) elastic properties of the substrate and the film are isotropic, (iii) Young moduli of elasticity of the substrate and the film are equal approximately, (iv) the thickness of the film does not amount to more than a few percent of the thickness of the substrate, (v) the residual stresses are plane, biaxial, and isotropic (that means $\sigma_{xx} = \sigma_{yy} = \sigma$ and $\sigma_{zz} = 0$ (Fig. 15) since the film is free of any deformation or constraint along the z axis), (vi) during bending, the transverse cross sections remain plane and, (vii) the adherence of the film to the substrate surface is ideal; i.e., the film–substrate interface is free of any delamination.

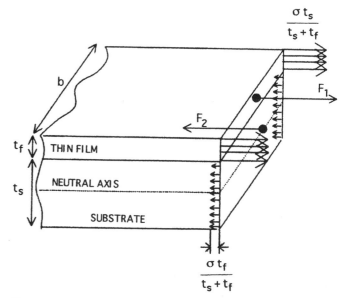

Fig. 16. Distribution of stresses in the structure after lateral contraction of the substrate (the sum of forces is equal to zero).

In an initial approach, the film is deposited on a substrate rigidly held, i.e., neither contraction nor bending of the strip can occur. Since the substrate has not been allowed to deform, there is no resultant stress in the substrate as represented in Figure 15. The condition for mechanical equilibrium of the structure corresponds to the sum of forces equal to zero, $\sum F = 0$. Then, the constraints applied on the substrate are partially removed so that the strip is allowed to shorten but not to curve. The film also contracts and its stress is reduced; the stress in the substrate is increased by the same amount. The situation is schematically represented in Figure 16. A couple of longitudinal and opposite forces with equal intensities acting at a distance of $(t_s + t_f)/2$ appears in the structure. The bending moment is given by

$$\begin{aligned}
M &= F_1\left(\frac{t_s + t_f}{2}\right) \\
&= \sigma\left(\frac{t_s}{t_s + t_f}\right)t_f b\left(\frac{t_s + t_f}{2}\right) \\
&= \sigma\left(\frac{t_s t_f b}{2}\right) \tag{40}
\end{aligned}$$

Now, the substrate is allowed to bend; as a result, a new distribution of the stresses occurs (Fig. 17). The new position at equilibrium is reached when the bending moment of the stresses is equal and opposite to the bending moment calculated previously and given by Eq. (40). In this situation, if the transverse cross section is assumed to remain plane according to the previous hypothesis (vi), the strain of a fiber at a distance z from the neutral axis is given by the following expression [82]:

$$\varepsilon_c = \frac{z}{R} \tag{41}$$

where R is the radius of curvature of the substrate. For elastically isotropic materials (hypothesis (ii)) and a state of plane, biaxial, and isotropic stress in the film (hypothesis (v)), the

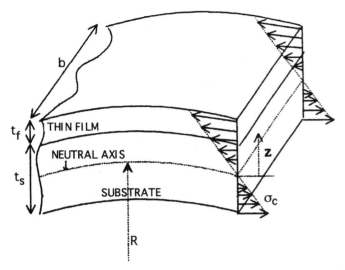

Fig. 17. Distribution of stresses in the structure after bending of the substrate (the sum of forces is still equal to zero).

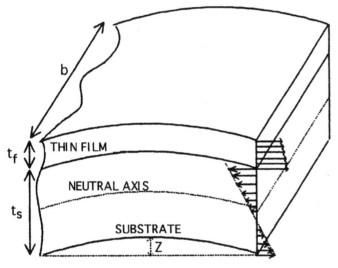

Fig. 18. Final distribution of stresses in the structure (the sum of forces and bending moments are equal to zero).

stress created in the structure by the curvature can be expressed by Hooke's law:

$$\sigma_c = \left(\frac{E}{1-\nu}\right)\varepsilon_c = \left(\frac{E}{1-\nu}\right)\frac{z}{R} \qquad (42)$$

The situation is schematically represented in Figure 17. The final distribution of stresses in the structure at equilibrium given in Figure 18 results from the superposition of stresses represented in Figures 16 and 17. According to previous hypotheses (iii) and (iv), i.e., $t_f \ll t_s$ and $E_f \approx E_s$, the neutral axis can be considered to be located at the midpoint of the substrate during bending. The bending moment corresponding to the stress, σ_c, is given by

$$M_c = b \int_{-t_s/2}^{(t_s/2)+t_f} \sigma_c z \, dz \approx b \int_{-t_s/2}^{t_s/2} \sigma_c z \, dz \qquad (43a)$$

or

$$M_c = \frac{E_s b}{(1-\nu_s)R} \int_{-t_s/2}^{t_s/2} z^2 \, dz = \frac{E_s b t_s^3}{12(1-\nu_s)R} \qquad (43b)$$

The bending moment given by Eq. (40) can be equated to the bending moment in Eq. (43b) since the structure is at mechanical equilibrium so that

$$\sigma = \frac{1}{6}\left(\frac{E_s}{1-\nu_s}\right)\left(\frac{t_s^2}{t_f R}\right) \qquad (44)$$

This equation corresponds to the Stoney equation established for electroplated coatings in 1909 [83]. When the substrate is not absolutely flat before deposition of the film, i.e., the initial radius of curvature of the substrate is R_1, the stress is given by

$$\sigma = \frac{1}{6}\left(\frac{E_s}{1-\nu_s}\right)\left(\frac{t_s^2}{t_f}\right)\left(\frac{1}{R_2} - \frac{1}{R_1}\right) \qquad (45)$$

where R_2 is the radius of curvature of the film–substrate structure.

The derivation of Eq. (44) or the Stoney equation is based on various assumptions, in particular, the thickness of the film is considered to be infinitesimal compared to the thickness of the substrate. However, the approximate equation can be used without much error when the thickness of the film does not amount to more than a few percent of the thickness of the substrate. Usually, the error on stress values calculated from Eq. (44) is less than the experimental error which is about 5–10%. The error involved in using Eq. (44) begins to exceed the experimental error for films with thicknesses greater than 5% of the substrate thickness.

Since the pioneer work of Stoney in 1909 for the determination of residual stresses in electroplated coatings, more rigorous derivations of the stress equation have been reported in the literature. A detailed review of stress equations proposed in the literature and more rigorous formulas covering all the common experimental arrangements were presented by Brenner and Senderoff [84]; the stress formulas for electroplated coatings were derived from the fundamentals of the theory of elasticity for three different methods of measurement or preparation of samples, namely, (i) the deposit is plated on a basis metal which is so rigidly held that neither contraction nor bending of the plated strip can occur, (ii) the deposit is plated on a strip that is constrained from bending but not from undergoing contraction, and (iii) the deposit is plated on a strip that is allowed to bend continuously during plating.

The stress calculation from Eqs. (44) and (45) is valid for samples in which the moduli of the substrate and the film were the same. A negligible effect of the difference in moduli is observed when the thickness of the film amounts to only a few percent of the thickness of the substrate but this effect may be significant with thicker films [84]. Stress equations were also derived and discussed when Young's moduli of the film and the substrate are different [84, 85].

Thus, stress calculations from Eq. (44) or others can be performed only after a cautious verification of the validity of assumptions used for the derivation of formulas. Major conditions must be fulfilled before calculations of residual stresses from Eq. (44). In particular, the ratio between the thickness of films and the thickness of substrates would be less than 10^{-3}. Precise calculation of the stress requires knowledge of Young's modulus, E_s, and Poisson's ratio, ν_s, of substrates, in particular for specific orientations within the crystallographic plane defining the surface of substrates. The isotropic elasticity theory was found to be exact for all directions within $\langle 111 \rangle$ planes for single crystal Si and Ge substrates [86]. The isotropy of residual stresses in the plane defining the substrate surface can be verified using film–substrate samples for which beam type substrates were perpendicular in the chamber during deposition of films. Using these various recommendations, the accuracy on the stress value would be essentially dependent on the experimental error on the measurement of the radius of curvature of film–substrate structures.

3.1.2. Measurement of the Radius of Curvature of Substrates

When the initial substrate is absolutely flat, the residual stresses in the film can be calculated from Eq. (44) after measurement of the radius of curvature, R, of the substrate covered with the film. This radius can be determined from measurement of either the camber of the curved strip or the deflection of the end of the strip. Initially, the flat strip is located at a position represented by the straight line ABC in Figure 19 whereas the position of

the strip covered with the film corresponds to the arc AOA'. The radius of curvature, R, is determined by measuring the sagitta Z of the arc AOA'. The curved strip is set on a flat surface with the convex side up; the camber of the arc can be measured with a microscope or other methods [73, 74, 76].

The expression of the radius of curvature, R, of the substrate as a function of the sagitta Z can be established from triangles AON and AIN in Figure 19,

$$(R - Z)^2 = R^2 - \frac{B^2}{4} \tag{46}$$

$$\frac{1}{R} = \frac{8Z}{B^2 + 4Z^2} \tag{47}$$

$$\frac{1}{R} = \frac{8Z}{Q^2} \tag{48}$$

The length, Q, of the broken line AOA' is approximately equal to the length, L, of the curve strip AOA', and Eq. (48) can be written as

$$\frac{1}{R} = \frac{8Z}{L^2} \tag{49}$$

The replacement of Q by the arc length, L, leads to an error of less than about 2% provided that the sagitta, Z, of the strip type substrate is not more than 10% of the length of the substrate. Therefore, Eq. (44) for stress calculations becomes

$$\sigma = \frac{4}{3}\left(\frac{E_s}{1 - \nu_s}\right)\left(\frac{t_s^2}{t_f L^2}\right) Z \tag{50}$$

The deflection, Z', of the end A' of the strip can be measured when the end A is firmly maintained in a fixed position (Fig. 19). The radius of curvature of the strip type substrate can also be expressed as a function of the deflection, Z'. The area of the trapezoid $AIA'B$ is related to the areas of triangles $AA'B$ and AIN: (area $AIA'B$) = (area $AA'B$) + 2(area AIN). From these areas, it can be demonstrated that $NH = AB/2$. In addition, since the area of the triangle AIA' is equal to two times the area of the triangle AIN, it can be established that

$$NH = \frac{B}{2R}(R - Z) \tag{51}$$

From triangle ABA', it follows:

$$B^2 = Z'^2 + AB^2 = Z'^2 + 4NH^2 = Z'^2 + \frac{B^2}{R^2}(R - Z)^2 \tag{52}$$

Then, the relationship between the radius of curvature, R, and the deflection, Z', of the strip is given by

$$\frac{1}{R} = \frac{2Z'}{B^2} \tag{53}$$

In this equation, the length, L, of the arc AOA' can be substituted for the chord without much error:

$$\frac{1}{R} = \frac{2Z'}{L^2} \tag{54}$$

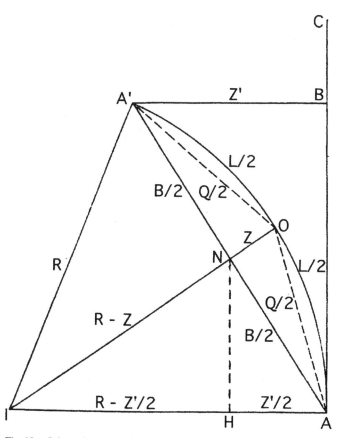

Fig. 19. Schematic representing the position of a curved strip with respect to the position of the flat strip.

Table I.   Estimates of Typical Experimental Errors Involved in the Stress
Calculation from Sagitta Measurements by Interferometry

Experimental parameter	Relative error	Observations
$\dfrac{\Delta t_s}{t_s}$	0.01	The thickness of substrate is measured with a mechanical gauge.
$\dfrac{\Delta L}{L}$	0.005	Using substrates of 50 mm in length.
$\dfrac{\Delta(E_s/(1-\nu_s))}{E_s/(1-\nu_s)}$	0.005	For example, biaxial moduli for $\langle 111\rangle$-oriented single crystal Si and Ge substrates equal to 229 and 183.7 GPa, respectively, are reported in the literature [86].
$\dfrac{\Delta t_f}{t_f}$	0.01	The thickness of films can be measured by spectrophotometry or ellipsometry. The thickness uniformity of films is about 0.5%.
$\dfrac{\Delta Z}{Z}$	0.02 to 0.5 depending on the sagitta value	The error on Z is about 0.05λ where λ is the wavelength used for interferometry measurements of the sagitta. Depending on the substrate thickness, relatively high values of Z can be obtained for a film with specific thickness and stress values.

The expression for the residual stresses as a function of the deflection can be derived from Eqs. (44) and (54):

$$\sigma = \frac{1}{3}\left(\frac{E_s}{1-\nu_s}\right)\left(\frac{t_s^2}{t_f L^2}\right)Z' \qquad (55)$$

Moreover, from Eqs. (49) and (54), it can be noted that the deflection, Z', is four times higher than the sagitta, Z ($Z' = 4Z$). The error on the value of R calculated from Eq. (54) is less than 2% if the deflection, Z', is not greater than 20% of the length of the strip type substrate. Finally, for equal values of the radius of curvature of substrates, the error involved in the calculation of residual stresses from Eq. (55) is about twice as large as that coming from Eq. (50).

The relative error on the residual stresses determined from Eq. (50), i.e., from measurements of the sagitta of substrates is given by

$$\frac{\Delta\sigma}{\sigma} = \frac{\Delta(E_s/(1-\nu_s))}{E_s/(1-\nu_s)} + 2\frac{\Delta t_s}{t_s} + 2\frac{\Delta L}{L} + \frac{\Delta t_f}{t_f} + \frac{\Delta Z}{Z} \quad (56)$$

where the symbol Δ is used to designate the absolute error of the variables. An estimate of the maximum absolute error on variables involved in Eq. (50) is given in Table I for PVD thin films deposited on $\langle 111\rangle$-oriented single crystal Si and Ge substrates, i.e., strip type substrates of (50×5) mm^2 with a thickness in the range 0.6–2 mm. For given values of the thickness of films and residual stresses, the thickness of substrates was chosen so that the sagitta to be measured was maximum. The relative accuracy on the stress values determined under these conditions was about $\pm 10\%$ [49].

3.2. Determination of Residual Stresses Using X-Ray Diffraction Techniques

X-ray diffraction techniques are powerful tools for the characterization of the microstructure of polycrystalline films, in particular for measuring changes in lattice spacing. The strain in the crystal lattice can be deduced from X-ray diffraction data and residual stresses producing the strain are calculated assuming a linear elastic distortion of the crystal lattice. In fact, residual stresses in thin films are an extrinsic property not directly measurable. All methods of stress determination require measurements of an intrinsic property such as strain, and then the associated stress is calculated.

3.2.1. Basic Principles of Stress Measurements by X-Ray Diffraction Techniques

The details of the theory and the interpretation of residual stress measurements by X-ray diffraction techniques are thoroughly described in the literature [75, 87–91]. The technique is based on measurements of the interatomic spacing of lattice planes near the surface as the gauge length for determining strain. The diffraction of a monochromatic beam of X-rays at a high diffraction angle (2θ) from the surface of a stressed sample is schematically represented in Figure 20 for two orientations of the sample relative to the X-ray beam. The orientation of the sample surface is defined by the angle ψ between the normal of the surface and the incident and the diffracted beam bisector; this angle ψ is also the angle between the normal to the diffracting lattice planes and the surface of samples.

The position of the sample for an angle $\psi = 0$ is represented in Figure 20a. The incident X-ray beam diffracts from a set of (hkl) lattice planes that satisfy Bragg's law in grains with these planes parallel to the sample surface. If biaxial compressive stresses develop in the film, the interatomic spacing of lattice planes is larger than in the stress-free state because of Poisson's effect. After the sample is rotated through a known angle ψ while the X-ray source and the detector are maintained in fixed positions (Fig. 20b), other grains in the film are responsible for the diffraction of the incident X-ray beam and diffraction occurs from the same set of (hkl) lattice planes; however, in new grains which diffract, the orientation of the diffracting planes with Miller indices (hkl) is more nearly perpendicular to the stress direction. These planes are less separated than in grains which were responsible for the X-ray diffraction when the sample was in position corresponding to $\psi = 0$ (Fig. 20a). The X-ray diffraction peak corresponding to this lattice plane family moves toward higher 2θ values. In other words, with the tilt of the sample, the lattice spacing decreases and the angle 2θ increases (Fig. 20). The calculation of the stress in the sample lying in the plane of diffraction which contains the incident and diffracted X-ray beams is possible by measuring the change in the angular position of the diffraction peak for at least two orientations of the sample defined by the angle ψ. To determine the stress in different directions at the same point, the sample is rotated about its surface normal to coincide the direction of

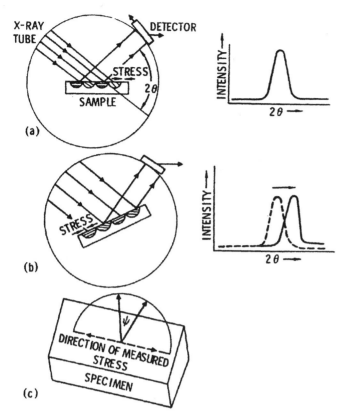

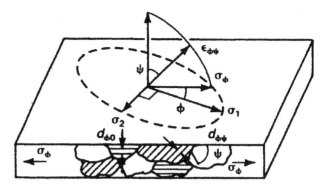

Fig. 21. Plane-stress elastic model [89]. Reprinted from P. S. Prevey, in "Metals Handbook—Materials Characterization" (R. E. Whan et al., Eds.), Vol. 10, p. 380. Am. Soc. Metals, Metals Park, OH, 1986, with permission.

Fig. 20. Principles of stress measurements by X-ray diffraction. (a) Schematic of a diffractometer. The incident beam diffracts X-rays of wavelength λ from planes that satisfy Bragg's law, in crystals with these planes parallel to the sample's surface. The diffracted beam is recorded as intensity versus scattering angle by a detector moving with respect to the specimen. If the surface is in compression, because of Poisson's effect these planes are further apart than in the stress-free state. Their spacing, d, is obtained from the peak in intensity versus scattering angle 2θ and Bragg's law, $\lambda = 2d \sin\theta$. (b) After the specimen is tilted, diffraction occurs from other grains, but from the same planes, and these are more nearly perpendicular to the stress. These planes are less separated than in (a). The peak occurs at higher angles 2θ. (c) After the specimen is tilted, the stress is measured in a direction which is the intersection of the circle of tilt and the surface of the specimen [87]. Reprinted from M. R. James and J. B. Cohen, in "Treatise on Materials Science and Technology, Experimental Methods" (H. Herman, Ed.), Vol. 19, Part A, p. 1. Academic Press, New York, 1980, with permission.

interest with the diffraction plane. The residual stresses determined by X-ray diffraction is the arithmetic average stress value in a volume of material defined by the irradiated area varying from square centimeters to square millimeters. The volume of material investigated also depends on the penetration depth of the X-ray beam.

Any interplanar spacing may be used to measure strain in the crystal lattice. However, the availability of the wavelengths produced by commercial X-ray tubes limits the choice to a few possible planes. The accuracy of the method depends on the diffraction peak selected for measurements. In practice, diffraction peaks corresponding to diffraction angles, 2θ, greater than $120°$ are currently selected since the higher the diffraction angle, the greater the precision.

3.2.2. Plane-Stress Elastic Model

The stress distribution is described by principal stresses σ_1 and σ_2 in the plane of the film surface whereas no stress is assumed to exist in the direction perpendicular to the film surface, i.e., $\sigma_3 = 0$. However, a strain component ε_3 exists as a result of Poisson's ratio contractions caused by the two principal stresses (Fig. 21).

The strain in the direction defined by the angles ϕ and ψ is given by

$$\varepsilon_{\phi\psi} = \left[\left(\frac{1+\nu_f}{E_f}\right)(\sigma_1\alpha_1^2 + \sigma_2\alpha_2^2)\right] - \left[\left(\frac{\nu_f}{E_f}\right)(\sigma_1 + \sigma_2)\right] \quad (57)$$

where E_f and ν_f are Young's modulus and Poisson's ratio of the deposited material, respectively; α_1 and α_2 are the angle cosines of the strain vector given by

$$\alpha_1 = \cos\phi \sin\psi \quad (58a)$$
$$\alpha_2 = \sin\phi \sin\psi \quad (58b)$$

As a result, Eq. (57) can be written as

$$\varepsilon_{\phi\psi} = \left[\left(\frac{1+\nu_f}{E_f}\right)(\sigma_1\cos^2\phi + \sigma_2\sin^2\phi)\sin^2\psi\right] - \left[\left(\frac{\nu_f}{E_f}\right)(\sigma_1 + \sigma_2)\right] \quad (59)$$

With the angle $\psi = 90°$, the strain vector lies in the plane of the surface, and the surface stress component, σ_ϕ, is given by

$$\sigma_\phi = \sigma_1\cos^2\phi + \sigma_2\sin^2\phi \quad (60)$$

From Eqs. (59) and (60), the strain in the sample surface at an angle ϕ from the principal stress σ_1 can be written as

$$\varepsilon_{\phi\psi} = \left[\left(\frac{1+\nu_f}{E_f}\right)\sigma_\phi\sin^2\psi\right] - \left[\left(\frac{\nu_f}{E_f}\right)(\sigma_1 + \sigma_2)\right] \quad (61)$$

Equation (61) relates the stress σ_ϕ in any direction defined by the angle ψ to the strain, $\varepsilon_{\phi\psi}$, in the direction (ϕ, ψ) defined by the angles ϕ and ψ and the principal stresses, σ_1 and σ_2, in the film.

The strain can also be expressed in terms of changes in the linear dimensions of the crystal lattice,

$$\varepsilon_{\phi\psi} = \frac{\Delta d}{d_0} = \frac{d_{\phi\psi} - d_0}{d_0} \quad (62)$$

where $d_{\phi\psi}$ and d_0 are the interplanar spacing measured in the direction (ϕ, ψ) and the stress-free lattice spacing, respectively. Equations (61) and (62) lead to

$$\frac{d_{\phi\psi} - d_0}{d_0} = \left[\left(\frac{1 + \nu_f}{E_f}\right)_{hkl} \sigma_\phi \sin^2 \psi\right] - \left[\left(\frac{\nu_f}{E_f}\right)_{hkl} (\sigma_1 + \sigma_2)\right] \quad (63)$$

where $((1 + \nu_f)/E_f)_{hkl}$ and $(\nu_f/E_f)_{hkl}$ are the elastic constant values for the crystallographic direction normal to the lattice planes in which the stress is measured as specified by the Miller indices (hkl). Because of elastic anisotropy, the elastic constants in the (hkl) direction vary significantly from the bulk mechanical values which are an average over all possible directions in the crystal lattice. Finally, the lattice spacing, $d_{\phi\psi}$, in the direction (ϕ, ψ) for a given set of (hkl) lattice planes is given by

$$d_{\phi\psi} = \left[\left(\frac{1 + \nu_f}{E_f}\right)_{hkl} \sigma_\phi d_0 \sin^2 \psi\right] - \left[\left(\frac{\nu_f}{E_f}\right)_{hkl} d_0(\sigma_1 + \sigma_2) + d_0\right] \quad (64)$$

Equation (64) is the fundamental relationship between the lattice spacing and the biaxial stresses in thin films or coatings. The lattice spacing $d_{\phi\psi}$ which is a linear function of $\sin^2 \psi$ can be plotted versus $\sin^2 \psi$. The intercept of the curve at $\sin^2 \psi = 0$ is given by

$$d_{\phi 0} = d_0\left[1 - \left(\frac{\nu_f}{E_f}\right)_{hkl} (\sigma_1 + \sigma_2)\right] \quad (65)$$

The slope of the curve is expressed as

$$\frac{\partial d_{\phi\psi}}{\partial \sin^2 \psi} = \left(\frac{1 + \nu_f}{E_f}\right)_{hkl} \sigma_\phi d_0 \quad (66)$$

As a result, the stress σ_ϕ is given by

$$\sigma_\phi = \left(\frac{E_f}{1 + \nu_f}\right)_{hkl} \frac{1}{d_0}\left(\frac{\partial d_{\phi\psi}}{\partial \sin^2 \psi}\right) \quad (67)$$

The elastic constants can be determined empirically; however, the stress-free lattice spacing, d_0, can be unknown. In general, since Young's modulus, E_f, is significantly higher than the sum $(\sigma_1 + \sigma_2)$, the value of $d_{\phi 0}$ given by Eq. (65) differs from d_0 by not more than $\pm 1\%$. As a result, Eq. (67) can be approximated to

$$\sigma_\phi = \left(\frac{E_f}{1 + \nu_f}\right)_{hkl} \frac{1}{d_{\phi 0}}\left(\frac{\partial d_{\phi\psi}}{\partial \sin^2 \psi}\right) \quad (68)$$

Finally, the stress value can be calculated from the intercept and the slope of the curve obtained by plotting lattice spacing versus $\sin^2 \psi$.

3.2.3. *Various Techniques of Measurement*

The three most common X-ray diffraction methods for determination of residual stresses, namely, the single-angle, two-angle, and $\sin^2 \psi$ techniques assume plane stress in films and are based on the fundamental relationship given by Eq. (64).

3.2.3.1. *The Single-Angle Technique*

The basic geometry for stress determination by the single-angle technique is represented in Figure 22. A collimated X-ray beam is inclined at a known angle, β, from the film surface normal. X-rays diffracted from the sample form a cone originating at point O. The diffracted X-rays are recorded using position-sensitive detectors placed on either side of the incident beam. With residual stresses in the sample, the lattice spacing varies slightly between the diffracting crystallites in positions 1 and 2 (Fig. 22). As a result, the diffraction angles are slightly different on either side of the X-ray beam. The average residual stress in the sample is given by [89]

$$\sigma_\phi = \left(\frac{E_f}{1 + \nu_f}\right)_{hkl} \left(\frac{S_1 - S_2}{2R}\right)\left(\frac{\cot \theta}{\sin^2 \psi_1 - \sin^2 \psi_2}\right) \quad (69)$$

where S_1 and S_2 are the arc lengths along the surface of the detector at a radius R from the sample surface, respectively; the angles ψ_1 and ψ_2 are related to the Bragg diffraction angles θ_1, θ_2, and the angle of inclination of the instrument, β, by

$$\psi_1 = \beta + \theta_1 - \frac{\pi}{2} \quad (70a)$$

$$\psi_2 = \beta + \theta_2 - \frac{\pi}{2} \quad (70b)$$

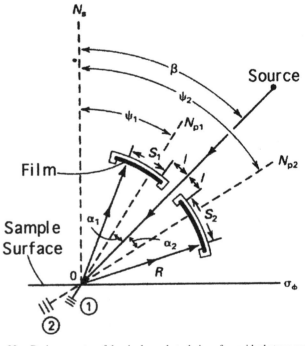

Fig. 22. Basic geometry of the single-angle technique for residual stress measurement by X-ray diffraction. N_p, normal to the lattice planes; N_s, normal to the surface [89]. Reprinted from P. S. Prevey, in "Metals Handbook—Materials Characterization" (R. E. Whan et al., Eds.), Vol. 10, p. 380. Am. Soc. Metals, Metals Park, OH, 1986, with permission.

The precision of this method is limited. Indeed, an increase in the diffraction angle 2θ to achieve precision in the determination of lattice spacing reduces the possible range of $\sin^2 \psi$ and, thereby reduces the sensitivity; generally, two-angle and $\sin^2 \psi$ techniques exhibit higher sensitivities.

3.2.3.2. The Two-angle Technique

According to Eq. (64), the lattice spacing, $d_{\phi\psi}$, is a linear function of $\sin^2 \psi$. Therefore, the stress can be determined by measuring the lattice spacing for any two ψ angles. The ψ angles are selected to provide a range of $\sin^2 \psi$ as large as possible within the limitations imposed by the diffraction angle 2θ. In practice, the lattice spacing is determined at two extreme values of ψ, typically 0 and 45°, and the stress is calculated using Eq. (68).

3.2.3.3. The $\sin^2 \psi$ Technique

The lattice spacing is determined for multiple ψ tilts instead of for two ψ angles as in the two-angle technique. The lattice spacing, $d_{\phi\psi}$, is plotted versus $\sin^2 \psi$ and a straight line is fitted by least-squares regression. The stress is calculated from the slope of the best fit line using Eq. (68). A sufficient number of measurements in various directions defined by the angle ψ should be conducted to improve the accuracy of the fit line. In addition, the significance of this distribution should be verified through a number of measurements at different ϕ angles. If a nonlinear dependence of the lattice spacing on $\sin^2 \psi$ is observed, then the film being measured is inhomogeneous. Therefore, the $\sin^2 \psi$ technique is no longer valid and applicable.

A possible explanation for a nonlinear dependence of the lattice spacing on $\sin^2 \psi$ is that stress exists normal to the film surface. An expression for the lattice spacing can be established as a function of the angles ϕ and ψ, assuming stresses normal to the surface. In principle, this full-tensor method can be used to determine stresses nondestructively in the presence of large subsurface stress gradients [92, 93]. Since extensive data collection is required for the full-tensor method, this approach is not acceptable for routine testing and is rather used for research applications. The $\sin^2 \psi$ method can be applied to films or coatings with a thickness ranging from 0.5 to 350 μm. However, it is difficult to determine residual stresses in extremely thin films and, in some cases, in highly textured films.

3.3. Magnitude of Residual Stresses in Multilayer Structures

The technological importance of multilayer structures has increased considerably over many years in many diverse fields, in particular for the production of optical coatings, integrated circuits, and mechanical components. The mechanical stability of these multilayer structures depends on the interactions between individual films, and between the stacked films and the substrate. Residual stresses in each individual film contribute to substrate bending. The description of the stress distribution and the state of elastic strain in multilayer structures was examined in various articles [94–96].

A general solution for the elastic relationships in a multilayer structure consisting of films with different elastic properties and/or thermal properties was proposed by Townsend, Barnett, and Brunner [95]. In this analysis, the multilayer structure is constructed from a set of individual films in a detached state. The components of the structure are assumed to be unstrained with relaxed lengths of d_i for the film i. The first step in the construction of the multilayer structure is to apply axial end forces to the films such that each film has the final length d_0 in the structure. This deformation leads to an elastic strain in a given film i equal to

$$\varepsilon_{i,0} = \ln\left(\frac{d_0}{d_i}\right) \tag{71}$$

Then, solutions were obtained for the case of equivalent elastic properties and for the case of different elastic properties among the films involved in the structure. The relationships and the formulas derived on the basis of this theoretical approach were approximated for the case of thin films stacked on a thick substrate of thickness t_s. The principal approximation is to assume that the thickness of one of the films in the structure is much greater than the thickness of the remainder of the multilayer structure, i.e., $t_s \gg \sum_{i=1}^{N} t_i$, where t_s is the thickness of this particular film which is the substrate and the summation is now over the set of thin films. The stress in the substrate, σ_s, and the stress in film i, σ_i, are now expressed as

$$\sigma_s = \left(\frac{E_s}{1 - v_s}\right)\left[\left(\frac{t_s}{2} - z\right)\frac{1}{R}\right.$$
$$\left. + \sum_{i=1}^{N} \frac{E_i(1 - v_s)t_i}{E_s(1 - v_i)t_s} \ln\left(\frac{d_i}{d_s}\right)\right] \tag{72a}$$

$$\sigma_i = \left(\frac{E_i}{1 - v_i}\right)\left[\ln\left(\frac{d_s}{d_i}\right)\right] \tag{72b}$$

where R is the radius of curvature of the substrate; z is the total thickness of the structure. The terms E_s, E_i, v_s, and v_i are Young's moduli and Poisson's ratios of the substrate and film, respectively. The terms in the square brackets in Eqs. (72a) and (72b) describe the elastic strains in the substrate and in each film, respectively.

The elastic strain in the substrate, in Eq. (72a), is composed of two terms. The first term corresponds to a bending strain proportional to the reciprocal radius of curvature; the second term is a planar distortion proportional to the product of the t_i / t_s ratio and the logarithm mismatch of the relaxed film and substrate planar dimensions. The thickness ratio, t_i / t_s, is much less than unity. Therefore, the planar dimension of the substrate is very slightly distorted from its relaxed configuration. The planar dimension of the multilayer structure is essentially determined by the relaxed substrate dimension.

The elastic strain in film i, given by Eq. (72b), is proportional to the relative mismatch in the relaxed film and substrate planar dimensions. In addition, the interaction between films is negligible. In other words, each individual film independently interacts with the substrate, and the presence of adjacent films or the stacking sequence of films in the multilayer structure can be totally disregarded.

A planar disparity between films can arise from differences in thermal expansion coefficients of films as the temperature of the multilayer structure is changed. A thermal strain, ε_{th}, appears in each individual film according to

$$\varepsilon_{\text{th}} = \ln\left(\frac{d_i}{d_{i0}}\right) = \alpha_i \Delta T \qquad (73)$$

where d_{i0} is the planar dimension of film i before the temperature change; ΔT is the deviation from the temperature at which the relaxed lengths were equal and α_i is the linear thermal expansion coefficient of film i. The planar dimension of a given film i depends on the temperature change ΔT as

$$\ln(d_i) = \alpha_i \Delta T + \ln(d_{i0}) \qquad (74)$$

A difference between the thermal expansion coefficients of the substrate and film i leads to a thermal strain in the structure. The thermal stress resulting from film i is obtained by combining Eq. (72b) with Eq. (74):

$$\sigma_i = \left(\frac{E_i}{1 - \nu_i}\right)\left[(\alpha_s - \alpha_i)\right]\Delta T \qquad (75)$$

From the theoretical development of Townsend, Barnett, and Brunner [95], for several thin films stacked on a substrate, the contribution of film i to the reciprocal radius of curvature of the multilayer structure is given by

$$\frac{1}{R_i} = 6\left[\frac{E_i(1 - \nu_s)}{E_s(1 - \nu_i)}\right]\left(\frac{t_i}{t_s}\right)\left[\ln\left(\frac{d_s}{d_i}\right)\right] \qquad (76)$$

The total value of the reciprocal radius of curvature depends on the contribution of each elementary film:

$$\frac{1}{R} = \sum_{i=1}^{N} \frac{1}{R_i} \qquad (77)$$

According to Eq. (77), the total substrate curvature consists of a linear superposition of the bending effects resulting from each of the individual films.

The stress in the substrate and each film, σ_s and σ_i, can be derived from Eqs. (72) and (75)–(77),

$$\sigma_s = \left(\frac{E_s}{1 - \nu_s}\right)\left(\frac{t_s}{3} - z\right)\frac{1}{R} \qquad (78a)$$

$$\sigma_i = \frac{1}{6}\left(\frac{E_s}{1 - \nu_s}\right)\left(\frac{t_s^2}{t_i R_i}\right) \qquad (78b)$$

Equation (78b) is quite similar to Eq. (44) established for the stress calculation in a simple thin film deposited on the substrate. The stress in each elementary film of a multilayer structure is proportional to the partial reciprocal radius of curvature of the substrate caused by the particular film. Finally, the individual films interact independently with the substrate and do not interact with each other. Experimental measurements of the relation between the plane strain and the bending strain in

the substrate performed for single crystal Si substrates covered with sputtered Ti-W films have demonstrated the correctness of the theoretical development and the validity of expressions derived for stress calculations in multilayer structures [95].

3.4. Mechanical Stability of PVD Thin Films

Elastic stresses in thin films are only sustained by the mechanical resistance of the substrate. Excessive residual stresses in thin films deposited on rigid substrates may cause crack formation or fracture of films, delamination of films from the substrate surface, and occasionally fracture of the substrate. The interfacial stresses are responsible for film blistering and peeling. These film failures correspond to various modes of elastic stress relaxation. The elastic energy, U, stored in the films which depend on the magnitude of residual stresses is an important criterion for the mechanical stability of film–substrate structures. Some mechanisms for the adhesion, the delamination, and the fracture of thin films in film–substrate structures with identical elastic properties of the film and the substrate were discussed in detail [2, 97].

3.4.1. Interfacial Stresses in Film–Substrate Structures

The major objective of the analysis of thermally induced stresses in single and multilayered structures proposed by Suhir [98–100] was to develop an engineering method for an approximate evaluation of the interfacial stresses in structures fabricated on thick substrates. The magnitude and the distribution of the shearing and peeling stresses in the interfaces, as well as the normal stresses in the films for film–substrate structures of finite size can be determined from this approach based on the concept of the interface compliance [98].

For the simplest case of a single layer structure fabricated at an elevated temperature, T_s, and subsequently cooled, the normal thermal stress, σ_f, in the film which is parallel to the film–substrate interface is given by [99]

$$\sigma_f = \left(\frac{E_f}{1 - \nu_f}\right)(\alpha_s - \alpha_f)(T_r - T_s)\chi_0(x) \qquad (79)$$

with

$$\chi_0(x) = \left(1 - \exp[-k(L - x)]\right) \qquad (80)$$

where α_s and α_f are the thermal expansion coefficients of the substrate and the film, respectively; T_s and T_r are the deposition temperature and the final temperature of the structure, respectively. The parameter L is the half of the film length and the x coordinate is the distance between the edge of the film and the point where the stress is equal to σ_f along the x axis parallel to the film–substrate interface. The origin O of the rectangular coordinates (x, y) is in the middle of the structure on the film–substrate interface.

The term k in Eq. (80) is a material-dependent parameter which depends on both the substrate and the film, i.e.,

$$k = \sqrt{\frac{\lambda}{\kappa}} \qquad (81)$$

with

$$\lambda = \lambda_f + \lambda_s + \frac{(t_f + t_s)^2}{(E_f/(1 - \nu_f))t_f^3 + (E_s/(1 - \nu_s))t_s^3}$$

$$\approx \lambda_f = \frac{1}{(E_f/(1 - \nu_f))t_f} \tag{82}$$

where λ_f and λ_s are the coefficients of axial compliance for the film and the substrate, respectively. The coefficient λ_s is given by [99]

$$\lambda_s = \frac{1}{(E_s/(1 - \nu_s))t_s} \tag{83}$$

For films and substrates with similar elastic properties (Young's modulus and Poisson's ratio), and with thin films deposited on relatively thick substrates ($t_f \ll t_s$), the coefficient of axial compliance, λ_s, for the substrate is negligible with respect to λ_f. Furthermore, the term, κ, in Eq. (81) is given by [99]

$$\kappa = \kappa_f + \kappa_s$$

$$= \frac{2}{3}\left[\left(\frac{1 + \nu_f}{1 - \nu_f}\right)\left(\frac{t_f}{E_f/(1 - \nu_f)}\right)\right.$$

$$\left. + \left(\frac{1 + \nu_s}{1 - \nu_s}\right)\left(\frac{t_s}{E_s/(1 - \nu_s)}\right)\right] \tag{84}$$

where κ_f and κ_s are the coefficients of interfacial compliance for the film and the substrate, respectively.

For the cross sections sufficiently remote from the film ends, the factor, $\chi_0(x)$, given by Eq. (80) is close to unity; hence, the stress, σ_f, in the film is independent of the location of the given cross section along the film length (x axis). Near the edges, where the x coordinate is of the order of half the film length, L, the stress drops rapidly and turns to zero at the edges of the film.

The magnitude and the distribution of the shearing stress, τ_f, along the film–substrate interface are given by [99]

$$\tau_f = -\left(\frac{E_f}{1 - \nu_f}\right)(\alpha_s - \alpha_f)(T_r - T_s)t_f k \exp[-k(L - x)]$$

$$= \tau_{\max} \exp[-k(L - x)] \tag{85}$$

for film–substrate structures with very great values of the parameter k. The shearing stress, τ_f, depends on the film thickness, t_f, and drops exponentially with the decrease in x; i.e., the shearing stress is concentrated near the film ends.

The magnitude and distribution of the peeling stress, p_f, along the film–substrate interface can be expressed as follows [99]:

$$p_f = p_{\max} \exp[-k(L - x)] \tag{86}$$

with

$$p_{\max} = -\frac{1}{2}\left(\frac{E_f}{1 - \nu_f}\right)(\alpha_s - \alpha_f)(T_r - T_s)t_f^2 k^2$$

$$= \frac{1}{2}k t_f \tau_{\max} \tag{87}$$

Hence, the distribution of the peeling stress is similar to the distribution of the shearing stress, i.e., the peeling stress is also concentrated at the film edges. In addition, the peeling stress depends on the square of the film thickness.

For large values of the material-dependent parameter, k, the normal thermal stress, σ_f, is independent of the film thickness (Eqs. (79) and (80)) while the interfacial stresses depend on the parameter k of the interfacial compliance which, in its turn, depends on the biaxial modulus, $E_s/(1 - \nu_s)$, of the substrate and the film thickness, t_s. The interfacial shearing and peeling stresses increase with increasing film thickness. For smaller interfacial stresses (responsible for the delamination of films), the thickness of films can be reduced in the zone of high stresses by tapering the film edges.

3.4.2. Modes of Failure

The system consisting of a residually stressed thin film deposited on a rigid substrate may fail through various mechanisms. A tensile stress in a brittle film may lead to crack formation and film fracture (Fig. 23a). This type of failure is often referred to as mud cracking since the crack pattern across the film surface looks like that observed in drying mud [101, 102]. Further failure may be initiated by these cracks. Depending on the relative fracture resistance of the film–substrate interface and substrate, interfacial delamination, or propagation of cracks into the substrate can occur (Fig. 23b and c). A tensile stress in a ductile film can be responsible for failure initiated from discontinuities such as free surfaces or inclusions within the film [102–104]. The location of cracks also depends on the fracture resistance of the substrate and the interface (Fig. 23d and e). Usually, excessive compressive residual stresses in films cause film buckling above an intrinsic defect at the film–substrate interface (Fig. 23f). The compressive stress

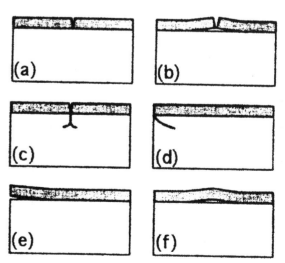

Fig. 23. Typical failures for thin films bonded to a substrate: (a) cracking of the film, (b) cracking of the film followed by delamination of the interface, (c) cracking of the film followed by failure of the substrate, (d) cracking of the substrate initiated from a free surface, (e) delamination initiated from a free surface, and (f) delamination and possible spalling resulting from buckling of the film [97]. Reprinted from M. D. Thouless, *Thin Solid Films* 181, 397 (1989), with permission.

is partially released by the film buckling. The released elastic strain energy is then available to drive the delamination of the film from the substrate [101, 102]. A compressive stress in a brittle film may initiate the buckling and the delamination which can be accompanied by spalling since the crack leaves the interface to propagate through the film.

No external loads are involved in the modes of failure described in Figure 23a–f. In fact, the crack-driving force or strain-energy release rate can be defined as

$$G = \left(\frac{\partial U}{\partial A}\right) \tag{88}$$

where U is the elastic energy stored in the film and A is the crack area. The criterion for crack advance is that G exceeds the toughness or the fracture resistance which is the energy needed to advance the crack by a unit area [97].

3.4.3. Elastic Energy and Surface Free Energy

The fracture and the delamination of films are governed by the elastic energy stored in the film which was initially firmly adherent to the substrate surface. This elastic energy depends on the film thickness and the magnitude of residual stresses in the film.

For residual stresses, σ, assumed to be biaxial and isotropic in the plane of films, the elastic energy, U, stored in a film of unit surface area is given by [105],

$$U = \left(\frac{1 - \nu_f}{E_f}\right) t_f \sigma^2 \tag{89}$$

where t_f is the film thickness and the term in brackets is the reciprocal biaxial modulus of the film. The stress value, σ, is often found to be independent of the film thickness; as a result, the elastic energy, U, increases linearly with increasing film thickness. The energy U is indicated by the shaded section at the right of Figure 24. This available energy can be consumed to produce mechanical failure of the film and the substrate. In particular, the mechanical instability of film–substrate structures can be observed when the elastic energy, U, exceeds a critical value, U_c, which is determined by the surface free energy, γ, required for film failure. As illustrated in Figure 24, the two new surfaces formed by the fracture correspond to an increase of 2γ in the total surface free energy of the film–substrate system.

For a perfect film–substrate interface having maximum film–substrate adhesion, the increase in surface free energy upon delamination of the film from the substrate is given by [2]

$$\gamma_d = \gamma_f + \gamma_s \tag{90}$$

where the subscripts f and s refer to film and substrate, respectively. In fact, γ_d is the work of adhesion for a given film–substrate system in which the film–substrate interface is perfect; i.e., the film and the substrate form a contiguous and perfect joint at the interface. For a film and a substrate totally disjointed or separated, the work of adhesion, γ_d, is equal to zero.

For not perfect film–substrate adhesion, a free energy, γ_i, is already expended and corresponds to a measure of the degree of

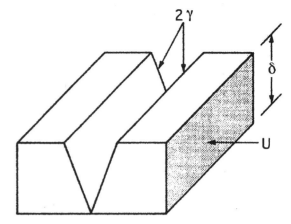

Fig. 24.   A section of film of thickness δ and of unit surface area, i.e., of unit length and width [2]. Reprinted from E. Klokholm, *IBM J. Res. Develop.* 31, 585 (1987), with permission.

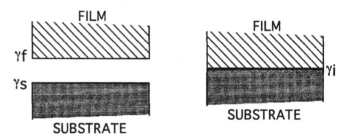

Fig. 25.   Free-energy change when materials F and S are joined.

adhesion [2]. This free energy, γ_i, is nil for a perfect adhesion between the film and the substrate. For a film totally separated from the substrate, the work of adhesion, γ_d, is equal to zero and the free energy, γ_i, is given by the sum $(\gamma_f + \gamma_s)$.

As a result, for the intermediate case in which an imperfect adhesion exists between the film and the substrate, the work of adhesion is given by [106]

$$\gamma_d = \gamma_f + \gamma_s - \gamma_i \tag{91}$$

The surface free energies, γ_f and γ_s, are essentially known quantities while γ_i, the free energy of the formed interface as illustrated in Figure 25, is an unknown quantity. In practice, γ_i depends not only upon the chemistry of the interface, but also upon its morphology and structural integrity. However, by minimizing γ_i, the work of adhesion, γ_d, may increase significantly for many systems.

3.4.4. Fracture of Films

Tensile residual stresses in a film bonded to a substrate may lead to crack formation in the film or the fracture of the film as shown in Figure 23a. In this case, the film is tightly adherent to the substrate. The crack is perpendicular to the film plane; in addition, the crack does not penetrate the substrate, and the film is not separated from the substrate at the intersection of the crack and the substrate. For this type of crack caused by a tensile fracture of the film, the Griffith fracture theory can be used [107].

The critical stress, σ_c, for fracture of the film which is normal to the crack plane (Fig. 24) can be approximated to

$$\sigma_c \approx \left(\frac{2E_f\gamma}{h}\right)^2 \tag{92}$$

where E_f is Young's modulus for the film, γ is the surface free energy for film fracture, and h is Griffith crack length [107]. In this derivation, for simplicity, Poisson's ratio, ν_f, of the film material was omitted. This ratio can be neglected without significantly affecting the results [2]. Therefore, the increase in total surface energy, 2γ, caused by the fracture of the film is given by

$$2\gamma = \frac{\sigma_c^2 h}{E_f} \tag{93}$$

The critical energy for crack formation in the film can be expressed as

$$U_c = \frac{\sigma_c^2 h}{E_f} \approx 2\gamma \tag{94}$$

By substituting the film thickness, t_f, for h in Eq. (92) or (94), the expression of the critical energy for film fracture can be written as

$$U_c = \frac{\sigma_c^2 t_{fc}}{E_f} \approx 2\gamma \tag{95}$$

where t_{fc} is a critical film thickness at which the elastic energy stored in the film exceeds the critical energy, U_c, for film fracture. The fracture of the film may occur when the elastic energy, U, given by Eq. (89) exceeds U_c,

$$U = \frac{\sigma_c^2 t_f}{E_f} \geq U_c \tag{96}$$

where Poisson's ratio, ν_f, for the film material is omitted. The value of U_c is governed by the magnitude of the surface energy, γ, for film fracture while the elastic energy, U, stored in the film depends on the product $(\sigma^2 t_f)$. For a given residual stress value, σ, independent of the film thickness, the condition for film fracture is $U \approx U_c$. As a result, the critical film thickness, t_{fc}, for film fracture is given by

$$t_{fc} = \frac{2E_f\gamma}{\sigma^2} \tag{97}$$

3.4.5. Delamination of Films

For films with compressive residual stresses, a common mechanical instability is the delamination of films from the substrate and the blister formation at the film–substrate interface as illustrated in Figure 23e and f. The separation of the film from the substrate as shown in Figure 23e is also a fracture phenomenon. The planar fracture model proposed by Barenblatt [108] is more appropriate than the Griffith fracture theory for modeling this type of instability [2].

The basic parameter in the Barenblatt model is a modulus of cohesion, K, defined by the energy, U_K, required for the separation of the two surfaces, i.e., for delamination of the film

from the substrate or the crack formation at the film–substrate interface. The energy U_K is given by

$$U_K = \left(\frac{1-\nu_f}{E_f}\right)\frac{K^2}{\pi} \tag{98}$$

The modulus of cohesion, K, can be approximated to [108]

$$K^2 \approx \left(\frac{E_f}{1-\nu_f^2}\right)\pi\gamma \tag{99}$$

By neglecting Poisson's ratio, ν_f, in Eqs. (98) and (99), the energy U_K is approximately equal to the surface free energy, γ, for film fracture. Furthermore, for delamination of the film from the substrate, the surface free energy, γ, can be equated to the work of adhesion, γ_d, and the critical energy for delamination of the film is given by

$$U_{cd} = \gamma_d \tag{100}$$

Equation (100) for delamination is quite similar to Eq. (94) established for film fracture. Thus, the condition for delamination of the film in terms of elastic energy, U, stored in the film is

$$U = \frac{\sigma^2 t_f}{E_f} \geq U_{cd} \tag{101}$$

Equation (101) for film delamination is similar to Eq. (96) valid for film fracture. The critical thickness of the film for delamination can be deduced from Eqs. (100) and (101):

$$t_{fcd} \approx \frac{E_f\gamma_d}{\sigma^2} \tag{102}$$

Hence, the critical residual stress value for delamination of the film can be written as

$$\sigma_{cd} \approx \left(\frac{E_f\gamma_d}{t_f}\right)^{1/2} \tag{103}$$

Finally, from the simplest aspects of fracture theory, it can be demonstrated that the condition for mechanical stability of film–substrate structures is governed by four major parameters (γ, γ_d, σ, and t_f). In addition, Eqs. (96) and (97) describing the criteria for film fracture exhibit the same form as Eqs. (101) and (102) which correspond to the criteria for delamination of a film from the substrate. Catastrophic film failure can be avoided by minimizing elastic energy, U, stored in the film, i.e., by reducing the product $(\sigma^2 t_f)$. The minimum value of the film thickness is usually fixed by various aspects or requirements for a given application. Therefore, the mechanical stability of film–substrate structures may be insured essentially by reducing residual stress value, σ. A significant reduction of residual stresses in thin films for a given film–substrate structure is not always possible for practical reasons. When fracture or delamination of films cannot by avoided, substitute materials for the film and/or the substrate or alternative deposition techniques must be considered for production of mechanically stable film–substrate structures.

4. ORIGIN OF RESIDUAL STRESSES IN PVD THIN FILMS

According to Eq. (1), three major contributions in residual stresses in PVD thin films bonded to rigid substrates can be schematically distinguished. There is no single model accounting for the origin of all the residual stresses observed for a wide variety of thin films. Major models invoked to explain the origin of tensile and compressive intrinsic stresses have already been reviewed extensively [73, 74, 109–114]. The intent of the present treatment is to concentrate on models most often used to explain the origin of residual stresses in amorphous and polycrystalline PVD thin films that have found many proponents. Then, investigations and results that have substantiated the prediction of these models and added new insights on the origin of stresses are described in more detail in Section 6 of this chapter.

4.1. Thermal Stresses

The thermal stress, σ_{th}, in a film bonded to a substrate results from the difference between thermal expansion coefficients of the film and the substrate when the film is deposited on the substrate at a temperature, T_s, which is higher or lower than the temperature of samples, T_r, during determination of residual stresses; in general, the residual stresses are determined at room temperature. The biaxial strain, ε, in the film is given by

$$\varepsilon = (\alpha_s - \alpha_f)(T_r - T_s) \tag{104}$$

where α_s and α_f are the thermal expansion coefficients of the substrate and the film, respectively. Without any plastic deformation in the film–substrate structure during temperature change, the elastic strain can be directly related to the thermal stress in the film through Hooke's law:

$$\sigma_{\text{th}} = \left(\frac{E_f}{1 - \nu_f}\right)(\alpha_s - \alpha_f)(T_r - T_s) \tag{105}$$

In general, the deposition temperature is higher than room temperature used for residual stress determination, i.e., $(T_r - T_s) < 0$; as a result, the thermal stress developed during temperature change is tensile ($\sigma_{\text{th}} > 0$) or compressive ($\sigma_{\text{th}} < 0$) depending on the relative values of α_s and α_f. For a film with $\alpha_f > \alpha_s$, during cooling down the film is trying to contract much more than the substrate, and the thermal stress is positive according to Eq. (105), i.e., the thermal stress is tensile.

In multilayer structures, Eq. (104) gives the biaxial strain valid for each elementary film and independent of the other films. The net thermal stress in the multilayer structure is the algebraic sum of thermal stresses developed in each elementary film. Furthermore, the net thermal stress is independent of the order of elementary films as long as the substrate temperature, T_s, is constant during deposition of elementary films. The major factor involved in the development of thermal stresses in multilayer structures is the thermal expansion coefficient value of individual films with respect to that of the substrate.

Generally, three types of stresses develop in PVD thin films and the value of residual stresses determined by one of techniques presented previously is the algebraic sum of the thermal, intrinsic and extrinsic stresses. The extrinsic stress, σ_e, is negligible when any type of interactions of the film–substrate structure and environment can be avoided between deposition of the film and the residual stress determination. In this case very frequently observed, the residual stress value is the result of the intrinsic stress built up during deposition of the film and the thermal stress which may develop during temperature change between deposition of the film and the residual stress determination. The determination of thermal and intrinsic stress contributions in the residual stress value is often required for elucidation of growth mechanisms of films and origin of stresses; however, it is not easy to achieve a clear separation of these respective contributions. The more straightforward method is to investigate the annealing temperature effect on the residual stresses in films deposited on two types of substrates having different thermal expansion coefficients. As an example, this approach was adopted to determine the contribution of the thermal stress in residual stresses developed in stoichiometric SiO$_2$ films deposited by direct evaporation; the stress variations were investigated as a function of the annealing temperature of film–substrate structures (Fig. 26). These SiO$_2$ films have been deposited simultaneously on ⟨111⟩-oriented single crystal Si and Ge substrates 1×5 cm^2 at room temperature and 200°C under a base pressure of 2×10^{-5} mbar. For films deposited at room temperature, the thermal stress was obviously nil; the residual stresses determined by interferometry immediately after deposition of films from samples under vacuum were essentially equal to the intrinsic stress developed in the films during deposition. The intrinsic stress was found to be independent of the nature of the substrate; the maximum intrinsic stress difference was about 10 MPa which is less than the accuracy on stress values. On the basis of these results, the intrinsic stress in SiO$_2$ films deposited at 200°C may be assumed to be independent of the nature of the substrate.

For determination of residual stresses versus annealing temperature in SiO$_2$ films deposited on Si and Ge substrates at 200°C, the samples were under vacuum to eliminate possible interactions with environment or generation of extrinsic stress and were heated up or cooled down with a linear temperature change of 20°C min^{-1} (Fig. 26). The residual stresses ($\sigma = \sigma_{\text{th}} + \sigma_i$) measured at room temperature for films deposited on Ge substrates were compressive (about -45 MPa) while those in films deposited on Si substrates were tensile (about $+10$ MPa). Since the intrinsic stress, σ_i, in these films is independent of the nature of the substrate, this difference in residual stresses probably reflects the disparity of thermal stresses developed in films deposited on Ge and Si substrates. Indeed, after deposition of SiO$_2$ films at 200°C, the samples were so rapidly cooled down to room temperature that the microstructure of films and, therefore, the intrinsic stress cannot be significantly modified during this quenching step. Hence, the difference in residual stresses, $\Delta\sigma$, of about 55 MPa can be related to the difference in thermal stresses which arises from the disparity of thermal expansion coefficients of Ge and Si substrates. The difference in residual or thermal stresses calculated

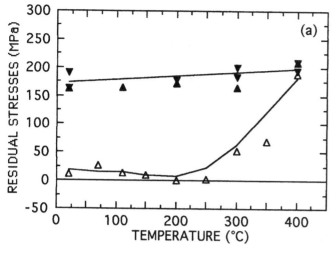

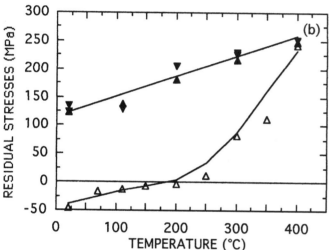

Fig. 26. Residual stresses in SiO$_2$ films versus annealing temperature for films deposited on (a) Si substrate and (b) Ge substrate during the first ($\triangle$) heating and ($\blacktriangledown$) cooling cycles, and subsequent ($\blacktriangle$) heating and ($\blacktriangledown$) cooling cycles [49]. Reprinted from H. Leplan, Ph.D. Thesis, National Polytechnic Institute of Grenoble, France, 1995.

Table II. Residual Stresses in SiO$_2$ Films Deposited by Evaporation on Si and Ge Substrates at 200°C under Various Oxygen Pressures

Oxygen pressure (mbar)	Stresses in SiO$_2$/Si, σ_{Si} (MPa)	Stresses in SiO$_2$/Ge, σ_{Ge} (MPa)	$\Delta\sigma = (\sigma_{Ge} - \sigma_{Si})$ (MPa)
2×10^{-5}	-177	-240	-63
5×10^{-5}	-117	-185	-68
2×10^{-4}	-33	-80	-47

Source: H. Leplan, National Polytechnic Institute of Grenoble, 1995.

temperature and 200°C were found to be dependent on the oxygen pressure in the deposition chamber [115, 116]. For SiO$_2$ films exhibiting similar intrinsic and extrinsic stresses, i.e., for films comparable in terms of deposition parameters and aging time in room air, the residual stresses were determined by interferometry at room temperature (Table II); the difference in residual stresses, $\Delta\sigma$, between films deposited simultaneously on Si and Ge substrates at 200°C corresponds approximately to the difference in thermal stresses developed in these films. The values of $\Delta\sigma$ are in good concordance with the value of -54 MPa estimated from Eq. (106) using the biaxial modulus of bulk SiO$_2$. This result suggests that the biaxial modulus of these SiO$_2$ films is similar to that of bulk SiO$_2$.

As the SiO$_2$-substrate samples were heated up to about 200°C, the tensile residual stresses in SiO$_2$ films deposited on Si substrates remained essentially unchanged while in films deposited on Ge substrates, the compressive stresses were reduced progressively to zero. In the range of annealing temperature above 250°C, tensile residual stresses developed in SiO$_2$ films, and at 400°C, the tensile stress values in SiO$_2$-Si and SiO$_2$-Ge samples were stable at $+200$ and $+250$ MPa, respectively. During cooling down of samples, the residual stresses decreased linearly with decreasing annealing temperature. At room temperature, the residual stresses in SiO$_2$-Si and SiO$_2$-Ge samples were approximately equal to $+160$ and $+125$ MPa, respectively. For subsequent annealing cycles, the tensile stresses in SiO$_2$ films increased or decreased linearly with a reversible manner during heating up or cooling down of samples (Fig. 26).

From these data, the variation of residual stresses above 250°C during the first annealing cycle may be attributed to the change of the thermal stress which increases with increasing annealing temperature and the release of the compressive intrinsic stress caused by the annealing temperature effect on the microstructure of amorphous SiO$_2$ films. After stabilization of stresses at 400°C, the linear variation of residual stresses as the annealing temperature was varied may arise from the change of the residual stresses in films deposited on substrates with different thermal expansion coefficients. For stabilized samples, the residual stresses, σ, can be expressed as

from Eq. (105) is given by

$$\Delta\sigma = (\sigma_{Ge} - \sigma_{Si}) = \Delta\sigma_{th}$$
$$= \left(\frac{E_{SiO_2}}{1 - \nu_{SiO_2}}\right)(\alpha_{Ge} - \alpha_{Si})(T_r - T_s) \quad (106)$$

where the difference, $(T_r - T_s)$, for SiO$_2$ films deposited at 200°C, is about -180; α_{Ge} and α_{Si} are the thermal expansion coefficients of Ge and Si substrates, i.e., 6×10^{-6} and 2.5×10^{-6} °C^{-1}, respectively. The value of the biaxial modulus of SiO$_2$ films, $[E_{SiO_2}/(1 - \nu_{SiO_2})]$, needed to calculate the difference, $\Delta\sigma$, is unknown and cannot be determined from these data. However, the difference in thermal stresses, $\Delta\sigma_{th}$, can be evaluated using the biaxial modulus of bulk SiO$_2$, i.e., $[E_{SiO_2}/(1 - \nu_{SiO_2})] = 86$ GPa; Eq. (106) leads to an estimate value of the difference in thermal stresses of -54 MPa, in good agreement with the experimental value reported in Figure 26.

Moreover, the residual stresses in stoichiometric SiO$_2$ films deposited by direct evaporation on Si and Ge substrates at room

$$\sigma = \sigma_i + \sigma_{th} = \sigma_i + \left(\frac{E_f}{1 - \nu_f}\right)(\alpha_s - \alpha_f)(T_r - T_s) \quad (107)$$

Table III. Thermal Expansion Coefficient and Biaxial Modulus of SiO$_2$ Films Deposited by Evaporation on Si and Ge Substrates Deduced from Curves Shown in Figure 26; the Values for Bulk Glass SiO$_2$ Are Also Given for Comparison

Material	Thermal expansion coefficient, α (per °C)	Biaxial modulus, $E/(1-\nu)$ (GPa)
SiO$_2$ films	1.4×10^{-6}	72
Bulk glass SiO$_2$	0.55×10^{-6}	86

Source: H. Leplan, National Polytechnic Institute of Grenoble, 1995.

where the intrinsic stress, σ_i, is independent of the annealing temperature and the nature of the substrate. The biaxial modulus and the thermal expansion coefficients of SiO$_2$ films may change during the first annealing step as the intrinsic stress together with the microstructure of films were modified. However, after stabilization of films at 400°C, these mechanical constants may be assumed to be stable and independent of the annealing temperature between room temperature and 400°C. Therefore, the slope of the straight line corresponding to the change of residual stresses for subsequent annealing cycles of stabilized samples can be derived from Eq. (107):

$$\frac{d\sigma}{dT} = \left(\frac{E_f}{1-\nu_f}\right)(\alpha_s - \alpha_f) \qquad (108)$$

The slopes of straight lines for SiO$_2$-Si and SiO$_2$-Ge samples are equal to 7.8×10^4 and 3.3×10^5 MPa °C^{-1}, respectively. The values of the thermal expansion coefficients and the biaxial modulus of SiO$_2$ films deduced from these data are given in Table III with an estimate relative accuracy of 10–20%. The biaxial modulus of the SiO$_2$ films is not very different from that of bulk SiO$_2$. By contrast, the difference between the thermal expansion coefficients of the films and the bulk SiO$_2$ is relatively large.

Not much data on thermomechanical properties of thin films have been reported in the literature. On the basis of a similar experimental approach, Blech and Cohen [117] observed a high thermal expansion coefficient (with respect to that of bulk material) for SiO$_2$ films produced by chemical vapor deposition (CVD), i.e., $\alpha = 4.1 \times 10^{-6}$ °C^{-1}. In addition, the biaxial modulus of these CVD SiO$_2$ films of about 46 GPa was relatively low. The disparity of mechanical properties with respect to the bulk material values was attributed to the porous or not fully dense structure of films produced by CVD. Indeed, the biaxial Young's modulus of a porous material is found to vary linearly with the packing density of the material, i.e., $E_f/(1-\nu_f) = p[E_{\text{bulk}}/(1-\nu_{\text{bulk}})]$ where the packing density, p, is the ratio between the mass densities of the porous film and the bulk material [118]. For the PVD SiO$_2$ films investigated, the mass density was 1.92 g cm^{-3} while the mass density of SiO$_2$ films produced by thermal oxidation is 2.2 g cm^{-3}, i.e., $p = 87$. Therefore, the biaxial modulus of not fully dense SiO$_2$ films is equal to 75 GPa; this value is in good agreement with

the value of 72 GPa deduced from these data (Table III). The relatively high value of the thermal expansion coefficient of SiO$_2$ films may be attributed to the effect of the hydration of films by water vapor contained in room air [119]. During hydration, OH radicals penetrated in the SiO$_2$ lattice, modified the atomic interactions and the thermal expansion coefficient which depends on the interatomic potential. A similar explanation was proposed for the thermal expansion coefficient of hydrogenated amorphous silicon films [120].

Finally, it is worthy to note that the experimental approach adopted to investigate the thermal stress in films allows us to determine mechanical constants of stabilized films in a given temperature range. In other words, the values of the biaxial modulus and the thermal expansion coefficient of SiO$_2$ films deduced from the variation of residual stresses as a function of the annealing temperature are valid for stabilized films in which the intrinsic stress was totally or partially released and was independent of the annealing temperature. However, the biaxial modulus and the thermal expansion coefficient of as-deposited films (just after quenching of samples) cannot be deduced from these experimental data and may be different from those for annealed films.

4.2. Intrinsic Stresses

Intrinsic stresses develop in films while the growth of films is in progress. The magnitude of intrinsic stresses is observed to be related to the microstructure (morphology) or the packing density of films which, in turn, depends on the kinetic energy of atoms condensed on the substrate or the film surface and other energetic species impinging on the surface of growing films. According to the findings of Targove and Macleod [62] described previously, the relevant parameter affecting the microstructure is the normalized momentum transfer rather than the kinetic energy. The variation of the intrinsic stress as the normalized momentum increases can be depicted by an idealized curve (Fig. 27). The intrinsic stress reported in this diagram is the average stress value which is independent of the thickness of films. The idealized diagram represents the state of continuous, amorphous, or polycrystalline films deposited at low temperatures, i.e., at a normalized growth temperature, T_s/T_m, lower than about 0.2 where thermally driven diffusion-induced strain relief is negligible. In addition, the magnitude of compensating impurity-induced stresses as might be produced by the incorporation of hydrogen [121–126], oxygen, and oxygen-based gases [127, 128] in the films is assumed to be nil. In other words, the composition of films and the impurity content are independent of the normalized momentum given by Eqs. (28)–(30) and Eqs. (33)–(35). The calculation of the normalized momentum is not always straightforward or possible since the determination of the kinetic energy and the flux of particles striking the film surface during deposition can be problematic. In addition, the normalized momentum depends on various process parameters such as pressure, target shape, target–substrate distance, and target atom mass to sputtering gas atom mass ratio for a sputter-deposition process which can be difficult to control

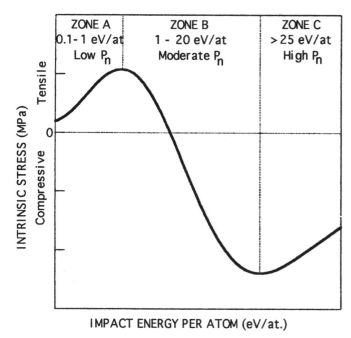

Fig. 27. Idealized intrinsic stress versus normalized momentum which is related to the kinetic energy of particles impinging on the film surface per condensing atom.

Table IV. Major Models Proposed to Explain the Origin of Tensile Intrinsic Stresses in Films of Various Microstructures

Film structure	Model
Films which consist of small isolated crystals (islands)	Surface tension
Polycrystalline films	Surface tension
	Phase change
	Buried layer
	Thermal gradient
	Grain boundary relaxation
Single crystal films	Misfit dislocation

Source: R. W. Hoffman, Academic Press, 1966.

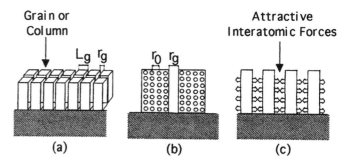

Fig. 28. Schematic of the grain boundary relaxation model.

or to determine and not always clearly mentioned in the literature. Hence, the normalized momentum is given in a qualitative manner while the impact energy value can be more easily indicated in the stress diagram represented in Figure 27.

Three distinct zones or regimes appear in this diagram depending on the normalized momentum or the impact energy which is the total kinetic energy of species colliding with the film surface per atom condensed and incorporated in the film. At a low normalized momentum, tensile intrinsic stresses are negligible since the microstructure of films (zone 1 of structure-zone diagrams in Figs. 2–4) is not sufficiently dense to support stress. As the normalized momentum increases in zone A (Fig. 27), the microstructure of films changes progressively from zone 1 to zone T (Figs. 3 and 4), the film density increases and the tensile intrinsic stress reaches a maximum value. For moderate normalized momenta (zone B in Fig. 27), the tensile intrinsic stress decreases with increasing momentum as the compensating compressive stress which is caused by the atomic peening mechanism becomes operative [129]. When the normalized momentum increases in zone B, a transition from tensile to compressive stress is observed as a result of the mass density improvement or densification of films and a maximum value of compressive stresses can be reached if the films remain tightly adherent to substrates. At high normalized momentum values (zone C in Fig. 27), intrinsic stresses decrease as the yield strength of the deposited material is attained. There is no model able to explain all the observations. Various models can be invoked for the origin of intrinsic stress, each of them is valid in a given zone of the stress-momentum diagram. Since tensile stress develops in films produced from nonenergetic particles and compressive stress appears in films obtained from energetic

particles, it is convenient to separate the presentation of models for stresses according to the kinetic energy of incident particles.

4.2.1. Intrinsic Stresses in Films Produced from Nonenergetic Particles

The models proposed for generation of tensile intrinsic stresses can be classified according to the film microstructure (Table IV). Models applicable to polycrystalline films have already been reviewed more or less thoroughly in the literature [111–114]. One of the most often used is the grain boundary relaxation (GBR) model proposed by Hoffman and co-workers [74, 109, 130–134]. This model developed for polycrystalline films is also adaptable to amorphous films exhibiting a columnar or fibrous microstructure very often observed for films produced by PVD. In these films, voids and intercolumnar spaces are assumed to play a role similar to grain boundaries in polycrystalline films.

4.2.1.1. Grain Boundary Relaxation Model

This GBR model developed for polycrystalline metal films is based on various physical features, arguments, and hypotheses. As the film grows from isolated atomic clusters, interatomic attractive forces acting across grain boundaries, gaps between grains, or intercolumnar spaces cause an elastic deformation (or relaxation) of the grain walls (Fig. 28). These attractive forces are counterbalanced by the intragrain or intracolumn tensile forces imposed by the constraint caused by the adhesion of the film to the substrate surface. The adhesion forces exceed the tensile forces developed between adjacent grains or columns. The intragrain tensile forces are prone to reduce the volume of the film as the film is detached from the substrate. The induced stress is tensile since the film is trying to contract. The intragrain strain energy can be related to the difference in the surface energy of adjacent crystallites and the energy of the resultant grain boundaries [134]. The change in energy, $\Delta\gamma$, corresponding to the formation of a grain boundary (with a grain boundary energy γ_{gb}) from the close approach of two free surfaces (of surface energy γ_s) is given by

$$\Delta\gamma = 2\gamma_s - \gamma_{gb} \quad (109)$$

For large angle grain boundaries, the average grain boundary energy is about $\frac{1}{3}\gamma_s$ [134]; thus, the change in energy, $\Delta\gamma$, is equal to $\frac{5}{3}\gamma_s$. A fraction of the energy produces a constrained relaxation of the lattice in the grain. The elastic strain is given by

$$\varepsilon = \frac{x - a}{a} = \frac{\Delta}{L_g} \quad (110)$$

where a is the unstrained lattice constant and $(x - a)$ is the variation of the lattice constant; Δ and L_g are the grain boundary relaxation distance and the final grain size, respectively. The average value of Δ can be calculated using the method of the interaction potential between two atoms [135],

$$\Delta \approx \frac{\bar{a}r_i - r_i^2/2}{2\bar{a} - r_i} \quad (111)$$

where $\bar{a}$ is the average bulk lattice constant and r_i is the ionic radius; the distance of closest approach is the sum of the ionic radii. The elastic deformation is responsible for the macroscopically observed tensile intrinsic stress, σ_i, according to Hooke's law:

$$\sigma_i = \left(\frac{E_f}{1 - \nu_f}\right)\frac{\Delta}{L_g} \quad (112)$$

The grain boundary relaxation distance can be computed from the grain separation potential for each grain diameter, L_g. The intrinsic stress in nickel films was obtained by inserting the values of Δ and L_g in Eq. (112); the calculated stress values were found to reproduce the form of the measured stress versus the substrate temperature curve satisfactorily and numerical agreement was within roughly 30% [132].

This approach can be simplified if the grain boundary relaxation distance is approximated to the ionic radius, r_i, i.e., the tensile intrinsic stress is given by

$$\sigma_i \approx \left(\frac{E_f}{1 - \nu_f}\right)\frac{r_i}{L_g} \quad (113)$$

The inverse grain size dependence of the intrinsic stress predicted by the GBR model has been observed for many film–substrate structures. This simplified approach leads to a reasonably good agreement between calculated and experimental values of tensile intrinsic stresses in chromium [134, 136],

nickel [132], diamond [137], and magnesium fluoride [138] films.

4.2.1.2. Adaptation of the GBR Model to Sputter-deposited Tungsten Films

The GBR model was adapted by Itoh, Hori, and Nadahara [139] to the calculation of intrinsic stresses developed in tungsten films sputter deposited on Si substrates at room temperature under an argon pressure in the range of 0.25–2.5 Pa. A Morse potential was chosen to calculate the force acting across grain boundaries. The average distance between the grain boundary faces, r_g, was calculated from the mass density, D_f, of the film consisting of rectangular grains (Fig. 28); the distance, r_g, is given by

$$r_g = \frac{L_g(1 - D_f)}{2D_b} \quad (114)$$

where L_g is the grain size and D_b, the bulk density of W, is equal to $19.2\,\mathrm{g\,cm^{-3}}$. The interatomic potential, $P(r)$, given by the Morse potential is expressed as

$$P(r) = A\big(\exp[-2B(r - r_0)] - 2\exp[-B(r - r_0)]\big) \quad (115)$$

where r is the interatomic distance between two grains or two columns and r_0 is the interatomic distance in the grain or the column. The other terms for tungsten are the following: $A = 0.9906$ eV, $r_0 = 0.3032$ nm, and $B = 0.14116\,\mathrm{nm^{-1}}$. The interatomic force, $F(r)$, acting across grain boundaries is given by

$$F(r) = 2AB\big(\exp[-B(r - r_0)] - \exp[-2B(r - r_0)]\big) \quad (116)$$

The Morse potential and the interatomic force versus the interatomic distance, r, are plotted in Figure 29. The average distance, r_g, between two adjacent grains or columns is usually higher than the interatomic distance, r_0 (Fig. 29). Therefore, interatomic attractive forces exist between two adjacent grain faces.

The tensile stresses in sputter-deposited W films were determined using Eq. (116) and the calculation method proposed by Itoh, Hori, and Nadahara [139]. The theoretical stress values are in good agreement with stress values determined experimentally for films deposited under argon pressures higher than 5 mTorr or 0.66 Pa (Fig. 30). In this pressure range, a thermalization of W atoms ejected from the target may occur between the target and the substrate by collisions with argon atoms. These films were probably grown from nonenergetic W atoms. Below 5 mTorr, compressive intrinsic stresses developed in W films since under low argon pressures energetic W atoms may condense on the film surface.

The origin of tensile intrinsic stresses in metal films and the existence of a long range attractive component in the interatomic potential were reexamined in detail by Machlin [114]. Various considerations and physical arguments based on the relation provided by Rose, Ferrante, and Smith [140] for the energy between metal surfaces are presented to determine the

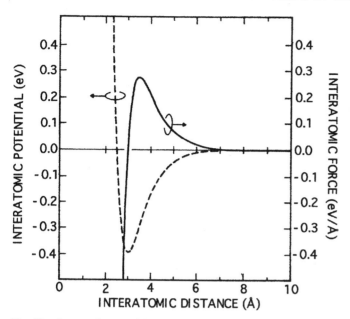

Fig. 29. Interatomic potential and interatomic force for tungsten [139]. Reprinted from M. Itoh, M. Hori, and S. Nadahara, *J. Vac. Sci. Technol. B* 9, 149 (1991), with permission.

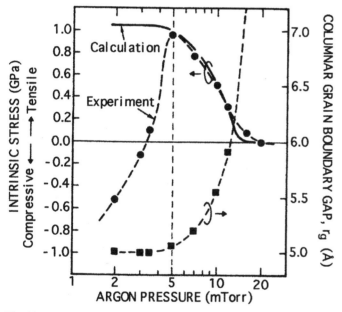

Fig. 30. Average distance between columnar grain boundary gaps (r_g) and stresses, calculated and measured [139]. Reprinted from M. Itoh, M. Hori, and S. Nadahara, *J. Vac. Sci. Technol. B* 9, 149 (1991), with permission.

1.7 Å are nil and not sufficient to account for the magnitude of the observed tensile intrinsic stresses. For grains or columns separated by less than 1.7 Å, the attractive interaction leading to a decrease in volume can result in a tensile strain. The maximum value of the tensile intrinsic stress is the yield strength of the deposited material. Thus, the GBR model is consistent with tensile intrinsic stress data in various thin films.

4.2.2. Intrinsic Stresses in Films Produced from Energetic Particles

The adatom surface mobility may be enhanced by increasing either the substrate temperature or the kinetic energy of particles impinging on the film surface which condense on the surface or simply reflect at the film surface. The adequate surface diffusion of adatoms leads to the removal of voids in the films and the microstructure or morphology of films is modified progressively; these films with an increased packing density exhibit a microstructure represented by zone T in the structure-zone diagrams (Figs. 3 and 4). The volume reduction of voids, grain boundaries, and intercolumnar spaces removes the origin of the tensile intrinsic stress; the magnitude of the stress decreases and tends to be equal to zero even though the yield strength of the deposited material is greater than zero. In addition, the removal of voids also may result in an acceleration of secondary recrystallization process and films exhibiting a bimodal grain size distribution can be obtained (Fig. 4). Therefore, the yield strength of these films may be markedly reduced. This phenomenon is likely responsible for the sudden drop in tensile intrinsic stress as the impact energy or normalized momentum increases (Fig. 27).

This type of sharp decrease in tensile intrinsic stress is observed in Figure 30 for sputter-deposited W films as the argon pressure decreases; i.e., the kinetic energy of sputtered W atoms which condense on the film surface increases. Beyond this sudden drop in tensile intrinsic stress, a transition between tensile and compressive intrinsic stress is observed and the magnitude of the compressive intrinsic stress increases progressively up to a maximum value with increasing normalized momentum (zone B in Fig. 27). Two major models have been used to account for compressive intrinsic stresses in PVD thin films produced from energetic particles with relatively high normalized momentum values [54, 113, 141].

4.2.2.1. Forward Sputtering Model Proposed by Windischmann

The forward sputtering model proposed by Windischmann [54, 113] corresponds to a quantitative development of the peening mechanism discussed by d'Heurle [52] who suggested that the compressive stress in sputter-deposited films was caused by the bombardment of the film surface by energetic species in a process akin to "shot peening" [129]. Quantitative modeling of the atom peening effect refers to the knock-on linear cascade theory of forward sputtering proposed by Sigmund [53].

maximum separation of metal surfaces across which an adequate tensile force can be generated to account for the observed intrinsic stresses in thin films. A tensile intrinsic stress can be induced either by a crack narrower than 1.7 times the spacing between close-packed planes or by a region that has less density than the equilibrium density, but in which the atomic distance is less than 1.4 times the equilibrium spacing. In other words, it has been demonstrated that the atomic interactions across voids (grain boundaries or intercolumnar spaces) thicker than about

The model proposed by Windischmann is based on three major assumptions: (i) energetic bombardment of the film surface causes displacements of atoms in the film from their equilibrium positions through a series of primary and recoil collisions, producing a volume distortion, (ii) for films deposited at low substrate temperatures ($T_s/T_m < 0.25$), mass transport and defect mobility are sufficiently low to freeze the volumetric distortion in place, and (iii) the relative volumetric distortion, d, which corresponds to a strain is proportional to the fractional number of atoms, n/N, displaced from equilibrium positions; i.e., d is given by Eq. (12). The calculation developed from Eqs. (12) to (19) leads to the expression of the volumetric distortion or the elastic strain,

$$d = 4.79 \left(\frac{K\phi_p}{N}\right)\delta\sqrt{E_p} \qquad (21)$$

with

$$\delta = \frac{M_t^{1/2}(Z_p Z_t)^{1/2}}{U_0 (M_p + M_t)^{1/2}\left(Z_p^{2/3} + Z_t^{2/3}\right)^{3/4}} \qquad (22)$$

The compressive intrinsic stress, σ_i, in the film is obtained by Hooke's law:

$$\sigma_i = \left(\frac{E_f}{1 - \nu_f}\right)d = \left(\frac{E_f}{1 - \nu_f}\right)4.79\left(\frac{K\phi_p}{N}\right)\delta\sqrt{E_p} \quad (117)$$

The atom number density, N, is given by

$$N = \frac{N_{Av}D_f}{M_t} \qquad (118)$$

where N_{Av} is Avogadro's number, D_f is the mass density of the film, and M_t is the target atomic mass (i.e., the mass of atoms incorporated in the film). By substituting Eq. (118) in Eq. (117), the expression of the compressive intrinsic stress becomes

$$\sigma_i = \frac{E_f M_t}{(1 - \nu_f)D_f} 4.79\left(\frac{K\phi_p}{N_{Av}}\right)\delta\sqrt{E_p} \qquad (119)$$

or

$$\sigma_i = k\phi_p\sqrt{E_p}Q \qquad (120)$$

where ϕ_p and E_p are the flux and the kinetic energy of energetic particles (projectiles). The constant k is equal to $4.79K\delta/N_{Av}$ and the term Q representing the stored elastic energy per mole is given by

$$Q = \left(\frac{E_f}{1 - \nu_f}\right)\left(\frac{M_t}{D_f}\right) \qquad (121)$$

The units for Q are $\mathrm{erg\,mol^{-1}}$; E_f is expressed in $\mathrm{dyn\,cm^{-2}}$, M_t in atomic mass unit (amu), and D_f in $\mathrm{g\,cm^{-3}}$. To convert $\mathrm{erg\,mol^{-1}}$ to $\mathrm{eV\,atom^{-1}}$, the value of Q is divided by 10^{12}.

The compressive intrinsic stress depends on the atomic volume of the film material, M_t/D_f. This dependence means that the interaction between the energetic particles (projectiles) and the deposited material (target) gives rise to a variable strain depending upon the atomic arrangement in the film. The linear relationship between the compressive intrinsic stress and the

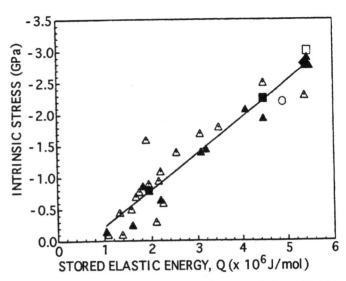

Fig. 31. Variation of the compressive intrinsic stress with the modified biaxial Young's modulus, $Q = EM/(1 - \nu)D$, for a variety of films prepared by different sputtering techniques. See [54] for identification of symbols. Reprinted from H. Windischmann, *J. Appl. Phys.* 62, 1800 (1987), with permission.

factor Q has been observed for films deposited by various sputtering techniques (Fig. 31).

The forward sputtering model is not applicable to films produced with very light energetic particle bombardment, i.e., for $Z_p/Z_t \ll 1$, since the assumptions involved in the sputtering theory of Sigmund are not valid. In addition, the model cannot be invoked to explain the intrinsic stress when the film surface is bombarded with particles having very low or very high kinetic energies. The lower energy limit of applicability corresponds approximately to the energy for atomic displacement in the films which is about 18–30 eV for most materials [142, 143]. With low energy projectiles, the assumption of isotropic cascades in the target (film) involved in the sputtering theory is no longer valid. Bombardment of growing films with ions having a normalized energy, E_n, of less than a few electron volts per atom condensed on the film surface leads to films containing a significant fraction of voids; as a result, tensile intrinsic stress develops in the films rather than compressive intrinsic stress [144]. The upper energy limit of applicability is related to resputtering of the deposited material and possible mechanical damage of the film such as void formation and plastic flow [145]. For instance, under krypton ion bombardment, the fraction of platinum films resputtered increases linearly with increasing normalized energy, E_n, and can reach 50% at $E_n = 50$ eV atom^{-1}. Moreover, the magnitude of the compressive intrinsic stress predicted by this model increases continuously with increasing projectile energy, E_p. However, the maximum compressive stress observed in PVD films is usually less than the yield strength. In addition, beyond this maximum, the decrease in compressive stress as the normalized momentum increases (zone C in Fig. 27) is not predicted by the forward sputtering model.

4.2.2.2. Model Proposed by Davis

This model proposed to explain the formation of compressive intrinsic stress in thin films deposited with simultaneous bombardment by energetic ions or atoms is also applicable to dense films. The energetic particles impinging on the surface of growing films cause atoms to be incorporated into spaces which are smaller than the usual atomic volume [146]. Therefore, the films expand outward from the substrate; however, in the plane of films, the films are not free to expand and the entrapped atoms are responsible for development of compressive intrinsic stresses.

The model proposed by Davis [141] is based on two major assumptions. The compressive intrinsic stress is assumed to be caused by film atoms implanted below the surface of the film by knock-on processes in accordance with the model proposed by Windischmann. In addition, thermal spikes are assumed to reduce the stress by causing displacement of the implanted atoms. The rate, n_i, per unit area with which atoms are implanted below the film surface deduced from the forward sputtering model is given by

$$n_i = \left(\frac{13.36}{U_0}\right)\phi_p\sqrt{E_p} \qquad (122)$$

where U_0 is the sublimation energy of the deposited material; ϕ_p and E_p are the flux and the kinetic energy of ions (projectiles) impinging on the surface of films grown by ion beam-assisted deposition, respectively.

Furthermore, the implanted atoms in metastable position which acquire more than some excitation energy, E_d, will escape from their metastable position to the surface of the film. Intense local heating resulting from the energy of bombarding particles transferred to film atoms in the very small area of the impact; i.e., thermal spike is supposed to provide the energy required for releasing implanted atoms from their metastable position within the film [147]. The number, n_a, of atoms which will receive more than the excitation energy when an energy, q, is deposited into the thermal spike is given by

$$n_a = 0.016\rho\left(\frac{q}{E_d}\right)^{5/3} \qquad (123)$$

where ρ is a material-dependent parameter which is of the order of unity [141]. For energies lower than 1 keV which are generally used for ion-assisted deposition processes, each energetic impact will produce only one thermal spike and the energy, q, can be approximated to the kinetic energy of ions, E_p. Assuming that every implanted atom receiving more than the excitation energy, q, will migrate to the film surface, the rate, n_R, per unit area with which atoms will experience relaxation can be expressed as

$$n_R = 0.016\rho\frac{n}{N}\phi_p\left(\frac{E_p}{E_d}\right)^{5/3} \qquad (124)$$

where n and N are the number of atoms displaced by energetic particle bombardment per unit volume and the atomic number density, respectively. Assuming that there is a balance between implantation and relaxation processes, the density, n, of implanted atoms (or atoms displaced by energetic particle bombardment) is constant with time. The rate per unit area, R_i, with which implanted atoms are incorporated into the film is given by

$$R_i = (n_i - n_R) = \phi_c\frac{n}{N} \qquad (125)$$

where ϕ_c is the flux of neutral atoms condensed on the film surface and incorporated in the growing film.

The expression of the ratio, n/N, calculated from Eqs. (122), (124), and (125) can be written as

$$\frac{n}{N} = \left(\frac{13.36}{U_0}\right)\left[\frac{(E_p)^{1/2}}{\phi_c/\phi_p + k_a(E_d)^{5/3}}\right] \qquad (126)$$

where the factor k_a is equal to $0.016\rho(E_d)^{-5/3}$. According to Hooke's law, the compressive intrinsic stress in the film can be expressed as

$$\sigma_i = \left(\frac{E_f}{1 - \nu_f}\right)\left(\frac{Kn}{N}\right)$$
$$= \left[\left(\frac{13.36K}{U_0}\right)\left(\frac{E_f}{1 - \nu_f}\right)\right]\left[\frac{(E_p)^{1/2}}{\phi_c/\phi_p + k_a(E_d)^{5/3}}\right] \qquad (127)$$

Equation (127) can also be written as [141]

$$\sigma_i = \kappa\left[\frac{(E_p)^{1/2}}{\phi_c/\phi_p + k_a(E_d)^{5/3}}\right] \qquad (128)$$

This model predicts that the magnitude of the compressive intrinsic stress is strongly dependent on the energy of ions used for ion-assisted deposition of films. The parameters κ and k_a in Eq. (128) are generally unknown. One method for determining these parameters is to use them as fitting parameters for a least-squares fit to data for which the kinetic energy, E_p, ion flux, ϕ_p, and deposition rate of films or flux of condensing atoms, ϕ_c, are known.

The applicability of this model was demonstrated for AlN films produced by ion-assisted deposition and amorphous carbon (a-C) films prepared by filtered-arc deposition. The experimental data may be fitted by Eq. (128) assuming that $\rho = 1$ and with an excitation energy, E_d, of 11 and 3 eV for AlN and a-C films, respectively, (Fig. 32); the normalized flux, ϕ_p/ϕ_c, was approximated to unity.

When the normalized flux, ϕ_p/ϕ_c, is low, the reciprocal term ϕ_c/ϕ_p in Eq. (127) is large with respect to $k_a(E_p)^{5/3}$ and the compressive intrinsic stress can be approximated to

$$\sigma_i \approx \left(\frac{E_f}{1 - \nu_f}\right)\left(\frac{13.36K}{U_0}\right)\frac{\phi_p(E_p)^{1/2}}{\phi_c} \qquad (129)$$

This expression is comparable with Eq. (120) which gives the compressive intrinsic stress predicted from the forward sputtering model. In the theory developed by Davis, the compressive intrinsic stress is found to be proportional to ϕ_p/ϕ_c whereas the forward sputtering model leads to a compressive intrinsic stress

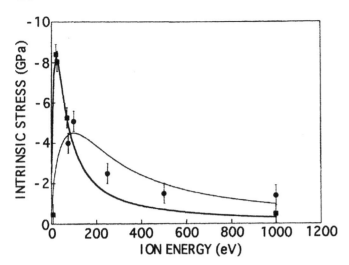

Fig. 32. Experimental measurements of the compressive stress in tetrahedral amorphous carbon films deposited by filtered cathodic arc (■) and AlN films prepared by ion-assisted deposition (●). The data were fitted by Eq. (127) assuming that $E_d = 3$ eV (bold solid line) for the amorphous carbon and $E_d = 11$ eV (light solid line) for the AlN, with $\rho = 1$ and $\phi_C/\phi_i = 1$ for both curves [141]. Reprinted from C. A. Davis, *Thin Solid Films* 226, 30 (1993), with permission.

intensity linearly depends on ϕ_p only. For a very large normalized flux, ϕ_p/ϕ_c, Eq. (127) can be approximated to

$$\sigma_i \approx \left(\frac{E_f}{1 - \nu_f} \right) \left(\frac{13.36K}{U_0} \right) \frac{1}{k_a (E_p)^{7/6}} \qquad (130)$$

Two different strategies can be adopted to reduce the magnitude of compressive intrinsic stresses in films produced by ion-assisted deposition techniques. The first possibility is to produce PVD films with low normalized fluxes, ϕ_p/ϕ_c, i.e., at high deposition rates and low ion fluxes. The compressive intrinsic stress given by Eq. (129) is also reduced with decreasing ion energy. However, in this case, the benefits of ion-assisted deposition may be considerably diminished. The alternative approach is to prepare PVD films with high normalized fluxes, ϕ_p/ϕ_c, or at low deposition rates with high ion fluxes. The magnitude of compressive intrinsic stress expressed by Eq. (130) may be reduced with increasing ion energy. However, too high ion energies may lead to resputtering phenomena and significant damage with void formation in the films. Eventually, films produced under these conditions exhibit reduced mass densities and tensile intrinsic stress.

4.3. Extrinsic Stresses

In addition to the thermal stress and intrinsic stress, a third term, named extrinsic stress can be distinguished in the overall residual stresses in PVD films (Eq. (1)). This type of stress originates from interactions between the deposited material and the chemical agents present in the film environment during or after deposition. As-deposited films exhibit more or less dense microstructure and tensile or compressive intrinsic stress depending upon process parameters adopted for film deposition.

Impurities such as oxygen, hydrogen, inert gas atoms can be incorporated in PVD films during or after deposition by evaporation or sputtering. A lattice distortion may result from incorporation of atoms of a size different from the host, reaction such as oxidation or hydrogenation which produces a new phase with a different molar volume, and grain surface energy reduction. These phase transformations lead generally to volume expansion and compressive stress can be observed [148]. Consequently, for not fully dense films exhibiting tensile intrinsic stress, an impurity-induced compensating compressive stress may develop during deposition. This compressive stress related to impurity incorporation may reduce the tensile stress in a nonovert manner or may produce a net compressive stress if the tensile intrinsic stress is totally compensated and overwhelmed even though the atomic peening mechanism is ineffective. Therefore, the interpretation of stress data must be performed cautiously for PVD films with compositions or impurity contents which vary with process parameters as well as for PVD films produced by reactive evaporation and reactive sputtering processes which may involve the formation of a new phase with a different molar volume by surface reactions such as oxidation, nitridation, or hydrogenation.

4.3.1. Effect of Various Impurities

Oxygen or oxygen-bearing species were recognized to produce compressive stresses in PVD films [73, 109]. Various examples have been described by Windischmann [113]. The decrease in high tensile stress in evaporated Cr films caused by oxygen and oxygen-bearing species was ascribed to the formation of chromium oxide at grain boundaries that suppresses grain coalescence [149]. This mechanism corresponds to the inverse GBR model; the oxidation of Cr leads to a decrease in the relaxation distance, Δ, contained in Eq. (112) which gives the magnitude of tensile intrinsic stress since the surface energy of neighboring free surfaces is modified [134].

Hydrogen atoms can also be incorporated in Si and C films produced by PVD. Compressive stresses in Si:H [121, 122] and amorphous C:H [126] films may be associated with modification of bond length or orientation and lattice distortion.

Inert gas atoms can be entrapped in thin films produced by sputtering under appropriate deposition conditions. The compressive stress in sputter-deposited films may be attributed to the effect of inert gas entrapment [150–152]. However, for various metal films produced from a cylindrical postcathode magnetron source, the compressive strain and the entrapped argon content were found to be totally independent [153–155]. Thornton, Tabock, and Hoffman [155] conclude that the compressive stress was caused by the energetic particle bombardment of the metal films which produces lattice damage by direct and recoil atomic displacement according to the forward sputtering mechanism proposed by Sigmund [53] and the atomic peening mechanism [52]. In spite of a number of articles attributing the compressive stress in sputter-deposited films to the incorporation of sputtering inert gas atoms in the crystal lattice, an exact mechanism to explain the origin of the stress in impurity containing films has not been clearly identified.

4.3.2. Adsorption Effect of Water Molecules or Other Polar Species in Porous Films

Various molecules can penetrate open voids or pores present in not fully dense films and adsorb on pore walls thereby interaction forces between adsorbed species, in particular, between polar species such as water molecules can act to modify residual stresses. These interaction forces may be responsible for extrinsic stress generation in porous films which were exposed to room air or chemical agents contained in various environments. Depending on the relative orientation of adsorbed polar molecules, nature of adsorbed species and composition of pore walls, attractive or repellent interaction forces may develop and the resulting extrinsic stress in the films can be tensile or compressive. A model to explain the origin of extrinsic stress based on the adsorption of polar species on pore walls was proposed by Hirsch [156]. This model is not of great numerical precision; however, dipole interactions between adsorbed molecules are demonstrated to be responsible for the observed forces and stresses produced by realistic amounts of adsorbate.

For films exhibiting columnar grains, the structure is similar to an aggregate of closely packed cylindrical columns of radius R. Voids of cross-sectional area, $(\sqrt{3} - \pi/2)R^2$, are formed between any three adjacent cylinders. For simplicity, this voided structure is replaced by equivalent cylindrical pores of radius, a, yielding the same cross-sectional area; the relationship between the radii, a and R, can be established from geometrical considerations [156]:

$$a = \left(\frac{\sqrt{3}}{\pi} - \frac{1}{2}\right)^{1/2} \qquad R = 0.23R \qquad (131)$$

In addition, the spacing, s, between centers of adjacent cylindrical pores is given by

$$s = \frac{2(\sqrt{3} - 1/\sqrt{3})a}{(\sqrt{3}/\pi - 1/2)^{1/2}} \approx 10a \qquad (132)$$

Hirsch assumes that the adsorbed molecular dipoles are arranged on the cylindrical pore wall with their axes normal to the surface and with charges of the same sign pointing inward and outward, respectively. In addition, the circumference of the adsorbing surface is assumed to remain at the constant value of $2\pi a$, irrespective of the amount of material adsorbed on pore walls.

A pore of length equal to the film thickness, t_f, containing two polar species adsorbed on the wall is represented in Figure 33. The z axis coincides with the pore axis. The x and y axes lie in the center plane of the film, with the x axis pointing along the length of the substrate, i.e., in the direction of the stress. A dipole of moment $\mathbf{M}_1$ aligned along the x axis in the x–y plane interacts with a second dipole of moment $\mathbf{M}_2$ located at a vertical height, h, above the center plane and displaced azimuthally from $\mathbf{M}_1$ by angle ϕ. The force along vector, $\mathbf{r}$,

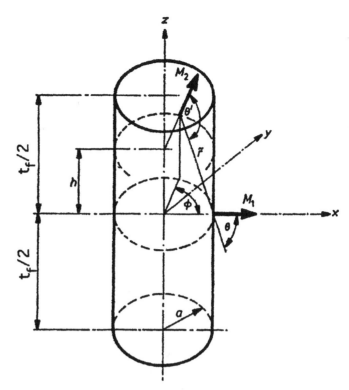

Fig. 33. Schematic of the model geometry with two dipoles in a circular cylindrical pore [156]. Reprinted from E. H. Hirsch, *J. Phys. D: Appl. Phys.* 13, 2081 (1980), with permission.

connecting $\mathbf{M}_1$ and $\mathbf{M}_2$ can be expressed from elementary electrostatic theory as [157]

$$F_r = \frac{3\mathbf{M}_1\mathbf{M}_2}{4\pi\varepsilon|\mathbf{r}|^4}\left(\sin\theta\sin\theta'\cos\psi - 2\cos\theta\cos\theta'\right) \qquad (133)$$

where ε is the dielectric constant of the free space, ψ is the angle between the planes intersecting in $\mathbf{r}$ that contain the dipole vectors $\mathbf{M}_1$ and $\mathbf{M}_2$. The angles, θ and θ', are angles between the vector $\mathbf{r}$ and the two dipole axes, $\mathbf{M}_1$ and $\mathbf{M}_2$, respectively. The terms in brackets on the right-hand side member of Eq. (133) can be expressed as functions of the normalized vertical height, $y = h/a$, and angle ϕ [156]. In addition, the absolute value of the vector $\mathbf{r}$ is given by

$$|\mathbf{r}| = a\left[4\sin^2\left(\frac{\phi}{2}\right) + y^2\right]^{1/2} \qquad (134)$$

Therefore, making the appropriate substitutions, the interaction force acting along the vector $\mathbf{r}$ can be expressed as a function of variables y and ϕ only. Multiplication by $\cos\theta$ yields the component, $_xF_{\mathbf{M}_1}$, of the interaction force on $\mathbf{M}_1$ in the direction of the x axis; this component represents the contribution of the interaction force to the stress,

$$_xF_{\mathbf{M}_1} = \frac{3\mathbf{M}_1\mathbf{M}_2}{4\pi\varepsilon a^4}g(\phi, y) \qquad (135)$$

where g is a function of ϕ and y only.

A summation over the entire set of discrete molecular dipoles is required for calculation of the interaction force on $\mathbf{M}_1$ due to all the polar molecules adsorbed on the pore wall. An alternate calculation procedure adopted by Hirsch consists of replacing the set of discrete dipoles by an equivalent continuous dipole layer of moment β per unit area and an integration can be performed by replacing $\mathbf{M}_1$ and $\mathbf{M}_2$ by dipole elements $\beta\, dS_1$ and $\beta\, dS_2$ of area dS_1 and dS_2, respectively. Indeed, the potential at a point in a dipole array is insensitive to a rearrangement of the individual dipoles provided that the change is made in a manner leaving the average dipole moment per unit area constant [158]. The area of the dipole element is given by

$$dS_2 = a^2\, d\phi\, dy \tag{136}$$

The total component of the interaction force on the dipole element, $d\mathbf{M}_1$, becomes

$$\begin{aligned} {}_x\overline{F_{\mathbf{M}_1}} &= \frac{3\beta^2\, dS_2}{4\pi\varepsilon a^2} \int_{y_0}^{y_{\max}} \int_{\phi_0}^{\pi} g(\phi, y)\, d\phi\, dy \\ &= \frac{3\beta^2\, dS_1}{4\pi\varepsilon a^2} I \end{aligned} \tag{137}$$

where the value of the interaction integral, I, depends on the integration limits of the variables ϕ and y which remain to be determined. On the basis of various arguments developed by Hirsch for the choice of integration limit values, Eq. (137) can be written as

$$\begin{aligned} {}_x\overline{F_{\mathbf{M}_1}} &= \frac{3\beta^2\, dS_2}{2\pi\varepsilon a^2} \int_0^{d_0} \int_{d_0/a}^{\pi} g(\phi, y)\, d\phi\, dy \\ &\quad + \int_{d_0}^{\infty} \int_0^{\pi} g(\phi, y)\, d\phi\, dy \\ &= \frac{3\beta^2\, dS_1}{2\pi\varepsilon a^2} I(a, d_0) \end{aligned} \tag{138}$$

where d_0 is the minimum distance between two neighboring molecular dipoles which also represents the effective diameter of polar molecules adsorbed on the pore wall. A factor of 2 has been introduced to take into account the contribution from points both above and below the center plane.

For a dipole element $d\mathbf{M}_1$ not oriented along the x axis, but inclined to it at an angle φ, the contribution to the interaction force acting on the film cross section is obtained by multiplying Eq. (138) by $\cos\varphi$. For the entire pore wall, the component of the interaction force along the x axis is given by

$$\begin{aligned} {}_x\overline{F_{\text{pore}}} &= \frac{3\beta^2 I(a, d_0)}{\pi\varepsilon a} \int_{-t_f/2}^{t_f/2} \int_{-\pi/2}^{\pi/2} \cos\varphi\, d\varphi\, dh \\ &= \frac{6\beta^2 I(a, d_0) t_f}{\pi\varepsilon a} \end{aligned} \tag{139}$$

The number of pores per unit length is deduced from Eq. (132), i.e., $1/s \approx 1/10a$; hence, the total interaction force acting on the cross section of the film per unit width is given by

$$F_{\text{total}} = \frac{6\beta^2 I(a, d_0) t_f}{10\pi\varepsilon a^2} \tag{140}$$

In addition, the moment β of the dipole layer per unit area is due to n dipoles per unit area, each having a dipole moment α, i.e., $\beta = n\alpha$. Substituting β in Eq. (140), the total interaction

force can be written as

$$F_{\text{total}} = \frac{3n^2\alpha^2 I(a, d_0) t_f}{5\pi\varepsilon a^2} \tag{141}$$

The interaction integral, $I(a, d_0)$, can be numerically computed. Hirsch has observed that for pore radius, a, of 10 Å, changes of 10–20% in diameter, d_0, of polar molecules do not significantly affect the interaction integral value; in addition, for molecular diameter, d_0, in the range 3–6 Å, the interaction integral can be expressed as

$$I(a, d_0) \approx 5.8 \times 10^8 a \tag{142}$$

where the pore radius, a, is expressed in meters. Substituting $I(a, d_0)$ by Eq. (142) in Eq. (141), the total interaction force becomes

$$F_{\text{total}} = \frac{1.74 \times 10^9 n^2\alpha^2 t_f}{5\pi\varepsilon a} \tag{143}$$

According to Eq. (143), for very porous films with relatively large pore radii, a, the extrinsic stress resulting from adsorption of molar molecules would be rather low. In these films, the distance between adsorbed polar molecules on opposite pore walls is too large for efficient attractive or repellent interactions. For dense films, the pore radii tend to be nil and the interaction force would be infinite. This situation is unrealistic since for too low pore sizes, the coverage of the pore walls by adsorbed polar molecules is expected to be nonuniform or eventually molecules cannot be adsorbed if the pore diameter is less than the polar molecule diameter. As the pore diameter decreases, the number of dipoles per unit area, n, adsorbed on the pore walls decreases concomitantly and the interaction force tends to be negligible. The magnitude of extrinsic stress originating from adsorption of polar molecules in not fully dense films would be maximum for pore sizes comparable to the diameter of polar molecules.

Furthermore, the magnitude of extrinsic stress may depend on the aging time of films when the interaction between adsorbed species and film material leads to chemical changes or to structural transformations of the deposited material. This type of slow evolution of residual stresses with increasing aging time was observed for SiO_2 films deposited by direct thermal evaporation [119].

5. EFFECT OF MAJOR PROCESS PARAMETERS ON THE INTRINSIC STRESS

On the basis of experimental data and major mechanisms proposed to explain the origin of intrinsic stress in PVD films, the magnitude and the sign (tensile or compressive) of intrinsic stress depend essentially on the normalized momentum. This relevant physical factor is related to: (i) the kinetic energy and the mass of atoms condensed on the substrate and other species striking the film surface during deposition, (ii) the flux of atoms condensed on the substrate or the deposition rate of films, and (iii) the flux of energetic particles impinging on the surface, reflected at the film surface and backscattered to the gas phase. These physical parameters affect the microstructure and the morphology of films as well as intrinsic stress and other physical characteristics of films such as packing density and microhardness which exhibit a strong microstructure dependence. The values of physical factors affecting the magnitude of intrinsic stress depend on various process parameters, in particular, pressure, P, in the deposition chamber and the substrate temperature, T_s, as well as substrate bias voltage, V_B, for films produced by sputtering, ion plating, or activated reactive evaporation (ARE). In ion plating and ARE processes, the gas phase consists of atoms evolved from an evaporation source and additional gases (inert gas or reactive gas such as nitrogen for deposition of nitride films). The gas phase is partially ionized by various techniques and positive ions contained in the plasma are accelerated to the growing film surface biased to an appropriate negative voltage. The main process parameters (P, T_s, and V_B) may vary in relatively wide ranges independently. Therefore, a large variety of deposition conditions can be adopted to produce PVD films and to control the magnitude of intrinsic stress. Additional process parameters may affect the magnitude of intrinsic stress in films deposited by sputtering, namely, the target–gas mass ratio, substrate orientation, cathode geometry, cathode–substrate distance, angle of deposition incidence, and concentration of sputtering gas atoms entrapped in the films. The literature data about the effects of these various process parameters or factors on intrinsic stresses in sputter-deposited films have been reviewed [113].

5.1. Pressure Effect

The magnitude of intrinsic stress in evaporated and sputter-deposited films is commonly observed to vary with the gas pressure in the deposition chamber as shown by typical curves in Figures 34 and 35. At low pressures, energetic particles evolved from the evaporation source or coming from the target as sputtered atoms or backscattered neutral atoms can strike the film surface with substantial kinetic energy. Compressive intrinsic stresses develop in thin films since the normalized momentum of particles arriving at the film surface is relatively high. The microstructure of films produced under these conditions corresponds to that of zones 2 and T in structure-zone diagrams (Figs. 3 and 4). The mechanisms developed by Windischmann [54, 113] and by Davis [141] based on the knock-on linear cascade theory of forward sputtering can be invoked to explain the origin of compressive intrinsic stresses in these dense films.

As the pressure in the deposition chamber is increased, particle scattering phenomena caused by collisions with residual gas atoms or molecules contribute to reduce both the kinetic energy and the normal flux component of species arriving at the film surface. The microstructure of films changes progressively from zone T to zone 1 as the pressure increases (Fig. 3). This evolution of the microstructure is accompanied by a reduction in compressive intrinsic stress and tensile stress component built up in these not fully dense films. The stress transition

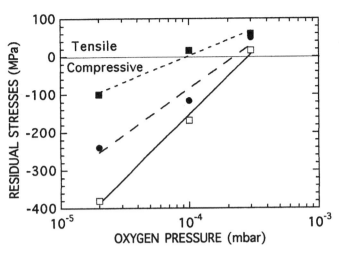

Fig. 34. Residual stresses versus oxygen pressure for SiO_2 films deposited by thermal evaporation on Si substrates at various temperatures: (■) ambient temperature, (●) 200°C, and (□) 250°C [49]. Reprinted from H. Leplan, Ph.D. Thesis, National Polytechnic Institute of Grenoble, France, 1995.

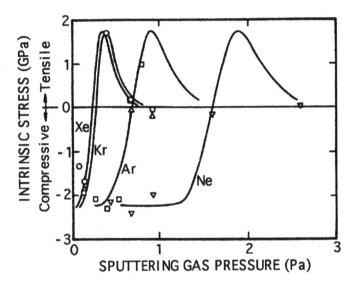

Fig. 35. Intrinsic stress versus sputtering pressure for molybdenum films sputter deposited from a cylindrical-postmagnetron source in Ne, Ar, Kr, and Xe at 1 nm s^{-1} [159]. Reprinted from D. W. Hoffman and R. C. McCune, in "Handbook of Plasma Processing Technology" (S. M. Rossnagel, J. J. Cuomo, and W. D. Westwood, Eds.), p. 483. Noyes, Park Ridge, NJ, 1990, with permission.

from compressive to tensile is observed as the tensile intrinsic stress caused by grain boundary relaxation exceeds the compressive intrinsic stress generated by atomic peening phenomena.

At high pressures, gas scattering phenomena are very efficient at reducing the kinetic energy of species impinging on the film surface. The microstructure of films is columnar or fibrous and a rough surface topography caused by a shadowing effect can be observed [160]. Tensile intrinsic stresses originating from grain boundary relaxation develop in these films. For highly porous films with large voids or intercolumnar spaces, zero intrinsic stress can be observed since lateral interaction forces between adjacent grains become too low to produce any strain in the films.

The sputtering gas pressure corresponding to the tensile-compressive stress transition depends on the atomic masses, M_t and M_i, of the target and the sputtering gas, respectively. For molybdenum films produced by magnetron sputtering, the inert gas pressure at the tensile–compressive stress transition decreases significantly with the increase in atomic mass of the sputtering inert gas (Fig. 35). When light element films are sputter-deposited by a high mass inert gas, the intrinsic stress in the films can remain tensile even for very low sputtering gas pressures. The kinetic energy and the flux of particles reflected at the target surface depend on the atom mass ratio, M_t/M_i, as shown in Figures 13 and 14. Using target–sputtering gas combinations corresponding to a low atom mass ratio, the reflected energy and the flux may be too low to induce atom peening phenomena and knock-on linear collision cascades involved in generation of compressive intrinsic stresses.

5.2. Substrate Bias Voltage Effect

The effect of the substrate bias voltage on residual stresses in sputter-deposited molybdenum films is clearly illustrated by data reported in Figure 36. The kinetic energy of positive ions striking the film surface is controlled by the negative substrate bias voltage. The variation of stress with increasing negative bias voltage is very similar to the stress variation observed as the sputtering gas pressure decreases. The nature of energetic particles impinging on the growing films may depend on the process parameters; however, the similarities in the pressure and the bias voltage data suggest that a common mechanism involved in the generation of intrinsic stress can be invoked.

Furthermore, the similarities observed between pressure and bias voltage effects also extend to the microstructure of films. The intrinsic stress for sputter-deposited tungsten films was investigated as a function of the substrate bias voltage, V_B, varying in the range $+30$ to -60 V [162, 163]. Three series of films were produced with various sputtering powers (1.25, 2, and 3.5 W cm^{-2}). The microstructure of films was examined by scanning electron microscopy [163]. The porosity of metal films deposited on grounded substrates ($V_B = 0$) was found to decrease with increasing sputtering power. More energetic particles collide the film surface with a relatively high sputtering power. In addition, the tensile-to-compressive stress transition and the maximum tensile stress occurred to a lower negative bias voltage as the sputtering power increased (Fig. 37). Using high sputtering powers, the impact energy on the film surface is relatively high; hence, lower substrate bias voltages are required for the tensile–compressive stress transition, i.e., to be under deposition conditions at which the atomic peening mechanism for compressive stress generation is operative. As shown in Figure 37, the intrinsic stress remains tensile using positive substrate bias voltage. In addition, metal films deposited under these conditions exhibited a porous microstructure corresponding to zone 1 of the structure-zone diagram (Fig. 3). With a positive bias voltage, the positive ions contained in the argon plasma were repelled from the substrate. The particles impinging on the film surface were tungsten atoms coming from the

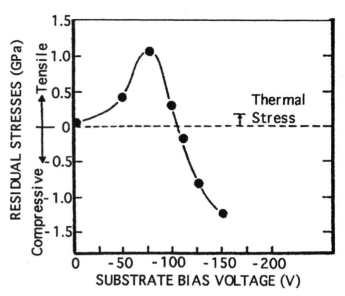

Fig. 36. Variation of the residual stresses with substrate bias voltage for molybdenum films of 280–350 nm in thickness deposited by RF planar diode sputtering on substrates at about 100°C [161]. Reprinted from J. A. Thornton, in "Semiconductor Materials and Process Technology Handbook" (G. E. McGuire, Ed.), p. 329. Noyes, Park Ridge, NJ, 1988, with permission.

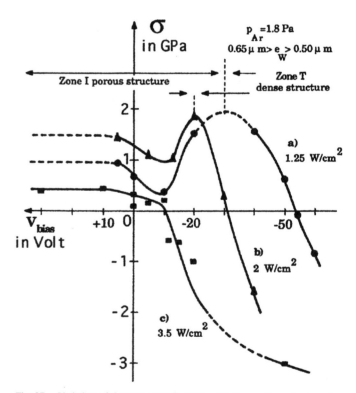

Fig. 37. Variation of the stress as a function of the bias voltage for tungsten films deposited at different sputtering powers. The various microstructure zones indicated in the figure correspond to a zero bias. At $V_B = 0$ V, (a) $T_s/T_m = 0.100$, zone 1 of the structure-zone diagram, (b) $T_s/T_m = 0.107$, zone 1/zone T transition structure, (c) $T_s/T_m = 0.117$, zone T of the structure-zone diagram [162]. Reprinted from A. M. Haghiri-Goisnet, F. R. Ladan, C. Mayeux, and H. Launois, *Appl. Surf. Sci.* 38, 295 (1989), with permission.

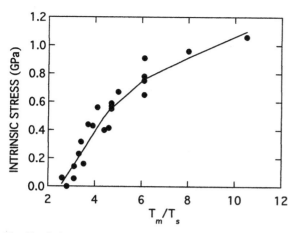

Fig. 38. Tensile intrinsic stress in 100-nm-thick nickel films as a function of the reduced growth temperature [164]. Reprinted from E. Klokholm, *J. Vac. Sci. Technol.* 6, 138 (1969), with permission.

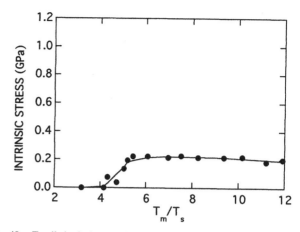

Fig. 40. Tensile intrinsic stress in 100-nm-thick aluminum films as a function of the reduced growth temperature [164]. Reprinted from E. Klokholm, *J. Vac. Sci. Technol.* 6, 138 (1969), with permission.

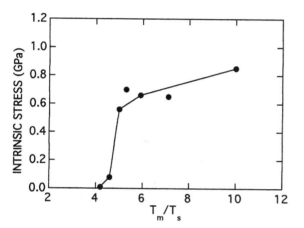

Fig. 39. Tensile intrinsic stress in 100-nm-thick copper films as a function of the reduced growth temperature [164]. Reprinted from E. Klokholm, *J. Vac. Sci. Technol.* 6, 138 (1969), with permission.

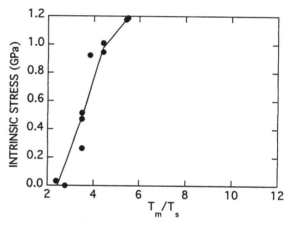

Fig. 41. Tensile intrinsic stress in 100-nm-thick platinum films as a function of the reduced growth temperature [131]. Reprinted from R. E. Rottmayer and R. W. Hoffman, *J. Vac. Sci. Technol.* 8, 151 (1971), with permission.

target and the electrons collected by the substrate. The condensation of these nonenergetic particles on the substrate occurring in the absence of recoil and forward sputtering phenomena produced not fully dense metal films. These results demonstrate that the kinetic energy of impact particles on the film surface, the microstructure of films, and the intrinsic stress developed in the films are closely correlated.

5.3. Substrate Temperature Effect

As illustrated by structure-zone diagrams (Figs. 3 and 4), the microstructure of PVD films depends on the normalized growth temperature, T_s/T_m; hence, the magnitude of intrinsic stresses is expected to be dependent on the deposition temperature, T_s. The effect of the substrate temperature on the intrinsic stress was investigated essentially for metal films deposited by evaporation (Figs. 38 to 41). The intrinsic stress determined by *in situ* measurements remained tensile for reduced growth temperatures, T_m/T_s, lower than 3 to 4. A sudden decreases in tensile intrinsic stress to zero was observed with the de-

crease in reduced growth temperature from 4 to 3 depending on the nature of the metal films. Similarly, a sudden increase in grain size is also observed at a reduced growth temperature of about 3 to 4 for a variety of metal films (Fig. 42). In addition, this value of the reduced growth temperature corresponds to the transition temperature between zone T and zone 2 in the structure-zone diagram for vapor-deposited metal films (Fig. 4). According to the zone model proposed by Grovenor, Hentzell, and Smith [40], the void network disappears in films deposited at a reduced growth temperature lower than 3 (normalized growth temperature, T_s/T_m, above 0.33). In these films deposited at moderated temperatures, the adatom diffusion becomes sufficiently rapid to fill some of the voids before the arrival of the next monolayer on the film surface. The partial disappearance of voids allows some grain boundaries to migrate and to produce an array of much larger grains by a secondary recrystallization process which contributes to significantly lower the yield strength of films. Therefore, the sudden increase in grain size associated to the sudden decrease in tensile stress at the reduced growth temperature of about 3 to 4 may be at-

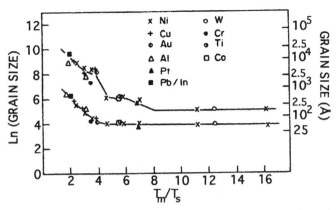

Fig. 42. Grain size in various metal films as a function of the reduced growth temperature. The upper and lower curves represent the maximum and the minimum observed grain sizes, respectively [40]. Reprinted from C. R. M. Grovenor, H. T. G. Hentzell, and D. A. Smith, *Acta Metall.* 32, 773 (1984), with permission.

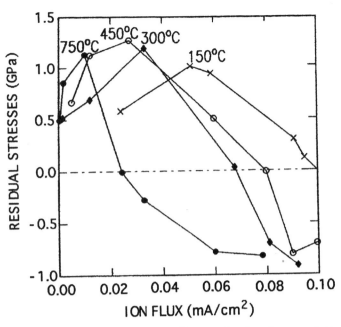

Fig. 43. Residual stresses in tungsten films as a function of concurrent argon ion bombardment at 400 eV for various deposition temperatures [169]. Reprinted from R. A. Roy, R. Petkie, and A. Boulding, *J. Mater. Res.* 6, 80 (1991), with permission.

tributed to the absence of the open voided structure of films deposited at moderate temperatures.

Klokholm and Berry [41] proposed the buried layer model to explain the sign and the magnitude of the intrinsic stresses observed in metal films deposited by evaporation. In this model, the intrinsic stress is believed to result from the annealing and the shrinkage of disordered material behind the surface of the growing film. The rearrangement of the disordered material involves a thermally activated process which could be inhibited (and the stress prevented from development) by deposition on a substrate maintained at a sufficiently low temperature. The magnitude of the intrinsic stress reflects the amount of disorder initially present on a surface layer before it becomes buried by the condensation of the succeeding layer. In other words, the magnitude of the stress in films deposited at a given substrate temperature, T_s, depends on the relative values of the deposition rate, R_0, and the annealing rate, Γ, which is assumed to be given by an Arrhenius relationship such as [165]

$$\Gamma = \nu_0 \exp\left(-\frac{Q}{RT_s}\right) \qquad (144)$$

where ν_0 is a frequency factor, Q is an activation energy, R is the ideal gas constant, and T_s is the absolute substrate temperature. On this basis, when $\Gamma \gg R_0$, i.e., at high deposition temperatures, the intrinsic stress as well as the amount of disorder are low. A reverse situation is observed at low deposition temperatures where $\Gamma \ll R_0$. The order of magnitude of the frequency, ν_0, is about 10^{14} s^{-1}, thereby the effective transition temperature corresponding to $\Gamma = R_0$ occurs when

$$\frac{Q}{RT_s} = 32 \qquad (145)$$

By analogy with an empirical rule for self-diffusion [165], the activation energy is assumed to be proportional to the melting point, T_m:

$$Q = KT_m \qquad (146)$$

Therefore, Eq. (145) can be written as

$$\frac{KT_m}{RT_s} \approx 32 \qquad (147)$$

As already observed for metal films deposited by evaporation, the intrinsic stress decreases suddenly for a reduced temperature, T_m/T_s, higher than about 3–4. Then, according to Eq. (147), the value of the proportionality factor K may vary in the range 15 to 20 cal mol^{-1} K^{-1}. These values of K represent about half the K value necessary to fit Eq. (146) to self-diffusion data for bulk materials. These reduced activation energies deduced from Eq. (146) are consistent with a mechanism involving surface diffusion processes.

The buried layer model was invoked to explain the relationship between intrinsic stress and deposition temperature for films produced by evaporation [164, 166] and sputtering [167, 168]. The successful applicability of the model indicates that thermally activation strain relief at moderate or elevated deposition temperatures is a general phenomenon independent of the nature of the film material, state of stress, and method of deposition.

The magnitude of intrinsic stresses in films produced by ion-assisted deposition depends on the normalized momentum as illustrated in Figure 27. For a constant value of the impact energy, the stress may vary as a function of the normalized flux, ϕ_p/ϕ_c, which represents the ratio between the fluxes of ions striking the film surface and atoms condensed on the film surface. The intrinsic stresses in tungsten films prepared by evaporation at various substrate temperatures were investigated as a function of the normalized flux; the growing films were subjected to an argon ion bombardment of 400 eV (Fig. 43).

The normalized flux for the tensile–compressive stress transition decreased with increasing substrate temperature. The stress reversal required 4 $(eV)^{1/2} atom^{-1}$ for W films deposited at 150°C ($T_s/T_m = 0.11$) and only 0.8 $(eV)^{1/2} atom^{-1}$ for films prepared at 750°C ($T_s/T_m = 0.28$). A lower void fraction or a higher packing density is obtained for films prepared at high temperatures. Therefore, the densification of films by ion-peening phenomena becomes operative at relatively low normalized momentum values as the substrate temperature increases.

6. DATA ON RESIDUAL STRESSES IN PVD THIN FILMS

The mechanical integrity of optical thin films and the surface flatness of substrates are recognized to be critical points for ensuring stable properties, performance, and reliability of optical devices operating under various conditions in different environments (vacuum, low or moderate temperatures, moisture, etc.). The mechanical characteristics of optical devices are essentially governed by the nature and the magnitude of residual stresses developed during physical or chemical vapor deposition of optical thin films and exposure of films to reactive atmosphere. Similarly, the tribological behavior and the mechanical damage of solid lubricant thin films may be governed by the magnitude of compressive residual stresses in the films. Excessive residual stresses may affect the mechanical resistance and the integrity of films, i.e., brittleness, fracture resistance, or fragility, and also the adherence of films to substrates. The mechanical defects lead to formation of debris in the wear track and rapid increase in the wear rate of films. The performance of optical and tribological films may greatly depend on the control of residual stresses. Therefore, comprehensive investigation of the residual stress dependence on the deposition parameters of films, of the origin of stresses in PVD films, and of the stress evolution during postdeposition exposure of films to various environments is needed to find ways to reduce and to control the magnitude of residual stresses which limits the optical or tribological performance of thin films.

Residual stresses were determined in silicon dioxide and silicon oxynitride films prepared by direct electron beam evaporation, ion-assisted deposition, and dual ion beam sputtering for optical applications. Stress determination was also performed for amorphous carbon films deposited by conventional and unbalanced magnetron sputtering for lubrication of mechanical assemblies and tribological applications in hostile environments. These stress data are presented as examples illustrating the applicability, the validity, and the limits of models described in previous sections which were proposed in the literature to explain the origin of residual stresses in PVD thin films.

6.1. Residual Stresses in Silicon Dioxide Films Prepared by Thermal Evaporation

Stoichiometric silicon dioxide films have been deposited on $\langle 111 \rangle$-oriented single crystal Si and Ge substrates ($50 \times 5 mm^2$, 0.6- or 1-mm thick), and glass plates (BK7 Schott) by electron-gun evaporation of a SiO_2 source under either an oxygen pressure varying from 2×10^{-5} mbar (base pressure during evaporation) to 4×10^{-4} mbar or an argon pressure maintained at 2×10^{-4} mbar [115, 116]. The deposition temperature was varied between 20 and 285°C. The refractive index and the thickness of films were measured by spectroscopic ellipsometry in room air. The composition of films was determined by Rutherford backscattering spectroscopy (RBS) measurements. The mass density of films was deduced from RBS data and film thickness. The concentration of hydrogen in these films was measured by elastic recoil detection analysis (ERDA) immediately after deposition and after one month of storage of samples in clean-room environment. The residual stresses in these amorphous films were deduced from the change of the radius of curvature of Si and Ge substrates measured by interferometry. The magnitude of stresses was calculated using Eq. (45). For measurements of the radius of curvature of film–substrate structures, the samples were placed either in room air or in an airtight chamber for analysis under vacuum (5×10^{-5} mbar).

6.1.1. Characteristics of Evaporated Silicon Dioxide Films

Stoichiometric SiO_2 films were produced independently of the deposition temperature and the oxygen pressure in the deposition chamber. The residual stresses in films of various thicknesses deposited on Si substrates at 200°C are given in Table V. The average compressive residual stresses in films deposited under an oxygen pressure of 2×10^{-4} mbar and under the base pressure (2×10^{-5} mbar) were equal to $-(31 \pm 6)$

Table V. Residual Stresses in SiO_2 Films of Various Thicknesses Deposited on Si Substrates Mounted at Various Positions on the Substrate Holder of the Evaporation Chamber (Position 1: Top; 2: Middle; and 3: Bottom of the Hemispherical Substrate Holder)

Oxygen pressure	2×10^{-4} mbar					2×10^{-5} mbar				
Position	1	2	3	1	1	1	2	3	1	1
Thickness (nm)	693.6	684	670.6	342.3	117.1	324.8	318.7	313.8	567	247.7
Stress (MPa)	−27	−25	−37	−33	−30	−177	−184	−192	−180	−191

Source: H. Leplan et al., *J. Appl. Phys.* 78, 962, 1995.

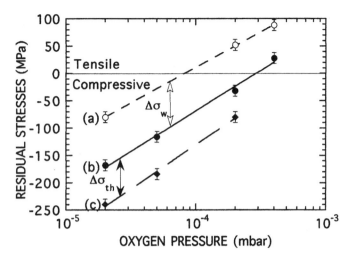

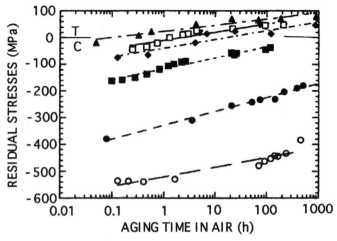

Fig. 44. Residual stresses versus oxygen pressure for SiO_2 films deposited at 200°C. The stress measurements were performed under vacuum (curve a) or in air (curve b) for films deposited on Si substrates, and in air (curve c) for films deposited on Ge substrates. The stress values are those obtained from films exposed for 20 days to clean-room environment [116]. Reprinted from H. Leplan, B. Geenen, J. Y. Robic, and Y. Pauleau, *J. Appl. Phys.* 78, 962 (1995), with permission.

Fig. 45. Residual stresses in SiO_2 films deposited on Si substrates under various oxygen (P_{O_2}) or argon (P_{Ar}) pressures at different substrate temperatures (T_s) as a function of time after air admission in the deposition chamber: (○) $T_s = 285°C$ under the base pressure; (●) $T_s = 285°C$, $P_{O_2} = 1 \times 10^{-4}$ mbar; (■) $T_s = 200°C$ under the base pressure; (♦) $T_s = 200°C$, $P_{Ar} = 2 \times 10^{-4}$ mbar; (□) $T_s = 200°C$, $P_{O_2} = 2 \times 10^{-4}$ mbar; (▼) ambient temperature under the base pressure; (▲) ambient temperature, $P_{O_2} = 2 \times 10^{-4}$ mbar [116]. Reprinted from H. Leplan, B. Geenen, J. Y. Robic, and Y. Pauleau, *J. Appl. Phys.* 78, 962 (1995), with permission.

and $-(185 \pm 7)$ MPa, respectively. These stress values lay in the range of experimental accuracy; hence, the residual stresses are essentially independent of both film thickness and incidence angle of the vapor flux on the substrate surface.

The residual stresses in SiO_2 films deposited on Si and Ge substrates at 200°C measured in room air were compressive for films produced at low oxygen pressures and decreased with increasing oxygen pressure in the deposition chamber (curves b and c, Fig. 44). Tensile residual stresses were observed in films deposited at relatively high oxygen pressures. The residual stresses in films deposited on Si substrates at 200°C measured under vacuum were less compressive than in room air. In addition, the variation of residual stresses observed when the samples were moved from air to vacuum was independent of the oxygen pressure in the deposition chamber (curves a and b, Fig. 44).

The residual stresses were found to vary with the time elapsed from air admission in the deposition chamber or aging time (Fig. 45). It can be noticed that an increase in substrate temperature or a decrease in oxygen pressure affects the initial stress value in a similar manner; i.e., reduced compressive stresses were obtained when either the deposition temperature was decreased or the oxygen pressure was increased. The initial magnitude of residual stresses increased considerably from -180 MPa to -550 MPa as the substrate temperature was varied from 200 to 285°C. The lowest stress value was obtained from SiO_2 films deposited at ambient temperature under high oxygen pressures (2×10^{-4} mbar). In addition, the stress value was nearly independent of the nature of the gas (oxygen or argon) in the deposition chamber. The compressive stresses in samples exposed to a conventional clean-room environment were found to decrease progressively whatever the initial stress value. The slopes of aging curves are very similar for all sam-

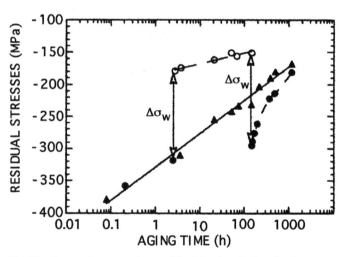

Fig. 46. Stress aging curves for two SiO_2 films deposited on Si substrates at 285°C under an oxygen pressure of 1×10^{-4} mbar during the same evaporation run. Sample 1 (▲) was exposed to room air; sample 2 was first exposed to room air for 2 h (●), then was placed under vacuum for 100 h (○), and finally was exposed again to room air (●) [116]. Reprinted from H. Leplan, B. Geenen, J. Y. Robic, and Y. Pauleau, *J. Appl. Phys.* 78, 962 (1995), with permission.

ples; hence, a unique mechanism is probably involved in the aging process of films.

To get a better insight into this aging mechanism, two SiO_2 films deposited during the same evaporation run were stored under different conditions. Sample 1 was kept in a conventional clean-room environment with its stress value periodically measured in air whereas sample 2 was placed in a vacuum chamber 2 h after sample exposure to room air and its stress value was measured *in situ* as a function of time (Fig. 46). The slope of the aging curve of sample 2 under vacuum was clearly smaller

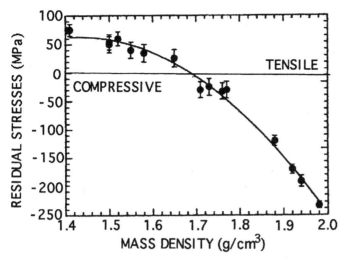

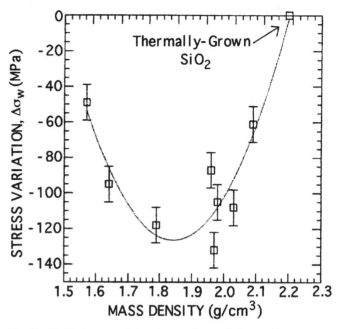

Fig. 47. Residual stresses, σ, as a function of the mass density of SiO_2 films deposited on Si substrates by evaporation at various temperatures under different pressures and using two different evaporation equipments. The stress values are those obtained from films exposed for 20 days to clean-room environment [116]. Reprinted from H. Leplan, B. Geenen, J. Y. Robic, and Y. Pauleau, *J. Appl. Phys.* 78, 962 (1995), with permission.

Fig. 48. Residual stress variation, $\Delta\sigma_w$, as the sample is moved from vacuum to air as a function of the mass density of SiO_2 films [49]. Reprinted from H. Leplan, Ph.D. Thesis, National Polytechnic Institute of Grenoble, France, 1995, with permission.

than that of sample 1 in room air; thus, the rate of the aging process decreased as the sample was maintained under vacuum (in a water vapor free environment). The final stress variation for sample 2 occurring when the sample was moved from vacuum to room air, after 100 h of aging in vacuum was nearly equal to the stress variation observed initially, i.e., $\Delta\sigma_w = +100$ MPa. In addition, these sudden and instantaneous stress variations were independent of the film thickness. Moreover, as sample 2 was moved from vacuum to room air, the rate of the aging process was rapidly very similar to the aging rate of sample 1 (about 1.45 MPa h^{-1}).

The depth concentration profiles of hydrogen in SiO_2 films deposited at 200°C under an oxygen pressure of 2×10^{-4} mbar were determined by ERDA just after deposition and after one month of sample storage in room air. The hydrogen concentration increased from 7 to 9 at.% during this aging time in room air. The samples were placed under vacuum during ERDA analyses and weakly bound water molecules (if any) can desorb readily from the films. Therefore, hydrogen atoms detected by ERDA were chemically fixed in SiO_2 films during sample exposure to room air.

The residual stresses, σ, were dependent on the mass density of films (Fig. 47). Films with reduced mass density exhibited tensile residual stresses. The magnitude of the tensile residual stresses decreased with increasing mass density; the tensile-to-compressive stress transition was observed for a mass density of about 1.7 g cm^{-3}. Compressive residual stresses were found in relatively dense SiO_2 films. The stress component, $\Delta\sigma_w$, was minimum for films with a low mass density and reached a maximum value of 120–140 MPa as the film density varied from 1.8 to 1.9 g cm^{-3} (Fig. 48). Then, this stress component decreased and was negligible for films with mass densities closed to the bulk SiO_2 density.

6.1.2. Origin of Residual Stresses in Evaporated Silicon Dioxide Films

The stress difference, $\Delta\sigma_{th}$, observed between films deposited on Si and Ge substrates arises from the disparity between the thermal stresses which is related to the difference of thermal expansion coefficients of substrates. This point was already discussed in a previous section on the origin of thermal stresses in SiO_2 films. The experimental value of $\Delta\sigma_{th}$ (about 53 MPa) was found to be in very good agreement with the value calculated using Eq. (106). The biaxial modulus of these evaporated films was very close to the bulk fused silica value. However, these films exhibited a much higher thermal expansion coefficient than the bulk material. The porous structure of evaporated films and the presence of water molecules adsorbed on pore walls are probably responsible for this relatively high thermal expansion coefficient. Indeed, Si—OH groups in the structure disrupt the local bonding symmetry of the SiO_2 network and, in turn, may increase the anharmonicity of the local potential since the thermal expansion coefficient depends on the anharmonicity of the interatomic potential [120]. The thermal stresses in evaporated silica films deposited on Si and Ge substrates represent only a small fraction of the high compressive residual stresses in films deposited at low oxygen pressures and high substrate temperatures.

Evaporated silica films may exhibit very low mass densities reflecting their porous microstructure (Fig. 7). Water molecules might adsorb in the pores; as a result, increased refractive indices and compressive stresses are noticed [115]. These water molecules incorporated in SiO_2 films can be detected by infrared (IR) absorption spectroscopy. The intensity of the IR

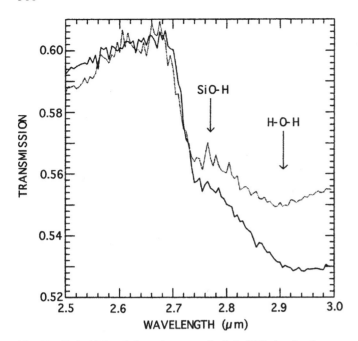

Fig. 49. Typical infrared absorption spectra in air (solid line) and under vacuum or a base pressure of 2×10^{-2} mbar (dotted line) for SiO_2 films deposited on Ge substrates at 250°C under an oxygen pressure of 1×10^{-4} mbar [49]. Reprinted from H. Leplan, Ph.D. Thesis, National Polytechnic Institute of Grenoble, France, 1995.

absorption band at 2.9 μm which characterizes H—O—H bonds was measured in room air and under vacuum for SiO_2 films deposited at 250°C under an oxygen pressure of 1×10^{-4} mbar (Fig. 49). The mass density of these not fully dense films was equal to 1.98 g cm^{-3}. The intensity of the absorption band at 2.9 μm was considerably reduced as the samples were analyzed under vacuum. This decrease in intensity confirms that water molecules desorb from the porous films when the samples are placed under vacuum. In addition, when the films were exposed anew to room air, the intensity of the absorption band increased rapidly to reach the initial value; i.e., the phenomenon responsible for the intensity variation is rapid and reversible. The IR absorption band at 2.76 μm characterizes SiO—H bonds (Fig. 49). These OH groups in SiO_2 films may result from the reaction between H_2O molecules and silicon dioxide. The intensity of this absorption band is nearly independent of the sample environment (air or vacuum). The model proposed by Hirsch [156] involving the interaction forces between the permanent electric dipole moments associated with H_2O molecules adsorbed on the pore walls is well designed to explain the origin of the stress difference $\Delta\sigma_w$. According to this model, the electrostatic dipole interaction force scales up inversely with the pore radius (Eq. (143)) so that negligible extrinsic stresses caused by water adsorption are found in films of coarse porosity with large pore radii. In fact, the water adsorption induced stress component was very low or negligible when the mass density of evaporated films was less than 1.5 g cm^{-3} (Fig. 48). This low extrinsic stress results from the decrease in repulsion efficiency as the pore radius and the resulting interaction distance between adsorbed dipoles increase in the films.

For intermediate mass density values (1.8 to 1.9 g cm^{-3}), the pore radii were probably well adapted to the diameter of water molecules for development of efficient interaction forces between adsorbed polar species and maximum value of the stress component, $\Delta\sigma_w$ (Fig. 48). For films with mass densities higher than 2.1 g cm^{-3}, the pore radii may become too small for penetration and adsorption of water molecules. As a result, the values of $\Delta\sigma_w$ were considerably reduced.

This extrinsic stress caused by adsorbed water molecules in porous films is greatly responsible for the mechanical instabilities of evaporated films since its magnitude will change with degree of humidity in ambient air, temperature, pressure, and dry gas flow. In addition, a subsequent reaction of water molecules with silica may play a significant role in the aging mechanism and stress evolution encountered in the evaporated SiO_2 films. The kinetics of the stress evolution was found to be considerably slower for samples stored under vacuum since most of water molecules desorb from the films (Fig. 46). The increase in hydrogen concentration in films as the storage time in room air increases appears as an additional evidence for a subsequent reaction between water molecules and SiO_2 which leads to a modification of the magnitude of residual stresses with increasing aging time.

As mentioned in previous sections, the fundamental factors governing the microstructure, the mass density, and the physical properties of evaporated films are the kinetic energy of condensing atoms and thermal energy supplied to the growing film. These arguments are consistent with various findings in particular residual stresses are independent of the film thickness, the nature of substrates, and the angle of incidence of the vapor flux since these parameters are not expected to greatly modify the kinetic energy of atoms condensed on the substrate. The residual stresses, i.e., ($\sigma_i + \sigma_{th} + \sigma_w$), after 20 days of aging in room air, varied from a moderate tensile to a highly compressive state as the film density increased (Fig. 47). The thermal and water adsorption induced stress contributions can be estimated as functions of the film density and can be deducted from the residual stress values. In addition, the stress variation during 20 days of sample aging in room air can be reasonably evaluated from the stress-aging curves given in Figure 45. By combining these data, the intrinsic stress can be calculated as a function of the film density. The general trend was an increase in the compressive intrinsic stress level as the film density increased; however, because of various uncertainties in the intrinsic stress calculations, it is not clear whether a real tensile intrinsic stress was built up during the deposition of low density SiO_2 films or if the compressive intrinsic stress simply tended to zero. This later assumption would be consistent with the fact that an amorphous and highly void-rich film will not develop intrinsic stress during deposition because of its low structural integrity or resistance.

6.1.3. Extrinsic Stresses Induced by Adsorption of Polar Molecules in Evaporated Silicon Dioxide Films

According to data obtained from IR spectroscopic analyses and stress measurements in room air and under vacuum, the extrin-

Table VI. Dipole Moment, α, Mass Density, ρ, Molar Mass, M, Effective Surface Area, S_{eff}, of a Molecule, and
Dipole Moment per Unit Area, α/S_{eff}, for Various Compounds

Molecule	Dipole moment of molecules, α ($\times 10^{-30}$ C m)	Mass density, ρ (g cm^{-3})	Molar mass, M (amu)	Effective surface area, S_{eff} (Å^2)	Dipole moment per unit area, α/S_{eff} ($\times 10^{-10}$ C m^{-1})
H_2O	6.17	1	18	10.5	58.76
Propanol	5.54	0.7855	60.1	27.6	20.07
Acetone	9.61	0.7899	58.08	26.86	35.78
N_2	0				0

Source: H. Leplan, National Polytechnic Institute of Grenoble, 1995.

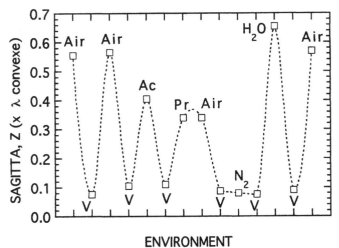

Fig. 50. Variation of the sagitta measured at the center of an Si substrate covered with a 560-nm-thick SiO$_2$ film deposited by thermal evaporation at a substrate temperature of 250°C under an oxygen pressure of 1×10^{-4} mbar as the sample was exposed to various environments: air, nitrogen (N$_2$), vacuum (V), water vapor (H$_2$O), acetone (Ac), and propanol (Pr) [49]. Reprinted from H. Leplan, Ph.D. Thesis, National Polytechnic Institute of Grenoble, France, 1995.

sic stress, $\Delta\sigma_w$, is caused by the adsorption of water molecules in porous SiO$_2$ films. On the basis of the model proposed by Hirsch [156], the extrinsic stress is expected to vary with the nature of adsorbed molecules, in particular with the square of the dipole moment, α, of molecules (Eq. (143)).

Silicon dioxide films of 1.98 g cm^{-3} in mass density were placed under vacuum (for H$_2$O desorption) and then exposed to various gas phase environment such as propanol-2 (CH$_3$−CHOH−CH$_3$), acetone, water vapor, and nitrogen [49]. The sagitta, Z, of the Si substrate, 28×1 cm^2 and 0.6-mm thick, covered with an evaporated SiO$_2$ film of 560 nm in thickness, was determined by interferometry and was investigated as a function of the sample environment (Fig. 50). The initial sagitta value was measured in room air. Two air-vacuum cycles were performed before exposure of the sample to acetone vapor. The first four data points in Figure 50 corresponding to a sagitta variation of 0.5λ (where λ is the wavelength used for interferometry measurements) reveal a perfect reversibility of phenomena responsible for sagitta variations. When the sam-

ple was exposed to acetone vapor, the sagitta variation was smaller (0.3λ); however, the phenomenon was always perfectly reversible. The propanol exposure of the sample led to a sagitta variation of 0.2λ. Since the sagitta remained constant when air was introduced in the chamber, the adsorption sites in SiO$_2$ films were probably totally saturated or occupied by adsorbed propanol molecules. The initial sagitta value was obtained anew when the sample was under vacuum. Therefore, the adsorption–desorption phenomena of propanol molecules were reversible. The variation of sagitta was nil when the sample under vacuum was then exposed to nitrogen. As a result, N$_2$ molecules either were not adsorbed in the silicon dioxide film or the nitrogen adsorption has no effect on the stress. The maximum sagitta variation of 0.65λ obtained from the sample exposed to water vapor suggests that the relative humidity of the environment may significantly affect the magnitude of the extrinsic stress. After the series of experiments, the curvature of the SiO$_2$-Si sample in air or under vacuum returned to its initial value. Major findings have been obtained from these experiments. First, the exposure of evaporated SiO$_2$ films to nonpolar species such as nitrogen molecules does not modify the residual stresses. The adsorption and desorption phenomena of polar molecules investigated are totally reversible at room temperature. In addition, the magnitude of the extrinsic stress depends on the nature of the polar molecules in agreement with the prediction of the model proposed by Hirsch.

The dipole moment, α, of molecules and the number of dipoles, n, per unit area are needed to verify the applicability of the model developed by Hirsch [156]. The dipole moment values of vapor molecules investigated are given in Table VI. The number of dipoles adsorbed per unit area may be approximated to be inversely proportional to the area of the polar molecules. The effective surface area, S_{eff}, of a molecule is given by [170]

$$S_{eff} = 2\sqrt{3}\left(\frac{M}{4\sqrt{2}N_{Av}\rho}\right)^{2/3} \qquad (148)$$

where M and ρ are the molar mass and the mass density of the molecules, respectively; N_{Av} is Avogadro's number. The values of S_{eff} and those of the dipole moment per unit area, α/S_{eff}, calculated from Eq. (148) are given in Table VI. According to Eq. (143), the square root of the extrinsic stress is

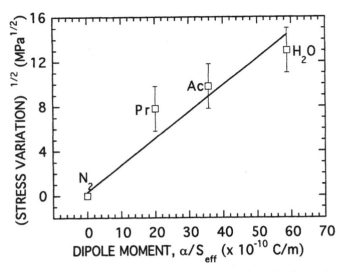

Fig. 51. Residual stress variation induced by adsorption of various polar molecules (H_2O, acetone, propanol) in porous SiO_2 films as a function of the dipole moment per unit area [49]. Reprinted from H. Leplan, Ph.D. Thesis, National Polytechnic Institute of Grenoble, France, 1995, with permission.

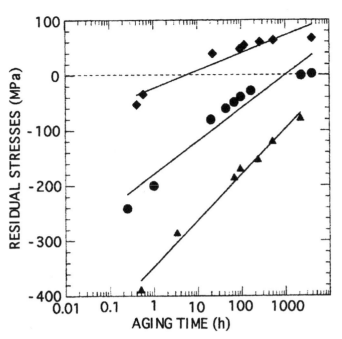

Fig. 52. Aging time effect on the residual stresses in SiO_2 films deposited on Si substrates at ambient temperature under a base pressure of 2×10^{-5} mbar (▲) or under an oxygen pressure of 1×10^{-4} mbar (●), and 3×10^{-4} mbar (♦); the mass density of these films was 1.97, 1.79, and 1.49 g cm^{-3}, respectively [119]. Reprinted from H. Leplan, J. Y. Robic, and Y. Pauleau, *J. Appl. Phys.* 79, 6926 (1996), with permission.

expected to be proportional to the dipole moment per unit area, α/S_{eff}. The diagram in Figure 51 was established from data given in Table VI. The linear dependence of $(\Delta\sigma_w)^{1/2}$ versus the dipole moment per unit area expected from Eq. (143) is not absolutely observed. The number of data points is probably too low and the values of α/S_{eff} are not sufficiently accurate to establish a better correlation. Nevertheless, the general trend illustrated by these data is in accordance with the prediction of the model.

The validation of the model would be completed from the investigation of the extrinsic stress as a function of the pore radius in SiO_2 films. According to Eq. (143), the extrinsic stress is expected to be inversely proportional to the pore radius, a. The radii of pores or the size distribution in the films were not measured directly; however, the radius of pores may be assumed to be dependent on the mass density of films. The extrinsic stress, $\Delta\sigma_w$, measured as the sample was moved from air to vacuum versus mass density of films is plotted in Figure 48. The compressive extrinsic stress is maximum in SiO_2 films having an intermediate mass density of about 1.9 g cm^{-3}. The extrinsic stress decreases either with decreasing mass density from 1.9 to 1.5 g cm^{-3} or with increasing mass density up to the bulk density value of 2.2 g cm^{-3}. Assuming that the pore size decreases progressively as the mass density of films increases, the dependence of the extrinsic stress induced by adsorption of water molecules on the film density is quite compatible with the prediction of the model. This dependence results from the superposition of two phenomena. For films with mass densities in the range 1.5 to 1.9 g cm^{-3}, the variation of the extrinsic stress is in agreement with the $1/a$ dependence predicted by Eq. (143). The magnitude of the compressive extrinsic stress increases as the interaction distance between adsorbed water molecules decreases. For dense SiO_2 films, the assumption corresponding to a uniform coverage of the pore surface by water

molecules is likely not valid. The extreme case corresponds to films in which the pore size is lower than the volume of water molecules or the pores are totally closed thereby the extrinsic stress tends to be zero. For dense films, the magnitude of the extrinsic stress is governed by the number of water molecules which can be accommodated on the pore wall and also by the number of pores open and sensitive to humidity. The magnitude of the extrinsic stress is maximum in films with a packing density of about 0.87 for which a closed porosity develops probably in the film.

6.1.4. Kinetics of Residual Stress Evolution in Evaporated Silicon Dioxide Films Exposed to Room Air

The magnitude of residual stresses in evaporated SiO_2 films of about 500-nm thick was investigated as a function of the time elapsed from air admission in the deposition chamber (aging time) [119]. For films deposited under various conditions, the residual stresses varied linearly with the logarithm of the aging time (Figs. 52–54). The difference between residual stress value after an aging time of 1000 h (about 40 days) and initial value (after an aging time of 0.1 h) with respect to the initial stress value was dependent on the mass density of films (Fig. 55). The value of this relative variation, $[\sigma(t = 1000\text{ h}) - \sigma(t = 0.1\text{ h})]/\sigma(t = 0.1\text{ h})$, decreased linearly with increasing mass density and was approximately nil for films exhibiting a mass density close to that of thermally grown silicon dioxide films.

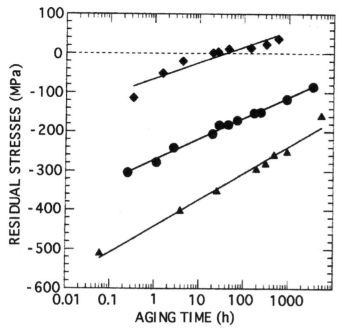

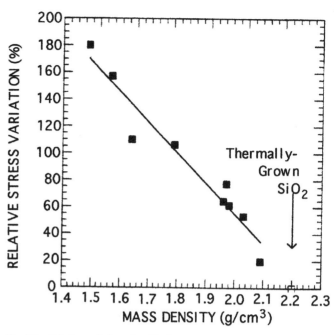

Fig. 53. Aging time effect on the residual stresses in SiO$_2$ films deposited on Si substrates at 200°C under a base pressure of 2×10^{-5} mbar (▲) or under an oxygen pressure of 1×10^{-4} mbar (●), and 3×10^{-4} mbar (◆); the mass density of these films was 2.03, 1.96, and 1.57 g cm^{-3}, respectively [119]. Reprinted from H. Leplan, J. Y. Robic, and Y. Pauleau, *J. Appl. Phys.* 79, 6926 (1996), with permission.

Fig. 55. Relative variation of residual stresses versus mass density of SiO$_2$ films after exposure of samples to room air for 1000 h [119]. Reprinted from H. Leplan, J. Y. Robic, and Y. Pauleau, *J. Appl. Phys.* 79, 6926 (1996), with permission.

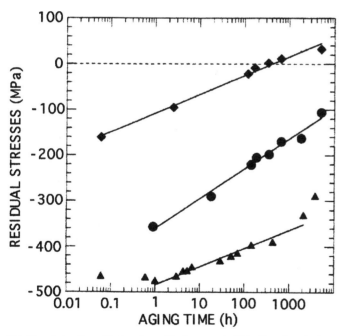

Fig. 54. Aging time effect on the residual stresses in SiO$_2$ films deposited on Si substrates at 250°C under a base pressure of 2×10^{-5} mbar (▲) or under an oxygen pressure of 1×10^{-4} mbar (●), and 3×10^{-4} mbar (◆); the mass density of these films was 2.09, 1.98, and 1.64 g cm^{-3}, respectively [119]. Reprinted from H. Leplan, J. Y. Robic, and Y. Pauleau, *J. Appl. Phys.* 79, 6926 (1996), with permission.

The composition of SiO$_2$ films deposited on Ge substrates was determined by IR spectroscopy after an aging time in room air of 15 min, 3 h, 3 and 4 days (Fig. 56a and b). As

the aging time increased, a progressive shift toward shorter wavelengths was observed for the absorption band at 9.3 μm (1080 cm^{-1}) corresponding to the stretching vibration of Si—O—Si bonds. In addition, the IR spectra of films after an aging time longer than 15 min exhibited an absorption band at 10.7 μm (935 cm^{-1}) ascribable to Si—OH groups [171, 172]. The intensity of this absorption band increased with increasing aging time (Fig. 56b).

The variation of residual stresses as the samples move from vacuum to air, $\Delta\sigma_w$, and the time evolution of stresses resulting from exposure of films to room air constitute two distinct parts of the extrinsic stress. The first part was attributed to the effect of water molecules adsorbed in voids and pores of not fully dense SiO$_2$ films. The magnitude of this reversible compressive stress component was discussed in previous sections on the basis of the model proposed by Hirsch. The second part of the extrinsic stress resulting from exposure of samples to room air was found to vary irreversibly and may be ascribed to the instability of films to water vapor in room air.

In addition to the increase in hydrogen concentration in films exposed to room air and detected by ERDA, the IR spectroscopic analyses revealed that the concentration of Si—OH atomic bonds increased with increasing aging time. A chemical reaction between water molecules and silicon dioxide occurring progressively as the aging time increases can be postulated on the basis of these experimental results. Porous SiO$_2$ films deposited by evaporation can behave similarly to silica gel employed as a drying agent. The hydration of silica leads to the formation of various silicic acids such as Si(OH)$_4$ or H$_8$Si$_4$O$_{12}$. Monosilicic acid, H$_4$SiO$_4$, would be the ultimate acidic com-

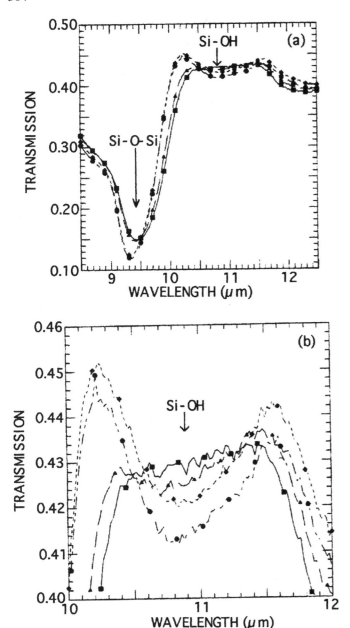

Fig. 56. (a) Typical infrared absorption spectra of SiO_2 films deposited on Ge substrates at ambient temperature under a base pressure of 2×10^{-5} mbar after an aging time of 15 min (■, solid line), 3 h (▲, dashed line), 3 days (●, dashed and dotted line), and 4 days (♦, dotted line); the mass density of films was 1.97 g cm^{-3}. (b) Typical infrared absorption spectra in the wavelength range 10–12 μm for SiO_2 films described in (a) [119]. Reprinted from H. Leplan, J. Y. Robic, and Y. Pauleau, *J. Appl. Phys.* 79, 6926 (1996), with permission.

pound formed by hydration of silica films ($SiO_2 + 2H_2O \rightarrow H_4SiO_4$). The fixation of OH groups in the SiO_2 lattice can occur via silicon dangling bonds but also may proceed by dissociation of Si—O—Si bonds [119]. The hydration of evaporated silica films leads to fragmentation and reduction of the degree of polymerization which result in the shrinkage of the SiO_2 lattice and the reduction in residual stresses as the aging time increases. The hydration of the SiO_2 lattice does not lead only to a stress relaxation but tensile residual stresses develop progres-

sively in SiO_2 films with increasing aging time (Figs. 52–54). New types of interactions appear in the material after partial hydration. Permanent electric dipole moments are associated with HO—Si—OH groups in silicic acids. As a result, the dipole interaction force may be invoked to explain the existence of tensile residual stresses in hydrated SiO_2 films.

The kinetics of hydration of films was discussed on the basis of a reaction mechanism composed of two successive elementary steps [119]:

(i) Adsorption of water vapor molecules on the SiO_2 surface,

$$H_2O(g) + s \underset{k_{-1}}{\overset{k_1}{\leftrightarrow}} H_2O{-}s \qquad (I)$$

(ii) Hydration of SiO_2 with formation of silanol radicals,

$$H_2O{-}s + Si{-}O{-}Si \overset{k_2}{\rightarrow} 2Si{-}OH + s \qquad (II)$$

where k_1 and k_{-1} are the rate constants of the direct and reverse reactions in step (I), respectively; k_2 is the rate constant of step (II). In these elementary steps, the species, s, represents free adsorption sites at the SiO_2 surface which can be Si or O atoms. Adsorbed water molecules, $H_2O{-}s$, disappear via reaction (II), i.e., react with two neighboring species, Si and O—Si, belonging to an Si—O—Si group in which one of the Si—O atomic bonds is weakened. In fact, Si and O—Si species at the SiO_2 surface may act as either adsorption sites or reactive species. The concentrations of Si and O—Si species decrease progressively as silica is converted into silicic acid.

The adsorption of water molecules on the SiO_2 surface (elementary step I) was assumed to be a rapid step while the formation of Si—OH radicals (elementary step II) was considered to be the rate-limiting step for the overall reaction. As a result, step (I) is considered to be at equilibrium and the concentration of adsorbed water molecules, q_a, on the SiO_2 surface is given by

$$q_a = K_1 PS \qquad (149)$$

where K_1 is the equilibrium constant of step (I), P is the partial pressure of water vapor, and S is the concentration of free adsorption sites at the SiO_2 surface. Moreover, the concentration of Si—O—Si reactive species on the SiO_2 surface is equal to S at a given aging time since these species may act as either adsorption sites in step (I) or reactive species in step (II). Assuming that step (II) is the rate-limiting step, the hydration rate of silica can be written as

$$-\frac{dS}{dt} = k_2 q_a S \qquad (150)$$

The rate of the elementary step (I) is given by

$$\frac{dq_a}{dt} = k_1 PS - k_{-1} q_a \qquad (151)$$

The probability of desorption of water molecules from the oxide surface is believed to be very small at a given aging time;

hence, the rate of formation of adspecies via step (I) can be approximated to

$$\frac{dq_a}{dt} = k_1 P S \qquad (152)$$

Therefore, Eqs. (149), (150), and (152) lead to

$$-\frac{dS}{S} = \frac{k_2}{k_{-1}} dq_a \qquad (153)$$

which integrates to

$$S = S_0 \exp\left(-\frac{k_2}{k_{-1}} q_a\right) \qquad (154)$$

since initially (at $t = 0$), $q_a = 0$ and $S = S_0$. Consequently, the concentration of free adsorption sites on the oxide surface is found to decrease exponentially as the quantity of water molecules adsorbed on the surface increases. Equation (154) corresponds to the concentration of free adsorption sites given by the Elovich equation which can be derived for a uniform or a nonuniform surface when the activation energy of the adsorption step varies with the quantity of adspecies [173].

At the steady state, the rate of consumption of free adsorption sites is equal to the rate of hydration of silica which can be expressed as

$$-\frac{dS}{dt} = \frac{dq_a}{dt} = k_1 P S = k_1 P S_0 \exp\left(-\frac{k_2}{k_{-1}} q_a\right) \qquad (155)$$

which integrates to

$$q_a = \frac{k_{-1}}{k_2} \ln(k_2 K_1 S_0 P t + 1) \qquad (156)$$

Equation (156) provides the amount of water molecules adsorbed on the SiO_2 surface and consumed in step (II) for hydration of silica. In other words, this equation gives the hydration rate of silica. This logarithmic kinetic law may account for the linear decrease in extrinsic stress in evaporated SiO_2

films with increasing the logarithm of the aging time. Furthermore, Eq. (156) predicts that the hydration rate decreases with decreasing water vapor pressure. This result is in good concordance with the reduced variation of the residual stresses observed when the samples were maintained under vacuum (Fig. 45).

6.2. Residual Stresses in Silicon Dioxide Films Produced by Ion-Assisted Deposition

Silicon dioxide films have been deposited on various substrates at room temperature by direct electron beam evaporation of SiO_2 using a high vacuum evaporation system equipped with a gridless ion source to produce high ion fluxes at low energy [174]. The kinetic energy and the flux of oxygen or argon ions can be adjusted by the source anode voltage, V_a, and the current, I_a. The energy distribution and the current density of ions were measured at the substrate surface under experimental conditions similar to those used for ion-assisted deposition of films. A broad distribution of the ion energy was obtained around the nominal value fixed by the anode voltage, $E_a = eV_a$. The maximum current density, J_a, measured at the center of the beam for each operating condition is listed in Table VII along with the experimental deposition conditions. The pressure in the deposition chamber was in the range 10^{-4} mbar and was dependent on the gas flow rate required for the ion source to monitor the anode current. The optical properties (refractive index and extinction coefficient) and the thickness of SiO_2 films deposited on glass substrates (BK7 Schott) were deduced from spectrophotometric measurements [175].

6.2.1. Characteristics of Silicon Dioxide Films Produced by Ion-Assisted Deposition

The refractive index and the mass density of SiO_2 films produced by IAD under various conditions are given in Table VII.

Table VII. Ion-assisted Deposition Parameters and Physical Properties for SiO_2 Films Prepared in an Alcatel EVA 700 System and Assisted with a Gridless Ion Source

Gas	R (nm s^{-1})	V_a (V)	I_a (A)	J_a (mA cm^{-2})	P_n (g eV mol^{-1})$^{1/2}$	ρ (g cm^{-3})	n at 600 nm	σ (MPa)	$\Delta\sigma_w$ (MPa)
O_2	0.72	0	0	0	0	1.56	1.44	−10	−64
O_2	0.76	150	1	0.37	117	1.80	1.46	−230	−108
O_2	0.52	150	2	0.56	228	2.03	1.48	−276	−40
O_2	0.52	150	3	0.75	284	2.21	1.48	−320	−18
O_2	0.54	150	4	0.93	346	2.13		−364	0
O_2	0.67	100	4	0.93			1.47	−230	−98
O_2	0.78	80	4	0.93			1.45	−100	−72
Ar	0.56	150	2	0.39		2.19		−470	0

Source: J. Y. Robic et al., *Thin Solid Films* 290–291, 34 (1996).

The gas is the type of assistance gas; R is the deposition rate; V_a is the anode voltage; I_a is the anode current; J_a is the maximum current density; P_n is the normalized momentum; n is the refractive index at 600-nm wavelength; ρ is the film mass density; s is the residual stress value in air; $\Delta\sigma_w$ is the variation of residual stress value as the sample is moved from vacuum to air.

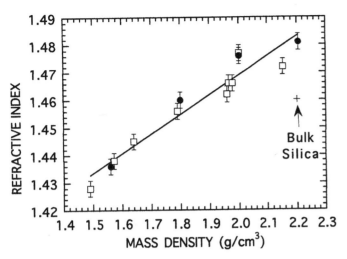

Fig. 57. Refractive index at 600 nm versus mass density of SiO$_2$ films deposited by evaporation in equipment I ($\square$), and prepared either under ion beam bombardment or without any ion beam assistance in equipment II ($\bullet$) [174]. Reprinted from J. Y. Robic et al., *Thin Solid Films* 290–291, 34 (1996), with permission.

The extinction coefficient for these films was below 10^{-4}. The refractive index of SiO$_2$ films produced by direct evaporation exhibited the same linear correlation with the mass density as that of films prepared by ion-assisted deposition (Fig. 57).

The residual stresses in IAD SiO$_2$ films are also given in Table VII. The magnitude of residual stresses, σ, measured in room air and the stress variation, $\Delta\sigma_w$, caused by adsorption of water molecules in the films were determined 1 h after deposition and exposure of films to room air. The compressive residual stresses in the films increased with increasing current and anode voltage. Films produced under argon ion bombardment exhibited more compressive stresses than films obtained with oxygen ion bombardment, i.e., −470 MPa and −276 MPa, respectively. The compressive stress variation, $\Delta\sigma_w$, corresponding to the transfer of samples from vacuum to air was reproducible and reversible; the value of $\Delta\sigma_w$ was reduced to zero as the current or anode voltage was increased. This stress variation versus mass density of SiO$_2$ films produced by direct evaporation and ion-assisted deposition is plotted in Figure 58. For both types of SiO$_2$ films, the stress variation increases with increasing mass density, reaches a maximum value of about −130 MPa as the mass density varies from 1.8 and 1.9 g cm^{-3}, and decreases to zero for higher mass densities.

6.2.2. Normalized Momentum Effect on the Mass Density and the Residual Stresses of Silicon Dioxide Films Produced by Ion-Assisted Deposition

The normalized momentum is the relevant process parameter which governs the properties of films produced by ion-assisted deposition. For IAD of SiO$_2$ films, the normalized momentum, P_n, is expressed as

$$P_n = \frac{\phi_i}{\phi_c}\sqrt{M_i E_i} \qquad (157)$$

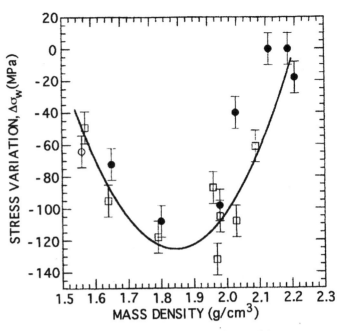

Fig. 58. Residual stress variation as the sample is moved from vacuum to air as a function of the mass density of SiO$_2$ films deposited on Si substrates using two different equipments: ($\square$) films prepared by thermal evaporation in equipment I and ($\bullet$) films obtained under ion beam bombardment or without any ion beam assistance in equipment II [49]. Reprinted from H. Leplan, Ph.D. Thesis, National Polytechnic Institute of Grenoble, France, 1995.

where ϕ_i/ϕ_c is the ratio between the ion flux and the film atom flux at the substrate surface; M_i and E_i are the mass and the kinetic energy of incident ions, respectively. It is difficult to accurately estimate the value of P_n; hence, a functional parameter P^* proportional to P_n was substituted for P_n,

$$P^* = \frac{\phi_i^*}{\phi_c^*}\sqrt{M_i E_i^*} = \frac{K J_a/e}{D_R \rho N_{Av}/M_c}\sqrt{M_i \frac{e V_a}{K}} \qquad (158)$$

where D_R is the deposition rate of films, ρ is the mass density of films, N_{Av} is Avogadro's number, M_c is the molar mass of silica, e is the electron charge, J_a and V_a are the maximum current density and the anode voltage of the ion source, respectively. The factor K is equal to unity and 2 for argon and oxygen ion assistance, respectively. The factor of 2 for oxygen is due to the dissociation of oxygen ions, O$^+$, at the substrate surface [176]. A linear dependence of the mass density and the refractive index of IAD SiO$_2$ films on the functional parameter, P^*, is observed in Figure 59. This result is in good agreement with the prediction of models for film densification proposed in the literature. As mentioned in the model proposed by Windischmann [113], the densification of films can be achieved only by ions with a kinetic energy higher than the displacement energy of atoms in a crystal lattice, i.e., about 25 eV. Because of the broad energy scattering of the gridless ion source, it is difficult to evaluate the number of ions which have an energy above this threshold energy. Therefore, the comparison of results is possible only for films obtained with the same anode voltage, i.e., with an anode voltage of 150 V for data points reported in Figure 59.

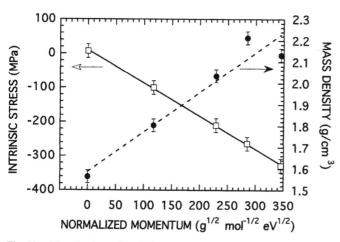

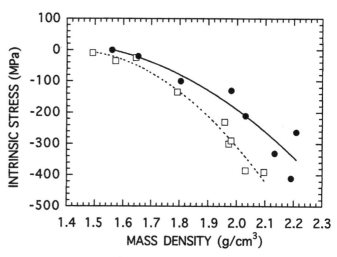

Fig. 59. Mass density and intrinsic stress of SiO$_2$ films produced by IAD as a function of the normalized momentum. The films have been deposited under oxygen ion bombardment with $V_a = 150$ V [174]. Reprinted from J. Y. Robic et al., *Thin Solid Films* 290–291, 34 (1996), with permission.

Fig. 60. Intrinsic stress as a function of the mass density of SiO$_2$ films deposited on Si substrates by: (□) thermal evaporation and (●) IAD [174]. Reprinted from J. Y. Robic et al., *Thin Solid Films* 290–291, 34 (1996), with permission.

The thermal stress in these IAD SiO$_2$ films deposited on substrates at room temperature are essentially nil. The extrinsic stress resulting from water adsorption in the films can be interpreted on the basis of the model proposed by Hirsch. For films with reduced mass densities (less than 1.5 g cm^{-3}) and large pore sizes, the interaction forces between adsorbed dipoles are negligible and the extrinsic stress tends to be zero. For films of intermediate mass density values (between 1.5 and 1.9 g cm^{-3}), the pore size is the major factor governing the water adsorption induced stress. For denser films with a mass density in the range 1.9 to 2.2 g cm^{-3}, the number of open pores is the factor limiting the extrinsic stress caused by water adsorption. No water molecules can be adsorbed in films with the bulk density and the extrinsic stress becomes negligible.

The values of the intrinsic stress in IAD SiO$_2$ films were calculated from Eq. (1) using data given in Table VII. The intrinsic stress versus the functional parameter, P^*, is plotted in Figure 59 for SiO$_2$ films produced under oxygen ion bombardment. Similar to the mass density, a linear dependence of the intrinsic stress on P^* is observed. A clear correlation appears between the mass density and the intrinsic stress in SiO$_2$ films produced by direct evaporation and IAD (Fig. 60). The compressive intrinsic stress in these films increases with increasing mass density. For films produced by direct evaporation (without any ion assistance of the growth), the intrinsic stress was compressive although the kinetic energy of species condensed on the substrate was very low. Thermodynamic calculations for the system composed of SiO$_2$ at equilibrium with gas species such as SiO, O$_2$, O, SiO$_2$, and Si at various temperatures under various oxygen pressures have shown that SiO molecules are the major species in the gas phase [49]. Therefore, the oxidation of the SiO species on the film surface is needed to produce stoichiometric SiO$_2$ films by direct evaporation and by IAD as well. This oxidation process at the growing surface may generate compressive stress in the deposited material since oxidation results in an increase in molecular volume. On the basis of this possible oxidation which may be responsible for compressive

intrinsic stress generation, it can be assumed that in films with a mass density lower than the bulk density the bombardment induced defects and atom displacements are not responsible for the compressive stress generation. Furthermore, for a given film density, the magnitude of the compressive intrinsic stress is lower in IAD films than in films deposited by direct evaporation (Fig. 60). This result suggests that a stress relief may be induced by ion bombardment of growing films.

6.3. Residual Stresses in Silicon Oxynitride Films Produced by Dual Ion Beam Sputtering

The control of the refractive index of optical films between the refractive indices of silica and silicon nitride (1.46 and 2.05, respectively) can be achieved by silicon oxynitride films for optical applications which need discrete films as well as continuously variable films like rugate. Silicon oxynitride, SiO$_x$N$_y$, films of 0.5- to 1-μm thick have been deposited using a high vacuum dual ion beam sputtering (DIBS) system equipped with two ion sources for sputtering and ion-assisted growth of films [177]. A silicon target of 35 cm in diameter was sputtered by ion beams using a Kaufman-type ion source of 15 cm in diameter operating with a mixture of argon and nitrogen of various compositions. A Kaufman-type ion source of 7.5 cm in diameter for ion beam-assisted deposition of films fed with nitrogen was focused on the rotating substrates which were bombarded with nitrogen ions in a discontinuous manner. Nitrogen was introduced into the deposition chamber through the ion sources while oxygen for oxidation of the deposited material was introduced directly into the chamber.

The deposition rate of films was maintained at about 0.1 nm s^{-1}. The substrate temperature and the pressure in the deposition chamber were fixed at 100°C and 2×10^{-4} mbar, respectively. Oxynitride films of various nitrogen and oxygen contents were prepared under constant parameters of the ion sources (ion current, ion energy, gas flow rate, Ar/N$_2$ ratio, etc.) while the flow rate of oxygen introduced in the chamber was varied in the range $0-22 \text{ cm}^3 \text{ min}^{-1}$. Two series of SiO_xN_y films (referred to as conditions 1 and 2) corresponding to two different sets of parameters for the ion sources were investigated. The second set of deposition conditions was investigated to improve nitridation of films by reducing the sputtering rate of the silicon target and by increasing the nitrogen ion flux on the film surface. The refractive index, the extinction coefficient, and the thickness of films deposited on glass substrates (BK7 Schott or silica glass) were deduced from spectrophotometric measurements [175]. The mass density and the composition of films were determined by Rutherford backscattering spectroscopy (RBS) and nuclear reaction analysis (NRA). In addition, the residual stresses in films deposited on ⟨111⟩-oriented Si substrates were determined from the change of the radius of curvature of substrates measured by interferometry.

6.3.1. Characteristics of Silicon Oxynitride Films Deposited by Dual Ion Beam Sputtering

The composition of SiO_xN_y films produced with two sets of deposition conditions was found to vary with the oxygen flow rate introduced in the chamber as shown in Figure 61 in which the oxygen to silicon and nitrogen to silicon atom number ratios versus oxygen flow rate to deposition rate ratio, D_{O_2}/R are plotted. The variation of the composition with D_{O_2}/R is independent of the deposition conditions (conditions 1 or 2). The O/Si atom number ratio increases while the N/Si atom number ratio decreases with increasing D_{O_2}/R. For films deposited without oxygen in the deposition chamber, the N/Si atom number ratio in films obtained under conditions 2 is higher than that in films deposited under conditions 1 (1.4 and 1.1, respectively) while the N/Si atom number ratio is 1.33 in stoichiometric

Si_3N_4 films. The mass density of silicon oxynitride films was linearly dependent on the N/Si atom number ratio and varied from the mass density of silicon dioxide to the mass density of silicon nitride (Fig. 62).

The refractive index of these films decreased from 2.09 to 1.54 with increasing D_{O_2}/R; the extinction coefficient of films in the range 10^{-4} decreased slowly with increasing D_{O_2}/R [177]. The optical properties of films were independent of the deposition conditions used (conditions 1 or 2) except for very low D_{O_2}/R values. The residual stresses in these silicon oxynitride films measured in room air were highly compressive (Fig. 63). A large decrease in the magnitude of the compressive residual stresses from -1480 to -650 MPa and -1840 to -640 MPa was observed in films deposited under conditions 1

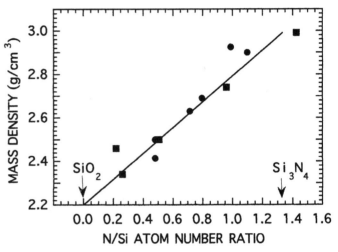

Fig. 62. Mass density of silicon oxynitride films versus N/Si atom number ratio. The films have been deposited under conditions 1 (●) and 2 (■) [177]. Reprinted from J. Y. Robic et al., in "Developments in Optical Component Coatings," Proceedings SPIE (I. Reid, Ed.), Vol. 2776, p. 381. SPIE, Bellingham, WA, 1994, with permission.

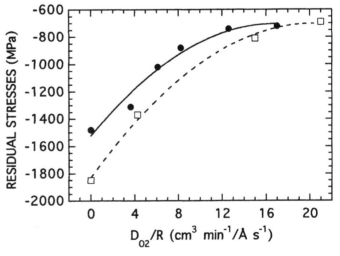

Fig. 63. Residual stresses in silicon oxynitride films versus ratio between the oxygen flow rate and deposition rate of films produced under deposition conditions 1 (●) and 2 (□) [177]. Reprinted from J. Y. Robic et al., in "Developments in Optical Component Coatings," Proceedings SPIE (I. Reid, Ed.), Vol. 2776, p. 381. SPIE, Bellingham, WA, 1994, with permission.

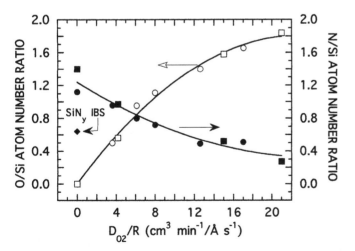

Fig. 61. O/Si (open symbols) and N/Si (closed symbols) atom number ratios in silicon oxynitride films versus ratio between the oxygen flow rate and deposition rate of films produced under deposition conditions 1 (○,●) and 2 (□, ■). The data point (◆) corresponds to the N/Si atom number ratio in SiN_y films deposited by IBS under conditions 1 without any ion bombardment and oxygen in the gas phase [177]. Reprinted from J. Y. Robic et al., in "Developments in Optical Component Coatings," Proceedings SPIE (I. Reid, Ed.), Vol. 2776, p. 381. SPIE, Bellingham, WA, 1994, with permission.

and 2, respectively. The residual stresses were independent of the environment. In other words, the residual stress value was unchanged when the samples were moved from room air to vacuum and no aging effect or evolution of stresses was observed for silicon oxynitride films stored for a long time (one year) in room air.

6.3.2. Origin of Residual Stresses in Silicon Oxynitride Films Deposited by Dual Ion Beam Sputtering

The oxidation process of silicon atoms or silicon–nitrogen species at the film surface is expected to depend on the flux ratio of oxygen and silicon atoms arriving at the surface. The value of this ratio is assumed to be proportional to the D_{O_2}/R ratio which was used as a process parameter. The data reported in Figure 61 suggest that the oxidation of silicon is the major reaction which controls the growth of silicon oxynitride films. Under deposition conditions (1 or 2) investigated, the nitrogen flux through ion sources was maintained constant for various D_{O_2}/R values; however, the nitrogen content in the films decreased with increasing D_{O_2}/R. Furthermore, except for low D_{O_2}/R values, the nitrogen content was similar in films deposited under conditions 1 or 2 although the nitrogen flux was higher for deposition of films under conditions 2. Therefore, silicon atoms are more or less oxidized depending on the D_{O_2}/R ratio and, then nitridation occurred on silicon sites unoccupied by oxygen up to saturation of the free silicon bonds. This dependence between nitrogen and oxygen contents in SiO_xN_y films is shown in Figure 64. The N/Si atom number ratio decreases from 1.33 (stoichiometric ratio in Si_3N_4) as the O/Si atom number ratio increases up to 2 (stoichiometric ratio in SiO_2), except for low O/Si values under conditions 1 where the nitrogen flux is not sufficient for total nitridation of silicon atoms. Stoichiometric SiO_xN_y films were obtained by varying the D_{O_2}/R ratio (Figs. 62 and 64). The relative high extinction coefficient of films prepared under conditions 1 without oxygen in the deposition chamber resulted from a large deviation of stoichiometry (with respect to Si_3N_4) in these films.

The mass density of silicon oxynitride films produced by DIBS was sufficiently high to avoid penetration or adsorption of water vapor molecules in the films and, thereby generation of extrinsic stress. The deposition temperature was 100°C; as a result, thermal stress can be generated in the films during cooling down of samples and can be calculated using Eq. (105). The physical parameters such as Young's modulus, Poisson's ratio, and thermal expansion coefficient for these films are not known over the entire range of composition. To estimate the value of the thermal stress in oxynitride films, the calculation was performed for two limit compositions corresponding to SiO_2 and Si_3N_4. Since the film density was nearly equal to the bulk density, the mechanical properties of bulk amorphous silica were adopted for calculation of thermal stress in SiO_2 films. For Si_3N_4 films, relatively scattered data can be found for thermomechanical properties in the literature; the values proposed for dense silicon nitride films deposited by CVD at 800°C appear to be well adapted for dense films produced by DIBS. The thermal stress values for SiO_2 and Si_3N_4 films calculated on the

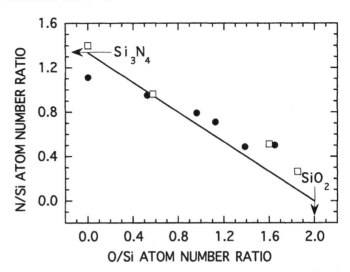

Fig. 64. Nitrogen and oxygen contents of silicon oxynitride films produced by DIBS under deposition conditions 1 (●) and 2 (□) [177]. Reprinted from J. Y. Robic et al., in "Developments in Optical Component Coatings," Proceedings SPIE (I. Reid, Ed.), Vol. 2776, p. 381. SPIE, Bellingham, WA, 1994, with permission.

Table VIII. Evaluation of Thermal Stress Values in SiO_2 and Si_3N_4 Films Deposited at 100°C on Si Substrates

Film	Biaxial modulus $E_f/(1-\nu_f)$ (GPa)	Thermal expansion coefficient, α (per °C)	Thermal stress (MPa)
SiO_2	86	0.55×10^{-6}	−14
Si_3N_4	370	1.6×10^{-6}	−27

Source: J. Y. Robic et al., SPIE, 1994.

basis of these data are given in Table VIII. The calculated values for silicon oxide and silicon nitride films are relatively low, i.e., −14 MPa and −24 MPa, respectively, so that the contribution of the thermal stress in the residual stresses for silicon oxynitride films deposited by DIBS can be neglected.

On the basis of data obtained from the models proposed by Windischmann [113] and by Davis [141], the intrinsic stress in silicon oxynitride films was expressed as

$$\sigma_i = \left(\frac{E_f}{1-\nu_f}\right)\left(\frac{M_t}{D_f}\right)\sum_i \frac{\phi_i}{\phi_c}\sqrt{M_i E_i} \qquad (159)$$

where the first term in brackets is the biaxial modulus of the film material; M_t and D_f are the molar mass and the mass density of the deposited material, respectively. The M_t/D_f ratio is equal to $1/N$ where N is the atom number density of the deposited material. The term, ϕ_i/ϕ_c, is the ratio between the ion flux and the film atom flux at the substrate surface; M_i and E_i are the mass and the kinetic energy of incident ions, respectively. Two terms, Q and P_n, can be distinguished in Eq. (159),

$$Q = \left(\frac{E_f}{1-\nu_f}\right)\left(\frac{M_t}{D_f}\right) = \left(\frac{E_f}{1-\nu_f}\right)\frac{1}{N} \qquad (160)$$

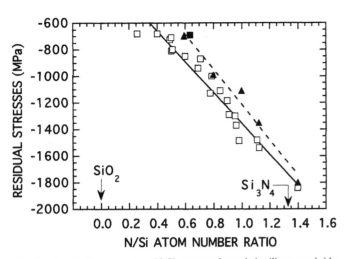

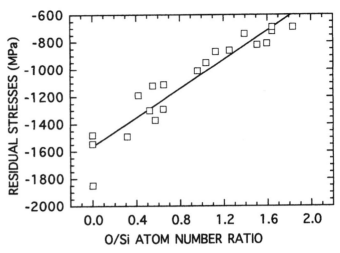

Fig. 65. Residual stresses versus N/Si atom number ratio in silicon oxynitride films produced by DIBS (□), or IBS (■) [49], and in SiN_y films prepared by IBS (▲) under similar deposition conditions according to the results reported by Bosseboeuf [178]. Reprinted from H. Leplan, Ph.D. Thesis, National Polytechnic Institute of Grenoble, France, 1995.

Fig. 66. Residual stresses in silicon oxynitride films produced by DIBS versus O/Si atom number ratio [49]. Reprinted from H. Leplan, Ph.D. Thesis, National Polytechnic Institute of Grenoble, France, 1995.

which represents the stored elastic energy per mole and corresponds to the elastic response of the film to a volumetric distortion induced by defects (interstitial atoms) and

$$P_n = \sum_i \frac{\phi_i}{\phi_c} \sqrt{M_i E_i} \qquad (161)$$

which is the normalized momentum.

Since the pressure in the deposition chamber was always less than 2×10^{-4} mbar and the path length of particles to reach the substrates was around 30 cm, the thermalization of energetic particles can be neglected. Thus, the normalized momentum was nearly constant for all experiments corresponding to data reported in Figure 63. As a result, in accordance with Eq. (159), the stress variation may originate essentially from the variation of the parameter Q which depends on the composition of films. The validity of this assumption seems to be confirmed by the direct correlation between residual stresses (or intrinsic stress since thermal and extrinsic stresses are negligible in these films) and the N/Si or O/Si atom number ratio (Figs. 65 and 66). The stress-nitrogen content correlation appeared more accurate than that with oxygen content and was also found for SiN_y films produced by ion beam sputtering (IBS) under similar experimental conditions [178]. To compare the stress values predicted by the models and the experimental stress values, the parameter Q must be calculated using Eq. (160) for various SiO_xN_y films. Since the elastic constants for these films are not determined, the comparison can be only qualitative. For instance, for a $SiO_{1.82}N_{0.27}$ film produced by DIBS and an $SiN_{1.4}$ film deposited by IBS, the intrinsic stress varied from -690 MPa to -1840 MPa and, the atom number density, N, of these films determined by RBS was 2.6×10^{22} and 3.5×10^{22} at cm^{-3}, respectively. Therefore, the variation of the atom number density is very small and, the stress variation with the composition of films reflects the variation of the biaxial modulus of films. It can be noted that this stress variation is compatible with the biax-

ial modulus of silica and silicon nitride (Table VIII). The stress variation with the composition of films depends only on the parameter Q which is the elastic response of the film whereas the absolute value of the intrinsic stress for each composition is dependent on both Q and P_n factors according to Eq. (159). The effect of the normalized momentum is illustrated in Figure 65 where the intrinsic stress in silicon oxynitride films obtained by IBS (without any ion assistance) is also reported. The stress variation between IBS and DIBS films with identical N/Si atom number ratios may be attributed to the effect of the difference in the normalized momentum values, P_n, used in IBS and DIBS processes.

6.4. Amorphous Carbon Films Deposited by Conventional Magnetron Sputtering on Grounded Substrates

Experiments were designed to estimate the flux and the kinetic energy of particles impinging on the surface of amorphous carbon (a-C) films produced by direct current (dc) magnetron sputtering from a graphite target in pure argon discharges [179]. The major objective of these experiments was to assess the contribution of various species to the energy deposited on the surface of growing films and to correlate the magnitude of residual stresses in a-C films to the kinetic energy and the flux of these species. The characteristics, in particular the plasma potential, of the argon discharge, the flux, and the kinetic energy of predominant energetic particles (neutral carbon atoms and argon ions) impinging on the growing film surface as well as residual stresses in a-C films were investigated as functions of the sputtering gas pressure at various sputtering powers. The films have been deposited on ⟨100⟩-oriented Si substrates of 6×1 cm^2 at ambient temperature. The substrate temperature measured by a thermocouple inserted in the substrate holder during sputter deposition of films was less than 60°C under the experimental conditions investigated. The substrates were mounted on a grounded substrate holder located at 7 cm from the surface of the graphite target of 21×9 cm^2. The argon pressure was varied in the range 0.1–2 Pa. The sputtering power was fixed at 0.5 or 2 kW.

The thickness of a-C films (up to 1.8-μm thick) was determined by profilometer measurements. The flux of carbon atoms, ϕ_C, condensed on Si substrates was deduced from RBS data. The mass density of films was calculated from RBS data and film thickness values. The residual stresses in the films were determined from the change of the radius of curvature of Si substrates measured before and after deposition; the stress values were calculated using Eq. (45).

The plasma potential, V_P, deduced from $I(V)$ characteristics of a cylindrical Langmuir probe located at 6 cm from the target surface was investigated as a function of the argon pressure at a sputtering power of 0.5 and 2 kW. The flux of argon ions, ϕ_{Ar}, impinging on the grounded substrates was calculated from the saturation ion current, I_{is}, collected by the probe assuming that the flux of carbon ions was negligible with respect to the ion argon flux, i.e., $\phi_{Ar} = I_{is}/(eS)$ where e is the charge of Ar^+ ions and S is the surface of the grid electrode. Since the surface of the collector electrode was readily covered with carbon films, the current of secondary electrons emitted by the probe and resulting from the ion bombardment with a maximum energy of 100 eV was neglected for the calculation of the argon ion flux [180].

6.4.1. Characteristics of the Argon Discharge and the Amorphous Carbon Films

The plasma potential reached a maximum value of 2.6 V under an argon pressure of 0.1 Pa and decreased progressively as the argon pressure increased up to 2 Pa before stabilization at about 1 V for high argon pressures [179]. The plasma potential was found to be independent of the sputtering power. The floating potential was equal to −0.5 V under the experimental conditions investigated. The electron temperature and the electron density of 0.5 eV and about 2×10^9 cm^{-3}, respectively, were in good agreement with values for argon magnetron discharges reported in the literature [181].

The flux of argon ions striking the surface of grounded substrates deduced from the saturation ion current was proportional to the sputtering power and was dependent on the argon pressure. The maximum value of this ion flux was reached under an argon pressure of 0.25 Pa. The uniformity of the film thickness was better than 8%. As a result, the flux of carbon atoms condensed on grounded Si substrates could be determined accurately by RBS measurements performed in the center of samples. The flux of carbon atoms was also proportional to the sputtering power [182] and varied with the argon pressure [179].

The ratio of the argon ion flux to the carbon atom flux, ϕ_{Ar}/ϕ_C, was found to vary from 0.10 to 0.05 as the argon pressure increased (Fig. 67). This flux ratio was nearly independent of the sputtering power, in particular under argon pressures higher than 1 Pa. Moreover, the flux of carbon atoms being independent of the sputtering gas pressure, the argon pressure dependence of the flux ratio was very similar to the dependence of the argon ion flux.

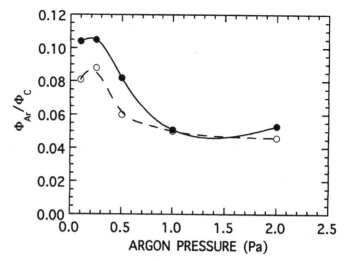

Fig. 67. Argon pressure effect on the ratio of the argon ion flux to the carbon atom flux for a sputtering power of 0.5 kW (●) and 2 kW (○) [179]. Reprinted from E. Mounier and Y. Pauleau, *J. Vac. Sci. Technol. A* 14, 2535 (1996), with permission.

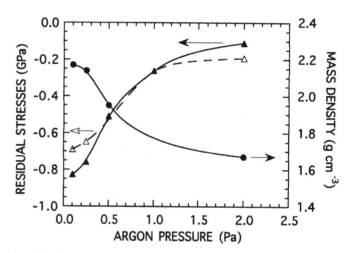

Fig. 68. Argon pressure effect on the mass density of a-C films prepared with a sputtering power of 0.5 kW and residual stresses developed in films obtained with a sputtering power of 0.5 kW (▲) and 2 kW (△) [179]. Reprinted from E. Mounier and Y. Pauleau, *J. Vac. Sci. Technol. A* 14, 2535 (1996), with permission.

The argon content in 30-nm-thick a-C films was less than the RBS detection limit evaluated to be about 0.5 at.%. The hydrogen content was less than 1 at.% with a base pressure in the deposition chamber lower than 4×10^{-4} Pa which was currently reached prior to sputter deposition of films. The growth rate of films was in the range 10–60 nm min^{-1} depending on the deposition conditions. Examinations of the cross section of a-C films by scanning electron microscopy revealed a densely packed structure for films deposited at low argon pressures while films produced at an argon pressure higher than 1 Pa exhibited columnar microstructures.

The mass density of films depended on the sputtering gas pressure (Fig. 68); the maximum value was close to 2.2 g cm^{-3}, i.e., nearly equal to the mass density of bulk graphite (2.25 g cm^{-3}). The residual stresses were found to be compressive and the magnitude of stresses decreased progressively with increasing argon pressure (Fig. 68). In addition, the effect of the sputtering power on the stress value was negligible. The magnitude of residual stresses was also determined in a-C films sputter deposited with various substrate–target distances, d, at an argon pressure value fixed in the range 0.1–0.5 Pa. The compressive residual stress values decreased progressively as the distance d increased under a given argon pressure.

6.4.2. Energy of Carbon Atoms Ejected from the Graphite Target

The kinetic energy of carbon atoms condensed on Si substrates depends on both the energy of atoms ejected from the graphite target and the argon pressure. The experimental determination of the kinetic energy of sputtered atoms can be performed using relatively sophisticated techniques; however, this energy can also be estimated more readily from various models. In particular, the energy distribution of atoms ejected by ion sputtering can be determined from the theoretical model developed by Thompson [183] in which the ejection of atoms results principally from the generation of atomic collision cascades by the bombarding ions. This model was used to establish the velocity distribution of iron atoms sputtered from a magnetron target. The theoretical velocity distribution was found to agree with the distribution determined by laser-induced fluorescence measurements for a substrate-target distance, d, lower than the mean free path of sputtered atoms, λ_{Ar}, colliding with sputtering gas atoms [184]; for a ratio, $d/\lambda_{Ar} > 1$, the thermalization of sputtered iron atoms caused by collisions in the gas phase became efficient.

The kinetic energy of argon ions impinging on the surface of the graphite target is required for calculation of the energy distribution of carbon atoms ejected from the target surface. The argon ions experience an acceleration in the vicinity of the target surface caused by the potential drop in the cathode zone. The thickness of the cathode zone depends on the target voltage and the argon pressure; its value can be estimated to be less than about 1 cm under the experimental conditions investigated [185]. A number of accelerated argon ions may collide with argon atoms and a charge transfer process may occur with formation of a nonenergetic argon ion. This new argon ion experiences a lower acceleration in the cathode zone since the potential drop decreases as the distance from the target surface decreases [186]. The mean free path of argon atoms (in centimeters) at room temperature is given by $\lambda_{Ar} = 0.7/P$ where P is the argon pressure expressed in Pascals. The value of λ_{Ar} varied from 7 to 0.35 cm as the argon pressure increased from 0.1 to 2 Pa. Furthermore, the mean free path for the charge transfer process, λ_{CT}, can be calculated from the following equation [187]

$$\lambda_{CT} = \frac{kT}{P\sigma(E)} \tag{162}$$

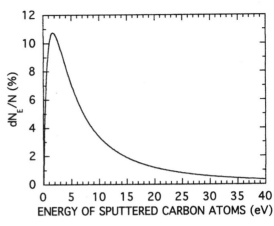

Fig. 69. Energy distribution of carbon atoms ejected from the graphite target biased to −680 V [179]. Reprinted from E. Mounier and Y. Pauleau, *J. Vac. Sci. Technol. A* 14, 2535 (1996), with permission.

where T and P are the absolute temperature and the pressure of the gas phase, respectively; k is Boltzmann's constant, and the charge transfer cross section, $\sigma(E)$, is equal to 3×10^{-15} cm^2 for argon ions accelerated by a potential drop of 600–700 V [187]. Assuming that the gas temperature in the discharge is about 500 K, the value of the mean free path for the charge transfer process, λ_{CT} (in centimeters), at an argon pressure P (in Pascals) calculated from $\lambda_{CT} = 2.3/P$ was found to vary in the range 23–1.1 cm as the argon pressure increased from 0.1 to 2 Pa. As a result, the thickness of the cathode zone was always lower than the mean free path for the charge transfer process and the mean free path of argon atoms. Therefore, the energy of argon ions colliding with the graphite target can be calculated directly from the target voltage since the energy loss caused by collisions in the cathode zone is expected to be negligible.

According to the model developed by Thompson [183], the number of carbon atoms, dN_C, ejected from the target with a kinetic energy ranging from E_C to $(E_C + dE_C)$ is proportional to

$$\frac{1 - [(M_{Ar} + M_C)^2 (E_B + E_C)/(4M_{Ar}M_C E_{Ar})]^{1/2}}{E_C^2 (1 + E_B/E_C)^3} \tag{163}$$

where E_{Ar} is the kinetic energy of incident ions and $E_B = 3.5$ eV is the graphite surface binding energy [188]; M_{Ar} and M_C are the masses of incident ions and target atoms, respectively. The energy distribution of Thompson for carbon atoms sputtered by argon ions from the magnetron discharge was determined for a target voltage of 680 V which corresponds to an average value of target voltages investigated (Fig. 69).

The average value of the kinetic energy of carbon atoms deduced from the energy distribution up to 40 eV was found to be 10.1 eV. For this calculation, the analytical expression of the energy distribution was integrated up to 40 eV. Indeed, atoms with energies higher than 40 eV can be neglected since their number accounts for only a small fraction of the total population [189]. This limitation is justified on the basis of the comparison between the average kinetic energy of Cu atoms ejected from a magnetron target determined experimentally and the average kinetic energy deduced from Thompson's distribution which revealed that the calculated energy value is overestimated [190].

6.4.3. Energy of Carbon Atoms Condensed on Silicon Substrates

The kinetic energy of carbon atoms traveling from the target to the substrates decreases progressively because of collisions in the gas phase. The scattering of sputtered atoms by the sputtering gas and the phenomena leading to kinetic energy losses of sputtered atoms in the gas phase have been modeled by Westwood [48]. A sputtered atom loses kinetic energy and changes direction on each elastic collision with an argon atom. The energy ratio, E_2/E_1, for carbon atoms is given by Eq. (9b) where E_1 and E_2 are the kinetic energies of a carbon atom before and after elastic collision with an argon atom, respectively. M is the atomic mass ratio, $M_{Ar}/M_C = 3.33$. As a result, the terms w and E_2/E_1 are equal to -0.494 and 0.610, respectively. After n elastic collisions between a carbon atom and argon atoms, the energy ratio can be written as

$$\frac{E_n}{E_1} = \exp(nw) \qquad (164)$$

The average diffusion angle of a carbon atom due to the collision with an argon atom is $\langle\theta\rangle = 78.46°$ and the average distance traveled by the carbon atom making n collisions is given by [48]

$$d = n\lambda\sqrt{2}\cos\left(\frac{\langle\theta\rangle}{2}\right) \qquad (165)$$

where λ is the mean free path of a carbon atom with mass M_C traveling through a gas consisting of a mixture of carbon atoms and argon atoms of mass M_{Ar}; the expression of λ can be derived from the kinetic gas theory [191]. The value of λ (in centimeters) can be calculated from $\lambda = 2.2/P$ where P is the argon pressure in Pascals. Therefore, the energy ratio, E_n/E_1, after n elastic collisions is given by

$$\frac{E_s}{E_1} = \exp\left(\frac{wd}{\lambda\sqrt{2}\cos(\langle\theta\rangle/2)}\right) = \exp(-0.205\,dP) \qquad (166)$$

where d and P are expressed in centimeters and Pascals, respectively. This ratio depends on the product $(d \times P)$, i.e., [(target–substrate distance) × (sputtering gas pressure)] as shown in Figure 70. As a result, with a target–substrate distance of 7 cm and an average energy of sputtered carbon atoms, E_1, of 10.1 eV, the average kinetic energy of carbon atoms condensed on the substrate surface is approximately equal to the kinetic energy of argon atoms under an argon pressure higher than 2 Pa. In other words, above 2 Pa, the carbon atoms condensed on the substrate surface are thermalized because of elastic collisions with sputtering gas atoms. This thermalization state is reached after a number of collisions, n_{th}, in the gas phase given by [48]

$$n_{th} = \frac{\ln(E_n/E_1)}{\ln(E_2/E_1)} = \frac{1}{w}\left(\frac{E_{th}}{E_1}\right) \qquad (167)$$

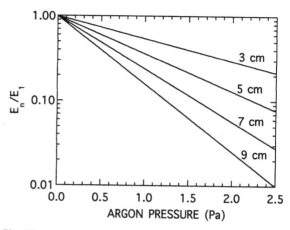

Fig. 70. Kinetic energy ratio, E_n/E_1, after n collisions versus argon pressure for various target–substrate distances [179]. Reprinted from E. Mounier and Y. Pauleau, *J. Vac. Sci. Technol. A* 14, 2535 (1996), with permission.

with $E_1 = 10.1$ eV and an average gas phase temperature of 500 K, i.e., $E_{th} = 0.065$ eV, this number of collisions, n_{th}, is equal to 10.

6.4.4. Energy of Argon Ions Impinging on the Surface of Growing Films

The thickness of the sheath formed on the surface of grounded substrates can be calculated from the Child–Langmuir equation [192]; its value is much lower than 0.1 mm, i.e., much lower than the mean free path of argon atoms and the mean free path for the charge transfer process. In other words, the collision frequency of argon ions traveling through this cathode sheath was negligible. Therefore, the energy of argon ions colliding with the surface of a-C films grown on grounded substrates could be deduced directly from the value of the plasma potential. The kinetic energy of argon ions impinging on the surface of growing films was varied in the range 2.6–1 eV as the argon pressure increased from 0.1 to 2 Pa.

6.4.5. Origin of Residual Stresses in Amorphous Carbon Films Sputter Deposited on Grounded Substrates

The residual stresses in a-C films sputter deposited on Si substrates may essentially result from thermal stress and intrinsic stress developed during the growth of films. The maximum substrate temperature for these sputter depositions of a-C films was less than 60°C. As a result, the contribution of the thermal stress to the residual stresses may be neglected and the magnitude of intrinsic stress is approximately equal to that of residual stresses which was determined as a function of the argon pressure (Fig. 68).

Intrinsic stresses are generated by energetic particle bombardment of the surface of growing films. The surface of films deposited by magnetron sputtering may sense four types of energetic particles, namely, (i) neutral target atoms sputtered from the target surface and condensed on the substrate, (ii) neutral sputtering gas atoms previously implanted in the target as

Table IX.   Residual Stresses in Amorphous Carbon (a-C) Films, Plasma Potential, and Characteristics of the Energetic Particles under
Various Experimental Conditions

Sputtering power (kW)	Argon pressure (Pa)	Residual stresses (GPa)	Plasma potential (V)	Flux ratio ϕ_{Ar}/ϕ_c	Average energy of C atoms (eV)
0.5	0.1	−0.83	2.6	0.104	8.7
0.5	0.25	−0.76	2	0.105	7
0.5	0.5	−0.51	1.7	0.082	4.9
0.5	1	−0.26	1.5	0.051	2.4
0.5	2	−0.11	1.1	0.053	0.6
2	0.1	−0.69	2.7	0.081	8.7
2	0.25	−0.65	1.9	0.088	7
2	0.5	—	1.7	0.06	4.9
2	1	−0.26	1.4	0.05	2.4
2	2	−0.19	1.2	0.046	0.6

Source: E. Mounier and Y. Pauleau, *J. Vac. Sci. Technol. A* 14, 2535, 1996.

sputtering ions which are ejected from the target surface together with neutral target atoms, (iii) ions of the sputtering
gas coming from the discharge, and (iv) neutralized sputtering
ions backscattered from the target surface by elastic collisions.
The flux of the second type of energetic particles is negligible
with respect to the flux of carbon atoms ejected from the target. The reflection coefficients, R_0 and γ_0, corresponding to the
flux and the energy of sputtering ions neutralized and backscattered from the target surface (fourth type of energetic particles)
are dependent on the atomic mass ratio, M_C/M_{Ar} (Figs. 13
and 14). For the carbon–argon system, the atomic mass ratio is
equal to 0.3; hence, the reflection coefficients R_0 and γ_0 are less
than 10^{-3} and 0.1, respectively. Therefore, for this system, the
contribution of backscattered neutral argon atoms to the energy
deposited on the surface of growing films is negligible with respect to the contribution of neutral carbon atoms and argon ions
originating from the plasma.

According to the model proposed by Windischmann [54],
the magnitude of the compressive intrinsic stress in films produced by ion beam sputtering was material specific and scaled
with elastic energy per mole defined by the parameter Q expressed by Eq. (121). Stress data reported in the literature for
films produced by other deposition techniques involving ion-
peening-induced stress show good correlation with Q for a
wide range of materials (Fig. 71). The Q parameter value was
found to be equal to 16×10^5 J mol^{-1} for a-C films sputter
deposited on grounded Si substrates using a biaxial modulus,
$E_f/(1 - \nu_f)$, of 200 GPa [193]. These a-C films deposited at
an argon pressure of 0.25 Pa and a sputtering power of 0.5 kW
exhibited a mass density of 2.15 g cm^{-3} and a compressive
residual (intrinsic) stress, σ, of −0.78 GPa. The values of Q
and σ for these a-C films represented in Figure 71 are in good
agreement with the values reported in the literature for various materials. On the basis of this excellent correlation, the
argon pressure effect on the intrinsic stress developed in sputter-
deposited a-C films was analyzed using the forward sputtering
model [54, 113]. The compressive intrinsic stress in a-C films

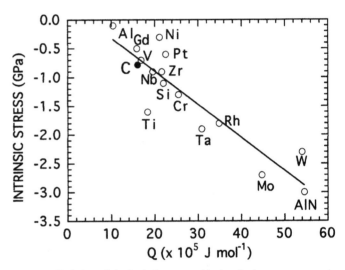

Fig. 71.   Variation of the intrinsic stress with the elastic energy per mole
factor Q for films prepared by magnetron sputtering [179]. Reprinted from
E. Mounier and Y. Pauleau, *J. Vac. Sci. Technol. A* 14, 2535 (1996), with permission.

is given by [179]

$$\sigma = 4.79 \left(\frac{E_f}{1 - \nu_f} \right) \left(\frac{K}{N} \right)$$
$$\times \left[2.79(E_{Ar})^{1/2} \phi_{Ar} + 2.64(E_C)^{1/2} \phi_C \right] \quad (168)$$

where E_{Ar} and E_C are the argon ion and the carbon atom energies (in electron volts) on the surface of a-C films, respectively;
ϕ_{Ar} and ϕ_C are the argon ion and the carbon atom fluxes (in particles cm^{-2} s^{-1}) on the surface of growing films, respectively.
K is a proportionality factor and N is the atom number density
in the deposited material.

The intrinsic stress in a-C films and the average kinetic energy of carbon atoms condensed on the surface of films grown
under various conditions are given in Table IX. Moreover, the
flux and the kinetic energy of argon ions impinging on the sur-

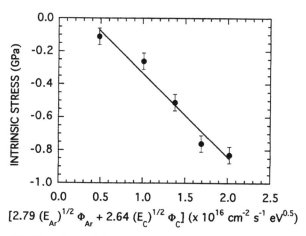

Fig. 72. Dependence of the intrinsic stress in a-C films on the energetics of the deposition process; the solid line was deduced from a least-squares fit to the experimental data [179]. Reprinted from E. Mounier and Y. Pauleau, *J. Vac. Sci. Technol. A* 14, 2535 (1996), with permission.

Table X. Effect of the product
(Target–substrate Distance) × (Argon Pressure) on the Residual Stresses in a-C Films Produced under Various Deposition Conditions

Argon pressure, P (Pa)	Target–substrate distance, d (cm)	$d \times P$ (cm Pa)	Growth rate of films (nm min^{-1})	Residual stresses (GPa)
0.125	14	1.75	4.36	−0.75
0.25	7	1.75	12.2	−0.75
0.5	3.5	1.75	27.6	−0.75

Source: E. Mounier and Y. Pauleau, *J. Vac. Sci. Technol. A* 14, 2535, 1996.

face of grounded substrates were evaluated from the ion current and the plasma potential measurements [179]. These data are reported in the diagram intrinsic stress versus flux and energy of projectile particles plotted in Figure 72. The dependence of the intrinsic stress on energetics of the deposition process is in good agreement with the prediction of the model proposed by Windischmann.

In this model, the projectile particles (argon ions and carbon atoms) are supposed to be sufficiently energetic to penetrate the surface and randomly displace carbon atoms from their equilibrium positions in the films through a series of primary and recoil collisions producing a volumetric distortion. The atom displacement energy in bulk materials is currently higher than 10 eV, i.e., higher than the average energy of carbon atoms condensed on Si substrates at low argon pressures and much higher than the energy of argon ions originating from the plasma. However, the displacement of carbon atoms located near the surface of the deposited material is expected to need less energy than the displacement of atoms in the crystal lattice of the bulk material. In addition, the carbon atoms displaced on the surface or in the outermost atomic layers are rapidly covered with other carbon atoms and are isolated from the flux of incident particles. As a result, these carbon atoms are frozen in nonequilibrium positions, leading to volumetric distortions and compressive intrinsic stress development. Furthermore, the energy distribution of carbon atoms ejected from the target shows that a small fraction of these atoms (in the distribution tail) can be sufficiently energetic to displace carbon atoms in the growing films from their equilibrium positions. These energetic carbon atoms may also contribute to the development of the compressive intrinsic stress in the films.

The contribution of argon ions to the energy deposited on the film surface is relatively small compared to that of neutral carbon atoms, in particular, at low argon pressures when the thermalization phenomena of sputtered carbon atoms are not very efficient. In other words, in Eq. (168), at low argon pressures, the term $2.79(E_{Ar})^{1/2}\phi_{Ar}$ is negligible with respect

to the term $2.64(E_C)^{1/2}\phi_C$. These thermalization phenomena involving collisions between sputtered carbon atoms and sputtering gas atoms play a major role in the deposition process. The number of collisions of carbon atoms in the gas phase and the thermalization of sputtered carbon atoms depend on the product ($d \times P$) between the target–substrate distance and the argon pressure. This product is a relevant factor affecting the development of the intrinsic stress and governing their magnitude in sputter-deposited films as shown in Table X. The compressive intrinsic stress was found to remain at a constant value of −0.75 GPa in a-C films produced under various experimental conditions with the same value of the product ($d \times P$).

6.5. Amorphous Carbon Films Deposited by Conventional and Unbalanced Magnetron Sputtering on Biased Substrates

The magnetron target was placed in the center of a cylindrical magnetic coil of 23 cm in inner diameter and the unbalanced magnetron sputtering mode was operated by varying the current in the coil [194]. The water-cooled substrate holder was either grounded or biased to negative voltages up to 300 V.

The characteristics of the deposition process such as substrate temperature, ion flux, ϕ_{Ar}, and carbon atom flux, ϕ_C, on the film surface were determined as functions of the substrate bias voltage for sputtering using conventional and unbalanced magnetron modes. The properties of sputter-deposited a-C films, in particular argon content, mass density, and residual stresses were correlated to the deposition parameters. The physical characteristics of the argon discharge (plasma potential, floating potential, electron density, and electron temperature) were deduced from $I(V)$ characteristics of a cylindrical Langmuir probe placed 6 cm from the target surface. A grid probe was placed at the substrate holder position to determine the flux of ions impinging on the surface of growing films.

6.5.1. Characteristics of the Deposition Process

The major characteristics of the argon discharge for sputter deposition of a-C films using the conventional and unbalanced magnetron modes have previously been reported [179, 182, 195]. The kinetic energy of ions striking the surface of growing films deposited on grounded substrates directly corresponds to the plasma potential which was independent of the sputtering power and the current in the magnetic coil. The plasma potential was found to vary in the range 2.5–1 V as the argon pressure was increased from 0.1 to 3.5 Pa. The kinetic energy of ions collected at the surface of films deposited on negatively biased substrates corresponds to the difference between the plasma potential and the substrate bias voltage. However, the value of the plasma potential was always negligible with respect to that of the substrate bias voltage.

The flux of positive ions impinging on the film surface increased rapidly as the negative substrate bias voltage was varied from 0 to about 40–50 V [194]. In addition, the ion flux decreased with increasing argon pressure for both magnetron modes and increased with increasing current in the coil, i.e., with increasing unbalanced level of the magnetron target. The ion flux value was in the range $(1-4) \times 10^{14}$ ions cm^{-2} s^{-1} with the conventional magnetron mode and could be 20 times higher using the unbalanced magnetron mode [194].

The growth rate of a-C films up to 1.8-μm thick was in the range 10–160 nm min^{-1} depending upon the deposition parameters and the magnetron modes used. The flux of carbon atoms condensed on Si substrates found in the range 10^{15}–10^{16} atoms cm^{-2} s^{-1} was proportional to the sputtering power and was essentially independent of the argon pressure and the negative substrate bias voltage [195]. In fact, the dependence of the growth rate of films (expressed in nanometers per minute) on the argon pressure resulted from the decrease in mass density of films with increasing argon pressure. Using the conventional magnetron mode at a sputtering power of 0.5 kW, the ratio of the ion flux to the carbon flux, ϕ_{Ar}/ϕ_C, was observed to vary from 0.05 to 0.17 with increasing negative substrate bias voltage [194]. The dependence of this ratio on the bias voltage applied to substrates was very similar to that of the ion flux since the carbon flux was approximately independent of the bias voltage. Using the unbalanced magnetron mode, the ratio, ϕ_{Ar}/ϕ_C, also increased with increasing negative substrate bias voltage; depending on the current in the coil and the substrate bias voltage, the flux ratio value could be higher by a factor of 10 to 20 than the ratio value obtained from the conventional magnetron mode [194].

The energetic particle bombardment of the surface of growing films is known to be a major factor affecting the compressive intrinsic stress in the films but also simultaneously the energy deposited on the film surface leads to a temperature rise which depends on the sputter-deposition conditions. Using the conventional magnetron mode with a sputtering power of 0.5 kW, the maximum substrate temperature was about 60°C. The maximum temperature of grounded substrates decreased with increasing argon pressure. The energy deposited on the surface of grounded substrates was carried by carbon atoms condensed on the surface. The energy deposited on negatively biased substrates was carried by positive ions accelerated by the bias voltage applied to substrates. Using the unbalanced magnetron mode, the substrate temperature was stabilized more rapidly; however, the maximum substrate temperature was de-

pendent on the current in the coil and the substrate bias voltage. With a positive substrate bias voltage of +20 V, the substrate temperature can reach a relatively high value of 350°C [195]. Consequently, a water-cooled substrate holder must be utilized in particular with the unbalanced magnetron mode for sputter deposition of thermal stress-free a-C films under the experimental conditions investigated.

6.5.2. Characteristics of Amorphous Carbon Films

Argon atoms were found as major impurities in the films. For a-C films sputter deposited on negatively biased substrates using the conventional magnetron mode at an argon pressure of 0.25 and 2 Pa, the argon concentration was proportional to the square of the bias voltage and became constant at a given bias voltage value which was dependent on the argon pressure [194]. The maximum amount of argon atoms incorporated in the films was 2 and 4 at.% under an argon pressure of 2 and 0.25 Pa, respectively, i.e., the maximum argon concentration increased with decreasing argon pressure. This dependence of the argon concentration on the negative bias voltage and the argon pressure is similar to that already observed for sputter-deposited metal films [196]. Argon atoms were also found as impurities in a-C films produced by unbalanced magnetron sputtering [194]. The argon concentration increased rapidly with increasing substrate bias voltage before stabilization at a value ranging from 1 to 1.4 at.%. However, the amount of argon atoms in the films was significantly lower than that incorporated in a-C films deposited by conventional magnetron sputtering. In addition, in this case, the linear dependence of the argon concentration on the square of the bias voltage was not observed.

The mass density of argon containing a-C films deposited on negatively biased substrates by conventional magnetron sputtering at an argon pressure of 0.25 and 2 Pa increased linearly with increasing bias voltage up to 90–100 V; beyond this bias voltage value, the mass density of films was approximately constant (Fig. 73). The mass density of a-C films deposited on grounded substrates at an argon pressure of 0.25 Pa was nearly equal to the bulk graphite density (2.25 g cm^{-3}). The variation of the mass density of a-C films with the substrate bias voltage is quite similar to the variation of the argon content; i.e., the mass density and the argon concentration were maxima for similar values of the bias voltage.

Since the substrates were maintained at room temperature by the water-cooled substrate holder, the thermal stress contribution in the residual stress value can be neglected. As a result, the residual stress values deduced from measurements of the radius of curvature of Si substrates are the values of the intrinsic stress built up during sputter deposition of a-C films. The compressive intrinsic stress in films sputter deposited by conventional magnetron mode at an argon pressure of 0.25 Pa and a sputtering power of 0.5 kW increased progressively up to a maximum value of −2.8 GPa with increasing bias voltage (Fig. 73). This maximum value is about three times higher than that obtained from films sputter deposited on grounded substrates. By contrast, a-C films with a reduced mass density sputter deposited

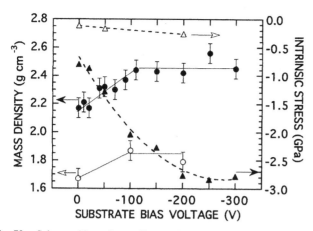

Fig. 73. Substrate bias voltage effect on the mass density and the intrinsic stress for a-C films sputter deposited using the conventional magnetron mode at a sputtering power of 0.5 kW under an argon pressure of: 0.25 Pa (closed symbols) and 2 Pa (open symbols) [194]. Reprinted from E. Mounier and Y. Pauleau, *Diamond Relat. Mater.* 6, 1182 (1997), with permission.

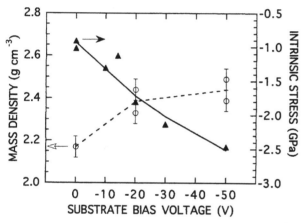

Fig. 74. Substrate bias voltage effect on the mass density and the intrinsic stress for a-C films sputter-deposited using the unbalanced magnetron mode at a sputtering power of 0.5 kW under an argon pressure of 0.25 Pa with a current intensity in the coil of 6 A [194]. Reprinted from E. Mounier and Y. Pauleau, *Diamond Relat. Mater.* 6, 1182 (1997), with permission.

at an argon pressure of 2 Pa exhibited negligible compressive intrinsic stress whatever the bias voltage value although the argon concentration in these films was dependent on the substrate bias voltage.

The mass density of a-C films produced by unbalanced magnetron sputtering increased from 2.2 to 2.4–2.5 g cm^{-3} with increasing negative substrate bias voltage before stabilization (Fig. 74). As already observed for films produced by conventional magnetron sputtering, the a-C films deposited on grounded substrates exhibited a mass density close to the bulk graphite density. In other words, the a-C films deposited on grounded substrates by conventional and unbalanced magnetron modes are quite similar in terms of mass density. The compressive intrinsic stress in a-C films produced by unbalanced magnetron sputtering also increased with increasing substrate bias voltage (Fig. 74). The films produced with a negative bias voltage higher than 50 V were not adherent to Si

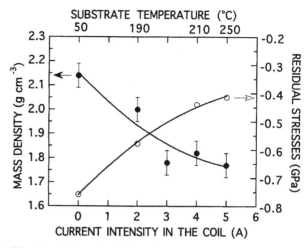

Fig. 75. Mass density of a-C films and magnitude of residual stresses in a-C films versus current density in the coil or the substrate temperature for sputter deposition with the unbalanced magnetron mode at a sputtering power of 0.5 kW under an argon pressure of 0.25 Pa [194]. Reprinted from E. Mounier and Y. Pauleau, *Diamond Relat. Mater.* 6, 1182 (1997), with permission.

substrates since the compressive intrinsic stress in these films was excessive, i.e., probably higher than 2.5 GPa. The value of the compressive intrinsic stress in a-C films produced on grounded substrates using the unbalanced magnetron mode was of the same order of magnitude as that for a-C films deposited by conventional magnetron sputtering, i.e., in the range −0.8 to −1 GPa. The compressive intrinsic stress in a-C films produced by unbalanced magnetron mode increased more rapidly with increasing bias voltage than the intrinsic stress in films deposited by conventional magnetron sputtering; however, the reverse trend was observed for the increase in argon concentration in films produced by unbalanced and conventional magnetron modes.

The mass density and the residual stresses for a-C films deposited on grounded and nonwater-cooled Si substrates by unbalanced magnetron sputtering is plotted versus current in the coil in Figure 75; the substrate temperature increased with increasing coil current. The contribution of thermal stress in the residual stress value cannot be neglected for these films. The mass density and the residual stresses decrease with increasing substrate temperature. The characteristics of these a-C films are similar to those of films deposited on substrates heated at moderate temperatures by infrared lamps using the conventional magnetron sputtering process [197].

6.5.3. Origin of Residual Stresses in Amorphous Carbon Films Sputter Deposited on Biased Substrates

The mass density of films and the concentration of argon atoms entrapped in the films are maximum for similar bias voltage values. This similarity suggests that the increase in mass density may be caused by the amount of argon atoms incorporated in the films rather than by the formation of really dense a-C films. In fact, the energy deposited on the film surface which increases with increasing bias voltage seems to have a negligible effect on the mass density of the deposited material.

The energy available on the film surface essentially originates from the kinetic energy of positive ions created in the argon discharge and accelerated by the negative substrate bias voltage. The maximum value of the average kinetic energy of neutral carbon atoms condensed on the substrate is about 10 eV whereas the kinetic energy of positive ions is in the range 10–300 eV. The energy deposited on the growing films and resulting from the positive ion bombardment of the surface appears to be the relevant factor affecting the magnitude of the compressive intrinsic stress in a-C films deposited by conventional and unbalanced magnetron sputtering. Since the flux ratio, ϕ_{Ar}/ϕ_C, is also dependent on the negative substrate bias voltage or energy of positive ions impinging on the film surface, this flux ratio can be considered as an additional factor affecting the value of the intrinsic stress in a-C films produced under the experimental conditions investigated.

6.5.3.1. Applicability of the Forward Sputtering Model Proposed by Windischmann

According to Windischmann's model, the compressive intrinsic stress is predicted to be proportional to the product of the particle flux, ϕ_p, and the square root of the particle energy, E_p (Eq. (120)); i.e., ϕ_p and E_p represent the flux and the kinetic energy of positive ions striking the surface of growing a-C films. For a-C films sputter deposited on Si substrates biased to a negative voltage in the range 20–300 V under an argon pressure of 0.25 Pa with a sputtering power of 0.5 kW, the intrinsic stress versus product $[\phi_p(E_p)^{1/2}]$ is plotted in Figure 76. The experimental values of the compressive intrinsic stress are in good agreement with the prediction of Windischmann's model as the product $[\phi_p \times (E_p)^{1/2}]$ is less than 5×10^{15} ions cm^{-2} s^{-1} eV$^{1/2}$. Beyond this value, a large deviation between experimental and predicted results can be observed. As a result, additional phenomena resulting in stress relaxation must be considered for interpreting and modeling the effect of the ion flux and energy on the intrinsic stress in a-C films deposited by sputtering under intense energetic particle bombardment.

6.5.3.2. Applicability of the Model Proposed by Davis

On the basis of the model proposed by Davis [141], the compressive intrinsic stress in films for which the film surface is subjected to energetic particle bombardment during deposition can be calculated from Eq. (128). For sputter deposition of a-C films, the flux ratio ϕ_C/ϕ_i was determined as a function of the energy E_i of Ar$^+$ ions striking the surface of negatively biased substrates. For a-C films sputter deposited with the conventional magnetron mode under an argon pressure of 0.25 Pa at a sputtering power of 0.5 kW, the expression of ϕ_C/ϕ_i deduced from experimental data [194] is given by

$$\frac{\phi_C}{\phi_i} = \frac{22.75}{2.4 + 0.1(E_f)^{1/2}} \tag{169}$$

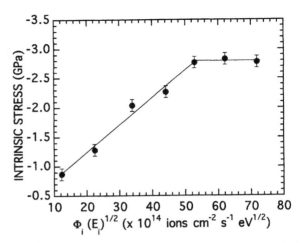

Fig. 76. Effect of the flux and the energy of positive ions on the intrinsic stress in a-C films deposited with the conventional magnetron mode at a sputtering power of 0.5 kW under an argon pressure of 0.25 Pa; the substrate bias voltage values for sputter deposition of these films were in the range −20 to −300 V [194]. Reprinted from E. Mounier and Y. Pauleau, *Diamond Relat. Mater.* 6, 1182 (1997), with permission.

For a-C films produced with the unbalanced magnetron mode under an argon pressure of 0.25 Pa at a sputtering power of 0.5 kW with a current in the coil of 5 A, the ratio ϕ_C/ϕ_i deduced from experimental data [194] is given by

$$\frac{\phi_C}{\phi_i} = \frac{22.75}{26.6 + 5.72(E_f)^{1/2}} \tag{170}$$

In Eqs. (169) and (170), the ion energy E_i in electron volts is equal to the absolute value of the substrate bias voltage expressed in volts. Finally, the compressive intrinsic stress in a-C films is related to the ion energy by

$$\sigma = \kappa \left[\frac{(E_f)^{1/2}}{(22.75/(2.4 + 0.1(E_f)^{1/2})) + k_a(E_f)^{5/3}} \right] \tag{171}$$

$$\sigma = \kappa \left[\frac{(E_f)^{1/2}}{(22.75/(26.6 + 5.72(E_f)^{1/2})) + k_a(E_f)^{5/3}} \right] \tag{172}$$

Equations (171) and (172) based on the model proposed by Davis [141] are valid for a-C films sputter deposited by conventional and unbalanced magnetron modes, respectively.

These equations contain two undetermined parameters, κ and k_a, which were used as fitting parameters for a least-squares fit to experimental results. The theoretical curves and the experimental values of the compressive intrinsic stress in the a-C films sputter deposited by conventional and unbalanced magnetron modes are represented in Figure 77. In addition, the values of parameters κ and k_a are given in Table XI. The experimental results and the predicted values of the intrinsic stress are in very good agreement for a-C films sputter deposited by conventional magnetron mode. For a-C films produced by an unbalanced magnetron mode, the theoretical curve also fits the experimental values; however, these values correspond to relatively low ion energies since the films produced under more energetic ion fluxes were found to be not adherent to Si substrates.

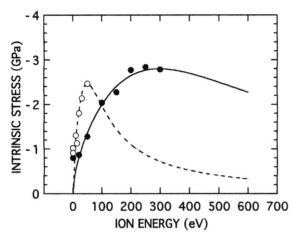

Fig. 77. Effect of the positive ion energy on the intrinsic stress in a-C films deposited at a sputtering power of 0.5 kW under an argon pressure of 0.25 Pa using: (●) the conventional magnetron sputtering mode, and (○) the unbalanced magnetron sputtering mode with a current intensity in the coil of 5 A; the curves in solid and dashed lines represent the intrinsic stress values predicted from the model proposed by Davis [194]. Reprinted from E. Mounier and Y. Pauleau, *Diamond Relat. Mater.* 6, 1182 (1997), with permission.

Table XI. Experimental Values of Fitting Parameters, κ and k_a, Included in Eqs. (171) and (172) Based on the Model Proposed by Davis

Fitting parameter	Conventional magnetron mode	Unbalanced magnetron mode
κ	1.44	0.2
k_a	2.54×10^{-4}	3.45×10^{-4}

Source: E. Mounier and Y. Pauleau, *Diamond Relat. Mater.* 6, 1182, 1997.

The value of E_d which is an excitation energy required for release of the stress or for displacement of one carbon atom from a metastable position on the surface of the film can be deduced from the expression of k_a, i.e., $k_a = 0.016\rho(E_d)^{-5/3}$ assuming that $\rho = 1$ (Eq. (126)). This energy was equal to 12 and 10 eV for films deposited by conventional and unbalanced magnetron modes, respectively. These excitation energy values are in concordance with those corresponding to various other materials [141]. The value of κ are of the same order of magnitude for a-C films sputter deposited by conventional and unbalanced magnetron modes (Table XI). It is more difficult to verify the concordance between these values and those calculated from the literature data of various physical constants (Young's modulus, Poisson's ratio, and sublimation energy of the deposited material) included in the expression of κ since these physical parameters are rather unknown for sputter-deposited a-C films. Nevertheless, the agreement between the experimental results and the predicted data as shown in Figure 77 contributes to assess the validity and to demonstrate the applicability of the mechanism proposed by Davis to explain the origin of the compressive intrinsic stress developed in a-C films deposited by magnetron sputtering.

7. SUMMARY AND CONCLUSION

Residual stresses in PVD films are the sum of: (1) thermal stress arising from the different thermal expansion coefficients of films and substrates combined with the difference between the deposition temperature and room temperature, (2) intrinsic stress developed as the films grow, and (3) extrinsic stress resulting from interactions between the deposited material and the reactive agents present in film environment. Excessive residual stresses may cause mechanical damage which has a detrimental effect on the performance of films and devices utilized for various advanced technologies. The control of residual stresses together with a clear understanding of mechanisms involved in the development of stresses are required for reliable manufacturing techniques such as evaporation, sputtering, or ion plating currently used for preparation of thin films.

The nucleation steps and the growth mechanisms of PVD films determine the microstructure or the morphology of the deposited materials which, in turn, affect the physical properties of films, in particular the mass or the packing density, the surface roughness, the microhardness, and the residual stresses. Therefore, it is important to investigate the microstructure and the growth mechanisms of films in combination with the nature and the magnitude of residual stresses as functions of experimental parameters related to the deposition equipment and the process. The major features of the microstructure of films which depend on the normalized growth temperature, T_s/T_m (where T_s is the absolute substrate temperature and T_m is the absolute melting point of the deposited material), are thoroughly described by structure-zone diagrams proposed for films deposited by evaporation and sputtering. The microstructure of films is related to the surface mobility of adatoms involved in the growth of films. The adatom mobility depends on the substrate temperature and the kinetic energy of species striking the growing film surface. Hence, the normalized growth temperature and the gas pressure in the deposition chamber are two major experimental parameters which affect the microstructure of PVD films and thereby the residual stresses.

A wide variety of methods can be used for measuring the magnitude of residual stresses in PVD films. The residual stresses in amorphous, nanocrystalline, or polycrystalline films can be determined by measurements of the deformation of film–substrate structures. For thin films deposited on relatively thick substrates (beams), the nature and the average values of stresses are obtained from measurements of the radius of curvature of beams, the deflection of the free end of the beam, or by observing the displacement (sagitta) of the center of the beam or a circular disk type substrate. Alternatively, the residual stresses in polycrystalline and single crystal films can be determined by X-ray diffraction techniques.

Various models can be invoked to explain the development of residual stresses, in particular the origin of intrinsic stress in PVD films. The intrinsic stress can be tensile or compressive depending on the deposition parameters. Tensile intrinsic stress is normally observed in films deposited on substrates at low temperatures by thermal evaporation in the absence of any energetic particle bombardment. These not fully dense films exhibit a zone 1 type microstructure represented in the structure-zone diagram and the grain boundary relaxation (GBR) model can explain the origin of the stress. Compressive intrinsic stress develops in films deposited at low temperatures under energetic particle bombardment. These films exhibit dense microstructures corresponding to zone T in the structure-zone diagram and their compressive intrinsic stress results from the atomic peening mechanism which can be associated with thermal spikes causing displacement of the implanted atoms at relatively high energetic bombardment. The sudden tensile-to-compressive stress transition may arise from a threshold phenomenon occurring at approximately the atomic displacement energy. Reduced intrinsic stress can be observed in PVD films deposited at relatively high temperatures since thermally activated atomic diffusion and stress relief phenomena may occur during deposition of films.

In addition to thermal and intrinsic stresses, extrinsic stress may develop in thin films subsequently to the deposition process. Water vapor molecules may penetrate in porous films exposed to room air. Adsorption phenomena and adsorbed dipole interactions can be invoked to explain the origin of extrinsic stress. Furthermore, interactions between the deposited material and the adsorbed species can be responsible for a slow evolution of the extrinsic stress as the exposure time of samples to room air (aging time) increases.

Stress data illustrating the applicability of models proposed for the origin of stresses have been obtained, in particular for silicon dioxide films prepared by thermal evaporation and ion-assisted deposition, silicon oxynitride films produced by dual ion beam sputtering, and amorphous carbon films deposited by conventional and unbalanced magnetron sputtering. To assess the applicability of models, in particular to sputter-deposited films, numerous data regarding the plasma discharge, the flux, and the kinetic energy of particles impinging on the film surface must be collected using more or less complex experimental techniques. The magnitude of compressive intrinsic stress predicted from the models proposed by Windischmann and by Davis are clearly substantiated by experimental results obtained from SiO_2, SiO_xN_y and a-C films produced by energetic deposition techniques.

REFERENCES

1. M. J. O'Keefe and J. M. Rigsbee, in "Materials and Processes for Surface and Interface Engineering," NATO-ASI Series, Series E: Applied Sciences (Y. Pauleau, Ed.), Vol. 290, p. 151. Kluwer Academic, Dordrecht, The Netherlands, 1995.
2. E. Klokholm, *IBM J. Res. Develop.* 31, 585 (1987).
3. M. D. Drory, M. D. Thouless, and A. G. Evans, *Acta Metall.* 36, 2019 (1988).
4. L. van den Hove, J. Vanhellemont, R. Wolters, W. Claassen, R. De Keersmaecker, and G. Declerk, in "Proceedings of the 1st Symposium on Advanced Materials for VLSI" (M. Scott, Y. Akasaka, and R. Reif, Eds.), The Electrochemical Soc., Pennington, NJ, 1988.
5. J. Vanhellemont, S. Amelinckx, and C. Claeys, *J. Appl. Phys.* 61, 2176 (1987).

6. H. Awano and T. Sato, *Japan. J. Appl. Phys.* 27, L880 (1988).
7. K. Kolkholm and J. F. Freedman, *J. Appl. Phys.* 38, 1354 (1967).
8. H. C. W. Huang, P. Chaudhari, and C. J. Kircher, *Philos. Mag. A* 54, 583 (1986).
9. C.-Y. Li, R. D. Black, and W. R. La Fontaine, *Appl. Phys. Lett.* 53, 31 (1988).
10. J. E. Greene, in "Multicomponent and Multilayered Thin Films for Advanced Microtechnologies: Techniques, Fundamentals and Devices," NATO-ASI Series, Series E: Applied Sciences (O. Auciello and J. Engemann, Eds.), Vol. 234, p. 39. Kluwer Academic, Dordrecht, The Netherlands, 1993.
11. J. K. Hirvonen, in "Materials and Processes for Surface and Interface Engineering," NATO-ASI Series, Series E: Applied Sciences (Y. Pauleau, Ed.), Vol. 290, p. 307. Kluwer Academic, Dordrecht, The Netherlands, 1995.
12. P. B. Barna and M. Adamik, in "Protective Coatings and Thin Films: Synthesis, Characterization and Applications," NATO-ASI Series, Partnership Sub-Series 3: High Technology (Y. Pauleau and P. B. Barna, Eds.), Vol. 21, p. 279. Kluwer Academic, Dordrecht, The Netherlands, 1997.
13. E. S. Machlin, "Materials Science in Microelectronics, The Relationships between Thin Film Processing and Structure," Chap. III, p. 49. Giro Press, Croton-on-Hudson, NY, 1995.
14. Y. W. Lee and J. M. Rigsbee, *Surf. Sci.* 173, 3 (1986).
15. J. P. Hirth and K. L. Moazed, "Thin Films Physics" (G. Hass and R. E. Thun, Eds.), Vol. 4, p. 97. Academic Press, New York, 1967.
16. C. A. Neugebauer, in "Handbook of Thin Film Technology" (L. I. Maissel and R. Glang, Eds.), Chap. 8. McGraw-Hill, New York, 1970.
17. R. S. Wagner and R. J. H. Voorhoeve, *J. Appl. Phys.* 43, 3948 (1971).
18. E. Krikorian and R. J. Sneed, *Astrophys. Space Sci.* 65, 129 (1979).
19. M. Marinov, *Thin Solid Films* 46, 267 (1977).
20. M.-A. Hasan, S. A. Barnett, J.-E. Sundgren, and J. E. Greene, *J. Vac. Sci. Technol. A* 5, 1883 (1987).
21. T. Narusawa, S. Shimizu, and S. Komiya, *J. Vac. Sci. Technol.* 16, 366 (1979).
22. K. Yagi, S. Tamura, and T. Tokuyama, *Japan. J. Appl. Phys.* 16, 245 (1977).
23. T. Tokuyama, K. Yagi, K. Miyaki, M. Tamura, N. Natsuaki, and S. Tachi, *Nucl. Instrum. Methods* 182–183, 241 (1981).
24. K.-H. Müller, *Phys. Rev. B* 35, 7906 (1987).
25. K.-H. Müller, *Surf. Sci.* 184, L375 (1987).
26. C.-H. Choi and S. A. Barnett, *Appl. Phys. Lett.* 55, 2319 (1989).
27. C.-H. Choi, L. Hultman, and S. A. Barnett, *J. Vac. Sci. Technol. A* 8, 1587 (1990).
28. C.-H. Choi, L. Hultman, R. Ai, and S. A. Barnett, *Appl. Phys. Lett.* 57, 2931 (1990).
29. S. A. Barnett, C.-H. Choi, and R. Kaspi, *Mater. Res. Soc. Symp. Proc.* 201, 43 (1991).
30. D. K. Brice, J. Y. Tsao, and S. T. Picraux, *Nucl. Instrum. Methods B* 44, 68 (1989).
31. B. A. Movchan and A. V. Demchishin, *Phys. Met. Metallogr.* 28, 83 (1969).
32. H. T. G. Hentzell, B. Andersson, and S.-E. Karlsson, *Acta Metall.* 31, 2103 (1983).
33. A. G. Dirks and H. J. Leamy, *Thin Solid Films* 47, 219 (1977).
34. J. A. Venables and G. L. Price, in "Epitaxial Growth" (J. W. Matthews, Eds.). Academic Press, New York, 1971.
35. B. A. Movchan, A. V. Demchishin, and L. D. Kooluck, *J. Vac. Sci. Technol.* 11, 869 (1974).
36. J. A. Thornton, *J. Vac. Sci. Technol.* 11, 666 (1974).
37. J. A. Thornton, *J. Vac. Sci. Technol.* 12, 830 (1975).
38. J. A. Thornton and D. W. Hoffman, *Thin Solid Films* 171, 5 (1989).
39. J. A. Thornton, *Annu. Rev. Mater. Sci.* 7, 239 (1977).
40. C. R. M. Grovenor, H. T. G. Hentzell, and D. A. Smith, *Acta Metall.* 32, 773 (1984).
41. E. Klokholm and B. S. Berry, *J. Electrochem. Soc.* 115, 823 (1968).

42. C. C. Wong, H. I. Smith, and C. V. Thompson, *Appl. Phys. Lett.* 48, 335 (1986).

43. E. S. Machlin, "Materials Science in Microelectronics, The Relationships between Thin Film Processing and Structure," Chap. III, p. 57. Giro Press, Croton-on-Hudson, NY, 1995.

44. W. L. Brown and A. Ourmazd, *MRS Bull.* 6, 30 (1992).

45. F. Jona, *J. Phys. Chem. Solids* 28, 2155 (1967).

46. S. D. Dahlgren, *J. Vac. Sci. Technol.* 11, 832 (1974).

47. L. Eckertova, "Physics of Thin Films" Chap. 2, p. 3. Plenum, New York, 1977.

48. W. D. Westwood, *J. Vac. Sci. Technol.* 15, 1 (1978).

49. H. Leplan, Ph.D. Thesis, National Polytechnic Institute of Grenoble, France, 1995.

50. M. Xia, X. Liu, and P. Gu, *Appl. Opt.* 32, 5443 (1993).

51. P. Sigmund, *Phys. Rev.* 184, 383 (1969).

52. F. M. d'Heurle, *Metall. Trans.* 1, 725 (1970).

53. P. Sigmund, in "Topics in Applied Physics: Sputtering by Particle Bombardment I" (R. Behrisch, Ed.), Vol. 47, Chap. 2. Springer-Verlag, Berlin, 1981.

54. H. Windischmann, *J. Appl. Phys.* 62, 1800 (1987).

55. G. H. Kinchin and R. S. Pease, *Rep. Prog. Phys.* 18, 1 (1955).

56. T. C. Huang, G. Lim, F. Parmigiani, and E. Kay, *J. Vac. Sci. Technol. A* 3, 2161 (1985).

57. E. Kay, F. Parmigiani, and P. Parrish, *J. Vac. Sci. Technol. A* 5, 44 (1987).

58. H. H. Andersen and H. L. Bay, in "Topics in Applied Physics: Sputtering by Particle Bombardment I" (R. Behrisch, Ed.), Vol. 47, p. 145. Springer-Verlag, Berlin, 1981.

59. W. D. Wilson, L. G. Haggmark, and J. P. Biersack, *Phys. Rev. B* 15, 2458 (1977).

60. K.-H. Müller, *J. Vac. Sci. Technol. A* 4, 184 (1986).

61. K.-H. Müller, *J. Appl. Phys.* 59, 2803 (1986).

62. J. D. Targove and H. A. Macleod, *Appl. Opt.* 27, 3779 (1988).

63. H. A. Macleod, *J. Vac. Sci. Technol. A* 4, 418 (1986).

64. F. Seitz and J. S. Koehler, *Solid State Phys.* 2, 307 (1956).

65. G. K. Wehner and G. S. Anderson, in "Handbook of Thin Film Technology" (L. I. Maissel and R. Glang, Eds.), p. 3–23. McGraw-Hill, New York, 1970.

66. J. Bottiger, J. A. Davies, P. Sigmund, and K. B. Winterbon, *Radiat. Effects* 11, 69 (1971).

67. C. Brunneé, *Z. Phys.* 147, 161 (1957).

68. N. N. Petrov, *Sov. Phys. Solid State* 2, 857 (1960).

69. P. Sigmund, *Can. J. Phys.* 46, 731 (1968).

70. H. F. Winters and D. Horne, *Phys. Rev. B* 10, 55 (1974).

71. W. R. Gesang, H. Oechsner, and H. Schoof, *Nucl. Instrum. Methods* 132, 687 (1976).

72. J. P. Biersack and L. G. Haggmark, *Nucl. Instrum. Methods* 174, 257 (1980).

73. D. S. Campbell, in "Handbook of Thin Film Technology" (L. I. Maissel and R. Glang, Eds.), p. 12–21. McGraw-Hill, New York, 1970.

74. R. W. Hoffman, in "Physics of Non-Metallic Thin Films," NATO-ASI Series, Series B: Physics (C. H. Dupuy and A. Cachard, Eds.), Vol. 14, p. 273. Plenum, New York, 1976.

75. M. R. James and O. Buck, *Crit. Rev. Solid State Mater. Sci.* 9, 61 (1980).

76. S. Tamulevicius, L. Augulis, G. Laukaitis, and L. Puodziukynas, *Medziagotyra* 2, 40 (1998).

77. M. Murakami, T.-S. Kuan, and I. A. Blech, *Treatise Mater. Sci. Technol.* 24, 163 (1982).

78. B. Borie, C. J. Sparcks, and J. V. Cathcart, *Acta Metall.* 10, 691 (1962).

79. C. R. Houska, *J. Appl. Phys.* 41, 69 (1970).

80. J. Unnam, J. A. Carpenter, and C. R. Houska, *J. Appl. Phys.* 44, 1957 (1973).

81. M. Murakami, *Acta Metall.* 26, 175 (1978).

82. S. Timoshenko, "Résistance des Matériaux," Vol. 1, p. 89. Librairie Polytechnique Béranger, Paris.

83. G. G. Stoney, *Proc. Roy. Soc. London Ser. A* 82, 172 (1909).

84. A. Brenner and S. Senderoff, *J. Res. Natl. Bur. Stand.* 42, 105 (1949).

85. J. D. Wilcock and D. S. Campbell, *Thin Solid Films* 3, 3 (1969).

86. W. A. Brantley, *J. Appl. Phys.* 44, 534 (1973).

87. M. R. James and J. B. Cohen, in "Treatise on Materials Science and Technology, Experimental Methods" (H. Herman, Ed.), Vol. 19, Part A, p. 1. Academic Press, New York, 1980.

88. V. M. Hauk, *Adv. X-Ray Anal.* 27, 81 (1984).

89. P. S. Prevey, in "Metals Handbook—Materials Characterization" (R. E. Whan et al., Eds.), Vol. 10, p. 380. Am. Soc. Metals, Metals Park, OH, 1986.

90. I. C. Noyan and J. B. Cohen, in "Residual Stress Measurement by Diffraction and Interpretation," Materials Research and Engineering Series (B. Ilschner and N. J. Grant, Eds.). Springer-Verlag, Berlin, 1987.

91. R. E. Cuthrell, D. M. Mattox, C. R. Peeples, P. L. Dreike, and K. P. Lamppa, *J. Vac. Sci. Technol. A* 6, 2914 (1988).

92. H. Dölle, *J. Appl. Cryst.* 12, 4598 (1979).

93. H. Dölle and J. B. Cohen, *Metall. Trans. Mater. A* 11, 159 (1980).

94. Z.-C. Feng and H.-D. Liu, *J. Appl. Phys.* 54, 83 (1983).

95. P. H. Townsend, D. M. Barnett, and T. A. Brunner, *J. Appl. Phys.* 62, 4438 (1987).

96. E. Suhir, *ASME J. Appl. Mech.* 55, 143 (1988).

97. M. D. Thouless, *Thin Solid Films* 181, 397 (1989).

98. E. Suhir, *ASME J. Appl. Mech.* 53, 657 (1986).

99. E. Suhir, *ASME J. Appl. Mech.* 55, 143 (1988).

100. E. Suhir, *ASME J. Appl. Mech.* 56, 595 (1989).

101. G. Gille, in "Current Topics in Materials Science" (E. Kaldis, Ed.), Vol. 12. North-Holland, Amsterdam, 1985.

102. M. S. Hu and A. G. Evans, *Acta Metall.* 37, 917 (1989).

103. M. D. Thouless, H. C. Cao, and P. A. Mataga, *J. Mater. Sci.* 24, 1406 (1989).

104. M. Y. He and J. W. Hutchinson, *ASME J. Appl. Mech.* 56, 270 (1989).

105. J. J. Prescott, "Applied Elasticity," p. 187. Dover, New York, 1961.

106. J. E. E. Baglin, in "Materials and Processes for Surface and Interface Engineering," NATO-ASI Series, Series E: Applied Sciences (Y. Pauleau, Ed.), Vol. 290, p. 111. Kluwer Academic, Dordrecht, The Netherlands, 1995.

107. A. Kelly, "Strong Solids." Clarendon, Oxford, 1976.

108. G. I. Barenblatt, *Adv. Appl. Mech.* 7, 55 (1962).

109. R. W. Hoffman, in "Physics of Thin Films—Advances in Research and Development" (G. Hass and R. E. Thun, Eds.), Vol. 3, p. 211. Academic Press, New York, 1966.

110. D. S. Campbell, in "Basic Problems in Thin Film Physics" (R. Niedermayer and H. Mayer, Eds.), p. 223. Vandenhoeck and Ruprecht, Gottingen, 1966.

111. K. Kinosita, *Thin Solid Films* 12, 17 (1972).

112. M. F. Doerner and W. D. Nix, *Crit. Rev. Solid State Mater. Sci.* 14, 225 (1988).

113. H. Windischmann, *Crit. Rev. Solid State Mater. Sci.* 17, 547 (1992).

114. E. S. Machlin, "Materials Science in Microelectronics, The Relationships between Thin Film Processing and Structure," Chap. 6, p. 157. Giro Press, Croton-on-Hudson, NY, 1995.

115. H. Leplan, B. Geenen, J. Y. Robic, and Y. Pauleau, in "Optical Interference Coatings," Proceedings SPIE (F. Abelès, Ed.), Vol. 2253, p. 1263. SPIE, Bellingham, WA, 1994.

116. H. Leplan, B. Geenen, J. Y. Robic, and Y. Pauleau, *J. Appl. Phys.* 78, 962 (1995).

117. I. Blech and U. Cohen, *J. Appl. Phys.* 53, 4202 (1982).

118. M. F. Ashby and D. R. H. Jones, "Matériaux-1. Propriétés et Applications," p. 53. Editions Dunod, Paris, 1991.

119. H. Leplan, J. Y. Robic, and Y. Pauleau, *J. Appl. Phys.* 79, 6926 (1996).

120. F. Jansen, M. A. Machonkin, N. Palmieri, and D. Kuhman, *J. Appl. Phys.* 62, 4732 (1987).

121. P. Paducsheck, P. Eichinger, G. Kristen, and H. Mitlehner, *Nucl. Instrum. Methods* 199, 421 (1982).

122. P. Paducsheck, C. Hopfl, and H. Mitlehner, *Thin Solid Films* 110, 291 (1983).

123. J. C. Knight, in "Topics in Applied Physics: The Physics of Hydrogenated Amorphous Silicon I—Structure, Preparation, and Devices"

(D. Joannopoulos and G. Lucovsky, Eds.), Vol. 55, p. 5. Springer-Verlag, Berlin, 1984.

124. J. P. Harbison, A. J. Williams, and D. V. Lang, *J. Appl. Phys.* 55, 946 (1984).

125. H. Windischmann, R. W. Collins, and J. M. Cavese, *J. Non-Cryst. Solids* 85, 261 (1986).

126. J. C. Angus, C. C. Hayman, and R. W. Hoffman, "Proceedings SPIE," Vol. 969, p. 2. SPIE, Bellingham, WA, 1988.

127. H. Sankur and W. Gunning, *J. Appl. Phys.* 66, 807 (1989).

128. G. Thurner and R. Abermann, *Thin Solid Films* 192, 277 (1990).

129. F. M. d'Heurle and J. M. E. Harper, *Thin Solid Films* 171, 81 (1989).

130. J. D. Finegan and R. W. Hoffman, *J. Appl. Phys.* 30, 597 (1959).

131. R. E. Rottmayer and R. W. Hoffman, *J. Vac. Sci. Technol.* 8, 151 (1971).

132. F. A. Doljack and R. W. Hoffman, *Thin Solid Films* 12, 71 (1972).

133. R. W. Hoffman, *Thin Solid Films* 34, 185 (1976).

134. H. K. Pulker, *Thin Solid Films* 89, 191 (1982).

135. R. W. Hoffman, J. D. Finegan, F. A. Doljack, and R. W. Springer, AEC Technical Report, Atomic Energy Commission, Case Western Reserve University, Cleveland, OH, 1961, p. 18; 1970, p. 64; 1971, p. 76; 1972, p. 79; 1975, pp. 82–83.

136. R. Berger and H. K. Pulker, "Proceedings SPIE," Vol. 401, p. 69. SPIE, Bellingham, WA, 1983.

137. H. Windischmann, G. F. Epps, Y. Cong, and R. W. Collins, *J. Appl. Phys.* 69, 2231 (1991).

138. H. K. Pulker and J. Mäser, *Thin Solid Films* 59, 65 (1979).

139. M. Itoh, M. Hori, and S. Nadahara, *J. Vac. Sci. Technol. B* 9, 149 (1991).

140. J. H. Rose, J. Ferrante, and J. R. Smith, *Phys. Rev. Lett.* 47, 675 (1981).

141. C. A. Davis, *Thin Solid Films* 226, 30 (1993).

142. E. A. Kenick and T. E. Mitchell, *Philos. Mag.* 32, 815 (1975).

143. M. J. Makin, S. N. Buckley, and G. P. Walters, *J. Nucl. Mater.* 68, 161 (1977).

144. J. E. Yehoda, B. Yang, K. Vedam, and R. Messier, *J. Vac. Sci. Technol. A* 6, 1631 (1988).

145. B. Window and K.-H. Müller, *Thin Solid Films* 171, 183 (1989).

146. O. Knotek, R. Elsing, G. Kramer, and F. Jungblut, *Surf. Coat. Technol.* 46, 265 (1991).

147. K.-H. Müller, *J. Vac. Sci. Technol. A* 4, 209 (1986).

148. E. S. Machlin, "Materials Science in Microelectronics, The Relationships between Thin Film Processing and Structure," Chap. 6, p. 179. Giro Press, Croton-on-Hudson, NY, 1995.

149. H. P. Martinez and R. Abermann, *Thin Solid Films* 89, 133 (1982).

150. A. Blachman, *Metall. Trans.* 2, 699 (1971).

151. R. S. Wagner, A. K. Sinha, T. T. Sheng, H. J. Levinstein, and F. B. Alexander, *J. Vac. Sci. Technol.* 11, 582 (1974).

152. R. D. Bland, G. J. Kominiak, and D. M. Mattox, *J. Vac. Sci. Technol.* 11, 671 (1974).

153. D. W. Hoffman and J. A. Thornton, *Thin Solid Films* 40, 355 (1977).

154. D. W. Hoffman and J. A. Thornton, *Thin Solid Films* 45, 387 (1977).

155. J. A. Thornton, J. Tabock, and D. W. Hoffman, *Thin Solid Films* 64, 111 (1979).

156. E. H. Hirsch, *J. Phys. D: Appl. Phys.* 13, 2081 (1980).

157. W. R. Smythe, "Static and Dynamic Electricity," 3rd ed., p. 7. McGraw-Hill, New York, 1968.

158. J. Topping, *Proc. Roy. Soc. London Ser. A* 114, 67 (1927).

159. D. W. Hoffman and R. C. McCune, in "Handbook of Plasma Processing Technology" (S. M. Rossnagel, J. J. Cuomo and W. D. Westwood, Eds.), p. 483. Noyes, Park Ridge, NJ, 1990.

160. E. S. Machlin, "Materials Science in Microelectronics, The Relationships between Thin Film Processing and Structure," Chap. 2, p. 21. Giro Press, Croton-on-Hudson, NY, 1995.

161. J. A. Thornton, in "Semiconductor Materials and Process Technology Handbook" (G. E. McGuire, Ed.), p. 329. Noyes, Park Ridge, NJ, 1988.

162. A. M. Haghiri-Goisnet, F. R. Ladan, C. Mayeux, and H. Launois, *Appl. Surf. Sci.* 38, 295 (1989).

163. A. M. Haghiri-Goisnet, F. R. Ladan, C. Mayeux, H. Launois, and M. C. Joncour, *J. Vac. Sci. Technol. A* 7, 2663 (1989).

164. E. Klokholm, *J. Vac. Sci. Technol.* 6, 138 (1969).

165. P. G. Shewmon, "Diffusion in Solids," p. 65. McGraw-Hill, New York, 1963.

166. R. A. Holmwood and R. Glang, *J. Electrochem. Soc.* 112, 827 (1965).

167. R. C. Sun, T. C. Tisone, and P. D. Cruzan, *J. Appl. Phys.* 46, 112 (1975).

168. H. Windischmann, *J. Vac. Sci. Technol. A* 7, 2247 (1989).

169. R. A. Roy, R. Petkie, and A. Boulding, *J. Mater. Res.* 6, 80 (1991).

170. S. Dushman, "Scientific Foundations of Vacuum Technique," 2nd ed., p. 460. Wiley, New York, 1962.

171. W. A. Pliskin, *J. Vac. Sci. Technol. A* 4, 418 (1986).

172. K. Ramkumar and A. N. Saxena, *J. Electrochem. Soc.* 139, 1437 (1992).

173. D. O. Hayward and B. M. W. Trapnell, "Chemisorption," 2nd ed., p. 93. Butterworth, London, 1964.

174. J. Y. Robic, H. Leplan, Y. Pauleau, and B. Rafin, *Thin Solid Films* 290–291, 34 (1996).

175. M. Ida, P. Chaton, and B. Rafin, in "Optical Interference Coatings," Proceedings SPIE (F. Abelès, Ed.), Vol. 2253, p. 404. SPIE, Bellingham, WA, 1994.

176. D. Van Vechten, G. K. Hubler, E. P. Donavan, and F. D. Correl, *J. Vac. Sci. Technol. A* 8, 821 (1990).

177. J. Y. Robic, H. Leplan, M. Berger, P. Chaton, E. Quesnel, O. Lartigue, C. Pelé, Y. Pauleau, and F. Pierre, in "Developments in Optical Component Coatings," Proceedings SPIE (I. Reid, Ed.), Vol. 2776, p. 381. SPIE, Bellingham, WA, 1994.

178. A. Bosseboeuf, Ph.D. Thesis, University of Paris-Sud, Orsay, France, 1989.

179. E. Mounier and Y. Pauleau, *J. Vac. Sci. Technol. A* 14, 2535 (1996).

180. K. Bewilogua and D. Wagner, *Vacuum* 42, 473 (1991).

181. S. M. Rossnagel, M. A. Russak, and J. J. Cuomo, *J. Vac. Sci. Technol. A* 5, 2150 (1987).

182. E. Mounier, E. Quesnel, and Y. Pauleau, in "New Diamond and Diamond-Like Films," Proceedings of the Topical Symposium II on New Diamond and Diamond-Like Films of the 8th CIMTEC-World Ceramic Congress and Forum on New Materials, Advances in Science and Technology 6 (P. Vincenzini, Ed.), Vol. 6, p. 183. Techna srl, Faenza, Italy, 1995.

183. M. W. Thompson, *Philos. Mag.* 18, 377 (1968).

184. W. Z. Park, T. Eguchi, C. Honda, K. Muraoka, Y. Yamagata, B. W. James, M. Maeda, and M. Akasaki, *Appl. Phys. Lett.* 58, 2564 (1991).

185. J. R. Roth, "Industrial Plasma Engineering," Vol. 1, Chap. 8, p. 281. Institute of Physics, Bristol, U.K., 1995.

186. W. D. Davis and T. A. Vanderslice, *Phys. Rev.* 131, 219 (1963).

187. R. S. Robinson, *J. Vac. Sci. Technol.* 16, 185 (1979).

188. Y. Lifshitz, S. R. Kasi, J. W. Rabalais, and W. Eckstein, *Phys. Rev. B* 41, 10,468 (1990).

189. K. Meyer, I. K. Schuller, and C. M. Falco, *J. Appl. Phys.* 52, 5803 (1981).

190. L. T. Ball, I. S. Falconer, D. R. McKenzie, and J. M. Smelt, *J. Appl. Phys.* 59, 720 (1986).

191. E. W. McDaniel, "Collision Phenomena in Ionized Gases," Wiley, New York, 1964.

192. J. R. Roth, "Industrial Plasma Engineering," Vol. 1, Chap. 9, p. 329. Institute of Physics, Bristol, U.K., 1995.

193. A. Rouzaud, E. Barbier, J. Ernoult, and E. Quesnel, *Thin Solid Films* 270, 270 (1995).

194. E. Mounier and Y. Pauleau, *Diamond Relat. Mater.* 6, 1182 (1997).

195. Y. Pauleau, E. Mounier, and P. Juliet, in "Protective Coatings and Thin Films: Synthesis, Characterization and Applications," NATO-ASI Series, Partnership Sub-Series 3: High Technology (Y. Pauleau and P. B. Barna, Eds.), Vol. 21, p. 197. Kluwer Academic, Dordrecht, The Netherlands, 1997.

196. J. J. Cuomo and R. J. Gambino, *J. Vac. Sci. Technol.* 14, 152 (1977).

197. E. Mounier, P. Juliet, E. Quesnel, Y. Pauleau, and M. Dubus, *Mater. Res. Soc. Symp. Proc.* 383, 465 (1995).

Chapter 10

LANGMUIR–BLODGETT FILMS OF BIOLOGICAL MOLECULES

Victor Erokhin

Fondazione El.B.A., Corso Europa 30, Genoa, 16132 Italy

Contents

1. INTRODUCTION

One of the essential parts of living organisms is the membrane—a thin layer that separates the internal part of living unit from the surrounding space. Therefore, the membrane can be considered as the boundary between life and death. Moreover, it not only serves as a passive boundary, but several extremely important processes take place in it: photosynthesis, ion exchange and other biochemical and biophysical processes.

However, the aim of this chapter is not to illustrate the state of the art in the field of biological membrane investigations, but to show possible ways of reconstructing membrane structures with artificial methods, namely, by the Langmuir–Blodgett (LB) technique.

The ancients knew of the ability of oillike compounds to form floating layers at an air–water interface. It was Lord Rayleigh who demonstrated for the first time that such layers can be one molecule thick [1]. Systematic investigations of floating monolayers began with work of Irving Langmuir

at the beginning of the twentieth century [2, 3]. The behavior of insoluble floating monolayers at the air–water interface was investigated. From the fundamental point of view, the concepts of the two-dimensional gas, two-dimensional liquid, and two-dimensional crystal and of phase transitions between them were introduced for the first time. The Nobel Prize was given to Irving Langmuir in 1937 as a result of these works.

The second stage of the investigations in this field was performed in the 1930s, when the collaboration of Irving Langmuir with Katrine Blodgett resulted in the development of the technique for monolayer transfer from an air–water interface to solid supports [4–8]. This invention opened more possibilities both from fundamental and applied points of view. On the one hand, more precise and quantitative methods of the investigation were available for the study of such objects. On the other hand, possibility of the manipulation with structures at the molecular level of resolution was established for the first time.

The third stage in LB film studies was started in 1960s with the work of Hans Kuhn's group. In this period the LB technique

Handbook of Thin Film Materials, edited by H.S. Nalwa
Volume 1: Deposition and Processing of Thin Films

ISBN 0-12-512909-2/$35.00

was applied for the realization of molecular architectures [9–19]. Complicated structures, with designed alternation of different functional layers, were realized and investigated. Exciting work on energy transfer in artificial structures composed of donor, acceptor, and spacer layers attracted the interest of numerous research groups to the molecular manipulation possibilities provided by the LB technique.

It is not surprising that the technique was applied for the investigation of biological objects practically as soon as it was discovered. In fact, a floating monolayer at an air–water interface can be considered as a model of half of a biological membrane. The surfactant molecules composing the layer are rather similar to lipid molecules—the main part of natural membranes. Thus, Langmuir and Schaefer published in 1938 the first work where the technique was applied for studying protein monolayers and multilayers [20]. This work was also important in that it introduced the horizontal deposition technique, which is widely used now for protein film deposition.

Summarizing, the LB approach is very useful for studying membrane-like structures. First of all, several processes can be studied by injecting active molecules under the monolayer into the volume of the liquid subphase. The interaction of those molecules with the model membrane (monolayer) provides insight into the processes in biological membranes. Second, the technique allows one to deposit structures similar to the model membranes onto solid supports, where one can apply various modern and powerful tools for investigation of their structure and properties. Third, it allows one to realize complicated layered structures, which can be used for sensors, biocatalytical reactors, drug design, etc.

The aim of this chapter is to give some ideas about the application of the LB technique for studying model systems of biological objects. The structure of the chapter is as follows.

Section 2 gives some details on the LB method itself.

Section 3 introduces several experimental techniques frequently used for studying monolayers at an air–water interface and LB films. This section gives some features of the techniques for subsequent reference. It does not give complete descriptions of the techniques, but describes their basic principles, ranges of application, and degrees of applicability for studying biological LB films.

One graphic representation will be useful for illustrating the subject of the chapter. Figure 1 shows schematic representation of a biological membrane. It contains a lipid bilayer (object 1)—the main part of the membrane. Characteristic features of lipid LB films are considered in Sections 3 and 4. Lipid film studies have been performed for understanding fundamental mechanisms of the membrane structure, its phase transitions, and its interactions with ions and bioactive components.

The main part of the chapter, Section 4, deals with proteins. As is seen in Figure 1, proteins in nature exist as integral membrane proteins (object 2), having large hydrophobic areas; partial membrane proteins (object 3), attached to the membrane by hydrophobic interactions of their restricted hydrophobic areas; and circulating hydrophilic proteins (object 4). Section 4 gives

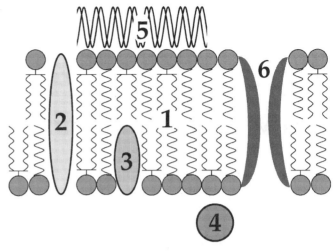

Fig. 1. Schematic representation of the biological membrane: 1, lipid bilayer; 2, integral membrane protein; 3, partial membrane protein; 4, circulating hydrophilic protein; 5, DNA molecule; 6, valinomycin.

advice on how to proceed with film formation in each of these cases.

In the conclusion (Section 5) are summarized the most important features of the application of the LB technique to lipid and protein layers. Some remarks on the applicability of the technique for studying some other biological objects, such as DNA (object 5 in Fig. 1) and valinomycine (object 6 in Fig. 1), which are not considered in detail in the text, will be presented.

2. PRINCIPLES OF THE LANGMUIR–BLODGETT TECHNIQUE

2.1. Monolayers at the Air–Water Interface

Salts of fatty acids are classical objects of the LB technique. The general structure of these molecules is

$$CH_3(CH_2)_n COOH$$

Fatty acids that form stable monolayers, include stearic ($n = 16$), arachidic ($n = 18$), and behenic ($n = 20$) acid.

Placed at an air–water interface, these molecules arrange themselves in such a way that the hydrophilic part (COOH) penetrates the water due to its electrostatic interactions with water molecules, which can be considered as electric dipoles. The hydrophobic part (aliphatic chain) faces the air, because it cannot penetrate water for entropy reasons. Therefore, if few enough molecules of this type are placed on the water surface, they form a two-dimensional system.

Let us consider what will happen when the layer is compressed with some kind of barrier. We consider the surface pressure as a parameter describing the monolayer behavior. It is determined as

$$\pi = \sigma_{H_2O} - \sigma_{ml}$$

where σ_{H_2O} is the surface tension of pure water and σ_{ml} is the surface tension of the monolayer-covered water surface. In

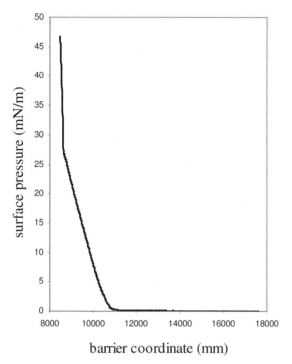

Fig. 2. Pressure–area isotherm of a stearic acid monolayer.

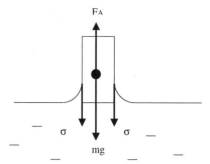

Fig. 3. Measurement principle and force balance of the Wilhelmy balance.

other words, the surface pressure can be considered as the decrease of the water surface tension due to the presence of the monolayer on it.

A compression isotherm of a stearic acid monolayer is presented in Figure 2. This important characteristic represents the dependence of the surface pressure upon the area per molecule, obtained at constant temperature. This dependence is usually called the $\pi - A$ isotherm [21]. Measurements of such isotherms are practically always performed for studying the behavior and phase transitions of the monolayers at the air–water interface [22–55].

Let us now consider the isotherm in detail. Initially, the compression does not result in surface pressure variations. Molecules at the air–water interface are rather far from each other and do not interact. This state of the monolayer is referred to as a *two-dimensional gas*. Further compression results in an increase of the surface pressure. Molecules come closer one to another and begin to interact. This state of the monolayer is referred to as a *two-dimensional liquid*. For some compounds it is possible to distinguish also liquid-expanded and liquid-condensed phases. Continuation of the compression results in the appearance of a *two-dimensional solid state* phase, characterized by a sharp increase in the surface pressure for even a small decrease in the area per molecule. Dense packing of molecules in the monolayer is reached. Further compression results in the collapse of the monolayer. 2D structure no longer exists. Uncontrollable multilayers are formed at the water surface.

Two instruments are usually considered for surface pressure measurements, namely, the Langmuir balance [56] and the Wilhelmy balance [57]. The Langmuir balance measures the surface tension directly. A barrier, separating the clean water sur-

face from that covered with the monolayer, is the sensitive element. This balance is used mainly when precise measurements are needed, and practically never when the monolayer is to be transferred onto solid substrates. There are several reasons for its restricted application. First of all, the utilization of the Langmuir balance requires that the compression of the monolayer be from one direction only. This can result in a gradient of the monolayer density (and thus the surface pressure), which is unacceptable in some cases. Second, the measurement of the surface pressure is not performed where the deposition takes place. This can result in weak control of the monolayer state during its transfer onto solid substrates. Third, there is rather large area of the monolayer that cannot be used. The first and the second drawbacks are critical when working with rigid monolayers. The third drawback is very important when working with expansive substances, as a significant part of the layer must be wasted.

The Wilhelmy balance has found more applications, even though it does not provide a direct measurement of the surface pressure. It allows one to avoid all three of the mentioned drawbacks of the Langmuir balance [58–61].

The sensitive element of the Wilhelmy balance is a plate. The measurement principle is illustrated in Figure 3. The forces acting to the plate are:

mg, the weight of the plate;
F_A, the Archimedean force;
F_s, the surface-tension-induced force.

The last force is just the product of the surface tension and the plate perimeter. The weight of the plate is constant, and Wilhelmy balances are now equipped with systems for maintaining the plate immersion depth in the water at the same level, and thereby keeping the buoyant force constant. Thus it is possible to attribute zero force to the clean water surface, and the differences from this value will directly yield the surface pressure.

The construction of the balance allows one to perform measurements at a point exactly corresponding to the deposition point with respect to the barrier position. It provides precise control of the surface pressure value, maintained by the feedback system. Another advantage of the Wilhelmy balance is the possibility of compressing the monolayer from both sides, achieving better homogeneity of the monolayer.

An important consideration in using the Wilhelmy balance is connected with the necessity of maintaining constant plate position with respect to the water surface level, in order to avoid variations in the buoyant force. Usually, it this is achieved by special construction of the balance. The Wilhelmy plate is connected to a magnet, which is inserted into a solenoid. Variations in the surface pressure displace the magnet, and electronics provides the current to the solenoid, which restores the initial magnet position. The position can be reset by optical methods. The value of the current is then proportional to the surface pressure value.

Another parameter that can be controlled when working with monolayers at an air–water interface is the surface potential [62–87, 89]. This potential results from the orientation of molecular charges and dipoles during the compression of the monolayer. Three different regions are usually considered for its interpretation. The first one is due to the orientation of the C−H bonds in the hydrocarbon chains of the amphiphilic molecules during monolayer formation. The second one is connected with the regular arrangement of the polar head groups. And the third one is due to the dipole orientation of water molecules in the region just under the monolayer. The relative configuration of each of these regions can vary according to the nature of the molecules forming the monolayer.

There are differences in behavior between the surface potential and the surface pressure. The variation of the surface potential begins long before the surface pressure begins to increase significantly. This behavior is due to the fact that molecules begin to aggregate, forming dimers, trimers, and small domains, at the initial stage of the monolayer formation. Being aggregated, molecules tend to orient themselves in energetically favorable positions, giving rise to variation in the surface potential. This happens while there is practically no increase of the surface pressure. In the later stages of the monolayer compression the variation of the surface potential is mainly due to the increase of the monolayer density.

The Kelvin probe is the tool [90–93] that is usually used for surface potential measurements. The instrument is equipped with a vibrating electrode, placed near the water surface. The reference electrode is inserted into the water subphase. The electrode and water surface form a capacitor. The vibration of the electrode modulates the capacitance, resulting in an alternating current proportional to the surface potential.

The surface potential of the monolayer can be either positive or negative; the sign is determined by the nature of the molecules in the monolayer.

2.2. Monolayer Transfer onto Solid Substrates

A floating monolayer can be transferred onto the surface of a solid support. Two main techniques are usually considered for monolayer deposition, namely, the Langmuir–Blodgett (or vertical lift) [4–8], and the Langmuir–Schaefer (or horizontal lift) [20, 94–97].

The scheme for Langmuir–Blodgett deposition is illustrated in Figure 4. A specially prepared substrate is passed vertically through the monolayer. The monolayer is transferred onto the

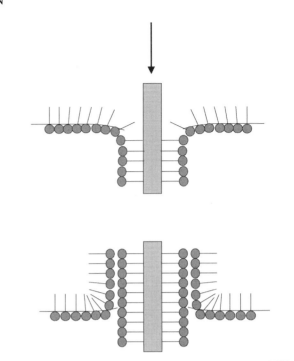

Fig. 4.   Scheme of the Langmuir–Blodgett deposition (vertical lift).

substrate surface during this passage. The important requirement for such deposition is to have the monolayer electrically neutral. If any charges in the monolayer molecule head groups are uncompensated, the deposition will not be performed because the electrostatic interaction of these charges with water molecules will be larger than the hydrophobic interactions of chains with the hydrophobized substrate surface.

Let us consider again a monolayer of fatty acids in order to demonstrate the necessity of head-group neutrality. If the monolayer is formed at the surface of distilled water (pH about 6.0), it cannot be transferred onto a solid substrate. Its head group is dissociated and contains negative charge (COO^-). There are two ways to enable deposition. The first one is protonation of the head groups. It requires a decrease of the pH of the subphase. In fact, the deposition begins to take place when the pH is less than 4.0. However, a monolayer of pure fatty acid is very rigid, and its transfer usually results in a defective LB film on a solid substrate. Therefore, usually fatty acid salt monolayers are deposited instead of fatty acids. In this case, bivalent metal ions are added to the water subphase. Normally, their concentration is of the order of magnitude of 10^{-4} M. These ions attach themselves electrostatically to the dissociated fatty acid head groups, rendering them electrically neutral. In the deposited layer, the bivalent metal ion coordinates four oxygen atoms in two fatty acid molecules in adjacent monolayers. This coordination is illustrated schematically in Figure 5. The metal atom is at the center of the tetrahedron formed by the oxygen atoms. Such coordination implies that the metal ions are bound to the fatty acid molecules in the adjacent layers and their attachment is very likely to take place when the substrate passes through the meniscus during its upward motion.

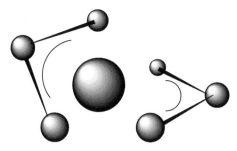

Fig. 5. Tetrahedral coordination of oxigen atoms by a bivalent metal ion in LB films of fatty acid salts.

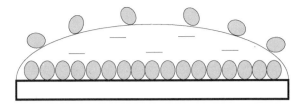

Fig. 7. Situation on the solid substrate surface after touching the protein monolayer.

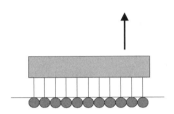

Fig. 6. Scheme of the Langmuir–Schaefer (horizontal lift) deposition.

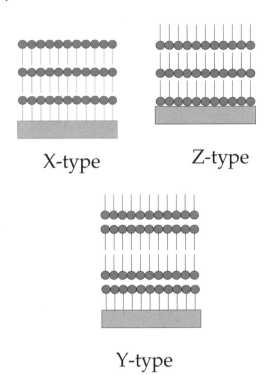

Fig. 8. Types of Langmuir–Blodgett films.

The other method of monolayer transfer from the air–water interface onto solid substrates is illustrated in Figure 6. It is called the Langmuir–Schaefer (LS) or horizontal lift technique. It was developed in 1938 by Langmuir and Schaefer for the deposition of protein layers. A specially prepared substrate touches the monolayer horizontally, and the layer transfers itself onto its surface. The method is often used for the deposition of rigid and protein monolayers. In both cases the application of LB method is not desirable, as it results in defective films.

In the application of the LS method to rigid monolayers special care must be taken. The monolayer at the air–water interface must be divided into parts after reaching the desired surface pressure [98]. This must be done with a special grid with windows corresponding to the solid support sizes. The main reason for the grid is the following. If the monolayer is rigid, the removal of some part of it will result in the formation of empty regions in the monolayer. As the layer is rigid, these empty zones will be maintained for a very long time. Repeating the deposition will result in the formation of many defects in the monolayer, and the resulting transferred layer will be very inhomogeneous. The use of the grid also assures that only one monolayer is transferred during one touch.

In the case of proteins the monolayer is soft. Therefore, the problems mentioned do not exist in this case and the use of the grid can be avoided [99]. In fact, the monolayer structure in the case of protein layers is practically amorphous, as it is easy to reveal by Brewster angle microscopy (to be considered later). Therefore, the removal of some monolayer regions can be rapidly compensated by the feedback system without the loss of the monolayer homogeneity. However, there is another problem when applying the LS technique for protein monolayer transfer. The situation on the solid support after the touching of the monolayer is schematically shown in Figure 7. A regular close-packed monolayer is on the surface of the solid support. Some amount of water, transferred together with the monolayer,

forms a drop on the substrate. Some protein molecules form an irregular layer at the top of this drop. If the sample is dried in a usual way, these molecules will form an inhomogeneous layer in an uncontrollable way. Therefore, these additional molecules must be removed before the sample drying. The most effective way is to use a strong jet of inert gas, such as nitrogen or argon. It removes the water drop together with the randomly distributed protein molecules on its top, leaving only a regular layer facing to the substrate surface [100].

Deposited films are usually divided into three types, schematically shown in the Figure 8, namely, X, Y, and Z types [101]. As is clear from the figure, the Y type is a centrosymmetric one, while X and Z types are polar ones and they differ one from the other only by the orientation of the head groups and hydrocarbon chains with respect to the substrate surface. This difference is due to the fact that in some cases there is no monolayer transfer during upward or downward motion of the substrate in the case of LB deposition. In the case of the LS deposition, moreover, the layers seem to be always transferred in a polar manner. However, the X and Z types are practically never realized in practice. Even when some nonlin-

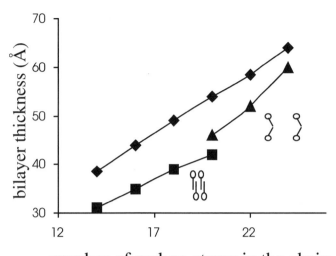

number of carbon atoms in the chain

Fig. 9. Dependence of the spacing of the LB films of fatty acid complexes upon the length of hydrocarbon chains (expressed as the number of carbon atoms in the chain).

ear properties, such as pyroelectricity, realizable only in polar structures, were observed, and the structures were considered as polar ones [102–106], detailed investigations revealed that the films are of Y type but with unequal densities of odd and even layers [107]. Moreover, in the case of LS deposition of fatty acid films, it was shown that the last three transferred monolayers are involved in the structural reorganization during the passage of the meniscus, so as to realize the thermodynamically stable Y-type packing [108]. This reorganization provides the orientation of the hydrocarbon chains toward the air in the last-deposited monolayer.

Normal packing of amphiphiles in LB films is characterized by head-to-head and tail-to-tail packing. The tilt angle of the hydrophobic chains with respect to the film plane can vary, but it must provide close packing of the chains. However, several examples of other packings with interdigitation of the hydrocarbon chains of adjacent layers were observed.

Initially, such packing was observed on two objects, namely, substituted anthracene [109] and amphiphilic TCNQ complex [110]. It is interesting to note that in both cases the head-group complex was rather large, and this fact was the first indication that the ratio of the head-group size to the hydrocarbon chain length is critical for the realization of such packing. However, this hypothesis was confirmed only after systematic study increasing the length of the aliphatic chain of a molecule with the same large hydrophilic head group.

The head group used in this case was a fluorine metal complex, attached to the fatty acid molecule from the water subphase during the monolayer formation [111]. The complex layers were transferred onto solid substrates, and their structure was studied by small-angle X-ray scattering. The results of the study are presented in Figure 9. It is interesting to note that the spacings of all the films were less than those of LB films of the salts of the same acids with bivalent metals, even if the size of the head group was larger. It is clear from Figure 9 that there are

two linear branches in the dependence, with different slopes. In the case of arachidic acid, the X-ray pattern contained two systems of reflections, corresponding to different spacing values. Calculations of the electron density profile determined that for the short-chain fatty acids packing with interdigitation of hydrocarbon chains takes place, while for the longer chains, they are just packed tilted.

The ratio of the head size to the hydrocarbon chain length was taken as the parameter indicating the realization of tilted or interdigitated packing. The value of the parameter for the arachidic acid layer was taken as the transition value (for coexistence of both phases).

Application of the mentioned criterion for interdigitated packing confirmed its validity.

Great interest in the LB method appeared after the works published by Kuhn's group [10–12, 14, 16, 18, 19, 112–121]. They had shown the possibility of realizing molecular architectures with this method. Complicated structures with alternation of layers of different molecules were realized. Such systems were used for studying energy transfer processes. Layers of donors and acceptors were separated by spacer layers of different thickness.

3. TECHNIQUES FOR STUDYING MONOLAYERS AND LB FILMS

3.1. Monolayers at the Air–Water Interface

The main technique for studying monolayers at an air–water interface is π–A isotherm measurement. It reveals the monolayer state and its phase transitions. The method was briefly described in the previous section, together with surface potential measurements. Therefore, let us here consider other methods useful for studying the monolayer at the air–water interface.

3.1.1. Fluorescence Microscopy

This technique is based on the fluorescent light from a monolayer containing fluorescent molecules. It reveals the morphology of the layer. In most cases, small dye molecules are added to the monolayer. They cannot penetrate the regions with close packing of the monolayer molecules, and therefore reveal the variation of the domain structure during compression [122–136]. However, the technique cannot be considered absolutely clean, as the monolayer under investigation contains additional dye molecules, the presence of which can vary the domain structure of the layer.

Microscopy can also be used for revealing the attachment of protein or other fluorescently labeled molecules to surfactant monolayers from the water subphase [137–140]. The technique has been used for imaging the variation in domain structure, in particular, the growth of 2D streptavidin and enzyme crystals. It has been applied for studying phase transitions and chiral discrimination in monolayers. It has also been widely used for studying biorecognition processes between components with high affinity in monolayers.

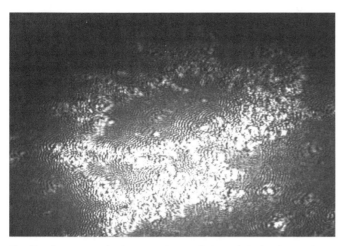

Fig. 10. Brewster angle microscopy image of a bacteriorhodopsin monolayer.

3.1.2. Brewster Angle Microscopy

This method is based on the fact that p-polarized light does not reflect from an interface when it is incident at the Brewster angle φ determined by the equation

$$\tan \varphi = \frac{n_1}{n_2}$$

where n_1 and n_2 are the refractive indices of the two media at the interface. The value of this angle for an air–water interface is 53.1°. Therefore, it is possible to adjust the analyzer's position in such a way that it will have a dark field when imaging the interface. Spreading of the monolayer varies the Brewster conditions for both air–monolayer and monolayer–water interfaces, making visible the morphology of the monolayer [141, 142]. A typical Brewster angle image of the monolayer at the air–water interface is shown in Figure 10.

To some extent, the applications of fluorescence and Brewster angle microscopies overlap. Where they do, the application of Brewster angle microscopy is preferable as it does not demand adding anything to the monolayer, with the risk of disturbing it.

Brewster angle microscopy was successfully applied for studying the morphology of monolayers [143–151], their phase transitions [131, 152, 153], their thickness and optical properties [154, 155], their 2D crystallization [156, 157] and 2D–3D transformations [158], and the miscibility of multicomponent monolayers [159–162].

3.1.3. X-Ray Reflectivity

X-ray methods are very powerful for studying the structure of LB films. Mainly, they are used for investigation of films on solid supports and will be discussed later. However, there are some applications of the method for studying monolayers at the air–water interface [163–190].

Such experiments are rather complicated to perform. First, the Langmuir trough must be placed on a support that prevents vibrations and surface wave formation. Second, the X-ray

source must be powerful, as few molecules (one monolayer only) are involved in the scattering. For this reason synchrotron sources must be used. Third, the source of the incident beam must be attached to the tool so as to provide for scanning of the incident angle (when working with films on solid supports, this is achieved by rotation of the sample with respect to the beam).

Modeling of the electron density profile in the direction perpendicular to the water surface is used for revealing the monolayer structure. The monolayer is divided into regions of fixed electron density. The experimental data are fitted by varying the length and relative positions of these regions. Usually the following regions are considered: bulk water; water just under the monolayer, which can be different from the bulk water due to the presence of ions or molecules attached to the monolayer by electrostatic or Van der Walls interactions; head-group region; hydrocarbon chain packing region; interface of the monolayer with air.

Such measurements reveal differences in the monolayer packing upon compression as well as interactions of the monolayer with ions and active molecules distributed in the water subphase.

3.2. LB Films on Solid Supports

3.2.1. Electron Diffraction and Microscopy

Electron diffraction is the most powerful technique for studying the in-plane structure of the layers. The technique was applied for the structural investigations of LB films at the end of 1930s [191, 192]. It is possible to apply the technique in both transmission and reflection modes. In the case of transmission the sample preparation is rather critical. The support must not significantly disturb the resulting diffraction pattern. In the most cases the supporting substrate is a thin organic layer [193]. However, some other methods have been used for the preparation of the supporting substrate. For example, a thin aluminum layer was anodically oxidized and used as a substrate [194, 195]. The layer was then chemically etched with weak HF solution. Cellulose acetate is another possibility for the supporting substrate [196]. The diffraction of this material is very diffuse and weak, and practically does not disturb the diffraction pattern of the layer under investigation.

When the electron spot is large, the resultant diffraction pattern contains several rings. This is because several domains with different orientations are involved in the diffraction and the situation is rather similar to that of diffraction from a polycrystal powder. Such a diffraction pattern allows only a determination of the elementary cell size, giving no information about the symmetry. Good, reliable conclusions about the symmetry can be reached when the spot diameter is less than the domain sizes.

In the case of reflected diffraction the supporting substrates are not so critical. Therefore, a larger variety of substrates have been used.

Electron diffraction allows determining the type of symmetry and repetition units of the film elementary cell. LB films of fatty acid salts have been intensively studied [197–211]. Initially, their structure was attributed to hexagonal packing [193]. However, precise study using a decreased spot diameter revealed that this apparent hexagonal packing has nothing to do with the real packing, but is due only to the different orientations of different film domains. Within the same domain the packing can be assigned to orthorhombic, monoclinic, or triclinic symmetry [212].

For electron microscopy the layers can be decorated with heavy atoms. This allows determination of the true structure (including defects) of LB films [213–216].

Both electron diffraction and electron microscopy have been applied even to single monomolecular layers [217–225].

The techniques have been widely applied for the investigation of the structure of lipid [226–231] and lipid–protein LB films [232–241].

3.2.2. Scanning Tunneling Microscopy

STM is based on the phenomenon of electron tunneling. A sharp metal tip is connected to a piezo mover, which provides scanning of the sample surface. A feedback system maintains constant the tunneling current between the tip and the sample. As the tunneling current depends exponentially upon the tunneling distance, Z-motion of the tip reproduces the relief of the sample surface. STM allows one to obtain even atomic resolution when the sample is highly conductive and atomically flat. Freshly cut pirolytic graphite is an example of such samples.

However, the technique has restricted application. The substrate must be conductive, and if the layer is insulating, its thickness must be not more than several angstroms. Otherwise, the tunneling current will be too small for effective measurements. Nevertheless, in some cases STM has been successfully applied for studying LB films of rather large molecules. In particular, good-quality images were obtained on LB films containing charge transfer salts, due to their rather high electrical conductivity [240–244]. Other successful imaging was performed on LB films of organic semiconductors [245–248] and inorganic clusters [249–253]. There are also publications on the successful STM imaging of LB films of large insulating molecules [97, 254–272]. One possible explanation of this success is the penetration of the tip into the layer during scanning. Image formation, in this case, can result from periodic disturbance of the tip by the film molecules. Finally, STM was also applied for protein LB film imaging [273–277]. Mechanisms of the image formation process in this case are still under discussion. However, it is clear that internal water can form passage ways for electrons in the protein films.

Wide use of STM among research groups is due to the fact that the instrument allows not only imaging of surfaces and thin layers, but also manipulations of them [278–281]. In fact, it has been used for lithography at nanometer resolution. Moreover, the tip can be used as an ultrafine electrode for study and realization of point contacts for junctions involving single molecules. Such utilization of STM has made possible room-temperature single-electron junctions, single-molecule switches, molecular memory, etc.

3.2.3. Atomic Force Microscopy

AFM is rather similar to STM. The difference is that in AFM the atomic forces are the parameters under control. A feedback system maintains constant the attractive or repulsive Van der Waals force according to the relative position of the tip and surface. The main advantage of AFM is the ability to study any kind of the surface regardless of its conductivity. However, it provides less resolution than STM. In fact, it is very hard to reach atomic resolution with AFM, due to the curvature of the tip. This is not the case for STM, due to the exponential dependence of the tunneling current upon tunneling distance.

The AFM has been widely applied for the investigation of different kinds of LB films [200, 221, 261, 266, 271, 281–289, 289–398].

The *taping mode* gives even more ability to the AFM for studying biological samples, as it is less affected by noise. In this case the tip is in vibration during the surface scanning [399–402].

Like the STM, the AFM can be used as nanometer-scale instrument for realizing patterns in thin layers [403, 404].

3.2.4. X-Ray Scattering

Several methods based on the utilization of X-rays are used for studying the structure of the LB films. These methods were used even in the initial stages of LB film investigation [405–412]. Diffractometry and reflectometry are the ones most often used.

Diffractometry is mainly used when a well-ordered periodic structure is under the investigation and there are several Bragg reflections in the X-ray pattern [190, 209, 212, 282, 287, 413–469]. The position of these reflections is determined by the Bragg equation

$$2D \sin \Theta = n\lambda$$

where D is the thickness of the periodic unit (period or spacing), λ is the wavelength of the incident X-ray beam, Θ is an incident angle, and n is the order of the reflection. Therefore, the thickness of the periodic unit (usually a bilayer) can be that obtained directly from the angular position of Bragg reflections.

The other information that can be directly obtained from the X-ray pattern is the correlation length L. This parameter can be considered as the thickness up to which the film can be still considered as an ordered one, and it is determined by the following formula:

$$L = \frac{\lambda}{2\Delta\Theta}$$

where $\Delta\Theta$ is the half-width of the Bragg reflections.

More information can be obtained if several reflections are registered. Each Bragg reflection can be considered as a term in the Fourier series representing the electron density on the repeating unit of the film. Therefore, the electron density on a period of the film can be expressed as

$$\rho(l) \sim \sum_n A_n \cos \varphi_n$$

where A_n is proportional to the intensity of the Bragg reflection and φ_n is the phase of the harmonic. Fortunately, the most LB films have Y-type structure, and therefore the phases for them can be only 0 or π, so that the typical crystallographic phase problem becomes a simpler sign problem. This problem is usually solved by fitting the electron density in some known regions—most often a region where close packing of hydrocarbon chains takes place.

The resolution in the determination of the electron density profile in the case of diffractometry (R) is connected with the number of registered reflections and is determined by the formula

$$R = \frac{D}{2n}$$

where D is the period and n is the number of registered reflection. Therefore, the resolution is half the wavelength of the harmonic determined by the last registered reflection.

Reflectometry is the other approach for studying LB films with X-ray techniques [181, 326, 470–500]. It allows one to study not very well ordered films and even monolayers. In reflectometry the whole registered scattering curve is considered. It can contain Bragg reflections and also Kiessig fringes. The latter correspond to the interference of the X-ray beams reflected from the air–film and film–substrate interfaces, and their angular position gives information about the total thickness of the LB film. The electron density profile in the case of reflectometry is calculated from models, fitting the whole experimental curve.

3.2.5. Infrared Spectroscopy

Infrared spectroscopy is among the most powerful tools for studying the structure of LB films, both at the air–water interface and after deposition onto solid substrates. It allows one to determine the orientation of molecular chains [501–511], the structure and composition of layers [512–536], and the attachment and secondary structure of proteins [537–541]. It also allows one to visualize reorientation of the layers due to temperature [358–360, 542–547], substrate influence [548–551], and electron beam action [552]. The technique has also been used for studying water transport through layers, in particular, hydration of DNA containing LB films [455, 553].

The main difficulty with IR experimental analysis of the LB films is connected with the small amount of the material in the sample. However, the good signal-to-noise ratio can be obtained if attenuated total reflection is used for the measurements of LB films [224, 554, 555].

3.2.6. Gravimetry

A very useful technique is based on the fact that a piezoelectric oscillator changes its resonant frequency when additional

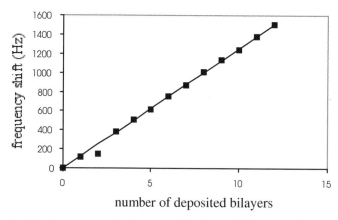

Fig. 11. Dependence of the frequency shift of a quartz balance upon the number of bilayers of cadmium arachidate deposited on its surface.

mass is deposited onto its surface. Usually, quartz oscillators are used. The phenomenon is described by the Sauerbrey equation:

$$\frac{\Delta f}{f_0} = -\frac{\Delta m}{A \rho l} l$$

where f_0 is the initial resonant frequency, Δf is the frequency shift, ρ is the density of the quartz, A is the electrode area covered by the film, and l is the oscillator thickness [566].

Calibration of the quartz balance can be performed by successive deposition of layers of fatty acid salts [567]. The typical dependence of the frequency shift upon the deposited layers is a straight line as presented in Figure 11.

In many practical cases it is better to calibrate the balance in terms of surface density instead of mass changes. In this case it is also necessary to cover the quartz surface with layers of a material with known surface density (again, fatty acid salts, for example). Such calibration is useful for studying transferred layers of substances with unknown surface density. In the case of monolayer or bilayer transfer, it will be possible to determine the surface density of the film and, knowing the molecular weight of the compound, to determine the area per molecule in the layer [568]. This possibility is important for protein layers, as the area per molecule for them can almost never be found from $\pi-A$ isotherms, for reasons that will be discussed later.

The technique was also applied for the investigation of the hydration of LB films [569–573], for the investigation of the attachment of ions [574, 575] and of other molecules [576–579] to layers, for studying chemical reactions in LB films [580, 581], and for studying viscoelastic properties [582] and phase transitions [583, 584]. The technique also finds wide application in the field of gas sensors [585–592] and biosensors [593–598].

3.2.7. Ellipsomerty

This technique is based on the fact that linearly polarized light becomes elliptically polarized after reflection from the interface. The measured parameters are ψ (the amplitudes of oscillation of the electric field vector in the plane of incidence and perpendicular to it) and Δ (the difference of phase between them). They allow one to calculate with high accuracy the thickness and refractive index of the layer.

The first time the technique was applied for LB film study was in 1934 for the investigation of properties of fatty acid layers on a mercury surface [599]. Since then, the technique has been successfully applied for studying films with increasing numbers of layers. In some cases, a three-layer model, including an intermediate layer (usually oxide) between the substrate (usually metal) and the LB film, is considered [600, 601].

Initial works considered the LB film as an isotropic system [602–617]. However, later the theory was extended to anisotropic LB films [618–620]. In this case, the difference in the refractive index in different directions with respect to the sample plane was measured. Such measurements made it possible to show that the dipping direction in the case of vertical lift deposition orients the layers, resulting in slight anisotropy of the LB films.

Ellipsometric studies were also performed at the air–water interface, allowing study of the variation of the monolayer thickness during compression, and thereby revealing phase transitions of the monolayer [621–625].

Ellipsometry at the air–water interface is also very useful for studying the interactions of a monolayer with different compounds dissolved in the water subphase [626]. As an example, we can mention the successful application of the technique for studying lipid–protein interactions [138, 627].

Recently, microscopic ellipsometry was established, allowing one to determine the distribution of thickness and refractive index in 2D images [628, 629].

3.2.8. Interference Methods

The interference method of film thickness determination has been applied since the beginning of LB film investigations. For the normal incidence, the film thickness h can be determined from the formula

$$h = m\lambda/4n$$

where λ is the wavelength, n is the film refractive index, and m is an integer. This simple technique is still used effectively for thickness determination [453, 630–635].

It is interesting to note that one of the first practical applications of LB films was as thickness standards, using the high precision in the thickness of the resultant layers [8]. Another application of the technique is in the field of sensors [636]. In fact, any change that varies the thickness or refractive index of the layer can be revealed by interferometric methods.

Finally, interferometry gave rise to very interesting and powerful method of thin-layer investigations, namely, scanning optical microscopy, where the thickness and refractive index patterning of the layer can be determined with rather high resolution [637].

3.2.9. Neutron Scattering

Neutron scattering is a complicated technique, demanding expensive, complicated equipment, which allows one to determine structural parameters of layers. There is one fundamental difference between neutron scattering and that of X-rays or electrons. In the case of neutrons, it is possible to observe so-called anomalous scattering, where the scattered beam changes its phase. Fortunately from the application point of view, one of the most widely distributed elements in nature, hydrogen, provides such anomalous scattering. Therefore, it is possible to vary the contrast of the measurements by using a mixture of ordinary and deuterated water (ordinaral water yields anomalous scattering; heavy water does not).

Neutron scattering was used for studying the LB film structure at an air–water interface [166, 170, 638–667] and after deposition onto solid substrates [473, 481, 484, 485, 668–672].

At the air–water interface, the technique was used for the visualization of phase transitions, in particular, by registering the difference in the head-group hydration in different phases [27, 646, 673, 674], the attachment of protein molecules to the monolayer [667, 675–678], and the behavior of multicomponent mixtures [679, 670–687].

In the solid phase, the technique allows performing experiments on the lateral and interlayer motion of molecules in LB films, which cannot be determined with any other technique [490, 497, 688]. Let us consider one example illustrating this statement [689, 690].

A multilayer barium stearate film was deposited. Each odd layer was deposited with ordinary stearic acid, while each even layer was deposited with deuterated stearic acid. The X-ray pattern revealed periodicity of about 5 nm, as is typical for stearic acid salt LB films. The neutron diffraction pattern, however, had a system of reflections with double that period. In fact, in the case of X-ray scattering measurements, there is no difference between normal and deuterated stearic acid, as their electron density is practically the same, while for the neutron scattering these layers are different and therefore the period was doubled. Quantitative analysis of the scattering curves showed that the layers are not uniform. Each deuterated layer contains about 70% of the deuterated and about 30% of the normal stearic acid. In the normal acid layer the situation was reversed. Therefore, this experiment showed that there is molecular exchange between adjacent layers in LB films. Very likely, this exchange takes place during deposition of the layers, when the solid substrate passes through the meniscus. In fact, there are other, indirect indications that monolayer molecules can be involved in flip-flop motion while passing a meniscus. This fact indicates once more that the resulting structure of the LB film can be different from the desired one, and that special structural investigations must be carried out to verify the real structure of realized films.

Similar investigations were performed for analyzing the temperature and temporal effects on the LB film structure [417, 493, 691].

4. PROTEIN FILMS

4.1. Protein Monolayers at the Air–Water Interface

Protein molecules are complicated systems, composed of polypeptide chains organized into globules. Usually, interior of the globule is hydrophobic, while its surface is mainly hydrophilic. This fact about proteins leads to their instability at an air–water interface. In fact, the surface tension of the water (72 mN/m) is enough to exert significant forces on protein molecules, resulting in most cases in denaturation. The time necessary for the denaturation depends upon the protein. In some cases it can be minutes, and in others, hours. Immunoglobulin, in particular IgG, is an example of the more stable proteins. The structure of this molecule is stabilized by covalent disulfide bridges. The presence of such strong bonds provides improved resistance to the action of the surface tension at the air–water interface. However, long exposition of IgG to surface tension leads to significant changes even in such a stable protein.

There are some examples where protein solutions can be spread directly on the air–water interface. That is mainly true of membrane proteins, and it happens at interfaces even in nature. We will consider two examples of monolayers of such proteins. Other proteins demand the utilization of some variation of the LB technique for realization of films without denaturation.

4.1.1. Monolayers of Bacteriorhodopsin

Bacteriorhodopsin (BR) is a light protein (12.5 kDa) providing light-induced transmembrane proton pumping. It is a main part (80%) of purple membranes. BR is very unstable in isolated form, while it is very stable in membranes. Therefore, a spreading solution is composed of membrane fragments. This makes it difficult to maintain a significant part of the protein at the air–water interface. In fact, the membrane fragments are about a micron in sizes. The top and bottom parts of these objects are hydrophilic, while only small lateral parts, about 5 nm high, are hydrophobic. Therefore, after the spreading on a pure water surface, most of the fragments penetrate the water volume and do not form a monolayer. In order to avoid this difficulty, the solubility of the membrane fragments must be decreased. This can be achieved by using concentrated salt solutions as a subphase. Usually, 1.5–2.0 M NaCl or KCl is enough to prevent leakage of the membranes from the surface to the volume of the liquid subphase [454]. A comparison of π–A isotherms of membrane fragments at the surface of pure water and on 1.5 M KCl clearly indicates that the presence of salt in the subphase allows one to maintain a significant amount of the sample in the monolayer.

Usually, BR films are deposited onto the solid substrate by the horizontal lift technique. Some amount of water subphase is also transferred together with the monolayer. As the subphase contains concentrated salt solution, a significant amount of salt can remain at the substrate in the case of BR films. However, the salt can be removed from the sample. There are two ways to do so. The first one requires the use of special salts, which can be decomposed into gas compounds. Ammonium acetate is an

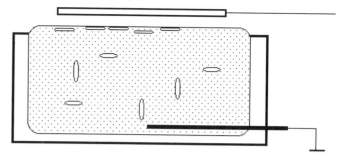

Fig. 12. Scheme for electric-field-assisted formation of a BR monolayer.

example. After the deposition, the BR LB film can be exposed to low vacuum, and the salt will be decomposed and removed from the sample. The second approach does not demand the use of special salts. The sample is simply washed with water after depositing each successive layer [454].

Many applications of BR films demand strong of the membrane fragments, with proton transfer vectors oriented in the same direction. However, that is not simple to realize in practice. The top and bottom parts of the fragment have nearly the same hydrophilic properties. Therefore, fragments are oriented in opposite directions, resulting in zero photopotential and photocurrent, as the proton displacements compensate each other in the adjacent fragments. This difficulty can be overcome by taking into consideration the charge distribution in the purple membranes: their interior is charged more positively than their exterior.

The following modification of the LB technique was realized for obtaining highly oriented BR LB films [692]. The scheme of the modified procedure is shown in Figure 12. One electrode is immersed in the water subphase, containing the concentrated salt solution. The other electrode, with area about 70% or more of that of the opening in the barrier, is placed about 2 mm above the subphase surface. A voltage of 30–50 V is applied between the electrodes (the upper one is biased positively). Purple membrane solution is injected into the subphase volume. Membrane fragments begin self-assembling at the air–water interface, orienting themselves in the electric field in such a way that all proton pathways are oriented in the same direction. The process is analogous to electrochemical sedimentation in solution [693, 694]. The monolayer formation can be monitored by registering the increase of the surface pressure in time. Comparison of these dependences in the presence and absence of the electric field makes it clear that self-assembling of fragments into the monolayer takes place only if the electric field is applied. Moreover, if the polarity of the applied field is opposite (upper electrode is biased negatively), the dependence of the surface pressure upon the time is similar to that in the absence of the electric field.

When the increase of the surface pressure with time becomes saturated (at surface pressure about 12 mN/m), the electric field can be switched off, the upper electrode removed, and

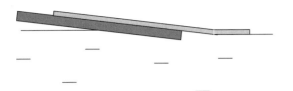

Fig. 13. Scheme for the spreading of RC solution on the air–water interface using a hydrophilic plate.

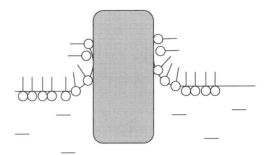

Fig. 14. Rearrangment of the RC–detergent complex at the air–water interface.

the monolayer further compressed (25 mN/m) for higher density and transferred onto solid substrates.

The proposed technique allows one to form much better-oriented mono- and multilayer BR films than the usual LB deposition. It also has significant advantages over the usual electrosedimentation technique, as it allows realizing structures at molecular resolution, while electrosedimented films are of about micron thickness.

Investigations of BR monolayers at the air–water interface have allowed determining the role of the subphase composition [695–697], electrical and optical properties [698, 699], and the structure of the monolayer [539, 700–702].

4.1.2. Films of Photosynthetic Reaction Centers

A photosynthetic reaction center (RC) is a large protein (about 100 kDa), containing three subunits and performing light-induced transmembrane electron transport. The electron transport chain of the protein contains a dimer of bacteriochlorophyll (primary donor), bacteriopheophetin, and two quinons. In nature the protein works together with multiheme cytochrome, which provides necessary electrons to the dimer of bacteriochlorophyll.

Spreading solutions contain separated RC molecules. Each molecule is surrounded by a detergent, attached to a hydrophobic area that is embedded naturally in the membrane. The detergent appears during the extraction of RCs from the membrane. Therefore, the resulting complex is hydrophilic, and special care must be taken in placing such molecules on the air–water interface. There are two common methods to do this. The first method is based on the use of a hydrophilic glass plate inserted into the water subphase (Fig. 13). RC solution is dropped onto the plate and transfers itself to the subphase surface [698]. The second method is based on dropping small drops of RC solution onto the air–water interface [100, 703, 704]. The spreading is good (small leakage into the subphase volume) if the drops remain for some time at the subphase surface, diminishing in size.

Being placed on the air–water interface, the detergent–RC complex rearranges itself. Interaction of the detergent head groups with water results in their detachment from the RC molecules and transfer onto the water surface (Fig. 14). As a result, detergent-free RC molecules form a monolayer, and the space between them is filled by detergent molecules. Usually, detergent molecules have rather short hydrocarbon chains and cannot form a stable monolayer at the air–water interface themselves. Therefore, dense RC packing in the monolayer with a reduced amount of detergent can be achieved by compressing the monolayer to high surface pressure. In this case, the detergent molecules can be pushed into the subphase volume in the form of micelles. They can also form local collapses. Therefore, at high surface pressure the monolayer is mainly composed of RC molecules.

The compression of the RC monolayer must be rather fast in order to minimize the denaturing action of high surface tension on the RC molecules.

In this regard, the chief weakness of the molecule is that it is composed of subunits. Therefore, it is easy to suppose that the surface tension will result first of all in the destruction of the quaternary structure of the protein. This suggestion is easy to check, as the two bacteriochlorophyll molecules in the dimer are attached to different subunits [705]. Absorption spectra of RC LB films deposited at different surface tensions have confirmed the splitting of the protein subunits. At surface pressures below 25 mN/m it is possible to see the absorbance of the monomer of bacteriochlorophyll (802 nm) and no absorbance of the dimer (860 nm). Increase of the deposition surface pressure results in the appearance of the dimer absorption peak, finally amounting to half the monomer peak intensity. Such behavior of the absorbance indicates that high surface tension at the initial stages of the monolayer compression splits the RC molecule into subunits. When the surface tension is reduced by monolayer compression, the spliting does not take place.

As in the BR case, there is also great difficulty in orienting all RC molecules in the same direction in the monolayer, as the hydrophilic areas at the top and bottom of the molecule have similar properties. However, the approach used in the case of BR cannot be applied to the RC monolayers. In fact, the membrane fragments used in the case of BR monolayers cannot overturn after the switching off of the electric field. Their sizes are large, and overturning would demand significant energy. Instead, RC molecules, being separated one from the other, will reorient themselves, resulting in mutually compensating orientation of the electron pathways in the adjacent molecules.

The following method was suggested to overcome this difficulty (Fig. 15). The monolayers were transferred onto the substrate, which was biased with respect to the liquid subphase [706–708]. Of course, the solid substrate must be conductive in this case. However, when the photoelectric proper-

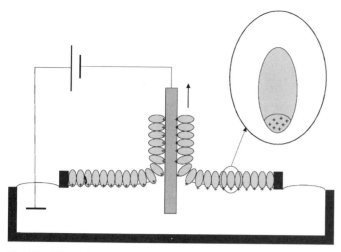

Fig. 15. Scheme for the deposition of RC monolayers onto electrically biased substrates.

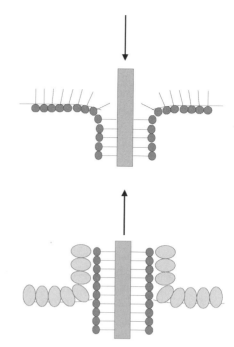

Fig. 16. Scheme for the deposition of an RC–arachidic acid heterostructure.

ties of the RC monolayers are to be used, the substrates must be conductive ones anyway. Therefore, that is not a big disadvantage of the method. A more important problem is the prevention of electrical current flow between the substrate and water subphase, because such a current can have undesirable effects on the film. Therefore, it is best to deposit an insulating layer between the substrate surface and RC layers. The presence of such a layer will not create additional difficulties if optoelectronic measurements are used to probe the electron displacement in the RC molecules, but it will make practically impossible any measurements for which electron exchange must take place between the RC molecules and the substrate electrode.

Another method, allowing better orientation of RC molecules in mono- and multilayers, was demonstrated on RC from *Chromatium minutissimum* [709, 710]. The cytochrome molecule is strongly attached to the RC molecule in these bacteria and is not lost during extraction from membranes. Therefore, the RC–cytochrome complex can be considered as a four-subunit aggregate in this case. The cytochrome subunit is charged positively. It is interesting to note that this charge on one subunit made it impossible to transfer the RC monolayer from the water surface to the solid substrate by the usual vertical or horizontal deposition techniques. The electrostatic interactions with the water were much stronger than that with the solid substrate surface. However, the presence of such a charged subunit is very useful for the realization of monolayer anisotropy. The cytochrome subunit is oriented into the water subphase. In order to keep this anisotropy in the deposited film, the RC monolayers must be transferred onto negatively charged substrates. Such deposition was realized by depositing heterostructures, with alternation of RC monolayers with arachidic acid monolayers (Fig. 16). A hydrophobized solid substrate was passed downward through an arachidic acid monolayer, formed at the subphase with pH 7.0. Head groups of the arachidic acid are mainly dissociated at this pH value and are charged negatively. During the subsequent upward motion, the substrate was passed through the RC monolayer. Electrostatic interaction of

the arachidic acid head groups with cytochrome subunit of RC took place at this stage. The process was repeated several times.

An X-ray diffraction study performed on such hererostructures revealed the presence of seven Bragg reflections, indicating rather high ordering of the resultant film. Normally, monocomponent protein films do not allow registering more than one Bragg reflection [99].

The suggested approach seems to be rather general for obtaining highly ordered protein layers. Electrostatic interactions provide anisotropy of the film, while the presence of fatty acid interlayers results in improved layer packing.

4.1.3. Complex Lipid–Protein Monolayers

The most natural way to eliminate the denaturing action of surface tension on protein molecules is to diminish it by spreading a lipid monolayer at the air–water interface and to attach proteins to it [237, 711–716]. A widely used example of such complex monolayers is based on biotin–streptavidin interactions, as the affinity of the resulting complex is extremely high [678, 717–727]. The streptavidin molecule has four binding sites where biotin can be attached.

Several synthetic lipids with biotin molecules in the head group were synthesized, and their monolayer behavior at the air–water interface was studied. Streptavidin molecules were injected into the water subphase. When strike conditions of the biotinated lipid were chosen in the correct way, growth of 2D streptavidin crystals was observed by different methods.

The proposed approach is very important for the realization of layer manipulation at the molecular level, called also molecular architecture. In fact, 2D crystals of streptavidin are attached with two biotin binding sites to the lipid monolayer, and two other binding sites are still available. The resulting 2D crystal of streptavidin can be used as a template for the next layer formation. This layer can be composed of any functional molecules, conjugated with biotin. Regular distribution of the binding in the template layer will make it possible to distribute desirable molecules in a quasi-crystallic manner, while attachment of the biotin at different positions will guarantee their preferential orientation. Therefore, combination of the LB technique with self-assembling, based on complexes with high affinity, allow one to form regular molecular systems with high ordering not only between monolayers, but also in each monolayer plane.

The suggested procedure can be repeated to realize the organization of complicated structures of different layers. In this case, functional molecules must be conjugated with two biotin molecules. One of them will be attached to the template layer, while the other will provide the binding for the formation of the next streptavidin layer. The attached streptavidin will form a template layer for the formation of the next active layer, formed from different molecules.

Another method to attach protein molecules to the lipid monolayer is based on the utilization of Coulomb interactions. From the realization point of view, the method can be considered as a simplified version of that based on specific affinity. Formation of cytochrome C monolayers can illustrate this method [442, 728–731].

Cytochrome C is a protein with molecular weight 30 kDa. It has charge +8 at neutral pH (7.0). Fatty acid molecules are dissociated at this pH value and contain negative charge. Therefore, a monolayer of stearic acid was formed and cytochrome C molecule solution was injected under it. Polarized absorbance measurements revealed the attachment of protein molecules to the monolayer with preferential orientation of heme groups in them parallel to the monolayer plane.

A special tool for producing lipid–protein interactions at the air–water interface and transfer the resultant complex monolayers onto solid substrate is called the *Fromherz trough* [238, 732, 733].

It is shown schematically in Figure 17. It is a circular trough divided into several sections (four are shown in the scheme, but their number can be varied). Two moving barriers allow one to compress monolayers in each section and also to transfer them from one section to another, maintaining the internal distances. The process of monolayer transfer from one section to another is schematically shown in Figure 18.

Initially, the monolayer of desired lipid molecules is formed in section I by compressing it with the barriers up to the chosen surface pressure. Then, the monolayer can be transferred to section II by synchronous motion of both barriers. Section II can contain protein molecules with affinity (opposite charge) to the head groups of the formed monolayer. Then, the monolayer with attached protein molecules can be transferred to section III, containing reagents for detaching protein molecules adsorbed nonspecifically on the monolayer. Section IV can contain other protein molecules, which can be attached to the formed complex monolayer.

Fig. 17.   Scheme of the Fromhertz trough.

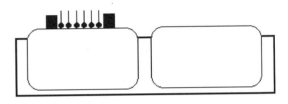

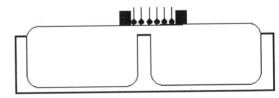

Fig. 18.   Scheme for monolayer transfer from one section to another in a Fromhertz trough.

An important feature of the instrument is its ability to form the monolayer at the surface of one subphase and to produce the lipid–protein interaction, at the other. In fact, different conditions, such as composition, pH, or ionic strength, can be optimized for forming the monolayer and providing the interaction attachment. Moreover, in some cases the interaction can result in different protein attachment (density and orientation) to monolayers compressed to different surface pressure values. Therefore, the ability provided by Fromherz trough to perform protein attachment when the monolayer compression is finished is very important.

Another instrument allowing the effective deposition of complex lipid–protein monolayers is based on a specially constructed sample holder [734]. The sample holder is equipped with a protective plate, located about 1 mm from the sample surface (Fig. 19). The plate is fabricated from hydrophilic materials. It can be in two positions: up (the sample surface is not protected by the plate) and down (the plate is near the sample surface, keeping a water layer between them by capillary forces).

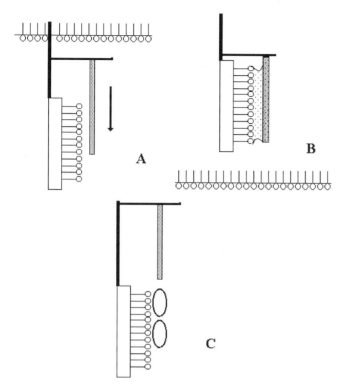

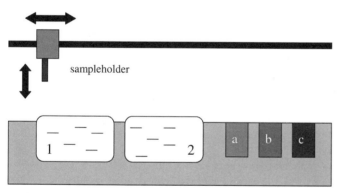

Fig. 19. Scheme of the sample holder with protective plate for deposition of complex lipid–protein layers.

Fig. 20. Scheme of the trough for deposition of complex lipid–protein films.

The trough contains also two separated Langmuir sections and several sections where lipid–protein interactions take place. A scheme of the whole instrument is presented in Figure 20. The main task of the protective plate is to prevent the exposure of the monolayer to air during its transfer from one section to another. That is very important, as monolayers are not fixed systems, and they can reorient themselves for minimum energy in the environment where they are placed. In practice, it has been shown that last three monolayers in multilayer lipid LB films can perform flip-flop reorientation on passing the meniscus of the air–water interface [108].

The instrument works in the following way. Monolayers of two different lipids are formed in sections I and II. The hydrophobized substrate passes downward through the monolayer in section I with the protecting plate in the up position. The

monolayer is transferred to its surface with active head groups exposed to water. At the lowest position of the substrate, the protective plates moved to the down position, and the sample holder begins to move up. The protective plate prevents the deposition of the next monolayer and preserves the orientation of the already deposited monolayer. After coming to the top, the sample holder can move horizontally to section a, containing the solution of protein molecules. After immersion in the section, the plate moves to the up position, allowing the attachment of the protein molecules to the monolayer. When the incubation is finished, the plate moves to the down position, and the sample holder transfers the sample to section b, containing the solution providing the detachment of nonspecifically attached proteins. After the washing, the sample can be transferred to section II. It passes the monolayer downward with the protective plate in the down position. At the bottom, the plate is moved to the up position and the sample passes the monolayer in upward motion. The monolayer in section II is transferred to the substrate. Finally, the protein molecules are encapsulated between two lipid monolayers. Such a structure is very useful for long-time storage of proteins without decrease of their activity.

Lipid–protein interactions at the air–water interface have been studied for 2D protein crystal growth [235, 436, 735–745], and for investigation of the interactions of model membranes with enzymes [746–763], with hydrophobic (membrane) proteins [140, 764–782], and with antibodies [579, 783–788].

4.1.4. Reversed Micelle Monolayers

The main difficulty in spreading proteins at the air–water interface, resulting in the penetration of a significant amount of protein molecules into the subphase volume, is connected with the fact that spreading solutions are aqueous rather than organic ones, and water is miscible with the subphase. This difficulty can be effectively overcome by using reversed micelles as spreading solutions [789–791].

Reversed micelles are complexes of proteins with surfactants, where proteins are incorporated into a vesicle formed by surfactant molecules. Proteins are in contact with polar head groups. The exterior of the complex is formed by hydrocarbon chains and is completely hydrophobic. This last fact makes the complex soluble in organic solvents. Usually, protein molecules enter the complex due to electrostatic interactions.

As an example of this approach, let us consider micelles formed by cytochrome C and aerosol OT (AOT) [790, 791]. All stages of the monolayer formation are shown in Figure 21.

Dropping of the micelle solution onto the air–water interface results in good spreading without penetration of the solution into the water subphase. Interaction with the water surface results in micelle reorganization such that surfactant molecules form a monolayer and cytochrome molecules remain attached to their charged head groups by electrostatic interactions. In fact, hydrocarbon chains of AOT molecules cannot penetrate the water volume, for entropy reasons. Therefore, the micelles

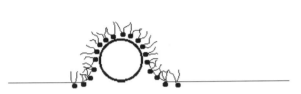

Fig. 21. Scheme for monolayer formation using the reversed micelle approach.

must be *opened* at the air–water interface. Just after the spreading, surfactant molecules cover all the trough surface, and protein molecules do not form a close packing; they are distant one from the other. This fact can be understood by simple geometrical consideration of the size of cytochrome C and the area of the monolayer after opening of the spherical surfactant neighborhood.

The short hydrocarbon chains of AOT do not allow the formation of a stable monolayer of these molecules themselves. Therefore, compression of the monolayer to high surface pressure will push these molecules from the interface into the subphase volume in the form of normal micelles (their interior is hydrophobic, while the head groups are exposed to water). An AOT monolayer with cytochrome C molecules associated with it is more stable. Compression of such a monolayer to high surface pressure will result in a situation where the surfactant molecules form a densely packed layer at the air–water interface and cytochrome C molecules are also densely packed under it.

Application of compression–expansion cycles to this complex monolayer can result in the formation of a densely packed layer. The first compression–expansion cycles result in a significant reduction of the area occupied by the monolayer during recompression to the same surface pressure, indicating pushing of the excess surfactant molecules from the monolayer into the water subphase. Further compression–expansion cycles do not change the monolayer properties, and the π–A isotherms are rather reproducible.

X-ray patterns of LB films obtained from reversed micelles reveal Bragg reflections, indicating good ordering of the films [790].

Similar studies were performed using vesicles instead of micelles [559, 792–799].

4.2. Monolayer Transfer

In the most cases when dealing with pure protein monolayers, the area contacting the substrate surface is hydrophilic [99]. Therefore, the substrate surface must be hydrophilic also. Carefully washed glass or silicon surfaces are good for protein monolayer transfer. However, experience shows that fresh surfaces treated in weak oxygen-containing plasma are most suitable for protein monolayer transfer.

In contrast, complex lipid–protein monolayers and those obtained using reversed micelles can be better transferred onto hydrophobic surfaces.

The horizontal deposition technique is often used for protein monolayer transfer. The technique was used for the first time in 1937 by Langmuir and Shaefer for depositing urease layers, and therefore it is also referred to as the Langmuir–Shaefer (LS) technique [20].

Generally speaking, the LS technique can disturb the monolayer at the air–water interface during deposition. In fact, when a part of the monolayer is taken for transfer onto a solid substrate, a void can be formed in the place where the substrate touched the layer. Multilayer deposition will result in the formation of extremely defective monolayers, where voids are distributed in a random way. Such behavior is typical of rigid monolayers, where 2D quasi-crystals are formed. In the case of protein layers this situation is not frequent. When it takes place, however, special care must be taken. A simple and effective procedure is based on a special grid, whose cells must correspond to the substrate dimensions. The grid is placed over the compressed monolayer, dividing it into separate areas. After that, the feedback of the trough must be switched off, and the monolayer is transferred by touching with the substrate the layer in each grid cell. Such deposition results in the transfer of the monolayer in amounts corresponding to the substrate surface area, and protects the monolayer in all other cells from disturbance.

When the monolayer is not rigid, as is true for the most protein films, the use of the grid can be avoided, as the transfer does not form serious defects, in view of the amorphous monolayer structure.

4.3. Protein Layers on Solid Substrates

Most of the powerful techniques available for structure and property investigation can be applied to layers transferred onto solid substrates. Moreover, most practical applications of protein films also depend on their transfer onto solid substrates. Therefore, some examples of LB films of different important classes of proteins will be considered in this subsection. Their structure, properties, and possible practical applications will be described.

4.3.1. Light-Sensitive Protein Films

Among light-sensitive proteins, bacteriorhodopsin (BR) and photosynthetic reaction centers (RC) have been widely studied by the LB technique.

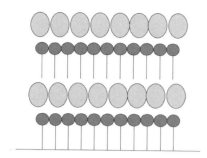

Fig. 22. Scheme of the RC–arachidic acid heterostructure.

The layered structure of BR LB films was determined by X-ray diffraction measurements [454, 800]. Practically always the pattern contains at least one Bragg reflection, corresponding to a spacing of 4.8 nm. This spacing value corresponds well to the thickness of the purple membrane. This fact confirms that the BR structure is determined by the membrane fragment packing, which is the main component of the spreading solutions.

In the case of RC the situation is rather different. Usually, LB films of RC do not display any Bragg reflections. In a few cases one Bragg reflection was registered [100, 661]. However, the ordering even in those cases was rather weak. The half-width of the reflection corresponded to only three monolayers within the ordering length. The ordering can be significantly improved by realizing heterostructures containing alternating layers of RC and fatty acid [709]. The resulting film structure is schematically shown in Figure 22. The X-ray pattern from such a film contained seven orders of Bragg reflections. Highly oriented interlayers of fatty acid maintained the order of the whole film.

The in-plane structure of BR [274] and RC [801] films was investigated by STM. In both cases molecular resolution was obtained during measurements. It is still not very clear what is the mechanism of the STM image formation for such large molecules. However, very likely the electron pathways in these molecules are formed by internal water, associated with the protein molecules.

It is interesting to mention that in the case of RC molecules effective STM imaging was obtained only in dark conditions, in light it was practically impossible to obtain any meaningful image. This behavior was attributed to the fact that an RC molecule transforms itself into an electrical dipole in light conditions, due to charge separation. This dipole interacts with the STM tip during scanning. This interaction results in molecular displacement during scanning, making structural investigations impossible. In order to check this suggestion, RC molecules in the monolayer were fixed with glutaraldehyde. This treatment provides intermolecular cross-linking, creating a rather rigid structure with fixed molecular positions. After the treatment, STM imaging was possible in both light and dark conditions, giving practically the same film structure.

In the case of BR films the mentioned problem does not exist. The reason is that BR molecules are already rigidly fixed in 2D lattices in membrane fragments and cannot be moved by interaction with the STM tip.

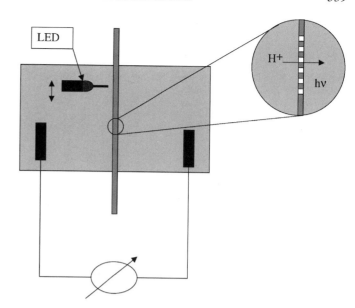

Fig. 23. Scheme of a light-addressable element based on an oriented BR layer.

Comparison of the behavior of RC and BR LB films under the STM allows us to reach rather general conclusions about the utilization of scanning tunnelling microscopy for the investigation of protein LB films. Large dimensions of the molecules do not create great problems, as there are always electron pathways formed by internal water associated with protein molecules. However, charges or dipoles of the protein molecules can interact with the STM tip during scanning, and this interaction can disturb the resultant image, and in some cases even make it impossible to obtain meaningful images. Such difficulty will be encountered when molecules in the layer do not form a rigid crystal-like structure. Therefore, special fixing techniques must be applied in such cases.

The electro-optical behavior of BR and RC LB films is the most interesting property of these objects. It is determined by the charge transfer properties: electron in the case of RC, and proton in the case of BR. Of course, utilization of these properties demands uniaxial orientation of charge pathways of these molecules in the film. Methods of orientation were considered in Sections 4.1.1 and 4.1.2.

The simplest setup for the photoresponse measurements is based on sandwich structures where RC [802–812] or BR [398, 813–828] LB films are placed between two electrodes, one of which is transparent or semitransparent. Measurements of the potential or current were performed during illumination of the layers by light of different wavelengths.

Another interesting application of BR LB films is based on local light-driven proton transfer [829]. The element is based on an ultrathin film (ideally, one monolayer) of oriented BR molecules. The element must have similar properties to a light-addressable potentiometric system (LAPS). A schematic representation of the element is shown in Figure 23. A BR LB film is deposited on a porous membrane, with holes of a size allowing unbroken film over it. The membrane separates two electrolyte solutions. Electrodes are inserted into these solutions. The cur-

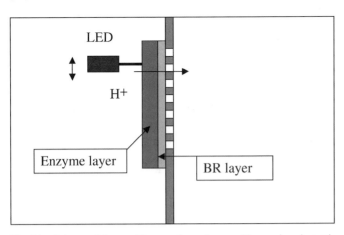

Fig. 24. Scheme of the sensitive membrane for a multisensor based on oriented BR layers.

rent in each section is determined by the proton and ion conductivity. In the ideal case, there is no possibility for the electric current to pass from one section to the other, though in real conditions that situation cannot be achieved—there are always pathways for, at least, protons. Therefore, current will appear in the system only after the illumination of some BR film area with light of suitable wavelength. In this case, the BR molecules will begin proton pumping, providing the electric current between two sections. The value of the current will depend on the light intensity. However, this dependence will come to saturation when the number of incoming photons is sufficient to provide continuous proton transfer to each BR molecule. The other parameters responsible for the current value are the absolute value of the pH in each section and the pH gradient between sections.

The system can be considered as a transducer for revealing pH variations. In fact, if the pH value in one section is maintained constant and the light intensity is also constant, the variation of the pH in the other section will be monitored as the variation of the current at the constant bias voltage. Such a scheme can be used for the construction of enzymatic sensors when the enzyme functioning results in the production or consumption of additional protons in the subphase volume (fortunately, most enzymes work in such a way). Moreover, more complicated devices, acting as artificial noses, can be based on the described principle.

The scheme of the sensitive membrane of such a device is shown in Figure 24. The porous membrane is covered by the BR monolayer. A complex enzyme film is deposited on top of the BR monolayer. The enzyme monolayer contains different types of enzymes, covering different areas of the BR monolayer (pixels), providing a 2D matrix of sensitive units. The presence of different analytes in the solution will result in a pH gradient map in the area adjacent to the BR layer. This map will be revealed by registering the current variation on scanning a light beam over the matrix. The resolution and the degree of integration of such a sensitive matrix can be very high. In fact, its spatial resolution will be determined only by the dimensions of the light beam, and can be a micron scale.

Other applications of BR layers are connected with the optical bistability of this material [830–835]. For natural BR there are two stable states, one with the absorbance at 570 nm and the other at 410 nm. It is possible to switch one state to the other by optical illumination, by thermal action, or by pH variation. Moreover, there are already some mutants of the BR [836], providing increased stability of the excited state (410 nm). These properties are very useful for designing optical memory and Fourier optics elements based on BR layers. However, such applications do not demand molecular thickness of the layers, and therefore the application of the LB technique is not required in this case. Therefore, we will not consider these applications in this chapter.

4.3.2. Antibody Films

Antibodies are a class of proteins responsible for the recognition and binding of specific antigens. There are several classes of antibodies. Here we will restrict ourselves to immunoglobulin G (IgG) antibodies, as most work using the LB technique has been performed on this class of antibodies. The IgG molecule is composed of two light chains and two heavy chains. These chains form one Fc fragment and two Fab fragments. IgG molecules are nearly identical. The only difference between them is in the Fab fragment, where there is a variable area, responsible for specificity to special groups in antigen molecules. Monoclonal antibodies have specificity of the variable area to a restricted group of antigen molecules, while polyclonal antibodies have specificity to different groups of antigens.

The presence of numerous covalent S—S bridges makes the molecule very stable. In fact, it forms stable monolayers at the air–water interface [837]. Significant denaturation of IgG molecules in the monolayer at the air–water interface takes place only after some hours of exposure to the high surface tension [838]. Therefore, if the monolayer compression is performed immediately after spreading with high velocity, practically all the IgG molecules maintain their native structure and functional activity in LB films.

The specific function of antibodies determines their field of applications. Their extremely high specificity is very important for sensor applications. The LB technique is very useful for their organization, one need have a layer only one molecule thick in order to provide sensing properties.

Several types of transducers can be used for immunosensor fabrication. The easiest one, basing on the binding properties, is the gravimetric one. IgG LB films (monolayers) are deposited onto a quartz crystal resonator [593, 838–843]. Binding of the antigen will vary the frequency of the resonator, loaded into the driving circuit. Therefore, the element will monitor the binding event directly as an increase of the mass attached to the resonator surface.

However, the technique has several limitations. First of all, the antigen must be a rather large molecule, in order to provide significant mass variation after the binding. Second, and

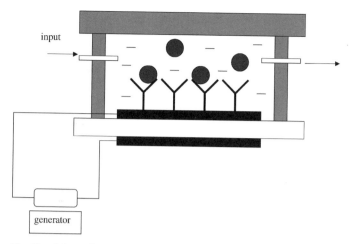

Fig. 25. Scheme for gravimetric measurements in liquid using only one electrode of the resonator.

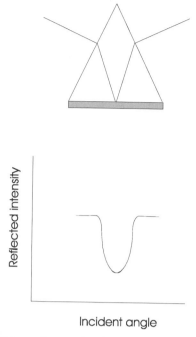

Fig. 26. Scheme of a sensor based on surface plasmon resonance.

probably more important, there is always nonspecific physical adsorption of the antigen molecules to the film surface. This adsorption must be taken into account and subtracted from the binding signal in order to have an exact specifically bound mass value. The nonspecific binding can be taken into account by performing two crystal differential scheme measurements. For this, two quartz resonators must be covered with IgG LB film. The first (measuring resonator) must be covered with IgG molecules specific to the analyte under the investigation, while the second (reference resonator) must be covered with IgG molecules with no affinity to the analyte. The similarity of the two covering molecules will lead to similar physical adsorption properties. Therefore, the registering electronics will monitor a differential signal, which will be proportional to the specifically attached mass. However, when the physical adsorption is intense, special washing must be performed after the binding.

First studies of such immunosensors were performed in the air phase, drying the sensitive elements after each measurement in the liquid phase. That is not surprising, in that dipping of the resonator in liquid will provide contact of electrodes with the aqueous solution. That can be considered as the appearance of a resistor between the two electrodes of the resonator—a parasitic resistor in the vibrating circuit, which causes a decrease of the quality factor of the circuit. It means that the accuracy of the measurements will be significantly decreased.

However, this difficulty was successfully overcome by using specially designed working chambers. These chambers provide the isolation of the working electrodes from each other [844]. In the simplest case (Fig. 25) only one electrode of the resonator is in contact with analyte solution and performs measurements. A simple modification will make both electrodes working, providing also inflow measurements.

Other transducers are based on surface acoustic wave propagation [845]. The measuring parameters in this case will be not the frequency shift, but the difference in sound velocity and/or phases of the signals, tunelling through a surface covered by the sensitive layer and without it. There are several descriptions of successful application of such transducers, but they will hardly

be used for industrial applications. The main reason is the extremely high price of the surface acoustic wave generators and receivers without significant advantages in performance over other techniques.

Different optical techniques can be used for fabrication of an immunosensor transducer. Up to now, fluorescence techniques are still the most widely used for medical applications [785, 846–858, 849]. The application of such a technique demands the initiation of a two-step reaction. The sensitive unit is the substrate covered by the antibody layer. This unit is exposed to the analyte solution. After incubation, the unit is immersed in a solution containing fluorescently labeled antibodies specific to the analyte under the investigation. Analysis of the fluorescence enables quantitation of the concentration of antigens in the analyte.

Among the optical techniques for immunosensing registration, surface plasmon resonance is the more widely used in recent times [786]. The technique is based on the following principle. There is a particular angle where the energy of the incident light is practically all transferred into vibration of the electron plasma of the thin metal layer on which the light impinges. This angle is strongly dependent on the condition of the metal layer surface. Attachment of several molecules can vary it over several degrees. Therefore, this phenomenon can be used for an immunosensor transducer. A scheme of the sensor with typical binding curves is shown in Figure 26. Note that the sensor can be also used in the liquid phase, performing continuous measurements of the binding events.

The technique can be used not only for transducer fabrication, but also as a measuring tool for the investigation of interactions of model membranes with different membrane-active compounds.

There are some other optical techniques for investigating antigen–antibody interactions. However, all of them have currently rather restricted application and are rather far from incorporation into real sensors.

4.3.3. Enzyme Films

The first work on protein films was performed by Langmuir and Schaefer on enzymes—pepsin and urease [20]. This work is interesting not only as the first use of the horizontal deposition technique, but also because it shows the possibility of preserving enzymatic activity in deposited layers.

Wide interest in enzyme films is due to their specific properties, which can be used for the realization of biosensors with high specificity and of enzymatic bioreactors. The LB technique allows one to work with very small quantities of the proteins. Therefore, it is considered as a useful tool for realizing enzyme-active layers, as the cost of some enzymes is extremely high.

The simpler and cheaper sensors are the enzymatic ones. Glucose oxidase (GOD) has been intensively studied by application of the LB technique, as its function can be easily detected, and GOD-based glucose sensors find wide application in medicine [860–874]. They are based on the following reaction:

$$\text{Glucose} \xrightarrow{\text{GOD}} \text{gluconic acid} + H_2O_2$$

A sensitive layer of GOD is deposited onto a platinum electrode. The measuring chamber contains also an Ag–AgCl reference electrode. Voltage must be applied between the enzymatic and reference electrodes. The current in the circuit will be proportional to the concentration of H_2O_2 produced by the enzymatic oxidation of the glucose, and therefore proportional to the glucose concentration. In order to improve the sensor performance, GOD molecules can be associated with conductive polymer layers. Such a complex layer provides better a charge drain between electrode and reactive media.

In all the cases, GOD layers are transferred to a solid substrate together with some amphiphilic molecules. However, selective removal of the lipid molecules after deposition has been shown to be possible. In particular, behenic acid molecules were removed from a film by isopropanol treatment of complex behenic acid–GOD LB films, deposited onto solid supports. Successful removal was confirmed by IR and AFM measurements. The possibility of the removal of fatty acid molecules (not salts) is well known, and is called *skeletonization* of the fatty acid salts [875]. Treatment of the fatty acid salt LB films with organic solvents results in the removal of fatty acid molecules not bound to metal ions, leaving only salt molecules, with reduced density of the layer.

Another interesting example enzyme for LB films is luciferase [796, 876]. The main reason for the academic interest to this protein is its clear function, which is practically free from mistreatment and artifacts. The enzyme catalyzes oxidation of luciferin (L) by molecular oxygen in the presence of the ATP

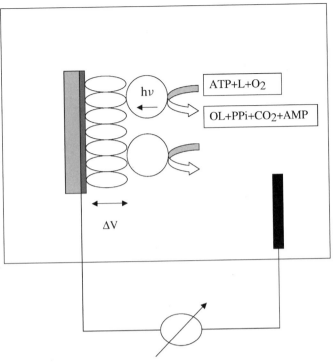

Fig. 27.   Scheme of a sensor based on a complex luciferase and BR layer.

complex according to the following reaction:

$$\text{ATP} + L + O_2 \xrightarrow{\text{luciferase}} \text{OL} + \text{PPi} + CO_2 + \text{AMP} \\ + h\nu(\lambda_{\max} = 562 \text{ nm})$$

where OL is oxyluciferin and Ppi is pyrophosphate. The activity of firefly luciferase, measured as light emission, was found to be preserved in LB films, but not stable in time.

The position of the emission maximum in the case of the luciferase reaction corresponds well to that of the maximum absorbance of BR in its ground state (570 nm). Therefore, we consider a complex layer, containing a BR film covered with luciferase, deposited on a potential-sensitive support. A possible scheme of such a structure is shown in Figure 27. The functioning of luciferase will result in the emission of light that will be absorbed by BR molecules. Light absorption will result in the appearance of a photopotential, which will be registered by the potential-sensitive substrate.

Another example of an enzyme successfully deposited with the LB technique is alcohol dehydrogenase (ADH) [877]. LB films of stearic acids with ADH were deposited and used as sensitive layers for ethanol monitoring. Detection of alcohol is based on the following reaction:

$$\text{CH}_3\text{CH}_2\text{OH} + \beta\text{-NAD}^+ \xrightarrow{\text{ADH}} \text{CH}_3\text{CHO} + \text{NADH} + H^+$$

The detectable substance in this case is the coenzyme NADH, which is reoxidized into β-NAD$^+$ at the electrode.

Like those of GOD, the properties of biosensors based on ADH LB films can be improved if the enzyme layers are associated with polymer sublayers. In this case, the polymer will play a role of a mediator, transferring the electron from the enzymatic reaction to the electrode. Such complex ADH–polymer layers were deposited and investigated by AFM.

In some cases the enzymatic activity in LB films can be registered by optical methods, if the absorbance of the reaction product is in a different wavelength range from that of the substrate and enzyme. Such measurements were performed for glutatione-S-transferase (GST) LB films [878–880]. GST catalyzes the conjugation of glutathione to electrophilic acceptors, such as quinones and organic peroxides. It has been shown that GST activity can be preserved in LB films. Moreover, thermal and temporal stability of the protein in the film were observed.

Some work on LB films of polymers have been performed with enzymes. Enzyme molecules were added into the volume of the water subphase under the monolayer of monomers, performing enzyme-catalyzed polymerization at the air–water interface [881–884].

4.4. Thermal Stability of Proteins in LB Films

Among the most interesting properties of the proteins organized in densely packed structures is their increased thermal stability. This property was described in 1993 for RC in LB films, in studying secondary structure variations by circular dichroism measurements [885, 886], and for BR self-assembled layers [887], in studying the structure by the X-ray method.

Variations of the CD spectra of RC in solution, LB films, and layers produced by drying solution drops demonstrate that RC molecules begin to denature in the solution at temperatures of about 50°C and are denatured practically completely at about 70°C.

In the films prepared by solution drop drying the complete denaturation takes place at more elevated temperatures. However, the CD spectrum even at room temperature is different from that in the solution. Very likely, these variations are due to the mutual redistribution of the protein–detergent complex during drying, which can result in some penetration of the detergent molecules into the proteins, resulting in partial denaturation of RC molecules.

In the case of LB film the situation was completely different. The initial spectrum is similar to that in the solution, indicating the preservation of the secondary structure of the protein after deposition onto solid substrates. The CD spectrum of the RC LB film is practically the same after heating to 150°C. Small variations begin to take place only at about 200°C.

It is interesting to compare the RC data with the behavior of CD spectra of BR also in solution, LB films, and layers obtained by drop drying. The behavior in solution is rather similar to that of the RC. The same is true for the LB films. However, the situation is completely different in the case of films obtained by drop drying. The effect of heating in this case is very similar to that in the case of LB films.

Comparison of the CD spectra allows one to state immediately that the water plays an important role in the thermal stability of proteins. In fact, variations of the secondary structure in

solutions begin at rather low temperatures. However, the comparison of the CD spectra in LB films and dried drops for RC and BR allows one to conclude that the absence of water is not the only condition responsible for the thermal stability of the secondary structure.

In the case of BR, the microscopic organization of molecules in both LB films and dried drop layers is practically the same. As was described in Section 4.1.1 the elementary unit in the case of BR layers is a membrane fragment. Therefore, the difference between the LB film and the dried drop layer is only in the arrangement of these fragments. In the case of LB it is slightly more regular. However, inside these fragments the BR molecular packing is the same—regular close packing, as in the membrane.

The situation is different in the case of RC films. The LB technique allows one to deposit regular close-packed layers. Moreover, detergent molecules are removed from the layer during the monolayer formation at the air–water interface, as is described in Section 4.1.2. In contrast, simple drying of the solution drop does not provide either close packing of molecules or regularity of the layer. Comparison of the behavior of CD spectra for RC and BR films [888] allows one to suggest that the molecular close packing and regularity of the layers are the other conditions responsible for the increased thermal stability of protein secondary structure in LB films.

However, the presence of the detergent molecules in the case of dried RC layers leaves some doubts about the last conclusion. In fact, the presence of detergent can vary the structure of the protein molecules during drying so significantly that it becomes impossible to make any comparison of such a sample with LB films. In this situation, our conclusion about the role of close packing and regularity in thermal stability cannot be strongly grounded, and the absence of water may be the only condition responsible for this phenomenon.

In order to check this possibility, experiments on antibody layers were performed [889]. The first test was performed on the heating of the lyophilized powder. Heating of this powder to 100°C resulted in a significant variations of the CD spectrum, while the LB film was practically the same up to 150°C, as with RC and BR LB films. Moreover, even boiling the LB film in water did not lead to so much denaturation as in the case of solution heating. The other interesting feature of the boiling of the LB sample is that there is practically no shift of the CD spectrum, but only a decrease of the peak values. This can be explained by the denaturation of the top layer (or layers), which forms (after the denaturation) a protective layer, preventing the penetration of water into the deep layers of the film.

The above considerations allow us to conclude that, in fact, the absence of water is not the only condition responsible for the thermal stability of proteins in thin films, but also the dense packing of molecules and regularity of the layer must be taken into consideration.

However, the preservation of the protein secondary structure does not always mean that the protein was not affected by the thermal treatment. Functional stability must be considered, as it

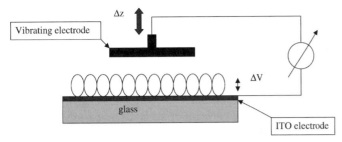

Fig. 28.   Scheme of the functional test of RC molecules using a Kelvin probe.

is the most important characteristic from the application point of view.

Let us consider RC films as the example of the stability of functional properties [890]. This choice is determined by the fact that the functioning of the protein, namely, light-induced electron transfer, can be rather easily detected. In the case of LB films, the detection was performed by light-induced electric potential measurements, using the Kelvin probe [889]. The scheme of the measurements is presented in Figure 28. The potential was measured at room temperature after heating to a fixed temperature for 10 min. The results of the experiment show that the functional properties of the RC in LB films are more affected by heating than by the secondary structure. Deactivation of the functional activity begins before significant changes in secondary structure, but later than in solution. It is also interesting to compare the stability of the functional properties of RC in LB films with that in dried solution samples. Unfortunately, the photopotential cannot be directly measured in this case. The unordered nature of the sample does not allow the preferential orientation of the protein molecules; therefore, electron displacements in adjacent molecules will compensate each other, giving no net potential. Therefore, functional activity was analyzed in this case by detecting the optical absorbance at 860 nm (dimer of bacteriochlorophyll). Though the presence of the bacteriochlorophyll dimer cannot be presumed to be absolutely without effect on the RC, we can use this method of investigation for comparison with LB films, as the disappearance of the peak will not take place before the deactivation of the protein. The results of this study show that the stability of the functional properties of RC in dried drops is much less than that in LB films, confirming once more the importance of close molecular packing and regularity of the layer for this property.

However, the question still remains—why is the functional activity destroyed if the secondary structure is preserved? The answer to this question can depend on the protein. As an example, we can consider RC again [891]. As was mentioned before, the protein contains three subunits. Heating to 110°C results in the splitting of the globule into its separate subunits. This fact was shown by STM investigation—the size of the elementary unit of the film was reduced to that of the subunit. Therefore, the quaternary structure of the protein was the first element that was affected by the thermal treatment. The functional properties of the RC are strongly connected with its quaternary structure. The

two bacteriochlorophyll molecules of the dimer—the primary donor of the electron—belong to different subunits. Therefore, the loss of the quaternary structure destroys the initial electron donor, eliminating completely the functional properties of the RC molecule.

It is interesting to note that STM images reveal not only the splitting of the RC into subunits, but also a significant improvement of the layer ordering after heating. The fact allows us to consider the heating as a rather general procedure (so long as it does not induce, for example, the destruction of the quaternary structure) for improvement of the protein LB film ordering. That statement was checked on different protein LB films, and in most cases was confirmed. The best results were obtained if the thermal treatment was performed not just on the finished multilayer film, but after each monolayer transfer [800]. Evaluation of the ordering-length improvement after thermal treatment of different protein LB films has been performed for antibodies, RC, BR, and cytochrome P450scc [800].

In some cases, such improvement of the film ordering resulted in improvement of the functional properties. Antibody films illustrate this statement [889, 891]. Increased antigen-binding ability was found for the thermally treated LB antibody monolayers, as represented both by the absolute value of the attached antigen (about 20% of the untreated value) and by the velocity of the reaction (about 5 times faster). It is difficult to suppose that this improvement is due to the variation of the individual protein molecule properties, caused by heating. More likely, it was due to the new organization of the layer. This suggestion was checked and confirmed by surface potential measurements of the antibody film before and after the heating. The increased value of the surface potential of the layer after the thermal treatment indicates improvement in the preferential orientation of molecules in the layer. This orientation can explain the improvement of the antigen-binding properties. Moreover, an electric potential can accelerate the reaction, providing additional driving force at its initial stage, when antigen must come close to the antibody layer surface.

The nature of the thermal stability of the protein molecules in LB films is still not completely clear. Of course, the absence of water is important, but it is not the only condition. Close molecular packing and regularity of the layer play also an important role. One analogy can show this behavior very roughly. Let us consider some fragile objects, matches, for example. If they are in a closely packed regular group, it is very hard to break them. Probably, something similar takes place also in the case of protein LB films. Close packing and regularity demand some additional energy to be applied to the system in order to break the molecular packing and only then to act on individual molecules.

5. CONCLUSIONS

The aim of this chapter has been to demonstrate the application of the LB technique for studying biological objects. Main attention was directed to protein LB films, as they are very important from the application point of view and it is rather difficult to work with them due to their fragile nature. Some biological objects were not considered in detail. However, we must mention the successful application of the LB technique for the formation of DNA-containing layers [553, 892–912] and for the incorporation of channel-forming molecules, such as valinomycin [679, 913–927].

The aim of the work, however, is to demonstrate the applicability of the technique for the fundamental investigation of the structure and processes in biological objects and for applied studies of thin active biological layers for applications in biosensors, transducers, and reactors.

ACKNOWLEDGMENT

The author would like to thank Svetlana and Konstantin for their help in the figure preparation.

REFERENCES

1. Lord Rayleigh, *Proc. Lond. Math. Soc.* 10, 4 (1879).
2. I. Langmuir, *J. Am. Chem. Soc.* 39, 1848 (1917).
3. I. Langmuir, *Trans. Faraday Soc.* 15, 62 (1920).
4. K. B. Blodgett, *J. Am. Chem. Soc.* 57, 1007 (1935).
5. K. B. Blodgett and I. Langmuir, *Phys. Rev.* 51, 964 (1937).
6. K. B. Blodgett, *J. Am. Chem. Soc.* 56, 495 (1934).
7. K. B. Blodgett, *J. Phys. Chem.* 41, 975 (1937).
8. K. B. Blodgett, *Phys. Rev.* 55, 391 (1939).
9. H. Kuhn and D. Möbius, *Angew. Chem. Int. Ed. Engl.* 10, 620 (1971).
10. H. Kuhn, *J. Chem. Phys.* 53, 101 (1970).
11. H. Kuhn, *Pure Appl. Chem.* 51, 341 (1979).
12. H. Kuhn, *Chem. Phys. Lipids* 8, 401 (1972).
13. H. Kuhn, *Thin Solid Films* 99, 1 (1983).
14. H. Kuhn, *Pure Appl. Chem.* 53, 2105 (1981).
15. H. Kuhn, *Thin Solid Films* 178, 1 (1989).
16. H. Kuhn, *Biosensors and Bioelectronics* 9/10, 707 (1994).
17. H. Kuhn, *J. Photochem.* 10, 111 (1979).
18. H. Kuhn, D. Möbius, and H. Bücher, in "Physical Methods of Chemistry" (A. Weissberger and B. W. Rossiter, Eds.), Part III B, Chap. VII. Wiley, New York, 1972.
19. H. Kuhn, D. Möbius, and H. Bücher, in "Techniques of Chemistry" (A. Weissberger and B. W. Rossiter, Eds.), Vol. I, Part IIIB, p. 577. Wiley, New York, 1972.
20. I. Langmuir and V. J. Schaefer, *J. Am. Chem. Soc.* 60, 1351 (1938).
21. G. L. Gaines, Jr., "Insoluble Monolayers at Liquid–Gas Interface." Wiley–Interscience, New York, 1966.
22. R. C. Ahuja, P.-L. Caruso, and D. Möbius, *Thin Solid Films* 242, 195 (1994).
23. R. C. Ahuja and D. Möbius, *Thin Solid Films* 243, 547 (1994).
24. E. Ando, M.-A. Suzuki, K. Moriyama, and K. Morimoto, *Thin Solid Films* 178, 103 (1989).
25. T. R. Baekmark, T. Wiesenthal, P. Kuhn, T. B. Bayerl, O. Nuyken, and R. Merkel, *Langmuir* 13, 5521 (1997).
26. J. F. Baret, H. Hasmonay, J. L. Firpo, J. J. Dupin, and M. Dupeyrat, *Chem. Phys. Lipids* 30, 177 (1982).
27. T. M. Bayerl, R. K. Thomas, J. Penfold, A. Rennie, and E. Sackmann, *Biophys. J.* 57, 1095 (1990).
28. G. M. Bommarito, W. J. Foster, P. S. Pershan, and M. L. Schlossman, *J. Phys. Chem.* 105, 5265 (1996).
29. G. Brezesinski, C. Böhm, A. Dietrich, and H. Möhwald, *Physica B* 198, 146 (1994).
30. G. Brezesinski, F. Bringezu, G. Weidemann, P. B. Howes, K. Kjaer, and H. Möhwald, *Thin Solid Films* 327–329, 256 (1998).
31. G. Brezesinski, M. Thoma, B. Struth, and H. Möhwald, *J. Phys. Chem.* 100, 3126 (1996).
32. F. Bringezu, G. Brezesinski, P. Nuhn, and H. Möhwald, *Thin Solid Films* 327–329, 28 (1998).
33. R. S. Cantor and P. M. McIlroy, *J. Chem. Phys.* 90, 4423 (1989).
34. M. K. Durbin, A. Malik, A. G. Richter, C.-J. Yu, R. Eisenhower, and P. Dutta, *Langmuir* 14, 899 (1998).
35. C. Duschl, D. Kemper, W. Frey, P. Meller, H. Ringsdorf, and W. Knoll, *J. Phys. Chem.* 93, 4587 (1989).
36. A. Fischer, M. Lösche, H. Möhwald, and E. Sackmann, *J. Phys. (Paris) Lett.* 45, L785 (1984).
37. A. Georgallas and D. A. Pink, *Can. J. Phys.* 60, 1678 (1982).
38. A. Jyoti, R. M. Prokop, and A. W. Neumann, *Colloids Surf. B: Biointerf.* 8, 115 (1997).
39. A. Jyoti, R. M. Prokop, J. Li, D. Vollhardt, D. Y. Kwok, R. Miller, H. Möhwald, and A. W. Neumann, *Colloids Surf. A* 116, 173 (1996).
40. C. M. Knobler and R. C. Desai, *Annu. Rev. Phys. Chem.* 43, 207 (1992).
41. D. Kramer and A. Ben-Shaul, *Physica A* 195, 12 (1993).
42. G. Lieser, B. Tieke, and G. Wegner, *Thin Solid Films* 68, 77 (1980).
43. M. Lösche, H.-P. Duwe, and H. Möhwald, *J. Colloid Interf. Sci.* 126, 432 (1988).
44. V. Melzer, D. Vollhardt, G. Brezesinski, and H. Möhwald, *J. Phys. Chem. B* 102, 591 (1998).
45. V. Melzer, D. Vollhardt, G. Weidemann, G. Brezesinski, R. Wagner, and H. Möhwald, *Phys. Rev. E* 57, 901 (1998).
46. R. M. Mendenhall and A. L. Mendenhall, Jr., *Rev. Sci. Instrum.* 34, 1350 (1963).
47. R. M. Mendenhall, *Rev. Sci. Instrum.* 42, 878 (1971).
48. R. M. Mendenhall, A. L. Mendenhall, Jr., and J. H. Tucker, *Ann. N. Y. Acad. Sci.* 130, 902 (1966).
49. K. Miyano, B. M. Abraham, J. B. Ketterson, and S. Q. Xu, *J. Chem. Phys.* 78, 1776 (1983).
50. G. A. Overbeck and D. Möbius, *J. Phys. Chem.* 97, 7999 (1993).
51. I. R. Peterson, *Phys. Scripta* T45, 245 (1992).
52. N. Sadrzadeh, H. Yu, and G. Zografi, *Langmuir* 14, 151 (1998).
53. E. ten Grotenhuis, R. A. Demel, M. Ponec, D. R. Boer, J. C. van Miltenburg, and J. A. Bouwstra, *Biophys. J.* 71, 1389 (1996).
54. V. Von Tscharner and H. M. McConnell, *Biophys. J.* 36, 409 (1981).
55. B. D. Casson and C. D. Bain, *J. Phys. Chem. B* 103, 4678 (1999).
56. I. Langmuir, *J. Am. Chem. Soc.* 39, 1848 (1917).
57. L. Wilhelmy, *Ann. Phys. Chem.* 119, 177 (1863).
58. G. L. Gains, Jr., *J. Colloid Interf. Sci.* 62, 191 (1977).
59. S. Sato and H. Kishimoto, *J. Colloid Interf. Sci.* 69, 188 (1979).
60. B. Sun, T. Curstedt, and B. Robertson, *Reproduction Fertility Devel.* 8, 173 (1996).
61. P. B. Welzel, I. Weis, and G. Schwartz, *Colloids Surf. A* 144, 229 (1998).
62. N. K. Adam, J. F. Danielli, and J. B. Harding, *Proc. Roy. Soc. London Ser. A* 147, 491 (1934).
63. M. Blank, L. Soo, R. E. Abbott, and U. Cogan, *J. Colloid Interf. Sci.* 73, 279 (1980).
64. H. Bümer, M. Winterhalter, and R. Benz, *J. Colloid Interf. Sci.* 168, 183 (1994).
65. G. Collacicco, *Chem. Phys. Lipids* 10, 66 (1973).
66. W. M. Heckl, H. Baumgärtner, and H. Möhwald, *Thin Solid Films* 173, 269 (1989).
67. C. M. Jones and H. L. Brockman, *Chem. Phys. Lipids* 60, 281 (1992).
68. S. V. Mello, R. M. Faria, L. H. C. Mattoso, A. Riul, Jr., and O. N. Oliveira, Jr., *Synth. Met.* 84, 773 (1997).
69. J. Mingins and B. A. Pethica, *Trans. Faraday Soc.* 59, 1892 (1963).
70. D. Möbius, W. Cordroch, R. Loschek, L. F. Chi, A. Dhathathreyan, and V. Vogel, *Thin Solid Films* 178, 53 (1989).
71. A. Noblet, H. Ridelaire, and G. Sylin, *J. Phys. E* 17, 234 (1984).
72. O. N. Oliveira, Jr., A. Riul, Jr., and G. F. Leal Ferreira, *Thin Solid Films* 242, 239 (1994).

73. O. N. Oliveira, Jr., D. M. Taylor, C. J. M. Stirling, S. Tripathi, and B. Z. Guo, *Langmuir* 8, 1619 (1992).

74. O. N. Oliveira, Jr., D. M. Taylor, T. J. Lewis, S. Salvagno, and C. J. M. Stirling, *J. Chem. Soc. Faraday Trans.* 85, 1009 (1989).

75. J. G. Petrov, E. E. Polymeropoulos, and H. Möhwald, *J. Phys. Chem.* 100, 9860 (1996).

76. F. Rustichelli, S. Dante, P. Mariani, I. V. Miagkov, and V. I. Troitsky, *Thin Solid Films* 242, 267 (1994).

77. D. M. Taylor, O. N. De Oliveira, Jr., and H. Morgan, *J. Colloid Interf. Sci.* 139, 508 (1990).

78. D. M. Taylor, O. N. Oliveira, Jr., and H. Morgan, *Chem. Phys. Lett.* 161, 147 (1989).

79. P. C. Tchoreloff, M. M. Boissonnade, A. W. Coleman, and A. Baszkin, *Langmuir* 11, 191 (1995).

80. R. H. Tredgold and G. W. Smith, *J. Phys. D* 14, L193 (1981).

81. R. H. Tredgold and G. W. Smith, *Thin Solid Films* 99, 215 (1983).

82. R. H. Tredgold, P. Hodge, Z. Ali-Adib, and S. D. Evans, *Thin Solid Films* 210/211, 4 (1992).

83. M. Winterhalter, H. Bürner, S. Marzinka, R. Benz, and J. J. Kasianowicz, *Biophys. J.* 69, 1372 (1995).

84. A. V. Hughes, D. M. Taylor, and A. E. Underhill, *Langmuir* 15, 2477 (1999).

85. A. Dhanabalan, L. Gaffo, A. M. Barros, W. C. Moreira, and O. N. Oliveira, Jr., *Langmuir* 15, 3944 (1999).

86. H. Kokubo, Y. Oyama, Y. Majima, and M. Iwamoto, *J. Appl. Phys.* 86, 3848 (1999).

87. M. Takeo, *J. Colloid Interf. Sci.* 157, 291 (1993).

88. B. Neys and P. Joos, *Colloids Surf. A* 143, 467 (1998).

89. D. M. Taylor and G. F. Bayers, *Mater. Sci. Eng. C* 8–9, 65 (1999).

90. N. A. Surplice and R. J. D'Arcy, *J. Phys. E* 3, 477 (1970).

91. M. Yasutake, D. Aoki, and M. Fujihira, *Thin Solid Films* 273, 279 (1996).

92. C. Goletti, A. Sgarlata, N. Motta, P. Chiaradia, R. Paolesse, A. Angelaccio, M. Drago, C. Di Natale, A. D'Amico, M. Cocco, and V. I. Troitsky, *Appl. Phys. Lett.* 75, 1237 (1999).

93. C. Di Natale, *Sensors Actuators B* 57, 183 (1999).

94. S. Lee, J. A. Virtanen, S. A. Virtanen, and R. M. Penner, *Langmuir* 8, 1243 (1992).

95. Z. Lu, S. Xiao, and Y. Wei, *Thin Solid Films* 210/211, 484 (1992).

96. Y. Okahata, K. Ariga, and K. Tanaka, *Thin Solid Films* 210/211, 702 (1992).

97. R. M. Stiger, J. A. Virtanen, S. Lee, S. A. Virtanen, and R. M. Penner, *Anal. Chim. Acta* 307, 377 (1995).

98. O. A. Aktsipetrov, E. D. Mishina, T. V. Murzina, V. R. Novak, and N. N. Akhmediv, *Thin Solid Films* 256, 176 (1995).

99. Yu. Lvov, V. Erokhin, and S. Zaitsev, *Biol. Membr.* 4(9), 1477 (1991).

100. V. Erokhin, L. Feigin, R. Kayushina, Yu. Lvov, A. Kononenko, P. Knox, and N. Zakharova, *Stud. Biophys.* 122, 231 (1987).

101. G. G. Roberts, "Langmuir–Blodgett Films." Plenum Press, New York, 1990.

102. L. M. Blinov, A. V. Ivanschenko, and S. G. Yudin, *Thin Solid Films* 160, 271 (1988).

103. L. M. Blinov, L. V. Mikhnev, E. B. Sokolova, and S. G. Yudin, *Sov. Tech. Phys. Lett.* 9, 640 (1983).

104. L. M. Blinov, N. N. Davydova, V. V. Lazarev, and S. G. Yudin, *Sov. Phys. Solid State* 24, 1523 (1983).

105. L. M. Blinov, N. V. Dubinin, and S. G. Yudin, *Opt. Spectr.* 56, 280 (1984).

106. L. M. Blinov, N. V. Dubinin, L. V. Mikhnev, and S. G. Yudin, *Thin Solid Films* 120, 161 (1984).

107. G. G. Roberts, *Ferroelectrics* 91, 21 (1989).

108. T. Kato, *Japan. J. Appl. Phys.* 26, L1377 (1987).

109. P. Vincett and W. Barlow, *Thin Solid Films* 71, 305 (1980).

110. B. Belbouch, M. Routliay, and M. Tournaric, *Thin Solid Films* 134, 89 (1985).

111. V. Erokhin, Y. Lvov, L. Mogilevsky, A. Zozulin, and E. Ilyin, *Thin Solid Films* 178, 111 (1989).

112. H. Kuhn and D. Mobius, *Angew. Chem. Int. Ed. Engl.* 10, 620 (1971).

113. H. Kuhn, *Thin Solid Films* 99, 1 (1983).

114. H. Kuhn, *Thin Solid Films* 178, 1 (1989).

115. H. Kuhn, *J. Photochem.* 10, 111 (1979).

116. B. Mann and H. Kuhn, *J. Appl. Phys.* 42, 4398 (1971).

117. B. Mann, H. Kuhn, and L. v. Szenpály, *Chem. Phys. Lett.* 8, 82 (1971).

118. E. E. Polymeropoulos, D. Möbius, and H. Kuhn, *Thin Solid Films* 68, 173 (1980).

119. U. Schoeler, K. H. Tews, and H. Kuhn, *J. Chem. Phys.* 61, 5009 (1974).

120. M. Sugi, K. Nembach, D. Möbius, and H. Kuhn, *Solid State Commun.* 15, 1867 (1974).

121. M. Sugi, K. Nembach, D. Möbius, and H. Kuhn, *Solid State Commun.* 13, 603 (1973).

122. L. F. Chi, R. R. Johnston, and H. Ringsdorf, *Langmuir* 7, 2323 (1991).

123. A. K. Dutta and C. Salesse, *Langmuir* 13, 5401 (1997).

124. K. Fujita, S. Kimura, Y. Imanishi, E. Rump, and H. Ringsdorf, *Langmuir* 11, 253 (1995).

125. L. F. Chi, R. R. Johnston, and H. Ringsdorf, *J. Am. Chem. Soc.* 109, 788 (1987).

126. K. Y. C. Lee, M. M. Lipp, D. Y. Takamoto, E. Ter-Ovanesyan, J. A. Zasadzinski, and A. J. Waring, *Langmuir* 14, 2567 (1998).

127. M. Lösche and H. Möhwald, *Rev. Sci. Instrum.* 55, 1968 (1984).

128. M. Lösche, E. Sackmann, and H. Möhwald, *Ber. Bunsenges. Phys. Chem.* 87, 848 (1983).

129. M. H. P. Moers, H. E. Gaub, and N. F. van Hulst, *Langmuir* 10, 2774 (1994).

130. F. Mojtabai, *Thin Solid Films* 178, 115 (1989).

131. S. Riviere, S. Hénon, J. Meunier, D. K. Schwartz, M.-W. Tsao, and C. M. Knobler, *J. Chem. Phys.* 101, 10045 (1994).

132. D. K. Schwartz and C. M. Knobler, *J. Phys. Chem.* 97, 8849 (1993).

133. K. J. Stine and C. M. Knobler, *Ultramicroscopy* 47, 23 (1992).

134. K. J. Stine and D. T. Stratmann, *Langmuir* 8, 2509 (1992).

135. K. J. Stine, *Microsc. Res. Technique* 27, 439 (1994).

136. R. M. Weis, *Chem. Phys. Lipids* 57, 227 (1991).

137. D. W. Grainger, A. Reichert, H. Ringsdorf, and C. Salesse, *Biochim. Biophys. Acta* 1023, 365 (1990).

138. J. N. Herron, W. Muller, M. Paudler, H. Riegler, H. Ringsdorf, and P. A. Suci, *Langmuir* 8, 1413 (1992).

139. T. Kondo, T. Kakiuchi, and M. Shimomura, *Thin Solid Films* 244, 887 (1994).

140. A. von Nahmen, A. Post, H.-J. Galla, and M. Sieber, *Eur. Biophys. J.* 26, 359 (1997).

141. D. Hönig and D. Möbius, *J. Phys. Chem.* 95, 4590 (1991).

142. D. Hönig and D. Möbius, *Thin Solid Films* 210/211, 64 (1992).

143. R. Castillo, S. Ramos, and J. Ruiz-Garcia, *J. Phys. Chem.* 100, 15235 (1996).

144. W. J. Foster, M. C. Shih, and P. S. Pershan, *J. Chem. Phys.* 105, 3307 (1996).

145. C. Lautz and T. M. Fischer, *J. Phys. Chem. B* 101, 8790 (1997).

146. G. A. Overbeck, D. Hönig, L. Wolthaus, M. Gnade, and D. Möbius, *Thin Solid Films* 242, 26 (1994).

147. T. Seki, H. Sekizawa, and K. Ichimura, *Polymer* 38, 725 (1997).

148. M.-W. Tsao, T. M. Fischer, and C. M. Knobler, *Langmuir* 11, 3184 (1995).

149. D. Vollhardt and T. Gutberlet, *Colloids Surf. A: Physicochem. Eng. Aspects* 102, 257 (1995).

150. J. M. Rodriguez Patino, C. C. Sánchez, and M. R. R. Niño, *Langmuir* 15, 2484 (1999).

151. L. Wolthaus, A. Schaper, and D. Möbius, *J. Phys. Chem.* 98, 10809 (1994).

152. W. J. Foster, M. C. Shih, and P. S. Pershan, in "Applications of Synchrotron Radiation Techniques to Material Science II" (D. L. Perry, G. Ice, N. Shinn, K. D'Amico, and L. Terminello, Eds.), Vol. 375, p. 187. Mater. Res. Soc. Proc., Pittsburgh, 1994.

153. E. Teer, C. M. Knobler, C. Lautz, S. Wurlitzer, J. Kildae, and T. M. Fischer, *J. Chem. Phys.* 106, 1913 (1997).

154. M. N. G. de Mul and J. A. Mann, Jr., *Langmuir* 14, 2455 (1998).

155. Y. Tabe and H. Yokoyama, *Langmuir* 11, 699 (1995).

156. W. Frey, W. R. Schief, Jr., and V. Vogel, *Langmuir* 12, 1312 (1996).

157. C. Lheveder, J. Meunier, and S. Hénon, in "Physical Chemistry of Biological Interfaces" (A. Baszkin and W. Norde, Eds.). Marcel Dekker, New York, 1999.

158. A. Angelova, D. Vollhardt, and R. Ionov, *J. Phys. Chem.* 100, 10710 (1996).

159. A. Angelova, M. Van der Auweraer, R. Ionov, D. Volhardt, and F. C. De Schryver, *Langmuir* 11, 3167 (1995).

160. Y. S. Kang, S. Risbud, J. Rabolt, and P. Stroeve, *Langmuir* 12, 4345 (1996).

161. M. M. Lipp, K. Y. C. Lee, A. J. Waring, and J. A. Zasadzinski, *Biophys. J.* 72, 2783 (1997).

162. J. M. R. Patino, C. C. Sánchez, and M. R. R. Niño, *Langmuir* 15, 4777 (1999).

163. P.-A. Albouy and P. Valerio, *Supramol. Sci.* 4, 191 (1997).

164. J. Als-Nielsen and K. Kjaer, in "Phase Transitions in Soft Condensed Matter" (T. Riste and D. Sherrington, Eds.), p. 113. Plenum Press, New York, 1989.

165. A. S. Brown, A. S. Holt, T. Dam, M. Trau, and J. W. White, *Langmuir* 13, 6363 (1997).

166. N. Dent, M. J. Grundy, R. M. Richardson, S. J. Roser, N. B. McKeown, and M. J. Cook, *J. Chim. Phys.* 85, 1003 (1988).

167. M. K. Durbin, M. C. Shih, A. Malik, P. Zschack, and P. Dutta, *Colloids Surf. A: Physicochem. Eng. Aspects* 102, 173 (1995).

168. M. Fukuto, K. Penanen, R. K. Heilmann, P. P. Pershan, and D. Vaknin, *J. Chem. Phys.* 107, 5531 (1997).

169. B. W. Gregory, D. Vaknin, T. M. Cotton, and W. S. Struve, *Thin Solid Films* 284–285, 849 (1996).

170. M. J. Grundy, R. M. Richardson, S. T. Roser, J. Penfold, and R. C. Ward, *Thin Solid Films* 159, 43 (1988).

171. H. Haas, G. Brezesinski, and H. Möhwald, *Biophys. J.* 68, 312 (1995).

172. C. A. Helm, H. Möhwald, K. Kjaer, and J. Als-Nielsen, *Europhys. Lett.* 4, 697 (1987).

173. M. Ibnelhaj, H. Riegler, H. Möhwald, M. Schwendler, and C. A. Helm, *Phys. Rev. A* 56, 1844 (1997).

174. K. Kago, H. Matsuoka, H. Endo, J. Eckelt, and H. Yamaoka, *Supramol. Sci.* 5, 349 (1998).

175. K. Kjaer, J. Als-Nielsen, C. A. Helm, P. Tippmann-Krayer, and H. Möhwald, *Thin Solid Films* 159, 17 (1988).

176. Z. Li, W. Zhao, J. Quinn, M. H. Rafallovich, J. Sokolov, R. B. Lennox, A. Eisenberg, X. Z. Wu, M. W. Kim, S. K. Sinha, and M. Tolan, *Langmuir* 11, 4785 (1995).

177. J. Majewski, T. L. Kuhl, K. Kjaer, M. C. Gerstenberg, J. Asl-nielsen, J. N. Israelachvili, and G. S. Smith, *J. Am. Chem. Soc.* 120, 1469 (1998).

178. M. L. Schlossman, D. K. Schwartz, E. H. Kawamoto, G. J. Kellogg, P. S. Pershan, M. W. Kim, and T. C. Chung, *J. Phys. Chem.* 95, 6628 (1991).

179. M. L. Schlossman, D. K. Schwartz, E. H. Kawamoto, G. J. Kellogg, P. S. Pershan, B. M. Ocko, M. W. Kim, and T. C. Chung, *Mat. Res. Soc. Symp. Proc.* 177, 351 (1990).

180. V. Skita, R. A. Butler, D. W. Chester, and R. F. Fishcetti, *Biophys. J.* 64, A259 (1993).

181. D. A. Styrkas, R. K. Thomas, and A. V. Sukhorukov, *Thin Solid Films* 243, 437 (1994).

182. D. A. Styrkas, R. K. Thomas, Z. A. Adib, F. Davis, P. Hodge, and X. H. Liu, *Macromolecules* 27, 5504 (1994).

183. U. Vierl and G. Cevc, *Biochim. Biophys. Acta Biomemb.* 1325, 165 (1997).

184. U. Vierl, G. Cevc, and H. Metzger, *Biochim. Biophys. Acta Biomemb.* 1234, 139 (1995).

185. H. Yamaoka, H. Matsuoka, K. Kago, H. Endo, J. Eckelt, and R. Yoshitome, *Chem. Phys. Lett.* 295, 245 (1998).

186. C. Fradin, A. Braslau, D. Luzet, M. Alba, F. Muller, J. Daillant, J. M. Petit, and F. Rieutord, *Langmuir* 14, 7327 (1998).

187. R. Steitz, J. B. Peng, I. R. Peterson, I. R. Gentle, R. M. Kenn, M. Goldmann, and G. T. Barnes, *Langmuir* 14, 7245 (1998).

188. K. Kago, M. Fürst, H. Matsuoka, H. Yamaoka, and T. Seki, *Langmuir* 15, 2237 (1999).

189. V. M. Kaganer, G. Brezesinski, H. Möhwald, P. B. Howes, and K. Kjaer, *Phys. Rev. E* 59, 2141 (1999).

190. S. Leporatti, S. Akari, F. Bringezu, G. Brezesinski, and H. Möhwald, *Appl. Phys. A* 66, S1245 (1998).

191. E. Havinga and J. De Wael, *Recl. Trav. Chim. Pays-Bas.* 56, 375 (1937).

192. J. De Wael and E. Havinga, *Recl. Trav. Chim. Pays-Bas.* 59, 770 (1940).

193. L. H. Germer and K. H. Storks, *J. Chem. Phys.* 6, 280 (1938).

194. W. Walkenhorst, *Naturwissenschaften* 34, 375 (1947).

195. H. P. Zingsheim, *Scanning Electron Microsc.* 1, 357 (1977).

196. D. Day and J. B. Lando, *J. Polym. Sci. Polym. Chem. Ed.* 16, 1431 (1978).

197. C. Böhm, R. Steitz, and H. Riegler, *Thin Solid Films* 178, 511 (1989).

198. T. Hayashi, T. Ito, M. Yamamoto, Y. Tsujii, M. Matsumoto, and T. Miyamoto, *Langmuir* 10, 4142 (1994).

199. T. Kato, K. Iriyama, and T. Araki, *Thin Solid Films* 210/211, 79 (1992).

200. T. Kato, N. Matsumoto, M. Kawano, N. Suzuki, T. Araki, and K. Iriyama, *Thin Solid Films* 242, 223 (1994).

201. V. V. Klechkovskaya, M. Anderle, R. Antolini, R. Canteri, L. Feigin, E. Rakova, and N. Sriopina, *Thin Solid Films* 284–285, 208 (1996).

202. F. Kopp, U. P. Fringeli, K. Mühlethaler, and H. H. Günthard, *Biophys. Struct. Mech.* 1, 75 (1975).

203. I. R. Peterson and G. J. Russell, *Phil. Mag. A* 49, 463 (1984).

204. H. E. Ries, Jr., and W. A. Kimball, *Nature* 181, 901 (1958).

205. I. Robinson, J. R. Sambles, and I. R. Peterson, *Thin Solid Films* 189, 149 (1989).

206. E. Sheppard, R. P. Bronson, and N. Tcheurekdjian, *J. Colloid Sci.* 19, 833 (1964).

207. R. Steitz, E. Mitchell, and I. R. Peterson, *Makromol. Chem. Macromol. Symp.* 46, 265 (1991).

208. I. R. Peterson, R. Steitz, H. Krug, and I. Voigt-Martin, *J. Phys. France* 51, 1003 (1990).

209. Yu. M. Lvov, V. I. Troitsky, and L. A. Feigin, *Mol. Cryst. Liq. Cryst.* 172, 89 (1989).

210. V. I. Troitsky, V. S. Bannikov, and T. S. Berzina, *J. Mol. Electron.* 5, 147 (1989).

211. V. I. Troitsky, in "Methods of Structure Analysis" (L. Feigin, Ed.), Vol. 272. Moscow, 1989.

212. L. A. Feigin, Y. M. Lvov, and V. I. Troitsky, *Sov. Sci. Rev. A. Phys.* 11, 287 (1989).

213. R. V. Sudiwala, C. Cheng, E. G. Wilson, and D. N. Batchelder, *Thin Solid Films* 210/211, 452 (1992).

214. J. D. Jontes and R. A. Milligan, *J. Mol. Biol.* 266, 331 (1997).

215. G. Lieser, S. Mittler-Neher, J. Spinke, and W. Knoll, *Biochim. Biophys. Acta* 1192, 14 (1994).

216. A. Tanaka, M. Yamaguchi, T. Iwasaki, and K. Iriyama, *Chem. Lett.* 1219 (1989).

217. C. Flament, K. Graf, F. Gallet, and H. Riegler, *Thin Solid Films* 243, 411 (1994).

218. E. W. Kubalek, R. D. Kornberg, and S. A. Darst, *Ultramicroscopy* 35, 295 (1991).

219. J. Majewski, L. Margulis, D. Jacquemain, F. Leveiller, C. Böhm, T. Arad, Y. Talmon, M. Lahav, and L. Leiserowitz, *Science* 261, 899 (1993).

220. H. E. Ries, Jr., and W. A. Kimball, *J. Phys. Chem.* 59, 94 (1955).

221. R. Rolandi, O. Cavalleri, C. Toneatto, and D. Ricci, *Thin Solid Films* 243, 431 (1994).

222. E. Sheppard, R. P. Bronson, and N. Tcheurekdjian, *J. Colloid Sci.* 20, 755 (1965).

223. R. Steitz and I. R. Peterson, in "Electron Crystallography of Organic Molecules" (J. R. Fryer and D. L. Dorset, Eds.), NATO ASI, Vol. C328, p. 365. Kluwer, Dordrecht, 1990.

224. F. Takeda, M. Matsumoto, T. Takenaka, and Y. Fujiyoshi, *J. Colloid Interf. Sci.* 84, 220 (1981).

225. J. R. Fryer, R. A. Hann, and B. L. Eyres, *Nature* 313, 382 (1985).

226. W. Chiu, A. J. Avila-Sakar, and M. F. Schmid, *Adv. Biophys.* 34, 161 (1997).

227. W. M. Heckl, M. Thompson, and H. Möhwald, *Langmuir* 5, 390 (1989).

228. S. W. Hui, M. Cowden, P. Papahadjopoulos, and D. F. Parsons, *Biochim. Biophys. Acta* 382, 265 (1975).

229. M. Li, E. Zhou, X. Wang, and J. Xu, *Thin Solid Films* 264, 94 (1995).

230. E. Okamura, J. Umemura, K. Iriyama, and T. Araki, *Chem. Phys. Lipids* 66, 219 (1993).

231. S.-X. Wang and S. Sui, *Supramol. Sci.* 5, 803 (1998).

232. A. J. Avila-Sakar and W. Chiu, *Biophys. J.* 70, 57 (1996).

233. W. Baumeister and M. Hahn, *Cytobiologie* 7, 244 (1973).

234. N. Braun, J. Tack, L. Bachmann, and S. Weinkauf, *Thin Solid Films* 284–285, 703 (1996).

235. A. Brisson, A. Olofsson, P. Ringler, M. Schmutz, and S. Stoylova, *Biol. Cell* 80, 221 (1994).

236. A. Brisson, W. Bergsma-Schutter, F. Oling, O. Lambert, and I. Reviakine, *J. Cryst. Growth* 196, 456 (1999).

237. D. G. Cornell and R. J. Carroll, *Colloids Surf.* 6, 385 (1983).

238. P. Fromherz, *Nature* 231, 267 (1971).

239. P. Helme and P. M. Vignais, *J. Mol. Biol.* 274, 687 (1997).

240. M. Mule, E. Stussi, D. De Rossi, T. S. Berzina, and V. I. Troitsky, *Thin Solid Films* 237, 225 (1994).

241. T. Sawaguchi, F. Mizutani, and I. Taniguchi, *Langmuir* 14, 3565 (1998).

242. P. Wang, J. L. Singleton, X.-L. Wu, M. Shamsuzzoha, R. M. Metzger, C. A. Panetta, J. W. Kim, and N. E. Heimer, *Synth. Met.* 57, 3824 (1993).

243. P. Wang, M. Shamsuzzoha, X.-L. Wu, W.-J. Lee, and R. M. Metzger, *J. Phys. Chem.* 96, 9025 (1992).

244. X.-L. Wu, M. Shamsuzzoha, R. M. Metzger, and G. J. Ashwell, *Synth. Met.* 57 (1993).

245. P. Quint, M. Hara, W. Knoll, H. Sasabe, and R. S. Duran, *Macromolecules* 28, 4019 (1995).

246. A. Zlatkin, S. Yudin, J. Simon, M. Hanack, and H. Lehman, *Adv. Mater. Opt. Electron.* 5, 259 (1995).

247. S. Alexandre, A. W. Coleman, A. Kasselouri, and J. M. Valleton, *Thin Solid Films* 284–285, 765 (1996).

248. O. V. Kolomytkin, A. O. Golubok, D. N. Davydov, V. A. Timofeev, S. A. Vinogradova, and S. Y. Tipisev, *Biophys. J.* 59, 889 (1991).

249. P. Facci, A. Diaspro, and R. Rolandi, *Thin Solid Films* 327–329, 532 (1998).

250. M. Hara, H. Sasabe, and W. Knoll, *Thin Solid Films* 273, 66 (1996).

251. H. J. Schmitt, R. Zhu, G. Min, and Y. Wei, *J. Phys. Chem.* 96, 8210 (1992).

252. P. Facci, V. Erokhin, A. Tronin, and C. Nicolini *J. Phys. Chem.* 98, 13323 (1994).

253. P. Facci, V. Erokhin, S. Carrara, and C. Nicolini, *Proc. Natl. Acad. Sci. U.S.A.* 93, 10556 (1996).

254. N. Elbel, W. Roth, E. Günther, and H. von Seggern, *Surf. Sci.* 303, 424 (1994).

255. Y. Zhang, Q. Li, Z. Xie, B. Mao, B. Hua, Y. Chen, and Z. Tian, *Thin Solid Films* 274, 150 (1996).

256. C. Zhu, J. Shen, Z. Ma, S. Pang, A. Wang, and K. Hu, *J. Chinese Electron Microsc. Soc.* 10, 167 (1991).

257. H.-J. Butt, R. Guckenberger, and J. P. Rabe, *Ultramicroscopy* 46, 375 (1992).

258. H. C. Day, D. R. Allee, R. George, and V. A. Burrows, *Appl. Phys. Lett.* 62, 1629 (1993).

259. L. Eng, H. R. Hidber, L. Rosenthaler, U. Staufer, R. Wiesendanger, and H.-J. Guentherodt, *J. Vac. Sci. Technol.* 6, 358 (1988).

260. J. Y. Fang, Z. H. Lu, G. W. Ming, Z. M. Ai, Y. Wei, and P. Stroeve, *Phys. Rev. A* 46, 4963 (1992).

261. W. M. Heckl, *Thin Solid Films* 210/211, 640 (1992).

262. M. Hibino, A. Sumi, and I. Hatta, *Thin Solid Films* 273, 272 (1996).

263. M. Hibino, A. Sumi, H. Tsuchiya, and I. Hatta, *J. Phys. Chem. B* 102, 4544 (1998).

264. J.-J. Kim, S.-D. Jung, H.-S. Roh, and J.-S. Ha, *Thin Solid Films* 244, 700 (1994).

265. C. A. Lang, J. K. H. Hörber, T. W. Hänsch, W. M. Heckl, and H. Möhwald, *J. Vac. Sci. Technol.* A6, 368 (1988).

266. R. M. Leblanc and J. A. DeRose, *Surf. Sci. Rep.* 22, 73 (1995).

267. R. M. Metzger and C. A. Panetta, *Synth. Met.* 41–43, 1407 (1991).

268. W. M. Sigmund, T. S. Bailey, M. Hara, H. Sasabe, W. Knoll, and R. S. Duran, *Langmuir* 11, 3153 (1995).

269. A. Stabel, L. Dasaradhi, D. O'Hagan, and J. P. Rabe, *Langmuir* 11, 1427 (1995).

270. C. Goletti, A. Sgarlata, N. Motta, P. Chiaradia, R. Paolesse, A. Angelaccio, M. Drago, C. Di Natale, A. D'Amico, M. Cocco, and V. I. Troitsky, *Appl. Phys. Lett.* 75, 1237 (1999).

271. L. C. Giancarlo and G. W. Flynn, *Annu. Rev. Phys. Chem.* 49, 297 (1998).

272. P. Facci, V. Erokhin, and C. Nicolini, *Thin Solid Films* 243, 403 (1994).

273. R. García, J. Tamayo, J. M. Soler, and C. Bustamante, *Langmuir* 11, 2109 (1995).

274. H. E.-M. Niemi, M. Ikonen, M. Levlin, and H. Lemmetyinen, *Langmuir* 9, 2436 (1993).

275. A. Tazi, S. Boussaad, J. A. DeRose, and R. M. Leblanc, *J. Vac. Sci. Technol. B, Microelectron. Nanometer Struct.* 14, 1476 (1996).

276. E. P. Friis, J. E. T. Andersen, Yu. I. Kharkatz, A. M. Kuznetsov, R. J. Nichols, J.-D. Zhang, and J. Ulstrup, *Proc. Natl. Acad. Sci. U.S.A.* 96, 1379 (1999).

277. D. Erts, J. Dzelme, and H. Olin, *Latv. J. Phys. Techn. Sci.* 6, 57 (1996).

278. L. Stockman, G. Neuttiens, C. Van Haesendonck, and Y. Bruynseraede, *Appl. Phys. Lett.* 62, 2935 (1993).

279. K. Takimoto, H. Kawada, E. Kishi, K. Yano, K. Sakai, K. Hatanaka, K. Eguchi, and T. Nakagiri, *Appl. Phys. Lett.* 61, 3032 (1992).

280. R. Yang, D. F. Evans, and W. A. Hendrickson, *Langmuir* 11, 211 (1995).

281. K. Yano, R. Kuroda, Y. Shimada, S. Shido, M. Kyogaku, H. Matsuda, K. Takimoto, K. Eguchi, and T. Nakagiri, *J. Vac. Sci. Technol. B, Microelectron. Nanometer Struct.* 14, 1353 (1996).

282. Z. Ali-Adib, P. Hodge, R. H. Tredgold, M. Woolley, and A. J. Pidduck, *Thin Solid Films* 242, 157 (1994).

283. S. Arisawa, T. Fujii, T. Okane, and R. Yamamoto, *Appl. Surf. Sci.* 60/61, 321 (1992).

284. C. E. H. Berger, K. O. van der Werf, R. P. H. Kooyman, B. G. de Grooth, and J. Greve, *Langmuir* 11, 4188 (1995).

285. E. Györvary, J. Peltonen, M. Lindén, and J. B. Rosenholm, *Thin Solid Films* 284–285, 368 (1996).

286. T. Imae and Y. Ikeo, *Supramol. Sci.* 5, 61 (1998).

287. N. Rozlosnik, G. Antal, T. Pusztai, and Gy. Faigel, *Supramol. Sci.* 4, 215 (1997).

288. A. K. Dutta, P. Vanoppen, K. Jeuris, P. C. M. Grim, D. Pevenage, C. Salesse, and F. C. De Schryver, *Langmuir* 15, 607 (1999).

289. R. Lu, K. Xu, J. Gu, and Z. Lu, *Phys. Lett. A* 260, 417 (1999).

290. K. Ekelund, E. Sparr, J. Engblom, H. Wennerström, and S. Engström, *Langmuir* 15, 6946 (1999).

291. E. Sparr, K. Ekelund, J. Engblom, S. Engström, and H. Wennerström, *Langmuir* 15, 6950 (1999).

292. H. Sugimura and N. Nakagiri, *J. Phys. A: Mater. Sci. Process.* 66, S427 (1998).

293. S. Morita, H. Wang, Y. Wang, K. Iriyama, and Y. Ozaki, *Mol. Cryst. Liq. Cryst.* 337, 325 (1998).

294. B. Lemkadem, P. Saulnier, F. Boury, and J. E. Proust, *Colloids Surf. B* 13, 233 (1999).

295. W. Barger, D. Koleske, K. Feldma, D. Krüger, and R. Colton, *Polymer Preprints* 37, 606 (1996).

296. G. K. Zhavnerko, V. N. Staroverov, V. E. Agabekov, M. O. Gallyamov, and I. V. Yaminsky, *Thin Solid Films* 359, 98 (2000).

297. K. S. Birdi and D. T. Vu, *Langmuir* 10, 623 (1994).

298. K. S. Birdi, D. T. Vu, L. Moesby, K. B. Andersen, and D. Kristensen, *Surf. Coating Technol.* 67, 183 (1994).

299. L. F. Chi and H. Fuchs, NATO-Series, Workshop "Manipulations of Atoms in High Fields and Temperatures: Applications" (B. V. Thien, Ed.), p. 287. Kluwer Academic, 1993.

300. L. F. Chi, H. Fuchs, R. R. Johnston, and H. Ringsdorf, *J. Vac. Sci. Technol. B* 12, 1967 (1994).

301. L. F. Chi, M. Anders, H. Fuchs, R. R. Johnston, and H. Ringsdorf, *Science* 259, 213 (1993).

302. Y. Chunbo, L. Xinmin, D. Desheng, L. Bin, Z. Hongjie, L. Zuhong, L. Juzeng, and N. Jiazuan, *Surf. Sci. Lett.* 366, L729 (1996).

303. Y. Chunbo, W. Ying, S. Yueming, L. Zuhong, and L. Juzheng, *Surf. Sci.* 392, L1 (1997).

304. Y. Chunbo, W. Ying, Y. Xiaomin, L. Zuhong, and L. Juzheng, *Appl. Surf. Sci.* 103, 531 (1996).

305. L. Daehne, J. Tao, and G. Mao, *Langmuir* 14, 565 (1998).

306. J. A. DeRose and R. M. Leblanc, *Surf. Sci. Rep.* 22, 73 (1995).

307. L. Dziri, S. Boussaad, N. Tao, and R. M. Leblanc, *Thin Solid Films* 327–329, 56 (1998).

308. J. Y. Fang, R. A. Uphaus, and P. Stroeve, *Thin Solid Films* 243, 450 (1994).

309. M. Flörsheimer, A. J. Steinford, and P. Günter, *Surf. Sci. Lett.* 297, L39 (1993).

310. M. Flörsheimer, A. J. Steinford, and P. Günter, *Thin Solid Films* 247, 190 (1994).

311. M. Flörsheimer, A. J. Steinfort, and P. Günter, *Thin Solid Films* 244, 1078 (1994).

312. R. A. Frazier, M. C. Davies, G. Matthijs, C. J. Roberts, E. Schatcht, S. J. Tendler, and P. M. Williams, *Langmuir* 13, 4795 (1997).

313. M. Fujihira and H. Takano, *Thin Solid Films* 243, 446 (1994).

314. I. Fujiwara, M. Ohnishi, and J. Seto, *Langmuir* 8, 2219 (1992).

315. J. Garnaes, D. K. Schwartz, R. Viswanathan, and J. A. N. Zasadzinski, *Synth. Met.* 57, 3795 (1993).

316. U.-W. Grummt, K.-H. Feller, F. Lehmann, R. Colditz, R. Gadonas, and A. Pugzlys, *Thin Solid Films* 284–285, 904 (1996).

317. U.-W. Grummt, K.-H. Feller, F. Lehmann, R. Colditz, R. Gadonas, and A. Pugzlys, *Thin Solid Films* 273, 76 (1996).

318. D.-F. Gu, C. Rosenblatt, and Z. Li, *Liq. Cryst.* 19, 489 (1995).

319. A. P. Gunning, P. J. Wilde, D. C. Clark, V. J. Morris, M. L. Parker, and P. A. Gunning, *J. Colloid Interf. Sci.* 183, 600 (1996).

320. J. Hu, M. Wang, H.-U. G. Weier, P. Frantz, W. Kolbe, D. F. Ogletree, and M. Salmeron, *Langmuir* 12, 1697 (1996).

321. S. W. Hui, R. Viswanathan, J. A. Zasadzinski, and J. N. Israelachvili, *Biophys. J.* 68, 171 (1995).

322. T. Imae and K. Aoki, *Langmuir* 14, 1196 (1998).

323. J. Y. Josefowicz, N. C. Maliszewskyj, S. H. J. Idziak, P. A. Heiney, J. P. McCauley, Jr., and A. B. Smith III, *Science* 260, 323 (1993).

324. T. Kajiyama, Y. Oishi, F. Hirose, K. Shuto, and T. Kuri, *Langmuir* 10, 1297 (1994).

325. T. Kato, M. Kameyama, and M. Kawano, *Thin Solid Films* 272, 232 (1996).

326. E. A. Kondrashkina, K. Hagedorn, D. Vollhardt, M. Schmidbauer, and R. Köhler, *Langmuir* 12, 5148 (1996).

327. B. L. Kropman, D. H. A. Blank, and H. Rogalla, *Thin Solid Films* 327–329, 185 (1998).

328. N. B. Larsen, T. Bjornholm, J. Garnaes, J. Larsen, and K. Schaumburg, *NATO ASI Ser. Ser. E* 292, 205 (1995).

329. M. Vélez, S. Vieira, I. Chambrier, and M. J. Cook, *Langmuir* 14, 4227 (1998).

330. N. C. Maliszewskyj, J. Y. Josefowicz, P. A. Heiney, J. P. McCauley, Jr., and A. B. Smith III, *Mater. Res. Soc. Proc.* 355, 147 (1995).

331. N. C. Maliszewskyj, P. A. Heiney, J. Y. Josefowicz, J. P. McCauley, Jr., and A. B. Smith III, *Science* 264, 77 (1994).

332. E. Mayer, L. Howald, R. M. Overney, H. Heinzelmann, J. Frommer, H.-J. Güntherodt, T. Wagner, H. Schler, and S. Roth, *Nature* 349, 398 (1991).

333. K. Meine, D. Vollhardt, and G. Weidemann, *Langmuir* 14, 1815 (1998).

334. O. Mori and T. Imae, *Langmuir* 11, 4779 (1995).

335. T. Nakagawa, K. Ogawa, and T. Kurumizawa, *Langmuir* 10, 525 (1994).

336. G. Nechev, M. Hibino, and I. Hatta, *Japan. J. Appl. Phys.* 36, L580 (1997).

337. N. Peachey and C. Eckhardt, *Micron* 25, 271 (1994).

338. J. Peltonen, M. Linden, H. Fagerholm, E. Györvary, and F. Eriksson, *Thin Solid Films* 242, 88 (1994).

339. J. B. Peng and G. T. Barnes, *Thin Solid Films* 284–285, 444 (1996).

340. B. Pignataro, C. Consalvo, G. Compagnini, and L. Licciardello, *Chem. Phys. Lett.* 299, 430 (1999).

341. X. Qian, Z. Tai, X. Sun, S. Xiao, H. Wu, Z. Lu, and Y. Wei, *Thin Solid Films* 284–285, 432 (1996).

342. H. Sato, Y. Ozaki, Y. Oishi, M. Kuramori, K. Suehiro, K. Nakashima, K. Uehara, and K. Iriyama, *Langmuir* 13, 4676 (1997).

343. P. Sawunyama, L. Jiang, A. Fujishima, and K. Hashimoto *J. Phys. Chem. B* 101, 11000 (1997).

344. F.-J. Schmitt, A. L. Weisenhorn, P. K. Hansma, and W. Knoll, *Makromol. Chem. Macromol. Symp.* 46, 133 (1991).

345. F.-J. Schmitt, A. L. Weisenhorn, P. K. Hansma, and W. Knoll, *Thin Solid Films* 210/211, 666 (1992).

346. D. K. Schwartz, J. Garnaes, R. Viswanathan, and J. A. N. Zasadzinski, *Scanning* 14, II-3 (1992).

347. D. K. Schwartz, J. Garnaes, R. Viswanathan, S. Chiruvolu, and J. A. N. Zasadzinski, *Phys. Rev. E* 47, 452 (1993).

348. D. K. Schwartz, R. Viswanathan, and J. A. N. Zasadzinski, *Science* 263, 1158 (1994).

349. N. Sigiyama, A. Shimizu, M. Nakamura, Y. Nakagawa, Y. Nagasawa, and H. Ishid, *Thin Solid Films* 331, 170 (1999).

350. J. M. Solletti, M. Botreau, F. Sommer, Tran Minh Duc, and M. R. Celio, *J. Vac. Sci. Technol. B, Microelectron. Nanometer Struct.* 14, 1492 (1996).

351. S. M. Stephens and R. A. Dluhy, *Thin Solid Films* 284–285, 381 (1996).

352. D. Y. Takamoto, E. TerOvanesyan, D. K. Schwartz, R. Viswanathan, A. J. Waring, and J. A. Zasadzinski, *Acta Phys. Polon.* 93, 373 (1998).

353. E. ten Grotenhuis, J. C. van Miltenburg, and J. P. van der Eerden, *Coll. Surf. A* 105, 309 (1995).

354. M. Velez, S. Mukhopadhyay, I. Muzikante, G. Matisova, and S. Vieira, *Langmuir* 13, 870 (1997).

355. M. Velez, S. Vieira, I. Chambrier, and M. J. Cook, *Langmuir* 14, 4227 (1998).

356. R. Viswanathan, D. K. Schwarz, J. Garnaes, and J. A. N. Zasadzinski, *Langmuir* 8, 54 (1992).

357. D. Vollhardt, T. Kato, and M. Kawano, *J. Phys. Chem.* 100, 4141 (1996).

358. Y. Wang, K. Nichogi, K. Iriyama, and Y. Ozaki, *J. Phys. Chem.* 100, 374 (1996).

359. Y. Wang, K. Nichogi, K. Iriyama, and Y. Ozaki, *J. Phys. Chem.* 100, 17238 (1996).

360. Y. Wang, K. Nichogi, S. Terashita, K. Iriyama, and Y. Ozaki, *J. Phys. Chem.* 100, 368 (1996).

361. A. L. Weisenhom, D. U. Roemer, and G. P. Lorenzi, *Langmuir* 8, 3145 (1992).

362. A. L. Weisenhorn, F.-J. Schmitt, W. Knoll, and P. K. Hansma, *Ultramicroscopy* 42–44, 1125 (1992).

363. A. L. Weisenhorn, H. E. Gaub, H. G. Hansma, R. L. Kelderman, and P. K. Hansma, *Scanning Microscopy* 4, 511 (1990).

364. A. L. Weisenhorn, M. Egger, F. Ohnesorge, S. A. C. Gould, S.-P. Heyn, H. G. Hansma, R. L. Sinsheimer, H. E. Gaub, and P. K. Hansma, *Langmuir* 7, 8 (1991).

365. T. M. Winger and E. L. Chaikof, *Langmuir* 14, 4148 (1998).

366. H.-M. Wu and S.-J. Xiao, *Jpn. J. Appl. Phys.* 35, L161 (1996).

367. S.-J. Xiao, H.-M. Wu, X.-M. Yang, N.-H. Li, Y. Wei, X.-Z. Sun, and Z.-H. Tai, *Thin Solid Films* 256, 210 (1995).

368. H. Yamada, Y. Hirata, M. Hara, and J. Miyake, *Thin Solid Films* 243, 455 (1994).

369. K. Yamanaka, H. Takano, E. Tomia, and M. Fujihira, *Jpn. J. Appl. Phys.* 35, 5421 (1996).

370. X.-M. Yang, D. Xiao, Z.-H. Lu, and Y. Wei, *Appl. Surf. Sci.* 90, 175 (1995).

371. X. M. Yang, G. M. Wang, and Z. H. Lu, *Supramol. Sci.* 5, 549 (1998).

372. K. Yano, M. Kyogaku, R. Kuroda, Y. Shimada, S. Shido, H. Matsuda, K. Takimoto, O. Albrecht, K. Eguchi, and T. Nakagiri, *Appl. Phys. Lett.* 68, 188 (1996).

373. S. Yokoyama, M. Kakimoto, and Y. Imai, *Synth. Met.* 81, 265 (1996).

374. C. Yuan, X. Yang, J. Liu, and Z. Lu, *Surf. Sci. Lett.* 355, L381 (1996).

375. M. Yumura, T. Nakamura, M. Matsumoto, S. Ohshima, Y. Kuriki, K. Honda, M. Kurahashi, and Y. F. Miura, *Synth. Met.* 57 (1993).

376. J. A. Zasadzinski, R. Viswanathan, D. K. Schwartz, J. Garnaes, L. Madsen, S. Chiruvolu, J. T. Woodward, and M. L. Longo, *Colloids Surf. A* 93, 305 (1994).

377. S. Terreyyaz, H. Yachibana, and M. Matsumoto, *Langmuir* 14, 7511 (1998).

378. B. Cappella, P. Baschieri, M. Ruffa, C. Ascoli, A. Relini, and R. Rolandi, *Langmuir* 15, 2152 (1999).

379. O. I. Kiselyova, O. L. Guryev, A. V. Krivosheev, S. A. Usanov, and I. V. Yaminsky, *Langmuir* 15, 1353 (1999).

380. A. Komolov, K. Schaumburg, P. J. Møller, and V. Monakhov, *Appl. Surf. Sci.* 142, 591 (1999).

381. C. Yuan, D. Ding, Z. Lu, and J. Liu, *Colloids Surf. A* 150, 1 (1999).

382. T. Geue, M. Schultz, U. Englisch, R. Stömmer, U. Pietsch, K. Meine, and D. Vollhardt, *J. Chem. Phys.* 110, 8104 (1999).

383. P. Qian, H. Nanjo, N. Sanada, T. Yokoyama, O. Itabashi, H. Hayashi, T. Miyashita, and T. M. Suzuki, *Thin Solid Films* 349, 250 (1999).

384. A. P. Gunning, A. R. Mackie, P. J. Wilde, and V. J. Morris, *Langmuir* 15, 4636 (1999).

385. F. Boury, P. Saulnier, J. E. Proust, I. Panaiotov, T. Ivanova, C. Postel, and O. Abillon, *Colloids Surf. A* 155, 117 (1999).

386. A. Tazi, S. Boussaad, and R. M. Leblanc, *Thin Solid Films* 353, 233 (1999).

387. S. A. Evenson, J. P. S. Badyal, C. Pearson, and M. C. Petty, *Adv. Mater.* 9, 58 (1997).

388. C. Consalvo, S. Panebianco, B. Pignataro, G. Compagnini, and O. Puglisi, *J. Phys. Chem. B* 103, 4687 (1999).

389. N. Kato, K. Saito, H. Aida, and Y. Uesu, *Chem. Phys. Lett.* 312, 115 (1999).

390. Y. Chunbo, W. Ying, L. Zuhong, and L. Juzheng, *J. Mater. Sci. Lett.* 16, 227 (1997).

391. J. Yang, L. K. Tamm, T. W. Tillack, and Z. Shao, *J. Mol. Biol.* 229, 286 (1993).

392. R. Viswanathan, D. K. Schwartz, J. Garnaes, and J. A. N. Zasadzinski, *Langmuir* 8, 1603 (1992).

393. D. K. Schwartz, *Curr. Opin. Colloid Interface Sci.* 3, 131 (1998).

394. T. Mikayama, K. Uehara, A. Sugimoto, H. Maruyama, K. Mizuno, and N. Inoue, *Mol. Cryst. Liq. Cryst.* 310, 137 (1998).

395. Y. Wang, K. Nichogi, K. Iriyama, and Y. Ozaki, *J. Phys. Chem. B* 101, 6379 (1997).

396. H. Sato, Y. Oishi, M. Kuramori, K. Suehiro, M. Kobayashi, K. Nakashima, K. Uehara, and K. Iriyama, *J. Chem. Soc. Faraday Trans.* 93, 621 (1997).

397. J. P. K. Peltonen, P. He, and J. B. Rosenholm, *J. Am. Chem. Soc.* 114, 7637 (1992).

398. P. Hartley, M. Matsumoto, and P. Mulvaney, *Langmuir* 14, 5203 (1998).

399. A. Ikai, *Surf. Sci. Rep.* 26, 261 (1997).

400. B. B. Sauer, R. S. McLean, and R. R. Thomas, *Langmuir* 14, 3045 (1998).

401. J. Yang, L. K. Tamm, A. P. Somlyo, and Z. Shao, *J. Microscopy* 171, 183 (1993).

402. H. Takano and M. Fujihira, *Thin Solid Films* 272, 312 (1996).

403. J. C. Kim, Y. M. Lee, E. R. Kim, H. Lee, Y. W. Shin, and S. W. Park, *Thin Solid Films* 327–329, 690 (1998).

404. H. F. Knapp, W. Wiegräbe, M. Heim, R. Eschrich, and R. Guckenberger, *Biophys. J.* 69, 708 (1995).

405. C. L. Andrews, *Rev. Sci. Instrum.* 11, 111 (1940).

406. W. T. Astbury, F. O. Bell, E. Gorter, and J. Van Ormondt, *Nature* 142, 33 (1938).

407. D. C. Bisset and J. Iball, *Proc. Phys. Soc. London A* 67, 315 (1954).

408. G. L. Clark and P. W. Leppla, *J. Am. Chem. Soc.* 58, 2199 (1936).

409. G. L. Clark, R. R. Sterrett, and P. W. Leppla, *J. Am. Chem. Soc.* 52, 330 (1935).

410. C. Holley and S. Bernstein, *Phys. Rev.* 49, 403 (1936).

411. C. Holley, *Phys. Rev.* 51, 1000 (1937).

412. D. S. Kapp and N. Wainfan, *Phys. Rev.* 138, 1490 (1965).

413. T. Arndt, A. J. Schouten, G. F. Schmidt, and G. Wegner, *Macromol. Chem.* 192, 2219 (1991).

414. T. A. Barberka, U. Höhne, U. Pietsch, and T. H. Metzger, *Thin Solid Films* 244, 1061 (1994).

415. S. W. Barton, B. N. Thomas, E. B. Flom, S. A. Rice, B. Lin, J. B. Peng, J. B. Ketterson, and P. Dutta, *J. Chem. Phys.* 89, 2257 (1988).

416. A. S. Belal, M. M. Salleh, and M. Yahaya, *Supramol. Sci.* 4, 535 (1997).

417. M. R. Buhaenko, M. J. Grundy, R. M. Richardson, and S. J. Roser, *Thin Solid Films* 159, 253 (1988).

418. M. W. Charles, *J. Colloid Interf. Sci.* 35, 167 (1971).

419. M. W. Charles, *J. Appl. Phys* 42, 3329 (1971).

420. J. Claudius, T. Gerber, J. Weigelt, and M. Kinzler, *Thin Solid Films* 287, 225 (1996).

421. S. Dante, M. De Rosa, E. Maccioni, A. Morana, C. Nicolini, F. Rustichelli, V. I. Troitsky, and B. Yang, *Mol. Cryst. Liq. Cryst.* 262, 191 (1995).

422. S. Dante, M. De Rosa, O. Francescangeli, C. Nicolini, F. Rustichelli, and V. I. Troitsky, *Thin Solid Films* 284–285, 459 (1996).

423. M. K. Durbin, A. Malik, A. G. Richter, K. G. Huang, and P. Dutta, *Langmuir* 13, 6547 (1997).

424. V. Erokhin, L. Feigin, G. Ivakin, V. Klechkovskaya, Yu. Lvov, and N. Stiopina, *Macromol. Chem. Macromol. Symp.* 46, 359 (1991).

425. N. P. Franks and W. R. Lieb, *J. Mol. Biol.* 133, 469 (1979).

426. P. Ganguly, M. Sastry, S. Choudhury, and D. V. Paranjape, *Langmuir* 13, 6582 (1997).

427. R. Geer, R. Shashidhar, A. Thibodeaux, and R. S. Duran, *Liq. Cryst.* 16, 869 (1994).

428. D. B. Hammond, T. Rayment, D. Dunne, P. Hodge, Z. Ali-Adib, and A. Dent, *Langmuir* 14, 5896 (1998).

429. W. Jark, G. Comelli, T. P. Russel, and J. Stöhr, *Thin Solid Films* 170, 309 (1989).

430. P. Lesieur, *J. Mol. Struct.* 383, 125 (1996).

431. W. Lesslauer, *Acta Crystallogr. Sect. B* 30, 1932 (1974).

432. H. Li, Z. Wang, B. Zhao, H. Xiong, X. Zhang, and J. Shen, *Langmuir* 14, 423 (1998).

433. Y. M. Lvov, D. Svergun, L. A. Feigin, C. Pearson, and M. C. Petty, *Phil. Mag. Lett.* 59, 317 (1989).

434. Yu. M. Lvov and L. A. Feigin, *Stud. Biophys.* 112, 221 (1986).

435. E. Maccioni, P. Mariani, F. Rustichelli, H. Delacroix, V. Troitsky, A. Riccio, A. Gambacorta, and M. De Rosa, *Thin Solid Films* 265, 74 (1995).

436. W. MacNaughton, K. A. Snook, E. Caspi, and N. P. Franks, *Biochim. Biophys. Acta* 818, 132 (1985).

437. A. Malik, M. K. Durbin, A. G. Richter, K. G. Huang, and P. Dutta, *Thin Solid Films* 284–285, 144 (1996).

438. P. Mariani, E. Maccioni, F. Rustichelli, H. Delacroix, V. Troitsky, A. Riccio, A. Gambacorta, and M. De Rosa, *Thin Solid Films* 265, 74 (1995).

439. H. J. Merle, Y. M. Lvov, and I. R. Peterson, *Makromol. Chem. Macromol. Symp.* 46, 271 (1991).

440. E. Milella, C. Giannini, and L. Tapfer, *Thin Solid Films* 293, 291 (1997).

441. H. Ohnuki, M. Izumi, K. Kitamura, H. Yamaguchi, H. Oyanagi, and P. Delhaes, *Thin Solid Films* 243, 415 (1994).

442. J. M. Pachence and J. K. Blasie, *Biochem. J.* 59, 894 (1991).

443. J. B. Peng, G. J. Foran, G. T. Barnes, and I. R. Gentle, *Langmuir* 13, 1602 (1997).

444. M. Pomerantz and A. Segmüller, *Thin Solid Films* 68, 33 (1980).

445. M. Pomerantz, A. Segmüller, L. Netzer, and J. Sagiv, *Thin Solid Films* 132, 153 (1985).

446. M. Pomerantz, F. H. Dacol, and A. Segmüller, *Bull. Am. Phys. Soc.* 20, 477 (1975).

447. M. Prakash, P. Dutta, J. B. Ketterson, and B. M. Abraham, *Chem. Phys. Lett.* 111, 395 (1984).

448. M. Prakash, J. B. Ketterson, and P. Dutta, *Thin Solid Films* 134, 1 (1985).

449. K. M. Robinson, C. Hernandez, and J. Adin Mann, Jr., *Thin Solid Films* 210/211, 73 (1992).

450. Y. Sasanuma and H. Nakahara, *Thin Solid Films* 261, 280 (1995).

451. M. Seul, P. Eisenburger, and H. McConnell, *Proc. Nat. Acad. Sci. U.S.A.* 80, 5795 (1983).

452. V. Skita, M. Filipkowski, A. F. Garito, and J. K. Blasie, *Phys. Rev. B.* 34, 5826 (1986).

453. V. K. Srivastava and A. R. Verma, *Solid State Commun.* 4, 367 (1966).

454. G. Sukhorukov, V. Lobyshev, and V. Erokhin, *Mol. Mat.* 1, 91 (1992).

455. G. B. Sukhorukov, L. A. Feigin, M. M. Montrel, and B. I. Sukhorukov, *Thin Solid Films* 259, 79 (1995).

456. T. Takamura, K. Matsushita, and Y. Shimoyama, *Japan. J. Appl. Phys.* 35, 5831 (1996).

457. R. H. Tredgold, A. J. Vickers, A. Hoorfar, P. Hodge, and E. Khoshdel, *J. Phys. D* 18, 1139 (1985).

458. M. Von Frieling, H. Bradaczek, and W. S. Durfee, *Thin Solid Films* 159, 451 (1988).

459. B. Yang and Y. Qiao, *Thin Solid Films* 284–285, 377 (1996).

460. B. Yang and Y. Qiao, *Thin Solid Films* 330, 157 (1998).

461. J. Cha, Y. Park, K.-B. Lee, and T. Chang, *Langmuir* 15, 1383 (1999).

462. V. M. Kaganer, G. Brezesinski, H. Möhwald, P. B. Howes, and K. Kjaer, *Phys. Rev. E* 59, 2141 (1999).

463. W. MacNaughtan, K. A. Snook, E. Caspi, and N. P. Franks, *Biochim. Biophys. Acta* 818, 132 (1985).

464. Y. Shufang, Z. Huilin, and H. Pingsheng, *J. Mater. Sci.* 34, 3149 (1999).

465. P. Dutta, *Cur. Opin. Solid State Mater. Sci.* 2, 557 (1997).

466. R. A. Uphaus, J. Y. Fang, R. Picorel, G. Chumanov, J. Y. Wang, T. M. Cotton, and M. Seibert, *Photochem. Photobiol.* 65, 673 (1997).

467. M. K. Durbin, A. Malik, R. Ghaskadvi, M. C. Shih, P. Zschack, and P. Dutta, *J. Phys. Chem.* 98, 1753 (1994).

468. S. Arisawa, S. Hara, and R. Yamamoto, *Mol. Cryst. Liq. Cryst.* 227, 29 (1993).

469. B. Lin, J. B. Peng, J. B. Ketterson, P. Dutta, B. N. Thomas, S. W. Barton, and S. A. Rice, *J. Chem. Phys.* 90, 2393 (1989).

470. J. M. Chen, X. Q. Yang, D. Chapman, M. Nelson, T. A. Skotheim, P. D. Hale, and Y. Okamoto, *Mol. Cryst. Liq. Cryst.* 190, 145 (1990).

471. J. B. Peng, G. J. Foran, G. T. Barnes, M. J. Crossley, and I. R. Gentle, *Langmuir* 16, 607 (2000).

472. A. Asmussen and H. Riegler, *J. Chem. Phys.* 104, 8151 (1996).

473. A. S. Brown, A. S. Holt, P. M. Saville, and J. W. White, *Austral. J. Phys.* 50, 391 (1997).

474. H. Franz, S. Dante, T. Wappmannsberger, W. Petry, M. de Rosa, and F. Rustichelli, *Thin Solid Films* 327–329, 52 (1998).

475. V. Gacem, J. Speakman, A. Gibaud, T. Richardson, and N. Cowlam, *Supramol. Sci.* 4, 275 (1997).

476. C. Giannini, L. Tapfer, M. Sauvage-Simkin, Y. Garreau, N. Jedrecy, M. B. Véron, R. Pinchaux, M. Burghard, and S. Roth, *Thin Solid Films* 288, 272 (1996).

477. C. Giannini, L. Tapfer, M. Sauvage-Simkin, Y. Garreau, N. Jedrecy, M. B. Véron, R. Pinchaux, M. Burghard, and S. Roth, *Nuovo Cimento D* 19D, 411 (1997).

478. B. W. Gregory, D. Vaknin, J. D. Gray, B. M. Ocko, P. Stroeve, T. M. Cotton, and W. S. Struve, *J. Phys. Chem.* 101, 2006 (1997).

479. H. Kepa, L. J. Kleinwaks, N. F. Berk, C. F. Majkrzak, T. S. Berzina, V. I. Troitsky, R. Antolini, and L. A. Feigin, *Physica B* 241–243, 1 (1998).

480. A. P. Kirilyuk and V. N. Bliznyuk, *Thin Solid Films* 269, 90 (1995).

481. O. V. Konovalov, I. I. Samoilenko, L. A. Feigin, and B. M. Shchedrin, *Crystallogr. Rep.* 41, 598 (1996).

482. O. V. Konovalov, L. A. Feigin, and B. M. Shchedrin, *Crystallogr. Rep.* 41, 592 (1996).

483. O. V. Konovalov, L. A. Feigin, and B. M. Shchedrin, *Crystallogr. Rep.* 41, 629 (1996).

484. O. V. Konovalov, L. A. Feigin, and B. M. Shchedrin, *Crystallogr. Rep.* 41, 640 (1996).

485. O. V. Konovalov, L. A. Feigin, and B. M. Shchedrin, *Crystallogr. Rep.* 41, 603 (1996).

486. F. Leveiller, C. Böhm, D. Jacquemain, H. Möhwald, L. Leiserowitz, K. Kjaer, and J. Als-Nielsen, *Langmuir* 10, 819 (1994).

487. A. Momose, Y. Hirai, I. Waki, S. Imazeki, Y. Tomioka, K. Hayakawa, and M. Naito, *Thin Solid Films* 178, 519 (1989).

488. G. Reiter, C. Bubeck, and M. Stamm, *Langmuir* 8, 1881 (1992).

489. M. Schaub, K. Mathauer, S. Schwiegk, P.-A. Albouy, G. Wenz, and G. Wegner, *Thin Solid Films* 210/211, 397 (1992).

490. A. Schmidt, K. Mathauer, G. Reiter, M. D. Foster, M. Stamm, G. Wegner, and W. Knoll, *Langmuir* 10, 3820 (1994).

491. J. Souto, F. Penacorada, M. L. Rodriguez-Mendez, and J. Reiche, *Mater. Sci. Eng. C* 5, 59 (1997).

492. A. Henderson and H. Ringsdorf, *Langmuir* 14, 5250 (1998).

493. T. R. Vierheller, M. D. Foster, H. Wu, A. Schmidt, W. Knoll, S. Satija, and C. F. Majkrzak, *Langmuir* 12, 5156 (1996).

494. H.-Y. Wang, J. A. Mann, Jr., J. B. Lando, T. R. Clark, and M. E. Kenney, *Langmuir* 11, 4549 (1995).

495. D. G. Wiesler, L. A. Feigin, C. F. Majkrzak, J. F. Ankner, T. S. Berzina, and V. I. Troitsky, *Thin Solid Films* 266, 69 (1995).

496. S. Yu. Zaitsev, and Y. M. Lvov, *Thin Solid Films* 254, 257 (1995).

497. U. Englisch, F. Peñacorada, L. Brehmer, and U. Pietsch, *Langmuir* 15, 1833 (1999).

498. R. K. Thomas and J. Penfold, *Curr. Opin. Colloid Interface Sci.* 1, 23 (1996).

499. K. N. Stoev and K. Sakurai, *Spectrochim. Acta B* 54, 41 (1999).

500. Y.-K. See, J. Cha, T. Chang, and M. Ree, *Langmuir* 16, 2351 (2000).

501. D. J. Ahn and E. I. Franses, *J. Phys. Chem.* 96, 9952 (1992).

502. D. Blaudez, T. Buffeteau, B. Desbat, N. Escafre, and J. M. Turlet, *Thin Solid Films* 243, 559 (1994).

503. F. Kimura, J. Umemura, and T. Takenaka, *Langmuir* 2, 96 (1986).

504. P. A. Chollet and J. Messier, *Chem. Phys.* 73, 235 (1982).

505. P. A. Chollet, *Thin Solid Films* 52, 343 (1978).

506. P. A. Chollet, J. Messier, and C. Rosilio, *J. Chem. Phys.* 64, 1042 (1976).

507. Y. Fujimoto, Y. Ozaki, and K. Iriyama, *J. Chem. Soc. Faraday Trans.* 92, 419 (1996).

508. D. D. Popenoe, S. M. Stole, and M. D. Porter, *Appl. Spectrosc.* 46, 79 (1992).

509. L. H. Sharpe, *Proc. Chem. Soc.* 461 (1961).

510. T. Takenaka, K. Nogami, and H. Gotoh, *J. Colloid Interf. Sci.* 40, 409 (1972).

511. T. Takenaka, K. Nogami, H. Gotoh, and R. Gotoh, *J. Colloid Interf. Sci.* 35, 395 (1971).

512. T. Wiesenthal, T. R. Baekmark, and R. Merkel, *Langmuir* 15, 6837 (1999).

513. T. L. Marshbanks, H. K. Jugduth, W. N. Delgass, and E. I. Franses, *Thin Solid Films* 232, 126 (1993).

514. Y. P. Song, A. S. Dhindsa, M. R. Bryce, M. C. Petty, and J. Yarwood, *Thin Solid Films* 210/211, 589 (1992).

515. X. Du and Y. Liang, *Chem. Phys. Lett.* 313, 565 (1999).

516. N. Katayama, Y. Ozaki, T. Seki, T. Tamaki, and K. Iriyama, *Langmuir* 10, 1898 (1994).

517. T. Kawai, J. Umemura, and T. Takenaka, *Langmuir* 6, 672 (1990).

518. R. Azumi, M. Matsumoto, S. Kuroda, and M. J. Crossley, *Langmuir* 11, 4495 (1995).

519. V. A. Burrows, *Solid State Electron.* 35, 231 (1991).

520. M. A. Chesters, M. J. Cook, S. L. Gallivan, J. M. Simmons, and D. A. Slater, *Thin Solid Films* 210/211, 538 (1992).

521. B. Desbat, L. Mao, and A. M. Ritcey, *Langmuir* 12, 4754 (1996).

522. J. W. Ellis and J. L. Pauley, *J. Colloid Sci.* 19, 755 (1964).

523. S. A. Francis and A. H. Ellison, *J. Opt. Soc. Am.* 49, 131 (1959).

524. T. Hasegawa, J. Umemura, and T. Takenaka, *J. Phys. Chem.* 97, 9009 (1993).

525. H. Hui-Litwin, L. Servant, M. J. Dignam, and M. Moskovits, *Langmuir* 13, 7211 (1997).

526. T. Kamata, J. Umemura, T. Takenaka, K. Takehara, K. Isomura, and H. Taniguchi, *J. Mol. Struct.* 240, 187 (1990).

527. L. Mao, A. M. Ritcey, and B. Desbat, *Langmuir* 12, 4754 (1996).

528. N. Nakahara and K. Fukuda, *J. Colloid Interf. Sci.* 69, 24 (1979).

529. J. F. Rabolt, F. C. Burns, W. E. Schlotter, and J. D. Swalen, *J. Chem. Phys.* 78, 946 (1983).

530. H. Sakai and J. Umemura, *Langmuir* 14, 6249 (1998).

531. T. Takenaka, K. Harda, and M. Matsumoto, *J. Colloid Interf. Sci.* 73, 569 (1980).

532. Y. Urai, C. Ohe, and K. Itoh, *Langmuir* 14, 4873 (1998).

533. D. A. Myrzakozha, T. Hasegawa, J. Nashijo, T. Imae, and Y. Ozaki, *Langmuir* 15, 3595 (1999).

534. D. A. Myrzakozha, T. Hasegawa, J. Nashijo, T. Imae, and Y. Ozaki, *Langmuir* 15, 3601 (1999).

535. Ya. I. Rabinovich, D. A. Guzanas, and R.-H. Yoon, *J. Colloid Interf. Sci.* 155, 221 (1993).

536. J. Yang, X.-G. Peng, Y. Zhang, H. Wang, and T.-S. Li, *J. Phys. Chem.* 97, 4484 (1993).

537. J. Buijs, W. Norde, and J. W. T. Lichtenbelt, *Langmuir* 12, 1605 (1996).

538. H. Ancelin, D. G. Zhu, M. C. Petty, and J. Yarwood, *Langmuir* 6, 1068 (1990).

539. M. Méthot, F. Boucher, C. Salesse, M. Subirade, and M. Pézolet, *Thin Solid Films* 284–285, 627 (1996).

540. I. M. Pepe, M. K. Ram, S. Paddeu, and C. Nicolini, *Thin Solid Films* 327–329, 118 (1998).

541. E. Bramanti, E. Benedetti, C. Nicolini, T. Berzina, V. Erokhin, A. D'Alessio, and E. Benedetti, *Biopolymers* 42, 227–237 (1997).

542. Y. Tian, X. Xu, Y. Zhao, X. Tang, T. Li, J. Sun, C. Li, and A. Pan, *Thin Solid Films* 284–285, 603 (1996).

543. J. Umemura, S. Takeda, T. Hasegawa, and T.Takenaka, *J. Mol. Struct.* 297, 57 (1993).

544. T. Hasegawa, T. Kamata, J. Umemura, and T. Takenaka, *Chem. Lett.* 1543 (1990).

545. F. Hoffmann, H. Huhnerfuss, and K.J. Stine, *Langmuir* 14, 4525 (1998).

546. Y. Wang, K. Nichogi, K. Iriyama, and Y. Ozaki, *J. Phys. Chem.* 100, 17232 (1996).

547. Z. Zhang, A. L. Verma, K. Nakashima, M. Yoneyama, K. Iriyama, and Y. Ozaki, *Thin Solid Films* 326, 211 (1998).

548. Y. Sakata, K. Domen, and T. Onishi, *Langmuir* 10, 2847 (1994).

549. T. Hasegawa, J. Nishijo, Y. Kobayashi, and J. Umemura, *Bull. Chem. Soc. Jpn.* 70, 525 (1997).

550. N. Katayama, Y. Fujimoto, Y. Ozaki, T. Araki, and K. Iriyama, *Langmuir* 8, 2758 (1992).

551. J. L. Dote and R. L. Mowery, *J. Phys. Chem.* 92, 1571 (1988).

552. F. Yano, V. A. Burrows, M. N. Kozicki, and J. Ryan, *J. Vac. Sci. Technol.* A11, 219 (1993).

553. B. I. Sukhorukov, M. M. Montrel', G. B. Sukhorukov, and L. I. Shabarchina, *Biophysics* 39, 273 (1994).

554. N. Katayama, Y. Ozaki, T. Araki, and K. Iriyama *J. Mol. Struct.* 242, 27 (1991).

555. T. Hasegawa, J. Umemura, and T. Takenaka, *Thin Solid Films* 210/211, 583 (1992).

556. T. Kamata, A. Kato, J. Umemura, and T. Takenaka, *Langmuir* 3, 1150 (1987).

557. T. L. Marshbanks, D. J. Ahn, and E. I. Franses, *Langmuir* 10, 276 (1994).

558. E. Okamura, J. Umemura, and T. Takenaka, *Biochim. Biophys. Acta* 812, 139 (1985).

559. P. Wenzl, M. Fringeli, J. Goette, and U. P. Fringeli, *Langmuir* 10, 4253 (1994).

560. D. J. Ahn, P. Sutandar, and E. I. Franses, *Macromolecules* 27, 7316 (1994).

561. T. I. Lotta, L. J. Laakkonen, J. A. Virtanen, and P. K. J. Kinnunen, *Chem. Phys. Lipids* 46, 1 (1988).

562. T. Hasegawa, J. Umemura, and T. Takenaka, *Appl. Spectrosc.* 47, 379 (1993).

563. T. Kamata, J. Umemura, and T. Takenaka, *Bull. Inst. Chem. Res. Kyoto Univ.* 65, 170 (1987).

564. T. Kamata, J. Umemura, and T. Takenaka, *Bull. Inst. Chem. Res. Kyoto Univ.* 65, 179 (1987).

565. F. Kopp, U. P. Fringeli, K. Mühlethaler, and H. H. Güthard, *Z. Naturforsch.* 30c, 711 (1975).

566. G. Z. Sauerbrey, *Z. Phys.* 178, 457 (1964).

567. P. Facci, V. Erokhin, and C. Nicolini, *Thin Solid Films* 230, 86 (1993).

568. C. M. Hanley, J. A. Quinn, and T. K. Vanderlick, *Langmuir* 10, 1524 (1994).

569. K. Ariga and Y. Okahata, *Langmuir* 10, 2272 (1994).

570. K. Ariga and Y. Okahata, *Langmuir* 10, 3255 (1994).

571. Y. Ebara, K. Itakura, and Y. Okahata, *Langmuir* 12, 5165 (1996).

572. Z. Lin, R. M. Hill, H. T. Davis, and M. D. Ward, *Langmuir* 10, 4060 (1994).

573. Y. Okahata and K. Ariga, *J. Chem. Soc. Chem. Commun.* 1535 (1987).

574. Y. Ebara, H. Ebato, K. Ariga, and Y. Okahata, *Langmuir* 10, 2267 (1994).

575. S. J. Roser and M. R. Lovell, *J. Chem. Soc. Faraday Trans.* 91, 1783 (1995).

576. T. Hasegawa, J. Nishijo, and J. Umemura, *J. Phys. Chem. B* 102, 8498 (1998).

577. H. J. Kim, S. Kwak, Y. S. Kim, B. I. Seo, E. R. Kim, and H. Lee, *Thin Solid Films* 327–329, 191 (1998).

578. Y. Okahata, X. Ye, A. Shimuzi, and H. Ebato, *Thin Solid Films* 180, 51 (1989).

579. I. Vikholm, W. M. Albers, H. Välimäki, and H. Helle, *Thin Solid Films* 327–329, 643 (1998).

580. D. N. Furlong, R. S. Urquhart, H. Mansur, F. Grieser, K. Tanaka, and Y. Okahata, *Langmuir* 10, 899 (1994).

581. P. Facci, V. Erokhin, A. Tronin, and C. Nicolini, *J. Phys. Chem.* 98, 13323 (1994).

582. D. Johannsmann, F. Embs, C. G. Willson, G. Wegner, and W. Knoll, *Makromol. Chem. Macromol. Symp.* 46, 247 (1991).

583. F. J. B. Kremer, H. Ringsdorf, A. Schuster, M. Seitz, and R. Weberskirch, *Thin Solid Films* 284–285, 436 (1996).

584. Y. Okahata, K. Kimura, and K. Ariga, *J. Amer. Chem. Soc.* 111, 9190 (1989).

585. Z.-K. Chen, S.-C. Ng, S. F. Y. Li, L. Zhong, L. Xu, and H. S. O. Chan, *Synt. Met.* 87, 201 (1997).

586. M. Furuki and L. S. Pu, *Thin Solid Films* 210/211, 471 (1992).

587. J.-D. Kim, S.-R. Kim, K. H. Choi, and Y. H. Chang, *Sensors Actuators B: Chem.* 40, 39 (1997).

588. S.-R. Kim, S.-A. Choi, J.-D. Kim, K. J. Kim, C. Lee, and S. B. Rhee, *Synth. Met.* 71, 2027 (1995).

589. K. Matsuura, Y. Ebara, and Y. Okahata, *Langmuir* 13, 814 (1997).

590. S. C. Ng, X. C. Zhou, Z. K. Chen, P. Miao, H. S. O. Chan, S. F. Y. Li, and P. Fu, *Langmuir* 14, 1748 (1998).

591. Y. Okahata, K. Matsuura, and Y. Ebara, *Supramol. Sci.* 3, 165 (1996).

592. S. Zhang, Z. K. Chen, G. W. Bao, and S. F. Y. Li, *Talanta* 45, 727 (1998).

593. T. Dubrovsky, V. Erokhin, and R. Kayushina, *Biol. Mem.* 6(1), 130 (1992).

594. V. Erokhin, R. Kayushina, and P. Gorkin, *Biochemistry* 45, 1049 (1990).

595. V. Erokhin, R. Kayushina, P. Gorkin, I. Kurochkin, B. Popov, and S. Chernov, *J. Anal. Chem.* 45, 1446 (1990) (in Russian).

596. C. Nicolini, M. Adami, T. Dubrovsky, V. Erokhin, P. Facci, P. Paschkevitch, and M. Sartore, *Sensors Actuators B. Chem.* 24–25, 121 (1995).

597. F. Patolsky, M. Zayats, E. Katz, and I. Willner, *Anal. Chem.* 71, 3171 (1999).

598. C. Nicolini, V. Erokhin, P. Facci, S. Guerzoni, A. Rossi, and P. Paschkevitch, *Biosensors Bioelectron.* 12, 613 (1997).

599. C. G. P. Feachem and L. Tronstad, *Proc. Roy. Soc. London Ser. A* 145, 127 (1934).

600. A. Tronin and V. Shapovalov, *Thin Solid Films* 313–314, 786 (1998).

601. A. Yu. Tronin and A. F. Konstantinova, *Phys. Chem. Mech. Surf.* 8, 722 (1993).

602. J. P. Cresswell, *Langmuir* 10, 3727 (1994).

603. D. den Engelsen and B. de Koning, *J. Chem. Soc. Faraday Trans. 1* 70, 2100 (1974).

604. D. den Engelsen, *J. Phys. Chem.* 76, 3390 (1972).

605. D. den Engelson and B. de Koning, *J. Chem. Soc. Faraday Trans.* 70, 1603 (1974).

606. M. J. Dignam, M. Moskovits, and R. W. Stobie, *Trans. Faraday Soc.* 67, 3306 (1971).

607. E. P. Honig and B. R. de Koning, *Surf. Sci.* 56, 454 (1976).

608. H. Knobloch, F. Penacorada, and L. Brehmer, *Thin Solid Films* 295, 210 (1997).

609. W. Knoll, J. Rabe, M. R. Philpott, and J. D. Swalen, *Thin Solid Films* 99, 173 (1983).

610. V. N. Kruchinin, S. M. Repinsky, L. L. Sveshnikova, E. M. Auvinen, I. N. Domnin, and N. P. Sysoeva, *Phys. Chem. Mech. Surf.* 7, 2899 (1992).

611. H. Motschmann, R. Reiter, R. Lawall, G. Duda, M. Stamm, G. Wegner, and W. Knoll, *Langmuir* 7, 2743 (1991).

612. M. Paudler, J. Ruths, and H. Riegler, *Langmuir* 8, 184 (1992).

613. T. Smith, *J. Colloid Interf. Sci.* 26, 509 (1968).

614. T. Smith, *J. Opt. Soc. Am.* 58, 1069 (1968).

615. R. Steiger, *Hevl. Chim. Acta* 54, 2645 (1971).

616. M. S. Tomar and V. K. Srivastava, *Thin Solid Films* 12, S29 (1972).

617. J. G. Petrov, T. Pfohl, and H. Möhwald, *J. Phys. Chem. B* 103, 3417 (1999).

618. D. den Engelsen, *J. Opt. Soc. Am.* 61, 1460 (1971).

619. B. Lecourt, D. Blaudez, and J. M. Turlet, *Thin Solid Films* 313–314, 791 (1998).

620. M. S. Tomar and V. K. Srivastava, *Thin Solid Films* 15, 207 (1973).

621. D. den Engelsen and B. de Koning, *J. Chem. Soc. Faraday Trans. 1* 70, 1603 (1974).

622. D. Ducharme, C. Salesse, and R. M. Leblanc, *Thin Solid Films* 132, 83 (1985).

623. D. Ducharme, J.-J. Max, C. Salesse, and R. Leblanc, *J. Phys. Chem.* 94, 1925 (1990).

624. L. Gambut, J.-P. Chauvet, C. Garcia, B. Berge, A. Renault, S. Rivière, J. Meunier, and A. Collet, *Langmuir* 12, 5407 (1996).

625. S. R. Goates, D. A. Schofield, and C. D. Bain, *Langmuir* 15, 1400 (1999).

626. M. W. Kim, B. B. Sauer, H. Yu, M. Yazdanian, and G. Zografi, *Langmuir* 6, 236 (1990).

627. S. W. H. Eijt, M. M. Wittebrood, M. A. C. Devillers, and T. Rasing, *Langmuir* 10, 4498 (1994).

628. M. Harke, M. Stelzle, and H. R. Motschmann, *Thin Solid Films* 284–285, 412 (1996).

629. R. Reiter, H. Motschmann, H. Orendi, A. Nemetz, and W. Knoll, *Langmuir* 8, 1784 (1992).

630. F. Kajzar, J. Messier, J. Zyss, and I. Leboux, *Opt. Commun.* 45, 133 (1983).

631. R. D. Mattuck, *J. Opt. Soc. Am.* 46, 782 (1956).

632. R. D. Mattuck, *J. Opt. Soc. Am.* 46, 621 (1956).

633. G. D. Scott, T. A. McLauchlan, and R. S. Sennet, *J. Appl. Phys.* 21, 843 (1950).

634. V. K. Srivastava and A. R. Verma, *Proc. Phys. Soc.* 80, 222 (1962).

635. S. Tolansky, "Multiple-Beam Interferometry of Surfaces and Films." Clarendon Press, London, 1948.

636. G. Gauglitz, A. Brecht, G. Kraus, and W. Nahm, *Sensors Actuators B* 11, 21 (1993).

637. H. Shiku, J. R. Krogmeier, and R. C. Dunn, *Langmuir* 15, 2162 (1999).

638. G. J. Ashwell, G. M. S. Wong, D. G. Bucknall, G. S. Bahra, and C. R. Brown, *Langmuir* 13, 1629 (1997).

639. H. D. Bijsterbosch, V. O. de Haan, A. W. de Graaf, M. Mellema, F. A. M. Leermakers, M. A. Cohen Stuart, and A. A. van Well, *Langmuir* 11, 4467 (1995).

640. A. Eaglesham, T. M. Herrington, and J. Penfold, *Colloids Surf.* 65, 9 (1992).

641. S. K. Gissing, B. R. Rochford, and R. W. Richards, *Colloids Surf. A* 86, 171 (1994).

642. J. A. Henderson, R. W. Richards, J. Penfold, and R. K. Thomas, *Macromolecules* 26, 65 (1993).

643. J. A. Henderson, R. W. Richards, J. Penfold, and R. K. Thomas, *Acta Polymer.* 44, 184 (1993).

644. J. A. Henderson, R. W. Richards, J. Penfold, C. Shackleton, and R. K. Thomas, *Polymer* 32, 3284 (1991).

645. P. Hodge, C. R. Towns, R. K. Thomas, and C. Shackleton, *Langmuir* 8, 585 (1992).

646. M. E. Lee, R. K. Thomas, J. Penfold, and R. C. Ward, *J. Phys. Chem.* 93, 381 (1989).

647. Z. X. Li, J. R. Lu, and R. K. Thomas, *Langmuir* 13, 3681 (1997).

648. Z. X. Li, J. R. Lu, R. K. Thomas, and J. Penfold, *Progr. Colloid Polymer Sci.* 98, 243 (1995).

649. M. Lösche, M. Piepenstock, D. Vaknin, and J. Als-Nielsen, *Thin Solid Films* 210/211, 659 (1992).

650. J. R. Lu, E. A. Simister, R. K. Thomas, and J. Penfold, *J. Phys. Condens. Matter* 6, A403 (1994).

651. J. R. Lu, E. A. Simister, R. K. Thomas, and J. Penfold, *J. Phys. Chem.* 97, 6024 (1993).

652. J. R. Lu, E. M. Lee, R. K. Thomas, J. Penfold, and S. L. Flitsch, *Langmuir* 9, 1352 (1993).

653. J. R. Lu, M. Hromadova, E. A. Simister, R. K. Thomas, and J. Penfold, *J. Phys. Chem.* 98, 11519 (1994).

654. J. R. Lu, Z. X. Li, J. Smallwood, R. K. Thomas, and J. Penfold, *J. Phys. Chem.* 94, 8233 (1995).

655. J. R. Lu, Z. X. Li, R. K. Thomas, and J. Penfold, *J. Chem. Soc. Faraday Trans.* 92, 403 (1996).

656. D. J. Lyttle, J. R. Lu, T. J. Su, R. K. Thomas, and J. Penfold, *Langmuir* 11, 1001 (1995).

657. G. Ma, D. J. Barlow, J. R. P. Webster, J. Penfold, and M. J. Lawrence, *J. Pharm. Pharmacol.* 47, 1071 (1995).

658. K. B. Migler and C. C. Han, *Macromolecules* 31, 360 (1998).

659. S. K. Peace, R. M. Richards, M. R. Taylor, J. R. P. Webster, and N. Williams, *Macromolecules* 31, 1261 (1998).

660. J. Penfold, E. M. Lee, and R. K. Thomas, *Mol. Phys.* 68, 33 (1989).

661. J. Penfold, R. M. Richardson, A. Zarbakhsh, J. R. P. Webster, D. G. Bucknall, A. R. Rennie, R. A. L. Jones, T. Cosgrove, R. K. Thomas, J. S. Higgins, P. D. I. Fletcher, E. Dicknson, S. J. Roser, I. A. McLure, A. R. Hillman, R. W. Richards, E. J. Staples, A. N. Burgess, E. A. Simister, and J. W. White, *J. Chem. Soc. Faraday Trans.* 93, 3899 (1997).

662. I. Reynolds, R. W. Richards, and J. R. P. Webster, *Macromolecules* 28, 7845 (1995).

663. R. W. Richards, B. R. Rochford, and J. R. P. Webster, *Faraday Discuss.* 98, 263 (1994).

664. P. M. Saville, I. R. Gentle, J. W. White, J. Penfold, and J. R. P. Webster, *J. Phys. Chem.* 98, 5935 (1994).

665. E. A. Simister, E. M. Lee, R. K. Thomas, and J. Penfold, *J. Phys. Chem.* 96, 1373 (1992).

666. E. A. Simister, R. K. Thomas, J. Penfold, L. Aveyard, B. P. Binks, P. Cooper, P. D. I. Fletcher, J. Rin, and A. Sokolowski, *J. Phys. Chem.* 96, 1383 (1992).

667. J. R. Lu and R. K. Thomas, in "Physical Chemistry of Biological Interfaces" (A. Baszkin and W. Norde, Eds.). Marcel Dekker, New York, 1999.

668. U. Englisch, T. A. Barberka, U. Pietsch, and U. Hohne, *Thin Solid Films* 266 (1995).

669. R. R. Highfield, R. K. Thomas, P. G. Gummins, D. P. Gregory, J. Mingins, J. B. Hayter, and O. Schärpf, *Thin Solid Films* 99, 165 (1983).

670. H. Kepa, L. J. Kleinwaks, N. F. Berk, C. F. Majkrzak, T. S. Berzina, V. I. Troitsky, R. Antolini, and L. A. Feigin, *Physica B* 241–243, 1 (1998).

671. R. M. Nicklow, M. Pomerantz, and A. Segmüller, *Phys. Rev. B* 23, 1081 (1981).

672. R. Stommer, U. Englisch, U. Pietsch, and V. Holy, *Physica B* 221, 284 (1996).

673. J. R. Lu, E. A. Simister, E. M. Lee, R. K. Thomas, A. R. Rennie, and J. Penfold, *Langmuir* 8, 1837 (1992).

674. C. Naumann, T. Brumm, A. R. Rennie, J. Penfold, and T. M. Bayeri, *Langmuir* 11, 3948 (1995).

675. P. J. Atkinson, E. Dickinson, D. S. Horne, and R. M. Richardson, *J. Chem. Soc. Faraday Trans.* 91, 2847 (1995).

676. P. J. Atkinson, E. Dickinson, D. S. Horne, and R. M. Richardson, *Am. Chem. Soc. Symp. Ser.* 602, 311 (1995).

677. S. Petrash, A. Liebmann-Vinson, M. D. Foster, L. M. Lander, and W. J. Brittain, *Biotechnol. Prog.* 13, 635 (1997).

678. A. Schmidt, J. Spinke, T. Bayerl, E. Sackmann, and W. Knoll, *Biophys. J.* 63, 1185 (1992).

679. A. Eaglesham and T. M. Herrington, *J. Colloid Interface Sci.* 171, 1 (1995).

680. J. D. Hines, G. Fragneto, R. K. Thomas, P. R. Garrett, G. K. Rennie, and A. R. Rennie, *J. Colloid Interface Sci.* 189, 259 (1997).

681. L. T. Lee, E. K. Mann, O. Guiselin, D. Langevin, B. Farnoux, and J. Penfold, *Macromolecules* 26, 7046 (1993).

682. J. R. Lu, J. A. K. Blondel, D. J. Cooke, R. K. Thomas, and J. Penfold, *Progr. Colloid Polym. Sci.* 100, 311 (1996).

683. J. R. Lu, R. K. Thomas, B. P. Binks, P. D. I. Fletcher, and J. Penfold, *J. Phys. Chem.* 99, 4113 (1995).

684. J. R. Lu, R. K. Thomas, R. Aveyard, B. P. Binks, P. Cooper, P. D. I. Fletcher, A. Sokolowski, and J. Penfold, *J. Phys. Chem.* 96, 10971 (1992).

685. J. Penfold, E. Staples, L. Thompson, and I. Tucker, *Colloids Surf. A: Physicochem. Eng. Aspects* 102, 127 (1995).

686. J. Y. Wang, D. Vaknin, R. A. Uphaus, K. Kjaer, and M. Lösche, *Thin Solid Films* 242, 40 (1994).

687. P. C. Griffiths, M. L. Whatton, R. J. Abbott, W. Kwan, A. R. Pitt, A. M. Howe, S. M. King, and R. K. Heenan, *J. Colloid Interf. Sci.* 215, 114 (1999).

688. U. Englisch, S. Katholy, F. Peñacorada, J. Reiche, and U. Pietsch, *Mater. Sci. Eng. C* 8–9, 99 (1999).

689. L. Feigin, O. Konovalov, D. G. Wiesler, C. F. Majkrzak, T. Berzina, and V. Troitsky, *Physica B* 221, 185 (1996).

690. D. G. Wiesler, L. A. Feigin, C. F. Majkrzak, J. F. Ankner, T. S. Berzina, and V. I. Troitsky, *Thin Solid Films* 266, 69 (1995).

691. A. Bolm, U. Englisch, F. Penacorada, M. Gerstenberg, and U. Pietsch, *Supramol. Sci.* 4, 229 (1997).

692. C. Nicolini, V. Erokhin, S. Paddeu, and M. Sartóre, *Nanotechnology* 9, 223 (1998).

693. A. V. Maximychev, A. S. Kholmansky, E. V. Levin, N. G. Rambidi, S. K. Chamorovsky, A. A. Kononenko, V. Erokhin, and L. N. Checulaeva, *Adv. Mater. Opt. Electron.* 1, 105 (1992).

694. A. V. Maximychev, S. K. Chamorovsky, A. S. Kholmanskii, V. V. Erokhin, E. V. Levin, L. N. Chekulaeva, A. A. Kononenko, and N. G. Rambidi, *Biol. Membr.* 8, 15 (1996).

695. B. F. Li, Y. C. Song, J. R. Li, J. A. Tang, and L. Jiang, *Colloids Surf. B: Biointerf.* 3, 317 (1995).

696. H. Lavoie and C. Salesse, *Mater. Sci. Eng. C* 10, 147 (1999).

697. A. Shibata, J. Kohara, S. Ueno, I. Uchida, T. Mashimo, and T. Yamashita, *Thin Solid Films* 244, 736 (1994).

698. D. M. Tiede, *Biochim. Biophys. Acta* 811, 357 (1985).

699. T. Miyasaka and K. Koyama, *Chem. Lett.* 1645 (1991).

700. S. B. Hwang, J. I. Korenbrot, and W. Stoeckenius, *J. Membr. Biol.* 36, 115 (1977).

701. G. Dencher and M. Lösche, *J. Mol. Biol.* 287, 837 (1999).

702. Y. Sugiyama, T. Inoue, M. Ikematsu, M. Iseki, and T. Sekiguchi, *Thin Solid Films* 310, 102 (1997).

703. V. Erokhin, R. Kayushina, Yu. Lvov, N. Zakharova, A. Kononenko, P. Knox, and A. Rubin, *Dokl. Akad. Nauk SSSR* 299, 1262 (1988).

704. J. Y. Fang, D. F. Gaul, G. Chumanov, and T. M. Cotton, *Langmuir* 11, 4366 (1995).

705. P. Facci, V. Erokhin, S. Paddeu, and C. Nicolini, *Langmuir* 14, 193 (1998).

706. Y. Yasuda, H. Sugino, H. Toyotama, Y. Hirata, M. Hara, and J. Miyake, *Bioelectrochem. Bioenerget.* 34, 135 (1994).

707. Y. Hirata and J. Miyake, *Thin Solid Films* 244, 865 (1994).

708. Y. Yasuda, H. Toyotama, M. Hara, and J. Miyake, *Thin Solid Films* 327–329, 800 (1998).

709. V. Erokhin, J. Sabo, N. Zakharova, R. Kayushina, A. Kononenko, Yu. Lvov, E. Lukashev, and P. Knox, *Biol. Membr.* 2(6), 1125 (1989).

710. V. Erokhin, R. Kayushina, A. Dembo, J. Sabo, P. Knox, and A. Kononenko, *Mol. Cryst. Liq. Cryst.* 221, 1 (1992).

711. M. Ahlers, W. Müller, A. Reichert, H. Ringsdorf, and J. Venzmer, *Angew. Chem. Int. Ed. Engl.* 29, 1269 (1990).

712. H. Mohwald, *Annu. Rev. Phys. Chem.* 41, 441 (1990).

713. R. Verger and F. Pattus, *Chem. Phys. Lipids* 30, 189 (1982).

714. H. Brockman, *Curr. Opin. Struct. Biol.* 9, 438 (1999).

715. G. Colacicco, *J. Colloid Interf. Sci.* 29, 345 (1969).

716. M. Saint-Pierre-Chazalet, C. Fressigné, F. Billoudet, and M. P. Pileni, *Thin Solid Films* 210/211, 743 (1992).

717. R. Blankenburg, P. Meller, H. Ringsdorf, and C. Salesse, *Biochemistry* 28, 8214 (1989).

718. L. A. Samuelson, Y. Yang, K. A. Marx, J. Kumar, S. K. Tripathy, and D. L. Kaplan, *Biomimetics* 1, 51 (1992).

719. S. A. Darst, M. Ahlers, P. H. Meller, E. W. Kubalek, R. Blankenburg, H. O. Ribi, H. Ringsdorf, and R. D. Kornberg, *Biophys. J.* 59, 387 (1991).

720. Z. Liu, H. Qin, C. Xiao, S. Wang, and S. F. Sui, *Eur. Biophys. J.* 24, 31 (1995).

721. L. A. Samuelson, P. Miller, D. M. Galotti, K. A. Marx, J. Kumar, S. K. Tripathy, and D. L. Kaplan, *Langmuir* 8, 604 (1992).

722. J. N. Herron, W. Muller, M. Paudler, H. Riegler, H. Ringsdorf, and P. A. Suci, *Langmuir* 8, 1413 (1992).

723. S. Koppenol, L. A. Klumb, V. Vogel, and P. S. Stayton, *Langmuir* 15, 7125 (1999).

724. V. Vogel, W. R. Schief, Jr., and W. Frey, *Supramol. Sci.* 4, 163 (1997).

725. L. Samuelson, P. Miller, D. Galotti, K. A. Marx, J. Kumar, S. Tripathy, and D. Kaplan, *Thin Solid Films* 210/211, 796 (1992).

726. L. A. Samuelson, D. L. Kaplan, J. O. Lim, M. Kamath, K. A. Marx, and S. K. Tripathy, *Thin Solid Films* 242, 50 (1994).

727. K. A. Marx, L. A. Samuelson, M. Kamath, S. Sengupta, D. Kaplan, J. Kumar, and S. K. Tripathy, in "Molecular and Biomolecular Electronics" (R. R. Birge, Ed.), Vol. 240, Chap. 15. American Chemical Society, 1994.

728. J. M. Pachence, S. M. Amador, G. Maniara, J. Vanderkooi, P. L. Dutton, and J. K. Blasie, *Biophys. J.* 58, 379 (1990).

729. P. L. Edmiston and S. S. Saavedra, *Biophys. J.* 74, 999 (1998).

730. J. Peschke and H. Möhwald, *Colloids Surf.* 27, 305 (1987).

731. S. Boussaad, L. Dziri, R. Arechabaleta, N. J. Tao, and R. M. Leblanc, *Langmuir* 14, 6215 (1998).

732. P. Fromhertz, *Biochim. Biophys. Acta* 225, 382 (1971).

733. P. Fromhertz and D. Marcheva, *FEBS Lett.* 49, 329 (1975).

734. V. I. Troitsky, M. Sartore, T. S. Berzina, D. Nardelli, and C. Nicolini, *Rev. Sci. Instrum.* 67, 4216 (1996).

735. E. E. Uzgiris, *Biochem. Biophys. Res. Commun.* 134, 819 (1986).

736. E. E. Uzgiris, *Biochem. J.* 242, 293 (1987).

737. M. Egger, S. P. Heyn, and H. E. Gaub, *Biophys. J* 57, 669 (1990).

738. L. Lebeau, S. Olland, P. Oudet, and C. Mioskowski, *Chem. Phys. Lipids* 62, 93 (1992).

739. N. Dubreuil, S. Alexandre, D. Lair, F. Sommer, T. Duc, and J. Valleton, *Colloids Surf. B: Biointerf.* 11, 95 (1998).

740. L. Lebeau, E. Regnier, P. Schultz, J. C. Wang, C. Mioskowski, and P. Oudet, *FEBS Lett.* 267, 38 (1990).

741. M. A. Requero, M. Gonzalez, F. M. Goni, A. Alonso, and G. Fidelio, *FEBS Lett.* 357, 75 (1995).

742. E. E. Uzgiris, *J. Cell. Biochem.* 29, 239 (1985).

743. A. Brisson, W. Bergsma-Schutter, F. Oling, O. Lambert, and I. Reviakine, *J. Cryst. Growth* 196, 456 (1999).

744. R. H. Newman and P. S. Freemont, *Thin Solid Films* 284–285, 18 (1996).

745. R. D. Kornberg and H. O. Ribi, in "Protein Structure, Folding and Design" (D. L. Oxender and A. R. Liss, Eds.), p. 175. Plenum Press, New York, 1987.

746. K. M. Maloney, M. Grandbois, D. W. Grainger, C. Salesse, K. A. Lewis, and M. F. Roberts, *Biochim. Biophys. Acta Biomemb.* 1235, 395 (1995).

747. J. P. Slotte, *Biochim. Biophys. Acta Lipids Lipid Metab.* 1259, 180 (1995).

748. G. Zografi, R. Verger, and G. H. de Haas, *Chem. Phys. Lipids* 7, 185 (1971).

749. D. M. Goodman, E. M. Nemoto, R. W. Evans, and P. M. Winter, *Chem. Phys. Lipids* 84, 57 (1996).

750. I. Arimoto, M. Fujita, H. Saito, T. Handa, and K. Miyajima, *Colloid Polymer Sci.* 275, 60 (1997).

751. T. Z. Ivanova, I. Panaiotov, F. Boury, J. E. Proust, and R. Verger, *Colloid Polymer Sci.* 275, 449 (1997).

752. V. Raneva, T. Ivanova, R. Verger, and I. Panaiotov, *Colloids Surf. B: Biointerf.* 3, 357 (1995).

753. A. Alsina, O. Valls, G. Piéroni, R. Verger, and S. Garcia, *Colloid Polymer Sci.* 261, 923 (1983).

754. G. Piéroni and R. Verger, *Eur. J. Biochem.* 132, 639 (1983).

755. T. O. Wieloch, B. Borgström, G. Piéroni, F. Pattus, and R. Verger, *FEBS Lett.* 128, 217 (1981).

756. G. Piéroni and R. Verger, *J. Biol. Chem.* 254, 10090 (1979).

757. T. Wieloch, B. Borgström, G. Piéroni, F. Pattus, and R. Verger, *J. Biol. Chem.* 257, 11523 (1982).

758. A. Verger, *J. Biol. Chem.* 258, 5477 (1983).

759. C. Souvignet, J. M. Pelosin, S. Daniel, E. M. Chambaz, S. Ransac, and R. Verger, *J. Biol. Chem.* 266, 40 (1991).

760. S. Yokoyama and F. J. Kezdu, *J. Biol. Chem.* 266, 4303 (1991).

761. M. Grandbois, B. Desbat, D. Blaudez, and C. Salesse, *Langmuir* 15, 6594 (1999).

762. M. Grandbois, J. Dufourcq, and C. Salesse, *Thin Solid Films* 284–285, 743 (1996).

763. M. C. E. van Dam-Mierac, A. J. Slotboom, H. M. Verheij, R. Verger, and G. H. de Haas, in "Structure of Biological Membranes" (S. Abrahmsson and I. Pascher, Eds.), p. 177. Plenum Press, New York, 1977.

764. S. Taneva, T. McEachren, J. Stewart, and K. M. W. Keough, *Biochemistry* 34, 10279 (1995).

765. M. C. Phillips, H. Hauser, R. B. Leslie, and D. Oldani, *Biochim. Biophys. Acta* 406, 402 (1975).

766. W. M. Heckl, M. Lösche, H. Scheer, and H. Möhwald, *Biochim. Biophys. Acta* 810, 73 (1985).

767. W. M. Heckl, B. N. Zaba, and M. H. Mohwald, *Biochim. Biophys. Acta* 903, 166 (1987).

768. S. G. Taneva and K. M. W. Keough, *Biochim. Biophys. Acta Biomembr.* 1236, 185 (1995).

769. P. Kernen, W. I. Gruszecki, M. Matuta, P. Wagner, U. Ziegler, and Z. Krupa, *Biochim. Biophys. Acta Biomembr.* 1373, 289 (1998).

770. C. R. Flach, A. Gericke, K. M. W. Keough, and R. Mendelsohn, *Biochim. Biophys. Acta Biomembr.* 1416, 11 (1999).

771. J. Pérez-Gil, K. Nag, S. Taneva, and K. M. W. Keough, *Biophys. J.* 63, 197 (1992).

772. K. Y. C. Lee, M. M. Lipp, J. A. Zasadzinski, and A. J. Waring, *Biophys. J.* 72, 368 (1997).

773. K. Nag, S. Taneva, J. Pérez-Gil, A. Cruz, and K. M. W. Keough, *Biophys. J.* 72, 2638 (1997).

774. M. L. F. Ruano, K. Nag, L.-A. Worthman, C. Casals, J. Pérez-Gil, and K. M. W. Keough, *Biophys. J.* 74, 1101 (1998).

775. K. Y. C. Lee, M. M. Lipp, J. A. Zasadzinski, and A. J. Waring, *Colloids Surf. A* 128, 225 (1997).

776. R. B. Weinberg, J. A. Ibdah, and M. C. Phillips, *J. Biol. Chem.* 267, 8977 (1992).

777. A. Chávez, M. Pujol, I. Haro, M. A. Alsina, and Y. Cajal, *Langmuir* 15, 1101 (1999).

778. M. M. Lipp, K. Y. C. Lee, J. A. Zasadzinski, and A. J. Waring, *Science* 273, 1196 (1996).

779. E. Ladanyi, I. R. Miller, D. Möbius, R. Popovits-Biro, Y. Marikovsky, P. von Wichert, B. Müller, and K. Stalder, *Thin Solid Films* 180, 15 (1989).

780. B. Lu, Y. Bai, and Y. Wei, *Thin Solid Films* 240, 138 (1994).

781. X. Caide, Z. Liu, Q. Gao, Q. Zhou, and S. Sui, *Thin Solid Films* 284–285, 793 (1996).

782. R. Maget-Dana, Ch. Hetru, and M. Ptak, *Thin Solid Films* 284–285, 841 (1996).

783. E. E. Uzgiris, *Biochem. J.* 272, 45 (1990).

784. M. M. Timbs, C. L. Poglitsch, M. L. Pisarchick, M. T. Sumner, and N. L. Thompson, *Biochim. Biophys. Acta* 1064, 219 (1991).

785. S. Tanimoto and H. Kitano, *Colloids Surf. B: Biointerf.* 4, 259 (1995).

786. B. Fischer, S. P. Heyn, M. Egger, and H. Gaub, *Langmuir* 9, 136 (1993).

787. I. Vikholm, E. Györvary, and G. Peltonen, *Langmuir* 12, 3276 (1996).

788. M. Piepenstock and M. Lösche, *Thin Solid Films* 210/211, 793 (1992).

789. V. Erokhin, S. Vakula, and C. Nicolini, *Progr. Colloid Polymer Sci.* 93, 229 (1993).

790. V. Erokhin, S. Vakula, and C. Nicolini, *Thin Solid Films* 238, 88 (1994).

791. L. V. Belovolova, T. V. Konforkina, V. V. Savransky, H. Lemmetyinen, and E. Vuorimaa, *Mol. Mater.* 6, 189 (1996).

792. S. Wang and S. Sui, *Acta Biophys. Sinica* 8, 148 (1992).

793. I. Vikholm, J. Peltonen, and O. Teleman, *Biochim. Biophys. Acta Biomembr.* 1233, 111 (1995).

794. K. Nag, J. Perez-Gil, A. Cruz, N. H. Rich, and K. M. W. Keough, *Biophys. J.* 71, 1356 (1996).

795. V. Marchi-Artzner, J.-M. Lehn, and T. Kunitake, *Langmuir* 14, 6470 (1998).

796. L. Marron-Brignone, R. M. Morélis, J.-P. Chauvet, and P. R. Coulet, *Langmuir* 16, 498 (2000).

797. S. Sui and S. Wang, *Thin Solid Films* 210/211, 57 (1992).

798. E. Kalb and L. K. Tamm, *Thin Solid Films* 210/211, 763 (1992).

799. V. S. Kulkarni and R. E. Brown, *Thin Solid Films* 244, 869 (1994).

800. V. Erokhin, S. Carrara, S. Guerzoni, P. Ghisellini, and C. Nicolini, *Thin Solid Films* 327–329, 636 (1998).

801. P. Facci, V. Erokhin, and C. Nicolini, *Thin Solid Films* 243, 403 (1994).

802. G. Alegria and P. L. Dutton, *Biochim. Biophys. Acta* 1057, 239 (1991).

803. G. Alegria and P. L. Dutton, *Biochim. Biophys. Acta* 1057, 258 (1991).

804. C. C. Moser, R. J. Sension, A. Z. Szarka, S. T. Repinec, R. M. Hochstrasser, and P. L. Dutton, *Chem. Phys.* 197, 343 (1995).

805. K. Iida, A. Kashiwada, and M. Nango, *Colloids Surf. A* 169, 199 (2000).

806. Y. Yasuda, M. Hara, J. Miyake, and H. Toyotama, *Japan. J. Appl. Phys. Lett.* 36, L557 (1997).

807. J. Goc, M. Hara, T. Tateishi, J. Miyake, A. Planner, and D. Frackowiak, *J. Photochem. Photobiol. A* 104, 123 (1997).

808. J. Miyake, T. Majima, K. Namba, M. Hara, Y. Asada, H. Sugino, S. Ajiki, and H. Toyotama, *Mater. Sci. Eng. C* 1, 63 (1994).

809. T. Ueno, J. Miyake, T. Fuju, M. Shirai, T. Arai, Y. Yasuda, and M. Hara, *Supramol. Sci.* 5, 783 (1998).

810. M. Fujihira, M. Sakomura, and T. Kamei, *Thin Solid Films* 180, 43 (1989).

811. Y. Yasuda, Y. Hirata, H. Sugino, M. Kumei, M. Hara, J. Miyake, and M. Fujihira, *Thin Solid Films* 210/211, 733 (1992).

812. Y. Yasuda, Y. Kawakami, and H. Toyotama, *Thin Solid Films* 292, 189 (1997).

813. M. Ikonen, A. Sharonov, N. Tkachenko, and H. Lemmetyinen, *Adv. Mater. Opt. Electron.* 2, 115 (1993).

814. M. Ikonen, A. Sharonov, N. Tkachenko, and H. Lemmetyinen, *Adv. Mater. Opt. Electron.* 2, 211 (1993).

815. J. P. Wang, J. R. Li, P. D. Tao, X. C. Li, and L. Jiang, *Adv. Mater. Opt. Electron.* 4, 219 (1994).

816. H. H. Weetall, A. B. Druzhko, L. A. Samuelson, A. R. de Lera, and R. Alvarez, *Bioelectrochem. Bioenerget.* 44, 37 (1997).

817. A. A. Kononenko, Y. P. Lukashev, D. S. Broun, S. K. Chamorovskii, B. A. Kruming, and S. A. Yakushev, *Biophysics* 38, 1025 (1993).

818. C. Steinem, A. Janshoff, F. Hohn, M. Sieber, and H. Galla, *Chem. Phys. Lipids* 89, 141 (1997).

819. B.-F. Li, J.-R. Li, Y.-C. Song, and L. Jiang, *Mater. Sci. Eng.* 3, 219 (1995).

820. F. T. Hong, *Mater. Sci. Eng. C* 4, 267 (1997).

821. E. P. Lukashev, S. Yu. Zaitsev, A. A. Kononenko, and V. P. Zubov, *Stud. Biophys.* 132, 111 (1989).

822. T. Miyasaka and K. Koyama, *Thin Solid Films* 210/211, 146 (1992).

823. J. R. Li, J. P. Wang, T. F. Chen, L. Jiang, K. S. Hu, A. J. Wang, and M. Q. Tan, *Thin Solid Films* 210/211, 760 (1992).

824. H. H. Weetall and L. A. Samuelson, *Thin Solid Films* 312, 314 (1998).

825. J. Min, H.-G. Choi, J.-W. Choi, W. H. Lee, and U. R. Kim, *Thin Solid Films* 327–329, 698 (1998).

826. L. Yuan, B. Li, and L. Jiang, *Thin Solid Films* 340, 262 (1999).

827. H. Lemmetyinen and M. Ikonen, *Trends Photochem. Photobiol. Res. Trends* 3, 413 (1994).

828. H. Lemmetyinen, M. Ikonen, A. Sharanov, and N. Tkachenko, "Laser Spectroscopy of Biomolecules," *Proc. SPIE* 1921, 209 (1993).

829. C. Nicolini, V. Erokhin, S. Paddeu, and M. Sartore, *Nanotechnology* 9, 223 (1998).

830. V. Volkov, Y. P. Svirko, V. F. Kamalov, L. Song, and M. A. El-Sayed, *Biophys. J.* 73, 3164 (1997).

831. A. S. Alexeev, S. I. Valiansky, and V. V. Savransky, *ICPAS Mater.* 38, 209 (1992).

832. A. V. Kir'yanov, I. A. Maslyanitsyn, V. V. Savranskii, V. D. Shigorin, and H. Lemmetyinen, *Quantum Electron.* 29, 85 (1999).

833. T. Miyasaka, K. Koyama, and I. Itoh, *Science* 255, 342 (1991).

834. Y. Okada-Shudo, I. Yamaguchi, H. Tomioka, and H. Sasabe, *Synth. Met.* 81, 147 (1996).

835. J. Yang and G. Wang, *Thin Solid Films* 324, 281 (1998).

836. T. V. Dyukova and E. P. Lukashev, *Thin Solid Films* 283, 1 (1996).

837. V. Erokhin, in "Thin Protein Films" (Yu. Lvov and H. Möhwald, Eds.). Marcel Dekker, New York, 1999.

838. T. Dubrovsky, A. Tronin, and C. Nicolini, *Thin Solid Films* 257, 130 (1995).

839. T. Dubrovsky, A. Tronin, S. Dubrovskaya, O. Guryev, and C. Nicolini, *Thin Solid Films* 284–285, 698 (1996).

840. T. Dubrovsky, A. Tronin, S. Dubrovskaya, S. Vakula, and C. Nicolini, *Sensors Actuators B: Chem.* 23, 1 (1995).

841. H. Ebato, C. A. Gentry, J. N. Herron, W. Müller, Y. Okahata, H. Ringsdorf, and P. A. Suci, *Anal. Chem.* 66, 1683 (1994).

842. V. Erokhin, R. Kayushina, Yu. Lvov, and L. Feigin, *Nuovo Cimento* 12D, 1253 (1990).

843. I. Turko, I. Pikuleva, and V. Erokhin, *Biol. Membr.* 4(10), 1745 (1991).

844. Z. Lin, R. M. Hill, H. T. Davis, and M. D. Ward, *Langmuir* 10, 4060 (1994).

845. V. Chernov, P. Monceau, M. Saint-Paul, L. Galchenkov, S. Ivanov, and I. Pyataikin, *Solid State Commun.* 97, 49 (1996).

846. A. Ahluwalia, D. De Rossi, and A. Schirone, *Thin Solid Films* 210/211, 726 (1992).

847. A. Ahluwalia, D. De Rossi, C. Ristori, A. Schirone, and G. Serra, *Biosensors Bioelectron.* 7, 207 (1992).

848. A. Ahluwalia, D. De Rossi, M. Monici, and A. Schirone, *Biosensors Bioelectron.* 6, 133 (1991).

849. T. B. Dubrovsky, M. V. Demecheva, A. P. Savitsky, E. Yu. Mantrova, A. I. Yaropolov, V. V. Savransky, and L. V. Belovolova, *Biosensors Bioelectron.* 8, 377 (1993).

850. D. Izhaky and L. Addadi, *Adv. Mater.* 10, 1009 (1998).

851. I. Vikholm, T. Viitala, W. M. Albers, and J. Peltonen, *Biochim. Biophys. Acta Biomembr.* 1421, 39 (1999).

852. G. K. Chudinova, A. V. Chudinov, V. V. Savransky, and A. M. Prokhorov, *Thin Solid Films* 307, 294 (1997).

853. K. Owaku, M. Goto, Y. Ikariyama, and M. Aizawa, *Anal. Chem.* 67, 1613 (1995).

854. A. Ahluwalia, G. Giusto, and D. De Rossi, *Mater. Sci. Eng. C* 3, 267 (1995).

855. I. V. Turko, G. I. Lepesheva, and V. L. Chashchin, *Thin Solid Films* 230, 70 (1993).

856. I. V. Turko, I. S. Yurkevich, and V. L. Chashchin, *Thin Solid Films* 205, 113 (1991).

857. I. V. Turko, I. S. Yurkevich, and V. L. Chashchin, *Thin Solid Films* 210/211, 710 (1992).

858. G. I. Lepesheva, I. V. Turko, I. A. Ges', and V. L. Chashchin, *Biokhimiya* 59, 939 (1994).

859. G. I. Lepesheva, T. N. Azeva, V. N. Knyukshto, V. L. Chashchin, and S. A. Usanov, *Sensors Actuators B* 68, 27 (2000).

860. Y. Okawa, H. Tsuzuki, S. Yoshida, and T. Watanabe, *Anal. Sci.* 5, 507 (1989).

861. H. Tsuzuki, T. Watanabe, Y. Okawa, S. Yoshida, S. Yano, K. Koumoto, M. Komiyama, and Y. Nihei, *Chem. Lett.* 1265 (1988).

862. F. Q. Tang, J. S. Yuan, and L. Jiang, *Chem. Sensors* 10, 26 (1990).

863. G. Dai, J. Li, and L. Jiang, *Colloids Surf. B* 13, 105 (1999).

864. Anicet, A. Anne, J. Moiroux, and J.-M. Saveant, *J. Am. Chem. Soc.* 120, 7115 (1998).

865. N. Dubreuil, S. Alexandre, C. Fiol, and J. M. Valleton, *J. Colloid Interf. Sci.* 181, 393 (1996).

866. A. Baszkin, M. M. Boissonnade, V. Rosilio, A. Kamyshny, and S. Magdassi, *J. Colloid Interf. Sci.* 209, 302 (1999).

867. V. Rosilio, M.-M. Boissonnade, J. Zhang, L. Jiang, and A. Baszkin, *Langmuir* 13, 4669 (1997).

868. S. Y. Zaitsev, *Sensors and Actuators B: Chem.* 24, 177 (1995).

869. A. J. Guiomar, S. D. Evans, and J. T. Guthrie, *Supramol. Sci.* 4, 279 (1997).

870. P. Zhu, Y. Li, L. Wang, W. Zhang, X. Ma, and Z. Zhu, *Thin Film Sci. Technol.* 4, 70 (1991).

871. Y. Okahata, T. Tsuruta, K. Ijiro, and K. Ariga, *Thin Solid Films* 180, 65 (1989).

872. C. Fiol, J.-M. Valleton, N. Delpire, G. Barbey, A. Barraud, and A. Ruaudel-Teixier, *Thin Solid Films* 210/211, 489 (1992).

873. C. Fiol, S. Alexandre, N. Dubreuil, and J. M. Valleton, *Thin Solid Films* 261, 287 (1995).

874. J.-r. Li, Y.-k. Du, P. Boullanger, and L. Jiang, *Thin Solid Films* 352, 212 (1999).

875. Skeletization.

876. L. Marron-Brignone, R. M. Morelis, and P. R. Coulet, *Langmuir* 12, 5674 (1996).

877. P. Pal, D. Nandi, and T. N. Misra, *Thin Solid Films* 239, 138 (1994).

878. F. Antolini, S. Paddeu, and C. Nicolini, *Langmuir* 11, 2719 (1995).

879. S. Paddeu, F. Antolini, T. Dubrovsky, and C. Nicolini, *Thin Solid Films* 268, 108 (1995).

880. S. Paddeu, V. Erokhin, and C. Nicolini, *Thin Solid Films* 284–285, 854 (1996).

881. F. Bruno, K. A. Marx, S. K. Tripathy, J. A. Akkara, L. A. Samuelson, and D. L. Kaplan, *J. Intelligent Mater. Syst. Struct.* 5, 631 (1994).

882. F. F. Bruno, J. A. Akkara, L. A. Samuelson, D. L. Kaplan, B. K. Mandal, K. A. Marx, J. Kumar, and S. K. Tripathy, *Langmuir* 11, 889 (1995).

883. F. F. Bruno, J. A. Akkara, L. A. Samuelson, D. L. Kaplan, K. A. Marx, and S. K. Tripathy, *MRS Proc.* 292, 147 (1993).

884. F. Bruno, J. Akkara, L. A. Samuelson, B. K. Mandal, D. K. Kaplan, K. A. Marx, and S. Tripathy, *Polym. Prep. (Am. Chem. Soc., Div. Polym. Chem.)* 32, 232 (1991).

885. V. Erokhin, F. Antolini, P. Facci, and C. Nicolini, *Progr. Colloid Polymer Sci.* 93, 228 (1993).

886. C. Nicolini, V. Erokhin, F. Antolini, P. Catasti, and P. Facci, *Biochim. Biophys. Acta* 1158, 273 (1993).

887. Y. Shen, C. R. Safinya, K. S. Liang, A. F. Ruppert, and K. J. Rothshild, *Nature* 336, 48 (1993).

888. V. Erokhin, P. Facci, A. Kononenko, G. Radicchi, and C. Nicolini, *Thin Solid Films* 284–285, 805 (1996).

889. V. Erokhin, P. Facci, and C. Nicolini, *Biosensors Bioelectron.* 10, 25 (1995).

890. F. Antolini, C. Nicolini, and M. Trotta, *Thin Solid Films* 254, 252 (1995).

891. P. Facci, V. Erokhin, F. Antolini, and C. Nicolini, *Thin Solid Films* 237, 19 (1994).

892. V. Erokhin, B. Popov, B. Samori, and A. Yakovlev, *Mol. Cryst. Liq. Cryst.* 215, 213 (1992).

893. Y. Fang and J. Yang, *J. Phys. Chem.* 101, 441 (1997).

894. M. A. Frommer, I. R. Miller, and A. Khaïat, *Adv. Exp. Med. Biol.* 7, 119 (1970).

895. K. Ijiro, M. Shimomura, M. Tanaka, H. Nakamura, and K. Hasebe, *Thin Solid Films* 284–285, 780 (1996).

896. R. Ionov, J. De Coninck, and A. Angelova, *Thin Solid Films* 284–285, 347 (1996).

897. M. A. Karymov, A. A. Kruchinin, A. Yu, A. Tarantov, L. A. Balova, L. A. Remisova, N. G. Sukhodolov, A. I. Yanklovich, and A. M. Yorkin, *Sensors Actuators B* 6, 208 (1992).

898. F. Nakamura, K. Ijiro, and M. Shimomura, *Thin Solid Films* 327–329, 603 (1998).

899. C. Nicolini, V. Erokhin, P. Facci, S. Guerzoni, A. Rossi, and P. Paschkevitsch, *Biosensors Bioelectron.* 12, 613 (1997).

900. Y. Okahata and K. Tanaka, *Thin Solid Films* 284–285, 6 (1996).

901. Y. Okahata, T. Kobayashi, and K. Tanaka, *Langmuir* 12, 1326 (1996).

902. M. Shimomura, F. Nakamura, K. Ijiro, H. Taketsuna, M. Tanaka, H. Nakamura, and K. Hasebe, *J. Am. Chem. Soc.* 119, 2341 (1997).

903. C. Xiao, M. Yang, and S. Sui, *Thin Solid Films* 327–329, 647 (1998).

904. R. Vijayalakshmi, A. Dhathathreyan, M. Kanthimathi, V. Subramanian, B. U. Nair, and T. Ramasami, *Langmuir* 15, 2898 (1999).

905. S. Huebner, E. Politsch, U. Vierl, and G. Ceve, *Biochim. Biophys. Acta Biomembr.* 1421, 1 (1999).

906. K. Kago, H. Matsuoka, R. Yoshitome, H. Yamaoka, K. Ijiro, and M. Shimomura, *Langmuir* 15, 5193 (1999).

907. M. Liu, J. Lang, and H. Nakahara, *Colloids Surf. A* 175, 153 (2000).

908. M. Shimomura, O. Karthaus, N. Maruyama, K. Ijiro, T. Sawadaishi, S. Tokura, and N. Nishi, *Rep. Prog. Polym. Phys. Jpn.* 40, 523 (1997).

909. M. M. Montrel, A. I. Petrov, L. I. Shabarchina, B. I. Sukhorukov, and G. B. Sukhorukov, *Sensors Actuators B: Chem.* 42, 225 (1997).

910. G. Sukhorukov, G. Ivakin, V. Erokhin, and V. Klechkovskaya, *Physica B* 198, 136 (1994).

911. G. Sukhorukov, V. Erokhin, and A. Tronin, *Biophysics* 38, 243 (1993).

912. G. B. Sukhorukov, M. M. Montrel, A. I. Petrov, L. I. Shabarchina, and B. I. Sukhorukov, *Biosensors Bioelectron.* 11, 913 (1996).

913. B. M. Abraham and J. B. Ketterson, *Langmuir* 1, 461 (1985).

914. T. Berzina, V. Troitsky, S. Vakula, A. Riccio, A. Morana, M. De Rosa, L. Gobbi, F. Rustichelli, V. Erokhin, and C. Nicolini, *Mater. Sci. Eng. C* 3, 33 (1995).

915. T. L. Fare, K. M. Rusin, and P. P. Bey, Jr., *Powder Technol.* 3, 51 (1991).

916. V. A. Howarth, M. C. Petty, G. H. Davies, and J. Yarwood, *Langmuir* 5, 330 (1989).

917. V. A. Howarth, M. C. Petty, G. H. Davies, and J. Yarwood, *Thin Solid Films* 160, 483 (1988).

918. C. Nicolini, F. Rustichelli, T. S. Berzina, V. I. Troitsky, V. V. Erokhin, S. Vakula, A. Riccio, A. Morana, M. De Rosa, and L. Gobbi, *Mater. Sci. Eng. C* 3, 33 (1995).

919. C. Nicolini, T. S. Berzina, V. I. Troitsky, A. Riccio, S. Vakula, and M. De Rosa, *Mater. Sci. Eng. C* 5, 1 (1997).

920. S. Pathirana, L. J. Myers, V. Vodyanoy, and W. C. Neely, *Supramol. Sci.* 2, 149 (1995).

921. J. B. Peng, B. M. Abraham, P. Dutta, J. B. Ketterson, and H. F. Gibbard, *Langmuir* 3, 104 (1987).

922. H. E. Ries, Jr., and H. S. Swift, *J. Colloid Interf. Sci.* 64, 111 (1978).

923. M. Shimomura, E. Shinihara, S. Kondo, N. Tajima, and K. Koshiishi, *Supramol. Chem.* 3, 23 (1993).

924. V. Vodyanoy, S. Pathirana, and W. C. Neely, *Langmuir* 10, 1354 (1994).

925. S. Y. Zaitsev, V. P. Zubov, and D. Möbius, *Colloids Surf. A: Physicochem. Eng. Aspects* 94, 75 (1995).

926. S. Yu. Zaitsev, V. P. Vereschetin, V. P. Zubov, W. Zeiss, and D. Möbius, *Thin Solid Films* 284–285, 667 (1996).

927. S. M. Amador, S. D. Myers, V. Vodyanoy, and W. C. Neely, *Biophys. J.* 74, A371 (1998).

Chapter 11

STRUCTURE FORMATION DURING ELECTROCRYSTALLIZATION OF METAL FILMS

V. M. Kozlov

Department of Physics, National Metallurgical Academy of Ukraine, Dniepropetrovsk, Ukraine

L. Peraldo Bicelli

Dipartimento di Chimica Fisica Applicata del Politecnico, Centro di Studio sui Processi Elettrodici del CNR, Milan, 20131 Italy

Contents

1. INTRODUCTION

Metal deposits are largely employed in microelectronics, for optical devices, in cosmic and atomic technology, in galvanotechnique, and in other fields. Their numerous practical applications are based on the possibility to deposit metal films with a favorable combination of their functional properties. In addition, there is also the possibility to prepare metal deposits with particular physical and mechanical properties which substantially differ from those of the same metals in their usual massive state. These possibilities offer the opportunity for improving the physical and technical parameters of materials and devices and for promoting new technical branches.

It may well be understood that the problem to obtain metal deposits with a foreseeable set of their properties becomes even more of a present interest. A solution to this important practical problem can be found on the basis of a deep and complex investigation of a whole series of fundamental questions, those related to the structure of deposits and to the mechanism of its formation included. Indeed, by means of the structure it is possible to realize the connection between the electrodeposition conditions of metals and the properties of the obtained deposits.

The need to investigate the structure of the deposits is suggested not only for practical reasons, but also for theoretical ones. As a matter of fact, the deposit structure is the end product of the crystallization process. Therefore, the experimental results of a structural research contain the information which may be utilized to deepen and to enlarge our knowledge of the details of the real crystal growth process. It may also be useful

Handbook of Thin Film Materials, edited by H.S. Nalwa
Volume 1: Deposition and Processing of Thin Films

ISBN 0-12-512909-2/$35.00

to create physical models of the formation of a deposit with either a single-crystal structure or a polycrystalline structure and to elaborate the mechanism of formation of the different types of crystallographic defects in deposits.

To more or less reproduce the overall mechanism of formation of the structure of the deposits, it is necessary to examine all the consecutive steps of their growth. In the case of the heterogeneous crystallization on a foreign substrate, two main growth steps of the deposits may be considered in connection with the formation of the different structural defects. They are: the initial nucleation step and the subsequent deposit growth step "in thickness."

The initial crystallization step includes the formation of the nuclei on the substrate, their lateral growth until a thin compact deposit layer is produced. The laws regulating the initial step as well as the type and the quantity of the structural defects which are formed during this step are first of all determined by the degree of interaction at the interphase between the substrate and the deposit. For this step, two main growth mechanisms of the deposit may be distinguished: the Volmer–Weber [1, 2] and the Frank–van der Merwe [3, 4] mechanism. The former describes the case of a weak interaction between the substrate and an atom of the deposit. It implies the formation of the deposit through the nucleation of three-dimensional isolated nuclei, their subsequent growth tangentially to the substrate surface until their coalescence. The Frank–van der Merwe mechanism occurs owing to a strong interaction of the atoms of the deposit with the substrate. In this case, the layer-by-layer growth takes place by a two-dimensional nucleation as a consequence of the oriented influence of the crystalline substrate on the depositing metal layer (epitaxial growth).

In agreement with the numerous experimental results, the main types of crystal defects which are formed during the initial step of heterogeneous crystallization are: low- and high-angle grain boundaries, twin boundaries, dislocations, and stacking faults. It is acknowledged that their main mechanism of formation is related to the process of coalescence of the isolated nuclei of the condensed phase during their tangential growth. Furthermore, it is important to point out the mechanism of formation of the mismatch dislocations during the epitaxial growth of the deposit. Such formation occurs when at a critical thickness of the depositing metal layer a relaxation of the accumulated elastic deformation takes place through the formation of dislocations at the substrate-deposit interphase to compensate the mismatch between the parameters of the two coordinating crystalline lattices.

At the end of the formation of a thin and complete layer of the condensating phase, the further growth of the deposit occurs in thickness, i.e., in the direction perpendicular to the substrate surface (the second growth step). During this step, the substrate does practically not influence the crystallization process any more and, correspondingly, the formation of the structural defects of the deposit. Hence, during the second growth step, the mechanism of formation of crystal defects has to depend on the typical characteristics of the elementary steps of the deposit growth process in thickness.

It may be concluded that the structural aspect plays an important role in the general problem of crystallization and growth of metal deposits. In the present chapter, the structural problem is discussed taking as an example the electrocrystallization of metals, although several of the problems which are examined may completely be applied to deposits obtained in other ways, first of all by deposition from the vapor phase. Furthermore, the main attention is devoted to the formation of structural defects during the second step of the growth of the electrolytic deposits.

2. CLASSIFICATION OF THE STRUCTURAL DEFECTS IN ELECTRODEPOSITS

According to the modern ideas of solid-state physics, the defects in crystals (structural defects) are classified on the basis of their geometric characteristics, i.e., of their extent in the three-dimensional space. Among them, the main role from the point of view of the influence of these defects in the majority of the metals properties is played from the two-dimensional (surface) defects such as the boundaries among near fragments of a crystal which are disoriented the one with respect to the other. This type of structural defect was investigated in detail for the first time in metals of metallurgical origin.

In a simple case, when such a boundary among parts of the grain is a screw boundary, its crystallographic characteristic is the azimuth angle of disorientation among the single parts, θ, while its energetic characteristic is the free energy per unit area of the boundary, γ, which depends in the general case on the value of the angle θ. The value of γ may be determined from the well-known Read–Shockley relationship [5],

$$\gamma = E_0\theta(1 + \ln\theta/\theta_m) \qquad (1)$$

where

$$E_0 = Gb/4\pi(1 - \nu)$$

G is the shear modulus, b is the absolute value of the Burgers vector of the dislocation, and ν is the Poisson's ratio. This relationship gives the variation of γ in the range of the θ values between zero and the maximum value θ_m, which for the greatest part of the metals has a value in the order of 10 to 15° [6]. Figure 1 shows the dependence of γ on θ evaluated for three metals which substantially differ by their physicomechanical parameters, in particular, heat of sublimation, melting point, and elastic constants.

In agreement with the theory of dislocations [7], the boundaries between the different fragments of the crystal are classified as low-angle boundary (of dislocations), if the disorientation angle $\theta < \theta_m$, and high-angle boundary, if $\theta > \theta_m$. In the former case, γ increases to its maximum value, γ_m, when the value of θ increases from zero to θ_m and each part is called subgrain; in the latter case, γ is equal to γ_m, and generally does not depend on the value of θ while each part is called grain.

In addition to these types of boundaries, it is necessary to consider the twin boundaries which are characterized by the well-defined twinning plane typical of each crystalline lattice and by the value of the specific free energy, γ_{tw}. In the case of two parallel twin boundaries which are at an interatomic distance, a stacking fault is formed and its specific free energy is more than two times γ_{tw}.

According to the results of numerous investigations obtained by means of the modern methods of metal structure examination, such as optical microscopy, X-ray diffraction, and transmission electron microscopy (TEM), the different types of structural surface defects previously mentioned have been recognized in metal electrodeposits. The quantity and the distribution of the structural defects in the bulk of the deposit depend on both the nature of the deposited metal as well as on the electrocrystallization conditions, in particular, on the current density, the acidity of the electrolyte, and the concentration of the surface-active additives which are present in the solution [8–14].

First of all, it may be noted that as a rule polycrystalline metal deposits are formed during electrodeposition. The single-crystal structure of the deposits is formed only during the deposition of a metal on its own single-crystal substrate or even on foreign substrates but in particular electrolysis conditions. The quantitative parameter which characterizes the polycrystalline microstructure of the deposits and strongly influences their properties is the average grain size, D. Its value is inversely proportional to the extension of the high-angle boundaries in the volume unit of the polycrystalline deposit.

As to their dependence on the D value, the electrodeposits are conventionally divided into three types: large-crystalline (when $D > 10 \mu m$), thin-crystalline ($1 \mu m < D < 10 \mu m$), and ultrathin-crystalline deposits ($D < 1 \mu m$). The microstructure of the deposits cross section of the first and second crystal types was investigated in detail by Fischer [8, 15, 16] through optical microscopy. Particular attention was devoted to the character of the electrodeposits microstructure from the point of view of the shape of the grains. According to the experimental results, a classification was performed of the different types of the deposits microstructure.

Here, we mention the two main types of the microstructure of compact electrodeposits which differ in principle one from the other. The first type of microstructure (columnar or FT-type according to Fischer's classification) is characterized by the columnar shape of the grains, as shown in Figure 2a [17]. As a rule, such type of microstructure is formed when electrodeposition is carried out at a constant current from simple aqueous electrolytes without additives. The second type (homogeneous or RD-type) is characterized from the size of the grains which is nearly equal in the different directions, Figure 2b. The microstructure of these types of deposits is usually observed when electrodeposition occurs from electrolytes containing surface-active addition agents in a sufficiently high concentration and when the processes of surface adsorption of the foreign molecules strongly influence the growth of the electrodeposits.

It is easy to understand that the possibilities offered from the optical microscopy methods to study the deposits structure are rather limited and that such methodology does not allow to obtain experimental data on the bulk structure of the grains. On the other hand, these data are extremely necessary for the

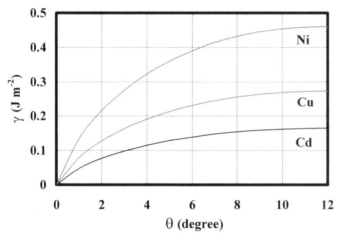

Fig. 1. Specific free energy of the dislocation boundary, γ, as a function of the azimuth disorientation angle, θ. Reprinted from [70] with kind permission from Elsevier Science.

(a)

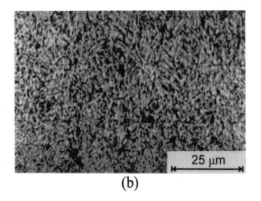

(b)

Fig. 2. Microstructure of the cross section of Cu electrodeposits obtained (a) in the absence and (b) in the presence of thiourea at a concentration of 15 mg L^{-1}. Reprinted from [17] with kind permission from Elsevier Science.

Table I.   Electrolyte Composition, Electrodeposition Conditions, and
Deposit Thickness

Metal	Electrolyte composition	pH	Current density (A dm^{-2})	t (°C)	Thickness (μm)
Ni	NiSO$_4$-400 g L^{-1}	4.0	1–5	45–70	80
Cu	CuSO$_4$-500 g L^{-1} H$_2$SO$_4$-50 g L^{-1}	—	1–3	30–60	120
Fe	FeSO$_4$-500 g L^{-1}	2.0	1–6	50–70	100
Zn	ZnSO$_4$-400 g L^{-1}	3.0	1–6	30–50	80

development of a theory of the growth of the deposits in real conditions. More complete information on the structural defects in electrolytic deposits has been obtained by the methods of the transmission electron microscopy (TEM) [18–29]. In particular, there is the possibility to investigate such structural defects as dislocations and twins.

As generalized results of this topic, we first of all consider our experimental data relevant to electrodeposits with different types of crystalline lattice (Ni, Cu—face-centered cubic (fcc), Fe—body-centered cubic (bcc), Zn—hexagonal close packed (hcp)). Electrodeposition of the former metals has been carried out from concentrated acid sulfate baths not containing organic addition agents, at relatively low overvoltages (see Table I). These are the favorable conditions to obtain deposits which are practically free from macrostresses known to be at the origin of plastic deformation and to produce deformation defects (in particular dislocations and twins). Therefore, the structural defects of these deposits observed during our TEM investigation may be classified from the point of view of their nature as defects formed during the growth process of the same deposits, only.

Moreover, care was taken to exclude the formation of structural defects due to the direct influence of the substrate. To reach this goal, first, the electrodeposition of these metals was performed on an indifferent substrate which had to assure a weak adhesion and, therefore, an insignificant influence on the formation of the deposits structure. For this purpose, electrodeposition was carried out on a mechanically polished stainless steel substrate from which the deposit could easily be detached.

Second, the deposits were thick (see Table I) and were analyzed following the TEM methodology. The samples to be submitted to TEM analysis were obtained by anodic polishing of the deposits in orthophosphoric acid. In this way, we investigated the layers of the deposited metals which were at a relatively high distance from the substrate. So, it could be assumed that the structural defects observed in these layers had been formed according to the mechanism relevant to the growth process of the deposits.

The investigation of the TEM micrographs showed that the deposits had the polycrystalline structure and that from the point of view of the average size of their grains those of Ni, Cu,

and Fe were thin crystalline, while those of Zn were large crystalline. As to the details of the internal structure of the grains, TEM investigation showed that they were composed of subgrains set as layers. Figure 3a and b shows the substructure of a Cu and Ni deposit when the planes of the subgrain boundaries were perpendicular to the deposit surface. The type of these boundaries was determined by a crystallographic analysis of the electron diffraction pattern of the regions containing them (e.g., the M and N areas in Fig. 3a and b, respectively). It was observed that in some cases the subgrain boundaries were twin boundaries, the octahedral (111) plane being the twin plane (Fig. 3c and e). In other cases, the angular separation of the diffraction spots (Fig. 3d and f) allowed us to evaluate the azimuth disorientation angle between neighboring subgrains. For example, such an angle was small, around 5 to 6°, as shown in Figure 3d. Consequently, the subgrain boundary in N (Fig. 3b) is a low-angle boundary, i.e., a dislocation boundary.

The presence of dislocation boundaries in the bulk of the grains was confirmed by direct observation when the boundary surface was parallel to, or only a little tilted toward, the deposit surface. In this case, it was possible to observe the dislocation boundaries as typical dislocation networks (Fig. 4). The dislocation density was proportional to the mean disorientation angle between the subgrains. Such an angle increased with the current density and at decreasing the electrodeposition temperature, that is, it increased with the cathodic overvoltage. For example, in the case of Cu deposits, such an angle increased from 1–2° to 5–6°, increasing the current density from 1 to 3 A dm^{-2}.

It was found that growth twins were present in large quantities in Ni and Cu deposits (fcc lattice, (111) twin plane, and [11$\bar{2}$] shear direction) but in small quantities in Fe deposits (bcc, (112), and [11$\bar{1}$], respectively) and in Zn deposits (hcp, (10$\bar{1}$2), and [$\bar{1}$011]) and only when they were obtained at current densities higher than 4 A dm^{-2}.

Moreover, it resulted that the deposits with an fcc lattice (Cu and Ni) had the specific structure for twinning: in addition to twins inside the same grain with parallel {111} planes, twinning with intersecting octahedral planes also took place. This was confirmed by direct observation of the grains with the ⟨110⟩ zone axis in TEM micrographs where the points of intersection of some twin boundaries were observed which were perpendicular to the surface of the deposit (Fig. 5).

In addition to dislocation boundaries and twinning, stacking faults (Fig. 6) were observed in Cu and Ni deposits obtained at high current densities (3 A dm^{-2} and 5 A dm^{-2}, respectively).

In conclusion, the experimental results of the investigation of the electrodeposits structure are the basis for stating that the main types of structural defects which are formed during the growth step when the substrate does not influence the process any more are high-angle boundaries of the grains, dislocation boundaries, twins, and stacking faults. As already stated, their presence and quantity depend on the nature of the deposited metal and on the electrocrystallization conditions.

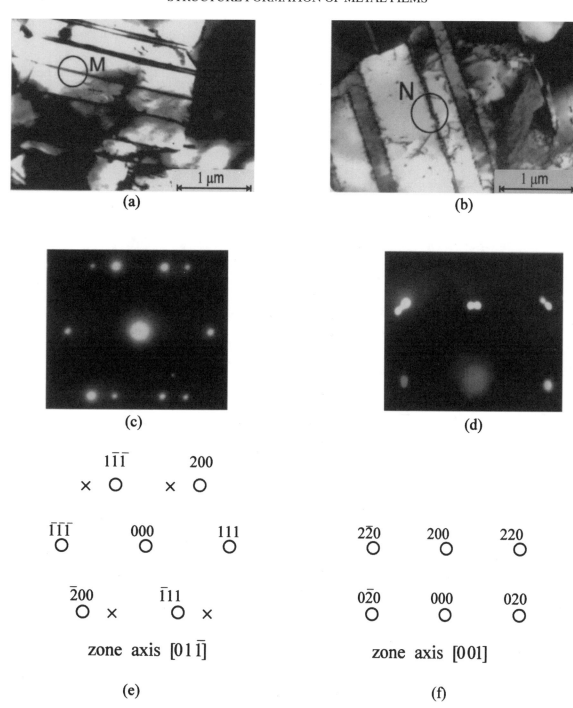

Fig. 3. TEM micrographs of (a) Cu and (b) Ni electrodeposits; (c) and (d) diffraction patterns of the M and N regions; (e) and (f) related schemes. (○) Reflexes of the matrix; (×) reflexes of the twins. The deposition conditions are: current density 2 A dm^{-2}; $t = 60°$C. Reprinted from [29] with kind permission from Elsevier Science.

3. MECHANISM OF FORMATION OF STRUCTURAL DEFECTS DURING NONCOHERENT NUCLEATION

The mechanism of formation of structural defects in electrodeposits is based on the one side on the generalized experimental data resulting from the deposits structure investigation, which was considered in the previous section and, on the other side,

is based on the theoretical data resulting from crystal growth science.

Regarding the results of structural research, it is worth underlining that the main types of crystallographic defects in electrodeposits (high-angle grain boundaries, subgrain boundaries of dislocations, and twins) have the same general character from the geometrical point of view; i.e., they belong to the group of two-dimensional surface defects. They sepa-

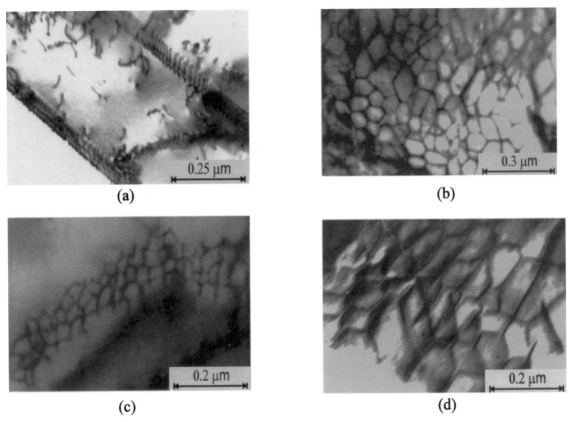

Fig. 4.   TEM micrographs showing the dislocation networks in the grains of (a) Ni, (b) Cu, (c) Fe, and (d) Zn electrodeposits. The deposition conditions are: Ni has a current density 2 A dm^{-2}; $t = 60°$C; Cu has a current density 1 A dm^{-2}; $t = 45°$C; Fe has a current density 3 A dm^{-2}; $t = 60°$C; Zn has a current density 3 A dm^{-2}; $t = 30°$C. Reprinted from [29] with kind permission from Elsevier Science.

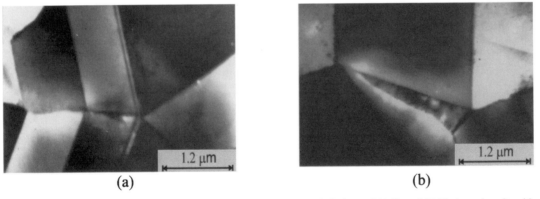

Fig. 5.   TEM micrographs showing the multitwinning of intersecting octahedral planes of (a) Cu and (b) Ni electrodeposits with a $\langle 110 \rangle$ orientation axis. The deposition conditions are: Cu has a current density 2 A dm^{-2}; $t = 30°$C; Ni has a current density 3 A dm^{-2}; $t = 60°$C.

rate nearby parts of a deposit which are in a determined different relative crystallographic orientation one with respect to the other. Hence, it may be expected that even the mechanism of their formation has a general unique character and has to be related to a determined elementary step of the electrodeposit growth process in thickness. Here and in the following, only structural defects are considered which are formed when the substrate does not influence the process any more.

Moreover, it must be evidenced that the considered structural defects in electrodeposits are not in thermodynamic equilibrium and, therefore, that their process of formation during crystal growth is expected to present typical fluctuations which are also typical of the nucleation step, one of the elementary step of metal electrocrystallization.

Taking account of the previously mentioned (geometrical and thermodynamic) factors relevant to defects in deposits, it may be assumed that the mechanism of formation of the main

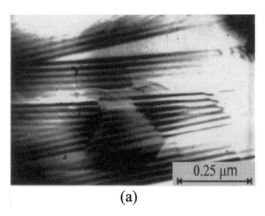

Fig. 6.   TEM micrographs showing the stacking faults of (a) Cu and (b) Ni electrodeposits. The deposition conditions are: Cu has a current density 3 A dm^{-2}; $t = 30°$C; Ni has a current density 5 A dm^{-2}; $t = 45°$C.

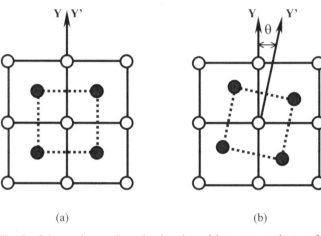

(a)   (b)

Fig. 7.   Scheme of a two-dimensional nucleus of four atoms on the top of a {100} plane of an fcc metal (a) in a normal position, and (b) in a disoriented position.

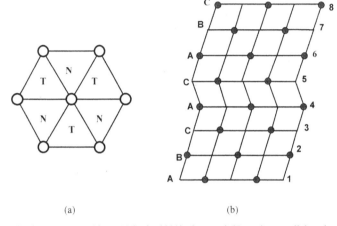

(a)   (b)

Fig. 8.   Atoms positions (a) in the {111} plane and (b) section parallel to the {110} plane for an fcc crystal.

structural defects is unique and is connected with the step of nuclei formation.

In agreement with the classical theory, crystal growth occurs through the continuous formation of two-dimensional nuclei [1, 4, 30–32]. It is necessary to stress that the classical theory assumes that the two-dimensional nuclei which are formed on their own substrate plane are in a normal (regular) position from the geometrical point of view. It is obvious that in this case, i.e., in the case of normal nucleation, no crystal defects are formed on the plane of the interface between the nucleus and the substrate.

Another possible situation may occur if we assume that, in addition to normal nuclei with a defined probability, noncoherent nuclei are also formed. These nuclei lie in an irregular crystallographic position of the growing surface and their growth forms a disordered layer which is bound to the underlying plane by the two-dimensional (surface) structural defect. Figure 7a shows the disposition of a two-dimensional normal nucleus of four atoms on the top of a {100} plane of an fcc crystal and Figure 7b the simplest model of a noncoherent nucleus having an azimuth disorientation angle, θ, with respect to its own substrate plane.

The actual type of a two-dimensional defect which is formed owing to noncoherent nucleation depends on the value of the disorientation angle. More precisely:

1. if the value of θ is smaller than 10–15°, a dislocation boundary is formed;
2. if it is greater than 10–15°, the usual high-angle grain boundary is formed.

It is necessary to draw attention to the third case which may occur, that is when nucleation takes place on a crystallographic plane which is the typical twinning plane of the crystal. As an example, an fcc crystal may be considered whose twinning plane is the octahedral {111} plane. This plane typically has two stable equilibrium positions for the growth of the next atomic plane, the normal (regular) N and the twin positions T (Fig. 8a). If this plane grows in the direction perpendicular to its surface through a subsequent normal nucleation, i.e., when the atoms of the nuclei occupy exactly the normal positions N, the usual stacking of the octahedral planes without defects ABCABCA... takes place. A different situation occurs when the atoms of one of the nuclei occupy the twin position T. In this case, a twin is formed and the stacking of the close-packed planes will be ABCACBA....

A particular case is observed when the nuclei are formed two times in sequence in the twin position and as a result a stacking fault is formed. This process will be examined considering as an example the case of an fcc crystal which has been formed through a continuous nucleation on octahedral planes (Fig. 8b). The figure shows that the octahedral atomic planes 1 to 4, 7, 8 are formed through normal nucleation, while planes 5 and 6 are formed by nuclei which have occupied twin positions. Hence, the stacking of the close-packed planes will be ABC$\underline{AC}$ABC. . . and the crystal will contain a stacking fault.

From the crystallographic point of view, while noncoherent nuclei of types (1) and (2) may be formed on any plane, twin nuclei may be formed on a twin plane, only. Therefore, the quantity of growth twins in the electrodeposits must depend not only on the thermodynamic factor (oversaturation value) but also on the crystallographic habitus of the growing crystals.

In agreement with literature data, Kern [33] was the first who suggested a twinning mechanism for the nucleation of two-dimensional nuclei on a twin plane of the growing crystal. Applying classical thermodynamics for phase transformation, he determined the nucleation work of the normal and twin nucleus, analyzing the influence of the electrodeposition conditions on their relative rate of formation.

Kern's theory had an evolution by Pangarov [34] who followed the method of the average work of separation [35]. The theoretical analysis of the equations he obtained for the work of formation of nuclei in normal and twin positions showed that the twin nuclei may be formed only when the value of the oversaturation is higher than a threshold value. Then, increasing the oversaturation values strongly increases the relative probability of the formation of the twin nuclei and for relatively high oversaturation values it is already equal to the probability of the formation of normal nuclei.

The fundamental ideas of the previous theory of twin formation during deposits growth were confirmed by the numerous research works investigating fcc metals (Ag, Ni, and Cu) [36–43].

As to the formation of dislocation boundaries in deposits, Sears was the first who suggested that such formation is connected to that of noncoherent two-dimensional nuclei [44]. The determination of the probability of formation of noncoherent nuclei was performed on the basis of the classical theory of nucleation.

The mechanism of formation of structural defects in the deposits which is also connected to the process of noncoherent nucleation was considered in Refs. [45–47]. It may occur when two or more nuclei are formed on a growing crystal plane which are casually disoriented with respect to this plane. These same nuclei will be disoriented among them. Therefore, as a result of their lateral growth and coalescence, structural defects between near atomic layers must be formed. The actual type of these defects (high-angle grain boundary, dislocation boundary, twin boundary) will depend on the value of the disorientation angle between the nuclei.

Hence, the formation of the main crystal defects in the deposits may be examined from the point of view of the noncoher-

ent nucleation process which may take place during the growth of the deposit crystals. Therefore, the step of formation of noncoherent nuclei must be analyzed in detail to determine the general laws concerning the influence of the electrocrystallization conditions on the structure of metal deposits.

4. CLASSICAL THEORY OF NONCOHERENT NUCLEATION

Since the formation of the main types of structural defects in deposits is related to the nucleation step during crystal growth, the process of formation of noncoherent nuclei will first of all be examined according to classical theory of heterogeneous nucleation [48, 49]. It will be assumed that the noncoherent nucleus presenting an azimuth disorientation angle with respect to the substrate plane of its same nature is three-dimensional with a well-defined shape, e.g., it is a cylinder.

For the thermodynamic analysis of noncoherent nucleation, as an example, a close-packed plane of an fcc lattice will be examined (Fig. 8a). A noncoherent nucleus may be obtained by rotation of the normal nucleus around the axis perpendicular to the octahedral plane by an angle θ. During the rotation of the nucleus by an angle in the range $0° < \theta < 60°$ and also $60° < \theta < 120°$, each atom of the nucleus crosses a series of unstable positions characterized by an excess free energy which depends on the azimuth disorientation angle of the same nucleus. The value $\theta = 60°$ corresponds to the stable equilibrium position of the twin nucleus and the value $\theta = 120°$ to the case of a normal nucleus occupying the regular stable position.

Contrary to the normal and twin nucleus, the noncoherent nucleus is not stable and, therefore, tends to decrease its free energy by spontaneously reorienting itself to the normal or twin position. However, in the actual electrocrystallization conditions, foreign atoms and/or molecules adsorbed on the surface of the growing crystal hinder such reorientation.

For the thermodynamic evaluation of the work of nucleation of a noncoherent nucleus, the excess free energy of the nucleus, E_s, must be considered, which depends on the azimuth disorientation angle, θ,

$$E_s = S\gamma(\theta) \qquad (2)$$

where S is the contact area of the nucleus with the surface of the underlying plane and $\gamma(\theta)$ is the free energy per unit area of the noncoherent bond between nucleus and substrate.

According to the dislocation theory, the azimuth disorientation of contiguous crystal layers with an angle not greater than the value $\theta_m = 10$–$15°$ causes the formation of a dislocation boundary in the plane in between these layers. Taking into account that the disoriented layers of the deposits are produced by noncoherent nucleation, in first approximation, it may be assumed that $\gamma(\theta)$ is also equal to the free energy per unit area of the dislocation boundary which may be evaluated by the Read–Shockley relationship [Eq. (1)]. When the θ values are greater than θ_m, $\gamma(\theta)$ may be assumed to be constant and equal to the free energy per unit area of the usual grain boundary, γ_m. Let us

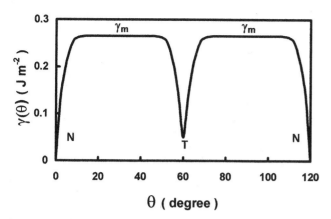

Fig. 9. Free energy per unit area, γ, of the noncoherent bond between the nucleus and the octahedral plane of an fcc lattice as a function of the azimuth disorientation angle, θ. The data refer to Cu.

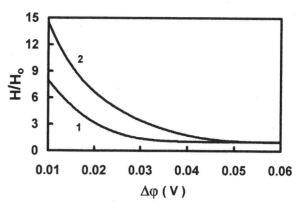

Fig. 10. Relative height of the noncoherent Cu nucleus, H/H_o, as a function of the crystallization overvoltage, $\Delta\varphi$. Disorientation angle of the nucleus: $3°$ (curve 1) and $10°$ (curve 2). $\sigma = 1\,\mathrm{J\,m^{-2}}$.

take these considerations into account and also that the atoms of the noncoherent nucleus periodically occupy the normal, N, and the twin, T, positions during their rotation around the axis perpendicular to the octahedral plane (Fig. 8a). Then, the dependence from the azimuth disorientation angle of the free energy per unit area of the noncoherent bond between the nucleus and the {111} plane of an fcc lattice is expected to follow the trend shown in Figure 9. The data employed for the calculations refer to copper.

Hence, it may be underlined once again that the types of structural defects which are formed during the metal deposition process owing to noncoherent nucleation depend on the azimuth disorientation angle of the nucleus. More precisely, as already stated:

1. if the value of θ is smaller than 10–15°, a dislocation boundary is formed;
2. if it is greater than 10–15°, the usual grain boundary is formed;
3. if the nucleus is in a favorable position for twinning, a twin boundary is formed.

To analyze the formation of three-dimensional noncoherent nuclei, we utilize the general Gibbs equation for heterogeneous nucleation taking into account that the nucleus has a cylindrical shape, with radius R and height H, and also Eq. (2),

$$\Delta G_3 = -\frac{\pi R^2 H}{V_o}\Delta\mu + 2\pi R H \sigma + \pi R^2 \gamma(\theta) \tag{3}$$

where ΔG_3 is the Gibbs free energy of formation of the noncoherent nucleus, V_o is the atom volume, σ is the (lateral) surface free energy per unit area of the nucleus, and $\Delta\mu$ is the variation of the chemical potential due to the phase transition which depends on the oversaturation of the system [50],

$$\Delta\mu = kT \ln C/C_o \tag{4}$$

k is the Bolzmann constant, T is the absolute temperature, C and C_o are the adatoms concentration in the oversaturation and equilibrium state, respectively.

In the case of metals electrocrystallization, $\Delta\mu$ is equal to $z e_o \Delta\varphi$, z is the number of charges transferred during the charge-transfer process, e_o is the electron absolute charge, and $\Delta\varphi$ is the absolute value of the crystallization overvoltage characterizing the deviation of the system from the equilibrium state.

Considering the minimum of Eq. (3), that is the conditions owing to which the nucleus assumes its equilibrium size, we obtain the dimensions and the nucleation work of the noncoherent three-dimensional critical nucleus,

$$R_3 = \frac{2V_o\sigma}{ze_o\Delta\varphi} \tag{5}$$

$$H_3 = \frac{2V_o\gamma(\theta)}{ze_o\Delta\varphi} \tag{6}$$

$$A_3 = \frac{4\pi\sigma^2 V_o^2 \gamma(\theta)}{(ze_o\Delta\varphi)^2} \tag{7}$$

According to Eqs. (5) and (6), the nucleus radius and the height decrease at increasing overvoltage. However, contrary to the R_3 value, the H_3 value depends on the disorientation angle of the noncoherent nucleus and not on the value of the surface free energy per unit area of the nucleus. Moreover, in any case, the nucleus height has a minimum value, H_o, equal to the atom diameter (Fig. 10). This means that the noncoherent nucleation process may be analyzed from the point of view of a three-dimensional nucleation process only in the limit of the H_3 values not smaller than $2H_o$. In the opposite case, it is necessary to utilize the Gibbs equation for the heterogeneous two-dimensional nucleation,

$$\Delta G_2 = -\frac{\pi R^2 H_o}{V_o}\Delta\mu + 2\pi R \chi + \pi R^2 \gamma(\theta) \tag{8}$$

where ΔG_2 is the Gibbs free energy of formation of the noncoherent two-dimensional nucleus and χ is the linear free energy per unit length of the nucleus boundary, equal to $H_o\sigma$.

From the minimum of Eq. (8), the radius and nucleation work of the noncoherent two-dimensional critical nucleus are

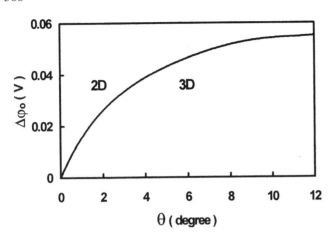

Fig. 11. Value of the critical crystallization overvoltage, $\Delta\varphi_o$, as a function of the azimuth disorientation angle, θ, of the noncoherent nucleus of Cu. $\sigma = 1\,\mathrm{J\,m^{-2}}$.

obtained:

$$R_2 = \frac{(V_o/H_o)\chi}{ze_o\Delta\varphi - (V_o/H_o)\gamma(\theta)} \quad (9)$$

$$A_2 = \frac{\pi\chi^2(V_o/H_o)}{ze_o\Delta\varphi - (V_o/H_o)\gamma(\theta)} \quad (10)$$

Since crystal growth occurs through the formation not only of noncoherent (two-dimensional or three-dimensional) but also of normal (two-dimensional) nuclei, introducing the condition $\gamma(\theta) = 0$ into Eqs. (9) and (10), the radius and nucleation work of the normal two-dimensional critical nucleus are obtained:

$$R_o = \frac{(V_o/H_o)\chi}{ze_o\Delta\varphi} \quad (11)$$

$$A_o = \frac{\pi\chi^2(V_o/H_o)}{ze_o\Delta\varphi} \quad (12)$$

Hence, the overvoltage values may be divided into two ranges, from 0 to $\Delta\varphi_o$ (the critical value) where noncoherent three-dimensional nucleation takes place, and for values greater than $\Delta\varphi_o$ where noncoherent two-dimensional nucleation occurs. The critical overvoltage may be obtained either from Eq. (6) assuming $H_3 = 2H_o$, or from Eq. (10) assuming the denominator equal to zero, i.e., $[ze_o\Delta\varphi - (V_o/H_o)\gamma(\theta)] = 0$,

$$\Delta\varphi_o = \frac{V_o\gamma(\theta)}{ze_oH_o} \quad (13)$$

In agreement with Eq. (13), increasing the disorientation angle of the noncoherent nucleus, $\Delta\varphi_o$ increases reaching its maximum value when $\theta = \theta_m$, that is when $\gamma(\theta) = \gamma_m$ (Fig. 11). Such curve $\Delta\varphi_o$ as a function of θ divides the field of existence of the two-dimensional and three-dimensional noncoherent nuclei. So, increasing the crystallization overvoltage, the range of θ values for the two-dimensional nuclei increases and, vice versa, that for the three-dimensional nuclei decreases. For example, as shown in Figure 11 when the crystallization process occurs with overvoltages greater than 0.055 V, the noncoherent nucleation process with the different θ values occurs through two-dimensional nucleation.

Table II. Estimated Values of the Work of Formation ($\times 10^{-19}$ J) of a Noncoherent, A_{nc}, and Normal, A_o, Cu Nucleus and of Their Difference, $A_{nc} - A_o$, for Different Values of the Crystallization Overvoltage, $\Delta\varphi$, and of the Disorientation Angle, θ

θ	$2°$	$4°$	$6°$	$8°$	$10°$
		$\Delta\varphi = 0.02$ V			
A_{nc}	106	159	192	213	224
A_o	21	21	21	21	21
$A_{nc} - A_o$	85	138	171	192	203
		$\Delta\varphi = 0.04$ V			
A_{nc}	26	38	46	52	56
A_o	10	10	10	10	10
$A_{nc} - A_o$	16	28	36	42	46
		$\Delta\varphi = 0.06$ V			
A_{nc}	12	16	19	21	22
A_o	7	7	7	7	7
$A_{nc} - A_o$	5	9	12	14	15
		$\Delta\varphi = 0.08$ V			
A_{nc}	7.7	9.9	11.6	13.1	14.1
A_o	5.2	5.2	5.2	5.2	5.2
$A_{nc} - A_o$	2.5	4.7	6.4	7.9	8.9
		$\Delta\varphi = 0.1$ V			
A_{nc}	5.8	6.9	7.7	8.4	8.9
A_o	4.1	4.1	4.1	4.1	4.1
$A_{nc} - A_o$	1.7	2.8	3.6	4.3	4.8

$\sigma = 1\,\mathrm{J\,m^{-2}}$.

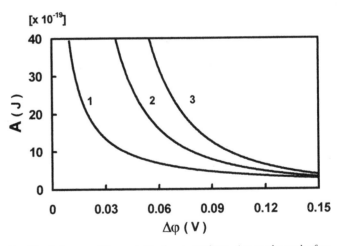

Fig. 12. Influence of the crystallization overvoltage, $\Delta\varphi$, on the work of nucleation, A, of the Cu critical nucleus, for different disorientation angles: $0°$ (curve 1), $3°$ (curve 2), and $10°$ (curve 3). $\sigma = 1\,\mathrm{J\,m^{-2}}$.

With the aid of Eqs. (7), (10), and (12), it is possible to analyze the influence of the crystallization overvoltage on the work of the formation of the normal two-dimensional and noncoherent (two-dimensional or three-dimensional) nuclei, A_o and A_{nc}, respectively. The results of the calculations in the considered example, relevant to copper, are reported in Table II and also in Figure 12. They show that increasing the value of the disorien-

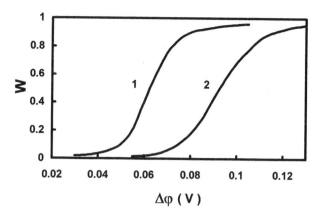

Fig. 13. Influence of the crystallization overvoltage, $\Delta\varphi$, on the relative probability of noncoherent Cu nucleation, W, for different values of the surface free energy per unit area of the nucleus: $0.6\,\mathrm{J\,m^{-2}}$ (curve 1) and $1\,\mathrm{J\,m^{-2}}$ (curve 2). $\theta = 1°$.

Table III. Estimated Values of the Work of Formation ($\times 10^{-19}$ J) of a Nucleus in Twinning, A_{tw}, and in Normal, A_o, Position, of Their Difference, $A_{\mathrm{tw}} - A_o$, and of the Relative Probability of the Formation of the Nucleus in the Twinning Position, W, for Different Values of the Crystallization Overvoltage, $\Delta\varphi$, and of the Free Energy per Unit Area of the Twin Boundary, γ_{tw}*

γ_{tw} (J m^{-2})	A_{tw}	A_o	$A_{\mathrm{tw}} - A_o$	W
	$\Delta\varphi = 0.02$ V			
0.02	26	21	5	0
0.05	42	21	21	0
0.15	125	21	104	0
	$\Delta\varphi = 0.04$ V			
0.02	12	10	2	10^{-12}
0.05	14	10	4	0
0.15	43	10	33	0
	$\Delta\varphi = 0.06$ V			
0.02	7.4	7	0.4	10^{-5}
0.05	8.3	7	1.3	10^{-15}
0.15	13.9	7	6.9	0
	$\Delta\varphi = 0.08$ V			
0.02	5.43	5.15	0.28	10^{-3}
0.05	5.90	5.15	0.75	10^{-8}
0.15	8.30	5.15	3.15	0
	$\Delta\varphi = 0.1$ V			
0.02	4.29	4.12	0.17	10^{-2}
0.05	4.58	4.12	0.46	10^{-5}
0.15	5.92	4.12	1.80	0

*$\sigma = 1\,\mathrm{J\,m^{-2}}$.

tation angle, at constant $\Delta\varphi$ values, the work of formation of the noncoherent nuclei increases as well as the difference between the A_{nc} and A_o values. Moreover, increasing the crystallization overvoltage, at constant θ values, the work of formation of both the noncoherent and normal nuclei decreases. Hence, increasing the $\Delta\varphi$ values, the difference between A_{nc} and A_o rapidly decreases so that the A_{nc} values become closer to that of A_o at sufficiently high $\Delta\varphi$ values (Fig. 12).

These results may be utilized to evaluate the influence of the electrocrystallization conditions on the degree of crystal imperfection of the deposits, which may be characterized by the relative probability of formation of the noncoherent nuclei,

$$W = \frac{W_{\mathrm{nc}}}{W_o} \tag{14}$$

where W_o is the probability of formation of the normal nuclei, proportional to $\exp(-A_o/kT)$, whereas W_{nc} is that of the noncoherent nuclei, proportional to $\exp(-A_{\mathrm{nc}}/kT)$ [1]. So, the relative probability may be written as

$$W = \exp\left(-\frac{A_{\mathrm{nc}} - A_o}{kT}\right) \tag{15}$$

Introduced in Eq. (15) A_o taken from (12) and A_{nc} from either (10) or (7), depending on whether noncoherent two-dimensional or three-dimensional nuclei are formed, we investigated the influence of the two main factors, $\Delta\varphi$ and σ, which depend on the electrolysis conditions, on the relative probability of formation of the noncoherent nuclei, and thus on the crystal imperfection of the deposits. Figure 13 depicts the influence of the crystallization overvoltage on W for two different values of the surface free energy per unit area of the nucleus.

Each curve shows that increasing the crystallization overvoltage, the relative probability of formation of the noncoherent nuclei is initially practically equal to zero, but then it increases very steeply and, at sufficiently high $\Delta\varphi$ values, the process of noncoherent nucleation has practically the same probability as that of the formation of the normal nuclei. Reminded that by increasing θ, the difference, $A_{\mathrm{nc}} - A_o$, increases (see Table II),

it may be concluded that, at constant overvoltage, the relative probability of formation of noncoherent nuclei decreases by increasing the disorientation angle.

In addition to the crystallization overvoltage, the relative probability of noncoherent nucleation also depends on the value of the surface free energy per unit area, which, in its turn, depends on the presence of foreign atoms and/or molecules in the electrolyte. Sometimes, surface-active organic addition agents are intentionally added to the electrolyte during metal electrocrystallization. They are adsorbed on the surface of the growing crystals of the electrodeposit and, therefore, decrease the surface free energy of the cathode. It may be assumed that they reduce the σ value of the nuclei. The comparison of curves 1 and 2 (Fig. 13) pertaining to different σ values (0.6 and 1 J m^{-2}, respectively) shows that the abrupt increase of the relative probability of formation of noncoherent nuclei with the same θ value occurs at lower overpotentials in the case of the lower σ value.

Similar to what was previously done, the process of twins formation in the electrodeposits will be analyzed. To do this, it is necessary to introduce in Eqs. (3) and (6)–(10) the value of γ_{tw}, the free energy per unit area of the twin boundary, instead of $\gamma(\theta)$. Table III shows the estimated values of the work of formation of the nuclei in twinning, A_{tw}, and in normal position, A_o, of their difference, $A_{\mathrm{tw}} - A_o$, and of the relative probability

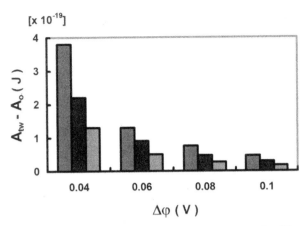

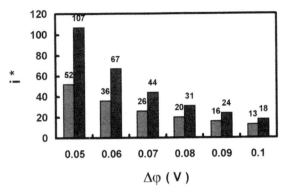

Fig. 14. Influence of the crystallization overvoltage, $\Delta\varphi$, on the difference between the work of formation of a nucleus in twinning and in normal position, $A_{\text{tw}} - A_o$, for different values of the surface free energy per unit area of the nucleus: $1\,\text{J m}^{-2}$ (left column); $0.8\,\text{J m}^{-2}$ (central column) and $0.6\,\text{J m}^{-2}$ (right column). $\gamma_{\text{tw}} = 0.05\,\text{J m}^{-2}$.

Fig. 15. Influence of the crystallization overvoltage, $\Delta\varphi$, on the number of atoms, i^*, of a normal (left column) and a noncoherent (right column) critical nucleus of Cu. $\theta = 1°$, $\sigma = 1\,\text{J m}^{-1}$.

of formation of nuclei in twinning positions, W, as a function of both the overvoltage and the γ_{tw}. The latter quantity depends first of all on the nature of the deposited metal. For example, the γ_{tw} value is very small for silver, is very high for iron and nickel, and is intermediate for copper.

As shown from the data in Table III, decreasing the free energy per unit area of the twin boundary in the deposit (at a constant $\Delta\varphi$ value) as well as increasing the overvoltage (at a constant γ_{tw} value), the work of formation of the nuclei in twinning positions and the difference, $A_{\text{tw}} - A_o$, decrease and the relative probability of the formation of the nuclei in twinning positions increases, consequently [Eq. (15)]. It is also worth noting that the W value remains practically equal to zero even at the high crystallization overvoltages when γ_{tw} has a high value. This means that twins are not expected to be formed in the case of deposits of metals having a sufficiently high value of γ_{tw}, as, for example, nickel. However, this result is in disagreement with the real situation. Indeed, the experimental results show that nickel electrodeposits contain a relatively high quantity of twins, and this although they are prepared in conditions of low overvoltage values.

Such behavior may be explained considering that hydrogen, too, discharges on the cathode during nickel electrodeposition and is adsorbed on the growing crystal planes, therefore, decreasing the value of the surface free energy per unit area. Figure 14 depicts the results of the calculation of the influence of the σ value on the difference $A_{\text{tw}} - A_o$ for the various overvoltage values. It shows that decreasing σ, the $A_{\text{tw}} - A_o$ value also decreases and, correspondingly, the relative probability of formation of nuclei in the twinning position has to increase, more especially at low $\Delta\varphi$ values. Therefore, it may be expected that in the case of nickel electrodeposition, as well as in the case of the electrodeposition of other metals with high γ_{tw} value, twins may be formed when the process of adsorption of atoms and foreign molecules occurs during their electrocrystallization.

Hence, the classical thermodynamic analysis of the noncoherent nucleation step has shown that there is a unique mechanism of formation of the main structural defects during metal electrocrystallization. On the basis of the calculations of the work of formation of noncoherent and normal nuclei, the influence was investigated for the crystallization overvoltage and for the surface free energy per unit area on the relative probability of noncoherent nucleation which qualitatively characterizes the degree of crystal imperfection of the electrodeposits.

On the other hand, it is worth underlining that the classical Gibbs–Volmer theory of nucleation is correct when the oversaturation (crystallization overvoltage) values are low enough that the nuclei consist of numerous atoms and a macroscopic formation may be considered. However, in many cases, in particular during both vapor phase condensation on a cold substrate and electrocrystallization of several metals, the deposition process occurs far from the equilibrium conditions while the nuclei consist of few atoms (Fig. 15). In these cases, it is necessary to investigate the noncoherent nucleation step from the atomistic point of view.

5. ATOMISTIC ANALYSIS OF NONCOHERENT NUCLEATION

Let us consider a (complex) nucleus of i atoms on its own crystalline plane (substrate) which has an azimuth disorientation angle with respect to the same plane, θ. This is the model of the noncoherent nucleus. In the general case, this complex is composed by i atoms and is a polylayer (i.e., three-dimensional). i_s is the number of the atoms of the nucleus of its lowest atomic layer, that is, it is the number of atoms which are in direct contact with the substrate. According to the atomistic theory of the heterogeneous nucleation [51–53], the Gibbs free energy of formation of this noncoherent nucleus, ΔG, depends on:

(a) the variation of the chemical potential, $\Delta\mu$, due to the phase transition which in turn depends on the oversaturation of the system;

(b) the excess free energy of a normal nucleus due to the nonsaturated bonds of the external atoms of the same nucleus;

(c) the excess free energy, ε, of each of the i_s atoms of the noncoherent nucleus due to its disorientation.

Hence, it results [54],

$$\Delta G = -i\Delta\mu + \left[i\varepsilon_{1/2} - E_o(i) - i_s\varepsilon_s\right] + i_s\varepsilon \qquad (16)$$

where $\varepsilon_{1/2}$ is the well-known work of separation of an atom from the "half-crystal position" [55], $E_o(i)$ is the bond energy among the i atoms of the nucleus, and ε_s is the work of separation of an isolated atom from the plane.

This equation is similar to that obtained according to the Gibbs–Volmer theory. However, while the atomistic theory refers to the individual atoms of a nucleus and to their bond energies, the classical theory refers to a macroscopic cylindrical nucleus and to its surface (lateral) free energy.

The dependence of ε from the azimuth disorientation angle of the noncoherent nucleus may be obtained from

$$\varepsilon = \gamma(\theta)/C \qquad (17)$$

where C is the surface density of the atoms in the plane where the nucleus is formed and $\gamma(\theta)$ is the free energy per unit area of the noncoherent bond between the nucleus and the substrate.

As previously done, we take into consideration that $\gamma(\theta)$ may be estimated between 0 and θ_m from the Read–Shockley relationship [see Eq. (1)] for the specific free energy of the dislocation boundary. In this case, $\gamma(\theta)$ increases with θ (Fig. 1); its maximum value is equal to the specific free energy of the usual grain boundary which is independent of the disorientation angle among the grains.

The values of $\varepsilon_{1/2}$, ε_s, and $E_o(i)$ in Eq. (16) may be determined from the work of separation of two neighbor atoms at the distance of the first, second, and third order (ε_1, ε_2, and ε_3, respectively), as

$$\varepsilon_{1/2} = N_1\varepsilon_1 + N_2\varepsilon_2 + N_3\varepsilon_3 \qquad (18)$$

$$\varepsilon_s = N_1'\varepsilon_1 + N_2'\varepsilon_2 + N_3'\varepsilon_3 \qquad (19)$$

$$E_o(i) = N_1''\varepsilon_1 + N_2''\varepsilon_2 + N_3''\varepsilon_3 \qquad (20)$$

where N_1, N_2, and N_3 are the number of the first, second, and third neighbors, respectively, of an atom in a half-crystal position, N_1', N_2', and N_3' are those of an isolated atom on the substrate plane, N_1'', N_2'', and N_3'' are those of the atoms in the nucleus. Reminded that the work needed for breaking the bonds between the atoms in the lattice is inversely proportional to the sixth power of their distance, it results: $\varepsilon_2 = 0.125\varepsilon_1$ and $\varepsilon_3 = 0.037\varepsilon_1$. The ε_1 value may be determined from the heat of sublimation of the considered metal.

For the subsequent application to noncoherent nucleation, Eq. (16) is represented as

$$\Delta G = -i\Delta\mu + iE_1 \qquad (21)$$

where E_1 is the total (sum of) excess free energy of an atom of the noncoherent nucleus due to both the nonsaturated bonds

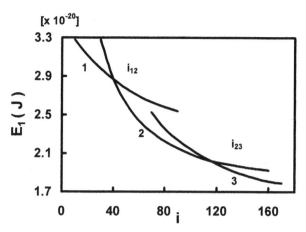

Fig. 16. Dependence of the total excess free energy per atom, E_1, of a noncoherent Zn nucleus of the monolayer (curve 1), bilayer (curve 2), and trilayer (curve 3) type on the number of its atoms, i. $\theta = 10°$.

of the atoms of the nucleus and to the nucleus disorientation which, according to Eqs. (16) and (17), is equal to

$$E_1 = \frac{i\varepsilon_{1/2} - E_o(i) - i_s(\varepsilon_s - \gamma(\theta)/C)}{i} \qquad (22)$$

Initially, we analyze how the type of the nucleus (monolayer or polylayer) influences the ΔG value for a determined $\Delta\mu$ value, recalling that in the case of metal electrocrystallization, $\Delta\mu$ is directly proportional to the crystallization overvoltage. With this aim, we compare the dependence of the two terms in the right side of Eq. (21) on the number of atoms for the different types of noncoherent nuclei: monolayer (ML), bilayer (BL), and trilayer (TL) nuclei. From Eq. (21), it results that the first term does not depend on the nucleus type: it decreases with a linear law by increasing the number of atoms in the nucleus (and the rate of such decrease increases with the $\Delta\mu$ value).

The second term directly depends on the value of the energetic factor E_1 which depends in a complicate way on the value of i and this dependence may be different for the various types of nuclei. As an example, we examine the nuclei on the close-packed plane of an fcc or an hcp lattice, the {111} and {0001} planes, respectively.

As shown by the calculations, the energetic factor E_1 of the different nucleus types (monolayer and polylayer) decreases with the increased number of atoms in the same nucleus but the curves $E_1 = f(i)$ are different for ML, BL, and TL nuclei. If we consider the normal nuclei [$\theta = 0°$, $\gamma(\theta) = 0$], it results that E_1 is independent from the number of atoms. Moreover, the E_1 values of the monolayer nuclei are lower than those of the polylayer nuclei, but their difference tends to decrease when the i value increases. This means that at the same number of atoms, ΔG always has the lowest value for the monolayer than for the polylayer nucleus.

Another situation occurs when the nuclei are noncoherent and, therefore, the value of $\gamma(\theta)$ for these nuclei is not equal to zero. As an example, in Figure 16 the curves are depicted showing the dependence of E_1 on the number of atoms in Zn nuclei of different types (ML, BL, and TL) which are formed on

Table IV.   Values of N_1 to N_3' for Fcc and Hcp Lattices

Lattice	N_1	N_2	N_3	N_1'	N_2'	N_3'
Fcc	6	3	12	3	3	9
Hcp	6	3	1	3	3	1

Reprinted from [54] with kind permission from Elsevier Science.

Table V.   Values of N_1'' and N_3'' as a Function of the Number of Atoms, i, of a Monolayer Nucleus on the Close-Packed Plane

i	1	2	3	4	5	6	7	8	9	10
N_1''	0	1	3	5	7	9	12	14	16	19
N_3''	0	0	0	1	2	4	6	7	9	11

Reprinted from [54] with kind permission from Elsevier Science.

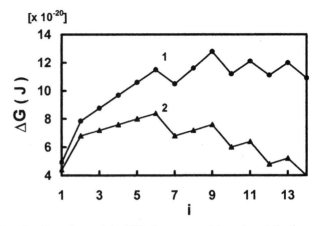

Fig. 17.   Dependence of the Gibbs free energy of formation, ΔG, of a noncoherent monolayer nucleus of Zn with the azimuth disorientation angle of $1°$ (broken curve 1) and of a normal nucleus (broken curve 2) on the number of its atoms, i. $\Delta \varphi = 0.05$ V.

the close-packed plane and are disoriented by an angle of $10°$. The curves intersect at consecutive critical values of the number of atoms, i_{12} and i_{23}. In the first range, where the number of atoms of the nuclei i is not higher than i_{12}, the monolayer nucleus has the lowest value of E_1. When the number of atoms lies in the range between i_{12} and i_{23}, the bilayer nucleus has the lowest value of E_1, whereas when i is higher than i_{23}, the trilayer nucleus has the lowest E_1 value.

Considering these results, we may note that the Gibbs free energy of formation of a noncoherent nucleus depends on its type and that, increasing the number of atoms, ΔG has its minimum value initially for the monolayer, then for the bilayer, and finally for the trilayer nucleus.

In the following, the principles of calculation of the work of formation of the critical nucleus are considered to evaluate the main parameters of the deposit structure (dislocation density and average grain size).

First of all, we examine the formation of the monolayer nuclei on the close-packed plane of an fcc or an hcp lattice. In this case, i_s is equal to i and, taking Eq. (17) into account, Eq. (16) becomes:

$$\Delta G = -i\Delta\mu + \left[i(\varepsilon_{1/2} - \varepsilon_s) - E_o(i) \right] + i\gamma(\theta)/C \quad (23)$$

The values of $\varepsilon_{1/2}$, ε_s, and $E_o(i)$ may be estimated on the basis of Eqs. (18)–(20). Table IV collects the values of N_1 to N_3' and Table V collects those of N_1'' and N_3'' as a function of the number of atoms in the monolayer nucleus (the second-order neighbors are missing).

According to these data, the difference $(\varepsilon_{1/2} - \varepsilon_s)$ is equal to $(3\varepsilon_1 + 3\varepsilon_3)$ and to $(3\varepsilon_1)$, for an fcc and an hcp lattice, respectively. However, since ε_3 is much smaller than ε_1, we may assume for both lattices,

$$\varepsilon_{1/2} - \varepsilon_s = 3\varepsilon_1 \quad (24)$$

Moreover, $N_3''\varepsilon_3$ may be neglected in comparison to $N_1''\varepsilon_1$ in Eq. (20), because the N_3'' values are lower than the N_1'' values

for any value of the nucleus atoms, except for $i = 1$. The calculations show that the error in the determination of $E_o(i)$ is not higher than 2–3% when the value of $N_3''\varepsilon_3$ is not taken into account.

So, in the case of metal electrocrystallization, the Gibbs free energy of formation of a noncoherent monolayer nucleus on a close-packed plane is

$$\Delta G = -ize_o\Delta\varphi + \left[3i\varepsilon_1 - E_o(i) \right] + i\gamma(\theta)/C \quad (25)$$

By using Eq. (25), it is possible to evaluate the work of formation of the critical noncoherent nucleus A_n with a disorientation angle, θ_n, for a definite crystallization overvoltage, $\Delta\varphi$. To do this, it is necessary to determine the value of ΔG as a function of the number of atoms in the nucleus. In the case of the normal nucleus, the value $\gamma(\theta) = 0$ has to be inserted in Eq. (25). Figure 17 shows the ΔG dependence on i for the formation of a (both normal and noncoherent) monolayer nucleus of Zn when the crystallization overvoltage is 0.05 V. A_n is equal to the maximum value of ΔG. In this same case, the work of the formation of the normal critical nucleus is equal to 8.4×10^{-20} J and the number of atoms in this nucleus is equal to 6, the corresponding values for the noncoherent nucleus are 12.8×10^{-20} J and 9, respectively.

In the same way, we calculated the value of A_n of nuclei of the polylayer type. Figure 18 illustrates the dependence on the crystallization overvoltage of the work of the formation of a noncoherent critical nucleus ($\theta_n = 10°$) of the different types: monolayer (curve 1), bilayer (curve 2), and trilayer (curve 3). It shows that, in the range of the low values of $\Delta\varphi$, the value of A_n is the lowest for the noncoherent trilayer nuclei, in the range of the medium overvoltage values, it is the lowest for the bilayer ones, and in the range of the relatively high $\Delta\varphi$ values, it is the lowest for the noncoherent monolayer nuclei. The first two overvoltage ranges decrease with the disorientation angle of the noncoherent nuclei. In the limiting case, when θ is equal to zero (normal nucleus), the work of formation of this nucleus has the minimum value for all the crystallization overvoltages.

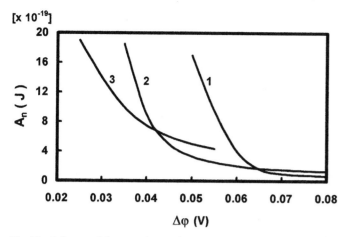

Fig. 18. Influence of the overvoltage value, $\Delta\varphi$, on the work of formation, A_n, of the critical noncoherent nucleus of Zn with the azimuth disorientation angle of $10°$ and of the monolayer (curve 1), bilayer (curve 2), and trilayer (curve 3) type.

Considering such dependence of the work of formation of noncoherent nuclei on their different types, our computation program automatically estimates the values for the formation of nuclei of the ML, BL, and TL types. Then, it selects the lowest one for further calculations. So, A_n was evaluated for the θ_n values from 0 to $60°$, at intervals of $1°$, taking account of the dependence of $\gamma(\theta)$ on the value of θ for a noncoherent nucleus on the close-packed plane of an fcc or an hcp lattice (see Fig. 9). It is worth recalling that, in agreement with our model of formation of the main structural defects in deposits, the noncoherent nuclei with disorientation angles in the range from 0 to $12°$ ($12°$ is the value we choose for θ_m) and also from 48 to $60°$ generate the dislocation boundaries of the subgrains, whereas the noncoherent nuclei with disorientation angles from 12 to $48°$ generate the usual grain boundaries.

The normalized probability of formation of the nucleus is

$$W_n = \frac{\exp(-A_n/kT)}{\sum_0^{60} \exp(-A_n/kT)} \qquad (26)$$

For each overvoltage, W_n is a decreasing function of θ_n between 0 and $12°$ and becomes constant between 12 and $48°$, where γ is independent of θ. Then, W_n becomes an increasing function of θ_n, between 48 and $60°$.

The average disorientation angle of the noncoherent nucleus was obtained from the statistical distribution of W_n for the θ_n values in the first and third range,

$$\langle\theta\rangle = \frac{\sum_0^{12}(W_n\theta_n) + \sum_{48}^{60}(W_n\theta_n)}{\sum_0^{12} W_n + \sum_{48}^{60} W_n} \qquad (27)$$

Moreover, assuming $\langle\theta\rangle$ also as the average disorientation angle among the subgrains of the deposit, the dislocation density in the dislocation boundaries was estimated [56],

$$\rho = \frac{\langle\theta\rangle^2}{b^2} \qquad (28)$$

b is the absolute value of the Burgers vector of the dislocation.

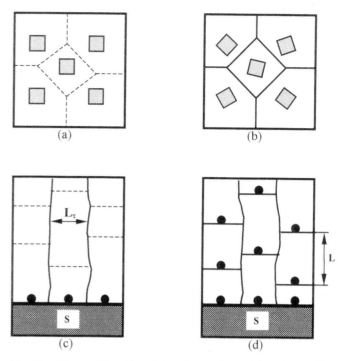

Fig. 19. Scheme of formation of the (a) subgrainy and (b) polycrystalline structure during the initial step of metal deposition and the scheme of the cross section of two different types of microstructure of deposits on an indifferent substrate (S): (c) columnar and (d) homogeneous type. Dots represent noncoherent nuclei with a great disorientation angle. The dashed and the continuous lines evidence the subgrain and the grain boundaries, respectively.

To estimate the average grain size, $\langle L \rangle$, the statistical distribution of W_n for the θ_n values in the second range was applied. Since the frequency of formation of the grain boundaries is equal to the overall probability of formation of the noncoherent nuclei with $12° \leq \theta \leq 48°$, it may be seen that

$$\langle L \rangle = \frac{d}{\sum_{12}^{48} W_n} \qquad (29)$$

where d is the distance among the close-packed planes.

Before analyzing the influence of the crystallization conditions on the structural parameters according to Eqs. (28) and (29), the growth model of a deposit will be considered, taking account of the previous mechanism of the formation of the main structural defects during noncoherent nucleation. First of all, the deposition of metals will be considered on the surface of a single crystal of the same metal or of a foreign one. If the crystallization conditions are such that isooriented nuclei are formed during the initial deposition step then, after their lateral growth and coalescence, a complete and compact layer having monocrystalline structure is formed. However, if the initial nuclei are disoriented one with respect to the other with a sufficiently small angle (not more than 10 to $15°$), dislocation boundaries are formed after their coalescence, that is, the subgrainy structure is formed (Fig. 19a).

The situation is different when the crystallization conditions are such that nuclei are initially formed which are disoriented among themselves with a relatively high angle (more than 10 to 15°). It is clear that the usual grain boundaries will be formed after the coalescence of the nuclei and that the structure of the deposited layer will be polycrystalline (Fig. 19b). The same mechanism of formation of a thin layer with the polycrystalline structure will also occur in the case of deposition on a polycrystalline substrate or, in the general case, when crystallization occurs on an indifferent substrate.

We will now consider how the polycrystalline structure of the deposit already formed during the initial step develops further. After formation of the initial layer, the growth process in thickness of the deposit takes place and during this step the main role in the formation of the crystalline defects is played by noncoherent nucleation. Fom the point of view of our ideas, if the probability of formation of the noncoherent nuclei with great disorientation angles ($\theta > \theta_m$) is practically equal to zero, so that no grain boundaries are formed in the whole deposit thickness, the dislocation boundaries, only, will be observed in the whole deposit thickness (Fig. 19c). Therefore, the final microstructure of the deposit cross section will be of the columnar type (FT-type according to Fischer's classification [8]). The polycrystalline structure of this same deposit will be characterized by the dislocation density, ρ, and by the average grain size parallel to the substrate surface, $\langle L_\tau \rangle$ (Fig. 19c).

Conversely, if the probability of formation of noncoherent nuclei with $\theta > \theta_m$ is different from zero, not only the dislocation boundaries will be formed in the whole deposit thickness but also the grain boundaries will be formed (Fig. 19d). The final microstructure of the deposit cross section will be of the homogeneous type (RD-type [8]) and will be characterized, besides by the ρ and $\langle L_\tau \rangle$ values, also by the average grain size in the growth direction of the deposit, $\langle L \rangle$ (Fig. 19d).

Of course, the different situations considered here and the values of the structural parameters of the deposits, ρ and $\langle L \rangle$, depend on the electrocrystallization conditions, which are examined in the next section.

6. FACTORS INFLUENCING THE STRUCTURE OF ELECTRODEPOSITS (THEORETICAL AND EXPERIMENTAL RESULTS)

6.1. Influence of the Crystallization Overvoltage on the Structure of Electrodeposits

The proposed method of calculation of the main structural parameters of the deposits (the dislocation density, ρ, and the average grain size, $\langle L \rangle$) is based on the model of noncoherent nucleation which allows us to investigate the influence of the conditions of electrocrystallization on the values of ρ and $\langle L \rangle$. To determine which factors directly influence the structural parameters, Eq. (21) will be considered, representing in a general way the work of formation of the noncoherent nucleus during electrocrystallization. In the case of formation of the critical nucleus, such an equation becomes

$$A = \Delta G_{\max} = -i^*(ze_o \Delta\varphi - E_1) \qquad (30)$$

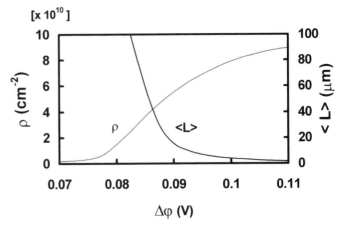

Fig. 20. Theoretical dependence of the dislocation density, ρ, and of the average grain size, $\langle L \rangle$, of Zn electrodeposits on the overvoltage, $\Delta\varphi$. Reprinted from [54] with kind permission from Elsevier Science.

where i^* is the number of atoms in the critical nucleus, $\Delta\varphi$ is the value of the crystallization overvoltage, and E_1 is the total excess free energy of an atom of the noncoherent nucleus, defined by Eq. (22).

It may be noted that the overvoltage is the first factor influencing A. Therefore, first of all the influence of the value of $\Delta\varphi$ on ρ and $\langle L \rangle$ will be examined. To compare the theoretical with the experimental results, we considered the electrodeposition of Zn and the nucleation on the (0001) plane.

Figure 20 shows the values of the dislocation density and of the average grain size of the Zn electrodeposits, calculated according to Eqs. (28) and (29), respectively, as a function of the overvoltage value.

The former remains practically constant (around 10^7–10^8 cm^{-2}) at relatively low overvoltages, significantly increases between 0.08 and 0.1 V, and remains again nearly constant at higher values. The latter rapidly decreases in the quite low overvoltage range from 0.08 to 0.09 V. These results may be explained according to previous considerations on the mechanism of growth of the electrodeposit and of the formation of the structural defects during noncoherent nucleation. Two cases of electrodeposition of metals will be considered, i.e., electrodeposition on a single-crystal substrate and on an indifferent substrate.

In the first case, the data in Figure 20 may be interpreted as in the following: increasing the crystallization overvoltage up to the critical value, the deposit is monocrystalline while the dislocation density is initially nearly constant and then rapidly increases. At sufficiently high overvoltages, a gradual transition to the polycrystalline microstructure occurs. Under these conditions, the dislocation density does not increase very much while the grain size decreases rapidly.

In the second case, it is necessary to take into account that the electrodeposit will have a polycrystalline microstructure with relatively large grains already at low overvoltages. Increasing the value of $\Delta\varphi$, the grain size of the deposit initially does not change very much while the dislocation density in-

Fig. 21. (a, c) Laue backreflection diffraction patterns and (b, d) optical micrographs of the surface of Zn electrodeposits obtained on (0001)-oriented Zn single crystals at overvoltages of 0.04 V (a, b) and 0.06 V (c, d). Reprinted from [54] with kind permission from Elsevier Science.

creases. Then, at relatively high overvoltages, the $\langle L \rangle$ value starts rapidly decreasing.

To verify the theoretical results, the experimental investigation was performed under specific conditions suggested by the theoretical treatment, also to simplify the process of defects formation during crystal growth. More precisely, to avoid the twinning process which introduces additional defects, a close-packed plane not being a twinning plane, that is, the basal plane of an hcp metal, was chosen as the substrate.

So, the influence of the cathodic overvoltage on the microstructure of Zn deposits on (0001)-oriented Zn single crystals was investigated. The Zn single crystals were cylinders of 8-mm diameter and were prepared from 99.99% pure Zn following the method by Bridgman and Stockbarger. They were cut along the basal plane in 2- to 3-mm thick disks at the temperature of liquid nitrogen to avoid plastic deformation. Immediately before electrodeposition, their surface was submitted to an anodic treatment. Zn electrodeposition was performed in potentiostatic conditions at overvoltages from 0.02 to 0.08 V, from an acid zinc sulfate bath (400 g L^{-1}), pH 2.5, at 25°C. The thickness of the Zn deposits was around 10 μm.

The microstructure of the deposits was investigated by X-ray diffraction (XRD) according to the Laue backreflection method, and their surface morphology by optical microscopy. The dislocation density was evaluated from the mean size of the domains with coherent diffraction [57] which was obtained from the width of the (0002) XRD line according to the Seliakov–Scherrer relationship. The average grain size was determined from the electrodeposit morphology.

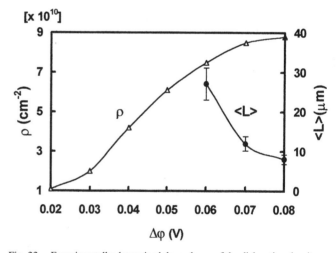

Fig. 22. Experimentally determined dependence of the dislocation density, ρ, and of the average grain size, $\langle L \rangle$, of Zn electrodeposits on (0001)-oriented Zn single crystals on the overvoltage, $\Delta\varphi$. Reprinted from [54] with kind permission from Elsevier Science.

The Laue backreflection patterns showed that the Zn deposits obtained potentiostatically at cathodic overvoltages from 0.02 to 0.05 V were monocrystalline, oriented according to the (0001) plane (Fig. 21a), as also confirmed by their surface morphology (Fig. 21b). On the contrary, those obtained from 0.06 to 0.08 V had the typical polycrystalline structure and morphology (Fig. 21c and d, respectively). Their grain size strongly decreased (Fig. 22). Furthermore, the most effective increase of the dislocation density occurred between 0.03 and 0.06 V (Fig. 22). So, the experimental results confirmed the theoretical ones, the trend of their curves being similar (Figs. 20 and 22, respectively).

Moreover, since overvoltage increases with current density, a similar behavior is also expected during galvanostatic electrodeposition of the metals on the single crystals, as experimentally observed by several authors [58, 59]. However, the considered mechanism of the transition from the monocrystalline to the polycrystalline microstructure is probably not the unique one. Indeed, if electrodeposition conditions are favorable to twinning, twinning may also be at the origin of the formation of the polycrystalline electrodeposits [41, 60].

The theoretical results obtained about the influence of the overvoltage on the structural parameters of the electrodeposits were also confirmed by our experimental results on the electrodeposition of some other metals (zinc, copper, and nickel) on a mechanically polished stainless steel substrate. The deposits were obtained following the galvanostatic method while the value of the electrodeposition overvoltage was determined as the difference between the electrode equilibrium potential and the electrode potential when the current was flowing. The electrolyte compositions and the pH values as well as the thicknesses of Zn, Cu, and Ni electrodeposits are reported in Table I. The electrocrystallization temperature was 30°C, 45°C, and 70°C, respectively, for Zn, Cu, and Ni deposition.

The method of determination of the values of the dislocation density and of the average grain size of the Zn electrodeposits was the same as in the case of Zn electrodeposition on the Zn single crystal. Instead, the $\langle L \rangle$ value of the deposits of Cu and Ni was determined from the micrographs obtained by transmission electron microscopy. By the way, the method of preparation of the thin samples for TEM analysis was considered in Section 2. The ρ value of the deposits of Cu and Ni, as well as of Zn, was evaluated from the mean size of the domains with coherent diffraction, D [57]. However, the D value of the Cu and Ni deposits was determined following the method of the Fourier analysis of the shape of the diffraction lines obtained during X-ray investigation. For the Cu deposits, the (111) and (222) lines were recorded, for the Ni ones, the (200) and (400) lines were recorded.

Figure 23 shows the influence of the current density, J, on the ρ and $\langle L \rangle$ values of the Zn electrodeposits on the stainless steel substrate. It is necessary to add that, according to the electrochemical experimental results, by increasing the value of J from 2 to 6 A dm^{-2}, the cathodic overvoltage increases from 0.058 to 0.095 V.

Table VI shows the influence of the current density on the $\Delta\varphi$, ρ, and $\langle L \rangle$ values of the Cu and Ni deposits on the stainless steel substrate.

The curves in Figure 23 and the data of Table VI show that increasing the current density increases the overvoltage and with it increases the dislocation density and decreases the average grain size of the Zn, Cu, and Ni electrodeposits, which is in agreement with our theoretical results.

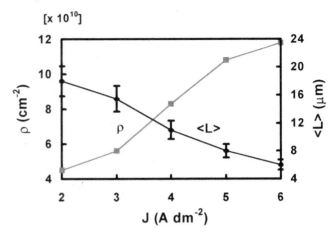

Fig. 23. Influence of the current density, J, on the dislocation density, ρ, and on the average grain size, $\langle L \rangle$, of Zn electrodeposits on the stainless steel substrate.

Table VI. Influence of the Current Density, J, on the Cathodic Overvoltage, $\Delta\varphi$, the Dislocation Density, ρ, and the Average Grain Size, $\langle L \rangle$, of Cu and Ni Electrodeposits on the Stainless Steel Substrate

J (A dm^{-2})	1	2	3	4	5
			Cu		
$\Delta\varphi$ (V)	0.072	0.088	0.102	0.111	0.118
$\rho \times 10^{10}$ (cm^{-2})	4	7	11	16	24
$\langle L \rangle$ (μm)	8.8	7.5	6.5	5.1	4.2
			Ni		
$\Delta\varphi$ (V)	0.168	0.207	0.236	0.258	0.275
$\rho \times 10^{10}$ (cm^{-2})	8	17	42	86	143
$\langle L \rangle$ (μm)	6.5	5.3	4.1	3.4	2.8

6.2. Influence of the Foreign Particle Adsorption on the Structure of Electrodeposits

According to Eq. (30), the work of formation of the noncoherent nucleus depends not only on the overvoltage but also on the value of the total excess free energy of an atom of the noncoherent nucleus, E_1, defined by Eq. (22).

Therefore, it will be analyzed in detail how the conditions of metals electrocrystallization may influence the value of E_1, and the structural parameters of the deposits, consequently. To solve this problem, the model of the noncoherent nucleus will be considered in a real situation. So far, it has been assumed that the noncoherent nucleus was formed on an ideal pure plane whereas in real conditions surface adsorption occurs for foreign particles (atoms and/or molecules) which are present in the electrolytic solution. In particular, in the practice of metal electrodeposition, electrolytes are rather often employed containing additives, i.e., surface-active agents which adsorb on the surface of the growing crystal planes.

Owing to the adsorption of foreign particles, the physicochemical state of the surface of the growing crystal planes changes and the degree of this change depends on the nature and the surface concentration of the foreign particles. Of course, it

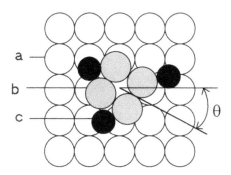

Fig. 24. Scheme of a two-dimensional noncoherent nucleus of four atoms with a disorientation angle, θ, in contact with surface-adsorbed foreign particles on the top of a {100} plane of an fcc metal. (a) and (b) represent the atoms of the {100} plane and of the nucleus, respectively; (c) the foreign particles.

is expected that such an adsorption process influences all the growth steps of the electrodeposits, the step of noncoherent nucleation included.

Therefore, a model of the noncoherent nucleus will be considered which takes the presence of foreign particles on the crystallographic plane into account, for example, on the {100} plane of an fcc lattice (Fig. 24). It is evident that the foreign particles in contact with the nucleus boundary decrease the number of the nonsaturated bonds of the external atoms of the nucleus. As a consequence, a decrease of the lateral surface free energy of the nucleus must occur, in agreement with the classical theory of nucleation.

In our considerations which follow the principles of the atomistic theory of nucleation, the presence of the surface-adsorbed foreign particles in contact with a nucleus may be taken into account by introducing the additional term, $-N\varepsilon_{ad}$, in the expression of the total excess free energy of an atom in the noncoherent nucleus [see Eq. (22)]. Hence, considering the adsorption of foreign particles and the formation of the critical nucleus, Eq. (22) may be written as

$$E_1^{ad} = \frac{i^*\varepsilon_{1/2} - E_o(i^*) - i_s(\varepsilon_s - \gamma(\theta)/C) - N\varepsilon_{ad}}{i^*} \quad (31)$$

where N is the number of foreign particles in contact with the nucleus and ε_{ad} is the bond energy of one particle with the nucleus.

Of course, N depends on the surface coverage with adsorbed foreign particles, ϑ, and in a first approximation may be assumed to be proportional to ϑ, ε_{ad} will depend on the nature of the foreign particles and its value may be estimated from their heat of adsorption, Q_{ad}.

So, the term representing the work of formation of the noncoherent nucleus when surface adsorption of foreign particles takes place will be

$$A^{ad} = -i^*\left(ze_o\Delta\varphi - E_1^{ad}\right) \quad (32)$$

Then, from Eq. (31) it results that, by increasing the value of ϑ as well as of Q_{ad}, the total excess free energy of an atom of the noncoherent nucleus must decrease and, as a consequence of Eq. (32), its work of formation must decrease, too.

Figure 25 shows the theoretical influence of the surface coverage with adsorbed foreign particles on the value of A.

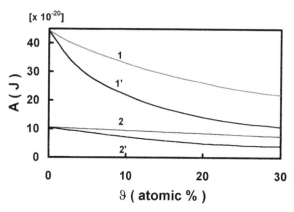

Fig. 25. Influence of the surface coverage, ϑ, with adsorbed foreign particles on the work of formation, A, of a noncoherent Ni nucleus with a disorientation angle of $10°$. Heat of adsorption: 5 (curves 1 and 2) and 15 kcal mol^{-1} (curves 1' and 2'). $\Delta\varphi = 0.1$ V (curves 1 and 1') and 0.2 V (curves 2 and 2').

These calculations were performed for noncoherent Ni nuclei on the {100} plane having a disorientation angle of $10°$ and for two different values of the overvoltage and of the heat of adsorption. The latter were 5 and 15 kcal mol^{-1} (around 21 and 63 kJ mol^{-1}) corresponding to physical and chemical adsorption, respectively. From Figure 25, it is clear that the particles which are more strongly adsorbed on the growing crystal planes and, therefore, have a higher value of the heat of adsorption, more strongly influence the value of A.

Following our method of determination of the structural parameters of the deposits, the values were determined for the work of formation of noncoherent Ni nuclei on the {100} plane with a disorientation angle from 0 to $90°$ ($\theta_m = 12°$), at intervals of $1°$, assuming a constant overvoltage of 0.2 V. Subsequently, the values were determined for the dislocation density and for the average grain size of the electrodeposits as a function of the surface coverage with adsorbed foreign particles. Again, two different values were assumed for the heat of adsorption, 5 and 15 kcal mol^{-1} [61].

Figure 26 shows that by increasing the surface coverage with adsorbed foreign particles, e.g., increasing the concentration of surface-active agents in the electrolyte, the structural defects in the electrodeposits dramatically increase. Indeed, the dislocation density increases while the grain size decreases. Such decrease in $\langle L \rangle$ means, in the case of a primary polycrystalline deposit, that its microstructure becomes ever finer. In the case of a primary monocrystalline structure (which occurs by deposition on a single-crystal substrate) it means a transition to a polycrystalline structure initially having relatively large grains which, subsequently, become ever smaller. Figure 26 also shows that by physical adsorption, foreign particles actually influence the deposit structure at relatively high surface coverages, whereas by chemical adsorption this still occurs at very low coverages.

To verify the theoretical results, the influence of phosphorus adsorption on the structure of Ni–P electrodeposits with different phosphorus content was first of all investigated and, subsequently, the influence of the surface-active agent (thiourea) with

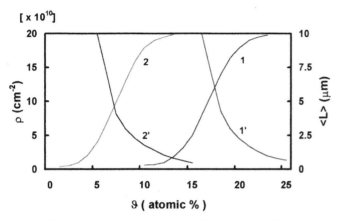

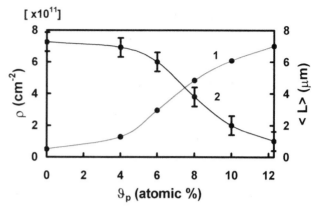

Fig. 26. Influence of the surface coverage, ϑ, with adsorbed foreign particles on the dislocation density (curves 1 and 2) and on the average grain size (curves 1′ and 2′) of Ni electrodeposits. Heat of adsorption: 5 (curves 1 and 1′) and 15 kcal mol^{-1} (curves 2 and 2′). $\Delta\varphi = 0.2$ V. Reprinted from [61] with kind permission from Elsevier Science.

Fig. 27. Influence of the surface coverage with adsorbed phosphorus atoms, ϑ_P, on the dislocation density, ρ (curve 1) and on the average grain size, $\langle L \rangle$ (curve 2) of Ni–P electrodeposits. $\Delta\varphi = 0.2$ V. Reprinted from [61] with kind permission from Elsevier Science.

different concentrations in the electrolyte on the structure of Cu electrodeposits.

Fifty-micrometer-thick Ni–P electrodeposits were prepared potentiostatically (the cathodic overvoltage, $\Delta\varphi$, was 0.2 V) from a Watts bath to which sodium hypophosphite (NaH$_2$PO$_2$) was added in various concentrations from 0 to 3 g L^{-1} at intervals of 0.5 g L^{-1}. The pH was 4, the temperature was 50°C. Sixty–micrometer-thick Cu deposits were also prepared potentiostatically ($\Delta\varphi = 0.15$ V) from an acid solution containing CuSO$_4$·5H$_2$O (250 g L^{-1}), H$_2$SO$_4$ (50 g L^{-1}), and thiourea in concentrations from 0 to 30 mg L^{-1}, at intervals of 5 mg L^{-1}. The temperature was 25°C. Electrodeposition of Ni–P and Cu was performed on a mechanically polished stainless steel substrate from which the deposit could easily be detached.

The microstructure of the cross section of the Ni–P and Cu deposits was investigated by optical microscopy. The average grain size of the deposits was determined from micrographs obtained by transmission electron microscopy. The dislocation density was evaluated from the average size of the domains with coherent diffraction, $\langle D \rangle$. The $\langle D \rangle$ value of the Ni–P deposits was obtained from the Fourier analysis of the (200) and (400) X-ray diffraction lines; that of the Cu deposits was obtained from the width of the (220) X-ray diffraction line, according to the Seliakov–Scherrer relationship.

The phosphorus content of the Ni–P deposits was determined from the already known [62] dependence of the phosphorus concentration in Ni–P deposits on the hypophosphite concentration in the electrolyte. According to this dependence, the volume concentration of phosphorus in the deposits, C_P, ranged from 0 to 5 at.% when the hypophosphite concentration varied from 0 to 3 g L^{-1}. Assuming a uniform distribution of phosphorus in the entire volume of the deposits, the surface concentration of P was evaluated from C_P and it was assumed as the surface coverage with the adsorbed P atoms, ϑ_P, during the Ni electrocrystallization process.

First, the experimental results relevant to the Ni–P electrodeposits will be considered. These results (Fig. 27) show the same

trend as the theoretical ones. Increasing ϑ_P, the degree of imperfection of the Ni–P deposits structure increases, starting from relatively low coverages, thus evidencing a relatively high energy of adsorption on the growing planes (compare the curves in Fig. 27 with curves 2 and 2′ of Fig. 26). Note that the {100} plane was selected for the theoretical calculations since Ni electrodeposits from Watts baths mainly show a ⟨100⟩ preferred orientation perpendicular to the substrate surface.

As observed by optical microscopy (Fig. 28a), Ni electrodeposits not containing P present a typical columnar microstructure of their cross section with a ⟨100⟩ preferred orientation. According to [63, 64], it may be assumed that such a polycrystalline structure is due to the independent formation of highly disoriented nuclei on the indifferent substrate during the earliest deposition stage (see the scheme of Fig. 19b). During the subsequent crystal growth process perpendicular to the substrate surface through a continuous nucleation, these crystals retain their original monocrystalline structure since the probability of formation of noncoherent nuclei with great disorientation angles is practically equal to zero in the experimented conditions when foreign particles adsorption is absent (see the scheme of Fig. 19c).

The columnar character of the microstructure does not change when the deposits contain phosphorus and the surface coverage with adsorbed P atoms is not higher than 4–6 at.%, but the width of the columns (grains), i.e., the size parallel to the substrate surface, decreases (Fig. 28b) as well as the preferentially oriented volume fraction. The latter decreases from 80 to 60%. Hence, phosphorus surface adsorption up to these percentages mainly influences the initial step of Ni electrocrystallization. However, with the increase of the ϑ_P values, the microstructure of the electrodeposits changes from columnar to homogeneous (Fig. 28c) while the preferentially oriented volume fraction of the Ni–P deposits decreases from 60 to 35%. So, phosphorus adsorption in these conditions causes the regular, periodic formation of nuclei with a great disorientation angle even during the outward growth process of Ni crystals, in agreement with the considered mechanism of formation

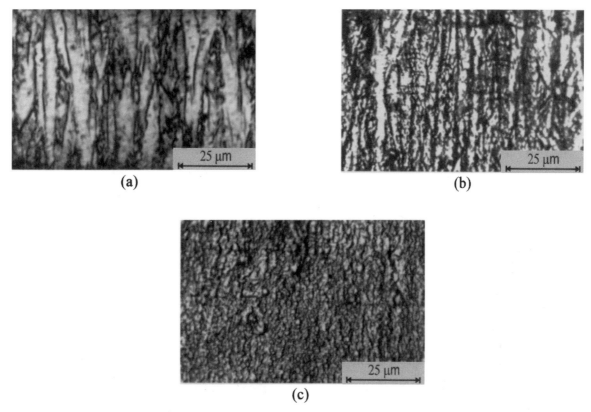

(a)

(b)

(c)

Fig. 28. Microstructure of the cross section of Ni–P electrodeposits. Surface coverage with adsorbed P atoms: (a) 0, (b) 6, and (c) 10 at.%. Reprinted from [61] with kind permission from Elsevier Science.

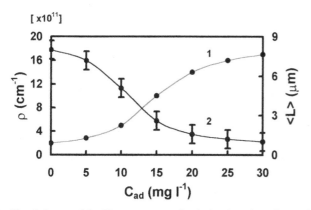

Fig. 29. Influence of the thiourea concentration in the electrolyte, C_{ad}, on the dislocation density, ρ, (curve 1) and on the average grain size, $\langle L \rangle$, (curve 2) of Cu electrodeposits. $\Delta\varphi = 0.15$ V. Reprinted from [17] with kind permission from Elsevier Science.

of structural defects in the electrodeposits (see the scheme of Fig. 19d).

As to the experimental results relevant to Cu electrocrystallization in the presence of thiourea [17], Figure 29 shows the dependence of the structural parameters, ρ and $\langle L \rangle$, of Cu electrodeposits on the thiourea concentration in the electrolyte, C_{ad}. It is evident that this dependence is similar to that observed for the structural parameters of the Ni–P deposits on the value of ϑ_P (Fig. 27). Assuming that increasing the value of C_{ad} increases

the surface coverage of the planes of the growing Cu crystals with adsorbed molecules of thiourea, it is possible to verify the agreement between the experimental and the theoretically predicted trend.

Experimental data concerning the microstructure of the Cu deposits cross section showed that it was columnar in the absence of the thiourea additive (Fig. 2a). However, increasing the thiourea concentration in the electrolyte, such microstructure gradually changed to the homogeneous type and was completely homogeneous for C_{ad} values greater than 10 mg L^{-1} (Fig. 2b). Hence, the same trend as in the case of Ni–P electrodeposits was observed in the change of the microstructure.

Summarizing the results, it may be concluded that the experimental findings agree with the considered mechanism of the influence of foreign particles adsorption on the process of noncoherent nucleation and of the formation of structural defects (grain and dislocation boundaries) in the electrodeposits.

As to the influence of the adsorbed foreign particles (phosphorus atoms and thiourea molecules) on the formation of growth twins in Ni–P and Cu electrodeposits, TEM results showed that increasing the values of ϑ_P and C_{ad}, respectively, the quantity of twins decreased (Fig. 30). This finding can be explained from our point of view assuming that, owing to the increased selective adsorption of phosphorus atoms and thiourea molecules, the habitus of the growing crystals changed so that the overall surface area of the octahedral-type planes decreased. This is the only crystallographic plane of fcc metals on which

Fig. 30. TEM micrographs of Ni–P electrodeposits (a, b) obtained in different conditions as to the surface coverage with adsorbed P atoms: (a) 0 and (b) 6 at.% and of Cu electrodeposits (c, d) obtained in the absence (c) and in the presence (d) of thiourea at a concentration of 15 mg L^{-1}. Micrographs (c, d) reprinted from [17] with kind permission from Elsevier Science.

the formation of nuclei in twin position is possible (Kern's twinning mechanism [33]).

6.3. Influence of the Nature of Metals on the Formation of the Polycrystalline Structure of the Deposit during Electrocrystallization

On the basis of the atomistic theory of nucleation, a quantitative relationship was obtained between the main structural parameters of the electrodeposits (average grain size and dislocation density) and the crystallization conditions (crystallization overvoltage and foreign particles adsorption). Another important aspect which is considered here is the influence of the nature of metals on noncoherent nucleation and on the formation of the polycrystalline structure of the electrodeposits.

First of all, it is important to note that several years ago Piontelli [65] divided the different metals into three classes according to their general electrokinetic behavior in aqueous solutions of their simple salts:

Class I, so-called normal metals: Pb, Sn, Tl, Cd, Hg, and Ag which behave in a practically reversible way for metal ion exchanges up to very high current densities and present a high surface diffusion coefficient for adatoms; instead, they show high overvoltages for hydrogen discharge;

Class II, intermediate metals: Zn, Bi, Cu, Au, Sb, and As which show lower reversibility and lower hydrogen overvoltages;

Class III, inert metals: Fe, Co, Ni, Mn, Cr, and Pt which show high overvoltages during the transfer of their ions (and a low surface diffusion coefficient for adatoms) and very low hydrogen overvoltages, to which correspond their well-known catalytic properties.

A substantially similar classification was later presented by other authors [66–69]. A so-called normality-inertia parameter has been introduced, obtained from the enthalpy contribution to the different steps of the Born–Haber cycle for redox reactions occurring at a metal surface. Such a parameter is low or high for class I or III metals, respectively [67, 68].

The logarithm of the metal and hydrogen ion exchange current density (argument in A cm^{-2}) is approximately greater than -2 and smaller than -8, respectively, for normal metals, ranges from -2 to -4 and from -8 to -6, respectively, for intermediate metals, and is smaller than -4 and greater than -6, respectively, for inert metals [67–69].

Class I–III metals present an increasing tendency to adsorb inhibitors; in addition, a so-called secondary inhibition due to an adsorbed layer of metal hydroxide is usually observed in class III metals [69].

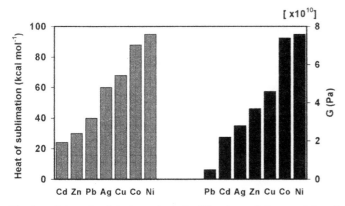

Fig. 31. Schematic of the latent heat of sublimation and shear modulus of classes I (Cd, Pb, Ag), II (Zn, Cu), and III (Co, Ni) metals. Reprinted from [70] with kind permission from Elsevier Science.

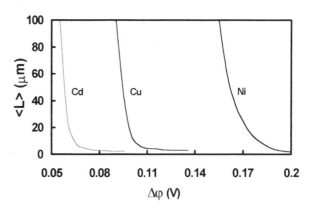

Fig. 32. Influence of the overvoltage, $\Delta\varphi$, on the average grain size, $\langle L \rangle$, of the electrodeposits of classes I (Cd), II (Cu), and III (Ni) metals on their close-packed plane. Reprinted from [70] with kind permission from Elsevier Science.

Table VII. Most Significant Overvoltages (See Text) of Classes I (Cd), II (Cu), and III (Ni) Deposits on Their Close-packed Plane

Metal class	I	II	III
$\Delta\varphi^*$, V	0.055	0.090	0.155
$\Delta\varphi^{**}$, V ($\langle L \rangle = 10\ \mu m$)	0.065	0.100	0.180

Reprinted from [70] with kind permission from Elsevier Science.

The three groups of metals may be distinguished according to their physical and mechanical properties, too. Indeed, in contrast with class III, class I metals show low values of melting point, latent heat of sublimation, hardness as well as of Young's and shear modulus. They are usually electrodeposited as large grains from solutions of their simple salts, while class III metals are electrodeposited as small crystallites. Typical examples of metals having a close-packed crystalline lattice (fcc or hcp) but belonging to the three different classes are Pb, Cd, Ag (class I), Zn, Cu (class II), and Co, Ni (class III). Some of their physicomechanical properties are compared in Figure 31 [70]. The observed trend follows the preceding considerations. To reach our goal, it is necessary to first consider the parameters depending on the metal nature and appearing in the theoretical expression of the work of nucleation, A, thus influencing its value. From Eq. (30), it results that the A value depends, besides on $\Delta\varphi$, also on the total excess free energy of an atom of the noncoherent nucleus, E_1, defined (in the case of the absence of foreign particles adsorption) by Eq. (22), that is,

$$E_1 = \frac{i^* \varepsilon_{1/2} - E_o(i^*) - i_s(\varepsilon_s - \gamma(\theta)/C)}{i^*} \tag{33}$$

where i^* is the number of atoms in the critical nucleus.

It is worth recalling that the values of $\varepsilon_{1/2}$, ε_s, and $E_o(i^*)$ depend on the work of separation of two neighbor atoms of the crystal of the considered metal which is proportional to the heat of sublimation. Therefore, they increase from class I to III metals. In addition, at constant θ values, $\gamma(\theta)$ increases from metals of the first group to those of the third group. This may be seen in Figure 1 which shows the dependence of γ on θ for metals belonging to the three classes, that is Cd, Cu, and Ni. The cause of this influence of the nature of metals on $\gamma(\theta)$ lies in the increase of the value of the shear modulus going from metals of the first group to those of the third group (Fig. 31). It is worth being reminded that $\gamma(\theta)$ has been estimated from the Read–Shockley relationship for the specific free energy of the dislocation boundary.

Equation (30) was applied to evaluate the influence of the overvoltage on the work of the formation of the noncoherent nu-

cleus of Cd, Cu, and Ni. Performing a statistical analysis of the data, as previously done, the dependence of the average grain size, $\langle L \rangle$, of the electrodeposits on the overvoltage was determined.

Figure 32 depicts the theoretical curves $\langle L \rangle$ versus $\Delta\varphi$ for Cd, Cu, and Ni which are regularly shifted toward higher overvoltages going from class I to III electrodeposits. In their upper part (not shown in Fig. 32) $\langle L \rangle$ is practically independent of the overvoltage. So, the change in the slope of the curves occurring close to 100 μm allows us to determine the overvoltage, $\Delta\varphi^*$, where $\langle L \rangle$ starts sharply decreasing.

When electrodeposition takes place on a single-crystal substrate, $\Delta\varphi^*$ represents the transition overvoltage from the monocrystalline to the polycrystalline structure. In the case of a polycrystalline substrate, the overvoltage, $\Delta\varphi^{**}$, for the typical value $\langle L \rangle = 10\ \mu$m, may also be considered. The theoretically estimated values of $\Delta\varphi^*$ and $\Delta\varphi^{**}$ are listed in Table VII. As expected, they increase from Cd to Cu and Ni.

The question now arises how to confirm the theoretical results. The best method would be to deposit the previously considered metals on a single-crystal electrode oriented according to the close-packed plane and to determine the overvoltage where the transition from the monocrystalline to the polycrystalline structure occurs. Another method would be to prepare polycrystalline deposits of the class I to III metals having nearly the same average grain size and to compare their deposition overvoltage. This is the method we followed here.

Cd, Cu, and Ni were electrodeposited in potentiostatic conditions on a mechanically polished stainless steel substrate

Table VIII. Electrolyte Composition, Electrodeposition Conditions, and $\Delta\varphi^{**}$ Overvoltage (See Text)

Metal class	Metal	Electrolyte composition	pH	Temperature, °C	$\Delta\varphi^{**}$, V $\langle L \rangle = 10\ \mu m$
I	Cd	CdSO$_4$-400 g L^{-1}	1.5	20	0.020
II	Cu	CuSO$_4$-500 g L^{-1}		45	0.070
		H$_2$SO$_4$-50 g L^{-1}			
III	Ni	NiSO$_4$-400 g L^{-1}	2.0	80	0.160

Reprinted from [70] with kind permission from Elsevier Science.

Fig. 33. Scanning electron microscopy micrograph of the Pb crystals obtained by electrocrystallization.

wherefrom they could easily be detached. The deposits were sufficiently thick (60–100 μm) to avoid the influence of the substrate on their structure. Electrodeposition was performed at different overvoltages and the overvoltage, $\Delta\varphi^{**}$, was determined at which the average grain size of the deposits was around 10 ± 2 μm. The average grain size of the Cd deposits was determined from their morphology obtained by scanning electron microscopy, that of the Cu and Ni deposits by transmission electron microscopy.

The overvoltages, $\Delta\varphi^{**}$, the other electrodeposition conditions, as well as the electrolyte composition are reported in Table VIII.

The experimental results show the increasing overvoltage at which grains with $\langle L \rangle$ around 10 μm are formed, passing from Cd to Cu and Ni deposits, in substantial agreement with the trend expected on the basis of theoretical calculations.

Nevertheless, further theoretical estimates in conditions closer to the experimental ones (e.g., for substrates with less packed planes) are necessary to verify the experimental findings quantitatively. Moreover, another aspect has to be considered. The experimental results as well as the traditional metal classification are based on the overvoltage of the entire electrodeposition process, whereas the calculations refer to the overvoltage of the crystallization step, only. Generally, the steps preceding crystallization (charge transfer, diffusion, and reactions occurring in the electrolyte) may influence the local distribution of the current density as well as the equilibrium concentration in adatoms and, therefore, their overvoltages should also be taken into account [69]. Moreover, the influence of the kinetic factors on the grain size, but those due to the crystallization step, was completely ignored. Owing to theoretical difficulties, a detailed analysis of this type is usually lacking. Instead, experimental conditions may be selected able to minimize the overvoltages of the steps preceding crystallization, as performed in the present case. The substantial similarity of the trend of the estimated and measured overvoltages gives a theoretical though qualitative support to the traditional metal classification. In any case, these interesting aspects deserve a deep look and along these lines it is worth developing further research.

7. MECHANISM OF MULTITWINNING

As already underlined in Section 2, in agreement with the experimental results, the deposits with an fcc lattice (Cu and Ni) have a specific twinning structure. Indeed, in addition to parallel {111} twin planes, twinning with intersecting octahedral planes also occurred. This behavior has been confirmed by the observation of the grains with the zone axis ⟨110⟩ in TEM micrographs which showed the points of intersection of some twin boundaries perpendicular to the surface of the deposit (Fig. 5). It is particularly important to pay attention to the points where five boundaries intersect. In this case, they are called pentagonal crystals with a geometrical "pseudo-five" symmetry. The ideal five symmetry of the crystals does not correspond to the laws of the crystallographic stacking because such a structure does not have the translational invariance.

The pentagonal crystals (particles) of the different sizes (by some tens of angstroms to some microns) have been observed during the initial step of the deposition process from the vapor phase and also during electrocrystallization of the different fcc metals (Au, Pd, Pt, Ni, Co, Ag, Cu, and Pb) [71–88]. As a rule, the pentagonal particles had the shape of a decahedron (ten planes with one axis of symmetry of the fifth order) or of an icosahedron (20 planes with six axes of symmetry of the fifth order). Figure 33 shows an example of the decahedral shape of the Pb crystals obtained by electrocrystallization on the mechanically polished stainless steel substrate.

From the point of view of the internal structure, there may be two types of pentagonal crystals of the fcc metals. If we consider five tetrahedra (each tetrahedron has four {111} planes) (Fig. 34a) and connect tetrahedron 1 to tetrahedron 2 through an octahedral plane and consecutively connect tetrahedron 2 to 3, 3 to 4, and 4 to 5, then each tetrahedron will be in twin position with respect to the neighbor tetrahedra but this is not the case for the first and the last tetrahedron. Since the angle between the {111} planes is equal to about 70.5°, but not to 360°/5, a "corner split" must be formed having the aspect of a dihedral angle by about 7.5° between the {111} plane of the first and the fifth tetrahedron. This means that these tetrahedra are mutually disoriented with respect to the ⟨110⟩ axis by 78°.

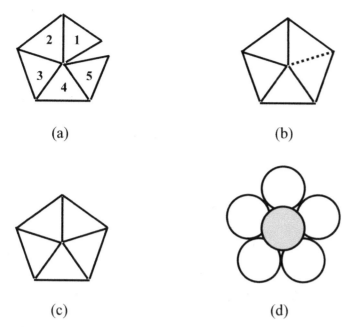

(a)

(b)

(c)

(d)

Fig. 34. Schematic showing (a) the connection of the five tetrahedra (b) with the formation of the pseudo-five and (c) of the veritable-five symmetry, and (d) a model of the cluster of seven atoms with "five" symmetry. The plane of the drawing is the {110} plane.

Therefore, with the filling of this corner split with the atoms of the deposits a connection takes place between the two extreme tetrahedra with the formation of a high-angle boundary of the grains [89–94] (the dashed line in Fig. 34b). Sometimes, during the morphological investigation of the pentagonal crystals, the corner split filled with the atoms of the metal has been observed and it had the aspect of a wedgelike "seam" [72, 75, 79, 82]. Hence, the pentagonal crystals of the first type (Fig. 34b) having four twin boundaries and one high-angle boundary present the pseudo-five symmetry.

We now consider the pentagonal crystals of the second type. To do this, we image that the system of the five tetrahedra previously considered (Fig. 34a) deforms itself homogeneously so that the corner split of 7.5° has been eliminated. In this case, a crystal with "veritable-five" symmetry (Fig. 34c) is formed containing five twin boundaries. However, it is necessary to note that such a pentagon will have an excess of internal energy owing to the elastic deformation of the lattice.

There are two alternative points of view as to the mechanism of formation of the pentagonal crystals.

According to one of them, the pentagonal particles with veritable-five symmetry are already formed during the nucleation step (cluster's mechanism). This point of view has been confirmed by the experimental results owing to the observation of pentagonal particles with a very small size (from 20 to 30 Å) having the shape of the decahedron and the icosahedron [78, 85, 93]. With this result, the deformed elastic state of the pentagonal particles has also been confirmed [85, 93, 94]. These results mean that the nuclei with the veritable-five symmetry are more stable than those with the usual close-packed structure which may be related to the condition that the internal energy of the

nucleus with the veritable-five symmetry, E_5, is lower than the internal energy of the nucleus with the close-packed structure, E_o. This fact may be connected to the relatively higher saturation of the bonds among the atoms of the pentagonal nucleus. The example reported in Figure 34d illustrates the stacking of the atoms in a pentagonal cluster formed by seven atoms.

In the articles [82, 93, 95–97], the stability of the pentagonal clusters with veritable-five symmetry having different sizes has theoretically been examined. Taking the energy of cohesion into consideration as well as the surface energy, the energy of the five twin boundaries and the energy of the elastic stress of the particles having the decahedral and icosahedral shape, the value of E_5 has been determined. This value was smaller than the value of E_o when the dimension of the cluster was not more than 2000–3000 Å. In the opposite case, $E_5 > E_o$. On the basis of the results of these calculations, the following conclusions were drawn:

(1) the cluster's mechanism of formation of pentagonal nuclei with veritable-five symmetry may occur when the conditions of crystallization favor the three-dimensional nucleation [86, 87];

(2) when during its growth the dimension of the pentagonal nucleus reaches a few thousands of angstroms, its structure has to change either into the usual close-packed structure [78, 82, 93] or into the structure with pseudo-five symmetry (four twin boundaries and one high-angle boundary) [94].

According to the second point of view relevant to the mechanism of formation of the pentagonal crystals, their formation is explained by a repeated twinning process during the growth step of the crystals [72, 75]. Pangarov and his co-workers [34, 79, 98] assumed this mechanism to interpret their experimental results regarding the electrocrystallization of fcc metals. The substantial points of the mechanism of a repeated twinning process may be explained as in the following. Let us suppose that during the electrodeposition of an fcc metal there is a separate crystal having a tetrahedral shape (tetrahedron 1 in Fig. 34a). On its octahedral plane, the formation of a nucleus in the twin position may occur which then increases forming a second tetrahedron. As a result of this process, the first twin boundary between tetrahedra 1 and 2 is formed (Fig. 34a).

Then, on the octahedral plane of tetrahedron 2, a nucleus is formed in the twin position and, as a consequence, tetrahedron 3 is formed and, correspondingly, the twin boundary between tetrahedra 2 and 3. Tetrahedra 4 and 5 are subsequently formed with the same mechanism. It is evident that the first and the last tetrahedron (1 and 5 in Fig. 34a) are not in the twin position between them. Therefore, with the filling of this corner split between tetrahedra 1 and 5 by the atoms of the deposit, the high-angle boundary of the grains is formed (Fig. 34b). Hence, as a result of the successively repeated twinning process, a crystal with pseudo-five symmetry is obtained presenting the shape of the decahedron (Fig. 33). In the same way, it is possible to explain the formation of the pentagonal crystal having the shape of the icosahedron.

(a)

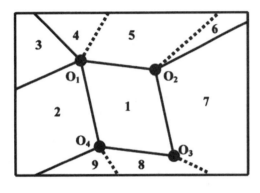

(b)

Fig. 35. (a) TEM micrograph showing the multitwinning of intersecting octahedral planes of Cu electrodeposits with a ⟨110⟩ orientation axis and (b) schematic where the twin boundaries (continuous lines) and the high-angle boundaries of the grains (dashed lines) are reported.

In the real case, during the inital deposition step of the fcc metals, both mechanisms of the formation of the pentagonal crystals may occur. We have a particular situation when the multitwinning of intersecting octahedral planes of electrodeposited fcc metals is observed not in the initial layers but in the layers which are formed at a distance of some tens of microns from the substrate (as shown in Fig. 5). In this case, the points of multitwinning are observed where two or more twin boundaries intersect. The crystallographic analysis shows that, if in the point of multitwinning n boundaries intersect (with $n = 3$, 4, 5), then $(n - 1)$ of them will be twin boundaries while one boundary will be the high-angle boundary of the grains.

We analyze this last case in detail considering as an example the TEM micrograph shown in Figure 35a. Figure 35b reports the scheme of this TEM micrograph. We observe the four points of the multiplying twinning, O_1, O_2, O_3, and O_4, where 5, 4, 3, and 4 boundaries intersect, respectively. In the figure, the twin boundaries are represented by the continuous lines whereas the high-angle boundaries of the grains are represented by the dashed lines.

Similar configurations of intersecting twin boundaries have also been observed in [89, 91, 92]. In agreement with the opinion of the authors of the article reported in [99], the formation of the points of multitwinning where several twin boundaries intersect (from two to four) may be explained by a repeated twinning process during the growth step of the crystals. In this case, the so-called cluster's mechanism can explain the formation of the points of one type, only; i.e., the point where four twin boundaries and one high-angle boundary of the grains (the pseudo-five symmetry point O_1, in Fig. 35b) intersect but not the formation of the points where three or two twin boundaries (O_2, O_3, and O_4, in Fig. 35b) intersect.

Let us now consider the formation of all the points of multitwinning in agreement with the schematic shown in Figure 35b. For example, crystal 1 is first formed which has the four octahedral planes perpendicular to the {110} plane, that is the plane of the drawing. Then, nuclei are formed on the four octahedral planes in the twin positions and fragments 2, 5, 7, and 8 will be in the twin position relative to fragment 1. Subsequently, fragments 3, 4, 6, and 9 may be formed in the same way with the formation of twin boundaries. After the connection of fragments 4 and 5, 6 and 5, 7 and 8, 9 and 8, the usual high-angle boundaries of the grains are formed.

So, it may be concluded that the separated crystals of the fcc metals with the pentagonal shape which are observed during the initial step of heterogeneous electrocrystallization are probably formed by the cluster's mechanism when the conditions of crystallization favor three-dimensional nucleation. This is related to the lower internal energy of the nuclei having the decahedral and icosahedral shape. On the other hand, the formation during the subsequent growth step of the different configurations of the intersecting twin boundaries of the fcc metals is probably connected to a repeated twinning process.

8. CONCLUSIONS

The elaboration of the mechanism of formation of the different types of crystallographic defects in metal deposits is important both for practical and theoretical reasons. In the present chapter, the problem of the structure formation of deposits was discussed taking the electrocrystallization of metals as an example, although several aspects which have been examined may completely be applied to metal deposits obtained in other ways, first of all by deposition from the vapor phase.

As the basis of this investigation, we considered the model of noncoherent (disoriented) nuclei, which are formed with a defined probability besides the normal nuclei during the growth of the deposit in thickness at the end of the initial crystallization step including the formation of the nuclei on the substrate, their lateral growth, and the formation of a thin compact deposit layer.

The application of this model allowed us to explain from a unique point of view the formation of the main types of structural defects in electrodeposits (high-angle grain boundaries, subgrain boundaries of dislocations and twins). In agreement with our ideas, the effective type of crystallographic two-dimensional defect which is formed owing to noncoherent nucleation depends on the value of the azimuth disorientation angle, θ, with respect to its own substrate plane. More precisely:

1. if the value of θ is smaller than 10–15°, a dislocation boundary is formed;
2. if it is greater than 10–15°, the usual high-angle grain boundary is formed.

It is necessary to draw attention to the third case which may occur, that is when nucleation takes place on a crystallographic plane being the typical twinning plane of the crystal (for example, the octahedral {111} plane of fcc crystals). In this case, if the atoms of the nucleus occupy twin positions; i.e., the electrodeposition conditions are favorable to twinning, then a twin is formed in the deposit (Kern's mechanism).

The classical thermodynamic analysis of the noncoherent nucleation step performed during our research has confirmed that there is a unique mechanism of formation of the main structural defects during metal electrocrystallization. On the basis of the calculations of the work of formation of noncoherent and normal nuclei, the influence of the crystallization overvoltage (oversaturation) and of the surface free energy per unit area on the relative probability of noncoherent nucleation which qualitatively characterizes the degree of crystal imperfection of the electrodeposits was investigated.

In more detail, the mechanism of formation of the main structural defects in the deposits was investigated according to a thermodynamic-statistical analysis of the noncoherent nucleation step from the atomistic point of view. This allowed us to determine the quantitative relationship between the main structural parameters of the deposits (dislocation density, ρ, and average grain size, $\langle L \rangle$) and the conditions of electrocrystallization.

First of all, we theoretically analyzed the influence of the crystallization overvoltage, $\Delta\varphi$, and of the surface coverage with adsorbed foreign particles, ϑ, on the values of ρ and $\langle L \rangle$ of the deposits. These results were interpreted in connection with the growth in thickness of the electrodeposits initially having either a polycrystalline structure or a monocrystalline structure and they were confirmed by a structural research (principally following the TEM method) of electrodeposited metals with different types of crystalline lattice (Cu, Ni—fcc, Fe—bcc, Zn—hcp).

The proposed model of noncoherent nucleation explained not only the mechanism of formation of the main structural defects but also the structural changes occurring by modifying the electrolysis conditions. In particular, it was possible to explain the well-known transition from the single-crystalline to the polycrystalline structure observed in metals deposited on single crystals at increasing cathodic overvoltage, current density, concentration of surface-active agents, or deposit thickness. Moreover, the transition of the microstructure of the polycrystalline electrodeposits from the columnar to the homogeneous type when the concentration of surface-active agents in the electrolyte was increased was also explained.

A thermodynamic-statistical analysis of the noncoherent nucleation step was also carried out to investigate the influence of the nature of metals on the formation of the electrodeposits polycrystalline structure. Considering the traditional classification of metals into three groups according to their increasingly slower kinetics during electrodeposition from aqueous solutions, we theoretically analyzed Cd, Cu, and Ni, they being typically representative metals of each group. The theoretical results were confirmed by the experimental results obtained during Cd, Cu, and Ni electrodeposition.

Particular attention was devoted to the phenomenon of multitwinning with intersecting octahedral planes in the fcc electrodeposited metals and in particular to the formation of the pentagonal crystals with a geometrical pseudo-five symmetry. In agreement with the results of the crystallographic analysis and of the generalization of the numerous literature data, some conclusions could be drawn regarding the mechanism of the multitwinning process in electrodeposits.

The proposed mechanism of the formation of the structure of electrodeposits certainly does not exclude other possible mechanisms, however, from our point of view, the major contribution to the formation of the main crystallographic defects during the growth in thickness of the electrodeposits is due to the noncoherent nucleation step. In any case, the series of problems connected to the structure of the electrodeposits could be explained just considering noncoherent nucleation.

ACKNOWLEDGMENT

V. M. Kozlov kindly acknowledges the Cariplo Foundation, "A. Volta" Center for Scientific Culture, for financial support.

REFERENCES

1. M. Volmer and A. Weber, *Z. Phys. Chem.* 119, 277 (1926).
2. E. Bauer, *Z. Kristallogr.* 110, 372 (1958).
3. F. C. Frank and J. H. van der Merwe, *Proc. R. Soc. A* 198, 205 (1949).
4. J. H. van der Merwe, *Cryst. Interfaces* 34, 117 (1963).
5. W. T. Read and W. Shockley, *Phys. Rev.* 78, 275 (1950).
6. H. Gleiter and B. Chalmers, "High-Angle Grain Boundaries," Pergamon, New York, 1972.
7. J. P. Hirth and J. Lothe, "Theory of Dislocation," Plenum, New York, 1966.
8. H. Fischer, "Elektrolytische Abscheidung und Elektrokristallisation von Metallen," Springer-Verlag, Berlin, 1954.
9. S. Steinemann and H. E. Hintermann, *Schweiz. Archiv Angew. Wiss. Tech.* 26, 202 (1960).
10. Y. M. Polukarov and Y. D. Gamburg, *Elektrokhimiya* 2, 487 (1966).
11. K. R. Lawless, *Phys. Thin Films* 4, 191 (1967).
12. V. M. Kozlov, L. Peraldo Bicelli, and E. A. Mamontov, *Ann. Chim. (Italy)* 63, 405 (1973).
13. M. Froment and G. Maurin, *Electrodep. Surface Treat.* 3, 245 (1975).
14. E. Budevski, G. Staikov, and W. J. Lorenz, "Electrochemical Phase Formation and Growth," VCH, Weinheim, New York, 1996.
15. H. Fischer, *Z. Elektrochem.* 54, 459 (1950).
16. H. Fischer, *Electrodep. Surface Treat.* 1, 319 (1973).
17. V. M. Kozlov and L. Peraldo Bicelli, *Mater. Chem. Phys.* 62, 158 (2000).
18. L. Reimer, *Z. Metallkd.* 47, 631 (1956).
19. S. Ogawa, J. Mitzumo, D. Watanabe, and F. E. Fujita, *J. Phys. Soc. Jpn.* 12, 999 (1957).

20. R. Weil and H. C. Cook, *J. Electrochem. Soc.* 109, 295 (1962).
21. E. M. Hofer and Ph. Javet, *Microtecnic* 17, 168 (1963).
22. T. G. Stoebe, F. H. Hammad, and M. L. Rudee, *Electrochim. Acta* 9, 925 (1964).
23. H. J. Read and E. J. Oles, *Plating* 52, 860 (1965).
24. E. M. Hofer, L. F. Chollet, and H. E. Hintermann, *J. Electrochem. Soc.* 112, 1145 (1965).
25. M. Froment and A. Ostrowiecki, *Metaux* 41, 83 (1966).
26. G. Maurin and M. Froment, *Metaux* 41, 102 (1966).
27. E. A. Mamontov and V. M. Kozlov, *Elektrokhimiya* 5, 1158 (1969).
28. V. M. Kozlov, *Elektrokhimiya* 18, 1353 (1982).
29. V. M. Kozlov and L. Peraldo Bicelli, *J. Cryst. Growth* 165, 421 (1996).
30. H. Brandes, *Z. Phys. Chem.* 126, 196 (1927).
31. R. Kaischev, *Izv. Bulg. Akad. Nauk, Ser. Fis.* 1, 100 (1950).
32. I. Markov and R. Kaischev, *Thin Solid Films* 32, 163 (1976).
33. R. Kern, *Bull. Soc. Franc. Miner. Cryst.* 84, 292 (1961).
34. N. A. Pangarov, *Phys. Status Solidi* 20, 371 (1967).
35. I. N. Stranski and R. Kaischev, *Ann. Phys.* 23, 330 (1935).
36. T. H. V. Setty and H. Wilman, *Trans. Faraday Soc.* 51, 984 (1955).
37. Th. H. Orem, *J. Res. Natl. Bur. Stand.* 60, 597 (1958).
38. S. C. Barnes, *Acta Metall.* 7, 700 (1959).
39. G. Poli and L. Peraldo Bicelli, *Met. Ital.* 51, 548 (1959).
40. Y. M. Polukarov and Z. V. Semionova, *Elektrokhimiya* 2, 184 (1966).
41. M. Lazzari and B. Rivolta, "Proceedings of the Symposium on Sulfamic Acid," Milan, 1966.
42. N. A. Pangarov and V. Velinov, *Electrochim. Acta* 13, 1909 (1968).
43. J. B. Cusminsky and H. Wilman, *Electrochim. Acta* 17, 237 (1972).
44. G. W. Sears, *J. Chem. Phys.* 31, 157 (1959).
45. G. A. Bassett, "Proceedings of the European Regional Conference on Electron Microscopy," Delft, 1960.
46. J. W. Matthews and D. L. Allinson, *Philos. Mag.* 8, 1283 (1963).
47. R. Weil and J. B. C. Wu, *Plating* 60, 622 (1973).
48. I. N. Stranski, *Z. Phys. Chem.* 136, 259 (1928).
49. M. Volmer, "Kinetik der Phasenbildung," Steinkopf, Dresden-Leipzig, 1939.
50. T. Erdey-Gruz and M. Volmer, *Z. Phys. Chem.* 157A, 165 (1931).
51. D. Walton, *J. Chem. Phys.* 37, 2188 (1962).
52. D. Walton, T. N. Rhodin, and R. W. Rollins, *J. Chem. Phys.* 38, 2698 (1963).
53. S. Stoyanov, *Thin Solid Films* 18, 91 (1973).
54. V. M. Kozlov and L. Peraldo Bicelli, *J. Cryst. Growth* 177, 289 (1997).
55. I. N. Stranski and R. Kaischev, *Z. Phys. Chem.* 35B, 27 (1937).
56. P. Hirsh, *Prog. Metal. Phys.* 6, 236 (1956).
57. G. Williamson and R. Smallman, *Philos. Mag.* 1, 34 (1956).
58. S. C. Barnes, *Electrochim. Acta* 5, 79 (1961).
59. G. Poli and L. Peraldo Bicelli, *Met. Ital.* 54, 497 (1962).
60. L. Peraldo Bicelli and G. Poli, *Electrochim. Acta* 11, 289 (1966).
61. V. M. Kozlov, L. Peraldo Bicelli, and V. N. Timoshenko, *J. Cryst. Growth* 183, 456 (1998).
62. V. V. Povetkin, *Izv. Akad. Nauk SSSR Met.* 3, 187 (1985).
63. S. Nakahara and R. Weil, *J. Electrochem. Soc.* 120, 1462 (1973).
64. R. Weil, G. J. Stanko, and D. E. Moser, *Plating Surf. Finishing* 63, 34 (1976).
65. R. Piontelli, "Proceedings IX Congress IUPAC," London, 1947, p. 785.
66. T. Erdey-Gruz, "Kinetics of Electrode Processes," Higler, London, 1972.
67. P. L. Cavallotti, D. Colombo, U. Ducati, and A. Piotti, *Proc. Electrochem. Soc. (Electrodeposition Technol. Theory Practice)* 87-17, 429 (1987).
68. P. L. Cavallotti, B. Bozzini, L. Nobili, and G. Zangari, *Electrochim. Acta* 39, 1123 (1994).
69. R. Winand, *Hydrometallurgy* 29, 567 (1992).
70. V. M. Kozlov and L. Peraldo Bicelli, *J. Cryst. Growth* 203, 255 (1999).
71. A. J. Melmed and D. O. Hayward, *J. Chem. Phys.* 31, 545 (1959).
72. H. Schlotterer, *Metalloberfläche* 18, 33 (1964).
73. M. A. Gedwill and C. J. Altstetter, *J. Appl. Phys.* 35, 2266 (1964).
74. F. Ogburn, B. Paretzkin, and H. S. Peiser, *Acta Crystallogr.* 17, 774 (1964).
75. R. W. De Blois, *J. Appl. Phys.* 36, 1647 (1965).
76. R. L. Schwoebel, *J. Appl. Phys.* 37, 2515 (1966).
77. J. G. Allpress and J. V. Sanders, *Surface Sci.* 7, 1 (1967).
78. S. Ino and S. Ogawa, *J. Phys. Soc. Jpn.* 22, 1365 (1967).
79. N. A. Pangarov and V. Velinov, *Electrochim. Acta* 13, 1641 (1968).
80. E. Gillet and M. Gillet, *Thin Solid Films* 4, 171 (1969).
81. I. Epelboin, M. Froment, and G. Maurin, *Plating* 56, 1356 (1969).
82. Y. Fukano and C. M. Wayman, *J. Appl. Phys.* 40, 1656 (1969).
83. E. Gillet and M. Gillet, *Thin Solid Films* 15, 249 (1973).
84. S. Kamasaki and Y. Tanabe, *Met. Finish. Soc. Jpn.* 25, 75 (1974).
85. K. Yagi, K. Takayanagi, K. Kobayashi, and G. Honjo, *J. Cryst. Growth* 28, 117 (1975).
86. M. Froment and J. Thevenin, *Metaux* N594, 43 (1975).
87. J. Thevenin, *J. Microsc. Spectrosc. Electron.* 1, 7 (1976).
88. C. Digard, G. Maurin, and J. Robert, *Metaux* N611, 255 (1976).
89. J. W. Faust and H. F. John, *J. Phys. Chem. Solids* 25, 1407 (1964).
90. S. Ino, *J. Phys. Soc. Jpn.* 21, 346 (1966).
91. J. W. Faust, F. Ogburn, D. Kahan, and A. W. Ruff, *J. Electrochem. Soc.* 114, 1311 (1967).
92. J. Smit, F. Ogburn, and C. J. Bechtoldt, *J. Electrochem. Soc.* 115, 371 (1968).
93. T. Komoda, *Jpn. J. Appl. Phys.* 7, 27 (1968).
94. E. Gillet, M. Gillet, and A. Renou, *Thin Solid Films* 29, 217 (1975).
95. S. Ino, *J. Phys. Soc. Jpn.* 26, 1559 (1969).
96. S. Ino, *J. Phys. Soc. Jpn.* 27, 941 (1969).
97. M. R. Hoare and P. Pal, *J. Cryst. Growth* 17, 77 (1972).
98. V. Velinov and N. Pangarov, *Izv. Otd. Him. Nauk Bulg. Akad. Nauk* 5, 207 (1972).
99. E. A. Mamontov, V. M. Kozlov, and L. A. Kurbatova, *Elektrokhimiya* 15, 257 (1979).

Chapter 12

EPITAXIAL THIN FILMS OF INTERMETALLIC COMPOUNDS

Michael Huth

Institute for Physics, Johannes Gutenberg-University Mainz, 55099 Mainz, Germany

Contents

1. INTRODUCTION

The potential of epitaxial thin films of intermetallic compounds in basic and applied research is only just emerging. Although the growth of semiconductor heterostructures and compounds based on molecular beam epitaxy (MBE) and related methods has come through a 30-year history of ongoing refinement and sophistication, still much has to be learned concerning the growth and characterization of even moderately complex metallic thin film structures. One of the main reasons for this used to be the lack of strong driving forces for possible industrial applications in metals epitaxy that is surely present in semiconductor science and technology. (It should be stressed here that investigations focused on the optimization of metallization layers based on silicides for semiconductor devices will not be considered in this chapter.) Nevertheless, this is changing now. At present, various intermetallic compounds are being investigated with respect to their applicability as magnetooptically active layers, to give just one example. In this context, the large variety of MBE techniques and *in situ* characterization methods, even though originally set up mainly for research on semiconductors, establishes a sound basis for the preparation of epitaxial metallic structures. Many guiding principles can be taken from the growth in the semiconductor knowledge base, but there are also salient differences in metallic systems. To give only two examples: first, the maintenance of epitaxial strain in metallic systems is difficult because there is no appreciable activation barrier for dislocation formation, whereas in semiconductors the Peierls barrier, due to directed covalent bonding, helps to sustain strain far beyond the critical film thickness. Particularly in magnetic systems exhibiting sufficient magnetoelastic coupling effects, "strain tailoring" is often necessary to generate desired magnetic properties, such as the orientation of the axis of easy magnetization. Second, the influence of impurity lev-

Handbook of Thin Film Materials, edited by H.S. Nalwa
Volume 1: Deposition and Processing of Thin Films

ISBN 0-12-512909-2/$35.00

els on the characteristic properties (e.g. the electronic structure) of metals is far less pronounced because of strong screening effects. This is in obvious contrast to semiconductors, the electronic structure of which can be appreciably changed, even by residual gas trace impurities introduced during the deposition process in a vacuum of 10^{-11} mbar. This means, in some instances, that reasonably decent metallic layers can be grown in a less stringent vacuum of 10^{-8} mbar.

MBE represents a well-defined crystallization technique based on the reactions between molecular or atomic beams of the constituent elements on a substrate or template at elevated temperatures in an ultrahigh vacuum (UHV) environment. Owing to the UHV environment, various surface sensitive characterization techniques, like electron diffraction in grazing and normal incidence, can be used to monitor and/or control the growth process. The appreciably improved control of beam fluxes and growth conditions, as compared with conventional physical vapor deposition techniques, allows the realization of synthetic structures on the atomic layer scale. This includes superlattice structures as well as metastable (ordered) alloys or intermetallic compounds. The growth process is much more driven than in near-equilibrium techniques because of an enhanced supersaturation that may be as large as 10 to 40. As a consequence, the growth of any compound, even if the constituent elements have vastly differing vapor pressures, is possible in principle.

In the following some general considerations concerning the growth of intermetallic compounds as epitaxial thin films will be given. This will comprise a short overview of different systems that show specific aspects of metallic compound growth employing MBE techniques. A detailed account of the technical side of MBE growth itself lies beyond the scope of this chapter. Only a short overview is given concerning specific instrumental aspects of MBE growth of metals. For recent publications concerning conceptional details of MBE and related techniques, the growth process, and relevant characterization techniques, the reader is referred to [1].

An appreciable part of this review will be devoted to the applications of thin films of intermetallic compounds in basic and applied research, which, more than anything else, spurs the growth of the field. This overview will certainly not be complete. It rather represents a subjective selection of investigations on intermetallic compounds and ordered alloys that either highlight specific aspects relevant to the field or are examples that are highly interesting in their own right.

2. MBE GROWTH OF INTERMETALLIC COMPOUNDS

2.1. Intermetallic Compounds: Definition of Terms

When should an ordered intermetallic alloy be designated an intermetallic compound? The physical reason for the formation of an ordered phase can give some guidance in the formulation of an answer to this question.

In the Helmholtz free energy of an alloy it is the entropy of mixing that favors the formation of mixed alloys. Only when the pair formation of unequal next neighbors results in an energy gain does the formation of an ordered phase tend to become advantageous. Consider, for example, the phase diagram of a binary alloy. With increasing negative energy of mixing the phase diagram will change, eventually showing the formation of an intermetallic phase out of the melt. Such a compound can be formed at high temperatures, with the liquidus and solidus curves forming a common maximum, resulting in a congruently melting compound. The compound can also form peritectoidically. For moderate negative mixing energies the formation of an ordered phase can also take place out of a disordered solid alloy phase. Cu_3Au can be considered one of the best-studied examples. Cu_3Au forms from the disordered phase by a congruent transformation at the stoichiometric composition and at the Au-rich limits by an eutectoid transformation. In this respect, one cannot discriminate between the terms "ordered alloy" and "intermetallic compound," at least below the order–disorder transition temperature. We therefore also include aspects of ordered alloys in this review. These are from growth studies that deal with the influence of interfaces and film morphology on the onset of long-range chemical order. We will also discuss some recent investigations concerning stabilization techniques for specific stacking variants in intermetallic compounds with fcc structure grown in (111) orientation.

2.2. Equipment

2.2.1. Preparation

Many aspects of modern MBE systems for metals epitaxy are analogous to the technology used in semiconductor growth. Figure 1 shows a state-of-the-art MBE system custom designed by Omicron Vacuumsystems.

The general setup of such a system is based on a modular concept that helps to sustain excellent vacuum conditions in the growth and analytic chambers. A small load lock with multiple sample cassette and heating capabilities permits the necessary preconditioning of the substrates. This can comprise outgassing, ion beam cleaning, and cleaving. The commonly used sample size is in the millimeter to centimeter range. For growth at high substrate temperatures, substrate carriers made from refractory metals (Mo, Ta) are often used, to which the substrate is attached by various clamping techniques or tungsten wires spot-welded over the edges, or with silver paint or ceramic adhesives. If the substrate is glued, care should be taken to degas the carrier and substrate before transfer into the MBE or analytic chambers because organic solvents can interfere with delicate components, like hair filaments of electron guns. The distribution chamber forms the central part of a modular system design, allowing the delivery of samples to the different attached chambers. The MBE chamber itself can be equipped with conventional Knudsen cells, high-temperature cells for refractory metals fitted with special liners (usually made from W, Ta, Mo, and pyrolytic graphite), and electron beam or electron

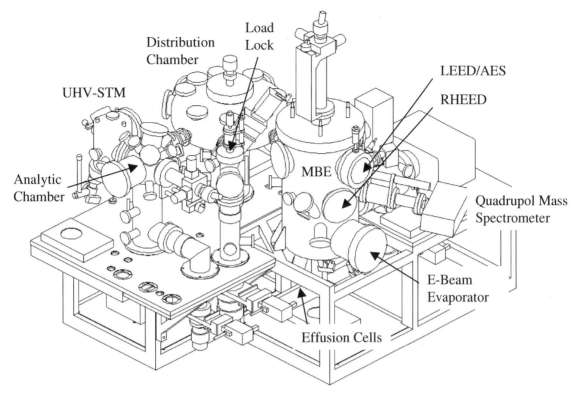

Fig. 1.   MBE system consisting of load lock, distribution chamber, growth chamber, and analytic chamber with UHV-STM as indicated.

impact evaporators. The use of high-temperature effusion cells relies on an open cell design to reach sufficient flux at acceptable cell temperatures, which can be as high as 2100 K. The heat transfer is then completely governed by radiation, which has to be taken into account for cell design and thermocouple placement. An accurate measurement of the source material's temperature can be difficult. However, the important parameters for compound growth are the rate ratios of the constituent components, rate stability, and the overal growth rate. Therefore, in some instances an exact knowledge of the absolute temperature of the source material is not mandatory.

Because of the high mobility of metal adatoms, the MBE sample heater should be able to sustain temperatures in the low-temperature regime (N_2 lq. or ^{4}He lq.). On the other hand, for the growth of refractory metals, often used as templates or buffer layers, temperatures as high as 1400 K can be needed. Such high temperatures are also mandatory for the preconditioning of some substrate materials, like sapphire. High-temperature annealing is necessary here to form highly ordered surfaces with a well-defined step-edge distribution. These demands can result in a complex sample stage design.

Typical growth rates in metals epitaxy vary from about 0.1 ML/s to 5 ML/s (ML: monolayer) but can be much slower in dedicated growth studies focusing on interface effects, pseudomorphism, or surface diffusion. To estimate the residual gas incorporation risk for a given background base pressure, we assume a growth rate of 0.1 ML/s. Especially for reactive metals with high oxygen affinity, like the rare earths or actinides, one would want to keep the maximum oxygen impurity concentration below the 100 ppm level. Because the maximum growth rate of adsorbants from the residual gas is about 0.5 ML/s at 10^{-6} mbar, this puts the upper limit for the base pressure at 5×10^{-11} mbar. On the other hand, in many instances even the incorporation of 0.1% gas impurities in metals is only weakly deteriorating of the characteristic electronic properties of the layers, because of the excellent screening characteristics in metals. Moreover, comparable impurity levels can be present in the source material. These can be enriched or depleted in the growing epilayer, depending on their respective vapor pressure, as compared with the base source material. As a result, even at a higher base pressure, metal layers can be grown that are of high quality structure and electronic properties. Nevertheless, for nucleation, surface diffusion, and morphology evolution, the residual gas pressure will in general exert a strong influence. Furthermore, in semimetallic systems screening is reduced, and the growth at elevated background pressure can result in severe changes in the electronic properties of the epilayers, as compared with bulk crystals.

The most common procedure for achieving base pressures in the low 10^{-11} mbar range is to bake the systems at about 170°C for about 24 h if the interior was exposed to air. Water readily desorbs from the walls at this temperature. An accurate monitoring of the residual gas composition during the bakeout process by mass spectroscopy is desirable. To sustain a low level of background pressure during the growth process, the desorption of adsorbants from the chamber surfaces has to be reduced.

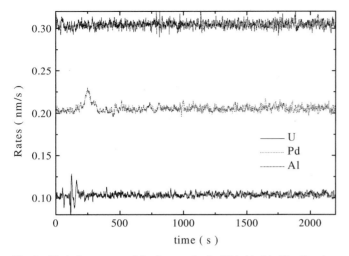

Fig. 2. Typical rate protocol for the growth of a UPd$_2$Al$_3$ thin film. For clarity, the rates for Pd and Al are vertically displaced by 0.1 nm/s and 0.2 nm/s, respectively.

This can be achieved by removing excessive heat loads created by the evaporation sources and sample heater, with the use of water cooling shrouds for the evaporation sources and cryopanels for liquid nitrogen cooling, predominantly in the sample region. The individual evaporation sources have to be properly screened from each other. Cross-talk prevention is especially important if materials with vastly different vapor pressures are used.

Evaporation control can be realized with various degrees of sophistication, depending on the desired precision in rate control. Oscillating quartz monitors are commonly used with feedback control for electron beam evaporators. Accuracies in the 0.01 nm/s range at overall growth rates of 0.1 nm/s can easily be achieved. Nevertheless, cross-talk of the evaporation sources in compound growth has to be avoided by proper shielding of the individual rate monitors. More flexibility and rate precision can be realized if a quadrupole mass spectrometer with cross-beam source and channel multiplexing for multiple source control is used. In either case, the monitoring of the actual growth rates over time is advantageous because it offers a direct means of judging the rate stability with regard to short-term fluctuations and long-term drift. Short-term rate instabilities are a drawback of electron-beam evaporators, whereas long-term drift is likely to happen when effusion cells are employed. In Figure 2 the rate protocol for the deposition of a UPd$_2$Al$_3$ thin film is presented. The rate control was performed with the use of three shielded, independent vibrating quartz monitors for the constituent elements U, Pd, and Al. On a short time scale the rates fluctuate by about 7%. With a typical deposition rate of 0.2 ML/s this reduces to about a 2.5% standard deviation averaged over the deposition time for one complete monolayer. The reproducibility of the respective rates can be judged by taking the average over 10 film depositions. Standard deviations below 0.7% can easily be obtained. For a given thicknesses d_i of an individual component, as measured by the vibrating quartz monitors, the resulting thickness of the compound layer d is given by the following dependence:

$$d = \mathcal{T} \cdot d_i \cdot \frac{\rho_i}{\rho} \cdot \frac{\sum_j c_j M_j}{c_i M_i} \qquad (1)$$

ρ_i, M_i, and c_i denote the density, molar mass, and concentration of the components. ρ is the density of the compound. $\mathcal{T}$ stands for the tooling factor, that is, the ratio of arrival rates of evaporant at the substrate and at the thickness monitors (assumed here to be the same for all components). For sticking coefficients below unity this relationship has to be adapted accordingly. To ensure an accurate average rate and stoichiometry calibration, independent measurements of the film thickness and composition should be peformed. This can be achieved by X-ray diffraction in grazing incidence and Rutherford backscattering (RBS), respectively. Because of backscattering from the substrate the application of energy-dispersive X-ray analysis for composition analysis is less straightforward for thin films.

Analytic equipment integrated with the MBE process mainly comprises electron reflection in grazing (RHEED) or normal incidence (LEED); the latter is often combined with Auger electron spectroscopy (AES) in retarding field mode. This can be augmented by various spectroscopic methodes in dedicated analytic chambers, such as photoelectron spectroscopy (PES) and scanning probe microscopy (SPM). RHEED has to be regarded as especially helpful for *in situ* growth analysis. It readily yields information about the various growth stages, such as nucleation and phase formation processes, epilayer orientation, strain evolution, and morphological changes.

In situ patterning capabilities based on shadow-mask techniques can be of practical use for electronic transport and tunneling investigations, thus avoiding air exposure of the interfaces. In this context, portable detachable sample chambers for transfer to other experiments under ultra-high-vacuum conditions may be needed. All of these requirements put some demands on the construction of the sample holder stages, heaters, and transfer system with respect to reliability, heat stability, and compatibility.

In metals epitaxy the substrates used cover a broad range of materials, ranging from $3d$ and refractory metals to different semiconductors and insulators, depending on the aim of the research. Dedicated growth studies tend to rely on highly oriented single-crystal metal substrates. But insulating substrates selected for minimal misfit and chemical inertness are also used. In many instances epitaxial-grade sapphire in various crystallographic orientations is preferred, on which excellent epitaxial-quality refractory metal buffer layers can be grown.

2.2.2. *Characterization*

In the following, two valuable tools for structural and chemical characterization are briefly outlined: RHEED and RBS. RHEED offers a very direct means for *in situ* studies of the growing epilayer with respect to phase formation and crystallographic orientation. Moreover, strain relaxation phenomena and the influence of annealing processes on the films' crystalline coherence and morphology can be analyzed in great detail. However, the interaction of the electrons with the crystal surface is strong and nonlinear. As a consequence, the diffraction patterns are often rather complex, with diffuse background, Kikuchi lines, diffraction spots, and streaks all of significant intensity. Even with advanced computational methods the calculation of the RHEED pattern for a given surface is a difficult task [2]. We will therefore limit the discussion to the kinematic approach to RHEED and refer to the literature for a more complete account [3]. RBS offers in many respects a simple means of analyzing sample composition in conjunction with depth profiling. Segregation and interface reaction processes can thus be identified. It is an especially valuable tool for composition calibration of thin films of intermetallic compounds.

2.2.2.1. RHEED

The information obtained in a RHEED experiment represents an averaging process over the longitudinal and transversal coherence length of the electron beam. For an estimate of these lengths, the influence of the energy spread of the electrons and the spatial extent of the electron emitter have to be considered. For a thermoelectric emitter operating at about 2600 K, the resulting energy spread amounts to $\Delta E = 2.45 k_B T = 0.59$ eV, assuming a Maxwell distribution for the emitted electrons. After acceleration of the electrons to an energy E, the resulting spatial extent Δx_t of the wave packet due to the finite temporal coherence is given by

$$E = \frac{\hbar^2 k^2}{2 m_e} \quad \Rightarrow \quad \Delta k = k \frac{\Delta E}{2E}; \quad \Delta x_t = \frac{2\pi}{\Delta k} \quad (2)$$

For k the component parallel to the film surface has to be used which amounts to

$$k = k_\parallel = \frac{2\pi}{\lambda} \cos\phi \quad (3)$$

where ϕ is the incidence angle of the electron beam on the film surface. Assuming a typical energy of 9 keV and an angle of incidence of 1°, the resulting longitudinal extent of the wave packet is about $x_t \simeq 400$ nm. The spatial coherence length is determined by the area of the virtual electron source (crossover region) as seen from the film surface. A sketch of the geometry appears in Figure 3. Any wave packet emitted from a given point of the virtual electron source is coherent and will result in an interference pattern on the screen. Figure 3 shows the travel distance of two wave packets being emitted from the same point of the virtual electron source for a film region of lateral extension Δx. Consider now the two most distant points on the electron source. The interference pattern on the screen will disappear if the difference in the travel distances amounts to $\lambda/2$. According to Figure 3 one obtains

$$\Delta x \leq \frac{\lambda}{4 \sin\phi \sin\beta} \quad (4)$$

β is given by the spatial extent of the virtual electron source and might be estimated to be 0.03° for a typical setup. The resulting longitudinal spatial coherence length at 9 keV is then $\Delta x \simeq 370$ nm. A longitudinal coherence length of about 400 nm might therefore be considered typical for a RHEED setup operating at 9 keV.

The semiquantitative description of electron diffraction in the RHEED geometry is based on the model of the limited penetration depth [4]. Elastic and inelastic scattering limit the penetration of electrons into a solid. According to Humphreys [5] the elastic scattering cross section for electrons off atoms follows an approximate dependence according to

$$\sigma_e = A \cdot \lambda^2 \cdot Z^{4/3} \cdot \frac{1}{\pi (1 - v^2/c^2)} \quad (5)$$

in which v and Z are the electron velocity and atomic number of the atom, respectively. The parameter A is of the order of unity, depending on the atomic wave function used. The resulting elastic scattering length $\Lambda_e = 1/N\sigma_e$ (N, number density of scattering centers) is about 20 nm for aluminum at 10 keV. For the inelastic scattering several contributions have to be considered. These comprise one-electron scattering, plasmon scattering, and core-level excitation. Seah and Dench give the following empirical formula [6]:

$$\Lambda_i = 0.41 \cdot \left(a[\text{Å}]\right)^{3/2} \cdot \sqrt{E[\text{keV}]} \quad (6)$$

with the atomic radius a. Again for aluminum at 10 keV the resulting inelastic scattering length is $\Lambda_i \simeq 5.5$ nm. With a typical given angle of incidence of 1°, the resulting penetration length with regard to elastic scattering is 0.5 nm. Consequently, RHEED can be considered as a manifestation of two-dimensional Bragg scattering. The respective expression of the scattered intensity,

$$I \propto F_{hkl}^2 \prod_{i=1}^{3} \frac{\sin^2(N_i k_i a_i/2)}{\sin^2(k_i a_i/2)} \quad (7)$$

with the structure factor F_{hkl}^2, the lattice constants a_i, and the number of scatterers N_i, is simplified by setting $N_3 \rightarrow 1$. The reciprocal lattice points thus degenerate into lattice rods perpendicular to the plane of scatterers. In the Ewald construction these reciprocal lattice rods intersect the Ewald sphere along circular bows that define the Laue zones. Analogously, one-dimensional regular defect structures, such as step edges, result in reciprocal lattice planes that intersect the Ewald sphere along a line. Three-dimensional structures on the film surface cause

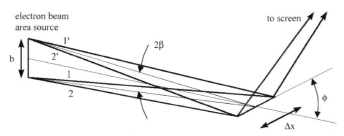

electron beam area source

to screen

Fig. 3. Schematics representing the travel distance for two wave packets emitted from the circumference of an area source. The distances are not to scale: $\Delta l = \Delta l_{1'2'} - \Delta l_{12} = \Delta x \cdot (\cos(\phi + \beta) - \cos(\phi - \beta)) = 2 \sin\phi \sin\beta \leq \lambda/2$.

three-dimensional Bragg scattering, which results in point reflections.

Many examples of RHEED images will be presented later on in this chapter. The surface net lattice constants can be determined from a simple geometric construction. Even with a simple setup, an accuracy of about 0.5% can be achieved. This is mainly limited by the resolution of the digitized image. An important application of real-time RHEED studies is the study of strain relaxation processes during growth. By taking a line through the reflections of the zeroth-order Laue zone, changes in the stripe distance as a function of film thickness can be monitored within the given resolution of the digitized image. Furthermore, the integration of the specular reflections or any other reflection during growth is commonly employed in so-called RHEED intensity oscillation studies. An example of this is shown for the growth of Ti in a separate section.

2.2.2.2. Rutherford Backscattering

RBS is based on the elastic scattering of nuclei. The dependence of the scattered particle's energy on the mass of the scatterer is used to discriminate the different atomic species in the target. The important aspects in RBS are the kinematics of the scattering process, the Rutherford scattering cross section, and the energy loss the projectile suffers when passing through matter. The last point is used to deduce information about the depth profile of the atomic species in the target. In the following a brief discourse on RBS is given that might be sufficient to convey a sense of the underlying principles of this method. For a detailed account the reader is referred to [7].

The kinematic factor K is the parameter for selecting the optimal scattering configuration in an RBS experiment with a given projectile-to-target mass ratio m_p/m_t and projectile energy E_0. K is defined as the ratio of the kinetic energy of the scattered projectile E_p and its initial kinetic energy, $K = E_p/E_0$. For energies below the Coulomb barrier the kinematic factor in the elastic Coulomb scattering process is given by

$$K = \left[\frac{\sqrt{1 - [(m_p/m_t)\sin\theta]^2} + (m_p/m_t)\cos\theta}{1 + (m_p/m_t)} \right]^2 \quad (8)$$

The optimal scattering angle θ for a given m_p/m_t ratio and E_0 will result in the lowest possible K. This is realized in backscattering geometry with $\theta = 180°$ if $m_p < m_t$ is assumed. This will also result in the best possible separation of two atomic species in the target with different masses, because the K over m_t/m_p curve shows the largest slope for $\theta = 180°$. In an experiment the scattering angle is typically 170° because the beam entrance region has to be left open.

The differential cross section for Rutherford scattering in the laboratory and the center-of-mass frame coincide in the limit $m_p \ll m_t$:

$$\frac{d\sigma}{d\Omega} = \left(\frac{Z_p Z_t e^2}{16\pi\epsilon_0 E} \right)^2 \frac{1}{\sin^4\theta/2} \quad (9)$$

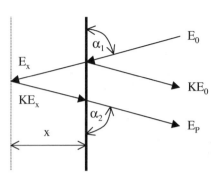

Fig. 4. Energies involved in the scattering of a particle at the film surface and from the interior of the film a distance x below the surface in an RBS experiment.

Here Z_p and Z_t denote the atomic numbers of the projectile and target atoms, respectively. Because of the quadratic dependence on Z_p, projectiles with a larger atomic number are usually preferred (^{4}He instead of H). One has also to take into account that the scattering rate falls off like E^{-2} with larger projectile energy.

The depth information in RBS relies on the energy loss that the charged projectile experiences when traveling through matter. The dominating process for energies above 1 MeV per nucleon is given by the interaction of the projectile with the electrons in the target, resulting in core excitation or ionization. In this energy range the specific energy loss is given by

$$-\frac{dE}{dx} \propto \frac{1}{v^2} \propto \frac{1}{E} \quad (10)$$

If the target material is composed of different atomic species, the additivity of the density-normalized energy losses is assumed according to the Bragg–Kleeman rule. For a binary alloy $A_a B_b$ this reads

$$\frac{dE}{dx}(A_a B_b) = n(A_a B_b)\left[\frac{a}{n(A)}\frac{dE}{dx}(A) + \frac{b}{n(B)}\frac{dE}{dx}(B) \right] \quad (11)$$

with the number density of target atoms $n(A)$ and $n(B)$ for the two species A and B, respectively.

The value of RBS for thin film studies becomes evident if the energy loss of the projectile as a function of the depth from which the backscattering occured is analyzed. A schematic representation appears in Figure 4. Assuming a constant energy loss dE/dx for the incoming and outgoing projectile trajectories (this is a reasonable approximation, because the energy loss varies slowly with energy), the resulting energy difference of the projectiles that are scattered at the surface and in a depth x depends linearly on x:

$$\Delta E = \left[\frac{K}{\sin\alpha_1}\frac{dE}{dx}\bigg|_{E_0} + \frac{1}{\sin\alpha_2}\frac{dE}{dx}\bigg|_{KE_0} \right]x \quad (12)$$

As an example, the RBS spectrum of a UPt$_3$ thin film sample grown on SrTiO$_3$ (111) is shown in Figure 5. The fit was performed by use of the algorithm implemented in RUMP [8].

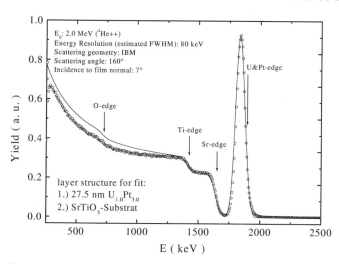

Fig. 5. RBS spectrum (open circles) and fit (solid line) of a UPt$_3$ (001) epitaxial film grown on SrTiO$_3$ (111). Parameters used for the fit are as indicated.

2.3. General Considerations in Compound Growth

Two main issues are at the core of growing thin films of intermetallic compounds. First, the nucleation problem has to be solved for the desired phase in a specific orientation. Second, depending on the energetics of the phase formation, exact compliance with the stoichiometry of the compound can be mandatory. Many parameters enter this equation, such as substrate material and orientation and the possible use of buffer layers for orientation selection, strain control, or as a diffusion barrier. Furthermore, the growth conditions influence the phase formation. These comprise, to mention just a few, growth rates, the surface mobility of the adatoms governed by the energy of the impinging atoms as well as the substrate temperature, the surface morphology of the template, and the background pressure and its residual gas composition. The phase nucleation and orientation selection problems are discussed in the following section. Here, the emphasis is on the influence of the sample composition on the phase formation.

There are various growth techniques for initiating compound formation. These comprise coevaporation of the constituent elements, layered growth followed by a postannealing step, layered growth at elevated temperature, and phase-spread alloying techniques to facilitate a systematic search for compounds with specific properties, like superconductivity [9]. The phase formation process will then be governed by surface or bulk diffusion, which will, in general, define quite different time scales. In either case, the common problem to be solved, once the desired phase formation is realized, is the accurate control of the composition of the growing layer to the greatest possible degree.

As most commonly expressed in the form of temperature–composition phase diagrams, in thermal equilibrium the phase under consideration might form a congruently or incongruently melting line compound, or, alternatively, it might have a large homogeneity region. Even though MBE is in general a nonequilibrium process, on a local scale the processes that drive the

phase formation can still be fast enough to establish thermodynamic equilibrium. The existence of a template in a specific orientation can then tip the energy balance toward an easier phase formation than would otherwise be obtained. However, materials that already exhibit a difficult metallurgy in bulk form represent yet another challenge in thin film growth. Heavy-fermion systems are a class of intermetallics for which sample homogeneity, composition, and crystallographic perfection influence very strongly the electronic properties. In these compounds strong f-electron correlations result in a pronounced increase in the effective charge carrier mass. Structural imperfections of any kind interfere with the formation of the low-temperature heavy-fermion state, which is the consequence of a cooperative Kondo singlet formation balanced against magnetic interactions of the RKKY type [10]. In these materials even a complete change of the ground state can be caused by slight deviations from the exact stoichiometry [11, 12].

One means of guaranteeing the correct stoichiometry is the systematic variation of the sample composition in a sequence of many samples, provided that all other process parameters are already reasonably optimized. Based on the rate stabilities achievable with various evaporation sources and feedback control techniques, this allows a reproducible adjustment of the composition in the 0.1% to 1% range. Nevertheless, this approach is very time and material consuming. A more efficient method is the growth of samples with a composition gradient. To give an example, TbFe$_2$ is a peritectoid line compound with no volatile components, and it is therefore necessary to provide the components in the exact stoichiometric proportions during the deposition process. To secure a high degree of structural perfection by avoiding nonstoichiometry, a collimated Fe beam can be established with the use of a cylindrical liner with a large length-to-diameter ratio (typically 4 : 1). The composition of the resulting Tb-Fe alloy then varies linearly as a function of position along the sample (see Fig. 6). In the investigation presented here [13], this was chosen as the long axis of rectangular substrates $17 \times 4 \times 0.5$ mm^3 in size. The compositions were determined by RBS. By adjusting the filling factor of the crucible from 1/3 to 1/2, the composition gradient could be changed from $\pm 15\%$ to $\pm 6\%$. This approach allows the investigator to eliminate off-stoichiometry to any required degree. For all further investigations the as-grown sample can simply be cut into several pieces. Only that part of the sample with the stoichiometric composition is then used to avoid impurity-phase contributions in the measured properties.

The growth with graded composition made it possible to establish a phase diagram for the epitaxial growth of TbFe$_2$ (111) [13]. This is shown in Figure 7. The shading in the center of the phase diagram represents the structural quality of the TbFe$_2$ films, with dark areas marking higher structural quality, as determined by X-ray diffraction. Impurity phases were observed to form in the off-stoichiometric regions. In particular, for Fe-rich growth TbFe$_3$ is found to grow epitaxially in the (0001) orientation. At elevated growth temperatures, increased interdiffusion occurs at the Mo–TbFe$_2$ interface, and this sets an upper limit to the growth temperature for TbFe$_2$ on Mo templates.

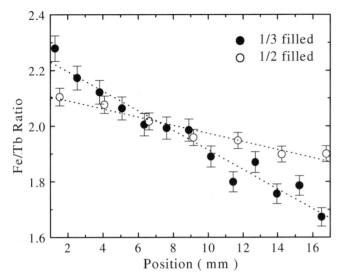

Fig. 6. Compositional gradient in Fe : Tb ratio as determined by RBS for two different filling ratios of the Fe liner as indicated. The horizontal axis represents the position along the long axis of a 17×4 mm^2 film. Reprinted with permission from [13], © 1998, American Physical Society.

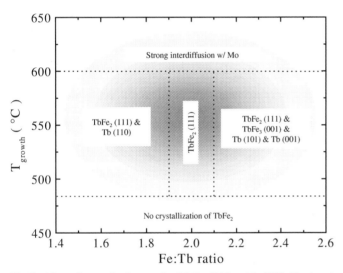

Fig. 7. Phase diagram for the growth of TbFe$_2$ (111) on Mo (110). The dotted lines indicate different growth regimes. Reprinted with permission from [13], © 1998, American Physical Society.

To some degree this corresponds to the binary phase diagram of Tb-Fe, thus suggesting that the phase formation proceeds close to equilibrium. Nevertheless, during the initial growth stage up to film thicknesses of about 1 nm, the epilayer is found to be under appreciable anisotropic biaxial strain induced by the large lattice misfit with respect to the Mo template. This results in an elastic energy contribution of about 0.1 eV per unit cell, which renders a direct comparison with the binary phase diagram difficult. It certainly highlights an important aspect in heteroepitaxial compound growth, namely the influence of misfit strain. This is discussed later in more detail.

2.4. Phase Stabilization and Orientation Selection

Original assumptions that the capillary theory of nucleation should be applicable to MBE growth proved to be wrong. Epitaxy in MBE is no classical nucleation phenomenon. Because of the tendency toward supersaturation, even collections of very few atoms are stable. The mechanism that favors large nucleation rates of favorably oriented nuclei as compared with differently oriented nuclei, which is responsible for phase selection in near-equilibrium growth, is less relevant for epitaxy. On the other hand, the orientation of larger islands during the island stage of growth is important. Provided that sufficiently glissile dislocations are present, for weakly interacting epilayer–substrate systems rotational jumps of larger islands can occur until the best epitaxial arrangement on the template is achieved [14].

The orientation selection itself is controlled by many parameters. One of these parameters is the symmetry of the underlying template. This was demostrated, for example, by the growth of Ti in six different crystallographic directions on sapphire with refractory bcc buffer layers carefully chosen to supply the needed symmetry-adapted template for the growth of Ti in the desired orientation [15]. Nevertheless, supplying a symmetry-adapted template with exact lattice parameter matching between film and substrate is not always sufficient to stabilize a desired orientation. For instance, a change in preferential orientation created by a change in only the lattice parameter of the template, with fixed symmetry and chemistry, was shown for the growth of TbFe$_2$ on Ta (110), Nb (110), and Mo (110) templates [13, 16, 17]. Furthermore, for the growth of UPd$_2$Al$_3$ and UNi$_2$Al$_3$ on Al$_2$O$_3$ (11$\bar{2}$0), a change in preferential orientation is observed as a consequence of the slightly different lattice constants of the epilayers for a fixed template [18]. The growth of CeSb provides yet another example for the various parameters that influence the orientation selection process. Here, epitaxial growth is controlled by the local adatom arrangement and the arrival rate of adatoms rather than the surface net of the underlying substrate. This is discussed in more detail later.

Epitaxy is a complex phenomenon. Different phase and orientation selection mechanisms prevail for different systems and different process conditions. Predictions of how to obtain a specific phase and orientation in thin film compound growth are difficult or impossible to make. There are, however, some guiding principles that are touched upon in what follows.

2.4.1. The Epitaxial Temperature

Already in 1963 Sloope and Tiller established for the growth of Ge (111) on CaF$_2$ (111) under UHV conditions a linear relationship between the logarithm of the evaporation rate and the inverse of the epitaxial temperature T_e [19]. The same relation also holds for the amorphous-to-crystalline transition temperature T^*. (For a given evaporation rate, T_e denotes the substrate temperature above which epitaxial growth can be observed. Accordingly, T^* represents the amorphous-to-crystalline transi-

Table I. The Relationship between the Epitaxial Temperature and the Melting Point of Simple Metal Layers and Intermetallic Compounds Related to This Work

System	T_m (K)	T_e (K)	T_e/T_m	Remarks
Ti (0001)/MgO (111)	1943	≤ 610	≤ 0.32	Anomalous diffusor
Nb (110)/Al$_2$O$_3$ (11$\bar{2}$0)	2742	1160	0.42	
Mo (110)/Al$_2$O$_3$ (11$\bar{2}$0)	2896	1050	0.36	
Ta (110)/Al$_2$O$_3$ (11$\bar{2}$0)	3293	1310	0.40	
			0.38 ± 0.04	avg.
UPd$_2$Al$_3$ (0001)/LaAlO$_3$ (111)	$\simeq 2100$	890	0.42	T_m poorly known
UNi$_2$Al$_3$ (1$\bar{1}$00)/Al$_2$O$_3$ (11$\bar{2}$0)	$\simeq 2100$	920	0.44	T_m poorly known
UPt$_3$ (0001)/SrTiO$_3$ (111)	1973	940	0.48	
CeSb (001)/Al$_2$O$_3$ (11$\bar{2}$0)	2083	1070–1320	0.51–0.63	Sb vapor pressure large
TbFe$_2$ (111)/Mo (110)	1460	820	0.56	
DyFe$_2$ (110)/Ta (110)	1543	820	0.53	
TiFe$_2$ (110)/Al$_2$O$_3$ (0001)	1700	890	0.52	
			0.49 ± 0.05	avg.

Only MBE-grown films with comparable growth rates of 0.1 to 0.5 monolayers/s are included. Vacua ranged from 2×10^{-11} to 10^{-7} mbar during the deposition process.

tion temperature.) The proposed relation was derived from nucleation theory for small clusters. Even though nucleation theory is in most instances not applicable to MBE growth, and it is conceded that Ge forms covalent bonds, in contrast to the metallic systems under discussion here, the presented relation nevertheless hints at an interesting relationship between the melting point of a metal and its epitaxial temperature [20, 21].

For crystalline order to occur, the surface diffusion rate of the epilayer adatoms $\dot{r}$ has to be larger than the effective adsorption rate from the vapor $\dot{r}_0$. In most instances $\dot{r}_0$ is solely given by the evaporation rate, with negligible contributions from desorbing adatoms. With ν denoting an attempt frequency and G_d denoting the Gibbs free energy for surface diffusion, this translates to [19]

$$\dot{r} = \nu \exp\left(-\frac{G_d}{k_B T}\right) = \nu \exp\left(-\frac{H_d}{k_B T}\right) \exp\left(\frac{S_d}{k_B}\right)$$
$$= D_0 \exp\left(-\frac{H_d}{k_B T}\right) > \dot{r}_0 \qquad (13)$$

T^* and T_e are then determined by the following condition:

$$\dot{r} = c^* \dot{r}_0 \quad \text{or} \quad \dot{r} = c_e \dot{r}_0 \quad \text{with}$$
$$c^* \quad \text{and} \quad c_e = \text{const.} > 1 \qquad (14)$$

Rearranging the terms leads to

$$\ln\left(\frac{c\dot{r}_0}{D_0}\right) = -\frac{H_d}{k_B T_c} \quad \text{with}$$
$$c = (c^*, c_e) \quad \text{and} \quad T_c = (T^*, T_e) \qquad (15)$$

Quite accurately, for self-diffusion in metals the diffusion coefficient $D(T_m)$ at the metal's melting point amounts to 10^{-8} cm^2/s [22]. For surface diffusion the activation barrier for jump processes is reduced as compared with bulk diffusion. This is due to the reduced binding energy at a surface site and

the availability of vacant nearest-neighbor sites. Nevertheless, assuming that

$$D(T_m) = D_0 \exp\left(-\frac{H_d}{k_B T_m}\right) \simeq \text{const.} \qquad (16)$$

still holds, inserting Eq. (16) into (15) leads to the following linear relationship between T^* or T_e and T_m:

$$\frac{T_c}{T_m} = -\frac{\ln(D(T_m)/D_0)}{\ln(c\dot{r}_0/D_0)} \simeq \text{const.} \qquad (17)$$

Inspection of Table I shows that such a T_e-to-T_m relationship can indeed be identified for all simple metal layers and intermetallic compounds related to the author's work. As can be expected, the epitaxial temperature for intermetallic compounds is enhanced as compared with simple metal films, because for compounds diffusion is slower because of inequivalent adatom sites. For a new metallic system to be grown, the relationship given here might be used to choose an appropriate substrate temperature for epitaxy before growth.

2.4.2. Early Growth Stages

As a recent example of *in situ* growth studies in the early stages of film growth we discuss the near-interface alloy formation process of the Gd-Fe system. Employing MBE growth and scanning tunneling microscopy (STM) with atomic resolution, Pascal et al. investigated the influence of the W (110) substrate on the formation and crystallographic structure of the cubic Laves phase compound GdFe$_2$ [23, 24]. Alloys and compounds of 3d transition metals and the rare earths are of high technological interest because they offer a wide variety of magnetic properties, ranging from strong magneto-optical active systems to magnetostrictive materials with record strictive distortions at room temperature.

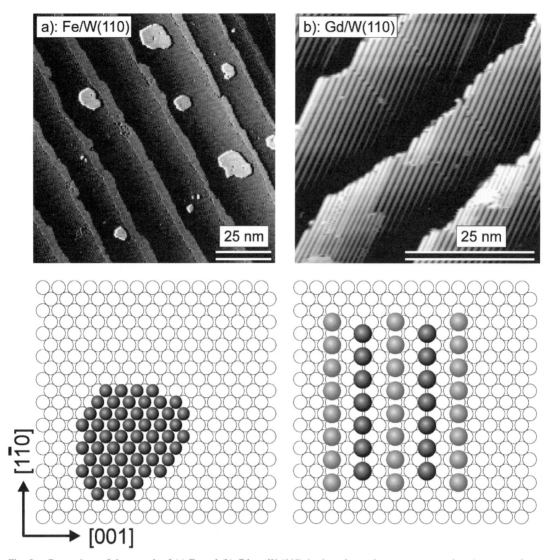

Fig. 8. Comparison of the growth of (a) Fe and (b) Gd on W (110) in the submonolayer coverage regime (coverage about 0.25 ML). Scan area 70×70 nm^2. Structure models below the STM images highlight the different growth modes. Image courtesy of M. Getzlaff. Reprinted with permission from [24], © 1999, American Physical Society.

The experiments were carried out in a two-chamber UHV system. The background pressure during growth did not exceed 4×10^{-10} mbar. The STM investigations were performed in a separate analysis chamber used to maintain a base pressure below 10^{-11} mbar. All measurements were performed in the constant-current mode. Because of the different ionic radii of Gd and Fe, the coverages are given in substrate units. In these units, the first closed Gd layer holds 0.64 monolayers and the first closed Fe layer holds 1 monolayer because of the pseudomorphic growth of Fe on W (110).

The first important aspect of the growth of Gd and Fe on W is their widely differing growth morphologies for coverages below one monolayer. Fe forms one-monolayer-high patches on the terraces and stripes along the substrate's step edges, indicating step-flow growth. The Gd atoms, on the other hand, tend to cover the substrate by forming quasi-one-dimensional fingered superstructures [25, 26]. The formation of dipole moments on the Gd atoms induced by a charge transfer from the Gd atoms

to the substrate results in repulsive dipolar interactions. These trigger the formation of the stripe pattern. The surface morphologies for submonolayer Fe and Gd growth are compared in Figure 8.

As a consequence of the difference in the layer morphology of the constituent elements in the early growth stages, the formation of GdFe$_2$ is not initiated in the C15 bulk structure of this compound. In a first step the Gd-to-Fe ratio was selected to be approximately 1 : 1. The deposition was performed at room temperature, followed by a 5-min annealing step at 700 K. The resulting film morphology appears in Figure 9. It consists of two regions formed by smooth areas and the Gd-typical one-dimensional finger structure. With the Fe content increased to the stoichiometric mixing ratio for GdFe$_2$, the nominal deposition of one monolayer does indeed result in a completely closed and smooth first GdFe$_2$ monolayer on W (110). The corresponding STM image is shown in Figure 10.

Subsequent LEED and STM analysis with atomic resolution led Pascal et al. to suggest the following structure model. The Fe and Gd atoms are assumed to occupy the W (110) bridge sites, which were previously shown to be the energetically preferred sites for Fe atoms on top of W (110) [27]. The resulting

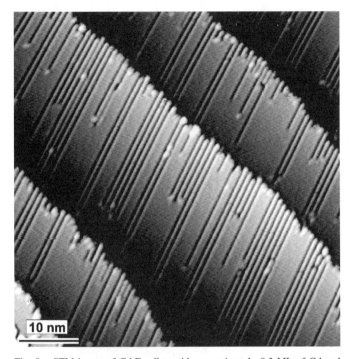

Fig. 9. STM image of Gd-Fe alloy with approximately 0.3 ML of Gd and 0.4 ML of Fe. The image was taken in constant current mode. The striped areas represent the Gd superstructure. The stripes are aligned along the [001] direction of the W substrate. The smooth areas correspond to an alloy of $GdFe_2$. The scan range is 70×70 nm^2, sample bias $U = 0.2$ V, tunneling current $I = 0.3$ nA. Image courtesy of M. Getzlaff. Reprinted with permission from [24], © 1999, American Physical Society.

arrangement of atoms is reproduced in Figure 11. Proceeding to the second monolayer, this structure is repeated as shown in Figure 12. As compared with the geometrical arrangement of the Fe and Gd atoms in the bulk C15 structure, the observed layer structure resembles the (111) plane being laterally compressed by 14% in the [1$\bar{1}$0] direction of the W (110) substrate and strained by 5.3% in the [001] direction. Nevertheless, the atomic arrangement of the Fe atoms with respect to the Gd atoms does not correspond to the C15 (111) lattice plane. Because the growth of the isostructural $TbFe_2$ on Mo (110) to larger film thicknesses (to be discussed next) shows the formation of the bulk C15 structure, a structural phase transition as a function of the film thickness in $GdFe_2$ on W (110) is likely to take place.

These STM studies are very well corroborated by RHEED studies of the growth of $TbFe_2$ on Mo (110) [13]. Although RHEED cannot give a comparable degree of information concerning the atomic arrangement of the growing layers, it is very well suited for the study of time-dependent phenomena and relaxation effects. Because of the large magnetostriction in $TbFe_2$, monitoring the strain evolution during growth is essential for assessing the reasons for residual strain present after film growth. This residual strain directly influences the overall magnetic ansiotropy, as will be discussed in the second part of this chapter.

The film deposition was performed in a Perkin-Elmer 430 MBE system. After the growth of the Mo buffer layer on sapphire (11$\bar{2}$0), the samples were cooled to a temperature range of 450°C to 680°C for deposition of $TbFe_2$ (111). This was accomplished at a growth rate of 0.02 nm/s for film thicknesses of 40 nm to 150 nm, with a background pressure of 1.1×10^{-10} mbar. For the RHEED investigation a 10-kV beam with an incidence angle of about 0.5° was used.

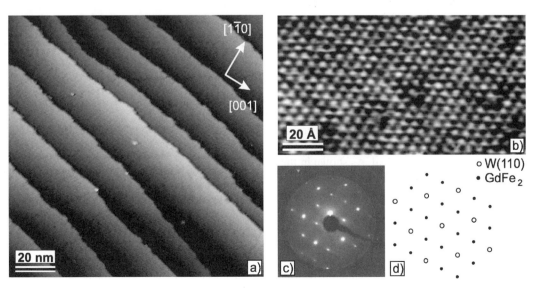

Fig. 10. (a) Completely closed and smooth first ML of $GdFe_2$ on W (110). (b) Atomic resolution obtained on this sample ($U = 0.18$ V, $I = 3$ nA). (c) Photograph and (d) sketch of the (2, 1; 1, 2) LEED pattern. Image courtesy of M. Getzlaff. Reprinted with permission from [24], © 1999, American Physical Society.

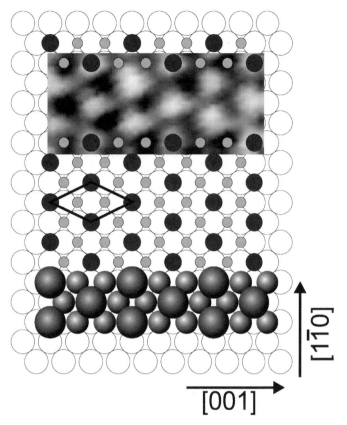

Fig. 11.   Structure model for the first ML of GdFe$_2$ on W (110). Gd is represented by the large, Fe by the small balls. In the upper part atomic diameters are scaled down by a factor of 2. An atomically resolved STM image is overlaid. In the structure model shown in the lower part the atomic diameters are to scale. Image courtesy of M. Getzlaff. Reprinted with permission from [24], © 1999, American Physical Society.

The initial growth forms well-defined diffraction streaks. The spacing with the electron beam aligned along [2$\bar{1}\bar{1}$] indicates that the films are under $-(1.5 \pm 1.0)\%$ compressive strain, whereas with the beam alignment along [0$\bar{1}$1] a compressive strain of $-(12.0 \pm 1.5)\%$ is observed. The orthorhombic distortion of the TbFe$_2$ (111) surface amounts to $+0.5\%$ and -9.1%, respectively, assuming a perfect accommodation of the Mo (110) template at a typical growth temperature of 550°C. It can then be concluded that the TbFe$_2$ surface is not fully pseudomorphic, most probably because of the large elastic strains involved. The epitaxial relationship and registry of the TbFe$_2$ (111) surface on the Mo (110) surface in the strained and relaxed states are depicted in Figure 13d. As shown in the STM studies on GdFe$_2$, the established Fe-to-Gd atomic arrangement is in fact different from the C15 (111) arrangement of the bulk.

As the thickness is increased to the 0.8–1.0-nm range, specular intensity shifts into the diffuse background, indicating a loss of long range coherence of the film surface. The RHEED image eventually recovers to show spots along the original diffraction streaks. This evidence for 3D scattering is accompanied by a relaxation of the streak spacing toward the lattice constants of unstrained TbFe$_2$ (111). This roughening transition most probably

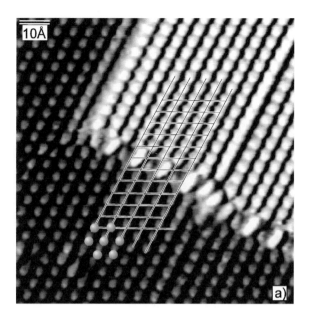

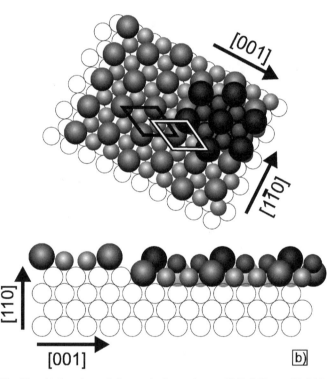

Fig. 12.   (a) Atomic resolution on the first and second ML GdFe$_2$ on W (110) ($U = 55$ mV, $I = 3$ nA). The grating and balls are used to determine the registry between the atoms of the first and second ML. (b) Structure model (top and side views) of the first and second ML of GdFe$_2$ deduced from the STM images. Image courtesy of M. Getzlaff. Reprinted with permission from [24], © 1999, American Physical Society.

occurs at the onset of dislocation formation. However, a structural phase transition as a function of film thickness cannot be excluded, as was suggested by Pascal et al. [24]. With increasing film thickness the diffraction pattern evolves back to streaks, which become well-defined at 8 nm. For film thick-

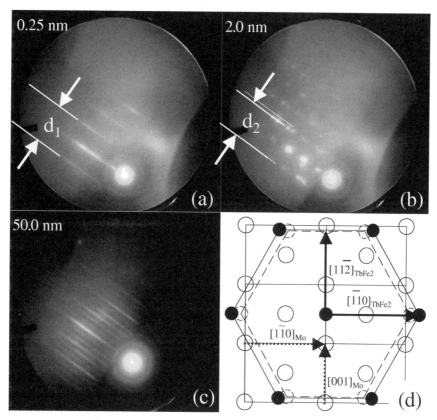

Fig. 13. (a–c) RHEED study of the TbFe$_2$ (111) growth on Mo (110) for the film thickness indicated. The beam was aligned along the [01̄1] direction. (d) Schematic diagram showing the orthorhombic lattice distortion of the TbFe$_2$ (111) surface in the initial pseudomorphic growth stage (dashed lines). Reprinted with permission from [13], © 1998, American Physical Society.

nesses of 40 nm and more, the diffraction pattern reveals that the surface is flat on the length scale of the longitudinal coherence of the electron beam, namely several hundred nanometers. The RHEED patterns corresponding to the three growth stages are shown in Figure 13a–c.

The TbFe$_2$ surface reconstructs through the sequence of 1×1, 3×3, and then 2×2 reconstructions as the temperature is increased from 520°C to 580°C. The surface also exhibits (110) faceting, as made evident by RHEED and atomic force microscopy (AFM) measurements. A discussion of these morphological aspects is deferred to the next section.

2.4.3. High Vapor Pressure Materials

From the technological point of view, semimetallic pnictides, particularly MnBi, hold some promise for applications in magnetooptical data storage because of their large Kerr rotations [28]. Very large Kerr rotations of up to 90° at low temperatures and in large magnetic fields were also reported for the Kondo semimetal CeSb [29, 30]. However, the main interest here lies in the investigation of the interplay between crystal field effects and Kondo-like electronic correlations despite the low charge carrier density of 0.02 holes per Ce ion. This results in a complex magnetic phase diagram including at least 15 different phases [31, 32]. Thin film investigations currently con-

centrate on the low charge carrier density aspect [33, 34]. The modulation of the charge carrier concentration in a well-defined fashion, employing an electric field effect structure with thin CeSb layers, represents a unique opportunity to vary the typical energy scales, namely the Kondo temperature T_K and the Néel temperature T_N, without disturbing the coherent nature of the Kondo lattice.

Investigations of the growth characteristics presented here for CeSb show striking similarities to III–V semiconductor growth. This is mostly related to the vastly different vapor pressures of the constituent elements. In the following, some aspects of the phase formation and orientation selection mechanisms in MBE-grown CeSb layers are discussed.

The growth was carried out by coevaporation of Sb from a Knudsen cell with a p-BN liner and Ce by means of electron impact evaporation from a tungsten crucible onto heated Al$_2$O$_3$ (112̄0). Ce was supplied as monomers from the liquid with a fixed evaporation rate. The availability of adsorbed Sb in various molecular forms on top of the growing epilayer represents the main controlling parameter in the formation of CeSb. At Sb cell temperatures of about 820 K, which resulted in a stoichiometric Sb : Ce ratio on the growing epilayer, the Sb vapor is predominantly formed by Sb$_4$ [35]. With increasing temperature, thermal dissociation tends to increase the dimer fraction in the Sb flux with equal vapor pressures of the tetramer and dimer molecules at cell temperatures of about 1160 K [35]. At these cell temperatures the Sb : Ce ratio is about 100.

Before the details of the phase formation and orientation selection processes for CeSb are discussed, a short review of homoepitaxial III–V semiconductor growth is given, with emphasis on GaAs. If the supply of As consists primarily of dimers, the growth proceeds through a physisorbed precursor state with a large desorption probability of the dimers. In the presence of Ga, dissociative chemisorption results in the formation of Ga–As bonds. Stoichiometric GaAs growth is ensured because excess As_2 readily desorbs from the surface, even at substrate temperatures as low as 200°C. For a dominantly tetramer supply, a transition to a chemisorbed As_4 pair state was proposed before the dissociation process sets in on adjacent Ga sites. Desorption of excess As results again in the formation of stoichiometric GaAs. This transient behavior in the phase formation process was studied in great detail by means of modulated beam mass spectrometry [36].

For CeSb some similarities can be noted with regard to the following observations. If Ce-Sb intermetallics form on Al_2O_3 (11$\bar{2}$0), regardless of the Sb : Ce flux ratio, only CeSb is stabilized analogously to III–V compound growth. This was checked by RBS with $^4He^+$ ions at 2 MeV. Nevertheless, the influence of the Sb molecular state and evaporation rate on the phase formation is more complex than in III–V materials. First, for the heteroepitaxial growth on Al_2O_3, CeO_x formation appears to be a competitive process. In contrast to Ga, Ce is a highly reactive metal. Second, the preferential CeSb orientation for stoichiometric Sb supply is (111), despite the lack of symmetry adaption to the Al_2O_3 (11$\bar{2}$0) surface. The (111) orientation is suppressed and (100) growth is favored for increased Sb dimer supply. As a result, the phase formation–orientation diagram shows re-entrant-like behavior for the CeSb phase formation and a change in the preferential orientation, depending on the Sb molecular state. This is presented in Figure 14.

As a first step in analyzing this growth behavior, the following model for the phase formation process is suggested. (In all cases discussed here the Ce evaporation rate was fixed and the substrate temperature was held constant at 1200 K.) The re-entrant behavior for the CeSb phase formation and the orientational preferences are governed by two factors that are inseparably connected for the Sb supply process employed here. These factors are the Sb tetramer-to-dimer ratio and the respective overall Sb evaporation rate. For pBN liner material a catalytic Sb_4-to-Sb_2 dissociation can be neglected [35]. The molecular distribution is therefore determined by the cell temperature alone, as indicated in Figure 14. The rock-salt structure of CeSb implies a tetrahedral next-nearest-neighbor configuration of equivalent atoms. The geometrical arrangement of adatoms in (111) orientation then suggests that Sb_4 can transfer from a precursor state (or even a Sb_4 pair state, as proposed for As in GaAs) to a chemisorbed state in the CeSb surface. The triangular base layer of Sb_4 can align parallel to the surface, leaving the pyramidal top atom for dissociation as an Sb monomer. Alternatively, two or four next-neighbor pyramidal atoms might form dimerized or tetramerized states, which will then desorb or occupy an available nearby triple site. Ce monomers fill in accordingly to complete the (111) stacking sequence, provided that the Ce-to-Sb ratio is close to 1. This will then proceed without significant CeSb (100) and CeO_x formation and represents the dominant growth process in the low-flux and large tetramer-to-dimer ratio region of the phase diagram.

The other limiting case is given by Sb_2-to-Sb_4 ratios close to 1 in conjunction with an Sb arrival rate that exceeds the Ce rate by about two orders of magnitude. In this case, Sb dimers in the precursor state can proceed to a chemisorbed state on the (100) surface. The formation of CeO_x is suppressed because Ce adatoms will very probably find Sb next neighbors because of the large arrival rate of Sb on the surface. Moreover, the overall growth rate for CeSb is reduced because the inclusion of tetramers in the formation of the (100) surface is kinetically limited as a competitive process to direct inclusion of dimers. Two-thirds of Sb adatoms will therefore tend to desorb. Furthermore, RBS experiments showed that the Ce-to-Sb ratio in these films is still 1. This implies that the sticking coefficient of Ce is also significantly reduced. As a result, the growth rate is reduced to about one third of the growth rate observed for (111) growth. This corresponds very well to the experimental observation as presented in Figure 14b. The geometrical adatom arrangements for the two limiting cases described above are shown in Figure 15 for reference.

In the high-flux regime the crystal quality tends to degrade because the large Sb arrival rate hinders surface diffusion. Between these limiting cases the overall Sb_x supply is too large to allow sufficient formation probability for the (111) stacking sequence, which necessarily contains pure Ce layers. No short-range ordered state of either (111) or (100) character can readily be formed. As a result, an increased tendency toward CeO_x formation is observed.

For further investigations, and to evaluate the present model, which is based mainly on geometrical considerations, the sep-

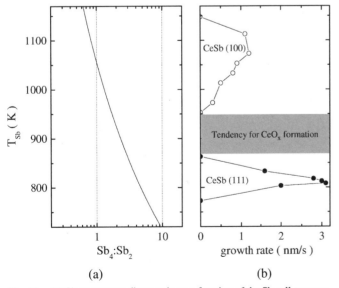

Fig. 14. (a) Sb tetramer-to-dimer ratio as a function of the Sb cell temperature. (b) Comparison of CeSb growth rates in the different growth regimes as indicated. The Ce evaporation rate and substrate temperature were fixed.

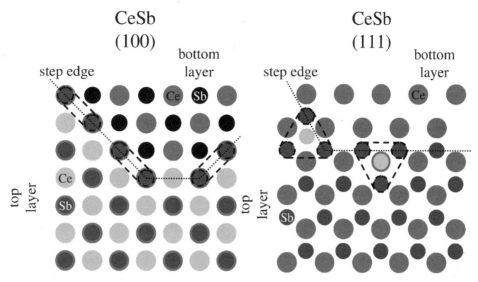

Fig. 15.   Scenario for the geometrical adatom arrangement on the CeSb (100) surface (left) and CeSb (111) surface (right).

aration of the tetramer-to-dimer ratio from the overall Sb evaporation rate is desirable. This suggests the use of a cracker cell, thus allowing the increase of the dimer contribution to the flux by catalytic Sb_4 dissociation in a separate hot zone. Keeping the overall Sb evaporation rate fixed, this could be used to tailor the orientation of the growing phase to be either (111) or (100). Very likely, these observations are relevant for the growth of manganese pnictides as well. They might offer a means of controlling epitaxy in these technologically promising materials.

2.5. Epitaxial Strain

Strain exerts several influences during epilayer growth. It can influence the energy balance in the phase formation process so that the growth of metastable phases becomes possible. The growth of metastable Fe layers on various substrate materials and in a variety of orientations represents a prominent example [37]. Strain, furthermore, can drive morphological instabilities that result in uniaxial or bidirectional corrugation effects (mound formation) that allow the relaxation of strain energy balanced against an increase in surface energy [38]. It can cause phase segregation effects that result in regular stripe domain patterns, as was recently demonstrated for the Fe/Co-Ag system [39]. The natural limit for these phenomena is given by the onset of dislocation formation. Isotropic metallic bonding results in very small activation barriers for dislocation formation. Consequently, for homogeneous epilayers strain can hardly be sustained beyond the critical film thickness beyond which strain relaxation sets in by the formation of interfacial dislocations [40]. Dislocation formation will generally result in an increase in rotational disorder; for example, small-angle grain boundaries are formed when glissile dislocations arrange on top of each other. The mosaicity and long-range coherence of the crystal are correlated with its strain state.

In semiconductor and metal epitaxy the establishment of a specific strain state can be desirable. Most often, this is not so with regard to the morphological changes strain can cause, because these frequently result in an unwanted increase of roughness. Nevertheless, the better our understanding of strain-induced roughening phenomena becomes, the better are the chances of finding specific applications of these effects in a controlled way. This can be considered a means of patterning on the nanometer scale based on self-organization effects.

Below are several examples of strain relaxation phenomena. RHEED and X-ray diffraction studies of the growth of Ti (0001) layers on large and small misfit substrates will serve to discuss some aspects related to strain relaxation phenomena and the critical thickness. We also briefly discuss the correlation between rotational disorder and the film's strain state.

2.5.1. Growth of Ti Layers on Large and Small Misfit Substrates

This study focuses on hcp Ti (0001) grown by MBE on symmetry-compatible MgO (111) with the rock-salt structure, and on Al_2O_3 (0001), which is quasi-hexagonal [41]; these represent small (-1%) and large ($+6.8\%$) misfit substrates, respectively. The initial stages of the film growth were monitored by RHEED, and the characterization was augmented by X-ray diffraction and AFM. The morphological aspects are postponed to the next section.

The thin films were deposited by electron beam evaporation of Ti in a Perkin-Elmer 430 MBE system with a base pressure of 2.5×10^{-11} mbar, which increased to about 6.5×10^{-10} mbar during growth, mainly because of hydrogen release from the Ti melt. Deposition was maintained at 0.03 nm/s to overall film thicknesses of 30 nm or 80 nm. Before deposition the substrates were heated to 1000°C in 1 h, kept there for 30 min, and then cooled to the growth temperature. RHEED measurements were

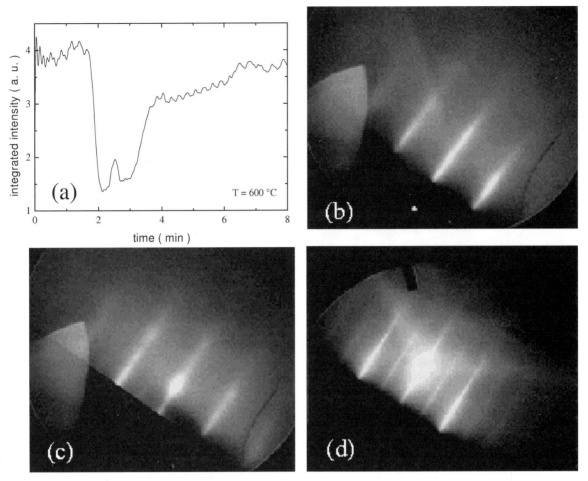

Fig. 16. (a) RHEED intensity oscillations for a Ti thin film sample grown on MgO at 600°C. The intensity was integrated over a rectangular region around the specular spot. RHEED images were taken after deposition of one monolayer of Ti on (b) MgO and (c) Al$_2$O$_3$. (d) RHEED image taken after growth, showing the 2 × 2 surface reconstruction (beam ∥[11$\bar{2}$0]). Reprinted with permission from [41], © 1997, American Institute of Physics.

performed at 9 keV with an incidence angle of about 0.5°. Film growth was monitored by RHEED imaging and intensity measurements. A computerized data acquisition system was used to integrate the intensity of the specular beam over a predefined integration window. X-ray measurements were later performed ex situ with the use of a two-circle diffractometer with Cu-K_α radiation, and an in-plane resolution $\Delta q_\parallel / q_\parallel = 1.6 \times 10^{-3}$.

RHEED measurements following the completion of one monolayer Ti on MgO reveal well-defined diffraction streaks characteristic of two-dimensional growth, as shown in Figure 16b. Layer-by-layer growth is also suggested by the appearance of intensity oscillations of the specular reflection for substrate temperatures in the range 400°C ≤ T ≤ 600°C (see Fig. 16a). These oscillations are caused by periodic variations in the step-edge density and are believed to occur only in regimes of strong surface diffusion between terraces [42]. Throughout the entire deposition period no damping of the oscillation amplitude was observed. The oscillation period corresponds to the full c axis spacing of hcp-titanium, as determined by comparison with an independent thickness calibration by RBS. After about six monolayers the intensity breaks down, but it recov-

ers after another two monolayers. For lower growth temperatures the amplitude of the intensity oscillations is reduced, most probably as a consequence of reduced interterrace adatom diffusion. For the highest growth temperature of 650°C, a step edge flow growth mode with constant step edge density can account for the observation that the intensity oscillations are suppressed [43]. Within the instrumental resolution of the RHEED experiment, no lattice parameter relaxation of the growing Ti film was observed. However, X-ray diffraction studies showed that the Ti lattice constants were completely relaxed to their bulk values for the 30-nm- and 80-nm-thick films, regardless of growth temperature from 340°C to 650°C. Based on these RHEED results it can be concluded that the Ti growth on the low-misfit substrate MgO (111) starts in a layer-by-layer mode until a critical thickness is reached. This is followed by the nucleation of dislocations that allow the Ti lattice constants to relax to their bulk values. With the use of the Matthews and Blakeslee model [44–46], the critical film thickness t_c for dislocation formation based on the $\langle 11\bar{2}0 \rangle$ {0001} slip system of Ti for perfect edge dislocations is given by

$$t_c = \left(a_{Ti}/8\pi(1+\nu)\epsilon \right) \ln(\alpha t_c/a_{Ti}) \qquad (18)$$

where α is the dislocation core energy parameter, $\epsilon = -0.01$ is the lattice misfit at growth temperature, $\nu = 0.33$ Poisson's ratio, and the basal plane lattice parameter of Ti is a_{Ti}. Assuming that the break in the specular reflection at 2.8 nm signifies the critical thickness, this results in a deduced dislocation core energy parameter of $\alpha = 2.5$. This value represents rather an upper limit because the activation energy for formation of dislocations can result in a sustained supercritical state of strain.

In contrast, Ti growth on Al$_2$O$_3$ (0001) was observed to proceed through three-dimensional nucleation, as deduced from the spot-like intensity enhancements on the RHEED diffraction streaks after completion of one monolayer, as shown in Figure 16c. No RHEED oscillations were observed, nor were any pronounced damping or recovery effects in the specular beam intensity, so no evidence exists for a crossover regime in the surface morphology in this case.

Independently of the substrate material, RHEED images taken after growth at room temperature revealed no appreciable differences for the different growth conditions. In each case well-pronounced surface resonance features in the form of Kikuchi lines indicate a well-ordered surface (see Fig. 16d). At the highest growth temperature of 650°C, or with heating of the films above 620°C after growth, a stable 2×2 surface reconstruction was observed. The high intensity of the superstructure reflections tends to indicate a reconstruction of the Ti top layer itself, as opposed to a possible regular arrangement of a surface adsorbate. However, the high adsorption probabilities for the residual gases N$_2$, O$_2$, and H$_2$ on the Ti surface suggest, to the contrary, an adsorbate-induced reconstruction. This question must remain for future resolution.

According to the X-ray diffraction analysis the epitaxial relationship between Ti (0001) and the substrates MgO (111) and Al$_2$O$_3$ (0001) is compatible with an adatom geometric arrangement of minimal misfit with Ti $[11\bar{2}0]\|$MgO $[10\bar{1}]$ and Ti $[11\bar{2}0]\|$Al$_2$O$_3$ $[1\bar{1}00]$.

Films of various thicknesses showed a sharp specular and a broad diffusive component in transverse scans (rocking curve) at the Bragg positions (0002) and (0004) for films of 30-nm thickness. This is shown for Ti grown on MgO and Al$_2$O$_3$ in Figure 17. The respective rocking curve widths of the sharp components are resolution limited. The broad components are characteristic of rotational disorder, with fixed angular width independent of the magnitude of the perpendicular component of the scattering vector. The width of the broad component for films grown on sapphire is independent of growth temperature with FWHM $\simeq 0.45°$, whereas for films on MgO the width follows a nearly linear dependence on the growth temperature and decreases from 0.25° to 0.16° for growth temperatures ranging from 340°C to 650°C.

As can be seen in the inset of Figure 17b, the Bragg scan of the (0002) reflection of Ti on sapphire is shifted to a lower scattering angle. Thus a 1.7% increase in the c lattice parameter of Ti is caused by clamping to the substrate. The asymmetric shape of the thin film oscillations visible on the high-angle side of the (0002) reflection might indicate that a strain field gradient exists along the Ti c axis as a consequence of incomplete strain

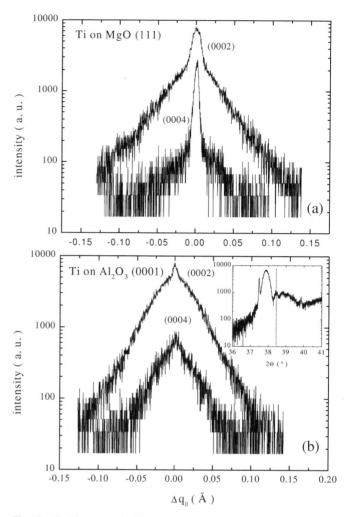

Fig. 17. Rocking curves for 30-nm-thick Ti on (a) MgO and (b) Al$_2$O$_3$, grown at 650°C. The inset in (b) is a Bragg scan through the Ti (0002) reflection. The dashed vertical line corresponds to the peak position for the c lattice constant of bulk Ti. Reprinted with permission from [41], © 1997, American Institute of Physics.

relaxation toward decreasing c at the film surface. For films of 80-nm thickness the peak position corresponds to a fully relaxed film. No evidence for residual strain was found for films of 30-nm and 80-nm thickness on MgO.

The occurrence of a two-component line shape in transverse scans of metal [47–49] and semiconducting [50] films with weak disorder was observed only recently. In the present case it can be inferred that long-range structural coherence in the film plane gives rise to the narrow component; the broad background component is characteristic of rotational disorder or mosaicity [51]. Both components can be described by introducing a displacement–difference correlation function,

$$2\sigma^2(\mathbf{r}) = \langle [u_z(\mathbf{r}) - u_z(\mathbf{0})]^2 \rangle \quad (19)$$

that describes displacements u_z along the film normal $\hat{z}$ for any given in-plane distance $\mathbf{r}$. The resulting differential cross section separates into two Debye–Waller-like contributions that represent the narrow and broad components [51, 52]. Because of an exponential damping the narrow component is visible only for weak disorder and is most pronounced for small momentum transfer q_z. If the reduction of the in-plane structural coherence is ascribed to dislocations in the film, the more pronounced sharp component for films grown on MgO is a consequence of the small lattice misfit. The curves shown are not corrected for q_z-dependent instrumental resolution.

2.5.2. Strain and Crystalline Coherence in Epitaxial UPt₃ Layers on SrTiO₃

Strain is an important parameter for heavy-fermion systems, because the hybridization of f-orbitals with itinerant states sensitively determines the electronic properties of the material [10]. In heavy-fermion thin films strain can be used to alter the heavy-fermion characteristics in a controlled way, but it might also be an unwanted effect. The investigation of the strain state of a given thin film sample is therefore an integral part of the characterization process. As an example, the correlation between biaxial strain and the long-range crystalline coherence in UPt₃ (0001)-oriented films on symmetry-adapted SrTiO₃ (111) is briefly discussed.

As was already shown for the growth of Ti (0001) on MgO and Al₂O₃, the concept of a simple mosaic crystal for epitaxial thin film samples has to be revised. The correlation length for rotational disorder in thin films is reduced as compared with bulk samples. This is due to the fact that reduced interfacial displacements are a consequence of epitaxy. A sharp component in transverse X-ray scans then arises from a structurally coherent contribution that is not present in the simple mosaic model. This narrow component is overlaid by a broad mosaic contribution. The more dislocations that are formed during the strain relaxation process, the more pronounced the mosaic contribution becomes. An example of this is shown in Figure 18

for c axis-oriented UPt₃ layers on SrTiO₃ as the relative normal lattice contraction versus the area ratio of the narrow and broad component of the (0002) reflection of UPt₃. Despite the strong scatter in the data, owing to the different defect structures present in different samples, the correlation is obvious. Larger strain or, equivalently, less dislocation formation results in a more pronounced long-range coherence of the lattice. The functional relationship between residual strain fields and the mosaic structure is very specific to the system under investigation. It is analyzed in more detail by lateral mosaicity studies that allow the identification of the dominant relaxation process [53].

2.6. Morphological Aspects

The evolution of surface morphology is governed by several processes, which are often interrelated. The substrate or template surface microstructure (miscut, step-edge distribution, defect types, impurities, etc.) strongly determines the onset of dislocation formation and its lateral distribution. The resulting surface morphology of the growing epilayer will therefore vary, depending on the substrate preprocessing [54]. Furthermore, kinetically driven surface instabilities, like mound formation, can result if there is a diffusion barrier (Ehrlich–Schwoebel barrier) at the step edges. Moreover, the visitation frequency of the diffusing adatoms at the step edges has to be low to obtain an increased probability for the formation of stable islands on the terraces [38]. Mound instabilities can therefore be avoided if substrates with a large miscut (i.e., a high step edge density) are employed or, alternatively, surfactant assisted growth can be realized. Substrate temperature increase, although an additional possible path to the stabilization of step-flow growth instead of mound formation, may very often be impracticable because of interface alloying or other degradation effects. Because, loosely speaking, compressive strain tends to enhance surface mobility whereas tensile strain lowers diffusion, the epilayer's strain state and the appearance of surface instabilities will in general be mutually dependent [38].

These aspects of surface morphology evolution are common to all metal layers. Specific to alloys and intermetallic compound thin films are segregation phenomena that can create regular chemical and morphological patterns [39, 55]. In the following a small selection of surface morphologies that can be found in metal epitaxy is given. First, the influence of the substrate's microstructure on the film morphology is discussed. The influence of strain and its relaxation is presented next. Finally, a new promising pathway for the selection of one specific stacking variant of (111) oriented fcc layers is outlined.

2.6.1. The Influence of the Substrate and Template Morphology

One of the most often employed insulating substrate materials is Al₂O₃. Epitaxial-grade sapphire is widely used in rare earth epitaxy in conjunction with adquate buffering [56, 57], but it is also an indispensable substrate material for the semiconductor industry. The necessary mechanochemical polishing

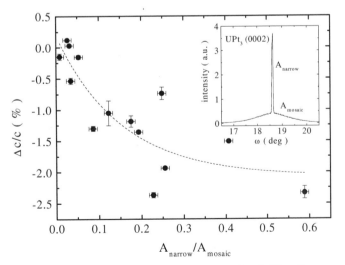

Fig. 18. Strain–coherence correlation of (0001)-oriented UPt₃ films on SrTiO₃ (111). The normal contraction is shown versus the area ratio of the coherent and mosaic component of the (0002) rocking curve of UPt₃. Inset: A typical (0002) rocking curve of UPt₃ on SrTiO₃ (111) for reference.

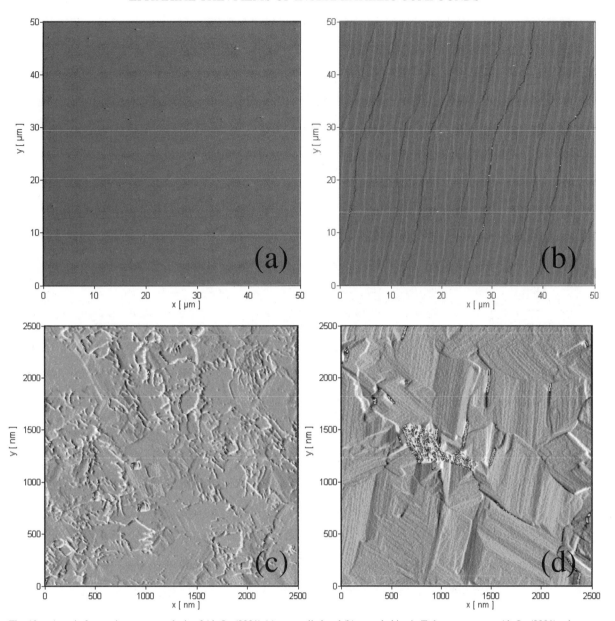

Fig. 19. Atomic force microscopy analysis of Al_2O_3 (0001) (a) as supplied and (b) annealed in air. Ta layers grown on Al_2O_3 (0001) substrates (c) as supplied and (d) annealed in air. Data were acquired in contact mode.

process results in a topmost layer with an irregular corrugation and crystallographic defects. To enhance the epitaxial growth on these substrates, flatness on the atomic scale and a well-ordered step edge distribution on vicinal surfaces are desirable. Vacuum annealing above 1000°C is known to improve the crystallinity of (0001), (11$\bar{2}$0), and (1$\bar{1}$02) surfaces [56], whereas it results in faceting of Al_2O_3 (1$\bar{1}$00) [58]. Moreover, annealing in air at temperatures between 1000°C and 1400°C results in the formation of atomically flat terraces separated by a regular arrangement of step edges [59]. The step edge height is uniformly 0.22 nm, showing a tendency toward step bunching at elevated annealing temperatures. An AFM analysis in air, presented in Figure 19, shows the result of 12 h of annealing of Al_2O_3 (0001) in air at 1200°C. Ta (110) template layers grown

on as-supplied and on annealed Al_2O_3 (0001) exhibit a pronounced change in surface morphology (see Fig. 19c and d).

Because the epilayer morphology strongly depends on the template, the controlled preparation of the substrate surface before growth offers one pathway to the realization of different surface morphologies. As an example, the microstructure of $TbFe_2$ (111) surfaces on Mo (110) templates can be varied by changing the molybdenum morphology. This can be achieved by inserting an additional Ta (110) buffer layer before the Mo growth. The resulting alteration of the $TbFe_2$ morphology is presented in Figure 20. The percolation network in which Mo grows on sapphire is reproduced in the magnetic layer. When the magnetic layer is grown on the much smoother Mo–Ta template, however, the structure is dominated by largely isolated

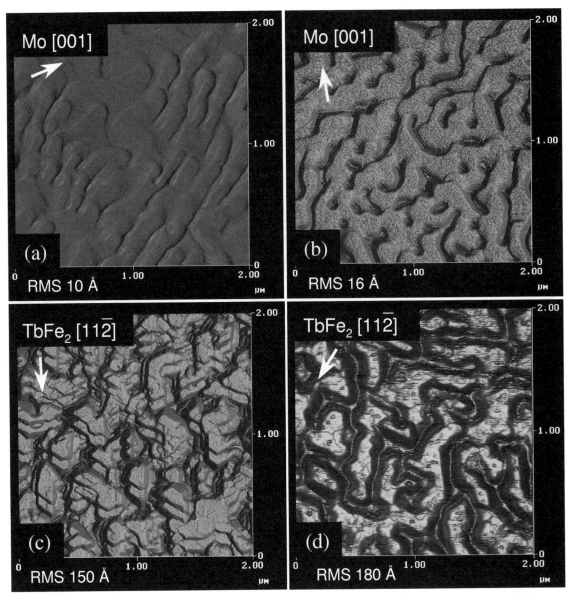

Fig. 20. AFM images of (a) 40 nm of Mo (110) grown on 30 nm of Ta (110) on sapphire (11$\bar{2}$0); and (b) 40 nm of Mo (110) grown directly on sapphire (11$\bar{2}$0). The arrow indicates the Mo [001] direction. (c) and (d) show the surface morphologies of TbFe$_2$ (111) grown on these templates. The film thickness is 50 nm in both cases. The arrows indicate the TbFe$_2$ [11$\bar{2}$] direction. Reprinted with permission from [13], © 1998, American Physical Society.

islands that reflect the symmetry of the TbFe$_2$ (111) surface. In both cases the grooving between TbFe$_2$ islands is pronounced, with a depth that is 60% of the nominal film thickness. The microstructure shows no dependence on the miscut of the sapphire substrates, which varied in the present case from 1.5° to 0.2°. The pronounced island structure of TbFe$_2$ films grown on Mo–Ta templates exerts an important influence on the magnetic hysteresis, as will become evident in the next section.

2.6.2. The Influence of Temperature and Strain

The sequence of AFM images shown in Figure 21 reveals a pronounced morphological change as Ti (0001) is grown on MgO (111) at various substrate temperatures. The strong in-

fluence of substrate defects blocking the propagation of the Ti growth fronts, causing step edges to bend, is strongly reduced with increasing substrate temperature. The direction of the step edges corresponds to the directions of substrate miscut. At the highest temperature the step edges follow the sixfold symmetry of the Ti (0001) surface. The growth is dominated by strong interlayer diffusion, which points to a reduced importance of the Schwoebel–Ehrlich barrier [60, 61] at the step edges as the thermal energy increases with substrate temperature. The surface free energy of the films is reduced by the formation of well-defined facets with step edges along the hexagonal symmetry directions. Unlike these results for MgO, the surface morphology for Ti grown on Al$_2$O$_3$ (0001) is stable against changes in the substrate temperature and exhibits reduced island diameters

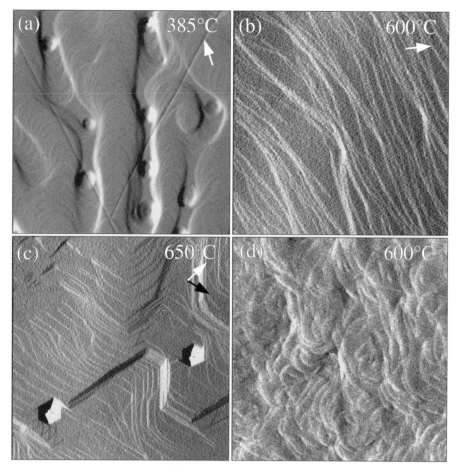

Fig. 21. Sequence (a–d) of AFM images for Ti on MgO (a–c) and Al$_2$O$_3$ (d) grown at the different substrate temperatures shown. The substrate miscut was 0.6–0.8° for MgO in directions shown by white arrows and <0.1° for Al$_2$O$_3$. The black arrow in (c) indicates the [1000] direction of Ti. Scan range 2100 nm. Reprinted with permission from [41], © 1997, American Institute of Physics.

due to the larger misfit. The rms roughness is found to be independent of the substrate at 0.2 nm to 0.5 nm and, consequently, of the lattice misfit.

The lateral length scales on which the equilibrium surface configuration can be achieved for a given temperature depend, among other things, on the step–step interaction. The merging step edges will show a tendency to form those facets with the lowest free energies. A corresponding detailed analysis of the step edge arrangement of a refractory metal was recently performed on Nb (110) layers grown on vicinal Al$_2$O$_3$ (11$\bar{2}$0) [62].

2.6.3. Heteroepitaxial Stacking Controlled by Substrate Miscut

Recent investigations by Bonham and Flynn point to a heteroepitaxial template effect for the growth of the ordered alloy Cu$_3$Au on Nb (110) [63]. They find a strong suppression, by a factor of 10^3, of one of the two close-packed stacking variants ABCA... or ACBA... of the fcc epilayer, which, by symmetry arguments, cannot be explained by the properties of the surface nets. The perfect bcc (110) surface, as supplied by the Nb template, cannot discriminate between the two fcc stacking

sequences, which are formed by successive lattice planes displaced by $\pm\frac{1}{6}[112]$. ABC and ACB stacked fcc domains are structural twins that are mutual mirror images in any of the {112} planes perpendicular to the (111) surface. The bcc (110) surface has two mirror planes perpendicular to the surface. Of these mirror planes the (1$\bar{1}$0) representative is parallel to the fcc (11$\bar{2}$) twin plane. This coincidence of one fcc twin plane and a bcc structural mirror plane implies that it is not possible for the bcc surface to discriminate between the two fcc stacking variants. Consequently, the strong suppression of one stacking variant has to be caused by a different feature of the surface structure, which was identified to be the vicinal miscut of the substrate in conjunction with the mesoscopic structures that emerge on the Nb surface. The Nb (110) surface was shown to develop faceted structures with a pronounced fingering along the [001] direction [64]. The preference of noble metals and Nb for neighbors of their own kind results in different interaction energies with the Nb step edges for the two possible second layers of the fcc stacking sequence [64]. This indicates one possible reason for the observed stacking preference.

Another example of stacking selection is the growth of GaAs (001) on Si (001), for which, for geometrical reasons, successive growth terraces on a miscut surface exhibit different dimer row reconstructions. In conjunction with their different interactions with the step edges, the miscut Si substrate can provide a template for the single-domain growth of GaAs (001) [65, 66].

Finally, recent investigations of the growth of Co–Pt heterostructures on MgO (111) reveal a suppression of one fcc stacking variant of the Pt (111) layers on miscut surfaces [67, 68]. Because MgO crystallizes in the rock-salt structure, surface net symmetry arguments cannot explain the stacking preference of the epilayers. At present, it remains an open question which microscopic mechanism on the MgO (111) surface is responsible for this growth selection phenomenon. Very likely the reason is related to the microfacet formation tendency of MgO (111) [69–71] in conjunction with the vicinal miscut of the substrate. In any event, a more detailed knowledge of the different mechanisms that induce the selection of one pure stacking variant is valuable because it represents one more step toward the goal of growing true single crystals.

3. SELECTED APPLICATIONS IN BASIC AND APPLIED RESEARCH

In the this part of the chapter we are concerned with applications of thin films of intermetallic compounds in basic and applied research. The selection of topics presented below is by no means complete. It rather reflects the author's own preferences. Nevertheless, by giving examples from such a diverse collection of fields as superconductivity, magnetism, and order–disorder phenomena, it is hoped that the reader will find something related to his or her own area of interest. Specifically, we discuss superconductivity in heavy-fermion materials as a prominent example of cooperative phenomena in systems with strong electronic correlations. In the framework of magnetism four examples are given. Magnetoelastic coupling effects are discussed to underline the importance of strain evolution with regard to the magnetic properties of epitaxial films. The influence of the substrate-induced microstructure on the magnetization dynamics of thin ferromagnetic layers is presented next. We then proceed to a nucleation study of the magnetic order parameter in an antiferromagnetic thin film based on resonant magnetic X-ray scattering. The magnetic systems section is closed by some remarks concerning the properties of ordered and disordered alloys for magnetic storage applications. We end by giving a short discourse on recent investigations of order–disorder phenomena in systems with interfaces, namely thin films.

3.1. Superconductivity in UPd$_2$Al$_3$

One of the greatest challenges in current solid-state physics is gaining a deeper insight into the implications of strong electron correlations. The materials studied in most detail in this context are heavy-fermion systems [10]. In heavy-fermion materials the character of cooperative phenomena, like superconductivity and magnetic order, is strongly affected by these correlations. Phenomenologically, the Landau–Fermi liquid ap-

proach adequately describes the low-lying excitations of most heavy-fermion systems in the low-temperature limit. The question remains, what are the residual interactions of the quasi-particles that induce a phase transition into a superconducting state? In conventional superconductors the pair-coupling mechanism is based on the exchange of virtual phonons. It can therefore be considered extrinsic to the electron system. In heavy-fermion superconductors, on the other hand, early theoretical investigations have already suggested an intrinsic pairing mechanism based on the polarization properties of the Fermi sea of Landau quasi-particles [72]. Because of the strong f-electron correlations, the realization of pairing states with nodes at the pair center was considered likely for energetic reasons. In all known heavy-fermion superconductors either antiferromagnetism and superconductivity coexist, or there are at least strong antiferromagnetic correlations present. Consequently, the exchange of antiferromagnetic spin excitations has been considered theoretically as a possible cause for the Cooper pair formation. This picture recently received some experimental support: the resistivity behavior under pressure of two cerium-based heavy-fermion compounds was shown to be consistent with a magnetically mediated pairing mechanism [73]. An analogous intimate relationship also appears to be central to the high-T_c superconductors [74, 75].

Spectroscopic investigations of the superconducting state of heavy-fermion materials promise to offer a direct means of obtaining detailed information about the pairing state. Special characteristics, such as the spatial anisotropy of the order parameter or the origin of the pairing interaction, are then, in principle, directly observable. Consequently, spectroscopic investigations of the superconducting state of heavy-fermion materials have a long history [76]. However, the application of established methods based on the formation of point contacts or break junctions proved to be difficult or impossible. This is mainly due to the inherent sensitivity of the superconducting order parameter to structural changes caused by local pressure, defects, or surface oxidation [77].

It was only very recently that clear-cut experimental evidence for unconventional superconductivity based on spectroscopic means was obtained for the heavy-fermion superconductor UPd$_2$Al$_3$. This resulted from two different approaches: a detailed analysis of the magnetic excitation spectrum of UPd$_2$Al$_3$ single crystals, applying inelastic magnetic neutron scattering in the superconducting state [78, 79], and tunneling spectroscopy on thin film tunnel junctions [80].

In the heavy-fermion system UPd$_2$Al$_3$ superconductivity below $T_c = 2$ K coexists with antiferromagnetic order setting in at $T_N = 14.5$ K [81]. The magnetic state of this hexagonal compound is formed by ferromagnetic easy planes of the U moments stacked antiferromagnetically along the crystallographic c axis [82]. The ordered moment amounts to $0.85\mu_B$. Such a large moment cannot be described within the scenario of small-moment band magnetism mostly observed in the other uranium-based heavy-fermion superconductors [10, 83].

Recent measurements of the characteristic neutron energy losses for momentum transfer at and in the vicinity of the mag-

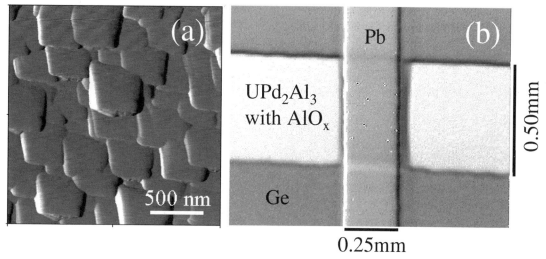

Fig. 22.   (a) Atomic force microscopy image of the UPd$_2$Al$_3$ (0001) surface taken in contact mode. (b) Optical image of the tunnel junction region in phase-contrast mode.

netic Bragg vector (0 0 1/2) showed two noticeable features: a strongly damped spin-wave excitation with an excitation energy of about 1.5 meV [84–86] and an energy loss peak in the superconducting state corresponding to the opening of an energy gap [78, 79, 87, 88]. The latter finding is compatible with a superconducting order parameter displaying a sign inversion when translated by the antiferromagnetic reciprocal lattice vector [89]. These neutron scattering results were corroborated by tunneling spectroscopy on cross-type UPd$_2$Al$_3$/AlO$_x$/Pb tunnel junctions based on UPd$_2$Al$_3$ thin films [80, 90–93] as is detailed below.

3.1.1. Experimental Details: Junction Preparation

The availability of well-defined uncontaminated surfaces of UPd$_2$Al$_3$ proved to be mandatory for preparing tunnel junctions. Because of the rather short Ginzburg–Landau coherence length in heavy-fermion superconductors (typically 10 nm), the conditions that the interfaces of tunneling devices have to fulfill are more stringent than for conventional superconductors. First, the surface of the base electrode formed by the heavy-fermion material has to be sufficiently smooth. This can be accomplished with the use of thin films. The growth was performed in a multichamber MBE system by coevaporation of the constituent elements onto heated LaAlO$_3$ (111) substrates, as detailed elsewhere [94–96]. Second, the individual steps in the contact preparation sequence have to be performed *in vacuo* to avoid a degradation of the respective interfaces. This was checked by AFM investigations, as shown in Figure 22a. Third, the insulating barrier has to be formed by a material that shows sufficient wetting properties on the base electrode.

After the UPd$_2$Al$_3$ layer growth the samples were transferred in vacuum to a dedicated contact preparation chamber [96]. Al layers that were 4.0 nm to 6.0 nm thick were sputter-deposited in an 8×10^{-2} mbar Ar atmosphere on top of the UPd$_2$Al$_3$ thin film. The Al layer was then oxidized in

an oxygen glow discharge. In the next step the whole AlO$_x$ layer, except for a stripe, was covered with 200 nm of amorphous Ge. Finally, again with the use of a shadow mask, the Pb counterelectrode was deposited so as to cross the stripe previously uncovered by the amorphous Ge. The resulting cross-type junction is shown in Figure 22b. The junction area typically amounted to about 0.13 mm^2 with area resistances in the range of 25–130 Ω mm^2. The differential conductivity was measured by a standard ac-current modulation technique in a ^{3}He cryostat. Further details can be found in [96].

3.1.2. Tunneling Spectroscopy: Results

In the simplest case, the differential conductivity in the gap region is directly proportional to the superconductor's density of states. Nevertheless, even for conventional superconductors the influence of the barrier can cause deviations from this ideal behavior. A well-known example is the appearance of zero-bias anomalies due to the presence of magnetic impurities in the barrier. In the present case the Pb counterelectrode gives a valuable means of judging the quality of the junction in this respect. The differential conductivity follows the well-known characteristic of Pb tunnel junctions exhibiting a Bardeen–Cooper–Schrieffer (BCS)-like density of states augmented by strong-coupling features due to the electron–phonon coupling mechanism. This is demonstrated in Figure 23. In a small applied magnetic field ($H_c^{Pb} < H < H_{c2}^{UPd_2Al_3}$) superconductivity in Pb is suppressed, thus allowing a direct investigation of the tunneling density of states of UPd$_2$Al$_3$. The temperature dependence of the energy gap of UPd$_2$Al$_3$ was then obtained with the use of Dynes' extension to the normalized BCS density of states [97],

$$\widetilde{N}(\epsilon, \Gamma) = \mathrm{Re}\left[\frac{\epsilon - i\Gamma}{\sqrt{(\epsilon - i\Gamma)^2 - \Delta^2}} \right] \tag{20}$$

which introduces a broadening parameter Γ that was originally intended to account for quasi-particle lifetime effects in strong-

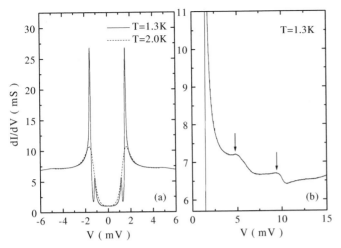

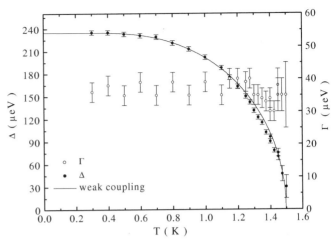

Fig. 23. (a) Differential conductivity of a UPd$_2$Al$_3$ junction at the temperatures indicated. (b) Detail: strong-coupling features of the Pb electrode.

Fig. 25. Temperature dependence of the energy gap in the c axis direction of UPd$_2$Al$_3$ as deduced from a Dynes fit (see text for details).

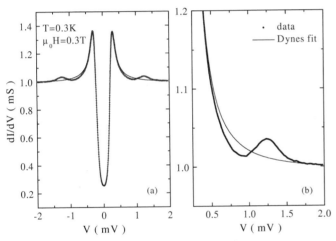

Fig. 24. (a) Differential conductivity normalized to the spectrum taken at 1.7 K in $\mu_0 H = 0.3$ T. The solid line represents a Dynes fit to the data as explained in the text. (b) Detail of the strong-coupling feature originating from the magnetic pairing interaction in UPd$_2$Al$_3$.

coupling superconductors. The measured differential conductivity can then be fitted with proper inclusion of thermal broadening and a fixed leakage conductivity $\sigma_\ell = 0.12\sigma_{NN}$ according to

$$\sigma_{SN}(eV) = \sigma_{NN} \int_{-\infty}^{\infty} \widetilde{N}(\epsilon, \Gamma)\left[-\frac{\partial f}{\partial \epsilon}(\epsilon + eV)\right]d\epsilon + \sigma_\ell \quad (21)$$

In Figure 24 the measured normalized conductivities and the corresponding Dynes fit are shown for $T = 0.3$ K and $\mu_0 H = 0.3$ T. The temperature-dependent fit parameters Δ and Γ are presented in Figure 25. The result for a BCS weak-coupling superconductor is also shown for reference.

The BCS ratio amounts to $2\Delta_0/k_B T_c = 3.62 \pm 0.05$ in an applied field of 0.3 T. The rather weak deviations in $\Delta(T)$ of UPd$_2$Al$_3$ from the universal weak-coupling curve are comparable to the deviations that are typically obtained for strong-coupling superconductors, like Pb or Hg. Furthermore, an anisotropic gap function can lead to these deviations if the tun-

neling process probes an extended part of the Fermi surface. In the present case, the tunneling process is highly directional, probing only the energy gap along the crystallographic c axis.

All of this seems to indicate that UPd$_2$Al$_3$ is a weak-coupling superconductor. However, the most important observation to be made is the systematic deviation of the data from the fit at about 1.2 meV. This deviation signifies a strong-coupling effect which can only be associated with the magnetic excitations in UPd$_2$Al$_3$ [80]. This will be briefly discussed next.

3.1.3. Antiferromagnetism and the Superconducting Order Parameter

Strong-coupling effects in the tunneling density of states of conventional superconductors, like Pb and Hg, are caused by phonon modes that contribute the dominant part to the pairing interaction. In analogy, the strong-coupling effect observed in the UPd$_2$Al$_3$ tunnel spectra lends strong support to the assumption that superconductivity in UPd$_2$Al$_3$ is caused by the exchange of antiferromagnetic spin excitations. This is consistent with the observed pronounced Pauli limiting in the upper critical field that indicates the formation of a singlet pairing state [98, 99].

Assuming such a pairing interaction, quite general arguments can be given concerning the structure of the superconducting order parameter based on the known properties of UPd$_2$Al$_3$. Because of a pronounced magnetic anisotropy, the spins tend to fluctuate in the basal planes [79]. It then follows that the pair partners cannot reside in the same plane because of a strong pair-breaking effect caused by the ferromagnetic spin alignment. The pairs might be formed by quasi-particles in neighboring planes. This necessarily leads to a node in the spatial pair function for vanishing quasi-particle distance. Accordingly, the order parameter in k-space is anisotropic, thus reducing the effective Coulomb repulsion in the pair state of the correlated electrons.

These general arguments can be substantiated by analyzing the node structure of the superconducting order parameter on the multisheeted Fermi surface of UPd_2Al_3. This has to be performed in two steps. First, the k-dependence of the superconducting order paramter of UPd_2Al_3 has to be analyzed within a weak-coupling scheme asuming a magnetic pairing interaction. Next, this analysis has to be extended to the strong-coupling limit within the Eliashberg formalism of strong-coupling superconductors. For further details the interested reader is refered to [89, 92, 93, 100].

3.2. Magnetoelastic Coupling Effects in RFe_2

As an example of the potential of intermetallic compound thin films, magnetoelastic coupling effects in highly strictive Laves phases are discussed next. The prospect for application of these materials in thin film form is only one important motivation for detailed investigations [101, 102]. More importantly, Laves phase compounds possess model character with regard to their magnetoelasticity and the corresponding interplay of microstructure and magnetism in the presence of epitaxial constraints.

The cubic RFe_2 (R: rare earth) Laves phase compounds with C15 structure are known for their pronounced magnetic anisotropy and extremely large magnetoelastic constants. These are mainly determined by single-ion effects exerted by the rare earth component. The most thoroughly studied and best understood example of this series of intermetallic compounds is $Tb_{0.35}Dy_{0.65}Fe_2$, also known as Terfenol-D. In Terfenol-D the magnetic anisotropy is greatly reduced as compared with the terminal compounds $TbFe_2$ and $DyFe_2$. This happens because in $TbFe_2$ the axes of easy magnetization point in the $\langle 111 \rangle$ directions, whereas in $DyFe_2$ these axes are aligned along $\langle 100 \rangle$ [103]. Despite the reduced magnetic anisotropy in Terfenol-D, a large magnetostriction along the $\langle 111 \rangle$ directions is preserved.

For thin films of the magnetic Laves phase compounds, such as $DyFe_2$, $ErFe_2$, or $TbFe_2$, epitaxial strain and clamping add further complexity to the interplay of various energy contributions that govern key magnetic parameters in single crystals. Strain changes the magnetic properties, as does the presence of an interface that breaks the translational symmetry of the lattice. Interface effects are less important in thicker films because of their inverse thickness dependence. In contrast, strain can be sustained through a thick film in several ways. For example, magnetic layers can be sandwiched between nonmagnetic layers that have slightly different lattice constants. Epitaxial strain can then be maintained throughout the magnetic layers by repeating the sandwich as a multilayer, provided that the individual layer thicknesses do not exceed the critical value for the onset of strain relaxation by dislocation mechanisms. This has been studied for rare earth and especially Dy c axis epilayers [58, 104].

In many instances adhesion of the epilayer to the substrate is sufficient to clamp the films, which then follow the elastic response of the substrate to external parameters like temperature. Therefore strains usually develop in cooling from an elevated growth temperature because of the difference between the thermal expansion coefficients of the substrate and epilayer. In epi-

taxial $TbFe_2$ (111) thin films, in-plane tensile strain results in anomalous magnetic hysteresis behavior [13, 16, 17].

In the following the relationship of the magnetic properties to the film microstructure and strain state is analyzed in more detail. This discussion might help to elucidate several important aspects of the strain–magnetism interplay.

3.2.1. Magnetic Properties of $TbFe_2$ with In-Plane Magnetic Anisotropy

The axis of easy magnetization of epitaxial (111)-oriented $TbFe_2$ thin films on Mo (110) templates grown on sapphire ($11\bar{2}0$) lies in the plane of the films, regardless of thickness, in the range from 40 nm to 150 nm. (The magnetization measurements presented here were performed in a SQUID magnetometer.) An in-plane easy axis is of course favored because of the shape anisotropy. It is also favored by the 0.5% tensile strain developed in the samples as they cool from the growth temperature to 300 K and below [13]. This strain lowers the free energy for those magnetic domains whose moments are aligned close to {111} crystal axes that lie at a 19° angle to the film plane.

A common feature observed for all samples is a pronounced drop of the magnetization close to or at zero applied field. This occurs in magnetic hysteresis loops measured at various temperatures between 300 K and 5 K. Figure 26 gives results for two 50-nm-thick $TbFe_2$ films, the AFM images of which were shown in Figure 20.

Wang et al. [105] inferred for sputtered $TbFe_2$ on Mo that this drop in the magnetization is solely caused by the formation of a magnetic alloy layer in the interfacial region. They studied the dependence of the saturation magnetization on film thickness and found the thickness of the soft magnetic alloy at the interface to be 10 nm. On the other hand, a 10-nm-thick alloy region can be responsible for the pronounced magnetization drop

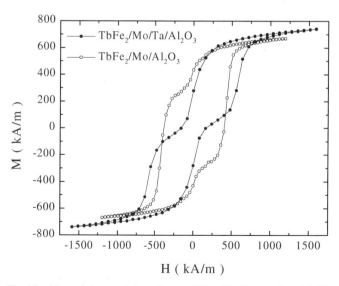

Fig. 26. Magnetic hysteresis loops for two $TbFe_2$ thin film samples with different morphologies shown in Figure 16. The field was aligned parallel to the $TbFe_2$ [$1\bar{1}0$] direction. Reprinted with permission from [13], © 1998, American Physical Society.

for samples as thick as 150 nm only if the effective saturated moment of the alloy surpasses the saturated moment of TbFe$_2$ by at least an order of magnitude. Furthermore, the saturation magnetization observed in the hysteresis loops taken at room temperature corresponds well to the expected 800 kA/m known for TbFe$_2$ single crystals. Finally, over an extended region of applied field after field reversal the magnetization remains close to zero, as can be seen in Figure 26. In what follows an alternative mechanism for magnetization reversal is presented that provides a more satisfactory account of the observed behavior [13].

To begin with, individual magnetic domains are assumed not to be coupled. Then the easy axis in zero field must be aligned perpendicular to the field direction, to account for the observed drop in magnetization close to zero field. An in-plane strain anisotropy can single out one easy axis of magnetization due to the strong magnetoelastic coupling. The analysis of the influence of pure one-domain effects on the magnetic hysteresis can be performed numerically. As a result, the magnetic hysteresis for a single-domain particle with cubic anisotropy, subject to a biaxial tensile strain in the (111) plane, can be calculated. The results of this calculation represent a necessary prerequisite to understanding the magnetization process and can later be augmented by consideration of domain coupling effects.

The calculation is based on the energy landscape for a single-domain magnetic particle whose moment is rotated with respect to the cubic crystal axes $x = [100]$, $y = [010]$, and $z = [001]$ (the extended Stoner–Wohlfarth model [106, 107]). The orientation of the component magnetizations M_i are specified by polar and azimuthal angles θ and ϕ. Relevant contributions to the energy come from the magnetic anisotropy energy $E_a(\theta, \phi)$, the magnetoelastic energy $E_{me}(\theta, \phi, \{\eta^{\alpha j}\})$, the Zeeman energy $E_z(\theta, \phi, \{H_i\})$, and the demagnetization energy $E_d(\theta, \phi, \{N_{ij}\})$ [108]. They may be written as

$$E = E_a(\theta, \phi) + E_{me}(\theta, \phi, \{\eta^{\alpha j}\})$$
$$+ E_z(\theta, \phi, \{H_i\}) + E_d(\theta, \phi, \{N_{ij}\}) \quad (22)$$

$$E_a(\theta, \phi) = K_1(a_1^2 a_2^2 + a_2^2 a_3^2 + a_1^2 a_3^2)$$
$$+ K_2 a_1^2 a_2^2 a_3^2 \quad (23)$$

$$E_{me}(\theta, \phi, \{\eta^{\alpha j}\}) = -\frac{3}{2} c^\gamma \lambda^{100} \times \left(\frac{2}{3} \left(a_3^2 - \frac{1}{2}(a_1^2 + a_2^2) \right) \eta^{\gamma 1} \right.$$
$$+ \frac{1}{2} (a_1^2 - a_2^2) \eta^{\gamma 2} \Bigg)$$
$$- \frac{3}{\sqrt{2}} c^\epsilon \lambda^{111} \times (a_1 a_2 \eta^{\epsilon 1} + a_2 a_3 \eta^{\epsilon 2}$$
$$+ a_3 a_1 \eta^{\epsilon 3}) \quad (24)$$

$$E_z(\theta, \phi, \{H_i\}) = -\mu_0 \sum_{i=1}^{3} M_i H_i \quad (25)$$

$$E_d(\theta, \phi, \{N_{ij}\}) = \frac{1}{2} \mu_0 \sum_{i,j=1}^{3} M_i N_{ij} M_j \quad (26)$$

Here the anisotropy constants are K_1 and K_2; the Lagrangian strain variables are $\{\eta^{\alpha j}\}$; the irreducible elastic constants are c^α and c^ϵ; the magnetostriction coefficients are λ^{100} and λ^{111} along the [100] and [111] directions, respectively; the components of the demagnetization tensor are N_{ij}; and the applied field is (H_1, H_2, H_3). The a_i are direction cosines with respect to the cube axes.

To simulate a typical hysteresis cycle the system is assumed to be in the global energy minimum at zero applied field. This defines the initial direction of the domain moment. The field is then increased to its maximum value, followed by one complete hysteresis cycle. During any field change the magnetization direction is determined by keeping the system adiabatically in a local energy minimum that remains accessible by rotating the moment without increasing the energy. Thermal excitations into neighboring energy minima that would correspond to thermally induced coherent domain rotation are not taken into account in the present approximation. Because TbFe$_2$ has a high Curie temperature of 698 K [109], this approach may be adequate even at room temperature.

In Figure 27 results are presented on the basis of the parameters as given in Table II. The Langrangian strain tensors are calulated from X-ray results with the use of the elastic constants of bulk Terfenol-D.

First, an isotropic biaxial strain is assumed based on the material parameters of bulk TbFe$_2$. The resulting hysteresis loop, given by the dashed line in Figure 27a, shows a significantly larger coercive field than that observed experimentally and lacks a magnetization drop close to zero field. To improve the correspondence between the calculation and measured hysteresis loops, K_1 is slightly reduced next (dash-dotted line in Figure 27a). The easy axes of magnetization are now essentially collinear with the in-plane $\langle 112 \rangle$ directions, pointing about $10°$ out of the plane toward the closest [111] direction.

A magnetization drop that takes place close to zero applied field is reproduced once the magnetoelastic energy contribution $E_{me}(\theta, \phi, \{\eta^{\alpha j}\})$ is allowed to break the sixfold symmetry of the anisotropy energy in the (111) plane, by introducing an anisotropic biaxial strain. Because the easy axis of magnetization under isotropic strain is close to the $\langle 112 \rangle$ directions, a small elongation along the [11$\bar{2}$] axis is selected. The result shown as the solid line in Figure 27a predicts a hysteresis loop that is in good agreement with the experimental data. In a field reversal close to zero applied field the magnetization discontinuously changes its orientation from close to the field direction along [1$\bar{1}$0] to the easy axis essentially collinear with the [11$\bar{2}$] direction. The projection of the moment direction in the film plane as it varies during the hysteresis cycle is shown in Figure 27b.

3.2.2. Further Discussion

Domain coupling effects must be taken into account in an improved model because measurements of in-plane hysteresis curves show sixfold symmetry with only weak anisotropy between hysteresis data taken along the $\langle 110 \rangle$ and $\langle 112 \rangle$ directions. For any given field direction there must then be an appreciable number of domains whose easy axes are aligned

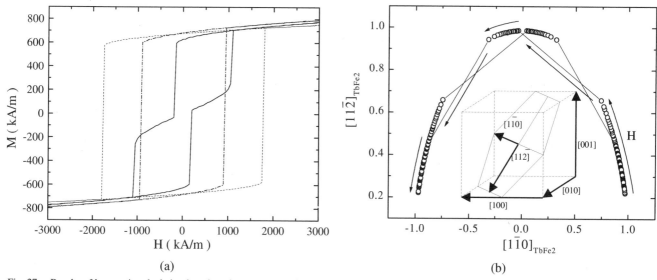

(a)

(b)

Fig. 27. Results of hysteresis calculation based on the parameters given in Table II (see text for details). (a) The dashed and dash-dotted lines were drawn assuming isotropic tensile strain in the (111) plane for two different values of the anisotropy constants (parameter sets 1 and 2 in Table II). The solid line is obtained when a biaxial strain anisotropy is assumed (parameter set 3 in Table II). (b) The orientation of the moment in the (111) plane for parameter set 3 during a hysteresis loop. The field is aligned parallel to the TbFe$_2$ [1$\bar{1}$0] direction. The relevant directions are depicted in the inset. Reprinted with permission from [13], © 1998, American Physical Society.

Table II. Parameters for Hysteresis Calculation

#	K_1 (J/m^3)	K_2 (J/m^3)	M_s (A/m)	λ^{111}	λ^{100}	$\eta^{\epsilon 1} = \eta^{\epsilon 2}$	$\eta^{\epsilon 3}$	c^ϵ (GPa)
1	-7.6×10^6	0	8×10^5	2.4×10^{-3}	0	-3.0×10^{-3}	-3.0×10^{-3}	97.4
2	-5.0×10^6	0	8×10^5	2.4×10^{-3}	0	-3.0×10^{-3}	-3.0×10^{-3}	97.4
3	-5.0×10^6	0	8×10^5	2.4×10^{-3}	0	-2.3×10^{-3}	-3.3×10^{-3}	97.4

See text for details.

perpendicular to the field. This can be accomplished by the following mechanism.

During the cool-down process after growth the sample remains in a demagnetized state below the Curie temperature. In this demagnetized state the individual magnetic domains are subject to biaxial strain because the magnetostrictive energy drives a rhombohedral lattice distortion along easy axes that lie close to the in-plane ⟨112⟩ directions. Minimization of the dipolar coupling energy then results in equal populations for all easy axis directions, as dictated by the sixfold structural symmetry in the film plane. On a macroscopic scale the strictions of the individual domains are therefore mutually compensated. The dislocation movements needed to accommodate the microscopic strain fields of any single domain involve one dislocation every 500 lattice sites, assuming about 0.2% magnetostrictve distortion along the easy axes. Given the in-plane lattice constant of 0.52 nm for TbFe$_2$, this results in one dislocation every 260 nm. High-resolution X-ray studies of reflections with an in-plane component of the scattering vector would then show a threefold splitting. Such investigations have not yet been performed on TbFe$_2$ epitaxial films. However, X-ray studies in zero applied field of c axis Dy thin films in the ferromagnetic state show precisely this splitting [110]. With its sixfold symmetry

in the basal plane, which is also the plane of easy magnetization, Dy provides a valuable system for comparison.

Suppose that when an external field is applied, the moments begin to align mainly parallel to the field direction. This causes a rotating stress field in the respective domains. The system can respond with a corresponding rotation of the biaxial strain anisotropy or may instead be clamped, thereby preventing a redistribution of the strains. In subsequent field reversal of the hysteresis cycle, the minimization of dipolar coupling must tend to restore the demagnetized state. When the applied field falls close to zero most domains that point in the original field direction are discontinuously depopulated in favor of symmetry-equivalent easy directions. This discontinuity is supported by the anisotropic strain fields, which break the sixfold symmetry each domain would otherwise possess. Only movement and not dislocation creation is necessary to accommodate the local variations of the stress, so the rhombohedral distortion is likely to follow changes of magnetization directions in the domains. However, dislocation creation and motion may in part be thermally activated. It is therefore possible that a crossover could take place from a plastic to a clamped state of the domains as a function of temperature. High-resolution X-ray studies in magnetic fields at various temperatures might possibly clarify this unresolved issue.

The magnetization process of (112) single crystals of TbFe$_2$ under tensile strain along the [112] direction shows comparable switching behavior, which was investigated in detail by Jiles and Thoelke [111].

In highly magnetostrictive films the elastic moduli change under the influence of an applied magnetic field. To mention just one technical application, this can serve as a means of reducing the speed of propagation of surface acoustic waves [112, 113]. Optimal behavior can only be achieved if the axis of easy magnetization of the films can be aligned perpendicular to the film surface. This represents the only direction in which magnetostriction can freely strain the film. Furthermore, the use of a substrate material that allows the direct coupling of the surface acoustic wave into the delay device is advantageous. The use of piezoelectric substrate materials with large thermal expansion coefficients can account for both conditions. In the clamped state the films are put under compressive strain in the cooling process after growth, which then can overcome the shape anisotropy. Investigations along these lines are therefore presented next.

3.2.3. TbFe$_2$ with Perpendicular Magnetic Anisotropy

Because of their large magnetostrictive response and moderate magnetic anisotropy, Tb$_{0.35}$Dy$_{0.65}$Fe$_2$ or Terfenol-D thin films hold some promise for the realization of surface-sensitive magnetostrictive devices, such as surface acoustic wave (SAW) delay lines, or the development of microactuators [114, 115]. Early attempts to realize SAW delay lines with amorphous TbFe$_2$ layers proved the feasibility but suffered from a severe reduction of the achievable magnetostrictive distortion [112, 116]. This reduced response is due to the inherent isotropic nature of amorphous and polycrystalline films, which results in

an averaging over all crystallographic directions. The magnetostriction of TbFe$_2$, however, is highly anisotropic, with a peak value of 0.3% along the crystallographic $\langle 111 \rangle$ directions. Because of clamping, which couples the epilayer elastically to the substrate, the films are free to expand and contract only along the normal direction. Perpendicular magnetization is therefore needed to maximize the achievable magnetostrictive response. However, the shape anisotropy or demagnetization energy of the thin film geometry favors an in-plane axis of easy magnetization.

The demagnetization effect can be overcome by exploiting the strong magnetoelastic coupling in the clamped epilayer, with the use of an adequate substrate material. This was shown for TbFe$_2$ by Wang et al. [105], who used CaF$_2$ as a substrate material. Alternatively, the use of LiNbO$_3$ (111) as a substrate material is beneficial in two respects. First, it exhibits a large thermal expansion coefficient in the (111) plane, which causes an in-plane compressive strain in the films during cool-down from the growth temperature. This in-plane compressive strain stabilizes perpendicular magnetization. Second, the piezoelectric character of LiNbO$_3$ allows a coupling of the magnetostrictive device to radar waves that propagate in a surface layer. Their propagation velocity can then be tuned by applying a magnetic field to the magnetostrictive layer. Unfortunately the growth of rare earth cubic Laves phases on a niobate substrate is not straightforward. These material difficulties can be overcome with Ti and Mo buffer layers [117]. In the following the magnetic properties of the layers are briefly summarized.

A comparison of the magnetic hysteresis in the temperature range of 5 K to 300 K for samples grown on sapphire and LiNbO$_3$ clearly shows in-plane magnetization for the former and a net perpendicular intrinsic magnetic anisotropy for the latter. This is established unambiguously by data taken at room temperature and shown in Figure 28. An in-plane easy axis of magnetization is favored for films grown on Al$_2$O$_3$ substrates.

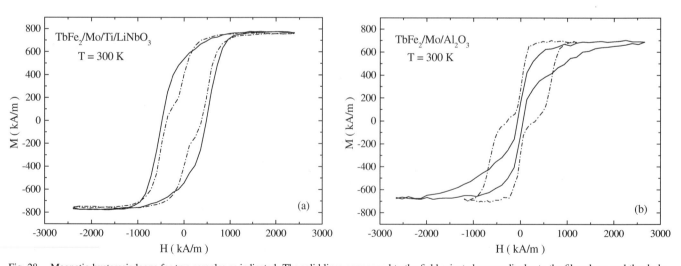

Fig. 28.   Magnetic hysteresis loops for two samples as indicated. The solid lines correspond to the field oriented perpendicular to the film plane, and the dash-dotted lines, to an in-plane field alignment along the TbFe$_2$ [11$\bar{2}$] direction. (a) Perpendicular magnetic anisotropy for the film grown on LiNbO$_3$. (b) In-plane easy axis of magnetization for the film grown on Al$_2$O$_3$. Reprinted with permission from [117], © 1999, Elsevier Science.

The smaller thermal expansion of sapphire as compared with $LiNbO_3$ (111) results in an in-plane tensile strain of the films, and this stabilizes the magnetic moment along those ⟨111⟩ directions that lie close to the film plane. As a consequence, magnetic saturation for fields aligned parallel to the [111] direction perpendicular to the film plane can only be reached above 2500 kA/m. The hysteresis curve shows a wasp-like shape close to zero field that characterizes this axis as magnetically hard. In contrast, films grown on $LiNbO_3$ (111) can be magnetically saturated in a perpendicular field orientation at a strongly reduced saturation field of 1200 kA/m. Furthermore, the hysteresis loop is more rectangular, which indicates an axis of easy magnetization in this case. Nevertheless, the onset of magnetization reversal is appreciably rounded, indicating the absence of nucleation of reversed domains. The shape of the hysteresis curve rather points toward domain wall movement initiated at edge domains as the dominant magnetization process. Note that no demagnetization corrections were performed to show the applied fields actually needed to magnetically saturate the films.

The magnetoelastic coupling energy contribution E_{me} and the demagnetization energy E_d can be used to estimate the compressive strain necessary to stabilize perpendicular magnetization, employing the condition $E_{me} - E_d < 0$. This leads to the following condition for the change of the (111) lattice plane spacing $(\delta d/d)_{111}$ (see, e.g., [108]):

$$\left(\frac{\delta d}{d}\right)_{111} > \frac{1}{3\lambda_{111}} \mu_0 M^2 \frac{|c^\epsilon - c^\alpha|}{c^\alpha c^\epsilon} \qquad (27)$$

With the magnetostrictive coefficient $\lambda_{111} = 2.5 \times 10^{-3}$ for fields along (111), the room temperature magnetization $M = 800$ kA/m, and the irreducible elastic constants $c^\epsilon = 97.4$ GPa and $c^\alpha = 271$ GPa [103, 109], it can be predicted in this way that perpendicular magnetization is energetically favored for $(\delta d/d)_{111} > 0.07\%$. For the observed strain of 0.4% the perpendicular magnetization is therefore expected to be fully stabilized. Specifically, the observed coercive fields in the perpendicular magnetized samples (see Fig. 28a) cannot be explained by the magneto-crystalline anisotropy in $TbFe_2$. Given the relevant energy contributions comprising anisotropy, magnetoelastic coupling, and shape anisotropy, the coercive field for coherent domain rotation can be estimated to exceed 8 MA/m in the present case. This further underlines the importance of magnetization reversal through domain-wall movement. It is in contradistinction to the coherent domain rotation that was observed for $TbFe_2$ (111) thin films grown on sapphire with in-plane magnetization, as was shown in the previous section.

The observed saturation magnetization approaches 90% of the expected value of 800 kA/m reported for single crystals. The deficiency can be attributed to the formation of an interface alloy region with reduced magnetic moments, as was suggested by comparative thickness studies of Wang et al. for films grown on Mo (110) templates on sapphire [105]. These structural defects are likely to contribute to the formation of a complicated domain structure partly masking the intrinsic perpendicular anisotropy of the films.

3.2.4. Concluding Remarks

The properties of epitaxially grown $TbFe_2$ reveal an intimate relationship between the magnetism and the microstructure of the films. Further studies are needed to clarify possible mechanisms of in-plane clamping and plastic evolution of magnetostrictive strain fields during the magnetization process. Nevertheless, the results obtained so far suggest how to proceed to possible technological applications that are briefly addressed in the text. To proceed in this way a further reduction of the coercive field is essential. This seems feasible by the partial replacement of Tb with Dy, following the path that proved succesful for bulk Terfenol-D. Investigations of the performance of delay lines for surface acoustic waves based on these samples will then be necessary to judge their technological relevance.

On the other hand, the influence of the microstructure on magnetic behavior is very difficult to quantify in films of the discussed thickness range. A much more direct approach to the study of microstructural aspects is given by the investigation of ultrathin magnetic layers for which dipolar coupling energies are strongly reduced. In the following, the magnetization reversal process of ultrathin Co–Pt heterostructures is discussed based on magnetooptical investigations in polar configuration. These films were grown on substrates with a predefined step edge distribution. The relationship to intermetallic compounds is given by realizing that the perpendicular magnetic anisotropy of ultrathin Co–Pt heterostructures is in fact mainly due to the formation of an ordered alloy at the Co–Pt interface [118, 119]. This point is taken up again later with regard to the application of ordered alloys in magnetooptics.

3.3. Magnetization Reversal of Ultrathin Co–Pt Heterostructures

The magnetization reversal process in thin films occurs either by nucleation of reversed domains or by the displacement of magnetic domain walls [120]. In the latter case, inhomogeneities in the film act as pinning sites for the domain walls and, consequently, have a strong influence on the magnetic domain shape and the dynamics of the magnetization reversal. These inhomogeneities are formed by crystallographic defects or deviations of the film morphology from an ideal atomically flat surface. In general, the extent to which the films' microstructures as opposed to their intrinsic properties are affecting the domain structure is difficult to quantify. This is due to the strong interrelationship of these effects. A precise picture of the importance of the microstructure can be developed by introducing well-controlled defects that allow the subsequent identification of their effect on the domain structure. As a model system one can use an epitaxial thin magnetic film on a substrate with a predefined defect structure (e.g., a regular array of step edges). To visualize the domain structure by means of the magnetooptical Kerr effect, the use of Co-based ultrathin

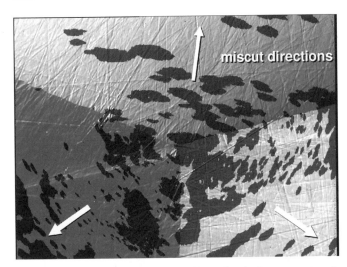

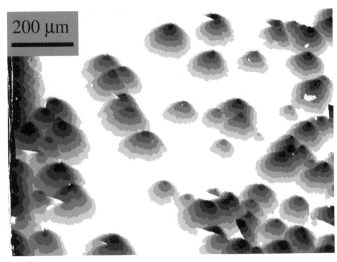

Fig. 29. Optical image of the Pt–Co–Pt film surface in phase-contrast mode and superimposed magnetic domain pattern (polarization mode). The magnetic domain pattern was taken during a magnetization reversal process. The miscuts are along the ⟨112⟩ directions. Reprinted with permission from [121], © 2000, American Physical Society.

Fig. 30. Polar Kerr image of the magnetic domain pattern on one miscut facet. The miscut direction is [11$\bar{2}$]. Magnetization reversal was in a field of −750 Oe. Five images are superimposed to demonstrate the time evolution of the domain growth. The time elapsed between two subsequent images is 0.25 s. Reprinted with permission from [121], © 2000, American Physical Society.

magnetic films with perpendicular anisotropy and large domain sizes is advantageous. This is due to the fact that the reduced stray field energy in ultrathin magnetic layers favors the formation of magnetic domains with large lateral dimensions. Because of its characteristic step-edge arrangement, vicinal MgO (111) as a substrate material is very suitable for studies of the influence of the substrate-induced microstructure on magnetic properties [121].

The Pt–Co–Pt trilayers used in this investigation consist of a 4-nm-thick Pt buffer followed by the magnetic Co layer and a Pt cap layer with a typical thickness of $t_{Co} = 0.3$ nm and $t_{Pt} = 1.8$ nm, respectively. Before growth, a high-pressure polishing procedure was applied, yielding a highly ordered substrate surface and giving rise to long-range ordered film growth with lateral crystalline coherence lengths above 450 nm [67]. The rms roughness of the individual layers, as determined by X-ray reflectivity measurements, was $\sigma_{Pt} = 0.05$ nm and $\sigma_{Co} = 0.2$ nm, respectively. To obtain a well-defined tilt of the substrate surface with respect to the MgO {111} planes, a precise tolerance of the substrate holder during the polishing procedure was introduced. The miscut values that can thereby be achieved vary between 0.5° and 1.5°. As a further consequence of the substrate treatment, the surface shows a pyramidal shape with three facets of equal tilt with respect to the (111) plane. These facets are oriented parallel to the ⟨11$\bar{2}$⟩ directions of the MgO, as shown in Figure 29.

The magnetization reversal of the samples was analyzed with the use of an optical polarization microscope in the polar configuration. The dark regions in Figure 29 represent the reversed magnetic domains generated in a field of −750 Oe, which was applied for 1 s. The domains reveal an anisotropic shape whose actual form depends on the miscut facet the domains are nucleating on. In contrast, for trilayers grown on MgO with negligible miscut, the domains reveal a circular

(isotropic) shape. This is in full accordance with previous results for Co–Pt alloy films [122, 123]. The origin of the systematic anisotropic domain shape is given by the microstructure of the vicinal substrate surface. The time evolution of the magnetization reversal in any single facet, as shown in Figure 30, further elucidates this mechanism. Starting from the nucleation sites, which appear to be hard nucleation centers, the domains expand only along the [$\bar{2}$11] and the [1$\bar{2}$1] directions, giving rise to a typical triangular domain shape throughout the whole facet. In contrast, samples with thicker Co layers on miscut MgO exhibit circular domain shapes. It is apparent that the substrate's surface microstructure cannot exert its influence on the domain wall evolution to arbitrary film thicknesses. This can be related to observations made for Co growth on Pt (111) [124]. For Co thicknesses larger than three monolayers, strain relaxation causes three-dimensional growth. Accordingly, anisotropic domain-wall movement was only observed for Co coverages below three monolayers, up to which Co grows layer by layer.

To formulate a microscopic description for the influence of the microstructure on the magnetic domain-wall movement, the surface miscut-induced formation of terraces separated by step edges has to be taken into account. The density of step edges is determined by the magnitude of the miscut. The step height can be mono-atomic or, in the case of step bunching, can comprise several unit cells of the substrate. The presence of terraces results in a surface potential that affects the diffusion of atoms deposited on the surface. As a result, the Ehrlich–Schwoebel barrier [60] leads to an increased reflexion of the diffusing atoms on downward step edges, resulting in a net flux toward the upward step edges. Thus, for coverages in the submonolayer regime and sufficient mobility, this potential gives rise to the accumulation of adatoms in the vicinity of the step edges and a depletion toward the middle of the terraces. In the present case, a magnetic Co monolayer is deposited on a single-crystalline Pt buffer, the surface of which ideally is a replica of the substrate's surface microstructure. Following the arguments given above, the concentration of Co atoms is significantly enhanced at the step edge positions, as is schematically shown in Figure 31. This modulation of the Co layer thickness results in a decrease of the effective anisotropy constant K_{eff} of the magnetic film in the vicinity of the steps. Because the spins at the center of a Bloch-like domain wall are aligned parallel to the film plane, the systems energy is minimized if the walls are lo-

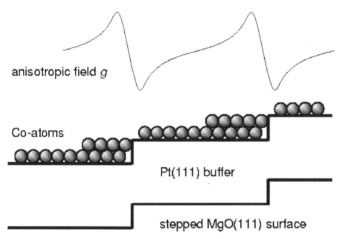

anisotropic field g

Co-atoms

Pt(111) buffer

stepped MgO(111) surface

Fig. 31. Schematic of step edge-induced thickness variation of Co layer. The upper curves depict the corresponding variation in the magnetic anisotropy energy. This results in a preferential pinning of the magnetic domain walls at the step edges and favors a domain-wall movement to the right. Reprinted with permission from [121], © 2000, American Physical Society.

cated at the step edges. Consequently, the thickness gradient of the Co layer results in an energy gradient for the domain-wall position across the terraces. This explains qualitatively the observed anisotropic movement of the domain walls antiparallel to the miscut direction. To explain the observed triangular shape of the magnetic domains, the lateral arrangement of the step edges on vicinal MgO (111) has to be considered. The step edges are aligned in a zigzag pattern along the $\langle \bar{2}11 \rangle$ directions. The evolution of this pattern is driven by surface energy minimization processes acting on the polar MgO (111) surface [71]. At the positions where two step edges running in different directions join, two different corner geometries will result. A schematic representation appears in Figure 32. The corner tip points in the miscut direction or in the opposite direction. Once a reversed magnetic domain is nucleated at a corner tip, it will move only along the growth terraces, because domain wall pinning is more effective at the step edges. A linearly extended domain shape results, causing a large increase in the effective domain wall circumference. To reduce the wall energy the domain wall will eventually be driven across the step edge, starting at the corner tip. For the corner configuration in Figure 32a this is supported by the Co thickness gradient, whereas it is hindered by the Co thickness gradient for the corner configuration shown in Figure 32b. As a result, the domain walls tend to move against the miscut direction, evolving into a triangular shaped form.

A more quantitative approach can be based on numerical simulations that describe the perpendicular spins of the ferromagnetic layer by a mesoscopic nonconserved order parameter $\phi(\mathbf{r}, t)$. Details can be found in [121].

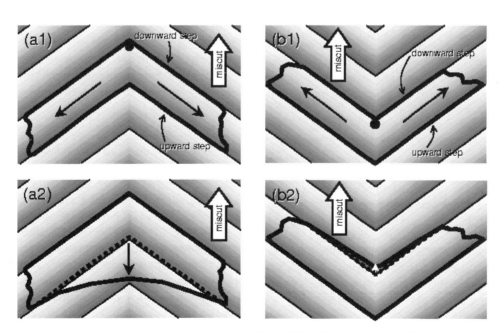

Fig. 32. Schematics of the step-edge arrangement on the MgO (111) surface in conjunction with the preferential magnetic domain wall movement (solid black lines). The MgO (111) step edges form a zigzag pattern, which results in an easier movement of the Pt–Co–Pt magnetic domain walls antiparallel to the miscut direction. Reprinted with permission from [121], © 2000, American Physical Society.

3.4. Intermetallic Compounds for Magnetooptics

Within the class of intermetallic compounds holding some promise for magnetooptical storage based on Curie point writing, MnBi can be considered one of the prime candidates. The goal of ever higher lateral storage density and the parallel development of blue light-emitting lasers based on GaN technology render systems with maximum polar Kerr rotation Θ_K in the blue light regime very attractive. Among these, MnBi has the largest Kerr response at room temperature. This represents the most important part in the figure of merit $\sqrt{R}\Theta_K$, where R is the reflectivity. Moreover, the large magnetocrystalline anisotropy of MnBi and the achievable uniaxial textured growth favor magnetooptical applications in the polar geometry [125]. The dominant obstacle that hinders the development of MnBi for rewritable optical recording is the fact that the first-order paramagnetic to ferromagnetic phase transition at 360°C is accompanied by a first-order lattice transition (hexagonal at low temperature to orthorhombic at high temperature [126]). This leads to significant media noise and results in nonreversible behavior close to the Curie point. Moreover, the coercive field increases with temperature [127]. As one possible solution for the decoupling of the structural and magnetic phase transitions, the substitution of Cr for Mn is currently under investigation [128]. Furthermore, it would be desirable to reduce the magnetic transition temperature below the eutectic temperature of about 270°C found at the Bi-rich side of the phase diagram. The apparent difficulty in completely eliminating the Bi inclusions in MnBi films would otherwise result in local melting effects during the Curie point writing process. As was shown by Bandaru et al., the substitution of Cr for Mn might serve both purposes [128]. The Curie temperature of $Mn_{0.9}Cr_{0.1}Bi$ films is reduced to about 255°C without any significant loss in the figure of merit. This reduction of T_c is thought to be caused by a weakening of the ferromagnetic exchange coupling due to the propensity for antiferromagnetic interactions between Cr ions. A strongly textured growth can be easily achieved on Bi-seeded glass and silicon nitride–silicon substrates. It remains to be shown that the magnetic phase transition at the reduced temperature is indeed decoupled from the first-order structural phase transition.

As an alternative to MnBi, Pt-Mn or more general Pt-T (T: V, Cr, Mn, Fe, Co) alloys attracted some interest. In alloys and compounds of Mn and Pt one can hope to combine the large d-band exchange splitting characteristic of Mn in various chemical environments and the strong spin-orbit interaction of Pt to form materials with outstanding magnetooptical attributes and sufficient chemical stability. Promising results have been obtained so far for the intermetallic compound $MnPt_3$. Here the extreme sensitivity of the polar Kerr effect on the preconditioning of the sample surface renders investigations of freshly prepared films particularly valuable [129].

3.5. Exchange Anisotropy with Metallic Antiferromagnets

Modern read heads in magnetic hard drives employ the giant magnetoresistance effect (GMR) in a spin-valve config-

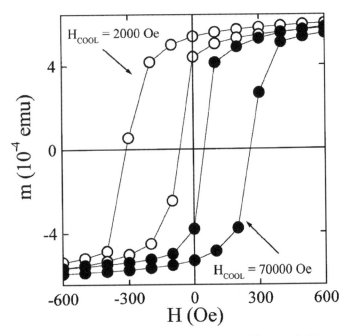

Fig. 33. Magnetic hysteresis loops for an FeF_2–Fe bilayer cooled in an applied field of 2000 Oe (open symbols) and 70,000 Oe (filled symbols). Reprinted with permission from [130], © 1999, Elsevier Science.

uration. GMR can be observed in magnetic heterostructures with antiparallel spin alignment of ferromagnetic layers weakly coupled over paramagnetic or antiferromagnetic spacer layers. In spin-valve devices the ferromagnetic base layer is strongly exchange-coupled to an underlying antiferromagnetic layer, whereas the ferromagnetic top layer can easily be switched from an antiparallel to a parallel spin alignment with regard to the ferromagnetic base layer by the application of a small magnetic field. The exchange coupling at the interface of the antiferromagnet (AFM) and ferromagnet (FM) leads to an undirectional anisotropy that is commonly refered to as the "exchange-bias effect" [130, 131]. In a simple AFM–FM sandwich the exchange-bias effect manifests itself as a unidirectional shift of the magnetic hysteresis loop along the applied field axis, as shown for a Fe–FeF$_2$ bilayer system in Figure 33.

Exchange-bias effects were first observed in magnetic cluster materials in which surface oxidation of the magnetic particles leads to the formation of an oxidic antiferromagnetic surface layer on a ferromagnetic core [132]. Nevertheless, for storage purposes virtually all applications are based on metallic AFMs [130, 131]. For several reasons this poses severe difficulties in the theoretical understanding of the origin of exchange-bias effects. Complex interface microstructures, as often observed in the growth of multicomponentmetallic heterostructures, render the discrimination of the various influences on the exchange-bias coupling very difficult. Interfacial stress and roughness, interdiffusion, and grain boundaries result in very ill-defined surface configurations. In conjunction with the magnetic character of the terminal layer of the AFM (degree of spin compensation), this leads to much less understanding of the basic phenomena than one would expect from the wealth

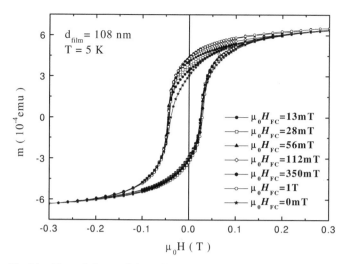

Fig. 34. Magnetic hysteresis loop shift of a TiFe$_2$ (110) thin film after cooling in different fields $\mu_0 H_{FC}$ as indicated.

of investigations performed so far. Within the class of metallic antiferromagnetic layers Mn alloys are dominant by far. First investigations on FeMn date back to the late 1970s [134]. Antiferromagnetic γ-Mn$_x$Fe$_{1-x}$ with $x = 0.3$–0.55 (fcc structure) are currently in use in most exchange-biased magnetoresistive read heads [130, 131]. In these disordered alloys the spin alignment is along the $\langle 111 \rangle$ directions. Other disordered and ordered alloys based on Mn with improved corrosion resistance are currently being investigated.

To ultimately identify the relevant influences that control the exchange-bias effects, the separation of microstructural and magnetic aspects at the AFM–FM interface are desirable. In this respect investigations on the hexagonal Laves phase compound TiFe$_2$ might help to resolve this issue. TiFe$_2$ has a wide homogeneity region ($-0.056 < x < 0.172$), showing ferromagnetic order for $x > 0$ and antiferromagnetic order for $x < 0$ [135, 136]. The magnetic ordering temperature varies with x, reaching a maximum Néel temperature of $T_N = 270$ K in the antiferromagnetic region and a maximum Curie temperature of $T_c = 385$ K in the ferromagnetic region. Recent investigations on (110)-oriented textured thin films with a composition close to $x = 0$ showed evidence for a FM–AFM phase separation process resulting in a large effective FM–AFM surface area [137]. Exchange-bias effects were observed in these films at low temperature, as presented in Figure 34. Proceeding now to the preparation of epitaxial FM–AFM heterostructures on sapphire substrates with varying step edge density, it might be possible to identify the individual control parameters for the exchange-bias effect in a systematic fashion. Neither interface strain nor chemical incompatibilities should play a role in this layer system.

3.6. Antiferromagnetic Order Parameter Nucleation on a Thin Film Surface

Information about phase transitions on a microscopic scale can be gained from scattering experiments. This is due to the fact

that the scattering cross section is proportional to a two-site correlation function [138]. In this and the following section two examples are given that might help to illustrate this fact.

For the interpretation of diffraction experiments on synchrotron light sources the probe coherence volume has to be taken into account. In particular, when the probe coherence volume approaches that of microscopically ordered regions in the sample, information about the location of the source of scattering below the sample surface can be obtained. In an experiment performed by Bernhoeft and collaborators [139] on the beamline ID20 at the European Synchrotron Radiation Facility, coherence effects in conventional high-resolution diffraction with partially coherent illumination were studied. The experiment utilized the fact that close to an absorption edge the absorption length can be made significantly smaller than the dimension of the coherently diffracting volume. In such circumstances the beam attenuation has to be taken into account at the level of the scattering amplitude instead of the scattered intensity.

A technique of resonant magnetic X-ray scattering was employed that relies on the enhancement of the magnetic scattering cross section as the photon energy is tuned to an absorption edge. To be more specific, (001)-oriented thin film samples of the heavy-fermion compound UPd$_2$Al$_3$ were analyzed. UPd$_2$Al$_3$ shows antiferromagnetic order below $T_N = 14$ K. The uranium moments are ferromagnetically aligned in the hexagonal basal plane, and they are stacked antiferromagnetically in the c axis direction. The beam line energy was tuned to the M$_4$ absorption edge of uranium at 3.73 keV. At the M$_4$ edge the absorption length is about 200 nm. Consequently, the change of the probe wave amplitude within the diffracting volume has to be taken into account because the longitudinal coherence length of the beam is generally larger than 1 μm. As a result, the line shape of an energy scan $I(E)$ about the absorption edge at the magnetic Bragg peak (0 0 1/2) depends on the number N of coherently scattering lattice planes. Basically, the observed broadening in the $I(E)$ line shape with increasing N can be understood to arise from a reduction in the effective scattering volume at the resonant energy. This was studied systematically by analyzing the $I(E)$ profile on a series of UPd$_2$Al$_3$ films of varying thicknesses ranging from 10 nm to 160 nm. In a quantitative analysis the necessity to include the absorption in the scattering on the amplitude level could be beautifully verified. This is shown in Figure 35.

Furthermore, a microscopic scenario for the nucleation of the antiferromagnetic order parameter in the films could be derived from an analysis of the temperature dependence of the magnetic Bragg peaks [140]. Transverse θ-scans and longitudinal L-scans were performed at the magnetic Bragg peaks (0 0 1/2) and (0 0 3/2). From longitudinal scans the magnetic correlation length $c/\Delta L$ in the c axis direction could be deduced from the observed peak width ΔL. At sufficiently low temperatures the longitudinal magnetic correlation length was shown to be identical to the structural correlation length as deduced from the respective scan at the (001) reflection. However, transverse scans revealed an enhanced peak width $\Delta \theta$ as compared with the structural reflection at all temperatures be-

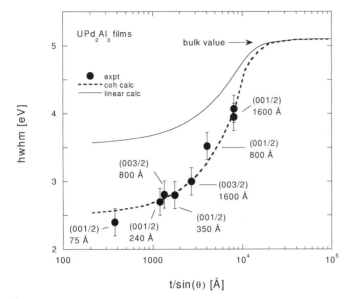

Fig. 35. HWHM of energy profiles vs the optical path of the antiferromagnetic peaks of various UPd_2Al_3 thin films taken at $T = 4$ K (t: film thickness, θ: Bragg angle for a reflection). The scans were performed at the uranium M_4 absorption edge. The dashed line is the numerical simulation in the coherent approximation, and the solid line corresponds to a classical (incoherent) summation, as discussed in detail in [139]. Image courtesy of N. Bernhoeft.

low T_N. This deduced lateral magnetic correlation length was shown to be considerably smaller then the structural correlation length of about 1 μm. Taking the width $\Delta\theta_{ch}$ of the (001) reflection as a measure of the mosaic spread of the films and the diffractometer resolution, a simple quadratic deconvolution was used to obtain the lateral magnetic correlation length $2c/\Delta\theta$. At low temperatures $2c/\Delta\theta$ corresponded to the film thickness. With increasing temperature the energy width ΔE, $2c/\Delta\theta$, and $c/\Delta L$ increased, whereas the ratio $\Delta L/\Delta\theta$ remained constant. This behavior is consistent with the assumption that the magnetic order initially develops on the film surface and penetrates into the bulk as the temperature is lowered.

In the same framework we consider next order–disorder phenomena on surfaces of thin films of ordered alloys.

3.7. Order–Disorder Phenomena

Surfaces modify the symmetry and dimensionality of order–disorder phenomena and can significantly alter phase transitions and their associated critical behavior [141, 142]. In a discussion of critical behavior, the appearance of novel universality classes can be parameterized by an additional scaling field that quantifies the extent to which the interactions at the surface are changed with regard to their value in the bulk. If this interaction is reduced, order at the surface sets in at a temperature below the bulk critical temperature $T_c^{(b)}$. In contrast, for enhanced interactions an ordered surface layer will be established above $T_c^{(b)}$, which is then floating on the disordered bulk. This surface layer, being a natural two-dimensional system, will show a 2D phase transition whose critical exponents will, in general, be very different from the 3D transition in the bulk.

Without going into too much detail, two examples are given below that highlight order–disorder phenomena on surfaces of ordered alloys grown *in situ* by an MBE process. Thin film studies of these phenomena offer the advantage of freshly prepared crystal surfaces available before to the investigation. Furthermore, the combination of surface sensitive diffraction techniques can be applied to study the order–disorder transformation in great detail. As a first example, resonant RHEED studies of the surface order of Cu_3Au (111) films performed by Bonham and Flynn are briefly reviewed [143]. Second, a recent investigation of critical phenomena at FeCo (100) surfaces by Krimmel et al. is taken up, which employed an advanced X-ray scattering technique [145, 146].

3.7.1. Surface Order of Cu_3Au (111) Films

The order–disorder transformation in bulk Cu_3Au takes place at about $T_c^{(b)} = 395°C$, above which the crystal exhibits an fcc order, with the Cu and Au atoms occupying the sites at random. Below $T_c^{(b)}$ the Au atoms segregate to one of the four possible fcc sites. Consequently, the formation of a single-domain ordered crystal must proceed through a two-step process. The fast nucleation of four equivalent ordered domains is followed by a slower ripening process in which one domain grows at the expense of others [143]. The order parameter, ranging from zero to unity as the order increases, can be specified as follows [147]:

$$\mathcal{O} = \frac{p - p_d}{p_0 - p_d} \tag{28}$$

in which p signifies the actual propability of finding the Au atom at the correct site. p_d is the occupation probability in the disordered state, and p_0 is the probability in the ordered state. To be able to measure the order with surface sensitive probes, the selected surface orientation should not exhibit strong segregation of one constituent element. In this respect, Cu_3Au (111) is ideally suited, because the unperturbed (111) planes have the same stoichiometry as the bulk crystal. Inherent composition changes during the phase transformation are therefore not to be expected.

RHEED investigations of order phenomena are advantageous because of their extreme surface sensitivity with regard to the average periodicity of the surface net. The average penetration depth at about 10 keV under grazing incidence, as was discussed earlier, is only a few atomic layers. Nevertheless, highly complex diffraction patterns comprising contributions from the main Laue reflections, surface resonance features (Kikuchi lines), diffraction spots, and diffuse background scattering render any quantitative analysis very difficult. Furthermore, the interaction of electrons with the crystal surface are strong and nonlinear. In some instances, as is the case in the work reviewed here, a quasi-kinematical approach is sufficient to model and interpret the results. In the quasi-kinematical approach the kinematical description of RHEED is augmented by a temperature-independent correction that takes dynamical phenomena, such as surface resonance effects, into

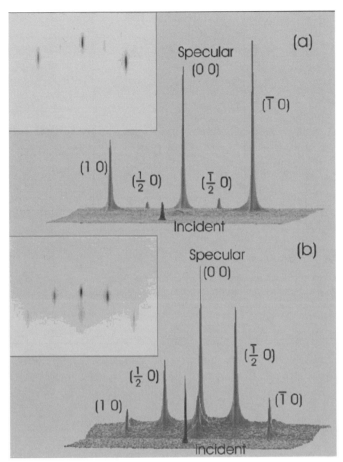

Fig. 36. RHEED diffraction patterns taken under surface resonance conditions, as a gray scale image and surface plot. The beam is aligned along the [1$\bar{1}$0] axis. (a) ±2 resonance. (b) ±3/2 resonance. Reprinted with permission from [143], © 1998, American Physical Society.

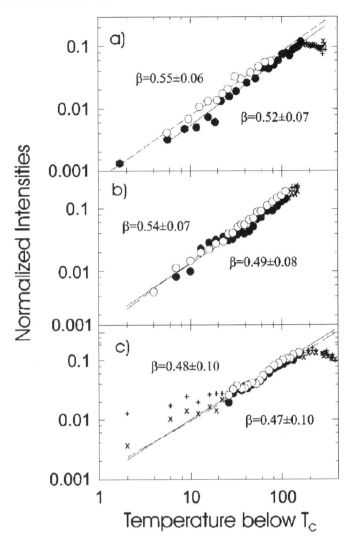

Fig. 37. Normalized RHEED superstructure peaks as a function of temperature for three different data sets. The intensities are proportional to the square of the surface order parameter. The empty and filled circles are taken from the (1/2 0) and ($\overline{1/2}$ 0) diffraction spots, respectively. The value of the critical exponent β is determined from the fits. + and × represent the data from the diffraction spots not used in the fit. The surface order reaches saturation at about 150 K below the transition temperature. Reprinted with permission from [143], © 1998, American Physical Society.

account. For details we refer to the orginal work by Bonham and Flynn [143].

The samples used in this research were grown by MBE onto epitaxial-grade Al_2O_3 (11$\bar{2}$0) with Nb (110) buffer layers of 50 nm thickness that act as a template for the nucleation of Cu_3Au (111). The RHEED investigations were performed with a Perkin-Elmer 10-keV electron gun and a computerized data acquisition system. The electron beam was aligned parallel to the [1$\bar{1}$0] direction. Independently of the order parameter, the structural reflections for the (111) surface are (00), (10), and ($\bar{1}$0). If the surface is chemically ordered, the most prominent superstructure reflexions are (1/2 0) and ($\overline{1/2}$ 0), which were used for the order–disorder analysis. The intensities of these superstructure reflections are proportional to the square of the surface order parameter $\mathcal{O}_s$. A typical diffraction pattern appears in Figure 36.

As the central result of this work, the critical exponent for the temperature dependence of the order parameter was deduced from the temperature-dependent evolution of the superstructure scattering intensity at and below the ordering temperature T_c. For a second-order transition (i.e., a continuous evolution of the order parameter), $\mathcal{O}_s(T)$ is predicted to follow a power law behavior below the critical temperature:

$$\mathcal{O}_s(T) \propto \left(\frac{T_c - T}{T_c} \right)^{\beta} \qquad (29)$$

The bulk order–disorder transformation of Cu_3Au has to be of first order for symmtry reasons [148]. However, the crystal symmetry is broken at the surface, which can result in a different order of the phase transition. Indeed, the research reviewed here unambiguously finds a critical exponent of 0.51 ± 0.03, which is in excellent correspondence with the mean field value 0.5 (see Fig. 37).

3.7.2. Critical Phenomena at FeCo (100) Surfaces

In the work by Krimmel et al. [144, 145], to be discussed next, X-ray scattering using synchrotron radiation was employed to

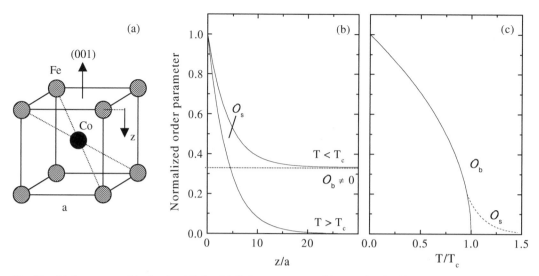

Fig. 38. (a) Structure model of ordered FeCo. (b) Order parameter profiles associated with surface-enhanced order for $T > T_c$ and $T < T_c$ (surface order parameter $\mathcal{O}_s$, bulk order parameter $\mathcal{O}_b$, lattice constant a). (c) Schematics of the temperature dependence of the bulk and surface order in the presence of a surface field h_s. See text for details.

study the continuous B2-A2 order–disorder transition of the FeCo (100) surface. In this work the authors could clarify two aspects specific to this surface phase transition. First, the surface layer undergoes a order–disorder transition at a temperature $T_c > T_c^{(b)}$. Second, the critical behavior of this surface layer is not caused by an enhanced surface coupling but is induced by surface segregation.

The binary alloy FeCo undergoes a continuous order–disorder transition from the B2 (CsCl) structure below $T_c^{(b)} \simeq$ 920 K to the A2 (bcc) structure above $T_c^{(b)}$ in a wide composition range [149]. In the B2 structure the normal direction of FeCo (100) consists of alternating A- or B-type layers. Because Fe-Co nearest-neighbor configurations are favored by the internal interactions, the surface segregation of either Fe or Co will induce an alternating layering into the bulk, starting with a second layer of the other type. This corresponds to a non-zero order parameter. According to Monte Carlo simulations [150] and field-theoretical studies [151], an alternative scenario, besides enhanced surface interactions, can induce the surface order transformation before bulk ordering sets in. This scenario is based on a surface field h_s induced by the segregation of one atomic species, which stabilizes an order parameter profile as indicated in Figure 38. As a result, the temperature dependence of the order parameter should not follow the behavior indicated in Eq. (29), but should show a leading temperature dependence of the form

$$\mathcal{O}_s(T) = \sqrt{1 - \left(\frac{T_c - T}{T_c}\right)^\gamma} \qquad (30)$$

with $\gamma = 0.33$.

Experimentally, the onset of surface order can be observed in X-ray scattering by the appearance of a broad feature related to the associated asymptotic (00ℓ) Bragg rod at the position of the superlattice reflection. This is indicated schematically in

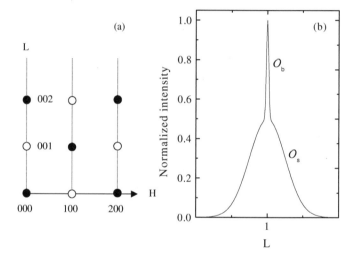

Fig. 39. (a) Reciprocal lattice map of FeCo with bcc points (full symbols) and superlattice points (empty symbols). The dashed vertical lines indicate the asymptotic Bragg scattering from the free (001) surface. (b) Schematic of asymptotic Bragg scattering intensity along the (001) reflection for surface-enhanced order.

Figure 39. Performing a detailed analysis of the Bragg intensity as a function of temperature, Krimmel et al. could show that even though the order is driven by a segregation-induced surface field, the order parameter in a 15-nm thin surface sheet still belongs to the ordinary universality class. It follows the temperature dependence given in Eq. (29) with $\beta = 0.79 \pm 0.10$. The crossover to extraordinary behavior, according to Eq. (30), may only occur very close to T_c. A selection of asymptotic Bragg profiles as function of temperature appears in Figure 40.

The short review of investigations concering order–disorder phenomena presented here represents only a small fraction of the research going on in this field. The interested reader is referred to recent reviews on this topic [141].

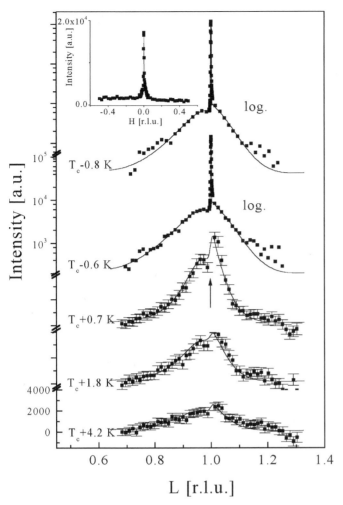

Fig. 40. Selected asymptotic Bragg profiles along (001) as observed at various temperatures below and above T_c. The two upper curves for $T < T_c$ are shown on a logarithmic scale. The full lines are theoretical curves. The inset shows a typical H-scan profile taken at $L = 0.98$ for $T = T_c + 0.7$ K (arrow). Image courtesy of W. Donner [145].

4. OUTLOOK

It should be apparrent from the foregoing sections that thin films of intermetallic compounds offer the possibility to tackle problems in basic and applied research in new and sometimes unique ways. Because of the consequences of the nonequilibrium phase formation process and various epitaxial constraints, the properties of such grown layers need not be in full correspondence with what is known from the respective bulk single crystals. This concerns purely crystallographic aspects, the defect structure, strain, and morphological variations, as well as the electronic and magnetic properties of the films. On the one hand, this can be an annoyance if the investigator wishes to study specific aspects known to be present in bulk crystals. An example is the strong strain dependence of the electronic properties of heavy-fermion materials, which have to be taken into account if thin films are to be employed. Very recent electron spectroscopy investigations $CeNi_2Ge_2$ and $CePd_2Si_2$ grown *in situ* show the wealth of electronic structural informa-

tion that can be gained with the use of thin films [152]. Yet, the interpretation of the results suffered from the lack of corresponding structure and strain information. On the other hand, in particular, the constraints imposed by the thin film form are exactly what are needed and can be used to tailor specific material aspects, as was shown for magnetostrictive layers. Moreover, patterning or the preparation of heterostructures is only feasible in thin film form. In any event, materials science has to go hand in hand with any research concerning the electronic and magnetic structure of the films under investigation. In this respect detailed studies of the microstructure evolution down to the atomic scale during the growth of alloy and compound films can give valuable insight into the early stages of nucleation and strain relaxation. This has to be augmented by theoretical studies of the growth of alloys and compounds in the form of thin films. Ultimately, it would be highly desirable to be able to judge the compatibility of certain substrate/template-epilayer combinations before film growth. Accompanying the current rise of designer materials, the understanding of the nature of film growth from first principles presents an important challenge for basic and applied research. Such theoretical studies are still in their infancy. The current main stream of research seems to go along the lines of kinetic Monte Carlo studies of alloy formation, mostly in binary form [153]. These investigations, combined with first-princicple calculations of the diffusion coefficients deduced from all of the relevant atomic processes, such as terrace diffusion, corner crossing, and king breaking, might eventually lead to a more detailed understanding of surface morphology evolution and the formation of ordered alloys and compounds. In this respect recent density–functional calculations of the self-diffusion of Al on Al (111) have to be mentioned [154].

The scope of issues that are accessible to thin film investigation on intermetallic compounds and ordered alloys is wide. The availability of topics of broad scientific interest is not the limiting factor in this field. Rather, the development and adaptation of adequate real-time rate monitoring and control for compound growth and an improved understanding of phase nucleation and orientation selection phenomena are needed. The elucidation of mechanisms for selecting the stacking variants in (111) growth of fcc metals is but one example of the research still to be done.

Acknowledgments

I cannot conclude without thanking all of the colleagues and students I had the pleasure to work with over the last few years. Specifically, I thank Hermann Adrian for his continuous support and C. P. Flynn, from whose knowledge I benefited during my time at the Physics Department and Materials Research Laboratory at the University of Illinois at Urbana-Champaign. I am also especially grateful to Martin Jourdan, Patrick Haibach, and Holger Meffert. Part of their work is presented in this chapter.

REFERENCES

1. M. A. Herman and H. Sitter, "Molecular Beam Epitaxy," 2nd ed. Springer-Verlag, Berlin, 1996.
2. Z. Mitura and P. A. Maksym, *Phys. Rev. Lett.* 70, 2904 (1993).
3. W. Braun, "Applied RHEED," Springer Tracts in Modern Physics, Vol. 154. Springer-Verlag, Berlin, 1999.
4. P. J. Dobson, "Surface and Interface Characterization by Electron Optical Methods" (A. Howie and U. Valdrè, Eds.), NATO ASI Serie B 191, 1987.
5. C. J. Humphreys, *Rep. Prog. Phys.* 42, 1825 (1979).
6. M. P. Seah and W. A. Dench, *Surf. Interface Anal.* 1, 2 (1979).
7. L. C. Feldman and J. W. Mayer, "Fundamentals of Surface and Thin Film Analysis." Appleton & Lange, New York, 1986.
8. Rutherford Backscattering Data Analysis, Plotting and Simulation Package. Computer Graphics Service, Ithaca, NY.
9. D. Lederman, D. C. Vier, D. Mendoza, J. Santamaria, S. Schultz, and I. K. Schuller, *Appl. Phys. Lett.* 66, 3677 (1995).
10. N. Grewe and F. Steglich, "Handbook on the Physics and Chemistry of Rare Earths," Vol. 14 (K. A. Gschneidner, Jr., and L. Eyring, Eds.), p. 343. North-Holland, Amsterdam, 1991.
11. G. Bruls, B. Wolf, D. Finsterbusch, P. Thalmeier, I. Kouroudis, W. Sun, W. Assmus, B. Lüthi, M. Lang, K. Gloos, F. Steglich, and R. Modler, *Phys. Rev. Lett.* 72, 1754 (1994).
12. K. Ishida, Y. Kawasaki, K. Tabuchi, K. Kashima, Y. Kitaoka, K. Asayama, C. Geibel, and F. Steglich, *Phys. Rev. Lett.* 82, 5353 (1999).
13. M. Huth and C. P. Flynn, *Phys. Rev. B: Condens. Matter* 58, 11526 (1998).
14. A. Masson, J. J. Metois, and R. Kern, "Advances in Epitaxy and Endotaxy" (H. G. Schneider and V. Ruth, Eds.), p. 103. VEB Deutscher Verlag für Grundstoffindustrie, Leipzig, 1971.
15. J. C. A. Huang, R. R. Du, and C. P. Flynn, *Phys. Rev. Lett.* 66, 341 (1991).
16. V. Oderno, C. Dufour, K. Dumesnil, Ph. Mangin, and G. Marchal, *J. Cryst. Growth* 165, 175 (1996).
17. M. Huth and C. P. Flynn, *J. Appl. Phys.* 83, 7261 (1998).
18. M. Jourdan, Preparation and Thermoelectric Power of Thin Films of the Heavy-Fermion Superconductors UPd$_2$Al$_3$ and UNi$_2$Al$_3$, Diploma Thesis, TH Darmstadt, 1995.
19. B. W. Sloope and C. O. Tiller, *J. Appl. Phys.* 36, 3174 (1963).
20. J. E. Cunningham, J. A. Dura, and C. P. Flynn, "Metallic Multilayers and Epitaxy" (M. Hong, S. Wolfe, and D. Gubser, Eds.), p. 75. Publics Metall Soc. Inc., 1988.
21. C. P. Flynn, I. Borchers, R. T. Demers, R. Du, J. A. Dura, M. V. Klein, S. H. Kong, M. B. Salamon, F. Tsui, S. Yadavaki, X. Zhu, H. Zabel, J. E. Cunningham, R. W. Erwin, J. J. Rhyne, "MRS International Meeting on Advanced Materials," 1989, Vol. 10, p. 275.
22. C. P. Flynn, "Point Defects and Diffusion." Clarendon, Oxford, 1972.
23. M. Getzlaff, R. Pascal, H. Tödter, M. Bode, and R. Wiesendanger, *Appl. Surf. Sci.* 142, 543 (1999).
24. R. Pascal, M. Getzlaff, H. Tödter, M. Bode, and R. Wiesendanger, *Phys. Rev. B: Condens. Matter* 60, 16109 (1999).
25. J. Kolaczkiewicz and E. Bauer, *Surf. Sci.* 175, 487 (1986).
26. R. Pascal, Ch. Zarnitz, M. Bode, and R. Wiesendanger, *Phys. Rev. B: Condens. Matter* 56, 3636 (1997).
27. E. D. Tober, R. X. Ynzunza, C. Westphal, and C. S. Fadley, *Phys. Rev. B: Condens. Matter* 53, 5444 (1996).
28. U. Rudiger, J. Kohler, A. D. Kent, T. Legero, J. Kubler, P. Fumagalli, and G. Güntherodt, *J. Magn. Magn. Mater.* 199, 131 (1999).
29. F. Salghetti-Drioli, K. Mattenberger, P. Wachter, and L. Degiorgi, *Solid State Commun.* 109, 687 (1999).
30. F. Salghetti-Drioli, P. Wachter, and L. Degiorgi, *Solid State Commun.* 109, 773 (1999).
31. J. Rossat-Mignod, P. Burlet, S. Quezel, J. M. Effantin, D. Delacôte, H. Bartholin, O. Vogt, and D. Ravot, *J. Magn. Magn. Mater.* 31–34, 398 (1983).
32. T. Chattopadhyay, P. Burlet, J. Rossat-Mignot, H. Bartholin, C. Vettier, and O. Vogt, *Phys. Rev. B: Condens. Matter* 49, 15096 (1994).
33. H. Meffert, J. Oster, P. Haibach, M. Huth, and H. Adrian, *Physica B* 259–261, 298 (1999).
34. H. Meffert, M. Huth, J. Oster, P. Haibach, and H. Adrian, *Physica B* 281–282, 447 (2000).
35. Landolt-Börnstein, *II/2a*, 6th ed. Springer-Verlag, Berlin/Göttingen/Heidelberg, 1960.
36. C. T. Foxon, in "Principles of Molecular Beam Epitaxy," Handbook of Crystal Growth 3a (D. T. J. Hurle, Ed.), p. 155. North-Holland, Amsterdam, 1994.
37. M. T. Kief and W. Egelhoff, *Phys. Rev. B: Condens. Matter* 47, 10785 (1993).
38. F. Family, *Physica A* 266, 173 (1999).
39. E. D. Tober, R. F. C. Farrow, R. F. Marks, G. Witte, K. Kalki, and D. D. Chambliss, *Phys. Rev. Lett.* 81, 1897 (1998).
40. C. P. Flynn, *J. Phys. Chem. Solids* 55, 1059 (1994).
41. M. Huth and C. P. Flynn, *Appl. Phys. Lett.* 71, 2466 (1997).
42. P. I. Cohen, G. S. Petrich, P. R. Pukite, G. J. Whaley, and A. S. Arrott, *Surf. Sci.* 216, 222 (1988).
43. P. J. Dobson, B. A. Joyce, J. H. Neave, and J. Zhang, *J. Cryst. Growth* 81, 1 (1987).
44. J. W. Matthews and A. E. Blakeslee, *J. Cryst. Growth* 27, 118 (1974).
45. J. W. Matthews, *J. Vac. Sci. Technol.* 12, 126 (1975).
46. J. W. Matthews and A. E. Blakeslee, *J. Cryst. Growth* 32, 265 (1976).
47. P. M. Reimer, H. Zabel, C. P. Flynn, and J. A. Dura, *Phys. Rev. B: Condens. Matter* 45, 11426 (1992).
48. A. Gibaud, R. A. Cowley, D. F. McMorrow, R. C. C. Ward, and M. R. Wells, *Phys. Rev. B: Condens. Matter* 48, 14463 (1993).
49. A. Stierle, A. Abromeit, N. Metoki, and H. Zabel, *J. Appl. Phys.* 73, 4808 (1993).
50. P. F. Miceli, C. J. Palmstrøm, and K. W. Moyers, *Appl. Phys. Lett.* 58, 1602 (1991).
51. P. F. Miceli and C. J. Palmstrøm, *Phys. Rev. B: Condens. Matter* 51, 5506 (1995).
52. P. F. Miceli, J. Weatherwax, T. Krentsel, and C. J. Palmstrøm, *Physica B* 221, 230 (1996).
53. V. Srikant, J. S. Speck, and D. R. Clarke, *J. Appl. Phys.* 82, 4286 (1997).
54. G. L. Zhou and C. P. Flynn, *Phys. Rev. B: Condens. Matter* 59, 7860 (1999).
55. G. L. Zhou and C. P. Flynn, *Appl. Phys. Lett.* 72, 34 (1998).
56. S. M. Durbin, J. E. Cunningham, M. E. Mochel, and C. P. Flynn, *J. Phys. F* 11, L223 (1981).
57. J. Kwo, E. M. Gyorgy, D. B. McWhan, M. Hong, F. J. DiSalvo, C. Vettier, and J. E. Bower, *Phys. Rev. Lett.* 55, 1402 (1985).
58. K. A. Ritley and C. P. Flynn, *Appl. Phys. Lett.* 72, 170 (1998).
59. M. Yoshimoto, T. Maeda, T. Ohnishi, H. Koinuma, O. Ishiyama, M. Shinohara, M. Kubo, R. Miura, and A. Miyamoto, *Appl. Phys. Lett.* 67, 2615 (1995).
60. G. Ehrlich and F. G. Hudda, *J. Chem. Phys.* 44, 1039 (1966).
61. R. L. Schwoebel, *J. Appl. Phys.* 40, 614 (1969).
62. C. P. Flynn and W. Święch, *Phys. Rev. Lett.* 83, 3482 (1999).
63. S. W. Bonham and C. P. Flynn, *Phys. Rev. B: Condens. Matter* 58, 10875 (1998).
64. G. L. Zhou, S. W. Bonham, and C. P. Flynn, *J. Phys.: Cond. Matter* 9, 671 (1997).
65. R. J. Fisher, N. Chand, W. F. Kopp, H. Morkoç, L. P. Erikson, and P. Youngman, *Appl. Phys. Lett.* 47, 397 (1985).
66. H. Kroemer, *J. Cryst. Growth* 81, 193 (1987).
67. P. Haibach, J. Köble, M. Huth, and H. Adrian, *Thin Solid Films* 336, 168 (1998).
68. P. Haibach, J. Köble, M. Huth, and H. Adrian, *J. Magn. Magn. Mater.* 198–199, 752 (1999).
69. P. W. Tasker, *J. Phys. C: Solid State Phys.* 112, 4977 (1979).
70. D. Wolf, *Phys. Rev. Lett.* 68, 3315 (1992).
71. A. Pojani, F. Finocchi, J. Goniakowski, and C. Noguera, *Surf. Sci.* 387, 354 (1997).
72. R. H. Heffner and M. R. Norman, *Comm. Cond. Matter Phys.* 17, 361 (1996).

73. N. D. Mathur, F. M. Grosche, S. R. Julian, I. R. Walker, D. M. Freye, R. K. W. Haselwimmer, and G. G. Lonzarich, *Nature (London)* 394, 39 (1998).

74. S. C. Zhang, *Science* 275, 1089 (1997).

75. W. Hanke, R. Eder, and E. Arrigoni, *Phys. Bl.* 54, 436 (1998).

76. H. v. Löhneysen, *Physica B* 218, 148 (1996).

77. K. Gloos, C. Geibel, R. Mueller-Reisener, and C. Schank, *Physica B* 218, 169 (1996).

78. N. Metoki, Y. Haga, Y. Koike, and Y. Onuki, *Phys. Rev. Lett.* 80, 5417 (1998).

79. N. Bernhoeft, B. Roessli, N. Sato, N. Aso, A. Hiess, G. H. Lander, Y. Endoh, and T. Komatsubara, *Phys. Rev. Lett.* 81, 4244 (1998).

80. M. Jourdan, M. Huth, and H. Adrian, *Nature (London)* 398, 47 (1999).

81. C. Geibel, C. Schank, S. Thies, H. Kitazawa, C. D. Bredl, A. Böhm, M. Rau, A. Grauel, R. Caspary, R. Helfrich, U. Ahlheim, G. Weber, and F. Steglich, *Z. Phys. B: Condens. Matter* 84, 1 (1991).

82. A. Krimmel, P. Fischer, B. Roessli, H. Maletta, C. Geibel, C. Schank, A. Grauel, A. Loidl, and F. Steglich, *Z. Phys. B: Condens. Matter* 86, 161 (1992).

83. H. Ikeda and Y. Ohashi, *Phys. Rev. Lett.* 81, 3723 (1998).

84. T. Petersen, T. E. Mason, G. Aeppli, A. P. Ramirez, E. Bucher, and R. N. Kleinman, *Physica B* 199–200, 151 (1994).

85. N. Sato, N. Aso, G. H. Lander, B. Roessli, T. Komatsubara, and Y. Endoh, *J. Phys. Soc. Jpn.* 66, 1884 (1997).

86. N. Metoki, Y. Haga, Y. Koike, N. Aso, and Y. Onuki, *J. Phys. Soc. Jpn.* 66, 2560 (1997).

87. N. Bernhoeft, B. Roessli, N. Sato, N. Aso, A. Hiess, G. H. Lander, Y. Endoh, and T. Komatsubara, in "Itinerant Electron Magnetism: Fluctuation Effects" (D. Wagner, W. Brauneck, and A. Solontsov, Eds.). Kluwer, Dordrecht, 1998.

88. N. Bernhoeft, B. Roessli, N. Sato, N. Aso, A. Hiess, G. H. Lander, Y. Endoh, and T. Komatsubara, *Physica B* 259–261, 614 (1999).

89. N. Bernhoeft, *Eur. Phys. J. B* 13, 685 (2000).

90. M. Jourdan, M. Huth, and H. Adrian, *Physica B* 259–261, 621 (1999).

91. M. Huth and M. Jourdan, in "Advances in Solid State Physics," Vol. 39, p. 351. Vieweg, Braunschweig/Wiesbaden, Germany, 1999.

92. M. Huth, M. Jourdan, and H. Adrian, *Eur. Phys. J. B* 13, 695 (2000).

93. M. Huth, M. Jourdan, and H. Adrian, *Physica B* 281–282, 882 (2000).

94. M. Huth, A. Kaldowski, J. Hessert, Th. Steinborn, and H. Adrian, *Solid State Commun.* 87, 1133 (1993).

95. M. Huth, Transport Phenomena and Coherence in Epitaxially-Grown Heavy-Fermion Thin Films, Ph.D. Thesis, TH Darmstadt, 1995.

96. M. Jourdan, Tunneling Spectroscopy of the Heavy-Fermion Superconductor UPd_2Al_3, Ph.D. Thesis, University of Mainz, 1999.

97. R. C. Dynes, V. Narayanamurti, and J. P. Garno, *Phys. Rev. Lett.* 41, 1509 (1978).

98. K. Gloos, R. Modler, H. Schimanski, C. D. Bredl, C. Geibel, F. Steglich, A. I. Buzdin, N. Sato, and T. Komatsubara, *Phys. Rev. Lett.* 70, 501 (1993).

99. J. Hessert, M. Huth, M. Jourdan, H. Adrian, C. T. Rieck, and K. Scharnberg, *Physica B* 230–232, 373 (1997).

100. G. Varelogiannis, *Z. Phys. B: Condens. Matter* 104, 411 (1997).

101. D. C. Jiles, "New Materials and their Applications" (D. Holland, Ed.). IOP, Bristol, 1990.

102. V. Koeninger, Y. Matsumara, H. H. Uchida, and H. Uchida, *J. Alloys Compounds* 211–212, 581 (1994).

103. A. E. Clark, "Handbook of the Physics and Chemistry of Rare Earth" (K. A. Gschneidner and L. Eyring, Eds.), Vol. 2, reprint. North-Holland, Amsterdam, 1982.

104. F. Tsui and C. P. Flynn, *Phys. Rev. Lett.* 71, 1462 (1993).

105. C. T. Wang, B. M. Clemens, and R. L. White, *IEEE Trans. Magn.* 32, 4752 (1996).

106. E. C. Stoner and E. P. Wohlfarth, *Philos. Trans. R. Soc. London, Ser. A* 240, 599 (1948).

107. E. W. Lee and J. E. L. Bishop, *Proc. Phys. Soc., London* 89, 661 (1966).

108. Etienne du Trémolet de Lacheisserie, "Magnetostriction." CRC Press, Boca Raton/Ann Arbor/London/Tokyo, 1993.

109. A. E. Clark, in "Ferromagnetic Materials" (E. P. Wohlfarth, Ed.), Vol. 1. North Holland, Amsterdam/New York/Oxford, 1980.

110. R. S. Beach, A. Matheny, M. B. Salamon, C. P. Flynn, J. A. Borchers, R. W. Erwin, and J. J. Rhyne, *J. Appl. Phys.* 73, 6901 (1993).

111. D. C. Jiles and J. B. Thoelke, *J. Magn. Magn. Mater.* 134, 143 (1994).

112. H. Uchida, M. Wada, K. Koike, H. H. Uchida, V. Koeninger, Y. Matsumura, H. Kaneko, and T. Kurino, *J. Alloys Comp.* 211–212, 576 (1994).

113. V. Koeninger, Y. Matsumura, H. H. Uchida, and H. Uchida, *J. Alloys Comp.* 211–212, 581 (1994).

114. F. Schatz, M. Hirscher, M. Schnell, G. Flik, and H. Kronmüller, *J. Appl. Phys.* 76, 5380 (1994).

115. S. F. Fischer, M. Kelsch, and H. Kronmüller, *J. Magn. Magn. Mater.* 195, 545 (1999).

116. H. Yamamoto, S. Matsumoto, M. Naoe, and S. Yamanaka, *Trans. Inst. Electrical Eng. Jpn.* 98, 31 (1978).

117. M. Huth and C. P. Flynn, *J. Magn. Magn. Mater.* 204, 203 (1999).

118. W. B. Zeper, F. J. A. M. Greidanus, P. F. Garcia, and C. R. Fincher, *J. Appl. Phys.* 65, 4971 (1989).

119. L. Uba, S. Uba, A. N. Yaresko, A. Ya. Perlow, V. N. Antonov, and R. Gontarz, *J. Magn. Magn. Mater.* 193, 159 (1999).

120. J. Pommier, P. Meyer, G. Pénissard, J. Ferré, P. Bruno, and D. Renard, *Phys. Rev. Lett.* 65, 2054 (1990).

121. P. Haibach, M. Huth, and H. Adrian, *Phys. Rev. Lett.* 84, 1312 (2000).

122. U. Nowak, J. Heimel, T. Kleinefeld, and D. Weller, *Phys. Rev. B: Condens. Matter* 56, 8143 (1997).

123. M. Jost, J. Heimel, and T. Kleinefeld, *Phys. Rev. B: Condens. Matter* 57, 5316 (1998).

124. P. Grütter and U. T. Dürig, *Phys. Rev. B: Condens. Matter* 49, 2021 (1994).

125. R. S. Tebble and D. J. Craik, "Magnetic Materials." Wiley-Interscience, London, 1969.

126. A. F. Andresen, W. Halg, P. Fischer, and E. Stoll, *Acta Chem. Scand.* 21, 1543 (1967).

127. X. Guo, X. Chen, Z. Altounian, and J. O. Strom-Olsen, *J. Appl. Phys.* 73, 6275 (1993).

128. P. R. Bandaru, T. D. Sands, Y. Kubota, and E. E. Marinero, *Appl. Phys. Lett.* 72, 2337 (1998).

129. A. Borgschulte, D. Menzel, T. Widmer, H. Bremers, U. Barkow, and J. Schoenes, *J. Magn. Magn. Mater.* 205, 151 (1999).

130. J. Nogués and I. K. Schuller, *J. Magn. Magn. Mater.* 192, 203 (1999).

131. A. E. Berkowitz and K. Takano, *J. Magn. Magn. Mater.* 200, 552 (1999).

132. W. H. Meiklejohn and C. P. Bean, *Phys. Rev.* 102, 1413 (1956).

133. W. H. Meiklejohn and C. P. Bean, *Phys. Rev.* 105, 904 (1957).

134. R. D. Hempstead, S. Krongelb, and D. A. Thompson, *IEEE Trans. Mag.* 14, 521 (1978).

135. T. Nakamichi, *J. Phys. Soc. Jpn.* 25, 1189 (1968).

136. E. F. Wassermann, B. Rellinghaus, Th. Roessel, and W. Pepperhoff, *J. Magn. Magn. Mater.* 190, 289 (1998).

137. J. Koeble and M. Huth, "European Conference on Magnetisma and Magnetic Applications (EMMA)," Kiev, 2000, Materials Science Forum. Transtech Publications, Zurich, Switzerland.

138. L. van Hove, *Phys. Rev.* 95, 249 (1954).

139. N. Bernhoeft, A. Hiess, S. Langridge, A. Stunault, D. Wermeille, C. Vettier, G. H. Lander, M. Huth, M. Jourdan, and H. Adrian, *Phys. Rev. Lett.* 81, 3419 (1998).

140. A. Hiess, N. Bernhoeft, S. Langridge, C. Vettier, M. Jourdan, M. Huth, H. Adrian, and G. H. Lander, *Physica B* 259–261, 631 (1999).

141. K. Binder, "Phase Transitions and Critical Phenomena" (C. Domb and J. L. Lebowitz, Eds.), p. 1. Academic Press, London, 1983.

142. H. Dosch, "Critical Phenomena at Surfaces and Interfaces," Springer Tracts in Modern Physics, Vol. 126. Springer-Verlag, Heidelberg, 1992.

143. S. W. Bonham and C. P. Flynn, *Phys. Rev. B: Condens. Matter* 57, 4099 (1998).

144. B. Nickel, W. Donner, H. Dosch, C. Detlefs, and D. Grübel, *Phys. Rev. Lett.* 85, 134 (2000).

145. S. Krimmel, W. Donner, B. Nickel, H. Dosch, C. Sutter, and G. Grubel, *Phys. Rev. Lett.* 78, 3880 (1997).

146. C. Ern, W. Donner, A. Ruhm, H. Dosch, B. P. Toperverg, and R. L. Johnson, *Appl. Phys. A* 64, 383 (1997).

147. B. E. Warren, "X-Ray Diffraction." Addison-Wesley, Reading, MA, 1969.

148. E. Domany, Y. Shnidman, and D. Mukamel, *J. Phys. C: Solid State Phys.* 15, L495 (1982).

149. M. F. Collins and J. B. Forsyth, *Philos. Mag.* 8, 401 (1963).

150. F. Schmid, *Z. Phys. B: Condens. Matter* 91, 77 (1993).

151. A. Drewitz, R. Leidl, T. W. Burkhardt, and H. W. Diehl, *Phys. Rev. Lett.* 77, 1090 (1997).

152. G. H. Fecher, B. Schmied, and G. Schönhense, *J. Electron. Spectrosc. Relat. Phenom.* 103, 771 (1999).

153. Y. Shim, D. P. Landau, and S. Pal, *J. Phys.: Condens. Matter* 11, 10007 (1999).

154. A. Bogicevic, J. Strömquist, and B. I. Lundqvist, *Phys. Rev. Lett.* 81, 637 (1997).

Chapter 13

PULSED LASER DEPOSITION OF THIN FILMS: EXPECTATIONS AND REALITY

Leonid R. Shaginyan

Institute for Problems of Materials Science, Kiev, 03142 Ukraine

Contents

1. INTRODUCTION

Three constituents of the deposition process (composition of the film-forming species (FFS), composition of the of medium where the FFS propagate, and conditions on the condensation surface) determine the composition, structure, and properties of the resulting films.

The composition of FFS generated by evaporation or sputtering of any substance during the preparation of the film by physical vapor deposition (PVD) methods depends primarily on

Handbook of Thin Film Materials, edited by H.S. Nalwa
Volume 1: Deposition and Processing of Thin Films

ISBN 0-12-512909-2/$35.00

the composition of the initial substance (target). However, the transfer of the substance from the target surface to the substrate includes several processes, which can substantially change its composition. These changes can start on the target surface, depending on the method of generation of the film-forming flux.

The next step, in which the composition of the generated film-forming flux can be changed, is its transfer from the surface of the target to the condensation surface. This can be affected by interaction between FFS in a gas phase and interaction between FFS and particles of the medium, where FFS propagate. Another mechanism that can influence the composition of evaporated (sputtered) species during their transfer from target to substrate is a scattering of these species by atoms or molecules of the medium. The scattering depends on several factors. The greater the working gas pressure during the deposition, the more probable are the collisions between the propagating species and the gas particles; and the greater the difference between the atomic masses of FFS and medium particles, the more probable is the deviation of the FFS from the initial direction of their movement. Both processes can substantially change the composition of the FFS deposited on the substrate (and hence the film composition) from the initial composition of the target.

The composition of the FFS can also be changed directly at the condensation surface. There are several reasons for this. The first is the difference in the sticking coefficients for different atoms. Metal atoms have sticking coefficients of ~ 1 at substrate temperatures in the range of $-195°C$ to several hundred degrees Celcius; the sticking coefficients of gas atoms at room temperature are between 0.1 and 0.5 [1]. Another possible reason is the ion or energetic particle bombardment of the condensation surface. If this is the case, the film composition can change drastically, especially if atoms of volatile component are present.

Qualitative and quantitative composition of FFS can influence not only the composition of the film, but its structure as well. A striking example of the influence of quantitative FFS composition on the structure of the resulting condensate can be the fabrication of diamond-like films. A diamond-like carbon (DLC) film with a high (up to 90%) content of sp^3 bonds can be obtained only from the atomic flux of carbon atoms with a certain amount of carbon ions with a certain energy. If the flux of sputtering from graphite target species contains clusters of several carbon atoms, the resulting condensate exhibits graphite-like properties [2].

All of the above considered stages (FFS formation, their transport, and condensation) of the pulsed laser deposition of thin films exhibit significant differences from those for other PVD methods.

The interaction of the high power pulse of laser irradiation with the target surface can heat it to such a temperature that all of the target constituents can be evaporated simultaneously and with the same velocity. This peculiarity of pulsed laser deposition (PLD) allows to assume that the composition of the generated vapor and the resulting condensate will be similar to that of the initial target.

Another peculiarity of the interaction of a high power density laser pulse with the condensed phase is the generation of high-density vapor just above the target surface. The supersonic expansion of the vapors during the initial hydrodynamic flow accompanies the effect of segregation of the atomic, molecular and micro-sized particulates due to their collisions in the laser plume. The clusters arising in such a process may contain several or hundreds of atoms [3, 4]. If the evaporation occurs in a gas medium the clusterization effect in PLD is substantially enhanced [5]. In this case, along with the clusters formed from the evaporated species, molecular clusters synthesized from the gas atoms and the evaporated species appear [6].

The other important peculiarities of the species generated by pulsed laser evaporation are their high kinetic energy and the presence of charged species among them. These factors result in the bombardment of the condensation surface by these species.

Each of the above considered peculiarities make the PLD method substantially different from other methods of physical vapor deposition.

Conventional methods of physical vapor deposition that use ion sputtering, electron bombardment, or resistive heating of a substance to genere the vapor phase yield a much lower vapor pressure compared with the instantaneous pressure of the vapor generated by laser pulse. The low vapor pressure almost excludes the interaction between the species in a vapor phase. The energy of sputtered and, in particular, evaporated particles is lower by orders of magnitude than that of the species generated by laser radiation. The amount of charged species in the film-forming flux generated by other PVD methods is also much smaller.

So it is natural to expect the properties of films fabricated by laser evaporation to be different from those obtained by other methods of physical deposition. It is assumed that, owing to the above considered peculiarities of pulsed laser evaporation, the target composition has to be easily reproduced in the film. It is expected that the structure of PLD film is more perfect than that deposited by any other PVD technique at lower substrate temperatures. The absence of the heating elements and discharging electrodes inside the vacuum chamber and the short time pulse duration contribute to the high purity of the deposited film.

The history of PLD started in 1964, when the first papers devoted to the application of laser to film deposition appeared. Even at that time discrepancies between the actual and predicted properties of condensates had been detected. The composition of the films of different compounds deviated from their initial target composition. Macroscopic defects (macrodefects) like solidified droplets and solid particles were present in the majority of condensates. Many investigations directed toward the development of the PLD method as well as the elimination of the aforementioned drawbacks have been carried out since that time. More attempts have been made to improve PLD than that for other film deposition techniques.

Major efforts have been made to eliminate the presence of macrodefects in a condensate. Different approaches had been used for the solution of this problem. One of them was related to improvements of the laser itself. The main directions here have been the improvement of the spatial and temporal homogeneity of the laser pulse [7, 8] and the search for optimal pulse duration and the optimal frequency of the pulse variation of radiation wavelength [9].

Other approach was to manipulate the target properties to improve the effectiveness of absorption of laser radiation energy and to enhance the vapor phase fraction in the products ejected from the target. For this purpose the evaporation of different types targets, such as alloys, single crystals, and free-falling and pressed powders has been investigated.

One more method, widely used in PLD technology to lower the number of macrodefects in films is the mechanical separation of film-forming particles [10–12].

According to the majority of publications related to PLD, the most important advantage of this method is the possibility of obtaining films of complex compounds with a composition identical to that of a target (see, e.g., [12, 13]). One may notice that the variety of the films fabricated by PLD was much wider in the early history of the method. Many investigations related to the fabrication of nitride, boride, carbide, and oxide films were carried out during the 1970s and the beginning of the 1980s. However, many of them were not successfully deposited with the correct stoichiometric composition. By the middle of the 1980s, after all merits and drawbacks of the method had been clarified, the activity in this field had declined. By the end of the 1980s PLD had been given new stimulus due to success in the laser deposition of high-temperature superconducting (HTSC) films. At the moment most research and development in PLD is concerned with the deposition of oxides and high-temperature superconductors. This is due to the fact that the composition of the films deposited by pulsed laser evaporation of these materials is very close to that of the initial target. Another reason for the rebirth of PLD is the development of accessible and relevant excimer lasers.

In recent years a number of systematic studies giving very detailed accounts of the ablation mechanism under typical PLD experimental conditions have appeared. This is related to the inspiration of the successful PLD deposition of HTSC films as well as the necessity to search for a solution to long-standing PLD problems, such as the presence of macrodefects and the nonstoichiometry of condensates composition. The effect of ambient background gases on YBCO plume propagation under film growth conditions was studied with the use of various analytical techniques, such as optical spectroscopy, ion probing, time-of-flight analysis, and gated high-speed CCD photography [4]. This study reveals features of laser–target interaction and plume transport that cannot be predicted otherwise.

Our analysis of the publications regarding PLD shows that despite of extensive past experience with PLD deposition of films from a wide variety of materials, the mechanisms of film formation and structure are still not completely clear; the main drawbacks of the method are still present. Furthermore, there are no definite answers to the following questions:

- Which materials are most promising for the PLD fabrication of films?
- Which material properties provide effective evaporation with identical (i.e., without loss of stoichiometry) transfer of composition to the film?
- Why, despite the prerequisites for fabrication of PLD films with more perfect structure at a lower substrate temperature (that is, high energy of evaporated FFS) usually not occur? Rather, the substrate temperature turns out to be even higher for the fabrication of laser condensates of the same structural perfection as those obtained by other methods.
- What properties of the target material determine the portion of vapor phase in the products ejected from the target under the laser pulse action? Let us mention that this parameter determines the effectiveness of absorption of laser radiation. The presence of the macrodefects in the film also depends on it.
- Which physical mechanisms govern the evaporation of targets of different types? What targets are optimal for the deposition of films with the required characteristics?

The answers to some of the above questions can, in principle, be found in the literature. However, we were not able to find a systematic survey of all of the aforementioned problems, despite a huge literature devoted to PLD. Hence, the aim of this chapter is to give a systematic survey of aforementioned problems, based on literature sources and the author's own investigations.

2. COMPOSITION OF PULSED LASER-DEPOSITED FILMS

One of the most important problems of film materials science is the fabrication of films with a specific composition. A short pulse of high-density optical radiation generated by a laser gives so great a thermal spike on the irradiated surface of the solid that all of its components should be evaporated similarly. If the sticking coefficients of each component are equal to each other, one can expect that obtained from such a flux condensate should have a composition similar to that of the initial target. This physical idea inspired the appearance of many experimental articles dedicated to the investigation of rules governing the formation of the aforementioned laser condensate composition. Analysis of those papers has shown that there are many factors influencing the composition. These factors can be divided into two groups: those related to the laser operation parameters and those related to the properties of both the evaporated substance and the medium where this evaporation occurs.

The density of laser radiation power, the radiation wavelength, the duration of a pulse and its spatial and temporal homogeneity, and the diameter of the radiation spot on the target surface can be related to the first group of factors.

The chemical composition of a target, its thermophysical characteristics, and the medium where evaporation occurs (vacuum, gas, its chemical properties) can be related to the second group of factors.

The method of laser deposition determines a great number of technological parameters influencing films properties. Because experimental investigations of laser deposition of the films were carried out by different scientists under different deposition conditions and with different sources of laser radiation, the analysis of the results of such investigations is quite complicated. This is why the conclusions provided here, while of a pretty general nature, cannot be regarded as final.

Let us consider the influence of the aforementioned groups of factors on the composition of laser condensates.

2.1. Dependence of the Composition of PLD Films on Laser and Deposition Processing Parameters

Film composition depends on the composition of the FFS. There are two main differences between species fluxes generated by laser radiation and those generated by other methods. These peculiarities are complex composition of the flux (the flux contains different particles, from multiply charged ions to multiatomic clusters and larger particles) and angular distribution of particles in the flux [14–16]. Both of these parameters depend essentially on the laser operation regime as well as on the deposition conditions (ambient medium, distance to target, etc.) [7]. It is obvious that the more inhomogeneous the composition and spatial distribution of the flux of FFS are, the less satisfactory is the quality of the resulting film.

Laser fluence (q), wavelength (λ), pulse duration (τ), target-to-substrate distance, laser spot size, and ambient medium (vacuum, gas) have been investigated for their influence on the plume angular distribution. Few general conclusions are possible, because even the laser fluence, the best studied of these parameters, has been examined in detail for only a few materials. Another consideration is the complication of data interpretation by the fact that variation of a single parameter may influence several aspects of the ablation process simultaneously. For example, a change in laser wavelength while the other laser parameters are kept constant may change both the initial density and energy of the plume, because of wavelength-dependent changes in the target and laser plume optical properties. Experimental data and theoretical analysis show that there are two regions of laser power density with quite different compositions of ejecta. For flux densities of $10^6 < q < 10^8$ W/cm^2, the ejecta consist mainly of the vapor in the form of neutral species and liquid phase. If $q > 10^9$ to 10^{10} W/cm^2, the amount of the ejecta is much less, but the charged particle fraction appears and increases rapidly with increasing laser power density [7, 14].

The angular distribution of ionized and neutral components of the laser plume defines the uniformity and physical properties of the film-forming flux. It is known that the angular, energy, and velocity distributions of different components of the laser plume are quite different [17, 18]. The angular distribution

of the laser generated flux is often much more strongly forward peaked than the flux obtainable from effusive sources. This forward peaking phenomenon for depositions in vacuum is now generally agreed to arise from collisions of the plume species among themselves. The axis of the laser generated plume is always oriented along the target surface normal for normal and nonnormal angles of laser incidence [19].

In addition to spatial separation, the time separation of the species in laser plumes has also been observed. The fast electrons are moving ahead of other species, and ions of maximal charge follow them. Ions of minimal charge are at the tail of the ion component of the flux. The slowest part of the flux is neutral. The ions of maximal charge have the narrowest angular distribution ($\theta_{max} \approx 30°$). As the ion charge decreases, the angular distribution widens so that neutral particles have the widest angular distribution. The energy distribution of laser plasma components follows their angular distribution. The ions with a maximal charge have the largest kinetic energy (up to tens of kiloelectron volts at a laser radiation density $q \approx 10^{11}$ to 10^{12} W/cm^2). The neutral atoms has the smallest kinetic energy (tens of electron volts) [18].

A general examination shows that the absorption coefficients of insulators and semiconductors tend to increase as one moves to the short wavelength (200–400 nm), whereas metals absorb more or less efficiently in a wide range of radiation wavelength (10 μm to 100 nm). The influence of wavelength on ion energy distribution was investigated in [18], where Ti was evaporated by excimer laser (308 nm, $\tau = 20$ ns) and by CO_2 laser (10.6 μm, $\tau = 100$ ns) at a constant energy density $q = 15$ J/cm^2. The energy spectrum of the ions was in the region of 0 eV to 2000 eV for a CO_2 laser, whereas for an excimer laser the spectrum was narrower and expanded up to ~1000 eV. The mean ion energy for both cases was 100–400 eV. These results correlate with the data obtained for the plume angular distribution for Se [18]. These data suggest that the angular distributions produced with IR wavelengths are significantly broader than those produced with visible and UV wavelengths [19].

The laser pulse length also has great influence on the plume composition and the angular distribution of its components. At a relatively long laser pulses (~ms) and low fluences ($q \approx 10^5$ W/cm^2) the energy of evaporated particles is much less than the dissociation energy of the molecules, and these conditions are suitable for obtaining stoichiometric films of polycomponent materials with a high energy of dissociation. At a short pulse duration (~ns) the laser-produced plasma contains a significant amount of high-energy particles, especially ions [18], which can greatly influence film properties, particularly their composition, because of the ion bombardment effect [20–23]. The influence of the pulse duration on the composition of Y-Ba-Cu-O films was observed when the films deposited at $\tau = 150$ ns were more compositionally uniform then those deposited at $\tau = 15$ ns [24].

Angular distribution also affects to such deposition conditions as the target–substrate distance, laser spot dimensions, and ambient gas presence. Noninteracting particles emitted in a vacuum from a small-area source are expected to have an angular distribution that does not change with distance from the source. Contrary to expectations, measurements for the deposition of some materials in a vacuum suggest that the angular distribution of the source may change with the distance from the source. The carbon data suggested a broader angular distribution close to the target that ceased at distances far from the target. These findings were attributed to scattering of the incident flux, with particles reflected and resputtered from the substrate [16].

The frequently observed effect of laser spot dimensions on the angular distribution of the evaporated species is that the smaller the spot dimensions are the more uniform is the film thickness. These effects arise because the number of intraplume collisions increases with the laser spot dimensions.

The angular distribution of evaporated material in a vacuum is determined by collisions among the plume species. Most of these collisions take place close to the target, while the plume is small. The angular distribution of the evaporated material changes if the evaporation occurs in ambient gas. Under these conditions the evaporated species experience additional collisions with the gas molecules that scatter the plume particles from their original trajectories and broaden the angular distribution. For a given background gas mass and plume species velocity, one qualitatively expects the dispersive effects of collisions to vary with the mass of the species: the higher the mass of the evaporated species, the lower the collision effect, and vice versa. Hence the background gas pressure at which this effect may be observed depends on the ratio of masses of evaporated species to a mass of the gas particles.

Because the different plume species have their own angular distributions, one may expect that each of the considered laser and deposition parameters may influence the composition of the film.

Measurements of laser film composition confirm that the film stoichiometry varies with the deposition angle [16]. The reason for this effect (see also above) is the different angular distributions of the plume component species. The latter may be associated with the difference in species charge (ions or neutrals) as well in their mass (heavy or light). For instance, the ions and neutrals in the plume have been observed to have distinctly different angular distributions.

Measurements of film composition produced by PLD in a vacuum from multicomponent targets sometimes indicate contradictory results. Higher on-axis light-to-heavy ratios were observed both for YBCO films [25] and for $PbZr_xTi_{1-x}O_3$ films [26]. However, for YBCO films this effect became less pronounced at higher fluences, whereas in contrast to YBCO films, composition inhomogeneities in $PbZr_xTi_{1-x}O_3$ films were enhanced as the fluence increased from 0.2 to 5 J/cm^2.

In [27] film composition homogeneity was investigated with variation in laser fluence, spot size, oxygen pressure, and the substrate-to-target distance. Slight on-axis enrichment of Cu and Ba was found in the films deposited in a vacuum, whereas the films deposited in oxygen were on-axis enriched in Y because of the gas-scattering effect. The last effect was increased with increasing of substrate-to-target distance. The authors concluded that the deposition of compositionally uniform films at

intermediate substrate-to-target distances results from cancellation of the intrinsic (in a vacuum) composition inhomogeneity by gas scattering effects.

Studies of laser plume composition established a very important effect that can significantly influence the film composition. It is the existence of two distinct components in the laser-produced fragments at the target surface [12, 25, 28]. One of the components with a weaker angular dependence was identified as a nonstoichiometric thermal evaporation. The other component, highly peaked in the forward direction, was identified as a stoichiometric component produced in a nonlinear evaporation process, such as a shock wave or highly excited dense plasma. The ratio of material emitted in the two different components depended upon the laser energy density. At lower energy densities (typically for Y-Ba-Cu-O below 0.9 J/cm^2, 248 nm, 30-ns laser pulses) the thermal evaporation component dominated. As a result the composition of deposited films deviated from the proper stoichiometry. At higher laser energy densities the propensity for deposition of laser-produced debris increased so that optimal energy density was required.

Our brief examination shows that PLD film composition is sensitive to the laser and deposition processing parameters. The sensitivity may be different for different evaporating materials. The main reasons for these effects are the multicomponent composition of the laser plume and the nonuniformity of the spatial distributions of each of the components.

2.2. Dependence of the Composition of PLD Films on the Evaporating Material

The main reason for the deviation of film stoichiometry during the evaporation of complex compounds by conventional methods is the difference in the vapor pressures of constituents of the compound. This is why the deposition of the film of a complex compound with the correct stoichiometry by conventional evaporation is possible in three cases: in the case of molecular evaporation of the compound; when the compound constituents evaporate congruently; when the volatilities of the constituents are similar. However, if the energy input to the evaporating compound is sufficient to instantly raise its temperature significantly above the temperature of evaporation of its less volatile component, one can expect that the composition of the condensate will be close to that of the initial compound. The flashing method (method of discrete evaporation) of thin compound film deposition is based on this principle [1].

Concentration of the laser radiation energy in a small volume for a short time interval gives rise to conditions suitable for the fabrication of complex compound films. The main aim of the majority of studies since the inception of the PLD method was to establish the proper laser operation regime and deposition conditions to obtain a film with a composition identical to that of the target.

More than 100 compounds have been investigated [16] over more than 30 years of PLD film deposition and investigation. These compounds fall into four groups: low-melting-point semiconductor compounds of $A^{II}B^{VI}$, $A^{III}B^{V}$, and $A^{IV}B^{IV}$

types; oxides and perovskites, including HTSC; and refractory compounds—carbides, silicides, and nitrides. The majority of compounds had been investigated during the first 15 years of the application of PLD. During these years a large number of papers dedicated to PLD had appeared. The peculiarity of recent investigations of PLD is the reduction of the number of materials deposited by this method. These materials are mainly binary and multicomponent oxides, including HTSC and ferroelectrics, as well as an amorphous carbon and carbon nitride. At the same time, much attention had been paid to different forms of reactive PLD [6] because of the necessity to correct the oxygen deficiency in oxide condensates.

2.2.1. Mechanisms of Formation of Low-Melting-Point Semiconductor Compounds and Oxides

To understand the mechanisms of laser condensate formation, let us consider the main results from investigation of the composition of the films, obtained by a pulse evaporation of compounds. Some data are reported in Table I. Although they are from quite a small number of articles about laser film fabrication, these data reflect a wide variety of film materials, obtained by PLD. Their analysis allows some conclusions about the properties of the compounds that can be obtained as films with proper composition by laser evaporation in a vacuum. Note that here we intentionally avoid discussion of HTSC films (like $YBa_2Cu_3O_{7-x}$) obtained by PLD, because much attention has been paid to these materials in the literature (see, e.g., [16 and references therein]).

A group of materials that can be deposited by PLD without serious deviation of the stoichiometry is low-melting-point semiconductors of $A^{II}B^{VI}$, $A^{III}B^{V}$, and $A^{IV}B^{IV}$ types. These substances incorporate components with a low melting temperature and a high pressure of saturation vapor at relatively low temperatures. This means that one may expect to obtain a film with the correct stoichiometry if the temperature achieved on the evaporation surface is substantially higher than those necessary for intensive evaporation of the compound components.

Because the temperature in a laser spot reaches several thousand degrees, it is sufficient for the complete evaporation of all components of low-melting-point compounds. It is essential to note that because of such thermophysical properties of components, the majority of considered semiconductor compounds can also be obtained as a films of stoichiometric composition by conventional evaporation [1].

Another group of materials that can be obtained by PLD with stoichiometric composition, includes some two-component (ZrO_2, Al_2O_3 [44], SiO_2 [31]) and multicomponent ($BaTiO_3$, $SrTiO_3$ [30], $CaTiO_3$ [31]) oxides. The main feature of the evaporation of many two-component oxides (BaO, BeO, Al_2O_3, SnO_2, In_2O_3, TiO_2, ZrO_2, SiO, SiO_2, and some others) is their preferential evaporation in a molecular form [1]. This peculiarity makes it possible, to obtain them in a stoichiometric form by both conventional and laser evaporation at relatively low power densities ($q = 10^5$ to 10^6 W/cm^2) [18]. The principal

mechanism of composition formation of multicomponent oxides films is most probably the same for conventional evaporation and PLD. During their evaporation, multicomponent oxides dissociate to the molecules of the MeO_x (Me = Ba, Ca, Sr) and TiO_y types, which during their condensation on the substrate form the stoichiometric compound. Most probably, such a mechanism of evaporation makes it possible to obtain the films of $BaTiO_3$, $CaTiO_3$, and $SrTiO_3$ not only by PLD but also by a flashing method (for instance, $BaTiO_3$ [1]). The other evaporation mechanism (e.g., complete dissociation of a compound with subsequent condensation and reaction between its components on a substrate) is less likely to occur because oxygen atoms have a sticking coefficient several times smaller than that of metal atoms [1], and the deposition occurs in a vacuum (see, e.g., [30]). In our view, the formation of the films of complex HTSC oxides exhibits the elements of a mechanism discussed for $BaTiO_3$, $CaTiO_3$, and $SrTiO_3$. That is, the action of a laser beam on ceramics does not lead to complete dissociation of a compound to separate chemical elements, but rather to the molecules of stable oxides of Y_xO_y and Ba_xO_y types. The second element of formation of the films of these compounds is a reaction between free metal atoms and oxygen in a gas phase [6]. This is due to the fact that the process takes place in an oxygen atmosphere at a relatively high pressure. The two aforementioned mechanisms make it quite easy to obtain the stoichiometric HTSC films by PLD.

The oxides of the Me_xO_y type can be attributed to the third group. The films of these oxides have been obtained by laser evaporation of their powders or sintered powders. Some of these oxide films were reduced completely (Cu_2O [29], ZnO [31], PbO, Fe_2O_3 [43]) during laser evaporation in vacuum or had an oxygen deficiency (TiO_2 [46], $In_2O_3 + SnO_2$ [47], RuO_2 [48]), which have been corrected by evaporation in an oxygen ambient. At the same time, in [18] the successful deposition of a whole series of stoichiometric oxide films by pulsed laser evaporation of BeO, Al_2O_3, SnO_2, TiO_2, and ZrO_2 powders in a vacuum has been reported. The author emphasizes that this was possible because of the use of relatively long laser pulses ($\sim$ms) and low fluences ($q \approx 10^5$ W/cm^2). These conditions are suitable for obtaining stoichiometric films of multicomponent materials with high dissociation energies and for oxides in particular. The data from Table I show that the oxides in [46–48] had been evaporated with higher power density, by nanosecond pulses and by short wavelength radiation. These conditions are favorable for the dissociation of oxides due to the higher absorption of laser radiation by the target material and by the generated vapor phase.

The latter leads to an oxygen deficiency in the condensate. To eliminate this deficiency, the target evaporation should be carried out in an oxygen medium. The evaporation in an oxygen ambient provides the interaction of evaporated species with gas particles and the formation of oxide molecules in a gas phase. The other mechanism is the reaction between oxygen and metal atoms on the substrate. The stoichiometric composition of the PLD deposited films is due, presumably, to the aforementioned mechanisms. Let us note that the formation of oxide molecules

Table I.   Deposition Conditions, Composition, and Structure of Thin Film Materials Deposited by PLD

Target (composition, type)	Laser processing parameters (q, J/cm^2; τ; λ)			Film composition	Film structure	Ambient gas pressure (torr)	Reference
CdTe, ZnTe, PbTe, InAs, ZnO, Cu$_2$O (P)	10^6	ms	Ruby	CdTe, ZnTe stoichiometric; PbTe(?); As/In = 35; Cu$_2$O reduction to Cu	P/C	Vac. 10^{-4}	[29]
ZnS, Sb$_2$S$_3$, SrTiO$_3$, BaTiO$_3$ (fine P)	(?)	ms	Nd:YAG	Stoichiometric	P/C	Vac. 10^{-7}	[30]
GaP, GaSb, GaAs, InAs; ZnS, ZnSe, ZnTe, ZnO, CdTe, CdSe; Al$_2$O$_3$, SiO$_2$, CaTiO$_3$, MgAl$_2$O$_4$, CdCrS$_4$ (P)	10^7–10^8	ms	Ruby	GaP, GaSb, GaAs, InAs; ZnS, ZnSe, ZnTe, CdTe, CdSe, CaTiO$_3$ stoichiometric; ZnO reduction to Zn	A^{iii}B^V, A^{ii}B^V P/C; residuary A	Vac. 10^{-6}	[31]
In$_2$Te$_3$, Cu$_2$Te (crystals)	10^8	ns	Nd:YAG	In$_2$Te$_3$, Cu$_2$Te; depending on T_s	A; P/C; Epi; depending on T_s	Vac. 10^{-6}	[32, 33]
GaAs, CdS, PbS, PbSe, PbTe, Pb-Cd-Se (crystals)	10^5–10^6	ms	Nd:YAG	Stoichiometric	P/C; Epi depending on T_s	Vac. 10^{-5}	[18]
AlN, GaN, InN (PP)	10^8	ns	KrF	AlN stoichiom.; GaN, InN partial reduction to Ga and In	Epi $\sim$500°C	Vac. 10^{-7}; $P_{N_2} = 5 \times 10^{-3}$	[34]
AlN (C)	$\sim 10^8$	ns	Nd:YAG, 266 nm	$0.95 < N/Al < 1.2$ depending on P_{N_2}	Epi 1000°C	$5 \times 10^{-5} \leq P_{N_2} \leq 10^{-1}$	[35]
h-BN	$\sim 10^8$	ns	KrF	Stoichiometric in N$_2$	P/C c-BN (main)	$P_{N_2} = 5 \times 10^{-2}$	[36]
TiN (SP)	$\sim 10^8$	ns	XeCl	Stoichiometric in N$_2$	P/C	$5 \times 10^{-4} \leq P_{N_2} \leq 10^{-3}$	[37]
TiC (SP)	$\sim 10^8$	ns	XeCl	Stoichiometric	Nanocrystal ($T_s = 300$°C)	Vac. 10^{-5}	[38]
SiC (SP)	$\sim 10^8$	ns	XeCl	Stoichiometric	P/C; 3C-SiC ($T_s = 800$°C)	Vac. 10^{-5}	[39]
WC (SP)	$\sim 10^7$	ms	ruby	(?)	A ($T_s = 400$°C)	Vac. 10^{-5}	[40]
MoSi$_2$ (free-falling P)	$\sim 10^8$	ns	CO$_2$	Si/Mo = 11	(?)	Vac. 10^{-5}	[41]
SiO$_2$, HfO$_2$, ZrO$_2$ (C)	$\sim 10^8$	ns	CO$_2$, KrF	SiO$_2$ stoichiom. (CO$_2$; in O$_2$) HfO$_2$, ZrO$_2$ stoichiom. under O$_2^+$ bombardment	(?)	$P_{O_2} = 10^{-4}$	[42]
PbO$_2$, Fe$_2$O$_3$ (PP)	$\sim 10^7$	ms	Nd:YAG	Reduction to metals	(?)	Vac. 10^{-5}	[43]
ZrO$_2$, Al$_2$O$_3$ (C)	$\sim 10^7$	Ns	CO$_2$	Stoichiometric ($T_s = 300$°C)	A ($T_s = $ RT); P/C	Vac. 10^{-5}	[44]
ZnO (SP)	$\sim 10^7$	ns	KrF	Stoichiometric in O$_2$	Epi ($T_s = 800$°C)	$P_{O_2} = 10^{-4}$	[45]
TiO$_2$ (SP)	$\sim 10^8$	ns	Nd:YAG	Stoichiometric in O$_2$	A ($T_s = $ RT); P/C ($T_s \geq 500$°C);	$10^{-4} \leq P_{O_2} \leq 10^{-1}$	[46]
95% In$_2$O$_3$ + 5% SnO$_2$ (SP)	$\sim 10^8$	ns	KrF	Stoichiometric in O$_2$ (resistivity, optical measurements)	A ($T_s = $ RT); P/C ($T_s \geq 200$°C);	$10^3 \leq P_{O_2} \leq 5 \times 10^{-2}$	[47]
RuO$_2$ (SP)	$\sim 10^8$	ns	XeCl	Stoichiometric in O$_2$	P/C ($T_s < 500$°C); Epi ($T_s \geq 500$°C);	$0.2 \leq P_{O_2} \leq 5 \times 10^{-4}$	[48]

Abbreviations: (P), powder; (PP), pressed powder; (C), ceramics; (SP), sintered powder; A, amorphous; P/C, polycrystalline; Epi, epitaxial; (?), no data.

in a gas phase is not peculiar to the composition formation of the films deposited by other PVD techniques.

2.2.2. Mechanisms of Formation of Refractory Materials Films

The next large group of materials that can be deposited as films by pulsed laser evaporation are the refractory nitrides, carbides, borides, and silicides. To discuss a mechanism of formation of the aforementioned condensates, it is appropriate to consider separately the materials with volatile components, such as nitrogen.

The fabrication of stoichiometric nitride films is difficult because of their dissociation during heating to high temperatures and the large difference in the sticking coefficients of nitrogen and the solid component. The majority of nitrides dissociate under evaporation. This may be due to the smaller binding energy of Me$-$N in the nitrides than of in Me$-$O in oxides. Examples are films of gallium and indium nitrides, where a noticeable number of droplets of indium and gallium metals have been detected after the evaporation of their pressed powders. These droplets were present both for evaporation in a vacuum and in nitrogen [34]. At the same time, stoichiometric films of aluminum [34, 35], boron [36], and titanium nitrides [37] could be obtained with substantially lower amounts of metal impurities from their laser evaporation in nitrogen. The authors of [34] mention a correlation between the growth of a number of metal droplets in AlN, GaN, and InN films and decrease in Me$-$N binding energy. AlN is stable up to its melting point of 2150°C, as opposed to GaN, which sublimes at 800°C, or especially to InN, which dissociates at $\sim$600°C [34]. Note that some nitrides, like oxides (e.g., aluminum nitride), can be obtained by conventional evaporation without essential dissociation [1]. This means that the mechanism of nitride film formation during laser evaporation of a nitride target is principally similar to that of oxides (see above). This is why the stoichiometric films of aluminum nitride could be (similar to some oxides) obtained by evaporation without pronounced deviation from stoichiometry and even without adding nitrogen as a working gas [34]. It should be noted that laser operating parameters can influence the probability of compound dissociation. This factor may lead to a drastic change in the final condensate composition. This phenomenon will be discussed below in more detail for the example of aluminum nitride.

It is of value to discuss the possibility of the fabrication of stoichiometric films from the other refractory compounds (carbides, silicides) with respect to the difference in the temperature of intensive evaporation (the temperature at which the vapor pressure is about 10^{-2} to 10^{-1} torr) of their constituents. If the volatilities of the constituents differ greatly, the film might be enriched by an element with a higher volatility. For instance, molybdenum silicide ($MoSi_2$) films contained 11 times more Si than the initial powder [41]. The reason for such silicon enrichment is the large difference in the volatilities of silicon and molybdenum vapors. For example, whereas the vapor pressure

of Si achieves the value of 10^{-2} torr at $\sim$1900 K, the same vapor pressure of Mo is reached at $\sim$2800 K only.

Films of titanium and silicon carbides had a stoichiometric composition during the laser evaporation of these targets in a vacuum [38, 39] despite the large difference in their components' volatilities. It is probable in such cases that the important role belongs to the molecular nature of the evaporation of these carbides. It is known that vapors generated during the conventional evaporation of silicon carbide contain a large number of SiC_2 and Si_2C molecules [49], wereas titanium carbide evaporates coherently [1].

The regularities of the formation of PLD films had been investigated in [50]. This paper covers the deposition and investigation of a broad spectrum of materials, including alloys and compounds. The results of the paper are of some interest because the deposition and investigation of all films were carried out under similar conditions. A more detailed consideration of the results of [50], along with those from Table I is of value with regard to getting a clearer idea of the mechanisms of the formation of PLD condensates.

2.3. Experimental Details of Pilyankevich et al.

A Nd:YAG laser ($\lambda = 1064$ nm, pulse duration $\sim$1 ms, power 1000 J per pulse) was used by Pilyankevich et al. [50]. The scheme of a PLD system used in [50] is presented in Figure 1. The dependence of the film composition on the laser power density was investigated for two values, $q = 1.5 \times 10^6$ W/cm^2 (defocused regime) and $q = 5 \times 10^7$ W/cm^2 (focused regime). The power density was varied by variation of the laser spot size at a constant power of 1000 J/pulse. The vacuum condiions were maintained by diffusion pump, and the process was carried out at 5×10^{-6} torr. Polished ceramic plates based on (SiO_2-TiO_2-Al_2O_3) oxides and fresh cleaves of NaCl or KCl crystals were used as substrates. The substrates were located in two positions relative to the target. One substrate was clamped to the substrate holder that was parallel to the target (line-of-sight position), and the other was placed on the holder positioned at the periphery of the target (off-axis position) (Fig. 1). The substrate–target distance for both substrates was $\sim$50 mm. The temperature of the substrates placed on the holder in the off-axis position can be varied from the liquid nitrogen temperature to 450°C. The upper limit of the substrate temperature was defined by the sublimation temperature of the NaCl or KCl single crystals used as a substrates. Such substrates are convenient for further investigation of films by transmission electron microscopy (TEM). Such an arrangement of the substrates relative to the target made it possible to estimate roughly the influence of the angular distribution of different species in the laser plume on the composition, structure, and macrodefects of films. Discs of alloys, ceramics (hot pressed powders of compounds), and single crystals were the targets for the evaporation. The composition of the films that did not contain the light constituent was investigated by electron probe microanalysis (EPMA). Because the analysis of light elements by the methods used was impossible, the conclusion about the

Table II. Composition and Structure of Films Deposited by PLD of Alloys and Compounds

Target composition	Film composition	Film structure: initial; after annealing in electron diffraction column	Total mass of ejected products (g); average film thickness (nm) ($q = 1.5 \times 10^6$ W/cm^2)	Total mass of ejected products (g); average film thickness (nm) ($q = 5 \times 10^7$ W/cm^2)
50Fe-50Co	Coincides with target comp.	α-Fe		
80Fe-17Ni-3Cr	Coincides with target comp.	Coincides with target structure		
50Fe-50Si	Coincides with target comp.	As-deposited amorphous; coincides with target structure at $T_s = 300°$C		
50Zr-50Ni	Film—Ni, droplets—Zr	Ni	0.1406; ~110	0.1399; ~70
80Fe-20B	Not investigated	As-deposited amorphous; α-Fe at $T_s = 300°$C		
75Nb-25Si	Si/Nb ~ 2.5–3	Amorphous at $T_s = 300°$C; transition to NbSi$_2$ after heating to 800°C		
LaB$_6$	Not investigated	As-deposited amorphous; transition to LaB$_6$ after heating to 700°C		
WSe$_2$	Coincides with target comp.	Not identified		
SiC	Not investigated	As-deposited amorphous; transition to 3C-SiC after heating to 850°C		
CdSe	Coincides with target comp.	As-deposited polycrystalline CdSe		
CrSi$_2$	Not investigated	As-deposited amorphous; transition to CrSi$_2$ after heating to 550°C	0.0591; 200	0.0779; 70
FeSi$_2$	Not investigated	As-deposited amorphous; transition to FeSi$_2$ after heating to 450°C		
TiSi$_2$	Not investigated	As-deposited amorphous; transition to TiSi$_2$ after heating to 750°C	0.1237; 160	0.1359; 90
TaSi$_2$	Coincides with target comp.	As-deposited amorphous; transition to TaSi$_2$ after heating to 750°C	0.7834; 130	0.3267; 50
WSi$_2$	Coincides with target comp.	As-deposited amorphous; transition to WSi$_2$ after heating to 800°C	0.1324; 170	0.1094; 70
BN	Not investigated	Main h-BN; locally c-BN		
AlN	Not investigated	Al		
Si$_3$N$_4$	Not investigated	As-deposited amorphous; transition to Si after heating to 850°C	0.0412; 220	0.0191; 170
90W-10Cu (cold-pressed)	Strongly enriched with Cu	Not established	0.2484; 130	0.1624; 90
90W-10Cu (hot-pressed)	Strongly enriched with Cu	Not established	0.0220; 20	0.0340; 20

Deposition conditions: substrate–target distance 5 cm; substrate is at off-axis position; $T_s = $ RT. (After [50, 86].)

composition of several compounds is based on their structural analysis (LaB$_6$, SiC, BN, AlN, Si$_3$N$_4$). The structure of the films was studied by high-energy electron diffraction (HEED) and selected area electron diffraction (SAED), and the target structure was investigated by X-ray diffraction (XRD). Comparison of these data made it possible to come to a conclusion about the correlation between film structure and composition. The amorphous condensates were annealed *in situ* in the electron diffraction column until the initiation of the transition from the amorphous to the crystalline state, and the results obtained made it possible to judge their composition and structure.

Such an approach to the investigation of a wide class of laser condensates, deposited under similar conditions, permits us to regard the results obtained to be fair indicators of the mechanisms of their composition formation.

The materials under investigation were divided into three groups. The first group consisted of two types of alloys. The alloys of the first type, namely, 50Fe-50Co, 70Fe-17Ni-3Cr, and 50Fe-50Si, have components with close melting temperatures and temperatures of the intensive evaporation (temperature at which $P_s = 10^{-1}$ torr) (Table III). For the alloys of the second type, namely, 50Ni-50Zr, 80Fe-20B, and 75Nb-25Si, these characteristics differ strongly from each other.

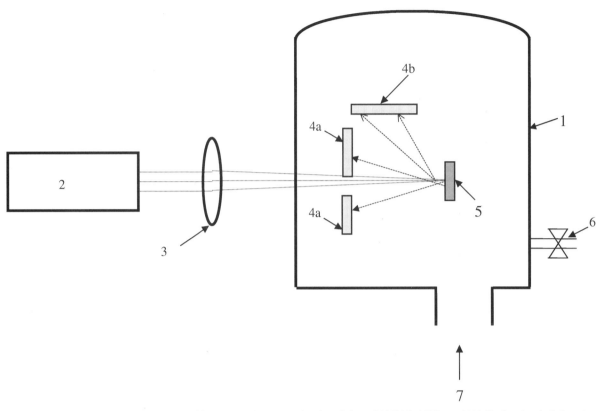

Fig. 1. A schematic of a pulsed laser deposition system. 1, vacuum chamber; 2, laser (Nd:YAG, 1064 nm, 1000 J/pulse, 1 ms); 3, focusing system; 4a, line-of-sight substrates; 4b, off-axis substrates; 5, target; 6, gas inlet; 7, pumping. (After [50].)

Table III. Melting Point and Temperatures at Which Saturated Vapor Pressure is $P_s = 10^{-1}$ torr

Element	Melting point (K)	Temperature at which $P_s = 10^{-1}$ torr
B	2360	2520
Al	933	1640
In	430	1355
Cd	594	593
Sn	505	1685
Co	1768	1960
Cr	2176	1820
Cu	1300	1690
Fe	1809	1920
La	1193	2200
Mo	2890	3060
Nb	2770	3170
Ni	1725	1970
Se	490	570
Si	1685	2090
Ti	1850	2210
W	3650	3810
Zr	2128	2930

After [1].

The second group incorporates refractory compounds, chosen by the same principle as alloys (CdSe, CrSi$_2$, FeSi$_2$, TiSi$_2$—compounds with close component melting temperature and volatility, whereas WSi$_2$, TaSi$_2$, WSe$_2$, LaB$_6$, and SiC are compounds with very different component melting temperatures and volatility). The third group contains the nitrides BN, AlN, and Si$_3$N$_4$ with a volatile nitrogen component.

The results of composition and structure investigations of obtained condensates are reported in Table II; the micrographs represent the characteristic structures of condensates.

2.3.1. Results: Alloy Evaporation

The majority of the results of investigations of composition and microstructure have been obtained for films deposited on the substrates located in an off-axis position. This is due to the fact that the much smaller number of large particles present in them made composition investigations difficult. The point is that EPMA gives the film composition as an average over the area under analysis. The presence of droplets with a mass larger than that of the entire film in this area shifts the results of such an analysis toward the excess of the droplet component in the film. In the cases where analysis was possible for the films deposited in both positions, no difference in their composition has been detected.

The mean thickness of the films of alloys, deposited by one laser "shot" on the off-axis substrate, was 30–140 nm/pulse

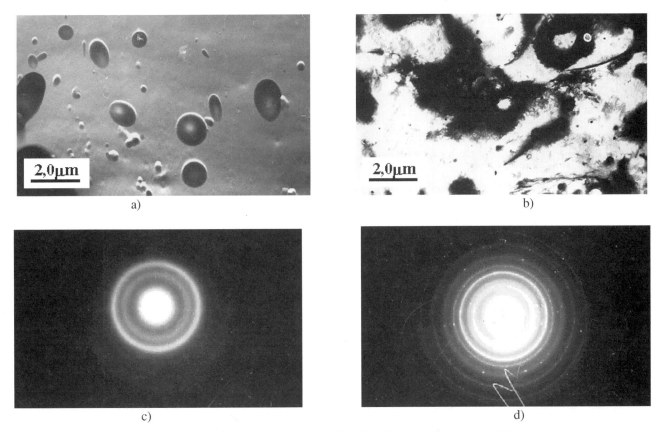

Fig. 2. Microstructure and corresponding SAED patterns of 50Zr-50Ni alloy films deposited at two different power densities. (a) $q = 1.5 \times 10^6$ W/cm^2 (defocused regime). (b) $q = 5 \times 10^7$ W/cm^2 (focused regime). (c, d) Corresponding SAED patterns.

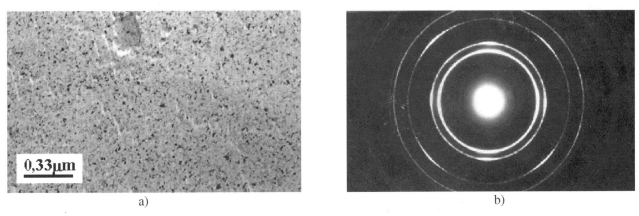

Fig. 3. Microstructure (a) and diffraction pattern (b) of the 70Fe-17Ni-3Cr alloy film.

when a laser was operated in the defocused regime, whereas the thickness of the films was ~10–50 nm/pulse in the case of a laser operating in the focused regime. The role of power density can be demonstrated by the example of films obtained by the evaporation of the alloy 50Zr-50Ni (Fig. 2a–c). For evaporation in the defocused regime when the laser spot diameter on the target was ~6 mm, the crater depth was less than its diameter and the number of defects (droplets and punctures) was much smaller than that for the case of focused regime evaporation. When the evaporation was carried out in the focused regime the depth of the crater was 1.2–1.5 times larger than its

diameter. In this case the traces of solidified melt that splashed out of the crater were present at the crater edges.

The micrographs of the films obtained by the evaporation of alloys 70Fe-17Ni-3Cr, 50Fe-50Co, and 80Fe-20B are presented in Figures 3–5. Such a microstructure was inherent to the alloy films deposited on the substrate at room temperature. The main difference in the morphology of the films of alloy deposited at room temperature was the different mean grain size.

The micrographs and electron diffraction patterns of Zr film deposited simultaneously on line-of-sight and off-axis substrates are shown in Figure 6a,b. It is seen that the film deposited directly from the laser plume (line-of-sight substrate, Fig. 6b) has a large number of defects, like punctures and droplets. The structure of the latter film was always polycrystalline (Fig. 6c), whereas the regions with quasi-amorphous (or fine-grained) structure (Fig. 6d) were present in the film deposited on the off-axis substrate.

The films deposited by evaporation of the alloys with silicon or boron constituents (50Fe-50Si, 80Fe-20B, 75Nb-25Si) on the substrate at room temperature were amorphous.

The EPMA investigations of the films deposited by the evaporation of 80Fe-17Ni-3Cr, 50Fe-50Co, and 50Fe-50Si alloys, containing the elements with close thermophysical parameters discussed above, have shown that the film composition coincided with that of corresponding targets (Table II). The structural analysis of targets and films confirmed the similarity of their structures.

If the components with strongly different volatilities were contained in an alloy, then the corresponding condensates were enriched in the element with the larger volatility. The films obtained by the evaporation of the 75Nb-25Si alloy were amorphous regardless of their deposition on cold or heated (300°C) substrates. Heating of the films to 800°C in an electron diffrac-

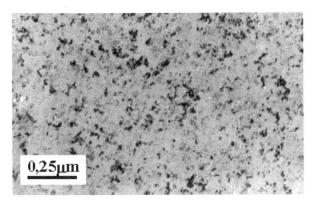

Fig. 4.   Microstructure of the film deposited from 50Fe-50Co alloy target.

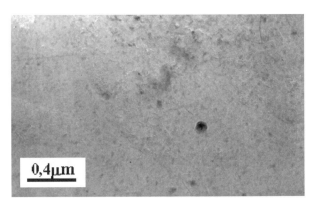

Fig. 5.   Microstructure of the film deposited from 80Fe-20B alloy target.

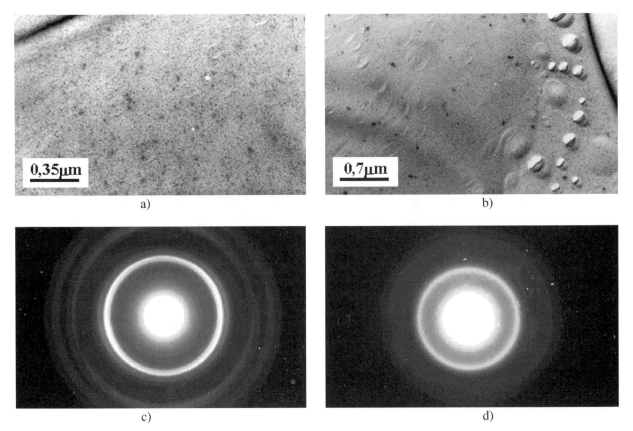

a)

b)

c)

d)

Fig. 6.   Microstructure of Zr films deposited on substrates located at (a) off-axis position and (b) line-of-sight position. (c, d) Corresponding SAED patterns.

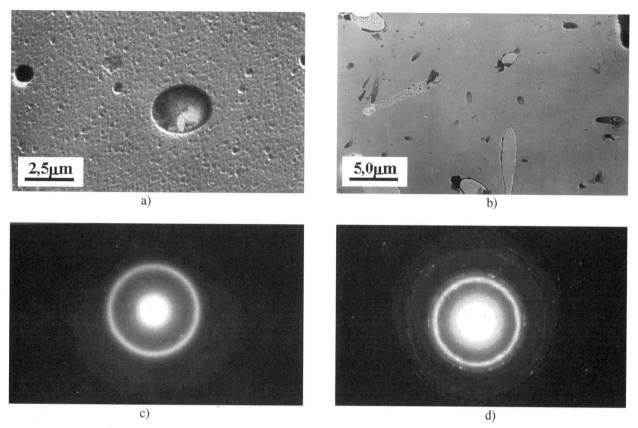

Fig. 7. Microstructure of CrSi$_2$ films deposited at two different power densities. (a) $q = 1.5 \times 10^6$ W/cm^2. (b) $q = 5 \times 10^7$ W/cm^2. (c, d) SAED patterns from the film presented in a and b, respectively.

tion column promoted the formation of the crystalline structure of the film with the NbSi$_2$ lattice. The analysis of the composition of as-deposited films by EPMA confirmed that the silicon amount in the film was 2.5–3 times higher than that in the target.

The structure of condensates obtained by the evaporation of the 80Fe-20B target was similar to that of α-Fe, whereas the target material consisted of two phases, Fe$_2$B and Fe.

The SAED investigation of the structure of films, obtained by the evaporation of the 50Zr-50Ni alloy, was hindered by the presence of many Zr droplets of different sizes (Fig. 2a, b). But in a few cases when it was possible to take the diffraction pattern from a film area free of droplets, its structure corresponded mainly to that of nickel.

2.3.2. Results: Evaporation of Compounds with Nonvolatile Components

The micro- and crystalline structures of the films obtained by pulsed laser evaporation of the refractory materials targets (LaB$_6$, YbB$_6$, SiC, CrSi$_2$, TiSi$_2$, TaSi$_2$, WSi$_2$) exhibit peculiarities different from those of films deposited by alloy evaporation. These films have significantly smaller amounts of macrodefects in the form of droplets and punctures. Those films were amorphous for the deposition on unheated substrates. The variation in laser power density has a weaker influence on the amount of macrodefects in these films, in contrast to that deposited

from the alloy targets. It is seen from a comparison of the microstructure of CrSi$_2$ and WSi$_2$ films deposited by the irradiation of corresponding targets in focused and defocused regimes (Figs. 7a–d and 8a, b). The large average thickness of the film deposited per laser pulse (100–200 nm/pulse) is a noticeable peculiarity of laser evaporation of these materials. Another important peculiarity is that the average thickness of the film deposited per pulse of focused radiation was 1.5–3 times smaller than that deposited by defocused radiation. This regularity was not observed for the average mass of the products ejected under the focused and defocused laser radiation what was revealed by the target weighting before and after laser "shot" (Table II). The depth of the crater left on these targets by the laser "shot" was usually less than its diameter, even for the case of focused radiation. No traces of molten materials have been detected on the crater edges.

The peculiar defects of films deposited by pulsed laser evaporation of the refractory materials were rarely encountered large droplets (0.1–10 μm), punctures, and traces of reflected droplets. The composition of droplets depends on the evaporated compound composition. For instance, the droplets on rare-earth metal hexaborides usually had the structure of the corresponding hexaboride (Figs. 9a–c, and 10). At the same time the droplets on the films obtained by the evaporation of hot pressed powders of silicides and single crystals of silicon

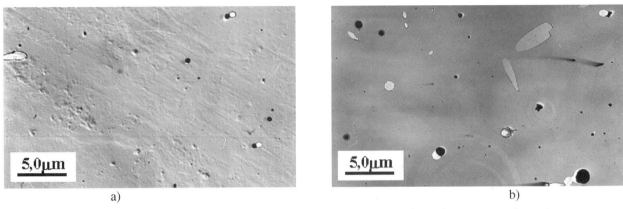

Fig. 8. Microstructure of WSi$_2$ films deposited at two different power densities. (a) $q = 1.5 \times 10^6$ W/cm^2. (b) $q = 5 \times 10^7$ W/cm^2. Solidified silicon droplets, pinholes and traces of reflected droplets are clearly seen in b.

Fig. 9. Microstructure and corresponding SAED patterns of LaB$_6$ film. The right side of micrograph (a) corresponds to the film microstructure. (b) SAED pattern from film area. The left side of micrograph (a) corresponds to the microstructure of a solidified LaB$_6$ droplet on the film. (c) Corresponding SAED pattern from the droplet.

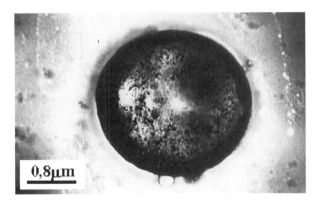

Fig. 10. Solidified droplet of YbB$_6$ on a film deposited by pulsed laser evaporation of YbB$_6$.

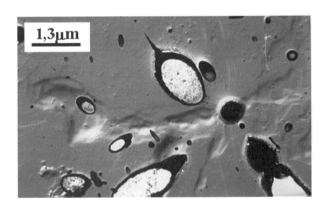

Fig. 11. TiS$_2$ film with silicon droplets and pinholes. The film was deposited at a laser power density $q = 1.5 \times 10^6$ W/cm^2.

carbide (6H-SiC) were the droplets of pure silicon (Figs. 11 and 12). The film structure in the close vicinity of a droplet was crystalline. The heat produced during freezing of the molten droplet was sufficient to crystallize the amorphous film in the droplet vicinity. This is seen from Figs. 9b and c, where SAED patterns, taken from the droplet itself and from the film area remote from the droplet, are presented. The structure of the droplet and the film in its close vicinity (the region of the film

adjacent to the droplet, see Fig. 9a) corresponds to LaB$_6$, and the structure of the film far from the droplet is amorphous.

The films of LaB$_6$ and other hexaborides (YbB$_6$, DyB$_6$) deposited on the substrates at a temperature below 400°C were amorphous. The *in situ* annealing of these films in the electron-diffraction column up to $\sim$700°C leads to their crystallization. The composition of the condensates was similar to the initial hexaboride in all of the cases considered.

The structure of condensates deposited on substrates at a temperature below 400°C by the evaporation of single-crystal 6H-SiC was amorphous. The heating of these condensates in the electron diffraction column to ~850°C promoted their crystallization in the 3C-SiC phase. Electron microscopy and selected area electron diffraction studies of as-deposited amorphous silicon carbide films reveal the presence in the amorphous matrix of small areas with a clearly seen crystalline structure (Fig. 13) corresponding to the 3C-SiC phase.

Fig. 12. Solidified silicon droplet on the surface of SiC film, deposited by PLD.

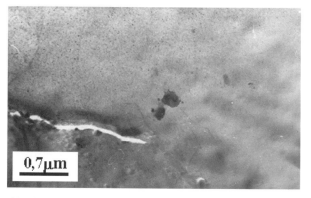

Fig. 13. Microstructure of as-deposited SiC film. The structure of the upper left side of the film corresponds to polycrystalline 3C-SiC. Most of the film area is amorphous.

Droplets have also been detected in films obtained by the evaporation of silicide ($CrSi_2$, $TiSi_2$, $TaSi_2$, WSi_2) targets. But the mean size of the droplets was smaller and their number was larger than those for hexaborides and carbides (Figs. 7 and 8). The structural analysis of the droplets was hindered because of their small size. But in a few cases where the analysis was possible, it was revealed from SAED patterns that the droplets have a structure peculiar to that of silicon. At the same time the silicide condensates in some cases were slightly enriched by a silicon, as was shown by EPMA. The as-deposited amorphous silicide films transited into the crystalline state after their annealing in the electron-diffraction column, and the structure of annealed films coincided with that of the target.

The study of the temperature of the transition from the amorphous to the crystalline state for different silicide films deposited under similar conditions shows that it is different for different silicides. For instance, the transition from the amorphous to the crystalline state for a $CrSi_2$ film occurred at a temperature of ~420°C, whereas for WSi_2 it was ~650°C. The temperature of crystallization initiation of amorphous silicide films also depends on the radiation power density. On average this temperature was ~100–150°C less when the film was deposited under a larger radiation power density (in the focused regime). For example, the crystallization of WSi_2 films deposited at $q = 5 \times 10^7$ W/cm², was initiated at ~650°C, whereas the crystallization of those deposited at $q = 1.5 \times 10^6$ W/cm² was initiated at ~750°C (Table IV).

The films obtained by the evaporation of a cold-pressed powder of low-melting-point CdSe are thick (~200 nm/pulse) and structurally and morphologically inhomogeneous. When deposited on a substrate at room temperature, they have a polycrystalline structure with frequently observed single crystal inclusions (Fig. 14). The composition and the structure of these films coincides with those of the target. The condensation of the film did not occur on the substrates heated to over ~400°C becuse of its reevaporation.

The films obtained by laser evaporation of a WSe_2 target were also thick (~150–200 nm/pulse) and on average had more macrodefects than the other films of this group of materials (Fig. 15). Their structure consists of two phases, amorphous

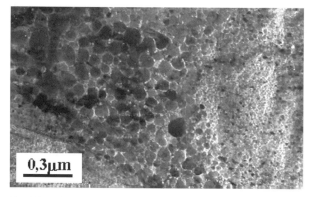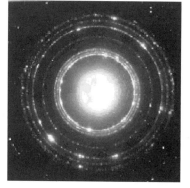

Fig. 14. Microstructure of CdSe film. Regions with different grain sizes are clearly seen. The SAED pattern confirms the film microstructure inhomogeneity.

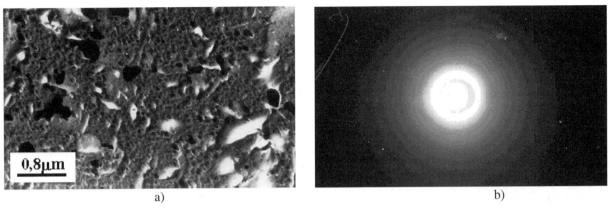

Fig. 15.   (a) Microstructure of a WSe$_2$ film. (b) Halos on the SAED patterns are the evidence of an amorphous film structure.

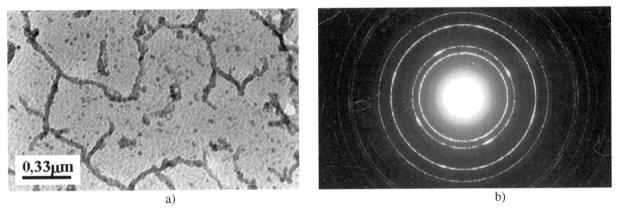

Fig. 16.   (a) Microstructure of a film deposited by laser ablation of an AlN ceramic target. (b) The crystal structure of the film corresponds to that of aliuminum, as follows from the SAED pattern.

Table IV.   Temperature of Amorphous to Crystalline Phase Transition for Silicide Films, Deposited at Two Power Densities by PLD and Obtained by Magnetron Sputtering

Silicide film	Temperature of amorphous to crystalline transition (°C)		
	PLD; $q = 10^6$ W/cm^2	PLD; $q = 5 \times 10^7$ W/cm^2	Sputtered [71]
CrSi$_2$	∼560	∼420	∼300
WSi$_2$	∼750	∼650	—
TaSi$_2$	∼750	∼600	∼380
TiSi$_2$	∼750	∼600	∼300

and polycrystalline. These phases were revealed in the films deposited on the substrate at room temperature and on that heated to 400°C. Unambiguous identification of the structure of the crystalline phase of this film was impossible. However, interplanar distances neither of WSe$_2$ nor of W and Se were revealed. EPMA analysis of these films showed the coincidence of the composition of the film and that of the target.

The general feature of all condensates deposited by pulsed laser evaporation of refractory compounds is the coincidence, to first order, of the film composition with that of the target.

2.3.3. Results: Evaporation of Compounds with Volatile Components

Micro- and crystalline structures of the films obtained by the evaporation of nitride (AlN, Si$_3$N$_4$, and BN) targets were different. The films deposited in the temperature range from room temperature to 500°C by the evaporation of an AlN target were homogeneous and coarse-grained. Droplets and other macrodefects were almost absent in these films. These films have a crystalline structure with a lattice constant peculiar to aluminum. The microstructure of a film deposited by the evaporation of an AlN target on a substrate heated to $T_s \leq 500$°C is shown on the micrograph in Figure 16. The patterns on the condensate surface are the consequence of decoration by aluminum of thermally etched regions of a single-crystal NaCl substrate surface during its preheating before film deposition. The variation of laser power density and substrate temperature has no influence on the film composition.

The thermal etching patterns of the single-crystal NaCl surface have also been observed on the microstructure of films deposited by evaporation of a Si$_3$N$_4$ target on substrates heated to over 450°C (Fig. 17a). However, the films remained amorphous, even at this substrate temperature (Fig. 17b). Droplets of solidified silicon are present on the surface of such films (Fig. 18a). When the droplet was large, then the structure of

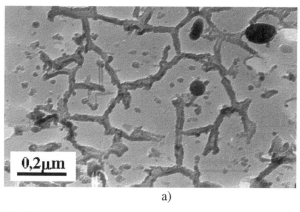

a)

b)

Fig. 17. (a) Microstructure of the film deposited by laser ablation of a Si_3N_4 ceramic target. (b) The film structure is amorphous, as is evident from the SAED pattern.

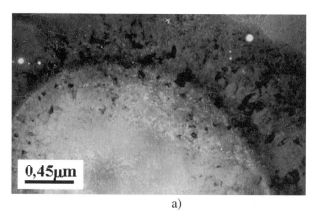

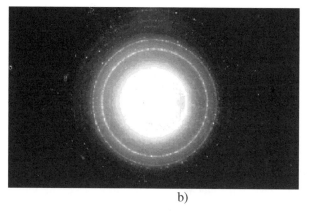

a)

b)

Fig. 18. (a) Solidified silicon droplet on the surface of a film deposited by laser ablation of a Si_3N_4 ceramic target. The structure of the film adjoining the droplet corresponds to silicon (spotty rings on SAED pattern b), whereas the main film area is amorphous (halo in SAED pattern b).

the film in the vicinity of the droplet was crystalline and its structure coincided with that of silicon (Fig. 18b). The density of the laser pulse power had no essential influence on the film structure. The *in situ* annealing of amorphous films obtained by Si_3N_4 evaporation provides the transition of amorphous film to the crystalline state, where the structure coincided with that of silicon. It should be noted that the crystallization temperature (850°C) for these films is higher than that for amorphous silicon films of the same thickness. This fact was established by pulsed laser evaporation of single-crystal silicon targets under similar conditions. The silicon films obtained by this method were also amorphous. Their crystallization during *in situ* annealing occurs at 650°C. The most probable reason for the higher thermal stability of the amorphous phase in the films deposited from a Si_3N_4 target as compared with that deposited from a Si target is the nitrogen impurity they contain.

Films obtained by laser evaporation of the graphite-like boron nitride had a substantially higher density of macrodefects and were inhomogeneous as compared with those of the above considered nitrides (Fig. 19a). The HEED studies of these films had shown that they are amorphous if deposited on substrates heated to 450°C (Fig. 19c). The SAED investigations revealed small areas of the film with polycrystalline structure. These

polycrystalline areas consisted mainly of graphite-like h-BN, although sometimes the areas with c-BN structure have also been observed (Fig. 19b).

The irradiation of a h-BN target at a lower power density ($q = 1.5 \times 10^6$ W/cm^2) resulted in the formation of BN film with larger areas of the polycrystalline h-BN phase together with amorphous areas. In this case the polycrystalline phase was revealed not only by SAED but also during the HEED investigations.

The amorphous phase in boron nitride films (similar to that of silicon nitride) consists, most probably, of boron with nitrogen admixture. This phase evidently arises because of incomplete boron nitride dissociation under the laser impact.

The composition of the films deposited by laser evaporation of AlN and Si_3N_4 targets in nitrogen gas ambient at $P_{N_2} = 10^{-2}$ torr was similar to that deposited in a vacuum.

2.4. Formation of PLD Film Composition

2.4.1. Role of the Target Type and Chemical Bonding

The above studies have shown that the film composition depends on the chemical bonding of target constituents (alloys or compounds), on the target type (cast or pressed powder), and

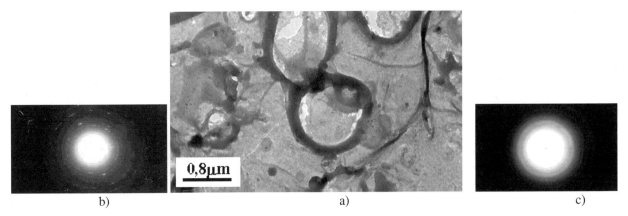

Fig. 19.   (a) Microstructure of a BN film obtained by laser ablation of a h-BN target. The main part of the film has an amorphous structure (SAED in c). Rarely observed small polycrystalline film areas correspond to the c-BN phase (SAED in b).

on the laser radiation power density [50]. If target components are not chemically bonded or are weakly bonded (as in some alloys), the resulting film composition depends greatly on the thermophysical properties of each of the components. If the target material is a compound in which the constituents are chemically bound, the resulting film composition does not depend for the most part on the thermophysical properties of each of the components. The film composition weakly depends on the radiation power density for the case of pulsed laser evaporation of the compound targets with nonvolatile constituents.

Let us consider the mechanism of the influence of the above noted target features on laser film composition formation. Physically the difference between the target types is determined by their thermal conductivity. The thermal conductivity of a cast target is substantially larger than that of a nonpressed or even a hot-pressed powder. The thermal conductivity of a material is one of the most important characteristics defining the effectiveness of laser evaporation. The thermal conductivity coefficient linearly enters the expression for the effectiveness of substance evaporation by laser radiation. This expression gives the relation between the thermophysical and optical characteristics of a substance and radiation energy [3]:

$$B = \frac{K \alpha H_{\mathrm{v}}}{I(1-R)C_{\mathrm{p}}}$$

where B is a dimensionless parameter, K is the thermal conductivity, α is the absorption coefficient, H_{v} is the heat of evaporation, I is the laser power density, R is the reflectivity, and C_{p} is the heat capacity. The dimensionless absorption parameter B can be regarded as a ratio of the heat (per unit area) required to evaporate a layer with a certain thickness and the deposited heat (per unit area) from laser irradiation. In general, for a smaller value of B ($\ll 1$), the material removal process is predominantly evaporation, whereas for a higher value of B ($\gg 1$), particulates are the dominant form of material removal. The material dependence of the B factor is reflected in various physical parameters: K, α, R, C_{p}. Under similar conditions the share of the vapor phase depends on the target material thermal conductivity, which in turn depends on the target type. The

lower the thermal conductivity, the higher is the share of vapor in ablated products. From this point of view the maximal temperature will be achieved at the surface of free (nonpressed) powder, and the temperature at the surface of a cast alloy target (at the same laser power density) will be minimal because of its high thermal conductivity. This means that the probability of total evaporation of all target constituents will be higher in the case of a target with low thermal conductivity.

The second reason for the difference in the composition of laser films deposited by evaporation of alloy and compound targets is the difference between the binding energies of the atoms in an alloy or compound. The atoms in an alloy are weakly bound or are not bound at all, whereas the binding energy in a compound is on the order of electron volts. Therefore during the conventional evaporation of the alloy the resulting film composition has to depend on the thermodynamic characteristics of each of the components because the evaporation occurs as an independent vaporization of each component according to Raul's law [1]. The atoms in compounds are bound by chemical interaction with the energy, depending on the particular properties of the compound components. This energy value and crystalline lattice symmetry determines whether the evaporation of a compound is accompanied by its complete dissociation or partial dissociation or occurs without dissociation [1]. It is clear that if the vaporized compound does not dissociate, then it is quite possible to obtain a condensate with a composition similar to that of the initial substance. In the case of evaporation with dissociation, the condensate composition depends on the volatilities of the compound components. The composition is similar or close to the initial composition in the case of close volatilities of compound components. If the initial compound involves a gas component as a constituent and the evaporation is accompanied by dissociation, it is impossible to obtain a stoichiometric condensate from its evaporation in a vacuum.

2.4.2. Target Effects during Excimer Laser Evaporation

Some problems related to the target effects influencing the film composition can arise during the use of an excimer laser operated in the frequency regime. One of the important problems related to target surface effects of low-power and low-pulse-length frequency laser (excimer laser) ablation was the texturing of the target by the incident laser beam, which makes a non-normal incidence angle with the target. Because of the shadowing effects, as the target is irradiated, surface features are enhanced via cone formation [51]. The cone formation leads to a shift in the angle of the emitted evaporant toward the laser beam. Because the composition and the thickness of the film are optimal when the film is deposited from the peak of the plume emitted from the target surface, a shift of the plume direction is a serious problem for the deposition of stoichiometric films.

Another problem related to target surface changes under low-power frequency irradiation is surface segregation [52]. Surface segregation is the preferential enrichment at a surface or grain boundary of one or more components of a multi-component target. The model of this process incorporates both the propagation into and withdrawal from the bulk target of a melt front. As resolidification begins, higher-melting-point components of the liquid freeze first, driving lower-melting-point liquid components toward the surface. This type of process is used to explain the compositional changes in YBCO PLD films in [52]. As Cu-rich material segregates to the top of the melt, it is removed by evaporation during laser pulses, leaving behind an yttrium-enriched surface. As Y enrichment proceeds, the laser begins interacting with increasingly transparent material—Y_2O_3 is transparent enough to allow a greater melt depth. However, the process is self-limiting, because the distance over which Cu must diffuse toward the surface becomes large. As a result, changes in the PLD (1 J/cm^2, KrF excimer laser) YBCO films were observed. These films were initially copper and barium rich until (at 40 shots/site) a steady-state composition was reached [53].

One more effect that is able to influence PLD film composition is the postevaporation from the hot target after termination of the pulse. This effect was experimentally established by high-speed photography of the film thickness during laser evaporation of a ZnS target with a millisecond pulse duration [54]. The duration of the postevaporation time can vary from one to hundreds of milliseconds, depending on the thermal conductivity of the target and melting point of the target constituents. Evidently the postevaporation mechanism in this case still has the same drawbacks of composition formation that are intrinsic to conventional evaporation.

The results described in [50] were obtained by the evaporation of the target with a high power laser pulse (1000 J/pulse) of sufficiently large duration ($\sim$1 ms). The thickness of the film deposited per pulse under these conditions can be sufficient for its potential applications, so the second "shot" has to be aimed at another location on the target. Therefore the above described problems inherent to frequency excimer laser evaporation are not urgent for lasers operating in a high pulse power regime.

2.4.3. Target Effects during High-Power–Long-Pulse Laser Evaporation

Our investigations of metal and alloy target surfaces had shown that two significantly differing zones appear after laser irradiation. The first zone (zone I) is a zone of direct laser influence. This zone is the bottom of the crater and appears as a smooth surface with rarely encountered solidified "bubbles" of metal. The bubbles are probably the result of subsurface boiling [10]. The second zone (zone II) adjoins the first and has a typical dendritic pattern formed during cooling of the melt of metals or alloys. The second zone is separated from the first by an intermediate region combining the features of the first and second zones (Fig. 20). It is clear that the temperature in zones I and II is quite different during the laser pulse action.

The heating of zone I (up to the highest temperature) occurs by direct laser power transfer to the target. The size of zone I (zone of complete evaporation) is approximately equal to the laser spot diameter. The temperature in zone II is lower than that in zone I because of losses of radiation power to the heating and melting of the target. Whereas the size of zone I depends mainly on the laser spot diameter, the zone II dimensions depend on the temperature in zone I and the thermal conductivity of the target material.

The composition of the film can be identical to that of the target in the case of the saturation vapor pressure balance of different target constituents. Vapor pressure balance is possible in two cases: when the saturation vapor pressure of different target constituents is equal at different temperatures, or when the temperature of the evaporation surface is so high that the

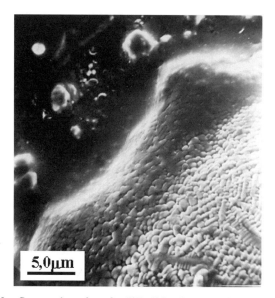

Fig. 20. Crater on the surface of an 50Fe-50Co alloy target after a laser "shot." Three temperature zones are clearly seen. The higher temperature zone is the smooth area with few bubbles at the bottom of the crater (the upper left corner of the picture). The minimum temperature zone is the edge of the crater with dendritic patterns formed during the melt splashing cooling (the lower right corner of the picture). The zone with the intermediate temperature is situated between the above two zones and corresponds to crater walls with partially crystallized melt.

difference between the saturation vapor pressures of different constituents disappears. The highest temperature of an evaporation surface can be achieved in zone I. Therefore one can expect that the composition of a PLD film will be identical or close to that of the target when the film is formed only by the vapors from zone I. However, the effect of postevaporation [54] from zone I can somehow change the composition of the film deposited from vapors of zone I. The deviation of the film composition depends on the thermophysical properties of constituents of the target and its thermal conductivity.

The evaporation mechanism from zone II is quite different from that of zone I, and it is similar to that at evaporation by conventional heating. Therefore the composition of the deposit from the vapors from zone II can deviate significantly from that of the evaporant [1].

Thus the effectiveness of PLD (with respect to the film composition formation) depends on the ratio of the sizes of zones I and II. The larger zone I is in comparison with zone II, the closer the film composition is to that of the target.

This argumentation confirms the results of our investigations of the ablated targets and correlates with the data of [12, 25, 28]. The authors of these papers described the appearance of two components in the laser plume. One component has a stoichiometric composition of vapors and is formed by a nonlinear evaporation process at optical irradiation, and the other component is the result of conventional evaporation. It was shown in [28] that at lower energy densities the thermal evaporation component dominates. The consequence of this is the deviation of the composition of the deposited film from the proper stoichiometry.

The ratio of zone I and zone II sizes depends on the ratio of parts of laser power consumed both by the evaporation and by the heating of the target material by thermal conductivity. One can increase the part of the laser power consumed by the evaporation by decreasing the time of its input (decreasing the pulse duration) and by increasing the absorption coefficient. The latter parameter can be optimized by using a laser with the relevant wavelength. Excimer lasers provide the possibility of variation of the wavelength and a pulse duration a few orders of magnitude shorter than that of lasers working in a free-running regime. However, at a nanosecond pulse length the power density of an excimer laser achieves values of 10^{10} to 10^{11} W/cm^2. Moreover, in this operating regime the absorption (by vapors) of laser radiation becomes significant [7]. In this case the deposition rate drops because of decreased evaporation effectiveness. Simultaneously the properties of the FFS change significantly. The number of charged species in the FFS increases drastically, and the process of molecule dissociation is also enhanced [18]. These effects can negatively influence the composition and the structure of the laser condensates.

Now let us consider the mechanisms of composition formation of laser films. This will be done according to the papers from Table I and [50]. For the sake of simplicity, we will base the subsequent discussion mainly on our results regarding laser alloy and compound films [50]. This is because all experiments in this work were carried out on a wide class of materials and

under identical deposition conditions. The latter means that the rate of laser power input is similar for different targets, and the rate of its dissipation depends on the individual target material properties, such as the absorption coefficient and thermal conductivity.

2.4.4. Mechanism of Composition Formation of Alloy Films

The diameter of the craters appearing on alloy targets after a laser "shot" was 4–5 times larger than the laser spot. One could observe the signs of molten and crystallized substance (Fig. 20) on the edges of the craters. This indicates that a zone heated to high temperatures because of thermal conductivity (zone II) significantly exceeds a zone of direct influence of laser radiation (zone I). We should remember here that because the temperature in zone I is much higher than the evaporation point of most refractory alloy components, all initial alloy components in this zone are evaporated simultaneously. But, in zone II, the temperature is much lower (than that in zone I), so that the actual evaporation point of each alloy component becomes important from the point of view of the resulting film composition. Hence, one may conclude that the film composition is mainly due to the vapor pressure of the alloy components (because the film is formed from almost all of the vapors from zone II). This means that the mechanism of evaporation of alloys by a laser operating in free-running mode is close to the conventional evaporation of alloys and obeys the Raul law. That is, the alloy constituents evaporate independently of each other like pure metals and predominantly as separate atoms [1]. In fact, if the alloy is composed of elements with equal vapor pressures (e.g., 10^{-1} torr) at close temperature, the composition of the condensates coincides with that of the target as it is observed for the alloys 80Fe-17Ni-3Cr, 50Fe-50Co, and 50Fe-50Si (see Table II). When these thermophysical characteristics of alloy constituents are significantly different, the film is enriched in elements with lower temperatures of intensive evaporation (more volatile element), as was observed for 75Nb-25Si and 80Fe-20B alloys.

An important effect for the laser evaporation of alloys was revealed during an investigation of a film obtained by laser ablation of the 50Zr-50Ni alloy. The large amount of droplets was revealed in the films deposited both in focused and defocused evaporation regimes. The films deposited in the focused regime contain relatively large amount of droplets (Fig. 2a, b). EPMA investigations had shown that the films are enriched in zirconium. Simultaneously, as shown by SAED investigations, the main phase in these films was nickel. Taking into account that the mass of the droplets per film area is larger than the mass of the film of the same area, and keeping in mind the peculiarities of EPMA and SAED methods, one can suppose that the droplets in the film are mainly composed of zirconium. This effect can be explained as follows. The difference between the melting point and the temperature at which the vapor pressure $P = 10^{-1}$ torr is significantly lower for nickel than that for zirconium. At the same time, the temperature at which $P = 10^{-1}$ torr for nickel is close to the melting point of zirconium.

Therefore the nickel on the evaporation surface is mainly present as a vapor, whereas the zirconium is present mainly as a melt. The recoil pressure of nickel vapors ejects the molten zirconium from the evaporation zone, and zirconium is transported to the condensation surface in the form of droplets. In the focused evaporation regime, when the narrow and deep crater forms, the vapors accumulated in the crater absorb laser irradiation and transfer the excessive heat energy to the crater walls. Because the molten evaporation area in this regime is larger, one can expect an increased number of zirconium droplets in the condensate deposited in this regime. Just this effect is observed in the micrographs (Fig. 2a, b), reflecting the microstructure of Zr-Ni condensates obtained under two deposition regimes.

The EPMA investigation of the surface of the crater on the 50Zr-50Ni alloy target showed that it was slightly enriched in zirconium compared with the proper composition. This is quite understandable, because the evaporation rate of nickel is higher than that of zirconium. The results obtained during the study of laser evaporation of the 50Zr-50Ni alloy confirm the above discussed laser evaporation mechanism of the alloys and simultaneously reveal the thermophysical parameters of the target material responsible for the droplets in PLD films. A more detailed study is made below of the influence of thermophysical parameters of metals on the particulates in PLD films.

The difference in the structure of the films deposited on the substrates at the same temperature in the range of 20–350°C by laser evaporation of alloys is striking. If the alloys are composed only of metals (e.g., 80Fe-17Ni-3Cr, 50Fe-50Co), the deposited films have a polycrystalline structure, whereas the films deposited from the alloys with nonmetal constituents like silicon or boron (e.g., 50Fe-50Si, 80Fe-20B, 75Nb-25Si) were amorphous. Most likely this effect has something to do with the partial formation of silicide or boride molecules in a gas phase and their condensation at the growth surface. The formation of molecules in a gas phase in a dense laser plasma is firmly established [3, 5]. At the same time the probability of the formation of metal molecules or clusters in a gas phase is low because of weak chemical bonding of metal atoms [5]. The surface mobility of molecules is lower than that of the atoms, and the crystallization temperature of silicides or borides is higher than that of the corresponding metals. These two factors hinder the formation of the crystalline phase in the films of such alloys on a substrate at a low temperature. Note that such an effect is typical for films of high-melting-point materials deposited by other PVD methods.

This observation permits us to suggest yet another probable composition formation mechanism for PLD films of alloys containing elements with higher (than metals) chemical bonding energies. This issue is discussed below.

2.4.5. Mechanisms of Composition Formation of Compound Films

The thermal conductivity of compound targets (AlN, BN, Si_3N_4, LaB_6, YbB_6, CdSe, WSe_2, SiC, $CrSi_2$, $TiSi_2$, $TaSi_2$, WSi_2) in the form of hot-pressed powders (ceramic targets) was significantly lower than that of the melting-cast alloy targets. Note that such targets have been used in many of the works cited in Table I.

The craters that appeared on such targets after a laser "shot" had dimensions close to the laser spot diameter and showed no signs of melting. There is one more important effect observed during laser evaporation of ceramic targets. This effect demonstrates the difference between the evaporation mechanism of alloys and compounds and consists of the difference in the mean thickness of the film deposited per pulse. The mean thickness of the film deposited per laser pulse during evaporation of the ceramic target was 100–200 nm, and this value was 30–140 nm for alloy targets. These facts indicate that the effectiveness of laser evaporation of compound targets is higher. This means that the share of laser power expended on the evaporation is larger for this type of target. The temperature in a zone of direct laser pulse action in the case of a ceramic target is significantly higher, and the evaporation was mainly from this zone (zone I). Therefore under these conditions the evaporation mechanism approaches that expected for laser evaporation. That is, it was expected that the temperature of the target in a zone of direct pulse action is so high that the difference in saturation vapor pressures for compound constituents disappears. In this case the composition of the condensate coincides with that of the target. Analysis of BN, CdSe, LaB_6, YbB_6, WSe_2, SiC, $CrSi_2$, $TiSi_2$, $TaSi_2$, and WSi_2 films confirmed the coincidence or similarity of the target and film composition despite the fact that these compounds have significantly different dissociation energies and are formed of components with substantially different volatilities.

At the same time, some silicide films (e.g., $TaSi_2$, $CrSi_2$) were slightly enriched in silicon, and an excess of silicon was also revealed on the surface of the crater of a $CrSi_2$ target. The surplus of the silicon in these films was concentrated in solidified silicon droplets on the film surface. This implies that despite a more effective input of laser optical radiation (due to lower thermal conductivity of the target), some deviation of the film composition from stoichiometry still takes place. The reason for this effect is the difference in the thermophysical properties of the target constituents. The influence of this factor is similar to that proposed above in the discussion of film composition formation by laser evaporation of a 50Zr-50Ni alloy. That is, the difference between the temperature at which the vapor pressure $P = 10^{-1}$ torr and the melting point of chromium is negative (-506 K), whereas the same difference for silicon is substantially larger (265 K). This means that when the temperature of the surface becomes so high that chromium begins to evaporate intensively, silicon only melts. The recoil pressure of chromium vapors ejects the molten silicon from the evaporation zone, so that silicon is transported to the condensation surface as droplets. This process can be realized at the very beginning of the pulse action, when the temperature at the target surface is not sufficiently high as well as after the pulse action during the cooling of the target. Therefore abundant silicon was revealed both on the surface of the crater on the CrSi$_2$ target and as droplets in the corresponding film (Fig. 7a).

The evaporation of a single-crystal 6H-SiC target also occurred with some dissociation that was revealed by the presence of solidified silicon droplets on the film surface (Fig. 12). The other indication of the dissociation of silicon carbide under laser irradiation was the structure of annealed SiC films that corresponds to the 3C-SiC phase. It is known that the presence of an over-stoichiometric silicon in silicon carbide stabilizes the cubic 3C-SiC phase [49]. Similar results were obtained in [39]. It should be noted that SiC target types were different in [39] and [50]. The SiC target in [39] was ceramic (sintered SiC powder); the target in [50] was 6H-SiC single crystal. The thermal conductivity of a single-crystal target is much larger than that of ceramics. Therefore if the evaporation mechanism of SiC was mainly dependent on the thermal conductivity of the target but not on the atomic bonding, the composition and structure of the films deposited in [39] and [50] would be different.

The above consideration shows that the mechanism of conventional low-temperature evaporation has also been realized under laser evaporation of compounds, but to a lesser degree than in the case of alloys. Aforementioned investigations of the composition of laser plasma during YBCO target evaporation with an excimer laser ($\lambda = 248$ nm, $q \approx 1$ J/cm^2, $\tau = 30$ ns) confirm this conclusion. The presence of two components in laser plasma has been revealed in these investigations. The composition of one of these components corresponds to the vapor generated under conventional evaporation. This was the reason for the deviation of the film composition from stoichiometric [28]. Ultrafast photography techniques have been used to observe the temporal evolution of the fragments of laser ablation of YBCO in an oxygen background gas [55]. It was established that the time interval over which material is evaporated from the target surface may be significantly longer than the laser pulse length. The authors also established that the temperature of the target surface was close to the melting point of YBCO (~ 1400 K) for a rather long time after the laser pulse ceased. These results correlate well with the data obtained in [54] for ZnS laser evaporation with a millisecond pulse length. The coincidence of the results obtained for different compounds deposited by laser evaporation with quite different power densities indicates that the processes occurring under such different conditions have the same nature and tend to suggest the important role of the mechanism of conventional evaporation in film composition formation during PLD.

Indirect confirmation of this conclusion is provided by the fact that most of the films deposited by PLD with stoichiometric composition can also be obtained by conventional evaporation of the same compounds. Among them are the compounds A^{II}B^{VI}, Al$_2$O$_3$, ZrO$_2$, In$_2$O$_3$, SnO$_2$, SiO, SiO$_2$, TiO$_2$, BaTiO$_3$ [1], SiC [49], LaB$_6$, and YbB$_6$ [56, 57], the evaporation of which occurs congruently, coherently, or molecularly. From Table I it is seen that the films of just these compounds with stoichiometric composition were successfully deposited by PLD.

2.4.6. Composition Formation Mechanism of PLD Films Compounds with Volatile Components

The mechanism of formation of a film with complex composition includes at least two processes. One is film composition formation from separate chemical elements on the growth surface, and the second is formation of a film from the molecules of a compound. The generation of FFS by conventional PVD methods occurs during the evaporation or sputtering of the target. In this case the source of separate atoms and molecules is the condensed matter (target). It is clear that the composition of the film is closer to that of the target in the case of molecular sputtering or evaporation of the condensed phase. For pulse laser evaporation there is one more source of molecules, which is the gas phase. As was already mentioned, the formation in a gas phase of clusters, containing two or more atoms is one of the significant peculiarities of PLD. Let us consider the role of this effect in the formation of the composition of PLD films.

Formation of molecules or polyatomic clusters in a gas phase can occur during pulse laser evaporation in a vacuum and in a chemically active background gas. The ability of the evaporant to aggregate is one of the main conditions of gas-phase cluster generation under direct laser evaporation of a material in a high vacuum. Some chemical elements such as sulfur, silicon, and carbon can easily aggregate, and their clusters have been produced by direct laser ablation of a solid target under high vacuum conditions. For typical metals the laser plasma is composed mostly of atomic species. Because of their weak bonding forces, metal clusters are not as easily formed by direct laser ablation as are silicon and carbon clusters. Aluminum is the exception to this rule. Aluminum clusters ranging from Al$_3^-$ to Al$_{50}^-$ were generated by direct laser ablation of an aluminum disk in a high vacuum [5]. The other metal clusters, like Fe, Ni, and Ti, were produced by laser ablation of these metals only in argon or helium background gas in a pressure range of 1–1500 torr. The decrease in the ambient gas pressure resulted in a decrease in size and a narrower cluster size distribution [3].

The study of laser ablation of YBCO in a vacuum has shown that clusters of nearly every combination of Y, Ba, Cu and O were observed, except for Y$_1$Ba$_2$Cu$_3$O$_7$. The majority corresponded to Y$_2$O$_3$ or YBa. On the basis of experimental results it was assumed that gas-phase condensation dominated over the direct ejection of these clusters [58].

The interaction of the laser plume with a reactive background gas plays an important role in the formation of compound thin films like oxides and nitrides because of the concomitant production of atomic and molecular precursors required for the growth of the compound phase. Note that such an effect is not typical for other PVD methods.

To facilitate the growth of an oxide film it is usually necessary to maintain an oxidizing environment during the deposition process. The laser ablation of some pure metals in an ambient of reactive gas can change the film composition because of the gas-phase formation of molecules containing the atoms of ablated material and gas atoms. The effect of formation of YO$_2$ molecules in the gas phase and Y$_2$O$_3$ film deposition was observed during the laser evaporation of Y into O_2 [59]. Nonetheless there are some oxides that are extremely stable even at high temperatures, and their molecules can be easily generated in a vacuum by direct ablation of an oxide target [6]. Although generally successful, the effectiveness of molecular oxygen as an oxidizing agent is somewhat limited because its low activity requires the use of relatively high pressures and deposition temperatures for the growth of high-quality HTSC films. For example, PLD deposition of YBCO films usually requires the use of 0.1–0.2 torr background O_2 and a deposition temperature of 700–750°C [12, 78]. The fact that a relatively high background O_2 pressure is required to obtain HTSC films with good quality shows that the oxidation during the growth of YBCO is kinetically limited by the instantaneous flux of species impinging onto the substrate [6]. The use of atomic species like O or N, which are extremely reactive and form much more efficient corresponding molecules, adsorbed on the surface, makes it possible to deposit the stoichiometric oxide and nitride films at lower background pressures. But it should be kept in mind that the efficiency of producing N atoms is significantly lower than that of O atoms, all other conditions being equal. Maybe this is why there are only a few reports of reactive deposition of nitride films by direct laser evaporation of metal in nitrogen [6].

One more element of the composition formation mechanism of PLD condensates distinguishing this film deposition method from other PVD methods has already been mentioned: noncontrollable postevaporation after cessation of the laser pulse. Confirmation of an established effect of postevaporation from a ZnS target after the and of pulse action [54] was recently obtained for YBCO and BN targets. The appearance of visible emission from laser-ablated particulates after 2.5 J/cm^2 248-nm irradiation of YBCO in a vacuum was observed. The emission is very weak, begins at the hot target surface at early times, and continues long after emission (to 500 μs) from the plasma has ceased. The same effect was also observed for a BN target [4].

The effect of postevaporation can play a negative role in the composition formation of the film. If the flux of postevaporated species is of a molecular nature, as was observed for a BN target, the film composition will not be deteriorated. But in the case of incongruent postevaporation, the film composition will deviate from the stoichiometric. The relative contribution of this effect in the film composition formation depends on many factors, including the properties of the compound, the thermal conductivity of the target, laser operation parameters, the deposition duration time, and the working gas pressure.

The existing experimental data about the conditions of cluster generation are very poor, despite the great importance of this detail in the mechanism of PLD film formation. It is known that the number of clusters that presumably form in the gas phase increases with increasing laser power density. The increasing of cluster amount with increasing laser power density was experimentally observed during mass spectrometric investigation of the laser ablation of YBCO in a vacuum [58]. Another general observation is that the probability of gas-phase molecule generation increases with increasing vapor density during the laser evaporation of the solid and with increasing pressure of

the ambient gas [5, 6]. Existing experimental results allow us to consider that the gas-phase formation of two- or few-atom molecules, all other conditions being equal, is more probable than the appearance of larger clusters. Obtaining such experimental results is hindered by the difficulties in distinguishing between the clusters ejected from the target and those generated in a gas phase.

During the deposition of films with a volatile component the most important role plays the formation of film-forming species (molecules or clusters) in a gas phase. It is known that the sticking coefficient of gas atoms on the substrate at room temperature ranges from 0.1 to 0.5 whereas this coefficient for most molecules containing metal and gas atoms is close to 1 [1]. Therefore the probability of formation of a film with the correct stoichiometry is substantially higher in the case of film-forming species containing oxide or nitride molecules or clusters, compared with the film forming flux containing individual gas and metal atoms. This is especially important for PLD, because the desorption probability of condensed gas adatoms increases sharply because of the increased energy of FFS bombarding the growth surface.

However, molecules and clusters are not generated under any deposition conditions of PLD. For example, in [50] it was found that the films deposited by pulsed laser evaporation (1064 nm, 1 ms, 10^6 to 10^7 W/cm^2) of an AlN ceramic target consisted completely of aluminum, and their composition did not depend both on the deposition conditions and on the ambient medium (vacuum or nitrogen background gas at $P_N = 10^{-2}$ torr). At the same time in [34, 35, 70] AlN films with the correct stoichiometry were produced by frequency laser (248, 266 nm, 3–10 ns, 1–3.5 J/cm^2) evaporation of an AlN ceramic target both in a vacuum and in nitrogen background gas. These films had a structure with parameters close to that of bulk AlN, and only small amount of little (<0.1 μm) metal droplets was observed on their surfaces. The authors related these droplets to clustering of aluminum atoms in a gas phase.

Considering these results on the basis of the above treated mechanism of composition formation of PLD films, one may conclude that the evaporation of a ceramic AlN target by high-power laser irradiation with a longer wavelength and a millisecond pulse duration occurs with the complete dissociation of nitride. Because no phases other than aluminum were revealed in the deposited film, one can assume that the interatomic gas phase interaction of evaporated N and Al atoms and of evaporated Al atoms with background nitrogen atoms did not take place. Therefore it can be concluded that the density of the vapor generated during the irradiation of an AlN ceramic by a laser with such parameters is not high, although the thickness of the film deposited per pulse was rather large, at 150–200 nm [50]. In fact, as cited above [5, 6] the probability of cluster formation increases with increasing the vapor density.

The evaporation of an AlN ceramic target by low-power laser irradiation with shorter wavelength and nanosecond pulse duration occurs with a low dissociation level. The dissociation of AlN is confirmed by the presence of a small amount of Al droplets on the surface of the film, as noted in [34, 35, 70]. The authors relate the appearance of Al droplets in AlN films, in view of their round shape and small dimensions, to the clustering of Al atoms in a gas phase. The deposition of AlN films with the correct composition is an indication of a low level of AlN dissociation.

The influence of nitrogen pressure on AlN film composition was revealed only in [35], whereas no indications of this effect were found in [34, 70]. A relative nitrogen content of N/Al = 0.95 was in the film deposited in vacuum, whereas a value of $1.25 > N/Al > 1$ was found in the film deposited in a nitrogen pressure range of $10^{-2} > P_{N_2} > 10^{-4}$ torr. Increased the nitrogen pressure to over 10^{-4} torr was accompanied by the appearance of over-stoichiometric nitrogen in the film and by deterioration of the film structure. Authors associate both of these effects with the formation of nitride clusters in a gas phase, which in turn is the result of enhancement of the interaction between ablated species and ambient gas at higher nitrogen pressures. Thus the third constituent of the composition formation mechanism of PLD films can be realized where an excimer laser is used for thin film deposition. Namely, apart from the growth of a film directly from molecules evaporated from a target and the formation of the film composition from the individual atoms on the growth surface, the third component is the formation of a film of molecules and larger clusters generated in a gas phase.

At the same time we would like to note the negative role of gas phase clustering during AlN film formation. In [34, 70] the presence of a small amount of little Al droplets on the film surface was explained by clustering of Al atoms in a gas phase. This means that Al atoms that appear because of dissociation of AlN under laser impact interact more easily with each other than with nitrogen atoms. This conclusion follows directly from the fact that the influence of the nitrogen ambient gas on (i) the film composition and (ii) the amount of Al droplets was not revealed in [34, 70]. A possible reason for this effect is the low concentration of atomic nitrogen compared with the concentration of molecular nitrogen, the chemical activity of which is significantly lower than that of atomic nitrogen [6]. Thus, the clustering of Al atoms in a gas phase on one hand weakens the microstructure of the film because of the presence of the droplets in it, and, on the other hand, it reduces the growth rate of the film.

Unsuccessful attempts to deposit silicon nitride film by laser evaporation (1064 nm, 1 ms, 10^7 W/cm^2) of a Si$_3$N$_4$ ceramic target also relate to the almost complete dissociation of nitride and the very low probability of nitride molecule formation in the gas phase under these conditions [50]. However, the dissociation of silicon nitride was not complete, as in the case of an aluminum nitride target. The higher crystallization temperature of these films compared to that for silicon films deposited under the same conditions indicates the presence of a small amount of nitrogen in these films. The less complete dissociation of silicon nitride compared to aluminum nitride during their laser evaporation under similar conditions [50] can be related to the lower binding energy in the AlN molecule (3.7 eV) compared to that of the SiN molecule (5.2 eV) [60].

The degree of dissociation of the boron nitride target under the laser impact was lower than that of aluminum nitride and silicon nitride. This indicates the presence of crystalline h-BN and c-BN phases along with an amorphous phase in the films deposited in these experiments [50]. The amorphous phase relates to the boron, with a low nitrogen content. The degree of dissociation of boron nitride decreases with decreasing laser power density (evaporation in defocused regime, $q = 10^6$ W/cm^2). This conclusion follows from the observation of decreasing amorphous phase areas in the films deposited at a lower power density in comparison with that obtained at $q = 5 \times 10^7$ W/cm^2.

Partial dissociation of boron nitride was also observed during evaporation by excimer laser (248 nm, 2.7 J/cm^2). This conclusion follows from the deposition conditions of these films, which included the presence of a nitrogen background gas ($P_{N_2} = 0.05$ torr) as a necessary condition [36]. An investigation of the film morphology indicated the presence of two types of particulates in an otherwise smooth matrix. Spherical particles with diameters up to 1 μm were composed of boron. Irregularly shaped particles about 350 nm in size consisted of h-BN and were presumed to be ejecta [36]. If one assumes that the boron particles in these films are the result of gas-phase clustering of boron atoms, as was observed for aluminum atoms during the evaporation of AlN targets in [34, 35, 74], then it is necessary to agree that boron atoms more readily interact with each other than with the nitrogen.

In [61] c-BN films were deposited by laser ablation (248 nm, 1.5 J/pulse) of a h-BN target in the presence of Ar-N$_2$ ambient gas. The films were deposited on Si substrates that had been heated to ~450°C and which were electrically biased. The laser plume was further excited by an electric discharge ignited in the nitrogen background gas. This mode of deposition was chosen to produce additional nitrogen atoms in the substrate region because the chemical efficiency of nitrogen molecules is quite low. The authors do not mention finding any particulates in their films.

The above results indicate that boron nitride is also dissociated under a laser impact, although the degree of dissociation of this nitride is much lower than that of aluminum nitride or silicon nitride. The degree of dissociation of boron nitride depends on the deposition conditions and laser parameters.

2.5. Conclusions

The above problems of composition formation of PLD films and related questions led us to the following conclusions.

1. The composition formation mechanism of the films deposited by pulse laser evaporation of targets with complex chemical composition includes at least three constituents: (i) the growth of the film due to chemical reaction of monoatomic components on the condensation surface being evaporated from the target; (ii) the formation of the film directly from molecules or clusters evaporated from the target; (iii) the formation of the film from molecules or clusters appearing in the gas phase due to the reaction of evaporated atoms with the background gas

atoms. The portion of each constituent in the composition formation mechanism depends on the properties of the evaporant, the deposition medium (vacuum, background gas), and the laser parameters.

2. If the evaporation of the compound by conventional methods occurs without dissociation, congruently or coherently, then it is reasonable to obtain a film of stoichiometric composition by PLD. Nevertheless there is the probability of compound dissociation, depending on the laser operation regime.

3. The clustering effect of atoms in a gas phase, which is possible under the proper laser operation parameters and at certain deposition conditions, is able to influence the film composition, especially for compounds with volatile components.

4. The mechanisms of composition formation of the films deposited by pulsed laser evaporation of alloys and compounds are quite different.

The laser evaporation of alloys composed of metals only (i) occurs mainly in monatomic form. (ii) The evaporated atoms do not interact with each other in the gas phase. (iii) The evaporation of cast alloys occurs both from the spot of direct laser impact and from the indirectly heated zone. These peculiarities of pulsed laser evaporation of alloys approach wthat is typical of conventional evaporation. Let us recall that the composition formation of alloy films deposited by conventional evaporation methods obeys the Raul law [1].

The laser evaporation of compounds can occur (i) directly in molecular form. (ii) The formation of molecules of compound can take place in a gas phase. (iii) The effectiveness of laser evaporation of compounds is higher than that of alloys because of the lower thermal conductivity of compound targets and because of their higher atomic binding energy. These differences in the evaporation mechanisms of alloys and compounds promote the laser deposition of compound films with compositions close to that of the target more frequently than that of the alloys.

3. STRUCTURE OF PLD FILMS

Perfection of the crystalline structure of a film depends on the nucleation and growth conditions. The most important of these are the arrival rate of FFS on the growth surface, substrate temperature, the energy and the composition (atoms, molecules, or clusters) of FFS, and the crystalline structure of the substrate.

3.1. Factors Influencing the PLD Film Structure

A high deposition rate leads to a high rate of nucleation and, as a result, the formation of small islands with high density. On one hand these processes lead to the formation of continuous films with a lower average thickness, but, on the other hand, such films have a smaller grain size. Therefore, to obtain a film with more perfect crystalline structure under such deposition conditions, a higher substrate temperature is required [62].

The increased energy of the species arriving at the substrate increases their sticking coefficient and surface mobility. These factors make it possible to deposit a film with greater structural perfection. However, if the energy of condensed species exceeds the optimal value (5–10 eV [63]), they promote the surface and bulk defects in the film, the creation of additional nucleation sites [64] and a decrease in the grain size of the film [65]. The presence of charged species also contributes to the increase in the number of nucleation sites (i.e., nuclei density) [62]. All of these effects negatively influence the film structure.

Each of the factors considered is present in the PLD process. Actually, instantaneous deposition rates during laser evaporation are very high and can be varied from 10^{14} to 10^{22} cm^{-2} s^{-1}; at a pulse duration of a nanosecond, the energy spectrum of the ions is in the region of 0–2000 eV, with a mean energy of 100–400 eV, depending on the experimental conditions. The energy of the neutral component of the laser plasma flux is about 10 eV; the ionization degree of the plasma is between 10% and 70% and depends strongly on the energetic and spectral laser parameters [18]. The composition of the flux of FFS is very complex and includes particles from ions to polyatomic clusters [5, 16, 18]. All of these factors are present simultaneously during the growth of the film.

3.2. Role of Molecules and Larger Clusters

Nucleation sites of different dimensions and different compositions, in the form of molecules or clusters, can be a noticeable limiting factor of the growth of crystallites in a film, even at a rather high surface mobility of adatoms. This means that the structure of a film deposited from the atomic flux onto the substrate with the same temperature has a more perfect crystalline structure than that formed of mixed film-forming particles (atoms, molecules, clusters). For example, whereas the structure of a film formed by atomic flux is polycrystalline, a film grown from a mixed film-forming flux has a fine-grained or amorphous structure at the same substrate temperature. The epitaxial growth of a film grown from mixed film-forming flux has to start at a higher substrate temperature than that of a film deposited from atomic flux. The role of the composition of film-forming flux in the formation of film structure has received little attention in the literature. Evidently this is related to the experimental difficulties encountered in creating a film-forming flux with a controlled composition. We presume, however, that the composition of the FFS (namely, atoms, molecules, clusters) plays an important role in the formation of the film structure.

In [68] the correlation between the composition of the vapor phase and the structure of simultaneously deposited films was studied. The composition of vapors generated by conventional and laser evaporation of a wide class of compounds (CdS, CdSe, ZnSe, SnTe, GeTe, GeSe, GeS, GaSe, InSe, Sb$_2$S$_3$, As$_2$S$_3$) was investigated by mass spectrometry. Structural studies of the corresponding films show that the presence of $\geq 10\%$ polyatomic clusters in a vapor phase results in the formation of a film with an amorphous structure. If the vapor is predominantly composed of atomic species, as is the case of evaporation of CdS, CdSe, ZnSe, and SnTe, then the deposited films are polycrystalline. The authors also noticed an unexpected similarity in the composition of vapors generated by conventional evaporation and pulsed laser evaporation.

Another example of the role of the composition of FFS in film structure formation are the results of investigations carried out in [69]. In this paper the authors also reported on a mass spectrometric investigation of vapor generated by pulse laser evaporation of GaAs single crystal and the structure of the films deposited from this vapor. It was revealed that the vapor contained charged GaAs, As_3, As_4, and Ga_2 molecules and neutral clusters containing three to five atoms of different types. The deposited films had a quasi-amorphous structure and were composed of very fine GaAs crystallites. The authors believe that such a superfine structure of the film is determined by a large number of nucleation sites created by condensed molecules and clusters.

Our investigations of the influence of laser power density on silicide film structure also indirectly demonstrate the dependence of the film structure on the properties of FFS. $CrSi_2$, WSi_2, $TaSi_2$, and $TiSi_2$ silicide films were deposited by PLD (1064 nm, ms, 1000 J/pulse) in two power density regimes, $q = 10^6$ and $q = 5 \times 10^7$ W/cm^2 which were varied by focusing and defocusing of the laser beam. The films were deposited on the substrates at room temperature and had amorphous structure. During their *in situ* annealing in an electron diffraction column it was established that each silicide underwent its proper temperature transition into the crystalline state. At the same time there a common feature in the structures of these films was revealed. That is, the silicide film deposited in the defocused regime had a $\sim$100–150°C higher phase transition temperature than that deposited in the focused regime (Table IV). Note that the transition from the amorphous to the crystalline state for similar silicide films with close thickness, deposited by magnetron sputtering, occurs at a temperature that is 150–200°C lower [71] than that for PLD films deposited in the focused regime (Table IV).

The results obtained for PLD silicide films suggest that the structure of the films deposited in the focused regime was more "perfect" than that of films deposited in the defocused regime. There are two possible reasons for this effect. One of them is that the effective substrate temperature was higher during the deposition in the focused regime. In other words, the arrival energy of the FFS generated in this regime was higher what is reasonable for such power density irradiation. Simultaneously, the dissociation of the silicide is more probable for higher power density, and thus the film deposited in this regime could be grown mostly from atoms. This effect can also provide a more "perfect" film structure [66, 68]. On the other hand, the FFS generated during low power density evaporation of materials with strong chemical bonding were mostly composed of molecules [18], the mobility of which is lower than that of the atoms [66, 67]. So it is reasonable to expect the film formed of molecules to be more amorphous than that grown from atomic flux. The lower transition temperature for similar silicide films deposited by sputtering supports this idea, because the flux of

species generated by sputtering is compositionally uniform and is mainly composed of atoms [13], and this is the reason for the more "perfect" structure of these films.

One more observation indirectly confirms the idea of a leading role for FFS composition in the structure formation of PLD films. The structure of Zr films was investigated with respect to the location of the substrate. Zr films were deposited in the defocused regime simultaneously on the substrates; one of these was placed at an off-axis position, and the other at a line-of-sight position (see Fig. 1). The structure of the main part of the film deposited on the substrate at the off-axis position was fine-grained, and the film deposited on the substrate placed at the line-of-sight position had a normal polycrystalline structure, as seen from diffraction patterns (Figs. 6c and d). Because the formation of molecules or clusters in a gas phase is not peculiar for the PLD evaporation of metals [5], the reason for the more "perfect" structure of Zr films deposited from the central region of the laser plume is the higher kinetic energy of Zr atoms arriving at the substrate from this region of the plume. The higher energy of the species from the central zone of the laser plume is an experimentally established fact [17, 18]. Higher energy provides a higher mobility to Zr adatoms, and, correspondingly, Zr film deposited on the substrate at the line-of-sight position had a more perfect structure. Because the silicide films were deposited on the substrates located at the line-of-sight position for both focused and defocused regimes, one can assume that the energy upon arrival at the substrate species is high. Therefore it is reasonable to expect to obtain the films with similar structures deposited in the two regimes. As this was not the case, it is reasonable to assume that the difference in the structural quality was conditioned by the lower surface mobility of FFS generated in the defocused regime. Because the energies of FFS generated in the two regimes are nearly equal, it is valid to associate the lower surface mobility of species generated in the defocused regime with their larger dimensions. In other words, it is reasonable to assert that the FFS generated in the defocused regime consist mainly of molecules and larger clusters, in contrast to mainly atomic flux generated in the focused regime. The difference in the compositions of fluxes generated in the two laser operation regimes is the main reason for the more disordered structure of the films deposited from the species of larger dimension.

3.3. Gas-Phase Clustering

When PLD is used for the deposition of films containing volatile constituents such as oxygen or nitrogen, the process is typically performed in a chamber with a background gas pressure of a few tenths of a torr. This means that interaction between the gas particles and evaporated species is inevitable. The interaction between these species has few effects that are expected to influence film formation. It reduces the energy of FFS and excludes ions. For example, during magnetron sputtering of WTi alloy, the concentration of Ar^+ ions with an energy $E_i > 80$ eV showed more than an order of magnitude decrease, and the average energy of sputtered atoms decreased from ~4 to 1.5 eV when P_{Ar} was increased from 10^{-4} to 10^{-1} torr [100]. Thus in the case of PLD in an ambient of gas, factors that negatively influence the film structure, like noncontrollable bombardment of the growth surface by energetic particles and charged species, are excluded. But the probability of formation of large-scale FFS (clusters) in gas phase increases in the presence of background gas. Because the mobility of clusters is low, one may expect that the substrate temperature during PLD in a gas ambient has to be higher for deposition of a film with high structural quality.

The last assertion can be confirmed by the results obtained in [34, 35, 70], where epitaxial AlN films of similar or close structural perfection were produced by PLD in a vacuum or in nitrogen as the background gas. These AlN films were produced by frequency pulsed excimer laser evaporation (248 and 266 nm, 30 ns) of an AlN ceramic target. The films were deposited on sapphire single crystal substrates and had a highly oriented structure with a basal plane parallel to the substrate.

However, such universal parameters as the substrate temperature and working gas pressure were quite different for different deposition methods. In [35] the films were deposited on the substrate at $T_s = 1000°C$ ($q = 1$ J/cm^2, target–substrate distance 3 cm, repetition rate 10 Hz), and the dependence of the film structure on the nitrogen pressure in the chamber was investigated. It was established that the epitaxial films with the most perfect structure (X-ray rocking curves with minimum full width at half-maximum, FWHM) could be deposited only at $P_{N_2} \leq 10^{-4}$ torr. The increase in nitrogen pressure in the chamber promoted rapid deterioration of the film structure (fast increase of FWHM). The authors relate the degradation of film crystallinity with increasing P_{N_2} to the increased number of collisions between the evaporated species and the gas particles in a gas phase. They assume the result of interaction is that the vapor species lose energy as they migrate on the substrate and form clusters with low surface mobility.

In [70] the substrate temperature needed for epitaxial growth of AlN films in a vacuum ($q = 2$ J/cm^2, target–substrate distance 4 cm, growth rate per pulse 0.025 nm/pulse, average growth rate $V_g \approx 1.25$ nm/s, repetition rate 50 Hz) was $T_s = 670°C$, which is substantially lower than that determined in former investigations. If the process was carried out in nitrogen background gas at $P_{N_2} = 5 \times 10^{-2}$ torr then films of similar crystallographic orientation and structural perfection were obtained at $T_s = 500°C$. As the nitrogen pressure was increased to $P_{N_2} = 0.4$ torr the structure of the film deteriorated, which was indicated by the simultaneous appearance of two different crystallographic orientations of film crystallites.

The most probable reason for increased epitaxial temperature during the deposition in a vacuum in comparison with that in nitrogen ($P_{N_2} = 5 \times 10^{-2}$ torr) is the high instantaneous rate of vapor condensation and the high laser repetition rate, which was noted in particular. As already mentioned, the nuclei are smaller, and their densities are higher at high deposition rates [62]. Therefore a higher substrate temperature is needed for the formation of epitaxial films with perfect structures. If the

deposition is carried out in a gas background, then the instantaneous condensation rate is reduced because of the scattering of evaporated species on the gas particles. The reduced condensation rate makes it possible to deposit epitaxial films of the same perfection at lower substrate temperatures. At the same time, the deterioration of the film structure with increasing nitrogen pressure was noted both in [35] and [70]. In the latter, the indicator of film structure deterioration was the appearance of crystallites with one more crystallographic orientation.

In [34] ($q = 3.5$ J/cm^2, target–substrate distance 3 cm, average growth rate $V_g = 0.015$ nm/s, laser repetition rate 1 Hz) epitaxial AlN films were deposited on a substrate in a vacuum at $T_s = 500°C$. Such a reduction of the epitaxial temperature in comparison with that used in [70] can be explained by the substantially lower deposition rate and lower pulse repetition rate used in [34]. However, it is difficult to explain the fact that the perfection of AlN films deposited on a substrate at $T_s = 500°C$ in a nitrogen background ($P_{N_2} = 5 \times 10^{-3}$ torr) was the same as that of films deposited in a vacuum in [70]. It is also impossible to understand why the growth rate of AlN film deposited in [70] was almost two orders of magnitude higher than that in [34], whereas the power density was higher and the target-to-substrate distance was lower in the latter case.

While considering the structure formation of AlN films in [34, 35, 70], we assumed that the main FFS are the particles that are not smaller than molecules. The following discussion provides basis for this assumption. In [35] the dependence of the composition of AlN films on nitrogen pressure was investigated. It was shown that the relative nitrogen content in films deposited in a vacuum is N/Al ≈ 0.95. At the same time, it is known that there are such laser operation regimes in which the evaporation of AlN occurs with complete dissociation of the compound, and as a result a pure aluminum film is deposited [50]. It is also known that it is possible to deposit stoichiometric AlN films through the evaporation of AlN by conventional methods [1]. Keeping in mind these experimental facts, one can assume that the evaporation of AlN by excimer lasers with the operation parameters used in [34, 35, 70] occurs almost without dissociation. Therefore the main FFS in these experiments were AlN molecules and possibly even larger molecular clusters. With this in mind, it is natural to expect that the epitaxial temperature of these films has to be higher than that of the film deposited from atomic aluminum and nitrogen fluxes. The latter assumption might be confirmed by the data presented in Table V. Information on the structures of different materials films deposited by different PVD methods is presented in this table.

3.4. Crystallization Temperature as an Index of Film Structure "Perfection"

The preceding discussion shows that one of the significant factors influencing the structure of films is the composition of the film-forming flux, which may include atoms, molecules, and larger clusters. We have already seen that the film-forming flux

Table V.   Temperatures of Crystallization and Epitaxial Growth of Films Deposited by PLD and Other PVD Techniques

Film composition	Deposition technique	Film structure; substrate temperature (°C)	Working gas pressure (torr)	Reference
AlN	PLD; excimer	SC; 500; 670; 1000;	N_2, 10^{-4}; vac.	[34, 35, 70]
AlN	PA MBE	SC; 500	Vac.	[72]
AlN	S	SC; 350	Ar-N_2, $\sim10^{-1}$	[73]
GaN	PLD; excimer	T/PC; 500	Vac.	[34]
GaN	PA MBE	SC; 500	Vac.	[72]
SiC	PLD, excimer	PC; 800	Vac.	[39]
SiC	PLD, Nd:YAG, ms	A; 450	Vac.	[50]
SiC	e-beam evaporation	Fine-grained PC; unheated	Vac.	[74]
LaB$_6$	PLD, Nd:YAG, ms	A; 450	Vac.	[50]
LaB$_6$	S	PC; 200	Ar; $\sim10^{-2}$	[75]
LaB$_6$	e-beam evaporation	PC; 350	Vac.	[76]
In-Sn-O	PLD; excimer	PC; 200	O_2, $\sim10^{-2}$	[47]
In-Sn-O	S	PC; RT	O_2, $\sim10^{-2}$	[77]
YBCO	PLD; excimer	T/PC; 700–750	O_2, $\sim10^{-1}$	[12, 78]
YBCO	S	SC; 550–730	Ar-O_2, $\sim10^3$–10^{-1}	[99]
GaAs	PLD, Nd glass, ns	SC; 600	Vac.	[101]
GaAs	RF ion-beam epitaxy	SC; 400	Vac.	[102]

Abbreviations: PA MBE, plasma activated molecular beam epitaxy; S, sputtering; SC, single crystal; PC, polycrystalline; T/PC, textured polycrystalline; A, amorphous.

generated by pulsed laser evaporation of compounds frequently has a complex composition, including ions, atoms, molecules, and larger particles. At the same time it is known that the surface mobility of large particles is lower than that of atoms. It is also known that the film-forming flux generated by such PVD techniques as molecular beam epitaxy or sputtering is mostly homogeneous and mainly contains atomic species. Therefore it is reasonable to assume that the substrate temperature for the deposition of films of identical composition and with similar structural perfection has to be lower in the case of PVD methods in comparison with PLD.

The particular experimental conditions can noticeably influence the temperature of crystallization or of epitaxial growth, as was seen, for instance, from a comparison of the results obtained in [34] and [70]. Nevertheless there is a certain temperature interval typical for the deposition of a film of a certain composition and of a certain structural perfection by a certain technique.

With this in mind, let us try to compare the structural quality of the films deposited by PLD and by other PVD methods. Let us choose substrate temperature as a performance criterion of the method. The lower is the substrate temperature for deposition of a film with higher structural perfection, the more optimal are the deposition conditions and deposition method for film production. The important condition for a comparison of deposition techniques is the equality or the closeness of the pressures of working gas during the film deposition.

Table V contains information about the structure of the films of some compounds deposited by different PVD methods and the values of the substrate temperature at which the films were obtained. It follows from these data that the substrate temperature at which films of the same composition and structural perfection were deposited by PLD was not lower and frequently was even higher compared with the temperature of deposition by other PVD methods.

One more criterion of the degree of perfection of the structure of as-deposited films may be the temperature of transition of initially amorphous film to the crystalline state (crystallization temperature). The lower is the transition temperature for the film of the same composition from the amorphous to the crystalline state, the closer is the structure of the film to the crystalline state. In other words, if the crystallization temperature of the amorphous film deposited by one of the PVD methods is lower than that for the film deposited by PLD, it is of value to consider that the former film structure has a greater degree of perfection. As an example, the crystallization temperatures of some amorphous silicide films deposited by PLD and by magnetron sputtering are listed in Table IV. From these data it follows that the temperature of transition from the amorphous to the crystalline state for silicide films deposited by PLD in our investigations was, on average, noticeably higher than that for similar films of close thickness deposited by magnetron sputtering.

All examples given indicate that the crystallization of films grown from the film-forming flux generated by laser pulse is impeded in comparison with that for films deposited by other deposition techniques.

3.5. Influence of Laser Parameters and Substrate Temperature on Epitaxial Growth of PLD Films

Analysis of the investigations devoted to the peculiarities of epitaxial growth of PLD condensates presented below confirms the conclusions of the previous section.

The epitaxial growth of the film is determined by the following conditions: (i) The substrate temperature cannot be lower than a certain value. This requirement provides the advantage of epitaxial nucleation and growth compared with nonoriented nucleation. (ii) The supersaturation of the condensed vapor has to be of a low value, at which the nucleation occurs only on the preferred centers, providing epitaxial growth. (iii) Increased supersaturation of the vapor enhances the possibility of chaotic nucleation exponentially.

A high substrate temperature activates the attachment of adatoms at a particular site of crystal lattice of the substrate and enhances the surface and bulk diffusion. The surface diffusion smooths the mismatch during the coalescence of nuclei. The temperature of film epitaxial growth is linearly related to the rate of FFS condensation: the logarithm of the condensation rate is proportional to the inverse substrate temperature [62]. The upper limit of the deposition rate is determined by the kinetics of atomic rearrangement. A certain time is needed for adatoms migrating along the substrate surface to reach thermodynamically stable sites. The kinetics of epitaxial growth is determined by the substrate temperature and by the kinetic energy of fast atoms and ions supplied to the growth surface. Epitaxial growth is an energetically limited process and is highly material dependent. For example, in molecular beam epitaxy of GaAs, the typical rate of deposition is about 1 μm/h. The temperature is chosen to be high enough to provide the surface atom mobility needed to form the film layer by layer. The epitaxial film could be grown faster by raising the temperature to accelerate the rearrangement process, but this results in the decomposition of GaAs by the evaporation of As and bulk diffusion of Ga or As into the substrate. In general, the temperature window will vary according to the material properties, including the surface mobility of the constituents of the compound, their bulk diffusivity, and the thermal stability of the compound.

As was mentioned several times, the characteristic features of pulsed laser evaporation are the pulses of instantaneous vapor fluxes of high density (which are separated by periods of the absence of vapor flux in the case of frequency deposition) and the relatively high arrival energy of vapor species. These peculiarities influence the epitaxial growth of PLD films.

Special in-depth investigations of the epitaxial growth of PLD films of different materials were carried out at the Moscow Engineering Physical Institute [101]. Because of the great importance of these investigations for the comprehension of the structure formation of PLD films, we consider the most interesting of them in more detail.

The influence of laser parameters and substrate temperature on perfection of the crystal structure of GaAs and PbTe films was investigated by both *ex situ* and *in situ* depositions. The *ex situ* depositions were carried out in a vacuum chamber, with subsequent structural investigations by RHEED (reflection high energy electron diffraction), HEED, and TEM. Depositions *in situ* were carried out in an electron diffraction or electron microscope column with simultaneous observation of the structural evolution during the film growth. The laser systems used for evaporation were based on Nd glass ($\lambda = 1064$ nm) and could operate in two different irradiation modes: in an ordered running regime with $q \approx 10^5$ to 10^6 and in Q-switch mode, with $q \approx 10^8$ to 10^9 W/cm^2. Pulse duration in the ordered pulse regime was 0.5 ms, while in Q-switch mode it was 30 ns. Single-crystal plates of cleaved NaCl and polished GaAs were used as substrates.

Ex situ investigations of the temperature dependence of the structure of the films with thickness of about 30 nm deposited in two regimes revealed the following results. The structure of the films deposited at $q \approx 10^5$ W/cm^2 on a NaCl substrate at room temperature was amorphous. The films deposited at 250°C had structures composed of amorphous and polycrystalline phases. The increase in the substrate temperature to 300°C promoted the refinement of the film structure up to textured polycrystalline, but a further increase of the substrate temperature to 350°C led to a drastic degradation of the structure.

The structure of the films deposited at $q \approx 10^8$ W/cm^2 in the same temperature interval was more perfect than that of films deposited at lower q. Films deposited at 250–350°C had a mosaic crystal structure with a different range of ordering. The films with the best crystalline structure were grown at 300°C and had highly oriented crystallites of a large size.

In situ monitoring of the growth of a GaAs film on a GaAs substrate, with respect to the film thickness per pulse d_i at different substrate temperatures, was carried out at $q \approx 10^8$ W/cm^2. It was revealed that the films deposited at $T_s \leq 300$°C and $d_i \approx 0.8$–1.0 nm/pulse were polycrystalline. The oriented crystallization of the film began at $T_s = 350$°C, and epitaxial films with a highly oriented structure could be grown at $T_s = 500$–600°C. Increasing d_i to ~4 nm at $T_s = 600$°C revealed the effect of the structure evolution for a few minutes after the pulse decay. Immediately after the pulse decay slightly diffuse diffraction rings were observed, indicating the formation of a fine-grained polycrystalline structure. One minute later diffraction spots appeared on the diffraction rings, the intensity of which decreased simultaneously with the appearance and increased intensity of the spots. At the end of the third minute only diffraction spots were seen on the electron diffraction screen. Highly oriented epitaxial layers up to ~50 nm in thickness could be grown at $d_i \approx 4$ nm only over an interval larger than 3 min between the depositing laser pulses. However, the structural perfection of these films was lower than that of films deposited at the same substrate temperature but with $d_i \approx$ 0.8–1.0 nm/pulse. This indicated the Kikuchi lines on the diffraction patterns of the GaAs films deposited at $d_i \approx 0.8$–1.0 nm and $T_s = 600$°C.

After the structural transformations ceased, the diffraction patterns of the films deposited at $d_i \approx 6$–7 nm and $T_s = 600$°C contained diffusive diffraction rings with barely visible spots. The films deposited at $d_i > 10$ nm/pulse and the same substrate temperature were amorphous. Special investigations had shown that orienting substrate influence extended no farther than 4–5 nm over the interface in the case of $d_i > 10$ nm/pulse. The deposition of a GaAs film at the substrate with $T_s \geq 650°C$ promoted the re-evaporation at first of As and later the Ga, which was indicated by structural transformation of the film surface.

To discuss the results described above, we will start with a comparison of the structures of films deposited at two different laser power densities, $q \approx 10^5$ W/cm^2 and $q \approx 10^8$ W/cm^2, keeping in mind that the power densities were changed by variation of the laser irradiation mode, not by simply changing the laser spot size.

The structural perfection of the films deposited at $q \approx 10^5$ W/cm^2 and millisecond pulse duration was lower compared with that deposited at $q \approx 10^8$ W/cm^2 and nanosecond pulse duration while other conditions were equal. The authors attributed this to the presence in the ejected products of large amounts of molecular clusters. Such clusters were found by the mass spectrometric investigation of laser plasma during the evaporation of GaAs by laser [69] operated in a regime ($\sim 10^5$ W/cm^2, ms pulse) close to that used here. Clusters, because of their lower mobility at the growth surface, hinder the ordering of the nuclei for epitaxial growth. The postdeposition structural transformations of these films also occurred more slowly, and such films did not achieve as great a structural perfection as the films deposited at $q \approx 10^8$ W/cm^2. These findings confirm two important ideas: (i) the laser operation mode can sufficiently influence the composition of FFS, and (ii) the presence of molecules or molecular clusters at the condensation surface can notably hinder the structural ordering of the film.

The dependence of the structured perfection of the film on the thickness of the layer deposited per pulse, d_i, was explained in the following way. The layer formed at the substrate from highly supersaturated vapors and from laser plasma may exist in a nonequilibrium state for a certain time. The processes developing in the layer during this period depend on two factors: the thermal motion of the atoms in the layer and the orienting influence of the substrate surface energy. As the temperature decreases, the solid–melt interface moves from the interface to the outer film surface. The crystallization proceeds fast and completely if the layer is thin, as was observed for $d_i \approx 0.8$–1.0 nm/pulse. In the case of a layer with a large thickness, the accumulation of structural defects in the bulk of the film takes place during the movement of the solid–melt interface. The stored structural defects inside the layer disrupt the oriented crystallization of the film and freeze the amorphous structure, as was observed for $d_i > 10$ nm/pulse. Estimations based on RHEED investigations have shown that the orienting influence of the substrate ($T_s = 600°C$) extended no farther than 4–5 nm over the substrate. So one of the factors discussed earlier that hinders epitaxial growth is readily seen from this series of experiments, that is, the high level of supersaturation of vapors and laser plasma. Taking this factor into account, the authors came to the conclusion that high-quality epitaxial GaAs films can be obtained only under conditions adequate for crystallization: (i) thickness per pulse $d_i \leq 0.8$–1.0 nm/pulse,

(ii) time between laser pulses ≥ 1 s, and (iii) substrate temperature $T_s = 600°C$.

Let us now compare the results of the deposition of epitaxial GaAs films by PLD with that deposited by rf ion beam epitaxy [102]. In this work the structures of GaAs films prepared by an ion beam epitaxial system with two ion sources and an rf coil have been studied. The films were deposited on cleavage surfaces of rock salt by the separate evaporation of gallium and arsenic from two ion sources. The deposition rate was about 1 nm/s, and the mean film thickness was 300 nm. It was established that epitaxial GaAs films with high structural quality could be grown on NaCl and GaAs substrates at $T_s = 400°C$, whereas the substrate temperature and growth rate needed for the deposition by conventional MBE of GaAs films of the same structural quality are about 550–570°C and 0.3 nm/s. Murayama [102] emphasizes that the main factors needed to obtain epitaxial films with good structure at a decreased growth temperature and increased growth rate are the ionization of the vapor beam, rf excitation, and dc bias of the substrate, with strong control of the bias voltage. This means that the film deposited under such conditions during the growth was exposed to continuous bombardment with ions at a certain energy. In this case the ion bombardment facilitated the epitaxial growth because of the enhancement of adatoms surface mobility.

A reasonable question arises: Why didn't the presence of the ions and the ion bombardment of the growth surface during PLD of GaAs films facilitate the epitaxial growth of this film?

The answer to this question can be found by considering the next series of experiments, devoted to the study of the epitaxial growth of PLD PbTe and (Bi,Sb)$_2$Te$_3$ films in [101]. In this series a Nd:glass laser operating in Q-switch mode (30-ns pulse duration) was used. The power density was varied in the range of $q = 2 \times 10^{11}$ to 10^9 W/cm^2 by variation of the pulse energy. The substrate–target distance was 3 cm, the rock salt-cleaved surfaces served as substrates. Simultaneously with the deposition, mass spectrometric investigations of the laser plasma composition were carried out. Oilless pumping of the vacuum chamber provided a vacuum better than 10^{-6} torr.

The mass spectrometric investigations of the energy distribution of charged particles inside the laser plume have shown that it is quite inhomogeneous spatially, both energetically and in number of charged particles. The species with maximal energy (~ 5 keV) and maximal charge ($Z_{max} = 3$) were distributed in a narrow ($\sim 30°$) spatial angle along the axis of the laser plume. The mass spectrometric analysis showed a fast decrease in the numbers of the energetic and charged species as the analysis angle was increased. The decrease in the power density from 10^{11} to 10^9 W/cm^2 promoted a decrease of the maximal charge of the ions from $Z_{max} = 3$ to $Z_{max} = 1$. An estimation showed that the total ion current to the substrate at $q = 10^{11}$ W/cm^2 and a 3-cm target–substrate distance can reach $\sim 10^3$ A/cm^2.

The TEM investigations of the films deposited on unheated substrates at $q = 2 \times 10^{11}$ W/cm^2 revealed a zone with the traces of re-evaporation and/or re-sputtering of the film. The center of this zone was located on the axis of the laser plume,

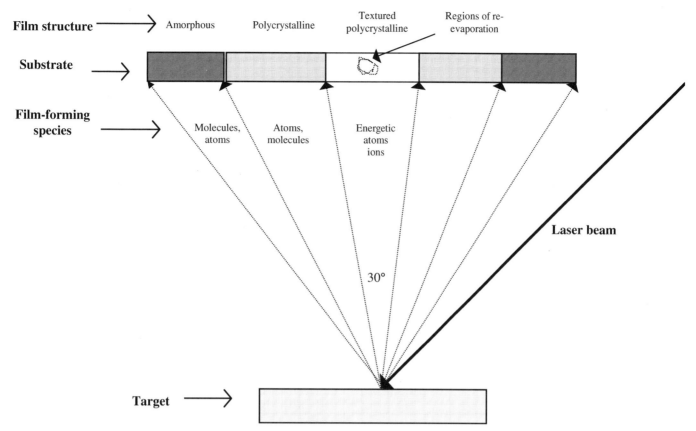

Fig. 21.   Scheme of the structure formation of PbTe and $(Bi,Sb)_2Te_3$ films deposited by pulsed laser evaporation. Deposition conditions: $\lambda = 1064$ nm, $\tau = 30$ ns, $q = 10^{11}$ W/cm^2, laser spot dimensions 4 mm^2; target-to-substrate distance 3 cm. The scheme is drawn according to the results of [101].

where the densities of most energetic and charged species were maximal. Both PbTe and $(Bi,Sb)_2Te_3$ films deposited at this zone had structures with maximal perfection. PbTe films were highly textured, whereas the $(Bi,Sb)_2Te_3$ films had a polycrystalline structure with a large grain size. The film structure became less ordered further from this zone and was completely amorphous at the periphery of the film. The films deposited at lower power densities ($q \approx 10^{10}$ W/cm^2) had less ordered structure and were completely amorphous at $q = 10^9$ W/cm^2 for both materials. A scheme of the structural formation of PbTe and $(Bi,Sb)_2Te_3$ films drawn according to these results is depicted in Figure 21.

The tendency of the structure to deteriorate toward the periphery of the film remained constant with the reduction in laser power density. The authors emphasized that the most important factor in the formation of the film structure is not the film thickness but the distance from the central zone. The thickness of the film was the same for regions with different structures.

An investigation of the influence of the specific energy of the laser plume on the film structure was carried out by depositing double-layered films of both materials. The upper layer of one of the double-layered films was deposited at a lower ($q = 7 \times 10^{10}$ W/cm^2) power density, and the lower layer was deposited at a higher power density ($q = 2 \times 10^{11}$ W/cm^2). The sequence order of the layers in one double-layered film was opposite that of the other. For both materials it was revealed that if

the upper layer was deposited at lower q, its structure was more ordered than that in the opposite case, where the upper layer was deposited at higher q.

At the same time the structure of a single-layered film deposited at lower q (thickness ≈ 7 nm) was more ordered than that of the layer deposited at higher q (thickness ≈ 20 nm), as was already observed for GaAs films. The structure of a three-layered film with layers of equal thickness (with a total thickness of 20 nm) was less ordered than that of a single-layered film.

In discussing the results described above, the authors stated that the main reason for the structural perfection of the films is the transformation of the specific energy of laser plasma into the crystallization heat. Undoubtedly this effect plays an important role in film structure formation. The deterioration of the film structure away from the center of the film and toward its periphery might be explained just by this effect. In fact, as follows from mass spectrometric investigations, the energy of FFS deposited at a peripheral regions of the film is lower than that of species deposited in the central zone. Another argument supporting this idea is the deterioration of the film structure as the laser power density decreases which results in a decrease in the energy of ejected species.

Nevertheless there may be another reasonable explanation for these effects. As is known from [62], the gas phase generated by conventional evaporation of PbTe mainly consists of

PbTe molecules. In the case of pulsed laser evaporation of PbTe, the composition of generated species depends on the opening angle of laser plume and heavy species with low kinetic energy usually observed at larger angles (see the results of mass spectrometric investigations described above and [16]). So it could be expected that the more distant regions of the film are formed mostly of condensed PbTe molecules. Because the mobility of molecules is significantly lower than that of atoms, it might be concluded that the latter is the reason for the low-order structure of the peripheral regions of the film (Fig. 21).

However, it is difficult to explain other results by discussing investigations based only on these aspects of laser plasma. For example, one could assume that the structure of upper layer in double-layered film is more ordered in the case of its deposition at a higher power density of $q = 2 \times 10^{11}$ W/cm^2. This is reasonable on one hand, because of the orienting effect of the lower layer, which has an ordered structure, and on the other hand, because of the higher specific energy of the laser plume at higher q. However, the result is just the opposite of what is expected. The more ordered structure has the upper layer deposited at a lower power density ($q = 7 \times 10^{10}$ W/cm^2) on the layer with the lower structural quality. This effect can be explained if one takes into account another aspect of laser plasma, namely the intense and noncontrollable bombardment of the condensation surface by energetic neutrals and ions. The higher the intensity of the bombardment, the higher the damage of the growth surface and growing layer, which prevents the ordering of the film structure. If the upper layer is deposited from a flux of species with lower energy, then the possibility of damaging to the growing film is lower. Because the thickness of the layer deposited at lower q is small, the possibility of annealing of the structural defects is higher. Another cause for the more ordered structure of the upper layer is the lower level of the supersaturation of the vapors generated at lower power density. This effect was also observed during GaAs film deposition. All of these factors provide the more ordered structure of the layer deposited at lower q.

Because the deposition conditions of GaAs were close to that for PbTe and the growing film and growth surface were bombarded by energetic neutrals and ions, it is reasonable to assume that the higher epitaxial temperature of GaAs films (600°C) compared to that deposited by rf ion beam epitaxy (400°C) relates to the noncontrollable bombardment. In fact, despite the decreased vapor supersaturation resulting from the reduced GaAs layer thickness deposited per pulse and the increased time interval between pulses, the optimal epitaxial temperature remained as high as 600°C. This hints at some other factor hindering the epitaxial growth of GaAs film under these conditions. In the case of deposition of GaAs films by ion beam epitaxy, the energy of ions bombarding the growing film could easily be controlled by the substrate bias voltage [101], whereas the energy distribution of neutral and ion species bombarding the growing film during PLD is rather wide and can hardly be controlled. Therefore a species with an energy that exceeds the permissible limit of 5–10 eV can damage the film structure [63].

3.6. Ways to Control the PLD Film Structure

After the preceding discussion it could be concluded that such peculiarities of the PLD as (i) inhomogeneity of the FFS, (ii) the wide energy spectrum of a deposited species with a high average energy per particle, and (iii) a high instantaneous supersaturation of condensed vapors (deposition rate) significantly affect the structure of PLD condensates. As was shown, these factors may frequently negatively influence the film structure.

Efficient control of parameters and properties of FFS by the variation of deposition conditions is difficult task. The most easily controlled parameter is the deposition rate, which can be changed by variation of the laser power density and pulse duration. It is possible to decrease the amount of the evaporated volume and thus decrease the instantaneous vapor supersaturation by decreasing the power density and shortening the pulse length.

The control of the composition of FFS is a much more complicated problem because atomic clusters and molecules are generated by at least two mechanisms. One is the direct evaporation of molecules from the target, and the second is the creation of the molecules and clusters in a gas phase. Moreover, the two of mechanisms are independent of each other and depend on the deposition conditions and the laser operation regime in different ways. For example, the evaporation of the target by a lower power laser pulse can yield mostly molecules in the case where the dissociation energy of the molecules of the target material is higher than the thermal energy of the evaporated particles [18]. At the same time, increasing the power density per pulse and/or carrying out deposition in a gas ambient will promote the clustering of evaporated species in the gas phase [3, 5]. The laser wavelength can also influence the composition of the evaporated particles. For example, the evaporation of a ceramic AlN target by a Nd:YAG laser (1064 nm) is accompanied by the formation of pure aluminum film [50], whereas the use of excimer lasers (248 and 266 nm) makes it possible to deposit films of pure aluminum nitride both in vacuum and in a nitrogen ambient [34, 35, 70].

Regulation of the energy of FFS bombarding a condensation surface is also a very complex problem because of the wide energy spectrum of the particles and because of their nature (neutrals and ions). The most efficient methods for the reduction of the energy of fast particles are to decrease the laser power density and to carry out the deposition in a gas ambient [6]. However, deposition in a gas ambient promotes clustering in the gas phase, which again can negatively influence the film structure.

Evidently the negative influence of noncontrollable bombardment of a growing film during PLD is greater than its positive effect, which is frequently defined as one of the important advantages of PLD. In fact, most of above considered investigations of PLD films show that the substrate temperature has a key influence upon their structure. As was shown earlier, the temperature of crystallization or epitaxial growth of PLD films is frequently higher than that for films obtained by other deposition techniques (Table IV). At the same time, if the positive effect of transferring tens of electron volts per atom by energetic particles to the growing film were high, then the energetic influence of the substrate temperature ($\sim 10^{-2}$ eV/atom) upon the film structure would be negligible against the background of the energy transferred to the growing film by fast particles. In practice, the role of the substrate temperature in the case of PLD is found to be even larger than that for the other PVD methods.

In general, separation and hence the control of the influence of simultaneously acting factors during PLD, such as the inhomogeneity of FFS, the bombardment of the growth surface by energetic particles, and a high instantaneous deposition rate, are impossible.

3.7. Conclusions

The crystalline structure of PLD films on average is less perfect than that of the films deposited at the same substrate temperature by the other methods of PVD. The reasons for this effect are weakly controlled factors: the inhomogeneity of the composition of FFS and bombardment of the growing film by energetic particles. Both of these factors are the result of an intrinsic feature of pulsed laser evaporation, which is the high nonequilibrium of the processes taking place during the interaction of a high-power short laser pulse with the condensed matter. These peculiarities of PLD are sufficient to limit the possibility of deposition of high-quality epitaxial films.

4. POLYMORPHISM IN PLD FILMS

The interaction of a high-power laser pulse with condensed matter supposes the simultaneous appearance of an instantaneous pulse of high temperature and high pressure on the target surface. Such conditions are favorable for the origin of metastable structural states at the point of interaction of the laser beam with the target. The bombardment of the condensation surface by energetic particles generated under a laser pulse impact with the target can also be the reason for the generation of metastable structural phases. In the latter case the nucleation of the metastable phase occurs on the growth surface. Most frequently the effect of the formation of metastable phases is observed in the case of laser evaporation of substances like carbon or polymorphous compounds.

4.1. Polymorphism of PLD Carbon Films

The properties of amorphous carbon films depend on the type of atomic bonding of carbon atoms. If sp^2 bonding prevails, then the film has graphite-like properties. When the prevailing bonding type is sp^3, then the hardness and elasticity of the film are similar to those of a diamond. DLC films with $\sim 90\%$ sp^3 bonding can be produced only with the use of a mass- and energy-filtered carbon ion or atomic source adjusted to give the required energy (about 100 eV per carbon atom) and when the substrate temperature $T_s < 100°C$ [2]. Because the laser plume generated by pulsed laser evaporation of carbon materials contains carbon ions and atoms with an energy of hundreds of electron volts, it is possible to obtain DLC films by choosing the proper laser operation regime (power density, wavelength, pulse duration) and other technological conditions [79–81].

The analysis of many publications devoted to the deposition of carbon films by pulsed laser evaporation of graphite led Voevodin et al. [80] to conclude that the probability of DLC film formation increases with decreasing laser wavelength and with increasing power density per pulse. Simultaneously, the shorter the laser wavelength is, the lower the power density can be in the production of an amorphous carbon film with diamond-like properties. An increase in the laser wavelength and a reduction of the power density promote the formation of amorphous graphite-like or graphite condensates. These observations can be supplemented by the results obtained in [81, 82]. In [81] it was shown that at a given laser wavelength there exists an optimal power density of laser irradiation at which the DLC films obtained had a maximal concentration of sp^3 bonding. If the power density deviates from a certain value, the proportion of sp^3 bonded carbon atoms decreases. At the same time, from the results of [82] it follows that the sp^3 content in the film may be increased by using a laser with a shorter wavelength.

The role of laser wavelength and power density in the deposition of carbon film with a high concentration of sp^3 bonded carbon atoms can be understood on the base of the preceding discussion. It is known that evaporated graphite can exist in the vapor phase in the form of separate atoms and as atomic clusters, because of the high anisotropy of atomic bonding in graphite [1]. It is also known that the substance absorption coefficient tends to increase as one moves to the short wavelength end, and the depth of penetration into the target materials is correspondingly reduced [9]. This means that the amount of laser power absorbed by a target volume unit increases with decreasing laser wavelength. In other words, the amount of laser power per evaporated atom increases as the laser wavelength decreases. The probability of graphite evaporation in graphite-like cluster form decreases, and the energy of evaporated carbon atoms and ions increases because of this effect. Hence the probability of DLC film formation has to be enhanced by a decrease in the laser wavelength, which was experimentally observed in [84]. The existence of an optimal combination of power density and laser wavelength ($q-\lambda$) for the formation of a carbon film with maximal sp^3 content is explained as follows. The amount and energy of evaporated carbon atoms and ions are not sufficient for the formation of a DLC film at low q. If q exceeds the optimal level, the energy of evaporated species is also enhanced, which negatively influences the sp^3 concentration in the condensate [2].

The carbon films deposited at relatively lower power density and larger wavelength (1000 J/pulse, 10^7 W/cm^2, $\lambda = 1064$ nm) consist of two phases, amorphous carbon and crystalline graphite (Fig. 22), in contrast to DLC films, which are deposited by the irradiation of a graphite target by an excimer laser with lower power (0.1–0.3 J/pulse) but a higher power density per pulse ($q = 10^8$ to 10^{11} W/cm^2, $\lambda = 248$ nm) [79–81]. This means that the concentration of carbon atoms and ions of a certain energy in a laser plume is insufficient for formation of a film with any sp^3 concentration in the former case.

Sometimes crystalline particles of different sizes were found embedded in amorphous carbon film (Fig. 23). Selected area electron diffraction investigations showed that these particles are the crystals of metastable carbon phase carbine [83]. Carbine phase was recently revealed in graphite subjected to laser irradiation and in simultaneously deposited condensate [84]. The appearance of this metastable carbon phase can be explained by the simultaneous action of an instantaneous pulse of high pressure and temperature induced by laser pulse on the graphite surface. Part of the carbine crystals that originated on the target were transported to the condensation surface by the recoil pressure of evaporated carbon. It is unlikely that the formation of such large particles could occur in the gas phase.

4.2. Polymorphism of Boron Nitride Films

Currently the growth of boron nitride (BN) films is one of the most attractive areas for the application of different PVD techniques. It is known that BN is stable in three crystallographic structures: a hexagonal graphite-like structure (h-BN), a cubic zincblende structure (c-BN), and a wurtzite structure analogous to hexagonal diamond (w-BN) [103]. Up to now BN films have been obtained only in two structure modifications, h-BN and c-BN, both by PLD and by other PVD methods. There is no information on how to obtain w-BN films. A short review of some properties of PLD BN films appears in [104]. In a majority of the works devoted to the PLD of BN films, excimer lasers operating in a frequency regime with a low power density per pulse ($\sim$3–4 J/cm^2) are used. The depositions were carried out in a nitrogen gas ambient. c-BN films could be successfully deposited by additional excitation of nitrogen by the electric discharge conditions [61], or from a separate source of nitrogen ions and atoms [104]. In a survey devoted to the review of deposition methods and the mechanisms of formation of c-BN films, it is emphasized that the necessary condition for c-BN film production is the bombardment of the growing film by ions with energies in a narrow range (80–120 eV for PVD methods and 200–400 eV for plasma-activated CVD) [105]. For the successful production of c-BN films by PLD, either deposition with electrical biasing of the substrate [61], or employment of an ion source to bombarding the substrate during the deposition [104] was also necessary. Hence it is reasonable to assume that the formation mechanism of BN films with cubic structure during their deposition under the above-described conditions by excimer laser is close to that realized during the deposition of c-BN films by other PVD techniques. The distinguishing feature of this mechanism is the formation of the c-BN phase on the condensation surface. It should be emphasized that such a mechanism can be realized during the PLD of BN films in the case of excimer lasers operating in frequency mode with a low power density per pulse.

However, a quite different mechanism of the formation of the c-BN phase during pulsed laser evaporation of an h-BN target is also possible. This mechanism was realized during the evaporation of an h-BN target with the use of a Nd:YAG laser ($\lambda = 1064$ nm) with high power per pulse (1000 J/pulse) and a millisecond pulse length [50]. The areas with a c-BN phase that were revealed in the films deposited under these conditions most possibly are the products of the phase transition that took place on the h-BN target surface under the influence of a laser pulse and were then transferred in the film. Such a mechanism is similar to that of carbine phase origination in a carbon film during pulsed laser evaporation of graphite under the same ex-

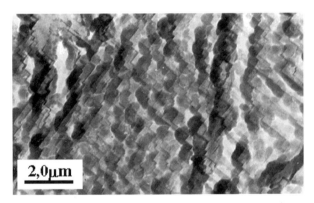

Fig. 22. Microstructure of graphite film deposited by laser ablation of a graphite target. (Nd:YAG laser, $q = 5 \times 10^7$ W/cm^2, 1 ms).

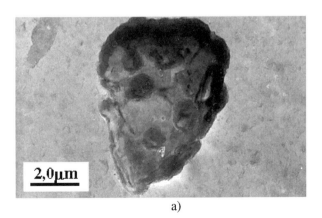

a)

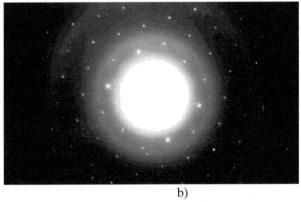

b)

Fig. 23. (a) Carbine particle embedded in an amorphous carbon film deposited by laser ablation of a graphite target. (b) Spots on the SAED pattern correspond to the particle, and halos relate to the amorphous carbon film.

perimental conditions. That is, the appearance of a c-BN phase is related to the simultaneous action of a pulse of high temperature and high pressure on a h-BN surface induced by laser pulse. It is known that the phase transition of h-BN to c-BN in bulk occurs only under conditions of simultaneous high temperature and high pressures [103]. However, the creation of c-BN nuclei in the gas phase due to the interaction of nitrogen and boron atoms evaporated from the target cannot be excluded.

4.3. Polymorphism of Silicon Carbide Films

The formation of a 3C-SiC phase in films deposited during the evaporation of a ceramic silicon carbide (SiC) target or a 6H-SiC single crystal has quite different causes. This result was obtained both in a majority of the works in which SiC films were obtained by the evaporation of a ceramic SiC target by excimer lasers [39], and by evaporation of a 6H-SiC single crystal with a Nd:YAG laser with a high power per pulse [50].

It is worth mentioning that ceramic SiC targets mostly have a structure of cubic SiC (3C-SiC) that relates to the peculiarities of the fabrication of this material. SiC ceramic material always contains an excess of silicon. And it is known that even an extremely small excess of the silicon in SiC stabilizes the 3C-SiC phase both in a single crystal [85] and in polycrystalline ceramic material.

Solidified droplets of silicon were revealed in SiC films deposited by pulsed laser evaporation of 6H-SiC single crystal during TEM investigations (Fig. 12). This fact indicates that during pulsed laser evaporation of SiC partial dissociation of the composition occurs. At the same time it is known that the volatility of silicon is noticeably higher than that of carbon, so the appearance of some excess of silicon in the film is not surprising. Evidently just this effect plays the main role in the formation of the cubic 3C-SiC phase in SiC films deposited by pulse laser evaporation of 6H-SiC single crystal or a ceramic SiC target. The same mechanism of 3C-SiC phase formation is typical for the other deposition methods of SiC films.

As follows from the above consideration of polymorphism in PLD films, the mechanisms of formation of polymorphous phases are different for different compounds. However, the mechanisms of formation of metastable phases during the deposition of films by excimer laser with a short pulse length and low power per pulse are similar to that realized in other deposition methods.

The mechanism of formation of such phases, however, can differ from what is typical for other deposition techniques, with the use of laser irradiation with high power per pulse. In this case the simultaneous action of pulses of high temperature and pressure that arise at the target surface under the influence of a laser pulse can induce new phases. Nuclei and particles of a new structural phase formed by this means then are transferred to the condensation surface. The formation of new phases apart from this mechanism can occur directly on the condensation surface because of its bombardment by particles with high energy. Each of these kinetic factors (pulses of high temperature

and pressure bombarding a growing film) can shift the thermodynamic balance on the condensation surface or on the target in the direction of the formation of a metastable or new phase.

4.4. Conclusions

The phase transitions in bulk polymorphic substances can occur under the influence of high temperatures and pressures (carbon, BN, and some others) or may be related to the concentration variations in the compound. In the latter case the phase transition can be observed even at minor (almost at stoichiometry limits) variations of the concentration of one of the constituents ($A^{II}B^{VI}$ compounds, SiC, and some others). The mechanisms of origination of structural phases in PLD films that are different from that intrinsic to the target depend on the peculiarities of the target substance and on the laser operation regimes. The influence of the laser irradiation directly on the target surface and indirectly on the growing surface can induce a new structure phase in the film due to the concentration changes in the film composition and under the simultaneous action of a pulse of high temperature and pressure. The compositional changes in the film inducing the formation of a new phase can take place on the target surface and on the growth surface, and this mechanism can be realized in other deposition methods. The induction of a new phase in the film by the transfer of the nuclei and particles of the new phase from the target to the growth surface is inherent only in the PLD method.

5. MACRODEFECTS IN PLD FILMS

One of the major drawbacks of PLD is the macrodefects in PLD films. We use the term "macrodefect" to designate a microstructure inhomogeneity in a PLD film that appears not due to condensation of the vapor phase. The origin of macrodefects in PLD films is an intrinsic problem of pulse laser interaction with condensed matter; therefore it is much more difficult or may even be impossible to overcome.

5.1. Mechanisms of Splashing

There are two major sources of macrodefects in PLD films. One source is the splashing that occurs during interaction of the laser pulse with the target, and this effect has many causes. The second source of macrodefects is the generation of particulates due to condensation of vapor species in the gas phase.

The size of particulates formed from the vapor phase tends to be in the nanometer range, whereas their counterparts are in the micron and submicron ranges. Particulates formed by splashing effects tend to be spherical, whereas the particulates formed from the solid ejecta tend to be irregularly shaped. Particulates formed from vaporized species mainly have a spherical shape.

Splashing includes a few mechanisms that lead to the deposition of particulates on the substrate [10]. For a given material splashing can include any one or a combination of these mechanisms.

One of these splashing mechanisms is subsurface boiling. It occurs if the time required to transform laser energy into heat is shorter than that needed to evaporate a surface layer with a thickness on the order of a skin depth. Under this condition the subsurface layer is superheated before the surface itself reaches the vapor state. Under this process micron-sized molten ejecta will be removed to the substrate.

In the second mechanism of splashing the force that causes expulsion of liquid droplets comes from above the liquid layer in the form of recoil pressure exerted by the shock wave of the plume. Because the droplets originating from this process also contain micron-sized balls, it is impossible to distinguish the last mechanism from the first. The magnitude of this effect can be reduced as in the previous case by lowering of the laser power density.

The third type of macrodefect observed in PLD films is the solid ejecta originating more during the evaporation of brittle material systems, particularly sintered ceramic targets. The mechanism of their generation relates to the formation of long, needle-shaped microstructures a few microns in dimension on the target surface as a result of erosion during laser beam forcing. These microdendritic structures are very fragile and can be broken by thermal shock induced during intense laser irradiation. This debris is carried toward the substrate by the rapidly expanding plume and condense on the growth surface.

5.2. Elimination of Particulates

There are a few approaches to the problem of particle deposition. One concerns the improvement of laser systems. It includes the improvement of the spatial and temporal homogeneity of the pulse structure [7, 8]; searching for optimal wavelength and pulse duration [9], and using the frequency evaporation regime. As a result of many years of investigations of laser development for film deposition, an ideal laser system was proposed [12]. This system is likely to have a pulse energy of ≤ 1 J at a repetition rate of 200 Hz with a beam homogeneity and pulse-to-pulse energy stability of better than 10%. The lower laser pulse energy makes it possible to reduce the particulate generation, because of the lowered condensation of vapor species in the gas phase and the decreased amount of liquid phase on the target and the lower magnitude of recoil pressure. Evaporation of a nonmetal target by irradiation with a shorter wavelength also decreases the possibility of clustering of evaporated species because of more efficient conversion of energy of shorter electromagnetic waves into heat for nonmetal materials [3].

Another approach to reducing the macrodefects in PLD films is the manipulation of the thermophysical properties of the targets to increase the efficiency of conversion of laser power into a heating–evaporating process and to increase by this means the portion of vapor phase in the ejecta. For this purpose the evaporation of different types of the targets like cast metals and alloys, single crystals, cold pressed and hot pressed powders [86], and free-falling powders [41] was investigated. These effects are considered in detail in Section 6.

One more widely used PLD technique for lowering the number of microparticles in films is the mechanical separation of the particles ejected from the target. These methods are widely treated elsewhere in surveys [3, 10, 12], and we will omit this question.

A quite different approach to enhancing the efficiency of laser irradiation in material evaporation is to establish the correlation between the thermophysical properties of the evaporant and the portion of a vapor phase in laser ablation products. In this case the efficiency of laser evaporation can be determined as the ratio of vapor mass to the total mass of the products, produced by laser irradiation of the target. One can enhance the evaporation efficiency and decrease or avoid the particulates in the film either by changing the thermophysical properties of the target (e.g., its thermal conductivity) or by choosing a relevant target composition.

In this context the problem of the efficiency of laser evaporation and splashing is tightly bound up with the problem of the determination of the portion of the vapor phase in laser-ablated products. The larger is the vapor portion in products ejected from the target by laser pulse, the greater is the film thickness and the more efficient is the utilization of laser radiation. The portion of the vapor in the ablated products depends on the laser parameters and on the thermophysical properties of the target material. Because there are very few publications devoted to this problem, we consider it in more detail in view of its importance for PLD.

5.3. Determination of Vapor Portion in Products of Laser-Ablated Metals

A few works have been devoted to the determination of the vapor portion in products ejected from the metal target during pulsed laser evaporation [87–90]. In [88] the efficiency of metal evaporation was estimated from the depth of the crater on the metal target that appeared after the impact of the laser pulse and from the mass of the products ejected from the crater. Using these criteria, the authors concluded that there is no correlation between the thermophysical properties of the metals and the amount of vapor produced by laser pulse impact. However, because the mass of the ejected products mainly consists of the liquid phase [7], the mass of the ejected products and the depth of the crater are not relevant parameters for the characterization of the efficiency of the evaporation of the metal by laser irradiation.

The ratio of vapor to liquid components in the ejected products during laser pulse impact with some metals was estimated and a correlation between the amount of vapor and the value λ/Q, where λ and Q are the specific energies of melting and vaporization, accordingly, was established in [87]. The authors determined the portion of the vapor component m_v/m (m_v, mass of the vapor, m, total mass of the products ejected by the pulse laser irradiation at $q = 5 \times 10^7$ W/cm^2) by two methods. In the first method m_v was calculated from the equation of the energy balance, where the total mass of ejecta m was the single experimentally determined parameter. However, in the equation of the energy balance the laser power was considered to be consumed only by the evaporation, and the unproductive losses of optical power on heating and melting of the metal were not taken into account. In the second method m_v was determined from the recoil momentum, inasmuch as the liquid fraction momentum in the total recoil momentum is negligible, because the velocity of the liquid droplets is $v_l \ll v_v$, where v_l and v_v are the velocity of the droplets and the vapor, accordingly. However, such an equation is acceptable for large droplets, the number of which is considerably lower than that of small droplets in the ejecta [10]. Second, the results of Kirillov and Ulyakov [87] show that the liquid fraction mass in the ejected products is much higher than the mass of the vapor. Therefore the neglect of the liquid fraction momentum in the total recoil momentum is not valid. Therefore the correlation established in [87] between the λ/Q value and the portion of vapor in the products ablated from the metal by the laser pulse is questionable.

In [89], based on film thickness measurements, the authors assert that the amount of vapor in the products ejected from the metal under a laser pulse depends on the heat of sublimation of the metal. The authors associate the laser power needed for the evaporation of a mole of metal with the thickness of the film deposited by laser evaporation of the metal. Because the thickness of the film deposited from a vapor generated by laser pulse does not linearly depend on the mass of the vapor, the results obtained in this investigation are not convincing.

In [90] the authors also experimentally determined the portion of the vapor in the ejected products generated by the impact of the laser pulse with different metals. The experiment was carried out for two different laser power densities, which were achieved by varying the laser round spot diameter ($\lambda = 1064$ nm, $\tau = 1$ ms, $q = 1.5 \times 10^6$ W/cm^2 and $q = 5 \times 10^7$ W/cm^2 at 1000 J/pulse). During these experiments the authors took into account uncertainties and the errors of the aforementioned works and obtained results different from those cited in [87–89]. On the base of the results obtained, a correlation between the laser power density, thermophysical properties of the metals, and the amount of vapor in the ejected products was established. A model of pulsed laser evaporation of metals was suggested. Let us briefly consider the results of this investigation.

The effectiveness of the laser irradiation for film deposition was estimated from the coefficient of evaporation. This coefficient was calculated as the ratio of the experimentally defined mass of the vapor generated by pulse laser evaporation of the metal to the total mass of the ejected products, m_v/m. The mass of the vapors m_v was calculated from the measurements of the thickness of the film, deposited by pulsed laser evaporation of the metal. The total mass m of the ejected products was defined by weighing the target before and after its laser irradiation. The procedure for the calculation of m_v calculation was as follows. It is known that the thickness of the film deposited by pulsed laser evaporation is not uniform and can be defined by the equation [91]

$$t = t_0 \cdot \left[1 + (l/r)^2 \right]^{2/3} \tag{1}$$

where t_0 is the film thickness at $l = 0$ (i.e., in the center of the substrate, where the film thickness is maximal); l is the distance from the center of the substrate (i.e., from the axis of the laser plume); and r is the target-to-substrate distance. Then m_v can be described by

$$m_v = \rho \int_0^{2\pi} d\varphi \int_0^{\pi/2} t(\theta) r^2 \sin\theta \, d\theta \tag{2}$$

where ϕ is a horizontal angle, $\sin\theta = l/r$, and ρ is the relative density of the metal.

When m, l, r, and ρ are known, it is easy to define t_0 and t for any l and r. It is suggested that the distribution of the film thickness is defined by (1) regardless of the type of evaporant and the laser operation regime. The distribution of the film thickness also does not depend on the horizontal angle ϕ; the film density is equal to the bulk density of the metal. It is of value to note that the change of the power in Eq. (1) in the range from 2 to 3 results in a change in the film thickness not higher than $\sim 30\%$.

The variation in the laser power density has significantly influenced the measured characteristics of the films and the targets. Thus, the crater depth was equal to or a little bit larger than its diameter $h \geq d$ in the case of lower power density, whereas the ratio $h \gg d$ was found for the focused laser beam (for the highest power density). The mass of ablated products was higher in the case of higher laser power density. Although the above-mentioned effects are not surprising, it was not expected that the thickness of the films deposited in the focused regime in a number of cases would be significantly lower than that of the films deposited at lower power density (see Table VI). The last result is also correlated with the dependence of the coefficient of vaporization m_v/m on the laser power density. This means that the efficiency of laser evaporation of metals by a defocused laser beam is significantly higher.

The next important result was that the correlation between the mass thickness of the film $t \cdot \rho$ (where t is the experimentally determined film thickness, ρ is the density of metal) and such thermodynamic parameters as the heat of evaporation or the λ/Q ratio (see above) was not found in our investigations (Table VII) [90], but it was established in [89] and [87], correspondingly. The most probable reason for the disagreement of the results of [90] and those obtained in [87, 89] are discussed above with regard to some uncertainties in cited works. A comparison of the results of [90] with those of [87, 89] shows that there is no simple relation between one or two thermophysical properties of a metal and the share of the vapor generated by laser irradiation of the metal.

The results obtained in our investigation [90] can be explained by a model that associates the portion of the vapor m_v/m in the ejected products simultaneously with several thermophysical properties of the evaporated metal. These properties are the melting point, the temperature at which the saturated vapor pressure is relatively high (e.g., 10^{-1} torr, Table III) and the thermal conductivity of the metal. A suggested mechanism for the pulsed laser ablation of metals follows.

Table VI. Total Mass of Ejected Products, Film Thickness, and Vapor Portion m_{v}/m in Ablated Products During Laser Ablation of Metal Targets with Different Power Densities

Metal	Mass of ablated metal (g)		Film thickness per pulse (nm)		Vapor portion in ablated products, m_{v}/m (%)	
	$q = 1.5 \times 10^6$ W/cm^2	$q = 5 \times 10^7$ W/cm^2	$q = 1.5 \times 10^6$ W/cm^2	$q = 5 \times 10^7$ W/cm^2	$q = 1.5 \times 10^6$ W/cm^2	$q = 5 \times 10^7$ W/cm^2
Cr	0.0079	0.0253	70.0	35.0	82.0	11.0
Fe	0.0290	0.0402	80.0	40.0	26.0	8.5
Ti	0.0249	0.0361	55.0	30.0	13.0	4.5
Zr	0.0571	0.0670	35.0	20.0	5.5	2.3
In	0.2477	0.2948	115.0	95.0	4.4	2.8
Sn	0.3033	0.2976	135.0	45.0	3.3	1.0
Mo	0.0002	0.0368	0.0	15.0	—	5.4
W	0.0005	0.0541	0.0	10.0	—	5.0
Al	0.0000	0.0303	0.0	35.0	—	3.7
Nb	0.0005	0.0456	0.0	15.0	—	3.0
Cu	0.0000	0.0558	0.0	10.0	—	2.3

After [90].

Table VII. Mass Thickness ($t \cdot \rho$) of Metal Films Deposited by Laser Ablation of Metal Targets at Two Power Densities and Some Thermophysical Properties of Metals Used as a Targets

Metal	$(t \cdot \rho) \cdot 10^{-5}$ (g/cm^2) ($q = 1.5 \times 10^6$ W/cm^2)	$(t \cdot \rho) \cdot 10^{-5}$ (g/cm^2) ($q = 5 \times 10^7$ W/cm^2)	Evaporation heat (kcal/g at 300 K)	Thermal conductivity (W/m · K at 273 K)	λ/Q (arb. un.)
Cr	5.0	2.4	1.82	88.6	0.034
Fe	6.3	2.9	1.78	75.8	0.033
W	0.0	2.5	1.11	160.0	0.041
Mo	0.0	1.6	1.65	138.0	0.055
Ti	2.5	1.4	2.35	21.9	0.036
Cu	0.0	1.0	1.27	405.0	0.038
Nb	0.8	1.2	1.86	55.0	0.038
Al	0.0	1.0	2.92	221.5	0.032
Zr	2.4	1.3	1.57	29.5	0.034
In	8.4	7.0	0.49	87.0	0.014
Sn	7.8	2.5	0.61	59.8	0.023

After [90].

5.3.1. Metal Ablation under Focused Irradiation

The narrow and deep crater on the surface of a metal target under focused laser pulse action appears and develops because of intensive melting and evaporation processes. An increasing amount of vapors appears in the crater volume and begins to absorb the laser irradiation during the development of the laser pulse. Overheated vapors transfer a portion of their heat to the crater walls, melt them, and throw the melt out of the crater. The redistribution of the laser power with the overheating of the vapors and with the melting of the target (crater walls) hinders the effective evaporation of the target by laser irradiation. Because the laser power in these conditions is consumed to large extent not only by evaporation, it is difficult to establish the correlation between the thermophysical properties of the metal and the generated vapor mass. The negative results obtained in [88] support this conclusion.

5.3.2. Metal Ablation under Defocused Irradiation

The crater formed under the influence of a defocused (lower power density) laser beam is wide and shallow. Therefore the density of the vapors in such a crater is low, and thus the portion of energy absorbed by the vapors laser irradiation is low. Under these conditions the laser power to a large extent is consumed by the evaporation, the melting, and the heating of the metal, whereas the overheating of the vapors is almost excluded. So it is reasonable to expect to reveal a correlation between the thermophysical properties of the metal and the amount of absorbed optical power.

The ratio of liquid to vapor masses in the ablated products depends on the ratio of speed of input of laser power into the metal to outflow speed from the zone of laser beam action of converted laser power. At a certain power input speed (certain laser operation regime) the outflow speed depends on the metal thermal conductivity and on the cooling speed of the melt surface due to evaporation.

Therefore if the vaporization of the metal is intensive at the melting point (i.e., the saturation vapor pressure P_{s} is high) and its thermal conductivity is low, these factors hinder the growth of a liquid layer thickness on the target surface. However, if the evaporation of the metal at the melting temperature is weak (i.e., the saturation vapor pressure P_{s} is low), then the laser power is consumed mostly by the further melting of the metal. The amount of liquid phase on the target surface under these conditions increases up to the moment when the surface reaches the temperature of intensive evaporation (i.e., at which $P_s = 10^{-1}$ torr). The lower is the difference between the temperature of intensive evaporation and melting point $\Delta T = T_v - T_m$ and the lower is the thermal conductivity of the metal, the faster the surface reaches the temperature of intensive evaporation. When the temperature of the melt on the target surface is lower than the temperature of intensive evaporation, the amount of vapor (and hence the vapor pressure) is not sufficient to throw out the melt from the crater. In other words, the presence of a noticeable amount of the liquid phase in the ejected products is possible only in the case of intensive evaporation. The substantially higher amount of vapor generated at the beginning of the laser pulse action than that at the end is related partly to the latter effect, as mentioned in [87]. The increased amount of the liquid phase in the ejected products at the end of the pulse action relates to the redistribution of the laser power by the thermal conductivity, which directs the power to the heating and melting of the target.

The following examples can illustrate the above suggested mechanism of evaporation of metals by defocused laser irradiation. The surfaces of Al, W, and Mo targets are only melted but not evaporated (Al) or are barely evaporated (W, Mo). The ejection of the substance from the target during the laser pulse impact with these metals was almost not observed; it was established by weighing of the targets before and after the laser "shot" (see Table VI). The latter could happen for the following reasons. First, the high thermal conductivity of these metals compared with the others resulted in the dissipation of the laser pulse power over the large volume of the metal; second, the temperature of the intensive evaporation for these metals is much higher than their melting temperature. Under these conditions the power of the laser pulse is insufficient for the generation of the amount of vapor necessary to eject the melt from the melted surface. As a result, no change in the weight of these metals was observed after their irradiation by defocused laser beam.

In contrast to this, a large mass of ejected melt and a low mass of vapor are observed for metals like indium and tin, which have low melting points, high temperature of intensive evaporation, and low thermal conductivity (see Table VI). In accordance with the suggested mechanism, this effect can be explained as follows. The low melting point and low thermal conductivity provide an accumulation of a large amount of liquid metal in the crater, and the high temperature of the intensive evaporation prevents the generation of large amounts of vapors. So, only at the end of the pulse action does the low vapor pressure generated up to that moment eject the melt from the crater volume. Therefore the thickness of the films deposited by laser evaporation of these metals is low, the amount of macrodefects in them is high, and, as a result, the effectiveness of PLD for the production of the films of these metals is very poor.

A striking example confirming the proposed mechanism of pulsed laser evaporation of metals is the evaporation of chromium and zirconium, which have close melting tempera-

tures and quite different temperatures of intensive evaporation and thermal conductivity. The temperature of intensive evaporation for chromium (temperature at which $P_s = 10^{-1}$ torr) is even lower than the melting point; therefore the evaporation process prevails over the melting. Hence for chromium the total mass of ejected products is relatively low, whereas the vapor portion in the ejecta is high (see Table VI). At the same time, the temperature of intensive evaporation for zirconium is significantly higher than that of its melting point, whereas the zirconium melting point is close to that for chromium. Therefore the amount of liquid zirconium in the crater will be relatively large up to the moment when it reaches the temperature of intensive evaporation. And it is reasonable to expect to obtain a large mass of liquid phase in the ejected products from a zirconium target irradiated by a laser pulse. And just this effect is observed experimentally for zirconium and 50Zr–50Ni alloy (see Table VI and Figs. 2b and 6b).

The surface of the copper target had not even a sign of melting under the action of a defocused laser pulse. This is related to the dissipation of the optical power due to the extremely high thermal conductivity of the copper. The same cause provided the effect that the mass of ejected products from the copper target almost completely consisted of the liquid phase ablated by focused laser irradiation. This means that the copper surface does not reach the temperature of intensive evaporation during the whole time of the pulse action (Table VI).

In Fig. 24 the plot illustrating the correlation between the portion of vapor in the ejected products (m_v/m) and the difference between the temperature of intensive evaporation (at which the saturation pressure is $P_s = 10^{-1}$ torr) and the melting point ($\Delta T = T_v - T_m$) is displayed. The metal targets were irradiated by defocused (lower power density) laser beam.

The above described effects were feebly marked during the use of focused irradiation because of the redistribution of the laser power. Actually, in this case the optical irradiation to a large extent is consumed by the superheating of the vapor, generated in a narrow crater. Superheated vapors in turn transfer the heat into the melting and ablation of the crater walls. In other words, under these conditions the laser power is mainly directed to the heating–melting processes but not to evaporation. Therefore the correlation between the vapor portion in the ejecta and the above discussed thermophysical properties of the metal is not so clear. And this is revealed by the plot in the Fig. 25, where the dependence of the vapor portion m_v/m in the ejecta on $\Delta T = T_v - T_m$ is not as pronounced as in the case of defocused laser irradiation. Despite this it seems that the presented model of laser evaporation of the metals is correct because it holds for the different laser operation regimes and for a wide class of the metals.

The discussed mechanism has allowed us to understand the reasons for the deviation of film composition from that of the target during laser evaporation of 50Zr–50Ni alloy, for the enhanced amount of macrodefects in these films in comparison with other PLD alloy films, and for the strong influence of laser power density on the amount of macrodefects in these condensates (see Section 2.4.4). The relationships revealed in this in-

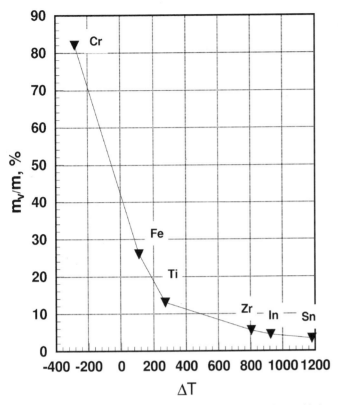

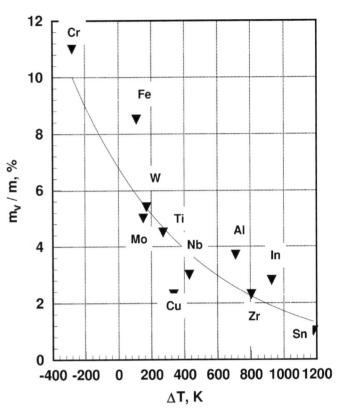

Fig. 24. The dependence of vapor portion m_v/m in products of laser ablation of metals on the temperature difference $\Delta T = T_v - T_m$, where T_v is the temperature at which the vapor saturation pressure is $P_s = 10^{-1}$ torr, and T_m is the melting point. Power density of laser irradiation 1.5×10^5 W/cm^2 (defocused regime).

Fig. 25. The dependence of the vapor portion m_v/m in products of laser ablation of metals on the temperature difference $\Delta T = T_s - T_m$, where T_s is the temperature at which vapor saturation pressure is $P_s = 10^{-1}$ torr, and T_m is the melting point. Power density of laser irradiation 5×10^7 W/cm^2 (focused regime).

vestigation may be used in PLD for the correct selection of a metal or alloy, which can be evaporated by laser pulse with a minor amount of droplets in deposited film.

An examination of the results obtained in [87–90] shows that to decrease the amount of macrodefects in PLD films the following conditions have to be taken into account:

- The difference in the temperatures of the intensive evaporation and the melting point $\Delta T = T_v - T_m$ and the thermal conductivity are the most important thermophysical properties of the material, defining the appearance and the amount of the droplets and punctures in PLD films.
- While a metal or alloy film is deposited by PLD, it is necessary to select an operation laser regime providing a crater depth that is smaller than the crater diameter.

5.4. Conclusions

Macrodefects in the films deposited by PLD are an intrinsic problem of this method. The main physical cause of this problem is the ratio of the speed of laser energy input into the target to the speed of its dissipation. The latter depends on the thermophysical properties of the target material. The main thermophysical properties determining the portion of vapor in the products of laser ablation of metals are the melting temperature,

T_m; the temperature of intensive evaporation, T_v; and the thermal conductivity. The optimal combination of thermophysical characteristics of a metal for PLD of the film with minimum particulates is minimal values of $\Delta T = T_v - T_m$ and thermal conductivity. Supposedly the last requirements are also relevant for pulsed laser evaporation of other materials. One of the most important requirements for the laser operation regime is that the crater appearing on the target surface during laser impact has to be shallow and wide. The amount of particulates in the film deposited in this regime can be expected to be minimal.

6. INFLUENCE OF TARGET PROPERTIES ON SOME FEATURES OF PLD COMPOUND FILMS

In the previous section the role of thermophysical properties of metal targets in the generation of particulates by pulsed laser ablation of metals was discussed. The role of the target properties in the composition formation of PLD films, their growth rate, and the appearance of macrodefects in these films was recently examined, say, on the "microscopic" level. This is due mainly to the fact that widely used excimer lasers provide a low power per pulse (0.1–1 J/pulse) and short-length pulses (~tens ns). The amount of ablated material from the target surface under these conditions is very small (the growth rate of the YBCO film normally deposited by an excimer laser is <0.1 nm/pulse). The thermal conductivity of the target material does not play as significant a role in this case as it does in the case of large laser power per pulse and longer ($\sim 10^3$ times) pulse duration ($\sim$ms pulses). This is due to the fact that the heat conduction over the large target volume is limited by the short time of pulse action and by the low power input per pulse. More important roles are played by such target properties as the roughness of the target surface [10], cone development and other kinds of the surface features, and segregation of higher melting point constituents on grain boundaries under laser irradiation [52] in the case of excimer lasers. These effects have a negative impact on the composition of laser films, decreasing of their growth rate and promoting the appearance of macrodefects in the films. These problems were widely treated in the recent literature (see, for example, [10, 56, 92]) and hence we will not concentrate on these questions here. Let us mention only that the generally accepted opinion about the target for PLD by excimer laser is that targets of high density can minimize the amount of particulates in the film [3].

In our opinion the contemporary direction of PLD method development, namely, the use of excimer lasers with low pulse energy ($\sim$1 J) operating in frequency regime, must be seen in perspective. The thickness of films depositing per pulse by excimer laser is many times lower in comparison with that of films deposited by laser with high power per pulse. Apart from this, such advantages of pulsed laser deposition as the high velocity of the evaporated species and the high level of their ionization practically disappear in the case of deposition in a gas ambient [6]. This modification of PLD becomes more prevalent in the case of deposition of films with one volatile component. Although the main drawbacks of PLD revealed at the earlier stages of the method development (particulates; the lack of uniformity, over a large area, of such PLD film properties as a thickness, composition, and structure; poor step coverage) have been diminished, they have not been eliminated.

At the same time it is reasonable to assume that the use of pulsed lasers with high power density distributed over a large area of the target can provide substantial improvement of the film properties. In fact, the ablation of a large area of the target without crater formation makes it possible: (i) to increase the deposition rate many times; (ii) to decrease substantially the nonuniformity of the composition and thickness of the film; and (iii) to increase the area of the film with uniform properties. When the laser spot size is larger than the target dimensions, the amount of droplets in the film is decreased. This is related to the effect that the melt appearing on the flat (without crater) surface of the target moves along it and normal to the laser plume under the recoil vapor pressure. The absence of the crater on the target surface makes it possible to use the same target many times, "cutting" it layer by layer with each laser "shot" [101]. The creators of this method managed to deposit a $(Bi,Sb)_2Te_3$ film of 1000-nm thickness with low particulate density by evaporation of a 6-mm target over 10 pulses. Thermophysical properties of the target begin to play an important role in PLD film formation in the deposition mode with large laser spot and large power

density. The major target material parameter in this case is the thermal conductivity that can be affected by changing the bulk density of the target material.

6.1. Role of Target Thermal Conductivity in Compound Film Property Formation

Different target types for the deposition of films by pulse laser evaporation were tested. Free-falling powders, dulk powders, pressed powders, hot-pressed (or sintered) powders, cast materials, and single crystals were tested as targets during PLD development (see references from Table I). However, there are very few works in which systematic investigations of film properties deposited by pulsed laser evaporation of a set of materials in different states have been carried out.

One such work is devoted to the investigation of the influence of target density (and hence the target's thermal conductivity) on PLD films ($\lambda = 1064$ nm, $\tau = 1$ ms, $q = 1.5 \times 10^6$ W/cm^2, and $q = 5 \times 10^7$ W/cm^2 at 1000 J/pulse) [86]. The methods of cold and hot pressing of powders were used to produce targets with different densities. Cold-pressed targets were prepared by pressing powders in open air at room temperature. Hot-pressed targets (ceramics) were fabricated by sintering of the powders under pressure in the shielding medium (normally an argon atmosphere) for a certain time. The density of the targets fabricated by hot-pressing of $CrSi_2$, WSi_2, $TaSi_2$, $TiSi_2$, TiC, SiC, and LaB_6 powders was $\sim$85–93% of that of the bulk. The density of the targets prepared by cold-pressing of WSe_2, CdSe, Nb_3Ge, and 80W-20Cu powders was 50–60% of that of the bulk. Considering the thermal conductivity of the target, it is worth remembering that at low temperatures for dispersed pores in a solid its thermal conductivity decreases linearly with increasing porosity [3].

The films deposited by laser evaporation of hot-pressed targets had large thickness (100–200 nm) per pulse. The films had a quite low density of particulates per film square, which increased slightly with increasing laser power density (Figs. 8a and b).

The thickness of films deposited by evaporation of cold-pressed targets was close to or sometimes higher than that obtained during the evaporation of hot-pressed targets. A comparative investigation of films deposited by the evaporation of different types of tungsten targets by lasers operating under similar conditions clearly demonstrates this effect. For this experiment single crystal of tungsten of 99.999% purity, cast tungsten of 95.0% purity, and hot-pressed tungsten powder were chosen as targets. The initial mass of the targets was about 5–7 g. The mass Δm of the products ablated by defocused ($q = 1.5 \times 10^6$ W/cm^2) laser pulse was, correspondingly, 0.0001 g and 0.0005 g. Because of the destruction of the target produced from cold-pressed tungsten powder, the Δm mass measurement failed. The thickness t of the films deposited in these experiments was, correspondingly, 0.0 nm, 35 nm, and 140 nm. Maximal efficiency of evaporation was achieved in the case of cold-pressed tungsten powder that had minimal thermal

Fig. 26. Microstructure of a film deposited by laser ablation of a hot-pressed 90W-10Cu powder target.

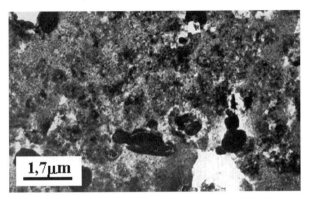

Fig. 27. Microstructure of a film deposited by laser ablation of a cold-pressed 90W-10Cu powder target.

conductivity. Tungsten single crystal exhibited maximal conductivity, which was the reason for the total consumption of laser power during heating of the metal without its evaporation.

The dependence of laser evaporation effectiveness on the target thermal conductivity for cold-pressed and hot-pressed 90W-10Cu powders was similar to that revealed for different types of tungsten targets (see Table II). Actually, the effectiveness of evaporation of more denser hot-pressed targets (hence with higher thermal conductivity) was lower than that of cold-pressed 90W-10Cu targets with lower thermal conductivities. The investigation of the composition of the copper concentration in condensates obtained during the evaporation of these targets was ~160 times higher than that in the targets, as revealed by EPMA investigations. Copper was mainly presented in these condensates in the form of solidified liquid droplets and powder particles expelled from the target (Figs. 26, 27). Because the composition of particulates differs from that of the film deposited from copper and tungsten vapors, the results of composition analysis do not reflect the real concentrations of these elements in the film.

The efficiency of cold-pressed target evaporation of other materials was also higher compared with that of hot-pressed targets. However, the amount of macrodefects in the form of droplets, punctures, and powder particles in films deposited by evaporation of cold-pressed targets was noticeably larger than

that in the films deposited from hot-pressed targets. The micrographs in Figs. 26 and 27 of the films deposited by evaporation of 90W-10Cu hot-pressed targets and cold-pressed targets confirm this assertion.

The composition of all of the films deposited by evaporation of cold-pressed targets, excluding 90W-10Cu targets, coincided with or was close to that of the target.

Although laser power density did not influence the film composition, it substantially affected the efficiency of evaporation. As seen from Table II, the increase in power density, due to a decrease in laser spot size, is accompanied by a decrease in evaporation efficiency. This means that the mass of ablated products was increased decreasing laser spot size, whereas the thickness of the deposited film was decreased.

6.2. Powder Targets: Mechanisms of Particulate Generation

The above results show that a decrease in target thermal conductivity due to its preparation method positively influences the thickness of films deposited from either cold-pressed targets or hot-pressed targets. The reason for this effect is more effective consumption of the laser power, which is due to the decrease in the portion of the power expended on heating–melting of the target.

However, a reasonable question arises: Why, then, is the density of macrodefects in films deposited from cold-pressed targets higher than that in the films deposited from hot-pressed targets? To answer this question, scanning electron microscope investigations of the surfaces of cold-pressed targets and hot-pressed targets after the laser "shots" were carried out. Micrographs of cold-pressed 75Nb-25Ge powder mixture, hot-pressed TiC, and cast Nb targets, subjected to a laser pulse ($q = 1.5 \times 10^6$ W/cm^2) are presented in Figs. 28a, b, and c, respectively. Many small pinholes on a solidified crater surface on a 75Nb-25Ge cold-pressed target are clearly seen (Fig. 28a). These pinholes are the tracks of outlets of gases occluded in the volume of the porous cold-pressed powder. The vapors generated by the laser pulse together with gases emerging through the melted layer on the target catch the liquid and solid particles and transport them on the surface of the growing film.

The tracks of outlet of gases occluded in the volume of the hot-pressed targets are less noticeable than those in the case of cold-pressed targets (Fig. 28b). It is quite clear that this is due to the fact that this type of target contains a substantially lower volume of pores and voids with occluded gases. As a result, the number of macrodefects in the films deposited from these targets is lower than that deposited from cold-pressed targets.

Only isolated and quite large bubbles that rise because of subsurface boiling are observed on the surface of the cast Nb target (Fig. 28c). The small pinholes typical of powder targets are absent on the cast targets.

Films deposited by the evaporation of cold-pressed targets and hot-pressed targets fabricated from 90W-10Cu powder had

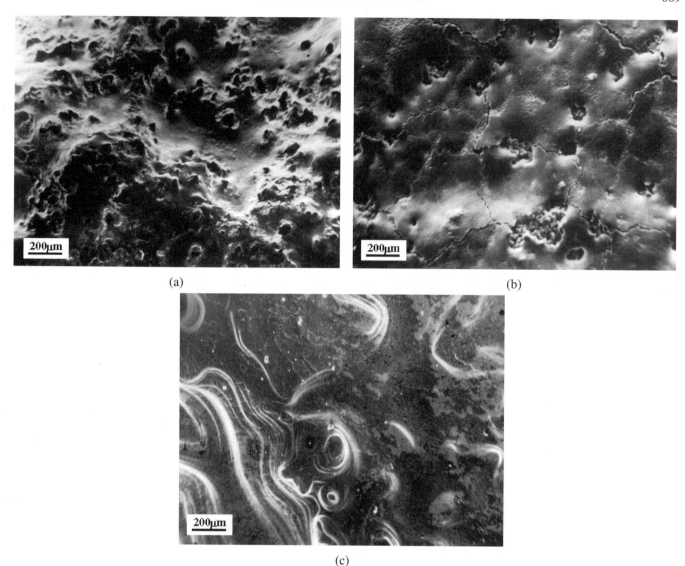

Fig. 28. (a) SEM micrograph of the bottom of a crater that appeared at the surface of a cold-pressed 50Nb-50Ge powder target after a laser "shot." A number of tracks (fumaroles) of gas outlet from the target are clearly seen. (b) SEM micrograph of the bottom of a crater that appeared at the surface of a hot-pressed (ceramic) TiC target after a laser "shot." The number of gas outlet tracks is much smaller than in a, a cold-pressed powder target. (c) SEM micrograph of the bottom of a creater that appeared on a cast Nb target after a laser "shot." The few solidified bubbles that are seen on the smooth crater surface are a result of subsurface boiling.

an anomalously high density of macrodefects in comparison with films deposited from cold-pressed targets and hot-pressed targets fabricated from other powder materials.

The 90W-10Cu targets are prepared from a cold- or hot-pressed mixture of tungsten and copper powders. These metals are insoluble in each other. Quite different optical and thermo-physical properties of these elements result in the development of high mechanical stress on the boundaries between the copper and tungsten particles due to the high temperature gradients during the heating by laser pulse. These mechanical stresses promote the enhanced destruction of such targets, which results in an increased density of macrodefects in the film and in deviation of the film composition from the stoichiometric. Experiments on the laser evaporation of cold-pressed tungsten powder confirm this argumentation. The films deposited by evap-

oration of this target in the same laser operation regime were almost free of macrodefects (Fig. 29). At the same time, the microstructure of the film deposited by evaporation of cold-pressed tungsten powder is close to that of the film deposited by evaporation of cold-pressed 90W-10Cu powder in the same operation regime. The main difference in the microstructures of these films is the presence of small copper droplets and particles in the condensate obtained by evaporation of cold-pressed targets from 90W-10Cu powder (compare Figs. 26 and 29). The mechanism of origination of copper particles in this film can be described in the following way. The melting point of copper is lower than that of tungsten; hence the amount of liquid copper on the target surface has to be larger than the amount of the liquid tungsten. Consequently, the droplets of liquid copper are transported by the vapors onto the condensation sur-

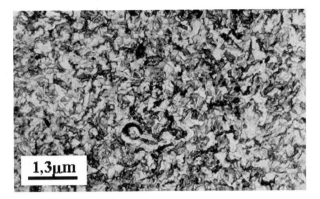

Fig. 29. TEM micrograph of tungsten film deposited by laser ablation of a cold-pressed tungsten powder target.

face. The size of the droplets is smaller and their density per film square area is higher in the case of hot-pressed 90W-10Cu powder evaporation. This is the result of enhanced thermal conductivity of hot-pressed targets compared with cold-pressed targets and, consequently, the decreased amount of liquid copper on the hot-pressed target surface during the laser "shot."

The mechanisms of formation of macrodefects in films deposited from the pressed powder targets are different from those of films that deposited from cast targets. The main role in macrodefect formation in the case of powder target evaporation is played by gases occluded in the pores and voids and on the boundaries of the powder particles in the target. The gases heated during the laser pulse action effluence through a thin melted layer with high speed and entrain the droplets and hard powder particles. The mechanisms of macrodefect formation during the evaporation of cast targets (metals and alloys) are discussed above in Sections 2.4.4 and 5.3. The difference in the evaporation mechanisms of cast and powder targets also results in the different average thickness of the films deposited per pulse under similar deposition conditions. The mean thickness of the film deposited per pulse by evaporation of cast targets is about 30–140 nm, and the mean film thickness in the case of evaporation of hot-pressed targets or cold-pressed targets of different materials reaches 100–250 nm per pulse when the laser operates in a regime with 1000 J/pulse and ∼1 ms pulse length.

From the point of view of the optimal consumption of laser power, the ideal target is that which can be completely evaporated in one laser pulse. In this case one may expect to obtain a film without macrodefects and with a composition coinciding with that of the target. Free-falling powder with particles that can be completely evaporated by one laser pulse may be used as such a target.

Such targets, consisting of free-falling $MoSi_2$ powder, were used by Schrag et al. [41]. It could be expected that the evaporation of the powder particles in this case has to be complete. However, this was not the case, and the composition of the film deposited by this means was highly enriched in silicon, whereas the powder particles after the laser impact were almost completely composed of molybdenum. Evidently the reason for the failure of this experiment is insufficient laser power density

employed for the evaporation of powder particles of the size used. In fact, it is known that the flashing method, based on the evaporation of powder particles falling on a heated surface, was successfully used for the deposition of films of quite different compounds including silicide films [1]. The more complete is the evaporation of a powder particle as it approaches the heated surface, the closer is the composition of the film deposited by flashing to stoichiometric.

6.3. Conclusions

First of all, it should be emphasized that the requirements for the target depend on the laser operation regime. The thermal conductivity of the target material does not play a significant role in the case of excimer lasers with a short pulse length and low power per pulse. The optimal target properties for excimer laser evaporation are a smooth surface, high density, and the uniformity of target composition and density over the whole volume [3]. Requirements that are valid for excimer laser evaporation lose their importance in the case of lasers with high power per pulse and relatively long pulse duration (∼ms). The most important role in this case is played by the thermal conductivity of the target. Thermal conductivity is a channel for nonefficient laser power that directs it to the heating and melting of the target. The artificial decrease in the thermal conductivity of the target by its fabrication from pressed powders increases the effectiveness of consumption of laser power directed to the evaporation. The decrease in the target thermal conductivity results in a noticeable increase of the thickness of the film deposited per pulse and frequently in a decrease in macrodefects in the film in comparison with films deposited by evaporation of cast targets. The probability of deposition of the film with a composition that is close or similar to that of the target also increases with the use of powder targets. The most appropriate targets for pulsed laser deposition of films with a minimum of composition and microstructure defects are ceramic targets fabricated by hot-pressing of powders.

7. GENERAL CONCLUSIONS

The above review of the contemporary state of the art of pulsed laser deposition of thin films shows that the hopes expressed for this method in its infancy are, only partially justified, even though significant effort has been expended on its development. The quotations listed below may confirm this assertion.

Schrag et al. [41] note that there are two principal difficulties historically associated with laser ablation that hinder its widespread use: "1) laser–target interaction produces large amounts of 'splatter' which is incorporated into resulting films making them useless for most applications; 2) stoichiometric deposition of film constituents is generally not possible for compound targets comprising elements having widely different enthalpies of vaporization." Another paper listed the following main disadvantages of PLD: "the deposition of the droplets; small-area deposition; narrow angle of stoichiometric transfer of target material" [93]. In a review chapter devoted to a consideration of approaches to the reduction of "splashing," the author points out that "There are two major drawbacks in PLD. One is the splashing and the other one is the lack of uniformity over a large area due to narrow angular distribution of the plume" [10]. In a "Handbook of Deposition Technologies for Films and Coatings" [13], while discussing PLD the author emphasizes that among the other limitations of the method, there are "(i) splashing effect, which involves the production of microparticles, diminishes film quality; (ii) the size of deposited films is small (10 to 20 mm, diameter), resulting from the small size of the laser impact spot."

It is difficult to argue with the authors of these citations, because the quoted drawbacks of pulsed laser deposition of thin films remain a matter of fact. Unceasing searches for methods of decreasing the macrodefects in PLD films [10, 12, 94–96] and ways of developing PLD for production of thin films with stoichiometric composition (e.g., [6]) that began early in the history of this method and have continued up to the present justify this conclusion.

The stated problems of PLD of thin films are principally caused by intrinsic particularities of this method. As was already mentioned, the main features of PLD are extremely nonequilibrium deposition conditions (extremely high deposition rates) and the high nonuniformity of film-forming flux. The instantaneous deposition rate during PLD can achieve the values of ~km/s, and the composition of the FFS includes a broad spectrum of particles, from photons, electrons, and ions up to clusters containing hundreds of atoms and even larger particulates of micron size. The distribution of different species in the target-to-substrate space is also not uniform and takes place at different angles.

These particularities of PLD result in the appearance of three levels of nonuniformity of laser films: (i) inhomogeneity on the atomic level, where the chemical composition of the film can change along the substrate; (ii) nonuniformity on the crystal structure level, where both amorphous and crystalline areas in the same film can be revealed; and (iii) inhomogeneity on the film microstructure level, where both macrodefects in the form of large droplets, particles, and punctures and an atomically smooth surface in the same film can be observed.

At the same time, it is known that equilibrium deposition conditions and a flux with an easily controlled energy of the uniform FFS are necessary to produce a film with perfect composition and structure. With an appropriate ratio of chemically different atomic species in film-forming flux and an appropriate substrate temperature, a film with perfect composition and structure can be deposited. Examples of such a technological approach are the molecular beam epitaxy [97] and ion beam epitaxy [98] methods. The main peculiarities of these deposition techniques are the uniformity of the FFS (either atoms or molecules), low supersaturation of the film-forming vapor, controllable ion concentration, and the energy of the FFS [98]. These conditions provide easy and efficient control of the growing layer structure by variation of the substrate structure, substrate temperature, and cleanliness of the condensation surface.

From this point of view, PLD in a vacuum with a laser with high energy and long pulse duration (~1 ms) is the complete opposite of molecular beam epitaxy deposition. At the same time, the deposition conditions for the production of epitaxial films by PLD with an excimer laser are close to what is peculiar to molecular beam epitaxy. Actually, the growth rate of epitaxial PLD films is very low, the substrate temperature is high, the deposition time is long, and the film thickness is small. These conditions are realized because of the short pulse length and low (~1 J/pulse) power. Under these evaporation conditions the vapor composition is more or less uniform, and its density is low. Because the process is frequently carried out in a gas ambient at relatively high pressures [6, 34, 35, 48], this promotes a noticeable decrease in the energy of evaporated species and the amount of ions among them. On one hand, these deposition conditions are favorable for the growing of films with low defect concentrations, but on the other hand, PLD is missing advantages mentioned by many authors such as high energy of FFS, the presence of charged species, and a high deposition rate, that are not peculiar to other PVD techniques.

While in its infancy PLD held out the hope of growing into a versatile method with unique possibilities for producing films with stoichiometric composition at high growth rates. It was supposed that this method would make possible the deposition of crystalline films on unheated amorphous substrates and the deposition of epitaxial films at substantially lower substrate temperatures [11, 14]. As follows from our review, this hope did not come true.

At present it is not anticipated that PLD will be able to play a substantial role in thin film deposition, even if the major problems associated with PLD (particulates, large area deposits, nonuniformity) were to be solved. The most realistic expectation for PLD among the other PVD methods is its use for the deposition of oxide, multicomponent oxide films, or films with any exotic composition, and mainly for laboratory research.

ACKNOWLEDGMENTS

I am grateful to a number of my colleagues, whose contributions have enriched the above investigations. In particular, I thank Dr. V. Yu. Kulikovsky for carrying out the depositions, V. P. Smirnov for the help with EPMA and SEM investigations, and Professor Stefanovich for his fruitful notes.

REFERENCES

1. R. Glang, "Handbook of Thin Film Technology" (L. Maissel and R. Glang, Eds.), pp. 9–174. McGraw-Hill, New York, 1970.
2. S. Neuville and A. Mathews, *MRS Bull.* September, 22 (1997).
3. L. C. Chen, in "Pulsed Laser Deposition of Thin Films" (D. B. Chirsey and G. K. Hubler, Eds.), pp. 167–178. Wiley–Interscience, New York, 1994.
4. D. Geohegan, in "Pulsed Laser Deposition of Thin Films" (D. B. Chirsey and G. K. Hubler, Eds.), pp. 115–166. Wiley–Interscience, New York, 1994.

5. R. Hettich and C. Jin, "Laser Ablation" (J. C. Miller, Ed.), pp. 128–156. Springer-Verlag, Berlin/New York, 1994.

6. A. Gupta, in "Pulsed Laser Deposition of Thin Films" (D. B. Chirsey and G. K. Hubler, Eds.), pp. 765–782. Wiley–Interscience, New York, 1994.

7. S. I. Anisimov, Ya. A. Imas, G. S. Romanov, and Yu. V. Khodyko, "Influence of High Power Radiation on Metals," p. 276. Nauka, Moscow, 1970.

8. Yu. A. Bykovsky, A. G. Dudoladov, and V. P. Kozlenkov, *Fizika i Khimiya Obrabotki Materialov* 2, 133 (1976) (in Russian).

9. S. M. Greene, A. Pique, K. S. Harshavardhan, and J. Bernstein, in "Pulsed Laser Deposition of Thin Films" (D. B. Chirsey and G. K. Hubler, Eds.), pp. 23–54. Wiley–Interscience, New York, 1994.

10. J. T. Cheung, in "Pulsed Laser Deposition of Thin Films" (D. B. Chirsey and G. K. Hubler, Eds.), pp. 1–22. Wiley–Interscience, New York, 1994.

11. V. G. Dneprovsky and V. N. Bankov, *Zarubezhnaya Radioelektronika* 9, 133 (1976) (in Russian).

12. T. V. Venkatesan, "Laser Ablation" (J. C. Miller, Ed.), pp. 85–106. Springer-Verlag, Berlin/New York, 1994.

13. R. F. Bunshah, "Handbook of Deposition Technologies for Films and Coatings" (R. F. Bunshah, Ed.), pp. 740–762. Noyes, Park Ridge, NJ, 1994.

14. A. A. Schuka, V. G. Dneprovsky, and A. G. Dudoladov, *Zarubezhnaya Radioelektronika* 72, 24, 38 (1973) (in Russian).

15. Yu. A. Bykovsky, and Yu. P. Kozyirev, *Priroda* 5, 54 (1977) (in Russian).

16. K. L. Saenger, in "Pulsed Laser Deposition of Thin Films" (D. B. Chirsey and G. K. Hubler, Eds.), pp. 199–238. Wiley–Interscience, New York, 1994.

17. D. B. Geohegan and D. N. Mashburn, *Appl. Phys. Lett.* 55, 2766 (1989).

18. S. Metev, in "Pulsed Laser Deposition of Thin Films" (D. B. Chirsey and G. K. Hubler, Eds.). Wiley–Interscience, New York, 1994.

19. S. G. Hansen and T. E. Robitaillle, *Appl. Phys. Lett.* 50, 359 (1987).

20. H. Ramarotafika and G. Lemperier, *Thin Solid Films* 266, 267 (1995).

21. A. N. Pilyankevich, V. Yu. Kulikovsky, and L. R. Shaginyan, *Poverkhnost* 10, 97 (1986).

22. L. R Shaginyan, *J. Chemical Vapor Deposition* 6, 219 (1998).

23. L. Jastrabik, L. Soukup, L. R. Shaginyan, and A. A. Onoprienko, *Surf. Coat. Technol.* 123, 261 (2000).

24. O. B. Vorob'ev, V. I. Kozlov, V. V. Korneev, V. N. Okhrimenko, B. V. Seleznev, and O. Yu. Tkachenko, *Supercond. Phys. Chem. Technol.* 4, 1935 (1991).

25. T. V. Venkatesan, X. D. Wu, A. Inam, and J. B. Wachtman, *Appl. Phys. Lett.* 52, 1193 (1988).

26. S. K. Hou, K. H. Wong, P. W. Chan, C. L. Choy, and H. K. Wong, *J. Mater. Sci. Lett.* 11, 1266 (1992).

27. M. C. Foote, B. B. Jones, B. D. Hunt, J. B. Barner, and L. J. Bajuk, *Physica C* 201, 176 (1992).

28. E. Fogarassy, C. Fuchs, J. P. Stoquert, P. Siffert, J. Perreire, and F. Rochet, *J. Less-Common Met.* 151, 249 (1989).

29. H. M. Smith and A. T. Turner, *Appl. Opt.* 4, 147 (1965).

30. H. Schwarz and H. A. Tourtellote, *J. Vac. Sci. Technol.* 6, 373 (1969).

31. V. S. Ban and D. A. Cramer, *J. Mater. Sci.* 5, 978 (1970).

32. A. A. Sokol, V. M. Kosevich, L. D. Barvinok, and E. A. Lubchenko, *Fizika i Khimiya Obrabotki Materialov* 3, 25 (1982) (in Russian).

33. V. M. Kosevich, A. A. Sokol, and L. D. Barvinok, *Fizika i Khimiya Obrabotki Materialov* 3, 66 (1982).

34. D. Feiler, R. S. Williams, A. A. Talin, H. Yoon, and M. S. Goorsky, *J. Cryst. Growth* 171, 12 (1997).

35. M. Okamoto, Y. Mori, and T. Sasaki, *Japan. J. Appl. Phys.* 38, 2114 (1999).

36. G. L. Doll, J. A. Sell, C. A. Taylor, and R. Clarke, *Phys. Rev. B* 43, 6816 (1991).

37. N. Biunno, J. Narayan, and A. R. Srivastava, *Appl. Phys. Lett.* 54, 151 (1989).

38. M. S. Donley, J. S. Zabinsky, W. J. Sessler, V. J. Dyhouse, and N. T. McDevitt, *Mater. Res. Soc. Symp. Proc.* 236, 461 (1992).

39. M. Balooch, R. J. Tench, W. J. Siekhaus, M. J. Allen, and A. L. Connor, *Appl. Phys. Lett.* 57, 1540 (1990).

40. S. Ghaisas, *J. Appl. Phys.* 70, 7626 (1991).

41. G. H. Schrag, S. O. Colgate, and P. H. Holloway, *Thin Solid Films* 199, 231 (1991).

42. G. Reisse, B. Keiper, S. Weissmantel, and H. Johansen, in "E-MRS Spring Meeting, Book of Abstracts," C-IX/P46, 1993. Europ. Materials Research Society.

43. Yu. A. Bykovsky, V. M. Boiyakov, Yu. P. Kozyrev, and P. A. Leont'ev, *Zhurnal Tekhnicheskoi Fiziki* 48, 991 (1978) (in Russian).

44. H. Sankur, *Mater. Res. Soc. Symp. Proc.* 29, 373 (1984).

45. J. F. Muth, R. M. Kolbas, A. K. Sharma, S. Oktyabrsky, and J. Narayan, *J. Appl. Phys.* 85, 7884 (1999).

46. M. Y. Chen and P. T. Murray, *Mater. Res. Soc. Symp. Proc.* 191, 43 (1991).

47. F. O. Adurodija, H. Izumi, T. Ishihara, H. Yoshioka, and M. Motoyama, *Japan. J. Appl. Phys.* 38, 2710 (1999).

48. Q. X. Jia, X. D. Wu, S. R. Foltyn, P. Tiwari, J. P. Zheng, and T. R. Jow, *Appl. Phys. Lett.* 67, 1677 (1995).

49. R. C. DeVries, in "Diamond and Diamond-Like Films and Coatings" (R. E. Clausing et al., Eds.), pp. 151–172. Plenum, New York, 1991.

50. A. N. Pilyankevich, V. Yu. Kulikovsky, and L. R. Shaginyan, *Elektronnaya Tekhnika Ser. Materialy (6)* 4, 71 (1981).

51. S. R. Foltyn, R. C. Dye, K. C. Ott, K. M. Hubbard, W. Hutchinson, R. E. Muenchausen, R. C. Estler, and X. D. Wu, *Appl. Phys. Lett.* 59, 594 (1991).

52. S. R. Foltyn, in "Pulsed Laser Deposition of Thin Films" (D. B. Chirsey and G. K. Hubler, Eds.), pp. 89–114. Wiley–Interscience, New York, 1994.

53. M. C. Foote, B. B. Jones, B. D. Hunt, J. B. Barner, R. P. Vasques, and L. J. Bajuk, *Physica C* 201, 176 (1992).

54. C. Cali, V. Daneu, A. Orioli, and S. Rava-Sanseverino, *Appl. Opt.* 15, 5, 1327 (1976).

55. A. Gupta, B. Braren, K. G. Kasey, B. W. Hussey, and R. Kelly, *Appl. Phys. Lett.* 59, 1302 (1991).

56. E. M. Dudnik, V. I. Bessaraba, and Yu. B. Paderno, *Poroshkovaya Metallurgiya* 12, 60 (1976) (in Russian).

57. V. N. Chernyaev, A. M. Vasil'ev, V. G. Blokhin, V. I. Bessaraba, and L. R. Shaginyan, *Elektronnaya Tekhnika Ser. Materialy (6)* 7, 3 (1982) (in Russian).

58. C. H. Becker and J. B. Pallix, *J. Appl. Phys.* 64, 5152 (1988).

59. C. Girault, D. Domiani, J. Auberton, and A. Catherinot, *Appl. Phys. Lett.* 55, 182 (1989).

60. L. V. Gurvich, G. V. Karachavcev, and V. N. Kondrat'ev, "Bonding Energy, Ionization Potentials, Electron Affinity." Nauka, Moscow, 1974.

61. R. W. Pryor, Z. L. Wu, K. R. Padmanabhan, S. Vallinueva, and R. L. Thomas, *Thin Solid Films* 253, 243 (1994).

62. I. Kh. Khan, in "Handbook of Thin Film Technology" (L. Maissel and R. Glang, Eds.), Vol. 2, pp. 97–125. McGraw-Hill, New York, 1970.

63. G. K. Hubler, in "Pulsed Laser Deposition of Thin Films" (D. B. Chirsey and G. K. Hubler, Eds.), pp. 327–356. Wiley–Interscience, New York, 1994.

64. J. S. Colligon, *J. Vac. Sci. Technol. A* 13, 1649 (1995).

65. A. N. Pilyankevich, V. Yu. Kulikovsky, and L. R. Shaginyan, *Thin Solid Films* 137, 215 (1986).

66. L. S. Palatnik, M. Ya. Fux, and V. M. Kosevich, "Mechanism of Formation and Substructure of Condensed Films," 320 p. Nauka, Moscow, 1972.

67. A. T. Pugachev, N. P. Churakov, N. I. Gorbenko, Kh. Saadly, and A. A. Solodovnik, *Low Temperature Phys.* 25, 3, 298 (1999).

68. V. P. Zakharov and I. M. Protas, *Dokl. Akad. Nauk SSSR* 215, 3, 562 (1974).

69. Yu. G. Poltavtsev and N. M. Zakharov, *Kristallografiya* 17, 1, 203 (1972) (in Russian).

70. M. G. Norton, P. G. Cotula, and C. B. Carter, *J. Appl. Phys.* 70, 2871 (1991).

71. L. A. Dvorina, I. V. Kud', G. Beddis, V. Bretschneider, and H. Helms, *Poroshkovaya Metallurgiya* 1, 81 (1987) (in Russian).

72. G. Ferro, H. Okumura, and S. Yoshida, *Japan. J. Appl. Phys.* 38, 3634 (1999).

73. H. L. Kao, P. J. Shih, and C.-H. Lai, *Japan. J. Appl. Phys.* 38, 1526 (1999).

74. M. Dubey and G. Singh, *J. Phys. D: Appl. Phys.* 7, 1482 (1974).

75. S. Winsztal, H. Majewska, M. Wisznewska, and T. Niemyski, *Mater. Res. Bull.* 8, 1329 (1973).

76. V. I. Bessaraba, L. A. Ivanchenko, and Yu. B. Paderno, *J. Less-Common Met.* 67, 505 (1979).

77. P. K. Song, Y. Shigesato, M. Kamei, and I. Yasui, *Japan. J. Appl. Phys.* 38, 2921 (1999).

78. J. S. Horwitz and J. A. Sprague, in "Pulsed Laser Deposition of Thin Films" (D. B. Chirsey and G. K. Hubler, Eds.), pp. 229–254. Wiley–Interscience, New York, 1994.

79. M. Buril, V. Elinek, D. Vorlicek, L. Chvostova, and J. Soukup, *J. Non-Cryst. Solids* 188, 118 (1995).

80. A. A. Voevodin, S. J. Laube, S. D. Walk, J. S. Solomon, M. S. Donley, and J. S. Zabinski, *J. Appl. Phys.* 78, 4123 (1995).

81. M. Tabbal, P. Merel, M. Chalker, M. A. El Khakani, E. G. Herbert, B. N. Lucas, and M. E. O'Hern, *J. Appl. Phys.* 85, 3860 (1999).

82. A. A. Puretzky, D. B. Geohegan, G. E. Jelinson, Jr., and M. M. McGibbon, *Appl. Surf. Sci.* 96–98, 859 (1996).

83. V. I. Kasatochkin, *Dokl. Akad. Nauk SSSR* 201, 5, 1104 (1971).

84. V. M. Babina, M. B. Guseva, and V. G. Babaev, in "Amorphous and Nanocrystalline Semiconductors," All-Russian Symposium Proceedings, St.-Petersburg, pp. 60–61. 1998.

85. Yu. A. Vodakov, G. A. Lomakina, and E. N. Mokhov, *Fizika Tverdogo Tela* 24, 1377 (1982) (in Russian).

86. V. Yu. Kulikovsky, L. R. Shaginyan, V. P. Smirnov, and T. V. Kukhtareva, *Elektronnaya Tekhnika Ser. Materialy (6)* 11, 29 (1981).

87. V. M. Kirillov and P. I. Ulyakov, *Fizika i Khimiya Obrabotki Materialov* 1, 8 (1971) (in Russian).

88. L. I. Mirkin, "Fundamentals of Physics of Material Treatment by Laser Irradiation." Moscow State Univ., Moscow, 1975.

89. Yu. A. Bykovsky and V. M. Boyakov, *Zhurnal Tekhnicheskoi Fiziki* 48, 991 (1978) (in Russian).

90. A. N. Pilyankevich, V. Yu. Kulikovskii, and L. R. Shaginyan, *Fizika i Khimiya Obrabotki Materialov* 18, 2, 134 (1984) (in Russian).

91. B. A. Osadin, *Elektronnaya Tekhnika Ser. Mikroelektronika* 4, 408 (1976) (in Russian).

92. J. C. S. Kools, in "Pulsed Laser Deposition of Thin Films" (D. B. Chirsey and G. K. Hubler, Eds.), pp. 465–469. Wiley–Interscience, New York, 1994.

93. M. Jelinek, L. Jastrabik, L. Soukup, J. Bulir and V. Trtik, *Jemna Mechanika Optika* 5–6, 177 (1995).

94. Z. Trajanovic, S. Choopun, P. R. Sharma, and T. Venkatesan, *Appl. Phys. Lett.* 70, 3461 (1997).

95. M. Tachiki, M. Noda, K. Yamada, and T. Kobayashi, *J. Appl. Phys.* 83, 5351 (1998).

96. M. Tachiki and T. Kobayashi, *Japan. J. Appl. Phys.* 38, 3642 (1999).

97. J. R. Arthur, *J. Vac. Sci. Technol.* 16, 2, 273 (1979).

98. A. S. Lutovich, *Crystal Growth Science* 14 (1983) (in Russian).

99. Yu. M. Boguslavsky, and A. P. Shapovalov, *Supercond. Sci. Technol.* 4, 149 (1991).

100. L. R. Shaginyan, M. Mišina, A. Mackova, V. Peřina, S. Vadlec, and L. Yestabik, *J. Vac. Sci. Technol. A* (2001), in print.

101. Yu. A. Bykovsky, Investigation of Epitaxial Growth of PLD Films, Report. Moscow Engineering Physical Institute, Solid State Physics Department, 1977.

102. Y. Murayama, *J. Vac. Sci. Technol.* 2, 4, 876 (1975).

103. A. S. Golubev, A. V. Kurdyumov, and A. N. Pilyankevich, "Boron Nitride: Structure, Properties, Fabrication." Naukova Dumka, Kiev, 1987.

104. M. S. Donley and J. S. Zabinski, in "Pulsed Laser Deposition of Thin Films" (D. B. Chirsey and G. K. Hubler, Eds.). Wiley–Interscience, New York, 1994.

105. T. Yoshida, *Diamond Relat. Mater.* 5, 501 (1996).

Chapter 14

SINGLE-CRYSTAL β''-ALUMINA FILMS

Chu Kun Kuo, Patrick S. Nicholson

Ceramic Engineering Research Group, Department of Materials Science and Engineering,
McMaster University, Hamilton, Ontario L8S 4L7, Canada

Contents

1. INTRODUCTION

Na-β''-Al$_2$O$_3$ single-crystal films have been grown on (001) sapphire substrates with a view to optical applications. They have identical c axis direction but are rotated 30° (a axis) vis-à-vis the substrate. The film "fits" on the closed-packed oxygen array of the substrate, resulting in low interfacial strains and stresses. High-level, vertical strain (and thus stress) limits film growth to shallow steps that vanish in the fully developed, β''-Al$_2$O$_3$ layers. β''-Al$_2$O$_3$ forms at the expense of the sapphire substrate via chemical reaction with alkali vapors generated from a vapor source of Na- and Li-β''-Al$_2$O$_3$ powders. The equilibrium alkali partial pressure values are maintained in an appropriate range for Li-stabilized Na-β''-Al$_2$O$_3$. One-square-inch, ≤ 40 μm single-crystal films have been prepared, and the growth rate determined dominated by a diffusion process.

The vapor–substrate film growth process was extended to prepare K-β''-Al$_2$O$_3$ single-crystal films and a β''-Al$_2$O$_3$ coating on monocrystalline α-Al$_2$O$_3$ platelets. The latter was veri-

Handbook of Thin Film Materials, edited by H.S. Nalwa
Volume 1: Deposition and Processing of Thin Films

ISBN 0-12-512909-2/$35.00

fied to constitute a mechanically weak interface that promotes platelet/crack deflection in Al_2O_3/ZrO_2 structural composites.

Na-β''-Al_2O_3 single-crystal films were ion-exchanged to the Li, K, Ag, and Ca isomorphs. Their optical refractivity was calculated via measured refractive indices and lattice parameters, and the results were employed to evaluate oxide ionicity and ion polarizability in the isomorphs.

Cu(I)-doped Na-, K-, Ba-, and Ag/Na-β''-Al_2O_3 films were prepared with the use of ion exchange, and emission spectra were measured under ultraviolet excitation. Emission band shifts and contributions are explained by the variation in polarizability of the latter. Among the host ions, large polarizability favors blue emission.

Luminescence patterning was achieved in β''-Al_2O_3 single-crystal films by charging in an aqueous electrochemical cell with a designed electrode configuration. Na- and Ag-rich β''-Al_2O_3 single-crystal film domains were formed that luminesce green and purple, respectively.

(The Appendix includes compilations of the β''-Al_2O_3 isomorphs prepared by ion exchange, their optical refractivity, and luminescent data.)

β''-Alumina is a superionic solid electrolyte with interesting optical properties [1, 2]. Its applications to optical devices have been demonstrated. The β''-Al_2O_3 crystal lattice accommodates exchanged or doped optically active ions at different and high concentration levels. Luminescence of transition-metal and rare-earth ions has been verified in the β''- and β-Al_2O_3 hosts over a wide range of emission wavelengths (Table XIX, Appendix). Among these luminescent β''- and β-aluminas, the emission of Cu^+-activated crystals [3–5] is sensitive to the environmental field of coexisting, conducting ions. This suggested tunable phosphors and lasers. Eu^{2+}-polyaluminate, with a β-Al_2O_3-type structure, is an attractive blue-emitting phosphor used in tricolor fluorescent lamps to attain high efficiency and good color rendition [6–9]. The laser function of Nd^{3+} was examined in β''-Al_2O_3 [10] and demonstrated large oscillating strength and a long fluorescence lifetime at high doping levels. This laser operates in both pulse and continuous wave (cw) modes. Degenerate four-wave mixing and optical memory [11] functions were also demonstrated in β''-Al_2O_3 crystals doped with Nd^{3+} and Cu^+ ions, respectively. It was determined that photoaggregation of Cu^+ dimers associated with ion mobility is the basis for the optical write/read characteristics of the doped material.

$\beta''(\beta)$-Alumina compounds have a heterogeneous structure composed of alternatively stacked Al-O blocks (spinel blocks) and conduction planes [12–15]. The former consists of a close-packed oxygen array with Al^{3+} located in a fraction of the octahedral and tetrahedral interstices. These close-packed Al-O blocks are interconnected via conduction planes of low atomic packing density. The cations and oxygen ions are weakly bonded; consequently, the cations exhibit fast diffusivity and are exchangeable with other cations in concentration gradients or electrical fields [16]. This behavior makes β''-Al_2O_3 properties adjustable via ion replacement in both the spinel-block and the conduction-plane units. In addition, the ion mobility in this material suggests application to field-adjusted optical devices.

Large β''-alumina crystals for optical planar devices are unavailable. This is due to the fact that Na-β''-Al_2O_3 crystal used as a precursor is grown by the flux evaporation method. The growth size and optical homogeneity are limited by microstresses due to structural asymmetry and the weak bonding within the conduction planes. To improve optical performance and increase the active area, a procedure for growing β''-Al_2O_3 single-crystal films was developed by Nicholson and co-workers [17, 18]. Large-area (up to 1 square inch), $\leq$40 μm Na-β''-Al_2O_3 single-crystal films have been grown on sapphire substrates. The films were successfully ion-exchanged and doped to luminescent isomorphs.

The present paper reviews the work on β''-Al_2O_3 single-crystal films and their preparation, characterization, luminescence, and mechanical composite studies. The first section concerns film growth technology. Film preparation involves the generation of alkali vapors and their chemical reactions with sapphire substrates. Vaporization and reaction thermodynamics are discussed. Na-β''-Al_2O_3 forms at the expense of the sapphire substrate via chemical reaction with alkali rather than vapor–vapor deposition. The latter has a minor influence on film growth at high temperatures. The section following is devoted to film characterization, including X-ray crystallography, microscopy, and growth-rate measurement. The conduction planes of the Na-β''-Al_2O_3 single-crystal films were determined parallel to the film surface, leading to a low interfacial misfit with the substrate. The mechanism of lattice transformation from sapphire to β''-Al_2O_3 is interpreted via restacking of the oxygen layers, redistribution of the aluminum ions, and insertion of the Na-O component. Dimension changes associated with the latter induce high vertical stresses that limit crystal growth to layer-by-layer, shallow steps. Film thickness is dominated by a diffusion process. The measured high activation energy for diffusion suggests diffusion up the c axis. Sections 3 and 4 cover extension of the Na-β''-Al_2O_3 film growth procedure to the synthesis of a single-crystal coating of β-Al_2O_3 on α-Al_2O_3 microplatelets for increased toughness of mechanical composites and to the synthesis of optical K-β''-Al_2O_3 single-crystal films. Ion exchange and isomorph preparation are dealt with in Section 5. Unit-cell constants and refractive indices of the film isomorphs are given, and optical refractivity is evaluated. The latter was employed to derive the polarizability of the component oxides and ions. An aqueous ion exchange approach is explored for the Na-Ag ion exchanges for purposes of ion pattering. The last section considers Cu^+ luminescence in the β''-Al_2O_3 single-crystal films hosts. Variations in luminescence versus codoping and conducting ions are explained by variation of ion polarizability. Luminescence patterning is achieved via ion patterning with the use of electrochemical cell in which the electrodes are designed to control ion activities in the electrolyte. Gu^+/Na and Cu^+/Ag domains are built up in a Na$^+$-β''-Al_2O_3 film and emit green and purple luminescence, respectively.

2. REVIEW OF THE LITERATURE ON LARGE-AREA THIN FILMS OF β''/β-Al$_2$O$_3$

Early attempts to fabricate β-Al$_2$O$_3$ films or coating materials for electrolyte application failed because of the evaporation of sodium [19]. It was found, however, that an α-alumina film was transformed to β-alumina when it was heated over Na-β-alumina powder.

Amorphous Na$_2$O·xAl$_2$O$_3$ ($1.5 \leq x \leq 20$) thin film on polyimide was prepared by Nobugai and Kanamaru by sputtering from sodium aluminate/alumina targets [20]. Postannealing amorphous material with $4.4 < x < 20$, at 700°C and 1050°, crystallized a metastable phase (called the λ or m phase), with a mullite-derived structure. The latter converted to Na-β-Al$_2$O$_3$ at temperatures above 1050°C. The authors found that the sodium content of the sputtered films increased with increasing Na$_2$O content in the target and increasing target temperature. Preferred (001)-textured coating parallel to the growing surface was obtained in films with thicknesses of ≤ 5 μm. The films produced had an ionic conductivity comparable to that of Na-β-alumina crystal.

Saalfeld [21] reported epitaxial overgrowths of β-Al$_2$O$_3$ on corundum and Al$_2$O$_3$-rich spinel (MgO·3.5Al$_2$O$_3$) substrates when these were heated over a cryolite/Al$_2$O$_3$ eutectic melt. He showed the overgrowth relations $(001)_{\beta\text{-alumina}} \| (001)_{\text{corundum}}$ or $(111)_{\text{spinel}}$, indicating that growth was taking place on the oxygen close-packing planes. Recent phase-equilibrium calculations by Guo et al. [22] revealed β''- and β-Al$_2$O$_3$ deposits in the fluoride system in the presence of H$_2$O. The fluorides hydrolyze to oxides and deposition arises therefrom.

The crystallography of α-Al$_2$O$_3$ formed on a β-Al$_2$O$_3$ substrate was explored by Udagawa et al. [23] and Segnit [24]. On thermal decomposition of Na-β-Al$_2$O$_3$, Udagawa et al. identified the crystallographic relationships of the oriented products, i.e., $[001]_{\beta\text{-Al}_2\text{O}_3} \| [001]_{\alpha\text{-Al}_2\text{O}_3}$, and $[100]_{\beta''\text{-Al}_2\text{O}_3} \| [110]_{\alpha''\text{-Al}_2\text{O}_3}$, with no intermediate phase formed, or $[001]_{\beta''\text{-Al}_2\text{O}_3} \| [001]_{\text{intermediate phase}} \| [001]_{\alpha\text{-Al}_2\text{O}_3}$ and $[100]_{\beta''\text{-Al}_2\text{O}_3} \| [100]_{\text{intermediate phase}} \| [110]_{\beta''\text{-Al}_2\text{O}_3}$, with the intermediate phase. Segnit conducted a microscope study on a slagged refractory lining from a kiln. He showed that when the overgrown hematite 001 structure (an α-Al$_2$O$_3$-type lattice) was rotated 30° and superimposed on the 001 close packing of a β-Al$_2$O$_3$, the crystals had identical [001] and coincident [100] and [110] directions.

Metal-organic chemical vapor deposition (MOCVD) β-Al$_2$O$_3$ coatings on silicon ribbon and sapphire substrates for composite applications were reported by Brown [25, 26]. The vapor was supplied by Al-diketonate derivatives and Na- and Rb-nitrate precursors at 900°C for 60 min. Coatings were a few to tens of micrometers in thickness. A mullite-alumina (m phase) of composition Na$_2$O·xAl$_2$O$_3$ ($x = 3$ to 12) and an amorphous or nanocrystalline phase were obtained in the Na and Rb systems, respectively. The intermediate phase converted to a β-Al$_2$O$_3$-type phase after heat treatment. LaAl$_{11}$O$_{18}$ coatings formed on c axis sapphire and YAG fibers, with the use

of Al- and La-diketonate precursors. Postdeposition heat treatment increased texturing in the MOCVDed coatings. The basal plane of the coating crystals aligned with the substrate surface.

Rare-earth hexa-aluminate (Nd-, Gd-Al$_{11}$O$_{18}$) films were grown on sapphire substrates by Lange and co-workers [27] by spin-coating pyrolyzed nitrate precursors. Single-crystal, magnetoplumbite films ($\approx$80 nm) were formed by chemical reaction between a perovskite intermediate phase and the sapphire. The film growth was matched on the close-packed oxygen layers of the substrate. The Nd- and Gd-polyaluminate films were found to be thermally and chemically compatible with Al$_2$O$_3$ at high temperatures in an oxidizing atmosphere.

No work on β''-Al$_2$O$_3$ single-crystal films has been reported. The present work thereon has been declared as a patent.

3. Na-β''-Al$_2$O$_3$ SINGLE-CRYSTAL FILM GROWTH

In the early 1960s, Na-β''-Al$_2$O$_3$ was recognized as an independent, intermediate compound in the Na$_2$O-Al$_2$O$_3$ system. The chemical formula proposed was Na$_2$O·5Al$_2$O$_3$ [28, 29] or Na$_2$O·6Al$_2$O$_3$ [30]. Théry and Briançon [28, 29] synthesized this compound via a reaction between Al$_2$O$_3$ and Na$_2$O·Al$_2$O$_3$ at 1050° to 1200°C. They related its lattice parameters to Na-β-Al$_2$O$_3$ as $a_0(\beta''\text{-Al}_2\text{O}_3) \approx a_0(\beta\text{-Al}_2\text{O}_3)$ and $c_0(\beta''\text{-Al}_2\text{O}_3) \approx 1.5c_0(\beta\text{-Al}_2\text{O}_3)$. The crystal structure of Na-β''-Al$_2$O$_3$ was determined by Yamaguchi and Suzuki [31] and Bettman and Peters [15] with the use of powder and single-crystal X-ray specimens, respectively. An idealized composition of Na$_2$MgAl$_{10}$O$_{17}$ was chosen in the single-crystal investigation, based on the replacement of one Al^{3+} from NaAl$_{11}$O$_{17}$ (Na-β-Al$_2$O$_3$) by one Na$^+$ and one Mg^{2+}. The authors ascertained that the β''-Al$_2$O$_3$ structure consists of Al-O blocks and Na-O planes, however, but with a oxygen-layer sequences different from those of β-Al$_2$O$_3$. In β''-Al$_2$O$_3$, the positions of the upper Al-O blocks relative to the lower ones are governed by a threefold screw axis, instead of the mirror plane of Na-β-Al$_2$O$_3$ between the upper and lower blocks. As a result, β''-Al$_2$O$_3$ has rhombohedral symmetry, with the unit cell height increased by 1.5 compared with hexagonal β-Al$_2$O$_3$. Excess Na$^+$ are introduced onto the conduction plane, enhancing the ionic conductivity and ion diffusivity.

According to the Na$_2$O-Al$_2$O$_3$ phase diagram of Rolin and Thanh [30], β-Al$_2$O$_3$ and β''-Al$_2$O$_3$ phases melt congruently at 1580° and 2000°C, respectively. Weber and Yereno [32] redetermined this diagram and showed that Na-β-Al$_2$O$_3$ is the only stable phase between Na$_2$O·Al$_2$O$_3$ and Al$_2$O$_3$. Na-β''-Al$_2$O$_3$ is plotted on the binary diagram as a metastable phase between 1000° and 1500°C. The metastability of binary Na-β''-Al$_2$O$_3$ was confirmed by Hodge's kinetic study [33] of the β''-to-β transformation. He found that once β''-Al$_2$O$_3$ forms, its decomposition to stable β-Al$_2$O$_3$ and sodium aluminate is slow. This dilatoriness was assumed to be the reason for the contradictory results obtained [30, 32, 34]. Comparison of the free energies of β''-Al$_2$O$_3$ (Na$_2$O·5Al$_2$O$_3$, by Róg and Kozłowska-Róg [35])

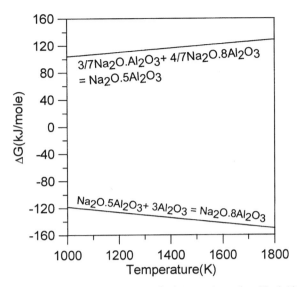

Fig. 1. Free energy increments of the reactions for Na-β-Al$_2$O$_3$ (Na$_2$O·8Al$_2$O$_3$) and Na-β''-Al$_2$O$_3$ (Na$_2$O·5Al$_2$O$_3$) formation.

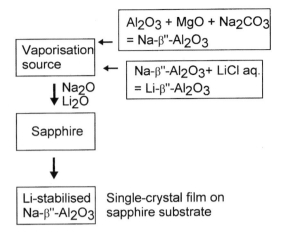

Fig. 2.   Flow chart for the growth of Na-β''-Al$_2$O$_3$ single-crystal films on sapphire substrates.

and β-Al$_2$O$_3$ (Na$_2$O·8Al$_2$O$_3$ and Na$_2$O·11Al$_2$O$_3$ by Kummer [36]), shows a positive free-energy result when β-Al$_2$O$_3$ and Na$_2$O·Al$_2$O$_3$ react to form Na-β''-Al$_2$O$_3$, whereas the reaction of Na-β''-Al$_2$O$_3$ and Al$_2$O$_3$ to form β-Al$_2$O$_3$ has a negative free energy. These calculation results are plotted in Figure 1, indicating that Na-β''-Al$_2$O$_3$ is thermodynamically unstable.

Na-β''-Al$_2$O$_3$ can be synthesized with aliovalent additives for structure stress stabilization. These result in conductivity enhancement. Boilot and Théry [37] concluded that cations of radius <0.97 are accommodated in the oxygen-polyhedra interstices of the spinel block. Transition ions, however, were excluded from the additives in most electrolyte studies because their variable oxidation states introduce electronic conductivity. Thus, the common stabilizers used are Li$_2$O and MgO. Ions of lower valence and higher polarizability than Al^{3+} introduce extra Na$^+$ to compensate for charge and release the stresses associated with structure asymmetry. Investigations of the ternary systems Li$_2$O- and MgO-Na$_2$O-Al$_2$O$_3$ [38–40] suggest that stabilized β''-Al$_2$O$_3$ coexists with the β-phase. Single-phase Na-β''-Al$_2$O$_3$ appears in a narrow region extending into the ternary compositions.

β''- and β-Al$_2$O$_3$ are nonstoichiometric phases. In the phase diagram compiled by De Vries and Roth [41], their compositions are located, respectively, in the ranges of 1 : 9–1 : 11 and 1 : 5–1 : 7 ratios of Na$_2$O : Al$_2$O$_3$. The deviation of Na content from the idealized $\beta''(\beta)$-Al$_2$O$_3$ compounds can be compensated for by Al and O point defects and, in ternary or multicomponent systems, by aliovalent ion substitution (Li$^+$ or Mg^{2+}). In the latter case, the ternary formula for Na-β''-Al$_2$O$_3$ is always written as Na$_{1+x}$Mg$_x$Al$_{10-x}$O$_{17}$ or Na$_{1+2x}$Li$_x$Al$_{10-x}$O$_{17}$ in which excess Na is balanced by replacing Al^{3+} with lower valence cations. The maximum Na content in β-Al$_2$O$_3$ estimated via the number of vacancies required to maintain electrostatic interaction is 1.57(Na$_2$O)·11Al$_2$O$_3$ and 1.67Na$_2$O· 1.33MgO·10.33Al$_2$O$_3$ for the binary and ternary β''-Al$_2$O$_3$

phases, respectively [42, 43]. Bettman and Peters [15] suggested the idealized formula Na$_2$O·MgO·5Al$_2$O$_3$ for Na-β''-Al$_2$O$_3$. Two sodium ions (on equivalent sites) reside on each conduction plane in the unit cell. This increases the sodium content in the O$_{17}$ formula to two. The nonstoichiometry of Na-β''-Al$_2$O$_3$ influences its ionic transport properties.

The block diagram in Figure 2 shows the synergy for the growth of an Na-β''-Al$_2$O$_3$ single-crystal film on a sapphire substrate in the Li ternary system. Na and Li vapors are derived from β''-Al$_2$O$_3$ powders, and these react with the sapphire substrate to form the Na-β''-Al$_2$O$_3$ film.

3.1. Sapphire Substrate

Substrate selection is critical for chemical epitaxy because it involves matching chemical proportionality and lattice geometry. Sapphire and Mg- and Li-Al spinels have close-packed oxygen lattices and contain the chemical elements required for β''-Al$_2$O$_3$ formation. Thus, they are candidates for single-crystal film growth. Low interfacial misfits on an a–b plane can be expected, as the close-packed oxygen array in the sapphire and spinel fit that in the spinel block of β''-Al$_2$O$_3$. The misfit on the a–b interface (estimated via lattice parameters) is as low as 2% for the Na-β''-Al$_2$O$_3$/sapphire or Mg-Al spinel, and 0.18% for Na-β''-Al$_2$O$_3$/Li$_2$O·5Al$_2$O$_3$ partners. The spinel-type substrate was excluded because it contains excess MgO or Li$_2$O, which result in secondary products (e.g., Mg-β''-Al$_2$O$_3$ or Li-β''-Al$_2$O$_3$).

Commercial 001-cut sapphire plates (0.5″ × 0.5″ and 1″ × 1″ × 0.5-mm-thick squares, obtained from ESPI, Oregon, USA, and Crystar, British Columbia, Canada) were used as substrates. The growing surface was optically polished for epitaxy purposes. The quality of polishing was found to be decisive in minimizing the film nucleation rate and guaranteeing uniform growth energy over the entire surface, thus inducing all nuclei to form in the same crystallographic direction and merge into a single film on contact. Misoriented, poorly polished, or ceramic substrates induced polycrystallization.

3.2. Vaporization Source

Alkali vapors were produced from mixed 90–95 mol% Na- and 5–10 mol% Li-MgO-stabilized β''-Al$_2$O$_3$ powders. The alkali vapor pressure derived therefrom was controlled as appropriate for Na-β''-Al$_2$O$_3$ formation but not sodium aluminate. At equilibrium, single-phase Na-β''-Al$_2$O$_3$ can be produced over a range of sodium pressures at a given temperature. This is due to the fact that Na-β''-Al$_2$O$_3$ is nonstoichiometric. Its vapor pressure varies with the solid–solution composition. If the sodium pressure is higher than the upper limit of the equilibrium pressure for the solid solutions, sodium aluminate results, giving a metastable phase assembly, α-Al$_2$O$_3$-Na-β''-Al$_2$O$_3$-sodium aluminate. When the sodium pressure is below the lower limit, α-Al$_2$O$_3$ forms via selective evaporation of sodium from the forming Na-β''-Al$_2$O$_3$. This suggests that vaporization is a desirable source of optimum surface area with minimum leakage.

Vaporization-source Na-β''-Al$_2$O$_3$ powders were synthesized from alumina, magnesia, and sodium carbonate. The weight ratios employed were 119.7 : 4.4 : 22.3. Powders were dried, weighed, and blended in a ball mill, then calcined at 1250°C for 5 h in a corundum crucible. The only weight loss was from carbonate decomposition. X-rays identified the principal phase as Na-β''-Al$_2$O$_3$ plus minor β-Al$_2$O$_3$. Crystallite dimensions were 0.06–0.1 μm.

Li-β''-Al$_2$O$_3$ was prepared by ion exchange of Na-β''-Al$_2$O$_3$-precursor powder at 90°C in a LiCl aqueous solution. Exchange was monitored via disappearance of the energy-dispersive X-ray analysis (EDX) peak for Na.

The Li concentration in the reaction atmosphere significantly influences the phase composition and growth rate of Na-β''-Al$_2$O$_3$. The optimum molecular ratio for Li$_2$O to Na$_2$O + Li$_2$O content was 0.05 in the β''-Al$_2$O$_3$ vaporization powders [17].

3.3. Alkali Vapor–Sapphire Substrate Reactions

The reaction chamber was either platinum or Na-β''-Al$_2$O$_3$-coated corundum crucibles with well-matched covers. The sapphire substrate was suspended over the Na- and Li-β''-Al$_2$O$_3$ vaporization bed. To suppress loss of volatiles at high temperatures, the assembly was placed in double crucibles, with the vaporization powder in the reaction crucible and between the internal and external crucibles. The whole assembly was placed in a furnace and heated to 1250–1450°C, at heating and cooling rates of 5–10°C/min, for several to tens of hours.

Calculation of the evaporation rate from a free surface with the Langmuir equation indicated that the sodium was consumed rapidly at film growth temperatures, whereas the evaporation rates of Al and Li were $\approx 10^{-8}$ to 10^{-6} and 10^{-2} to 10^{-1} g/h, respectively [44]. Weight loss during 72–96-h film growth experiments was <0.5% (0.05 g/$\approx$10-g of source powder). X-ray analysis of the powder detected a slight increase in the β-Al$_2$O$_3$ level but no α-Al$_2$O$_3$. Thus, the films grow under equilibrium or near-equilibrium conditions.

Figure 3 shows plots of partial pressure for MgO-stabilized Li- and Na-β''-Al$_2$O$_3$, calculated by the multicomponent equi-

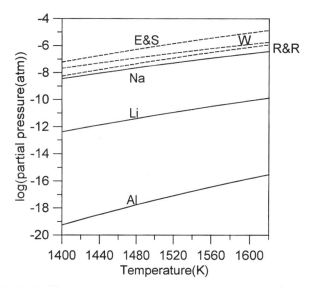

Fig. 3. Equilibrium vapor pressures on Mg-stabilized Na- and Li-β''-Al$_2$O$_3$, in the systems Na$_2$O- and Li$_2$O-0.64MgO-9.55Al$_2$O$_3$ at $p_{O_2} = 0.21$ atm, and comparisons with those for binary Na-β- and β''-Al$_2$O$_3$ calculated from the measurements of Weber [36] (W), Elfrie and Smeltzer [65] (E&S), and Rog and Rog [35] (R&R).

librium method [44, 45]. Binary Na-β- and β''-Al$_2$O$_3$ are also plotted for comparison. The high vapor pressure of binary Na-β''-Al$_2$O$_3$ is due to its metastability. The major gaseous species are Li, LiO, Na, NaO, and Al oxides, with distributions as in Figure 4 [46]. High ratios of Al and Li oxide species are due to the higher affinity of Al and Li than Na at high temperatures. MgO has a very low volatility and is neglected. Calculations reveal that oxygen suppresses the vaporization of sodium but enhances that of lithium and aluminum. Water vapor always promotes the decomposition of β''-aluminas and so increases the vapor pressure. Gaseous hydroxide molecules were the major species in H$_2$O-containing atmospheres. The extremely low partial pressure of Al species versus alkalies suggests that in the formation of β''-Al$_2$O$_3$, the vapor–vapor reaction is negligible.

Figure 5 shows simulations of the Na-β''-Al$_2$O$_3$ single-crystal film thickness that results from the different reaction models [47]. Mode 1 is a CVD process wherein Na-β''-Al$_2$O$_3$ is deposited on the sapphire substrate by chemical reaction between alkali and Al gaseous molecules. The weight gain is directly proportional to the film thickness. Modes 2 and 3 model the vapor–substrate reactions. The Na-O component was introduced from the vapor phase with or without Al-O loss, respectively. Both modes require rearrangement of the oxygen layer sequence and redistribution of the Al ions in the resulting newly arranged oxygen array. The second mode leads to a weight decrease, owing to loss of aluminum and oxygen. The third provides a slow weight increase because Na$_2$O must be introduced for β''-Al$_2$O$_3$ formation. A comparison of the measured weight changes for given film thicknesses indicates that mode 3 decisively contributes to the film growth process, i.e., the film grows by chemical reaction of gaseous Na species with the sapphire substrate. The positive deviation of the experi-

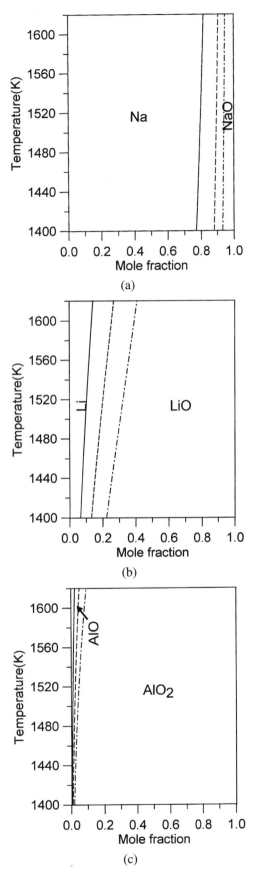

Fig. 4. Phase diagrams of the distribution of (a) Na, (b) Li, and (c) Al species in the system Li_2O-0.64MgO-9.55Al_2O_3 at $p_{O_2} = 1$ (—), 0.21 (– – –) and 0.06 atm (– · – · –).

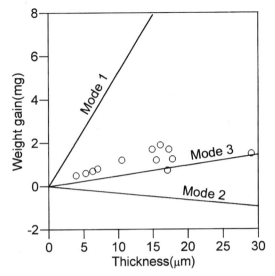

Fig. 5. Simulated and observed weight gain (o) versus thickness calculated via different reaction models.

mental data versus mode 3 is attributed to the chemical vapor reaction and deposition.

The following reactions are suggested, based on the thermodynamic analysis of (Na,Li)-MgO-stabilized β''-Al_2O_3-Al_2O_3-O_2.

Major reactions—contributions to film growth:

$$[Na(g), NaO(g)] + [LiO(g), Li(g)] + Al_2O_3(sapphire)$$
$$\rightarrow Na\text{-}(Li_2O\ stabilized)\text{-}\beta''\text{-}Al_2O_3$$

Minor reactions—contributions to film growth:

$$[Na(g), NaO(g)] + [LiO(g), Li(g)] + [AlO(g), AlO_2(g)]$$
$$\rightarrow Na\text{-}(Li_2O\ stabilized)\text{-}\beta''\text{-}Al_2O_3$$

((g) indicates gaseous species). The chemical formula of Na-β''-Al_2O_3 is 1.6Na_2O·0.3Li_2O·10.7Al_2O_3 [48].

The free energies for the Na_2O reaction, calculated by taking equilibrium partial pressures from the vaporization source, are given in Table I. The total free energy for the formation of Na-β''-Al_2O_3 from α-Al_2O_3 involves three subreactions: (1) free energy needed to produce Na-O planes by the alkali vapor reaction, (2) the free energy of the hexagonal-to-cubic oxygen lattice transition and the associated Al^{3+} redistribution, and (3) the free energy associated with the introduction of Li^+ ions onto the $Al(IV)^{3+}$ sites in the spinel blocks. The free energy terms related to the latter two processes were estimated from the polymorphic transition between α- and γ-Al_2O_3 and the formation of Li_2O·Al_2O_3 from α-Al_2O_3. Both γ-Al_2O_3 and the high-temperature modification of Li_2O·5Al_2O_3 have a disordered spinel structure wherein Al^{3+} and Li^+ ions are randomly distributed in a cubic oxygen array. The free energy increment for this disorder–order transition is calculated approximately from the mixing entropy, because negligible latent heat accompanies the ordering reaction.

The positive ΔG for the α- to γ-Al_2O_3 transformation is attributed to the metastability of the γ-Al_2O_3 modification

Table I. Free Energy Increments (kJ/mol) for the Sapphire (α-Al$_2$O$_3$) to Na-β''-Al$_2$O$_3$ Transformation (based on 1 mole of Al$_2$O$_3$)

Temperature (°C)	$\Delta G_{Na_2O \text{ reaction}}$	ΔG_{H-C}[a]	$\Delta G_{Li_2O \cdot 5Al_2O_3}$	$\Delta G_{disorder-order}$
1200	−51.66 to −51.81	9.50	−7.15	2.38
1300	−50.17 to −50.32	8.99	−7.18	2.55
1400	−48.69 to −48.83	8.48	−7.21	2.71

[a]Free energy for the hexagonal-cubic oxygen lattice transition estimated via α- to γ-Al$_2$O$_3$ polymorphic transformation.

Table II. Thicknesses of Na-β''-Al$_2$O$_3$ Single-Crystal Films and Calculated Diffusion Coefficients

Temperature (°C)	Time (h)	Thickness (μm)	Diffusivity (cm^3/s)
1235	48	3.9	1.4×10^{-12}
1250	48	6.2	3.8×10^{-12}
	96	10.5	4.2×10^{-12}
1270	24	6.3	7.2×10^{-12}
1280	48	16.1	2.0×10^{-11}
1300	48	18.5	3.0×10^{-11}
	96	28.0	3.2×10^{-11}
1350	48	24.9	8.0×10^{-11}

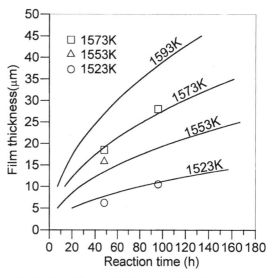

Fig. 6. Simulated and observed thicknesses of Na-β''-Al$_2$O$_3$ single-crystal films grown on sapphire substrates.

(Table I). The spinel lattice, on the other hand, is stabilized by Li$^+$ ions. Negative free energy increments were obtained for the chemical reaction of α-Al$_2$O$_3$ to L$_2$O·5Al$_2$O$_3$. The Na-O term contributes to the major part of the free energy increment of Na-β''-Al$_2$O$_3$ formation.

3.4. β''-Al$_2$O$_3$ Film Growth Kinetics

The velocity of single-crystal film growth was determined to be 2–3 times lower for polycrystalline films and is diffusion-dominated. Film thickness and diffusion coefficients versus temperature are listed in Table II. The activation energy determined therefrom is 718 kJ/mol.

Previous investigators reported activation energies of diffusion-controlled growth of 189 and 470 kJ/mol for Na-β- and β''-Al$_2$O$_3$ on α-Al$_2$O$_3$ ceramic and crystal in contact with sodium aluminate. The reaction mechanism was explained via the oxygen [49] or, alternatively, the Na-Al counter-diffusion [50]. The low activation energy for the ceramic system is believed to be due to the presence of grain boundaries therein with high oxygen diffusivity [51, 52]. The counter-diffusion mechanism assumes that the reaction at the Al$_2$O$_3$ takes place by the removal of Al and introduction of Na. This would lead to two reaction zones forming β- and β''-Al$_2$O$_3$ surface products. If Al diffusion is upward, in addition to chemical vapor deposition, the surface morphology can be modified. There are few cases showing such surface overgrowths, and these were explained via vapor–vapor reaction and deposition in Section 2.2.

The high activation energy determined for single-crystal film growth is similar to that for oxygen and aluminum diffusion in α-Al$_2$O$_3$ crystals [53–55]. This observation seems to suggest that the film growth process is controlled by diffusion across the close-packed oxygen spinel blocks of β''-Al$_2$O$_3$. The misfit calculations showed that creation of new Na-O conduction planes in sapphire will induce lattice stresses. The strain energies that developed because of c axis misfit could contribute to the slow diffusion rate and thus limit growth to small steps.

Figure 6 shows simulations of film thickness using the known diffusion coefficients. Increased temperature is required to prepare thick films and maintain a reasonable growth rate. High temperature also promotes material vaporization and disturbs crystal growing conditions. Reduction of leakage losses is critical to high-temperature processing.

4. SINGLE-CRYSTAL FILM CHARACTERIZATION

Na-β''-Al$_2$O$_3$ single-crystal films are transparent. A single-crystal film grown on a $1'' \times 1''$ substrate is shown in Figure 7. The crystallinity and thickness of the films were determined by scanning electron microscopy and quantitative X-ray diffraction. EDX was carried out on selected specimens.

Fig. 7. A one-inch-square Na-β''-Al$_2$O$_3$ single-crystal film grown on a sapphire substrate.

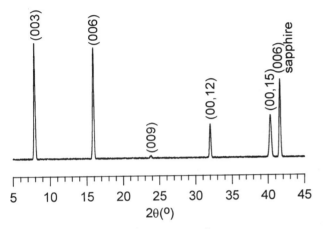

Fig. 8. An X-ray diffraction pattern of a Na-β''-Al$_2$O$_3$ single-crystal film, grown on a (001) sapphire substrate, taken on the film surface, showing (001) reflections.

4.1. X-Ray Diffraction

An X-ray diffraction pattern of the film, taken on the surface, consisted exclusively of (001) reflections of Na-β''-Al$_2$O$_3$ and the sapphire substrate (Fig. 8), indicating that the c axis of β''-Al$_2$O$_3$ is parallel to the sapphire and perpendicular to the film surface. The strong (003) and (006) peaks of β''-Al$_2$O$_3$ are sensitive indicators of film thickness. The diffraction background is low and flat, indicating high crystallinity and low stress levels.

The intensity of diffraction from β''-Al$_2$O$_3$ increases steeply with increasing film thickness in the range 10–25 μm, whereas that for the sapphire decreases and vanishes at film thicknesses of >30 μm. The evolution of intensity ratios for reflections (003), (00,12), and (00,15) of β''-Al$_2$O$_3$ to the (006) of α-Al$_2$O$_3$ were used for quantitative analysis of the film thickness. These ratios are plotted in Figure 9; they satisfy an exponential equation attributable to X-ray absorption in the β''-Al$_2$O$_3$ layer.

The rhombohedral symmetry of β''-Al$_2$O$_3$ was exhibited by X-ray precession and electron diffraction photographs. Figure 10 is an X-ray precession photograph for a 15-μm Na-β''-Al$_2$O$_3$ single-crystal film on a 0.5-mm sapphire substrate. Diffraction was collected from the surface in the vertical direction. There are three sets of diffraction patterns from the [001]$_{\text{Na-}\beta''\text{-Al}_2\text{O}_3}$, [001]$_{\alpha\text{-Al}_2\text{O}_3}$, and [110]$_{\alpha\text{-Al}_2\text{O}_3}$ directions. In-

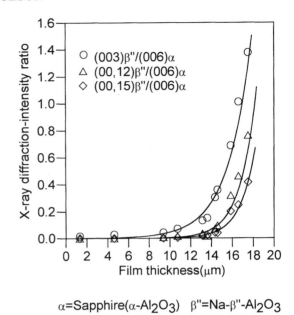

α=Sapphire(α-Al$_2$O$_3$) β''=Na-β''-Al$_2$O$_3$

Fig. 9. Variation of the intensity ratio of given diffraction pairs from a grown Na-β''-Al$_2$O$_3$ single-crystal film and sapphire substrate with film thickness.

dices were derived from the lattice parameters of α-Al$_2$O$_3$ ($a = 4.758$ Å and $c = 12.991$ Å) and Na-β''-Al$_2$O$_3$ ($a = 5.61$ Å and $c = 33.565$ Å). Analysis revealed that the angle between the [001] patterns of α- and β''-Al$_2$O$_3$ is $\pi/6$, whereas that between the [001] of β''-Al$_2$O$_3$ and the [110] of α-Al$_2$O$_3$ ($\pi/2$) indicates the epitaxial relationships, i.e.,

$$[001]_{\text{Na-}\beta''\text{-Al}_2\text{O}_3} \parallel [001]_{\alpha\text{-Al}_2\text{O}_3}$$
$$[100]_{\text{Na-}\beta''\text{-Al}_2\text{O}_3} \wedge [100]_{\alpha\text{-Al}_2\text{O}_3} = \pi/6$$

4.2. Microscopy

Scanning electron micrographs of as-grown single-crystal films are shown in Figure 11a and b. The growth steps show 120° with triangular intersections, characteristic of rhombohedral symmetry. The heights of the growth steps measured with an atomic force microscope were 3–4 nm, approximately equal to the c axis length of the Na-β''-Al$_2$O$_3$ unit cell. Low stripes were 0.2–0.4 nm high, i.e., one or two monolayers of the close-packed oxygen arrays. Figure 11c shows a multilayer growth feature that was tracked by etching in fused borax. These results indicate a layer-by-layer growth mechanism. The low step heights are explained by lattice expansion along the c axis during the transformation of sapphire to β''-Al$_2$O$_3$. The latter induces high vertical stresses.

A few triangular islands on the surface of the film grown at high temperatures are attributed to deposits from vapor–vapor reactions. Chemical vapor deposition will become appreciable versus the vapor reaction, as the concentration of gaseous Al species increases. Chemical vapor deposition contributed <5% of the total film thickness (calculated from Fig. 3) because of limited gaseous AlO$_x$ species from the alkali-β''-Al$_2$O$_3$ vaporization source.

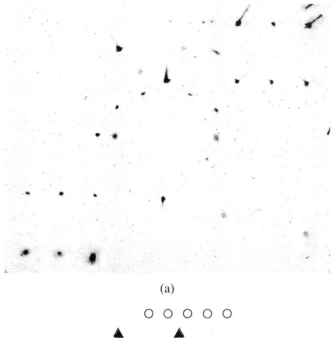

(a)

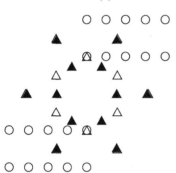

○ α-Al$_2$O$_3$ [110] diffraction pattern
△ α-Al$_2$O$_3$ [001] diffraction pattern
▲ Na-β"-Al$_2$O$_3$ [001] diffraction pattern

(b)

Fig. 10. (a) An X-ray precession photograph and (b) its index map, interpreted by Na-β''-Al$_2$O$_3$ [001] and α-Al$_2$O$_3$ [001] and [110] diffraction patterns.

Improper film growth conditions led to oriented polycrystalline deposits (Fig. 12). In contrast to the single-crystal deposit, the oriented polycrystals exposed the edge of the conduction planes to the film surface, with the following crystallographic relations:

$$[001]_{\beta''\text{-alumina(overgrowth)}} \parallel [100]_{\beta''\text{-alumina(film)}} \quad \text{and}$$
$$[100]_{\beta''\text{-alumina(overgrowth)}} \parallel [001]_{\beta''\text{-alumina(film)}}$$

or

$$[001]_{\beta''\text{-alumina(overgrowth)}} \parallel [100]_{\beta''\text{-alumina(film)}} \quad \text{and}$$
$$[110]_{\beta''\text{-alumina(overgrowth)}} \parallel [001]_{\beta''\text{-alumina(film)}}$$

X-ray diffraction patterns taken from polycrystalline films exhibited abnormally high intensities of (h00) reflections versus (hk0), indicating the higher probability of the former alignment.

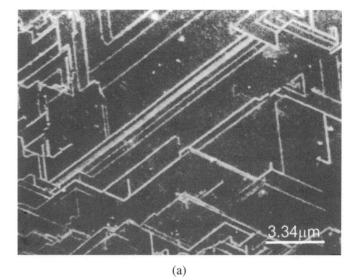

(a)

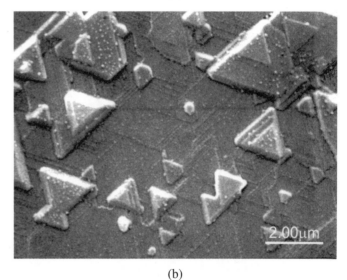

(b)

(c)

Fig. 11. SEM of Na-β''-Al$_2$O$_3$ single-crystal films showing (a) the white growth lines intersecting at 60° and 120° angles; (b) the triangle-shaped overgrowths on the β''-Al$_2$O$_3$ single-crystal background; and (c) the multiple-layered feature after etching with fused borax.

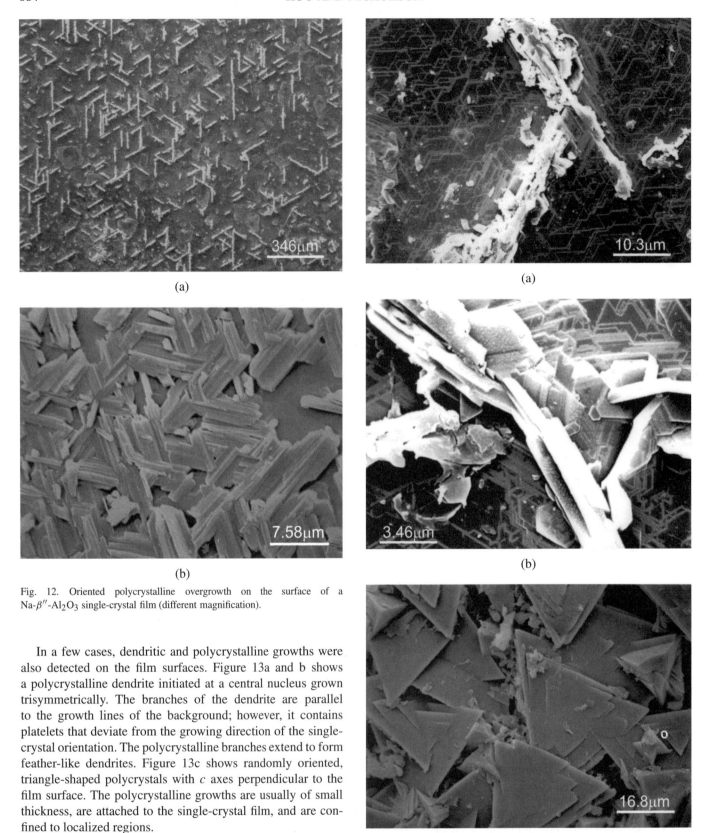

Fig. 12. Oriented polycrystalline overgrowth on the surface of a Na-β''-Al$_2$O$_3$ single-crystal film (different magnification).

In a few cases, dendritic and polycrystalline growths were also detected on the film surfaces. Figure 13a and b shows a polycrystalline dendrite initiated at a central nucleus grown trisymmetrically. The branches of the dendrite are parallel to the growth lines of the background; however, it contains platelets that deviate from the growing direction of the single-crystal orientation. The polycrystalline branches extend to form feather-like dendrites. Figure 13c shows randomly oriented, triangle-shaped polycrystals with c axes perpendicular to the film surface. The polycrystalline growths are usually of small thickness, are attached to the single-crystal film, and are confined to localized regions.

4.3. Structural Transformation from α- to β''-Al$_2$O$_3$

As indicated, α-Al$_2$O$_3$ and the Al-O spinel block in the β''-Al$_2$O$_3$ crystal have similar close-packed oxygen layers with

Fig. 13. Growth defects showing (a) a dendrite initiated at a central nucleus and growing trisymmetrically; (b) the details of the central nucleus area, containing microplatelets; and (c) randomly oriented polycrystals with their c axis perpendicular to the film surface.

Table III.   The Distribution of Al^{3+} in the Interstices between Oxygen Layers in Sapphire and Na-β''-Al$_2$O$_3$

Oxygen layers		Sapphire	Na-β''-Al$_2$O$_3$
Sapphire	Na-β''-Al$_2$O$_3$		
B, A	B, C	C2, C4, C3	(A1), (A1)
A, B	C, B	C4, C1	A4, A3, A2
B, A	B, A	C2, C3, C1	C2, (A1), (B3)
A, B	A, C	C2, C4, C3	B1, B4, B2
B, A	C, A	C4, C1	(B3), (B3)
A, B	A, C	C2, C3, C1	B1, B4, B2
B, A	C, B	C2, C4, C3	(C2), (B3), A1
A, B	B, A	C4, C1	C4, C1, C3
B, A	A, B	C2, C3, C1	(C2), (C2)
A, B	B, A	C2, C4, C3	C1, C4, C3
B, A	A, C	C4, C1	B3, (A1), (C2)
A, B	C, B	C2, C3, C1	A4, A2, A3
B, A	B, C	C2, C4, C3	(A1), (A1)

The site positions of A^{3+} refer to Figure 15, and those in parentheses are IV-coordinated.

different stacking sequences. β''-Al$_2$O$_3$ contains extra "loose-packed" Na-O planes alternately placed between Al-O blocks, so it has a less dense structure than α-Al$_2$O$_3$. The oxygen occupation volume ratio of these two crystals is 0.79 : 1. Based on these structural features, the transformation of sapphire to Na-β''-Al$_2$O$_3$ crystal can be imagined as follows:

1. Na-O planes are introduced. Sapphire has a homogeneous close-packed oxygen lattice. The introduction of loosely packed Na-O planes lowers the lattice density and makes the structure heterogeneous. The dimensional expansion is large along the c direction in comparison with changes in the a–b plane.

2. The restacking of the close-packed oxygen layers. The oxygen layers in α-Al$_2$O$_3$ are stacked ABAB..., in the hexagonal manner. Al^{3+} is in two-thirds of the octahedral interstices. In contrast, the oxygen of the spinel block of β''-Al$_2$O$_3$ is packed ABCABC..., in the cubic manner. Because of the Na-O conduction plane inserted between two spinel blocks, the oxygen layers separate into four-layer blocks as BCAB, ABCA, and CABC in sequence or, to account for the loose-packed Na-O layers, A$'$BCABC$'$ABCAB$'$-CABCA$'$..., where the prime denotes the loosely packed Na-O planes.

3. The Al^{3+} ions are redistributed in the Al-O sublattice of the β''-Al$_2$O$_3$ structure. Table III lists the difference in Al^{3+} distribution between the sapphire and β''-Al$_2$O$_3$ lattices.

The α-to-β''-Al$_2$O$_3$ lattice transformation is schematically illustrated in Figure 14, with an inset showing the transition of

Table IV.   Lattice Misfits in Sapphire and in Na-β''-Al$_2$O$_3$ Based on the Unit-Cell Constants

Temperature (°C)	Sapphire		Na-β''-Al$_2$O$_3$	
	a axis	c axis	a axis	c axis
25	0.0211	0.292	−0.0207	−0.226
1200	0.0221	0.290	−0.0216	−0.225
1300	0.0221	0.290	−0.0217	−0.225
1400	0.0222	0.290	−0.0217	−0.225

Al^{3+} distribution. Table IV lists the lattice misfit calculated via Figure 14 and the unit-cell constants. Dimension changes in the a–b plane are

$$a_{0,\text{sapphire}} - a_{0,\text{Na-Al}_2\text{O}_3} \times \sin(\pi/3)$$

The thermal mismatch in the single-crystal film–substrate assembly is low because the differences in thermal expansion coefficients for α-Al$_2$O$_3$ and Na-β''-Al$_2$O$_3$ are 0.8×10^{-6}/K and 1.1×10^{-6}/K for the a and c axes, respectively.

Stresses generated during the growth of Na-β''-Al$_2$O$_3$ from α-Al$_2$O$_3$ can be estimated from lattice misfit via the elastic stress–strain relations,

$$T_j = \sum C_{ji} S_i \qquad j, i = 1, 2, 3, 4, 5, \text{ or } 6$$

where T_j are the six components of the stress tensor, S_i are the six components of the strain matrix, and C_{ji} are the 6×6 arrays of the elastic stiffness constant. The subscripts j and i for the stress and strain orientation are $i = 1, 2, 3$ for the longitudinal stress and strain along the x, y, and z axes and j, $i = 4, 5, 6$ for the shear about each axis. Substituting the elastic constants for sapphire and Na-β''-Al$_2$O$_3$ into the strain–stress equation, the maximum stresses in the x, y, and z directions in the transforming and transformed layers are summarized in Table V. As expected, the film–substrate interface temporarily suffers a large vertical stress due to elongation of the c axis during the chemical phase transformation. These stresses will be completely released for a fully transformed layer because no lattice mismatch results.

From a thermodynamic point of view, when sapphire is exposed to alkali vapor, it reacts to form sodium aluminate, β''- or β-Al$_2$O$_3$, depending on the concentration of alkali and stabilizer. The presence of β''-Al$_2$O$_3$ was ensured by reduction of the alkali concentration and introduction of the Li stabilizer. The Na vapor pressure derived from the mixed β''-Al$_2$O$_3$ vaporization source equilibrates with the β''-Al$_2$O$_3$ product. As a consequence, sodium aluminate can coexist with α-Al$_2$O$_3$ only under meta-equilibrium conditions, or it forms on the surface of Na-β''-Al$_2$O$_3$ when the sodium dissipation rate is too low to transfer the extra sodium in contact with the sapphire substrate. The growth of single-crystal films from the substrate is envisaged as involving the following transition steps (rather than

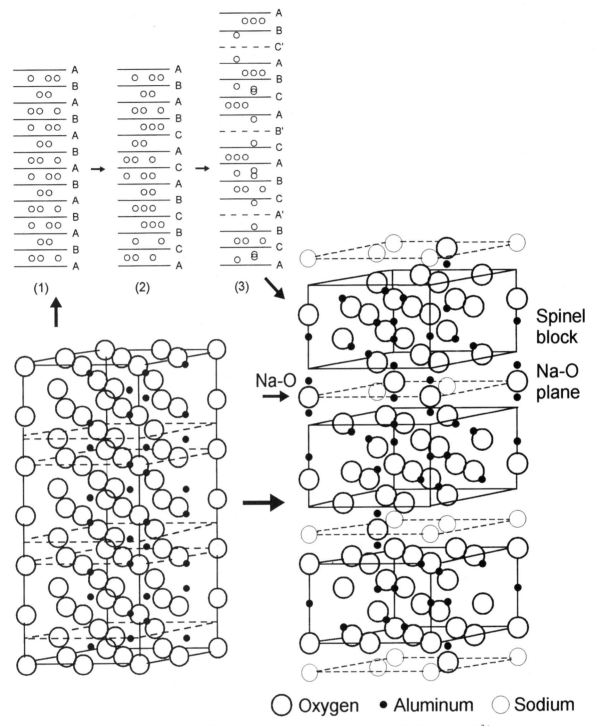

Fig. 14. A schematic drawing of the α-to-β''-Al$_2$O$_3$ transformation on Na-O insertion. The inset shows Al^{3+} rearrangement in the close-packed oxygen array. A, B, and C indicate the close-packed oxygen layers, and the prime indicates the loose-packed Na-O plane.

chemical disintegration and reconstruction, which would lead to polycrystallization):

1. The chemical reaction between sapphire and sodium vapor is initiated by the adsorption of Na and NaO species onto the sapphire. Absorbed oxygen is most likely located in the central interstices of the oxygen triangles at the surface, resulting in maximum contact with the oxygen network. The adsorbed Na$^+$ would occupy sites equivalent to "BR," "aBR," and "MO" sites, the first two being equivalent to those it occupies in the β''-Al$_2$O$_3$ structure.

2. Thus adsorption of Na-O induces β''-Al$_2$O$_3$ formation, followed by a change from a "hexagonal" oxygen lattice to a cubic one. This stacking sequence change proceeds if the oxy-

Table V. Stress (GPa) Calculation in Sapphire and Na-β''-Al$_2$O$_3$ with and without the c Axis Misfit Considered

Temperature (°C)	c misfit[a]	Sapphire			Na-β''-Al$_2$O$_3$		
		x	y	z	x	y	z
25	(1)	47.2	46.2	150.6	−24.3	−23.7	−64.4
	(2)	13.5	12.5	4.5	−9.6	−9.0	−2.5
1200	(1)	43.6	42.8	135.6	−21.8	−21.2	−56.7
	(2)	13.2	12.4	4.3	−8.9	−8.3	−2.3
1300	(1)	43.3	42.5	133.7	−21.6	−21.0	−56.0
	(2)	13.2	12.4	4.3	−8.8	−8.2	−2.3
1400	(1)	43.0	42.2	132.4	−21.4	−20.8	−55.4
	(2)	13.2	12.4	4.3	−8.8	−8.2	−2.3

[a] (1) With the c axis misfit considered. (2) Without the c axis misfit considered.

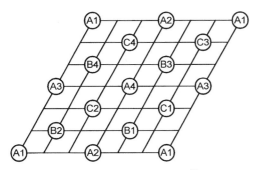

Fig. 15. A key pattern for Table III explaining the Al^{3+} rearrangement for the α-to-β''-Al$_2$O$_3$ structural transformation.

gen layer glides per the lower, energy construction process in an oxide; i.e., structural reconstruction is avoided.

3. The Al^{3+} ions rearrange to match the distribution pattern of the oxygen ions in the spinel as required for the β''-Al$_2$O$_3$ lattice. Four of the 11 aluminums decrease their coordination number from VI to IV. Such a coordination change is well known for aluminates, because the ionic radius of Al^{3+} is between the cation-to-anion ratio values appropriate for Al-O octahedral and tetrahedral coordination. This rearrangement of Al^{3+} is shown in Table III and Figures 14 and 15 and was discussed in detail in [18].

4. Further adsorption of Na-O species induces new Na-deficient, loose-packed oxygen planes built up, either by movement of the adsorbed O^{2-} downward through the previously reacted region or via oxygen exchange with the close-packed layers. The latter process will be accompanied by an Al^{3+} displacement upward.

5. The newly formed, open oxygen slabs inside the substrate fill with Na$^+$ ions, either from the vapor phase or from the surface, via lateral diffusion. c axis diffusion of Na$^+$ in Na-β''-Al$_2$O$_3$ is known to be much slower than that along the a–b plane. As a result, β''-Al$_2$O$_3$ growth occurs by downward movement of the open-structured oxygen layer into the sapphire. Excess Na adsorbed at the surface may disperse by surface diffusion or rebond with oxygen from the atmosphere.

6. Gaseous Li species enter the Al-O block to occupy Al(IV)$^{3+}$ sites. A few Li$^+$ ions will possibly reside on the "conduction" plane. The resultant mixed Na$^+$-Li$^+$ combination on the conduction plane is thought to be the source of reduction of the ionic conductivity of β''-Al$_2$O$_3$ single-crystal films [17].

5. NA-β''-Al$_2$O$_3$-COATED, α-Al$_2$O$_3$ SINGLE-CRYSTAL PLATELETS

β''-Al$_2$O$_3$-coated, monocrystalline α-Al$_2$O$_3$ microplatelets were introduced into alumina-reinforced, zirconia-based composites as a weak interface [56, 57]. The surfaces of the α-Al$_2$O$_3$ platelets are basal planes, thus providing a perfect substrate for the formation of epitaxial β''-Al$_2$O$_3$ films with c axis orientation identical to that of single-crystal films. The strong anisotropy of the elastic properties and fracture toughness of β''-Al$_2$O$_3$ provides a layered (intervening) phase at the α-Al$_2$O$_3$ platelet/zirconia–matrix interface, which promotes debonding, pull-out, and propagating-crack deflection.

α-Al$_2$O$_3$ platelets (Atochem, Paris la Defense cedex 42, France) were vapor-phase-reacted to give them β''-Al$_2$O$_3$ coatings. The monocrystalline α-Al$_2$O$_3$ platelets are hexagonal with equivalent diameter, 5–15 μm, and a thickness of 0.5–1 μm. They were subjected to Na/Li-β''-Al$_2$O$_3$ powder vaporization from 1200° to 1280°C. The β''-Al$_2$O$_3$ coatings were analyzed by X-ray diffraction and electron microscopy. They are single crystal. Figure 16 is a scanning electron micrograph of a partially transformed α-Al$_2$O$_3$ platelet. Electron diffraction shows that the β''-Al$_2$O$_3$ coating and the α-Al$_2$O$_3$ partner have identical epitaxial relationships, per the single-crystal films grown on sapphire (Section 2.2). The single-crystal coating exhibited trigonal symmetry and was thickened by progressive joining of growth steps. The α-Al$_2$O$_3$ platelets were consumed during excessive growth of β''-Al$_2$O$_3$.

The fractions of α-Al$_2$O$_3$ transformed to Na-β''-Al$_2$O$_3$ were estimated from the relative intensities of the X-ray diffraction

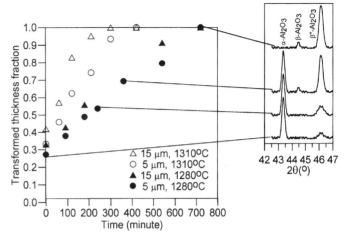

Fig. 18.   The transformation of α-Al$_2$O$_3$ platelets to Na-β''-Al$_2$O$_3$.

Fig. 16.   Na-β''-Al$_2$O$_3$ single-crystal coating on monocrystalline α-Al$_2$O$_3$ microplatelets.

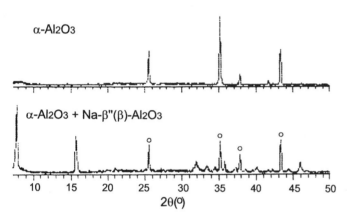

Fig. 17.   The X-ray pattern of starting and Na-β''-Al$_2$O$_3$-coated (1250°C, 4 h) α-Al$_2$O$_3$ platelets.

peaks, $(02,10)_{\text{Na-}\beta''\text{-Al}_2\text{O}_3}$ to $(113)_{\alpha\text{-Al}_2\text{O}_3}$ at $2\theta = 46.1°$ and $43.4°$ at CuK$_\alpha$ radiation, respectively, as follows:

$$\%\beta''\text{-Al}_2\text{O}_3 = \frac{I(02,10)_{\beta''\text{-Al}_2\text{O}_3}}{I(02,10)_{\beta''\text{-Al}_2\text{O}_3} + I(113)_{\alpha\text{-Al}_2\text{O}_3}}$$

The optimum coating was grown in 4 h at 1250°C; i.e., the α-Al$_2$O$_3$ platelets were uniformly covered with a thin layer of single-crystal Na-β''-Al$_2$O$_3$. Figure 17 is an X-ray pattern indicating a $\approx$0.27-μm film thickness.

The Na-β''-Al$_2$O$_3$ layer was unstable at zirconia composite processing temperatures (1550°C) and decomposed to α-Al$_2$O$_3$ because of the volatility of the sodium. Thus the Na$^+$ was ion-exchanged to Ca$^+$ (in a CaCl$_2$ melt) to improve the thermal stability [59]. The Ca-β''-Al$_2$O$_3$ surface films were unaffected on heating to 1550°C, and X-ray diffraction indicated a magnetoplumbite structure similar to that of Ca-β''-Al$_2$O$_3$.

Ca-β''-Al$_2$O$_3$-coated (25 vol%) platelets were incorporated into the zirconia composites. Chemical compatibility was verified by microscopic investigation and free-energy calculations. Experiments on α-Al$_2$O$_3$/zirconia composites with and without β''-Al$_2$O$_3$ surface modification revealed a higher degree of

debonding and pullout for the surface-modified platelets and increased fracture toughness (4.6 MPa$\sqrt{m}$ to 8.6 MPa$\sqrt{m}$).

The rate of conversion of the α-Al$_2$O$_3$ microplatelet to Na-β''-Al$_2$O$_3$ was determined to be $\leq$1310°C [58]. The starting platelets were loosely packed to a 2–3-mm depth in a Na-β''-Al$_2$O$_3$-coated alumina boat and exposed to the vapor generated by a (Na$_{0.95}$Li$_{0.05}$)- or (K$_{0.9}$Li$_{0.1}$)-β''-Al$_2$O$_3$ powder. The conversion thickness versus time is plotted in Figure 18. Complete transformation was attained after >5 h. The large particles reacted faster than small ones because of the higher gas permeability in the packing of large particles. The transformation is also accelerated by sequential heating with intermediate mixing to refresh platelet contact with the reaction atmosphere. Completely converted K-β''-Al$_2$O$_3$ platelets were also prepared. The latter platelets frequently showed a multilayered structure that was attributed to the increased reaction rate at the higher vapor pressure of potassium and the reactivity of thereof.

The vapor process is also of interest for the preparation of ultrapure, single-phase, monocrystalline β''-Al$_2$O$_3$ powders. The advantages of such synthesis are as follows: (1) The epitaxial transformation retains the original platelet shape and single crystallinity. (2) Because the phase conversion takes place isothermally in an atmosphere saturated or nearly saturated with sodium vapor over β''-Al$_2$O$_3$ powders, no by-products are produced and a high-purity product results. (3) No sodium aluminate or transient, unstable phases form, thus avoiding transient or partial melting, grain agglomeration, and exaggerated grain growth.

6. THE GROWTH OF K-β''-Al$_2$O$_3$ SINGLE-CRYSTAL FILMS

The K$_2$O-Al$_2$O$_3$ binary system is known to contain four intermediate compounds: 5K$_2$O·Al$_2$O$_3$, 3K$_2$O·Al$_2$O$_3$, K$_2$O·Al$_2$O$_3$, and K-β-Al$_2$O$_3$ solid solutions. This is more complex than the Na$_2$O system. K-β''-Al$_2$O$_3$ is metastable in the binary system;

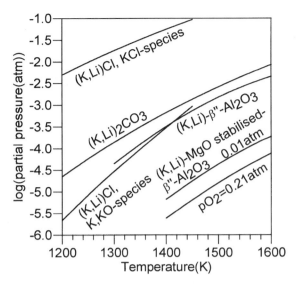

Fig. 19. Equilibrium vapor pressures in the $(K_{0.9}Li_{0.1})_2CO_3$, $(K_{0.9}Li_{0.1})Cl$, and $(K_{0.9}Li_{0.1})$-β''-Al_2O_3 systems in air and low oxygen partial pressure.

however, the upper K_2O limit of K-β-Al_2O_3 solid solutions extends as far as $K_2O·4.9Al_2O_3$, i.e., over the solid solution compositions generally accepted for the β''-Al_2O_3 phase.

The growth of K-β''-Al_2O_3 on Al_2O_3 was pursued with the use of chloride (KCl + LiCl), carbonate (K_2CO_3 + Li_2CO_3), and β''-Al_2O_3 (K-β''-Al_2O_3 + Li-β''-Al_2O_3) vaporization sources. Ceramic substrates were used instead of sapphire in the preliminary tests. The results can be summarized as follows:

1. K-β''-Al_2O_3 forms on Al_2O_3 with carbonate, chloride, and (K,Li)-β''-Al_2O_3 sources. The equilibrium vapor pressures of the K species are plotted in Figure 19.
2. The chloride and carbonate systems generated no Al species, so formation of β''- or β-Al_2O_3 must be from vapor–substrate reactions. β''-Al_2O_3 first appeared at temperatures as low as 950–1000°C because of the high chemical reactivity and vapor pressures of the K_2O species. The yield of β''-Al_2O_3, however, was low in the chloride system because of the high potassium pressure and a deficiency of oxide.
3. A K_2O-rich aluminate phase (in most cases $K_2O·Al_2O_3$) was found to coexist with β''- or β-Al_2O_3. This was assumed to be a reaction product of the high K vapor pressure or to be produced by a reaction between the K species and the β''- or β-Al_2O_3 formed. Potassium aluminate was very hygroscopic and severely corroded the ceramic substrate.
4. K-β''-Al_2O_3 single-crystal films only grew on sapphire substrates. Growth started at 1040°C, $\approx$50°C lower than that of Na-β''-Al_2O_3. A continuous film formed above 1100°C.

The K-β''-Al_2O_3/sapphire epitaxial partnership has a crystallography identical to that of the Na isomorph.

7. ION EXCHANGE PREPARATION OF OTHER β''-Al_2O_3 ISOMORPHS IN SINGLE-CRYSTAL FILM FORM

7.1. Ion Exchange

β''-Al_2O_3 single-crystal films ion exchange like bulk crystals [48]. The ion exchange ability of β- and β''-Al_2O_3 arises from the loose-packed crystal structure. Cations on the conduction planes are exchangeable with ions of different sizes and valences. The extent of the exchange depends on the chemical potentials of the exchanging and exchanged ions in the solid and the ion exchanging medium. It therefore varies with the temperature of the exchange and the concentration and nature of the anion associated with the exchanging cation. The rate of exchange is diffusion-controlled and can be accelerated by applied electrical or ultrasonic fields. In some cases it is decelerated by the "mixed-cation effect."

The facility of ion exchange of the $\beta(\beta'')$-Al_2O_3 family provides a number of chemical and electrochemical paths to new isomorphs. In general, the temperature of ion exchange is lower than that of solid synthesis (or crystal growth), facilitating the preparation of isomorphs of (or doping with) cations unstable at higher processing temperatures. This facility offers control of the valence state of optically active, transition-metal and rare-earth ions. Over 50 isomorphs of β- and β''-Al_2O_3 have been prepared by ion exchange (Appendix, Tables X–XV). Na-β''-Al_2O_3 is the commonly used precursor because of its high diffusivity (conductivity) and availability. β''-Al_2O_3 single-crystal films have been exchanged into Li, K, Cu, Ag, and Ca isomorphs. Because the isomorphs have similar a_0 lattice constant values, the film–substrate interface is subjected to low stress. The film is free to extend vertically, so stresses associated with c axis dimension changes are released. This is not true for polycrystalline ceramics, and the stresses developed during ion exchange destroy them.

The conditions for the ion exchange of β''-Al_2O_3 single-crystal films are summarized in Table VI. Partial or complete exchange was identified by EDX (Fig. 20). The associated X-ray patterns are given in Figure 21. The calculated diffraction intensities for the exchanged isomorphs, based on the atomic occupation of bulk crystals, are presented in Table VII. A distinct intensity change is observed for silver-β''-Al_2O_3. The intensity of reflections (001) was simulated and found to be sensitive to the atomic scattering factor but insensitive to the ionic occupation site.

Conventional ion exchange is carried out in fused salts. The melting temperature of solid mixed salts is high enough for ionic diffusion and substitution, and ion replacement is complete in a few hours. An aqueous ion exchange method [60] was designed to avoid sealing corrosion problems associated with fused-salt media at high temperatures. Aqueous electrochemical cells were constructed to create ion distribution profiles in β''-Al_2O_3 single-crystal films for luminescence patterning (Section 6.2). Aqueous-solution ion exchange was conducted on a single-crystal Na-β''-Al_2O_3 film in a 0.2–0.3 M $AgNO_3$

Table VI.   Conditions for Ion Exchange and Cu^+ Doping in β''-Al_2O_3 Single-Crystal Films

Starting phase	Medium	Temperature (°C)	Time (h)
Ag-β''-Al_2O_3	LiCl (exchanging)	650	10.5
Na-β''-Al_2O_3	KNO_3 (exchanging)	405	87
Na-β''-Al_2O_3	$AgNO_3$ (exchanging)	270	46
Na-β''-Al_2O_3	$CaCl_2$ (exchanging)	800	24
Na-β''-Al_2O_3	0.75CuCl-0.25NaCl (doping)	410–450	2–4
Na-β''-Al_2O_3	0.70CuCl-0.30KCl (doping)	420–470	2–4
Na-β''-Al_2O_3	0.80CuCl-0.20LiCl (doping)	470	3–6
Na-β''-Al_2O_3	0.67CuCl-0.33RbCl (doping)	450	4–6
Na-β''-Al_2O_3	0.936CuCl-0.064$CaCl_2$ (doping)	485	7
Na-β''-Al_2O_3	0.70CuCl-0.30$BaCl_2$ (doping)	650	10.5

Table VII.   Observed and Calculated X-ray Diffraction Intensities of β''-Al_2O_3 Isomorphs

(hkl)	Na-β''-Al_2O_3 0.42(6c), 1.18(18h)		Li-β''-Al_2O_3 0.42(6c), 1.18(18h)		K-β''-Al_2O_3 0.56(6c), 1.04(18h)		Ag-β''-Al_2O_3 1.60(6c), 0.0(18h)	
	Obs.	Calc.	Obs.	Calc.	Obs.	Calc.	Obs.	Calc.
(003)	100	100	100	100	100	100	11	84
(006)	79	59	45	58	67	67	3	1
(009)	2	3	4	5	3	12	86	93
(00,12)	23	17	20	9	54	30	100	100
(00,15)	36	13	21	9	66	26	49	77

The figures indicate the occupation on 6c and 18h sites used for the diffraction-intensity calculation.

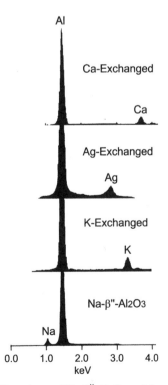

Fig. 20.   The EDX spectrum of Na-β''-Al_2O_3 and ion-exchanged single-crystal films.

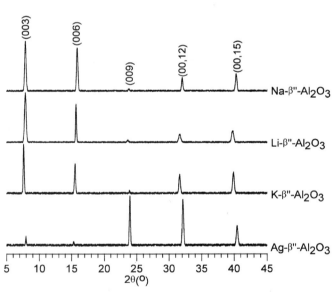

Fig. 21.   The X-ray diffraction pattern of Na-β''-Al_2O_3 and ion-exchanged single-crystal films.

solution. Na^+ and Ag^+ have high diffusivity and mobility in β''-Al_2O_3. Exchanged fractions were determined via weight gain, microbeam, and X-ray diffraction analyses. The increases in relative intensities of reflections (009) and (00,12) are asso-

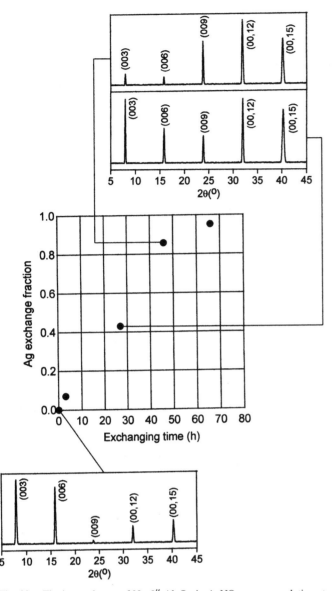

Fig. 22. The ion exchange of Na-β''-Al$_2$O$_3$ in AgNO$_3$ aqueous solution at 92°C (the X-ray diffraction patterns were taken at 0.00, 0.43, and 0.86 exchange fractions). The relative intensity changes are distinct.

ciated with a decrease in intensity of (003) and (006) as Ag$^+$ replaces Na$^+$ (Fig. 22). Nearly complete Ag exchange was obtained below 90°C. The reaction rate decreases with time, as expected for a diffusion-controlled process.

7.2. The Optical Refractivity of β''-Al$_2$O$_3$ Isomorphs

The Lorentz–Lorenz molar refractivity (R) was determined from the measured lattice constants and refractive indices for β''-Al$_2$O$_3$ single-crystal films and crystals, i.e.,

$$R = V_M(n^2 - 1)/(n^2 + 2)$$

$$V_M = W_M/d \quad \text{or} \quad N_A/N_f\{a_0^2 c_0 \sin(2\pi \cdot 3)\}$$

$$n = (n_0^2 n_e)^{1/3}$$

where V_M and W_M are the mole-volume and mole-weight; N_A and N_f are Avogadro's number and the number of formulae per unit cell, respectively; d is the density; and n is the geometric mean refractive index from the ordinary and extraordinary refractive indices. a_0 and c_0 are the lattice constants of the unit cell. The molar refractivity of Na-, Li-, K- and Ca-β''-Al$_2$O$_3$ is given in Table VIII. Refractivity increases with increasing size and charge of the exchanged cation.

Previous investigations have shown that the refractivities of β''- and β-Al$_2$O$_3$ compounds obey the additivity principle [61] and are correlated with the bond ionicity [62]. The refractivity of Al$_2$O$_3$ in β''- and β-Al$_2$O$_3$ is estimated to be approximately equal to that of γ-Al$_2$O$_3$ rather than α-Al$_2$O$_3$. The polarizability (α) is

$$\alpha = \frac{3R}{4\pi N_A}$$

$$R(\beta''\text{-Al}_2\text{O}_3) = \sum R(\text{oxide}) = \sum R(\text{cation}) + \sum R(\text{O}^{2-})$$

$$\alpha(\text{cation}) = \frac{3R(\text{cation})}{4\pi N_A}$$

$$\alpha(\text{O}^2) = \frac{3R(\text{O}^{-2})}{4\pi N_A}$$

where the molar refractivity is summed from the refractivity and polarizability of the cation and oxygen anion components as correlated with bond ionicity. The calculated polarizability is included in Table VIII. This suggests a heterogeneous bonding pattern for the β''-Al$_2$O$_3$ crystals. There are two types of chemical bond, (conduction ion)$-$(oxygen), and Al$-$O. As given in [62], the ionicities of the two oxide components are estimated as 0.63–0.65, and $\geq$0.90, respectively. The low ionicity of the Al$_2$O$_3$ component is attributable to the strong polarizing power of trivalent Al^{3+}, which increases the proximity of the valence electrons shared with the oxygen. The conducting cations have low field strengths, so their valence electrons are less attracted, thus increasing the oxide ionicity and the oxide-ion polarizability.

The additivity principle was used to estimate the mean refractive index of β''- and β-Al$_2$O$_3$ via molar volume. The results are summarized in the Appendix, Tables XVII and XVIII.

8. LUMINESCENCE INVESTIGATION OF Cu$^+$-DOPED, SINGLE-CRYSTAL β''-Al$_2$O$_3$ FILMS

8.1. Luminescence

Luminescence in activated β''- and β-Al$_2$O$_3$ hosts has been verified. Active ions can be doped into the spinel block or onto the conduction plane. Cations on the conduction plane are mobile and weakly bonded and are believed to be responsible for environmental effects.

Cu$^+$ was doped into single-crystal films via ion exchange in CuCl melts in dry nitrogen atmospheres (see Table VI). The luminescence and spectra for Cu$^+$-doped β''-Al$_2$O$_3$ single-crystal films excited by ultraviolet radiation (254 nm) are shown

Table VIII.   Optical Refractivity of β''-Al_2O_3 and Ion Polarizability of Ions Derived via Single-Crystal Film Data

Composition	a_0 (Å)	c_0 (Å)	n_0	n_e	R	$R_{basic\ oxide}$	α_{cation} (Å)^3	$\alpha_{oxygen\ anion}$ (Å)^3
$Na_{1.6}Li_{0.3}Al_{10.7}O_{17}$	5.61	33.565	1.687	1.637	68.6	8.95	0.192	2.79
$Li_{1.6}Li_{0.3}Al_{10.7}O_{17}$	5.605	33.67	1.662	1.627	67.1	7.05	0.069	2.62
$K_{1.6}Li_{0.3}Al_{10.7}O_{17}$	5.60	34.126	1.697	1.642	70.2	10.92	0.647	2.67
$Ag_{1.6}Li_{0.3}Al_{10.7}O_{17}$	5.604	33.46						
$Ca_{0.8}Li_{0.3}Al_{10.7}O_{17}$	5.607	33.47	1.702	1.672	70.1	9.143	0.594	2.97

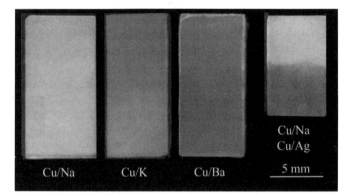

Fig. 23.   The luminescence of Cu^+-doped β''-Al_2O_3 single-crystal films irradiated with ultraviolet light (254 nm).

Table IX.   Emission Bands of the Cu^+-Doped β''-Al_2O_3 Spectrum

Cation	Emission (nm) (fraction)		
	Blue	Green	Red
Na^+	438 (0.100)	535 (0.847)	600 (0.053)
K^+	440 (0.266)	495 (0.734)	
Ba^{2+}	440 (0.539)	485 (0.461)	
Na^+	442 (0.053)	540 (0.776)	625 (0.172)
Ag^+	442 (0.255)	505 (0.046)	606 (0.699)

in Figures 23 and 24. The emission bands and their contributions are summarized in Table IX. Activated Na^+-, K^+-, and Ba-β''-Al_2O_3 exhibited green-blue luminescence. The spectra were resolved into three emitting bands at 440, 485–540, and 600–625 nm, respectively, i.e., in blue, green, and red visible. The transition in the Cu^+-doped β''-Al_2O_3's can be assigned as $3d^{10} \rightarrow 3d^9 4s$, where the blue and green emissions are assumed to arise, respectively, from cuprous ion monomers and dimers, as compared with the spectral data for Cu^+ in various phosphors [4, 63–65]. The band shifts and the efficiency vary with site occupation and crystal field. The wide shift of the emitting band observed in the Cu^+-activated β''-Al_2O_3 bulk crystals is explained by the crystal field changes associated with the expansion and contraction of the host lattice. However, the crystal field explanation is inapplicable to the blue and red shifts for the Ba and Ag hosts, where the emission energy increased with increasing lattice constant (c_0) and decreasing crystal field. The blue band activated in the single-crystal film host emits an identical $\approx$440-nm wavelength in the different isomorphs, whereas the green band varies in the wavelength range 485–540 nm, shifting to short wavelengths from Na- to K- and to the Ba-β''-Al_2O_3 hosts. This abnormality in the crystal field effect is explained by the Cu^+ and O^{2-} ligand interaction with the codoped, matrix cations. The O^{2-} is less deformed with increasing polarizability of the matrix ions and so contributes more to the bond with Cu^{2+} ions and enhances the environmental field, increasing the emission energy. As a result, green luminescence shifts to shorter wavelengths from sodium to potassium to barium. This suggests that the Cu^+

dimers are more flexibly affected by the host environment as compared with the stable blue emission.

The contribution of the blue emission was enhanced in the same sequence of increasing polarizability from Na^+ to K^+ to Ba^{2+}. The increase in polarizability of the matrix cations strengthens the bonding between Cu^{2+} and O^{2-} and suppresses the tendency of ions to cluster. These results suggest that cations of high polarizability are preferred for blue luminescence in a β''-Al_2O_3 host.

The red emission band of the Na-β''-Al_2O_3 host is enhanced in Ag/Na-β''-Al_2O_3. Polarizability (or refractivity) data are unavailable for silver β''-Al_2O_3, so further calculations are not possible. However, the high electronic polarizability of Ag^+ is expected because of the non-inert gas electronic configuration. The increased covalency of the bonding in the Ag-β''-Al_2O_3 isomorph may account for the red-band enhancement.

8.2. Luminescence Patterning of Single-Crystal β''-Al_2O_3 Films

Single-crystal films patterned into green and purple luminescent domains (Fig. 23) were obtained via ion exchange in an aqueous, electrochemical cell. A Cu^+-doped Na-β''-Al_2O_3 single-crystal film was the electrolyte separator, and Na- and Ag-nitrate solutions were the electrodes [64]:

$$(-)\ NaCl\ aq.\ |\ Na\text{-}\beta''\text{-}Al_2O_3\ |\ AgNO_3\ aq.\ (+)$$

An electrical potential was imposed on the cell to accelerate ion migration. The chemical potential of Na^+ and Ag^+ ions was dominated by the $NaNO_3$ and $AgNO_3$ solutions. The solution electrodes are reversible to Na- and Ag-β''-Al_2O_3, respectively, under low current density at moderate temperatures. During

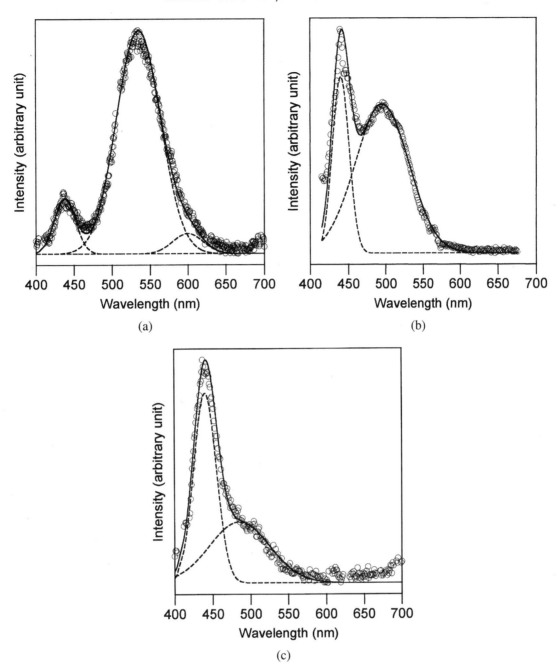

Fig. 24.   Emission spectra excited by 254-nm radiation. (a) Cu^+-activated Na-β''-Al$_2$O$_3$. (b) Cu^+-activated K-β''-Al$_2$O$_3$. (c) Cu^+-activated Ba-β''-Al$_2$O$_3$.

charging with the AgNO$_3$ compartment positively charged against NaNO$_3$, Ag$^+$ ions are forced toward the β''-Al$_2$O$_3$, and Na$^+$ ions exit therefrom into the NaNO$_3$ solution. The charged Ag$^+$ are expelled from the β''-Al$_2$O$_3$ electrolyte into the NaNO$_3$ solution because of the extremely low Ag$^+$ concentration therein. On the other hand, Ag$^+$ ions are strongly held in the electrolyte soaked in the AgNO$_3$, as [Ag$^+$] is much higher than [Na$^+$]. The difference in chemical potential in the electrode compartments determines the Na$^+$ and Ag$^+$ concentrations in the β''-Al$_2$O$_3$ electrolyte. Figure 25 shows the emission spectrum for the two halves of the luminescent β''-Al$_2$O$_3$ single-crystal film electrolyte and the EDX analysis

thereof. The latter indicates the Na and Na/Ag content of the separate domains.

The emission spectra of the ion-patterned β''-Al$_2$O$_3$ film consists of blue, green, and red emission bands. The green and purple luminescences (Fig. 23) are the result of the ionic distribution. For the Na-rich part, a strong spectral peak is positioned at 540 nm, emitting green luminescence. The green emission is greatly depressed in the Ag-rich part of the film, and the red band is intensified. The Ag domain appears purple.

A sharp green-to-purple luminescence boundary builds in the Na/Ag-β''-Al$_2$O$_3$ single-crystal film. The chemical potential of the solution electrodes dominates the Na$^+$ and Ag$^+$

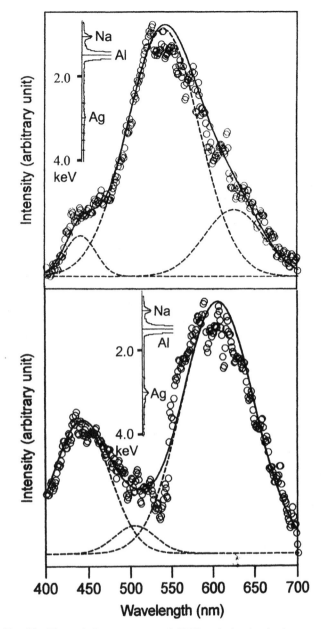

Fig. 25. The emission spectrum and EDX analysis of a luminescence-patterned β''-Al$_2$O$_3$ single-crystal film.

Table X. Monovalent Ion Exchange in β-Al$_2$O$_3$

Ion	Starting material	Preparation conditions	Reference
Li	Ag-β-Al$_2$O$_3$	LiCl saturated LiNO$_3$ melt	[1]
K	Na-β-Al$_2$O$_3$	KNO$_3$ melt	[1]
Rb	Na-β-Al$_2$O$_3$	RbNO$_3$ melt	[1]
	Ag-β-Al$_2$O$_3$	RbCl-RbNO$_3$ melt	[1]
Cs	Ag-β-Al$_2$O$_3$	CsCl melt, 54% exchange	[1]
Ag	Na-β-Al$_2$O$_3$	AgNO$_3$ melt	[1]
Cu	Na-β-Al$_2$O$_3$	CuCl + Cu, 527°C, 24 h, Ar	[2]
	Na-β-Al$_2$O$_3$	CuCl melt + Cu, electrolysis	[2]
Tl	Na-β-Al$_2$O$_3$	TlNO$_3$ melt	[1]
Ga	Ag-β-Al$_2$O$_3$	(Ga iodide + Ga) melt	[1]
In	Na-β-Al$_2$O$_3$	InI, 400°C, no exchange	[1]
	Ag-β-Al$_2$O$_3$	In metal, 350°C, 3 days	[1]
NO	Ag-β-Al$_2$O$_3$	NOCl-AlCl$_3$ melt, 197°C, 23 h	[4]
H	Na-β-Al$_2$O$_3$	H$_2$, 300°C	[5]
H$_3$O	Na-β-Al$_2$O$_3$	Boiling in concentrated H$_2$SO$_4$	[6]
NH$_4$	Na-β-Al$_2$O$_3$	NH$_4$NO$_3$ melt, 170°C	[1]

2. Ion-exchanged β''-Al$_2$O$_3$ single-crystal films doped with the cuprous ion have been prepared and employed as luminescent hosts. The emission spectrum observed includes blue-, green-, and red-emitting bands. The band position and emission contribution are associated with the polarizability of codoped cations. Blue emission is enhanced by codoping with ions of high polarizability.

3. Luminescence patterning was achieved with the use of an electrochemical cell. A Cu$^+$-doped, Na-β''-Al$_2$O$_3$ single-crystal film was used as an electrolyte separator, and NaNO$_3$ (aq.) and AgNO$_3$ (aq.) were used as the solution electrodes. Green and purple luminescent emission areas were patterned into the Na- and Ag-rich domains of the electrolyte.

concentrations in the luminescence-patterned β''-Al$_2$O$_3$ single-crystal films.

9. SUMMARY

1. Na-β''-Al$_2$O$_3$ single-crystal films have been grown on (001) sapphire substrates via chemical-epitaxial reaction between sapphire and alkali vapors. The two phases have identical c axes and 30°-rotated a axes. The film area achieved is 6.5 cm^2, the thickness, ≤ 40 μm. The epitaxial process was extended to K-β''-Al$_2$O$_3$ single-crystal films and β''-Al$_2$O$_3$-coated α-Al$_2$O$_3$ platelets. The latter was employed as a weak-interface dispersant to toughen alumina/zirconia composites.

10. APPENDIX

10.1. β''- and β-Al$_2$O$_3$ Isomorphs

A great variety of β''- and β-Al$_2$O$_3$ isomorphs are prepared via ion exchange in fused salts and, in a few cases, via vapors or aqueous solutions. The isomorphs and preparation conditions are compiled in Tables X–XV. It is known that the ion exchange rate and efficiency vary with the starting material and the exchanging medium. The preparation conditions can be considerably different in ceramic samples because of grain boundary and grain size effects. In addition, because the stresses induced by anisotropic dimension changes in the polycrystals destroy the material structure, ion exchange of ceramics involving volume changes mostly requires special compositional, structural, and procedure designs. Such β''-Al$_2$O$_3$ isomorph ceramics are beyond the scope of single-crystal films. The following tables include only the data from single crystals.

Table XI.   Divalent Ion Exchange in β-Al$_2$O$_3$

Ion	Starting material	Preparation conditions	Reference
Ca	Li-β-Al$_2$O$_3$	LiNO$_3$-Ca(NO$_3$)$_2$, 400°C, 4 days	[1]
		1.21Ca-0.11Li*	
	Na-β-Al$_2$O$_3$	CaCl$_2$ melt, 2 h	[7]
Sr	Na-β-Al$_2$O$_3$	SrI$_2$, 600°C, 3 days, 1.33Sr-0.22Na*	[1]
	Na-β-Al$_2$O$_3$	SrCl$_2$ melt, 2 h	[7]
Ba	Li-β-Al$_2$O$_3$	LiNO$_3$-Ba(NO$_3$)$_2$, 495°C, 4 days, 0.3Ba*	[1]
	Na-β-Al$_2$O$_3$	BaCl$_2$ melt, 2 h	[7]
Cu	Na-β-Al$_2$O$_3$	CuCl$_2$, 550°C, 6 days, 10% exchange	[1]
Zn	Na-β-Al$_2$O$_3$	ZnCl$_2$, 300°C, 6 days, no exchange	[1]
	Ag-β-Al$_2$O$_3$	Zn, 700°C, 1 day	[1]
Cd	Na-β-Al$_2$O$_3$	CdI$_2$, 450°C, 3 days, no exchange	[1]
Hg	Na-β-Al$_2$O$_3$	HgCl$_2$, 350°C, 14 days, no exchange	[1]
	Ag-β-Al$_2$O$_3$	Hg, 300°C, 3 days, no exchange	[1]
Pb	Na-β-Al$_2$O$_3$	PbCl$_2$, 550°C, 70 h, 1.20Pb-0.40Na*	[1]
Fe	Na-β-Al$_2$O$_3$	FeCl$_2$, 700°C, 3 days, 0.40Fe*	[1]
Mn	Na-β-Al$_2$O$_3$	MnCl$_2$, 550°C, 1 day, 0.44Mn*	[1]
Sn	Na-β-Al$_2$O$_3$	SnCl$_2$, 300°C, 6 days, $\approx$37% exchange	[1]

Table XII.   Trivalent Ion Exchange in β-Al$_2$O$_3$

Ion	Starting material	Preparation conditions	Reference
Pr	Na-β-Al$_2$O$_3$	PrCl$_3$, 800°C, 23.5 h, 90.3% exchange	[8]
Nd	Na-β-Al$_2$O$_3$	NdCl$_3$, 760°C, 25 h, 95.2% exchange	[8]
Ho	Na-β-Al$_2$O$_3$	HoCl$_3$, 720°C, 4.5 h, 95.6% exchange	[8]

Table XIII.   Monovalent Ion Exchange in β''-Al$_2$O$_3$

Ion	Starting material	Preparation conditions	Reference
Li	Ag-β''-Al$_2$O$_3$	LiNO$_3$ saturated LiCl, 400°C	[9]
	Na-β''-Al$_2$O$_3$	LiNO$_3$, 350°C	[9]
	Na-β''-Al$_2$O$_3$	LiCl, 700°C	[9]
K	Na-β''-Al$_2$O$_3$	KNO$_3$, 400°C, 6 days	[9]
Rb	Na-β''-Al$_2$O$_3$	RbNO$_3$ melt	[10]
Cu	Na-β''-Al$_2$O$_3$	75CuCl-25NaCl, 460°–600°C,	[11]
		0.1% exchange	
	La-β''-Al$_2$O$_3$	CuCl, 600°C, 1.5 h	[12]
Ag	Na-β''-Al$_2$O$_3$	AgNO$_3$, 250°C, 3 days	[9]
H$_3$O	Na-β''-Al$_2$O$_3$	Concentrated H$_2$SO$_4$, 240°C,	[13]
		>95% exchange	
NH$_4$	Na-β''-Al$_2$O$_3$	NH$_4$NO$_3$, 200°C, 14–43 days,	[9]
		>93% exchange	

Table XIV.   Divalent Ion Exchange in β''-Al$_2$O$_3$

Ion	Starting material	Preparation conditions	Reference
Ca	Ag-β''-Al$_2$O$_3$	CaCl$_2$, 770°C, 23 h	[14]
Sr	Na-β''-Al$_2$O$_3$	47Sr(NO$_3$)$_2$-53SrCl$_2$, 550°C, 20 h	[14]
Ba	Na-β''-Al$_2$O$_3$	63Ba(NO$_3$)$_2$-38BaCl$_2$, 550°C, 20 h	[14]
Zn	Ag-β''-Al$_2$O$_3$	ZnCl$_2$, 500°C, 24 h	[14]
Cd	Na-β''-Al$_2$O$_3$	CdCl$_2$, 600°C, 20 h	[14]
Hg	Na-β''-Al$_2$O$_3$	HgCl$_2$, 300°C, 14 days	[14]
Pb	Na-β''-Al$_2$O$_3$	PbCl$_2$, 535°C, 20 h	[15]
Mn	Na-β''-Al$_2$O$_3$	MnCl$_2$, 600°C, 5 days, vacuum	[15]
	Na-β''-Al$_2$O$_3$	MnCl$_2$, 650°C, 15 h	[14]
Ni	Na-β''-Al$_2$O$_3$	32NiCl$_2$-NaCl, 700°C, 12.5 h,	[16]
		25% exchange	
	Na-β''-Al$_2$O$_3$	NiCl$_2$ vapor, 35 h, 90% exchange	[16]
	Na-β''-Al$_2$O$_3$	NiI$_2$ vapor, 600°C, 6 days,	[15]
		32% exchange	
Co	Na-β''-Al$_2$O$_3$	CoCl$_2$, 750°C, 2 h	[16]
	Na-β''-Al$_2$O$_3$	CoBr$_2$, 500°–550°C, 18 days, vacuum,	[15]
		2 steps	
Sn	Na-β''-Al$_2$O$_3$	SnCl$_2$, 400°C, 1 min, 10% exchange,	[15]
		severely cracked	
Eu	Na-β''-Al$_2$O$_3$	EuI$_2$ vapor, 500°C, 5 days, vacuum	[15]

Table XV.   Trivalent Ion Exchange in β''-Al$_2$O$_3$

Ion	Starting material	Preparation conditions	Reference
Bi	Na-β''-Al$_2$O$_3$	BiCl$_3$, 270°C, 12 h, 70% exchange	[17]
Cr	Na-β''-Al$_2$O$_3$	CrCl$_3$, 530°–550°C, 14 days, 2 steps,	[15]
		75% exchange	
	Na-β''-Al$_2$O$_3$	31CrCl$_3$-69NaCl, 70°C, 1 h,	[16]
		30% exchange	
	Na-β''-Al$_2$O$_3$	CrCl$_3$ vapor, 12 h, 90% exchange	[16]
Ce	Na-β''-Al$_2$O$_3$	CeCl$_3$-NaCl eutectic melt, 650°C,	[18]
		3.6–70% exchange	
Pr	Na-β''-Al$_2$O$_3$	37PrCl$_3$-63NaCl, 600°C, 54 h,	[17]
		43% exchange	
Nd	Na-β''-Al$_2$O$_3$	45NdCl$_3$-55NaCl, 600°C, 12 h,	[17]
		53% exchange	
	Na-β''-Al$_2$O$_3$	NdBr$_2$, 720°C, 0.5 h, 95% exchange	[17]
Sm	Na-β''-Al$_2$O$_3$	SmCl$_3$, 700°C, 20 h, 95% exchange	[17]
Eu	Na-β''-Al$_2$O$_3$	EuCl$_3$, 870°C, 20 h, 95% exchange	[17]
	Na-β''-Al$_2$O$_3$	EuCl$_3$, 600°C, 5 days, Cl$_2$,	[15]
		92% exchange	
Tb	Na-β''-Al$_2$O$_3$	TbCl$_3$, 740°C, 48 h, 90% exchange	[17]
Dy	Na-β''-Al$_2$O$_3$	DyCl$_3$, 1000°C, 0.5 h, 70% exchange	[17]
Er	Na-β''-Al$_2$O$_3$	ErCl$_3$, 600°C, 8 days, 96% exchange	[15]
Yb	Na-β''-Al$_2$O$_3$	TbCl$_3$, 740°C, 24 h, 90% exchange	[17]

10.2. Optical Refractivity of β''- and β-Al$_2$O$_3$

Table XVI lists the Lorentz–Lorenz refractivities of β''- and β-Al$_2$O$_3$ compounds calculated via known molar volume and refractive indices. Tables XVII and XVIII list refractivities calculated via summation of the oxide components as veri-

fied by the additivity principle. The mean refractive index was evaluated with the use of the calculated refractivity from additivity and known lattice constants for the unit cell or the density. The refractivity component of α-Al$_2$O$_3$ was taken as

Table XVI.   Molar Refractivity of β''- and β-Al_2O_3 Evaluated via Refractive Indices and Lattice Constants

Composition	a_0 (Å)	c_0 (Å)	n_0	n_e	R (cm^3)
$Na_{1.6}Li_{0.3}Al_{10.7}O_{17}$ [19]	5.61	33.565	1.687	1.637	68.6
$Li_{1.6}Li_{0.3}Al_{10.7}O_{17}$ [19]	5.605	33.67	1.662	1.627	67.1
$K_{1.6}Li_{0.3}Al_{10.7}O_{17}$ [19]	5.60	34.126	1.697	1.642	70.2
$Ca_{1.6}Li_{0.3}Al_{10.7}O_{17}$ [20]	5.607	33.47	1.702	1.672	70.1
$Na_{2.58}Al_{21.81}O_{34}$ [21]	5.594	22.53	1.6655	1.6254	134.5
$Na_{1.67}Mg_{0.67}Al_{10.33}O_{17}$ [14, 22]	5.61	33.54	1.673	1.637	67.8
$Ca_{0.835}Mg_{0.67}Al_{10.33}O_{17}$ [14, 22]	5.613	33.27	1.681	1.655	68.3
$Ba_{0.835}Mg_{0.67}Al_{10.33}O_{17}$ [14, 22]	5.619	34.084	1.687	1.682	71.2

Table XVII.   Molar Refractivity and Mean Refractive Index of β-Al_2O_3 Calculated via the Refractivity Additive Principle

Composition	a_0	c_0	Density (g/cm^3)	R_{calc} (cm^3)	n_{mean}
$Li_{2.58}Al_{12.81}O_{34}$ [1]	5.596	22.642	3.111	132.7	1.64
$Na_{2.58}Al_{12.81}O_{34}$ [1, 21]	5.594	22.53	3.241	134.5	1.65
$K_{2.58}Al_{12.81}O_{34}$ [1]	5.596	22.729	3.322	139.8	1.68
$Na_{1.29}K_{1.29}Al_{12.81}O_{34}$ [1]	5.595	22.606	3.285	137.1	1.67
$Na_{0.5}K_{1.95}Al_{12.81}O_{34}$ [23]	5.592	22.651	3.306	138.2	1.67
$Rb_{2.58}Al_{21.81}O_{34}$ [1]	5.597	22.877	3.620	143.6	1.69
$Cs_{2.58}Al_{21.81}O_{34}$ [24]	5.584	22.83	3.780	146.1	1.72
$Ag_{2.58}Al_{21.81}O_{34}$ [1]	5.594	22.498	3.842	146.8	1.73
$(NH_4)_{2.58}Al_{21.81}O_{34}$ [1]	5.5961	22.888	3.154	144.0	1.70
$(H_3O)_2Al_{22}O_{34}$ [1]	55.602	22.677	3.168	137.3	1.66
$Ca_{1.097}Al_{21.94}O_{34}$ [25]	5.5902	22.455	3.224	134.7	1.66
$SrMgAl_{20}O_{34}$ [26]	5.61	22.334	3.261	129.8	1.63
$Ba_{1.567}Al_{21.62}O_{34}$ [27]	5.587	22.72	3.630	144.7	1.71
$La_{1.88}Al_{20.78}O_{34}$ [28]	5.556	22.03	3.770	144.5	1.73
$Na_{0.12}Nd_{0.81}Al_{21.81}O_{34}$ [8]	5.577	22.53	3.426	136.2	1.68
$Na_{1.79}Ho_{0.39}Al_{21.67}O_{34}$ [8]	5.587	22.54	3.363	131.0	1.63
$Na_{0.12}Ho_{0.98}Al_{21.74}O_{34}$ [8]	5.588	22.52	3.491	135.4	1.66

Table XVIII.   Molar Refractivity and Mean Refractive Index of β''-Al_2O_3 Calculated via the Refractivity Additive Principle

Composition	a_0	c_0	Density (g/cm^3)	R_{calc} (cm^3)	n_{mean}
$Na_2MgAl_{10}O_{17}$ [29]	5.614	33.85	3.300	69.0	1.67
$Na_{1.76}Mg_{0.6}Al_{10.35}O_{17}$ [10]	5.623	33.591	3.284	68.4	1.66
$Na_{1.71}Mg_{0.71}Al_{10.29}O_{17}$ [31]	5.61	33.54	3.303	68.3	1.67
$K_{1.6}Mg_{0.6}Al_{10.4}O_{17}$ [31]	5.63	34.01	3.360	71.4	1.69
$Rb_{1.72}Li_{0.3}Al_{10.66}O_{17}$ [30]	5.613	34.344	3.766	74.1	1.72
$Ag_{1.65}Mg_{0.68}Al_{10.33}O_{17}$ [10]	5.6295	33.4210	4.049	76.1	1.76
$Na_{0.12}K_{1.55}Mg_{0.67}Al_{10.33}O_{17}$ [23]	5.621	33.644	3.411	71.4	1.70
$Na_{1.67}Mg_{0.67}Al_{10.33}O_{17}$ [14]	5.61	33.54	3.299	68.2	1.67
$Rb_{1.67}Mg_{0.67}Al_{10.33}O_{17}$ [14]	5.613	33.344	3.773	74.1	1.72
$Ca_{0.835}Mg_{0.67}Al_{10.33}O_{17}$ [14]	5.613	33.27	3.295	68.8	1.68
$Sr_{0.835}Mg_{0.67}Al_{10.33}O_{17}$ [14]	5.61	33.72	3.469	70.9	1.69
$Ba_{0.835}Mg_{0.67}Al_{10.33}O_{17}$ [14]	5.619	34.084	3.643	73.2	1.71
$Pb_{0.835}Mg_{0.67}Al_{10.33}O_{17}$ [14]	5.61	33.967	3.980	75.8	1.75
$Nd_{0.556}Mg_{0.67}Al_{10.33}O_{17}$ [32]	5.628	33.259	3.537	69.5	1.68
$Eu_{0.556}Mg_{0.67}Al_{10.33}O_{17}$ [32]	5.627	33.19	3.566	69.1	1.68
$Gd_{0.556}Mg_{0.67}Al_{10.33}O_{17}$ [32]	5.612	33.134	3.590	69.0	1.68

Table XIX. The Luminescence of Transition-Metal and Rare-Earth Ions in β''-Al_2O_3 and Structure-Related Hosts

Activator ion	Host or phosphor material, state	Emission (nm)	Reference
Ce^{3+}	$CaAl_{12}O_{19}$, polycrystal	310	[33]
Ce^{3+}	$SrAl_{12}O_{19}$, polycrystal	300	[33]
Ce^{3+}	$BaAl_{12}O_{19}$, polycrystal	340	[33]
Ce^{3+}	$Sr_{0.95}Ce_{0.05}Mg_{0.05}Al_{11.95}O_{19}$, polycrystal	305	[34]
Ce^{3+}	$La_{0.9}Ce_{0.1}MgAl_{11}O_{17}$, polycrystal	330	[34]
Ce^{3+}	$CeMgAl_{11}O_{19}$, polycrystal	370	[34]
Cr^{3+}	Na-β''-Al_2O_3, single crystal	690–900	[16]
Cr^{3+}	Na-β''-Al_2O_3, single crystal	700, 802	[35]
Cu^{2+}	Na-β''-Al_2O_3, single crystal	535	[12]
Cu^{2+}	K-β''-Al_2O_3, single crystal	552^a	[12]
Cu^{2+}	Ag-β''-Al_2O_3, single crystal	605^a	[12]
Cu^{2+}	Ba-β''-Al_2O_3, single crystal	480, 575	[12]
Cu^{2+}	Cd-β''-Al_2O_3, single crystal	505^a	[12]
Eu^{2+}	$BaMgAl_{10}O_{17}$, polycrystal	450	[34, 36]
Eu^{2+}	$CaAl_{12}O_{19}$, polycrystal	410	[12]
Eu^{2+}	$SrAl_{12}O_{19}$, polycrystal	395	[33, 37]
Eu^{2+}	$BaAl_{12}O_{19}$, polycrystal	440	[33]
Eu^{2+}	$SrMg_2Al_{10}O_{18}$, polycrystal	470	[33]
Eu^{2+}	$BaAl_{10.5}Mg_{1.5}Al_{10.5}O_{18.25}$, polycrystal	445	[33]
Eu^{2+}	$Ba_{0.86}Eu_{0.14}Mg_2Al_{16}O_{27}$, polycrystal	450	[34]
Eu^{2+}	$Ba_{0.86}Eu_{0.14}Mg_{2-y}Mn_yAl_{16}O_{27}$, polycrystal	450, 515	[34]
Eu^{2+}	$Sr_5Eu_{0.5}Mg_6Al_{55}O_{94}$, polycrystal	465	[34]
Eu^{2+}	$Ba_{0.86}Eu_{0.14}Mg_2Al_{16}O_{27}$, polycrystal	450	[34]
Eu^{2+}	$Ba_{0.82}Al_{12}O_{18.82}$, polycrystal	445	[38, 39]
Eu^{2+}	$Ba_{1.29}Al_{12}O_{19.29}$, polycrystal	445, 500	[38, 39]
Eu^{2+}	$(Ba,Ca)_{1.29}Al_{12}O_{19.29}$, polycrystal	445	[39]
Eu^{2+}	Ba aluminate phase I, polycrystal	450^b	[36]
Eu^{2+}	$BaAl_{11}O_{16}N$, polycrystal	450^b	[36]
Pb^{2+}	$PbAl_{12}O_{19}$, polycrystal	340, 390	[40]
Pb^{2+}	$SrAl_{12}O_{19}$, polycrystal	3745, 410	[40]
Tb^{3+}	$Ce_{0.67}Tl_{0.33}MgAl_{11}O_{19}$, polycrystal	545	[34, 41]
Tl^+	$K_{0.9}Tl_{0.1}Al_{11}O_{17}$, polycrystal	385	[34]
Tl^+	$Rb_{0.9}Tl_{0.1}Al_{11}O_{17}$, polycrystal	385	[34]
Tl^+, Mn^{2+}	$Na_{0.6}Tl_{0.1}Ba_{0.3}Al_{10.7}Mn_{0.3}O_{17}$, polycrystal	510	[34]

^{a} Measured on the spectrum.

^{b} With long-wave shoulder.

11.53 cm^3 from the experimental data for Na-β-Al_2O_3. For the sake of clarity and comparison, the chemical composition of the β''-Al_2O_3 compounds is recalculated from the O_{17} and O_{34} formula.

10.3. Luminescence of Activated $\beta''(\beta)$-Al_2O_3

The luminescence of activated β''- and β-Al_2O_3 has not been extensively studied. This collection (Table XIX) includes some data for activated magnetoplumbite compounds. The structure of magnetoplumbite is similar to that of β-Al_2O_3, and these two phases have overlapping stability and metastability regions. The emission data collected were excited by ultraviolet radiation and measured at room or low temperatures.

Acknowledgment

The authors wish to a acknowledge the unfailing support of Dr. Peter McGreer of the Ontario Centre for Materials Research. Without his support, this work would never have been completed.

REFERENCES

1. G. C. Farrington, B. Dunn, and J. O. Thomas, *Appl. Phys. A* 32, 159 (1983).
2. B. Dunn and G. C. Farrington, *Solid State Ionics* 18/19, 31 (1986).
3. J. D. Barrie, B. Dunn, O. M. Stafsudd, and P. Nelson, *J. Lumin.* 37, 303 (1987).

4. J. D. Barrie, B. Dunn, G. Hollingsworth, and J. I. Zink, *J. Phys. Chem.* 93, 3958 (1989).
5. C. K. Kuo and P. S. Nicholson, *J. Can. Ceram. Soc.* 65, 139 (1996).
6. J. M. P. J. Verstegen, D. Radielović, and L. E. Vrenken, *J. Electrochem. Soc.* 121, 1627 (1974).
7. C. R. Rondo and B. M. J. Smets, *J. Electrochem. Soc.* 136, 570 (1989).
8. A. L. N. Stevels, *J. Lumin.* 17, 121 (1978).
9. S. R. Jansen, J. M. Migchels, H. T. Hintzen, and R. Metselaar, *J. Electrochem. Soc.* 146, 800 (1999).
10. M. Jansen, A. Alfrey, O. M. Stafsudd, B. Dunn, D. L. Yang, and G. C. Farrington, *Opt. Lett.* 9, 119 (1984).
11. G. Hollingsworth, J. D. Barrie, B. Dunn, and J. I. Zink, *J. Am. Chem. Soc.* 110, 6569 (1988).
12. W. L. Bragg, C. Gottfried, and J. West, *Z. Kristallogr.* 77, 255 (1931).
13. C. A. Beevers and M. A. S. Ross, *Z. Kristallogr.* 97, 59 (1937).
14. C. R. Peters, M. Bettman, J. W. Moore, and M. D. Glick, *Acta Crystallogr.* B27, 1826 (1971).
15. M. Bettman and C. R. Peters, *J. Phys. Chem.* 73, 1774 (1969).
16. Y.-F. Yu Yao and J. T. Rummer, *J. Inorg. Nucl. Chem.* 29, 2453 (1967).
17. A. Tan, C. K. Kuo, and P. S. Nicholson, *Solid State Ionics* 67, 131 (1993).
18. C. K. Kuo and P. S. Nicholson, *Solid State Ionics* 67, 157 (1993).
19. J. H. Kennedy, *Thin Solid Films* 43, 41 (1977).
20. K. Nobugai and F. Kanamaru, *Mater. Sci. Monogr.* 28B (*React. Solids*, Pt. B), 811 (1985).
21. H. Saalfeld, *Z. Anorg. Allgem. Chem.* 286, 174 (1956).
22. Y. Y. Guo, C. K. Kuo, and P. S. Nicholson, *Solid State Ionics* 99, 95 (1997).
23. S. Udagawa, H. Ikawa, M. Yamamoto, and N. Otsuka, *Nippon Kagaku Kaishi* No. 3, 297 (1980).
24. E. R. Segnit, *Miner. Mag.* 34, 416 (1965).
25. P. W. Brown and S. Sambasivan, *Ceram. Eng. Sci. Proc.* 15, 729 (1994).
26. P. W. Brown, *Ceram. Eng. Sci. Proc.* 16, 385 (1995).
27. K. J. Vaidya, C. Y. Yang, M. DeGraf, and F. F. Lange. *J. Mater. Res.* 9, 410 (1994).
28. J. Théry and D. Briançon, *C. R. Acad. Sci.* 254, 2782 (1962).
29. J. Théry and D. Briançon, *Rev. Hautes Témper. Réfract.* 1, 221 (1964).
30. M. Rolin and P. H. Thanh, *Rev. Hautes Témper. Réfract.* 2, 175 (1965).
31. G. Yamaguchi and K. Suzuki, *Bull. Chem. Soc. Jpn.* 41, 93 (1968).
32. N. Weber and A. F. Venero, *Am. Ceram. Soc. Bull.* 49, 491 (1970).
33. J. D. Hodge, *J. Am. Ceram. Soc.* 66, 166 (1983).
34. Y. Le Cars, J. Théry, and R. Collongues, *C. R. Acad. Sci.* C274, 4 (1972).
35. G. Róg and A. Kozłowska-Róg, *Solid State Ionics* 7, 291 (1982).
36. J. T. Kummer, *Prog. Solid State Chem.* 7, 141 (1972).
37. J. P. Boilot and J. Théry, *Mater. Res. Bull.* 11, 407 (1976).
38. N. Weber and A. F. Venero, *Am. Ceram. Soc. Bull.* 49, 498 (1970).
39. M. Aldén, *Solid State Ionics* 20, 17 (1986).
40. G. K. Duncan and A. R. West, *Solid State Ionics* 9/10, 259 (1983).
41. R. C. De Vries and W. L. Roth, *J. Am. Ceram. Soc.* 52, 364 (1969).
42. B. C. Tofield, in "Intercalation Chemistry," p. 181. Academic Press, New York, 1982.
43. P. T. Moseley, in "The Sodium Sulfur Battery" (J. L. Sudworth and A. R. Tilley, Eds.), pp. 19–77. Chapman and Hall, London, 1985.
44. C. K. Kuo, Y. M. Yan, and P. S. Nicholson, *Solid State Ionics* 92, 45 (1996).
45. Y. M. Yan, C. K. Kuo, and P. S. Nicholson, *Solid State Ionics* 68, 85 (1994).
46. Y. M. Yan, *Solid State Ionics* 53/56, 733 (1992).
47. C. K. Kuo and P. S. Nicholson, *Solid State Ionics* 84, 41 (1996).
48. C. K. Kuo and P. S. Nicholson, *Solid State Ionics* 69, 163 (1994).
49. J. D. Hodge, *J. Am. Ceram. Soc.* 66, C-154 (1983).
50. A. Ya. Neiman, I. E. Tigieva, L. A. Litvinov, and V. V. Pishchik, *Russ. J. Inorg. Chem. (Engl. Transl.)* 31, 490 (1986).
51. R. O. Ansell, A. Gilmour, and R. J. Cole, *J. Electroanal. Chem.* 244, 123 (1988).
52. L. C. De Jonghe and A. Buechele, *J. Mater. Sci.* 17, 885 (1982).
53. Y. Oishi and W. D. Kingery, *J. Chem. Phys.* 33, 480 (1960).
54. D. J. Reed and B. J. Wuensch, *J. Am. Ceram. Soc.* 63, 88 (1980).
55. H. Haneda and C. Monty, *J. Am. Ceram. Soc.* 72, 1153 (1989).
56. X. N. Huang, C. K. Kuo, and P. S. Nicholson, *J. Am. Ceram. Soc.* 78, 892 (1995).
57. C. K. Kuo, X. N. Huang, and P. S. Nicholson, *J. Am. Ceram. Soc.* 78, 824 (1995).
58. C. K. Kuo and P. S. Nicholson, *Solid State Ionics* 83, 225 (1996).
59. C. K. Kuo and P. S. Nicholson, *Solid State Ionics* 73, 297 (1994).
60. C. K. Kuo and P. S. Nicholson, *Solid State Ionics* 91, 81 (1996).
61. C. K. Kuo and P. S. Nicholson, *Solid State Ionics* 58, 173 (1992).
62. Y. Y. Guo, C. K. Kuo, and P. S. Nicholson, *Solid State Ionics* 110, 327 (1998).
63. G. Blasse, *Struct. Bonding (Berlin)* 76, 153 (1991).
64. Y. Shimizu, C. K. Kuo, and P. S. Nicholson, *Solid State Ionics* 110, 69 (1998).
65. F. A. Elrefaie and W. W. Smeltzer, *J. Electrochem. Soc.* 128, 1443 (1981).

Index